BUSINESS/SCIENCE/TECHNOLOGY DIVISION
CHICAGO PUBLIC LIBRARY
400 SOUTH STATE STREET
CHICAGO, ILLINOIS 60605

REF
TK
1001
.E398
2001

HWLCTC

Chicago Public Library

R0175439851
The electric power engineering handbook

THE ELECTRIC POWER ENGINEERING HANDBOOK

The Electrical Engineering Handbook Series

Series Editor
Richard C. Dorf
University of California, Davis

Titles Included in the Series

The Avionics Handbook, Cary R. Spitzer

The Biomedical Engineering Handbook, 2nd Edition, Joseph D. Bronzino

The Circuits and Filters Handbook, Wai-Kai Chen

The Communications Handbook, Jerry D. Gibson

The Control Handbook, William S. Levine

The Digital Signal Processing Handbook, Vijay K. Madisetti & Douglas Williams

The Electrical Engineering Handbook, 2nd Edition, Richard C. Dorf

The Electric Power Engineering Handbook, L.L. Grigsby

The Electronics Handbook, Jerry C. Whitaker

The Engineering Handbook, Richard C. Dorf

The Handbook of Formulas and Tables for Signal Processing, Alexander D. Poularikas

The Industrial Electronics Handbook, J. David Irwin

Measurements, Instrumentation, and Sensors Handbook, John Webster

The Mechanical Systems Design Handbook, Osita D.I. Nwokah

The RF and Microwave Handbook, J. Michael Golio

The Mobile Communications Handbook, 2nd Edition, Jerry D. Gibson

The Ocean Engineering Handbook, Ferial El-Hawary

The Technology Management Handbook, Richard C. Dorf

The Transforms and Applications Handbook, 2nd Edition, Alexander D. Poularikas

The VLSI Handbook, Wai-Kai Chen

The Electromagnetics Handbook, Aziz Inan and Umran Inan

The Mechatronics Handbook, Robert Bishop

THE ELECTRIC POWER ENGINEERING HANDBOOK

EDITOR-IN-CHIEF

L.L. GRIGSBY

Auburn University
Auburn, Alabama

 CRC PRESS   IEEE PRESS

A CRC Handbook Published in Cooperation with IEEE Press

> **Library of Congress Cataloging-in-Publication Data**
>
> The electric power engineering handbook / editor-in-chief L.L. Grigsby.
>   p. cm. -- (The electrical engineering handbook series)
>   Includes bibliographical references and index.
>   ISBN 0-8493-8578-4 (alk.)
>   1. Electric power production. I. Grigsby, Leonard L. II. Series.
>
> TK1001 .E398 2000
> 621.31′2--dc21   00-030425

This book contains information obtained from authentic and highly regarded sources. Reprinted material is quoted with permission, and sources are indicated. A wide variety of references are listed. Reasonable efforts have been made to publish reliable data and information, but the author and the publisher cannot assume responsibility for the validity of all materials or for the consequences of their use.

Neither this book nor any part may be reproduced or transmitted in any form or by any means, electronic or mechanical, including photocopying, microfilming, and recording, or by any information storage or retrieval system, without prior permission in writing from the publisher.

All rights reserved. Authorization to photocopy items for internal or personal use, or the personal or internal use of specific clients, may be granted by CRC Press LLC, provided that $.50 per page photocopied is paid directly to Copyright Clearance Center, 222 Rosewood Drive, Danvers, MA 01923 USA. The fee code for users of the Transactional Reporting Service is ISBN 0-8493-8578-4/01/$0.00+$.50. The fee is subject to change without notice. For organizations that have been granted a photocopy license by the CCC, a separate system of payment has been arranged.

The consent of CRC Press LLC does not extend to copying for general distribution, for promotion, for creating new works, or for resale. Specific permission must be obtained in writing from CRC Press LLC for such copying.

Direct all inquiries to CRC Press LLC, 2000 N.W. Corporate Blvd., Boca Raton, Florida 33431.

Trademark Notice: Product or corporate names may be trademarks or registered trademarks, and are used only for identification and explanation, without intent to infringe.

© 2001 by CRC Press LLC

No claim to original U.S. Government works
International Standard Book Number 0-8493-8578-4
Library of Congress Card Number 00-030425
Printed in the United States of America 1 2 3 4 5 6 7 8 9 0
Printed on acid-free paper

Preface

The generation, delivery, and utilization of electric power and energy remain among the most challenging and exciting fields of electrical engineering. The astounding technological developments of our age are highly dependent upon a safe, reliable, and economic supply of electric power. The objective of *The Electric Power Engineering Handbook* is to provide a contemporary overview of this far-reaching field as well as a useful guide and educational resource for its study. It is intended to define electric power engineering by bringing together the core of knowledge from all of the many topics encompassed by the field. The articles are written primarily for the electric power engineering professional who is seeking factual information and secondarily for the professional from other engineering disciplines who wants an overview of the entire field or specific information on one aspect of it.

The book is organized into 15 sections in an attempt to provide comprehensive coverage of the generation, transformation, transmission, distribution, and utilization of electric power and energy as well as the modeling, analysis, planning, design, monitoring, and control of electric power systems. The individual articles within the 15 sections are different from most technical publications. They are not journal type articles nor are they textbook in nature. They are intended to be tutorials or overviews providing ready access to needed information, while at the same time providing sufficient references to more in-depth coverage of the topic. This work is a member of the Electrical Engineering Handbook Series published by CRC Press. Since its inception in 1993, this series has been dedicated to the concept that when readers refer to a handbook on a particular topic they should be able to find what they need to know about the subject at least 80% of the time. That has indeed been the goal of this handbook.

In reading the individual articles of this handbook, I have been most favorably impressed by how well the authors have accomplished the goals that were set. Their contributions are, of course, most key to the success of the work. I gratefully acknowledge their outstanding efforts. Likewise, the expertise and dedication of the editorial board and section editors have been critical in making this handbook possible. To all of them I express my profound thanks. I also wish to thank the personnel at CRC Press who have been involved in the production of this book, with a special word of thanks to Nora Konopka and Ron Powers. Their patience and perseverance have made this task most pleasant.

Leo Grigsby
Editor-in-Chief

Editor-in-Chief

Leonard L. ("Leo") Grigsby received BSEE and MSEE degrees from Texas Tech University and a Ph.D. from Oklahoma State University. He has taught electrical engineering at Texas Tech, Oklahoma State University, and Virginia Tech. He has been at Auburn University since 1984, first as the Georgia Power Distinguished Professor, later as the Alabama Power Distinguished Professor, and currently as Professor Emeritus of Electrical Engineering. He also spent nine months during 1990 at the University of Tokyo as the Tokyo Electric Power Company Endowed Chair of Electrical Engineering. His teaching interests are in network analysis, control systems, and power engineering.

During his teaching career, Professor Grigsby has received 12 awards for teaching excellence. These include his selection for the university-wide William E. Wine Award for Teaching Excellence at Virginia Tech in 1980, his selection for the ASEE AT&T Award for Teaching Excellence in 1986, the 1988 Edison Electric Institute Power Engineering Educator Award, the 1990–91 Distinguished Graduate Lectureship at Auburn University, the 1995 IEEE Region 3 Joseph M. Beidenbach Outstanding Engineering Educator Award, and the 1996 Birdsong Superior Teaching Award at Auburn University.

Dr. Grigsby is a Fellow of IEEE. During 1998–99 he was a member of the Board of Directors as Director of Div. VII for power and energy. He has served the Institute in 27 different offices at the chapter, section, region, or national level. For this service, he has received seven distinguished service awards, the IEEE Centennial Medal in 1984, and the Power Engineering Society Meritorious Service Award in 1994.

During his academic career, Professor Grigsby has conducted research in a variety of projects related to the application of network and control theory to modeling, simulation, optimization and control of electric power systems. He has been the major advisor for 35 M.S. and 21 Ph.D. graduates. With his students and colleagues, he has published over 120 technical papers and a textbook on introductory network theory. He is currently Editor for CRC Press for a book series on electric power engineering. In 1993 he was inducted into the Electrical Engineering Academy at Texas Tech University for distinguished contributions to electrical engineering.

Editorial Board

Pritindra Chowdhuri
Tennessee Technological University
Cookeville, Tennessee

Richard G. Farmer
Arizona State University
Tempe, Arizona

L.L. Grigsby
Auburn University
Auburn, Alabama

S.M. Halpin
Mississippi State University
Mississippi State, Mississippi

Andrew Hanson
ABB Power T&D Company
Raleigh, North Carolina

James H. Harlow
Harlow Engineering Associates
Largo, Florida

George G. Karady
Arizona State University
Tempe, Arizona

William H. Kersting
New Mexico State University
Las Cruces, New Mexico

John D. McDonald
KEMA Consulting
Norcross, Georgia

Mark Nelms
Auburn University
Auburn, Alabama

Arun Phadke
Virginia Polytechnic Institute
Blacksburg, Virginia

Saifur Rahman
Virginia Tech
Alexandria, Virginia

Rama Ramakumar
Oklahoma State University
Stillwater, Oklahoma

Gerald B. Sheblé
Iowa State University
Ames, Iowa

Robert Waters
Alabama Power Company
Birmingham, Alabama

Bruce F. Wollenberg
University of Minnesota
Minneapolis, Minnesota

Contributors

Rambabu Adapa
Electric Power Research Institute
Palo Alto, California

Bajarang L. Agrawal
Arizona Public Service Co.
Phoenix, Arizona

Hirofumi Akagi
Tokyo Institute of Technology
Tokyo, Japan

Alex Apostolov
Alstom T&D
Los Angeles, California

John Appleyard
S&C Electric Company
Sauk City, Wisconsin

Miroslav Begovic
Georgia Institute of Technology
Atlanta, Georgia

Gabriel Benmouyal
Schweitzer Engineering
  Laboratories, Ltd.
Boucherville, Quebec, Canada

Michael J. Bio
Power Resources, Inc.
Pelham, Alabama

Al Bolger
BC Hydro
Burnaby, British Columbia,
  Canada

Philip Bolin
Mitsubishi Electric Power
  Products Inc.
Warrendale, Pennsylvania

M.H.J. Bollen
Chalmers University of
  Technology
Gothenburg, Sweden

Anjan Bose
Washington State University
Pullman, Washington

Simon W. Bowen
Alabama Power Company
Birmingham, Alabama

John R. Boyle
Power System Analysis
Signal Mountain, Tennessee

Wolfgang Breuer
Maschinenfabrik Reinhausen
  GmbH
Regensburg, Germany

Steven R. Brockschink
Pacific Engineering Corporation
Portland, Oregon

Richard E. Brown
ABB Power T&D Company
Raleigh, North Carolina

Kristine Buchholz
Pacific Gas and Electric
San Francisco, California

Antonio Castanheira
Trench Ltd.
Scarborough, Ontario, Canada

Wilford Caulkins
Sherman & Reilly, Inc.
Chattanooga, Tennessee

William Chisholm
Ontario Hydro Technologies
Toronto, Ontario, Canada

Pritindra Chowdhuri
Tennessee Technological
  University
Cookeville, Tennessee

George L. Clark
Alabama Power Company
Birmingham, Alabama

Patrick Coleman
Alabama Power Company
Birmingham, Alabama

Craig A. Colopy
Cooper Power Systems
Waukesha, Wisconsin

Robert C. Degeneff
Rensselaer Polytechnic Institute
Troy, New York

Don Delcourt
BC Hydro
Burnaby, British Columbia, Canada

Scott H. Digby
Waukesha Electric Systems
Goldsboro, North Carolina

Dieter Dohnal
Maschinenfabrik Reinhausen GmbH
Regensburg, Germany

M.K. Donnelly
Pacific Northwest National Laboratory
Richland, Washington

D.A. Douglass
Power Delivery Consultants, Inc.
Niskayuna, New York

Richard Dudley
Trench Ltd.
Scarborough, Ontario, Canada

M.E. El-Hawary
Dalhousie University
Halifax, Nova Scotia, Canada

Ahmed Elneweihi
BC Hydro
Burnaby, British Columbia, Canada

James W. Evans
Detroit Edison Company
Detroit, Michigan

Richard G. Farmer
Arizona State University
Tempe, Arizona

James W. Feltes
Power Technologies
Schenectady, New York

Shelia Frasier
Southern Engineering
Atlanta, Georgia

Rulon Fronk
Fronk Consulting
Cerritos, California

Dudley L. Galloway
ABB Power T & D Company
Jefferson City, Missouri

Michael G. Giesselmann
Texas Tech University
Lubbock, Texas

Jay C. Giri
ALSTOM ESCA Corporation
Bellevue, Washington

L.L. Grigsby
Auburn University
Auburn, Alabama

Charles A. Gross
Auburn University
Auburn, Alabama

John V. Grubbs
Alabama Power Company
Birmingham, Alabama

James H. Gurney
BC Hydro
Burnaby, British Columbia, Canada

Nouredine Hadjsaid
Institut National Polytechnique de Grenoble (INPG)
France

S.M. Halpin
Mississippi State University
Mississippi State, Mississippi

Andrew Hanson
ABB Power T & D Company
Raleigh, North Carolina

James H. Harlow
Harlow Engineering Associates
Largo, Florida

David L. Harris
Waukesha Electric Systems
Waukesha, Wisconsin

Tim A. Haskew
The University of Alabama
Tuscaloosa, Alabama

Robert Haas
Haas Engineering
Villa Hills, Kentucky

J.F. Hauer
Pacific Northwest National Laboratory
Richland, Washington

Ted Haupert
TJ/H2b Analytical Services
Sacramento, California

William R. Henning
Waukesha Electric Systems
Waukesha, Wisconsin

Felimón Hernandez
Arizona Public Service Company
Phoenix, Arizona

Philip J. Hopkinson
Square D Company
Monroe, North Carolina

Stan H. Horowitz
Consultant
Columbus, Ohio

Gary L. Johnson
Kansas State University
Manhattan, Kansas

Anthony J. Jonnatti
Loci Engineering
Palm Harbor, Florida

Gerhard Juette
Siemens
Munich, Germany

Danny Julian
ABB Power T & D Company
Raleigh, North Carolina

Tonia Jurbin
BC Hydro
Burnaby, British Columbia, Canada

John G. Kappenman
Metatech Corporation
Duluth, Minnesota

George G. Karady
Arizona State University
Tempe, Arizona

Richard P. Keil
Dayton Power & Light Company
Dayton, Ohio

John R. Kennedy
Georgia Power Company
Atlanta, Georgia

William H. Kersting
New Mexico State University
Las Cruces, New Mexico

Tibor Kertesz
Hydro One Networks, Inc.
Toronto, Ontario, Canada

Alireza Khotanzad
Southern Methodist University
Dallas, Texas

Prabha Kundur
Powertech Labs, Inc.
Surrey, British Columbia, Canada

Stephen R. Lambert
Shawnee Power Consulting, LLC
Williamsburg, Virginia

Einar Larsen
GE Power Systems
Schenectady, New York

W.H. Litzenberger
Bonneville Power Administration
Portland, Oregon

Andre Lux
ABB Power T&D Company
Raleigh, North Carolina

Yakout Mansour
BC Hydro
Burnaby, British Columbia, Canada

Juan A. Martinez-Velasco
Universitat Politecnica de Catalunya
Barcelona, Spain

John D. McDonald
KEMA Consulting
Norcross, Georgia

Shirish P. Mehta
Waukesha Electric Systems
Waukesha, Wisconsin

Christopher J. Melhorn
EPRI PEAC Corporation
Knoxville, Tennessee

Hyde M. Merrill
Merrill Energy, LLC
Schenectady, New York

Roger A. Messenger
Florida Atlantic University
Boca Raton, Florida

William A. Mittelstadt
Bonneville Power Adminstration
Portland, Oregon

Harold Moore
H. Moore & Associates
Niceville, Florida

Kip Morrison
Powertech Labs Inc.
Surrey, British Columbia, Canada

Dan Mulkey
Pacific Gas & Electric Co.
Petaluma, California

Randy Mullikin
Kuhlman Electric Corp.
Versailles, Kentucky

Paul I. Nippes
Magnetic Product and Services, Inc.
Holmdel, New Jersey

Robert S. Nowell
Georgia Power Company
Atlanta, Georgia

Carlos V. Núñez-Noriega
Glendale Community College
Glendale, Arizona

Alan Oswalt
Waukesha Electric Systems
Waukesha, Wisconsin

John Paserba
Mitsubishi Electric Power Products Inc.
Warrendale, Pennsylvania

Paulette A. Payne
Potomac Electric Power Company
Washington, DC

Dan D. Perco
Perco Transformer Engineering
Stoney Creek, Ontario, Canada

Joe C. Pohlman
Consultant
Pittsburgh, Pennsylvania

William W. Price
GE Power Systems
Schenectady, New York

Jeewan Puri
Square D Company
Monroe, North Carolina

Saifur Rahman
Virginia Tech
Falls Church, Virginia

Kaushik Rajashekara
Delphi Automotive Systems
Kokomo, Indiana

N. Dag Reppen
Niskayuna Power Consultants, LLC
Niskayuna, New York

Manuel Reta-Hernández
Arizona State University
Tempe, Arizona

Charles W. Richter
ALSTOM ESCA Corporation
Bellevue, Washington

Francisco de la Rosa
DLR Electric Power Reliability
Houston, Texas

Anne-Marie Sahazizian
Hydro One Networks, Inc.
Toronto, Ontario, Canada

Juan Sanchez-Gasca
GE Power Systems
Schenectady, New York

Peter W. Sauer
University of Illinois
Urbana, Illinois

Leo J. Savio
ADAPT Corporation
Kennett Square, Pennsylvania

Kenneth H. Sebra
Baltimore Gas & Electric Company
Baltimore, Maryland

Douglas B. Seely
Pacific Engineering Corporation
Portland, Oregon

Michael Sharp
Trench Ltd.
Scarborough, Ontario, Canada

Gerald B. Sheblé
Iowa State University
Ames, Iowa

Raymond R. Shoults
University of Texas at Arlington
Arlington, Texas

H. Jin Sim
Waukesha Electric Systems
Goldsboro, North Carolina

James H. Sosinski
Consumers Energy
Jackson, Mississippi

K. Neil Stanton
Stanton Associates
Bellevue, Washington

Robert P. Stewart
BC Hydro
Burnaby, British Columbia, Canada

C.M. Mike Stine
Raychem Corporation
Menlo Park, California

Mahesh M. Swamy
Yaskawa Electric America
Waukegan, Illinois

Glenn W. Swift
APT Power Technologies
Winnipeg, Manitoba, Canada

Larry D. Swift
University of Texas at Arlington
Arlington, Texas

Carson W. Taylor
Carson Taylor Seminars
Portland, Oregon

Rao S. Thallam
Salt River Project
Phoenix, Arizona

James S. Thorp
Cornell University
Ithaca, New York

Ridley Thrash
Southwire Company
Carrollton, Georgia

Robert F. Tillman, Jr.
Alabama Power Company
Birmingham, Alabama

Giao N. Trinh, Jr.
Log-In
Boucherville, Quebec, Canada

Vijay Vittal
Iowa State University
Ames, Iowa

Loren B. Wagenaar
America Electric Power
Pickerington, Ohio

Contents

1 Electric Power Generation: Non-Conventional Methods *Saifur Rahman*
  1.1 Wind Power *Gary L. Johnson* 1-1
  1.2 Advanced Energy Technologies *Saifur Rahman* 1-7
  1.3 Photovoltaics *Roger A. Messenger* 1-14

2 Electric Power Generation: Conventional Methods *Rama Ramakumar*
  2.1 Hydroelectric Power Generation *Steven R. Brockschink, James H. Gurney, and Douglas B. Seely* 2-1
  2.2 Synchronous Machinery *Paul I. Nippes* 2-12
  2.3 Thermal Generating Plants *Kenneth H. Sebra* 2-20
  2.4 Distributed Utilities *John R. Kennedy* 2-27

3 Transformers *James H. Harlow*
  3.1 Theory and Principles *Harold Moore* 3-3
  3.2 Power Transformers *H. Jin Sim and Scott H. Digby* 3-11
  3.3 Distribution Transformers *Dudley L. Galloway* 3-30
  3.4 Underground Distribution Transformers *Dan Mulkey* 3-52
  3.5 Dry Type Transformers *Paulette A. Payne* 3-63
  3.6 Step-Voltage Regulators *Craig A. Colopy* 3-68
  3.7 Reactors *Richard Dudley, Antonio Castanheira, and Michael Sharp* 3-81
  3.8 Instrument Transformers *Randy Mullikin and Anthony J. Jonnatti* 3-104
  3.9 Transformer Connections *Dan D. Perco* 3-130
  3.10 LTC Control and Transformer Paralleling *James H. Harlow* 3-135
  3.11 Loading Power Transformers *Robert F. Tillman, Jr.* 3-149
  3.12 Causes and Effects of Transformer Sound Levels *Jeewan Puri* 3-159
  3.13 Electrical Bushings *Loren B. Wagenaar* 3-171
  3.14 Load Tap Changers (LTCs) *Dieter Dohnal and Wolfgang Breuer* 3-184
  3.15 Insulating Media *Leo J. Savio and Ted Haupert* 3-204
  3.16 Transformer Testing *Shirish P. Mehta and William R. Henning* 3-209
  3.17 Transformer Installation and Maintenance *Alan Oswalt* 3-234
  3.18 Problem and Failure Investigations *Harold Moore* 3-242
  3.19 The United States Power Transformer Equipment Standards and Processes *Philip J. Hopkinson* 3-249
  3.20 On-Line Monitoring of Liquid-Immersed Transformers *Andre Lux* 3-268

4 Transmission System *George G. Karady*
4.1 Concept of Energy Transmission and Distribution *George G. Karady* 4-2
4.2 Transmission Line Structures *Joe C. Pohlman* 4-13
4.3 Insulators and Accessories *George G. Karady and R.G. Farmer* 4-23
4.4 Transmission Line Construction and Maintenance *Wilford Caulkins and Kristine Buchholz* 4-42
4.5 Insulated Power Cables for High Voltage Applications *Carlos V. Núñez-Noriega and Felimón Hernandez* 4-48
4.6 Transmission Line Parameters *Manuel Reta-Hernández* 4-62
4.7 Sag and Tension of Conductor *D.A. Douglass and Ridley Thrash* 4-89
4.8 Corona and Noise *Giao N. Trinh* 4-129
4.9 Geomagnetic Disturbances and Impacts upon Power System Operation *John G. Kappenman* 4-150
4.10 Lightning Protection *William A Chisholm* 4-165
4.11 Reactive Power Compensation *Rao S. Thallam* 4-169

5 Substations *John D. McDonald*
5.1 Gas Insulated Substations *Philip Bolin* 5-2
5.2 Air Insulated Substations — Bus/Switching Configurations *Michael J. Bio* 5-18
5.3 High Voltage Switching Equipment *David L. Harris* 5-23
5.4 High Voltage Power Electronics Substations *Gerhard Juette* 5-29
5.5 Considerations in Applying Automation Systems to Electric Utility Substations *James W. Evans* 5-41
5.6 Substation Automation *John D. McDonald* 5-53
5.7 Oil Containment *Anne-Marie Sahazizian and Tibor Kertesz* 5-62
5.8 Community Considerations *James H. Sosinski* 5-76
5.9 Animal Deterrents/Security *C.M. Mike Stine and Sheila Frasier* 5-88
5.10 Substation Grounding *Richard P. Keil* 5-92
5.11 Grounding and Lightning *Robert S. Nowell* 5-104
5.12 Seismic Considerations *R.P. Stewart, Rulon Frank, and Tonia Jurbin* 5-124
5.13 Substation Fire Protection *Al Bolger and Don Delcourt* 5-134

6 Distribution Systems *William H. Kersting*
6.1 Power System Loads *Raymond R. Shoults and Larry D. Swift* 6-1
6.2 Distribution System Modeling and Analysis *William H. Kersting* 6-11
6.3 Power System Operation and Control *George L. Clark and Simon W. Bowen* 6-67

7 Electric Power Utilization *Andrew Hanson*
7.1 Metering of Electric Power and Energy *John V. Grubbs* 7-1
7.2 Basic Electric Power Utilization — Loads, Load Characterization and Load Modeling *Andrew Hanson* 7-12
7.3 Electric Power Utilization: Motors *Charles A. Gross* 7-18

8 Power System Analysis and Simulation *L.L. Grigsby and Andrew Hanson*
8.1 The Per-Unit System *Charles A. Gross* 8-1
8.2 Symmetrical Components for Power System Analysis *Tim A. Haskew* 8-14
8.3 Power Flow Analysis *L. L. Grigsby and Andrew Hanson* 8-34
8.4 Fault Analysis in Power Systems *Charles A. Gross* 8-44

9 Power System Protection *Arun Phadke*
9.1 Transformer Protection *Alex Apostolov, John Appleyard, Ahmed Elneweihi, Robert Haas, and Glenn W. Swift* 9-1
9.2 The Protection of Synchronous Generators *Gabriel Benmouyal* 9-11
9.3 Transmission Line Protection *Stanley H. Horowitz* 9-30
9.4 System Protection *Miroslav Begovic* 9-39
9.5 Digital Relaying *James S. Thorp* 9-44
9.6 Use of Oscillograph Records to Analyze System Performance *John R. Boyle* 9-62

10 Power System Transients *Pritindra Chowdhuri*
10.1 Characteristics of Lightning Strokes *Francisco de la Rosa* 10-2
10.2 Overvoltages Caused by Direct Lightning Strokes *Pritindra Chowdhuri* 10-8
10.3 Overvoltages Caused by Indirect Lightning Strokes *Pritindra Chowdhuri* 10-21
10.4 Switching Surges *Stephen R. Lambert* 10-36
10.5 Very Fast Transients *Juan A. Martinez-Velasco* 10-41
10.6 Transient Voltage Response of Coils and Windings *Robert C. Degeneff* 10-54
10.7 Transmission System Transients — Grounding *William Chisholm* 10-80
10.8 Insulation Coordination *Stephen R. Lambert* 10-93

11 Power System Dynamics and Stability *Richard G. Farmer*
11.1 Power System Stability — Overview *Prabha Kundur* 11-2
11.2 Transient Stability *Kip Morrison* 11-10
11.3 Small Signal Stability and Power System Oscillations *John Paserba, Prabha Kundur, Juan Sanchez-Gasca, and Einar Larsen* 11-20
11.4 Voltage Stability *Yakout Mansour* 11-34
11.5 Direct Stability Methods *Vijay Vittal* 11-42
11.6 Power System Stability Controls *Carson W. Taylor* 11-55
11.7 Power System Dynamic Modeling *William W. Price* 11-72
11.8 Direct Analysis of Wide Area Dynamics *J. F. Hauer, W. A. Mittelstadt, M. K. Donnelly, W. H. Litzenberger, and Rambabu Adapa* 11-82
11.9 Power System Dynamic Security Assessment *Peter W. Sauer* 11-120
11.10 Power System Dynamic Interaction with Turbine-Generators *Richard G. Farmer and Bajarang L. Agrawal* 11-126

12 Power System Operation and Control Bruce F. Wollenberg
12.1 Energy Management *K. Neil Stanton, Jay C. Giri, and Anjan Bose* 12-1
12.2 Generation Control: Economic Dispatch and Unit Commitment
       Charles W. Richter, Jr. 12-10
12.3 State Estimation *Danny Julian* 12-27
12.4 Optimal Power Flow *M. E. El-Hawary* 12-38
12.5 Security Analysis *Nouredine Hadjsaid* 12-53

13 Power System Planning (Reliability) Gerald B. Sheblé
13.1 Planning *Gerald B. Sheblé* 13-1
13.2 Short-Term Load and Price Forecasting with Artificial Neural Networks
       Alireza Khotanzad 13-16
13.3 Transmission Plan Evaluation — Assessment of System Reliability
       N. Dag Reppen and James W. Feltes 13-26
13.4 Power System Planning *Hyde M. Merrill* 13-40
13.5 Power System Reliability *Richard E. Brown* 13-51

14 Power Electronics Mark Nelms
14.1 Power Semiconductor Devices *Kaushik Rajashekara* 14-1
14.2 Uncontrolled and Controlled Rectifiers *Mahesh M. Swamy* 14-8
14.3 Inverters *Michael Giesselmann* 14-37
14.4 Active Filters for Power Conditioning *Hirofumi Akagi* 14-44

15 Power Quality S.M. Halpin
15.1 Introduction *S.M. Halpin* 15-1
15.2 Wiring and Grounding for Power Quality *Christopher J. Melhorn* 15-2
15.3 Harmonics in Power Systems *S.M. Halpin* 15-16
15.4 Voltage Sags *M. H. J. Bollen* 15-24
15.5 Voltage Fluctuations and Lamp Flicker in Power Systems *S.M. Halpin* ... 15-42
15.6 Power Quality Monitoring *Patrick Coleman* 15-49

Index I-1

Electric Power Generation: Non-Conventional Methods

Saifur Rahman
Virginia Tech

1.1   **Wind Power**   *Gary L. Johnson* 1-1

1.2   **Advanced Energy Technologies**   *Saifur Rahman* 1-7

1.3   **Photovoltaics**   *Roger A. Messenger* 1-14

1
Electric Power Generation: Non-Conventional Methods

Gary L. Johnson
Kansas State University

Saifur Rahman
Virginia Tech

Roger A. Messenger
Florida Atlantic University

1.1 Wind Power 1-1
   Applications • Wind Variability
1.2 Advanced Energy Technologies 1-7
   Storage Systems • Fuel Cells • Summary
1.3 Photovoltaics 1-14
   Types of PV Cells • PV Applications

1.1 Wind Power

Gary L. Johnson

The wind is a free, clean, and inexhaustible energy source. It has served humankind well for many centuries by propelling ships and driving wind turbines to grind grain and pump water. Denmark was the first country to use wind for generation of electricity. The Danes were using a 23-m diameter wind turbine in 1890 to generate electricity. By 1910, several hundred units with capacities of 5 to 25 kW were in operation in Denmark (Johnson, 1985). By about 1925, commercial wind-electric plants using two- and three-bladed propellers appeared on the American market. The most common brands were Wincharger (200 to 1200 W) and Jacobs (1.5 to 3 kW). These were used on farms to charge storage batteries which were then used to operate radios, lights, and small appliances with voltage ratings of 12, 32, or 110 volts. A good selection of 32-VDC appliances was developed by the industry to meet this demand.

In addition to home wind-electric generation, a number of utilities around the world have built larger wind turbines to supply power to their customers. The largest wind turbine built before the late 1970s was a 1250-kW machine built on Grandpa's Knob, near Rutland, Vermont, in 1941. This turbine, called the Smith-Putnam machine, had a tower that was 34 m high and a rotor 53 m in diameter. The rotor turned an ac synchronous generator that produced 1250 kW of electrical power at wind speeds above 13 m/s.

After World War II, we entered the era of cheap oil imported from the Middle East. Interest in wind energy died and companies making small turbines folded. The oil embargo of 1973 served as a wakeup call, and oil-importing nations around the world started looking at wind again. The two most important countries in wind power development since then have been the U.S. and Denmark (Brower et al., 1993).

The U.S. immediately started to develop utility-scale turbines. It was understood that large turbines had the potential for producing cheaper electricity than smaller turbines, so that was a reasonable decision. The strategy of getting large turbines in place was poorly chosen, however. The Department of

TABLE 1.1 Wind Power Installed Capacity

Canada	83
China	224
Denmark	1450
India	968
Ireland	63
Italy	180
Germany	2874
Netherlands	363
Portugal	60
Spain	834
Sweden	150
U.K.	334
U.S.	1952
Other	304
Total	9839

Energy decided that only large aerospace companies had the manufacturing and engineering capability to build utility-scale turbines. This meant that small companies with good ideas would not have the revenue stream necessary for survival. The problem with the aerospace firms was that they had no desire to manufacture utility-scale wind turbines. They gladly took the government's money to build test turbines, but when the money ran out, they were looking for other research projects. The government funded a number of test turbines, from the 100 kW MOD-0 to the 2500 kW MOD-2. These ran for brief periods of time, a few years at most. Once it was obvious that a particular design would never be cost competitive, the turbine was quickly salvaged.

Denmark, on the other hand, established a plan whereby a landowner could buy a turbine and sell the electricity to the local utility at a price where there was at least some hope of making money. The early turbines were larger than what a farmer would need for himself, but not what we would consider utility scale. This provided a revenue stream for small companies. They could try new ideas and learn from their mistakes. Many people jumped into this new market. In 1986, there were 25 wind turbine manufacturers in Denmark. The Danish market gave them a base from which they could also sell to other countries. It was said that Denmark led the world in exports of two products: wind turbines and butter cookies! There has been consolidation in the Danish industry since 1986, but some of the companies have grown large. Vestas, for example, has more installed wind turbine capacity worldwide than any other manufacturer.

Prices have dropped substantially since 1973, as performance has improved. It is now commonplace for wind power plants (collections of utility-scale turbines) to be able to sell electricity for under four cents per kilowatt hour.

Total installed worldwide capacity at the start of 1999 was almost 10,000 MW, according to the trade magazine *Wind Power Monthly* (1999). The countries with over 50 MW of installed capacity at that time are shown in Table 1.1.

Applications

There are perhaps four distinct categories of wind power which should be discussed. These are

1. small, non-grid connected
2. small, grid connected
3. large, non-grid connected
4. large, grid connected

By small, we mean a size appropriate for an individual to own, up to a few tens of kilowatts. Large refers to utility scale.

Small, Non-Grid Connected

If one wants electricity in a location not serviced by a utility, one of the options is a wind turbine, with batteries to level out supply and demand. This might be a vacation home, a remote antenna and transmitter site, or a Third-World village. The costs will be high, on the order of $0.50/kWh, but if the total energy usage is small, this might be acceptable. The alternatives, photovoltaics, microhydro, and diesel generators, are not cheap either, so a careful economic study needs to be done for each situation.

Small, Grid Connected

The small, grid connected turbine is usually not economically feasible. The cost of wind-generated electricity is less because the utility is used for storage rather than a battery bank, but is still not competitive.

In order for the small, grid connected turbine to have any hope of financial breakeven, the turbine owner needs to get something close to the retail price for the wind-generated electricity. One way this is done is for the owner to have an arrangement with the utility called net metering. With this system, the meter runs backward when the turbine is generating more than the owner is consuming at the moment. The owner pays a monthly charge for the wires to his home, but it is conceivable that the utility will sometimes write a check to the owner at the end of the month, rather than the other way around. The utilities do not like this arrangement. They want to buy at wholesale and sell at retail. They feel it is unfair to be used as a storage system without remuneration.

For most of the twentieth century, utilities simply refused to connect the grid to wind turbines. The utility had the right to generate electricity in a given service territory, and they would not tolerate competition. Then a law was passed that utilities had to hook up wind turbines and pay them the avoided cost for energy. Unless the state mandated net metering, the utility typically required the installation of a second meter, one measuring energy consumption by the home and the other energy production by the turbine. The owner would pay the regular retail rate, and the utility would pay their estimate of avoided cost, usually the fuel cost of some base load generator. The owner might pay $0.08 to $0.15 per kWh, and receive $0.02 per kWh for the wind-generated electricity. This was far from enough to economically justify a wind turbine, and had the effect of killing the small wind turbine business.

Large, Non-Grid Connected

These machines would be installed on islands or in native villages in the far north where it is virtually impossible to connect to a large grid. Such places are typically supplied by diesel generators, and have a substantial cost just for the imported fuel. One or more wind turbines would be installed in parallel with the diesel generators, and act as fuel savers when the wind was blowing.

This concept has been studied carefully and appears to be quite feasible technically. One would expect the market to develop after a few turbines have been shown to work for an extended period in hostile environments. It would be helpful if the diesel maintenance companies would also carry a line of wind turbines so the people in remote locations would not need to teach another group of maintenance people about the realities of life at places far away from the nearest hardware store.

Large, Grid Connected

We might ask if the utilities should be forced to buy wind-generated electricity from these small machines at a premium price which reflects their environmental value. Many have argued this over the years. A better question might be whether the small or the large turbines will result in a lower net cost to society. Given that we want the environmental benefits of wind generation, should we get the electricity from the wind with many thousands of individually owned small turbines, or should we use a much smaller number of utility-scale machines?

If we could make the argument that a dollar spent on wind turbines is a dollar not spent on hospitals, schools, and the like, then it follows that wind turbines should be as efficient as possible. Economies of scale and costs of operation and maintenance are such that the small, grid connected turbine will always need to receive substantially more per kilowatt hour than the utility-scale turbines in order to break even. There is obviously a niche market for turbines that are not connected to the grid, but small, grid connected turbines will probably not develop a thriving market. Most of the action will be from the utility-scale machines.

Sizes of these turbines have been increasing rapidly. Turbines with ratings near 1 MW are now common, with prototypes of 2 MW and more being tested. This is still small compared to the needs of a utility, so clusters of turbines are placed together to form wind power plants with total ratings of 10 to 100 MW.

Wind Variability

One of the most critical features of wind generation is the variability of wind. Wind speeds vary with time of day, time of year, height above ground, and location on the earth's surface. This makes wind generators into what might be called energy producers rather than power producers. That is, it is easier to estimate the energy production for the next month or year than it is to estimate the power that will be produced at 4:00 PM next Tuesday. Wind power is not dispatchable in the same manner as a gas turbine. A gas turbine can be scheduled to come on at a given time and to be turned off at a later time, with full power production in between. A wind turbine produces only when the wind is available. At a good site, the power output will be zero (or very small) for perhaps 10% of the time, rated for perhaps another 10% of the time, and at some intermediate value the remaining 80% of the time.

This variability means that some sort of storage is necessary for a utility to meet the demands of its customers, when wind turbines are supplying part of the energy. This is not a problem for penetrations of wind turbines less than a few percent of the utility peak demand. In small concentrations, wind turbines act like negative load. That is, an increase in wind speed is no different in its effect than a customer turning off load. The control systems on the other utility generation sense that generation is greater than load, and decrease the fuel supply to bring generation into equilibrium with load. In this case, storage is in the form of coal in the pile or natural gas in the well.

An excellent form of storage is water in a hydroelectric lake. Most hydroelectric plants are sized large enough to not be able to operate full-time at peak power. They therefore must cut back part of the time because of the lack of water. A combination hydro and wind plant can conserve water when the wind is blowing, and use the water later, when the wind is not blowing.

When high-temperature superconductors become a little less expensive, energy storage in a magnetic field will be an exciting possibility. Each wind turbine can have its own superconducting coil storage unit. This immediately converts the wind generator from an energy producer to a peak power producer, fully dispatchable. Dispatchable peak power is always worth more than the fuel cost savings of an energy producer. Utilities with adequate base load generation (at low fuel costs) would become more interested in wind power if it were a dispatchable peak power generator.

The variation of wind speed with time of day is called the diurnal cycle. Near the earth's surface, winds are usually greater during the middle of the day and decrease at night. This is due to solar heating, which causes "bubbles" of warm air to rise. The rising air is replaced by cooler air from above. This thermal mixing causes wind speeds to have only a slight increase with height for the first hundred meters or so above the earth. At night, however, the mixing stops, the air near the earth slows to a stop, and the winds above some height (usually 30 to 100 m) actually increase over the daytime value. A turbine on a short tower will produce a greater proportion of its energy during daylight hours, while a turbine on a very tall tower will produce a greater proportion at night.

As tower height is increased, a given generator will produce substantially more energy. However, most of the extra energy will be produced at night, when it is not worth very much. Standard heights have been increasing in recent years, from 50 to 65 m or even more. A taller tower gets the blades into less turbulent air, a definite advantage. The disadvantages are extra cost and more danger from overturning in high winds. A very careful look should be given the economics before buying a tower that is significantly taller than whatever is sold as a standard height for a given turbine.

Wind speeds also vary strongly with time of year. In the southern Great Plains (Kansas, Oklahoma, and Texas), the winds are strongest in the spring (March and April) and weakest in the summer (July and August). Utilities here are summer peaking, and hence need the most power when winds are the lowest and the least power when winds are highest. The diurnal variation of wind power is thus a fairly good match to utility needs, while the yearly variation is not.

Electric Power Generation: Non-Conventional Methods

TABLE 1.2 Monthly Average Wind Speed in MPH and Projected Energy Production at 65 m, at a Good Site in Southern Kansas

Month	10 m Speed	60 m Speed	Energy (MWh)	Month	10 m Speed	60 m Speed	Energy (MWh)
1/96	14.9	20.3	256	1/97	15.8	21.2	269
2/96	16.2	22.4	290	2/97	14.7	19.0	207
3/96	17.6	22.3	281	3/97	17.4	22.8	291
4/96	19.8	25.2	322	4/97	15.9	20.4	242
5/96	18.4	23.1	297	5/97	15.2	19.8	236
6/96	13.5	18.2	203	6/97	11.9	16.3	167
7/96	12.5	16.5	169	7/97	13.3	18.5	212
8/96	11.6	16.0	156	8/97	11.7	16.9	176
9/96	12.4	17.2	182	9/97	13.6	19.0	211
10/96	17.1	23.3	320	10/97	15.0	21.1	265
11/96	15.3	20.0	235	11/97	14.3	19.7	239
12/96	15.1	20.1	247	12/97	13.6	19.5	235

The variability of wind with month of year and height above ground is illustrated in Table 1.2. These are actual wind speed data for a good site in Kansas, and projected electrical generation of a Vestas turbine (V47-660) at that site. Anemometers were located at 10, 40, and 60 m above ground. Wind speeds at 40 and 60 m were used to estimate the wind speed at 65 m (the nominal tower height of the V47-660) and to calculate the expected energy production from this turbine at this height. Data have been normalized for a 30-day month.

There can be a factor of two between a poor month and an excellent month (156 MWh in 8/96 to 322 MWh in 4/96). There will not be as much variation from one year to the next, perhaps 10 to 20%. A wind power plant developer would like to have as long a data set as possible, with an absolute minimum of one year. If the one year of data happens to be for the best year in the decade, followed by several below average years, a developer could easily get into financial trouble. The risk gets smaller if the data set is at least two years long.

One would think that long-term airport data could be used to predict whether a given data set was collected in a high or low wind period for a given part of the country, but this is not always true. One study showed that the correlation between average annual wind speeds at Russell, Kansas, and Dodge City, Kansas, was 0.596 while the correlation between Russell and Wichita was 0.115. The terrain around Russell is very similar to that around Wichita, and there is no obvious reason why wind speeds should be high at one site and low at the other for one year, and then swap roles the next year.

There is also concern about long-term variation in wind speeds. There appears to be an increase in global temperatures over the past decade or so, which would probably have an impact on wind speeds. It also appears that wind speeds have been somewhat lower as temperatures have risen, at least in Kansas. It appears that wind speeds can vary significantly over relatively short distances. A good data set at one location may underpredict or overpredict the winds at a site a few miles away by as much as 10 to 20%. Airport data collected on a 7-m tower in a flat river valley may underestimate the true surrounding hilltop winds by a factor of two. If economics are critical, a wind power plant developer needs to acquire rights to a site and collect wind speed data for at least one or two years before committing to actually constructing turbines there.

Land Rights

Spacing of turbines can vary widely with the type of wind resource. In a tradewind or a mountain pass environment where there are only one or two prevailing wind directions, the turbines can be located "shoulder to shoulder" crossways to the wind direction. A downwind spacing of ten times the rotor diameter is usually assumed to be adequate to give the wind space to recover its speed. In open areas, a crosswind spacing of four rotor diameters is usually considered a minimum. In the Great Plains, the prevailing winds are from the south (Kansas, Oklahoma, and Texas) or north (the Dakotas). The energy in the winds from east and west may not be more than 10% of the total energy. In this situation, a spacing of ten rotor diameters north–south and four rotor diameters east–west would be minimal. Adjustments would be made to avoid roads, pipelines, power lines, houses, ponds, and creeks.

The results of a detailed site layout will probably not predict much more than 20 MW of installed capacity per square mile (640 acres). This figure can be used for initial estimates without great error. That is, if a developer is considering installing a 100-MW wind plant, rights to at least five square miles should be acquired.

One issue that has not received much attention in the wind power community is that of a fair compensation to the land owner for the privilege of installing wind turbines. The developer could buy the land, hopefully with a small premium. The original deal could be an option to buy at some agreed upon price, if two years of wind data were satisfactory. The developer might lease the land back to the original landowner, since the agricultural production capability is only slightly affected by the presence of wind turbines. Outright purchase between a willing and knowledgeable buyer and seller would be as fair an arrangement as could be made.

But what about the case where the landowner does not want to sell? Rights have been acquired by a large variety of mechanisms, including a large one-time payment for lease signing, a fixed yearly fee, a royalty payment based on energy produced, and combinations of the above. The one-time payment has been standard utility practice for right-of-way acquisitions, and hence will be preferred by at least some utilities. A key difference is that wind turbines require more attention than a transmission line. Roads are not usually built to transmission line towers, while they are built to wind turbines. Roads and maintenance operations around wind turbines provide considerably more hassle to the landowner. The original owner got the lease payment, and 20 years later the new owner gets the nuisance. There is no incentive for the new landowner to be cooperative or to lobby county or state officials on behalf of the developer.

A one-time payment also increases the risk to the developer. If the project does not get developed, there has been a significant outlay of cash which will have no return on it. These disadvantages mean that the one-time payment with no yearly fees or royalties will probably not be the long-term norm in the industry.

To discuss what might be a fair price for a lease, it will be helpful to use an example. We will assume the following:

- 20 MW per square mile
- Land fair-market value $500/acre
- Plant factor 0.4
- Developer desired internal rate of return 0.2
- Electricity value $0.04/kWh
- Installed cost of wind turbine $1000/kW

A developer that purchased the land at $500/acre would therefore want a return of $(500)(0.2) = $100/acre. America's cheap food policy means that production agriculture typically gets a much smaller return on investment than the developer wants. Actual cash rent on grassland might be $15/acre, or a return of 0.03 on investment. We see an immediate opportunity for disagreement, even hypocrisy. The developer might offer the landowner $15/acre when the developer would want $100/acre if he bought the land. This hardly seems equitable.

The gross income per acre is

$$I = \frac{(20{,}000 \text{ kW})(0.4)(8760 \text{ hours/year})(\$0.04)}{640 \text{ acres}} = \$4380/\text{acre}/\text{year} \tag{1.1}$$

The cost of wind turbines per acre is

$$CT_a = \frac{(20{,}000 \text{ kW})(\$1000/\text{kW})}{640 \text{ acres}} = \$31{,}250/\text{acre} \tag{1.2}$$

We see that the present fair-market value for the land is tiny compared with the installed cost of the wind turbines. A lease payment of $100/acre/year is slightly over 2% of the gross income. It is hard to imagine financial arrangements so tight that they would collapse if the landowner (either rancher or developer) were paid this yearly fee. That is, it seems entirely reasonable for a figure like 2% of gross income to be a starting point for negotiations.

There is another factor that might result in an even higher percentage. Landowners throughout the Great Plains are accustomed to royalty payments of 12.5% of wholesale price for oil and gas leases. This is determined independently of any agricultural value for the land. The most worthless mesquite in Texas gets the same terms as the best irrigated corn ground in Kansas. We might ask if this rate is too high. A royalty of 12.5% of wholesale amounts to perhaps 6% of retail. Cutting the royalty in half would have the potential of reducing the price of gasoline about 3%. In a market where gasoline prices swing by 20%, this reduction is lost in the noise. If a law were passed which cut royalty payments in half, it is hard to argue that it would have much impact on our gasoline buying habits, the size of vehicles we buy, or the general welfare of the nation.

One feature of the 12.5% royalty is that it is high enough to get most oil and gas producing land under lease. Would 6.25% have been enough to get the same amount of land leased? If we assumed that some people would sign a lease for 12.5% that would not sign if the offer were 6.25%, then we have the interesting possibility that the supply would be less. If we assume the law of supply and demand to apply, the price of gasoline and natural gas would increase. The possible increase is shear speculation, but could easily be more than the 6.25% that was "saved" by cutting the royalty payment in half.

The point is that the royalty needs to be high enough to get the very best sites under lease. If the best site produces 10% more energy than the next best, it makes no economic sense to pay a 2% royalty for the second best when a 6% royalty would get the best site. In this example, the developer would get 10% more energy for 4% more royalty. The developer could either pocket the difference or reduce the price of electricity a proportionate amount.

References

Brower, M. C., Tennis, M. W., Denzler, E. W., and Kaplan, M. M., *Powering the Midwest*, A Report by the Union of Concerned Scientists, 1993.

Johnson, G. L., *Wind Energy Systems*, Prentice-Hall, New York, 1985.

Wind Power Monthly, 15(6), June, 1999.

1.2 Advanced Energy Technologies

Saifur Rahman

Storage Systems

Energy storage technologies are of great interest to electric utilities, energy service companies, and automobile manufacturers (for electric vehicle application). The ability to store large amounts of energy would allow electric utilities to have greater flexibility in their operation because with this option the supply and demand do not have to be matched instantaneously. The availability of the proper battery at the right price will make the electric vehicle a reality, a goal that has eluded the automotive industry thus far. Four types of storage technologies (listed below) are discussed in this section, but most emphasis is placed on storage batteries because it is now closest to being commercially viable. The other storage technology widely used by the electric power industry, pumped-storage power plants, is not discussed as this has been in commercial operation for more than 60 years in various countries around the world.

- Flywheel storage
- Compressed air energy storage
- Superconducting magnetic energy storage
- Battery storage

Flywheel Storage

Flywheels store their energy in their rotating mass, which rotates at very high speeds (approaching 75,000 rotations per minute), and are made of composite materials instead of steel because of the composite's ability to withstand the rotating forces exerted on the flywheel. In order to store enegy the flywheel is placed in a sealed container which is then placed in a vacuum to reduce air resistance. Magnets embedded in the flywheel pass near pickup coils. The magnet induces a current in the coil changing the rotational energy into electrical energy. Flywheels are still in research and development, and commercial products are several years away.

Compressed Air Energy Storage

As the name implies, the compressed air energy storage (CAES) plant uses electricity to compress air which is stored in underground reservoirs. When electricity is needed, this compressed air is withdrawn, heated with gas or oil, and run through an expansion turbine to drive a generator. The compressed air can be stored in several types of underground structures, including caverns in salt or rock formations, aquifers, and depleted natural gas fields. Typically the compressed air in a CAES plant uses about one third of the premium fuel needed to produce the same amount of electricity as in a conventional plant. A 290-MW CAES plant has been in operation in Germany since the early 1980s with 90% availability and 99% starting reliability. In the U.S., the Alabama Electric Cooperative runs a CAES plant that stores compressed air in a 19-million cubic foot cavern mined from a salt dome. This 110-MW plant has a storage capacity of 26 h. The fixed-price turnkey cost for this first-of-a-kind plant is about $400/kW in constant 1988 dollars.

The turbomachinery of the CAES plant is like a combustion turbine, but the compressor and the expander operate independently. In a combustion turbine, the air that is used to drive the turbine is compressed just prior to combustion and expansion and, as a result, the compressor and the expander must operate at the same time and must have the same air mass flow rate. In the case of a CAES plant, the compressor and the expander can be sized independently to provide the utility-selected "optimal" MW charge and discharge rate which determines the ratio of hours of compression required for each hour of turbine-generator operation. The MW ratings and time ratio are influenced by the utility's load curve, and the price of off-peak power. For example, the CAES plant in Germany requires 4 h of compression per hour of generation. On the other hand, the Alabama plant requires 1.7 h of compression for each hour of generation. At 110-MW net output, the power ratio is 0.818 kW output for each kilowatt input. The heat rate (LHV) is 4122 BTU/kWh with natural gas fuel and 4089 BTU/kWh with fuel oil. Due to the storage option, a partial-load operation of the CAES plant is also very flexible. For example, the heat rate of the expander increases only by 5%, and the airflow decreases nearly linearly when the plant output is turned down to 45% of full load. However, CAES plants have not reached commercial viability beyond some prototypes.

Superconducting Magnetic Energy Storage

A third type of advanced energy storage technology is superconducting magnetic energy storage (SMES), which may someday allow electric utilities to store electricity with unparalled efficiency (90% or more). A simple description of SMES operation follows.

The electricity storage medium is a doughnut-shaped electromagnetic coil of superconducting wire. This coil could be about 1000 m in diameter, installed in a trench, and kept at superconducting temperature by a refrigeration system. Off-peak electricity, converted to direct current (DC), would be fed into this coil and stored for retrieval at any moment. The coil would be kept at a low-temperature superconducting state using liquid helium. The time between charging and discharging could be as little as 20 ms with a round-trip AC–AC efficiency of over 90%.

Developing a commercial-scale SMES plant presents both economic and technical challenges. Due to the high cost of liquiud helium, only plants with 1000-MW, 5-h capacity are economically attractive. Even then the plant capital cost can exceed several thousand dollars per kilowatt. As ceramic superconductors, which become superconducting at higher temperatures (maintained by less expensive liquid nitrogen), become more widely available, it may be possible to develop smaller scale SMES plants at a lower price.

Battery Storage

Even though battery storage is the oldest and most familiar energy storage device, significant advances have been made in this technology in recent years to deserve more attention. There has been renewed interest in this technology due to its potential application in non-polluting electric vehicles. Battery systems are quiet and non-polluting, and can be installed near load centers and existing suburban substations. These have round-trip AC–AC efficiencies in the range of 85%, and can respond to load changes within 20 ms. Several U.S., European, and Japanese utilities have demonstrated the application of lead–acid batteries for load-following applications. Some of them have been as large as 10 MW with 4 h of storage.

The other player in battery development is the automotive industry for electric vehicle application. In 1991, General Motors, Ford, Chrysler, Electric Power Research Institute (EPRI), several utilities, and the U.S. Department of Energy (DOE) formed the U.S. Advanced Battery Consortium (USABC) to develop better batteries for electric vehicle (EV) applications. A brief introduction to some of the available battery technologies as well some that are under study is presented in the following (Source:http://www.eren.doe.gov/consumerinfo/refbriefs/fa1/html).

Battery Types

Chemical batteries are individual cells filled with a conducting medium-electrolyte that, when connected together, form a battery. Multiple batteries connected together form a battery bank. At present, there are two main types of batteries: primary batteries (non-rechargeable) and secondary batteries (rechargeable). Secondary batteries are further divided into two categories based on the operating temperature of the electrolyte. Ambient operating temperature batteries have either aqueous (flooded) or nonaqueous electrolytes. High operating temperature batteries (molten electrodes) have either solid or molten electrolytes. Batteries in EVs are the secondary-rechargeable-type and are in either of the two sub-categories. A battery for an EV must meet certain performance goals. These goals include quick discharge and recharge capability, long cycle life (the number of discharges before becoming unserviceable), low cost, recyclability, high specific energy (amount of usable energy, measured in watt-hours per pound [lb] or kilogram [kg]), high energy density (amount of energy stored per unit volume), specific power (determines the potential for acceleration), and the ability to work in extreme heat or cold. No battery currently available meets all these criteria.

Lead–Acid Batteries

Lead–acid starting batteries (shallow-cycle lead–acid secondary batteries) are the most common battery used in vehicles today. This battery is an ambient temperature, aqueous electrolyte battery. A cousin to this battery is the deep-cycle lead–acid battery, now widely used in golf carts and forklifts. The first electric cars built also used this technology. Although the lead–acid battery is relatively inexpensive, it is very heavy, with a limited usable energy by weight (specific energy). The battery's low specific energy and poor energy density make for a very large and heavy battery pack, which cannot power a vehicle as far as an equivalent gas-powered vehicle. Lead–acid batteries should not be discharged by more than 80% of their rated capacity or depth of discharge (DOD). Exceeding the 80% DOD shortens the life of the battery. Lead–acid batteries are inexpensive, readily available, and are highly recyclable, using the elaborate recycling system already in place. Research continues to try to improve these batteries.

A lead–acid nonaqueous (gelled lead acid) battery uses an electrolyte paste instead of a liquid. These batteries do not have to be mounted in an upright position. There is no electrolyte to spill in an accident. Nonaqueous lead–acid batteries typically do not have as high a life cycle and are more expensive than flooded deep-cycle lead–acid batteries.

Nickel Iron and Nickel Cadmium Batteries

Nickel iron (Edison cells) and nickel cadmium (nicad) pocket and sintered plate batteries have been in use for many years. Both of these batteries have a specific energy of around 25 Wh/lb (55 Wh/kg), which is higher than advanced lead–acid batteries. These batteries also have a long cycle life. Both of these batteries are recyclable. Nickel iron batteries are non-toxic, while nicads are toxic. They can also be discharged to 100% DOD without damage. The biggest drawback to these batteries is their cost. Depending on the size of battery bank in the vehicle, it may cost between $20,000 and $60,000 for the batteries. The batteries should last at least 100,000 mi (160,900 km) in normal service.

Nickel Metal Hydride Batteries

Nickel metal hydride batteries are offered as the best of the next generation of batteries. They have a high specific energy: around 40.8 Wh/lb (90 Wh/kg). According to a U.S. DOE report, the batteries are benign to the environment and are recyclable. They also are reported to have a very long cycle life. Nickel metal hydride batteries have a high self-discharge rate: they lose their charge when stored for long periods of time. They are already commercially available as "AA" and "C" cell batteries, for small consumer appliances and toys. Manufacturing of larger batteries for EV applications is only available to EV manufacturers. Honda is using these batteries in the EV Plus, which is available for lease in California.

Sodium Sulfur Batteries

This battery is a high-temperature battery, with the electrolyte operating at temperatures of 572°F (300°C). The sodium component of this battery explodes on contact with water, which raises certain safety concerns. The materials of the battery must be capable of withstanding the high internal temperatures they create, as well as freezing and thawing cycles. This battery has a very high specific energy: 50 Wh/lb (110 Wh/kg). The Ford Motor Company uses sodium sulfur batteries in their Ecostar, a converted delivery minivan that is currently sold in Europe. Sodium sulfur batteries are only available to EV manufacturers.

Lithium Iron and Lithium Polymer Batteries

The USABC considers lithium iron batteries to be the long-term battery solution for EVs. The batteries have a very high specific energy: 68 Wh/lb (150 Wh/kg). They have a molten-salt electrolyte and share many features of a sealed bipolar battery. Lithium iron batteries are also reported to have a very long cycle life. These are widely used in laptop computers. These batteries will allow a vehicle to travel distances and accelerate at a rate comparable to conventional gasoline-powered vehicles. Lithium polymer batteries eliminate liquid electrolytes. They are thin and flexible, and can be molded into a variety of shapes and sizes. Neither type will be ready for EV commercial applications until early in the 21st century.

Zinc and Aluminum Air Batteries

Zinc air batteries are currently being tested in postal trucks in Germany. These batteries use either aluminum or zinc as a sacrificial anode. As the battery produces electricity, the anode dissolves into the electrolyte. When the anode is completely dissolved, a new anode is placed in the vehicle. The aluminum or zinc and the electrolyte are removed and sent to a recycling facility. These batteries have a specific energy of over 97 Wh/lb (200 Wh/kg). The German postal vans currently carry 80 kWh of energy in their battery, giving them about the same range as 13 gallons (49.2 liters) of gasoline. In their tests, the vans have achieved a range of 615 mi (990 km) at 25 miles per hour (40 km/h).

Fuel Cells

In 1839, a British Jurist and an amateur physicist named William Grove first discovered the principle of the fuel cell. Grove utilized four large cells, each containing hydrogen and oxygen, to produce electricity and water which was then used to split water in a different container to produce hydrogen and oxygen. However, it took another 120 years until NASA demonstrated its use to provide electricity and water for some early space flights. Today the fuel cell is the primary source of electricity on the space shuttle. As a result of these successes, industry slowly began to appreciate the commercial value of fuel cells. In addition to stationary power generation applications, there is now a strong push to develop fuel cells for automotive use. Even though fuel cells provide high performance characterisitics, reliability, durability, and environmental benefits, a very high investment cost is still the major barrier against large-scale deployments.

Basic Principles

The fuel cell works by processing a hydrogen-rich fuel — usually natural gas or methanol — into hydrogen, which, when combined with oxygen, produces electricity and water. This is the reverse electrolysis process. Rather than burning the fuel, however, the fuel cell converts the fuel to electricity using a highly efficient electrochemical process. A fuel cell has few moving parts, and produces very little waste heat or gas.

A fuel cell power plant is basically made up of three subsystems or sections. In the fuel-processing section, the natural gas or other hydrocarbon fuel is converted to a hydrogen-rich fuel. This is normally accomplished through what is called a steam catalytic reforming process. The fuel is then fed to the power section, where it reacts with oxygen from the air in a large number of individual fuel cells to produce direct current (DC) electricity, and by-product heat in the form of usable steam or hot water. For a power plant, the number of fuel cells can vary from several hundred (for a 40-kW plant) to several thousand (for a multi-megawatt plant). In the final, or third stage, the DC electricity is converted in the power conditioning subsystem to electric utility-grade alternating current (AC).

In the power section of the fuel cell, which contains the electrodes and the electrolyte, two separate electrochemical reactions take place: an oxidation half-reaction occurring at the anode and a reduction half-reaction occurring at the cathode. The anode and the cathode are separated from each other by the electrolyte. In the oxidation half-reaction at the anode, gaseous hydrogen produces hydrogen ions, which travel through the ionically conducting membrane to the cathode. At the same time, electrons travel through an external circuit to the cathode. In the reduction half-reaction at the cathode, oxygen supplied from air combines with the hydrogen ions and electrons to form water and excess heat. Thus, the final products of the overall reaction are electricity, water, and excess heat.

Types of Fuel Cells

The electrolyte defines the key properties, particularly the operating temperature, of the fuel cell. Consequently, fuel cells are classified based on the types of electrolyte used as described below.

1. Polymer Electrolyte Membrane (PEM)
2. Alkaline Fuel Cell (AFC)
3. Phosphoric Acid Fuel Cell (PAFC)
4. Molten Carbonate Fuel Cell (MCFC)
5. Solid Oxide Fuel Cell (SOFC)

These fuel cells operate at different temperatures and each is best suited to particular applications. The main features of the five types of fuel cells are summarized in Table 1.3.

Fuel Cell Operation

Basic operational characteristics of the four most common types of fuel cells are discussed in the following.

Polymer Electrolyte Membrane (PEM)

The PEM cell is one in a family of fuel cells that are in various stages of development. It is being considered as an alternative power source for automotive application for electric vehicles. The electrolyte in a PEM cell is a type of polymer and is usually referred to as a membrane, hence the name. Polymer electrolyte membranes are somewhat unusual electrolytes in that, in the presence of water, which the membrane readily absorbs, the negative ions are rigidly held within their structure. Only the positive (H) ions contained within the membrane are mobile and are free to carry positive charges through the membrane in one direction only, from anode to cathode. At the same time, the organic nature of the polymer electrolyte membrane structure makes it an electron insulator, forcing it to travel through the outside circuit providing electric power to the load. Each of the two electrodes consists of porous carbon to which very small platinum (Pt) particles are bonded. The electrodes are somewhat porous so that the gases can diffuse through them to reach the catalyst. Moreover, as both platinum and carbon conduct

TABLE 1.3 Comparison of Five Fuel Cell Technologies

Type	Electrolyte	Operating Temperature (°C)	Applications	Advantages
Polymer Electrolyte Membrane (PEM)	Solid organic polymer poly-perflouro-sulfonic acid	60–100	Electric utility, transportation, portable power	Solid electrolyte reduces corrosion, low temperature, quick start-up
Alkaline (AFC)	Aqueous solution of potassium hydroxide soaked in a matrix	90–100	Military, space	Cathode reaction faster in alkaline electrolyte; therefore high performance
Phosphoric Acid (PAFC)	Liquid phosphoric acid soaked in a matrix	175–200	Electric utility, transportation, and heat	Up to 85% efficiency in co-generation of electricity
Molten Carbonate (MCFC)	Liquid solution of lithium, sodium, and/or potassium carbonates soaked in a matrix	600–1000	Electric utility	Higher efficiency, fuel flexibility, inexpensive catalysts
Solid Oxide (SOFC)	Solid zirconium oxide to which a small amount of yttria is added	600–1000	Electric utility	Higher efficiency, fuel flexibility, inexpensive catalysts. Solid electrolyte advantages like PEM

electrons well, they are able to move freely through the electrodes. Chemical reactions that take place inside a PEM fuel cell are presented in the following.

Anode

$$2H_2 \rightarrow 4H^+ + 4e^-$$

Cathode

$$O_2 + 4H^+ + 4e^- \rightarrow 2H_2O$$

Net reaction: $2H_2 + O_2 = 2H_2O$

Hydrogen gas diffuses through the polymer electrolyte until it encounters a Pt particle in the anode. The Pt catalyzes the dissociation of the hydrogen molecule into two hydrogen atoms (H) bonded to two neighboring Pt atoms. Only then can each H atom release an electron to form a hydrogen ion (H$^+$) which travels to the cathode through the electrolyte. At the same time, the free electron travels from the anode to the cathode through the outer circuit. At the cathode the oxygen molecule interacts with the hydrogen ion and the electron from the outside circuit to form water. The performance of the PEM fuel cell is limited primarily by the slow rate of the oxygen reduction half-reaction at the cathode, which is 100 times slower than the hydrogen oxidation half-reaction at the anode.

Phosphoric Acid Fuel Cell (PAFC)

Phosphoric acid technology has moved from the laboratory research and development to the first stages of commercial application. Turnkey 200-kW plants are now available and have been installed at more than 70 sites in the U.S., Japan, and Europe. Operating at about 200°C, the PAFC plant also produces heat for domestic hot water and space heating, and its electrical efficiency approaches 40%. The principal obstacle against widespread commercial acceptance is cost. Capital costs of about $2500 to $4000/kW must be reduced to $1000 to $1500/kW if the technology is to be accepted in the electric power markets.

The chemical reactions occurring at two electrodes are written as follows:

At anode: $\quad 2H_2 \rightarrow 4H^+ + 4e^-$

At cathode: $\quad O_2 + 4H^+ + 4e^- \rightarrow 2H_2O$

Molten Carbonate Fuel Cell (MCFC)

Molten carbonate technology is attractive because it offers several potential advantages over PAFC. Carbon monoxide, which poisons the PAFC, is indirectly used as a fuel in the MCFC. The higher operating temperature of approximately 650°C makes the MCFC a better candidate for combined cycle applications whereby the fuel cell exhaust can be used as input to the intake of a gas turbine or the boiler of a steam turbine. The total thermal efficiency can approach 85%. This technology is at the stage of prototype commercial demonstrations and is estimated to enter the commercial market by 2003 using natural gas, and by 2010 with gas made from coal. Capital costs are expected to be lower than PAFC. MCFCs are now being tested in full-scale demonstration plants. The following equations illustrate the chemical reactions that take place inside the cell.

At anode: $\quad 2H_2 + 2CO_3^{2-} \rightarrow 2H_2O + 2CO_2 + 4e^-$

and $\quad 2CO + 2CO_3^{2-} \rightarrow 4CO_2 + 4e^-$

At cathode: $\quad O_2 + 2CO_2 + 4e^- \rightarrow 2O_3^{2-}$

Solid Oxide Fuel Cell (SOFC)

A solid oxide fuel cell is currently being demonstrated at a 100-kW plant. Solid oxide technology requires very significant changes in the structure of the cell. As the name implies, the SOFC uses a solid electrolyte, a ceramic material, so the electrolyte does not need to be replenished during the operational life of the cell. This simplifies design, operation, and maintenance, as well as having the potential to reduce costs. This offers the stability and reliability of all solid-state construction and allows higher temperature operation. The ceramic make-up of the cell lends itself to cost-effective fabrication techniques. The tolerance to impure fuel streams make SOFC systems especially attractive for utilizing H_2 and CO from natural gas steam-reforming and coal gasification plants. The chemical reactions inside the cell may be written as follows:

At anode: $\quad 2H_2 + 2O^{2-} + 2H_2O + 4e^-$

and $\quad 2CO + 2O^{2-} \rightarrow 2CO_2 + 4e^-$

At cathode: $\quad O_2 + 4e^- \rightarrow 2O^{2-}$

Summary

Fuel cells can convert a remarkably high proportion of the chemical energy in a fuel to electricity. With the efficiencies approaching 60%, even without co-generation, fuel cell power plants are nearly twice as efficient as conventional power plants. Unlike large steam plants, the efficiency is not a function of the plant size for fuel cell power plants. Small-scale fuel cell plants are just as efficient as the large ones, whether they operate at full load or not. Fuel cells contribute significantly to the cleaner environment; they produce dramtically fewer emissions, and their by-products are primarily hot water and carbon dioxide in small amounts. Because of their modular nature, fuel cells can be placed at or near load centers, resulting in savings of transmission network expansion.

1.3 Photovoltaics

Roger A. Messenger

Types of PV Cells

Silicon Cells

Silicon PV cells come in several varieties. The most common cell is the single-crystal silicon cell. Other variations include multicrystalline (polycrystalline), thin silicon (buried contact) cells, and amorphous silicon cells.

Single-Crystal Silicon Cells
While single crystal silicon cells are still the most common cells, the fabrication process of these cells is relatively energy intensive, resulting in limits to cost reduction for these cells. Since single-crystal silicon is an indirect bandgap semiconductor ($E_g = 1.1$ eV), its absorption constant is smaller than that of direct bandgap materials. This means that single-crystal silicon cells need to be thicker than other cells in order to absorb a sufficient percentage of incident radiation. This results in the need for more material and correspondingly more energy involved in cell processing, especially since the cells are still produced mostly by sawing of single-crystal silicon ingots into wafers that are about 200 µm thick. To achieve maximum fill of the module, round ingots are first sawed to achieve closer to a square cross-section prior to wafering.

After chemical etching to repair surface damage from sawing, the junction is diffused into the wafers. Improved cell efficiency can then be achieved by using a preferential etch on the cell surfaces to produce textured surfaces. The textured surfaces reflect photons back toward the junction at an angle, thus increasing the path length and increasing the probability of the photon being absorbed within a minority carrier diffusion length of the junction. Following the chemical etch, contacts, usually aluminum, are evaporated and annealed and the front surface is covered with an antireflective coating.

The cells are then assembled into modules, consisting of approximately 33 to 36 individual cells connected in series. Since the open-circuit output voltage of an individual silicon cell typically ranges from 0.5 to 0.6 V, depending upon irradiance level and cell temperature, this results in a module open-circuit voltage between 18 and 21.6 V. The cell current is directly proportional to the irradiance and the cell area. A 4-ft^2 (0.372-m^2) module (active cell area) under full sun will typically produce a maximum power close to 55 W at approximately 17 V and 3.2 A.

Multicrystalline Silicon Cells
By pouring molten silicon into a crucible and controlling the cooling rate, it is possible to grow multi-crystalline silicon with a rectangular cross-section. This eliminates the "squaring-up" process and the associated loss of material. The ingot must still be sawed into wafers, but the resulting wafers completely fill the module. The remaining processing follows the steps of single-crystal silicon, and cell efficiencies in excess of 15% have been achieved for relatively large area cells. Multicrystalline material still maintains the basic properties of single-crystal silicon, including the indirect bandgap. Hence, relatively thick cells with textured surfaces have the highest conversion efficiencies. Multicrystalline silicon modules are commercially available and are recognized by their "speckled" surface appearance.

Thin Silicon (Buried Contact) Cells
The current flow direction in most PV cells is between the front surface and the back surface. In the thin silicon cell, a dielectric layer is deposited on an insulating substrate, followed by alternating layers of n-type and p-type silicon, forming multiple pn junctions. Channels are then cut with lasers and contacts are buried in the channels, so the current flow is parallel to the cell surfaces in multiple parallel conduction paths. These cells minimize resistance from junction to contact with the multiple parallel conduction paths and minimize blocking of incident radiation by the front contact. Although the material is not single crystal, grain boundaries cause minimal degradation of cell efficiency. The collection efficiency is

very high, since essentially all photon-generated carriers are generated within a diffusion length of a pn junction. This technology is relatively new, but has already been licensed to a number of firms worldwide (Green and Wenham, 1994).

Amorphous Silicon Cells

Amorphous silicon has no predictable crystal structure. As a result, the uniform covalent bond structure of single-crystal silicon is replaced with a random bonding pattern with many open covalent bonds. These bonds significantly degrade the performance of amorphous silicon by reducing carrier mobilities and the corresponding diffusion lengths. However, if hydrogen is introduced into the material, its electron will pair up with the dangling bonds of the silicon, thus passivating the material. The result is a direct bandgap material with a relatively high absorption constant. A film with a thickness of a few micrometers will absorb nearly all incident photons with energies higher than the 1.75 eV bandgap energy.

Maximum collection efficiency for a-Si:H is achieved by fabricating the cell with a pin junction. Early work on the cells revealed, however, that if the intrinsic region is too thick, cell performance will degrade over time. This problem has now been overcome by the manufacture of multi-layer cells with thinner pin junctions. In fact, it is possible to further increase cell efficiency by stacking cells of a-SiC:H on top, a-Si:H in the center, and a-SiGe:H on the bottom. Each successive layer from the top has a smaller bandgap, so the high-energy photons can be captured soon after entering the material, followed by middle-energy photons and then lower energy photons.

While the theoretical maximum efficiency of a-Si:H is 27% (Zweibel, 1990), small-area lab cells have been fabricated with efficiencies of 14% and large-scale devices have efficiencies in the 10% range (Yang et al., 1997).

Amorphous silicon cells have been adapted to the building integrated PV (BIPV) market by fabricating the cells on stainless steel (Guha et al., 1997) and polymide substrates (Huang et al., 1997). The "solar shingle" is now commercially available, and amorphous silicon cells are commonly used in solar calculators and solar watches.

Gallium Arsenide Cells

Gallium arsenide (GaAs), with its 1.43 eV direct bandgap, is a nearly optimal PV cell material. The only problem is that it is very costly to fabricate cells. GaAs cells have been fabricated with conversion efficiencies above 30% and with their relative insensitivity to severe temperature cycling and radiation exposure, they are the preferred material for extraterrestrial applications, where performance and weight are the dominating factors.

Gallium and arsenic react exothermically when combined, so formation of the host material is more complicated than formation of pure, single-crystal silicon. Modern GaAs cells are generally fabricated by growth of a GaAs film on a suitable substrate, such as Ge. A typical GaAs cell has a Ge substrate with a layer of n-GaAs followed by a layer of p-GaAs and then a thin layer of p-GaAlAs between the p-GaAs and the top contacts. The p-GaAlAs has a wider bandgap (1.8 eV) than the GaAs, so the higher energy photons are not absorbed at the surface, but are transmitted through to the GaAs pn junction, where they are then absorbed.

Recent advances in III-V technology have produced tandem cells similar to the a-Si:H tandem cell. One cell consists of two tandem GaAs cells, separated by thin tunnel junctions of GaInP, followed by a third tandem GaInP cell, separated by AlInP tunnel junctions (Lammasniemi et al., 1997). The tunnel junctions mitigate voltage drop of the otherwise forward-biased pn junction that would appear between any two tandem pn junctions in opposition to the photon-induced cell voltage. Cells have also been fabricated of InP (Hoffman et al., 1997).

Copper Indium (Gallium) Diselenide Cells

Another promising thin film material is copper indium (gallium) diselenide (CIGS). While the basic copper indium diselenide cell has a bandgap of 1.0 eV, the addition of gallium increases the bandgap to closer to 1.4 eV, resulting in more efficient collection of photons near the peak of the solar spectrum. CIGS has a high absorption constant and essentially all incident photons are absorbed within a distance of 2 µm, as in a-Si:H. Indium is the most difficult component to obtain, but the quantity needed for a module is relatively minimal.

The CIGS cell is fabricated on a soda glass substrate by first applying a thin layer of molybdenum as the back contact, since the CIGS will form an ohmic contact with Mo. The next layer is p-type CIGS, followed by a layer of n-type CdS, rather than n-type CIGS, because the pn homojunction in CIGS is neither stable nor efficient. While the cells discussed thus far have required metals to obtain ohmic front contacts, it is possible to obtain an ohmic contact on CdS with a transparent conducting oxide (TCO) such as ZnO. The top surface is first passivated with a thin layer (50 nm) of intrinsic ZnO to prevent minority carrier surface recombination. Then a thicker layer (350 nm) of n^+ ZnO is added, followed by an MgF_2 antireflective coating.

Efficiencies of laboratory cells are now near 18% (Tuttle et al., 1996), with a module efficiency of 11.1% reported in 1998 (Tarrant and Gay, 1998). Although at the time of this writing, CIGS modules were not commercially available, the technology has been under field tests for nearly 10 years. It has been projected that the cells may be manufactured on a large scale for $1/W or less. At this cost level, area-related costs become significant, so that it becomes important to increase cell efficiency to maximize power output for a given cell area.

Cadmium Telluride Cells

Of the II-VI semiconductor materials, CdTe has a theoretical maximum efficiency of near 25%. The material has a favorable direct bandgap (1.44 eV) and a large absorption constant. As in the other thin film materials, a 2-µm thickness is adequate for the absorption of most of the incident photons. Small laboratory cells have been fabricated with efficiencies near 15% and module efficiencies close to 10% have been achieved (Ullal et al., 1997). Some concern has been expressed about the Cd content of the cells, particularly in the event of fire dispersing the Cd. It has been determined that anyone endangered by Cd in a fire would be far more endangered by the fire itself, due to the small quantity of Cd in the cells. Decommissioning of the module has also been analyzed and it has been concluded that the cost to recycle module components is pennies per watt (Fthenakis and Moskowitz, 1997).

The CdTe cell is fabricated on a glass superstrate covered with a thin TCO (1 µm). The next layer is n-type CdS with a thickness of approximately 100 nm, followed by a 2-µm thick CdTe layer and a back contact of an appropriate metal for ohmic contact, such as Au, Cu/Au, Ni, Ni/Al, ZnTe:Cu or (Cu, HgTe). The back contact is then covered with a layer of ethylene vinyl acetate (EVA) or other suitable encapsulant and another layer of glass. The front glass is coated with an antireflective coating.

Experimental CdTe arrays up to 25 kW have been under test for several years with no reports of degradation. It has been estimated that the cost for large-scale production can be reduced to below $1/W. Once again, as in the CIGS case, module efficiency needs to be increased to reduce the area-related costs.

Emerging Technologies

The PV field is moving so quickly that by the time information appears in print, it is generally outdated. Reliability of cells, modules, and system components continues to improve. Efficiencies of cells and modules continue to increase, and new materials and cell fabrication techniques continue to evolve.

One might think that Si cells will soon become historical artifacts. This may not be the case. Efforts are underway to produce Si cells that have good charge carrier transport properties while improving photon absorption and reducing the energy for cell production. Ceramic and graphite substrates have been used with thinner layers of Si. Processing steps have been doubled up. Metal insulator semiconductor inversion layer (MIS-IL) cells have been produced in which the diffused junction is replaced with a Schottky junction. By use of clever geometry of the back electrode to reduce the rear surface recombination velocity along with front surface passivation, an efficiency of 18.5% has been achieved for a laboratory MIS-IL cell. Research continues on ribbon growth in an effort to eliminate wafering, and combining crystalline and amorphous Si in a tandem cell to take advantage of the two different bandgaps for increasing photon collection efficiency has been investigated.

Electric Power Generation: Non-Conventional Methods

At least eight different CIS-based materials have been proposed for cells. The materials have direct bandgaps ranging from 1.05 to 2.56 eV. A number of III-V materials have also emerged that have favorable photon absorption properties. In addition, quantum well cells have been proposed that have theoretical efficiencies in excess of 40% under concentrating conditions.

The PV market seems to have taken a strong foothold, with the likelihood that annual PV module shipments will exceed 200 MW before the end of the century and continue to increase by approximately 15% annually as new markets open as cost continues to decline and reliability continues to improve.

PV Applications

PV cells were first used to power satellites. Through the middle of the 1990s the most common terrestrial PV applications were stand-alone systems located where connection to the utility grid was impractical. By the end of the 1990s, PV electrical generation was cost-competitive with the marginal cost of central station power when it replaced gas turbine peaking in areas with high afternoon irradiance levels. Encouraged by consumer approval, a number of utilities have introduced utility-interactive PV systems to supply a portion of their total customer demand. Some of these systems have been residential and commercial rooftop systems and other systems have been larger ground-mounted systems. PV systems are generally classified as utility interactive (grid connected) or stand-alone.

Orientation of the PV modules for optimal energy collection is an important design consideration, whether for a utility interactive system or for a stand-alone system. Best overall energy collection on an annual basis is generally obtained with a south-facing collector having a tilt at an angle with the horizontal approximately 90% of the latitude of the site. For optimal winter performance, a tilt of latitude +15° is best and for optimal summer performance a tilt of latitude −15° is best. In some cases, when it is desired to have the PV output track utility peaking requirements, a west-facing array may be preferred, since its maximum output will occur during summer afternoon utility peaking hours. Monthly peak sun tables for many geographical locations are available from the National Renewable Energy Laboratory (Sandia National Laboratories, 1996; Florida Solar Energy Center).

Utility-Interactive PV systems

Utility-interactive PV systems are classified by IEEE Standard 929 as small, medium, or large (ANSI/IEEE, 1999). Small systems are less than 10 kW, medium systems range from 10 to 500 kW, and large systems are larger than 500 kW. Each size range requires different consideration for the utility interconnect. In addition to being able to offset utility peak power, the distributed nature of PV systems also results in the reduction of load on transmission and distribution lines. Normally, utility-interactive systems do not incorporate any form of energy storage — they simply supply power to the grid when they are operating. In some instances, however, where grid power may not be as reliable as the user may desire, battery backup is incorporated to ensure uninterrupted power.

Since the output of PV modules is DC, it is necessary to convert the module output to AC before connecting it to the grid. This is done with an inverter, also known as a power conditioning unit (PCU). Modern PCUs must meet the standards set by IEEE 929. If the PCU is connected on the customer side of the revenue meter, the PV system must meet the requirements of the *National Electrical Code® (NEC®)* (National Fire Protection Association, 1998). For a system to meet *NEC* requirements, it must consist of UL listed components. In particular, the PCU must be tested under UL 1741 (Underwriters Laboratories, 1997). But UL 1741 has been set up to test for compliance with IEEE 929, so any PCU that passes the UL 1741 test is automatically qualified under the requirements of the *NEC*.

Utility-interactive PCUs are generally pulse code modulated (PCM) units with nearly all *NEC*-required components, such as fusing of PV output circuits, DC and AC disconnects, and automatic utility disconnect in the event of loss of utility voltage. They also often contain surge protectors on input and output, ground fault protection circuitry, and maximum power tracking circuitry to ensure that the PV array is loaded at its maximum power point. The PCUs act as current sources, synchronized by the utility voltage. Since the PCUs are electronic, they can sample the line voltage at a high rate and readily shut down under conditions of utility voltage or frequency as specified by IEEE 929.

The typical small utility-interactive system of a few kilowatts consists of an array of modules selected by either a total cost criterion or, perhaps, by an available roof area criterion. The modules are connected to produce an output voltage ranging from 48 V to 300 V, depending upon the DC input requirements of the PCU. One or two PCUs are used to interface the PV output to the utility at 120 V or, perhaps, 120/240 V. The point of utility connection is typically the load side of a circuit breaker in the distribution panel of the occupancy if the PV system is connected on the customer side of the revenue meter. Connections on the utility side of the meter will normally be with double lugs on the line side of the meter. Section 690 of the *NEC* provides the connection and installation requirements for systems connected on the customer side of the revenue meter. Utility-side interconnects are regulated by the local utility.

Since the cost of PCUs is essentially proportional to their power handling capability, to date there has been no particular economy of scale for PV system size. As a result, systems are often modular. One form of modularity is the AC module. The AC module incorporates a small PCU ($\approx$300 W) mounted on the module itself so the output of the module is 120 V AC. This simplifies the hook-up of the PV system, since *NEC* requirements for PV output circuits are avoided and only the requirements for PCU output circuits need to be met.

Medium- and large-scale utility-interactive systems differ from small-scale systems only in the possibility that the utility may require different interfacing conditions relating to power quality and/or conditions for disconnect. Since medium- and large-scale systems require more area than is typically available on the rooftop of a residential occupancy, they are more typically found either on commercial or industrial rooftops or, in the case of large systems, are typically ground-mounted. Rooftop mounts are attractive since they require no additional space other than what is already available on the rooftop. The disadvantage is when roof repair is needed, the PV system may need to be temporarily removed and then reinstalled. Canopies for parking lots present attractive possibilities for large utility-interactive PV systems.

Stand-Alone PV Systems

Stand-alone PV systems are used when it is impractical to connect to the utility grid. Common stand-alone systems include PV-powered fans, water pumping systems, portable highway signs, and power systems for remote installations, such as cabins, communications repeater stations, and marker buoys. The design criteria for stand-alone systems is generally more complex than the design criteria for utility-interactive systems, where most of the critical system components are incorporated in the PCU. The PV modules must supply all the energy required unless another form of backup power, such as a gasoline generator, is also incorporated into the system. Stand-alone systems also often incorporate battery storage to run the system under low sun or no sun conditions.

PV-Powered Fans

Perhaps the simplest of all PV systems is the connection of the output of a PV module directly to a DC fan. When the module output is adequate, the fan operates. When the sun goes down, the fan stops. Such an installation is reasonable for use in remote bathrooms or other locations where it is desirable to have air circulation while the sun is shining, but not necessarily when the sun goes down. The advantage of such a system is its simplicity. The disadvantage is that it does not run when the sun is down, and under low sun conditions, the system operates very inefficiently due to a mismatch between the fan I-V characteristic and the module I-V characteristic that results in operation far from the module maximum power point.

If the fan is to run continuously, or beyond normal sunlight hours, then battery storage will be needed. The PV array must then be sized to provide the daily ampere-hour (Ah) load of the fan, plus any system losses. A battery system must be selected to store sufficient energy to last for several days of low sun, depending upon whether the need for the fan is critical, and an electronic controller is normally provided to prevent overcharge or overdischarge of the batteries.

PV-Powered Water Pumping System

If the water reservoir is adequate to provide a supply of water at the desired rate of pumping, then a water pumping system may not require battery storage. Instead, the water pumped can be stored in a storage tank for availability during low sun times. If this is the case, then the PV array needs to be sized to meet the power requirements of the water pump plus any system losses. If the reservoir provides water at a limited rate, the pumping rate may be limited by the reservoir replenishment rate, and battery storage may be required to extend the pumping time.

While it is possible to connect the PV array output directly to the pump, it is generally better to employ the use of an electronic maximum power tracker (MPT) to better match the pump to the PV array output. The MPT is a DC–DC converter that either increases or decreases pump voltage as needed to maximize pump power. This generally results in pumping approximately 20% more water in a day. Alternatively, it allows for the use of a smaller pump with a smaller array to pump the same amount of water, since the system is being used more efficiently.

PV-Powered Highway Information Sign

The PV-powered highway information sign is now a familiar sight to most motorists. The simpler signs simply employ bidirectional arrows to direct traffic to change lanes. The more complex signs display a message. The array size for a PV-powered highway information sign is limited by how it can be mounted without becoming a target for vandalism. Generally this means the modules must be mounted on the top of the sign itself to get them sufficiently above grade level to reduce temptation. This limits the array dimensions to the width of the trailer (about 8 ft) and the length of the modules (about 4 ft). At full sun, such a 32-ft^2 array, if 15% efficient, can produce approximately 450 W. Depending on location and time of year, about 5 h of full sun is typically available on an average day. This means the production of approximately 2250 Wh of energy on the typical day. Taking into account system losses in the batteries, the control circuitry, and degraded module performance due to dirty surfaces, about 70 to 75% of this energy can be delivered to the display, or about 1600 Wh/d. Hence, the average power available to the display over a 24-h period is 67 W. While this may not seem to be very much power, it is adequate for efficient display technology to deliver a respectable message.

If the system is a 12 V DC system, a set of deep discharge batteries will need to have a capacity of 185 Ah for each day of battery back-up (day of autonomy). For 3 d of autonomy, a total of 555 Ah of storage will be needed, which equates to eight batteries rated at 70 Ah each.

Hybrid PV-Powered Single Family Dwelling

In areas where winter sunlight is significantly less than summer sunlight, and/or where winter electrical loads are higher than summer electrical loads, if sufficient PV is deployed to meet winter needs, then the system produces excess power for many months of the year. If this power is not used, then the additional capacity of the system is wasted. Thus, for such cases, it often makes sense to size the PV system to completely meet the system needs during the month(s) with the most sunlight, and then provide backup generation of another type, such as a gasoline generator, to provide the difference in energy during the remaining months.

Such a system poses an interesting challenge for the system controller. It needs to be designed to make maximum use of PV power before starting the generator. Since generators operate most efficiently at about 90% of full load, the controller must provide for battery charging by the generator at the appropriate rate to maximize generator efficiency. Typically the generator will be sized to charge the batteries from 20 to 70% charge in about 5 h. When the batteries have reached 70% charge, the generator shuts down to allow available sunlight to complete the charging cycle. If the sunlight is not available, the batteries discharge to 20% and the cycle is repeated.

Figure 1.1 shows schematic diagrams of a few typical PV applications.

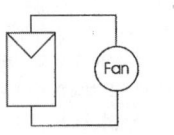
a. Simple PV-powered fan

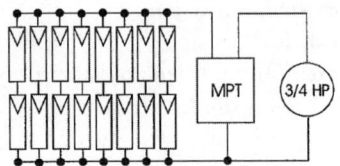
b. Water pump with maximum power tracking.

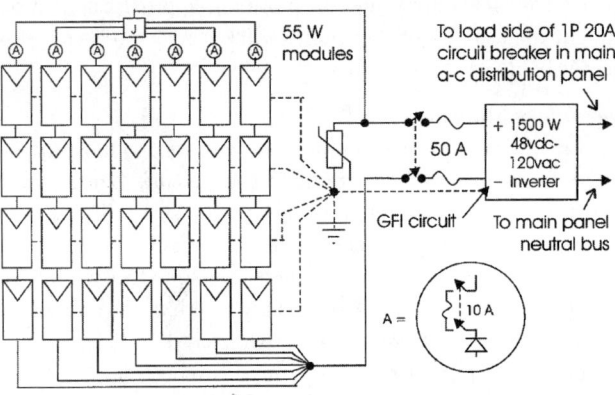
c. A 1.5 kW residential rooftop utility interactive system connected on customer side of revenue meter.

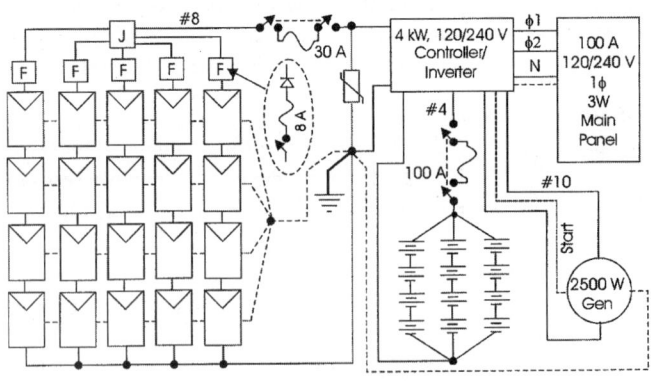
d. A hybrid residential installation.

FIGURE 1.1 Examples of PV systems.

References

ANSI/IEEE P929, IEEE Recommended Practice for Utility Interface of Residential and Intermediate Photovoltaic (PV) Systems, IEEE Standards Coordinating Committee 21, Photovoltaics, Draft 10, February 1999.

Florida Solar Energy Center site with extensive links to other sites, including solar radiation tables). http://alpha.fsec.ucf.edu/~pv/inforesource/links/

Fthenakis, V. M. and Moskowitz, P. D., Emerging photovoltaic technologies: environmental and health issues update, *NREL/SNL Photovoltaics Program Review*, AIP Press, New York, 1997.

Green, M. A. and Wenham, S. R., Novel parallel multijunction solar cell, *Appl. Phys. Lett.*, 65, 2907, 1994.

Guha, S., Yang, J., et al., *Proc. 26th IEEE PV Spec. Conf.*, 607-610, 1997.

Hoffman, R., et al., *Proc. 26th IEEE PV Spec. Conf.*, 815-818, 1997.

Huang, J., Lee, Y., et al., *Proc. 26th IEEE PV Spec. Conf.*, 699-702, 1997.

Lammasniemi, J., et al., *Proc. 26th IEEE PV Spec. Conf.*, 823-826, 1997.

Messenger, R., and Ventre, J., *Photovoltaic Systems Engineering*, CRC Press, Boca Raton, FL, 1999.

NFPA 70 National Electrical Code, 1999 Edition, National Fire Protection Association, Quincy, MA, 1998.

Stand-Alone Photovoltaic Systems: A Handbook of Recommended Design Practices, Sandia National Laboratories, Albuquerque, NM, 1996.

Tarrant, D. E. and Gay, R. R., *CIS-Based Thin Film PV Technology, Phase 2 Technical Report, October 1996 — October 1997*, NREL, Golden, CO, May 1998.

Tuttle, J.R., et al., *Proc. 14th NREL PV Program Review*, AIP Conf. Proceedings 394, Lakewood, CO, 1996, 83-105.

UL Subject 1741, Standard for Power Conditioning Units for Use in Residential Photovoltaic Power Systems, Underwriters Laboratories Inc., 1997.

Ullal, H. S., Zweibel, K., and von Roedern, B., *Proc. 26th IEEE PV Spec. Conf.*, 301-305, 1997.

Yang, J., Banerjee, A., et al., *Proc. 26th IEEE PV Spec Conf.*, 563-568, 1997.

Zweibel, K., *Harnessing Solar Power*, Plenum Press, New York, 1990.

2
Electric Power Generation: Conventional Methods

Rama Ramakumar
Oklahoma State University

2.1   **Hydroelectric Power Generation**   *Steven R. Brockschink, James H. Gurney, and Douglas B. Seely* 2-1

2.2   **Syncrhonous Machinery**   *Paul I. Nippes* 2-12

2.3   **Thermal Generating Plants**   *Kenneth H. Sebra* 2-20

2.4   **Distributed Utilities**   *John R. Kennedy* 2-27

2
Electric Power Generation: Conventional Methods

Steven R. Brockschink
Pacific Engineering Corporation

James H. Gurney
BC Hydro

Douglas B. Seely
Pacific Engineering Corporation

Paul I. Nippes
Magnetic Products and Services, Inc.

Kenneth H. Sebra
Baltimore Gas and Electric Company

John R. Kennedy
Georgia Power Company

2.1 Hydroelectric Power Generation 2-1
Planning of Hydroelectric Facilities • Hydroelectric Plant Features • Special Considerations Affecting Pumped Storage Plants • Commissioning of Hydroelectric Plants

2.2 Synchronous Machinery 2-12
General • Construction • Performance

2.3 Thermal Generating Plants 2-20
Plant Auxiliary System • Plant One-Line Diagram • Plant Equipment Voltage Ratings • Grounded vs. Ungrounded Systems • Miscellaneous Circuits • DC Systems • Power Plant Switchgear • Auxiliary Transformers • Motors • Main Generator • Cable • Electrical Analysis • Maintenance and Testing • Start-Up

2.4 Distributed Utilities 2-27
Available Technologies • Fuel Cells • Microturbines • Combustion Turbines • Storage Technologies • Interface Issues • Applications

2.1 Hydroelectric Power Generation

Steven R. Brockschink, James H. Gurney, and Douglas B. Seely

Hydroelectric power generation involves the storage of a hydraulic fluid, normally water, conversion of the hydraulic energy of the fluid into mechanical energy in a hydraulic turbine, and conversion of the mechanical energy to electrical energy in an electric generator.

The first hydroelectric power plants came into service in the 1880s and now comprise approximately 22% (660 GW) of the world's installed generation capacity of 3000 GW (Electric Power Research Institute, 1999). Hydroelectricity is an important source of renewable energy and provides significant flexibility in base loading, peaking, and energy storage applications. While initial capital costs are high, the inherent simplicity of hydroelectric plants, coupled with their low operating and maintenance costs, long service life, and high reliability, make them a very cost-effective and flexible source of electricity generation. Especially valuable is their operating characteristic of fast response for start-up, loading, unloading, and following of system load variations. Other useful features include their ability to start without the availability of power system voltage ("black start capability"), ability to transfer rapidly from generation mode to synchronous condenser mode, and pumped storage application.

Hydroelectric units have been installed in capacities ranging from a few kilowatts to nearly 1 GW. Multi-unit plant sizes range from a few kilowatts to a maximum of 18 GW.

Planning of Hydroelectric Facilities

Siting

Hydroelectric plants are located in geographic areas where they will make economic use of hydraulic energy sources. Hydraulic energy is available wherever there is a flow of liquid and head. Head represents potential energy and is the vertical distance through which the fluid falls in the energy conversion process. The majority of sites utilize the head developed by fresh water; however, other liquids such as salt water and treated sewage have been utilized. The siting of a prospective hydroelectric plant requires careful evaluation of technical, economic, environmental, and social factors. A significant portion of the project cost may be required for mitigation of environmental effects on fish and wildlife and re-location of infrastructure and population from flood plains.

Hydroelectric Plant Schemes

There are three main types of hydroelectric plant arrangements, classified according to the method of controlling the hydraulic flow at the site:

1. Run-of-the-river plants, having small amounts of water storage and thus little control of the flow through the plant.
2. Storage plants, having the ability to store water and thus control the flow through the plant on a daily or seasonal basis.
3. Pumped storage plants, in which the direction of rotation of the turbines is reversed during off-peak hours, pumping water from a lower reservoir to an upper reservoir, thus "storing energy" for later production of electricity during peak hours.

Selection of Plant Capacity, Energy, and Other Design Features

The generating capacity of a hydroelectric plant is a function of the head and flow rate of water discharged through the hydraulic turbines, as shown in Eq. (2.1).

$$P = 9.8\, \eta\, Q\, H \qquad (2.1)$$

where P = power (kilowatts)
η = plant efficiency
Q = discharge flow rate (meter^3/s)
H = head (meter)

Flow rate and head are influenced by reservoir inflow, storage characteristics, plant and equipment design features, and flow restrictions imposed by irrigation, minimum downstream releases, or flood control requirements. Historical daily, seasonal, maximum (flood), and minimum (drought) flow conditions are carefully studied in the planning stages of a new development. Plant capacity, energy, and physical features such as the dam and spillway structures are optimized through complex economic studies that consider the hydrological data, planned reservoir operation, performance characteristics of plant equipment, construction costs, the value of capacity and energy, and discount rates. The costs of substation, transmission, telecommunications, and remote control facilities are also important considerations in the economic analysis. If the plant has storage capability, then societal benefits from flood control may be included in the economic analysis.

Another important planning consideration is the selection of the number and size of generating units installed to achieve the desired plant capacity and energy, taking into account installed unit costs, unit availability, and efficiencies at various unit power outputs (American Society of Mechanical Engineers Hydro Power Technical Committee, 1996).

Hydroelectric Plant Features

Figures 2.1 and 2.2 illustrate the main components of a hydroelectric generating unit. The generating unit may have its shaft oriented in a vertical, horizontal, or inclined direction depending on the physical

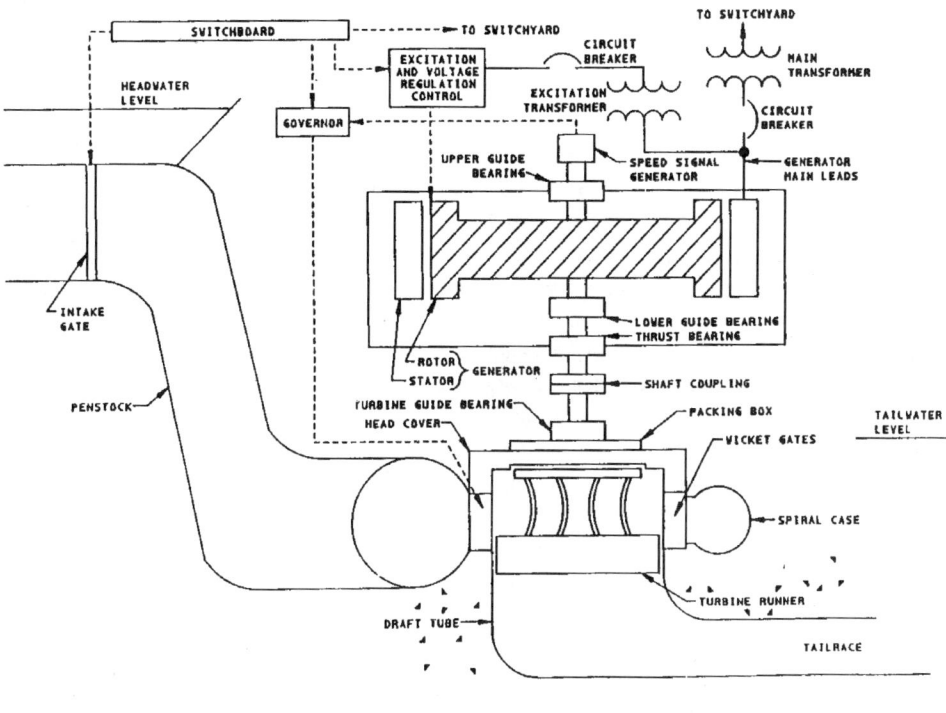

FIGURE 2.1 Vertical Francis unit arrangement. (*Source:* IEEE Standard 1020-1988 (Reaff 1994), *IEEE Guide for Control of Small Hydroelectric Power Plants,* 12. Copyright 1988 IEEE. All rights reserved.)

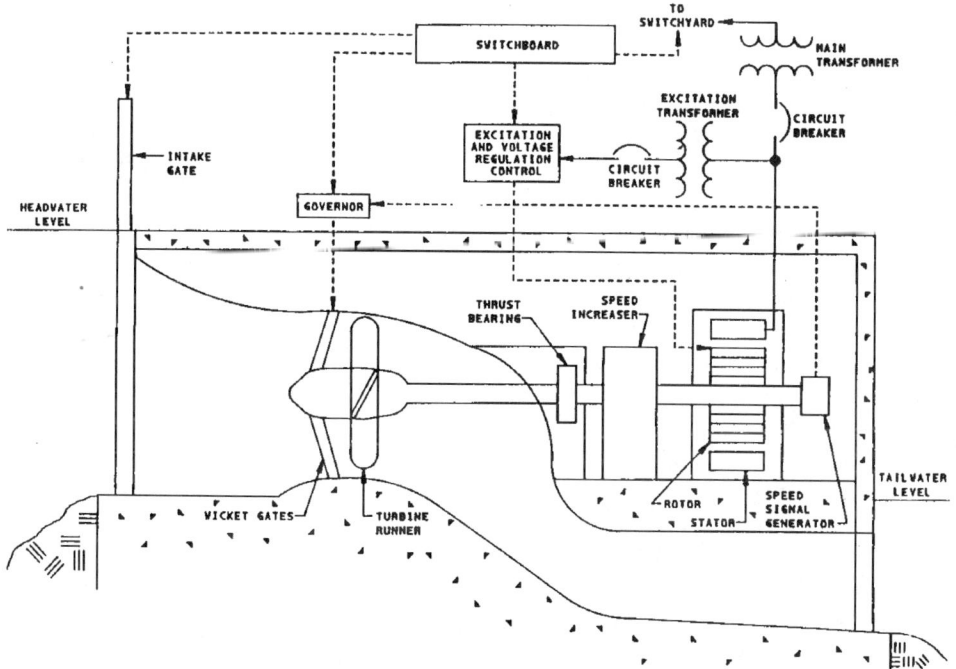

FIGURE 2.2 Horizontal axial-flow unit arrangement. (*Source:* IEEE Standard 1020-1988 (Reaff 1994), *IEEE Guide for Control of Small Hydroelectric Power Plants,* 13. Copyright 1988 IEEE. All rights reserved.)

conditions of the site and the type of turbine applied. Figure 2.1 shows a typical vertical shaft Francis turbine unit and Fig. 2.2 shows a horizontal shaft propeller turbine unit. The following sections will describe the main components such as the turbine, generator, switchgear, and generator transformer, as well as the governor, excitation system, and control systems.

Turbine

The type of turbine selected for a particular application is influenced by the head and flow rate. There are two classifications of hydraulic turbines: impulse and reaction.

The impulse turbine is used for high heads — approximately 300 m or greater. High-velocity jets of water strike spoon-shaped buckets on the runner which is at atmospheric pressure. Impulse turbines may be mounted horizontally or vertically and include perpendicular jets (known as a Pelton type), diagonal jets (known as a Turgo type) or cross-flow types.

In a reaction turbine, the water passes from a spiral casing through stationary radial guide vanes, through control gates and onto the runner blades at pressures above atmospheric. There are two categories of reaction turbine — Francis and propeller. In the Francis turbine, installed at heads up to approximately 360 m, the water impacts the runner blades tangentially and exits axially. The propeller turbine uses a propeller-type runner and is used at low heads — below approximately 45 m. The propeller runner may use fixed blades or variable pitch blades (known as a Kaplan or double regulated type) which allows control of the blade angle to maximize turbine efficiency at various hydraulic heads and generation levels. Francis and propeller turbines may also be arranged in slant, tubular, bulb, and rim generator configurations.

Water discharged from the turbine is directed into a draft tube where it exits to a tailrace channel, lower reservoir, or directly to the river.

Flow Control Equipment

The flow through the turbine is controlled by wicket gates on reaction turbines and by needle nozzles on impulse turbines. A turbine inlet valve or penstock intake gate is provided for isolation of the turbine during shutdown and maintenance.

Spillways and additional control valves and outlet tunnels are provided in the dam structure to pass flows that normally cannot be routed through the turbines.

Generator

Synchronous generators and induction generators are used to convert the mechanical energy output of the turbine to electrical energy. Induction generators are used in small hydroelectric applications (less than 5 MVA) due to their lower cost which results from elimination of the exciter, voltage regulator, and synchronizer associated with synchronous generators. The induction generator draws its excitation current from the electrical system and thus cannot be used in an isolated power system. Also, it cannot provide controllable reactive power or voltage control and thus its application is relatively limited.

The majority of hydroelectric installations utilize salient pole synchronous generators. Salient pole machines are used because the hydraulic turbine operates at low speeds, requiring a relatively large number of field poles to produce the rated frequency. A rotor with salient poles is mechanically better suited for low-speed operation, compared to round rotor machines which are applied in horizontal axis high-speed turbo-generators.

Generally, hydroelectric generators are rated on a continuous-duty basis to deliver net kVA output at a rated speed, frequency, voltage, and power factor and under specified service conditions including the temperature of the cooling medium (air or direct water). Industry standards specify the allowable temperature rise of generator components (above the coolant temperature) that are dependent on the voltage rating and class of insulation of the windings (ANSI, C50.12-1982; IEC, 60034-1). The generator capability curve, Fig. 2.3, describes the maximum real and reactive power output limits at rated voltage within which the generator rating will not be exceeded with respect to stator and rotor heating and other limits. Standards also provide guidance on short circuit capabilities and continuous and short-time current unbalance requirements (ANSI, C50.12-1982; IEEE, 492-1999).

Electric Power Generation: Conventional Methods

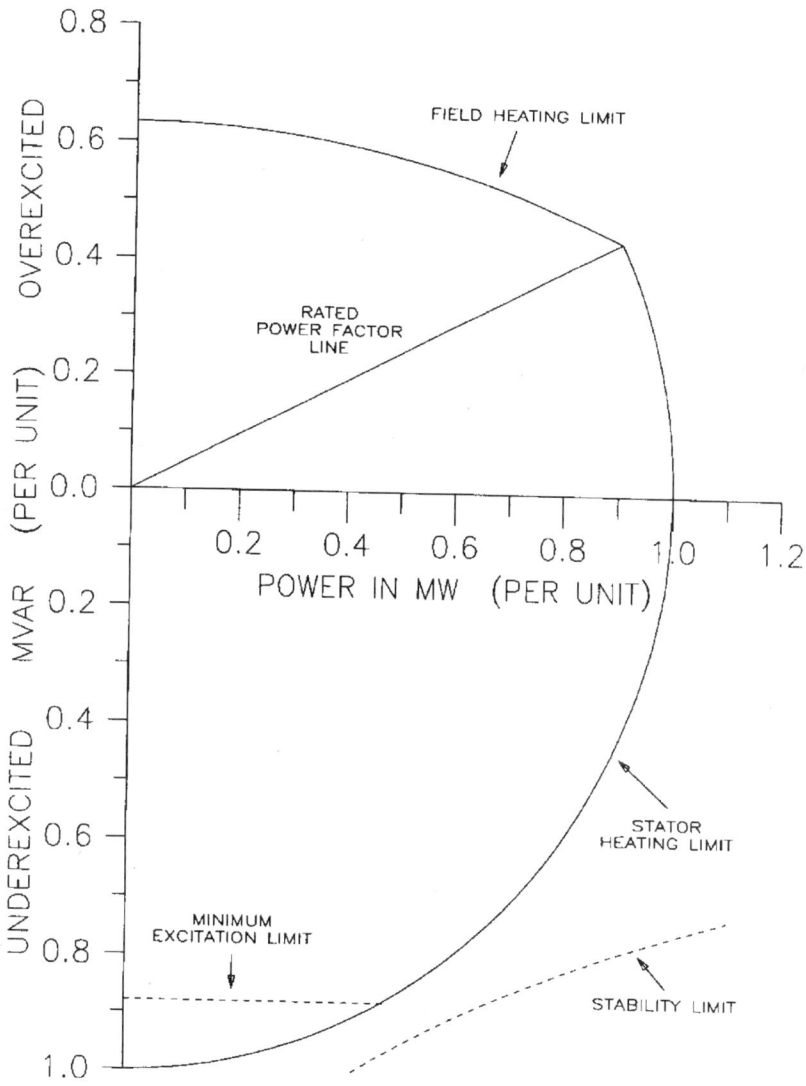

FIGURE 2.3 Typical hydro-generator capability curve (0.9 power factor, rated voltage). (*Source:* IEEE Standard 492-1999, *IEEE Guide for Operation and Maintenance of Hydro-Generators,* 16. Copyright 1999 IEEE All rights reserved.)

Synchronous generators require direct current field excitation to the rotor, provided by the excitation system described in Section entitled "Excitation System". The generator saturation curve, Fig. 2.4, describes the relationship of terminal voltage, stator current, and field current.

While the generator may be vertical or horizontal, the majority of new installations are vertical. The basic components of a vertical generator are the stator (frame, magnetic core, and windings), rotor (shaft, thrust block, spider, rim, and field poles with windings), thrust bearing, one or two guide bearings, upper and lower brackets for the support of bearings and other components, and sole plates which are bolted to the foundation. Other components may include a direct connected exciter, speed signal generator, rotor brakes, rotor jacks, and ventilation systems with surface air coolers (IEEE, 1095-1989).

The stator core is composed of stacked steel laminations attached to the stator frame. The stator winding may consist of single turn or multi-turn coils or half-turn bars, connected in series to form a three phase circuit. Double layer windings, consisting of two coils per slot, are most common. One or more circuits are connected in parallel to form a complete phase winding. The stator winding is normally

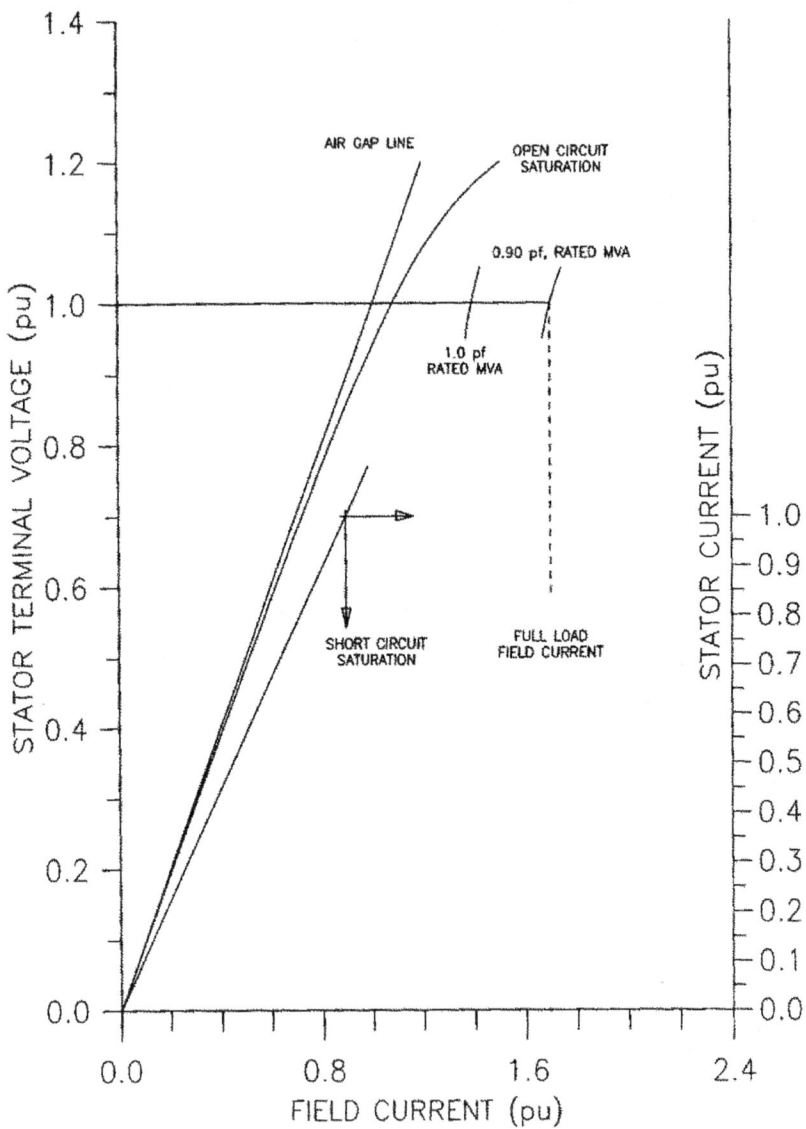

FIGURE 2.4 Typical hydro-generator saturation curves. (*Source:* IEEE Standard 492-1999, *IEEE Guide for Operation and Maintenance of Hydro-Generators,* 14. Copyright 1999 IEEE. All rights reserved.)

connected in wye configuration, with the neutral grounded through one of a number of alternative methods which depend on the amount of phase-to-ground fault current that is permitted to flow (IEEE, C62.92.2-1989; C37.101-1993). Generator output voltages range from approximately 480 VAC to 22 kVAC line-to-line, depending on the MVA rating of the unit. Temperature detectors are installed between coils in a number of stator slots.

The rotor is normally comprised of a spider attached to the shaft, a rim constructed of solid steel or laminated rings, and field poles attached to the rim. The rotor construction will vary significantly depending on the shaft and bearing system, unit speed, ventilation type, rotor dimensions, and characteristics of the driving hydraulic turbine. Damper windings or amortisseurs in the form of copper or brass rods are embedded in the pole faces, for damping rotor speed oscillations.

Electric Power Generation: Conventional Methods

The thrust bearing supports the mass of both the generator and turbine plus the hydraulic thrust imposed on the turbine runner and is located either above the rotor ("suspended unit") or below the rotor ("umbrella unit"). Thrust bearings are constructed of oil-lubricated, segmented, babbit-lined shoes. One or two oil lubricated generator guide bearings are used to restrain the radial movement of the shaft.

Fire protection systems are normally installed to detect combustion products in the generator enclosure, initiate rapid de-energization of the generator and release extinguishing material. Carbon dioxide and water are commonly used as the fire quenching medium.

Excessive unit vibrations may result from mechanical or magnetic unbalance. Vibration monitoring devices such as proximity probes to detect shaft run-out are provided to initiate alarms and unit shutdown.

The choice of generator inertia is an important consideration in the design of a hydroelectric plant. The speed rise of the turbine-generator unit under load rejection conditions, caused by the instantaneous disconnection of electrical load, is inversely proportional to the combined inertia of the generator and turbine. Turbine inertia is normally about 5% of the generator inertia. During design of the plant, unit inertia, effective wicket gate or nozzle closing and opening times, and penstock dimensions are optimized to control the pressure fluctuations in the penstock and speed variations of the turbine-generator during load rejection and load acceptance. Speed variations may be reduced by increasing the generator inertia at added cost. Inertia can be added by increasing the mass of the generator, adjusting the rotor diameter, or by adding a flywheel. The unit inertia also has a significant effect on the transient stability of the electrical system, as this factor influences the rate at which energy can be moved in or out of the generator to control the rotor angle acceleration during system fault conditions [see Chapter 11 — Power System Dynamics and Stability and (Kundur, 1994)].

Generator Terminal Equipment

The generator output is connected to terminal equipment via cable, busbar, or isolated phase bus. The terminal equipment comprises current transformers (CTs), voltage transformers (VTs), and surge suppression devices. The CTs and VTs are used for unit protection, metering and synchronizing, and for governor and excitation system functions. The surge protection devices, consisting of surge arresters and capacitors, protect the generator and low-voltage windings of the step-up transformer from lightning and switching-induced surges.

Generator Switchgear

The generator circuit breaker and associated isolating disconnect switches are used to connect and disconnect the generator to and from the power system. The generator circuit breaker may be located on either the low-voltage or high-voltage side of the generator step-up transformer. In some cases, the generator is connected to the system by means of circuit breakers located in the switchyard of the generating plant. The generator circuit breaker may be of the oil filled, air-magnetic, air blast, or compressed gas insulated type, depending on the specific application. The circuit breaker is closed as part of the generator synchronizing sequence and is opened (tripped) either by operator control, as part of the automatic unit stopping sequence, or by operation of protective relay devices in the event of unit fault conditions.

Generator Step-Up Transformer

The generator transformer steps up the generator terminal voltage to the voltage of the power system or plant switchyard. Generator transformers are generally specified and operated in accordance with international standards for power transformers, with the additional consideration that the transformer will be operated close to its maximum rating for the majority of its operating life. Various types of cooling systems are specified depending on the transformer rating and physical constraints of the specific application. In some applications, dual low-voltage windings are provided to connect two generating units to a single bank of step-up transformers. Also, transformer tertiary windings are sometimes provided to serve the AC station service requirements of the power plant.

Excitation System

The excitation system fulfills two main functions:

1. It produces DC voltage (and power) to force current to flow in the field windings of the generator. There is a direct relationship between the generator terminal voltage and the quantity of current flowing in the field windings as described in Fig. 2.4.
2. It provides a means for regulating the terminal voltage of the generator to match a desired set point and to provide damping for power system oscillations.

Prior to the 1960s, generators were generally provided with rotating exciters that fed the generator field through a slip ring arrangement, a rotating pilot exciter feeding the main exciter field, and a regulator controlling the pilot exciter output. Since the 1960s, the most common arrangement is thyristor bridge rectifiers fed from a transformer connected to the generator terminals, referred to as a "potential source controlled rectifier high initial response exciter" or "bus-fed static exciter" (IEEE, 421.1-1986; 421.2-1990; 421.4-1990; 421.5-1992). Another system used for smaller high-speed units is a brushless exciter with a rotating AC generator and rotating rectifiers.

Modern static exciters have the advantage of providing extremely fast response times and high field ceiling voltages for forcing rapid changes in the generator terminal voltage during system faults. This is necessary to overcome the inherent large time constant in the response between terminal voltage and field voltage (referred to as T'_{do}, typically in the range of 5 to 10 sec). Rapid terminal voltage forcing is necessary to maintain transient stability of the power system during and immediately after system faults. Power system stabilizers are also applied to static exciters to cause the generator terminal voltage to vary in phase with the speed deviations of the machine, for damping power system dynamic oscillations [see Chapter 11 — Power System Dynamics and Stability and (Kundur, 1994)].

Various auxiliary devices are applied to the static exciter to allow remote setting of the generator voltage and to limit the field current within rotor thermal and under excited limits. Field flashing equipment is provided to build up generator terminal voltage during starting to the point at which the thyristors can begin gating. Power for field flashing is provided either from the station battery or alternating current station service.

Governor System

The governor system is the key element of the unit speed and power control system (IEEE, 125-1988; IEC, 61362 [1998-03]; ASME, 29-1980). It consists of control and actuating equipment for regulating the flow of water through the turbine, for starting and stopping the unit, and for regulating the speed and power output of the turbine generator. The governor system includes set point and sensing equipment for speed, power and actuator position, compensation circuits, and hydraulic power actuators which convert governor control signals to mechanical movement of the wicket gates (Francis and Kaplan turbines), runner blades (Kaplan turbine), and nozzle jets (Pelton turbine). The hydraulic power actuator system includes high-pressure oil pumps, pressure tanks, oil sump, actuating valves, and servomotors.

Older governors are of the mechanical-hydraulic type, consisting of ballhead speed sensing, mechanical dashpot and compensation, gate limit, and speed droop adjustments. Modern governors are of the electro-hydraulic type where the majority of the sensing, compensation, and control functions are performed by electronic or microprocessor circuits. Compensation circuits utilize proportional plus integral (PI) or proportional plus integral plus derivative (PID) controllers to compensate for the phase lags in the penstock — turbine — generator — governor control loop. PID settings are normally adjusted to ensure that the hydroelectric unit remains stable when serving an isolated electrical load. These settings ensure that the unit contributes to the damping of system frequency disturbances when connected to an integrated power system. Various techniques are available for modeling and tuning the governor (Working Group, 1992; Wozniak, 1990).

A number of auxiliary devices are provided for remote setting of power, speed, and actuator limits and for electrical protection, control, alarming, and indication. Various solenoids are installed in the hydraulic actuators for controlling the manual and automatic start-up and shutdown of the turbine-generator unit.

TABLE 2.1 Summary of Control Hierarchy for Hydroelectric Plants

Control Category	Sub-Category	Remarks
Location	Local	Control is local at the controlled equipment or within sight of the equipment.
	Centralized	Control is remote from the controlled equipment, but within the plant.
	Off Site	Control location is remote from the project.
Mode	Manual	Each operation needs a separate and discrete initiation; could be applicable to any of the three locations.
	Automatic	Several operations are precipitated by a single initiation; could be applicable to any of the three locations.
Operation (supervision)	Attended	Operator is available at all times to initiate control action.
	Unattended	Operation staff is not normally available at the project site.

Source: IEEE Standard 1249-1996, *IEEE Guide for Computer-Based Control for Hydroelectric Power Plant Automation*, 6. Copyright 1997. All rights reserved.

Control Systems

Detailed information on the control of hydroelectric power plants is available in industry standards (IEEE, 1010-1987; 1020-1988; 1249-1996). A general hierarchy of control is illustrated in Table 2.1. Manual controls, normally installed adjacent to the device being controlled, are used during testing and maintenance, and as a backup to the automatic control systems. Figure 2.5 illustrates the relationship of control locations and typical functions available at each location. Details of the control functions available at each location are described in (IEEE, 1249-1996). Automatic sequences implemented for starting, synchronizing, and shutdown of hydroelectric units are detailed in (IEEE, 1010-1987).

Modern hydroelectric plants and plants undergoing rehabilitation and life extension are utilizing increasing levels of computer automation (IEEE, 1249-1996; 1147-1991). The relative simplicity of hydroelectric plant control allows most plants to be operated in an unattended mode from remote control centers.

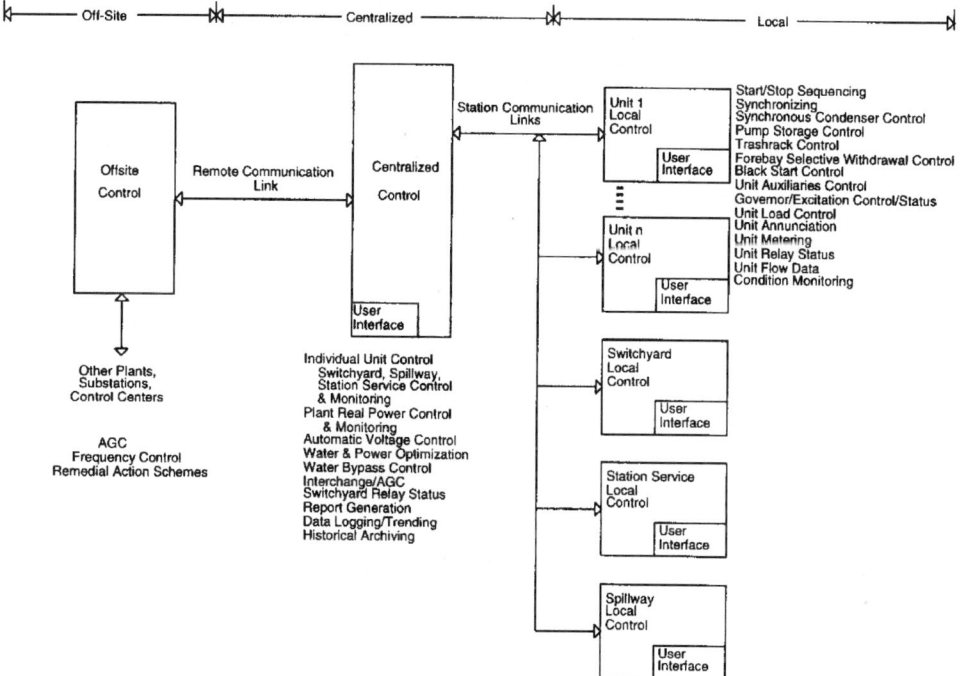

FIGURE 2.5 Relationship of local, centralized and off-site control. (*Source:* IEEE Standard 1249-1996, *IEEE Guide for Computer-Based Control for Hydroelectric Power Plant Automation*, 7. With permission.)

An emerging trend is the application of automated condition monitoring systems for hydroelectric plant equipment. Condition monitoring systems, coupled with expert system computer programs, allow plant owners and operators to more fully utilize the capacity of plant equipment and water resources, make better maintenance and replacement decisions, and maximize the value of installed assets.

Protection Systems

The turbine-generator unit and related equipment are protected against mechanical, electrical, hydraulic, and thermal damage that may occur as a result of abnormal conditions within the plant or on the power system to which the plant is connected. Abnormal conditions are detected automatically by means of protective relays and other devices and measures are taken to isolate the faulty equipment as quickly as possible while maintaining the maximum amount of equipment in service. Typical protective devices include electrical fault detecting relays, temperature, pressure, level, speed, and fire sensors, and vibration monitors associated with the turbine, generator, and related auxiliaries. The protective devices operate in various isolation and unit shutdown sequences, depending on the severity of the fault.

The type and extent of protection will vary depending on the size of the unit, manufacturer's recommendations, owner's practices, and industry standards.

Specific guidance on application of protection systems for hydroelectric plants is provided in (IEEE, 1010-1987; 1020-1988; C37.102-1995; C37.91-1985).

Plant Auxiliary Equipment

A number of auxiliary systems and related controls are provided throughout the hydroelectric plant to support the operation of the generating units (IEEE, 1010-1987; 1020-1988). These include:

1. Switchyard systems (see Chapter 5 — Substations).
2. Alternating current (AC) station service. Depending on the size and criticality of the plant, multiple sources are often supplied, with emergency backup provided by a diesel generator.
3. Direct current (DC) station service, normally provided by one or more battery banks, for supply of protection, control, emergency lighting, and exciter field flashing.
4. Lubrication systems, particularly for supply to generator and turbine bearings and bushings.
5. Drainage pumps, for removing leakage water from the plant.
6. Air compressors, for supply to the governors, generator brakes, and other systems.
7. Cooling water systems, for supply to the generator air coolers, generator and turbine bearings, and step-up transformer.
8. Fire detection and extinguishing systems.
9. Intake gate or isolation valve systems.
10. Draft tube gate systems.
11. Reservoir and tailrace water level monitoring.
12. Synchronous condenser equipment, for dewatering the draft tube to allow the runner to spin in air during synchronous condenser operation. In this case, the generator acts as a synchronous motor, supplying or absorbing reactive power.
13. Service water systems.
14. Overhead crane.
15. Heating, ventilation, and air conditioning.
16. Environmental systems.

Special Considerations Affecting Pumped Storage Plants

A pumped storage unit is one in which the turbine and generator are operated in the reverse direction to pump water from the lower reservoir to the upper reservoir. The generator becomes a motor, drawing its energy from the power system, and supplies mechanical power to the turbine which acts as a pump. The motor is started with the wicket gates closed and the draft tube water depressed with compressed air. The motor is accelerated in the pump direction and when at full speed and connected to the power system, the depression air is expelled, the pump is primed, and the wicket gates are opened to commence pumping action.

Pump Motor Starting

Various methods are utilized to accelerate the generator/motor in the pump direction during starting (IEEE, 1010-1987). These include:

1. Full voltage, across the line starting. Used primarily on smaller units, the unit breaker is closed and the unit is started as an induction generator. Excitation is applied near rated speed and machine reverts to synchronous motor operation.
2. Reduced voltage, across the line starting. A circuit breaker connects the unit to a starting bus tapped from the unit step-up transformer at one third to one half rated voltage. Excitation is applied near rated speed and the unit is connected to the system by means of the generator circuit breaker. Alternative methods include use of a series reactor during starting and energization of partial circuits on multiple circuit machines.
3. Pony motor starting. A variable speed wound-rotor motor attached to the AC station service and coupled to the motor/generator shaft is used to accelerate the machine to synchronous speed.
4. Synchronous starting. A smaller generator, isolated from the power system, is used to start the motor by connecting the two in parallel on a starting bus, applying excitation to both units, and opening the wicket gates on the smaller generator. When the units reach synchronous speed, the motor unit is disconnected from the starting bus and connected to the power system.
5. Semi-synchronous (reduced frequency, reduced voltage) starting. An isolated generator is accelerated to about 80% rated speed and paralleled with the motor unit by means of a starting bus. Excitation is applied to the generating unit and the motor unit starts as an induction motor. When the speed of the two units is approximately equal, excitation is applied to the motor unit, bringing it into synchronism with the generating unit. The generating unit is then used to accelerate both units to rated speed and the motor unit is connected to the power system.
6. Static starting. A static converter/inverter connected to the AC station service is used to provide variable frequency power to accelerate the motor unit. Excitation is applied to the motor unit at the beginning of the start sequence and the unit is connected to the power system when it reaches synchronous speed. The static starting system can be used for dynamic braking of the motor unit after disconnection from the power system, thus extending the life of the unit's mechanical brakes.

Phase Reversing of the Generator/Motor

It is necessary to reverse the direction of rotation of the generator/motor by interchanging any two of the three phases. This is achieved with multi-pole motor operated switches or with circuit breakers.

Draft Tube Water Depression

Water depression systems using compressed air are provided to lower the level of the draft tube water below the runner to minimize the power required to accelerate the motor unit during the transition to pumping mode. Water depression systems are also used during motoring operation of a conventional hydroelectric unit while in synchronous condenser mode. Synchronous condenser operation is used to provide voltage support for the power system and to provide spinning reserve for rapid loading response when required by the power system.

Commissioning of Hydroelectric Plants

The commissioning of a new hydroelectric plant, rehabilitation of an existing plant, or replacement of existing equipment requires a rigorous plan for inspection and testing of equipment and systems and for organizing, developing, and documenting the commissioning program (IEEE, 1248-1998).

References

American Society of Mechanical Engineers Hydro Power Technical Committee, *The Guide to Hydropower Mechanical Design*, HCI Publications, Kansas City, KA, 1996.

ANSI Standard C50.12-1982 (Reaff 1989), *Synchronous Generators and Generator/Motors for Hydraulic Turbine Applications*.

ASME PTC 29-1980 (R1985), *Speed Governing Systems for Hydraulic Turbine Generator Units*.

Electricity Technology Roadmap — 1999 Summary and Synthesis, Report C1-112677-V1, Electric Power Research Institute, Palo Alto, July, 1999, pp. 74, 83.

IEC Standard 60034-1 (1996-12), *Rotating Electrical Machines — Part 1: Rating and Performance*.

IEC Standard 61362 (1998-03), *Guide to Specification of Hydraulic Turbine Control Systems*.

IEEE Standard C37.91-1985 (Reaff 1990), *IEEE Guide for Protective Relay Applications to Power Transformers*.

IEEE Standard 421.1-1986 (Reaff 1996), *IEEE Standard Definitions for Excitation Systems for Synchronous Machines*.

IEEE Standard 1010-1987 (Reaff 1992), *IEEE Guide for Control of Hydroelectric Power Plants*.

IEEE Standard 125-1988 (Reaff 1996), *IEEE Recommended Practice for Preparation of Equipment Specifications for Speed-Governing of Hydraulic Turbines Intended to Drive Electric Generators*.

IEEE Standard 1020-1988 (Reaff 1994), *IEEE Guide for Control of Small Hydroelectric Power Plants*.

IEEE Standard C62.92.2-1989 (Reaff 1993), *IEEE Guide for the Application of Neutral Grounding in Electrical Utility Systems, Part II — Grounding of Synchronous Generator Systems*.

IEEE Standard 1095-1989 (Reaff 1994), *IEEE Guide for Installation of Vertical Generators and Generator/Motors for Hydroelectric Applications*.

IEEE Standard 421.2-1990, *IEEE Guide for Identification, Testing and Evaluation of the Dynamic Performance of Excitation Control Systems*.

IEEE Standard 421.4-1990, *IEEE Guide for the Preparation of Excitation System Specifications*.

IEEE Standard 1147-1991 (Reaff 1996), *IEEE Guide for the Rehabilitation of Hydroelectric Power Plants*.

IEEE Standard 421.5-1992, *IEEE Recommended Practice for Excitation Systems for Power Stability Studies*.

IEEE Standard C37.101-1993, *IEEE Guide for Generator Ground Protection*.

IEEE Standard C37.102-1995, *IEEE Guide for AC Generator Protection*.

IEEE Standard 1249-1996, *IEEE Guide for Computer-Based Control for Hydroelectric Power Plant Automation*.

IEEE Standard 1248-1998, *IEEE Guide for the Commissioning of Electrical Systems in Hydroelectric Power Plants*.

IEEE Standard 492-1999, *IEEE Guide for Operation and Maintenance of Hydro-Generators*.

Kundur, P., *Power System Stability and Control*, McGraw-Hill, New York, 1994.

Working Group on Prime Mover and Energy Supply Models for System Dynamic Performance Studies, Hydraulic turbine and turbine control models for system dynamic studies, *IEEE Trans. Power Syst.*, 7(1), February 1992.

Wozniak, L., Graphical Approach to Hydrogenerator Governor Tuning, *IEEE Trans. Energy Conv.*, 5(3), September 1990.

2.2 Synchronous Machinery

Paul I. Nippes

General

Synchronous motors convert electrical power to mechanical power; synchronous generators convert mechanical power to electrical power; and synchronous condensers supply only reactive power to stabilize system voltages.

Electric Power Generation: Conventional Methods

Synchronous motors, generators, and condensers perform similarly, except for a heavy cage winding on the rotor of motors and condensers for self-starting.

A rotor has physical magnetic poles, arranged to have alternating north and south poles around the rotor diameter which are excited by electric current, or uses permanent magnets, having the same number of poles as the stator electromagnetic poles.

The rotor RPM = 120 × Electrical System Frequency/Poles.

The stator winding, fed from external AC multi-phase electrical power, creates rotating electromagnetic poles.

At speed, rotor poles turn in synchronism with the stator rotating electromagnetic poles, torque being transmitted magnetically across the "air gap" power angle, lagging in generators and leading in motors.

Synchronous machine sizes range from fractional watts, as in servomotors, to 1500 MW, as in large generators.

Voltages vary, up to 25,000 V AC stator and 1500 V DC rotor.

Installed horizontal or vertical at speed ranges up to 130,000 RPM, normally from 40 RPM (waterwheel generators) to 3600 RPM (turbine generators).

Frequency at 60 or 50 Hz mostly, 400 Hz military; however, synthesized variable frequency electrical supplies are increasingly common and provide variable motor speeds to improve process efficiency.

Typical synchronous machinery construction and performance are described; variations may exist on special smaller units.

This document is intentionally general in nature. Should the reader want specific application information, refer to standards: NEMA MG-1; IEEE 115, C50-10 and C50-13; IEC 600034: 1-11,14-16,18, 20, 44, 72, and 136, plus other applicable specifications.

Construction (See Fig. 2.6)

Stator

Frame

The exterior frame, made of steel, either cast or a weldment, supports the laminated stator core and has feet, or flanges, for mounting to the foundation. Frame vibration from core magnetic forcing or rotor unbalance is minimized by resilient mounting the core and/or by designing to avoid frame resonance with forcing frequencies. If bracket type bearings are employed, the frame must support the bearings, oil seals, and gas seals when cooled with hydrogen or gas other than air. The frame also provides protection from the elements and channels cooling air, or gas, into and out of the core, stator windings, and rotor. When the unit is cooled by gas contained within the frame, heat from losses is removed by coolers having water circulating through finned pipes of a heat exchanger mounted within the frame. Where cooling water is unavailable and outside air cannot circulate through the frame because of its dirty or toxic condition, large air-to-air heat exchangers are employed, the outside air being forced through the cooler by an externally shaft-mounted blower.

Stator Core Assembly

The stator core assembly of a synchronous machine is almost identical to that of an induction motor. A major component of the stator core assembly is the core itself, providing a high permeability path for magnetism. The stator core is comprised of thin silicon steel laminations and insulated by a surface coating minimizing eddy current and hysteresis losses generated by alternating magnetism. The laminations are stacked as full rings or segments, in accurate alignment, either in a fixture or in the stator frame, having ventilation spacers inserted periodically along the core length. The completed core is compressed and clamped axially to about 10 kg/cm^2 using end fingers and heavy clamping plates. Core end heating from stray magnetism is minimized, especially on larger machines, by using non-magnetic materials at the core end or by installing a flux shield of either tapered laminations or copper shielding.

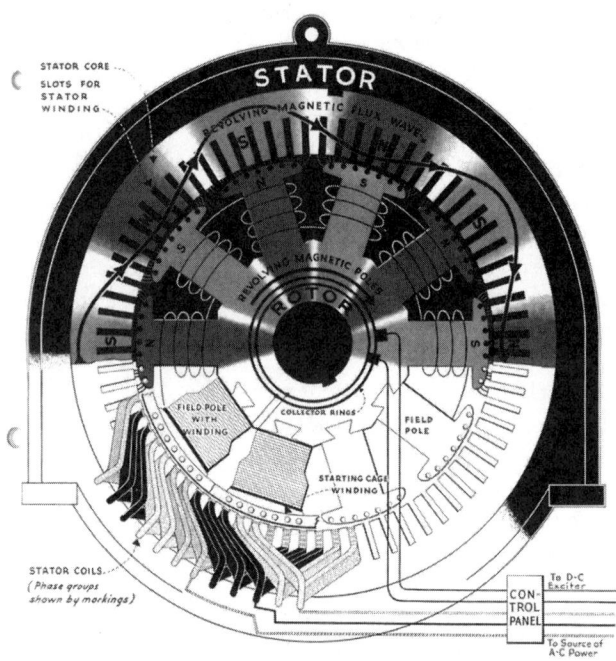

FIGURE 2.6 Magnetic "skeleton" (upper half) and structural parts (lower half) of a ten-pole (720 rpm at 60 cycles) synchronous motor. (From *The ABC's of Synchronous Motors*, 7(1), 5, 1944. The Electric Machinery Company, Inc. With permission.)

A second major component is the stator winding made up of insulated coils placed in axial slots of the stator core inside diameter. The coil make-up, pitch, and connections are designed to produce rotating stator electromagnetic poles in synchronism with the rotor magnetic poles. The stator coils are retained into the slots by slot wedges driven into grooves in the top of the stator slots. Coil end windings are bound together and to core-end support brackets. If the synchronous machine is a generator, the rotating rotor pole magnetism generates voltage in the stator winding which delivers power to an electric load. If the synchronous machine is a motor, its electrically powered stator winding generates rotating electromagnetic poles and the attraction of the rotor magnets, operating in synchronism, produces torque and delivery of mechanical power to the drive shaft.

Rotor

The Rotor Assembly

The rotor of a synchronous machine is a highly engineered unitized assembly capable of rotating satisfactorily at synchronous speed continuously according to standards or as necessary for the application. The central element is the shaft, having journals to support the rotor assembly in bearings. Located at the rotor assembly axial mid-section is the rotor core embodying magnetic poles. When the rotor is round it is called "non-salient pole", or turbine generator type construction and when the rotor has protruding pole assemblies, it is called "salient pole" construction.

The non-salient pole construction, used mainly on turbine generators (and also as wind tunnel fan drive motors), has two or four magnetic poles created by direct current in coils located in slots at the rotor outside diameter. Coils are restrained in the slots by slot wedges and at the ends by retaining rings on large high-speed rotors, and fiberglass tape on other units where stresses permit. This construction is not suited for use on a motor requiring self-starting as the rotor surface, wedges, and retaining rings overheat and melt from high currents of self-starting.

A single piece forging is sometimes used on salient pole machines, usually with four or six poles. Salient poles can also be integral with the rotor lamination and can be mounted directly to the shaft or fastened to an intermediate rotor spider. Each distinct pole has an exciting coil around it carrying

Electric Power Generation: Conventional Methods

excitation current or else it employs permanent magnets. In a generator, a moderate cage winding in the face of the rotor poles, usually with pole-to-pole connections, is employed to dampen shaft torsional oscillation and to suppress harmonic variation in the magnetic waveform. In a motor, heavy bars and end connections are required in the pole face to minimize and withstand the high heat of starting duty.

Direct current excites the rotor windings of salient, and non-salient pole motors and generators, except when permanent magnets are employed. The excitation current is supplied to the rotor from either an external DC supply through collector rings or a shaft-mounted brushless exciter. Positive and negative polarity bus bars or cables pass along and through the shaft as required to supply excitation current to the windings of the field poles.

When supplied through collector rings, the DC current could come from a shaft-driven DC or AC exciter rectified output, from an AC-DC motor-generator set, or from plant power. DC current supplied by a shaft-mounted AC generator is rectified by a shaft-mounted rectifier assembly.

As a generator, excitation current level is controlled by the voltage regulator. As a motor, excitation current is either set at a fixed value, or is controlled to regulate power factor, motor current, or system stability.

In addition, the rotor also has shaft-mounted fans or blowers for cooling and heat removal from the unit plus provision for making balance weight additions or corrections.

Bearings and Couplings

Bearings on synchronous machinery are anti-friction, grease, or oil-lubricated on smaller machines, journal type oil-lubricated on large machines, and tilt-pad type on more sophisticated machines, especially where rotor dynamics are critical. Successful performance of magnetic bearings, proving to be successful on turbo-machinery, may also come to be used on synchronous machinery as well.

As with bearings on all large electrical machinery, precautions are taken with synchronous machines to prevent bearing damage from stray electrical shaft currents. An elementary measure is the application of insulation on the outboard bearing, if a single-shaft end unit, and on both bearing and coupling at the same shaft end for double-shaft end drive units. Damage can occur to bearings even with properly applied insulation, when solid-state controllers of variable frequency drives, or excitation, cause currents at high frequencies to pass through the bearing insulation as if it were a capacitor. Shaft grounding and shaft voltage and grounding current monitoring can be employed to predict and prevent bearing and other problems.

Performance

Synchronous Machines, in General

This section covers performance common to synchronous motors, generators, and condensers.

Saturation curves (Fig. 2.7) are either calculated or obtained from test and are the basic indicators of machine design suitability. From these the full load field, or excitation, amperes for either motors or generators are determined as shown, on the rated voltage line, as "*Rated Load.*" For synchronous condensers, the field current is at the crossing of the zero P.F. saturation line at 1.0 V. As an approximate magnetic figure of merit, the no-load saturation curve should not exceed its extrapolated straight line by more than 25%, unless of a special design. From these criteria, and the knowledge of the stator current and cooling system effectiveness, the manufacturer can project the motor component heating, and thus insulation life, and the efficiency of the machine at different loads.

Vee curves (Fig. 2.8) show overall loading performance of a synchronous machine for different loads and power factors, but more importantly show how heating and stability limit loads. For increased hydrogen pressures in a generator frame, the load capability increases markedly.

The characteristics of all synchronous machines when their stator terminals are short-circuited are similar (see Fig. 2.9). There is an initial subtransient period of current increase of 8 to 10 times rated, with one phase offsetting an equal amount. These decay in a matter of milliseconds to a transient value of 3 to 5 times rated, decaying in tenths of a second to a relatively steady value. Coincident with this, the field current increases suddenly by 3 to 5 times, decaying in tenths of a second. The stator voltage on the shorted phases drops to zero and remains so until the short circuit is cleared.

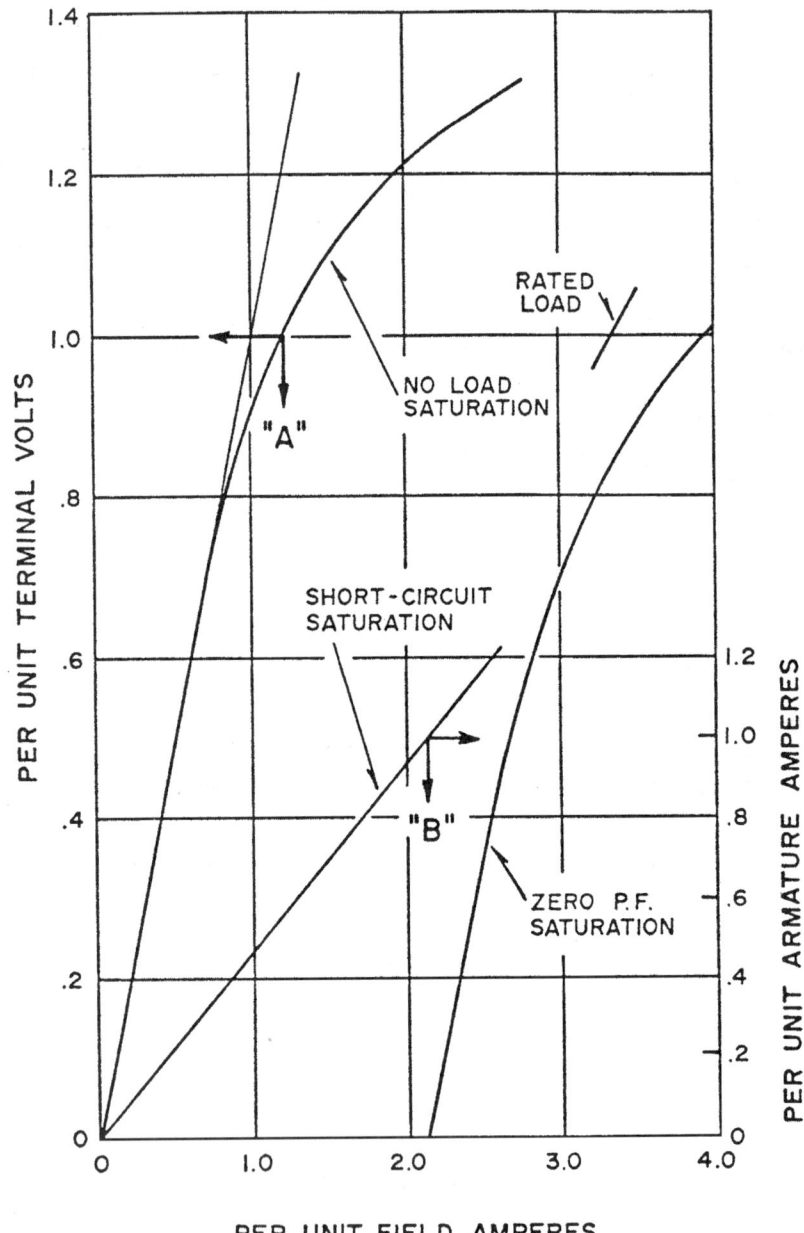

FIGURE 2.7 Saturation curves.

Synchronous Generator Capability

The synchronous generator normally has easy starting duty as it is brought up to speed by a prime mover. Then the rotor excitation winding is powered with DC current, adjusted to rated voltage, and transferred to voltage regulator control. It is then synchronized to the power system, closing the interconnecting circuit breaker as the prime mover speed is advancing, at a snail's pace, leading the electric system. Once on line, its speed is synchronized with the power system and KW is raised by increasing the prime mover KW input. The voltage regulator adjusts excitation current to hold voltage. Increasing the voltage regulator

Electric Power Generation: Conventional Methods 2-17

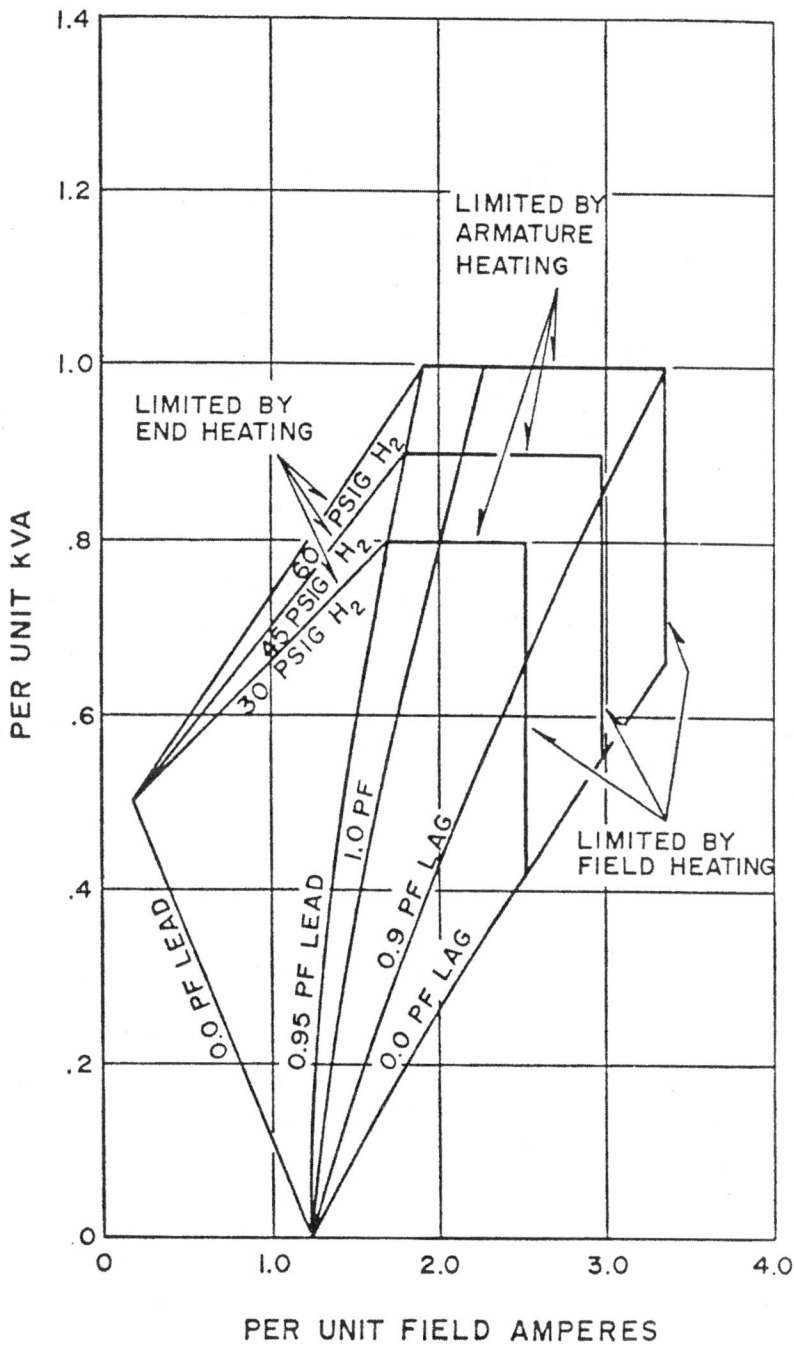

FIGURE 2.8 Vee curves.

set point increases KVAR input to the system, reducing the power factor toward lagging and vice versa. Steady operating limits are provided by its Reactive Capability Curve (see Fig. 2.10). This curve shows the possible kVA reactive loading, lagging, or leading, for given KW loading. Limitations consist of field heating, armature heating, stator core end heating, and operating stability over different regions of the reactive capability curve.

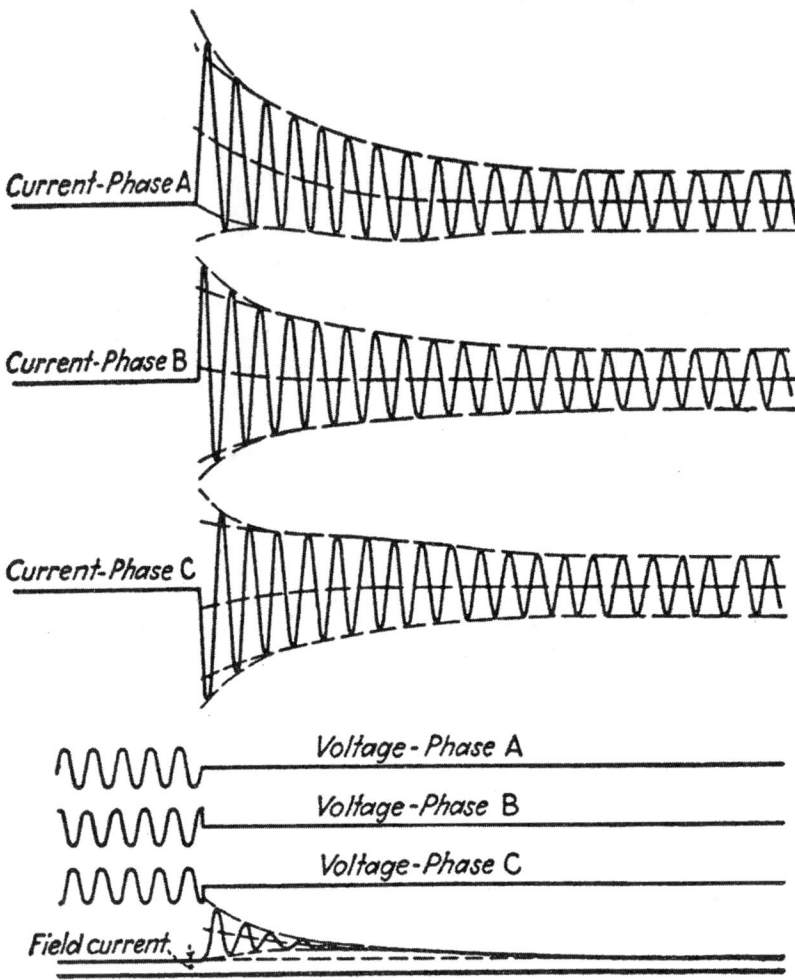

FIGURE 2.9 Typical oscillogram of a sudden three-phase short circuit.

Synchronous Motor and Condenser Starting

The duty on self-starting synchronous motors and condensors is severe, as there are large induction currents in the starting cage winding once the stator winding is energized (see Fig. 2.11). These persist as the motor comes up to speed, similar to but not identical to starting an induction motor. Similarities exist to the extent that extremely high torque impacts the rotor initially and decays rapidly to an average value, increasing with time. Different from the induction motor is the presence of a large oscillating torque. The oscillating torque decreases in frequency as the rotor speed increases. This oscillating frequency is caused by the saliency effect of the protruding poles on the rotor. Meanwhile, the stator current remains constant until 80% speed is reached. The oscillating torque at decaying frequency may excite train torsional natural frequencies during acceleration, a serious train design consideration. An anomaly occurs at half speed as a dip in torque and current due to the coincidence of line frequency torque with oscillating torque frequency. Once the rotor is close to rated speed, excitation is applied to the field coils and the rotor pulls into synchronism with the rotating electromagnetic poles. At this point, stable steady-state operation begins.

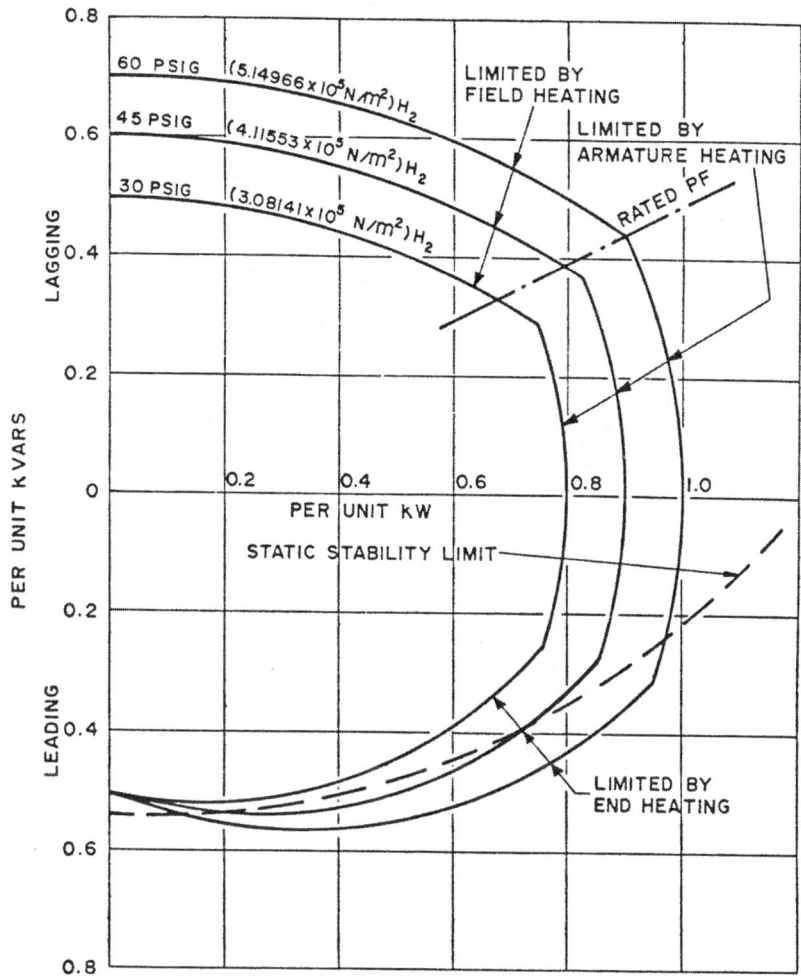

FIGURE 2.10 Typical reactive capability curve.

Increasingly, variable frequency power is supplied to synchronous machinery primarily to deliver the optimum motor speed to meet load requirements, improving the process efficiency. It can also be used for soft-starting the synchronous motor or condenser. Special design and control are employed to avert problems imposed, such as excitation of train torsional natural frequencies and extra heating from harmonics of the supply power.

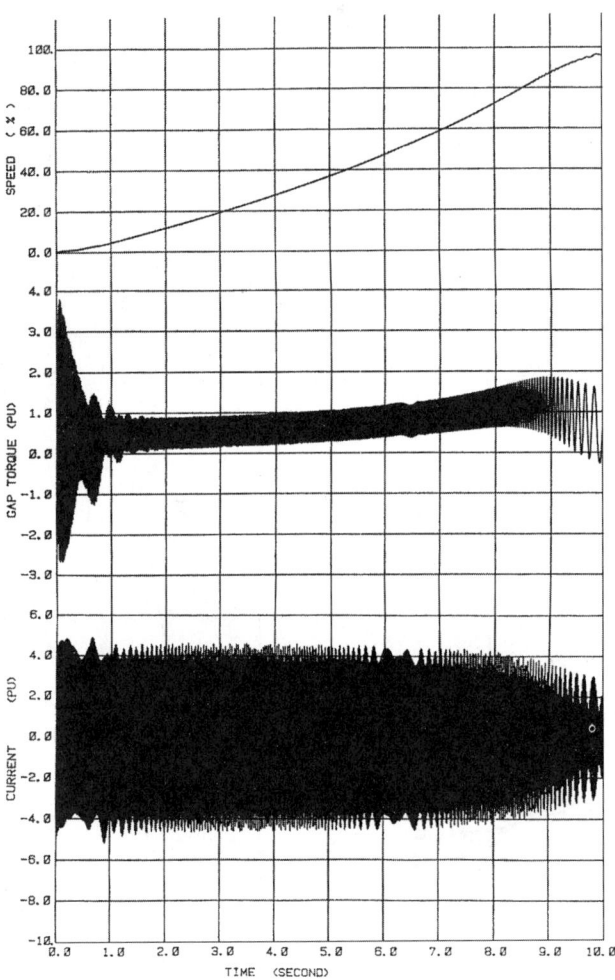

FIGURE 2.11 Synchronous motor and condensor starting.

2.3 Thermal Generating Plants

Kenneth H. Sebra

Thermal generating plants are designed and constructed to convert energy from fuel (coal, oil, gas, or radiation) into electric power. The actual conversion is accomplished by a turbine-driven generator. Thermal generating plants differ from industrial plants in that the nature of the product never changes. The plant will always produce electric energy. The things that may change are the fuel used (coal, oil, or gas) and environmental requirements. Many plants that were originally designed for coal were later converted to oil, converted back to coal, and then converted to gas. Environmental requirements have changed, which has required the construction of air and water emissions control systems. Plant electrical systems should be designed to allow for further growth. Sizing of transformers and buses is at best a matter of guesswork. The plant electrical system should be sized at 5 to 10% the size of the generating unit depending on the plant configuration and number of units at the plant site.

Plant Auxiliary System

Selection of Auxiliary System Voltages

The most common plant auxiliary system voltages are 13,800 V, 6900 V, 4160 V, 2400 V, and 480 V. The highest voltage is determined by the largest motor. If motors of 4000 hp or larger are required, one should consider using 13,800 V. If the largest motor required is less than 4000 hp, then 4160 V should be satisfactory.

Auxiliary System Loads

Auxiliary load consists of motors and transformers. Transformers supply lower level buses which supply smaller motors and transformers which supply lower voltage buses. Generation plants built before 1950 may have an auxiliary generator that is connected to the main generator shaft. The auxiliary generator will supply plant loads when the plant is up and running.

Auxiliary System Power Sources

The power sources for a generating plant consist of one or more off-site sources and one or more on-site sources. The on-site sources are the generator and, in some cases, a black start diesel generator or a gas turbine generator which may be used as a peaker.

Auxiliary System Voltage Regulation Requirements

Most plants will not require voltage regulation. A load flow study will indicate if voltage regulation is required. Transformers with tap changers, static var compensators, or induction regulators may be used to keep plant bus voltages within acceptable limits. Switched capacitor banks and overexcited synchronous motors may also be used to regulate bus voltage.

Plant One-Line Diagram

The one-line diagram is the most important document you will use. Start with a conceptual one-line and add detail as it becomes available. The one-line diagram will help you think about your design and make it easier to discuss with others. Do not be afraid to get something on paper very early and modify as you get more information about the design. Consider how the plant will be operated. Will there be a start-up source and a running source? Are there on-site power sources?

Plant Equipment Voltage Ratings

Establish at least one bus for each voltage rating in the plant. Two or more buses may be required depending on how the plant will be operated

Grounded vs. Ungrounded Systems

A method of grounding must be determined for each voltage level in the plant.

Ungrounded

Most systems will be grounded in some manner with the exception for special cases of 120-V control systems which may be operated ungrounded for reliability reasons. An ungrounded system may be allowed to continue to operate with a single ground on the system. Ungrounded systems are undesirable because ground faults are difficult to locate. Also, ground faults can result in system overvoltage, which can damage equipment that is connected to the ungrounded system.

Grounded

Most systems 480 V and lower will be solidly grounded.

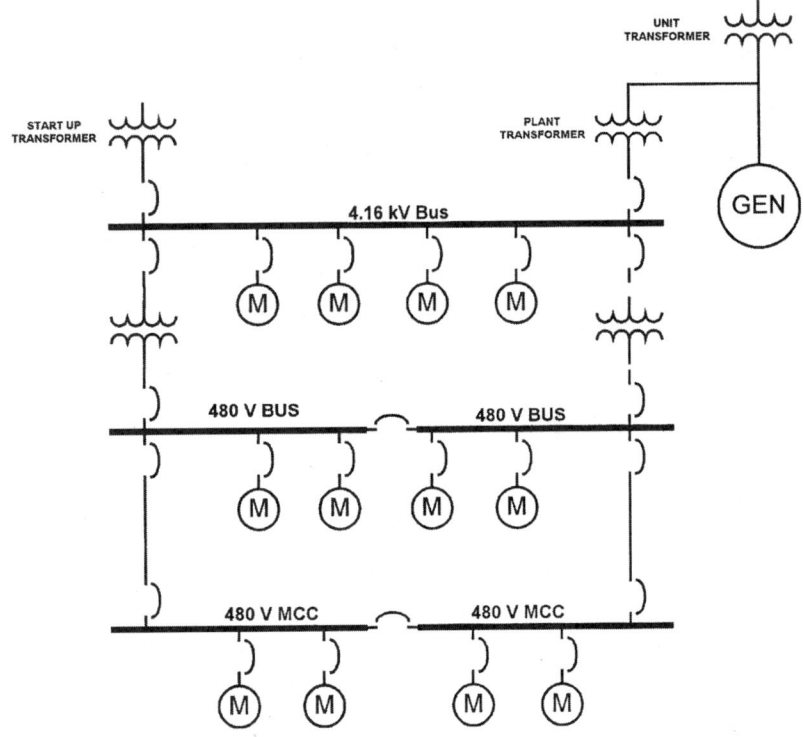

FIGURE 2.12 Typical plant layout.

Low-Resistance Grounding

Low-resistance grounding systems are used at 2400 V and above. This system provides enough ground fault current to allow relay coordination and limits ground fault current to a value low enough to prevent equipment damage.

High-Resistance Grounding

High-resistance grounding systems limit ground fault current to a very low value but make relay coordination for ground faults difficult.

Miscellaneous Circuits

Essential Services

Essential services such as critical control required for plant shutdown, fire protection, and emergency lighting should be supplied by a battery-backed inverter. This is equipment that must continue to operate after a loss of off-site power. Some of these loads may be supplied by an on-site diesel generator or gas turbine if a delay after loss of off-site power is acceptable.

Lighting Supply

Lighting circuits should be designed with consideration to emergency lighting to the control room and other vital areas of the plant. Consideration should be given to egress lighting and lighting requirements for plant maintenance.

DC Systems

The plant will require at least one DC system for control and operation of essential systems when off-site power is lost. The required operating time for the emergency equipment that will be operated from the DC systems must be established in order to size the batteries. The installation of a diesel generator may reduce the size of the battery.

125-V DC

A 125-V DC system is supplied for circuit breaker and protective relaying. The system voltage may collapse to close to zero during fault conditions and would not be capable of supplying relay control and breaker trip current when it is needed to operate.

250-V DC

A 250-V DC system may be required to supply turbine generator emergency motors such as turning gear motors and emergency lube oil motors.

Power Plant Switchgear

High-Voltage Circuit Breakers

High-voltage circuit breakers of 34.5 kV and above may be used in the switchyard associated with the generating plant, but are rarely used in a generating plant.

Medium-Voltage Switchgear

Medium-voltage breakers are 2.4 to 13.8 kV. Breakers in this range are used for large motors in the plant. The most prevalent is 4.16 kV.

Medium-Voltage Air Circuit Breakers

Air circuit breakers were the most common type of breaker until about 1995. Due to large size and high maintenance requirements of air circuit breakers, they have been replaced by vacuum breakers.

Medium-Voltage Vacuum Circuit Breakers

Vacuum circuit breakers are the most common type of circuit breaker used in new installations. Vacuum circuit breakers are being used to replace air circuit breakers. Vacuum breakers are smaller and can provide additional space if the plant needs to be expanded to meet new requirements. Before using vacuum circuit breakers, a transient analysis study should be performed to determine if there is a need for surge protection. If required, surge protection can be supplied by the installation of capacitors and/or surge suppressors can be used to eliminate voltage surge problems.

Medium-Voltage SF6 Circuit Breakers

SF6 circuit breakers have the same advantages as vacuum circuit breakers but there is some environmental concern with the SF6 gas.

Low-Voltage Switchgear

Low voltage is 600 V and below. The most common voltage used is 480 V.

Low-Voltage Air Circuit Breakers

Air circuit breakers are used in load centers that may include a power transformer. Air circuit breakers are used for motors greater than 200 hp and less than about 600 hp. Low-voltage circuit breakers are self-contained in that fault protection is an integral part of the breaker. Low-voltage devices, which do not contain fault protection devices, are referred to as low-voltage switches. Low-voltage breakers may be obtained with various combinations of trip elements. Long time, short time, and ground trip elements may be obtained in various combinations.

Low-voltage breakers manufactured before 1970 will contain oil dashpot time delay trip elements. Breakers manufactured after the mid-1970s until about 1990 will contain solid-state analog trip elements. Breakers manufactured after 1990 will contain digital trip elements. The digital elements provide much more flexibility.

A circuit that may be large enough for a load center circuit breaker but is operated several times a day should not be put on a load center circuit breaker. The circuit breaker would be put through its useful life in a very short time. A motor starter would be a better choice.

Motor Control Centers

Motor control centers are self-contained and may include molded case breakers or combination starters. Molded case breakers are available as either magnetic or thermal-magnetic. The magnetic trip breakers are instantaneous trip only and the thermal-magnetic trip breakers are time delay with instantaneous trip. Magnetic breakers can be used with a contactor to make a combination starter. Time delay trip is provided by overload relays mounted on the contactor. Solid-state equipment is available to use in motor control centers and allows much greater flexibility

Circuit Interruption

The purpose of a circuit breaker is to provide a method of interrupting the circuit either to turn the load on and off or to interrupt fault current. The first requirement is based on the full load current of the load. The second requirement is based on the maximum fault current as determined by the fault current study. There is no relationship between the load current and the fault current. If modifications are made to the electric power system, the fault interrupting current requirement may increase. Failure to recognize this could result in the catastrophic failure of a breaker.

Auxiliary Transformers

Selection of Percent Impedance

The transformer impedance is always compromised. High transformer impedance will limit fault current and reduce the required interrupting capability of switchgear and, therefore, reduce the cost. Low impedance will reduce the voltage drop through the transformer and therefore improve voltage regulation. A careful analysis using a load flow study will help in arriving at the best compromise.

Rating of Voltage Taps

Transformers should be supplied with taps to allow adjustment in bus voltage. Optimum tap settings can be determined using a load flow study.

Motors

Selection of Motors

Many motors are required in a thermal generating plant and range in size from fractional horsepower to several thousand horsepower. These motors may be supplied with the equipment they drive or they may be specified by the electrical engineer and purchased separately. The small motors are usually supplied by the equipment supplier and the large motors specified by the electrical engineer. How this will be handled must be resolved very early in the project. The horsepower cut-off point for each voltage level must be decided. The maximum plant voltage level must be established. A voltage of 13.8 kV may be required if very large horsepower motors are to be used. This must be established very early in the plant design so that a preliminary one-line diagram may be developed.

Types of Motors

Squirrel Cage Induction Motors

The squirrel cage induction motor is the most common type of large motor used in a thermal generating plant. Squirrel cage induction motors are very rugged and require very little maintenance.

Electric Power Generation: Conventional Methods

Wound Rotor Induction Motors

The wound rotor induction motor has a rotor winding which is brought out of the motor through slip rings and brushes. While more flexible than a squire cage induction motor, the slip rings and brushes are an additional maintenance item. Wound rotor motors are only used in special applications in a power plant.

Synchronous Motors

Synchronous motors may be required in some applications. Large slow-speed, 1800 rpm or less may require a synchronous motor. A synchronous motor may used to supply VARs and improve voltage regulation. If the synchronous motor is going to be used as a VAR source, the field supply must be sized large enough to over-excite the field.

Direct Current Motors

Direct current motors are used primarily on emergency systems such as turbine lube oil and turbine turning gear. Direct current motors may also be used on some control valves.

Single-Phase Motors

Single-phase motors are fractional horsepower motors and are usually supplied with the equipment.

Motor Starting Limitations

The starting current for induction motors is about 6 times full load current. This must be taken into account when sizing transformers and should be part of the load flow analysis. If the terminal voltage is allowed to drop too low, below 80%, the motor will stall. Methods of reduced voltage starting are available, but should be avoided if possible. The most reliable designs are the simplest.

Main Generator

The turbine generator will be supplied as a unit. The size and characteristics are usually determined by the system planners as a result of system load requirements and system stability requirements.

Associated Equipment

Exciters and Excitation Equipment

The excitation system will normally be supplied with the generator.

Electronic exciters — Modern excitation systems are solid state and, in recent years, most have digital control systems.

Generator Neutral Grounding

The generator neutral is never connected directly to ground. The method used to limit the phase to ground fault current to a value equal to or less than the three-phase fault current is determined by the way the generator is connected to the power system. If the generator is connected directly to the power system, a resistor or inductor connected between the neutral of the generator and ground will be used to limit the ground fault current. If the generator is connected to the power system through a transformer in a unit configuration, the neutral of the generator may be connected to ground through a distribution transformer with a resistor connected across the secondary of the transformer. The phase-to-ground fault current can be limited to 5 to 10 A using this method.

Isolated Phase Bus

The generator is usually connected to the step-up transformer through an isolated phase bus. This separated phase greatly limits the possibility of a phase-to-phase fault at the terminals of the generator.

Cable

Large amounts of cable are required in a thermal generating plant. Power, control, and instrumentation cable should be selected carefully with consideration given to long life. Great care should be given in the installation of all cable. Cable replacement can be very expensive.

Electrical Analysis

All electrical studies should be well-documented for use in plant modifications. These studies will be of great value in evaluating plant problems.

Load Flow

A load flow study should be performed as early in the design as possible even if the exact equipment is not known. The load flow study will help in getting an idea of transformer size and potential voltage drop problems.

A final load flow study should be performed to document the final design and will be very helpful if modifications are required in the future.

Short-Circuit Analysis

Short-circuit studies must be performed to determine the requirements for circuit breaker interrupting capability. Relay coordination should be studied as well.

Surge Protection

Surge protection may be required to limit transient overvoltage caused by lightning and circuit switching. A surge study should be performed to determine the needs of each plant configuration. Surge arrestors and/or capacitors may be required to limit transient voltages to acceptable levels.

Phasing

A phasing diagram should be made to determine correct transformer connections. An error here could prevent busses from being paralleled.

Relay Coordination Studies

Relay coordination studies should be performed to ensure proper coordination of the relay protection system. The protective relay system may include overcurrent relays, bus differential relays, transformer differential relays, voltage relays, and various special function relays.

Maintenance and Testing

A good plant design will take into account maintenance and testing requirements. Equipment must be accessible for maintenance and provisions should be made for test connections.

Start-Up

A start-up plan should be developed to ensure equipment will perform as expected. This plan should include insulation testing. Motor starting current magnitude and duration should be recorded and relay coordination studies verified. Voltage level and load current readings should be taken to verify load flow studies. This information will also be very helpful when evaluating plant operating conditions and problems.

References

General
Beeman, D., Ed., *Industrial Power Systems Handbook,* McGraw-Hill, New York.
Electrical Transmission and Distribution Reference Book, Westinghouse Electric Corporation.
IEEE Standard 666-1991, IEEE Design Guide for Electric Power Service Systems for Generating Stations.

Grounding
IEEE 665-1955, IEEE Guide for Generating Station Grounding.
IEEE 1050-1996, IEEE Guide for Instrumentation and control Grounding in Generating Stations.

DC Systems
IEEE 485-1997, IEEE Recommended Practice for Sizing Lead-Acid Batteries for Station Applications.
IEEE 946-1992, IEEE Recommended Practice for the Design of DC Auxiliary Power Systems for Generating Stations.

Switchgear
IEEE Standards Collection: Power and Energy-Switchgear, 1998 Edition.

Auxiliary Transformers
IEEE Distribution, Power and Regulating Transformers Standards Collection, 1998 Edition.

Motors
IEEE Electric Machinery Standards Collection, 1997 Edition.

Cable
IEEE 835-1994, IEEE Standard Power Cable Ampacity Tables.

Electrical Analysis
Clarke, E., *Circuit Analysis of AC Power Systems,* General Electric Company, 1961.
Stevenson, W. D., *Elements of Power Systems Analysis,* McGraw-Hill, New York, 1962.
Wager, C. F. and Evans, R. D., *Symmetrical Components,* McGraw-Hill, New York, 1933.

2.4 Distributed Utilities

John R. Kennedy

Distributed utilities (sometimes referred to as DU) is the current term used in describing distributed generation and storage devices operating separately and in parallel with the utility grid. In most cases, these devices are small in comparison to traditional utility base or peaking generation, but can range up to several megawatts. For the purposes of this section, DU will be limited to devices 5 MW and below applied at either the secondary voltage level, 120 V single phase to 480 V three phase, and at the medium voltage level, 2.4 kV to 25 kV, although many of the issues discussed would apply to the larger units as well.

In this section, we will give an overview of the different issues associated with DU, including available technologies, interfacing, a short discussion on economics and possible regulatory treatment, applications, and some practical examples. Emerging technologies discussed will include fuel cells, microturbines, and small turbines. A brief discussion of storage technologies is also included. Interfacing issues include general protection, overcurrent protection, islanding issues, communication and control, voltage regulation, frequency control, fault detection, safety issues, and synchronization. In the applications section, deferred investment, demand reduction, peak shaving, ancillary services, reliability, and power quality will be discussed. Economics and possible regulatory treatment will be discussed briefly.

Available Technologies

Many of the "new" technologies have been around for several years, but the relative cost per kilowatt of small generators compared to conventional power plants has made their use limited. Utility rules and interconnect requirements have also limited the use of small generators and storage devices to mostly emergency, standby, and power quality applications. The prospect of deregulation has changed all that. Utilities are no longer assured that they can recover the costs of large base generation plants, and stranded investment of transmission and distribution facilities is a subject of debate. This, coupled with improvements in the cost and reliability of DU technologies, has opened an emerging market for small power plants. In the near future, these new technologies should be competitive with conventional plants, providing high reliability with less investment risk. Some of the technologies are listed below. All of the

Technology	Size	Fuel Sources	AC Interface Type	Applications
Fuel Cells	.5Kw – Larger units With Stacking	Natural Gas Hydrogen Petroleum Products	Inverter type	Continuous
Microturbines	10Kw-100Kw Larger sizes	Natural Gas Petroleum Products	Inverter type	Continuous Standby
Batteries	.1Kw-2Mw+	Storage	Inverter type	PQ, Peaking
Flywheel	>.1Kw - .5Kw	Storage	Inverter type	PQ, Peaking
PV	>.1Kw- 1Kw	Sunlight	Inverter type	Peaking
Gas Turbine	10Kw – 5Mw+	Natural Gas Petroleum Products	Rotary type	Continuous, Peaking Standby

FIGURE 2.13 Distributed generation technology chart.

energy storage devices and many of the small emerging generation devices are inverter/converter based. Figure 2.13 is a listing of different technologies, their size ranges, fuel sources, and AC interface type, and most likely applications.

Fuel Cells

Fuel cell technology has been around since its invention by William Grove in 1839. From the 1960s to the present, fuel cells have been the power source used for space flight missions. Unlike other generation technologies, fuel cells act like continuously fueled batteries, producing direct current (DC) by using an electrochemical process. The basic design of all fuel cells consists of an anode, electrolyte, and cathode. Hydrogen or a hydrogen-rich fuel gas is passed over the anode, and oxygen or air is passed over the cathode. A chemical combination then takes place producing a constant supply of electrons (DC current) with by-products of water, carbon dioxide, and heat. The DC power can be used directly or it can be fed to a power conditioner and converted to AC power (see Fig. 2.14).

Most of the present technologies have a fuel reformer or processor that can take most hydrocarbon-based fuels, separate out the hydrogen, and produce high-quality power with negligible emissions. This would include gasoline, natural gas, coal, methanol, light oil, or even landfill gas. In addition, fuel cells can be more efficient than conventional generators. Theoretically they can obtain efficiencies as high as 85% when the excess heat produced in the reaction is used in a combined cycle mode. These features, along with relative size and weight, have also made the fuel cell attractive to the automotive industry as an alternative to battery power for electric vehicles. The major differences in fuel cell technology concern the electrolyte composition. The major types are the Proton Exchange Membrane Fuel Cell (PEFC) also called the PEM, the Phosphoric Acid Fuel Cell (PAFC), the Molten Carbonate Fuel Cell (MCFC), and the Solid Oxide Fuel Cell (SOFC) (Fig. 2.15).

Fuel cell power plants can come in sizes ranging from a few watts to several megawatts with stacking. The main disadvantage to the fuel cell is the initial high cost of installation. With the interest in efficient and environmentally friendly generation, coupled with the automotive interest in an EV alternative power source, improvements in the technology and lower costs are expected. As with all new technologies, volume of sales should also lower the unit price.

Microturbines

Experiments with microturbine technology have been around for many decades, with the earliest attempts of wide-scale applications being targeted at the automotive and transportation markets. These experiments

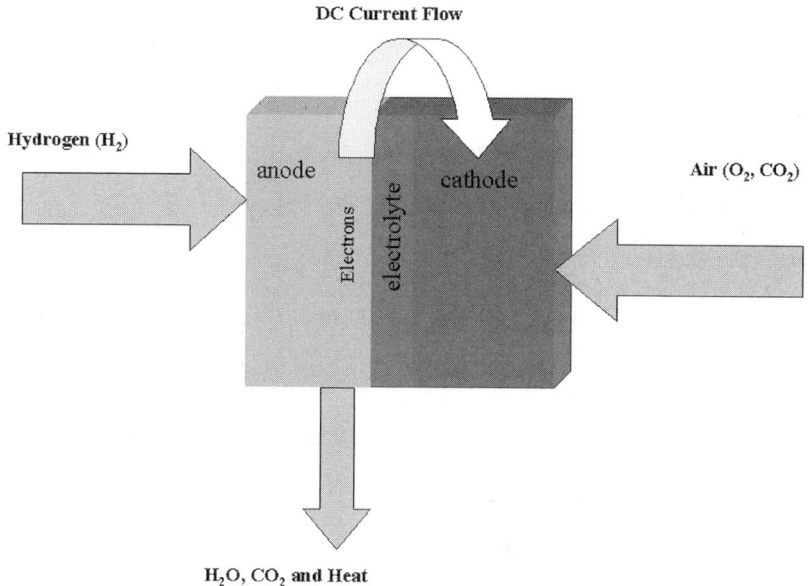

FIGURE 2.14 Basic fuel cell operation.

	PAFC	MCFC	SOFC	PEMFC
Electrolyte	Phosphoric acid	Molten carbonate salt	Ceramic	Polymer
Operating Temperature	375°F (190°C)	1200°F (650°C)	1830°F (1000°C)	175°F (80°C)
Fuels	Hydrogen (H_2)	H_2/CO	H_2/CO_2/CH_4	H_2
	Reformate	Reformate	Reformate	Reformate
Reforming	External	External	External	External
Oxidant	O_2/Air	CO_2/O_2/Air	O_2/Air	O_2/Air
Efficiency (HHV)	40–50%	50–60%	45–55%	40–50%

FIGURE 2.15 Comparison of fuel cell types. (*Source:* DoD Website, www.dodfuelcell.com/fcdescriptions.html.)

later expanded into markets associated with military and commercial aircraft and mobile systems. Microturbines are typically defined as systems with an output power rating of between 10 kW up to a few hundred kilowatts. As shown in Fig. 2.16, these systems are usually a single-shaft design with compressor, turbine, and generator all on the common shaft, although some companies are engineering dual-shaft systems. Like the large combustion turbines, the microturbines are Brayton Cycle systems, and will usually have a recuperator in the system.

The recuperator is incorporated as a means of increasing efficiency by taking the hot turbine exhaust through a heavy (and relatively expensive) metallic heat exchanger and transferring the heat to the input air, which is also passed through parallel ducts of the recuperator. This increase in inlet air temperature helps reduce the amount of fuel needed to raise the temperature of the gaseous mixture during combustion to levels required for total expansion in the turbine. A recuperated Brayton Cycle microturbine can operate at efficiencies of approximately 30%, while these aeroderivative systems operating without a recuperator would have efficiencies in the mid-teens.

Another requirement of microturbine systems is that the shaft must spin at very high speeds, in excess of 50,000 RPM and in some cases doubling that rate, due to the low inertia of the shaft and connected components. This high speed is used to keep the weight of the system low and increase the power density

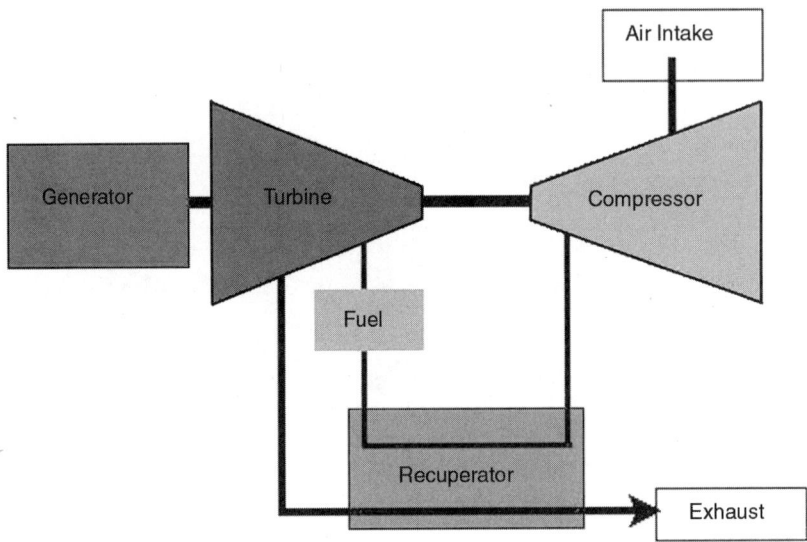

FIGURE 2.16 Turbine block diagram configuration with recuperator.

over other generating technologies. Although many of the microturbines are touted as having only a single moving part, there are numerous ancillary devices required that do incorporate moving parts such as cooling fans, fuel compressors, and pumps.

Since the turbine requires extremely high speeds for optimal performance, the generator cannot operate as a synchronous generator. Typical microturbines have a permanent magnet motor/generator incorporated onto the shaft of the system. The high rotational speed gives an AC output in excess of 1000 Hz, depending on the number of poles and actual rotational speed of the microturbine. This high-frequency AC source is rectified, forming a common DC bus voltage that is then converted to a 60-Hz AC output by an onboard inverter.

The onboard electronics are also used to start the microturbine, either in a stand-alone mode or in grid parallel applications. Typically, the utility voltage will be rectified and the electronics are used to convert this DC voltage into a variable frequency AC source. This variable frequency drive will power the permanent magnet motor/generator (which is operating as a motor), and will ramp the turbine speed up to a preset RPM, a point where stabile combustion and control can be maintained. Once this preset speed is obtained and stabile combustion is taking place, the drive shuts down and the turbine speed increases until the operating point is maintained and the system operates as a generator. The time from a "Shaft Stop" to full load condition is anywhere from 30 sec to 3 min, depending on manufacturer recommendations and experiences.

Although things are in the early stages of commercialization of the microturbine products, there are cost targets that have been announced from all of the major manufacturers of these products. The early market entry price of these systems is in excess of $600 per kW, more than comparably sized units of alternative generation technologies, but all of the major suppliers have indicated that costs will fall as the number of units being put into the field increases.

The microturbine family has a very good environmental rating, due to natural gas being a primary choice for fuel and the inherent operating characteristics, which puts these units at an advantage over diesel generation systems.

Combustion Turbines

There are two basic types of combustion turbines (CTs) other than the microturbines: the heavy frame industrial turbines and the aeroderivative turbines. The heavy frame systems are derived from similar

	Heavy Frame	Aeroderivative
Size (Same General Rating)	Large	Compact
Shaft Speed	Synchronous	Higher Speed (coupled through a gear box)
Air Flow	High (lower compression)	Lower (high compression)
Start-up Time	15 Minutes	2-3 minutes

FIGURE 2.17 Basic combustion turbine operating characteristics.

models that were steam turbine designs. As can be identified from the name, they are of very heavy construction. The aeroderivative systems have a design history from the air flight industry, and are of a much lighter and higher speed design. These types of turbines, although similar in operation, do have some significant design differences in areas other than physical size. These include areas such as turbine design, combustion areas, rotational speed, and air flows.

Although these units were not originally designed as a "distributed generation" technology, but more so for central station and large co-generation applications, the technology is beginning to economically produce units with ratings in the hundreds of kilowatts and single-digit megawatts. These turbines operate as Brayton Cycle systems and are capable of operating with various fuel sources. Most applications of the turbines as distributed generation will operate on either natural gas or fuel oil. The operating characteristics between the two systems can best be described in tabular form as shown in Fig. 2.17.

The combustion turbine unit consists of three major mechanical components: a compressor, a combustor, and a turbine. The compressor takes the input air and compresses it, which will increase the temperature and decrease the volume per the Brayton Cycle. The fuel is then added and the combustion takes place in the combustor, which increases both the temperature and volume of the gaseous mixture, but leaves the pressure as a constant. This gas is then expanded through the turbine where the power is extracted through the decrease in pressure and temperature and the increase in volume.

If efficiency is the driving concern, and the capital required for the increased efficiency is available, the Brayton Cycle systems can have either co-generation systems, heat recovery steam generators, or simple recuperators added to the combustion turbine unit. Other equipment modifications and improvements can be incorporated into these types of combustion turbines such as multistage turbines with fuel re-injection, inter-cooler between multistage compressors, and steam/water injection.

Typical heat rates for simple cycle combustion turbines vary across manufacturers, but are in a range from 11,000 to 20,000 BTU/kWh. However, these numbers decrease as recuperation and co-generation are added. CTs typically have a starting reliability in the 99% range and operating reliability approaching 98%. The operating environment has a major effect on the performance of combustion turbines. The elevation at which the CT is operating has a degradation factor of around 3.5% per 1000 ft of increased elevation and the ambient temperature has a similar degradation per 10° increase.

Figure 2.18 shows a block diagram of a simple cycle combustion turbine with a recuperator (left) and a combustion turbine with multistage turbine and fuel re-injection (right).

Storage Technologies

Storage technologies include batteries, flywheels, ultra-capacitors, and to some extent photovoltaics. Most of these technologies are best suited for power quality and reliability enhancement applications, due to their relative energy storage capabilities and power density characteristics, although some large battery installations could be used for peak shaving. All of the storage technologies have a power electronic converter interface and can be used in conjunction with other DU technologies to provide "seamless" transitions when power quality is a requirement.

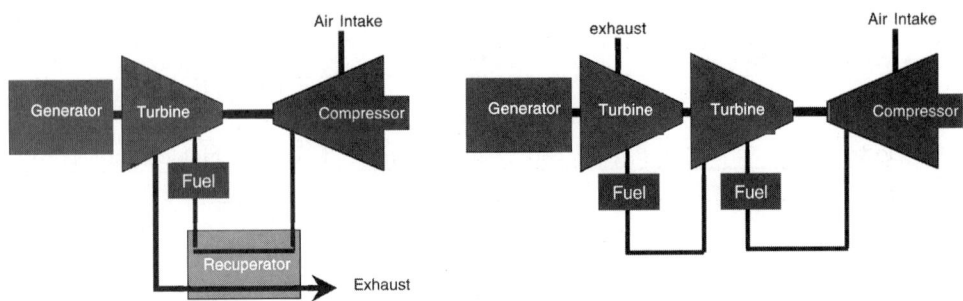

FIGURE 2.18 Basic combustion turbine designs.

Interface Issues

A whole chapter could be written just about interface issues, but this discussion will touch on the highlights. Most of the issues revolve around safety and quality of service. We will discuss some general guidelines and the general utility requirements and include examples of different considerations. In addition to the interface issues, the DU installation must also provide self-protection to prevent short circuit or other damage to the unit. Self-protection will not be discussed here. The most important issues are listed in Table 2.2.

In addition to the interface issues identified in Table 2.2, there are also operating limits that must be considered. These are listed in Table 2.3.

Utility requirements vary but generally depend on the application of a distributed source. If the unit is being used strictly for emergency operation, open transition peak shaving, or any other stand-alone type operation, the interface requirements are usually fairly simple, since the units will not be operating in parallel with the utility system. When parallel operation is anticipated or required, the interface requirements become more complex. Protection, safety, power quality, and system coordination become issues that must be addressed. In the case of parallel operation, there are generally three major factors that determine the degree of protection required. These would include the size and type of the generation, the location on the system, and how the installation will operate (one-way vs. two-way). Generator sizes are generally classified as:

Large: Greater than 3 MVA or possibility of "islanding" a portion of the system
Small: Between large and extremely small
Extremely small: Generation less than 100 kVA

TABLE 2.2 Interface Issues

Issue	Definition	Concern
Automatic reclosing	Utility circuit breakers can test the line after a fault.	If a generator is still connected to the system, it may not be in synchronization, thus damaging the generator or causing another trip.
Faults	Short circuit condition on the utility system.	Generator may contribute additional current to the fault, causing a miss operation of relay equipment.
Islanding	A condition where a portion of the system continues to operate isolated from the utility system.	Power quality, safety, and protection may be compromised in addition to possible synchronization problems.
Protection	Relays, instrument transformers, circuit breakers.	Devices must be utility grade rather than industrial grade for better accuracy. Devices must also be maintained on a regular schedule by trained technicians.
Communication	Devices necessary for utility control during emergency conditions.	Without control of the devices, islanding and other undesirable operation of devices.

TABLE 2.3 Operating Limits

1. Voltage — The operating range for voltage must maintain a level of ±5% of nominal for service voltage (ANSI C84.1), and have a means of automatic separation if the level gets out of the acceptable range within a specified time.
2. Flicker — Flicker must be within the limits as specified by the connecting utility. Methods of controlling flicker are discussed in IEEE Std. 519-1992, 10.5.
3. Frequency — Frequency must be maintained within ±0.5 Hz of 60 Hz and have an automatic means of disconnecting if this is not maintained. If the system is small and isolated, there might be a larger frequency window. Larger units may require an adjustable frequency range to allow for clock synchronizaton.
4. Power factor — The power factor should be within 0.85 lagging or leading for normal operation. Some systems that are designed for compensation may operate outside these limits.
5. Harmonics — Both voltage and current harmonics must comply with the values for generators as specified in IEEE Std. 519-1992 for both total and individual harmonics.

Location on the system and individual system characteristics determine impedance of a distribution line, which in turn determines the available fault current and other load characteristics that influence "islanding" and make circuit protection an issue. This will be discussed in more detail later.

The type of operation is the other main issue and is one of the main determinants in the amount of protection required. One-way power flow where power will not flow back into the utility has a fairly simple interface, but is dependent on the other two factors, while two-way interfaces can be quite complex. An example is shown in Fig. 2.19. Smaller generators and "line-commutated" units would have less stringent requirements. Commutation methods will be discussed later. Reciprocating engines such as diesel and turbines with mass, and "self-commutating" units which could include microturbines and fuel cells, would require more stringent control packages due to their islanding and reverse power capabilities.

Most of the new developing technologies are inverter based and there are efforts now in IEEE to revise the old Standard P929 *Recommended Practice for Utility Interface of Photovoltaic (PV) Systems* to include other inverter-based devices. The standards committee is looking at the issues with inverter-based devices in an effort to develop a standard interface design that will simplify and reduce the cost, while not sacrificing the safety and operational concerns. Inverter interfaces generally fall into two classes: line-commutated inverters and self-commutated inverters.

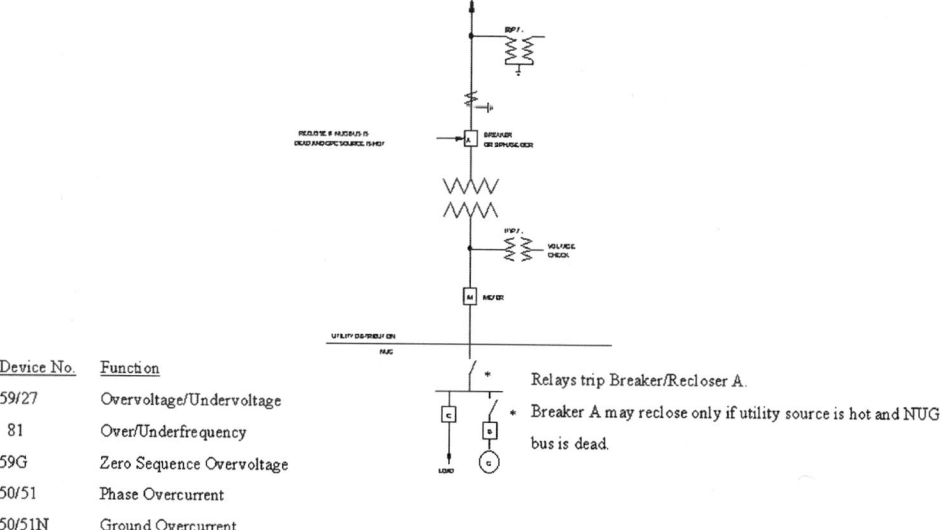

FIGURE 2.19 Example of large generator interface requirements for distribution. (*Source: Georgia Power Bulletin*, 18-8, generator interface requirements.)

Line-Commutated Inverters

These inverters require a switching signal from the line voltage in order to operate. Therefore, they will cease operation if the line signal, i.e., utility voltage, is abnormal or interrupted. These are not as popular today for single-phase devices due to the filtering elements required to meet the harmonic distortion requirements, but are appearing in some of the three-phase devices where phase cancellation minimizes the use of the additional components.

Self-Commutated Inverters

These inverters, as implied by the name, are self-commutating. All stand-alone units are self-commutated, but not all self-commutated inverters are stand-alone. They can be designed as either voltage or current sources and most that are now being designed to be connected to the utility system are designed to be current sources. These units still use the utility voltage signal as a comparison and produce current at that voltage and frequency. A great deal of effort has gone into the development of non-islanding inverters that are of this type.

Applications

Applications vary and will become more diverse as utilities unbundle. Listed below are some examples of the most likely.

Ancillary Services

Ancillary services support the basic electrical services and are essential for the reliability and operation of the electric power system. The electrical services that are supported include generating capacity, energy supply, and the power delivery system. FERC requires six ancillary services, including system control, regulation (frequency), contingency reserves (both spinning and supplemental), voltage control, and energy imbalance. In addition, load following, backup supply, network stability, system "black-start", loss replacement, and dynamic scheduling are necessary for the operation of the system. Utilities have been performing these functions for decades, but as vertically integrated regulated monopoly organizations. As these begin to disappear, and a new structure with multiple competing parties emerges, distributed utilities might be able to supply several of these.

The distributed utilities providing these services could be owned by the former traditional utility, customers, or third-party brokers, depending on the application. The main obstacles to this approach are aggregation and communication when dealing with many small resources rather than large central station sources.

"Traditional Utility" Applications

Traditional utilities may find the use of DU a practical way to solve loading and reliability problems if each case is evaluated on a stand-alone individual basis. Deferring investment is one likely way that DU can be applied. In many areas, substations and lines have seasonal peaks that are substantially higher than the rest of the year. In these cases, the traditional approach has been to increase the capacity to meet the demand. Based on the individual situation, delaying the upgrade for 2 to 5 years with a DU system could be a more economical solution. This would be especially true if different areas had different seasonal peaks and the DU system was portable, thus deferring two upgrades. DU could also be used instead of conventional facilities when backup feeds are required or to improve reliability or power quality.

In addition, peak shaving and generation reserve could be provided with strategically placed DU systems that take advantage of reducing system losses as well as offsetting base generation. Again, these have to be evaluated on an individual case basis and not a system average basis as is done in many economic studies. The type of technology used will depend on the particular requirements. In general, storage devices such as flywheels and batteries are better for power quality applications due to their fast response time, in many cases half a cycle. Generation devices are better suited for applications that require more than 30 min of supply, such as backup systems, alternate feeds, peak shaving, and demand deferrals. Generation sources can also be used instead of conventional facilities in certain cases.

Customer Applications

Individual customers with special requirements may find DU technologies that meet their needs. Customers who require "enhanced" power quality and reliability of service already utilize UPS systems with battery backup to condition the power to sensitive equipment, and many hospitals, waste treatment plants, and other emergency services providers have emergency backup systems supplied by standby generator systems. As barriers go down and technologies improve, customer-sited DU facilities could provide many of the ancillary services as well as sell excess power into the grid. Fuel cell and even diesel generators could be especially attractive for customers with requirements of heat and steam. Many of the fuel cell technologies are now looking at the residential market with small units that would be connected to the grid but supply the additional requirements for customers with special power quality needs.

Third-Party Service Providers

Third-party service providers could provide all the services listed above for the utilities and customers, in addition to selling power across the grid. In many cases, an end user does not have the expertise to operate and maintain generation systems and would prefer to purchase the services.

Conclusions

Disbursed generation will be a part of the distribution utility system of the future. Economics, regulatory requirements, and technology improvements will determine the speed at which they are integrated.

References

ANSI/IEEE Std. 1001-1998, *IEEE Guide for Interfacing Dispersed Storage and Generation Facilities with Electric Utility Systems*, IEEE Standards Coordinating Committee 23, Feb. 9, 1989.

Davis, M. W. *Microturbines — An Economic and Reliability Evaluation for Commercial, Residential, and Remote Load Applications*, IEEE Transactions PE-480-PWRS-0-10-1998.

Delmerico, R. W., Miller, N. W., and Owen, E. L. *Power System Integration Strategies for Distributed Generation*, Power Systems Energy Consulting GE International, Inc., Distributed Electricity Generation Conference, Denver, CO, Jan. 25, 1999.

Department of Defense Website, www.dodfuelcell.com/fcdescriptions.html.

Goldstein, H. L. *Small Turbines in Distributed Utility Application Natural Gas Pressure Supply Requirements*, NREL/SP-461-21073, May, 1996.

Hirschenhofer, J. H. DOE Forum on Fuel Cell Technologies, IEEE Winter Power Meeting, Parsons Corporation Presentation, Feb. 4, 1999.

Kirby, B. *Distributed Generation: A Natural for Ancillary Services*, Distributed Electric Generation Conference, Denver CO, Jan. 25, 1999.

Oplinger, J. L. *Methodology to Assess the Market Potential of Distributed Generation*, Power Systems Energy Consulting GE International, Inc., Distributed Electric Generation Conference, Denver, CO, Jan. 25, 1999.

Recommended Practice for Utility Interface of Photovoltaic (PV) Systems, IEEE Standard P929, Draft 10, Feb. 1999.

Southern Company Parallel Operation Requirements, Protection and Control Committee, Aug. 4, 1998.

Technology Overviews, DOE Forum on Fuel Cell Technology, IEEE Winter Power Meeting, Feb. 4, 1999.

3
Transformers

James H. Harlow
Harlow Engineering Associates

3.1 Theory and Principles *Harold Moore* 3-3

3.2 Power Transformers *H. Jin Sim and Scott H. Digby* 3-11

3.3 Distribution Transformers *Dudley L. Galloway* 3-30

3.4 Underground Distribution Transformers *Dan Mulkey* 3-52

3.5 Dry Type Transformers *Paulette A. Payne* 3-63

3.6 Step-Voltage Regulators *Craig A. Colopy* 3-68

3.7 Reactors *Richard Dudley, Antonio Castanheira, and Michael Sharp* 3-81

3.8 Instrument Transformers *Randy Mullikin and Anthony J. Jonnatti* 3-104

3.9 Transformer Connections *Dan D. Perco* 3-130

3.10 LTC Control and Transformer Paralleling *James H. Harlow* 3-135

3.11 Loading Power Transformers *Robert F. Tillman, Jr.* 3-149

3.12 Causes and Effects of Transformer Sound Levels *Jeewan Puri* 3-159

3.13 Electrical Bushings *Loren B. Wagenaar* 3-171

3.14 Load Tap Changers (LTCs) *Dieter Dohnal and Wolfgang Breuer* 3-184

3.15 Insulating Media *Leo J. Savio and Ted Haupert* 3-204

3.16 Transformer Testing *Shirish P. Mehta and William R. Henning* 3-209

3.17 Transformer Installation and Maintenance *Alan Oswalt* 3-234

3.18 Problem and Failure Investigations *Harold Moore* 3-242

3.19 The United States Power Transformer Equipment Standards and Processes
  Philip J. Hopkinson 3-249

3.20 On-Line Monitoring of Liquid-Immersed Transformers *Andre Lux* 3-268

3
Transformers

Harold Moore
H. Moore & Associates

H. Jin Sim
Waukesha Electric Systems

Scott H. Digby
Waukesha Electric Systems

Dudley L. Galloway
ABB Power T&D Company

Dan Mulkey
Pacific Gas & Electric Co.

Paulette A. Payne
Potomac Electric Power Company

Craig A. Colopy
Cooper Power Systems

Richard Dudley
Trench Ltd.

Antonio Castanheira
Trench Ltd.

Michael Sharp
Trench Ltd.

Randy Mulliken
Kuhlman Electric Corp.

Anthony J. Jonnatti
Loci Engineering

Dan D. Perco
Perco Transformer Engineering

James H. Harlow
Harlow Engineering Associates

Robert F. Tillman, Jr.
Alabama Power Company

Jeewan Puri
Square D Company

3.1 Theory and Principles...3-3
 Air Core Transformer • Iron or Steel Core Transformer • Equivalent Circuit of an Iron Core Transformer • The Practical Transformer • Thermal Considerations • Voltage Considerations

3.2 Power Transformers 3-11
 Rating and Classifications • Short Circuit Duty • Efficiency and Losses • Construction • Accessory Equipment • Inrush Current • Modern and Future Developments

3.3 Distribution Transformers..3-30
 Historical Background • Construction • Modern Processing • General Transformer Design • Transformer Locations • Transformer Losses • Performance • Transformer Loading • Special Tests • Protection • Economic Application

3.4 Underground Distribution Transformers..............3-52
 Vault Installations • Surface Operable Installations • Pad-Mounted Distribution Transformers

3.5 Dry Type Transformers..3-63
 Dry Type Transformers

3.6 Step-Voltage Regulators 3-68
 Power Systems Applications • Theory • Regulator Control

3.7 Reactors..3-81
 Background and Historical Perspective • Applications of Reactors • Some Important Application Considerations

3.8 Instrument Transformers..3-104
 Scope • Overview • Transformer Basics • Core Design • Burdens • Relative Polarity • Industry Standards • Accuracy Classes • Insulation Systems • Thermal Ratings • Primary Winding • Overvoltage Ratings • VT Compensation • Short-Circuit Operation • VT Connections • Ferroresonance • VT Construction • Capacitive Coupled Voltage Transformer (CCVT) • Current Transformer • Saturation Curve • CT Rating Factor • Open-Circuit Conditions • Overvoltage Protection • Residual Magnetism • CT Connections • Construction • Proximity Effects • Linear Coupler • Direct Current Transformer • CT Installations • Combination Metering Units • New Horizons

3.9 Transformer Connections......................................3-130
 Polarity of Single-Phase Transformers • Angular Displacement of Three-Phase Transformers • Three-Phase Transformer Connections • Three-Phase to Six-Phase Connections • Paralleling of Transformers

Loren B. Wagenaar
America Electric Power

Dieter Dohnal
Maschinenfabrik Reinhausen GmbH

Wolfgang Breuer
Maschinenfabrik Reinhausen GmbH

Leo J. Savio
ADAPT Corporation

Ted Haupert
TJ/H2b Analytical Services, Inc.

Shirish P. Mehta
Waukesha Electric Systems

William R. Henning
Waukesha Electric Systems

Alan Oswalt
Waukesha Electric Systems

Philip J. Hopkinson
Square D Company

Andre Lux
ABB Power T&D Company, Inc.

3.10 LTC Control and Transformer Paralleling 3-135
  System Perspective, Single Transformer • Control Inputs • The Need for Voltage Regulation • LTC Control with Power Factor Correction Capacitors • Extended Control of LTC Transformers and Step-Voltage Regulators • Introduction to Control for Parallel Operation of LTC Transformers and Step-Voltage Regulators • Defined Paralleling Procedures • Characteristics Important for LTC Transformer Paralleling • Paralleling Transformers with Mismatched Impedance

3.11 Loading Power Transformers 3-149
  Design Criteria • Nameplate Ratings • Other Thermal Characteristics • Thermal Profiles • Temperature Measurements • Predicting Thermal Response • Load Cyclicality • Science of Transformer Loading • Water in Transformers Under Load • Voltage Regulation • Loading Recommendations

3.12 Causes and Effects of Transformer Sound Levels 3-159
  Transformer Sound Levels • Sound Energy Measurement Techniques • Sources of Sound in Transformers • Sound Level and Measurement Standards for Transformers • Factors Affecting Sound Levels in Field Installations

3.13 Electrical Bushings 3-171
  Types of Bushings • Bushing Standards • Important Design Parameters • Other Features on Bushings • Tests on Bushings

3.14 Load Tap Changers (LTCs) 3-184
  Principle Design • Applications of Load Tap Changers • Rated Characteristics and Requirements for Load Tap Changers • Selection of Load Tap Changers • Maintenance of Load Tap Changers • Refurbishment/Replacement of Old LTC Types • Future Aspects

3.15 Insulating Media 3-204
  Solid Insulation — Paper • Liquid Insulation — Oil • Sources of Contamination

3.16 Transformer Testing 3-209
  Standards • Classification of Tests • Sequence of Tests • Voltage Ratio and Proper Connections • Insulation Condition • Control Devices and Control Wiring • Dielectric Withstand • Performance Characteristics • Other Tests

3.17 Transformer Installation and Maintenance 3-234
  Transformer Installation • Transformer Maintenance

3.18 Problem and Failure Investigations 3-242
  Background Investigation • Problem Analysis Where No Failure is Involved • Failure Investigations • Analysis of Information

3.19 The United States Power Transformer Equipment Standards and Processes 3-249
  Major Standards Organizations • Process for Acceptance of American National Standards • Relevant Power Transformer Standards Documents

3.20 On-Line Monitoring of Liquid-Immersed Transformers 3-268
  Benefits • On-Line Monitoring Systems • On-Line Monitoring Applications

3.1 Theory and Principles

Harold Moore

Transformers are devices that transfer energy from one circuit to another by means of a common magnetic field. In all cases except autotransformers, there is no direct electrical connection from one circuit to the other.

When an alternating current flows in a conductor, a magnetic field exists around the conductor as illustrated in Fig. 3.1. If another conductor is placed in the field created by the first conductor as shown in Fig. 3.2, such that the flux lines link the second conductor, then a voltage is induced into the second conductor. The use of a magnetic field from one coil to induce a voltage into a second coil is the principle on which transformer theory and application is based.

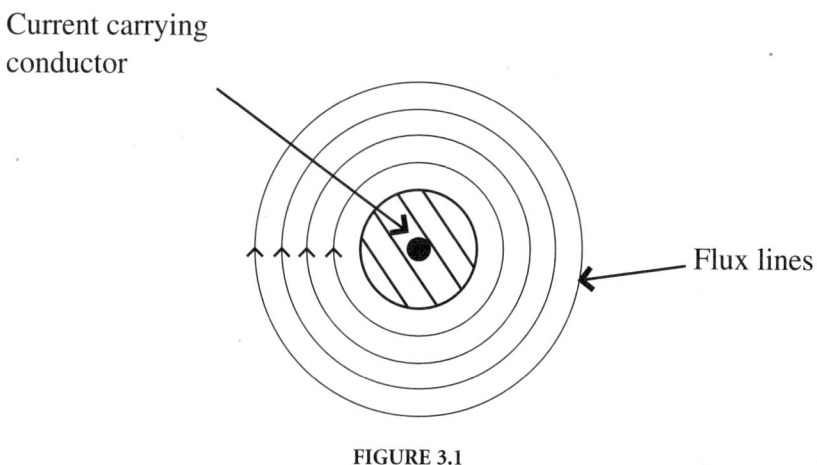

FIGURE 3.1

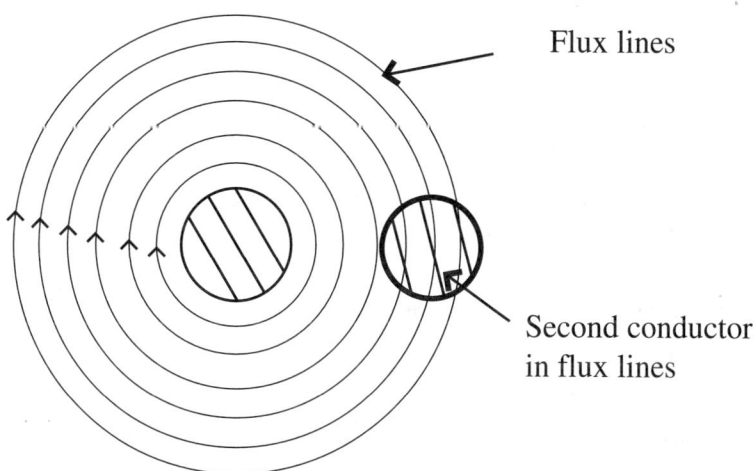

FIGURE 3.2

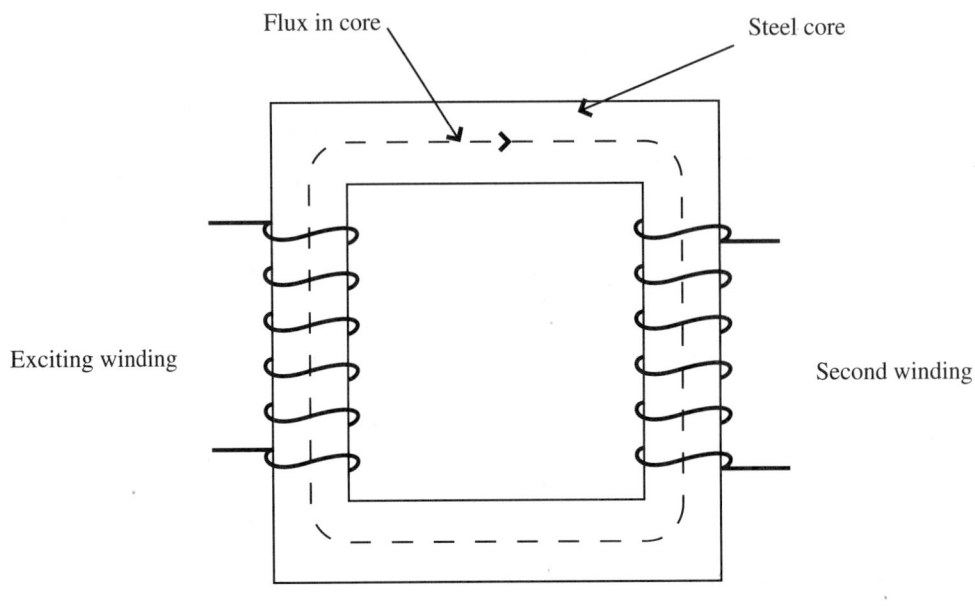

FIGURE 3.3

Air Core Transformer

Some small transformers for low power applications are constructed with air between the two coils. Such transformers are inefficient because the percentage of the flux from the first coil that links the second coil is small. The voltage induced in the second coil is determined as follows.

$$E = N \, d\emptyset/dt] 10]^8$$

where N = number of turns in the coil
 $d\emptyset/dt$ = time rate of change of flux linking the coil

Since the amount of flux ∅ linking the second coil is a small percentage of the flux from coil 1, the voltage induced into the second coil is small. The number of turns can be increased to increase the voltage output, but this will increase costs.

The need then is to increase the amount of flux from the first coil that links the second coil.

Iron or Steel Core Transformer

The ability of iron or steel to carry magnetic flux is much greater than air. This ability to carry flux is called permeability. Modern electrical steels have permeabilities on the order of 1500 compared to 1.0 for air. This means that the ability of a steel core to carry magnetic flux is 1500 times that of air. Steel cores were used in power transformers when alternating current circuits for distribution of electrical energy were first introduced. When two coils are applied on a steel core as illustrated in Fig. 3.3, almost 100% of the flux from coil 1 circulates in the iron core so that the voltage induced into coil 2 is equal to the coil 1 voltage if the number of turns in the two coils are equal.

The equation for the flux in the steel core is as follows:

$$\emptyset = \frac{3.19 \, N A u I}{d} \qquad (3.1)$$

Transformers

where

    0 = core flux in lines
    N = number of turns in the coil
    u = permeability
    I = maximum current in amperes
    d = mean length of the core

Since the permeability of the steel is very high compared to air, all of the flux can be considered as flowing in the steel and is essentially of equal magnitude in all parts of the core. The equation for the flux in the core can be written as follows:

$$0 = \frac{349\, E\, A}{f\, N} \qquad (3.2)$$

where

    A = area of the core in square inches
    E = applied alternating voltage
    f = frequency in cycles/second
    N = number of turns in the winding

It is useful in transformer design to use flux density so that Eq. (3.2) can be written as follows:

$$B = \frac{0}{A} = \frac{349\, E}{f\, A\, N} \qquad (3.3)$$

where B = flux density in Tesla.

Equivalent Circuit of an Iron Core Transformer

When voltage is applied to the exciting or primary winding of the transformer, a magnetizing current flows in the primary winding. This current produces the flux in the core. The flow of flux in magnetic circuits is analogous to the flow of current in electrical circuits.

When flux flows in the steel core, losses occur in the steel. There are two components of this loss which are termed "eddy" and "hystersis" losses. An explanation of these losses would require a full chapter. For the purpose of this text, it can be stated that the hysteresis loss is caused by the cyclic reversal of flux in the magnetic circuit . The eddy loss is caused by the flow of flux normal to the width of the core. Eddy loss can be expressed as follows:

$$W = K [w]^2 [B]^2 \qquad (3.4)$$

where

    K = constant
    w = width of the material normal to the flux
    B = flux density

If a solid core were used in a power transformer, the losses would be very high and the temperature would be excessive. For this reason, cores are laminated from very thin sheets such as 0.23 mm and 0.28 mm to reduce the losses. Each sheet is coated with a very thin material to prevent shorts between the laminations. Improvements made in electrical steels over the past 50 years have been the major contributor to smaller and more efficient transformers. Some of the more dramatic improvements are as follows:

- Development of grain-oriented electrical steels in the mid-1940s.
- Introduction of thin coatings with good mechanical properties.
- Improved chemistry of the steels.
- Introduction of laser scribed steels.
- Further improvement in the orientation of the grains.
- Continued reduction in the thickness of the laminations to reduce the eddy loss component of the core loss.

The combination of these improvements has resulted in electrical steels having less than 50% of the no load loss and 30% of the exciting current that was possible in the late 1940s.

The current to cause rated flux to exist in the core is called the magnetizing current. The magnetizing circuit of the transformer can be represented by one branch in the equivalent circuit shown in Fig. 3.4. The core losses are represented by [Xr], and the excitation characteristics by [Xm].

When the magnetizing current, which is about 0.5% of the load current, flows in the primary winding, there is a small voltage drop across the resistance of the winding and a small inductive drop across the inductance of the winding. We can represent these voltage drops as Rl and Xl in the equivalent circuit. However, these drops are very small and can be neglected in the practical case.

Since the flux flowing in all parts of the core is essentially equal, the voltage induced in any turn placed around the core will be the same. This results in the unique characteristics of transformers with steel cores. Multiple secondary windings can be placed on the core to obtain different output voltages. Each turn in each winding will have the same voltage induced in it. Refer to Fig. 3.5.

The ratio of the voltages at the output to the input at no load will be equal to the ratio of the turns. The voltage drops in the resistance and reactance at no load are very small with only magnetizing current flowing in the windings so that the voltage appearing at A can be considered to be the input voltage. The relationship $E1/N1 = E2/N2$ is important in transformer design and application.

A steel core has a nonlinear magnetizing characteristic as shown in Fig. 3.6. As shown, greater ampere turns are required as the flux density B is increased. Above the knee of the curve as the flux approaches saturation, a small increase in the flux density requires a large increase in the ampere turns. When the core saturates, the circuit behaves much the same as an air core.

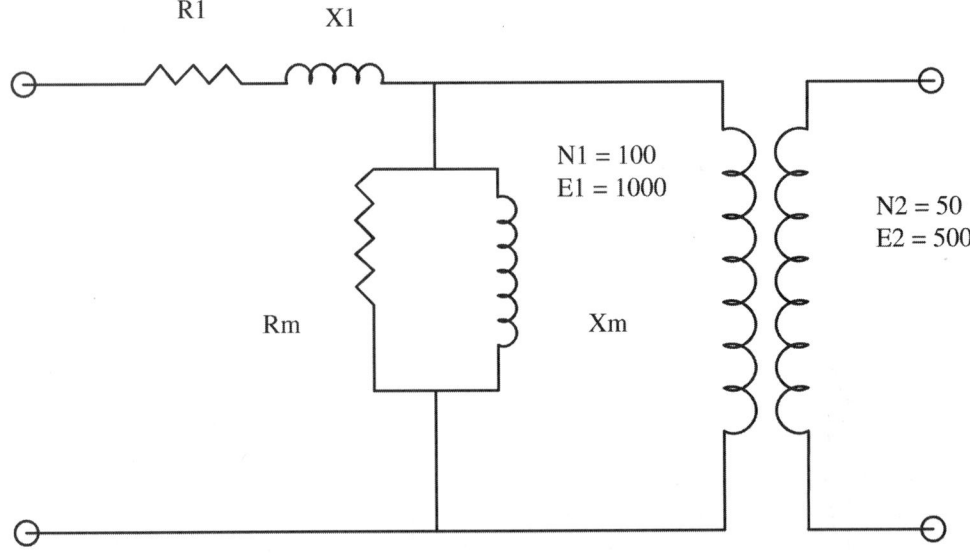

FIGURE 3.4

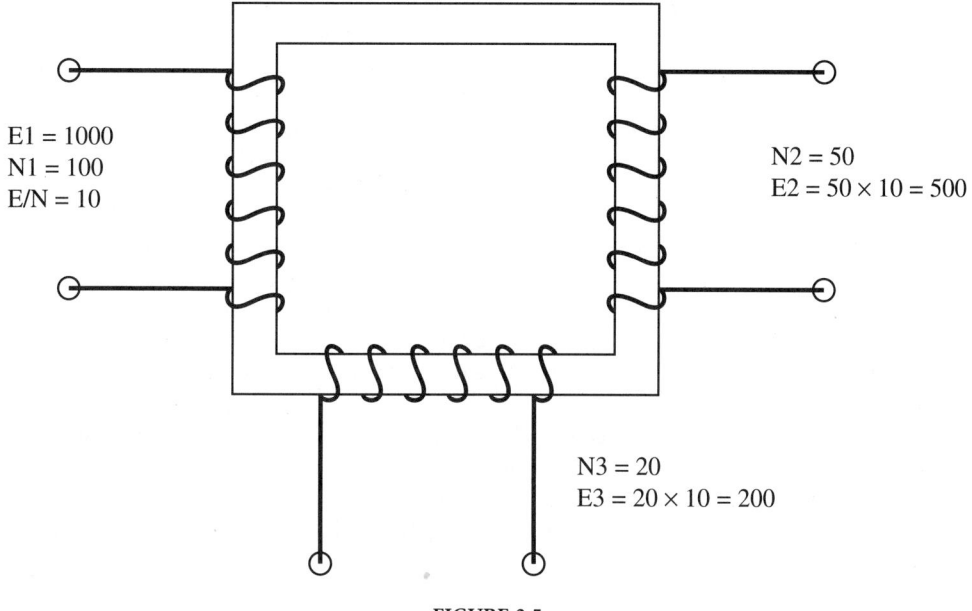

FIGURE 3.5

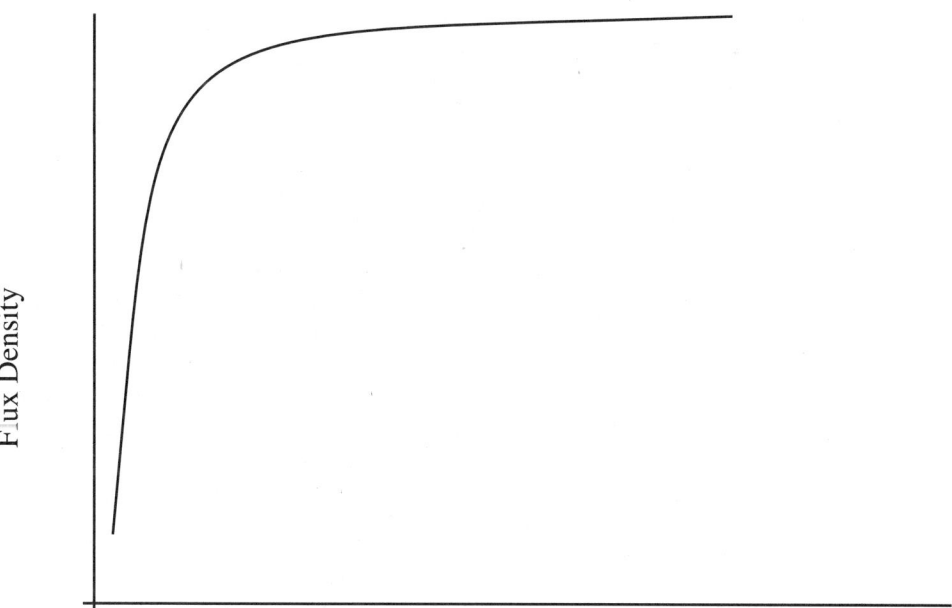

Ampere Turns

FIGURE 3.6

The Practical Transformer

Magnetic Circuit

In actual transformer design, the constants for the ideal circuit are determined from tests on materials and on transformers. For example, the resistance component of the core loss, usually called no load loss, is determined from curves derived from tests on samples of electrical steel and measured transformer no load losses. The designer will have curves for the different electrical steel grades as a function of induction. In the same manner, curves have been made available for the exciting current as a function of induction.

A very important relationship is derived from Eq. (3.4). It can be written in the following form.

$$B = \frac{349 [E/N]}{f A} \tag{3.5}$$

The term [E/N] is called "volts per turn". It determines the number of turns in the windings, the flux density in the core, and is a variable in the leakage reactance which will be discussed below. In fact, when the designer starts to make a design for an operating transformer, one of the first things selected is the volts per turn.

The no load loss in the magnetic circuit is a guaranteed value in most designs. The designer must select an induction level that will allow him to meet the guarantee. The design curves or tables usually show the loss/# or loss/kg as a function of the material and the induction.

The induction must also be selected so that the core will be below saturation under specified overvoltage conditions. Saturation is around 2.0 T.

Leakage Reactance

When the practical transformer is considered, additional concepts must be introduced. For example, the flow of load current in the windings results in high magnetic fields around the windings. These fields are termed leakage flux fields. The term is believed to have started in the early days of transformer theory when it was thought that this flux "leaked" out of the core. This flux exists in the spaces between windings and in the spaces occupied by the windings. See Fig. 3.7. These flux lines effectively result in an impedance between the windings, which is termed "leakage reactance" in the industry. The magnitude of this reactance is a function of the number of turns in the windings, the current in the windings, the leakage field, and the geometry of the core and windings. The magnitude of the leakage reactance is usually in the range of 4 to 10% at the base rating of power transformers. The load current through this reactance results in a considerable voltage drop. Leakage reactance is termed "percent leakage reactance" or "percent reactance". Percent reactance is the ratio of the reactance voltage drop to the winding voltage × 100. It is calculated by designers using the number of turns, the magnitude of the current and the leakage field, and the geometry of the transformer. It is measured by short circuiting one winding of the transformer and increasing the voltage on the other winding until rated current flows in the windings. This voltage divided by the rated winding voltage times 100 is the percent reactance voltage or percent reactance. The voltage drop across this reactance results in the voltage at the load being less than the value determined by the turns ratio. The percentage decrease in the voltage is termed "regulation". Regulation is a function of the power factor of the load, and it can be determined using the following equation for inductive loads:

$$\% \text{Reg.} = \% R [\cos 0] + \% X [\sin 0] + \frac{[\% X (\cos 0) - \% R (\sin 0)]^2}{200}$$

where

    % Reg. = percentage voltage drop across the resistance and the leakage reactance
    % R = % resistance = kilowatts of load loss/kVA of transformer × 100
    % X = % leakage reactance
    0 = angle corresponding to the power factor of the load. If the power factor is 0.9, the angle is 36.87°.

For capacitance loads, change the sign of the sin terms.

Transformers

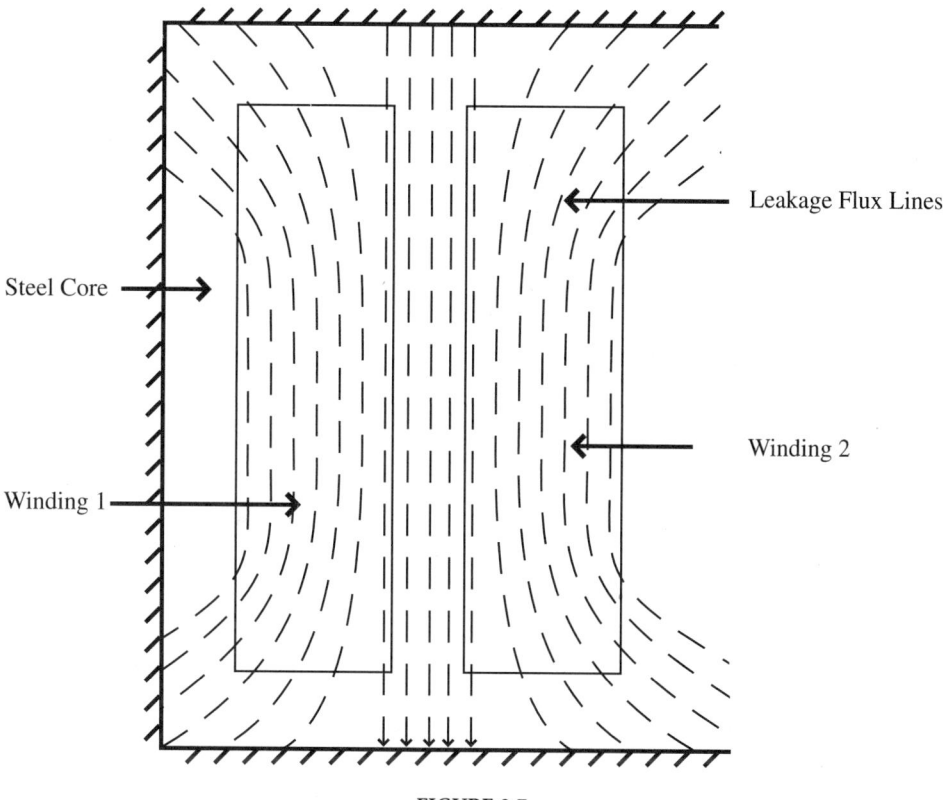

FIGURE 3.7

In order to compensate for these voltage drops, taps are usually added in the windings. The unique volts/turn feature of steel core transformers makes it possible to add or subtract turns to change the voltage outputs of windings. A simple illustration is shown in Fig. 3.8.

Load Losses

This term represents the losses in the transformer that result from the flow of load current in the windings. Load losses are composed of the following elements.

- Resistance losses as the current flows through the resistance of the conductors and leads.
- Eddy losses. These losses are caused by the leakage field, and they are a function of the second power of the leakage field density and the second power of the conductor dimensions normal to the field.
- Stray losses. The leakage field exists in parts of the core, steel structural members, and tank walls. Losses result in these members.

Again, the leakage field caused by flow of the load current in the windings is involved and the eddy and stray losses can be appreciable in large transformers.

Short Circuit Forces

Forces exist between current-carrying conductors when they are in an alternating current field. These forces are determined using the following equation:

$$F = B\ I\ \sin 0$$

where

F = force density
0 = angle between the flux and the current. (In transformers, sin 0 is almost always equal to 1.)

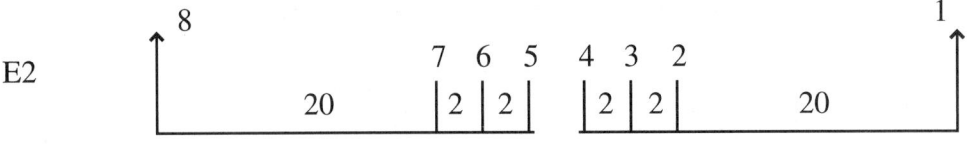

$$E1 = 100$$
$$N1 = 10$$
$$E/N = 10$$

$$E2 = E/N \times N2$$

N2	E2
4 to 5 = 48	E2 = 10 × 48 = 480 Volts
4 to 6 = 46	E2 = 10 × 46 = 460 Volts
3 to 6 = 44	E2 = 10 × 44 = 440 Volts
3 to 7 = 42	E2 = 10 × 42 = 420 Volts
2 to 7 = 40	E2 = 10 × 40 = 400 Volts

FIGURE 3.8

Since the leakage flux field is between windings and has a rather high density, the forces can be quite high. This is a special area of transformer design. Complex programs are needed to get a reasonable representation of the field in different parts of the windings. Much effort has gone into the study of stresses in the windings and the withstand criteria for different types of conductors and support systems. This subject is obviously very broad and beyond the scope of this section.

Thermal Considerations

The losses in the windings and the core cause temperature rises in the materials. This is another important area in which the temperatures must be limited to the long-term capability of the insulating materials. Refined paper is still used as the primary solid insulation in power transformers. Highly refined mineral oil is still used as the cooling and insulating medium in power transformers. Gases and vapors have been introduced in a limited number of special designs. The temperatures must be limited to the thermal capability of these materials. Again, this subject is quite broad and involved. It includes the calculation of the temperature rise of the cooling medium, the average and hottest spot rise of the conductors and leads, and the heat exchanger equipment.

Voltage Considerations

A transformer must withstand a number of different voltage stresses over its expected life. These voltages include:

- The operating voltages at the rated frequency
- Rated frequency overvoltages

- Natural lightning impulses that may strike the transformer or transmission lines
- Switching surges that result from opening and closing breakers and switches
- Combinations of the above voltages

This is a very specialized field in which the resulting voltage stresses must be calculated in the windings and withstand criteria must be established for the different voltages and combinations of voltages. The designer must design the insulation system so that it will withstand these various stresses.

3.2 Power Transformers

H. Jin Sim and Scott H. Digby

A transformer has been defined by ANSI/IEEE as a static electrical device, involving no continuously moving parts, used in electric power systems to transfer power between circuits through the use of electromagnetic induction. The term *power* transformer is used to refer to those transformers used between the generator and the distribution circuits and are usually rated at 500 kVA and above. Power systems typically consist of a large number of generation locations, distribution points, and interconnections within the system or with nearby systems, such as a neighboring utility. The complexity of the system leads to a variety of transmission and distribution voltages. Power transformers must be used at each of these points where there is a transition between voltage levels.

Power transformers are selected based on the application, with the emphasis towards custom design being more apparent the larger the unit. Power transformers are available for step-up operation, primarily used at the generator and referred to as generator step-up (GSU) transformers, and for step-down operation, mainly used to feed distribution circuits. Power transformers are available as a single phase or three phase apparatus.

The construction of a transformer depends upon the application, with transformers intended for indoor use primarily dry-type but also as liquid immersed and for outdoor use usually liquid immersed. This section will focus on the outdoor, liquid-immersed transformers, such as those shown in Fig. 3.9.

FIGURE 3.9 20 MVA, 161:26.4 × 13.2 kV with LTC, three-phase transformers.

TABLE 3.1 Standard Limits for Temperature Rises Above Ambient

Average winding temperature rise	65°C[a]
Hot spot temperature rise	80°C
Top liquid temperature rise	65°C

[a] The base rating is frequently specified and tested as a 55°C rise.

Rating and Classifications

Rating

In the U.S., transformers are rated based on the power output they are capable of delivering continuously at a specified rated voltage and frequency under "usual" operating conditions without exceeding prescribed internal temperature limitations. Insulation is known to deteriorate, among other factors, with increases in temperature, so insulation used in transformers is based on how long it can be expected to last by limiting operating temperatures.

The temperature that insulation is allowed to reach under operating conditions essentially determines the output rating of the transformer, called the kVA rating. Standardization has led to temperatures within a transformer being expressed in terms of the rise above ambient temperature, since the ambient temperature can vary under operating or test conditions. Transformers are designed to limit the temperature based on the desired load, including the average temperature rise of a winding, the hottest spot temperature rise of a winding, and, in the case of liquid-filled units, the top liquid temperature rise. To obtain absolute temperatures from these values, simply add the ambient temperature. Standard temperature limits for liquid-immersed power transformers are listed in Table 3.1.

The normal life expectancy of power transformers is generally assumed to be about 30 years of service when operated within their ratings; however, they may be operated beyond their ratings, overloaded, under certain conditions with moderately predictable "loss of life". Situations that may involve operation beyond rating are emergency re-routing of load or through-faults prior to clearing.

Outside the U.S., the transformer rating may have a slightly different meaning. Based on some standards, the kVA rating can refer to the power that can be input to a transformer, the rated output being equal to the input minus the transformer losses.

Power transformers have been loosely grouped into three market segments based upon size ranges. These three segments are:

1. Small power transformers   500 to 7500[1] kVA
2. Medium power transformers   7500[1] to 100 MVA
3. Large power transformers   100 MVA and above

It was noted that the transformer rating is based on "usual" service conditions, as prescribed by standards. Unusual service conditions may be identified by those specifying a transformer so that the desired performance will correspond to the actual operating conditions. Unusual service conditions include, but are not limited to, the following: high (above 40°C) or low (below −20°C) ambient temperatures; altitudes above 3300 ft above sea level; seismic conditions; and loads with harmonic content above 0.05 per unit.

Insulation Classes

The insulation class of a transformer is determined based on the test levels that it is capable of withstanding. Transformer insulation is rated by the BIL, or Basic Insulation Impulse Level, in conjunction with the voltage rating. Internally, a transformer is considered to be a non-self-restoring insulation system, mostly consisting

---

[1] The upper range of small power and the lower range of medium power can vary between 2500 and 10,000 kVA throughout the industry.

of porous, cellulose material impregnated by the liquid insulating medium. Externally, the transformer's bushings and, more importantly, the surge protection equipment must coordinate with the transformer rating to protect the transformer from transient overvoltages and surges. Standard insulation classes have been established by standards organizations stating the parameters by which tests are to be performed.

Wye connected transformers will typically have the common point brought out of the tank through a neutral bushing. Depending on the application, for example in the case of a solidly grounded neutral vs. a neutral grounded through a resistor or reactor or even an ungrounded neutral, the neutral may have a lower insulation class than the line terminals. There are standard guidelines for rating the neutral based on the situation. It is important to note that the insulation class of the neutral may limit the test levels of the line terminals for certain tests, such as the applied potential, or hi-pot, test where the entire circuit is brought up to the same voltage level. A reduced rating for the neutral can significantly reduce the cost of larger units and autotransformers as opposed to a fully rated neutral.

Cooling Classes

Since no transformer is truly an "ideal" transformer, each will incur a certain amount of energy loss, mainly that which is converted to heat. Methods of removing this heat can depend on the application, the size of the unit, and the amount of heat that needs to be dissipated.

The insulating medium inside a transformer, usually oil, serves multiple purposes, first to act as an insulator, and second to provide a good medium through which to remove heat.

The windings and core are the primary sources of heat; however, internal metallic structures can act as a heat source as well. It is imperative to have proper cooling ducts and passages in proximity to the heat sources through which the cooling medium can flow such that the heat can be effectively removed from the transformer. The natural circulation of oil through a transformer through convection has been referred to as a "thermosiphon" effect. The heat is carried by the insulating medium until it is transferred through the transformer tank wall to the external environment. Radiators, typically detachable, provide an increase in the convective surface area without increasing the size of the tank. In smaller transformers, integral tubular sides or fins are used to provide this increase in surface area. Fans can be installed to increase the volume of air moving across the cooling surfaces thus increasing the rate of heat dissipation. Larger transformers that cannot be effectively cooled using radiators and fans rely on pumps that circulate oil through the transformer and through external heat exchangers, or coolers, which can use air or water as a secondary cooling medium.

Allowing liquid to flow through the transformer windings by natural convection is also identified as non-directed flow. In cases where pumps are used, and even some instances where only fans and radiators are being used, the liquid is often guided into and through some or all of the windings. This is called directed flow in that there is some degree of control of the flow of the liquid through the windings. The difference between directed and non-directed flow through the winding in regard to winding arrangement will be discussed further with the description of winding types.

The use of auxiliary equipment such as fans and pumps with coolers, called forced circulation, increases the cooling and thereby the rating of the transformer without increasing the unit's physical size. Ratings are determined based on the temperature of the unit as it coordinates with the cooling equipment that is operating. Usually, a transformer will have multiple ratings corresponding to multiple stages of cooling, as equipment can be set to run only at increased loads.

Methods of cooling for liquid-immersed transformers have been arranged into cooling classes identified by a four-letter designation as follows.

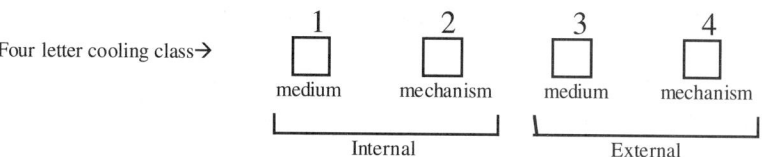

Table 3.2 lists the code letters that are used to make up the four-letter designation.

TABLE 3.2 Cooling Class Letter Descriptions

		Code Letter	Description
Internal	First letter (Cooling medium)	O	Liquid with flash point less than or equal to 300°C
		K	Liquid with flash point greater than 300°C
		L	Liquid with no measurable flash point
	Second letter (Cooling mechanism)	N	Natural convection through cooling equipment and windings
		F	Forced circulation through cooling equipment, natural convection in windings
		D	Forced circulation through cooling equipment, directed flow in main windings
External	Third letter (Cooling medium)	A	Air
		W	Water
	Fourth letter (Cooling medium)	N	Natural convection
		F	Forced circulation

This system of identification has come about through standardization between different international standards organizations and represents a change from what has traditionally been used in the U.S. Where OA classified a transformer as liquid-immersed self-cooled in the past, it is designated by the above system as ONAN. Similarly, the previous FA classification is identified as ONAF. FOA could be OFAF or ODAF, depending on whether directed oil flow is employed or not. In some cases, there are transformers with directed flow in windings without forced circulation through cooling equipment.

An example of multiple ratings would be ONAN/ONAF/ONAF, where the transformer has a base rating where it is cooled by natural convection and two supplemental ratings where groups of fans are turned on to provide additional cooling so the transformer will be capable of supplying additional kVA. This rating would have been designated OA/FA/FA per past standards.

Short Circuit Duty

A transformer supplying a load current will have a complicated network of internal forces acting on and stressing the conductors, support structures, and insulation structures. These forces are fundamental to the interaction of current-carrying conductors within magnetic fields involving an alternating current source. Increases in current result in increases in the magnitude of the forces proportional to the square of the current. Severe overloads, particularly through-fault currents resulting from external short circuit events, involve significant increases in the current above rated current and can result in tremendous forces inside the transformer.

Since the fault current is a transient event, it will have the offset sinusoidal waveshape decaying with time based on the time constant of the equivalent circuit that is characteristic of switching events. The amplitude of the basic sine wave, the symmetrical component, is determined from the formula

$$I_{sc} = I_{rated} / (Z_{xfmr} + Z_{sys}) \tag{3.6}$$

where Z_{xfmr} and Z_{sys} are the transformer and system impedances, respectively, expressed in per unit, and I_{sc} and I_{rated} are the short circuit and rated currents. An offset factor, K, determines the magnitude of the first peak, the asymmetrical peak, of the transient current when multiplied by the I_{sc} found above and the square root of 2 to convert from r.m.s. value. This offset factor is derived from the equivalent transient circuit; however, standards give values that must be used based upon the ratio of the effective inductance (x) and resistance (r), x/r.

As indicated by Eq. (3.6), the short circuit current is primarily limited by the internal impedance of the transformer, but may be further reduced by impedances of adjacent equipment, such as current limiting reactors, or by system power delivery limitations. Existing standards define the magnitude and duration of the fault current based on the rating of the transformer.

The transformer must be capable of withstanding the maximum forces experienced at the first peak of the transient current as well as the repeated pulses at each of the subsequent peaks until the fault is cleared or the transformer is disconnected. The current will experience two peaks per cycle, so the forces will pulsate at 120 Hz, twice the power frequency, acting as a dynamic load. Magnitudes of forces during these situations can range from several thousand pounds to millions of pounds in large power transformers. For analysis, the forces acting on the windings are generally broken up into two subsets, radial and axial forces, based on their apparent effect on the windings. Figure 3.10 illustrates the difference between radial and axial forces in a pair of circular windings.

The high currents experienced during through-fault events will also cause elevated temperatures in the windings. Limitations are also placed on the calculated temperature the conductor may reach during fault conditions. These high temperatures are rarely a problem due to the short time span of these events, but the transformer may experience an associated "loss of life" increase. This "loss of life" can become more prevalent, even critical, based on the duration of the fault conditions and how often such events occur. It is also possible for the conductor to experience changes in mechanical strength due to annealing that can occur at high temperatures. The temperature at which this can occur will depend on the properties and composition of the conductor material, such as the hardness, which is sometimes increased through cold-working processes, or the presence of silver in certain alloys.

Efficiency and Losses

Efficiency

Power transformers are very efficient pieces of equipment with efficiencies typically above 99%. The efficiency is derived from the rated output and the losses incurred in the transformer. The basic relationship for efficiency is the output over the input, which according to U.S. standards translates to

$$\text{Efficiency} = \left[\text{kVA rating} \big/ \left(\text{kVA rating} + \text{Total losses}\right)\right] * 100\% \quad (3.7)$$

and will generally decrease slightly with increases in load. Total losses are the sum of the no-load and load losses.

Losses

The no-load losses are essentially the power required to keep the core energized, and so are many times referred to as the core losses. They exist whenever the unit is energized. No-load losses depend primarily upon the voltage and frequency, so under operational conditions it will only vary slightly with system variations. Load losses, as the terminology might suggest, result from load currents flowing through the transformer. The two components of the load losses are the I^2R losses and the stray losses. I^2R losses are based on the measured DC resistance, the bulk of which is due to the winding conductors, and the current at a given load. The stray losses are a term given to the accumulation of the additional losses experienced by the transformer, which includes winding eddy losses and losses due to the effects of leakage flux entering internal metallic structures. Auxiliary losses refer to the power required to run auxiliary cooling equipment, such as fans and pumps, and are not typically included in the total losses as defined above.

Economic Evaluation of Losses

Transformer losses represent power that cannot be delivered to customers and therefore have an associated economic cost to the transformer user/owner. A reduction in transformer losses generally results in an increase in the transformer's cost. Depending on the application, there may be an economic benefit to a transformer with reduced losses and high price (initial cost), and vice versa. This process is typically

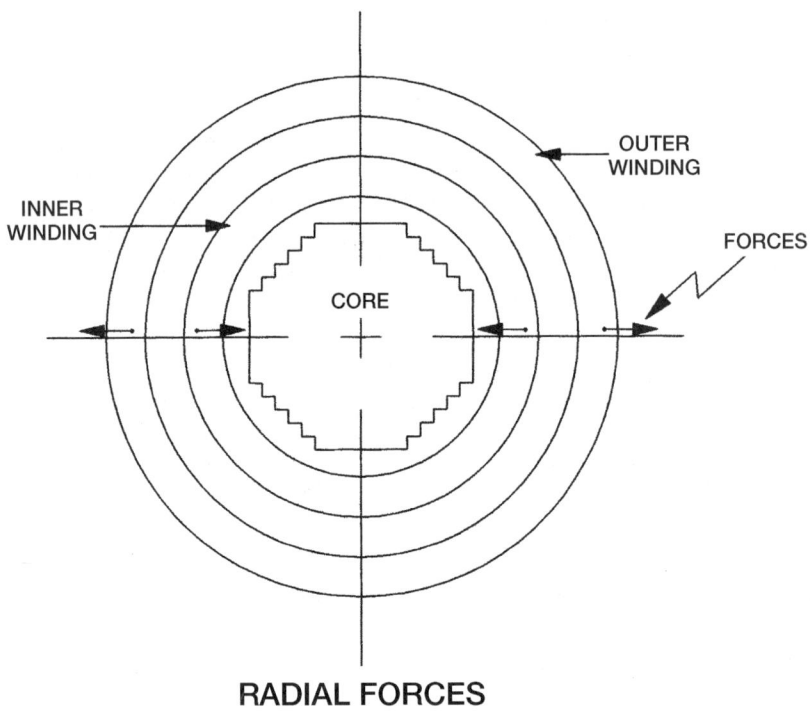

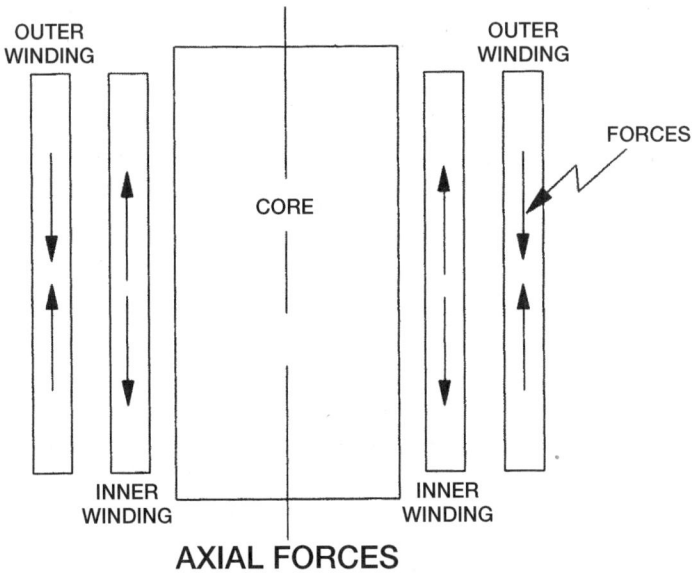

FIGURE 3.10 Radial and axial forces in a transformer winding.

dealt with through the use of "loss evaluations", which place a dollar value on the transformer losses to calculate a total owning cost that is a combination of the price and the losses. Typically, each of the transformer's individual loss parameters, no-load losses, load losses, and auxiliary losses, are assigned a dollar value per kilowatt ($/kW). Information obtained from such an analysis can be used to compare prices from different manufacturers or to decide on the optimum time to replace existing transformers. There are guides available, through standards organizations, for the estimation of the cost associated with transformer losses. Loss evaluation values can range from about $500/kW upwards of $12000/kW for the no-load losses and from a few hundred dollars per kilowatt to about $6000 to 8000/kW for load losses and auxiliary losses. Values will depend upon the application.

Construction

The construction of a power transformer will vary throughout the industry to a certain degree. The basic arrangement is essentially the same and has seen little significant change in recent years, so some of the variations may be discussed here.

The Core

The core, which provides the magnetic path to channel the flux, consists of thin strips of high-grade steel, called laminations, which are electrically separated by a thin coating of insulating material. The strips can be stacked or wound, with the windings either built integrally around the core or built separately and assembled around the core sections. Core steel may be hot or cold rolled, grain oriented or non-grain oriented, and even laser-scribed for additional performance. Thickness ranges from 9 mils (1 mil = 1 thousandth of an inch) upwards of 14 mils. The core cross-section may be circular or rectangular, with circular cores commonly referred to as cruciform construction. Rectangular cores are used for smaller ratings and as auxiliary transformers used within a power transformer. Rectangular cores, obviously, use a single width of strip steel, while circular cores use a combination of different strip widths to approximate a circular cross-section. The type of steel and arrangement will depend on the transformer rating as related to cost factors such as labor and performance.

Just like other components in the transformer, the heat generated by the core must be adequately dissipated. While the steel and coating may be capable of withstanding higher temperatures, it will come in contact with insulating materials with limited temperature capabilities. In larger units, cooling ducts are used inside the core for additional convective surface area and sections of laminations may be split to reduce localized losses.

The core will be held together by, but insulated from, mechanical structures and will be grounded to a single point, usually some readily accessible point inside the tank, but may also be brought through a bushing on the tank wall or top for external access. This grounding point should be removable for testing purposes, such as checking for unintentional core grounds.

The maximum flux density of the core steel is normally designed as close to the knee of the saturation curve as practical, accounting for required over-excitations and tolerances that exist due to materials and manufacturing processes. For power transformers, the flux density is typically between 13 and 18 kG with the saturation point for magnetic steel being around 20.3 to 20.5 kG.

The two basic types of core construction used in power transformers are called core-form and shell-form.

In core-form construction, there is a single path for the magnetic circuit. Figure 3.11 shows a schematic of a single-phase core with the arrows showing the magnetic path. For single-phase applications, the windings are typically divided on both core legs as shown, whereas in three-phase applications, the windings of a particular phase are typically on the same core leg, as illustrated in Fig. 3.12. Windings are constructed separate of the core and placed on their respective core legs during core assembly. Figure 3.13 shows what is referred to as the "E"-assembly of a three-phase core-form core during assembly.

In shell-form construction, the core provides multiple paths for the magnetic circuit. A schematic of a single-phase shell-form core is shown in Fig. 3.14, with the two magnetic paths illustrated. The core is typically stacked directly around the windings, which are usually "pancake" type windings, although some applications are such that the core and windings are assembled similar to core form. Due to advantages in short circuit and transient voltage performance, shell forms tend to be used more frequently in larger transformers where conditions can be more severe. There are variations of three-phase shell-form construction that include five- and seven-legged cores, depending on size and application.

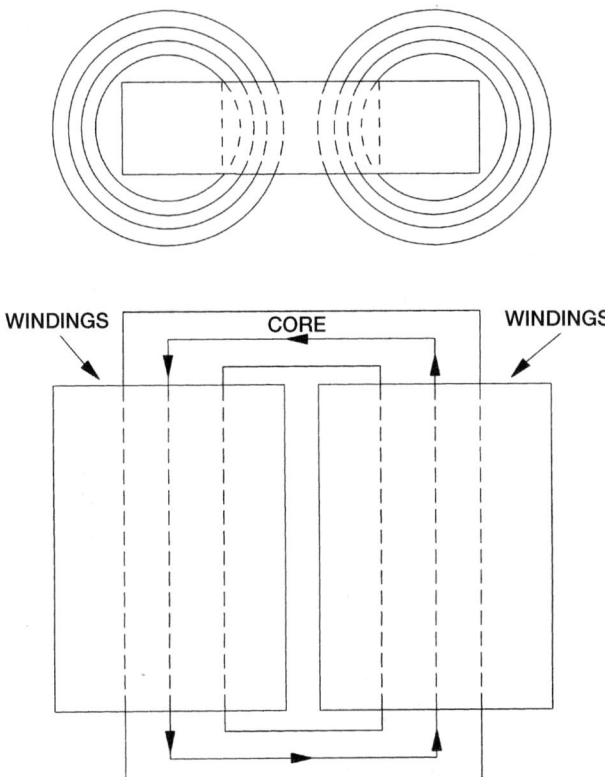

FIGURE 3.11 Schematic of single-phase core-form construction.

The Windings

The windings consist of the current carrying conductors wound around the sections of the core and must be properly insulated, supported, and cooled to withstand operational and test conditions. The terms winding and coil are used interchangeably in this discussion.

Copper and aluminum are the primary materials used as conductors in power transformer windings. While aluminum is lighter and generally less expensive than copper, a larger cross-section of aluminum conductor must be used to carry a current with similar performance as copper. Copper has higher mechanical strength and is used almost exclusively in all but the smaller size ranges, where aluminum conductors may be perfectly acceptable. In cases where extreme forces are encountered, materials such as silver-bearing copper may be used for even greater strength. The conductors used in power transformers will typically be stranded with a rectangular cross-section, although some transformers at the lowest ratings may use sheet or foil conductors. A variation involving many rectangular conductor strands combined into a cable is called continuously transposed cable (CTC), as shown in Fig. 3.15.

In core-form transformers, the windings are usually arranged concentrically around the core leg, as illustrated by Fig. 3.16 of a winding being lowered over another winding already on the core leg of a three-phase transformer. A schematic of coils arranged in this three-phase application was also shown in Fig. 3.12. Shell-form transformers may use a similar concentric arrangement or windings may be stacked into sections or groups as illustrated by Fig. 3.17 and as seen in the picture in Fig. 3.21.

When considering concentric windings, it is generally understood that circular windings have inherently higher mechanical strength than rectangular windings, whereas rectangular coils can have lower associated material and labor costs. Rectangular windings permit a more efficient use of space, but their use is limited to small power transformers and the lower range of medium power transformers where

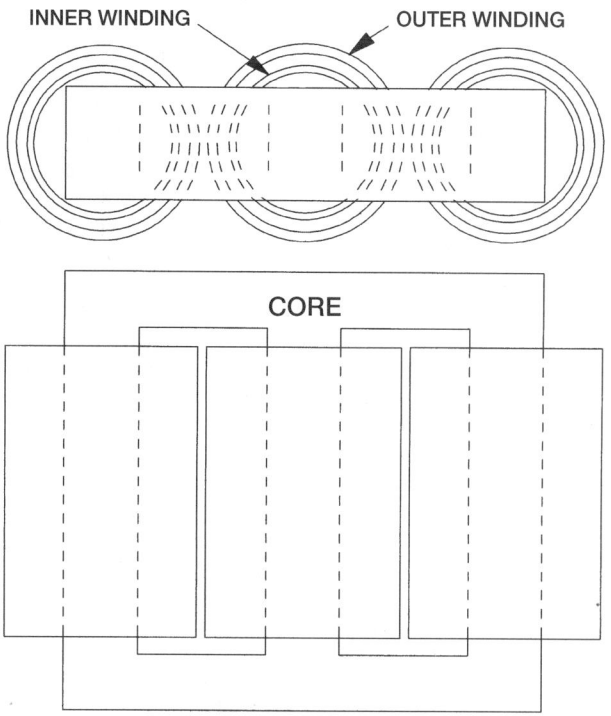

FIGURE 3.12 Schematic of three-phase core-form construction.

FIGURE 3.13 "E"-assembly, prior to insertion of top yoke.

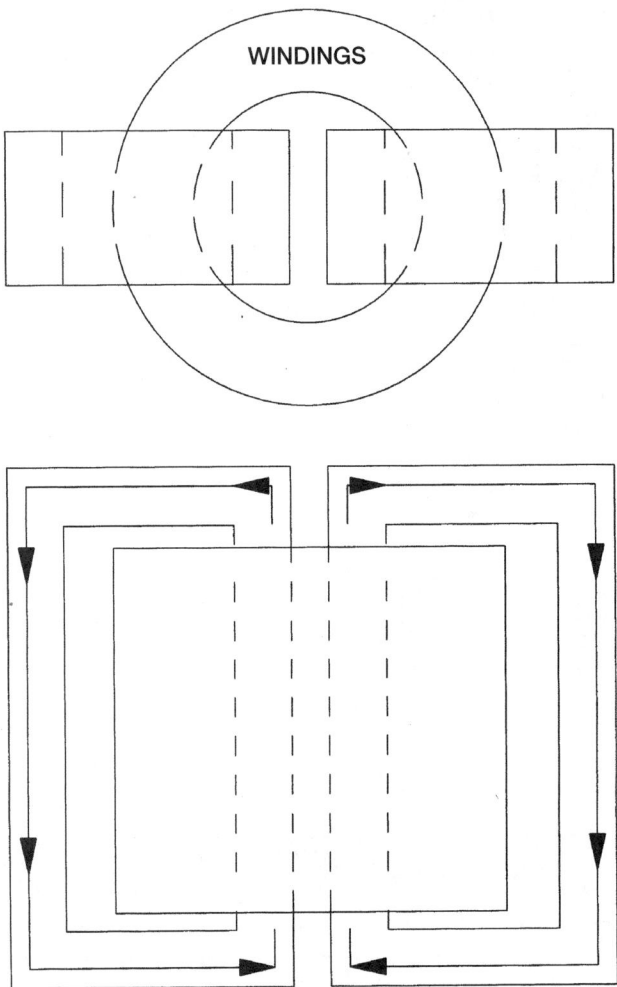

FIGURE 3.14 Schematic of single-phase shell-form construction.

the internal forces are not extremely high. As the rating increases, the forces significantly increase and there is need for added strength in the windings, so circular coils, or shell-form construction, are used. In some special cases, elliptical-shaped windings can even be used.

Concentric coils will typically be wound over cylinders with spacers attached so as to form a duct between the conductors and the cylinder. As previously mentioned, the flow of liquid through the windings can be based solely on natural convection or the flow can be somewhat controlled through the use of strategically placed barriers within the winding. Figures 3.18 and 3.19 show winding arrangements comparing non-directed and directed flow. This concept is sometimes referred to as guided liquid flow.

There are a variety of different types of windings that have been used in power transformers through the years. Coils can be wound in an upright, vertical orientation, as is necessary with larger, heavier coils, or can be wound horizontally and uprighted upon completion. As mentioned before, the type of winding will depend on the transformer rating as well as the core construction. Several of the more common winding types are discussed below.

While it is recognized that several types of windings are sometimes referred to as "pancake" windings due to the arrangement of conductors into discs, the term most often refers to the type of coil that is almost exclusively used in shell-form transformers. The conductors are wound around a rectangular form, with the widest face of the conductor either oriented horizontally or vertically, with layers of

Transformers

FIGURE 3.15 Continuously transposed cable (CTC).

conductors stacked on top of one another and separated by spacers. Figure 3.20 illustrates how these coils are typically wound. This type of winding lends itself to grouping different windings along the same axial space, as previously shown in Fig. 3.17 and further illustrated in Fig. 3.21.

Layer, or barrel, windings are among the simplest of windings in that the insulated conductors are wound directly next to each other around the cylinder and spacers. Several layers may be wound on top of one another, with the layers separated by solid insulation, ducts, or a combination of both. Several strands may be wound in parallel if the current dictates. Variations of this winding are often used for applications such as tap windings used in load tap changing transformers and for tertiary windings used for, among other things, third harmonic suppression. Figure 3.22 shows a layer winding during assembly that will be used as a regulating winding in an LTC transformer.

Helical windings are also referred to as screw or spiral windings with each term accurately characterizing the coil's construction. A helical winding will consist of anywhere from a few to more than 100 insulated strands wound in parallel continuously along the length of the cylinder, with spacers inserted between adjacent turns or discs and suitable transpositions to minimize circulating currents between strands. The manner of construction is such that the coil will somewhat resemble a corkscrew. Figure 3.23 shows a helical winding during the winding process. Helical windings are used for relatively higher current applications frequently encountered in the lower voltage classes.

A disc winding can involve a single strand or several strands of insulated conductors wound in a series of parallel discs of horizontal orientation with the discs connected at either the inside or outside as a "cross-over" point. Each disc will be comprised of multiple turns wound over other turns with the crossovers alternating between inside and outside. Figure 3.24 outlines the basic concept with Fig. 3.25 showing typical crossovers during the winding process. Most windings 25 kV class and above used in

FIGURE 3.16 Concentric arrangement, outer coil being lowered onto core leg over top of inner coil.

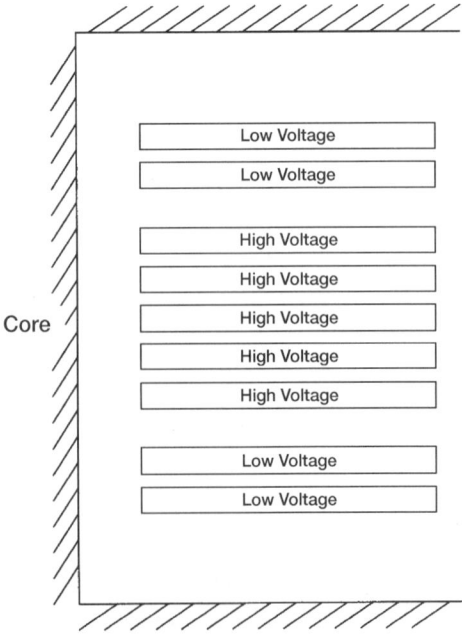

FIGURE 3.17 Example of stacking arrangement of windings in shell-form construction.

Transformers

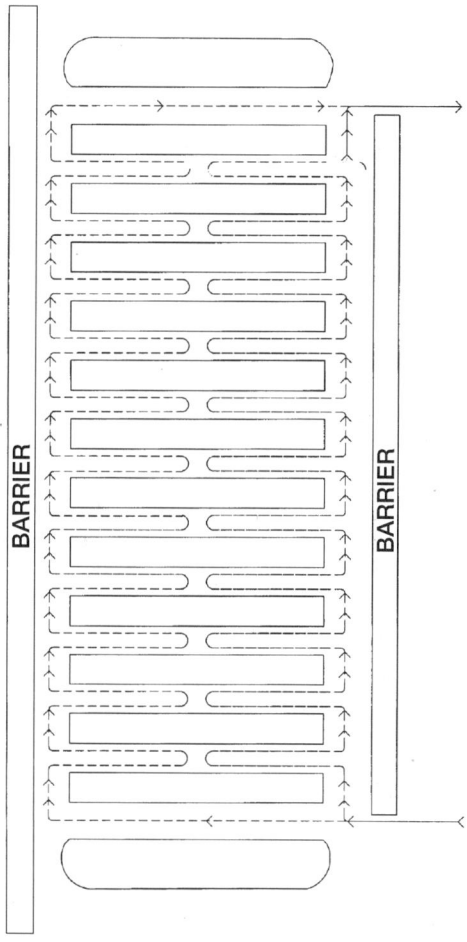

FIGURE 3.18 Non-directed flow.

core-form transformers are disc type so, due to the high voltages involved in test and operation, particular attention must be given to avoid high stresses between discs and turns near the end of the winding when subjected to transient voltage surges. Numerous techniques have been developed to ensure an acceptable voltage distribution along the winding under these conditions.

Taps — Turns Ratio Adjustment

The capability of adjusting the turns ratio of a transformer is oftentimes desirable to compensate for variations in voltage that occur due to loading cycles, and there are several means by which the task can be accomplished. There is a significant difference in a transformer that is capable of changing the ratio while the unit is on-line, referred to as a Load Tap Changing (LTC) transformer, and one that must be taken off-line, or de-energized, to perform a tap change.

Currently, most transformers are provided with a means to change the number of turns in the high-voltage circuit, whereby part may be tapped out of the circuit. In many transformers this is done using one of the main windings and tapping out a section or sections, whereas with larger units a dedicated tap winding may be necessary to avoid ampere-turn voids along the length of the winding that occur in the former case. Use and placement of tap windings vary with the application and among manufacturers. A manually operated switching mechanism, a DETC (De-energized Tap Changer), is normally provided accessible external to the transformer to change the tap position. When LTC capabilities are desired,

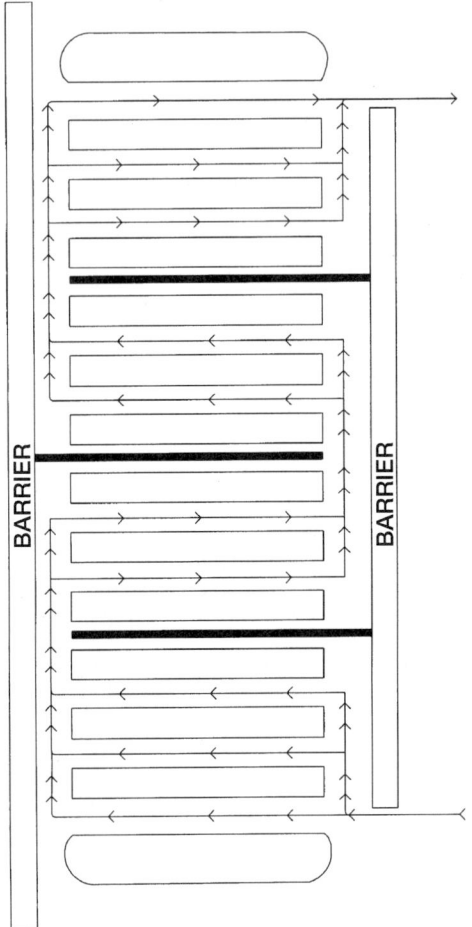

FIGURE 3.19 Directed flow.

FIGURE 3.20 Pancake winding during winding process.

Transformers

FIGURE 3.21 Stacked pancake windings.

additional windings and equipment are required that significantly increase the size and cost of the transformer.

It is also possible for a transformer to have dual voltage ratings, as is popular in spare and mobile transformers. While there is no physical limit to the ratio between the dual ratings, even ratios (for example, 24.94 × 12.47 kV or 138 × 69 kV) are easier for manufacturers to accommodate.

Accessory Equipment

Accessories

There are a great many different accessories used for the purpose of monitoring and protecting power transformers, some of which are considered standard features and others which are used based on miscellaneous requirements. A few of the basic accessories will be discussed briefly.

Liquid Level Indicator

A liquid level indicator is a standard feature on liquid-filled transformer tanks because the liquid medium is critical for cooling and insulation. This indicator is typically a round-faced gauge on the side of the tank with a float and float arm that moves a dial pointer as the liquid level changes.

Pressure Relief Devices

Pressure relief devices are mounted on transformer tanks to relieve excess internal pressures that might build up during operating conditions to avoid damage to the tank itself. On larger transformers, several pressure relief devices may be required due to the large quantities of oil.

FIGURE 3.22 Layer windings (single layer with two strands wound in parallel).

Liquid Temperature Indicator

A liquid temperature indicator will measure the temperature of the internal liquid at a point near the top of the liquid through a probe inserted in a well mounted through the side of the transformer tank.

Winding Temperature Indicator

A winding temperature simulation method is used to approximate the hottest spot in the winding because of the difficulties involved in direct winding temperature measurements. The method applied to power transformers involves a current transformer which will be located to incur a current proportional to the load current through the transformer. The current transformer feeds a circuit that essentially adds heat to the top liquid temperature reading to give an approximate reading of the winding temperature. This method relies on design or test data of the temperature differential between the liquid and the windings, called the winding gradient.

Sudden Pressure Relay

A sudden (or rapid) pressure relay is intended to indicate a quick increase in internal pressure that can occur when there is an internal fault. Varieties are used that can be mounted on the top or side of the transformer or that operate in liquid or gas space.

Desiccant (Dehydrating) Breathers

Desiccant breathers use a material such as silica gel to allow air to enter and exit the tank which removes moisture as the air passes through. Most tanks will be somewhat free-breathing and such a device, if properly maintained, allows a degree of control over the quality of air entering the transformer.

Liquid Preservation Systems

There are several methods in practice to preserve the properties of the transformer liquid and associated insulation structures that it penetrates. Preservation systems attempt to isolate the transformer's internal environment from the external environment (atmosphere) while understanding that a certain degree of

Transformers

FIGURE 3.23 Helical winding during assembly.

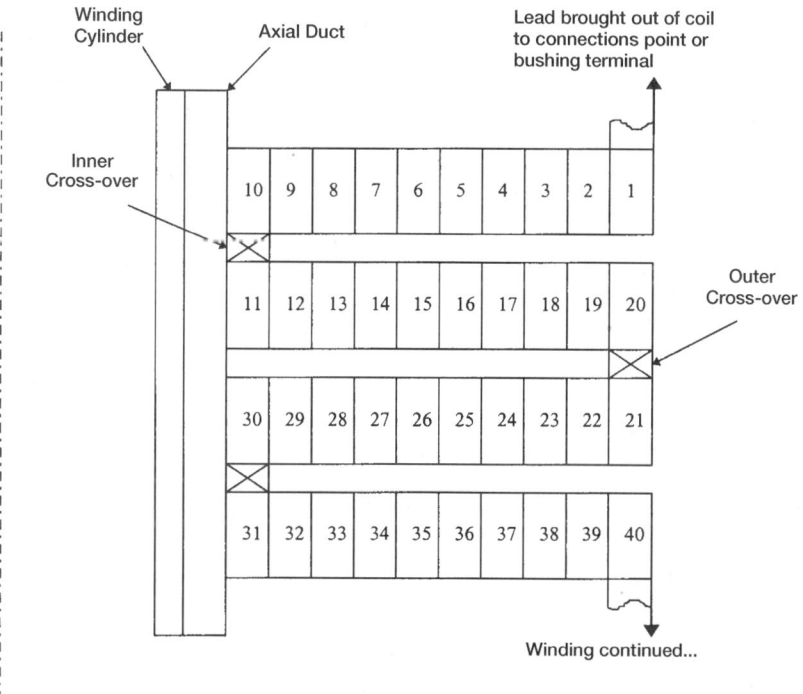

FIGURE 3.24 Basic disc winding layout.

FIGURE 3.25 Disc winding inner and outer crossovers.

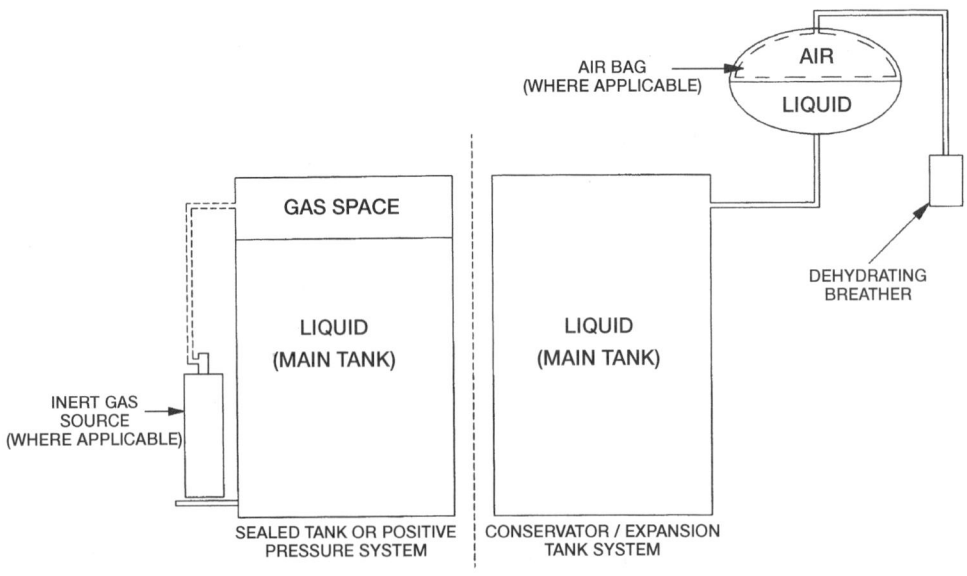

FIGURE 3.26 General arrangements of liquid preservation systems.

interaction or "breathing" is required due to variations in pressure that occur under operational conditions, such as expansion and contraction of liquid with temperature. The most commonly used methods are outlined as follows and illustrated in Fig. 3.26.

1. Sealed tank systems have the tank interior sealed from the atmosphere and will maintain a layer of gas, a gas space or cushion, that will sit above the liquid. The gas plus liquid volume will remain constant. Negative internal pressures can exist in sealed tank systems at lower loads or temperatures with positive pressures existing as load and temperatures increase. A pressure-vacuum bleeder is used to limit operating pressures in transformers over a certain size.
2. Positive pressure systems involve the use of inert gases to maintain a positive pressure in the gas space. A source of inert gas, typically a bottle of compressed nitrogen, will be injected incrementally into the gas space when the internal pressure falls out of range.
3. Conservator (expansion tank) systems are used both with and without air bags, also called bladders or diaphragms, which involve a separate auxiliary tank. The main transformer tank will be completely filled with liquid, the auxiliary tank will be partially filled, and the liquid will expand and contract within the auxiliary tank. The auxiliary tank will be allowed to "breath", usually through a dehydrating breather. The use of an air bag in the auxiliary tank can provide further separation from the atmosphere.

Inrush Current

When a transformer is taken off-line, there will be a certain amount of residual flux that can remain in the core due to the properties of the magnetic core material. The residual flux can be as much as 50 to 90% of the maximum operating flux, depending on the type of core steel. When voltage is reapplied to the transformer, the flux introduced by this source voltage will build upon that which already exists in the core. In order to maintain this level of flux in the core, which can be well into the saturation range of the core steel, the transformer can draw current well in excess of the transformer's rated full load current. Depending on the transformer design, the magnitude of this current inrush can be anywhere from 3.5 to 40 times the rated full load current. The waveform of the inrush current will be similar to a sine wave, but largely skewed towards the positive or negative direction. This inrush current will experience a decay, partially due to losses, which will provide a dampening effect; however, the current can remain well above rated current for many cycles.

This inrush current can have an effect on the operation of relays and fuses located in the system near the transformer. Decent approximations of the inrush current require detailed information regarding the transformer design which may be available from the manufacturer but is not typically available to the user. Actual inrush currents will also depend upon where in the source voltage wave the switching operations occur, the moment of opening effecting the residual flux magnitude, and the moment of closing effecting the new flux.

Modern and Future Developments

High-Voltage Generator (Powerformer)

Because electricity is currently generated at voltage levels that are too low to be efficiently transmitted across the great distances that the power grid typically spans, step-up transformers are required at the generator. With developments in high-voltage cable technology, a high-voltage generator, called the powerformer, has been developed that will eliminate the need for this GSU transformer and associated equipment. This powerformer can reportedly be designed to generate power at voltage levels between 20 and 400 kV to directly feed the transmission network.

High-Temperature Superconducting (HTS) Transformer

Superconducting technologies are being applied to power transformers in the development of what are being referred to as high-temperature superconducting (HTS) transformers. In HTS transformers, the copper and aluminum in the windings would be replaced by superconductors. In the field of superconductors, high temperatures are considered to be in the range of −250 to −200°F, which represents quite a significant deviation in the operating temperatures of conventional transformers. At these temperatures, insulation of the type currently used in transformers would not degrade in the same manner. Using superconducting conductors in transformers requires advances in cooling, specifically refrigeration technology directed toward use in transformers. The predominant cooling medium in HTS development has been liquid nitrogen, but some other mediums have been investigated as well. Transformers built using HTS technology would reportedly be of reduced size and weight and capable of overloads without experiencing "loss of life" due to insulation degradation, instead using an increased amount of the replaceable coolant. An additional benefit would be an increase in efficiency of HTS transformers over conventional transformers due to the fact that resistance in superconductors is virtually zero, thus eliminating the I^2R loss component of the load losses.

References

American National Standard for Transformers — 230 kV and Below 833/958 through 8333/10417 kVA, Single-Phase, and 750/862 through 60000/80000/100000 kVA, Three-Phase Without Load Tap Changing; and 3750/4687 through 60000/80000/100000 kVA with Load Tap Changing — Safety Requirements, ANSI C57.12.10-1997, National Electrical Manufacturers Association, 1998.

Bean, R. L., Chackan, N. Jr., Moore, H. R., and Wentz, E. C., *Transformers for the Electric Power Industry,* McGraw-Hill, New York, 1959.

Goldman, A. W. and Pebler, C. G., *Power Transformers,* Vol. 2, Electrical Power Research Institute, Inc., Palo Alto, CA, 1987.

Hobson, J. E. and Witzke, R. L., Power transformers and reactors, in *ElectricalTransmission and Distribution Reference Book,* 4th ed., Central Station Engineers of the Westinghouse Electric Corporation, Westinghouse Electric Corporation, East Pittsburgh, PA, 1950, chap. 5.

IEEE Standard General Requirements for Liquid-Immersed Distribution, Power, and Regulating Transformers, IEEE C57.12.00-1993, The Institute of Electrical and Electronics Engineers, Inc., 1993.

IEEE Standard Terminology for Power and Distribution Transformers, ANSI/IEEE C57.12.80-1978, The Institute of Electrical and Electronics Engineers, Inc., 1998.

Mehta, S. P., Aversa, N., and Walker, M. S., Transforming transformers, *IEEE Spectrum,* p. 43, 1997.

3.3 Distribution Transformers[2]

Dudley L. Galloway

Historical Background

Long-Distance Power

In 1886, George Westinghouse built the first long-distance alternating current electric lighting system in Great Barrington, Massachusetts. The power source was a 25-hp steam engine driving an alternator with

---

[2]Figures 3.31, 3.32, 3.33, and 3.35 adapted or reprinted from IEEE Std. C57.105-1978 "IEEE Guide for Application of Transformer Connections in Three-Phase Distribution Systems", Copyright © 1978 by the Institute of Electrical and Electronics Engineers, Inc. The IEEE disclaims any responsibility or liability resulting from the placement and use in the described manner. Information is reprinted with the permission of the IEEE.

Figure 3.36 adapted from IEEE Std. C57.12.90-1993 "IEEE Standard Test Code for Liquid-Immersed Distribution, Power, and Regulating Transformers and IEEE Guide for Short Circuit Testing of Distribution and Power Transformers", Copyright © 1993 by the Institute of Electrical and Electronics Engineers, Inc. The IEEE disclaims any responsibility or liability resulting from the placement and use in the described manner. Information is reprinted with the permission of the IEEE.

Figure 3.37 adapted from IEEE Std. C57.12.00-1993 "IEEE Standard General Requirements for Liquid-Immersed Distribution, Power, and Regulating Transformers", Copyright © 1993 by the Institute of Electrical and Electronics Engineers, Inc. The IEEE disclaims any responsibility or liability resulting from the placement and use in the described manner. Information is reprinted with the permission of the IEEE.

All others, by permission of ABB Power T&D Company Inc., Raleigh, NC, and Jefferson City, MO.

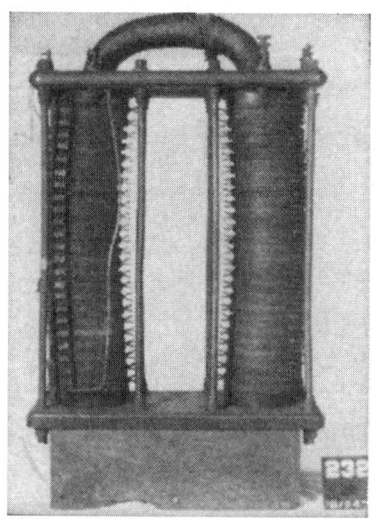 

FIGURE 3.27 (a) The Gaulard and Gibbs transformer. (b) William Stanley's early transformer.

an output of 500 V and 12 A. In the middle of town, 4000 ft away, transformers were used to reduce the voltage to serve light bulbs located in nearby stores and offices (Powel, 1997).

The First Transformers

Westinghouse realized that electric power could only be delivered over distances by transmitting at a higher voltage and then reducing the voltage at the location of the load. He purchased U.S. patent rights to the transformer from Gaulard and Gibbs, shown in Fig. 3.27a. William Stanley, Westinghouse's electrical expert, designed and built the transformers to reduce the voltage from 500 to 100 V on the Great Barrington system. See Fig. 3.27b.

What Is a Distribution Transformer?

Just like the transformers in the Great Barrington system, any transformer that takes voltage from a primary distribution circuit and "steps down" or reduces it to a secondary distribution circuit or a consumer's service circuit is a distribution transformer. Although many industry standards tend to limit this definition by kVA rating (e.g., 5 to 500), distribution transformers can have lower ratings and can have ratings of 5000 kVA or even higher, so the use of kVA ratings to define transformer types is being discouraged (IEEE, 1978).

Construction

Early Transformer Materials

From the pictures, the Gaulard–Gibbs transformer seems to have used a coil of many turns of iron wire to create a ferromagnetic loop. The Stanley model, however, appears to have used flat sheets of iron, stacked together and clamped with wooden blocks and steel bolts. Winding conductors were most likely made of copper from the very beginning. Several methods of insulating the conductor were used in the early days. Varnish dipping was often used and is used for some applications today. Paper tape wrapping of conductors has been used extensively, but has now been almost completely replaced by other methods.

Oil Immersion

In 1887, the year after Stanley designed and built the first transformers in the U.S., Elihu Thompson patented the idea of using mineral oil as a transformer cooling and insulating medium (Myers et al., 1981). Although materials have improved dramatically, the basic concept of an oil-immersed cellulosic insulating system has changed very little in well over a century.

Core Improvements

The major improvement in core materials was the introduction of silicon steel in 1932. Over the years, the performance of electrical steels has been improved by grain orientation (1933) and continued improvement of the chemistry and surface insulating properties. The thinner and more effective the insulating coatings are, the more efficient a particular core material will be. The thinner the laminations of electrical steel, the lower the losses in the core due to circulating currents. Mass production of distribution transformers has made it feasible to replace stacked cores with wound cores. 'C' cores were first used in distribution transformers around 1940. A 'C' core is made from a continuous strip of steel, wrapped and formed into a rectangular shape, then annealed and bonded together. The core is then sawed in half to form two 'C'-shaped sections that are machine-faced and reassembled around the coil. In the mid-1950s, various manufacturers developed wound cores that were die-formed into a rectangular shape and then annealed to relieve their mechanical stresses. The cores of almost all distribution transformers are made this way today, with each turn cut where it laps over itself. This allows the core to be disassembled and put back together around the coil structures while creating a minimum of energy loss in the core. Electrical steel manufacturers now produce steel for wound cores that is from 14 to 7 mils thick in various grades. The recent introduction of an amorphous core steel having reduced losses has offered another choice in the marketplace for utilities that have very high energy costs.

Winding Materials

Conductors for low-voltage windings were originally made from small rectangular copper bars, referred to as "strap." Larger ratings could require as many as 16 of these strap conductors in parallel to make one winding with the proper cross-section. A substantial improvement was gained by using copper strip, which could be much thinner than a strap but the same width as the coil itself. In the early 1960s, instability in the copper market encouraged the use of aluminum strip conductors. The use of aluminum round wire in the primary windings followed in the early 1970s (Palmer, 1983). Today, both aluminum and copper conductors are used in distribution transformers, and the choice is largely dictated by economics. Round wire separated by paper insulation between layers has several disadvantages. The wire tends to "gutter," that is, fall into the troughs in the layer below. Also, the contact between the wire and paper occurs only along two lines on either side of the conductor. This is a significant disadvantage when an adhesive is used to bind the wire and paper together. To prevent these problems, manufacturers often flatten the wire into an oval or rectangular shape in the process of winding the coil. This allows more conductor to be wound into a given size of coil and improves the mechanical and electrical integrity of the coil.

Conductor Insulation

The most common insulation today for high-voltage windings is an enamel coating on the wire, with kraft paper used between layers. Low-voltage strip can be bare with paper insulation between layers. The use of paper wrapping on the strap conductor is slowly being replaced by synthetic polymer coatings or wrapping with synthetic cloth.

Thermally Upgraded Paper

In 1958, manufacturers introduced insulating paper that was chemically treated to resist breakdown due to thermal aging. At the same time, testing programs throughout the industry were showing that the estimates of transformer life being used at the time were extremely conservative. By the early 1960s, citing the functional life testing results, the industry began to change the standard average winding temperature rise for distribution transformers, first to a dual rating of 55/65°C, and then to a single 65°C rating (IEEE, 1995). In some parts of the world, the distribution transformer standard remains at 55°C rise using non-upgraded paper.

Conductor Joining

The introduction of aluminum wire, strap, and strip conductors, and enamel coatings presented a number of challenges to distribution transformer manufacturers. Aluminum spontaneously forms an insulating oxide coating when exposed to air. This oxide coating must be removed or avoided whenever an electrical connection is desired. Also, electrical conductor grades of aluminum are quite soft, and are subject to cold flow and differential expansion problems when mechanical clamping is attempted. Some methods of splicing aluminum wires include soldering, and crimping — with special crimps that penetrate enamel and oxide coatings and seal out oxygen at the contact areas. Aluminum strap or strip conductors can be TIG welded. Aluminum strip can also be cold-welded or crimped to other copper or aluminum connectors. Bolted connections can be made to soft aluminum if the joint area is properly cleaned. 'Belleville' spring washers and proper torquing are used to control the clamping forces and contain the metal that wants to flow out of the joint. Aluminum joining problems are sometimes mitigated by using hard alloy tabs with tin plating to make bolted joints with standard hardware.

Coolants

Mineral Oil

Mineral oil surrounding a transformer core-coil assembly enhances the dielectric strength of the winding and prevents oxidation of the core. Dielectric improvement occurs because oil has a greater electrical withstand than air and because the dielectric constant of oil (2.2) is closer to that of the insulation. As a result, the stress on the insulation is lessened when oil replaces air in a dielectric system. Oil also picks up heat while it is in contact with the conductors and carries the heat out to the tank surface by self-convection. Thus, a transformer immersed in oil can have smaller electrical clearances and smaller conductors for the same voltage and kVA ratings.

Askarels

Beginning about 1932, a class of liquids called askarels or polychlorinated biphenyls (PCBs) was used as a substitute for mineral oil where flammability was a major concern. Askarel-filled transformers could be placed inside or next to a building where only dry-types were used previously. Although these coolants were considered non-flammable, as used in electrical equipment they could decompose when exposed to electric arcs or fires to form hydrochloric acid, and toxic furans and dioxins. The compounds were further undesirable because of their persistence in the environment and their ability to accumulate in higher animals including humans. The use of askarels in new transformers was outlawed in 1977 (Claiborne, 1999). Work still continues to retire and properly dispose of transformers containing askarels or askarel-contaminated mineral oil.

HTHCs

Among the coolants used to take the place of askarels in distribution transformers are high-temperature hydrocarbons or high-molecular-weight hydrocarbons. These coolants are classified by the National Electric Code as "less flammable" if they have a fire point above 300°C. The disadvantages of HTHCs are a diminished cooling capacity from the higher viscosity that accompanies the higher molecular weight, plus increased cost.

Silicones

Another coolant that meets the National Electric Code requirements for a less flammable liquid is a silicone, chemically known as polydimethylsiloxane. Silicones are only occasionally used because they exhibit biological persistence if spilled and are more expensive than mineral oil and HTHCs.

Halogenated Fluids

Mixtures of tetrachloroethane and mineral oil were tried as an oil substitute for a few years. This and other chlorine-based compounds are no longer used because of a lack of biodegradability, the tendency to produce toxic by-products, and possible effects on the earth's ozone layer.

Esters

Synthetic esters are being used in Europe where high-temperature capability and biodegradability are most important and their high cost can be justified, for example, in traction transformers. Transformer manufacturers in the U.S. are now investigating the use of natural esters obtained from vegetable seed

FIGURE 3.28 Typical three-phase padmount distribution transformer.

oils. It is possible that agricultural esters will provide the best combination of high-temperature properties, stability, biodegradability, and cost as an alternate for mineral oil in distribution transformers (Oommen and Claiborne, 1996).

Tank and Cabinet Materials

A distribution transformer is expected to operate satisfactorily for a minimum of 30 years in an outdoor environment while extremes of loading work to weaken the insulation systems on the inside. This high expectation demands the best in state-of-the-art design, metal processing, and coating technologies. See a typical three-phase padmount in Fig. 3.28.

Mild Steel

Almost all overhead and padmount transformers have a tank and cabinet parts made from mild carbon steel. In recent years, major manufacturers have started using coatings applied by electrophoretic methods (aqueous deposition) and by powder coating. These new methods have replaced the traditional flow-coating and solvent spray applications.

Stainless Steel

Since the mid-1960s, single-phase submersibles have almost exclusively used AISI 400 series stainless steel. These grades of stainless were selected for good welding properties and the tendency not to pit-corrode. Both 400 series and the more expensive 304L (low carbon chromium-nickel) stainless have been used for padmounts and pole-types where severe environments justify the added cost. Transformer users with severe coastal environments have observed that padmounts show the worst corrosion damage where the cabinet sill and lower areas of the tank contact the pad. This is easily explained by the tendency for moisture, leaves, grass clippings, lawn chemicals, etc., to collect on the pad surface. Higher areas of a tank and cabinet are warmed and dried by the operating transformer, but the lowest areas in contact with the pad remain cool. Also, the sill and tank surfaces in contact with the pad are most likely to have

Transformers

FIGURE 3.29 Single-phase transformer with composite hood.

the paint scratched. To address this, manufacturers sometimes offer hybrid transformers, where the cabinet sill, hood, or the tank base may be selectively made from stainless.

Composites

There have been many attempts to conquer the corrosion tendencies of transformers by replacing metal structures with reinforced plastics. One of the more successful is a one-piece composite hood for single-phase padmount transformers. See Fig. 3.29.

Modern Processing

Adhesive Bonding

Today's distribution transformers almost universally use a kraft insulating paper that has a diamond pattern of epoxy adhesive on each side. Each finished coil is heated prior to assembly. The heating drives out any moisture that might be absorbed in the insulation. Bringing the entire coil to the elevated temperature also causes the epoxy adhesive to bond and cure, making the coil into a solid mass, more capable of sustaining high thermal and mechanical stresses, such as the transformer might see under short-circuit current conditions while in service.

Vacuum Processing

With the coil still warm from the bonding process, transformers are held at a high vacuum while oil flows into the tank. The combination of heat and vacuum assures that all moisture and all air bubbles have been removed from the coil, ensuring electrical integrity and a long service life. Factory processing with heat and vacuum is impossible to duplicate in the field or in most service facilities. Transformers, if opened, should be exposed to the atmosphere for minimal amounts of time, and oil levels should never be taken down below the tops of the coils. All efforts must be taken to keep air bubbles out of the insulation structure.

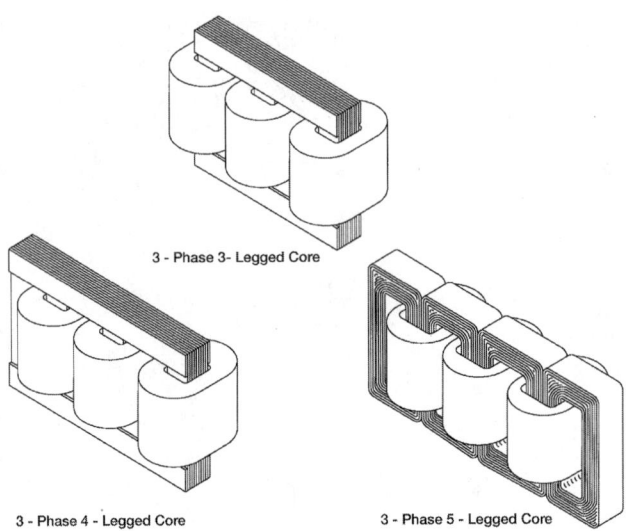

FIGURE 3.30 Three- and four-legged stacked cores, and five-legged wound core.

General Transformer Design

Liquid Filled vs. Dry Type

The vast majority of distribution transformers on utility systems today are liquid filled. Liquid-filled transformers offer the advantages of smaller size, lower cost, and greater overload capabilities compared to dry type.

Stacked vs. Wound Cores

Stacked-core construction favors the manufacturer that makes a small quantity of widely varying special designs in his facility. A manufacturer that builds large quantities of identical designs will benefit from the automated fabrication and processing of wound cores. See Fig. 3.30.

Single Phase

The vast majority of distribution transformers used in North America are single phase, usually serving a single residence or as many as 14 to 16, depending on the characteristics of the residential load. Single-phase transformers can be connected into banks of two or three separate units. Each unit in a bank should have the same voltage ratings, but need not supply the same kVA load.

Core Form
A single core loop linking two identical winding coils is referred to as "core-form" construction. See Fig. 3.31.

Shell Form
A single winding structure linking two core loops is referred to as "shell-form" construction. See Fig. 3.32.

Winding Configuration
Most distribution transformers for residential service are built as "shell form" with the secondary winding split into two sections with the primary winding in between. This so-called LO-HI-LO configuration results in a lower impedance than if the secondary is contiguous. The LO-HI configuration is used where the higher impedance is desired and especially on higher kVA ratings where higher impedances are mandated by standards to limit short-circuit current. "Core-form" transformers are always built LO-HI because the two coils must always carry the same currents. A 120/240 service using "core form" in the

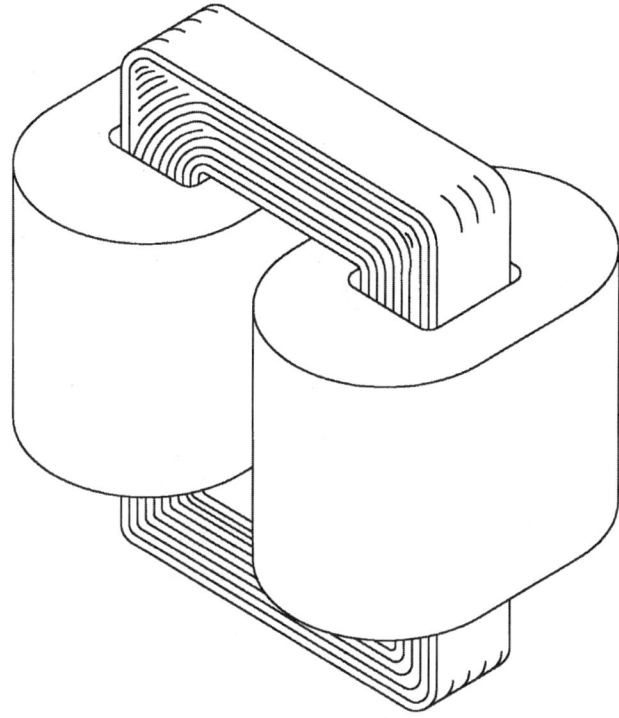

SINGLE - PHASE CORE - TYPE TRANSFORMER

FIGURE 3.31 Core-form construction.

LO-HI-LO configuration would need eight interconnected coil sections. This is considered too complicated to be commercially practical. See Fig. 3.33.

Three Phase

Most distribution transformers built and used outside North America are three phase, even for residential service. In North America, three-phase transformers serve commercial and industrial sites only. All three-phase distribution transformers are said to be of core-form construction although the definitions above do not hold. Three-phase transformers will have one coaxial coil for each phase encircling a vertical leg of the core structure. Stacked cores will have three or possibly four vertical legs, while wound cores will have a total of four loops creating five legs or vertical paths, three down through the center of the three coils and one on the end of each outside coil. The use of three vs. four or five legs in the core structure will have a bearing on which electrical connections and loads can be used by a particular transformer. The advantage of three-phase electrical systems in general is the economy gained by having the phases share common conductors and other components. This is especially true of three-phase transformers using common core structures.

Duplex and Triplex Construction

Occasionally, utilities will require a single tank that contains two completely separate core-coil assemblies. Such a design is sometimes called a duplex, and may have any size combination of single-phase core-coils inside. The effect is the same as constructing a two-unit bank with the advantage of having only one tank to place. Similarly, a utility may request a transformer with three completely separate and distinct core structures mounted inside one tank.

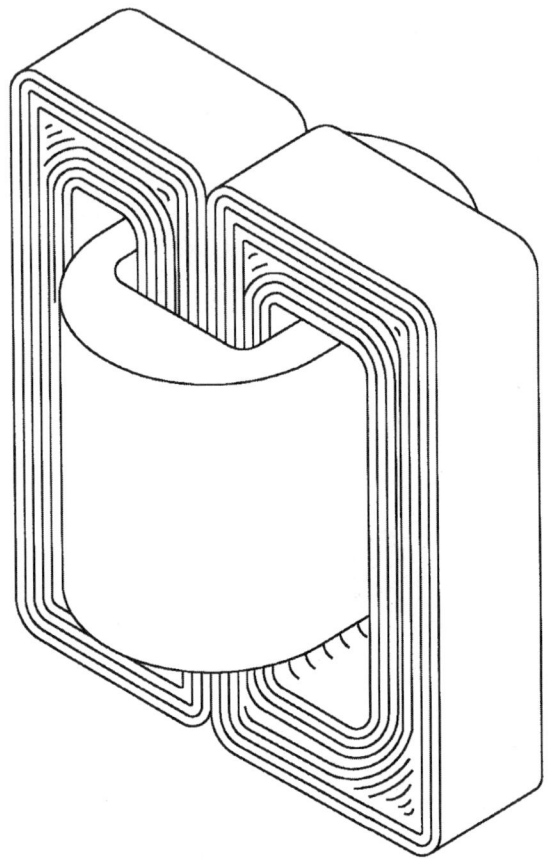

SINGLE - PHASE SHELL - TYPE TRANSFORMER

FIGURE 3.32 Shell-form construction.

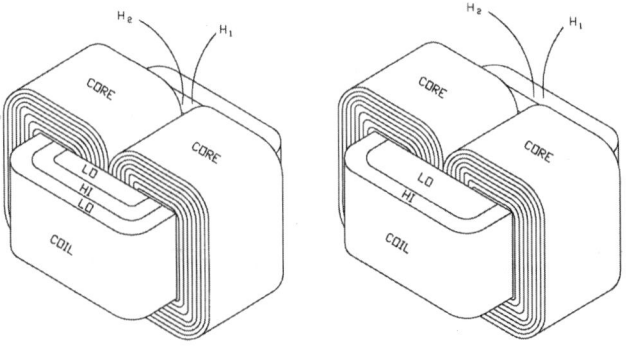

FIGURE 3.33 LO-HI-LO and LO-HI configurations.

Serving Mixed Single- and Three-Phase Loads

The utility engineer has a number of transformer configurations to choose from and it is important to match the transformer to the load being served. A load that is mostly single phase with a small amount

Transformers

of three phase is best served by a bank of single-phase units, or a duplex pair, one of which is larger to serve the single-phase load. A balanced three-phase load is best served by a three-phase unit, with each phase's coil identically loaded. See (ABB, 1995).

Transformer Connections

Single-Phase Primary Connections

The primary winding of a single-phase transformer may be connected between a phase conductor and ground or between two phase conductors.

Grounded Wye Connection — Those units that must be grounded on one side of the primary are usually only provided with one primary connection bushing. The primary circuit is completed by grounding of the transformer tank to the system neutral. Thus, it is imperative that proper grounding procedure be followed when the transformer is installed, so that the tank never becomes "hot". Since one end of the primary winding is always grounded, the manufacturer can economize the design and grade the high-voltage insulation. Grading provides less insulation at the end of the winding closest to ground. A transformer with graded insulation usually cannot be converted to operate phase-to-phase. The primary voltage designation on the nameplate of a graded insulation transformer will include the letters "GRDY", as in "12470 GRDY/7200," indicating that it must be connected phase-to-ground on a grounded wye system.

Fully Insulated Connection — Single-phase transformers supplied with fully insulated coils and two separate primary connection bushings may be connected phase-to-phase on a three-phase system or phase-to-ground on a grounded wye system as long as the proper voltage is applied to the coil of the transformer. The primary voltage designation on the nameplate of a fully insulated transformer will look like "7200/12470Y" where 7200 is the coil voltage. If the primary voltage shows only the coil voltage, as in "7200", then the bushings can sustain only a limited potential from the system ground and the transformer can *only* be connected phase-to-phase.

Single-Phase Secondary Connections

Distribution transformers will usually have 2, 3, or 4 secondary bushings, and common voltage ratings are 240 and 480, with and without a mid-tap connection. See Fig. 3.34.

Two Secondary Bushings — A transformer with two bushings can supply only a single voltage to the load.

Three Secondary Bushings — A transformer with three bushings supplies a single voltage with a tap at the mid-point of that voltage. This is the common three-wire residential service used in North America. For example, a 120/240 secondary can supply load at either 120 or 240 V as long as neither 120-V coil section is overloaded. Transformers with handholes or removable covers can be internally reconnected from three to two bushings in order to serve full kVA from the parallel connection of coil sections. These are designated 120/240 or 240/480 with the smaller value first. Most padmount distribution transformers

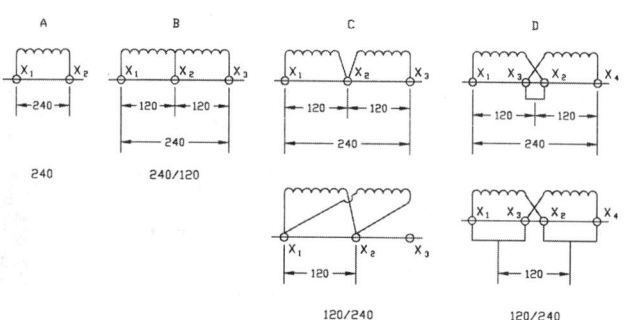

FIGURE 3.34 Single-phase secondary connections.

are permanently and completely sealed and therefore cannot be reconnected from three to two bushings. The secondary voltage for permanently sealed transformers with three bushings is 240/120 or 480/240.

Four Secondary Bushings — Secondaries with four bushings can be connected together external to the transformer to create a mid-tap connection with one bushing in common, or a two-bushing connection where the internal coil sections are paralleled. The four-bushing secondary will be designated as 120/240 or 240/480, indicating that a full kVA load can be served at the lower voltage. The distinction between 120/240 and 240/120 must be carefully followed when padmounted transformers are being specified.

Three-Phase Connections

When discussing three-phase distribution transformer connections, it is good to remember that you can be speaking about a single three-phase transformer or about single-phase transformers interconnected to create a three-phase bank. For either an integrated transformer or a bank, either the primary or secondary can be connected in either delta or wye connection. The wye connections can be either grounded or ungrounded. However, not all combinations will operate satisfactorily, but will depend upon the transformer construction, characteristics of the load, and the source system. Detailed information on three-phase connections can be found in (ANSI, 1978) and (ABB, 1995). Some connections that are of special concern are listed below.

Ungrounded Wye–Grounded Wye — A wye-wye connection where the primary neutral is left floating produces an unstable neutral where high third harmonic voltages are likely to appear. In some Asian systems, the primary neutral is stabilized by using a three-legged core and by limiting current imbalance on the feeder at the substation.

Grounded Wye–Delta — This connection is called a grounding transformer. Unbalanced primary voltages will create high currents in the delta circuit. Unless the transformer is specifically designed to handle these circulating currents, the secondary windings can be overloaded and burn out. Use of the ungrounded wye–delta is suggested instead.

Grounded Wye–Grounded Wye — This connection will sustain unbalanced voltages, but must use a four- or five-legged core to provide a return path for zero sequence flux.

Preferred Connections — In the earliest days of electric utility systems, it was found that induction motors drew currents that had substantial amounts of third harmonics. In addition, transformers on the system that were operating close to the saturation point of their cores had third harmonics in the exciting current. One way to keep these harmonic currents from spreading over an entire system was to use delta-connected windings in transformers. Third harmonic currents add up in-phase in a delta loop and flow around the loop, dissipating themselves as heat in the windings, but minimizing the harmonic voltage distortion that might be seen elsewhere on the utility's system. With the advent of suburban underground systems in the 1960s, it was found that a transformer with a delta-connected primary was more prone to ferroresonance problems. An acceptable preventive was to go to grounded wye–grounded wye transformers on all but the heaviest industrial applications.

Ferroresonance

Ferroresonance is an overvoltage phenomenon that occurs when charging current for a long underground cable or other capacitive reactance saturates the core of a transformer. Such a resonance can result in voltages as high as 5 times the rated system voltage, damaging lightning arresters and other equipment, possibly even the transformer itself. When ferroresonance is occurring, the transformer is likely to produce loud squeals and groans, and has been likened to the sound of steel roofing being dragged across a concrete surface. A typical ferroresonance situation is shown in Fig. 3.35 and consists of long underground cables feeding a transformer with a delta-connected primary. The transformer is unloaded or very lightly loaded and switching or fusing for the circuit operates one phase at a time. Ferroresonance can occur when energizing the transformer as the first switch is closed, or can occur if one or more distant fuses open and the load is very light. Ferroresonance is more likely to occur on systems with

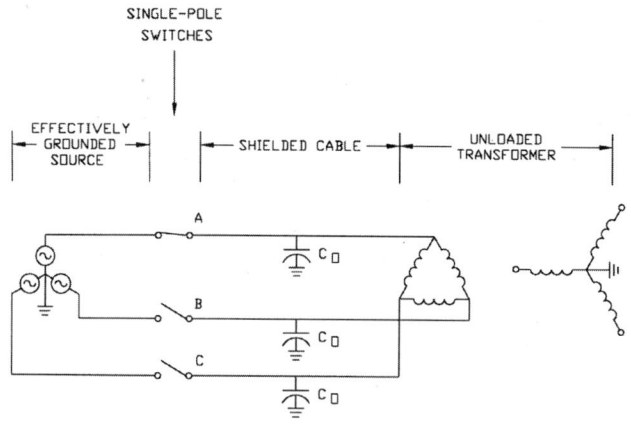

FIGURE 3.35 Typical ferroresonance situation.

higher primary voltage and has been observed even when there is no cable present. All of the contributing factors, delta connection, cable length, voltage, load, and single-phase switching, must be considered together. Attempts to set precise limits for prevention of the phenomenon have been frustrating.

Tank Heating
Another phenomenon that can occur to three-phase transformers because of the common core structure between phases is tank heating. Wye–wye connected transformers that are built on four- or five-legged cores are likely to saturate the return legs when zero-sequence voltage exceeds about 33% of the normal line-to-neutral voltage. This can happen, for example, if two phases of an overhead line wrap together and are energized by a single electrical phase. When the return legs are saturated, magnetic flux is then forced out of the core and finds a return path through the tank walls. Eddy currents produced by magnetic flux in the ferromagnetic tank steel will produce tremendous localized heating, occasionally burning the tank paint and boiling the oil inside. For most utilities, the probability of this happening is so low that it is not economically feasible to take steps to prevent it, other than keeping trees trimmed. A few, with a higher level of concern, purchase only triplex transformers, having three separate core-coil assemblies in one tank.

Three-Phase Secondary Connections — Delta
Three-phase transformers or banks with delta secondaries will have simple nameplate designations such as '240' or '480'. If one winding has a mid-tap, say for lighting, then the nameplate will say '240/120' or '480/240', similar to a single-phase transformer with a center tap. Delta secondaries can be grounded at the mid-tap or any corner.

Three-Phase Secondary Connections — Wye
Popular voltages for wye secondaries are 208Y/120, 480Y/277, and 600Y/347

Duplex Connections
Two single-phase transformers can be connected into a bank having an "open wye" or "open delta" primary and an "open delta" secondary. Such banks are used to serve loads that are predominantly single phase but with some three phase. The secondary leg serving the single-phase load can have a mid-tap, which may be grounded.

Other Connections
For details on other connections such as T-T and Zig-Zag, consult the listed references (IEEE, 1978a, b; ABB, 1995).

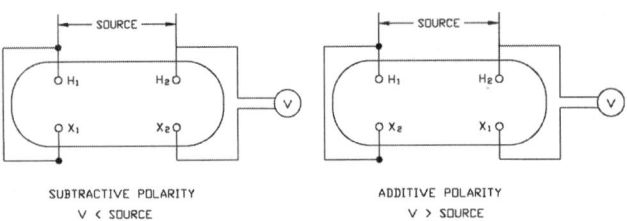

FIGURE 3.36 Single-phase polarity.

Polarity and Angular Displacement

The phase relationship of single-phase voltages is described as "polarity". The term for voltage phasing on three-phase transformers is "angular displacement".

Single-Phase Polarity — The "polarity" of a transformer can either be "additive" or "subtractive". These terms describe the voltage that may appear on adjacent terminals if the remaining terminals are jumpered together. The origin of the "polarity" concept is obscure, but apparently early transformers having lower primary voltages and smaller kVA sizes were first built with "additive" polarity. When the range of kVA's and voltages was extended, a decision was made to switch to subtractive polarity so that voltages between adjacent bushings could never be higher than the primary voltage already present. Thus, the transformers built to ANSI standards today are "additive" if the voltage is 8660 or below and the kVA is 200 or less, but "subtractive" otherwise. This differentiation is strictly a U.S. phenomenon. Transformers built to Canadian standards are all "additive" and those built to Mexican standards are all "subtractive". Although the technical definition of polarity involves the relative position of primary and secondary bushings, the position of primary bushings is always the same according to standards. Therefore, when facing the secondary bushings of an "additive" transformer, the X1 bushing is located to the right (of X3), while for a "subtractive" transformer, X1 is farthest to the left. To complicate this definition, a single-phase padmount built to ANSI Standard Type 2 will always have the X2 mid-tap bushing on the lowest right-hand side of the low-voltage slant pattern. Polarity has nothing to do with the internal construction of the transformer windings, but only with the routing of leads to the bushings. Polarity only becomes important when transformers are being paralleled or banked. (See Fig. 3.36.)

Three-Phase Angular Displacement — The phase relation of voltage between H1 and X1 bushings on a three-phase distribution transformer is referred to as angular displacement. ANSI standards require that wye–wye and delta–delta transformers have 0° displacement. Wye–delta and delta–wye transformers will have X1 lagging H1 by 30°. This difference in angular displacement means that care must be taken when three-phase transformers are paralleled to serve large loads. Sometimes the phase difference is used to advantage when supplying power to 12-pulse rectifiers and other specialized loads. European standards permit a wide variety of displacements, the most common being "Dy11". This IEC designation is interpreted as Delta primary–wye secondary, with X1 lagging H1 by 11 × 30° = 330°, or leading by 30°, 60° different than the ANSI standard angular displacement. (See Fig. 3.37.)

Transformer Locations

Overhead

With electric wires being strung on poles, it is obvious that transformers would be hung on the same poles, as close as possible to the high-voltage source. Larger units are often placed on overhead platforms in alleyways or along side of buildings or on ground-level pads, protected by fencing.

Underground

Distribution transformers can be installed in ventilated underground or above-ground vaults, away from public access. Depending on the type of distribution system and the vault's exposure to standing water, the transformer may be referred to as a vault-type, network, or submersible.

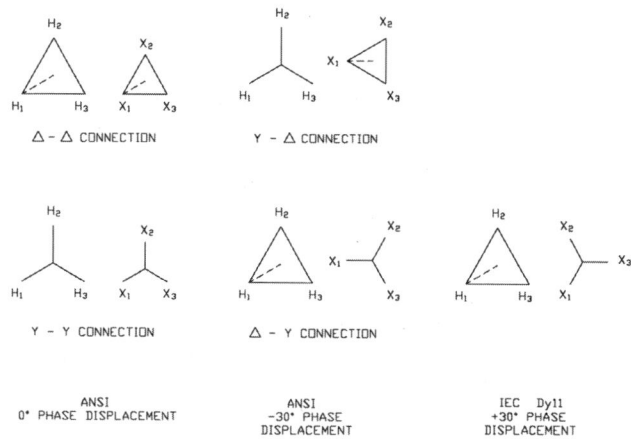

FIGURE 3.37 Three-phase angular displacement.

Direct Buried

Through the years, attempts have been made to place distribution transformers directly in the ground. A direct-buried installation is desirable because it is completely out of sight and cannot be damaged by windstorms, automobiles, or lawn mowers. There are three major challenges when direct-buried installations are considered: (1) the limited operational accessibility, (2) the corrosive environment, and (3) the challenge of dissipating heat from the transformer. The overall experience has been that heat from a buried transformer tends to dry out earth that surrounds it, causing the earth to shrink and create gaps in the heat conduction paths to the ambient soil. If a site is found that is always moist, then heat conduction may be assured but corrosion of the tank or of cable shields is still a major concern. The cost of preventing corrosion damage usually makes direct burial impractical.

Padmounts

The most common site for transformers serving new residential and light industrial and commercial distribution is on a pad at grade level. The problems of heat dissipation and corrosion are only slightly more severe than overheads, but substantially better than transformers confined in below-grade ventilated vaults. Padmounts are intended to be placed in locations that are frequented by the general public. Therefore, the operating utility has to be concerned about security of the locked cabinet covering the primary and secondary connections to the transformer. The industry has established standards for security against unauthorized entry and vandalism of the cabinet and locking provisions (ANSI, 1988). The minimization of sharp corners or edges that may be hazardous to children at play is another concern that has been addressed by standards. The fact that padmount transformers can operate with surface temperatures near the boiling point of water is another concern that is voiced from time to time. Justification is often found in comparing a hot transformer to the hood of an automobile on a sunny day. From a scientific standpoint, research has shown that people will pull away after touching a hot object in a much shorter time than it takes to sustain a burn injury. The point above which persons might be burned is about 150°C (Hayman, 1973).

Interior Installations

Building codes generally prohibit the installation of a distribution transformer containing mineral oil inside or immediately adjacent to an occupied building. The options available include use of a dry-type transformer and the replacement of mineral oil with a less flammable coolant. See the section entitled "Coolants".

Transformer Losses

No-Load Loss and Exciting Current

When alternating voltage is applied to a transformer winding, an alternating magnetic flux is induced in the core. The alternating flux produces hysteresis and eddy currents within the electrical steel, resulting in heat generated in the core. Heating of the core due to applied voltage is called no-load loss. Other names are iron loss or core loss. The term "no-load" is descriptive because the core is heated regardless of the amount of load on the transformer. If the applied voltage is varied, the no-load loss is roughly proportional to the square of the peak voltage, as long as the core is not taken into saturation. The current that flows when a winding is energized is called the "exciting current" or "magnetizing current", consisting of a real component and a reactive component. The real component delivers power for no-load losses in the core. The reactive current delivers no power but represents energy momentarily stored in the winding inductance. Typically, the exciting current of a distribution transformer is less than 0.5% of the rated current.

Load Loss

A transformer supplying load has current flowing in both the primary and secondary windings that will produce heat in those windings. Load loss is divided into two parts: (1) I^2R loss and (2) eddy loss.

I^2R Loss

Each transformer winding has an electrical resistance that produces heat when load currents flow. Resistance of a winding is measured by passing DC current through the winding to eliminate inductive effects.

Eddy Losses

When alternating current flows through a winding, the losses created are always greater than the I^2R losses measured with DC current. The additional AC losses are called "eddy losses" and are caused by currents circulating within the winding conductors. This phenomenon is sometimes called "skin effect" because eddy currents tend to flow close to the surface of the conductor. Additional losses occur in transformer frame parts and in the tank walls, especially in the areas close to secondary bushings. The stray losses external to the windings are called "other strays". Stray losses are greater in larger transformers because of the higher currents and larger conductors. Stray losses are proportional to current frequency and thus can increase dramatically when loads with harmonic currents are served. The effects can be reduced by subdividing large conductors and by using stainless steel or other non-ferrous materials for frame parts and bushing plates.

Harmonics and DC Effects

Rectifier and discharge lighting loads cause currents to flow in the distribution transformer that are not pure power frequency sine waves. Using Fourier analysis, distorted load currents can be resolved into components that are integer multiples of the power frequency, and thus are referred to as harmonics. Distorted load currents are expected to be high in the 3rd, 5th, 7th, and sometimes the 11th and 13th harmonics, depending on the character of the load.

Odd-Ordered Harmonics

Load currents that contain the odd-numbered harmonics will increase both the eddy losses and other stray losses within a transformer. If the harmonics are substantial, then the transformer must be derated to prevent localized and general overheating. ANSI standards suggest that any transformer with load containing more than 5% total harmonic distortion should be loaded according to the appropriate ANSI Guide (IEEE, 1998).

Even-Ordered Harmonics

Analysis of most harmonic currents will show very low amounts of even harmonics, 2nd, 4th, 6th, etc. Components that are even multiples of the fundamental frequency generally cause the waveform to be

Transformers

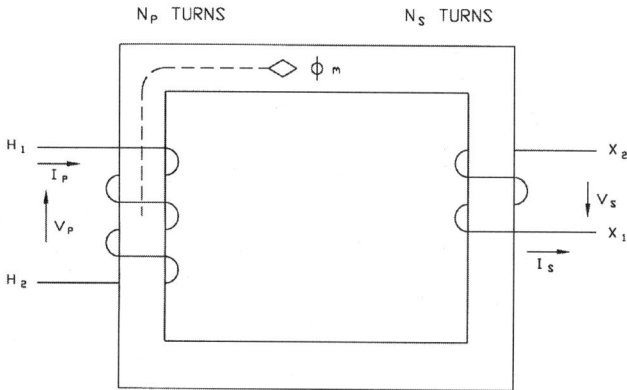

FIGURE 3.38 Two-winding transformer schematic.

non-symmetrical about the zero current axis. The current therefore has a 0th harmonic or DC offset component. The cause of a DC offset is usually found to be half-wave rectification due to a defective rectifier or other component. The effect of a DC current offset is to drive the transformer core into saturation on alternate half-cycles. When the core saturates, exciting current can be extremely high, which can burn out the primary winding in a very short time. Transformers that are experiencing DC offset problems are usually noticed because of objectionably loud noise coming from the core structure. Industry standards are not clear regarding the limits of DC offset on a transformer. A recommended value is a DC current no larger than the normal exciting current, which is usually 1% or less of a winding's rated current (Galloway, 1993).

Performance

Transformer Model

To help explain performance characteristics of a distribution transformer, namely impedance, short-circuit current, regulation, and efficiency, a simple model will be developed.

Schematic

A simple two-winding transformer is shown above in the schematic diagram of Fig. 3.38. A primary winding of N_p turns is on one side of a ferromagnetic core loop and a similar coil having N_s turns is on the other. Both coils are wound in the same direction with the starts of the coils at H1 and X1, respectively. When an alternating voltage V_p is applied from H2 to H1, an alternating magnetizing flux, ϕ_m, flows around the closed core loop. A secondary voltage $V_s = V_p \times N_p/N_s$ is induced in the secondary winding and appears from X2 to X1 and very nearly in phase with V_p. With no load connected to X1 – X2, I_p consists of only a small current called the magnetizing current. When load is applied, current I_s flows out of terminal X1 and results in a current $I_p = I_s \times N_p/N_s$ flowing into H1 in addition to magnetizing current. The ampere-turns of flux due to current $I_p \times N_p$ cancel the ampere-turns of flux due to current $I_s \times N_s$ so that only the magnetizing flux exists in the core for all the time the transformer is operating normally.

Complete Equivalent Circuit

Figure 3.39 shows a complete equivalent circuit of the transformer. An ideal transformer is inserted to represent the current and voltage transformations. A parallel resistance and inductance representing the magnetizing impedance are placed across the primary of the ideal transformer. Resistance and inductance of the two windings are placed in the H1 and X1 legs, respectively.

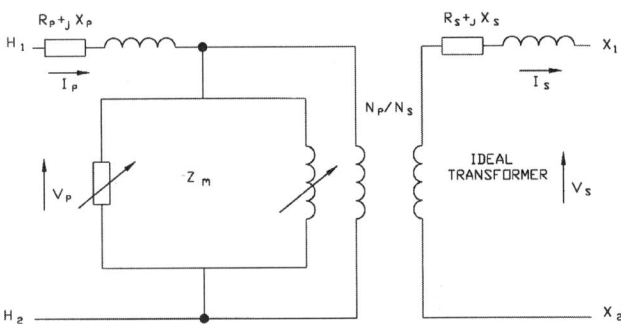

FIGURE 3.39 Complete equivalent circuit.

Simplified Model

To create a simplified model, the magnetizing impedance has been removed, acknowledging that no-load loss is still generated and magnetizing current still flows, but is so small that it can be ignored when compared to the rated currents. The R and X values in either winding can be translated to the other side by using percent values or by converting ohmic values with a factor equal to the turns ratio squared $(N_p/N_s)^2$. To convert losses or ohmic values of R and X to percent use:

$$\%R = \frac{Load\ Loss}{10 \cdot kVA} = \frac{\Omega_{(R)} \cdot kVA}{kV^2} \tag{3.8}$$

or

$$\%X = \frac{AW}{10 \cdot kVA} = \frac{\Omega_{(L)} \cdot kVA}{kV^2} \tag{3.9}$$

where AW is apparent watts or the scalar product of applied voltage and exciting current. Once the resistances and inductances are translated to the same side of the transformer, the ideal transformer can be eliminated and the percent values of R and X, respectively, combined. The result is the simple model shown in Fig. 3.40.

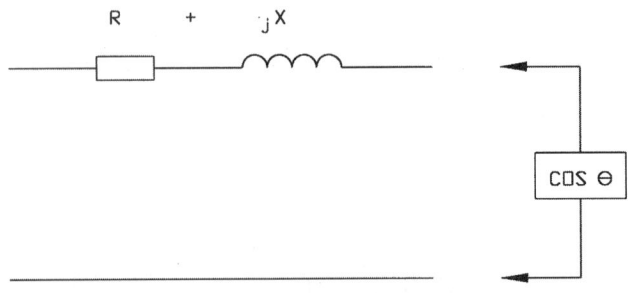

FIGURE 3.40 Simplified transformer model.

Transformers

Impedance

The values of %R and %X form the legs of what is known as the "impedance triangle". The hypotenuse of the triangle is called the transformer's impedance and can be calculated using:

$$\%Z = \sqrt{\%R^2 + \%X^2} \qquad (3.10)$$

A transformer's impedance is sometimes called "impedance volts" because it can be measured by shorting the secondary terminals and applying sufficient voltage to the primary so that rated current flows in each winding. The ratio of applied voltage to rated voltage is equal to the percent impedance.

Short-Circuit Current

If the load (right) side of the model of Fig. 3.40 is shorted and rated voltage from an infinite source is applied to the left side, the current I_{SC} will be limited only by the transformer impedance:

$$I_{SC} = 100 \times I_R / \%Z \qquad (3.11)$$

For example, if the rated current, I_R, is 100 amperes and the impedance is 2.0%, the short-circuit current will be 100 × 100/2 = 5000 amperes.

Percent Regulation

When a transformer is energized with no load, the secondary voltage will be exactly the primary voltage divided by the turns ratio (N_P/N_S). When the transformer is loaded, the secondary voltage will be diminished by an amount determined by the impedance and the power factor of the load. This change in voltage is called regulation and is defined as the *rise* in voltage when the load is removed. The primary voltage is assumed to be held constant at the rated value during this process. One result of the definition of regulation is that it is always a positive number.

$$\%Reg = \left[\%R^2 + \%X^2 + 200 \cdot \left(\%X \cdot \sin\theta + \%R \cdot \cos\theta\right) + 10000\right]^{0.5} - 100 \qquad (3.12)$$

where $\cos\theta$ is the power factor of the load. The most significant portion of this equation is the cross-products, and since %X predominates over %R in the transformer impedance and $\cos\theta$ predominates over $\sin\theta$ in the load, the percent regulation is usually less than the impedance. It may be necessary to consider the regulation at a load other than rated kVA. When the load is L per unit, then the equation becomes:

$$\%Reg = \left[L^2\left(\%R^2 + \%X^2\right) + 200 \cdot L \cdot \left(\%X \cdot \sin\theta + \%R \cdot \cos\theta\right) + 10000\right]^{0.5} - 100 \qquad (3.13)$$

Percent Efficiency

As with any other energy conversion device, the efficiency of a transformer is the ratio of energy delivered to the load divided by the total energy drawn from the source. Percent efficiency is expressed as:

$$\%Efficiency = \frac{L \cdot kVA \cdot \cos\theta \cdot 10^5}{L \cdot kVA \cdot \cos\theta \cdot 10^3 + NL + L^2 \cdot LL} \qquad (3.14)$$

where $\cos\theta$ is again the power factor of the load, and therefore kVA· $\cos\theta$ is real energy delivered to the load. **NL** is the no-load loss and **LL** is the load loss of the transformer. Most distribution transformers serving residential or light industrial loads are not fully loaded all the time. It is assumed that such transformers are loaded to about 50% of nameplate rating on the average. Thus, efficiency is often calculated at L = 0.5, where the load loss is about 25% of the value at full load. Since a typical transformer will have no-load loss of around 25% of load loss at 100% load, at L = 0.5, the no-load loss will equal the load loss and the efficiency will be at a maximum.

Transformer Loading

Temperature Limits

According to ANSI standards, modern distribution transformers are to operate at a maximum of 65°C average winding rise over a 30°C ambient air temperature at rated kVA. One exception to this is submersible distribution transformers where a 55°C rise over a 40°C ambient is specified. The bulk oil temperature near the top of the tank is called the "top oil temperature" and cannot be more than 65°C over ambient but will typically be about 55°C over ambient, or about 10°C less than the average winding rise.

Hottest Spot Rise

The location in the transformer windings that has the highest temperature is called the "hottest spot". Standards require that the hottest spot temperature not exceed 80°C rise over a 30°C ambient, or 110°C. These are steady-state temperatures at rated kVA. The hottest spot is of great interest because, presumably, this is where the greatest thermal degradation of the transformer's insulation system will take place. For calculation of thermal transients, the top oil rise over ambient air and the hottest spot rise over top oil are the parameters used.

Load Cycles

If all distribution loads were constant, then determining the proper loading of transformers would be a simple task. Loads on transformers, however, vary through the hours of a day, the days of a week, and through the seasons of the year. Insulation aging is a highly nonlinear function of temperature that accumulates over time. The best use of a transformer, then, is to balance brief periods of hottest spot temperatures slightly above 110°C with extended periods at hottest spots well below 110°C. Methods for calculating the transformer loss-of-life for a given daily cycle are found in the ANSI Guide for Loading (IEEE, 1995). Parameters needed to make this calculation are the no-load and load losses, the top oil rise and hottest spot rise, and the thermal time constant.

Thermal Time Constant

Liquid-filled distribution transformers can sustain substantial short-time overloads because the mass of oil, steel, and conductor takes time to come up to a steady state operating temperature. Time constant values can vary from 2 to 6 h, mainly due to the differences in oil volume vs. tank surface for different products.

Special Tests

Design Tests

Tests that manufacturers perform on prototypes or production samples are referred to as "design tests". These tests may include sound level tests, temperature rise tests, and short-circuit current tests. The purpose of a design test is to establish a design limit that can be applied by calculation to every transformer built. In particular, short-circuit tests are destructive and may result in some invisible damage to the sample, even if the test is passed successfully. The ANSI standard calls for a transformer to sustain six tests, four with symmetrical fault currents and two with asymmetrical currents. One of the symmetrical shots is to be of long duration, up to 2 sec, depending on the impedance for lower ratings. The remaining 5 shots are to be 0.25 sec in duration. The long shot duration for distribution transformers 750 kVA and above is 1 sec. The design has passed the short-circuit test if the transformer has sustained no internal or external damage as determined by visual inspection and minimal impedance changes. The transformer also has to pass production dielectric tests and experience no more than a 25% change in exciting current (Bean et al., 1959).

Production Tests

Production tests are given to and passed by each transformer made. Tests such as ratio, polarity or phase-relation, iron loss, load loss, and impedance are done to verify that the nameplate information is correct. Dielectric tests specified by industry standards are intended to prove that the transformer is capable of sustaining unusual but anticipated electrical stresses that may be encountered in service.

Impulse Test

Distribution lines are routinely disturbed by voltage surges caused by lightning strokes and switching transients. A standard 1.2×50 μsec impulse wave with a peak equal to the BIL of the primary system (60 to 150 kV) is applied to verify that each transformer will withstand these surges.

Applied Voltage Test

Standards require application of a voltage of (very roughly) twice the normal line-to-line voltage to each winding for 1 min. This checks the ability of one phase to withstand voltage it may see when another phase is faulted to ground and transients are reflected and doubled.

Induced Voltage Test

The original applied voltage test is now supplemented with an induced test. Voltage at higher frequency (usually 400 Hz) is applied at twice the rated value of the winding. This induces the higher voltage in each winding simultaneously. If a winding is permanently grounded on one end, the applied voltage test cannot be performed. In this case, many ANSI product standards specify that the induced test voltage is raised to 1000 plus 3.46 times the rated winding voltage (Bean et al., 1959).

Protection

Goals of Protection

Distribution transformers are often provided with fuses or circuit breakers to interrupt excessive currents that may otherwise harm the transformer itself or cause damage and service interruptions on the remainder of the system. Lightning arresters may be used to limit voltage surges that may otherwise damage the transformer or the nearby system. Protective devices may be located either internal or external to the transformer.

Protective Links

Distribution transformers that have no other protection are often supplied with a small high-voltage fuse. The protective link is sized to melt at from 6 to 10 times the rated current of the transformer. Thus, it will not protect against long-time overloads and will permit short-time overloads that may occur during inrush or cold-load-pickup phenomena. For this reason they are often referred to as "fault-sensing" links. Depending upon the system voltage, protective links can safely interrupt faults of 1000 to 3000 A.

Dual Sensing or Eutectic Links

High-voltage fuses made from a low-melting-point tin alloy will melt at 145°C and so will protect a transformer by detecting the combination of overload current and high oil temperature. A eutectic link, therefore, prevents long-time overloads, but allows high inrush and cold-load-pickup currents.

Current-Limiting Fuses

If available fault currents on the primary system exceed the interrupting ratings of protective links, then current-limiting fuses can be used. Current-limiting fuses can typically interrupt 40,000 to 50,000 A faults and do so in less than one half of a cycle. The interruption of a high current internal fault in such a short time will prevent severe damage to the transformer and avoid damage to surrounding property or hazard to personnel that might otherwise occur. Current-limiting fuses cannot prevent long-time overloads.

Bayonets

Padmounts and submersibles may use a primary link that is mounted internally in the transformer oil, but that can be withdrawn for inspection of the fuse element or to interrupt the primary feed. This device is called a bayonet and consists of a probe with a cartridge on the end that contains the replaceable fuse element. Fuses for bayonets may be either fault sensing or dual sensing.

Bayonet and Partial Range Current-Limiting Fuse

The most common method of protection for padmounted distribution transformers is the combination of a bayonet fuse (usually dual sensing) and a partial range current-limiting fuse (PRCL). The PRCL only responds to a high fault current, while the bayonet fuse is only capable of interrupting low fault currents. These fuses must be coordinated in such a way that any secondary fault will melt the bayonet fuse. Fault currents above the bolted secondary fault level are assumed to be due to internal faults. Thus the PRCL, which is mounted inside the tank, will operate only when the transformer must be removed from service anyway.

Lightning Arresters

Overhead transformers may be supplied with primary lightning arresters mounted nearby on the pole structure, on the transformer itself, directly adjacent to the primary bushing, or within the tank. Pad-mounted transformers may have arresters too, especially those at the end of a radial line, and they may be inside the tank, plugged into dead-front bushings, or at a nearby riser pole, where primary lines transition from overhead to underground.

Internal Secondary Circuit Breakers

Secondary breakers that are placed in the bulk oil of a transformer can protect against overloads that might otherwise cause thermal damage to the conductor-insulation system. Some breakers also have magnetically actuated trip mechanisms that rapidly interrupt the secondary load in case of secondary faults. When properly applied, secondary breakers should limit the top oil temperature of a transformer to about 110°C during a typical residential load cycle. Breakers on overhead transformers are often equipped with a red signal light. When this light is on, it signifies that the transformer has come close to tripping the breaker. The light will not go off until a lineman climbs the pole and resets the breaker. The lineman may also set the breaker on its "emergency" position, which allows the transformer to supply a higher overload without tripping, until the utility replaces the unit with one having a larger kVA capacity.

CSP®[1] Transformers

Overhead transformers that are built with the combination of secondary breaker, primary protective link, and external lightning arrester are referred to generically as "CSPs" or completely self-protected transformers. This protection package is expected to prevent failures caused by excessive loads, external voltage surges, and to protect the system from internal faults. The breaker will be furnished with a signal light and emergency control described above. See Fig. 3.41.

Protection Philosophy

CSP transformers are still in use, especially in rural areas, but the trend is away from secondary breakers to prevent transformer burnouts. Continued growth of residential load is not a foregone conclusion, as it once was. Utilities are becoming more sophisticated in their initial transformer sizing and are using computerized billing data to detect a transformer that is being overloaded. Experience is showing that modern distribution transformers can sustain more temporary overload than a breaker would allow. Most utilities would rather have service to their customers maintained than trip a breaker unnecessarily.

---

[1]CSP is a Registered Trademark of ABB Power T&D Company Inc., Raleigh, NC.

FIGURE 3.41 CSP® transformer.

Economic Application

Historical Perspective

Serious consideration of the economics of transformer ownership did not begin until the oil embargo of the early 1970s. With large increases in the cost of all fuels, utilities could no longer just pass along these increases to their customers without demonstrating fiscal responsibility by controlling losses on their distribution systems.

Evaluation Methodology

An understanding soon developed that the total cost of owning a transformer consisted of two major parts: the purchase price and the cost of supplying thermal losses of the transformer over an assumed life, which might be 20 to 30 years. To be consistent, the future costs of losses would have to be brought back to the present so that the two costs would both be on a present-worth basis. The calculation methodologies were first published by Edison Electric Institute, and recently updated in the form of a proposed ANSI standard (IEEE, 1998). The essential part of the evaluation method is the derivation of 'A' and 'B' factors, which are the utility's present worth cost for supplying no-load and load losses, respectively, in the transformer, measured in $/watt.

References

ABB Distribution Transformer Guide, Distribution Transformer Division, ABB Power T&D Company Inc., 1995, 40-70.
ANSI C57.12.28 — 1988, Pad-Mounted Equipment — Enclosure Integrity.
ANSI/IEEE C57.105 — 1978, IEEE Guide for Application of Transformer Connections in Three-Phase Distribution System, Section 2, 1978b.
ANSI/IEEE C57.12.80 — 1978, IEEE Standard Terminology for Power and Distribution Transformers, Clause 2.3, 1978a.
Bean, R. L., Chackan, Jr., N., Moore, H. R., and Wentz, E. C., Transformers for the Electric Power Industry, Westinghouse Electric Corp. Power Transformer Division, McGraw-Hill, New York, 1959, 338-340.
Claiborne, C. C., ABB Electric systems Technology Institute, personal communication, 1999.
Galloway, D. L., Harmonic and DC Currents in Distribution Transformers, 46th Annual Power Distribution Conference, Austin, TX, October 27, 1993.
Hayman, J. L., E. I. duPont de Nemours & Co., Letter to Betty Jane Palmer, Westinghouse Electric Corp., Jefferson City, MO, October 11, 1973.
IEEE C57.110 — 1998, IEEE Recommended Practice for Establishing Transformer Capability When Supplying Nonsinusoidal Load Currents.
IEEE C57.12.33, Draft 6 — 1998, Guide for Distribution Transformer Loss Evaluation.
IEEE Std C57.91 — 1995, IEEE Guide for Loading Mineral-Oil-Immersed Transformers, Introduction, iii.
Myers, S. D., Kelly, J. J., and Parrish, R. H., *A Guide to Transformer Maintenance,* Transformer Maintenance Division, S. D. Myers, Inc, Akron, OH, 1981, 12-footnote.
Oommen, T. V. and Claiborne, C. C., Natural and Synthetic High Temperature Fluids for Transformer Use, Internal Report, ABB Electric systems Technology Institute, 1996.
Palmer, B. J., History of Distribution Transformer Core/Coil Design, Distribution Transformer Engineering Report # 83-17, Westinghouse Electric Corporation, Jefferson City, MO, 1983.
Powel, C. A., General Considerations of Transmission, *Electrical Transmission and Distribution Reference Book,* ABB Power T&D Company Inc., Raleigh, NC, 1997, 1.

3.4 Underground Distribution Transformers

Dan Mulkey

Underground transformers are self-cooled, liquid-filled, sealed units designed for step-down operation from an underground primary cable supply. They are available in both single- and three-phase designs. Underground transformers can be separated into three subgroups: (1) those designed for installation in room-like vaults, (2) those designed for installation in surface operable enclosures, and (3) those designed for installation on a pad at ground level.

Vault Installations

The vault provides the required ventilation, access for operation, maintenance, and replacement, while at the same time providing protection against unauthorized entry. Vaults used for transformer installations are large enough to allow personnel to enter the enclosure, typically through a manhole and down a ladder. Vaults have been used for many decades and it is not uncommon to find installations that date back to the days when only paper-and-lead insulated primary cable was available. Transformers for vault installations are typically designed for radial application and have a separate fuse installation on the source side.

Vaults can incorporate many features:

- Removable top sections for transformer replacement
- Automatic sump pumps to keep water levels down
- Chimneys to increase natural air flow
- Forced air circulation

Transformers

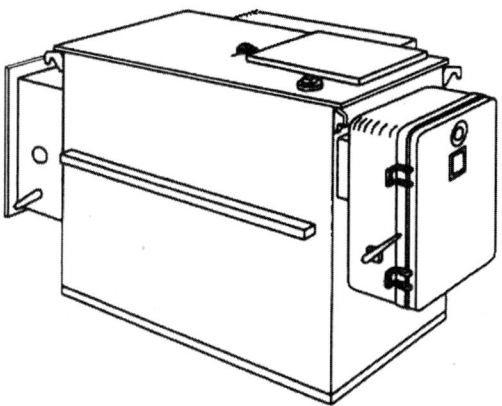

FIGURE 3.42 Network transformer with protector.

Transformers designed for vault installation are sometimes installed in a room inside a building. This, of course, requires a specially designed room to limit exposure to fire and access by unauthorized personnel and to provide sufficient ventilation. Both mineral-oil-filled units and units with one of the less-flammable insulating oils are used in these installations. These installations are also made using dry-type or pad-mounted transformers.

Transformers for vault installation are manufactured as either *subway transformers* or as *vault-type transformers*, which according to ANSI C57.12.40 are defined as:

- *Vault-type transformers* are suitable for occasional submerged operation.
- *Subway-type transformers* are suitable for frequent or continuous submerged operation.

Transformers for Vault Installation

Network Transformers (see Fig. 3.42): As defined in ANSI/IEEE C57.12.80, it is a transformer designed for use in a vault to feed a variable-capacity system of interconnected secondaries. They are three-phase transformers that are designed to connect through a network protector to a secondary network system. Network transformers are typically applied to serve loads in the downtown areas of major cities. National standard ANSI C57.12.40 details network transformers. The standard kVA ratings are 300, 500, 750, 1000, 1500, 2000, and 2500. The primary voltages range from 2400 to 34500 V. The secondary voltages are 216Y/125 or 480Y/277.

Network transformers are built as either vault-type or subway-type. They incorporate a primary switch with open, closed, and ground positions. Primary cable entrances are made by one of the following methods:

- Wiping sleeves or entrance fittings for connecting to lead cables — either one three-conductor or three single-conductor fittings or sleeves.
- Bushing wells or integral bushings for connecting to plastic cables — three wells or three bushings.

Single-Phase Subway or Vault (see Fig. 3.43): These are round single-phase transformers designed to be installed in a vault and capable of being banked together to provide three-phase service. These can be manufactured as either subway-type or vault-type. They are typically applied to serve small- to medium-sized commercial three-phase loads. The standard kVA ratings are 25, 37.5,

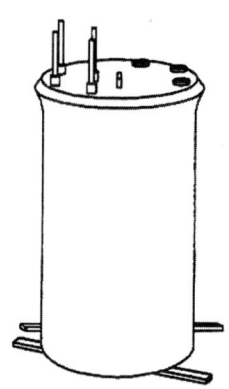

FIGURE 3.43 Single-phase subway.

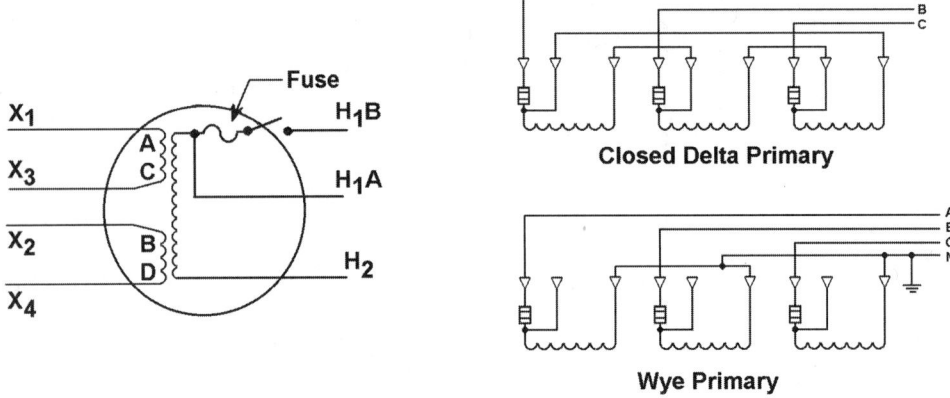

FIGURE 3.44 Three-bushing subway.

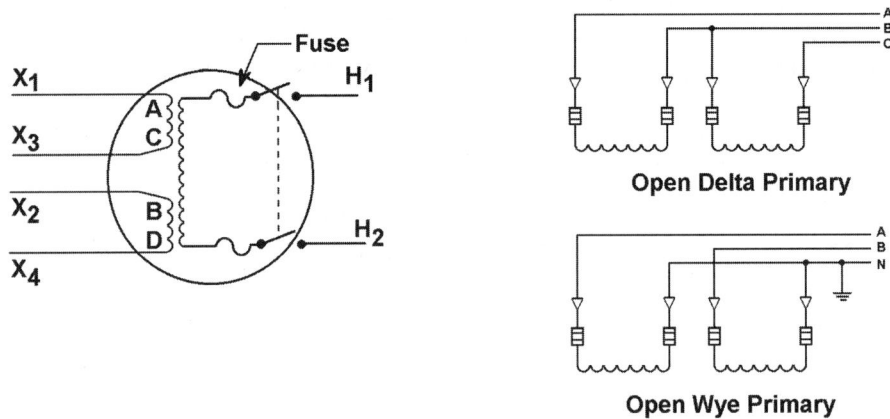

FIGURE 3.45 Two-bushing subway.

50, 75, 100, 167, and 250. Primary voltages range from 2400 to 34500 V, with the secondary voltage usually being 120/240. Four secondary bushings allow the secondary windings to be connected in parallel for wye connections or in series for delta connections. The secondary can be either insulated cables or spades. The units are designed to fit through a 36-in. manhole. They are not specifically covered by a national standard; however, they are very similar to the units in IEEE C57.12.23.

Units with three primary bushings or wells, and with an internal primary fuse (see Fig. 3.44), allow for connection in closed-delta, wye, or open-wye banks. They can also be used for single-phase phase-to-ground connections.

Units with two primary bushings or wells, and with two internal primary fuses (see Fig. 3.45), allow for connection in an open-delta or an open-wye bank. It also allows for single-phase line-to-line connection.

Three-Phase Subway or Vault (see Fig. 3.46): These are rectangular shaped three-phase transformers, and can be manufactured as either subway-type or vault-type. These are used to supply large three-phase commercial loads. Typically they have primary bushing well terminations on one of the small sides and the secondary bushings with spades are on the opposite end. These are also designed for radial installation and require external fusing. They can be manufactured in any of the standard three-phase kVA sizes and voltages. They are not detailed in a national standard.

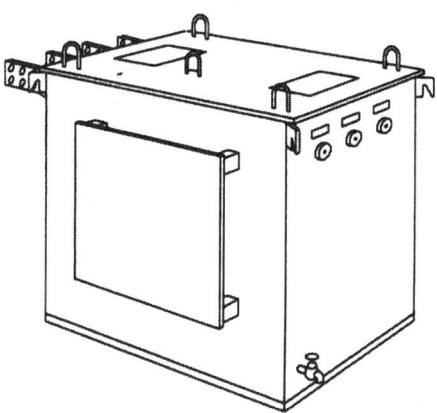

FIGURE 3.46 Three-phase vault.

Surface Operable Installations

The subsurface enclosure provides the required ventilation, access for operation, maintenance, and replacement, while at the same time providing protection against unauthorized entry. Surface operable enclosures have grade-level covers that can be removed to gain access to the equipment. The enclosures typically are just large enough to accommodate the largest size of transformer and allow for proper cable bending. Submersible transformers are designed to be connected to an underground distribution system that utilizes 200-A class equipment. The primary is most often #2 or 1/0 cables with 200-A elbows. While larger cables such as 4/0 can be used with the 200-A elbows, it is not recommended. The extra stiffness of 4/0 cable makes it very difficult to avoid putting strain on the elbow-bushing interface, which leads to early failure.

The operating points of the transformer are arranged on or near the cover. The installation is designed to be hot-stick operable by a person standing at ground level at the edge of the enclosure. Transformers for installation in surface operable enclosures are manufactured as submersible transformers, which is defined in IEEE C57.12.80 as "… so constructed as to be successfully operable when submerged in water under predetermined conditions of pressure and time." These transformers are designed for loop application and thus require internal protection. There are three typical variations of submersible transformers.

1. **Single-Phase Round Submersible:** Single-phase round transformers (see Fig. 3.47) have been used since the early 1960s. These transformers are typically applied to serve residential single-phase loads. These units are covered by IEEE C57.12.23 — Underground-Type, Self-Cooled, Single-Phase Distribution Transformers With Separable, Insulated, High-Voltage Connectors; High Voltage (24940 GrdY/14400 V and Below) and Low Voltage (240/120 V, 167 kVA and Smaller). They are manufactured in the normal single-phase kVA ratings of 25, 37.5, 50, 75, 100, and 167 kVA. Primary voltages are available from 2400 through 24940 GrdY/14400 and the secondary is 240/120 V. They are designed for loop-feed operation with a 200-A internal bus connecting the two bushings. Three low-voltage cable leads are provided through 100 kVA, while the 167-kVA size has six.

    They commonly come in two versions — a two-primary-bushing unit (see Fig. 3.48), and a four-primary-bushing unit (see Fig. 3.49) — although only the first is detailed in the IEEE

FIGURE 3.47 Single-phase round.

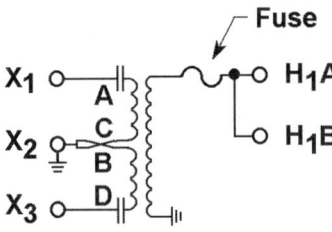
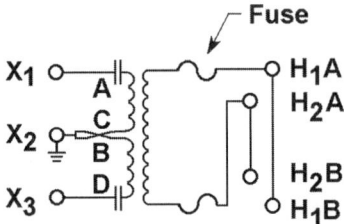

FIGURE 3.48 Two-primary-bushing.   FIGURE 3.49 Four-primary-bushing.

standard. The two-bushing unit is for phase-to-ground connected transformers, while the four-bushing unit is for phase-to-phase connected transformers.

As these are designed for application where the primary continues on after feeding through the transformer, the transformers require internal protection. The most common method is to use a secondary breaker and an internal non-replaceable primary expulsion fuse element.

These units are designed for installation in a 36-in. round enclosure. Enclosures have been made out of fiberglass and concrete. Installations have been made with and without a solid bottom. Those without a solid bottom simply rest on a gravel base.

2. **Single-Phase Horizontal Submersible:** Functionally these are the same as the round single-phase. However, they are designed to be installed in a rectangular enclosure as shown in Fig. 3.50. Three low-voltage cable leads are provided through 100 kVA, while the 167-kVA size has six. They are manufactured in both four-primary-bushing designs (see Fig. 3.51) and in six-primary-bushing designs (see Fig. 3.52). As well as the normal single-phase versions, there is also a duplex version. This is used to supply four-wire, three-phase, 120/240 volt services from two core-coil assemblies connected open-delta on the secondary side. The primary can be either open-delta or open-wye. Horizontal transformers have also been in use since the early 1960s. These units are not specifically covered by a national standard. The enclosures used have included treated plywood, fiberglass,

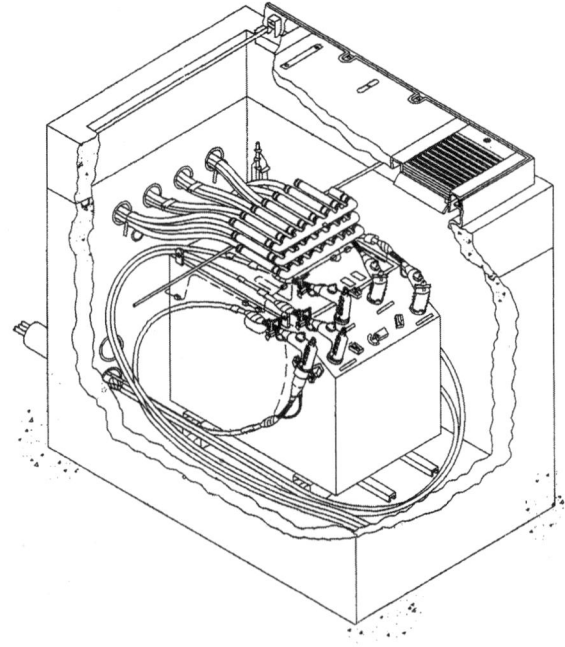

FIGURE 3.50 Four-bushing horizontal installed.

Transformers

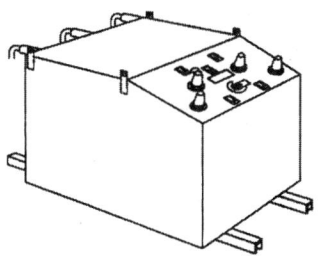

FIGURE 3.51 Four-bushing horizontal.

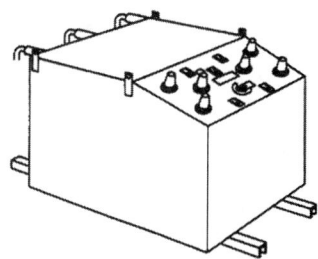

FIGURE 3.52 Six-bushing horizontal.

and concrete. The plywood and fiberglass enclosures are typically bottomless with the transformer resting on a gravel base.

3. **Three-Phase Submersible** (see Fig. 3.53): The three-phase surface operable units are detailed in ANSI C57.12.24 — Underground-Type Three-Phase Distribution Transformers, 2500 kVA and Smaller; High-Voltage, 34500 GrdY/19920 Volts and Below; Low Voltage, 480 volts and Below — Requirements. Typical application for these transformers is to serve three-phase commercial loads from loop-fed primary underground cables. Primary voltages are available from 2400 V through 34,500. The standard three-phase kVA ratings from 75 to 1000 kVA are available with secondary voltage of 208Y/120 V. With a 480Y/277 volt secondary, the available sizes are 75 to 2500 kVA.

Protection options include:

- Dry-well current-limiting fuses with an interlocked switch to prevent the fuses from being removed while energized.
- Submersible bayonet fuses with backup, under-oil, partial-range current-limiting fuses, or with backup internal non-replaceable primary expulsion fuse elements.

These are commonly installed in concrete rectangular-shaped boxes with removable cover sections.

Vault and Subsurface Common Elements

Tank Material: The substrate and coating should meet the requirements detailed in ANSI C57.12.32, Submersible Equipment — Enclosure Integrity. The smaller units can be constructed out of 400 series or 300 series stainless steels, or out of mild carbon steel. In general, 300 series stainless steel will outperform 400 series stainless steel, which significantly outperforms mild carbon steel. Most of the small

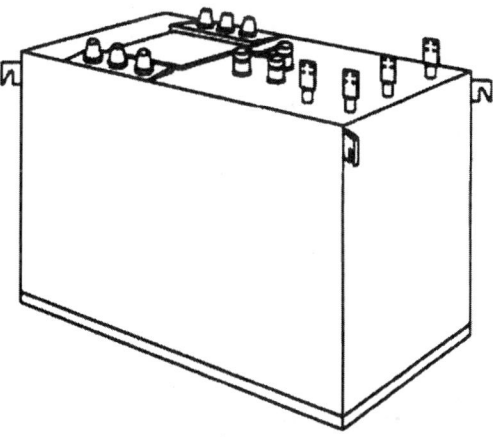

FIGURE 3.53 Three-phase submersible.

units are manufactured out of 400 series stainless steel since it is significantly less expensive than 300 series. 400 stainless with a good coating has been found to give satisfactory field performance. Due to material availability, many of the larger units cannot be manufactured from 400 series stainless. With the choice being limited to mild carbon steel or 300 series stainless, most of the large units are constructed out of mild carbon steel.

Temperature Rating: Kilovoltampere ratings are based on not exceeding an average winding temperature rise of 55°C and a hottest-spot temperature rise of 70°C. However, they are constructed with the same 65°C rise insulation systems used in overhead and pad mounted transformers. This allows for continuous operation at rated kVA, provided that the enclosure ambient air temperature does not exceed 50°C and the average temperature does not exceed 40°C. Utilities commonly restrict loading on underground units to a lower limit than they do with pad-mounted or overhead units.

Siting: Subsurface units should not be installed when any of the following exist:

- Soil is severely corrosive.
- Heavy soil erosion occurs.
- High water table exists causing repeated flooding of the enclosures.
- Heavy snowfall occurs.
- A severe mosquito problem exists.

Maintenance: Maintenance mainly consists of keeping the enclosure and the air vents free of foreign material. Dirt allowed to stay packed against the tank has led to accelerated anaerobic corrosion resulting in tank puncture and loss of mineral oil.

Emerging Issues

Water Pumping: Pumping of water from subsurface enclosures has been increasingly regulated. In some areas, water with any oily residue or turbidity must be collected for hazardous waste disposal. Subsurface and vault enclosures are often subject to runoff water from streets. This water can include oily residue from vehicles. So even without a leak from the equipment, water collected in the enclosure may be judged a hazardous waste.

Solid Insulation: Transformers with solid insulation are just being made commercially available in limited designs for subsurface application. The total encapsulation of what is essentially a dry-type transformer allows it to be applied in a subsurface environment.

Pad-Mounted Distribution Transformers

Pad-mounted transformers are the most used type of transformer for serving loads from underground distribution systems. They offer many advantages over subsurface, vault, or subway transformers.

- Installation: Less expensive to purchase and easier to install.
- Maintenance: Easier to maintain.
- Operability: Easier to find, less time to open and operate.
- Loading: Has better cooling so has greater loadability.

Many users and suppliers break distribution transformers into just two major categories — overhead and underground, with pad-mounted transformers included in underground. The IEEE standards, however, divide distribution transformers into three categories — overhead, underground, and pad-mounted.

Pad-mounted transformers are manufactured as either:

- *Single-Phase or Three-Phase:* Single-phase units are designed to transform only one phase. Three phase units transform all three phases.
- *Loop or Radial:* Loop-style units have the capability of terminating two primary conductors per phase. Radial-style units can only terminate one primary cable per phase. The primary must end if at a radial-style unit, but from a loop style it can continue on to serve other units.

- *Live-Front or Dead-Front:* Live-front units have the primary cables terminated in a stress cone supported by a bushing. Thus, the primary has exposed energized metal, or "live", parts. Dead-front units use primary cables that are terminated with high-voltage separable insulated connectors. Thus, the primary has all "dead" parts — no exposed energized metal.

Single-Phase Pad-Mounted Transformers

Single-phase pad-mounted transformers are usually applied to serve residential subdivisions. Most single-phase transformers are manufactured as clamshell, dead-front, loop-type with an internal 200-A primary bus designed to allow the primary to loop through and continue on to feed the next transformer. There is one national standard that details the requirements of single-phase pad-mounted transformers: ANSI C57.12.25 — Requirements for Pad-Mounted, Compartmental-Type, Self-Cooled, Single-Phase Distribution Transformers with Separable Insulated High-Voltage Connectors: High-Voltage, 34 500 GrdY/19 920 Volts and Below; Low Voltage 240/120 Volts; 167 kVA and Smaller.

The standard assumes that the residential subdivision is served by a one-wire primary extension. It details two terminal arrangements for loop feed systems — Type 1 (see Fig. 3.54) and Type 2 (see Fig. 3.55). Both have two primary bushings and three secondary bushings. The primary is always on the left as you face the transformer and the secondary is on the right. There is no barrier or division between the primary and secondary. In Type 1 units both primary and secondary cables rise directly up from the pad. In Type 2 units the primary rises from the right and crosses the secondary cables that rise from the left. Type 2 units can be shorter than Type 1 units because the crossed cable configuration gives enough free cable length to operate the elbow without requiring the bushing to be placed as high.

Although not detailed in the national standard, there are units built with four and six primary bushings. The four-bushing unit is used for single-phase lines with the transformers connected phase-to-phase. The six primary bushing units are used to supply single-phase loads from three-phase taps. Terminating all of the phases in the transformer allows all of the phases to be sectionalized at the same location. The internal single-phase transformer can be connected either phase-to-phase or phase-to-ground. The six

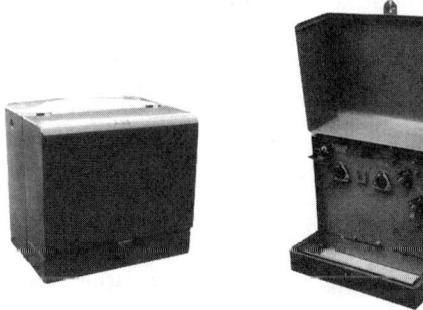

FIGURE 3.54   Typical Type 1.

FIGURE 3.55   Typical Type 2.

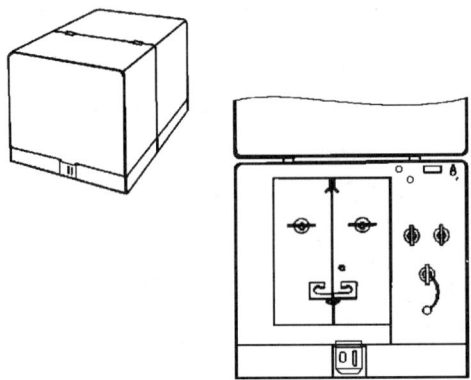

FIGURE 3.56 Single-phase live-front.

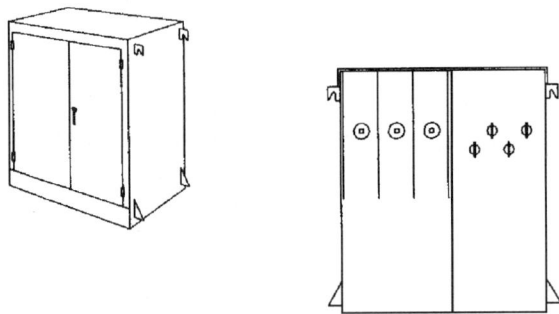

FIGURE 3.57 Radial-style live-front.

bushing units also allow the construction of duplex pad-mount units that can be used to supply small three-phase loads along with the normal single-phase residential load. In those cases, the service voltage is four-wire, three-phase, 120/240 V.

Cabinets for single-phase transformers are typically built in the clamshell configuration with one large door that swings up, as shown in Figs. 3.54 and 3.55. Older units were manufactured with two doors similar to the three-phase cabinets. New installations are almost universally dead-front; however, live-front units (see Fig. 3.56) are still purchased for replacements. These units are also built with clamshell cabinets but have an internal box-shaped insulating barrier constructed around the primary connections.

Three-Phase Pad-Mounted Transformers

Three-phase pad-mounted transformers are typically applied to serve commercial and industrial three-phase loads from underground distribution systems. There are two national standards that detail requirements for pad-mounted transformers — one for live-front and one for dead-front.

Live-Front: Live-front pad-mounted transformers are detailed in ANSI C57.12.22 — Requirements for Pad-Mounted, Compartmental-Type, Self-Cooled, Three-Phase Distribution Transformers with High-Voltage Bushings, 2500 kVA and Smaller: High Voltage 34 500 GrdY/19 920 Volts and Below; Low Voltage 480 Volts and Below. These live-front transformers are specified as radial units and thus do not come with any fuse protection (see Fig. 3.57). The primary compartment is on the left and the secondary compartment is on the right. A rigid barrier separates them. The secondary door must be opened before the primary door can be opened. Stress cone terminated primary cables rise vertically and connect to the terminals on the end of the high-voltage bushings. Secondary cables rise vertically and are terminated on spades connected to the secondary bushings. Units with a secondary of 208Y/120 V are available up to 1000 kVA. Units with a secondary of 480Y/277 V are available up to 2500 kVA.

Transformers

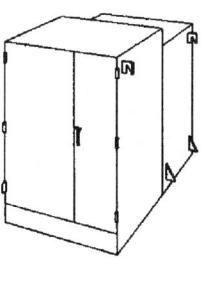 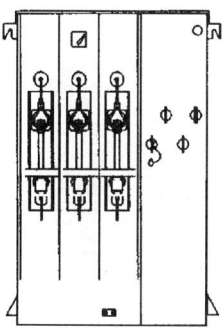

FIGURE 3.58 Loop-style live-front.

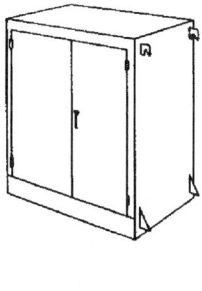 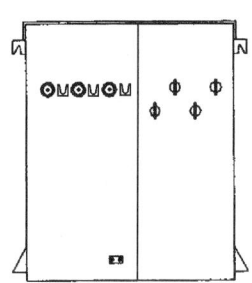

FIGURE 3.59 Radial-style dead-front.

Although not detailed in a national standard, there are many similar types available. A loop-style live front (see Fig. 3.58) can be constructed by adding fuses mounted below the primary bushings. Two primary cables are then both connected to the bottom of the fuse. The loop is then made at the terminal of the high-voltage bushing, external to the transformer but within its primary compartment.

Dead-Front: Dead-front pad-mounted transformers are detailed in IEEE C57.12.26 — Standard for Transformers — Pad-Mounted, Compartmental-Type, Self-Cooled, Three-Phase Distribution Transformers for Use with Separable Insulated High-Voltage Connectors, High-Voltage, 34 500 GrdY/19 920 Volts and Below; 2500 kVA and Smaller. Both radial and loop designs are detailed in the standard. Radial-style units have three primary bushings arranged horizontally (see Fig. 3.59). Loop-style units have six primary bushings arranged in a "V" pattern (see Figs. 3.60 and 3.61). In both, the primary

FIGURE 3.60 Small loop-style dead-front.

FIGURE 3.61 Large loop-style dead-front.

FIGURE 3.62 Mini three-phase in clamshell cabinet.

compartment is on the left and the secondary compartment is on the right with a rigid barrier between them. The secondary door must be opened before the primary door can be opened. The primary cables are terminated with separable insulated high-voltage connectors, commonly referred to as 200-A elbows. These plug onto the primary bushings, which can be either bushing wells with an insert or they can be integral bushings. Bushing wells with inserts are recommended as they allow the insert and elbow to both be easily replaced. Units with a secondary of 208Y/120 V are available up to 1000 kVA. Units with a secondary of 480Y/277 V are available up to 2500 kVA.

In addition to what is shown in the national standards, there are other variations available. The smallest size in the national standards is the 75 kVA; however, 45-kVA units are also manufactured in the normal secondary voltages. Units with higher secondary voltages such as 2400 and 4160Y/2400 are manufactured in sizes up to 3750 kVA. There is a new style being produced that is a cross between single and three-phase units. A small three-phase transformer is placed in a six-bushing loop-style clamshell cabinet (see Fig. 3.62). These are presently available from 45 to 150 kVA in both 208Y/120 and 480Y/277 V secondaries.

Pad-Mount Common Elements

Protection: Units designed for radial installation require the installation of an external protective device such as a fuse. Units designed for loop installation usually incorporate internal protective devices. The options include:

- Secondary breaker and internal non-replaceable primary expulsion fuse element
- Dry-well current-limiting fuses with an interlocked switch to prevent the fuses from being removed while energized
- Submersible bayonet fuses with backup, under-oil, partial-range current-limiting fuses, or with backup internal non-replaceable primary expulsion fuse elements

Primary Conductor: Pad-mounted transformers are designed to be connected to an underground distribution system that utilizes 200-A class equipment. The primary is most often #2 or 1/0 cables with 200-A elbows or stress cones. While larger cables such as 4/0 can be used with the 200-A elbows, it is not recommended. The extra stiffness of 4/0 cable makes it very difficult to avoid putting strain on the elbow-bushing interface, which has led to early failure.

Pad: Pads are made out of various materials. The most common is concrete and can be either poured in place or precast. Concrete is suitable for any size pad. Pads for single-phase transformers are also commonly made out of fiberglass or polymer-concrete.

Enclosure: There are two national standards that specify the requirements for enclosure integrity for pad-mounted equipment: ANSI C57.12.28 for normal environments and ANSI C57.12.29 for coastal environments. The tank and cabinet of pad-mounted transformers is commonly manufactured out of mild carbon steel. When applied in corrosive areas, such as near the ocean, they are commonly made out of 300 or 400 series stainless steel. In general, 300 series stainless steel will outperform 400 series stainless steel, which significantly outperforms mild carbon steel in corrosive applications.

Maintenance: Maintenance mainly consists of keeping the enclosure in good repair so that it remains tamper-resistant and rust-free.

Temperature Rating: The normal temperature ratings are used — the kilovoltampere ratings are based on not exceeding an average winding temperature rise of 65°C and a hottest-spot temperature rise of 80°C.

3.5 Dry Type Transformers

Paulette A. Payne

A dry type transformer is one in which the insulating medium surrounding the winding assembly is a gas or dry compound. This section covers single- and three-phase, ventilated, non-ventilated, and sealed dry type transformers with voltage in excess of 600 V in the highest voltage winding.

Dry type transformers compared to oil-immersed are lighter and non-flammable. Increased experience with thermal behavior of materials, continued development of materials and transformer design has improved transformer thermal capability. Upper limits of voltage and kVA have increased. Winding insulation materials have advanced from protection against moisture to protection under more adverse conditions (e.g., abrasive dust and corrosive environments).

Transformer Taps

Transformers may be furnished with voltage taps in the high-voltage winding. Typically two taps above and two taps below rated voltage are provided, yielding a 10% total tap voltage range (ANSI/IEEE, 1981 (R1989); ANSI/IEEE C57.12.52-1981 (R1989)).

Cooling Classes for Dry Type Transformers

American and European cooling class designations are indicated in Table 3.3. Cooling classes for dry type transformers are as follows (IEEE, 100, 1996; ANSI/IEEE, C57.94-1982 (R-1987)):

Ventilated — Ambient air may circulate, cooling the transformer core and windings.
Non-ventilated — No intentional circulation of external air through the transformer.
Sealed — Self-cooled transformer with hermetically sealed tank.
Self-cooled — Cooled by natural circulation of air.

TABLE 3.3 Cooling Class Designation

Cooling Class	IEEE Designation (ANSI/IEEE 57.12.01-1989 (R1998))	IEC Designation (IEC 60726-1982 [Amend. 1-1986])
Ventilated self-cooled	AA	AN
Ventilated forced-air cooled	AFA	AF
Ventilated self-cooled/forced-air cooled	AA/FA	ANAF
Non-ventilated self-cooled	ANV	ANAN
Sealed self-cooled	GA	GNAN

Force-air cooled — Cooled by forced circulation of air.

Self-cooled/forced air-cooled — A rating with cooling by natural circulation of air and a rating with cooling by forced circulation of air.

Winding Insulation System

General practice is to seal or coat dry type transformer windings with resin or varnish to provide protection against adverse environmental conditions that can cause degradation of transformer windings. Insulating mediums for primary and secondary windings are categorized as follows:

Cast coil — The winding is reinforced or placed in a mold and cast in a resin under vacuum pressure. Lower sound levels are realized as the winding is encased in solid insulation. Filling the winding with resin under vacuum pressure eliminates voids that can cause corona. With a solid insulation system, the winding has superior mechanical and short-circuit strength and is impervious to moisture and contaminants.

Vacuum pressure encapsulated — The winding is embedded in a resin under vacuum pressure. Encapsulating the winding with resin under vacuum pressure eliminates voids that can cause corona. The winding has excellent mechanical and short-circuit strength, and provides protection against moisture and contaminants.

Vacuum pressure impregnated — The winding is permeated in a varnish under vacuum pressure. An impregnated winding provides protection against moisture and contaminants.

Coated — The winding is dipped in a varnish or resin. A coated winding provides some protection against moisture and contaminants for application in moderate environments.

Application

Non-ventilated and sealed dry type transformers are suitable for indoor and outdoor applications (ANSI/IEEE, 57.94-1982 (R-1987)). As the winding is not in contact with the external air, it is suitable for applications, e.g., exposure to fumes, vapors, dust, steam, salt spray, moisture, dripping water, rain, and snow.

Ventilated dry type transformers are recommended only for dry environments unless designed with additional environmental protection. External air carrying contaminants or excessive moisture could degrade winding insulation. Dust and dirt accumulation can reduce air circulation through the windings (ANSI/IEEE, 57.94-1982 (R-1987)). Table 3.4 indicates transformer applications based upon the process employed to protect the winding insulation system from environmental conditions.

Enclosures

All energized parts should be enclosed to prevent contact. Ventilated openings should be covered with baffles, grills, or barriers to prevent entry of water, rain, snow, etc. The enclosure should be tamper resistant. A means for effective grounding should be provided (ANSI/IEEE, C2-1997). The enclosure should provide protection suitable for the application, e.g., a weather- and corrosion-resistant enclosure for outdoor installations.

If not designed moisture resistant, ventilated and non-ventilated dry type transformers operating in a high moisture or high humidity environment when de-energized should be kept dry to prevent moisture

Transformers

TABLE 3.4 Transformer Applications

Winding Insulation System	Cast Coil	Encapsulated	Impregnated or Coated	Sealed Gas
Harsh environments[a]	Yes	Yes	Yes	Yes
Severe climates[b]	Yes	Yes		Yes
Load cycling	Yes	Yes	Yes	Yes
Short circuit	Yes	Yes	Yes	Yes
Non-flammability	Yes	Yes	Yes	Yes
Outdoor	Yes	Yes	Yes[c]	Yes
Indoor	Yes	Yes	Yes	Yes

[a] Fumes, vapors, excessive or abrasive dust, steam, salt spray, moisture, or dripping water.
[b] Extreme heat or cold, moisture.
[c] If designed for installation in dry environments.

TABLE 3.5 Usual Operating Conditions for Transformers (ANSI/IEEE, C57.12.01-1989 (R1998))

Temperature of cooling air	$\leq 40°C$
24hr average temperature of cooling air	$\leq 30°C$
Minimum ambient temperature	$\geq -30°C$
Load current[a]	Harmonic factor ≤ 0.05 per unit
Altitude[b]	≤ 3300 ft (1000 m)
Voltage[c] (without exceeding limiting temperature rise)	• Rated output KVA at 105% rated secondary voltage, power factor ≥ 0.80 • 110% rated secondary voltage at no load

[a] Any unusual load duty should be specified to the manufacturer.
[b] At higher altitudes, the reduced air density decreases dielectric strength; it also increases temperature rise reducing capability to dissipate heat losses (ANSI/IEEE, C57.12.01-1989 (R1998)).
[c] Operating voltage in excess of rating may cause core saturation and excessive stray losses, which could result in overheating and excessive noise levels (ANSI/IEEE, C57.94-1982 (R1987), C57.12.01-1989 (R1998)).

ingress. Strip heaters can be installed to switch on manually or automatically when the transformer is de-energized for maintaining temperature after shutdown to a few degrees above ambient temperature.

Operating Conditions

The specifier should inform the manufacturer of any unusual conditions for which the transformer will be subjected. Dry type transformers are designed for application under usual operating conditions indicated in Table 3.5.

Gas may condense in a gas-sealed transformer left de-energized for a significant period of time at low ambient temperature. Supplemental heating may be required to vaporize the gas before energizing the transformer (ANSI/IEEE, C57.94-1982 (R1987)).

Limits of Temperature Rise

Winding temperature rise limits are chosen so that the transformer will experience normal life expectancy for the given winding insulation system under usual operating conditions. Table 3.6 indicates the limits of temperature rise for the thermal insulation systems most commonly applied. Operation at rated load and loading above nameplate will result in normal life expectancy. A lower average winding temperature rise, 80°C rise for 180°C temperature class and 80°C or 115°C rise for 220°C temperature class, may be designed providing increased life expectancy and additional capacity for loading above nameplate rating.

Accessories

The winding temperature indicator can be furnished with contacts to provide indication and/or alarm of winding temperature approaching or in excess of maximum operating limits.

For sealed dry type transformers, a gas pressure switch can be furnished with contacts to provide indication and/or alarm of gas pressure deviation from recommended range of operating pressure.

TABLE 3.6 Limits of Temperature Rise for Commonly Applied Thermal Insulation Systems

Insulation System Temperature Class (°C)	Winding Hottest-Spot Temperature Rise (°C) for Normal Life Expectancy		Average Winding Temperature Rise (°C)[a]
	Continuous Operation at Rated Load[a]	Loading above Nameplate Rating[b]	
150	110	140	80
180	140	170	115
220	180	210	150

[a] ANSI/IEEE Standard C57.12.01-1989 (R1998).
[b] ANSI/IEEE Standard C57.96-1989.

Surge Protection

For transformers with exposure to lightning or other voltage surges, protective surge arresters should be coordinated with transformer Basic Lightning Impulse Insulation Level, BIL. The lead length connecting from transformer bushing to arrester, and from arrester ground to neutral should be minimum length to eliminate inductive voltage drop in the ground lead and ground current (ANSI-IEEE, C62.2-1987 (R1994)). Table 3.7 provides transformer BIL levels corresponding to nominal system voltage. Lower BIL levels may be applied where surge arresters provide appropriate protection. At 25 kV and above, higher BIL levels may be required due to exposure to overvoltage or for a higher protective margin (ANSI/IEEE, C57.12.01-1989 (R1998)).

TABLE 3.7 Transformer BIL Levels (ANSI/IEEE, 1981a,b)

Nominal System Voltage	BIL (kV)
1200 and lower	10
2500	20
5000	30
8700	45
15000	60
25000	110
34500	150

Harmonics

When the harmonic content of the load current exceeds 0.05 per unit, eddy current loss can cause excessive winding loss from the high-frequency heating effects. This results in higher winding temperature rise, which can shorten transformer life expectancy (IEEE, 519-1992). Transformer capability to supply non-sinusoidal loading as a percent of 60-Hz sinusoidal load current can be calculated given the harmonic composition of the load current. The maximum allowable non-sinusoidal rms load current per unit of rated rms load current is (ANSI/IEEE, C57.110-1986 (R1993)):

$$I_{max}(pu) = \left[\frac{P_{W,R}(pu)}{1 + \left[\sum_{h=1}^{h=max} f_h^2 h^2 \bigg/ \sum_{h=1}^{h=max} f_h^2 \cdot P_{E,R}(pu) \right]} \right]^{1/2} \qquad (3.15)$$

where

$P_{W,R}(pu)$ = load loss density at rated conditions in per unit of rated load I^2R loss density
h = harmonic order
f_h = harmonic current distribution factor for harmonic "h" equivalent to the harmonic component of current divided by the fundamental 60-Hz component of current
$P_{E,R}(pu)$ = winding eddy current loss at rated conditions in per unit of rated load I^2R loss density

K-rated transformers are designed for operation under non-sinusoidal loading. The K-factor indicates transformer ability to withstand non-sinusoidal loading without exceeding the limits of insulation temperature rise for normal life expectancy. Short-circuit calculations should be performed and overcurrent and transient protection sized accordingly. The K-factor is calculated as follows:

$$K = \sum_{h=1}^{h=max} \left(I_h(pu)\right)^2 h^2 \quad (3.16)$$

where

$I_h(pu)$ = rms current at harmonic "h" per unit of rated rms load current
h = the harmonic order

Efficiency

Life cycle cost derives the economic value of electrical losses over transformer useful life. Total Owning Cost on an equivalent first cost basis, over useful life, or defined period of time is a commonly used methodology:

$$TOC = T_{FC} + C_{NL} \cdot P_{NL} + C_{LL} \cdot P_{LL} \quad (3.17)$$

where

T_{FC} = transformer first cost
C_{NL} = equivalent cost of no load losses
P_{NL} = no load power loss watts
C_{LL} = equivalent cost of load losses
P_{LL} = load loss power loss watts

An alternative to evaluation of Total Owning Costs is to use energy efficiency tables (NEMA, TP1-1996).

References

ANSI/IEEE Standard C57.12.50-1981 (R1989), Requirements for Ventilated Dry-Type Distribution Transformers, 1 to 500 kVA, Single-Phase, and 15 to 500 kVA, Three-Phase, with High-Voltage 601 to 34500 Volts, Low-Voltage 120 to 600 Volts.
ANSI/IEEE Standard C57.12.52-1981 (R1989), Requirements for Sealed Dry-Type Power Transformers, 501 kVA and Larger, Three-Phase, with High-Voltage 601 to 34500 Volts, Low-Voltage 208Y/120 to 4160 Volts.
ANSI/IEEE Standard C57.94-1982 (R-1987), IEEE Recommended Practice for Installation, Application, Operation, and Maintenance of Dry-Type General Purpose Distribution and Power Transformers.
ANSI/IEEE Standard C57.12.01-1989 (R1998), Standard General Requirements for Dry-Type Distribution and Power Transformers Including Those with Solid Cast and/or Resin-Encapsulated Windings.
ANSI/IEEE Standard C57.96-1989, Guide for Loading Dry-Type Distribution and Power Transformers.
ANSI/IEEE C57.110-1986 (R1993), Recommended Practice for Establishing Transformer Compatibility When Supplying Non-Sinusoidal Currents.
ANSI/IEEE C62.2-1987 (R1994), Guide for the Application of Gapped Silicon-Carbide Surge Arresters for Alternating Current Systems.
ANSI/IEEE C2-1997, National Electrical Safety Code.
IEC Standard 60726-1982 (Amendment 1-1986), Dry-Type Power Transformers, 1st ed.
IEEE 519-1992, Recommended Practices and Requirements for Harmonic Control in Electrical Power Systems.
IEEE Standard 100-1996, Standard Dictionary of Electrical and Electronics Terms, 6th ed.
NEMA Standard TP1-1996, Guide for Determining Energy Efficiency for Distribution Transformers.

3.6 Step-Voltage Regulators

Craig A. Colopy

Requirements for electrical power become stronger every day, both in quality and quantity. Quality means that consumers need stable voltage without distortions and interruptions. Quantity means that they can have as much load as needed without reducing the quality of the supply. These requirements come from industry as well as from domestic consumers. They have influence on each other. Maintaining voltage magnitude in a specified range is an important component of power quality. Power transformers and feeders pose their own impedance, and voltage drops depend on the loads and, consequently, the currents that flow through them. Feeder voltage magnitudes decrease along the feeder, which means that consumers at the feeder end will have the lowest voltage.

Distribution systems must be designed in such a way that voltage magnitudes always remain within a specified range. This is achieved by voltage control equipment and system design. Step-voltage regulators, as shown in Fig. 3.63, are one of the methods used for improving the voltage "profile" used for power systems.

Figure 3.64 shows an example of a location in a power system where the voltage regulators are commonly applied.

Step-voltage regulators are applied in the substations and out on distribution feeders to maintain a constant voltage by on-load voltage regulation wherever the voltage magnitude is beyond specified upper and lower limits. The combination of step-voltage regulators and power transformers are sometimes used in lieu of Load-Tap-Changing power transformers in the substation. To obtain constant voltage at some distance (load center) from the step-voltage regulator bank, a line drop compensation feature in the control can be utilized.

Step-voltage regulators are designed, manufactured, and tested in accordance with IEEE Standard C57.15–IEEE Standard Requirements, Technology, and Test Code for Step-Voltage Regulators. A step-voltage regulator is defined as "an induction device having one or more windings in shunt with and excited from the primary circuit, and having one or more windings in series between the primary circuit and the regulated circuit, all suitably adapted and arranged for the control of the voltage, or of the phase angle, or of both, of the regulated circuit in steps by means of taps without interrupting the load."

The most common step-voltage regulators manufactured today are single-phase, using reactive switching resulting in $32/5$% voltage steps (16 boosting and 16 bucking the applied voltage) providing an overall ±10% regulation. They are oil-immersed and typically use ANSI Type II insulating oil in accordance with ANSI/ASTM D-3487.

Although not required by IEEE standards, most manufacturers today design and manufacture step voltage regulators rated 55°C average winding rise over a 30°C average ambient and use a sealed tank-type construction. The gases generated from arcing in oil, as a result of normal operation of the internal load tap changer, are vented through a pressure relief device located on the tank side. Standard paper insulation designed for an average winding rise of 65°C, along with the use of a sealed tank-type system, allows for a 12% increase in load over the nameplate 55°C rise kVA rating.

Many regulators with a continuous current rating of 668 A and below can be loaded in excess of their 55°C rise rated ampere load if the range of voltage regulation is limited at a value less than the normal ±10% value. Table 3.8 shows the percent increase in ampere load permitted on each regulator when the percent regulation range is limited to values of ±10%, ±8¾%, ±7½%, ±6¼%, and ±5%.

Some regulators have limitations in this increased current capacity due to tap changer ampacity limitations. In those cases, the maximum tap changer capacity is shown on the regulator nameplate. Limiting the percent regulation range is accomplished by setting limit switches in the position indicator of the regulator, to prevent the tap changer from traveling beyond a set position in either raise or lower directions. It should be recognized, however, that although the regulators can be loaded with these additional amounts, when the percent regulation is decreased, without affecting the regulator's normal coil insulation longevity, the life of the tap changer contacts will decrease proportionately to the inverse square of the percent increase in current.

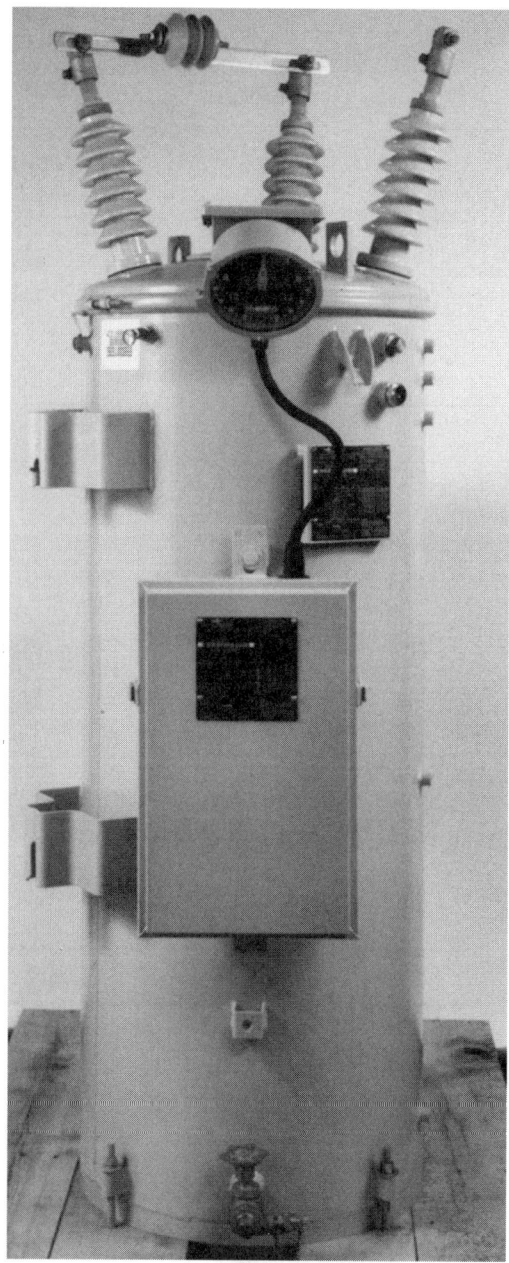

FIGURE 3.63 Step-voltage regulator.

Power Systems Applications

The following common types of circuits can be regulated:

- single-phase circuit
- three-phase, four-wire, multi-grounded wye circuit with three regulators
- three-phase, three-wire circuit with two regulators
- three-phase, three-wire circuit with three regulators

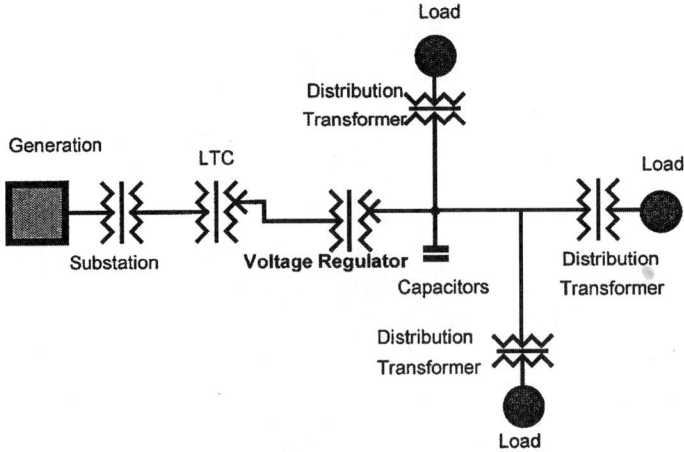

FIGURE 3.64 Power system.

TABLE 3.8

Range of Voltage Regulation (%)	Continuous Current Rating (%)
10.0	100
8.75	110
7.5	120
6.25	135
5.0	160

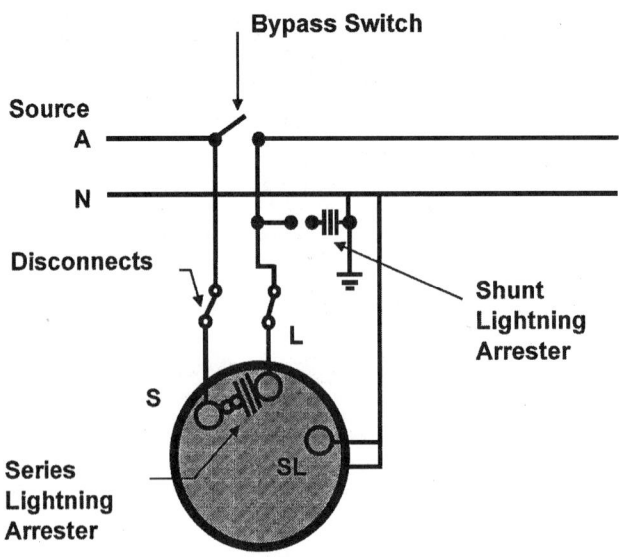

FIGURE 3.65 Voltage regulator connection in a single-phase circuit.

Figure 3.65 shows a regulator that can be used to maintain voltage on a single-phase circuit or lateral off of a main feeder.

Regulators are designed to withstand severe fault currents and frequent switching or lightning surges that may be found in substations or out on a main feeder. Three-phase power can be regulated by a bank of three single-phase step-voltage regulators connected in a Wye or two or three single-phase regulators

Transformers

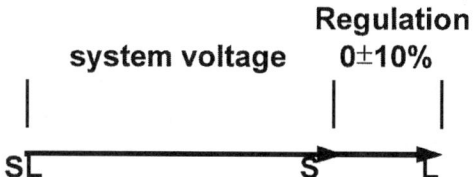

FIGURE 3.66 Vector diagram of a voltage regulator regulating a single-phase circuit.

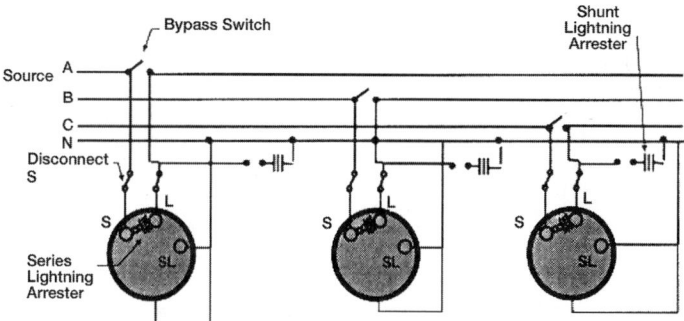

FIGURE 3.67 Connection of three regulators regulating a three-phase, four-wire, multi-grounded Wye circuit.

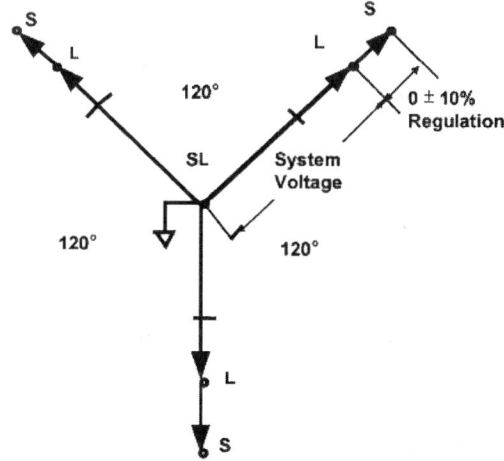

FIGURE 3.68 Vector diagram of three regulators regulating a three-phase, four-wire, multi-grounded wye circuit.

connected in an open or closed Delta configuration to keep voltages within specified limits for these sectors of the distribution system.

Wye-connected regulators (Fig. 3.67) work independently from each other. Regulators will regulate the voltage between the phases and neutral.

It is not necessary for loads on each phase to be balanced. Unbalanced current will flow in the neutral wire keeping the neutral reference from floating. In the case of a three-wire Wye, the neutral can shift, so the regulator has no stable reference point from which to excite the controls of the regulator. This can cause overstressing of insulation and erratic regulator operation. Therefore, three regulators cannot be connected in ungrounded Wye on a three-phase, three-wire circuit. But, if for some reason it is necessary to connect regulators in Wye configuration, a special stabilizing provision is required. Two methods are

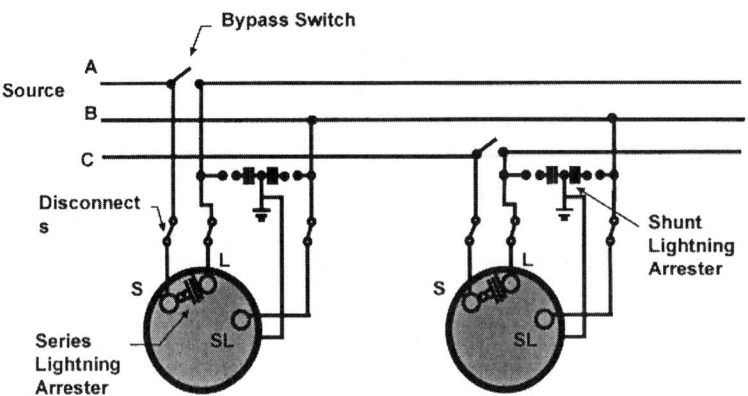

FIGURE 3.69 Connection of two voltage regulators regulating a three-phase, three-wire Wye or Delta circuit.

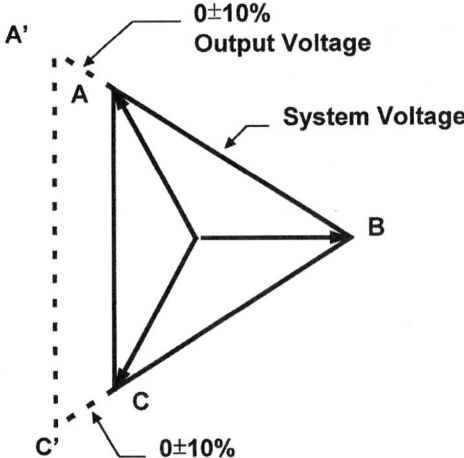

FIGURE 3.70 Vector diagram of two voltage regulators regulating a three-phase circuit.

available to provide this stabilization. One method is to link the common SL connections back to the impedance grounded secondary neutral of a substation transformer that is located in the same vicinity. If a substation transformer is not available, another method is to install a small grounding bank consisting of three transformers, each from one third to two thirds the kVA rating of the individual regulators. The rating within the range depends on the expected unbalance in load. Either method allows a path for the unbalanced current to flow.

Two single-phase regulators connected in an open Delta bank as shown in Fig. 3.69 allow for two of the phases to be regulated independent of each other with the third phase tending to read the average of the other two. A 30° phase displacement between the regulator current and voltage is a result of an open Delta connection. This is shown by the vector diagram in Fig. 3.70. Depending on the phase rotation, one regulator has its current lagging the voltage while the other has its current leading the voltage.

When a three-phase, three-wire circuit incorporates three single-phase voltage regulators in a closed-Delta-connected bank as shown in Fig. 3.71, the overall range of regulation of each phase is dependent on the range of regulation of each regulator. This type of connection will give approximately 50% more regulation than is obtained with two regulators in an open Delta configuration. A 30° phase displacement is also realized between the regulator current and voltage as a result of the Delta connection. This is shown by the vector diagram in Fig. 3.72. Depending on the phase rotation, all three regulators have their currents lagging or leading the voltage.

Transformers

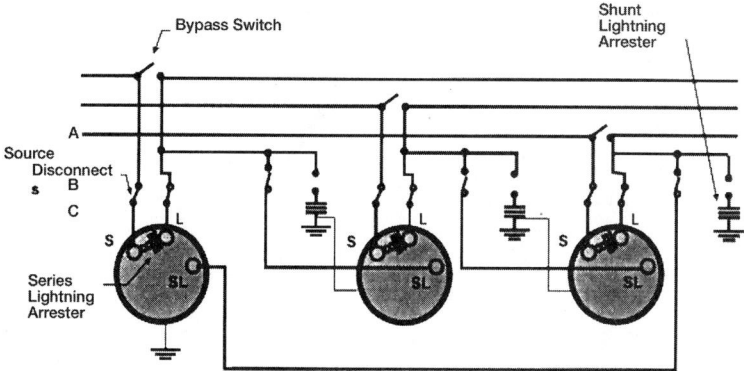

FIGURE 3.71 Connection of three voltage regulators in a three-phase, three-wire Delta circuit.

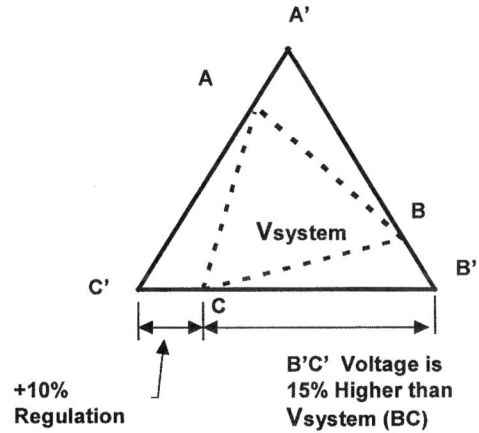

FIGURE 3.72 Vector diagram of closed-Delta-connected regulators.

Contributing to the effect is that the phase angle increases as the individual range of regulation of each regulator changes. A 4 to 6% increase in the phase angle will be recognized with the regulators set at the same extreme tap position.

A voltage improvement of 10% in the phase obtained with the regulator operation causes a 5% voltage improvement in the adjacent phase. If all three regulators operate to the same extreme position, the overall effect is increasing the overall range of regulation to ±15%, as shown in the vector diagram of Fig. 3.72.

Figure 3.73 reflects the power connection of a closed delta bank of regulators. Current within the windings will reach a maximum of, approximately, minus or plus 5% of the line current as the regulators approach the full raise or lower positions, respectively. Due to the fact that operation of any regulator changes the voltage across the other two, it may be necessary for the other two to make additional tap changes to restore voltage balance. The tap changer contact life will be affected by these additional steps. The 30° phase displacement between the regulator current and voltage for open and closed Delta connections will also affect the resulting interrupting power and corresponding life of the tap changer load breaking contacts.

The rated load current of a step-voltage regulator is determined by the following equation:

$$I_{rated} = \frac{3\phi \, load}{V_{L-L} \times \sqrt{3}}$$

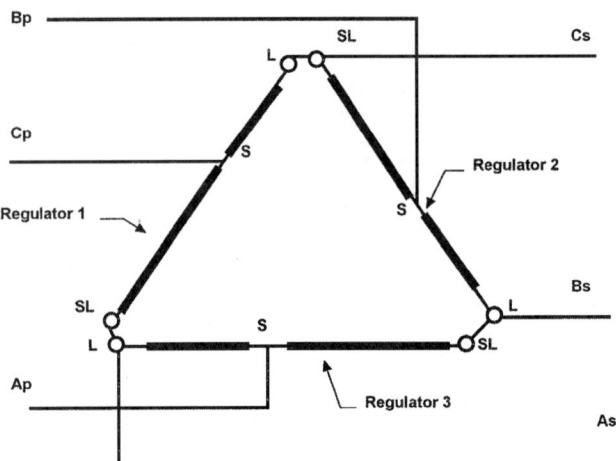

FIGURE 3.73 Power connection of closed-Delta-connected voltage regulators.

If the regulators are used in a single-phase circuit, four wire-grounded Wye circuit or connected in a Wye configuration in a three-wire system, the rated voltage of the regulator would be $V_{L-L}/\sqrt{3}$. If the regulators are connected in an open or closed Delta configuration for a three-wire system, the rated voltage of the regulator would be V_{L-L}. As a result, the kVA rating of the regulator would be determined by the following equation:

$$kVA = \frac{V_{rated} \times I_{rated} \times \text{range of regulation}}{1000}$$

Single-phase voltage regulators are available in the common ratings shown in Table 3.9.

Theory

A step-voltage regulator is a tapped autotransformer. To understand how a regulator operates, one can start by comparing it to a two-winding transformer.

Figure 3.74 is a basic diagram of a transformer with a 10:1 turns ratio. If the primary winding has 1000 V applied, the secondary winding will have an output of 100 V (10%). These two independent windings can be connected so that their voltages may aid or oppose one another. A voltmeter connected across the output terminals will measure either the sum of the two voltages or the difference between them. The transformer becomes an autotransformer with the ability to raise (Fig. 3.75) or lower (Fig. 3.76) the primary or system voltage by 10%.

In a voltage regulator, the equivalent of the high-voltage winding in a two-winding transformer would be referred to as the shunt winding. The low voltage winding would be referred to as the series winding. The series winding voltage is 10% of the regulator applied voltage. The polarity of its connection to the shunt winding to either boost or buck the regulator applied voltage is accomplished by the use of a reversing switch on the internal motor driven tap changer. Eight 1¼% taps are added to the series winding to provide small voltage adjustment increments. To go even further to provide fine voltage adjustment, such as sixteen 5/8% tap steps, a center tapped bridging reactor used in conjunction with two movable contacts on the motorized tap changer is utilized.

The process of moving from one voltage regulator tap to the adjacent voltage regulator tap consists of closing the circuit at one tap before opening the circuit at the other tap. The tap changer movable contacts move through stationary taps alternating in eight bridging and eight non-bridging (symmetrical) positions. An asymmetrical position is realized when one tap connection is open before transferring the

TABLE 3.9

Rated Volts	BIL (kV)	Rated kVa	Load Current (amps)
2500	60	50	200
		75	300
		100	400
		125	500
		167	668
		250	1000
		333	1332
		416	1665
5000	75	50	100
		75	150
		100	200
		125	250
		167	334
		250	500
		333	668
		416	833
7620	95	38.1	50
		57.2	75
		76.2	100
		114.3	150
		167	219
		250	328
		333	438
		416	546
		500	656
		667	875
		833	1093
13800	95	69	50
		138	100
		207	150
		276	200
		414	300
		552	400
14400	150	72	50
		144	100
		216	150
		288	200
		333	231
		432	300
		576	400
		667	463
		833	578
19920	150	100	50
		200	100
		333	167
		400	201
		667	334
		833	418

load to the adjacent tap. At this juncture, all the load current flows through one half of the reactor, magnetizing the reactor, and the reactance voltage is introduced into the circuit.

Figure 3.77 shows the two movable tap changer contacts on a symmetrical position; center tap of the reactor is at the same potential.

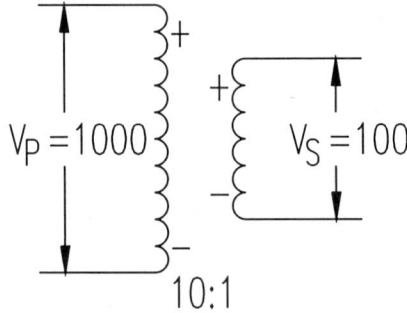

FIGURE 3.74 Transformer with 10:1 turns ratio.

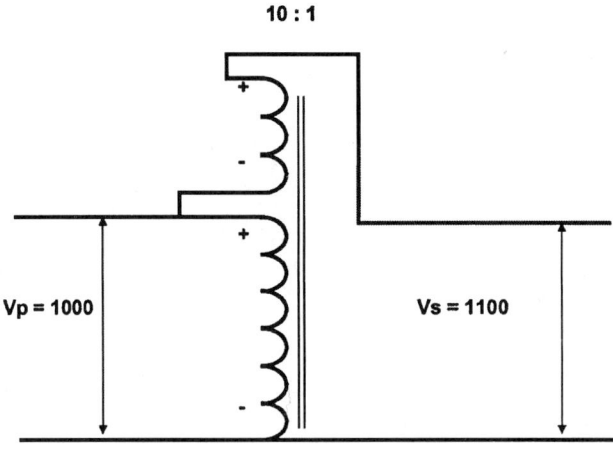

FIGURE 3.75 Step-up autotransformer.

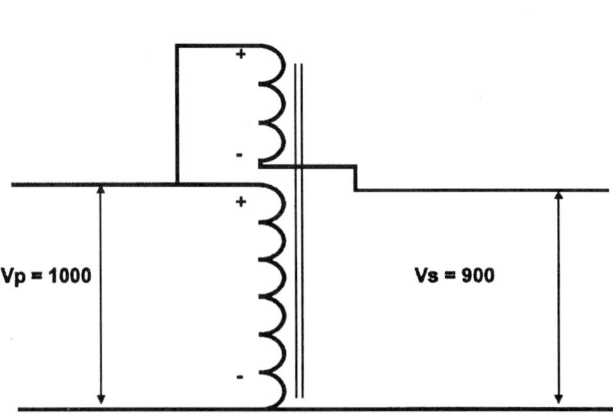

FIGURE 3.76 Step-down autotransformer.

Figure 3.78 shows the movable contacts in a bridging position; voltage change is one half the 1¼% tap voltage of the series winding because of its center tap and movable contacts located on adjacent stationary contacts.

A circulating current caused by the two contacts being at different positions (reactor energized with 1¼% tap voltage) is limited by the reactive impedance of this circuit. Two opposing requirements must

Transformers

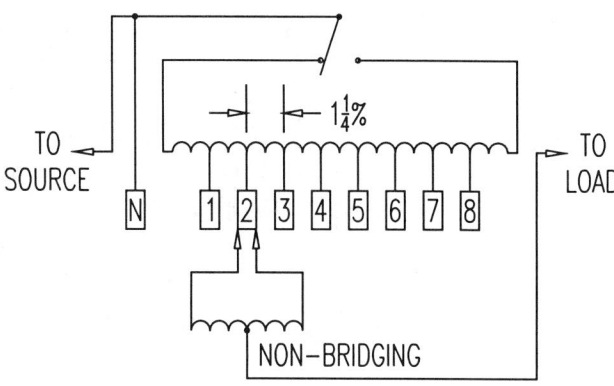

FIGURE 3.77 Two movable contacts on the same stationary contact (symmetrical position).

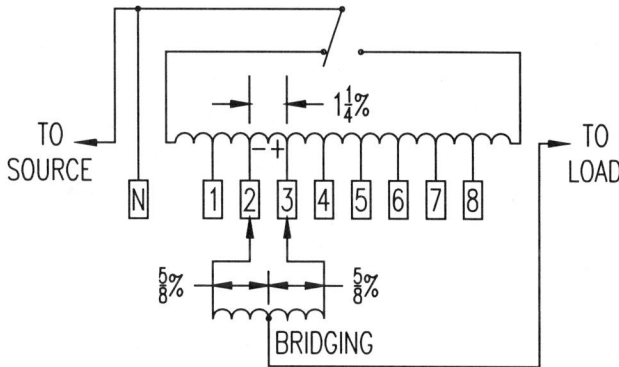

FIGURE 3.78 Two movable contacts on adjacent stationary contacts (bridging position).

be kept in mind in the design of the amount of reactance to which the value of the circulating current is set. First, the circulating current must not be excessive, and second, the variation of reactance during the switching cycle should not be so large as to introduce undesirable fluctuations in the line voltage. The reactor has an iron core with gaps in the magnetic circuit to set this magnetizing circulating current between 25 and 60%, inclusively, of full load current allowing for an equitable compromise between no-load and load conditions. The value of this circulating current also has a very decided effect on switching ability and contact life. The ideal reactor from an arcing standpoint would be one that has a closed magnetic circuit at no-load with an air gap that would increase in direct proportion to increase in load.

The voltage at the center tap is 5/8%, one half of the 1¼% tap voltage of the series winding taps. Some regulators, depending upon the rating, use an additional winding called an equalizer winding in the bridging reactor circuit. The equalizer winding is a 5/8% voltage winding on the same magnetic circuit (core) as the shunt and series winding. The equalizer winding is connected into the reactor circuit opposite in polarity to the tap voltage so that the reactor is excited at 5/8% of line voltage on both the symmetrical and bridging positions. Figure 3.79 shows an equalizer winding incorporated into the main coil of a regulator.

Voltage regulators are designed and manufactured in two basic different types of construction. These are defined by IEEE standards as Type A and Type B.

Type A step-voltage regulators have the primary circuit (source voltage) connected directly to the shunt winding of the regulator. The series winding is connected to the load side of the regulator and, by adjusting taps, changes the output voltage. With Type A construction, the core excitation varies because the shunt winding is connected across the primary circuit. See the schematic diagram in Fig. 3.80.

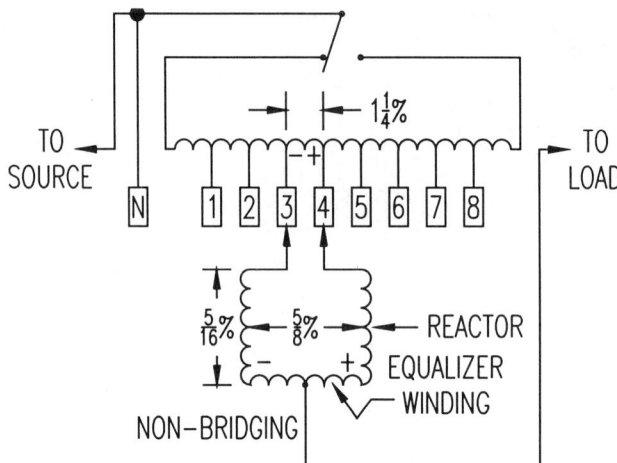

FIGURE 3.79 An equalizer winding incorporated into the main coil of a regulator.

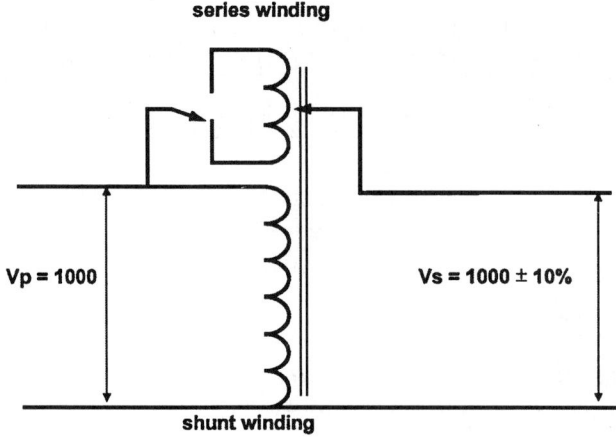

FIGURE 3.80 Voltage regulation on the load side (Type A).

Type B step-voltage regulators are constructed so that the primary circuit (source voltage) is applied by way of taps to the series winding of the regulator, which is connected to the source side of the regulator. With Type B construction, the core excitation is constant because the shunt winding is connected across the regulated circuit. See Fig. 3.81.

Regulator Control

The purpose of the regulator control is to provide an output action as a result of changing input conditions, in accordance with preset values which are selected by the regulator user. The output action is the energization of the motorized tap changer to change taps to maintain the correct regulator output voltage. The changing input is the output load voltage and current from an internal PT and CT, respectively, as shown in Fig. 3.82. The preset values are the values the regulator user has selected as control parameters for the regulated voltage. Basic regulator control settings are:

- Set Voltage
- Bandwidth

Transformers

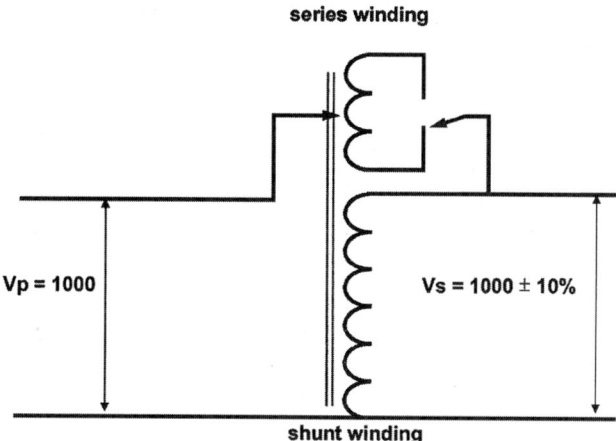

FIGURE 3.81 Voltage regulation on the source side (Type B).

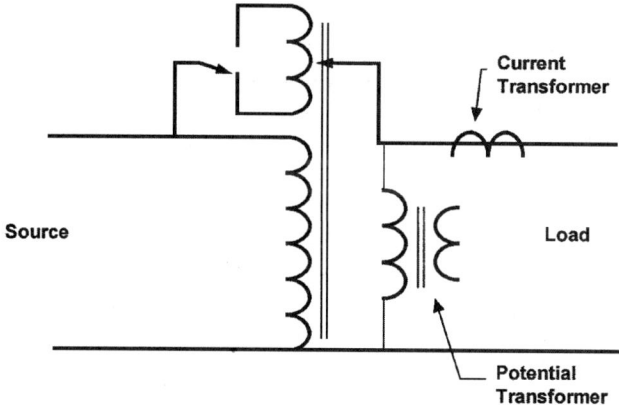

FIGURE 3.82 Location of current and potential transformers in a voltage regulator.

- Time Delay
- Line Drop Resistive and Reactive Compensation

The control set voltage is dependent upon the regulator rating and the system voltage on which it is installed. The regulator nameplate shows the potential transformer ratio that corresponds to the system voltage. The regulator load voltage is the product of the PT ratio and the control set voltage.

The bandwidth is the total voltage range around the set voltage value which the control will consider as a satisfied condition. For example, a 2-V bandwidth on a 120-V setting means that the control will not activate a tap change until the voltage is above 121 V or below 119 V.

The time delay is the period of time in seconds that the control waits from the time the voltage goes out of band to when power is applied to the tap changer motor to make a tap change.

Quite often, regulators are installed some distance from a theoretical load center or the location at which the voltage is attempted to be regulated. This means the load will not be served at the desired voltage level due to the losses (voltage drop) on the line between the regulator and the load. Furthermore, as the load increases, line losses also increase, causing the lowest voltage condition to occur during the time of heaviest loading. This is the least desirable time for this to occur.

To provide the regulator with the capability to regulate at a "projected" load center, a line drop compensation feature is incorporated in the control. This circuitry consists of a secondary supply from an internal current transformer, proportional to the load current, and resistive and reactive components through which this current flows. As the load current increases, the resulting secondary current flowing through these elements produces voltage drops which simulate the voltage drops on the primary line. This causes the "sensed" voltage to be altered correspondingly; therefore, the control responds by operating upon this pseudo load center voltage.

To select the proper resistive and reactive values, the user must take into account several factors about the line being regulated. A number of publications are available from manufacturers to assist in determining the variables needed and the resulting calculations.

When line drop compensation is used, the correct polarity of the resistance and reactance components is necessary for proper regulation. On four-wire, Wye-connected systems, the polarity selector is always set for +X and +R values. On Delta connected systems, however, the line current is 30° displaced from the line-to-line voltage (assuming 100% power factor). On open-Delta-connected regulator banks, one regulator is 30° leading, the other is 30° lagging. On closed Delta regulator banks, all regulators will be either leading or lagging. As a result of this displacement, the polarity of the appropriate resistive or reactive element must be reversed. The setting of the selector switch would be set on the +X+R, –X+R, or +X–R setting.

Unique Applications

Most step-voltage regulators are installed in circuits with a well-defined power flow from source to load. However, some circuits have interconnections or loops in which the direction of power flow through the regulator may change. For optimum utility system performance, step-voltage regulators, installed on such circuits, have the capability of detecting this reverse power flow, and sensing and controlling the voltage, regardless of the power flow direction.

In other systems, increasing levels of embedded (dispersed) generation pose new challenges to utilities in their use of step-voltage regulators. Traditionally, distribution networks have been used purely to transport energy from the transmission system down to lower voltage levels. A generator delivering electricity directly to the distribution network reverses the normal power flow direction. Options in the electronic control of the step-voltage regulator are available for handling different types of scenarios that give rise to reverse power flow conditions.

References

ANSI/IEEE C57.95 — 1987, IEEE Guide for Loading Liquid-Immersed Step-Voltage and Induction-Voltage Regulators.

Colopy, C. A., Grimes, S., and Foster, J. D., Proper Operation of Step Voltage Regulators in the Presence of Embedded Generation, *CIRED Conference*, June 1999.

Cooper Power Systems Bulletin 77006 2/93, How Step Voltage Regulators Operate.

Cooper Power Systems Distribution Voltage Regulation and Voltage Regulator Service Workshops.

Day, T. R. and Down, D. A., The Effects of Open-Delta Line Regulation on Sensitive Earth Fault Protection of 3-Wire Medium Voltage Distribution Systems, Cooper Power Systems.

Foster, J. D., Bishop, M. T., and Down, D. A., The Application of Single-Phase Voltage Regulators on Three-Phase Distribution Systems, *IEEE Conf. Paper* 94 C2, June 1994.

IEEE C57.15 — 1999, Standard Requirements, Terminology, and Test Code for Step-Voltage Regulators.

Kojovic, L. A., McCall, J. C., and Colopy, C. A., Voltage Regulator and Capacitor Application Considerations in Distribution Systems for Voltage Improvements, *Syst. Eng. Ref. Bull.*, SE9701, January 1997.

Voltage Regulator Instruction and Operating Manuals and Product Catalogs published by Cooper Power Systems, Siemens Energy and Automation, Inc. and General Electric Company.

3.7 Reactors

Richard Dudley, Antonio Castanheira, and Michael Sharp

Reactors, like capacitors, are basic to and an integral part of both distribution and transmission power systems. Depending on their function, reactors are either connected in shunt or in series with the network; singularly (current limiting reactors, shunt reactors) or in conjunction with other basic components such as power capacitors (shunt capacitor switching reactors, capacitor discharge reactors, filter reactors).

Reactors are utilized to provide inductive reactance in power circuits for a wide variety of purposes. These include fault current limiting, inrush current limiting for capacitors and motors, harmonic filtering, VAR compensation, reduction of ripple currents, blocking of power line carrier signals, neutral grounding, damping of switching transients, flicker reduction for arc furnace applications, circuit de-tuning, load balancing, and power conditioning.

Reactors may be installed at any industrial, distribution, or transmission voltage level and may be rated for any current duty from a few amperes to tens of thousands of amperes and fault current levels of up to hundreds of thousands of amperes.

Background and Historical Perspective

Reactors can be either dry type or oil immersed. Dry type reactors may be of air core or iron core construction. In the past, dry type air core reactors were only available in open style construction (Fig. 3.83), their windings held in place by a mechanical clamping system and the basic insulation provided by the air space between turns. Modern dry type air core reactors (Fig. 3.84) feature fully encapsulated windings with the turns insulation provided by film, fiber, or enamel dielectric. Oil immersed reactors may be of gapped iron core (Fig. 3.85) or magnetically shielded construction. The application range for the different reactor technologies has undergone a major realignment from historical usage. In the past, dry type air core reactors (open style winding technology) were limited to applications at distribution voltage class. Modern dry type air core reactors (fully encapsulated with solid dielectric insulated windings) are employed over the full range of distribution and transmission voltages, including EHV (high voltage series reactors) and (filter reactors, smoothing reactors) HVDC. Oil immersed reactors are primarily used for EHV shunt reactor and for some HVDC smoothing reactor applications. Dry type iron core reactors (Fig. 3.86) are usually used at low voltage and indoors, for applications such as harmonic filtering and power conditioning (di/dt, smoothing, etc.). Applicable IEEE standards, such as IEEE C57.21-1990 (R1995), IEEE C57.16-1996, and IEEE 1277, reflect these practices.

These standards provide considerable information not only concerning critical reactor ratings, operational characteristics, tolerances, and test code, but also guidance for installation and important application specific considerations.

Applications of Reactors

General Overview

Reactors have always been an integral part of power systems. The type of technology employed for the various applications has changed over the years based on design evolution and construction/materials breakthroughs. Dry type air core reactors have traditionally been used for current limiting applications due to their inherent linearity of inductance vs. current. For this application, fully encapsulated construction usually became the design of choice because its improved mechanical characteristics enabled the reactors to withstand higher fault currents. Conversely, high voltage series reactors were initially oil immersed shielded core designs. However, beginning in the early 1970s, the requirements of these applications were also met by fully encapsulated dry type air core designs. Due to such developments, the latest revision of IEEE C57.16-1996, the series reactor standard, is now a dry type air core reactor standard only.

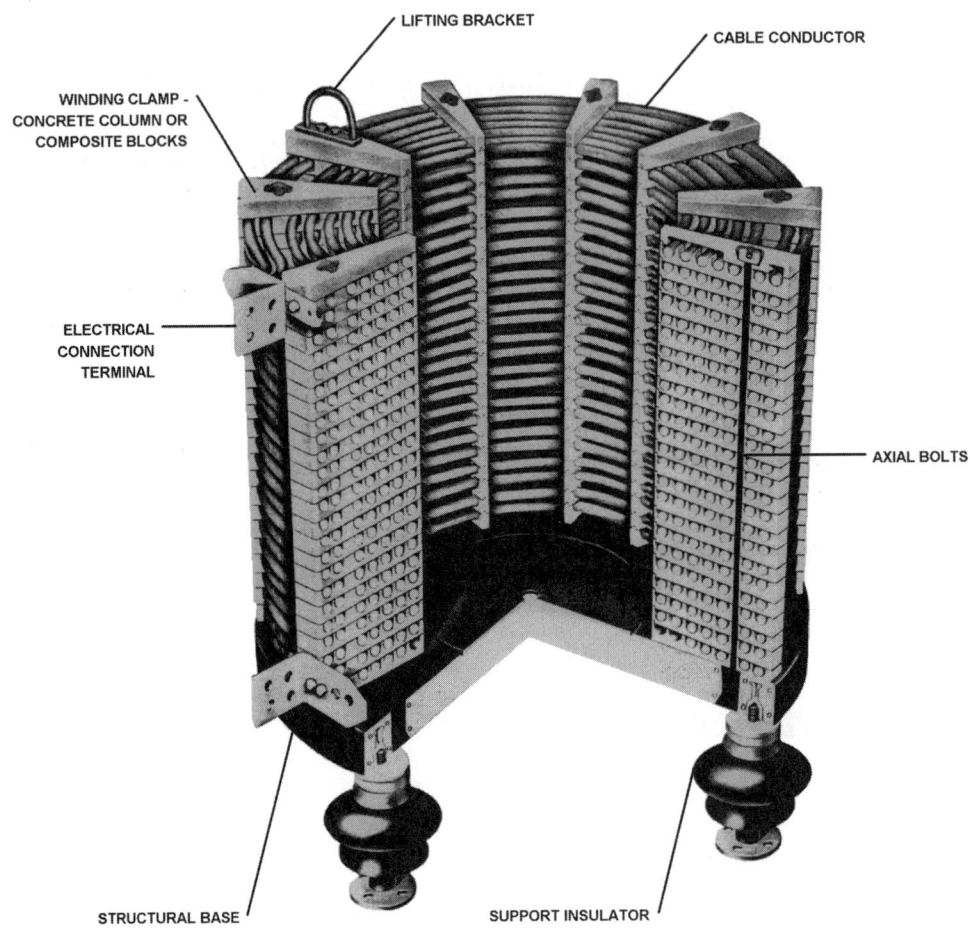

FIGURE 3.83 Open style reactor.

The construction technology employed for modern shunt reactors, on the other hand, is more dependent on the applied voltage. Transmission class shunt reactors are of oil immersed construction, whereas tertiary connected, or lower voltage direct connect shunt reactors, utilize either dry type air core or oil immersed construction. Hence, ANSI/IEEE C57.21-1990 (R1995) covers both oil immersed and dry type air core shunt reactors.

A review of modern reactor applications is provided below.

Current Limiting Reactors

Current limiting reactors are now used to control short circuit levels in electrical power systems covering the range from large industrial power complexes to utility distribution networks and to bulk HV/EHV transmission systems.

Current limiting reactors are primarily installed to reduce the short circuit current to levels consistent with the electro-mechanical withstand level of circuit components (especially transformers and circuit breakers) and to reduce the short circuit voltage drop on bus sections to levels that are consistent with insulation coordination practice. High fault currents on distribution or transmission systems, if not limited, can cause catastrophic failure of distribution equipment and present a serious threat to the safety of operating crews. In summary, current limiting reactors are installed to reduce the magnitude of short circuit currents in order to achieve one or more of the following benefits:

Transformers

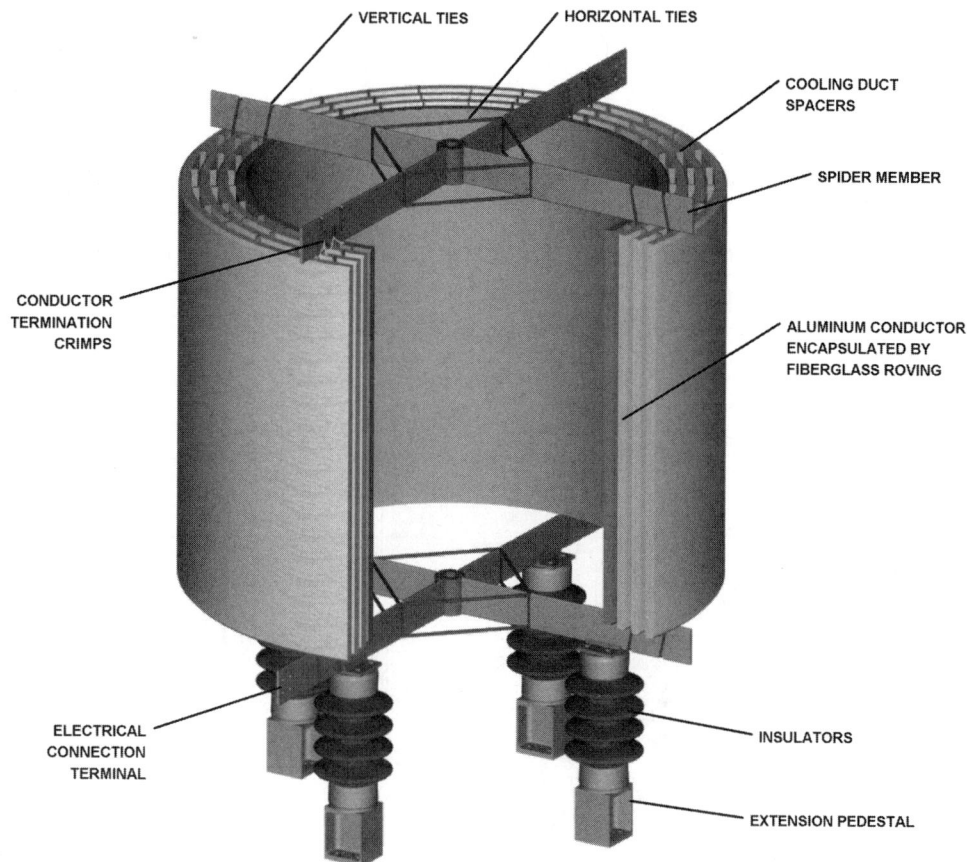

FIGURE 3.84 Modern fully encapsulated reactor.

- Reduction of electro-mechanical loading and thermal stresses on transformer windings, thus extending the service life of transformers and associated equipment.
- Improvement of the stability of primary bus voltage during a fault event on a feeder.
- Reduction of current-interrupting duty of feeder circuit breakers.
- Reduction of line-to-line fault current to levels below those of line-to-ground faults or vice versa.
- Protection of distribution transformers and all other downstream power equipment and devices from the propagation of initial fast front voltage transients due to faults and/or circuit breaker operations.
- Reduction of the requirement for downstream protection devices such as reclosers, sectionalizers, and current limiting fuses.
- Allowance of complete control over the steady-state losses by meeting any specified Q-factor for any desired frequency; this feature is particularly important for networks where high harmonic currents are to be damped without increasing the fundamental frequency loss.
- Increase in system reliability.

Current limiting reactors may be installed at different points in the power network and as such they are normally referred to by a name that reflects their location. The most common nomenclatures are:

Phase Reactors — installed in series with incoming or outgoing lines or feeders
Bus Tie Reactors — used to tie together two otherwise independent buses
Neutral Grounding Reactors — installed between the neutral of a transformer and ground
Duplex Reactors — installed between a single source and two buses

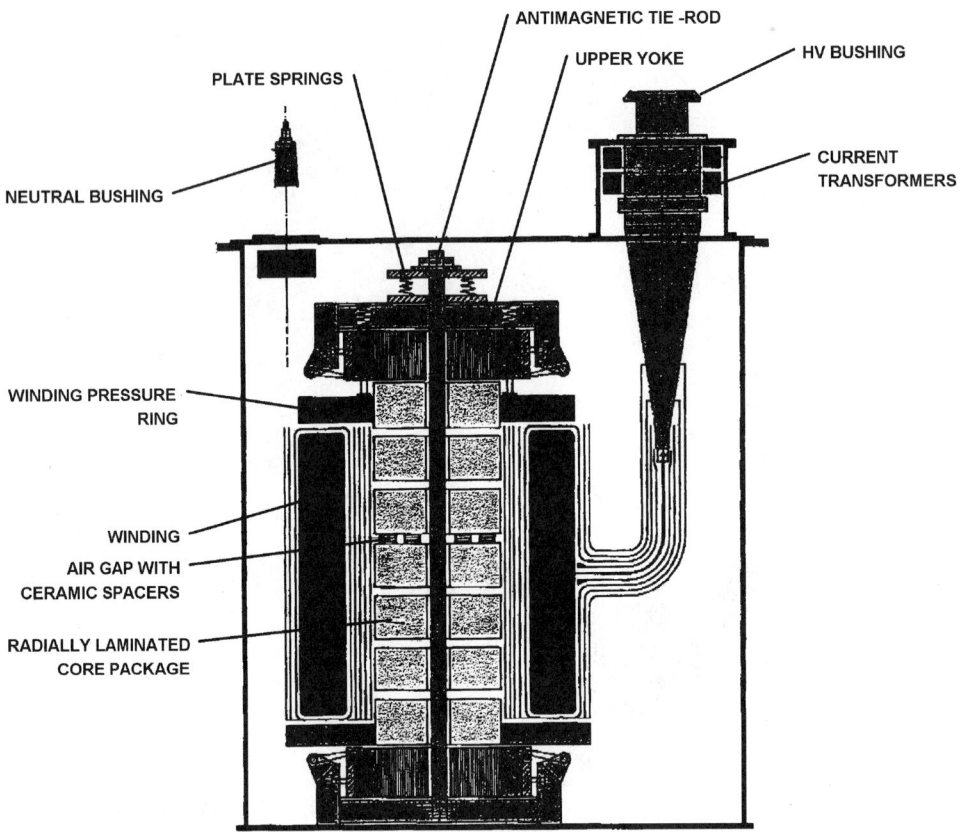

FIGURE 3.85 Gapped iron core oil immersed reactor.

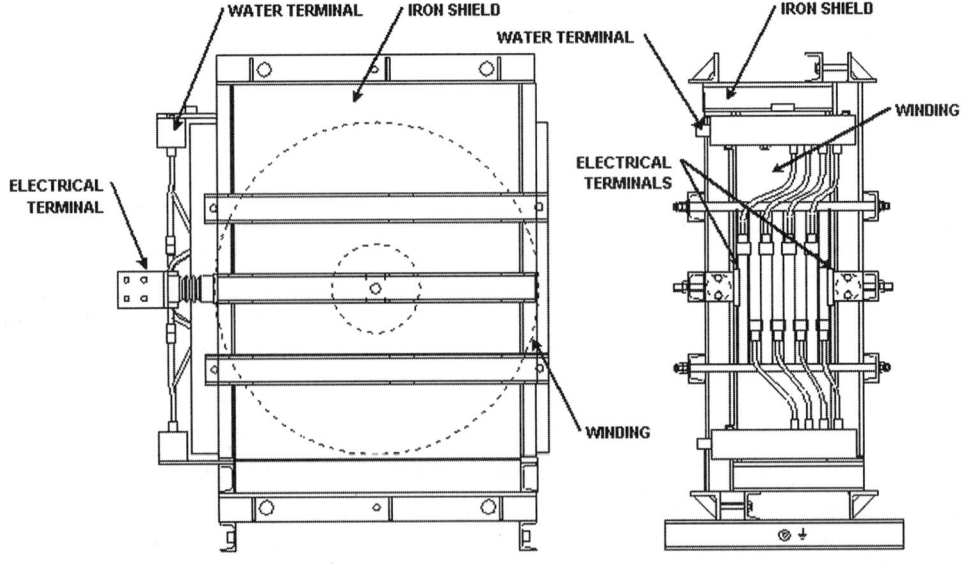

FIGURE 3.86 Dry type iron core reactor (water cooled).

Transformers

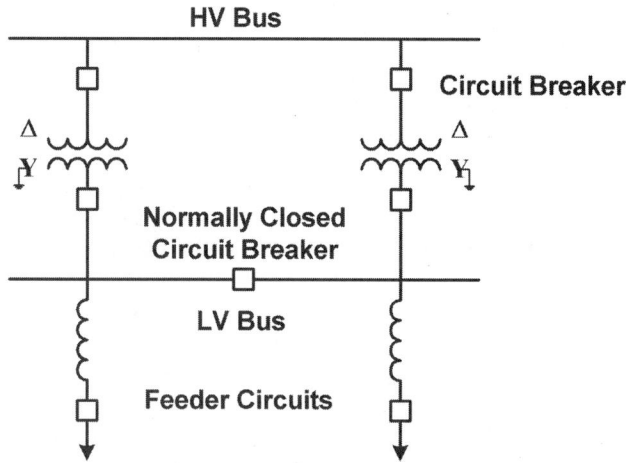

FIGURE 3.87 Typical phase reactor connection.

FIGURE 3.88 345-kV phase reactors.

Phase Reactors

Figure 3.87 shows the location of phase reactors in the feeder circuits of a distribution system. Current limiting reactors may also be applied at transmission voltage levels as shown in Fig. 3.88, depicting a 345 kV, 1100 A, 94.7 mH phase reactor installed in a U.S. utility substation.

The main advantage of the illustrated configuration is that it reduces line-to-line *and* phase-to-ground short-circuit current to any desired level at a strategic location in the distribution or transmission system.

Phase reactors are one of the most versatile manifestations of series connected reactors in that they can be installed in one feeder of a distribution system to protect one circuit or at the output of a generator to limit fault contribution to the entire power grid or anywhere in between. Any of the benefits listed in the section entitled "Current Limiting Reactors" can be achieved with this type of current limiting reactor. The impedance of the phase reactor required to limit the 3Φ short-circuit current to a given value, may be calculated as:

$$X_{CLR} = V_{LL} * \left[\left(1/I_{SCA}\right) - \left(1/I_{SCB}\right) \right] / \sqrt{3} \quad [\Omega] \tag{3.18}$$

$$X_{CLR} = V_{LL}^2 * \left[\left(1/MVA_{SCA}\right) - \left(1/MVA_{SCB}\right) \right] \quad [\Omega] \tag{3.19}$$

where V_{LL} is the rated system voltage, in [kV]

I_{SCA} or MVA_{SCA} is the required value of the short-circuit current or power after the installation of the phase reactor, in [kA] or [MVA]

I_{SCB} or MVA_{SCB} is the available value of the short circuit current or power before the installation of the phase reactor, in [kA] or [MVA]

Bus Tie Reactors

Bus tie reactors are used when two or more feeders and/or power sources are connected to a single bus and it is desirable to sectionalize the bus due to high fault levels, without losing operational flexibility. Figure 3.89 illustrates the arrangement.

As in the case of phase reactors, bus tie reactors may be applied at any voltage level. Figure 3.90 shows an example of 230 kV, 2000 A, 26.54 mH bus tie reactors installed in a U.S. utility substation.

The advantages of this configuration are similar to those associated with the use of phase reactors. An added benefit is that if the load is essentially balanced on both sides of the reactor under normal operating conditions, the reactor has negligible effect on voltage regulation or system losses.

The required reactor impedance is calculated using either Eq. (3.18) or (3.19) for 3Φ faults or Eq. (3.20) for single line to ground faults.

A method to evaluate the merits of using either phase reactors or bus tie reactors is presented in the section entitled "Phase Reactors vs. Bus Tie Reactors".

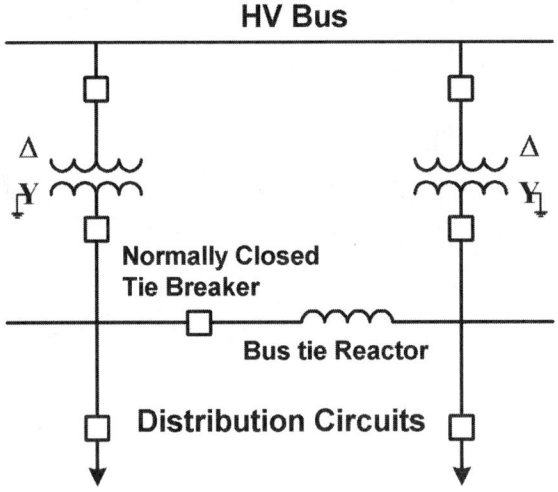

FIGURE 3.89 Typical bus tie reactor connection.

FIGURE 3.90 230-kV bus tie reactors.

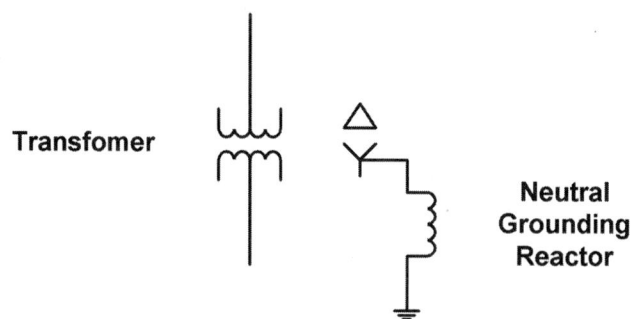

FIGURE 3.91 Typical neutral grounding reactor connection.

Neutral Grounding Reactor

Neutral grounding reactors (NGRs) are used to control single line to ground faults only. (They do not limit line to line fault current levels). They are particularly useful at transmission voltage levels, when autotransformers with a delta tertiary are employed.

These transmission station transformers can be a strong source of zero-sequence currents and, as a result, the ground fault current may substantially exceed the 3Φ fault current.

Figure 3.91 shows a typical neutral grounding reactor arrangement.

These devices, normally installed between the transformer or generator neutral and ground, are effective in controlling single line to ground faults since, in general, the system short circuit impedance is largely reactive. NGRs reduce short circuit stresses on station transformers for the most prevalent type of fault in an electrical system.

If the objective is to reduce the single line to ground (1Φ) fault, then Eqs. (3.30) and (3.30a) must be used and, after algebraic manipulations, the required phase reactor impedance, in ohms, is calculated as:

$$X_{CLR} = \sqrt{3} * V_{LL} * \left[\left(1/I_{SCA} \right) - \left(1/I_{SCB} \right) \right] / 3 \quad [\Omega] \qquad (3.20)$$

where the parameters are defined as before, with the exception that the short-circuit currents in question are the *single line-to-ground fault currents, in kA*.

A factor to be taken into consideration when applying NGRs is that the resulting X_0/X_1 may exceed a critical value and, as a result, give rise to transient overvoltages on the unfaulted phases ($X_0 > 10X_1$) (See IEEE Std. 142-1991, Green Book).

Because only one NGR is required per three-phase transformer and their continuous current is the system unbalance current, the cost of installing NGRs is lower than that for phase CLRs. Operating losses are also lower than for phase CLRs and steady state voltage regulation need not be considered with their application.

The impedance rating of a neutral grounding reactor may be calculated using Eq. (3.18), provided both the short circuit currents before and after the NGR installation are the single line to ground faults.

Although NGRs do not have any direct effect on line-to-line faults, they are of significant benefit since most faults start from line-to-ground, some progressing quickly to a line-to-line fault if fault site energy is high and the fault current is not interrupted in time. Therefore, the NGR can contribute indirectly to a reduction in the number of occurrences of line-to-line faults by reducing the energy available at the location of the line to ground fault.

Duplex Reactors

Duplex reactors are usually installed at the point where a large source of power is split into two simultaneously and equally loaded buses (Fig. 3.92). They are designed to provide low rated reactance under normal operating conditions and full rated or higher reactance under fault conditions. A duplex reactor consists of two magnetically coupled coils per phase. This magnetic coupling, which is dependent upon the geometric proximity of the two coils, determines the properties of a duplex reactor under steady state and short circuit operating conditions. During steady-state operation, the magnetic fields produced by the two windings are in opposition, and the effective reactance between the power source and each bus is a minimum. Under short-circuit condition, the linking magnetic flux between the two coils becomes unbalanced, resulting in higher impedance on the faulted bus, thus restricting the fault current. The voltage on the unfaulted bus is supported significantly until the fault is cleared; both by the effect of the reactor impedance between the faulted and unfaulted bus and also by the "voltage boosting" effect caused by the coupling of the faulted leg with the unfaulted leg of duplex reactors.

The impedance of a duplex reactor can be calculated using Eqs. (3.18) and (3.19), the same as those used for phase reactors.

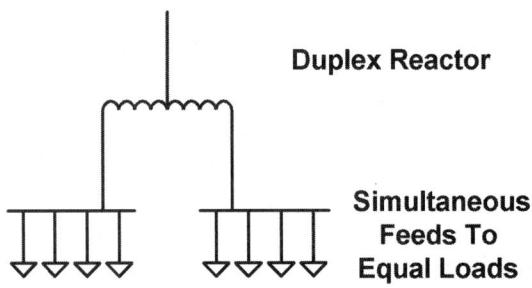

FIGURE 3.92 Typical duplex reactor connection.

Capacitor Inrush/Outrush Reactors

Capacitor switching can cause significant transients at both the switched capacitor and remote locations. The most common transients are:

- overvoltage on the switched capacitor during energization
- voltage magnification at lower voltage capacitors
- transformer phase-to-phase overvoltages at line termination
- inrush current from another capacitor during back-to-back switching
- current outrush from a capacitor into a nearby fault
- dynamic overvoltage when switching a capacitor and transformer simultaneously

Capacitor inrush/outrush reactors (Fig. 3.93) are used to reduce the severity of some of the transients listed above in order to minimize dielectric stresses on breakers, capacitors, transformers, surge arresters, and associated station electrical equipment. High frequency transient interference in nearby control and communication equipment is also reduced.

Reactors are effective in reducing all transients associated with capacitor switching since they not only limit the magnitude of the transient current, but also significantly reduce the transient frequency, as indicated by Eqs. (3.21) and (3.22).

$$I_{peak} = V_s * \sqrt{(C_{eq}/L_{eq})} \quad [\text{Amps}] \tag{3.21}$$

$$f = 1 / \left[2\pi \sqrt{L_{eq} C_{eq}} \right] \quad [\text{Hz}] \tag{3.22}$$

where C_{eq} and L_{eq} are the equivalent capacitance and inductance of the circuit, respectively, and V_s is the system line to line voltage.

Therefore, reflecting the information presented in the preceding discussion, the standard IEEE 1036-1992, Guide for Application of Shunt Power Capacitors, calls for the installation of series reactors in series with each capacitor bank, especially when switching back-to-back capacitor banks.

Transformers

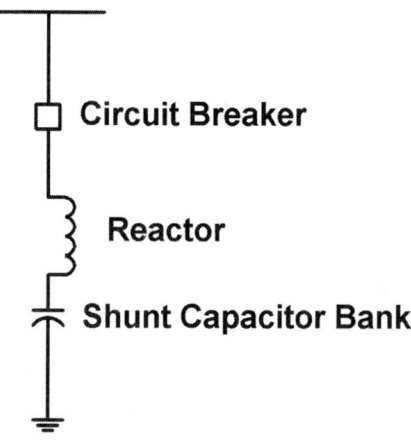

FIGURE 3.93 Typical capacitor inrush/outrush reactor connection.

FIGURE 3.94 550-kV capacitor inrush/outrush reactors.

A typical EHV shunt capacitor installation, utilizing reactors rated 550 kV/1550 kV BIL, 600 A, 3.0 mH, is shown in Fig. 3.94.

Discharge Current Limiting Reactors

High voltage series capacitor banks are utilized in transmission systems to improve stability operating limits. Series capacitor banks may be supplied with a number of discrete steps, insertion or bypass being achieved using a switching device. For contingencies, a bypass gap is also provided for fast bypass of the capacitors. In both cases, bypass switch closed or bypass gap activated, a discharge of the capacitor occurs, and the energy associated with the discharge must be limited by a damping circuit. A discharge current limiting reactor is an integral part of this damping circuit. Therefore, the discharge reactor must be designed to withstand the high frequency discharge current superimposed on the system power frequency current. The damping characteristic of this reactor is a critical parameter of the discharge circuit. Sufficient damping may be provided as an integral component of the reactor design (de-Q'ing) or can be supplied as a separate element (resistor). See Fig. 3.95.

Power Flow Control Reactors

A more recent application of series reactors in transmission systems is that of power flow control (Fig. 3.96) or its variant, overload mitigation.

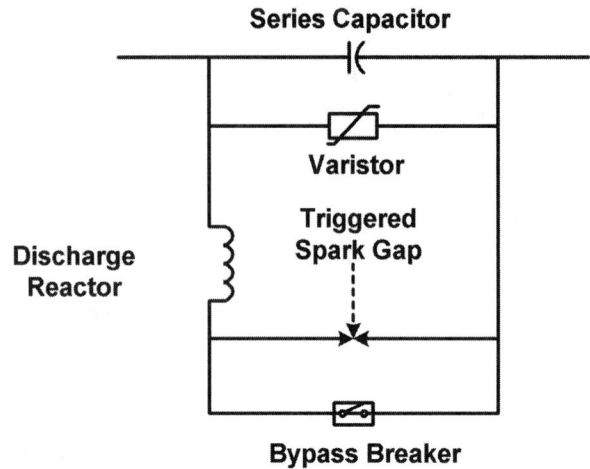

FIGURE 3.95 Tyical discharge current limiting reactor connection.

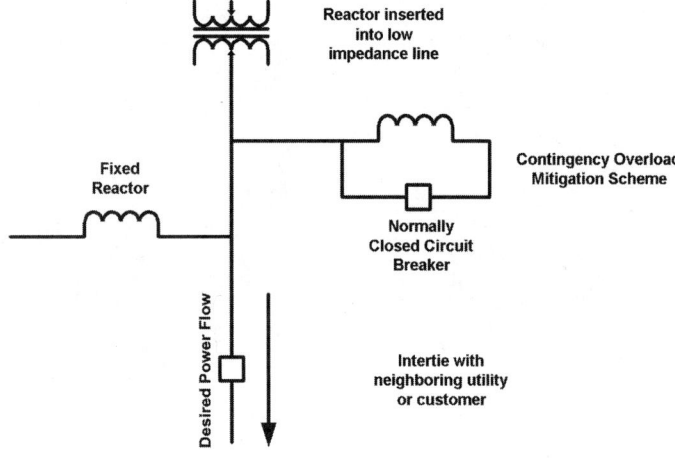

FIGURE 3.96 Typical high voltage power flow control reactor connections.

The flow of power through a transmission system is a function of the path impedance and the complex voltage (magnitude and phase) at the ends of the line. In interconnected systems, the control of power flow is a major concern for the utilities, because unscheduled power flow may give rise to a number of problems, such as:

- overloading of lines
- increased system losses
- reduction in security margins
- contractual violations concerning power import/export
- increase in fault levels beyond equipment rating

Typical power flow inefficiencies and limitations encountered in modern power systems may be the result of one or more of the following:

- Non-optimized parallel line impedances resulting in one line reaching its thermal limit well before the other line, thereby limiting peak power transfer.
- Parallel lines having different X/R ratios: a significant reactive component will flow in the *opposite* direction to that of the active power flow.

- High loss line more heaviliy loaded than lower loss parallel line, resulting in higher power transfer losses.
- "Loop flow" (the difference between scheduled and actual power flow): although inherent to interconnected systems, "loop flows" may be so severe as to adversely affect the system reliability.

Power flow control reactors are used to optimize power flow on transmission lines through a modification of the transfer impedance. As utility systems grow and the number of interties increase, parallel operation of AC transmission lines is becoming more common in order to provide adequate power to load centers. In addition, the complexity of contemporary power grids results in situations where the power flow experienced (by a given line of one utility) can be affected by switching, loading, and outage conditions occurring in another service area. Strategic placement of power flow reactors may serve to increase peak power transfer, reduce power transfer loss, and improve system reliability. The insertion of high voltage power flow control reactors in a low impedance circuit allows parallel lines to reach their thermal limits simultaneously and hence optimize peak power transfer at reduced overall losses. Optimum system performance may be achieved by insertion of one reactor rating to minimize line losses during periods of off peak power transfer and one of an alternative rating to achieve simultaneous peak power transfer on parallel lines during peak load periods or contingency conditions.

Contingency overload mitigation reactor schemes are used when the removal of generation sources and/or lines in one area affects the loading of other lines feeding the same load center. This contingency may overload one or more of the remaining lines. The insertion of series reactors, shunted by a normally closed breaker, in the potentially overloaded line(s), keeps the line current below thermal limits. The parallel breaker carries the line current under normal line loading conditions and the reactor is switched into the circuit only under contingency situations.

Shunt Reactors (Steady State Reactive Compensation)

High voltage transmission lines, particularly long ones, generate a substantial amount of leading reactive power when lightly loaded. Conversely, they absorb a large amount of lagging reactive power when heavily loaded. In other words, unless the transmission line is operating under reactive power balance, the voltage on the system cannot be maintained at rated values.

$$\text{Reactive Power Balance} = \text{Total Line Charging} - \text{Line Reactive Losses}$$

If the Power Balance $\neq 0$, the line must be compensated for a given operational condition. Under heavy load, the power balance is negative and capacitive compensation (voltage support) is required. Conversely, under light load, the power balance is positive and inductive compensation is required. For details of power balance definition, please refer to the section entitled "Power Line Balance".

Shunt reactors absorb reactive power (Mvar's) and thus lower the voltage on a system. They are therefore used when lightly loaded transmission lines are producing Mvar's resulting in higher operating voltages. Shunt reactors are used to provide inductive reactive compensation to mitigate the effects of the high charging current of transmission lines and cable systems; therefore, switching surges during energization or reclosing of high voltage transmission lines are limited.

Shunt reactors may be connected to the transmission system through the tertiary winding of a power transformer connected to the transmission line being compensated; typically 13.8 kV, 34.5 kV, and 69 kV. Tertiary connected shunt reactors (Fig. 3.97) may be of dry type air core single phase per unit construction, or oil-immersed three phase or oil immersed single phase per unit construction.

Alternatively, shunt reactors can be connected directly to the transmission line to be compensated. Connection may be at the end of a transmission line or at an intermediate point, depending on voltage profile considerations. Directly connected shunt reactors are usually of oil immersed construction; dry type air core shunt reactors are presently available only for voltages up to 235 kV.

Oil immersed shunt reactors are available in two design configurations: coreless and iron core (and either self cooled or force cooled). Coreless oil immersed shunt reactor designs utilize a magnetic circuit or shield which surrounds the coil to contain the flux within the reactor tank. The steel core that normally

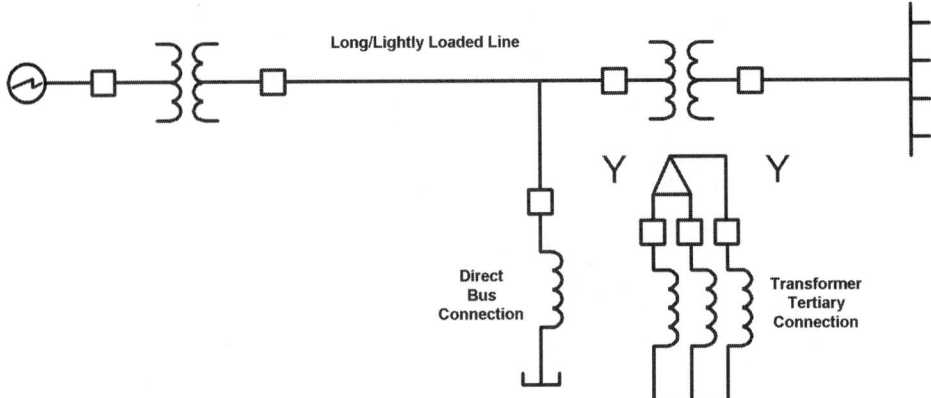

FIGURE 3.97 Typical shunt reactor connections.

FIGURE 3.98 20 kV, 20 MVA (per phase) tertiary connected shunt reactors.

provides a magnetic flux path through the primary and secondary windings of a power transformer is replaced by insulating support structures resulting in an inductor that is nearly linear with respect to applied voltage. Conversely, the magnetic circuit of an oil immersed iron core shunt reactor is constructed in a manner similar to that used for power transformers with the exception that an air gap or distributed air gap is introduced to provide the desired reluctance. Both types of oil immersed shunt reactors can be constructed as single phase or three phase units and are similar in appearance to conventional power transformers. (See Fig. 3.98.)

Thyristor Controlled Reactors (Dynamic Reactive Compensation)

As the network operating characteristics approach system limits, such as dynamic or voltage stability, or in the case of large dynamic industrial loads, such as arc furnaces, the need for dynamic compensation arises.

Typically, static VAR compensators (SVCs) are used to provide dynamic compensation at a receiving end bus, through microprocessor control, for maintaining a dynamic reserve of capacitive support when there is a sudden need.

Transformers

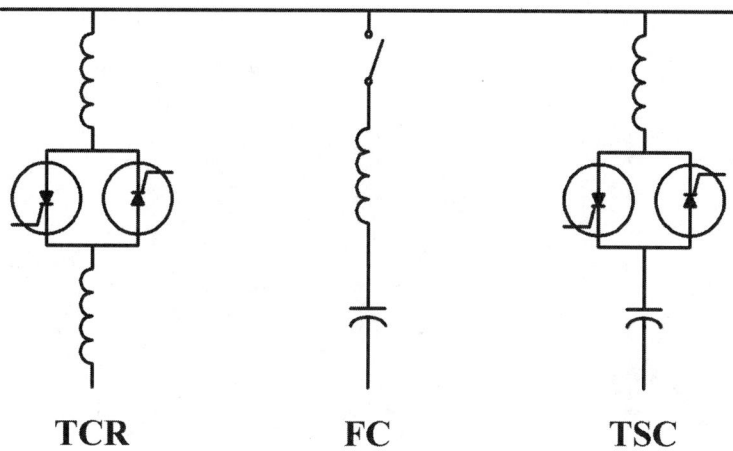

FIGURE 3.99 Static VAR compensator.

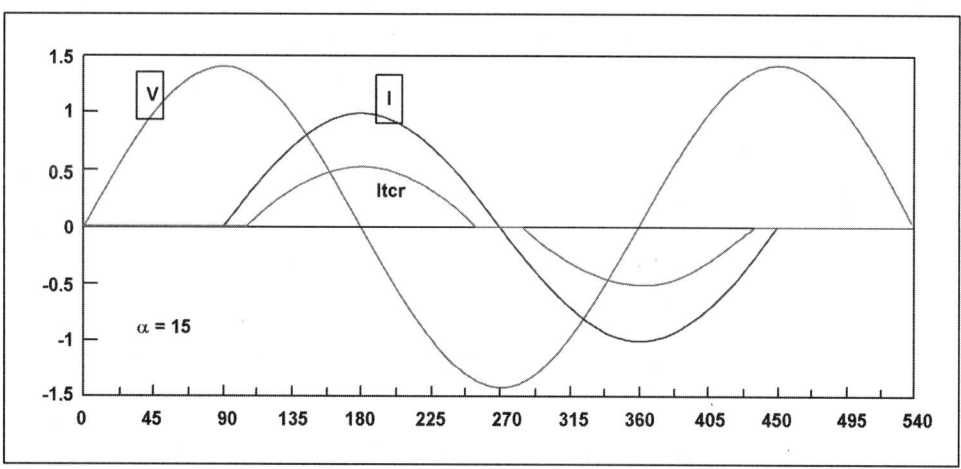

FIGURE 3.100 SVC voltage and current waveforms.

Figures 3.99 and 3.100 illustrate a typical configuration for an SVC and its basic operation and methodology, respectively.

By varying the firing angle, a, of the thyristor controlled reactor (TCR), the amount of current absorbed by the reactor can be continuously varied. The reactor then behaves as an infinitely variable inductance. Consequently, the capacitive support provided by the fixed capacitor (FC) and/or by the thyristor switched capacitor (TSC) can be adjusted to the specific need of the system.

The efficiency, as well as voltage control and stability, of power systems is greatly enhanced with the installation of SVCs.

The use of SVCs is also well established in industrial power systems. Demands for increased production and more strict regulations regarding both the consumption of reactive power by and disturbance mitigation on the power system may require the installation of SVCs. A typical example of an industrial load which can cause annoyance to consumers, usually in the form of flicker, is the extreme load fluctuations of electrical arc furnaces in steel works.

A typical installation at a steel mill is shown in Fig. 3.101. The thyristor controlled reactors are rated at 34 kV, 710 A, and 25 MVAR per phase.

FIGURE 3.101 34 kV, 25 MVAR (per phase) thyristor controlled reactors.

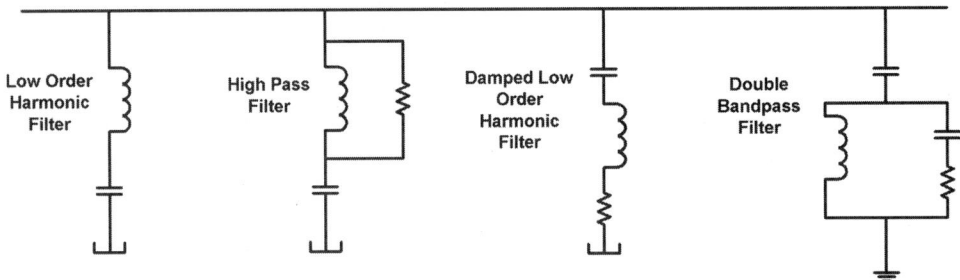

FIGURE 3.102 Typical filter reactor connections.

Filter Reactors

The increasing presence of nonlinear loads and the widespread use of power electronic switching devices in industrial power systems is causing an increase of harmonics in the power system.

Major sources of harmonics are industrial arcing loads (arc furnaces, welding devices), power converters for variable speed motor drives, distributed arc lightning for roads, fluorescent lightning, residential sources such as TV-sets and home computers, etc.

Power electronic switching devices are also applied in modern power transmission systems and include HVDC (High Voltage Direct Current) converters as well as FACTS (Flexible AC Transmission Systems) devices, such as SVCs.

Harmonics can have detrimental effects on equipment such as transformers, motors, switchgear, capacitor banks, fuses, and protective relays. Transformers, motors, and switchgear may experience increased losses and excessive heating. Capacitors may fail prematurely from increased heating and higher dielectric stress. If distribution feeders and telephone lines have the same "right-of-way", harmonics may also cause telephone interference problems.

In order to minimize the propagation of harmonics into the connected power distribution or transmission system, shunt filters are often applied close to the origin of the harmonics. Such shunt filters in their simplest embodiment consist of a series inductance (filter reactor) and capacitance (filter capacitor). If more than one harmonic is to be filtered, several sets of filters of different rating are applied to the same bus. More complex filters are also used to filter multiple harmonics.

Reactors for HVDC Application

In an HVDC system, reactors are used for various functions as shown, in principle, in Fig. 3.103. The HVDC smoothing reactors are connected in series with an HVDC transmission line or inserted in the

Transformers

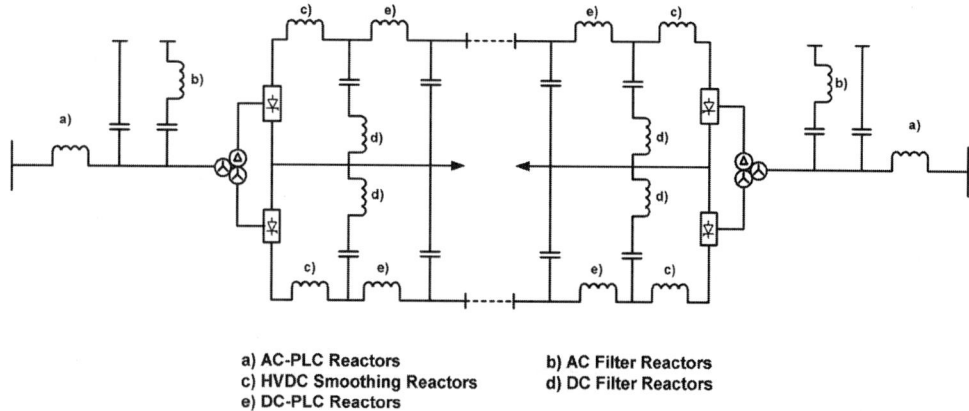

a) AC-PLC Reactors
c) HVDC Smoothing Reactors
e) DC-PLC Reactors

b) AC Filter Reactors
d) DC Filter Reactors

FIGURE 3.103 One line diagram of a typical HVDC bipole link illustrating reactor applications.

intermediate DC circuit of a back-to-back link to reduce the harmonics on the DC side, to reduce the current rise caused by faults in the DC system, and to improve the dynamic stability of the HVDC transmission system.

Filter reactors are installed for harmonic filtering on the AC and DC sides of the converters. AC filters serve two purposes simultaneously: the supply of reactive power and the reduction of harmonic currents. AC filter reactors are utilized in three types of filter configurations employing combinations of resistors and capacitors, namely single tuned filters, double tuned filters, and high pass filters. A single tuned filter is normally designed to filter the low order harmonics on the AC side of the converter. A double tuned filter is designed to filter multiple discrete frequencies using a single combined filter circuit. A high pass filter is essentially a single tuned damped filter. Damping flattens and extends the filter response to more effectively cover high order harmonics. DC filter reactors are installed in shunt with the DC line, on the line side of the smoothing reactors. The function of these DC filter banks is to further reduce the harmonic currents on the DC line. (See Figs. 3.102 and 3.103.)

PLC (Power Line Carrier) and RI (Radio Interference) filter reactors are employed on the AC and/or DC side of the HVDC converter to reduce high frequency noise propagation in the lines.

Series Reactors for Electric Arc Furnace Application

Series reactors may be installed in the medium voltage feeder (high voltage side of the furnace transformer) of an AC electric arc furnace in order to improve efficiency, reduce furnace electrode consumption, and limit short circuit current (thus reducing mechanical forces on the furnace electrodes). Such reactors may be either "built into" the furnace transformer or are separate, stand-alone units of oil immersed or dry type air core construction. Usually, the reactors are equipped with taps to facilitate optimization of the furnace performance. (See Figs. 3.104 and 3.105.)

Arc Suppression Reactors (Petersen Coils)

An arc suppression coil is a single phase, variable inductance, oil immersed, iron core reactor that is connected between the neutral of a transformer and ground for the purpose of achieving a resonant neutral ground. The zero sequence impedance of the transformer is taken into consideration in rating the inductance of the arc suppression coil. The adjustment of inductance is achieved in steps by means of taps on the winding or can be continuously adjusted by varying the reluctance of the magnetic circuit; the length of an air gap is adjusted by means of a central moveable portion of the core (usually motor driven). See Fig. 3.106. The inductance is adjusted, in particular, during non-ground fault conditions, to achieve cancellation of the capacitive ground current, so that in the case of a single line-to-ground fault, cancellation of the capacitive fault current is achieved with an inductive current of equal magnitude. Current injection by an active component (power converter) into the neutral, usually through an auxiliary

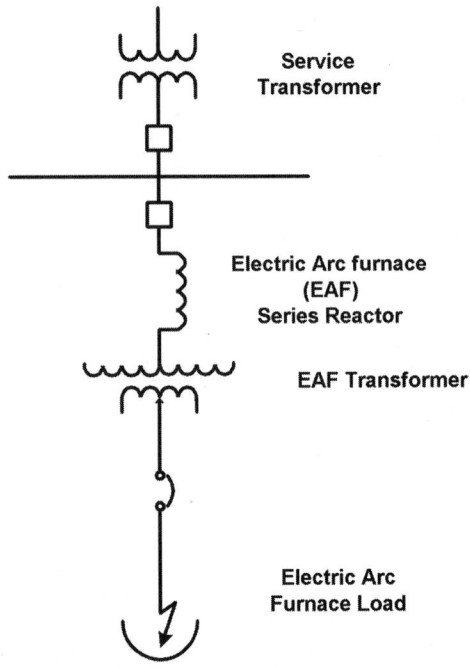

FIGURE 3.104 Typical electric arc furnace series reactor connection.

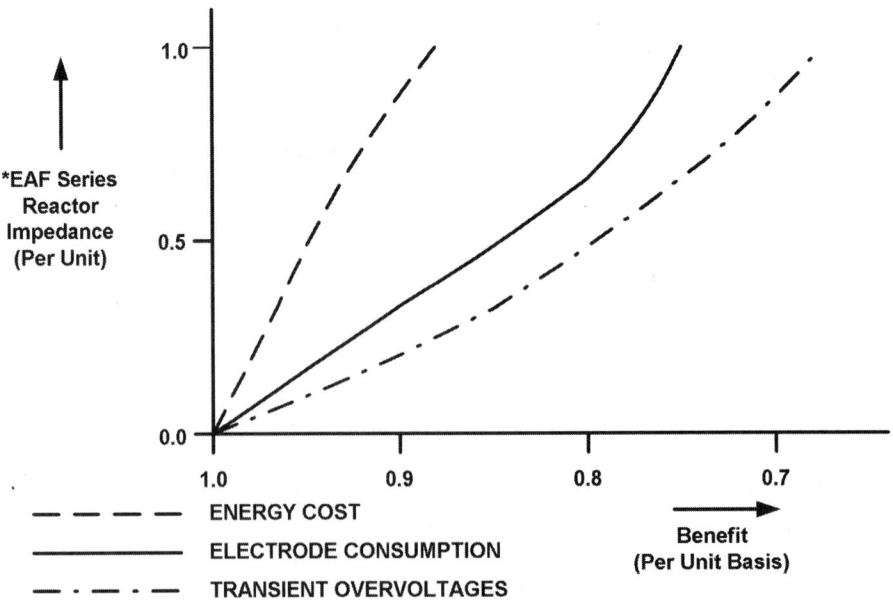

* 1.0 PU. - Series Reactor impedance corresponding to optimum arc length with respect to furnace refractories.

FIGURE 3.105 EAF series reactor benefits.

Transformers

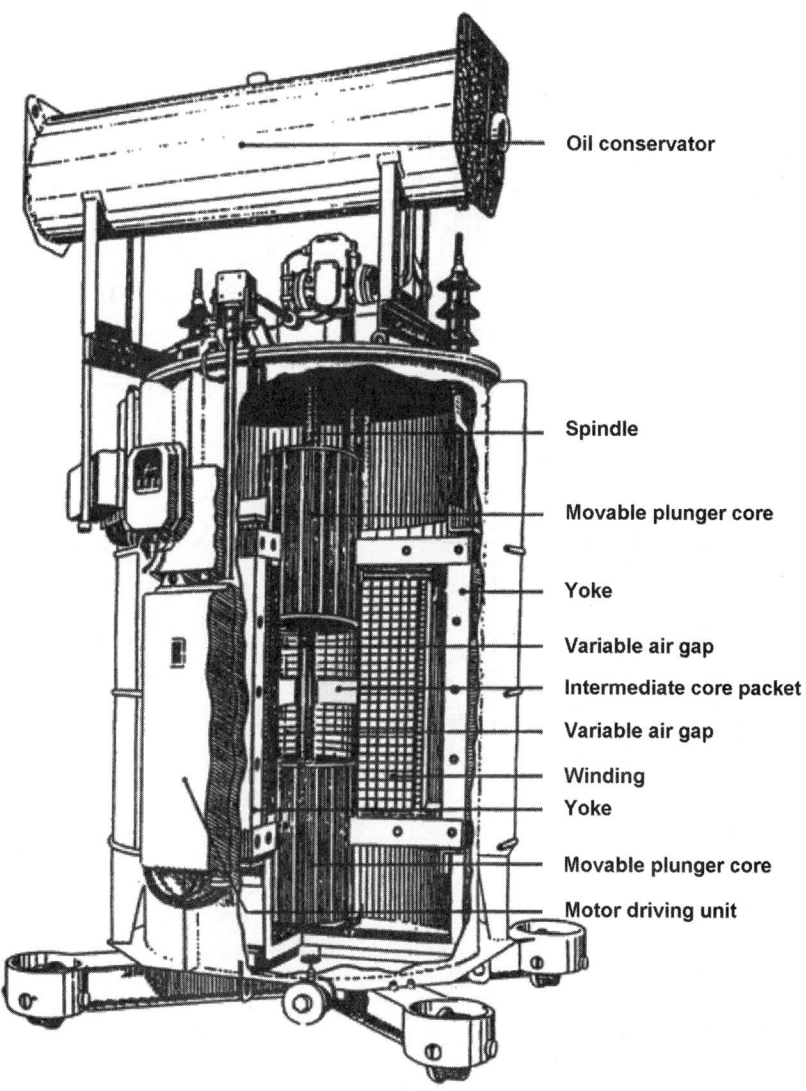

FIGURE 3.106 Arc suppression reactor.

winding of the arc suppression coil, can also provide cancellation of the resistance component of the fault current. See Figs. 3.107 and 3.108, which illustrate this principle. Resonant grounding is used in distribution systems in Europe, parts of Asia, and in a few areas of the U.S. The type of system ground employed is a complex function of system design, safety considerations, contingency (fault) operating practices, and legislation. Arc suppression reactors are typically used to the best advantage on distribution systems with overhead lines, to reduce intermittent arcing type single line-to-ground faults, which may otherwise occur on ungrounded systems.

Other Reactors

Reactors are also used in such diverse applications as motor starting (current limiting), test laboratory circuits (current limiting, dv/dt control, di/dt control), and insertion impedance (circuit switchers). Design considerations, insulation system, conductor design, cooling method, construction concept (dry type, oil immersed), and subcomponent/subassembly variants (mounting and installation considerations) are selected based on the application requirements.

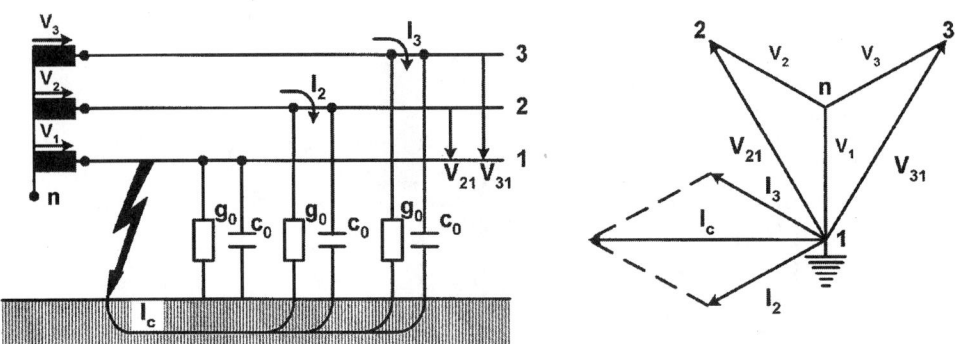

FIGURE 3.107 Single line-to-ground fault in non-grounded system.

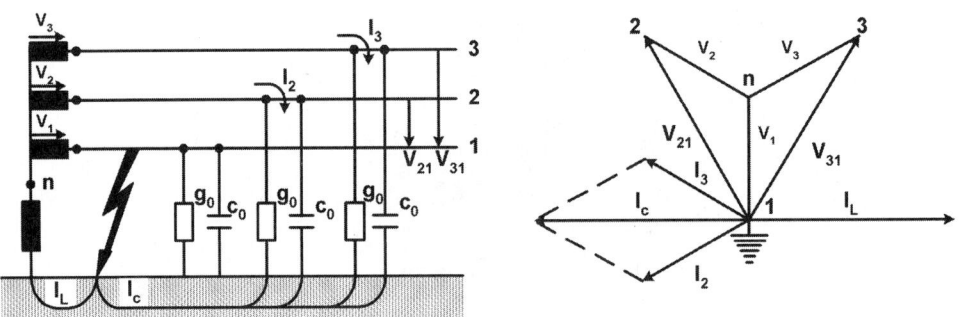

FIGURE 3.108 Single line-to-ground fault in resonant-grounded system.

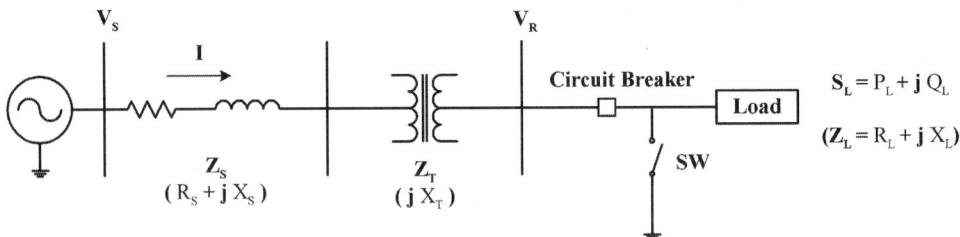

FIGURE 3.109 Radial system.

Some Important Application Considerations

Short Circuit: Basic Concepts

Figure 3.109 represents a radial system in which the sending end bus is connected to an infinite source. From inspection, the following equations may be established:

$$V_S = V_R + \Delta V \tag{3.23}$$

$$I = (V_R + \Delta V)/(Z_S + Z_T + Z_L) \tag{3.24}$$

Under steady state, the load impedance[2] (Z_L) essentially controls the current **I** since both Z_S and Z_T are small. Also, typically $X_S \gg R_S$; $X_T \gg X_S$ and the load $P_L > Q_L$.

Therefore,

$$\Delta V = V_S - V_R \cong \left[P * R_S + Q * \left(X_S + X_T\right)\right] \quad (3.25)$$

in **per unit**, is small.

When a short circuit occurs (closing the switch **SW**), $V_R \to 0$ and $Z_L = 0$ (bolted fault) and Eq. (3.24) can be rewritten as:

$$I_{SC} = \Delta V \Big/ \left[R_S + j\left(X_S + X_T\right)\right] \quad (3.26)$$

$$\cong \Delta V \angle -90° \Big/ \left(X_S + X_T\right) \quad (3.26a)$$

Since $|(X_S + X_T)|$ is small, the short circuit current (I_{SC}) may become very large. The total transmitted power then equals the available power from the source (**MVA**$_{SC}$),

$$S = V_S * I_{SC} \quad (3.27)$$

$$S = V_S * \Delta V \angle -90° \Big/ \left(X_S + X_T\right) \quad (3.27a)$$

which is essentially reactive in nature.

Therefore, the system voltage is shared, as voltage drops, between the system impedance (transmission lines) and the transformer impedance:

$$V_S = \Delta V \cong Q_S * \left(X_S + X_T\right) \quad (3.28)$$

Since X_T is typically much larger than X_S, the voltage drop across the transformer almost equals the system voltage. Two major concerns arise from this scenario:

- The mechanical **stresses in the transformer**; the windings will experience a force **proportional to the square of the current**.
- The ability of the **circuit breaker** to successfully interrupt the fault current.

Therefore, it is imperative to limit the short circuit current so that it will not exceed the ratings of equipment exposed to it.

Basic formulas are as follows:

Three Phase Fault:

$$I_{3\Phi}[kA] = 100[MVA] \Big/ \left[\sqrt{3}\, V_{LL} * Z_1\right] \quad (3.29)$$

1Φ to Ground Fault:

$$I_{SLG}[kA] = 3*100[MVA] \Big/ \left[\sqrt{3}\, V_{LL} * Z_T\right] \quad (3.30)$$

$$Z_T = Z_1 + Z_2 + Z_0 + 3*Z_N \quad (3.30a)$$

---

[2]The load may in fact be a rotating machine, in which case the **back e.m.f.** generated by the motor is the controlling factor in limiting the current. Nevertheless, this **e.m.f.** may be related to an impedance.

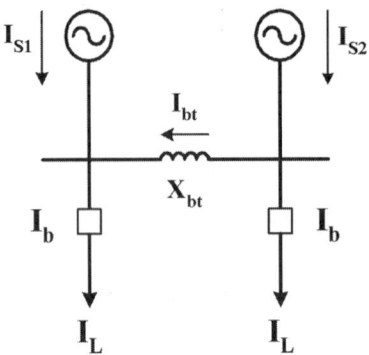

FIGURE 3.110 Bus tie reactor connection.

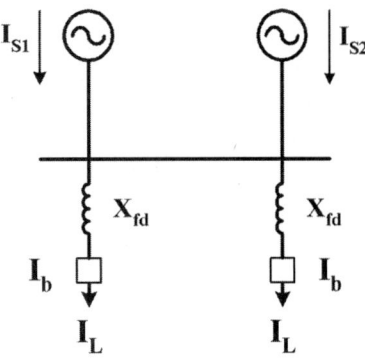

FIGURE 3.111 Phase reactor connection.

where V_{LL} [kV], is the line to line base voltage;

Z_T [pu], is the total equivalent system impedance seen from the fault;

Z_1, Z_2, and Z_0, are the equivalent system positive, negative and zero sequence impedance seen from the fault (in PU @ 100 MVA base);

Z_N [pu] is any impedance intentionally connected to ground, in the path of the fault current.

Phase Reactors vs. Bus Tie Reactors

A method to evaluate the merits of using either phase reactors vs. bus tie reactors is indicated below (see also Figs. 3.110 and 3.111).

I_{Si} is the available fault contribution from the sources, I_L is the rated current of each feeder, and **n** is the number of feeders in the bus.

Define: $I_b = \alpha\, I_{S2}$, the interrupting rating of the feeder CBs

$I_{S2} = K*I_{S1}$ (I_{S2} assumed $<I_{S1}$, K < 1)

$I_{bt} = \beta*n*I_L$, the rated current of the bus tie reactor

β = split of feeders on either side of the reactor

Therefore, the required reactor impedance in each configuration, their ratio, and the ratio of the rated power of the reactors are given by the equations that follow:

$$X_{bt} = \frac{V_{LL}}{\sqrt{3}\, I_{S2}} \left[\frac{K(1-\alpha)+1}{\alpha K - 1} \right] \qquad (3.31a)$$

$$X_{fd} = \frac{V_{LL}}{\sqrt{3}\, I_{S2}} \left[\frac{K(1-\alpha)+1}{\alpha(1+K)} \right] \qquad (3.31b)$$

$$\frac{X_{bt}}{X_{fd}} = \frac{\alpha(1+K)}{\alpha K - 1} \qquad (3.31c)$$

$$\frac{MVA_{bt}}{MVA_{fd}} = n\beta^2 \left[\frac{\alpha(1+K)}{\alpha K - 1} \right] \qquad (3.31d)$$

Total

Transformers

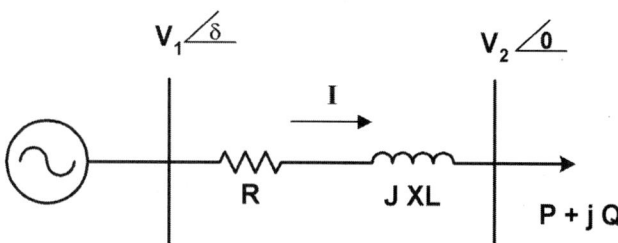

FIGURE 3.112 Simplified radial system.

From the above it is apparent that bus tie reactors are a good solution where a relatively small reduction in fault level is required on a number of downstream feeders.

Also, bus tie reactors grow rapidly in size and cost when (1) the fault contributions on either side of the reactor are significantly different (i.e., as **K** moves away from 1.0) and (2) when the largest fault contribution (I_{S1}) approaches the breaker rating I_b.

Conversely, bus tie reactors decrease rapidly in size and cost if the reactor can be given a low continuous rating due to low normal power transfer across the tie.

Power Line Balance

Consider the radial system shown in Fig. 3.112, in which the sending end bus is fed from an infinite power source.

By inspection, the following equations may be written:

$$\Delta V = V_1 - V_2 \tag{3.32}$$

$$\Delta V = Z * I \tag{3.33}$$

$$\Delta V = \left[(R*I*\cos\Theta + X*I*\sin\Theta)\right] + j\left[(X*I*\cos\Theta - R*I*\sin\Theta)\right] \tag{3.34}$$

where $\Theta = \tan^{-1}$ (**Q/P**)

Since $P = V_2 * I * \cos\Theta$, and
$\quad\quad\; Q = V_2 * I * \sin\Theta$

then,

$$\Delta V = \left[(P*R + Q*X) + j(P*X - Q*R)\right]/V_2 \tag{3.35}$$

$$\Delta V = \Delta V_f + j\,\Delta V_q \tag{3.36}$$

The vector diagram, shown in Fig. 3.113, illustrates the meaning of Eqs. (3.20) and (3.21). By inspection from Fig. 3.113,

$\sin\delta = \Delta V_q/V_1$, therefore,

$$P = \left[V_1 * V_2 * \sin\delta\right]/X + Q*R/X \tag{3.37}$$

and

$$\delta = \tan^{-1}\left\{\left[P*X - Q*R\right]/\left[V_2^2 + (P*R + Q*X)\right]\right\} \tag{3.38}$$

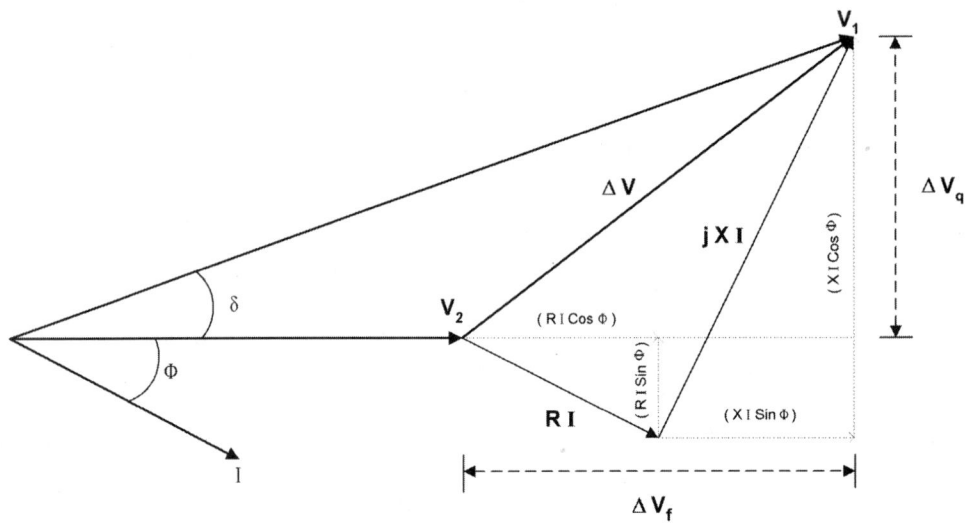

FIGURE 3.113 Power line balance vector diagram.

In transmission systems, the X/R ratio is large and the grid usually operates with power factor close to unity.

Thus, assuming R = 0 and Q = 0,

$$\Delta V = j\, P * X / V_2 \tag{3.39}$$

$$P = V_1 * V_2 * \sin(\delta) / X \tag{3.40}$$

and

$$\delta = \tan^{-1}(P * X / V_2) \tag{3.41}$$

Therefore, apart from the voltage magnitude, which must be kept within regulated limits, control of power flow can only be achieved by variation of the line reactance (X), the transmission angle (δ), or both.

Reactive Power Balance

Figure 3.114 shows a transmission system represented by its π equivalent.

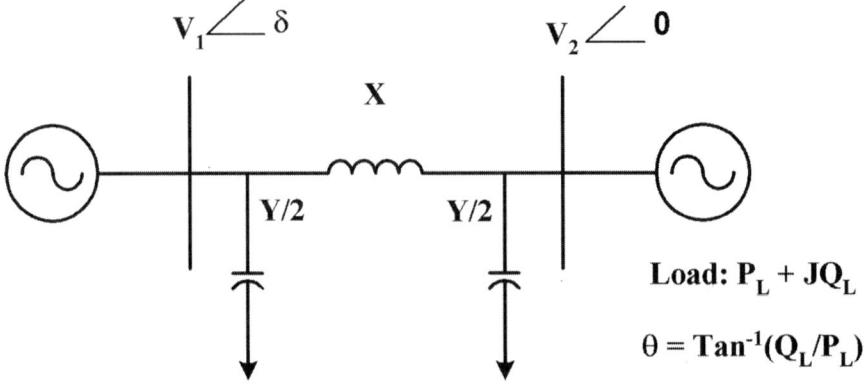

FIGURE 3.114 Transmission system π equivalent circuit.

Transformers

By inspection, the following expressions can be derived:

$$V_1 = V_2 + \Delta V \tag{3.42}$$

$$V_1 * \cos\delta + j V_1 * \sin\delta = V_2 + X * I * \sin\theta + j X * I * \cos\theta \tag{3.43}$$

$$X * I * \cos\theta = P_L * X/V_2 = V1 \sin\delta \tag{3.44}$$

$$P_L = V1 * V2 * \sin(\delta)/X \tag{3.45}$$

$$X * I * \sin\theta = Q_L * X/V_2 = V_1 * \cos\delta - V_2 \tag{3.46}$$

$$Q_L = V1 * V2 * \cos\delta - V_2^2/X \tag{3.47}$$

$$S_i \sqrt{\left(P_i^2 + Q_i^2\right)} \tag{3.48}$$

Therefore,

$$\text{Line reactive losses} = \left(S_i/V_i\right)^2 * X \tag{3.49}$$

$$\text{Line charging (at each end)} = V_i^2 * Y/2, \quad i = 1, 2 \tag{3.50}$$

$$\text{Line surge impedance, } Z_S = \sqrt{X/Y} = \sqrt{L/C} \tag{3.51}$$

$$\text{Surge impedance loading, } SIL = V_{LL}^2/Z_S \tag{3.52}$$

Power balance = Total line charging − Line reactive losses

References

Bonheimer, D., Lim, E., Dudley, R.F., and Castanheira, A., A modern alternative for power flow control.

Bonner, J.A., Hurst, W., Rocamora, R.G., Dudley, R.F., Sharp, M.R., and Twiss, J.A., Selecting ratings for capacitors and reactors in applications involving multiple single tuned filters, *IEEE Trans. Power Delivery*, 10(1), Jan. 1995.

IEEE C57.21-1990 (Reaff. 1995), IEEE Standard Requirements, Terminology and Test Code for Shunt Reactors Rated over 500 kVA.

IEEE Standard C37.015-1993, Application Guide for Shunt Reactor Switching.

IEEE Standard C57.16-1996 Standard Requirements, Terminology and Test Code for Dry-Type Air Core Series Connected Reactors.

IEEE 1277 — Trial Use General Requirements and Test Code for Dry Type and Oil-Immersed Smoothing Reactors for DC Power Transmission.

Peelo, D.F. and Ross, E.M., A new IEEE application guide for shunt reactor switching, *IEEE Trans. Power Delivery*, 11(2), April 1996.

Shunt Reactor Protection Working Group, Shunt reactor protection practices, A Power System Relaying Committee Report.

3.8 Instrument Transformers

Randy Mullikin and Anthony J. Jonnatti

Scope

It is the purpose of this section to inform the reader of the fundamental basics and theory of operation of instrument transformers. Common types and construction highlights will be discussed. Application features emphasizing characteristics associated with instrument transformers will be covered without detail of three-phase circuit fundamentals, fault analysis, or the operation and selection of protective devices and measuring instruments. Though not complete, it will cover some of the more common practices used in industry over the last 30 years. Information contained within this section may be found in more detail in other sections of this handbook.

Overview

Instrument transformers are primarily used to provide isolation between the main primary circuit and the secondary control and measuring devices. This isolation is achieved by magnetically coupling the two circuits. In addition to isolation, levels in magnitude are reduced to safer levels.

Instrument transformers are divided into two categories: (1) voltage transformers (VT) and (2) current transformers (CT). The primary winding of the VT is connected in parallel with the monitoring circuit, while the primary winding of the CT is connected in series (see Fig. 3.115). The secondary windings will proportionally transform the primary levels to typical values of 120 V and 5 A, respectively. Monitoring devices such as wattmeters, power factor meters, voltmeters, ammeters, and relays are often connected to the secondary circuits.

Transformer Basics

The ideal transformer (see Fig. 3.116) will magnetically induce from the primary circuit, a level exactly proportional to the turns ratio into the secondary circuit, and exactly opposite in phase regardless of the changes occurring in the primary circuit. A review of the general relationships of the ideal case yields

$V_p/V_s = N_p/N_s$ $V_p I_p = V_s I_s$ $I_p N_p = I_s N_s$

Transformation ratio   From the law of conservation of energy

Core Design

In practice, the use of steel core material is a major factor in forcing the transformer to deviate from the ideal. There is steel available that offers different properties upon which to design around. The most common type used is electrical-grade silicon-iron. It offers low losses at high flux densities, but has low initial permeability. Exotic materials, such as nickel-iron, offer high initial permeability and low losses,

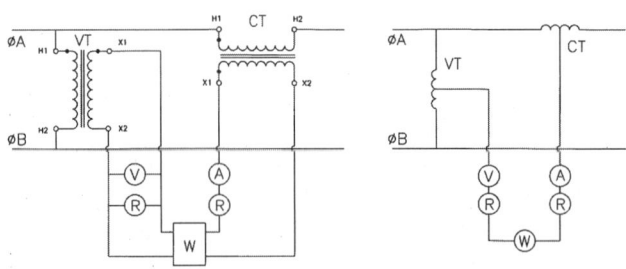

FIGURE 3.115 Typical wiring and single-line diagram.

Transformers

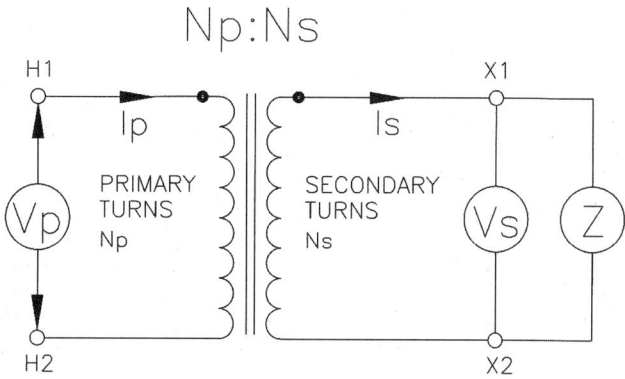

FIGURE 3.116 Ideal transformer.

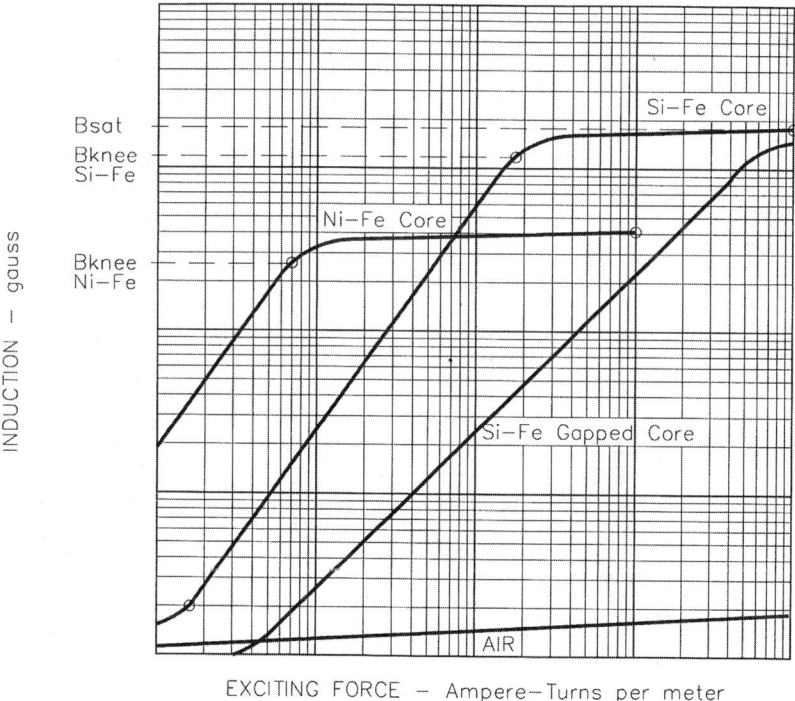

FIGURE 3.117 Typical excitation characteristic.

but have much lower saturation levels. These are often used when extremely high accuracy is desired, but is cost prohibitive in standard products. A typical excitation characteristic is shown in Fig. 3.117. There are three areas of concern: (1) the ankle point, (2) the knee point, and (3) saturation. The ankle region is at the lowest permeability and flux levels. The knee represents the maximum permeability and is the beginning of the saturation zone. Saturation is the point at which no more flux is entering the core. There are occasions when the designer would like the benefits of silicon-iron and nickel-iron. This type is called a composite core. The ratio of each is dependent on the properties desired and overall cost. Quality of steel is improving, as are processing and manufacturing techniques, but there are some inherent properties that must be overcome, such as:

1. Some portion of the primary energy will be required to establish the magnetic flux to induce the secondary winding.
2. The magnetization of the core is nonlinear in nature.

From statement 1, the primary energy required to magnetize the core is a product of the flux in the core and the magnetic reluctance of the core. This energy is called the magnetomotiveforce, mmf, and is defined as

$$\text{mmf} = \Phi \mathfrak{R} = k_1 \left[\frac{Zs\ Is}{Ns\ f} \right] k_2 \left[\frac{\text{mmp}}{Ac\ \mu c} \right] = k_1 k_2 \left[\frac{Zs\ Is\ \text{mmp}}{Ns\ f\ Ac\ \mu c} \right] \qquad (3.53)$$

where Φ = flux in the core $\qquad\mathfrak{R}$ = magnetic reluctance
$\qquad k_1$ = constants of proportionality $\qquad k_2$ = constants of proportionality
$\qquad Zs$ = secondary impedance $\qquad$ mmp = core mean magnetic path
$\qquad Is$ = secondary current $\qquad Ac$ = core cross-sectional area
$\qquad Ns$ = secondary turns $\qquad \mu c$ = permeability of core material
$\qquad f$ = frequency, Hz

The magnetic reluctance, in terms of Ohm's law, would be analogous to resistance. It is a function of the core type used. An annular or toroidal core, one that is a continuous tape wound core, has the least amount of reluctance. A cut-core with a straight cut through all the laminations, has the highest reluctance. Minimizing gaps in core constructions will reduce reluctance. Generally after the steel material is cut, stamped, or wound, it undergoes a stress-relief anneal to restore the magnetic properties that may have been altered during fabrication. At this point, the core is constructed and insulated.

From statement 2, the permeability, μc, changes with flux density, Φ/Ac. Neglecting leakage flux we can now see the error-producing elements. From Eq. (3.53), an increase in any of the elements in the denominator will decrease errors, while an increase in Zs and mmp will increase errors.

There are also other contributing factors which, based on the construction of the instrument transformer, can introduce errors. The windings, typically of copper wire and/or foil, have resistance, which introduce volt drops (see Fig. 3.118). The physical geometry and arrangement of the windings with respect to each other and the core can introduce inductance, and sometimes capacitance, which has an effect on magnetic leakage reducing the flux linkage from the primary circuit effecting performance. A winding utilizing all of the magnetic path will have the lowest reactance.

Those factors not directly related to construction would be elements introduced from the primary circuit, such as harmonics, which account for hysteresis and eddy current losses in the core material, and fault conditions which can cause magnetic saturation.

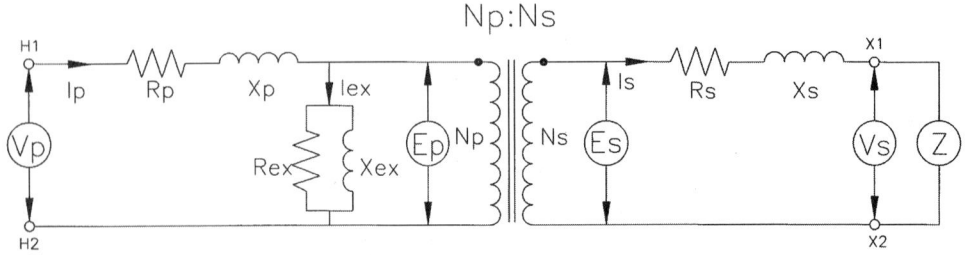

FIGURE 3.118 Equivalent transformer circuit. Where Vp = primary terminal voltage; Vs = secondary terminal voltge; Ep = primary induceded voltage; Es = secondary induced voltage; Ip = primary current; Is = secondary current; Np = primary turns; Ns = secondary turns; Rp = primary winding resistance; Rs = secondary winding resistance; Xp = primary winding reactance; Xs = secondary winding reactance; Rex = wattfull magnetizing component; Z = secondary burden (load); Xex = wattless magnetizing component; Iex = magnetizing current

TABLE 3.10 Typical Burden Values for Common Devices

Device	Voltage Transformers		Current Transformers	
	Burden, VA	P.F.	Burden, VA	P.F.
Voltmeter	0.1–20	0.7–1.0	—	—
Ammeter	—	—	0.1–15	0.4–1.0
Wattmeter	1–20	0.3–1.0	0.5–25	0.2–1.0
P.F. meter	3–25	0.8–1.0	2–6	0.5–0.95
Freq. meter	1–50	0.7–1.0	—	—
KWH meter	2–50	0.5–1.0	0.25–3	0.4–0.95
Relays	0.1–50	0.3–1.0	0.1–150	0.3–1.0
Regulator	50–100	0.5–0.9	10–180	0.5–0.95

Almost all can be compensated for by careful selection of core and winding type. It is also possible to offset error by adjusting turns on one of the windings, preferably the one with the higher turns — it will provide better control. It is also possible to compensate using external means, such as RCL networks, but this will limit the transformer for operation to a specific load, and can have adverse effects over the operating range.

Burdens

The burden of the instrument transformer is considered to be everything connected externally to its terminals. This will include monitoring devices, relays, and pilot wiring. The impedance values of each component, which can be obtained from manufacturers data sheets, should be added algebraically to determine the total load. The units should be the same, and in the rectangular form R + jX. Table 3.10 shows typical ranges of burdens for various devices used.

For the purpose of establishing a uniform basis of test, a series of standard burdens has been defined for calibrating VTs and CTs. The burdens are inductive and designated in terms of VA. All are based on 120 V and 5 A, respectively, at 60 Hz.

Relative Polarity

The instantaneous relative polarity of instrument transformers may be critical for proper operation in metering and protection schemes. These will be discussed later. The basic convention is as current flows into the H1 terminal, it flows out of the X1 terminal, making this polarity subtractive. These terminals are identified on the transformer by the letters and/or a white dot.

Industry Standards

In the U.S., the utility industry relies heavily on IEEE Standard C57.13 — Requirements for Instrument Transformers. This standard establishes the basis for test and manufacture of all instrument transformers used in this country. It defines the parameters for insulation class and accuracy class. The burdens listed in Table 3.11 are defined in C57.13. Often, standards for other electrical apparatus that may use instrument transformers have adopted their own criteria based on C57.13. These standards, along with utility practices and the National Electric Code, are used in conjunction with each other to assure maximum safety and system reliability. The industrial market may also coordinate with Underwriters Laboratories. As the marketplace becomes global, there is a drive for standard harmonization with the International Electrotechnical Commission (IEC), but we are not quite there yet. One must be aware of the international standards in use. Most major countries had, at one time, developed their own standards. More and more are beginning to adopt IEC to supercede their own. (See Table 3.12.)

TABLE 3.11 Standard Metering and Relaying Class Burdens

	Voltage Transformers			Current Transformers		
Type Used	Burden	VA	P.F.	Burden	VA	P.F.
Metering	W	12.5	0.1	B0.1	2.5	0.9
Metering	X	25	0.7	B0.2	5	0.9
Metering	M	35	0.2	B0.5	12.5	0.9
Metering	—	—	—	B0.9	22.5	0.9
Metering	—	—	—	B1.8	45	0.9
Relaying	Y	75	0.85	B1.0	25	0.5
Relaying	Z	200	0.85	B2.0	50	0.5
Relaying	Zz	400	0.85	B4.0	100	0.5
Relaying	—	—	—	B8.0	200	0.5

TABLE 3.12 Instrument Transformer Standards

Country	CT Standard	VT Standard
U.S.	C57.13	C57.13
Canada	CAN-C13-M83	CAN-C13-M83
IEC	60044-1 (formerly 185)	60044-2 (formerly 186)
U.K.	BS 3938	BS 3941
Australia	AS 1675	AS 1243
Japan	JIS C 1731	JIS C 1731

Accuracy Classes

Instrument transformers are rated by performance in conjunction with a secondary burden. As the burden increases, the accuracy class, may in fact, decrease. For revenue metering use, the coordinates of ratio error and phase error must lie within a prescribed parallelogram. This parallelogram is based on a 0.6 system power factor. The ratio error (R.E.) is converted into a Ratio Correction Factor (RCF) which is simply

$$RCF = 1 - (R.E./100) \tag{3.54}$$

The total error component is the Transformer Correction Factor (TCF), which is the combined ratio and phase angle error. The limits of phase angle error were determined from the following relationship

$$TCF = RCF \pm \left[\frac{PA \tan\theta}{3438} \right] \tag{3.55}$$

where TCF = transformer correction factor
 RCF = ratio correction factor
 PA = phase angle error, in minutes
 θ = supply system PF angle
 + = for VTs only (see Fig. 3.119)
 − = for CTs only (see Fig. 3.120)
 3438 = minutes of angle in 1 radian

Therefore, using 0.6 system power factor ($\theta = 53°$) and substituting in Eq. (3.55),

$$\text{for VTs,} \qquad TCF = RCF + \left[PA/2600 \right] \tag{3.56}$$

Transformers

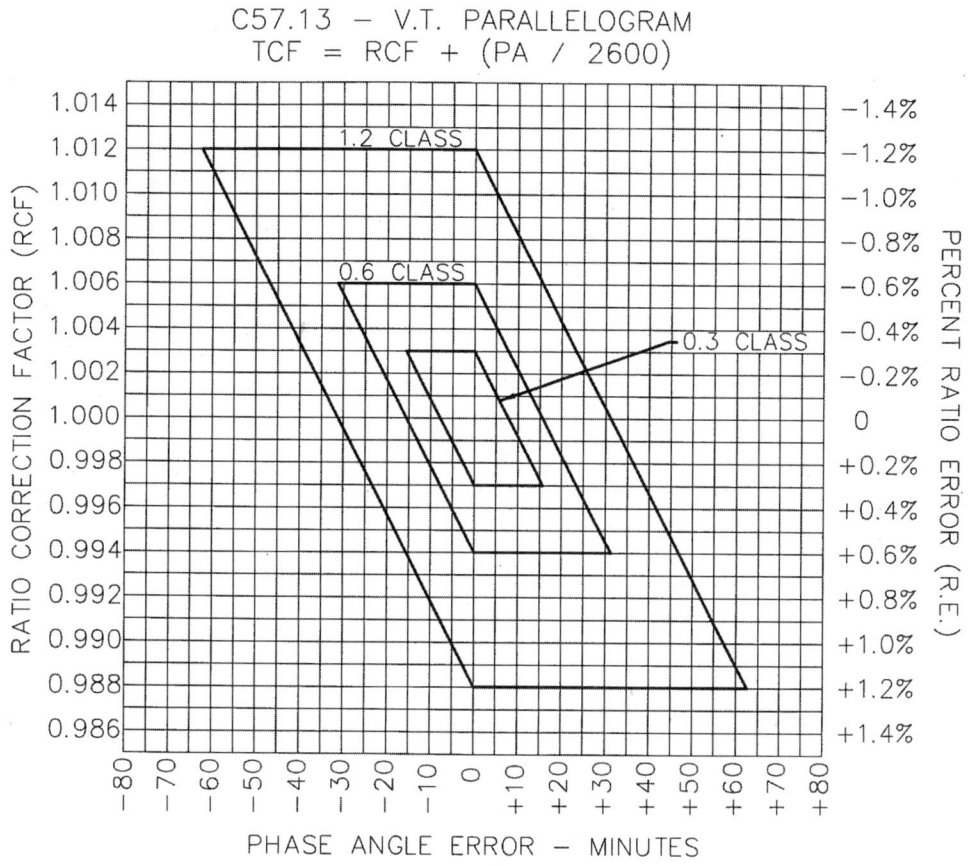

FIGURE 3.119 Accuracy coordinates for VTs.

and

$$\text{for CTs,} \quad \text{TCF} = \text{RCF} - [\text{PA}/2600]. \tag{3.57}$$

The TCF is mainly applied when the instrument transformer is being used to measure energy usage. The limits of TCF are also the same as RCF in Table 3.13. A negative R.E. will yield an RCF > 1, while a positive R.E. will yield an RCF < 1. The adopted class in Table 3.13 is extrapolated from these relationships, and recognized in industry.

The accuracy class limits apply to the errors at 100% of rated current up through the rating factor of the CTs. At 10% of rated current, the error limits permitted are twice that of the 100% class. There is no defined requirement for the current range between 10 and 100%, nor is their any requirement below 10%. There are certain instances in which the user is concerned about the errors at 5% and will rely on the manufacturer's guidance. Because of the non-linearity in the core and the ankle region, the errors are exponential. At some point the errors will become linear up until it drives the core into saturation, at which point the errors increase at a tremendous rate (see Fig. 3.121).

In the case of the VT, the accuracy class range is between 90 and 110% of rated voltage for each designated burden. Unlike the CT, the accuracy class is maintained through the entire range. The

TABLE 3.13 Accuracy Classes

	Class	RCF Range
New	0.15	1.0015–0.9985
C57.13	0.3	1.003–0.997
C57.13	0.6	1.006–0.994
C57.13	1.2	1.012–0.988
C57.13	2.4	1.024–0.976
Adopted	4.8	1.048–0.952

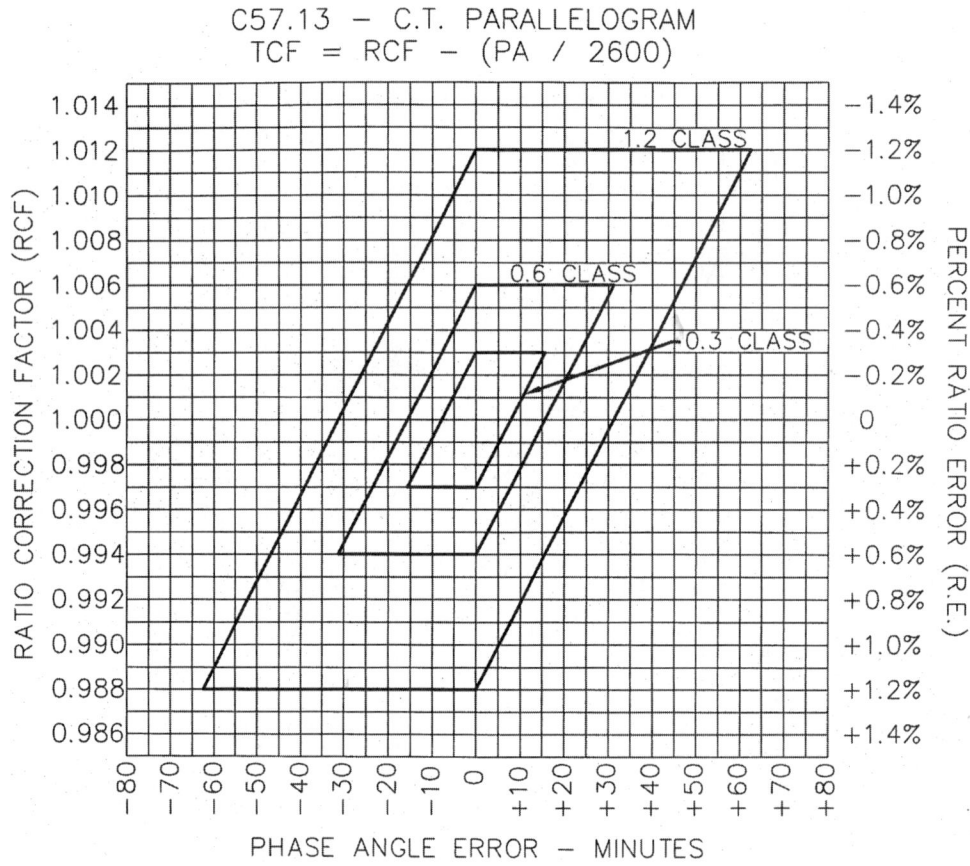

FIGURE 3.120 Accuracy coordinates for CTs.

manufacturer will provide test data at 100% rated voltage, but can furnish at other levels if required by the end user. The response is somewhat similar to that of the CT being linear over a long range below 90%. But since the normal operating flux densities are much higher than in the CT, saturation will occur much sooner at voltages above 110%, depending on the overvoltage rating.

Protection, or relay class, is based on the instrument transformer's performance at some defined fault level. In VTs it may also be associated with an under and overvoltage condition. In this case, the VT may have errors as high as 5% at levels as low as 5% of rated voltage and at the VT overvoltage rating. In CTs, the accuracy is based on a terminal voltage developed at 20 times nominal rated current. The limits of RCF are 0.90 to 1.10, or 10% R.E. from nominal through 20 times. This applies to rated burden or any burden less than rated burden.

Insulation Systems

This is one of the most important features of the instrument transformer. This will define the construction, insulation medium used, and the overall size. Generally, the insulation system is defined by the source voltage to which the instrument transformer is to be directly connected. This source must also have a BIL rating. Finally, where the instrument transformer is to be installed will help determine the insulation medium and temperature class. In indoor applications, the instrument transformer is protected from external weather elements. However, the ambient temperature may be significantly higher, thus affecting the insulation used. In outdoor installations, the transformer must endure all weather conditions from extremely low temperatures to UV radiation, and must be impervious to moisture penetration.

Transformers

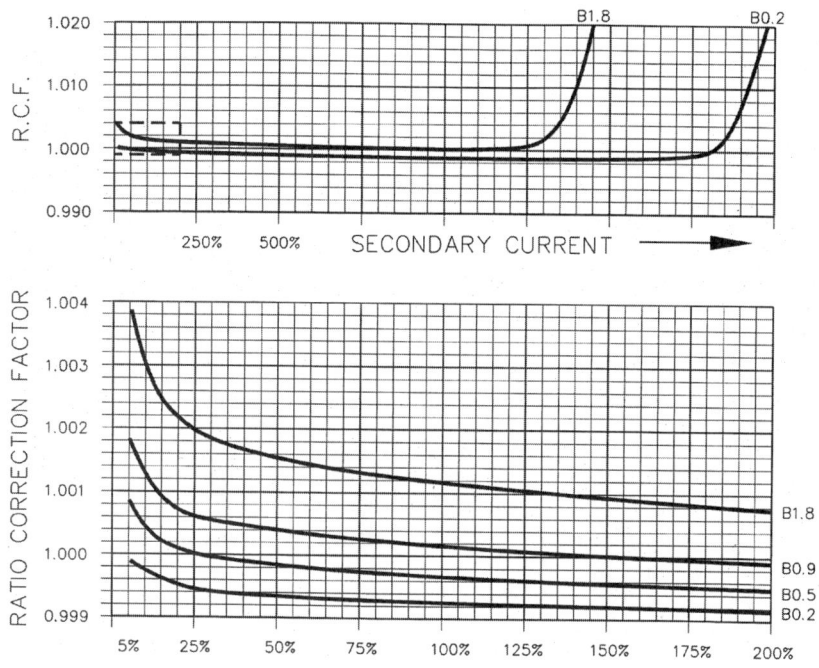

FIGURE 3.121 CT RCF characteristic.

TABLE 3.14 Low and Medium Voltage Classes

Class (kV)	BIL (kV)	Indoor Applications Materials/Construction	Outdoor Applications Materials/Construction
0.6	10	Tape, varnished, plastic, cast	Cast
5.0	60/45	Plastic, cast	Cast
8.7	75/60	Cast	Cast
15	110/95	Cast	Cast or tank/oil/porcelain
25	150/125	Cast	Cast or tank/oil/porcelain
34.5	200/150	Cast	Cast or tank/oil/porcelain
46	250	Not available	Cast or tank/oil/porcelain
69	350	Not available	Cast or tank/oil/porcelain

Note: Cast can imply any polymeric material, e.g., Butyl rubber, epoxy, resin, etc.

The outer protection can range from tape, varnish treatment, or plastic housings, to molding compounds, porcelain, or steel enclosures. Table 3.14 identifies by voltage rating where materials and construction types are commonly found.

Some voltage classes have several BIL classes. Often in indoor applications, this requirement is lower than in the outdoor applications. And it is not uncommon to require higher BIL class for use in a highly polluted environment. All installations above 69 kV are for outdoor service only, and are of the tank/oil/porcelain construction type. The classes and BIL ratings are shown in Table 3.15.

Thermal Ratings

An important part of the insulation system is the temperature class. Generally for instrument transformers, only three classes are defined in the standard and are listed in Table 3.16. Of course, others can be used to fit the application. These apply to the instrument transformer under the most extreme continuous

TABLE 3.15 High-Voltage Classes

Class, kV	BIL, kV
92	450
115	550
138	650
161	750
230	1050/900
345	1300
500	1800/1675
765	2050

TABLE 3.16 Temperature Class

Temperature Class	30°C Ambient		55°C Ambient
	Temperature Rise	Hot-Spot Temp. Rise	Temperature Rise
105°C	55°C	65°C	30°C
120°C	65°C	80°C	40°C
150°C	80°C	110°C	55°C

conditions for which it is rated. The insulation system used must be coordinated within the designated temperature class.

Primary Winding

The primary winding is subjected to the same dynamic and thermal stresses as the rest of the primary system when large short-circuit currents are present. It must be sized to safely carry the maximum continuous current without exceeding the insulation system's temperature class.

Voltage Transformer (VT)

The voltage transformer is connected in parallel with the circuit to be monitored. It operates under the same principles as power transformers.

The significant differences are power capability, size, operating flux levels, and compensation. VTs are not typically used to supply raw power; however, they do have limited power ratings. They can often be used to supply temporary 120-V service for light-duty maintenance purposes where supply voltage normally would not be available. In switchgear compartments they may be used to drive motors that open and close circuit breakers. In voltage regulators they may drive a tap-changing motor. The power ranges are from 500 VA and less for low-voltage VT, and 1 to 3 kVA for medium-voltage VT. Since they have such low power ratings, their physical size is much smaller. The performance characteristics of the VT are based on standard burdens and power factors, which are not always the same as the actual connected burden. It is possible to predict, graphically, the anticipated performance when given at least two reference points. Manufacturers typically provide this data with each VT produced. From that, one can construct what is often referred to as the VT circle diagram, shown in Fig. 3.122. Knowing the ratio error and phase error coordinates, and the values of standard burdens, the graph can be produced to scale in terms of VA and power factor. Other power factor lines can be inserted to pinpoint actual circuit conditions.

Performance can also be calculated using the same vector concept by the following relationships, provided that the unknown burden is of less value than the known burden. Two coordinates must be known, at zero and one other standard burden value.

Transformers

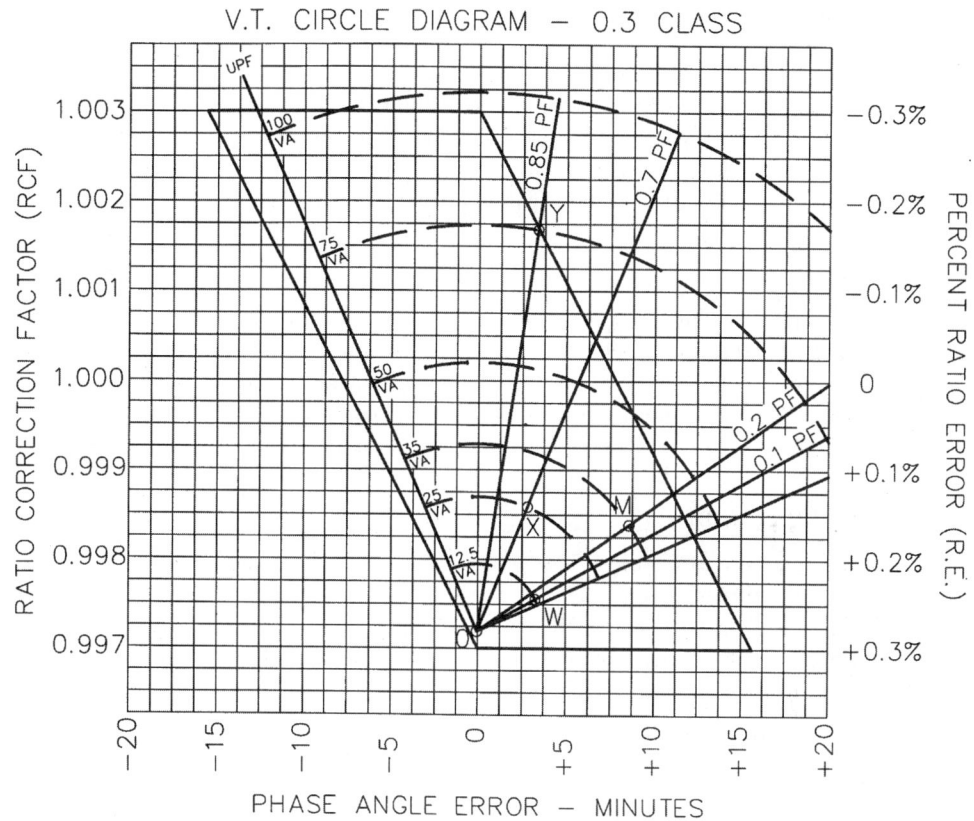

FIGURE 3.122 VT circle diagram.

$$RCF_x = \left[\frac{B_x}{B_t}\right]\left[(RCF_t - RCF_0)\cos(\theta_t - \theta_x) + (\gamma_t - \gamma_0)\sin(\theta_t - \theta_x)\right] \quad (3.58)$$

$$\gamma_x = \left[\frac{B_x}{B_t}\right]\left[(\gamma_t - \gamma_0)\cos(\theta_t - \theta_x) - (RCF_t - RCF_0)\sin(\theta_t - \theta_x)\right] \quad (3.59)$$

where

RCF_x = RCF of new burden
RCF_t = RCF of known burden
RCF_0 = RCF at zero burden
B_x = new burden
B_t = known burden
γ_x => in minutes, multiply value from Eq. (3.59) by 3438

γ_x = phase error of new burden, radians
γ_t = phase error of known burden, radians
γ_0 = phase error at zero burden, radians
θ_x = new burden PF angle, radians
θ_t = known burden PF angle, radians

Overvoltage Ratings

The operating flux density is much lower. This is to help minimize the losses and to prevent the VT from possible overheating during overvoltage conditions. They are normally designed to withstand 110% rated voltage continuously, unless otherwise designated. C57.13 divide VTs into groups based on voltage and application. Group 1 includes those intended for L-L or L-G connection, and are rated 125%. Group 3 is for L-G connection only, and have two secondary windings. They are designed to withstand 173% of rated voltage for 1 min except those rated 230 kV and above must withstand 140% for the same duration. Group 4 is L-G connections with 125% in emergency conditions. Group 5 is L-G connections with 140% rating for 1 min. Other standards have more stringent requirements such as the Canadian Standard, which defines its Group 3 VTs for L-G connection on ungrounded systems to withstand 190% for 30 sec to 8 h, depending on ground-fault protection. This also falls in line with the IEC standard.

VT Compensation

The high-voltage windings are always compensated to provide the widest range of performance within an accuracy class. Since there is compensation, the actual turns ratio will vary from the rated voltage ratio. For example, say a 7200:120 V, 60:1 is required to meet 0.3 class. The designer may desire to adjust the primary turns by 0.3% by removing them from the nominal turns, thus reducing the actual turns ratio to say 59.82:1. Adjustment of turns has little to no effect on the phase angle error.

Short-Circuit Operation

Under no normal circumstances is the VT secondary to be short-circuited. The VT must be able to withstand mechanical and thermal stresses for 1 sec, with full voltage applied to the primary terminals, and showing no damage. In most situations this condition would cause some protective device to operate and remove the applied voltage, hopefully in less than 1 sec. If prolonged, the temperature rise would far exceed the insulation limits and the axial and radial forces on the windings would cause severe destruction to the VT.

VT Connections

VTs are provided in two arrangements: (1) dual or two bushing type and (2) single bushing type. Two bushing types are designed for L-L connection, but in most cases can be connected L-G with reduced output voltage. Single bushing types are strictly for L-G connection. The VT should never be connected to a system that is higher than its rated terminal voltage. As for the connection between phases, polarity must always be observed. Low- and medium-voltage VTs may be configured in delta or wye. As the system voltages exceed 69 kV, only single-bushing types are available. Precautions must be taken when connecting VT primaries in wye on an ungrounded system (this will be discussed further in the next section). Primary fusing is always recommended. Indoor switchgear types are often available with fuse holders mounted directly on the VT body.

Ferroresonance

VT with Y-connected primaries on three-wire systems that are ungrounded can oscillate with the distributed line-to-ground capacitance (refer to Fig. 3.123). Under balanced conditions, line-to-ground voltages are normal. Momentary ground faults or switching surges can upset the balance and raise the line-to-ground voltage above normal. This condition can initiate a resonant oscillation between the primary windings and the system capacitance to ground. Higher current flows in the primary windings due to fluctuating saturation that can result in overheating. The current levels may not be high enough to blow the primary fuses since they are generally for short-circuit protection and not thermal protection of the VT. Not every disturbance will cause ferroresonance. This phenomenon depends on several factors:

- initial state of magnetic flux in the cores
- saturation characteristics (magnetizing impedance) of the VT
- air-core inductance of the primary winding
- system circuit capacitance

Transformers

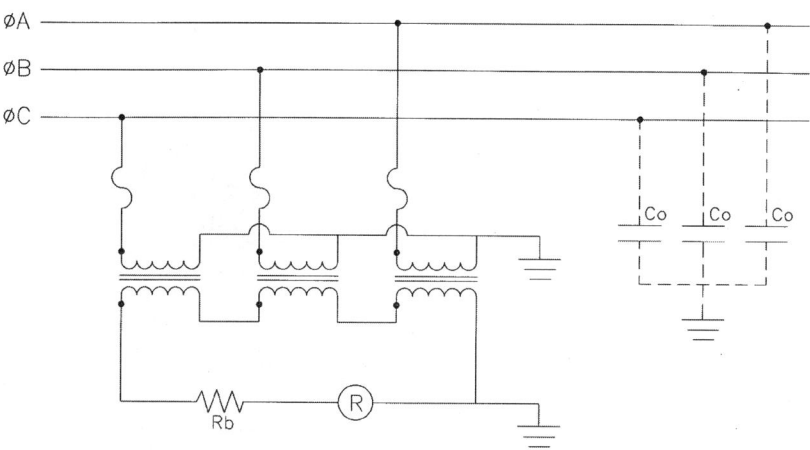

FIGURE 3.123 VTs Y-connected on an ungrounded system.

One method often used to protect the VT is to connect a resistive load to each of the secondaries individually, or connect them in a Δ-configuration and insert a load resistance in one corner of the Δ. This resistance can be emperically approximated by

$$\Delta\text{res} = (100 * L_A)/N^2$$

where

    Δres = loading resistance in ohms
    L_A = VT primary inductance during saturation, in mH
    N = VT turns ratio

This is not a fix-all solution as ferroresonance may still occur, but this may reduce the chances of it happening. The loading will have an effect on VT errors and may cause it to exceed 0.3%, but that is not critical for this scheme since it is seldom used for metering.

VT Construction

The electromagnetic wound-type VT is similar in construction to that of the power transformer. The magnetic circuit is a core-type or shell-type arrangement with the windings concentrically wound on one leg of the core. A barrier is placed between the primary and secondary winding(s) to provide adequate insulation for its voltage class. In low-voltage applications, it is usually a two-winding arrangement. But in medium- and high-voltage transformers, a third (tertiary) winding is often added, isolated from the other windings. This provides more flexibility for using the same VT for metering and protective purposes simultaneously. As mentioned previously, the VT is available in single- or dual-bushing arrangements. A single bushing has one lead accessible for connection to the high-voltage conductor, while the other side of the winding is grounded. The grounded terminal (H2) may be accessible somewhere on the VT body near the base plate. There is usually a grounding strap connected from it to the base, and can be removed to conduct field power factor tests. In service, the strap must always be connected to ground. Some medium-voltage transformers are solidly grounded and have no H2 terminal access. The dual-bushing arrangement has two live terminal connections, and both are fully rated for the voltage to which it is to be connected. (See Fig. 3.124.)

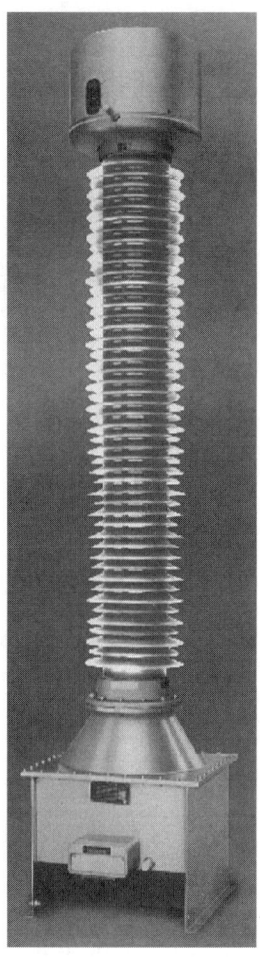

Medium Voltage Dual-Bushing VT. (Photos Courtesy of Kuhlman Electric Corp)

High Voltage single-bushing VT.

Capacitive Coupled Voltage Transformer (CCVT)

The CCVT is primarily a capacitance potential divider and electromagnetic VT combined. Developed in the early 1920s, it was used to couple telephone carrier current with the high voltage transmission lines. The next decade brought a capacitive tap extending its use for indication and relaying. To provide sufficient energy, the divider output had to be relatively high, typically 11 kV. This created the need for an electromagnetic VT to step the voltage down to 120 V. A tuning reactor was used to increase energy transfer (see Fig. 3.124). As transmission voltage levels increased, so did the use of CCVTs. Its traditional low-cost vs. the conventional VT, and that it was nearly impervious to ferroresonance due to its low flux density, made it an ideal choice. It proved to be quite stable for protective purposes, but was not adequate for revenue metering. In fact, the accuracy has been known to drift over time and temperature ranges. This would often warrant the need for routine maintenance. They are commonly used in the 345 to 500 kV systems. Improvements have been made to better stabilize the output, but its popularity has declined.

Another consideration with CCVTs is its transient response. When a fault reduces the line voltage, the secondary output does not respond instantaneously due to the energy storing elements. The higher the

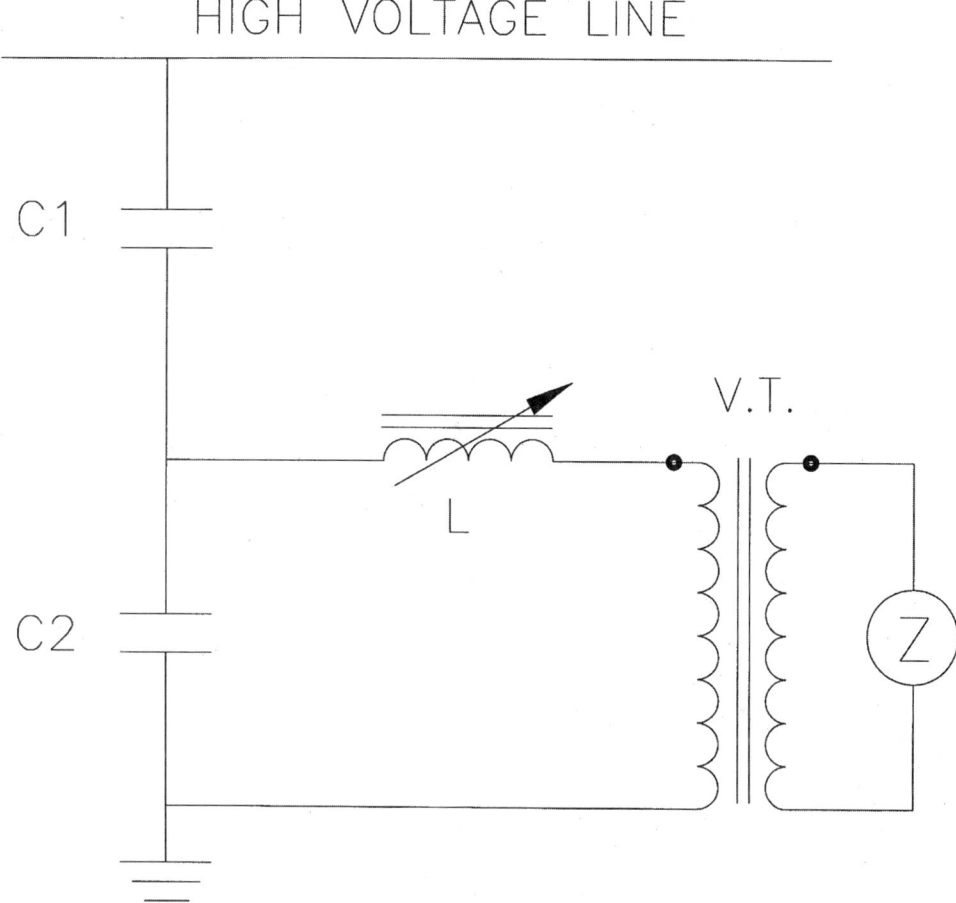

FIGURE 3.124 CCVT simplified circuit.

capacitance, the lower the transient magnitude. Another element is the ferroresonance suppression circuit, usually on the secondary side of the VT. There are two types, active and passive. Active circuits, which also contain energy storing components, add to the transient. Passive circuits have little effect on transients. The concern of the transient response is with distance relaying and high speed line protection. This transient may cause out-of-zone tripping, which is not tolerable.

Current Transformer (CT)

The current transformer is often treated as a "black box". It is a transformer that is governed by the laws of electromagnetic induction:

$$\varepsilon = k\, \beta\, Ac\, N\, f$$

As previously stated, the CT is connected in series with the circuit to be monitored, and it is this difference that leads to its ambiguous description. The primary winding is to offer a constant current source of supply through a low impedance loop. Because of this low impedance, current passes through it with very little regulation. The CT operates on the ampere-turn principle (Faraday's Law):

$$\text{Ip Np} = \text{Is Ns}.$$

Primary Ampere-Turns = Secondary Ampere-Turns

Since there is energy loss during transformation, this loss can be expressed in ampere-turns. Thus;

$$\text{Ip Np} - \text{Iex Np} = \text{Is Ns}.$$

Primary Ampere-Turns − Magnetizing Ampere-Turns = Secondary Ampere-Turns

The CT is not voltage dependant, but is voltage limited. As current passes through an impedance, a voltage is developed (Ohm's Law, $V = I \times Z$). As this occurs, energy is depleted from the primary supply, thus acting like a shunt. This depletion of energy results in the CT errors. As the secondary impedance increases, the voltage proportionally increases. Thus, the limit of the CT is magnetic saturation, a condition when the core flux can no longer support the increased voltage demand. At this point, nearly all of the available energy is going into the core, leaving none left to support the secondary circuit.

Saturation Curve

The saturation curve, often called secondary excitation curve, is a plot of secondary exciting voltage vs. secondary exciting current drawn on log–log paper. The units are in RMS with the understanding that the applied voltage is sinusoidal. This characteristic defines the core properties after the stress relief annealing process. It can be demonstrated by test that cores processed in the same manor will always follow this characteristic within the specified tolerances. Figure 3.125 shows a typical characteristic of a

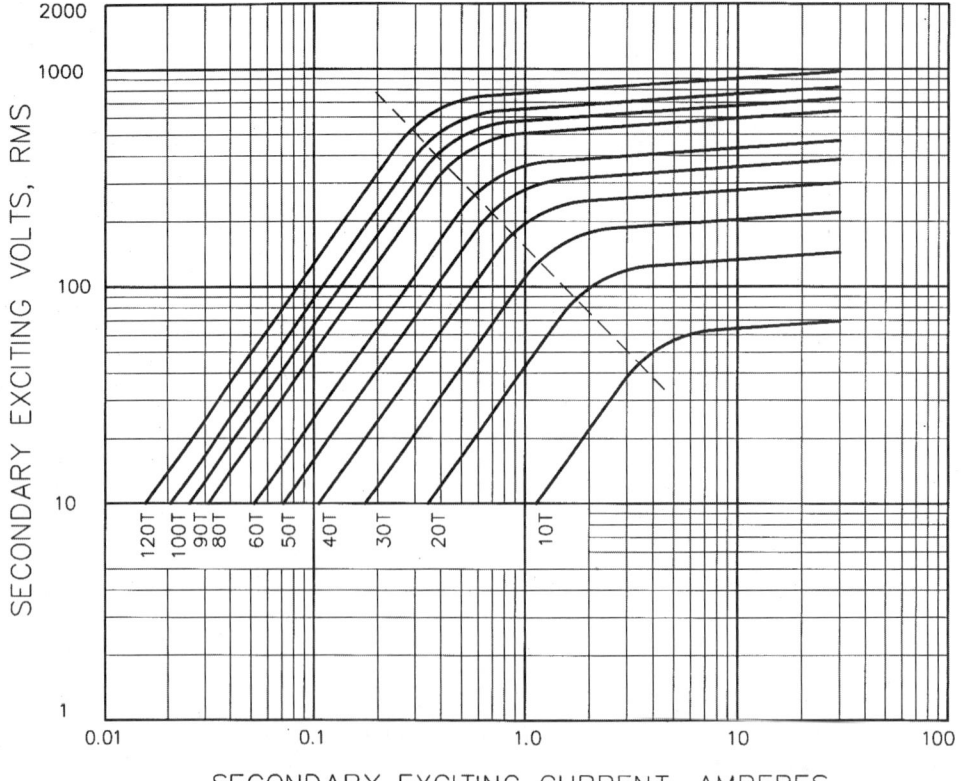

FIGURE 3.125 Saturation curve for a multi-ratio CT.

Transformers

600:5 multi-ratio CT. The knee point is indicated by the dashed line. Since the voltage is proportional to the turns, the V/T at the knee is constant. The tolerances are 95% of saturation voltage for any exciting current above the knee, and 125% of exciting current for any voltage below the knee. Knowing the secondary winding resistance and the excitation characteristic, the user can calculate the expected RCF under fault conditions. Using this type of curve is only valid for non-metering applications. The required voltage needed from the CT must be calculated using the total circuit impedance and the anticipated secondary current level. The corresponding exciting current is read from the curve and used to approximate the anticipated errors.

$$Vex = Is\, Zt = Is\sqrt{(R_S + R_B)^2 + X_B} \tag{3.60}$$

$$RCF = (Is + Ie)/Is \tag{3.61}$$

$$\%R.E. = Ie/Is \times 100 \tag{3.62}$$

where

Zt = total circuit impedance, in ohms
Is = secondary nominal current at desired magnitude
Ie = secondary exciting current at Vex, obtained from curve

CT Rating Factor

The continuous current rating factor is given at a reference ambient temperature, usually 30°C. The standard convention is that the average temperature rise will not exceed 55°C for general purpose use, but can be any rise shown in Table 3.16. From this rating factor, a given CT can be de-rated for use in higher ambient temperatures from the following relationship

$$\frac{RF_{NEW}}{RF_{STD^2}} = \frac{85°C - AMB_{NEW}}{85°C - 30°C} \tag{3.63}$$

This expression is only valid for 55°C rise ratings, and maximum ambient less than 85°C. The rating factor will assure that the CT will not exceed its insulation class rating. If a higher temperature class insulation system is provided, then the rise must be in compliance with Table 3.16.

Open Circuit Conditions

The CT functions best with the minimum burden possible, which would be its own internal impedance. This can only be accomplished by applying a short circuit across the secondary terminals. Since the core mmf acts like a shunt, with no load connected to its secondary, the mmf becomes the primary current, thus driving the CT into hard saturation. With no load on the secondary to control the voltage, the winding develops an extremely high peak voltage. This voltage can be in the thousands, or even tens of thousands of volts. This situation puts the winding under incredible stress, thus ultimately failing. This could result in damage to other equipment, or even to personnel. It is for this reason that the secondary circuit should never be open. It must always have a load connected. If it is installed to the primary, but not in use, then the terminals should be shorted until it is to be used. Most manufacturers ship CTs with a shorting strap or wire across the secondary terminals. The CT must be able to withstand 3500 Vpeak for 1 min under open circuit conditions.

Overvoltage Protection

Under load the CT is voltage limited. The level of this voltage is dependant upon the turns and core area. The user must evaluate limits of the burdens connected to assure equipment safety. Sometimes protective devices are used on the secondary side to maintain safe levels of voltage. These devices are also incorporated to protect the CT during an open circuit condition. In metering applications, it is possible for such a device to introduce a DC current across the winding, which could saturate the core or leave it in some state of residual flux. In high voltage equipment, arrestors may be used to protect the primary winding from high voltage spikes produced by switching transients or lightning.

Residual Magnetism

Residual magnetism, residual flux, or remanence, is referred to the amount of stored or trapped flux in the core. This can be introduced during heavy saturation or with the presence of some DC component. Figure 3.126 shows a typical B-H curve for silicon-iron driven into hard saturation. The point at which the curve crosses zero force, identified by +Bres, represents the residual flux. If at some point the CT was disconnected from the source, this flux would remain in the magnetic core until a source becomes present. If a fault current drove the CT into saturation, when the supply current resumed normal levels, the core will contain some residual component. Residual flux does not gradually decay but remains constant once steady-state equilibrium has been reached. Under normal conditions the minor B-H loop must be high enough to remove the residual component. If it is not, then it will remain present until another fault occurrence takes place. The result could be effectively reducing the saturation flux. However, if a transient of opposite polarity occurs, saturation will be reduced with the assistance of the residual. Conversely, the magnitude of residual is also based on polarity of the transient and the phase relationship of the flux and current. Whatever the outcome, the result could cause a delayed response to the connected relay.

It has been observed that in a tape wound toroidal core, as much as 85% of saturation flux could be left in the core as residual component. The best way to remove residual flux is to demagnetize the core. This is not always practical. The user could select a CT with a relay class that is twice the required amount.

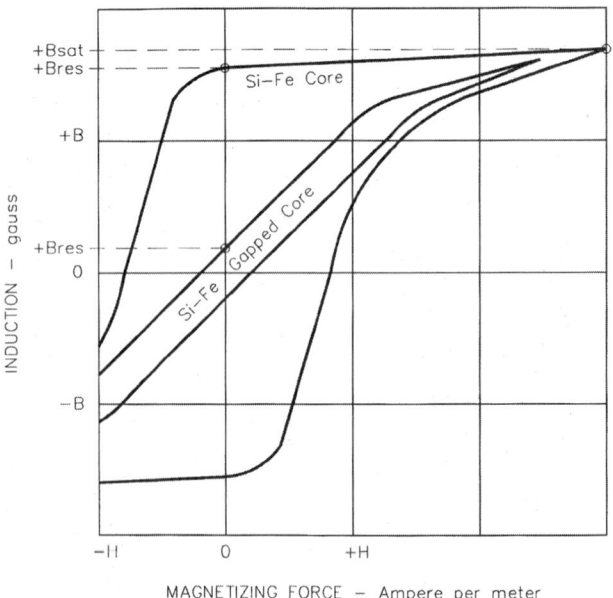

FIGURE 3.126 Typical B-H curve for Si-Fe steel.

Transformers

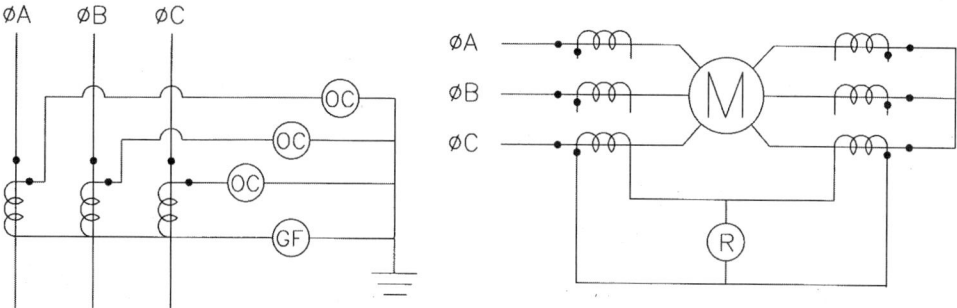

FIGURE 3.127 (a) Over-current and ground fault protection scheme. (b) Differential protection scheme.

This may not eliminate residual flux, but it will certainly reduce the magnitude. The use of hot-rolled steel may inherently reduce the residual component to 40 to 50% of saturation flux. Another way is to use an air-gapped core. Normally, the introduction of a gap that is say, 0.0001 in. per inch of core circumference, could limit the residual flux to about 10% of the saturation flux. Referring to Fig. 3.126, a typical B-H loop for an air-gapped core is shown. The drawback is significantly higher exciting current and lower saturation levels, as can be seen in Fig. 3.117. To overcome the high exciting current, the core would be made larger. That, coupled with the gap, will increase its overall cost. This type of core construction is often referred to as a linearized core.

CT Connections

As previously mentioned, some devices are sensitive to the direction of current flow. It is often critical in three-phase schemes to maintain proper phase shifting. Residually connected CTs in three-phase ground fault scheme sum to zero when the phases are balanced. Reversed polarity of a CT could cause a ground fault relay to trip under a normal balanced condition. In differential protection schemes, current sources are compared. Reverse polarity of a CT could effectively double the phase current flowing into the relay, thus causing a nuisance tripping of a relay. When two CTs are driving a three-phase ammeter through a switch, a reversed CT could show 1.73 times the monitored current flowing in the unmonitored circuit. (See Fig. 3.127.)

Y-connected secondary circuit is the most commonly used. The CT will reproduce positive, negative, and zero sequence elements as they occur in the primary circuit. In the Δ-connected secondary, the zero sequence components are filtered and left to circulate in the delta. This is a common scheme for differential protection of Δ-Y transformer. A general rule of thumb is to connect the CT secondaries in Y when they are on the Δ winding of a transformer and, conversely, connect the secondaries in Δ when they are on the Y winding of the transformer. (See Table 3.17.)

TABLE 3.17 Total Burden on CT in Fault Conditions

CT Secondary Connections	3-Phase or Ph-to-Ph Fault	Phase-to-GND Fault
Y-connected at CT	$R_S + R_L + Z_R$	$R_S + 2R_L + Z_R$
Y-connected at switchboard	$R_S + 2R_L + Z_R$	$R_S + 2R_L + Z_R$
Δ-connected at CT	$R_S + 2R_L + 3Z_R$	$R_S + 2R_L + 2Z_R$
Δ-connected at switchboard	$R_S + 3R_L + 3Z_R$	$R_S + 2R_L + 2Z_R$

Note: where R_S = CT secondary winding resistance + CT lead resistance; R_L = one-way circuit lead resistance; Z_R = relay impedance in secondary current path.

Construction

There are four major types of CT: (1) window-type, which includes bushing-type (BCT); (2) bar-type; (3) split-core type; and (4) wound-type, which has both a primary and secondary winding. There is ongoing development and limited use of optical-type current transformers (OCT) which rely on the principles of light deflection.

Window-Type

The window-type CT is the simplest form of instrument transformer. It is considered to be an incomplete transformer assembly because it consists of a secondary winding wound on its core. The most common type is that wound on a toroidal core. The secondary winding is fully distributed around the periphery of the core. In special cases when taps are employed, they are distributed such that any connection made would utilize the entire core periphery. Windings in this manner assure optimum flux linkage and distribution. Coupling is almost impervious to primary conductor position provided that the return path is sufficiently distanced from the outer periphery of the secondary winding. The effects of stray flux are negligible, thus making this type of winding a low reactance design. As for the primary winding, in most cases it is a single conductor centrally located in the window. A common application is to position the CT over a high voltage bushing, hence the name BCT. Nearly all window-type CTs manufactured today are rated 600-V class. In practice they are intended to be used over insulated conductors when the conductor voltage exceeds 600 V. It is common practice to utilize a 600-V class window-type in conjunction with air space between the window and the conductor on higher voltage systems. Such use may be seen in isolated-phase bus compartments. There are some window-types which may be rated for higher voltages as stand alone units, or with the use of an integral sleeve or tube made of porcelain or some polymeric material. Window-types generally have a round window opening but are also available with rectangular openings. This is sometimes provided to fit a specific bus arrangement which could be found in the rear of switchgear panels or on draw-out type circuit breakers. This type is used for general purpose monitoring, revenue metering and billing, and protective relaying.

Bar-Type

The bar-type CT is, for all practical purposes, a window-type CT with a primary bar inserted straight through the window. This bar assembly can be permanently attached or held in place with brackets. Either way, the primary conductor is a single turn through the window, fully insulated from the secondary winding. The bar must be sized to handle the continuous current to be passed through it, and mechanically secured to handle high-level short-circuit currents without incurring damage. Uses are the same as for the window-type.

Split-Core Type

The split-core type CT is a special case of window-type CT. Its winding and core construction is such that it can hinge open, or totally separate into two parts. This arrangement is ideal for use in cases where the primary conductor cannot be opened or broken. However, because of this cut, the winding is not fully distributed. Often, only 50% of the effective mmp is used. Use of this type should be with discretion since this construction results in higher than normal errors. There is also some uncertainty in repeatability of performance from installation to installation. The re-assembly of the core halves is critical. It is intended for general-purpose monitoring and temporary installations.

Wound-Type

The wound-type CT is that which has a primary winding that is fully insulated from the secondary winding. Both windings are permanently assembled on the core. The insulation medium used, whether polymeric, oil, or even air, in conjunction with its rated voltage class dictates the core and coil construction. There are several core types used from a low reactance toroid to high reactance cut-cores and laminations. The distribution of the cut(s) and gap(s) helps control the magnetizing losses. The manner

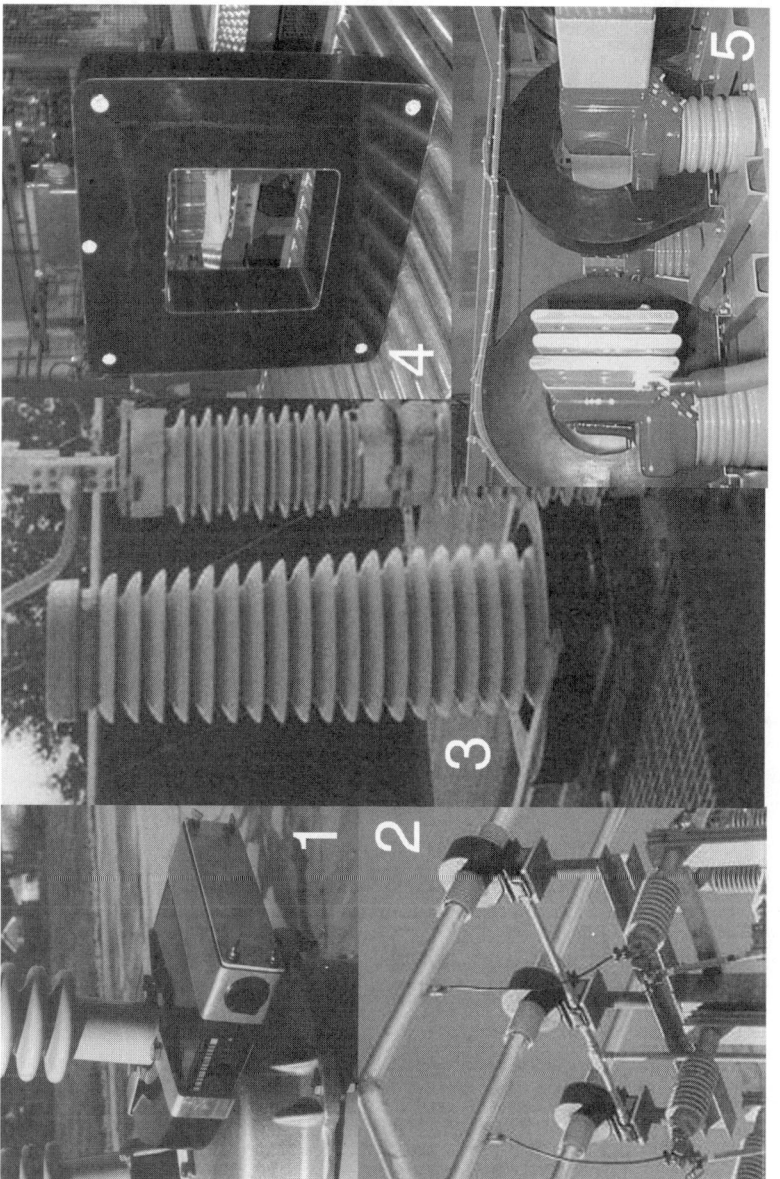

1. Small 600-V class window-type CT mounted over a bushing on 15-kV voltage regulator. 2. 15-kV class window-type CTs with porcelain tube mounted on sub-station structure. 3. Large 600-V class window-type (slipover) CT mounted over a high voltage bushing. 4. 8.7-kV class window-type CT with rectangular opening. 5. 600-V class window-type CTs mounted over 15-kV bus inside a metal enclosure. (All photos courtesy of Kuhlman Electric Corp.)

Bar-type CT. (Courtesy of Kuhlman Electric Corp.)

Split-core type CT with secondary winding on three legs of core. (Courtesy of Kuhlman Electric Corp.)

Medium-voltage wound-type CT cast in epoxy resin. (Photos Courtesy of Kuhlman Electric Corp.)

High-voltage wound-type CT in combination steel tank, oil, and porcelain construction.

in which the windings are arranged on the core will affect the reactance since it is a geometric function. Generally, the windings do not utilize the mmp efficiently. The proper combination of core type and coil arrangement can greatly reduce the total reactance, thus reducing errors.

Transformers

600-V indoor class auxiliary CT. (Courtesy of Kuhlman Electric Corp.)

The auxiliary CT is a wound-type used in secondary circuits for totalizing, summation, phase shifting, or to change ratios. They are typically 600-V class since they are used in the low voltage circuit. When applying auxiliary CTs, the user must be aware of its reflected impedance on the main line CT.

Proximity Effects

Current flowing in a conductor will induce magnetic flux through the air. This flux is inversely proportional to the distance squared, $B \Leftrightarrow 1/d^2$. As current increases, the flux increases. Considering the case in a three-phase bus compartment, each bus is equally spaced from the other. If CTs are mounted on each phase, then it is possible that the flux from adjacent conductor fields links to the CTs. Often the distance is sufficient enough that stray flux linkage is almost negligible. But at higher current levels it can cause problems, especially in differential protection schemes. This stray flux can cause localized saturation in segments of the core. This saturation can cause heating in the winding, and increase the CT errors. This same effect can be seen when return paths are also in close proximity to a CT. In the case of draw-out circuit breakers rated 2000 A and above, the phase distances and return paths are close together, causing problems with CTs mounted on the stabs. Sometimes magnetic shunts or special winding arrangements are incorporated to offset the effects. Another concern is CTs mounted over large generator bushings. The distances are typically adequate for normal operation, but under overcurrent situations may lead to mis-operation of protective devices. It is for that reason CTs used in this application that are rated above 10,000 A are shielded. The shield may be external, or internal as an integral part of the secondary winding.

Linear Coupler

The linear coupler (LC) utilizes a non-magnetic core, e.g., wood, plastic, or paper, which acts as a former for the winding. It is typically a window-type construction. With an air-core we remove the magnetizing components of error, thus offering linear response. They do not produce secondary currents that would be provided by an ideal CT. If the magnetizing impedance approached infinity, the secondary current would approach the ideal, $i_s = -i_p$ (Np/Ns). The low permeability of the core prevents a high magnetizing inductance, thus considerable divergence of performance from that of the conventional CT.

Consequently, protective equipment must be designed to present infinite impedance to the LC and operate as mutual inductors. Closely matching the LC impedance will provide maximum power transfer to the device. LC outputs are typically defined by Vs/Ip, e.g., 5 V per 1000 A. Since the load is high impedance, the LC can be safely open-circuited, unlike the conventional iron-core CT.

Because coupling is important, the window size is made as small as possible to accommodate the primary conductor. Positioning is critical, and the return conductor and adjacent phases must be far enough away from the outer diameter of the LC so that stray flux is not introduced into the winding.

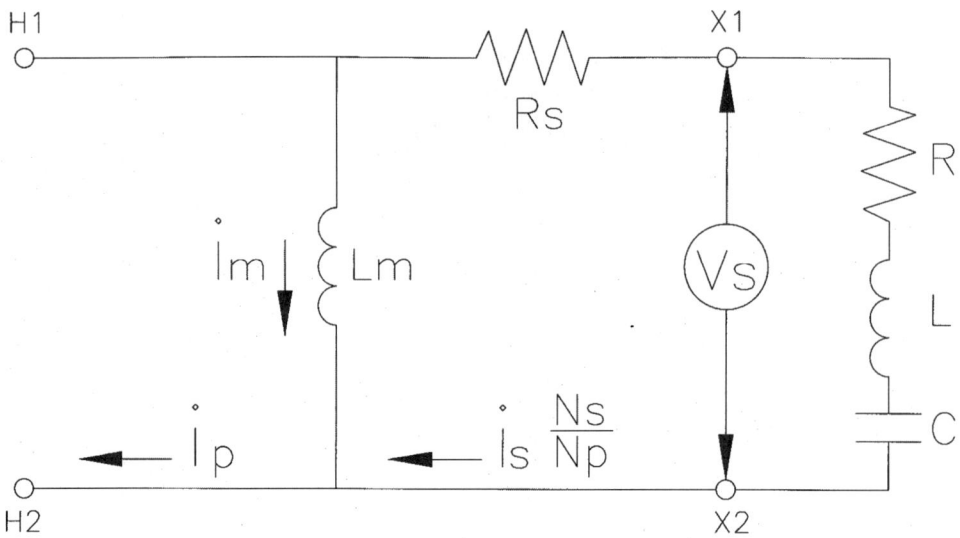

FIGURE 3.128 Linear coupler equivalent circuit.

Direct Current Transformer

The basic direct current transformer (DCT) utilizes two coils, referred to as elements, that require external AC excitation. The elements are window-type CTs that fit over the DC bus. The elements are connected in opposition such that the instantaneous AC polarity of one is always in opposition with the other. The AC flux in one element opposes the DC flux in the primary bus and de-saturates the core, while in the other element the AC flux will aid the DC flux and further saturate the core. This cycle is repeated during the other half AC cycle. The need to rectify is due to the square wave output. The direction of current flow in the primary is not important.

Proximity of the return conductor may have an effect on accuracy due to local saturation. For best results, all external influences should be kept at a minimum.

The output waveform contains a commutation notch at each half-cycle of the applied exciting voltage. These notches contribute to the errors and can interfere with the operation of fast-acting devices. Ideally, the core material should have a square B-H characteristic, which will minimize the notches. There are several connection schemes that can eliminate the notches, but will run up the overall cost.

One way is to use two additional elements and a Scott-connected transformer. The supply voltage of one element is shifted 90° in phase with the other element, such that one carries secondary current proportional to the primary current. Six elements can be used and connected to a different phase of the three-phase supply. This will reduce the need for the Scott transformer, but requires even more elements.

A simpler approach could be the three-element connection, where the third element acts as a smoothing choke to the notches. This also increases overall frequency response by providing AC coupling between the primary and output circuits.

Finally, there is the Hall-effect device. This solid-state chip is inserted into the gap of an iron core whose area is much larger than the device itself. The core has no secondary winding. The Hall effect requires a low voltage DC source to power, and will provide an output proportional to the DC primary current and the flux linked to the core and gap.

CT Installations

CTs are often installed on existing systems as power requirements increase. One of the most common retrofit installations is the application of a window-type CT mounted over the high voltage bushing of

Transformers

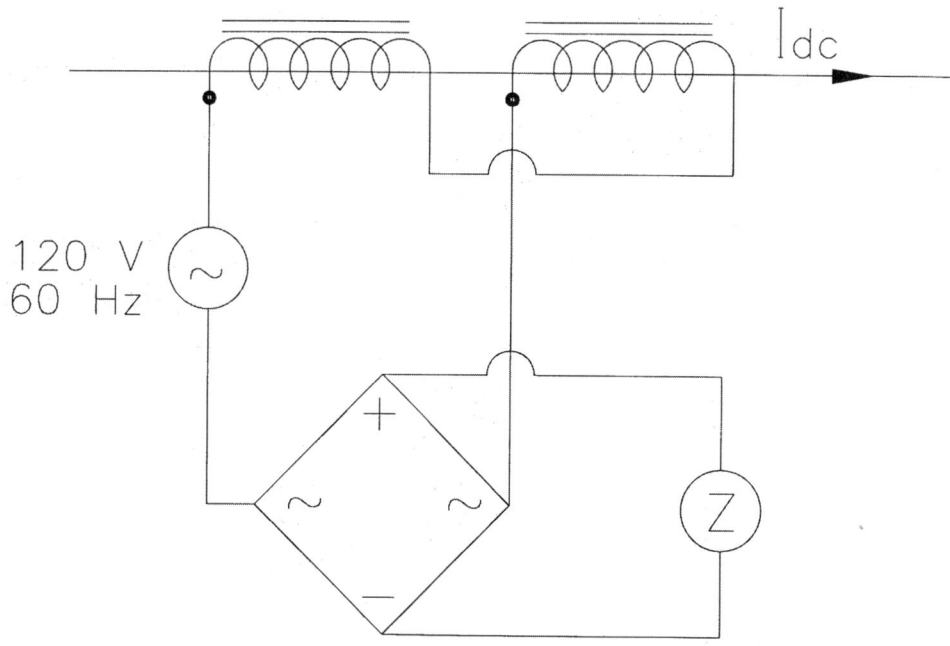

FIGURE 3.129 Standard two-element connection.

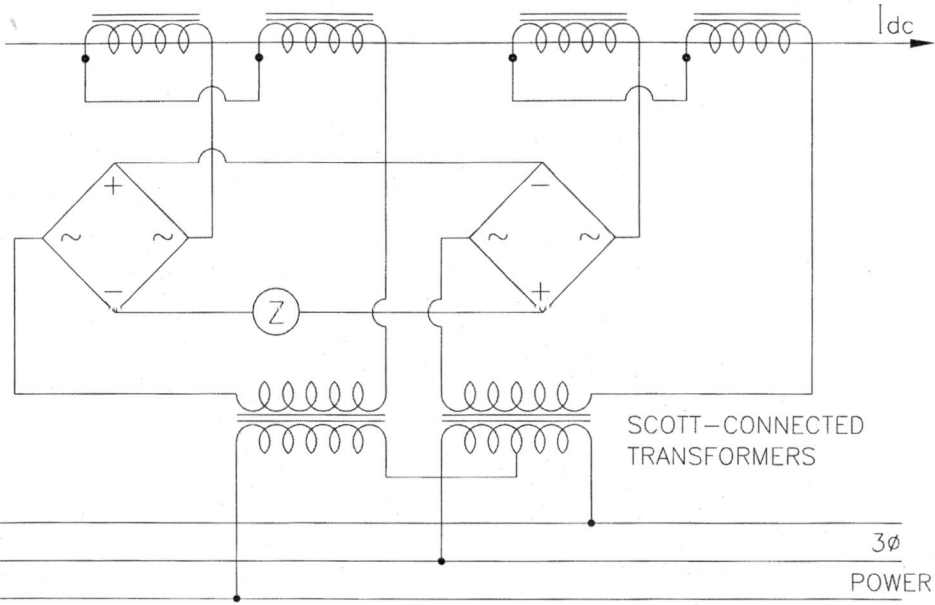

FIGURE 3.130 Four-element Scott-transformer connection.

a power transformer. To maintain the insulation system integrity, adequate strike clearances must be observed. It is also important to protect the CT from a high voltage flash-over to ground. This is done by placing the CT body below the ground plane. When using a molded CT, a ground shield may be placed on top of the unit and connected to ground. When making this connection it is important that

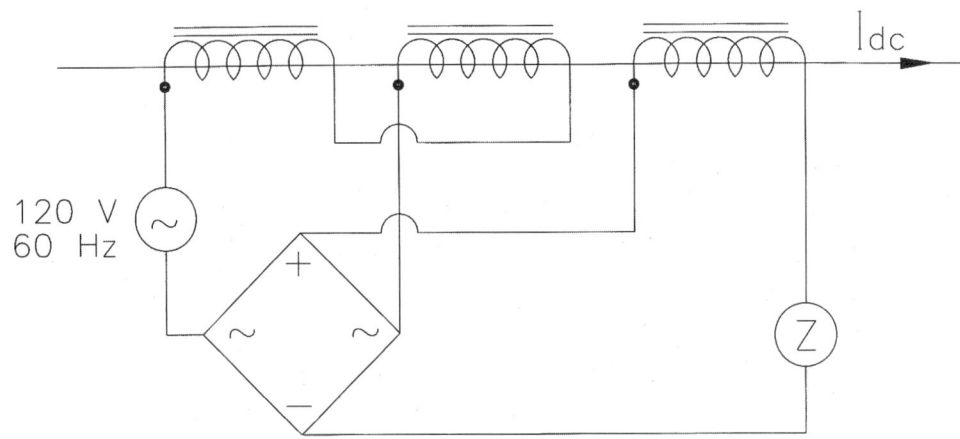

FIGURE 3.131 Three-element connection.

600V class slip-over CT installed on high voltage bushing with ground shield. (Courtesy of Kuhlman Electric Corp.)

High voltage circuit breaker with BCTs mounted underneath the metallic cover.

3 VT and 3 CT elements on pole mounted structure. (Courtesy of Kuhlman Electric Corp.)

the lead is routed such that it does not make a shorted electrical turn around the CT. In high voltage switchgear, BCTs are fitted over the high voltage bushing and enclosed inside a metallic cover, protecting it from the weather. In some cases it is possible to encapsulate a CT coil assembly inside a metallic housing which also shields the CT.

Combination Metering Units

The last major assembly is the metering unit, which consists of VT and CT elements in single phase or three phase arrangements. Metering units can be single molded elements mounted on a structure that is bolted to a utility pole, or in a pad-mount compartment. The elements can be in any combination needed to provide accurate energy measurement.

Metering units are also available in three-phase tanks with the elements submerged in oil. In high voltage systems, a single- phase combination CT/VT unit is available that looks much like a high voltage CT. The VT element is L-G and is common to the H1 terminal. The elements are independent of each other. The CT element is mounted on top and can house several cores, which can be used for metering and relaying applications. The VT element is located in the bottom.

When measuring energy usage for the purposes of revenue billing, and knowing the RCF and phase angle readings of each element, you can correct the watts or watthours by multiplying the reading by the product of $[TCF_{CT}][TCF_{VT}]$.

3-Phase metering unit. (Courtesy of Kuhlman Electric Corp.)

New Horizons

With de-regulation of the utility industry, buying and selling of power, leasing power lines, etc., the need for monitoring power at the transmission and distribution level will increase. There will be a need to add more metering points within existing systems starting at the generator. The utilities will want to add this feature at the most economical cost. The use of window-type slipover CTs for revenue metering will drive the industry towards improved performance. The need for higher accuracies at all levels will be desired, and can be obtained easily with the use of low burden solid-state devices. Products will become more environmentally safe and smaller in size. To help the transformer industry fulfill these needs, steel producers will need to make improvements by lowering losses, increasing initial permeability, and developing new composites.

References

Burke, H. E. *Handbook of Magnetic Phenomena*. New York: Van Nostrand Reinhold Co., 1986.

Cosse, R. E. Jr. et al., Eds. The practice of ground differential relaying. *I.A.S. Trans.* 30, 1472-1479, 1994.

Funk, D. G. and Beaty, H. W. *Standard Handbook for Electrical Engineers, 12th ed.* New York: McGraw-Hill, 1987.

Hague, B. *Instrument Transformers, Their Theory, Characteristics, and Testing.* London: Sir Isaac Pitman & Sons, LTD, 1936.

Hou, D. and Roberts, J. *Capacitive Voltage Transformer Transient Overreach Concerns and Solutions for Distance Relaying.* May 1-3. Atlanta, GA: Paper presented to 50th Annual Georgia Tech. Protective Relaying Conference, 1996.

IEEE Standard C57.13 — *Requirements for instrument Transformers.* New York: IEEE, 1993.

IEEE Standard C37.110 –*Guide for the Application of Current Transformers Used for Protective Relaying Purposes.* New York: IEEE, 1996.

Jenkins, B. D. *Introduction to Instrument Transformers.* London: George Newnes LTD, 1967.

Karlicek, R. F. and Taylor, Jr. E. R. Ferroresonance of grounded potential transformers on ungrounded power systems. *AIEE Trans.* 607-618, 1959.

Moreton, S. D. A simple method for determination of bushing current transformer characteristics. *AIEE Trans.* 62, 581-585, 1943.

Pagon, J. *Fundamentals of Instrument Transformers.* April 21-25. Gainsville, FL: Electromagnetic Industries, Inc., 1975.

Settles, J. L. et al., Eds. The analytical and graphical determination of complete potential transformer characteristics. *AIEE Trans.* 1213-1219, 1961.

Swindler, D. L. et al., Eds. Modified differential ground fault protection for systems having multiple sources and grounds. *I.A.S. Trans.* 30, 1490-1505, 1994.

Wentz, E. C. A simple method for determination of ratio error and phase angle in current transformers. *AIEE Trans.* 60, 949-954, 1941.

West, D. J. Current transformer application guidelines. *I.A.S. Annual*, 110-126, 1977.

Working Group — System Relaying Committee, IEEE. Relay performance considerations with low ratio current transformers and high fault currents. *I.A.S. Trans.* 31, 392-404, 1995.

Yarbrough, R. B. *Electrical Engineering Reference Manual, 5th ed.* Belmont, CA: Professional Publications, Inc., 1990.

3.9 Transformer Connections

Dan D. Perco

In deciding the transformer connections required in a particular application, there are so many considerations to be taken into account that the final solution must necessarily be a compromise. It is therefore necessary to study in detail the various features of the transformer connections together with the local requirements under which the transformer will be operated.

This section describes the common connections for Distribution, Power, HVDC Converter and Rectifier Transformers. Space does not permit a detailed discussion of each type of transformer connection or other uncommon connections. The information presented in this section is primarily directed at transformer users. Additional information can be obtained from the IEEE transformer standards. In particular, reference is made to IEEE Standards C57.12.70, "Terminal Markings and Connections for Distribution and Power Transformers", C57.105, " Application of Transformer Connections in Three-Phase Distribution Systems", C57.129 " General Requirements and Test Code for Oil-Immersed HVDC Converter Transformers", and C57.18.10 " Practices and Requirements for Semiconductor Power and Rectifier Transformers".

Polarity of Single-Phase Transformers

Transformers can be either subtractive or additive polarity. Most of the standards require subtractive polarity. In either case, the polarity of the transformer is identified by the terminal markings as shown in Fig. 3.132. Subtractive polarity has correspondingly marked terminals for the primary and secondary windings opposite each other. For additive polarity, the terminal markings of the secondary winding are reversed.

Angular Displacement of Three-Phase Transformers

Connection of three-phase transformers or three single-phase transformers in a three-phase bank can create angular displacement between the primary and secondary terminals. The convention for the direction of rotation of the voltage phasors is taken as counterclockwise. The standard angular displacement for two winding transformers is shown in Fig. 3.133. The references for the angular displacement are shown as dashed lines. The angular displacement is the angle between the lines drawn from the neutral to H1 and from the neutral to X1 in a clockwise direction from H1 to X1. The angular displacement between the primary and secondary terminals can be changed from 0° to 330° in 30° steps simply by altering the three-phase connections of the transformer. Therefore, systems with difference angular displacements can be connected by selecting the appropriate three-phase transformer connections.

Figure 3.133 shows angular displacement for common, double wound three-phase transformers. Multiwinding and autotransformers are similarly connected.

Transformers

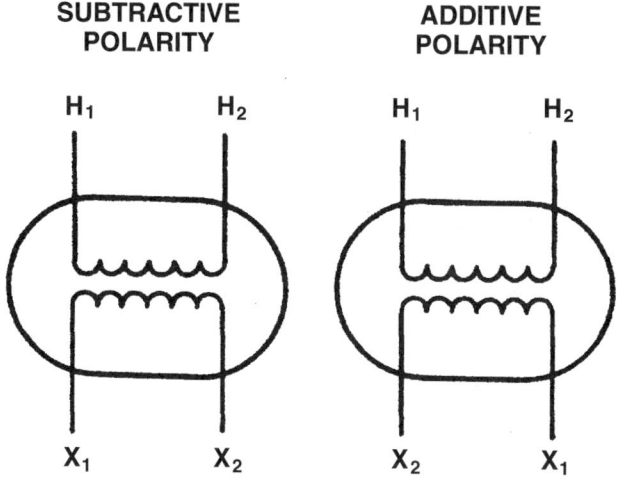

FIGURE 3.132 Single-phase transformer terminal markings.

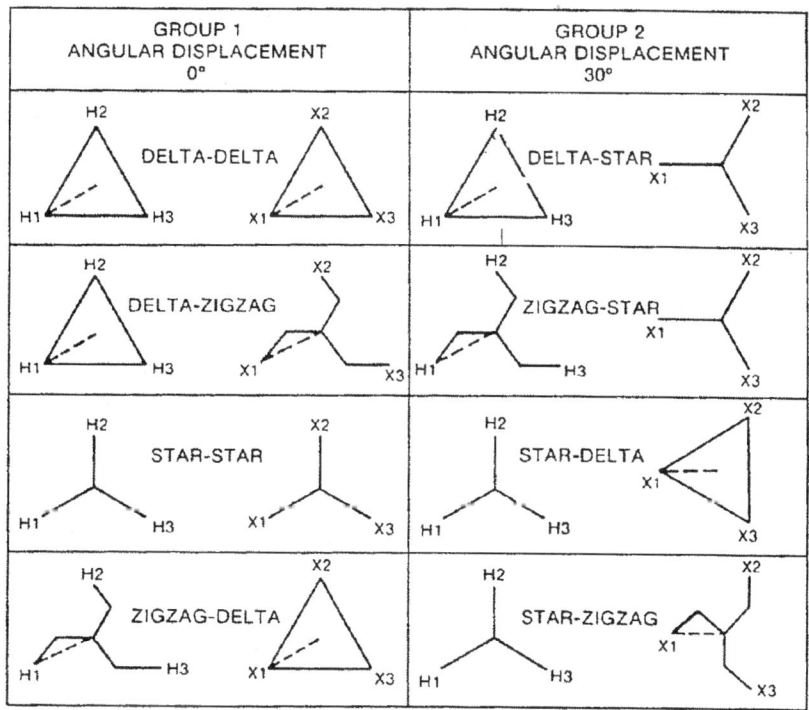

FIGURE 3.133 Standard angular displacement for three-phase transformers.

Three-Phase Transformer Connections

Three-phase transformer connections can be compared with each other with respect to:

1. Ratio of kVA output to the kVA rating of the bank
2. Degree of voltage symmetry with unbalanced phase loads
3. Voltage and current harmonics

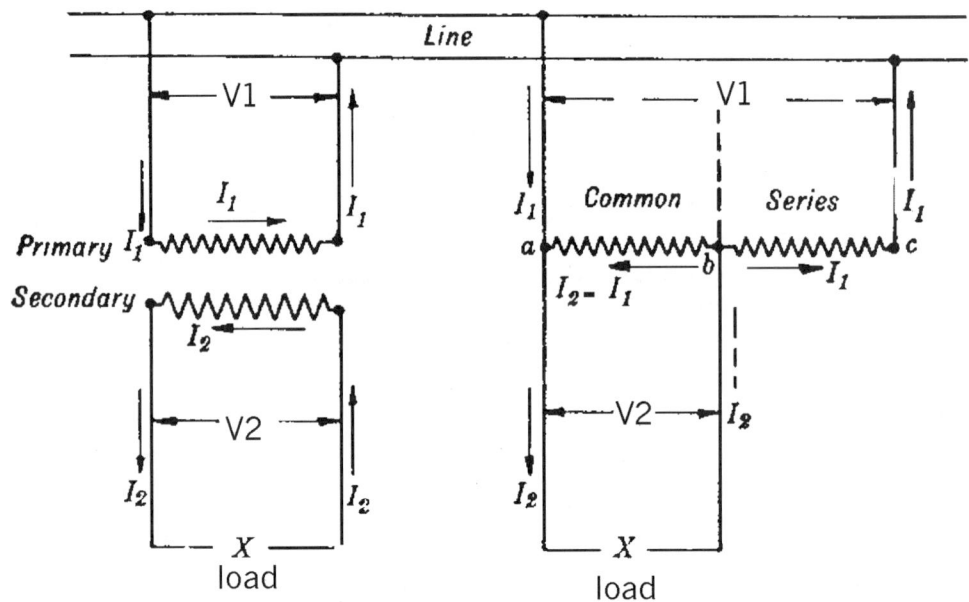

FIGURE 3.134 Current flow in double wound and autotransformers.

4. Transformer ground availability
5. System fault or transient voltages

and other operating characteristics to determine the most suitable connection for each application.

Double Wound Transformers

The majority of three-phase transformer connections are made by connecting single phase or the phases of a three-phase transformer, either between the power system lines, thus forming a delta connection, or one end of each phase together and the other ends to the lines thus forming a Y connection. For these connections, the internal winding rating is equal to the through load rating. This accounts for the popularity of these connections. For all other double wound transformer connections, the ratio is less than unity. For example, in the interconnected star or zigzag connection, the transformer is capable of delivering a load equal to only 86.6% of the internal winding rating. These types of three-phase connections are shown in Fig. 3.133.

Voltage and current symmetry, both with respect to the three lines and also lines to neutral are obtained only in the delta and zigzag connections. The Y connection is symmetrical as far as the lines are concerned, but introduce third-harmonic voltage and current dissymmetry between lines and neutral. If the transformer and generator neutrals are grounded, third harmonic current will flow that can create interference in telephone circuits. Third harmonic voltages are also created on the lines that can subject the power system to dangerous overvoltages due to resonance with the line capacitance. This is particularly true for shell type three-phase transformers, five-limb coreform and three-phase banks of single phase transformers. For any three-phase connection of three-limb coreform transformers, the impedance to third harmonic flux is relatively high on account of the magnet coupling between the three phases, resulting in a more stabilized neutral voltage. A delta tertiary winding can be added on Y–Y transformers to provide a path for third harmonic and zero sequence currents and stabilize the neutral voltage.

Delta-connected transformers do not introduce third harmonics or their multiples into the power lines. However, this type of three-phase transformer connection will have higher ground voltages during system fault or transient voltages. Supplying an artificial neutral to the system with a grounding transformer can control these voltages. The delta connection is also more costly to manufacture for high voltages and is generally limited to 345-kV systems and below.

In the Y-delta or delta-Y connections, complete voltage and current symmetry is maintained by the presence of the delta. These connections are more free from objectionable features than any other connections.

Multiwinding Transformers

Transformers having more than two windings are frequently used in power and distribution systems. The arrangement of windings can be varied to change the value of leakage reactance between winding pairs. In this way, the voltage regulation and the short circuit requirements are optimized. The application of multiwinding transformers permits:

1. Interconnection of several power systems operating at different voltages.
2. Use of a delta-connected stabilizing winding, which can also be used to supply external loads.
3. Regulation and control of reactive power compensation.

A disadvantage of multiwinding transformers is that all the windings are magnetically coupled and are affected by the loading of the other windings. It is therefore essential to understand the leakage impedance behavior of this type of transformer to be able to calculate the voltage regulation of each winding. As an example, for three winding transformers, the leakage reactance between each pair of windings must be converted into a star equivalent circuit. After the loading of each winding is determined, the regulation can be calculated for each impedance branch.

Autotransformers Connections

Autotransformers will deliver more than the internal winding ratings, depending on the voltage ratios of the primary and secondary voltages as shown in Fig. 3.134 and the following formula:

$$\text{Output/Internal rating} = V1/(V1 - V2)$$

where $V1$ = the voltage of the highest voltage winding
$V2$ = the voltage of the lower voltage winding

Consequently, the internal rating, size, cost, and efficiency of an autotransformer is better than a double wound transformer. This also shows that the greatest benefit of the autotransformer is achieved when the system voltages are close to each other.

A disadvantage of the autotransformer is that the short circuit current and forces are increased because of the reduced leakage reactance. In addition, most three-phase autotransformers are Y-Y connected. This form of connection has the same limitations as for the Y-Y double wound transformers.

Often this type of transformer has a delta-connected tertiary winding to reduce third harmonic voltages, permit the transformation of unbalanced three-phase loads, and supply station auxiliary load or power factor improvement equipment. The tertiary winding must be designed to accept all these external loads as well as the severe short circuit currents and forces associated with three-phase faults on its own terminals or single line-to-ground faults on any other terminal. If no external loading is required, the tertiary winding terminals should not be brought out except for one terminal to ground the delta during service operation.

The problem of transformer insulation stresses and system transient protection is more complicated for autotransformers, particularly when tapping windings are also required. Transients can also be more easily transferred between the power systems with the autotransformer connection.

Interconnected Star and Grounding Transformers

The star/interconnected star connections have the advantages of the star/delta connections with the additional advantage of the neutral. The interconnected star or zigzag connection allows unbalanced phase load currents without creating severe neutral voltages. This connection also provides a path for third harmonic currents created by the non-linearity of the magnetic core material. As a result, interconnected

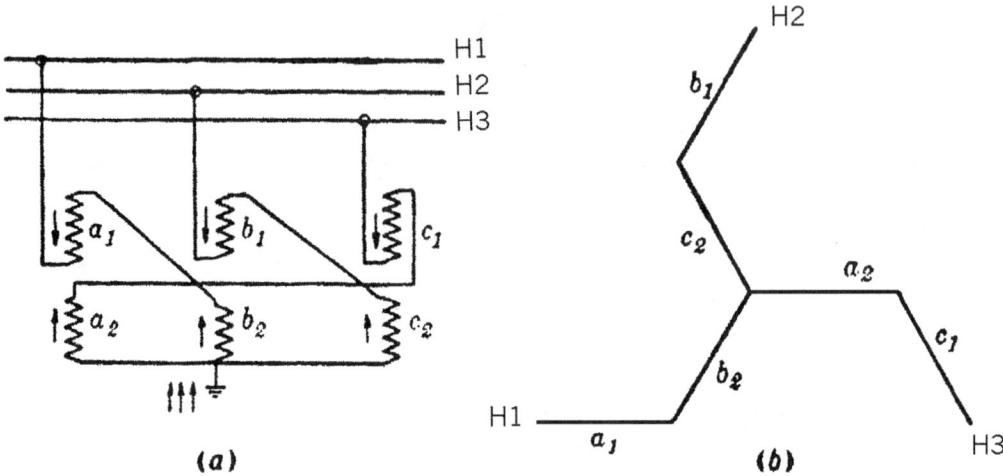

FIGURE 3.135 Interconnected star grounding transformer. (a) Current distribution in the coils for a line-to-ground fault. (b) Normal operating voltages in the coils.

star neutral voltages are essentially eliminated. However, the zero sequence impedance of interconnected star windings is often so low that high third harmonic and zero sequence currents will result when the neutral is directly grounded. These can be limited to an acceptable level by connecting a reactor between the neutral and ground.

The stable neutral inherent in the interconnected star or zigzag connection has made its use possible as a grounding transformer for otherwise isolated systems. This is shown in Fig. 3.135. The connections to the second set of windings can also be reversed to produce the winding angular displacements shown in Fig. 3.133.

For a line-to-neutral load or a line-to-ground fault on the system, the current is limited by the leakage reactance between the two coils on each phase of the grounding transformer.

Three-Phase to Six-Phase Connections

For six-pulse rectifier systems, the three-phase connections discussed above are commonly used. However, for 12-pulse rectifier systems, three-phase to six-phase transformations are required. For low voltage DC applications, there are numerous practical connection arrangements possible to achieve this. However, for high voltage DC applications, there are few practical arrangements. The most commonly used connections are either a delta or Y primary with two secondaries, one Y and one Delta connected.

Paralleling of Transformers

Transformers having terminals marked in the manner shown in the section entitled "Polarity of Single-Phase Transformers" may be operated in parallel by connecting similarly marked terminals together, provided their ratios, voltages, angular displacement, resistances, reactances, and ground connections are such as to permit parallel operation.

The difference in the no load terminal voltages of the transformers will cause a circulating current to flow between the transformers when paralleled. This current will flow at any load. The impedance of the circuit, which is usually the sum of the impedances of the transformers paralleled, limits the circulating current. The circulating current adds vectorially to the load current to establish the total current in the transformer. As a result, the capacity of the transformer to carry load current is reduced by the circulating current when the transformers are paralleled.

The load currents in the paralleled transformers will divide inversely to the impedances of the transformers paralleled. Generally, the difference in resistance has an insignificant effect on the circulating current because the leakage reactance of the transformers involved is much larger than the resistance. Transformers having different impedance values can be made to divide their load in proportion to their load ratings by placing a reactor in series with one transformer so that the resultant impedance of the two branches will create the required load sharing.

When delta–delta connected transformer banks are paralleled, the voltages are completely determined by the external circuit, but the division of current among the phases depends upon internal characteristics of the transformers. Considerable care must be taken in the selection of transformers, particularly single-phase transformers in three-phase banks, if the full capacity of the banks is to be used when the ratios of transformation on all phases are not alike. In the delta-Y connection, the division of current is indifferent to the differences in the characteristics of individual transformers.

3.10 LTC Control and Transformer Paralleling

James H. Harlow

Tap changing under load (TCUL), be it with load tap changing power transformers or step-voltage regulators, is the primary means of dynamically regulating the voltage on utility power systems. Switched shunt capacitors may also be used for this purpose, but in the context of this discussion, shunt capacitors are presumed to be applied with the objective of improving the system power factor.

The control of a tap changer is much more involved than simply responding to a voltage excursion at the transformer secondary. Modern digital versions of LTC control include so many ancillary functions and calculated parameters that it is often used to serve as the means for system condition monitoring.

The control of the tap changer in a transformer or step-voltage regulator is essentially the same. Unless stated otherwise, the use of the term "transformer" in this section applies equally to step-voltage regulators. It should be recognized that either type of product may be constructed as a single-phase or three-phase apparatus, but that transformers are most often three-phase while step-voltage regulators are most commonly single-phase.

System Perspective, Single Transformer

This discussion is patterned to a typical utility distribution system, the substation and the feeder, although much of the material is also applicable to transmission applications. This first discussion of the control considers that the control operates only one LTC in isolation; that is, there is no opportunity for routine operation of transformers in parallel.

The system may be configured in any of several ways, according to the preference of the user. Figure 3.136 depicts two common implementations. In the illustrations of Fig. 3.136, the dashed line box depicts a substation enclosing a transformer or step-voltage regulators. The implementations illustrated accomplish bus voltage regulation on a three-phase or single-phase basis. Another common application is to use voltage regulators on the distribution feeders. A principal argument for the use of single-phase regulators is that the voltages of the three phases are controlled independently, whereas a three-phase transformer or regulator will control the voltage of all phases based on knowledge of the voltage of only one phase. For the figure:

S = source, the utility network at transmission or sub-transmission voltage, usually 69 kV or greater.
Z = impedance of the source as "seen" looking back from the substation.
L = loads, distributed on the feeders, most often at 15 to 34.5 kV.

The dominance of load tap changing apparatus involves either 33 voltage steps of 5/8% voltage change per step, or 17 voltage steps of 1¼% voltage change per step. In either case, the range of voltage regulation is ±10%.

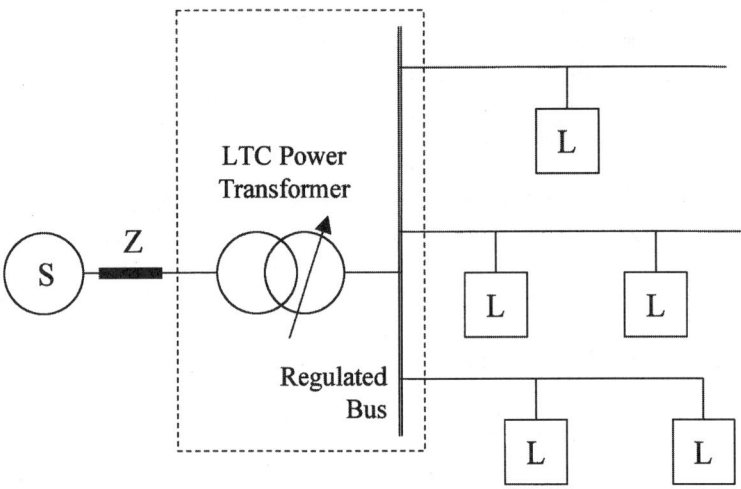

FIGURE 3.136(a) Three-phase bus regulation, three-phase LTC transformer.

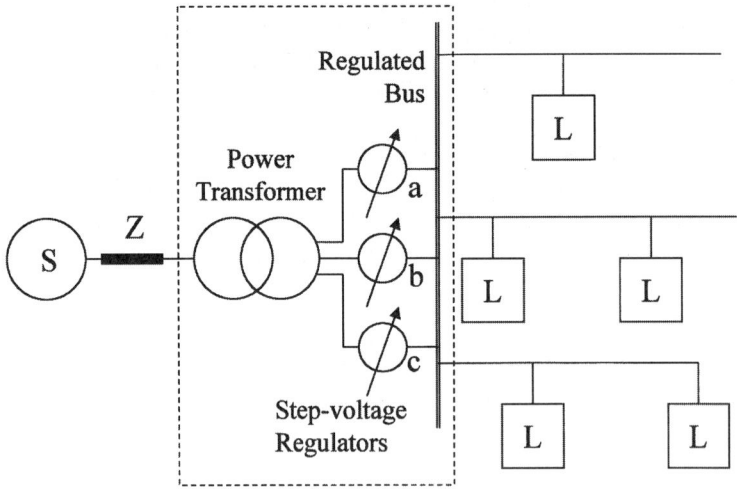

FIGURE 3.136(b) Independent phase bus regulation, three single-phase step-voltage regulators.

Control Inputs

Voltage Input

The voltage class of the primary system is unimportant to the LTC control. The system will always include a voltage transformer (VT) or other means to drop the system voltage to a nominal 120 V for use by the control. Because of this, the control is calibrated in terms of 120 V and it is common to speak of the voltage as being, say, "118 V", or "124 V", it being understood that the true system voltage is the value stated times the VT ratio. Presuming a single tap step change represents 5/8% voltage, it is easily seen that a single tap step change will result in 0.00625 pu × 120 V = 0.75 V change at the control.

The control will receive only a single 120-V signal from the voltage transformer, with it tracking the line-ground voltage of one phase, or a line-line voltage of two phases. For the case of three-phase apparatus, the user must exhibit great care, as later described, in selecting the phase(s) for connecting the VT.

Transformers

Current Input

The current transformer(s) (CT) is provided by the transformer manufacturer so as to deliver control current of "…not less than 0.15 A and not more than 0.20 A … when the transformer is operating at the maximum continuous current for which it is designed…" (This is per ANSI/IEEE standards where the nominal current is 0.2 A. Other systems may be based on a different nominal current, such as 5.0 A.) As with the voltage, the control will receive only one current signal, but it may be that of one phase or the cross-connection (the paralleling of two CTs) of two phases.

Phasing of Voltage and Current Inputs

In order for the control to perform all of its functions properly, it is essential that the voltage and current input signals be in-phase for a unity power factor load, or, if not, that appropriate recognition and corrective action be made for the expected phasing error.

Figure 3.137(a,b,c) depicts the three possible CT and VT orientations for three-phase apparatus. Note that for each of the schemes the instrument transformers could be consistently shifted to different phases from that illustrated without changing the objective. The first scheme, involving only a line to ground rated VT and a single CT, is clearly the simplest and least expensive; however, it causes all control action to be taken solely on the basis of knowledge of conditions of one phase. Some may prefer the second scheme as it gives reference to all three phases, one for voltage and the two not used for voltage to current. The third scheme is often found with a delta connected transformer secondary. Note that the current signal derived in this case is $\sqrt{3}$ times the magnitude of the individual CT secondary currents. This must be scaled before the signal is delivered to the control.

The Need for Voltage Regulation

Referring to the circuits of Fig. 3.136, it will be recognized that system conditions will change over time with the result that the voltage at the substation bus, and as delivered to the load, will change. From Fig. 3.136, the source voltage, the source impedance, and the load conditions will be expected to change with time. The most notable of these, the load, must be recognized to consist of two factors: (1) the magnitude and the power factor, or what is the same point, the real (watt) and (2) reactive (var) components.

Regulation of the Voltage at the Bus

Many times the object of the LTC is to simply hold the substation bus voltage at the desired level. If this is the sole objective, it is sufficient to bring only a VT signal that is representative of the bus voltage into the control. The secondary of the VT is usually 120 V at the nominal bus voltage, but other VT secondary voltages, especially 125 V, 115 V, and 110V, are used. Figure 3.138 shows the circuit. The figure shows a motor (M) on the LTC which is driven in either the Raise or Lower direction by the appropriate output (M_R or M_L) of the control. This first control, Fig. 3.138, is provided with only three settings, these being those required for the objective of regulating the substation secondary bus voltage:

- **Voltage Set Point.** This, on the control voltage base, as 120 V, is the voltage desired to be held *at the load*. (The load location for this first case is the substation secondary bus because line drop compensation is not yet considered and is therefore zero). This characteristic is also commonly spoken of as "Voltage Band Center", this being illustrative of the point that there is a band of acceptable voltage, and this is the midpoint of that band. If line drop compensation (LDC, as detailed later) is not used, the set point will often be somewhat higher than 120 V, perhaps 123 V to 125 V; with use of LDC the setting will be lower, perhaps 118 V.
- **Bandwidth.** The bandwidth describes the voltage range, or band, which is considered acceptable, i.e., in which there is no need for any LTC corrective action. The bandwidth is defined in the ANSI/IEEE standard as a voltage, with one-half of the value above and one-half below the voltage set point. Some other controls adjust the bandwidth as a percentage of the voltage set point and the value represents the band on each side of the bandcenter. The bandwidth voltage selected is

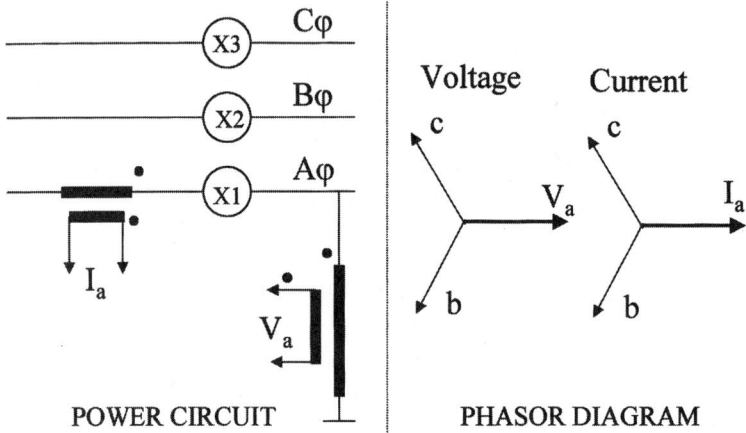

VT (L-G), 1 CT on same phase

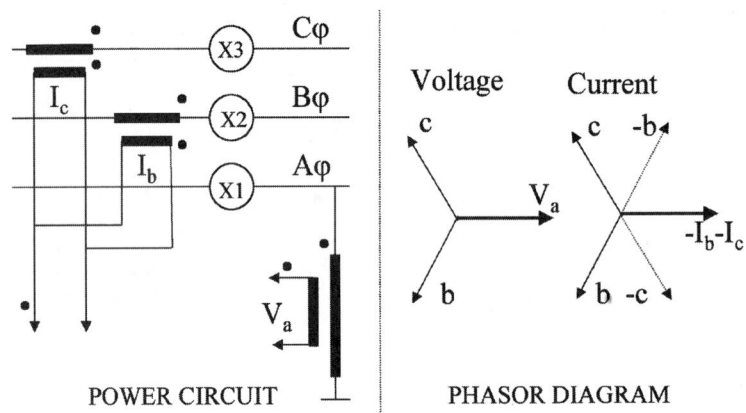

VT (L-G), 2 CTs on other phases

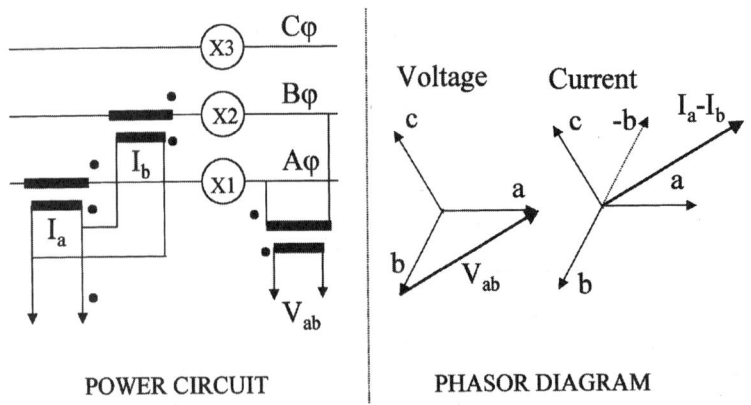

VT (L-L), 2 CTs on same phases

FIGURE 3.137(a,b,c) Phasing of voltage and current inputs.

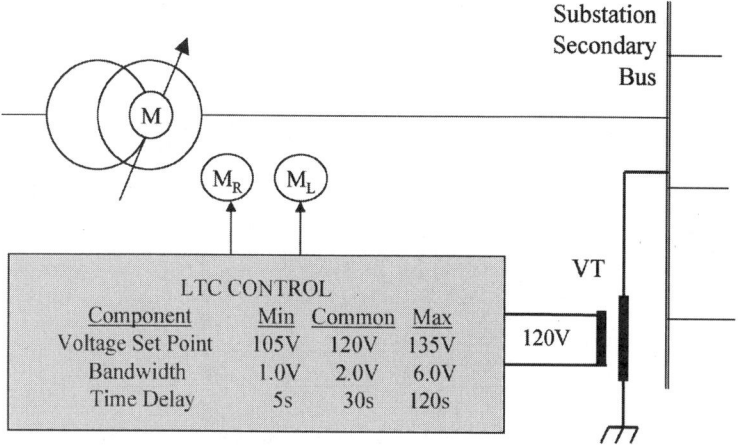

FIGURE 3.138 Control for voltage regulation of the bus.

basically determined by the LTC voltage change per step. Consider a transformer where the voltage change per step is nominally 0.75 V (5/8% of 120 V). Often this is only the average; the actual voltage change per step may differ appreciably at different steps. Clearly, the bandwidth must be somewhat greater than the maximum step change voltage as if the bandwidth were less than the voltage change per step, the voltage could pass fully across the band with a single step, causing a severe hunting condition. The minimum suggested bandwidth setting is twice the nominal step change voltage, plus 0.5 V, for a 2.0-V minimum setting for the most common 5/8% systems. Many users use somewhat higher bandwidths when the voltage is not critical and there is a desire to reduce the number of daily tap changes.

- **Time Delay.** All LTC controls incorporate an intentional time delay from the time the voltage is "out-of-band" until a command is given for tap changer action. Were it not for the control delay, the LTC would respond to short-lived system voltage sags and swells causing many unwanted and unnecessary tap changes. Most applications use a linear time delay characteristic, usually set in the range of 30 to 60 sec, although controls with inverse time delay characteristics are also available, where the delay is related inversely to the voltage digression from the set point.

Regulation of the Voltage at the Load

It is recognized that if it were easy to do so, the preferred objective would be to regulate the voltage at the load, rather than at the substation bus. The difficulty is that the voltage at the load is not commonly measured and communicated to the control; therefore, it must be calculated in the control using system parameters calculated by the user. Basically, the calculation involves determining the line impedance $(R + jX_L)$ between the substation and the load, the location of which is itself usually very nebulous.

The procedure used is that of Line Drop Compensation (LDC), that is the boosting of the voltage at the substation in order to compensate for the voltage drop on the line. The validity of the method is subject to much debate because of (1) the uncertainty of where to consider the load to be when it is in fact distributed, and (2) the inaccuracies encountered in determining the feeder line resistance and reactance.

The principle upon which LDC is based is that there is one concentrated load located a sufficient distance from the LTC transformer for the voltage drop in the line to be meaningful. Consider Fig. 3.139 which is similar to Fig. 3.138 with the addition of a load located remote from the substation and a current signal input to the control. The distance from the substation to the load must be defined in terms of the electrical distance, the resistance (R) and inductive reactance (X) of the feeder. The means of determining the line R and X values is available in the literature of most producers of the control equipment.

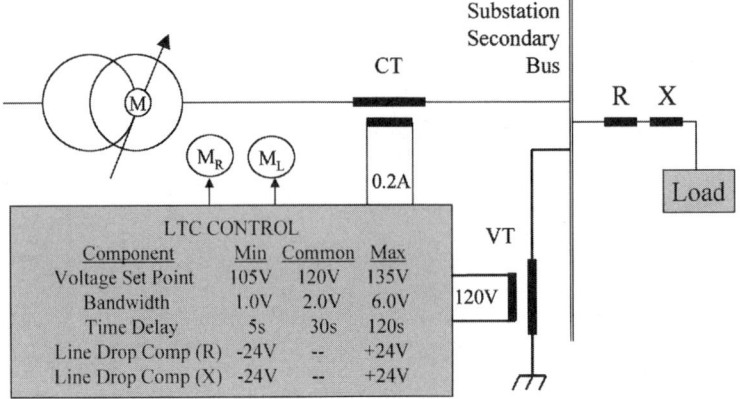

FIGURE 3.139 Control for voltage regulation at the load.

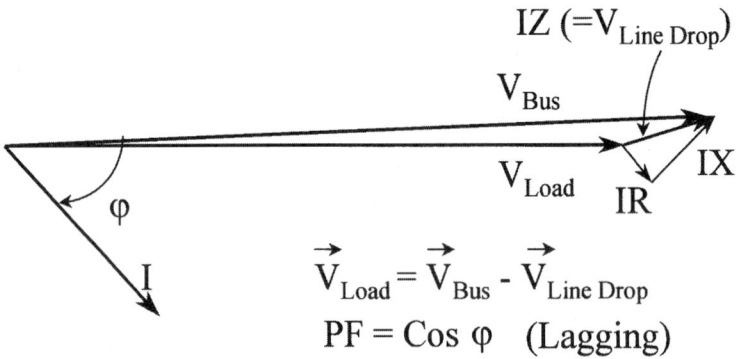

FIGURE 3.140(a) Load phasor diagram, normal load, lagging power factor.

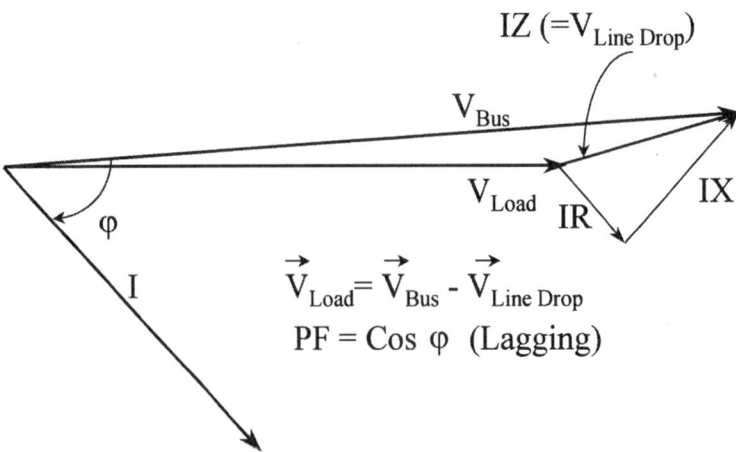

FIGURE 3.140(b) Load phasor diagram, heavy load, lagging power factor.

The LDC resistance and reactance settings are expressed as a value of volts on the 120-V base. These voltage values are the voltage drop on the line (R = in-phase component; X = quadrature component) when the line current magnitude is the CT primary rated current.

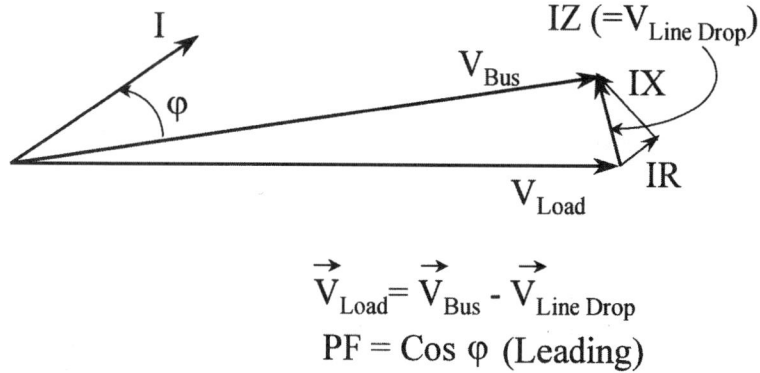

FIGURE 3.140(c) Load phasor diagram, normal load, leading power factor.

The manner in which the control accounts for the line voltage drop is illustrated with phasor diagrams. Figure 3.140 presents three illustrations showing the applicable phasor diagram as it changes by virtue of the load magnitude and power factor. In the illustrations, the voltage desired at the load, V_{Load}, is the reference phasor; its magnitude does not change. All of the other phasors will change when the load current changes in magnitude or phase angle. For all of the diagrams:

IR = voltage drop on the line due to line resistance; in-phase with the current.
IX = voltage drop on the line due to line inductive reactance; leads the current by 90°.
IZ = total line voltage drop, the phasor sum of IR and IX.

In the first illustration, the power factor angle, φ, is about 45°, lagging, for an illustration of an exaggerated power factor of about 0.7. It is seen that the voltage at the bus will need to be boosted to the value V_{Bus} in order to overcome the IR and IX voltage drops on the line. The second illustration simply shows that if the line current doubles, with no change of phase angle, the IR and IX phasors also double and a commensurately greater boost of V_{Bus} is required to hold the V_{Load} constant. The third illustration considers that the line current magnitude is the same as the first case, but the angle is now leading. The IZ phasor simply pivots to reflect the new phase angle. It is interesting to note that in this case, the V_{Load} magnitude *exceeds* V_{Bus}. This is modeling the real system: too much shunt capacitance on the feeder (excessive leading power factor) will result in a voltage rise along the feeder. The message for the user is that LDC accurately models the line drop, both in magnitude and phase.

No "typical" set point values for LDC can be given, unless it is zero, as the values are so specific to the application. Perhaps due to the difficulty in calculating reasonable values, line drop compensation is not used in many applications. An alternative to line drop compensation, Z-compensation is sometimes preferred for its simplicity and essential duplication of LDC. To use Z-comp, the control is programmed to simply raise the output voltage as a linear function of the load current, to some maximum voltage boost. This method is not concerned with the location of the load, but also does not accurately compensate for changes in the power factor of the load.

LTC Control with Power Factor Correction Capacitors

Many utility distribution systems include shunt capacitors to improve the load power factor as seen from the substation, and reduce the overall losses by minimizing the need for volt-amperes reactive (vars) from the utility source. This practice is often implemented both with capacitors that are fixed and others that are switched in response to some user selected criteria. Further, the capacitors may be located at the substation, at the load, or at any intermediate point.

The capacitors, located at the load and source, are illustrated in Fig. 3.141. The position of the capacitors is important to the LTC control when LDC is used.

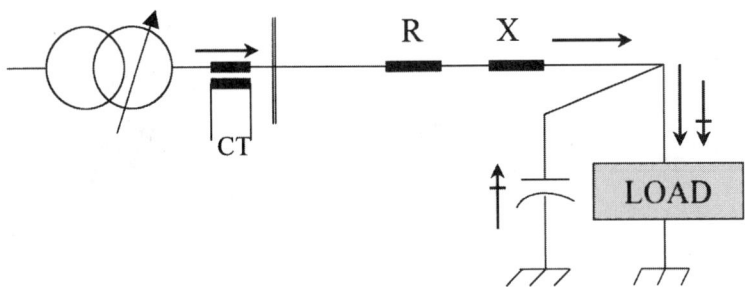

FIGURE 3.141(a)   Feeder with power factor correction capacitors, capacitors at the load.

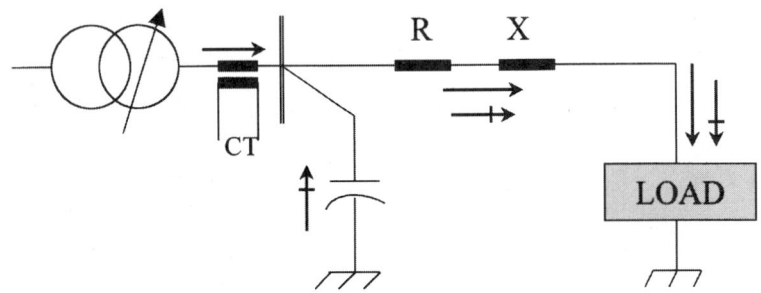

FIGURE 3.141(b)   Feeder with power factor correction capacitors, capacitors at the source bus.

With the capacitors located at the load, Fig. 3.141a, the transformer CT current is exactly the current of the line. Here LDC is correctly calculated in the control because the control LDC circuit accurately represents the current causing the line voltage drop.

In the second illustration, Fig. 3.141b, the capacitor is at the substation bus. There is a voltage drop in the line due to the reactive current in the line. This reactive current is not measured by the LDC CT. In this case, the line drop voltage is not correctly calculated by the control. To be accurate for this case, it is necessary to determine the voltage drop on the feeder due to the capacitor produced portion of the load current, and add that voltage to the control set point voltage to account for the drop not recognized by the LDC circuit.

The matter is further confused when the capacitor bank is switched. With the capacitor bank in the substation it is possible devise a control change based on the presence of the bank. No realistic procedure is recognized for the case where the bank is not in close physical proximity to the control, keeping in mind that the need is lessened to zero as the capacitor location approaches the load location.

Extended Control of LTC Transformers and Step-Voltage Regulators

An LTC control often includes much more functionality than is afforded by the five basic set points. Much of the additional functionality can be provided for analog controls with supplemental hardware packages, or is provided as standard equipment with the newer digital controls. Some functionality, most notably serial communications, is available only with digital controls.

Voltage Limit Control

Perhaps the most requested supplemental LTC control function is voltage limit control, also commonly known as "first house (or first customer) protection". This feature may be important with the use of line drop compensation, where excessive line current could result in the voltage at the substation bus becoming excessive. System realities are such that this means that the voltage at the "first house", or the load immediately outside of the substation, is also exposed to this high voltage condition.

Transformers

Review again Fig. 3.139 where LDC is used and it is desired to hold 118 V at the load. The voltage at the source (the secondary substation bus) is boosted as the load increases in order to hold the load at 118 V. If the load continues to grow, the voltage at the source will rise accordingly. At some point the load could increase to the point where the first customer is receiving power at excessive voltage. Recognize that the control is performing properly; it is because of the unanticipated excessive load on the system that the bus voltage is too high.

The Voltage Limit Control functions only in response to the actual local bus voltage, opening the Raise circuit to the drive motor at a user selected voltage, thereby defeating the basic control Raise signal when the bus voltage becomes excessive. In this way, the LTC action will not be responsible for first customer overvoltage.

These controls provide additional control if the bus voltage should further exceed the selected voltage set point cutoff voltage, as may be due to a sudden loss of load or other system condition unrelated to the LTC. If the voltage exceeds a second value, usually about 2 V higher than the selected voltage set point cutoff voltage, the voltage limit control will, of itself, command LTC Lower action.

Most of these controls provide a third capability, that of undervoltage LTC blocking so that the LTC will not run Lower if the voltage is already below a set point value. This function mirrors the Raise block function described above.

The voltage limit control functionality described is built into all of the digital controls. This is good in that it is conveniently available to the user, but caution needs to be made regarding its expected benefit. The function cannot be called a backup control unless it is provided as a physically separate control. Consider that the bus voltage is rising, not correctly because of LDC action, but because of a failure of the LTC control. The failure mode of the control may cause the LTC to run high and the bus voltage to rise accordingly. This high voltage will not be stopped if the Voltage Limit Control function used is that which is integral to the defective digital control. Only a supplemental device will stop this, and thereby qualify as a backup control.

Voltage Reduction Control

Numerous studies have reported that, for the short term, a load reduction essentially proportional to a voltage reduction can be a useful tool to reduce load and conserve generation during critical periods of supply shortage. It is very logical that this can be implemented using the LTC control. Many utilities prepare for this with the voltage reduction capability of the control.

With analog controls, voltage reduction is usually implemented using a "fooler" transformer at the sensing voltage input of the control. This transformer could be switched into the circuit using SCADA to boost the voltage at the control input by a given percentage, or often up to three different percentages using different taps on the fooler transformer. Having the sensed voltage boosted by, say, 5% without changing the voltage set point of the control will cause the control to run the tap position down by 5% voltage, accomplishing the desired voltage reduction. The percentage of three reduction steps, most commonly 2.5, 5.0, and 7.5%, is pre-established by the design of the fooler transformer.

Digital controls do the same function much more conveniently, more accurately, and more quickly. The voltage reduction applicable to steps 1, 2, and 3 are individually programmed, and upon implementation effectively lower the voltage set point. Control based on the new set point is implemented without intentional time delay, reverting to panel time delay after the voltage reduction has been implemented. Most often controls provide for three steps of voltage reduction, with each step individually programmed up to 10.0%.

Reverse Power Flow

Voltage regulators as used on distribution feeders are sometimes subjected to reverse power flow due to system switching, a situation treated in the step-voltage regulator section of this chapter.

The apparatus and procedures defined for step-voltage regulators are not correct for most transformer applications where reverse power flow can occur. The basic difference is that the feeder regulator application remains a radial system after the line switching is complete. Reverse power in transformers is more likely to occur on a system where the reverse power is due to a remote generator, which is operating continuously in parallel with the utility. The proper operation of the LTC in this case must be evaluated for the system. Perhaps the preferred operation would be to control the LTC so as to minimize the var exchange between the systems. Some systems are simply operated with the LTC control turned off of automatic operation during RPF so that the LTC stays fixed on position until FPF resumes.

Introduction to Control for Parallel Operation of LTC Transformers and Step-Voltage Regulators

For a variety of reasons, it may be desirable to operate LTC transformers or regulators in parallel with each other. This may be done simply to add additional load handling capability to an existing overloaded transformer, or it may be by initial design to afford additional system reliability anticipating that there may be a failure of one transformer.

Most common paralleling schemes have the end objective of having the load tap changers operate on the same, or on nearly the same, tap position at all times. For more complex schemes this may not be the objective. A knowledge of the system is required in order to assess the merits of the various techniques.

The Need for Special Control Considerations

To understand why special control consideration needs to be given to paralleling, consider two LTC transformers operating in parallel, i.e., the primary and secondary of the transformers are bussed together as in Fig. 3.142. If the transformers are identical, they will evenly divide the load between them at all times while they are operating on the same tap position. Consider that the voltage at the secondary bus goes out-of-band. Even if the controls are set the same, they are not *identical* and one will command tap change operation before the other. Later, when the voltage again changes so as to require a second voltage correction, the same control which operated more quickly the first time will be expected to do so again. This can continue indefinitely with one LTC doing all of the operation. As the tap positions of the LTC transformers digress, the current that circulates in the substation increases. This current simply circulates around the loop formed by the busses and the two transformers doing no useful work, but causing an increase in losses and perhaps causing one or both of the transformers to overheat. To put the matter in perspective, consider the case of a distribution substation where there are two 5/8% steps, 15 MVA LTC transformers of 8.7% impedance in parallel. The secondary voltage is 12.47 kV. For a tap difference of only 1 step, a circulating current of 25 A flows in the substation LV bus. This current magnitude increases linearly with tap position difference.

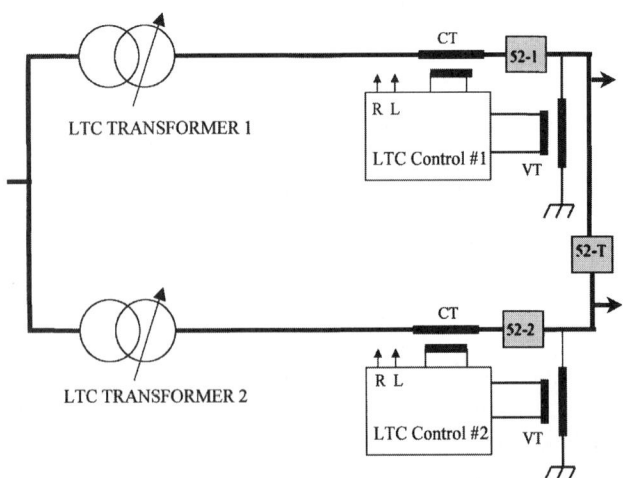

FIGURE 3.142 LTC transformers in parallel with no interconnections.

Transformers

In the illustration above for LTC transformers, the circulating current was limited by the impedances of the transformers. It is very important to recognize that the same procedure cannot be done with step-voltage regulators as the impedance of a regulator is very low at even the extreme tap positions, and may be essentially zero at the Neutral tap position. In this case, if one regulator is on Neutral and the other moves to position 1R, the circulating current will be expected to be sufficient to cause catastrophic failure of the regulators. *Step-voltage regulators can only be operated in parallel when there is adequate supplemental impedance included in the current loop.* This is most often the impedance of the transformers which are in series with each regulator bank, or may be a series current limiting reactor.

Instrument Transformer Considerations

Refer again to Fig. 3.137. Most techniques for paralleling use the voltage and current signals derived for line drop compensation; it is essential that the transformers which will be paralleled deliver voltage and current signals that are in phase with each other, when the system has no circulating current. This means that the instrument transformers must deliver signals from equivalent phases. Further, the ratios of the instrument transformers must be the same, except as very special conditions will require otherwise.

Defined Paralleling Procedures

Numerous procedures have been identified over the years to accomplish LTC transformer paralleling using electronic control. These are listed with some limited description, with alternative names sometimes heard:

1. Negative Reactance (Reverse Reactance): Seldom used today except in some network applications, this is one of the oldest procedures accomplished by other than mechanical means. This means of paralleling is the reason LTC controls are required by the standard to provide negative X capability on the line drop compensation.
    Advantages: Simplicity of installation. The system requires no apparatus other than the basic control, with LDC X set as a negative value. There is no control interconnection wiring so transformers may be distant from each other.
    Disadvantages: Operation is with a usually high −X LDC set point, meaning that the bus voltage will be *lowered* as the load increases. Attempts to compensate with +R LDC settings are only marginally successful.
2. Cross-connected Current Transformers: Unknown in practice today. The system operates on precepts similar to the Negative Reactance method. The LDC circuits of two controls are fed from the line CT of the opposite transformer.
    Advantages: System requires no apparatus other than the basic control, but does require CT circuits to pass between the transformers.
    Disadvantages: Operation may need to be with a value of +X LDC much higher than desired for LDC purposes, thereby boosting the voltage too much. The system may be used on two transformers only.
3. Circulating Current (Current Balance): The most common method in use in the U.S. today; about 90% of new installations use this procedure. It has been implemented with technical variations by several sources.
    Advantages: Generally reliable operation for any reasonable number of paralleled transformers. Uses the same CT as that provided for LDC, but operates independently of the line drop compensation.
    Disadvantages: Control circuits can be confusing, and must be accurate as to instrument transformer polarities, etc. Proper operation is predicated on the system being such that any significant difference in CT currents must be due only to circulating current. Matched transformers will, at times, operate unbalanced under normal conditions.

4. **Master/Follower (Master/Slave, Electrical Interlock, Lock-in-step):** Used by the 10% of new installations in the U.S. which do not use circulating current. It is much more commonly used worldwide than in the U.S.
   Advantages: Matched transformers will always be balanced, resulting in minimum system losses.
   Disadvantages: As usually implemented, involves numerous auxiliary relays which may fail, locking out the system.
5. **Reactive Current Balance (Delta VAR):** Generally used only when special system circumstances require it. Operates so as to balance the reactive current in the transformers.
   Advantages: Can be made to parallel transformers in many more complex systems where other methods do not work.
   Disadvantages: May be more expensive than more common means.

The two most common paralleling procedures are Master/Follower and Circulating Current Minimization.

Master/Follower

The Master/Follower method operates on the simple premise: Designate one control as the Master, any other units are Followers. Only the Master needs to know the voltage and need for a tap change. Upon recognition of such a need, the Master so commands a tap change of the LTC on the first transformer. Tap changer action of #1 activates contacts and relays, which make the circuit for LTC following action in all of the Followers, and temporarily lock out further action by the Master. The LTC action of the Followers in turn activates additional contacts and relays which free the Master to make a subsequent tap change when required.

Circulating Current Method

All of the common analog implementations of the circulating current method provide an electronic means (a "paralleling balancer") of extracting (1) the load current and (2) the circulating current from the total transformer current. These currents are then used for their own purposes, the load current to be the basis for line drop compensation and the circulating current to bias the control to favor the next LTC action which will tend to keep the circulating current to a minimum.

Consider Fig. 3.143. The balancers receive the scaled transformer current and divide it into the components of load current and circulating current. The load current portion of the transformer current is that required for line drop compensation. The circulating current portion is essentially totally reactive

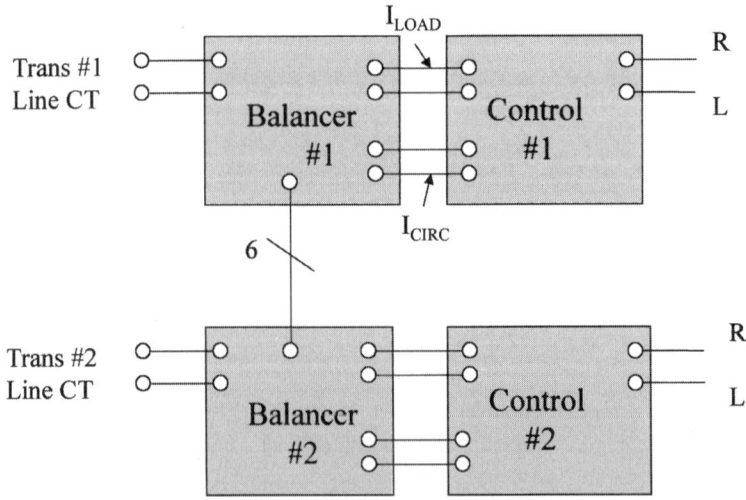

FIGURE 3.143 Control block diagram, paralleling by usual circulating current method.

Transformers

and is the same in magnitude in the two controls, but of opposite polarity. Presuming a lagging power factor load, the control monitoring the transformer on the higher tap position will "see" a more lagging current; that on the lower tap position a less lagging, or perhaps leading, current. This circulating current is injected into the controls. The polarity difference serves to bias them differently. The LTC which next operates is that which will correct the voltage, while tending to reduce the circulating current, i.e., bring the tap positions into closer relation to each other.

Characteristics Important for LTC Transformer Paralleling

There are many transformer characteristics which must be known and evaluated when it is planned to parallel LTC transformers. Some of the more notable follow:

1. Impedance and MVA. "Impedance" as a criteria for paralleling is more correctly stated as the percent impedance referred to as a common MVA base. Two transformers of 10% impedance but one of 10 MVA and the other of 15 MVA are not the same. Two other transformers, one 10% impedance and 10 MVA and another of 15% impedance and 15 MVA are suitably matched per this criteria. There is no definitive difference in the impedances which will be the limit of acceptability, but a difference of no greater than 7.5% is realistic.
2. Voltage rating and turns ratio. It may not be essential that the voltage ratings and turns ratios be identical. If one transformer is 69-13.8 kV and the other 69-12.47 kV, the difference may be tolerated by recognizing and accepting a fixed step tap discrepancy, or it may be that the ratios may be made more nearly the same using the de-energized tap changers.
3. Winding configuration. The winding configuration, as delta–wye, wye–wye, etc., is critical, yet transformers of different configurations may be paralleled if care is taken to assure that the phase shift through the transformers is the same.
4. Instrument transformers. The transformers must have VTs and CTs which produce in-phase signals of the correct ratio to the control, and must be measuring the same phase in the different transformers.

Paralleling Transformers with Mismatched Impedance

Very often it is desired to use two existing transformers in parallel where it is recognized that the impedance mismatch is greater than that recommended for proper operation. This can usually be accomplished, although some capacity of one transformer will be sacrificed.

If the impedances of the transformers in parallel are not equal, the current will divide inversely as the impedances in order for the same voltage to appear across both impedances.

The impedances are effectively the transformer impedance as read from the nameplate, which may be taken to be wholly reactive. The problem when dealing with mismatched transformers in parallel is that the current will divide per the impedances, but the control, if operating using the conventional circulating current method, is attempting to match the currents.

Realize that the LTC control and the associated paralleling equipment really has no knowledge of the actual line current; it knows only a current on its scaled base which could represent anything, say 100 A to 3000 A. The objective of the special considerations to permit the mismatched impedance transformers to be paralleled is to supply equal current signals to the controls when the transformers are carrying load current in inverse proportion to their impedances.

To illustrate, consider the paralleling of two 20 MVA transformers of 9% and 11% impedance, which is much more than the 7.5% difference criteria stated earlier. We establish that $Z_1 = k \times Z_2$, so:

$$k = Z_1/Z_2 = 9/11 = 0.818.$$

And since $I_1 = I_2/k$:

$$I_1 = I_2/k = 1.222\ I_2.$$

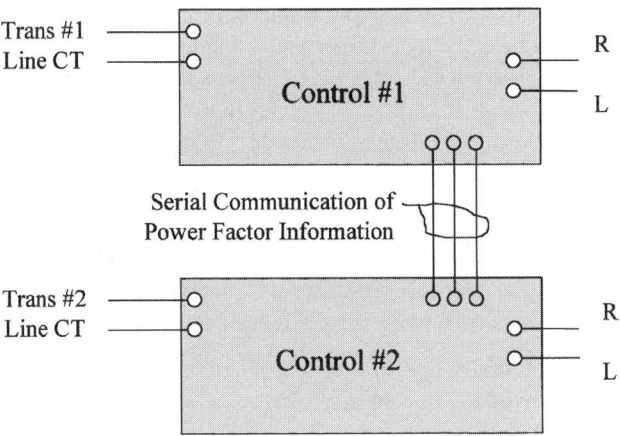

FIGURE 3.144 Control block diagram, paralleling by equal power factor, circulating current method.

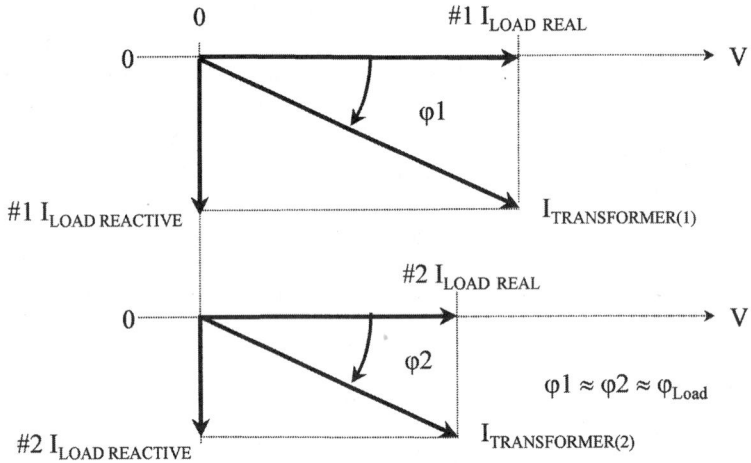

FIGURE 3.145 Phasor diagram — mismatched transformers in parallel by equal power factor method.

Transformer T1 will carry 22.2% more load than transformer T2, even though they are of the same MVA rating. A resolution is to fool the controls to act as though the current is balanced when, in fact, it is mismatched by 22%.

A solution is found by placing a special ratio auxiliary CT in the control current path of T2, which boosts that current by 22%. This will cause both controls to see the same current when, in fact, T1 is carrying 22% more current.

In this way the controls are fooled into thinking that the load is balanced when it is actually unbalanced due to the impedance mismatch. Transformer T2 has effectively been derated by 22% in order to have the percent impedances match.

The circulating current paralleling described above is that commonly used in the U.S. Another basis for implementing circulating current paralleling is now available. The new scheme does not require the breakout of the circulating current and the load current from the total transformer current. Rather, the procedure is to recognize the apparent power factors as seen by the transformers and act so as to make the power factors be equal. The control configuration used is as shown in Fig. 3.144, where the phasor diagram, Fig. 3.145, shows typical loading for mismatched transformers. The fundamental difference in this manner of circulating current paralleling is that the principle involves the equalization of the apparent power factors as seen by the transformers, i.e., the control acts to make $\varphi 1 \approx \varphi 2$. The benefit of this subtle difference is that it is more amenable to use where the transformers exhibit mismatched impedance.

3.11 Loading Power Transformers
Robert F. Tillman, Jr.

Design Criteria

ANSI Standards Collection C57.12.99-1993 sets forth the general requirements for the design of power transformers. With respect to loading, the requirements of concern are those that define the transformer's thermal characteristics. The average ambient temperature of the air in contact with the cooling equipment for a 24-h period shall not exceed 30°C (86°F) and the maximum shall not exceed 40°C (104°F) (IEEE, 1994). For example, a day with a high temperature of 96°F and a low of 76°F has an approximate average ambient temperature of 86°F.

The average winding temperature rise above ambient shall not exceed 65°C for a specified continuous load current (IEEE, 1994). The industry uses this criteria because manufacturers can obtain it by measuring the resistance of the windings during temperature rise tests. The manufacturer must guarantee to meet this requirement.

The hottest conductor temperature rise (hot spot) in the winding shall not exceed 80°C under these same conditions (IEEE, 1994). This 80°C limit is no assurance that the hot spot rise is 15°C higher than the average winding temperature rise. In the past, using the 15°C adder to determine the hot spot rise produced very conservative test results. The greatest level of thermal degradation occurs at this location. The hot spot location is near the top of the high or low voltage winding. The most common reason for hot spots to occur is that these regions have higher localized eddy losses because the leakage flux fringes radially at the winding ends.

ANSI Standards Collection C57.12.99-1993 paragraph 4.1.6 specifies requirements for operation above rated voltage. At no load, the voltage shall not exceed 110% (IEEE, 1994). At full load, the voltage shall not exceed 105% (IEEE, 1994). Modern transformers are usually capable of excitation beyond these limits without causing saturation of the core. However, this causes the core to contribute greater than predicted heating. Consequently, all temperature rises will increase. While not critical at rated load, this becomes important when loading transformers past their rating criteria.

Nameplate Ratings

The transformer nameplate provides the voltage (excitation) rating for all taps. If a transformer has a 4-step de-energized high voltage tap changer and a 32-step LTC, then it has 5 primary voltage ratings and 33 secondary voltage ratings. The excitation limits apply to all ratings.

The cooling class affects the design of the cooling package. Different cooling classes have different thermal profiles. Therefore, gauge readings can be different for equivalently rated transformers under the same loading conditions. Consequently, the operator must understand the cooling classes and their thermal profiles in order to confirm that a given transformer is responding thermally as it should.

OA cooling uses natural air flow through the radiators to dissipate heat. FA cooling uses fans to force air through radiators in order to substantially increase the rate of cooling. There is natural convective flow of the oil in the radiators from top to bottom with both OA and FA cooling. FOA non-directed flow (NDF) cooling uses fans, as with FA, along with pumps to draw the hot oil out of the top of the transformer and force it through the radiators into the bottom of the tank. The most common modern transformer found in service with this cooling class is General Electric medium power transformers (12 MVA and larger) manufactured in Rome, Georgia. FOA directed flow also uses fans and pumps. However, this design incorporates a cooling duct that directs the oil from the radiator outlet directly through the winding from bottom to top (Pierce, 1993).

Medium power transformers, 12 MVA and larger, have three stages of cooling, OA/FA/FA or OA/FOA/FOA. Most are OA/FA/FA. Large power transformers are usually OA/FOA/FOA directed flow cooling. Some large power transformers, especially GSU transformers, are FOA only. They have no rating unless the auxiliary cooling equipment is operating.

The nameplate kVA (or MVA) rating is the continuous load at rated voltage that produces an average winding temperature rise within the 65°C limit and hot spot temperature rise within the 80°C limit (IEEE, 1994). The rating of each cooling stage is given. The transformer design must meet temperature rise guarantees for each cooling stage. Operators and planners generally express loading capability as a percent (or per unit) of the maximum MVA rating.

The windings contribute most of the heat that produces the temperature rises. The winding heat results from the "I squared r" loss in the conductor. When users express loading as a percent on an MVA basis, they are assuming that the transformer is operating at rated voltage, both primary and secondary. Operators should express the load for thermal calculations as a percent (or per unit) of the full load current for a given set of primary and secondary tap positions (Tillman, 1998).

Other Thermal Characteristics

Other thermal characteristics are dependent on the design, loss optimization, and cooling class. Top oil temperature is the temperature of the bulk oil in the top of the tank. Users can directly measure this with a gauge mounted on the transformer tank. The temperature of the oil exiting the oil duct in the top of the winding is actually the important characteristic (Chew, 1978; Pierce, 1993). Users normally do not have a direct measurement of this in practice. Manufacturers can measure this during heat run tests but normally they do not.

Bottom oil rise is the temperature rise above ambient of the oil entering the bottom of the core and coil assembly. In all cooling classes, the bottom oil temperature is approximately equal to the temperature of the oil exiting the radiators (Chew, 1978; Pierce, 1993). Manufacturers can easily measure it during heat run tests and provide the information on certified test reports. Users should require manufacturers to provide this information. In service, users can measure the bottom oil temperature with an infrared camera.

Average oil rise is the key oil thermal characteristic needed to calculate the winding gradient (Chew, 1978). It is approximately equal to the average of the temperature rise of oil entering the bottom of the core and coil assembly and the oil exiting the top. The location of the average oil is approximately equal to the location of the average winding.

The winding gradient is the difference between the average winding and average oil temperature rises (Chew, 1978). Designers assume that this gradient is the same from the bottom to the top of the winding except at the ends where the hot spot exists. The hot spot gradient is somewhat higher than the winding gradient. Designers calculate this gradient by determining the effects of localized eddy losses at the end of the winding. Computers allow designers to calculate this very accurately. An engineer can obtain an estimate of the hot spot gradient by multiplying the winding gradient by 1.1 (Chew, 1978).

Thermal Profiles

Thermal profiles show the relationships between different temperatures inside a transformer. Operators must understand these thermal profiles for given cooling classes in order to understand a transformer's thermal response to a given load. A plot of winding temperature vs. its axial position in the tank is a thermal profile.

For FA cooling, the oil temperature outside the winding is approximately equal to the oil temperature inside the winding from bottom to top (Chew, 1978). This is due to the oil movement by natural convection both inside and outside the core and coil assembly. The oil temperature exiting the bottom of the radiators approximately equals that entering the bottom of the core and coil assembly (Chew, 1978; Pierce, 1993). The top bulk oil temperature approximately equals the oil temperature at the top of the winding (IEEE, 1995). It is relatively easy for engineers to determine the gradients for this cooling class. They can verify that a transformer is responding properly to a given load by obtaining load information, reading the top oil temperature gauge, measuring the bottom oil temperature by use of an infrared camera, and comparing with calculations and data from the manufacturer's test reports.

For FOA NDF cooling, the oil pumps force oil into the bulk oil at the bottom of the tank. The higher rate of flow produces a small bottom to top temperature differential in the coolers. The oil flow in the winding is by natural convection. The pumps circulate cool oil throughout the tank and around the core and coil assembly while only a small portion passes through the windings. Therefore, the temperature profile of the oil internal to the windings does not match that of the external profile. The top winding duct oil, the key temperature, is much hotter than the top bulk oil (Pierce, 1992; 1993).

For FOA directed flow cooling, the oil pumps force oil through a sealed oil duct directly into the bottom of the core and coil assembly. The oil flows within the winding and exits the top of the winding into the bulk oil. The temperature profile of the oil external to the winding approximately equals that of the oil internal to the winding. The bottom to top oil temperature rise is very small. Top oil temperature rises for this design are lower than equivalent FA designs. The hot spot temperature is not lower because the hot spot gradient is much larger (Pierce, 1992; 1993).

Temperature Measurements

The user specifies the gauges and monitors installed by the manufacturer. The top oil temperature gauge directly measures the temperature of the top bulk oil. Most oil gauges in service are mechanical and provide instantaneous and maximum temperature readings. They also have contacts for controlling cooling and providing alarms. Some of these gauges are electronic and provide analog outputs for SCADA. This gauge reading is approximately equal to the temperature of the top winding duct oil for FA and FOA directed flow cooling and is much lower for FOA non-directed flow cooling (IEEE, 1995; Chew, 1978; Pierce, 1992; 1993).

The winding temperature gauge is actually an assembly that simulates the hot spot winding temperature. This hot spot temperature simulator measures top bulk oil and adds to it another temperature increment (hot spot gradient) proportional to the square of the load current. This incremental temperature is actually a simulation of the hot spot gradient. An oil probe inserted in a resistive well and a current obtained from a BTCT in the main winding is a method used for simulating the hot spot temperature. Designers calibrate the resistor and the current to the hot spot gradient. For FA and FOA directed flow cooling, this simulation is valid because the measured top oil temperature is approximately equal to the top winding duct oil temperature (IEEE, 1995; Chew, 1978). However, this simulation is not valid for FOA non-directed flow cooling because the measured top oil temperature is less than the top winding duct oil temperature (Pierce, 1992). It is more difficult to predict and verify the thermal response of a transformer with FOA non-directed flow cooling.

The hot spot simulator is accurate only for steady state constant loads because there is a thermal lag between the top oil and hot spot winding temperatures. A given load takes several hours to heat up the bulk oil to its ultimate temperature. On the other hand, the same load takes only a few minutes to heat up the winding conductors to their ultimate temperature. Consequently, a transformer must have an applied constant load for many hours for the hot spot simulator to indicate an accurate measurement. For transient loading, the instantaneous indication is meaningless to an operator. The actual hot spot temperature is greater for increasing loads than the simulator indicates and lower for decreasing loads (IEEE, 1995).

Users also specify electronic gauges. These have many features built into one box. They include many adjustable alarm and control contacts, both top oil and simulated hot spot temperatures, and analog outputs for SCADA. This monitor creates the hot spot simulation by routing the current directly to the monitor. Designers calibrate a microprocessor in the monitor to calculate the hot spot gradient. This provides a more reliable indication than mechanical gauges but is still a simulation with the same drawbacks discussed above.

Direct measurements of the hot spot temperature are possible through application of fiber optics. A sensor inserted in the oil duct between two winding discs replaces a key-spacer. Design engineers determine the locations to install the sensors. Fiber optic probes transmit the temperature from the sensors to a storage device (PC). These applications are relatively unreliable. Designers typically call for many probes in order to have a few that work properly. The application of fiber optics also produces additional risk of internal dielectric failure in the transformer. While fiber optics is a good research tool and appropriate for special applications, its general application is usually not worth the added risk and cost (Pierce, 1993).

Predicting Thermal Response

The IEEE Standards C57.91-1995 Guide for Loading Mineral-Oil-Immersed Transformers gives detailed formulas for calculating oil and winding temperatures. It gives formulas for constant steady state and transient loading. Users apply the transient loading equations in computer programs. Computer simulations are good for gaining understanding but are not necessary for most situations.

Users must calculate oil temperatures in order to predict a transformer's thermal response. The total losses in a transformer cause the oil temperature to rise for a given load (IEEE, 1995). Total losses include core, load, stray, and eddy losses. The oil temperatures vary directly with the ratio of the total losses raised to an exponent. The industry designation for the oil rise exponent is "n" (IEEE, 1995; Chew, 1978; Pierce, 1993).

$$TO_2 = TO_1(TL_2/TL_1)^n \tag{3.64}$$

where:

TO_2 = ultimate top oil rise
TO_1 = initial or known top oil rise
TL_2 = total losses at ultimate load
TL_1 = total losses at known load

Users can calculate a conservative approximation by neglecting the core, stray, and eddy losses and considering only the "i^2r" load losses:

$$TO_2 = TO_1(i_2^2 r/i_1^2 r)^n = TO_1(i_2^2/i_1^2)^n$$

$$TO_2 = TO_1(i_2/i_1)^{2n} \tag{3.65}$$

The value of "n" varies between transformer designs. The industry generally accepts an approximate value of 0.9 for FA class transformers (IEEE, 1995; Chew, 1978; Pierce, 1993). This is a conservative value.

$$TO_2 = TO_1(i_2/i_1)^{1.8} \tag{3.66}$$

If i_1 is the full load current rating, then i_2/i_1 is the per unit current loading. As an example, consider a transformer with FA cooling that has a top oil temperature rise of 55°C at full load current. Calculate the ultimate steady state top oil temperature rise due to a constant 120% load:

$$TO_2 = 55(1.2)^{1.8} = 76°C$$

If the maximum ambient temperature is 95°F (35°C), then the ultimate top oil temperature is 111°C. The user can find detailed transient equations in IEEE C57.91-1995. They appear complex but are only expanded forms of the above equations. The same basic principles apply. However, their use requires many incremental calculations summed over given time periods. There are computer programs available that apply these equations.

Transient analysis has the most value when actual temperature profiles are available. Except for special applications, this is often not the case. Usually, the temperature available is the maximum for a given time period. Furthermore, engineers do not always have time available to analyze and compare computer outputs vs. field data. The key issue is that the engineer must have the ability to predict a transformer's thermal response to a given load (steady state or transient) in order to ensure that the transformer is responding properly. They can use the above simplified steady state equations to calculate maximum temperatures reached for a given load cycle. Using a load equal to 90% of a peak cyclical load in the steady state equations yields a good approximation of the peak temperature for an operator to expect. In the above example, the resultant expected top oil temperature for a cyclical 120% load is 98°C.

The relationships for winding temperature equations are similar to the oil temperature equations. The key element is the hot spot gradient. Once an engineer calculates this gradient, he/she adds it to the calculated top oil temperature to obtain the hot spot temperature. Similar to Eq. (3.65), the hot spot gradient varies directly with the ratio of the current squared raised to an exponent "m" (IEEE, 1995; Chew, 1978; Pierce, 1993).

$$HSG_2 = HSG_1(i_2/i_1)^{2m} \qquad (3.67)$$

where:

HSG_2 = ultimate hot spot gradient
HSG_1 = initial or known hot spot gradient
i_2 = current at the ultimate load
i_1 = current at the known or initial load

As with "n", the values of "m" vary between transformers. However, the variance for "m" is much greater. It is very dependent on the current density and loss optimization used to design a particular transformer. High losses and current densities produce large values of "m". Low losses and current densities produce small values for "m". For example, a high loss transformer design may have a lower hot spot temperature at rated load than a similar low loss transformer design. However, the low loss transformer is likely more suitable for loading past its nameplate criteria. The industry generally accepts an approximate value of 0.8 for FA class transformers (IEEE, 1995; Chew, 1978; Pierce, 1993). Again, this is a conservative value:

$$HSG_2 = HSG_1(i_2/i_1)^{1.6} \qquad (3.68)$$

Engineers can calculate the "m" and "n" constants by requiring manufacturers to perform overload temperature tests (i.e., 125%). This, along with temperature data from standard temperature tests, provides two points for calculating these values.

There are limiting elements in a transformer other than the active part. These include bushings, leads, tap changers, and BTCTs. Sometimes it is difficult to determine the bushing rating in older transformers. Infrared scanning can help determine if a bushing is heating excessively. In modern transformers, manufacturers choose bushings with ample margin for overloading the transformer. Insulated leads are also subject to overheating, especially if the manufacturer applies too much insulation. LTC contact life can accelerate. For arcing-under-oil tap changers, oil contamination increases. It is also important to know if a transformer has a series (booster) winding. The user will probably not know the ratio of the series transformer for older transformers.

Load Cyclicality

Past practices called for loading power transformers to a percentage of their nameplate kVA or MVA rating. A typical loading criteria was 100%. The measured load used to determine the percent loading was the annual peak 15-min integrated output demand in kVA or MVA. Few users considered the cyclicality of the load or the ambient temperature. These thermal cycles allow operators to load transformers beyond their nameplate rating criteria for temporary peak loads with little or no increased loss of life or probability of failure.

To completely define the load profile, the user must know the current for thermal calculations and the power factor for voltage regulation calculations as a function of time, usually in hourly increments. Cyclical loads generally vary with temperature due to air conditioning and heating. However, summer and winter profiles are very different. Summer peak loads occur when peak temperatures occur. The reverse occurs in winter. In fall and spring, not only is the ambient temperature mild but there is little or no air conditioning or heating load. The thermal stress on a transformer with a cyclical load during fall and spring is negligible. It is also usually negligible during winter peak loading.

If ambient conditions differ from the nameplate criteria, then the user must adjust the transformer capability accordingly. IEEE C57.91-1995 provides tables and equations for making these adjustments. A good approximation is an adjustment of 1% of the maximum nameplate rating for every degree C above or below the nameplate rating (IEEE, 1995). If the transformer operates in 40°C (104°F) average ambient, then the user must de-rate the nameplate kVA by 10% in order to meet the nameplate thermal rating criteria. Conversely, operating in a 0°C (32°F) average ambient environment allows the user to up-rate the transformer by approximately 30%.

Twenty-four hour summer load cycles resemble a sine wave. The load peak occurs approximately when the temperature peak occurs, 3 to 5 PM. The load peak lasts less than 1 h. The load valley occurs during predawn hours when ambient temperature minimums occur, 4 to 6 AM. The magnitude of the valley load is 50 to 60% of the peak load. The user can calculate the top oil temperature at the valley in the same manner as at the peak. The value of making these calculations is to verify that a particular loaded transformer response is correct. If the transformer responds properly, then it meets one of the criteria for recommending overload capability.

Science of Transformer Loading

The thermal rating of a power transformer differs from the thermal rating of other current carrying elements in a substation. Examples of other elements are conductor, bus-work, connectors, disconnects, circuit breakers, etc. The insulation system for these elements is air and solid support insulators. The cooling system is passive (ambient air). The thermal limits depend on the properties of the conductor itself. These elements are maximum rated devices. In a power transformer, the cooling system is active. The thermal limits depend on the dielectric and mechanical properties of the cellulose and oil insulation system. As a maximum rated device, the transformer capability is 200% of the maximum nameplate rating (IEEE, 1995).

With overly conservative loading practices, cellulose insulation life due to thermal stress is practically limitless, theoretically more than 1000 years. In practice, deterioration of accessories and non-active parts limits the practical life of the transformer. These elements include the tank, gauges, valves, fans, radiators, bushings, LTC, etc. The average, practical life of a transformer is probably 30 to 50 years. Many fail beyond economical repair before 30 years. Therefore, it is reasonable and responsible to allow some thermal aging of the cellulose insulation system. The key is to identify the risk. Loss of insulation life is seldom the real risk. Most of the time, risks other than loss of insulation life are the limiting characteristics.

Paper insulation must have mechanical and dielectric strength. The mechanical strength allows it to withstand forces caused by through faults. When through faults occur, the winding conductors try to move. If the paper has sufficient mechanical strength, the coil assemblies will also have sufficient strength to withstand these forces. When aged paper loses its mechanical strength, through faults will cause excessive winding movement. Physical damage due to excessive winding movement during through fault conditions reduces the dielectric withstand of the paper insulation. At this point, the risk of dielectric failure is relatively high.

Arrhenius reaction equation is the basic principle for thermal aging calculations (Kelly et al., 1988a; Dakin, 1948):

$$L = Ae^{(B/T)} \tag{3.69}$$

"L" is the calculated insulation life in hours. "A" and "B" are constants that depend on the aging rate and end-of-life definition. They also depend on the condition of the insulation system. "T" is the absolute temperature in degrees Kelvin (degrees C + 273) (Kelly et al., 1988a). There is a limited amount of functional life test data available. Most available data is over 20 years old. Past calculations of life expectancy used extremely conservative numbers for the "A" and "B" constants (Kelly et al., 1988a).

Normal life of cellulose insulation is the time in years for a transformer operated with a constant 110°C hot spot winding temperature to reach its defined end of life criteria. DP (degree of polymerization) and tensile strength are properties that quantify aging of cellulose paper insulation (Kelly et al., 1988b). In the past, the industry accepted approximately 7½ years for normal life. This is a misleadingly low value. Actual practice and more recent studies show that normal life under these conditions is somewhere between 20 years and infinity (IEEE, 1995). It is not important for users to master these relationships and equations. However, the user should understand the following realities:

1. The relationship between life expectancy and temperature is logarithmic (McNut, 1995). As the hottest spot conductor temperature moves below 110°C, the life expectancy increases rapidly and vice versa. An accepted rule of thumb is the life doubles for every 8°C decrease in operating temperature. It halves for every 8°C increase in operating temperature (Kelly et al., 1988a).
2. Users seldom operate transformers such that the hot spot winding temperature is above, at, or anywhere near 110°C. Even when peak loads cause 140°C hot spot winding temperatures, the cumulative time that it operates above 110°C is relatively short; it is probably less than 200 to 400 h per year.
3. Cellulose aging is a chemical reaction. As in all reactions, heat, water, and oxygen act as a catalyst. In the Arrhenius equation, the "A" and "B" constants are highly dependent on the presence of moisture and oxygen in the system (Kelly et al., 1988b). The expected life of the paper halves when the water content doubles (Bassetto et al., 1997; Kelly et al., 1988b). Also, tests indicate that the aging rate increases by a factor of 2.5 when oxygen content is high (Bassetto et al., 1997; Kelly et al., 1988b).

Water in Transformers Under Load

The action of water in the insulation system of power transformers poses one of the major risks in loading transformers past their nameplate rating criteria. The insulation system inside a transformer consists of cellulose paper and oil working together. Water always exists in this system. The paper must have some moisture content in order to maintain its tensile strength (Kelly et al., 1988b). However, the distribution of the water in this system is uneven. The paper attracts much more water than the oil (Kelly et al., 1988c). As the transformer cycles thermally throughout its life, the water redistributes itself. The water will collect in the coldest part of the winding (bottom disks) and in the area of highest electrical intensity (Kelly et al., 1988c). This redistribution of moisture is very unpredictable.

When a transformer is hot due to a heavy load, water moves from the paper to the oil. The heat in the conductor pushes the water out of the paper insulation while the solubility of water in oil increases with temperature (Kelly et al., 1988b). When the transformer cools, water moves back to the paper. However, the water goes back into the paper much more slowly than it is driven out (Kelly et al., 1988b). As the oil cools, the water in the oil can approach saturation.

As long as changes take place gradually and temperatures do not reach extreme levels, the insulation system can tolerate the existence of a significant amount of water (Kelly et al., 1988b). However, loading transformers to a higher level causes higher temperatures and greater changes during the thermal cycling. Emergency loading causes greater temperature swings than normal loading. The key is to understand what is an excessive temperature and when it will occur. That is the point at which elevated temperatures introduce the risk of failure. That temperature level depends on the moisture content in the insulation system. There are two basic risks associated with the relationship between loading and water in the insulation system; they are reduction in dielectric strength due to saturation of moisture in oil, and bubble evolution.

Dielectric Effects of Moisture in Oil

The dielectric strength of oil is a function of the average oil temperature and percent saturation of water in the oil (Moore, 1997). As long as the oil is hot, the water solubility is high (Kelly et al., 1988b). Therefore, a given amount of water in the oil produces a lower percent saturation at high temperature than at low temperature. As an example, consider a transformer with 1.5% water in the paper. A 100% load on a hot summer day produces a 70°C average oil temperature. The water content of the oil reaches 20 ppm as the moisture is driven from the paper into the oil (Kelly et al., 1988c). The water saturation at this point is 220 ppm, resulting in less than 10% saturation (Kelly et al., 1988b). The dielectric breakdown of the oil as measured by ASTM D1816 method is quite high, approximately 50 kV (Moore, 1997). In the evening, the load and temperature decrease, producing an average oil temperature of 50°C. The saturation level of water in the oil reduces to approximately 120 ppm (Kelly et al., 1988b). Since the oil goes back into the paper slowly, the water in oil is still 20 ppm resulting in almost 20% saturation, which is still quite low. The dielectric breakdown is somewhat lower, approximately 45 kV (Moore, 1997). This is still quite high and poses no real problem.

Now consider the same example where the peak load increases to 120%. The resultant peak average oil temperature is 90°C causing the moisture content of the oil to reach 60 ppm (Kelly et al., 1988c). The percent saturation at peak temperature is still less than 10% (Kelly et al., 1988b). However, during the evening the average oil temperature is 55°C. The water saturation level is 140 ppm, greater than 40% saturation (Kelly et al., 1988b). The dielectric breakdown reduces significantly to approximately 35 kV (Moore, 1997).

The conditions in this example should not cause problems, especially if the transformer in question is relatively new. However, the insulation systems in service aged transformers have properties that magnify the risk illustrated in this example (Oommen et al., 1995). Severe emergency loading can produce higher risk levels, especially if they occur in winter. At 0°C, oil saturates at approximately 20 ppm (Kelly et al., 1988b). There is a point where the probability of failure increases to an unacceptable level.

Bubble Evolution

Bubble evolution is a function of conductor temperature, water content by dry weight of the paper insulation, and gas content of the oil (Oommen et al., 1995). Gas bubbles in a transformer insulation system are of concern because the dielectric strength of the gases is significantly lower than that of the cellulose insulation and oil. When bubbles evolve, they replace the liquid insulation. At this point, the dielectric strength decreases and the risk of dielectric failure of the major and minor insulation systems increases.

Bubbles result from a sudden thermal change in the insulation at the hottest conductor. In a paper insulation system, there is an equilibrium state between the partial pressures of the dissolved gases in the paper. As temperature increases, the vapor pressure increases exponentially. When the equilibrium balance tips, the water vapor pressure causes the cellulose insulation to suddenly release the water vapor as bubbles (Oommen et al., 1995).

According to past studies, moisture content is the most important factor influencing bubble evolution. The temperature at which bubbles evolve decreases exponentially as the moisture content in the cellulose insulation increases. Increasing content of other gases also significantly influences bubble evolution when high moisture content exists. Increasing content of other gases does not significantly influence bubble evolution at low moisture content (McNut, 1995). Data show that in a dry transformer (less than 0.5% moisture by dry weight) bubble evolution from overload may not occur below 200°C. A service aged transformer with 2.0% moisture may evolve bubbles at 140°C or less (Oommen et al., 1995).

Voltage Regulation

Voltage regulation is the voltage drop through a transformer's impedance. It determines the output voltage at the load side terminals of a power transformer. If output voltage is too low, then regulating equipment cannot provide adequate voltage control to the system it serves. Voltage regulation also affects the thermal response of a transformer (Tillman, 1998).

A transformer load flowing through its impedance produces voltage drop. In order to calculate this drop, an engineer must know the load current and power factor in addition to the impedance of the transformer. The impedance is "r + jx" (usually expressed in percent at OA rating and rated voltage). The load is "I∠ – θ" where "I" is the magnitude of the load current and "θ" is the inverse cosine of the power factor (p). Voltage regulation is the magnitude of the input voltage (V_{in}) minus the magnitude of the output voltage (V_{out}) divided by the magnitude of the output voltage (Bean et al., 1959):

$$\text{Regulation} = \{|V_{in}| - |V_{out}|\}/|V_{out}| \quad (3.70)$$

To calculate the voltage drop, one must solve the following equation (Bean et al., 1959):

$$V_{in} = \{I\angle - \theta\}\{r + jx\} + V_{out} \quad (3.71)$$

Letting V_{out} equal 1∠0, solving Eq. (3.71) for V_{in}, and plugging into Eq. (3.70), one derives the following regulation equation in per unit values:

$$\text{Regulation} = \text{SQRT}\left[I^2(r^2 + x^2) + 2I(pr + qx) + 1\right] - 1 \quad (3.72)$$

where

    I = load current in per unit of the OA load current
    p = load power factor in per unit
    q = load reactive factor in per unit; positive for lagging and negative for leading
    r = load loss resistance in per unit at the OA rating
    x = leakage reactance in per unit at the OA rating

For illustration, consider an example where the load is 120% of the maximum nameplate rating (2 per unit of the OA rating), power factor is .90 lagging (p = .90 and q = .45), and impedance is .005 + j.105 per unit. The calculated voltage drop through the transformer is approximately 12%. Tap changing equipment will have to operate at or near maximum raise in order to compensate for this large voltage drop. Further loading in this example is impossible unless users improve the power factor by providing Varr support.

Planners use measured output MVA and a transformer's maximum nameplate MVA rating in order to calculate the percent (or per unit) loading. However, the loading used in Eq. (3.72) is current. The per unit MVA load only equals the per unit current load when the actual output voltage equals the rated output voltage (Tillman, 1998). This almost never happens. In the previous example, consider an input voltage equal to 100% (or 1.00 per unit) of the transformer HV rating. The calculated output voltage due to the 12% voltage drop is .88 per unit of the output voltage rating. In order to deliver nameplate MVA at .88 per unit voltage, the actual current has to be the rated voltage divided by the actual voltage times the rated current:

$$I_{actual} = I_{rated}(1/0.88) = 1.136(I_{rated})$$

In the previous example, the 120%, of MVA, load used to calculate voltage drop was actually a 136% (120% × 1.136), of current, load. The actual voltage drop is approximately 13½% by recalculating Eq. (3.68) using per unit current rather than per unit MVA. Also, Eq. (3.66) calls for per unit current to calculate the top oil temperature rise. The calculated top oil temperature rise is 7 to 10°C higher using current instead of MVA (Tillman, 1998).

Loading Recommendations

The IEEE loading guide and many published papers give guidelines and equations for calculating transformer loading. In theory, operators can load modern transformers temporarily up to 200% of nameplate rating or 180°C hottest conductor temperature. These guidelines provide a tool to help a user understand the relationship between loading and the design limitations of a power transformer. However, understanding the relationship between loading and the general condition of and moisture in the insulation system along with voltage regulation is also important. Unless conditions are ideal, these factors will limit a transformer's loading capability before quantified loss of life considerations.

The following loading recommendations assume the load is summer and winter peaking, the fall and spring peak is approximately 60% of summer peak, average ambient temperature on a normal summer day is 30°C, the daily load profile is similar to the cyclical load profile described earlier, the winter peak is less than 120% of the summer peak, and the non-cyclical load is constant except for downtime:

Type of Load	FA Max Top Oil Temp (°C)	NDFOA Max Top Oil Temp (°C)	Max Winding Temp (°C)	Max % Load
Normal summer load	105	95	135	130
Normal winter load	80	70	115	140
Emergency summer load	115	105	150	140
Emergency winter load	90	80	130	150
Non-cyclical load	95	85	115	110

Alarm Settings	FA 65°C Rise	NDFOA 65°C Rise
Top Oil	105°C	95°C
Hot Spot	135°C	135°C
Load Amps	130%	130%

1. The normal summer loading accounts for periods when temperatures are abnormally high. These might occur every three to five years. For every degree C that the normal ambient temperature during the hottest month of the year exceeds 30°C, de-rate the transformer 1% (i.e., 129% loading for 31°C average ambient).
2. The percent load is given on the basis of the current rating. For MVA loading, multiply by the per unit output voltage. If the output voltage is .92 per unit, the recommended normal summer MVA loading is 120%.
3. Exercise caution if the load power factor is less than 0.95 lagging. If the power factor is less than 0.92 lagging, then lower the recommended loading by 10% (i.e., 130% to 120%).
4. Verify that cooling fans and pumps are in good working order and oil levels are correct.
5. Verify that the oil condition is good: moisture is less than 1.5% (1.0% preferred) by dry weight, oxygen is less than 2.0%, acidity is less than 0.5, and CO gas increases after heavy load seasons are not excessive.
6. Verify that the gauges are reading correctly when transformer loads are heavy. If correct field measurements differ from manufacturer's test report data, then investigate further before loading past nameplate criteria.
7. Verify with infrared camera or RTD during heavy load periods that the LTC top oil temperature relative to the main tank top oil temperature is correct. For normal LTC operation, the LTC top oil is cooler than the main tank top oil. A significant deviation from this indicates LTC abnormalities.
8. If the load current exceeds the bushing rating, do not exceed 110°C top oil temperature (IEEE, 1995). If bushing size is not known, perform an infrared scan of the bushing terminal during heavy load periods. Investigate further if the temperature of the top terminal cap is excessive.

9. Use winding power factor tests as a measure to confirm the integrity of a transformer's insulation system. This gives an indication of moisture and other contaminants in the system. High BIL transformers require low winding power factors (<0.5%), while low BIL transformers can tolerate higher winding power factors (<1.5%).
10. If the transformer is extremely dry (less than 0.5% by dry weight) and the load power factor is extremely good (0.99 lag to 0.99 lead), then add 10% to the above recommendations.

References

Bassetto, A., Mak, J., Batista, R. P., and de Faria, T. A. Economic assessment of power transformer loading, Doble Conference Proceedings 64PAIC97, pp 8-10.1 — 8-10.11, 1997.

Bean, R. L., Chackan, Jr., N., Moore, H. R., and Wentz, E. C. *Transformers for the Electric Power Industry*, New York: McGraw-Hill, 1959, 115-120.

Chew, O. Operation of transformers at loads in excess of nameplate ratings, IEEE-Power Engineering Society Transformer Seminar, New Orleans, LA, March 9, 1978.

Dakin, T. W. Electrical insulation deterioration treated as a chemical reaction rate phenomenon, *AIEE Trans.*, 67, 113-122, 1948.

IEEE Standard General Requirements for Liquid-Immersed Distribution, Power, and Regulating Transformers C57.12.00-1993, IEEE Standards Collection Distribution, Power and Regulating Transformers C57, 1994.

IEEE Guide for Loading Mineral-Oil-Immersed Transformers C57.91-1995, IEEE Standard C57, 1995.

IEEE, Adaptive Transformer Thermal Overload Protection, IEEE PSRC WG K3 Transaction Paper Draft 1.03, March 4, 1997.

Kelly, J. J., Myers, S. D., and Parrish, R. H., *A Guide to Transformer Maintenance*, S. D. Myers, Inc., 200-209, 1988a.

Kelly, J. J., Myers, S. D., and Parrish, R. H. *A Guide to Transformer Maintenance*, S. D. Myers, Inc., 211-258, 1988b.

Kelly, J. J., Myers, S. D., and Parrish, R. H. *A Guide to Transformer Maintenance*, S. D. Myers, Inc., 297-322, 1988c.

McNut, W. J. Discussion of the T. V. Oommen, E. M. Petrie, and S. R. Lindgren paper, bubble generation in transformer windings under overload conditions, Doble Conference Proceedings 62PAIC95 pp 8-5A.1–8-5A.3, 1995.

Moore, H. R. Water in transformers, Presentation to Southern Company, Atlanta, GA, 1997.

Oommen, T. V., Petrie, E. M., and Lindgren, S R. Bubble generation in transformer windings under overload conditions, Doble Conference Proceedings 62PAIC95 pp 8-5.1–8-5.5, 1995.

Pierce, L. W. An investigation of the thermal performance of an oil-filled transformer winding, *IEEE Trans. Power Delivery*, 7,(3), 1347-1358, 1992.

Pierce, L. W. Current developments for predicting transformer loading capability, 1993 Minnesota Power Systems Conference, October 5-7, 1993.

Tillman, R. F. Relationships between power factor, voltage regulation, and power transformer loading, Doble Conference Proceedings 65PAIC98, pp 8-9.1–8-9.9, 1998.

3.12 Causes and Effects of Transformer Sound Levels

Jeewan Puri

In modern communities, there is an increasing prevalence of local ordinances specifying sound levels at commercial and residential property lines. Consequently, the sound energy radiated from transformers has become a factor of increasing importance to the neighboring residential areas. It is therefore appropriate that a good understanding of sound power radiation and its measurement principles be developed for appropriately specifying sound levels in transformers. A good understanding of these principles can be helpful in minimizing community complaints regarding the present and future installations of transformers.

Transformer Sound Levels

In order to evaluate a sound source, we must understand the following basic principles used for the quantification of sound energy.

Sound Pressure Level

The main quantity used to describe a sound is the size or amplitude of the pressure fluctuations at a human ear. The weakest sound a healthy human ear can detect has an amplitude of 20 millionths of a Pascal (20 µPa). A pressure change of 20 µPa is so small that it causes the eardrum to deflect a distance less than the diameter of a single hydrogen molecule. Amazingly, the ear can tolerate sound pressures more than one million times higher. Thus, if we measured sound in Pa, we would end up with some quite large, unmanageable numbers. To avoid this, another scale is used — the decibel or dB scale.

The decibel is not an absolute unit of measurement. It is a ratio between a measured quantity and an agreed reference level. The dB scale is logarithmic and uses the hearing threshold of 20 µPa as the reference level. This is defined as 0 dB. A sound pressure level L_p may therefore be defined as:

$$L_p \ dB = 10 \log (P/P_o)^2$$

where P_o = reference level = 20 µPa

One useful aspect of the decibel scale is that it gives a much better approximation of the human perception of relative loudness than the Pascal scale.

Perceived Loudness

We have already defined sound as a pressure variation, which can be heard by a human ear. A healthy human ear of a young person can hear frequencies ranging from 20 Hz to 20 kHz. In terms of sound pressure level, audible sounds range from the threshold of hearing at 0 dB to the threshold of pain which can be over 130 dB.

Although an increase of 6 dB represents a doubling of the sound pressure, in actuality an increase of about 10 dB is required before the sound subjectively appears to be twice as loud. The smallest change in sound level we can perceive is about 3 dB.

The subjective or perceived loudness of a sound is determined by several complex factors. One such factor is that the human ear is not equally sensitive at all frequencies. It is most sensitive to sounds between 2 and 5 kHz and less sensitive at higher and lower frequencies.

Sound Power

A source sound radiates energy and this results in a sound pressure. Sound energy is the cause. Sound pressure is the effect. Sound power is the rate at which energy is radiated (energy per unit time). The sound pressure that we hear or measure with a microphone is dependent on the distance from the source and the acoustic environment (or sound field) in which sound waves are present. This in turn depends on the size of the room and the sound absorption characteristics of its wall surfaces. Therefore by measuring sound pressure, we cannot necessarily quantify how much noise a machine makes. We have to find the sound power because this quantity is more or less independent of the environment and is the unique descriptor of the noisiness of a sound source.

Sound Intensity Level

Sound intensity describes the rate of energy flow through a unit area. The units for sound intensity are watts per square meter (W/m^2).

Sound intensity also gives a measure of direction, as there will be energy flow in some directions but not in others. Therefore, sound intensity is a vector quantity as it has both magnitude and direction. On the other hand, pressure is a scalar quantity as it has magnitude only. Usually we measure the intensity in a direction normal (at 90°) to a specified unit area through which the sound energy is flowing.

Sound intensity is measured as the time-averaged rate of energy flow per unit area. At some points of measurements, energy may be traveling back and forth. If there is no net energy flow in the direction of measurement, there will be no net recorded intensity.

Like sound pressure, sound intensity level L_I is also quantified using a dB scale where the measured intensity I in W/m² is expressed as ratio to a reference intensity level I_o as follows:

$$L_I \text{ dB} = 10 \log (I/I_o)$$

where I_o = reference level = 10^{-12} W/m²

Sound Intensity and Sound Pressure Level Relationship

For any free progressive wave there is a unique relation between the mean-square sound pressure and the intensity. This relation at a particular point and in the direction of the wave propagation is described as follows:

$$I = P^2_{rms}/\rho c \text{ W/m}^2$$

where

- I = intensity, W/m²
- P^2_{rms} = mean-square sound pressure, (N/m²)², measured at that particular point where I is desired in the free progressive wave
- ρc = characteristic resistance, mks rayls

Note that for air, at T = 20°C and atmospheric pressure = 0.751 m of Hg, ρc = 406 mks rayls.
As described earlier, sound intensity level in decibels is:

$$L_I = 10 \log (I/I_o) \text{ dB}$$

where I = sound intensity (power passing in a specified direction through a unit area), W/m²
Combining the above equations, the sound intensity level may be expressed as:

$$L_I = 10 \log((P^2_{rms}/\rho c)/I_o)$$

$$= 10 \log(P_{rms}/P_o)^2 + 10 \log ((P^2_o/\rho c)/I_o)$$

From this expression, L_I may be defined as follows:

$$L_I = L_p - 10 \log K$$

where K = constant = $I_o * \rho c/P^2_o$, which is dependent upon ambient pressure and temperature.
By definition,

$$P^2_o/I_o = (20*10^{-6})^2/10^{-12} = 400 \text{ mks rayls}$$

Note that the quantity 10 log K will equal zero, when K = 1 or when ρc equals 400.
As described earlier, under commonly encountered temperature and atmospheric conditions, ρc is ~400.

Therefore, in free field measurements $L_p \approx L_I$.

That is, noise pressure and noise intensity measurement in free space yield the same numerical value.

Sound Energy Measurement Techniques

Sound level of a source may be measured by directly measuring sound pressure or sound intensity at a known distance. Both of these measurement techniques are quite equivalent and acceptable. In most of the industry worldwide, sound pressure measurements have been used for quantifying sound levels in transformers. As a result of the recent work completed by CIGRE, sound intensity measurements are now being incorporated as an alternative in the IEC Standard 60076-10.

Sound Pressure Level Measurement

A sound level meter is an instrument designed to respond to sound in approximately the same way as the human ear and to give objective, reproducible measurements of sound pressure level. There are many different sound measuring systems available. Although different in detail, each system consists of a microphone, a processing section, and a read-out unit.

The microphone converts the sound signal to an equivalent electrical signal. The most suitable type of microphone for sound level meters is the condenser microphone, which combines precision with stability and reliability. The electrical signal produced by the microphone is quite small. It is therefore amplified by a preamplifier before being processed.

Several different types of processing may be performed on the signal. The signal may pass through a weighting network of filters. It is relatively simple to build an electronic circuit whose sensitivity varies with frequency in the same way as the human ear, thus simulating the equal loudness contours. This has resulted in three different internationally standardized characteristics termed the "A", "B", and "C" weightings.

Nowadays the "A" weighting network is the most widely used since the "B" and "C" weightings do not correlate well with subjective tests.

Sound Intensity Measurements

Until recently, we could only measure sound pressure that was dependent on the sound field. Sound power can be related to sound pressure only under carefully controlled conditions where special assumptions are made about the sound field. Therefore, a noise source had to be placed in specially constructed rooms such as anechoic or reverberant chambers to measure sound power levels with the desired accuracy.

Sound intensity, however, can be measured in any sound field. This property allows all the measurements to be done directly in situations where a plurality of sound sources are present. Measurements on any sound source can be made even when all the others are radiating noise simultaneously. It should be noted that steady background noise makes no contribution to the sound power of the source determined with sound intensity measurements.

Sound intensity is the time-averaged product of the pressure and particle velocity. A single microphone can measure pressure. However, measuring particle velocity is not as simple. With Euler's linearized equation, the particle velocity can be related to the *pressure gradient* (i.e., the rate at which the instantaneous pressure changes with distance).

Euler's equation is essentially Newton's second law applied to a fluid. Newton's Second Law relates the acceleration given to a mass to the force acting on it. If we know the force and the mass, we can find the acceleration and then integrate it with respect to time to find the velocity.

With Euler's equation, it is the pressure gradient that accelerates a fluid of density ρ.

With the knowledge of pressure gradient and density of the fluid, the particle acceleration (or deceleration) can be calculated as follows:

$$a = -1/\rho \; \partial P/\partial r$$

where a = particle deceleration due to a pressure change ∂P in a fluid of density ρ across a distance ∂r.

Integrating the above gives the particle velocity 'u' as follows:

$$u = -\int 1/\rho \; \partial P/\partial r \; dt$$

Transformers

It is possible to measure the pressure gradient with two closely spaced microphones facing each other and relate it to the particle velocity using the above equation.

With two closely spaced microphones 'A' and 'B' separated by a distance Δr, it is possible to obtain a straight line approximation to the pressure gradient by taking the difference in their measured pressures P_A and P_B and dividing it by the distance Δr between them. This is called a finite difference approximation.

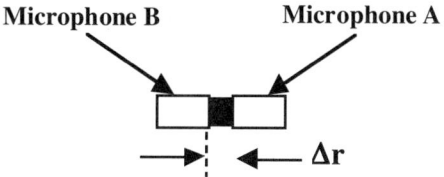

The pressure gradient signal must now be integrated to give the particle velocity 'u' as follows:

$$u = -1/\rho \int ((P_A - P_B)/\Delta r) dt$$

Since intensity I is the time averaged product of pressure P and particle velocity u

$$I = -P/\rho \int ((P_A - P_B)/\Delta r) dt$$

where $P = (P_A + P_B)/2$

This is the basic principle of signal processing in sound intensity measuring equipment.

Sources of Sound in Transformers

Unlike cooling fan or pump noise, the sound radiated from a transformer is tonal in nature, consisting of even harmonics of the power frequency. It is generally recognized that the predominant source of transformer noise is the core. The low frequency, tonal nature of this noise makes it harder to mitigate than the broad band higher frequency noise that comes from the other sources. This is because low frequencies propagate farther with less attenuation. Also, tonal noise can be perceived more acutely than broad band levels, even with high background noise levels. This combination of low attenuation and high perception makes tonal noise the dominant problem in the neighboring communities around transformers. To address this problem, most noise ordinances impose penalties or stricter requirements for tonal noise.

Even though the core is the principal noise source in transformers, the load noise, which is principally caused by the electromagnetic forces in the windings, can also be a significant influence in low sound level transformers. The cooling equipment (fans and pumps) noise typically dominates the very low and very high frequency ends of the sound spectrum, whereas the core noise dominates in the intermediate range of frequencies between 100 and 600 Hz.

These sound producing mechanisms may be further characterized as follows:

Core Noise — When a strip of iron is magnetized, it undergoes a very small change in its dimensions (usually only a few parts in a million). This phenomenon is called magnetostriction. This change in dimension is independent of the direction of magnetic flux; therefore, it occurs at twice the line frequency. Because the magnetostriction curve is nonlinear, even higher harmonics also appear in the resulting core vibration at higher induction levels (above 1.4 T).

Flux density, core material, core geometry, and the waveform of excitation voltage are the factors that influence the magnitude and frequency components of the transformer core sound levels. The mechanical resonance in transformer mounting structure, core, and tank walls can also have a significant influence on the magnitude of transformer vibrations and consequently on the acoustic noise generated.

Load Noise — Load noise is caused by vibrations in tank walls, magnetic shields, and transformer windings due to the electromagnetic forces resulting from leakage fields produced by load currents. These electromagnetic forces are proportional to the square of the load currents.

The load noise is predominantly produced by axial and radial vibration of transformer windings. However, marginally designed magnetic shielding can also be a significant source of sound in transformers. A rigid design for laminated magnetic shields with firm anchoring to the tank walls can greatly reduce their influence on the overall load sound levels. The frequency of load noise is usually twice the power frequency. An appropriate mechanical design for laminated magnetic shields can be helpful in avoiding resonance in the tank walls. The design of the magnetic shields should take into account the effects of overloads to avoid saturation, which would cause higher sound levels during such operating conditions.

Studies have shown that except in very large coils, radial vibrations do not make any significant contribution to the winding noise. The compressive electromagnetic forces produce axial vibrations and thus can be a major source of sound in poorly supported windings. In some cases, the natural mechanical frequency of winding clamping systems may tend to resonate with electromagnetic forces, thereby severely intensifying the load noise. In such cases, damping of the winding system may be required to minimize this effect.

The presence of harmonics in load current and voltage (e.g., in rectifier transformers) can produce vibrations at twice the harmonic frequencies and thus a sizeable increase in the overall sound level of a transformer.

Through several decades the contribution of the load noise to the total transformer noise has remained moderate. However, in transformers designed with low induction levels and improved core designs for complying with low sound level specifications, the load-dependent winding noise of electromagnetic origin can become a significant contributor to the overall sound level of the transformer. In many such cases, the sound power of the winding noise is only a few dB below that of the core noise.

Fan and Pump Sound — Power transformers generate considerable heat because of the losses in the core, coils, and other metallic structural components of the transformer. This heat is removed by fans which blow air over radiators or coolers. Noise produced by the cooling fans is usually broad band in nature. Cooling fans usually contribute more to the total noise for transformers of smaller rating and for low-induction transformers. Factors that affect the total fan noise output include tip speed, blade design, number of fans, and the arrangement of the radiators.

Sound Level and Measurement Standards for Transformers

NEMA Publication TR-1, Tables 02 thorough 04, lists standard sound levels for liquid filled power, liquid filled distribution, and dry type transformers. These sound level requirements must be met unless special lower sound levels are specified by the customer.

The present sound level measurement procedures as described in IEEE Standards C57.12.90 and C57.12.91 specify that the sound level measurements on a transformer shall be made under no load conditions. Sound pressure measurements shall be made to quantify the total sound energy radiated by a transformer. Sound intensity measurements have already been incorporated into IEC Sound Level Measurement Standard 60076-10 as an acceptable alternative. It is anticipated that this method will be adopted in the IEEE standards in the near future. The following is a brief description of the procedures used for this determination.

Transformer Connections During Test

This test is performed by exciting one of the transformer windings at rated voltage of sinusoidal wave shape at rated frequency while all the other windings are open circuited. The tap-changer (if any) is at the rated voltage tap position. In some cases (e.g., transformers equipped with reactor type on-load-tap-changers), a tap position other than the rated may be used if the transformer produces maximum sound levels at this position.

Transformers

Principal Radiating Surface for Measurements

This is the surface from which the sound energy is emanating toward the receiver locations. The location of the radiating surface is determined based on the proximity of the cooling equipment to the transformer.

For transformers with no cooling equipment or with cooling equipment mounted less than 3 m from the transformer tank or dry type transformers with enclosures provided with cooling equipment (if any) inside the enclosure, the principal radiating surface is obtained by taking the vertical projection of a string contour surrounding the transformer and its cooling equipment (if any) as shown in Fig. 3.146 (taken from IEC 60076-10). The vertical projection begins at the tank cover and terminates at the base of the transformer.

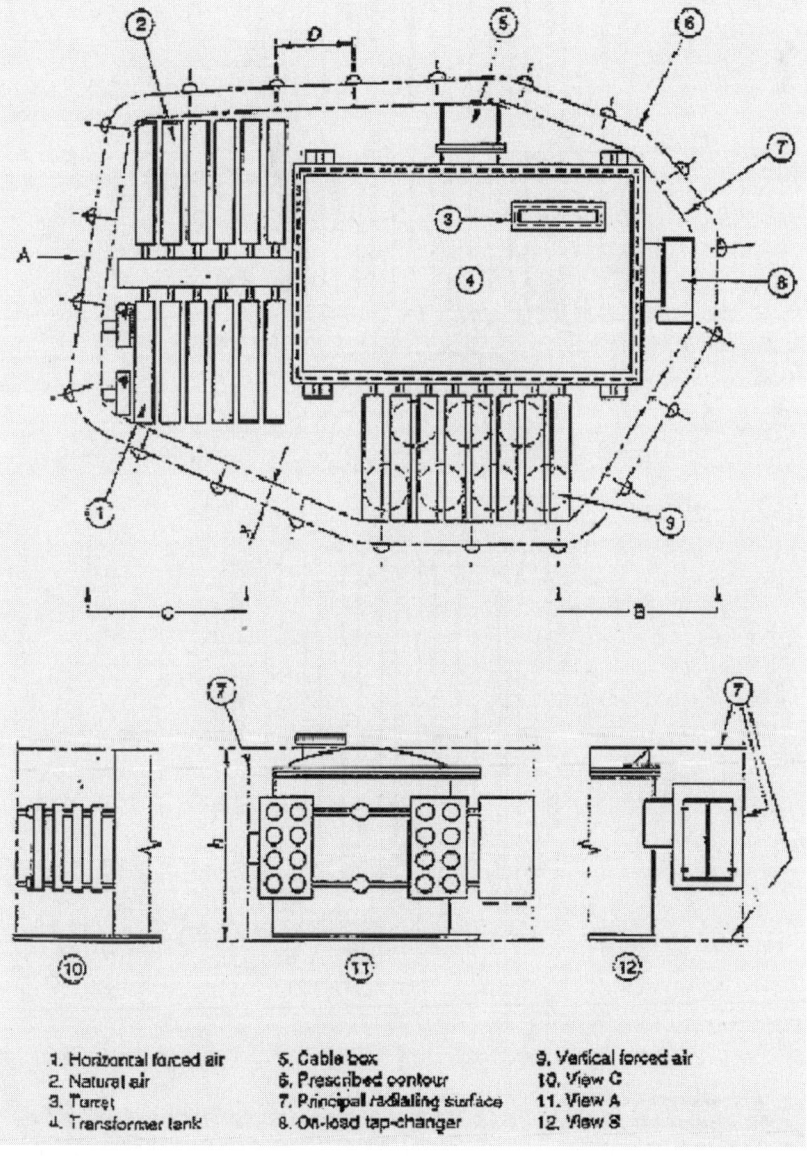

1. Horizontal forced air
2. Natural air
3. Turret
4. Transformer tank
5. Cable box
6. Prescribed contour
7. Principal radiating surface
8. On-load tap-changer
9. Vertical forced air
10. View C
11. View A
12. View B

FIGURE 3.146 Typical microphone positions for sound measurement on transformers having cooling auxiliaries mounted either directly on the tank or on a separate structure spaced <3 m away from the principal radiating surface of the main tank.

Separate radiating surfaces for the transformer and its cooling equipment are determined if the cooling equipment is mounted more than 3 m from the transformer tank. The principal radiating surface for the cooling equipment is determined by taking the vertical projection of the string perimeter surrounding the cooling equipment as shown in Fig. 3.147 (taken from IEC 60076-10). The vertical projection begins at the top of the cooling structure and terminates at its base.

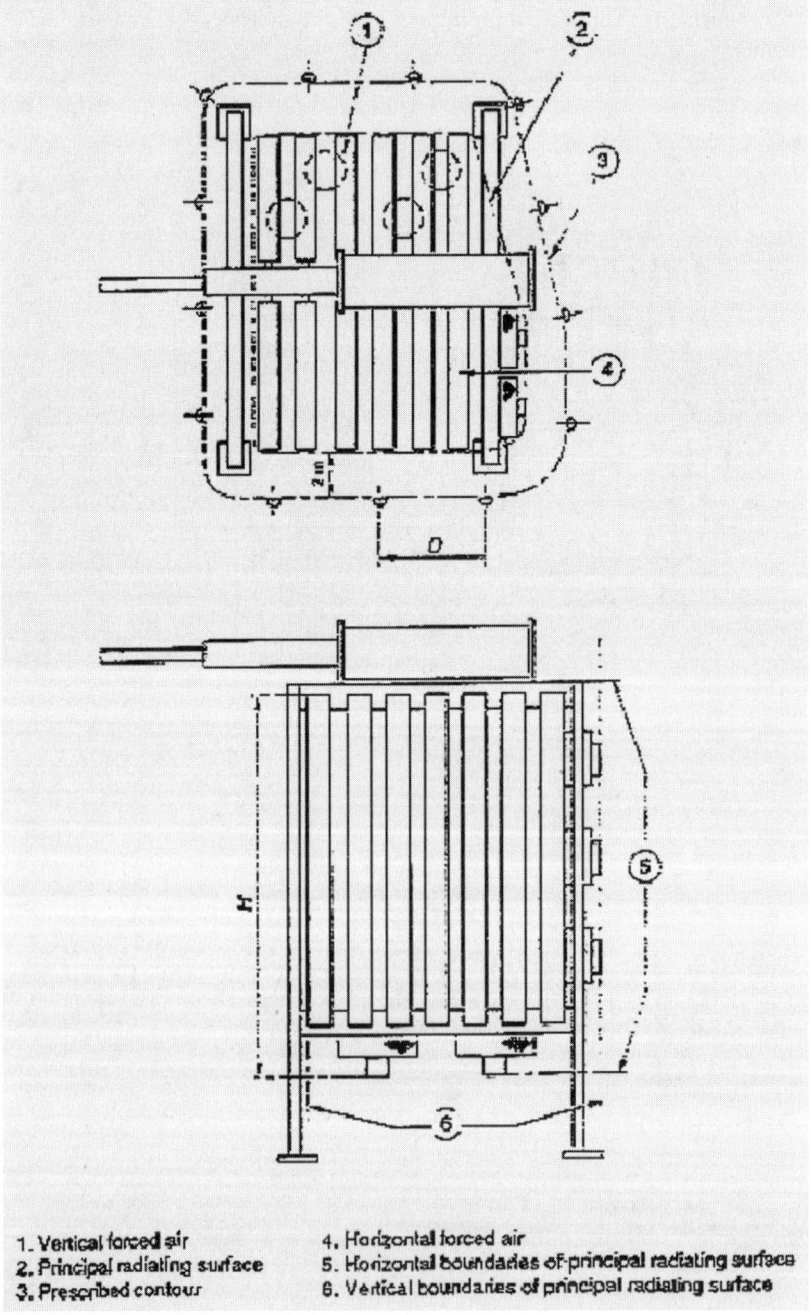

FIGURE 3.147 Typical microphone positions for sound measurement on cooling auxiliaries mounted on a separate structure spaced ≥3 m away from the principal radiating surface of the transformer.

Transformers

Prescribed Contour Location for Measurements

All sound level measurements are made on a prescribed contour located 0.3 m away from the radiating surface. The location of this contour will depend on the radiating surface as determined by the proximity of the cooling equipment to the transformer as shown in Figs. 3.146 and 3.147.

The location of the prescribed contours above the base of the transformer shall be at half the tank height for transformer tanks < 2.5 m high or at one-third and two-thirds the tank height for transformer tanks <2.5 m high.

Measuring Positions on Prescribed Contour

The first microphone position is located on the prescribed contour opposite the main tank drain valve. Proceeding in a clockwise direction (as viewed from the top of the transformer), additional measuring positions on the prescribed contour are located no more than 1 m apart.

The minimum number of measurements as stipulated in IEEE C57.12.90 or IEEE C57.12.91 for North American practices is taken on each prescribed contour. These standards specify that sound level measurements shall be made with and without the cooling equipment in operation. IEC 60076-10 standard should be consulted for European practices, which are slightly different.

Sound Pressure Level Measurements

A-weighted sound pressure level measurements are the most commonly used method for determining sound levels in transformers.

Sound pressure measurements are quite sensitive to the ambient sound levels on the test floor. Therefore, appropriate corrections for the ambient sound level and reflected sound from the surrounding surfaces must also be quantified for determining the true sound level of the transformer.

It is recommended that acceptable ambient sound level conditions should be met for obtaining reliable measurements on transformers. For this reason, industry standards specify that A-weighted ambient sound pressure levels must be measured immediately before and after the measurements on the transformer. The ambient noise level readings are taken at each microphone position on the prescribed contours with the transformer and cooling equipment (if any) de-energized. These measurements are used for correcting the measurements made on the transformer. The magnitude of this correction depends upon the difference between the ambient and the transformer sound levels. This difference should not be less than 5 dB for a valid measurement. No correction is necessary if the ambient sound level is more than 10 dB lower than the transformer sound level.

From the measured sound pressure levels, L_{pAi}, at each microphone position on the prescribed contour, an A-weighted average sound pressure level, L_{pA}, may be calculated using the following equation:

$$L_{pA} = 10 \log \left[1/N \sum_{i=1}^{N} 10^{0.1 L_{pAi}} \right] - K \text{ dB}$$

where

L_{pA} = A-weighted average sound pressure level, in decibels — Reference: 20 µPa
L_{pAi} = A-weighted sound pressure level measured at the ith position and corrected for the ambient noise level, in decibels — Reference: 20 µPa
N = total number of measuring positions on the prescribed contour
K = environmental correction for the influence of reflected sound and ambient sound level, in decibels. IEEE and IEC standards should be consulted for details.

Sound Intensity Measurements

The equipment for these measurements has only recently emerged in the industry. By definition, the A-weighted sound intensity measurements provide a measure of sound power radiated in watts through a unit area per unit time. This type of measurement yields a vector quantity that represents the sound energy radiated in a direction normal to the principal radiating surface of the transformer.

The noise intensity measuring probes use two matched microphones that respond to sound pressure so that the readings taken by them does not differ by more than 1.5 dB at any location.

From the measured sound intensity levels, L_{IAi}, at each position of the prescribed contour, an A-weighted average sound intensity level in decibels, L_{IA}, may be calculated using the following equation:

$$L_{IA} = 10 \log \left[1/N \sum_{i=1}^{N} 10^{0.1 L_{ia}} \right] dB$$

where

L_{IA} = A-weighted average sound intensity level, in decibels — Reference: 10^{-12} W/m²
L_{ia} = A-weighted sound intensity level measured at the ith position for the ambient noise level — Reference: 10^{-12} W/m²
N = total number of measuring positions on the prescribed contour

Unlike sound pressure, the noise intensity measurements are not influenced by the ambient noise level provided the ambient noise remains constant as the measurements are taken around the prescribed contour. Under such conditions, noise intensity measurements can be made in ambient sound level even higher than the sound level of the transformer. For this reason, these types of measurements are especially suitable for transformers designed for very low noise levels.

It should be recognized that at this time the transformer industry experience in sound intensity measurement is rather limited. Actual measurements published in CIGRE publications have demonstrated that the reliability of the sound intensity measurements depends upon the difference ΔL between the average sound intensity and pressure measurements made by the same probes. This work suggests that for optimum results, ΔL should not be more than 8 dB.

Calculation of Sound Power Level

For demonstrating compliance with the local ordinances, it becomes necessary to calculate sound pressure levels at property lines located away from the transformer.

Sound power level provides a measure of the total sound energy radiated by a transformer. With this quantity, the sound pressure levels at any desired distance (outside the prescribed contour) from the transformer may be calculated.

Since sound pressure or sound intensity measurements yield the same numerical result, the following equations may be used for calculating sound power levels in decibels from the measured sound pressure or sound intensity levels.

$$L_{WA} = L_{IA} + 10 \log (S) \text{ dB}$$

or

$$L_{WA} = L_{pA} + 10 \log (S) \text{ dB}$$

where

S = the area of the radiating surface in square meters
$L_{WA} = 10 \log(W/W_0)$ dB = A-weighted sound power level in decibels, ($W_0 = 10^{-12}$ W)
L_{pA} = A-weighted average sound pressure level, in decibels — Reference: 20 μPa
L_{IA} = A-weighted average sound intensity level, in decibels — Reference: 10^{-12} W/m²

The calculation of effective radiating surface area S is also a function of the distance at which the sound measurements were made. IEEE or IEC standards may be consulted for details.

Transformers

Sound Pressure Level Calculations at Far Field Receiver Locations

Sound level requirements at a specific receiver location in the far field play a major role in specifying sound levels for transformers.

The following basic method may be considered for estimating transformer sound pressure levels for simple installations. These calculations provide an accurate estimate of transformer noise emissions at distances roughly greater than twice the largest dimension of the transformer.

Given the sound pressure level requirement of L_{pD} dB at a distance D from the transformer, the maximum allowable sound power L_W for the transformer may be estimated as follows:

$$L_W = L_{pD} + 10 \, \text{Log} \, (2 \prod D^2) \, W$$

Therefore, the maximum allowable sound pressure L_{pT} level at the measurement surface near the transformer will be:

$$L_{pT} = L_W - 10 \, \text{Log} \, (\text{Measurement Surface Area})$$

Measurement Surface Areas = (1.25 * Transformer tank height in meters * Measurement contour length in meters)

Factors Affecting Sound Levels in Field Installations

In order to assure repeatability, the factory measurements are made under controlled conditions specified by sound measurement standards. Every effort is made to maintain core excitation voltage constant and sinusoidal for assuring known core induction while making sound level evaluations in transformers.

The effects of ambient sound levels and reflections must be subtracted from the overall sound level measurements in order to determine the true sound power of a transformer. It is therefore assured that the ambient sound levels are constant and are measured accurately. Many times it becomes necessary to make these measurements in a sound chamber where ambient sound level is very low and the effects of reflecting surfaces can also be eliminated.

The sound level measurements at the transformer site can be drastically different depending upon its operating conditions. The effects of the following factors should therefore be considered while defining the sound level requirements for transformers.

Load Power Factor

In the factory, core and winding sound levels are measured separately at rated voltage and full load current, respectively. These sound levels are produced by core and winding vibrations of twice the power frequency and its even harmonics. It is assumed that these vibrations are in phase with each other and therefore their power levels are added to predict the overall sound level of the transformer. However, this assumption applies only when the transformer is carrying purely resistive load.

Under actual operating conditions, depending upon the power factor of the load, the phase angle between voltage and load current may induce a change in the factory predicted transformer sound level.

Internal Regulation

The magnitude and the phase angle of the load currents also change the internal voltage drop in the transformer windings. The transformer loading conditions therefore can change the core induction level and significantly influence the core sound levels.

Load Current and Voltage Harmonics

During factory tests, only sinusoidal load current is simulated for measuring winding noise. This noise is produced by magnetic forces that are proportional to the square of the load current. However, harmonic content in the load current has a larger impact on the sound level than might be expected from the amplitude of the harmonic currents since they interact with the power frequency load current. In such cases, the magnetic force is proportional to the cross product between the power frequency current and the harmonic current in addition to the force that is proportional to the square of load current and the square of the harmonic current. Thus, the highest contribution to the sound level due to the harmonic current occurs when the product of the load current and the harmonic current reaches the maximum. The resulting audible tones are made up of frequencies of the harmonic current ± the fundamental power frequency.

Current harmonics are a major source of increase in sound levels in HVDC and rectifier transformers.

Non-linear loads cause harmonics in the excitation voltage, resulting in an increase in core sound levels. This influence must be considered while specifying sound level for a transformer.

DC Magnetization

Even a moderate DC magnetization of a transformer core will result in a significant increase in the transformer audible sound level. In addition to increasing the power level of the normal harmonics in the transformer vibrations (i.e., even harmonics of the power frequencies), DC magnetization will add odd harmonic tones to the overall sound level of the transformer.

Modern cores have high remnant flux density. Upon energization, the core sound levels may be as much as 20 dB higher than the factory test value. It is therefore recommended that a transformer should be energized for approximately 6 h before evaluating its sound levels.

Traditionally, circuits like DC feeders to the transportation systems have been a source of DC fields in transformers. However, with the increased use of power electronic equipment in power transmission systems and in the industry, the number of possible sources for DC magnetization is increasing. Geomagnetic storms may also cause severe DC magnetization in transformers connected to long transmission lines.

Acoustical Resonance

Dry type transformers are most frequently applied inside buildings. In a room with walls of low sound absorption coefficient, the sound from the transformer will reflect back and forth between walls, resulting in a build-up of sound level in the room.

The number of dBs by which the sound level at the transformer will increase may be approximated as follows:

$$\text{dB build-up} = 10 \log \left(1 + 4(1-a) A_T / (a A_U)\right)$$

where

A_T = surface area of the transformer
A_U = area of the reflecting surface
a = average absorption coefficient of the surfaces

In a room with concrete walls (with an absorption coefficient of 0.01) and with sound reflecting surface area four times that of the transformer ($A_U/A_T = 4$), the increase in sound level at the transformer can be 20 dB. However, covering the reflecting surfaces of this room with sound absorbing material with absorption coefficient of 0.3 will reduce this build-up to 5.5 dB.

Sound propagation is affected by many factors such as atmospheric absorption, interceding barriers, and reflective surfaces. An explanation of these factors is beyond the scope of this text; however, they are mentioned to make the reader aware of their potential influence. If the site conditions that will influence the sound propagation exist, the reader is advised to use reference textbooks dealing with the subject of acoustic propagation or consult an expert in conducting more accurate sound propagation calculations.

References

Beranek, L., Ed., *Noise and Vibration Control,* Institute of Noise Control Engineering, Washington, D.C.
Electra No. 144 October 1992, Transformer noise: determination of sound power level using the sound intensity measurement method, IEC SC 12- WG 12 (April 1990). Convenor: J.P Fanton.
Empire State Electric Energy Research Corporation Report EP 9-14, Power transformer noise abatement, Westinghouse Electric Corporation and Bolt Beranek and Newman Inc., October 1981.
Gade, S., *Sound Intensity Instrumentation and Applications,* Technical Review No 4, Bruel & Kjaer, 1982.
Gade, S., *Sound Intensity Theory,* Technical Review No 3, Bruel & Kjaer, 1982.
IEEE C57.156 (Draft 9), Guide for sound level abatement and detrermination for liquid immersed power transformers and shunt reactors rated obe 500 kVA, Audible Sound and Vibration Subcommittee Working Group.
Measuring Sound, Bruel & Kjaer, 1984.
Replinger, E., Study of noise emitted by power transformers based on today's viewpoint, Cigre Paper 12-08, Session Aug. 28–Sept. 3 1988, Siemens AG, Transformatorenwerk, Nurnberg, Germany.
Sound Intensity, Bruel & Kjaer, 1993.
Specht, T. R., Consulting Engineer, Noise levels of indoor transformers, October 1955, Transformer Division, Westinghouse Corporation.

3.13 Electrical Bushings

Loren B. Wagenaar

ANSI/IEEE Std. C57.19.00 (1997) defines an electrical bushing as "an insulating structure, including a through conductor or providing a central passage for such a conductor, with provision for mounting a barrier, conducting or otherwise, for the purpose of insulating the conductor from the barrier and conducting current from one side of the barrier to the other." As a less formal explanation, the purpose of an electrical bushing is simply to transmit electrical power in or out of enclosures, i.e., barriers, of an electrical apparatus such as transformers, circuit breakers, shunt reactors, and power capacitors. The bushing conductor may take the form of a conductor built directly as a part of the bushing, or alternately, as a separate conductor which is drawn through, usually through the center of, the bushing.

Since electrical power is the product of voltage and current, insulation in a bushing must be capable of withstanding the voltage at which it is applied, and its current carrying conductor must be capable of carrying rated current without overheating the adjacent insulation. For practical reasons, bushings are not rated by the power transmitted through them; rather, they are rated by the maximum voltage and current for which they are designed.

Types of Bushings

There are many methods to classify the types of bushings. These classifications are based on practical reasons, which will become apparent in the following.

According to Insulating Media on Ends

One method is to designate the types of insulating media at the ends of the bushing. This classification depends primarily on the final application of the bushing.

An air-to-oil bushing has air insulation at one end of the bushing and oil insulation at the other. Since oil is more than twice as strong dielectrically as air at atmospheric pressure, the oil end is approximately half as long or less than the air end. This type of bushing is quite common for usage between atmospheric air and any oil-filled apparatus.

An air-to-air bushing has air insulation on both ends and is normally used in building applications where one end is exposed to outdoor atmospheric conditions, and the other end is exposed to indoor

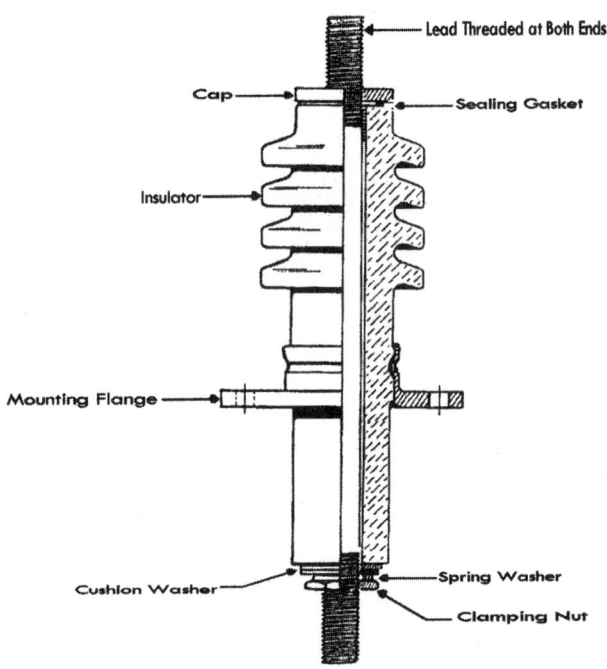

FIGURE 3.148 Solid type bushing.

conditions. The outer end may have higher creep distances in order to withstand higher pollution environments and possibly higher strike distances in order to withstand transient voltages during adverse weather conditions such as rainstorms.

Special application bushings have limited usage and include: air-to-SF_6 bushings, usually used in SF_6 insulated circuit breakers; SF_6-to-oil bushings, used as transitions between SF_6 bus ducts and oil filled apparatus; and oil-to-oil bushing, used between oil bus ducts and oil filled apparatus.

According to Construction

There are basically two types of construction, the solid or bulk type, and the capacitance-graded or condenser type.

Solid Bushing

The solid type bushing, depicted in Fig. 3.148, is typically made with a central conductor and porcelain or epoxy insulators at either end, and is primarily used at the lower voltages through 25 kV. This construction is generally relatively simple compared to the capacitance-graded type, which was used for the original bushings, and its present usage is quite versatile with respect to size. It is commonly used for applications ranging from small distribution transformers and circuit switchers to large generator step-up transformers and hydrogen cooled power generators.

At the lower end of the applicable voltage range, the central conductor may be a small diameter lead connected directly to the transformer winding, and such a lead typically passes through an arbitrarily shaped bore of an outer and inner porcelain or epoxy insulator(s). Between the two insulators is typically located a mounting flange for mounting the bushing to the transformer or other apparatus. In one rather unique design, only one porcelain insulator was used, and the flange was assembled onto the porcelain after the porcelain had been fired. At higher voltages, particularly at 25 kV, more care is taken to make certain that the lead and bore of the insulator(s) are circular and concentric, so that the electric stresses in the gap between these two items are more predictable and uniform. For higher current bushings, typically up to 20 kA, large diameter circular copper leads or several copper bars arranged in a circle and brazed to copper end plates may be used.

The space between the lead and insulator may consist of only air on lower voltage solid type bushings, or this space may be filled with electric grade mineral oil or some other special compound on higher voltage bushings. The oil may be self-contained within the bushing, or it may be oil from the apparatus in which the bushing is installed. Special compounds are typically self-contained. Oil and compounds are used for three reasons. First, they enable better cooling of the conductor than air does. Second, they have higher dielectric constants (about 2.2 for oil) than air, and therefore, when used with materials with higher dielectric constants, such as porcelain or epoxy, they endure a smaller share of the voltage than an equally sized gap occupied by air. The result is that oil and compounds withstand higher voltages than air does. Third, oil and other compounds display higher breakdown strengths than air.

The primary limitation of the solid bushing is its ability to withstand 60-Hz voltages above 90 kV. Hence, its applications are limited to 25-kV equipment ratings, which have test voltages of 70 kV. Recent applications require low partial discharge limits on the 25-kV terminals during transformer test and have caused further restrictions on the use of this type of bushing. In these cases, either a specially designed solid bushing, with unique grading shielding that enables low inherent partial discharge levels, or a more expensive capacitance-graded bushing must be used.

Capacitance-Graded Bushings

Technical literature dating back to the early twentieth century describes the principles of the capacitance-graded bushing (Easly and Stockum, 1984). R. Nagel of Siemens published a German paper (Nagel, 1906) in 1906 describing an analysis and general principles of condenser bushings, and Reynders of Westinghouse published a U.S. paper (Reynders, 1909) which described the principles of the capacitance-graded bushing and compared the characteristics of these bushings with those of solid type construction. Thereafter, several additional papers were published, including those by individuals from Micafil of Switzerland and ASEA of Sweden.

The value of the capacitance-graded bushing was quickly demonstrated, and this bushing type was produced extensively by those companies possessing the required patents. Currently, this construction is used for virtually all voltage ratings above 25 kV system voltage and has been used for bushings through 1500 kV system voltage. This construction uses conducting layers at predetermined radial intervals within oil-impregnated paper or some other insulation material which is located in the space between the central conductor and the insulator. Different manufacturers have used a variety of materials and methods for making capacitance-graded bushings. Early methods were to insert concentric porcelain cylinders with metallized surfaces or laminated pressboard tubes with embedded conductive layers. Later designs used conductive foils, typically aluminum or copper, in oil-impregnated kraft paper. An alternative method is to print semi-conductive ink (different manufacturers have used different conductivities) on all or some of the oil-impregnated kraft paper wraps.

Figure 3.149 shows the general construction of an oil-filled, capacitance-graded bushing. The principal elements are the central circular conductor, onto which the capacitance-graded core is wound, the top and lower insulators, the mounting flange, the oil and an oil expansion cap, and the top and bottom terminals. Figure 3.150 is a representation of the equipotential lines in a simplified capacitance-graded bushing in which neither the expansion cap nor the sheds on either insulator are shown. The bold lines within the capacitance-graded core depict the voltage grading elements. The contours of the equipotential lines show the influence of the grading elements, both radially within the core and axially along the length of the insulators.

The mathematical equation for the radial voltage distribution as a function of diameter between two concentric conducting cylinders is:

$$V(d) = V\left(\ln(D_2/d)/\ln(D_2/D_1)\right) \tag{3.73}$$

where V is the voltage between the two cylinders

D_1 and D_2 are the diameters of the inner and outer cylinders, respectively.

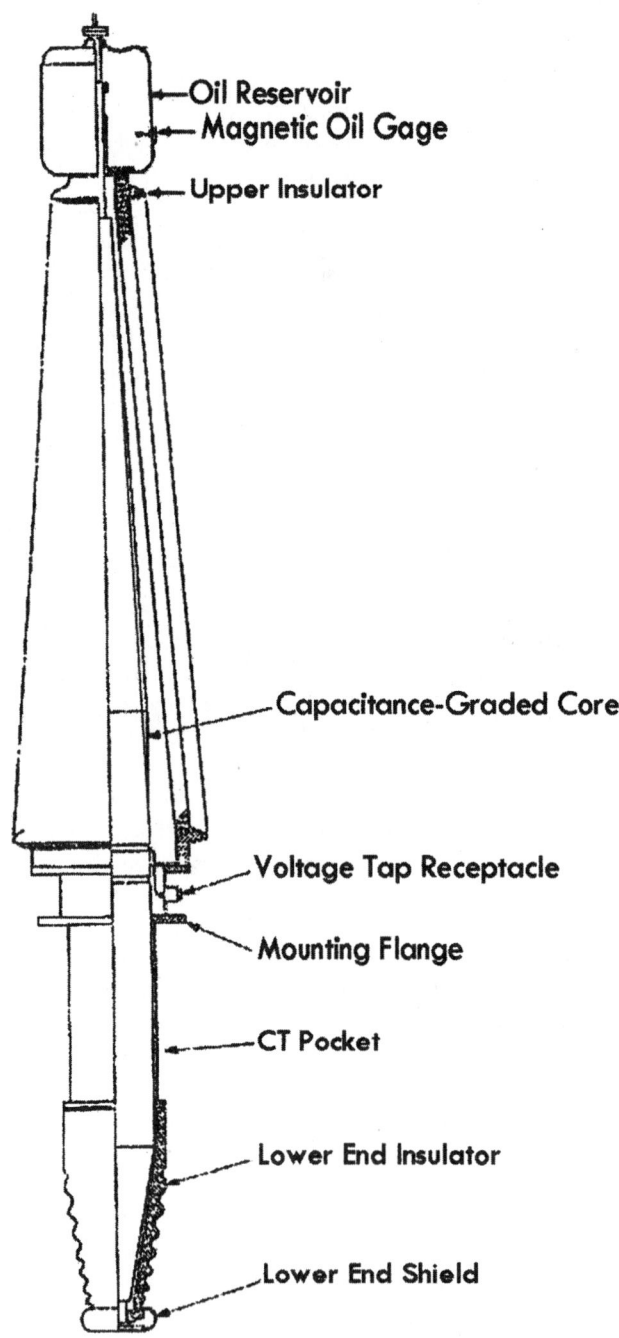

FIGURE 3.149 Capacitance-graded bushing.

Since this is a logarithmic function, the voltage is non-linear, concentrating around the central conductor and decreasing near the outer cylinder. Likewise, the associated radial electric stress, calculated by

$$E(d) = 2V / \left(d \ln(D_2/D_1) \right) \tag{3.74}$$

Transformers

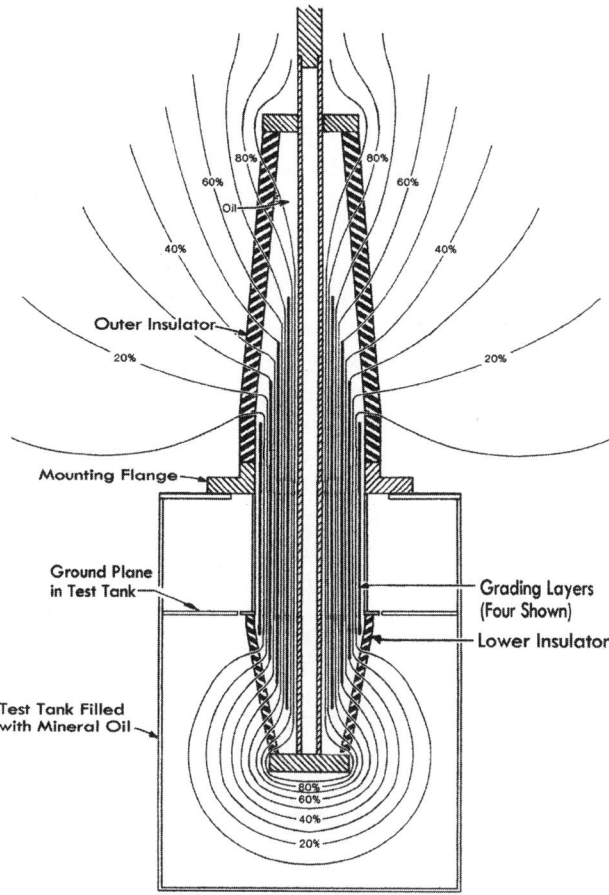

FIGURE 3.150 Equipotential plot of capacitance graded bushing.

will be the greatest at $d = D_1$. The lengths of grading elements and the diameters at which they are positioned are such as to create a more uniform radial voltage distribution than found in a solid type bushing.

As seen from Fig. 3.150, the axial voltage distribution along the inner and outer insulators is almost linear when the proper capacitance grading is employed. Therefore, both insulators on capacitance-graded bushings can be shorter than their solid bushing counterparts.

Capacitance-graded bushings involve many more technical and manufacturing details than solid bushings and are therefore more expensive. These details include the insulation/conducting layer system, equipment to wind the capacitor core, and the oil to impregnate the paper insulation. However, it should be noted that the radial dimension required for the capacitance-graded bushing is much less than the solid construction, and this saves on material within the bushing as well as in the apparatus in which the bushing is used. Also, from a practical standpoint, higher voltage bushings could not possibly be manufactured with a solid construction.

According to Insulation Inside Bushing

Still another classification relates to the insulating material used inside the bushing. In general, these materials can be used in either the solid or capacitance-graded construction, and more than one of these insulating materials can be used in conjunction. The following text gives a brief description of these types:

1. **Air-insulated bushings** generally are used only with air-insulated apparatus and of the solid construction that employs air at atmospheric pressure between the conductor and the insulators.
2. **Oil-insulated or oil-filled bushings** have electrical grade, mineral oil between the conductor and the insulators in solid type bushings. This oil may be contained within the bushing or may be shared with the apparatus in which the bushing is used. Capacitance-graded bushings also use mineral oil, usually contained within the bushing, between the insulating material and the insulators for the purposes of impregnating the kraft paper and transferring heat from the conducting lead.
3. **Oil-impregnated paper insulated bushings** use the dielectric synergy of mineral oil and electric grades of kraft paper to produce a composite material with superior dielectric withstand characteristics. This material has been extensively used as the insulating material in capacitance-graded cores for approximately the last 50 years.
4. **Resin-bonded paper insulated bushings** use a resin-coated kraft paper to fabricate the capacitance-graded core, whereas **resin-impregnated paper insulated bushings** use paper impregnated with resin which is then used to fabricate the capacitance-graded core. The latter type of bushings has superior dielectric characteristics, comparable with oil-impregnated paper insulated bushings.
5. **Cast insulation bushings** are constructed of a solid cast material with or without an inorganic filler. These bushings may be either of the solid or capacitance-graded types, although the former type is more representative of present technology.
6. **Gas-insulated bushings** (Spindle) use pressurized gas, such as SF_6 gas, to insulate between the central conductor and the flange. The bushing shown in Fig. 3.151 is one of the simpler designs and is typically used with circuit breakers. It uses the same pressurized gas as the circuit breaker, has no capacitance grading and uses the dimensions and placement of the ground shield to control the electric fields. Other designs use a lower insulator to enclose the bushing, which permits the gas pressure to be different than the circuit breaker. Still other designs use capacitance-graded cores made of plastic film material that is compatible with SF_6 gas.

Bushing Standards

Several bushing standards exist in the various countries around the world. The major standards have been established by the Transformers Committee within the IEEE Power Engineering Society and by IEC Committee 37. Five important standards established by these committees include the following:

1. ANSI/IEEE Std. C57.19.00, Standard Performance Characteristics and Test Procedure for Outdoor Power Apparatus Bushings (1997). This general standard is widely used by countries in the western hemisphere and contains definitions, service conditions, ratings, general electrical and mechanical requirements, and detailed descriptions of routine and design test procedures for outdoor power apparatus bushings.
2. IEEE Std. C57.19.01, Standard General Requirements and Test Procedure for Outdoor Power Apparatus Bushings (IEEE, 1997a). This standard lists the electrical insulation and test voltage requirements for power apparatus bushings rated from 15 kV through 800 kV maximum system voltages. It also lists dimensions for standard dimensioned bushings, cantilever test requirements for bushings rated through 345 kV system voltage and partial discharge limits, as well as limits for power factor and capacitance change from before to after the standard electrical tests.
3. IEEE Std. C57.19.03, Standard Requirements, Terminology and Test Procedures for Bushings for DC Applications (IEEE, 1996). This standard gives the same type of information as ANSI/IEEE Std. C57.19.00 for bushings for direct current equipment, including oil-filled converter transformers and smoothing reactors. It also covers air-to-air DC bushings.
4. IEEE Std. C57.19.100, Guide for Application of Power Apparatus Bushings (IEEE, 1997b). This guide recommends practices to be used for thermal loading above nameplate rating for bushings applied on power transformers and circuit breakers, and for bushings connected to isolated-phase

Transformers

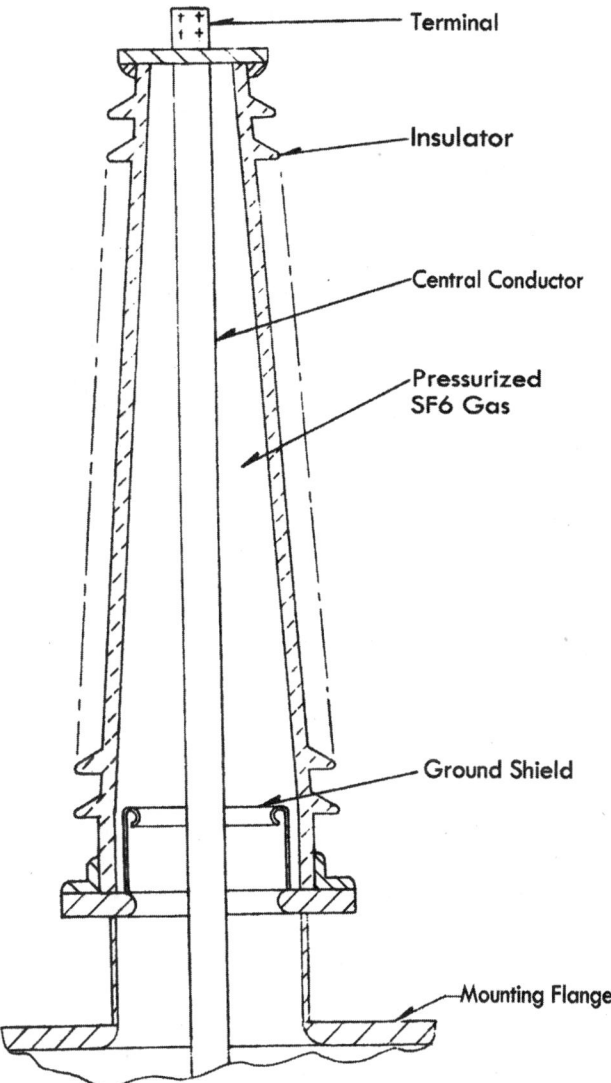

FIGURE 3.151 Pressurized SF$_6$ gas bushing.

bus. It also recommends practices for allowable cantilever loading caused by the pull of the line connected to the bushing, applications for contaminated environments and high altitudes, and maintenance practices.

5. IEC Publication 137, Bushings for Alternating Voltages above 1000 V. This standard is the IEC equivalent to the first standard listed above and is used widely in European and Asian countries.

Important Design Parameters

Conductor Size and Material

The conductor diameter is determined primarily by the current rating. There are two factors at work here. First, the skin depth of copper material at 60 Hz is about 1.3 cm and that of aluminum is about 1.6 cm. This means that most of the current will flow in the region from the outer portion of the conductor and radially inward to a depth of the skin depth δ. Second, the losses generated within a conductor will be:

$$P_{loss} = I^2 R = I^2 \rho L / A = 4 I^2 \rho L / \pi \left(D_1^2 - D_0^2 \right) \tag{3.75}$$

where I = rated current
ρ = resistivity of the conductor material, ohm m
L = length of conductor, m
A = cross-section of conductor = $\pi(D_1^2 - D_0^2)/4$
D_1 = outside diameter of conductor, m
$D_0 = D_1 - \delta$, m

It can be seen from Eq. (3.75) that P_{loss} decreases as D_1 increases. Hence, design practice is to increase the outside diameter of the conductor for higher current ratings and to limit the wall thickness to near the skin depth. There are other technical advantages to increasing the outside conductor diameter. First, from Eq. (3.74), observe that electric field stress reduces as $d = D_1$ increases. Therefore, a larger diameter conductor will have higher partial discharge inception and withstand voltages. Second, the mechanical strength of the conductor is dependent on the total cross-sectional area of the conductor, so that a larger diameter is sometimes used to achieve higher withstand forces in the conductor.

Insulators

Insulators must have sufficient length to withstand the steady state and transient voltages that the bushing will experience. Adequate lengths depend on the insulating media in which the insulator is used and on whether the bushing is capacitively graded. In cases where there are two different insulating media on either side of an insulator, the medium with the inferior dielectric characteristics determines the length of the insulator.

Air Insulators

Primary factors that determine the required length of insulators used in air at atmospheric pressure are lightning impulse voltage under dry conditions and power frequency and switching impulse voltages under wet conditions. Standard dry conditions are based on 760 mm Hg atmospheric pressure and 20°C, and wet conditions are discussed in the section entitled "Low Frequency Tests".

Bushings are normally designed to be adequate for altitudes up to 1000 m (3300 ft). Beyond 1000 m, longer insulators must be used in order to accommodate the lower reduced air density present at the higher altitudes.

Insulators exposed to pollution must have adequate creep distance, measured along the external contour of the insulator, for purposes of withstanding detrimental insulating effects of contamination on the insulator surface. Figure 3.149 shows the undulations on the weather sheds, and additional creep distance is obtained by adding undulations or increasing their depth. Recommendations (IEEE, 1997b) for creep distance are shown in Table 3.18, according to four different classifications of contamination.

For example, a 345-kV bushing has a maximum line to ground voltage of 220 kV, so that the minimum creep is 220 × 28 = 6160 mm for a light contamination level and 220 × 44 = 9680 mm for a heavy contamination level.

The term ESDD (Equivalent Salt Density Deposit) used in Table 3.18 is the conductivity of the water-soluble deposits on the insulator surface. It is expressed in terms of the density of sodium chloride

TABLE 3.18 Recommended Creep Distances for Four Contamination Levels

Contamination Level	(ESDD) (mg/cm²)	Recommended Minimum Creep Distance (mm/kV)
Light	0.03–0.08	28
Medium	0.08–0.25	35
Heavy	0.25–0.6	44
Extra heavy	Above 0.6	54

Transformers

deposited on the insulator surface that will produce the same conductivity. Following are typical environments for the four contamination levels listed (IEEE, 1997b):

Light contamination areas include areas without industry and with low density emission producing residential heating systems, and areas with some industrial areas or residential density but with frequent winds and/or precipitation. These areas are not exposed to sea winds or located near the sea.

Medium contamination areas include areas with industries not producing highly polluted smoke and/or with average density of emission producing residential heating systems, areas with high industrial and/or residential density but subject to frequent winds and/or precipitation, and areas exposed to sea winds but not located near the sea coast.

Heavy contamination areas include those areas with high industrial density and large city suburbs with a high density emission producing residential heating systems, and areas close to the sea or exposed to strong sea winds.

Extra heavy contamination areas include those areas subject to industrial smoke producing thick conductive deposits and small coastal areas exposed to very strong and polluting sea winds.

Oil Insulators

Since mineral oil is dielectrically stronger than air, the length of insulators immersed in oil is typically 30 to 40% the length of air insulators. In equipment having oil with low contamination levels, no sheds are required on oil-immersed insulators. In situations where some contamination exists in the oil, such as carbon particles in oil insulated circuit breakers, small ripples are generally cast on the outer insulator surface exposed to the oil.

Pressurized SF_6 Gas Insulators

Since various pressures may be used for this application, the length of the insulator may be equal or less than an insulator immersed in oil. Since particles are harmful to the dielectric strength of any pressurized gas, precautions are generally taken to keep the SF_6 gas free of particles, and therefore, in such cases no sheds are required on the insulators.

Flange

The flange has two purposes: First, to mount the bushing to the apparatus on which it is utilized; second, to contain the gaskets located on the extreme ends of the flange, as described in the section entitled "Clamping System". Flange material may be cast aluminum for high activity bushings where the casting mold can be economically justified. In cases where production activities are not so high, flanges may be fabricated from steel or aluminum plate material. A further consideration for high current bushings is that aluminum or some other non-magnetic material is used in order to eliminate magnetic losses caused by currents induced in the flange by the central conductor.

Oil Reservoir

An oil reservoir, often called the expansion cap, is required on larger bushings with self-contained oil for at least one and often two related reasons. First, mineral oil expands and contracts with temperature, and the oil reservoir is required to contain the oil expansion at high oil temperatures. Second, oil-impregnated insulating paper must be totally submerged in oil in order to keep its insulating qualities. Hence, the reservoir must have sufficient oil in it to maintain oil over the insulating paper at the lowest anticipated temperatures. Since oil is an incompressible fluid, the reservoir must also contain a sufficient volume of gas, such as nitrogen, so that excessive pressures are not created within the bushing at highest temperatures. Excessive pressures within a bushing may cause oil leakage.

On bushings for mounting at angles up to about 30° from vertical, the reservoir is mounted on the top end of the bushing. On smaller, lower voltage bushings, the reservoir may be within the top end of the upper insulator. Oil-filled bushings that are horizontally mounted usually have an oil reservoir mounted on the flange, but some have bellows, either inside or outside the bushing, which expand and contract with the temperature of the oil.

For the purpose of checking the oil level in the bushing, an oil level gage is often incorporated into the reservoir. There are two basic types of oil gages, the clear glass type and the magnetic type. The former type is cast from colored or clear glass such that the oil level can be seen from any angle of rotation around the bushing. The second type is two-piece gage; the part inside the reservoir being a float attached to a magnet that rotates on an axis perpendicular to the reservoir wall. The part outside the reservoir is then a gage dial attached to a magnet that follows the rotation of the magnet mounted inside the reservoir. This type of gage suffers the disadvantage of only being able to viewed at an angle of approximately 120° around the bushing. For this reason, bushings with this type of gage are normally rotated on the apparatus such that the gage can be seen from ground level.

Clamping System

Two types of clamping systems are generally used on bushings. The first uses an external flange on the end of each insulator, and bolts are used to fasten them to mating parts, i.e., the mounting flange and the top and bottom terminals. A grading ring is often placed over this area so as to shield the bolts from electric fields. The pressurized gas bushing shown in Fig. 3.151 uses this type of clamping system.

The second system is to place a compression type spring assembly in the reservoir located at the top of the bushing, thereby placing the central conductor in tension when the spring assembly is released. This action simultaneously places the insulators, flange, and gaskets between these members, and the terminals at the extreme ends of the insulators in compression, thereby sealing the gaskets.

Whatever method is used for the clamping system, the clamping force must be adequate to withstand the cantilever forces that will be exerted on the ends of a bushing during its service life. The major mechanical force to which the top end of an outdoor bushing is subjected during service is the cantilever force applied to the top terminal by the line pull of the connecting lead. This force is comprised of the static force exerted during normal conditions plus those forces exerted due to wind loading and/or icing on the connecting lead. In addition, bushings mounted at an angle from vertical exert a force equivalent to a static cantilever force at the top of the bushing, and this force must be accounted for in the design.

In addition to the static forces, bushings must also withstand short-time dynamic forces created by short-circuit currents and seismic shocks. In particular, the lower end of bushings mounted in circuit breakers must also withstand the forces created by the interruption devices within the breaker.

In order to give the user some guidance for allowable line pull, standards (IEEE, 1997b) recommend permissible loading levels: the static line loading should not exceed 50% of the test loading, as defined later in the section entitled "Mechanical Tests", and the short-time, dynamic loading should not exceed 85% of the same test loading.

Other Features on Bushings

Voltage Taps

It is possible within capacitance-graded bushings to create a capacitance divider arrangement wherein a small voltage, on the order of 5 kV, appears at the "voltage tap" when the bushing is operated at normal voltage. The voltage tap is created by attaching to one of the grading elements just to the inside of the grounded element. This tap, shown in Fig. 3.149, can be used during the testing operation of the bushing and the apparatus into which it is installed, as well as during field operation. In the former application, it is used to measure partial discharge within the bushing tested by itself or within the transformer. It is used during field operation to provide voltage to relays, which monitor phase voltages and instruct the circuit breakers to operate under certain conditions.

Bushing Current Transformer Pockets

The bushing flange creates a very convenient site to locate bushing current transformers (BCTs). The flange is extended on its inner end, and the BCTs, having 500 to 5000 turns in the windings, are placed around the flange. This location is called the BCT pocket and is shown in Fig. 3.149. In this case, the bushing central conductor forms the single turn primary of the BCT, and the turns in the windings form the secondary.

Lower End Shield

It can be seen from Fig. 3.150 that all regions of the lower end of air-to-oil bushings experience high dielectric stresses. In particular, the areas near the corners of the lower terminal are very highly stressed. Therefore, electrostatic shields with large radii, such as the one shown on Fig. 3.149, are attached to the lower end of these bushings in order to reduce the electric fields that appear in this area. Such shields also serve the purpose of shielding the bolted connections used to connect the lead to the bushing. Since shields with a thin dielectric barrier are somewhat stronger dielectrically, crepe paper is wrapped or molded pressboard is placed on the outer surfaces of the shield.

Tests on Bushings

Categories of Tests

Standards (ANSI/IEEE, 1997) designate three types of tests to be applied to bushings, as follows:

Design Tests

Design, or type, tests are only made on prototype bushings, i.e., the first of a design. The purpose of design tests is to ascertain that the bushing design is adequate to meet its assigned ratings, to ensure that the bushing can operate satisfactorily under usual or special service conditions, and to demonstrate compliance with industry standards. These tests need not be repeated unless the customer deems it necessary to have them performed on a routine basis.

Test levels at which bushings are tested during design tests are higher than the levels encountered during normal service so as to establish margins that take into account dielectric aging of insulation as well as material and manufacturing variations in successive bushings. Bushings must withstand these tests without evidence of partial or full failure, and incipient damage which initiates during the dielectric tests is usually detected by comparing values of power factor, capacitance, and partial discharge before and after the testing program.

Standards (ANSI/IEEE, 1997) prescribe the following design tests:

a. Low-frequency wet withstand voltage on bushings rated 242 kV maximum system voltage and less
b. Full-wave lightning impulse withstand voltage
c. Chopped-wave lightning impulse withstand voltage
d. Wet switching impulse withstand voltage on bushings rated 345 kV maximum system voltage and greater
e. Draw-lead bushing cap pressure test
f. Cantilever withstand test
g. Temperature test at rated current

Routine Tests

Routine, or production, tests are made on every bushing produced, and their purpose is to check the quality of the workmanship and the materials used in the manufacture. Standards (ANSI/IEEE, 1997) prescribe the following routine tests:

a. Capacitance and power factor measurements at 10 kV
b. Low-frequency dry withstand test with partial discharge measurements
c. Tap withstand voltage test
d. Internal hydraulic pressure test

Special Tests

Special tests are for establishing the characteristics of a design practice and are not part of routine or design tests. The only special test currently included in standards (ANSI/IEEE, 1997) is the thermal stability test, only applicable to EHV bushings, but other tests could be added in the future. These include short time, short-circuit withstand, and seismic capabilities.

Dielectric Tests

Low Frequency Tests

There are two low frequency tests: The **low-frequency wet-withstand voltage test** is applied on bushings rated 242 kV and below while a waterfall at a particular precipitation rate and conductivity is applied. The values of precipitation rate, water resistivity, and the time of application vary in different countries. American standard practice is a precipitation rate of 5 mm/min, a resistivity of 178 ohm-m, and a test duration of 10 sec, whereas European practice is 3 mm/min, 100 ohm-m, and 60 sec, respectively (IEEE, 1995). If the bushing flashes over externally during the test, it is allowed to apply the test one additional time. If this attempt also flashes over, then the test fails and something must be done to modify the bushing design or test set up so that the capability can be established.

The **low frequency dry withstand test** was until recently made for a 1-min duration without the aid of partial discharge measurements to detect incipient failures, but standards (ANSI/IEEE, 1997) currently specify a 1-h duration for the design test, in addition to partial discharge measurements. The present test procedure is:

1. Partial discharge (either radio influence voltage or apparent charge) shall be measured at 1.5 times the maximum line-ground voltage. Maximum limits for partial discharge vary for different bushing constructions and range from 10 to 100 μV or pC.
2. A 1-min test at the dry withstand level, approximately 1.7 times the maximum line-ground voltage, is applied. If an external flashover occurs, it is allowed to make another attempt; but if this one also flashes over, the bushing fails the test. No partial discharge tests are required for this test.
3. Partial discharge measurements are repeated every 5 min during the 1-h test duration at 1.5 maximum line-ground voltage required for the design test. Routine tests specify only a measurement of partial discharge at 1.5 maximum line-ground voltage, after which the test is considered complete.

Bushing standards were changed in the early 1990s to align with the transformer practice, which started to use the 1-h test with partial discharge measurements in the late 1970s. Experience with this new approach has been good in that incipient failures were uncovered in the factory test laboratory, rather than in service, and it was decided to add this procedure to the bushing test procedure. Also, from a more practical standpoint, bushings are applied to every transformer, and transformer manufacturers require that these tests be applied to the bushings prior to application so as to reduce the number of bushing failures during the transformer tests.

Wet Switching Impulse Withstand Voltage

This test is required on bushings rated 345 kV systems and above. The test waveshape is 250 μs time to crest and 2500 μs time to half value with tolerance of ±30% on the time to crest and ±20% on the time to half value. This is the standard waveshape for testing insulation systems without magnetic core steel present in the test object and is different than the waveshape for transformers.

Three different standard test procedures are commonly used for establishing the wet switching impulse withstand voltage of the external insulation:

1. Fifteen impulses of each polarity are applied with no more than two flashovers.
2. Three impulses of each polarity are applied. If a flashover occurs, then it is permitted to apply three additional impulses. If no flashovers occur at either polarity, then the bushing passes the test. Otherwise, the bushing fails the test.
3. The 90% (1.3 σ) level is established from the 50% flashover tests.

Lightning Impulse Tests

The same waveshapes are used for establishing the lightning impulse capability of bushings and transformers. The waveshape for the full wave is 1.2 μs for the wavefront and 50 μs for the time to half value, and the chopped wave flashes over at a minimum of 3.0 μs. One of the same procedures as described

Transformers

above for the wet switching impulse tests is followed to establish the full wave capability for both polarities. The chopped wave capability is established by applying a minimum of three chopped impulses at each polarity.

Mechanical Tests

IEEE Std. C57.19.01 (IEEE, 1997a) specifies the static cantilever withstand forces to be applied separately to the top and bottom ends of outdoor apparatus bushings. The forces applied to the top end range from 150 lb for the smaller, lower voltage bushings to 1200 lb for the larger, higher voltage or current bushings, and the forces applied to the lower end are generally about twice the top end forces.

The test procedure is to apply the specified forces perpendicular to the bushing axis, first at one end and then at the other, each application of force lasting 1 min. Permanent deflection, measured at the bottom end, shall not exceed 0.76 mm, and there shall be no oil leakage at either end at any time during the test or within 10 min after removing the force.

Thermal Tests

There are two thermal tests. The first is the thermal test at rated current, and it is applied to all bushing designs. The second test is the thermal stability test and it is applied for only EHV bushings:

Thermal Test at Rated Current

This test demonstrates a bottom-connected bushing's ability to carry rated current. The bushing is first equipped with a sufficient number of thermocouples, usually placed inside the inner diameter of the hollow tube conductor, to measure the hottest spot temperature of the conductor. The bushing is then placed in an oil-filled tank, the oil is heated to a temperature rise above ambient air of 55°C for transformers and 40°C for circuit breakers, and rated current is passed through the central conductor until thermal equilibrium is reached. The bushing passes the test if the hottest spot temperature rise above ambient air does not exceed 65°C.

Thermal Stability Test (Wagenaar, 1994)

Capacitive leakage currents in the insulating material within bushings cause dielectric losses. Dielectric losses within a bushing can be calculated by the following equation using data directly from the nameplate or test report:

$$P_d = 2\pi f C V^2 \tan\delta \qquad (3.76)$$

where P_d = dielectric losses, W
 f = applied frequency, Hz
 C = capacitance of bushing (C_1), Farads
 V = operating voltage, rms V
 $\tan\delta$ = dissipation factor, p.u.

A bushing operating at rated voltage and current generates both ohmic and dielectric losses within the conductor and insulation, respectively. Since these losses, which both appear in the form of heat, are generated at different locations within the bushing, they are not directly additive. However, heat generated in the conductor influences the quantity of heat which escapes from within the core. A significant amount of heat generated in the conductor will raise the conductor temperature and prevent losses from escaping from the inner surface of the core. This causes the dielectric losses to escape from only the outer surface of the core, and consequently raises the hottest spot temperature within the core. Most insulating materials display an increasing dissipation factor, $\tan\delta$, with higher temperatures, such that as the temperature rises, $\tan\delta$ also raises, which in turn, raises the temperature even more. If this cycle does not stabilize, then $\tan\delta$ increases rapidly and total failure of the insulation system ensues.

Bushing failures due to thermal stability have occurred both on the test floor and in service. One of the classic symptoms of a thermal stability failure is the high internal pressure caused by the gases generated from the deteriorating insulation. These high pressures cause an insulator, usually the outer one because of its larger size, either to lift off the flange or to explode. If the latter event occurs with a porcelain insulator, shards of porcelain saturated with oil become flaming projectiles that may endanger the lives of personnel and cause damage to nearby substation equipment.

Note from Eq. (3.76) that the operating voltage, V, particularly influences the losses generated within the insulating material. It has been found from testing experience that thermal stability only becomes a factor at operating voltages 500 kV and above.

The test procedure given in (ANSI/IEEE, 1997) is to first immerse the lower end of the bushing in oil at a temperature of 95°C tan δ and then pass rated current through the bushing. When the bushing comes to thermal equilibrium, a test voltage equal to 1.2 times the maximum line-ground voltage is applied, and tan δ is measured at regular, normally hourly, intervals. These conditions are maintained until tan δ rises no more than 0.02% over a period of 5 h. The bushing is considered to have passed the test if it has reached thermal stability at this time and it withstands all of the routine dielectric tests without significant change from the previous results.

References

ANSI/IEEE Std. C57.19.00 00-1991(R1997), Standard Performance Characteristics and Test Procedure for Outdoor Power Apparatus Bushings, 1997.
Easley, J. K. and Stockum, F. R., Bushings, IEEE Tutorial on Transformers, IEEE Publication EH0209-7/84/0000-0032.
IEEE Std. C57.19.01-1991 (R1997), Standard General Requirements and Test Procedure for Outdoor Power Apparatus Bushings, 1997a.
IEEE Std. C57.19.03-1996, Standard Requirements, Terminology and Test Procedures for Bushings for DC Applications, 1996.
IEEE Std. C57.19.100-1995 (R1997), Guide for Application of Power Apparatus Bushings, 1997b.
IEC Publication 137, Bushings for Alternating Voltages above 1000 V.
IEEE Std. 4-1995, Standard Techniques for High Voltage Testing.
Nagel, R., Uber Eine Neuerung An Hochspannungstransformer Der Siemens- Schuckertwerke, Elektrische Bahnen Und Betriebe, 4, 275-278, May 23, 1906.
Reynders, A. B., Condenser type of insulation for high-tension terminals, *AIEE Trans.*, XXVIII, Part I, 209; 220, 1909.
Spindle, H. E., Project Manager, Evaluation, design and development of a 1200 kV prototype termination, U.S. Dept. of Energy Report DOE/ET/29068-T8 (DE86005473).
Wagenaar, L. B., The significance of thermal stability tests in EHV bushings and current transformers, 1994 Doble Conference Paper 4-6.

3.14 Load Tap Changers (LTCs)

Dieter Dohnal and Wolfgang Breuer

For many decades power transformers equipped with LTCs have been the main components of electrical networks and industry. The LTC allows voltage regulation and/or phase shifting by varying the transformer ratio under load without interruption.

From the beginning of LTC development, two switching principles have been used for the load transfer operation, the high-speed resistance-type and the reactance-type. Over the decades both principles have been developed into reliable transformer components available in a broad range of current and voltage applications to cover the needs of today's network and industrial process transformers as well as ensuring optimum system and process control (Goosen, 1996).

This section refers to LTCs immersed in transformer mineral oil. The use of other insulating fluids or gas insulation requires the approval of the LTC's manufacturer and may lead to a different LTC design.

Transformers

Principle Design

The LTC changes the ratio of a transformer by adding turns to or subtracting turns from either the primary or the secondary winding. The main components of an LTC are contact systems for make and break currents as well as carrying currents, transition impedances, gearings, spring energy accumulators, and a drive mechanism.

The transition impedance in the form of a resistor or reactor consists of one or more units that are bridging adjacent taps for the purpose of transferring load from one tap to the other without interruption or appreciable change in the load current. At the same time they are limiting the circulating current for the period when both taps are used. Normally, reactance-type LTCs use the bridging position as a service position and, therefore, the reactor is designed for continuous loading.

The voltage between the mentioned taps is the step voltage. It normally lies between 0.8% and 2.5% of the rated voltage of the transformer.

The majority of resistance-type LTCs are installed inside the transformer tank (in-tank LTCs), whereas the reactance-type LTCs are in a separate compartment that is normally welded to the transformer tank.

Resistance-Type Load Tap Changer

The LTC design that is normally applied to larger powers and higher voltages is comprised of an arcing switch and a tap selector. For lower ratings LTC designs are used where the functions of the arcing switch and the tap selector are combined in a so-called arcing tap switch.

With an LTC comprising an arcing switch and a tap selector (Fig. 3.152), the tap change takes place in two steps (Fig. 3.153). First the next tap is preselected by the tap selector at no load (Fig. 3.153, position a–c). Then the arcing switch transfers the load current from the tap in operation to the preselected tap (Fig. 3.153, position c–g). The LTC is operated by means of a drive mechanism. The tap selector is operated by a gearing directly from the drive mechanism. At the same time, a spring energy accumulator

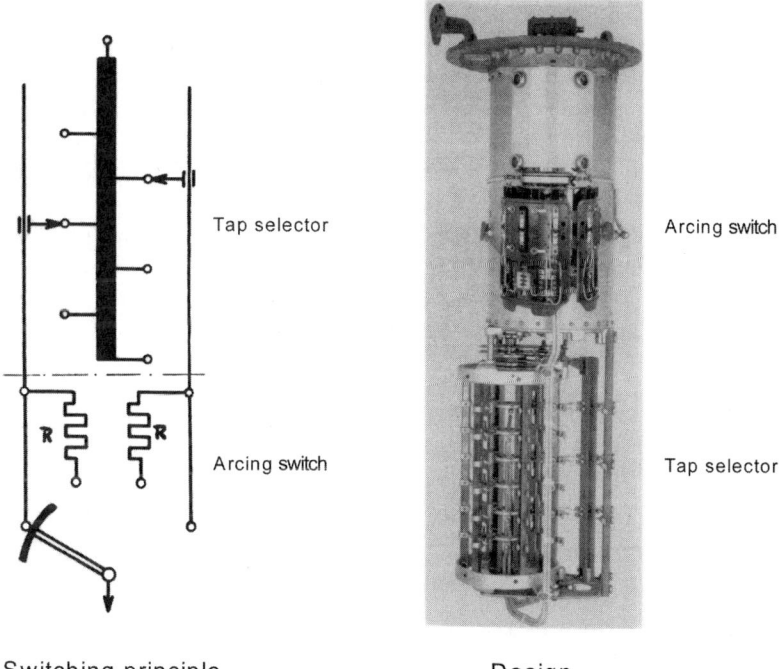

FIGURE 3.152 Design principle — arcing switch with tap selector.

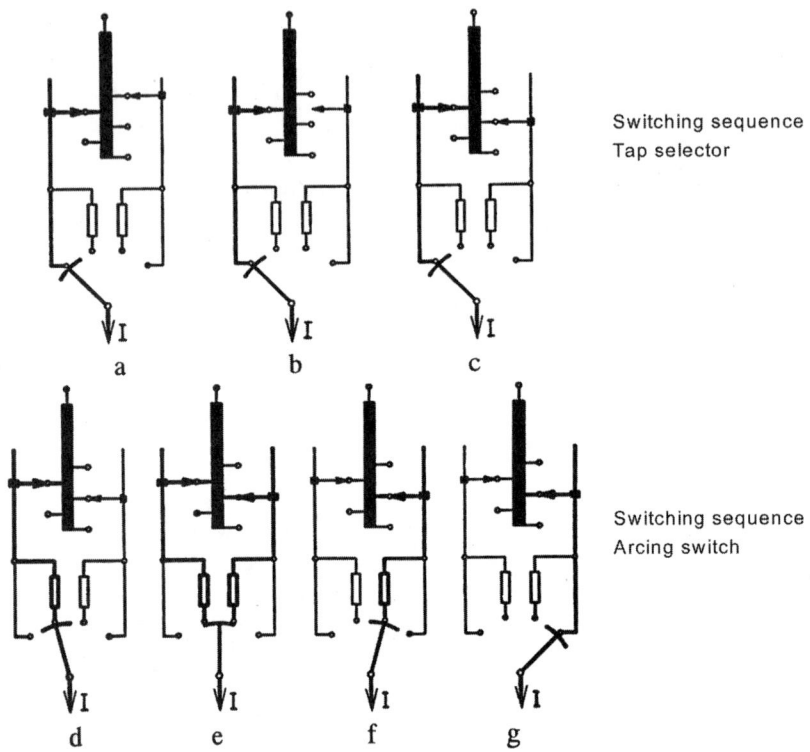

FIGURE 3.153 Switching sequence of tap selector — arcing switch.

is tensioned. This operates the arcing switch — after releasing in a very short time — independently of the motion of the drive mechanism. The gearing ensures that this arcing switch operation always takes place after the tap preselection operation has been finished. The switching time of an arcing switch lies between 40 and 60 ms with today's designs.

During the arcing switch operation, transition resistors are inserted (Fig. 3.153, position d–f) which are loaded for 20 to 30 ms, i.e., the resistors can be designed for short-term loading. The amount of resistor material required is therefore relatively small. The total operation time of an LTC is between 3 and 10 sec, depending on the respective design.

An arcing tap switch (Fig. 3.154) carries out the tap change in one step from the tap in service to the adjacent tap (Fig. 3.155). The spring energy accumulator, wound up by the drive mechanism actuates the arcing tap switch sharply after releasing. For switching time and resistor loading (Fig. 3.155, position b–d), the above statements are valid.

The details of switching duty including phasor diagrams are described in Annex A of (IEEE, 1995).

Reactance-Type Load Tap Changer

For reactance-type LTCs, the following types of switching are used [Annex B of (IEEE, 1995)]:

- Arcing tap switch
- Arcing switch with tap selector
- Vacuum interrupter with tap selector

Today the greater part of arcing tap switches is produced for voltage regulators, whereas the vacuum-type LTC is going to be the state of the art in the field of power transformers.

Transformers

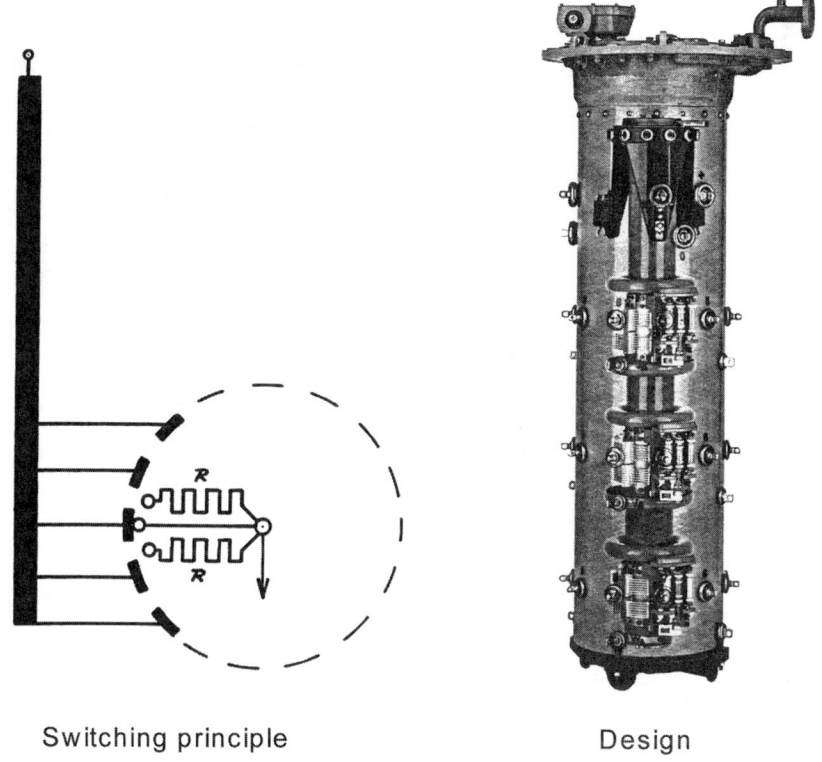

FIGURE 3.154 Design principle — arcing tap switch.

Figure 3.156 shows a three-phase vacuum-type LTC with full insulation between phases and to ground (nominal voltage level 69 kV). It consists of an oil compartment containing tap selector and reversing/coarse change-over selector, vacuum interrupters, and bypass switches.

A typical winding layout and the operating sequence of the said LTC is shown in Fig. 3.157. The operating sequence is divided into three major functions:

1. Current transfer from the tap selector part preselecting the next tap to the part remaining in position by means of the vacuum interrupter in conjunction with the associated bypass switch (Fig. 3.157-b, position A–C).
2. Selection of the next tap position by the tap selector in proper sequence with the reclosing of the vacuum interrupter and bypass switch (Fig. 3.157-b, position C–F). Contrary to resistance-type LTCs, the bridging position in which the moving selector contacts p1 and p4 are on neighboring fixed selector contacts (in Fig. 3.157-b, contacts 4 and 5, position F) is a service position and therefore the preventive autotransformer/reactor (normally produced by the transformer manufacturer) is designed for continuous loading; i.e., the number of tap positions is twice the number of steps of the tap winding. In other words, the preventive autotransformer works as a voltage divider for step voltage of the tap winding in the bridging position. In comparison with the resistance-type LTC, the reactance-type LTC requires only half the number of taps of the tap winding for the equivalent number of service tap positions.
3. Operation of reversing or coarse change-over selector in order to double the number of positions; for this operation the moving selector contacts p1 and p2 have to be on the fixed selector contact M (Fig. 3.157-a).

For more detailed information about switching duty and phasor diagrams, see Annex B of (IEEE, 1995).

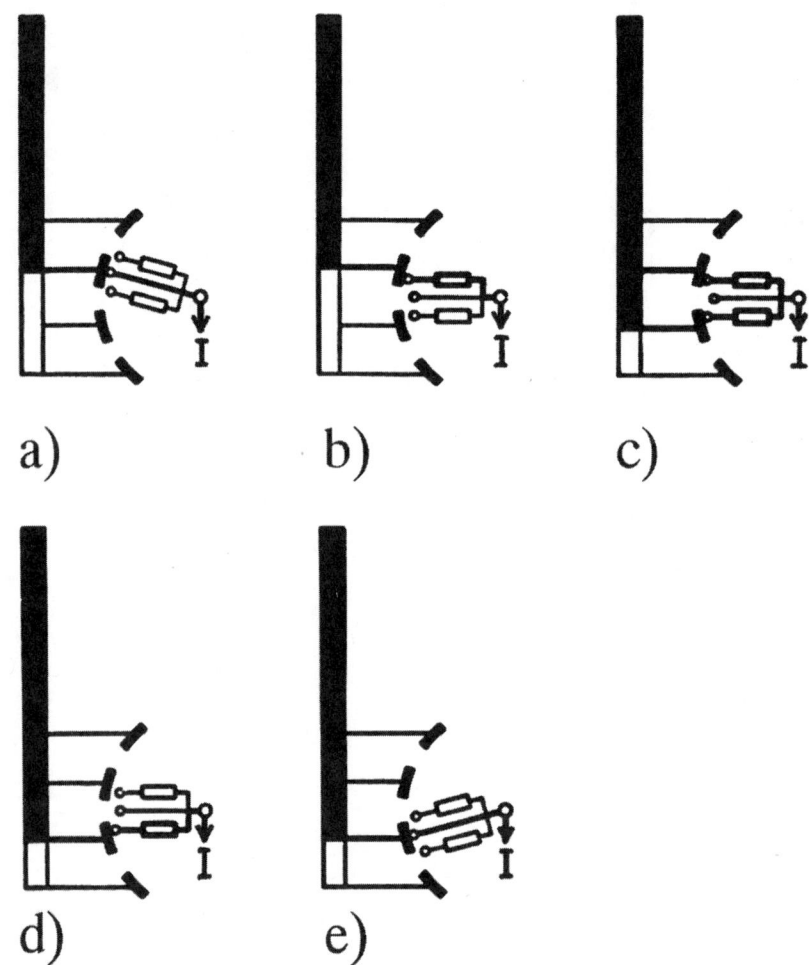

FIGURE 3.155 Switching sequence of arcing tap switch.

Applications of Load Tap Changers

Basic Arrangements of Regulating Transformers

The following basic arrangements of tap windings are used (Fig. 3.158).

Linear arrangement (Fig. 3.158-a), is generally used on power transformers with moderate regulating ranges up to a maximum of 20%.

With a **reversing change-over selector** (Fig. 3.158-b), the tap winding is added to or subtracted from the main winding so that the regulating range can be doubled or the number of taps be reduced. During this operation the tap winding is disconnected from the main winding (for a discussion on problems arising from this disconnection, see the section entitled "Potential Connection of Tap Winding During Change-Over Operaiton"). The greatest copper losses occur, however, in the position with the minimum number of effective turns. This reversing operation is realized with the help of a change-over selector which is part of the tap selector or of the arcing tap switch. The **double reversing change-over selector** (Fig. 3.158-c) avoids the disconnection of tap winding during the change-over operation. In phase-shifting transformers (PST), this apparatus is called an **advance-retard switch (ARS)**.

By means of a **coarse change-over selector** (Fig. 3.158-d) the tap winding is either connected to the plus or minus tapping of the coarse winding. Also during coarse selector operation, the tap winding is

FIGURE 3.156 Reactance-type LTC, inside view showing vacuum interrupter assembly.

disconnected from the main winding (special winding arrangements can cause same disconnection problems as above, the series impedance of coarse winding/tap winding has to be checked, see the section entitled "Effects of the Leakage Impedance of Tap Winding/Coarse Winding During Operation of the Arcing Switch When Passing the Mid-Position of the Resistance-Type LTC"). In this case, the copper losses are lowest in the position of the lowest effective number of turns. This advantage, however, puts higher demands on insulation material and requires a larger number of windings. The **multiple coarse change-over selector** (Fig. 3.158-e) allows a multiplication of the regulating range. It is mainly applied for industrial process transformers (rectifier/furnace transformers). The coarse change-over selector is also part of the LTC.

An arrangement of **LTC for tap winding with tickler coil** is applied when the tap winding has, for example, 8 steps and where the regulation should be carried out in ±16 steps. In this case, a tickler coil, whose voltage is half the step voltage of the tap winding, is used. The tickler coil is electrically separated from the tap winding and is looped into the arcing switch/tap selector connection. Figure 3.159 shows the switching sequence; in positions 1,3,5,..., the tickler coil is out of circuit; in positions 2,4,6,..., the tickler coil is inserted.

It depends on the system and the operating requirements which of these basic winding arrangements is used in the individual case. These arrangements are applicable to two winding transformers as well as to voltage and phase-shifting transformers (PST).

The position in which the tap winding and therefore the LTC is inserted in the windings depends on the transformer design.

Examples of Commonly Used Winding Arrangements

Two winding transformers with **wye connected windings** have the regulation applied to the neutral end as shown in Fig. 3.160. This results in relatively simple and compact solutions for LTCs and tap windings.

Regulation of **delta connected windings** (Fig. 3.161) requires a three-phase LTC whose three phases are insulated according to the highest system voltage applied (Fig. 3.161a), or 3 single-phase LTCs, or 1 single-phase and 1 two-phase LTC (Fig. 3.161b). Today, the design limit for three-phase LTCs with phase-to-phase insulation is a nominal voltage level of 138 kV (BIL 650 kV). To reduce the phase-to-phase stresses on the delta-LTC, the three pole mid-winding arrangement (Fig. 3.161c) can be used.

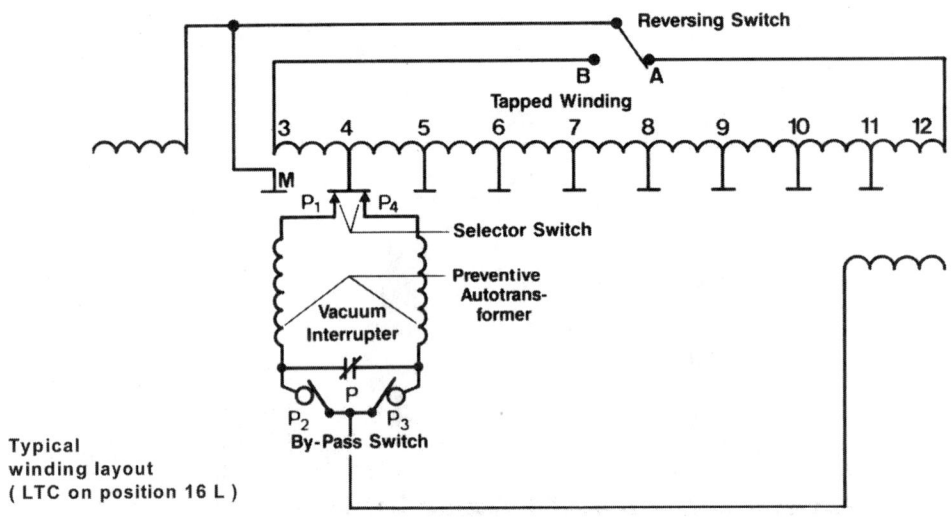

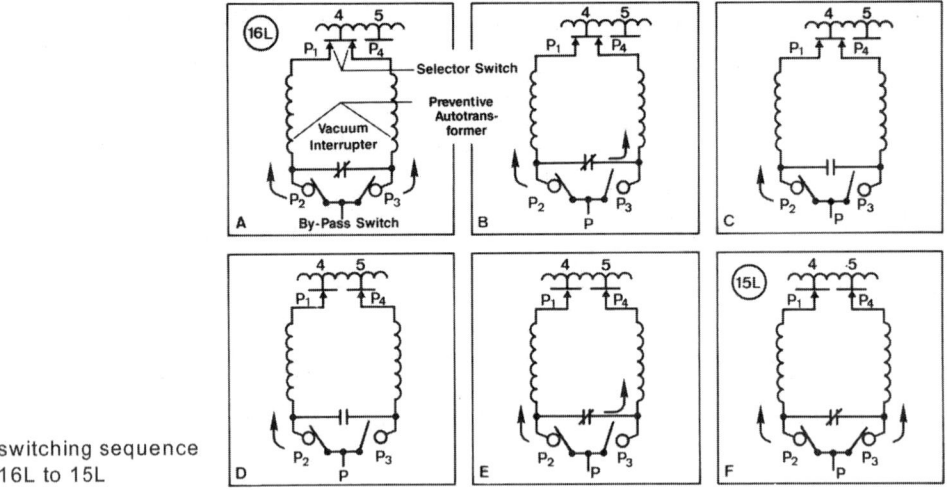

FIGURE 3.157 Reactance-type LTC. (a) Typical winding layout, LTC in position 16L. (b) Switching sequence, position 16L to 15L.

For regulated **autotransformers**, Fig. 3.162 shows various schemes. Depending on their regulating range, system conditions and/or requirements, weight and size restrictions during transportation, the most appropriate scheme is chosen. Autotransformers are always wye-connected.

- Neutral end regulation (Fig. 3.162a) may be applied with a ratio above 1:2 and a moderate regulating range up to 15%. It operates with variable flux.
- A scheme shown in Fig. 3.162c is used for regulation of high voltage U1.
- For regulation of low voltage U2 the circuits Fig. 3.162b,d,e,f are applicable. The arrangements Fig. 3.162e and 3.162f are two core solutions. Circuit Fig. 3.162f is operating with variable flux in the series transformer, but it has the advantage that a neutral end LTC can be used. In case of arrangement according to Fig. 3.162e, main and regulating transformers are often placed in separate tanks to reduce transport weight. At the same time this solution allows some degree of phase shifting by changing the excitation-connections within the intermediate circuit.

Transformers

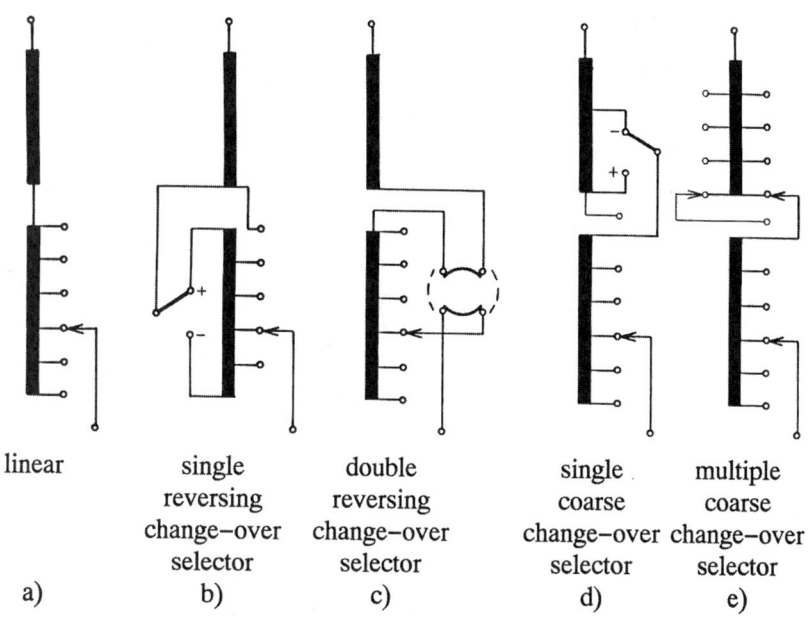

FIGURE 3.158 Basic arrangements of tap windings.

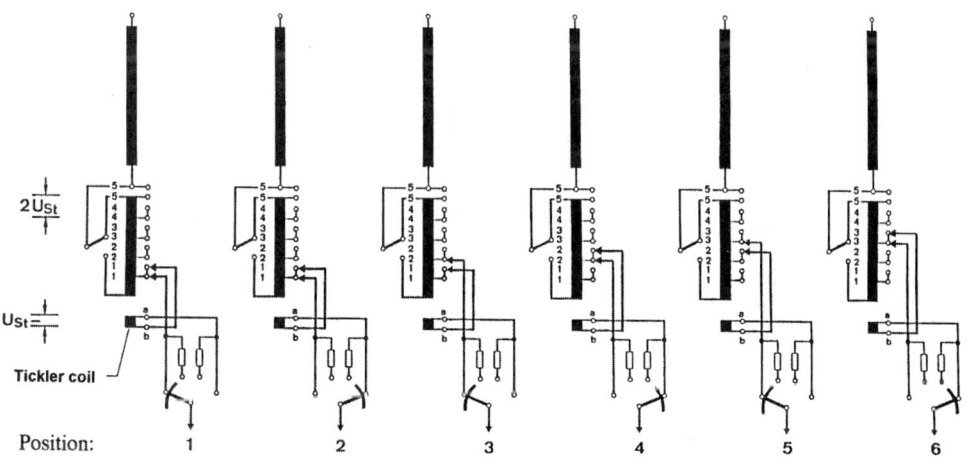

FIGURE 3.159 LTC for tap windings with tickler coil.

Phase-Shifting Transformers (PST)

In recent years, the importance of PSTs used to control the power flow on transmission lines in meshed networks has steadily been increasing (Krämer and Ruff, 1998). The fact that IEEE is preparing a "Guide for the Application, Specification and Testing of Phase-Shifting Transformers" proves the demand for PSTs. These transformers often require regulating ranges which exceed those normally used. To reach such regulating ranges, special circuit arrangements are necessary. Two examples are given in Figs. 3.163 and 3.164. Figure 3.163 shows a circuit with direct line-end regulation (single-core design). Fig. 3.164 shows an intermediate circuit arrangement (two-core design).

Figure 3.163 illustrates very clearly how the phase-angle between the voltages of the source- and load-system can be varied by the LTC position. Various other circuit arrangements have been realized.

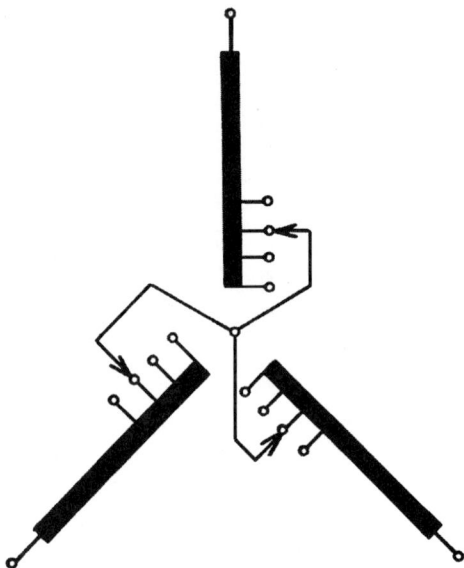

FIGURE 3.160 LTC with wye connected windings.

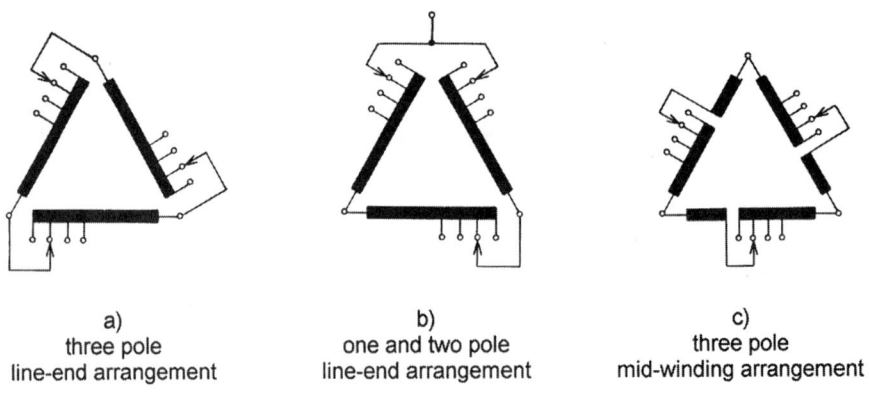

a)
three pole
line-end arrangement

b)
one and two pole
line-end arrangement

c)
three pole
mid-winding arrangement

FIGURE 3.161 LTC with delta connected windings.

The number of LTC operations of PSTs is much higher than that of other regulating transformers in networks (10 to 15 times higher). In some cases, according to regulating ranges — especially for line-end arrangements (Fig. 3.163) — the transient overvoltage stresses over tapping ranges have to be limited by the application of non-linear resistors. Furthermore, the short-circuit current ability of the LTC must be checked, as the short-circuit power of the network determines the said current. The remaining features of LTCs for such transformers can be selected according to the usual rules (see the section entitled "Selection of Load Tap Changers").

Significant benefits resulting from the use of a PST are:

- Reduction of overall system losses by elimination of circulating currents.
- Improvement of circuit capability by proper load management.
- Improvement of circuit power factor.
- Control of power flow to meet contractual requirements.

Transformers

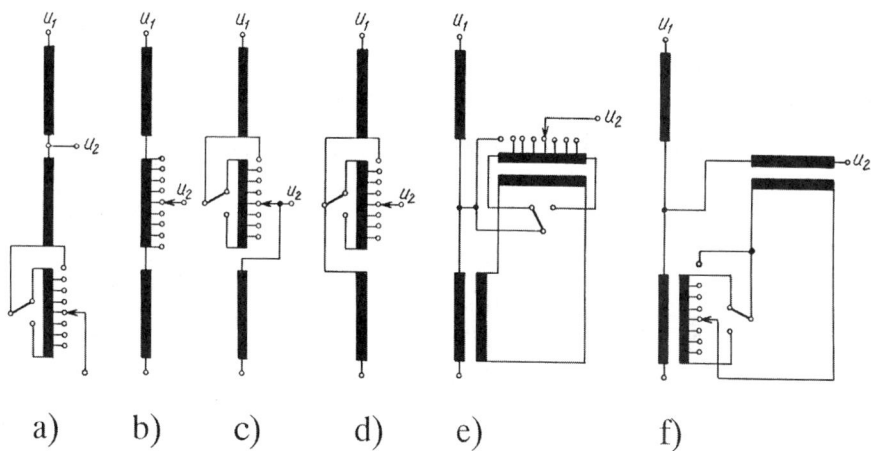

FIGURE 3.162 LTC in autotransformers.

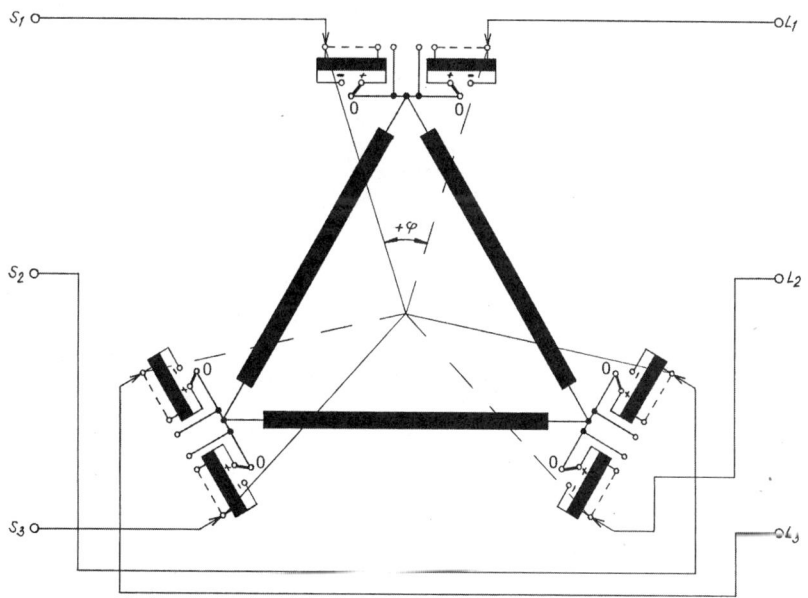

FIGURE 3.163 Phase-shifting transformer — direct circuit arrangement.

Rated Characteristics and Requirements for Load Tap Changers

The rated characteristics of an LTC are as follows:

- Rated through current[1]
- Maximum rated through current[1]
- Rated step voltage[1]
- Maximum rated step voltage[1]

---

[1]Within the maximum rated through current of the LTC there may be different combinations of values of rated through current and corresponding rated step voltage. Figure 3.165 shows that relationship. When a value of rated step voltage is referred to as a specific value of rated through current, it is called the relevant rated step voltage.

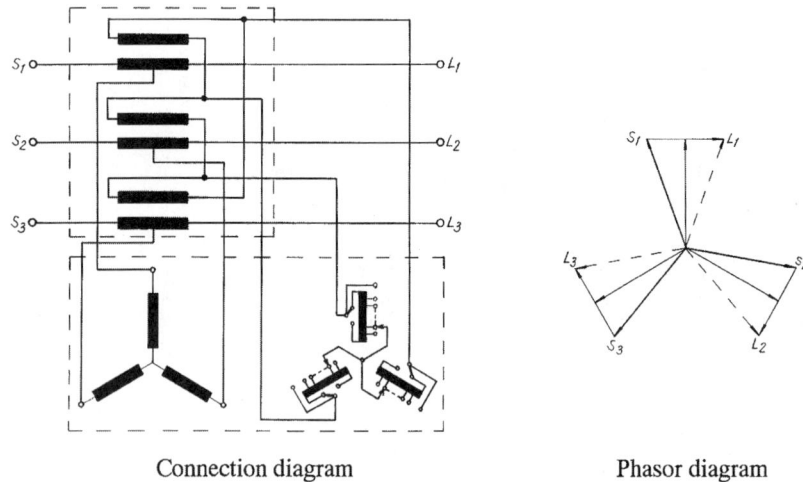

FIGURE 3.164 Phase-shifting transformer — intermediate circuit arrangement.

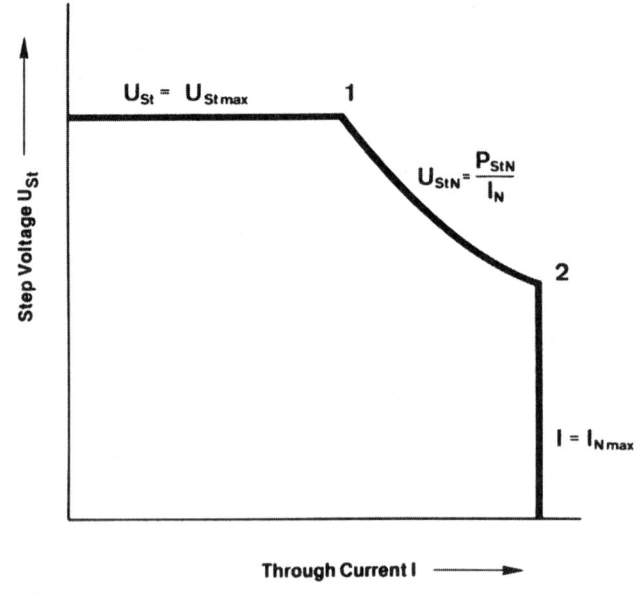

Correlation of step voltage U_{St} and through-current I at rated switching capacity

1 - Upper point of rated load
2 - Lower point of rated load

FIGURE 3.165 Characteristics of the arching switch — through current, step voltage, switching capacity.

- Rated frequency
- Rated insulation level

The basic requirements for LTCs are laid down in standards (IEEE Std. C57.131-1995 (1995); IEC 60214 (1989); and IEC 60542 (1988)).

The main features to be tested during design tests are:

- Contact life: IEEE Std. C57.131-1995-6.2.1.1/IEC 60214-8.2.1
  50,000 operations at the max rated through-current and the relevant rated step voltage shall be performed. The result of these tests may be used by the manufacturer to demonstrate that the contacts used for making and breaking current are capable of performing, without replacement of the contacts, the number of operations guaranteed by the manufacturer at the rated through current and the relevant rated step voltage.
- Temporary overload: IEEE Std. C57.131-1995-6.1.3 and 6.2.2/IEC 60214-8.1.1 and 8.2.2
  At 1.2 times maximum rated through current, temperature rise tests of each type of contact carrying current continuously shall be performed to verify that the steady-state temperature rise does not exceed 20°C above the temperature of the insulating fluid surrounding the contacts. In addition, breaking capacity tests shall be performed with 40 operations at a current up to twice the maximum rated through-current and at the relevant rated step voltage.
  LTCs that comply with the said design tests, and when installed and properly applied to the transformer, can be loaded in accordance with the applicable IEEE or IEC loading guides.
- Mechanical life: IEEE Std. C57.131-1995-6.5.1/IEC 60214-8.5.1
  A mechanical endurance test of 500,000 tap-change operations without load has to be performed. During this test the LTC shall be assembled and filled with insulating fluid or immersed in a test tank filled with clean insulating fluid, and operated as for normal service conditions.
  Compared with the actual number of tap-change operations in various fields of application (Table 3.19) it can be seen that the mechanical endurance test covers the service requirements.
- Short-circuit current strength: IEEE Std. C57.131-1995-6.3/IEC 60214-8.4
  All contacts of different design that carry current continuously shall be subjected to three short-circuit currents of 10 times maximum rated through current (valid for I > 400A) with an initial peak current of 2.5 times the rms value of the short-circuit current, each current application of at least 2-s duration.
- Dielectric requirements: IEEE Std. C57.131-1995-6.6/IEC 60214-8.6
  The dielectric requirements of an LTC depend on the transformer winding to which it is to be connected.
  The transformer manufacturer shall be responsible not only for selecting an LTC of the appropriate insulation level, but also for the insulation level of the connecting leads between the LTC and the winding of the transformer.
  The insulation level of the LTC is demonstrated by dielectric tests — in accordance with the standards (applied voltage, basic lightning impulse, switching impulse, partial discharge, if applicable) — on all relevant insulation spaces of the LTC.
  The test and service voltages of the insulation between phases and to ground shall be in accordance with the standards. The values of the withstand voltages of all other relevant insulation spaces of an LTC shall be declared by the manufacturer of the LTC.

TABLE 3.19 Number of LTC Switching Operations in Various Fields of Application

Transformer	Transformer Data			Number of On-Load Tap Changer Operations Per Year		
	Power MVA	Voltage kV	Current A	Min.	Medium	Max.
Power station	100–1300	110–765	100–2000	500	3,000	10,000
Interconnected	200–1500	110–765	300–3000	300	5,000	25,000
Network	15–400	60–525	50–1600	2,000	7,000	20,000
Electrolysis	10–100	20–110	50–3000	10,000	30,000	150,000
Chemistry	1.5–80	20–110	50–1000	1,000	20,000	70,000
Arc furnace	2.5–150	20–230	50–1000	20,000	50,000	300,000

- Oil tightness of arcing switch oil compartment: IEC 60214-7.1
  The analysis of gases dissolved in the transformer oil is an important, very sensitive, and commonly used indication for the operational behavior of a power transformer. To avoid any influence of the switching gases produced by each operation of the arcing switch, on the results of the said gas-in-oil analysis, the arcing switch oil compartment has to be oil tight. Furthermore, the arcing switch conservator tank must be completely separated from the transformer conservator tank on the oil and on the gas side.
  Vacuum and pressure withstand values of the oil compartment shall be declared by the manufacturer of the LTC.

Selection of Load Tap Changers

The selection of a particular LTC will render optimum technical and economical efficiency if requirements due to operation and testing of all conditions of the associated transformer windings are met. In general, usual safety margins may be neglected as LTCs designed, tested, selected, and operated in accordance with IEEE (1995) and IEC standards (1976; 1989) are most reliable.

To select the appropriate LTC, the following important data of associated transformer windings should be known:

- MVA rating
- Connection of tap winding (for wye, delta, or single-phase connection)
- Rated voltage and regulating range
- Number of service tap positions
- Insulation level to ground
- Lightning impulse and power frequency voltage of the internal insulation

The following LTC operating data may be derived from this information:

- Maximum through-current: I_{max}
- Step voltage: U_{st}
- Switching capacity: $P_{st} = U_{st} \times I_{max}$

and the appropriate tap changer can be determined:

- LTC type
- Number of poles
- Nominal voltage level of LTC
- Tap selector size/insulation level
- Basis connection diagram

If necessary, the following characteristics of the tap changer should be checked:

- Breaking capacity
- Overload capability
- Short-circuit current (especially to be checked in case of Fig. 3.163 applications)
- Contact life

In addition to that, the following two important LTC stresses resulting from the arrangement and application of the transformer design have to be checked.

Potential Connection of Tap Winding During Change-Over Operation

During the operation of the reversing or coarse change-over selector, the tap winding is disconnected momentarily from the main winding. It thereby takes a potential that is determined by the voltages of the adjacent windings as well as by the coupling capacities to these windings and to grounded parts. In general, this potential is different from the potential of the tap winding before the change-over selector

Transformers

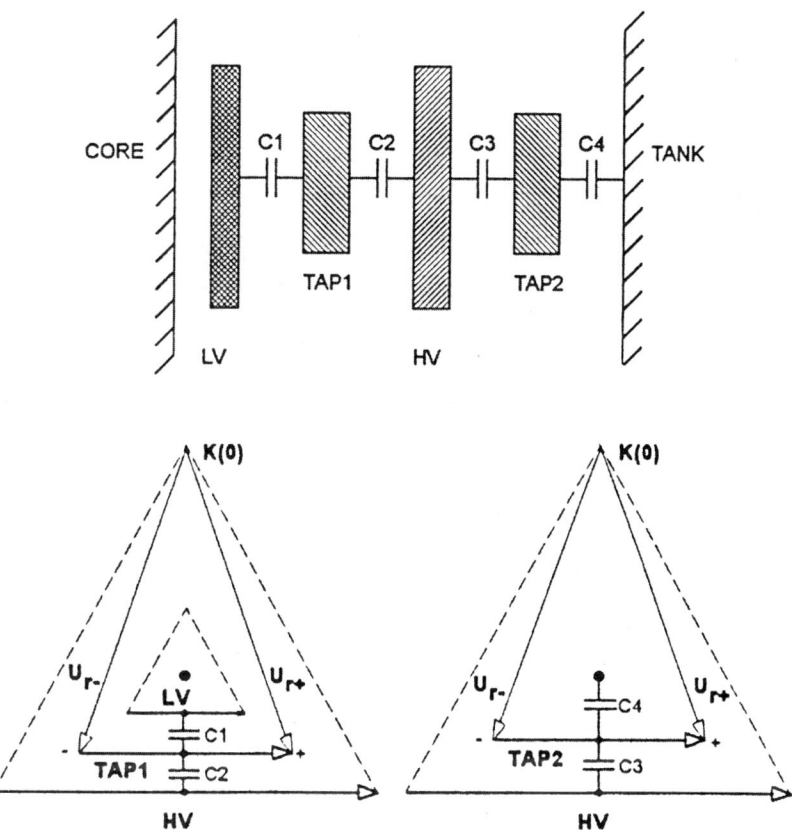

FIGURE 3.166 Phase-shifting transformer, circuit (as shown in Fig. 3.163). (a) Typical winding arrangement with two tap windings. (b) Recovery voltages (Ur+, Ur−) for tap windings 1 and 2 (phasor diagram).

operation. The differential voltages are the recovering voltages at the opening contacts of the change-over selector and, when reaching a critical level, they are liable to cause inadmissible discharges on the change-over selector. If these voltages exceed a certain limit value (for special product series, said limit voltages are in the range of 15 to 35 kV), measures regarding potential control of the tap winding must be taken.

Especially in case of PSTs with regulation at the line end (e.g., Fig. 3.163), high recovery voltages can occur due to the winding arrangement. Figure 3.166a illustrates a typical winding arrangement of PST according to Fig. 3.163. Figure 3.166b gives the phasor diagram of that arrangement without limiting measures. As it can be seen, the recovery voltages appearing at the change-over selector contacts are in the range of the system voltages on the source and the load side.

It is sure that an LTC cannot be operated under such conditions. This fact has already been taken into account during the planning stage of the PST design. There are three ways to solve the above-mentioned problem:

1. Install screens between the windings. These screens must have the potential of the movable change-over selector contact 0 (Fig. 3.163). See Figs. 3.167a and b.
2. Connect the tap winding to a fixed potential by a fixed ohmic resistor (tie-in resistor) or by an ohmic resistor which is only inserted during change-over selector operation by means of a potential switch. This ohmic resistor is usually connected to the middle of the tap winding and to the current take-off terminal of the LTC (Fig. 3.168).
3. Use an Advanced Retard Switch (ARS) as change-over selector (Fig. 3.169). This additional unit allows the change-over operation to be carried out in two steps without interruption. With this

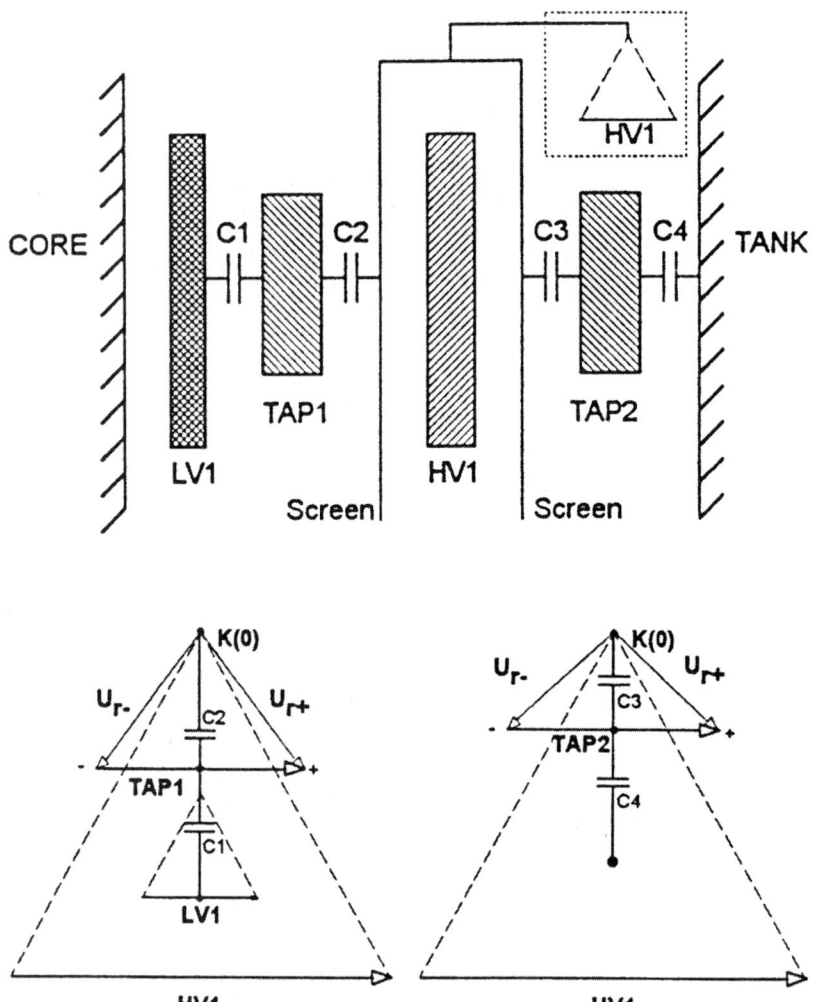

FIGURE 3.167 Phase-shifting transformer, circuit (as shown in Fig. 3.163). (a) Winding arrangement with two windings and screens. (b) Recovery voltages (U_{r+}, U_{r-}) for tap windings 1 and 2 (phasor diagram).

arrangement, the tap winding is connected to the desired potential during the whole change-over operation. As this method is relatively complicated, it is only used for high power PSTs.

The common method for the potential connection of tap windings is to use tie-in resistors. The following information is required to dimension tie-in resistors:

- all characteristic data of the transformer such as power, high and low voltages with regulating range, winding connection, insulation levels
- design of the winding, i.e., location of the tap winding in relation to the adjacent windings or winding parts (in case of layer windings)
- voltages across the windings and electrical position of the windings within the winding arrangement of the transformer which is adjacent to the tap winding
- capacity between tap winding and adjacent windings or winding parts
- capacity between tap winding and ground or, if existing, grounded adjacent windings
- surge stress across half of tap winding
- service and test power-frequency voltages across half of the tap winding

Transformers

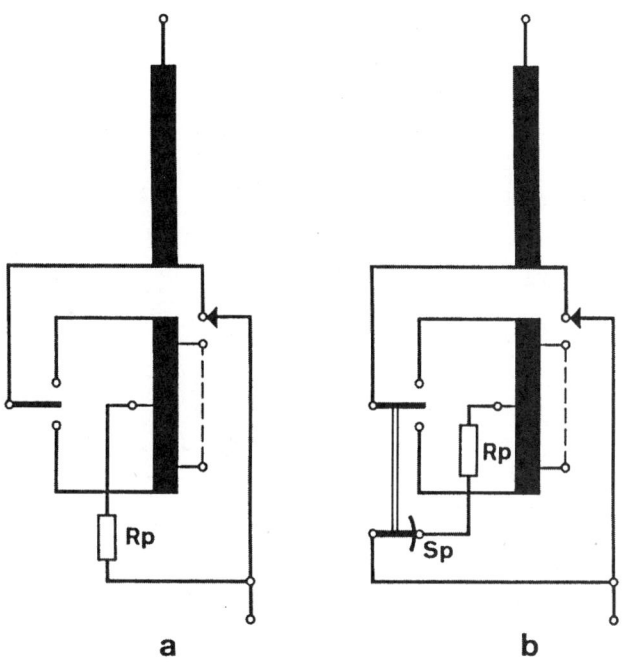

FIGURE 3.168 Methods of potential connection (reversing change-over selector in mid-position). (a) Fixed tie-in resistor Rp. (b) With potential switch Sp and tie-in resistor Rp.

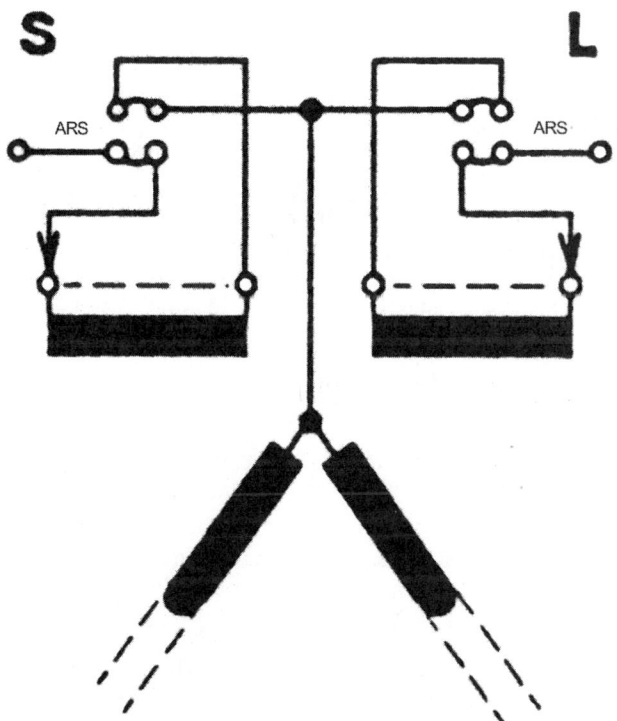

FIGURE 3.169 Phase-shifting transformer — change-over operation by means of an Advanced Retard Switch (ARS).

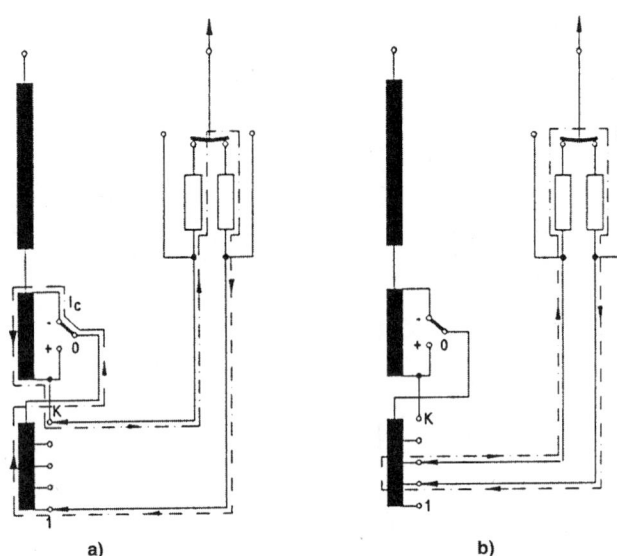

FIGURE 3.170 Effect of the leakage impedance of coarse winding/tap winding arrangement. (a) Operation through mid-position. (b) Operation through any tap poisiton beside mid-position.

Effects of the Leakage Impedance of Tap Winding/Coarse Winding During the Operation of the Arcing Switch When Passing the Mid-Position of the Resistance-Type LTC

During the operation of the arcing switch from the end of the tap winding to the end of the coarse winding and vice versa (passing mid-position, see Fig. 3.170a), all turns of the whole tap winding and coarse winding are inserted in the circuit.

This results in a leakage impedance value which is substantially higher than during operation within the tap winding where only negligible leakage impedance of one step is relevant (Fig. 3.170b). The higher impedance value in series with the transition resistors has an effect on the circulating current which is flowing in the opposite direction through coarse winding and tap winding during the arcing switch operation.

Consequently, a phase shift between switched current and recovery voltage takes place at the transition contacts of the arcing switch and may result in an extended arcing time.

In order to ensure optimum selection and adaptation of the LTC to these operating conditions, it is necessary to specify the leakage impedance of coarse winding and tap winding connected in series.

Protective Devices for Load Tap Changers

The protective devices for LTCs are designed to limit or prevent the effect of the following stresses: inadmissible increase of pressure within the arcing switch compartment or the separate compartment of the reactance-type LTC, respectively, operation of LTCs with overcurrents above certain values, operation of LTCs at oil temperatures below the limit laid down in the standards (−25°C) (IEEE, 1995; IEC, 1989), inadmissible voltage stresses of the insulation in the arcing switch caused by transient overvoltages. The following control and protective equipment are in use:

- Oil-flow relays inserted into the pipe between LTC head and conservator are mostly used for in-tank LTCs (Fig. 3.171). They respond to disturbances in the arcing switch compartment of relatively low-energy up to high-energy dissipation within a reasonable time, avoiding damages to the LTC and the transformer. The oil-flow relay has to disconnect the transformer. To give alarm only, as it is practiced in some users' systems, is not allowed because it is dangerous and could lead to severe faults.

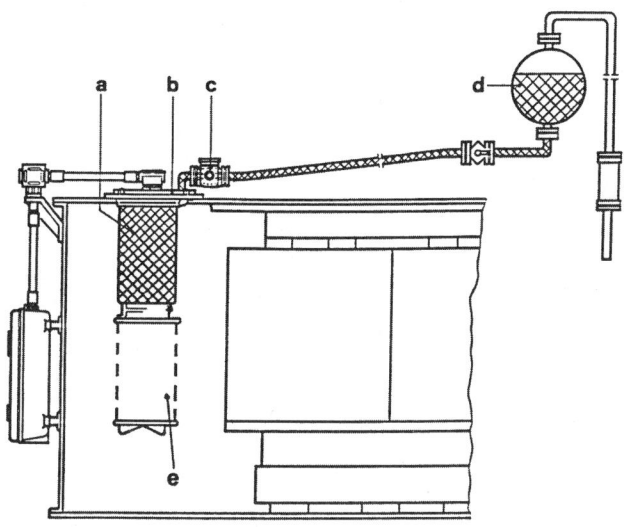

(a): Arcing switch oil compartment (d): Oil conservator
(b): Integrated pressure relief diaphragm (e): Tap selector
(c): Oil-flow relay

FIGURE 3.171 Arrangements of overpressure protection of LTCs in the transformer tank.

- Pressure sensing and/or pressure releasing relays are also often used parallel to the oil-flow relay or alone. Their response time is a little bit shorter than that of the oil-flow relay. But decrease in the response time is of minor importance because the complete disconnecting time of the transformer is determined by the total response time of the control circuit which trips the circuit breakers of the transformer and which is much longer than the response time of the oil-flow relay.
- At oil temperatures below −25°C, it may be necessary to provide special devices, e.g., blocking the drive mechanism to obtain reliable service behavior of the LTC.
- An overcurrent blocking device which stops the LTC's drive mechanism during an overload is used in many utilities as a standard device. It is normally set at 1.5 times the rated current of the transformer.
- In transformers with regulation on the high voltage side and coarse winding arrangements, extremely high voltage stresses can occur at the inner insulation of the arcing switch of the resistance-type LTC during impulse testing when the LTC is in mid-position (Fig. 3.172). Up to 25% of the incoming wave for BIL-tests or 40% for chopped wave tests can appear over the said distance. Critical values could be reached above a BIL of 550 kV.
- There are two principles to protect the arcing switch from undue overvoltages. Spark gaps or non-linear resistors could be installed in series to the transition resistors as shown in Fig. 3.173. The spark gap is a safe overvoltage protection for applications in medium-size power transformers. Non-linear resistors are solutions for high power transformers and for all transformers where the service conditions would cause the spark gaps to respond frequently.

In the early stages, silicon-carbide (SiC) elements were installed. The specific characteristics of this material did not allow full range application.

After high power zinc-oxide (ZnO) varistors had come on the market, the application of these elements for over-voltage protection were studied in detail with best results. For more than 20 years, ZnO varistors have solely been in use.

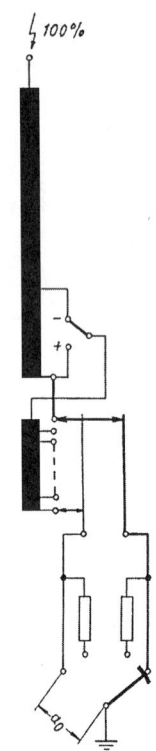

Voltage stress at "a_0":

wave shape	stress in % due to imput impulse
1.2/50 µs	10% – 20% max. 25%
chopped wave	15% – 30% max 40%

Coarse/fine winding arrangement
OLTC in mid–position

FIGURE 3.172 Voltage stress between selected and preselected tap of coarse/tap winding arrangement, LTC in mid-position.

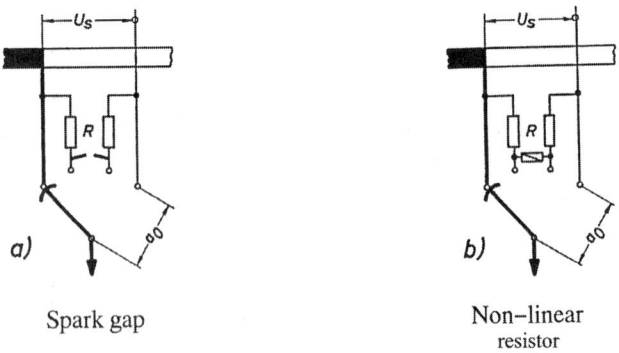

a) Spark gap

b) Non–linear resistor

FIGURE 3.173 Overvoltage protection devices arranged within the arching switch.

Maintenance of Load Tap Changers

LTC maintenance is the basis for the regulating transformer's high level of reliability. The background for maintenance recommendations is as follows:

For LTCs where oil is used for arc-quenching, the arcing at the arcing switch or arcing tap switch contacts causes contact erosion and carbonization of the arcing switch oil. The degree of contamination depends upon the operating current of the LTC, the number of operation, and to some degree the quality of the insulating oil. For LTCs using vacuum interrupters for arc-quenching, contact life of the vacuum interrupters and the mechanically stressed parts of the device are the key indicators for the maintenance recommendations.

Maintenance and inspection intervals depend on the type of LTC, the LTC rated through current, the field experience, and the individual operating conditions. They are suggested as periodical measures with respect to a certain number of operations or after a certain operating time, whichever comes first.

The recommended maintenance intervals for an individual LTC type are given in the operating and inspection manuals available for each LTC type.

Normally, maintenance of an LTC can be performed within a few hours by qualified and experienced personnel, provided that it has been properly planned and organized. In countries with tropical or subtropical climate, the humidity must also be taken into consideration. In some countries customers decide to start maintenance work only if the relative humidity is less than 75%.

Economical factors are taken more and more into consideration by users of large power transformers in distribution networks when assessing the operating parameters for cost-intensive operating equipment. While users are aiming at cost reduction for transformer maintenance, they are also demanding higher system reliability. Modern supervisory concepts on LTCs (LTC monitoring) offer a solution for the control of these divergent development tendencies.

Today a few products are on the market which differ significantly in their performance.

A state-of-the-art LTC on-line monitoring system should include an early fault detection function and information on condition-based maintenance which requires an expert-system of the LTC manufacturer. The data processing and visualization should provide information about status signal messages, trend analyses, and prognoses. Monitoring application is a judgment of transformer size and importance, and maintenance and equipment costs.

Refurbishment/Replacement of Old LTC Types

With regard to system planning of power utilities, the lifetime of regulating transformers is normally assumed to be 25 to 30 years. The actual lifetime is, however, much longer. Due to economic aspects and aging networks, as well as the requirement to improve reliability, refurbishment/replacement is becoming a major policy issue for utility companies.

Refurbishment includes a complete overhaul of the regulating transformer plus other improvements regarding loading capability, an increase in insulation levels, a decrease in noise levels, and the possible replacement of the bushings and of the LTC or a complete overhaul of the LTC. This overhaul should be performed by specialists from the LTC manufacturer in order to avoid any risk when judging the condition of the LTC components, when deciding which components have to be replaced, with regard to the disassembly and the reassembly as well as the cleaning of insulation material.

The replacement of an old risky LTC (for which neither maintenance work nor spare parts are available) by a new LTC may economically be justified, compared to the expenses for a new regulating transformer, even if the transformer design has to be modified for that reason. The manufacturer of the new LTC must, of course, guarantee maintenance work and spare parts for the foreseeable future.

Future Aspects

For the time being, no alternative to regulating transformers is expected. The LTC will therefore continue to play an essential part in the optimum operation of electrical networks and industrial processes in the foreseeable future.

With regard to the future of LTC systems, one can say that a static LTC, without any mechanical system consisting only of power electronics, leads to extremely uneconomical solutions and this will not change in the near future. Therefore, the mechanical LTC will still be used.

Conventional LTC technology has reached a very high level and is capable of meeting most requirements of the transformer manufacturer. This applies to the complete voltage and power fields of today which will probably remain unchanged in the foreseeable future. It is very unlikely that, due to new impulses given to development, greater power and higher voltages will be required.

Today the main concern goes to service behavior as well as reliability of LTCs and how to keep this reliability at a consistently high level during the regulating transformer's life cycle.

Another target of development is the insulation and cooling media with low or no flammability for regulating transformers, mainly relevant in the field of medium power transformers (<100 MVA). Such media are: several synthetic/organic fluids provided as PCB-replacement, SF_6-gas, and air. The application of these transformers is preferred in areas where mineral oil would be too dangerous, considering pollution and fire. The application of these media may require a different LTC design. For many years SF_6-regulating transformers have been in service in Japan and a few other countries.

As an alternative, dry-type distribution transformers with regulation have been available for several years. The LTC is operating in air with vacuum interrupters. These transformers are used for indoor application with extreme fire hazard and/or pollution requirements, existing in metropolitan and special industrial areas.

References

Goosen, P.V. Transformer accessories, (On behalf of Study Committee 12), *CIGRE,* 12-104, 1996.
IEC Standard Publication 60214, On-Load Tap Changers, 3rd ed., 1989.
IEC Standard Publication 60542, Application Guide for On-Load Tap Changers, 1976, First Amendment 1988.
IEEE Guide for the Application, Specification and Testing of Phase-Shifting Transformers, PC57.135 (Draft).
IEEE Std C57.131-1995, IEEE Standard Requirements for Load Tap Changers.
Krämer, A. and Ruff, J. Transformers for phase angle regulation, considering the selection of on-load tap changers, *IEEE Trans. Power Delivery,* 13(2), April 1998.

3.15 Insulating Media

Leo J. Savio and Ted Haupert

Insulating media in high voltage transformers consists of paper wrapped around the conductors in the transformer coils plus mineral oil and pressboard to insulate the coils from ground. From the moment a transformer is placed in service, both the solid and liquid insulation begin a slow but irreversible process of degradation.

Solid Insulation — Paper

Composition of Paper — Cellulose

Paper and pressboard are composed primarily of cellulose, which is a naturally occurring polymer of plant origin. From a chemical perspective, cellulose is a naturally occurring polymer. Each cellulose molecule is initially composed of approximately 1000 repeating units of a monomer that is very similar to glucose. As the cellulose molecule degrades, the polymer chain ruptures and the average number of repeating units in each cellulose molecule decreases. With this reduction in the degree of polymerization of cellulose, there is a decrease in the mechanical strength of the cellulose as well as a change in brittleness and color. As a consequence of this degradation, cellulose will reach a point at which it will no longer properly function as an insulator separating conductors. When cellulose will reach its end of life as an insulator depends greatly on the rate at which it degrades.

Parameters that Affect Degradation of Cellulose

Heat
Several chemical reactions contribute to the degradation of cellulose. Oxidation and hydrolysis are the most significant reactions that occur in oil-filled electrical equipment. These reactions are dependent on the amounts of oxygen, water, and acids that are in contact with the cellulose. In general, the greater the level of these components, the faster the degradation reactions. Also, the rates of the degradation reactions are greatly dependent on temperature. As the temperature rises, the rates of chemical reactions increase.

For every 10° (Celsius) rise in temperature, reaction rates double!

Consequently, the useful life of cellulose and oil is markedly reduced at higher temperatures. Paper and oil subjected to an increased temperature of 10° will have their lives reduced by a factor of one half. Elevations in temperature can result from voluntary events such as increased loading or they can result from a large number of involuntary events such as the occurrence of fault processes such as partial discharge and arcing.

Oxygen

The cellulose that is present in paper, pressboard, and wood oxidizes readily and directly to carbon oxides. The carbon oxides (carbon dioxide and carbon monoxide) that are found in oil-filled electrical equipment result primarily from the cellulose material. This has very important consequences since the useful life of major electrical devices such as power transformers is generally limited by the integrity of the solid insulation — the paper. It is now possible to determine more closely the extent and the rate of degradation of the cellulose by observing the levels of the carbon oxides in the oil as a function of time.

As cellulose reacts with oxygen, carbon dioxide, water, and possibly carbon monoxide, are produced. Carbon monoxide will be produced if there is an insufficient supply of oxygen to meet the demands of the oxidation reaction. The levels of these products in the oil will continue to increase as oxidation continues. However, they will never exceed concentrations in the oil that are referred to as their solubility limits (which are temperature and pressure dependent). After the solubility limit of each has been reached, all further production will result in no increased concentration in the oil. If carbon monoxide and carbon dioxide were to ever exceed their solubility limits, they would form bubbles that would be lost to the atmosphere or to a gas blanket; this rarely happens. For water, free water (drops) will form that can fall to the bottom of the tank or be adsorbed into the solid insulation (the cellulose).

Moisture

Cellulose has a great affinity for holding water (notice how well paper towels work!) Water that is held in the paper can migrate into the oil as the temperature of the system increases or the reverse as the temperature of the system decreases. In a typical large power transformer, the quantity of cellulose in the solid insulation can be several thousand pounds.

For new transformers, the moisture content of the cellulose is generally recommended to be no more than 0.5%. Water will distribute between the oil and the paper in a ratio that is a constant and dependent on the temperature of the system. As the temperature increases, water will move from the paper into the oil until the distribution ratio for the new temperature is achieved. Likewise, as the temperature decreases, water will move in the opposite direction.

In addition to the water that is in the paper and the oil at the time a transformer is put into service, there is also water introduced into the system because of the ongoing oxidation of the cellulose. Water is a product of the oxidation of cellulose and it is therefore always increasing in concentration with time. Even if the transformer were perfectly sealed, the moisture concentration of the paper would continue to increase. The rate of generation of water is determined primarily by the oxygen content of the oil and the temperature of the system. Increases in each of these factors increase the rate of water generation.

Acid

Cellulose can degrade by a chemical process referred to as hydrolysis. During hydrolysis, water is consumed in the breaking of the polymeric chains in the cellulose molecules. The process is catalyzed by acids. Acids are present in the oil that is in contact with the cellulose. Carboxylic acids are produced from the oil as a result of oxidation. The acid content of the oil increases as the oil oxidizes. With an increase in acidity, the degradation of the cellulose increases.

Liquid Insulation — Oil

The insulating fluid that has the greatest use in electrical equipment is mineral oil. There are insulating materials that may be superior to mineral oil with respect to both dielectric and thermal properties; however, to date, none has achieved the requisite combination of equal or better performance at an equal or better price. Consequently, mineral oil continues to serve as the major type of liquid insulation used in electrical equipment.

Composition of Oil

Types of Hydrocarbons and Properties of Each

Mineral oil can vary greatly in its composition. All mineral oils are mixtures of hydrocarbon compounds with about 25 carbon atoms per molecule. The blend of compounds that is present in a particular oil is dependent on several factors, such as the source of the crude oil and the refining process. Crude oils from different geographical areas will have different chemical structures (arrangement of the carbon atoms within the molecules). Crude oils from some sources are higher in *paraffinic* compounds, whereas others are higher in *naphthenic* compounds. Crude oils also contain significant amounts of aromatic and polyaromatic compounds. Some of the polyaromatic compounds are termed "heterocyclics" because besides carbon and hydrogen, they contain other atoms such as nitrogen, sulfur, and oxygen. Some heterocyclics are beneficial (e.g., oxidation inhibitors), but most are detrimental (e.g., oxidation initiators, electrical charge carriers).

The refining of crude oil for the production of dielectric fluids reduces the aromatic and polyaromatic content to enhance the dielectric properties and stability of the oil.

The terms paraffinic and naphthenic refer to the arrangement of carbon atoms in the oil molecule. Carbon atoms that are arranged in straight or branched chains, that is carbon atoms bonded to one another in straight or branched lines, are referred to as being paraffinic. Carbon atoms that are bonded to one another to form rings of generally five, six, or seven carbons are referred to as being naphthenic. Carbon atoms that are bonded as rings of benzene are referred to as being aromatic. Carbon atoms that are contained in "fused" benzene rings are referred to as being *polyaromatic*. These forms of bonded carbon atoms are depicted in Fig. 3.174. The straight lines represent the chemical bonds between carbon atoms that are present (but not depicted) at the ends and vertices of the straight lines.

Figure 3.175 illustrates a typical oil molecule. Remember that a particular oil will contain a mixture of many different molecular species and types of carbon atoms. Whether a particular oil is considered paraffinic or naphthenic is a question of degree. If the oil contains more paraffinic carbon atoms than naphthenic carbons, it is considered a paraffinic oil. If it contains more naphthenic carbons, it is considered naphthenic.

The differences in the chemical composition will result in differences in physical properties and in the chemical behavior of the oils after they are put in service. The chemical composition will have profound effects on the physical characteristics of the oil.

For electrical equipment, the main concerns are:

- Paraffinic oils tend to form waxes (solid compounds) at low temperature.
- Paraffinic oils have a lower thermal stability than that of naphthenic and aromatic oils.
- Paraffinic oils have a higher viscosity at low temperature than that of naphthenic and aromatic oils.

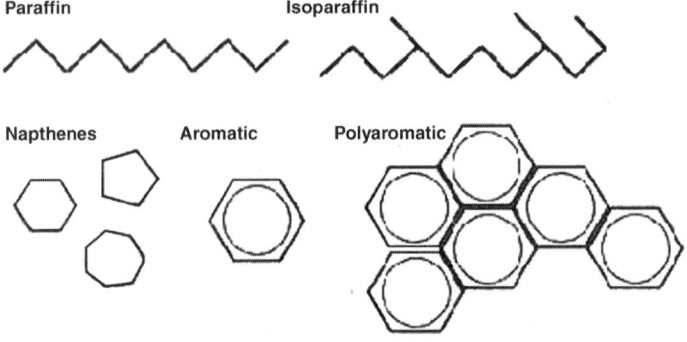

FIGURE 3.174 Carbon configurations in oil molecules.

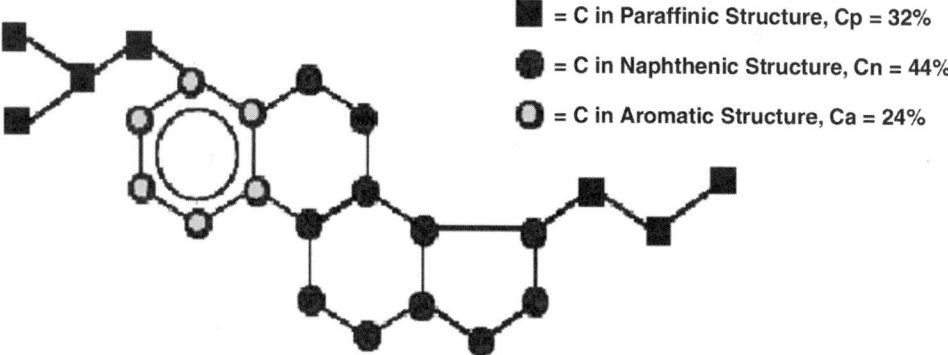

FIGURE 3.175 Typical oil molecule.

These factors can impair the performance of high-voltage electrical equipment. The first two factors have an unfavorable effect on the dielectric characteristics of the oil. The third factor unfavorably affects the heat dissipation ability of the oil. Unfortunately, the availability of insulating oil is limited. Therefore, electrical equipment owners have a choice of only a few producers and they produce only a very few different products.

Oxidation Inhibitors

Oxidation inhibitors, such as DBPC (di-tertiary butyl paracresol) and DBP (di-tertiary butylphenol), are often added to oil to retard the oxidation process. These compounds work by attracting oxygen molecules to themselves rather than allowing oxygen to bind with oil molecules. With time, the inhibitor gets consumed because of its preferential reaction with oxygen. As a result, the oil will then oxidize at a more rapid rate. The remedy is to add inhibitor to oil that has lost its anti-oxidant capabilities.

Functions of Oil

Electrical Insulation

The primary function of insulating oil is to provide a dielectric medium that acts as insulation surrounding various energized conductors.

Another function of the insulating oil is to provide a protective coating to the metal surfaces within the device. This coating protects against chemical reactions such as oxidation that can influence the integrity of connections, affect the formation of rust, and consequent contamination of the system.

Insulating oil, however, is *not* a good lubricant. Despite this fact, it is widely used in both load tap changers and circuit breakers in addition to transformers. Its use, therefore, in these devices presents a challenge to the mechanical design of the system.

Heat Dissipation

Second, the insulating fluid functions as a dissipater of heat. This is of particular importance in transformers where localized heating of the windings and core can be severe. The oil aids in the removal of heat from these areas and distributes the thermal energy over a generally large mass of oil and the tank of the device. Heat from the oil can then be transferred by means of conduction, convection, and radiation to the surrounding environment.

All mineral oils are comparable in their ability of the oil to conduct and dissipate heat. To ensure that a given oil will perform satisfactorily with respect to heat dissipation, a variety of specifications is placed on the oil. These specifications are based upon certain factors that influence the oil's ability to dissipate heat over a wide range of possible operating conditions. These factors include such properties as viscosity, pour point, and flash point.

Diagnostic Purposes

Another function of the insulating fluid that has gained importance is the role that insulating fluid plays as an indicator of the operational condition of the liquid-filled equipment. The condition (both chemical and electrical) of the insulating fluid reflects the operational condition of the electrical device. In a sense, the fluid can provide diagnostic information about the electrical device much like blood can provide diagnostic information about the human body. The condition of the blood is important as it relates to its primary function of transporting oxygen and other chemical substances to the various parts of the body. In addition, the condition of the blood is also symptomatic of the overall health of the body. For example, the analysis of the blood can be used to diagnose a wide variety of health problems related to abnormal organ function.

In much the same way, insulating fluid can be viewed as serving its primary functions as an insulator and heat dissipater. It can also be viewed as serving another (and perhaps equally important) function as a diagnostic indicator of the operational health of liquid-filled equipment. This is possible because when faults develop in liquid-filled equipment, they cause energy to be dissipated through the liquid. This energy can cause a chemical degradation of the liquid. An analysis for these degradation products can provide information about the type of fault that is present.

Parameters that Affect Oil Degradation

Heat

Just as temperature influences the rate of degradation of the solid insulation, so does it affect the rate of oil degradation. Although the rates of both processes are different, both are influenced by temperature in the same way. As the temperature rises, the rates of degradation reactions increase.

For every 10° (Celsius) rise in temperature, reaction rates double!

Oxygen

Hydrocarbon-based insulating oil, like all products of nature, is subject to the ongoing, relentless process of oxidation. Oxidation is often referred to as aging. The abundance of oxygen in the atmosphere provides the reactant for this most common degradation reaction. The ultimate products of oxidation of hydrocarbon materials are carbon dioxide and water. However, the process of oxidation can involve the production of other compounds that are formed by intermediate reactions such as alcohols, aldehydes, ketones, peroxides, and acids.

Partial Discharge and Thermal Faulting

Of all the oil degradation processes, hydrogen gas requires the lowest amount of energy to be produced. Hydrogen gas results from the breaking of carbon-hydrogen bonds in the oil molecules. All of the three fault processes (partial discharge, thermal faulting, and arcing) will produce hydrogen, but it is only with partial discharge or corona that hydrogen will be the only gas produced in significant quantity. In the presence of thermal faults, along with hydrogen will be the production of methane together with ethane and ethylene. The ratio of ethylene to ethane will increase as the temperature of the fault increases.

Arcing

With arcing, acetylene will be produced along with the other fault gases. Acetylene is characteristic of arcing. Because arcing can generally lead to failure over a much shorter time interval than faults of other types, even trace levels of acetylene (a few parts per million) must be taken seriously and be a cause for concern.

Acid

High levels of acid (generally acid levels greater than 0.6 mg KOH/g of oil) cause sludge formation in the oil. Sludge is a solid product of complex chemical composition that can deposit throughout the transformer. The deposition of sludge can seriously and adversely affect heat dissipation and ultimately result in equipment failure.

Sources of Contamination

External

External sources of contamination can be minimized generally by maintaining a sealed system, but on some types of equipment (e.g., free-breathing devices) this is not possible. Examples of external sources of contamination are moisture, oxygen, and solid debris introduced during maintenance of the equipment or during oil processing.

Internal

Internal sources of contamination can be controlled only to a limited extent because these sources of contamination are generally chemical reactions (like the oxidation of cellulose and the oxidation of oil) that are constantly ongoing. They cannot be stopped, but their rates are determined by factors that are well understood and often controllable. Examples of these factors are temperature and the oxygen content of the system.

Internal sources of contamination are:

- Non-metallic particles such as cellulose particles from the paper and pressboard
- Metal particles from mechanical or electrical wear
- Moisture from the chemical degradation of cellulose (paper insulation and pressboard)
- Chemical degradation products of the oil that result from its oxidation (e.g., acids, aldehydes, ketones)

3.16 Transformer Testing

Shirish P. Mehta and William R. Henning

Reliable delivery of electric power is, in great part, dependent on the reliable operation of power transformers in the electric power system. Power transformer reliability is enhanced considerably by a well-written Test Plan, which should include specifications for transformer tests. Developing a test plan with effective test specifications is a joint effort between manufacturers and users of power transformers. The written test plan and specifications should take into consideration the anticipated operating environment of the transformer, including factors such as atmospheric conditions, types of grounding, and exposure to lightning and switching transients. In addition to nominal rating information, special ratings for impedance, sound level, or other requirements should be considered in the test plan and included in the specifications. Selection of appropriate tests and the specification of correct test levels, which ensure transformer reliability in service, are important parts of this joint effort.

Transformers may be subjected to a wide variety of tests for a number of reasons, including:

1. Compliance with user specifications
2. Assessment of quality and reliability
3. Verification of design calculations
4. Compliance with applicable industry standards

Standards

ANSI/IEEE standards for power transformers are given in the C57 series of standards. Requirements that apply generally to all power and distribution transformers are given in the following two standards. These standards are particularly relevant and useful for those needing information on transformer testing.

1. ANSI/IEEE C57.12.00, IEEE Standard General Requirements for Liquid-Immersed Distribution, Power, and Regulating Transformers [1].
2. ANSI/IEEE C57.12.90, IEEE Standard Test Code for Liquid-Immersed Distribution, Power, and Regulating Transformers; and Guide for Short-Circuit Testing of Distribution and Power Transformers [2].

TABLE 3.20 Transformer Tests By Category

Dielectric Tests Transients	Performance Characteristics	Thermal Tests	Other Tests
1. Lightning impulse • Full-wave • Chopped-wave • Steep-wave 2. Switching impulse	1. No-load loss 2. % Excitation current 3. Load loss 4. % Impedance 5. Zero sequence impedance 6. Ratio test 7. Short circuit test	1. Winding resistance 2. Heat run test • Oil rise • Winding rise • Hottest-spot rise 3. Overload heat run 4. Gas in oil 5. Thermal scan	1. Insulation capacitance and dissipation factor[a] 2. Sound level tests 3. 10-kV Exc. current 4. Megger 5. Core ground 6. Electrical center[a] 7. Recurrent surge 8. Dew point 9. Core loss before impulse 10. Control circuit test[a] 11. Test on series transformer[a] 12. LTC tests[a] 13. Preliminary ratio tests[a] 14. Test on bushing CT[a] 15. Oil preservation system[a]

Dielectric Tests Low (Power) Frequency
1. Applied voltage
2. Single-phase induced
3. Three-phase induced
4. Partial discharge

[a] Quality control tests.

These two ANSI and IEEE standards, and others, will be cited frequently in this section. There are many other standards in the ANSI C57 series for transformers, covering the requirements for many specific products, product ranges, and special applications. They also include guides, tutorials, and recommended practices. These documents will be of interest to readers wanting to find out detailed, authoritative information on testing power and distribution transformers.

The standards cited above are very important documents because they facilitate precise communication and understanding between manufacturers and users. They identify critical features, provide minimum requirements for safe and reliable operation, and serve as valuable references of technical information.

Classification of Tests

According to ANSI and IEEE standards (IEEE, 1993a), all tests on power transformers fall into one of three categories: (1) routine tests, (2) design tests, and (3) other tests. The manufacturer may perform additional testing to ensure the quality of the transformer at various stages of the production cycle. For this discussion, tests on power transformers are categorized as shown in Table 3.20.

Some of these tests are performed before the transformer core and coil assembly is placed in the tank, while other tests are performed after the transformer is completely assembled and ready for "final testing". However, what are sometimes called "final tests" are not really final. Additional tests are performed just before transformer shipment, and still others are carried out at the customer site during installation and commissioning. All of these tests, the test levels, and the accept/reject criteria represent an important aspect of the joint test plan development effort made between the manufacturer and the purchaser of the transformer.

Sequence of Tests

The sequence in which the various tests are performed is also specified. An example of test sequence is as follows:

(A) Tests before tanking
  • Preliminary ratio, polarity, and connection of the transformer windings
  • Core insulation tests
  • Ratio and polarity tests of bushing current transformers

(B) Tests after tanking (final tests)
- Final ratio, polarity, and phase rotation
- Insulation capacitance and dissipation factor
- Insulation resistance
- Control wiring tests
- Lightning voltage impulse tests
- Applied potential tests
- Induced potential tests and partial discharge measurements
- No-load loss and excitation current measurements
- Winding resistance measurements
- Load loss and impedance voltage measurements
- Temperature rise tests (heat runs)
- Tests on gauges, accessories, LTCs, etc.
- Sound level tests
- Other tests as required

(C) Tests before shipment
- Dew point of gas
- Core ground megger test
- Excitation frequency response test

(D) Commissioning tests
- Ratio, polarity, and phase rotation
- Capacitance, insulation dissipation factor, and megger tests
- LTC settings check
- Test on transformer oil
- Excitation frequency response test
- Space above the oil in the transformer tank

Scope of this Section

This section covers testing to verify or measure the following:

- Voltage ratio and proper connections
- Insulation condition
- Control devices and control wiring
- Dielectric withstand
- Performance characteristics
- Other tests

Voltage Ratio and Proper Connections

The Purpose of Ratio, Polarity, and Phase Relation Tests

Tests for checking the winding ratios, and for checking the polarity and phase relationships of winding connections are carried out on all transformers during factory tests. The purpose of these tests is to ensure that all windings have the correct number of turns according to the design, that they are assembled in the correct physical orientation, and that they are connected properly to provide the desired phase relationship for the case of polyphase transformers. If a transformer is equipped with either a deenergized tap changer (DETC) or a load tap changer (LTC), or both, then ratio tests are also carried out at the various positions of the tap changer(s). The objective of ratio tests at different tap positions is to ensure that all winding taps are made at the correct turns and that the tap connections are properly made to the tap changing devices.

The ANSI/IEEE test code defines three separate tests:

1. Ratio test
2. Polarity test
3. Phase relation test

However, test sets are available that, once all the connections are made, determine ratio and polarity simultaneously, and facilitate testing of three-phase transformers to determine phase relationship by switch selection of the required leads. In this sense, the three tests can be combined.

Ratio Test

ANSI/IEEE general standard [1] requires that the measured voltage ratio between any two windings be within ± 0.5% of the value indicated on the nameplate. To verify this requirement, ratio tests are performed in which the actual voltage ratio is determined through measurements. Ratio tests can be made by energizing the transformer with a low AC test voltage and measuring the voltage induced in other windings at various tap settings, etc. In each case the voltage ratio is calculated and compared to the voltage ratio indicated on the transformer nameplate. More commonly, Transformer Turns Ratio (TTR) test sets are used for making the tests. In this method the transformer to be tested is energized with a low AC voltage at power frequency in parallel with the high-turn winding of a standard reference transformer in the test set. The induced voltage in the LV winding of the transformer under test is compared against the induced voltage of the variable low-turn reference winding, in both magnitude and phase, to verify voltage ratio and polarity.

Polarity Test

Polarity is usually checked at the same time as turns ratio, using the test set. At balance, a TTR test set displays the voltage ratio between two windings and also indicates winding polarity.

There are three other ways to check polarity:

1. Inductive kick
2. Alternating voltage
3. Comparison with transformer of known polarity and same ratio

Information on these methods may be found in the ANSI/IEEE test code [2].

Phase Relationship Test

The transformer test sets used to determine voltage ratio and polarity are designed for single-phase operation, but supplemental switching arrangements are employed to facilitate testing of three-phase transformers. Based on the phasor diagram of the three-phase transformer, appropriate HV and LV voltages are selected by switches for determining the phase relationships.

The ANSI/IEEE test code [2] discusses additional methods that may be used to determine or verify phase relationship.

Insulation Condition

The Purpose of Insulation Condition Tests

Full-scale dielectric testing of the transformer insulation system will be discussed later. Here we will discuss insulation tests, performed at voltages about 10 kV, to verify the condition of the insulation system. Initial measurements at the factory can be recorded and compared to later measurements in the field to assess changes in the condition of the transformer insulation.

The quality of the transformer insulation and the efficacy of the insulation processing for moisture removal are evaluated through the results of Insulation Power Factor Tests and Insulation Resistance Tests.

Transformers

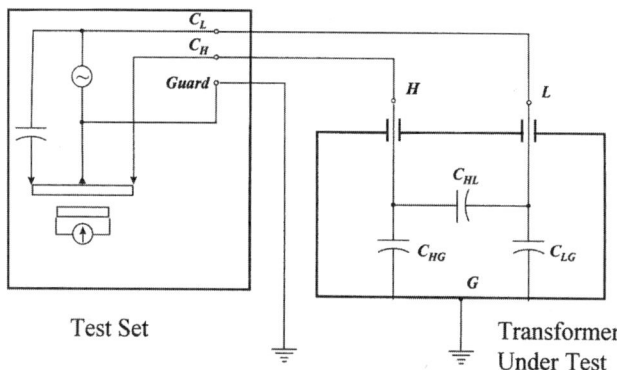

FIGURE 3.176 Measurement of capacitance and dissipation factor — high to low.

Insulation Power-Factor

Insulation Power Factor Tests are performed in one of two ways. In the first general method, a bridge circuit is used to measure the capacitance and dissipation factor (Tanδ) of the insulation system. In the second general method, a specially designed test set measures voltage, current, and power when tests are made at about 10 kV. From these measurements, the insulation power factor (Cosϕ) is determined. Insulation power factor (Cosϕ) and insulation dissipation factor (Tanδ) are approximately equal for a well-processed insulation system with low dielectric loss and small loss angle. Both methods are, therefore, equivalent.

Insulation Capacitance and Dissipation Factor (Tanδ) Test

The measurement of insulation capacitance and dissipation factor (Tanδ) is carried out using a capacitance bridge. A transformer ratio arm bridge or a Schearing bridge can be used for this purpose. In a two-winding transformer, there are three measurements of capacitance: (1) HV to Ground, (2) LV to Ground, and (3) HV to LV. These values of capacitance and their respective values of insulation dissipation factor (Tanδ) are to be measured.

For performing these tests, the following connections are made. All HV line terminals are connected together, labeled (H); all LV line terminals are connected together, labeled (L); and a connection is made to a Ground Terminal, usually a connection to the transformer tank, which is labeled (G). Leads from the measuring instrument or bridge are connected to one or both terminals and ground. Either Grounded Specimen Measurements or Guarded Measurements are possible so that all capacitance values and dissipation factor values can be determined. Figure 3.176 shows the measurement of capacitance and dissipation factor of the HV to LV capacitance, using guarded measurements. Figure 3.177 is for the low-

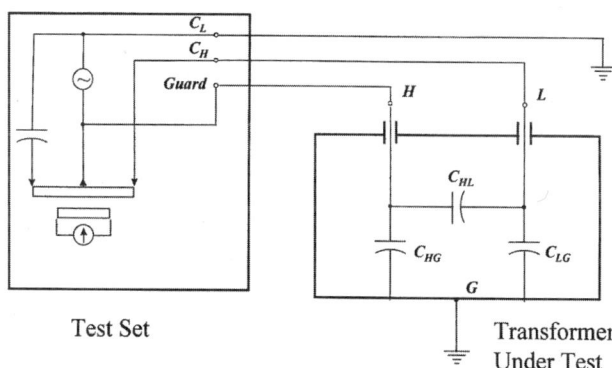

FIGURE 3.177 Measurement of capacitance and dissipation factor — low to ground.

to-ground capacitance. These measurements are usually made at voltages of 10 kV or less, at or near power frequency. In substations and factory test floors, interference control circuits may be required to achieve the desired sensitivity at balance.

The Volt-Ampere-Watt Test

Test sets, specifically designed for this purpose, determine the required parameters by measurements of voltage, current, and power when tests are made with about 10 kV excitation. Calculations are made to determine the appropriate values of capacitance and insulation power factor.

Interpretation of Results

Insulation power factor values of less than 0.5% are considered an indication of well-processed, dried-out transformer insulation. Although there is general recognition and agreement that power factor and dissipation factor values are dependent on temperatures, the exact relationship is not uniformly agreed upon.

Insulation Resistance

Insulation resistance of a two-winding transformer insulation system, HV to ground, LV to ground, and HV to LV is determined with a Megger type of instrument. Historically, insulation resistance measurements are also made to assess the amount of moisture in transformer insulation. However, the measurement of insulation dissipation factor has shown to be a better indicator of the overall condition of insulation in a power transformer.

Control Devices and Control Wiring

Modern power transformers often incorporate equipment condition monitoring devices and substation automation related devices as part of their control systems. Testing of the devices and the functional verification of the control system is a very critical aspect of the Controls Test Plan.

The Controls Test should include performance verification of control devices with various equipment protection, monitoring, and supervisory functions. Such devices may include: liquid temperature gauges, winding temperature simulator gauges, liquid level gauges, pressure-vacuum gauges, conservator Bucholz relays, sudden-pressure relays, pressure relief devices, etc. Many power transformers are equipped with Load Tap Changers (LTCs), and tests to verify operation of the LTC at rated current during factory tests are now included in the latest ANSI standards. LTC control settings for voltage level, bandwidth, time delay, first-customer protection, backup protection, etc. should also be verified during factory tests and at commissioning.

Bushing current transformer leads are usually routed to the control cabinet for connection by the customer. That these leads are connected to the proper terminals is usually checked during the load loss test on power transformers by monitoring the secondary current in various tap positions of the bushing CTs.

Hipot tests are performed to check the dielectric integrity of the control wiring. A 2.5-kV AC test is recommended for current transformer leads and a 1.5-kV AC test is required for other control circuit wiring.

Dielectric Withstand

In actual operation on a power system, a transformer is subjected to both normal and abnormal dielectric stresses. For example, a power transformer is required to operate continuously at 105% of rated voltage when delivering full load current and at 110% of rated voltage under no-load for an indefinite duration [1]. These are examples of conditions defined as *normal operating conditions* [1]. The voltage stresses associated with normal conditions as defined above, although higher than stresses at rated values, are nonetheless considered normal stresses.

A transformer may be subjected to *abnormal* dielectric stresses, arising out of various power system events or conditions. Sustained power frequency overvoltage can result from Ferranti rise, load rejection,

Transformers

and ferroresonance. These effects can produce abnormal turn-to-turn and phase-to-phase stresses. On the other hand, line-to-ground faults can result in unbalance and very high terminal-to-ground voltages, depending upon system grounding. Abnormal transient overvoltages of short duration arise out of lightning related phenomena, and longer duration transient overvoltages can result from line switching operations.

Even though these dielectric stresses are described as *abnormal*, the events causing them are expected to occur, and the transformer insulation system must be designed to withstand them. To verify the transformer capability to withstand these kinds of abnormal but expected transient and low frequency dielectric stresses, transient and low frequency dielectric tests are routinely performed on all transformers. The general ANSI/IEEE transformer standard [1] identifies the specific tests required. It also defines test levels for each test. The ANSI/IEEE test code [2] describes exactly how the tests are to be made; it defines pass-fail criteria; and it provides valid methods of corrections to the results.

Transient Dielectric Tests

The Purpose of Transient Dielectric Tests

Transient dielectric tests consist of Lighting Impulse Tests and Switching Impulse Tests. They demonstrate the strength of the transformer insulation system to withstand transient voltages impinged upon the transformer terminals during surge arrester discharges, line shielding flashovers, and line switching operations.

Power transformers are designed to have certain transient dielectric strength characteristics based on Basic Impulse Insulation Levels (BIL). The general ANSI/IEEE transformer standard [1] provides a table listing various system voltages, BIL, and test levels for selected insulation classes. The Transient Dielectric Tests demonstrate that the power transformer insulation system has the necessary dielectric strength to withstand the voltages indicated in the tables.

Lightning Impulse Test

Impulse tests are performed on all power transformers. In addition to verification of dielectric strength of the insulation system, impulse tests are excellent indicators of the quality of insulation, workmanship, and processing. The sequence of tests, test connections, and applicable standards will be described below.

Lightning Impulse Voltage Tests simulate traveling waves due to lightning strikes and line flashovers. The full-wave lightning impulse voltage wave shape is one where the voltage reaches crest magnitude in 1.2 μs, then decays to 50% of crest magnitude in 50 μs. This is shown in Fig. 3.178. Such a wave is said to have a *wave shape* of 1.2 × 50.0 μs. The term *wave shape* will be used in this section to refer to the test wave in a general way. The term *waveform* will be used when referring to detailed features of the test voltage or current records, such as oscillations, "mismatches", and chops. The difference in meaning of these two terms can be found in the IEEE dictionary [3].

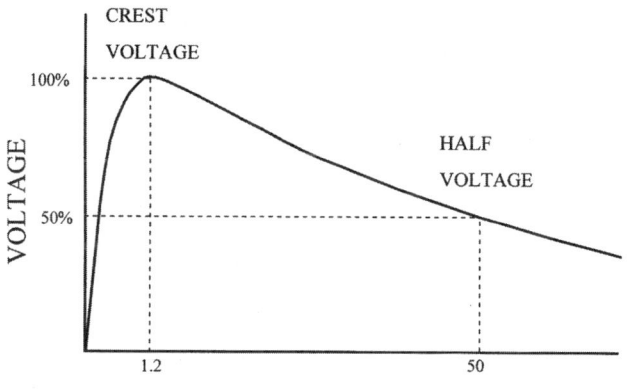

FIGURE 3.178 Standard lightning impulse full wave.

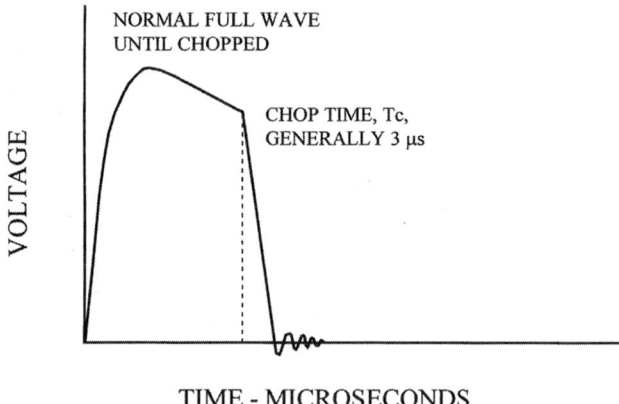

FIGURE 3.179 Standard lightning impulse chopped wave.

In addition to the standard impulse full wave, a second type of lightning impulse wave, known as the chopped wave, or sometimes called the tail-chopped wave, is used in transformer work. The chopped wave employs the same wave shape as a full-wave lightning impulse, except that its crest value is 10% greater than that of the full-wave and the wave is chopped at about 3 μs. The chop in the voltage wave is accomplished by the flashover of a rod gap or by using some other chopping device, connected in parallel with the transformer terminal being tested. This wave is shown in Fig. 3.179. The chopped-wave test simulates the sudden flashover of line insulation. Its significance in transformer work is that the high rate of change in voltage, impinged on the transformer terminals, results in internal oscillations that can produce high dielectric stresses in specific regions of the transformer winding.

In addition to the full-wave test and the chopped-wave test, a third type of test known as front-of-wave test is sometimes made. (The test is sometimes called the steepwave test or front-chopped test.) The front-of-wave test simulates a direct lightning strike on the transformer terminals. Although direct strokes to transformer terminals in substations of modern design have very low probabilities of occurrence, front-of-wave tests are often specified. The voltage wave for this test is chopped on the front of the wave before the prospective crest value is reached. The rate of rise of voltage of the wave is set to about 1000 kV/μs. Chopping is set to occur at a chop time corresponding to an assigned instantaneous crest value. Front-of-wave tests, when required, must be specified.

Lightning impulse tests, including full-wave impulse and chopped-wave impulse test waves, are made on each line terminal of power transformers. The recommended sequence is:

1. One reduced-voltage, full-wave impulse, with crest value of 50 to 70% of the required full-wave crest magnitude (BIL) to establish reference pattern waveforms (impulse voltage and current) for failure detection.
2. Two chopped-wave impulses, meeting the requirements of crest voltage value and time to chop.
3. One full-wave impulse with crest value corresponding to the BIL of the winding line terminal.

When front-of-wave tests are specified, impulse tests are carried out in the following sequence: one reduced full-wave impulse, followed by two front-of-wave impulses, two chopped-wave impulses, and one full-wave impulse.

Generally, impulse tests are made on line terminals of windings, one terminal at a time. Terminals not being tested are usually solidly grounded or grounded through resistors with values of resistance in the range of 300 to 450 ohms. The voltage on terminals not being tested should be limited to 80% of the terminal BIL. Details about connections, tolerances on wave shapes, voltage levels, and correction factors are given in the IEEE test code [2] and the IEEE impulse guide [4].

Transformers

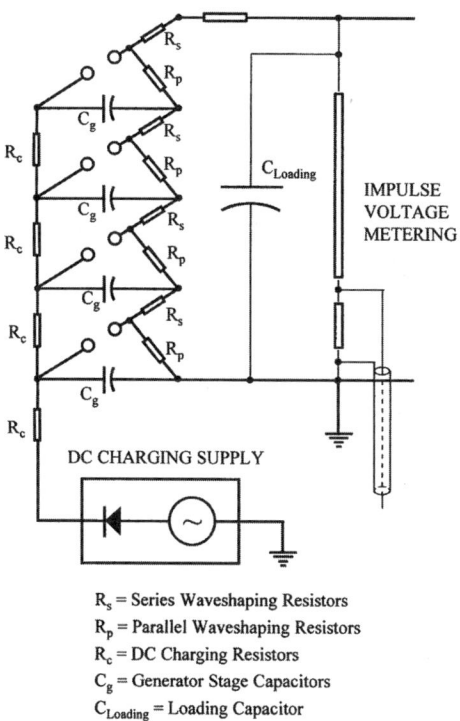

R_s = Series Waveshaping Resistors
R_p = Parallel Waveshaping Resistors
R_c = DC Charging Resistors
C_g = Generator Stage Capacitors
$C_{Loading}$ = Loading Capacitor

FIGURE 3.180 Marx generator with four stages.

Lightning Impulse Test Equipment
The generation, measurement, and control of impulse voltage waves is a very specialized subject. In this section only a very brief general introduction to the subject is provided. Most impulse generator designs are based on the Marx circuit. Figure 3.180 shows a schematic diagram of a typical Marx-circuit impulse generator with four stages. In principle, voltage multiplication is obtained by charging a set of parallel-connected capacitors in many stages of the impulse generator to a predetermined DC voltage, then momentarily reconnecting the capacitor stages in series to make the individual capacitor voltages add. The reconnection from parallel to series is accomplished through the controlled firing of a series of adjustable sphere gaps, adjusted to be near breakdown at the DC charging voltage. After the capacitors are charged to the proper DC voltage level, a sphere gap in the first stage is made to flash over by some means. This initiates a cascade flashover of all the sphere gaps in the impulse generator. The gaps function as switches, reconnecting the capacitor stages from parallel to series, producing a generator output voltage that is approximately equal to the voltage per stage times the number of stages.

The desired time to crest value on the front of the wave and the time to half crest value on the tail of the wave are controlled by wave-shaping circuit elements. These elements are indicated as R_s, R_p, and $C_{Loading}$ in Fig. 3.180. Generally, control of the time to crest on the front of the wave is realized by changing the values of series resistance, the impulse generator capacitance, and the load capacitance. Control of the time to 50% magnitude on the tail of the wave is realized by changing the values of parallel resistors and the load capacitance. Control of the voltage crest magnitude is provided by adjustment of the DC charging voltage and by changing the load on the impulse generator. The time of flashover for chopped waves is controlled by adjustment of gap spacings of the chopping gaps or the rod gaps.

The capacitor charging current path for the impulse generator is shown in Fig. 3.181. At steady state, each of the capacitors is charged to a voltage equal to the DC supply voltage. After the cascade firing of the sphere gaps, the main discharging current path becomes, in simplified form, that of Fig. 3.182. The RC time constants of the DC charging resistors, R_c as defined in Fig. 3.180, have values typically expressed

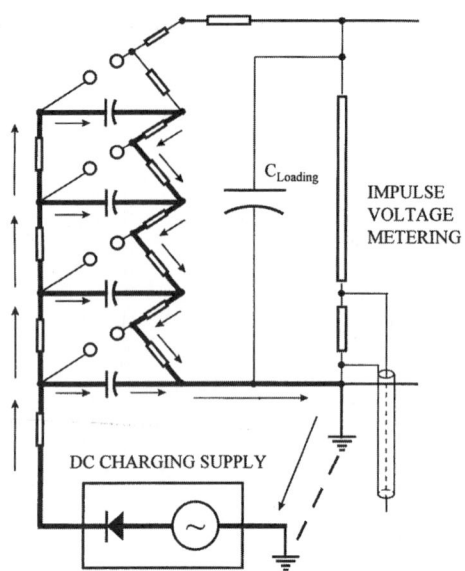

FIGURE 3.181 Charging the capacitors of an impulse generator.

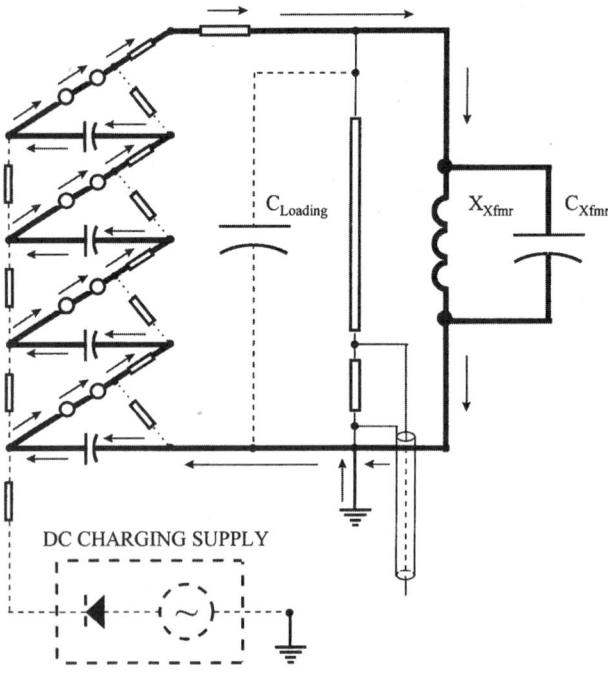

FIGURE 3.182 Discharging the capacitors of an impulse generator.

in seconds, while the wave shape control elements, R_p and R_s as defined in Fig. 3.180, have RC time constants typically expressed in microseconds. Hence, for the time period of the impulse generator discharge, the relatively high resistance values of the charging resistors represent open circuits for the relatively short time period of the generator discharge. This is indicated by dotted lines in Fig. 3.182. The discharge path shown in the figure is somewhat simplified for clarity: Significant currents do flow in the

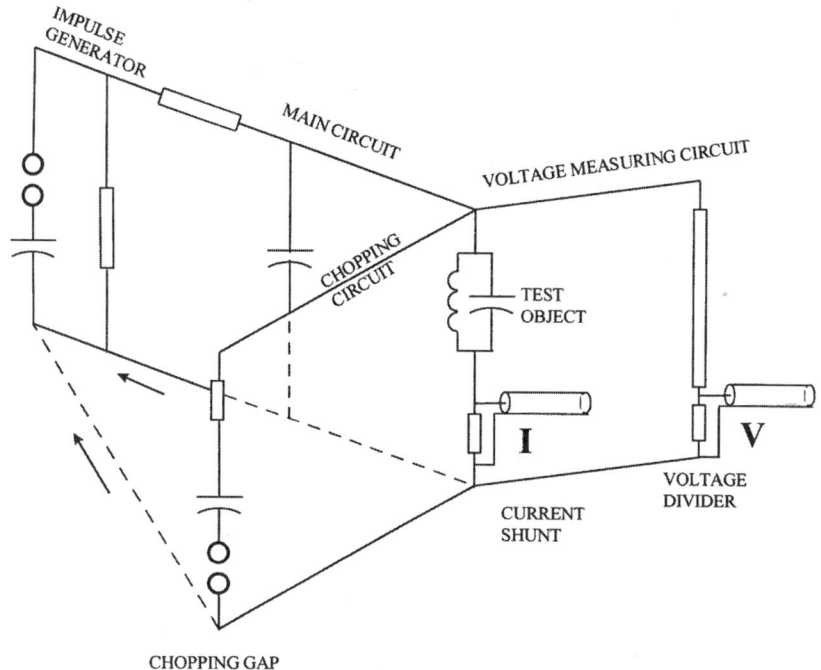

FIGURE 3.183 Impulse test setup.

shunt wave-shaping resistors, R_p, and significant current also flows in the loading capacitor, $C_{LOADING}$. These currents, which are significant in controlling the wave shape, are ignored in Fig. 3.182.

The measurement of impulse voltage in the range of a million volts in magnitude requires the use of voltage dividers. Depending upon requirements, either resistive, capacitive, or optimally damped (RC) types of dividers, having stable ratios and fast response times, are utilized to scale the high voltage impulses to provide a suitable input for instruments. Most impulse test facilities utilize specially designed impulse oscilloscopes or, more recently, specially designed transient digitizers, for accurate measurement of impulse voltages.

Measurement of the transient currents associated with impulse voltages is carried out with the aid of special non-inductive shunts or wide-band current transformers, included in the path of current flow. Usually, voltages proportional to impulse currents are measured with the impulse oscilloscopes or transient digitizers described earlier.

Impulse Test Setup

For consistent results it is important that the test setup be carefully made, especially with respect to grounding, external clearances, and induced voltages produced by impulse currents. Otherwise, impulse failure detection analysis could be flawed. One example of proper impulse test setup is shown in Fig. 3.183. This figure illustrates proper physical arrangement of the impulse generator, main circuit, chopping circuit, chopping gap, test object, current shunt, voltage measuring circuit, and voltage divider. High voltages and currents at high frequencies in the main circuit and the chopping circuit can produce rapidly changing electromagnetic fields, capable of inducing unwanted noise and error voltages in the low-voltage signal circuits connected to the impulse recorder inputs. The purpose of this physical arrangement is to minimize these effects.

Impulse Test Failure Detection

To accomplish failure detection or to verify the absence of a dielectric failure in the transformer insulation system, the impressed impulse voltage waveforms and the resulting current waveforms of the full-wave test are compared to the reduced full-wave test reference waveforms. The main idea behind failure detection in transformers is that if no dielectric breakdowns or partial discharges occur, then the final

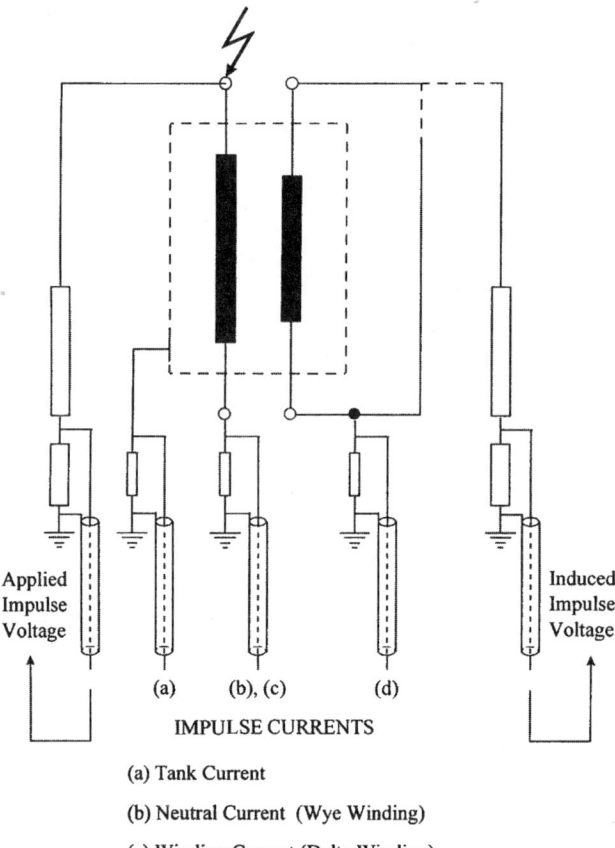

(a) Tank Current
(b) Neutral Current (Wye Winding)
(c) Winding Current (Delta Winding)
(d) Capacitively Transferred Current

FIGURE 3.184 Impulse current measurement locations.

full-wave test voltage and current waves will exhibit waveforms identical to the initial reduced full-wave reference tests when appropriately scaled. The occurrence of a dielectric breakdown would produce a sudden change in the inductance-capacitance network seen at the tested terminals of the transformer, causing a deviation in the test waveforms compared to the reference waveforms. The act of comparing the reduced full-wave records and the full-wave records is sometimes called "matching" the waves. If they are identical, the waves are said to be *matched*. Any differences in the waves, judged to be significant, are said to be *mismatches*. If there are mismatches, something is not correct, either in the test setup or in the dielectric system of the transformer. Various waveforms of voltages and currents associated with different types of defects are presented in great detail the IEEE impulse guide [4]. When digital recorders are employed, methods of waveform analysis using the frequency dependence of the transformer impedance, transformer transfer function, and other digital waveform analysis tools are now being developed and used to aid failure detection. Measurements of the voltages and currents in various parts of the transformer under test can aid in location of dielectric defects. These schemes are summarized in Fig. 3.184.

Switching Impulse Test

Man-made transients, as opposed to nature-made transients, are often the result of switching operations in power systems. Switching surges are relatively slow impulses. They are characterized by a wave that:

1. Rises to peak value in no less than 100 μs.
2. Falls to zero voltage in no less than 1000 μs.
3. Remains above 90% of peak value, before and after time of crest, for no less than 200 μs.

Transformers

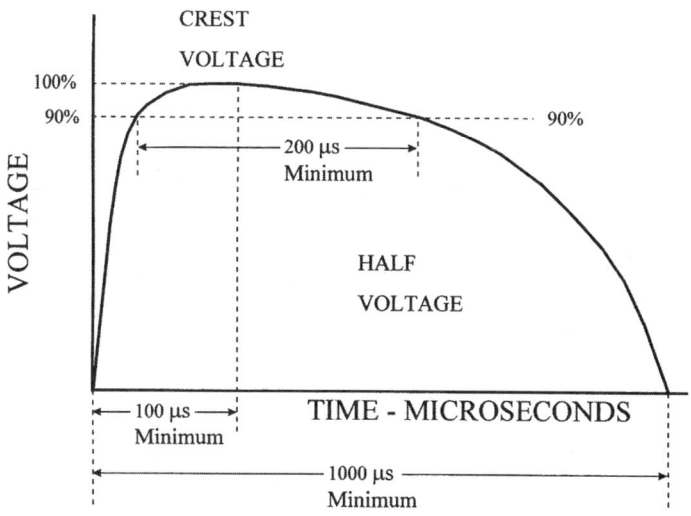

FIGURE 3.185 Standard switching impulse wave.

This is shown in Fig. 3.185. Generally, the peak value of the switching impulse voltage is approximately 83% of the BIL.

Voltages of significant magnitude are induced in all windings due to core flux buildup that results from the relatively long duration of the impressed voltage during the switching impulse test. The induced voltages are approximately proportional to the turns ratios between windings. Depending upon the transformer construction, shell-form vs. core-form, three-leg vs. five-leg construction, etc., many connections for tests are possible. Test voltages at the required levels can be applied directly to the winding under test, or they can be induced in the winding under test by application of switching impulse voltage of suitable magnitude across another winding, taking into consideration the turns ratio between the two windings. The magnitudes of voltages between windings and between different phases depend on the connections. This is discussed in great detail in the IEEE impulse guide [4].

Because of its long duration and high peak voltage magnitude, application of switching impulses on windings can result in saturation of the transformer core. When saturation of the core occurs, the resulting waves exhibit faster falling, shorter duration tails. By reversing polarity of the applied voltages between successive shots, the effects of core saturation can be reduced. Failures during switching impulse tests are readily visible on voltage wave oscillograms and are often accompanied by loud noises and external flashover.

Switching impulse tests are generally carried out with impulse generators having adequate energy capacity and appropriate wave-shaping resistors and loading capacitors.

Low-Frequency Dielectric Tests

The Purpose of Low-Frequency Dielectric Tests

When high-frequency voltages are applied to transformer terminals, the stress distributions within the windings are not linear but depend on the L-C characteristics of the windings. Also, the effects of oscillations penetrating the windings produce complex and changing voltage distributions. Low frequency stresses that result from power frequency overvoltage, on the other hand, result in stresses with a linear distribution along the winding. Because the insulation system is stressed differently at low frequency, a second set of tests is required to demonstrate dielectric withstand under power-frequency conditions. The Low-Frequency Dielectric Tests demonstrate that the power transformer insulation system has the necessary dielectric strength to withstand the voltages indicated in the tables of the standards for low frequency tests.

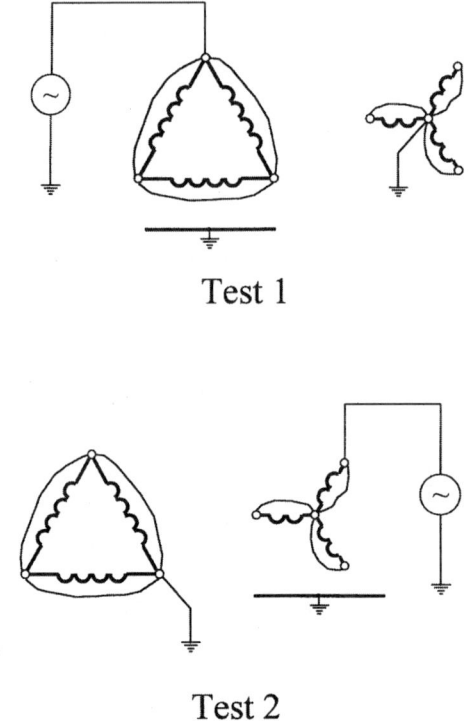

FIGURE 3.186 Applied voltage tests on two-winding transformers.

Applied Voltage Test

The applied voltage test is often called the hipot test. The purpose of this test is to verify the major insulation in transformers. More specifically, the purpose is to ensure that the insulation between windings and the insulation of windings to ground can withstand the required power-frequency voltages, applied for 1-min duration. For fully insulated windings, the test voltage levels are related to the BIL of the windings. For windings with graded insulation, test levels correspond to the winding terminal with the lowest BIL rating.

For two-winding transformers, there are two applied tests. Test 1 is the applied potential test of the HV winding, which is carried out by connecting all HV terminals together and connecting them to a high potential test set, as shown in Fig. 3.186. The LV terminals are connected together and connected to ground. Power frequency voltage of the correct level with respect to ground is applied to the HV terminals with LV terminals grounded. In this test the insulation system between HV to ground and between HV to LV is stressed. A second test, Test 2, is made at the correct level of test voltage for the LV winding, with the LV winding terminals connected together and connected to the high potential test set. In Test 2 the HV winding terminals are connected together and to ground. In the second test the insulation system LV to ground and LV to HV are stressed. This is also shown in Fig. 3.186.

A dielectric failure during this test is the result of internal or external flashover or tracking. Internal failures may be accompanied by a loud noise heard on the outside and "smoke and bubbles" seen in the oil.

Induced Voltage Test

The induced voltage test, with monitoring of partial discharges during the test, is one of the most significant tests to demonstrate the integrity of the transformer insulation system. During this test the turn-to-turn, disc-to-disc, and phase-to-phase insulation systems are stressed simultaneously at levels that are considerably higher than during normal operation. Weaknesses in dielectric design, processing,

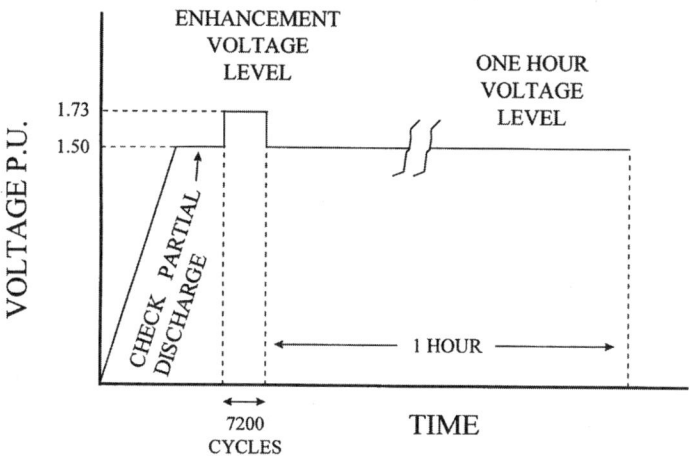

FIGURE 3.187 Induced voltage test for Class II power transformers.

or manufacture may cause partial discharge activity during this test. Partial discharges are generally monitored on all line terminals rated 115 kV or higher during the induced voltage test. This test produces the required voltages in the windings by magnetic induction. Because the test voltages are significantly higher than rated voltages, the core would ordinarily saturate if 60-Hz voltage were to be employed. To avoid core saturation, the test frequency is increased to a value normally in the range of 100 to 400 Hz.

Tests for Class II power transformers require an extended duration induced test. Class II is defined in the ANSI/IEEE general standard [1]. The test is usually made by raising the voltage on the LV windings to a value that induces 1.5 times maximum operating voltage in the HV windings. If no partial discharge activity is detected, the voltage is raised to the enhancement level (usually 1.73 p.u.) for a time period of 7200 cycles at the test frequency. Voltage is then reduced to the 1.5 p.u level and held for a period of 1 h. This test is described graphically in Fig. 3.187. In the U.S., PD level measurements are carried out with wide-band PD detectors or narrow-band RIV meters. The wide-band PD detectors read apparent charge in picocolumbs (pC). The narrow-band RIV meters read the radio influence voltage in microvolts (μV). Often both instruments are used, taking wide-band pC and narrow-band μV readings simultaneously. During the 1-h period, partial discharge readings are taken every 5 min. The criteria for a successful test are:

1. The magnitude of partial discharge activity does not exceed 100 μV (500 pC).
2. Any increase in partial discharge activity does not exceed 30 μV (150 pC).
3. The partial discharge levels during the 1-h period do not show an increasing trend.
4. There is no sustained increase in partial discharge level during the last 20 min of the test.

The test circuit for PD measurement during the induced voltage test is shown in Fig. 3.188. The bushing capacitance tap is connected to a coupling impedance unit, which provides PD signals to the PD detection unit. Methods for measurement and calibration are described in detail in the IEEE test code [2] and in the IEEE partial discharge measurement guide [5].

If there is significant level of PD activity during the test, it is desirable to identify the location of the PD source inside the transformer so that the problem can be corrected. This will ensure PD-free operation. Methods for location of PD sources inside the transformer are often based on triangulation. Ultrasonic acoustic waves arriving at transformer tank surfaces due to PD activity are monitored at various locations on the tank walls. Knowing wave propagation velocities and travel paths, it is possible to locate sources of PD activity within reasonable accuracy.

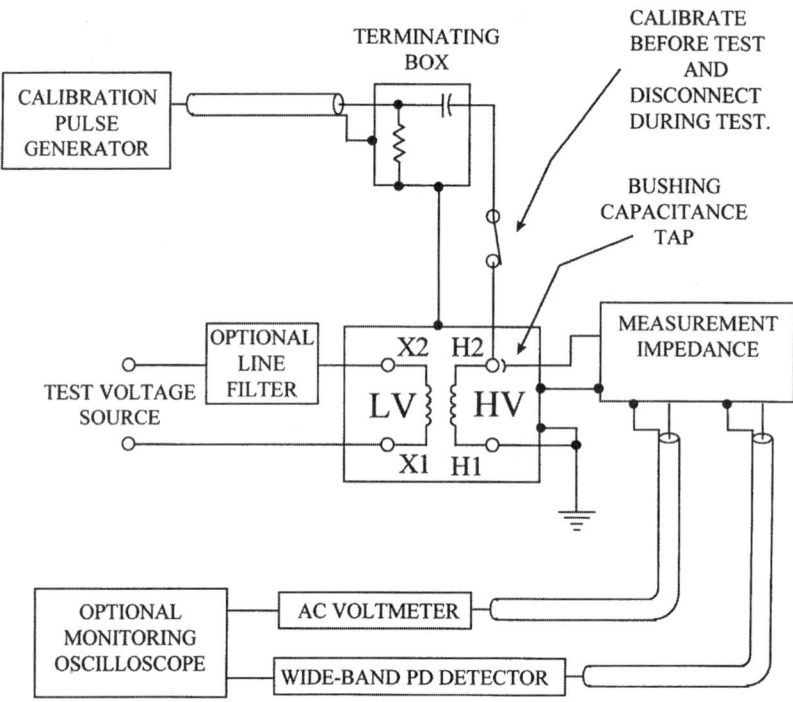

FIGURE 3.188 Test circuit for PD measurement during the induced test.

Performance Characteristics

No-Load Loss and Excitation Current Measurements

The Purpose of No-Load Loss Measurements

A transformer dissipates a constant no-load loss as long as it is energized at constant voltage, 24 hours a day, for all conditions of loading. This power loss represents a cost to the user during the lifetime of the transformer. Maximum values of the no-load loss of transformers are specified and often guaranteed by the manufacturer. No-load loss measurements are made to verify that the no-load loss does not exceed the specified or guaranteed value.

The Nature of the Quantity Being Measured

Transformer no-load loss, often called core loss or iron loss, is the power loss in a transformer excited at rated voltage and frequency but not supplying load. The no-load loss includes three sources of loss:

1. Core loss in the core material.
2. Dielectric loss in the insulation system.
3. I^2R loss due to excitation current in the energized winding.

The no-load loss of a transformer is primarily caused by losses in the core steel, Item 1 above. The remaining two sources are sometimes ignored. As a result, the terms *no load loss*, *core loss*, and *iron loss* are often used interchangeably. *Core loss* and *iron loss*, strictly speaking, refer only to the power loss that appears within the core material. The following discussion on no-load loss, or core loss, will explain why the average voltage voltmeter method, to be described later, is recommended. The magnitude of no load loss is a function of the magnitude, frequency, and waveform of the impressed voltage. These variables affect the magnitude and shape of the core magnetic flux waveform and hence affect the value of the core loss. Core loss also depends upon the temperature of the core. Two main components of the core

Transformers

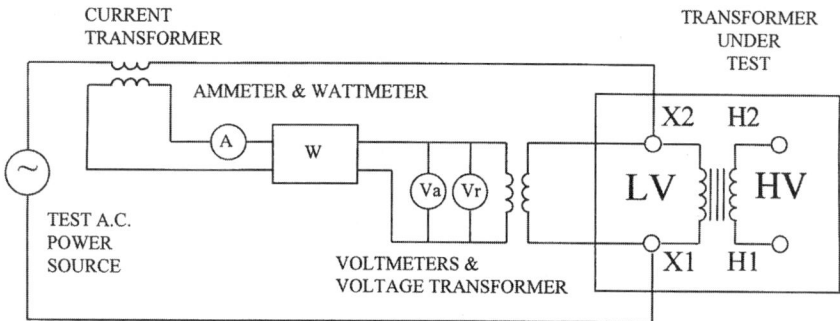

FIGURE 3.189 Test circuit for no-load loss measurement.

loss are hysteresis loss and eddy current loss. The hysteresis loss magnitude is a function of the peak flux density in the core flux waveform. When the impressed voltage waveform is distorted (not a pure sine wave), the resulting peak flux density in the flux waveform depends on the average absolute value of the impressed voltage wave. Eddy current loss is a function of the frequency of the power source and the thickness of the core steel laminations. Eddy loss is strongly influenced by harmonics in the impressed voltage. The IEEE transformer test code [2] recommends the Average Voltage Voltmeter Method, to be described below, for measuring no-load loss.

How No-Load Loss is Measured

The measurement of no-load loss, according to the Average Voltage Voltmeter Method, is illustrated in Fig. 3.189. Voltage and current transformers are required to scale the inputs for voltmeters, ammeters, and wattmeters. Three phase no-load loss measurements are carried out the same way except that three sets of instruments and instrument transformers are utilized. The test involves raising voltage on one winding, usually the low voltage winding, to its rated voltage, while the other windings are in open-circuit. Two voltmeters connected in parallel are employed. The voltmeter labeled V_a in Fig. 3.189 represents an average-responding, rms-calibrated voltmeter. The voltmeter labeled V_r represents a true rms-responding voltmeter. Harmonics in the impressed voltage will cause the rms value of the waveform to be different from the average-absolute (rms-scaled) value, and the two voltmeter readings will differ. When the voltage reading, as measured by the average-responding voltmeter, reaches a value corresponding to the rated voltage of the excited winding, readings are taken of the rms current, the rms voltage, and the no-load power. The ratio of the measured rms current to the rated load current of the excited winding, expressed in percent, is commonly referred to as the percent excitation current. The measured no-load loss is corrected to a sine-wave basis by a formula given in the IEEE test code [2], using the readings of the two voltmeters. The correction is shown below. The corrected value is reported as the no-load loss of the transformer.

$$P_c = \frac{P_m}{P1 + \left(\dfrac{V_r}{V_a}\right)^2 P2} \tag{3.77}$$

where

   P_c is the corrected (reported) value of no-load loss
   P_m is the measured value of no-load loss
   V_a is the reading of the average-responding, rms-calibrated voltmeter
   V_r is the reading of the true rms-responding voltmeter
   $P1$ and $P2$ are the per-unit hysteresis and per-unit eddy-current losses, respectively

Load Loss and Impedance Measurements

The Purpose of Load Loss Measurements

A transformer dissipates load loss that depends upon the transformer load current. Load loss is a cost to the user during the lifetime of the transformer. Maximum values of the load loss of transformers at rated current are specified and often guaranteed by the manufacturer. Load loss measurements are made to verify that the load loss does not exceed the specified or guaranteed value.

The Nature of the Quantity Being Measured

The magnitude of the load loss is a function of the transformer load current. Its magnitude is zero when there is no load on the transformer. Load loss is always given for a specified transformer load, usually at rated load. Transformer load loss, often called copper loss, includes I^2R losses due to load current in the winding conductors and stray losses in various metallic transformer parts due to eddy currents induced by leakage fields. Stray losses are produced in the winding conductors, in core clamps, in metallic structural parts, in magnetic shields, and in tank walls due to the presence of leakage fields. Stray losses also include power loss due to circulating currents in parallel windings and in parallel conductors within windings.

Because winding resistance varies with conductor temperature, and because the resistivities of the structural parts producing stray losses vary with temperature, the transformer load losses are a function of temperature. For this reason a standard reference temperature (usually 85°C) for reporting the load loss is established in the ANSI/IEEE general standard [1]. To correct the load loss measurements from the temperature at which they are measured to the standard reference temperature, a correction formula is provided in the ANSI/IEEE test code [2]. This correction involves the calculation of winding I^2R losses, where I is the rated current of the winding in amperes, and R is the measured DC resistance of the winding. The I^2R losses and the stray losses are separately corrected and combined in the formula given in the standard. The measurement of the DC winding resistance, R, is covered in the section entitled "Winding Resistance Measurements". Stray losses are determined by subtracting the I^2R losses from the measured load losses. All of this is covered in detail in the IEEE test code [2]. The formula for this conversion is stated in general form below:

$$W_{LL} = \frac{\left(I_p^2 R_p + I_s^2 R_s\right)\left(T_k + T_r\right)}{\left(T_k + T_m\right)} + \frac{\left(W_m - I_p^2 R_p - I_s^2 R_s\right)\left(T_k + T_m\right)}{\left(T_k + T_r\right)} \quad (3.78)$$

where

W_{LL} is the load loss (W), corrected to a reference temperature, T_r (°C).
T_m is the temperature at which the load loss is measured (°C).
T_k is a winding material constant; 234.5 for copper; 225 for aluminum.
R_p and R_s are the primary and secondary resistances in ohms.
I_p and I_s are the primary and secondary rated currents (A).
W_m is the load loss (W), as measured at temperature, T_m (°C).

It can be seen from the above equation that the measurement of load losses and correction to the reference temperature involves the measurement of five separate quantities:

1. Electric power (the load loss as measured).
2. Temperature (the temperature at time of test).
3. Resistance of the primary winding (at a known temperature).
4. Resistance of the secondary winding (at a known temperature).
5. Electric current (needed to adjust the current to the required values).

The discussion to follow focuses on the measurement of a single quantity: electric power at low power factor.

Transformers

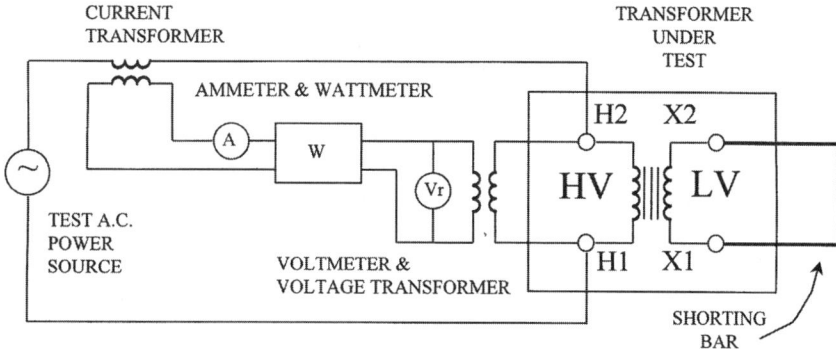

FIGURE 3.190 Test circuit for load loss measurement.

How Load Loss is Measured

Load losses are normally measured by connecting one winding, usually the low voltage winding, to a short circuit with adequately sized shorting bars and connecting the other winding, usually the high voltage winding, to a power-frequency voltage source. The source voltage is adjusted until the impressed voltage causes rated current to flow in both windings. Input rms voltage, rms current, and electric power are then measured. Figure 3.190 shows a circuit commonly used for measurement of load losses of a single-phase transformer. Three-phase measurement is carried out in the same way but with three sets of instruments and instrument transformers. Because of the high magnitudes of current, voltage, and power involved, precision scaling devices are usually required.

The applied test voltage when the transformer is connected as in Fig. 3.190, with rated currents in the windings, is equal to the impedance voltage of the transformer. Hence, impedance is also measured during this test. The ratio of this voltage to the rated voltage of the winding, expressed in percent, is the percent impedance or the percent impedance voltage of the transformer.

Discussion of the Measurement Process

The equivalent circuit of the transformer being tested in the load loss test and the phasor diagram of test voltage and current during the test are shown in Fig. 3.191. The load loss power factor for the load loss test is $\cos(\vartheta) = \frac{E_R}{E_Z}$. Because the transformer leakage impedance, consisting of R and X in Fig. 3.191, is mainly reactive, and more so the larger the transformer, the power factor during the load loss test is very low. In addition, there is a trend in modern transformers to create designs with lower losses due to increased demands for improved efficiency and transformer load loss evaluations for optimal life cycle costs. Transformer designs for low values of load loss lead to reductions in the equivalent resistance shown in Fig. 3.191, and hence to low values for the quantity, E_R, which translates to lower values for the load loss power factor in modern transformers. In fact, the load loss power factors for large modern transformers are often very low, in the range of from 0.01 to 0.05. Under circuit conditions with very low power factor, the accurate measurement of electric power requires special scaling devices, having very low phase-angle error and power measuring instruments having high accuracy at low power factor. In addition to the IEEE test code [2], a Transformer Loss Measurement Guide is being developed by a working group of the IEEE PES Transformers Committee. This guide is not yet approved or published, but when generally available, it will provide a complete background and the basis for carrying out measurements and calibrations to ensure accurate values of reported losses as required by the ANSI/IEEE general standard [1].

Because the magnitudes of load losses and impedance depend upon the tap positions of the Deenergized Tap Changer (DETC) and the Load Tap Changer (LTC), if present, load loss and impedance voltage measurements are usually carried out in the rated voltage connection and at the tap extremes. If the transformer under test has multiple MVA ratings that depend on the type of cooling, the tests are normally carried out at all ratings.

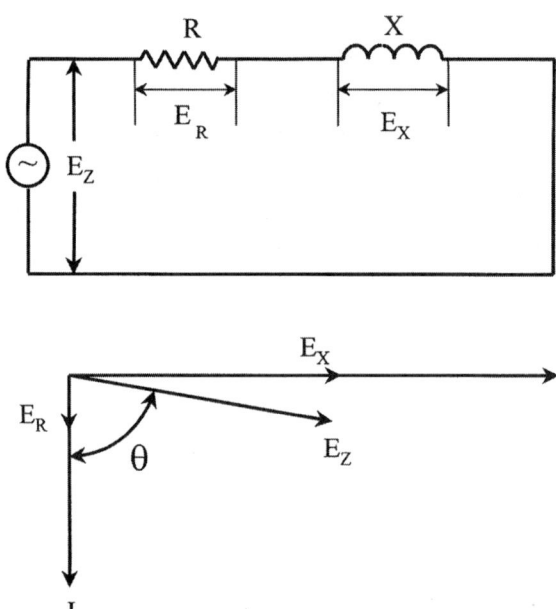

FIGURE 3.191 Equivalent circuit and phasor diagram for load loss test.

Winding Resistance Measurements

Purpose of Winding Resistance Measurements

Winding DC resistance measurements are of fundamental importance because they form the basis for determining the following:

1. Resistance measurements, taken at known temperatures, are used in the calculation of winding conductor I^2R losses. The I^2R losses at known temperatures are used to correct the measured load losses to a standard reference temperature. Correction of load losses is discussed in the previous section.
2. Resistance measurements, taken at known temperatures, provide the basis to determine the temperature of the same winding at a later time by measuring the resistance again. From the change in resistance, the change in temperature can be deduced. This measurement is employed to determine average winding temperatures at the end of heat run tests. Taking resistance measurements after a heat run test is discussed in the section entitled "Heat Run Tests".
3. Resistance measurements across the transformer terminals provide an assessment of the quality of internal connections made to the transformer windings. Loose or defective connections are indicated by unusually high or unstable resistance readings.

The Nature of the Quantity Being Measured

The DC winding resistance differs from the value of resistance indicated for the resistor shown in Fig. 3.191 or the resistors that appear in textbook illustrations of the *PI* or *T* equivalent circuits of transformers to represent the resistance of the windings. The resistors in the equivalent circuits include the effects of winding I^2R loss, eddy loss in the windings, stray losses in structural parts, and circulating currents in parallel conductors — namely, they represent the resistance of the load loss. The resistors shown in the equivalent circuits may be thought of as representing an equivalent AC resistance of the windings. The DC resistance of the windings is a different quantity, one that is relevant for calculating I^2R, for determining average winding temperature, and for evaluating electrical connections.

Transformers

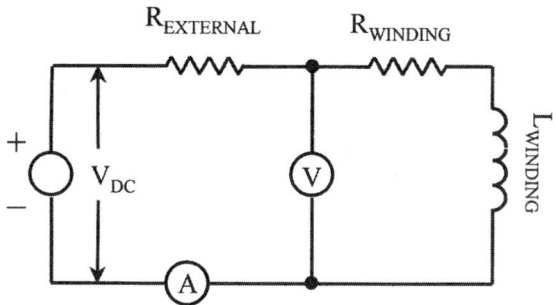

FIGURE 3.192 Circuit for measuring winding resistance.

How Winding Resistance Measurements are Made

The measurement of power transformer winding resistance is normally done using the voltmeter-ammeter method or using a ratiometric method to display the voltage-current ratio directly. A circuit for the measurement of winding resistance is shown in Fig. 3.192. A DC voltage or current source is used to establish the flow of steady direct current in the transformer winding to be measured. After the R-L transient has subsided, simultaneous readings are taken of the voltage across the winding and the current through the winding. The resistance of the winding is determined from these readings based on Ohm's law.

Discussion of the Measurement Process

If a DC voltage is applied to a series R-L circuit, the current will rise exponentially with a time constant of L/R. This is familiar for the case where both resistance and inductance remain constant during the transient period. For a transformer winding, however, it is possible for the true resistance, the apparent resistance, and the inductance of the winding to change with time. The true resistance may change if the direct current is of high enough magnitude and is applied long enough to heat the winding substantially, thereby changing its resistance during the measurement. The inductance changes with time because of the non-linear B-H curve of the core steel and varies in accordance with the slope of the core steel saturation curve. In addition there is an apparent resistance, Ra, during the transient period.

$$Ra = \frac{V}{I} = R + \frac{L}{I}\frac{dI}{dt} \tag{3.79}$$

Note that the apparent resistance, Ra, is higher than the true resistance, R, during the transient period and that the apparent resistance derived from the voltmeter and ammeter readings equals the true resistance only after the transient has subsided.

Resistance measurement error due to heating of the winding conductor is usually not a problem in testing transformers, but the possibility of this effect should be taken into consideration, especially for some low-current distribution transformer windings where the DC current can be significant compared to the rated current. It is more likely that errors will occur because of meter readings taken before core saturation is achieved. The process involved in core saturation is described below.

Compared to the exponential current vs. time relationship for the R-L circuit with constant R and constant L, the current in a transformer winding when a DC voltage is first applied rises slowly. The slow rate of rise comes about because of the high initial impedance of the winding. The high impedance results from the large effective inductance of the winding with a large iron core during this initial period. As the current slowly increases, the flux density in the core slowly rises until the core begins to saturate. At this point the winding no longer behaves like an iron-core coil and instead behaves like an air-core coil, with relatively low inductance. The rate of rise of the current increases for a period as the core saturates; then the current levels off at a steady-state value. Typical shapes for the voltage, current, and apparent

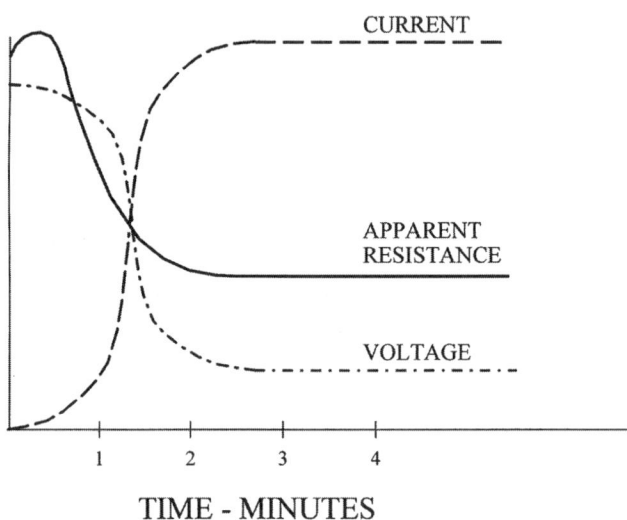

FIGURE 3.193 Current, voltage, and apparent resistance with time.

resistance are shown in Fig. 3.193. The magnitude of the DC voltage affects the rate at which flux builds up in the core since $V = N \frac{d\Phi}{dt}$. The higher the magnitude of the DC voltage, the shorter the time to saturation because of a higher value for $\frac{d\Phi}{dt}$. At the same time, though, the coil must be able to provide the required magnetomotive force in ampere turns, $N \cdot I$, needed to force the core into saturation, which leads to a minimum value for the DC current. Of course, there is an upper limit to the value for DC current, namely the point at which conductor heating would disturb the resistance measurement.

Note the time scale of the graph in Fig. 3.193. It is very important that the steady-state DC current be attained before meter readings are taken. If this is not done, errors in excess of 20% are easily realized.

Winding Resistance and Average Winding Temperature

Two of the three purposes listed above for measuring the DC resistance of a transformer winding inherently involve a concomitant measurement of temperature. When measuring resistance for the purpose of calculating I²R at a given temperature, the I²R value obtained will be used to determine the load loss value at a different temperature. When the winding resistance is measured before and during a heat run, the determination of average winding temperature at the end of a heat run test requires knowledge of winding resistance at two temperatures.

The winding DC resistance at two temperatures, $T1$ and $T2$, will have values of $R1$ and $R2$, respectively, at the two temperatures. The functional relationship between winding resistance and average temperature is shown below:

$$\frac{R1}{R2} = \frac{T1 + T_k}{T2 + T_k} \qquad (3.80)$$

where

$R1$ is the value of winding resistance, corresponding to average winding temperature of $T1$.
$R2$ is the value of winding resistance, corresponding to average winding temperature of $T2$.
Tk is 235.5 for copper; 225 for aluminum.

Correction of load loss for temperature is covered in the section titled "Load Loss and Impedance Measurements". Determination of average winding temperature in a heat run test is covered in the next section.

Heat Run Tests

Purpose of Heat Run Tests

The maximum allowable average and hottest-spot temperature rises of the windings over ambient temperature and the maximum allowable temperature rise of the top oil of the transformer are specified by ANSI/IEEE standards and are guaranteed by the manufacturer. The purpose of temperature rise tests is to demonstrate that the transformer will deliver rated load without exceeding the guaranteed values of the temperature rises of the windings and oil. According to the ANSI/IEEE standards, these tests are performed at the minimum and maximum load ratings of a transformer.

Test Methods

For factory testing, it is not practical to connect the transformer to a load impedance with full rated secondary voltage applied to the simulated load. Although this would most directly simulate service conditions, most of the total test input power would dissipate in the load impedance. The load power, which equals the load rating of the transformer, is much larger than the sum of no-load and load losses that dissipate in the transformer. The electrical heating of the load would not contribute to transformer heating. Electric power consumption for the test would be excessive and the test would not be practical for routine testing.

In factory tests according to methods specified in the ANSI/IEEE standards, several artificial loading schemes may be used to simulate heat dissipation caused by the load and no-load losses of the transformer. The back-to-back loading method, described in the IEEE test code [2], requires two identical transformers and is often used for heat runs of distribution transformers. For power transformers it is most common to use the short-circuit loading method, as specified in the IEEE test code [2]. The test setup is similar to that used for measurement of load loss and impedance voltage. One winding is connected to a short circuit, and sufficient voltage is applied to the other winding to result in currents in both windings that generate the required power loss to heat the oil and windings.

Determination of the Top, Bottom, and Average Oil Rises

This discussion applies to the short-circuit loading method. Initially, the test current is adjusted to provide an input power loss equal to the no-load loss plus the load loss. This may be called the total loss. The total loss is corrected to the guaranteed temperature rise plus ambient temperature. During this portion of the heat run, the windings provide a heat source for the oil and the oil cooling system. The winding temperatures will be higher than expected because higher-than-rated currents are applied to the windings, but here only the oil temperature rises are being determined. For an oil-filled power transformer, the total power loss for a given rating is maintained until the top oil temperature rise is stabilized. Stabilization is defined as no more than 1°C change in three consecutive 1-h periods.

After stabilization is achieved, the top and bottom oil readings are used to determine the top, bottom, and average oil temperature rises over ambient at the specified load.

Determination of the Average Winding Temperature Rise

After the top, bottom, and average oil temperature rises are determined, the currents in the windings are reduced to rated value for a period of 1 h. Immediately following this 1-h period, the AC power leads are disconnected, and resistance measurements are carried out on both windings. The total time from disconnection of power and the first resistance readings should be as short as possible, typically less than 2 min — certainly not more than 4 min. Repeated measurements of winding resistance are carried out for a period of 10 to 15 min as the windings cool down toward the surrounding average oil temperature. The average oil temperature is itself falling, but at a much slower rate (with a longer time constant.) Data for the changing resistance vs. time is then plotted and extrapolated back to the instant of shutdown. With computers the extrapolation can be done using a regression-based curve fitting approach. The extrapolated value of winding resistance at the instant of shutdown is used to calculate winding temperature, using the method discussed below, with some correction for the drop in top oil rise during the 1-h loading at rated current. The winding temperature, thus determined, minus ambient temperature, is equal to the winding temperature rise at a given loading.

Similar tests are repeated for all ratings for which temperature rise tests are required.

Determining the Average Temperature by Resistance

The average winding temperature, as determined by the following method, is sometimes called the Average Winding Temperature by Resistance. The word *measurement* is implied at the end of this phrase. The conversion of measured winding resistance to average winding temperature is accomplished as follows. Initial resistance measurements are made at some time before commencement of the heat run when the transformer is in thermal equilibrium. When in equilibrium, the assumption can be made that the temperature of the conductors is uniform and is equal to that of the transformer oil surrounding the coils. Initial resistance measurements are made and recorded, along with the oil temperature. This measurement is sometimes called the cold resistance test, so the winding resistance measured during the test will be called the cold resistance, R_c and the temperature will be called the cold temperature, T_c. At the end of the heat run, R_h, the hot resistance is determined from the time series of measured resistance values by extrapolation to the moment of shutdown. The formula given below is used to determine T_h, the hot temperature, knowing the hot resistance, cold resistance, and the cold temperature. This calculated temperature is the Average Winding Temperature by Resistance.

$$T_h = \frac{R_h}{R_c}\left[T_c + T_k\right] - T_k \tag{3.81}$$

where

T_h is the "hot" temperature
T_c is the "cold" temperature
R_h is the "hot" resistance
R_c is the "cold" resistance
T_k is a material constant; 234.5 for copper; 225 for aluminum

Accurate measurements of R_c and T_c during the cold resistance test, as well as accurate measurements of hot winding resistance, R_h, at the end of the heat run are extremely critical for accurate determination of average winding temperature rises by resistance. The reason for this will be evident by analyzing the above formula. The following discussion illustrates how measurement errors in the three measured quantities, R_h, R_c, and T_c, affect the computed quantity, T_h, via the functional relationship by which T_h is computed.

Shown in Table 3.21 are computed values for T_h and the resulting error, e_{Th}, in the computation of T_h for sample sets of the measured quantities R_h, R_c, and T_c, measured in error by the amounts e_{Rh}, e_{Rc}, and e_{Tc}. Let us examine this table row by row. The way that measurement error propagates in the calculation is shown in the formula below.

$$\left(T_h + e_{Th}\right) = \frac{\left(R_h + e_{Rh}\right)}{\left(R_c + e_{Rc}\right)}\left[\left(T_c + e_{Tc}\right) + 234.5\right] - 234.5. \tag{3.82}$$

TABLE 3.21 Effect of Measurement Error in Average Winding Temperature by Resistance

Row	$T_h + e_{Th}$ (°C)	$R_h + e_{Rh}$ (Ω)	$R_c + e_{Rc}$ (Ω)	$T_c + e_{Th}$ (°C)
1	87.375 + 0	0.030 + 0	0.024 + 0	23.0 + 0
2	87.375 − 3.187 = 84.188 (−3.6%)	0.030 + 0	0.024 + 0.00024 = 0.02424 (+1.0%)	23.0 + 0
3	87.375 + 3.218 = 90.594 (+3.7%)	0.030 + 0.0003 = 0.0303 (+1.0%)	0.024 + 0	23.0 + 0
4	87.375 + 1.25 = 88.625 (+1.43%)	0.030 + 0	0.024 + 0	23.0 + 1.0 = 24.0 (4.35%)

Transformers

In Row 1 of Table 3.21, there is no measurement error. The sample shows a set of typical measured values. The value for T_h in this row may be considered the "correct answer." In Row 2, the cold resistance measurement was 1% too high, causing the calculated value of T_h to be 3.6% too low. This amplification of the relative measurement error is due to the functional relationship employed to perform the calculated result. This example illustrates the importance of measuring the resistance very carefully and accurately. Similarly, in Row 3 a hot resistance reading 1% too high results in a calculated hot temperature which is 3.7% too high. Row 4 shows the result if the cold resistance reading is 1°C too high. The result is that the determined hot resistance is 1.25°C too high. In this case, while there is a reduction in the error expressed as percent, the absolute error in degrees Celsius is in fact greater than the original temperature error in the cold temperature reading. These examples show that all three measured quantities R_h, R_c, and T_c must be measured accurately to obtain an accurate determination of the average winding temperature.

Other methods for correction to the instant of shutdown based on W/kg, or W/kg and time, are given in the ANSI/IEEE test code [2]. The cooling curve method, however, is preferred.

Other Tests

Audible Sound Level Measurements

The Purpose of Audible Sound Level Measurements
Sound level tests demonstrate that the transformer audible sound level meets any local noise regulations and that it does not exceed the design guarantee.

The Nature of the Quantity Being Measured
A tacky joke in the transformer industry goes like this: "Why does a transformer hum?" "Because a transformer does not know the lyrics," is the reply. When operational, all transformers produce audible sound and vibration. The source of this "audible noise" is in part the effect of cyclic magnetostriction in the core steel laminations. For this reason, the sound that emanates from a transformer is, among other factors, a function of the flux density and the voltage applied to the transformer. If a higher than rated voltage is applied to the transformer, the sound level will be higher than as tested according to the method described below.

How the Audible Sound Level is Measured
The sound level test is normally made on an unloaded transformer excited at rated voltage and rated frequency. Sound pressure level measurements are usually made at a specified number of points, located at specified distances around and away from the string perimeter. The locus defining the set of required microphone locations is shown conceptually in Fig. 3.194. The ANSI/IEEE transformer test code [2] defines the exact locations for microphone placement. For the purpose of standardization, noise levels are measured with the A-weighted network, in an environment where the ambient sound level is preferably 10 dB lower than the level being measured. Usually, the average of measurements at specified locations and height are reported as the sound level of the transformer under test. Typically, the results of measurements at the self-cooled, forced-air cooled, and forced-oil cooled ratings are reported and compared to guaranteed values.

Octave-band measurements of sound levels at various frequencies and with various defined bandwidths are often specified to meet local regulations. Methods for conducting such tests as well as methods for sound power and sound intensity determinations are described in the ANSI/IEEE transformer test code [2].

Special Tests

Many additional tests are available to obtain certain information about the transformer, usually to address a specific application issue. These are listed below.

1. Overload heat run
2. Gas in oil sampling and analysis (in conjunction with other tests)
3. Extended-duration no-load loss tests
4. Zero sequence impedance measurements

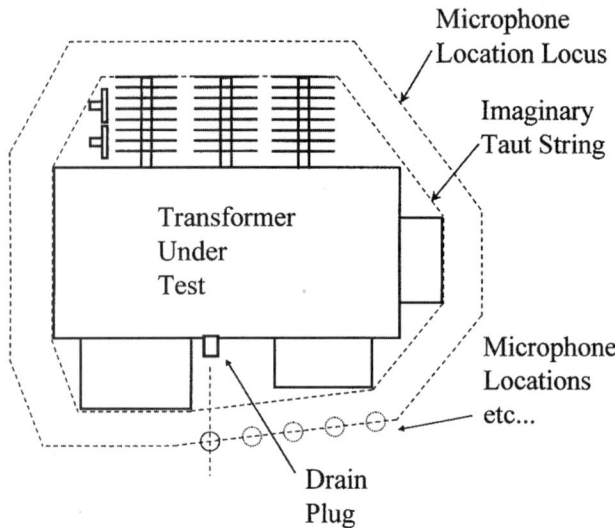

FIGURE 3.194 Microphone locations for audible sound test.

5. Tests of the load tap changer
6. Short-circuit withstand tests
7. Fault current capability of enclosures (overhead distribution transformers)
8. Telephone line voice frequency electrical noise (overhead distribution transformers)
9. Tests on controls

These tests and others, of a specific nature, are beyond the scope of this general section. Interested readers will find more information about these and other tests in the national and international standards dealing with transformers.

References

[1] ANSI/IEEE C57.12.00-1993, IEEE Standard General Requirements for Liquid-Immersed Distribution, Power, and Regulating Transformers.
[2] ANSI/IEEE C57.12.90-1992, IEEE Standard Test Code for Liquid-Immersed Distribution, Power, and Regulating Transformers; and Guide for Short-Circuit Testing of Distribution and Power Transformers (ANSI).
[3] ANSI/IEEE Std. 100-1992, The New IEEE Standard Dictionary of Electrical and Electronics Terms.
[4] IEEE C57.98-1993, IEEE Guide for Transformer Impulse Tests.
[5] IEEE C57.113-1991, IEEE Guide for Partial Discharge Measurement in Liquid-Filled Power Transformers and Shunt Reactors.

3.17 Transformer Installation and Maintenance

Alan Oswalt

Transformer Installation

The first priority is to hire a reliable contractor to move and assemble the transformer. There are many stories where the contractors, lacking experience or proper equipment, drop the transformer or do not assemble the components correctly. Accepting a low bid could cost your company more than securing a competent contractor.

Transformers

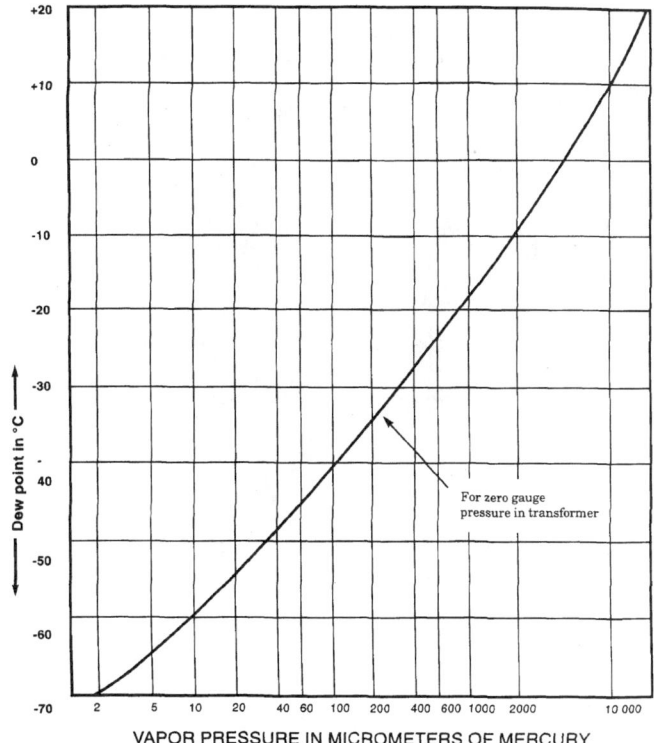

FIGURE 3.195(a) Conversion of dew point to vapor pressure. (Courtesy of Waukesha Electric Systems.)

Do not assume that all manufacturers have the same methods of installation. Your understanding of the manufacturer's transformer installation book and reviewing the complete set of drawings in advance will help you to understand "their" procedures. Some manufacturers have a toll-free number which allows you to clarify drawings and/or the assembly methods. Others have put together a series of videos and/or CDs that will assist you to understand the complete assembly. Then you should review all of the information with your assembly contractor.

Receiving Inspection

Prior to unloading a transformer and the accessories, a complete inspection is necessary. If any damage or problems are found, contact the transformer manufacturer *before* unloading. Freight damage should be resolved, as it may be required to return the damaged transformer or the damaged accessories. Photographs of the damage should be sent to the manufacturer. Good receiving records and photographs are important, should there be any legal problems.

Three important inspections checks are (1) loss of pressure on the transformer, (2) above zone 3 on the impact recorder, and (3) signs of movement by the transformer or its accessories. If any of the three inspection checks indicate a problem, an internal inspection is recommended.

A shorted core reading could also mean a bad transit ride. With a railroad shipment, if any of the checks indicate problems and an internal inspection does not reveal the problem, get an *exception report* filled out by a railroad representative. This report will assist you later if hidden damages are found.

Low core meggar readings (200 Megonms) could be an indication of moisture in the unit and require extra costs to remove.[1] The moisture could have entered the unit through a cracked weld caused by the bad transit ride. (See Figs. 3.195a and b.)

---

[1] A dew point test will determine the moisture in the transformer. A dew point reading should be used with the winding temperature value (insulation temperature) to determine the percentage of moisture. (See Figs. 3.195a and b.)

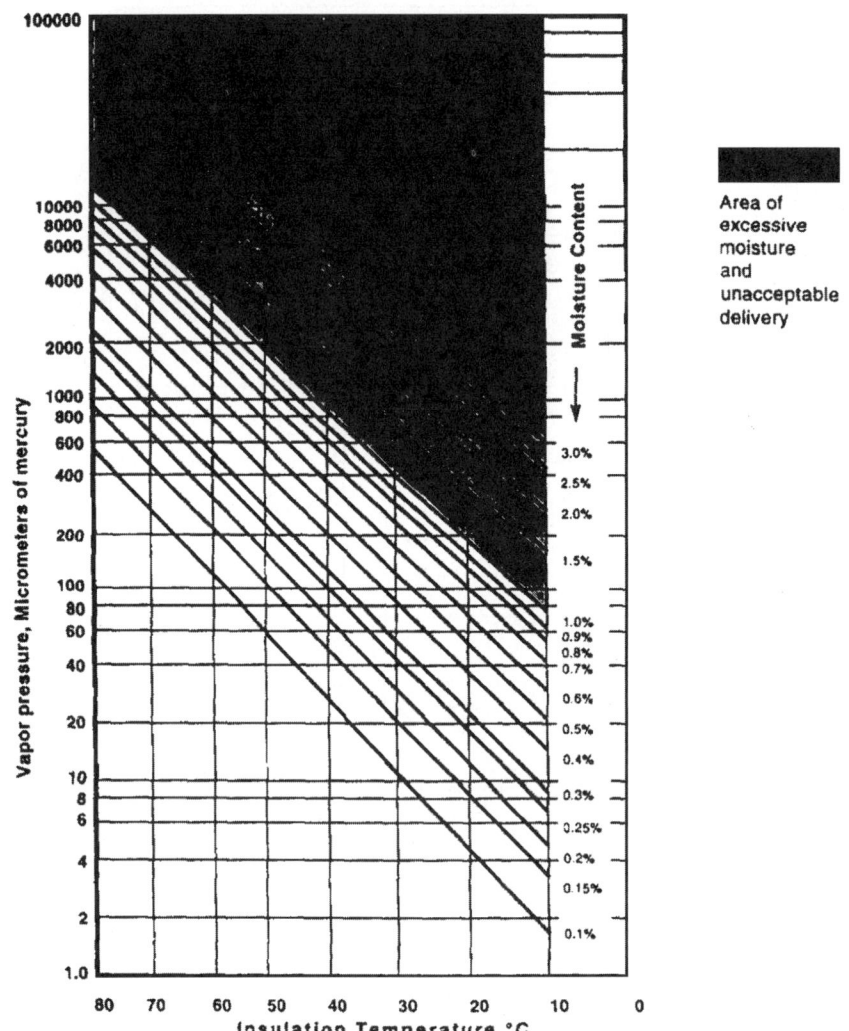

FIGURE 3.195(b) Moisture equilibrium chart (with moisture content in percent of dry weight of insulation). (Courtesy of Waukesha Electric Systems.)

Entering a unit requires good confined entry procedures and can be done after contacting the manufacture, as they may want to have a representative to do the inspection. Units shipped full of oil require a storage tanker and the costs should be agreed upon before starting.

Assuming that we now have a good transformer and it is setting on its substation pad, there are some items that are essential for assembly. First ground the transformer before starting the assembly. Static electricity can build up in the transformer and cause a problem for the assembly crew. A static discharge could cause a crew member to jump or move and lose their balance while assembling parts.

Another item is to have all accessories to be assembled set close to the unit, as this eliminates a lot of lost time moving parts closer or from a storage yard. With the contractor setting the accessories close to the unit, you can usually save a day of assembly time. Keep in mind that some transformer manufactures "match-mark" each item. This means that each part has a specific location on the unit. Do not try to interchange the parts. Some manufacturers do not have this requirement, which allows bushings, radiators, and other parts to be assembled at the contractor or customer's discretion.

Transformers

Weather is a major factor during the assembly of any transformer. Always have an ample supply of dry air flowing through the unit during the assembly. Be ready to seal the unit on positive pressure at the end of the day or if the weather turns bad. If the weather is questionable, keep the openings to a minimum and have everything ready to seal the unit.

There are many types of contaminants that can cause a transformer to fail. Foreign objects dropped into the windings, dirt brought into the unit on the assemblers' shoes, moisture left in the assembled parts, and misplaced or forgotten tools left inside are just a few items that could cause a failure. Take time to caution the assemblers about the preventions of contaminants and to follow good safety procedures. Again, an experienced contractor should have experienced assemblers and good assembly procedures in place.

If weather conditions are a problem, there are many other assembly operations that can be done during marginal weather. The following are a few:

1. Uncrate all items.
2. Locate and count all hardware.
3. Clean bushings with denatured alcohol and cover.
4. Count and replace gaskets — if possible.

Caution: Do not supply power to the control cabinet as it could back-feed into the inside current transformers, which could energize the primary and secondary bushings. A shock from the bushings could cause serious injury.

Bushings

You need to read the installation manual to understand the correct lifting and assembly methods. There are a variety of bushings so it will save time to have read this information. All bushing surfaces should be cleaned again with denatured alcohol. This also includes the inside tube (draw-through type bushing). During shipment, even though the bushings may be protected, contaminants, such as moisture, could be found inside the bushing tube. The draw-through bushings have a conductive cable, or a rod, that has to be pulled through the bushing while it is being installed. In some cases, the corona shield on HV bushings should be removed and cleaned. There are also bottom connected bushings that require copper bus, a terminal, and hardware to secure the connection to the bushing and winding. All connections should be cleaned and free of oxidation or corrosion and then wiped down with denatured alcohol.[2]

After the installation of all bushings and all internal connections are made, another inspection should be made for the following:

1. Lead clearances. During the internal assembly work, some leads may have been moved. Check the manufacturer's installation book for the necessary clearances. The information should include the basic installation level (bil) rating along with the clearances needed.
2. Bolted connections, done by the assemblers, should be inspected for proper clearances.
3. Wipe down and vacuum clean the inside of the unit around the assembly area to remove any dirt or oil smudges.
4. Operate the DLTC and check its mechanical operation.
5. Check for items, such as tools, that may have been left inside during the assembly.
6. Replace man-hole gaskets, if required.

Some units have conservators and require gas piping and oil piping connected to the transformer, after the man-hole covers are installed on the transformer and before pressure or vacuum cycles are started.

---

[2]Some manufacturers will require all bushing gaskets to be replaced. Others furnish a Buta-N O-Ring that, in most cases, will not need replacing.

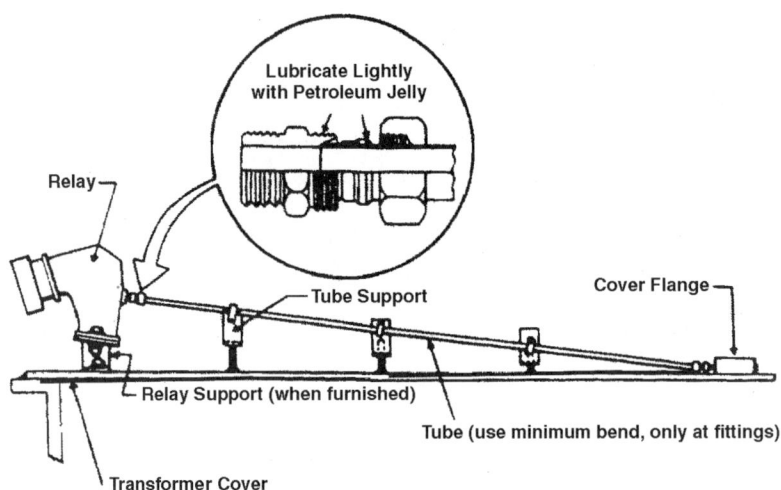

FIGURE 3.196 Outline of typical relay installation on transformer cover. (Courtesy of Waukesha Electric Systems.)

Oil Conservators

Conservators are usually mounted on one end of the transformer and well above the cover and bushings. Conservators normally have a rubber bladder inside. This bladder expands or retracts due to the temperature of the oil vs. the ambient temperature. The inside of the bladder is connected to external piping, and then to a silica gel breather. All exposure of the oil to the air is eliminated, yet the bladder can flex. (See Fig. 3.197.)

Gas Monitoring and Piping

The piping is used to bring any combustible gases to a monitor. The monitor is usually located on the cover where it is visible from ground level. All gas piping should be cleaned by blowing dry air through them, or cleaning with a rag and denatured alcohol. Gas pipes are usually not connected to the gas monitor until after the vacuum/oil filling. The gas monitor could have tubing running down the side of the unit to allow ground-level sampling or bleeding of the line. There are other types of oil/gas monitors than the one shown in Fig. 3.196.

The oil supply piping, from the conservator to the transformer, should have at least one valve. The valve(s) must be closed during the vacuum cycle as the vacuum will try to pull the rubber bladder through the piping. The oil piping should have been cleaned prior to installation and the valves inspected. The conservator should have an inspection cover and the inside bladder inspected. While making this inspection, also check the operation of the oil float. (See Fig. 3.197.)

Radiators

All radiators should be free of moisture and contaminants such as rust. If anything is found, the radiators should be cleaned and oil flushed with new transformer oil. The radiators may have to be replaced with new ones. Take time to inspect each radiator for bent fins or welding defects. If a problem is found, the repair should be made before installation. Touch-up painting, if needed, should be done, as it is difficult to reach all areas after the radiators are installed. During the radiator installation, all of the radiator valves need to be tested on at least 2 lb of pressure, or under oil, for a good seal.

Some gaskets for mounting the radiator/valve mounting flange may have to be replaced. Coating the outside of the gasket with petroleum jelly protects the surface of the gasket during the radiator assembly. The radiator surface will then slide without damaging the gasket.

Transformers

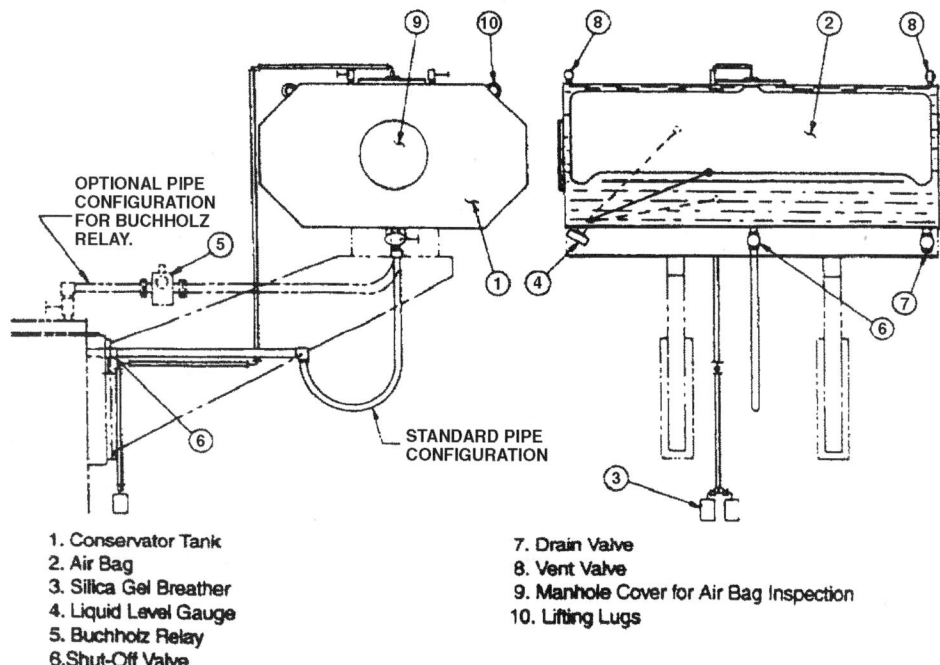

1. Conservator Tank
2. Air Bag
3. Silica Gel Breather
4. Liquid Level Gauge
5. Buchholz Relay
6. Shut-Off Valve
7. Drain Valve
8. Vent Valve
9. Manhole Cover for Air Bag Inspection
10. Lifting Lugs

FIGURE 3.197 Conservator tank construction. (Courtesy of Waukesha Electric Systems.)

Coolers

Coolers are oil to air heat exchangers which require oil pumps and usually less space around the transformer. Forced oil cooling can be controlled by a top oil or a winding temperature gauge, or both. All pumps, piping, and coolers should be inspected for contaminants before assembly. The correct pump rotation is an important checkpoint.

Load Tap Changers (LTC)

Some are shipped full of oil but you may want to make an internal inspection. After removing the oil you can inspect for problems. Check the manufacturer's installation book for information concerning vacuum oil filling of the unit. Some LTCs require a vacuum line to main tank for equalizing the pressure. *Do not* operate the LTC mechanism while the unit is on vacuum, as severe damage could occur to the mounting board.

If the LTC requires oil, *do not* add the oil while the main tank is on vacuum as the unequal pressure could damage the LTC.

Look for loose hardware or any misalignment of the contacts. Operate the LTC through all positions and check at each step.

Positive Pressure System

This system consists of a cabinet with regulating equipment and alarms with an attached nitrogen bottle. A positive supply of nitrogen is kept on the transformer. With positive pressure on the tank, the possibility of moisture entering is reduced. Loss of nitrogen pressure, without any oil spillage, is usually found in the transformer "gas space" or the nitrogen supply system.

Control Cabinet

All control equipment must be inspected for loose wiring or problems caused by the shipping. The fan, gauges, LTC controls, and monitoring equipment must be tested or calibrated. Information for the installation and/or the calibration should be supplied by the manufacturer.

Accessories

There are many items that could be required for your particular transformer. A few are listed below.

1. HV and LV arrestors
2. External current transformers
3. Discharge meters
4. Neutral grounding resistors
5. Bushing potential devices
6. Cooling fan (hi speed)
7. Gas monitors (various types)

Caution: All oil handling equipment, transformer bushings, and the transformer should be grounded before starting the vacuum oil cycle. Special requirements are needed for vacuum oil filling in cold weather. Check your manufacturer's manual.

Vacuum Cycle

Pulling vacuum on a transformer is usually done through the mechanical relief flange or a special vacuum valve located on the cover of the transformer. A vacuum sensor, to send a signal to the vacuum recording gauge, should be at the highest location on the transformer's cover. All readings from this gauge should be recorded at least every hour. *Note:* All radiator and cooler valves should be open prior to starting the vacuum cycle.

Vacuum Filling System

Manufacturers differ on the length of vacuum required and the method to add oil to the unit. It is important that the vacuum crew doing this process follow the correct procedure or the warranty can be invalid. Good record-keeping during this process is just as important for your information as it is for supplying the manufacturer with information that validates the warranty.

The length of time (pulling vacuum) will vary as to the exposure time to atmospheric air, the transformer rating, and the dew point/moisture calculations. Most of the needed information as to the vacuum cycle time, should be furnished in the installation book. (See Figs. 3.195a and b.)

Always consider that some contractors have used their equipment on older and/or failed transformers. The equipment needs to be thoroughly cleaned with new transformer oil and a new filter medium added to the oil filtering equipment. The vacuum oil pump should have new vacuum oil installed and it should be able to "pull-down" against a closed valve to below IMM of vacuum.

Transformer Oil

The oil supplied should be secured from an approved source and meet the IEEE C57 106-1991 guide for acceptance. When requested, an inhibitor can be added to the oil, to a level of 0.3% by weight.

Adding the Oil

All oil tankers should be field tested for acceptable dielectric level prior to pumping oil through the oil handling equipment. It is always desirable to heat the oil before pumping it into the transformer as this method also assists in driving the moisture out of the internal core and coil assembly. Oil filling a conservator transformer takes a lot more time as the piping and the conservator have to be slowly filled while air is "bled" out of the piping, bushings, CT turrets, and time gas monitor. Methods vary for adding the oil to the conservator because of the risk to the oil bladder. Weeks later, the air should be "bled" again. If this is not done, you could receive a false signal that may take the transformer out of service.

Field Test

After the vacuum oil filling cycle, the transformer should be field tested and compared to some of the factory tests. The field test also will give you a baseline record for future reference.

Transformers

These are a few of the tests that are routinely done:

1. Current transformer ratio
2. Turns ratio of the transformer
3. Power factor bushings
4. Power factor transformer
5. Winding resistance
6. Core ground
7. Lab test of the oil
8. Test all gauges
9. Test all pressure switches

A field test report may be required by the manufacturer in order to validate the warranty. Any questionable test values should be brought to the attention of the transformer manufacturer.

Transformer Maintenance

The present maintenance trend is to reduce cost, which in some cases means lengthening the intervals of time to do maintenance or eliminating the maintenance completely. The utility, or company, realizes some savings on manpower and material by lengthening the maintenances cycle, but by doing this, the risk factor is increased. A few thousand dollars for a maintenance program could save your utility or company a half-million dollar transformer. Consider the following:

1. The length of time to have a transformer rebuilt or replaced
2. The extra load on your system
3. Rigging costs to move the transformer
4. Freight costs for the repair, or buying a new one
5. Disassembly and reassembly costs
6. Costs to set up and use a mobile transformer
7. Costs for oil handling of a failed unit
8. Vacuum oil filling of the rebuilt or new transformer
9. Customer's dissatisfaction with outage
10. Labor costs, which usually cover a lot of overtime or employees pulled away from their normal work schedule

You will have to answer questions such as: Why did this happen? Could it have been prevented? Time spent on a scheduled maintenance program is well worth the expense.

There are many systems available to monitor the transformer which can assist you in scheduling your maintenance program. Many transformer manufacturers can supply monitoring equipment that alerts the owner to potential problems. However, relying solely on monitoring equipment may not give your notice or alert you to mechanical problems. Some of these problems can be: fans that fail, pressure switches that malfunction, or oil pumps that cease to function. You could also have oil leaks that need to be repaired. Annual inspections can provide a chance for correcting a minor repair before it becomes a major repair.

Maintenance Tests

There are two important tests that could prevent a field failure. Using an infaray scan on a transformer could locate "hot spots". The high temperature areas could be caused by a radiator valve closed, low oil in a bushing, or an LTC problem. Early detection could allow time to repair the problem.

Another important test is dissolved gas analysis test of the oil by a lab. A dissolved gas analysis lab test will let you know if high levels of gases are found and they will inform you as to the recommended action. Following the lab report could let you plan your course of action. If there seems to be a problem, it would be worthwhile to take a second dissolved gas in oil sample and send it to a different lab and compare the results (IEEE C57 104-1991). Maintenance inspection and tests can be divided into two sections: (1) minor and, after a set period of years, (2) major inspection. Annual tests are usually done while the transformer is in service, and consist of the following:

1. Check the operation of the LTC.
2. Take an oil sample from the LTC.
3. Change silica gel in breathers.
4. Inspect fan operation.
5. Take an oil sample from the main tank.
6. Check oil level in bushings.
7. Check tank and radiators for oil leaks.
8. Check for oil levels in main tank and the LTC.
9. Make sure all control heaters are operating.
10. Check all door gaskets.
11. Record the amount of LTC operations and operate through a couple of positions.
12. Most importantly, have your own check-off list and take time to do each check. This record (check-off list) can be used for future reference.

Major inspections require the transformer to be out of service. Both primary and secondary bushings should be grounded before doing the work. Besides the annual inspection checks that should be made, the following should also be done:

1. Power factor the bushings and compare to the values found during the installation tests.
2. Power factor the transformer and check these valves.
3. Make a complete inspection of the LTC and replace any questionable parts. If major repair is required during this inspection, a turns ratio test should be done.
4. Painting rusty areas may be necessary.
5. Test all pressure switches and alarms.
6. Check the tightness of all bolted connections.
7. Check and test the control cabinet components.

The lists for annual and major inspections may not be complete for your transformer and you may want to make a formal record for your company's reference. Some transformer installation/installation books furnished by the manufacturer may have a list of their inspections areas which you should utilize in making your own formal record.

3.18 Problem and Failure Investigations

Harold Moore

The investigation of transformer problems or failures in many respects is similar to medical exploratory procedures. Elements of transformer design, application, and operation are involved. The elements to be investigated depend on the nature and the severity of the problem. If a failure is involved, all of the elements are usually investigated. The analyses required can be quite complex and involved. It will be impossible to describe the many details involved in complex failure investigations in this one section, and a complete book could be written on this subject. It will only be possible to describe the processes involved in such investigations.

Excellent references are ANSI/IEEE C57.117, *IEEE Guide for Reporting Failure Data for Power Transformers and Shunt Reactors on Electric Power Systems* and ANSI/IEEE C57.125, *IEEE Guide for Failure Investigation, Documentation, and Analysis for Power Transformers and Shunt Reactors*.

The following steps are involved in problem and failure investigations.

- Collect background data that may be involved.
- Visit the site to obtain application and operation data.
- Interview all persons that may be involved.
- Inspect the transformer and perform a partial or complete dismantling if a failure is involved.
- Analyze the available information and background.
- Prepare a preliminary report and review with persons that are involved or that have a direct interest to generate additional inputs.
- Write a final report.

As a general statement, no details should be neglected in such investigations. Experience has indicated that what may appear to be minor details sometimes hold the essential clues for solutions.

Background Investigation

The background investigation usually involves the following steps.

How did the transformer problem manifest itself or how did the failure occur? — Unusual trends or events that occurred which indicated a possible problem should be recorded. Operation of relays or protective equipment that indicated a failure should be studied. Copies of oscillographic or computer records of the events surrounding the problem or failure should be obtained.

History of similar designs — The performance of transformers made by the manufacturer should be studied. Industry records of transformers of similar rating and voltage class may be helpful in establishing base data for the investigation. For example, there have been transformers made by certain manufacturers which have a history of short-circuit failures in service. At one time, the failure record of EHV transformers designed with three steps to reduce BILs had a higher failure rate than those with higher BILs.

Transformer design — The specifications for the transformer, instruction books and literature, nameplate, and drawings such as the outline and internal assembly should be examined. If failures involving the core and windings have occurred, the investigative process is much like a design review in which the details of the insulation design, winding configuration, lead configurations and clearances, short-circuit capability, and the core construction and induction must be studied. If components are involved, bushings and bushing clearances, tap changers, heat exchanger equipment, and control equipment should be investigated.

Manufacturing and testing of the transformer — The manufacturing and test records should be studied to determine if discrepancies occurred. Of particular importance are deviations from normal manufacturing specifications or practices. Such deviations could be involved in the problem or failure. All parts of the test data must be studied to determine if discrepancies or deviations existed. Partial discharge and impulse test data that have any deviation from good practices should be recorded. Approvals of deviations made by the manufacturer of dielectric and test standards are sometimes indications of difficulty during testing.

Application and operation data — The operation records of the transformer should be examined in detail. Records of field tests such as insulation resistance and power factor, gas in oil analyses, oil test data, and bushing tests should be studied. Any trends from normal such as increasing power factor, water in oil, or deterioration of oil properties should be noted. Any internal inspections that may have been performed, changes in oil, and any repairs or modifications are of interest. Maintenance records should be examined for evidence of either good or poor maintenance practices. Application of the transformer in the power system should be determined. Operational history such as switching events, number and type of system short circuits, known lightning strikes, overloads, etc. should be recorded. Effectiveness of arresters and other protective schemes should be determined.

Problem Analysis Where No Failure Is Involved

It is advisable to make some investigations after severe operating events such as direct lightning strikes (if known), high magnitude short-circuit faults, and inadvertent high magnitude overloads. Problem investigations are sometimes more difficult than failure analyses because some of the internal parts cannot be seen without dismantling the transformer. The same data collection process is recommended for problems as for failures. It is recommended that a plan be prepared including a checklist when the investigation is initiated to guide the study and to ensure that no important steps are omitted. A recommended checklist is in ANSI/IEEE C57.125. Specific steps that are recommended are as follows:

Communication with Persons Involved or Site Visit

- Obtain the background of the events indicating a problem.
- If there is any external evidence such as loss of oil, overheated parts, or other external indications, a site visit may be advisable to get first-hand information and to discuss the events with persons who operated the transformer.
- Determine if there have been any unusual events such as short circuits, overvoltages, or overloads.

Diagnostic Testing

If the utility has a comprehensive field testing program, obtain and study the following data. If the data are not available, arrange for tests to be made.

- Gas in oil analyses; obtain for several years prior to the event if possible
- Oil test data
  - Dielectric strength in accordance with ASTM D 1816
  - Water in oil
  - Power factor at 25 and 100°C if available
  - General characteristics such as color, IFT, etc.
- Turns ratio
- Resistance (such measurements must be made using suitable high-accuracy instruments; these measurements are not easy to make under service conditions)
- Insulation power factor, resistance, and capacitance.
- Low-voltage exciting current (if previous data are available)
- Short-circuit impedance if faults have been involved

Information on interpretation of the data could take volumes, and there is much information on this subject in the technical literature. Some simple guides are listed below for reference.

- The presence of high carbon monoxide and carbon dioxide are indications of thermal or oxidation damage to cellulose insulation. If there is high CO and there have been no overloads or previous indications of thermal problems, the problem may be excessive oxygen in the oil.
- High oxygen is usually an indication of inadequate oil processing, gasket leaks, or leak of air through the rubber bag in expansion tanks.
- Acetylene is an indication of arcing or very high temperatures.
- Deterioration of oil dielectric strength usually results from particulate contamination or excessive water.
- High power factor or low resistance between windings or from the windings to ground is usually the result of excessive water in the insulation.
- High water in oil may result from excessive water in the paper. Over 95% of the total water in the system is in the paper so that high water in oil is a reflection of the water in the paper.

- Turns ratio different than previous measurements is an indication of shorts in a winding. The shorts can be between turns or between parts of windings such as disk to disk.
- Measurable changes in the leakage impedance is an indication of winding movement or distortion.
- Open circuits result from major burning in the windings or possibly a tap changer malfunction.
- High hydrogen with methane being about 20% of the hydrogen is an indication of partial discharges. "Spitting" or "cracking" noises are sometimes indicators of high partial discharges.

If the problem cannot be identified from the test data and behavior analyses, an internal inspection may be necessary. In general, internal inspections should be avoided because the probability of failure increases after persons have entered transformers. The following items should be checked when the transformer is inspected.

- Is there evidence of carbon tracks indicating flashovers or severe partial discharges?
- Check leads for evidence of overheating. The insulation will be tan, brown, or black in extreme cases.
- Check for evidence of partial discharge trees or failure paths. Are there odors indicating burned insulation or oil?
- Check bolted connections for proper tightness.
- Are leads and bushing lower shields in position? Is the lead insulation tight?
- Inspect the windings for evidence of distortion or movement.
- Check the end insulation on core form for evidence of movement or looseness.
- Are the support members at the ends of the phases tight in shell form designs?
- Are coil clamping devices tight and in position?
- Check tap changers for contact deterioration. Is there evidence of problems in the operating mechanisms?
- In core form, check visible parts of the core for evidence of heating.
- After the inspection has been completed, look over everything again for evidence of "does anything look abnormal". Such final inspections frequently reveal something of value.

Design Review Process

An explanation of the design review process would require a separate chapter and is beyond the scope of this subject. The process involves a detailed study of the winding and insulation designs, short circuit capability, thermal design, magnetic circuit characteristics, leakage flux losses and heating analysis, materials used, oil preservation systems, etc. In some cases, it may be desirable to conduct part or all of a design review as a part of a problem investigation.

Determine if Similar Designs Experienced Problems or Failures

This can be difficult since there is no agency that collects and distributes data on all industry problems. It has been made even more difficult by the closing of several major transformer manufacturing plants. However, some information can usually be obtained by discussions with users having transformers made by the same manufacturer.

Dismantling Inspections

Complete dismantling is usually performed when a failure has occurred. In a few instances, such operations are performed to determine the cause for a problem such as excessive gassing that could not be explained by other investigations. This subject will be developed under the discussion on failures.

Failiure Investigations

The same processes are required in failure investigations as described for problem analyses. In fact, it is recommended that the approaches described for problem investigations be performed before dismantling of the transformer to determine possible causes for the failure. Performing this work in advance usually results in hypotheses for the failure and frequently indicates directions for the dismantling.

Dismantling Process

The following steps are recommended for this process.

- The process should be directed by a person or persons having knowledge of transformer design. If it is done in the original manufacturer's plant, the manufacturer's experts will usually be available. However, it is recommended that the user have experts available if the failure mechanism is in doubt. It is good to have two experts available because they will usually look into the failure from different perspectives and will provide the opportunity to discuss various aspects of the investigation from the viewpoints of experts.
- Inspection before untanking. Inspect the tank for distortion resulting from high internal pressures that sometimes result from failures. Check the position of leads and connections. Determine if there has been movement of bushings.
- Inspection after removal of core and windings:
  - Make a detailed inspection of the top ends of windings and cores.
  - Inspect the mechanical and electrical condition of the interphase insulation and the insulation between the windings and the tank.
  - Check the core ground.
  - Check lead entrances to windings for mechanical, thermal, and electrical condition.
  - Check the general condition of leads, connections, and tap changers.
  - Check the leakage flux shields on tank walls or on frames for heating or arcing.
  - Examine the wedging between phases and from the windings to the core on shell form and the winding clamping structures on core form to determine if there is still pressure on the windings.
- Detailed inspection of the windings:
  - Is there evidence of electrical failure paths from the windings over the major insulation to ground?
  - Is there evidence of distortion of the coils or windings resulting from short circuit failures?
  - Has electrical failure occurred between turns and has mechanical distortion of the turns occurred?
  - Have failures occurred between windings or between coils?
  - Are there weaknesses in non-failed portions that could affect the electrical strength? Such instances are damage to turn insulation, distorted disks or coils, insulation pieces not properly assembled.
  - Is there evidence of metallic contamination on the insulation and windings?
  - Is there evidence of hot spots at the ends of windings or in tap leads inside the disks or coils?
  - Is there any evidence of partial discharges or overheating in the leads or connections?
  - Were the windings properly supported for short circuit?
    - Spacers in alignment?
    - Insulation between windings and between the inner winding and the core tight?
- Examination of the magnetic circuit:
  - Have electrical arcs occurred to the core? If so, determine if the fault current flowed from the failure point to ground which has resulted in damage to laminations in this path which usually requires scrapping of the laminations.
  - Is there evidence of leakage flux heating in the outer laminations?
  - Has heating occurred as the result of large gaps at the joints, excessive burrs at slit or cut edges and joints?

- Does mechanical distortion exist in any parts of the core?
- Is there evidence of heating in the lock plates used for mechanical support of the frames?
- Is the core ground in good condition with no evidence of heating or burning?
- Mechanical components:
  - Is there distortion in the mechanical supports?
  - Is there evidence of leakage flux heating in the frames or frame shields?

Analysis of Information

The most important part of the process is analysis of the information gathered. The objective is determining the cause of the problem or failure, and adequate analysis is obviously necessary if problems are to be solved and failures are to be prevented. There is no one process that is best for all situations. However, it is suggested that there are two approaches which are helpful for reaching conclusions in such matters.

1. Make a systematic analysis of the data.
2. Compare data analyses to known problem and failure modes.

Analysis of Data

- Prepare a list of known facts and unknowns. If some of the unknowns appear to be of importance, attempt to find answers.
- Analyze known facts to determine if there is a pattern which indicates the nature of the problem.
- Prepare a spreadsheet of test data and observations, including inspections. Note items which appear to indicate the cause of the problem or failure.
- Use problem-solving techniques.

Known Problem and Failure Modes

There are a number of failure modes, and a few will be listed for examples and guidance.

1. Dielectric or Insulation Failures
   - Surface or creepage over long distances. This phenomenon is usually caused by contamination if the design is adequate. If the design is marginal, slight amounts of contamination may initiate the discharges or failure.
   - Oil space breakdown. This can occur in any part of the insulation since oil is the weak link in the insulation system. If the design is marginal, discharges can be initiated by particulate contamination or water in the oil. This type of breakdown usually occurs at interfaces with paper such as at the edge of a radial spacer in a disk to disk space or at the edge of a spacer in a high-voltage winding to low-voltage winding space.
   - Oil breakdown over long distances as from a bushing shield to tank wall or from a lead to ground. This problem type is usually caused by over-stress in the large oil gap. It can occur in marginal situations if particles or gas bubbles are present in the gap. The dielectric strength of oil is lower at low temperatures if there is an appreciable amount of water in the oil. If such breakdowns occur in very low temperature conditions, investigate the oil strength at the low temperature as a function of the water in the oil.
   - Turn to turn failures. If the design is adequate, such failures can result from mechanical weakness in the paper or from damage during short circuits if the paper is brittle due to thermal aging or oxidation. These failures usually are associated with fast transients such as lightning.
   - Extensive treeing in areas of high oil velocity such as the oil entrance to the windings in forced cooled designs. This can be associated with static electrification and usually occurs when the oil temperature is less than 40°C and all pumps are in operation.

- Discharges or failure originating from joints in leads. This type of failure usually results from the paper not being tight at the joint in the tape. Discharges start in the oil space at the surface of the cable and propagate out through the joint.

2. Thermal or Oxidation Failure Modes
    - Deteriorated insulation at the end turns of core form or the outer turns of line coils in shell form. Such deterioration is caused by local hot spots. The eddy losses are higher in these regions, and the designer may have used added insulation in these regions which have high electrical stress.
    - Overheated tap leads. This usually occurs because the designer has used added insulation on the leads. The leads may have added eddy loss because they are in a high leakage flux field.
    - Leads with brown or black paper at the surface of the conductor. This results from excessive paper insulation on the lead.
    - Joints with deteriorated paper. The resistance of the joint may be too high or there may be leakage flux heating if the connector is wide.
    - Damaged paper or pressboard adjacent to the core or core supports. This type of heating is usually the result of leakage flux heating in the laminations or core joints.
    - Paper has lost much of its strength, but there have been no thermal stresses. This is the result of excessive oxygen in the oil. In the initial stages of the process, the outer layers of paper will have more damage than the inner layers.

3. Magnetic Circuit Heating
    - Large gaps at joints can result in local saturation. The gap area will be black, and there may be low levels of methane, ethane, and some hydrogen.
    - Local heating on the surface or at joints. Such heating is caused by excessive burrs on the edges or at end cuts of the laminations.
    - Burning at the joint of outer laminations. This may be caused by circulating currents in the outer laminations of cores. It results from an imbalance in leakage flux.

4. Short Circuit Failure Modes
    - Failure at the ends of shell form coils. This type of failure results from inadequate support of the outer layers of the coils.
    - Beam bending between spacers in either shell or core form which is caused by stresses higher than the beam strength of the conductor. Missing spacers can also be involved.
    - Inner winding buckling in core form. This results from inadequate strength of the conductor or inadequate spacer support. The evidence is radial distortion of the winding.
    - Telescoping of conductors. This occurs in windings with a thin conductor where there is not adequate support by an inner cylinder. The layers slip over adjacent layers.
    - Lead distortion.
    - Turns or disks telescoping over the end insulation or supports. This results from high axial forces in combination with insulation that may not have been properly dried and compressed.

It is important to consider that many problems and failures are a combination of events or problems. Some examples will be used to illustrate the technique.

Excessive Water in the Solid Insulation of a Transformer

- Measured water in the oil is high particularly after the transformer has been loaded and the oil is hot. (At elevated temperatures, water comes from the paper to the oil rather quickly.)
- Power factor of the insulation is increasing.
- The hydrogen content of the oil is increasing.
- The dielectric strength of the oil is decreasing when tested in accordance with ASTM D-1816.

This analysis indicates that the insulation has excessive water and should be dried. This conclusion could have been made without the power factor measurements.

Excessive Oxygen in the System (usually recognized as >3000 ppm of oxygen in the oil)

- Generation of CO is high.
- There is no history of overloads.
- Design analysis indicated no excessive hot spots.
- Oxygen appears to vary as the CO increases. (The oxygen is being consumed by the process which forms CO.)
- Internal inspection indicated that the outer layers of paper on a taped cable had greater deterioration than the inner layers.

The importance of keeping good records of transformer operation and maintenance events and making a complete analysis of all data involved in problem-solving cannot be overemphasized. Many failures and problems result from multiple causes. The following example demonstrates the importance of diligent investigations.

- Transformer experienced a severe short circuit as the result of a through fault on the system. Transformer did not fail.
- The oxygen in the oil had been high — 4000 ppm or higher for years.
- Transformer had history of high CO generation.
- Failure occurred some months after a switching event.

The failure was at first attributed to the switching event alone. However, the investigation showed that it was initiated by damage to brittle insulation probably during the short circuit event. The brittle condition was caused by the high oxygen in the system. The overvoltage involved in the switching event caused the failure at the damaged paper location.

Another important factor in problem and failure analysis is to use two experienced persons when possible. Each can challenge the ideas expressed by the other and offer suggestions for different approaches in the investigation. Experience has shown that a better analysis results when using this approach.

3.19 The United States Power Transformer Equipment Standards and Processes

Philip J. Hopkinson

This section is devoted to a description of the Power Transformer Equipment Standards approval processes and a listing of the Standards that are in place in the U.S. in 1999. The section on Accredited Standards Approval Processes provides an abbreviated description of the methods that are employed in the U.S. for gaining American National Standards Institute approval and recognition. Similarly, an approval process is also shown for International Electrotechnical Commission (IEC) documents. Many thanks are offered by the author for assistance that has been rendered by representatives of the National Electrical Manufacturer's Association, American National Standards Institute, and individuals from IEEE, IEC, EL & P, UL, and especially by Ms. Purefoy, who arranged and documented these contents.

The U.S. uses a voluntary process for the development of nationally recognized power transformer equipment standards. This section describes the U.S. standards accreditation process and provides flowcharts to show how accredited standards are approved. With the International Electrotechnical Committee (IEC) taking on increased importance, the interaction of the U.S. technical experts with IEC is also described. Finally, relevant power transformer documents are listed for the key power transformer equipment standards that guide U.S. industry.

I. Major Standards Organizations

ANSI	American National Standards Institute
ANSI C57	Accredited Standards Committee "C57" for Power and Distribution Transformers
IEEE	Institute of Electrical and Electronic Engineers
NEMA	National Electrical Manufacturing Association
EL & P	Electric Light and Power Delegation
EEI	Edison Electric Institute
AEIC	Association of Edison Illuminating Companies
UL	Underwriter's Laboratory
IEC	International Electrotechnical Commission

II. Processes for Acceptance of American National Standards

The acceptance of a standard as an American National Standard requires that it be processed through one of three methods: by canvas list, accredited standards committee, or accredited standards organization action. All three methods share the common requirement that the process used has prior approval of the American National Standards Institute (ANSI), that the methodology incorporates due process, and that consensus among interests is achieved. Inherent in that approval is presentation of accepted operational procedures and/or a balloting group that is balanced among users, manufacturers, and general interest groups. Other considerations include (1) that the document is within the scope previously registered, (2) identified conflicts are resolved, and (3) that known national standards were examined to avoid duplication or conflict, any appeal has been completed, and the ANSI patent policy is met.

A. Canvas List (see Fig. 3.198)

The canvas list method provides procedures for seeking approval/acceptance of a document without the structure of a committee or an organization. Under the canvas list, the originator of a standard seeking its acceptance as an American National Standard (ANS) must assemble a balloting group consisting of a balance of interests and present that assemblage to ANSI for approval. Once the list is approved, the document is circulated for voting. Simultaneously, the manager of the canvas list completes and submits the Board for Standard Revision 8 (BSR-8) form for public notification of the undertaking and providing opportunity for comment from persons outside the balloted group. Once the balloting period ends, usually after 60 days, the manager completes Board for Standard Review 9 (BSR-9) to provide validation that in the balloting, the proposal received consensus approval and a report on how each of the participants voted. The BSR-9 is forwarded to ANSI for the Board of Standard Review action. Of particular concern is that the document review was completed under an open and fair procedure, and that a consensus of the voting group approved its acceptance.

Underwriter's Laboratories (UL) uses the Canvas List method to obtain ANSI recognition of UL documents.

B. Accredited Standards Committee (see Fig. 3.199)

The accredited standards committee is a second method for gaining "national" acceptance of a standard. Under current procedures, accredited standards committees (ASCs) are entities established through the coalescence of a balance of interest groups focused on a particular product area. An ASC is approved to address specific product areas. The product area and the committee's organization and organizational procedures are approved by ANSI. The ASC charter is subject to periodic review and reaffirmation, but, generally, it is unobstructed. The ASC has the option to develop and submit standards for acceptance as ANSs or to process documents that fall within their operational scope that originate in other bodies — trade associations, business groups, and the like. Documents submitted to ASCs are subjected to the

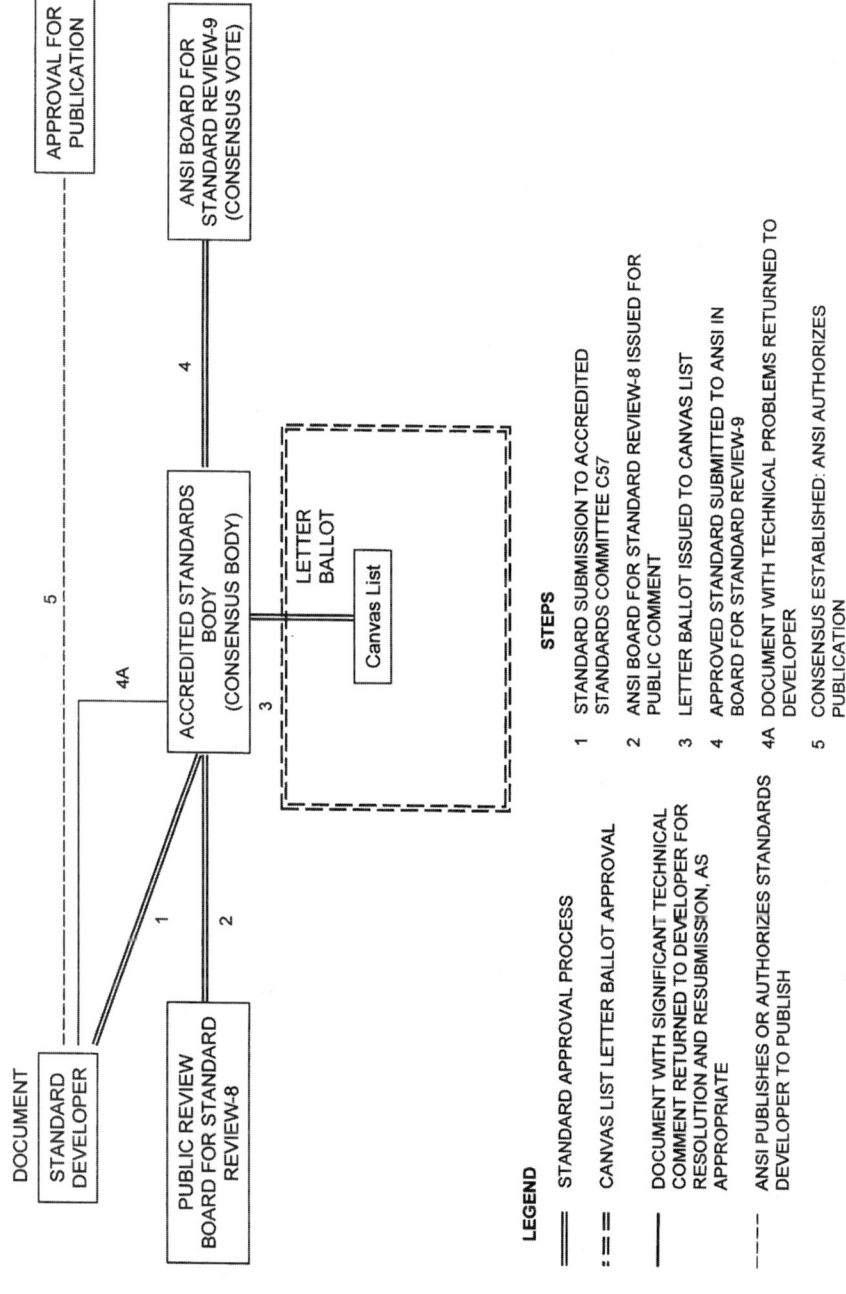

FIGURE 3.198 ANSI standards canvas list approval process.

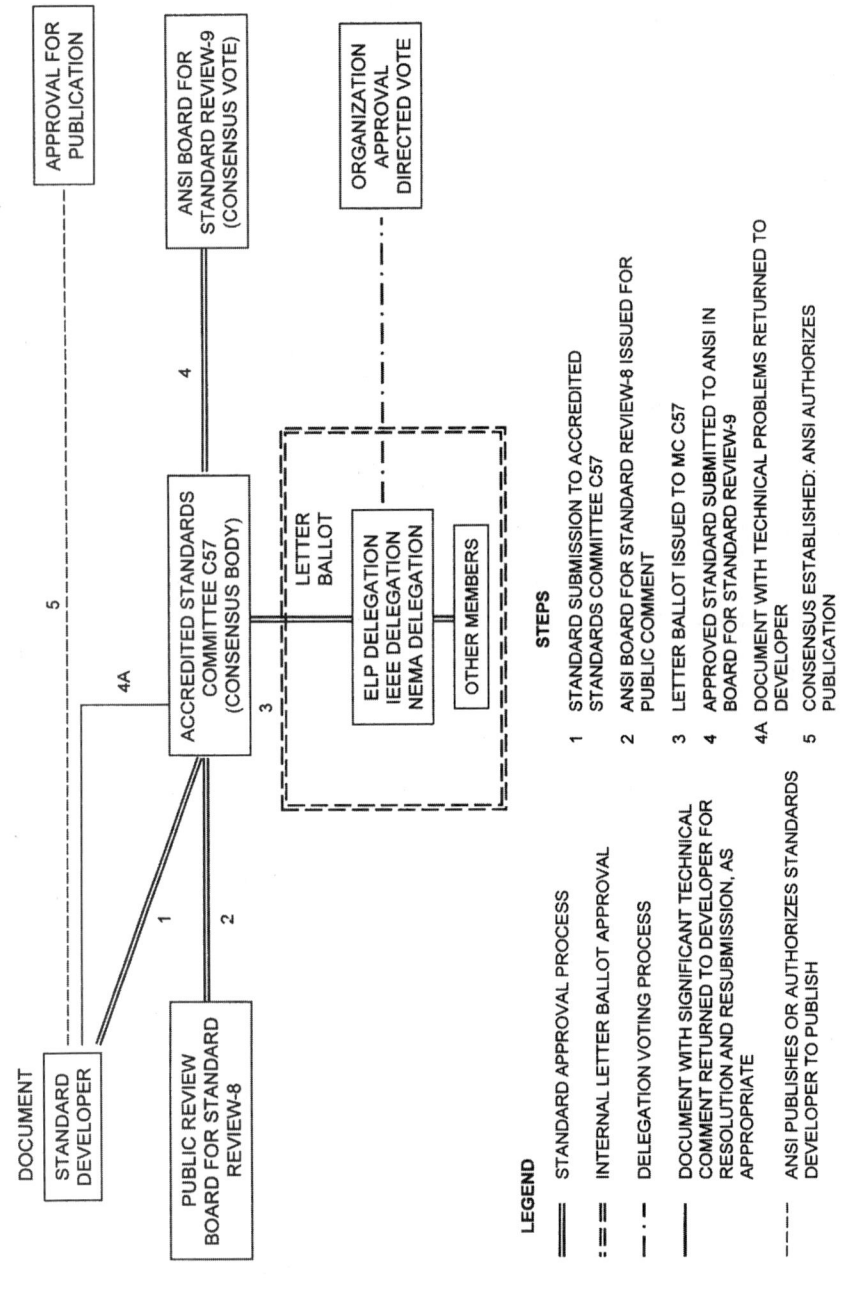

FIGURE 3.199 ANSI accredited standards committee approval process for C57.

Transformers

same procedures for consideration as in the Canvas List method. A BSR-8 is issued upon receipt of a document for acceptance and the initiation of committee review and vote. The BSR-8 provides for public notification of the undertaking and for public comment. Once the balloting period is ended, the BSR-9 report is sent to ANSI confirming the voting and the consensus. ANSI reviews the report and provides appropriate approval.

IEEE and NEMA use the Accredited Standards Committee to gain ANSI C57 document approvals.

The Electric Light and Power delegation (EL & P) represents the Edison Electric Institute (EEI) and the Association of Edison Illuminating Companies (AEIC). EL & P, predominantly through EEI, is well represented in the IEEE Transformers Committee. EL & P is not currently a standards development organization, but does vote as a delegation on documents that are submitted to ANSI C57 for approval.

C. Accredited Standards Organization (see Fig. 3.200)

The third method for acceptance is the accredited standards organization. Under these procedures, an organization demonstrates the openness and balance of its voting groups and its operating procedures. The ASO's purview or authority for processing documents for acceptance as an ANS may be restricted to particular products or a group of products, depending upon organizational interests and goals. The documents developed by the ASO and conforming to the interest balance and openness procedures are submitted to ANSI utilizing the BSR-8 and BSR-9 reports, in appropriate sequence. The ANSI BSR evaluates the documentation and makes its decision using the criteria as in the other methodologies.

The developer of a standard is presented with a range of options in pursuing the document's acceptance as an American National Standard. All methods require prior, or standing, approval until organization scope is changed from ANSI. The canvas method provides the greatest flexibility for the developer but places a greater involvement in assembling the necessary balloting group and, therefore, latitude in determining voting participants. For a developer outside an organization, the canvas list and the ASC provide the greatest and quickest access. The ASO route is not a normal venue for outsiders — or non-members — without special arrangements or agreements from the sponsoring ASO, particularly if the document falls outside the scope of the organization's accreditation.

III. The International Electrotechnical Commission (IEC) (see Fig. 3.201)

The International Electrotechnical Commission (IEC) is composed of a Central Office and approximately 100 Technical Committees and Subcommittees. The IEC Central Office is located in Geneva, Switzerland. All balloting of technical documents is conducted through the Central Office.

The National Committees of each country are responsible for establishing participation status on the various Technical Committees and Subcommittees as well as casting votes on the respective ballots.

Participation status by a national committee of a country can be either of three categories:

P-Member: To participate actively in the work, with an obligation to vote on all questions formally submitted for voting within the technical committee or subcommittee, on enquiry drafts and Final Draft International Standards, and whenever possible, to participate in meetings.
O-Member: To follow the work as an observer, and therefore to receive committee documents and to have the right to submit comments and to attend meetings.
Non P-Member and non O-member: Such a country will have neither the rights nor obligations described above for the work of a particular committee. Nevertheless, all national bodies irrespective of their status within a technical committee of subcommittee have the right to vote on enquiry drafts and on Final Draft International Standards. All ballots are cast by the national committees of the respective countries, as one vote per country.

The U.S. National Committee is established by the American National Standards Institute (ANSI), with headquarters in New York. Technical Advisors (TAs) are appointed by the U.S. National Committee to represent U.S. interests on the various Technical Committees and Subcommittees. TA appointments are based on technical experience and capability. TAs are responsible for establishing Technical Advisory Groups (TAGs) to assist the TA in technical representation. This includes the establishment of working group experts.

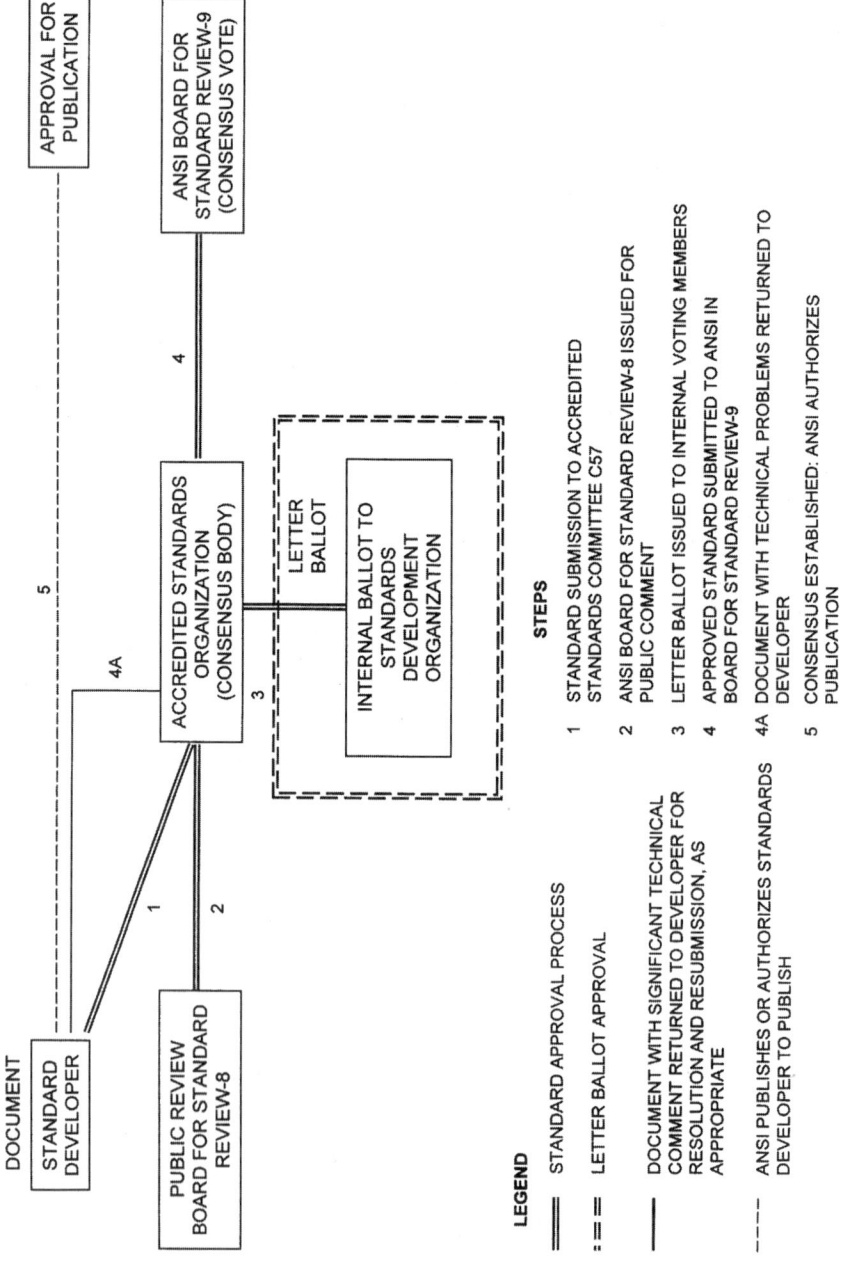

FIGURE 3.200 ANSI standards organization approval process.

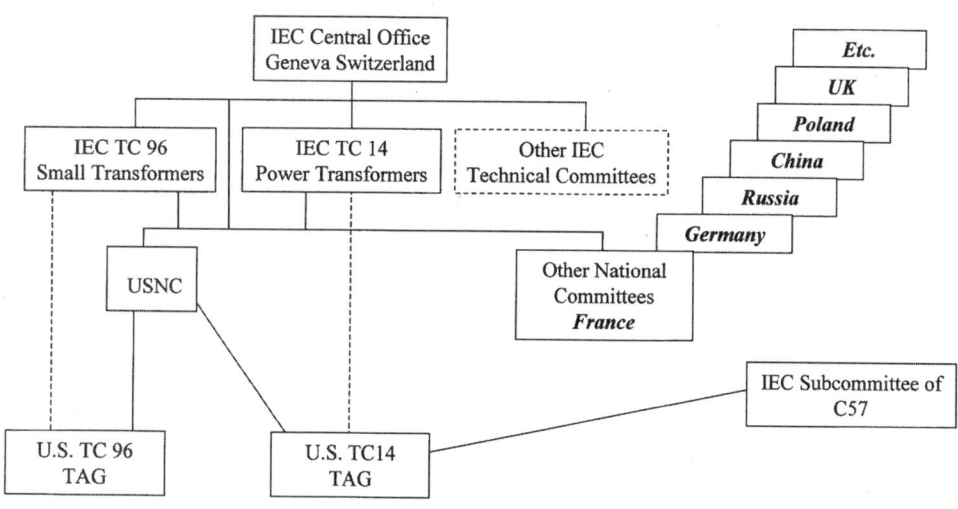

FIGURE 3.201 IEC technical committee document approval process.

As stated above, all balloting of technical documents is conducted through the Central Office of the IEC. The ballots are first distributed to the National Committees. The national committees next send the ballots to the appropriate technical expert for input. Within the U.S., ballots are submitted to the TA's office. The TA has the responsibility to distribute the ballots to the TAG for direction and/or comment. A consensus process is used to be certain that votes are truly representative of the U.S. position.

The TA sends all recommended actions to the Secretary of the USNC. The secretary then sends the official U.S. vote to the IEC Central Office. Figure 3.201 shows a flow diagram of the IEC ballot process.

IV. Relevant Power Transformer Standards Documents

There are numerous issued documents that apply to the specifications and performance requirements for the various power transformers in the industry today. This section organizes them in ascending order of power ratings, in the following categories:

Small Dry-Type Transformers
- NEMA
- UL
- IEC

Electronics Power Transformers
- IEEE

Low Voltage Medium Power Dry-Type Transformers
- NEMA
- UL
- IEC

Medium Voltage and Large Power Dry-Type Transformers
- ANSI C57/IEEE C57
- NEMA

- UL
- IEC

Liquid-Filled Transformers
- ANSI C57/IEEE C57
- NEMA
- IEC

A. Small Dry-Type Transformers

1. NEMA ST-1 "Specialty Transformers (Except General Purpose Type)"

Scope — This Standards Publication covers control transformers, industrial control transformers, Class 2 transformers, signaling transformers, ignition transformers, and luminous-tube transformers. The publication contains service conditions, tests, classifications, performance characteristics, and construction data for the transformers.

kVA Range: 0–5 kVA Single Phase

Voltage:
   Control Transformers Through 4800 V
   Ignition and Luminous Tube Transformers Through 15,000 V

2. ANSI/UL 506 "Standard for Safety for Specialty Transformers"

Scope — These requirements cover air-cooled transformers and reactors for general use, and ignition transformers for use with gas burners and oil burners. Transformers incorporating overcurrent or overtemperature protective devices, transient voltage surge protectors, or power factor correction capacitors are also covered by these requirements. These transformers are intended to be used in accordance with the National Electrical Code, NFPA 70.

These requirements do not cover liquid-immersed transformers, variable voltage autotransformers, transformers having a nominal primary rating of more than 600 V, transformers having overvoltage taps rated over 660 V, cord and plug connected transformers (other than gas-tube-sign transformers), garden light transformers, voltage regulators, swimming pool and spa transformers, or other special types of transformers covered in requirements for other electrical devices or appliances.

These requirements do not cover:

a) Autotransformers used in industrial control equipment, which are evaluated in accordance with the requirements for industrial control equipment, UL 508.
b) Class 2 or Class 3 transformers, which are evaluated in accordance with the Standard for Class 2 and Class 3 Transformers, UL 1585.
c) Toy transformers, which are evaluated in accordance with the Standard for Toy Transformers, UL 697.
d) Transformers for use with radio- and television-type appliances, which are evaluated in accordance with the requirements for transformers and motor transformers for use in audio-, radio-, and television-type appliances, UL 1411.
e) Transformers for use with high-intensity discharge lamps, which are evaluated in accordance with the Standard for High-Intensity-Discharge Lamp Ballasts, UL 1029.
f) Transformers for use with fluorescent lamps, which are evaluated in accordance with the Standard for Fluorescent-Lamp Ballasts, UL 935.
g) Ventilated transformers for general use or non-ventilated transformers for general use (other than compound filled or exposed core types), which are evaluated in accordance with the requirements for dry-type general purpose and power transformers, UL 1561.
h) Dry-type distribution transformers rated over 600 V, which are evaluated in accordance with the requirements for transformers, distribution, dry-type — over 600 V, UL 1562.
i) Transformers incorporating rectifying or waveshaping circuitry evaluated in accordance with the requirements for power units other than Class 2, UL 1012.

Transformers

j) Transformers of the direct plug-in type evaluated in accordance with the requirements for Class 2 power units, UL 1310.
k) Transformers for use with electric discharge and neon tubing, which are evaluated in accordance with the Standard for Neon Transformers and Power Supplies, UL 2161.

A product that contains features, characteristics, components, materials, or systems new or different from those in use when the Standard was developed, and that involves a risk of fire, electric shock, or injury to persons, shall be evaluated using the appropriate additional component and end-product requirements as determined necessary to maintain the level of safety for the user of the product as originally anticipated by the intent of this Standard.

3. ANSI/UL 1446 "Standard for Safety for Systems of Insulating Materials — General"

Scope — These requirements cover test procedures to be used in the evaluation of Class 120(E) or higher electrical insulation systems intended for connection to branch circuits rated 600 V or less. These requirements also cover the investigation of the substitution of minor components of insulation in a previously evaluated insulation system and also the test procedures to be used in the evaluation of magnet wire coatings, magnet wires, and varnishes.

These requirements do not cover a single insulating material or a simple combination of materials, such as a laminate or a varnished cloth.

These requirements do not cover insulation systems exposed to radiation or operating in oils, refrigerants, soaps, or other media that potentially degrade insulating materials.

These requirements shall be modified or supplemented as determined by the applicable requirements in the end-product standard covering the device, appliance, or equipment in which the insulation system is used.

Additional consideration shall be given to conducting tests for an insulating material, such as a coil encapsulant, that is used as the ultimate electrical enclosure.

Additional consideration shall be given to conducting tests for an insulating material or component that is a functional support of, or in direct contact with, a live part.

A product that contains features, characteristics, components, materials, or systems new or different from those in use when the Standard was developed, and that involves a risk of fire, electric shock, or injury to persons, shall be evaluated using the appropriate additional component and end-product requirements as determined necessary to maintain the level of safety for the user of the product as originally anticipated by the intent of this Standard.

4. IEC TC96

Scope — Standardization, in the field of safety of transformers, power supply units, and reactors with a rated voltage not exceeding 1000 V and a rated frequency not exceeding 1 MHz of the following kinds:

a) Power transformers and power supply units with a rated power less than 1 kVA single-phase and 5 kVA polyphase.
b) Special transformers and power supply units other than those intended to supply distribution networks, in particular transformers and power supply units intended to allow the application of protective measures against electric shock as defined by IEC TC 64. Electrical installations in buildings, with no limitation of rated power, but in certain cases including limitation of voltage.

Reactors with a rated power less than 2 kVAR single-phase and 10 kVAR polyphase.
Special reactors other than those covered by IEC 289.
Note: Excluded are switch mode power supplies (dealt with by SC 22E).

Safety group function — Special transformers and power supply units other than those intended to supply distribution networks, in particular transformers and power supply units intended to allow the application of protective measures against electric shock as defined by TC 64, with no limitation of rated power but in certain cases including limitation of voltage.

Relevant Documents

IEC TC96	Description
61558-1	Safety of power transformers, power supply units and similar
61558-2-1	Particular requirements for separating transformers for general use
61558-2-2	Particular requirements for separating control transformers
61558-2-3	Particular requirements for ignition transformers for gas or oil burners
61558-2-4	Particular requirements for isolating transformers for general use
61558-2-5	Particular requirements for shaver transformers and supply
61558-2-6	Particular requirements for safety isolating transformers for general use
61558-2-7	Particular requirements for toys
61558-2-8	Particular requirements for bell and chime
61558-2-9	Particular requirements for transformers for Class III handlamps (NP)
61558-2-10	Particular requirements for high insulation level transformers (NP)
61558-2-11	Particular requirements for stray field transformers
61558-2-12	Particular requirements for stabilizing transformers
61558-2-13	Particular requirements for autotransformers
61558-2-14	Particular requirements for variable transformers
61558-2-15	Particular requirements for insulating transformers for the supply of medical rooms (CDV)
61558-2-16	Particular requirements for power supply units and similar (NP)
61558-2-17	Particular requirements for transformers for switch mode power
61558-2-18	Particular requirements for medical appliances
61558-2-19	Particular requirements for mainsborne perturbation attenuation transformers wtih earthed midpoint (CDV)
61558-2-20	Particular requirements for small reactors (CDV)
61558-2-21	Particular requirements for transformers with special dielectric (liquid SF_6)
61558-2-22	Particular requirements for transformers with rated maximum temperature for luminaries (NP)
61558-23	Particular requirements for transformers for construction sites (CDV)

B. Electronics Power Transformers

1. IEEE 295 "Electronics Power Transformers"

Scope — This Standard pertains to power transformers and inductors that are used in electronic equipment and supplied by power lines or generators of essentially sine wave or polyphase voltage. Guides to application and test procedures are included. Appendices contain certain precautions, recommended practices, and guidelines for typical values. Provision is made for relating the characteristics of transformers to the associated rectifiers and circuits.

Certain pertinent definitions, which have not been found elsewhere, are included with appropriate discussion. Attempts are made to alert the industry and profession to factors that are commonly overlooked.

This Standard includes, but is not limited to, the following specific transformers and inductors.

- Rectifier supply transformers for either high- or low-voltage supplies.
- Filament and cathode heater transformers.
- Transformers for alternating current resonant charging circuits.
- Inductors used in rectifier filters.
- Autotransformers with fixed taps.

kVA Range: 0–1000+
Voltage: 0–15kV

C. Low Voltage Medium Power Dry-Type Transformers

1. NEMA ST-20 "Dry-Type Transformers for General Applications"

Scope — This Standards Publication applies to single-phase and polyphase dry-type transformers (including autotransformers and non-current-limiting reactors) for supplying energy to power, heating, and lighting circuits and designed to be installed and used in accordance with the *National Electrical Code*.

Transformers

It applies to transformers with and/or without accessories having the following ratings:

a) 1.2 kV class (600 V nominal and below), 0.25 kVA and up.
b) Above 1.2 kV class sound level limits are supplied. These limits are applicable to commercial, institutional, and industrial transformers.

This Standards Publication applies to transformers, commonly known as general-purpose transformers, for commercial, institutional, and industrial use in non-hazardous locations both indoors and outdoors. The publication includes ratings and information on the application, design, construction, installation, operation, inspection, and maintenance as an aid in obtaining a high level of safe performance. These standards, except for those for ratings, may be applicable to transformers having other than standard ratings. These standards, as well as applicable local codes and regulations, should be consulted to secure the safe installation, operation, and maintenance of dry-type transformers.

This publication does not apply to the following types of specialty transformers: control, industrial control, Class 2, signaling, oil- or gas-burner ignition, luminous tube, cold cathode lighting, incandescent, and mercury lamp. Also excluded are network transformers, unit substation transformers, and transformer distribution centers.

2. NEMA TP-1 "Guide for Determining Energy Efficiency for Distribution Transformers"

Scope — This Standard is intended for use as a basis for determining the energy efficiency performance of the equipment covered and to assist in the proper selection of such equipment.

This Standard covers single-phase and three-phase dry-type and liquid-filled distribution transformers as defined in the following table:

Voltage Class	Primary Voltage/ Secondary Voltage	34.5 kV and Below/ 600 V and Below
Liquid rating	Single phase	10–833 kVA
	Three phase	15–2500 kVA
Dry rating	Single phase	15–833 kVA
	Three phase	15–2500 kVA

Note: Includes all products at 1.2 kV and below.

Products excepted from this standard include:

a. Liquid-filled transformers below 10 kVA
b. Dry-type transformers below 15 kVA
c. Drives transformers, both AC and DC
d. All rectifier transformers and transformers designed for high harmonics
e. Autotransformers
f. Non-distribution transformers, such as UPS transformers
g. Special impedance and harmonic transformers
h. Regulating transformers
i. Sealed and non-ventilated transformers
j. Retrofit transformers
k. Machine tool transformers
l. Welding transformers
m. Transformers with tap ranges greater than 15%
n. Transformers with frequency other than 60 Hz
o. Grounding transformers
p. Testing transformers

3. NEMA TP-2 "Standard Test Method for Measuring the Energy Consumption of Distribution Transformers"

Scope — This Standard is intended for use as a basis for determining the energy efficiency performance of the equipment covered and to assist in the proper selection of such equipment.

This Standard covers single-phase and three-phase dry-type and liquid-immersed distribution transformers (transformers for transferring electrical energy from a primary distribution circuit to a secondary distribution circuit or within a secondary distribution circuit) as defined in the following table:

Transformer Type	No. Phases	Rating Range
Liquid-immersed	Single phase	10–833 kVA
	Three phase	15–2500 kVA
Dry-type	Single phase	15–833 kVA
	Three phase	15–2500 kVA

This standard addresses the test procedures for determining the efficiency performance of the transformers covered in NEMA Publication TP-1

Note: Includes all products at 1.2 kV and below.

Products excepted from this standard include:

a. Liquid-filled transformers below 10 kVA
b. Dry-type transformers below 15 kVA
c. Transformers connected to converter circuits
d. All rectifier transformers and transformers designed for high harmonics
e. Autotransformers
f. Non-distribution transformers, such as UPS transformers
g. Special impedance and harmonic transformers
h. Regulating transformers
i. Sealed and non-ventilated transformers
j. Retrofit transformers
k. Machine tool transformers
l. Welding transformers
m. Transformers with tap ranges greater than 15%
n. Transformers with frequency other than 60 Hz
o. Grounding transformers
p. Testing transformers

4. ANSI/UL 1561 "Standard for Safety for Dry-Type General Purpose and Power Transformers"

Scope — These requirements cover:

a. General purpose and power transformers of the air-cooled, dry, ventilated, and nonventilated types rated no more than 500 kVA single-phase or no more than 1500 kVA three-phase to be used in accordance with the National Electrical Code, NFPA 70. Constructions include step up, step down, insulating, and autotransformer type transformers as well as air-cooled and dry-type reactors; or
b. General purpose and power transformers of the exposed core, air-cooled, dry, and compound-filled types rated more than 10 kVA but no more than 333 kVA single-phase or no more than 1000 kVA three-phase to be used in accordance with the National Electrical Code, NFPA 70. Constructions include step up, step down, insulating, and autotransformer type transformers as well as air-cooled, dry, and compound-filled type reactors.

Transformers

These requirements do not cover ballasts for high intensity discharge (HID) lamps (metal halide, mercury vapor, and sodium types) or fluorescent lamps, exposed core transformers, compound-filled transformers, liquid-filled transformers, voltage regulators, general use or special types of transformers covered in requirements for other electrical equipment, autotransformers forming part of industrial control equipment, motor-starting autotransformers, variable voltage autotransformers, transformers having a nominal primary or secondary rating of more than 600 V, or overvoltage taps rated greater than 660 V.

These requirements do not cover transformers provided with waveshaping or rectifying circuitry. Waveshaping or rectifying circuits may include components such as diodes and transistors. Components such as capacitors, transient voltage surge suppressors, and surge arresters are not considered to be waveshaping or rectifying devices.

A product that contains features, characteristics, components, materials, or systems new or different from those in use when the Standard was developed, and that involves a risk of fire, electric shock, or injury to persons, shall be evaluated using the appropriate additional component and end-product requirements as determined necessary to maintain the level of safety for the user of the product as originally anticipated by the intent of this Standard.

5. ANSI/UL 1446 "Standard for Safety for Systems of Insulating Materials — General"

Scope — These requirements cover test procedures to be used in the evaluation of Class 120(E) or higher electrical insulation systems intended for connection to branch circuits rated 600 V or less. These requirements also cover the investigation of the substitution of minor components of insulation in a previously evaluated insulation system and also the test procedures to be used in the evaluation of magnet wire coatings, magnet wires, and varnishes.

These requirements do not cover a single insulating material or a simple combination of materials, such as a laminate or a varnished cloth.

These requirements do not cover insulation systems exposed to radiation or operating in oils, refrigerants, soaps, or other media that potentially degrade insulating materials.

These requirements shall be modified or supplemented as determined by the applicable requirements in the end-product standard covering the device, appliance, or equipment in which the insulation system is used.

Additional consideration shall be given to conducting tests for an insulating material, such as a coil encapsulant, that is used as the ultimate electrical enclosure.

Additional consideration shall be given to conducting tests for an insulating material or component that is a functional support of, or in direct contact with, a live part.

A product that contains features, characteristics, components, materials, or systems new or different from those in use when the Standard was developed, and that involves a risk of fire, electric shock, or injury to persons, shall be evaluated using the appropriate additional component and end-product requirements as determined necessary to maintain the level of safety for the user of the product as originally anticipated by the intent of this Standard.

6. IEC Technical Committee 14 Power Transformers

Scope — To prepare international standards for power transformers, on-load tapchangers and reactors, without limitation of voltage or power (instrument transformers, testing transformers, traction transformers mounted on rolling stock and welding transformers are not included).

Relevant Documents	
60076-1	General requirements
60076-2	Temperature rise
60076-3	Insulation levels dielectric tests and external clearances in air
60076-4	Guide for lighting impulse and switching impulse testing
60076-5	Ability to withstand short-circuit
60076-6	Reactors [IEC 289]
60076-8	Power transformers — Application guide
60076-9	Terminal and tapping markings [IEC 616]

Relevant Documents (continued)	
60076-10	Determination of transformer reactor sound levels
60076-11	Dry-type power transformers [IEC726]
60076-12	Loading guide for dry-type power transformers [IEC905]
61378	Converter transformers
61378-1	Transformers for industrial applications
61378-3	Applications guide

D. Medium Voltage and Large Power Dry-Type Transformers

1. ANSI C57/IEEE C57 Documents

Scope — These Standards are intended as a basis for the establishment of performance, interchangeability, and safety requirements of equipment described and for assistance in the proper selection of such equipment.

Electrical, mechanical, and safety requirements of ventilated, nonventilated, and sealed dry-type distribution and power transformers or autotransformers (single and polyphase, with a voltage of 601 V or higher in the highest voltage winding) are described. Instrument transformers and rectifier transformers are also included.

The information in these Standards apply to all dry-type transformers except as follows:

a. Arc-furnace transformers
b. Rectifier transformers
c. Specialty transformers
d. Mine transformers

When these Standards are used on a mandatory basis, the word *shall* indicates mandatory requirements; the words *should* and *may* refer to matters that are recommended or permissive, but not mandatory.

Note: The introduction of this voluntary consensus standard describes the circumstances under which the standard may be used on a mandatory basis.

Relevant Documents	
IEEE 259-1994	Standard Test Procedure for Evaluation of Systems of Insulation for Specialty Transformers (ANSI)
IEEE 638-1992	Standard for Qualification of Class 1E Transformers for Nuclear Generating Stations
ANSI C57.12.00-1993	IEEE Standard General Requirements for Liquid-Immersed Distribution, Power, and Regulating Transformers (ANSI)
ANSI C57.12.01-1998	IEEE Standard General Requirements for Dry-Type Distribution and Power Transformers Including Those with Solid Cast and/or Resin-Encapsulated Windings
IEEE C57.12.35-1996	Standard for Bar Coding for Distribution Transformers
ANSI C57.12.40-1994	American National Standard Requirements for Secondary Network Transformers — Subway and Vault Types (Liquid Immersed)
IEEE C57.12.44-1994	Standard Requirements for Secondary Network Protectors
ANSI C57.12.50-1981 (R1989)	American National Standard Requirements for Ventilated Dry-Type Distribution Transformers, 1 to 500 kVA, Single-Phase, and 15 to 500 kVA, Three-Phase, with High-Voltage 601 to 34 500 V, Low-Voltage 120 to 600 V
ANSI C57.12.51-1981(R1989)	American National Standard Requirements for Ventilated Dry-Type Power Transformers, 501 kVA and Larger, Three-Phase, with High-Voltage 601 to 34 500 V, Low-Voltage 208Y/120 to 4160 V
ANSI C57.12.52-1981 (R1989)	American National Standard Requirements for Sealed Dry-Type Power Transformers, 501 kVA and Larger, Three-Phase, with High-Voltage 601 to 34 500 V, Low-Voltage 208Y/120 to 4160 V
ANSI C57.12.55-1987	American National Standard for Transformers — Dry-Type Transformers Used in Unit Installations, Including Unit Substations — Conformance Standard
IEEE C57.12.56-1986	Standard Test Procedure for Thermal Evaluation of Insulation Systems for Ventilated Dry-Type Power and Distribution Transformers

Transformers 3-263

Relevant Documents (continued)

ANSI C57.12.57-1987 (R1992)	American National Standard for Transformers — Ventilated Dry-Type Network Transformers 2500 kVA and Below, Three-Phase, with High-Voltage 34 500 V and Below, Low-Voltage 216Y/125 and 480Y/277 Volts — Requirements
IEEE C57.12.58-1991 (R1996)	Guide for Conducting a Transient Voltage Analysis of a Dry-Type Transformer Coil
Draft C57.12.60-1998	Guide for Test Procedures for Thermal Evaluation of Insulation Systems for Solid-Cast and Resin-Encapsulated Power and Distribution Transformers
ANSI C57.12.70-1978 (R1992)	American National Standard Terminal Markings and Connections for Distribution and Power Transformers
IEEE C57.12.80-1978 (R1992)	Standard Terminology for Power and Distribution Transformers
IEEE C57.12.91-1995	Standard Test Code for Dry-Type Distribution and Power Transformers
IEEE C57.13-1993	Standard Requirements for Instrument Transformers
IEEE C57.13.1-1981 (R1992)	Guide for Field Testing of Relaying Current Transformers
IEEE C57.13.3-1983 (R1991)	Guide for the Grounding of Instrument Transformer Secondary Circuits and Cases
IEEE C57.15-1986 (R1992)	Standard Requirements, Terminology, and Test Code for Step-Voltage and Induction-Voltage Regulators
IEEE C57.16-1996	Standard Requirements, Terminology, and Test Code for Dry-Type Air-Core Series-Connected Reactors
IEEE C57.18.10-1998	Standard Practices and Requirements for Semiconductor Power Rectifier Transformers
IEEE C57.19.00-1991 (R1997)	Standard General Requirements and Test Procedures for Outdoor Power Apparatus Bushings
IEEE C57.19.01-1991 (R1997)	Standard Performance Characteristics and Dimensions for Outdoor Apparatus Bushings
IEEE C57.19.03-1996	Standard Requirements, Terminology, and Test Code for Bushings for DC Applications
IEEE C57.19.100-1995 (R1997)	Guide for Application of Power Apparatus Bushings
IEEE C57.21-1990 (R1995)	Standard Requirements, Terminology, and Test Code for Shunt Reactors Rated Over 500 kVA
*IEEE C57.94-1982 (R1987)	Recommended Practice for the Installation, Application, Operation, and Maintenance of Dry-Type General Purpose Distribution and Power Transformers
IEEE C57.96-1989	Guide for Loading Dry-Type Distribution and Power Transformers
IEEE C57.98-1993	Guide for Transformer Impulse Tests (An errata sheet is available in PDF format)
IEEE C57.105-1978 (R1999)	Guide for Application of Transformer Connections in Three-Phase Distribution Systems
IEEE C57.109-1993	Guide for Liquid-Immersed Transformer Through-Fault-Current Duration
IEEE C57.110-1998	Recommended Practice for Establishing Transformer Capability When Supplying Nonsinusoidal Load Currents
IEEE C57.116-1989 (R1994)	Guide for Transformers Directly Connected to Generators
IEEE C57.117-1986 (R1992)	Guide for Reporting Failure Data for Power Transformers and Shunt Reactors on Electric Utility Power Systems
IEEE C57.124-1991 (R1996)	Recommended Practice for the Detection of Partial Discharge and the Measurement of Apparent Charge in Dry-Type Transformers
IEEE C57.138-1998	Recommended Practice for Routine Impulse Test for Distribution Transformers

2. ANSI/UL 1562 "Standard for Safety for Transformers, Distribution, Dry-Type — Over 600 Volts"

Scope — These requirements cover single-phase or three-phase, dry-type, distribution transformers. The transformers are provided with either ventilated or non-ventilated enclosures and are rated for a primary or secondary voltage from 601 to 35000 V and from 1 to 5000 kVA.

These transformers are intended for installation in accordance with the National Electrical Code. These requirements do not cover the following transformers:

   a. Instrument transformers
   b. Step-voltage and induction voltage regulators
   c. Current regulators
   d. Arc furnace transformers
   e. Rectifier transformers
   f. Specialty transformers (such as rectifier, ignition, gas tube sign transformers, and the like)
   g. Mining transformers
   h. Motor-starting reactors and transformers

These requirements do not cover transformers under the exclusive control of electrical utilities utilized for communication, metering, generation, control, transformation, transmission, and distribution of electric energy regardless of whether such transformers are located indoors, in buildings and rooms used exclusively by utilities for such purposes; or outdoors on property owned, leased, established rights on private property, or on public rights of way (highways, streets, roads, and the like).

A product that contains features, characteristics, components, materials, or systems new or different from those in use when the Standard was developed, and that involves a risk of fire, electric shock, or injury to persons, shall be evaluated using the appropriate additional component and end-product requirements as determined necessary to maintain the level of safety for the user of the product as originally anticipated by the intent of this Standard.

3. IEC Technical Committee 14 Power Transformers

Scope — To prepare international standards for power transformers, on-load tapchangers, and reactors, without limitation of voltage or power (instrument transformers, testing transformers, traction transformers mounted on rolling stock and welding transformers are not included).

Relevant Documents	
60076-1	General requirements
60076-2	Temperature rise
60076-3	Insulation levels dielectric tests and external clearances in air
60076-4	Guide for lighting impulse and switching impulse testing
60076-5	Ability to withstand short-circuit
60076-6	Reactors [IEC 289]
60076-8	Power transformers — Application guide
60076-9	Terminal and tapping markings [IEC 616]
60076-10	Determination of transformer reactor sound levels
60076-11	Dry-type power transformers [IEC726]
60076-12	Loading guide for dry-type power transformers [IEC905]
60214-1	Tapchangers for power transformers
60214-2	Application guide for on-load tapchangers [IEC542]
61378	Converter transformers
61378-1	Transformers for industrial applications
61378-3	Applications guide

E. Liquid-Filled Transformers

1. ANSI/IEEE

Scope — These Standards are a basis for the establishment of performance, limited electrical and mechanical interchangeability, and safety requirements of equipment described. They are also a basis for assistance in the proper selection of such equipment.

These Standards describe electrical, mechanical, and safety requirements of liquid-immersed distribution and power transformers, and autotransformers and regulating transformers, single and polyphase, with voltages of 601 V or higher in the highest voltage winding. These Standards also cover instrument transformers and rectifier transformers.

These Standards apply to all liquid-immersed distribution, power, regulating, instrument, and rectifier transformers except as indicated below:

- Arc furnace transformers
- Specialty transformers
- Grounding transformers
- Mobile transformers
- Mine transformers

Relevant Documents

IEEE 62-1995	Guide for Diagnostic Field Testing of Electric Power Apparatus — Part 1: Oil Filled Power Transformers, Regulators, and Reactors
IEEE 259-1994	Standard Test Procedure for Evaluation of Systems of Insulation for Specialty Transformers (ANSI)
IEEE 637-1985	Guide for the Reclamation of Insulating Oil and Criteria for Its Use (ANSI)
IEEE 638-1992	Standard for Qualification of Class 1E Transformers for Nuclear Generating Stations
IEEE 799-1987	Guide for Handling and Disposal of Transformer Grade Insulating Liquids Containing PCBs (ANSI)
IEEE 1276-1997	Trial-Use Guide for the Application of High Temperature Insulation Materials in Liquid-Immersed Power Transformers
ANSI C57.12.00-1993	Standard General Requirements for Liquid-Immersed Distribution, Power, and Regulating Transformers (ANSI)
ANSI C57.12.10-1988	American National Standard for Transformers — 230 kV and Below 833/958 through 8333/10 417 kVA, Single-Phase, and 750/862 through 60 000/80 000/100 000 kVA, Three-Phase without Load Tap Changing; and 3750/4687 through 60 000/80 000/100 000 kVA with Load Tap Changing — Safety Requirements
ANSI C57.12.20-1997	American National Standard for Overhead Distribution Transformers, 500 kVA and Smaller: High Voltage, 34 500 V and below: Low Voltage 7970/13 800 Y V and below
ANSI C57.12.22-1989	American National Standard for Transformers — Pad-Mounted, Compartmental-Type, Self-Cooled, Three-Phase Distribution Transformers with High-Voltage Bushings, 2500 kVA and Smaller: High-Voltage, 34 500 GrdY/19 920 V and Below; Low Voltage, 480 V and Below
IEEE C57.12.23-1992	Standard for Transformers — Underground-Type, Self-Cooled, Single-Phase Distribution Transformers With Separable, Insulated, High-Voltage Connectors; High Voltage (24 940 GrdY/14 400 V and Below) and Low Voltage (240/120 V, 167 kVA and Smaller)
ANSI C57.12.24-1994	American National Standard Requirements for Transformers — Underground-Type, Three-Phase Distribution Transformers, 2500 kVA and Smaller; High Voltage, 34 500 GrdY/19 920 V and Below; Low Voltage, 480 V and Below — Requirements
ANSI C57.12.25-1990	American National Standard for Transformers — Pad-Mounted, Compartmental-Type, Self-Cooled, Single-Phase Distribution Transformers with Separable Insulated High-Voltage Connectors; High Voltage, 34 500 GrdY/19 920 V and Below; Low Voltage, 240/120 V; 167 kVA and Smaller
IEEE C57.12.26-1992	Standard for Pad-Mounted, Compartmental-Type, Self-Cooled, Three-Phase Distribution Transformers for Use with Separable Insulated High-Voltage Connectors (34 500 GrdY/19 920 V and Below; 2500 kVA and Smaller)
ANSI C57.12.29-1991	American National Standard Switchgear and Transformers — Pad-Mounted Equipment-Enclosure Integrity for Coastal Environments
ANSI C57.12.31-1996	American National Standard for Pole-Mounted Equipment — Enclosure Integrity
ANSI C57.12.32-1994	American National Standard for Submersible Equipment — Enclosure Integrity
IEEE C57.12.35-1996	Standard for Bar Coding for Distribution Transformers
ANSI C57.12.40-1994	American National Standard Requirements for Secondary Network Transformers — Subway and Vault Types (Liquid Immersed)
IEEE C57.12.44-1994	Standard Requirements for Secondary Network Protectors
ANSI C57.12.70-1978 (R1992)	American National Standard Terminal Markings and Connections for Distribution and Power Transformers
IEEE C57.12.80-1978 (R1992)	Standard Terminology for Power and Distribution Transformers
IEEE C57.12.90-1993	Standard Test Code for Liquid-Immersed Distribution, Power, and Regulating Transformers and Guide for Short Circuit Testing of Distribution and Power Transformers
IEEE C57.13-1993	Standard Requirements for Instrument Transformers
IEEE C57.13.1-1981 (R1992)	Guide for Field Testing of Relaying Current Transformers
IEEE C57.13.3-1983 (R1991)	Guide for the Grounding of Instrument Transformer Secondary Circuits and Cases
IEEE C57.15-1986 (R1992)	Standard Requirements, Terminology, and Test Code for Step-Voltage and Induction-Voltage Regulators
IEEE C57.16-1996	Standard Requirements, Terminology, and Test Code for Dry-Type Air-Core Series-Connected Reactors
IEEE C57.18.10-1998	Standard Practices and Requirements for Semiconductor Power Rectifier Transformers
IEEE C57.19.00-1991 (R1997)	Standard General Requirements and Test Procedures for Outdoor Power Apparatus Bushings

Relevant Documents (continued)

IEEE C57.19.01-1991 (R1997)	Standard Performance Characteristics and Dimensions for Outdoor Apparatus Bushings
IEEE C57.19.03-1996	Standard Requirements, Terminology, and Test Code for Bushings for DC Applications
IEEE C57.19.100-1995 (R1997)	Guide for Application of Power Apparatus Bushings
IEEE C57.21-1990 (R1995)	Standard Requirements, Terminology, and Test Code for Shunt Reactors Rated Over 500 kVA
IEEE C57.91-1995	Guide for Loading Mineral-Oil-Immersed Transformers
IEEE C57.93-1995	Guide for Installation of Liquid-Immersed Power Transformers
IEEE C57.98-1993	Guide for Transformer Impulse Tests (*An errata sheet is available in PDF format*)
IEEE C57.100-1986 (R1992)	Standard Test Procedures for Thermal Evaluation of Oil-Immersed Distribution Transformers
IEEE C57.104-1991	Guide for the Interpretation of Gases Generated in Oil-Immersed Transformers.
IEEE C57.105-1978 (R1999)	Guide for Application of Transformer Connections in Three-Phase Distribution Systems
IEEE C57.109-1993	Guide for Liquid-Immersed Transformer Through-Fault-Current Duration
IEEE C57.110-1998	Recommended Practice for Establishing Transformer Capability When Supplying Nonsinusoidal Load Currents
IEEE C57.111-1989 (R1995)	Guide for Acceptance of Silicone Insulating Fluid and Its Maintenance in Transformers
IEEE C57.113-1991	Guide for Partial Discharge Measurement in Liquid-Filled Power Transformers and Shunt Reactors
IEEE C57.116-1989 (R1994)	Guide for Transformers Directly Connected to Generators
IEEE C57.117-1986 (R1992)	Guide for Reporting Failure Data for Power Transformers and Shunt Reactors on Electric Utility Power Systems
IEEE C57.121-1998	Guide for Acceptance and Maintenance of Less Flammable Hydrocarbon Fluid in Transformers
IEEE C57.131-1995	Standard Requirements for Load Tap Changers
IEEE C57.138-1998	Recommended Practice for Routine Impulse Test for Distribution Transformers

1. NEMA TP-1 "Guide for Determining Energy Efficiency for Distribution Transformers"

Scope — This Standard is intended for use as a basis for determining the energy efficiency performance of the equipment covered and to assist in the proper selection of such equipment.

This Standard covers single-phase and three-phase dry-type and liquid-filled distribution transformers as defined in the following table:

Voltage Class	Primary Voltage/ Secondary Voltage	34.5 kV and Below/ 600 V and Below
Liquid rating	Single phase	10–833 kVA
	Three phase	15–2500 kVA
Dry rating	Single phase	15–833 kVA
	Three phase	15–2500 kVA

Note: Includes all products at 1.2 kV and below.

Products excepted from this standard include:

a. Liquid-filled transformers below 10 kVA
b. Dry-type transformers below 15 kVA
c. Drives transformers, both AC and DC
d. All rectifier transformers and transformers designed for high harmonics
e. Autotransformers
f. Non-distribution transformers, such as UPS transformers
g. Special impedance and harmonic transformers
h. Regulating transformers
i. Sealed and non-ventilated transformers
j. Retrofit transformers

Transformers

    k. Machine tool transformers
    l. Welding transformers
    m. Transformers with tap ranges greater than 15%
    n. Transformers with frequency other than 60 Hz
    o. Grounding transformers
    p. Testing transformers

2. NEMA TP-2 "Standard Test Method for Measuring the Energy Consumption of Distribution Transformers"

Scope — This Standard is intended for use as a basis for determining the energy efficiency performance of the equipment covered and to assist in the proper selection of such equipment.

This Standard covers single-phase and three-phase dry-type and liquid-immersed distribution transformers (transformers for transferring electrical energy from a primary distribution circuit to a secondary distribution circuit or within a secondary distribution circuit) as defined in the following table:

Transformer Type	No. Phases	Rating Range
Liquid-immersed	Single phase	10–833 kVA
	Three phase	15–2500 kVA
Dry-type	Single phase	15–833 kVA
	Three phase	15–2500 kVA

This standard addresses the test procedures for determining the efficiency performance of the transformers covered in NEMA Publication TP-1

Note: Includes all products at 1.2 kV and below.

Products excepted from this standard include:

    a. Liquid-filled transformers below 10 kVA
    b. Dry-type transformers below 15 kVA
    c. Transformers connected to converter circuits
    d. All rectifier transformers and transformers designed for high harmonics
    e. Autotransformers
    f. Non-distribution transformers, such as UPS transformers
    g. Special impedance and harmonic transformers
    h. Regulating transformers
    i. Sealed and non-ventilated transformers
    j. Retrofit transformers
    k. Machine tool transformers
    l. Welding transformers
    m. Transformers with tap ranges greater than 15%
    n. Transformers with frequency other than 60 Hz
    o. Grounding transformers
    p. Testing transformers

3. NEMA TR-1 "1993 (R-1999) Transformers, Regulators, and Reactors"

Scope — This publication provides a list of all ANSI C57 Standards that have been approved by NEMA. In addition, it includes certain NEMA Standard test methods, test codes, properties, etc. of liquid-immersed transformers, regulators, and reactors that are not American National Standards.

IEC 76-1

Scope — To prepare international standards for power transformers, on-load tapchangers and reactors, without limitation of voltage or power (instrument transformers, testing transformers, traction transformers mounted on rolling stock and welding transformers are not included).

Relevant Documents

60076-1	General requirements
60076-2	Temperature rise
60076-3	Insulation levels dielectric tests and external clearances in air
60076-4	Guide for lighting impulse and switching impulse testing
60076-5	Ability to withstand short-circuit
60076-6	Reactors [IEC 289]
60076-7	Loading guide for oil-immersed power transformer [IEC 354]
60076-8	Power transformers — Application guide
60076-9	Terminal and tapping markings [IEC 616]
60076-10	Determination of transformer reactor sound levels
60214-1	Tapchangers for power transformers
60214-2	Application guide for on-load tapchangers [IEC 542]
61378	Converter transformers
61378-1	Transformers for industrial applications
61378-2	Transformers for HVDC applications
61378-3	Applications guide

3.20 On-Line Monitoring of Liquid-Immersed Transformers

Andre Lux[3]

On-line monitoring of transformers and associated accessories (measuring certain parameters or conditions while energized) is an important consideration in their operation and maintenance. The justification for on-line monitoring is driven by the need to increase the availability of transformers, to facilitate the transition from time-based and/or operational-based maintenance to condition-based maintenance, to improve asset and life management, and to enhance failure cause analysis.

This discussion covers most of the on-line monitoring methods that are currently in common practice, including their benefits, system configurations, and application to the various operational parameters that can be monitored. For the purposes of this section, the term transformer refers, but is not limited, to: step-down power transformers, generator step-up transformers; autotransformers, phase-shifting transformers, regulating transformers, intertie transmission transformers; DC converter transformers; high-voltage instrument transformers, and shunt, series, and saturable reactors.

---

[3]Acknowledgment: This section was a project of the IEEE Working Group and Task Force on On-Line Monitoring of Liquid-Immersed Transformers of the IEEE Transformers Committee. The Working Group Chairman is F.N. Young, and the Task Force Chairman is D. Chu. The members of the Working Group and Task Force, and the co-authors of this chapter are: J.W. Abbott, P. A. Alex, D. Anderegg, R. L. Barker, M. F. Barnes, W. Bartley, T. Bengtsson, P. Boss, K. Carrander, C. Claiborne, D. Chu, J. Crouse, D. Dohnal, D. J. Fallon, N. Field, G. E. Forrest, S. Foss, M.A.Franchek, J. D. Fyvie, A.C. Hall, W.X. Hansen, J.W. Harley, J.H. Harlow, T. Haupert, D.A. Hoch, M. Horning, A.Q. Huang, A.F. Hueston, J. Ipser, R. James, J. Kelly, S.V. Kulkarni, S.B.; Ladd, B. Langan, J. LaSalle, S. Lindgren, A. Lux, A. Molden, J.M. Patton, D. Payne, M. Perkins, P. Pillitteri, T.A. Prevost, M. Rivers, B.D. Sparling, C. Spoorenberg, J. Smith, P. Stiller, R.W. Stoner, R.S. Thompson, T. Traub, K. Viereck, B.H. Ward, R.C. Wicks, F.N. Young, Y. Zhang, H. Zhu

Benefits

Various issues must be considered when determining whether or not the installation of an on-line monitoring system is appropriate. Prior to the installation of on-line monitoring equipment, cost-benefit and risk-benefit analyses are typically performed in order to determine the value of the monitoring system as applied to a particular transformer. For example, for an aging transformer, especially with critical functions, on-line monitoring of certain key parameters is appropriate and valuable. Monitoring equipment can also be justified for transformers with certain types of load tap changers that have a history of coking or other types of problems, or for transformers with symptoms of certain types of problems such as overheating, partial discharge, excessive aging, bushing problems, etc. However, for transformers that are operated normally without any overloading and have acceptable routine maintenance and dissolved gas analysis (DGA) test results, monitoring can probably not be justified economically.

Categories

Both direct and strategic benefits can arise from the installation of on-line monitoring equipment. Direct benefits are cost savings benefits obtained strictly from changing maintenance activities. They include reducing expenses by reduced frequency of equipment inspections, and reducing or delaying active interventions (repair, replacement, etc.) on the equipment. Strategic benefits are based on the ability to prevent (or mitigate) failures, or to avoid catastrophe. These benefits can be substantial since failures can be very damaging and costly. Benefits in this category include better safety (preventing injuries to workers or the public in the event of catastrophic failure), protection of the equipment, and avoiding the potentially large impact caused by system instability, loss of load, environmental cleanup, etc.

Direct Benefits

Maintenance Benefits: These benefits represent resources saved in maintenance activities by the application of on-line monitoring as a predictive maintenance technique. On-line monitoring can mitigate or eliminate the need for manual time-based or operation-based inspections by identifying problems early and allowing corrective actions to be implemented.

Equipment Usage Benefits: These benefits arise because additional reinforcement capacity may be deferred due to on-line monitoring and diagnostics allowing more effective utilization of existing equipment. On-line monitoring equipment can continuously provide real-time capability limits, both operationally and in terms of equipment life.

Strategic Benefits

Strategic benefits are those that accrue when the results of system failures can be mitigated, reduced, or eliminated. A key feature of on-line monitoring technology is its ability to anticipate and forestall catastrophic failures. The value of the technology is its ability to lessen the frequency of such failures.

Service Reestablishment Benefits: These benefits represent the need for reduced repair and/or replacement of damaged equipment because on-line monitoring has been able to identify a component failure in time for planned corrective action. Unscheduled repairs can be very costly in terms of equipment damage and its potential impact on worker safety and public relations.

System Operations Benefits: These benefits represent the avoidance of operational adjustments to the power system as a result of having identified the component failure prior to a general failure. System adjustments, in the face of a delivery system breakdown, can range from negligible to significant. An example of a negligible adjustment is when the failure is in a non-critical part of the network and adequate redundancy exists. Significant adjustments are necessary if the failure causes large, baseload generation to experience a forced outage, or if contractual obligations to independent generators cannot be met. These benefits are driven in part by the duration of the resulting circuit outage.

Outage Benefits: These benefits represent the impact of component failure and resulting system breakdown on end-use customers. A utility incurs direct revenue losses as a result of a system or component failure. A utility's customers, in turn, may also experience losses during failures. The magnitude and/or frequency of such losses may result in the customer's loss of significant revenues.

On-Line Monitoring Systems

The characteristics of transformer on-line monitoring equipment can vary depending on the number of parameters that are monitored and the desired accessibility of the data. An on-line monitoring system typically records data at regular intervals, alarms and reports when preset limits are exceeded. The equipment required for an on-line transformer monitoring system consists of sensors, data acquisition units (DAU), and a computer connected with a communication link.

Sensors

Sensors measure electrical, chemical, and physical signals. Individual sensor types and monitoring methods are discussed in the section "On-Line Monitoring Applications". Standard sensor output signal levels are 4 to 20 mA, 0 to 1 mA, and 0 to 10 V. The sensors may be directly connected to the data acquisition unit(s). Another category of sensors is represented by intelligent electronic devices (IEDs).

Information/data about a function or status that is being monitored is captured by a sensor that may be attached directly to the transformer or within the control house. Once captured, the data is transferred to a data acquisition unit (DAU) that may also be attached to the transformer or located elsewhere in the substation. The transfer is triggered by either a predefined event such as a motor operation, a signal reaching a threshold, or the changing state of a contact. The transfer can also be initiated by a time-based schedule such as an hourly measurement of a bushing's power factor.

The method of data collection depends on the characteristics of the on-line monitoring system. A common characteristic of all systems is the need to move information/data from the sensor level to the user. The following represent examples of possible components in a data collection system.

Data Acquisition Units

A data acquisition unit (DAU) collects signals from one or more sensors and performs signal conditioning and analog-to-digital conversions. The DAU also provides electrical isolation and insulation between the measured output signals and the DAU electronics. For example, a trigger could cause the DAU to start recording, store information about the event, and send it to a substation computer.

DAU to Computer Communications Line

The data collection process usually involves transferring the data to a computer. The computer could be located within the DAU, elsewhere in the substation, or off-site. The data may be transferred via a variety of communications networks such as permanent direct connection, manual direct connection, local area networks (LAN), or wide area networks (WAN).

Computer

At the computer, information is held resident for additional analysis. The computer may be an integral part of the DAU or it may be located separately in the station. The computer is based on standard technology. From a platform point-of-view, software functions of the substation computer program include support of the computer, users, and communications systems, storage of data, and communications to users or other systems, such as SCADA. The computer manages the DAUs and acts as the data and communication server to the user interface software. The computer facilitates expert system diagnostics and contains the basic platform for data acquisition and storage.

Data Processing

The first step in data processing is the extraction of sensor data. Some types of data can be used in the form in which it is acquired, while other types of data need to be processed further. For example, a transformer's top oil temperature can be directly used, while a bushing's sum current waveform requires additional processing to calculate the fundamental (50 or 60 Hz) phasor. The data is then compared with various reference values such as limits, nameplate values, and other measurements, depending on the user's application.

In situations where reference data is not available, a learning period may be used to generate a baseline for comparison. Data is accumulated during a specified period of time, and statistical evaluation is used to either accept or reject the data. In some applications, the rejected data is still saved, but is not used in the calculation of the initial benchmark. In other applications, the initial benchmark is determined using only the accepted data.

The next data processing step is to determine if variations suggest actual apparatus problems or if they are due to ambient fluctuations (such as weather effects), power system variables, or other effects. A combination of signal processing techniques and/or the correlation of the information obtained from measurements from locations on the same bus may be used to eliminate both the power system effects and temperature influences.

The next step in processing is dependent on the sophistication of the monitoring system; however, generally the data needs to be interpreted, and the resulting information communicated to the user. An approach that is often employed is that if the parameter measured does not change significantly from the most previous measurement, then no data is recorded, saved, or transmitted.

On-Line Monitoring Applications

Various basic parameters of power transformers, load tap changers, instrument transformers, and bushings can be monitored with available sensor technologies.

Power Transformers

Transformer problems can be characterized as those that arise from defects and develop into incipient faults, those that derive from deterioration processes, and those induced by operating conditions that exceed the capability of the transformer. These problems may take many years to gestate before developing into a problem or failure. However, in some cases undesirable consequences can be created quite precipitously.

Deterioration processes relating to aging are accelerated by thermal and voltage stresses. Increasing temperature, oxygen, moisture, and other contaminants significantly contribute to insulation degradation. The deterioration is particularly exaggerated in the presence of catalysts, and/or through faults, and by mechanical or electro-mechanical wear. Characteristics of the deterioration processes include sludge accumulation, weakened mechanical strength of insulation materials such as paper-wrapped conductor, shrinkage of materials that provide mechanical support, and improper alignment of tap changer mechanisms. Excessive moisture will accelerate the aging of insulation materials over many years of operation. During extreme thermal transients that may occur during some loading cycles, high moisture content can result in water vapor bubbles. The bubbles can cause serious reduction in dielectric strength of the insulating liquid, resulting in a dielectric failure.

The processes causing eventual problems (e.g., shrinkage of the insulation material or excessive moisture) may take many years to develop but the consequences can appear suddenly. Continuous monitoring permits timely remedial action: not too early, saving valuable maintenance resources and not too late, resulting in costly consequences. Higher loading can be tolerated, as continuous automated evaluation will alert users of conditions that could result in failure or excessive aging of critical insulation structures (Griffin, 1999).

Table 3.22 lists the major transformer components along with their associated problems and the parameters that can be monitored on-line to detect them.

Dissolved Gas-in-Oil Analysis

Monitored Parameters

Dissolved gas-in-oil analysis (DGA) has proven to be a valuable and reliable diagnostic technique for the detection of incipient fault conditions within liquid-immersed transformers by detecting certain key gases. DGA has been widely used throughout the industry as the primary diagnostic tool for transformer maintenance, and it is of major importance in a transformer owner's loss prevention program.

Data has been acquired from the analysis of samples from electrical equipment in the factory, laboratory, and field installations over the years. A large body of information relating certain fault conditions

TABLE 3.22 Main Tank Transformer Components, Failure Mechanisms, and Measured Signals

Component		Phenomenon Leading to Failure	Measured Signals
Non-current carrying metal components	Core	Overheating of laminations	Top and bottom temperatures Ambient temperature Line currents Voltage Hydrogen — minor overheating Multi-gas, particularly ethane, ethylene, and methane — moderate or severe overheating
	Frames, clamping, cleats, shielding, tank walls, etc.	Overheating due to circulating currents, leakage flux	Top and bottom temperatures Ambient temperature Line currents Voltage Multi-gas, particularly ethane, ethylene, and methane
Winding insulation	Core ground Magnetic shield	Floating core and shield grounds create discharge	Hydrogen or multi-gas Acoustic and electric PD
	Cellulose: Paper, pressboard, wood products	Local and general overheating and excessive aging	Top and bottom temperatures Ambient temperature Line currents RS moisture in oil Multi-gas, particularly carbon monoxide, carbon dioxide, and oxygen
		Severe hot spot Overheating	Top and bottom temperatures Ambient temperature Line currents Moisture in oil Multi-gas, particularly carbon monoxide, carbon dioxide, ethane, hydrogen, and oxygen
		Moisture contamination	Top and bottom temperatures Ambient temperature Relative saturation of moisture in oil
		Bubble generation	Top and bottom temperatures Ambient temperature Total percent dissolved gas-in-oil Line currents Relative saturation of moisture in oil Hydrogen Acoustic and electric PD
		Partial discharge	Hydrogen or multi-gas Acoustic and electric PD
Liquid insulation		Moisture contamination	Top and bottom temperatures Ambient temperature Relative saturation of moisture in oil
		Partial discharge	Hydrogen Acoustic and electric PD
		Arcing	Hydrogen and acetylene
		Local overheating	Ethylene, ethane, methane
Cooling system	Fans Pumps Temperature Measurement devices	Electrical failures of pumps and fans	Motor (fan, pump) currents Top oil temperature Line currents
		Failure or inaccuracy of top liquid or winding temperature indicators or alarms	Ambient temperature Top and bottom temperatures Line currents

TABLE 3.22 (continued) Main Tank Transformer Components, Failure Mechanisms, and Measured Signals

Component		Phenomenon Leading to Failure	Measured Signals
	Internal cooling path	Defects or physical damage in the directed flow system	Top and bottom temperatures
			Ambient temperature
		Localized hot spots	Line currents
			Carbon monoxide and carbon dioxide
	Radiators and coolers	Internal or external blocking of radiators resulting in poor heat exchange	Top and bottom temperatures
			Ambient temperature
			Line currents
Oil and winding temperature Forecasting		Overloading of transformer	Top and bottom temperatures
			Ambient temperature
			Line currents
			Moisture in oil
			Multi-gas, particularly carbon monoxide, carbon dioxide, and oxygen

to the various gases that can be detected and easily quantified by gas chromatography has been developed. The gases that are generally measured and their significance are shown in Table 3.23, based on (Griffin, 1999). Methods for interpreting fault conditions associated with various gas concentration levels and combinations of these gases are provided in (Griffin, 1999).

Laboratory-based DGA programs are typically conducted on a periodic basis dictated by the application or transformer type. Some problems with short gestation times may go undetected between normal laboratory test intervals. Installation of continuous gas-in-oil monitors may detect the start of incipient failure conditions to allow confirmation of the presence of a suspected fault through laboratory DGA testing. This early warning may allow the user to plan necessary steps required to identify the fault and implement corrective actions where possible. Present technology exists that can determine gas type, concentration, trending, and production rates of generated gases. The rate of change of gases dissolved in oil is a valuable diagnostic in terms of determining the severity of the developing fault. A conventional unscheduled gas-in-oil analysis is typically performed after an alarm condition has been reported. The application of on-line dissolved gas monitoring considerably reduces the risk of missing the detection or prolonged delay in detecting fault initialization due to long on-site oil sampling intervals.

Laboratory-based sampling and analysis of a frequency sufficient to obtain real-time feedback becomes impractical and too expensive. For critical transformers, on-line gas-in-oil monitors can provide timely and continuous information in a manner that permits load adjustments to prevent excessive gassing from

TABLE 3.23 Gases Typically Found in Transformer Insulating Liquid under Fault Conditions, based on (IEEE Guide C57.104)

Gas	Chemical Formula	Predominant Source
Nitrogen	N_2	Inert gas blanket, atmosphere
Oxygen	O_2	Atmosphere
Hydrogen[1]	H_2	Partial discharge
Carbon dioxide	CO_2	Overheated cellulose, atmosphere
Carbon monoxide[1]	CO	Overheated cellulose, air pollution
Methane[1]	CH_4	Overheated oil (Hot metal gas)
Ethane[1]	C_2H_6	Overheated oil
Ethylene[1]	C_2H_4	Very overheated oil (May have trace of C_2H_2)
Acetylene[1]	C_2H_2	Arcing in oil

[1] Denotes combustible gas. Overheating can be caused both by high temperatures and by unusual or abnormal electrical stress.

thermal-type faults. This can keep a transformer operating for many months while ensuring safety limits are observed.

Gas Sensor Development

Early attempts to identify and document the gases found in energized transformers date back to 1919. This analysis was conducted by liquid column chromatography (Myers et al., 1981). An early type of gas monitor, still in use in many locations, is a device similar to the Buchholz Relay which was developed in the late 1920s. This type of relay detects and measures the pressure of free gas generated in the transformer and indicates an alarm signal.

The gas chromatograph was first applied in the early 1960s to this area. Its ability to differentiate and quantify the various gases that are generated and found in the insulating oil of transformers and other electrical equipment has proven quite useful (Myers et al., 1981). Beginning in the late 1970s and continuing to the present, efforts have been made to develop a gas chromatograph for on-line applications. These efforts have been focused on analyzing gases in the gas space of transformers and on extracting the gases from transformer oil and injecting the gases into the gas chromatograph. Recently, on-site laboratory quality analyses have become available utilizing a portable gas chromatograph that is not permanently connected to the transformer.

In the 1980s and early 1990s, an alternate method to using gas chromatography was developed. Sensors based on fuel cell technology and thermal conductivity detection (TCD) were developed. Both methods use membrane technologies to separate dissolved gases from the transformer oil and produce voltage signals proportional to the amount of dissolved gases. The fuel-cell sensor senses hydrogen and carbon monoxide together with small amounts of other hydrocarbon gases. This method has been successful in providing an early warning of detecting incipient faults initiated by the dielectric breakdown of the insulating fluid and cellulose found in the solid insulation.

Successive efforts have been targeted toward measuring the other gases that can be produced inside the transformer that are detectable by gas chromatography. These efforts are designed to provide on-line access to data that can then be used to indicate the need for further sampling of the insulating oil. The oil is then analyzed in the laboratory to confirm the monitoring data.

During the mid-1990s, a multi-gas on-line DGA monitor that could detect and quantify the gas concentrations in parts per million (ppm) was developed (Chu et al., 1993; 1994). This monitor samples all seven key gases and was designed to provide sufficient dissolved gas data ensuring that analysis and interpretation of faults could take place on-line using the criteria provided in (IEEE Std. C57.104). The sampling approach is non-invasive; both the extraction and sensor systems are external to the transformer (Glodjo, 1998). The system takes multiple oil samples per day and senses changes in the absolute values of gas concentrations and gas ratios. This information is analyzed with the transformer load and temperature levels, environmental conditions, and known fault conditions from repair records and diagnostic software programs. A second system developed uses membrane extraction technology combined with infrared spectroscopy (FTIR) sensing for all gases except hydrogen. For hydrogen, this system uses fuel cell technology. It can also detect all seven key gases (per IEEE Std. C57.104).

Moisture in Oil

The measurement of moisture in oil is a routine test (in addition to other physical characteristics of the oil) performed in the laboratory on a sample taken from the transformer. The moisture level of the sample is evaluated with the sample temperature and the winding temperature of the transformer. This combination of data is vital in determining the relative saturation of moisture in the cellulose/liquid insulation complex that establishes the dielectric integrity of the transformer. Moisture in the transformer reduces the insulation strength by decreasing the dielectric strength of the transformer's insulation system. As the transformer warms up, moisture migrates from the solid insulation into the fluid. The rate of migration is dependent on the conductor temperature and the rate-of-change of the conductor temperature. As the transformer cools, the moisture returns to the solid insulation at a slower rate. The time constants for these migrations depend on the design of the transformer and the solid and liquid components in use. The combination of moisture, heat, and oxygen are the key conditions that indicate accelerated degradation of the cellulose. Excessive amounts of moisture can accelerate the degradation process of the cellulose and prematurely age the transformer's insulation system. The existence of a particular type of furanic compound in the oil is also an indication of moisture in the cellulose insulation.

Moisture-in-oil sensors were first successfully tested and used in the early 1990s (Oommen, 1991; 1993). The sensors measure the relative saturation of the water in oil, which is a more meaningful measure than the more familiar units of parts per million (ppm). Continuous measurements allow for detection of the true moisture content of the transformer insulation system and hazardous conditions which may occur during temperature cycling, thereby helping to prevent transformer failures.

Partial Discharge

One cause of transformer failures is dielectric breakdown. Failure of the dielectrics inside transformers is often preceded by partial discharge activity. A significant increase either in the partial discharge (PD) level or in the rate of increase of partial discharge level can provide an early indication that changes are evolving inside the transformer. Since partial discharge can deteriorate into complete breakdown, it is desirable to monitor this parameter on-line. Partial discharges in oil will produce hydrogen dissolved in the oil. However, the dissolved hydrogen may or may not be detected, depending on the location of the PD source and the time necessary for the oil to carry or transport the dissolved hydrogen to the location of the sensor. The PD sources most commonly encountered are moisture in the insulation, cavities in solid insulation, metallic particles, and gas bubbles generated due to some fault condition.

The interpretation of detected PD activity is not straightforward. No general rules exist that correlate the remaining life of a transformer to PD activity. As part of the routine factory acceptance tests, most transformers are tested to have a PD level below a specified value. From a monitoring and diagnostic view, detection of PD above this level is therefore cause for an alarm but not generally for a tripping action. These realities illustrate one of the many difficulties encountered in PD diagnosis. The results need to be interpreted with knowledge of the studied equipment. Two methods are used for measuring partial discharges: electrical and acoustic. Both of these have attracted considerable attention, but neither is able to yield an unambiguous PD measurement without additional procedures.

Electrical Method

The electrical signals from PD are in the form of a unipolar pulse with a risetime that can be as short as nanoseconds (Morshuis, 1995). They have a very wide frequency content. The high frequencies are attenuated when the signal propagates through the equipment and the network. The detected signal frequency is dependent both on the original signal and the measurement method.

Electrical PD detection methods are generally hampered by electrical interference signals from surrounding equipment and the network as illustrated in Fig. 3.202. Any on-line PD sensing method has to find a way to minimize the influence of such signals. One way is to use a directional high-frequency field sensor (Lemke, 1987). The high detection frequency limits the disturbance from PD sources at a distance and the directionality simplifies a remote scan of many objects. Therefore, this type of sensor seems most appropriate for periodic surveillance. It is not known whether this principle has been tried in a continuous monitoring system.

A popular method to interpret PD signals is to study their occurrence and amplitude as a function of the power phase position; this is called phase-resolved PD analysis (PRPDA). This method can give valuable insight into the type of PD problem present. In (Fruth and Fuhr, 1990), it is suggested that by identifying typical problem patterns in a PRPDA, one could minimize external influences. The conceptual difficulty with this method is that the problem type must be known beforehand, which is not always the case. Second, the relevant signals may be corrupted by an external disturbance.

There have been many attempts to use neural networks or adaptive digital filters (Wenzel et al., 1995), but it is not clear if this has led to a standard method. The problem with this approach is that the measured and the background signals are very similar and the variation within each of the groups may be much larger than the difference between them. Adaptive filters and neural networks can be useful to diminish other background sources such as medium-wave radio and rectifier pulses.

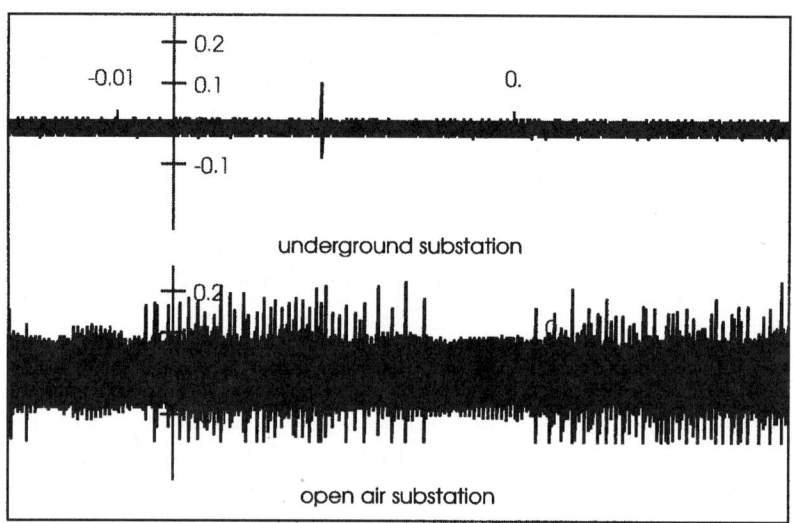

FIGURE 3.202 Electric PD measurements on transformers in underground and open-air substations. The overhead transmission lines cause a multitude of signals, making a PD measurement very insensitive. Underground stations are generally fed by cables that attenuate the high-frequency signals from the network and PD measurements are quite sensitive. Horizontal scale in seconds, vertical scale in mV.

These methods employ a single sensor for the PD measurement. If several sensors of different types or at different locations are employed, the possibilities to reduce external influences are greatly enhanced. Generally, the multi-sensor approach can be split into two branches: separate detection of external signals and energy flow measurements.

When there is a clear source for the disturbing signals, it is tempting to use a separate sensor as a pick-up for those and simply turn off the PD measurements when the external level is too large. Methods like this have the disadvantage of being insensitive some percentage of the measurement time. In addition, a very large signal from the equipment under study may be detected by the external pick-up as well and thus be rejected.

Energy-flow measurements use both an inductive and a capacitive sensor to measure current and voltage in the PD pulse (Eriksson et al., 1995; Wenzel et al., 1995). By careful tuning of the signals from the two sensors, they may be reliably multiplied and the polarity of the resulting energy pulse determines whether the signal originated inside the apparatus or outside. This approach seems to be the most promising for on-line electric PD detection.

Acoustic Method

Like electrical methods, acoustic methods have a long history of use for PD detection. The sensitivity can be shown to be comparable to electric sensing. Acoustic signals are generated from bubble formation and collapse during the PD event and have frequencies of approximately 100 kHz (Bengtsson et al., 1993). Like the electric signals, the high frequencies are generally attenuated during propagation. Due to the limited propagation velocity, acoustic signals are commonly used for location of PD sources.

The main advantage with acoustic detection is that disturbing signals from the electric network do not interfere with the measurement. As the acoustic signal propagates from the PD source to the sensor, it will generally encounter different materials. Some of these materials may attenuate the signal considerably; furthermore, each material interface further attenuates the propagated signal. Therefore, acoustic signals can only be detected within a limited distance from the source. Consequently, the sensitivity for PD inside transformer windings, for example, may be quite low. In typical applications, many acoustic sensors are carefully distributed around the tested equipment (Eleftherion, 1995; Bengtsson et al., 1997).

Though not disturbed by the electric network, external influences in the form of rain or wind and non-PD vibration sources like loose parts and cooling fans will generate acoustic signals that interfere with the PD detection. One way to decrease the external influence is to use acoustic wave-guides (Harrold, 1983) that detect signals from inside the transformer tank. This solution is typically only considered for permanent monitoring of important transformers. As an alternative, phase position analysis can be used to reject these disturbances (Bengtsson et al., 1997).

A transformer generates disturbing acoustic signals in the form of core noise, which may extend up to the 50 to 100 kHz region. To diminish this disturbance, acoustic sensors with sensitivity in the 150 kHz range are usually employed (Eleftherion, 1995). Such sensors may, however, have less sensitivity to PD signals as well (Bengtsson et al., 1997). The properties of these signals are such that it is relatively easy to distinguish them from PD signals; thus, their main effect is to limit sensitivity.

As a generalization of the electric multi-sensor systems discussed above, there are a few descriptions of combined electric and acoustic PD monitoring systems for transformers in the literature (Wang et al., 1997). Rather elaborate software must be employed to utilize the potential sensitivity of these systems. If both the acoustic and the electric parts are designed with the considerations above in mind and an effective software constructed, systems like this will become effective yet costly.

Oil Temperatures
Overheating or overloading can cause transformer failures. Continuous measurement of the top oil temperature is an important factor in maximizing the service life. Top oil temperature, ambient temperature, load (current), fan/pump operations, and direct reading winding temperatures (if available) can be combined in algorithms to determine hottest-spot temperature and manage the overall temperature conditions of the transformer.

Winding Temperatures
There is a direct correlation between winding temperature and normally expected service life of a transformer. The hottest spot temperature of the winding is one of a number of limiting factors for the load capability of transformers. Insulation materials lose their mechanical strength with prolonged exposure to excessive heat. This can result in tearing and displacement of the paper and dielectric breakdown that will result in premature failures. Conventional winding temperature measurements are not typically direct; the hot spot is indirectly calculated from oil temperature and load current measurements using a method described and recommended by (Domun, 1994; Duval and Lamarre, 1977; Feser et al., 1993; Fox, 1983; IEEE, 1995). Fiber optic temperature sensors can be installed in the winding only when the transformer is manufactured, rebuilt, or refurbished. Two sensor types are available: optical fibers that measure the temperature at one point, and distributed optical fibers that measure the temperature along the length of the winding. Since a distributed fiber optic temperature sensor is capable of measuring the temperature along the fiber as a function of distance, it can replace a large number of discrete sensors and allow a real-time measurement of the temperature distribution.

Load Current and Voltage
Maximum loading of transformers is restricted by the temperature to which the transformer and its accessories can be exposed without excessive loss of life. Continuous on-line monitoring of current and voltage coupled with temperature measurements can provide a means to gauge thermal performance. Load current and voltage monitoring can also automatically track the loading peaks of the transformer, increase the accuracy of simulated computer load flow programs, provide individual load profiles to assist in distribution system planning, and aid in dynamically loading the transformer.

Insulation Power Factor
The dielectric loss in any insulation system is the power dissipated by the insulation when an AC voltage is applied. All electrical insulation has a measurable quantity of dielectric loss, regardless of condition. Good insulation usually has a very low loss. Normal aging of an insulating material will cause the dielectric

loss to increase. Contamination of insulation by moisture or chemical substances may cause losses to be higher than normal. Physical damage from electrical stress or other outside forces also affects the level of losses.

When an AC voltage is applied to insulation, the leakage current flowing through the insulation has two components, one resistive and the other capacitive. This is depicted in Fig. 3.203. The power factor is a dimensionless ratio of the resistive current (I_R) to total current (I_T) flowing through the insulation, and is given by the cosine of the angle θ depicted in Fig. 3.203. The dissipation factor, also known as tan delta, is a dimensionless ratio of the resistive current to the reactive current flowing through the insulation, and is the tangent of the angle δ in Fig. 3.203. By convention, these factors are usually expressed in percent.

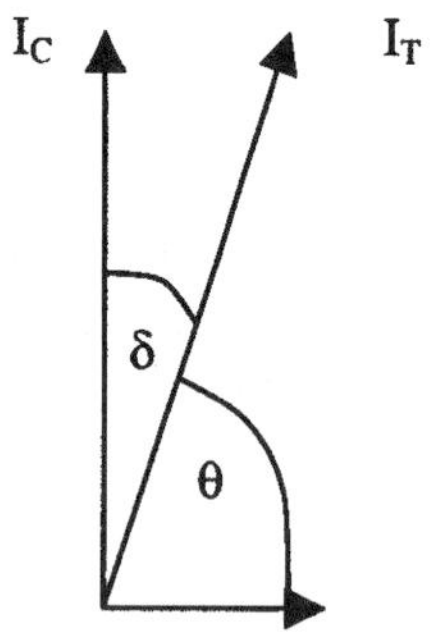

FIGURE 3.203 Power factor graphical representation.

Pump/Fan Operation

The most frequent failure mode of the cooling system is the failure of pumps and fans. The objective of continuous on-line analysis of pumps and fans is to determine if they are on when they are supposed to be on and are off when they are supposed to be off. This is accomplished by measuring the currents drawn by pumps and fans and correlating them with the measurement of the temperature that controls the cooling system. This can also be accomplished by measuring pump/fan current and top oil temperature. Mode of operation is verified based on current level. Normal operational modes may indicate rotation of fan blades and correct rotation of pump impeller. Abnormal operational modes are usually the result of improper control wiring to those devices.

Pump failures due to malfunctioning bearings could be a source of metallic particles. The particles could be a potential dielectric hazard. Sensors that detect bearing wear are available. The ultrasonic sensors are embedded in the pump bearings and measure the bearing length, thus determining if metal loss is occurring.

Furthermore, continuous on-line analysis should take into account that:

- The temperature that controls the cooling system can differ from the temperature measured by the diagnostic system.
- The initial monitoring parameters are set for the cooling stages based on the original transformer design. Any modifications to the cooling sequences or upgrades must be noted since this will change the monitoring system output.
- The sensitivity of the diagnostic system is influenced by the number of motors that are measured by each current sensor.

Instrument Transformers

The techniques available to monitor instrument transformers on-line may be focused on fewer possible degradation mechanisms than those that can be monitored on power transformers. However, the mechanisms by which instrument transformers fail are among the most difficult to detect on-line and are not easy to simulate or accelerate in the laboratory.

Failure Mechanisms Associated With Instrument Transformers

While the failure rates of instrument transformers around the world are generally low, the large numbers of installed instrument transformers has lead to the development of a database of failures and failure statistics. One problem associated with compiling a database of failures of porcelain-housed instrument transformers is that such failures are often catastrophic, leaving little evidence to determine the cause of the fault. Nevertheless, the following mechanisms have been observed and identified as probable causes of failure.

Moisture Ingress

Moisture ingress is commonly identified as a cause of failure of instrument transformers. The ingress of moisture into the instrument transformer may be through loss of integrity of a mechanical seal, e.g., gaskets. The moisture penetrates the oil and oil/paper insulation (which increases the losses in the insulating materials) and failure then follows. This would appear to be a particular problem if the moisture penetrates to certain high stress regions within the instrument transformer. The increase in the dielectric losses will be detected as a change in the power factor of the material and will also appear as increased moisture levels in oil quality tests.

Partial Discharge

The insulation of instrument transformers may have voids within it. Such voids will discharge if subjected to a high enough electric field. Such partial discharges may produce aggressive chemical by-products, which then enlarge the size of the void causing an increase in the energy of the discharge within the void. Eventually, these small partial discharges may degrade individual insulation layers resulting in the short-circuiting of stress grading layers. Such a developing fault may be detected in two ways. One is the observance of a change in the capacitance of the device (through the shorting of one stress grading layer) and which may also reflect as a change in tan delta; and the second is an increase in the partial discharge levels (in pC) associated with the failing item.

Overvoltages

Overvoltages produced by induced lightning surges are also a failure mechanism, particularly where thunderstorms occurred in the vicinity of the failure. More recently, the observance of fast risetime transients (Trise ~ 100 ns) in substations during disconnect switch operations has led to concerns that these transients may cause damage to the insulation of instrument transformers. There is significant speculation that instrument transformers do not perform well when exposed to a number of disconnect switch operations in quick succession. These disconnector-generated fast transients will remain a suspected cause of failure until more is understood about the stress distribution within the instrument transformer under these conditions. Switching overvoltages are a further source of overstressing that may lead to insulation failure.

Through-faults

In order to prevent failures due to the mechanisms outlined above, experience seems to indicate that slower forming faults are probably detectable and preventable, while fast forming faults due to damage caused by lightning strikes will be difficult to prevent.

Another possible mechanism may relate to mechanical damage to the insulation after a current transformer has been subjected to fault current through its primary. After current transformer failures, it is often stated that one to two weeks prior to the failure the CT had been subjected to a through-fault. Again, it is difficult to state that damage is caused to the CT under these conditions and additional information would be required before this mechanism may be considered a probable cause of failure.

Instrument Transformer On-Line Monitoring Methods

On-line techniques for the measurement of relative tan delta and relative capacitance by comparing individual units against a larger population of similar units have been installed by a number of utilities with reports of some success in identifying suspect units. On-line partial discharge measurement techniques may provide important additional information as to the condition of the insulation within the instrument transformer, but research and development work is still under way in order to address issues related to noise rejection vs. required sensitivity and on-site calibration. Other possible future developments may include on-line dissolved gas analyzers that will be able to detect all gases associated with the partial discharge degradation of oil/paper insulation. This section reviews applicable methods to on-line monitoring of instrument transformers.

Relative Tan Delta and Relative Capacitance Measurements

Off-line partial discharge and tan delta monitoring are well-established techniques. These may be supplemented by taking small samples of mineral oil from the instrument transformer for DGA. The development of on-line monitoring techniques is ongoing but significant progress has been made, particularly with respect to on-line tan delta and capacitance measurements. Laboratory-type tan delta and capacitance measurements usually require a standard low loss capacitor at the voltage rating of the

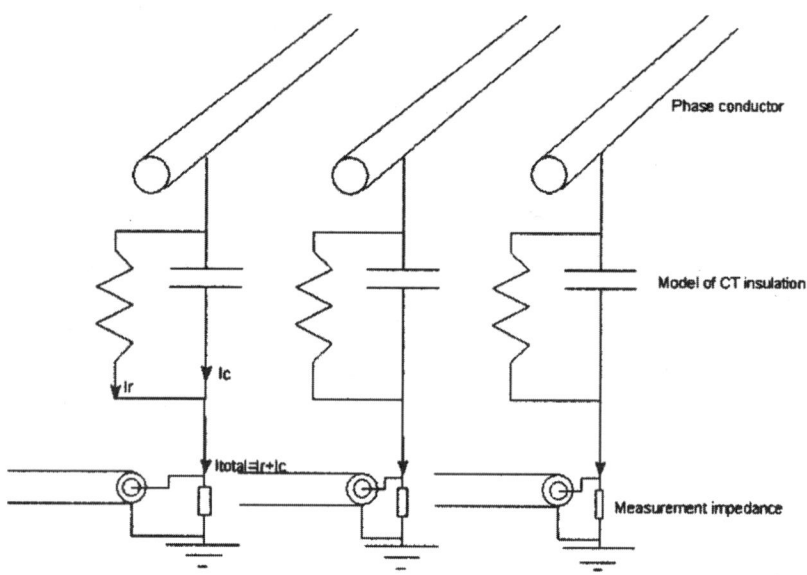

FIGURE 3.204 Schematic representation of relative tan delta measurements.

equipment under test such that a sensitive bridge technique may be used to determine the capacitance and the tan delta (also know as the insulation power factor) of the insulation. This is not practical for on-line measurements.

This problem is overcome by relying on relative measurements in which the insulation of one instrument transformer is compared to the insulation of the other instrument transformers that are installed in the same substation. By comparing sufficient numbers of instrument transformers to each other, changes in one unit when compared to other similar units (and not explained as normal statistical fluctuations due to changes in loading and ambient temperatures) may be identified. There are two commercially available units that monitor tan delta on-line. In the first, the ground currents from each of the three instrument transformers per phase are detected. This is done by isolating the base of the instrument transformer from its base except at one connection point which forms the only current path to ground. This current may then be measured using a suitable sensor. The current consists of two components: a capacitive component (the capacitance of a typical CT to earth being of the order of 0.5–1 nF) and a resistive component dependent upon the insulation loss factor or tan delta of the insulation within the instrument transformer. If each of the three instrument transformers are in similar condition and of similar design, then the sum of the three phase currents to earth should be zero. Any resistive component of current to earth will cause slight phase and magnitude shifts in these currents. If all three units on each phase have a low tan delta, then changes in one unit with respect to the other two may be readily detected. As the insulation deteriorates and possibly a grading layer is shorted out, a change in the capacitance of the unit will be reflected as a change in the capacitive current to earth. As the measurements are made with respect to other similar units, such measurements are referred to as relative tan delta and relative capacitance change measurements. Figure 3.204 shows this arrangement schematically.

Another technique involves comparing each instrument transformer to a number of different units, possibly on the same busbar or on each of three phases. The capacitive and resistive current flows to earth are monitored and the results for each instrument transformer may then be compared to those values measured on other units. Relative changes in tan delta and capacitance may then be determined and if these exceed established norms determined from software algorithms, then an alarm is raised.

These two techniques are currently in service and have achieved success in detecting instrument transformers behaving in a markedly different manner to other similar units. These two measurement tools are trending instruments as they detect changes of certain parameters of a large sample of units over a period of time and identify an individual unit or units performing outside the parameter variations seen for other units.

On-Line Gas Analysis
The fuel cell sensor–membrane technology that has been applied widely to power transformers in which the oil is in circulation may be applied to instrument transformers; however, in instrument transformers, the oil is confined, and this confinement can affect sensor operation. Installation can require factory modifications, depending on the type of sensor that is installed (Boisseau and Tantin, 1993). Typically, the hydrogen sensor is located in an area where the oil is stagnant, especially during periods of low ambient temperatures. This arrangement results in poor accuracy for low hydrogen concentration levels. For significant hydrogen concentrations (above 300 ppm) in stagnant oil, the accuracy has been determined to be acceptable (Cummings et al., 1988). These constraints may not apply to the TCD technology. In this case, the sensor is located externally to the apparatus and utilizes active oil circulation through the monitor and also provides continuous moisture level monitoring.

On-Line Partial Discharge Measurements
On-line partial discharge measurement techniques that were discussed in the section on power transformers are applicable to instrument transformers.

Pressure
Due to partial discharge activity inside the tank, gases can be formed which result in a pressure increase after the gases saturate the oil. A threshold pressure switch can be used to perform this measurement. The operation of this sensor is possible with an inflatable bellows that is placed between the expansion device and the enclosure. The installation of the device typically requires factory modification. In some applications, pressure sensors take a considerable amount of time (on the order of months) to detect any significant pressure change. The sensitivity of this type of measurement is less than that of hydrogen and partial discharge sensors (Boisseau and Tantin, 1993). Pressure sensors are also available that mount on the drain valve (Cummings et al., 1988).

Bushings

Bushings are subjected to high dielectric and thermal stresses, and bushing failures are one of the leading causes of forced outages and transformer failures. The methods of detecting deterioration of the bushing insulation have been well understood for decades, and traditionally off-line diagnostics are very effective at discovering problems if present. The challenge facing a maintenance engineer is that some problems have gestation time (i.e., going from good condition to failure) that is shorter than typical routine test intervals. Since on-line monitoring of power-factor and capacitance can be performed continuously, and with the same sensitivity as the off-line measurement, deciding whether to apply an on-line system is reduced to an economic exercise of weighing the direct and strategic benefits with the cost.

Failure Mechanisms Associated with Bushings
The two most common bushing failure mechanisms are moisture contamination and partial discharge. Moisture usually enters the bushing via deterioration of gasket material or cracks in terminal connections resulting in an increase in the dielectric loss and insulation power factor. Tracking over the surface or burning through the condenser core is typically associated with partial discharge. The first indication of this type of problem is an initial increase in power factor. As the deterioration progresses, increases in capacitance will be observed.

On-Line Bushing Power Factor and Capacitance Measurements
Measurement of power factor and capacitance is a very useful and reliable diagnostic indicator. A very sensitive method for obtaining these parameters on-line is the sum current method. The basic principle of the sum current method is based on the fact that the sums of the voltage and current phasors are zero in a symmetrical three-phase system. Therefore, analysis of bushing condition can be performed by

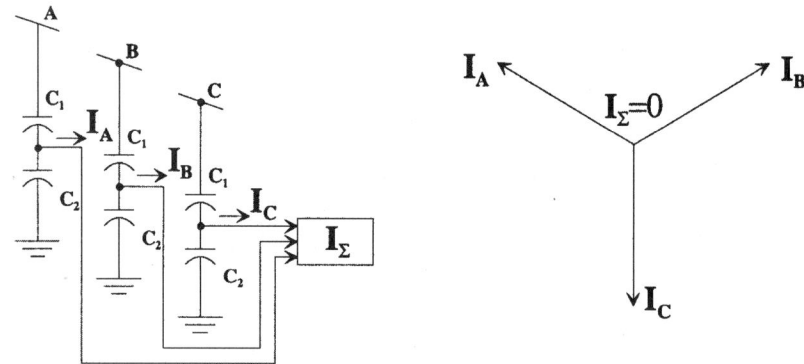

FIGURE 3.205 Bushing sum current measurements.

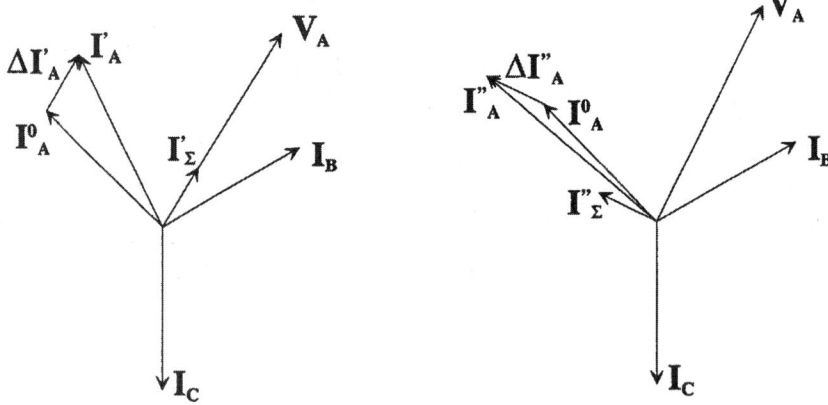

FIGURE 3.206 Analysis of bushing sum currents. (a) Change in current phasor due to change in power factor of bushing A. (b) Change in current phasor due to change in capacitance of bushing A.

vectorially adding currents from the capacitance or power factor taps, as depicted in Fig. 3.205. If the bushings are identical and system voltages are perfectly balanced, the sum current, I_Σ, will equal zero.

In reality, bushings are never identical and system voltages are never perfectly balanced. As a result, the sum current is a non-zero value and is unique for each set of bushings. The initial sum current can be learned and the condition of the bushings can be determined by evaluating changes in the sum current phasor. Incorporating software techniques and an expert system for analysis of changes to the sum current, changes in either the capacitance or power factor of any of the bushings being monitored can be detected, as shown in Fig. 3.206.

Figure 3.206a depicts a change that is purely resistive, i.e., only the in-phase component of current is changing. It is due to a change in C_1 insulation power factor and it results in the current phasor change $\Delta I'_A$ from I^0_A to I'_A. The change in current is in-phase with A phase line voltage, V_A, and it is equal to I'_Σ. This is then evidence of a power factor increase of the A phase bushing.

Figure 3.206b depicts a change that is purely capacitive, i.e., only a quadrature component of current is changing. In this case, the change is due to a change in C_1 insulation capacitance and it results in the current phasor change $\Delta I''_A$ from I^0_A to I''_A. The change in current leads the voltage V_A by 90° and it is equal to I'_Σ.

Expert systems are also used to determine if the sum current change is related to actual bushing deterioration or changes in environmental conditions such as fluctuations in system voltages, changes in bushing or ambient temperature, and changes in surface conditions (Lachman, 1999).

Load Tap Changers

High maintenance costs for load tap changers (LTC) result from several causes. The main reasons include: misalignment of contacts, poor design of the contacts, high loads, excessive number of tap changers, mechanical failures, and coking caused by contact heating. Load tap changer failures account for approximately 41% of substation transformer failures (Bengtsson, 1996; CIGRE, 1983).

LTC contact wear occurs as the LTC operates to maintain a constant voltage with varying loads. This mechanical erosion is a normal operating characteristic, but the rate can be accelerated by improper design, faulty installation, and high loads. If an excessive wear situation is undetected, the contacts may burn open or weld together. Monitoring a combination of parameters suitable for a particular LTC design can help avoid such failures.

LTC failures are either mechanical or electrical in nature. Faults that are mechanical in nature include failures of springs, bearings, shafts, and drive mechanisms. Faults that are electrical in nature can be attributed to coking of contacts, burning of transition resistors, and insulation problems (Bengtsson, 1996). This section discusses the various parameters that can be monitored on-line that will give an indication of tap changer condition.

Mechanical Diagnostics for On-Load Tap Changers

A variety of diagnostic algorithms for on-load tap changers can be implemented using drive motor torque or motor current information. Mechanical and control problems can be detected because additional friction, contact binding, extended changer operation times, and other anomalies significantly impact torque and current.

A signature, or event record, is captured each time the tap changer moves to a different tap. This event record can be either motor torque or as a vibro-acoustic pattern and motor current as a function of time. The signature can then be examined by several methods to detect mechanical and, in the case of vibro-acoustic patterns, electrical (arcing) problems. The following mechanical parameters can be monitored on-line:

Initial peak torque or current — Initial current inrush and starting torque are related to mechanical static friction and backlash in the linkages. Monitoring this peak value during the first 50 msec of the event provides a useful diagnostic. Increasing values are cause for concern.

Average torque or motor current — Running current or torque provides a measure of dynamic friction and also helps detect binding. Monitoring the average value after initial inrush/start-up is a useful diagnostic measure. Motor current measurement is most effective when the motor directly drives the mechanical linkages. Several common tap changer designs employ a motor to charge a spring. It is the spring that supplies energy to move the linkages during a tap change. In this case, motor current measurement is not very effective at detecting mechanical trouble. Torque or force sensors measuring drive force will yield the desired information.

A monitoring system is available that determines the torque curve by measuring the active power of the motor. Anomalies in the torque curve are detected by using an expert system that performs a separate assessment of the individual functions of a switching operation (Liebfried et al., 1998). Figure 3.207 is a sample torque curve for a resistance-type tap changer.

Motor current index — The area under the motor current curve is called the motor index and is usually given in ampere-cycles, based on the power frequency. A similar parameter based on torque can be used. This parameter characterizes the initial inrush, average running conditions, and total running time. Not all types of tap changer operations have similar index values. An operation through neutral can have a significantly higher index as the reversing switch is exercised. Similarly, tap changer raise operations can have different index values depending on whether the previous operation was also a raise or a lower. This is related primarily to linkage backlash. Figure 3.208 shows an example of the motor current curve for a load tap changer, and Fig. 3.209 shows an example of the motor current index.

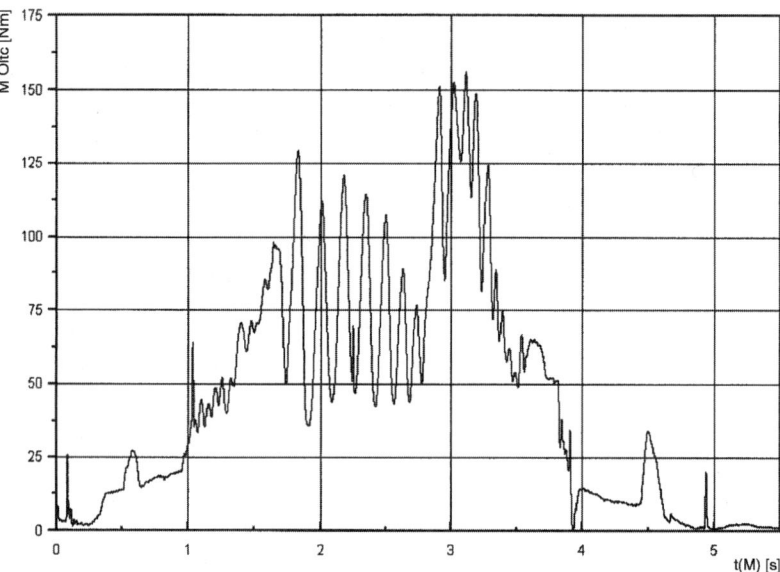

FIGURE 3.207 Sample torque curve.

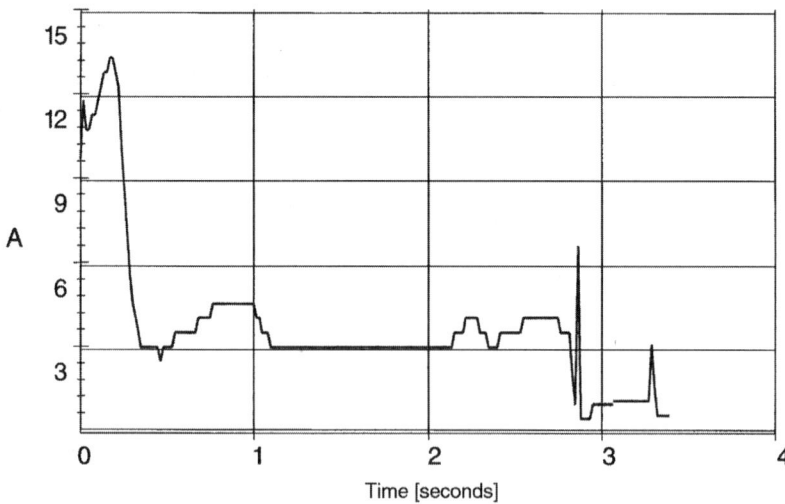

FIGURE 3.208 Load tap changer motor current during a tap changing event.

Sequential controls and other operational issues must also be considered. For example, the index will be very large if the tap changer moves more than one step during an operation. The index will be very small if the controls call for a tap change and then rescind the request before seal-in. All of these situations must be considered when performing diagnostics based on motor current or torque measurements.

Contact wear model — Monitoring systems are available in which an expert system is employed that calculates the total wear on the tap changer switch contacts and issues a recommendation con-

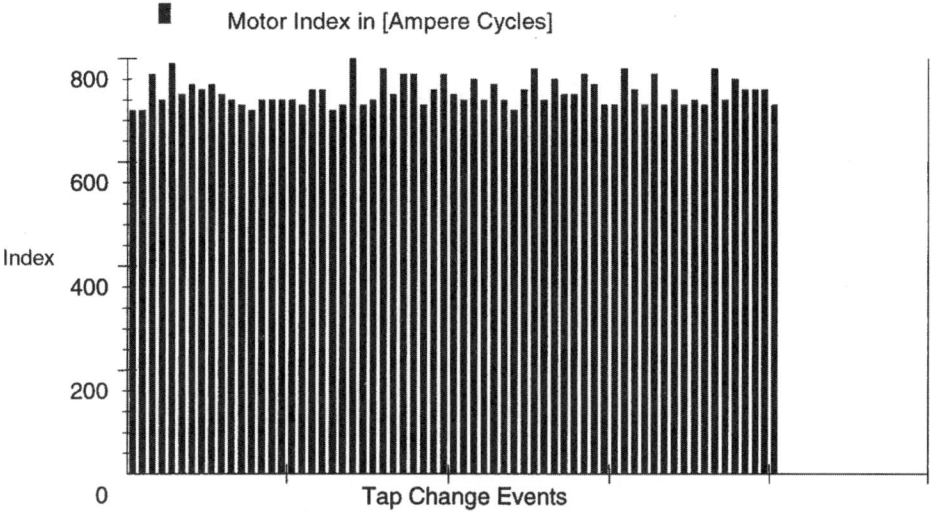

FIGURE 3.209 Sample motor current index curve.

cerning when the contacts should be replaced. The model used by the expert system is based on tap changer switch life tests and field experience.

Position determination — Monitoring systems are available that determine the exact position of the fine tap selector during a switching operation. This information is used by the system to correlate tap position with the motor torque. In this manner, the position of the tap changer after a completed switching operation is determined and the end position of the tap changer is monitored.

Thermal Diagnostics For On-Load Tap Changers

A variety of diagnostic algorithms for on-load tap changers can be implemented using temperature data. The heat transfer pattern resulting from energy losses results in a temperature profile that is easily measured with external temperature sensors. Temperature profiles are normally influenced by weather conditions, cooling bank status, and electrical load. However, abnormal sources of energy (losses) also impact the temperature profile, thus providing a method of detection. The following electrical/thermal parameters can be monitored on-line.

Temperature — The simplest temperature-related diagnostic involves monitoring the temperature level. Load tap changer temperature in excess of a certain level may be an indication of equipment trouble. However, there are also many factors that normally influence temperature level. One LTC monitoring system measures the temperature of the diverter switch oil and the main tank oil temperature as a way to estimate the overload capacity of the tap changer.

Simple differential temperature — Another simple algorithm involves monitoring the temperature difference between the main tank and load tap changer compartment for those tap changer designs in which the tap changer is in a compartment separate from the main tank. Under normal operating conditions, the main tank temperature will be higher than the tap changer compartment temperature. This result is expected, given the energy losses in the main tank and general flow of thermal energy from that point to other regions of the equipment. Differential temperature is most effective on external tap changer designs because this arrangement naturally results in larger temperature differences. Smaller differences are expected on tap changers that are physically located inside the main tank.

Many factors influence differential temperature. Excessive losses caused by bad contacts in the tap changer are detectable. However, load tap changer temperature can exceed main tank temperature periodically under normal conditions. Hourly variations in electrical load, weather conditions, and cooling bank activation can result in main tank temperatures below the tap changer. Reliable diag-

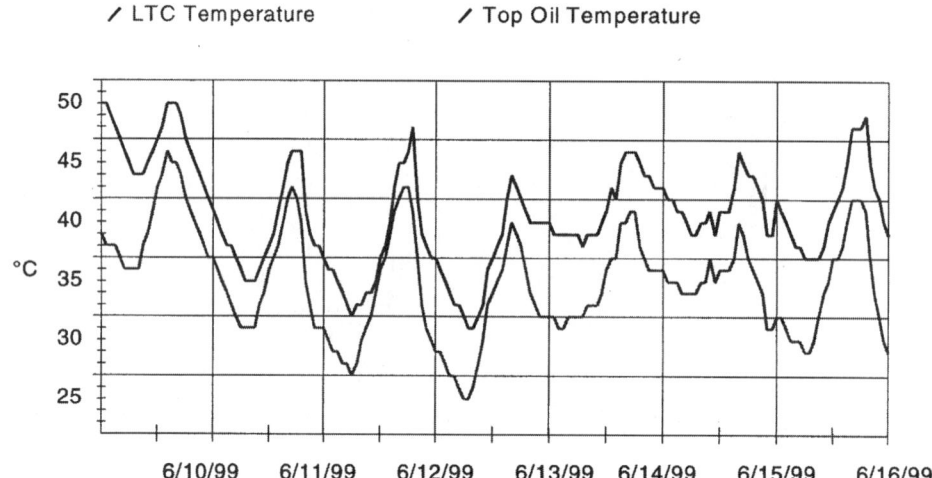

FIGURE 3.210 Sample differential temperature measurement. The top trace is the main tank top oil temperature, and the bottom trace is the LTC compartment temperature.

nostic algorithms must account for these normal variations in some way. Figure 3.210 is a graphical representation of the top oil temperature in the main tank and of the LTC compartment temperature.

Differential temperature with trending — One method used to distinguish between normal and abnormal differential temperature is trending. When the load tap changer temperature exceeds the main tank temperature, the temperature trends are examined. If the tap changer temperature is decreasing, this is deemed a normal condition. However, if the tap changer temperature exceeds the main tank temperature and is increasing, this may indicate an equipment problem.

Temperature index — Another method used to examine temperature differential involves computing the area between the two temperature curves over a rolling window of time (usually one week). This quantity is called the temperature index, and is usually expressed in degree-hours. Normal temperature difference (main tank above tap changer) is counted as "negative" area and the reverse is "positive" area. Therefore, over a period of 7 days, the index reflects the general relationship between the two measurements without changing significantly due to normal daily variations in temperature. Under abnormal conditions, the index will exhibit an increasing trend as the load tap changer tends to run hotter relative to the main tank. This method eliminates false alarms associated with simple differential monitoring but responds slowly to abnormal conditions. A change in tap changer temperature characteristics that takes place over the course of several hours will require several days to be reflected in the index.

Vibro-Acoustic Monitoring

The vibrations caused by various mechanical movements during a tap changing operation may be recorded and analyzed for signs of deterioration. This gives continuous control of the transition time as well as an indication of contact wear and detection of sudden mechanical ruptures faults (Bengtsson et al., 1998).

Acoustic monitoring of on-load tap changers has been under development. The LTC operation can be analyzed by recording the acoustic signature and comparing it with the running average representative of recent operations. The signal is analyzed in distinct frequency bands, which facilitates the distinction between problems with electrical causes and those with mechanical causes.

Every operation of the tap changer produces a characteristic acoustic wave, which propagates through the oil and structure of the transformer. Field measurements show that in the case of a properly functioning tap changer, for a given operation, this vibration pattern proves to be very repeatable in time.

The acoustic signal is split into two frequency bands. Experience has shown that electrical problems (arcing when there should not be any) are detected in a higher frequency band than those mechanical in nature (excessive wear or ruptured springs). This system has the intelligence to distinguish imminent failure conditions and normal wear of the LTC to allow for just-in-time maintenance (Foata et al., 1999).

Dissolved Gas Analysis

Analysis of gases dissolved in the oil in the load tap changer compartment is proving to be a useful diagnostic. Key gases for this analysis include acetylene and ethylene. However, correlation between dissolved gases and certain type faults are still in the infancy stage and depend, to a great extent, on the design and material used in the tap changer.

References

Bengtsson, C., Status and Trends in Transformer Monitoring, *IEEE Trans. Power Delivery*, 11(3), 1379–1384, July 1996.

Bengtsson, T., Kols, H., Foata, M., and Leonard, F., Monitoring Tap Changer Operations, *Int. Conf. Large High Voltage Electric Syst. (CIGRE)*, Paper 12.209, 1998 Session.

Bengtsson, T., Kols, H., and Jönsson, B., Transformer PD Diagnosis using Acoustic Emission Technique, *Proc. 10th ISH*, Montréal 1997. 4, 115.

Bengtsson, T., Leijon, M., and Ming L., Acoustic Frequencies Emitted by Partial Discharges in Oil, *Proc. 7th ISH*, Dresden, 1993, Paper no. 63.10.

Boisseau, C. and Tantin, P., Evaluation of Monitoring Methods Applied to Instrument Transformers, *Doble Conference*, 1993.

Boisseau, C., Tantin, P., Despiney, P., and Hasler, M., Instrument Transformers Monitoring, *CIGRE Diagnostics and Maintenance Techniques Symposium*, Berlin, April 19–21, 1993, Paper 110-13.

Canadian Electricity Association, On-Line Condition Monitoring of Substation Power Equipment and Utility Needs, CEA No. 485 T 1049, December 1996.

Chu, D., El Badaly, H., and Slemon, C., Development of an Automated Transformer Oil Monitor, *EPRI 2nd Conf. Substation Diagnostics*, November 1993.

Chu, D., El Badaly, H., and Slemon, C., Status Report on the Automated Transformer Oil Monitor, *EPRI 3rd Conf. Substation Diagnostics*, November 1994.

CIGRE Working Group 05, An International Survey on Failures in Large Power Transformers in Service, *Electra*, 88, May, 1983.

Cummings, H. B., et,. al., Continuous, On-Line Monitoring of Freestanding, Oil-Filled Current Transformers to Predict an Imminent Failure, *IEEE Trans. Power Delivery*, 3(4), 1776–1783, October 1988.

Domun, M. K., Condition Monitoring of Power Transformers by Oil Analysis Techniques, *Science, Education and Technology Division Colloquium on Condition Monitoring and Remanent Life Assessment in Power Transformers*, IEE Colloquium (Digest) n 075 Mar. 22, 1994.

Duval, M. and Lamarre, C., The Characterization of Electrical Insulating Oils by High Performance Liquid Chromatography, *IEEE Trans. Electrical Insulation*, 12(5), October 1977.

Eleftherion, P., Partial Discharge XXI: Acoustic Emission-Based PD Source Location in Transformers, *IEEE Electrical Insulation Magazine*, 11(6), 22, 1995.

Eriksson, T., Leijon, M., and Bengtsson, C., PD On-Line Monitoring of Power Transformers, *Stockholm Power Tech 1995*, p. 101.0, Paper SPT HV 03-08-0682.

Feser, K., Maier, H. A., Freund, H., Rosenow, U., Baur, A., and Mieske, H., On-Line Diagnostic System for Monitoring the Thermal Behaviour of Transformers, *CIGRE Diagnostics and Maintenance Techniques Symposium*, Berlin, April 19–21, 1993, Paper 110-08.

Foata, M., Aubin, J., and Rajotte, C., Field Experience with Acoustic Monitoring of On Load Tap Changers, *1999 Proc. Sixty Sixth Annu. Int. Conf. Doble Clients.*

Fox, R. J., Measurement of Peak Temperatures Along an Optical Fiber, *Applied Optics*, 22, April 1, 1983.

Fruth, B. and Fuhr, J., Partial Discharge Pattern Recognition — A Tool for Diagnostics and Monitoring of Aging, *International Conference on Large High Voltage Electric Systems (CIGRE) 1990*, Paper 15/33-12.
Glodjo, A., A Field Experience with Multi-Gas On-Line Monitors, *1998 Proc. Sixty Fifth Annu. Int. Conf. Doble Clients*.
Griffin, P., Continuous Condition Assessment and Rating of Transformers, *1999 Proc. Sixty Sixth Annu. Int. Conf. Doble Clients*, pp. 8-8.1–8-5.21.
Harrold, R.T., Acoustic Waveguides for Sensing and Locating Electric Discharges Within High Voltage Power Transformers and Other Apparatus, *IEEE Trans. Power Apparatus Systems*, Vol. PAS-102, April 1983.
IEEE Guide for the Interpretation of Gases Generated in Oil-Immersed Transformers, C57.104.
IEEE Std. C57.91-1995, Guide for Loading Mineral-Oil-Immersed Transformers.
Lachman, M. F., On-Line Diagnostics of High-Voltage Bushings and Current Transformers Using the Sum Current Method, *IEEE Trans Power Delivery*, PE-471-PWRD-0-02-1999.
Leibfried, T., Knorr, W., Viereck, D., Dohnal, D., Kosmata, A., Sundermann, U., and Breitenbauch, B., On-Line Monitoring of Power Transformers — Trends, New Developments, and First Experiences, *Intl. Conf. Large High Voltage Electric Syst. (CIGRE)*, Paper 12.211, 1998 Session.
Lemke, E., A New Procedure for Partial Discharge Measurements on the Basis of an Electromagnetic Sensor, *Proc. 5th ISH*, Braunschweig, August, 1987, Paper 41.02.
Morshuis, P. H. F., Partial Discharge Mechanisms in Voids Related to Dielectric Degradation, *IEE Proc.-Sci. Meas. Technol.*, 142(1), 62, 1995.
Myers, S. D., Kelly, J. J., and Parrish, R. H., *A Guide to Transformer Maintenance*, The Transformer Maintenance Institute, Akron, OH, 1981.
Oommen, T. V., On-Line Moisture Sensing in Transformers, *Proc. 20th Electrical/Electronics Insulation Conf.*, October 7–10, 1991, Boston, MA, p. 236–241.
Oommen, T. V., Further Experimentation on Bubble Generation During Transformer Overload, Electric Power Research Institute, Report EL-7291, Final Report, March 1992.
Oommen, T. V., On-Line Moisture Monitoring in Transformers and Oil Processing Systems, *CIGRE Diagnostics and Maintenance Techniques Symposium*, Berlin, April 19–21, 1993, Paper 110-03.
Sokolov, V. V. and Vanin, B. V., In-Service Assessment of Water Content in Power Transformers, *Doble Conference*, 1995.
Wang, C., Dong, X., Wang, Z., Jing, W., Jin, X., and Cheng, T.C., On-line Partial Discharge Monitoring System for Power Transformers, *Proc. 10th ISH*, Montréal 1997, 4, 379.
Wenzel, D., Borsi, H., and Glockenbach, E., Pulse Shaped Noise Reduction and Partial Disharge Localisation on Transformers using the Karhunen-Loéve-Transform, *Proc. 9th ISH*, Graz, August 1995, Paper 5627.
Wenzel, D., Schichler, U., Borsi, H., and Glockenbach, E., Recognition of Partial Discharges on Power Units by Directional Coupling, *Proc. 9th ISH*, Graz, August 1995, Paper 5626.
Zaretsky, M. C. et al., Moisture Sensing in Transformer Oil Using Thin-Film Microdielectrometry, *IEEE Trans. Electrical Insulation*, 24(6), December, 1989.

4
Transmission System

George G. Karady
Arizona State University

4.1	Concept of Energy Transmission and Distribution *George G. Karady*	4-2
4.2	Transmission Line Structures *Joe C. Pohlman*	4-13
4.3	Insulators and Accessories *George G. Karady and R.G. Farmer*	4-23
4.4	Transmission Line Construction and Maintenance *Wilford Caulkins and Kristine Buchholz*	4-42
4.5	Insulated Power Cables for High-Voltage Applications *Carlos V. Núñez-Noriega and Felimón Hernandez*	4-48
4.6	Transmission Line Parameters *Manuel Reta-Hernández*	4-62
4.7	Sag and Tension of Conductor *D.A. Douglass and Ridley Thrash*	4-89
4.8	Corona and Noise *Giao N. Trinh*	4-129
4.9	Geomagnetic Disturbances and Impacts upon Power System Operation *John G. Kappenman*	4-150
4.10	Lightning Protection *William A. Chisholm*	4-165
4.11	Reactive Power Compensation *Rao S. Thallam*	4-169

4
Transmission System

George G. Karady
Arizona State University

Joe C. Pohlman
Consultant

R.G. Farmer
Arizona State University

Wilford Caulkins
Sherman & Reilly

Kristine Buchholz
Pacific Gas & Electric Company

Carlos V. Núñez-Noriega
Glendale Community College

Felimón Hernandez
Arizona Public Service Company

Manuel Reta-Hernández
Arizona State University

D.A. Douglass
Power Delivery Consultants, Inc.

Ridley Thrash
Southwire Company

Giao N. Trinh
Log-In

John G. Kappenman
Metatech Corporation

William A. Chisholm
Ontario Hydro Technologies

Rao S. Thallam
Salt River Project

4.1 Concept of Energy Transmission and Distribution 4-2
Generation Stations • Switchgear • Control Devices • Concept of Energy Transmission and Distribution

4.2 Transmission Line Structures..4-13
Traditional Line Design Practice • Current Deterministic Design Practice • Improved Design Approaches

4.3 Insulators and Accessories 4-23
Electrical Stresses on External Insulation • Ceramic (Porcelain and Glass) Insulators • Nonceramic (Composite) Insulators • Insulator Failure Mechanism • Methods for Improving Insulator Performance

4.4 Transmission Line Construction and Maintenance 4-42
Tools • Equipment • Procedures • Helicopters

4.5 Insulated Power Cables for High-Voltage Applications....4-48
Typical Cable Description • Overview of Electric Parameters of Underground Power Cables • Underground Layout and Construction • Testing, Troubleshooting, and Fault Location

4.6 Transmission Line Parameters 4-62
Equivalent Circuit • Resistance • Current-Carrying Capacity (Ampacity) • Inductance and Inductive Reactance • Capacitance and Capacitive Reactance • Characteristics of Overhead Conductors

4.7 Sag and Tension of Conductor 4-89
Catenary Cables • Approximate Sag-Tension Calculations • Numerical Sag-Tension Calculations • Ruling Span Concept • Line Design Sag-Tension Parameters • Conductor Installation

4.8 Corona and Noise 4-129
Corona Modes • Main Effects of Discharges on Overhead Lines • Impact on the Selection of Line Conductors • Conclusions

4.9 Geomagnetic Disturbances and Impacts upon Power System Operation...4-150
Power System Reliability Threat • Transformer Impacts Due to GIC • Magneto-Telluric Climatology and the Dynamics of a Geomagnetic Superstorm • Satellite Monitoring and Forecast Models Advance Forecast Capabilities

4.10 Lightning Protection...4-165
Ground Flash Density • Stroke Incidence to Power Lines • Stroke Current Parameters • Calculation of Lightning Overvoltages on Shielded Lines • Insulation Strength • Conclusion

4.11 Reactive Power Compensation 4-169
   The Need for Reactive Power Compensation • Application of
   Shunt Capacitor Banks in Distribution Systems — A Utility
   Perspective • Static VAR Control (SVC) • Series
   Compensation • Series Capacitor Bank

4.1 Concept of Energy Transmission and Distribution

George G. Karady

The purpose of the electric transmission system is the interconnection of the electric energy producing power plants or generating stations with the loads. A three-phase AC system is used for most transmission lines. The operating frequency is 60 Hz in the U.S. and 50 Hz in Europe, Australia, and part of Asia. The three-phase system has three phase conductors. The system voltage is defined as the rms voltage between the conductors, also called line-to-line voltage. The voltage between the phase conductor and ground, called line-to-ground voltage, is equal to the line-to-line voltage divided by the square root of three. Figure 4.1 shows a typical system.

The figure shows the Phoenix area 230-kV system, which interconnects the local power plants and the substations supplying different areas of the city. The circles are the substations and the squares are the generating stations. The system contains loops that assure that each load substation is supplied by at least two lines. This assures that the outage of a single line does not cause loss of power to any customer. For example, the Aqua Fria generating station (marked: Power plant) has three outgoing lines. Three high-voltage cables supply the Country Club Substation (marked: Substation with cables). The Pinnacle Peak Substation (marked: Substation with transmission lines) is a terminal for six transmission lines. This example shows that the substations are the node points of the electric system. The system is

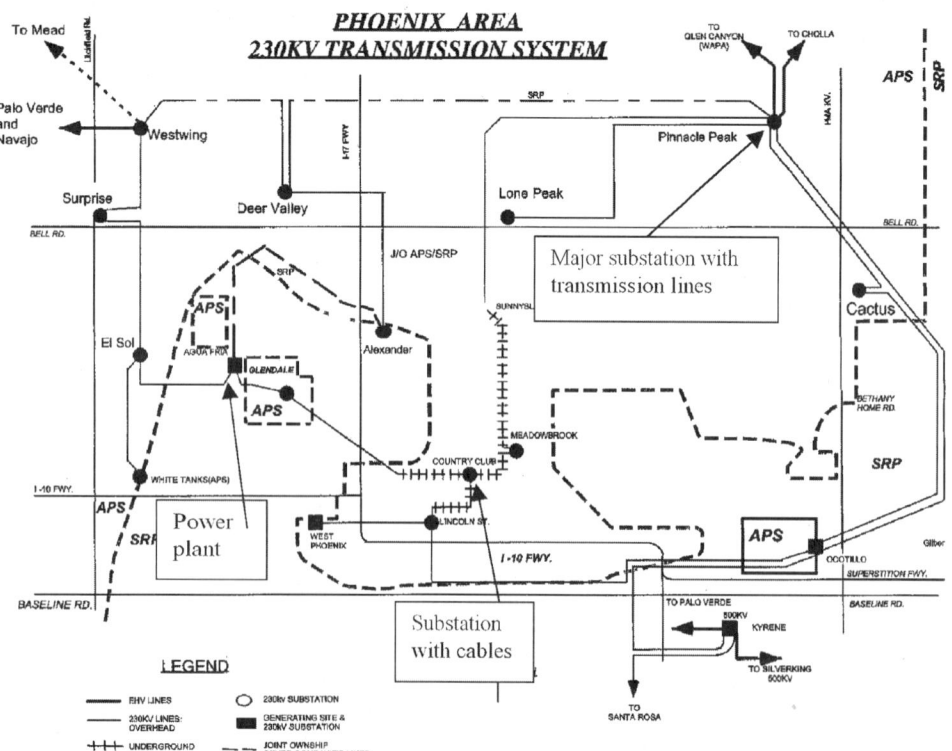

FIGURE 4.1 Typical electrical system.

interconnected with the neighboring systems. As an example, one line goes to Glen Canyon and the other to Cholla from the Pinnacle Peak substation.

In the middle of the system, which is in a congested urban area, high-voltage cables are used. In open areas, overhead transmission lines are used. The cost per mile of overhead transmission lines is 6 to 10% less than underground cables.

The major components of the electric system, the transmission lines, and cables are described briefly below.

Generation Stations

The generating station converts the stored energy of gas, oil, coal, nuclear fuel, or water position to electric energy. The most frequently used power plants are:

Thermal Power Plant. The fuel is pulverized coal or natural gas. Older plants may use oil. The fuel is mixed with air and burned in a boiler that generates steam. The high-pressure and high-temperature steam drives the turbine, which turns the generator that converts the mechanical energy to electric energy.

Nuclear Power Plant. Enriched uranium produces atomic fission that heats water and produces steam. The steam drives the turbine and generator.

Hydro Power Plants. A dam increases the water level on a river, which produces fast water flow to drive a hydro-turbine. The hydro-turbine drives a generator that produces electric energy.

Gas Turbine. Natural gas is mixed with air and burned. This generates a high-speed gas flow that drives the turbine, which turns the generator.

Combined Cycle Power Plant. This plant contains a gas turbine that generates electricity. The exhaust from the gas turbine is high-temperature gas. The gas supplies a heat exchanger to preheat the combustion air to the boiler of a thermal power plant. This process increases the efficiency of the combined cycle power plant. The steam drives a second turbine, which drives the second generator. This two-stage operation increases the efficiency of the plant.

Switchgear

The safe operation of the system requires switches to open lines automatically in case of a fault, or manually when the operation requires it. Figure 4.2 shows the simplified connection diagram of a generating station.

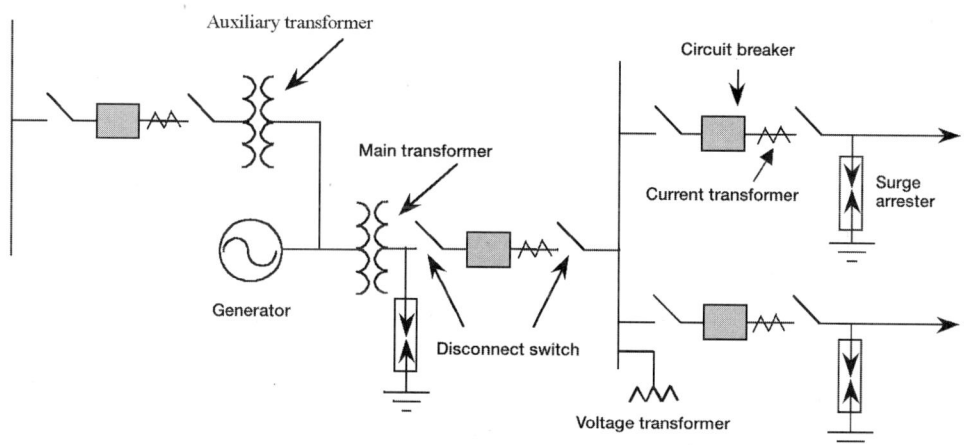

FIGURE 4.2 Simplified connection diagram of a generating station.

The generator is connected directly to the low-voltage winding of the main transformer. The transformer high-voltage winding is connected to the bus through a circuit breaker, disconnect switch, and current transformer. The generating station auxiliary power is supplied through an auxiliary transformer through a circuit breaker, disconnect switch, and current transformer. Generator circuit breakers, connected between the generator and transformer, are frequently used in Europe. These breakers have to interrupt the very large short-circuit current of the generators, which results in high cost.

The high-voltage bus supplies two outgoing lines. The station is protected from lightning and switching surges by a surge arrester.

Circuit breaker (CB) is a large switch that interrupts the load and fault current. Fault detection systems automatically open the CB, but it can be operated manually.

Disconnect switch provides visible circuit separation and permits CB maintenance. It can be operated only when the CB is open, in no-load condition.

Potential transformers (PT) and current transformers (CT) reduce the voltage to 120 V, the current to 5 A, and insulates the low-voltage circuit from the high-voltage. These quantities are used for metering and protective relays. The relays operate the appropriate CB in case of a fault.

Surge arresters are used for protection against lightning and switching overvoltages. They are voltage dependent, nonlinear resistors.

Control Devices

In an electric system the voltage and current can be controlled. The voltage control uses parallel connected devices, while the flow or current control requires devices connected in series with the lines.

Tap-changing transformers are frequently used to control the voltage. In this system, the turns-ratio of the transformer is regulated, which controls the voltage on the secondary side. The ordinary tap changer uses a mechanical switch. A thyristor-controlled tap changer has recently been introduced.

A shunt capacitor connected in parallel with the system through a switch is the most frequently used voltage control method. The capacitor reduces lagging-power-factor reactive power and improves the power factor. This increases voltage and reduces current and losses. Mechanical and thyristor switches are used to insert or remove the capacitor banks.

The frequently used Static Var Compensator (SVC) consists of a switched capacitor bank and a thyristor-controlled inductance. This permits continuous regulation of reactive power.

The current of a line can be controlled by a capacitor connected in series with the line. The capacitor reduces the inductance between the sending and receiving points of the line. The lower inductance increases the line current if a parallel path is available.

In recent years, electronically controlled series compensators have been installed in a few transmission lines. This compensator is connected in series with the line, and consists of several thyristor-controlled capacitors in series or parallel, and may include thyristor-controlled inductors.

Medium- and low-voltage systems use several other electronic control devices. The last part in this section gives an outline of the electronic control of the system.

Concept of Energy Transmission and Distribution

Figure 4.3 shows the concept of typical energy transmission and distribution systems. The generating station produces the electric energy. The generator voltage is around 15 to 25 kV. This relatively low voltage is not appropriate for the transmission of energy over long distances. At the generating station a transformer is used to increase the voltage and reduce the current. In Fig. 4.3 the voltage is increased to 500 kV and an extra-high-voltage (EHV) line transmits the generator-produced energy to a distant substation. Such substations are located on the outskirts of large cities or in the center of several large loads. As an example, in Arizona, a 500-kV transmission line connects the Palo Verde Nuclear Station to the Kyrene and Westwing substations, which supply a large part of the city of Phoenix.

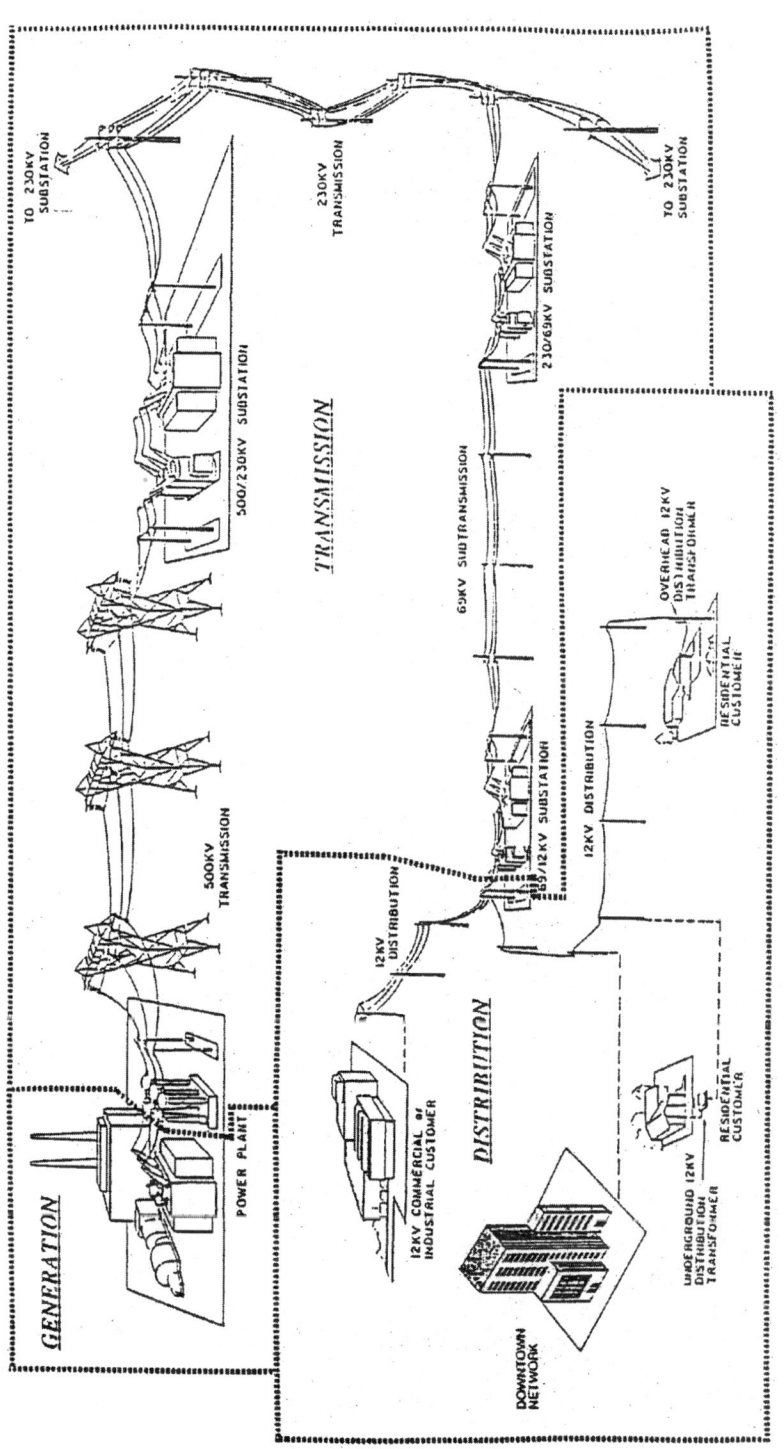

FIGURE 4.3 Concept of electric energy transmission.

The voltage is reduced at the 500 kV/220 kV EHV substation to the high-voltage level and high-voltage lines transmit the energy to high-voltage substations located within cities.

At the high-voltage substation the voltage is reduced to 69 kV. Sub-transmission lines connect the high-voltage substation to many local distribution stations located within cities. Sub-transmission lines are frequently located along major streets.

The voltage is reduced to 12 kV at the distribution substation. Several distribution lines emanate from each distribution substation as overhead or underground lines. Distribution lines distribute the energy along streets and alleys. Each line supplies several step-down transformers distributed along the line. The distribution transformer reduces the voltage to 230/115 V, which supplies houses, shopping centers, and other local loads. The large industrial plants and factories are supplied directly by a subtransmission line or a dedicated distribution line as shown in Fig. 4.3.

The overhead transmission lines are used in open areas such as interconnections between cities or along wide roads within the city. In congested areas within cities, underground cables are used for electric energy transmission. The underground transmission system is environmentally preferable but has a significantly higher cost. In Fig. 4.3 the 12-kV line is connected to a 12-kV cable which supplies commercial or industrial customers. The figure also shows 12-kV cable networks supplying downtown areas in a large city. Most newly developed residential areas are supplied by 12-kV cables through pad-mounted step-down transformers as shown in Fig. 4.3.

High-Voltage Transmission Lines

High-voltage and extra-high-voltage (EHV) transmission lines interconnect power plants and loads, and form an electric network. Figure 4.4 shows a typical high-voltage and EHV system.

This system contains 500-kV, 345-kV, 230-kV, and 115-kV lines. The figure also shows that the Arizona (AZ) system is interconnected with transmission systems in California, Utah, and New Mexico. These

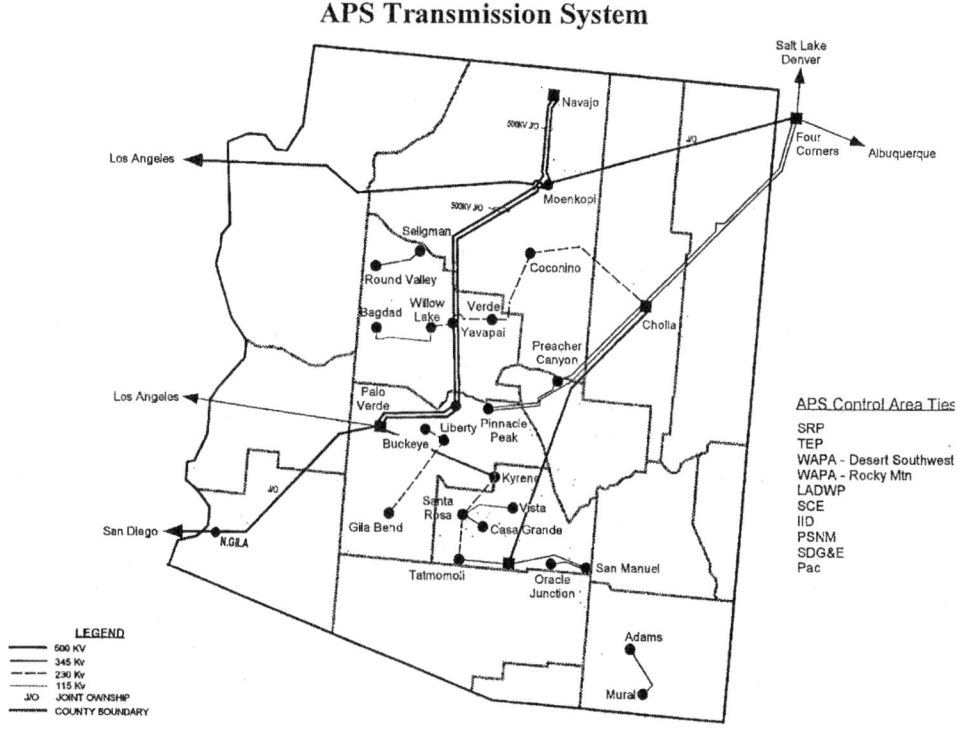

FIGURE 4.4 Typical high-voltage and EHV transmission system (Arizona Public Service, Phoenix area system).

Transmission System

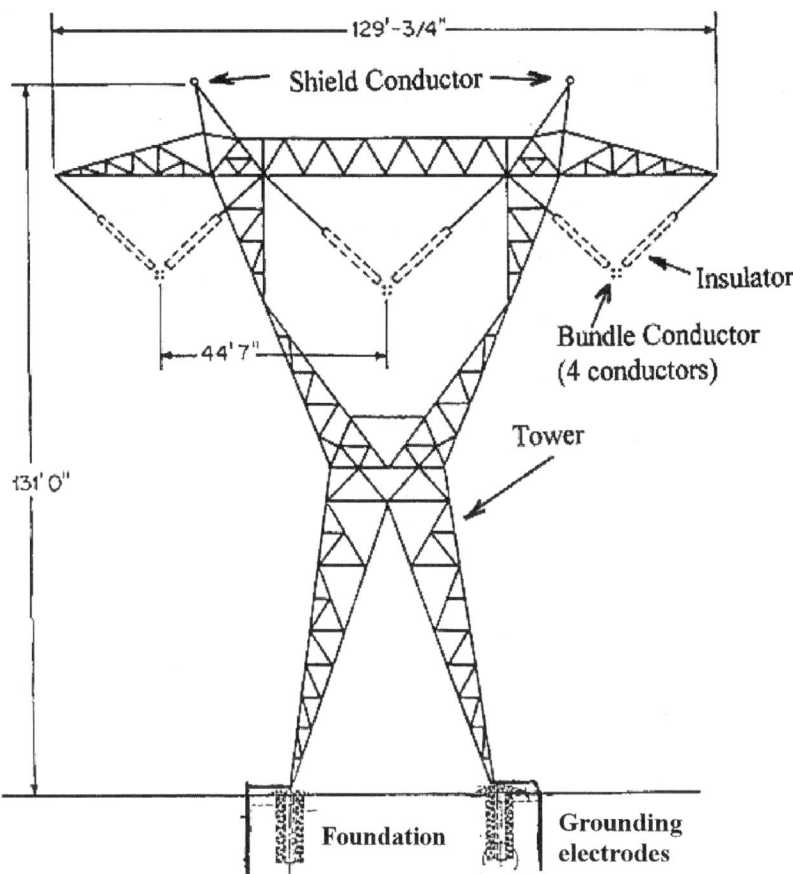

FIGURE 4.5 Typical high-voltage transmission line.

interconnections provide instantaneous help in case of lost generation in the AZ system. This also permits the export or import of energy, depending on the needs of the areas.

Presently, synchronous ties (AC lines) interconnect all networks in the eastern U.S. and Canada. Synchronous ties also (AC lines) interconnect all networks in the western U.S. and Canada. Several nonsynchronous ties (DC lines) connect the East and the West. These interconnections increase the reliability of the electric supply systems.

In the U.S., the nominal voltage of the high-voltage lines is between 100 kV and 230 kV. The voltage of the extra-high-voltage lines is above 230 kV and below 800 kV. The voltage of an ultra-high-voltage line is above 800 kV. The maximum length of high-voltage lines is around 200 miles. Extra-high-voltage transmission lines generally supply energy up to 400–500 miles without intermediate switching and var support. Transmission lines are terminated at the bus of a substation.

The physical arrangement of most extra-high-voltage (EHV) lines is similar. Figure 4.5 shows the major components of an EHV, which are:

1. Tower: The figure shows a lattice, steel tower.
2. Insulator: V strings hold four bundled conductors in each phase.
3. Conductor: Each conductor is stranded, steel reinforced aluminum cable.
4. Foundation and grounding: Steel-reinforced concrete foundation and grounding electrodes placed in the ground.
5. Shield conductors: Two grounded shield conductors protect the phase conductors from lightning.

FIGURE 4.6 Typical 230-kV constructions.

At lower voltages the appearance of lines can be improved by using more aesthetically pleasing steel tubular towers. Steel tubular towers are made out of a tapered steel tube equipped with banded arms. The arms hold the insulators and the conductors. Figure 4.6 shows typical 230-kV steel tubular and lattice double-circuit towers. Both lines carry two three-phase circuits and are built with two conductor bundles to reduce corona and radio and TV noise. Grounded shield conductors protect the phase conductors from lightning.

High-Voltage DC Lines

High-voltage DC lines are used to transmit large amounts of energy over long distances or through waterways. One of the best known is the Pacific HVDC Intertie, which interconnects southern California with Oregon. Another DC system is the ±400 kV Coal Creek-Dickenson lines. Another famous HVDC system is the interconnection between England and France, which uses underwater cables. In Canada, Vancouver Island is supplied through a DC cable.

In an HVDC system the AC voltage is rectified and a DC line transmits the energy. At the end of the line an inverter converts the DC voltage to AC. A typical example is the Pacific HVDC Intertie that operates with ±500 kV voltage and interconnects Southern California with the hydro stations in Oregon.

Figure 4.7 shows a guyed tower arrangement used on the Pacific HVDC Intertie. Four guy wires balance the lattice tower. The tower carries a pair of two-conductor bundles supported by suspension insulators.

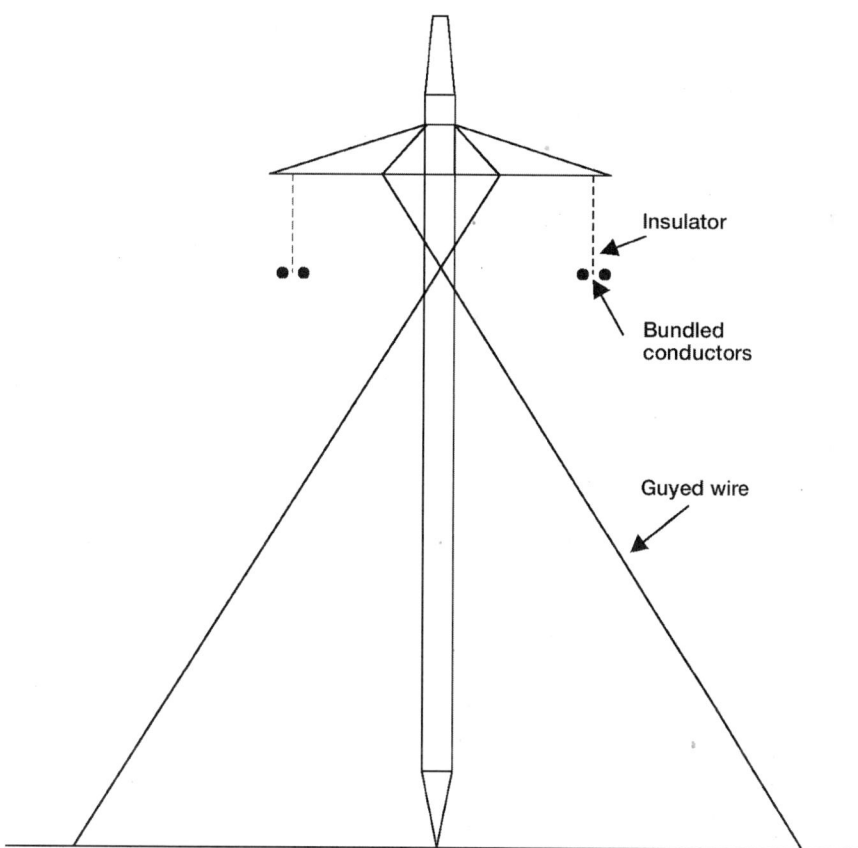

FIGURE 4.7 HVDC tower arrangement.

Sub-Transmission Lines

Typical sub-transmission lines interconnect the high-voltage substations with distribution stations within a city. The voltage of the subtransmission system is between 46 kV, 69 kV, and 115 kV. The maximum length of sub-transmission lines is in the range of 50–60 miles. Most subtransmission lines are located along streets and alleys. Figure 4.8 shows a typical sub-transmission system.

This system operates in a looped mode to enhance continuity of service. This arrangement assures that the failure of a line will not interrupt the customer's power.

Figure 4.9 shows a typical double-circuit sub-transmission line, with a wooden pole and post-type insulators. Steel tube or concrete towers are also used. The line has a single conductor in each phase. Post insulators hold the conductors without metal cross arms. One grounded shield conductor on the top of the tower shields the phase conductors from lightning. The shield conductor is grounded at each tower. Plate or vertical tube electrodes (ground rod) are used for grounding.

Distribution Lines

The distribution system is a radial system. Figure 4.10 shows the concept of a typical urban distribution system. In this system a main three-phase feeder goes through the main street. Single-phase subfeeders supply the crossroads. Secondary mains are supplied through transformers. The consumer's service drops supply the individual loads. The voltage of the distribution system is between 4.6 and 25 kV. Distribution feeders can supply loads up to 20–30 miles.

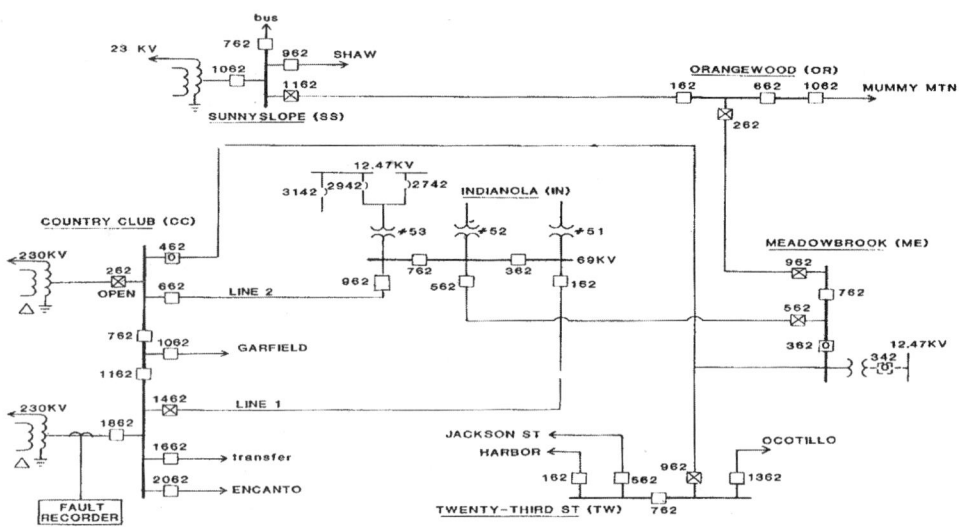

FIGURE 4.8 Subtransmission system.

FIGURE 4.9 Typical subtransmission line.

Many distribution lines in the U.S. have been built with a wood pole and cross arm. The wood is treated with an injection of creosote or other wood preservative that protects the wood from rotting and termites. Most poles are buried in a hole without foundation. Lines built recently may use a simple

Transmission System

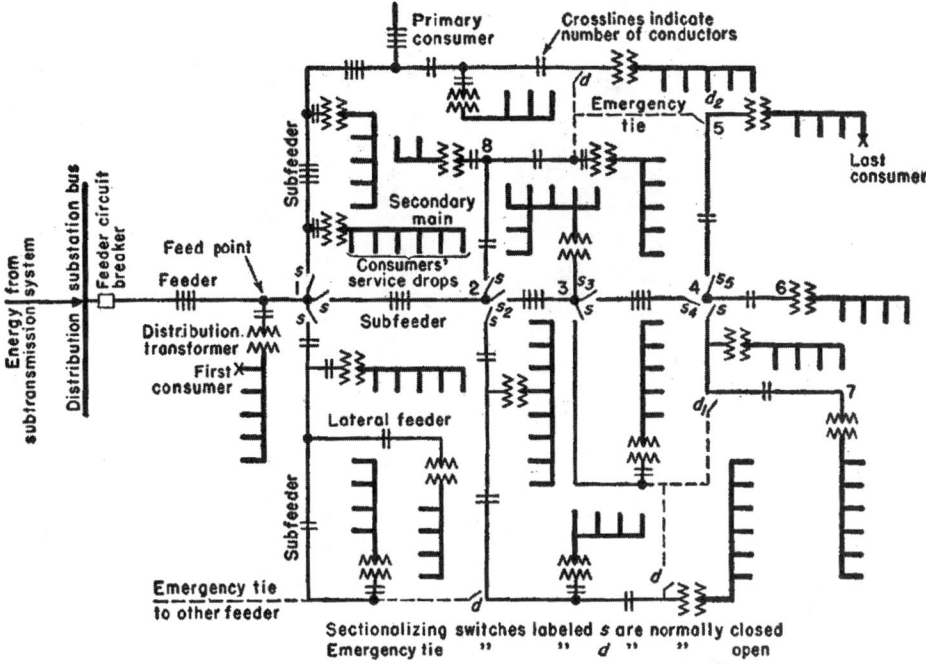

FIGURE 4.10 Concept of radial distribution system.

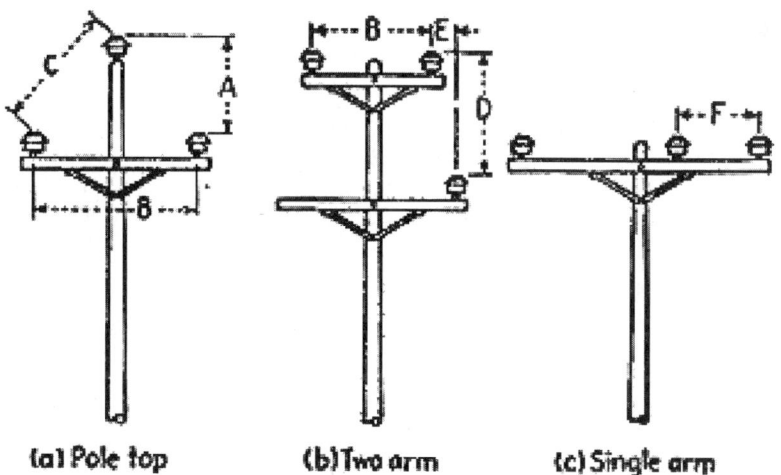

FIGURE 4.11 Distribution line arrangements.

concrete block foundation. Small porcelain or non-ceramic, pin-type insulators support the conductors. The insulator pin is grounded to eliminate leakage current, which can cause burning of the wood tower. A simple vertical copper rod is used for grounding. Shield conductors are seldom used. Figure 4.11 shows typical distribution line arrangements.

Because of the lack of space in urban areas, distribution lines are often installed on the subtransmission line towers. This is referred to as underbuild. A typical arrangement is shown in Figure 4.12.

The figure shows that small porcelain insulators support the conductors. The insulators are installed on metal brackets that are bolted onto the wood tower. This arrangement reduces the right-of-way requirement and saves space.

FIGURE 4.12 Distribution line installed under the subtransmission line.

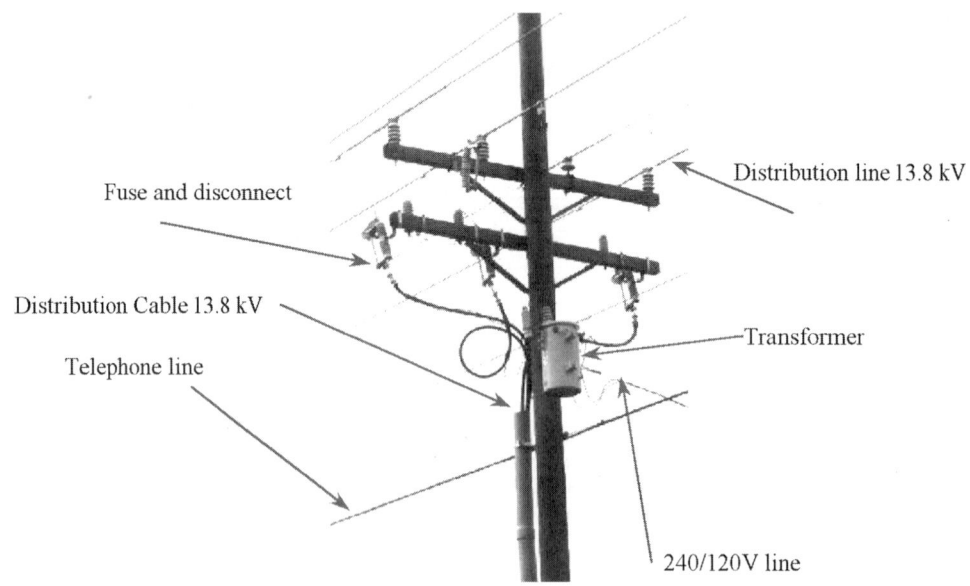

FIGURE 4.13 Service drop.

Transformers mounted on distribution poles frequently supply individual houses or groups of houses. Figure 4.13 shows a typical transformer pole, consisting of a transformer that supplies a 240/120-V service drop, and a 13.8-kV distribution cable. The latter supplies a nearby shopping center, located on the other side of the road. The 13.8-kV cable is protected by a cut-off switch that contains a fuse mounted on a pivoted insulator. The lineman can disconnect the cable by pulling the cut-off open with a long insulated rod (hot stick).

References

Electric Power Research Institute, *Transmission Line Reference Book, 345 kV and Above,* Electric Power Research Institute, Palo Alto, CA, 1987.

Fink, D.G. and Beaty, H.W., *Standard Hand Book for Electrical Engineering,* 11th ed., McGraw-Hill, New York, Sec. 18, 1978.

Gonen, T., *Electric Power Distribution System Engineering,* Wiley, New York, 1986.

Gonen, T., *Electric Power Transmission System Engineering,* Wiley, New York, 1986.

Zaborsky J.W. and Rittenhouse, *Electrical Power Transmission,* 3rd ed. The Rensselaer Bookstore, Troy, NY, 1969.

4.2 Transmission Line Structures

Joe C. Pohlman

An overhead transmission line (OHTL) is a very complex, continuous, electrical/mechanical system. Its function is to transport power safely from the circuit breaker on one end to the circuit breaker on the other. It is physically composed of many individual components made up of different materials having a wide variety of mechanical properties, such as:

- flexible vs. rigid
- ductile vs. brittle
- variant dispersions of strength
- wear and deterioration occurring at different rates when applied in different applications within one micro-environment or in the same application within different micro-environments

This discussion will address the nature of the structures which are required to provide the clearances between the current-carrying conductors, as well as their safe support above the earth. During this discussion, reference will be made to the following definitions:

Capability: Capacity (×) availability
Reliability level: Ability of a line (or component) to perform its expected capability
Security level: Ability of a line to restrict progressive damage after the failure of the first component
Safety level: Ability of a line to perform its function safely

Traditonal Line Design Practice

Present line design practice views the support structure as an isolated element supporting half span of conductors and overhead ground wires (OHGWs) on either side of the structure. Based on the voltage level of the line, the conductors and OHGWs are configured to provide, at least, the minimum clearances mandated by the National Electrical Safety Code (NESC) (IEEE, 1990), as well as other applicable codes. This configuration is designed to control the separation of:

- energized parts from other energized parts
- energized parts from the support structure of other objects located along the r-o-w
- energized parts above ground

The NESC divides the U.S. into three large global loading zones: heavy, medium, and light and specifies radial ice thickness/wind pressure/temperature relationships to define the minimum load levels that must be used within each loading zone. In addition, the Code introduces the concept of an Overload Capacity Factor (OCF) to cover uncertainties stemming from the:

- likelihood of occurrence of the specified load
- dispersion of structure strength

- grade of construction
- deterioration of strength during service life
- structure function (suspension, dead-end, angle)
- other line support components (guys, foundations, etc.)

Present line design practice normally consists of the following steps:

1. The owning utility prepares an agenda of loading events consisting of:
   - mandatory regulations from the NESC and other codes
   - climatic events believed to be representative of the line's specific location
   - contingency loading events of interest; i.e., broken conductor
   - special requirements and expectations

Each of these loading events is multiplied by its own OCF to cover uncertainties associated with it to produce an agenda of final ultimate design loads (see Fig. 4.14).

2. A ruling span is identified based on the sag/tension requirements for the preselected conductor.
3. A structure type is selected based on past experience or on recommendations of potential structure suppliers.
4. Ultimate design loads resulting from the ruling span are applied statically as components in the longitudinal, transverse, and vertical directions, and the structure deterministically designed.
5. Using the loads and structure configuration, ground line reactions are calculated and used to accomplish the foundation design.
6. The ruling span line configuration is adjusted to fit the actual r-o-w profile.
7. Structure/foundation designs are modified to account for variation in actual span lengths, changes in elevation, and running angles.
8. Since most utilities expect the tangent structure to be the weakest link in the line system, hardware, insulators, and other accessory components are selected to be stronger than the structure.

Inasmuch as structure types are available in a wide variety of concepts, materials, and costs, several iterations would normally be attempted in search of the most cost effective line design based on total installed costs (see Fig. 4.15).

While deterministic design using static loads applied in quadrature is a convenient mathematical approach, it is obviously not representative of the real-world exposure of the structural support system. OHTLs are tens of yards wide and miles long and usually extend over many widely variant microtopographical and microclimatic zones, each capable of delivering unique events consisting of magnitude of

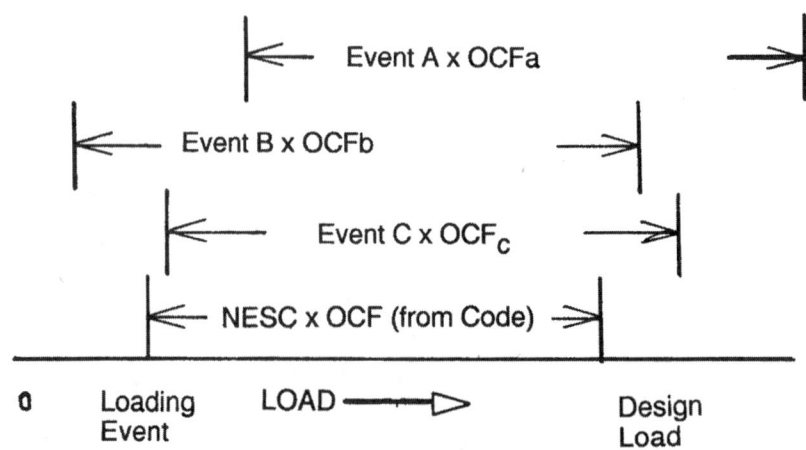

FIGURE 4.14 Development of a loading agenda.

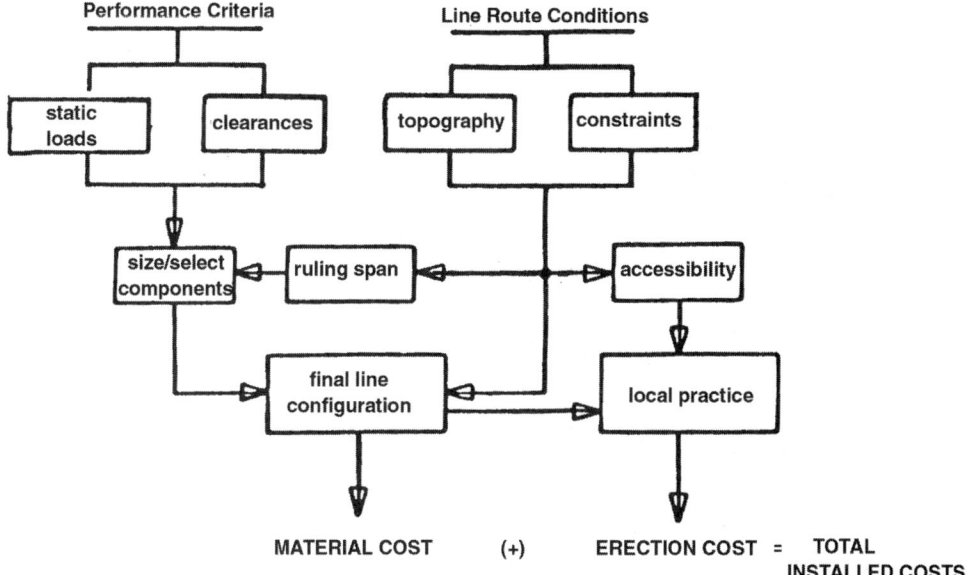

FIGURE 4.15 Search for cost effectiveness.

load at a probability-of-occurrence. That component along the r-o-w that has the highest probability of occurrence of failure from a loading event becomes the weak link in the structure design and establishes the reliability level for the total line section. Since different components are made from different materials that have different response characteristics and that wear, age, and deteriorate at different rates, it is to be expected that the weak link:

- will likely be different in different line designs
- will likely be different in different site locations within the same line
- can change from one component to another over time

Structure Types in Use

Structures come in a wide variety of styles:

- lattice towers
- cantilevered or guyed poles and masts
- framed structures
- combinations of the above

They are available in a wide variety of materials:

- Metal
    galvanized steel and aluminum rods, bars and rolled shapes
    fabricated plate
    tubes
- Concrete
    spun with pretensioned or post-tensioned reinforcing cable
    statically cast nontensioned reinforcing steel
    single or multiple piece

- Wood
    - as grown
    - glued laminar
- Plastics
- Composites
- Crossarms and braces
- Variations of all of the above

Depending on their style and material contents, structures vary considerably in how they respond to load. Some are rigid. Some are flexible. Those structures that can safely deflect under load and absorb energy while doing so, provide an ameliorating influence on progressive damage after the failure of the first element (Pohlman and Lummis, 1969).

Factors Affecting Structure Type Selection

There are usually many factors that impact on the selection of the structure type for use in an OHTL. Some of the more significant are briefly identified below.

Erection Technique: It is obvious that different structure types require different erection techniques. As an example, steel lattice towers consist of hundreds of individual members that must be bolted together, assembled, and erected onto the four previously installed foundations. A tapered steel pole, on the other hand, is likely to be produced in a single piece and erected directly on its previously installed foundation in one hoist. The lattice tower requires a large amount of labor to accomplish the considerable number of bolted joints, whereas the pole requires the installation of a few nuts applied to the foundation anchor bolts plus a few to install the crossarms. The steel pole requires a large-capacity crane with a high reach which would probably not be needed for the tower. Therefore, labor needs to be balanced against the need for large, special equipment and the site's accessibility for such equipment.

Public Concerns: Probably the most difficult factors to deal with arise as a result of the concerns of the general public living, working, or coming in proximity to the line. It is common practice to hold public hearings as part of the approval process for a new line. Such public hearings offer a platform for neighbors to express individual concerns that generally must be satisfactorily addressed before the required permit will be issued. A few comments demonstrate this problem.

The general public usually perceives transmission structures as "eyesores" and distractions in the local landscape. To combat this, an industry study was made in the late 1960s (Dreyfuss, 1968) sponsored by the Edison Electric Institute and accomplished by Henry Dreyfuss, the internationally recognized industrial designer. While the guidelines did not overcome all the objections, they did provide a means of satisfying certain very highly controversial installations (Pohlman and Harris, 1971).

Parents of small children and safety engineers often raise the issue of lattice masts, towers, and guys, constituting an "attractive challenge" to determined climbers, particularly youngsters.

Inspection, Assessment, and Maintenance: Depending on the owning utility, it is likely their in-house practices will influence the selection of the structure type for use in a specific line location. Inspections and assessment are usually made by human inspectors who use diagnostic technologies to augment their personal senses of sight and touch. The nature and location of the symptoms of critical interest are such that they can be most effectively examined from specific perspectives. Inspectors must work from the most advantageous location when making inspections. Methods can include observations from ground or fly-by patrol, climbing, bucket trucks, or helicopters. Likewise, there are certain maintenance activities that are known or believed to be required for particular structure types. The equipment necessary to maintain the structure should be taken into consideration during the structure type selection process to assure there will be no unexpected conflict between maintenance needs and r-o-w restrictions.

Future Upgrading or Uprating: Because of the difficulty of procuring r-o-w's and obtaining the necessary permits to build new lines, many utilities improve their future options by selecting structure types for current line projects that will permit future upgrading and/or uprating initiatives.

Transmission System

```
TANGENT AND LIGHT ANGLE SUSPENSION TOWER — 345 DOUBLE CIRCUIT
OHGW:          Two 7/16" diameter galvanized steel strand
Conductors:    Six twin conductor bundles of 1431 KCM 45/7 ACSR
Weight span:   1,650 feet
Wind span:     1,100 feet
Line angle:    0° to 2°
```

Load Case	Load Event	Radial Ice (")	Wind Pressure Wire (psf)	Wind Pressure Structure (psf)	Load Direction	OCF
1	NESC Heavy	1/2	4	5.1	T	2.54
					L	1.65
					V	1.27
2	One broken OHGW combined with wind and ice	1/2	8	13.0	T	1.0
					L	1.0
					V	1.0
3	One broken conductor bundle combined with wind and ice	1/2	8	13.0	T	1.0
					L	1.0
					V	1.0
4	Heavy wind	0	16	42.0	T	1.0
					L	1.0
					V	1.0
5	Wind on bare tower (no conductors or OHGW)	0	0	46.2	T	1.0
					L	1.0
					V	1.0
6	Vertical load at any OHGW support of 3780 lbs. (not simultaneously)	0	0	0	V	1.0
7	Vertical load at any conductor support of 17,790 lbs. (not simultaneously)	0	0	0	V	1.0

FIGURE 4.16   Example of loading agenda.

Current Deterministic Design Practice

Figure 4.16 shows a loading agenda for a double-circuit, 345-kV line built in the upper Midwest region of the U.S. on steel lattice towers. Over and above the requirements of the NESC, the utility had specified these loading events:

- a heavy wind condition (Pohlman and Harris, 1971)
- a wind on bare tower (Carton and Peyrot, 1992)
- two maximum vertical loads on the OHGW and conductor supports (Osterdorp, 1998; CIGRE, 1995)
- two broken wire contingencies (Pohlman and Lummis, 1969; Dreyfuss, 1968)

It was expected that this combination of loading events would result in a structural support design with the capability of sustaining 50-year recurrence loads likely to occur in the general area where the line was built. Figure 4.17 shows that different members of the structure, as designed, were under the control of different loading cases from this loading agenda. While interesting, this does not:

- provide a way to identify weak links in the support structure
- provide a means for predicting performance of the line system
- provide a framework for decision-making

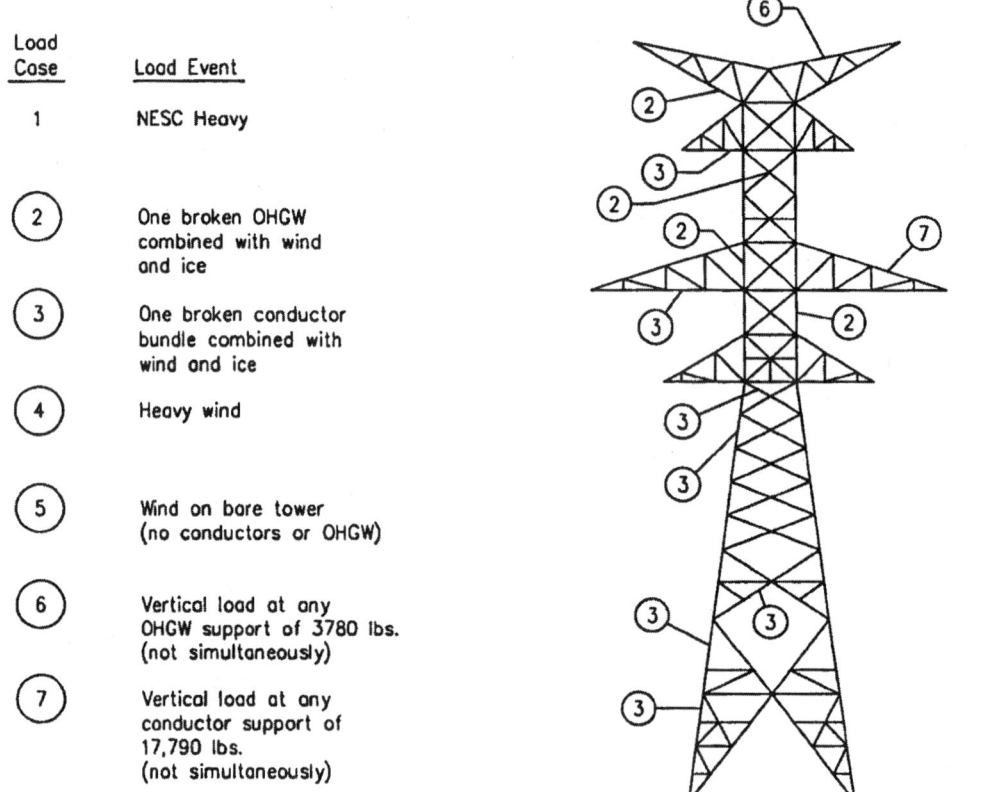

Load Case	Load Event
1	NESC Heavy
2	One broken OHGW combined with wind and ice
3	One broken conductor bundle combined with wind and ice
4	Heavy wind
5	Wind on bare tower (no conductors or OHGW)
6	Vertical load at any OHGW support of 3780 lbs. (not simultaneously)
7	Vertical load at any conductor support of 17,790 lbs. (not simultaneously)

FIGURE 4.17 Results of deterministic design.

Reliability Level

The shortcomings of deterministic design can be demonstrated by using 3D modeling/simulation technology which is in current use (Carton and Peyrot, 1992) in forensic investigation of line failures. The approach is outlined in Fig. 4.18. After the structure (as designed) is properly modeled, loading events of increasing magnitude are analytically applied from different directions until the actual critical capacity for each key member of interest is reached. The probability of occurrence for those specific loading events can then be predicted for the specific location of that structure within that line section by professionals skilled in the art of micrometerology.

Figure 4.19 shows a few of the key members in the example for Fig. 4.17:

- The legs had a probability of failure in that location of once in 115 years.
- Tension chords in the conductor arm and OHGW arm had probabilities of failure of 110 and 35 years, respectively.
- A certain wind condition at an angle was found to be critical for the foundation design with a probability of occurrence at that location of once in 25 years.

Some interesting observations can be drawn:

- The legs were conservatively designed.
- The loss of an OHGW is a more likely event than the loss of a conductor.
- The foundation was found to be the weak link.

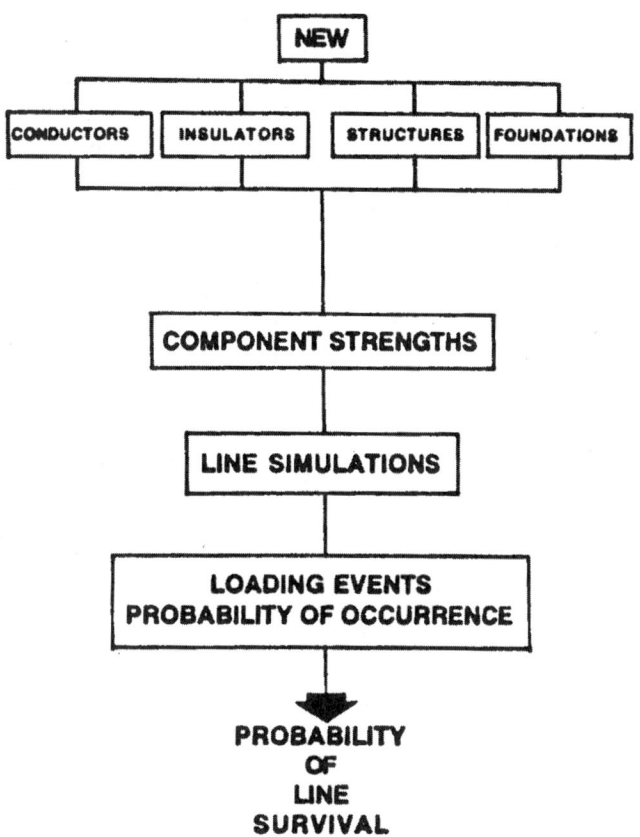

FIGURE 4.18 Line simulation study.

Member	Controlling Climatic Load Condition	Controlling Load Return Period (years)
Legs	Wind, no ice	115
Tension chord of conductor arm	Ice, no wind	110
Tension chord of OHGW arm	Ice, no wind	35
Foundation	Wind, no ice	25

Controlling Climatic Loads

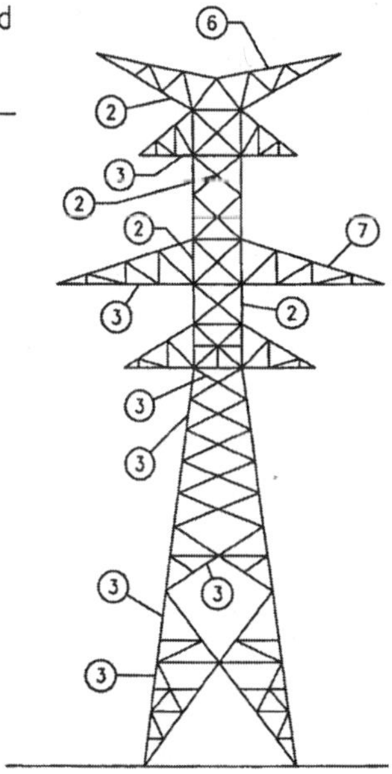

FIGURE 4.19 Simulation study output.

In addition to the interesting observations on relative reliability levels of different components within the structural support system, the output of the simulation study also provides the basis for a decision-making process which can be used to determine the cost effectiveness of management initiatives. Under the simple laws of statistics, when there are two independent outcomes to an event, the probability of the first outcome is equal to one minus the probability of the second. When these outcomes are survival and failure:

$$\text{Annual probability of survival} = 1 - \text{Annual probability of failure}$$
$$Ps = 1 - Pf \tag{4.1}$$

If it is desired to know what the probability of survival is over an extended length of time, i.e., n years of service life:

$$[Ps1 \times Ps2 \times Ps3 \times \ldots Psn] = (ps)n \tag{4.2}$$

Applying this principle to the components in the deterministic structure design and considering a 50-year service life as expected by the designers:

- the legs had a Ps of 65%
- the tension chord in the conductor arm had a Ps of 63%
- the tension chord of the OHGW arm had a Ps of 23%
- the foundation had a Ps of 13%

Security Level

It should be remembered, however, that the failure of every component does not necessarily progress into extensive damage. A comparison of the total risk that would result from the initial failure of components of interest can be accomplished by making a security-level check of the line design (Osterdorp, 1998).

Since the OHTL is a contiguous mechanical system, the forces from the conductors and OHGWs on one side of each tangent structure are balanced and restrained by those on the other side. When a critical component in the conductor/OHGW system fails, energy stored within the conductor system is released suddenly and sets up unbalanced transients that can cause failure of critical components at the next structure. This can set off a cascading effect that will continue to travel downline until it encounters a point in the line strong enough to withstand the unbalance. Unfortunately, a security check of the total line cannot be accomplished from the information describing the one structure in Fig. 4.17; but perhaps some generalized observations can be drawn for demonstration purposes.

Since the structure was designed for broken conductor bundle and broken OHGW contingencies, it appears the line would not be subjected to a cascade from a broken bare conductor, but what if the conductor was coated with ice at the time? Since ice increases the energy trapped within the conductor prior to release, it might be of interest to determine how much ice would be "enough." Three-dimensional modeling would be employed to simulate ice coating of increasing thicknesses until the critical amount is defined. A proper micrometerological study could then identify the probability of occurrence of a storm system capable of delivering that amount of ice at that specific location.

In the example, a wind condition with no ice was identified that would be capable of causing foundation failure once every 25 years. A security-level check would predict the amount of resulting losses and damages that would be expected from this initiating event compared to the broken-conductor-under-ice-load contingencies.

Improved Design Approaches

The above discussion indicates that technologies are available today for assessing the true capability of an OHTL that was created using the conventional practice of specifying ultimate static loads and designing a structure that would properly support them. Because there are many different structure types made from different materials, this was not always straightforward. Accordingly, many technical societies prepared guidelines on how to design the specific structure needed. These are listed in the accompanying references. The interested reader should realize that these documents are subject to periodic review and revision and should, therefore, seek the most current version.

While the technical fraternity recognizes that the mentioned technologies are useful for analyzing existing lines and determining management initiatives, something more direct for designing new lines is needed. There are many efforts under way. The most promising of these is *Improved Design Criteria of OHTLs Based on Reliability Concepts* (Ostendorp, 1998), currently under development by CIGRE Study Committee 22: Recommendations for Overhead Lines. Appendix A outlines the methodology involved in words and in a diagram. The technique is based on the premise that loads and strengths are stochastic variables and the combined reliability is computable if the statistical functions of loads and strength are known. The referenced report has been circulated internationally for trial use and comment. It is expected that the returned comments will be carefully considered, integrated into the report, and the final version submitted to the International Electrotechnical Commission (IEC) for consideration as an International Standard.

References

1. Carton, T. and Peyrot, A., Computer Aided Structural and Geometric Design of Power Lines, *IEEE Trans. on Power Line Syst.*, 7(1), 1992.
2. Dreyfuss, H., *Electric Transmission Structures*, Edison Electric Institute Publication No. 67-61, 1968.
3. Guide for the Design and Use of Concrete Poles, ASCE 596-6, 1987.
4. Guide for the Design of Prestressed Concrete Poles, ASCE/PCI Joint Commission on Concrete Poles, February, 1992. Draft.
5. Guide for the Design of Transmission Towers, *ASCE Manual on Engineering Practice*, 52, 1988.
6. Guide for the Design Steel Transmission Poles, *ASCE Manual on Engineering Practice*, 72, 1990.
7. IEEE Trial-Use Design Guide for Wood Transmission Structures, IEEE Std. 751, February, 1991.
8. *Improved Design Criteria of Overhead Transmission Lines Based on Reliability Concepts*, CIGRE SC-22 Report, October 1995.
9. *National Electrical Safety Code ANSI C-2*, IEEE, 1990.
10. Ostendorp, M., Longitudinal Loading and Cascading Failure Assessment for Transmission Line Upgrades, *ESMO Conference '98*, Orlando, Florida, April 26-30, 1998.
11. Pohlman, J. and Harris, W., Tapered Steel H-Frames Gain Acceptance Through Scenic Valley, *Electric Light and Power Magazine*, 48(vii), 55-58, 1971.
12. Pohlman, J. and Lummis, J., Flexible Structures Offer Broken Wire Integrity at Low Cost, *Electric Light and Power*, 46(V, 144-148.4), 1969.

Appendix A — General Design Criteria — Methodology

The recommended methodology for designing transmission line components is summarized in Fig. 4.20 and can be described as follows:

a) Gather preliminary line design data and available climatic data.[1]
b1) Select the reliability level in terms of return period of design loads. (Note: Some national regulations and/or codes of practice sometimes impose design requirements, directly or indirectly, that may restrict the choice offered to designers).
b2) Select the security requirements (failure containment).
b3) List safety requirements imposed by mandatory regulations and construction and maintenance loads.
c) Calculate climatic variables corresponding to selected return period of design loads.

---

[1] In some countries, design wind speed, such as the 50-year return period, is given in National Standards.

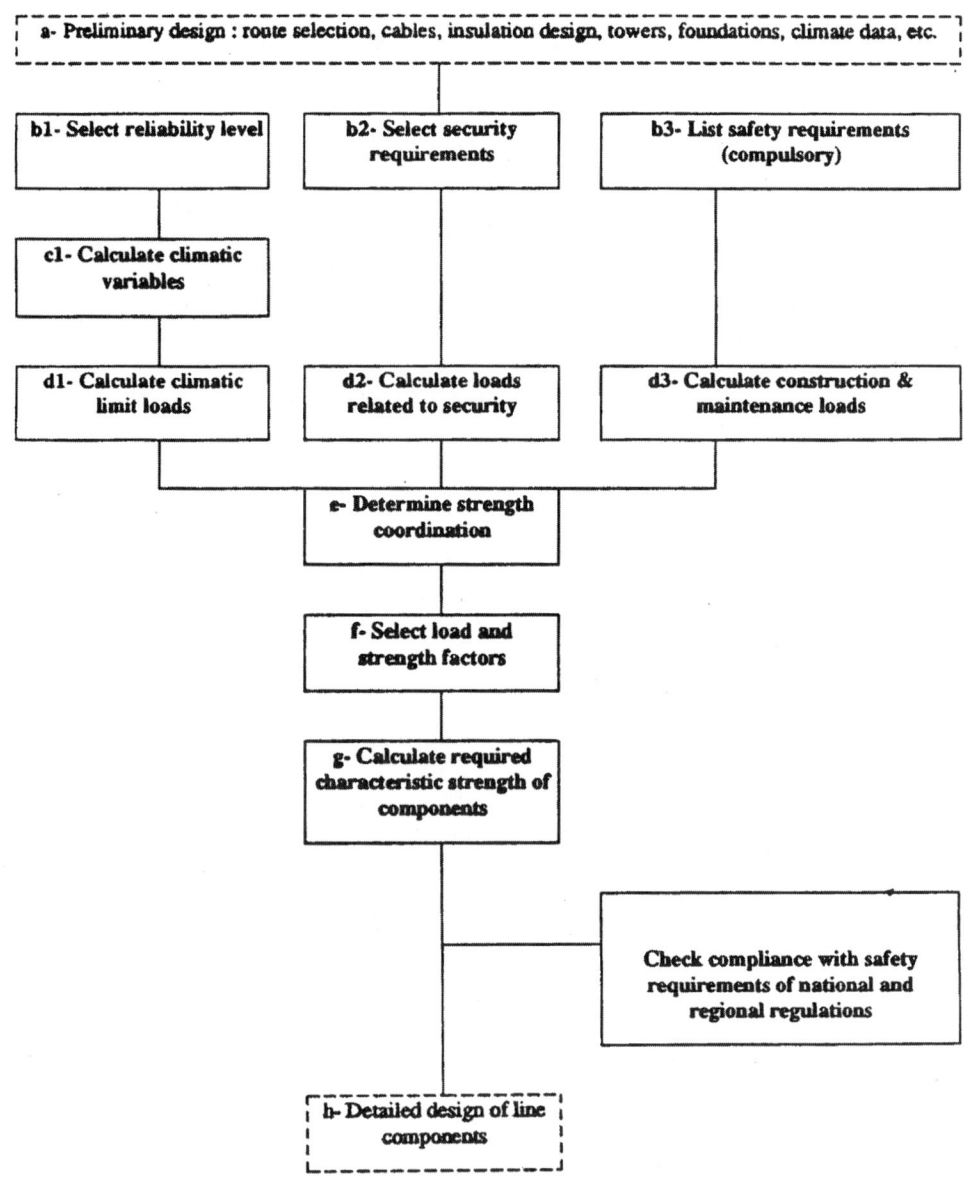

FIGURE 4.20 Methodology.

d1) Calculate climatic limit loadings on components
d2) Calculate loads corresponding to security requirements.
d3) Calculate loads related to safety requirements during construction and maintenance.
 e) Determine the suitable strength coordination between line components.
 f) Select appropriate load and strength factors applicable to load and strength equations.
 g) Calculate the characteristic strengths required for components.
 h) Design line components for the above strength requirements.

This document deals with items b) to g). Items a) and h) are not part of the scope of this document. They are identified by a dotted frame in Fig. 4.20.

Source: Improved design criteria of overhead transmission lines based on reliability concepts, *CIGRE SC22 Report,* October, 1995.

4.3 Insulators and Accessories

George G. Karady and R.G. Farmer

Electric insulation is a vital part of an electrical power system. Although the cost of insulation is only a small fraction of the apparatus or line cost, line performance is highly dependent on insulation integrity. Insulation failure may cause permanent equipment damage and long-term outages. As an example, a short circuit in a 500-kV system may result in a loss of power to a large area for several hours. The potential financial losses emphasize the importance of a reliable design of the insulation.

The insulation of an electric system is divided into two broad categories:

1. Internal insulation
2. External insulation

Apparatus or equipment has mostly internal insulation. The insulation is enclosed in a grounded housing which protects it from the environment. External insulation is exposed to the environment. A typical example of internal insulation is the insulation for a large transformer where insulation between turns and between coils consists of solid (paper) and liquid (oil) insulation protected by a steel tank. An overvoltage can produce internal insulation breakdown and a permanent fault.

External insulation is exposed to the environment. Typical external insulation is the porcelain insulators supporting transmission line conductors. An overvoltage caused by flashover produces only a temporary fault. The insulation is self-restoring.

This section discusses external insulation used for transmission lines and substations.

Electrical Stresses on External Insulation

The external insulation (transmission line or substation) is exposed to electrical, mechanical, and environmental stresses. The applied voltage of an operating power system produces electrical stresses. The weather and the surroundings (industry, rural dust, oceans, etc.) produce additional environmental stresses. The conductor weight, wind, and ice can generate mechanical stresses. The insulators must withstand these stresses for long periods of time. It is anticipated that a line or substation will operate for more than 20–30 years without changing the insulators. However, regular maintenance is needed to minimize the number of faults per year. A typical number of insulation failure-caused faults is 0.5–10 per year, per 100 mi of line.

Transmission Lines and Substations

Transmission line and substation insulation integrity is one of the most dominant factors in power system reliability. We will describe typical transmission lines and substations to demonstrate the basic concept of external insulation application.

Figures 4.21 shows a high-voltage transmission line. The major components of the line are:

1. Conductors
2. Insulators
3. Support structure tower

The insulators are attached to the tower and support the conductors. In a suspension tower, the insulators are in a vertical position or in a V-arrangement. In a dead-end tower, the insulators are in a horizontal position. The typical transmission line is divided into sections and two dead-end towers terminate each section. Between 6 and 15 suspension towers are installed between the two dead-end towers. This sectionalizing prevents the propagation of a catastrophic mechanical fault beyond each section. As an example, a tornado caused collapse of one or two towers could create a domino effect, resulting in the collapse of many miles of towers, if there are no dead ends.

FIGURE 4.21 A 500-kV suspension tower with V string insulators.

Figure 4.22 shows a lower voltage line with post-type insulators. The rigid, slanted insulator supports the conductor. A high-voltage substation may use both suspension and post-type insulators.

Electrical Stresses

The electrical stresses on insulation are created by:

1. Continuous power frequency voltages
2. Temporary overvoltages
3. Switching overvoltages
4. Lightning overvoltages

Continuous Power Frequency Voltages

The insulation has to withstand normal operating voltages. The operating voltage fluctuates from changing load. The normal range of fluctuation is around ±10%. The line-to-ground voltage causes the voltage stress on the insulators. As an example, the insulation requirement of a 220-kV line is at least:

$$1.1 \times \frac{220 \text{ kV}}{\sqrt{3}} \cong 140 \text{ kV} \tag{4.3}$$

FIGURE 4.22 69-kV transmission line with post insulators.

This voltage is used for the selection of the number of insulators when the line is designed. The insulation can be laboratory tested by measuring the dry flashover voltage of the insulators. Because the line insulators are self-restoring, flashover tests do not cause any damage. The flashover voltage must be larger than the operating voltage to avoid outages. For a porcelain insulator, the required dry flashover voltage is about 2.5–3 times the rated voltage. A significant number of the apparatus standards recommend dry withstand testing of every kind of insulation to be two (2) times the rated voltage plus 1 kV for 1 min of time. This severe test eliminates most of the deficient units.

Temporary Overvoltages

These include ground faults, switching, load rejection, line energization and resonance, cause power frequency, or close-to-power frequency, and relatively long duration overvoltages. The duration is from 5 sec to several minutes. The expected peak amplitudes and duration are listed in Table 4.1.

The base is the crest value of the rated voltage. The dry withstand test, with two times the maximum operating voltage plus 1 kV for 1 minute, is well-suited to test the performance of insulation under temporary overvoltages.

TABLE 4.1 Expected Amplitude of Temporary Overvoltages

Type of Overvoltage	Expected Amplitude	Duration
Fault overvoltages		
Effectively grounded	1.3 per unit	1 sec
Resonant grounded	1.73 per unit or greater	10 sec
Load rejection		
System substation	1.2 per unit	1–5 sec
Generator station	1.5 per unit	3 sec
Resonance	3 per unit	2–5 min
Transformer energization	1.5–2.0 per unit	1–20 sec

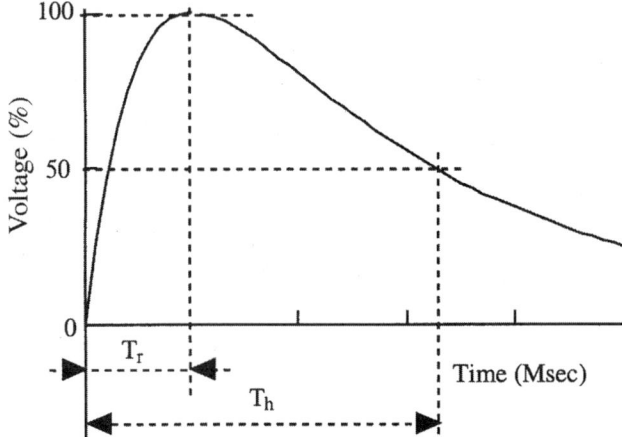

FIGURE 4.23 Switching overvoltages. T_r = 20–5000 μsec, T_h < 20,000 μsec, where T_r is the time-to-crest value and T_h is the time-to-half value.

Switching Overvoltages

The opening and closing of circuit breakers causes switching overvoltages. The most frequent causes of switching overvoltages are fault or ground fault clearing, line energization, load interruption, interruption of inductive current, and switching of capacitors.

Switching produces unidirectional or oscillatory impulses with durations of 5000–20,000 μsec. The amplitude of the overvoltage varies between 1.8 and 2.5 per unit. Some modern circuit breakers use pre-insertion resistance, which reduces the overvoltage amplitude to 1.5–1.8 per unit. The base is the crest value of the rated voltage.

Switching overvoltages are calculated from computer simulations that can provide the distribution and standard deviation of the switching overvoltages. Figure 4.23 shows typical switching impulse voltages. Switching surge performance of the insulators is determined by flashover tests. The test is performed by applying a standard impulse with a time to crest of 250 μsec and time to half value of 5000 μsec. The test is repeated 20 times at different voltage levels and the number of flashovers is counted at each voltage level. These represent the statistical distribution of the switching surge impulse flashover probability. The correlation of the flashover probability with the calculated switching impulse voltage distribution gives the probability, or risk, of failure. The measure of the risk of failure is the number of flashovers expected by switching surges per year.

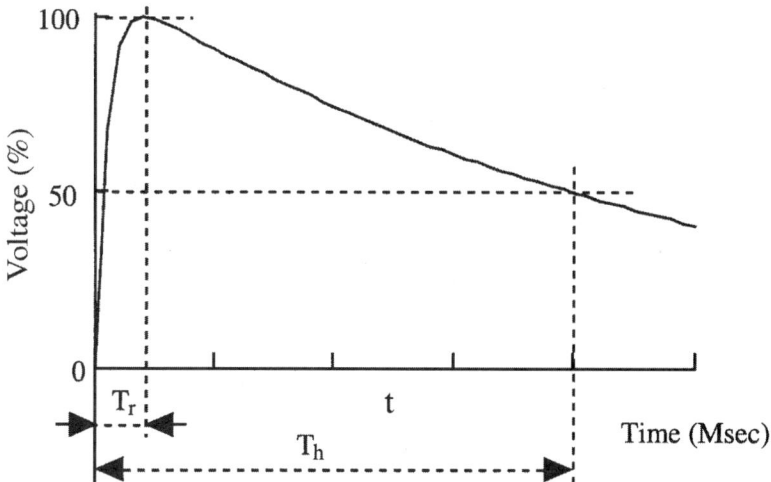

FIGURE 4.24 Lightning overvoltages. T_r = 0.1–20 μsec, T_h 20–200 μsec, where T_r is the time-to-crest value and T_h is the time-to-half value.

Lightning Overvoltages
Lightning overvoltages are caused by lightning strikes:

1. to the phase conductors
2. to the shield conductor (the large current-caused voltage drop in the grounding resistance may cause flashover to the conductors [back flash]).
3. to the ground close to the line (the large ground current induces voltages in the phase conductors).

Lighting strikes cause a fast-rising, short-duration, unidirectional voltage pulse. The time-to-crest is between 0.1–20 μsec. The time-to-half value is 20–200 μsec.

The peak amplitude of the overvoltage generated by a direct strike to the conductor is very high and is practically limited by the subsequent flashover of the insulation. Shielding failures and induced voltages cause somewhat less overvoltage. Shielding failure caused overvoltage is around 500 kV–2000 kV. The lightning-induced voltage is generally less than 400 kV. The actual stress on the insulators is equal to the impulse voltage.

The insulator BIL is determined by using standard lightning impulses with a time-to-crest value of 1.2 μsec and time-to-half value of 50 μsec. This is a measure of the insulation strength for lightning. Figure 4.24 shows a typical lightning pulse.

When an insulator is tested, peak voltage of the pulse is increased until the first flashover occurs. Starting from this voltage, the test is repeated 20 times at different voltage levels and the number of flashovers are counted at each voltage level. This provides the statistical distribution of the lightning impulse flashover probability of the tested insulator.

Environmental Stresses

Most environmental stress is caused by weather and by the surrounding environment, such as industry, sea, or dust in rural areas. The environmental stresses affect both mechanical and electrical performance of the line.

Temperature
The temperature in an outdoor station or line may fluctuate between −50°C and +50°C, depending upon the climate. The temperature change has no effect on the electrical performance of outdoor insulation. It is believed that high temperatures may accelerate aging. Temperature fluctuation causes an increase of mechanical stresses, however it is negligible when well-designed insulators are used.

UV Radiation

UV radiation accelerates the aging of nonceramic composite insulators, but has no effect on porcelain and glass insulators. Manufacturers use fillers and modified chemical structures of the insulating material to minimize the UV sensitivity.

Rain

Rain wets porcelain insulator surfaces and produces a thin conducting layer most of the time. This reduces the flashover voltage of the insulators. As an example, a 230-kV line may use an insulator string with 12 standard ball-and-socket-type insulators. Dry flashover voltage of this string is 665 kV and the wet flashover voltage is 502 kV. The percentage reduction is about 25%.

Nonceramic polymer insulators have a water-repellent hydrophobic surface that reduces the effects of rain. As an example, with a 230-kV composite insulator, dry flashover voltage is 735 kV and wet flashover voltage is 630 kV. The percentage reduction is about 15%. The insulator's wet flashover voltage must be higher than the maximum temporary overvoltage.

Icing

In industrialized areas, conducting water may form ice due to water-dissolved industrial pollution. An example is the ice formed from acid rain water. Ice deposits form bridges across the gaps in an insulator string that result in a solid surface. When the sun melts the ice, a conducting water layer will bridge the insulator and cause flashover at low voltages. Melting ice-caused flashover has been reported in the Quebec and Montreal areas.

Pollution

Wind drives contaminant particles into insulators. Insulators produce turbulence in airflow, which results in the deposition of particles on their surfaces. The continuous depositing of the particles increases the thickness of these deposits. However, the natural cleaning effect of wind, which blows loose particles away, limits the growth of deposits. Occasionally, rain washes part of the pollution away. The continuous depositing and cleaning produces a seasonal variation of the pollution on the insulator surfaces. However, after a long time (months, years), the deposits are stabilized and a thin layer of solid deposit will cover the insulator. Because of the cleaning effects of rain, deposits are lighter on the top of the insulators and heavier on the bottom. The development of a continuous pollution layer is compounded by chemical changes. As an example, in the vicinity of a cement factory, the interaction between the cement and water produces a tough, very sticky layer. Around highways, the wear of car tires produces a slick, tar-like carbon deposit on the insulator's surface.

Moisture, fog, and dew wet the pollution layer, dissolve the salt, and produce a conducting layer, which in turn reduces the flashover voltage. The pollution can reduce the flashover voltage of a standard insulator string by about 20–25%.

Near the ocean, wind drives salt water onto insulator surfaces, forming a conducting salt-water layer which reduces the flashover voltage. The sun dries the pollution during the day and forms a white salt layer. This layer is washed off even by light rain and produces a wide fluctuation in pollution levels.

The Equivalent Salt Deposit Density (ESDD) describes the level of contamination in an area. Equivalent Salt Deposit Density is measured by periodically washing down the pollution from selected insulators using distilled water. The resistivity of the water is measured and the amount of salt that produces the same resistivity is calculated. The obtained mg value of salt is divided by the surface area of the insulator. This number is the ESDD. The pollution severity of a site is described by the average ESDD value, which is determined by several measurements.

Table 4.2 shows the criteria for defining site severity.

The contamination level is light or very light in most parts of the U.S. and Canada. Only the seashores and heavily industrialized regions experience heavy pollution.

TABLE 4.2 Site Severity (IEEE Definitions)

Description	ESDD (mg/cm^2)
Very light	0–0.03
Light	0.03–0.06
Moderate	0.06–0.1
Heavy	<0.1

TABLE 4.3 Typical Sources of Pollution

Pollution Type	Source of Pollutant	Deposit Characteristics	Area
Rural areas	Soil dust	High resitivity layer, effective rain washing	Large areas
Desert	Sand	Low resistivity	Large areas
Coastal area	Sea salt	Very low resistivity, easily washed by rain	10–20 km from the sea
Industrial	Steel mill, coke plants, chemical plants, generating stations, quarries	High conductivity, extremely difficult to remove, insoluble	Localized to the plant area
Mixed	Industry, highway, desert	Very adhesive, medium resistivity	Localized to the plant area

Typically, the pollution level is very high in Florida and on the southern coast of California. Heavy industrial pollution occurs in the industrialized areas and near large highways. Table 4.3 gives a summary of the different sources of pollution.

The flashover voltage of polluted insulators has been measured in laboratories. The correlation between the laboratory results and field experience is weak. The test results provide guidance, but insulators are selected using practical experience.

Altitude

The insulator's flashover voltage is reduced as altitude increases. Above 1500 feet, an increase in the number of insulators should be considered. A practical rule is a 3% increase of clearance or insulator strings' length per 1000 ft as the elevation increases.

Mechanical Stresses

Suspension insulators need to carry the weight of the conductors and the weight of occasional ice and wind loading.

In northern areas and in higher elevations, insulators and lines are frequently covered by ice in the winter. The ice produces significant mechanical loads on the conductor and on the insulators. The transmission line insulators need to support the conductor's weight and the weight of the ice in the adjacent spans. This may increase the mechanical load by 20–50%.

The wind produces a horizontal force on the line conductors. This horizontal force increases the mechanical load on the line. The wind-force-produced load has to be added vectorially to the weight-produced forces. The design load will be the larger of the combined wind and weight, or ice and wind load.

The dead-end insulators must withstand the longitudinal load, which is higher than the simple weight of the conductor in the half span.

A sudden drop in the ice load from the conductor produces large amplitude mechanical oscillations, which cause periodic oscillatory insulator loading (stress changes from tension to compression and back).

The insulator's one-minute tension strength is measured and used for insulator selection. In addition, each cap-and-pin or ball-and-socket insulator is loaded mechanically for one minute and simultaneously energized. This mechanical and electrical (M&E) value indicates the quality of insulators. The maximum load should be around 50% of the M&E load.

The Bonneville Power Administration uses the following practical relation to determine the required M&E rating of the insulators.

1. M&E > 5∗ Bare conductor weight/span
2. M&E > Bare conductor weight + Weight of 3.81 cm (1.5 in) of ice on the conductor (3 lb/sq ft)
3. M&E > 2∗ (Bare conductor weight + Weight of 0.63 cm (1/4 in) of ice on the conductor and loading from a wind of 1.8 kg/sq ft (4 lb/sq ft)

The required M&E value is calculated from all equations above and the largest value is used.

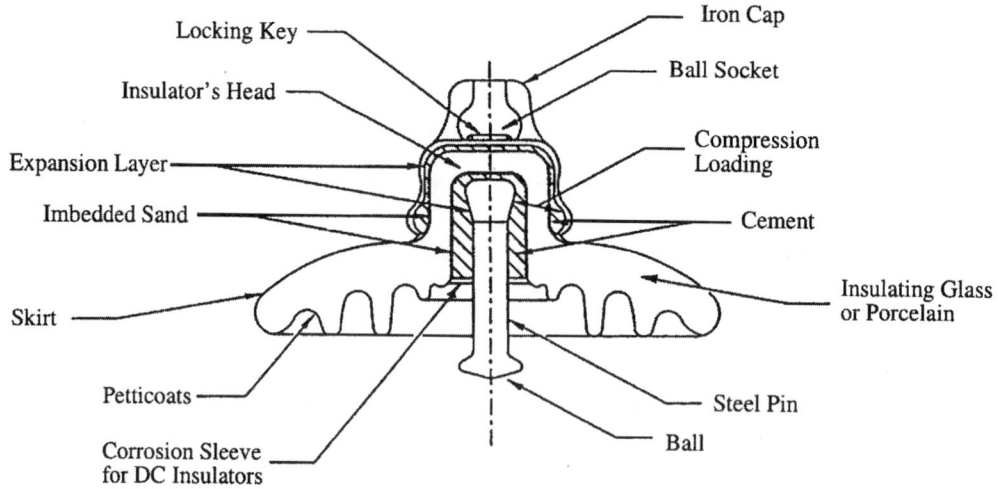

FIGURE 4.25 Cross-section of a standard ball-and-socket insulator.

Ceramic (Porcelain and Glass) Insulators

Materials

Porcelain is the most frequently used material for insulators. Insulators are made of wet, processed porcelain. The fundamental materials used are a mixture of feldspar (35%), china clay (28%), flint (25%), ball clay (10%), and talc (2%). The ingredients are mixed with water. The resulting mixture has the consistency of putty or paste and is pressed into a mold to form a shell of the desired shape. The alternative method is formation by extrusion bars that are machined into the desired shape. The shells are dried and dipped into a glaze material. After glazing, the shells are fired in a kiln at about 1200°C. The glaze improves the mechanical strength and provides a smooth, shiny surface. After a cooling-down period, metal fittings are attached to the porcelain with Portland cement.

Toughened glass is also frequently used for insulators. The melted glass is poured into a mold to form the shell. Dipping into hot and cold baths cools the shells. This thermal treatment shrinks the surface of the glass and produces pressure on the body, which increases the mechanical strength of the glass. Sudden mechanical stresses, such as a blow by a hammer or bullets, will break the glass into small pieces. The metal end-fitting is attached by alumina cement.

Insulator Strings

Most high-voltage lines use ball-and-socket-type porcelain or toughened glass insulators. These are also referred to as "cap and pin." The cross section of a ball-and-socket-type insulator is shown in Fig. 4.25. The porcelain skirt provides insulation between the iron cap and steel pin. The upper part of the porcelain is smooth to promote rain washing and cleaning of the surface. The lower part is corrugated, which prevents wetting and provides a longer protected leakage path. Portland cement attaches the cup and pin. Before the application of the cement, the porcelain is sandblasted to generate a rough surface. A thin expansion layer (e.g., bitumen) covers the metal surfaces. The loading compresses the cement and provides high mechanical strength. The basic technical data of a standard ball-and-socket insulator is as follows:

TABLE 4.4 Technical Data of a Standard Insulator

Diameter	25.4 cm	(10 in.)
Spacing	14.6 cm	(5-3/4 in.)
Leakage distance	305 cm	(12 ft)
Typical operating voltage	10 kV	
Mechanical strength	75 kN	(15 klb)

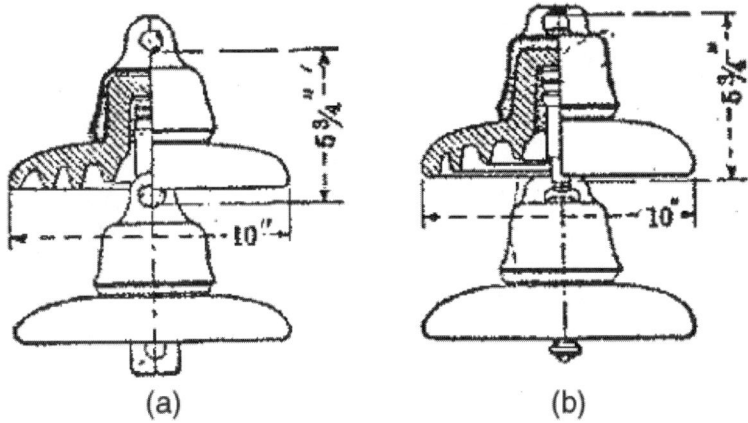

FIGURE 4.26 Insulator string: (a) clevis type, (b) ball-and-socket type.

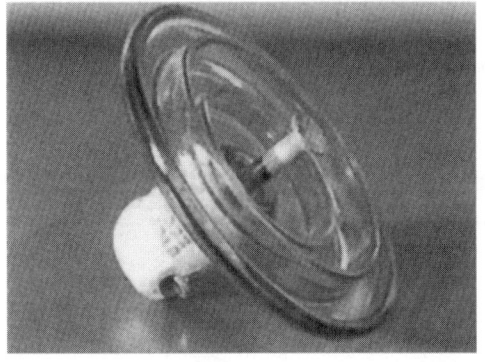 

FIGURE 4.27 Standard and fog-type insulators. (Courtesy of Sediver, Inc., Nanterre Cedex, France.)

The metal parts are designed to fail before the porcelain fails as the mechanical load increases. This acts as a mechanical fuse protecting the tower structure.

The ball-and-socket insulators are attached to each other by inserting the ball in the socket and securing the connection with a locking key. Several insulators are connected together to form an insulator string. Figure 4.26 shows a ball-and-socket insulator string and the clevis-type string, which is used less frequently for transmission lines.

Fog-type, long leakage distance insulators are used in polluted areas, close to the ocean, or in industrial environments. Figure 4.27 shows representative fog-type insulators, the mechanical strength of which is higher than standard insulator strength. As an example, a 6 1/2 × 12 1/2 fog-type insulator is rated to 180 kN (40 klb) and has a leakage distance of 50.1 cm (20 in.).

Insulator strings are used for high-voltage transmission lines and substations. They are arranged vertically on support towers and horizontally on dead-end towers. Table 4.5 shows the typical number of insulators used by utilities in the U.S. and Canada in lightly polluted areas.

TABLE 4.5 Typical Number of Standard (5-1/4 ft × 10 in.) Insulators at Different Voltage Levels

Line Voltage (kV)	Number of Standard Insulators
69	4–6
115	7–9
138	8–10
230	12
287	15
345	18
500	24
765	30–35

Post-Type Insulators

Post-type insulators are used for medium- and low-voltage transmission lines, where insulators replace the cross-arm (Fig. 4.23). However, the majority of post insulators are used in substations where insulators support conductors, bus bars, and equipment. A typical example is the interruption chamber of a live tank circuit breaker. Typical post-type insulators are shown in Fig. 4.28.

Older post insulators are built somewhat similar to cap-and-pin insulators, but with hardware that permits stacking of the insulators to form a high-voltage unit. These units can be found in older stations. Modern post insulators consist of a porcelain column, with weather skirts or corrugation on the outside surface to increase leakage distance. For indoor use, the outer surface is corrugated. For outdoor use, a deeper weather shed is used. The end-fitting seals the inner part of the tube to prevent water penetration. Figure 4.28 shows a representative unit used at a substation. Equipment manufacturers use the large post-type insulators to house capacitors, fiber-optic cables and electronics, current transformers, and operating mechanisms. In some cases, the insulator itself rotates and operates disconnect switches.

Post insulators are designed to carry large compression loads, smaller bending loads, and small tension stresses.

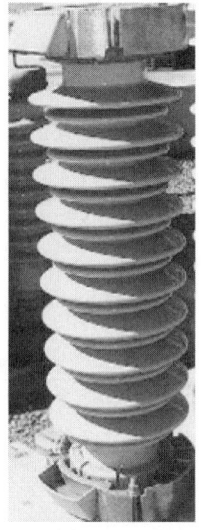

FIGURE 4.28 Post insulators.

Long Rod Insulators

The long rod insulator is a porcelain rod with an outside weather shed and metal end fittings. The long rod is designed for tension load and is applied on transmission lines in Europe. Figure 4.29 shows a typical long rod insulator. These insulators are not used in the U.S. because vandals may shoot the insulators, which will break and cause outages. The main advantage of the long rod design is the elimination of metal parts between the units, which reduces the insulator's length.

Nonceramic (Composite) Insulators

Nonceramic insulators use polymers instead of porcelain. High-voltage composite insulators are built with mechanical load-bearing fiberglass rods, which are covered by polymer weather sheds to assure high electrical strength.

The first insulators were built with bisphenol epoxy resin in the mid-1940s and are still used in indoor applications. Cycloaliphatic epoxy resin insulators were introduced in 1957. Rods with weather sheds were molded and cured to form solid insulators. These insulators were tested and used in England for several years. Most of them were exposed to harsh environmental stresses and failed. However, they have been successfully used indoors. The first composite insulators, with fiberglass rods and rubber weather sheds, appeared in the mid-1960s. The advantages of these insulators are:

- Lightweight, which lowers construction and transportation costs.
- More vandalism resistant.

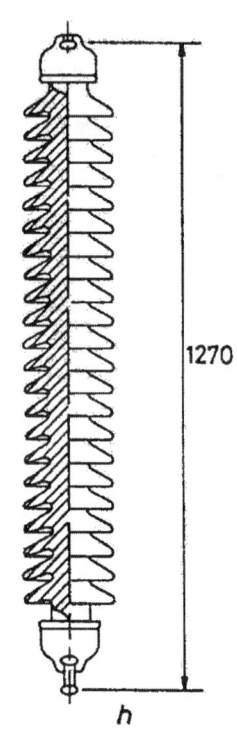

FIGURE 4.29 Long rod insulator.

- Higher strength-to-weight ratio, allowing longer design spans.
- Better contamination performance.
- Improved transmission line aesthetics, resulting in better public acceptance of a new line.

However, early experiences were discouraging because several failures were observed during operation. Typical failures experienced were:

- Tracking and erosion of the shed material, which led to pollution and caused flashover.
- Chalking and crazing of the insulator's surface, which resulted in increased contaminant collection, arcing, and flashover.
- Reduction of contamination flashover strength and subsequent increased contamination-induced flashover.
- Deterioration of mechanical strength, which resulted in confusion in the selection of mechanical line loading.
- Loosening of end fittings.
- Bonding failures and breakdowns along the rod-shed interface.
- Water penetration followed by electrical failure.

As a consequence of reported failures, an extensive research effort led to second- and third-generation nonceramic transmission line insulators. These improved units have tracking free sheds, better corona resistance, and slip-free end fittings. A better understanding of failure mechanisms and of mechanical strength-time dependency has resulted in newly designed insulators that are expected to last 20–30 years. Increased production quality control and automated manufacturing technology has further improved the quality of these third-generation nonceramic transmission line insulators.

Composite Suspension Insulators

A cross-section of a third-generation composite insulators is shown in Fig. 4.30. The major components of a composite insulator are:

- End fittings
- Corona ring(s)
- Fiberglass-reinforced plastic rod
- Interface between shed and sleeve
- Weather shed

End Fittings

End fittings connect the insulator to a tower or conductor. It is a heavy metal tube with an oval eye, socket, ball, tongue, and a clevis ending. The tube is attached to a fiberglass rod. The duty of the end fitting is to provide a reliable, non-slip attachment without localized stress in the fiberglass rod. Different manufacturers use different technologies. Some methods are:

1. The ductile galvanized iron-end fitting is wedged and glued with epoxy to the rod.
2. The galvanized forged steel-end fitting is swaged and compressed to the rod.
3. The malleable cast iron, galvanized forged steel, or aluminous bronze-end fitting is attached to the rod by controlled swaging. The material is selected according to the corrosion resistance requirement. The end fitting coupling zone serves as a mechanical fuse and determines the strength of the insulator.
4. High-grade forged steel or ductile iron is crimped to the rod with circumferential compression.

The interface between the end fitting and the shed material must be sealed to avoid water penetration. Another technique, used mostly in distribution insulators, involves the weather shed overlapping the end fitting.

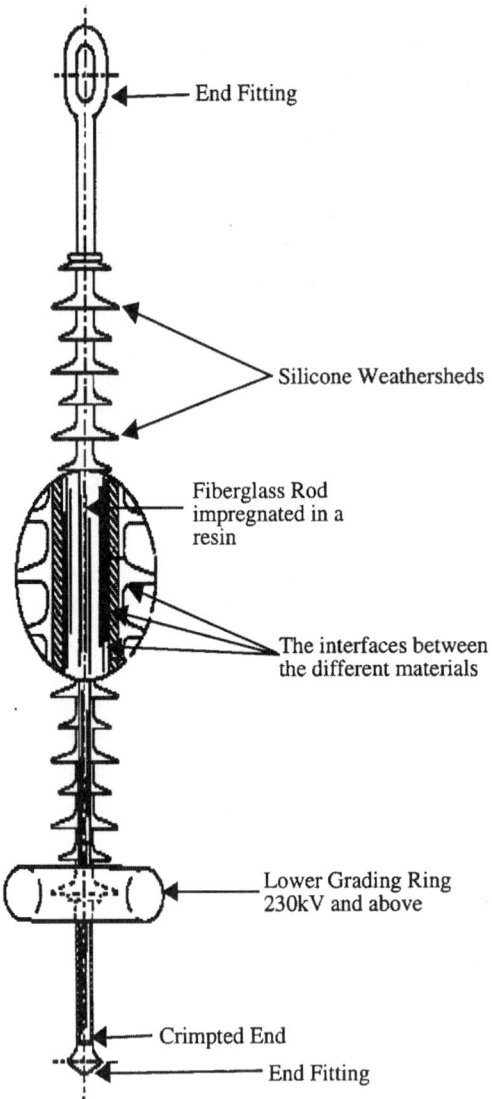

FIGURE 4.30 Cross-section of a typical composite insulator. (*Toughened Glass Insulators.* Sediver, Inc., Nanterre Cedex, France. With permission.)

Corona Ring(s)
Electrical field distribution along a nonceramic insulator is nonlinear and produces very high electric fields near the end of the insulator. High fields generate corona and surface discharges, which are the source of insulator aging. Above 230 kV, each manufacturer recommends aluminum corona rings be installed at the line end of the insulator. Corona rings are used at both ends at higher voltages (>500 kV).

Fiberglass-Reinforced Plastic Rod
The fiberglass is bound with epoxy or polyester resin. Epoxy produces better-quality rods but polyester is less expensive. The rods are manufactured in a continuous process or in a batch mode, producing the required length. The even distribution of the glass fibers assures equal loading, and the uniform impregnation assures good bonding between the fibers and the resin. To improve quality, some manufacturers use E-glass to avoid brittle fractures. Brittle fracture can cause sudden shattering of the rod.

Interfaces Between Shed and Fiberglass Rod

Interfaces between the fiberglass rod and weather shed should have no voids. This requires an appropriate interface material that assures bonding of the fiberglass rod and weather shed. The most frequently used techniques are:

1. The fiberglass rod is primed by an appropriate material to assure the bonding of the sheds.
2. Silicon rubber or ethylene propylene diene monomer (EPDM) sheets are extruded onto the fiberglass rod, forming a tube-like protective covering.
3. The gap between the rod and the weather shed is filled with silicon grease, which eliminates voids.

Weather Shed

All high-voltage insulators use rubber weather sheds installed on fiberglass rods. The interface between the weather shed, fiberglass rod, and the end fittings are carefully sealed to prevent water penetration. The most serious insulator failure is caused by water penetration to the interface.

The most frequently used weather shed technologies are:

1. Ethylene propylene copolymer (EPM) and silicon rubber alloys, where hydrated-alumina fillers are injected into a mold and cured to form the weather sheds. The sheds are threaded to the fiberglass rod under vacuum. The inner surface of the weather shed is equipped with O-ring type grooves filled with silicon grease that seals the rod-shed interface. The gap between the end-fittings and the sheds is sealed by axial pressure. The continuous slow leaking of the silicon at the weather shed junctions prevents water penetration.
2. High-temperature vulcanized silicon rubber (HTV) sleeves are extruded on the fiberglass surface to form an interface. The silicon rubber weather sheds are injection-molded under pressure and placed onto the sleeved rod at a predetermined distance. The complete subassembly is vulcanized at high temperatures in an oven. This technology permits the variation of the distance between the sheds.
3. The sheds are directly injection-molded under high pressure and high temperature onto the primed rod assembly. This assures simultaneous bonding to both the rod and the end-fittings. Both EPDM and silicon rubber are used. This one-piece molding assures reliable sealing against moisture penetration.
4. One piece of silicon or EPDM rubber shed is molded directly to the fiberglass rod. The rubber contains fillers and additive agents to prevent tracking and erosion.

Composite Post Insulators

The construction and manufacturing method of post insulators is similar to that of suspension insulators. The major difference is in the end fittings and the use of a larger diameter fiberglass rod. The latter is necessary because bending is the major load on these insulators. The insulators are flexible, which permits bending in case of sudden overload. A typical post-type insulator used for 69-kV lines is shown in Fig. 4.31.

Post-type insulators are frequently used on transmission lines. Development of station-type post insulators has just begun. The major problem is the fabrication of high strength, large diameter fiberglass tubes and sealing of the weather shed.

Insulator Failure Mechanism

Porcelain Insulators

Cap-and-pin porcelain insulators are occasionally destroyed by direct lightning strikes, which generate a very steep wave front. Steep-front waves break down the porcelain in the cap, cracking the porcelain. The penetration of moisture results in leakage currents and short circuits of the unit.

Mechanical failures also crack the insulator and produce short circuits. The most common cause is water absorption by the Portland cement used to attach the cap to the porcelain. Water absorption

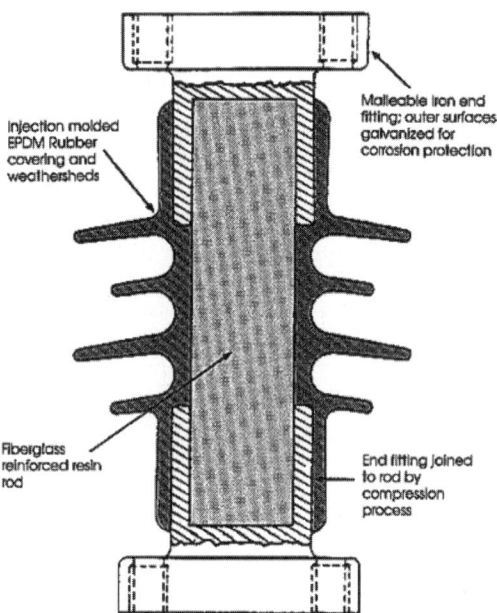

FIGURE 4.31 Post-type composite insulator. (*Toughened Glass Insulators*. Sediver, Inc., Nanterre Cedex, France. With permission.)

expands the cement, which in turn cracks the porcelain. This reduces the mechanical strength, which may cause separation and line dropping.

Short circuits of the units in an insulator string reduce the electrical strength of the string, which may cause flashover in polluted conditions.

Glass insulators use alumina cement, which reduces water penetration and the head-cracking problem. A great impact, such as a bullet, can shatter the shell, but will not reduce the mechanical strength of the unit.

The major problem with the porcelain insulators is pollution, which may reduce the flashover voltage under the rated voltages. Fortunately, most areas of the U.S. are lightly polluted. However, some areas with heavy pollution experience flashover regularly.

Insulator Pollution

Insulation pollution is a major cause of flashovers and of long-term service interruptions. Lightning-caused flashovers produce short circuits. The short circuit current is interrupted by the circuit breaker and the line is reclosed successfully. The line cannot be successfully reclosed after pollution-caused flashover because the contamination reduces the insulation's strength for a long time. Actually, the insulator must dry before the line can be reclosed.

Ceramic Insulators

Pollution-caused flashover is an involved process that begins with the pollution source. Some sources of pollution are: salt spray from an ocean, salt deposits in the winter, dust and rubber particles during the summer from highways and desert sand, industrial emissions, engine exhaust, fertilizer deposits, and generating station emissions. Contaminated particles are carried in the wind and deposited on the insulator's surface. The speed of accumulation is dependent upon wind speed, line orientation, particle size, material, and insulator shape. Most of the deposits lodge between the insulator's ribs and behind the cap because of turbulence in the airflow in these areas (Fig. 4.32).

The deposition is continuous, but is interrupted by occasional rain. Rain washes the pollution away and high winds clean the insulators. The top surface is cleaned more than the ribbed bottom. The

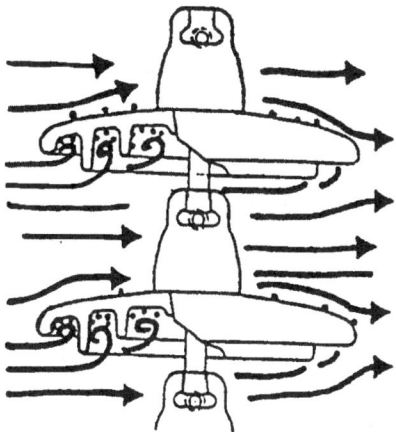

FIGURE 4.32 Deposit accumulation. (*Application Guide for Composite Suspension Insulators.* Sediver, Inc., York, SC, 1993. With permission.)

horizontal and V strings are cleaned better by the rain than the I strings. The deposit on the insulator forms a well-dispersed layer and stabilizes around an average value after longer exposure times. However, this average value varies with the changing of the seasons.

Fog, dew, mist, or light rain wets the pollution deposits and forms a conductive layer. Wetting is dependent upon the amount of dissolvable salt in the contaminant, the nature of the insoluble material, duration of wetting, surface conditions, and the temperature difference between the insulator and its surroundings. At night, the insulators cool down with the low night temperatures. In the early morning, the air temperature begins increasing, but the insulator's temperature remains constant. The temperature difference accelerates water condensation on the insulator's surface. Wetting of the contamination layer starts leakage currents.

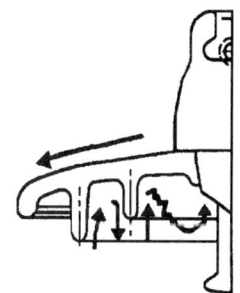

FIGURE 4.33 Dry-band arcing. (*Application Guide for Composite Suspension Insulators.* Sediver, Inc., York, SC, 1993. With permission.)

Leakage current density depends upon the shape of the insulator's surface. Generally, the highest current density is around the pin. The current heats the conductive layer and evaporates the water at the areas with high current density. This leads to the development of dry bands around the pin. The dry bands modify the voltage distribution along the surface. Because of the high resistance of the dry bands, it is across them that most of the voltages will appear. The high voltage produces local arcing. Short arcs (Fig. 4.33) will bridge the dry bands.

Leakage current flow will be determined by the voltage drop of the arcs and by the resistance of the wet layer in series with the dry bands. The arc length may increase or decrease, depending on the layer resistance. Because of the large layer resistance, the arc first extinguishes, but further wetting reduces the resistance, which leads to increases in arc length. In adverse conditions, the level of contamination is high and the layer resistance becomes low because of intensive wetting. After several arcing periods, the length of the dry band will increase and the arc will extend across the insulator. This contamination causes flashover.

In favorable conditions when the level of contamination is low, layer resistance is high and arcing continues until the sun or wind dries the layer and stops the arcing. Continuous arcing is harmless for ceramic insulators, but it ages nonceramic and composite insulators.

The mechanism described above shows that heavy contamination and wetting may cause insulator flashover and service interruptions. Contamination in dry conditions is harmless. Light contamination and wetting causes surface arcing and aging of nonceramic insulators.

Nonceramic Insulators

Nonceramic insulators have a dirt and water repellent (hydrophobic) surface that reduces pollution accumulation and wetting. The different surface properties slightly modify the flashover mechanism.

Contamination buildup is similar to that in porcelain insulators. However, nonceramic insulators tend to collect less pollution than ceramic insulators. The difference is that in a composite insulator, the diffusion of low-molecular-weight silicone oil covers the pollution layer after a few hours. Therefore, the pollution layer will be a mixture of the deposit (dust, salt) and silicone oil. A thin layer of silicone oil, which provides a hydrophobic surface, will also cover this surface.

Wetting produces droplets on the insulator's hydrophobic surface. Water slowly migrates to the pollution and partially dissolves the salt in the contamination. This process generates high resistivity in the wet region. The connection of these regions starts leakage current. The leakage current dries the surface and increases surface resistance. The increase of surface resistance is particularly strong on the shaft of the insulator where the current density is higher.

Electrical fields between the wet regions increase. These high electrical fields produce spot discharges on the insulator's surface. The strongest discharge can be observed at the shaft of the insulator. This discharge reduces hydrophobicity, which results in an increase of wet regions and an intensification of the discharge. At this stage, dry bands are formed at the shed region. In adverse conditions, this phenomenon leads to flashover. However, most cases of continuous arcing develop as the wet and dry regions move on the surface.

The presented flashover mechanism indicates that surface wetting is less intensive in nonceramic insulators. Partial wetting results in higher surface resistivity, which in turn leads to significantly higher flashover voltage. However, continuous arcing generates local hot spots, which cause aging of the insulators.

Effects of Pollution

The flashover mechanism indicates that pollution reduces flashover voltage. The severity of flashover voltage reduction is shown in Fig. 4.34. This figure shows the surface electrical stress (field), which causes flashover as a function of contamination, assuming that the insulators are wet. This means that the salt in the deposit is completely dissolved. The Equivalent Salt Deposit Density (ESDD) describes the level of contamination.

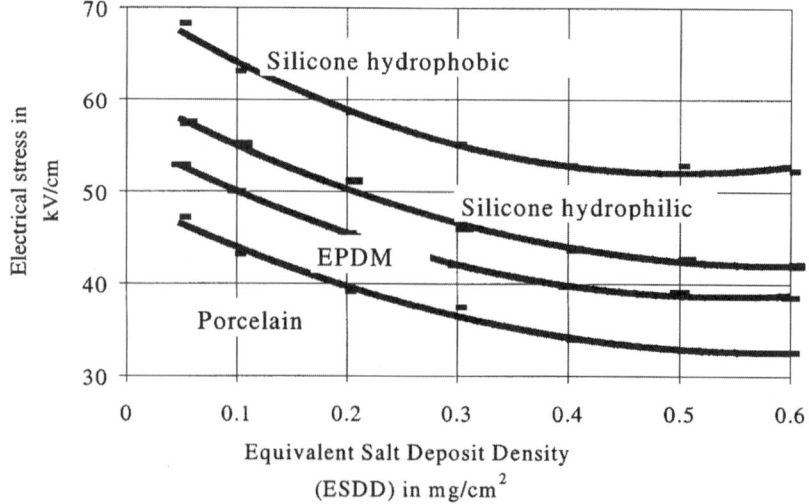

FIGURE 4.34 Surface electrical stress vs. ESDD of fully wetted insulators (laboratory test results). (*Application Guide for Composite Suspension Insulators.* Sediver, Inc., York, SC, 1993. With permission.)

TABLE 4.6 Number of Standard Insulators for Contaminated Areas

System Voltage kV	Level of Contamination			
	Very light	Light	Moderate	Heavy
138	6/6	8/7	9/7	11/8
230	11/10	14/12	16/13	19/15
345	16/15	21/17	24/19	29/22
500	25/22	32/27	37/29	44/33
765	36/32	47/39	53/42	64/48

Note: First number is for I-string; second number is for V-string.

These results show that the electrical stress, which causes flashover, decreases by increasing the level of pollution on all of the insulators. This figure also shows that nonceramic insulator performance is better than ceramic insulator performance. The comparison between EPDM and silicone shows that flashover performance is better for the latter.

Table 4.6 shows the number of standard insulators required in contaminated areas. This table can be used to select the number of insulators, if the level of contamination is known.

Pollution and wetting cause surface discharge arcing, which is harmless on ceramic insulators, but produces aging on composite insulators. Aging is a major problem and will be discussed in the next section.

Composite Insulators

The Electric Power Research Institute (EPRI) conducted a survey analyzing the cause of composite insulator failures and operating conditions. The survey was based on the statistical evaluation of failures reported by utilities.

Results show that a majority of insulators (48%) are subjected to very light pollution and only 7% operate in heavily polluted environments. Figure 4.35 shows the typical cause of composite insulator failures. The majority of failures are caused by deterioration and aging. Most electrical failures are caused by water penetration at the interface, which produces slow tracking in the fiberglass rod surface. This tracking produces a conduction path along the fiberglass surface and leads to internal breakdown of the insulator. Water penetration starts with corona or erosion-produced cuts, holes on the weather shed, or mechanical load-caused separation of the end-fitting and weather shed interface.

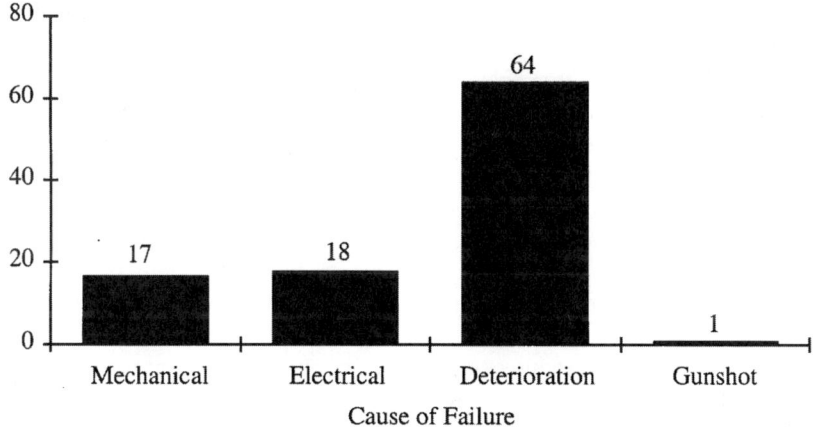

FIGURE 4.35 Cause of composite insulator failure. (Schneider et al., Nonceramic insulators for transmission lines, *IEEE Transaction on Power Delivery,* 4(4), 2214-2221, April, 1989.)

Most of the mechanical failures are caused by breakage of the fiberglass rods in the end fitting. This occurs because of local stresses caused by inappropriate crimping. Another cause of mechanical failures is brittle fracture. Brittle fracture is initiated by the penetration of water containing slight acid from pollution. The acid may be produced by electrical discharge and acts as a cathalizator, attacking the bonds and the glass fibers to produce a smooth fracture. The brittle fractures start at high mechanical stress points, many times in the end fitting.

Aging of Composite Insulators

Most technical work concentrates on the aging of nonceramic insulators and the development of test methods that simulate the aging process. Transmission lines operate in a polluted atmosphere. Inevitably, insulators will become polluted after several months in operation. Fog and dew cause wetting and produce uneven voltage distribution, which results in surface discharge. Observations of transmission lines at night by a light magnifier show that surface discharge occurs in nearly every line in wet conditions. UV radiation and surface discharge cause some level of deterioration after long-term operation. These are the major causes of aging in composite insulators which also lead to the uncertainty of an insulator's life span. If the deterioration process is slow, the insulator can perform well for a long period of time. This is true of most locations in the U.S. and Canada. However, in areas closer to the ocean or areas polluted by industry, deterioration may be accelerated and insulator failure may occur after a few years of exposure. Surveys indicate that some insulators operate well for 18–20 years and others fail after a few months. An analysis of laboratory data and literature surveys permit the formulation of the following aging hypothesis:

1. Wind drives dust and other pollutants into the composite insulator's water-repellent surface. The combined effects of mechanical forces and UV radiation produces slight erosion of the surface, increasing surface roughness and permitting the slow buildup of contamination.
2. Diffusion drives polymers out of the bulk skirt material and embeds the contamination. A thin layer of polymer will cover the contamination, assuring that the surface maintains hydrophobicity.
3. High humidity, fog, dew, or light rain produce droplets on the hydrophobic insulator surface. Droplets may roll down from steeper areas. In other areas, contaminants diffuse through the thin polymer layer and droplets become conductive.
4. Contamination between the droplets is wetted slowly by the migration of water into the dry contaminant. This generates a high resistance layer and changes the leakage current from capacitive to resistive.
5. The uneven distribution and wetting of the contaminant produces an uneven voltage stress distribution along the surface. Corona discharge starts around the droplets at the high stress areas. Additional discharge may occur between the droplets.
6. The discharge consumes the thin polymer layer around the droplets and destroys hydrophobicity.
7. The deterioration of surface hydrophobicity results in dispersion of droplets and the formation of a continuous conductive layer in the high stress areas. This increases leakage current.
8. Leakage current produces heating, which initiates local dry band formation.
9. At this stage, the surface consists of dry regions, highly resistant conducting surfaces, and hydrophobic surfaces with conducting droplets. The voltage stress distribution will be uneven on this surface.
10. Uneven voltage distribution produces arcing and discharges between the different dry bands. These cause further surface deterioration, loss of hydrophobicity, and the extension of the dry areas.
11. Discharge and local arcing produces surface erosion, which ages the insulator's surface.
12. A change in the weather, such as the sun rising, reduces the wetting. As the insulator dries, the discharge diminishes.
13. The insulator will regain hydrophobicity if the discharge-free dry period is long enough. Typically, silicon rubber insulators require 6–8 h; EPDM insulators require 12–15 h to regain hydrophobicity.

14. Repetition of the described procedure produces erosion on the surface. Surface roughness increases and contamination accumulation accelerates aging.
15. Erosion is due to discharge-initiated chemical reactions and a rise in local temperature. Surface temperature measurements, by temperature indicating point, show local hot-spot temperatures between 260°C and 400°C during heavy discharge.

The presented hypothesis is supported by the observation that the insulator life spans in dry areas are longer than in areas with a wetter climate. Increasing contamination levels reduce an insulator's life span. The hypothesis is also supported by observed beneficial effects of corona rings on insulator life.

DeTourreil et al. (1990) reported that aging reduces the insulator's contamination flashover voltage. Different types of insulators were exposed to light natural contamination for 36–42 months at two different sites. The flashover voltage of these insulators was measured using the "quick flashover salt fog" technique, before and after the natural aging. The quick flashover salt fog procedure subjects the insulators to salt fog (80 kg/m^3 salinity). The insulators are energized and flashed over 5–10 times. Flashover was obtained by increasing the voltage in 3% steps every 5 min from 90% of the estimated flashover value until flashover. The insulators were washed, without scrubbing, before the salt fog test. The results show that flashover voltage on the new insulators was around 210 kV and the aged insulators flashed over around 184–188 kV. The few years of exposure to light contamination caused a 10–15% reduction of salt fog flashover voltage.

Natural aging and a follow-up laboratory investigation indicated significant differences between the performance of insulators made by different manufacturers. Natural aging caused severe damage on some insulators and no damage at all on others.

Methods for Improving Insulator Performance

Contamination caused flashovers produce frequent outages in severely contaminated areas. Lines closer to the ocean are in more danger of becoming contaminated. Several countermeasures have been proposed to improve insulator performance. The most frequently used methods are:

1. **Increasing leakage distance by increasing the number of units or by using fog-type insulators.** The disadvantages of the larger number of insulators are that both the polluted and the impulse flashover voltages increase. The latter jeopardizes the effectiveness of insulation coordination because of the increased strike distance, which increases the overvoltages at substations.
2. **Application insulators are covered with a semiconducting glaze.** A constant leakage current flows through the semiconducting glaze. This current heats the insulator's surface and reduces the moisture of the pollution. In addition, the resistive glaze provides an alternative path when dry bands are formed. The glaze shunts the dry bands and reduces or eliminates surface arcing. The resistive glaze is exceptionally effective near the ocean.
3. **Periodic washing of the insulators with high-pressure water.** The transmission lines are washed by a large truck carrying water and pumping equipment. Trained personnel wash the insulators by aiming the water spray toward the strings. Substations are equipped with permanent washing systems. High-pressure nozzles are attached to the towers and water is supplied from a central pumping station. Safe washing requires spraying large amounts of water at the insulators in a short period of time. Fast washing prevents the formation of dry bands and pollution-caused flashover. However, major drawbacks of this method include high installation and operational costs.
4. **Periodic cleaning of the insulators by high pressure driven abrasive material, such as ground corn cobs or walnut shells.** This method provides effective cleaning, but cleaning of the residual from the ground is expensive and environmentally undesirable.
5. **Replacement of porcelain insulators with nonceramic insulators.** Nonceramic insulators have better pollution performance, which eliminates short-term pollution problems at most sites. However, insulator aging may affect the long-term performance.

6. **Covering the insulators with a thin layer of room-temperature vulcanized (RTV) silicon rubber coating.** This coating has a hydrophobic and dirt-repellent surface, with pollution performance similar to nonceramic insulators. Aging causes erosion damage to the thin layer after 5–10 years of operation. When damage occurs, it requires surface cleaning and a reapplication of the coating. Cleaning by hand is very labor intensive. The most advanced method is cleaning with high pressure driven abrasive materials like ground corn cobs or walnut shells. The coating is sprayed on the surface using standard painting techniques.
7. **Covering the insulators with a thin layer of petroleum or silicon grease.** Grease provides a hydrophobic surface and absorbs the pollution particles. After one or two years of operation, the grease saturates the particles and it must be replaced. This requires cleaning of the insulator and application of the grease, both by hand. Because of the high cost and short life span of the grease, it is not used anymore.

References

Application Guide for Composite Suspension Insulators, Sediver Inc., York, SC, 1993.

DeTourreil, C.H. and Lambeth, P.J., Aging of composite insulators: Simulation by electrical tests, *IEEE Trans. on Power Delivery,* 5(3), 1558-1567, July, 1990.

Fink, D.G. and Beaty, H.W., *Standard Handbook for Electrical Engineers,* 11th ed., McGraw-Hill, New York, 1978.

Gorur, R.S., Karady, G.G., Jagote, A., Shah, M., and Yates, A., Aging in silicon rubber used for outdoor insulation, *IEEE Transaction on Power Delivery,* 7(2), 525-532, March, 1992.

Hall, J.F., History and bibliography of polymeric insulators for outdoor application, *IEEE Transaction on Power Delivery,* 8(1), 376-385, January, 1993.

Karady, G.G., Outdoor insulation, *Proceedings of the Sixth International Symposium on High Voltage Engineering,* New Orleans, LA, September, 1989, 30.01-30.08.

Karady, G.G., Rizk, F.A.M., and Schneider, H.H., Review of CIGRE and IEEE Research into Pollution Performance of Nonceramic Insulators: Field Aging Effect and Laboratory Test Techniques, in *International Conference on Large Electric High Tension Systems (CIGRE), Group 33,* (33-103), Paris, 1-8, August, 1994.

Looms, J.S.T., *Insulators for High Voltages,* Peter Peregrinus Ltd., London, 1988.

Schneider, H., Hall, J.F., Karady, G., and Rendowden, J., Nonceramic insulators for transmission lines, *IEEE Transaction on Power Delivery,* 4(4), 2214-2221, April, 1989.

Toughened Glass Insulators. Sediver Inc., Nanterre Cedex, France.

Transmission Line Reference Book (345 kV and Above), 2nd ed., EL 2500 Electric Power Research Institute (EPRI), Palo Alto, CA, 1987.

4.4 Transmission Line Construction and Maintenance

Wilford Caulkins and Kristine Buchholz

The information herein was derived from personal observation and participation in the construction of overhead transmission lines for over 35 years. Detailed information, specific tools and equipment have been provided previously and are available in IEEE Standard 524-1992 and IEEE Standard 524A-1993.

The purpose of this section is to give a general overview of the steps that are necessary in the planning and construction of a typical overhead transmission line, to give newcomers to the trade a general format to follow, and assist transmission design engineers in understanding how such lines are built.

Stringing overhead conductors in transmission is a very specialized type of construction requiring years of experience, as well as equipment and tools that have been designed, tried, and proven to do the work. Because transmission of electrical current is normally at higher voltages (69 kV and above), conductors must be larger in diameter and span lengths must be longer than in normal distribution.

Although proximity to other energized lines may be limited on the right-of-way, extra care must be exercised to protect the conductor so that when energized, power loss and corona are not a problem.

There are four methods that can be used to install overhead transmission conductors:

1. Slack stringing
2. Semi-tension stringing
3. Full-tension stringing
4. Helicopter stringing

Slack stringing can only be utilized if it is not necessary to keep the conductor off of the ground, and if no energized lines lie beneath the line being strung. In this method the pulling lines are pulled out on the ground, threaded through the stringing blocks, and the conductor is pulled in with less tension than is required to keep it off the ground. This is not considered to be an acceptable method when demands involve maximum utilization of transmission requirements.

Semi-tension methods are merely an upgrading of slack stringing, but do not necessarily keep the conductor completely clear of the ground, or the lines used to pull.

Full-tension stringing is a method of installing the conductors and overhead groundwire in which sufficient pulling capabilities on one end and tension capabilities on the other, keep the wires clear of any obstacles during the movement of the conductor from the reel to its final sag position. This ensures that these current-carrying cables are "clipped" into the support clamps in the best possible condition, which is the ultimate goal of the work itself.

Stringing with helicopters, which is much more expensive per hour of work, can be much less expensive when extremely arduous terrain exists along the right-of-way and when proper pre-planning is utilized. Although pulling conductors themselves with a helicopter can be done, it is limited and normally not practical. Maximum efficiency can be achieved when structures are set and pilot lines are pulled with the helicopter, and then the conductor stringing is done in a conventional manner. Special tools (such as stringing blocks) are needed if helicopters are used.

So that maximum protection of the conductor is realized and maximum safety of personnel is attained, properly designed and constructed tools and equipment are tantamount to a successful job. Because the initial cost of these tools and equipment represent such a small percentage of the overall cost of the project, the highest quality should be used, thus minimizing "down time" and possible failure during the course of construction.

Tools

Basic tools needed to construct overhead transmission lines are as follows:

1. Conductor blocks
2. Overhead groundwire blocks
3. Catch-off blocks
4. Sagging blocks
5. Pulling lines
6. Pulling grips
7. Catch-off grips
8. Swivels
9. Running boards
10. Conductor lifting hooks
11. Hold-down blocks

Conductor blocks are made in the following configurations:

1. Single conductor
2. Multiple conductor
3. Multiversal type (can be converted from bundle to single, and vice versa)
4. Helicopter

Conductor blocks should be large enough to properly accommodate the conductor and be lined with a resilient liner such as neoprene or polyurethane and constructed of lightweight, high-strength materials. Sheaves should be mounted on anti-friction ball bearings to reduce the tension required in stringing and facilitate proper sagging. Conductor blocks are available for stringing single conductors or multiple conductors. Some are convertible, thus enhancing their versatility. When stringing multiple conductors, it is desirable to pull all conductors with a single pulling line so that all conductors in the bundle have identical tension history. The running board makes this possible. Pulling lines are divided into two categories:

1. Steel cable
2. Synthetic rope

Because of the extra high tension required in transmission line construction, steel pulling lines and pilot lines are most practical to use. Torque-resistant, stranded, and swagged cable are used so that ball bearing swivels can be utilized to prevent torque buildup from being transferred to the conductor. Some braided or woven steel cables are also used. If synthetic ropes are utilized, the most important features should include:

1. No torque
2. Very minimum elongation
3. No "kinking"
4. Easily spliced
5. High strength/small diameter
6. Excellent dielectric properties

Stringing overhead groundwires does not normally require the care of current-carrying conductors. Most overhead groundwires are stranded steel construction and the use of steel wire with a fiber-optic core for communications has become a common practice. Special care should be taken to ensure that excessive bending does not occur when erecting overhead groundwires with fiber-optic centers, such as OPT-GW (Optical Power Telecommunications — Ground Wire) and ADSS (All Dielectric Self-Supporting Cable). Special instructions are available from the manufacturer, which specify minimum sheave and bullwheel diameter for construction. OPT-GW should be strung using an antirotational device to prevent the cable from twisting.

Equipment

Pullers are used to bring in the main pulling line. Multidrum pullers, called pilot line winders, are used to tension string the heavy pulling cable.

Primary pullers are used to tension string the conductors. These pullers are either drum type or bullwheel type. The drum type is used more extensively in many areas of North America because the puller and pulling cable are stored on one piece of equipment, but it is not practical in other areas because it is too heavy. Thus, the bullwheel type is used allowing the puller and pulling cable to be separated onto two pieces of equipment. Also, the pulling cable can be separated into shorter lengths to allow easier handling, especially if manual labor is preferred.

Tensioners should be bullwheel type using multigroove wheels for more control. Although V groove machines are used on some lighter, smaller conductors, they are not recommended in transmission work because of the crushing effect on the conductor. Tensioners are either mounted on a truck or trailer.

Reel stands are used to carry the heavy reels of conductor and are equipped with brakes to hold "tailing tension" on the conductor as it is fed into the bullwheel tensioner. These stands are usually mounted on a trailer separated from the tensioner.

Helicopters are normally used to fly in a light line which can be used to pull in the heavier cable.

Procedures

Once the right-of-way has been cleared, the following are normal steps taken in construction:

1. Framing
2. Pulling
3. Pulling overhead groundwire up to sag and installation
4. Pulling in main line with pilot line
5. Stringing conductors
6. Sagging conductors
7. Clipping in conductors
8. Installing spacer or spacer dampers where applicable

Framing normally consists of erecting poles, towers, or other structures, including foundations and anchors on guyed structures. It is desirable for the stringing blocks to be installed, with finger lines, on the ground before structures are set, to eliminate an extra climb later. Helicopters are used to set structures, especially where rough terrain exists or right-of-way clearances are restricted.

Once structures are secure, overhead groundwire and pilot lines are pulled in together with a piece of equipment such as a caterpillar or other track vehicle. A helicopter is also used to fly in these lines. Once the overhead groundwires are in place, they are sagged and secured, thus giving the structures more stability for the stringing of the conductors. This is especially important for guyed structures.

Normally the three pilot lines (typically 3/8 in. diameter swagged steel cable) pull in the heavier pulling line (typically 3/4 in. diameter or 7/8 in. diameter swagged steel) under tension. The main pulling line is then attached to the conductor which is strung under full tension. Once the conductor is "caught off," the main pulling line is returned for pulling of the next phase.

Once the conductors are in place, they are then brought up to final sag and clipped into the conductor clamps provided. If the conductor is a part of a bundle per phase, the spacers or spacer dampers are installed, using a spacer cart which is either pulled along from the ground or self-propelled.

Coordination between design engineers and construction personnel is very important in the planning and design of transmission lines. Although it is sometimes impossible to accommodate the most efficient capabilities of the construction department (or line contractor), much time and money can be conserved if predesign meetings are held to discuss items such as the clearances needed for installing overhead groundwire blocks, hardware equipped with "work" holes to secure lifting hooks or blocks, conductor reel sizes compatible with existing reel stands, length of pull most desirable, or towers equipped to facilitate climbing.

For maximum safety of personnel constructing transmission lines, proper and effective grounding procedures should be utilized. Grounding can be accomplished by:

1. adequate grounding of conductors being strung and pulling cables being used, or
2. fully insulating equipment and operator, or
3. isolating equipment and personnel.

All equipment, conductors, anchors, and structures within a defined work area must be bonded together and connected to the ground source. The recommended procedures of personnel protection are the following:

1. Establish equipotential work zones.
2. Select grounding equipment for the worst-case fault.
3. Discontinue all work when the possibility of lightning exists which may affect the work site.

In addition to the grounding system, the best safety precaution is to treat all equipment as if it could become energized.

Helicopters

As already mentioned, the use of helicopters is another option that is being chosen more frequently for transmission system construction and maintenance. There are a wide variety of projects where helicopters become involved, making the projects easier, safer, or more economical. When choosing any construction or maintenance method, identify the work to be accomplished, analyze the potential safety aspects, list the possible alternatives, and calculate the economics. Helicopters add a new dimension to this analytical process by adding to the alternatives, frequently reducing the risks of accident or injury, and potentially reducing costs. The most critical consideration in the use of a helicopter is the ability to safely position the helicopter and line worker at the work location.

Conductor Stringing

Helicopters are used for conductor stringing on towers through the use of pilot lines. Special stringing blocks are installed at each tower and a helicopter is brought in and attached to a pilot line. The helicopter flies along the tower line and slips the pilot line in through each stringing block until it reaches the end of the set of towers for conductor pulling, where it disconnects and the pilot line is transferred to a ground crew. The ground crew then proceeds to pull the conductor in the conventional manner (Caulkins, 1987). The helicopter may also be used to monitor the conductor pulling and is readily available to assist if the conductor stalls at any tower location.

Structure and Material Setting

The most obvious use of helicopters is in the setting of new towers and structures. Helicopters are frequently used in rough terrain to fly in the actual tower to a location where a ground crew is waiting to spot the structure into a preconstructed foundation. In addition, heavy material can be transported to remote locations, as well as the construction crew.

The use of helicopters can be especially critical if the tower line is being replaced following a catastrophe or failure. Frequently, roads and even construction paths are impassable or destroyed following natural disasters. Helicopters can carry crews and materials with temporary structures that can be erected within hours to restore tower lines. Again, depending on the terrain and current conditions, whether the existing structure is repaired or temporary tower structures are utilized, the helicopter is invaluable to carry in the needed supplies and personnel.

Insulator Replacement

A frequent maintenance requirement on a transmission system is replacing insulators. This need is generated for various reasons, including line upgrading, gunshots, environmental damage, or defects in the original insulator manufacturing. With close coordinated crews, helicopters can maximize the efficiency of the replacement project.

Crews are located at several towers to perform the actual insulator removal and installation. The crews will do the required setup for a replacement, but the helicopter can be used to bring in the necessary tools and equipment. The crew removes the old insulator string and sets it to one side of the work location. When the crews are ready, the helicopter flies in the new insulator string to each tower. The crew on the tower detaches the new insulator string from the helicopter, positions it, and then attaches the old string to the helicopter, which removes the string to the staging area. With a well-coordinated team of helicopters and experienced line workers, it is not unusual to achieve a production rate of replacing all insulators on four three-phase structures per crew per day. Under ideal conditions, crews are able to replace the insulators on a structure in one hour (Buchholz, 1987).

Replacing Spacers

One of the first uses of helicopters in live-line work was the replacement of spacers in the early 1980s. This method was a historic step in live-line work since it circumvented the need for hot sticks or insulated aerial lift devices.

The first projects involved a particular spacer wearing into the conductor strands, causing the separation of the conductor. Traditionally, the transmission line would have been de-energized, grounded, and either a line worker would have utilized a spacer cart to move out on the line to replace the spacer, or the line would have been lowered and the spacer replaced and the conductor strengthened. The obvious safety dilemma was whether the conductor could support a line worker on a spacer cart or whether it was physically able to withstand the tensions of lowering it to the ground. By utilizing a helicopter and bare-hand work methods, the spacers were able to be replaced and the conductor strengthened where necessary with full-tension compression splices while providing total safety to the line workers and a continuous supply of energy over the transmission lines. One of the early projects achieved a replacement and installation of 25,000 spacers without a single accident or injury. A typical spacer replacement required about 45 sec, including the travel time between work locations (Buchholz, 1987).

Insulator Washing

Another common practice is to utilize helicopters for insulator washing. Again, this is a method that allows for the line to remain energized during the process. The helicopter carries a water tank that is refilled at a staging area near the work location. A hose and nozzle are attached to a structure on the helicopter and are operated by a qualified line worker who directs the water spray and adequately cleans the insulator string. Again, with the ease of access afforded by the helicopter, the speed of this operation can result in a typical three-phase tower being cleaned in a few minutes.

Inspections

Helicopters are invaluable for tower line and structure inspections. Due to the ease of the practice and the large number of inspections that can be accomplished, utilities have increased the amount of maintenance inspections being done, thus promoting system reliability.

Helicopters typically carry qualified line workers who utilize stabilizing binoculars to visually inspect the transmission tower for signs of rusting or weakness and the transmission hardware and conductor for damage and potential failure. Infrared inspections and photographic imaging can also be accomplished from the helicopter, either by mounting the cameras on the helicopter or through direct use by the crew. During these inspections, the helicopter provides a comfortable situation for accomplishing the necessary recording of specific information, tower locations, etc. In addition, inspections from helicopters are required following a catastrophic event or system failure. It is the only logical method of quickly inspecting a transmission system for the exact location and extent of damage.

Helicopter Method Considerations

The ability to safely position a helicopter and worker at the actual work site is the most critical consideration when deciding if a helicopter method can be utilized for construction or maintenance. The terrain and weather conditions are obvious factors, as well as the physical spacing needed to position the helicopter and worker in the proximity required for the work method. If live-line work methods are to be utilized, the minimum approach distance required for energized line work must be calculated very carefully for every situation. The geometry of each work structure, the geometry of the individual helicopter, and the positioning of the helicopter and worker for the specific work method must be analyzed. There are calculations that are available to analyze the approach distances (IEEE Task Force 15.07.05.05, 1999).

When choosing between construction and maintenance work methods, the safety of the line workers is the first consideration. Depending on circumstances, a helicopter method may be the safest work method. Terrain has always been a primary reason for choosing helicopters to assist with projects since the ability to drive to each work site may not be possible. However, helicopters may still be the easiest and most economic alternative when the terrain is open and flat, especially when there are many individual tower locations that will be contacted. Although helicopters may seem to be expensive on a per person basis, the ability to quickly position workers and easily move material can drastically reduce costs. When live-line methods can be utilized, the positioning of workers, material, and equipment becomes comparatively easier.

Finally, if the safe use of the helicopter allows the transmission systems to remain energized throughout the project, the helicopter may be the only possible alternative. Since the transmission system is a major link in the competitive energy markets, transmission operation will have reliability performance measures which must be achieved. Purchasing replacement energy through alternate transmission paths, as was done in the regulated world, is no longer an option. Transmission system managers are required to keep systems operational and will be fined if high levels of performance are not attained. The option of de-energizing systems for maintenance practices may be too costly in the deregulated world.

References

Buchholz, F., Helicopter application in transmission system maintenance and repair, in *IEEE/CSEE Joint Conference on High-Voltage Transmission Systems in China,* October, 1987.

Caulkins, III., W., Practical applications and experiences in the installation of overhead transmission line conductors, in *IEEE/CSEE Joint Conference on High-Voltage Transmission Systems in China,* October, 1987.

Guide to Grounding During the Installation of Overhead Transmission Line Conductors: Supplement to IEEE Guide to the Installation of Overhead Transmission Line Conductors, IEEE 524A-1993, 1998.

Guide to the Installation of Overhead Transmission Line Conductors, IEEE 524-1992, 1998.

IEEE Task Force 15.07.05.05, PE 046 PRD (04-99), Recommended Practices for Helicopter Bonding Procedures for Live Line Work.

4.5 Insulated Power Cables for High-Voltage Applications

Carlos V. Núñez-Noriega and Felimón Hernandez

The choice of transmitting and distributing electric power through underground vs. overhead systems requires consideration of economical, technical, and environmental issues. Underground systems have traditionally been favored when distributing electric power to densely populated areas and when reliability or aesthetics is important. Underground distribution systems provide superior reliability because they are not exposed to wind, lightning, vandalism, or vehicle damage. These factors are the main contributors to failures of overhead electric power distribution systems. When designed and installed properly, underground distribution systems also require less preventive maintenance. The main disadvantage of underground transmission and distribution systems as compared with overhead lines is its higher cost.

Underground transmission and distribution of electric power is done through the use of insulated power cables. Cable designs vary widely depending on many factors such as voltage, power rating, application, etc.; however, all cables have certain common components. This section presents an overview of insulated electric power cables, standard usage practices, and common formulas used to calculate electric parameters important for the design of underground electric power systems.

Typical Cable Description

The basic function of an insulated power cable is to transmit electric power at a predetermined current and voltage. A typical single-core insulated power cable consists of a copper or aluminum conductor and several layers of insulation. The constructional design of power cables is more complex for cables designed for higher voltages. A cable for voltage in the tens of kilovolts may also include a metallic shield to provide a moisture barrier, an extruded semiconductive screen, and a metallic mesh or sheath which provides a path for the return currents and a reference to ground for the voltage.

Since insulated cables present no voltage at the external protecting jacket while in operation, they can be installed either directly buried (in direct contact with ground) or in a duct bank. Power insulated

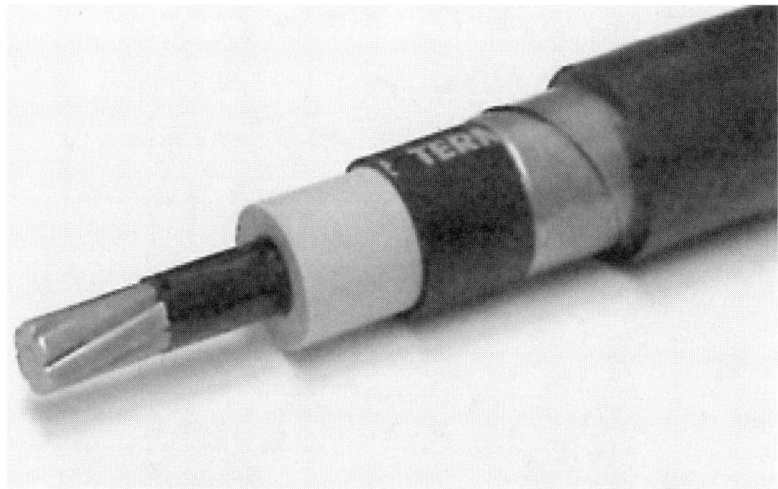

FIGURE 4.36 Elements of typical (a) single-core and (b) three-core, steel armored, insulated power cables.

cables are designed to endure different types of stresses such as electrical stress caused by the rated voltage and by transient overvoltages, mechanical stress due to tension and compression during the installation, thermal stress produced during normal operation, and chemical stress caused by the reaction with the environment that may occur when the cable is installed in aggressive soils or in the presence of some chemicals. An insulated power cable typically consists of the following elements: a) a central conductor (or conductors), through which the electric current flows, b) the insulation, which withstands the applied voltage to ground, c) the shielding, consisting of a semiconductor shield that uniformly distributes the electric field around the insulation, d) a metallic sheath that provides a reference for the voltage and a path for the return and short-circuit currents, and e) an external cover that provides mechanical protection. Figure 4.36(a) illustrates a single-core insulated power cable rated at 1 kV with several layers of insulation and a metallic shield. Figure 4.36(b) shows a typical three-core, steel wire armored cable designed for voltages up to 15 kV.

Conductor

The conductor of a typical power cable is made of copper or aluminum. A single-core cable with a *concentric* configuration consists of a single central wire around which concentric layers of wire are built. An alternative configuration is the *shaped compact* single-core cable. The compacted cable is obtained

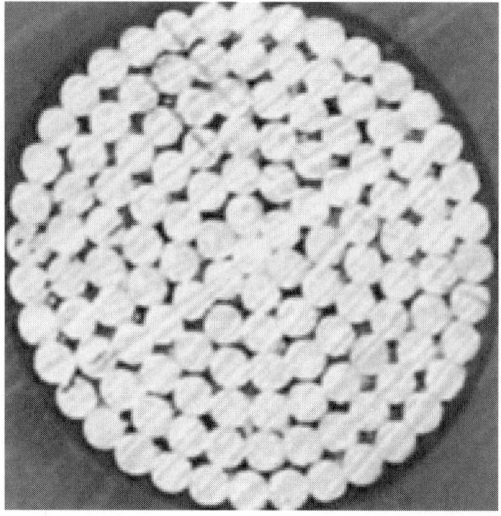
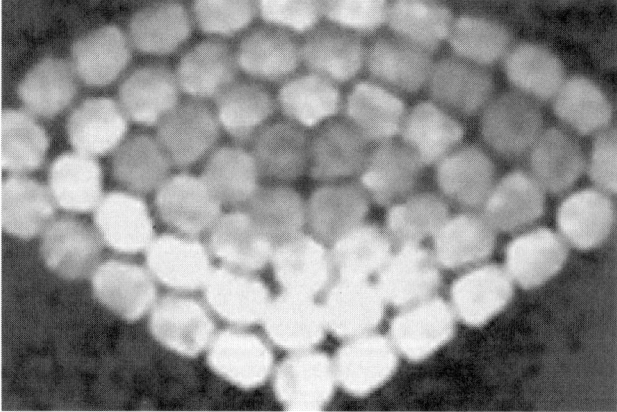

FIGURE 4.37 Cross-section view of (a) a concentric cable and (b) a single sector of a compact-shaped three-sector conductor.

by passing the cable through compacting machines to obtain its form. Because of compacting ratios of 85 to 90%, more current capacity per unit of transversal section is achieved with compact cables, and they are therefore sometimes preferred over regular concentric cables. Figure 4.37 illustrates (a) a cross-section view of a concentric cable and (b) a single section of a compact-shaped three-sector cable. For voltages of 1 kV or less, cables with solid conductors are sometimes used. Solid conductors have the disadvantage of reduced flexibility. This is partially overcome by using three or more sectors per core.

Insulation

Paper and natural rubbers were used for many years as insulating materials in underground power cables. Presently, synthetic materials are preferred for insulation of cables. The chemical composition of such materials can be altered to produce polymers with specific chemical, electrical, and mechanical properties. Although the list of materials used as cable insulation is extensive, ethylene propylene (*EP* or *EPR*) and cross-linked polyethylene (*XLP* or *XLPE*) are by far the most popular. EP and XLP insulation have similar insulating characteristics and expected long life under the same operational conditions. Some companies prefer the XLP to the EP because the former is transparent and phenomena like treeing, which is a common cause of power cable failures, can be easily analyzed under the microscope. Other insulating

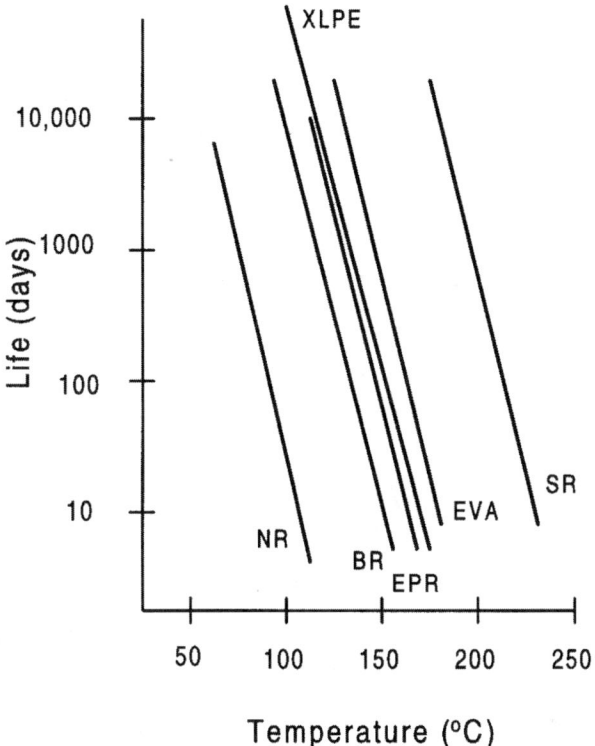

FIGURE 4.38 Life vs. temperature for different insulating materials.

materials used in cables include natural rubber (NR), silicone rubber (SR), ethyl vinyl acetate (EVA), and butyl rubber (BR).

An important consideration in the selection of insulating material is made by comparing aging performance as it relates to the maximum operating temperature of the cable, typically 60°C. This relation is obtained by using the Arrhenius reaction rate equation. A graph of life in days vs. temperature in °C is obtained. A typical comparison for different materials is illustrated in Fig. 4.38.

Semiconducting Shield

In a typical high-voltage cable, two layers of semiconductor material surround the metallic core. The first layer, placed directly around the conductor, has the following purposes:

1. To distribute the electric field uniformly around the conductor.
2. To prevent the formation of ionized voids in the conductor.
3. To dampen impulse currents traveling over the conductor surface.

The second layer of semiconductor material is placed around the first insulating layer and has the following purposes:

1. To reduce the surface voltage to zero.
2. To confine the electric field to the insulation, eliminating tangential stresses.
3. To offer a direct path to ground for short-circuit current if the shield is grounded.

Metallic Sheath

The metallic sheath surrounding insulated cables serves several purposes: as an electrostatic shield, as a ground fault current conductor, and as a neutral wire.

1. The metallic sheath designed for electrostatic purpose should be made of non-magnetic tape or non-magnetic wires. Copper is usually used for these sheaths.
2. When properly grounded, this sheath provides a path for short-circuit current.
3. With appropriate dimensions, the metallic sheath can be used as a system neutral, as in residential distribution cables where single-phase transformers are common. When single-core cables are used in three-phase systems, the grounded sheath provides a path for unbalanced currents.

When high-voltage underground cables require some type of moisture insulating barrier, a metallic pipe surrounds the current-carrying cable. The materials most commonly used as moisture barriers are lead and aluminum. The electric currents carried by the high-voltage cable will induce a voltage in the metallic shield that surrounds the conductor, and in the shields of other surrounding cables. These induced voltages, in turn, will generate an induced current flow with its associated heat losses. In the case of steel shields, the losses will include magnetization and hysteresis losses.

External Layer

The purpose of the external layer in insulated power cables is to provide mechanical protection against the environment during the installation and operation of the power cable. Currently, materials commonly used as the external layer for extruded power cables include PVC and polyethylene of low and high density. These materials are used for their ability to withstand the cable operating temperature, their resistance to excessive degradation when in contact with some chemicals typical of some operating environments, and their excellent mechanical properties for undergoing stresses during transportation, or compression and tension during installation and operation.

Overview of Electric Parameters of Underground Power Cables

Cable Electrical Resistance

The cable operation parameters determine the cable behavior under emergency and normal operating conditions. The cable impedance, $(R + jX_L)$, is useful for calculating regulation and losses under normal conditions and short-circuit current magnitude under short-circuit fault conditions. Alterations in parameters due to temperature and length variations include the change of resistance. In order to simplify the selection of a power cable for a given application, the maximum cable operating temperature is determined based on engineering calculations and laboratory tests. In the case of single wires, which are stranded together following a helical path to form a cable, a factor must be added to account for the extra length. In multi-core cables, the extra length due to the layout of the individual cores must also be accounted for. To simplify the calculation of electrical parameters, graphs of conductor size vs. resistance in ohm/kilometer can be developed and are provided by the manufacturer. Typically, manufacturers provide AC and DC resistance values which are temperature dependent. In general, the DC resistance is given by:

$$R_{t_c} = R_{20}\left[1+\alpha_{20}\left(t_c - 20\right)\right] \quad (4.4)$$

where

R_{tc} = conductor resistance at temperature t_c°C (in ohms)
R_{20} = conductor resistance at temperature 20°C (in ohms)
α_{20} = temperature coefficient of resistance of conductor material at 20°C
t_c = conductor temperature (in °C)

The AC resistance of cables is important because it affects the current-carrying capacity. The AC resistance of cables is affected mainly by skin effect and proximity effect. The analysis of these effects is complicated and graphs and tables provided by the cable manufacturer should be used to simplify any

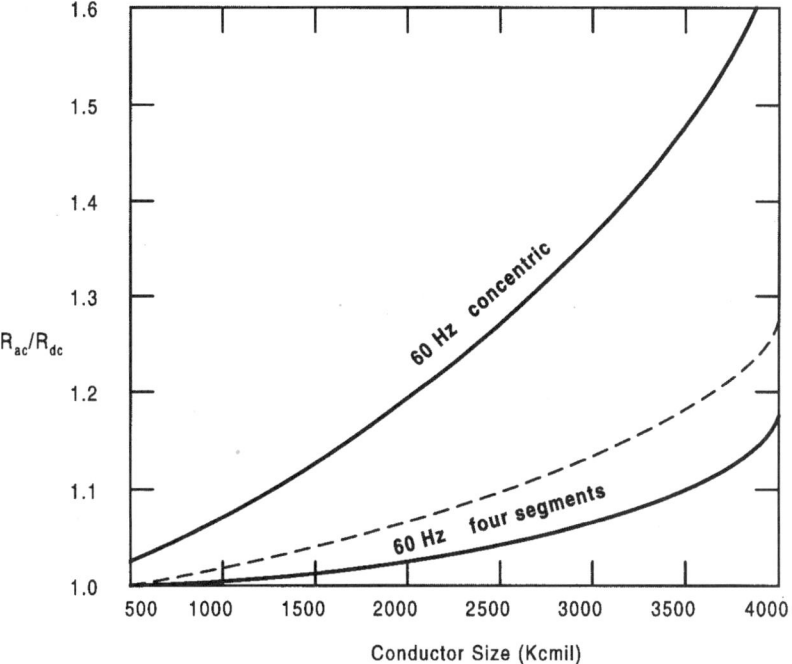

FIGURE 4.39 Conductor size vs. resistance.

calculations. Figure 4.39 illustrates a typical chart of AC/DC resistance vs. cable size for two different types of cable cores.

Cable Inductance

A variable magnetic field is created when an electric alternating current passes through a conductor. This field interacts with the magnetic field of other adjacent current-carrying conductors. The time-varying magnetic field divided by the time-varying current is called inductance. The total cable inductance is composed of the self (or internal) inductance and the mutual (or external) inductance. The mutual inductance is caused by the interaction of adjacent conductors carrying an alternating current. The inductance L per core of either a three-core cable or three single-core cables is obtained from

$$L = K + 0.2 \ln \frac{2S}{d} \quad (\text{in mH/km}) \tag{4.5}$$

where

K = a constant that accounts for the number of wires in the core (see Table 4.7)
S = distance between conductors in mm for trefoil spacing, or
 = 1.26 × spacing for single-core cables in flat configuration (in mm)
d = equivalent conductor diameter (in mm)

Cable Capacitance

The capacitance between two conductors is defined as the charge between the conductors divided by the difference in voltage between them. The capacitance of power cables is affected by several factors, including geometry of the construction (if single-core or triplex), existence of metallic shielding, and type and thickness of insulating material.

TABLE 4.7 Approximate Values of K for Stranded Conductors

Number of Wires in Conductor	K
3	0.078
7	0.064
19	0.055
37	0.053
61	0.051

For a single-core cable, the capacitance per meter is given by:

$$C = \frac{q}{V} \quad (4.6)$$

or

$$C = \frac{\varepsilon_r}{18 \ln(D/d)} \quad (4.7)$$

where

D = diameter over insulation (in meters)
d = diameter over conductor (in meters)
ε_r = relative permitivity

The relative permitivity is a function of the insulating material of the cables and can be safely neglected for cables operating at 60 Hz and at normal operating temperatures.

For three-core type cables, the capacitance between one conductor and the other conductors can be approximated using the previous equation where D becomes the diameter of one conductor plus the thickness of the insulation between conductors plus the thickness of insulation between any conductor and the metallic shield.

Shield Bonding Methods and Electric Parameters

Effect of Shield Bonding Method in the Electric Parameters

Metallic shields of cables can be grounded in several ways, depending on national or regional practices, safety issues, and practical considerations. The grounding method employed is important since induced and return currents may potentially flow through grounded shields, effectively derating the cable. A single current-carrying conductor inside its shield can be modeled as shown in Fig. 4.40. For simplicity it is assumed that the conductor and shield are of infinite length. Three shield bonding methods will be discussed in this section. In the first method of shield bonding, the shield is not grounded or it is grounded at one end only; in the second method, the shield is grounded at each end; and in the third method, the shields are grounded and transposed at multiple locations throughout the length of the cable. In each case, a three-phase system composed of three single-core cables is assumed.

Ungrounded and Single-Point Grounded Shields

When the shield is not grounded, the current flowing in the center conductor will induce electric and magnetic fields in the surrounding shield and in the conductors and shields of adjacent cables. Because induction is maximum in the shield wall closest to the inner conductor, and minimum in the outer wall of the shield, a current flow is driven by the difference in electromagnetic fields. The losses associated

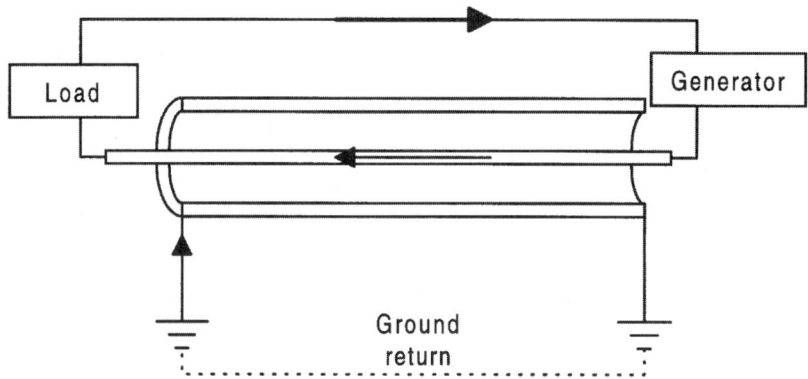

FIGURE 4.40 Model of a current carrying conductor within a metallic shield.

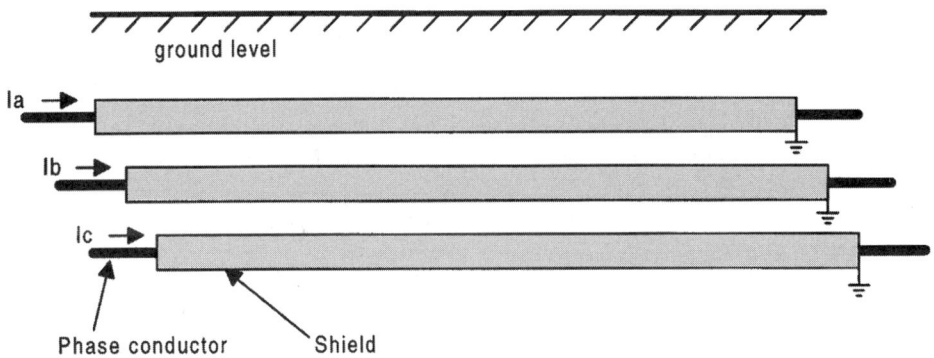

FIGURE 4.41 Single-point grounded shield system.

with these currents are described as *eddy current* losses. In single-core lead-shield conductors, eddy current losses are normally small when compared with conductor losses, and are therefore neglected. In the case of more than one conductor with aluminum shield, losses may become significant, especially when the shielded cables are in close proximity.

The single-point grounded shield system is the simplest form of bonding a cable shield. The single-point grounded system consists of connecting and grounding the shield of each cable at a single point along its length, usually at one end as shown in Fig. 4.41. In this condition, the shield circuit is not closed and therefore no shield current will flow. If the phase currents are unbalanced, a return current will flow through the earth (and the ground wires, if installed) but no return current will flow through the shield. In the ungrounded and single-point grounded shield systems, only eddy currents flow through the shields.

Multigrounded Shields

When the shields of a cable are connected at both ends, the system is said to be multigrounded (or multipoint grounded). A multigrounded system is illustrated in Fig. 4.42. In this system a closed path exists for the shield currents to flow through. The shield circulating currents produce a magnetic field that tends to cancel the magnetic field generated by the phase current. This cancellation occurs in single-phase underground systems such as those feeding individual homes in some residential areas. The reason for the cancellation is that phase current and return current flow in opposite directions. In multicircuit three-phase systems, however, the magnetic field cancellation rarely occurs, mainly due to mutual inductance and unpredicted current unbalances.

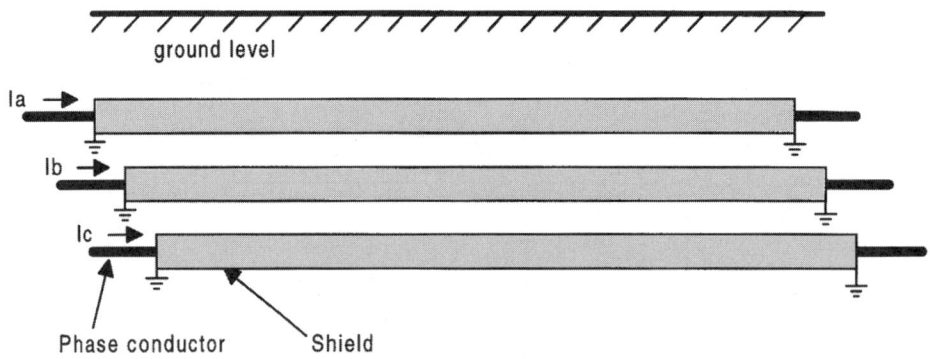

FIGURE 4.42 Multigrounded shield system.

It is important to note that multigrounded systems, such as the one depicted in Fig. 4.42, are rarely used since the shield circulating currents increases the joule losses, therefore decreasing the ampere rating of the cable.

Cross-Bonded with Transposition and Sectionalized with Transposition

A third method of bonding cable shields is known as cross-bonded with transposition and is illustrated in Fig. 4.43. In this method, the shields are divided into three or more sections and grounded at each end of the route. A route consists of three or a multiple of three sections. The conductor shields are then adequately transposed, minimizing the induced shield currents and consequent losses. The shields are isolated from ground using polyethylene or other plastic material. A variant of this method calls for sectionalizing the shields and grounding at both ends, each section of the shield further canceling the shield circulating currents. It is common practice to cross-bond cables at the splices. That is, the total length of cable that can normally be stored in a reel is the cross-bonding length. For long cable runs, the number of cross-bonding sections must be a multiple of three, making a ground connection at every third station. Cross-bonding with transposition and the sectionalizing with transposition techniques reduce the induced shield currents to zero or to values low enough to be safely neglected.

The most prevalent practice for sheath bonding in the U.S., Canada, and Great Britain has been either to use single-point grounded or sectionalized cross-bonded sheath to minimize sheath losses. A secondary benefit of these bonding practices associated with the reduction of sheath circulating currents is a drastic reduction of the magnetic field generated by unbalanced currents, especially at large distances from the lines.

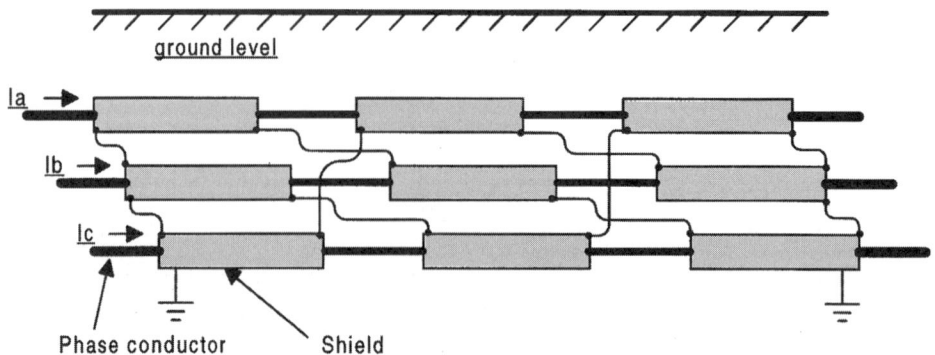

FIGURE 4.43 Multipoint grounded with transposition system.

Calculation of Losses

When the sheath is grounded or bonded to earth at more than one point, current flows through the sheath due to the emf induced in it by the alternating current in the central conductor. The mechanism of induction is the same as the mechanism of induction in a transformer. The induced voltage in the sheath of such conductor is:

$$E_s = I X_m \tag{4.8}$$

where I = conductor current in amperes and

$$X_m = 2\pi f M \, 10^{-3} \quad \text{(in ohms/km)} \tag{4.9}$$

where the mutual inductance between conductors and sheath is:

$$M = 0.2 \log_e \frac{2S}{d_m} \quad \text{(in mH/km)} \tag{4.10}$$

The impedance of the sheath is given by:

$$Z_s = \sqrt{R_s^2 + X_m^2} \quad \text{(in ohms/km)} \tag{4.11}$$

where R_s is the sheath resistance.

The sheath current is therefore:

$$I_s = \frac{E_s}{\sqrt{R_s^2 + X_m^2}} \tag{4.12}$$

or

$$I_s = \frac{I X_m}{\sqrt{R_s^2 + X_m^2}} \tag{4.13}$$

Finally, the sheath current losses per phase is given by:

$$I_s^2 R_s = \frac{I^2 X_m^2 R_s}{R_s^2 + X_m^2} \quad \text{(in watts/km)} \tag{4.14}$$

The calculations described above assume that voltage induction by nearby conductors and nearby sheaths is negligible.

Underground Layout and Construction

Residential Distribution Layout

In general, the design of underground distribution systems feeding residential areas is similar to overhead distribution systems. As in overhead systems, a primary main is installed from which distribution transformers supply consumers at a lower voltage (120–240 V). There are two widely used circuit

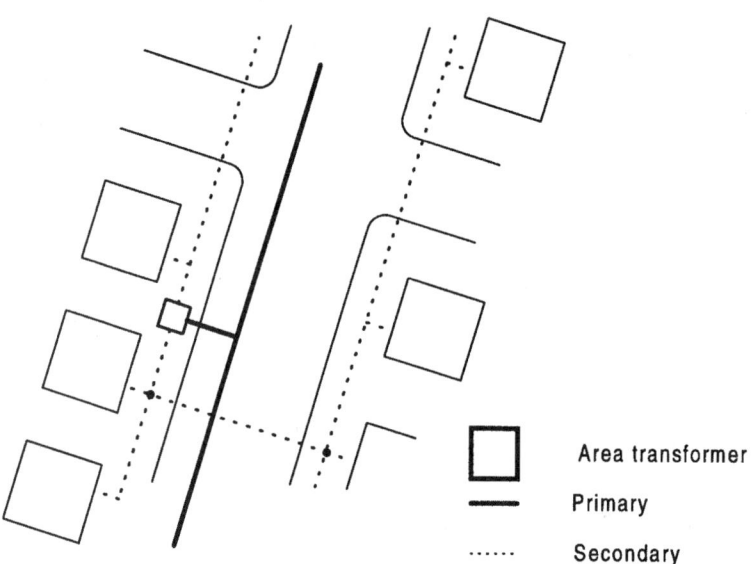

FIGURE 4.44 Underground layout with area transformer.

configurations preferred in underground systems for residential areas. The selection between these two is a matter of simple economics.

In the first case, the design is such that the primary main supplies a step-down transformer from which several secondary mains connect to consumers. This design configuration is illustrated in Fig. 4.44. In the second design, the primary main supplies the consumer directly through individual transformers placed next to the consumer service entrance. In this design pattern, the secondary mains are eliminated. This design configuration is illustrated in Fig. 4.45.

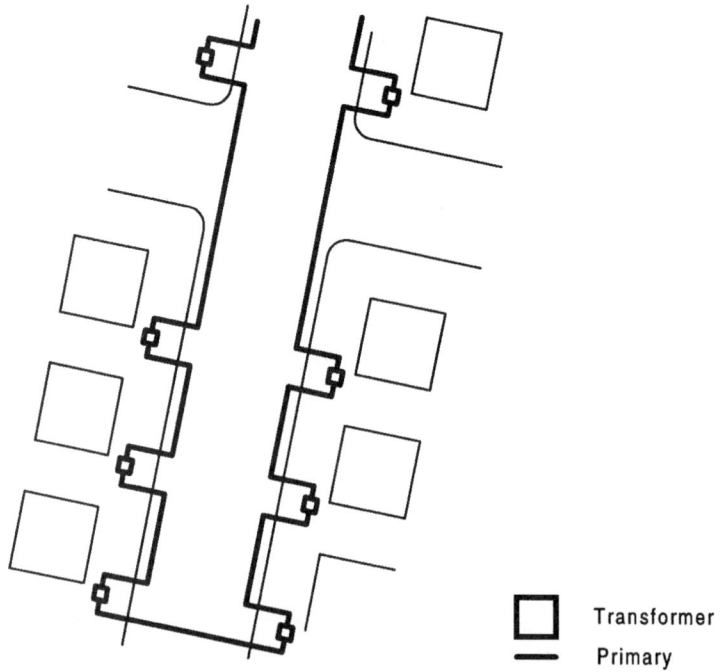

FIGURE 4.45 Underground layout with individual transformers.

Transmission System

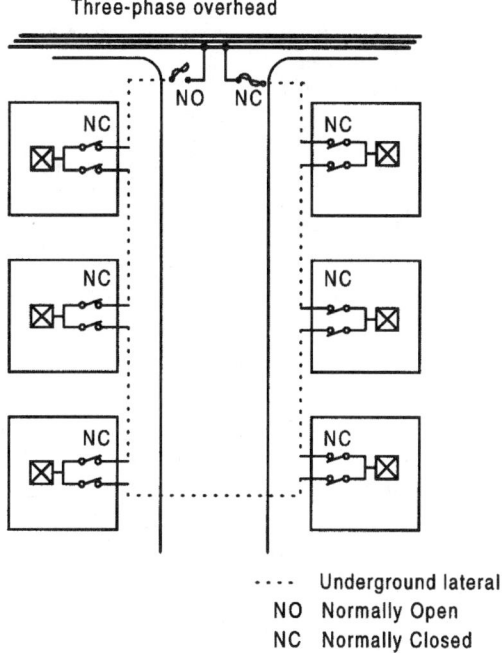

FIGURE 4.46 Open-loop design with NO and NC switches.

In each case, the primary feeder radiates from a substation, and laterals are connected, usually through adequate protective devices. The purpose of these devices is to protect the feeder from faults occurring in the lateral circuits.

Because faults in underground systems are often difficult to locate and repair, an open loop design is frequently used in these systems. As seen in Fig. 4.46, this design consists of a primary main, from which a lateral supplies a group of consumers through an open loop. In the case of a fault in the primary feeder, the faulted section is open at both ends and service to the lateral is re-established by closing the loop through the normally open switch.

Commercial and Industrial Layout

Commercial and industrial consumers are treated differently than residential. Commercial and industrial loads require three phase supply and a more reliable service. In distribution circuits serving these loads, a neutral wire is installed to carry the unbalanced current. Because digging in these areas may be restricted, the cables and equipment are installed in ducts and manholes, improving maintenance and repair time.

Economics and quality of service required are factors to consider during the design phase of underground distribution systems for commercial and industrial consumers. In general, two designs are favored. In the first design, illustrated in Fig. 4.47, the load is connected through a transformer, the primary of which is connected to an automatic throw-over device. In case of a fault in the main feeder, to which the load is connected, the device automatically switches over to an alternate main feeder.

The second design is more complex and expensive, but provides superior reliability. In this design, several primary feeders connect the primary of a group of transformers. The secondaries of the transformers are connected together through fuses and protectors, forming a low-voltage network. In the event of failure in one of the primary feeders, the appropriate protective device opens, and power distribution continues through the other primaries without interruption.

Direct Buried Cables

Underground distribution power cables can be installed directly in a trench (direct burial) or in a duct. In some instances, power cables can be installed with telephone, gas, water, or other facilities. Direct

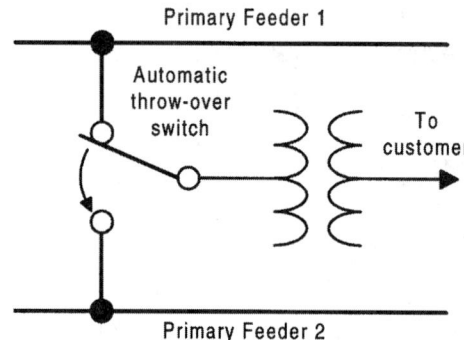

FIGURE 4.47 Diagram of throw-over switch and two supplies for extra reliability.

burial of power cables is commonly used in low-density residential areas. The main advantage of direct burial installation is cost, since conduits and manholes are eliminated. The National Electrical Safety Code (NESC) provides minimum requirements for width and depth of the trench as well as the number and types of facilities that can be installed together with the power cable. The direct burial of cables is divided into several stages, which commonly include surveying the land and digging trial holes to determine soil characteristics. In the U.S., mechanized equipment is commonly used for the installation of directly buried cable. Such equipment performs the trenching, cable lying, and backfilling in a single operation.

Cable in Ducts

Ducts or conduits are normally used under roadways, or in locations where mechanical or other types of damage may be expected. Conduit installation is expensive and complex and the conduit type should be selected carefully. The conduit materials most commonly used for underground cables are precast concrete, plastic, fiber, and iron. Because iron pipes are so costly, their use is normally limited to places where only very shallow digging is possible and mechanical rigidity is required. Conduits used under roads and railways are commonly encased in concrete.

All pipes and ducts must be installed and joined according to the manufacturer's specifications. It is common practice to seal the duct to prevent foreign materials from entering the conduit. Another practice consists of including a noncorroding draw line in the duct prior to sealing it. This is a beneficial provision, especially in long lengths between manholes. Finally, it is not uncommon to install spare ducts for future use during the construction stage.

Manholes

Manholes are typically built at the splices as a way for workers to install cables or other equipment, provide test points, and perform routine or emergency maintenance operations. The dimensions of the manholes should accommodate the conduits and cables entering the manhole and any equipment installed, such as transformers or protective equipment. Finally, the manhole should be big enough to provide adequate headroom for workers.

In less than perfect soils, manholes should be provided with adequate drainage. In extreme cases of wet soil, and if sewer connections are possible, appropriate connections should be made from the manholes to the sewer lines. If sewer connections are not possible, manholes should be constructed with waterproof concrete. This is a necessary requirement where the natural water table is higher than the bottom of the manhole.

Testing, Troubleshooting, and Fault Location

Testing of insulated cables is performed for a variety of reasons. When completed at the factory, tests are usually an integral part of the manufacturing process. These tests aim at measuring important parameters such as thickness of insulation, conductivity, etc. After installation, tests are performed almost exclusively to detect the location of failures. This section presents an overview of some commonly performed pre- and post-installation tests.

Four types of tests are performed on cables:

1. Routine tests performed on every cable to ensure compliance with specifications and the integrity of every cable.
2. Sample tests performed periodically on cable samples to ensure manufacturing consistency.
3. Experimental tests performed to determine characteristics of newly developed materials.
4. Site tests performed after installation in the field.

Tests 1 through 3 provide insight into cable dimensions, insulation resistance and capacitance. In addition, high voltage, impulse voltage, and measurement of partial discharge trials are performed as part of these tests.

Impulse Test

The purpose of the impulse voltage test is to check the cable under conditions of transient overvoltages or other temporary high-voltage surges normally associated with switching operations or atmospheric discharges. Such conditions are often encountered during the regular operation of a cable. During a typical impulse voltage test, the cable is first subjected to mechanical stresses typically found in a routine installation procedure, such as bending. A set of negative and positive impulses at the withstand level are applied to the cable which has been heated at the maximum conductor temperature. After the impulses are applied, the cable may be subjected to a high-voltage test.

High-Voltage Test

High-voltage tests are performed to locate defects in insulated cables. The selection of voltage level is extremely important: if the voltage is too high, it may cause incipient damage of the cable, affecting future service life. On the other hand, the breakdown of the insulation under high voltage is also time dependent and therefore a number of voltage/time to breakdown tests are performed. These tests are important for developmental purposes but have little use as predictors of service life.

Partial Discharge Test

Gas-filled voids are found in the insulation and at the juncture of dielectric and conductive sheaths of cables. The breakdown of this gas produces a phenomenon known as partial discharge. Partial discharge occurs at these voids because the breakdown strength of the gas within voids is much less than that of the typical cable insulation material. In fact, discharges may occur at voltages lower than the operating voltage of the cable. The level of voltage at which partial discharge occurs first is called the discharge inception voltage.

Incidence of Faults and Fault Location

The major causes of faults in underground cables (listed below) are an indication of some of the hazards to which cables are exposed: accidental contact, aging, faulty installation, lightning, and others. Depending on the location, accidental contact and aging account for more than 50% of all faults. Because accidental contacts with cables often produce only minor damage, there is a time lapse until the cable deteriorates enough to produce failure and operation of protective equipment. When there is accidental contact with cables, failures may occur as a result of prolonged exposure to moisture, loss of dielectric material, corrosion, etc.

An effective location of faults in an underground cable system requires a systematic approach. Typically, the following steps are followed to successfully locate a fault:

1. The existence and type of fault is first determined. This requires the determination of damage to the shield and or changes in continuity of the conductor.
2. Once the existence of the fault is assured, it is necessary to diagnose the cause of the fault. This can be determined in several ways, including ruling out faulty operation of protective equipment. A Megger may be used to check continuity.

3. In the case of intermittent faults, preconditioning of the cable may be required to produce a "stable" fault. This step may or may not be required, depending on the type of equipment available for fault location. The most common preconditioning technique used consists of allowing current to pass through the fault to carbonize the insulation.
4. Once a consistent fault is observed, procedures for fault prelocation are followed. This is known as pinpointing the fault location. This step assures that the location and repair of a faulted underground cable will be accomplished in one single excavation. Most modern instruments used to pinpoint faults are based on traveling waves theory.

References

Ametani, A., A general formulation of impedance and admittance of cables, *IEEE Trans. PAS-99*, 3, 902-910, May/June 1980.

Densley, R. J., An investigation into the growth of electrical trees in XLPE cable insulation, *IEEE Trans. EI-14*, 148-158, June 1979.

Gale, P. F., Cable fault location by impulse current method, *Proc. IEEE 122*, 403-408, April 1975.

Graneau, P., *Underground Power Transmission: The Science, Technology, and Economics of High Voltage Cables*, John Wiley & Sons, New York, 1979.

IEEE Guide for the Application of Sheath-Bonding Methods for Single-Conductor Cable and the Calculation of Induced Voltages and Currents in Cable Sheaths. *ANSI/IEEE Std. 575-1988*.

Kalsi, S. S. and Minnich, S. H., Calculation of circulating current losses in cable conductors, *IEEE Trans. PAS-99*, March/April, 1980.

Kojima, K. et al., Development and commercial use of 275 kV XLPE power cable, *IEEE Trans. PAS-100*, 1, 203-210.

McAllister, D., *Electric Cables Handbook*, Granada Publishing Ltd., Great Britain, 1982.

Milne, A. G. and Mochlinski, K., Characteristics of soil affecting cable ratings, *Proc. IEEE 111* (5), 1964.

Pansini, A., *Guide to Electrical Power Distribution Systems*, Pennwell Books, Tulsa, OK, 1996.

Transmission cable magnetic field research, *EPRI EL-6759-D, Project 7898-21, Final Report*, April 1990.

4.6 Transmission Line Parameters

Manuel Reta-Hernández

The power transmission line is one of the major components of an electric power system. Its major function is to transport electric energy, with minimal losses, from the power sources to the load centers, usually separated by long distances. The three basic electrical parameters of a transmission line are:

1. Series resistance
2. Series inductance
3. Shunt capacitance

Once evaluated, the parameters are used to model the line and to perform design calculations. The arrangement of the parameters (equivalent circuit) representing the line depends upon the length of the line.

Equivalent Circuit

A transmission line is defined as a short-length line if its length is less than 80 km (50 mi). In this case, the capacitive effect is negligible and only the resistance and inductive reactance are considered. Assuming balanced conditions, the line can be represented by the equivalent circuit of a single phase with resistance R, and inductive reactance X_L in series, as shown in Fig. 4.48.

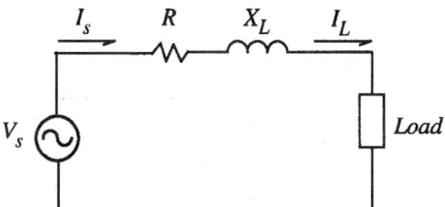

FIGURE 4.48 Equivalent circuit of a short-length transmission line.

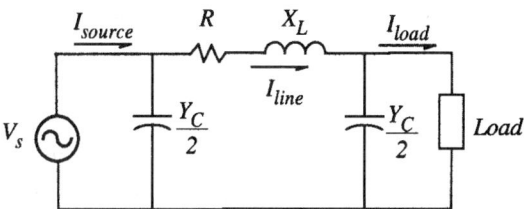

FIGURE 4.49 Equivalent circuit of a medium-length transmission line.

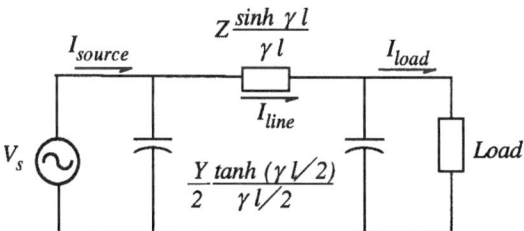

FIGURE 4.50 Equivalent circuit of a long-length transmission line.

If the line is between 80 km (50 mi) and 240 km (150 mi) long, the line is considered a medium-length line and its single-phase equivalent circuit can be represented in a nominal π circuit configuration. The shunt capacitance of the line is divided into two equal parts, each placed at the sending and receiving ends of the line. Figure 4.49 shows the equivalent circuit for a medium-length line.

Both short- and medium-length transmission lines use approximated lumped-parameter models. However, if the line is more than 240 km long, the model must consider parameters uniformly distributed along the line. The appropriate series impedance and shunt capacitance are found by solving the corresponding differential equations, where voltages and currents are described as a function of distance and time. Figure 4.50 shows the equivalent circuit for a long line,

where $Z = z\,l$ = equivalent total series impedance (Ω)
  $Y = y\,l$ = total shunt admittance (S)
  z = series impedance per unit length (Ω/m)
  y = shunt admittance per unit length (S/m)
  $\gamma = \sqrt{z\,y}$ = propagation constant

Detailed methods for calculating the three basic transmission line parameters are presented in the next sections.

Resistance

The AC resistance of a conductor in a transmission line is based on the calculation of its DC resistance. If DC is flowing along a round cylindrical conductor, the current is uniformly distributed over its cross-section area and the DC resistance is evaluated by:

$$R_{dc} = \frac{\rho l}{A} \quad [\Omega] \tag{4.15}$$

where ρ = conductor resistivity at a given temperature (Ω-m)
l = conductor length (m)
A = conductor cross-section area (m²)

If AC current is flowing, rather than DC current, the conductor effective resistance is higher due to the skin effect (presented in the next section).

Frequency Effect

The frequency of the AC voltage produces a second effect on the conductor resistance due to the nonuniform distribution of the current. This phenomenon is known as skin effect. As frequency increases, the current tends to go toward the surface of the conductor and the current density decreases at the center. Skin effect reduces the effective cross-section area used by the current and thus the effective resistance increases. Also, although in small amount, a further resistance increase occurs when other current-carrying conductors are present in the immediate vicinity. A skin correction factor k, obtained by differential equations and Bessel functions, is considered to reevaluate the AC resistance. For 60 Hz, k is estimated around 1.02:

$$R_{ac} = R_{ac} \, k \tag{4.16}$$

Other variations in resistance are caused by:

- temperature
- spiraling of stranded conductors
- bundle conductors arrangement

Temperature Effect

The resistivity of any metal varies linearly over an operating temperature, and therefore the resistance of any conductor suffers the same variations. As temperature rises, the resistance increases linearly, according to the following equation:

$$R_2 = R_1 \left(\frac{T + t_2}{T + t_1} \right) \tag{4.17}$$

where R_2 = resistance at second temperature t_2 (°C)
R_1 = resistance at initial temperature t_1 (°C)
T = temperature coefficient for the particular material (°C)

Resistivity (ρ) and temperature coefficient (T) constants depend on the particular conductor material. Table 4.8 lists resistivity and temperature coefficients of some typical conductor materials.

Spiraling and Bundle Conductor Effect

There are two types of transmission line conductors: overhead and underground. Overhead conductors, made of naked metal and suspended on insulators, are preferred over underground conductors because of the lower cost and ease of maintenance.

Transmission System

TABLE 4.8 Resistivity and Temperature Coefficient of Some Materials

Material	Resistivity at 20°C (Ω-m)	Temperature Coefficient (°C)
Silver	1.59×10^{-8}	243.0
Annealed copper	1.72×10^{-8}	234.5
Hard-drawn copper	1.77×10^{-8}	241.5
Aluminum	2.83×10^{-8}	228.1

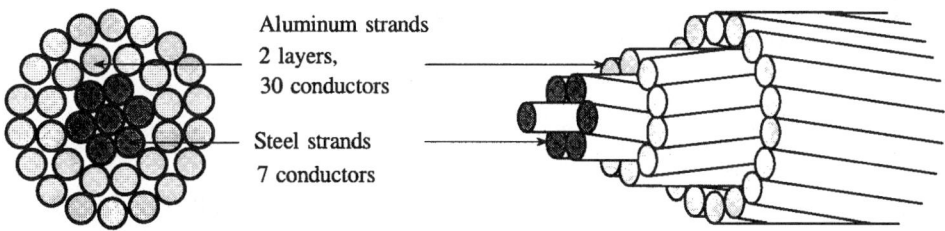

FIGURE 4.51 Stranded aluminum conductor with stranded steel core (ACSR).

In overhead transmission lines, aluminum is a common material because of the lower cost and lighter weight compared to copper, although more cross-section area is needed to conduct the same amount of current. The aluminum conductor, steel-reinforced (ACSR), is one of the most used conductors. It consists of alternate layers of stranded conductors, spiraled in opposite directions to hold the strands together, surrounding a core of steel strands as shown in Fig. 4.51. The purpose of introducing a steel core inside the stranded aluminum conductors is to obtain a high strength-to-weight ratio. A stranded conductor offers more flexibility and is easier to manufacture than a solid large conductor. However, the total resistance is increased because the outside strands are larger than the inside strands due to the spiraling.

The resistance of each wound conductor at any layer, per unit length, is based on its total length as follows:

$$R_{cond} = \frac{\rho}{A}\sqrt{1+\left(\pi\frac{1}{p}\right)^2} \quad [\Omega \text{ m}] \tag{4.18}$$

where R_{cond} = resistance of wound conductor (Ω)

$\sqrt{1+\left(\pi\frac{1}{p}\right)^2}$ = length of wound conductor (m)

$p_{cond} = \dfrac{l_{turn}}{2r_{layer}}$ = relative pitch of wound conductor

l_{turn} = length of one turn of the spiral (m)

$2r_{layer}$ = diameter of the layer (m)

The parallel combination of n conductors with the same diameter per layer gives the resistance per layer as follows:

$$R_{layer} = \frac{1}{\sum_{i=1}^{n}\frac{1}{R_i}} \quad [\Omega/\text{m}]. \tag{4.19}$$

Similarly, the total resistance of the stranded conductor is evaluated by the parallel combination of resistances per layer.

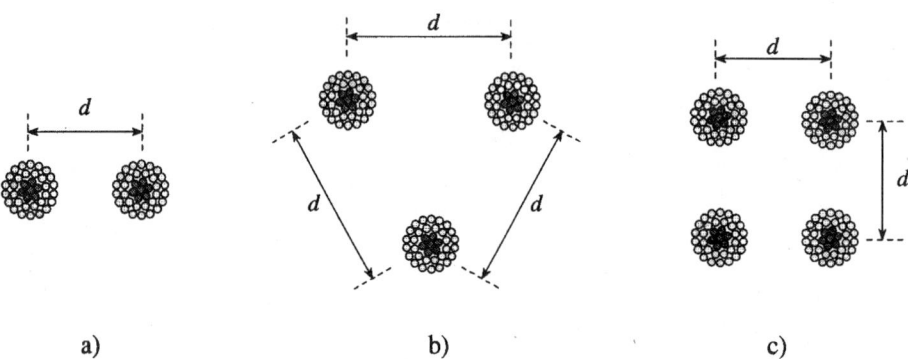

FIGURE 4.52 Stranded conductors arranged in bundles of (a) two, (b) three, and (c) four.

In high-voltage transmission lines, there may be more than one conductor per phase. This is a bundle configuration used to increase the current capability and to reduce corona discharge. By increasing the number of conductors per phase, the current capacity is increased, and the total AC resistance is proportionally decreased with respect to the number of conductors per bundle. Corona occurs when high electric field strength along the conductor surface causes ionization of the surrounding air, producing conducting atmosphere and thus producing corona losses, audible noise, and radio interference. Although corona losses depend on meteorological conditions, their evaluation takes into account the conductance between conductors and between conductors and ground. Conductor bundles may be applied to any voltage but are always used at 345 kV and above to limit corona. To maintain the distance between bundle conductors, spacers are used which are made of steel or aluminum bars. Figure 4.52 shows some arrangements of stranded bundle configurations.

Current-Carrying Capacity (Ampacity)

In overhead transmission lines, the current-carrying capacity is determined mostly by the conductor resistance and the heat dissipated from its surface. The heat generated in a conductor (I^2R) is dissipated from its surface area by convection and by radiation:

$$I^2 R = S(w_c + w_r) \quad [\text{W}] \tag{4.20}$$

where R = conductor resistance (Ω)
  I = conductor current-carrying (A)
  S = conductor surface area (sq. in.)
  w_c = convection heat loss (W/sq. in.)
  w_r = radiation heat loss (W/sq. in.)

Dissipation by convection is defined as:

$$w_c = \frac{0.0128 \sqrt{pv}}{T_{air}^{0.123} \sqrt{d_{cond}}} \Delta t \quad [\text{W}] \tag{4.21}$$

where p = atmospheric pressure (atm)
  v = wind velocity (ft/sec)
  d_{cond} = conductor diameter (in.)
  T_{air} = air temperature (Kelvin)
  Δt = $T_c - T_{air}$ = temperature rise of the conductor (°C)

Dissipation by radiation is obtained from the Stefan-Boltzmann law and is defined as:

$$w_r = 36.8\, E \left[\left(\frac{T_c}{1000}\right)^4 - \left(\frac{T_{air}}{1000}\right)^4 \right] \quad [\text{W/sq. in.}] \qquad (4.22)$$

where w_r = radiation heat loss (W/sq. in.)
$\quad\ E$ = emissivity constant (1 for the absolute black body and 0.5 for oxidized copper)
$\quad\ T_c$ = conductor temperature (°C)
$\quad\ T_{air}$ = ambient temperature (°C)

Substituting Eqs. (4.21) and (4.22) in (4.20) we can obtain the conductor ampacity at given temperatures.

$$I = \sqrt{\frac{(w_c + w_r) S}{R}} \quad [\text{A}] \qquad (4.23)$$

$$I = \sqrt{\frac{S}{R}\left(\left(\frac{\Delta t\left(0.0128\sqrt{pv}\right)}{T_{air}^{0.123}\sqrt{d_{cond}}}\right) + \left(36.8\, E\left(\frac{T_c^4 - T_{air}^4}{1000^4}\right)\right)\right)} \quad [\text{A}] \qquad (4.24)$$

Some approximated current-carrying capacity values for overhead aluminum and aluminum reinforced conductors are presented in the section "Characteristics of Overhead Conductors."

Inductance and Inductive Reactance

The magnetic flux generated by the current in transmission line conductors produces a total inductance whose magnitude depends on the line configuration. To determine the inductance of the line, it is necessary to calculate, as in any magnetic circuit with permeability μ:

1. the magnetic field intensity H,
2. the magnetic field density B, and
3. the flux linkage λ.

Inductance of a Solid, Round, Infinitely Long Conductor

Consider a long, solid, cylindrical conductor with radius r, carrying current I as shown in Fig. 4.53. If the conductor is a nonmagnetic material, and the current is assumed to be uniformly distributed (no skin effect), then the generated internal and external magnetic field lines are concentric circles around the conductor with direction defined by the right-hand rule.

Internal Inductance

To obtain the internal inductance, a magnetic field at radius x inside the conductor is chosen as shown in Fig. 4.54.

The fraction of the current I_x enclosed in the area of the circle is determined by:

$$I_x = \frac{\pi x^2}{\pi r^2}\, I \quad [\text{A}]. \qquad (4.25)$$

Ampere's law determines the magnetic field intensity Hx constant at any point along the circle contour:

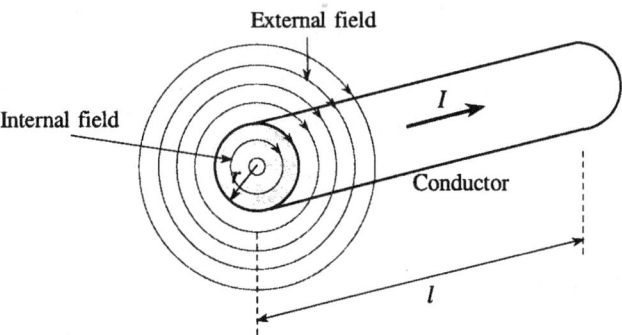

FIGURE 4.53 External and internal concentric magnetic flux lines around the conductor.

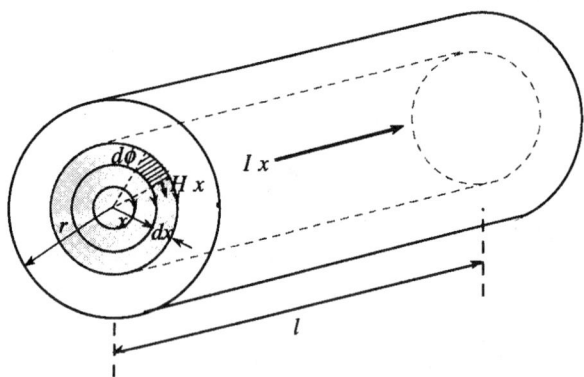

FIGURE 4.54 Internal magnetic flux.

$$H_x = \frac{I_x}{2\pi x} = \frac{I}{2\pi r^2} x \quad [\text{A/m}]. \tag{4.26}$$

The magnetic flux density B_x is obtained by

$$B_x = \mu H_x = \frac{\mu_0}{2\pi}\left(\frac{I x}{r^2}\right) \quad [\text{T}] \tag{4.27}$$

where $\mu = \mu_0 = 4\pi \times 10^{-7}$ (H/m) for a nonmagnetic material.

The differential flux $d\phi$ enclosed in a ring of thickness dx for a 1-m length of conductor, and the differential flux linkage $d\lambda$ in the respective area are

$$d\phi = B_x\, dx = \frac{\mu_0}{2\pi}\left(\frac{I x}{r^2}\right) dx \quad [\text{Wb/m}] \tag{4.28}$$

$$d\lambda = \frac{\pi x^2}{\pi r^2} d\phi = \frac{\mu_0}{2\pi}\left(\frac{I x^3}{r^4}\right) dx \quad [\text{Wb-turn/m}]. \tag{4.29}$$

Transmission System

The internal flux linkage is obtained by integrating the differential flux linkage from $x = 0$ to $x = r$

$$\lambda_{int} = \int_0^r d\lambda = \frac{\mu_0}{8\pi} I \quad [\text{Wb-turn}/\text{m}]. \tag{4.30}$$

The inductance due to internal flux linkage per-unit length becomes

$$L_{int} = \frac{\lambda_{int}}{I} = \frac{\mu_0}{8\pi} \quad [\text{H}/\text{m}]. \tag{4.31}$$

External Inductance

The external inductance is evaluated assuming that the total current I is concentrated at the conductor surface (maximum skin effect). At any point on an external magnetic field circle of radius y (Fig. 4.55), the magnetic field intensity H_y and the magnetic field density B_y are

$$H_y = \frac{I}{2\pi y} \quad [\text{A}/\text{m}] \quad \text{and} \tag{4.32}$$

$$B_y = \mu H_y = \frac{\mu_0}{2\pi} \frac{I}{y} \quad [\text{T}]. \tag{4.33}$$

The differential flux $d\phi$ enclosed in a ring of thickness dy, from point D_1 to point D_2, for a 1-m length of conductor is

$$d\phi = B_y \, dy = \frac{\mu_0}{2\pi} \frac{I}{y} dy \quad [\text{Wb}/\text{m}]. \tag{4.34}$$

As the total current I flows in the surface conductor, then the differential flux linkage $d\lambda$ has the same magnitude as the differential flux $d\phi$.

$$d\lambda = d\phi = \frac{\mu_0}{2\pi} \frac{I}{y} dy \quad [\text{Wb-turn}/\text{m}] \tag{4.35}$$

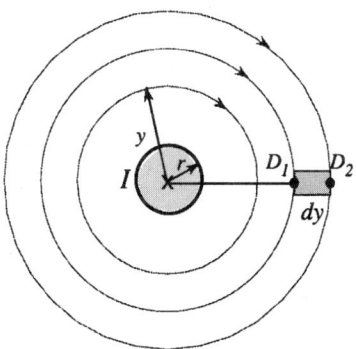

FIGURE 4.55 External magnetic flux.

The total external flux linkage enclosed by the ring is obtained by integrating from D_1 to D_2

$$\lambda_{1-2} = \int_{D_1}^{D_2} d\lambda = \frac{\mu_0}{2\pi} I \int_{D_1}^{D_2} \frac{dy}{y} = \frac{\mu_0}{2\pi} I \ln\left(\frac{D_1}{D_2}\right) \quad [\text{Wb-turn}/\text{m}]. \qquad (4.36)$$

In general, the external flux linkage from the surface of the conductor to any point D is

$$\lambda_{\text{ext}} = \int_{r}^{D} d\lambda = \frac{\mu_0}{2\pi} I \ln\left(\frac{D}{r}\right) \quad [\text{Wb-turn}/\text{m}]. \qquad (4.37)$$

The summation of the internal and external flux linkage at any point D permits evaluation of the total inductance of the conductor L_{tot} per-unit length as follows,

$$\lambda_{\text{intl}} + \lambda_{\text{ext}} = \frac{\mu_0}{2\pi} I \left[\frac{1}{4} + \ln\left(\frac{D}{r}\right)\right] = \frac{\mu_0}{2\pi} I \ln\left(\frac{D}{e^{-1/4} r}\right) \quad [\text{Wb-turn}/\text{m}] \qquad (4.38)$$

$$L_{\text{tot}} = \frac{\lambda_{\text{int}} + \lambda_{\text{ext}}}{I} = \frac{\mu_0}{2\pi} \ln\left(\frac{D}{\text{GMR}}\right) \quad [\text{H}/\text{m}] \qquad (4.39)$$

where GMR (geometric mean radius) = $e^{-1/4} r = 0.7788\, r$.

Inductance of a Two-Wire, Single-Phase Line

Consider a two-wire, single-phase line with conductors A and B with the same radius r, separated by a distance $D > r_A$ and r_B, and conducting the same current I as shown in Fig. 4.56. The current flows from the source to the load in conductor A and returns in conductor B ($I_A = -I_B$).

The magnetic flux generated by one conductor links the second conductor. The total flux linking conductor A, for instance, has two components: (a) the flux generated by conductor A, and (b) the flux generated by conductor B which links conductor A.

As shown in Fig. 4.57, the total flux linkage from conductors A and B at point P is

$$\lambda_{AP} = \lambda_{AAP} + \lambda_{ABP} \qquad (4.40)$$

$$\lambda_{BP} = \lambda_{BAP} + \lambda_{BBP} \qquad (4.41)$$

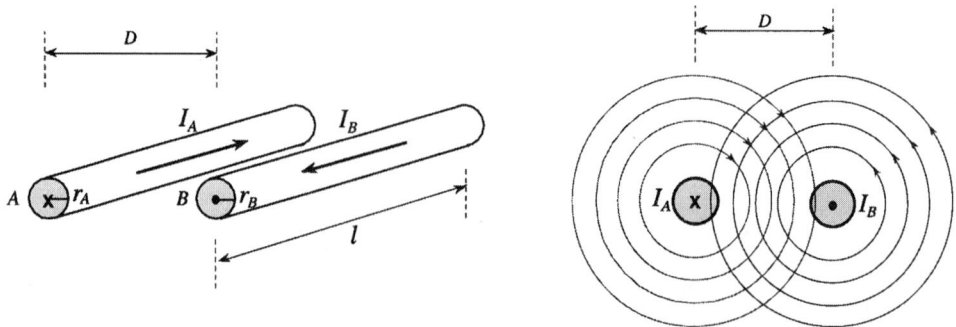

FIGURE 4.56 External magnetic flux around conductors in a two-wire, single-phase line.

Transmission System

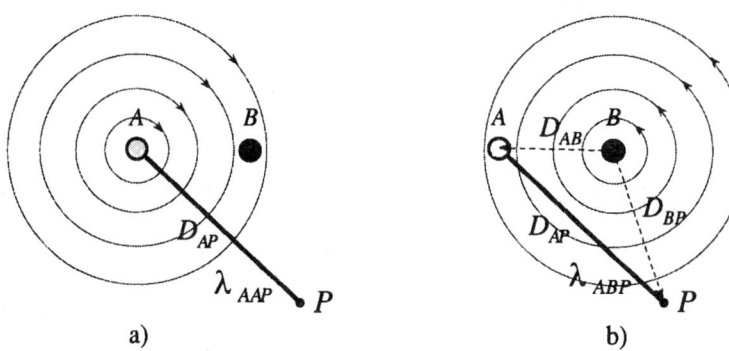

FIGURE 4.57 Flux linkage of (a) conductor A at point P, and (b) conductor B on conductor A at point P. Single-phase.

where λ_{AAP} = flux linkage from magnetic field of conductor A at point P
λ_{ABP} = flux linkage from magnetic field of conductor B on conductor A at point P
λ_{BAP} = flux linkage from magnetic field of conductor A on conductor B at point P
λ_{BBP} = flux linkage from magnetic field of conductor B at point P

A graphical description of flux linkages λ_{AAP} and λ_{ABP} is shown in Fig. 4.57.

The equations of the different flux linkage above are found by analyzing the flux linkage in a single conductor.

$$\lambda_{AAP} = \frac{\mu_0}{2\pi} I \ln\left(\frac{D_{AP}}{GMR_A}\right) \tag{4.42}$$

$$\lambda_{ABP} = \int_D^{D_{BP}} B_{BP}\, dP = -\frac{\mu_0}{2\pi} I \ln\left(\frac{D_{BP}}{D}\right) \tag{4.43}$$

$$\lambda_{BAP} = \int_D^{D_{AP}} B_{AP}\, dP = \frac{\mu_0}{2\pi} I \ln\left(\frac{D_{AP}}{D}\right) \tag{4.44}$$

$$\lambda_{BBP} = -\frac{\mu_0}{2\pi} I \ln\left(\frac{D_{BP}}{GMR_B}\right) \tag{4.45}$$

The total flux linkage of the system at point P is the algebraic summation of λ_{AP} and λ_{BP}

$$\lambda_P = \lambda_{AP} + \lambda_{BP} = \left(\lambda_{AAP} + \lambda_{ABP}\right) - \left(\lambda_{BAP} + \lambda_{BBP}\right) \tag{4.46}$$

$$\lambda_P = \frac{\mu_0}{2\pi} I \ln\left[\left(\frac{D_{AP}}{GMR_A}\right)\left(\frac{D}{D_{BP}}\right)\left(\frac{D_{BP}}{GMR_B}\right)\left(\frac{D}{D_{AP}}\right)\right] = \frac{\mu_0}{2\pi} I \ln\left(\frac{D^2}{GMR_A\, GMR_B}\right) \tag{4.47}$$

If the conductors have the same radius, $r_A = r_B = r$, and the point P is shifted to infinity, then the total flux leakage of the system is

$$\lambda = \frac{\mu_0}{\pi} I \ln\left(\frac{D}{GMR}\right) \;[\text{Wb-turn}/m], \tag{4.48}$$

and the total inductance per-unit length becomes

$$L_{\text{1-phase system}} = \frac{\lambda}{I} = \frac{\mu_0}{\pi} \ln\left(\frac{D}{GMR}\right) \quad [\text{H/m}]. \tag{4.49}$$

It can be seen that the inductance of the single-phase system is twice the inductance of a single conductor.

For a line with stranded conductors, the inductance is determined using a new GMR value (GMR_{stranded}) evaluated according to the number of conductors. Generally, the GMR_{stranded} for any particular cable can be found in conductors tables. Additionally, if the line is composed of bundle conductors, the inductance is reevaluated taking into account the number of bundle conductors and the separation among them. The GMR_{bundle} is introduced to determine the final inductance value.

Assuming the same separation among bundles, the equation for GMR_{bundle}, up to three conductors per bundle, is defined as

$$GMR_{n \text{ bundle conductors}} = \left(d^{n-1} GMR_{\text{stranded}}\right)^{1/n} \tag{4.50}$$

where n = number of conductors per bundle
GMR_{stranded} = GMR of the stranded conductor
d = distance between bundle conductors

Inductance of a Three-Phase Line

The derivations for the inductance in a single-phase system can be extended to obtain the inductance per phase in a three-phase system. Consider a three-phase, three-conductor system as shown in Fig. 4.58. The currents I_A, I_B, and I_C circulate along conductors with radius r_A, r_B, and r_C, and the separation between conductors are D_{AB}, D_{BC}, and D_{CA} (where $D > r$).

The flux linkage calculation of conductor A at point P, is calculated as

$$\lambda_{AP} = \lambda_{AAP} + \lambda_{ABP} + \lambda_{ACP} \tag{4.51}$$

$$\lambda_{AAP} = \frac{\mu_0}{2\pi} I_A \ln\left(\frac{D_{AP}}{GMR_A}\right) \tag{4.52}$$

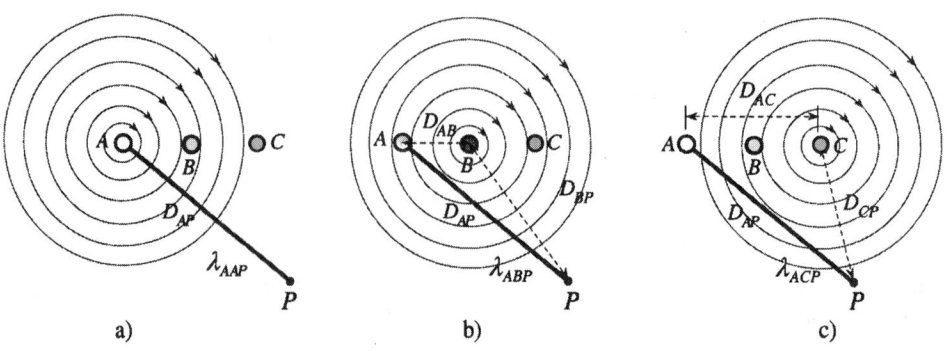

FIGURE 4.58 Flux linkage of (a) conductor A at point P, (b) conductor B on conductor A at point P, and (c) conductor C on conductor A at point P. Three-phase.

$$\lambda_{ABP} = \int_{D_{AB}}^{D_{BP}} B_{BP}\, dP = \frac{\mu_0}{2\pi} I_B \ln\left(\frac{D_{BP}}{D_{AB}}\right) \tag{4.53}$$

$$\lambda_{ACP} = \int_{D_{AC}}^{D_{CP}} B_{CP}\, dP = \frac{\mu_0}{2\pi} I_C \ln\left(\frac{D_{CP}}{D_{AC}}\right) \tag{4.54}$$

where λ_{AP} = total flux linkage of conductor A at point P
λ_{AAP} = flux linkage from magnetic field of conductor A at point P
λ_{ABP} = flux linkage from magnetic field of conductor B on conductor A at point P
λ_{ACP} = flux linkage from magnetic field of conductor C on conductor B at point P

The flux linkages of conductors B and C to point P have expressions similar to those for conductor A. All flux linkages expressions are

$$\lambda_{AP} = \frac{\mu_0}{2\pi}\left[I_A \ln\left(\frac{D_{AP}}{GMR_A}\right) + I_B \ln\left(\frac{D_{BP}}{D_{AB}}\right) + I_C \ln\left(\frac{D_{CP}}{D_{AC}}\right) \right] \tag{4.55}$$

$$\lambda_{BP} = \frac{\mu_0}{2\pi}\left[I_A \ln\left(\frac{D_{AP}}{D_{BA}}\right) + I_B \ln\left(\frac{D_{BP}}{GMR_B}\right) + I_C \ln\left(\frac{D_{CP}}{D_{BC}}\right) \right] \tag{4.56}$$

$$\lambda_{CP} = \frac{\mu_0}{2\pi}\left[I_A \ln\left(\frac{D_{AP}}{D_{CA}}\right) + I_B \ln\left(\frac{D_{BP}}{D_{CB}}\right) + I_C \ln\left(\frac{D_{CP}}{GMR_C}\right) \right] \tag{4.57}$$

Rearranging the expressions, we have

$$\begin{aligned}\lambda_{AP} &= \frac{\mu_0}{2\pi}\left[I_A \ln\left(\frac{1}{GMR_A}\right) + I_B \ln\left(\frac{1}{D_{AB}}\right) + I_C \ln\left(\frac{1}{D_{AC}}\right) \right] \\ &+ \frac{\mu_0}{2\pi}\left[I_A \ln(D_{AP}) + I_B \ln(D_{BP}) + I_C \ln(D_{CP}) \right]\end{aligned} \tag{4.58}$$

$$\begin{aligned}\lambda_{BP} &= \frac{\mu_0}{2\pi}\left[I_A \ln\left(\frac{1}{D_{BA}}\right) + I_B \ln\left(\frac{1}{GMR_B}\right) + I_C \ln\left(\frac{1}{D_{BC}}\right) \right] \\ &+ \frac{\mu_0}{2\pi}\left[I_A \ln(D_{AP}) + I_B \ln(D_{BP}) + I_C \ln(D_{CP}) \right]\end{aligned} \tag{4.59}$$

$$\begin{aligned}\lambda_{CP} &= \frac{\mu_0}{2\pi}\left[I_A \ln\left(\frac{1}{D_{CA}}\right) + I_B \ln\left(\frac{1}{D_{CB}}\right) + I_C \ln\left(\frac{1}{GMR_C}\right) \right] \\ &+ \frac{\mu_0}{2\pi}\left[I_A \ln(D_{AP}) + I_B \ln(D_{BP}) + I_C \ln(D_{CP}) \right]\end{aligned} \tag{4.60}$$

The second part of each equation is zero for practical conditions assuming that point P is shifted to infinity. Therefore, the expressions of the total flux linkage for all conductors are

$$\lambda_A = \frac{\mu_0}{2\pi}\left[I_A \ln\left(\frac{1}{GMR_A}\right) + I_B \ln\left(\frac{1}{D_{AB}}\right) + I_C \ln\left(\frac{1}{D_{AC}}\right)\right] \quad (4.61)$$

$$\lambda_B = \frac{\mu_0}{2\pi}\left[I_A \ln\left(\frac{1}{D_{BA}}\right) + I_B \ln\left(\frac{1}{GMR_B}\right) + I_C \ln\left(\frac{1}{D_{BC}}\right)\right] \quad (4.62)$$

$$\lambda_C = \frac{\mu_0}{2\pi}\left[I_A \ln\left(\frac{1}{D_{CA}}\right) + I_B \ln\left(\frac{1}{D_{CB}}\right) + I_C \ln\left(\frac{1}{GMR_C}\right)\right] \quad (4.63)$$

The flux linkage of each phase conductor depends on the three currents, and therefore the inductance per phase is not only one as in the single-phase system. Instead, three different inductances (self and mutual conductor inductances) exist. Calculating the inductance values from the equations above and arranging the equations in a matrix form, we can obtain the set of inductances in the system

$$\begin{bmatrix}\lambda_A \\ \lambda_B \\ \lambda_C\end{bmatrix} = \begin{bmatrix}L_{AA} & L_{AB} & L_{AC} \\ L_{BA} & L_{BB} & L_{BC} \\ L_{CA} & L_{CB} & L_{CC}\end{bmatrix}\begin{bmatrix}I_A \\ I_B \\ I_C\end{bmatrix} \quad (4.64)$$

where $\lambda_A, \lambda_B, \lambda_C$ = total flux linkage of conductor A, B, and C
L_{AA}, L_{BB}, L_{CC} = self-inductance of conductors A, B, and C field of conductor A at point P
$L_{AB}, L_{BC}, L_{CA}, L_{BA}, L_{CB}, L_{AC}$ = mutual inductance among conductors

With nine different inductances in a simple three-phase system, the analysis could be a little more complicated. A single inductance per phase can be obtained, however, if the three conductors are arranged with the same separation among them $D = D_{AB} = D_{BC} = D_{CA}$ (triangle configuration). In this case the flux linkage of conductor A per-unit length is

$$\lambda_A = \frac{\mu_0}{2\pi}\left[I_A \ln\left(\frac{1}{GMR_A}\right) + I_B \ln\left(\frac{1}{D}\right) + I_C \ln\left(\frac{1}{D}\right)\right]. \quad (4.65)$$

Assuming a balanced system ($I_A + I_B + I_C = 0$, or $I_A = -I_B - I_C$), then the flux linkage is

$$\lambda_A = \frac{\mu_0}{2\pi}I_A \ln\left(\frac{D}{GMR_A}\right) \quad [\text{Wb-turn}/\text{m}]. \quad (4.66)$$

If the GMR value is the same in all phase conductors, the total flux linkage expression is the same for all phases. Therefore, the equivalent inductance per phase is

$$L_{phase} = \frac{\mu_0}{2\pi}\ln\left(\frac{D}{GMR_{cond}}\right) \quad [\text{H}/\text{m}]. \quad (4.67)$$

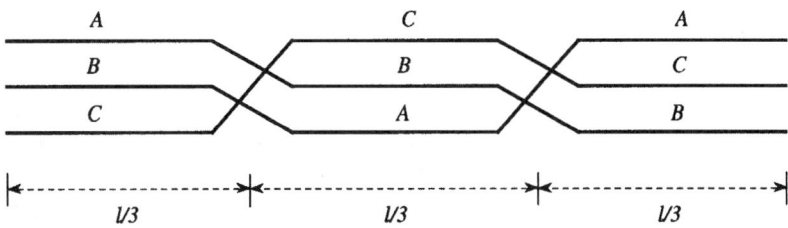

FIGURE 4.59 Arrangement of conductors in a transposed line.

Inductance of Transposed Three-Phase Transmission Lines

In actual transmission lines, the phase conductors generally do not have a symmetrical (triangular) arrangement. However, if the phase conductors are transposed, an average distance GMD (geometrical mean distance) is substituted for distance D, and the calculation of the phase inductance derived for symmetrical arrangement is still valid. In a transposed system, each phase conductor occupies the location of the other two phases for one third of the total line length as shown in Fig. 4.59.

The inductance per phase per unit length in a transmission line is

$$L_{phase} = \frac{\mu_0}{2\pi} \ln\left(\frac{GMD}{GMR_{cond}}\right) \quad [H/m] \tag{4.68}$$

where $GMD = \sqrt[3]{D_{AB} D_{BC} D_{CA}}$ = geometrical mean distance for a three-phase line.

Once the inductance per phase is obtained, the inductive reactance can be evaluated as

$$X_{L_{phase}} = 2\pi f L_{phase} = \mu_0 f \ln\left(\frac{GMD}{GMR_{cond}}\right) \quad [\Omega/m]. \tag{4.69}$$

For bundle conductors, the GMR_{bundle} value is determined, as in the single-phase transmission line case, by the number of conductors, and by the number of conductors per bundle and the separation among them. The expression for the total inductive reactance per phase is

$$X_{L_{phase}} = \mu_0 f \ln\left(\frac{GMD}{GMR_{bundle}}\right) \quad [\Omega/m] \tag{4.70}$$

where GMR_{bundle} = $(d^{n-1} GMR_{stranded})^{1/n}$ = geometric mean radius of bundle conductors
GMD = geometric mean distance
$GMR_{stranded}$ = geometric mean radius of stranded conductor
d = distance between bundle conductors
n = number of conductors per bundle
f = frequency (60 Hz)

Capacitance and Capacitive Reactance

To evaluate the capacitance between conductors in a surrounding medium with permitivity ε, it is necessary to first determine the voltage between the conductors, and the electric field strength of the surrounding.

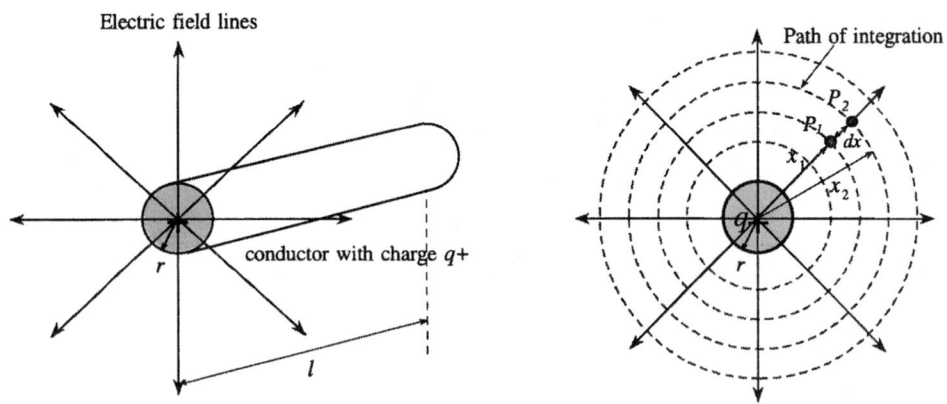

FIGURE 4.60 Electric field produced from a single conductor.

Capacitance of a Single Solid Conductor

Consider a solid, cylindrical, long conductor with radius r, in a free space with permitivity ε_0, and with a charge of $q+$ C per meter uniformly distributed on the surface. There is a constant electric field strength on the surface of cylinder (Fig. 4.60). The resistivity of the conductor is assumed to be zero (perfect conductor), which results in zero internal electric field due to the charge on the conductor.

The charge $q+$ produces an electric field radial to the conductor with equipotential surfaces concentric to the conductor. According to Gauss's law, the total electric flux leaving a closed surface is equal to the total charge inside the volume enclosed by the surface. Therefore, at an outside point P separated x meters from the center of the conductor, the electric field flux density, and the electric field intensity are

$$\text{Density}_P = \frac{q}{A} = \frac{q}{2\pi x} \quad [C] \quad \text{and} \tag{4.71}$$

$$E_P = \frac{\text{Density}_P}{\varepsilon} = \frac{q}{2\pi\varepsilon_0 x} \quad [V/m] \tag{4.72}$$

where Density_P = electric flux density at point P
E_P = electric field intensity at point P
A = surface of a concentric cylinder with one-meter length and radius x (m²)
ε = ε_0 = $\dfrac{10^{-9}}{36\pi}$ = permitivity of free space assumed for the conductor (F/m)

The potential difference or voltage difference between two outside points P_1 and P_2 with corresponding distances x_1 and x_2 from the conductor center is defined by integrating the electric field intensity from x_1 to x_2

$$V_{1-2} = \int_{x_1}^{x_2} E_P \frac{dx}{x} = \int_{x_1}^{x_2} \frac{q}{2\pi\varepsilon_0} \frac{dx}{x} = \frac{q}{2\pi\varepsilon_0} \ln\left[\frac{x_2}{x_1}\right] \quad [V]. \tag{4.73}$$

Then, the capacitance between points P_1 and P_2 is evaluated as

$$C_{1-2} = \frac{q}{V_{1-2}} = \frac{2\pi\varepsilon_0}{\ln\left[\dfrac{x_2}{x_1}\right]} \quad [\text{F}/\text{m}]. \tag{4.74}$$

If point P_1 is located at the conductor surface ($x_1 = r$), and point P_2 is located at ground surface below the conductor ($x_2 = H$), then the voltage of the conductor and the capacitance between the conductor and ground are

$$V_{cond} = \frac{q}{2\pi\varepsilon_0}\ln\left[\frac{H}{r}\right] \quad [\text{V}] \quad \text{and} \tag{4.75}$$

$$C_{cond\text{-}ground} = \frac{q}{V_{cond}} = \frac{2\pi\varepsilon_0}{\ln\left[\dfrac{H}{r}\right]} \quad [\text{F}/\text{m}]. \tag{4.76}$$

Capacitance of a Single-Phase Line with Two Wires

Consider a two-wire, single-phase line with conductors A and B with the same radius r, separated by a distance $D > r_A$ and r_B. The conductors are energized by a voltage source such that conductor A has a charge q^+ and conductor B a charge q^- as shown in Fig. 4.61.

The charge on each conductor generates independent electric fields. Charge q^+ on conductor A generates a voltage $V_{AB\text{-}A}$ between both conductors. Similarly, charge q^- on conductor B generates a voltage $V_{AB\text{-}B}$ between conductors.

$V_{AB\text{-}A}$ is calculated by integrating the electric field intensity, due to the charge on conductor A, on conductor B from r_A to D

$$V_{AB-A} = \int_{r_A}^{D} E_A \, dx = \frac{q}{2\pi\varepsilon_0}\ln\left[\frac{D}{r_A}\right]. \tag{4.77}$$

$V_{AB\text{-}B}$ is calculated by integrating the electric field intensity due to the charge on conductor B from D to r_B

$$V_{AB-B} = \int_{D}^{r_B} E_B \, dx = \frac{-q}{2\pi\varepsilon_0}\ln\left[\frac{r_B}{D}\right]. \tag{4.78}$$

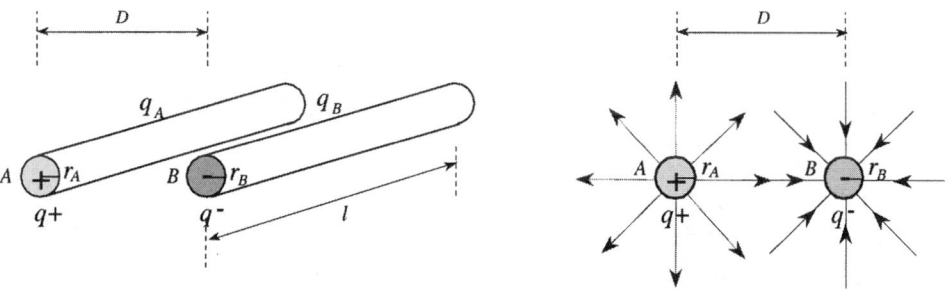

FIGURE 4.61 Electric field produced from a two-wire, single-phase system.

The total voltage is the sum of the generated voltages V_{AB-A} and V_{AB-B}

$$V_{AB} = V_{AB-A} + V_{AB-B} = \frac{q}{2\pi\varepsilon_0}\ln\left[\frac{D}{r_A}\right] - \frac{q}{2\pi\varepsilon_0}\ln\left[\frac{r_B}{D}\right] = \frac{q}{2\pi\varepsilon_0}\ln\left[\frac{D^2}{r_A r_B}\right]. \quad (4.79)$$

If the conductors have the same radius, $r_A = r_B = r$, then the voltage between conductors V_{AB}, and the capacitance between conductors C_{AB}, for a one-meter line length are

$$V_{AB} = \frac{q}{\pi\varepsilon_0}\ln\left[\frac{D}{r}\right] \quad [\text{V}] \quad \text{and} \quad (4.80)$$

$$C_{AB} = \frac{\pi\varepsilon_0}{\ln\left[\dfrac{D}{r}\right]} \quad [\text{F/m}]. \quad (4.81)$$

The voltage between each conductor and ground (Fig. 4.62) is one-half of the voltage between the two conductors. Therefore, the capacitance from either line to ground is twice the capacitance between lines

$$V_{AG} = V_{BG} = \frac{V_{AB}}{2} \quad [\text{V}] \quad (4.82)$$

$$C_{AG} = \frac{q}{V_{AG}} = \frac{2\pi\varepsilon_0}{\ln\left[\dfrac{D}{r}\right]} \quad [\text{F/m}]. \quad (4.83)$$

Capacitance of a Three-Phase Line

Consider a three-phase line with the same voltage magnitude between phases, and assume a balanced system with *abc* (positive) sequence such that $q_A + q_B + q_C = 0$. The conductors have radii r_A, r_B, and r_C, and the spaces between conductors are D_{AB}, D_{BC}, and D_{AC} (where D_{AB}, D_{BC}, and $D_{AC} > r_A$, r_B, and r_C). Also, the effect of earth and neutral conductors is neglected.

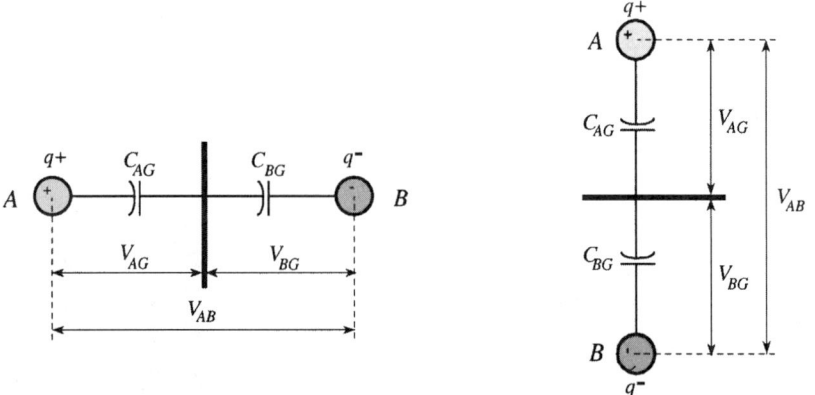

FIGURE 4.62 Capacitance between line-to-ground in a two-wire, single-phase line.

The expression for voltage between two conductors in a single-phase system can be extended to obtain the voltages between conductors in a three-phase system. The expressions for V_{AB} and V_{AC} are

$$V_{AB} = \frac{1}{2\pi\varepsilon_0}\left[q_A \ln\left[\frac{D_{AB}}{r_A}\right] + q_B \ln\left[\frac{r_B}{D_{AB}}\right] + q_C \ln\left[\frac{D_{BC}}{D_{AC}}\right]\right] \quad \text{and} \quad (4.84)$$

$$V_{AC} = \frac{1}{2\pi\varepsilon_0}\left[q_A \ln\left[\frac{D_{CA}}{r_A}\right] + q_B \ln\left[\frac{D_{BC}}{D_{AB}}\right] + q_C \ln\left[\frac{r_C}{D_{AC}}\right]\right] \quad (4.85)$$

If the three-phase system has a triangular arrangement with equidistant conductors such that $D_{AB} = D_{BC} = D_{AC} = D$, with the same radii for the conductors such that $r_A = r_B = r_C = r$ (where $D > r$), the expressions for V_{AB} and V_{AC} are

$$V_{AB} = \frac{1}{2\pi\varepsilon_0}\left[q_A \ln\left[\frac{D}{r}\right] + q_B \ln\left[\frac{r}{D}\right] + q_C \ln\left[\frac{D}{D}\right]\right]$$

$$= \frac{1}{2\pi\varepsilon_0}\left[q_A \ln\left[\frac{D}{r}\right] + q_B \ln\left[\frac{r}{D}\right]\right] \text{ [V]} \quad (4.86)$$

$$V_{AC} = \frac{1}{2\pi\varepsilon_0}\left[q_A \ln\left[\frac{D}{r}\right] + q_B \ln\left[\frac{D}{D}\right] + q_C \ln\left[\frac{r}{D}\right]\right]$$

$$= \frac{1}{2\pi\varepsilon_0}\left[q_A \ln\left[\frac{D}{r}\right] + q_C \ln\left[\frac{r}{D}\right]\right] \text{ [V]} \quad (4.87)$$

Balanced line-to-line voltages with sequence abc, expressed in terms of the line-to-neutral voltage are: $V_{AB} = \sqrt{3}\,V_{AN}\angle 30°$, and $V_{AC} = -V_{CA} = \sqrt{3}\,V_{AN}\angle -30°$; where V_{AN} is the line-to-neutral voltage. Therefore, V_{AN} can be expressed in terms of V_{AB} and V_{AC} as

$$V_{AN} = \frac{V_{AB} + V_{AC}}{3} \quad (4.88)$$

and thus, substituting V_{AB} and V_{AC} from Eqs. (4.81) and (4.82), we have

$$V_{AN} = \frac{1}{6\pi\varepsilon_0}\left[\left[q_A \ln\left[\frac{D}{r}\right] + q_B \ln\left[\frac{r}{D}\right]\right] + \left[q_A \ln\left[\frac{D}{r}\right] + q_C \ln\left[\frac{r}{D}\right]\right]\right]$$

$$= \frac{1}{6\pi\varepsilon_0}\left[2q_A \ln\left[\frac{D}{r}\right] + (q_B + q_C)\ln\left[\frac{r}{D}\right]\right]. \quad (4.89)$$

Under balanced conditions ($q_A + q_B + q_C = 0$, or $-q_A = (q_B + q_C)$) then, the final expression for the line-to-neutral voltage is

$$V_{AN} = \frac{1}{2\pi\varepsilon_0} q_A \ln\left[\frac{D}{r}\right] \quad [\text{V}]. \tag{4.90}$$

The positive sequence capacitance per unit length between phase A and neutral can now be obtained. The same result is obtained for capacitance between phases B and C to neutral.

$$C_{AN} = \frac{q_A}{V_{AN}} = \frac{2\pi\varepsilon_0}{\ln\left[\dfrac{D}{r}\right]} \quad [\text{F}/\text{m}] \tag{4.91}$$

Capacitance of Stranded Bundle Conductors

The calculation of the capacitance in the equation above is based on:

1. solid conductors with zero resistivity (zero internal electric field)
2. distributed charge uniformly
3. equilateral spacing of phase conductors

In actual transmission lines, the resistivity of the conductors produces a small internal electric field and, therefore, the electric field at the conductor surface is smaller than estimated. However, the difference is negligible for practical purposes. Because of the presence of other charged conductors, the charge distribution is nonuniform, and therefore the estimated capacitance is different. However, this effect is negligible for most practical calculations. In a line with stranded conductors, the capacitance is evaluated assuming a solid conductor with the same radius as the outside radius of the stranded conductor. This produces a negligible difference.

Most transmission lines do not have equilateral spacing of phase conductors. This causes differences between the line-to-neutral capacitances of the three phases. However, transposing the phase conductors balances the system, resulting in equal line-to-neutral capacitance for each phase and is developed in the following manner.

Consider a transposed three-phase line with conductors having the same radius r, and with space between conductors D_{AB}, D_{BC}, and D_{AC}, where D_{AB}, D_{BC}, and $D_{AC} > r$. Assuming abc positive sequence, the expressions for V_{AB} on the first, second, and third sections of the transposed line are

$$V_{AB\,\text{first}} = \frac{1}{2\pi\varepsilon_0}\left[q_A \ln\left[\frac{D_{AB}}{r}\right] + q_B \ln\left[\frac{r}{D_{AB}}\right] + q_C \ln\left[\frac{D_{AB}}{D_{AC}}\right]\right] \tag{4.92}$$

$$V_{AB\,\text{second}} = \frac{1}{2\pi\varepsilon_0}\left[q_A \ln\left[\frac{D_{BC}}{r}\right] + q_B \ln\left[\frac{r}{D_{BC}}\right] + q_C \ln\left[\frac{D_{AC}}{D_{AB}}\right]\right] \tag{4.93}$$

$$V_{AB\,\text{third}} = \frac{1}{2\pi\varepsilon_0}\left[q_A \ln\left[\frac{D_{AC}}{r}\right] + q_B \ln\left[\frac{r}{D_{AC}}\right] + q_C \ln\left[\frac{D_{AB}}{D_{BC}}\right]\right]. \tag{4.94}$$

Similarly, the expressions for V_{AC} on the first, second, and third sections of the transposed line are

$$V_{AC\,\text{first}} = \frac{1}{2\pi\varepsilon_0}\left[q_A \ln\left[\frac{D_{AC}}{r}\right] + q_B \ln\left[\frac{D_{BC}}{D_{AB}}\right] + q_C \ln\left[\frac{r}{D_{AC}}\right]\right] \tag{4.95}$$

$$V_{AC\,second} = \frac{1}{2\pi\varepsilon_0}\left[q_A \ln\left[\frac{D_{AB}}{r}\right] + q_B \ln\left[\frac{D_{AC}}{D_{BC}}\right] + q_C \ln\left[\frac{r}{D_{AB}}\right]\right] \quad (4.96)$$

$$V_{AC\,third} = \frac{1}{2\pi\varepsilon_0}\left[q_A \ln\left[\frac{D_{BC}}{r}\right] + q_B \ln\left[\frac{D_{AB}}{D_{AC}}\right] + q_C \ln\left[\frac{r}{D_{BC}}\right]\right] \quad (4.97)$$

Taking the average value of the three sections, we have the final expressions of V_{AB} and V_{AC} in the transposed line

$$V_{AB\,transp} = \frac{V_{AB\,first} + V_{AB\,second} + V_{AB\,third}}{3}$$
$$= \frac{1}{6\pi\varepsilon_0}\left[q_A \ln\left[\frac{D_{AB}D_{AC}D_{BC}}{r^3}\right] + q_B \ln\left[\frac{r^3}{D_{AB}D_{AC}D_{BC}}\right] + q_C \ln\left[\frac{D_{AC}D_{AC}D_{BC}}{D_{AC}D_{AC}D_{BC}}\right]\right] \quad (4.98)$$

$$V_{AB\,transp} = \frac{V_{AC\,first} + V_{AC\,second} + V_{AC\,third}}{3}$$
$$= \frac{1}{6\pi\varepsilon_0}\left[q_A \ln\left[\frac{D_{AB}D_{AC}D_{BC}}{r^3}\right] + q_B \ln\left[\frac{D_{AC}D_{AC}D_{BC}}{D_{AB}D_{AC}D_{BC}}\right] + q_C \ln\left[\frac{r^3}{D_{AC}D_{AC}D_{BC}}\right]\right]. \quad (4.99)$$

For a balanced system where $-q_A = (q_B + q_C)$, the phase-to-neutral voltage V_{AN} (phase voltage) is

$$V_{AN\,transp} = \frac{V_{AB\,transp} + V_{AC\,transp}}{3}$$
$$= \frac{1}{18\pi\varepsilon_0}\left[2q_A \ln\left[\frac{D_{AB}D_{AC}D_{BC}}{r^3}\right] + (q_B + q_C)\ln\left[\frac{r^3}{D_{AB}D_{AC}D_{BC}}\right]\right] \quad (4.100)$$
$$= \frac{1}{6\pi\varepsilon_0}q_A \ln\left[\frac{D_{AB}D_{AC}D_{BC}}{r^3}\right] = \frac{1}{2\pi\varepsilon_0}q_A \ln\left[\frac{GMD}{r}\right]\,[V]$$

where $GMD = \sqrt[3]{D_{AB}D_{BC}D_{CA}}$ = geometrical mean distance for a three-phase line.

For bundle conductors, an equivalent radius r_e replaces the radius r of a single conductor and is determined by the number of conductors per bundle and the spacing of conductors. The expression of r_e is similar to GMR_{bundle} used in the calculation of the inductance per phase, except that the actual outside radius of the conductor is used instead of the GMR_{cond}. Therefore, the expression for V_{AN} is

$$V_{AN\,transp} = \frac{1}{2\pi\varepsilon_0}q_A \ln\left[\frac{GMD}{r_e}\right]\,[V] \quad (4.101)$$

where $r_e = (d^{n-1}r)^{1/n}$ = equivalent radius for up to three conductors per bundle (m)
d = distance between bundle conductors (m)
n = number of conductors per bundle

Finally, the capacitance and capacitive reactance per unit length from phase to neutral can be evaluated as

$$C_{AN\,transp} = \frac{q_A}{V_{AN\,transp}} = \frac{2\pi\varepsilon_0}{\ln\left[\dfrac{GMD}{r_e}\right]} \quad [F/m] \tag{4.102}$$

$$X_{AN\,transp} = \frac{1}{2\pi f\,C_{AN\,transp}} = \frac{1}{4\pi f\varepsilon_0}\ln\left[\dfrac{GMD}{r_e}\right] \quad [\Omega/m]. \tag{4.103}$$

Characteristics of Overhead Conductors

Tables 4.9a and 4.9b present general characteristics of aluminum cable steel reinforced conductors (ACSR). The size of the conductors (cross-section area) is specified in square millimeters and kcmil, where a cmil is the cross-section area of a circular conductor with a diameter of 1/1000 inch. Typical values of resistance, inductive reactance, and capacitive reactance are listed. The approximate current-carrying capacity of the conductors is also included assuming 60 Hz, wind speed of 1.4 mi/h, and conductor and air temperatures of 75°C and 25°C, respectively. Similarly, Tables 4.10a and 4.10b present the corresponding characteristics of aluminum conductors (AAC).

References

Barnes, C. C., *Power Cables: Their Design and Installation*, 2nd ed., Chapman and Hall, Ltd., London, 1966.
Electric Power Research Institute, *Transmission Line Reference Book 345 kV and Above*, 2nd ed., Palo Alto, CA, 1987.
Glover J. D., Sarma, M., and Digby, G., *Power System Analysis and Design with Personal Computer Applications*, 2nd ed., PWS Publishing Company, 1994.
Gross, Ch. A., *Power System Analysis*, John Wiley & Sons, New York, 1979.
Gungor, B. R., *Power Systems*, Harcourt Brace Jovanovich, Florida, 1988.
Yamayee, Z. A., and Bala, J. L. Jr., *Electromechanical Energy Devices and Power Systems*, John Wiley & Sons, New York, 1994.
Zaborszky, J., and Rittenhouse, J. W., *Electric Power Transmission: The Power System in the Steady State*, The Ronald Press Company, New York, 1954.

Transmission System

TABLE 4.9a Characteristics of Aluminum Cable Steel Reinforced Conductors (ACSR)

Code	Cross-Section Area Total (mm²)	Aluminum (kcmil)	Aluminum (mm²)	Stranding Al/Steel	Diameter Conductor (mm)	Diameter Core (mm)	Layers	Approx. Current-Carrying Capacity (Amp.)*	Resistance (mΩ/km) DC 25°C	AC (60 Hz) 25°C	50°C	75°C	GMR (mm)	60-Hz Reactances (Dm = 1m) X_L (Ω/km)	X_C (MΩ/km)
—	1521	2776	1407	84/19	50.80	13.87	4		21.0	4.5	26.2	28.1	20.33	0.294	0.175
Joree	1344	2515	1274	76/19	47.75	10.80	4		22.7	26.0	28.0	30.0	18.93	0.299	0.178
Thrasher	1235	2312	1171	76/19	45.77	10.34	4		24.7	27.7	30.0	32.2	18.14	0.302	0.180
Kiwi	1146	2167	1098	72/7	44.07	8.81	4		26.4	29.4	31.9	34.2	17.37	0.306	0.182
Bluebird	1181	2156	1092	84/19	44.75	12.19	4		26.5	29.0	31.4	33.8	17.92	0.303	0.181
Chukar	9767	1781	902	84/19	40.69	11.10	4		32.1	34.1	37.2	40.1	16.28	0.311	0.186
Falcon	908	1590	806	54/19	39.24	13.08	3	1380	35.9	37.4	40.8	44.3	15.91	0.312	0.187
Lapwing	862	1590	806	45/7	38.20	9.95	3	1370	36.7	38.7	42.1	45.6	15.15	0.316	0.189
Parrot	862	1510	765	54/19	38.23	12.75	3	1340	37.8	39.2	42.8	46.5	15.48	0.314	0.189
Nuthatch	818	1510	765	45/7	37.21	9.30	3	1340	38.7	40.5	44.2	47.9	14.78	0.318	0.190
Plover	817	1431	725	54/19	37.21	12.42	3	1300	39.9	41.2	45.1	48.9	15.06	0.316	0.190
Bobolink	775	1431	725	45/7	36.25	9.07	3	1300	35.1	42.6	46.4	50.3	14.39	0.320	0.191
Martin	772	1351	685	54/19	36.17	12.07	3	1250	42.3	43.5	47.5	51.6	14.63	0.319	0.191
Dipper	732	1351	685	45/7	35.20	8.81	3	1250	43.2	44.9	49.0	53.1	13.99	0.322	0.193
Pheasant	726	1272	645	54/19	35.10	11.71	3	1200	44.9	46.1	50.4	54.8	14.20	0.321	0.193
Bittern	689	1272	644	45/7	34.16	8.53	3	1200	45.9	47.5	51.9	56.3	13.56	0.324	0.194
Grackle	681	1192	604	54/19	34.00	11.33	3	1160	47.9	49.0	53.6	58.3	13.75	0.323	0.194
Bunting	646	1193	604	45/7	33.07	8.28	3	1160	48.9	50.4	55.1	59.9	13.14	0.327	0.196
Finch	636	1114	564	54/19	32.84	10.95	3	1110	51.3	52.3	57.3	62.3	13.29	0.326	0.196
Bluejay	603	1113	564	45/7	31.95	8.00	3	1110	52.4	53.8	58.9	64.0	12.68	0.329	0.197
Curlew	591	1033	523	54/7	31.62	10.54	3	1060	56.5	57.4	63.0	68.4	12.80	0.329	0.198
Ortolan	560	1033	525	45/7	30.78	7.70	3	1060	56.5	57.8	63.3	68.7	12.22	0.332	0.199
Merganser	596	954	483	30/7	31.70	13.60	2	1010	61.3	61.8	67.9	73.9	13.11	0.327	0.198
Cardinal	546	954	483	54/7	30.38	10.13	3	1010	61.2	62.0	68.0	74.0	12.31	0.332	0.200
Rail	517	954	483	45/7	29.59	7.39	3	1010	61.2	62.4	68.3	74.3	11.73	0.335	0.201
Baldpate	562	900	456	30/7	30.78	13.21	2	960	65.0	65.5	71.8	78.2	12.71	0.329	0.199
Canary	515	900	456	54/7	29.51	9.83	3	970	64.8	65.5	72.0	78.3	11.95	0.334	0.201
Ruddy	478	900	456	45/7	28.73	7.19	3	970	64.8	66.0	72.3	78.6	11.40	0.337	0.202
Crane	501	875	443	54/7	29.11	9.70	3	950	66.7	67.5	74.0	80.5	11.80	0.335	0.202
Willet	474	874	443	45/7	28.32	7.09	3	950	66.7	67.9	74.3	80.9	11.25	0.338	0.203

TABLE 4.9a (continued) Characteristics of Aluminum Cable Steel Reinforced Conductors (ACSR)

Code	Cross-Section Area		Stranding Al/Steel	Diameter		Layers	Approx. Current-Carrying Capacity (Amp.)*	Resistance (mΩ/km)				GMR (mm)	60-Hz Reactances (Dm = 1m)	
	Total (mm²)	Aluminum		Conductor (mm)	Core (mm)			DC 25°C	AC (60 Hz)				X_L (Ω/km)	X_C (MΩ/km)
		(kcmil)	(mm²)							25°C	50°C	75°C			
Skimmer	479	795	403	30/7	29.00	12.40	2	940	73.5	74.0	81.2	88.4	11.95	0.334	0.202
Mallard	495	795	403	30/19	28.96	12.42	2	910	73.5	74.0	81.2	88.4	11.95	0.334	0.202
Drake	469	795	403	26/7	28.14	10.36	2	900	73.3	74.0	81.2	88.4	11.43	0.337	0.203
Condor	455	795	403	54/7	27.74	9.25	3	900	73.4	74.1	81.4	88.6	11.22	0.339	0.204
Cuckoo	455	795	403	24/7	27.74	9.25	2	900	73.4	74.1	81.4	88.5	11.16	0.339	0.204
Tern	431	795	403	45/7	27.00	6.76	3	900	73.4	74.4	81.6	88.8	10.73	0.342	0.205
Coot	414	795	403	36/1	26.42	3.78	3	910	73.0	74.4	81.5	88.6	10.27	0.345	0.206
Buteo	447	715	362	30/7	27.46	11.76	2	840	81.8	82.2	90.2	98.3	11.34	0.338	0.204
Redwing	445	715	362	30/19	27.46	11.76	2	840	81.8	82.2	90.2	98.3	11.34	0.338	0.204
Starling	422	716	363	26/7	26.7	9.82	2	840	81.5	82.1	90.1	98.1	10.82	0.341	0.206
Crow	409	715	362	54/7	26.31	8.76	3	840	81.5	82.2	90.2	98.2	10.67	0.342	0.206

Source: Transmission Line Reference Book 345 kV and Above, 2nd ed., Electronic Power Research Institute, Palo Alto, CA, 1987. With permission.

TABLE 4.9b Characteristics of Aluminum Cable Steel Reinforced Conductors (ACSR)

Code	Cross-Section Area Total (mm²)	Aluminum (kcmil)	Aluminum (mm²)	Stranding Al/Steel	Diameter Conductor (mm)	Diameter Core (mm)	Layers	Approx. Current-Carrying Capacity (Amp.)*	Resistance (mΩ/km) DC 25°C	AC (60 Hz) 25°C	50°C	75°C	GMR (mm)	60-Hz Reactances (Dm = 1m) X_L (Ω/km)	X_C (MΩ/km)
Stilt	410	716	363	24/7	26.31	8.76	2	840	81.5	82.2	90.2	98.1	10.58	0.343	0.206
Grebe	388	716	363	45/7	25.63	6.4	3	840	81.5	82.5	90.4	98.4	10.18	0.3446	0.208
Gannet	393	666	338	26/7	25.76	9.5	2	800	87.6	88.1	96.6	105.3	10.45	0.344	0.208
Gull	382	667	338	54/7	25.4	8.46	3	800	87.5	88.1	96.8	105.3	10.27	0.345	0.208
Flamingo	382	667	338	24/7	25.4	8.46	2	800	87.4	88.1	96.7	105.3	10.21	0.346	0.208
Scoter	397	636	322	30/7	25.88	11.1	2	800	91.9	92.3	101.4	110.4	10.70	0.342	0.207
Egret	396	636	322	30/19	25.88	11.1	2	780	91.9	92.3	101.4	110.4	10.70	0.342	0.207
Grosbeak	375	636	322	26/7	25.15	9.27	2	780	91.7	92.2	101.2	110.3	10.21	0.346	0.209
Goose	364	636	322	54/7	24.82	8.28	3	770	91.8	92.4	101.4	110.4	10.06	0.347	0.208
Rook	363	636	322	24/7	24.82	8.28	2	770	91.7	92.3	101.3	110.3	10.06	0.347	0.209
Kingbird	340	636	322	18/1	23.88	4.78	2	780	91.2	92.2	101.1	110.0	9.27	0.353	0.211
Swirl	331	636	322	36/1	23.62	3.38	3	780	91.3	92.4	101.3	110.3	9.20	0.353	0.212
Wood Duck	378	605	307	30/7	25.25	10.82	2	760	96.7	97.0	106.5	116.1	10.42	0.344	0.208
Teal	376	605	307	30/19	25.25	10.82	2	770	96.7	97.0	106.5	116.1	10.42	0.344	0.208
Squab	356	605	356	26/7	25.54	9.04	2	760	96.5	97.0	106.5	116.0	9.97	0.347	0.208
Peacock	346	605	307	24/7	24.21	8.08	2	760	96.4	97.0	106.4	115.9	9.72	0.349	0.210
Duck	347	606	307	54/7	24.21	8.08	3	750	96.3	97.0	106.3	115.8	9.81	0.349	0.210
Eagle	348	557	282	30/7	24.21	10.39	2	730	105.1	105.4	115.8	126.1	10.00	0.347	0.210
Dove	328	556	282	26/7	23.55	8.66	2	730	104.9	105.3	115.6	125.9	9.54	0.351	0.212
Parakeet	319	557	282	24/7	23.22	7.75	2	730	104.8	105.3	115.6	125.9	9.33	0.352	0.212
Osprey	298	556	282	18/1	22.33	4.47	2	740	104.4	105.2	115.4	125.7	8.66	0.358	0.214
Hen	298	477	242	30/7	22.43	9.6	2	670	122.6	122.9	134.9	147.0	9.27	0.353	0.214
Hawk	281	477	242	26/7	21.79	8.03	2	670	122.4	122.7	134.8	146.9	8.84	0.357	0.215
Flicker	273	477	273	24/7	21.49	7.16	2	670	122.2	122.7	134.7	146.8	8.63	0.358	0.216
Pelican	255	477	242	18/1	20.68	4.14	2	680	121.7	122.4	134.4	146.4	8.02	0.364	0.218
Lark	248	397	201	30/7	20.47	8.76	2	600	147.2	147.4	161.9	176.4	8.44	0.360	0.218
Ibis	234	397	201	26/7	19.89	7.32	2	590	146.9	147.2	161.7	176.1	8.08	0.363	0.220
Brant	228	398	201	24/7	19.61	6.53	2	590	146.7	147.1	161.6	176.1	7.89	0.365	0.221
Chickadee	213	397	201	18/1	18.87	3.78	2	590	146.1	146.7	161.0	175.4	7.32	0.371	0.222
Oriole	210	336	170	30/7	18.82	8.08	2	530	173.8	174.0	191.2	208.3	7.77	0.366	0.222

TABLE 4.9b (continued) Characteristics of Aluminum Cable Steel Reinforced Conductors (ACSR)

Code	Cross-Section Area			Stranding Al/Steel	Diameter		Layers	Approx. Current-Carrying Capacity (Amp.)*	Resistance (mΩ/km)				GMR (mm)	60-Hz Reactances (Dm = 1m)	
	Total (mm²)	Aluminum (kcmil)	(mm²)		Conductor (mm)	Core (mm)			DC 25°C	AC (60 Hz) 25°C	50°C	75°C		X_L (Ω/km)	X_C (MΩ/km)
Linnet	198	336	170	26/7	18.29	6.73	2	530	173.6	173.8	190.9	208.1	7.41	0.370	0.224
Widgeon	193	336	170	24/7	18.03	6.02	2	530	173.4	173.7	190.8	207.9	7.5	0.371	0.225
Merlin	180	336	170	18/1	16.46	3.48	2	530	173.0	173.1	190.1	207.1	6.74	0.377	0.220
Piper	187	300	152	30/7	17.78	7.62	2	500	195.0	195.1	214.4	233.6	7.35	0.370	0.225
Ostrich	177	300	152	26/7	17.27	6.38	2	490	194.5	194.8	214.0	233.1	7.01	0.374	0.227
Gadwall	172	300	152	24/7	17.04	5.69	2	490	194.5	194.8	213.9	233.1	6.86	0.376	0.227
Phoebe	160	300	152	18/1	16.41	3.28	2	490	193.5	194.0	213.1	232.1	6.37	0.381	0.229
Junco	167	267	135	30/7	16.76	7.19	2	570	219.2	219.4	241.1	262.6	6.92	0.375	0.228
Partridge	157	267	135	26/7	16.31	5.99	2	460	218.6	218.9	240.5	262.0	6.61	0.378	0.229
Waxwing	143	267	135	18/1	15.47	3.1	2	460	217.8	218.1	239.7	261.1	6.00	0.386	0.232

* For conductor temperature at 75°C, air at 25°C, wind speed at 1.4 mi/hr, frequency at 60 Hz.

Source: Transmission Line Reference Book 345 kV and Above, 2nd ed., Electronic Power Research Institute, Palo Alto, CA, 1987. With permission.

Transmission System

TABLE 4.10a Characteristics of All-Aluminum Conductors (AAC)

Code	Cross-Section Area (mm²)	Cross-Section Area (kcmil or AWG)	Stranding	Diameter (mm)	Layers	Approx. Current-Carrying Capacity (Amp.)*	Resistance (mΩ/km) DC 25°C	Resistance (mΩ/km) AC (60 Hz) 25°C	Resistance (mΩ/km) 50°C	Resistance (mΩ/km) 75°C	GMR (mm)	60-Hz Reactances (Dm = 1 m) X_L (Ω/km)	60-Hz Reactances (Dm = 1 m) X_C (MΩ/km)
Coreopsis	806.2	1591	61	36.93	4	1380	36.5	39.5	42.9	46.3	14.26	0.320	0.190
Gladiolus	765.8	1511	61	35.99	4	1340	38.4	41.3	44.9	48.5	13.90	0.322	0.192
Caranation	725.4	1432	61	35.03	4	1300	40.5	43.3	47.1	50.9	13.53	0.324	0.193
Columbine	665.3	1352	61	34.04	4	1250	42.9	45.6	49.6	53.6	13.14	0.327	0.196
Narcissus	644.5	1272	61	33.02	4	1200	45.5	48.1	52.5	46.7	12.74	0.329	0.194
Hawthorn	604.1	1192	61	31.95	4	1160	48.7	51.0	55.6	60.3	12.34	0.331	0.197
Marigold	564.2	1113	61	30.89	4	1110	52.1	54.3	59.3	64.3	11.92	0.334	0.199
Larkspur	524	1034	61	29.77	4	1060	56.1	58.2	63.6	69.0	11.49	0.337	0.201
Bluebell	524.1	1034	37	29.71	3	1060	56.1	58.2	63.5	68.9	11.40	0.337	0.201
Goldenrod	483.7	955	61	28.6	4	1010	60.8	62.7	68.6	74.4	11.03	0.340	0.203
Magnolia	483.6	954	37	28.55	3	1010	60.8	62.7	68.6	74.5	10.97	0.340	0.203
Crocus	443.6	875	61	27.38	4	950	66.3	68.1	74.5	80.9	10.58	0.343	0.205
Anemone	443.5	875	37	27.36	3	950	66.3	68.1	74.5	80.9	10.49	0.344	0.205
Lilac	403.1	796	61	26.11	4	900	73.0	74.6	81.7	88.6	10.09	0.347	0.207
Arbutus	402.9	795	37	26.06	3	900	73.0	74.6	81.7	88.6	10.00	0.347	0.207
Nasturtium	362.5	715	61	24.76	4	840	81.2	82.6	90.5	98.4	9.57	0.351	0.209
Violet	362.8	716	37	24.74	3	840	81.1	82.5	90.4	98.3	9.48	0.351	0.209
Orchid	322.2	636	37	23.32	3	780	91.3	92.6	101.5	110.4	8.96	0.356	0.212
Mistletoe	281.8	556	37	21.79	3	730	104.4	105.5	115.8	126.0	8.38	0.361	0.215
Dahlia	281.8	556	19	21.72	2	730	104.4	105.5	115.8	125.9	8.23	0.362	0.216
Syringa	241.5	477	37	20.19	3	670	121.8	122.7	134.7	146.7	7.74	0.367	0.219
Cosmos	241.9	477	19	20.14	2	670	121.6	122.6	134.5	146.5	7.62	0.368	0.219
Canna	201.6	398	19	18.36	2	600	145.9	146.7	161.1	175.5	6.95	0.376	0.224
Tulip	170.6	337	19	16.92	2	530	172.5	173.2	190.1	207.1	6.40	0.381	0.228
Laurel	135.2	267	19	15.06	2	460	217.6	218.1	239.6	261.0	5.70	0.390	0.233
Daisy	135.3	267	7	14.88	1	460	217.5	218	239.4	260.8	5.39	0.394	0.233
Oxlip	107.3	212 or (4/0)	7	13.26	1	340	274.3	274.7	301.7	328.8	4.82	0.402	0.239
Phlox	85	168 or (3/0)	7	11.79	1	300	346.4	346.4	380.6	414.7	4.27	0.411	0.245
Aster	67.5	133 or (2/0)	7	10.52	1	270	436.1	439.5	479.4	522.5	3.81	0.40	0.25
Poppy	53.5	106 or (1/0)	7	9.35	1	230	550	550.2	604.5	658.8	3.38	0.429	0.256
Pansy	42.4	#1 AWG	7	8.33	1	200	694.2	694.2	763.2	831.6	3.02	0.438	0.261
Iris	33.6	#2 AWG	7	7.42	1	180	874.5	874.5	960.8	1047.9	2.68	0.446	0.267
Rose	21.1	#3 AWG	7	5.89	1	160	1391.5	1391.5	1528.9	1666.3	2.13	0.464	0.278
Peachbell	13.3	#4 AWG	7	4.67	1	140	2214.4	2214.4	2443.2	2652	1.71	0.481	0.289

* For conductor at 75°C, air at 25°C, wind speed at 1.4 mi/hr, frequency at 60 Hz.

Source: Transmission Line Reference Book 345 kV and Above, 2nd ed., Electronic Power Research Institute, Palo Alto, CA, 1987. With permission.

TABLE 4.10b Characteristics of All-Aluminum Conductors (AAC)

Code	Cross-Section Area (mm²)	Cross-Section Area (kcmil or AWG)	Stranding	Diameter (mm)	Layers	Approx. Current-Carrying Capacity (Amp.)*	Resistance (mΩ/km) DC 25°C	Resistance (mΩ/km) AC (60 Hz) 25°C	Resistance (mΩ/km) 50°C	Resistance (mΩ/km) 75°C	GMR (mm)	60-Hz Reactances (Dm = 1 m) X_L (Ω/km)	60-Hz Reactances (Dm = 1 m) X_C (MΩ/km)
EVEN SIZES													
Bluebonnet	1773.3	3500	7	54.81	6		16.9	22.2	23.6	25.0	21.24	0.290	0.172
Trillium	1520.2	3000	127	50.75	6		19.7	24.6	26.2	27.9	19.69	0.296	0.175
Lupine	1266.0	2499	91	46.30	5		23.5	27.8	29.8	31.9	17.92	0.303	0.180
Cowslip	1012.7	1999	91	41.40	5		29.0	32.7	35.3	38.0	16.03	0.312	0.185
Jessamine	887.0	1750	61	38.74	4		33.2	36.5	39.5	42.5	14.94	0.317	0.188
Hawkweed	506.7	1000	37	29.24	3	1030	58.0	60.0	65.5	71.2	11.22	0.339	0.201
Camelia	506.4	999	61	29.26	4	1030	58.1	60.1	65.5	71.2	11.31	0.338	0.201
Snapdragon	456.3	900	61	27.79	4	970	64.4	66.3	72.5	78.7	10.73	0.342	0.204
Cockscomb	456.3	900	37	27.74	3	970	64.4	66.3	72.5	78.7	10.64	0.343	0.204
Cattail	380.1	750	61	25.35	4	870	77.4	78.9	86.4	93.9	9.78	0.349	0.208
Petunia	380.2	750	37	23.85	3	870	77.4	78.9	86.4	93.9	9.72	0.349	0.208
Flag	354.5	700	61	24.49	4	810	83.0	84.4	92.5	100.6	9.45	0.352	0.210
Verbena	354.5	700	37	24.43	3	810	83.0	84.4	92.5	100.6	9.39	0.352	0.210
Meadowswee	303.8	600	37	2.63	3	740	96.8	98.0	107.5	117.0	8.69	0.358	0.214
Hyacinth	253.1	500	37	20.65	3	690	116.2	117.2	128.5	140.0	7.92	0.365	0.218
Zinnia	253.3	500	19	20.60	2	690	116.2	117.2	128.5	139.9	7.80	0.366	0.218
Goldentuft	228.0	450	19	19.53	2	640	129.0	129.9	142.6	155.3	7.41	0.370	0.221
Daffodil	177.3	350	19	17.25	2	580	165.9	166.6	183.0	199.3	6.52	0.379	0.227
Peony	152.1	300	19	15.98	2	490	193.4	194.0	213.1	232.1	6.04	0.385	0.230
Valerian	126.7	250	19	14.55	2	420	232.3	232.8	255.6	278.6	5.52	0.392	0.235
Sneezewort	126.7	250	7	14.40	1	420	232.2	232.7	255.6	278.4	5.21	0.396	0.235

Source: *Transmission Line Reference Book 345 kV and Above*, 2nd ed., Electronic Power Research Institute, Palo Alto, CA, 1987. With permission.

4.7 Sag and Tension of Conductor

D.A. Douglass and Ridley Thrash

The energized conductors of transmission and distribution lines must be placed to totally eliminate the possibility of injury to people. Overhead conductors, however, elongate with time, temperature, and tension, thereby changing their original positions after installation. Despite the effects of weather and loading on a line, the conductors must remain at safe distances from buildings, objects, and people or vehicles passing beneath the line at all times. To ensure this safety, the shape of the terrain along the right-of-way, the height and lateral position of the conductor support points, and the position of the conductor between support points under all wind, ice, and temperature conditions must be known.

Bare overhead transmission or distribution conductors are typically quite flexible and uniform in weight along their length. Because of these characteristics, they take the form of a catenary (Ehrenberg, 1935; Winkelmann, 1959) between support points. The shape of the catenary changes with conductor temperature, ice and wind loading, and time. To ensure adequate vertical and horizontal clearance under all weather and electrical loadings, and to ensure that the breaking strength of the conductor is not exceeded, the behavior of the conductor catenary under all conditions must be known before the line is designed. The future behavior of the conductor is determined through calculations commonly referred to as sag-tension calculations.

Sag-tension calculations predict the behavior of conductors based on recommended tension limits under varying loading conditions. These tension limits specify certain percentages of the conductor's rated breaking strength that are not to be exceeded upon installation or during the life of the line. These conditions, along with the elastic and permanent elongation properties of the conductor, provide the basis for determinating the amount of resulting sag during installation and long-term operation of the line.

Accurately determined initial sag limits are essential in the line design process. Final sags and tensions depend on initial installed sags and tensions and on proper handling during installation. The final sag shape of conductors is used to select support point heights and span lengths so that the minimum clearances will be maintained over the life of the line. If the conductor is damaged or the initial sags are incorrect, the line clearances may be violated or the conductor may break during heavy ice or wind loadings.

Catenary Cables

A bare-stranded overhead conductor is normally held clear of objects, people, and other conductors by periodic attachment to insulators. The elevation differences between the supporting structures affect the shape of the conductor catenary. The catenary's shape has a distinct effect on the sag and tension of the conductor, and therefore, must be determined using well-defined mathematical equations.

Level Spans

The shape of a catenary is a function of the conductor weight per unit length, w, the horizontal component of tension, H, span length, S, and the maximum sag of the conductor, D. Conductor sag and span length are illustrated in Fig. 4.63 for a level span.

The exact catenary equation uses hyperbolic functions. Relative to the low point of the catenary curve shown in Fig. 4.63, the height of the conductor, $y(x)$, above this low point is given by the following equation:

$$y(x) = \frac{H}{w}\cosh\left(\left(\frac{w}{H}x\right) - 1\right) = \frac{w(x^2)}{2H} \tag{4.104}$$

Note that x is positive in either direction from the low point of the catenary. The expression to the right is an approximate parabolic equation based upon a MacLaurin expansion of the hyperbolic cosine.

For a level span, the low point is in the center, and the sag, D, is found by substituting $x = S/2$ in the preceding equations. The exact and approximate parabolic equations for sag become the following:

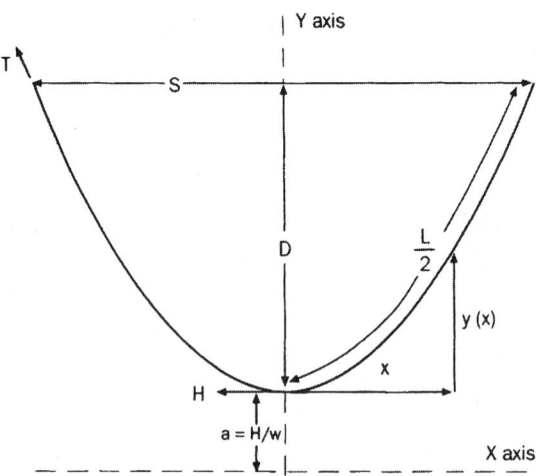

FIGURE 4.63 The catenary curve for level spans.

$$D = \frac{H}{w}\left(\cosh\left(\frac{wS}{2H}\right) - 1\right) = \frac{w(S^2)}{8H} \quad (4.105)$$

The ratio, H/w, which appears in all of the preceding equations, is commonly referred to as the catenary constant. An increase in the catenary constant, having the units of length, causes the catenary curve to become shallower and the sag to decrease. Although it varies with conductor temperature, ice and wind loading, and time, the catenary constant typically has a value in the range of several thousand feet for most transmission-line catenaries.

The approximate or parabolic expression is sufficiently accurate as long as the sag is less than 5% of the span length. As an example, consider a 1000-ft span of Drake conductor (w = 1.096 lb/ft) installed at a tension of 4500 lb. The catenary constant equals 4106 ft. The calculated sag is 30.48 ft and 30.44 ft using the hyperbolic and approximate equations, respectively. Both estimates indicate a sag-to-span ratio of 3.4% and a sag difference of only 0.5 in.

The horizontal component of tension, H, is equal to the conductor tension at the point in the catenary where the conductor slope is horizontal. For a level span, this is the midpoint of the span length. At the ends of the level span, the conductor tension, T, is equal to the horizontal component plus the conductor weight per unit length, w, multiplied by the sag, D, as shown in the following:

$$T = H + wD \quad (4.106)$$

Given the conditions in the preceding example calculation for a 1000-ft level span of Drake ACSR, the tension at the attachment points exceeds the horizontal component of tension by 33 lb. It is common to perform sag-tension calculations using the horizontal tension component, but the average of the horizontal and support point tension is usually listed in the output.

Conductor Length

Application of calculus to the catenary equation allows the calculation of the conductor length, $L(x)$, measured along the conductor from the low point of the catenary in either direction.

The resulting equation becomes:

$$L(x) = \frac{H}{w}\text{SINH}\left(\frac{wx}{H}\right) = x\left(1 + \frac{x^2(w^2)}{6H^2}\right) \quad (4.107)$$

Transmission System

For a level span, the conductor length corresponding to $x = S/2$ is half of the total conductor length and the total length, L, is:

$$L = \left(\frac{2H}{w}\right) \text{SINH}\left(\frac{Sw}{2H}\right) = S\left(1 + \frac{S^2(w^2)}{24H^2}\right) \quad (4.108)$$

The parabolic equation for conductor length can also be expressed as a function of sag, D, by substitution of the sag parabolic equation, giving:

$$L = S + \frac{8D^2}{3S} \quad (4.109)$$

Conductor Slack

The difference between the conductor length, L, and the span length, S, is called slack. The parabolic equations for slack may be found by combining the preceding parabolic equations for conductor length, L, and sag, D:

$$L - S = S^3\left(\frac{w^2}{24H^2}\right) = D^2\left(\frac{8}{3S}\right) \quad (4.110)$$

While slack has units of length, it is often expressed as the percentage of slack relative to the span length. Note that slack is related to the cube of span length for a given H/w ratio and to the square of sag for a given span. For a series of spans having the same H/w ratio, the total slack is largely determined by the longest spans. It is for this reason that the ruling span is nearly equal to the longest span rather than the average span in a series of suspension spans.

Equation (4.110) can be inverted to obtain a more interesting relationship showing the dependence of sag, D, upon slack, $L-S$:

$$D = \sqrt{\frac{3S(L-S)}{8}} \quad (4.111)$$

As can be seen from the preceding equation, small changes in slack typically yield large changes in conductor sag.

Inclined Spans

Inclined spans may be analyzed using essentially the same equations that were used for level spans. The catenary equation for the conductor height above the low point in the span is the same. However, the span is considered to consist of two separate sections, one to the right of the low point and the other to the left as shown in Fig. 4.64 (Winkelmann, 1959). The shape of the catenary relative to the low point is unaffected by the difference in suspension point elevation (span inclination).

In each direction from the low point, the conductor elevation, $y(x)$, relative to the low point is given by:

$$y(x) = \frac{H}{w}\cosh\left(\left(\frac{w}{H}x\right) - 1\right) = \frac{w(x^2)}{2H} \quad (4.112)$$

Note that x is considered positive in either direction from the low point.

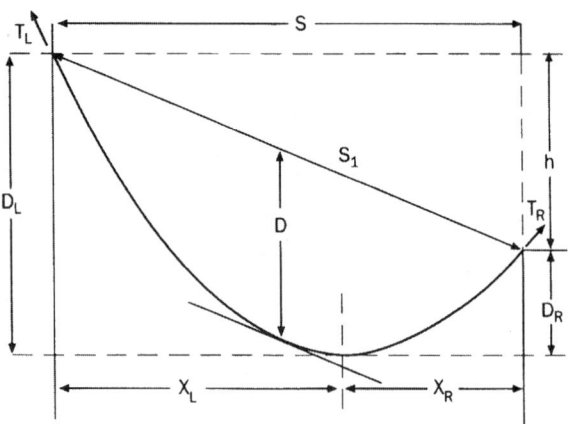

FIGURE 4.64 Inclined catenary span.

The horizontal distance, x_L, from the left support point to the low point in the catenary is:

$$x_L = \frac{S}{2}\left(1 + \frac{h}{4D}\right) \tag{4.113}$$

The horizontal distance, x_R, from the right support point to the low point of the catenary is:

$$x_R = \frac{S}{2}\left(1 - \frac{h}{4D}\right) \tag{4.114}$$

where

S = horizontal distance between support points.
h = vertical distance between support points.
S_1 = straight-line distance between support points.
D = sag measured vertically from a line through the points of conductor support to a line tangent to the conductor.

The midpoint sag, D, is approximately equal to the sag in a horizontal span equal in length to the inclined span, S_1.

Knowing the horizonal distance from the low point to the support point in each direction, the preceding equations for $y(x)$, L, D, and T can be applied to each side of the inclined span.

The total conductor length, L, in the inclined span is equal to the sum of the lengths in the x_R and x_L sub-span sections:

$$L = S + \left(x_R^3 + x_L^3\right)\left(\frac{w^2}{6H^2}\right) \tag{4.115}$$

In each sub-span, the sag is relative to the corresponding support point elevation:

$$D_R = \frac{wx_R^2}{2H} \quad D_L = \frac{wx_L^2}{2H} \tag{4.116}$$

Transmission System

or in terms of sag, D, and the vertical distance between support points:

$$D_R = D\left(1 - \frac{h}{4D}\right)^2 \quad D_L = D\left(1 + \frac{h}{4D}\right)^2 \tag{4.117}$$

and the maximum tension is:

$$T_R = H + wD_R \quad T_L = H + wD_L \tag{4.118}$$

or in terms of upper and lower support points:

$$T_u = T_l + wh \tag{4.119}$$

where

D_R = sag in right sub-span section
D_L = sag in left sub-span section
T_R = tension in right sub-span section
T_L = tension in left sub-span section
T_u = tension in conductor at upper support
T_l = tension in conductor at lower support

The horizontal conductor tension is equal at both supports. The vertical component of conductor tension is greater at the upper support and the resultant tension, T_u, is also greater.

Ice and Wind Conductor Loads

When a conductor is covered with ice and/or is exposed to wind, the effective conductor weight per unit length increases. During occasions of heavy ice and/or wind load, the conductor catenary tension increases dramatically along with the loads on angle and deadend structures. Both the conductor and its supports can fail unless these high-tension conditions are considered in the line design.

The National Electric Safety Code (NESC) suggests certain combinations of ice and wind corresponding to heavy, medium, and light loading regions of the United States. Figure 4.65 is a map of the U.S. indicating those areas (NESC, 1993). The combinations of ice and wind corresponding to loading region are listed in Table 4.11.

The NESC also suggests that increased conductor loads due to high wind loads without ice be considered. Figure 4.66 shows the suggested wind pressure as a function of geographical area for the United States (ASCE Std 7-88).

Certain utilities in very heavy ice areas use glaze ice thicknesses of as much as two inches to calculate iced conductor weight. Similarly, utilities in regions where hurricane winds occur may use wind loads as high as 34 lb/ft^2.

As the NESC indicates, the degree of ice and wind loads varies with the region. Some areas may have heavy icing, whereas some areas may have extremely high winds. The loads must be accounted for in the line design process so they do not have a detrimental effect on the line. Some of the effects of both the individual and combined components of ice and wind loads are discussed in the following.

Ice Loading

The formation of ice on overhead conductors may take several physical forms (glaze ice, rime ice, or wet snow). The impact of lower density ice formation is usually considered in the design of line sections at high altitudes.

The formation of ice on overhead conductors has the following influence on line design:

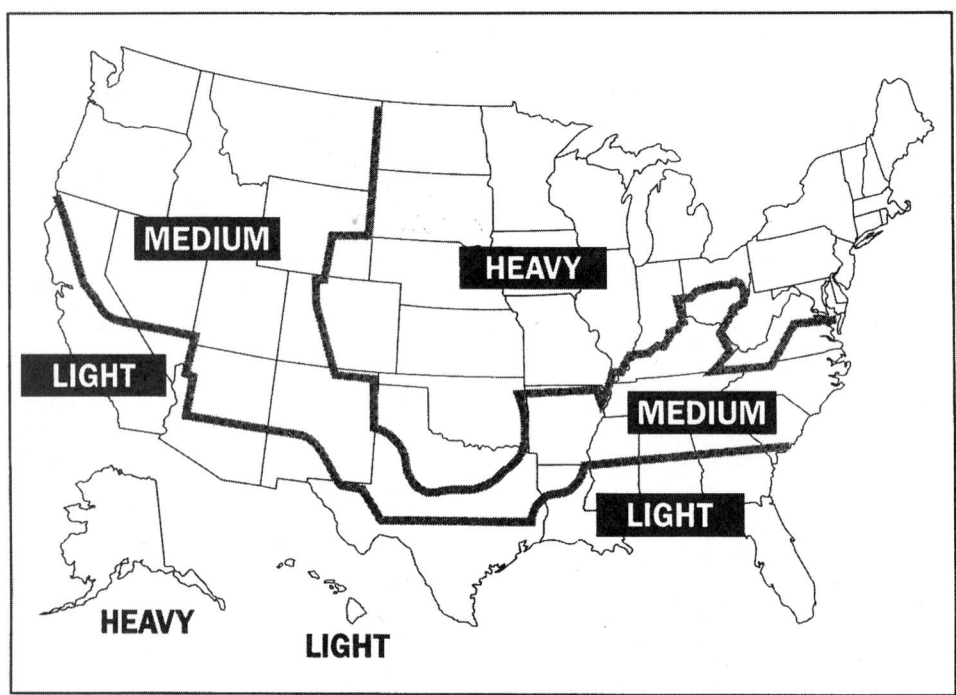

FIGURE 4.65 Ice and wind load areas of the U.S.

TABLE 4.11 Definitions of Ice and Wind Load for NESC Loading Areas

	Loading Districts			
	Heavy	Medium	Light	Extreme Wind Loading
Radial thickness of ice				
(in.)	0.50	0.25	0	0
(mm)	12.5	6.5	0	0
Horizontal wind pressure				
(lb/ft^2)	4	4	9	See Fig. 4.66
(Pa)	190	190	430	
Temperature				
(°F)	0	+15	+30	+60
(°C)	−20	−10	−1	+15
Constant to be added to the resultant for all conductors				
(lb/ft)	0.30	0.20	0.05	0.0
(N/m)	4.40	2.50	0.70	0.0

- Ice loads determine the maximum vertical conductor loads that structures and foundations must withstand.
- In combination with simultaneous wind loads, ice loads also determine the maximum transverse loads on structures.
- In regions of heavy ice loads, the maximum sags and the permanent increase in sag with time (difference between initial and final sags) may be due to ice loadings.

Ice loads for use in designing lines are normally derived on the basis of past experience, code requirements, state regulations, and analysis of historical weather data. Mean recurrence intervals for heavy ice loadings are a function of local conditions along various routings. The impact of varying assumptions concerning ice loading can be investigated with line design software.

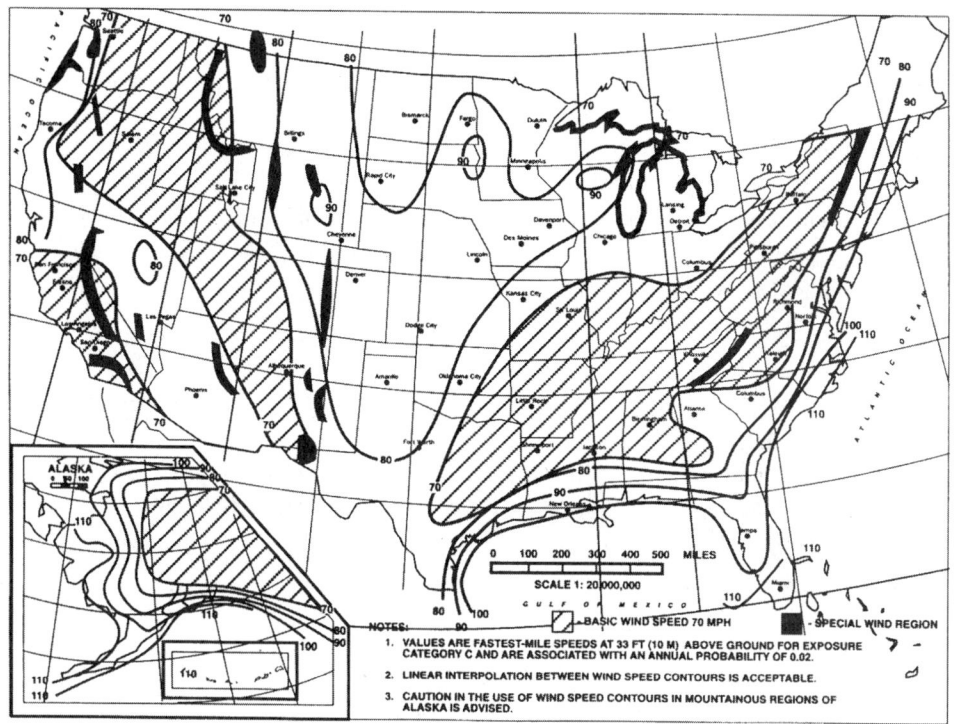

FIGURE 4.66 Wind pressure design values in the United States. (*Source:* Overend, P.R. and Smith, S., Impulse Time Method of Sag Measurement, American Society of Civil Engineers. With permission.)

TABLE 4.12 Ratio of Iced to Bare Conductor Weight

ACSR Conductor	D_c, in.	w_{bare}, lb/ft	w_{ice}, lb/ft	$\dfrac{w_{bare} + w_{ice}}{w_{bare}}$
#1/0 AWG -6/1 "Raven"	0.398	0.1451	0.559	4.8
477 kcmil-26/7 "Hawk"	0.858	0.6553	0.845	2.3
1590 kcmil-54/19 "Falcon"	1.545	2.042	1.272	1.6

The calculation of ice loads on conductors is normally done with an assumed glaze ice density of 57 lb/ft³. The weight of ice per unit length is calculated with the following equation:

$$w_{ice} = 1.244\, t \left(D_c + t \right) \tag{4.120}$$

where

t = thickness of ice, in.
D_c = conductor outside diameter, in.
w_{ice} = resultant weight of ice, lb/ft

The ratio of iced weight to bare weight depends strongly upon conductor diameter. As shown in Table 4.12 for three different conductors covered with 0.5-in radial glaze ice, this ratio ranges from 4.8 for #1/0 AWG to 1.6 for 1590-kcmil conductors. As a result, small diameter conductors may need to have a higher elastic modulus and higher tensile strength than large conductors in heavy ice and wind loading areas to limit sag.

Wind Loading

Wind loadings on overhead conductors influence line design in a number of ways:

- The maximum span between structures may be determined by the need for horizontal clearance to edge of right-of-way during moderate winds.
- The maximum transverse loads for tangent and small angle suspension structures are often determined by infrequent high wind-speed loadings.
- Permanent increases in conductor sag may be determined by wind loading in areas of light ice load.

Wind pressure load on conductors, P_w, is commonly specified in lb/ft². The relationship between P_w and wind velocity is given by the following equation:

$$P_w = 0.0025(V_w)^2 \qquad (4.121)$$

where V_w = the wind speed in miles per hour.

The wind load per unit length of conductor is equal to the wind pressure load, P_w, multiplied by the conductor diameter (including radial ice of thickness t, if any), is given by the following equation:

$$W_w = P_w \frac{(D_c + 2t)}{12} \qquad (4.122)$$

Combined Ice and Wind Loading

If the conductor weight is to include both ice and wind loading, the resultant magnitude of the loads must be determined vectorially. The weight of a conductor under both ice and wind loading is given by the following equation:

$$w_{w+i} = \sqrt{(w_b + w_i)^2 + (W_w)^2} \qquad (4.123)$$

where

w_b = bare conductor weight per unit length, lb/ft
w_i = weight of ice per unit length, lb/ft
W_w = wind load per unit length, lb/ft
w_{w+i} = resultant of ice and wind loads, lb/ft

The NESC prescribes a safety factor, K, in pounds per foot, dependent upon loading district, to be added to the resultant ice and wind loading when performing sag and tension calculations. Therefore, the total resultant conductor weight, w, is:

$$w = w_{w+i} + K \qquad (4.124)$$

Conductor Tension Limits

The NESC recommends limits on the tension of bare overhead conductors as a percentage of the conductor's rated breaking strength. The tension limits are: 60% under maximum ice and wind load, 33.3% initial unloaded (when installed) at 60°F, and 25% final unloaded (after maximum loading has occurred) at 60°F. It is common, however, for lower unloaded tension limits to be used. Except in areas experiencing severe ice loading, it is not unusual to find tension limits of 60% maximum, 25% unloaded initial, and 15% unloaded final. This set of specifications could easily result in an actual maximum tension on the order of only 35 to 40%, an initial tension of 20% and a final unloaded tension level of 15%. In this case, the 15% tension limit is said to govern.

Transmission-line conductors are normally not covered with ice, and winds on the conductor are usually much lower than those used in maximum load calculations. Under such everyday conditions, tension limits are specified to limit aeolian vibration to safe levels. Even with everyday lower tension levels of 15 to 20%, it is assumed that vibration control devices will be used in those sections of the line that are subject to severe vibration. Aeolian vibration levels, and thus appropriate unloaded tension limits, vary with the type of conductor, the terrain, span length, and the use of dampers. Special conductors, such as ACSS, SDC, and VR, exhibit high self-damping properties and may be installed to the full code limits, if desired.

Approximate Sag-Tension Calculations

Sag-tension calculations, using exacting equations, are usually performed with the aid of a computer; however, with certain simplifications, these calculations can be made with a handheld calculator. The latter approach allows greater insight into the calculation of sags and tensions than is possible with complex computer programs. Equations suitable for such calculations, as presented in the preceding section, can be applied to the following example:

It is desired to calculate the sag and slack for a 600-ft level span of 795 kcmil-26/7 ACSR "Drake" conductor. The bare conductor weight per unit length, w_b, is 1.094 lb/ft. The conductor is installed with a horizontal tension component, H, of 6300 lb, equal to 20% of its rated breaking strength of 31,500 lb.

By use of Eq. (4105), the sag for this level span is:

$$D = \frac{1.094(600^2)}{(8)6300} = 7.81 \text{ ft } (2.38 \text{ m})$$

The length of the conductor between the support points is determined using Eq. (4.109):

$$L = 600 + \frac{8(7.81)^2}{3(600)} = 600.27 \text{ ft } (182.96 \text{ m})$$

Note that the conductor length depends solely on span and sag. It is not directly dependent on conductor tension, weight, or temperature. The conductor slack is the conductor length minus the span length; in this example, it is 0.27 ft (0.0826 m).

Sag Change with Thermal Elongation

ACSR and AAC conductors elongate with increasing conductor temperature. The rate of linear thermal expansion for the composite ACSR conductor is less than that of the AAC conductor because the steel strands in the ACSR elongate at approximately half the rate of aluminum. The effective linear thermal expansion coefficient of a non-homogenous conductor, such as Drake ACSR, may be found from the following equations (Fink and Beatty):

$$E_{AS} = E_{AL}\left(\frac{A_{AL}}{A_{TOTAL}}\right) + E_{ST}\left(\frac{A_{ST}}{A_{TOTAL}}\right) \tag{4.125}$$

$$\alpha_{AS} = \alpha_{AL}\left(\frac{E_{AL}}{E_{AS}}\right)\left(\frac{A_{AL}}{A_{TOTAL}}\right) + \alpha_{ST}\left(\frac{E_{ST}}{E_{AS}}\right)\left(\frac{A_{ST}}{A_{TOTAL}}\right) \tag{4.126}$$

where

E_{AL} = Elastic modulus of aluminum, psi
E_{ST} = Elastic modulus of steel, psi
E_{AS} = Elastic modulus of aluminum-steel composite, psi
A_{AL} = Area of aluminum strands, square units
A_{ST} = Area of steel strands, square units
A_{TOTAL} = Total cross-sectional area, square units
α_{AL} = Aluminum coefficient of linear thermal expansion, per °F
α_{ST} = Steel coefficient of thermal elongation, per °F
α_{AS} = Composite aluminum-steel coefficient of thermal elongation, per °F

The elastic moduli for solid aluminum wire is 10 million psi and for steel wire is 30 million psi. The elastic moduli for stranded wire is reduced. The modulus for stranded aluminum is assumed to be 8.6 million psi for all strandings. The moduli for the steel core of ACSR conductors varies with stranding as follows:

- 27.5×10^6 for single-strand core
- 27.0×10^6 for 7-strand core
- 26.5×10^6 for 19-strand core

Using elastic moduli of 8.6 and 27.0 million psi for aluminum and steel, respectively, the elastic modulus for Drake ACSR is:

$$E_{AS} = \left(8.6 \times 10^6\right)\left(\frac{0.6247}{0.7264}\right) + \left(27.0 \times 10^6\right)\left(\frac{0.1017}{0.7264}\right) = 11.2 \times 10^6 \text{ psi}$$

and the coefficient of linear thermal expansion is:

$$\alpha_{AS} = 12.8 \times 10^{-6}\left(\frac{8.6 \times 10^6}{11.2 \times 10^6}\right)\left(\frac{0.6247}{0.7264}\right) + 6.4 \times 10^{-6}\left(\frac{27.0 \times 10^6}{11.2 \times 10^6}\right)\left(\frac{0.1017}{0.7264}\right)$$

$$= 10.6 \times 10^{-6} / °F$$

If the conductor temperature changes from a reference temperature, T_{REF}, to another temperature, T, the conductor length, L, changes in proportion to the product of the conductor's effective thermal elongation coefficient, α_{AS}, and the change in temperature, $T - T_{REF}$, as shown below:

$$L_T = L_{T_{REF}}\left(1 + \alpha_{AS}\left(T - T_{REF}\right)\right) \qquad (4.127)$$

For example, if the temperature of the Drake conductor in the preceding example increases from 60°F (15°C) to 167°F (75°C), then the length at 60°F increases by 0.68 ft (0.21 m) from 600.27 ft (182.96 m) to 600.95 ft (183.17 m):

$$L_{(167°F)} = 600.27\left(1 + \left(10.6 \times 10^{-6}\right)\left(167 - 60\right)\right) = 600.95 \text{ ft}$$

Ignoring for the moment any change in length due to change in tension, the sag at 167°F (75°C) may be calculated for the conductor length of 600.95 ft (183.17 m) using Eq. (4.111):

$$D = \sqrt{\frac{3(600)(0.95)}{8}} = 14.62 \text{ ft}$$

Using a rearrangement of Eq. (4.105), this increased sag is found to correspond to a decreased tension of:

$$H = \frac{w(S^2)}{8D} = \frac{1.094(600^2)}{8(14.62)} = 3367 \text{ lb}$$

If the conductor were inextensible, that is, if it had an infinite modulus of elasticity, then these values of sag and tension for a conductor temperature of 167°F would be correct. For any real conductor, however, the elastic modulus of the conductor is finite and changes in tension do change the conductor length. Use of the preceding calculation, therefore, will overstate the increase in sag.

Sag Change Due to Combined Thermal and Elastic Effects

With moduli of elasticity around the 8.6 million psi level, typical bare aluminum and ACSR conductors elongate about 0.01% for every 1000 psi change in tension. In the preceding example, the increase in temperature caused an increase in length and sag and a decrease in tension, but the effect of tension change on length was ignored.

As discussed later, concentric-lay stranded conductors, particularly non-homogenous conductors such as ACSR, are not inextensible. Rather, they exhibit quite complex elastic and plastic behavior. Initial loading of conductors results in elongation behavior substantially different from that caused by loading many years later. Also, high tension levels caused by heavy ice and wind loads cause a permanent increase in conductor length, affecting subsequent elongation under various conditions.

Accounting for such complex stress-strain behavior usually requires a sophisticated, computer-aided approach. For illustration purposes, however, the effect of permanent elongation of the conductor on sag and tension calculations will be ignored and a simplified elastic conductor assumed. This idealized conductor is assumed to elongate linearly with load and to undergo no permanent increase in length regardless of loading or temperature. For such a conductor, the relationship between tension and length is as follows:

$$L_H = L_{H_{REF}}\left(1 + \frac{H - H_{REF}}{E_C A}\right) \tag{4.128}$$

where

L_H = Length of conductor under horizontal tension H
$L_{H_{REF}}$ = Length of conductor under horizontal reference tension H_{REF}
E_C = Elastic modulus of elasticity of the conductor, psi
A = Cross-sectional area, in.^{2}

In calculating sag and tension for extensible conductors, it is useful to add a step to the preceding calculation of sag and tension for elevated temperature. This added step allows a separation of thermal elongation and elastic elongation effects, and involves the calculation of a zero tension length, ZTL, at the conductor temperature of interest, T_{cdr}.

This ZTL(T_{cdr}) is the conductor length attained if the conductor is taken down from its supports and laid on the ground with no tension. By reducing the initial tension in the conductor to zero, the elastic elongation is also reduced to zero, shortening the conductor. It is possible, then, for the zero tension length to be less than the span length.

Consider the preceding example for Drake ACSR in a 600-ft level span. The initial conductor temperature is 60°F, the conductor length is 600.27 ft, and E_{AS} is calculated to be 11.2 million psi. Using Eq. (4.128), the reduction of the initial tension from 6300 lb to zero yields a ZTL (60°F) of:

$$ZTL_{(60°F)} = 600.27\left(1 + \frac{0 - 6300}{(11.2 \times 10^6)0.7264}\right) = 599.81 \text{ ft}$$

Keeping the tension at zero and increasing the conductor temperature to 167°F yields a purely thermal elongation. The zero tension length at 167°F can be calculated using Eq. (4.127):

$$ZTL_{(167°F)} = 599.81\left(1 + \left(10.6 \times 10^{-6}\right)\left(167 - 60\right)\right) = 600.49 \text{ ft}$$

According to Eqs. (4.105) and (4.111), this length corresponds to a sag of 10.5 ft and a horizontal tension of 4689 lb. However, this length was calculated for zero tension and will elongate elastically under tension. The actual conductor sag-tension determination requires a process of iteration as follows:

1. As described above, the conductor's zero tension length, calculated at 167°F (75°C), is 600.49 ft, sag is 10.5 ft, and the horizontal tension is 4689 lb.
2. Because the conductor is elastic, application of Eq. (4.128) shows the tension of 4689 lb will increase the conductor length from 600.49 ft to:

$$L_{1(167°F)} = 600.49\left(1 + \frac{4689 - 0}{0.7264\left(11.2 \times 10^6\right)}\right) = 600.84 \text{ ft}$$

3. The sag, $D_{1(167°F)}$, corresponding to this length is calculated using Eq. (4.111):

$$D_{1(167°F)} = \sqrt{\frac{3(600)(0.84)}{8}} = 13.72 \text{ ft}$$

4. Using Eq. (4.105), this sag yields a new horizontal tension, $H_{1(167°F)}$, of:

$$H_1 = \frac{1.094\left(600^2\right)}{8(13.7)} = 3588 \text{ lb}$$

A new trial tension is taken as the average of H and H_1, and the process is repeated. The results are described in Table 4.13.

Note that the balance of thermal and elastic elongation of the conductor yields an equilibrium tension of approximately 3700 lbs and a sag of 13.3 ft. The calculations of the previous section, which ignored elastic effects, results in lower tension, 3440 lb, and a greater sag, 14.7 ft.

Slack is equal to the excess of conductor length over span length. The preceding table can be replaced by a plot of the catenary and elastic curves on a graph of slack vs tension. The solution occurs at the intersection of the two curves. Figure 4.67 shows the tension versus slack curves intersecting at a tension of 3700 lb, which agrees with the preceding calculations.

Sag Change Due to Ice Loading

As a final example of sag-tension calculation, calculate the sag and tension for the 600-ft Drake span with the addition of 0.5 inches of radial ice and a drop in conductor temperature to 0°F. Employing Eq. (4.120), the weight of the conductor increases by:

$$w_{ice} = 1.244\, t(D + t)$$

$$w_{ice} = 1.244(0.5)(1.108 + 0.5) = 1.000 \text{ lb/ft}$$

Transmission System   4-101

TABLE 4.13 Interative Solution for Increased Conductor Temperature

Iteration #	Length, L_n, ft	Sag, D_n, ft	Tension, H_n, lb	New Trial Tension, lb
ZTL	600.550	11.1	4435	—
1	600.836	13.7	3593	$\frac{4435+3593}{2}=4014$
2	600.809	13.5	3647	$\frac{3647+4014}{2}=3831$
3	600.797	13.4	3674	$\frac{3674+3831}{2}=3753$
4	600.792	13.3	3702	$\frac{3702+3753}{2}=3727$

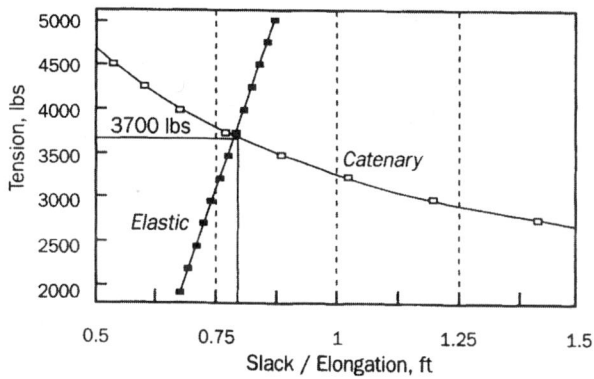

FIGURE 4.67 Sag-tension solution for 600-ft span of Drake at 167°F.

As in the previous example, the calculation uses the conductor's zero tension length at 60°F, which is the same as that found in the previous section, 599.81 ft. The ice loading is specified for a conductor temperature of 0°F, so the ZTL(0°F), using Eq. (4.127), is:

$$ZTL_{(0°F)} = 599.81\left[1+\left(10.6 \times 10^{-6}\right)(0-60)\right] = 599.43 \text{ ft}$$

As in the case of sag-tension at elevated temperatures, the conductor tension is a function of slack and elastic elongation. The conductor tension and the conductor length are found at the point of intersection of the catenary and elastic curves (Fig. 4.68). The intersection of the curves occurs at a horizontal tension component of 12,275 lb, not very far from the crude initial estimate of 12,050 lb that ignored elastic effects. The sag corresponding to this tension and the iced conductor weight per unit length is 9.2 ft.

In spite of doubling the conductor weight per unit length by adding 0.5 in. of ice, the sag of the conductor is much less than the sag at 167°F. This condition is generally true for transmission conductors where minimum ground clearance is determined by the high temperature rather than the heavy loading condition. Small distribution conductors, such as the 1/0 AWG ACSR in Table 4.11, experience a much larger ice-to-conductor weight ratio (4.8), and the conductor sag under maximum wind and ice load may exceed the sag at moderately higher temperatures.

The preceding approximate tension calculations could have been more accurate with the use of actual stress-strain curves and graphic sag-tension solutions, as described in detail in *Graphic Method for Sag*

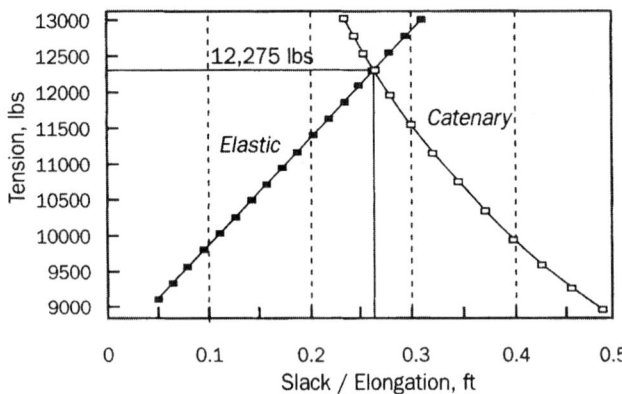

FIGURE 4.68 Sag-tension solution for 600-ft span of Drake at 0°F and 0.5 in. ice.

Tension Calculations for ACSR and Other Conductors (Aluminum Company of America, 1961). This method, although accurate, is very slow and has been replaced completely by computational methods.

Numerical Sag-Tension Calculations

Sag-tension calculations are normally done numerically and allow the user to enter many different loading and conductor temperature conditions. Both initial and final conditions are calculated and multiple tension constraints can be specified. The complex stress-strain behavior of ACSR-type conductors can be modeled numerically, including both temperature, and elastic and plastic effects.

Stress-Strain Curves

Stress-strain curves for bare overhead conductor include a minimum of an initial curve and a final curve over a range of elongations from 0 to 0.45%. For conductors consisting of two materials, an initial and final curve for each is included. Creep curves for various lengths of time are typically included as well.

Overhead conductors are not purely elastic. They stretch with tension, but when the tension is reduced to zero, they do not return to their initial length. That is, conductors are plastic; the change in conductor length cannot be expressed with a simple linear equation, as for the preceding hand calculations. The permanent length increase that occurs in overhead conductors yields the difference in initial and final sag-tension data found in most computer programs.

Figure 4.69 shows a typical stress-strain curve for a 26/7 ACSR conductor (Aluminum Association, 1974); the curve is valid for conductor sizes ranging from 266.8 to 795 kcmil. A 795 kcmil-26/7 ACSR "Drake" conductor has a breaking strength of 31,500 lb (14,000 kg) and an area of 0.7264 in.^{2} (46.9 mm^2) so that it fails at an average stress of 43,000 psi (30 kg/mm^2). The stress-strain curve illustrates that when the percent of elongation at a stress is equal to 50% of the conductor's breaking strength (21,500 psi), the elongation is less than 0.3% or 1.8 ft (0.55 m) in a 600-ft (180 m) span.

Note that the component curves for the steel core and the aluminum stranded outer layers are separated. This separation allows for changes in the relative curve locations as the temperature of the conductor changes.

For the preceding example, with the Drake conductor at a tension of 6300 lb (2860 kg), the length of the conductor in the 600-ft (180 m) span was found to be 0.27 ft longer than the span. This tension corresponds to a stress of 8600 psi (6.05 kg/mm^2). From the stress-strain curve in Fig. 4.69, this corresponds to an initial elongation of 0.105% (0.63 ft). As in the preceding hand calculation, if the conductor is reduced to zero tension, its unstressed length would be less than the span length.

Figure 4.70 is a stress-strain curve (Aluminum Association, 1974) for an all-aluminum 37-strand conductor ranging in size from 250 kcmil to 1033.5 kcmil. Because the conductor is made entirely of aluminum, there is only one initial and final curve.

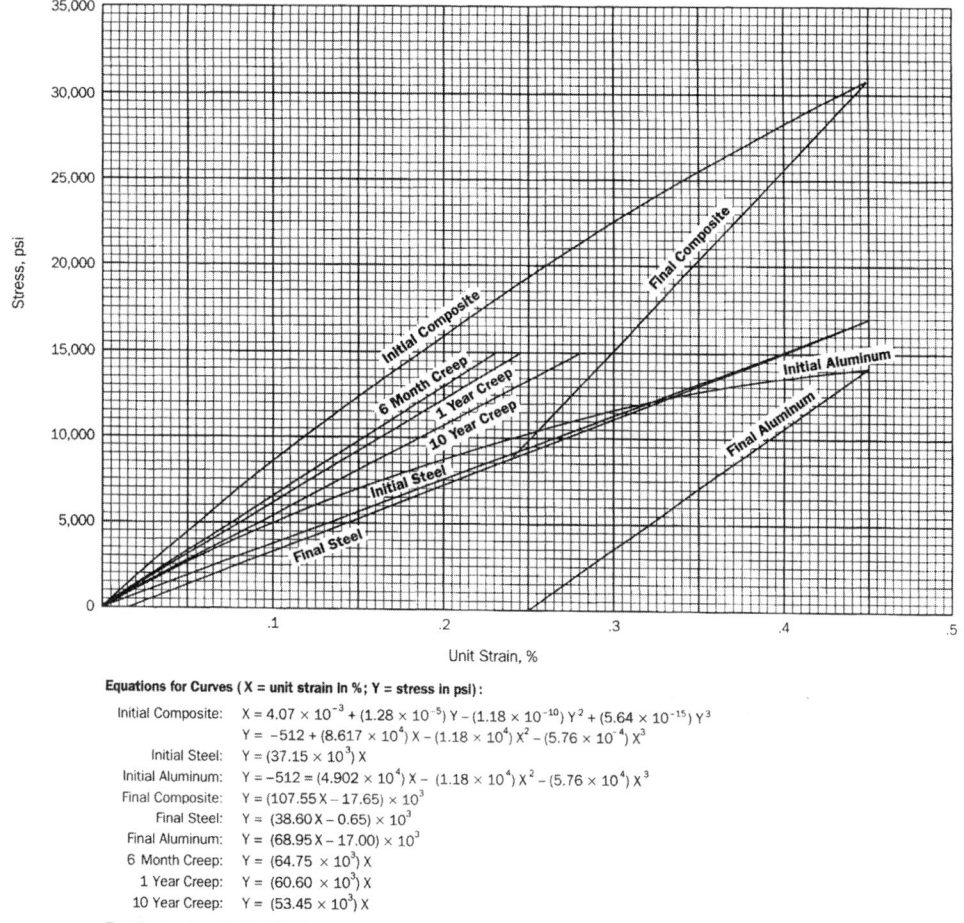

FIGURE 4.69 Stress-strain curves for 26/7 ACSR.

Permanent Elongation

Once a conductor has been installed at an initial tension, it can elongate further. Such elongation results from two phenomena: permanent elongation due to high tension levels resulting from ice and wind loads, and creep elongation under everyday tension levels. These types of conductor elongation are discussed in the following sections.

Permanent Elongation Due to Heavy Loading

Both Figs. 4.69 and 4.70 indicate that when the conductor is initially installed, it elongates following the initial curve that is not a straight line. If the conductor tension increases to a relatively high level under ice and wind loading, the conductor will elongate. When the wind and ice loads abate, the conductor elongation will reduce along a curve parallel to the final curve, but the conductor will never return to its original length.

For example, refer to Fig. 4.70 and assume that a newly strung 795 kcmil-37 strand AAC "Arbutus" conductor has an everyday tension of 2780 lb. The conductor area is 0.6245 in.^{2}, so the everyday stress is 4450 psi and the elongation is 0.062%. Following an extremely heavy ice and wind load event, assume that the conductor stress reaches 18,000 psi. When the conductor tension decreases back to everyday levels, the conductor elongation will be permanently increased by more than 0.2%. Also the sag under everyday conditions will be correspondingly higher, and the tension will be less. In most numerical sag-

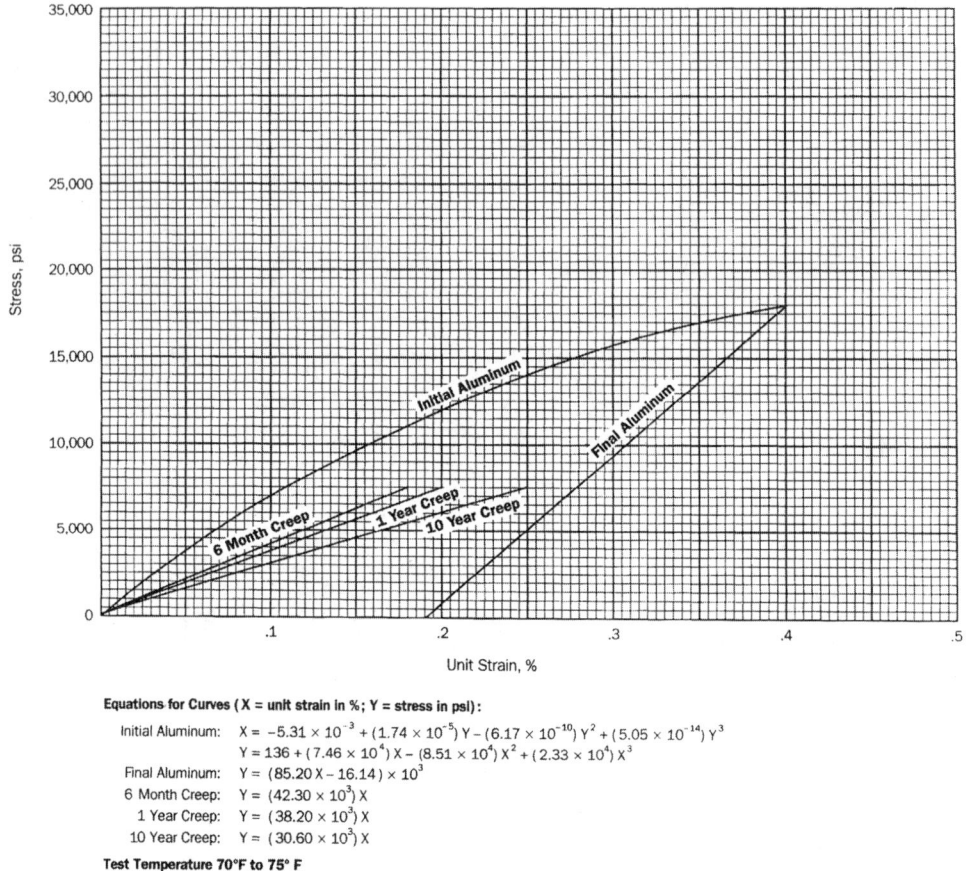

FIGURE 4.70 Stress-strain curves for 37-strand AAC.

tension methods, final sag-tensions are calculated for such permanent elongation due to heavy loading conditions.

Permanent Elongation at Everyday Tensions (Creep Elongation)

Conductors permanently elongate under tension even if the tension level never exceeds everyday levels. This permanent elongation caused by everyday tension levels is called creep (Aluminum Company of America, 1961). Creep can be determined by long-term laboratory creep tests, the results of which are used to generate creep curves. On stress-strain graphs, creep curves are usually shown for 6-mo, 1-yr, and 10-yr periods. Figure 4.70 shows these typical creep curves for a 37 strand 250.0 through 1033.5 kcmil AAC. In Figure 4.70 assume that the conductor tension remains constant at the initial stress of 4450 psi. At the intersection of this stress level and the initial elongation curve, 6-month, 1-year, and 10-year creep curves, the conductor elongation from the initial elongation of 0.062% increases to 0.11%, 0.12%, and 0.15%, respectively. Because of creep elongation, the resulting final sags are greater and the conductor tension is less than the initial values.

Creep elongation in aluminum conductors is quite predictable as a function of time and obeys a simple exponential relationship. Thus, the permanent elongation due to creep at everyday tension can be found for any period of time after initial installation. Creep elongation of copper and steel conductors is much less and is normally ignored.

Permanent increase in conductor length due to heavy load occurrences cannot be predicted at the time that a line is built. The reason for this unpredictability is that the occurrence of heavy ice and wind is random. A heavy ice storm may occur the day after the line is built or may never occur over the life of the line.

Transmission System

Sag-Tension Tables

To illustrate the result of typical sag-tension calculations, refer to Tables 4.14 through 4.19 showing initial and final sag-tension data for 795 kcmil-26/7 ACSR "Drake", 795 kcmil-37 strand AAC "Arbutus", and 795-kcmil Type 16 "Drake/SDC" conductors in NESC light and heavy loading areas for spans of 1000 and 300 ft. Typical tension constraints of 15% final unloaded at 60°F, 25% initial unloaded at 60°F, and 60% initial at maximum loading are used.

With most sag-tension calculation methods, final sags are calculated for both heavy ice/wind load and for creep elongation. The final sag-tension values reported to the user are those with the greatest increase in sag.

Initial vs. Final Sags and Tensions

Rather than calculate the line sag as a function of time, most sag-tension calculations are determined based on initial and final loading conditions. Initial sags and tensions are simply the sags and tensions at the time the line is built. Final sags and tensions are calculated if (1) the specified ice and wind loading has occurred, and (2) the conductor has experienced 10 years of creep elongation at a conductor temperature of 60°F at the user-specified initial tension.

TABLE 4.14 Sag and Tension Data for 795 kcmil-26/7 ACSR "Drake" Conductor

Span = 600 ft
NESC Heavy Loading District

Creep is **not** a factor

Temp, °F	Ice, in.	Wind, lb/ft²	K, lb/ft	Resultant Weight, lb/ft	Final Sag, ft	Final Tension, lb	Initial Sag, ft	Initial Tension, lb
0	0.50	4.00	0.30	2.509	11.14	10153 5415 Al 4738 St	11.14	10153 5415 Al 4738 St
32	0.50	0.00	0.00	2.094	44.54	8185 3819 Al 4366 St	11.09	8512 4343 Al 4169 St
−20	0.00	0.00	0.00	1.094	6.68	7372 3871 Al 3501 St	6.27	7855 4465 Al 3390 St
0	0.00	0.00	0.00	1.094	7.56	6517 3111 Al 3406 St	6.89	7147 3942 Al 3205 St
30	0.00	0.00	0.00	1.094	8.98	5490 2133 Al 3357 St	7.95	6197 3201 Al 2996 St
60	0.00	0.00	0.00	1.094	10.44	4725[a] 1321 Al 3404 St	9.12	5402 2526 Al 2875 St
90	0.00	0.00	0.00	1.094	11.87	4157 634 Al 3522 St	10.36	4759 1922 Al 2837 St
120	0.00	0.00	0.00	1.094	13.24	3727 35 Al 3692 St	11.61	4248 1379 Al 2869 St
167	0.00	0.00	0.00	1.094	14.29	3456 0 Al 3456 St	13.53	3649 626 Al 3022 St
212	0.00	0.00	0.00	1.094	15.24	3241 0 Al 3241 St	15.24	3241 0 Al 3239 St

[a] Design condition.

TABLE 4.15 Tension Differences in Adjacent Dead-End Spans

Conductor: Drake
795 kcmil-26/7 ACSR
Area = 0.7264 in.²
Creep *is* a factor

Span = 700 ft

NESC Heavy Loading District

Temp, °F	Ice, in.	Wind, lb/ft²	K, lb/ft	Resultant Weight, lb/ft	Final Sag, ft	Final Tension, lb	Initial Sag, ft	Initial Tension, lb
0	0.50	4.00	0.30	2.509	13.61	11318	13.55	11361
32	0.50	0.00	0.00	2.094	13.93	9224	13.33	9643
−20	0.00	0.00	0.00	1.094	8.22	8161	7.60	8824
0	0.00	0.00	0.00	1.094	9.19	7301	8.26	8115
30	0.00	0.00	0.00	1.094	10.75	6242	9.39	7142
60	0.00	0.00	0.00	1.094	12.36	5429	10.65	6300[a]
90	0.00	0.00	0.00	1.094	13.96	4809	11.99	5596
120	0.00	0.00	0.00	1.094	15.52	4330	13.37	5020
167	0.00	0.00	0.00	1.094	16.97	3960	15.53	4326
212	0.00	0.00	0.00	1.094	18.04	3728	17.52	3837

[a] Design condition.

Conductor: Drake
795 kcmil-26/7 ACSR
Area = 0.7264 in.²
Creep is *not* a factor

Span = 1000 ft

NESC Heavy Loading District

Temp, °F	Ice, in.	Wind, lb/ft²	K, lb/ft	Resultant Weight, lb/ft	Final Sag, ft	Final Tension, lb	Initial Sag, ft	Initial Tension, lb
0	0.50	4.00	0.30	2.509	25.98	12116	25.98	12116
32	0.50	0.00	0.00	2.094	26.30	9990	25.53	10290
−20	0.00	0.00	0.00	1.094	18.72	7318	17.25	7940
0	0.00	0.00	0.00	1.094	20.09	6821	18.34	7469
30	0.00	0.00	0.00	1.094	22.13	6197	20.04	6840
60	0.00	0.00	0.00	1.094	24.11	5689	21.76	6300[a]
90	0.00	0.00	0.00	1.094	26.04	5271	23.49	5839
120	0.00	0.00	0.00	1.094	27.89	4923	25.20	5444
167	0.00	0.00	0.00	1.094	30.14	4559	27.82	4935
212	0.00	0.00	0.00	1.094	31.47	4369	30.24	4544

[a] Design condition.

Special Aspects of ACSR Sag-Tension Calculations

Sag-tension calculations with ACSR conductors are more complex than such calculations with AAC, AAAC, or ACAR conductors. The complexity results from the different behavior of steel and aluminum strands in response to tension and temperature. Steel wires do not exhibit creep elongation or plastic elongation in response to high tensions. Aluminum wires do creep and respond plastically to high stress levels. Also, they elongate twice as much as steel wires do in response to changes in temperature.

Table 4.20 presents various initial and final sag-tension values for a 600-ft span of a Drake ACSR conductor under heavy loading conditions. Note that the tension in the aluminum and steel components is shown separately. In particular, some other useful observations are:

1. At 60°F, without ice or wind, the tension level in the aluminum strands decreases with time as the strands permanently elongate due to creep or heavy loading.
2. Both initially and finally, the tension level in the aluminum strands decreases with increasing temperature reaching zero tension at 212°F and 167°F for initial and final conditions, respectively.
3. At the highest temperature (212°F), where all the tension is in the steel core, the initial and final sag-tensions are nearly the same, illustrating that the steel core does not permanently elongate in response to time or high tension.

Transmission System

TABLE 4.16 Sag and Tension Data for 795 kcmil-26/7 ACSR "Drake" 600-ft Ruling Span

Conductor: Drake
795 kcmil-26/7 ACSR Span = 600 ft
Area = 0.7264 in.²
Creep is ***not*** a factor NESC Heavy Loading District

Temp, °F	Ice, in.	Wind, lb/ft²	K, lb/ft	Resultant Weight, lb/ft	Final Sag, ft	Final Tension, lb	Initial Sag, ft	Initial Tension, lb
0	0.50	4.00	0.30	2.509	11.14	10153	11.14	10153
32	0.50	0.00	0.00	2.094	11.54	8185	11.09	8512
−20	0.00	0.00	0.00	1.094	6.68	7372	6.27	7855
0	0.00	0.00	0.00	1.094	7.56	6517	6.89	7147
30	0.00	0.00	0.00	1.094	8.98	5490	7.95	6197
60	0.00	0.00	0.00	1.094	10.44	4725[a]	9.12	5402
90	0.00	0.00	0.00	1.094	11.87	4157	10.36	4759
120	0.00	0.00	0.00	1.094	13.24	3727	11.61	4248
167	0.00	0.00	0.00	1.094	14.29	3456	13.53	3649
212	0.00	0.00	0.00	1.094	15.24	3241	15.24	3241

[a] Design condition.

Ruling Span Concept

Transmission lines are normally designed in line sections with each end of the line section terminated by a strain structure that allows no longitudinal (along the line) movement of the conductor (Winkelman, 1959). Structures within each line section are typically suspension structures that support the conductor vertically, but allow free movement of the conductor attachment point either longitudinally or transversely.

Tension Differences for Adjacent Dead-End Spans

Table 4.21 contains initial and final sag-tension data for a 700-ft and a 1000-ft dead-end span when a Drake ACSR conductor is initially installed to the same 6300-lb tension limits at 60°F. Note that the difference between the initial and final limits at 60°F is approximately 460 lb. Even the initial tension (equal at 60°F) differs by almost 900 lb at −20°F and 600 lb at 167°F.

Tension Equalization by Suspension Insulators

At a typical suspension structure, the conductor is supported vertically by a suspension insulator assembly, but allowed to move freely in the direction of the conductor axis. This conductor movement is possible due to insulator swing along the conductor axis. Changes in conductor tension between spans, caused by changes in temperature, load, and time, are normally equalized by insulator swing, eliminating horizontal tension differences across suspension structures.

Ruling Span Calculation

Sag-tension can be found for a series of suspension spans in a line section by use of the ruling span concept (Ehrenberg, 1935; Winkelman, 1959). The ruling span (RS) for the line section is defined by the following equation:

$$RS \sqrt{\frac{S_1^3 \quad S_2^3 \quad \cdots \quad S_n^3}{S_1 \quad S_2 \quad \cdots \quad S_n}} \qquad (4.129)$$

where

RS = Ruling span for the line section containing n suspension spans
S_1 = Span length of first suspension span
S_2 = Span length of second suspension span
S_n = Span length of nth suspension span

TABLE 4.17 Stringing Sag Table for 795 kcmil-26/7 ACSR "Drake" 600-ft Ruling Span

600-ft Ruling Span

Controlling Design Condition:
15% RBS at 60°F, No Ice or Wind, Final

NESC Heavy Load District

Horizontal Tension, lb	6493	6193	5910	5645	5397	5166	4952	4753	4569
Temp, °F	20	30	40	50	60	70	80	90	100
Spans	Sag, ft-in.	Sag, ft-in.	Sag, ft-in.	Sag, ft-in.	Sag, ft-in.	Sag, ft-in.	Sag, ft-in.	Sag, ft-in.	Sag, ft-in.
400	3 - 4	3 - 6	3 - 8	3 - 11	4 - 1	4 - 3	4 - 5	4 - 7	4 - 9
410	3 - 6	3 - 9	3 - 11	4 - 1	4 - 3	4 - 5	4 - 8	4 - 10	5 - 0
420	3 - 9	3 - 11	4 - 1	4 - 3	4 - 6	4 - 8	4 - 10	5 - 1	5 - 3
430	3 - 11	4 - 1	4 - 3	4 - 6	4 - 8	4 - 11	5 - 1	5 - 4	5 - 6
440	4 - 1	4 - 3	4 - 6	4 - 8	4 - 11	5 - 2	5 - 4	5 - 7	5 - 10
450	4 - 3	4 - 6	4 - 8	4 - 11	5 - 2	5 - 4	5 - 7	5 - 10	6 - 1
460	4 - 5	4 - 8	4 - 11	5 - 2	5 - 4	5 - 7	5 - 10	6 - 1	6 - 4
470	4 - 8	4 - 11	5 - 1	5 - 4	5 - 7	5 - 10	6 - 1	6 - 4	6 - 7
480	4 - 10	5 - 1	5 - 4	5 - 7	5 - 10	6 - 1	6 - 4	6 - 8	6 - 11
490	5 - 1	5 - 4	5 - 7	5 - 10	6 - 1	6 - 4	6 - 8	6 - 11	7 - 2
500	5 - 3	5 - 6	5 - 9	6 - 1	6 - 4	6 - 7	6 - 11	7 - 2	7 - 6
510	5 - 6	5 - 9	6 - 0	6 - 4	6 - 7	6 - 11	7 - 2	7 - 6	7 - 9
520	5 - 8	6 - 0	6 - 3	6 - 7	6 - 10	7 - 2	7 - 6	7 - 9	8 - 1
530	5 - 11	6 - 2	6 - 6	6 - 10	7 - 1	7 - 5	7 - 9	8 - 1	8 - 5
540	6 - 2	6 - 5	6 - 9	7 - 1	7 - 5	7 - 9	8 - 1	8 - 5	8 - 9
550	6 - 4	6 - 8	7 - 0	7 - 4	7 - 8	8 - 0	8 - 4	8 - 8	9 - 1
560	6 - 7	6 - 11	7 - 3	7 - 7	7 - 11	8 - 4	8 - 8	9 - 0	9 - 5
570	6 - 10	7 - 2	7 - 6	7 - 10	8 - 3	8 - 7	9 - 0	9 - 4	9 - 9
580	7 - 1	7 - 5	7 - 9	8 - 2	8 - 6	8 - 11	9 - 4	9 - 8	10 - 1
590	7 - 4	7 - 8	8 - 1	8 - 5	8 - 10	9 - 3	9 - 7	10 - 0	10 - 5
600	7 - 7	7 - 11	8 - 4	8 - 9	9 - 1	9 - 6	9 - 11	10 - 4	10 - 9
610	7 - 1	8 - 3	8 - 7	9 - 0	9 - 5	9 - 10	10 - 3	10 - 9	11 - 2
620	8 - 1	8 - 6	8 - 11	9 - 4	9 - 9	10 - 2	10 - 7	11 - 1	11 - 6
630	8 -	8 - 9	9 - 2	9 - 7	10 - 1	10 - 6	11 - 0	11 - 5	11 - 11
640	8 - 8	9 - 1	9 - 6	9 - 11	10 - 5	10 - 10	11 - 4	11 - 9	12 - 3
650	8 - 11	9 - 4	9 - 9	10 - 3	10 - 9	11 - 2	11 - 8	12 - 2	12 - 8
660	9 - 2	9 - 7	10 - 1	10 - 7	11 - 1	11 - 6	12 - 0	12 - 6	13 - 1
670	9 - 5	9 - 11	10 - 5	10 - 11	11 - 5	11 - 11	12 - 5	12 - 11	13 - 5
680	9 - 9	10 - 3	10 - 8	11 - 2	11 - 9	12 - 3	12 - 9	13 - 4	13 - 10
690	10 - 0	10 - 6	11 - 0	11 - 6	12 - 1	12 - 7	13 - 2	13 - 8	14 - 3
700	10 - 4	10 - 10	11 - 4	11 - 11	12 - 5	13 - 0	13 - 6	14 - 1	14 - 8

Alternatively, a generally satisfactory method for estimating the ruling span is to take the sum of the average suspension span length plus two-thirds of the difference between the maximum span and the average span. However, some judgment must be exercised in using this method because a large difference between the average and maximum span may cause a substantial error in the ruling span value.

As discussed, suspension spans are supported by suspension insulators that are free to move in the direction of the conductor axis. This freedom of movement allows the tension in each suspension span to be assumed to be the same and equal to that calculated for the ruling span. This assumption is valid for the suspension spans and ruling span under the same conditions of temperature and load, for both initial and final sags. For level spans, sag in each suspension span is given by the parabolic sag equation:

$$D_i = \frac{w(S_i^2)}{8 H_{RS}} \tag{4.130}$$

TABLE 4.18 Time-Sag Table for Stopwatch Method

					Return of Wave					
Sag, in.	3rd Time, sec	5th Time, sec	Sag, in.	3rd Time, sec	5th Time, sec	Sag, in.	3rd Time, sec	5th Time, sec	Sag, in.	3rd Time, sec	5th Time, sec

Sag, in.	3rd Time, sec	5th Time, sec	Sag, in.	3rd Time, sec	5th Time, sec	Sag, in.	3rd Time, sec	5th Time, sec	Sag, in.	3rd Time, sec	5th Time, sec
5	1.9	3.2	55	6.4	10.7	105	8.8	14.7	155	10.7	17.9
6	2.1	3.5	56	6.5	10.8	106	8.9	14.8	156	10.8	18.0
7	2.3	3.8	57	6.5	10.9	107	8.9	14.9	157	10.8	18.0
8	2.4	4.1	58	6.6	11.1	109	9.0	15.0	158	10.9	18.1
9	2.6	4.3	59	6.6	11.1	109	9.0	15.0	159	10.9	18.1
10	2.7	4.6	60	6.7	11.1	110	9.1	15.1	160	10.9	18.2
11	2.9	4.8	61	6.7	11.2	111	9.1	15.2	161	11.0	18.2
12	3.0	5.0	62	6.8	11.3	112	9.1	15.2	162	11.0	18.2
13	3.1	5.2	63	6.9	11.4	113	9.2	15.3	163	11.0	18.4
14	3.2	5.4	64	6.9	11.5	114	9.2	15.4	164	11.1	18.4
15	3.3	5.6	65	7.0	11.6	115	9.3	15.4	165	11.1	18.5
16	3.5	5.8	66	7.0	11.7	116	9.3	15.5	166	11.1	18.5
17	3.6	5.9	67	7.1	11.8	117	9.3	15.6	167	11.2	18.6
18	3.7	6.1	68	7.1	11.9	118	9.4	15.6	168	11.2	18.7
19	3.8	6.3	69	7.2	12.0	119	9.4	15.7	169	11.2	18.7
20	3.9	6.4	70	7.2	12.0	120	9.5	15.8	170	11.3	18.8
21	4.0	6.6	71	7.3	12.1	121	9.5	15.8	171	11.3	18.8
22	4.0	6.7	72	7.3	12.2	122	9.5	15.9	172	11.3	18.9
23	4.1	6.9	73	7.4	12.3	123	9.6	16.0	173	11.4	18.9
24	4.2	7.0	74	7.4	12.4	124	9.6	16.0	174	11.4	19.0
25	4.3	7.2	75	7.5	12.5	125	9.7	16.1	175	11.4	19.0
26	4.4	7.3	76	7.5	12.5	126	9.7	16.2	176	11.4	19.1
27	4.5	7.5	77	7.6	12.6	127	9.7	16.2	177	11.5	19.1
28	4.6	7.6	78	7.6	12.7	128	9.8	16.3	178	11.5	19.2
29	4.6	7.7	79	7.7	12.8	129	9.8	16.3	179	11.5	19.3
30	4.7	7.9	80	7.7	12.9	130	9.8	16.4	180	11.6	19.3
31	4.8	8.0	81	7.8	13.0	131	9.9	16.5	181	11.6	19.4
32	4.9	8.1	82	7.8	13.0	132	9.9	16.5	182	11.6	19.4
33	5.0	8.3	83	7.9	13.1	133	10.0	16.6	183	11.7	19.5
34	5.0	8.4	84	7.9	13.2	134	10.0	16.7	184	11.7	19.5
35	5.1	8.5	85	8.0	13.3	135	10.0	16.7	185	11.7	19.6
36	5.2	8.6	86	8.0	13.3	136	10.1	16.8	186	11.8	19.6
37	5.3	8.8	87	8.1	13.4	137	10.1	16.8	187	11.8	19.7
38	5.3	8.9	88	8.1	13.5	138	10.1	16.9	188	11.8	19.7
39	5.4	9.0	89	8.1	13.6	139	10.2	17.0	189	11.9	19.8
40	5.5	9.1	90	8.2	13.7	140	10.2	17.0	190	11.9	19.8
41	5.5	9.2	91	8.2	13.7	141	10.3	17.1	191	11.9	19.9
42	5.6	9.3	92	8.3	13.8	142	10.3	17.1	192	12.0	19.9
43	5.7	9.4	93	8.3	13.9	143	10.3	17.2	193	12.0	20.0
44	5.7	9.5	94	8.4	14.0	144	10.4	17.3	194	12.0	20.0
45	5.8	9.7	95	8.4	14.0	145	10.4	17.3	195	12.1	20.1
46	5.9	9.8	96	8.5	14.1	146	10.4	17.4	196	12.1	20.1
47	5.9	9.9	97	8.5	14.2	147	10.5	17.4	197	12.1	20.2
48	6.0	10.0	98	8.5	14.2	148	10.5	17.5	198	12.1	20.0
49	6.0	10.1	99	8.6	14.3	149	10.5	17.6	199	12.2	20.3
50	6.1	10.2	100	8.6	14.4	150	10.6	17.6	200	12.2	20.3
51	6.2	10.3	101	8.7	14.5	151	10.6	17.7	201	12.2	20.4
52	6.2	10.4	102	8.7	14.5	152	10.6	17.7	202	12.3	20.5
53	6.3	10.5	103	8.8	14.6	153	10.7	17.8	203	12.3	20.5
54	6.3	10.6	104	8.8	14.7	154	10.7	17.9	204	12.3	20.6

Note: To calculate the time of return of other waves, multiply the time in seconds for one wave return by the number of wave returns or, more simply, select the combination of values from the table that represents the number of wave returns desired. For example, the time of return of the 8th wave is the sum of the 3rd and 5th, while for the 10th wave it is twice the time of the 5th.

The approximate formula giving the relationship between sag and time is given as:

$$D = 12.075 \left(\frac{T}{N}\right)^2 \text{ (inches)}$$

where

D = sag, in.
T = time, sec
N = number of return waves counted

TABLE 4.19 Typical Sag and Tension Data 795 kcmil-26/7 ACSR "Drake," 300- and 1000-ft Spans

Conductor: Drake
795 kcmil-26/7 ACSR
Area = 0.7264 in.²
Creep *is* a factor

Span = 300 ft

NESC Heavy Loading District

Temp, °F	Ice, in.	Wind, lb/ft²	K, lb/ft	Weight, lb/ft	Final Sag, ft	Final Tension, lb	Initial Sag, ft	Initial Tension, lb
30	0.00	9.00	0.05	1.424	2.37	6769	2.09	7664
30	0.00	0.00	0.00	1.094	1.93	6364	1.66	7404
60	0.00	0.00	0.00	1.094	2.61	4725[a]	2.04	6033
90	0.00	0.00	0.00	1.094	3.46	3556	2.57	4792
120	0.00	0.00	0.00	1.094	1.00	3077	3.25	3785
167	0.00	0.00	0.00	1.094	4.60	2678	4.49	2746
212	0.00	0.00	0.00	1.094	5.20	2371	5.20	2371

[a] Design condition.

Conductor: Drake
795 kcmil-26/7 ACSR
Area = 0.7264 in.²
Creep *is* a factor

Span = 1000 ft

NESC Heavy Loading District

Temp, °F	Ice, in.	Wind, lb/ft²	K, lb/ft	Weight, lb/ft	Final Sag, ft	Final Tension, lb	Initial Sag, ft	Initial Tension, lb
30	0.00	9.00	0.05	1.424	28.42	6290	27.25	6558
30	0.00	0.00	0.00	1.094	27.26	5036	25.70	5339
60	0.00	0.00	0.00	1.094	29.07	4725[a]	27.36	5018
90	0.00	0.00	0.00	1.094	30.82	4460	28.98	4740
120	0.00	0.00	0.00	1.094	32.50	4232	30.56	4498
167	0.00	0.00	0.00	1.094	34.49	3990	32.56	4175
212	0.00	0.00	0.00	1.094	35.75	3851	35.14	3917

[a] Design condition.

Note: **Calculations based on:** (1) NESC Light Loading District. (2) Tension Limits: a. Initial Loaded – 60% RBS @ 30°F; b. Initial Unloaded – 25% RBS @ 60°F; c. Final Unloaded – 15% RBS @ 60°F.

where

D_i = sag in the ith span
S_i = span length of the ith span
H_{RS} = tension from ruling span sag-tension calculations

The sag in level suspension spans may also be calculated using the ratio:

$$D_i = D_{RS} \left(\frac{S_i}{S_{RS}} \right)^2$$

where D_{RS} = sag in ruling span

Suspension spans vary in length, though typically not over a large range. Conductor temperature during sagging varies over a range considerably smaller than that used for line design purposes.

If the sag in any suspension span exceeds approximately 5% of the span length, a correction factor should be added to the sags obtained from the above equation or the sag should be calculated using catenary Eq. (4.132). This correction factor may be calculated as follows:

Transmission System

TABLE 4.20 Typical Sag and Tension Data 795 kcmil-26/7 ACSR "Drake," 300- and 1000-ft Spans

Conductor: Drake
795 kcmil-26/7 ACSR/SD
Area = 0.7264 in.²
Creep *is* a factor

Span = 300 ft

NESC Heavy Loading District

Temp, °F	Ice, in.	Wind, lb/ft²	K, lb/ft	Weight, lb/ft	Final Sag, ft	Final Tension, lb	Initial Sag, ft	Initial Tension, lb
0	0.50	4.00	0.30	2.509	2.91	9695	2.88	9802
32	0.50	0.00	0.00	2.094	3.13	7528	2.88	8188
−20	0.00	0.00	0.00	1.094	1.26	9733	1.26	9756
0	0.00	0.00	0.00	1.094	1.48	8327	1.40	8818
30	0.00	0.00	0.00	1.094	1.93	6364	1.66	7404
60	0.00	0.00	0.00	1.094	2.61	4725[a]	2.04	6033
90	0.00	0.00	0.00	1.094	3.46	3556	2.57	4792
120	0.00	0.00	0.00	1.094	4.00	3077	3.25	3785
167	0.00	0.00	0.00	1.094	4.60	2678	4.49	2746
212	0.00	0.00	0.00	1.094	5.20	2371	5.20	2371

[a] Design condition.

Conductor: Drake
795 kcmil-26/7 ACSR
Area = 0.7264 in.²
Creep is *not* a factor

Span = 1000 ft

NESC Heavy Loading District

Temp, °F	Ice, in.	Wind, lb/ft²	K, lb/ft	Weight, lb/ft	Final Sag, ft	Final Tension, lb	Initial Sag, ft	Initial Tension, lb
0	0.50	4.00	0.30	2.509	30.07	10479	30.07	10479
32	0.50	0.00	0.00	2.094	30.56	8607	29.94	8785
−20	0.00	0.00	0.00	1.094	24.09	5694	22.77	6023
0	0.00	0.00	0.00	1.094	25.38	5406	23.90	5738
30	0.00	0.00	0.00	1.094	27.26	5036	25.59	5362
60	0.00	0.00	0.00	1.094	29.07	4725[a]	27.25	5038
90	0.00	0.00	0.00	1.094	30.82	4460	28.87	4758
120	0.00	0.00	0.00	1.094	32.50	4232	30.45	4513
167	0.00	0.00	0.00	1.094	34.36	4005	32.85	4187
212	0.00	0.00	0.00	1.094	35.62	3865	35.05	3928

[a] Design condition.

Note: **Calculations based on:** (1) NESC Heavy Loading District. (2) Tension Limits: a. Initial Loaded – 60% RBS @ 0°F; b. Initial Unloaded – 25% RBS @ 60°F; c. Final Unloaded – 15% RBS @ 60°F.

$$Correction = D^2 \frac{w}{6H} \quad (4.131)$$

where

D = sag obtained from parabolic equation
w = weight of conductor, lb/ft
H = horizontal tension, lb

The catenary equation for calculating the sag in a suspension or stringing span is:

$$Sag = \frac{H}{w}\left(\cosh \frac{Sw}{2H} - 1\right) \quad (4.132)$$

TABLE 4.21 Typical Sag and Tension Data 795 kcmil-Type 16 ACSR/SD, 300- and 1000-ft Spans

Conductor: Drake
795 kcmil-Type 16 ACSR/SD
Area = 0.7261 in.²
Creep *is* a factor

Span = 300 ft

NESC Heavy Loading District

Temp, °F	Ice, in.	Wind, lb/ft²	K, lb/ft	Weight, lb/ft	Final Sag, ft	Final Tension, lb	Initial Sag, ft	Initial Tension, lb
30	0.00	9.00	0.05	1.409	1.59	9980	1.31	12373
30	0.00	0.00	0.00	1.093	1.26	9776	1.03	11976
60	0.00	0.00	0.00	1.093	1.60	7688	1.16	10589[a]
90	0.00	0.00	0.00	1.093	2.12	5806	1.34	9159
120	0.00	0.00	0.00	1.093	2.69	4572	1.59	7713
167	0.00	0.00	0.00	1.093	3.11	3957	2.22	5545
212	0.00	0.00	0.00	1.093	3.58	3435	3.17	3877

[a] Design condition.

Conductor: Drake
795 kcmil-Type 16 ACSR/SD
Area = 0.7261 in.²
Creep *is* a factor

Span = 1000 ft

NESC Heavy Loading District

Temp, °F	Ice, in.	Wind, lb/ft²	K, lb/ft	Weight, lb/ft	Final Sag, ft	Final Tension, lb	Initial Sag, ft	Initial Tension, lb
30	0.00	9.00	0.05	1.409	17.21	10250	15.10	11676
30	0.00	0.00	0.00	1.093	15.22	8988	12.69	10779
60	0.00	0.00	0.00	1.093	17.21	7950[a]	13.98	9780
90	0.00	0.00	0.00	1.093	19.26	7108	15.44	8861
120	0.00	0.00	0.00	1.093	21.31	6428	17.03	8037
167	0.00	0.00	0.00	1.093	24.27	5647	19.69	6954
212	0.00	0.00	0.00	1.093	25.62	5352	22.32	6136

[a] Design condition.

Note: **Calculations based on:** (1) NESC Light Loading District. (2) Tension Limits: a. Initial Loaded – 60% RBS @ 30°F; b. Initial Unloaded – 25% RBS @ 60°F; c. Final Unloaded – 15% RBS @ 60°F.

where

S = span length, ft
H = horizontal tension, lb
w = resultant weight, lb/ft

Stringing Sag Tables

Conductors are typically installed in line section lengths consisting of multiple spans. The conductor is pulled from the conductor reel at a point near one strain structure progressing through travelers attached to each suspension structure to a point near the next strain structure. After stringing, the conductor tension is increased until the sag in one or more suspension spans reaches the appropriate stringing sags based on the ruling span for the line section. The calculation of stringing sags is based on the preceding sag equation.

Table 4.23 shows a typical stringing sag table for a 600-ft ruling span of Drake ACSR with suspension spans ranging from 400 to 700 ft and conductor temperatures of 20–100°F. All values in this stringing table are calculated from ruling span initial tensions, shown in Table 4.22 using the parabolic sag equation.

Line Design Sag-Tension Parameters

In laying out a transmission line, the first step is to survey the route and draw up a plan-profile of the selected right-of-way. The plan-profile drawings serve an important function in linking together the

TABLE 4.22 Typical Sag and Tension Data 795 kcmil-Type 16 ACSR/SD, 300- and 1000-ft Span

Conductor: Drake
795 kcmil-Type 16 ACSR/SD   **Span = 300 ft**
Area = 0.7261 in.^{2}
Creep *is* a factor   *NESC Heavy Loading District*

Temp, °F	Ice, in.	Wind, lb/ft^2	K, lb/ft	Weight, lb/ft	Final Sag, ft	Final Tension, lb	Initial Sag, ft	Initial Tension, lb
0	0.50	4.00	0.30	2.486	2.19	12774	2.03	13757
32	0.50	0.00	0.00	2.074	2.25	10377	1.90	12256
−20	0.00	0.00	0.00	1.093	.91	13477	.87	14156
0	0.00	0.00	0.00	1.093	1.03	11962	.92	13305
30	0.00	0.00	0.00	1.093	1.26	9776	1.03	11976
60	0.00	0.00	0.00	1.093	1.60	7688	1.16	10589[a]
90	0.00	0.00	0.00	1.093	2.12	5806	1.34	9159
120	0.00	0.00	0.00	1.093	2.69	4572	1.59	7713
167	0.00	0.00	0.00	1.093	3.11	3957	2.22	5545
212	0.00	0.00	0.00	1.093	3.58	3435	3.17	3877

[a] Design Condition

Conductor: Drake
795 kcmil-Type 16 ACSR/SD   **Span = 1000 ft**
Area = 0.7261 in.^{2}
Creep *is* a factor   *NESC Heavy Loading District*

Temp, °F	Ice, in.	Wind, lb/ft^2	K, lb/ft	Weight, lb/ft	Final Sag, ft	Final Tension, lb	Initial Sag, ft	Initial Tension, lb
0	0.50	4.00	0.30	2.486	20.65	15089	20.36	15299
32	0.50	0.00	0.00	2.074	20.61	12607	19.32	13445
−20	0.00	0.00	0.00	1.093	12.20	11205	10.89	12552
0	0.00	0.00	0.00	1.093	13.35	10244	11.56	11832
30	0.00	0.00	0.00	1.093	15.22	8988	12.69	10779
60	0.00	0.00	0.00	1.093	17.21	7950[a]	13.98	9780
90	0.00	0.00	0.00	1.093	19.26	7108	15.44	8861
120	0.00	0.00	0.00	1.093	21.31	6428	17.03	8037
167	0.00	0.00	0.00	1.093	24.27	5647	19.69	6954
212	0.00	0.00	0.00	1.093	25.62	5352	22.32	6136

[a] Design condition.

Note: **Calculations based on:** (1) NESC Heavy Loading District. (2) Tension Limits: a. Initial Loaded − 60% RBS @ 0°F; b. Initial Unloaded − 25% RBS @ 60°F; Final Unloaded − 15% RBS @ 60°F.

various stages involved in the design and construction of the line. These drawings, prepared based on the route survey, show the location and elevation of all natural and man-made obstacles to be traversed by, or adjacent to, the proposed line. These plan-profiles are drawn to scale and provide the basis for tower spotting and line design work.

Once the plan-profile is completed, one or more estimated ruling spans for the line may be selected. Based on these estimated ruling spans and the maximum design tensions, sag-tension data may be calculated providing initial and final sag values. From this data, sag templates may be constructed to the same scale as the plan-profile for each ruling span, and used to graphically spot structures.

Catenary Constants

The sag in a ruling span is equal to the weight per unit length, w, times the span length, S, squared, divided by 8 times the horizontal component of the conductor tension, H. The ratio of conductor horizontal tension, H, to weight per unit length, w, is the catenary constant, H/w. For a ruling span sag-tension calculation using eight loading conditions, a total of 16 catenary constant values could be defined, one for initial and final tension under each loading condition.

TABLE 4.23 Typical Sag and Tension Data 795 kcmil-37 Strand AAC "Arbutus," 300- and 1000-ft Spans

Conductor: Arbutus
795 kcmil-37 Strands AAC **Span = 300 ft**
Area = 0.6245 in.^{2}
Creep *is* a factor NESC Light Loading District

Temp, °F	Ice, in.	Wind, lb/ft^2	K, lb/ft	Weight, lb/ft	Final Sag, ft	Final Tension, lb	Initial Sag, ft	Initial Tension, lb
30	0.00	9.00	0.05	1.122	3.56	3546	2.82	4479
30	0.00	0.00	0.00	0.746	2.91	2889	2.06	4075
60	0.00	0.00	0.00	0.746	4.03	2085[a]	2.80	2999
90	0.00	0.00	0.00	0.746	5.13	1638	3.79	2215
120	0.00	0.00	0.00	0.746	6.13	1372	4.86	1732
167	0.00	0.00	0.00	0.746	7.51	1122	6.38	1319
212	0.00	0.00	0.00	0.746	8.65	975	7.65	1101

[a] Design condition.

Conductor: Arbutus
795 kcmil-37 Strands AAC **Span = 1000 ft**
Area = 0.6245 in.^{2}
Creep *is* a factor NESC Light Loading District

Temp, °F	Ice, in.	Wind, lb/ft^2	K, lb/ft	Weight, lb/ft	Final Sag, ft	Final Tension, lb	Initial Sag, ft	Initial Tension, lb
30	0.00	9.00	0.05	1.122	44.50	3185	42.85	3305
30	0.00	0.00	0.00	0.746	43.66	2158	41.71	2258
60	0.00	0.00	0.00	0.746	45.24	2085[a]	43.32	2175
90	0.00	0.00	0.00	0.746	46.76	2018	44.89	2101
120	0.00	0.00	0.00	0.746	48.24	1958	46.42	2033
167	0.00	0.00	0.00	0.746	50.49	1873	48.72	1939
212	0.00	0.00	0.00	0.746	52.55	1801	50.84	1860

[a] Design condition.

Note: **Calculations based on:** (1) NESC Light Loading District. (2) Tension Limits: a. Initial Loaded – 60% RBS @ 30°F; b. Initial Unloaded – 25% RBS @ 60°F; c. Final Unloaded – 15% RBS @ 60°F.

Catenary constants can be defined for each loading condition of interest and are used in any attempt to locate structures. Some typical uses of catenary constants for locating structures are to avoid overloading, assure ground clearance is sufficient at all points along the right-of-way, and minimize blowout or uplift under cold weather conditions. To do this, catenary constants are typically found for: (1) the maximum line temperature; (2) heavy ice and wind loading; (3) wind blowout; and (4) minimum conductor temperature. Under any of these loading conditions, the catenary constant allows sag calculation at any point within the span.

Wind Span

The maximum wind span of any structure is equal to the distance measured from center to center of the two adjacent spans supported by a structure. The wind span is used to determine the maximum horizontal force a structure must be designed to withstand under high wind conditions. Wind span is not dependent on conductor sag or tension, only on horizontal span length.

Weight Span

The weight span of a structure is a measure of the maximum vertical force a structure must be designed to withstand. The weight span is equal to the horizontal distance between the low points and the vertex of two adjacent spans. The maximum weight span for a structure is dependent on the loading condition being a minimum for heavy ice and wind load. When the elevations of adjacent structures are the same, the wind and weight spans are equal.

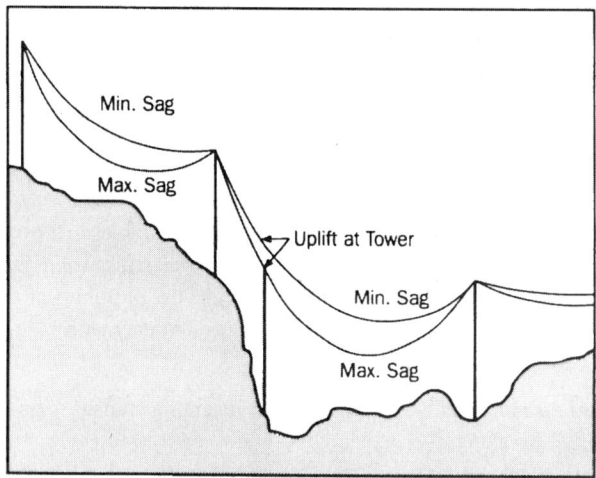

FIGURE 4.71 Conductor uplift.

Uplift at Suspension Structures

Uplift occurs when the weight span of a structure is negative. On steeply inclined spans, the low point of sag may fall beyond the lower support. This indicates that the conductor in the uphill span is exerting a negative or upward force on the lower tower. The amount of this upward force is equal to the weight of the conductor from the lower tower to the low point in the sag. If the upward pull of the uphill span is greater than the downward load of the next adjacent span, actual uplift will be caused and the conductor will swing free of the tower. This usually occurs under minimum temperature conditions and must be dealt with by adding weights to the insulator suspension string or using a strain structure (Fig. 4.71).

Tower Spotting

Given sufficiently detailed plan-profile drawings, structure heights, wind/weight spans, catenary constants, and minimum ground clearances, structure locations can be chosen such that ground clearance is maintained and structure loads are acceptable. This process can be done by hand using a sag template, plan-profile drawing, and structure heights, or numerically by one of several commercial programs.

Conductor Installation

Installation of a bare overhead conductor can present complex problems. Careful planning and a thorough understanding of stringing procedures are needed to prevent damage to the conductor during the stringing operations. The selection of stringing sheaves, tensioning method, and measurement techniques are critical factors in obtaining the desired conductors sagging results. Conductor stringing and sagging equipment and techniques are discussed in detail in the *IEEE Guide to the Installation of Overhead Transmission Line Conductors*, IEEE Std. 524-1992. Some basic factors concerning installation are covered in this section. Because the terminology used for equipment and installation procedures for overhead conductors varies throughout the utility industry, a limited glossary of terms and equipment definitions excerpted from IEEE Std. 524-1992 is provided in the chapter appendix. A complete glossary is presented in the *IEEE Guide to the Installation of Overhead Transmission Line Conductors*.

Conductor Stringing Methods

There are two basic methods of stringing conductors, categorized as either slack or tension stringing. There are as many variations of these methods as there are organizations installing conductors. The selected method, however, depends primarily on the terrain and conductor surface damage requirements.

Slack or Layout Stringing Method

Slack stringing of conductor is normally limited to lower voltage lines and smaller conductors. The conductor reel(s) is placed on reel stands or "jack stands" at the beginning of the stringing location. The conductor is unreeled from the shipping reel and dragged along the ground by means of a vehicle or pulling device. When the conductor is dragged past a supporting structure, pulling is stopped and the conductor placed in stringing sheaves attached to the structure. The conductor is then reattached to the pulling equipment and the pull continued to the next structure.

This stringing method is typically used during construction of new lines in areas where the right-of-way is readily accessible to vehicles used to pull the conductor. However, slack stringing may be used for repair or maintenance of transmission lines where rugged terrain limits use of pulling and tensioning equipment. It is seldom used in urban areas or where there is any danger of contact with high-voltage conductors.

Tension Stringing

A tension stringing method is normally employed when installing transmission conductors. Using this method, the conductor is unreeled under tension and is not allowed to contact the ground. In a typical tension stringing operation, travelers are attached to each structure. A pilot line is pulled through the travelers and is used, in turn, to pull in heavier pulling line. This pulling line is then used to pull the conductor from the reels and through the travelers. Tension is controlled on the conductor by the tension puller at the pulling end and the bullwheel tension retarder at the conductor payout end of the installation. Tension stringing is preferred for all transmission installations. This installation method keeps the conductor off the ground, minimizing the possibility of surface damage and limiting problems at roadway crossings. It also limits damage to the right-of-way by minimizing heavy vehicular traffic.

Tension Stringing Equipment and Setup

Stringing equipment typically includes bullwheel or drum pullers for back-tensioning the conductor during stringing and sagging; travelers (stringing blocks) attached to every phase conductor and shield wire attachment point on every structure; a bullwheel or crawler tractor for pulling the conductor through travelers; and various other special items of equipment. Figure 4.72 illustrates a typical stringing and sagging setup for a stringing section and the range of stringing equipment required. Provision for conductor splicing during stringing must be made at tension site or midspan sites to avoid pulling splices through the travelers.

During the stringing operation, it is necessary to use proper tools to grip the strands of the conductor evenly to avoid damaging the outer layer of wires. Two basic types or categories of grips are normally used in transmission construction. The first is a type of grip referred to as a pocketbook, suitcase, bolted, etc., that hinges to completely surround the conductor and incorporates a bail for attaching to the pulling line. The second type is similar to a Chinese finger grip and is often referred to as a basket or "Kellem" grip. Such a grip, shown in Fig. 4.73, is often used because of its flexibility and small size, making it easily pulled through sheaves during the stringing operation. Whatever type of gripping device is used, a swivel should be installed between the pulling grip and pulling line or running board to allow free rotation of both the conductor and the pulling line.

A traveler consists of a sheave or pulley wheel enclosed in a frame to allow it to be suspended from structures or insulator strings. The frame must have some type of latching mechanism to allow insertion and removal of the conductor during the stringing operation. Travelers are designed for a maximum safe working load. Always ensure that this safe working load will not be exceeded during the stringing operation. Sheaves are often lined with neoprene or urethane materials to prevent scratching of conductors in high-voltage applications; however, unlined sheaves are also available for special applications.

Travelers used in tension stringing must be free rolling and capable of withstanding high running or static loads without damage. Proper maintenance is essential. Very high longitudinal tension loads can develop on transmission structures if a traveler should "freeze" during tension stringing, possibly causing conductor and/or structure damage. Significant levels of rotation resistance will also yield tension differences between spans, resulting in incorrect sag.

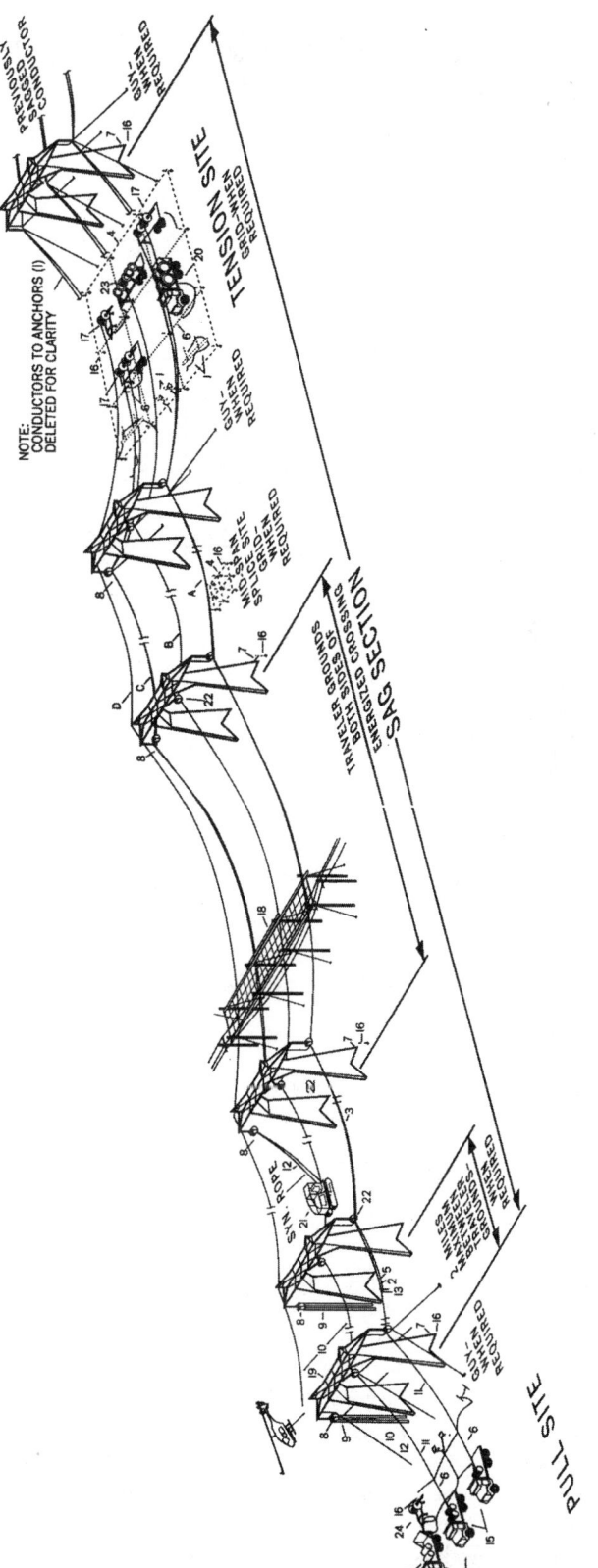

FIGURE 4.72 Tension stringing equipment setup.

FIGURE 4.73 Basket grip pulling device.

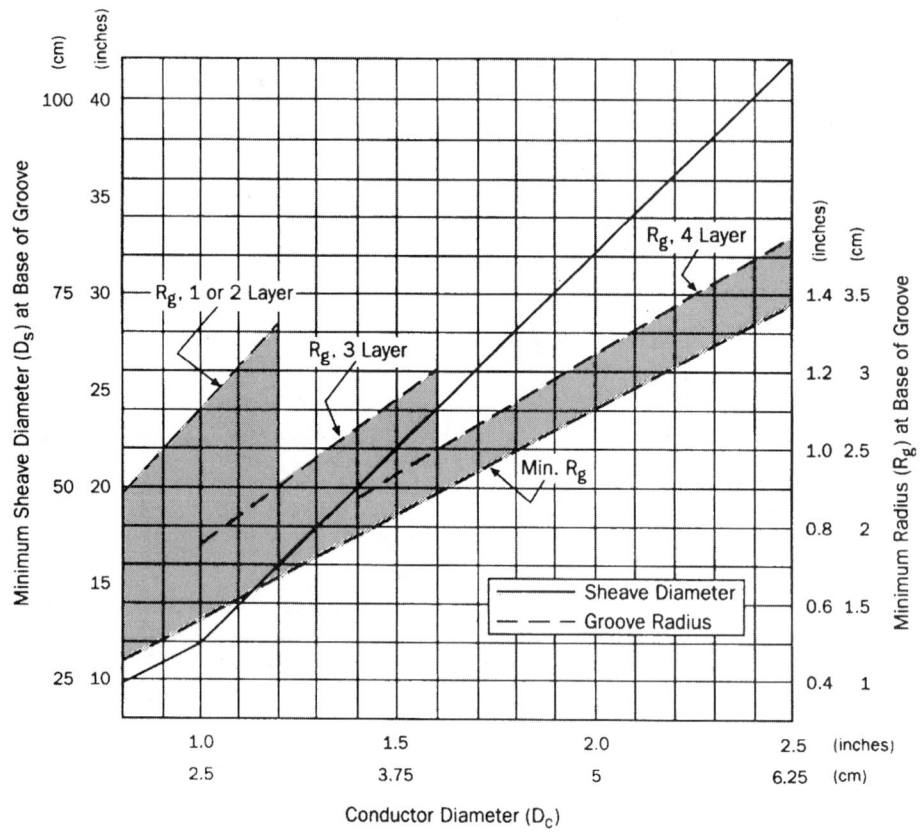

FIGURE 4.74 Recommended minimum sheave dimensions.

Proper selection of travelers is important to assure that travelers operate correctly during tension stringing and sagging. The sheave diameter and the groove radius must be matched to the conductor.

Figure 4.74 illustrates the minimum sheave diameter for typical stringing and sagging operations. Larger diameter sheaves may be required where particularly severe installation conditions exist.

Sagging Procedure

It is important that the conductors be properly sagged at the correct stringing tension for the design ruling span. A series of several spans, a line section, is usually sagged in one operation. To obtain the correct sags and to insure the suspension insulators hang vertically, the horizontal tension in all spans must be equal. Figures 4.76 through 4.81 depict typical parabolic methods and computations required for sagging conductors. Factors that must be considered when sagging conductors are creep elongation during stringing and prestressing of the conductor.

Creep elongation during stringing: Upon completion of conductor stringing, a time of up to several days may elapse before the conductor is tensioned to design sag. Since the conductor tension during the stringing process is normally well below the initial sagging tension, and because the conductor remains in the stringing sheaves for only a few days or less, any elongation due to creep is neglected. The conductor should be sagged to the initial stringing sags listed in the sag tables. However, if the conductor tension is excessively high during stringing, or the conductor is allowed to remain in the blocks for an extended period of time, then the creep elongation may become significant and the sagging tables should be corrected prior to sagging.

Creep is assumed exponential with time. Thus, conductor elongation during the first day under tension is equal to elongation over the next week. Using creep estimation formulas, the creep strain can be estimated and adjustments made to the stringing sag tables in terms of an equivalent temperature. Also, should this become a concern, Southwire's Wire and Cable Technology Group will be happy to work with you to solve the problem.

Prestressing conductor. Prestressing is sometimes used to stabilize the elongation of a conductor for some defined period of time. The prestressing tension is normally much higher than the unloaded design tension for a conductor. The degree of stabilization is dependent upon the time maintained at the prestress tension. After prestressing, the tension on the conductor is reduced to stringing or design tension limits. At this reduced tension, the creep or plastic elongation of the conductor has been slowed, reducing the permanent elongation due to strain and creep for a defined period of time. By tensioning a conductor to levels approaching 50% of its breaking strength for times on the order of a day, creep elongation will be temporarily halted (Cahill, 1973). This simplifies concerns about creep during subsequent installation but presents both equipment and safety problems.

Sagging by Stopwatch Method
A mechanical pulse imparted to a tensioned conductor moves at a speed proportional to the square root of tension divided by weight per unit length. By initiating a pulse on a tensioned conductor and measuring the time required for the pulse to move to the nearest termination, the tension, and thus the sag of the conductor, can be determined. This stopwatch method (Overend and Smith) has come into wide use even for long spans and large conductors.

The conductor is struck a sharp blow near one support and the stopwatch is started simultaneously. A mechanical wave moves from the point where the conductor was struck to the next support point at which it will be partially reflected. If the initiating blow is sharp, the wave will travel up and down the span many times before dying out. Time-sag tables such as the one shown in Table 4.24 are available from many sources. Specially designed sagging stopwatches are also available.

The reflected wave can be detected by lightly touching the conductor but the procedure is more likely to be accurate if the wave is both initiated and detected with a light rope over the conductor. Normally, the time for the return of the 3rd or 5th wave is monitored.

Traditionally, a transit sagging method has been considered to be more accurate for sagging than the stopwatch method. However, many transmission-line constructors use the stopwatch method exclusively, even with large conductors.

Sagging by Transit Methods
IEEE Guide Std. 524-1993 lists three methods of sagging conductor with a transit: "Calculated Angle of Sight," "Calculated Target Method," and "Horizontal Line of Sight." The method best suited to a particular line sagging situation may vary with terrain and line design.

Sagging Accuracy
Sagging a conductor during construction of a new line or in the reconductoring of a old line involves many variables that can lead to a small degree of error. IEEE Std. 524-1993 suggests that all sags be within 6 in. of the stringing sag values. However, aside from measurement errors during sagging, errors in terrain measurement and variations in conductor properties, loading conditions, and hardware installation have led some utilities to allow up to 3 ft of margin in addition to the required minimum ground clearance.

Clipping Offsets
If the conductor is to be sagged in a series of suspension spans where the span lengths are reasonably close and where the terrain is reasonably level, then the conductor is sagged using conventional stringing sag tables and the conductor is simply clipped into suspension clamps that replace the travelers. If the conductor is to be sagged in a series of suspension spans where span lengths vary widely or more

TABLE 4.24 Typical Sag and Tension Data 795 kcmil-37 Strand AAC "Arbutus," 300- and 1000-ft Spans

Conductor: Arbutus
795 kcmil-37 Strands AAC **Span = 300 ft**
Area = 0.6245 in.²
Creep *is* a factor

Temp, °F	Ice, in.	Wind, lb/ft²	K, lb/ft	Weight, lb/ft	Final Sag, ft	Final Tension, lb	Initial Sag, ft	Initial Tension, lb
0	0.50	4.00	0.30	2.125	3.97	6033	3.75	6383
32	0.50	0.00	0.00	1.696	4.35	4386	3.78	5053
−20	0.00	0.00	0.00	0.746	1.58	5319	1.39	6055
0	0.00	0.00	0.00	0.746	2.00	4208	1.59	5268
30	0.00	0.00	0.00	0.746	2.91	2889	2.06	4075
60	0.00	0.00	0.00	0.746	4.03	2085[a]	2.80	2999
90	0.00	0.00	0.00	0.746	5.13	1638	3.79	2215
120	0.00	0.00	0.00	0.746	6.13	1372	4.86	1732
167	0.00	0.00	0.00	0.746	7.51	1122	6.38	1319
212	0.00	0.00	0.00	0.746	8.65	975	7.65	1101

[a] Design condition.

Conductor: Arbutus
795 kcmil-37 Strands AAC **Span = 1000 ft**
Area = .6245 in.²
Creep *is* a factor   *NESC Heavy Loading District*

Temp, °F	Ice, in.	Wind, lb/ft²	K, lb/ft	Weight, lb/ft	Final Sag, ft	Final Tension, lb	Initial Sag, ft	Initial Tension, lb
0	0.50	4.00	0.30	2.125	45.11	59.53	44.50	6033
32	0.50	0.00	0.00	1.696	45.80	4679	44.68	4794
−20	0.00	0.00	0.00	0.746	40.93	2300	38.89	2418
0	0.00	0.00	0.00	0.746	42.04	2240	40.03	2350
30	0.00	0.00	0.00	0.746	43.66	2158	41.71	2258
60	0.00	0.00	0.00	0.746	45.24	2085[a]	43.32	2175
90	0.00	0.00	0.00	0.746	46.76	2018	44.89	2101
120	0.00	0.00	0.00	0.746	48.24	1958	46.42	2033
167	0.00	0.00	0.00	0.746	50.49	1873	48.72	1939
212	0.00	0.00	0.00	0.746	52.55	1801	50.84	1860

[a] Design condition.

Note: **Calculations based on:** (1) NESC Light Loading District. (2) Tension Limits: a. Initial Loaded – 60% RBS @ 0°F; b. Initial Unloaded – 25% RBS @ 60°F; c. Final Unloaded – 15% RBS @ 60°F.

commonly, where the terrain is steep, then clipping offsets may need to be employed in order to yield vertical suspension strings after installation.

Clipping offsets are illustrated in Fig. 4.75, showing a series of steeply inclined spans terminated in a "snub" structure at the bottom and a "deadend" structure at the top. The vector diagram illustrates a balance of total conductor tension in the travelers but an imbalance in the horizontal component of tension.

Defining Terms

Block: A device designed with one or more single sheaves, a wood or metal shell, and an attachment hook or shackle. When rope is reeved through two of these devices, the assembly is commonly referred to as a *block and tackle*. A *set of 4s* refers to a block and tackle arrangement utilizing two 4-inch double-sheave blocks to obtain four load-bearing lines. Similarly, a *set of 5s* or a *set of 6s* refers to the same number of load bearing lines obtained using two 5-inch or two 6-inch double-sheave blocks, respectively.

Synonyms: set of 4s, set of 5s, set of 6s.

Transmission System

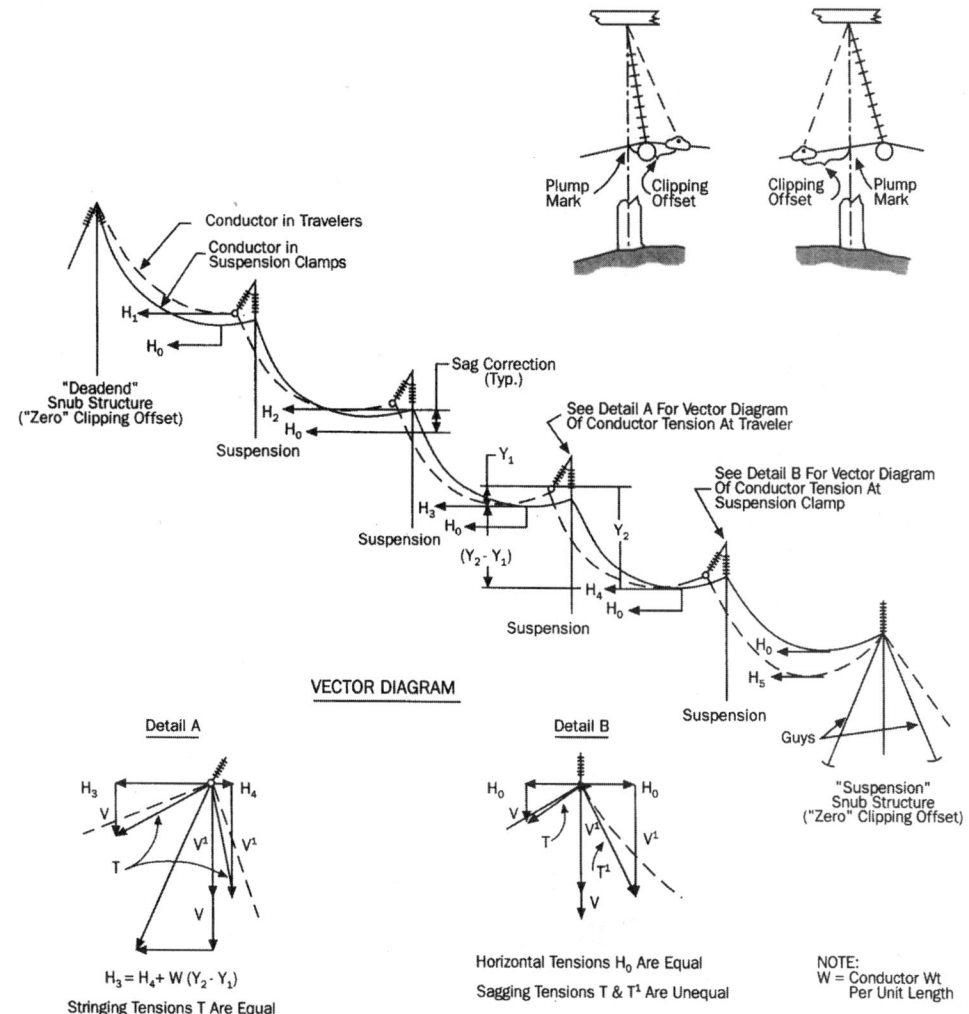

FIGURE 4.75 Clipping offset illustration.

Bullwheel: A wheel incorporated as an integral part of a bullwheel puller or tensioner to generate pulling or braking tension on conductors or pulling lines, or both, through friction. A puller or tensioner normally has one or more pairs arranged in tandem incorporated in its design. The physical size of the wheels will vary for different designs, but 17-in. (43 cm) face widths and diameters of 5 ft (150 cm) are common. The wheels are power driven or retarded and lined with single- or multiple-groove neoprene or urethane linings. Friction is accomplished by reeving the pulling line or conductor around the groove of each pair.

Clipping-in: The transferring of sagged conductors from the traveler to their permanent suspension positions and the installing of the permanent suspension clamps.

Synonyms: clamping, clipping.

Clipping offset: A calculated distance, measured along the conductor from the plum mark to a point on the conductor at which the center of the suspension clamp is to be placed. When stringing in rough terrain, clipping offset may be required to balance the horizontal forces on each suspension structure.

Grip, conductor: A device designed to permit the pulling of conductor without splicing on fittings, eyes, etc. It permits the pulling of a *continuous* conductor where threading is not possible. The

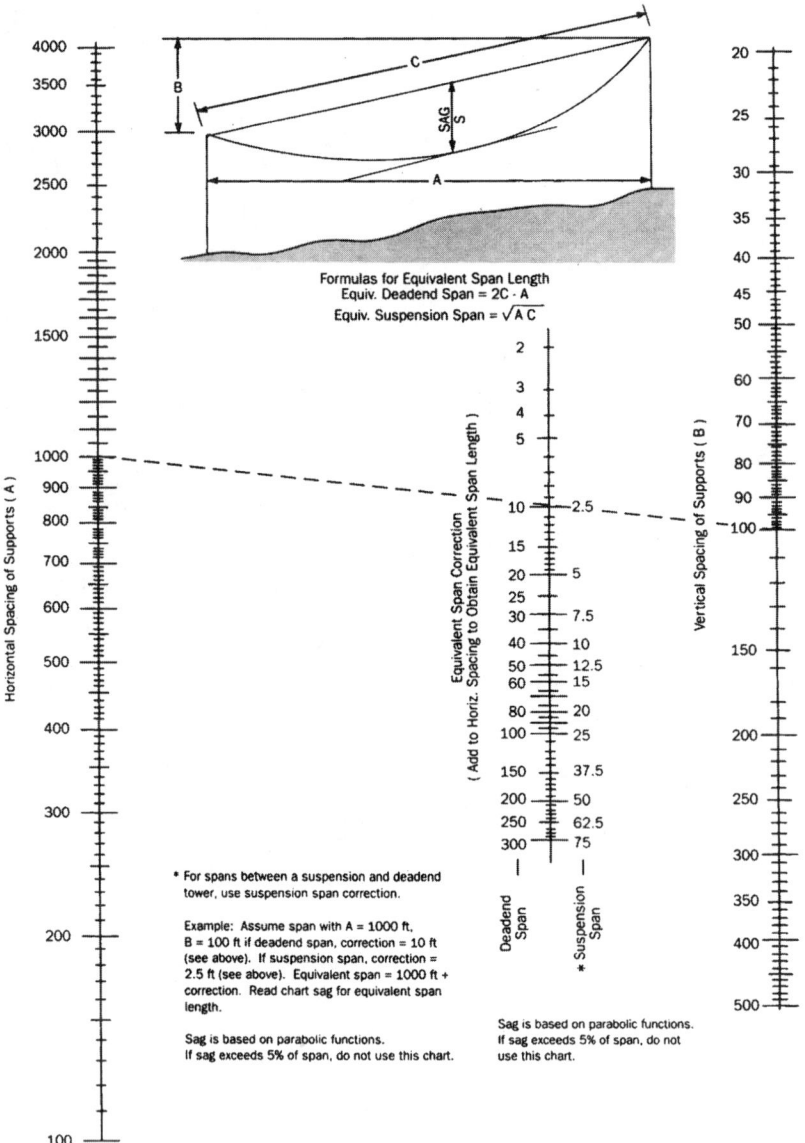

FIGURE 4.76 Nomograph for determining level span equivalents of non-level spans.

designs of these grips vary considerably. Grips such as the Klein (Chicago) and Crescent utilize an open-sided rigid body with opposing jaws and swing latch. In addition to pulling conductors, this type is commonly used to tension guys and, in some cases, pull wire rope. The design of the come-along (pocketbook, suitcase, four bolt, etc.) incorporates a bail attached to the body of a clamp which folds to completely surround and envelope the conductor. Bolts are then used to close the clamp and obtain a grip.

Synonyms: buffalo, Chicago grip, come-along, Crescent, four bolt, grip, Klein, pocketbook, seven bolt, six bolt, slip-grip, suitcase.

Line, pilot: A lightweight line, normally synthetic fiber rope, used to pull heavier pulling lines which in turn are used to pull the conductor. Pilot lines may be installed with the aid of finger lines or by helicopter when the insulators and travelers are hung.

Synonyms: lead line, leader, P-line, straw line.

Transmission System

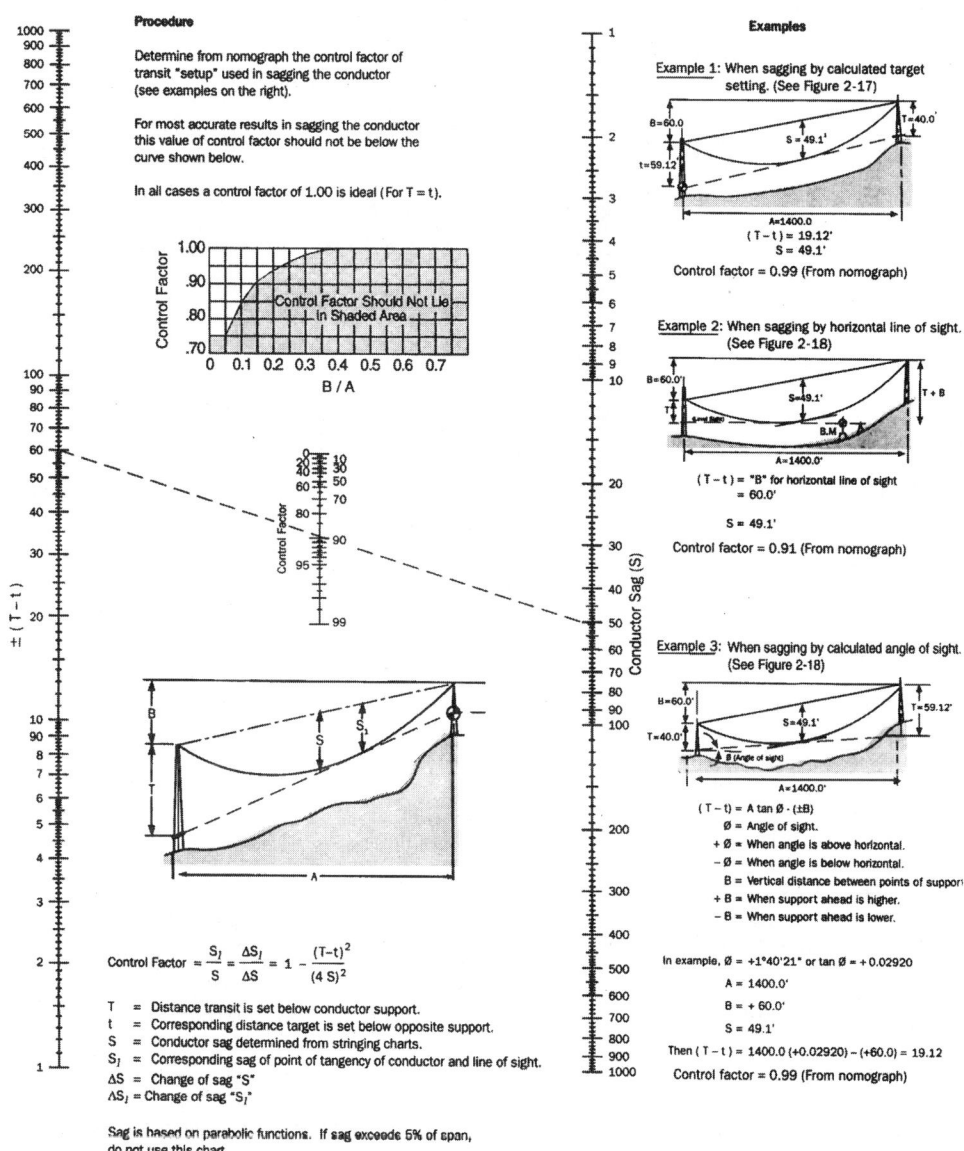

FIGURE 4.77 Nomograph for determining control factor for conductor sagging.

Line, pulling: A high-strength line, normally synthetic fiber rope or wire rope, used to pull the conductor. However, on reconstruction jobs where a conductor is being replaced, the old conductor often serves as the pulling line for the new conductor. In such cases, the old conductor must be closely examined for any damage prior to the pulling operations.
  Synonyms: bull line, hard line, light line, sock line.

Puller, bullwheel: A device designed to pull pulling lines and conductors during stringing operations. It normally incorporates one or more pairs of urethane- or neoprene-lined, power-driven, single- or multiple-groove bullwheels where each pair is arranged in tandem. Pulling is accomplished by friction generated against the pulling line which is reeved around the grooves of a pair of the bullwheels. The puller is usually equipped with its own engine which drives the bullwheels mechanically, hydraulically, or through a combination of both. Some of these devices function as either a puller or tensioner.
  Synonym: puller.

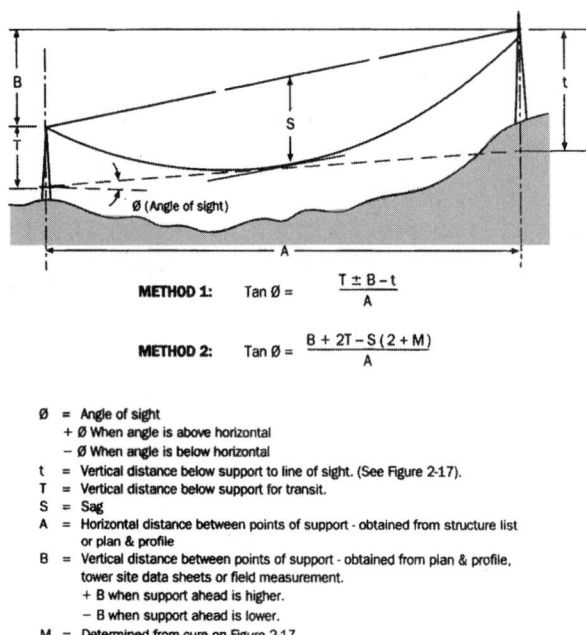

FIGURE 4.78 Conductor sagging by calculated angle of sight.

Puller, drum: A device designed to pull a conductor during stringing operations. It is normally equipped with its own engine which drives the drum mechanically, hydraulically, or through a combination of both. It may be equipped with synthetic fiber rope or wire rope to be used as the pulling line. The pulling line is payed out from the unit, pulled through the travelers in the sag section and attached to the conductor. The conductor is then pulled in by winding the pulling line back onto the drum. This unit is sometimes used with synthetic fiber rope acting as a pilot line to pull heavier pulling lines across canyons, rivers, etc.

Synonyms: hoist, single drum hoist, single drum winch, tugger.

Puller, reel: A device designed to pull a conductor during stringing operations. It is normally equipped with its own engine which drives the supporting shaft for the reel mechanically, hydraulically, or

Transmission System

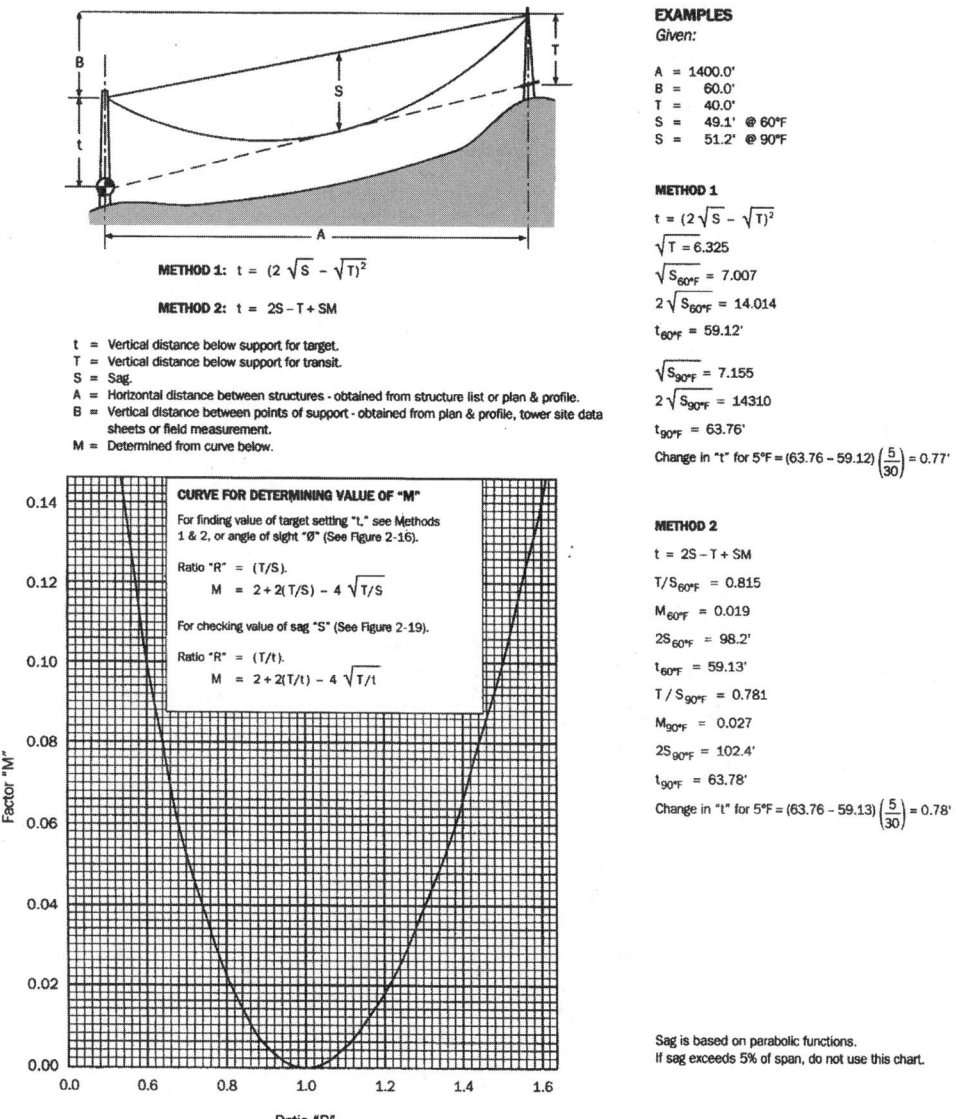

FIGURE 4.79 Conductor sagging by calculated target method.

through a combination of both. The shaft, in turn, drives the reel. The application of this unit is essentially the same as that for the drum puller previously described. Some of these devices function as either a puller or tensioner.

Reel stand: A device designed to support one or more reels and having the possibility of being skid, trailer, or truck mounted. These devices may accommodate rope or conductor reels of varying sizes and are usually equipped with reel brakes to prevent the reels from turning when pulling is stopped. They are used for either slack or tension stringing. The designation of reel trailer or reel truck implies that the trailer or truck has been equipped with a reel stand (jacks) and may serve as a reel transport or *payout* unit, or both, for stringing operations. Depending upon the sizes of the reels to be carried, the transporting vehicles may range from single-axle trailers to semi-trucks with trailers having multiple axles.

Synonyms: reel trailer, reel transporter, reel truck.

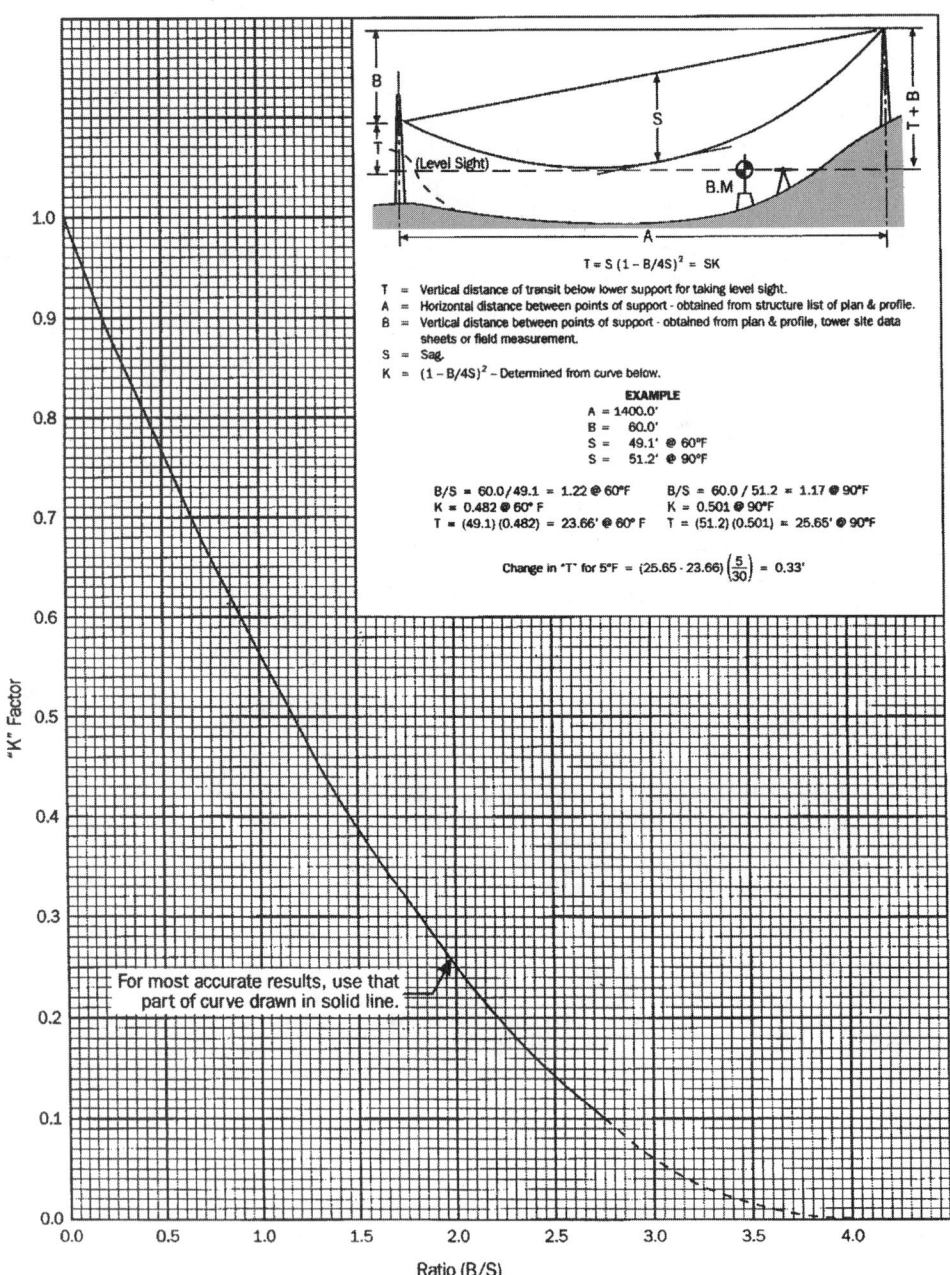

FIGURE 4.80 Conductor sagging by horizontal line of sight.

Running board: A pulling device designed to permit stringing more than one conductor simultaneously with a single pulling line. For distribution stringing, it is usually made of lightweight tubing with the forward end curved gently upward to provide smooth transition over pole cross-arm rollers. For transmission stringing, the device is either made of sections hinged transversely to the direction of pull or of a hard-nose rigid design, both having a flexible pendulum tail suspended from the

Transmission System

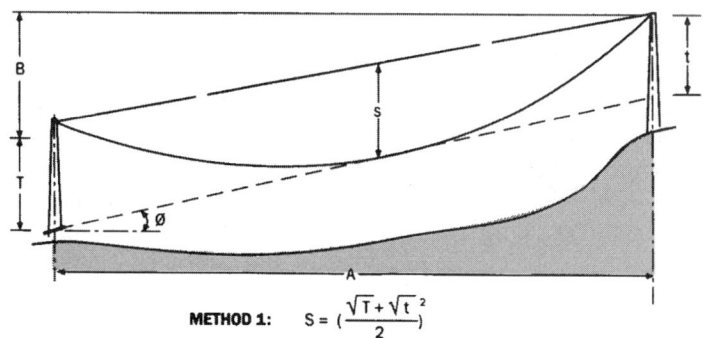

METHOD 1: $S = \left(\dfrac{\sqrt{T}+\sqrt{t}}{2}\right)^2$

METHOD 2: $S = \dfrac{B}{2} + \dfrac{t}{2} - \dfrac{tM}{8}$

- S = Sag
- t = Vertical distance below support to line of sight.
  - = T ± B − A tan Ø when angle Ø is above horizontal.
  - = T ± B + A tan Ø when angle Ø is below horizontal.
- T = Vertical distance below support for transit.
- B = Vertical distance between points of support - obtained from plan & profile, tower site data sheets or field measurement.
  - + B when support ahead is higher.
  - − B when support ahead is lower.
- A = Horizontal distance between points of support - obtained from structure list or plan & profile
- Ø = Angle of sight
- M = Determined from curve on Figure 2-17.

EXAMPLES
Given:

A = 1400.0'   T = 40.0'
B = 60.0'   Ø = +1° 40' 21" @ 60°F
              (Field Measured)

METHOD 1

$S = \left(\dfrac{\sqrt{T}+\sqrt{t}}{2}\right)^2$

t = 40.0 + 60.0 − 1400.0 tan 1° 40' 21"

  = 59.12'

$\sqrt{t}$ = 7.689

$\sqrt{T}$ = 6.325

$S_{60°F}$ = 49.1'

METHOD 2
Note: When using Method 2, value of "T" should lie between 3/4 "S" & 4/3 "S"

$S = \dfrac{B}{2} + \dfrac{t}{2} - \dfrac{tM}{8}$

t = 59.12'

t/2 = 29.56'

T/2 = 20.0"

M = 0.061

$S_{60°F} = 20.0 + 29.56 - \dfrac{(59.12)(0.061)}{8}$

$S_{60°F}$ = 49.1'

Sag is based on parabolic functions. If sag exceeds 5% of span, do not use this chart.

FIGURE 4.81 Conductor sagging for checking sag S.

rear. This configuration stops the conductors from twisting together and permits smooth transition over the sheaves of bundle travelers.

Synonyms: alligator, bird, birdie, monkey tail, sled.

Sag section: The section of line between snub structures. More than one sag section may be required in order to properly sag the actual length of conductor which has been strung.

Synonyms: pull, setting, stringing section.

Site, pull: The location on the line where the puller, reel winder, and anchors (snubs) are located. This site may also serve as the pull or tension site for the next sag section.
Synonyms: reel setup, tugger setup.

Site, tension: The location on the line where the tensioner, reel stands and anchors (snubs) are located. This site may also serve as the pull or tension site for the next sag section.
Synonyms: conductor payout station, payout site, reel setup.

Snub structure: A structure located at one end of a sag section and considered as a *zero* point for sagging and clipping offset calculations. The section of line between two such structures is the sag section, but more than one sag section may be required in order to sag properly the actual length of conductor which has been strung.
Synonyms: 0 structure, zero structure.

Tensioner, bullwheel: A device designed to hold tension against a pulling line or conductor during the stringing phase. Normally, it consists of one or more pairs of urethane- or neoprene-lined, power braked, single- or multiple-groove bullwheels where each pair is arranged in tandem. Tension is accomplished by friction generated against the conductor which is reeved around the grooves of a pair of the bullwheels. Some tensioners are equipped with their own engines which retard the bullwheels mechanically, hydraulically, or through a combination of both. Some of these devices function as either a puller or tensioner. Other tensioners are only equipped with friction-type retardation.
Synonyms: retarder, tensioner.

Tensioner, reel: A device designed to generate tension against a pulling line or conductor during the stringing phase. Some are equipped with their own engines which retard the supporting shaft for the reel mechanically, hydraulically, or through a combination of both. The shaft, in turn, retards the reel. Some of these devices function as either a puller or tensioner. Other tensioners are only equipped with friction type retardation.
Synonyms: retarder, tensioner.

Traveler: A sheave complete with suspension arm or frame used separately or in groups and suspended from structures to permit the stringing of conductors. These devices are sometimes bundled with a center drum or sheave, and another traveler, and used to string more than one conductor simultaneously. For protection of conductors that should not be nicked or scratched, the sheaves are often lined with nonconductive or semiconductive neoprene or with nonconductive urethane. Any one of these materials acts as a padding or cushion for the conductor as it passes over the sheave. Traveler grounds must be used with lined travelers in order to establish an electrical ground.
Synonyms: block, dolly, sheave, stringing block, stringing sheave, stringing traveler.

Winder reel: A device designed to serve as a recovery unit for a pulling line. It is normally equipped with its own engine which drives a supporting shaft for a reel mechanically, hydraulically, or through a combination of both. The shaft, in turn, drives the reel. It is normally used to rewind a pulling line as it leaves the bullwheel puller during stringing operations. This unit is not intended to serve as a puller, but sometimes serves this function where only low tensions are involved.
Synonyms: take-up reel.

References

Cahill, T., Development of Low-Creep ACSR Conductor, *Wire Journal*, July 1973.

Ehrenburg, D.O., Transmission Line Catenary Calculations, AIEE Paper, Committee on Power Transmission & Distribution, July 1935.

Fink, D.G. and Beaty, H.W., *Standard Handbook for Electrical Engineers*, 13th ed., McGraw-Hill.

IEEE Guide to the Installation of Overhead Transmission Line Conductors, IEEE Standard 524-1993, IEEE, New York, 1993.

Graphic Method for Sag Tension Calculations for ACSR and Other Conductors, Aluminum Company of America, 1961.

Minimum Design Loads for Buildings and Other Structures, American Society of Civil Engineers Standard, ASCE 7-88.

National Electrical Safety Code, 1993 edition.

Overend, P.R. and Smith, S., Impulse Time Method of Sag Measurement.

Stress-Strain-Creep Curves for Aluminum Overhead Electrical Conductors, Aluminum Association, 1974.

Winkelman, P.F., Sag-Tension Computations and Field Measurements of Bonneville Power Administration, AIEE Paper 59-900, June 1959.

4.8 Corona and Noise

Giao N. Trinh

Modern electric power systems are often characterized by generating stations located far away from the consumption centers, with long overhead transmission lines to transmit the energy from the generating sites to the load centers. From the few tens of kilovolts in the early years of the 20th century, the line voltage has reached the EHV levels of 800 kV AC (Lacroix and Charbonneau, 1968) and 500 kV DC (Bateman, 1969) in the 1970s, and touched the UHV levels of 1200 kV AC (Bortnik, 1988) and 600 kV DC (Krishnayya et al., 1988). Although overhead lines operating at high voltages are the most economical means of transmitting large amounts of energy over long distances, their exposure to atmospheric conditions constantly alters the surface conditions of the conductors and causes large variations in the corona activities on the line conductors.

Corona discharges follow an electron avalanche process whereby neutral molecules are ionized by electron impacts under the effect of the applied field (Raether, 1964). Since air is a particular mixture of nitrogen (79%), oxygen (20%), and various impurities, the discharge development is significantly conditioned by the electronegative nature of oxygen molecules, which can readily capture free electrons to form negative ions and thus hamper the electron avalanche process (Loeb, 1965). Several modes of corona discharge can be distinguished; and while all corona modes produce energy losses, the streamer discharges also generate electromagnetic interference, and aubible noise in the immediate vicinity of HV lines (Trinh et al., 1968; Trinh, 1995). These parameters are currently used to evaluate the corona performance of conductor bundles and to predict the energy losses and environmental impact of HV lines prior to their installation.

Adequate control of line corona is obtained by controlling the surface gradient at the line conductors. The introduction of bundled conductors by Whitehead in 1910 has greatly influenced the development of HV lines to today's EHV voltages (Whitehead, 1910). In effect, HV lines as we know them today would not exist without the bundled conductors. This section reviews the physical processes leading to the development of corona discharges on the line conductors and presents the current practices in selecting the line conductors.

Corona Modes (Trinh et al., 1968; Trinh, 1995)

In a nonuniform field gap in atmospheric air, corona discharges can develop over a whole range of voltages in a small region near the highly stressed electrode before the gap breaks down. Several criteria have been developed for the onset of corona discharge, the most familiar being the streamer criterion. They are all related to the development of an electron avalanche in the gas gap and can be expressed as

$$1 - \gamma \exp\left(\int \alpha' dx\right) = 0, \quad \text{with } \alpha' = (\alpha - \eta) \tag{4.133}$$

where α' is the net coefficient of ionization by electron impact of the gas, α and η are respectively the ionization and attachment coefficients in air, and γ is a coefficient representing the efficiency of secondary processes in maintaining the ionization activities in the gap. The net coefficient of ionization varies with the distance x from the highly stressed electrode and the integral is evaluated for values of x where α' is positive.

A physical meaning may be given to the above corona onset criteria. Rewriting the onset conditions as

$$\exp\left[\int(\alpha-\eta)dx\right]=\frac{1}{\gamma} \qquad (4.134)$$

the left-hand side represents the avalanche development from a single electron and $1/\gamma$, the critical size of the avalanche to assure the stable development of the discharge.

The nonuniform field necessary for the development of corona discharges and the electronegative nature of air favor the formation of negative ions during the discharge development. Due to their relatively slow mobility, ions of both polarities from several consecutive electron avalanches accumulate in the low field region of the gap and form ion space charges. To properly interpret the development of corona discharges, account must be taken of the active role of these ion space charges, which continuously modify the local field intensity and, hence, the development of corona discharges according to their relative build-up and removal from the region around the highly stressed electrode.

Negative Corona Modes

When the highly stressed electrode is at a negative potential, electron avalanches are initiated at the cathode and develop toward the anode in a continuously decreasing field. Referring to Fig. 4.82, the nonuniformity of the field distribution causes the electron avalanche to stop at the boundary surface S_0 where the net ionization coefficient is zero, i.e., $\alpha = \eta$. Since free electrons can move much faster than ions under the influence of the applied field, they concentrate at the avalanche head during its progression. A concentration of positive ions thus forms in the region of the gap between the cathode and the boundary surface, while free electrons continue to migrate across the gap. In air, free electrons rapidly attach themselves to oxygen molecules to form negative ions which, because of the slow drift velocity, start to

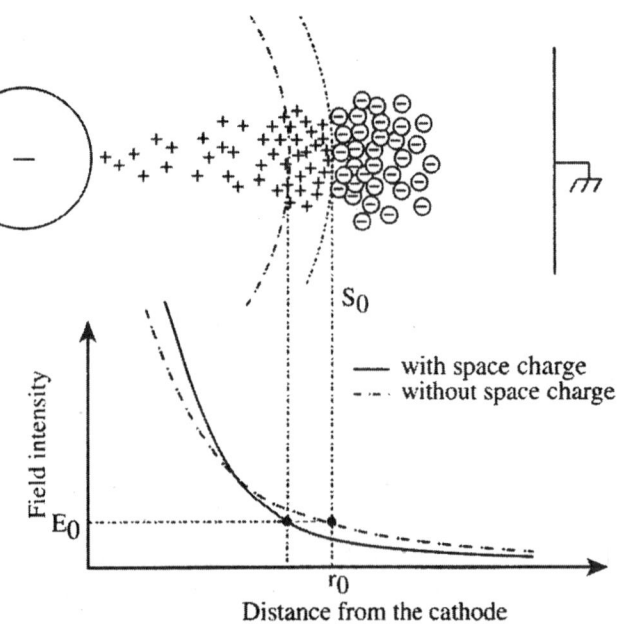

FIGURE 4.82 Development of an electron avalanche from the cathode. (Trinh, 1995.)

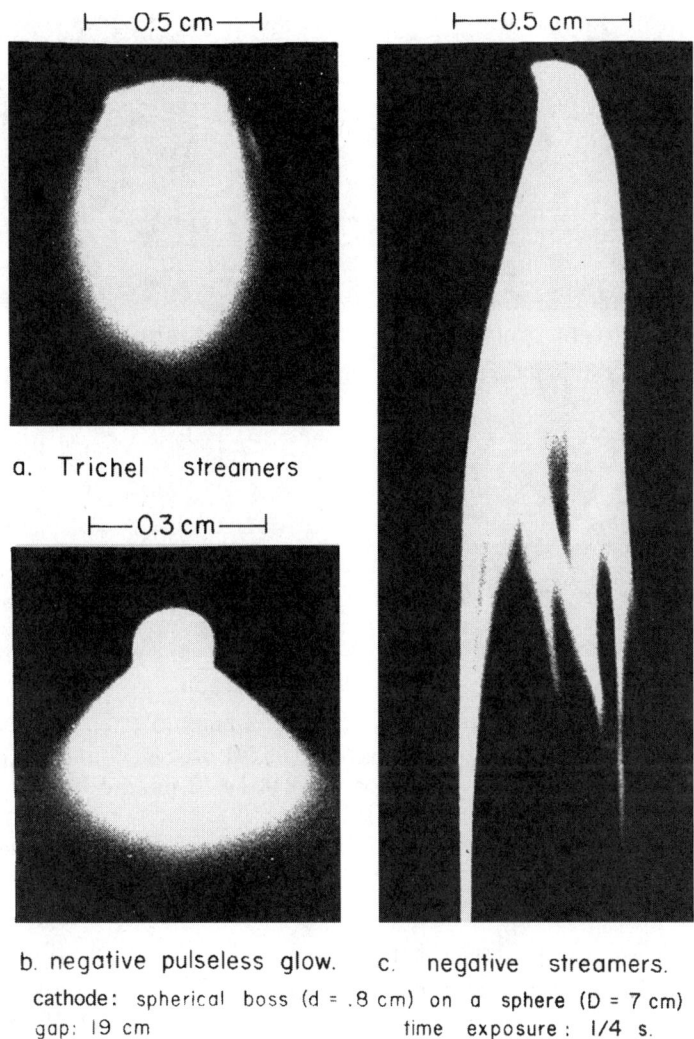

FIGURE 4.83 Corona modes at cathode: (a) trichel streamers; (b) negative pulseless glow; (c) negative streamers. (Trinh and Jordan, 1968; Trinh, 1995.) Cathode: spherical protrusion (d = 0.8 cm) on a sphere (D = 7 cm); gap 19 cm; time exposure 1/4 sec.

accumulate in the region of the gap beyond S_0. Thus, as soon as the first electron avalanche has developed, there are two ion space charges in the gap.

The presence of these space charges increases the field near the cathode but it reduces the field intensity at the anode end of the gap. The boundary surface of zero ionization activity is therefore displaced toward the cathode. The subsequent electron avalanche develops in a region of slightly higher field intensity but covers a shorter distance than its predecessor. The influence of the ion space charge is such that it actually conditions the development of the discharge at the highly stressed electrode, producing three modes of corona discharge with distinct electrical, physical, and visual characteristics (Fig. 4.83). These are, respectively, with increasing field intensity: Trichel streamer, negative pulseless glow, and negative streamer. An interpretation of the physical mechanism of different corona modes is given below.

Trichel Streamer
Figure 4.83a shows the visual aspect of the discharge; its current and light characteristics are shown in Fig. 4.84. The discharge develops along a narrow channel from the cathode and follows a regular pattern

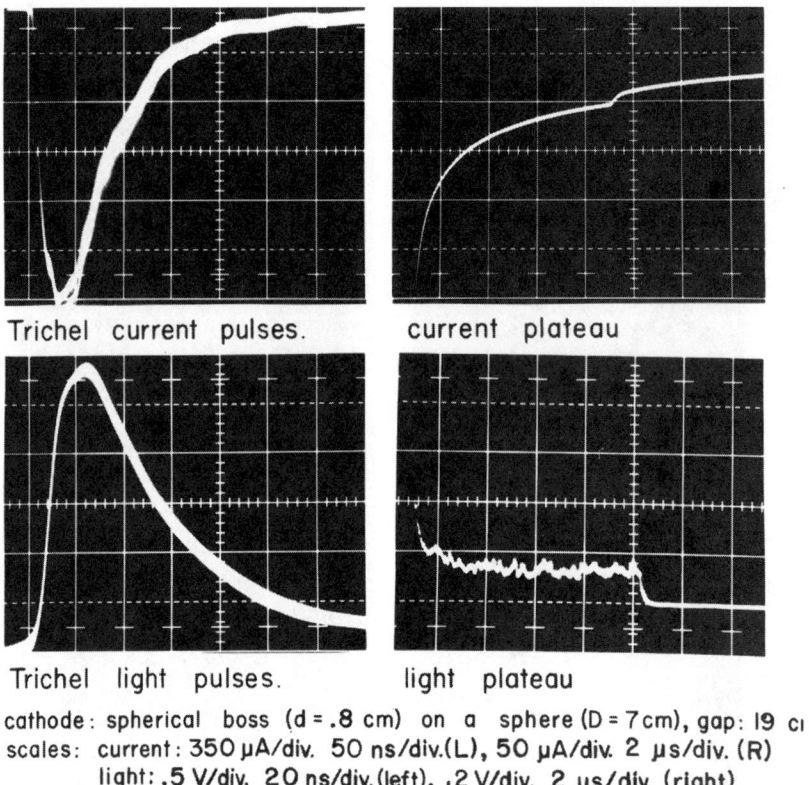

FIGURE 4.84 Current and light characteristics of Trichel streamer. (Trinh and Jordan, 1968; Trinh, 1995.) Cathode: spherical protrusion (d = 0.8 cm) on a sphere (D = 7 cm); gap 19 cm. Scales: current 350 μA/div., 50 ns/div. (left), 50 μA/div., 2 μs/div (right). Light: 0.5 V/div, 20 ns/div. (left), 0.2 V/div., 2 μs/div (right).

in which the streamer is initiated, develops, and is suppressed; a short dead time follows before the cycle is repeated. The duration of an individual streamer is very short, a few tens of nanoseconds, while the dead time varies from a few microseconds to a few milliseconds, or even longer. The resulting discharge current consists of regular negative pulses of small amplitude and short duration, succeeding one another at the rate of a few thousand pulses per second. A typical Trichel current pulse is shown in Fig. 4.84a where, it should be noted, the waveshape is somewhat influenced by the time constant of the measuring circuit. The discharge duration may be significantly shorter, as depicted by the light pulse shown in Fig. 4.84b.

The development of Trichel streamers cannot be explained without taking account of the active roles of the ion space charges and the applied field. The streamer is initiated from the cathode by a free electron. If the corona onset conditions are met, the secondary emissions are sufficient to trigger new electron avalanches from the cathode and maintain the discharge activity. During the streamer development, several generations of electron avalanches are initiated from the cathode and propagate along the streamer channel. The avalanche process also produces two ion space charges in the gap, which moves the boundary surface S_0 closer to the cathode. The positive ion cloud thus finds itself compressed at the cathode and, in addition, is partially neutralized at the cathode and by the negative ions produced in subsequent avalanches. This results in a net negative ion space charge, which eventually reduces the local field intensity at the cathode below the onset field and suppresses the discharge. The dead time is a period during which, the remaining ion space charges are dispersed by the applied field. A new streamer will develop when the space charges in the immediate surrounding of the cathode have been cleared to a sufficient extent.

This mechanism depends on a very active electron attachment process to suppress the ionization activity within a few tens of nanoseconds following the beginning of the discharge. The streamer repetition rate is essentially a function of the removal rate of ion space charges by the applied field, and generally shows a linear dependence on the applied voltage. However, at high fields a reduction in the pulse repetition rate may be observed, which corresponds to the transition to a new corona mode.

Negative Pulseless Glow

This corona mode is characterized by a pulseless discharge current. The discharge itself is particularly stable, as indicated by the well-defined visual aspect of the discharge (Fig. 4.83b) which shows the basic characteristics of a miniature glow discharge. Starting from the cathode, a cathode dark space can be distinguished, followed by a negative glow region, a Faraday dark space and, finally, a positive column of conical shape. As with low-pressure glow discharges, these features of the pulseless glow discharge result from very stable conditions of electron emission from the cathode by ionic bombardment. The electrons, emitted with very low kinetic energy, are first propelled through the cathode dark space where they acquire sufficient energy to ionize the gas, and intensive ionization occurs at the negative glow region. At the end of the negative glow region, the electrons lose most of their kinetic energy and are again accelerated across the Faraday dark space before they can ionize the gas atoms in the positive column. The conical shape of the positive column is attributed to the diffusion of the free electrons in the low-field region.

These stable discharge conditions may be explained by the greater efficiency of the applied field in removing the ion space charges at higher field intensities. Negative ion space charges cannot build up sufficiently close to the cathode to effectively reduce the cathode field and suppress the ionization activities there. This interpretation of the discharge mechanism is further supported by the existence of a plateau in the Trichel streamer current and light pulses (Fig. 4.84) which indicate that an equilibrium exists for a short time between the removal and the creation of the negative ion space charge. It has been shown (Trinh et al., 1970) that the transition from the Trichel streamer mode to the negative pulseless glow corresponds to an indefinite prolongation of one such current plateau.

Negative Streamer

If the applied voltage is increased still further, negative streamers may be observed, as illustrated in Fig. 4.83c. The discharge possesses essentially the same characteristics observed in the negative pulseless glow discharge but here the positive column of the glow discharge is constricted to form the streamer channel, which extends farther into the gap. The glow discharge characteristics observed at the cathode imply that this corona mode also depends largely on electron emissions from the cathode by ionic bombardment, while the formation of a streamer channel characterized by intensive ionization denotes an even more effective space charge removal action by the applied field. The streamer channel is fairly stable. It projects from the cathode into the gap and back again, giving rise to a pulsating fluctuation of relatively low frequency in the discharge current.

Positive Corona Modes

When the highly stressed electrode is of positive polarity, the electron avalanche is initiated at a point on the boundary surface S_0 of zero net ionization and develops toward the anode in a continuously increasing field (Fig. 4.85). As a result, the highest ionization activity is observed at the anode. Here again, due to the lower mobility of the ions, a positive ion space charge is left behind along the development path of the avalanche. However, because of the high field intensity at the anode, few electron attachments occur and the majority of free electrons created are neutralized at the anode. Negative ions are formed mainly in the low-field region farther in the gap. The following discharge behavior may be observed (Trinh and Jordan, 1968; Trinh, 1995):

- The incoming free electrons are highly energetic and cannot be immediately absorbed by the anode. As a result, they tend to spread over the anode surface where they lose their energy through ionization of the gas particles, until they are neutralized at the anode, thus contributing to the development of the discharge over the anode surface.

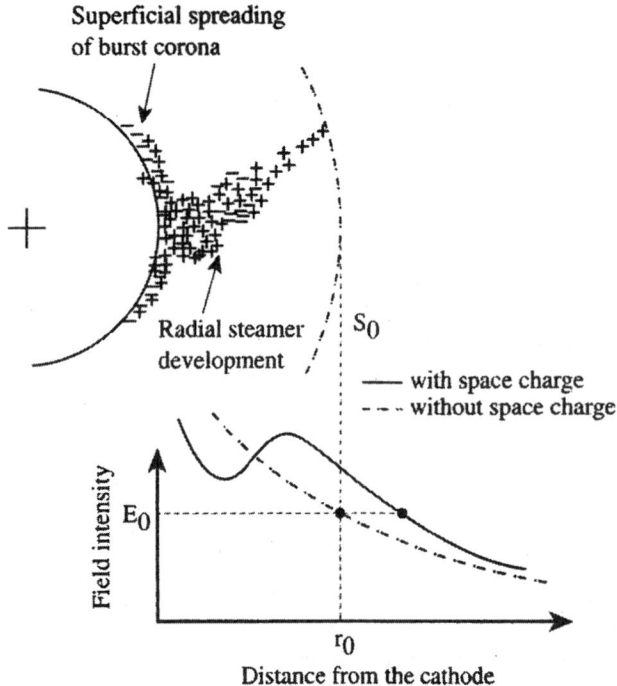

FIGURE 4.85 Development of an electron avalanche toward the anode. (Trinh, 1995.)

- Since the positive ions are concentrated immediately next to the anode surface, they may produce a field enhancement in the gap that attracts secondary electron avalanches and promotes the radial propagation of the discharge into the gap along a streamer channel. During streamer discharge, the ionization activity is observed to extend considerably into the low-field region of the gap via the formation of corona globules, which propagate owing to the action of the electric field generated by their own positive ion space charge. Dawson (1965) has shown that if a corona globule containing 10^8 positive ions within a spherical volume of $3\ 10^{-3}$ cm in radius is produced, the ion space charge field is such that it attracts sufficient new electron avalanches to create a new corona globule a short distance away. In the meantime, the initial corona globule is neutralized, causing the corona globule to effectively move ahead toward the cathode.

The presence of ion space charges of both polarities in the anode region greatly affects the local distribution of the field, and, consequently, the development of corona discharges at the anode. Four different corona discharge modes having distinct electrical, physical, and visual characteristics can be observed at a highly stressed anode, prior to flashover of the gap. These are, respectively, with increasing field intensity (Fig. 4.86): burst corona, onset streamers, positive-glow, and breakdown streamers. An interpretation of the physical mechanisms leading to the development of these corona modes is given below.

Burst Corona

This corona mode appears as a thin luminous sheath adhering closely to the anode surface (Fig. 4.86a). The discharge results from the spread of ionization activities at the anode surface, which allows the high-energy incoming electrons to lose their energy prior to neutralization at the anode. During this process, a number of positive ions are created in a small area over the anode which builds up a local positive space charge and suppresses the discharge. The spread of free electrons then moves to another part of the anode. The resulting discharge current consists of very small positive pulses (Fig. 4.87a), each corresponding to the ionization spreading over a small area at the anode and then being suppressed by the positive ion space charge produced.

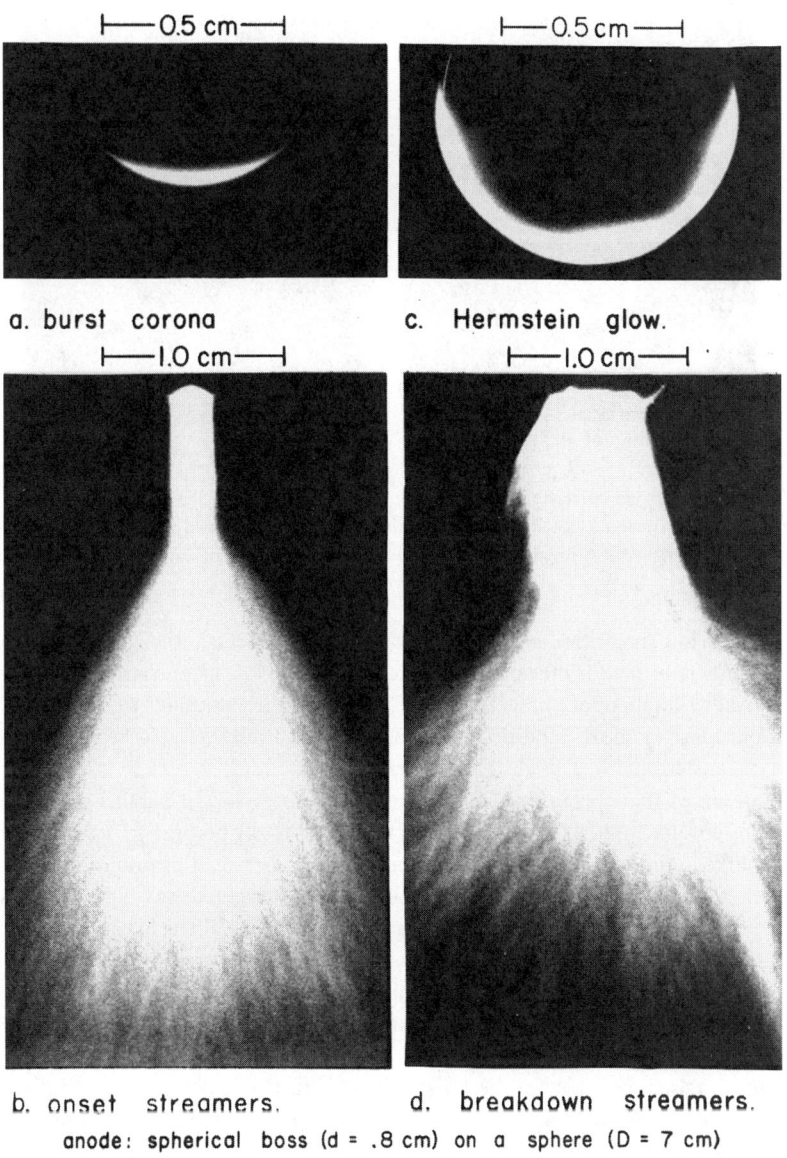

FIGURE 4.86 Corona modes at anode: (a) burst corona, (b) onset streamers; (c) Hernstein anode glow corona, and (d) breakdown streamers. (Trinh and Jordan, 1968; Trinh, 1995.) Anode spherical protrusion (d = 0.8 cm) on a sphere (D = 7cm); gap 35 cm; time exposure 1/4 sec.

Onset Streamer

The positive ion space charge formed adjacent to the anode surface causes a field enhancement in its immediate vicinity, which attracts subsequent electron avalanches and favors the radial development of onset streamers. This discharge mode is highly effective and the streamers are observed to extend farther into the low-field region of the gap along numerous filamentary channels, all originating from a common stem projecting from the anode (Fig. 4.86b). During this development of the streamers, a considerable number of positive ions are formed in the low-field region. As a result of the cumulative effect of the successive electron avalanches and the absorption at the anode of the free electrons created in the discharge, a net residual positive ion space charge forms in front of the anode. The local gradient at the

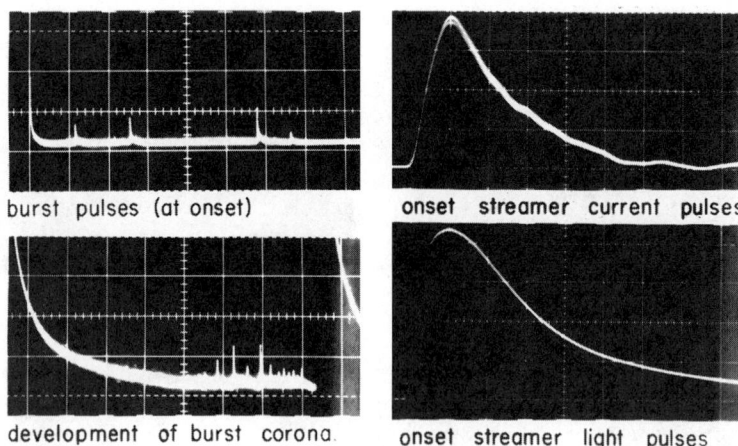

FIGURE 4.87 (a) Burst corona current pulse. Scales: 5 mA/div., 0.2 ms/div. (Trinh and Jordan, 1968; Trinh, 1995). (b) Development of burst corona following a streamer discharge. Scales: 5 mA/div., 0.2 ms/div. (c) Current characteristics of onset streamers. Scales: 7 mA/div., 50 ns/div. (d) Light characteristics of onset streamers. Scales: 1 V/div., 20 ns/div.

anode then drops below the critical value for ionization and suppresses the streamer discharge. A dead time is consequently required for the applied field to remove the ion space charge and restore the proper conditions for the development of a new streamer. The discharge develops in a pulsating mode, producing a positive current pulse of short duration, high amplitude, and relatively low repetition rate due to the large number of ions created in a single streamer (Figs. 4.87c and d).

It has been observed that these first two discharge modes develop in parallel over a small range of voltages following corona onset. As the voltage is increased, it rapidly becomes more effective in removing the ion space charge in the immediate vicinity of the electrode surface, thus promoting the lateral spread of burst corona at the anode. In fact, burst corona can be triggered just a few microseconds after suppression of the streamer (Fig. 4.87b). This behavior can be explained by the rapid clearing of the positive ion space charge at the anode region, while the incoming negative ions encounter a high enough gradient to shed their electrons and sustain the ionization activity over the anode surface in the form of burst corona. The latter will continue to develop until it is again suppressed by its own positive space charge.

As the voltage is raised even higher, the burst corona is further enhanced by a more effective space charge removal action of the field at the anode. During the development of the burst corona, positive ions are created and rapidly pushed away from the anode. The accumulation of positive ions in front of the anode results in the formation of a stable positive ion-space charge that prevents the radial development of the discharge into the gap. Consequently, the burst corona develops more readily, at the expense of the onset streamer, until the latter is completely suppressed. A new mode, the positive-glow discharge, is then established at the anode.

Positive Glow

A photograph of a positive glow discharge developing at a spherical protrusion is presented in Fig. 4.86. This discharge is due to the development of the ionization activity over the anode surface, which forms a thin luminous layer immediately adjacent to the anode surface, where intense ionization activity takes place. The discharge current consists of a direct current superimposed by a small pulsating component with a high repetition rate, in the hundreds of kilohertz range. A photomultiplier study of the light emitted reveals that the uniform ionizing sheath projects from a central region and back again, continuously, following the burst of ionization activity at the anode, which gives rise to the pulsating current component.

The development of the positive glow discharge may be interpreted as resulting from a particular combination of removal and creation of positive ions in the gap. The field is high enough for the positive ion space charge to be rapidly removed from the anode, thus promoting surface ionization activity. Meanwhile, the field intensity is not sufficient to allow radial development of the discharge and the formation of streamers. The main contribution of the negative ions is to supply the necessary triggering electrons to sustain ionization activity at the anode.

Breakdown Streamer

If the applied voltage is further increased, streamers are again observed and they eventually lead to breakdown of the gap. The development of breakdown streamers is preceded by local *streamer spots* of intense ionization activity which may be seen moving slowly over the anode surface. The development of streamer spots is not accompanied by any marked change in the current or the light signal. Only when the applied field becomes sufficiently high to rapidly clear the positive ion space charges from the anode region does radial development of the discharge become possible, resulting in breakdown streamers.

Positive breakdown streamers develop more and more intensively with higher applied voltage and eventually cause the gap to break down. The discharge is essentially the same as the onset streamer type but can extend much farther into the gap. The streamer current is more intense and may occur at a higher repetition rate. A streamer crossing the gap does not necessarily result in gap breakdown, which proves that the filamentary region of the streamer is not fully conducting.

AC Corona

When alternating voltage is used, the gradient at the highly stressed electrode varies continuously, both in intensity and in polarity. Different corona modes can be observed in the same cycle of the applied voltage. Figure 4.88 illustrates the development of different corona modes at a spherical protrusion as a function of the applied voltage. The corona modes can be readily identified by the discharge current. The following observations can be made:

- For short gaps, the ion space charges created in one half-cycle are absorbed by the electrodes in the same half-cycle. The same corona modes that develop near onset voltages can be observed, namely: negative Trichel streamers, positive onset streamers, and burst corona.
- For long gaps, the ion space charges created in one half-cycle are not completely absorbed by the electrodes. The residual space charges are drawn back to the region of high field intensity in the following half-cycle and can influence discharge development. Onset streamers are suppressed in favor of the positive glow discharge. The following corona modes can be distinguished: negative Trichel streamers, negative glow discharge, positive glow discharge, and positive breakdown streamers.
- Negative streamers are not observed under AC voltage, owing to the fact that their onset gradient is higher than the breakdown voltage that occurs during the positive half-cycle.

Main Effects of Corona Disharges on Overhead Lines (Trinh, 1995)

Impact of corona discharges on the design of high-voltage lines has been recognized since the early days of electric power transmission when the corona losses were the limiting factor. Even today, corona losses remain critical for HV lines below 300 kV. With the development of EHV lines operating at voltages between 300 and 800 kV, electromagnetic interferences become the designing parameters. For UHV lines operating at voltages above 800 kV, the audible noise appears to gain in importance over the other two parameters. The physical mechanisms of these effects — corona losses, electromagnetic interference, and audible noise — will be given and their current evaluation methods discussed below.

Corona Losses

The movement of ions of both polarities generated by corona discharges, and subjected to the applied field around the line conductors, is the main source of energy loss. For AC lines, the movement of the

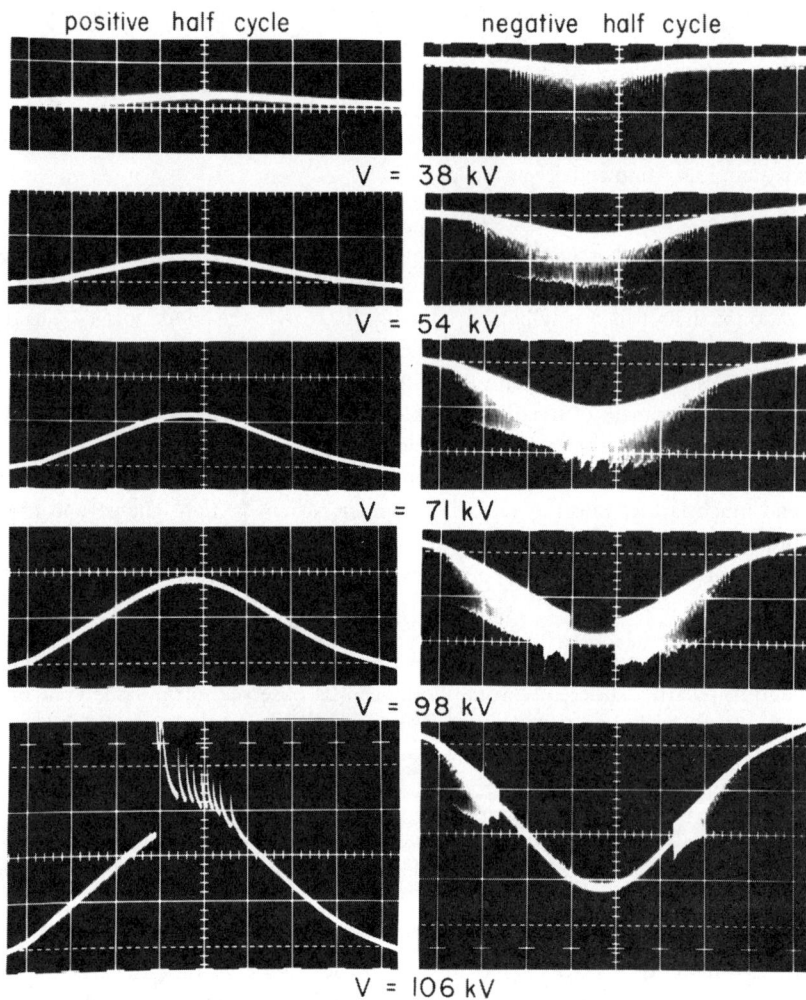

FIGURE 4.88 Corona modes under AC voltage. (Trinh and Jordan, 1968; Trinh, 1995.) Electrode: conical protrusion (θ = 30°) on a sphere (D = 7cm); gap 25 cm; R = 10 kOhm; Scales: 50 μA/div., 1.0 ms/div.

ion space charges is limited to the immediate vicinity of the line conductors, corresponding to their maximum displacement during one half-cycle, typically a few tens of centimeters, before the voltage changes polarity and reverses the ionic movement. For DC lines, the ion displacement covers the whole distance separating the line conductors, and between the conductors and the ground.

Corona losses are generally described in terms of the energy losses per kilometer of the line. They are generally negligible under fair-weather conditions but can reach values of several hundreds of kilowatts per kilometer of line during foul weather. Direct measurement of corona losses is relatively complex, but foul-weather losses can be readily evaluated in *test cages* under artificial rain conditions, which yield the highest energy loss. The results are expressed in terms of the *generated loss W*, a characteristic of the conductor to produce corona losses under given operating conditions.

Electromagnetic Interference

Electromagnetic interference is associated with streamer discharges that inject current pulses into the conductor. These steep-front, short-duration pulses have a high harmonic content, reaching the tens of

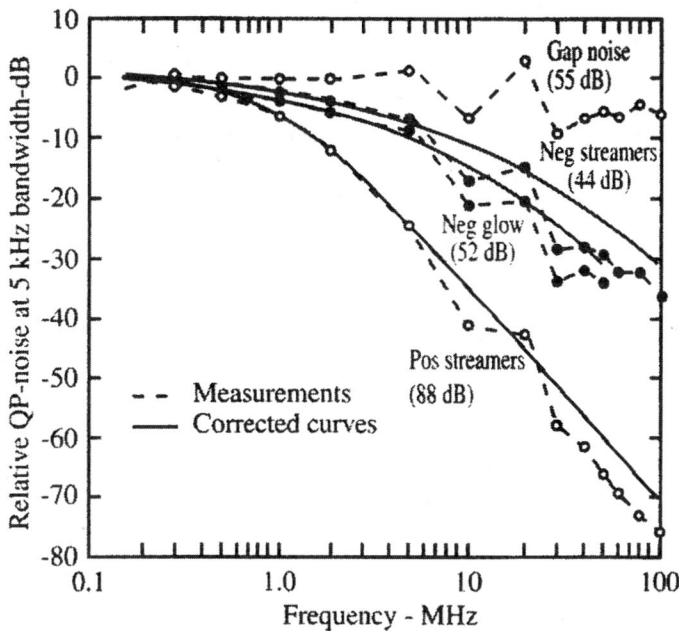

FIGURE 4.89 Relative frequency spectra for different noise types. (Trinh, 1995; Juette, 1972.)

megahertz range, as illustrated in Fig. 4.89, which shows the typical frequency spectra associated with various streamer modes (Juette, 1972). A tremendous research effort was devoted to the subject during the years 1950–80 in an effort to evaluate the electromagnetic interference from HV lines. The most comprehensive contributions were made by Moreau and Gary (1972) of Électricité de France, who introduced the concept of the *excitation function*, $\Gamma(\omega)$, which characterizes the ability of a line conductor to generate electromagnetic interference under given operating conditions.

Consider first the case of a single-phase line, where the contribution to the electromagnetic interference at the measuring frequency ω, from corona discharges developing at a section dx of the conductor is

$$i0(\omega)dx = C\,\Gamma(\omega)dx \tag{4.135}$$

where C is the capacitance per unit length of the line conductor to ground.

Upon injection, the discharge current pulse splits itself in two identical current pulses of half-amplitude propagating in opposite directions away from the discharge site. At a point of observation located at a distance x along the line from the discharge site, the noise current is distorted according to

$$i(\omega, x)dx = i0(\omega)\exp(-\gamma x)dx \tag{4.136}$$

where γ represents the propagation constant, which can be approximated by its real component α.

The total noise current circulating in the line conductor is the sum of all contributions from the corona discharges along the conductor and is given by

$$I(\omega) = \sqrt{\int_{-\infty}^{\infty}\bigl[i(\omega,x)\bigr]^2\,dx} = \frac{i0(\omega)}{\sqrt{\alpha}}. \tag{4.137}$$

Circulation of the noise current in the line conductor effectively generates an electromagnetic interference field around the conductors, which is readily picked up by any radio or TV receiver located in the vicinity of the HV line. The current practices characterize the interference field in terms of its electric component, $E(\omega)$, expressed in decibels (dB) above a reference level of 1 µV/m. Evaluation of the electromagntic interference is usually made by first calculating the magnetic interference field $H(\omega)$ at the measuring point

$$H(\omega) = \sum_j \frac{1}{2\pi r_j} I_j(\omega) ar. \qquad (4.138)$$

The summation was made with respect to the number of phase conductors of the lines and their images with respect to the magnetic ground. The electric interference field can next be related to the magnetic interference field according to

$$E(\omega) = \sqrt{\frac{\mu 0}{\varepsilon 0}} H(\omega). \qquad (4.139)$$

For a multi-phase line, because of the high-frequency nature of the noise current, the calculation of the interference field must take account of the mutual coupling among the conductors, which further complicates the process (Gary, 1972; Moreau et al., 1972). Modal analysis provides a convenient means of evaluating the noise currents on the line conductors. In this approach, the noise currents are first transposed into their modal components, which propagate without distortion along the line conductors at their own velocity according to the relation

$$[i0(\omega) dx] = [M][j0(\omega) dx]. \qquad (4.140)$$

Consequently,

$$[j0(\omega) dx] = [M]^{-1}[i0(\omega) dx] \qquad (4.141)$$

where [M] is the modal transposition matrix and $j0(\omega)$ are the modal components of the injected noise current. The modal current at the measuring point located at a distance x from the injection point is:

$$j(\omega, x) dx = j0(\omega) \exp(-\alpha x) dx, \qquad (4.142)$$

and the modal current component at the measuring point is

$$J(\omega) = \sqrt{\int_{-\infty}^{\infty} [j(\omega, x)]^2 dx} = \frac{j0(\omega)}{\sqrt{\alpha}}, \qquad (4.143)$$

or, in a general way

$$[J(\omega)] = \frac{1}{[\sqrt{\alpha}]}[j0(\omega)] = \frac{1}{[\sqrt{\alpha}]}[M]^{-1}[i0(\omega)]. \qquad (4.144)$$

Finally, the line current can be obtained from

$$[I(\omega)] = [M][J(\omega)] \tag{4.145}$$

The magnetic and electric fields produced by the noise currents in the line conductors can then be evaluated using Eqs. (4.138) and (4.139). Gary and Moreau (1972) obtained good agreement between calculated and experimental results with the symmetrical modes of Clarke for the modal transposition.

$$[M] = \begin{bmatrix} 1/\sqrt{6} & 1/2 & 1/\sqrt{3} \\ -2/\sqrt{6} & 0 & 1/\sqrt{3} \\ 1/\sqrt{6} & -1/2 & 1/\sqrt{3} \end{bmatrix}$$

The attenuation coefficients at 0.5 MHz are 11.1, 54, and 342 Np/m for the modal currents, and the magnetic ground was assumed to be located at a depth equal to the penetration depth of the magnetic field as defined by

$$P = \sqrt{\frac{2\rho}{\mu_0 \omega}}. \tag{4.146}$$

For a typical soil resistivity of 100 Ohm-m and a measuring frequency of 0.5 MHz, the depth of the magnetic ground is equal to 7.11 m.

TV Interference
The frequency spectrum of corona discharges has cut-off frequencies around a few tens of megahertz. As a result, the interference levels at the TV frequencies are very much attenuated. In fact, gap discharges, which generate sharp current pulse with nanosecond risetimes, are the principal discharges that effectively interfere with TV reception. These discharges are produced by loose connections, a problem common on low-voltage distribution lines but rarely observed on high-voltage transmission lines. Another source of interference is related to reflections of TV signals at high-voltage line towers, producing ghost images. However, the problem is not related in any way to corona activities on the line conductors (Juette, 1972).

Audible Noise
The high temperature in the discharge channel produced by the streamer creates a corresponding increase in the local air pressure. Consequently, a pulsating sound wave is generated from the discharge site, propagates through the surrounding ambient air, and is perfectly audible in the immediate vicinity of the HV lines. The typical octave-band frequency spectra of line corona in Fig. 4.90 contains discrete components corresponding to the second and higher harmonics of the line voltage superimposed on a relatively broad-band noise, extending well into the ultrasonic range (Ianna et al., 1974). The octave band measurements in this figure show a sharp drop at frequencies over 20 kHz, due principally to the limited frequency response of the microphone and associated sound-level meter.

Similar to the case of electromagnetic interference, the ability of the line conductors to produce audible noise is characterized by the *generated acoustic power density A*, defined as the acoustic power produced per unit length of the line conductor under specific operating conditions. The acoustic power generated by corona discharges developing in a portion dx of the conductor is then

$$dA = A\,dx. \tag{4.147}$$

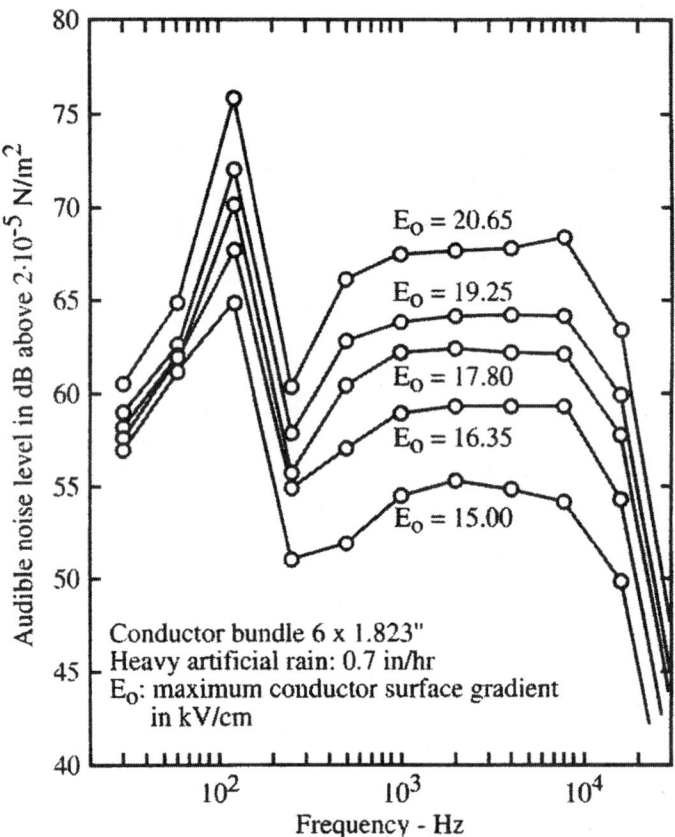

FIGURE 4.90 Octave-band frequency-spectrum of line corona audible noise at 10 m from the conductor. (Trinh, 1995; Trinh et al., 1977.)

Its contribution to the acoustic intensity at a measuring point located at a distance r from the discharge site is

$$dI = \frac{A}{4\pi r^2} dx. \qquad (4.148)$$

The acoustic intensity at the measuring point is the sum of all contributions from corona discharge distributed along the conductor:

$$I(R) = 2A \int_{-\infty}^{\infty} \frac{1}{4\pi(R^2 + x^2)} dx = \frac{A}{2R} \qquad (4.149)$$

where R is the distance from the measuring point to the conductor, and the integral is evaluated in terms of the longitudinal distance x along the conductor. Finally, the acoustic intensity at the measuring point is the sum of the contributions from the different phase conductors of the line

$$I(R) = \sum_j I_j(R). \qquad (4.150)$$

Transmission System

TABLE 4.25 Hydro-Québec 735-kV Line

	Center phase	Outer phase
Distance between phase (m)	13.7	
Height of conductors (m)	19.8	
Number of sub-conductors	4	
Diameter of sub-conductor (cm)	3.05	
Electric field at the conductor surface (kVrms/cm)	19.79	18.46
Capacitance per unit length (pF/m)	10.57	
Generated loss W (W/m)	59.77	33.92
RI excitation function Γ (dB above 1 $\mu A/\sqrt{m}$)	43.52	39.59
Subconductor generated acoustic power density A (dBA above 1 $\mu W/m$)	3.28	−0.24

The sound pressure, usually expressed in terms of decibel (dBA) above a reference level of 2×10^{-5} N/m², is

$$p(r) = \sqrt{\rho_0 C I} \qquad (4.151)$$

Example of Calculation

It is obvious from the preceding sections that the effects of corona discharges on HV lines — the corona losses, the electromagnetic interferences, and audible noise — can be readily evaluated from the generated loss W, the excitation function $\Gamma(\omega)$, and the generated acoustic power density A of the conductor. The latter parameters are characteristics of the bundle conductor and are usually derived from tests in a test cage or on experimental line. An example calculation of the corona performance of a HV line is given below for the case of the Hydro-Québec's 735-kV lines under conditions of heavy rain. The line parameters are given in Table 4.25, together with the various corona-generated parameters taken from Trinh et al. (1977). The calculation of the radio interference and audible noise levels will be made for a lateral distance of 15 m from the outer phase, i.e., at the limit of the right of way of the line.

Corona losses: The corona losses are the sum of the losses generated at the three phases of the line, which amount to 127.63 kW/km.

Radio interference: The calculation of the radio interference requires that the noise current be first transformed into its modal components. Consider a noise current of unit excitation function $\Gamma a (\omega) = 1.0 \ \mu A/\sqrt{m}$ circulating in phase A of the line. Because of the capacitive coupling, it induces currents to the other two phases of the line as well. For Hydro-Québec's 735-kV line, the capacitance matrix is

$$C = \begin{bmatrix} 11.204 & -2.241 & -0.73 \\ -2.241 & 11.605 & -2.241 \\ -0.73 & -2.241 & 11.204 \end{bmatrix} \quad pF/m,$$

and the noise current in phase A and its induced currents to phases B and C are

$$ia(\omega) = \begin{bmatrix} 11.204 \\ -2.241 \\ -0.73 \end{bmatrix} \quad \mu A.$$

The modal transformation using Eqs. (4.141–4.144) gives the following modal noise currents at the measuring point, taking into account of the different attenuations of the modal currents.

$$Ja(\omega) = \begin{bmatrix} 16.472 & 10.321 & 2.31 \\ -30.497 & 0 & 1.998 \\ 16.472 & -10.321 & 2.31 \end{bmatrix} \mu A$$

These modal currents, once transformed back to the current mode, Eq. (4.145), give the modal components of the noise currents flowing in the line conductors at the measuring point as related to the noise current injected to phase A.

$$Ia(\omega) = \begin{bmatrix} 6.725 & 7.298 & 1.333 \\ -13.449 & 0 & 1.333 \\ 6.725 & -7.298 & 1.333 \end{bmatrix} \mu A$$

These currents can then be used to calculate the magnetic and electric interference field using Eqs. (4.138) and (4.139)

$$Ha(\omega) = \begin{bmatrix} 0.0124 & 0.0449 & 0.0239 \end{bmatrix} \mu A/m$$

$$Ea(\omega) = \begin{bmatrix} 4.674 & 16.938 & 9.017 \end{bmatrix} \mu V/m$$

The corresponding electric interference level is 25.911 dB above 1µV/m.

The above electric interference field and interference level are obtained assuming a noise excitation function of 1.0 $\mu A/\sqrt{m}$. For the case of interest, the excitation function at phase A is 39.59 dB and the corresponding interference level is 64.98 dB. By repeating the same process for the noise currents injected in phases B and C, one obtains effectively three sets of magnetic and electric field components generated by the circulation of the noise currents on the line conductors.

$$Eb(\omega) = \begin{bmatrix} -8.653 & 0 & 7.80 \end{bmatrix} \mu V/m, \text{ and}$$

$$Ec(\omega) = \begin{bmatrix} 4.674 & -16.938 & 9.017 \end{bmatrix} \mu V/m$$

Their contributions to the noise level are, respectively, 64.26 dB and 64.98 dB, resulting in a total noise level of 69.53 dB at the measuring point. The measuring frequency is 0.5 MHz.

Audible noise: Calculation of the audible noise is straightforward, since each phase of the line can be considered as an independent noise source. Consider the audible noise generated from phase A. The subconductor generated acoustic power density is –0.24 dBA or 1.58 10^{-5} µW/m for the bundle conductor. The acoustic intensity at 15 m from the outer phase of the line as given by Eq. (4.149) is 3.19 10^{-7} W/m² and the noise level is 55.14 dBA above 2 10^{-5} N/m².

By repeating the process for the other two phases of the line, the contributions to the acoustic intensity at the measuring point from the phase B and C of the line are 2.64 10^{-7} and 1.69 10^{-7} W/m², respectively, and the corresponding noise levels are 54.33 dBA and 52.38 dBA. The total noise level is 58.87 dBA.

Impact on the Selection of Line Conductors

Corona Performance of HV Lines

Corona performance is a general term used to characterize the three main effects of corona discharges developing on the line conductors and their related hardware, namely corona losses (CL), electromagnetic interference (RI), and audible noise (AN). All are sensitive to weather conditions, which dictate the corona

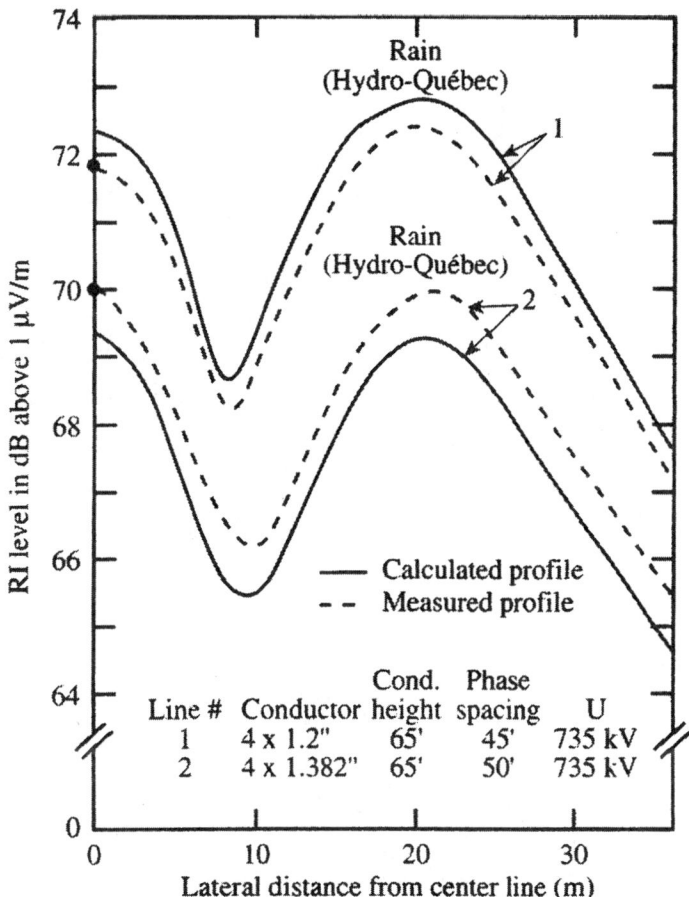

FIGURE 4.91 Comparison of calculated and measured RI performances of Hydro-Québec 735-kV lines at 1 MHz and using natural modes. (Trinh, 1995; Trinh et al., 1977.)

activities. Corona losses can be described by a lump figure, which is equal to the total energy losses per kilometer of the line. Both the electromagnetic interference and the audible noise levels vary with the distance from the line and are best described by lateral profiles, which show the variations in the RI and AN level with the lateral distance from the line. Typical lateral profiles are presented in Figs. 4.91 and 4.92 for a number of HV lines under foul-weather conditions. For convenience, the interference and noise levels at the edge of the right-of-way, typically 15 m from the outside phases of the line, are generally used to quantify the interference and noise level.

The time variations in the corona performance of HV lines is best described in terms of a statistical distribution, which shows the proportion of time that the energy losses, the electromagnetic interference, and audible noise exceed their specified levels. Figure 4.93 illustrates typical corona performances of Hydro-Québec's 735-kV lines as measured at the edge of the right-of-way. It can be seen that the RI and AN levels vary over wide ranges. In addition, the cumulative distribution curves show a typical inverted-S shape, indicating that the recorded data actually result from the combination of more than one population, usually associated with fair and foul weather conditions.

DC coronas are less noisy than AC coronas. In effect, although DC lines can become very lossy during foul weather, the radio interference and audible noises are significantly reduced. This behavior is related to the fact that water drops become elongated, remain stable, and produce glow corona modes rather than streamers in a DC field (Ianna et al., 1974).

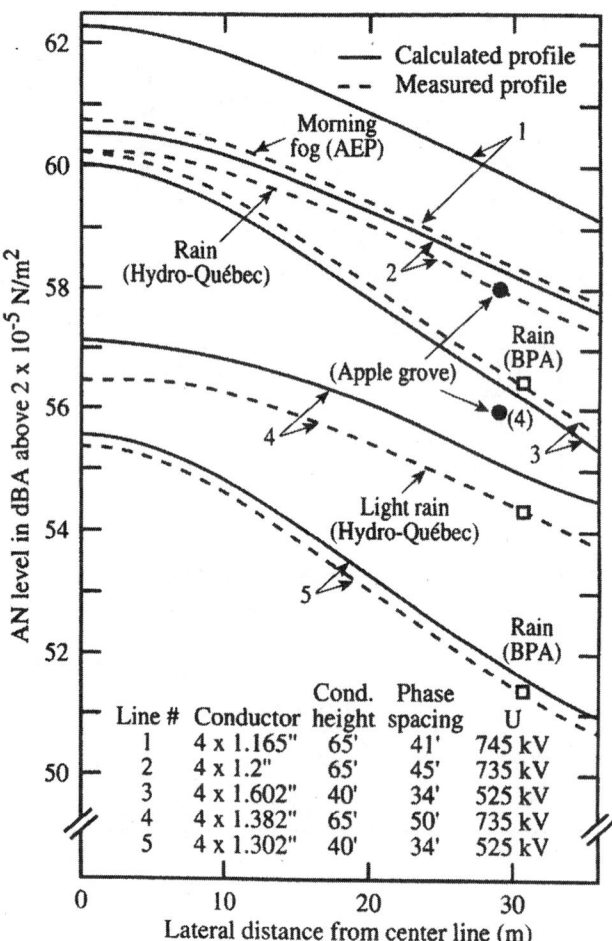

FIGURE 4.92 Comparison of calculated and measured AN performances of HV lines. (Trinh, 1995; Trinh et al., 1977.)

Approach to Control the Corona Performance

The occurrence of corona discharges on line conductors is dictated essentially by the local field intensity, which, in turn, is greatly affected by the surface conditions, e.g., rugosity, water drops, snow and ice particles, etc. For a smooth cylindrical conductor, the corona onset field is well described by the Peek's experimental law

$$Ec = 30m\delta\left(1 + \frac{0.301}{\sqrt{\delta a}}\right) \quad (kVp/cm) \qquad (4.152)$$

where Ec is the corona onset field, a is the radius of the conductor, and m is an experimental factor to take account of the surface conditions. Typical values of m are 0.8–0.9 for a dry aged conductor, 0.5–0.7 for a conductor under foul weather conditions, and δ is the relative air density factor.

The above corona onset condition emphasizes the great sensitivity of corona activities to the conductor surface condition and, hence, to changes in weather conditions. In effect, although the line voltage and the nominal conductor surface gradient remain constant, the surface condition factor varies continuously

Transmission System

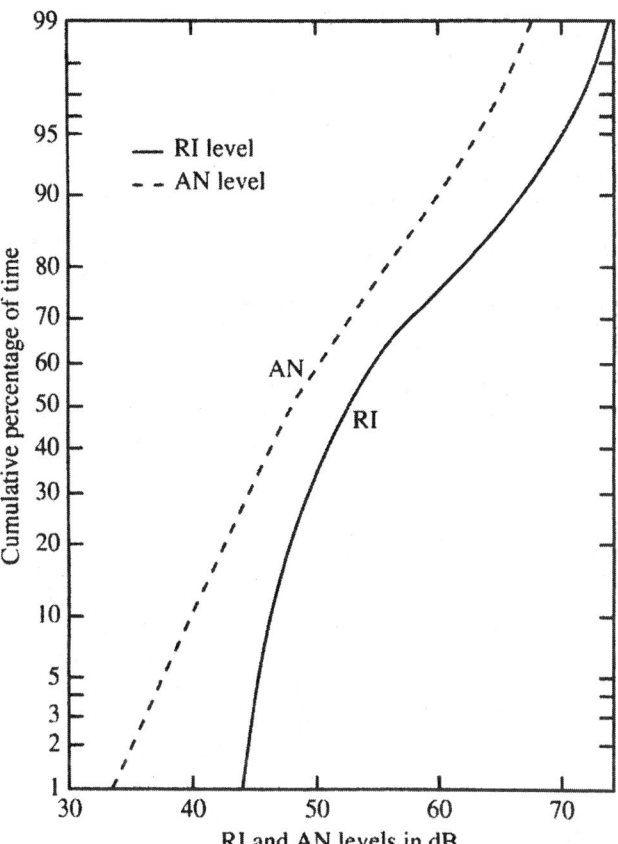

FIGURE 4.93 Cumulative distribution of RI and AN levels measured at 15 m from the outer phases of Hydro-Québec 735-kV lines. (Trinh, 1995.)

due to the exposure of line conductors to atmospheric conditions. The changes are particularly pronounced during foul weather as a result of the numerous discharge sites associated with water drops, snow, and ice particles deposited on the conductor surface.

Adequate corona performance of HV lines is generally achieved by a proper control of the field intensity at the surface of the conductor. It can be well-illustrated by the simple case of a single-phase, single-conductor line for which the field intensity at the conductor surface is

$$E0 = \frac{1}{\ln\left(\frac{2h}{a}\right)} \frac{U}{a} \leq E_c. \qquad (4.153)$$

It can be seen that the field intensity at the conductor surface is inversely proportional to its radius and, to a lesser extent, to the height of the conductor above ground. By properly dimensioning the conductor, the field intensity at its surface can be kept below the fair-weather corona-onset field for an adequate control of the corona activities and their undesirable effects.

With the single-conductor configuration, the size required for the conductor to be corona-free under fair weather conditions is roughly proportional to the line voltage, and consequently will reach unrealistic values when the latter exceeds some 400 kV. Introduced in 1910 by Whitehead to increase the transmission capability of overhead lines (1910), the concept of *bundled conductors* quickly revealed itself as an effective means of controlling the field intensity at the conductor surface, and hence, the line corona activities.

TABLE 4.26 Comparison of Single and Bundled Conducotrs Performances

Line voltage (kV)	400	735	1100
Distance between phases (m)	12	13.7	17
Number of subconductors	2	4	8
Bundle diameter (cm)	45	65	84
Conductor diameter (cm)	**3.2**	**3.05**	**3.2**
Corona onset gradient, m = 0.85, (kVrms/cm)	22.32	22.04	22.32
Maximum surface gradient (kVrms/cm)	16.3	19.79	17.3
Single conductor diameter of the same gradient (cm)	**4.7**	**8.5**	**13.8**
Transmission capability (GW)	0.5	2.0	4.9
Single conductor diameter of the same transmission capability (cm)	**8.5**	**22**	**64**

This is well-illustrated by the results in Table 4.26, which compare the single conductor design required to match the bundle performances in terms of power transmission capabilities, and the maximum conductor surface gradient for different line voltages. Bundled conductors are now used extensively in EHV lines rated 315 kV and higher; as a matter of fact, HV lines as we know them today would not exist without the introduction of conductor bundles.

Selection of Line Conductors

Even with the use of bundled conductors, it is not economically justifiable to design line conductors that would be corona-free under all weather conditions. The selection of line conductors is therefore made in terms of them being relatively corona-free under fair weather. While corona activities are tolerated under foul weather, their effects are controlled to acceptable levels at the edge of the rights-of-way of the line. For AC lines, the design levels of 70 dB for the radio interference and 60 dBA for the audible noise at the edge of the right-of-way are often used (Trinh et al., 1974). These levels may be reached during periods of foul weather, and for a specified annual proportion of time, typically 15–20%, depending on the local distribution of the weather pattern. The design process involves extensive field calculations and experimental testing to determine the number and size of the line conductors required to minimize the undesirable effects of corona discharges. Current practices in dimensioning HV line conductors usually involve two stages of selection according to their worst-case and long-term corona performances.

Worst-Case Performance

Several conductor configurations (number, spacing, and diameter of the subconductors) are selected with respect to their worst-case performances which, for AC lines, corresponds to foul-weather conditions, in particular heavy rain. Evaluation of the conductor worst-case performance is best done in *test cages* under artificial heavy rain conditions (Trinh and Maruvada, 1977). Test cages of square section, typically 3 m × 3 m, and a few tens of meters long, are adequate for evaluating full-size conductor bundles located along its central axis, for lines up to the 1500-kV class. The advantages of this experimental setup are the relatively modest test voltage required to reproduce the same field distribution on real-size bundled conductors, and the possibility of artificially producing the heavy rain conditions. The worst-case performance of various bundled conductors can then be determined over a wide range of surface gradients.

Under DC voltage, the worst-case corona performance is not directly related to foul-weather conditions. Although heavy rain was found to produce the highest losses, both the electromagnetic interference and the audible noise levels decrease under rain conditions. This behavior is related to the fact that under DC field conditions, the water droplets have an optimum shape, favorable to the development of stable glow-corona modes (Ianna et al., 1974). For this reason, test cage is less effective in evaluating the worst-case DC performance of bundled conductors.

A significant amount of data was gathered in cage tests at IREQ during the 1970s and provided the database for the development of a method to predict the worst-case performance of bundled conductors for AC voltage (Trinh and Maravuda, 1977). The results presented in Figs. 4.91 and 4.92, which compare the calculated and measured lateral RI and AN profiles of a number of HV lines, illustrate the good concordance of this approach. Commercial softwares exist that evaluate the wosrt-case performance of

HV line conductors using available experimental data obtained in cage tests under conditions of artificial heavy rain, making it possible to avoid undergoing tedious and expensive tests to help select the best configurations for line conductors for a given rating of the line.

Long-Term Corona Performance

Because of their wide range of variation in different weather conditions, representative corona performances of HV line are best evaluated in their natural environment. Test lines are generally used in this study which involves energizing the conductors for a sufficiently long period, usually one year to cover most of the weather conditions, and recording their corona performances together with the weather conditions. The higher cost of the long-term corona performance study usually limits its application to a small number of conductor configurations selected from their worst-case performance.

It should be noted that best results for the long-term corona performance evaluated on test lines are obtained when the weather pattern at the test site is similar to that existing along the actual HV line. A direct transposition of the results is then possible. If this condition is not met, some interpretation of the experimental data is needed. This is done by first decomposing the recorded long-term data into two groups, corresponding to the fair and foul weather conditions, then recombining these data according to the local weather pattern to predict the long-term corona performance along the line.

Conclusions

This section on transmission systems has reviewed the physics of corona discharges and discussed their impact on the design of high-voltage lines, specifically in the selection of the line conductors. The following conclusions can be drawn.

- Corona discharges can develop in different modes, depending on the equilibrium state existing under a given test condition, between the buildup and removal of ion space charges from the immediate vicinity of the highly stressed electrode. Three different corona modes, Trichel streamer, negative glow, and negative streamer, can be observed at the cathode with increasing applied field intensities. With positive polarity, four different corona modes are observed, namely burst corona, onset streamers, positive glow, and breakdown streamers.
- While all corona modes produce energy losses, the streamer discharges also generate electromagnetic interference and aubible noise in the immediate vicinity of HV lines. These parameters are currently used to evaluate the corona performance of conductor bundles and to predict the energy losses and environmental impact of HV lines prior to their installation.
- Adequate control of line corona is obtained by controlling the surface gradient at the line conductors. The introduction of bundled conductors in 1910 has greatly influenced the development of HV lines to today's EHV voltages.
- Commercial softwares are available to select the bundle configuration: number and size of the subconductors, with respect to corona performances, which can be verified in test cages and lines in the early stage of new HV-line projects.

References

Bateman, L.A., Haywood, R.W., and Brooks, R.F., Nelson River DC Transmission Project, *IEEE Trans.*, PAS-88, 688, 1969.

Bortnik, I.M., Belyakov, N.N., Djakov, A.F., Horoshev, M.I., Ilynichin, V.V., Kartashev, I.I., Nikitin, O.A., Rashkes, V.S., Tikhodeyev, N.N., and Volkova, O.V., *1200 kV Transmission Line in the USSR: The First Results of Operation,* in *CIGRE Report No. 38-09,* Paris, August 1988.

Dawson, G. A., A model for streamer propagation, *Zeitchrift fur Physic,* 183, 159, 1965.

Gary, C. H., The theory of the excitation function: A demonstration of its physical meaning, *IEEE Trans.*, PAS-91, 305, 1972.

Ianna, F., Wilson, G.L., and Bosak, D.J., Spectral characteristics of acoustic noise from metallic protrusion and water droplets in high electric fields, *IEEE Trans.*, PAS-93, 1787, 1974.
Juette, G. W., Evaluation of television interference from high-voltage transmission lines, *IEEE Trans.*, PAS-91, 865, 1972.
Krishnayya, P.C.S., Lambeth, P.J., Maruvada, P.S., Trinh, N.G., and Desilets, G., *An Evaluation of the R&D Requirements for developing HVDC Converter Stations for Voltages above ±600 kV*, in CIGRE Report No. 14-07, Paris, August 1988.
Lacroix, R. and Charbonneau, H., Radio Interference from the first 735-kV line of Hydro-Quebec, *IEEE Trans.*, PAS-87, 932, 1968.
Loeb, L.B., *Electrical Corona*, University of California Press, 1965.
Moreau, M. R. and Gary, C. H., Predetermination of the radio-interference level of high voltage transmission lines — I: Predetermination of the excitation function, *IEEE Trans.*, PAS-91, 284, 1972.
Moreau, M. R. and Gary, C. H., Predetermination of the radio-interference level of high voltage transmission lines — II: Field calculating method, *IEEE Trans.*, PAS-91, 292, 1972.
Raether, H., *Electron Avalanche*, Butterworth Co., 1964.
Trinh, N. G., Partial discharge XX: Partial discharges in air — Part II: Selection of line conductors, *IEEE Electrical Insulation Magazine*, 11, 5, 1995.
Trinh, N. G., Partial discharge XIX: Discharge in air — Part I: Physical mechanisms, *IEEE Electrical Insulation Magazine*, 11, 23, 1995.
Trinh, N. G. and Jordan, I. B., Modes of corona discharges in air, *IEEE Trans.*, PAS-87, 1207, 1968.
Trinh, N. G. and Jordan, I. B., Trichel streamers and their transition into the pulseless glow discharge, *J. Appl. Physics*, 41, 3991, 1970.
Trinh, N. G. and Maruvada, P. S., A method of predicting the corona performance of conductor bundles based on cage test results, *IEEE Trans.*, PAS-96, 312, 1977.
Trinh, N. G., Maruavada, P. S., and Poirier, B., A comparative study of the corona performance of conductor bundles for 1200-kV transmission lines, *IEEE Trans.*, PAS-93, 940, 1974.
Whitehead, J. B., *Systems of Electrical Transmission*, U.S. Patent No. 1,078,711, 1910.

4.9 Geomagnetic Disturbances and Impacts Upon Power System Operation

John G. Kappenman

Nearly all modern technolgy systems (power systems, communications, satellites, and navigation to name a few) are more susceptible to geomagnetic disturbances than their counterparts of previous solar cycles. This is certainly the pattern that has been witnessed in the electric power industry.

Geomagnetic disturbances can induce near DC($f < 0.01$ Hz) currents (i.e., Geomagnetically Induced Currents, GIC) to flow through the power system entering and exiting the many grounding points on a transmission network. This is generally of most concern at the latitudes of the northern U.S., Canada, and northern Europe, for example, but regions much farther south are affected during intense magnetic storms. GICs are caused when the auroral electrojet (a large multimillion ampere current structure in the conductive portion of the ionosphere at an approximate altitude of 100 km) subjects portions of the earth's nonhomogeneous, conductive surface to time-varying fluctuations in the planet's normally quiescent magnetic field. These field fluctuations induce electric fields in the earth which give rise to potential differences between grounding points. The resulting electric field can extend over large regions and essentially behave as an ideal voltage source applied between remote neutral ground connections of transformers in a power system. This voltage potential difference causes a GIC to flow through the transformers and associated power system lines and neutral ground points. Over 100 amps have been measured in the neutral leads of transformers in such areas, while only a few amps are sufficient to initiate disruption of transformer operation. Solar Cycle 22 (the prior 11-year sunspot cycle 1986–97) has been especially important because of the unprecedented impact that storms have had on electric power systems.

Power System Reliability Threat

Low-level and very small scale investigations of the impact of geomagnetic storms had been underway in the power industry for a number of years (even dating back to observed impacts from a storm in 1940). However, threats to power system integrity are no longer just academic speculation with the events that unfolded during the Great Geomagnetic Storm of March 13, 1989, when the entire Hydro-Québec system (a system serving more than 6 million customers) was plunged into a blackout, triggered by GIC, causing voltage collapse and equipment malfunction. The impact of this particular storm was simultaneously felt over the entire North American continent with most of Hydro-Québec's neighboring systems in the U.S. coming uncomfortably close to experiencing the same sort of voltage collapse/cascading outage scenario.

Additional, though less severe, storm events in September 1989, March 1991, and October 1991 reinforced, for utilities around the world, that geomagnetic disturbances can hamper reliable operation as voltage regulation is impacted, as undesired relay operations occur on important system equipment, and as new areas of vulnerability are exposed from the unintended consequences of interactions of GIC with various advanced technology apparatus and devices that have been added to the grid. In part, utility system impacts have been greater in recent years because of a more severe and active storm cycle than has been experienced over the prior 30 years. On the other hand, this previous benign era has had the effect of lulling designers into neglecting consideration of these possible influences in their design decisions.

Many portions of the North American power grid have all the elements that contribute to susceptibility to geomagnetic storms: located in northern geomagnetic latitudes, near the auroral electrojet current; located in broad areas of highly resistive igneous rock; and dependent upon remote generation sources linked by long transmission lines to deliver energy to load centers. In fact, the evolving growth of the North American transmission grid over the past few decades has made the grid, along with the geological formations occurring in much of North America, the equivalent of a large efficient antenna that is electromagnetically coupled to the disturbance signals produced by fluctuations of the earth's magnetosphere. GIC, when present in transformers on the system, will produce half-cycle saturation of numerous transformers simultaneously across the network. The large geographic scale coupled with the simultaneous and global impact of these storms produces voltage regulation and harmonic effects in each of these transformers in quantities that add in a cumulative fashion. The result is sufficient to overwhelm the voltage regulation capability and the protection margins of equipment over large regions of the network. Combinations of events such as these can rapidly lead to system-wide problems. For example, the Hydro-Québec outage was the end result of over 15 discrete protective-system operations linked into a chain of events. Further, from the initial event to complete blackout, there was a total elapsed time of a mere one and a half minutes — hardly enough time to even assess what was occurring, let alone provide any meaningful human intervention.

Power systems in areas of igneous rock, typical across the Laurentian shield, for instance, are the most vulnerable to the effects of intense geomagnetic activity because the relatively high resistance of igneous rock encourages more current to flow in alternative conductors such as power transmission lines situated above these geological formations. Research has been done to investigate devices to block GIC flow, but they continue to remain too complex and expensive to blanket such a large network/ground topology.

Operational strategies, and decisions on when to implement them, are presently based upon combinations of storm forecasts and alerts and are often confirmed by locally monitored impacts on power system operations that would be due to a storm (i.e., GIC in a transformer neutral). However, as previously noted, storm events can, at times, progress quickly and some operational changes such as generation redispatch can take up to an hour or more to implement. In situations such as this, it is not always possible to respond quickly enough after a storm is confirmed to prevent serious damage. Forecasts as a means of preparing operational strategies have been problematic in that they have, over prior sunspot cycles, been of low reliability. Since operating postures that would harden power systems to effects of geomagnetic disturbances can be risky, costly, and difficult to effectively maintain for extended durations, and since forecasting technology has been so poor, utilities find themselves caught in the paradox of

(1) either implementing response measures prior to a storm confirmation, or (2) awaiting local confirmation to avoid false alarms and hoping reaction time can be rapid enough.

Also important to electric system operators, though perhaps less obvious, is to know when to deactivate these procedures. The intermittent nature of the effects of geomagnetic storms makes it difficult to tell when the storm activity is over. There may be lulls in activity followed by additional, regionally severe activity. Most guidelines are held for a period of time, usually two to four hours, after the last observed indication of geomagnetic activity. The optimal choice of operating procedures depends on the prediction of the level of GIC in the system which, in turn, requires knowledge of the expected storm severity and local manifestation characteristics. Thus, the ability of system operators to maintain and manage grids under geomagnetically disturbed conditions can be significantly enhanced if the severity and duration of geomagnetic disturbances can be predicted accurately.

Transformer Impacts due to GIC

The primary concern with geomagnetically induced currents (GIC) is the effect that they have on the operation of a large power transformer. Under normal conditions the large power transformer is a very efficient device for converting one voltage level into another. Decades of design engineering and refinement have increased efficiencies and capabilities of these complex apparatus to the extent that only a few amperes of AC exciting current are necessary to provide the magnetic flux for the voltage transformation in even the largest modern power transformer.

However, in the presence of GIC, the near-direct current essentially biases the magnetic circuit of the transformer with resulting disruptions in performance. The three major effects produced by GIC in transformers are (1) the increased reactive power consumption of the affected transformer, (2) the increased even and odd harmonics generated by the half-cycle saturation, and (3) the possibilities of equipment damaging stray flux heating.

Transformers use steel in their cores to enhance their transformation capability and efficiency, but this core steel introduces nonlinearities into their performance. Common design practice minimizes the effect of the nonlinearity while also minimizing the amount of core steel. Therefore, the transformers are usually designed to operate over a predominantly linear range of the core steel characteristics (as shown in blue in Fig. 4.94a), with only slightly nonlinear conditions occurring at the voltage peaks. This produces a relatively small exciting current (blue in Fig. 4.94b). With GIC present, the normal operating point on the core steel saturation curve is offset and the system voltage variation that is still impressed on the transformer causes operation in an extremely nonlinear portion of the core steel characteristic for half of the AC cycle (red in Fig. 4.94a). Hence the term half-cycle saturation.

Because of the extreme saturation that occurs on half of the AC cycle, the transformer now draws an extremely large asymmetrical exciting current. The red waveform in Fig. 4.94b depicts a typical example from field tests of the exciting current from a three-phase 600 MVA power transformer that has 75 amps of GIC in the neutral (25 A per phase). Spectrum analysis reveals this distorted exciting current to be rich in even, as well as odd harmonics. As is well documented, the presence of even a small amount of GIC (3 to 4 amps per phase or less) will cause half-cycle saturation in a large transformer.

Since the exciting current lags the system voltage by 90°, it creates reactive-power loss in the transformer and the impacted power system. Under normal conditions, this reactive loss is very small. However, the several orders of magnitude increase in exciting current under half-cycle saturation also results in extreme reactive-power losses in the transformer. For example, the three-phase reactive power loss associated with the abnormal exciting current of Fig. 4.94b produces a reactive power loss of over 40 MVars for this transformer alone. The same transformer would draw less than 1 MVar under normal conditions. Figure 4.95 provides a comparison of reactive power loss for two core types of transformers as a function of the amount of GIC flow.

Under a geomagnetic storm condition in which a large number of transformers are experiencing a simultaneous flow of GIC and undergoing half-cycle saturation, the cumulative increase in reactive power demand can be significant enough to impact voltage regulation across the network, and in extreme

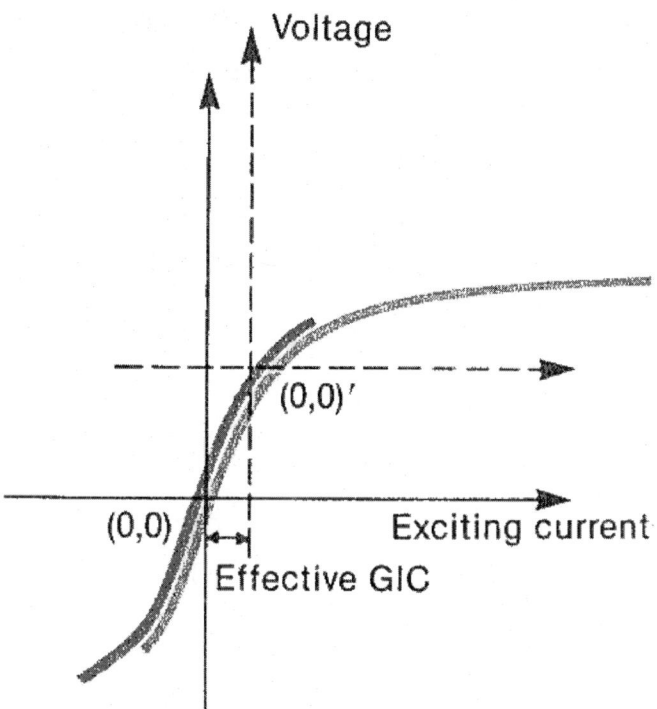

FIGURE 4.94a The presence of GIC causes the transformer magnetization characteristics to be biased or offset due to the DC. Therefore on one-half of the AC cycle, the transformer is driven into saturation by the combination of applied voltage and DC bias. Normal excitation operation is shown in the left curve, the biased operation in the right.

situations, lead to network voltage collapse. In the March 13, 1989, superstorm, Manitoba Hydro observed dramatic increases in reactive power output from synchronous condensers at one of their substations. The total reactive requirements at this station alone increased by 420 MVars during the course of the storm event within a few minutes time span. Studies of impacts across a system indicate probable and extreme storm events can cause system-wide reactive demand increases of several thousand MVars. A system GIC flow distribution model taking into account the transformer specific design and GIC flow can provide a means of evaluating the potential for system-wide GIC voltage regulation impacts.

The large and distorted exciting current drawn by the transformer under half-cycle saturation poses a hazard to operation of the network because of the rich source of even and odd harmonic currents this injects into the network and the undesired interactions that these harmonics may cause with relay and protective systems or other power system apparatus. Figure 4.96 is the spectrum analysis of the asymmetrical exciting current from Fig. 4.94b. Even and odd harmonics are present typically in the first ten-orders and the variation of harmonic current production varies somewhat with the level of GIC and the degree of half-cycle saturation. A larger GIC and resulting larger degree of saturation may actually decrease the total harmonic distortion produced by the transformer as more fundamental frequency current is drawn on each half-cycle of saturated operation An example of this relationship is shown in Fig. 4.97.

In addition to the power system effects of the harmonics and reactive power demands, the transformer itself can be severely stressed by this mode of operation. Measurements have shown that audible noise from the transformer can increase more than fivefold because the magnetostriction of saturated operation increases core steel vibration. Figure 4.98 provides an example of the spectral content of the transformer audible noise variation for various levels of DC excitation. Many anecdotal observations of power system impacts due to geomagnetic disturbances have been as a result of reported transformer audible noise increases.

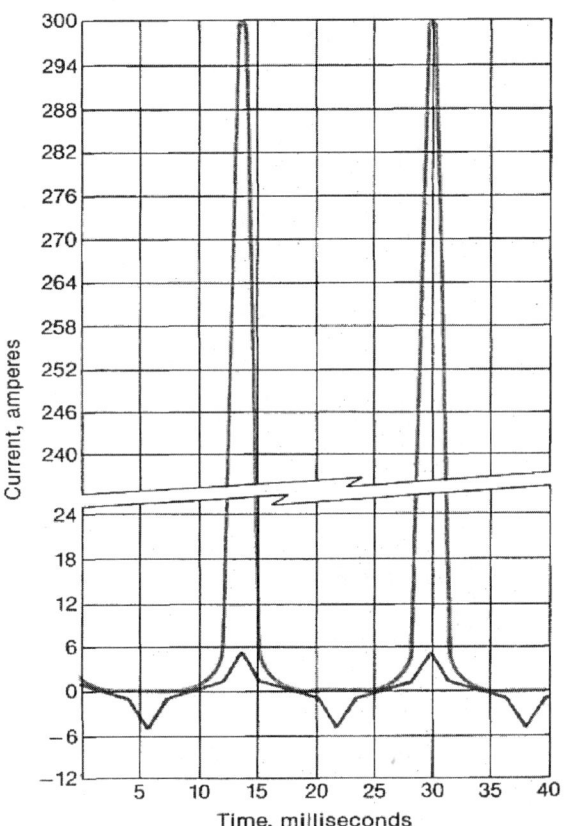

FIGURE 4.94b Under normal conditions, the excitation current of this 600 MVA 500/230 kV transformer is less than 1% of transformer rated current. However, with 25 amps/phase of GIC present, the excitation current drawn by the transformer (top curve) is highly distorted by the half-cycle saturation conditions and has a large peak magnitude rich in harmonics.

With the magnetic circuit of the core steel saturated, the magnetic core will no longer contain the flow of flux within the transformer. This stray flux will impinge upon or flow through adjacent paths such as the transformer tank or core clamping structures. The flux in these alternate paths can concentrate to the densities found in the heating elements of a kitchen stove. This abnormal operating regime can persist for extended periods as GIC flows from storm events can last for hours. The hot spots that may then form can severely damage the paper winding insulation, produce gassing and combustion of the transformer oil, or lead to other serious internal failures of the transformer. Such saturation and the unusual flux patterns which result, are not typically considered in the design process and, therefore, a risk of damage or loss of life is introduced. Further a transformer's vulnerability is extremely design dependent, so general conclusions are inappropriate.

One of the more thoroughly investigated incidents of transformer stray flux heating occurred in the Allegheny Power System on a 350 MVA 500/138 kV autotransformer at their Meadow Brook Substation near Winchester, Virginia. The transformer was first removed from service on March 14, 1989, because of high gas levels in the transformer oil which were a by-product of internal heating. The gas-in-oil analysis showed large increases in the amounts of hydrogen, methane, and acetylene, indicating core and tank heating. External inspection of the transformer indicated four areas of blistering or discolored paint due to tank surface heating. In the case of the Meadow Brook transformer, calculations estimate the flux densities were high enough in proximity to the tank to create hot spots approaching 400°C. Reviews made by Allegheny Power indicated that similar heating events (though less severe) occurred in several

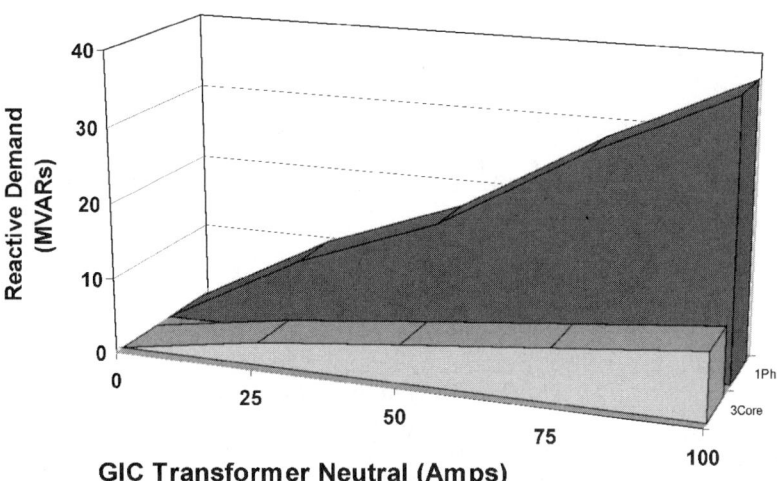

FIGURE 4.95 The exciting current drawn by half-cycle saturation conditions shown in Fig. 4.94b produces a reactive power loss in the transformer as shown in the top plot. This reactive loss varies with GIC flow as shown. This was measured from field tests of a 3-phase bank of single-phase 500/230 kV transformers. Also shown in the bottom curve is measured reactive demand vs. GIC from a 230/115 kV 3-phase 3-legged core-form transformer. Transformer core design is a significant factor in estimating GIC reactive power impact.

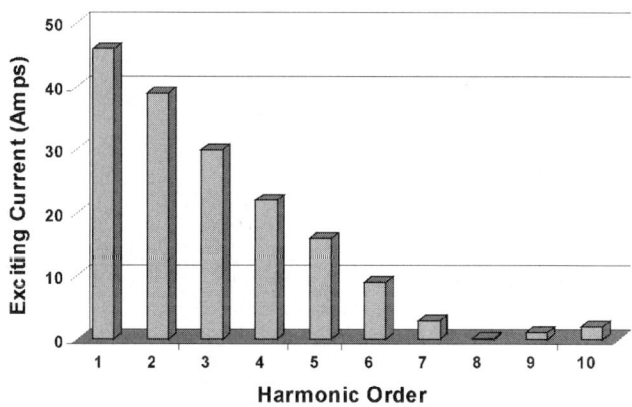

FIGURE 4.96 The distorted transformer exciting current shown in Fig. 4.94b has even and odd harmonic current distortion. This spectrum analysis was half-cycle saturation conditions resulting from a GIC flow of 25 amps per phase.

other large power transformers in their system due to the March 13 disturbance. Figure 4.99 is a recording that Allegheny Power made on their Meadow Brook transformer during a storm in 1992. This measurement shows an immediate transformer tank hot-spot developing in response to a surge in GIC entering the neutral of the transformer, while virtually no change is evident in the top oil readings. The manufacturer had not predicted or anticipated this mode of operation and therefore could not expect standard over-temperature sensors to be effective deterrents.

Designing a large transformer that would be immune to the near-DC geomagnetically induced current would be technically difficult and prohibitively costly. The ampere-turns of excitation (the product of

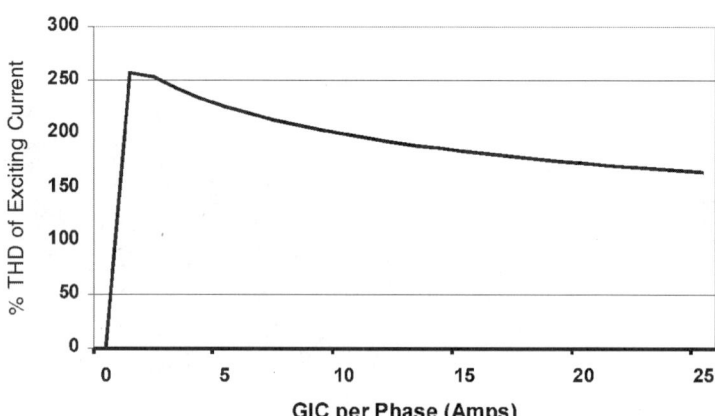

FIGURE 4.97 The total harmonic distortion (THD) of the transformer excitation current is shown above. While the excitation current magnitude increases substantially with increasing GIC flow, the THD percentage decreases as more fundamental frequency excitation current is drawn by the half-cycle saturated transformer.

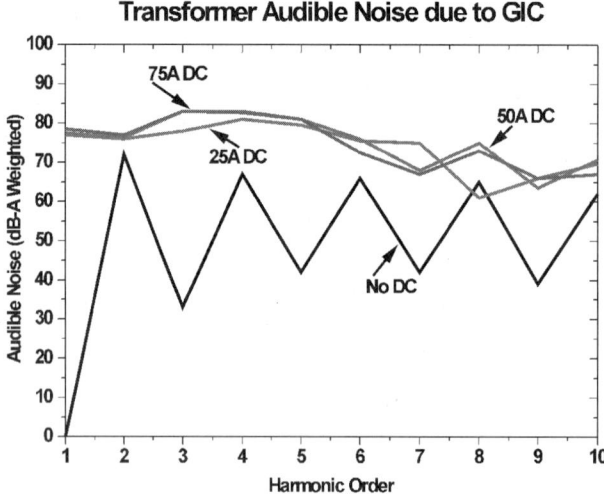

FIGURE 4.98 Half-cycle saturation produces substantial increases in audible noise emissions from a transformer due to core steel magnetostriction. Because the saturation is asymmetrical, the normal 120-Hz hum noise spectrum is replaced by 60 Hz and harmonic noise spectrum.

the normal exciting current and the number of winding turns) generally determine the core steel volume requirements of a transformer. However, designing for unsaturated operation with the high level of GIC present would require a core of excessive size. Blocking the flow of GIC into a transformer is an alternate approach and several design options have been developed for blocking capacitors to install in the transformer neutral. While effective in blocking the flow of GIC for single winding designs, in the case of an autotransformer, the flow of GIC can predominate in the series winding, and as a result, the transformer can still experience severe half-cycle saturation. Series capacitors at the high voltage level can also be employed for GIC blocking but efficacy of GIC mitigation in complex networks requires detailed storm and GIC flow modeling simulations.

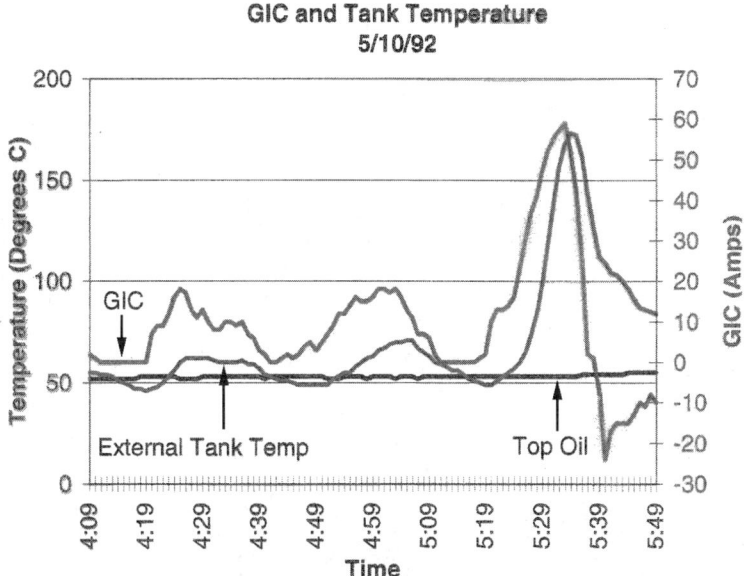

FIGURE 4.99 Transformer hot-spot heating due to stray flux can be a concern in operation of a transformer with GIC present. This transformer experienced stray flux heating that could be monitored with a thermocouple mounted on the tank exterior surface. This storm demonstrated that the GIC and resulting half-cycle saturation produced a rapid heating in the tank hot-spot. Notice also that transformer top-oil temperature did not show any significant change, indicating that the hot-spot was relatively localized.

Magneto-Telluric Climatology and the Dynamics of a Geomagnetic Superstorm

The sunspot cycle (a measure of variable sunspot observations) is the traditionally defined metric of solar activity. However, impacts at earth due to solar activity are separately defined by the geomagnetic storm cycle. These two cycles are not perfectly in synchronism as shown in Fig. 4.100. The geomagnetic storm cycle tends to have two or three peaks of activity during the course of a typical solar cycle and also presents a broader plateau of higher frequency of activity than implied by the narrower peak of the sunspot cycle. Further geomagnetic activity tends to peak during the declining phase of the sunspot cycle rather than coincident with the sunspot peak. The disconnect between the sunspot cycle and the geomagnetic storm cycle is primarily due to differing solar processes (coronal hole activity) that come into play during the latter half of the solar cycle. These coronal hole processes are not measured by the traditional sunspot count, yet can become the primary driver for geomagnetic storm activity during this stage in the cycle.

Long-term projections of solar cycle activity and expected terrestrial manifestations is more art than science at present owing to the fundamental data and knowledge gaps that exist in understandings of solar processes that drive the solar cycle. As a result, statistical data analysis and trending methods are one of the key inputs in developing projections for upcoming solar cycles, and this is still true for the consensus projections developed for Cycle 23. The consensus opinion holds that Cycle 23 will be a cycle similar in characteristics to Cycle 22, which was in the top quartile of the 22 solar cycles on record (Joselyn et al.). Therefore, this forecast implies an early ascent in activity with the majority of sunspot activity to occur by year 2002. Further, the frequency and severity of geomagnetic storm events is projected to place this solar cycle in the top quartile of severity, even for the lowest projection estimate. Figure 4.101 provides a summary of severe geomagnetic storms over the limited database of measured Ap index (1932 to present). Ap is a measure of the daily range of magnetic field variation at several globally distributed stations. While not directly a measure of GIC production at any specific location, its value stems from

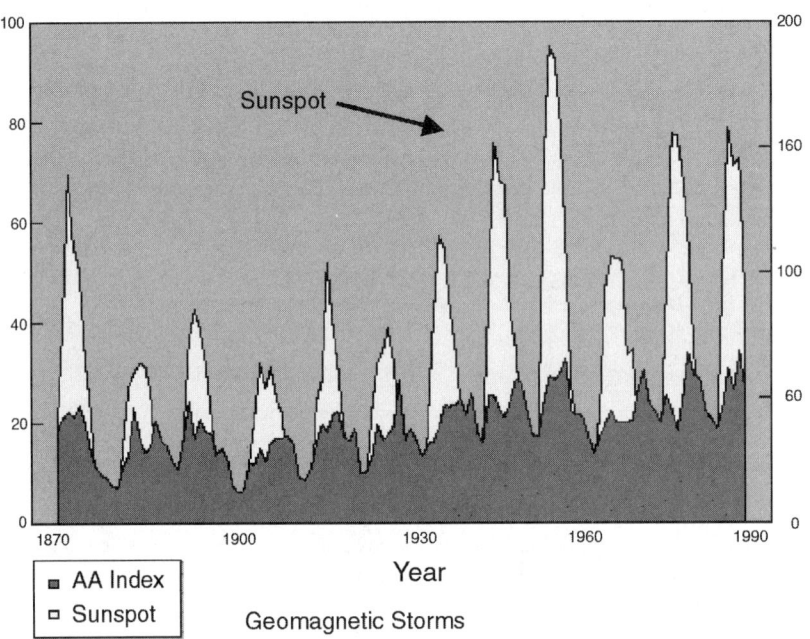

FIGURE 4.100 The sunspot cycle provides a measure of solar variability. The geomagnetic cycle is the measure of magnetic disturbances at the earth caused by solar activity. The two cycles have different peak and duration characteristics because of solar activity drivers that are not captured by the sunspot count.

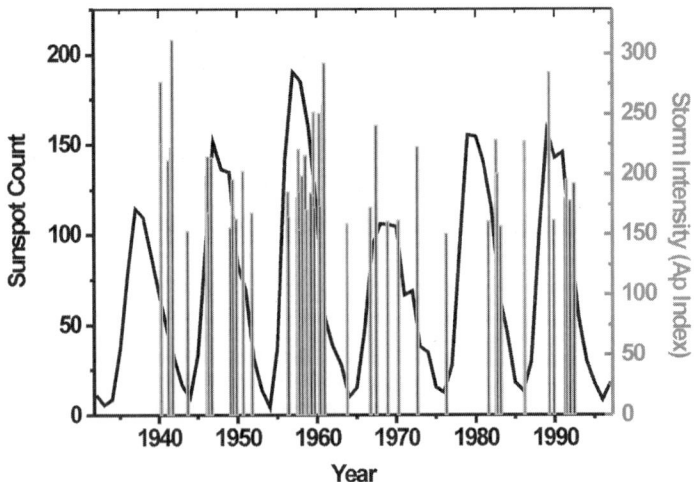

FIGURE 4.101 The climatology of large geomagnetic storms (Ap > 150) is shown over the last 68 years relative to the sunspot cycle. Storms this large have been sufficient to cause large GIC flows with resulting power system impacts. A storm of Ap > 150 occurs approximately 1.06 times per year. Large storms can also occur at any time during the solar cycle and are not confined to the peak sunspot count years.

the fact that it is one of the oldest measures of storm intensity for long-term climatology comparisons. Even this measure is relatively recent, dating back to only 1932, about six solar cycles. Each storm is unique in many aspects, especially in time- and region-specific geomagnetic fluctuations and intensities.

Transmission System

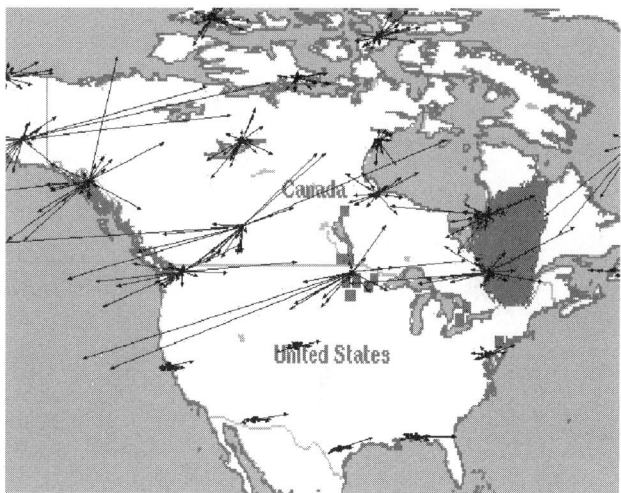

FIGURE 4.102 Disturbance conditions observed (dBh/dt in nT/min) at North American magnetic observatories on March 13, 1989, between 02:36–03:02 EST and resulting power system impacts reported to NERC. Glen Lea (near Winnipeg) reported the largest dB/dt of 869 nT/min. The Hydro-Québec system experienced a complete system collapse at approximately 02:45 EST.

Indices, by their nature, provide an averaging of the highly specific impacts over broad windows of time and regions of space and as such tend to blur the details of the spatial dynamics of the electrojet current driver in the ground induction manifestations of a geomagnetic storm.

While the Ap index is imperfect as an absolute measure of GIC impact, it generally takes a storm of Ap intensity of 150 or greater to trigger significant power system events. Given these limitations, a review of Ap index tendencies indicates that a storm of this intensity occurs on a planetary basis at a rate of 1.06 times per year. Further large storms, while more frequent during geomagnetic storm cycle peaks, can occur at any time as they only need one well-aimed eruption from the sun to be created (a solar process that still occurs several times per week even during solar quiet conditions). A good case in point is the 225 Ap storm that shook large parts of the North American power grid in February 1986, the absolute minimum between Sunspot Cycles 21 and 22. The March 13–14, 1989, superstorm was the third largest on record with an Ap of 285. While Cycle 22 produced several noteworthy storm events, the worst cycle of record for large storms was Cycle 19, which produced six storms of Ap 200 or greater versus only two in Cycle 22.

A geomagnetic disturbance produces the large ionospheric current structures that predominate in the nighttime regions of the planet. Spatial and intensity variations in the electrojet interact with the local geomagnetic field and cause intense and impulsive geomagnetic field fluctuations to drive the ground induction process. The electrojet current storm process can be exceedingly large, both in intensity and geographic breadth. In addition, severe and periodic substorms can extend for excessively long durations (several days is typical for large storms). The fluctuations in the million ampere plus electrojet structure produce comparably severe and sudden fluctuations in the ground-level magnetic field in proximity to the electrojet. The coupling of these magnetic field perturbations with the earth and overlaying transmission grid will trigger flows of geomagnetically induced currents (GIC) that can cause transformer half-cycle saturation and associated power system impacts. Figures 4.102 and 4.103 summarize the dynamic and widespread impact that the superstorm of March 13, 1989, presented across the North American continent. The time period from 02:30–07:00 EST produced the largest rate of change of ground horizontal magnetic field (dBh/dt variations measured in nano-Tesla per minute) in a region centered on the U.S.-Canadian border. Later substorm events as noted in the time period from 17:00–21:00 EST produced severe dBh/dt events further south, with the largest events occurring over a region from the Canadian border down to the Fredericksburg, VA, and Boulder, CO, observatories. Even

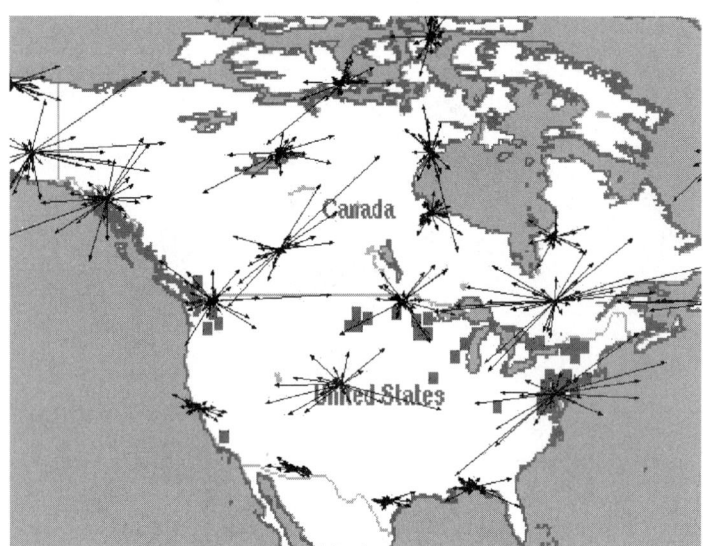

FIGURE 4.103 Disturbance conditions observed (dBh/dt in nT/min) at North American magnetic observatories on March 13, 1989, between 17:00–17:30 EST and resulting power system impacts reported to NERC. The extent of the storm electrojet structure extended further equatorward, resulting in system impacts through many midlatitude locations.

the Bay St. Louis observatory near the Gulf of Mexico experienced several dBh/dt fluctuations in excess of 300nT/min during this series of substorms. From the observation of large dBh/dt at various observatories around the world during this storm, a plot of geomagnetic field disturbance extrema can be projected on a world map showing equatorward extensions of large dBh/dt events. The projection provides a correction for the asymmetry between the geographic and geomagnetic poles and translates to the appropriate geographic coordinates that these impulsive shocks would extend worldwide.

Figures 4.104 and 4.105 characterize the dBh/dt variations and onsets that are the important drivers for ground-induction, these figures also denote significant region-specific power system impacts that occurred associated with the storm. As shown in Fig. 4.103, the onset of severe magnetic field fluctuations

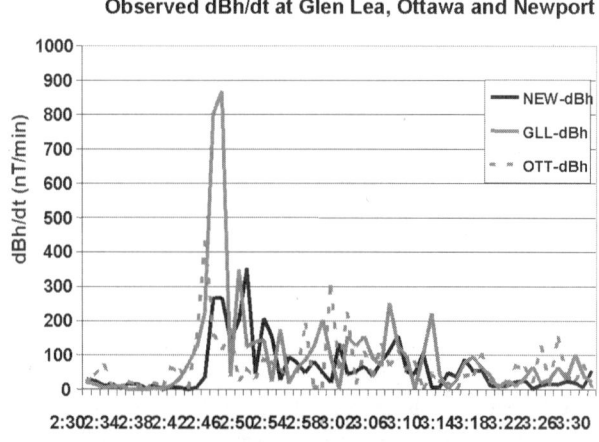

FIGURE 4.104 Observed dBh/dt (nT/min) at magnetic observatories near the U.S.-Canadian border during the time of the Hydro-Québec blackout and other noteworthy power system impacts.

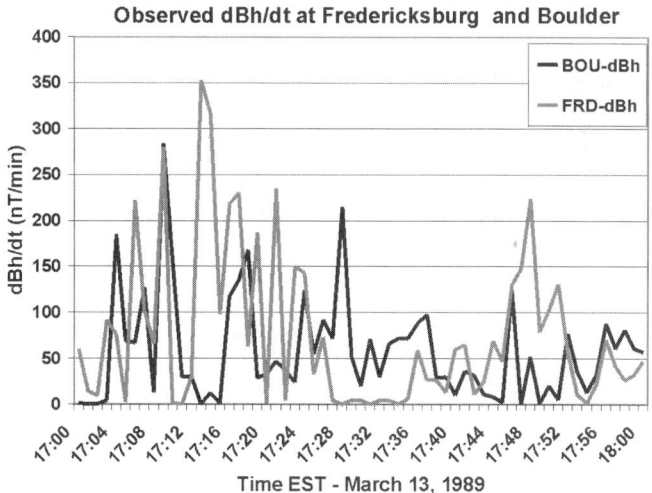

FIGURE 4.105 Observed dBh/dt (nT/min) at magnetic observatories at U.S. midlatitude sites during a substorm from 17:00–18:00 EST. A large number of transmission network events were reported coincident with these severe disturbances.

can be rapid and essentially allow no lead time for meaningful response measures. Therefore, reacting on the basis of locally observed confirmation of storm activity (such as a measured GIC) would not provide fail-safe lead time especially in the case of the large and important storm events. While NOAA and other governmental agency forecast products would continue to emphasize environmental assessment, impacted systems need additional translation of these broad environmental conditions into potential impacts on their respective systems. Advanced modeling techniques satisfy these industry-specific translation needs.

Satellite Monitoring and Forecast Models Advance Forecast Capabilities

In January 1998, a NASA satellite (Advanced Composition Explorer or ACE satellite) began providing continuous and real-time monitoring of the solar-wind conditions that are the primary drivers for a geomagnetic storm. The data is fundamental to enabling the formulation of highly accurate forecast techniques and the subsequent issuance of alerts and warnings of impending major geomagnetic disturbances. Because it takes a disturbance in the solar wind about an hour to travel from where ACE is, near the L1 point (about 1 million miles upstream in the solar wind) to earth, telemetry from ACE will allow alerts of imminent, severe geomagnetic storms to be issued nominally an hour in advance of their onset. Data from the ACE satellite and its array of instruments will provide virtually fail-safe certainty in the forecast of major disturbances on a planetary scale.

Unlike the terrestrial weather conditions that are monitored routinely at thousands of locations worldwide, the conditions in space are much more difficult to monitor; therefore, only a handful of space-based and ground-based monitoring stations are available. As a result, space weather forecasters are required to specify and to predict conditions in space and earth's magnetosphere using a minimum of guidance from actual measurements. The extreme under-sampling of the diverse, coupled regions of space demands that numerical models be utilized to provide continuous quantitative assessment and prediction of the geospace environment. As discussed in the last section, a geomagnetic disturbance can have a rapid and dynamic onset, which, if monitored locally at the earth, would not be able to provide impacted systems meaningful "lead time" for severe storm activity. As evidenced by the concern about system-collapse-type scenarios in operation of transmission networks, remedial measures applied in response to locally detected storm events cannot assure any degree of success in severe storm scenarios.

Therefore, forecast or predictive modeling of space weather is needed if an impacted system requires an advance warning of storm conditions in order to take preventive or mitigative actions. In order to perform the evaluation for the power industry, a number of model development and storm database review procedures were undertaken in order to extend forecasts from planetary-level quality to region-specific projections of severe geomagnetic field disturbances. Solar wind velocity, density, and direction and magnitude of the interplanetary magnetic field provide basic inputs to forecast models which in turn provide an equivalent "lead-time" of a storm event for the processes modeled. These efforts are developing capabilities to predict not only on a global scale, but also more importantly, for concerned transmission grid operators, an ability to provide a projection of region and time-specific meso-scale processes of concern. Further, these can be provided with the expectation that major events can be forecast with a low false alarm rate (Maynard, 1995; Kappenman, 1998).

NOAA and other governmental agencies provide "Environmental Assessments" of forecast storm conditions, whereas power-industry users of forecasts who are responsible for important operational functions during storm events need to have a "System Impact Assessment" of the storm potential. The primary focus of system impact analysis is the desire to quantify the region, system-specific severity, and impact of a storm. NOAA's forecasts, for example, primarily provide forecast products in index-style severity classification. The most familiar NOAA index is the "K Index," a logarithmic scale from 0 to 9 that classifies storm severity in a manner similar to the Richter scale for earthquakes. Index approaches are inherently difficult to apply for system impact analysis on power systems, in that the ground-induction process requires detailed knowledge of the electrojet current location and temporal variations. In contrast, indices are derived from averaging that highly detailed information over broad regions and time windows. However, as shown in Figs. 4.102–4.105, sudden and dramatic dB/dt variations and onsets are the important drivers for ground-induction. Therefore, the index approach only "blurs" the induction process details. While NOAA and other governmental agency forecast products will continue to emphasize environmental assessment, impacted systems need additional translation of these broad environmental conditions into potential impacts on their respective systems. Advanced modeling techniques satisfy these industry-specific translation needs.

Since preventing the flow of GIC in power systems is usually not a viable threat mitigation strategy, a management plan to prepare the system for the stress imposed by a resulting geomagnetic is the most prudent course of action. Decisions on when to implement operation measures have been problematic in the past because of the inherent low quality of forecasts that have been provided. Further storm onsets can develop suddenly and as a result, some operation changes cannot be implemented in time to address the paramount priority of system reliability. Forecasts prior to the deployment of the ACE satellite have been very unreliable (less than a 40% accuracy rate with numerous false alarms as well as missed disturbances). The recent deployment of the ACE satellite will provide highly accurate and reliable advance warnings of solar wind conditions that will trigger geomagnetic storm conditions.

Extremely large magnitude magnetic field disturbances can be produced during the course of a severe geomagnetic disturbance. For accurate impact assessment to operational power system forecast users, the forecast needs to provide for the following forecast aspects: (1) a lead time of onset of the most severe portion of the storm event, (2) an intensity prediction of ground level B field deviation and resulting E field, and (3) an expected duration of severe storm conditions. Further, to provide client-specific impact assessment, the definition of the positional and intensity definition of the electrojet current structure has to be specified to provide for a first-order assessment of the ground-induction coupling potential to nearby transmission grids.

The forecasting of the electrojet current structure is a highly refined specification of the environment resulting during the course of a storm event. This refined electrojet environment data can then be used in an electromagnetic coupling model to ground-based systems to provide a first-order assessment of the storm event impact on the modeled ground-based system. System impact forecast assessment capability would provide specific power systems with the ability (through detailed modeling of their respective systems) to evaluate the threat potential of various levels of geomagnetic storm activity and the potential impact that the threat poses to reliable operation. In this example, the magnetosphere and ionosphere

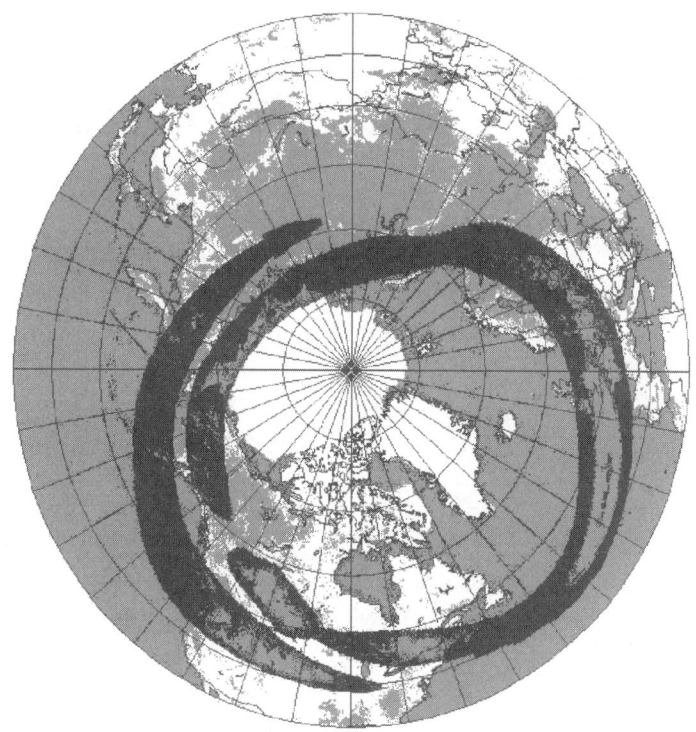

FIGURE 4.106 Ionospheric model output projecting 45 minutes in advance the electrojet current location and intensity for the May 4, 1998, storm at time UT 04:00. This model output is derived from a magnetospheric/ionospheric model that forecasts these storm patterns typically 45 minutes in advance. The model uses real-time solar wind data from the NASA ACE satellite and updates the model calculation in one-minute time steps.

modeling advances to derive the forecast electrojet current is then coupled with ground-induction modeling to the power system of interest to provide highly reliable and accurate calculations of GIC flows in the networks and resulting voltage regulation impacts with lead times of 45 min or more in advance. Figures 4.106 through 4.109 demonstrate the implementation of the "End-to-End" modeling of a storm-onset-to-power-system-impact forecast process.

This progression from solar wind inputs to forecast GIC flow and power system impact would allow power industry users to have specific impact ranges, magnitudes, and locations of the storm event. Then this forecast expected onset and duration could be incorporated into operational evaluation and state-estimation models. This will allow more precise implementation of power network storm operational measures (for example, transfer constraints and curtailments) when absolutely needed for network security, but will also prevent unnecessarily long periods of operational postures that are restrictive of the energy market functions (Kappenman et al., 1997; Albertson and Van Baelen, 1970).

References

Albertson, V.D. and Van Baelen, J.A., Electric and magnetic fields at the earth's surface due to auroral currents, *IEEE Trans. Power Apparatus Syst.*, PAS-89, 578-594, 1970.
Barnes, P.R. and Van Dyke, J.W., Potential economic costs from geomagnetic storms, *IEEE Special Publication*, 90th 0357-4-PWR, Geomagnetic Storm Cycle 22: Power System Problems on the Horizon, July 1990.
Boteler, D.H. and Jansen Van Beek, G., Mapping the March 13, 1989, magnetic disturbance and its consequences across North America, *Solar Terrestrial Predictions IV, Proceedings of a Workshop*, Ottawa, Canada, May 18-22, 1992, Volume 3, pages 57-70.

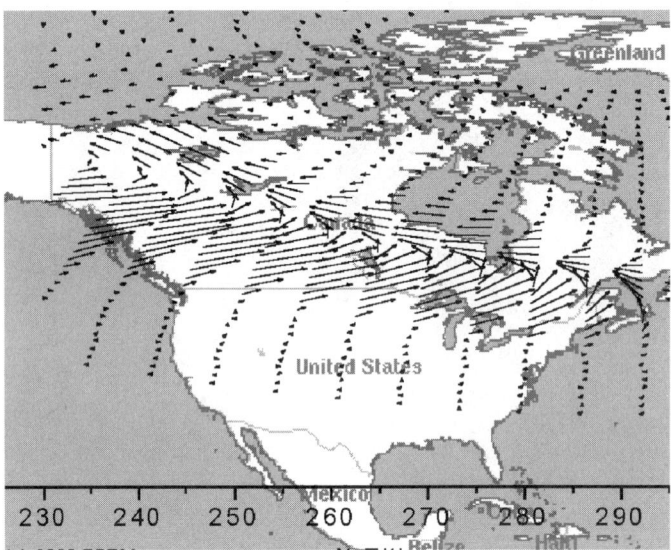

FIGURE 4.107 A regional view of the electrojet forecast example for May 4, 1998, at 05:00 UT. The electrojet current intensity and location as shown as vector equivalents that are used to electromagnetically couple to client-specific ground-based systems. This provides the first step in the calculation of estimated GIC and power system impact potential due to the geomagnetic storm.

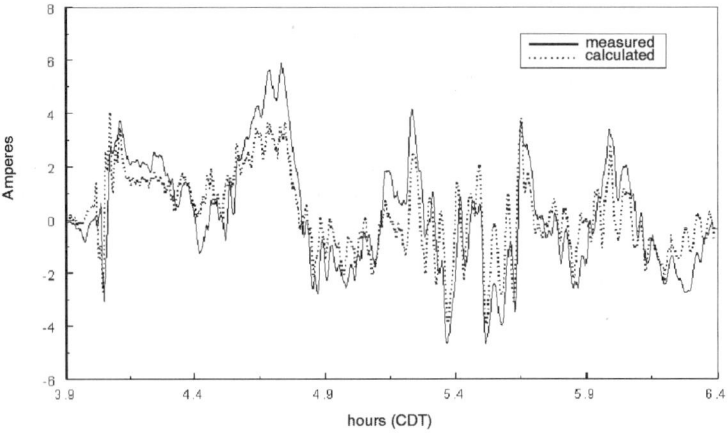

FIGURE 4.108 Validation of the ground-induction modeling accuracy has shown the ability to replicate GIC flow through transmission networks for storm events with reasonable accuracy over extended time history.

IEEE, The effects of GIC on protective relaying, IEEE Working Group K-11 Report, *IEEE Trans. Power Delivery,* 11(2), 725-739, 1996.

Joselyn J.A. et al., Panel achieves consensus prediction on Solar Cycle 23, EOS, *Trans. Am. Geophys. Union,* 78, 205, 211-212.

Kappenman, J.G., Geomagnetic storm forecasting mitigates power system impacts, *IEEE Power Eng. Rev.,* 4-7, November 1998.

Kappenman J.G. and Albertson, V.D., Bracing for the geomagnetic storms, *IEEE Spectrum Magazine,* March 1990.

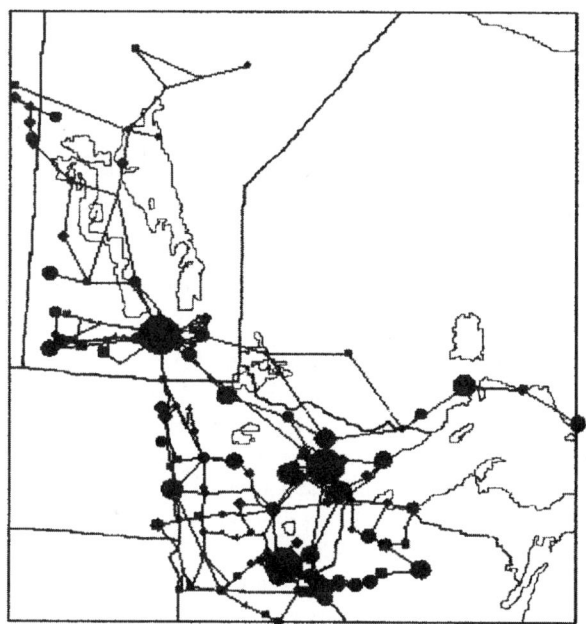

FIGURE 4.109 With the forecast of the electrojet location and intensity as described in Figs. 4.106 and 4.107, a model of the induction process can be used to calculate the flow of GIC in the power system of interest. With GIC flow calculated in network transformers, projections can be made on the number of transformers that will be driven into half-cycle saturation and the system reactive power losses (MVARs) that will occur.

Kappenman, J.G., Zanetti, L.J., and Radasky, W.A., Space weather from the users perspective: Geomagnetic storm forecasts and the power industry, EOS, *Trans. Am. Geophys. Union*, 78(4), 37, 41, 44-45, 1997.

Maynard, N.C., Space weather prediction, *Rev. Geophys.*, Supplement, U.S. National Report to International Union of Geodesy and Geophysics 1991-1994, pg. 547-557, July 1995.

4.10 Lightning Protection

William A. Chisholm

The study of lightning predates electric power systems by many centuries. Observations of thunder have been maintained in some areas for more than a millenium. Benjamin Franklin and others established the electrical nature of lightning and introduced the concepts of shielding and grounding to protect structures. Early power transmission lines used as many as six overhead shield wires, strung above the phase conductors and grounded at the towers for effective lightning protection. Later in the twentieth century, repeated strikes to tall towers, buildings, and power lines, contradicting the adage that "it never strikes twice," allowed systematic study of stroke current parameters. Improvements in electronics, computers, telecommunications, rocketry, and satellite technologies have all extended our knowledge about lightning, while at the same time exposing us to ever-increasing risks of economic damage from its consequences.

Ground Flash Density

The first negative, downward, cloud-to-ground lightning stroke is the dominant risk element to power system components. Positive first strokes, negative subsequent strokes, and continuing currents can also cause specific problems. A traditional indicator of cloud-to-ground lightning activity is given by thunder observations, collected to World Meteorological Organization standards and converted to Ground Flash Density (Anderson et al., 1984; MacGorman et al., 1984):

$$\text{GFD} = 0.04 \text{ TD}^{1.25} \tag{4.154}$$

$$\text{GFD} = 0.054 \text{ TH}^{1.1} \tag{4.155}$$

where

TD = number of days with thunder per year
TH = number of hours with thunder per year
GFD = number of first cloud-to-ground strokes per square kilometer per year

Long-term thunder data suggest that GFD has a relative standard deviation of 30%.

Electromagnetic signals from lightning are unique and have a high signal-to-noise ratio at large distances. Many single-station lightning flash counters have been developed and calibrated, each with good discrimination between cloud-flash and ground-flash activity using simple electronic circuits (Heydt, 1982). It has also been feasible for more than twenty years (Krider et al., 1976) to observe these signals with two or more stations and to triangulate lightning stroke locations on a continent-wide basis. Lightning location networks (Global Atmospherics, Inc., Website, 2000) have improved continuously to the point where multiple ground strikes from a single flash can be resolved with high spatial and temporal accuracy and high probability of detection. A GFD value from these data should be based on approximately 400 counts in each cell to reduce relative standard deviation of the observation process below 5%. In areas with moderate flash density, a minimum cell size of 20 × 20 km is appropriate.

In areas where there are presently no commercial lightning location networks, Optical Transient Detector (OTD) observations from low-earth orbit offer some merit. This class of instrument, first operated systematically in 1995 (GHCC Website, 2000), provides an unbiased quantitative sample of lightning activity over most of the globe. The OTD responds to both cloud and ground flashes, and has some blind areas at the poles and in the south Atlantic near Brazil. However, comparisons of orographic lightning features, such as an elevated flash density to the east of the Rocky Mountains in North America, confirm and extend important trends seen in limited ground-based network observations. The OTD technology is also improving as new satellites build on the successful experiences.

Stroke Incidence to Power Lines

The lightning leader, a thin column of electrically charged plasma, develops from cloud down to the ground in a series of step breakdowns (Uman, 1987). Near the ground, electric fields are high enough to satisfy the conditions for continuous positive leader inception upward from tall objects or conductors. Analysis of a single overhead conductor with this approach (Rizk, 1990) leads to:

$$N_S = 3.8 \text{ GFD } h^{0.45} \tag{4.156}$$

where

N_s = the number of strikes to the conductor per 100 km of line length per year
h = the average height of the conductor above ground, in meters

In areas of moderate to high ground flash density, one or more overhead shield wires are usually installed above the phase conductors. This shielding usually has a success rate of greater than 95%, but adds nearly 10% to the cost of line construction and also wastes energy from induced currents. The leader inception model (Rizk, 1990) has also been used to analyze shielding failures.

Stroke Current Parameters

Once the downward leader contacts a power system component through an upward-connecting leader, the stored charge will be impressed through a high channel impedance of 600 to 2000 Ω. With this high source impedance, an impulse current source model is suitable.

Berger made the most reliable direct measurements of negative downward cloud-to-ground lightning parameters on an instrumented tower from 1947 to 1977 (Berger, 1977). Additional observations have been provided by many researchers and then summarized (Anderson and Eriksson, 1980; CIGRE, 1991). The overall stroke current distribution can be approximated (CIGRE, 1991) as log-normal with a mean of 31 kA and a log standard deviation of 0.48. The waveshape rises with a concave front, giving the maximum steepness near the crest of the wave, then decays with a time to half-value of 50 μs or more. The median value of maximum steepness (CIGRE, 1991) is 24 kA/μs, with a log standard deviation of 0.60. Steepness has a positive correlation to the peak amplitude (CIGRE, 1991) that allows simplified modeling using a single equivalent front time (peak current divided by peak rate of rise). The mean equivalent front is 1.4 μs for the median 31-kA current, rising to 2.7 μs as peak stroke current increases to the 5% level of 100 kA (CIGRE, 1991). An equivalent front time of 2 μs is recommended for simplified analysis (IEEE, 1997).

Calculation of Lightning Overvoltages on Shielded Lines

The voltage rise V_R of the ground resistance R at each tower will be proportional to peak stroke current: $V_R = R\,I$. The relation between the tower base geometry and its resistance is:

$$R = \frac{\rho}{2\pi s} \ln\left(\frac{17 s^2}{A}\right) + \frac{\rho}{l} \quad (4.157)$$

where

ρ = soil resistivity (Ω-m)
s = the three-dimensional distance from the center to its outermost point (m)
A = the surface area (sides + base) of the hole needed to excavate the electrode (m²)
l = the length (m) of the wire used to make up the electrode (infinite for solid electrodes)

For large surge currents, local ionization will tend to reduce the second ρ/l contact resistance term but not the first geometric resistance term in Eq. (4.157).

The voltage rise V_L associated with conductor and tower series inductance L and the equivalent front time (dt ~ 2 μs) is V_L ~ L I/dt. The V_L term will add to, and sometimes dominate, V_R. Lumped inductance can be approximated from the expression:

$$L = Z\,\tau = 60 \ln\left(\frac{2h}{r}\right) \times \frac{l}{c} \quad (4.158)$$

where

L = the inductance (H)
Z = the element antenna impedance (Ω)
h = the wire height above conducting ground (m)
r = the wire radius (m)
l = the wire length (m)
c = the speed of light (3×10^8 m/s)

In numerical analyses, series and shunt impedance elements can be populated using the same procedure. Tall transmission towers have longer travel times and thus higher inductance, which further exacerbates the increase of stroke incidence with line height. The high electromagnetic fields surrounding any stricken conductor will induce currents and couple voltages in nearby, unstricken conductors through their mutual surge impedances. In the case where lightning strikes a grounded overhead shield wire, this coupling increases common-mode voltage and reduces differential voltage across insulators. Additional shield wires and corona (CIGRE, 1991; IEEE, 1997) can improve this desirable surge-impedance coupling to mitigate half of the total tower potential rise ($V_R + V_L$).

The strong electromagnetic fields from vertical lightning strokes can induce large overvoltages in nearby overhead lines without striking them directly. This is a particular concern only for MV and LV systems.

Insulation Strength

Power system insulation is designed to withstand all anticipated power system overvoltages. Unfortunately, even the weakest direct stroke from a shielding failure to a phase conductor will cause a lightning flashover. Once an arc appears across an insulator, the power system fault current keeps this arc alive until voltage is removed by protective relay action. Effective overhead shielding is essential on transmission lines in areas with moderate to high ground flash density.

When the overhead shield wire is struck, the potential difference on insulators is the sum of the resistive and inductive voltage rises on the tower, minus the coupled voltage on the phase conductors. The potential difference can lead to a "backflashover" from the tower to the phase conductor. Backflashover is more frequent when the stroke current is large (5% > 100 kA), when insulation strength is low (<1 m or 600 kV Basic Impulse Level) and/or when footing resistance is high (>30 Ω). Simplified models (CIGRE, 1991; IEEE, 1997) are available to carry out the overvoltage calculations and coordinate the results with insulator strength, giving lightning outage rates in units of interruptions per 100 km per year.

Mitigation Methods

Lightning mitigation methods need to be appropriate for the expected long-term ground flash density and power system reliability requirements. Table 4.27 summarizes typical practices at five different levels of lightning activity to achieve a reliability of one outage per 100 km of line per year on an HV line.

TABLE 4.27 Lightning Mitigation Methods

Ground Flash Density Range	Typical Design Approaches
0.1–0.3 ground flashes/km² per year	Unshielded, one- or three-pole reclosing
0.3–1 ground flashes/km² per year	Single overhead shield wire
1–3 ground flashes/km² per year	Two overhead shield wires
3–10 ground flashes/km² per year	Two overhead shield wires with good grounding or line surge arresters
10–30 ground flashes/km² per year	Three or more overhead and underbuilt shield wires with good grounding; line surge arresters; underground transmission cables

Note: Designs to achieve reliability of one outage per 100 km of line per year on an HV line.

Conclusion

Direct lightning strokes to any overhead transmission line are likely to cause impulse flashover of supporting insulation, leading to a circuit interruption. The use of overhead shield wires, located above the phase conductors and grounded adequately at each tower, can reduce the risk of flashover by 95–99.5%, depending on system voltage.

References

Anderson, R. B. and Eriksson, A. J., Lightning parameters for engineering applications, *Electra* No. 69, 65-102, 1980.

Anderson, R. B., Eriksson, A. J., Kroninger, H., Meal, D. V., Lightning and thunderstorm parameters, IEEE Conference Publication 236, *Lightning and Power Systems*, London, June 1984.

Berger, K., The earth flash, in *Lightning*, Golde, R., Ed., Academic Press, London, 119-190, 1977.

CIGRE Working Group 01 (Lightning) of Study Committee 33, Guide to Procedures for Estimating the Lightning Performance of Transmission Lines, *CIGRE Brochure 63*, Paris, October 1991.

Global Atmospherics, Inc., Global Atmospherics, Inc., April 10, 2000, ⟨http://www.glatmos.com⟩.

Heydt, G., Instrumentation, in *Handbook of Atmospherics*, Volume II, Volland, H., Ed., CRC Press, Boca Raton, FL, 203-256, 1982.

IEEE Guide for Improving the Lightning Performance of Transmission Lines, IEEE Standard 1243-1997, December 1997.

Krider, E.P., Noggle, R. C, and Uman, M. A., A gated, wideband direction finder for lightning return strokes, *J. Appl. Meteor.*, 15, 301, 1976.

Lightning and Atmospheric Electricity Research at the GHCC, Global Hydrology and Climate Center (GHCC), April 10, 2000, ⟨http://thunder.msfc.nasa.gov/⟩.

MacGorman, D. R., Maier, M. W., Rust, W. D., Lightning Strike Density for the Contiguous United States from Thunderstorm Duration Records. Report to U.S. Nuclear Regulatory Commission, NUREG/CR-3759, 1984.

Rizk, F. A. M., Modeling of transmission line exposure to direct lightning strokes, *IEEE Trans PWRD*, 5(4), 1983, 1990.

Uman, M. A., *The Lightning Discharge*, Academic Press, New York, 1987.

4.11 Reactive Power Compensation

Rao S. Thallam

The Need for Reactive Power Compensation

Except in a very few special situations, electrical energy is generated, transmitted, distributed, and utilized as alternating current (AC). However, alternating current has several distinct disadvantages. One of these is the necessity of reactive power that needs to be supplied along with active power. Reactive power can be leading or lagging. While it is the active power that contributes to the energy consumed, or transmitted, reactive power does not contribute to the energy. Reactive power is an inherent part of the "total power." Reactive power is either generated or consumed in almost every component of the system, generation, transmission, and distribution and eventually by the loads. The impedance of a branch of a circuit in an AC system consists of two components, resistance and reactance. Reactance can be either inductive or capacitive, which contribute to reactive power in the circuit. Most of the loads are inductive, and must be supplied with lagging reactive power. It is economical to supply this reactive power closer to the load in the distribution system.

In this section, reactive power compensation, mainly in transmission systems installed at substations, is discussed. Reactive power compensation in power systems can be either shunt or series. Both will be discussed.

Shunt Reactive Power Compensation

Since most loads are inductive and consume lagging reactive power, the compensation required is usually supplied by leading reactive power. Shunt compensation of reactive power can be employed either at load level, substation level, or at transmission level. It can be capacitive (leading) or inductive (lagging) reactive power, although in most cases as explained before, compensation is capacitive. The most common form of leading reactive power compensation is by connecting shunt capacitors to the line.

Shunt Capacitors

Shunt capacitors are employed at substation level for the following reasons:

1. Voltage regulation: The main reason that shunt capacitors are installed at substations is to control the voltage within required levels. Load varies over the day, with very low load from midnight to early morning and peak values occurring in the evening between 4 PM and 7 PM. Shape of the load curve also varies from weekday to weekend, with weekend load typically low. As the load varies, voltage at the substation bus and at the load bus varies. Since the load power factor is always lagging, a shunt connected capacitor bank at the substation can raise voltage when the load is high. The shunt capacitor banks can be permanently connected to the bus (fixed capacitor bank) or can be switched as needed. Switching can be based on time, if load variation is predictable, or can be based on voltage, power factor, or line current.
2. Reducing power losses: Compensating the load lagging power factor with the bus connected shunt capacitor bank improves the power factor and reduces current flow through the transmission lines, transformers, generators, etc. This will reduce power losses (I^2R losses) in this equipment.
3. Increased utilization of equipment: Shunt compensation with capacitor banks reduces kVA loading of lines, transformers, and generators, which means with compensation they can be used for delivering more power without overloading the equipment.

Reactive power compensation in a power system is of two types — shunt and series. Shunt compensation can be installed near the load, in a distribution substation, along the distribution feeder, or in a transmission substation. Each application has different purposes. Shunt reactive compensation can be inductive or capacitive. At load level, at the distribution substation, and along the distribution feeder, compensation is usually capacitive. In a transmission substation, both inductive and capacitve reactive compensation are installed.

Application of Shunt Capacitor Banks in Distribution Systems — A Utility Perspective

The Salt River Project (SRP) is a public power utility serving more than 720,000 (April 2000) customers in central Arizona. Thousands of capacitor banks are installed in the entire distribution system. The primary usage for capacitor banks in the distribution system is to maintain a certain power factor at peak loading conditions. The target power factor is .98 leading at system peak. This figure was set as an attempt to have a unity power factor on the 69-kV side of the substation transformer. The leading power factor compensates for the industrial substations that have no capacitors. The unity power factor maintains a balance with ties to other utilities.

The main purpose of the capacitors is not for voltage support, as the case may be at utilities with long distribution feeders. Most of the feeders in the SRP service area do not have long runs (substations are about two miles apart) and load tap changers on the substation transformers are used for voltage regulation.

The SRP system is a summer peaking system. After each summer peak, a capacitor study is performed to determine the capacitor requirements for the next summer. The input to the computer program for evaluating capacitor additions consists of three major components:

- Megawatts and megavars for each substation transformer at peak.
- A listing of the capacitor banks with size and operating status at time of peak.
- The next summer's projected loads.

By looking at the present peak MW and Mvars and comparing the results to the projected MW loads, Mvar deficiencies can be determined. The output of the program is reviewed and a listing of potential needs is developed. The system operations personnel also review the study results and their input is included in making final decisions about capacitor bank additions.

Once the list of additional reactive power requirements is finalized, determinations are made about the placement of each bank. The capacitor requirement is developed on a per-transformer basis. The

Transmission System

TABLE 4.28 Number and Size of Capacitor Banks in the SRP System

Kvar	Number of Banks	
	Line	Station
150	1	
300	140	
450	4	
600	758	2
900	519	
1200	835	581
Total	2257	583

TABLE 4.29 SRP Line Capacitors by Type of Control

Type of Control	Number of Banks
Current	4
Fixed	450
Time	1760
Temperature	38 (used as fixed)
Voltage	5

ratio of the kvar connected to kVA per feeder, the position on the feeder of existing capacitor banks, and any concentration of present or future load are all considered in determining the position of the new capacitor banks. All new capacitor banks are 1200 kvar. The feeder type at the location of the capacitor bank determines if the capacitor will be pole-mounted (overhead) or pad-mounted (underground).

Capacitor banks are also requested when new feeders are being proposed for master plan communities, large housing developments, or heavy commercial developments.

Table 4.28 shows the number and size of capacitor banks in the SRP system in 1998. Table 4.29 shows the number of line capacitors by type of control.

Substation capacitor banks (three or four per transformer) are usually staged to come on and go off at specific load levels.

Static VAR Control (SVC)

Static VAR compensators, commonly known as SVCs, are shunt connected devices, vary the reactive power output by controlling or switching the reactive impedance components by means of power electronics. This category includes the following equipment:

Thyristor controlled reactors (TCR) with fixed capacitors (FC)
Thyristor switched capacitors (TSC)
Thyristor controlled reactors in combination with mechanically or Thyristor switched capacitors

SVCs are installed to solve a variety of power system problems:

1. Voltage regulation
2. Reduce voltage flicker caused by varying loads like arc furnace, etc.
3. Increase power transfer capacity of transmission systems
4. Increase transient stability limits of a power system
5. Increase damping of power oscillations
6. Reduce temporary overvoltages
7. Damp subsynchronous oscillations

A view of an SVC installation is shown in Figure 4.110.

FIGURE 4.110 View of static VAR compensator (SVC) installation. (Photo courtesy of ABB.)

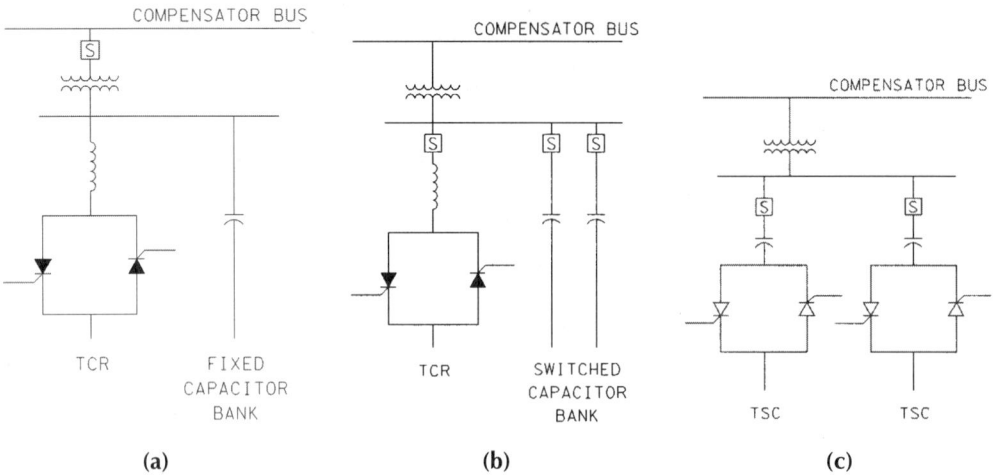

FIGURE 4.111 Three versions of SVC. (a) TCR with fixed capacitor bank; (b) TCR with switched capacitor banks; and (c) thyristor switched capacitor compensator.

Description of SVC

Figure 4.111 shows three basic versions of SVC. Figure 4.111a shows configuration of TCR with fixed capacitor banks. The main components of a SVC are thyristor valves, reactors, the control system, and the step-down transformer.

How Does SVC Work?

As the load varies in a distribution system, a variable voltage drop will occur in the system impedance, which is mainly reactive. Assuming the generator voltage remains constant, the voltage at the load bus

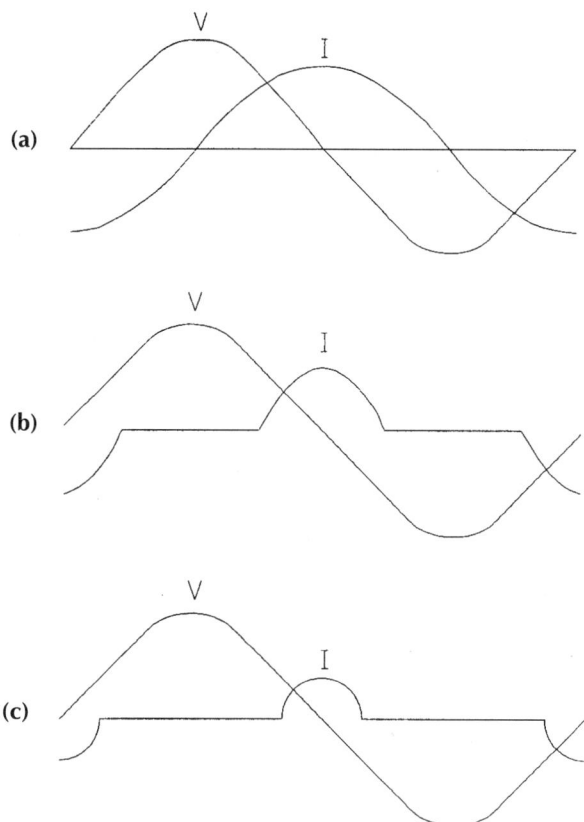

FIGURE 4.112 TCR voltage (V) and current (I) waveforms for three conduction levels. Thyristor gating angle = α; conduction angle = σ. (a) $\alpha = 90°$ and $\sigma = 180°$; (b) $\alpha = 120°$ and $\sigma = 120°$; and (c) $\alpha = 150°$ and $\sigma = 60°$.

will vary. The voltage drop is a function of the reactive component of the load current, and system and transformer reactance. When the loads change very rapidly, or fluctuate frequently, it may cause "voltage flicker" at the customers' loads. Voltage flicker can be annoying and irritating to customers because of the "lamp flicker" it causes. Some loads can also be sensitive to these rapid voltage fluctuations.

An SVC can compensate voltage drop for load variations and maintain constant voltage by controlling the duration of current flow in each cycle through the reactor. Current flow in the reactor can be controlled by controlling the gating of thyristors that control the conduction period of the thyristor in each cycle, from zero conduction (gate signal off) to full-cycle conduction. In Fig. 4.111a, for example, assume the MVA of the fixed capacitor bank is equal to the MVA of the reactor when the reactor branch is conducting for full cycle. Hence, when the reactor branch is conducting full cycle, the net reactive power drawn by the SVC (combination of capacitor bank and thyristor controlled reactor) will be zero. When the load reactive power (which is usually inductive) varies, the SVC reactive power will be varied to match the load reactive power by controlling the duration of the conduction of current in the thyristor controlled reactive power branch. Figure 4.112 shows current waveforms for three conduction levels, 60, 120 and 180°. Figure 4.112a shows waveforms for thyristor gating angle (α) of 90°, which gives a conduction angle (σ) of 180° for each thyristor. This is the case for full-cycle conduction, since the two back-to-back thyristors conduct in each half-cycle. This case is equivalent to shorting the thyristors. Figure 4.112b is the case when the gating signal is delayed for 30° after the voltage peak, and results in a conduction angle of 120°. Figure 4.112c is the case for $\alpha = 150°$ and $\sigma = 60°$.

With a fixed capacitor bank as shown in Figure 4.111a, it is possible to vary the net reactive power of the SVC from 0 to the full capacitive VAR only. This is sufficient for most applications of voltage regulation, as in most cases only capacitive VARs are required to compensate the inductive VARs of the load. If the

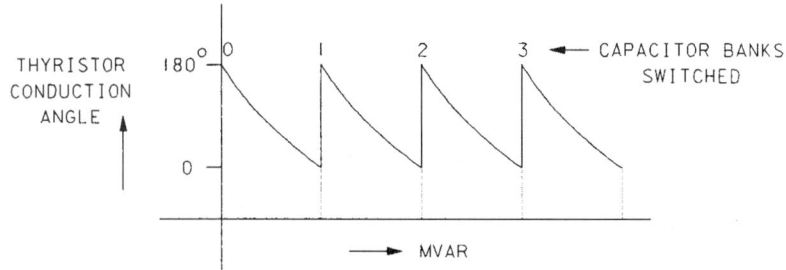

FIGURE 4.113 Reactive power variation of TCR with switched capacitor banks.

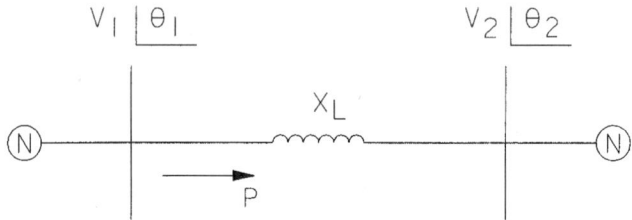

FIGURE 4.114 Power flow through transmission line.

capacitor can be switched on and off, the MVAR can be varied from full inductive to full capacitive, up to the rating of the inductive and capacitive branches. The capacitor bank can be switched by mechanical breakers (see Fig. 4.111b) if time delay (usually five to ten cycles) is not a consideration, or they can be switched fast (less than one cycle) by thyristor switches (see Fig. 4.111c).

Reactive power variation with switched capacitor banks for an SVC is shown in Fig. 4.113.

Series Compensation

Series compensation is commonly used in high-voltage AC transmission systems. They were first installed in that late 1940s. Series compensation increases power transmission capability, both steady state and transient, of a transmission line. Since there is increasing opposition from the public to construction of EHV transmission lines, series capacitors are attractive for increasing the capabilities of transmission lines. Series capacitors also introduce some additional problems for the power system. These will be discussed later.

Power transmitted through the transmission system (shown in Fig. 4.114) is given by:

$$P_2 = \frac{V_1 \cdot V_2 \cdot \sin\delta}{X_L} \qquad (4.159)$$

where

P_2 = Power transmitted through the transmission system
V_1 = Voltage at sending end of the line
V_2 = Voltage at receiving end of transmission line
X_L = Reactance of the transmission line
δ = Phase angle between V_1 and V_2

Equation (4.159) shows that if the total reactance of a transmission system is reduced by installing capacitance in series with the line, the power transmitted through the line can be increased.

With a series capacitor installed in the line, Eq. (4.159) can be written as

$$P_2 = \frac{V_1 \cdot V_2 \cdot \sin\delta}{X_L - X_C} \qquad (4.160)$$

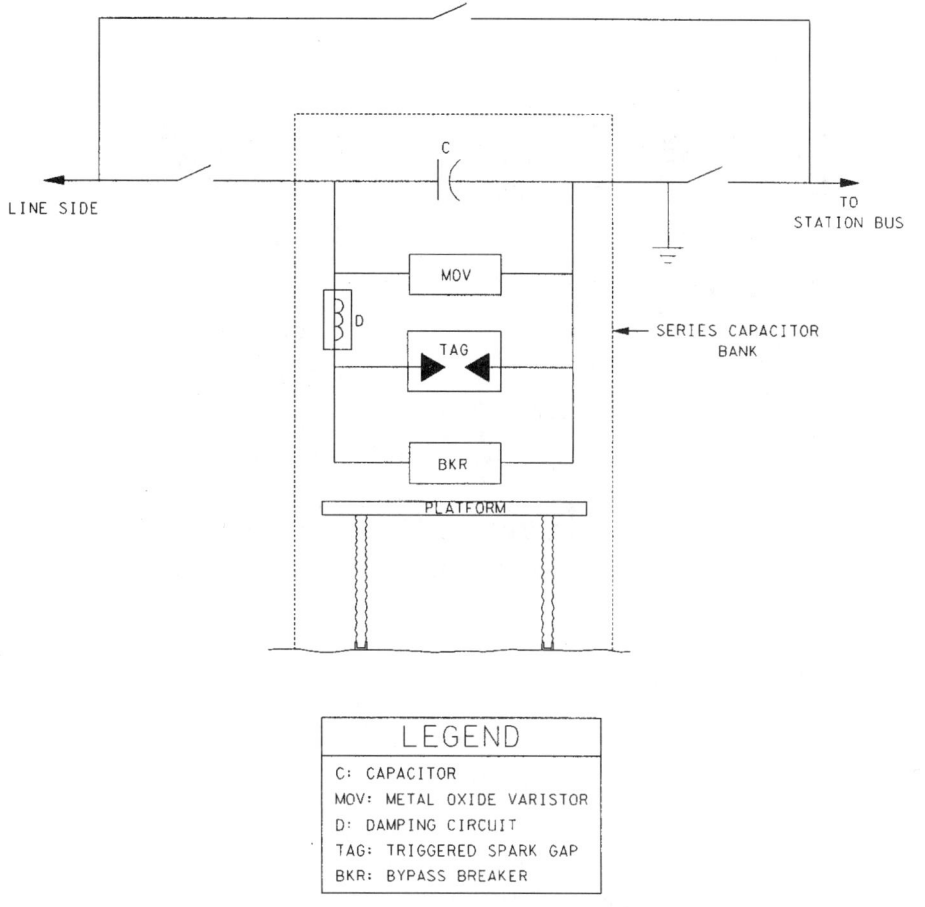

FIGURE 4.115 Schematic one-line diagram of series capacitor bank.

$$= \frac{V_1 \cdot V_2 \cdot \sin\delta}{X_L(1-K)} \qquad (4.161)$$

where $K = \dfrac{X_C}{X_L}$ is degree of the compensation, usually expressed in percent. A 70% series compensation means the value of the series capacitor in ohms is 70% of the line reactance.

Series Capacitor Bank

A series capacitor bank consists of a capacitor bank, overvoltage protection system, and a bypass breaker, all elevated on a platform, which is insulated for the line voltage. See Fig. 4.115. The overvoltage protection is comprised of a zinc oxide varistor and a triggered spark gap, which are connected in parallel to the capacitor bank, and a damping reactor. Prior to the development of the high-energy zinc oxide varistor in the 1970s, a silicon carbide nonlinear resistor was used for overvoltage protection. Silicon carbide resistors require a spark gap in series because the nonlinearity of the resistors is not high enough. The zinc oxide varistor has better nonlinear resistive characteristics, provides better protection, and has become the standard protection system for series capacitor banks.

The capacitor bank is usually rated to withstand the line current for normal power flow conditions and power swing conditions. It is not economical to design the capacitors to withstand the currents and

FIGURE 4.116 Aerial view of 500-kV series capacitor installation. (Photo courtesy of ABB.)

voltages associated with faults. Under these conditions capacitors are protected by a metal oxide varistor (MOV) bank. The MOV has a highly nonlinear resistive characteristic and conducts negligible current until the voltage across it reaches the protective level. For internal faults, which are defined as faults within the line section in which the series capacitor bank is located, fault currents can be very high. Under these conditions, both the capacitor bank and MOV will be bypassed by the "triggered spark gap." The damping reactor (D) will limit the capacitor discharge current and damps the oscillations caused by spark gap operation or when the bypass breaker is closed. The amplitude, frequency of oscillation, and rate of damping of the capacitor discharge current will be determined by the circuit parameters, C (series capacitor), L (damping inductor), and resistance in the circuit, which in most cases is losses in the damping reactor.

A view of series capacitor bank installation is shown in Fig. 4.116.

Description of Main Components

Capacitors

The capacitor bank for each phase consists of several capacitor units in series-parallel arrangement, to make up the required voltage, current, and Mvar rating of the bank. Each individual capacitor unit has one porcelain bushing. The other terminal is connected to the stainless steel casing. The capacitor unit usually has a built-in discharge resistor inside the case. Capacitors are usually all film design with insulating fluid that is non-PCB. Two types of fuses are used for individual capacitor units — internally fused or externally fused. Externally fused units are more commonly used in the U.S. Internally fused capacitors are prevalent in European installations.

Metal Oxide Varistor (MOV)

A metal oxide varistor is built from zinc oxide disks in series and parallel arrangement to achieve the required protective level and energy requirement. One to four columns of zinc oxide disks are installed in each sealed porcelain container, similar to a high-voltage surge arrester. A typical MOV protection system contains several porcelain containers, all connected in parallel. The number of parallel zinc oxide disk columns required depends on the amount of energy to be discharged through the MOV during the worst-case design scenario. Typical MOV protection system specifications are as follows.

The MOV protection system for the series capacitor bank is usually rated to withstand energy discharged for all faults in the system external to the line section in which the series capacitor bank is located.

Faults include single-phase, phase-to-phase, and three-phase faults. The user should also specify the fault duration. Most of the faults in EHV systems will be cleared by the primary protection system in 3 to 4 cycles. Back-up fault clearing can be from 12 to 16 cycles duration. The user should specify whether the MOV should be designed to withstand energy for back-up fault clearing times. Sometimes it is specified that the MOV be rated for all faults with primary protection clearing time, but for only single-phase faults for back-up fault clearing time. Statistically, most of the faults are single-phase faults.

The energy discharged through the MOV is continuously monitored and if it exceeds the rated value, the MOV will be protected by the firing of a triggered air gap, which will bypass the MOV.

Triggered Air Gap

The triggered air gap provides a fast means of bypassing the series capacitor bank and the MOV system when the trigger signal is issued under certain fault conditions (for example, internal faults) or when the energy discharged through the MOV exceeds the rated value. It typically consists of a gap assembly of two large electrodes with an air gap between them. Sometimes two or more air gaps in series can also be employed. The gap between the electrodes is set such that the gap assembly sparkover voltage without trigger signal will be substantially higher than the protective level of the MOV, even under the most unfavorable atmospheric conditions.

Damping Reactor

A damping reactor is usually an air-core design with parameters of resistance and inductance to meet the design goal of achieving the specified amplitude, frequency, and rate of damping. The capacitor discharge current when bypassed by a triggered air gap or a bypass breaker will be damped oscillation with amplitude, rate of damping, and frequency determined by circuit parameters.

Bypass Breaker

The bypass breaker is usually a standard line circuit breaker with a rated voltage based on voltage across the capacitor bank. In most of the installations, the bypass breaker is located separate from the capacitor bank platform and outside the safety fence. This makes maintenance easy. Both terminals of the breaker standing on insulator columns are insulated for the line voltage. It is usually a SF_6 puffer-type breaker, with controls at ground level.

Relay and Protection System

The relay and protection system for the capacitor bank is located at ground level, in the station control room, with information from and to the platform transmitted via fiber-optic cables. The present practice involves all measured quantities on the platform being transmitted to ground level, with all signal processing done at ground level.

Subsynchronous Resonance

Series capacitors, when radially connected to the transmission lines from the generation near by, can create a subsynchronous resonance (SSR) condition in the system under some circumstances. SSR can cause damage to the generator shaft and insulation failure of the windings of the generator. This phenomenon is well-described in several textbooks, given in the reference list at the end of this section.

Adjustable Series Compensation (ASC)

The ability to vary the series compensation will give more control of power flow through the line, and can improve the dynamic stability limit of the power system. If the series capacitor bank is installed in steps, bypassing one or more steps with bypass breakers can change the amount of series compensation of the line. For example, as shown in Fig. 4.117, if the bank consists of 33% and 67% of the total compensation, four steps, 0%, 33%, 67%, and 100%, can be obtained by bypassing both banks, smaller bank (33%), larger bank (67%), and not bypassing both banks, respectively.

Varying the series compensation by switching with mechanical breakers is slow, which is acceptable for control of steady-state power flow. However, for improving the dynamic stability of the system, series compensation has to be varied quickly. This can be accomplished by thyristor controlled series compensation (TCSC).

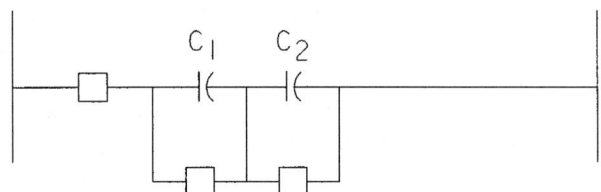

FIGURE 4.117 Breaker controlled variable series compensation.

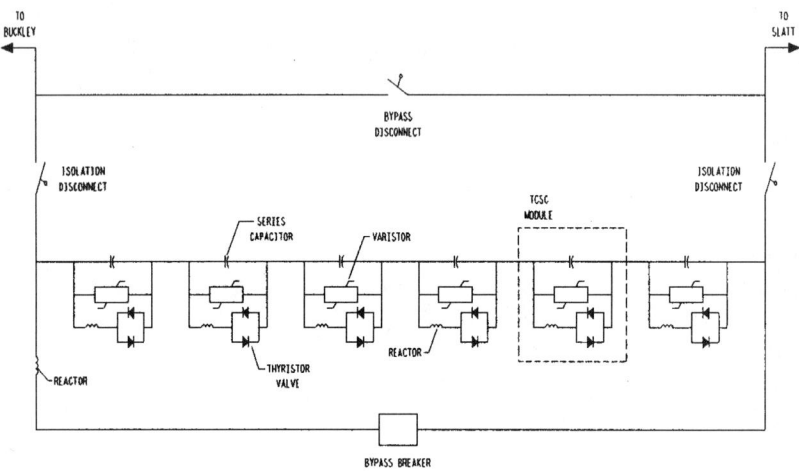

FIGURE 4.118 One-line diagram of TCSC installed at slatt substation.

Thyristor Controlled Series Compensation (TCSC)

Thyristor controlled series compensation provides fast control and variation of the impedance of the series capacitor bank. To date (1999), three prototype installations, one each by ABB, Siemens, and the General Electric Company (GE), have been installed in the U.S. TCSC is part of the Flexible AC Transmission System (FACTS), which is an application of power electronics for control of the AC system to improve the power flow, operation, and control of the AC system. TCSC improves the system performance for subsynchronous resonance damping, power swing damping, transient stability, and power flow control.

The latest of the three prototype installations is the one at the Slatt 500-kV substation in the Slatt-Buckley 500-kV line near the Oregon-Washington border in the U.S. This is jointly funded by the Electric Power Research Institute (EPRI), the Bonneville Power Administration (BPA), and the General Electric Company (GE). A one-line diagram of the Slatt TCSC is shown in Fig. 4.118. The capacitor bank (8 ohms) is divided into six identical TCSC modules. Each module consists of a capacitor (1.33 ohms), back-to-back thyristor valves controlling power flow in both directions, a reactor (0.2 ohms), and a varistor. The reactors in each module, in series with thyristor valves, limit the rate of change of current through the thyristors. The control of current flow through the reactor also varies the impedance of the combined capacitor-reactor combination, giving the variable impedance. When thyristor gating is blocked, complete line current flows through the capacitance only, and the impedance is 1.33 ohms capacitive (see Fig. 4.119a). When the thyristors are gated for full conduction (Fig. 119b), most of the line current flows through the reactor-thyristor branch (a small current flows through the capacitor) and the resulting impedance is 0.12 ohms inductive. If thyristors are gated for partial conduction only (Fig. 4.119c), circulating current will flow between capacitor and inductor, and the impedance can be varied from 1.33 ohms and 4.0 ohms, depending on the angle of conduction of the thyristor valves. The latter is called the vernier operating mode.

Transmission System

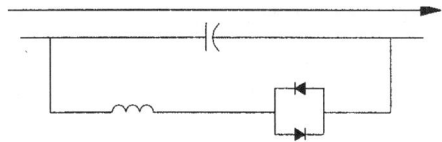

(a) No Thyristor Valve Current (Gating Blocked).

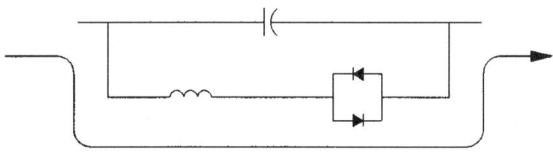

(b) Bypassed With Thyristor.

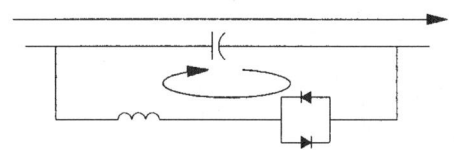

(c) Inserted With Vernier Control,
Circulating Some Current Through
Thyristor Valve.

FIGURE 4.119 Current flow during various operating modes of TCSC.

The complete capacitor bank with all six modules can be bypassed by the bypass breaker. This bypass breaker is located outside the main capacitor bank platform, similar to the case for the conventional series capacitor bank. There is also a reactor connected in series with the bypass breaker to limit the magnitude of capacitor discharge current through the breaker. All reactors are of air-core dry-type design and rated for the full line current rating. Metal oxide varistors (MOV) connected in parallel with the capacitors in each module provide overvoltage protection. The MOV for a TCSC requires significantly less energy absorption capability than is the case for a conventional series capacitor of comparable size, because gating of thyristor valves provides quick protection for faulted conditions.

STATic COMpensator (STATCOM)

STATCOM provides variable reactive power from lagging to leading, but with no inductors or capacitors for var generation. Reactive power generation is achieved by regulating the terminal voltage of the converter. The STATCOM consists of a voltage source inverter using gate turn-off thyristors (GTOs) which produces an alternating voltage source in phase with the transmission voltage, and is connected to the line through a series inductance which can be the transformer leakage inductance required to match the inverter voltage with line voltage. If the terminal voltage (V_t) of the voltage source inverter is higher than the bus voltage, STATCOM generates leading reactive power. If V_t is lower than the bus voltage, STATCOM generates lagging reactive power. The performance is similar to the performance of a synchronous condenser (unloaded synchronous motor with varying excitation).

Reactive power generated or absorbed by STATCOM is not a function of the capacitor on the DC bus side of the inverter. The capacitor is rated to limit only the ripple current, and hence the harmonics in the output voltage.

The first demonstration STATCOM of ±100 Mvar rating was installed at the Tennessee Valley Authority's Sullivan substation in 1994.

Defining Terms

Shunt capacitor bank: A large number of capacitor units connected in series and parallel arrangement to make up the required voltage and current rating, and connected between the high-voltage line and ground, between line and neutral, or between line-to-line.

Voltage flicker: Commonly known as "flicker" and "lamp flicker," this is a rapid and frequent fluctuation of supply voltage that causes lamps to flicker. Lamp flicker can be annoying, and some loads are sensitive to these frequent voltage fluctuations.

Subsynchronous resonance: Per IEEE, subsynchronous resonance is an electric power system condition where the electric network exchanges energy with a turbine generator at one or more of the natural frequencies of the combined system below the synchronous frequency of the system.

References

Anderson, P.M., Agrawal, B.L., and Van Ness, J.E., *Subsynchronous Resonance in Power Systems*, IEEE Press, 1990.

Anderson, P.M. and Farmer, R.G., *Series Compensation in Power Systems*, PBLSH! Inc. 1996.

Gyugyi, L., Otto, R.A., and Putman, T.H., Principles and application of thyristor-controlled shunt compensators, *IEEE Trans. on Power Appar. and Syst.*, 97, 1935-1945, Sept/Oct 1978.

Gyugyi, L. and Taylor, Jr., E.R., Characteristics of static thyristor-controlled shunt compensators for power transmission applications, *IEEE Trans. on Power Appar. and Syst.*, PAS-99, 1795-1804, 1980.

Hammad, A.E., Analysis of power system stability enhancement by static VAR compensators, *IEEE Trans. on Power Syst.*, 1, 222-227, 1986.

Miller, T.J.E., Ed., *Reactive Power Control in Electric Systems*, John Wiley & Sons, New York, 1982.

Miske, Jr., S.A. et al., Recent Series Capacitor Applications in North America, Paper presented at *CEA Electricity '95 Vancouver Conference*, March 1995.

Padiyar, K.R., *Analysis of Subsynchronous Resonance in Power Systems*, Kluwer Academic Publishers, 1999.

Schauder, C. et al., Development of a ±100 MVAR static condenser for voltage control of transmission systems, *IEEE Trans. on Power Delivery*, 10(3), 1486-1496, July 1995.

5
Substations

John D. McDonald
KEMA Consulting

5.1   Gas Insulated Substations   *Philip Bolin* 5-2

5.2   Air Insulated Substations — Bus/Switching Configurations   *Michael J. Bio* 5-18

5.3   High-Voltage Switching Equipment   *David L. Harris* 5-23

5.4   High-Voltage Power Electronics Substations   *Gerhard Juette* 5-29

5.5   Considerations in Applying Automation Systems to Electric Utility Substations
       James W. Evans 5-41

5.6   Substation Automation   *John D. McDonald* 5-53

5.7   Oil Containment   *Anne-Marie Sahazizian and Tibor Kertesz* 5-62

5.8   Community Considerations   *James H. Sosinski* 5-76

5.9   Animal Deterrents/Security   *C.M. Mike Stine and Sheila Frasier* 5-88

5.10   Substation Grounding   *Richard P. Keil* 5-92

5.11   Grounding and Lightning   *Robert S. Nowell* 5-104

5.12   Seismic Considerations   *R.P. Stewart, Rulon Fronk, and Tonia Jurbin* 5-124

5.13   Substation Fire Protection   *Al Bolger and Don Delcourt* 5-134

5
Substations

Philip Bolin
Mitsubishi Electric Power Products, Inc.

Michael J. Bio
Power Resources, Inc.

David L. Harris
Waukesha Electric Systems

Gerhard Juette
Siemens

James W. Evans
Detroit Edison Company

John D. McDonald
KEMA Consulting

Anne-Marie Sahazizian
Hydro One Networks, Inc.

Tibor Kertesz
Hydro One Networks, Inc.

James H. Sosinski
Consumers Energy

C. M. Mike Stine
Raychem Corporation

Sheila Frasier
Southern Engineering

Richard P. Keil
Dayton Power & Light Company

Robert S. Nowell
Georgia Power Company

Robert P. Stewart
BC Hydro

Rulon Fronk
Fronk Consulting

Tonia Jurbin
BC Hydro

Al Bolger
BC Hydro

Don Delcourt
BC Hydro

5.1 Gas Insulated Substations 5-2
   SF6 • Construction and Service Life • Economics of GIS

5.2 Air Insulated Substations — Bus/Switching
   Configurations 5-18
   Single Bus • Double Bus, Double Breaker • Main and
   Transfer Bus • Double Bus, Single Breaker • Ring Bus •
   Breaker-and-a-Half • Comparison of Configurations

5.3 High-Voltage Switching Equipment 5-23
   Ambient Conditions • Disconnect Switches • Load Break
   Switches • High-Speed Grounding Switchers • Power Fuses •
   Circuit Switchers • Circuit Breakers • GIS Substations •
   Environmental Concerns

5.4 High-Voltage Power Electronics Substations 5-29
   Types • Control • Losses and Cooling • Buildings •
   Interference • Reliability • Specifications • Training and
   Commissioning • The Future

5.5 Considerations in Applying Automation Systems to
   Electric Utility Substations 5-41
   Physical Considerations • Analog Data Acquisition • Status
   Monitoring • Control Functions

5.6 Substation Automation 5-53
   Definitions and Terminology • Open Systems • Substation
   Automation Technical Issues • IEEE Power Engineering
   Society Substations Committee • EPRI-Sponsored Utility
   Substation Communication Initiative

5.7 Oil Containment 5-62
   Oil-Filled Equipment in Substation • Spill Risk Assessment •
   Containment Selection Consideration • Oil Spill Prevention
   Techniques

5.8 Community Considerations 5-76
   Community Acceptance • Planning Strategies and Design •
   Permitting Process • Construction • Operations

5.9 Animal Deterrents/Security 5-88
   Animal Types • Mitigation Methods

5.10 Substation Grounding 5-92
   Accidental Ground Circuit • Permissible Body Current
   Limits • Tolerable Voltages • Design Criteria

5.11 Grounding and Lightning 5-104
   Lightning Stroke Protection • Lightning Parameters •
   Empirical Design Methods • The Electromagnetic Model •
   Calculation of Failure Probability • Active Lightning
   Terminals

        5.12 Seismic Considerations 5-124
              A Historical Perspective • Relationship Between Earthquakes
              and Substations • Applicable Documents • Decision Process
              for Seismic Design Consideration • Performance Levels and
              Desired Spectra • Qualification Process
        5.13 Substation Fire Protection 5-134
              Fire Hazards • Fire Protection Measures • Hazard
              Assessment • Risk Analysis • Conclusion

5.1 Gas Insulated Substations

Philip Bolin

A gas insulated substation (GIS) uses a superior dielectric gas, SF6, at moderate pressure for phase-to-phase and phase-to-ground insulation. The high voltage conductors, circuit breaker interrupters, switches, current transformers, and voltage transformers are in SF6 gas inside grounded metal enclosures. The atmospheric air insulation used in a conventional, air insulated substation (AIS) requires meters of air insulation to do what SF6 can do in centimeters. GIS can therefore be smaller than AIS by up to a factor of ten. A GIS is mostly used where space is expensive or not available. In a GIS the active parts are protected from the deterioration from exposure to atmospheric air, moisture, contamination, etc. As a result, GIS is more reliable and requires less maintenance than AIS.

GIS was first developed in various countries between 1968 and 1972. After about 5 years of experience, the use rate increased to about 20% of new substations in countries where space is limited. In other countries with space easily available, the higher cost of GIS relative to AIS has limited use to special cases. For example, in the U.S., only about 2% of new substations are GIS. International experience with GIS is described in a series of CIGRE papers (CIGRE, 1992; 1994; 1982). The IEEE (IEEE Std. C37. 122-1993; IEEE Std C37. 122.1-1993) and the IEC (IEC, 1990) have standards covering all aspects of the design, testing, and use of GIS. For the new user, there is a CIGRE application guide (Katchinski et al., 1998). IEEE has a guide for specifications for GIS (IEEE Std. C37.123-1996).

SF6

Sulfur hexaflouride is an inert, non-toxic, colorless, odorless, tasteless, and non-flammable gas consisting of a sulfur atom surrounded by and tightly bonded to six flourine atoms. It is about five times as dense as air. SF6 is used in GIS at pressures from 400 to 600 kPa absolute. The pressure is chosen so that the SF6 will not condense into a liquid at the lowest temperatures the equipment experiences. SF6 has two to three times the insulating ability of air at the same pressure. SF6 is about one hundred times better than air for interrupting arcs. It is the universally used interrupting medium for high voltage circuit breakers, replacing the older mediums of oil and air. SF6 decomposes in the high temperature of an electric arc, but the decomposed gas recombines back into SF6 so well that it is not necessary to replenish the SF6 in GIS. There are some reactive decomposition byproducts formed because of the trace presence of moisture, air, and other contaminants. The quantities formed are very small. Molecular sieve absorbants inside the GIS enclosure eliminate these reactive byproducts. SF6 is supplied in 50-kg gas cylinders in a liquid state at a pressure of about 6000 kPa for convenient storage and transport. Gas handling systems with filters, compressors, and vacuum pumps are commercially available. Best practices and the personnel safety aspects of SF6 gas handling are covered in international standards (IEC, 1995).

The SF6 in the equipment must be dry enough to avoid condensation of moisture as a liquid on the surfaces of the solid epoxy support insulators because liquid water on the surface can cause a dielectric breakdown. However, if the moisture condenses as ice, the breakdown voltage is not affected. So dew points in the gas in the equipment need to be below about –10°C. For additional margin, levels of less than 1000 ppmv of moisture are usually specified and easy to obtain with careful gas handling. Absorbants inside the GIS enclosure help keep the moisture level in the gas low, even though over time, moisture will evolve from the internal surfaces and out of the solid dielectric materials (IEEE Std. 1125-1993).

Small conducting particles of mm size significantly reduce the dielectric strength of SF6 gas. This effect becomes greater as the pressure is raised past about 600 kPa absolute (Cookson and Farish, 1973). The particles are moved by the electric field, possibly to the higher field regions inside the equipment or deposited along the surface of the solid epoxy support insulators, leading to dielectric breakdown at operating voltage levels. Cleanliness in assembly is therefore very important for GIS. Fortunately, during the factory and field power frequency high voltage tests, contaminating particles can be detected as they move and cause small electric discharges (partial discharge) and acoustic signals, so they can be removed by opening the equipment. Some GIS equipment is provided with internal "particle traps" that capture the particles before they move to a location where they might cause breakdown. Most GIS assemblies are of a shape that provides some "natural" low electric field regions where particles can rest without causing problems.

SF6 is a strong greenhouse gas that could contribute to global warming. At an international treaty conference in Kyoto in 1997, SF6 was listed as one of the six greenhouse gases whose emissions should be reduced. SF6 is a very minor contributor to the total amount of greenhouse gases due to human activity, but it has a very long life in the atmosphere (half-life is estimated at 3200 years), so the effect of SF6 released to the atmosphere is effectively cumulative and permanent. The major use of SF6 is in electrical power equipment. Fortunately, in GIS the SF6 is contained and can be recycled. By following the present international guidelines for use of SF6 in electrical equipment (Mauthe et al., 1997), the contribution of SF6 to global warming can be kept to less than 0.1% over a 100-year horizon. The emission rate from use in electrical equipment has been reduced over the last three years. Most of this effect has been due to simply adopting better handling and recycling practices. Standards now require GIS to leak less than 1% per year. The leakage rate is normally much lower. Field checks of GIS in service for many years indicate that the leak rate objective can be as low as 0.1% per year when GIS standards are revised.

Construction and Service Life

GIS is assembled of standard equipment modules (circuit breaker, current transformers, voltage transformers, disconnect and ground switches, interconnecting bus, surge arresters, and connections to the rest of the electric power system) to match the electrical one-line diagram of the substation. A cross-section view of a 242-kV GIS shows the construction and typical dimensions (Fig. 5.1). The modules are joined using bolted flanges with an "O" ring seal system for the enclosure and a sliding plug-in contact for the conductor. Internal parts of the GIS are supported by cast epoxy insulators. These support insulators provide a gas barrier between parts of the GIS, or are cast with holes in the epoxy to allow gas to pass from one side to the other.

Up to about 170 kV system voltage, all three phases are often in one enclosure (Fig. 5.2). Above 170 kV, the size of the enclosure for "three-phase enclosure," GIS becomes too large to be practical. So a "single-phase enclosure" design (Fig. 5.1) is used. There are no established performance differences between three-phase enclosure and single-phase enclosure GIS. Some manufacturers use the single-phase enclosure type for all voltage levels.

Enclosures today are mostly cast or welded aluminum, but steel is also used. Steel enclosures are painted inside and outside to prevent rusting. Aluminum enclosures do not need to be painted, but may be painted for ease of cleaning and a better appearance. The pressure vessel requirements for GIS enclosures are set by GIS standards (IEEE Std. C37.122-1993; IEC, 1990), with the actual design, manufacture, and test following an established pressure vessel standard of the country of manufacture. Because of the moderate pressures involved, and the classification of GIS as electrical equipment, third-party inspection and code stamping of the GIS enclosures are not required.

Conductors today are mostly aluminum. Copper is sometimes used. It is usual to silver plate surfaces that transfer current. Bolted joints and sliding electrical contacts are used to join conductor sections. There are many designs for the sliding contact element. In general, sliding contacts have many individually

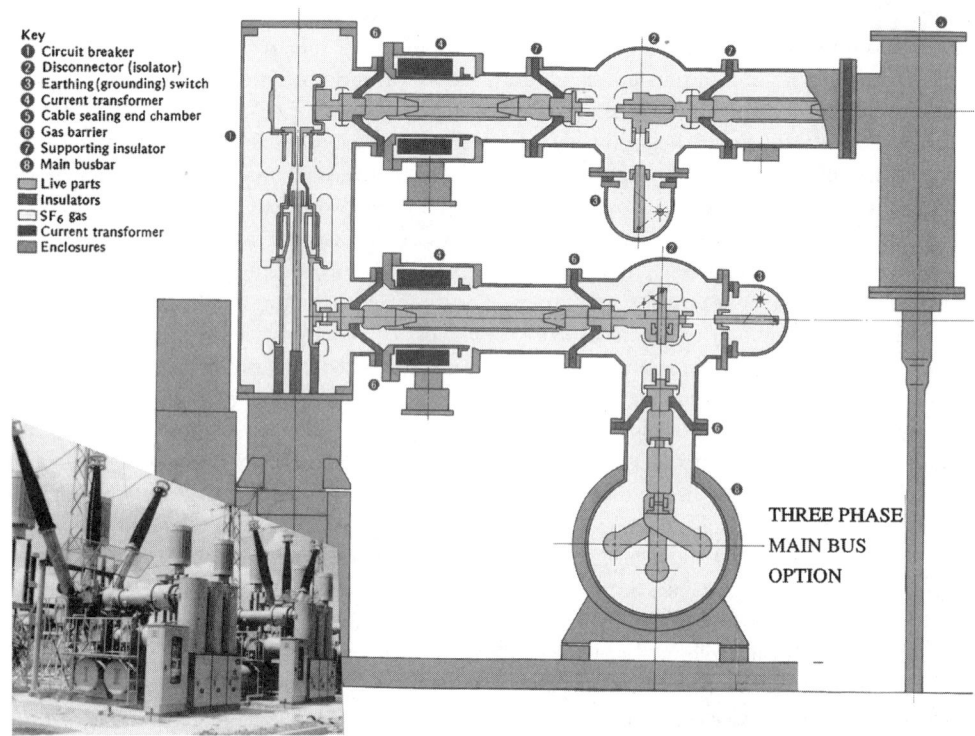

FIGURE 5.1 Single-phase eclosure GIS.

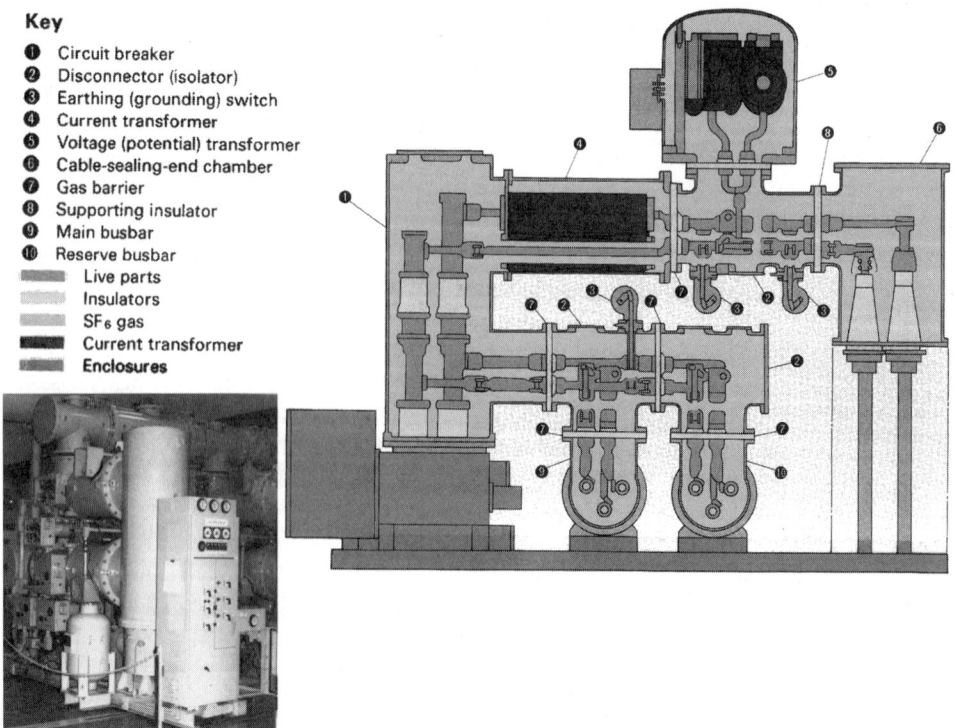

FIGURE 5.2 Three-phase enclosure GIS.

sprung copper contact fingers working in parallel. Usually the contact fingers are silver plated. A contact lubricant is used to ensure that the sliding contact surfaces do not generate particles or wear out over time. The sliding conductor contacts make assembly of the modules easy and also allow for conductor movement to accommodate the differential thermal expansion of the conductor relative to the enclosure. Sliding contact assemblies are also used in circuit breakers and switches to transfer current from the moving contact to the stationary contacts.

Support insulators are made of a highly filled epoxy resin cast very carefully to prevent formation of voids and/or cracks during curing. Each GIS manufacturer's material formulation and insulator shape has been developed to optimize the support insulator in terms of electric field distribution, mechanical strength, resistance to surface electric discharges, and convenience of manufacture and assembly. Post, disc, and cone type support insulators are used. Quality assurance programs for support insulators include a high voltage power frequency withstand test with sensitive partial discharge monitoring. Experience has shown that the electric field stress inside the cast epoxy insulator should be below a certain level to avoid aging of the solid dielectric material. The electrical stress limit for the cast epoxy support insulator is not a severe design constraint because the dimensions of the GIS are mainly set by the lightning impulse withstand level and the need for the conductor to have a fairly large diameter to carry to load current of several thousand amperes. The result is space between the conductor and enclosure for support insulators having low electrical stress.

Service life of GIS using the construction described above has been shown by experience to be more than 30 years. The condition of GIS examined after many years in service does not indicate any approaching limit in service life. Experience also shows no need for periodic internal inspection or maintenance. Inside the enclosure is a dry, inert gas that is itself not subject to aging. There is no exposure of any of the internal materials to sunlight. Even the "O" ring seals are found to be in excellent condition because there is almost always a "double seal" system — Fig. 5.3 shows one approach. The lack of aging has been found for GIS, whether installed indoors or outdoors.

Circuit Breaker

GIS uses essentially the same dead tank SF6 puffer circuit breakers used in AIS. Instead of SF6-to-air as connections into the substation as a whole, the nozzles on the circuit breaker enclosure are directly connected to the adjacent GIS module.

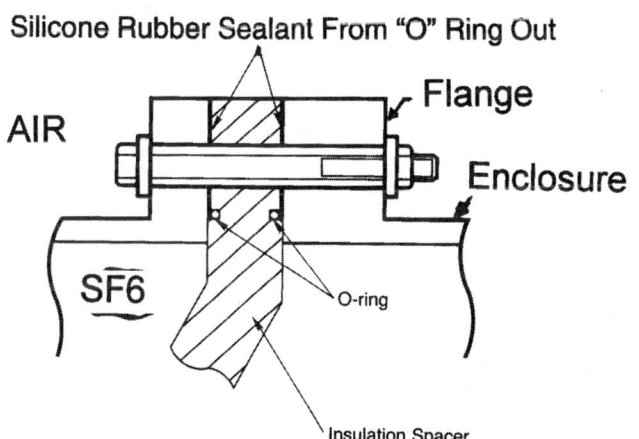

FIGURE 5.3 Gas seal for GIS enclosure.

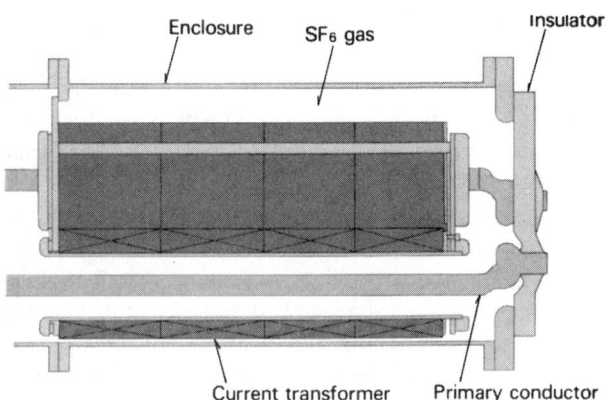

FIGURE 5.4 Current transformers for GIS.

Current Transformers

CTs are inductive ring type installed either inside the GIS enclosure or outside the GIS enclosure (Fig. 5.4). The GIS conductor is the single turn primary for the CT. CTs inside the enclosure must be shielded from the electric field produced by the high voltage conductor or high transient voltages can appear on the secondary through capacitive coupling. For CTs outside the enclosure, the enclosure itself must be provided with an insulating joint, and enclosure currents shunted around the CT. Both types of construction are in wide use.

Voltage Transformers

VTs are inductive type with an iron core. The primary winding is supported on an insulating plastic film immersed in SF6. The VT should have an electric field shield between the primary and secondary windings to prevent capacitive coupling of transient voltages. The VT is usually a sealed unit with a gas barrier insulator. The VT is either easily removable so the GIS can be high voltage tested without damaging the VT, or the VT is provided with a disconnect switch or removable link (Fig. 5.5).

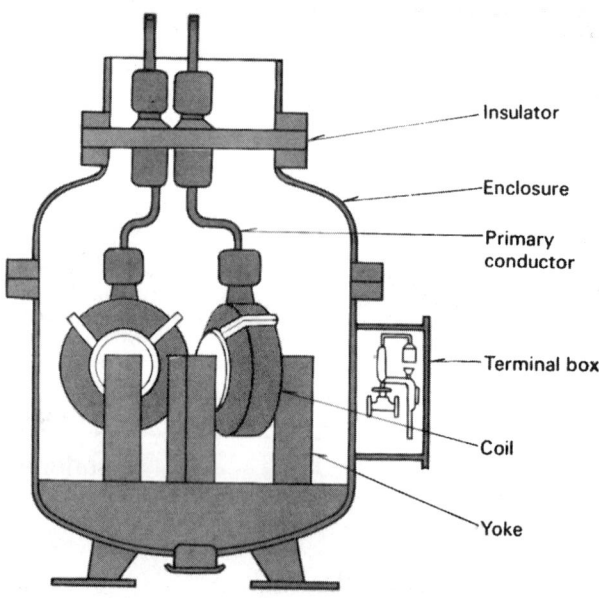

FIGURE 5.5 Voltage transformers for GIS.

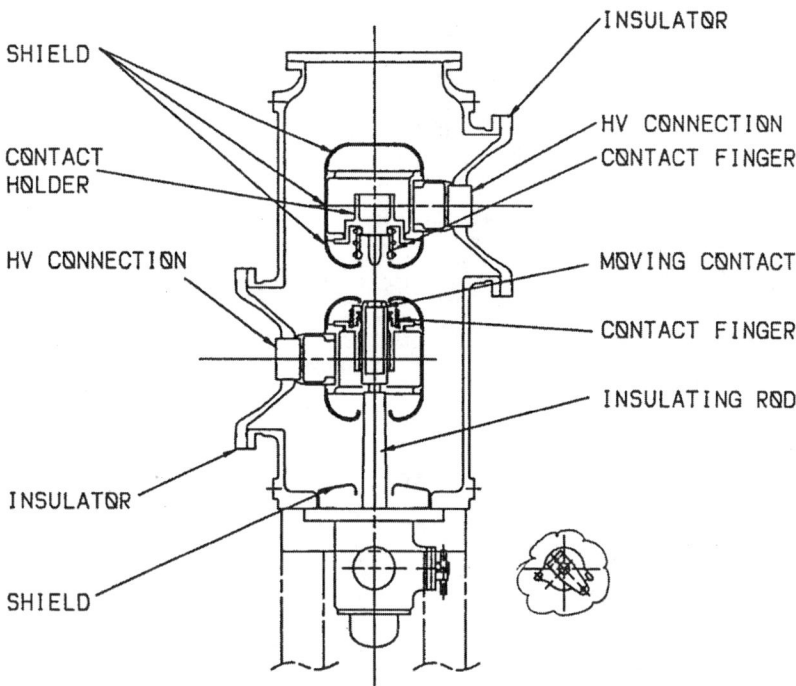

FIGURE 5.6 Disconnect switches for GIS.

Disconnect Switches

Disconnect switches (Fig. 5.6) have a moving contact that opens or closes a gap between stationary contacts when activated by a insulating operating rod that is itself moved by a sealed shaft coming through the enclosure wall. The stationary contacts have shields that provide the appropriate electric field distribution to avoid too high a surface stress. The moving contact velocity is relatively low (compared to a circuit breaker moving contact) and the disconnect switch can interrupt only low levels of capacitive current (for example, disconnecting a section of GIS bus) or small inductive currents (for example, transformer magnetizing current). Load break disconnect switches have been furnished in the past, but with improvements and cost reductions of circuit breakers, it is not practical to continue to furnish load break disconnect switches, and a circuit breaker should be used instead.

Ground Switches

Ground switches (Fig. 5.7) have a moving contact that opens or closes a gap between the high voltage conductor and the enclosure. Sliding contacts with appropriate electric field shields are provided at the enclosure and the conductor. A "maintenance" ground switch is operated either manually or by motor drive to close or open in several seconds and when fully closed to carry the rated short-circuit current for the specified time period (1 or 3 sec) without damage. A "fast acting" ground switch has a high speed drive, usually a spring, and contact materials that withstand arcing so it can be closed twice onto an energized conductor without significant damage to itself or adjacent parts. Fast-acting ground switches are frequently used at the connection point of the GIS to the rest of the electric power network, not only in case the connected line is energized, but also because the fast-acting ground switch is better able to handle discharge of trapped charge and breaking of capacitive or inductive coupled currents on the connected line.

Ground switches are almost always provided with an insulating mount or an insulating bushing for the ground connection. In normal operation the insulating element is bypassed with a bolted shunt to the GIS enclosure. During installation or maintenance, with the ground switch closed, the shunt can be removed and the ground switch used as a connection from test equipment to the GIS conductor. Voltage

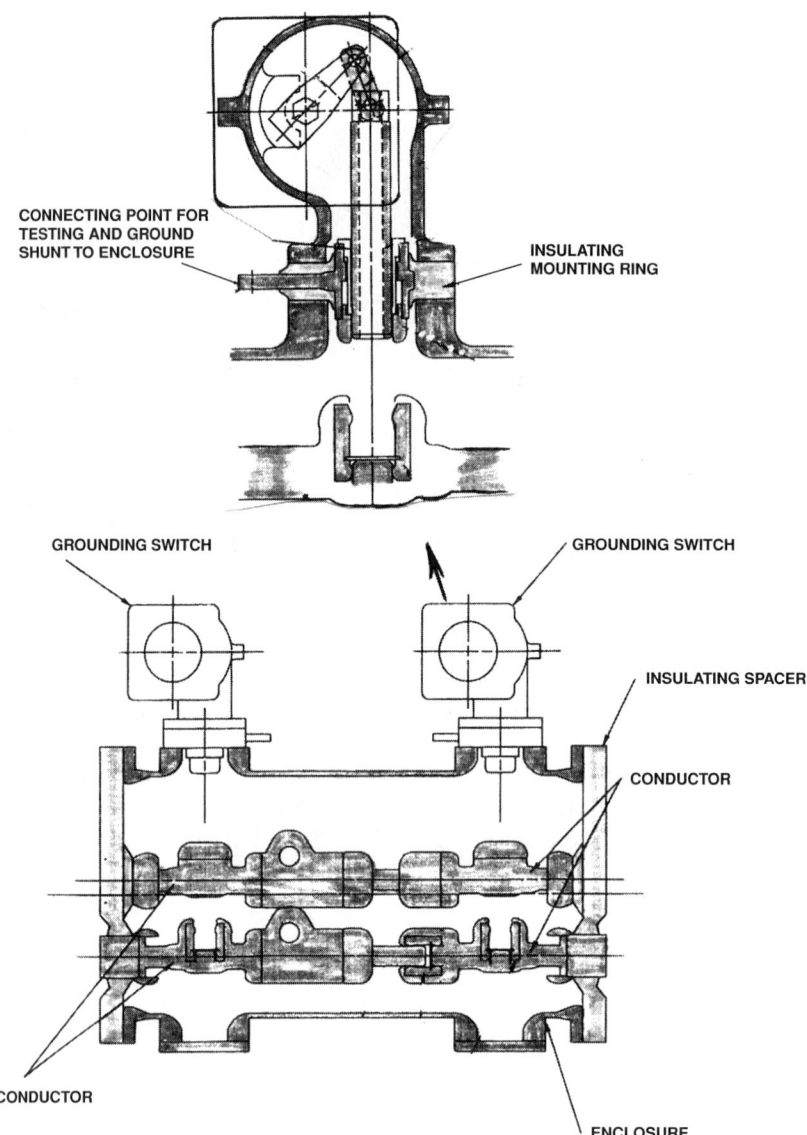

FIGURE 5.7 Ground switches for GIS.

and current testing of the internal parts of the GIS can then be done without removing SF6 gas or opening the enclosure. A typical test is measurement of contact resistance using two ground switches (Fig. 5.8).

Bus

To connect GIS modules that are not directly connected to each other, an SF6 bus consisting of an inner conductor and outer enclosure is used. Support insulators, sliding electrical contacts, and flanged enclosure joints are usually the same as for the GIS modules.

Air Connection

SF6-to-air bushings (Fig. 5.9) are made by attaching a hollow insulating cylinder to a flange on the end of a GIS enclosure. The insulating cylinder contains pressurized SF6 on the inside and is suitable for exposure to atmospheric air on the outside. The conductor continues up through the center of the insulating cylinder to a metal end plate. The outside of the end plate has provisions for bolting to an air

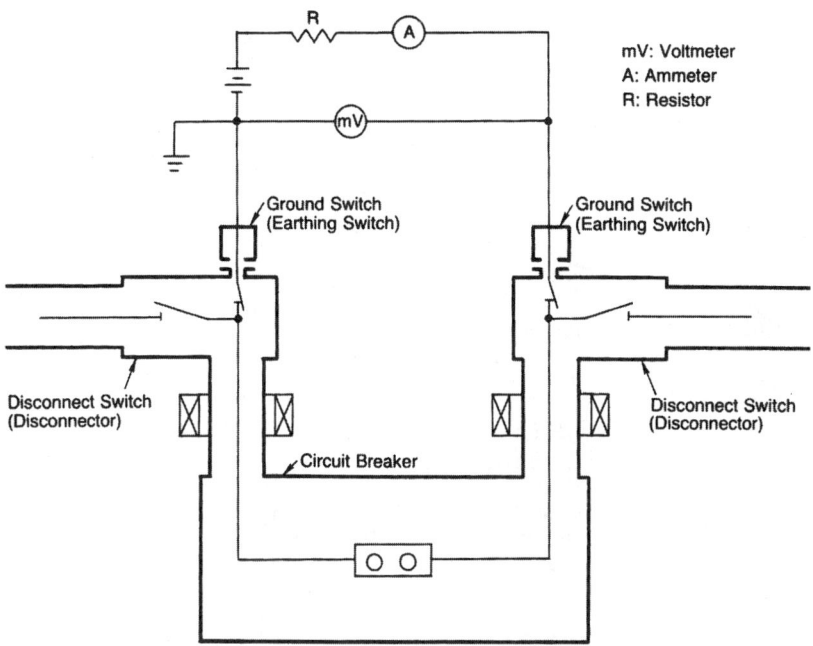

FIGURE 5.8 Contact resistance measured using ground switch.

insulated conductor. The insulating cylinder has a smooth interior. Sheds on the outside improve the performance in air under wet and/or contaminated conditions. Electric field distribution is controlled by internal metal shields. Higher voltage SF6-to-air bushings also use external shields. The SF6 gas inside the bushing is usually the same pressure as the rest of the GIS. The insulating cylinder has most often been porcelain in the past, but today many are a composite consisting of a fiberglass epoxy inner cylinder with an external weather shed of silicone rubber. The composite bushing has better contamination resistance and is inherently safer because it will not fracture as will porcelain.

Cable Connections

A cable connecting to a GIS is provided with a cable termination kit that is installed on the cable to provide a physical barrier between the cable dielectric and the SF6 gas in the GIS (Fig. 5.10). The cable termination kit also provides a suitable electric field distribution at the end of the cable. Because the cable termination will be in SF6 gas, the length is short and sheds are not needed. The cable conductor is connected with bolted or compression connectors to the end plate or cylinder of the cable termination kit. On the GIS side, a removable link or plug in contact transfers current from the cable to the GIS conductor. For high voltage testing of the GIS or the cable, the cable is disconnected from the GIS by removing the conductor link or plug-in contact. The GIS enclosure around the cable termination usually has an access port. This port can also be used for attaching a test bushing.

Direct Transformer Connections

To connect a GIS directly to a transformer, a special SF6-to-oil bushing that mounts on the transformer is used (Fig. 5.11). The bushing is connected under oil on one end to the transformer's high voltage leads. The other end is SF6 and has a removable link or sliding contact for connection to the GIS conductor. The bushing may be an oil-paper condenser type or more commonly today, a solid insulation type. Because leakage of SF6 into the transformer oil must be prevented, most SF6-to-oil bushings have a center section that allows any SF6 leakage to go to the atmosphere rather than into the transformer. For testing, the SF6 end of the bushing is disconnected from the GIS conductor after gaining access through an opening in the GIS enclosure. The GIS enclosure of the transformer can also be used for attaching a test bushing.

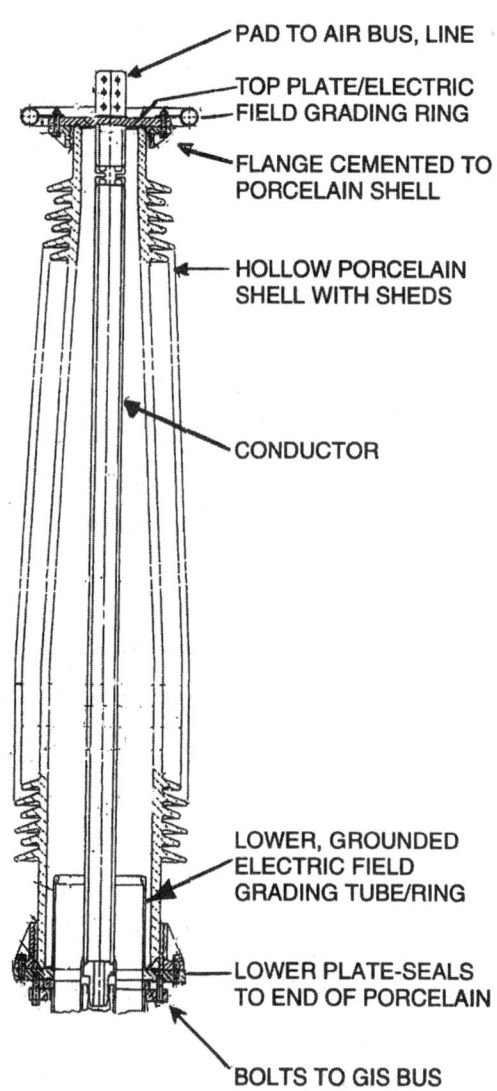

FIGURE 5.9 SF6-to-air bushing.

Surge Arrester

Zinc oxide surge arrester elements suitable for immersion in SF6 are supported by an insulating cylinder inside a GIS enclosure section to make a surge arrester for overvoltage control (Fig. 5.12). Because the GIS conductors are inside in a grounded metal enclosure, the only way for lightning impulse voltages to enter is through the connections of the GIS to the rest of the electrical system. Cable and direct transformer connections are not subject to lightning strikes, so only at SF6-to-air bushing connections is lightning a concern. Air insulated surge arresters in parallel with the SF6-to-air bushings usually provide adequate protection of the GIS from lightning impulse voltages at a much lower cost than SF6 insulated arresters. Switching surges are seldom a concern in GIS because with SF6 insulation the withstand voltages for switching surges are not much less than the lightning impulse voltage withstand. In AIS there is a significant decrease in withstand voltage for switching surges than for lightning impulse because the longer time span of the switching surge allows time for the discharge to completely bridge the long insulation distances in air. In the GIS, the short insulation distances can be bridged in the short time span of a lightning impulse so the longer time span of a switching surge does not significantly decrease

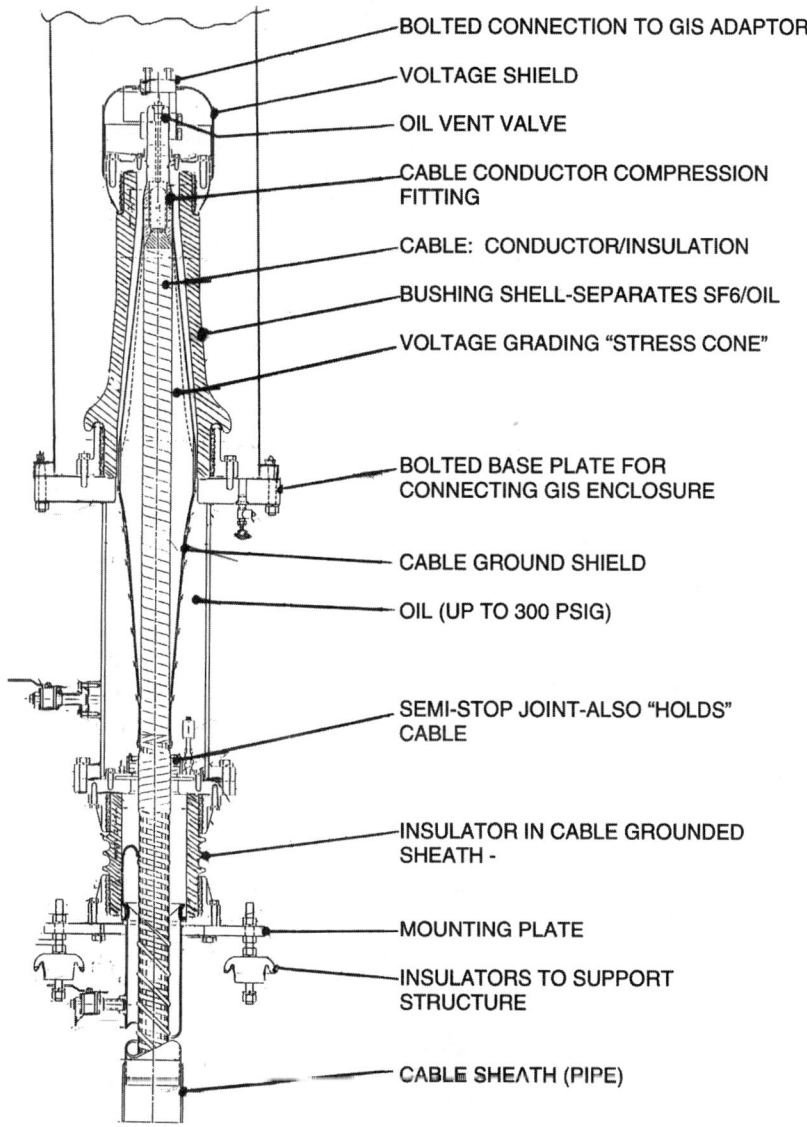

FIGURE 5.10 Power cable connection.

the breakdown voltage. Insulation coordination studies usually show there is no need for surge arresters in a GIS; however, many users specify surge arresters at transformers and cable connections as the most conservative approach.

Control System

For ease of operation and convenience in wiring the GIS back to the substation control room, a local control cabinet (LCC) is provided for each circuit breaker position (Fig. 5.13). The control and power wires for all the operating mechanisms, auxiliary switches, alarms, heaters, CTs, and VTs are brought from the GIS equipment modules to the LCC using shielded multiconductor control cables. In addition to providing terminals for all the GIS wiring, the LCC has a mimic diagram of the part of the GIS being controlled. Associated with the mimic diagram are control switches and position indicators for the circuit breaker and switches. Annunciation of alarms is also usually provided in the LCC. Electrical interlocking

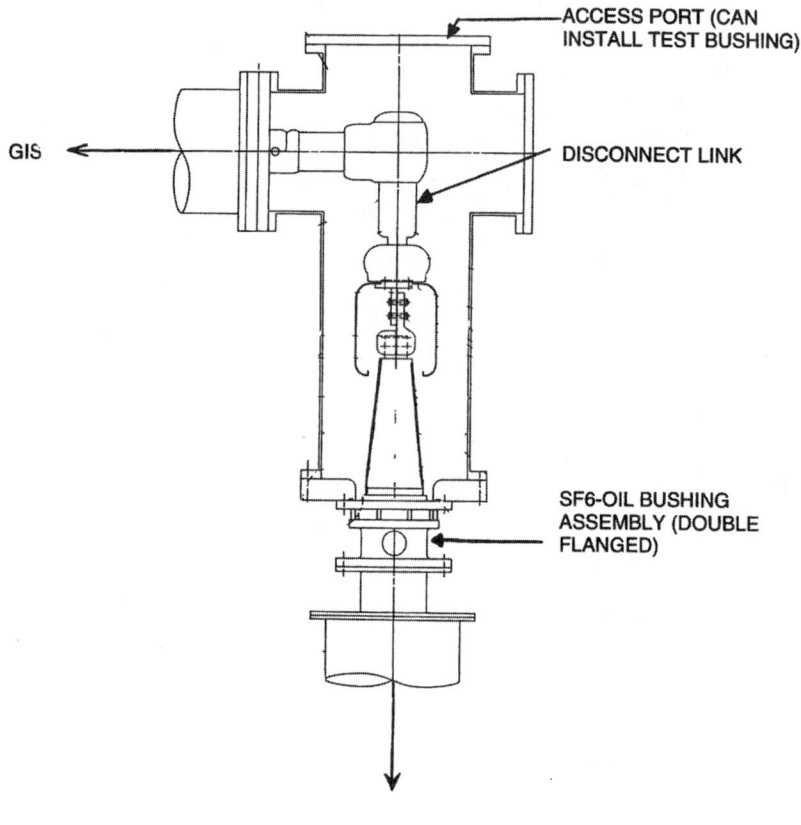

FIGURE 5.11 Direct SF6 bus connection to transfromer.

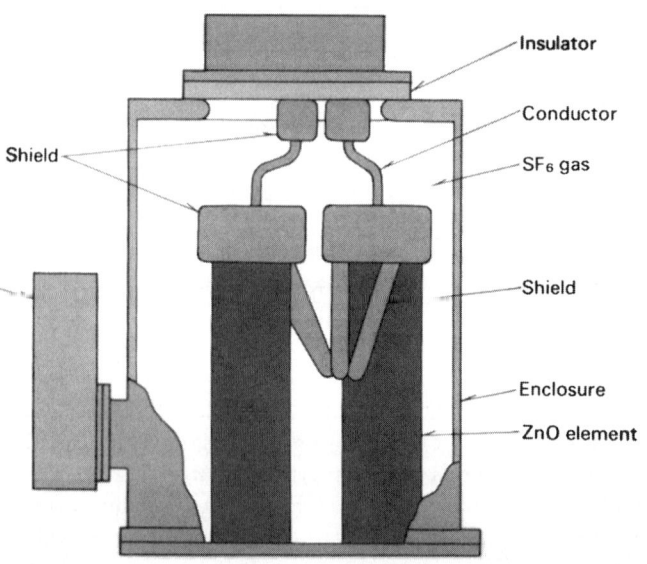

FIGURE 5.12 Surge arrester for GIS.

FIGURE 5.13 Local control cabinet for GIS.

and some other control functions can be conveniently implemented in the LCC. Although the LCC is an extra expense, with no equivalent in the typical AIS, it is so well established and popular that attempts to eliminate it to reduce cost have not succeeded. The LCC does have the advantage of providing a very clear division of responsibility between the GIS manufacturer and user in terms of scope of equipment supply.

Switching and circuit breaker operation in a GIS produces internal surge voltages with a very fast rise time on the order of nanoseconds and a peak voltage level of about 2 per unit. These "very fast transient overvoltages" are not a problem inside the GIS because the duration of this type of surge voltage is very short — much shorter than the lightning impulse voltage. However, a portion of the VFTO will emerge from the inside of the GIS at any place where there is a discontinuity of the metal enclosure — for example, at insulating enclosure joints for external CTs or at the SF6-to-air bushings. The resulting "transient ground rise voltage" on the outside of the enclosure may cause some small sparks across the insulating enclosure joint or to adjacent grounded parts. These may alarm nearby personnel but are not harmful to a person because the energy content is very low. However, if these VFT voltages enter the control wires, they could cause faulty operation of control devices. Solid-state controls can be particularly affected. The solution is thorough shielding and grounding of the control wires. For this reason, in a GIS, the control cable shield should be grounded at both the equipment and the LCC ends using either coaxial ground bushings or short connections to the cabinet walls at the location where the control cable first enters the cabinet.

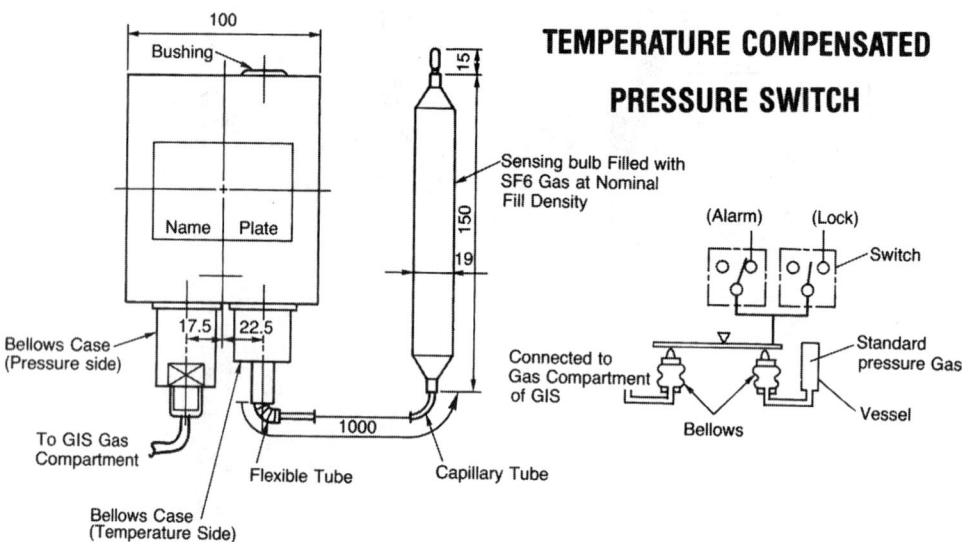

FIGURE 5.14 SF6 density monitor for GIS.

Gas Monitor System

The insulating and interrupting capability of the SF6 gas depends on the density of the SF6 gas being at a minimum level established by design tests. The pressure of the SF6 gas varies with temperature, so a mechanical temperature-compensated pressure switch is used to monitor the equivalent of gas density (Fig. 5.14). GIS is filled with SF6 to a density far enough above the minimum density for full dielectric and interrupting capability so that from 10% to 20% of the SF6 gas can be lost before the performance of the GIS deteriorates. The density alarms provide a warning of gas being lost, and can be used to operate the circuit breakers and switches to put a GIS that is losing gas into a condition selected by the user. Because it is much easier to measure pressure than density, the gas monitor system usually has a pressure gage. A chart is provided to convert pressure and temperature measurements into density. Microprocessor-based measurement systems are available that provide pressure, temperature, density, and even percentage of proper SF6 content. These can also calculate the rate at which SF6 is being lost. However, they are significantly more expensive than the mechanical temperature-compensated pressure switches, so they are supplied only when requested by the user.

Gas Compartments and Zones

A GIS is divided by gas barrier insulators into gas compartments for gas handling purposes. In some cases, the use of a higher gas pressure in the circuit breaker than is needed for the other devices, requires that the circuit breaker be a separate gas compartment. Gas handling systems are available to easily process and store about 1000 kg of SF6 at one time, but the length of time needed to do this is longer than most GIS users will accept. GIS is therefore divided into relatively small gas compartments of less than several hundred kg. These small compartments may be connected with external bypass piping to create a larger gas zone for density monitoring. The electrical functions of the GIS are all on a three-phase basis, so there is no electrical reason not to connect the parallel phases of a single-phase enclosure type of GIS into one gas zone for monitoring. Reasons for not connecting together many gas compartments into large gas zones include a concern with a fault in one gas compartment causing contamination in adjacent compartments and the greater amount of SF6 lost before a gas loss alarm. It is also easier to locate a leak if the alarms correspond to small gas zones, but a larger gas zone will, for the same size leak, give more time to add SF6 between the first alarm and second alarm. Each GIS manufacturer has a standard approach to gas compartments and gas zones, but will, of course, modify the approach to satisfy the concerns of individual GIS users.

Substations

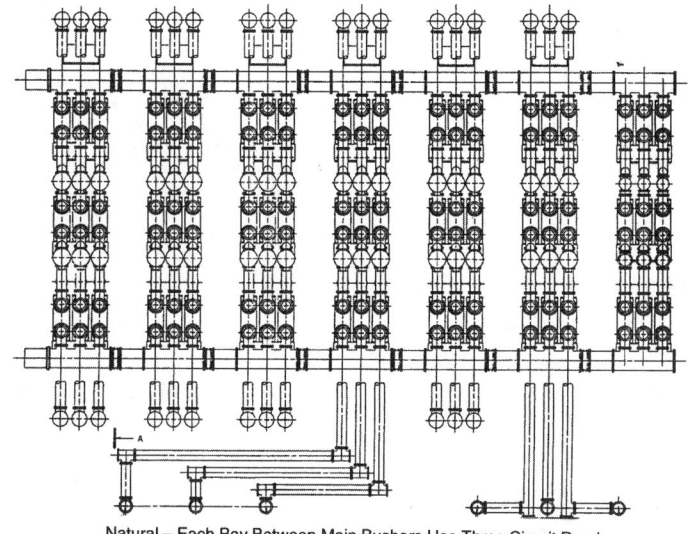

Natural – Each Bay Between Main Busbars Has Three Circuit Breakers

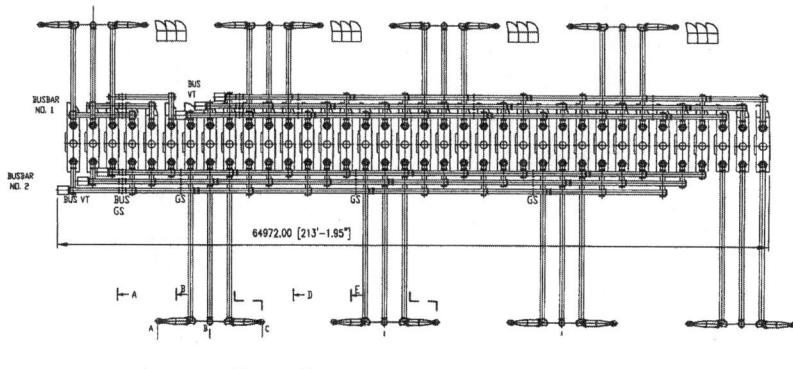

Linear – Circuit Breakers Are Side by Side

FIGURE 5.15   One-and-one-half circuit breaker layouts.

Electrical and Physical Arrangement

For any electrical one-line diagram there are usually several possible physical arrangements. The shape of the site for the GIS and the nature of connecting lines and/or cables should be considered. Figure 5.15 compares a "natural" physical arrangement for a breaker and a half GIS with a "linear" arrangement.

Most GIS designs were developed initially for a double bus, single breaker arrangement (Fig. 5.2). This widely used approach provides good reliability, simple operation, easy protective relaying, excellent economy, and a small footprint. By integrating several functions into each GIS module, the cost of the double bus, single breaker arrangement can be significantly reduced. An example is shown in Fig. 5.16. Disconnect and ground switches are combined into a "three-position switch" and made a part of each bus module connecting adjacent circuit breaker positions. The cable connection module includes the cable termination, disconnect switches, ground switches, a VT, and surge arresters.

Grounding

The individual metal enclosure sections of the GIS modules are made electrically continuous either by the flanged enclosure joint being a good electrical contact in itself or with external shunts bolted to the

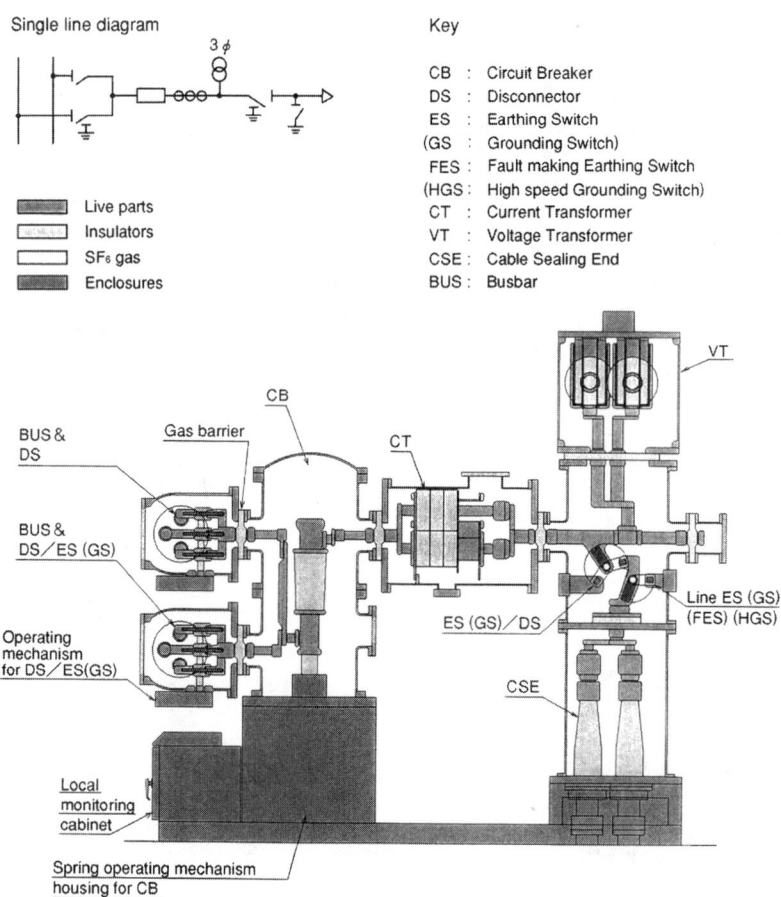

FIGURE 5.16 Integrated (combined function) GIS.

flanges or to grounding pads on the enclosure. While some early single-phase enclosure GIS were "single point grounded" to prevent circulating currents from flowing in the enclosures, today the universal practice is to use "multipoint grounding" even though this leads to some electrical losses in the enclosures due to circulating currents. The three enclosures of a single-phase GIS should be bonded to each other at the ends of the GIS to encourage circulating currents to flow. These circulating enclosure currents act to cancel the magnetic field that would otherwise exist outside the enclosure due to the conductor current. Three-phase enclosure GIS does not have circulating currents, but does have eddy currents in the enclosure, and should also be multipoint grounded. With multipoint grounding and the resulting many parallel paths for the current from an internal fault to flow to the substation ground grid, it is easy to keep the touch and step voltages for a GIS to the safe levels prescribed in IEEE 80.

Testing

Test requirements for circuit breakers, CTs, VTs, and surge arresters are not specific for GIS and will not be covered in detail here. Representative GIS assemblies having all of the parts of the GIS except for the circuit breaker are design tested to show that the GIS can withstand the rated lightning impulse voltage, switching impulse voltage, power frequency overvoltage, continuous current, and short-circuit current. Standards specify the test levels and how the tests must be done. Production tests of the factory-assembled GIS (including the circuit breaker) cover power frequency withstand voltage, conductor circuit resistance, leak checks, operational checks, and CT polarity checks. Components such as support insulators, VTs, and CTs are tested in accordance with the specific requirements for these items before assembly into the GIS. Field tests repeat the factory tests. The power frequency withstand voltage test is most important as a check of the cleanliness of the inside of the GIS in regard to contaminating conducting particles, as explained in the SF6 section above. Checking of interlocks is also very important. Other field tests may be done if the GIS is a very critical part of the electric power system, when, for example, a surge voltage test may be requested.

Installation

The GIS is usually installed on a monolithic concrete pad or the floor of a building. It is most often rigidly attached by bolting and/or welding the GIS support frames to embedded steel plates or beams. Chemical drill anchors can also be used. Expansion drill anchors are not recommended because dynamic loads may loosen expansion anchors when the circuit breaker operates. Large GIS installations may need bus expansion joints between various sections of the GIS to adjust to the fit-up in the field and, in some cases, provide for thermal expansion of the GIS. The GIS modules are shipped in the largest practical assemblies. At the lower voltage level, two or more circuit breaker positions can be delivered fully assembled. The physical assembly of the GIS modules to each other using the bolted flanged enclosure joints and sliding conductor contacts goes very quickly. More time is used for evacuation of air from gas compartments that have been opened, filling with SF6 gas, and control system wiring. The field tests are then done. For a high voltage GIS shipped as many separate modules, installation and testing takes about two weeks per circuit breaker position. Lower voltage systems shipped as complete bays, and mostly factory-wired, can be installed more quickly.

Operation and Interlocks

Operation of a GIS in terms of providing monitoring, control, and protection of the power system as a whole is the same as for an AIS except that internal faults are not self-clearing so reclosing should not be used for faults internal to the GIS. Special care should be taken for disconnect and ground switch operation because if these are opened with load current flowing, or closed into load or fault current, the arcing between the switch moving and stationary contacts will usually cause a phase-to-phase fault in three-phase enclosure GIS or to a phase-to-ground fault in single-phase enclosure GIS. The internal fault will cause severe damage inside the GIS. A GIS switch cannot be as easily or quickly replaced as an AIS switch. There will also be a pressure rise in the GIS gas compartment as the arc heats the gas. In extreme cases, the internal arc will cause a rupture disk to operate or may even cause a burn-through of the enclosure. The resulting release of hot, decomposed SF6 gas may cause serious injury to nearby personnel. For both the sake of the GIS and the safety of personnel, secure interlocks are provided so that the circuit breaker must be open before an associated disconnect switch can be opened or closed, and the disconnect switch must be open before the associated ground switch can be closed or opened.

Maintenance

Experience has shown that the internal parts of GIS are so well protected inside the metal enclosure that they do not age and as a result of proper material selection and lubricants, there is negligible wear of the switch contacts. Only the circuit breaker arcing contacts and the teflon nozzle of the interrupter experience wear proportional to the number of operations and the level of the load or fault currents being interrupted. Good contact and nozzle materials combined with the short interrupting time of modern circuit breakers provide, typically, for thousands of load current interruption operations and tens of full-rated fault current interruptions before there is any need for inspection or replacement. Except for circuit breakers in special use such as at a pumped storage plant, most circuit breakers will not be operated enough to ever require internal inspection. So most GIS will not need to be opened for maintenance. The external operating mechanisms and gas monitor systems should be visually inspected, with the frequency of inspection determined by experience.

Economics of GIS

The equipment cost of GIS is naturally higher than that of AIS due to the grounded metal enclosure, the provision of an LCC, and the high degree of factory assembly. A GIS is less expensive to install than an

AIS. The site development costs for a GIS will be much lower than for an AIS because of the much smaller area required for the GIS. The site development advantage of GIS increases as the system voltage increases because high voltage AIS take very large areas because of the long insulating distances in atmospheric air. Cost comparisons in the early days of GIS projected that, on a total installed cost basis, GIS costs would equal AIS costs at 345 kV. For higher voltages, GIS was expected to cost less than AIS. However, the cost of AIS has been reduced significantly by technical and manufacturing advances (especially for circuit breakers) over the last 30 years, but GIS equipment has not shown any cost reduction until very recently. Therefore, although GIS has been a well-established technology for a long time, with a proven high reliability and almost no need for maintenance, it is presently perceived as costing too much and is only applicable in special cases where space is the most important factor.

Currently, GIS costs are being reduced by integrating functions as described in the arrangement section above. As digital control systems become common in substations, the costly electromagnetic CTs and VTs of a GIS will be replaced by less-expensive sensors such as optical VTs and Rogowski coil CTs. These less-expensive sensors are also much smaller, reducing the size of the GIS and allowing more bays of GIS to be shipped fully assembled. Installation and site development costs are correspondingly lower. The GIS space advantage over AIS increases. GIS can now be considered for any new substation or the expansion of an existing substation without enlarging the area for the substation.

References

Cookson, A. H. and Farish, O., Particle-initiated breakdown between coaxial electrodes in compressed SF6, *IEEE Transactions on Power Appratus and Systems, Vol. PAS-92(3),* 871-876, May/June, 1973.
IEC 1634: 1995, IEC technical report: High voltage switchgear and controlgear — use and handling of sulphur hexafluoride (SF6) in high-voltage switchgear and controlgear.
IEEE Guide for Moisture Measurement and Control in SF6 Gas-Insulated Equipment, IEEE Std. 1125-1993.
IEEE Guide for Gas-Insulated Substations, IEEE Std. C37.122.1-1993.
IEEE Standard for Gas-Insulated Substations, IEEE Std. C37.122-1993.
IEEE Guide to Specifications for Gas-Insulated, Electric Power Substation Equipment, IEEE Std. C37.123-1996.
IEC 517: 1990, Gas-insulated metal-enclosed switchgear for rated voltages of 72.5 kV and above (3rd ed.).
Jones, D. J., Kopejtkova, D., Kobayashi, S., Molony, T., O'Connell, P., and Welch, I. M., GIS in service — experience and recommendations, *Paper 23-104 of CIGRE General Meeting,* Paris, 1994.
Katchinski, U., Boeck, W., Bolin, P. C., DeHeus, A., Hiesinger, H., Holt, P.-A., Murayama, Y., Jones, J., Knudsen, O., Kobayashi, S., Kopejtkova, D., Mazzoleni, B., Pryor, B., Sahni, A. S., Taillebois, J.-P., Tschannen, C., and Wester, P., User guide for the application of gas-insulated switchgear (GIS) for rated voltages of 72.5 kV and above, *CIGRE Report 125,* Paris, April 1998.
Kawamura, T., Ishi, T., Satoh, K., Hashimoto, Y., Tokoro, K., and Harumoto, Y., Operating experience of gas insulated switchgear (GIS) and its influence on the future substation design, *Paper 23-04 of CIGRE General Meeting,* Paris, 1982.
Kopejtkova, D., Malony, T., Kobayashi, S., and Welch, I. M., A twenty-five year review of experience with SF6 gas insulated substations (GIS), *Paper 23-101 of CIGRE General Meeting,* Paris, 1992.
Mauthe, G., Pryor, B. M., Neimeyer, L., Probst, R., Poblotzki, J., Bolin, P., O'Connell, P., and Henriot, J., SF6 recycling guide: Re-use of SF6 gas in electrical power equipment and final disposal, *CIGRE Report 117,* Paris, August, 1997.

5.2 Air Insulated Substations — Bus/Switching Configurations

Michael J. Bio

Various factors affect the reliability of a substation or switchyard, one of which is the arrangement of the buses and switching devices. In addition to reliability, arrangement of the buses/switching devices will impact maintenance, protection, initial substation development, and cost.

Substations

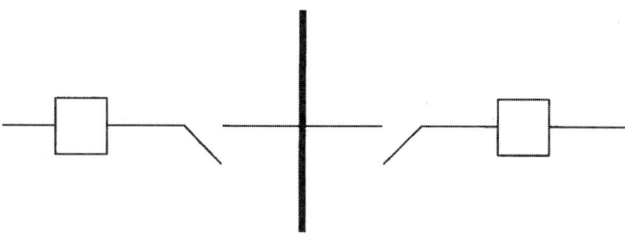

FIGURE 5.17 Single bus.

There are six types of substation bus/switching arrangements commonly used in air insulated substations:

1. Single bus
2. Double bus, double breaker
3. Main and transfer (inspection) bus
4. Double bus, single breaker
5. Ring bus
6. Breaker and a half

Single Bus (Fig. 5.17)

This arrangement involves one main bus with all circuits connected directly to the bus. The reliability of this type of an arrangement is very low. When properly protected by relaying, a single failure to the main bus or any circuit section between its circuit breaker and the main bus will cause an outage of the entire system. In addition, maintenance of devices on this system requires the de-energizing of the line connected to the device. Maintenance of the bus would require the outage of the total system, use of standby generation, or switching, if available.

Since the single bus arrangement is low in reliability, it is not recommended for heavily loaded substations or substations having a high availability requirement. Reliability of this arrangement can be improved by the addition of a bus tiebreaker to minimize the effect of a main bus failure.

Double Bus, Double Breaker (Fig. 5.18)

This scheme provides a very high level of reliability by having two separate breakers available to each circuit. In addition, with two separate buses, failure of a single bus will not impact either line. Maintenance of a bus or a circuit breaker in this arrangement can be accomplished without interrupting either of the circuits.

This arrangement allows various operating options as additional lines are added to the arrangement; loading on the system can be shifted by connecting lines to only one bus.

A double bus, double breaker scheme is a high-cost arrangement, since each line has two breakers and requires a larger area for the substation to accommodate the additional equipment. This is especially true in a low profile configuration. The protection scheme is also more involved than a single bus scheme.

Main and Transfer Bus (Fig. 5.19)

This scheme is arranged with all circuits connected between a main (operating) bus and a transfer bus (also referred to as an inspection bus). Some arrangements include a bus tie breaker that is connected between both buses with no circuits connected to it. Since all circuits are connected to the single, main bus, reliability of this system is not very high. However, with the transfer bus available during maintenance, de-energizing of the circuit can be avoided. Some systems are operated with the transfer bus normally de-energized.

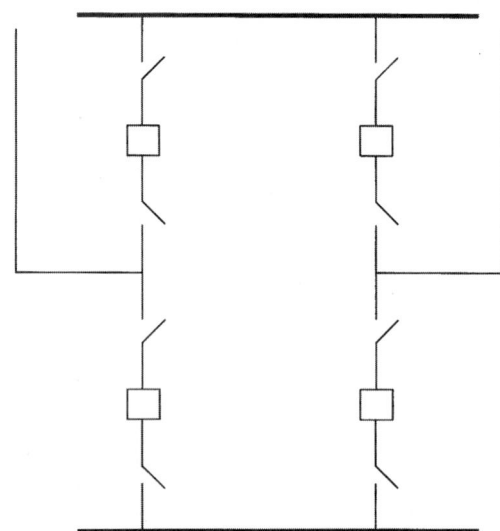

FIGURE 5.18 Double bus, double breaker.

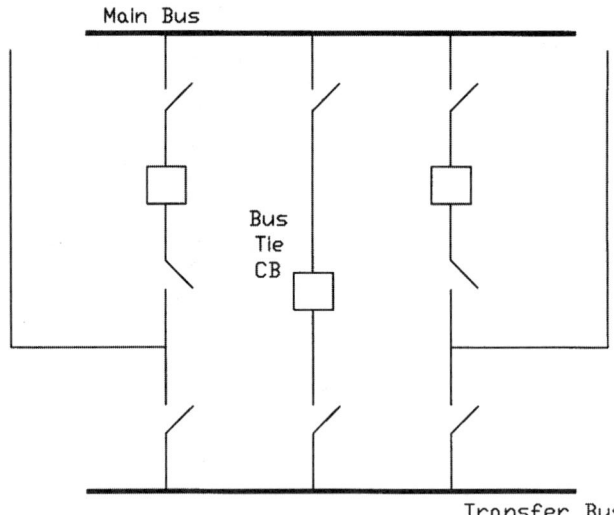

FIGURE 5.19 Main and transfer bus.

When maintenance work is necessary, the transfer bus is energized by either closing the tie breaker, or when a tie breaker is not installed, closing the switches connected to the transfer bus. With these switches closed, the breaker to be maintained can be opened along with its isolation switches. Then the breaker is taken out of service. The circuit remaining in service will now be connected to both circuits through the transfer bus. This way, both circuits remain energized during maintenance. Since each circuit may have a different circuit configuration, special relay settings may be used when operating in this abnormal arrangement. When a bus tie breaker is present, the bus tie breaker is the breaker used to replace the breaker being maintained, and the other breaker is not connected to the transfer bus.

A shortcoming of this scheme is that if the main bus is taken out of service, even though the circuits can remain energized through the transfer bus and its associated switches, there would be no relay protection for the circuits. Depending on the system arrangement, this concern can be minimized through the use of circuit protection devices (reclosure or fuses) on the lines outside the substation.

Substations

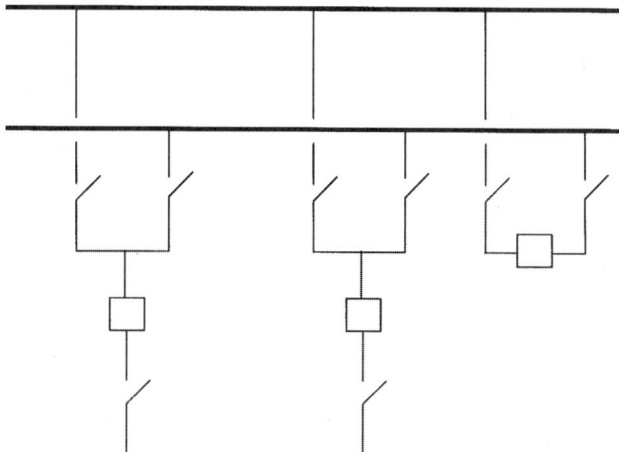

FIGURE 5.20 Double bus, single breaker.

This arrangement is slightly more expensive than the single bus arrangement, but does provide more flexibility during maintenance. Protection of this scheme is similar to that of the single bus arrangement. The area required for a low profile substation with a main and transfer bus scheme is also greater than that of the single bus, due to the additional switches and bus.

Double Bus, Single Breaker (Fig. 5.20)

This scheme has two main buses connected to each line circuit breaker and a bus tie breaker. Utilizing the bus tie breaker in the closed position allows the transfer of line circuits from bus to bus by means of the switches. This arrangement allows the operation of the circuits from either bus. In this arrangement, a failure on one bus will not affect the other bus. However, a bus tie breaker failure will cause the outage of the entire system.

Operating the bus tie breaker in the normally open position defeats the advantages of the two main buses. It arranges the system into two single bus systems, which as described previously, has very low reliability.

Relay protection for this scheme can be complex, depending on the system requirements, flexibility, and needs. With two buses and a bus tie available, there is some ease in doing maintenance, but maintenance on line breakers and switches would still require outside the substation switching to avoid outages.

Ring Bus (Fig. 5.21)

In this scheme, as indicated by the name, all breakers are arranged in a ring with circuits tapped between breakers. For a failure on a circuit, the two adjacent breakers will trip without affecting the rest of the system. Similarly, a single bus failure will only affect the adjacent breakers and allow the rest of the system to remain energized. However, a breaker failure or breakers that fail to trip will require adjacent breakers to be tripped to isolate the fault.

Maintenance on a circuit breaker in this scheme can be accomplished without interrupting any circuit, including the two circuits adjacent to the breaker being maintained. The breaker to be maintained is taken out of service by tripping the breaker, then opening its isolation switches. Since the other breakers adjacent to the breaker being maintained are in service, they will continue to supply the circuits.

In order to gain the highest reliability with a ring bus scheme, load and source circuits should be alternated when connecting to the scheme. Arranging the scheme in this manner will minimize the potential for the loss of the supply to the ring bus do to a breaker failure.

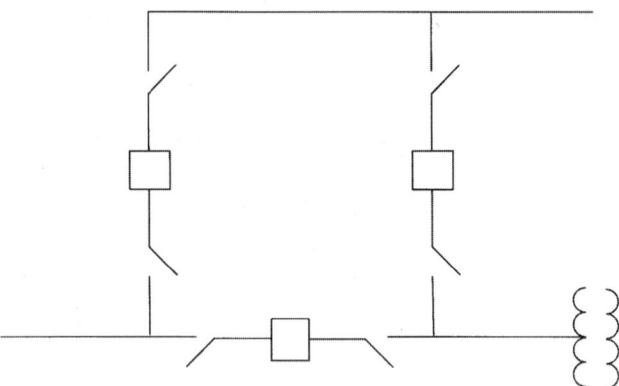

FIGURE 5.21 Ring bus.

Relaying is more complex in this scheme than some previously identified. Since there is only one bus in this scheme, the area required to develop this scheme is less than some of the previously discussed schemes. However, expansion of a ring bus is limited, due to the practical arrangement of circuits.

Breaker-and-a-Half (Fig. 5.22)

The breaker-and-a-half scheme can be developed from a ring bus arrangement as the number of circuits increase. In this scheme, each circuit is between two circuit breakers, and there are two main buses. The failure of a circuit will trip the two adjacent breakers and not interrupt any other circuit. With the three breaker arrangement for each bay, a center breaker failure will cause the loss of the two adjacent circuits. However, a breaker failure of the breaker adjacent to the bus will only interrupt one circuit.

Maintenance of a breaker on this scheme can be performed without an outage to any circuit. Furthermore, either bus can be taken out of service with no interruption to the service.

This is one of the most reliable arrangements, and it can continue to be expanded as required. Relaying is more involved than some schemes previously discussed. This scheme will require more area and is costly due to the additional components.

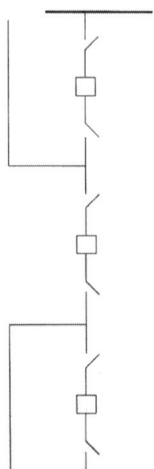

FIGURE 5.22 Breaker-and-a-half.

Substations

TABLE 5.1 Comparison of Configurations

Configuration	Reliability	Cost	Available Area
Single bus	Least reliable — single failure can cause complete outage	Least cost (1.0) — fewer components	Least area — fewer components
Double bus	Highly reliable — duplicated components; single failure normally isolates single component	High cost (1.8) — duplicated components	Greater area — twice as many components
Main bus and transfer	Least reliable — same as *Single bus*, but flexibility in operating and maintenance with transfer bus	Moderate cost (1.76) — fewer components	Low area requirement — fewer components
Double bus, single breaker	Moderately reliable — depends on arrangement of components and bus	Moderate cost (1.78) — more components	Moderate area — more components
Ring bus	High reliability — single failure isolates single component	Moderate cost (1.56) — more components	Moderate area — increases with number of circuits
Breaker-and-a-half	Highly reliable — single circuit failure isolates single circuit, bus failures do not affect circuits	Moderate cost (1.57) — breaker-and-a-half for each circuit	Greater area — more components per circuit

Note: The number shown in parenthesis is a per unit amount for comparison of configurations.

Comparison of Configurations

In planning an electrical substation or switchyard facility, one should consider major parameters as discussed above: reliability, cost, and available area. Table 5.1 has been developed to provide specific items for consideration.

In order to provide a complete evaluation of the configurations described, other circuit-related factors should also be considered. The arrangement of circuits entering the facility should be incorporated in the total scheme. This is especially true with the ring bus and breaker and a half schemes, since reliability in these schemes can be improved by not locating source circuits or load circuits adjacent to each other. Arrangement of the incoming circuits can add greatly to the cost and area required.

Second, the profile of the facility can add significant cost and area to the overall project. A high-profile facility can incorporate multiple components on fewer structures. Each component in a low-profile layout requires a single area, thus necessitating more area for an arrangement similar to a high-profile facility.

Therefore, a four-circuit, high-profile ring bus may require less area and be less expensive than a four-circuit, low-profile main and transfer bus arrangement.

5.3 High Voltage Switching Equipment

David L. Harris

The design of the high voltage substation must include consideration for the safe operation and maintenance of the equipment. Switching equipment is used to provide isolation, no load switching, load switching, and/or interruption of fault currents. The magnitude and duration of the load and fault currents will be significant in the selection of the equipment used.

System operations and maintenance must also be considered when equipment is selected. One significant choice is the decision of single-phase or three-phase operation. High voltage power systems are generally operated as a three-phase system, and the imbalance that will occur when operating equipment in a single-phase mode, must be considered.

Ambient Conditions

Air-insulated high voltage electrical equipment is generally covered by standards based on assumed ambient temperatures and altitudes. Ambient temperatures are generally rated over a range from –40°C to

+40°C for equipment that is air-insulated and dependent on ambient cooling. Altitudes above 1000 meters (3300 feet) may require derating.

At higher altitudes, air density decreases, hence the dielectric strength is also reduced and derating of the equipment is recommended. Operating (strike distances) clearances must be increased to compensate for the reduction in dielectric strength of the ambient air. Also, current ratings generally decrease at higher elevations due to the decreased density of the ambient air, which is the cooling medium used for dissipation of the heat generated by the load losses associated with load current levels.

Disconnect Switches

A disconnect switch is a mechanical device used to change connections within a circuit or isolate a circuit from its power source, and are normally used to provide isolation of the substation equipment for maintenance. Typically a disconnect switch would be installed on each side of a piece of equipment to provide a visible confirmation that the power conductors have been opened for personnel safety. Once the switches are placed in the open position, safety grounds can be attached to the de-energized equipment for worker protection. Switches can be equipped with grounding blades to perform the safety grounding function.

Disconnect switches are designed to continuously carry load currents and momentarily carry higher capacity for short-circuit currents for a specified duration (typically specified in seconds). They are designed for no load switching, opening or closing circuits where negligible currents are made or interrupted, or when there is no significant voltage across the open terminals of the switch. They are relatively slow-speed operating devices and therefore are not designed for arc interruption. Disconnect switches are also installed to bypass breakers or other equipment for maintenance and can also be used for bus sectionalizing. Interlocking equipment is available to prevent inadvertent operating sequence by inhibiting operation of the disconnect switch operation until the fault and/or load currents have been interrupted by the appropriate equipment.

Single-phase or three-phase operation is possible for some switches. Operating mechanisms are normally installed to permit operation of the disconnect switch by an operator standing at ground level. The operating mechanisms provide a swing arm or gearing to permit operation with reasonable effort by utility personnel. Motor operating mechanisms are also available and are applied when remote switching is necessary.

Disconnect switch operation can be designed for vertical or horizontal operating of the switch blades. Several configurations are frequently used for switch applications including:

- Vertical break
- Double break switches
- V switches
- Center-break switches
- Hook stick switches
- Vertical reach switches
- Grounding switches

Phase spacing is usually adjusted to satisfy the spacing of the bus system installed in the substation.

Load Break Switches

A load break switch is a disconnect switch that has been designed to provide making or breaking of specified currents. This is accomplished by addition of equipment that increases the operating speed of the disconnect switch blade and the addition of some type of equipment to alter the arcing phenomena and allow the safe interruption of the arc resulting when switching load currents.

Disconnect switches can be supplied with equipment to provide a limited load switching capability. Arcing horns, whips, and spring actuators are typical at lower voltages. These switches are used to de-energize or energize a circuit that possesses some limited amount of magnetic or capacitive current, such as transformer exciting current or line charging currents.

An air switch can be modified to include a series interrupter (typically vacuum or SF6) for higher voltage and current interrupting levels. These interrupters increase the load break capability of the disconnect switch and can be applied for switching load or fault currents of the associated equipment.

High Speed Grounding Switches

Automatic high-speed grounding switches are applied for protection of transformer banks when the cost of supplying other protective equipment is too costly. The switches are generally actuated by discharging a spring mechanism to provide the "high-speed" operation. The grounding switch operates to provide a deliberate ground on the high voltage bus supplying the equipment (generally a transformer bank), which is detected by protective relaying equipment remotely, and operates the transmission line breakers at the remote end of the line supplying the transformer. This scheme also imposes a voltage interruption to all other loads connected between the same remote breakers. A motor-operated disconnect switch is frequently installed along with a relay system to sense bus voltage and allow operation of a motor-operated disconnect switch when there is no voltage on the transmission line to provide automatic isolation of the faulted bank, and allow reclosing operation of the remote breaker to restore service to the transmission line.

The grounding switch scheme is dependent on the ability of the source transmission line relay protection scheme to recognize and clear the fault by opening the remote line circuit breaker. Clearing times are necessarily longer since the fault levels are not normally within the levels appropriate for an instantaneous trip response. The lengthening of the trip time also imposes additional stress on the equipment being protected and should be considered when selecting this method for bank protection. Grounding switches are usually considered when relative fault levels are low so that there is not the risk of significant damage to the equipment with the associated extended trip times.

Power Fuses

Power fuses are a generally accepted means of protecting power transformers in distribution substations. The primary purpose of a power fuse is to provide interruption of permanent faults. Fusing is an economical alternative to circuit switcher or circuit breaker protection. Fuse protection is generally limited to voltages from 34.5 kV through 69 kV, but has been applied for protection of 115-kV and 138-kV transformers.

To provide the greatest protective margin, it is necessary to use the smallest fuse rating possible. The advantage of close fusing is the ability of the fuse unit to provide backup protection for some secondary faults. For the common delta-wye connected transformer, a fusing ratio of 1.0 would provide backup protection for a phase-to-ground fault as low as 230% of the secondary full-load rating. Fusing ratio is defined as the ratio of the fuse rating to the transformer full load current rating. With low fusing ratios, the fuse may also provide backup protection for line-to-ground faults remote to the substation on the distribution network.

Fuse ratings also must consider parameters other than the full load current of the transformer being protected. Coordination with other overcurrent devices, accommodation of peak overloadings, and severe duty may require increased ratings of the fuse unit. The general purpose of the power transformer fuse is to accommodate, not interrupt, peak loads. Fuse ratings must consider the possibility of nuisance trips if the rating is selected too low for all possible operating conditions.

The concern of unbalanced voltages in a three-phase system must be considered when selecting fusing for power transformer protection. The possibility of one or two fuses blowing must be reviewed. Unbalanced voltages can cause tank heating in three-phase transformers and overheating and damage to three-phase motor loads. The potential for ferroresonance must be considered for some transformer configurations when using fusing.

Fuses are available in a number of tripping curves (standard, slow, and very slow) to provide coordination with other system protective equipment. Fuses are not voltage-critical; they may be applied at any voltage equal to or greater than their rated voltage. Fuses may not require additional structures, and are generally mounted on the incoming line structure, resulting in space savings in the substation layout.

Circuit Switchers

Circuit switchers have been developed to overcome some of the limitations of fusing for substation transformers. They are designed to provide three-phase interruption (solving the unbalanced voltage considerations) and provide protection for transient overvoltages and overloads at a competitive cost between the costs of fuses and circuit breakers. Additionally, they can provide protection from transformer faults based on differential, sudden pressure, and overcurrent relay schemes as well as critical operating constraints such as low oil level, high oil or winding temperature, pressure relief device operation, and others.

Circuit switchers are designed and supplied as a combination of a circuit breaking interrupter and an isolating disconnect switch. Later models have been designed with improved interrupters that have reduced the number of gaps and eliminated the necessity of the disconnect switch blades in series with the interrupter. Interrupters are now available in vertical or horizontal mounting configurations, with or without an integral disconnect switch. Circuit switchers have been developed for applications involving protection of power transformers, lines, capacitors, and line connected or tertiary connected shunt reactors.

Circuit switchers are an alternative to the application of circuit breakers for equipment protection. Fault duties may be lower and interrupting times longer than a circuit breaker. Some previous designs employed interrupters with multiple gaps and grading resistors and the integral disconnect switch as standard. The disconnect switch was required to provide open-circuit isolation in some earlier models of circuit switchers.

Circuit switchers originally were intended to be used for transformer primary protection. Advancements in the interrupter design have resulted in additional circuit switcher applications, including:

- Line and switching protection
- Cable switching and protection
- Single shunt capacitor bank switching and protection
- Shunt reactor switching and protection (line connected or tertiary connected reactors)

Circuit Breakers

A circuit breaker is defined as "a mechanical switching device capable of making, carrying and breaking currents under normal circuit conditions and also making, carrying and breaking for a specified time, and breaking currents under specified abnormal circuit conditions such as a short circuit" (IEEE Std. C37.100-1992).

Circuit breakers are generally classified according to the interrupting medium used to cool and elongate the electrical arc permitting interruption. The types are:

- Air magnetic
- Oil
- Air blast
- Vacuum
- SF6 gas

Air magnetic circuit breakers are limited to older switchgear and have generally been replaced by vacuum or SF6 for switchgear applications. Vacuum is used for switchgear applications and some outdoor breakers, generally 38 kV class and below. Air blast breakers, used for high voltages (≥765 kV), are no longer manufactured and have been replaced by breakers using SF6 technology.

Oil circuit breakers have been widely used in the utility industry in the past but have been replaced by other breaker technologies for newer installations. Two designs exist — bulk oil (dead tank designs) dominant in the U.S.; and oil minimum breaker technology (live tank design). Bulk oil circuit breakers were designed as single-tank or three-tank mechanisms; generally, at higher voltages, three-tank designs were dominant. Oil circuit breakers were large and required significant foundations to support the weight and impact loads occurring during operation. Environmental concerns forcing the necessity of oil retention systems, maintenance costs, and the development of the SF6 gas circuit breaker have led to the gradual replacement of the oil circuit breaker for new installations.

Oil circuit breaker development has been relatively static for many years. The design of the interrupter employs the arc caused when the contacts are parted and the breaker starts to operate. The electrical arc generates hydrogen gas due to the decomposition of the insulating mineral oil. The interrupter is designed to use the gas as a cooling mechanism to cool the arc and to use the pressure to elongate the arc through a grid (arc chutes), allowing extinguishing of the arc when the current passes through zero.

Vacuum circuit breakers use an interrupter that is a small cylinder enclosing the moving contacts under a high vacuum. When the contacts part, an arc is formed from contact erosion. The arc products are immediately forced to and deposited on a metallic shield surrounding the contacts. Without anything to sustain the arc, it is quickly extinguished.

Vacuum circuit breakers are widely employed for metal-clad switchgear up to 38 kV class. The small size of the breaker allows vertically stacked installations of breakers in a two-high configuration within one vertical section of switchgear, permitting significant savings in space and material compared to earlier designs employing air magnetic technology. When used in outdoor circuit breaker designs, the vacuum cylinder is housed in a metal cabinet or oil-filled tank for dead tank construction popular in the U.S. market.

Gas circuit breakers generally employ SF6 (sulfur hexaflouride) as an interrupting and sometimes as an insulating medium. In "single puffer" mechanisms, the interrupter is designed to compress the gas during the opening stroke and use the compressed gas as a transfer mechanism to cool the arc and to elongate the arc through a grid (arc chutes), allowing extinguishing of the arc when the current passes through zero. In other designs, the arc heats the SF6 gas and the resulting pressure is used for elongating and interrupting the arc. Some older two-pressure SF6 breakers employed a pump to provide the high-pressure SF6 gas for arc interruption.

Gas circuit breakers typically operate at pressures between six and seven atmospheres. The dielectric strength of SF6 gas reduces significantly at lower pressures, normally as a result of lower ambient temperatures. Monitoring of the density of the SF6 gas is critical and some designs will block operation of the circuit breaker in the event of low gas density.

Circuit breakers are available as live-tank or dead-tank designs. Dead-tank designs put the interrupter in a grounded metal enclosure. Interrupter maintenance is at ground level and seismic withstand is improved versus the live-tank designs. Bushings are used for line and load connections which permit installation of bushing current transformers for relaying and metering at a nominal cost. The dead-tank breaker does require additional insulating oil or gas to provide the insulation between the interrupter and the grounded tank enclosure.

Live-tank circuit breakers consist of an interrupter chamber that is mounted on insulators and is at line potential. This approach allows a modular design as interrupters can be connected in series to operate at higher voltage levels. Operation of the contacts is usually through an insulated operating rod or rotation of a porcelain insulator assembly by an operator at ground level. This design minimizes the quantity of oil or gas used for interrupting the arc as no additional quantity is required for insulation of a dead-tank enclosure. The design also readily adapts to the addition of pre-insertion resistors or grading capacitors when they are required. Seismic capability requires special consideration due to the high center of gravity of the interrupting chamber assembly.

Interrupting times are usually quoted in cycles and are defined as the maximum possible delay between energizing the trip circuit at rated control voltage and the interruption of the main contacts in all poles. This applies to all currents from 25 to 100% of the rated short-circuit current.

Circuit breaker ratings must be examined closely. Voltage and interrupting ratings are stated at a maximum operating voltage rating, i.e., 38 kV voltage rating for a breaker applied on a nominal 34.5-kV circuit. The breakers have an operating range designated as K factor per IEEE C37.06, (see Table 3 in the document's appendix). For a 72-kV breaker, the voltage range is 1.21, indicating that the breaker is capable of its full interrupting rating down to a voltage of 60 kV.

Breaker ratings need to be checked for some specific applications. Applications requiring reclosing operation should be reviewed to be sure that the duty cycle of the circuit breaker is not being exceeded. Some applications for out-of-phase switching or back-to-back switching of capacitor banks also require review and may require specific-duty circuit breakers to insure proper operation of the circuit breaker during fault interruption.

GIS Substations

Advancements in the use of SF6 as an insulating and interrupting medium have resulted in the development of gas insulated substations. Environmental and/or space limitations may require the consideration of GIS (gas insulated substation) equipment. This equipment utilizes SF6 as an insulating and interrupting medium and permits very compact installations.

Three-phase or single-phase bus configurations are normally available up to 145 kV class, and single-phase bus to 500 kV and higher, and all equipment (disconnect/isolating switches, grounding switches, circuit breakers, metering current, and potential transformers, etc.) are enclosed within an atmosphere of SF6 insulating gas. The superior insulating properties of SF6 allow very compact installations.

GIS installations are also used in contaminated environments and as a means of deterring animal intrusions. Although initial costs are higher than conventional substations, a smaller substation footprint can offset the increased initial costs by reducing the land area necessary for the substation.

Environmental Concerns

Environmental concerns will have an impact on the siting, design, installation, maintenance, and operation of substation equipment.

Sound levels, continuous as well as momentary, can cause objections. The operation of a disconnect switch, switching cables, or magnetizing currents of a transformer will result in an audible noise associated with the arc interruption in air. Interrupters can be installed to mitigate this noise. Closing and tripping of a circuit breaker will result in an audible momentary sound from the operating mechanism. Transformers and other magnetic equipment will emit continuous audible noise.

Oil insulated circuit breakers and power transformers may require the installation of systems to contain or control an accidental discharge of the insulating oil and prevent accidental migration beyond the substation site. Lubricating oils and hydraulic fluids should also to be considered in the control/containment decision.

References

American National Standard for Switchgear — AC High-Voltage, IEEE Std. C37.06-1997, Circuit Breakers Rated on a Symmetrical Current Basis — Preferred Ratings and Related Required Capabilities.
IEEE Guide for Animal Deterrents for Electric Power Supply Substations, IEEE Std. 1264-1993.
IEEE Guide for Containment and Control and Containment of Oil Spills in Substations, IEEE Std. 980-1994.
IEEE Guide for the Design, Construction and Operation of Safe and Reliable Substations for Environmental Acceptance, IEEE Std. 1127-1998.
IEEE Guide for Gas-Insulated Substations, IEEE Std. C37.122.1-1993.
IEEE Standards Collection: Power and Energy — Substations, 1998.
IEEE Standards Collection: Power and Energy — Switchgear, 1998.
IEEE Standard for Interrupter Switches for Alternating Current, Rated Above 1000 Volts, IEEE Std. 1247-1998.
IEEE Standard Definitions for Power Switchgear, IEEE Std. C37.100-1992.
IEEE Standard for Gas-Insulated Substations, IEEE Std. C37.122-1993.

5.4 High Voltage Power Electronics Substations

Gerhard Juette

Details on power electronics are provided in Chapter 15, whereas gas insulated and air insulated substations in general are covered in Sections 5.1 and 5.2 of this chapter. This section focuses on the specifics of power electronics as applied in substations for power transmission purposes.

The dramatic development of power electronics in the past few decades made significant progress in electric power transmission technology possible, resulting in special types of transmission substations.

The most important high voltage power electronics substations are frequency converters, above all for High Voltage Direct Current (HVDC) transmission, and controllers for Flexible AC Transmission Systems (FACTS), including electric energy storage, Static VAr Compensators (SVC), Static Compensator (STATCOM), Thyristor Controlled Series Compensation (TCSC), and the Unified Power Flow Controller (UPFC). Detailed descriptions of these circuits can be found in Chapter 15, Power Electronics.

In addition to the conventional substation elements covered in Sections 5.1 and 5.2, high voltage power electronic substations typically include harmonic filtering and reactive compensation equipment, as well as the main power electronic equipment with its dedicated transformers, buildings, coolers, and auxiliaries.

Most high voltage power electronics substations are air insulated, although some use combinations of air and gas insulation. Typically, passive filters and reactive compensation equipment are air insulated and outdoors, while power and control electronics, active filters, and most communication and auxiliary systems are air insulated, but indoors.

Basic community considerations, grounding, lightning protection, seismic, and fire protection considerations also apply. In addition, high voltage power electronic substations may emit characteristic electric and acoustic noise and may require extra fire protection.

The IEEE, CIGRE, and other international technical societies continue to develop technical standards, disseminate information, and facilitate the exchange of know-how in this high-tech power engineering field. Within the IEEE, the following group is concerned with high voltage power electronics substations: Power Engineering Society (PES) — Substations Committee — High Voltage Power Electronics Stations Subcommittee. The related Web page is: http://home.att.net/(gengmann/.

Types

The high voltage power electronic circuits covered here are frequency converters, with an emphasis on HVDC and FACTS controllers, including recent developments.

Frequency Converters (HVDC)

Frequency converters transmit power between systems with different constant or variable frequencies. Some examples are converters between variable speed machines and power grids, energy storage converters, converters for railroad systems, and, most importantly, HVDC (High Voltage Direct Current transmission).

HVDC converters convert AC power to DC power and vice versa. They terminate DC transmission lines and cables or form back-to-back asynchronous AC system couplings. When connected to DC transmission lines, the converter voltages can be on the order of a million volts (±500 kV) and power ratings can reach thousands of megawatts. With back-to-back converters, where DC line economies are not a consideration, the DC voltage and current are chosen so as to minimize converter cost. This choice results in DC voltages up to and exceeding 100 kV at power ratings up to several hundred megawatts.

Most HVDC converters are line commutated Graetz bridge converters that require substantial AC harmonic filtering and reactive compensation. Converters connected to DC lines have harmonic filters on the DC side as well. Traditionally, passive filters which consist of capacitors, reactors, and resistors have been used. Recently, self-commutated HVDC converters are being introduced (Torgerson et al., 1997; Asplund, 1998; EPRI, 1998), as are active (electronic) AC and DC harmonic filters (Pereira, 1995; Andersen, 1997), using GTO thyristors, IGBTs, and other devices with gate-turn-off capability.

The AC system or systems to which a converter station is connected significantly impact the station's design in many ways. This is true for harmonic filters, reactive compensation devices, fault duties, and insulation coordination. Weak AC systems (i.e., low short-circuit ratios), represent special challenges for the design of HVDC converters (IEEE Std. 1204-1997). Some stations include temporary overvoltage limiting devices, which consist of MOV arresters with forced cooling or fast switches (deLaneuville et al., 1991).

Many converter stations, HVDC stations in particular, require DC voltage insulation coordination. Internal equipment insulation, for example the insulation of transformers and bushings, must take the DC voltage gradient distribution in solid and mixed dielectrics into account. Substation clearances and creepage distances must have the proper dimensions. Standards for indoor and outdoor clearances and creepage distances are currently being promulgated (CIGRE Working Group 33-05, 1984). DC electric fields enhance the pollution of exposed surfaces. This pollution, particularly in combination with water, can adversely influence the conductivity, voltage distribution, and withstand capability of insulating surfaces. Therefore, it is especially important with converter stations to apply either grease, special compounds, and/or booster sheds, and to engage in adequate cleaning practices. Insulation problems with extra high voltage DC bushings continue to be a matter of concern and study (Schneider et al., 1991; Porrino et al., 1995).

A specific issue with DC transmission is the use of ground return. Used during contingencies, ground (and/or sea) return can increase the economy and availability of long distance HVDC transmission. The necessary electrodes are usually located at some distance from the station, with a neutral line that leads to them. The related neutral buswork, switching devices, and protection systems form part of the station. Electrode design depends on the local soil or water conditions (Tykeson et al., 1996; Holt et al., 1997). The National Electric Safety Code (NESC) does not allow the use of ground as the sole conductor. Monopolar HVDC operation is permitted only under emergencies and for a limited time. The IEEE-PES is working toward the introduction of changes to the code to better meet the needs of HVDC transmission while addressing potential side effects to other systems.

Disconnect switches are the only mechanical switching devices on the DC side of a typical HVDC converter station. No true DC breakers exist and DC fault currents are best and fastest interrupted by the converters themselves. Mechanical breaker systems with limited DC current interrupting capability have been developed (Vithayathil et al., 1985). They include commutation circuits, i.e., parallel L/C resonance circuits that create current zeroes across the breaker contacts. So far, only experimental "DC breakers" have ever been installed in actual HVDC stations.

Figures 5.23 through 5.26 show photos of different converter stations. The station shown in Fig. 5.23 is one of several asynchronous links between the western and eastern North American power grids. In the photo, one can recognize the control building (next to the communication tower), the valve hall attached to it, the converter transformers on both sides, the AC filter circuits (near the center line), as well as the AC buses (at the outer left and right) with major reactive power compensation and TOV suppression equipment which was used in this low-short-circuit-ratio installation. The valve groups shown in Fig. 5.24 are arranged back-to-back, i.e., across the aisle in the same room. Fig. 5.25 shows a 500 kV long-distance HVDC valve hall, with the valves suspended from the ceiling to withstand major seismic events. The converter station shown in Fig. 5.26 is the south terminal of the Nelson River ± 500 kV HVDC transmission sytem in Manitoba, Canada. It consists of two bipoles completed in 1973 and 1985. The DC yard and line connections are on the left side, while the 230 kV AC yard with harmonic filters and converter transformers is on the right side of the picture. In total, the station is rated 3670 MW.

FACTS Controllers

Different types of FACTS controllers and the theory underlying their function are covered in Chapter 15, Power Electronics. Typical ratings of these controllers range from about thirty to several hundred MVA. Normally, FACTS controllers are an integral part of AC substations. Like HVDC, they require controls, cooling systems, harmonic filters, fixed capacitors, transformers, and building facilities.

Substations

FIGURE 5.23 200 MW HVDC back-to-back converter station, at Sidney, Nebraska. (Photo courtesy Siemens.)

Static VAr Compensators (SVC) have been used successfully for many years, either for load (flicker) compensation of large industrial loads, or for transmission compensation in utility systems. Harmonic filter and capacitor banks, as well as (normally air core) reactors, step down transformers, a building, breakers and disconnect switches on the high voltage side, and heavy duty busbars on the medium voltage side characterize most SVC stations. The power electronic (thyristor) controllers can have air or liquid cooling. A new type of controlled shunt compensator, called STATCOM, uses voltage source inverters (Schander et al., 1995). STATCOM requires fewer harmonic filters and capacitors than an SVC, and no reactors at all. This fact makes the footprint of a STATCOM station significantly more compact than that of a conventional SVC.

Thyristor Controlled Series Capacitors (TCSC) (Piwko et al., 1994; Montoya et al., 1990) involve insulated platforms at phase potential, with weatherproof valve housings, as well as communication links between platform and ground. Liquid cooling requires ground-to-platform pipes made of insulating material. Auxiliary platform power, where needed, is extracted from the line current via CTs. As most conventional SCs, TCSCs are typically integrated into existing substations. Upgrading an existing SC to TCSC is generally possible.

While SVC and STATCOM controllers are shunt devices, and TCSC are series devices, the UPFC (Unified Power Flow Controller) is a combination of both (Mehraban et al., 1996). It uses a shunt connected transformer and a transformer with series connected line windings, both of which are interconnected to a DC capacitor via related power electronic circuits in the control building. The newest FACTS station project (Fardanesh et al., 1998) involves similar shunt and series elements as the UPFC, and can be reconfigured to meet changing system requirements.

The ease with which FACTS stations can be reconfigured or even relocated, may be important and can influence the substation design (Renz et al., 1994; Knight et al., 1998). Changes in generation and load patterns may make such flexibility desirable.

Figures 5.27 through 5.31 are photos of FACTS substations. In Fig. 5.27, one can distinguish the 500-kV feeder (on the left side), the transformers (three single-phase units plus one spare), the medium voltage bus work and three Thyristor Switched Capacitor (TSC) banks, as well as the building which houses the thyristor switches and controls. The SVC shown in Fig. 5.28 is also connected to 500 kV. It uses Thyristor

FIGURE 5.24 600 MW HVDC back-to-back converter valves. (Photo courtesy Siemens.)

Controlled Reactors (TCRs) and TSCs, which are visible together with the 11 kV high current buswork behind the building. The harmonic filters are before the building and are not visible in the photo. Figure 5.29 shows the three 500-kV platforms of one of the world's first commercial TCSC installations in Brazil. The platform-mounted valve housings are clearly visible. Figure 5.30 shows an SVC being relocated. The controls and valves are in container-like e-houses, which allow for faster relocation. Figure 5.31 is a photo of the world's first Unified Power Flow Controller (UPFC) connected to AEP's "Inez" 138-kV substation in eastern Kentucky, U.S. The main components are identified and clearly visible.

Control

HVDC and FACTS controllers allow steady-state, quasi-steady-state, dynamic, and transient control actions and they provide important equipment and system protection functions. Fault monitoring and

FIGURE 5.25 Valve hall of ±500 kV, 1000 MW long-distance HVDC converter station. (Photo courtesy Siemens.)

FIGURE 5.26 Dorsey Terminal of the Nelson River HVDC transmission system. (Photo courtesy Manitoba Hydro.)

sequence of event recording equipment are used in most power electronics stations. Typically, these stations are remotely controlled and offer full local controllability. Man-machine interfaces are often highly computerized, with extensive supervision and control being exercised via monitor and keyboard. All of these functions add to the basic substation secondary systems described in Section 5.4.

FIGURE 5.27 500 kV, 400 MVAr SVC at Adelanto, California. (Photo courtesy Siemens.)

FIGURE 5.28 500 kV, 400 MVAr SVC Chinu, I.S.A., Colombia. (Photo courtesy Siemens.)

One of the most complex control algorithms applies to HVDC converters. Real power, reactive power, AC bus frequency and voltage, start-up and shut-down sequences, contingency and fault recovery sequences, remedial action schemes, modulation schemes for system oscillation and SSR damping, and loss of communication are some of the applicable control parameters and conditions. Special v/i control characteristics are used for converters in multi-terminal HVDC systems to allow their safe operation even under a loss of inter-station communication. Furthermore, HVDC controls provide equipment and system protection, such as thyristor overcurrent, thyristor temperature, and DC line fault protection.

FIGURE 5.29 TCSC Serra da Mesa, FURNAS, Brazil, 500 kV, 107MVAr, (1...3)×13.17Ω. (Photo courtesy Siemens.)

FIGURE 5.30 Static Var compensator is relocated where the system needs it. (Photo courtesy ALSTOM T&D Power Electronic Systems.)

Other converter stations and FACTS stations use different specific controls in addition to the basic secondary systems described in this section. Industrial SVCs often have open-loop, direct load compensation control. Static shunt and series compensators in transmission systems provide closed-loop, steady-state, and dynamic reactive power and bus voltage control, some degree of load flow control, with

FIGURE 5.31 UPFC at Inez substation. (Photo courtesy American Electric Power.)

modulation loops for stability and SSR mitigation, as well as equipment and system protection functions. The complexity of the control for the UPFC may well rival that of HVDC converters.

Most controllers included here have the potential to provide power system damping, i.e., to improve system stability. By the same token, if not properly designed, they may add to, or even create, system undamping, especially of SSR. It is imperative to pay proper attention to SSR in the control design and function testing of power electronic stations, near existing or planned turbo generators in particular.

The following principal hard- and software structure applies to the control systems described above:

- Valve firing and monitoring circuits
- Main (closed-loop) control
- Open-loop control (sequences, interlocks, etc.)
- Protective functions
- Monitoring and alarms
- Diagnostic functions
- Operator interface and communications
- Data handling

Figures 5.32 and 5.33 show parts of HVDC control systems. The first photo depicts the local control interface of a back-to-back converter used for power transmission between non-synchronous grids. The second photo was taken during the real-time simulator function testing of the AC and DC control panel of a major long distance HVDC converter.

Losses and Cooling

Converter and controller valve losses are comparable in magnitude to those of the associated transformers. Standard procedures to determine and evaluate high voltage power electronic substation losses have been developed (IEC). As opposed to the relatively even distribution of the losses in transformers, there exist areas of extreme loss density in power electronic equipment. This and their location inside a building makes special cooling techniques necessary for the valves. Deionized water in a closed loop is often used as the prime valve coolant. Various types of dry or evaporative secondary coolers dissipate the heat, usually into the surrounding air.

Substations

FIGURE 5.32 Local control desk in 600 MW back-to-back converter station. (Photo courtesy Siemens.)

FIGURE 5.33 Control for ±500 kV, 1800MW HVDC; function test. (Photo courtesy Siemens.)

Buildings

High voltage power electronic substation buildings are special in that they house specific control equipment and the high voltage power electronic valves. Insulation clearance requirements can lead to large valve rooms (halls). The valves are connected to the yard through wall bushings. Converter transformers are sometimes placed adjacent to the valve building with the valve side bushings penetrating through the walls in order to save space.

Not only do the valves require controlled air temperature, humidity, and cleanliness, but they also generate losses that can be on the order of hundreds of kilowatts, and even up to several megawatts with HVDC. The major part of these losses is handled by the equipment cooling system, but some fraction of them is dissipated into the valve room and adds to its air conditioning load. The valves also generate electromagnetic and acoustic noise. Therefore, valve rooms are usually shielded electrically with wire mesh in walls and windows. Walls, windows, and doors must also assure that legal or contractual limits for acoustic noise inside and outside the building are met. Of course, national and local building codes also apply.

Extreme electric power flow densities and voltages in the valves create a certain risk of fire. Valve fires with more or less severe consequences have happened in the past (Krishnayya et al., 1994). Improved internal valve insulation materials and designs, as well as special fire protection, reduce this risk. In addition to firewalls and sprinkler systems around the transformers, several different fire detection and extinction systems are often used to protect the valves.

In addition to the actual valve and control rooms, power electronic substation buildings typically include rooms for coolant pumps and treatment devices, for auxiliary power distribution systems, and for air conditioning systems, battery rooms, and communication rooms.

Some HVDC schemes use outdoor valves. They avoid the cost of large valve buildings at the expense of more complicated valve maintenance.

Interference

The periodic fast switching of electronic converter and controller valves causes a wide spectrum of harmonic currents and electromagnetic fields as well as significant audible noise. Electric interference with radio, TV, and communication systems can usually be controlled with power line carrier filters and harmonic filters and by shielding the valve room. Audible noise from the valves, transformers, capacitors, reactors, and coolers must be controlled to conform with acceptable limits within the building and outside the station. This is not always easy and may require low-noise equipment, noise damping walls, barriers, and a special arrangement of equipment in the yard. The theory of audible noise propagation is well understood (Beranek, 1988) and analytical tools for audible noise design are available (Smede et al., 1995). Specified noise limits can thus be met, but doing so may impact the total station layout and cost.

Reliability

Together with all other components, power electronics and their controls and auxiliary circuits naturally add to substation unavailability. The complex structure and large number of interdependent parts within power electronic systems have an inherent probability for outages. On the other hand, they also allow the effective application of advanced reliability analysis and design. By means of built-in redundancy, detailed monitoring, self-supervision, segmentation, and automatic switch-over strategies, together with consistent quality control and prudent operation and maintenance, one can achieve almost any level of reliability. HVDC converters have enjoyed the highest level of scrutiny, systematic monitoring, and standardized international reporting of reliability design and performance. CIGRE has developed a reporting system (Protocol for Reporting the Operational Performance…, CIGRE) and publishes bi-annual HVDC station reliability reports (Christofersen et al., 1996). At least one publication discusses the importance of substation operation and maintenance practices on actual reliability (Cochrane et al., 1995). The IEEE has issued a guide for HVDC converter reliability (IEEE Std. 1240). Other high voltage power electronic technologies have benefited from these efforts as well. Power electronic systems in substations have reached levels of reliability comparable to the balance of substation components. Reliability, availability, and maintainability (RAM) are frequent terms used in major high voltage power electronic substations specifications (Vancers et al., 1993) and contracts.

Specifications

High voltage power electronic systems warrant detailed specifications to assure successful implementation. In addition to applicable industry and owner standards for conventional substations and equipment, many specific conditions and requirements need to be defined for high voltage power electronic substations. To facilitate the introduction of advanced power electronic technologies in substations, the IEEE

Training and Commissioning

Operation and maintenance training are important for the success of high voltage power electronic substation projects. A substantial part of this training is best performed during commissioning. The IEEE and other organizations have, to a large degree, standardized high voltage power electronic component and substation testing and commissioning procedures (IEEE Std. 857-1996; IEEE Std. 1378-1997; IEEE Std. 1303-2000). Real-time digital system simulators have become a major tool for the off-site function testing of all controls, thus reducing the amount of actual on-site testing. Nonetheless, staged fault tests are still performed with power electronic substations including, for example, with the Kayenta TCSC (Weiss et al., 1996).

The Future

In the future, one can expect increased application of power electronics in transmission systems. Innovations such as the voltage-source converter (Zhao and Iravani, 1994) or the capacitor-commutated converter (Bjorklund and Johnsson, 1995), active filters, outdoor valves (Asplund et al., 1995), or the transformer-less converter (Vithayathil et al., 1995) may reduce the complexity of HVDC converter stations (Carlsson et al., 1994). New and more economical FACTS technologies may be introduced. Self-commutated converters and active filters will change the footprint of high voltage power electronic substations. Eventually, STATCOM may replace rotating synchronous condensers and TCSC or UPFC may replace phase shifting transformers to some degree. New developments such as electronic transformer tap changers, semiconductor breakers, electronic fault current limiters, and arresters may even affect the "conventional" parts of the substation. As a result, the high voltage power electronics substations of the future will be more common, more effective, more compact, and easier to relocate.

References

The following list provides somewhat random access to the vast bibliography on high voltage power electronic substations. Even within its limited focus on the practical substation aspect, this list is far from complete. The reader may consult some of the other reference lists in this book, as well as ongoing conferences, proceedings, and publications provided by the IEEE-PES and other power engineering societies.

Andersen, N., Gunnarsson, S., Pereira, M., Fritz, P., Damstra, G.C., Enslin, J.H.R., and O'Lunelli, D., *Active filters in HVDC applications*, Proceedings, CIGRE International Colloquium on HVDC and FACTS, Johannesburg, South Africa, September 1997.

Application Guide for Insulation Coordination and Arrester Protection of HVDC Converter Stations, CIGRE Working Group 33-05, Electra No. 96, 101-156, October 1984.

Asplund, G. et al., Outdoor thyristor valve for HVDC, in *Proceedings, IEEE/Royal Institute of Technology Sockholm Power Tech; Power Electronics*, June 18–22, 1995.

Asplund, G., Eriksson, K., Svensson, K., Jiang, H., Lindberg, J., and Palsson, R., DC transmission based on voltage source converters, CIGRE, Paper 14-302, Paris, France, 1998.

Beranek, L.L., *Noise and Vibration Control*, New York, McGraw Hill, 1971, rev. ed., Institute of Noise Control Engineering, 1988.

Bjorklund, P.E. and Jonsson, T., Capacitor commutated converters for HVDC, in *Proceedings, IEEE/Royal Institute of Technology Sockholm Power Tech; Power Electronics*, June 18–22, 1995.

Canelhas, A., Peixoto, C.A.O., and Porangaba, H.D., Converter Station Specification Considering Sate of the Art Technology, in *Proceedings, IEEE/Royal Institute of Technology Stockholm Power Tech: Power Electronics*, June 18–22, 1995.

Carlsson, L. et al., Present trends in HVDC converter station design, in *Proceedings, Fourth Symposium of Specialists in Electrical Operation and Expansion Planning (IV SEPOPE)*, Brazil, May 23–27, 1994.

Christofersen, D.J., Elahi, H., and Bennett, M.G., A survey of the reliability of HVDC systems throughout the world during 1993–1994, Paper No. 14-101, CIGRE Session, August 26–31, Paris, France, 1996.

Cochrane, J.J., Emerson, M.P., Donahue, J.A., and Wolf, G., A survey of HVDC operating and maintenance practices and their impact on reliability and performance, *IEEE Trans. on Power Delivery*, 11(1), January, 1995.

Protocol for Reporting the Operational Performance of HVDC Transmission Systems, CIGRE report 14-97, (WG04) 1989, revised 1997.

Determination of Power Losses in HVDC Converter Stations, IEC Standard No. 61803.

Fardanesh, B., Henderson, M., Shperling, B., Zelingher, S., Gyugi, L., Schauder, C., Lam, B., Mountford, J., Adapa, R., and Edris, A., *Feasibility Studies for Application of a FACTS Device on the New York State Transmission System*, CIGRE, Paris, 1998.

Guide for Commissioning High Voltage Direct Current Converter Stations and Associated Transmission Systems, IEEE Standard No. 1378-1997.

Guide for a Detailed Functional Specification and Application of Static VAr Compensators, IEEE Standard No. 1031-2000.

Guide for Reliability of HVDC Converter Stations, IEEE Standard No. 1240.

Guide for Static VAr Compensator Field Tests, IEEE Standard No. 1303-2000.

Holt, R.J., Dabkowski, J., and Hauth, R.L., *HVDC Power Transmission Electrode Siting and Design*, Oak Ridge National Laboratory, ORNL/Sub/95-SR893/3, April 1997.

IEEE Guide for Planning DC Links Terminating at AC System Locations Having Low Short Circuit Capacities, Parts I and II, IEEE Std. No. 1204-1997.

Knight, R.C., Young, D.J., Trainer, D.R., Relocatable GTO-based Static Var Compensator for NGC Substations, Paper 14-106, CIGRE Session, 1998.

Krishnayya, P. C. S. et al., Fire aspects of HVDC thyristor valves and valve halls, in *Proceedings of CIGRE 35th International Conference on Large High Voltage Electric Systems*, Paris, August 28–September 3, 1994.

deLaneuville, H., Haidle, L., McKenna, S., Sanders, S., Torgerson, D., Klenk, E., Povh, D., Flairty, C.W., and Piwko, R., Miles City Converter Station and Virginia Smith Converter Station Operating Experiences, IEEE Summer Power Meeting, San Diego, CA, July 1991.

Mehraban, A. S. et al., *Application of the world's first UPFC on the AEP system*, EPRI Conference on the Future of Power Delivery, Washington, D.C., April 9–11, 1996.

Modeling Development of Converter Topologies and Control for BTB Voltage Source Converters, EPRI, Palo Alto, California and Western Area Power Administration, Golden, Colorado, TR-111182, 1998.

Montoya, A.H., Torgerson, D.R., Vossler, B.A., Feldmann, W., Juette, G., Sadek, K., and Schultz, A., *230 kV Advanced Series Compensation Kayenta Substation (Arizona), Project Overview*, EPRI FACTS Workshop, Cincinnati, Ohio, November 14–16, 1990.

Piwko, R.J. et al., The Slatt thyristor-controlled series capacitor..., Paper 14-104, in Proceedings of CIGRE 35th International Conference on Large High Voltage Electric Systems, Paris, August 28–September 3, 1994.

Pereira, M. and Sadek, K., *Application of power active filters for damping harmonics*, CIGRE Study Committee 14 International Colloquium on HVDC and FACTS, Montreal, Quebec, September 1995.

Porrino, A. et al., *Flashover in HVDC bushings under nonuniform rain*, Proceedings, 9th International Symposium on High Voltage Engineering, Vol. 3, Graz, Austria, 3204, August 28–September 1, 1995.

Recommended Practice for Test Procedures for High Voltage Direct Current Thyristor Valves, IEEE Standard No. 857-1996.

Renz, K.W. and Tyll, H.K., Design aspects of relocatable SVCs, in *Proceedings, VIII National Power Systems Conference*, New Delhi, India, December, 1994.

Schauder, C.D. et al., *TVA STATCON Installation*, CIGRE Study Committee 14 International Colloquium on HVDC and FACTS, Montreal, September 18–19, 1995.

Schneider H.M. and Lux, A.E., Mechanism of HVDC wall bushing flashover in nonuniform rain, *IEEE Trans. on Power Delivery*, 6(1), 448-455, January 1991.
Smede, J., Johansson, C.G., Winroth, O., and Schutt, H.P., Design of HVDC converter stations with respect to audible noise requirements, *IEEE Trans. on Power Delivery*, 10(2), 747-758, April 1995.
Torgerson, D.R., Rietman, T.R., Edris, A., Tang, L., Wong, W., Mathews, H., and Imece, A.F., A transmission application of back-to-back connected voltage source converters, in *EPRI Conference on the Future of Power Delivery in the 21st Century*, La Jolla, California, November 18-20, 1997.
Tykeson, K., Nyman, A., and Carlsson, H., Environmental and geographical aspects in HVDC electrode design, *IEEE Trans. on Power Delivery*, 11(4), 1948-1954, October 1996.
Vancers, I., Hormozi, F.J. et al., A summary of North American HVDC converter station reliability specifications, *IEEE Trans. on Power Delivery*, 8(3), 1114-1122, July 1993.
Vithayathil, J. J. et al., DC systems with transformerless converters, *IEEE Trans. on Power Delivery*, 10(3), 1497-1504, July 1995.
Vithayathil, J.J., Courts, A.L., Peterson, W.G., Hingorani, N.G., Nilson, S., and Porter, J.W., HVDC circuit breaker development and field tests, *IEEE Trans. on Power Appar. and Syst.*, PAS 104(10), 2693-2705, October 1985.
Weiss, S. et al., Kayenta staged fault tests, *EPRI Conference on the Future of Power Delivery*, Washington, D.C., April 9–11, 1996.
Zhao, Z. and Iravani, M.R., Application of GTO voltage source inverter in a hybrid HVDC link, *IEEE Trans. on Power Delivery*, 9(1), 369-377, January 1994.

5.5 Considerations in Applying Automation Systems to Electric Utility Substations

James W. Evans

An electric utility Substation Automation system depends on the interface between the substation and its associated equipment to provide and maintain the high level of confidence demanded for power system operation. It must also serve the needs of other corporate users to a level that justifies its existence. This section describes typical functions provided in utility Substation Automation (SA) systems and some important considerations in the interface between substation equipment and the automation system components.

Physical Considerations

Components of a Substation Automation System

The electric utility Substation Automation (SA) system uses a variety of devices integrated into a functional package by a communications technology for the purpose of monitoring and controlling the substation. SA systems incorporate microprocessor-based intelligent electronic devices (IEDs) that provide inputs and outputs to the system. Common IEDs are protective relays, load survey and/or operator indicating meters, revenue meters, programmable logic controllers (PLC), and power equipment controllers of various descriptions. Other devices may also be present, dedicated to specific functions for the SA system. These may include transducers, position monitors, and clusters of interposing relays. Dedicated devices may use a controller (SA controller) or interface equipment such as a conventional remote terminal unit as a means of integration into the SA system. The SA system typically has one or more communications connections to the outside world. Common communications connections include utility operations centers, maintenance offices, and/or engineering centers. Most SA systems connect to a traditional SCADA System Master Station serving the real-time needs for operating the utility network from an operations center. SA systems may also incorporate a variation of SCADA Remote Terminal Unit (RTU) for this purpose or the RTU functions may appear in an SA controller or substation host computer.

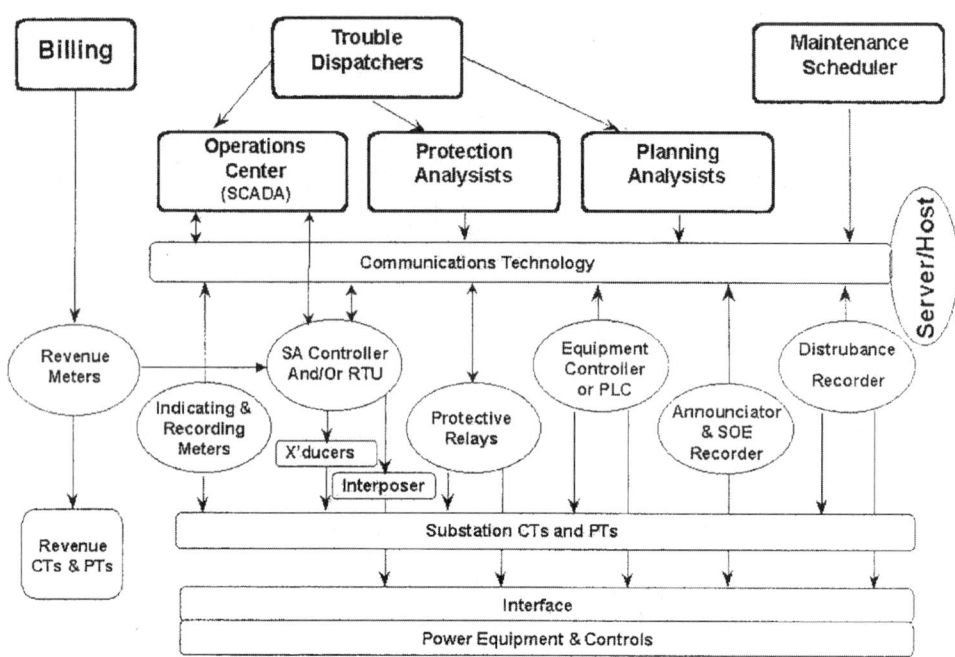

FIGURE 5.34 Power station SA system functional diagram.

Communications for other utility users are usually through a bridge, gateway, or network processor. The components described here are illustrated in Fig. 5.34.

Locating the Interface Equipment

The SA system interfaces to control station equipment through interposing relays and to measuring circuits through meters, protective relays, transducers, and other measuring devices as indicated in Fig. 5.34. These interfaces may be associated with, and integral to an IED, or dedicated interface devices for a specific automation purpose. The interfaces may be distributed throughout the station or centralized within one or two cabinets. Available panel space, layout of station control centers, as well as engineering and economic judgment are major factors in selecting a design.

The centralized interface simplifies installing a SA system in an existing substation since the placement of the interface equipment affects only one or two panels (the new SA controller and interface equipment panels). However, cabling will be required from each controlled and monitored equipment panel that meets station panel wiring standards for insulation, separations, conductor sizing, and interconnection termination. Centralizing the SA system/station equipment interface has the potential to adversely affect the security of the station, as many control and instrument transformer circuits become concentrated in a single panel or cabinet and can be seriously compromised by fire and invite human error.

Placing the interface equipment on each monitored or controlled panel is much less compromising, but is more costly and difficult to design. Each interface placement must be individually located, and more panels are affected. If a low energy interface (less than 50 volts) is used, a substantial savings in cable cost may be realized since interconnections between the SA controller and the interface devices may be made with less expensive cable and hardware.

The distributed approach is more logical when the SA system incorporates IEDs like protective relays, indicating meters, or PLCs. Protection engineers usually insist on separating protection devices into logical groups based on substation configuration for security. Similar concerns often dictate the placement of indicating meters and PLCs. The interface to the SA system becomes that of the IED on the substations side and a communications channel on the SA side. Depending on the communications capability of the

IEDs, the SA interface can be as simple as a shielded, twisted-pair cable routed between IEDs and the SA controllers. The communications interface can also be complex where short haul RS-232 connections to a communications controller are required.

As the cabling distances within the substation increase, system installation costs increase, particularly if additional cable trays, conduit, or ducts are required. Using SA communication technology and IEDs can reduce interconnection cost. Distributing multiple small SA "hubs" throughout the substation can reduce cabling to that needed for a communications link to the SA controller. Likewise, these "hubs" can be isolated using fiber-optic technology for improved security and reliability.

Environment

The environment of a substation is challenging for SA equipment. Substation control buildings are seldom heated or air conditioned. Ambient temperatures may range from below freezing to above 100°F (40°C). Metal-clad switchgear substations can reach ambient temperatures in excess of 140°F (50°C). Temperature changes stress the stability of measuring components in IEDs, RTUs, and transducers. Good temperature stability is important in SA system equipment and needs to be defined in the equipment purchase specifications. Designers of SA systems for substations need to pay careful attention to the temperature specifications of the equipment selected for SA. In many environments, self-contained heating or air conditioning may be advisable.

When equipment is installed in outdoor enclosures, not only is the temperature cycling problem aggravated, but moisture becomes a problem. Outdoor enclosures usually need heaters to control their temperature to prevent condensation. The placement of heaters should be reviewed carefully when designing an enclosure, as they can aggravate temperature stability and even create hot spots within the cabinet that can damage components and shorten life span. Incident solar radiation shields may also be required to keep enclosure temperature manageable. Specifications which identify the need for wide temperature range components, coated circuit boards, and corrosion-resistant hardware are part of specifying and selecting SA equipment for outdoor installation.

Environmental factors also include airborne contamination from dust, dirt, and corrosive atmospheres found at some sites. Special non-corrosive cabinets and air filters may be required for protection against the elements. In some regions, seismic requirements are important enough to be given special consideration.

Electrical Environment

The electrical environment of a substation is severe. High levels of electrical noise and transients are generated by the operation of power equipment and their controls. Operating high voltage disconnect switches can generate transients that appear throughout the station on current, potential, and control wiring entering or leaving the switch yard. Station controls for circuit breakers, capacitor, and tap changers can also generate transients found throughout the station on battery power and station service wiring. EHV stations also have high electrostatic field intensities which couple to station wiring. Finally, ground rise during faults or switching can damage electronic equipment in stations.

Effective grounding is critical to controlling the effects of substation electrical noise on electronic devices. IEDs need a solid ground system to make their internal suppression effective. Ground systems should be radial with signal and protective grounds separated. They require large conductors for "surge" grounds, making grounds as short as possible and establishing a single ground point for logical groupings of equipment. These measures help to suppress the introduction of noise and transients into measuring circuits. A discussion of this topic is usually found in the IED manufacturer's instruction book and their advice should be heeded.

The effects of electrical noise can be controlled with surge suppression, shielded and twisted pair cabling, as well as careful cable separation practices. However, suppressing surges with capacitors, Metal Oxide Varistors (MOV) and semi-conducting overvoltage "Transorbs" on substation instrument transformer and control wiring can protect IEDs. They can create reliability problems as well. Surge suppressors must have sufficient energy-absorbing capacity and be coordinated so that all suppressors clamp around the same voltage. Otherwise, the lowest dissipation, lowest voltage suppressor will become sacrificial.

Multiple failures of transient suppressors can short circuit important station signals to ground leading to blown potential fuses, shorted CTs, and shorted control wiring; even false tripping.

While every installation has a unique noise environment, some testing can help prevent noise problems from becoming unmanageable. The transients generated by operating high-voltage disconnect switches and the operation of electromechanical control devices are represented by the IEEE Surge Withstand Capability Test C37.90-1992. This test can be applied to devices in a laboratory or on the factory floor. It should be considered when specifying station interface equipment. Insulation resistance and high potential tests are also sometimes useful.

Analog Data Acquisition

Measured Quantities

Electric utility SA systems gather power system performance parameters (i.e., volts, amperes, watts, and vars) for system generators, transmission lines, transformer banks, and station buses and distribution feeders. Energy output and usage quantities (i.e., kilowatt-hours and kilovar-hours) are also important. Other quantities such as transformer temperatures, insulating gas pressures, fuel tank levels for on-site generation, or head level for hydro generation might also be measured and transmitted as analog values. Often, transformer tap positions, regulator positions, or other multiple position quantities are also transmitted as if they were analog values. These values enter the SA system through IEDs, transducers, and sensors.

Transducers and IEDs measure electrical quantities (watts, vars, volts, amps) with instrument transformers provided in power equipment as shown in Fig. 5.35. They convert instrument transformer outputs to digital values or DC voltages, or currents which can be readily accepted by a traditional SCADA RTU or SA controller.

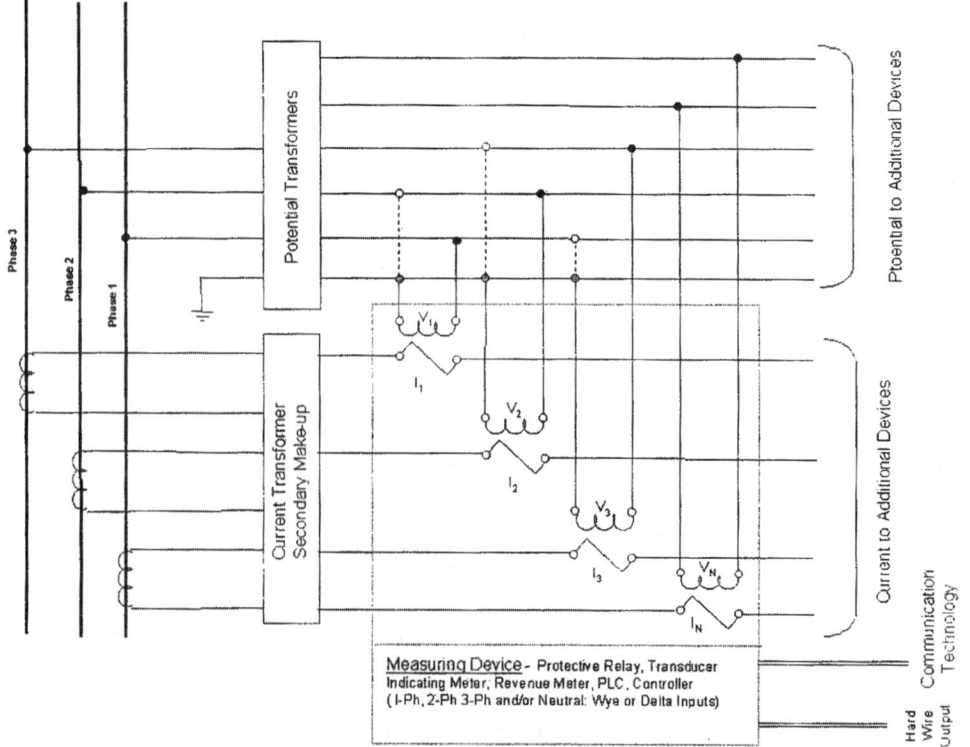

FIGURE 5.35 SA system measuring interface.

Substations

Analog values may also be collected by the SA system from substation meters, protective relays, and recloser controls as IEDs. Functionally, the process is equivalent but the IEDs perform signal processing and digital conversion directly as part of their primary function. IEDs use a communications channel for passing data to the SA controller instead of conventional analog signals.

Performance Requirements

In the initial planning stages of an SA system, the economic value of the data to be acquired needs to be weighed against the cost to measure it. A balance must be struck to achieve the data quality required to suit the purpose of the system. This affects the conceptual design of the measuring interface and provides input to the performance specifications for IEDs and transducers as well as the measuring practices applied. This step is important. Specifying a higher performance measuring system than required raises the overall system cost. Conversely, constructing a low-performance system adds costs when the measuring system must be upgraded.

The electrical relationship between measurements and the placement of available instrument transformer sources deserves careful attention to insure satisfactory performance. Many design compromises must be made when installing SA monitoring in an existing power station, because of the availability of measuring sources. This is especially true when using protective relays as load monitoring data sources (IEDs). Protection engineers often ignore current omissions or contributions at a measuring point as they may not affect the performance of protection during faults. These variances are often intolerable for power flow measurements. Measuring source placement may also result in measurements that include or exclude reactive contributions of a series or shunt reactor or capacitor; measurements that include reactive component contributions of a transformer bank; or measurements that are affected when a section breaker is open because the potential source is on an adjacent bus. Power system charging current and imbalances also influence measurement accuracy, especially at low load levels. The compromises are endless and each produces an unusual operating condition in some state. When deficiencies are recognized, the changes to correct them can be very costly, especially if instrument transformers must be installed or replaced to correct the problem.

The overall accuracy of measured quantities is affected by a number of factors. These include instrument transformer errors, IED or transducer performance, and analog-to-digital (A/D) conversion. Accuracy is not predictable based solely on the IED, transducer, or A/D converter specifications. Significant measuring errors often result from instrument transformer performance and errors induced in scaling.

Revenue metering accuracy is usually required for monitoring power system interconnections and feeding economic area interchange dispatch systems. High accuracy, revenue metering grade, instrument transformers, and 0.25% accuracy class IEDs or transducers can produce consistent real power measurements with accuracies of 1% or better at 0.5 to 1.0 power factor, and reactive power measurements with accuracies of 1% or better at 0 to 0.5 power factors.

When an SA system provides information for internal power flow telemetering, revenue-grade instrument transformers are not usually available. SA IEDs and transducers must often share lesser accuracy instrument transformers provided for protective relaying or load monitoring. Overall accuracy under these conditions can easily decrease to 2–3% for real power, voltage, and current measurements, and 5% or greater for reactive power.

Instrument Transformers

Current Transformers

Current transformers (CTs) of all sizes and types find their way into substations to provide the current replicas for metering, controls, and protective relaying. Some will perform well for SA applications and some may be marginal. CT performance is characterized by ratio correction factor (turns ratio error), saturation voltage, phase angle error, and rated burden. Bushing CTs installed in power equipment, as shown in Fig. 5.36, are the most common type found in medium- and high-voltage power equipment. They are toroidal, having a single primary turn (the power conductor), which passes through their center. The current transformation ratio results from the number of turns wound on the core to make up the

FIGURE 5.36 Bushing current transformer installation.

secondary. More than one ratio is often provided by tapping the secondary winding at multiple turns ratios. The core cross-sectional area, diameter, and magnetic properties determine the CTs performance. As the CT is operated over its current ranges, its deviations from specified turns ratio is characterized by its ratio correction curve, sometimes provided by the manufacturer. At low currents, the exciting current causes ratio errors which are predominant until sufficient primary flux overcomes the effects of core magnetizing. Thus, watt or Var measurements made at very low load may be substantially in error from both ratio error and phase shift. Exciting current errors are a function of individual CT construction. They are generally higher for protection CTs than metering CTs by design.

Metering CTs are designed with core cross-sections chosen to minimize exciting current effects and are allowed to saturate at fault currents. Larger cores are provided for protection CTs where high current saturation must be avoided for the CT to faithfully reproduce high currents for fault sensing. The exciting current of the larger core at low load is not considered important for protection. Core size and magnetic properties limit the ability of CTs to develop voltage to drive secondary current through the circuit load impedance (burden). This is an important consideration when adding SA IEDs or transducers to existing metering CT circuits, as added burden can affect accuracy. The added burden of SA devices is less likely to create metering problems with protection CTs at load levels, but could have undesirable effects on protective relaying at fault levels. In either case, CT burdens are an important consideration in the design. Experience with both protection and metering CTs has shown, however, that both perform comparably once the operating current sufficiently exceeds the exciting current.

Occasions arise where it is necessary to obtain current from more than one source by summing currents with auxiliary CTs. This will perform satisfactorily only if the auxiliaries used are adequate. If the core size is too small to drive the added circuit burden, the auxiliaries will introduce excessive ratio and phase angle errors which will degrade measurement accuracy. Using an auxiliary transformer must be approached with caution.

Potential Sources

The most common potential sources for power system measurements are wound transformers (potential transformers) and capacitive divider devices (capacitor voltage transformers or bushing potential devices). Some new applications of resistor dividers and exotic technologies are also becoming available. All provide scaled replicas of their high-voltage potential characterized by their ratio, load capability, and phase angle response. Wound potential transformers (PTs) provide the best performance with ratio and phase angle errors suitable for revenue metering. Even protection-type potential transformers can provide revenue metering performance. PTs are usually capable of supplying large potential circuit loads without degradation, provided their secondary wiring is of adequate size. For substation automation purposes, PTs are unaffected by changes in load or temperature. They are the preferred source for measuring potential.

Capacitor voltage transformers (CVTs) use a series stack of capacitors, connected as a divider, along with a low voltage transformer to obtain a secondary voltage replica. They are less expensive than wound transformers and can approximate wound transformer performance under controlled conditions. While revenue-grade CVTs are available, CVTs are less stable and less accurate than wound PTs. Older CVTs may be totally unsatisfactory. CVTs can be affected by secondary load and ambient temperature. CVTs must be individually calibrated in the field to bring their ratio errors within 1%, and must be recalibrated whenever the load is changed. Older CVTs can change ratio up to +5% with ambient temperature variation. In all, CVTs are a reluctant choice for SA system measuring. When CVTs are the only choice, consideration should be given to using the more modern devices for better performance and a periodic calibration program to maintain their performance at satisfactory levels.

Bushing capacitor potential devices use a tap made in the capacitive grading of a high-voltage bushing to provide the potential replica. They can supply only very limited secondary load and are very load sensitive. They can also be very temperature sensitive. As with CVTs, if BCPDs are the only choice, they should be individually calibrated and periodically checked.

Transducers

Transducers measure power system parameters with instrument transformers. They provide a scaled, low-energy signal which represents a power system quantity that the SA interface controller can easily accept. Transducers also isolate and buffer the SA interface controller from the power system and substation environments. Transducer outputs are DC voltages or currents in the range of a few tens of volts or milliamperes.

Transducers measuring power system electrical quantities are designed to be compatible with instrument transformer outputs. Potential inputs are based around 120 or 115 VAC and current inputs accept 0–5 amperes. Many transducers can operate at levels above their normal ranges with little degradation in accuracy provided their output limits are not exceeded. Transducer input circuits may share the same instrument transformers as the station metering and protection systems; thus, they must conform to the same standards as any switchboard component. Wiring standards for current and potential circuits vary between utilities, but generally 600 volt class wiring is required, and #12 AWG or larger wire is used. Special termination standards also apply in many utilities. Test switches for "in-service" testing of transducers are often provided to make it possible to test transducers without shutting down the monitored equipment. Transducers may also require an external power source to supply their power supply requirements. When this is the case, the reliability of this source must be considered.

Transducer outputs are voltage or current sources specified to supply a rated voltage or current into a specific load. For example, full output may correspond to 10 V at up to 1.0 mA or 1.0 mA into 10 kΩ, up to 10 V maximum. Some over-range capability is provided in transducers so long as the maximum current and/or voltage capability is not exceeded. The over-range may vary from 20 to 100%, depending on the transducer; however, accuracy is usually not specified for the over-range area.

Transducer outputs are usually wired with shielded, twisted pair cable to minimize stray signal pickup. In practice, #18 AWG conductors or smaller are satisfactory, but individual utility practices differ. It is common to allow transducer output circuits to remain isolated from ground to reduce the susceptibility to transient damage, although some SA controller suppliers provide a common ground for all analogs, often to accommodate electronic multiplexers. Some transducers may also have a ground reference associated with their outputs. Double grounds, where transducer and controller both have ground references, can cause major reliability problems. Practices also differ somewhat on shield grounding with some shields grounded at both ends; but it is also common practice to ground shields at the SA controller end only. When these signals must cross a switchyard, however, it is a good practice to not only provide the shielded twisted pairs, but to also provide a heavy-gauge overall cable shield. This shield should be grounded where it leaves a station control house to enter a switchyard and where it re-enters another control house. These grounds are terminated to the station ground mass, and not the SA analog grounds bus.

Scaling of Analog Values

In an SA system, the transition of power system measurements to database values or displays is a process that entails several steps of scaling, each with its own dynamic range. Power system parameters are first scaled by current and potential transformers, then by IEDs or transducers. In the process, an A/D conversion occurs as well. Each of these steps has its own proportionality constant which, when combined, relate the digital coding of the data value to the primary quantities. At the data receiver or master station, these are operated on by one or more constants to convert the data to user-acceptable values for databases and displays.

SA system measuring performance can be severely affected by data value scaling. Optimally, under normal power system conditions, each IED or transducer should be operating in its most linear range and utilize as much A/D conversion range as possible. Scaling should take into account the minimum, normal, and maximum value for the quantity, even under abnormal or emergency loading conditions. Optimum scaling balances the expected value at maximum, the current and potential transformer ratios, the IED or transducer range, and the A/D range, to utilize as much of the IED or transducer output and A/D range as possible under normal power system conditions without driving the conversion over its full scale at maximum value. This practice minimizes the quantizing error of the A/D conversion process and provides the best quantity resolution.

Conversely, scaling some IEDs locally makes their data difficult to use in an SA system. A solution to this problem is to set the IED scaling to unity in, and apply all the scale factors at, the data receivers. Under the practical restraints imposed when applying SA to an existing substation, scaling can be expected to be compromised by available instrument transformer ratios and A/D or IED scaling provisions. A reasonable selection of scale factor and range would provide half output or more under normal conditions, but not exceed 90% of full range under maximum load.

Intelligent Electronic Devices (IED) as Analog Data Sources

Technological advancements have made it practical to use electronic substation meters, protective relays, and even reclosers and regulators as sources of analog data. IED measurements are converted directly to digital form and passed to the SA system via a communications channel while the IED performs its primary function. In order to use IEDs effectively, it is necessary to assure that the performance characteristics of the IED fit the requirements of the system. Some IEDs designed for protection functions, where they must accurately measure fault currents, do not measure low load accurately. Others, where measuring is part of a control function, may lack overload capability or have insufficient resolution. Sampling rates and averaging techniques will affect the quality of data and should be evaluated as part of the system product selection process. With reclosers and regulators, the measuring CTs and PT are often contained within the equipment. They may not be accurate enough to meet the measuring standards set for the SA system. Regulators may only have a single-phase CT and PT. These issues challenge the SA system integrator to deliver a quality system.

The IED communications channel becomes an important data highway and needs attention to security, reliability, and most of all throughput. A communication interface is needed in the SA system to retrieve and convert the data to meet the requirement of the master or data receiver.

Integrated Analog Quantities — Pulse Accumulators (PA)

Some power system quantities of interest are energy transfer values derived from integrating instantaneous values over an arbitrary time period, usually fifteen-minute values for one hour. The most common of these is watt-hours, although var-hours and amp-squared-hours are not uncommon. They are usually associated with energy interchange over interconnecting tie lines, generator output, or the load of major customers. In most instances, they originate from a revenue metering package which includes revenue-grade instrument transformers, one or more watt-hour and var-hour meters, and a local recording device such as a magnetic tape recorder or remote meter reading device. They also can be interfaced to an SA system.

Integrated energy transfer values are traditionally recorded by counting the revolutions of the disc on an electromechanical watt-hour meter type device. Newer technology makes this concept obsolete but the integrated interchange value continues as a mainstay of energy interchange between utilities and users. In the old technology, a set of contacts opens and closes in direct relation to the disc rotation, either mechanically from a cam driven by the meter disc shaft, or through the use of opto-electronics and a light beam interrupted by or reflected off the disc. These contacts may be standard form "a", form "b", form "c", or a form "k", which is peculiar to watt-hour meters. Modern revenue meters often mimic this feature, as do some analog transducers. Each contact transfer ("pulse") represents an increment of energy transfer as measured by a watt-hour meter. Pulses are accumulated over a period of time in a register, and then the total is recorded on command from a clock.

When applied to SA systems, energy transfer quantities are processed by metering IEDs, PAs in an RTU, or SA controller. The PA receives contact closures from the metering package and accumulates them in a register. On command from the master station command, the pulse count is frozen, then transmitted. The register is sometimes reset to zero to begin the cycle for the next period. This command is synchronized to the master station clock, and all "frozen" accumulator quantities are polled some time later when time permits. Some RTUs can freeze and store their pulse accumulators from an internal or local external clock should the master "freeze-and-transmit command" be absent. These may be internally "time tagged" for transmission when commanded by the master station. Other options may include the capability to arithmetically process several demand quantities to derive a resultant. Software to "de-bounce" the demand contacts is also sometimes available.

Integrated energy transfer telemetering is almost always provided on tie lines between bordering utilities. The location of the measuring point is usually specified in the interconnection contractual agreement, along with a procedure to insure metering accuracy. Some utilities agree to share a common metering point at one end of a tie and electronically transfer the interchange reading. Others insist on having their own duplicate metering, sometimes specified to be a "backup" service. When a tie is metered at both ends, it is important to verify that the metering installations are in agreement. Even with high accuracy metering, however, some disagreement can be expected, and this is often a source of friction between utilities.

Status Monitoring

Status indications are an important function of SA systems for the electrical utility. Status monitoring is provided for power circuit breakers, circuit switchers, reclosers, motor-operated disconnect switches, and a variety of other on-off functions in a substation. Status points may be provided with status change memory for between-scan monitoring or time tagging to provide sequence of events. Status indications originate from auxiliary switch contacts which are mechanically actuated by the monitored device. Interposing relay contacts are also used for status points where the interposer is driven from auxiliary switches. This practice is common, depending on the utility and the availability of spare contacts. The exposure of status point wiring to the switchyard environment is a consideration in installing interposing relays.

Contact Performance

The mechanical response of either relay contacts or auxiliary switch contacts can create monitoring difficulties. Contacts mechanically "bounce" when they transfer from one position to another. The input point may interpret the bouncing of the status contact as multiple operations of the primary device. A "mercury wetted" contact is sometimes used to minimize contact bounce. Another technique used is to employ "c" form contacts for status indications so that status changes are recognized only when one contact closes preceded by the opening of its companion. Contact changes occurring on one contact only are ignored; "c" contact arrangements are more immune to noise pulses.

Event recording with high-speed resolution is particularly sensitive to contact bounce as each transition is recorded. When the primary device is subject to pumping or bouncing induced from mechanical characteristics, it may be difficult to prevent excessive status change reporting. Event contacts can also contain unwanted delays which can confuse interpretation of event sequences. While this may not be avoidable, it is important to know the response time of all event devices so that event sequences can be correctly interpreted.

IEDs often have "de-bounce" algorithms in their programming to filter contact bouncing. These algorithms allow the user to "tune" the de-bouncing to be tolerant of bouncing contacts.

Wetting Sources

Status points are usually isolated "dry" contacts wetted from the input point, but voltage signals from a station control circuit can also be monitored by SA controllers. Equipment suppliers can provide a variety of status point input options. When selecting between options, the choice balances circuit isolation against design convenience. The availability of spare isolated contacts often becomes an issue in choice. Voltage signals may eliminate the need for spare contacts, but can require circuits from various parts of the station and from different control circuits be brought to a common termination location. This compromises circuit isolation within the station and raises the possibility of test personnel causing circuit misoperations. Usually, switchboard wiring standards would be required for this type of installation, which could increase costs. Voltage inputs are often fused with small fuses at the source to minimize the risk that the exposed wiring will compromise the control circuits.

In installations using isolated "dry" contacts, the wetting voltage is sourced from the station battery, an SA controller, or IED supply. Each monitored control circuit must then provide an isolated contact for status monitoring. Circuit isolation occurs at each control panel, thus improving the overall security of the installation. It is common for many status points share a common supply, either station battery or a low voltage supply provided for this purpose. When status points are powered from the station battery, the monitored contacts have full control potential appearing across their surfaces and thus can be expected to be more immune to open circuit failures from contact surface contamination. Switchboard wiring standards would be required for this type of installation. An alternative source for status points is a low voltage wetting supply. Wiring for low voltage sourced status points may not need to be switchboard standard in this application, which may realize some economies. Usually, shielded, twisted pairs are used with low voltage status points to minimize noise effects. Concern over contact reliability due to the lower "wetting" voltage can be partially overcome by using contacts which are closed when the device is in its normal position, thereby maintaining a loop current through the contact. Some SA systems provide a means to detect wetting supply failure for improved reliability. Where multiple IEDs are status point sources, it can be difficult to detect a lost wetting supply.

In either approach, the status point loop current is determined by the monitoring device design. Generally, the loop current is 1.0 to 20 milliamperes. Filter networks and/or software filtering are usually provided to reduce noise effects and false changes resulting from bouncing contacts.

Wiring Practices

When wiring status points, it is important to make cable runs radially between the monitor and the monitored device. Circuits where status circuit loops are not parallel pairs are subject to induced currents that can cause false status changes. Circular loops most often occur when using spare existing conductors in multiple cables or when using a common return connection for several status points. Designers should be wary of this practice. The resistance of status loops can also be an important consideration. Shielded, twisted pairs make the best interconnection for status points, but this type of cable is not always readily available in switchboard standard sizes and insulation for use in control battery-powered status circuits.

Substations

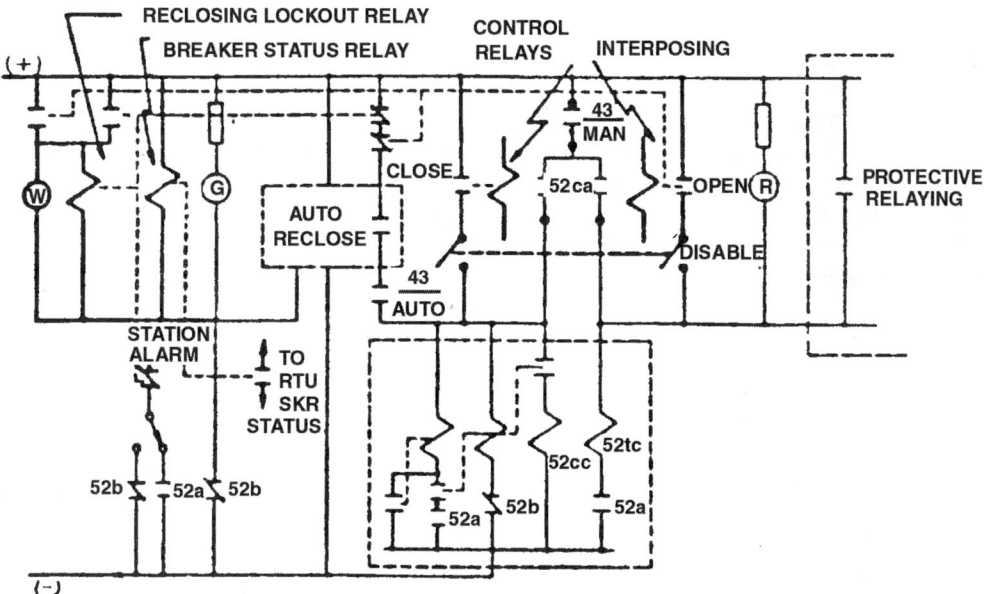

FIGURE 5.37 Schematic diagram of a breaker control interface.

Finally, it is important to provide for testing status circuits. Test switches or jumper locations for simulating open or closed status circuits are needed, as well as a means for isolating the circuit for testing.

Control Functions

The supervisory control functions of electric utility SA systems provide routine and emergency switching and operating capability for station equipment. SA controls are most often provided for circuit breakers, reclosers, and switchers. It is not uncommon to also include control for voltage regulators, tap 20 amperes to effect their action, which imposes constraints on the interposing devices, changing transformers, motor-operated disconnects, valves or even peaking units through an SA system. A variety of different control outputs are available from IEDs and SA controllers which can provide both momentary timed control outputs and latching-type interposing. Latching is commonly associated with blocking of automatic breaker reclosing or voltage controllers for capacitor switching. A typical interface application for controlling a circuit breaker is shown in Fig. 5.37.

Interposing Relays

Power station controls often require high power levels and operate in circuits powered from 48, 125, or 250-VDC station batteries or from 120 or 240-VAC station service. Control circuits often must switch 10 or 20 A to effect their action, which imposes constraints on the interposing devices. The interposing between an SA controller and station controls commonly use large electromechanical relays. Their coils are driven by the SA control system through static or pilot duty relay drivers, and their contacts switch the station control circuits. Interposing relays are often specified with 25 A, 240 VAC contact rating to insure adequate interrupting duty. Smaller interposing relays are also used, however, often with only 10 or 3 A contacts, where control circuits allow. When controlling DC circuits, the large relays may be required, not because of the current requirements, but to provide the long contact travel needed to interrupt the arc associated with interrupting a DC circuit. Note that most relays which would be considered for the interposing function do not carry DC interrupting ratings. "Magnetic Blowout" contacts, contacts fitted with small permanent magnets that lengthen the interruption arc to aid in extinguishing it, may also be used to improve interrupting duty. They are polarity sensitive, however, and work only if correctly wired. Correct current flow direction must be observed.

Control Circuit Designs

Many station control circuits can be designed so that the interrupting duty problem for interposing relays is minimized, thereby allowing smaller interposing relays to be used. These circuits are designed so that once they are initiated, some other device in the circuit interrupts control current in preference to the initiating device. These are easily driven from momentary outputs. The control logic is such that the initiating contact is bypassed once control action begins, and remains bypassed until control action is completed. The initiating circuit current is then interrupted, or at least greatly reduced, by a device in another portion of the control circuit. This eliminates the need for the interposing relay to interrupt heavy control circuit current. This is typical of modern circuit breaker closing circuits, motor-operated disconnects, and many circuit switchers. Other controls which "self-complete" are breaker tripping circuits, where the tripping current is interrupted by the breaker auxiliary switch contacts long before the initiating contact opens.

Redesigning control circuits often simplifies the application of supervisory control. The need for large interposing relay contacts can be eliminated in many cases by simple modifications to the controlled circuit to make them "self-completing." An example of this would be the addition of any auxiliary control relay to a breaker control circuit, which maintains the closing circuit until the breaker has fully closed and provides anti-pumping should it trip free. This type of revision is often desirable anyway, if a partially completed control action could result in some equipment malfunction.

Control circuits may also be revised to limit control circuit response to prevent more than one action from taking place while under supervisory control. This includes preventing a circuit breaker from "pumping" if it were closed into a fault or failed to latch. Another example is to limit tap changer travel to one tap per initiation.

Latching Devices

It is often necessary to modify control circuit behavior when supervisory control is used to operate station equipment. Control mode changes that would ordinarily accompany manual local operation must also occur when action occurs through supervisory control. Many of these require latched interposing relaying that modifies control behavior when supervisory control is exercised, and can be restored through supervisory or local control. The disabling of automatic circuit breaker reclosing when a breaker is opened through supervisory control action is an example. Automatic reclosing must also be restored and/or reset when a breaker is closed through supervisory control. This concept also applies to automatic capacitor switcher controls that must be disabled when supervisory control is used and can be restored to automatic control through local or supervisory control.

These types of control modifications generally require a latching-type interposing design. Solenoid-operated control switches have become available which can directly replace a manual switch on a switchboard and can closely mimic manual control action. These can be operated through supervisory control, and can frequently provide the proper control behavior.

Intelligent Electronic Devices (IEDs) for Control

IEDs are available that have control capability accessible through their communications ports. Protective relays, panel meters, reclosers controls, and regulators are common devices with control capability. They offer the opportunity to control substation equipment without a traditional interposing relay cluster for the interface — sometimes without any control circuit additions. Instead, the control interface is embedded in the IED. When using embedded control interfaces, the SA system designer needs to assess the security and capability of the interface provided. These requirements should not change just because the interface devices are within an IED. External interposing may be required to meet circuit loads or interrupting duty.

When controlling equipment with IEDs over a communications channel, the integrity of the channel and the security of the messaging system become important factors. Not all IEDs have select-before-operate capability common to RTUs and SCADA systems. Their protocols may also not have efficient error detection, which could lead to misoperations. In addition, the requirements to have supervisory control disabled for test and maintenance must not impact the IED's primary function.

Summary

The addition of SA systems control impacts station security and deserves a great deal of consideration. It should be recognized that SA control can concentrate station control in a small area and can increase the vulnerability of station control to human error and accident. This deserves careful attention to the control interface design for SA systems. The security of the equipment installed must insure freedom from false operation, and the design of operating and testing procedures must recognize these risks and minimize them.

5.6 Substation Automation

John D. McDonald

The integration of Intelligent Electronic Devices (IEDs) in the substation presents many challenges to electric utilities. For example, each IED can provide massive amounts of data, both instantaneous and historical, with limited filtering capabilities in the IED or in the IED protocol. The proprietary IED protocols have required interface modules, at additional cost, for protocol conversion and for functionality not supported by present IEDs. In addition, there is no standard local area network protocol for the substation. Finding a network protocol to handle data acquisition as well as high-speed protection requirements has been an ongoing industry activity. Various industry efforts have attempted to solve the standardization problems within the substation. This section defines and discusses substation automation and current standardization efforts and their status.

Definitions and Terminology

Substation integration and automation can be broken down into five levels, or layers. Two of these levels are interface levels: power system equipment (e.g., transformers, breakers) to IEDs, and substation to utility via a utility enterprise connection. The three middle levels are IED implementation, IED integration via data concentrator/substation host processor, and substation automation applications. The five-layer architecture is shown in Table 5.2.

TABLE 5.2 Five-Layer Architecture

Utility Enterprise Connection
Substation Automation Applications
IED Integration Via Data Concentrator/Substation Host Processor
IED Implementation
Power System Equipment (transformers, breakers)

Definitions of important terms are included below:

- Intelligent Electronic Device (IED): Any device incorporating one or more processors with the capability to receive or send data/control from or to an external source (e.g., electronic multifunction meters, digital relays, controllers) (IEEE Std. C37.1-1994; IEEE Std. 1000, 1997).
- IED Integration: Integration of protection, control, and data acquisition functions into a minimal number of platforms to reduce capital and operating costs, reduce panel and control room space, and eliminate redundant equipment and databases.
- Substation Automation: Deployment of substation and feeder operating functions and applications ranging from SCADA and alarm processing to integrated volt/Var control in order to optimize the management of capital assets and enhance operation and maintenance (O&M) efficiencies with minimal human intervention.

Open Systems

What is an open system? An open system is an evolutionary means for a control system, based on the use of non-proprietary and standard software and hardware interfaces. Open systems enable future upgrades to be available from multiple suppliers at lowered cost and integrated with relative ease and low risk. An open system is a computer system that embodies supplier-independent standards so that software can be applied on many different platforms and can interoperate with other applications on local and remote systems.

How do open systems apply to substation automation? First, the different *de jure* (by law) and *de facto* (in fact or actual) standards should be known and applied to eliminate proprietary approaches. An open systems approach allows the incremental upgrade of the automation system, without the need for complete replacement as in the past with proprietary systems. There is no longer the reliance on one supplier for complete implementation. The system and IEDs from competing suppliers are able to interchange and share information. The benefits of open systems include longer expected system life, investment protection, upgradeability and expandability, and readily available third-party components.

Substation Automation Technical Issues

There are many technical issues in substation automation. They will be discussed in this section in the areas of system responsibilities, system architecture, substation host processor, substation LAN requirements, substation LAN protocols, user interface, communication interfaces, and the data warehouse.

System Responsibilities

The system must interface to all the IEDs in the substation. This includes polling the IEDs for readings and event notifications. The data from all the IEDs must be sent to the utility enterprise to populate the data warehouse, or the location for storage of the substation data. The system processes data and control requests from users and from the data warehouse. The system must provide supplier isolation from the IEDs, or provide a generic interface to the IEDs. In other words, there should be a standard interface, regardless of the IED supplier. The system should be updated with a report-by-exception scheme, so that only status point changes, and analog point changes beyond their significant deadband, are reported. This reduces the load on the communications channel. In some systems, the data is reported in an unsolicited response mode. When the end device has something to report, it does not have to wait for a poll request from a master (master to slave). The device initiates the communication by grabbing the communication channel and transmitting its information.

Current substation automation systems perform protocol translation, converting all the IED protocols from the various IED suppliers. Even with the protocol standardization efforts going on in the industry, there will always be legacy protocols that will require protocol translation.

The system must manage the IEDs and devices in the substation. The system must be aware of the address of each IED, aware of alternate communication paths, and aware of IEDs that may be utilized to accomplish a specific function. The system must know the status of all connected IEDs at all times.

The system provides data exchange and control support for the data warehouse. It should use a standard messaging service in the interface (standard protocol). The interface should be independent of any IED protocol, and should use a report-by-exception scheme to reduce channel loading.

The system must provide an environment to support user applications. These user applications can be internally written by the utility, or purchased from a third party and integrated into the substation automation system (Fig. 5.38).

System Architecture

The types of data and control that the system will be expected to facilitate are dependent on the choice of IEDs and devices chosen to be implemented. This must be addressed on a substation-by-substation basis. The primary requirement is that the analog readings be obtained in a way that provides an accurate representation of their values.

The data concentrator shall store all analog and status information available at the substation. This information is required for both operational and nonoperational reasons (e.g., load forecasting, engineering studies, outage investigations).

FIGURE 5.38 Substation automatic system.

There are several levels of data exchange and requirements associated with the Substation Automation system:

Level 1 — Field Devices
Each electronic device (relay, meter, PLC, etc.) has internal memory to store some or all of the following data: analog values, status changes, sequence-of-events, and power quality. This data is typically stored in a FIFO queue and varies in the number of events, etc., maintained.

Level 2 — Substation Data Concentrator
The substation data concentrator will poll each device (both electronic and other) for analog values and status changes at data collection rates consistent with the utility's SCADA system (e.g., status points every 2 sec, tie-line and generator analogs every 2 sec, and remaining analog values every 2 to 10 sec). The substation data concentrator shall maintain a local database.

Level 3 — SCADA System, Data Warehouse
All data required for operational purposes shall be communicated to the SCADA system via a communication link from the data concentrator. All data required for nonoperational purposes shall be communicated to the data warehouse via a communication link from the data concentrator.

A data warehouse is necessary to support a mainframe or client/server architecture of data exchange between the system and corporate users over the corporate WAN. This provides many users with more up-to-date information without them having to wait for access using a single line of communication to the system, such as telephone dial-up through the modem.

The system's logical architecture is illustrated in Table 5.3. At the lowest level of the architecture are the IEDs that collect information from, and provide control capability for, power system field devices. The next level of the architecture is the substation data concentrator that has responsibility for determining the underlying IED configuration, interfacing with the IEDs and utility enterprise, and providing

TABLE 5.3 SCADA Architecture

User Interface
Applications
Application ↔ ECS Upgrade System Interface   Application ↔ Data Warehouse Interface
ECS Upgrade System   **Data Warehouse Server**
SCADA System ↔ Substation Data   Data Warehouse ↔ Substation Data
Concentrator Interface   Concentrator Interface
Substation Data Concentrator
Substation Data Concentrator ↔ IED Interface
Intra-Substation LAN
Intelligent Electronic Devices (IEDs)
Power System Field Devices

data to the SCADA system. The SCADA system is a database with security to different levels of user access. The data warehouse server is one or more (i.e., centralized or regional) repositories for fault/event oscillographic data from the IEDs. At the top layer of the architecture are utility applications that will interact with either the SCADA system or the data warehouse server (Fig. 5.39).

Each layer of the architecture communicates with only the adjacent layers. For example, if the SCADA system requires data from the field, it will communicate through the data concentrator that will in turn communicate with the IEDs over the substation LAN.

Substation Host Processor

The substation host processor must be based on industry standards and strong networking ability, such as Ethernet, X/Windows, Motif, TCP/IP, UNIX, Windows NT, etc. It must also support an open architecture, with no proprietary interfaces or products. An industry-accepted relational database (RDB) with

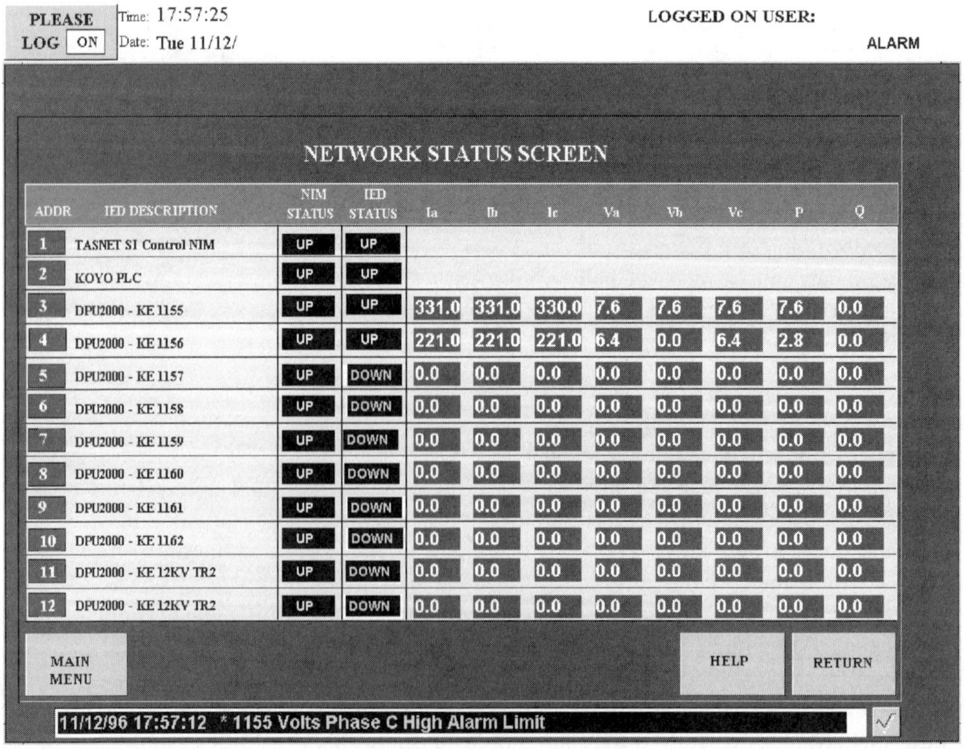

FIGURE 5.39 Network status display.

SQL capability and enterprise-wide computing must be supported. The RDB supplier must have implemented replication capabilities to provide the capability to support a redundant or backup database. A full graphics user interface (bit or pixel addressable) should be provided with Windows-type capability. There should be interfaces to Windows-type applications (i.e., Excel, DDE interface, OLE interface, etc.). The substation host processor should be flexible and expandable, and transportable to multiple hardware platforms (PCs, Power PC, DEC Alpha, Sun, HP, etc.).

Should the host processor be single or redundant, or distributed? For a smaller distribution substation, it may be a single processor. For a large transmission substation, there may be redundant processors with failover. If the supplier offers a distributed processor system with levels of redundancy, this may be more cost-effective for the larger substations. PLCs may be used as controllers, running special application programs at the substation level, coded in ladder logic. Smaller "slave" substations will have IEDs but may not have a substation automation system. The IED data from these "slave" substations will be sent upstream to a larger "master" substation that contains a complete substation automation system.

Substation Local Area Network (LAN) Requirements

The substation LAN must meet industry standards to allow interoperability and plug-and-play device capabilities. Open architecture principles should be followed, such as use of industry standard protocols (e.g., TCP/IP, IEEE 802.x [Ethernet], UCA 2.0). The LAN technology employed must be applicable to the substation environment, interfacing to process-level equipment (IEDs, PLCs) and incorporating substation noise immunity and isolation.

The LAN must have enough throughput and bandwidth to support integrated data acquisition, control, and protection requirements. Should the LAN utilize deterministic protocol technologies, such as token ring and token bus schemes? Response times for data transfer must be deterministic and repeatable. The definition of *deterministic* is: pertaining to a process, model, or variable whose outcome, result, or value does not depend on chance (IEEE Std. 100-1997).

The LAN should support peer-to-peer communications capability for high-speed protection functions, and file transfer support for IED configuration and programmable logic controller (PLC) programs. Priority data transfer would allow configuration files and other low priority data to be downloaded without affecting time-critical data transfer. The IED and peripheral interface should be a common bus for all input/output. If the LAN is compatible with the substation computer (e.g., Ethernet), a front-end processor may not be needed. There are stringent speed requirements for interlocking and intertripping data transfer, which the LAN must support. The LAN must be able to support bridges and routers for the utility enterprise wide area network (WAN) interface. Test equipment for the LAN must be readily available and economic. Implementation of the LAN technology must be competitive to drive the cost down. For example, Ethernet is more widely used than FDDI, and therefore Ethernet interface equipment is lower cost. *Peer-to-peer communication* is defined as communication between two or more network nodes in which either node can initiate sessions, and is able to poll or answer to polls (IEEE Std. 100-1997).

Substation LAN Protocols

A substation local area network (LAN) is a communications network, typically high speed, within the substation, and extending into the switchyard, which provides the ability to quickly transfer measurements, indications, control adjustments, configuration, and historic data between intelligent devices at the site. The benefits achievable using this architecture include: a reduction in the amount and complexity of the cabling currently required between devices; an increase in the available communications bandwidth to support faster updates and more advanced functions such as virtual connection, file transfer, peer-to-peer communications, and plug-and-play capabilities; and the less tangible benefit of an open LAN architecture including laying the foundation for future upgrades, access to third-party equipment, and increased interoperability.

The EPRI-sponsored Utility Substation Communication Initiative performed benchmark and simulation testing of different local area network technologies for the substation in late 1996. The initial substation configuration tested included 47 IEDs with these data types: analog, accumulator, control and

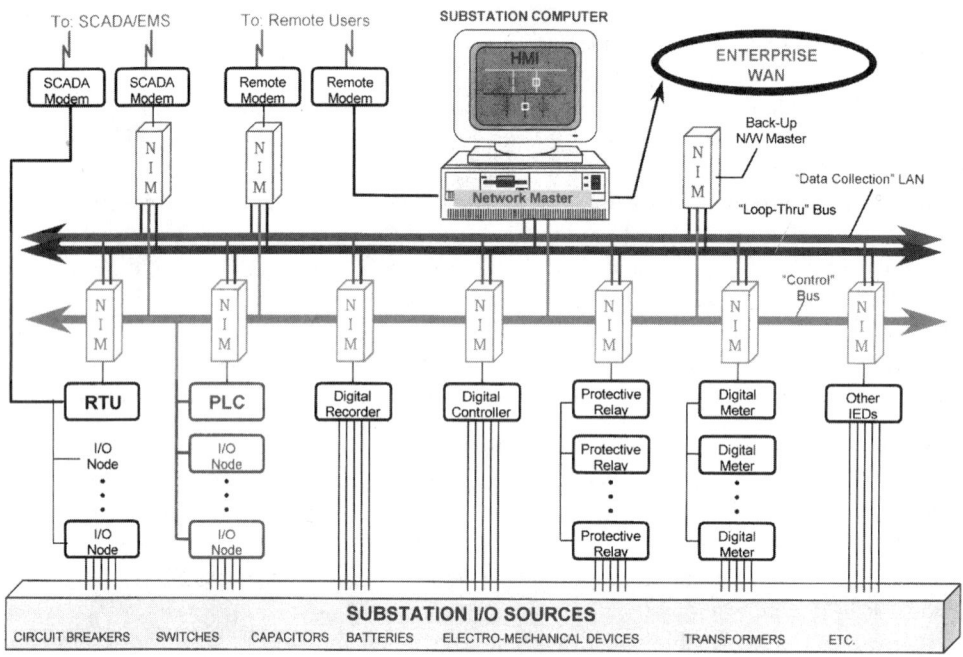

FIGURE 5.40 Substation automation system configuration. (*Automating a Distribution Cooperative, A to Z,* National Rural Electric Cooperative Association [NRECA], Cooperative Research Network [CRN], 1999. With permission.)

events, and fault records. The response requirements were 4 msec for a protection event, 111 transactions per second for SCADA traffic, and 600 sec to transmit a fault record. The communication profiles tested were FMS/Profibus at 12Mbps, MMS/Trim7/Ethernet at 10Mbps and 100Mbps, and Switched Ethernet. Initially, the testing was done with four test bed nodes, using four 133-MHz Pentium computers. The four nodes simulated 47 devices in the substation. Analysis of the preliminary results from this testing resulted in a more extensive test done with 20 nodes, using 20 133-MHz Pentium computers. The 20 nodes simulated a large substation issuing four trip signals each to simulate 80 trip signals from 80 different IEDs.

The tests determined that FMS/Profibus at 12Mbps (fast FMS implementation) would not meet the trip time requirements for protective devices. However, MMS/Ethernet would meet the requirements. In addition, it was found that varying the SCADA load does not impact transaction performance. Also, the transmission of oscillographic data and SCADA data does not impact transaction times.

User Interface

The user interface in the substation must be an intuitive design to ensure effective use of the system with minimal confusion. An efficient display hierarchy will result in all essential activities being performed from a few displays. It is critical that the amount of typing needed be minimized, or better yet, be eliminated. There should be a common "look and feel" established for all displays. A library of standard symbols should be used to represent substation power apparatus on graphical displays. In fact, this library should be established and used in all substations, and coordinated with other systems in the utility, such as the distribution SCADA system, the Energy Management System, the Geographic Information System (GIS), the Trouble Call Management System, etc. The field personnel, or the users of the system, should be involved in determining what information should be on the different displays. Multiple databases should be avoided, such as the database for the IED and the database associated with a third-party user interface software package (e.g., U.S. Data FactoryLink, BJ Systems RealFlex, Intellution, WonderWare, etc.).

Substations   5-59

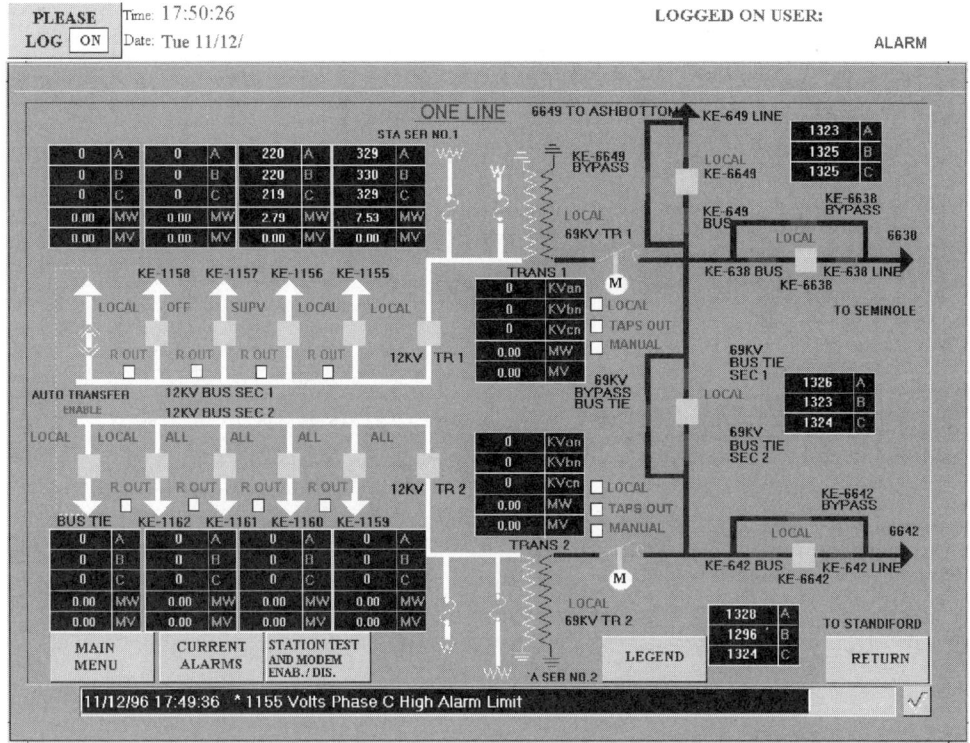

FIGURE 5.41 Substation one-line display.

The substation one-line displays (Fig. 5.41) may be similar in appearance to the displays on a Distribution Management System or an Energy Management System. Typically, analog panel meter functionality is integrated in the substation automation system user interface. The metered values can be viewed in a variety of formats. Alarm displays (Fig. 5.42) can have tabular and graphical display formats. There should be a convenient means to obtain detailed information about the alarm condition. There are two types of logs in the substation. The substation automation system has the capability to log any information in the system at a specified periodicity. It can log alarms and events with time tags. In other words, any information the system has, can be included in a log. A manual log is also in the substation for documenting all activities performed in the substation. The information in the manual log can be included in the system, but would have to be entered into the system using the keyboard, and typing is not something field personnel normally want to do.

Communication Interfaces

There are interfaces to substation IEDs to acquire data, determine the operating status of each IED, support all communication protocols used by the IEDs, and support standard protocols being developed. There may be an interface to the Energy Management System (EMS), where the system operators are able to monitor and control each substation, and the EMS is able to receive data from the substation automation system at different periodicities. There may be an interface to the Distribution Management System with the same capabilities as the EMS interface. There may also be an interface to a capacitor control system, to control switched capacitor banks located on feeders emanating from the substation, and monitoring Vars on all three phases for the decision to switch the capacitor bank and for verification that the capacitor bank did switch. There is always an interface for remote dial-in capabilities, providing authorized access to the system via dial-in telephone to obtain data and alarms, execute diagnostic programs, and retrieve results of diagnostic programs. Last, there is an interface to a time standard source.

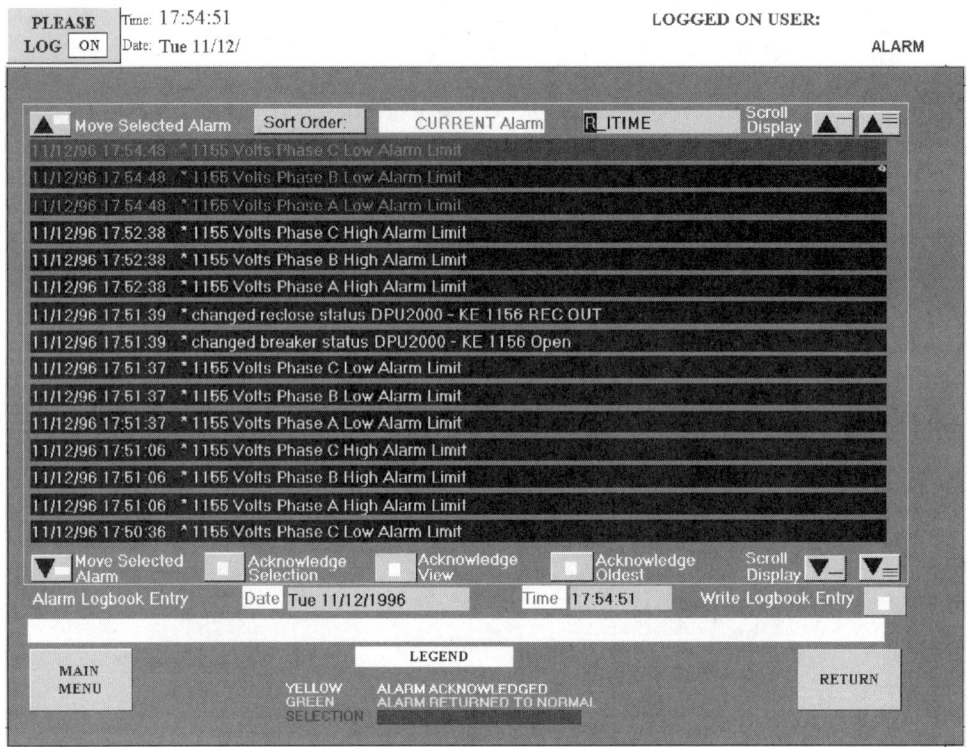

FIGURE 5.42 Substation alarm display.

A time reference unit (TRU) is typically provided at each substation for outputting time signals to the substation automation system IEDs, controllers, and host computers. The GPS time standard is used and synchronized to GPS satellite time. The GPS system includes an alphanumeric display for displaying time, satellite tracking status, and other setup parameters.

Data Warehouse

The data warehouse enables users to access substation data while maintaining a firewall to substation control and operation functions. Both operational and non-operational data (load forecasting, engineering studies, outage investigations) are needed in the data warehouse. The utility must determine who the users of the substation automation system data are, the nature of their application, the type of data needed, how often the data is needed, and the frequency of update required for each user. Examples of user groups within a utility are operations, planning, engineering, SCADA, protection, distribution automation, metering, substation maintenance, and information technology. A possible data warehouse is the "information management system" typically associated with an Energy Management System (EMS), populated by the EMS with both real-time and historical data, and employing firewalls to prevent external access beyond the information management system into the EMS. A *firewall* protects a trusted network from an outside, "untrusted" network — a difficult job to do quickly and efficiently, since the firewall must inspect each network communication and decide whether to allow it to cross over to the trusted network (Holzbaur, 1997).

IEEE Power Engineering Society Substations Committee

The Data Acquisition, Processing and Control Systems Subcommittee of the Institute of Electrical and Electronic Engineers (IEEE) Power Engineering Society (PES) Substations Committee recognized the need for a standard IED protocol in the late 1980s and formed a task force to examine existing protocols, and determine, based on two sets of screening criteria, the two best candidates. IEEE Standard 1379, *Trial Use Recommended Practice for Data Communications Between Intelligent Electronic Devices and Remote Terminal Units in a Substation*, was published in early 1998. This document does not establish a communication standard. To quickly achieve industry acceptance and use, it instead provides a specific implementation of two existing communication protocols in the public domain. The first is DNP 3.0, the Level 2-subset implementation as published by the DNP Users Group. The second is IEC 870-5-101, developed by IEC Technical Committee (TC) 57 Working Group (WG) 03, including the T101 companion standard (profile). The Task Force decided to use the IEEE "trial use recommended practice" designation for this work, with a limited lifetime, intending that these recommendations would fill a void on an interim basis until a longer term, more permanent solution (such as EPRI UCA 2.0) was ready to be implemented.

EPRI-Sponsored Utility Substation Communication Initiative

In mid-1996, American Electric Power (AEP) hosted the first Utility Substation Communication Initiative meeting as a continuation of the EPRI UCA/Substation Automation Project (RP3599). After 15 meetings in a little more than 2 years, there are approximately 25 utilities and 15 suppliers participating, having formed supplier/utility teams to define the supplier IED functionality, and to implement a standard IED protocol (UCA 2.0 profile) and LAN protocol (Ethernet). Prototype protocols and topologies have been staged and benchmarked to verify that the recommended approaches are based on sound technology (Burger et al., 1998).

Generic Object Models for Substation and Feeder Equipment (GOMSFE) are being developed to facilitate suppliers in implementing the UCA/Substation Automation Project substation and feeder elements of the Power System Object Model (PSOM). The GOMSFE work merges the UCA Forum Substation and Feeder Automation work with that of UCA 2.0 in order to produce common generic object models for implementation of UCA 2.0-compliant field devices in electric utilities (UCA, 1997; Harlow, 1998).

New IED products with this functionality were commercially available in 1999. The utilities began providing demonstration sites for the implementation of the new IED products in 1998, to demonstrate interoperability between IED equipment from different suppliers, and to evaluate and recommend a suitable UCA-compliant substation LAN (McDonald and Saxton, 1997).

References

Adamiak, M., Patterson, R., and Melcher, J., Inter and Intra Substation Communications: Requirements and Solutions, Paper SIPSEP-96-13, System Protection Seminar, Monterrey, Mexico, November 24–27, 1996.
Advancements in Microprocessor Based Protection and Communication, IEEE Tutorial Course 97TP120-0, 1997.
Automating a Distribution Cooperative, A to Z, National Rural Electric Cooperative Association (NRECA), Cooperative Research Network (CRN), 1999.
Borlase, S. H., Advancing to true station and distribution system integration in electric utilities, IEEE PES paper PE-675-PWRD-0-02-1997.
Burger, J., Melcher, J. C., and Robinson, J. T., Substation Communications and Protocols — EPRI UCA 2.0 Demonstration Initiative, in *Proceedings of the EPRI-Sponsored Substation Equipment Diagnostic Conference VI*, EPRI EV-108393, February 1998.
Communication Protocols, IEEE Tutorial 95 TP 103, 1995.
Draft Input for the Utility Communications Architecture Version 2.0, UCA Version 2.0 Profiles, prepared under the Auspices of the Profile Working Group of the MMS Forum, Editorial Draft 1.0, 1997.
An Enhanced Version of the IEEE Standard Dictionary of Electrical and Electronics Terms, (IEEE Std. 100) CD-ROM, 1997.
Enterprise-wide Data Integration in a Distribution Cooperative, Cornice Engineering, Inc., 1996.
EPRI UCA report Number EL-7547 (multi-volume), Electric Power Research Institute, 1991.
Fundamentals of Supervisory Systems, IEEE Tutorial 94 EH0392-1 PWR, 1994.

Harlow, J. H., Application of UCA Communications to Power Transformers, in *Proceedings of the EPRI-Sponsored Substation Equipment Diagnostic Conference VI,* EPRI EV-108393, February 1998.

Harmon, G. L., Substation protection, control and monitoring protocols: past, present and future, Paper SIPSEP-96-12, System Protection Seminar, Monterrey, Mexico, November 24–27, 1996.

Holzbaur, H., Firewalls — What every utility should know, *Information Technologies for Utilities,* 19–26, Fall 1997.

IEEE Standard Definition, Specification and Analysis of Systems Used for Supervisory Control, Data Acquisition, and Automatic Control, IEEE Std C37.1-1994.

IEEE Standard Electrical Power System Device Function Numbers and Contact Designations, IEEE Std. C37.2-1996.

Marks, J., Data that talks UCA, *Rural Electrification,* 28–30, December 1992.

McDonald, J. D. and Saxton, T. L., Understanding today's protocol standardization efforts, *Utility Automation,* 32-36, September/October 1997.

Spinney, A., Which LAN is best for substation automation?, *Synergy News,* GE Harris Energy Control Systems, 5–6, July 1997.

Profibus Technical Overview, Profibus Trade Organization (PTO), April 1996.

Marks, J., The last piece in the puzzle: The quest for a common communications protocol for utility sensing devices, *Rural Electrification,* 28-30, September 1995.

The New Telecommunications Environment: Opportunities for Electric Cooperatives, NRECA, 1996.

Substation Automation Communications Demonstration Initiative, Revised Statement of Work, Scope/Requirements Checklist, Review Draft Revision 0.7, September 12, 1997.

Substation LAN Performance Scenario, Version 0.6, January 12, 1997.

Substation UCA Protocol Reference Specification (Review Draft), KEMA Consulting, July 1997.

Trial Use Recommended Practice for Data Communications Between Intelligent Electronic Devices and Remote terminal Units in a Substation, IEEE Std. 1379-1997.

UCA 2.0 General Object Models for Substation and Feeder Field Devices, Draft Version 0.7, December 1997.

5.7 Oil Containment[1]

Anne-Marie Sahazizian and Tibor Kertesz

Containment and control of oil spills at electric supply substations is a concern for most electric utilities. The environmental impact of oil spills and their cleanup is governed by several federal, state, and local regulations, necessitating increased attention in substations to the need for secondary oil containment, and a Spill Prevention Control and Countermeasure (SPCC) plan. Beyond the threat to the environment, cleanup costs associated with oil spills could be significant, and the adverse community response to any spill is becoming increasingly unacceptable.

The probability of an oil spill occurring in a substation is very low. However, certain substations, due to their proximity to navigable waters or designated wetlands, the quantity of oil on site, surrounding topography, soil characteristics, etc., have or will have a higher potential for discharging harmful quantities of oil into the environment. At minimum, a SPCC plan will probably be required at these locations, and installation of secondary oil-containment facilities might be the right approach to mitigate the problem.

Before an adequate spill prevention plan is prepared and a containment system is devised, the engineer must first be thoroughly aware of the requirements included in the federal, state, and local regulations.

---

[1]Sections of this chapter reprinted from IEEE Std. 980-1994, *IEEE Guide for Containment and Control of Oil Spills in Substations,* 1995, Institute of Electrical and Electronics Engineers, Inc. (IEEE). The IEEE disclaims any responsibility or liability resulting from the placement and use in the described manner. Information is reprinted with permission of the IEEE.

The federal requirements of the U.S. for discharge, control, and countermeasure plans for oil spills are contained in the Code of Federal Regulations, Title 40 (40CFR), Parts 110 and 112. The above regulations only apply if the facility meets the following conditions:

1. Facilities with above-ground storage capacities greater than 2500 l (approximately 660 gal) in a single container or 5000 l (approximately 1320 gal) in aggregate storage, *or*
2. Facilities with a total storage capacity greater than 159,000 l (approximately 42,000 gal) of buried oil storage, *or*
3. *Any* facility which has spilled more than 3786 l (1000 gal) of oil in a single event or spilled oil in two events occurring within a 12-month period, and
4. Facilities which, due to their location, could reasonably be expected to discharge oil into or upon the navigable waters of the U.S. or its adjoining shorelines.

In other countries, applicable governmental regulations will cover the above requirements.

Oil-Filled Equipment in Substation (IEEE, 1994)

A number of electrical apparatus installed in substations are filled with oil that provides the necessary insulation characteristics and assures their required performance. Electrical faults in this power equipment can produce arcing and excessive temperatures that may vaporize insulating oil, creating excessive pressure that may rupture the electrical equipment tanks. In addition, operator errors, sabotage, or faulty equipment may also be responsible for oil releases.

The initial cause of an oil release or fire in electrical apparatus may not always be avoidable, but the extent of damage and the consequences for such an incident can be minimized or prevented by adequate planning in prevention and control.

Described below are various sources of oil spills within substations. Spills from any of these devices are possible. The user must evaluate the quantity of oil present, the potential impact of a spill, and the need for oil containment associated with each oil-filled device.

Large Oil-Filled Equipment

Power transformers, oil-filled reactors, large regulators, and circuit breakers are the greatest potential source of major oil spills in substations, since they typically contain the largest quantity of oil.

Power transformers, reactors, and regulators may contain anywhere from a few hundred to 100,000 l or more of oil (500 to approximately 30,000 gal), with 7500–38,000 l (approximately 2000–10,000 gal) being typical. Substations usually contain one to four power transformers, but may have more.

The higher voltage oil circuit breakers may have three independent tanks, each containing 400–15,000 l (approximately 100–4000 gal) of oil, depending on their rating. However, most circuit breaker tanks contain less than 4500 l (approximately 1200 gal) of oil. Substations may have 10–20 or more oil circuit breakers.

Cables

Substation pumping facilities and cable terminations (potheads) that maintain oil pressure in pipe-type cable installations are another source of oil spills. Depending on its length and rating, a pipe-type cable system may contain anywhere from 5000 l (approximately 1500 gal) up to 38,000 l (approximately 10,000 gal) or more of oil.

Mobile Equipment

Although mobile equipment and emergency facilities may be used infrequently, consideration should be given to the quantity of oil contained and associated risk of oil spill. Mobile equipment may contain up to 30,000 l (approximately 7500 gal) of oil.

Oil-Handling Equipment

Oil filling of transformers, circuit breakers, cables, etc. occurs when the equipment is initially installed. In addition, periodic reprocessing or replacement of the oil may be necessary to ensure that proper insulation qualities are maintained. Oil pumps, temporary storage facilities, hoses, etc. are brought in to accomplish this task. Although oil processing and handling activities are less common, spills from these devices can still occur.

Oil Storage Tanks

Some consideration must be given to the presence of bulk oil storage tanks (either above-ground or below-ground) in substations as these oil tanks could be responsible for an oil spill of significant magnitude. Also, the resulting applicability of the 40CFR, Part 112 rules for these storage tanks could require increased secondary oil containment for the entire substation facility. The user may want to reconsider storage of bulk oil at substation sites.

Other Sources

Station service, voltage, and current transformers, as well as smaller voltage regulators, oil circuit reclosers, capacitor banks, and other pieces of electrical equipment typically found in substations, contain small amounts of insulating oil, usually less than the 2500 l (approximately 660 gal) minimum for a single container.

Spill Risk Assessment

The risk of an oil spill caused by an electric equipment failure is dependent on many factors, including:

- Engineering and operating practices (i.e., electrical fault protection, loading practices, switching operations, testing, and maintenance).
- Quantities of oil contained within apparatus.
- Station layout (i.e., spatial arrangement, proximity to property lines, streams, and other bodies of water).
- Station topography and site preparation (i.e., slope, soil conditions, ground cover).
- Rate of flow of discharged oil.

Each facility must be evaluated to select the safeguards commensurate with the risk of a potential oil spill.

The engineer must first consider whether the quantities of oil contained in the station exceed the quantities of oil specified in the Regulations, and secondly, the likelihood of the oil reaching navigable waters if an oil spill or rupture occurs. If no likelihood exists, no SPCC plan is required.

SPCC plans must be prepared for each piece of portable equipment and mobile substations. These plans have to be general enough that the plan may be used at any and all substations or facility location.

Both the frequency and magnitude of oil spills in substations can be considered to be very low. The probability of an oil spill at any particular location depends on the number and volume of oil containers, and other site-specific conditions.

Based on the applicability of the latest regulatory requirements, or when an unacceptable level of oil spills has been experienced, it is recommended that a program be put in place to mitigate the problems. Typical criteria for implementing oil spill containment and control programs incorporate regulatory requirements, corporate policy, frequency and duration of occurrences, cost of occurrences, safety hazards, severity of damage, equipment type, potential impact on nearby customers, substation location, and quality-of-service requirements (IEEE, 1994).

The decision to install secondary containment at new substations (or to retrofit existing substations) is usually based on predetermined criteria. A 1992 IEEE survey addressed the factors used to determine where oil spill containment and control programs are needed. Based on the survey, the criteria in Table 5.4 are considered when evaluating the need for secondary oil containment.

TABLE 5.4 Secondary Oil-Containment Evaluation Criteria

Criteria	Utilities Responding That Apply This Criteria
Volume of oil in individual device	88%
Proximity to navigable waters	86%
Total volume of oil in substation	62%
Potential contamination of groundwater	61%
Soil characteristics of the station	42%
Location of substation (urban, rural, remote)	39%
Emergency response time if a spill occurs	30%
Failure probability of the equipment	21%
Age of station or equipment	10%

Source: IEEE, 1994.

TABLE 5.5 Secondary Oil-Containment Equipment Criteria

Equipment	Utilities Responding That Provide Secondary Containment
Power transformers	86%
Above-ground oil storage tanks	77%
Station service transformers	44%
Oil circuit breakers	43%
Three-phase regulators	34%
Below-ground oil storage tanks	28%
Shunt reactors	26%
Oil-filling equipment	22%
Oil-filled cables and terminal stations	22%
Single-phase regulators	19%
Oil circuit reclosers	15%

Source: IEEE, 1994.

The same 1992 IEEE survey provided no clear-cut limit for the proximity to navigable waters. Relatively, equal support was reported for several choices over the range of 45–450 m (150–1500 feet).

Rarely is all of the equipment within a given substation provided with secondary containment. Table 5.5 lists the 1992 IEEE survey results identifying the equipment for which secondary oil containment is provided.

Whatever the criteria, each substation has to be evaluated by considering the criteria to determine candidate substations for oil-containment systems (both new and retrofit). Substations with planned equipment change-outs and located in environmentally sensitive areas have to be considered for retrofits at the time of the change-out.

Containment Selection Consideration (IEEE, 1994)

Containment selection criteria have to be applied in the process of deciding the containment option to install in a given substation. Criteria to be considered include: operating history of the equipment, environmental sensitivity of the area, the solution's cost-benefit ratio, applicable governmental regulations, and community acceptance.

The anticipated cost of implementing the containment measures must be compared to the anticipated benefit. However, cost alone can no longer be considered a valid reason for not implementing containment and/or control measures, because any contamination of navigable waters may be prohibited by government regulations.

Economic aspects can be considered when determining which containment system or control method to employ. Factors such as proximity to waterways, volume of oil, response time following a spill, etc., can allow for the use of less effective methods at some locations.

Due to the dynamic nature of environmental regulations, some methods described in this section of the handbook could come in conflict with governmental regulations or overlapping jurisdictions. Therefore, determination of which containment system or control method to use must include research into applicable laws and regulations.

Community acceptance of the oil spill containment and control methods is also to be considered. Company policies, community acceptance, customer relations, etc. may dictate certain considerations. The impact on adjacent property owners must be addressed and, if needed, a demonstration of performance experiences could be made available.

Oil Spill Prevention Techniques

Upon an engineering determination that an oil spill prevention system is needed, the engineer must weigh the advantages and disadvantages that each oil retention system may have at the facility in question. The oil retention system chosen must balance the cost and sophistication of the system to the risk of the damage to the surrounding environment. The risks, and thus the safeguards, will depend on items such as soil, terrain, relative closeness to waterways, and potential size of discharge. Each of the systems that are described below may be considered based on their relative merits to the facility under consideration. Thus, one system will not always be the best choice for all situations and circumstances.

Containment Systems

The utility has to weigh the advantages and disadvantages that each oil retention system may have at the facility in question. Some of the systems that could be considered based on their relative merits to the facility under consideration are presented in the next paragraphs.

Yard Surfacing and Underlying Soil

100 to 150 mm (4 to 6 in.) of rock gravel surfacing are normally required in all electrical facility yards. This design feature benefits the operation and maintenance of the facility by providing proper site drainage, reducing step and touch potentials during short-circuit faults, eliminating weed growth, improving yard working conditions, and enhancing station aesthetics. In addition to these advantages, this gravel will aid in fire control and in reducing potential oil spill cleanup costs and penalties that may arise from federal and state environmental laws and regulations.

Yard surfacing is not to be designed to be the primary or only method of oil containment within the substation, but rather has to be considered as a backup or bonus in limiting the flow of oil in the event that the primary system does not function as anticipated.

Soil underlying power facilities usually consists of a non-homogeneous mass that varies in composition, porosity, and physical properties with depth.

Soils and their permeability characteristics have been adapted from typical references and can be generalized as in the following Table 5.6.

TABLE 5.6 Soil Permeability Characteristics

Permeability (cm/sec)	Degree of Permeability	Type of Soil
Over 10^{-1}	High	Stone, gravel, and coarse- to medium-grained sand
10^{-1} to 10^{-3}	Medium	Medium-grained sand to uniform, fine-grained sand
10^{-3} to 10^{-6}	Low	Uniform, fine-grained sand to silty sand or sandy clay
Less than 10^{-6}	Practically impermeable	Silty sand or sandy clay to clay

Source: IEEE, 1994.

Substations

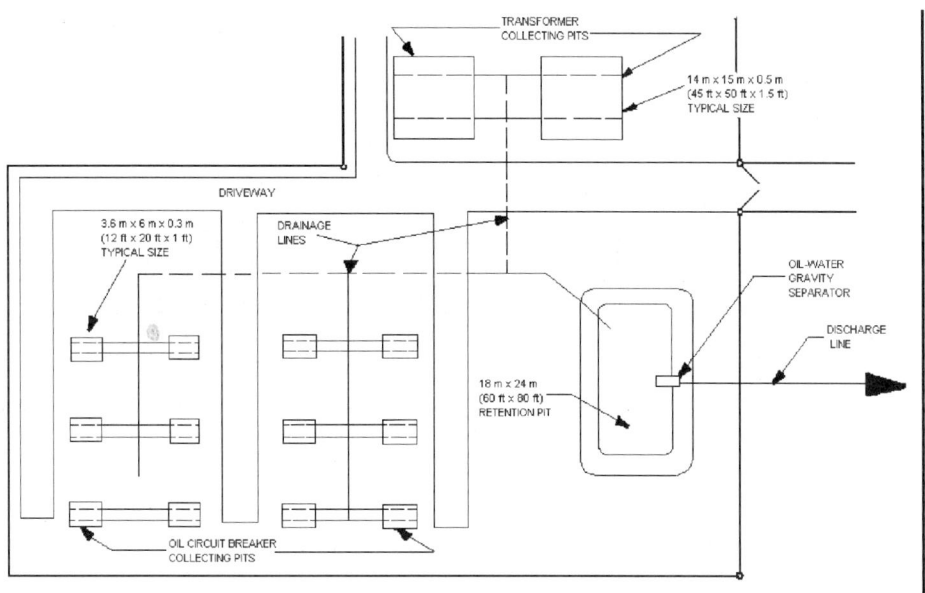

FIGURE 5.43 Typical containment system with retention and collection pits.

Substation Ditching
One of the simplest methods of providing total substation oil spill control is the construction of a ditch entirely around the outside periphery of the station. The ditch has to be of adequate size as to contain all surface run-offs due to rain and insulating oil. These ditches may be periodically drained by the use of valves.

Collecting Ponds with Traps
In this system, the complete design consists of a *collection pit* surrounding the protected equipment, drains connecting the collection pits to an open *containment pit* and an oil trap which is sometimes referred to as a skimming unit and the discharge drain. Figure 5.43 (IEEE, 1994) presents the general concept of such a containment solution. The collection pit surrounding the equipment is filled with rocks and designed only deep enough to extinguish burning oil. The bottom of this pit is sloped for good drainage to the drainpipe leading to an open containment pit. This latter pit is sized to handle all the oil of the largest piece of equipment in the station. To maintain a dry system in the collecting units, the invert of the intake pipe to the containment pit must be at least the maximum elevation of the oil level. In areas of the country subject to freezing temperatures, it is recommended that the trap (skimmer) be encased in concrete, or other similar means available, to eliminate heaving due to ice action.

Oil Containment Equipment Pits
Probably one of the most reliable but most expensive methods of preventing oil spills and insuring that oil will be contained on the substation property is by placing all major substation equipment on or in containment pits. This method of oil retention provides a permanent means of oil containment. These containment pits will confine the spilled oil to relatively small areas that in most cases will greatly reduce the cleanup costs.

One of the most important issues related to an equipment pit is to prevent escape of spilled oil into underlying soil layers. Pits with liners or sealers may be used as part of an oil containment system capable of retaining any discharged oil for an extended period of time. Any containment pit must be constructed with materials having medium to high permeability (above 10^{-3} cm/sec) and be sealed in order to prevent migration of spilled oil into underlying soil layers and groundwater. These surfaces may be sealed and/or lined with any of the following materials:

1. Plastic or rubber — Plastic or rubber liners may be purchased in various thickness and sizes. It is recommended that a liner be selected that is resistant to mechanical injury which may occur due to construction and installation, equipment, chemical attacks on surrounding media, and oil products.
2. Bentonite (clay) — Clay and Bentonite may also be used to seal electrical facility yards and containment pits. These materials can be placed directly in 100 to 150 mm (4 to 6 in.) layers or may be mixed with the existing subsoil to obtain an overall soil permeability of less than 10^{-3} cm/sec.
3. Spray-on fiberglass — Spray-on fiberglass is one of the most expensive pit liners available, but in some cases, the costs may be justifiable in areas which are environmentally sensitive. This material offers very good mechanical strength properties and provides excellent oil retention.
4. Reinforced concrete — 100 to 150 mm (4 to 6 in.) of reinforced concrete may also be used as a pit liner. This material has an advantage over other types of liners in that it is readily available at the site at the time of initial construction of the facility. Concrete has some disadvantages in that initial preparation is more expensive and materials are not as easily workable as some of the other materials.

If materials other than those listed above are used for an oil containment liner, careful consideration must be given to selecting materials, which will not dissolve or become soft with prolonged contact with oil, such as asphalt.

Fire Quenching Considerations (IEEE, 1994)
In places where the oil-filled device is installed in an open pit (not filled with stone), an eventual oil spill associated with fire will result in a pool fire around the affected piece of equipment. If a major fire occurs, the equipment will likely be destroyed. Most utilities address this concern by employing active or passive quenching systems, or drain the oil to a remote pit. Active systems include foam or water spray deluge systems.

Of the passive fire quenching measures, pits filled with crushed stone are the most effective. The stone-filled pit provides a fire quenching capability designed to extinguish flames in the event that a piece of oil-filled equipment catches on fire. An important point is that in sizing a stone-filled collecting or retention pit, the final oil level elevation (assuming a total discharge) has to be situated approximately 300 mm (12 in) below the top elevation of the stone.

All the materials used in construction of a containment pit have to be capable of withstanding the higher temperatures associated with an oil fire without melting. If any part of the containment (i.e., discharge pipes from containment to a sump) melts, the oil will be unable to drain away from the burning equipment, and the melted materials may pose an environmental hazard.

Volume Requirements
Before a substation oil-containment system can be designed, the volume of oil to be contained must be known. Since the probability of an oil spill occurring at a substation is very low, the probability of simultaneous spills is extremely low. Therefore, it would be unreasonable and expensive to design a containment system to hold the sum total of all of the oil contained in the numerous oil-filled pieces of equipment normally installed in a substation. In general, it is recommended that an oil-containment system be sized to contain the volume of oil in the single largest oil-filled piece of equipment, plus any accumulated water from sources such as rainwater, melted snow, and water spray discharge from fire protection systems. Interconnection of two or more pits to share the discharged oil volume may provide an opportunity to reduce the size requirements for each individual pit.

Typically, equipment containment pits are designed to extend 1.5–3 m (5–10 ft) beyond the edge of the tank in order to capture a majority of the leaking oil. A larger pit size is required to capture all of the oil contained in an arcing stream from a small puncture at the bottom of the tank (such as from a bullet hole). However, the low probability of the event and economic considerations govern the 1.5–3 m (5–10 ft) design criteria. For all of the oil to be contained, the pit or berm has to extend 7.5 m (25 ft) or more beyond the tank and radiators.

The volume of the pit surrounding each piece of equipment has to be sufficient to contain the spilled oil in the air voids between the aggregate of gravel fill or stone. A gravel gradation with a nominal size of 19–50 mm (3/4 to 2 in.) which results in a void volume between 30 and 40% of the pit volume is generally being used. The theoretical maximum amount of oil that can be contained in 1 ft³ or 1 m³ of stone is given by the following formulae:

$$\text{Oil Volume}\,[\text{gal}] = \frac{\text{void volume of stone}\,[\%]}{100 \times 0.1337\,\text{ft}^3} \tag{5.1}$$

$$\text{Oil Volume}\,[\text{l}] = \frac{\text{void volume of stones}\,[\%]}{100 \times 0.001\,\text{m}^3} \tag{5.2}$$

where 1 gallon = 0.1337 ft³ and 1 liter = 0.001 m³ = 1 dm³.

If the pits are not to be automatically drained of rainwater, then an additional allowance must be made for precipitation. The additional space required would depend on the precipitation for that area and the frequency at which the facility is periodically inspected. It is generally recommended that the pits have sufficient space to contain the amount of rainfall for this period plus a 20% safety margin.

Expected rain and snow accumulations can be determined from local weather records. A severe rainstorm is often considered to be the worst-case event when determining the maximum volume of short-term water accumulation (for design purposes). From data reported in a 1992 IEEE survey, the storm water event design criteria employed ranged from 50 to 200 mm (2 to 8 in.) of rainfall within a short period of time (1–24 h). Generally accepted design criteria is assuming a one in a 25-year storm event.

The area directly surrounding the pit must be graded to slope away from the pit to avoid filling the pit with water in times of rain.

Typical Equipment Containment Solutions

Figure 5.44 illustrates one method of pit construction that allows the equipment to be installed partially below ground. The sump pump can be manually operated during periods of heavy rain or automatically operated. If automatic operation is preferred, special precautions must be included to insure that oil is not pumped from the pits. This can be accomplished with either an oil-sensing probe or by having all major equipment provided with oil-limit switches (an option available from equipment suppliers). These limit switches are located just below the *minimum* top oil line in the equipment and will open when the oil level drops below this point.

A typical above-grade pit and/or berm, as shown in Fig. 5.45, has maintenance disadvantages but can be constructed relatively easily after the equipment is in place at new and existing electrical facilities. These pits may be emptied manually by gate valves or pumps depending on the facility terrain and layout, or automatically implemented by the use of equipment oil limit switches and dc-operated valves or sump pumps.

Another method of pit construction is shown in Fig. 5.46. The figure shows all-concrete containment pits installed around transformers. The sump and the control panel for the oil pump (located inside the sump) are visible and are located outside the containments. Underground piping provides the connection between the two adjacent containments and the sump. The containments are filled with fire-quenching stones.

Discharge Control Systems (IEEE, 1994)

An adequate and effective station drainage system is an essential part of any oil-containment design. Drains, swales, culverts, catch basins, etc., provide measures to ensure that water is diverted away from the substation. In addition, the liquid accumulated in the collecting pits or sumps of various electrical

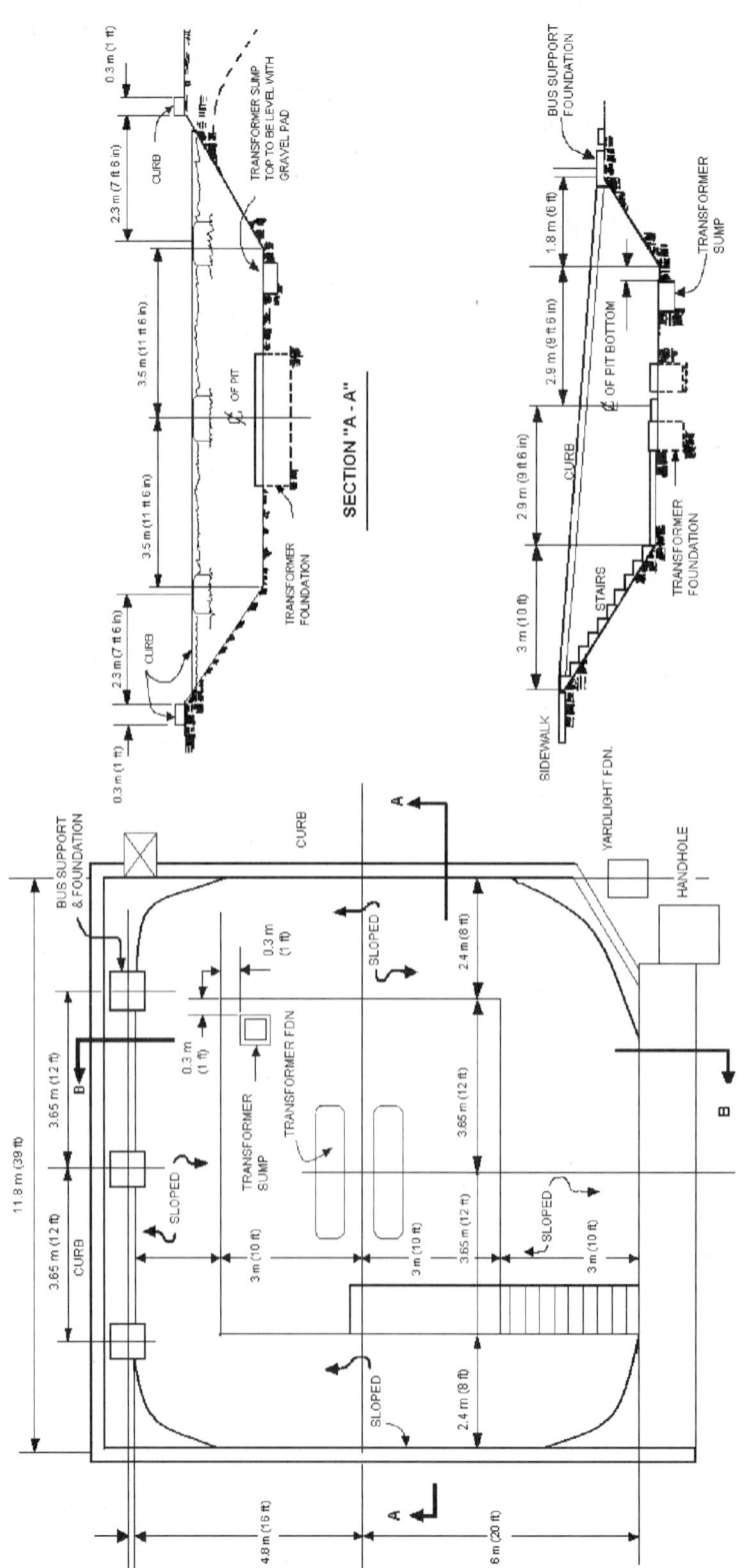

FIGURE 5.44 Typical below-grade containment pit.

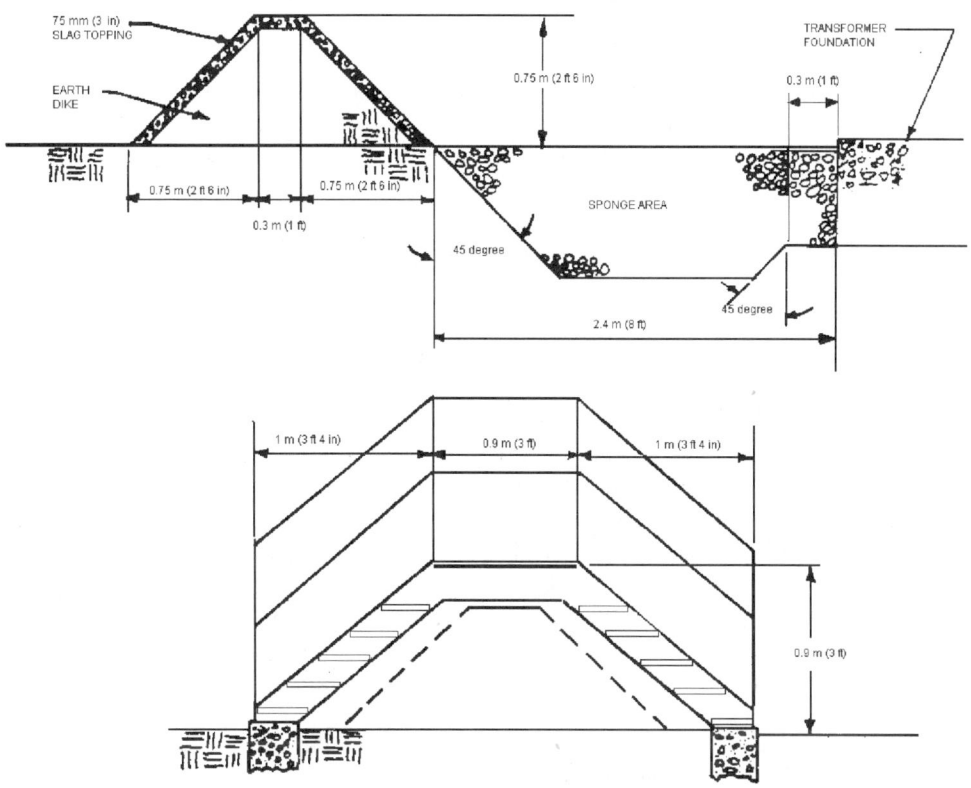

FIGURE 5.45 Typical above-grade berm/pit.

FIGURE 5.46 All-concrete containment pits.

equipment, or in the retention pit has to be discharged. This liquid consists mainly of water (rainwater, melted snow or ice, water spray system discharges, etc.). Oil will be present only in case of an equipment discharge. It is general practice to provide containment systems that discharge the accumulated water into the drainage system of the substation or outside the station perimeter with a discharge control system.

These systems, described below, provide methods to release the accumulated water from the containment system while blocking the flow of discharged oil for later cleanup. Any collected water has to be released as soon as possible so that the entire capacity of the containment system is available for oil containment in the event of a spill. Where the ambient temperatures are high enough, evaporation may eliminate much of the accumulated water. However, the system still should be designed to handle the worst-case event.

Oil-Water Separator Systems (IEEE, 1994)

Oil-water separator systems rely on the difference in specific gravity between oil and water. Because of that difference, the oil will naturally float on top of the water, allowing the water to act as a barrier and block the discharge of the oil.

Oil-water separator systems require the presence of water to operate effectively, and will allow water to continue flowing even when oil is present. The presence of emulsified oil in the water may, under some turbulent conditions, allow small quantities of oil to pass through an oil-water separator system.

Figure 5.47 (IEEE, 1994) illustrates the detail of an oil-water gravity separator that is designed to allow water to discharge from a collecting or retention pit, while at the same time retaining the discharged oil.

Figure 5.48 (IEEE, 1994) illustrates another type of oil-water separator. This separator consists of a concrete enclosure, located inside a collecting or retention pit and connected to it through an opening located at the bottom of the pit. The enclosure is also connected to the drainage system of the substation. The elevation of the top of the concrete weir in the enclosure is selected to be slightly above the maximum elevation of discharged oil in the pit. In this way, the level of liquid in the pit will be under a layer of fire quenching stones where a stone-filled pit is used. During heavy accumulation of water, the liquid will flow over the top of the weir into the drainage system of the station. A valve is incorporated in the weir. This normally closed, manually operated valve allows for a controlled discharge of water from the pit when the level of liquid in the pit and enclosure is below the top of the weir.

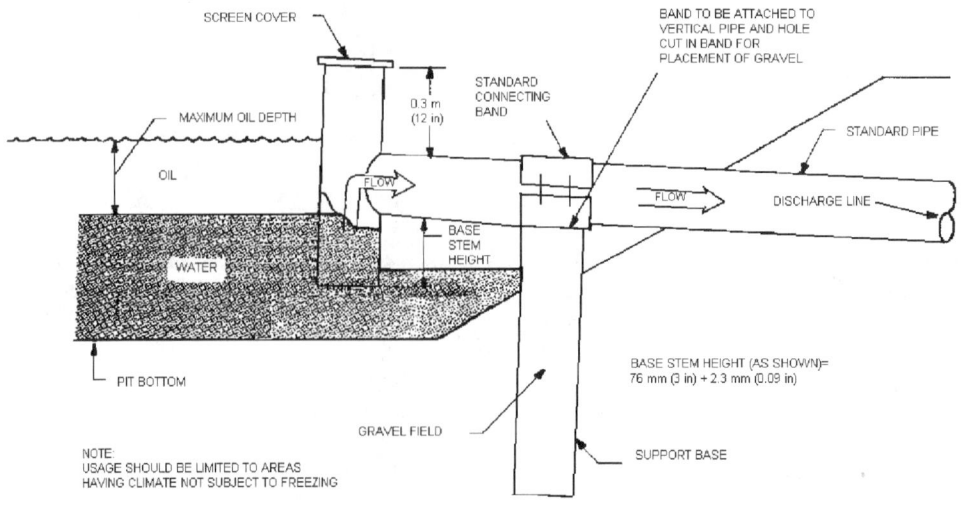

FIGURE 5.47 Oil-water gravity separator.

Substations   5-73

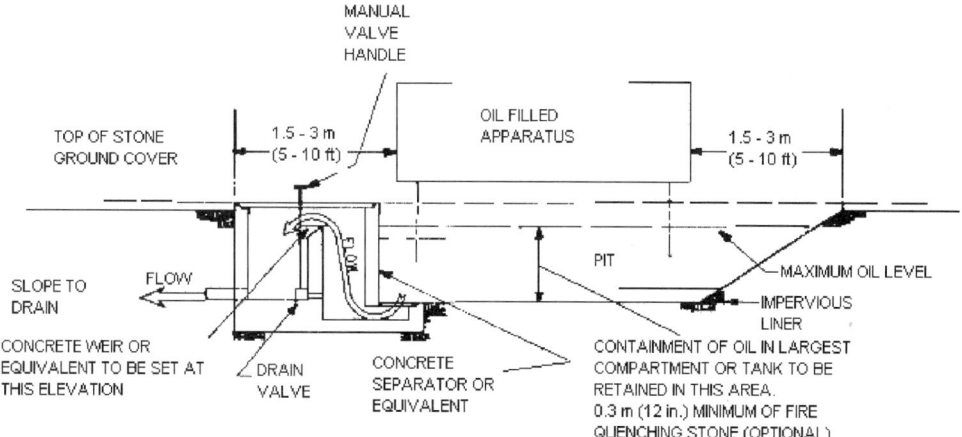

FIGURE 5.48 Equipment containment with oil-water separator.

Figure 5.49 (IEEE, 1994) provides typical detail of an oil trap type oil-water separator. In this system, the oil will remain on top of the water and not develop the head pressure necessary to reach the bottom of the inner vertical pipe. In order for this system to function properly, the water level in the manhole portion of the oil trap must be maintained at an elevation no lower than 0.6 m (2 ft) below the inlet elevation. This will ensure that an adequate amount of water is available to develop the necessary hydraulic head within the inner (smaller) vertical pipe, thereby preventing any discharged oil from leaving the site. It is important to note that the inner vertical pipe should be extended downward past the calculated water-oil interface elevation sufficiently to ensure that oil cannot discharge upward through the inner pipe. Likewise, the inner pipe must extend higher than the calculated oil level elevation in the manhole to ensure that oil does not drain downward into the inner pipe through the vented plug. The reason for venting the top plug is to maintain atmospheric pressure within the vertical pipe, thereby preventing any possible siphon effect.

Flow Blocking Systems (IEEE, 1994)
Described below are two oil flow blocking systems that do not require the presence of water to operate effectively. These systems detect the presence of oil and block all flow (both water and oil) through the discharge system. The best of these systems have been shown to be the most sensitive in detecting and blocking the flow of oil. However, they are generally of a more complex design and may require greater maintenance to ensure continued effectiveness.

Figure 5.50 illustrates an oil stop valve installed inside a manhole. The valve has only one moving part: a ballasted float set at a specific gravity between that of oil and water. When oil reaches the manhole, the float in the valve loses buoyancy and sinks as the oil level increases until it sits on the discharge opening of the valve and blocks any further discharge. When the oil level in the manhole decreases, the float will rise automatically and allow discharge of water from the manhole. Some of the oil stop valves have a weep hole in the bottom of the valve that allows the ballasted float to be released after the oil is removed. This can cause oil to discharge if the level of the oil is above the invert of the discharge pipe.

Figure 5.51 illustrates a discharge control system consisting of an oil-detecting device and a pump installed in a sump connected to the collecting or retention pits of the oil-containment system. The oil-detecting device may use different methods of oil sensing (e.g., capacitance probes, turbidimeters, and fluorescence meters). The capacitance probe shown detects the presence of oil on the surface of the water, based on the significant capacitance difference of these two liquids and, in combination with a logic of liquid level switches, stops the sump pump when the water-oil separation layer reaches a preset height

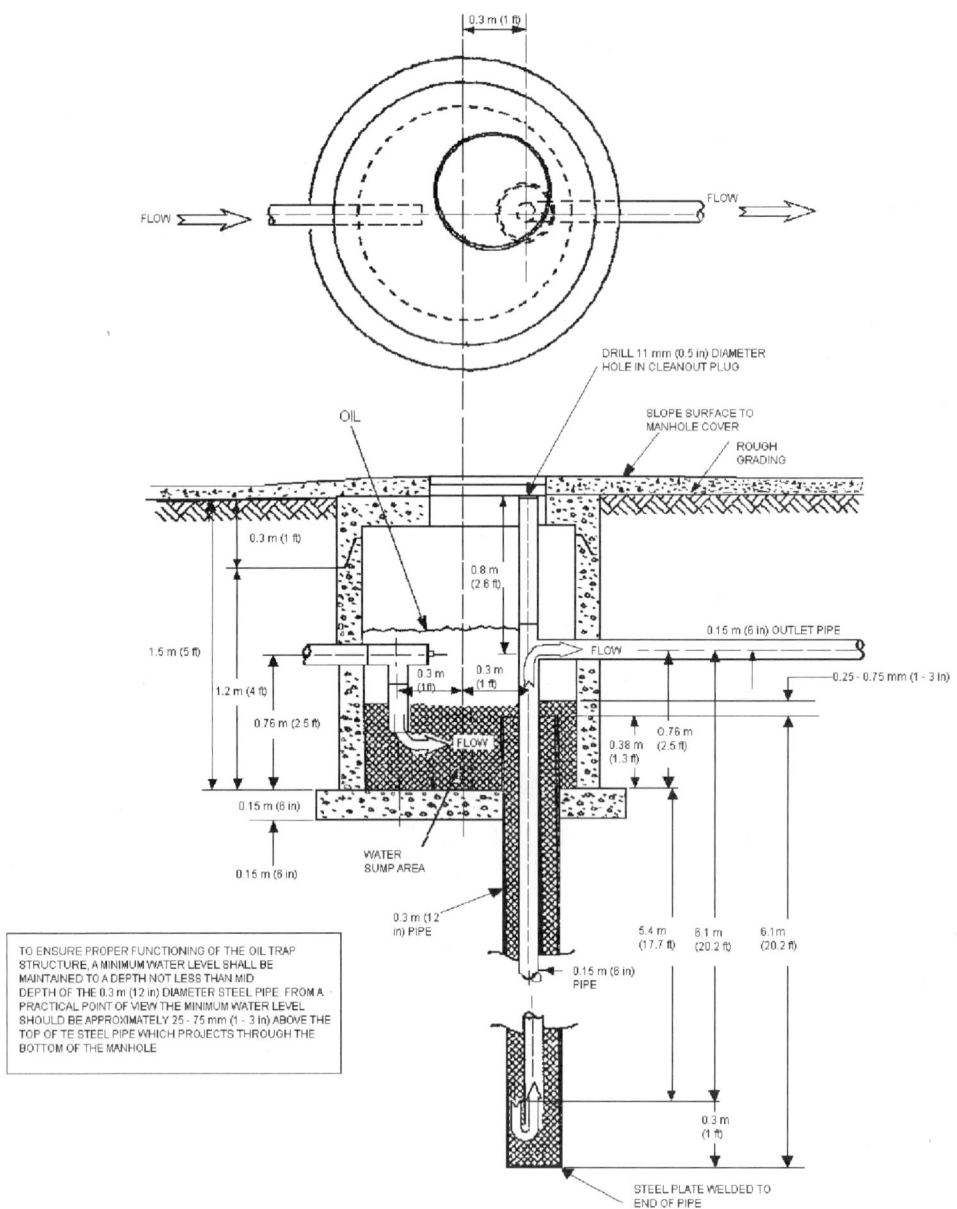

FIGURE 5.49 Oil trap type oil-water separator.

in the sump. Transformer low oil level or gas protection can be added into the control diagram of the pump in order to increase the reliability of the system during major spills.

Some containment systems consist of collecting pits connected to a retention pit or tank that have no link to the drainage system of the substation. Discharge of the liquid accumulated in these systems requires the use of permanently installed or portable pumps. However, should these probes become contaminated, they may cease to function properly. Operating personnel manually activate these pumps. This system requires periodic inspection to determine the level of water accumulation. Before pumping any accumulated liquid, an inspection is required to assess whether the liquid to be pumped out is contaminated.

Substations

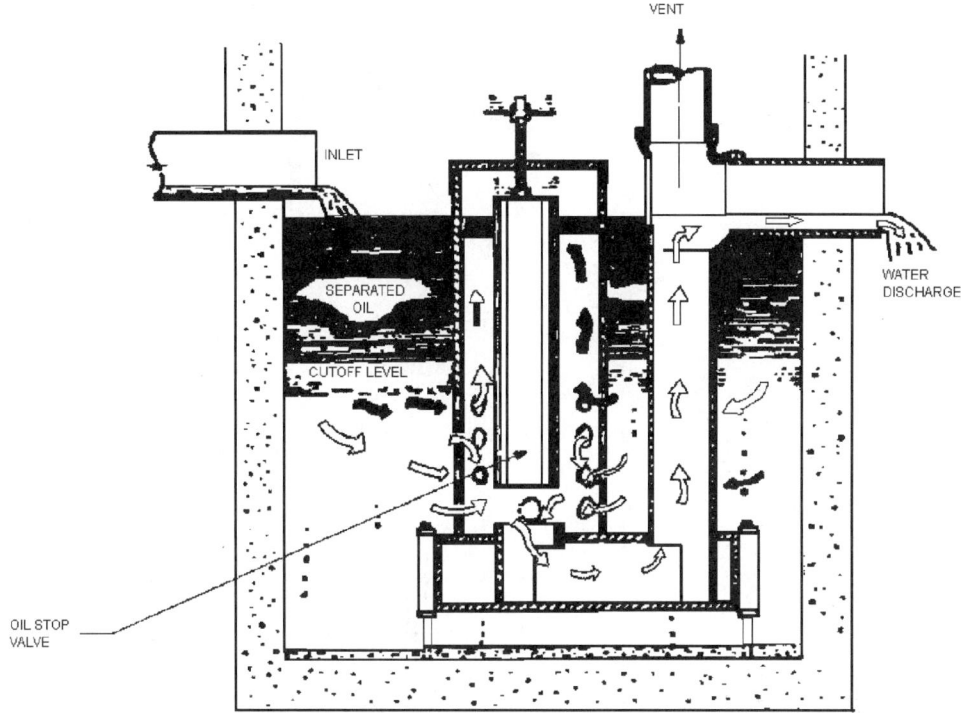

FIGURE 5.50 Oil stop valve installed in manhole.

Warning Alarms and Monitoring (IEEE, 1994)

In the event of an oil spill, it is imperative that cleanup operations and procedures be initiated as soon as possible to prevent the discharge of any oil, or to reduce the amount of oil reaching navigable waters. Hence, it may be desirable to install an early detection system for alerting responsible personnel of an oil spill. Some governmental regulations may require that the point of discharge (for accumulated water) from a substation be monitored and/or licensed.

The most effective alarms are the ones activated by the presence of oil in the containment system. A low oil level indicator within the oil-filled equipment can be used; however, it may not activate until 3–6% of the transformer oil has already discharged. In cases where time is critical, it may be worthwhile to also consider a faster operating alarm such as one linked to the transformer sudden gas pressure relay. Interlocks have to be considered as a backup to automatic pump or valve controls.

Alarms are transmitted via supervisory equipment or a remote alarm system to identify the specific problem. The appropriate personnel are then informed so that they can determine if a spill has occurred and implement the SPCC contingency plan.

References

Design Guide for Oil Spill Prevention and Control at Substations, U.S. Department of Agriculture, Rural Electrification Administration Bulletin 65-3, January, 1981.

IEEE Guide for Containment and Control of Oil Spills in Substations, IEEE Std. 980-1994.

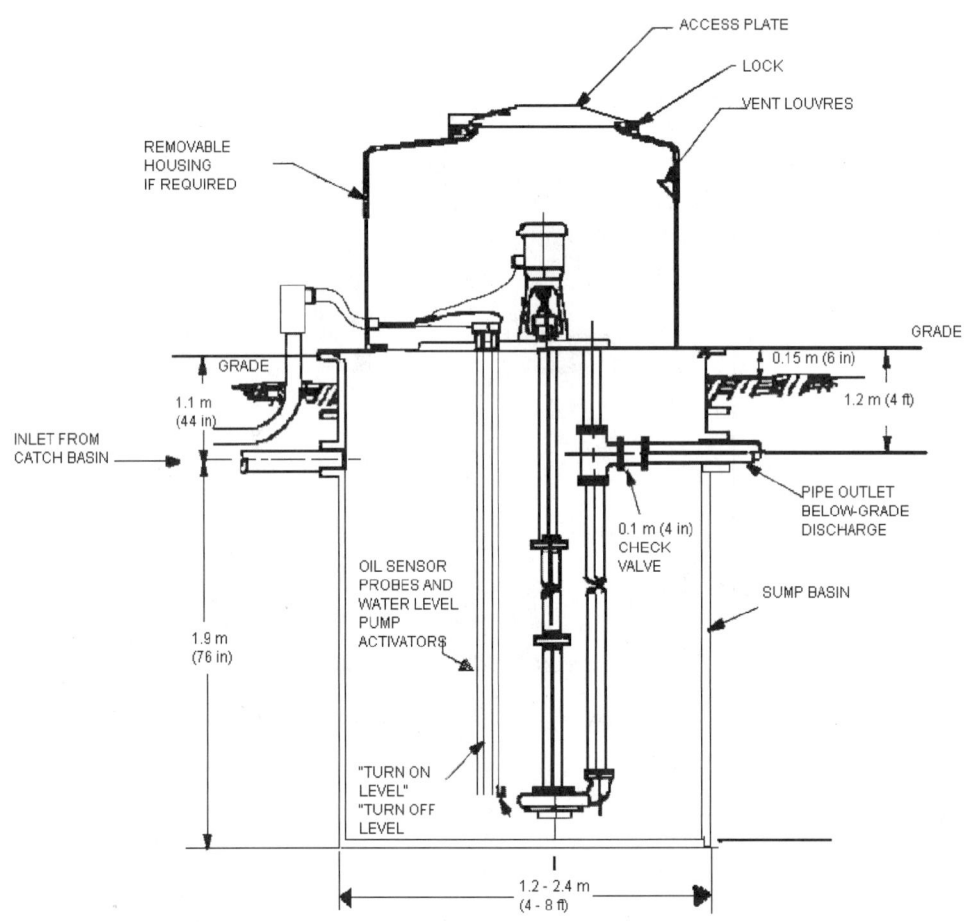

FIGURE 5.51 Sump pump water discharge (with oil sensing probe).

5.8 Community Considerations[1]

James H. Sosinski

Community Acceptance

Community acceptance generally encompasses the planning, design, and construction phases of a substation as well as the in-service operation of the substation. It takes into account those issues that could influence a community's willingness to accept building a substation at a specific site. New substations or expansions of existing facilities often require extensive review for community acceptance. Government bodies typically require a variety of permits before construction may begin.

---

[1]Sections 3, 4, 5, 6, and 7 (excluding sections 4.3.2.2, 4.3.5, 4.4.2.1, 4.4.2.2, 4.4.2.3, 4.4.3.1, 4.4.3.2, 4.4.3.3, 4.4.3.4, 4.4.3.5, 5.1, 5.2, 6.1.4, 6.4, 7.2.1., 7.2.2, tables 1 and 2, and figures 1 and 2) reprinted from IEEE Std. 1127-1998, "IEEE Guide for the Design, Construction, and Operation of Electric Power Substations for Community Acceptance and Environmental Compatibility" Copyright© 1998, by the Institute of Electrical and Electronics Engineers, Inc. (IEEE). The IEEE disclaims any responsibility or liability resulting from the placement and use in the described manner. Information is reprinted with the permission of the IEEE.

For community acceptance, several considerations should be satisfactorily addressed, including the following:

- Noise
- Site preparations
- Aesthetics
- Fire protection
- Potable water and sewage
- Hazardous materials
- Electric and magnetic fields
- Safety and security

This section on Community Considerations is essentially a condensed version of IEEE Standard 1127-1998.

Planning Strategies and Design

Planning is essential for the successful design, construction, and operation of a substation. The substation's location and proximity to wetlands, other sensitive areas, and contaminated soils; its aesthetic impact; and the concerns of nearby residents over noise and electric and magnetic fields (EMF) can significantly impact the ability to achieve community acceptance. Public perceptions and attitudes toward both real and perceived issues can affect the ability to obtain all necessary approvals and permits.

These issues can be addressed through presentations to governmental officials and the public. Failure to obtain community acceptance can delay the schedule or, in the extreme, stop a project completely.

Site Location and Selection, and Preparation

The station location (especially for new substations) is the key factor in determining the success of any substation project. Although the site location is based on electric system load growth studies, the final site location may ultimately depend upon satisfying the public and resolving potential community acceptance concerns. If necessary, a proactive public involvement program should be developed and implemented. The best substation site selection is influenced by several factors including, but not limited to, the following:

1. Community attitudes and perceptions
2. Location of nearby wetlands, bodies of water, or environmentally sensitive areas
3. Site contamination (obvious or hidden)
4. Commercial, industrial, and residential neighbors, including airports
5. Permit requirements and ordinances
6. Substation layout (including future expansions) and placement of noise sources
7. Levels of electric and magnetic fields
8. Availability and site clearing requirements for construction staging
9. Access to water and sewer
10. Drainage patterns and storm water management
11. Potential interference with radio, television, and other communication installations
12. Disturbance of archaeological, historical, or culturally significant sites
13. Underground services and geology
14. Accessibility
15. Aesthetic and screening considerations

Wetlands
A site-development plan is necessary for a substation project that borders wetlands. Such a plan for the site and its immediate surroundings should include the following:

1. Land-use description
2. Grades and contours

3. Locations of any wetland boundaries and stream-channel encroachment lines
4. Indication of flood-prone areas and vertical distance or access to ground water
5. Indication of existing wildlife habitats and migratory patterns

The plan should describe how site preparation will modify or otherwise impact these areas and what permanent control measures will be employed, including ground water protection.

Site Contamination

Soil borings should be taken on any proposed substation site to determine the potential presence of soil contaminants.

There are many substances that, if found on or under a substation site, would make the site unusable or require excessive funds to remediate the site before it would be usable. Some of the substances are as follows:

1. Polychlorinated biphenyls (PCBs)
2. Asbestos
3. Lead and other heavy metals
4. Pesticides and herbicides
5. Radioactive materials
6. Petrochemicals
7. Dioxin
8. Oil

Governmental guidelines for the levels of these substances should be used to determine if the substance is present in large enough quantities to be of concern.

The cost of removal and disposal of any contaminants should be considered before acquiring or developing the site. If a cleanup is needed, the acquisition of another site should be considered as governmental regulations can hold the current owner or user of a site responsible for cleanup of any contamination present, even if substances were deposited prior to acquisition. If a cleanup is initiated, all applicable governmental guidelines and procedures should be followed.

Potable Water and Sewage

The substation site may need potable water and sewage disposal facilities. Water may be obtained from municipal or cooperative water utilities or from private wells. Sewage may be disposed of by municipal services or septic systems, or the site could be routinely serviced by portable toilet facilities, which are often used during construction. Where municipal services are used for either water or sewer service, the requirements of that municipality must be met. Septic systems, when used, should meet all applicable local, state, and federal regulations.

Aesthetics

Aesthetics play a major role where community acceptance of a substation is an issue. Sites should be selected that fit into the context of present and future community patterns.

Community acceptability of a site can be influenced by:

1. Concerns about compatibility with present and future land uses
2. Building styles in the surrounding environment
3. Landscape of the site terrain
4. Allowance for buffer zones for effective blending, landscaping, and safety
5. Site access that harmonizes with the community

In addition, the site may need to be large enough to accommodate mobile emergency units and future expansions without becoming congested.

Visual Simulation

Traditionally, a site rendering was an artist's sketch, drawing, painting, or photomontage with airbrush retouching, preferably in color, and as accurate and realistic as possible. In recent years, these traditional techniques, although still employed, have given way to two- and three-dimensional computer-generated images, photorealism, modeling, and animation to simulate and predict the impact of proposed developments.

This has led to increased accuracy and speed of image generation in the portrayal of new facilities for multiple-viewing (observer) positions, allowing changes to be made early in the decision-making process while avoiding costly alterations that sometimes occur later during construction.

A slide library of several hundred slides of aesthetic design choices is available from the IEEE. It is a compilation of landscaping, decorative walls and enclosures, plantings, and site location choices that have been used by various utilities worldwide to ensure community acceptance and environmental compatibility.

Landscaping and Topography
Landscaping: Where buffer space exists, landscaping can be a very effective aesthetic treatment. On a site with little natural screening, plantings can be used in concert with architectural features to complement and soften the visual effect.

All plantings should be locally available and compatible types, and should require minimum maintenance. Their location near walls and fences should not compromise either substation grounding or the security against trespass by people or animals.

Topography: Topography or land form, whether shaped by nature or by man, can be one of the most useful elements of the site to solve aesthetic and functional site development problems.

Use of topography as a visual screen is often overlooked. Functionally, earth forms can be permanent, visual screens constructed from normal on-site excavating operations. When combined with plantings of grass, bushes, or evergreens and a planned setback of the substation, berms can effectively shield the substation from nearby roads and residents.

Fences and walls: The National Electrical Safety Code® [(NESC®) (Accredited Standards Committee C2-1997)] requires that fences, screens, partitions, or walls be employed to keep unauthorized persons away from substation equipment.

Chain-link fences: This type of fence is the least vulnerable to graffiti and is generally the lowest-cost option. Chain-link fences can be galvanized or painted in dark colors to minimize their visibility, or they can be obtained with vinyl cladding. They can also be installed with wooden slats or colored plastic strips woven into the fence fabric. Grounding and maintenance considerations should be reviewed before selecting such options.

Wood fences: This type of fence should be constructed using naturally rot-resistant or pressure-treated wood, in natural color or stained for durability and appearance. A wood fence can be visually overpowering in some settings. Wood fences should be applied with caution because wood is more susceptible to deterioration than masonry or metal.

Walls: Although metal panel and concrete block masonry walls cost considerably more than chain-link and wood fences, they deserve consideration where natural or landscaped screening does not provide a sufficient aesthetic treatment. Brick and precast concrete can also be used in solid walls, but these materials are typically more costly. These materials should be considered where necessary for architectural compatibility with neighboring facilities. Walls can offer noise reduction (discussed later) but can be subject to graffiti. All issues should be considered before selecting a particular wall or fence type.

Color
When substations are not well screened from the community, color can have an impact on the visual effect.

Above the skyline, the function of color is usually confined to eliminating reflective glare from bright metal surfaces. Because the sun's direction and the brightness of the background sky vary, no one color can soften the appearance of substation structures in the course of changing daylight. Below the skyline, color can be used in three aesthetic capacities. Drab coloring, using earth tones and achromatic hues, is a technique that masks the metallic sheen of such objects as chain-link fences and steel structures, and reduces visual contrast with the surrounding landscape. Such coloring should have very limited variation in hues, but contrast by varying paint saturation is often more effective than a monotone coating. Colors and screening can often be used synergistically. A second technique is to use color to direct visual attention to more aesthetically pleasing items such as decorative walls and enclosures. In this use, some brightness is warranted, but highly saturated or contrasting hues should be avoided. A third technique is to brightly color equipment and structures for intense visual impact.

Lighting
When attractive landscaping, decorative fences, enclosures, and colors have been used to enhance the appearance of a highly visible substation, it may also be appropriate to use lighting to highlight some of these features at night. Although all-night lighting can enhance substation security and access at night, it should be applied with due concern for nearby residences.

Structures
The importance of aesthetic structure design increases when structures extend into the skyline. The skyline profile typically ranges from 6 m to 10 m (20 ft to 35 ft) above ground. Transmission line termination structures are usually the tallest and most obvious. Use of underground line exits will have the greatest impact on the substation's skyline profile. Where underground exits are not feasible, low-profile station designs should be considered. Often, low-profile structures will result in the substation being below the nearby tree line profile.

For additional cost, the most efficient structure design can be modified to improve its appearance. The following design ideas may be used to improve the appearance of structures:

1. Tubular construction
2. Climbing devices not visible in profile
3. No splices in the skyline zone
4. Limiting member aspect ratio for slimmer appearance
5. Use splices other than pipe-flange type
6. Use of gusset plates with right-angle corners not visible in profile
7. Tapering ends of cantilevers
8. Equal length of truss panel
9. Making truss diagonals with an approximate 60° angle to chords
10. Use of short knee braces or moment-resistant connections instead of full-height diagonal braces
11. Use of lap splice plates only on the insides of H-section flanges

Enclosures
Total enclosure of a substation within a building is an option in urban settings where underground cables are used as supply and feeder lines. Enclosure by high walls, however, may be preferred if enclosure concealment is necessary for community acceptance.

A less costly design alternative in nonurban locales that are served by overhead power lines is to take advantage of equipment enclosures to modify visual impact. Relay and control equipment, station batteries, and indoor power switchgear all require enclosures. These enclosures can be aesthetically designed and strategically located to supplement landscape concealment of other substation equipment. The exterior appearance of these enclosures can also be designed (size, color, materials, shape) to match neighboring homes or buildings.

Industrial-type, pre-engineered metal enclosures are a versatile and economic choice for substation equipment enclosures. Concrete block construction is also a common choice for which special shaped and colored blocks may be selected to achieve a desired architectural effect. Brick, architectural metal panels, and precast concrete can also be used.

Substation equipment enclosures usually are not exempt from local building codes. Community acceptance, therefore, requires enclosure design, approval, and inspection in accordance with local regulation.

Bus Design
Substations can be constructed partly or entirely within aboveground or belowground enclosures. However, cost is high and complexity is increased by fire-protection and heat-removal needs. The following discussion deals with exposed aboveground substations.

Air-insulated substations: The bus and associated substation equipment are exposed and directly visible. An outdoor bus may be multitiered or spread out at one level. Metal or wood structures and insulators support such bus and power line terminations. Space permitting, a low-profile bus layout is generally best for aesthetics and is the easiest to conceal with landscaping, walls, and enclosures. Overhead transmission line terminating structures are taller and more difficult to conceal in such a layout. In dry climates, a low-profile bus can be achieved by excavating the earth area, within which outdoor bus facilities are then located for an even lower profile.

Switchgear: Metal-enclosed or metal-clad switchgear designs that house the bus and associated equipment in a metal enclosure are an alternative design for distribution voltages. These designs provide a compact low-profile installation that may be aesthetically acceptable.

Gas-insulated substation (GIS): Bus and associated equipment can be housed within pipe-type enclosures using sulfur hexafluoride or another similar gas for insulation. Not only can this achieve considerable compactness and reduced site preparation for higher voltages, but it can also be installed lower to the ground. A GIS can be an economically attractive design where space is at a premium, especially if a building-type enclosure will be used to house substation equipment (see IEEE Std. C37.123-1996).

Cable bus: Short sections of overhead or underground cables can be used at substations, although this use is normally limited to distribution voltages (e.g., for feeder getaways or transformer-to-switchgear connections). At higher voltages, underground cable can be used for line-entries or to resolve a specific connection problem.

Noise

Audible noise, particularly continuously radiated discrete tones (e.g., from power transformers), is the type of noise that the community may find unacceptable. Community guidelines to ensure that acceptable noise levels are maintained can take the form of governmental regulations or individual/community reaction (permit denial, threat of complaint to utility regulators, etc.). Where noise is a potential concern, field measurements of the area background noise levels and computer simulations predicting the impact of the substation may be required. The cost of implementing noise reduction solutions (low-noise equipment, barriers or walls, noise cancellation techniques, etc.) may become a significant factor when a site is selected.

Noise can be transmitted as a pressure wave either through the air or through solids. The majority of cases involving the observation and measurement of noise have dealt with noise being propagated through the air. However, there are reported, rare cases of audible transformer noise appearing at distant observation points by propagating through the transformer foundation and underground solid rock formations. It is best to avoid the situation by isolating the foundation from bedrock where the conditions are thought to favor transmission of vibrations.

Noise Sources

Continuous audible sources: The most noticeable audible noise generated by normal substation operation consists of continuously radiated audible discrete tones. Noise of this type is primarily generated by power transformers. Regulating transformers, reactors, and emergency generators, however, could also be sources. This type of noise is most likely to be subject to government regulations. Another source of audible noise in substations, particularly in extra high voltage (EHV) substations, is corona from the bus and conductors.

Continuous radio frequency (RF) sources: Another type of continuously radiated noise that can be generated during normal operation is RF noise. These emissions can be broadband and can cause interference to radio and television signal reception on properties adjacent to the substation site. Objectionable RF noise is generally a product of unintended sparking, but can also be produced by corona.

Impulse sources: While continuously radiated noise is generally the most noticeable to substation neighbors, significant values of impulse noise can also accompany normal operation. Switching operations will cause both impulse audible and RF noise with the magnitude varying with voltage, load, and operation speed. Circuit-breaker operations will cause audible noise, particularly operation of air-blast breakers.

Typical Noise Levels

Equipment noise levels: Equipment noise levels may be obtained from manufacturers, equipment tendering documents, or test results. The noise level of a substation power transformer is a function of the MVA and BIL rating of the high voltage winding. These transformers typically generate a noise level ranging from 60 to 80 dBA.

Transformer noise will "transmit" and attenuate at different rates depending on the transformer size, voltage rating, and design. Few complaints from nearby residents are typically received concerning substations with transformers of less than 10 MVA capacity, except in urban areas with little or no buffers. Complaints are more common at substations with transformer sizes of 20–150 MVA, especially within the first 170–200 m (500–600 ft). However, in very quiet rural areas where the nighttime ambient can reach 20–25 dBA, the noise from the transformers of this size can be audible at distances of 305 m (1000 ft) or more. In urban areas, substations at 345 kV and above rarely have many complaints because of the large parcels of land on which they are usually constructed.

Attenuation of noise with distance: The rate of attenuation of noise varies with distance for different types of sound sources depending on their characteristics. Point sound sources that radiate equally in all directions will decrease at a rate of 6 dB for each doubling of distance. Cylindrical sources vibrating uniformly in a radial direction will act like long source lines and the sound pressure will drop 3 dB for each doubling of distance. Flat planar surfaces will produce a sound wave with all parts of the wave tracking in the same direction (zero divergence). Hence, there will be no decay of the pressure level due to distance only. The designer must first identify the characteristics of the source before proceeding with a design that will take into account the effect of distance.

A transformer will exhibit combinations of all of the above sound sources, depending on the distance and location of the observation point. Because of its height and width, which can be one or more wavelengths, and its nonuniform configuration, the sound pressure waves will have directional characteristics with very complex patterns. Close to the transformer (near field), these vibrations will result in lobes with variable pressure levels. Hence, the attenuation of the noise level will be very small. If the width (W) and height (H) of the transformer are known, then the near field is defined, from observation, as any distance less than $2\sqrt{WH}$ from the transformer.

Further from the transformer (far field), the noise will attenuate in a manner similar to the noise emitted from a point source. The attenuation is approximately equal to 6 dB for every doubling of the distance. In addition, if a second adjacent transformer produces an identical noise level to the existing transformer (e.g., 75 dBA), the total sound will be 78 dBA for a net increase of only 3 dB. This is due to the logarithmic effect associated with a combination of noise sources.

Governmental Regulations

Governmental regulations may impose absolute limits on emissions, usually varying the limits with the zoning of the adjacent properties. Such limits are often enacted by cities, villages, and other incorporated urban areas where limited buffer zones exist between property owners. Typical noise limits at the substation property line used within the industry are as follows:

- Industrial zone   <75 dBA
- Commercial zone   <65 dBA
- Residential zone   <55 dBA

Additional governmental noise regulations address noise levels by limiting the increase above the existing ambient to less than 10 dB. Other regulations could limit prominent discrete tones, or set specific limits by octave bands.

Noise Abatement Methods

The likelihood of a noise complaint is dependent on several factors, mostly related to human perceptions. As a result, the preferred noise abatement method is time-dependent as well as site-specific.

Reduced transformer sound levels: Since power transformers, voltage regulators, and reactors are the primary sources of continuously radiated discrete tones in a substation, careful attention to equipment design can have a significant effect on controlling noise emissions at the substation property line. This equipment can be specified with noise emissions below manufacturer's standard levels, with values as much as 10 dB below those levels being typical.

In severely restrictive cases, transformers can be specified with noise emissions 20 dB less than the manufacturers' standard levels, but usually at a significant increase in cost. Also, inclusion of bid evaluation factor(s) for reduced losses in the specification can impact the noise level of the transformer. Low-loss transformers are generally quieter than standard designs.

Low-impulse noise equipment: Outdoor-type switching equipment is the cause of most impulse noise. Switchgear construction and the use of vacuum or puffer circuit breakers, where possible, are the most effective means of controlling impulse emissions. The use of circuit switchers or air-break switches with whips and/or vacuum bottles for transformer and line switching, may also provide impulse-emission reductions over standard air-break switches.

RF noise and corona-induced audible noise control: Continuously radiated RF noise and corona-induced audible noise can be controlled through the use of corona-free hardware and shielding for high-voltage conductors and equipment connections, and through attention to conductor shapes to avoid sharp corners. Angle and bar conductors have been used successfully up to 138 kV without objectionable corona if corners are rounded at the ends of the conductors and bolts are kept as short as possible.

Tubular shapes are typically required above this voltage. Pronounced edges, extended bolts, and abrupt ends on the conductors can cause significant RF noise to be radiated. The diameter of the conductor also has an effect on the generation of corona, particularly in wet weather. Increasing the size of single grading rings or conductor diameter may not necessarily solve the problem. In some cases it may be better to use multiple, smaller diameter grading rings.

Site location: For new substations to be placed in an area known to be sensitive to noise levels, proper choice of the site location can be effective as a noise abatement strategy. Also, locations in industrial parks or near airports, expressways, or commercial zones that can provide almost continuous background noise levels of 50 dB or higher will minimize the likelihood of a complaint.

Larger yard area: Noise intensity varies inversely with distance. An effective strategy for controlling noise of all types involves increasing the size of the parcel of real estate on which the substation is located.

Equipment placement: Within a given yard size, the effect of noise sources on the surroundings can be mitigated by careful siting of the noise sources within the confines of the substation property. In addition, making provisions for the installation of mobile transformers, emergency generators, etc. near the center of the property, rather than at the edges, will lessen the effect on the neighbors.

Barriers or walls: If adequate space is not available to dissipate the noise energy before it reaches the property line, structural elements might be required. These can consist of walls, sound-absorbing panels, or deflectors. In addition, earth berms or below-grade installation may be effective. It may be possible to deflect audible noises, especially the continuously radiated tones most noticeable to the public, to areas not expected to be troublesome. Foliage, despite the potential aesthetic benefit and psychological effect, is not particularly effective for noise reduction purposes.

Properly constructed sound barriers can provide several decibels of reduction in the noise level. An effective barrier involves a proper application of the basic physics of

1. Transmission loss through masses
2. Sound diffraction around obstacles
3. Standing waves behind reflectors
4. Absorption at surfaces

For a detailed analysis of wall sound barriers, refer to IEEE Std. 1127-1998.

Active noise cancellation techniques: Another solution to the problem of transformer noise involves use of active noise control technology to cancel unwanted noise at the source, and is based on advances in digital controller computer technology. Active noise cancellation systems can be tuned to specific problem frequencies or bands of frequencies achieving noise reduction of up to 20 dB.

Electric and Magnetic Fields

Electric substations produce electric and magnetic fields. In a substation, the strongest fields around the perimeter fence come from the transmission and distribution lines entering and leaving the substation. The strength of fields from equipment inside the fence decreases rapidly with distance, reaching very low levels at relatively short distances beyond substation fences.

In response to the public concerns with respect to EMF levels, whether perceived or real, and to governmental regulations, the substation designer may consider design measures to lower EMF levels or public exposure to fields while maintaining safe and reliable electric service.

Electric and Magnetic Field Sources in a Substation

Typical sources of electric and magnetic fields in substations include the following:

1. Transmission and distribution lines entering and exiting the substation
2. Buswork
3. Transformers
4. Air core reactors
5. Switchgear and cabling
6. Line traps
7. Circuit breakers
8. Ground grid
9. Capacitors
10. Battery chargers
11. Computers

Electric Fields

Electric fields are present whenever voltage exists on a conductor. Electric fields are not dependent on the current. The magnitude of the electric field is a function of the operating voltage and decreases with the square of the distance from the source. The strength of an electric field is measured in volts per meter. The most common unit for this application is kilovolts per meter. The electric field can be easily shielded (the strength can be reduced) by any conducting surface such as trees, fences, walls, buildings, and most structures. In substations, the electric field is extremely variable due to the screening effect provided by the presence of the grounded steel structures used for electric bus and equipment support.

Although the level of the electric fields could reach magnitudes of approximately 13 kV/m in the immediate vicinity of high-voltage apparatus, such as near 500-kV circuit beakers, the level of the electric field decreases significantly toward the fence line. At the fence line, which is at least 6.4 m (21 ft) from the nearest live 500-kV conductor (see the NESC), the level of the electric field approaches zero kV/m. If the incoming or outgoing lines are underground, the level of the electric field at the point of crossing the fence is negligible.

Magnetic Fields

Magnetic fields are present whenever current flows in a conductor, and are not voltage dependent. The level of these fields also decreases with distance from the source but these fields are not easily shielded. Unlike electric fields, conducting materials such as the earth, or most metals, have little shielding effect on magnetic fields.

Magnetic fields are measured in Webers per square meter (Tesla) or Maxwells per square centimeter (Gauss). One Gauss = 10^{-4} Tesla. The most common unit for this application is milliGauss (10^{-3} Gauss).

Various factors affect the levels of the fields, including the following:

1. Current magnitude
2. Phase spacing
3. Bus height
4. Phase configurations

Substations

5. Distance from the source
6. Phase unbalance (magnitude and angle)

Magnetic fields decrease with increasing distance (r) from the source. The rate is an inverse function and is dependent on the type of source. For point sources such as motors and reactors, the function is $1/r^2$; and for single-phase sources such as neutral or ground conductors the function is $1/r$. Besides distance, conductor spacing and phase balance have the largest effect on the magnetic field level because they control the rate at which the field changes.

Magnetic fields can sometimes be shielded by specially engineered enclosures. The application of these shielding techniques in a power system environment is minimal because of the substantial costs involved and the difficulty of obtaining practical designs.

Safety and Security

Fences and Walls
The primary means of ensuring public safety at substations is by the erection of a suitable barrier, such as a fence or a wall with warning signs. As a minimum, the barrier should meet the requirements of the NESC and other applicable electrical safety codes. Recommended clearances from substation live parts to the fence are specified in the NESC, and security methods are described in IEEE P1402/D8.

Lighting
Yard lighting may be used to enhance security and allow equipment status inspections. A yard-lighting system should provide adequate ground-level lighting intensity around equipment and the control-house area for security purposes without disruption to the surrounding community. High levels of nightly illumination will often result in complaints.

Grounding
Grounding should meet the requirement of IEEE Std. 80-1986 to ensure the design of a safe and adequate grounding system. All non-current-carrying metal objects in or exiting from substations should be grounded (generally to a buried metallic grid) to eliminate the possibility of unsafe touch or step potentials, which the general public might experience during fault conditions.

Fire Protection
The potential for fires exists throughout all stations. Although not a common occurrence, substation fires are an important concern because of potential for long-term outages, personnel injury or death, extensive property and environmental damage, and rapid uncontrolled spreading. Refer to IEEE Std. 979-1994 for detailed guidance and identification of accepted substation fire-protection design practices and applicable industry standards.

Permitting Process

A variety of permits may be required by the governing bodies before construction of a substation may begin. For the permitting process to be successful, the following factors may have to be considered:

1. Site location
2. Level of ground water
3. Location of wetlands
4. Possibility of existing hazardous materials
5. Need for potable water and sewage
6. Possible noise
7. Aesthetics
8. EMF

Timing for the permit application is a critical factor because the permit application may trigger opposition involvement. If it is determined that the situation requires public involvement, the preparation and implementation of a detailed plan using public participation can reduce the delays and costs associated with political controversy and litigation. In these situations, public involvement prior to permit application can help to build a positive relationship with those affected by the project, identify political and community concerns, obtain an informed consensus from project stakeholders, and provide a basis for the utility to increase its credibility and reputation as a good neighbor.

Construction
Site Preparation
Clearing, Grubbing, Excavation, and Grading
Concerns include the creation of dust, mud, water runoff, erosion, degraded water quality, and sedimentation. The stockpiling of excavated material and the disposal of excess soil, timber, brush, etc. are additional items that should be considered. Protective measures established during the design phase or committed to through the permitting process for ground water, wetlands, flood plains, streams, archeological sites, and endangered flora and fauna should be implemented during this period.

Site Access Roads
The preparation and usage of site access roads create concerns that include construction equipment traffic, dust, mud, water runoff, erosion, degraded water quality, and sedimentation. Access roads can also have an impact on agriculture, archaeological features, forest resources, wildlife, and vegetation.

Water Drainage
Runoff control is especially important during the construction process. Potential problems include flooding, erosion, sedimentation, and waste and trash carried off the site.

Noise
Noise control is important during construction in areas sensitive to this type of disturbance. An evaluation should be made prior to the start of construction to determine noise restrictions that may be imposed at the construction site.

Safety and Security
Safety and security procedures should be implemented at the outset of the construction process to protect the public and prevent unauthorized access to the site. These procedures should be developed in conformance with governmental agencies. See IEEE P1402/D8 for detailed descriptions of the security methods that can be employed. The safety and security program should be monitored continuously to ensure that it is functioning properly.

The following are suggestions for safety and security at the site:

1. Temporary or permanent fencing
2. Security guards
3. Security monitoring systems
4. Traffic control
5. Warning signs
6. Construction safety procedures
7. Temporary lighting

Site Housekeeping
During construction, debris and refuse should not be allowed to accumulate. Efforts should be made to properly store, remove, and prevent these materials from migrating beyond the construction site. Burning of refuse should be avoided. In many areas this activity is prohibited by law. Portable toilets that are routinely serviced should be provided.

Substations 5-87

Hazardous Material

The spillage of transformer and pipe cable insulating oils, paints, solvents, acids, fuels, and other similar materials can be detrimental to the environment as well as a disturbance to the neighborhood. Proper care should be taken in the storage and handling of such materials during construction.

Operations

Site Housekeeping

Water and sediment control: Routine inspection of control for water flows is important to maintain proper sediment control measures. Inspection should be made for basin failure and for gullies in all slopes. Inspection of all control measures is necessary to be sure that problems are corrected as they develop and should be made a part of regular substation inspection and maintenance.

Yard surface maintenance: Yard surfacing should be maintained as designed to prevent water runoffs and control dust. If unwanted vegetation is observed on the substation site, approved herbicides may be used with caution to prevent runoff from damaging surrounding vegetation. If runoffs occur, the affected area should be covered with stone to retard water runoff and to control dust.

Paint: When material surfaces are protected by paint, a regular inspection and repainting should be performed to maintain a neat appearance and to prevent corrosion damage.

Landscaping: Landscaping should be maintained to ensure perpetuation of design integrity and intent.

Storage: In some areas, zoning will not permit storage in substations. The local zoning must therefore be reviewed before storing equipment, supplies, etc. The appearance of the substation site should be considered so it will not become visually offensive to the surrounding community.

Noise: Inspection of all attributes of equipment designed to limit noise should be performed periodically.

Safety and security: All substations should be inspected regularly, following established and written procedures to ensure the safety and security of the station. Safe and secure operation of the substation requires adequate knowledge and proper use of each company's accident prevention manual. See IEEE P1402/D8 for detailed descriptions of the security methods that can be employed.

Routine inspections of the substation should be performed and recorded, and may include the following:

1. Fences
2. Gates
3. Padlocks
4. Signs
5. Access detection systems
6. Alarm systems
7. Lighting systems
8. Grounding systems
9. Fire-protection equipment
10. All oil-filled equipment
11. Spill-containment systems

Fire Protection

Refer to IEEE Std. 979-1994 for detailed guidance and identification of accepted substation fire-protection practices and applicable industry standards. Any fire-protection prevention system installed in the substation should be properly maintained.

Hazardous Material

A spill-prevention control and counter-measures plan should be in place for the substation site and should meet governmental requirements. For general guidance, see IEEE Std. 980-1994.

Defining Terms (IEEE, 1998)

A-weighted sound level: The representation of the sound pressure level that has as much as 40 dB of the sound below 100 Hz and a similar amount above 10,000 Hz filtered out. This level best approximates the response of the average young ear when listening to most ordinary, everyday sounds. Generally designated as dBA.

Commercial zone: A zone that includes offices, shops, hotels, motels, service establishments, or other retail/commercial facilities as defined by local ordinances.

Hazardous material: Any material that has been so designated by governmental agencies or adversely impacts human health or the environment.

Industrial zone: A zone that includes manufacturing plants where fabrication or original manufacturing is done, as defined by local ordinances.

Noise: Undesirable sound emissions or undesirable electromagnetic signals/emissions.

Residential zone: A zone that includes single-family and multifamily residential units, as defined by local ordinances.

Wetlands: Any land that has been so designated by governmental agencies. Characteristically, such land contains vegetation associated with saturated types of soil.

For additional definitions, see IEEE Std. 100-1996.

References

Guide for Electric Power Substation Physical and Electronic Security, IEEE P1402/D8, draft dated April 1997.[2]

IEEE Guide for the Design, Construction, and Operation of Electric Power Substations for Community Acceptance and Environmental Compatibility, IEEE Std. 1127-1998.

IEEE Guide for Safety in AC Substation Grounding, IEEE Std. 80-1991.

IEEE Guide to Specifications for Gas-Insulated, Electric Power Substation Equipment, IEEE Std C37.123-1996.

IEEE Guide for Substation Fire Protection, IEEE Std. 979-1994.

The IEEE Standard Dictionary of Electrical and Electronics Terms, IEEE Std. 100-1996.

IEEE Standard Procedures for Measurement of Power Frequency Electric and Magnetic Fields from AC Power Lines, IEEE Std. 644-1994.

National Electrical Safety Code® (NESC®) (ANSI), Accredited Standards Committee C2-1997, Institute of Electrical Electronics Engineers, Piscataway, NJ, 1997.

5.9 Animal Deterrents/Security

C.M. Mike Stine and Sheila Frasier

The vast majority of electrical utility substations designed to transform transmission voltages to distribution class voltages employ an open-air design. The configurations may vary, but usually consist of equipment that utilizes polymer or porcelain insulators or bushings to create electrically insulated creepage and dry arc distances between the potential voltage carried by the bus or conductor and the grounded portions of the equipment or structure. Although these insulators or bushings provide the proper insulation distance for normal operation voltages (AC, DC, and BIL), they do not provide sufficient distances to eliminate bridging of many animals from potential to ground. This animal bridging situation usually exists at the low side or distribution voltage portion of the substation (12 through 36 kV), but depending on the size and type of the animal, can also affect higher voltage equipment. Utilities have reported that animal-caused outages have become a major problem affecting the reliability and continuity of the electrical system and are actively taking steps to prevent it.

---

[2]This IEEE standards project was not approved by the IEEE Standards Board at the time this publication went to press. For information about obtaining a draft, contact the IEEE.

The effects of animal bridging ranges from nuisance trips of the electrical system which may be a momentary occurrence, to faults that may interrupt power for long periods of time. Aside from the inconvenience and reliability aspects of animal-induced outages, there can be damage to the substation equipment ranging from porcelain bushings and insulators that may cost as little as $20.00, to complete destruction of large transformers running into the millions of dollars. There can also be an environmental risk involved with catastrophic failure such as oil spillage from equipment that has ruptured due to electrical faults.

Damage from outages is not limited to the equipment owned by the electrical utility. Many heavy industrial plants such as pulp and paper, petrochemical, and car manufacturers employ processes that are sensitive to interruptions and may result in significant time and money to reestablish production. The proliferation of computers, programmable logic controllers (PLCs), and other electrically sensitive devices in the workplace is also a reliability concern.

In addition to the concern for protecting assets such as substation equipment, improving the reliability of the system, eliminating environmental risks, and ensuring customer satisfaction and loyalty, the conservation of endangered and protected animal species is an issue. It is important to be educated and informed about the species and types of animals that are protected in each individual area or location.

To evaluate the problem and its possible solutions, several aspects need to be investigated:

- Animal type, size, and tendencies
- Equipment voltage rating and clearance from electrical ground
- Natural surroundings
- Method animals enter substation
- Influences attracting the animals
- Barrier methods available to keep the animal out
- Deterrent methods to repel the animals
- Insulation options

Animal Types

Clearance Requirements

The following table has been developed to aid in establishing minimum phase-to-ground and phase-to-phase clearances for the associated animals. This table is for reference only.

TABLE 5.7 Typical Clearance Requirement by Animal

Animal Type	Phase-to-Phase	Phase-to-Ground
Squirrel	18″ (450 mm)	18″ (450 mm)
Opossum/Raccoon	30″ (750 mm)	30″ (750 mm)
Snake	36″ (900 mm)	36″ (900 mm)
Crow/Grackle	24″ (600 mm)	18″ (450 mm)
Migratory Large Bird	36″ (900 mm)	36″ (900 mm)
Frog	18″ (450 mm)	18″ (450 mm)
Cat	24″ (600 mm)	24″ (600 mm)

Squirrels

In North America, a common culprit causing bridging is the squirrel. Although there are many varieties of squirrels, it can be assumed that the nominal length of a squirrel is 18″ (450 mm). Using this dimension, you can evaluate equipment and clearances to determine areas where bridging could occur between potential and ground or phase-to-phase. Clearances for modern substation equipment rated 35 kV and above will normally be sufficient to eliminate squirrel-caused problems; however, distances between phases and between phase and grounded structures should be examined.

There are several schools of thought regarding the reason squirrels often enter substations. One explanation offered is the proximity of trees and vegetation near the substation site that may attract squirrels. Some utilities report that removal of this vegetation had no effect on the squirrel-caused outages. Experts have theorized that the animals path is predetermined and the construction of a structure will not deter a squirrel from following his intended route. Others believe that the animals are attracted by heat or vibration emitted from the electrical equipment. Regardless of the reason, squirrels are compelled toward intrusion.

The entry into the substation does not always occur over, under, or through the outer fence of the site. Squirrels are very adept at traveling along overhead conductors and often enter the substation in this manner. Because of this fact, perimeter barriers are often ineffectual in preventing squirrel entry.

Birds

Birds create several problems when entering an electrical substation. The first and most obvious is the bridging between phase-to-ground or phase-to-phase caused by the wingspan when flying into or exiting the structure. Another problem is the bridging caused by debris used to build nests. Many times material such as strands of conductors or magnetic recording tape may be readily available from the surrounding area and be utilized by the birds. This conductive debris is often dragged across the conductor/busbar and results in flashovers, trips, or faults. The third problem is contamination of insulators caused by regurgitation or defecation of the birds. When this residue is allowed to remain, it can result in flashovers from potential to ground across the surface of the porcelain or polymer insulator by essentially decreasing the insulated creepage distance. The fourth possibility is commonly known as a "streamer outage." Streamers are formed when a bird defecates upon exiting a nest that has been built above an insulator. The streamers may create a path between the structure and conductor/bus, resulting in a flashover. Birds will tend to make nests in substations in an effort to eliminate possible predators from attacking the nest for food. The construction of nests in substations can, in turn, attract other animals such as snakes, cats, and raccoons into the area searching for food.

Snakes

Snakes are a major contributor to substation outages. In some areas, snakes are responsible for virtually all substation wildlife outages. Because of their size and climbing ability, snakes can reach most parts of a substation without difficulty. Snake-proofing substations can sometimes create problems rather than solving them. Snakes typically enter substations hunting birds and eggs. Eliminating these predators can lead to an increase in the bird population inside the substation boundaries. This bird infestation can then lead to bird-induced problems unless additional measures are taken.

Raccoons

Raccoons are excellent climbers and can easily gain access to substations. Unlike snakes, raccoons will occasionally enter substations for no particular reason except curiosity. Because of their large size, raccoons can easily bridge phase-to-phase and phase-to-ground distances on equipment with voltage ratings up to 25 kV.

Mitigation Methods

Barriers

Some of the barrier methods available include cyclone fences, small mesh wire fences, smooth climbing guards, electric fences, solid wall barriers, and fences with unconventional geometries. Barrier methods can be very effective against certain animals. Some utilities report that the use of small mesh fencing along the lower 3–4 feet (1–1.3 meters) of the perimeter has prevented intrusion of certain types of snakes. Several substation owners have incorporated the use of a bare wire attached to a PVC pipe energized with a low voltage transformer creating an electric fence that surrounds the structure inside of the normal property fence. This method has also been proven effective for snakes. Although these barrier designs prevent snakes from entering substations, they do little or nothing to eliminate legged animal intrusions. Smooth climbing guards are also used on structures to prevent some animals from scaling the vertical framework. While these guards work for some legged animals such as dogs and foxes, more agile animals such as squirrels, opossums, and cats can easily circumnavigate the devices.

Deterrents

There are a myriad of commercially available deterrent devices on the market. Many of the devices have actually come from applications in the household market to repel pests such as squirrels and pigeons from property. Although numerous, most devices have a limited effect on wildlife. Some of these include ultrasonic devices, devices producing loud noises at intermittent periods, chemical repellents, sticky gels, predator urine, plastic owls or snakes, poisons, and spined perching deterrents for birds. Ultrasonic devices tend to have an initial impact on animals, but have reportedly become ineffective after a relatively short period of time either due to the animal adapting to the sounds or the need to maintain the devices. Loud noise devices, like ultrasonics, soon lose the ability to repel the animals as they become familiar with the sound and lack of consequence. Chemical repellents, sticky gels, and predator urine have been shown effective against some animals when reapplied at frequent intervals. Poisons have been used to curb infestations of pests such as pigeons, but will sometimes result in collateral effects on pets and other animals if the pest is allowed to die outside of the substation boundaries. Spined perching deterrents have proven very successful in preventing smaller birds from building nests or congregating above electrically sensitive areas, but can sometimes serve as a functional anchor for greater sized birds to secure large nests.

Insulation

Insulating live conductors and hardware can be very effective in eliminating animal outages. Insulation systems are available in several forms:

- Spray on RTV coatings
- Insulating tapes
- Heat-shrinkable tubings, tapes, and sheet materials
- Pre-formed insulating covers

Insulation systems should be used at locations where animals can possibly make contact phase-to-ground or phase-to-phase. Typical applications include:

- Equipment bushing hardware (i.e., circuit breakers, reclosers, transformers, potential transformer, capacitors, regulators, etc.).
- Bus support insulator connections to structure or bus.
- Hook switch insulator connections to switch base or bus.
- Any area where clearance between bus and grounded equipment or structure is insufficient to eliminate bridging.
- Busbar and conductors where phase-to-phase spacing is inadequate.

Because these products are used as insulation on bus, conductor, or hardware, it is critical that they be of a material that is designed for the rigors of the high voltage environment. Unlike barriers and deterrents, the insulating materials are subjected to the electric field and are sometimes applied to the leakage path of other insulating materials such as porcelain. Care should be taken to select products that will withstand the outdoor environment as well as the electrical stress to which they may be subjected.

Isolation Devices

Isolation devices are rigid insulating discs that are installed in the leakage path of porcelain insulators. These devices force animals to climb onto them, isolating them from ground. These discs are used on both support insulators as well as switch insulators. As with insulating covers, the insulating material must be designed for the outdoor high voltage environment.

5.10 Substation Grounding

Richard P. Keil

The substation grounding system is an essential part of the overall electrical system. The proper grounding of a substation is essential for the following two reasons:

1. It provides a means of dissipating electric current into the earth without exceeding the operating limits of the equipment.
2. It provides a safe environment to protect personnel in the vicinity of grounded facilities from the dangers of electric shock under fault conditions.

The grounding system includes all the interconnected grounding facilities in the substation area, including the ground grid, overhead ground wires, neutral conductors, underground cables, foundations, etc. The ground grid consists of the horizontal interconnected bare conductors, ground rods, foundations, deep well, etc.

The following information is mainly concerned with personnel safety. The information about the grounding system resistance, grid current, and ground potential rise can also be used to determine if the operating limits of the equipment will be exceeded.

To provide a safe condition for personnel within and around the substation area, the ground grid design limits the potential difference a person can come in contact with to safe levels. IEEE-80, *Guide for Safety in AC Substation Grounding*, provides general information about substation grounding and the specific design equations necessary to design a safe substation grounding system. The following discussion is a brief description of the information presented in IEEE-80.

The Guide's design is based on the permissible body current when a person becomes part of an accidental ground circuit. Permissible body current will not cause ventricular fibrillation, that is, stoppage of the heart. The design methodology will limit the voltages that produce the permissible body current to a safe level.

Accidental Ground Circuit

There are two conditions that a person within or around the substation can experience that can cause them to become part of the ground circuit. One of these conditions, touch voltage, is illustrated in Figs. 5.52 and 5.53. The other condition, step voltage, is illustrated in Figs. 5.54 and 5.55. Figure 5.52 shows the fault current being discharged to the earth by the substation grounding system and a person touching a grounded metallic structure, H. Figure 5.53 shows the Thevenin equivalent for the person's feet in parallel, Z_{th}, in series with the body resistance, R_B. V_{th} is the voltage between terminal H and F

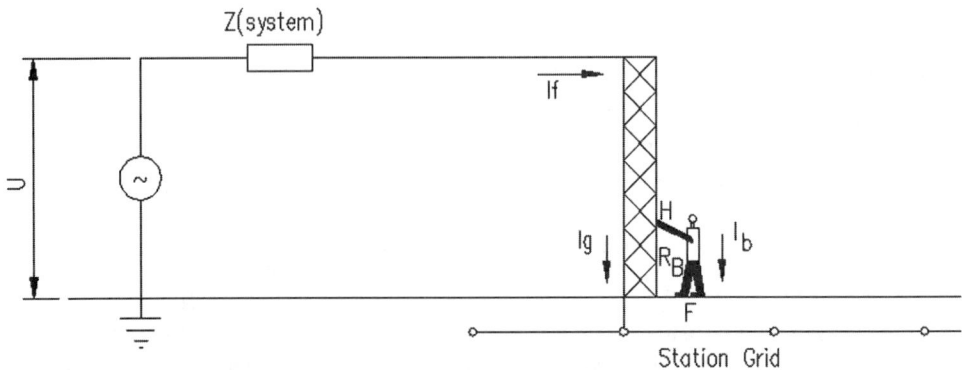

FIGURE 5.52 Exposure to touch voltage.

Substations

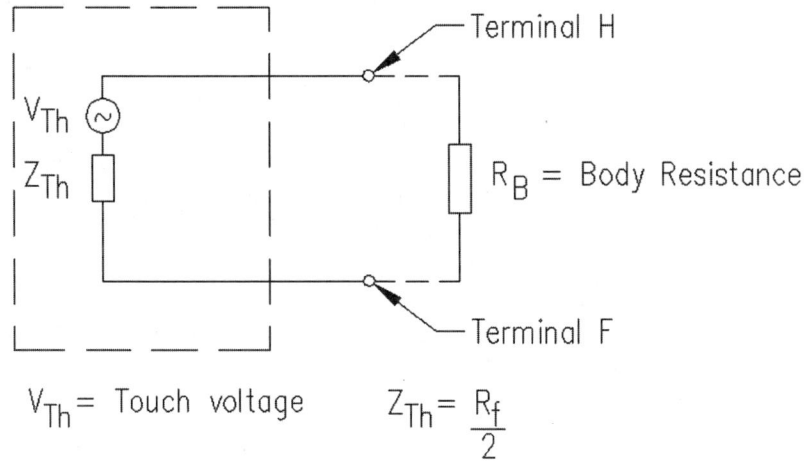

V_{Th} = Touch voltage $Z_{Th} = \dfrac{R_f}{2}$

FIGURE 5.53 Touch voltage circuit.

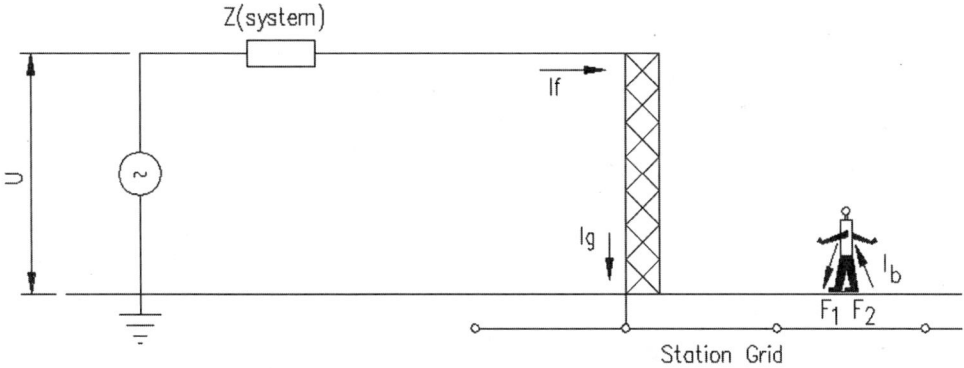

FIGURE 5.54 Exposure to step voltage.

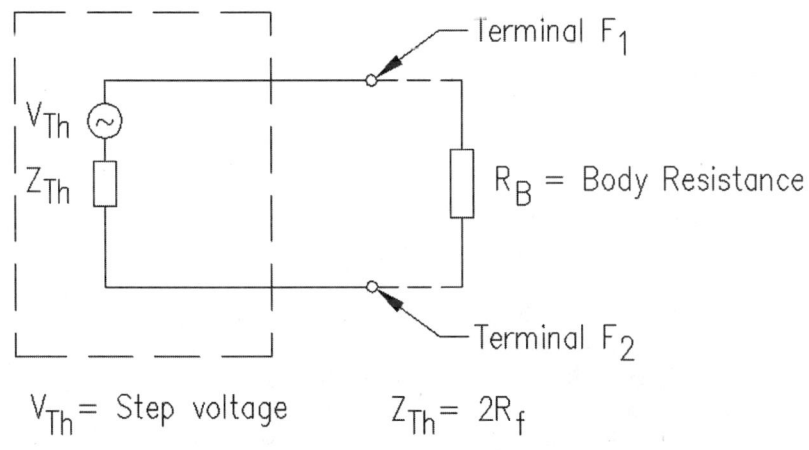

V_{Th} = Step voltage $Z_{Th} = 2R_f$

FIGURE 5.55 Step voltage circuit.

when the person is not present. I_b is the body current. Z_{th} is equal to the resistance of two feet in parallel, the touch voltage is

$$E_{touch} = I_b(R_B + Z_{th}). \tag{5.3}$$

Figures 5.54 and 5.55 show the conditions for step voltage. Z_{th} is the Thevenin equivalent for the person's feet in series and in series with the body. Based on the Thevenin equivalent, the step voltage is

$$E_{step} = I_b(R_B + Z_{th}). \tag{5.4}$$

The resistance of the foot is represented by a metal circular plate of radius 0.08 m in homogeneous earth of resistivity ρ (Ω-m) and is equal to:

$$R_f = 3\rho. \tag{5.5}$$

The Thevenin equivalent for two feet in parallel in the touch voltage, E_{touch}, equation is

$$Z_{th} = \frac{R_f}{2} = 1.5\rho. \tag{5.6}$$

The Thevenin equivalent for two feet in series in the step voltage, E_{step}, equation is

$$Z_{th} = 2R_f = 6\rho. \tag{5.7}$$

The above equations assume uniform soil resistivity. In a substation, a thin layer of high resistivity material is often spread over the earth surface to introduce a high-resistance contact between the soil and the feet, reducing the body current. The surface layer derating factor, C_s, increases the foot resistance and depends on the relative values of the resistivity of the soil, the surface material, and the thickness of the surface material. C_s is defined by the following equation:

$$C_s = 1 + \frac{16b}{\rho_s} \sum_{n=1}^{\infty} K^n R_{m(2nh_s)} \tag{5.8}$$

$$K = \frac{\rho - \rho_s}{\rho + \rho_s} \tag{5.9}$$

where

C_s is the surface layer derating factor
K is the reflection factor between different material resistivities
ρ_s is the surface material resistivity in Ω-m
ρ is the resistivity of the earth beneath the surface material in Ω-m
h_s is the thickness of the surface material in m
b is the radius of the circular metallic disc representing the foot in m
$R_{m(2nh_s)}$ is the mutual ground resistance between the two similar, parallel, coaxial plates, separated by a distance $(2nh_s)$, in an infinite medium of resistivity ρ_s in Ω

A series of curves has been developed based on $b = 0.08$ m and is shown in Fig. 5.56.

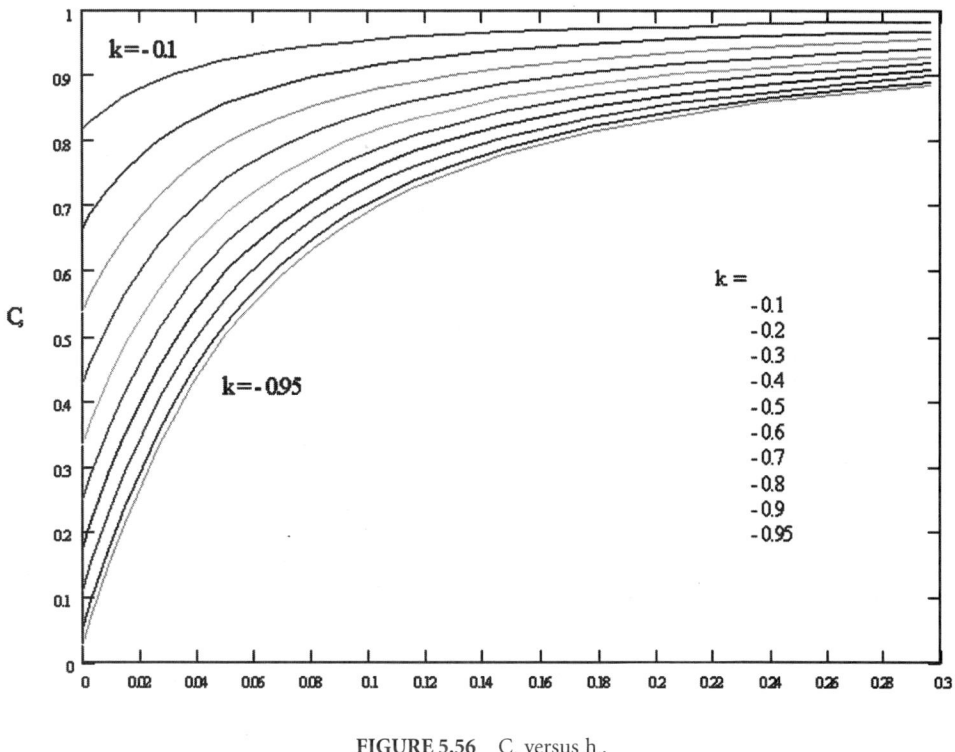

FIGURE 5.56 C_s versus h_s.

The following empirical equation by Sverak (1984) and later modified, gives the value of C_s. The values of C_s obtained using Eq. 5.10 are within 5% of the values obtained with the analytical method (Thaper et al., 1994).

$$C_s = 1 - \frac{0.09\left(1 - \frac{\rho}{\rho_s}\right)}{2h_s + 0.09} \tag{5.10}$$

Permissible Body Current Limits

Current duration, magnitude, and frequency impact the human body as it passes through it. The most dangerous impact on the body is a heart condition known as ventricular fibrillation, a stoppage of the heart resulting in immediate loss of blood circulation.

Humans are very susceptible to the effects of electric currents at 60 Hz. The most common physiological effect as the current increases are perception, muscular contraction, unconsciousness, fibrillation, respiratory nerve blockage, and burning (Dalziel and Lee, 1969). The threshold of perception, the detection of a slight tingling sensation, is generally recognized as 1 mA. The let-go current, the ability to control the muscles and release the source of current, is recognized as between 1 and 6 mA. Nine to 25 mA may cause the loss of muscular control, making it impossible to release the source of current. At slightly higher currents, breathing may become very difficult, caused by the muscular contractions of the chest muscles. Although very painful, these levels of current do not cause permanent damage to the body. In a range of 60 to 100 mA, ventricular fibrillation occurs. Ventricular fibrillation can be a fatal electric shock. The only way to restore the normal heartbeat is through another controlled electric shock called defibrillation. Larger currents will inflict nerve damage and burning, causing other life-threatening conditions.

The substation grounding system design must limit the electric current flow through the body to a value below the fibrillation current. In 1960, Charles Dalziel published a paper introducing an equation relating the flow of current through the body for a specific time that statistically, 99.5% of the population could survive before the onset of fibrillation. This equation determines the allowable body current.

$$I_B = \frac{k}{\sqrt{t_s}} \qquad (5.11)$$

where

I_B is the rms magnitude of the current through the body in A
t_s is the duration of the current exposure in s

$k = \sqrt{S_B}$

S_B is the empirical constant related to the electric shock energy tolerated by a certain percent of a given population

Dalziel found the value of $k = 0.116$ for persons weighing approximately 50 (110 lb) kg or $k = 0.157$ for a body weight of 70 kg (154 lb) (Dalziel, 1968).

Based on a 50 kg weight, the tolerable body current is

$$I_B = \frac{0.116}{\sqrt{t_s}}. \qquad (5.12)$$

The equation is based on tests limited to values of time in the 0.03–3.0 s range. It is not valid for other values of time.

Other researchers have suggested other limits (Biegelmeier and Lee, 1980). Their results have been similar to Dalziel's for the range of 0.03–3.0 s.

Tolerable Voltages

Figures 5.57 and 5.58 show the five voltages a person can be exposed to in a substation. The following definitions describe the voltages.

Ground potential rise (GPR): The maximum electrical potential that a substation grounding grid may attain relative to a distant grounding point assumed to be at the potential of remote earth. GPR is the product of the magnitude of the grid current, the portion of the fault current conducted to earth by the grounding system, and to the ground grid resistance.

Mesh voltage: The maximum touch voltage within a mesh of a ground grid.

Metal-to-metal touch voltage: The difference in potential between metallic objects or structures within the substation site that may be bridged by direct hand-to-hand or hand-to-feet contact. Note: The metal-to-metal touch voltage between metallic objects or structures bonded to the ground grid is assumed to be negligible in conventional substations. However, the metal-to-metal touch voltage between metallic objects or structures bonded to the ground grid, and metallic objects internal to the substation site, such as an isolated fence, but not bonded to the ground grid may be substantial. In the case of gas-insulated substations, the metal-to-metal touch voltage between metallic objects or structures bonded to the ground grid may be substantial because of internal faults or induced currents in the enclosures.

Step voltage: The difference in surface potential experienced by a person bridging a distance of 1 m with the feet without contacting any other grounded object.

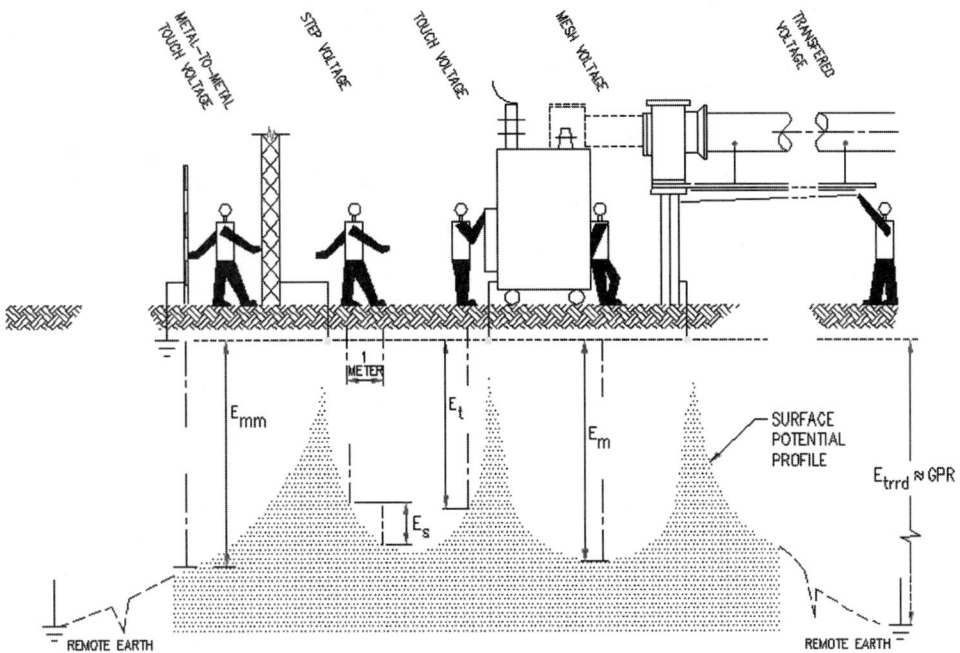

FIGURE 5.57 Basic shock situations.

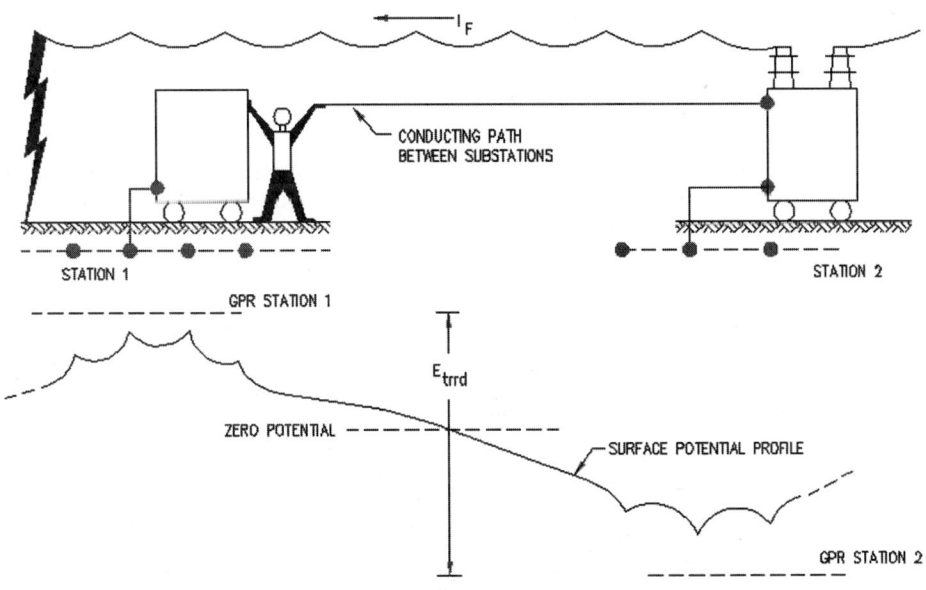

FIGURE 5.58 Typical situation of external transferred potential.

Touch voltage: The potential difference between the ground potential rise (GPR) and the surface potential at the point where a person is standing while at the same time having a hand in contact with a grounded structure.

Transferred voltage: A special case of the touch voltage where a voltage is transferred into or out of the substation, from or to a remote point external to the substation site.

The maximum voltage of any accidental circuit must not exceed the limit that would produce a current flow through the body that will cause fibrillation.

Assuming the more conservative body weight of 50 kg to determine the permissible body current and a body resistance of 1000 Ω, the tolerable touch voltage is

$$E_{touch50} = (1000 + 1.5 C_s \cdot \rho_s) \frac{0.116}{\sqrt{t_s}} \tag{5.13}$$

and the tolerable step voltage is

$$E_{step50} = (1000 + 6 C_s \cdot \rho_s) \frac{0.116}{\sqrt{t_s}}. \tag{5.14}$$

where

E_{step} is the step voltage in V
E_{touch} is the touch voltage in V
C_s is determined from Fig. 5.56 or Eq. 5.10
ρ_s is the resistivity of the surface material in Ω-m
t_s is the duration of shock current in seconds

Since the only resistance for the metal-to-metal touch voltage is the body resistance, the voltage limit is

$$E_{mm-touch50} = \frac{116}{\sqrt{t_s}}. \tag{5.15}$$

The shock duration is usually assumed to be equal to the fault duration. If reclosing of a circuit is planned, the fault duration time should be the sum of the individual faults and used as the shock duration time t_s.

Design Criteria

The design criteria for a substation grounding system are to limit the actual step and mesh voltages to levels below the tolerable step and touch voltages as determined by Eqs. 5.13 and 5.14. The worst-case touch voltage as shown in Fig. 5.57 is the mesh voltage.

Actual Touch and Step Voltages

Mesh Voltage (E_m)

The actual mesh voltage E_m (maximum touch voltage) is the product of the soil resistivity, ρ; the geometrical factor based on the configuration of the grid, K_m; a correction factor, K_i, that accounts for some of the error introduced by the assumptions made in deriving K_m; and the average current per unit of effective buried length of the conductor that makes up the grounding system.

Substations

$$E_m = \frac{\rho \cdot K_m \cdot K_i \cdot I_G}{L_m} \quad (5.16)$$

The geometrical factor K_m (Sverak, 1984), is as follows:

$$K_m = \frac{1}{2 \cdot \pi} \cdot \left[\ln\left[\frac{D^2}{16 \cdot h \cdot d} + \frac{(D+2\cdot h)^2}{8 \cdot D \cdot d} - \frac{h}{4 \cdot d}\right] + \frac{K_{ii}}{K_h} \cdot \ln\left[\frac{8}{\pi(2 \cdot n - 1)}\right] \right]. \quad (5.17)$$

For grids with ground rods along the perimeter, or for grids with ground rods in the grid corners, as well as both along the perimeter and throughout the grid area,

$$K_{ii} = 1.$$

For grids with no ground rods or grids with only a few ground rods, none located in the corners or on the perimeter,

$$K_{ii} = \frac{1}{(2 \cdot n)^{\frac{2}{n}}} \quad (5.18)$$

$$K_h = \sqrt{1 + \frac{h}{h_o}} \quad h_o = 1\text{m (grid reference depth)} \quad (5.19)$$

Using four grid shape components (Thaper et al., 1991), the effective number of parallel conductors in a given grid, n, can be made applicable to both rectangular and irregularly shaped grids that represent the number of parallel conductors of an equivalent rectangular grid:

$$n = n_a \cdot n_b \cdot n_c \cdot n_d \quad (5.20)$$

where

$$n_a = \frac{2 \cdot L_C}{L_p} \quad (5.21)$$

$n_b = 1$ for square grids
$n_c = 1$ for square and rectangular grids
$n_d = 1$ for square, rectangular, and L-shaped grids

Otherwise,

$$n_b = \sqrt{\frac{L_p}{4 \cdot \sqrt{A}}} \quad (5.22)$$

$$n_c = \left[\frac{L_x \cdot L_y}{A}\right]^{\frac{0.7 \cdot A}{L_x \cdot L_y}} \quad (5.23)$$

$$n_d = \frac{D_m}{\sqrt{L_x^2 + L_y^2}} \quad (5.24)$$

L_C is the total length of the conductor in the horizontal grid in m
L_p is the peripheral length of the grid in m
A is the area of the grid in m²
L_x is the maximum length of the grid in the x direction in m
L_y is the maximum length of the grid in the y direction in m
D_m is the maximum distance between any two points on the grid in m
D is the spacing between parallel conductors in m
h is the depth of the ground grid conductors in m
d is the diameter of the grid conductor in m

The irregularity factor, K_i, used in conjunction with the above defined n is:

$$K_i = 0.644 + 0.148 \cdot n. \quad (5.25)$$

For grids with no ground rods, or grids with only a few ground rods scattered throughout the grid, but none located in the corners or along the perimeter of the grid, the effective buried length, L_M, is:

$$L_M = L_C + L_R \quad (5.26)$$

where L_R is the total length of all ground rods in m.

For grids with ground rods in the corners, as well as along the perimeter and throughout the grid, the effective buried length, L_M, is:

$$L_M = L_C + \left[1.55 + 1.22\left(\frac{L_r}{\sqrt{L_x^2 + L_y^2}}\right)\right] L_R \quad (5.27)$$

where L_r is the length of each ground rod in m.

Step Voltage (E_s)

The maximum step voltage is assumed to occur over a distance of 1 m, beginning at and extending outside of the perimeter conductor at the angle bisecting the most extreme corner of the grid.

The step voltage values are obtained as the product of the soil resistivity (ρ), the geometrical factor K_s, the corrective factor K_i, and the average current per unit of buried length of grounding system conductor (I_G/L_S):

$$E_s = \frac{\rho \cdot K_s \cdot K_i \cdot I_G}{L_S} \quad (5.28)$$

For the usual burial depth of 0.25 m < h < 2.5 m (Sverak, 1984), K_s is:

Substations

$$K_s = \frac{1}{\pi}\left[\frac{1}{2\cdot h} + \frac{1}{D+h} + \frac{1}{D}\left(1 - 0.5^{n-2}\right)\right] \quad (5.29)$$

and K_i as defined in Eq. (5.25).

For grids with or without ground rods, the effective buried conductor length, L_S, is:

$$L_S = 0.75\cdot L_C + 0.85\cdot L_R \quad (5.30)$$

Evaluation of the Actual Touch and Step Voltage Equations

It is essential to determine the soil resistivity and maximum grid currents to design a substation grounding system. The touch and step voltages are directly proportional to these values. Overly conservative values will increase the cost dramatically. Underestimating them may cause the design to be unsafe.

Soil Resistivity

Soil resistivity investigations are necessary to determine the soil structure. There are a number of tables in the literature showing the ranges of resistivity based on soil types (clay, loam, sand, shale, etc.) (Rüdenberg, 1926; Sunde, 1968; Wenner, 1916). These tables give only very rough estimates. The soil resistivity can change dramatically with changes in moisture, temperature, and chemical content. To determine the resistivity of a particular site, resistivity measurements must be taken. Soil resistivity can vary both horizontally and vertically, making it necessary to take more than one set of measurements. A number of measuring techniques are described in detail in IEEE Std. 81-1983, *Guide for Measuring Earth Resistivity, Ground Impedance, and Earth Surface Potential of a Ground System*. The most widely used test for determining soil resistivity data was developed by Wenner and is called either the Wenner or four-pin method. Using four pins or electrodes driven in to the earth along a straight line at equal distances of a, to a depth of b, current is passed through the outer pins while a voltage reading is taken with the two inside pins. Based on the resistance as determined by the voltage and current, the apparent resistivity can be calculated using the following equation, assuming b is small compared to a:

$$\rho_a = 2\pi a R \quad (5.31)$$

where it is assumed the apparent resistivity, ρ_a, at depth a is given by the equation.

Interpretation of the apparent resistivity based on field measurements is difficult. Uniform and two-layer soil models are the most commonly used soil resistivity models. The objective of the soil model is to provide a good approximation to the actual soil conditions. Interpretation can be done either manually or by the use of computer analysis. There are commercially available programs that take the soil information and mathematically calculate the resistivity and give a confidence level based on the test. Sunde developed a graphical method to interpret the test results.

The equations in IEEE-80 require a uniform soil resistivity. Engineering judgment is required to interpret the soil resistivity measurements to determine the value of the soil resistivity, ρ, to use in the equations. IEEE-80 gives suggestions on how to interpret the data to choose a uniform soil resistivity.

Resistance

The grid resistance, that is, the resistance of the ground grid to remote earth without other metallic conductors connected, can be calculated based on the following Sverak (1984) equation:

$$R_g = \rho\left[\frac{1}{L_T} + \frac{1}{\sqrt{20A}}\left(1 + \frac{1}{1+h\sqrt{20/A}}\right)\right] \quad (5.32)$$

where

> R_g is the substation ground resistance in Ω
> ρ is the soil resistivity in Ω-m
> A is the area occupied by the ground grid in m²
> h is the depth of the grid in m
> L_T is the total buried length of conductors in m

Grid Current

The maximum grid current must be determined, since it is this current that will produce the greatest ground potential rise (GPR) and the largest local surface potential gradients in and around the substation area. It is the flow of the current from the ground grid system to remote earth that determines the GPR.

There are many types of faults that can occur on an electrical system. Therefore, it is difficult to determine what condition will produce the maximum fault current. In practice, single-line-to-ground and line-to-line-to-ground faults will produce the maximum grid current. Figures 5.59, 5.60, 5.61, and 5.62 show the maximum grid current, I_G, for various fault locations and system configurations.

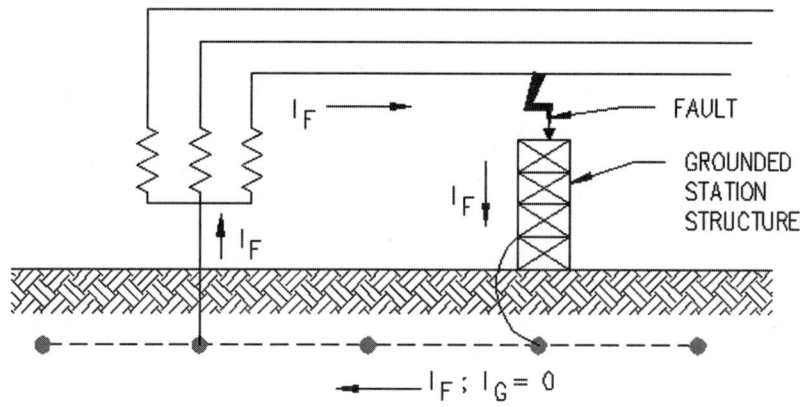

FIGURE 5.59 Fault within local substation; local neutral grounded.

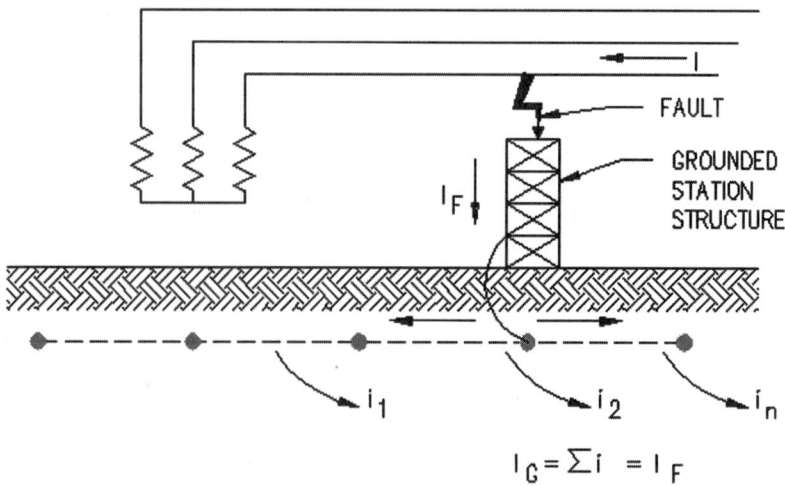

FIGURE 5.60 Fault within local substation; neutral grounded at remote location.

Substations

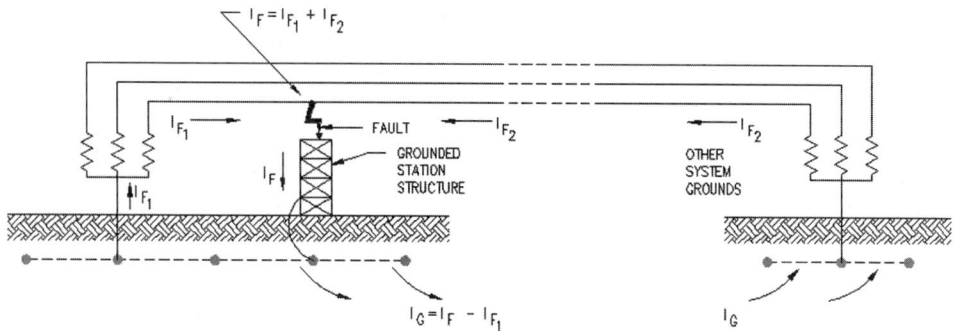

FIGURE 5.61 Fault in substation; system grounded at local station and also at other points.

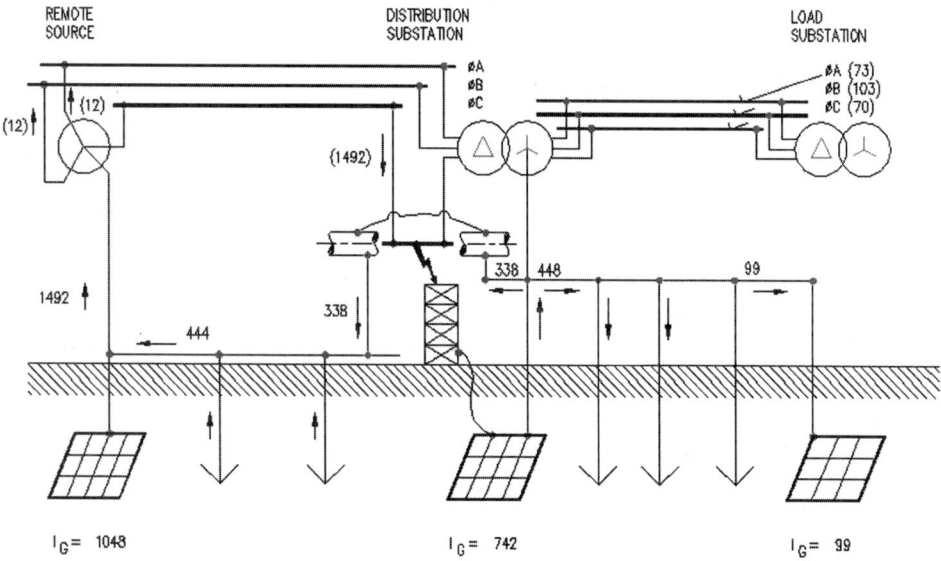

FIGURE 5.62 Typical current division for a fault on high side of distribution substation.

Overhead ground wires, neutral conductors, direct buried pipes, and cables conduct a portion of the ground fault current away from the substation ground grid and need to be considered when determining the maximum grid current. The effect of these other current paths in parallel with the ground grid is difficult to determine because of the complexities and uncertainties in the current flow. Computer programs are available to determine the split between the various current paths. There are many papers available to determine the effective impedance of a static wire as seen from the fault point.

Using the maximum grid current instead of the maximum fault current will reduce the overall cost of the ground grid system.

Use of the Design Equations

The design equations above are limited to a uniform soil resistivity, equal grid spacing, specific buried depths, and relatively simple geometric layouts of the grid system. It may be necessary to use more sophisticated computer techniques to design a substation ground grid system for nonuniform soils or complex geometric layouts. Commercially available computer programs can be used to optimize the layout and provide for unequal grid spacing and maximum grid current based on the actual system configuration, including overhead wires, neutral conductors, underground facilities, etc. Computer programs can also handle special problems associated with fences, interconnected substation grounding systems at power plants, customer substations, and other unique situations.

There are other elements of substation design that have not been discussed. These elements include the sizing of conductors and connectors, effects of direct buried pipes and cables, refinement of the design, special areas of concern including fence grounding, control and power cable grounding, surge arrester grounding, transferred potentials, and installation considerations.

References

Biegelmeier, U. G. and Lee, W. R., New considerations on the threshold of ventricular fibrillation for AC shocks at 50–60 Hz, *Proceedings of the IEEE*, 127, 103, 1980.
Dalziel, C. F., Threshold 60-Cycle fibrillating currents, *AIEE Trans. on Power Appar. and Syst.*, 79, part III, 667, 1960.
Dalziel, C. F. and Lee, R. W., Reevaluation of lethal electric currents, *IEEE Trans. on Ind. and Gen. Appl.*, IGA-4, no. 5, 467, October 1968.
Dalziel, C. F. and Lee, W. R., Lethal electric currents, *IEEE Spectrum*, 44, Feb. 1969.
Rüdenberg, R., Basic considerations concerning systems, *Electrotechnische Zeitschrift*, 11 and 12, 1926.
Sunde, E. D., *Earth Conduction Effects in Transmission Systems*, McMillan, New York, 1968.
Sverak, J. G., Simplified analysis of electrical gradients above a ground grid; Part I — How good is the present IEEE method?, *IEEE Trans. on Power Appar. and Syst.*, PAS-103, no. 1, 7, January 1984.
Thapar, B., Gerez, V., Balakrishnan, A., and Blank, D., Simplified equations for mesh and step voltages in an AC substation, *IEEE Trans. on Power Delivery*, 6, no. 2, 601, April 1991.
Thapar, B., Gerez, V., and Kejriwal, H., Reduction factor for the ground resistance of the foot in substation yards, *IEEE Trans. on Power Delivery*, 9, no. 1, 360, January 1994.
Wenner, F., A Method of Measuring Earth Resistances, *Bulletin of the Bureau of Standards*, Rep. No. 258, 12, No. 3, 469, February 1916.

5.11 Grounding and Lightning[3]

Robert S. Nowell

Lightning Stroke Protection

Substation design involves more than installing apparatus, protective devices, and equipment. The significant monetary investment and required reliable continuous operation of the facility requires detailed attention to preventing surges (transients) from entering the substation facility. These surges can be switching surges, lightning surges on connected transmission lines, or direct strokes to the substation facility. The origin and mechanics of these surges, including lightning, are discussed in detail in Chapter 10 of this handbook. This section focuses on the design process for providing *effective shielding* (that which permits lightning strokes no greater than those of critical amplitude [less design margin] to reach phase conductors [IEEE Std. 998-1996]) against direct lightning stroke in substations.

---

[3]A large portion of the text and all of the figures used in the following discussion were prepared by the Direct Stroke Shielding of Substations Working Group of the Substations Committee — IEEE Power Engineering Society, and published as IEEE Std. 998-1996, *IEEE Guide for Direct Lightning Stroke Shielding of Substations*, Institute of Electrical and Electronics Engineers, Inc., 1996. The IEEE disclaims any responsibility or liability resulting from the placement or use in the described manner. Information is reprinted with the permission of the IEEE. The author has been a member of the working group since 1987.

The Design Problem

The engineer who seeks to design a direct stroke shielding system for a substation or facility must contend with several elusive factors inherent in lightning phenomena, namely:

- The unpredictable, probabilistic nature of lightning
- The lack of data due to the infrequency of lightning strokes in substations
- The complexity and economics involved in analyzing a system in detail

There is no known method of providing 100% shielding short of enclosing the equipment in a solid metallic enclosure. The uncertainty, complexity, and cost of performing a detailed analysis of a shielding system has historically resulted in simple rules of thumb being utilized in the design of lower voltage facilities. Extra high voltage (EHV) facilities, with their critical and more costly equipment components, usually justify a more sophisticated study to establish the risk vs. cost benefit.

Because of the above factors, it is suggested that a four-step approach be utilized in the design of a protection system:

1. Evaluate the importance and value of the facility being protected.
2. Investigate the severity and frequency of thunderstorms in the area of the substation facility and the exposure of the substation.
3. Select an appropriate design method consistent with the above evaluation and then lay out an appropriate system of protection.
4. Evaluate the effectiveness and cost of the resulting design.

The following paragraphs and references will assist the engineer in performing these steps.

Lightning Parameters

Strike Distance

Return stroke current magnitude and strike distance (length of the last stepped leader) are interrelated. A number of equations have been proposed for determining the striking distance. The principal ones are as follows:

$$S = 2I + 30\left(1 - e^{-I/6.8}\right) \quad \text{Darveniza (1975)} \quad (5.33)$$

$$S = 10\,I^{0.65} \quad \text{Love (1987; 1993)} \quad (5.34)$$

$$S = 9.4\,I^{2/3} \quad \text{Whitehead (1974)} \quad (5.35)$$

$$S = 8\,I^{0.65} \quad \text{IEEE (1985)} \quad (5.36)$$

$$S = 3.3\,I^{0.78} \quad \text{Suzuki (1981)} \quad (5.37)$$

where

S is the strike distance in meters
I is the return stroke current in kiloamperes

It may be disconcerting to note that the above equations vary by as much as a factor of 2:1. However, lightning investigators now tend to favor the shorter strike distances given by Eq. (5.36). Anderson, for example, who adopted Eq. (5.34) in the 1975 edition of the *Transmission Line Reference Book* (1987), now feels that Eq. (5.36) is more accurate. Mousa (1988) also supports this form of the equation. The equation may also be stated as follows:

$$I = 0.041\, S^{1.54} \tag{5.38}$$

From this point on, the return stroke current will be referenced as the *stroke current*.

Stroke Current Magnitude

Since the stroke current and striking distance are related, it is of interest to know the distribution of stroke current magnitudes. The median value of strokes to OHGW, conductors, structures, and masts is usually taken to be 31 kA (Anderson, 1987). Anderson (1987) gave the probability that a certain peak current will be exceeded in any stroke as follows:

$$P(I) = 1 \Big/ \left[1 + (I/31)^{2.6} \right] \tag{5.39}$$

where

$P(I)$ is the probability that the peak current in any stroke will exceed I
I is the specified crest current of the stroke in kiloamperes

Mousa (1989) has shown that a median stroke current of 24 kA for strokes to flat ground produces the best correlation with available field observations to date. Using this median value of stroke current, the probability that a certain peak current will be exceeded in any stroke is given by the following equation:

$$P(I) = 1 \Big/ \left[1 + (I/24)^{2.6} \right] \tag{5.40}$$

where the symbols have the same meaning as above.

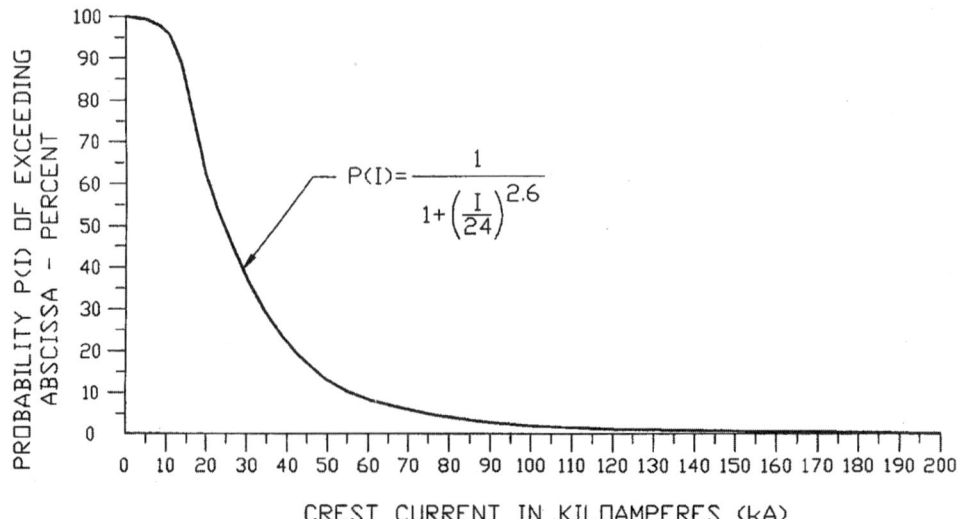

FIGURE 5.63 Probability of stroke current exceeding abscissa for strokes to flat ground. (IEEE Std. 998-1996. With permission.)

Substations

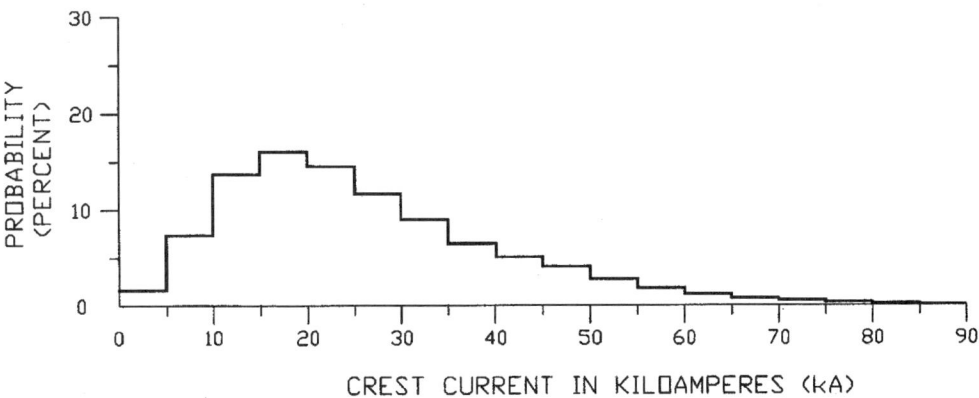

FIGURE 5.64 Stroke current range probability for strokes to flat ground. (IEEE Std. 998-1996. With permission.)

Figure 5.63 is a plot of Eq. (5.40), and Fig. 5.64 is a plot of the probability that a stroke will be within the ranges shown on the abscissa.

Keraunic Level

Keraunic level is defined as the average annual number of thunderstorm days or hours for a given locality. A daily keraunic level is called a thunderstorm-day and is the average number of days per year on which thunder will be heard during a 24-h period. By this definition, it makes no difference how many times thunder is heard during a 24-h period. In other words, if thunder is heard on any one day more than one time, the day is still classified as one thunder-day (or thunderstorm day). The average annual keraunic level for locations in the U.S. can be determined by referring to isokeraunic maps on which lines of equal keraunic level are plotted on a map of the country. Figure 5.65 gives the mean annual thunderstorm days for the U.S.

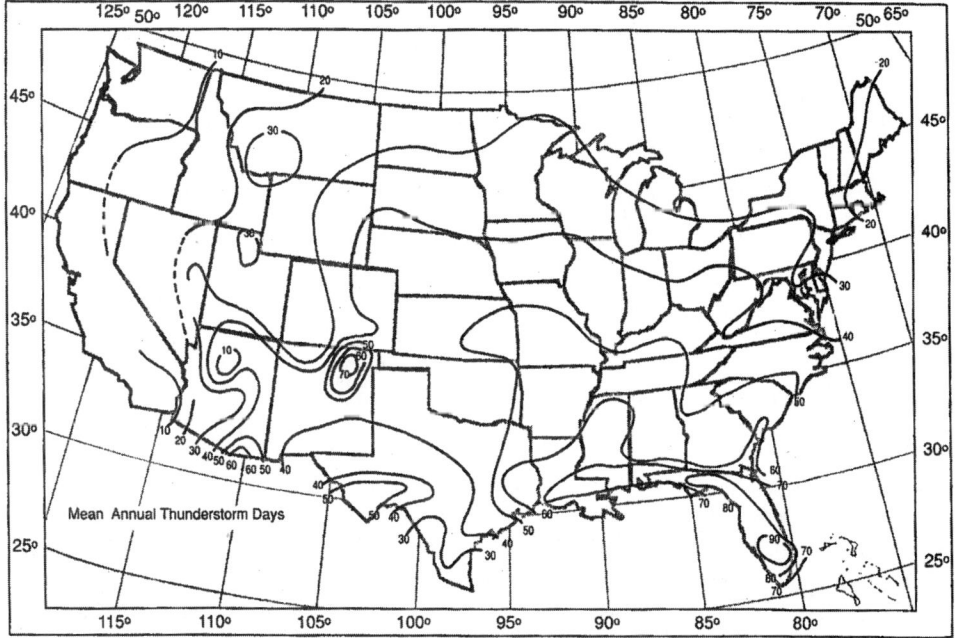

FIGURE 5.65 Mean annual thunderstorm days in the U.S. (IEEE Std. 998-1996. With permission.)

Ground Flash Density

Ground flash density (GFD) is defined as the average number of strokes per unit area per unit time at a particular location. It is usually assumed that the GFD to earth, a substation, or a transmission or distribution line is roughly proportional to the keraunic level at the locality. If thunderstorm days are to be used as a basis, it is suggested that the following equation be used (Anderson, 1987):

$$N_k = 0.12\, T_d \tag{5.41}$$

or

$$N_m = 0.31\, T_d \tag{5.42}$$

where

N_k is the number of flashes to earth per square kilometer per year
N_m is the number of flashes to earth per square mile per year
T_d is the average annual keraunic level, thunderstorm days

Lightning Detection Networks

A new technology is now being deployed in Canada and the U.S. that promises to provide more accurate information about ground flash density and lightning stroke characteristics. Mapping of lightning flashes to the earth has been in progress for over a decade in Europe, Africa, Australia, and Asia. Now a network of direction finding receiving stations has been installed across Canada and the U.S. By means of triangulation among the stations, and with computer processing of signals, it is possible to pinpoint the location of each lightning discharge. Hundreds of millions of strokes have been detected and plotted to date.

Ground flash density maps have already been prepared from this data, but with the variability in frequency and paths taken by thunderstorms from year to year, it will take a number of years to develop data that is statistically significant. Some electric utilities are, however, taking advantage of this technology to detect the approach of thunderstorms and to plot the location of strikes on their system. This information is very useful for dispatching crews to trouble spots and can result in shorter outages that result from lightning strikes.

Empirical Design Methods

Two classical design methods have historically been employed to protect substations from direct lightning strokes:

1. Fixed angles
2. Empirical curves

The two methods have generally provided acceptable protection.

Fixed Angles

The fixed-angle design method uses vertical angles to determine the number, position, and height of shielding wires or masts. Figure 5.66 illustrates the method for shielding wires, and Fig. 5.67 illustrates the method for shielding masts. The angles used are determined by the degree of lightning exposure, the importance of the substation being protected, and the physical area occupied by the substation. The value of the angle alpha that is commonly used is 45°. Both 30° and 45° are widely used for angle beta. (Sample calculations for low voltage and high voltage substations using fixed angles are given in annex B of IEEE Std. 998-1996.)

Substations   5-109

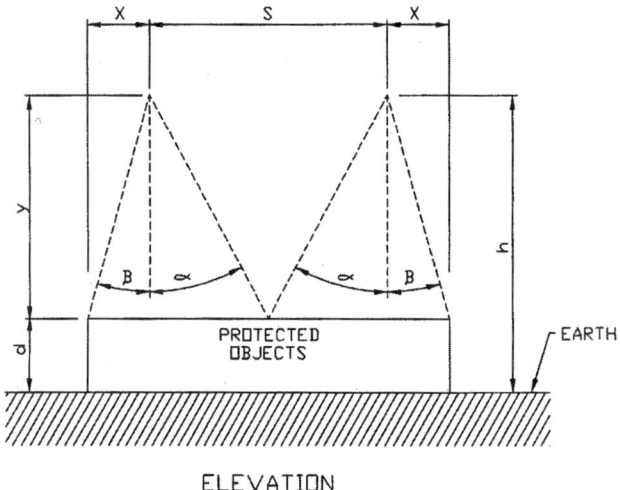

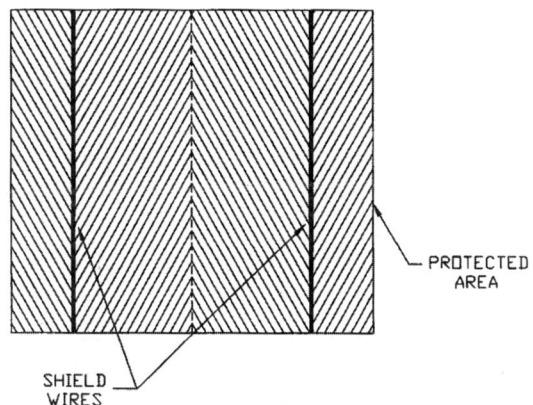

FIGURE 5.66 Fixed angles for shielding wires. (IEEE Std. 998-1996. With permission.)

Empirical Curves

From field studies of lightning and laboratory model tests, empirical curves have been developed to determine the number, position, and height of shielding wires and masts (Wagner et al., 1941; Wagner, 1942; Wagner, McCann, Beck, 1941). The curves were developed for shielding failure rates of 0.1, 1.0, 5.0, 10, and 15%. A failure rate of 0.1% is commonly used in design. Figures 5.68 and 5.69 have been developed for a variety of protected object heights, d. The empirical curve method has also been referred to as the Wagner method.

Areas Protected by Lightning Masts

Figures 5.70 and 5.71 illustrate the areas that can be protected by two or more shielding masts (Wagner et al., 1942). If two masts are used to protect an area, the data derived from the empirical curves give shielding information only for the point B, midway between the two masts, and for points on the semicircles drawn about the masts, with radius x, as shown in Fig. 5.70a. The locus shown in Fig. 5.70a, drawn by the semicircles around the masts, with radius x, and connecting the point B, represents an approximate limit for a selected exposure rate. Any single point falling within the cross-hatched area should have <0.1% exposure. Points outside the cross-hatched area will have >0.1% exposure. Figure 5.70b illustrates this phenomenon for four masts spaced at the distance s as in Fig. 5.70a.

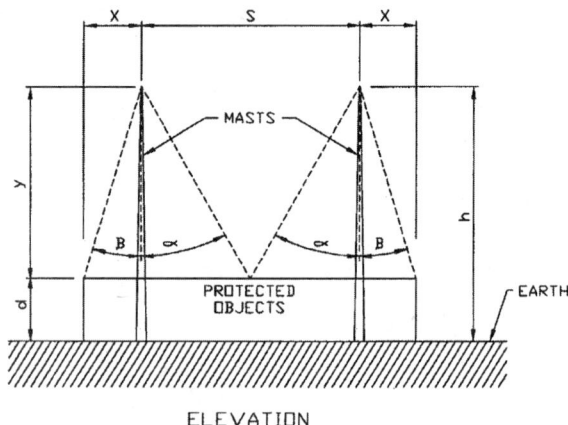

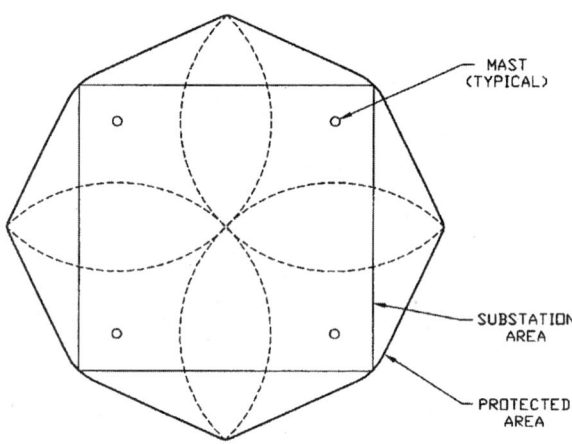

FIGURE 5.67 Fixed angles for masts. (IEEE Std. 998-1996. With permission.)

The protected area can be improved by moving the masts closer together, as illustrated in Fig. 5.71. In Fig. 5.71a, the protected areas are, at least, as good as the combined areas obtained by superimposing those of Fig. 5.70a. In Fig. 5.71a, the distance s' is one-half the distance s in Fig. 5.70a. To estimate the width of the overlap, x', first obtain a value of y corresponding to twice the distance, s' between the masts. Then use Fig. 5.68 to determine x' for this value of y. This value of x is used as an estimate of the width of overlap x' in Fig. 5.71. As illustrated in Fig. 5.71b, the size of the areas with an exposure greater than 0.1% has been significantly reduced. (Sample calculations for low voltage and high voltage substations using empirical curves are given in annex B of IEEE Std. 998-1996.)

The Electrogeometric Model (EGM)

Shielding systems developed using classical methods (fixed-angle and empirical curves) of determining the necessary shielding for direct stroke protection of substations have historically provided a fair degree of protection. However, as voltage levels (and therefore structure and conductor heights) have increased over the years, the classical methods of shielding design have proven less adequate. This led to the development of the electrogeometric model.

Substations

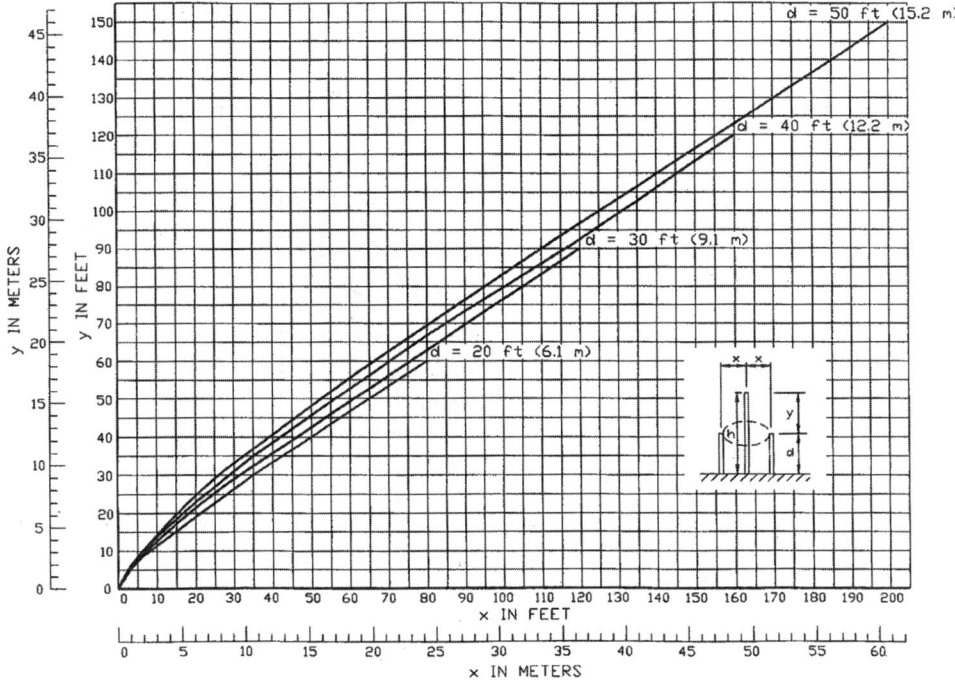

FIGURE 5.68 Single lightning mast protecting single ring of object — 0.1% exposure. Height of mast above protected object, y, as a function of horizontal separation, x, and height of protected object, d. (IEEE Std. 998-1996. With permission.)

Whitehead's EGM

In 1960, Anderson developed a computer program for calculation of transmission line lightning performance that uses the *Monte Carlo Method* (1961). This method showed good correlation with actual line performance. An early version of the EGM was developed in 1963 by Young et al., but continuing research soon led to new models. One extremely significant research project was performed by Whitehead (1971). Whitehead's work included a theoretical model of a transmission system subject to direct strokes, development of analytical expressions pertaining to performance of the line, and supporting field data that verified the theoretical model and analyses. The final version of this model was published by Gilman and Whitehead in 1973.

Recent Improvements in the EGM

Sargent made an important contribution with the *Monte Carlo Simulation* of lightning performance (1972) and his work on lightning strokes to tall structures (1972). Sargent showed that the frequency distribution of the amplitudes of strokes collected by a structure depends on the structure height as well as on its type (mast vs. wire). In 1976, Mousa extended the application of the EGM (which was developed for transmission lines) to substation facilities.

Criticism of the EGM

Work by Eriksson reported in 1978 and later work by Anderson and Eriksson reported in 1980 revealed apparent discrepancies in the EGM that tended to discredit it. Mousa (1988) has shown, however, that explanations do exist for the apparent discrepancies, and that many of them can be eliminated by adopting a revised electrogeometric model. Most investigators now accept the EGM as a valid approach for designing lightning shielding systems.

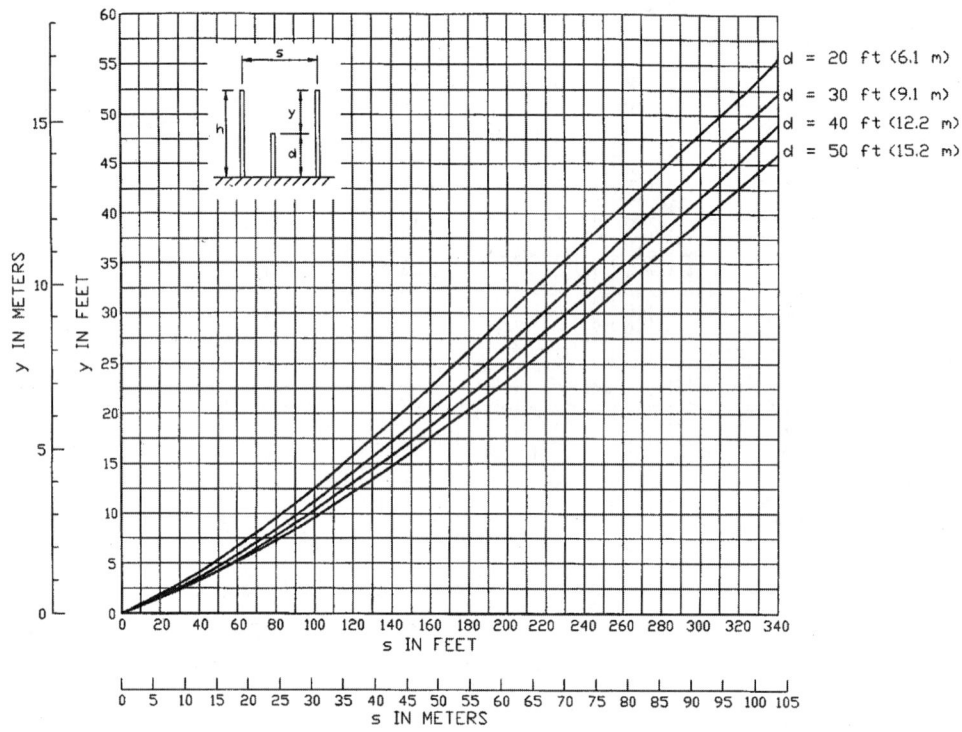

FIGURE 5.69 Two lightning masts protecting single object, no overlap — 0.1% exposure. Height of mast above protected object, *y*, as a function of horizontal separation, *s*, and height of protected object, *d*. (IEEE Std. 998-1996. With permission.)

A Revised EGM

The revised EGM was developed by Mousa and Srivastava (1986; 1988). Two methods of applying the EGM are the modified version of the rolling sphere method (Lee, 1979; Lee, 1978; Orell, 1988), and the method given by Mousa and Srivastava (1988; 1991).

The revised EGM model differs from Whitehead's model in the following respects:

1. The stroke is assumed to arrive in a vertical direction. (It has been found that Whitehead's assumption of the stroke arriving at random angles is an unnecessary complication [Mousa and Srivastava, 1988].)
2. The differing striking distances to masts, wires, and the ground plane are taken into consideration.
3. A value of 24 kA is used as the median stroke current (Mousa and Srivastava, 1989). This selection is based on the frequency distribution of the first negative stroke to flat ground. This value best reconciles the EGM with field observations.
4. The model is not tied to a specific form of the striking distance equations Eq. 5.33–5.38. Continued research is likely to result in further modification of this equation as it has in the past. The best available estimate of this parameter may be used.

Description of the Revised EGM

Previously, the concept that the final striking distance is related to the magnitude of the stroke current was introduced and Eq. (5.36) was selected as the best approximation of this relationship. A coefficient *k* accounts for the different striking distances to a mast, a shield wire, and to the ground. Eq. (5.36) is repeated here with this modification:

Substations

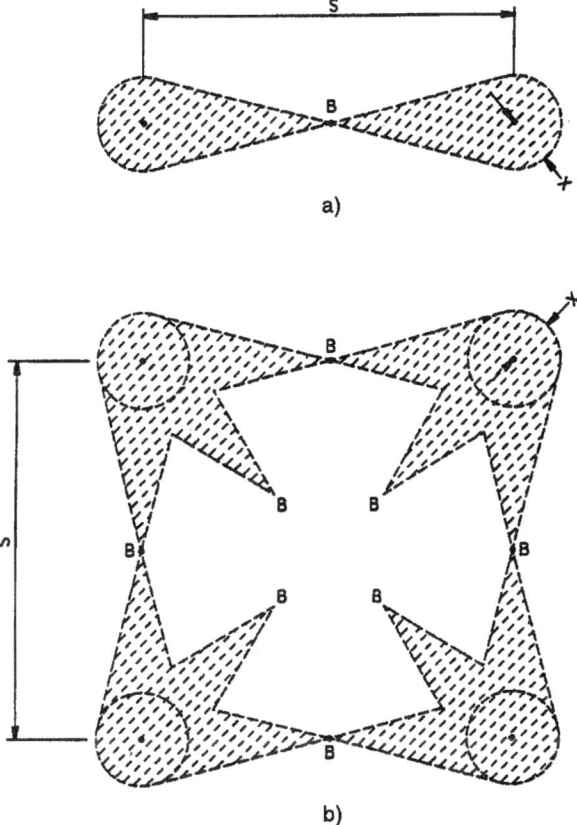

FIGURE 5.70 Areas protected by multiple masts for point exposures shown in Fig. 5.67a with two lightning masts, 5.67b with four lightning masts. (IEEE Std. 998-1996. With permission.)

$$S_m = 8kI^{0.65} \quad (5.43)$$

or

$$S_f = 26.25kI^{0.65} \quad (5.44)$$

where

S_m is the strike distance in meters
S_f is the strike distance in feet
I is the return stroke current in kiloamperes
k is a coefficient to account for different striking distances to a mast, a shield wire, or the ground plane

Mousa (1988) gives a value of $k = 1$ for strokes to wires or the ground plane and a value of $k = 1.2$ for strokes to a lightning mast.

Lightning strokes have a wide distribution of current magnitudes, as shown in Fig. 5.63. The EGM theory shows that the protective area of a shield wire or mast depends on the amplitude of the stroke current. If a shield wire protects a conductor for a stroke current I_s, it may not shield the conductor for a stroke current less than I_s that has a shorter striking distance. Conversely, the same shielding arrangement

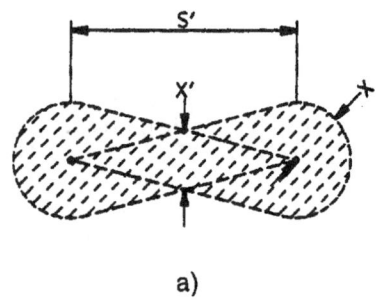

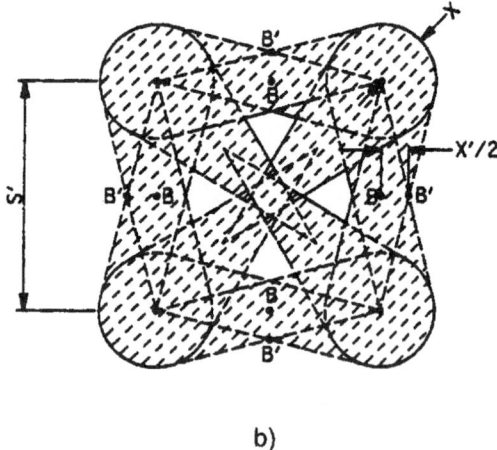

FIGURE 5.71 Areas protected by multiple masts for point exposures shown in Figures 5.67. (a) With two lightning masts; (b) with four lightning masts. (IEEE Std. 998-1996. With permission.)

will provide greater protection against stroke currents greater than I_s that have greater striking distances. Since strokes less than some critical value I_s can penetrate the shield system and terminate on the protected conductor, the insulation system must be able to withstand the resulting voltages without flashover. Stated another way, the shield system should intercept all strokes of magnitude I_s and greater so that flashover of the insulation will not occur.

Allowable Stroke Current

Some additional relationships need to be introduced before showing how the EGM is used to design a zone of protection for substation equipment. Bus insulators are usually selected to withstand a *basic lightning impulse level* (BIL). Insulators may also be chosen according to other electrical characteristics, including negative polarity *impulse critical flashover* (C.F.O.) voltage. Flashover occurs if the voltage produced by the lightning stroke current flowing through the surge impedance of the station bus exceeds the withstand value. This may be expressed by the Gilman & Whitehead equation (1973):

$$I_S = \text{BIL} \times 1.1 / (Z_S/2) = 2.2(\text{BIL})/Z_S \tag{5.45}$$

or

$$I_S = 0.94 \times \text{C.F.O.} \times 1.1 / (Z_S/2) = 2.068(\text{C.F.O.})/Z_S \tag{5.46}$$

Substations

where

- I_s is the allowable stroke current in kiloamperes
- BIL is the basic lightning impulse level in kilovolts
- C.F.O. is the negative polarity critical flashover voltage of the insulation being considered in kilovolts
- Z_s is the surge impedance of the conductor through which the surge is passing in ohms
- 1.1 is the factor to account for the reduction of stroke current terminating on a conductor as compared to zero impedance earth (Gilman and Whitehead, 1973)

In Eq. 5.46, the C.F.O. has been reduced by 6% to produce a withstand level roughly equivalent to the BIL rating for post insulators.

Withstand Voltage of Insulator Strings

BIL values of station post insulators can be found in vendor catalogs. A method is given below for calculating the withstand voltage of insulator strings. The withstand voltage in kV at 2 μs and 6 μs can be calculated as fo6

$$V_{I2} = 0.94 \times 820\, w \qquad (5.47)$$

$$V_{I6} = 0.94 \times 585\, w \qquad (5.48)$$

where

- w is the length of insulator string (or air gap) in meters
- 0.94 is the ratio of withstand voltage to C.F.O. voltage
- V_{I2} is the withstand voltage in kilovolts at 2 μs
- V_{I6} is the withstand voltage in kilovolts at 6 μs

Equation 5.48 is recommended for use with the EGM.

Application of the EGM by the Rolling Sphere Method

It was previously stated that it is only necessary to provide shielding for the equipment from all lightning strokes greater than I_s that would result in a flashover of the buswork. Strokes less than I_s are permitted to enter the protected zone since the equipment can withstand voltages below its BIL design level. This will be illustrated by considering three levels of stroke current: I_s, stroke currents greater than I_s, and stroke currents less than I_s. First, let us consider the stroke current I_s.

Protection Against Stroke Current I_s

I_s is calculated from Eq. (5.45) or Eq. (5.46) as the current producing a voltage the insulation will just withstand. Substituting this result in Eq. (5.43) or Eq. (5.44) gives the striking distance S for this stroke current. In 1977, Lee developed a simplified technique for applying the electromagnetic theory to the shielding of buildings and industrial plants (1982; 1979; 1978). Orrell extended the technique to specifically cover the protection of electric substations (1988). The technique developed by Lee has come to be known as the rolling sphere method. For the following illustration, the *rolling sphere* method will be used. This method employs the simplifying assumption that the striking distances to the ground, a mast, or a wire are the same. With this exception, the rolling sphere method has been updated in accordance with the revised EGM.

Use of the rolling sphere method involves rolling an imaginary sphere of radius S over the surface of a substation. The sphere rolls up and over (and is supported by) lightning masts, shield wires, substation fences, and other grounded metallic objects that can provide lightning shielding. A piece of equipment is said to be protected from a direct stroke if it remains below the curved surface of the sphere by virtue of the sphere being elevated by shield wires or other devices. Equipment that touches the sphere or penetrates its surface is not protected. The basic concept is illustrated in Fig. 5.72.

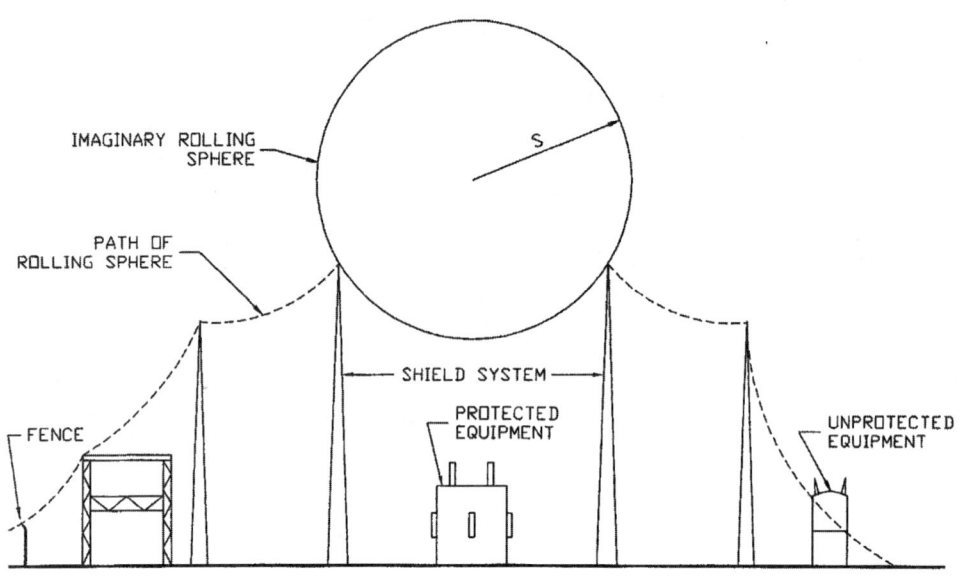

FIGURE 5.72 Principle of the rolling sphere. (IEEE Std. 998-1996. With permission.)

Continuing the discussion of protection against stroke current I_s, consider first a single mast. The geometrical model of a single substation shield mast, the ground plane, the striking distance, and the zone of protection are shown in Fig. 5.73. An arc of radius S that touches the shield mast and the ground plane is shown in Fig. 5.73. All points below this arc are protected against the stroke current I_s. This is the protected zone. The arc is constructed as follows (see Fig. 5.73). A dashed line is drawn parallel to the ground at a distance S (the striking distance as obtained from Eq. (5.43) or Eq. (5.44)) above the ground plane. An arc of radius S, with its center located on the dashed line, is drawn so the radius of the arc just touches the mast. Stepped leaders that result in stroke current I_s and that descend outside of the point where the arc is tangent to the ground will strike the ground. Stepped leaders that result in stroke current I_s and that descend inside the point where the arc is tangent to the ground will strike the shield mast, provided all other objects are within the protected zone. The height of the shield mast that will provide the maximum zone of protection for stroke currents equal to I_s is S. If the mast height is less than S, the zone of protection will be reduced. *Increasing the shield mast height greater than S will provide additional protection in the case of a single mast. This is not necessarily true in the case of multiple masts and shield wires.* The protection zone can be visualized as the surface of a sphere with radius S that is rolled toward the mast until touching the mast. As the sphere is rolled around the mast, a three-dimensional surface of protection is defined. It is this concept that has led to the name *rolling sphere* for simplified applications of the electrogeometric model.

Protection Against Stroke Currents Greater than I_s

A lightning stroke current has an infinite number of possible magnitudes, however, and the substation designer will want to know if the system provides protection at other levels of stroke current magnitude. Consider a stroke current I_{s1} with magnitude greater than I_s. Strike distance, determined from Eq. (5.43) or Eq. (5.44), is S1. The geometrical model for this condition is shown in Fig. 5.74. Arcs of protection for stroke current I_{s1} and for the previously discussed I_s are both shown. The figure shows that the zone of protection provided by the mast for stroke current I_{s1} is greater than the zone of protection provided by the mast for stroke current I_s. Stepped leaders that result in stroke current I_{s1} and that descend outside of the point where the arc is tangent to the ground will strike the ground. Stepped leaders that result in stroke current I_{s1} and that descend inside the point where the arc is tangent to the ground will strike the

Substations

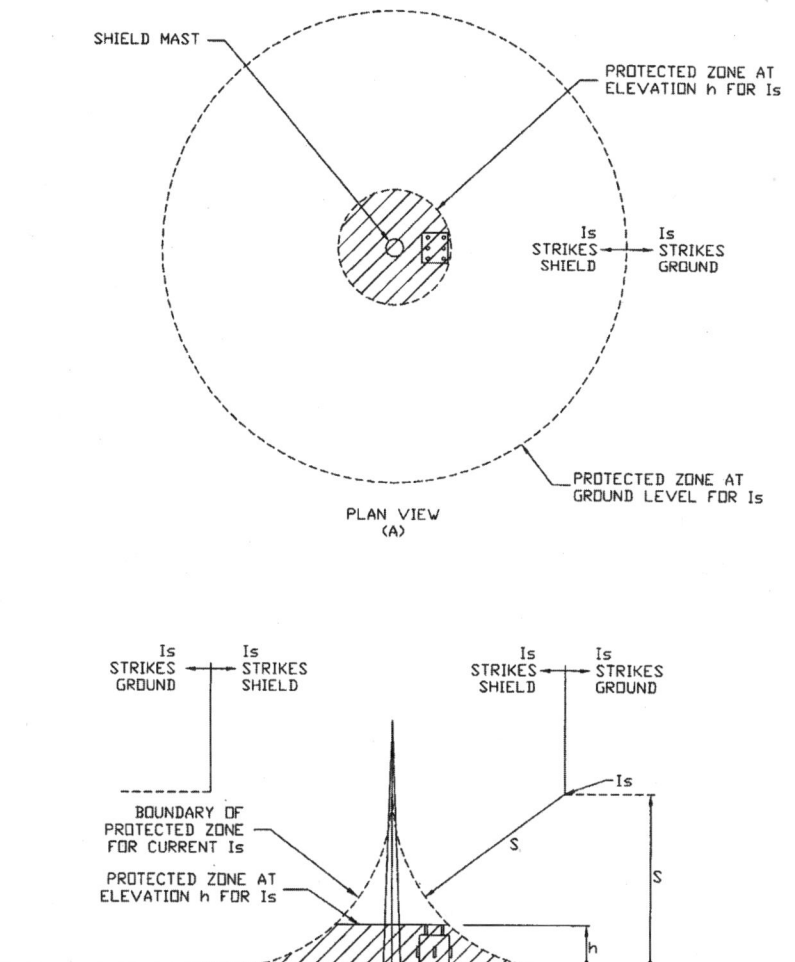

FIGURE 5.73 Shield mast protection for stroke current I_s. (IEEE Std. 998-1996. With permission.)

shield mast, provided all other objects are within the S1 protected zone. Again, the protective zone can be visualized as the surface of a sphere touching the mast. In this case, the sphere has a radius S1.

Protection Against Stroke Currents Less than I_s

It has been shown that a shielding system that provides protection at the stroke current level I_s provides even better protection for larger stroke currents. The remaining scenario to examine is the protection afforded when stroke currents are less than I_s. Consider a stroke current I_{so} with magnitude less than I_s. The striking distance, determined from Eq. (5.43) or Eq. (5.44), is S_0. The geometrical model for this condition is shown in Fig. 5.75. Arcs of protection for stroke current I_{so} and I_s are both shown. The figure shows that the zone of protection provided by the mast for stroke current I_{so} is less than the zone of protection provided by the mast for stroke current I_s. It is noted that a portion of the equipment protrudes above the dashed arc or zone of protection for stroke current I_{so}. Stepped leaders that result in stroke current I_{so} and that descend outside of the point where the arc is tangent to the ground will strike the ground. However, some stepped leaders that result in stroke current I_{so} and that descend inside the point where the arc is tangent to the ground could strike the equipment. This is best shown by observing the

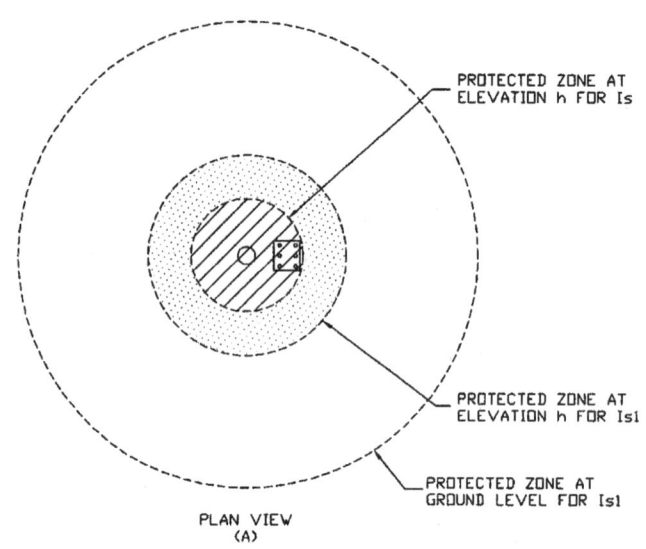

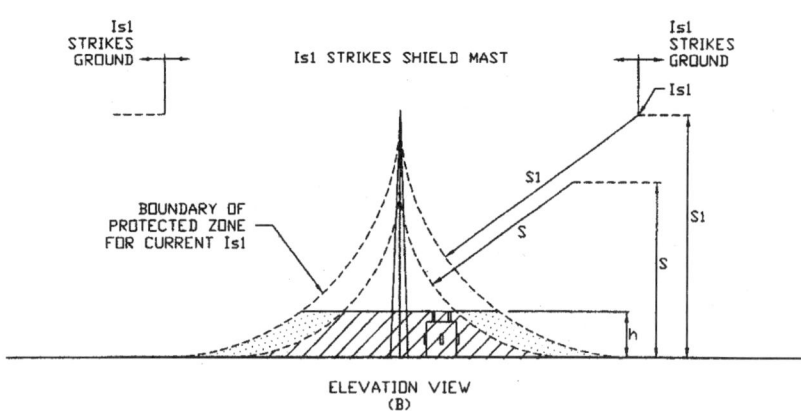

FIGURE 5.74 Shield mast protection for stroke current I_{s1}. (IEEE Std. 998-1996. With permission.)

plan view of protective zones shown in Fig. 5.75. Stepped leaders for stroke current I_{so} that descend inside the inner protective zone will strike the mast and protect equipment that is h in height. Stepped leaders for stroke current I_{so} that descend in the shaded unprotected zone will strike equipment of height h in the area. *If, however, the value of I_s was selected based on the withstand insulation level of equipment used in the substation, stroke current I_{so} should cause no damage to equipment.*

Multiple Shielding Electrodes

The electrogeometric modeling concept of direct stroke protection has been demonstrated for a single shield mast. A typical substation, however, is much more complex. It may contain several voltage levels and may utilize a combination of shield wires and lightning masts in a three-dimensional arrangement. The above concept can be applied to multiple shielding masts, horizontal shield wires, or a combination of the two. Figure 5.76 shows this application considering four shield masts in a multiple shield mast arrangement. The arc of protection for stroke current I_s is shown for each set of masts. The dashed arcs represent those points at which a descending stepped leader for stroke current I_s will be attracted to one of the four masts. The protected zone between the masts is defined by an arc of radius S with the center at the intersection of the two dashed arcs. The protective zone can again be visualized as the surface of

Substations

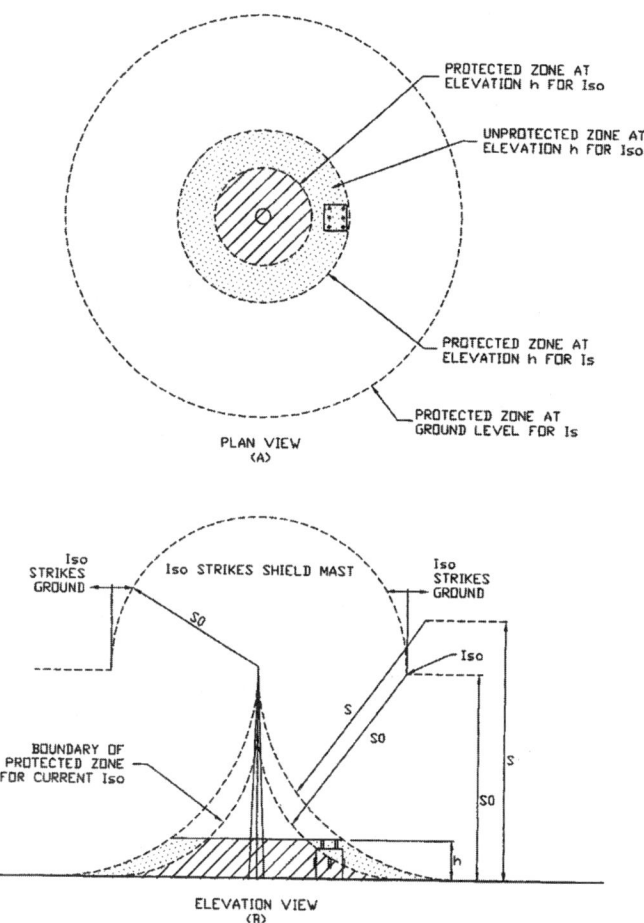

FIGURE 5.75 Shield mast protection for stroke current I_{s0}. (IEEE Std. 998-1996. With permission.)

a sphere with radius S, which is rolled toward a mast until touching the mast, then rolled up and over the mast such that it would be supported by the masts. The dashed lines would be the locus of the center of the sphere as it is rolled across the substation surface. Using the concept of rolling sphere of the proper radius, the protected area of an entire substation can be determined. This can be applied to any group of different height shield masts, shield wires, or a combination of the two. Figure 5.77 shows an application to a combination of masts and shield wires.

Changes in Voltage Level

Protection has been illustrated with the assumption of a single voltage level. Substations, however, have two or more voltage levels. The rolling sphere method is applied in the same manner in such cases, except that the sphere radius would increase or decrease appropriate to the change in voltage at a transformer. (Sample calculations for a substation with two voltage levels are given in annex B of IEEE Std. 998-1996.)

Minimum Stroke Current

The designer will find that shield spacing becomes quite close at voltages of 69 kV and below. It may be appropriate to select some minimum stroke current, perhaps 2 kA for shielding stations below 115 kV. Such an approach is justified by an examination of Figs. 5.63 and 5.64. It will be found that 99.8% of all strokes will exceed 2 kA. Therefore, this limit will result in very little exposure, but will make the shielding system more economical.

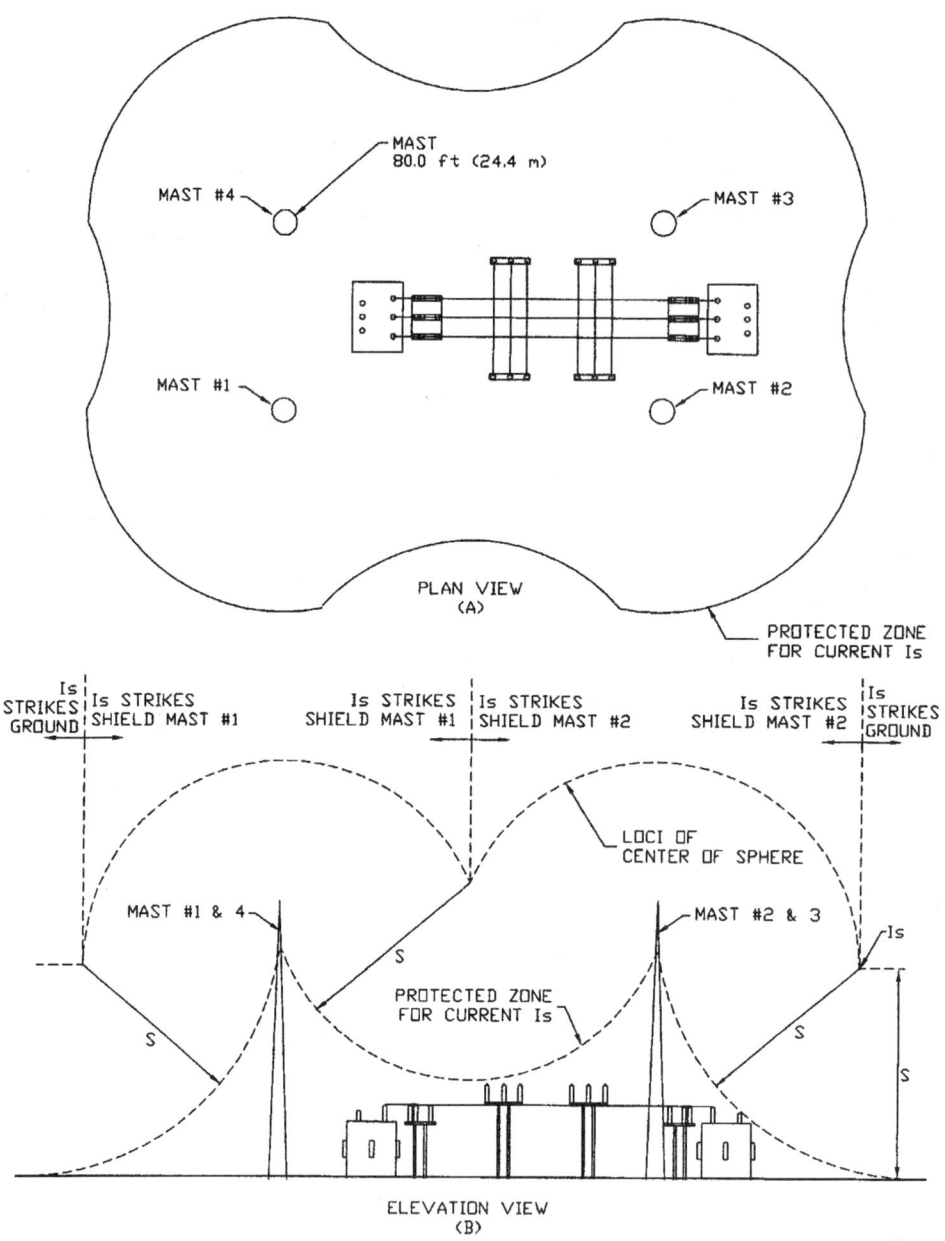

FIGURE 5.76 Multiple shield mast protection for stroke current I_s. IEEE Std. 998-1996, with permission.

Application of Revised EGM by Mousa and Srivastava Method

The rolling sphere method has been used in the preceding paragraphs to illustrate application of the EGM. Mousa describes the application of the revised EGM (1976). Figure 5.78 depicts two shield wires, G1, and G2, providing shielding for three conductors, W1, W2, and W3. S_c is the critical striking distance as determined by Eq. (5.43), but reduced by 10% to allow for the statistical distribution of strokes so as to preclude any failures. Arcs of radius S_c are drawn with centers at G1, G2, and W2 to determine if the shield wires are positioned to properly shield the conductors. The factor ψ is the horizontal separation of the outer conductor and shield wire, and b is the distance of the shield wires above the conductors. Figure 5.79 illustrates the shielding provided by four masts. The height h_{mid} at the center of the area is

Substations

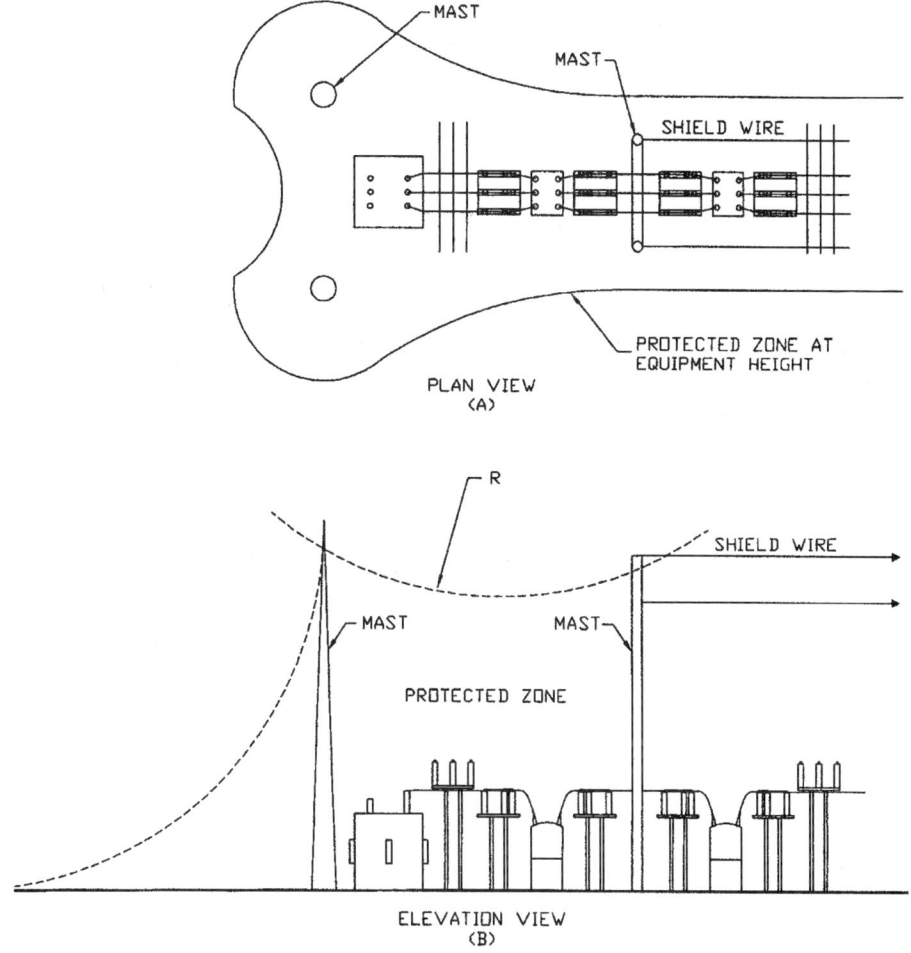

FIGURE 5.77 Protection by shield wires and masts. (IEEE Std. 998-1996. With permission.)

the point of minimum shielding height for the arrangement. For further details in the application of the method, see Mousa (1976). At least two computer programs have been developed that assist in the design of a shielding system. One of these programs (Mousa, 1991) uses the revised EGM to compute the surge impedance, stroke current, and striking distance for a given arrangement of conductors and shield systems, then advises the user whether or not effective shielding is provided. (Sample calculations are provided in annex B of IEEE Std. 998-1996 to further illustrate the application.)

Calculation of Failure Probability

In the revised EGM just presented, striking distance is reduced by a factor of 10% so as to exclude all strokes from the protected area that could cause damage. In the empirical design approach, on the other hand, a small failure rate is permitted, typically 0.1%. Linck (1975) also developed a method to provide partial shielding using statistical methods. It should be pointed out that for the statistical approach to be valid, the size of the sample needs to be large. For power lines that extend over large distances, the total exposure area is large and the above criterion is met. It is questionable, therefore, whether the statistical approach is as meaningful for substations that have very small exposure areas by comparison. Engineers do, however, design substation shielding that permits a small statistical failure rate. Orrell (1988) has developed a method of calculating failure rates for the EGM rolling sphere method. (This method is described with example calculations in annex D of IEEE Std. 998-1996.)

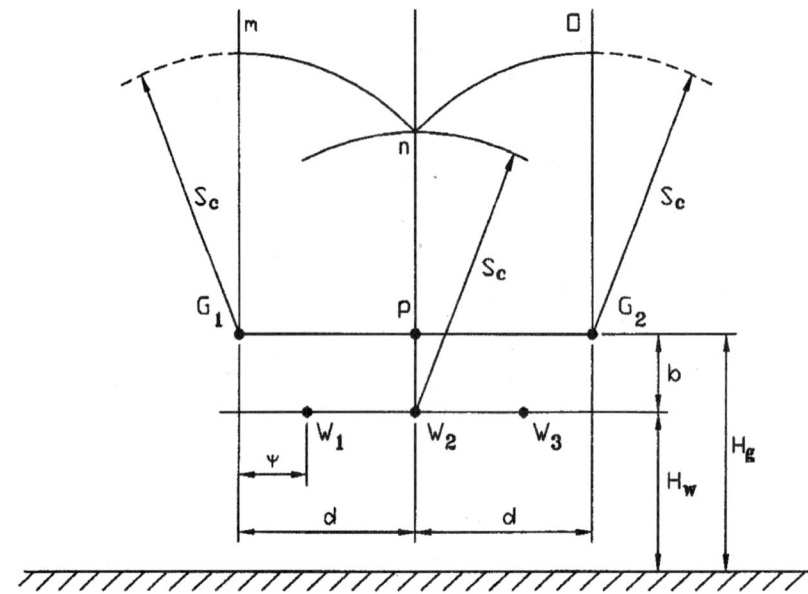

FIGURE 5.78 Shielding requirements regarding the strokes arriving between two shield wires. (IEEE Std. 998-1996. With permission.)

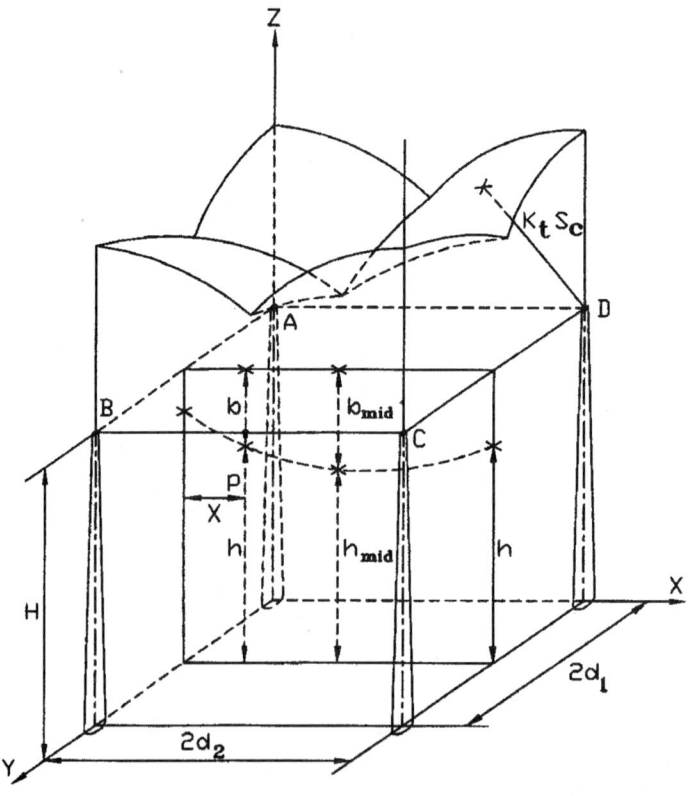

FIGURE 5.79 Shielding of an area bounded by four masts. (IEEE Std. 998-1996. With permission.)

Active Lightning Terminals

In the preceding methods, the lightning terminal is considered to be a *passive* element that intercepts the stroke merely by virtue of its position with respect to the live bus or equipment. Suggestions have been made that lightning protection can be improved by using what may be called *active* lightning terminals. Three types of such devices have been proposed over the years:

- *Lightning rods with radioactive tips* (Golde, 1973). These devices are said to extend the attractive range of the tip through ionization of the air.
- *Early Streamer Emission (ESM) lightning rods* (Berger and Floret, 1991). These devices contain a triggering mechanism that sends high-voltage pulses to the tip of the rod whenever charged clouds appear over the site. This process is said to generate an upward streamer that extends the attractive range of the rod.
- *Lightning prevention devices*. These devices enhance the *point discharge* phenomenon by using an array of needles instead of the single tip of the standard lightning rod. It is said that the space charge generated by the many needles of the array neutralize part of the charge in an approaching cloud and prevent a return stroke to the device, effectively extending the protected area (Carpenter, 1976).

Some of the latter devices have been installed on facilities (usually communications towers) that have experienced severe lightning problems. The owners of these facilities have reported no further lightning problems in many cases.

There has not been sufficient scientific investigation to demonstrate that the above devices are effective; and since these systems are proprietary, detailed design information is not available. It is left to the design engineer to determine the validity of the claimed performance for such systems.

References

This reference list is reprinted in part from IEEE Working Group D5, Substations Committee, *Guide for Direct Lightning Stroke Shielding of Substations*, IEEE Std. 998-1996.

Anderson, R. B. and Eriksson, A. J., Lightning parameters for engineering application, *Electra*, no. 69, 65–102, Mar. 1980.

Anderson, J. G., Monte Carlo computer calculation of transmission-line lightning performance, *AIEE Transactions*, 80, 414–420, Aug. 1961.

Anderson, J. G., *Transmission Line Reference Book 345 kV and Above*, 2nd ed. Rev. Palo Alto, CA: Electric Power Research Institute, 1987, chap. 12.

Berger, G. and Floret, N., Collaboration produces a new generation of lightning rods, *Power Technol. Int.*, London: Sterling Publications, 185–190, 1991.

Carpenter, R. B., Jr., Lightning Elimination. Paper PCI-76-16 given at the *23rd Annual Petroleum and Chemical Industry Conference* 76CH1109-8-IA, 1976.

Darveniza, M., Popolansky, F., and Whitehead, E. R., Lightning protection of UHV transmission lines, *Electra*, no. 41, 36–69, July 1975.

Eriksson, A. J., Lightning and tall structures, *Trans. South African IEE*, 69(8), 238–252, Aug. 1978. Discussion and closure published May 1979, vol. 70, no. 5, 12 pages.

Gilman D. W. and Whitehead, E. R., The mechanism of lightning flashover on high voltage and extra-high voltage transmission lines, *Electra*, no. 27, 65-96, Mar. 1973.

Golde, R. H., Radio-active lightning conductors, *Lightning Protection*, London: Edward Arnold Publishing Co., 37–40, 196–197, 1973.

IEEE Working Group, Estimating lightning performance of transmission lines. II. Updates to analytic models, *IEEE Trans. on Power Delivery*, 8(3), 1254–1267, July 1993.

Guide for Direct Lightning Stroke Shielding of Substations, IEEE Std. 998-1996, IEEE Working Group D5, Substations Committee.

IEEE Working Group, A simplified method for estimating lightning performance of transmission lines, *IEEE Trans. on Power Appar. and Syst.*, PAS-104, no. 4, 919–932, 1985.

Lee, R. H., Lightning protection of buildings, *IEEE Trans. on Ind. Appl.*, IA-15(3), 236–240, May/June 1979.

Lee, R. H., Protection zone for buildings against lightning strokes using transmission line protection practice, *IEEE Trans. on Ind. Appl.*, 1A-14(6), 465–470, 1978.

Lee, R. H., Protect your plant against lightning, *Instruments and Control Systems*, 55(2), 31–34, Feb. 1982.

Linck, H., Shielding of modern substations against direct lightning strokes, *IEEE Trans. on Power Appar. and Syst.*, PAS-90(5), 1674–1679, Sept./Oct. 1975.

Mousa, A. M., A computer program for designing the lightning shielding systems of substations, *IEEE Trans. on Power Delivery*, 6(1), 143–152, 1991.

Mousa, A. M., Shielding of high-voltage and extra-high-voltage substations, *IEEE Trans. on Power Appar. and Syst.*, PAS-95(4), 1303–1310, 1976.

Mousa, A. M., A Study of the Engineering Model of Lightning Strokes and its Application to Unshielded Transmission Lines, Ph.D. thesis, University of British Columbia, Vancouver, Canada, Aug. 1986.

Mousa, A. M. and Srivastava, K. D., The implications of the electrogeometric model regarding effect of height of structure on the median amplitudes of collected lightning strokes, *IEEE Trans. on Power Delivery*, 4(2), 1450–1460, 1989.

Mousa, A. M. and Srivastava, K. D., A revised electrogeometric model for the termination of lightning strokes on ground objects, in *Proceedings of International Aerospace and Ground Conference on Lightning and Static Electricity*, Oklahoma City, OK, Apr. 1988, 342–352.

Orrell, J. T., Direct stroke lightning protection, Paper presented at EEI Electrical System and Equipment Committee Meeting, Washington, D.C., 1988.

Sargent, M. A., The frequency distribution of current magnitudes of lightning strokes to tall structures, *IEEE Trans. on Power Appar. and Syst.*, PAS-91(5), 2224–2229, 1972.

Sargent, M. A., Monte Carlo simulation of the lightning performance of overhead shielding networks of high voltage stations, *IEEE Trans. on Power Appar. and Syst.*, PAS-91(4), 1651–1656, 1972.

Suzuki, T., Miyake, K., and Shindo, T., Discharge path model in model test of lightning strokes to tall mast, *IEEE Trans. on Power Appar. and Syst.*, PAS-100(7), 3553–3562, 1981.

Wagner, C. F., McCann, G. D., and Beck, E., Field investigations of lightning, *AIEE Trans.*, 60, 1222–1230, 1941.

Wagner C. F., McCann, G. D., and Lear, C. M., Shielding of substations, *AIEE Trans.*, 61, 96–100, 313,448, Feb. 1942.

Wagner, C. F., McCann, G. D., and MacLane, G. L., Shielding of transmission lines, *AIEE Trans.*, 60, 313-328, 612–614, 1941.

Whitehead, E. R., CIGRE survey of the lightning performance of extra-high-voltage transmission lines, *Electra*, 63–89, Mar. 1974.

Whitehead, E. R., Mechanism of lightning flashover, EEI Research Project RP 50, Illinois Institute of Technology, Pub 72-900, Feb. 1971.

Young, E. S., Clayton, J. M., and Hileman, A. R., Shielding of transmission lines, *IEEE Trans. on Power Appar. and Syst.*, S82, 132–154, 1963.

5.12 Seismic Considerations

R.P. Stewart, Rulon Fronk, and Tonia Jurbin

A Historical Perspective

Prior to 1970, seismic requirements for substation components were minimal. In the 1970s and 1980s, several large-magnitude earthquakes struck California, causing millions of dollars in damage to substation components and lost revenue. As a result of these losses, it became apparent to owners and operators of substation facilities in seismically active areas that the existing seismic requirements for substation components were inadequate. The 1997 version of the Institute of Electrical and Electronic Engineers (IEEE) Standard 693, *Seismic Design for Substations* (1997), and the document presently being produced by the American Society of Civil Engineers (ASCE), entitled *Substation Structure Design Guide*, have enhanced the current state of knowledge in this area and promote seismic standardization of substation power equipment in the electric power industry.

The requirements necessary to qualify power equipment developed from research into seismic activity and how it relates to substation equipment have proven to be complex. Because of the complexities, the IEEE 693 committee has attempted to simplify the application of the qualification process for the end user by providing a single set of requirements that can be applied by specifying a few simple instructions. These instructions will be discussed further, but briefly they are:

1. Note the equipment type, such as surge arresters or circuit breakers.
2. Select the qualification level — *Low, Moderate,* or *High.*
3. Note the equipment *in-situ* configuration, such as mounting information, etc.

Another IEEE 693 committee goal was to minimize testing costs, for by using one set of seismic qualifications the cost of the qualifications could be amortized over all purchasers, similar to what is done in the development of new equipment.

The purpose of this document is to guide the substation designer with little or no familiarity with substation seismic design considerations by illustrating the basic steps required for securing and protecting substation components within a given substation. It is only a guide and it is not intended to be all-inclusive or to provide all the necessary details to undertake this work. For further details and information on this topic, it is recommended that the documents listed at the end of this discussion be reviewed.

Relationship Between Earthquakes and Substations

To secure and protect substation equipment from a seismic event, the relationship between earthquakes and substation components must first be clarified. Earthquakes occur when there is a sudden rupture along a pre-existing geologic fault. Shock waves that radiate from the fracture zone amplify, and depending on the surficial geology, these waves will arrive at the surface as a complex set of multifrequency vibratory ground motions, with horizontal and vertical components.

The response of structures and buildings to this ground motion depends on their construction, ductility, dynamic properties, and design. Lightly damped structures that have one or more natural modes of oscillation within the frequency band of the ground motion excitation can experience considerable movement which can generate forces and deflections that the structures were not designed to accommodate, while mechanisms that absorb energy in a structure in response to its motion can provide damping of these forces. If two or more structures or pieces of equipment are rigidly linked, they will interact with one another, producing a modified response. If they are flexibly linked, an ideal situation, then no forces are transferred between the two components; however, the link must be designed with sufficient flexibility to accommodate the relative displacements.

For electrical reasons, many pieces of substation power equipment are interconnected and contain porcelain, which is a relatively brittle, low-strength, and low-damping material compared to steel. Furthermore, unless instructed to do otherwise, construction personnel may not install flexible bus conductors used to electrically connect this equipment with sufficient slack so that small differential motions of one piece of equipment can easily impact an adjacent piece of equipment. Substation equipment whose natural frequencies lie in the range of earthquake ground motions are especially vulnerable to this type of damage by seismic events.

Applicable Documents

Once the relationship between substation components and earthquakes is understood, the substation designer should become familiar with the standards and references currently available (see reference list below). It is important for the user to appreciate how the various documents interrelate. Although the title of IEEE 693 is *Recommended Practice for Seismic Design of Substations,* it was clear to the IEEE 693 committee that other documents had already addressed many of the aspects of seismic design substations. Therefore, IEEE 693 simply refers the users to the appropriate document if the information is not contained therein. It was also clear that a single set of seismic qualification requirements was needed; therefore, the IEEE 693 emphasizes those aspects associated with the seismic qualification of power equipment.

Special attention also needs to be given to the ASCE's *Substation Structure Design Guide*. This guide provides information for all of the structures within a substation, such as A-frames, buildings, racks, etc. Since these two documents, IEEE 693 and ASCE, were developed at about the same time, the two committees collaborated so that the two documents would complement each other. Simply stated, IEEE 693 addresses the equipment and its 'first' support structure and ASCE addresses all the other structures.

Decision Process for Seismic Design Consideration

Once document familiarization is complete, the designer can follow the steps as outlined in Fig. 5.80, which was created with the assumption that each substation component will be reviewed independently.

The first step in the decision-making process is to determine whether the substation component under consideration is classified as power equipment or not. Assuming the component is classified as nonpower equipment, the next step is to determine what type of nonpower equipment the component is. For example, a structure such as a bus support may require foundation modification or anchor design work. Once the component type is determined, the appropriate references can be accessed and the required engineering work carried out. The decision-making process for substation components classified as nonpower equipment is then complete.

If the substation component under consideration is found to be a piece of power equipment, the next step in the power equipment decision process stream is to determine if this equipment is classified as Class 1E, equipment for nuclear power generating stations. IEEE 693 does not cover Class 1E equipment, but this information is available in IEEE 344 (1993).

The next step of the power equipment stream is to determine if this equipment's voltage class is more or less than 15,000 volts. If it is less than 15,000 volts, there is a possibility that it could be seismically qualified by experience data as indicated in IEEE 693 and IEEE 344. If the power equipment's voltage class is greater than or equal to 15,000 volts, it must be determined if there are any secondary support, foundation, or anchor design issues, and if the equipment is existing or to be installed, for although IEEE 693 was written primarily for new installations, it can be used to assist designers in the analysis of seismic requirements for existing equipment as well.[4] Anchor design issues should be addressed as per the ASCE document and IEEE 693 as indicated in IEEE 693.

Performance Levels and Required Spectra

Following the voltage classification, determination of the appropriate performance level for seismic qualification of the site in question must be selected.

The performance level of earthquake motion is represented by response spectra that reasonably envelop response spectra from anticipated ground motions determined using historical earthquake records. The shape of the performance level is a broadband response spectrum that envelopes the effects of earthquakes in different areas for site conditions ranging from rock to soft soils as described in the National Earthquake Hazard Reduction Program (NEHRP) (1997).

The performance level, and the required response spectrum shapes, bracket the vast majority of substation site conditions, and in particular provide longer period coverage for soft sites, but sites with

---

[4]It is assumed that although IEEE 693 recommends testing or analysis, that testing of existing equipment would not be undertaken due to the high cost of transporting the piece of equipment to the test laboratory, its loss of availability during the testing, and the risk of damage or destroying the equipment during transportation or testing.

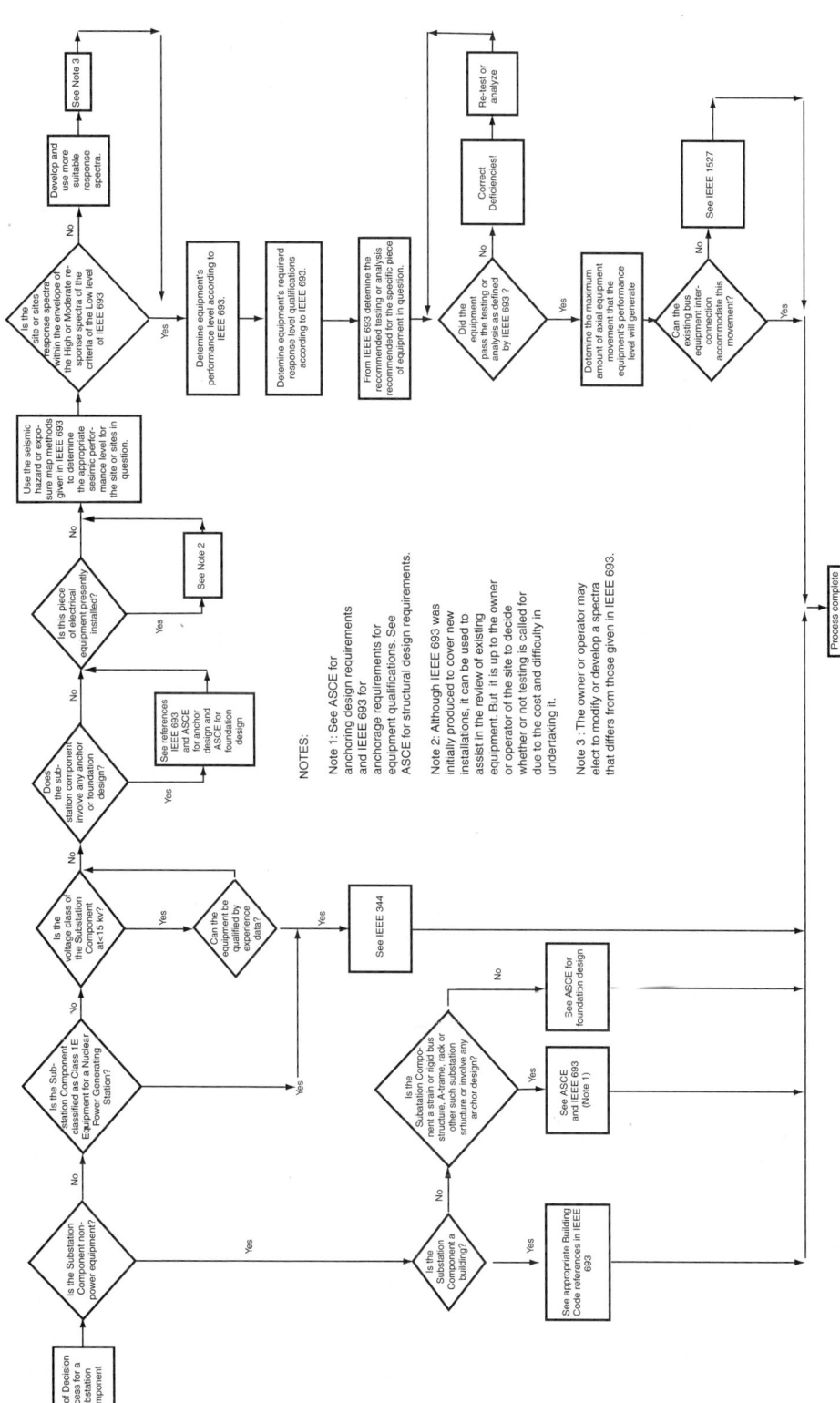

FIGURE 5.80 Decision process for seismic design considerations for substation components.

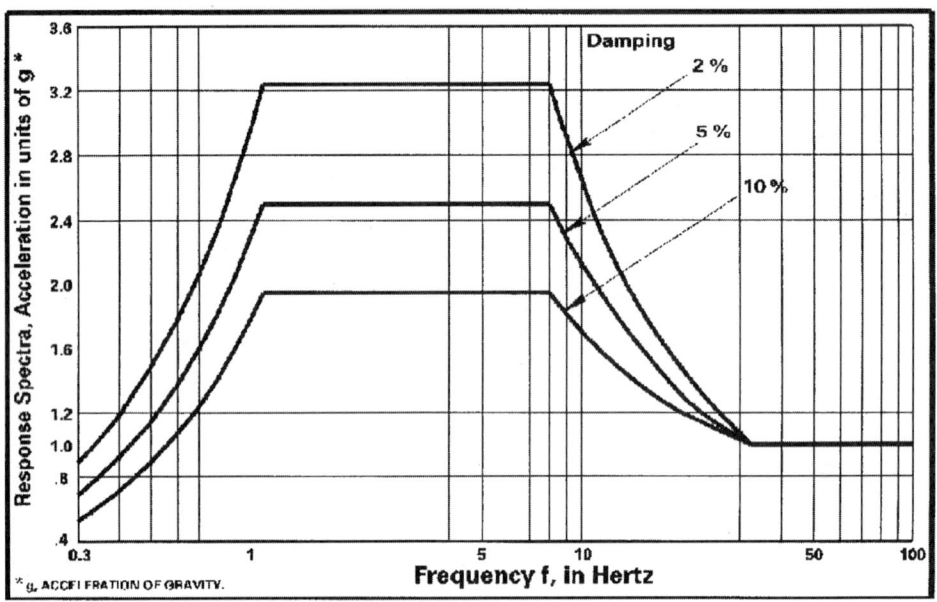

FIGURE 5.81 High seismic performance level.

very soft soils and sites located on moderate to steep slopes may not be adequately covered by these spectral shapes. Equipment that is shown by this practice to perform acceptably in ground shaking up to the *High* seismic performance level is understood to be seismically qualified to the *High* level. The *High* seismic performance level is shown in Fig. 5.81 with different damping percentages.

Equipment that has demonstrated acceptable performance during a *Moderate* event is said to be seismically qualified to the *Moderate* level. The *Moderate* seismic performance level is shown in Fig. 5.82

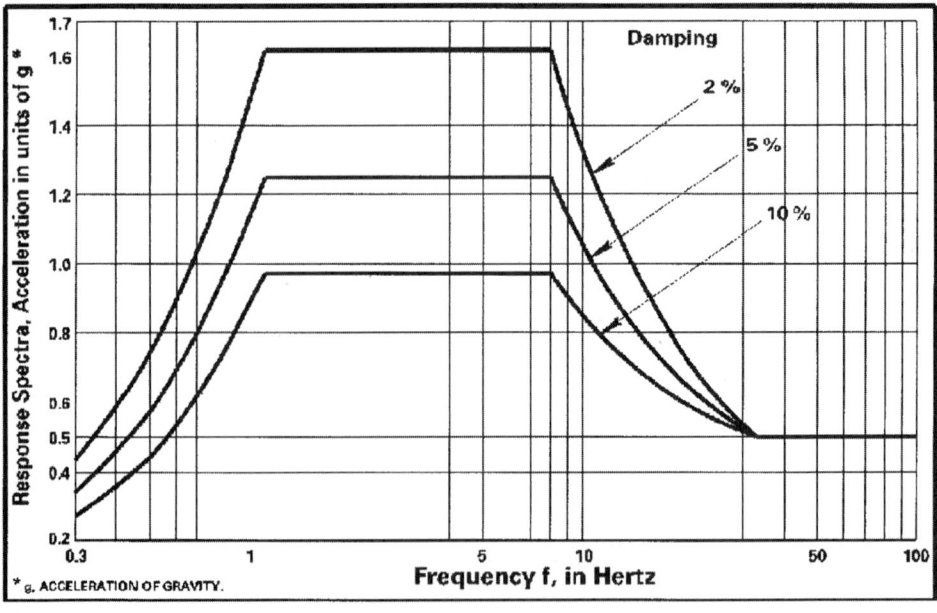

FIGURE 5.82 Moderate seismic performance level.

with different damping percentages. Finally, equipment that has demonstrated acceptable performance during a *Low* event is said to be seismically qualified to the *Low* level.

The *Low* seismic performance level represents the performance that can be expected when good construction practices are used and no special consideration is given to seismic performance. In general, it is expected that the majority of equipment will have acceptable performance at 0.1g or less. However, it should be noted that live tank circuit breakers in voltage classifications of 500,000 volts and higher that have not been designed for earthquake loading, but have been seismically qualified at higher performance levels by testing or analysis, may be damaged even at low acceleration levels.

The performance level for a site is determined by using either an earthquake hazard map or seismic exposure map for the appropriate part of North America as specified in IEEE 693. For example, in the U.S., the procedure to select the appropriate seismic qualification level for a site using the earthquake hazard map method consists of the following steps:

1. Establish the probabilistic earthquake hazard exposure of the site where the equipment will be placed. Use the site-specific peak ground acceleration developed in a study of the site's seismic hazard, selected at a 2% probability of exceedance in 50 years, modified for site soil conditions.
2. Compare the resulting site-specific peak acceleration value and spectral acceleration with the three seismic performance levels *High*, *Moderate*, or *Low* that best accommodates the expected ground motions. If the peak ground acceleration is equal to or less than 0.1g, the site is classified as *Low*. If the peak ground acceleration is greater than 0.1g, but equal or less than 0.5g, the site is classified as *Moderate*. If the peak ground acceleration is greater than 0.5g, the site is classified as *High*. This level then specifies the seismic qualification level used for procurement.

When selecting the qualification level based on performance levels, it should be remembered that performance levels represent levels of ruggedness based on testing at lower levels combined with factors of safety for material, or analysis combined with experience from previous earthquakes. These performance levels therefore have an inherent degree of uncertainty. For better assurance of structural performance during an earthquake, owners or operators may require that the qualification spectra be increased from *Low* to *Moderate* or from *Moderate* to *High* to better fit the equipment performance level that they desire. The owners or operators should carefully weigh the benefits of deviating from the criteria specified herein against the added costs.

The earthquake hazard method is the preferred approach and can be used at any site, but the seismic exposure map method can be undertaken utilizing the NEHRP-1997 seismic exposure maps available in the United States. The NEHRP-1997 maps provide spectral acceleration levels at several response spectra periods of ground motion vibration for several probability levels.

To select the appropriate seismic qualification level using this method, the following steps are suggested:

1. Determine the soil classification of the site (NEHRP-1997 soil type A, B, C, D, E, or F).
2. Locate the site on the NEHRP-1997 seismic hazard map showing peak acceleration (%g) with 2% probability of exceedance in 50 years on a soft rock site.
3. Estimate the site's peak soft rock acceleration from the 2% map.
4. Multiply that value by the Fa value for the site soil conditions as a function of the peak acceleration for soft rock.
5. Select the performance level from the resulting value, which is the site peak ground acceleration. If the peak ground acceleration is less than or equal to 0.1g, the site is classified as *Low*. If the peak ground acceleration is greater than 0.1g, but less than 0.5g, the site is classified as *Moderate*. If the peak ground acceleration is greater than 0.5g, the site is classified as *High*.

Similar methods for evaluating seismic qualification methods used in Canada and Mexico are also given in IEEE 693 with appropriate county specific references and maps as required. Other countries may use a method similar to those given or, preferably use the hazard method.

Judgment and experience must be exercised when selecting the performance level for seismic qualification as the site hazard may not fall directly on the *High*, *Moderate*, or *Low* seismic performance level and a strategy on accepting more or less risk will be required. It is recommended that large blocks of service areas be dedicated to a single performance level to increase post-event performance consistency, interchangeability and to help reduce costs through bulk purchases. For existing facilities it will mean increased efficiency in upgrade or repair design work that may be required. Additional operational requirements must also be considered when selecting equipment for an active inventory of an operating utility. The owner/operator must therefore evaluate all of the sites in the service territory and establish a master plan, designating the required (or desired, as the case may be) performance level of each site, and prioritizing those sites which need to be upgraded to meet current standards. Likewise, after a site for new electrical equipment has been identified, the owner or operator's agent must determine the appropriate seismic performance level.

If the seismic response spectra for a specific site falls outside of the response spectra indicated in Figs. 5.81 and 5.82, then a more appropriate response spectra will have to be developed for use by the owner or operator. If the new response spectra falls outside the ones defined in IEEE 693, then the basic procedure laid out in the rest of the decision-making process of Fig. 5.80 can still be followed. However, the *High*, *Moderate*, and *Low* levels specified in IEEE 693 should be used without deviation, unless it is very clear that one of the performance levels will not adequately represent the site or sites. Note that if the owner or operator elects to modify or develop a spectra that differs from those given, the user will lose the benefits of the standardization.

It is often not practical or cost effective to test to the *High* or *Moderate* performance level because:

1. Test laboratories may not be able to attain these acceleration levels, especially at low frequencies.
2. More importantly, the yield strength of the in-service ductile materials may be considered acceptable at the performance level, and testing to a higher performance level could lead to damage of components resulting in an unnecessary financial loss.

For these reasons, the equipment may be tested at 50% of the required performance level. For consistency, analysis will also be performed at 50% of the performance level. This reduced level is called the *required response spectra* (RRS). For the *High* level, compare Fig. 5.81 to Fig. 5.83 and for the *Moderate* level, compare Fig. 5.82 to Fig. 5.84.

The ratio of performance level (PL) to required response spectra (RRS) in this practice is 2.0. This factor is called the *Performance Factor (PF)*, that is, the performance factor is PF = PL/RRS. The performance factor does not apply to the *Low* seismic level.

Equipment that is tested or analyzed to the required response spectra is expected to perform acceptably at that performance level. This is achieved by measuring the stresses in the component obtained from the test or from the analysis at the required response spectra and by:

1. Comparing the measured or calculated stresses to 50% of the ultimate strength of the porcelain or cast aluminum component. The 50% is the inverse of the performance factor.
2. Using a lower factor of safety against yield combined with an allowance for ductility for steel and other ductile materials.
3. Considering that for most materials, the hysteresis damping capability increases at the higher levels of stress normally associated with higher levels of shaking.

Theoretically, for the reasons stated, components qualified using the *Moderate* or *High* required response spectra should be able to withstand ground shaking at the respective performance level. It is cautioned that this approach is dependent upon identifying the locations with the highest stresses within an individual piece of equipment and monitoring the stresses at these locations during testing or analysis. If the testing or analysis is not carried out in this manner, the critical locations within the equipment may fail prematurely during a seismic event. In addition to these considerations, the response of the equipment to the dynamic load may change between the required response spectra and the performance level. If this is not anticipated, premature failures may occur.

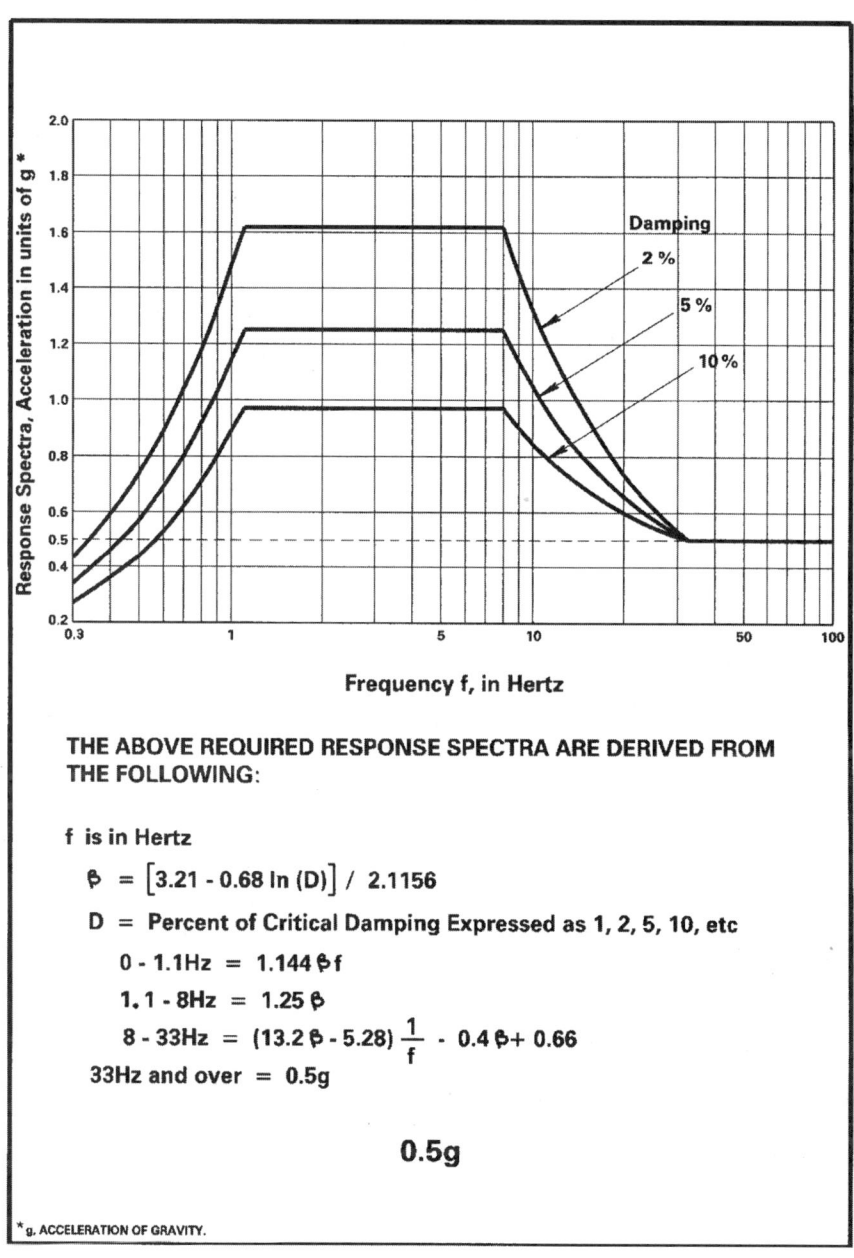

FIGURE 5.83 High required response level.

The above discussion pertains to the structural performance of the equipment. Qualification by analysis provides no assurance of electrical function. Shake-table testing provides assurance for only those electrical functions verified by electrical testing and only to the required response spectra level, not to the performance level.

Shake-table testing may be required for equipment that in previous years was qualified by dynamic analysis, but performed poorly during past earthquakes. However, static or static coefficient analysis may still be specified when past seismic performance of equipment qualified by such methods has led to acceptable performance.

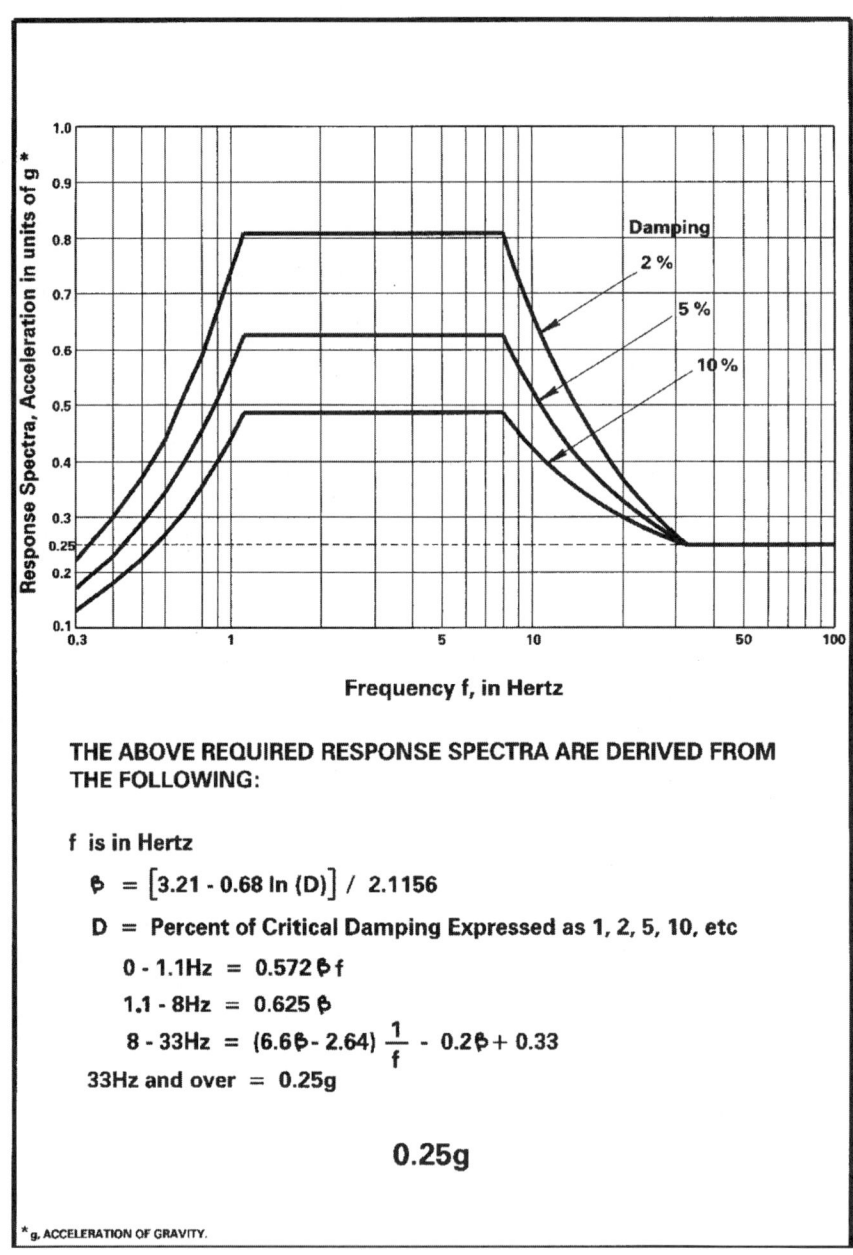

FIGURE 5.84 Moderate required response level.

High and Moderate Levels

The *High* and *Moderate* required response spectra are given in Figs. 5.83 and 5.84, respectively. The required response spectra spectral shape is the same spectral shape as is used in the performance, except at 50% of the performance level. The equations for the spectra are listed in Figs. 5.83 and 5.84.

Low Level

A rigorous seismic qualification, such as that required to meet the *High* and *Moderate* performance levels, is not required for equipment qualified to the *Low* performance level. That is, no required response spectrum or seismic report is required. However, the following criteria should be met:

1. Anchorage for the *Low* seismic performance level shall be capable of withstanding at least 0.2 times the equipment weight applied in one horizontal direction, combined with 0.16 times the weight applied in the vertical direction at the center of gravity of the equipment and support. The resultant load shall be combined with the maximum normal operating load and dead load to develop the greatest stress on the anchorage. The anchorage should be designed using the requirements specified in IEEE 693 and ASCE.
2. The equipment and its support structure shall have a well-defined load path. The determination of the load path shall be established so that it describes the transfer of loads generated by, or transmitted to, the equipment from the point of origin of the load to the anchorage of the supplied equipment. Among the forces that shall be considered are seismic (simultaneous triaxial loading — two horizontal and one vertical), gravitational, and normal operating loads. The load path shall not include:
   - Sacrificial collapse members
   - Materials that will undergo nonelastic deformations, unrestrained translation, or rotational degrees of freedom
   - Solely friction-dependent restraint (control energy dissipating devices excepted)

Qualification Process

Once the performance level has been established, the testing or analysis as required in the IEEE 693 document must be undertaken. For example, to qualify a 138,000-volt circuit breaker to meet the *Moderate* seismic qualification level, the following criteria, as specified in IEEE 693, must successfully demonstrate:

1. The seismic withstand capability shall be demonstrated by performing a dynamic analysis, and the analyzed equipment shall include the control cabinet, stored energy sources, and the associated current transformer assuming this equipment is on the same support structure.
2. The circuit breaker and the supporting structure shall be designed so that there will be no damage during and following the seismic event.
3. The response spectrum shown in Fig. 5.84 shall be used in the analysis.

The IEEE 693 document also provides guidance on the following for this piece of equipment:

1. General requirements for dynamic analysis.
2. General and detailed qualification procedures required and when the qualification is considered acceptable.
3. Equipment and support design.
4. A report analysis checklist.
5. Information on how to include base isolation and other damping systems in the analysis.
6. Recommendations on what seismic information should be listed on the equipment identification plate.

This IEEE 693 document also contains similar material for nearly all other substation components.

Because of the way IEEE 693 is written, the information necessary to be provided in the users tendering specifications is minimal. A few paragraphs are usually all that are necessary as the controlling language is contained in IEEE 693. Therefore, it is strongly suggested that rather than copying information from the IEEE 693 document into the user's specifications, that it is referred to in its entirety. This eliminates the possibility of a misunderstanding between the owner and the manufacturer.

Based on the results of the testing and analysis undertaken, corrective measures can be carried out to seismically upgrade the power equipment in question. The maximum amount of equipment displacement is also determined from these tests or dynamic analysis.

The final step of the decision process for power equipment stream is to determine the flexible bus interconnection required for the piece of equipment. IEEE 693 provides guidance in determining the minimum length of flexible bus, while IEEE 1527, which is still in a draft form, will provide a more detailed design procedure to follow. The basic decision-making process for substation components that are classified as power equipment is now complete.

Reference

ASCE, *Substation Structure Design Guide,* American Society of Civil Engineers, Reston, VA.
IEEE Recommended Practice for Seismic Design of Substations, IEEE Std. 693-1997.
IEEE Recommended Practice for Seismic Qualifications of Class 1E Equipment for Nuclear Power Generating Stations, IEEE Std. 344-1987 (Reaffirmed in 1993).
IEEE Recommended Practice for the Design of Flexible Buswork Located in Seismically Active Areas, IEEE Std. P1527 (Draft).
Recommended Provision for Seismic Regulations for New Building, NEHRP-1997 (National Earthquake Hazards Reduction Program), Federal Emergency Management Agency (FEMA), 1997.

5.13 Substation Fire Protection

Al Bolger and Don Delcourt

The occurrence and risk of fire in substations has been historically low, but the possible impacts of a fire can be catastrophic. Fires in substations can severely impact the supply of power to customers and the utility company's revenue and assets. These fires can also create fire risks to utility personnel, emergency personnel, and the general public. The recognition of the fire hazards, the risks involved and the appropriate fire protection mitigation measures are some of the key considerations for the design and operation of new or existing substations.

The purpose of this document is to act as an overview to the substation designer to help identify fire hazards within the substation, to identify remedial measures, and to evaluate the value of incorporating these measures. It is only an overview and is not intended to be all-inclusive or to provide the necessary details to do the work. For further details and information on this topic, it is recommended that the designer refer to IEEE Std. 979-1994, which identifies the standards and clauses that apply to each specific area.

Fire Hazards

The physical objects or conditions that create latent (undeveloped) demands for fire protection are called hazards. Every hazard has the following attributes:

- a probability that a fire will actually occur during a specified time interval
- the magnitude of a possible fire
- the consequence of the potential loss

For the purpose of this overview, substation fire hazards are be broken down into two areas of occurrence: the switchyard and control buildings, and indoor substations.

Switchyard and Control Buildings

Switchyard: The main components of most substation switchyards that create a fire hazard are the electrical cables and the mineral oil-insulated electrical equipment. However, in some switchyards there are a number of other components that must be considered, such as:

- Hydrogen-cooled synchronous condensers
- Service buildings
- Standby diesel generator buildings
- Breaker air storage buildings
- Miscellaneous storage

Control building: The components of most control buildings that create a fire hazard are cables, electrical panels, and station batteries. Some of the bigger substation control buildings include:

- Standby diesel generators
- Batteries
- Air compressors
- Service areas
- Miscellaneous storage

Indoor Substations

Indoor substations generally include the same hazards as outdoor switchyard stations, but because they are often enclosed in a building, they pose a much greater threat to personnel, equipment, and the building structure. Mineral oil-insulated equipment and oil-insulated cables in a building can create a major hazard.

Hazards

Cables are a major hazard because they are a combination of fuel supply and ignition source. A cable failure can result in sufficient heat to ignite the cable insulation that could continue to burn and produce high heat and large quantities of toxic smoke. Oil insulated cables are an even greater hazard since the oil increases the fuel load and spill potential.

The hazard created by mineral oil-insulated equipment such as transformer, reactors, and circuit breakers is that the oil is a significant fuel supply that can be ignited by an electrical failure within the equipment. Infiltration of water, failure of core insulation, exterior fault currents, and tap changer failures are some of the causes of internal arcing within the mineral insulating oil that can result in fire. This arcing can produce breakdown gases such as acetylene and hydrogen. Depending on the type of failure and its severity, the gases can build up sufficient pressure to cause the external shell of the transformer tank or ceramic bushings to fail or rupture. Once the tank or bushing fails, there is a strong likelihood that a fire will occur.

The other hazards such as diesel generators and air compressors are not present at all substations, but where they are present, they do contain a combination of significant fuel load combined with an ignition source.

Miscellaneous storage and shops are one of the more difficult hazards to deal with due to their transient nature. These hazards could be in the form of stored wooden crates, spare cables, oil, etc., along with various types of shops such as carpentry, electrical, and civil shops. It is difficult to deal with these types of hazards because of their changing nature and because staff often do not recognize the potential threat.

Fire Protection Measures

Fire protection measures are broken down into passive, active, and manual measures.

Passive Measures

Passive measures are measures that normally control the spread of fire. These measures are the most frequently used methods of protecting life and property in buildings from a fire. This protection confines a fire to a limited area or insures that the structure remains sound for a designated period of fire exposure. Its popularity is based on the reliability of this mode of protection since it does not require human intervention or equipment operation. Common types of passive protection include fire stopping, fire separations, equipment spacing, use of noncombustible construction materials, use of low flame spread/smoke developed rated materials, substation grading, provision of gravel or crushed rock around oil-filled equipment, etc.

The degree of passive protection for building structure would be based on the occupancy of the area and the required structural integrity. The structural integrity of a building is critical in order to preserve life and property. The premature structural failure of a building before the occupants can evacuate or the fire department can suppress the fire is a major concern.

IEEE 979 includes recommendations on these measures relative to substation design.

Active Measures

Active fire protection measures are automatic fire protection measures that warn occupants of the existence of fire, extinguish, or control the fire. These measures are designed to automatically extinguish or control a fire at its earliest stage, without risking life or sacrificing property. The benefits of these systems have been universally identified and accepted by building and insurance authorities. Insurance companies have found significant reductions in losses when automatic suppression systems have been installed.

An automatic suppression system consists of an extinguishing agent supply, control valves, a delivery system, and fire detection and control equipment. The agent supply may be virtually unlimited (such as with a city water supply for a sprinkler system) or of limited quantity (such as with water tank supply for a sprinkler system). Typical examples of agent control valves are deluge valves, sprinkler valves, and halon control valves. The agent delivery systems are a configuration of piping, nozzles, and generators that apply the agent in a suitable form and quantity to the hazard area (e.g., sprinkler piping and heads). Fire detection and control equipment may be either mechanical or electrical in operation.

Active systems include detection and alarm, wet, dry, and pre-action sprinklers, deluge systems, foam systems, halon systems, halon substitute systems (such as Inergen), and CO_2 systems.

Detailed descriptions of each of these systems, Code references, and recommendations on application are covered in IEEE 979.

Manual Measures

Manual measures include items such as the various types of fire extinguishers, fire hydrants, hose stations, etc. requiring active participation by staff or the fire department in order to control a fire.

Typical readily accessible portable fire equipment is provided for the extinguishment of incipient stage fires by building occupants. Since the majority of fires start small, it is an advantage to extinguish them during their incipient stage to ensure that potential losses are minimized.

Hazard Assessment

Hazards are generally split into life safety measures and investment measures. The hazards analysis consists of a field review for existing substations or a design review for new stations to identify the hazards and potential threats to personnel and equipment, and to recommend measures to adequately deal with these hazards. The availability of a trained fire fighting response definitely is a major consideration into the level of fire protection recommended.

Life Safety

Life safety items generally include the fire protection areas covered under Building, Fire, or Life Safety Codes. The main objective of these Codes is to ensure that:

- the occupants are able to leave the station without being subject to hazardous or untenable conditions (radiant exposure, carbon monoxide, carbon dioxide, soot, and other gases),
- firefighters are safely able to effect a rescue and prevent the spread of fire, and
- building collapse does not endanger people (including firefighters) who are likely to be near the building.

To meet this objective, fire safety systems provide the following mitigating measures to:

- detect a fire at its earliest stage
- signal the building occupants of the fire,
- provide adequate illumination to an exit,
- provide illuminated exit signs, and
- provide fire separated exits within reasonable travel distances from all areas of a building. These exits shall terminate at the exterior of the building.

These items are considered to be mandatory by Code and should be installed without any further financial analysis.

Investment Measures

Investment measures are fire protection measures that are recommended for the protection of the assets and revenue. These measures include passive, active, and manual measures such as sprinklers, gaseous systems, fire stopping, etc. The measures are generally recommended on the basis of asset protection and revenue only. A probabilistic risk analysis and financial evaluation is done as part of the risk analysis to show whether the fire protection measures should proceed; i.e., if the potential loss due to fire is greater than the cost of providing some means of fire protection, then the fire protection measure is recommended.

Design Considerations

As part of the hazard assessment, the various alternatives and designs should be considered. It is at this stage that the type of gravel or crushed rock to be distributed around the transformers, the impact of fire protection on oil containment and other environmental measures, the potential for freezing in cold climates, explosion venting requirements, cable trench limitations, lightning protection, grounding, fuel handling systems, alarm and detection choices, separation around transformers, etc. should be reviewed. IEEE 979 provides more insight into how to deal with these various choices.

Risk Analysis

The risk analysis is the review of the investment measures in relation to the probability of fire, the potential loss due to fire, and the cost of the fire protection measures. This part of the analysis requires a reasonable database of the probability of fires for the different hazard areas, an assessment of the effectiveness of the proposed fire protection measures, the potential loss due to a fire within the equipment, and a fair degree of engineering judgment. The potential loss usually includes the potential equipment loss as well as an assessment of the lost revenue due to the outage resulting from the loss of equipment.

Once the potential loss due to a fire has been calculated, the designer should assess the effectiveness of any proposed fire protection measure and consider this to be the benefit of the measure when doing a benefit-cost analysis of providing fire protection. It may be that the right decision is to do nothing.

An example would be transformer fire protection, where potential protection measures would be deluge protection ($50 K for deluge + $150 K for a water supply), spatial separation ($20 K), barriers ($50 K), locating the transformer so that the spilled oil would not flow toward any other equipment or building, and provision of gravel around the base of the transformer ($10 K).

Assuming the deluge protection is 50% effective in saving the adjacent transformers, and based on the annual probability of a distribution transformer fire to be in the order of .0005 and the replacement time to be 1 year plus, using a 25 MW transformer with a mill rate of $25/MW, the annual potential avoided loss would be .5*.0005*365d/y*24h/d*25MW/h*25$/MW = $13,679.

Based on a 25-year life expectation for the deluge system, the PV would be approximately $140,000. Since the cost of the deluge system will be approximately $200,000, and the PV of the benefit ($140,000) is considerably lower than the cost, the recommendation would be to consider other protection measures and to recommend against the installation of a deluge system.

Conclusion

The assessment of the hazards involved with an existing or planned substation and the selection of the most appropriate fire protection are the best ways to ensure that the power supply to customers, company revenue, and assets are protected from fire. A Substation, Switchyard and Substation Control Building fire protection review checklists are included in the appendix of this section to aid in the assessment process. The IEEE *Guide for Substation Fire Protection*, 979-1994, provides an excellent guide to the assessment process.

References

IEEE Guide for Substation Fire Protection, IEEE Std. 979-1994.

Appendix — Substation Control Building Fire Protection Assessment Process Checklist

Risk Assessment

- Review the criticality of the control room and building fire loss to the substation operation and asset base.
- Review the historical frequency of fire in control buildings.

Life Safety Assessment

- Review the control room layout to ensure that the room has a minimum of two outward swinging exit doors.
- Ensure that the travel distance from any area within the control building to an exit does not exceed 100 feet.
- Ensure that exit signs are installed at each exit door.
- Review that emergency lighting is provided that will provide a minimum lighting level of 10 lux at the floor, along the exit paths.
- Review the size and number of stories of the building to ensure proper exits are provided to ensure that maximum travel distances to the exits do not exceed 100 feet.
- Determine if there are any building or fire code requirements for the installation of a fire detection system.

Fire Protection Assessment

- Review the availability of fire department response to the site.
- Review the availability of fire fighting water supply at or adjacent to the site.
- Review the adequacy of any existing Control Building fire protection.
- Review criticality of control building equipment, hazards involved, response time of station personnel and the fire department. Determine the type of equipment that will provide an acceptable, very early detection (air sampling detection) of a fire (small electronic component failure — arcing), or at an early stage with smoke detection (photoelectric detection) of a fire at a smoldering or small flame stage. Determine the type of fire suppression system that will provide acceptable equipment losses and outages; i.e., gaseous suppression systems to suppress a fire at an early stage (component loss), or sprinkler protection to suppress a fire at the stage where the loss would be restricted to a single control cabinet.
- Review the hours during which the building is occupied and the ability of site personnel to safely extinguish a fire with portable fire equipment. Determine the levels of portable fire equipment required by the local fire code and that it is suitable for safe staff operation.

Hazard Assessment

- Review the other uses (shops, offices, storage, etc.) within the control building and their exposure to the critical substation equipment.
- Review the use of combustible construction in the control building (i.e., exterior surfaces and roofs).
- Review the use of combustible interior surface finishes in the Control Room and ensure that the surface finishes have a flame spread rating of less than 25.
- Review the combustibility of any exposed cable used in the building to ensure that it meets the requirements of IEEE 383.
- Review the control room separation walls to other occupied areas to ensure that they have a fire resistance rating of a minimum of 1 hour.

Substation Switchyard Fire Protection Assessment Process Checklist

- Determine the initial electric equipment layout and equipment types.

Risk Assessment

- Review the criticality of the various pieces of equipment.
- Review types of insulating fluid used and their flammability.
- Review the historical frequency of fire for the various types of equipment.
- Review the availability of a fire department response to the site.
- Review the availability of a fire fighting water supply at or adjacent to the site.
- Review the adequacy of any existing substation fire protection.

Radiant Exposure Assessment

- Review the spacing between individual single-phase transformers and breakers with IEEE 979 Table 5.1.
- Review the spacing between large three phase transformers, banks of single-phase transformers, or groups of breakers with IEEE 979 Table 5.1.
- Review the spacing of oil-filled equipment with respect to substation buildings with IEEE 979 Table 5.2. Note that the presence of combustible surfaces and unprotected windows on exposed surfaces of the buildings may require detailed thermal radiation calculations, or the application of safety factors to the table distances. The Society of Fire Protection Engineers' *Engineering Guide for Assessing Flame Radiation to External Targets from Pool Fires* can be used as a reference for detailed thermal radiation calculations.
- Review the distances between oil-filled equipment and the property line. Note that combustible vegetation and building structures beyond the property line of the substation may be exposed to high enough heat fluxes to ignite combustible surfaces. Detailed thermal radiation calculations should be considered.
- Review the use of the various methods of fire protection discussed in IEEE 979 that will address the hazard determined in the radiant exposure assessment such as changing the type of equipment and insulating fluid used, increased spacing, provision of gravel ground cover, oil containment, fire barriers, and automatic water deluge fire protection.

Fire Spread Assessment

- Is the surface around oil-filled equipment pervious (gravel) or impervious? Use of 12-inch-thick gravel ground covers will suppress the flames from a burning oil spill fire. Impervious surfaces can allow the burning oil to form a large pool fire, which will increase the heat flux to adjacent equipment and structures.
- Is there any oil containment in place around the oil-filled equipment? Oil containment can contain pool fires and prevent their spread.
- Does the grade surrounding the oil-filled equipment slope toward the equipment or away from the oil-filled equipment toward adjacent oil-filled equipment, cable trenches, drainage facilities, or buildings? The burning oil released from ruptured oil-filled equipment can spread for significant distances if the ground surrounding the equipment has a slope greater than 1%.
- Review the use of the various methods of fire protection discussed in IEEE 979 that address the hazard determined in the fire spread assessment. These methods include changing the type of equipment and insulating fluid used, increased spacing, use of gravel ground cover, provision of oil containment, changing the grade surrounding the equipment, use of liquid tight noncombustible cable trench cover adjacent to oil-filled equipment, fire stopping of cable trench entries into control buildings, and the use of automatic water deluge fire protection.

6
Distribution Systems

William H. Kersting
New Mexico State University

6.1 Power System Loads   *Raymond R. Shoults and Larry D. Swift* 6-1

6.2 Distribution System Modeling and Analysis   *William H. Kersting* 6-11

6.3 Power System Operation and Control   *George L. Clark and Simon W. Bowen* 6-67

6
Distribution Systems

Raymond R. Shoults
University of Texas at Arlington

Larry D. Swift
University of Texas at Arlington

William H. Kersting
New Mexico State University

George L. Clark
Alabama Power Company

Simon W. Bowen
Alabama Power Company

6.1 Power System Loads 6-1
    Load Classification • Modeling Applications • Load Modeling
    Concepts and Approaches • Load Characteristics and Models •
    Static Load Characteristics • Load Window Modeling

6.2 Distribution System Modeling and Analysis 6-11
    Modeling • Analysis

6.3 Power System Operation and Control 6-67
    Implementation of Distribution Automation • Distribution
    SCADA History • SCADA System Elements • Tactical and
    Strategic Implementation Issues • Distribution Management
    Platform • Trouble Management Platform • Practical
    Considerations

6.1 Power System Loads

Raymond R. Shoults and Larry D. Swift

The physical structure of most power systems consists of generation facilities feeding bulk power into a high-voltage bulk transmission network, that in turn serves any number of distribution substations. A typical distribution substation will serve from one to as many as ten feeder circuits. A typical feeder circuit may serve numerous loads of all types. A light to medium industrial customer may take service from the distribution feeder circuit primary, while a large industrial load complex may take service directly from the bulk transmission system. All other customers, including residential and commercial, are typically served from the secondary of distribution transformers that are in turn connected to a distribution feeder circuit. Figure 6.1 illustrates a representative portion of a typical configuration.

Load Classification

The most common classification of electrical loads follows the billing categories used by the utility companies. This classification includes residential, commercial, industrial, and other. Residential customers are domestic users, whereas commercial and industrial customers are obviously business and industrial users. Other customer classifications include municipalities, state and federal government agencies, electric cooperatives, educational institutions, etc.

Although these load classes are commonly used, they are often inadequately defined for certain types of power system studies. For example, some utilities meter apartments as individual residential customers, while others meter the entire apartment complex as a commercial customer. Thus, the common classifications overlap in the sense that characteristics of customers in one class are not unique to that class. For this reason some utilities define further subdivisions of the common classes.

A useful approach to classification of loads is by breaking down the broader classes into individual load components. This process may altogether eliminate the distinction of certain of the broader classes,

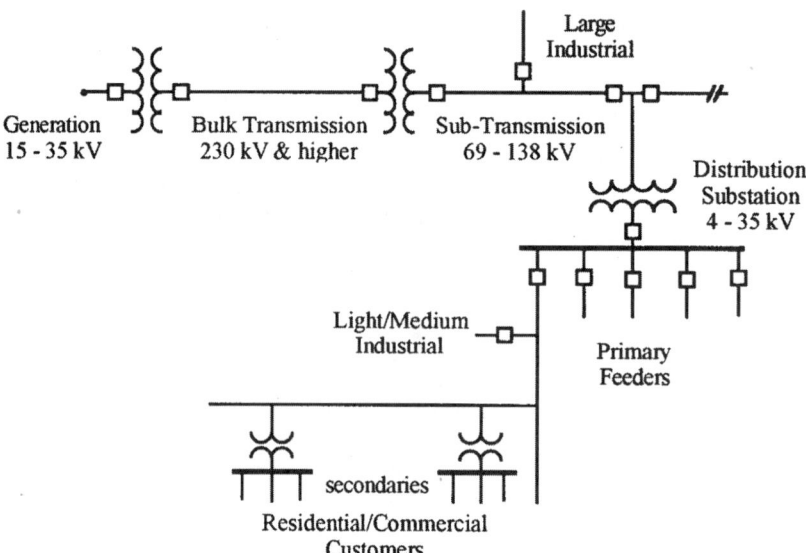

FIGURE 6.1 Representative portion of a typical power system configuration.

but it is a tried and proven technique for many applications. The components of a particular load, be it residential, commercial, or industrial, are individually defined and modeled. These load components as a whole constitute the composite load and can be defined as a "load window."

Modeling Applications

It is helpful to understand the applications of load modeling before discussing particular load characteristics. The applications are divided into two broad categories: static ("snap-shot" with respect to time) and dynamic (time varying). Static models are based on the steady-state method of representation in power flow networks. Thus, static load models represent load as a function of voltage magnitude. Dynamic models, on the other hand, involve an alternating solution sequence between a time-domain solution of the differential equations describing electromechanical behavior and a steady-state power flow solution based on the method of phasors. One of the important outcomes from the solution of dynamic models is the time variation of frequency. Therefore, it is altogether appropriate to include a component in the static load model that represents variation of load with frequency. The lists below include applications outside of Distribution Systems but are included because load modeling at the distribution level is the fundamental starting point.

Static applications: Models that incorporate only the voltage-dependent characteristic include the following.

- Power flow (PF)
  - Distribution power flow (DPF)
  - Harmonic power flow (HPF)
  - Transmission power flow (TPF)
- Voltage stability (VS)

Dynamic applications: Models that incorporate both the voltage- and frequency-dependent characteristics include the following.

- Transient stability (TS)
- Dynamic stability (DS)
- Operator training simulators (OTS)

Strictly power-flow based solutions utilize load models that include only voltage dependency characteristics. Both voltage and frequency dependency characteristics can be incorporated in load modeling for those hybrid methods that alternate between a time-domain solution and a power flow solution, such as found in Transient Stability and Dynamic Stability Analysis Programs, and Operator Training Simulators.

Load modeling in this section is confined to static representation of voltage and frequency dependencies. The effects of rotational inertia (electromechanical dynamics) for large rotating machines are discussed in Chapters 11 and 12. Static models are justified on the basis that the transient time response of most composite loads to voltage and frequency changes is fast enough so that a steady-state response is reached very quickly.

Load Modeling Concepts and Approaches

There are essentially two approaches to load modeling: component based and measurement based. Load modeling research over the years has included both approaches (EPRI, 1981; 1984; 1985). Of the two, the component-based approach lends itself more readily to model generalization. It is generally easier to control test procedures and apply wide variations in test voltage and frequency on individual components.

The component-based approach is a "bottom-up" approach in that the different load component types comprising load are identified. Each load component type is tested to determine the relationship between real and reactive power requirements versus applied voltage and frequency. A load model, typically in polynomial or exponential form, is then developed from the respective test data. The range of validity of each model is directly related to the range over which the component was tested. For convenience, the load model is expressed on a per-unit basis (i.e., normalized with respect to rated power, rated voltage, rated frequency, rated torque if applicable, and base temperature if applicable). A composite load is approximated by combining appropriate load model types in certain proportions based on load survey information. The resulting composition is referred to as a "load window."

The measurement approach is a "top-down" approach in that measurements are taken at either a substation level, feeder level, some load aggregation point along a feeder, or at some individual load point. Variation of frequency for this type of measurement is not usually performed unless special test arrangements can be made. Voltage is varied using a suitable means and the measured real and reactive power consumption recorded. Statistical methods are then used to determine load models. A load survey may be necessary to classify the models derived in this manner. The range of validity for this approach is directly related to the realistic range over which the tests can be conducted without damage to customers' equipment. Both the component and measurement methods were used in the EPRI research projects EL-2036 (1981) and EL-3591 (1984–85). The component test method was used to characterize a number of individual load components that were in turn used in simulation studies. The measurement method was applied to an aggregate of actual loads along a portion of a feeder to verify and validate the component method.

Load Characteristics and Models

Static load models for a number of typical load components appear in Tables 6.1 and 6.2 (EPRI 1984–85). The models for each component category were derived by computing a weighted composite from test results of two or more units per category. These component models express per-unit real power and reactive power as a function of per-unit incremental voltage and/or incremental temperature and/or per-unit incremental torque. The incremental form used and the corresponding definition of variables are outlined below:

$\Delta V = V_{act} - 1.0$ (incremental voltage in per unit)
$\Delta T = T_{act} - 95°F$ (incremental temperature for Air Conditioner model)
$\quad\quad = T_{act} - 47°F$ (incremental temperature for Heat Pump model)
$\Delta \tau = \tau_{act} - \tau_{rated}$ (incremental motor torque, per unit)

If ambient temperature is known, it can be used in the applicable models. If it is not known, the temperature difference, ΔT, can be set to zero. Likewise, if motor load torque is known, it can be used in the applicable models. If it is not known, the torque difference, $\Delta \tau$, can be set to zero.

Based on the test results of load components and the developed real and reactive power models as presented in these tables, the following comments on the reactive power models are important.

- The reactive power models vary significantly from manufacturer to manufacturer for the same component. For instance, four load models of single-phase central air-conditioners show a Q/P ratio that varies between 0 and 0.5 at 1.0 p.u. voltage. When the voltage changes, the $\Delta Q/\Delta V$ of each unit is quite different. This situation is also true for all other components, such as refrigerators, freezers, fluorescent lights, etc.
- It has been observed that the reactive power characteristic of fluorescent lights not only varies from manufacturer to manufacturer, from old to new, from long tube to short tube, but also varies from capacitive to inductive depending upon applied voltage and frequency. This variation makes it difficult to obtain a good representation of the reactive power of a composite system and also makes it difficult to estimate the $\Delta Q/\Delta V$ characteristic of a composite system.
- The relationship between reactive power and voltage is more non-linear than the relationship between real power and voltage, making Q more difficult to estimate than P.
- For some of the equipment or appliances, the amount of Q required at the nominal operating voltage is very small; but when the voltage changes, the change in Q with respect to the base Q can be very large.
- Many distribution systems have switchable capacitor banks either at the substations or along feeders. The composite Q characteristic of a distribution feeder is affected by the switching strategy used in these banks.

Static Load Characteristics

The component models appearing in Tables 6.1 and 6.2 can be combined and synthesized to create other more convenient models. These convenient models fall into two basic forms: exponential and polynomial.

Exponential Models

The exponential form for both real and reactive power is expressed in Eqs. (6.1) and (6.2) below as a function of voltage and frequency, relative to initial conditions or base values. Note that neither temperature nor torque appear in these forms. Assumptions must be made about temperature and/or torque values when synthesizing from component models to these exponential model forms.

$$P = P_o \left[\frac{V}{V_o}\right]^{\alpha_v} \left[\frac{f}{f_o}\right]^{\alpha_f} \tag{6.1}$$

$$Q = Q_o \left[\frac{V}{V_o}\right]^{\beta_v} \left[\frac{f}{f_o}\right]^{\beta_f} \tag{6.2}$$

The per-unit models of Eqs. (6.1) and (6.2) are as follows.

$$P_u = \frac{P}{P_o} = \left[\frac{V}{V_o}\right]^{\alpha_v} \left[\frac{f}{f_o}\right]^{\alpha_f} \tag{6.3}$$

Distribution Systems

TABLE 6.1 Static Models of Typical Load Components — AC, Heat Pump, and Appliances

Load Component	Static Component Model
1-φ Central Air Conditioner	$P = 1.0 + 0.4311 \ast \Delta V + 0.9507 \ast \Delta T + 2.070 \ast \Delta V^2 + 2.388 \ast \Delta T^2 - 0.900 \ast \Delta V \ast \Delta T$
	$Q = 0.3152 + 0.6636 \ast \Delta V + 0.543 \ast \Delta V^2 + 5.422 \ast \Delta V^3 + 0.839 \ast \Delta T^2 - 1.455 \ast \Delta V \ast \Delta T$
3-φ Central Air Conditioner	$P = 1.0 + 0.2693 \ast \Delta V + 0.4879 \ast \Delta T + 1.005 \ast \Delta V^2 - 0.188 \ast \Delta T^2 - 0.154 \ast \Delta V \ast \Delta T$
	$Q = 0.6957 + 2.3717 \ast \Delta V + 0.0585 \ast \Delta T + 5.81 \ast \Delta V^2 + 0.199 \ast \Delta T^2 - 0.597 \ast \Delta V \ast \Delta T$
Room Air Conditioner (115V Rating)	$P = 1.0 + 0.2876 \ast \Delta V + 0.6876 \ast \Delta T + 1.241 \ast \Delta V^2 + 0.089 \ast \Delta T^2 - 0.558 \ast \Delta V \ast \Delta T$
	$Q = 0.1485 + 0.3709 \ast \Delta V + 1.5773 \ast \Delta T + 1.286 \ast \Delta V^2 + 0.266 \ast \Delta T^2 - 0.438 \ast \Delta V \ast \Delta T$
Room Air Conditioner (208/230V Rating)	$P = 1.0 + 0.5953 \ast \Delta V + 0.5601 \ast \Delta T + 2.021 \ast \Delta V^2 + 0.145 \ast \Delta T^2 - 0.491 \ast \Delta V \ast \Delta T$
	$Q = 0.4968 + 2.4456 \ast \Delta V + 0.0737 \ast \Delta T + 8.604 \ast \Delta V^2 - 0.125 \ast \Delta T^2 - 1.293 \ast \Delta V \ast \Delta T$
3-φ Heat Pump (Heating Mode)	$P = 1.0 + 0.4539 \ast \Delta V + 0.2860 \ast \Delta T + 1.314 \ast \Delta V^2 - 0.024 \ast \Delta V \ast \Delta T$
	$Q = 0.9399 + 3.013 \ast \Delta V - 0.1501 \ast \Delta T + 7.460 \ast \Delta V^2 - 0.312 \ast \Delta T^2 - 0.216 \ast \Delta V \ast \Delta T$
3-φ Heat Pump (Cooling Mode)	$P = 1.0 + 0.2333 \ast \Delta V + 0.5915 \ast \Delta T + 1.362 \ast \Delta V^2 + 0.075 \ast \Delta T^2 - 0.093 \ast \Delta V \ast \Delta T$
	$Q = 0.8456 + 2.3404 \ast \Delta V - 0.1806 \ast \Delta T + 6.896 \ast \Delta V^2 + 0.029 \ast \Delta T^2 - 0.836 \ast \Delta V \ast \Delta T$
1-φ Heat Pump (Heating Mode)	$P = 1.0 + 0.3953 \ast \Delta V + 0.3563 \ast \Delta T + 1.679 \ast \Delta V^2 + 0.083 \ast \Delta V \ast \Delta T$
	$Q = 0.3427 + 1.9522 \ast \Delta V - 0.0958 \ast \Delta T + 6.458 \ast \Delta V^2 - 0.225 \ast \Delta T^2 - 0.246 \ast \Delta V \ast \Delta T$
1-φ Heat Pump (Cooling Mode)	$P = 1.0 + 0.3630 \ast \Delta V + 0.7673 \ast \Delta T + 2.101 \ast \Delta V^2 + 0.122 \ast \Delta T^2 - 0.759 \ast \Delta V \ast \Delta T$
	$Q = 0.3605 + 1.6873 \ast \Delta V + 0.2175 \ast \Delta T + 10.055 \ast \Delta V^2 - 0.170 \ast \Delta T^2 - 1.642 \ast \Delta V \ast \Delta T$
Refrigerator	$P = 1.0 + 1.3958 \ast \Delta V + 9.881 \ast \Delta V^2 + 84.72 \ast \Delta V^3 + 293 \ast \Delta V^4$
	$Q = 1.2507 + 4.387 \ast \Delta V + 23.801 \ast \Delta V^2 + 1540 \ast \Delta V^3 + 555 \ast \Delta V^4$
Freezer	$P = 1.0 + 1.3286 \ast \Delta V + 12.616 \ast \Delta V^2 + 133.6 \ast \Delta V^3 + 380 \ast \Delta V^4$
	$Q = 1.3810 + 4.6702 \ast \Delta V + 27.276 \ast \Delta V^2 + 293.0 \ast \Delta V^3 + 995 \ast \Delta V^4$
Washing Machine	$P = 1.0 + 1.2786 \ast \Delta V + 3.099 \ast \Delta V^2 + 5.939 \ast \Delta V^3$
	$Q = 1.6388 + 4.5733 \ast \Delta V + 12.948 \ast \Delta V^2 + 55.677 \ast \Delta V^3$
Clothes Dryer	$P = 1.0 - 0.1968 \ast \Delta V - 3.6372 \ast \Delta V^2 - 28.32 \ast \Delta V^3$
	$Q = 0.209 + 0.5180 \ast \Delta V + 0.363 \ast \Delta V^2 - 4.7574 \ast \Delta V^3$
Television	$P = 1.0 + 1.2471 \ast \Delta V + 0.562 \ast \Delta V^2$
	$Q = 0.2431 + 0.9830 \ast \Delta V + 1.647 \ast \Delta V^2$
Fluorescent Lamp	$P = 1.0 + 0.6534 \ast \Delta V - 1.65 \ast \Delta V^2$
	$Q = -0.1535 - 0.0403 \ast \Delta V + 2.734 \ast \Delta V^2$
Mercury Vapor Lamp	$P = 1.0 + 0.1309 \ast \Delta V + 0.504 \ast \Delta V^2$
	$Q = -0.2524 + 2.3329 \ast \Delta V + 7.811 \ast \Delta V^2$
Sodium Vapor Lamp	$P = 1.0 + 0.3409 \ast \Delta V - 2.389 \ast \Delta V^2$
	$Q = 0.060 + 2.2173 \ast \Delta V + 7.620 \ast \Delta V^2$
Incandescent	$P = 1.0 + 1.5209 \ast \Delta V + 0.223 \ast \Delta V^2$
	$Q = 0.0$
Range with Oven	$P = 1.0 + 2.1018 \ast \Delta V + 5.876 \ast \Delta V^2 + 1.236 \ast \Delta V^3$
	$Q = 0.0$
Microwave Oven	$P = 1.0 + 0.0974 \ast \Delta V + 2.071 \ast \Delta V^2$
	$Q = 0.2039 + 1.3130 \ast \Delta V + 8.738 \ast \Delta V^2$
Water Heater	$P = 1.0 + 0.3769 \ast \Delta V + 2.003 \ast \Delta V^2$
	$Q = 0.0$
Resistance Heating	$P = 1.0 + 2 \ast \Delta V + \Delta V^2$
	$Q = 0.0$

$$Q_u = \frac{Q}{P_o} = \frac{Q_o}{P_o} \left[\frac{V}{V_o}\right]^{\beta_v} \left[\frac{f}{f_o}\right]^{\beta_f} \qquad (6.4)$$

The ratio Q_o/P_o can be expressed as a function of power factor (pf) where ± indicates a lagging/leading power factor, respectively.

$$R = \frac{Q_o}{P_o} = \pm\sqrt{\frac{1}{pf^2} - 1}$$

TABLE 6.2 Static Models of Typical Load Components – Transformers and Induction Motors

Load Component	Static Component Model
Transformer Core Loss Model	$P = \dfrac{KVA(rating)}{KVA(system\ base)} \left[0.00267 V^2 + 0.73 \times 10^{-9} \times e^{13.5 V^2} \right]$
	$Q = \dfrac{KVA(rating)}{KVA(system\ base)} \left[0.00167 V^2 + 0.268 \times 10^{-13} \times e^{22.76 V^2} \right]$
	where V is voltage magnitude in per unit
1-φ Motor Constant Torque	$P = 1.0 + 0.5179^*\Delta V + 0.9122^*\Delta\tau + 3.721^*\Delta V^2 + 0.350^*\Delta\tau^2 - 1.326^*\Delta V^*\Delta\tau$
	$Q = 0.9853 + 2.7796^*\Delta V + 0.0859^*\Delta\tau + 7.368^*\Delta V^2 + 0.218^*\Delta\tau^2 - 1.799^*\Delta V^*\Delta\tau$
3-φ Motor (1-10HP) Const. Torque	$P = 1.0 + 0.2250^*\Delta V + 0.9281^*\Delta\tau + 0.970^*\Delta V^2 + 0.086^*\Delta\tau^2 - 0.329^*\Delta V^*\Delta\tau$
	$Q = 0.7810 + 2.3532^*\Delta V + 0.1023^*\Delta\tau - 5.951^*\Delta V^2 + 0.446^*\Delta\tau^2 - 1.48^*\Delta V^*\Delta\tau$
3-φ Motor (10HP/Above) Const. Torque	$P = 1.0 + 0.0199^*\Delta V + 1.0463^*\Delta\tau + 0.341^*\Delta V^2 + 0.116^*\Delta\tau^2 - 0.457^*\Delta V^*\Delta\tau$
	$Q = 0.6577 + 1.2078^*\Delta V + 0.3391^*\Delta\tau + 4.097^*\Delta V^2 + 0.289\Delta\tau^2 - 1.477^*\Delta V^*\Delta\tau$
1-φ Motor Variable Torque	$P = 1.0 + 0.7101^*\Delta V + 0.9073^*\Delta\tau + 2.13^*\Delta V^2 + 0.245^*\Delta\tau^2 - 0.310^*\Delta V^*\Delta\tau$
	$Q = 0.9727 + 2.7621^*\Delta V + 0.077^*\Delta\tau + 6.432^*\Delta V^2 + 0.174^*\Delta\tau^2 - 1.412^*\Delta V^*\Delta\tau$
3-φ Motor (1-10HP) Variable Torque	$P = 1.0 + 0.3122^*\Delta V + 0.9286^*\Delta\tau + 0.489^*\Delta V^2 + 0.081^*\Delta\tau^2 - 0.079^*\Delta V^*\Delta\tau$
	$Q = 0.7785 + 2.3648^*\Delta V + 0.1025^*\Delta\tau + 5.706^*\Delta V^2 + 0.13^*\Delta\tau^2 - 1.00^*\Delta V^*\Delta\tau$
3-φ Motor (10HP & Above) Variable Torque	$P = 1.0 + 0.1628^*\Delta V + 1.0514^*\Delta\tau \angle 0.099^*\Delta V^2 + 0.107^*\Delta\tau^2 + 0.061^*\Delta V^*\Delta\tau$
	$Q = 0.6569 + 1.2467^*\Delta V + 0.3354^*\Delta\tau + 3.685^*\Delta V^2 + 0.258^*\Delta\tau^2 - 1.235^*\Delta V^*\Delta\tau$

After substituting R for Q_o/P_o, Eq. (6.4) becomes the following.

$$Q_u = R \left[\frac{V}{V_o} \right]^{\beta_v} \left[\frac{f}{f_o} \right]^{\beta_f} \quad (6.5)$$

Eqs. (6.1) and (6.2) [or (6.3) and (6.5)] are valid over the voltage and frequency ranges associated with tests conducted on the individual components from which these exponential models are derived. These ranges are typically ±10% for voltage and ±2.5% for frequency. The accuracy of these models outside the test range is uncertain. However, one important factor to note is that in the extreme case of voltage approaching zero, both P and Q approach zero.

EPRI-sponsored research resulted in model parameters such as found in Table 6.3 (EPRI, 1987; Price et al., 1988). Eleven model parameters appear in this table, of which the exponents α and β and the power factor (pf) relate directly to Eqs. (6.3) and (6.5). The first six parameters relate to general load models, some of which include motors, and the remaining five parameters relate to nonmotor loads — typically resistive type loads. The first is load power factor (pf). Next in order (from left to right) are the exponents for the voltage (α_v, α_f) and frequency (β_v, β_f) dependencies associated with real and reactive power, respectively. N_m is the motor-load portion of the load. For example, both a refrigerator and a freezer are 80% motor load. Next in order are the power factor (pf_{nm}) and voltage (α_{vnm}, α_{fnm}) and frequency (β_{vnm}, β_{fnm}) parameters for the nonmotor portion of the load. Since the refrigerator and freezer are 80% motor loads (i.e., $N_m = 0.8$), the nonmotor portion of the load must be 20%.

Polynomial Models

A polynomial form is often used in a Transient Stability program. The voltage dependency portion of the model is typically second order. If the nonlinear nature with respect to voltage is significant, the order can be increased. The frequency portion is assumed to be first order. This model is expressed as follows.

$$P = P_o \left[a_o + a_1 \left(\frac{V}{V_o} \right) + a_2 \left(\frac{V}{V_o} \right)^2 \right] \left[1 + D_p \Delta f \right] \quad (6.6)$$

Distribution Systems

TABLE 6.3 Parameters for Voltage and Frequency Dependencies of Static Loads

Component/Parameters	pf	α_v	α_f	β_v	β_f	N_m	pf_{nm}	α_{vnm}	α_{fnm}	β_{vnm}	β_{fnm}
Resistance Space Heater	1.0	2.0	0.0	0.0	0.0	0.0	—	—	—	—	—
Heat Pump Space Heater	0.84	0.2	0.9	2.5	−1.3	0.9	1.0	2.0	0.0	0.0	0.0
Heat Pump/Central A/C	0.81	0.2	0.9	2.5	−2.7	1.0	—	—	—	—	—
Room Air Conditioner	0.75	0.5	0.6	2.5	−2.8	1.0	—	—	—	—	—
Water Heater & Range	1.0	2.0	0.0	0.0	0.0	0.0	—	—	—	—	—
Refrigerator & Freezer	0.84	0.8	0.5	2.5	−1.4	0.8	1.0	2.0	0.0	0.0	0.0
Dish Washer	0.99	1.8	0.0	3.5	−1.4	0.8	1.0	2.0	0.0	0.0	0.0
Clothes Washer	0.65	0.08	2.9	1.6	1.8	1.0	—	—	—	—	—
Incandescent Lighting	1.0	1.54	0.0	0.0	0.0	0.0	—	—	—	—	—
Clothes Dryer	0.99	2.0	0.0	3.3	−2.6	0.2	1.0	2.0	0.0	0.0	0.0
Colored Television	0.77	2.0	0.0	5.2	−4.6	0.0	—	—	—	—	—
Furnace Fan	0.73	0.08	2.9	1.6	1.8	1.0	—	—	—	—	—
Commercial Heat Pump	0.84	0.1	1.0	2.5	−1.3	0.9	1.0	2.0	0.0	0.0	0.0
Heat Pump Comm. A/C	0.81	0.1	1.0	2.5	−1.3	1.0	—	—	—	—	—
Commercial Central A/C	0.75	0.1	1.0	2.5	−1.3	1.0	—	—	—	—	—
Commercial Room A/C	0.75	0.5	0.6	2.5	−2.8	1.0	—	—	—	—	—
Fluorescent Lighting	0.90	0.08	1.0	3.0	−2.8	0.0	—	—	—	—	—
Pump, Fan, (Motors)	0.87	0.08	2.9	1.6	1.8	1.0	—	—	—	—	—
Electrolysis	0.90	1.8	−0.3	2.2	0.6	0.0	—	—	—	—	—
Arc Furnace	0.72	2.3	−1.0	1.61	−1.0	0.0	—	—	—	—	—
Small Industrial Motors	0.83	0.1	2.9	0.6	−1.8	1.0	—	—	—	—	—
Industrial Motors Large	0.89	0.05	1.9	0.5	1.2	1.0	—	—	—	—	—
Agricultural H$_2$O Pumps	0.85	1.4	5.6	1.4	4.2	1.0	—	—	—	—	—
Power Plant Auxiliaries	0.80	0.08	2.9	1.6	1.8	1.0	—	—	—	—	—

$$Q = Q_o \left[b_o + b_1 \left(\frac{V}{V_o} \right) + b_2 \left(\frac{V}{V_o} \right)^2 \right] \left[1 + D_q \, \Delta f \right] \tag{6.7}$$

where $a_o + a_1 + a_2 = 1$
$b_o + b_1 + b_2 = 1$
$D_p \equiv$ real power frequency damping coefficient, per unit
$D_q \equiv$ reactive power frequency damping coefficient, per unit
$\Delta f \equiv$ frequency deviation from scheduled value, per unit

The per-unit form of Eqs. (6.6) and (6.7) is the following.

$$P_u = \frac{P}{P_o} = \left[a_o + a_1 \left(\frac{V}{V_o} \right) + a_2 \left(\frac{V}{V_o} \right)^2 \right] \left[1 + D_p \, \Delta f \right] \tag{6.8}$$

$$Q_u = \frac{Q}{P_o} = \frac{Q_o}{P_o} \left[b_o + b_1 \left(\frac{V}{V_o} \right) + b_2 \left(\frac{V}{V_o} \right)^2 \right] \left[1 + D_q \, \Delta f \right] \tag{6.9}$$

Combined Exponential and Polynomial Models

The two previous kinds of models may be combined to form a synthesized static model that offers greater flexibility in representing various load characteristics (EPRI, 1987; Price et al., 1988). The mathematical expressions for these per-unit models are the following.

$$P_u = \frac{P_{poly} + P_{exp1} + P_{exp2}}{P_o} \quad (6.10)$$

$$Q_u = \frac{Q_{poly} + Q_{exp1} + Q_{exp2}}{P_o} \quad (6.11)$$

where

$$P_{poly} = a_0 + a_1\left(\frac{V}{V_o}\right) + a_3\left(\frac{V}{V_o}\right)^2 \quad (6.12)$$

$$P_{exp1} = a_4\left(\frac{V}{V_o}\right)^{\alpha_1}\left[1 + D_{p1}\,\Delta f\right] \quad (6.13)$$

$$P_{exp2} = a_5\left(\frac{V}{V_o}\right)^{\alpha_2}\left[1 + D_{p2}\,\Delta f\right] \quad (6.14)$$

The expressions for the reactive components have similar structures. Devices used for reactive power compensation are modeled separately.

The flexibility of the component models given here is sufficient to cover most modeling needs. Whenever possible, it is prudent to compare the computer model to measured data for the load.

Table 6.4 provides typical values for the frequency damping characteristic, D, that appears in Eqs. (6.6) through (6.9), (6.13), and (6.14) (EPRI, 1979). Note that nearly all of the damping coefficients for reactive power are negative. This means that as frequency declines, more reactive power is required which can cause an exacerbating effect for low-voltage conditions.

Comparison of Exponential and Polynomial Models

Both models provide good representation around rated or nominal voltage. The accuracy of the exponential form deteriorates when voltage significantly exceeds its nominal value, particularly with exponents (α) greater than 1.0. The accuracy of the polynomial form deteriorates when the voltage falls significantly below its nominal value when the coefficient a_o is non zero. A nonzero a_o coefficient represents some portion of the load as constant power. A scheme often used in practice is to use the polynomial form, but switch to the exponential form when the voltage falls below a predetermined value.

TABLE 6.4 Static Load Frequency Damping Characteristics

Component	Frequency Parameters	
	D_p	D_q
Three-Phase Central AC	1.09818	−0.663828
Single-Phase Central AC	0.994208	−0.307989
Window AC	0.702912	−1.89188
Duct Heater w/blowers	0.528878	−0.140006
Water Heater, Electric Cooking	0.0	0.0
Clothes Dryer	0.0	−0.311885
Refrigerator, Ice Machine	0.664158	−1.10252
Incandescent Lights	0.0	0.0
Florescent Lights	0.887964	−1.16844
Induction Motor Loads	1.6	−0.6

Distribution Systems

Devices Contributing to Modeling Difficulties

Some load components have time-dependent characteristics that must be considered if a sequence of studies using static models is performed that represents load changing over time. Examples of such a study include Voltage Stability and Transient Stability. The devices that affect load modeling by contributing abrupt changes in load over periods of time are listed below.

Protective Relays — Protective relays are notoriously difficult to model. The entire load of a substation can be tripped off line or the load on one of its distribution feeders can be tripped off line as a result of protective relay operations. At the utilization level, motors on air conditioner units and motors in many other residential, commercial, and industrial applications contain thermal and/or over-current relays whose operational behavior is difficult to predict.

Thermostatically Controlled Loads — Air conditioning units, space heaters, water heaters, refrigerators, and freezers are all controlled by thermostatic devices. The effects of such devices are especially troublesome to model when a distribution load is reenergized after an extended outage (cold-load pickup). The effect of such devices to cold-load pickup characteristics can be significant.

Voltage Regulation Devices — Voltage regulators, voltage controlled capacitor banks, and automatic LTCs on transformers exhibit time-dependent effects. These devices are present at both the bulk power and distribution system levels.

Discharge Lamps (Mercury Vapor, Sodium Vapor, and Fluorescent Lamps) — These devices exhibit time-dependent characteristics upon restart, after being extinguished by a low-voltage condition — usually about 70% to 80% of rated voltage.

Load Window Modeling

The static load models found in Tables 6.1 and 6.2 can be used to define a composite load referred to as the "load window" mentioned earlier. In this scheme, a distribution substation load or one of its feeder loads is defined in as much detail as desired for the model. Using the load window scheme, any number of load windows can be defined representing various composite loads, each having as many load components as deemed necessary for accurate representation of the load. Figure 6.2 illustrates the load window concept. The width of each subwindow denotes the percentage of each load component to the total composite load.

Construction of a load window requires certain load data be available. For example, load saturation and load diversity data are needed for various classes of customers. These data allow one to (1) identify the appropriate load components to be included in a particular load window, (2) assign their relative percentage of the total load, and (3) specify the diversified total amount of load for that window. If load modeling is being used for Transient Stability or Operator Training Simulator programs, frequency dependency can be added. Let P(V) and Q(V) represent the composite load models for P and Q,

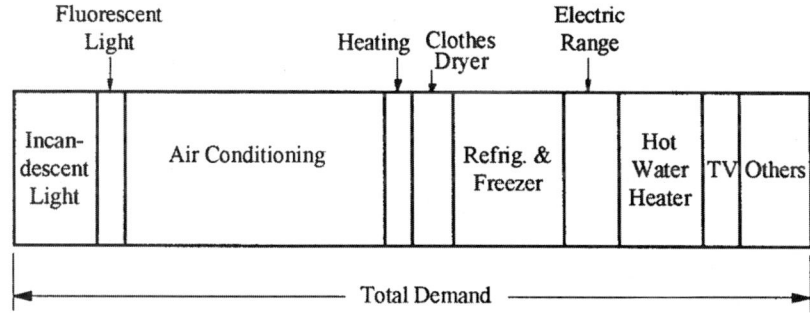

FIGURE 6.2 A typical load window with % composition of load components.

TABLE 6.5 Composition of Six Different Load Window Types

Load Window Type Load Component	LW 1 Res. 1 (%)	LW 2 Res. 2 (%)	LW 3 Res. 3 (%)	LW 4 Com 1 (%)	LW 5 Com 2 (%)	LW 6 Indust (%)
3-Phase Central AC	25	30	10	35	40	20
Window Type AC	5	0	20	0	0	0
Duct Heater with Blower	5	0	0	0	0	0
Water Heater, Range Top	10	10	10	0	0	0
Clothes Dryer	10	10	10	0	0	0
Refrigerator, Ice Machine	15	15	10	30	0	0
Incandescent Lights	10	5	10	0	0	0
Fluorescent Lights	20	30	30	25	30	10
Industrial (Induct. Motor)	0	0	0	10	30	70

respectively, with only voltage dependency (as developed using components taken from Tables 6.1 and 6.2). Frequency dependency is easily included as illustrated below.

$$P = P(V) \times (1 + D_p \Delta f)$$

$$Q = Q(V) \times (1 + D_q \Delta f)$$

Table 6.5 shows six different composite loads for a summer season in the southwestern portion of the U.S. This "window" serves as an example to illustrate the modeling process. Note that each column must add to 100%. The entries across from each component load for a given window type represent the percentage of that load making up the composite load.

References

EPRI User's Manual — Extended Transient/Midterm Stability Program Package, version 3.0, June 1992.

General Electric Company, Load modeling for power flow and transient stability computer studies, *EPRI Final Report EL-5003,* January 1987 (four volumes describing LOADSYN computer program).

Kundur, P., *Power System Stability and Control,* EPRI Power System Engineering Series, McGraw-Hill, Inc., 271–314, 1994.

Price, W. W., Wirgau, K. A., Murdoch, A., Mitsche, J. V., Vaahedi, E., and El-Kady, M. A., Load Modeling for Power Flow and Transient Stability Computer Studies, *IEEE Trans. on Power Syst.,* 3(1), 180–187, February 1988.

Taylor, C. W., *Power System Voltage Stability,* EPRI Power System Engineering Series, McGraw-Hill, Inc., 67–107, 1994.

University of Texas at Arlington, Determining Load Characteristics for Transient Performances, *EPRI Final Report EL-848,* May 1979 (three volumes).

University of Texas at Arlington, Effect of Reduced Voltage on the Operation and Efficiency of Electrical Loads, *EPRI Final Report EL-2036,* September 1981 (two volumes).

University of Texas at Arlington, Effect of Reduced Voltage on the Operation and Efficiency of Electrical Loads, *EPRI Final Report EL-3591,* June 1984 and July 1985 (three volumes).

Warnock, V. J. and Kirkpatrick, T. L., Impact of Voltage Reduction on Energy and Demand: Phase II, *IEEE Trans. on Power Syst.,* 3(2), 92–97, May 1986.

6.2 Distribution System Modeling and Analysis

William H. Kersting

Modeling

Radial distribution feeders are characterized by having only one path for power to flow from the source ("distribution substation") to each customer. A typical distribution system will consist of one or more distribution substations consisting of one or more "feeders". Components of the feeder may consist of the following:

- Three-phase primary "main" feeder
- Three-phase, two-phase ("V" phase), and single-phase laterals
- Step-type voltage regulators or load tap changing transformer (LTC)
- In-line transformers
- Shunt capacitor banks
- Three-phase, two-phase, and single-phase loads

The loading of a distribution feeder is inherently unbalanced because of the large number of unequal single-phase loads that must be served. An additional unbalance is introduced by the nonequilateral conductor spacings of the three-phase overhead and underground line segments.

Because of the nature of the distribution system, conventional power-flow and short-circuit programs used for transmission system studies are not adequate. Such programs display poor convergence characteristics for radial systems. The programs also assume a perfectly balanced system so that a single-phase equivalent system is used.

If a distribution engineer is to be able to perform accurate power-flow and short-circuit studies, it is imperative that the distribution feeder be modeled as accurately as possible. This means that three-phase models of the major components must be utilized. Three-phase models for the major components will be developed in the following sections. The models will be developed in the "phase frame" rather than applying the method of symmetrical components.

Figure 6.3 shows a simple one-line diagram of a three-phase feeder; it illustrates the major components of a distribution system. The connecting points of the components will be referred to as "nodes." Note in the figure that the phasing of the line segments is shown. This is important if the most accurate models are to be developed.

The following sections will present generalized three-phase models for the "series" components of a feeder (line segments, voltage regulators, transformer banks). Additionally, models are presented for the "shunt" components (loads, capacitor banks). Finally, the "ladder iterative technique" for power-flow studies using the models is presented along with a method for computing short-circuit currents for all types of faults.

Line Impedance

The determination of the impedances for overhead and underground lines is a critical step before analysis of distribution feeder can begin. Depending upon the degree of accuracy required, impedances can be calculated using Carson's equations where no assumptions are made, or the impedances can be determined from tables where a wide variety of assumptions are made. Between these two limits are other techniques, each with their own set of assumptions.

Carson's Equations

Since a distribution feeder is inherently unbalanced, the most accurate analysis should not make any assumptions regarding the spacing between conductors, conductor sizes, or transposition. In a classic paper, John Carson developed a technique in 1926 whereby the self and mutual impedances for **ncond**

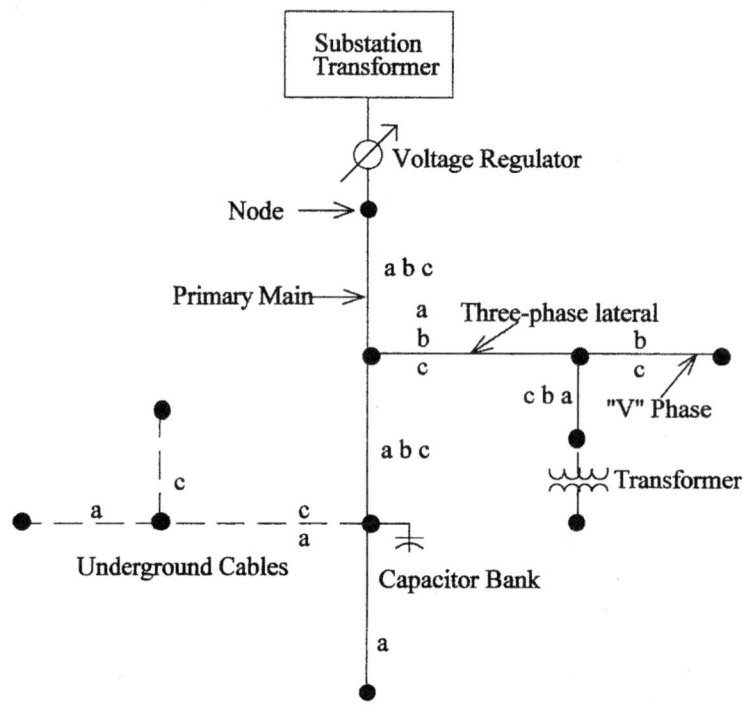

FIGURE 6.3 Distribution feeder.

overhead conductors can be determined. The equations can also be applied to underground cables. In 1926, this technique was not met with a lot of enthusiasm because of the tedious calculations that had to be done on the slide rule and by hand. With the advent of the digital computer, Carson's equations have now become widely used.

In his paper, Carson assumes the earth is an infinite, uniform solid, with a flat uniform upper surface and a constant resistivity. Any "end effects" introduced at the neutral grounding points are not large at power frequencies, and therefore are neglected. The original Carson equations are given in Eqs. (6.15) and (6.16).

Self-impedance:

$$\hat{z}_{ii} = r_i + 4\omega P_{ii} G + j\left(X_i + 2\omega G_{ii} \cdot \ln\frac{S_{ii}}{R_i} + 4\omega Q_{ii} G \right) \text{ Ohms/mile} \quad (6.15)$$

Mutual impedance:

$$\hat{z}_{ij} = 4\omega P_{ij} G + j\left(2\omega G \cdot \ln\frac{S_{ij}}{D_{ij}} + 4\omega Q_{ij} G \right) \text{ Ohms/mile} \quad (6.16)$$

where $\hat{z}_{ii}$ = self-impedance of conductor **i** in Ohms/mile
$\hat{z}_{ij}$ = mutual impedance between conductors **i** and **j** in ohms/mile
r_i = resistance of conductor **i** in Ohms/mile
ω = system angular frequency in radians per second
G = 0.1609347×10^{-7} Ohm-cm/abohm-mile
R_i = radius of conductor **i** in feet
GMR_i = geometric mean radius of conductor **i** in feet

Distribution Systems

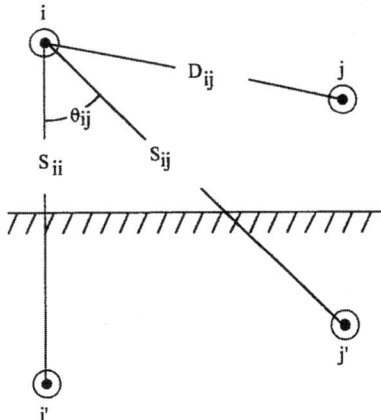

FIGURE 6.4 Conductors and images.

f = system frequency in Hertz
ρ = resistivity of earth in ohm-meters
D_{ij} = distance between conductors **i** and **j** in feet
S_{ij} = distance between conductor **i** and image **j** in feet
θ_{ij} = angle between a pair of lines drawn from conductor **i** to its own image and to the image of conductor **j**

$$X_i = 2\omega G \cdot \ln \frac{R_i}{GMR_i} \quad \text{Ohms/mile} \tag{6.17}$$

$$P_{ij} = \frac{\pi}{8} - \frac{1}{3\sqrt{2}} k_{ij} \cos(\theta_{ij}) + \frac{k_{ij}^2}{16} \cos(2\theta_{ij}) \cdot \left(0.6728 + \ln \frac{2}{k_{ij}}\right) \tag{6.18}$$

$$Q_{ij} = -0.0386 + \frac{1}{2} \ln \frac{2}{k_{ij}} + \frac{1}{3\sqrt{2}} k_{ij} \cos(\theta_{ij}) \tag{6.19}$$

$$k_{ij} = 8.565 \times 10^{-4} \cdot S_{ij} \cdot \sqrt{\frac{f}{\rho}} \tag{6.20}$$

As indicated above, Carson made use of conductor images; that is, every conductor at a given distance above ground has an image conductor the same distance below ground. This is illustrated in Fig. 6.4.

Modified Carson's Equations

Only two approximations are made in deriving the "Modified Carson Equations." These approximations involve the terms associated with P_{ij} and Q_{ij}. The approximations are shown below:

$$P_{ij} = \frac{\pi}{8} \tag{6.21}$$

$$Q_{ij} = -0.03860 + \frac{1}{2} \ln \frac{2}{k_{ij}} \tag{6.22}$$

It is also assumed:

f = frequency = 60 Hertz
ρ = resistivity = 100 Ohm-meter

Using these approximations and assumptions, Carson's equations reduce to:

$$\hat{z}_{ii} = r_i + 0.0953 + j0.12134\left(\ln\frac{1}{GMR_i} + 7.93402\right) \text{ Ohms/mile} \quad (6.23)$$

$$\hat{z}_{ij} = 0.0953 + j0.12134\left(\ln\frac{1}{D_{ij}} + 7.93402\right) \text{ Ohms/mile} \quad (6.24)$$

Overhead and Underground Lines

Equations (6.23) and (6.24) can be used to compute an **ncond × ncond** "primitive impedance" matrix. For an overhead four wire, grounded wye distribution line segment, this will result in a 4 × 4 matrix. For an underground grounded wye line segment consisting of three concentric neutral cables, the resulting matrix will be 6 × 6. The primitive impedance matrix for a three-phase line consisting of **m** neutrals will be of the form:

$$[z_{primitive}] = \begin{bmatrix} \hat{z}_{aa} & \hat{z}_{ab} & \hat{z}_{ac} & | & \hat{z}_{an1} & \cdot & \hat{z}_{anm} \\ \hat{z}_{ba} & \hat{z}_{bb} & \hat{z}_{bc} & | & \hat{z}_{bn1} & \cdot & \hat{z}_{bnm} \\ \hat{z}_{ca} & \hat{z}_{cb} & \hat{z}_{cc} & | & \hat{z}_{cn1} & \cdot & \hat{z}_{cnm} \\ --- & --- & --- & --- & --- & --- & --- \\ \hat{z}_{n1a} & \hat{z}_{n1b} & \hat{z}_{n1c} & | & \hat{z}_{n1n1} & \cdot & \hat{z}_{n1nm} \\ \cdot & \cdot & \cdot & | & \cdot & \cdot & \cdot \\ \hat{z}_{nma} & \hat{z}_{nmb} & \hat{z}_{nmc} & | & \hat{z}_{nmn1} & \cdot & \hat{z}_{nmnm} \end{bmatrix} \quad (6.25)$$

In partitioned form Eq. 6.11 becomes:

$$[z_{primitive}] = \begin{bmatrix} [\hat{z}_{ij}] & [\hat{z}_{in}] \\ [\hat{z}_{nj}] & [\hat{z}_{nn}] \end{bmatrix} \quad (6.26)$$

Phase Impedance Matrix

For most applications, the primitive impedance matrix needs to be reduced to a 3 × 3 "phase frame" matrix consisting of the self and mutual equivalent impedances for the three phases. One standard method of reduction is the "Kron" reduction (1952) where the assumption is made that the line has a multi-grounded neutral. The Kron reduction results in the "phase impedances matrix" determined by using Eq. 6.27 below:

$$[z_{abc}] = [\hat{z}_{ij}] - [\hat{z}_{in}] \cdot [\hat{z}_{nn}]^{-1} \cdot [\hat{z}_{nj}] \quad (6.27)$$

For two-phase (V-phase) and single-phase lines in grounded wye systems, the modified Carson equations can be applied, which will lead to initial 3 × 3 and 2 × 2 primitive impedance matrices. Kron reduction will reduce the matrices to 2 × 2 and a single element. These matrices can be expanded to 3 × 3 "phase frame" matrices by the addition of rows and columns consisting of zero elements for the missing phases.

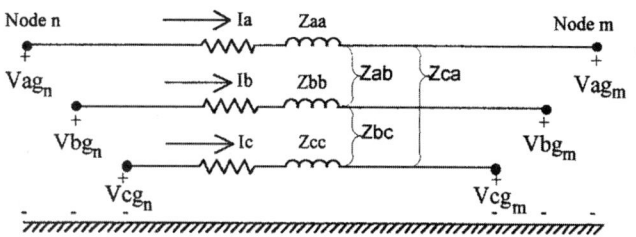

FIGURE 6.5 Three-phase line segment.

The phase frame matrix for a three-wire delta line is determined by the application of Carson's equations without the Kron reduction step.

The phase frame matrix can be used to accurately determine the voltage drops on the feeder line segments once the currents flowing have been determined. Since no approximations (transposition, for example) have been made regarding the spacing between conductors, the effect of the mutual coupling between phases is accurately taken into account. The application of Carson's equations and the phase frame matrix leads to the most accurate model of a line segment.

Figure 6.5 shows the equivalent circuit of a line segment.

The voltage equation in matrix form for the line segment is given by Eq. (6.28).

$$\begin{bmatrix} V_{ag} \\ V_{bg} \\ V_{cg} \end{bmatrix}_n = \begin{bmatrix} V_{ag} \\ V_{bg} \\ V_{cg} \end{bmatrix}_m + \begin{bmatrix} Z_{aa} & Z_{ab} & Z_{ac} \\ Z_{ba} & Z_{bb} & Z_{bc} \\ Z_{ca} & Z_{cb} & Z_{cc} \end{bmatrix} \cdot \begin{bmatrix} I_a \\ I_b \\ I_c \end{bmatrix} \qquad (6.28)$$

where $Z_{ij} = z_{ij} \cdot length$

The "phase impedance matrix" is defined in Eq. (6.29). The phase impedance matrix for single-phase and "V"-phase lines will have a row and column of zeros for each missing phase.

$$[Z_{abc}] = \begin{bmatrix} Z_{aa} & Z_{ab} & Z_{ac} \\ Z_{ba} & Z_{bb} & Z_{bc} \\ Z_{ca} & Z_{cb} & Z_{cc} \end{bmatrix} \qquad (6.29)$$

Equation (6.28) can be written in "condensed" form as:

$$[VLG_{abc}]_n = [VLG_{abc}]_m + [Z_{abc}] \cdot [I_{abc}] \qquad (6.30)$$

This condensed notation will be used throughout the document.

Sequence Impedances

Many times the analysis of a feeder will use the positive and zero sequence impedances for the line segments. There are basically two methods for obtaining these impedances. The first method incorporates the application of Carson's equations and the Kron reduction to obtain the phase frame impedance matrix. The 3 × 3 "sequence impedance matrix" can be obtained by:

$$[z_{012}] = [A_s]^{-1} \cdot [z_{abc}] \cdot [A_s] \quad \text{Ohms/mile} \qquad (6.31)$$

where

$$[A_s] = \begin{bmatrix} 1 & 1 & 1 \\ 1 & a^2 & a \\ 1 & a & a^2 \end{bmatrix} \quad (6.32)$$

$$a = 1.0\underline{/120} \quad a^2 = 1.0\underline{/240}$$

The resulting sequence impedance matrix is of the form:

$$[z_{012}] = \begin{bmatrix} z_{00} & z_{01} & z_{02} \\ z_{10} & z_{11} & z_{12} \\ z_{20} & z_{21} & z_{22} \end{bmatrix} \text{Ohms/mile} \quad (6.33)$$

where z_{00} = the zero sequence impedance
z_{11} = the positive sequence impedance
z_{22} = the negative sequence impedance

In the idealized state, the off diagonal terms of Eq. (6.33) would be zero. When the off diagonal terms of the phase impedance matrix are all equal, the off diagonal terms of the sequence impedance matrix will be zero. For high voltage transmission lines, this will generally be the case because these lines are transposed, which causes the mutual coupling between phases (off diagonal terms) to be equal. Distribution lines are rarely if ever transposed. This causes unequal mutual coupling between phases, which causes the off diagonal terms of the phase impedance matrix to be unequal. For the nontransposed line, the diagonal terms of the phase impedance matrix will also be unequal. In most cases, the off diagonal terms of the sequence impedance matrix are very small compared to the diagonal terms and errors made by ignoring the off diagonal terms are small.

Sometimes the phase impedance matrix is modified such that the three diagonal terms are equal and all of the off diagonal terms are equal. The usual procedure is to set the three diagonal terms of the phase impedance matrix equal to the average of the diagonal terms of Eq. (6.29) and the off diagonal terms equal to the average of the off diagonal terms of Eq. (6.29). When this is done, the self and mutual impedances are defined as:

$$z_s = \frac{1}{3} \cdot (z_{aa} + z_{bb} + z_{cc}) \quad (6.34)$$

$$z_m = \frac{1}{3} \cdot (z_{ab} + z_{bc} + z_{ca}) \quad (6.35)$$

The phase impedance matrix is now defined as:

$$[z_{abc}] = \begin{bmatrix} z_s & z_m & z_m \\ z_m & z_s & z_m \\ z_m & z_m & z_s \end{bmatrix} \text{Ohms/mile} \quad (6.36)$$

When Eq. (6.31) is used with this phase impedance matrix, the resulting sequence matrix is diagonal (off diagonal terms are zero). The sequence impedances can be determined directly as:

$$z_{00} = z_s + 2 \cdot z_m$$
$$z_{11} = z_{22} = z_s - z_m \quad (6.37)$$

Distribution Systems

A second method that is commonly used to determine the sequence impedances directly is to employ the concept of Geometric Mean Distances (GMD). The GMD between phases is defined as:

$$D_{ij} = GMD_{ij} = \sqrt[3]{D_{ab} \cdot D_{bc} \cdot D_{ca}} \tag{6.38}$$

The GMD between phases and neutral is defined as:

$$D_{in} = GMD_{in} = \sqrt[3]{D_{an} \cdot D_{bn} \cdot D_{cn}} \tag{6.39}$$

The GMDs as defined above are used in Eqs. (6.23) and (6.24) to determine the various self and mutual impedances of the line resulting in:

$$\hat{z}_{ii} = r_i + 0.0953 + j0.12134 \cdot \left[\ln\left(\frac{1}{GMR_i}\right) + 7.93402 \right] \tag{6.40}$$

$$\hat{z}_{nn} = r_n + 0.0953 + j0.12134 \cdot \left[\ln\left(\frac{1}{GMR_n}\right) + 7.93402 \right] \tag{6.41}$$

$$\hat{z}_{ij} = 0.0953 + j0.12134 \cdot \left[\ln\left(\frac{1}{D_{ij}}\right) + 7.93402 \right] \tag{6.42}$$

$$\hat{z}_{in} = 0.0953 + j0.12134 \cdot \left[\ln\left(\frac{1}{D_{in}} + 7.93402\right) \right] \tag{6.43}$$

Equations (6.40) through (6.43) will define a matrix of order **ncond** × **ncond**, where ncond is the number of conductors (phases plus neutrals) in the line segment. Application of the Kron reduction (Eq. 6.27) and the sequence impedance transformation [Eq. (6.37)] leads to the following expressions for the zero, positive, and negative sequence impedances:

$$z_{00} = \hat{z}_{ii} + 2 \cdot \hat{z}_{ij} - 3 \cdot \left(\frac{\hat{z}_{in}^2}{\hat{z}_{nn}}\right) \text{ Ohms/mile} \tag{6.44}$$

$$z_{11} = z_{22} = \hat{z}_{ii} - \hat{z}_{ij}$$

$$z_{11} = z_{22} = r_i + j0.12134 \cdot \ln\left(\frac{D_{ij}}{GMR_i}\right) \text{ Ohms/mile} \tag{6.45}$$

Eq. (6.45) is recognized as the standard equation for the calculation of the line impedances when a balanced three-phase system and transposition are assumed.

Example 1

The spacings for an overhead three-phase distribution line is constructed as shown in Fig. 6.6. The phase conductors are 336,400 26/7 ACSR (Linnet) and the neutral conductor is 4/0 6/1 ACSR.

a. Determine the phase impedance matrix.
b. Determine the positive and zero sequence impedances.

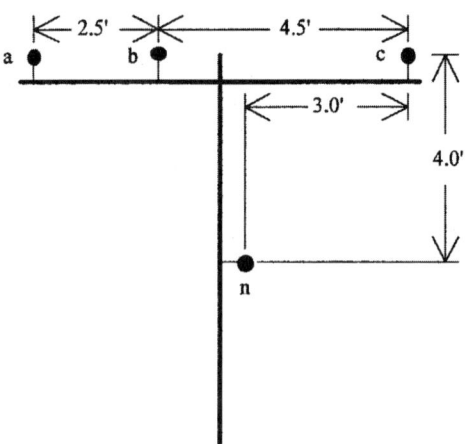

FIGURE 6.6 Three-phase distribution line spacings.

From the table of standard conductor data, it is found that:

336,400 26/7 ACSR: GMR = 0.0244 ft
                              Resistance = 0.306 Ohms/mile
4/0 6/1 ACSR: GMR = 0.00814 ft
                              Resistance = 0.5920 Ohms/mile

From Fig. 6.6 the following distances between conductors can be determined:

D_{ab} = 2.5 ft D_{bc} = 4.5 ft D_{ca} = 7.0 ft
D_{an} = 5.6569 ft D_{bn} = 4.272 ft D_{cn} = 5.0 ft

Applying Carson's modified equations [Eqs. (6.23) and (6.24)] results in the "primitive impedance matrix."

$$[\hat{z}] = \begin{bmatrix} 0.4013+j1.4133 & 0.0953+j0.8515 & 0.0953+j0.7266 & 0.0953+j0.7524 \\ 0.0953+j0.8515 & 0.4013+j1.4133 & 0.0953+j0.7802 & 0.0953+j0.7865 \\ 0.0953+j0.7266 & 0.0953+j0.7802 & 0.4013+j1.4133 & 0.0953+j0.7674 \\ 0.0953+j0.7524 & 0.0953+j0.7865 & 0.0953+j0.7674 & 0.6873+j1.5465 \end{bmatrix} \quad (6.46)$$

The "Kron" reduction of Eq. (6.27) results in the "phase impedance matrix."

$$[z_{abc}] = \begin{bmatrix} 0.4576+j1.0780 & 0.1560+j0.5017 & 0.1535+j0.3849 \\ 0.1560+j0.5017 & 0.4666+j1.0482 & 0.1580+j0.4236 \\ 0.1535+j0.3849 & 0.1580+j0.4236 & 0.4615+j1.0651 \end{bmatrix} \text{ Ohms/mile} \quad (6.47)$$

The phase impedance matrix of Eq. (6.47) can be transformed into the "sequence impedance matrix" with the application of Eq. (6.31).

$$[z_{012}] = \begin{bmatrix} 0.7735+j1.9373 & 0.0256+j0.0115 & -0.0321+j0.0159 \\ -0.0321+j0.0159 & 0.3061+j0.6270 & -0.0723-j0.0060 \\ 0.0256+j0.0115 & 0.0723-j0.0059 & 0.3061+j0.6270 \end{bmatrix} \text{ Ohms/mile} \quad (6.48)$$

In Eq. (6.48), the 1,1 term is the zero sequence impedance, the 2,2 term is the positive sequence impedance, and the 3,3 term is the negative sequence impedance. Note that the off-diagonal terms are not zero, which implies that there is mutual coupling between sequences. This is a result of the nonsymmetrical spacing between phases. With the off-diagonal terms nonzero, the three sequence networks representing the line will not be independent. However, it is noted that the off-diagonal terms are small relative to the diagonal terms.

In high voltage transmission lines, it is usually assumed that the lines are transposed and that the phase currents represent a balanced three-phase set. The transposition can be simulated in this example by replacing the diagonal terms of Eq. (6.47) with the average value of the diagonal terms (0.4619 + j1.0638) and replacing each off-diagonal term with the average of the off-diagonal terms (0.1558 + j0.4368). This modified phase impedance matrix becomes:

$$[z1_{abc}] = \begin{bmatrix} 0.3619+j1.0638 & 0.1558+j0.4368 & 0.1558+j0.4368 \\ 0.1558+j0.4368 & 0.3619+j1.0638 & 0.1558+j0.4368 \\ 0.1558+j0.4368 & 0.1558+j0.4368 & 0.3619+j1.0638 \end{bmatrix} \text{Ohms/mile} \quad (6.49)$$

Using this modified phase impedance matrix in the symmetrical component transformation, Eq. (6.31) results in the modified sequence impedance matrix.

$$[z1_{012}] = \begin{bmatrix} 0.7735+j1.9373 & 0 & 0 \\ 0 & 0.3061+j0.6270 & 0 \\ 0 & 0 & 0.3061+j0.6270 \end{bmatrix} \text{Ohms/mile} \quad (6.50)$$

Note now that the off-diagonal terms are all equal to zero, meaning that there is no mutual coupling between sequence networks. It should also be noted that the zero, positive, and negative sequence impedances of Eq. (6.50) are exactly equal to the same sequence impedances of Eq. (6.48).

The results of this example should not be interpreted to mean that a three-phase distribution line can be assumed to have been transposed. The original phase impedance matrix of Eq. (6.47) must be used if the correct effect of the mutual coupling between phases is to be modeled.

Underground Lines

Figure 6.7 shows the general configuration of three underground cables (concentric neutral, or tape shielded) with an additional neutral conductor.

Carson's equations can be applied to underground cables in much the same manner as for overhead lines. The circuit of Fig. 6.7 will result in a 7 × 7 primitive impedance matrix. For underground circuits that do not have the additional neutral conductor, the primitive impedance matrix will be 6 × 6.

Two popular types of underground cables in use today are the "concentric neutral cable" and the "tape shield cable." To apply Carson's equations, the resistance and GMR of the phase conductor and the equivalent neutral must be known.

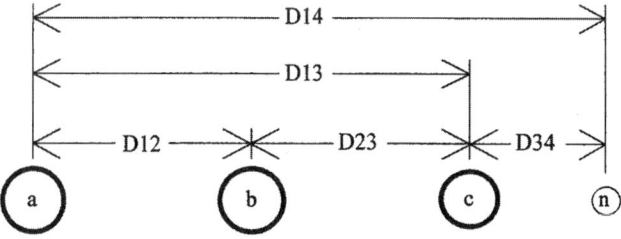

FIGURE 6.7 Three-phase underground with additional neutral.

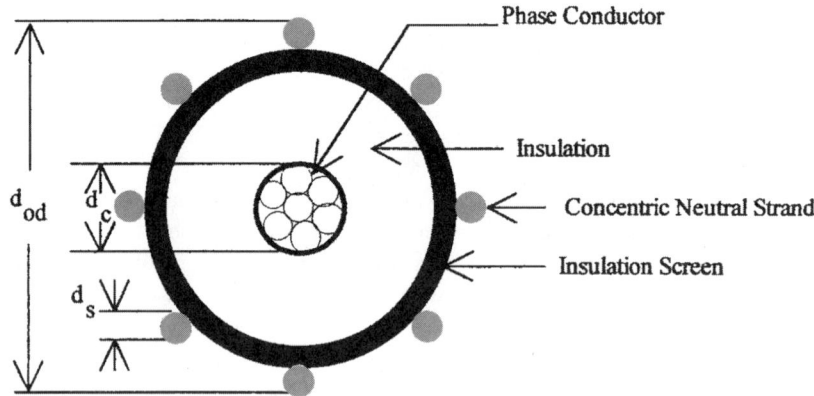

FIGURE 6.8 Concentric neutral cable.

Concentric Neutral Cable

Figure 6.8 shows a simple detail of a concentric neutral cable. The cable consists of a central "phase conductor" covered by a thin layer of nonmetallic semiconducting screen to which is bonded the insulating material. The insulation is then covered by a semiconducting insulation screen. The solid strands of concentric neutral are spiralled around the semiconducting screen with a uniform spacing between strands. Some cables will also have an insulating "jacket" encircling the neutral strands.

In order to apply Carson's equations to this cable, the following data needs to be extracted from a table of underground cables.

d_c = phase conductor diameter (inches)
d_{od} = nominal outside diameter of the cable (inches)
d_s = diameter of a concentric neutral strand (inches)
GMR_c = geometric mean radius of the phase conductor (ft)
GMR_s = geometric mean radius of a neutral strand (ft)
r_c = resistance of the phase conductor (Ohms/mile)
r_s = resistance of a solid neutral strand (Ohms/mile)
k = number of concentric neutral strands

The geometric mean radii of the phase conductor and a neutral strand are obtained from a standard table of conductor data. The equivalent geometric mean radius of the concentric neutral is given by:

$$GMR_{cn} = \sqrt[k]{GMR_s \cdot k \cdot R^{k-1}} \qquad (6.51)$$

where R = radius of a circle passing through the center of the concentric neutral strands

$$R = \frac{d_{od} - d_s}{24} \text{ ft} \qquad (6.52)$$

The equivalent resistance of the concentric neutral is:

$$r_{cn} = \frac{r_s}{k} \text{ Ohms/mile} \qquad (6.53)$$

The various spacings between a concentric neutral and the phase conductors and other concentric neutrals are as follows:

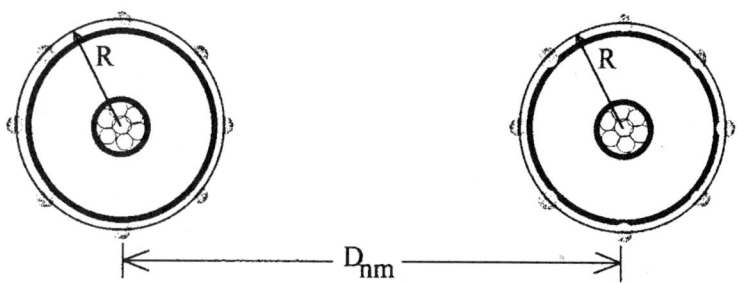

FIGURE 6.9 Distances between concentric neutral cables.

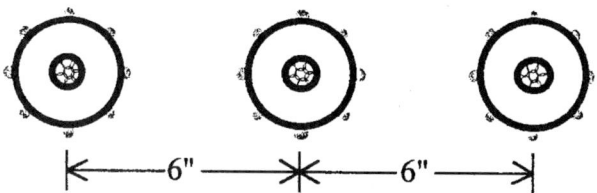

FIGURE 6.10 Three-phase concentric neutral cable spacing.

Concentric neutral to its own phase conductor

$D_{ij} = R$ [Eq. (6.52) above]

Concentric neutral to an adjacent concentric neutral

D_{ij} = center-to-center distance of the phase conductors

Concentric neutral to an adjacent phase conductor

Figure 6.9 shows the relationship between the distance between centers of concentric neutral cables and the radius of a circle passing through the centers of the neutral strands.

The geometric mean distance between a concentric neutral and an adjacent phase conductor is given by Eq. (6.54).

$$D_{ij} = \sqrt[k]{D_{nm}^k - R^k} \ \ (\text{ft}) \tag{6.54}$$

where D_{nm} = center-to-center distance between phase conductors

For cables buried in a trench, the distance between cables will be much greater than the radius R and therefore very little error is made if D_{ij} in Eq. (6.54) is set equal to D_{nm}. For cables in conduit, that assumption is not valid.

Example 2

Three concentric neutral cables are buried in a trench with spacings as shown in Fig. 6.10. The cables are 15 kV, 250,000 CM stranded all aluminum with 13 strands of #14 annealed coated copper wires (1/3 neutral). The data for the phase conductor and neutral strands from a conductor data table are:

250,000 AA phase conductor: GMR_p = 0.0171 ft, resistance = 0.4100 Ohms/mile
14 copper neutral strands: GMR_s = 0.00208 ft, resistance = 14.87 Ohms/mile
Diameter (d_s) = 0.0641 in.

The equivalent GMR of the concentric neutral [Eq. (6.51)] = 0.04864 ft
The radius of the circle passing through strands [Eq. (6.52)] = 0.0511 ft
The equivalent resistance of the concentric neutral [Eq. (6.53)] = 1.1440 Ohms/mile

Since R (0.0511 ft) is much less than D_{12} (0.5 ft) and D_{13} (1.0 ft), then the distances between concentric neutrals and adjacent phase conductors are the center-to-center distances of the cables.

Applying Carson's equations results in a 6 × 6 primitive impedance matrix. This matrix in partitioned form [Eq. (6.26)] is:

$$[Z_{ij}] = \begin{bmatrix} 0.5053+j1.4564 & 0.0953+j1.0468 & 0.0953+j0.9627 \\ 0.0953+j1.0468 & 0.5053+j1.4564 & 0.0953+j1.0468 \\ 0.0953+j0.9627 & 0.0953+j1.0468 & 0.5053+j1.4564 \end{bmatrix}$$

$$[Z_{in}] = \begin{bmatrix} 0.0953+j1.3236 & 0.0953+j1.0468 & 0.0953+j0.9627 \\ 0.0953+j1.0468 & 0.0953+j1.3236 & 0.0953+j1.0468 \\ 0.0953+j0.9627 & 0.0953+j1.0468 & 0.0953+j1.3236 \end{bmatrix}$$

$$[z_{nj}] = [z_{in}]$$

$$[Z_{nn}] = \begin{bmatrix} 1.2393+j1.3296 & 0.0953+j1.0468 & 0.0953+j0.9627 \\ 0.0953+j1.0468 & 1.2393+j1.3296 & 0.0953+j1.0468 \\ 0.0953+j0.9627 & 0.0953+j1.0468 & 1.2393+j1.3296 \end{bmatrix}$$

Using the Kron reduction [Eq. (6.27)] results in the phase impedance matrix:

$$[z_{abc}] = \begin{bmatrix} 0.7982+j0.4463 & 0.3192+j0.0328 & 0.2849-j0.0143 \\ 0.3192+j0.0328 & 0.7891+j0.4041 & 0.3192+j0.0328 \\ 0.28490-j0.0143 & 0.3192+j0.0328 & 0.7982+j0.4463 \end{bmatrix} \text{Ohms/mile}$$

The sequence impedance matrix for the concentric neutral three-phase line is determined using Eq. (6.17). The resulting sequence impedance matrix is:

$$[z_{012}] = \begin{bmatrix} 1.4106+j0.4665 & -0.0028-j0.0081 & -0.0056+j0.0065 \\ -0.0056+j0.0065 & 0.4874+j0.4151 & -0.0264+j0.0451 \\ -0.0028-j0.0081 & 0.0523+j0.0003 & 0.4867+j0.4151 \end{bmatrix} \text{Ohms/mile}$$

Tape Shielded Cables

Figure 6.11 shows a simple detail of a tape shielded cable.

Parameters of Fig. 6.11 are:

d_c = diameter of phase conductor (in.)
d_s = inside diameter of tape shield (in.)
d_{od} = outside diameter over jacket (in.)
T = thickness of copper tape shield in mils
 = 5 mils (standard)

Once again, Carson's equations will be applied to calculate the self-impedances of the phase conductor and the tape shield as well as the mutual impedance between the phase conductor and the tape shield. The resistance and GMR of the phase conductor are found in a standard table of conductor data.

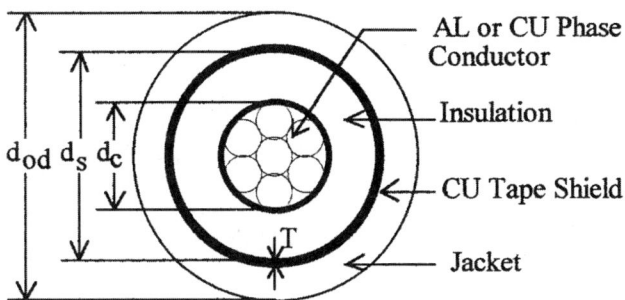

FIGURE 6.11 Taped shielded cable.

The resistance of the tape shield is given by:

$$r_{shield} = \frac{18.826}{d_s \cdot T} \quad \text{Ohms/mile} \tag{6.55}$$

The resistance of the tape shield given in Eq. (6.55) assumes a resistivity of 100 Ohm-meter and a temperature of 50°C. The diameter of the tape shield d_s is given in inches and the thickness of the tape shield T is in mils.

The GMR of the tape shield is given by:

$$GMR_{shield} = \frac{\frac{d_s}{2} - \frac{T}{2000}}{12} \quad \text{ft} \tag{6.56}$$

The various spacings between a tape shield and the conductors and other tape shields are as follows:
Tape shield to its own phase conductor

$$D_{ij} = GMR_{tape} = \text{radius to midpoint of the shield} \tag{6.57}$$

Tape shield to an adjacent tape shield

$$D_{ij} = \text{center-to-center distance of the phase conductors} \tag{6.58}$$

Tape shield to an adjacent phase or neutral conductor

$$D_{ij} = D_{nm} \tag{6.59}$$

where D_{nm} = center to center distance between phase conductors.

In applying Carson's equations for both concentric neutral and tape shielded cables, the numbering of conductors and neutrals is important. For example, a three-phase underground circuit with an additional neutral conductor must be numbered as:

1 = phase conductor #1
2 = phase conductor #2
3 = phase conductor #3
4 = neutral of conductor #1

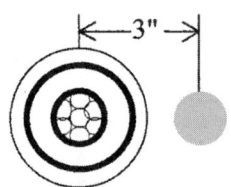

FIGURE 6.12 Single-phase tape shield with neutral.

5 = neutral of conductor #2
6 = neutral of conductor #3
7 = additional neutral conductor (if present)

Example 3

A single-phase circuit consists of a 1/0 AA tape shielded cable and a 1/0 CU neutral conductor as shown in Fig. 6.12.

Cable Data: 1/0 AA
Inside diameter of tape shield = d_s = 1.084 in.
Resistance = 0.97 Ohms/mi
GMR_p = 0.0111 ft
Tape shield thickness = T = 8 mils

Neutral Data: 1/0 Copper, 7 strand
Resistance = 0.607 Ohms/mi
GMR_n = 0.01113 ft

Distance between cable and neutral = D_{nm} = 3 in.
The resistance of the tape shield is computed according to Eq. (6.55):

$$r_{shield} = \frac{18.826}{d_s \cdot T} = \frac{18.826}{1.084 \cdot 8} = 2.1705 \text{ Ohms/mile}$$

The GMR of the tape shield is computed according to Eq. (6.56):

$$GMR_{shield} = \frac{\frac{d_s}{2} - \frac{T}{2000}}{12} = \frac{\frac{1.084}{2} - \frac{8}{2000}}{12} = 0.0455 \text{ ft}$$

Using the relations defined in Eqs. (6.57) to (6.59) and Carson's equations results in a 3 × 3 primitive impedance matrix:

$$z_{primitive} = \begin{bmatrix} 1.0653 + j1.5088 & 0.0953 + j1.3377 & 0.0953 + j1.1309 \\ 0.0953 + j1.3377 & 2.2658 + j1.3377 & 0.0953 + j1.1309 \\ 0.0953 + j1.1309 & 0.0953 + j1.1309 & 0.7023 + j1.5085 \end{bmatrix} \text{ Ohms/mile}$$

Applying Kron's reduction method will result in a single impedance which represents the equivalent single-phase impedance of the tape shield cable and the neutral conductor.

$$z_{1p} = 1.3368 + j0.6028 \quad \text{Ohms/mile}$$

Shunt Admittance

When a high-voltage transmission line is less than 50 miles in length, the shunt capacitance of the line is typically ignored. For lightly loaded distribution lines, particularly underground lines, the shunt capacitance should be modeled.

The basic equation for the relationship between the charge on a conductor to the voltage drop between the conductor and ground is given by:

$$Q_n = C_{ng} \cdot V_{ng} \qquad (6.60)$$

where Q_n = charge on the conductor
C_{ng} = capacitance between the conductor and ground
V_{ng} = voltage between the conductor and ground

For a line consisting of **ncond** (number of phase plus number of neutral) conductors, Eq. (6.60) can be written in condensed matrix form as:

$$[Q] = [C] \cdot [V] \qquad (6.61)$$

where $[Q]$ = column vector of order **ncond**
$[C]$ = **ncond** × **ncond** matrix
$[V]$ = column vector of order **ncond**

Equation (6.61) can be solved for the voltages:

$$[V] = [C]^{-1} \cdot [Q] = [P] \cdot [Q] \qquad (6.62)$$

where

$$[P] = [C]^{-1} = \text{"Potential Coefficient Matrix"} \qquad (6.63)$$

Overhead Lines

The determination of the shunt admittance of overhead lines starts with the calculation of the "potential coefficient matrix" (Glover and Sarma, 1994). The elements of the matrix are determined by:

$$P_{ii} = 11.17689 \ln \frac{S_{ii}}{RD_i} \qquad (6.64)$$

$$P_{ij} = 11.17689 \cdot \ln \frac{S_{ij}}{D_{ij}} \qquad (6.65)$$

See Fig. 6.4 for the following definitions.

S_{ii} = distance between a conductor and its image below ground in ft
S_{ij} = distance between conductor **i** and the image of conductor **j** below ground in ft
D_{ij} = overhead spacing between two conductors in ft
RDi = radius of conductor **i** in ft

The potential coefficient matrix will be an **ncond** × **ncond** matrix. If one or more of the conductors is a grounded neutral, then the matrix must be reduced using the "Kron" method to an **nphase** × **nphase** matrix $[P_{abc}]$.

The inverse of the potential coefficient matrix will give the **nphase × nphase** capacitance matrix $[C_{abc}]$. The shunt admittance matrix is given by:

$$[y_{abc}] = j \cdot \omega \cdot [C_{abc}] \quad \text{micro-S/mile} \qquad (6.66)$$

where $\omega = 2 \cdot \pi f = 376.9911$

Example 4

Determine the shunt admittance matrix for the overhead line of Example 1. Assume that the neutral conductor is 25 ft above ground.

For this configuration, the image spacing matrix is computed to be:

$$[S] = \begin{bmatrix} 58 & 58.0539 & 58.4209 & 54.1479 \\ 58.0539 & 58 & 58.1743 & 54.0208 \\ 58.4209 & 58.1743 & 58 & 54.0833 \\ 54.1479 & 54.0208 & 54.0835 & 58 \end{bmatrix} \text{ ft}$$

The primitive potential coefficient matrix is computed to be:

$$[P_{primitive}] = \begin{bmatrix} 84.56 & 35.1522 & 23.7147 & 25.2469 \\ 35.4522 & 84.56 & 28.6058 & 28.359 \\ 23.7147 & 28.6058 & 84.56 & 26.6131 \\ 25.2469 & 28.359 & 26.6131 & 85.6659 \end{bmatrix}$$

"Kron" reduce to a 3 × 3 matrix:

$$[P] = \begin{bmatrix} 77.1194 & 26.7944 & 15.8714 \\ 26.7944 & 75.172 & 19.7957 \\ 15.8714 & 19.7957 & 76.2923 \end{bmatrix}$$

Invert [P] to determine the shunt capacitance matrix:

$$[C_{abc}] = [P]^{-1} = \begin{bmatrix} 0.015 & -0.0049 & -0.0019 \\ -0.0019 & 0.0159 & -0.0031 \\ -0.0019 & -0.0031 & 0.0143 \end{bmatrix}$$

Multiply $[C_{abc}]$ by the radian frequency to determine the final three-phase shunt admittance matrix.

$$[Y_{abc}] = j \cdot 376.9911 \cdot [C_{abc}] = \begin{bmatrix} j5.6711 & -j1.8362 & -j0.7033 \\ -j1.8362 & j5.9774 & -j1.169 \\ -j0.7033 & -j1.169 & j5.391 \end{bmatrix} \mu\text{S/mile}$$

Underground Lines

Because the electric fields of underground cables are confined to the space between the phase conductor and its concentric neutral to tape shield, the calculation of the shunt admittance matrix requires only the determination of the "self" admittance terms.

Concentric Neutral

The self-admittance in micro-S/mile for a concentric neutral cable is given by:

$$Y_{cn} = j\frac{77.582}{\ln\left(\dfrac{R_b}{R_a}\right) - \dfrac{1}{k}\ln\left(\dfrac{k \cdot R_n}{R_b}\right)} \tag{6.67}$$

where R_b = radius of a circle to center of concentric neutral strands (ft)
R_a = radius of phase conductor (ft)
R_n = radius of concentric neutral strand (ft)
k = number of concentric neutral strands

Example 5

Determine the three-phase shunt admittance matrix for the concentric neutral line of Example 2.

$$R_b = R = 0.0511 \text{ ft}$$

Diameter of the 250,000 AA phase conductor = 0.567 in.

$$R_a = \frac{0.567}{24} = 0.0236 \text{ ft}$$

Diameter of the #14 CU concentric neutral strand = 0.0641 in.

$$R_n = \frac{0.0641}{24} = 0.0027 \text{ ft}$$

Substitute into Eq. (6.67):

$$Y_{cn} = j\frac{77.582}{\ln\left(\dfrac{R_b}{R_a}\right) - \dfrac{1}{k}\ln\left(\dfrac{k \cdot R_n}{R_b}\right)} = j\frac{77.582}{\ln\left(\dfrac{0.0511}{0.0236}\right) - \dfrac{1}{13}\ln\left(\dfrac{13 \cdot 0.0027}{0.0511}\right)} = j96.8847$$

The three-phase shunt admittance matrix is:

$$[Y_{abc}] = \begin{bmatrix} j96.8847 & 0 & 0 \\ 0 & j96.8847 & 0 \\ 0 & 0 & j96.8847 \end{bmatrix} \mu\text{S/mile}$$

Tape Shield Cable

The shunt admittance in micro-S/mile for tape shielded cables is given by:

$$Y_{ts} = j\frac{77.586}{\ln\left(\dfrac{R_b}{R_a}\right)} \; \mu\text{S/mile} \tag{6.68}$$

where R_b = inside radius of the tape shield
R_a = radius of phase conductor

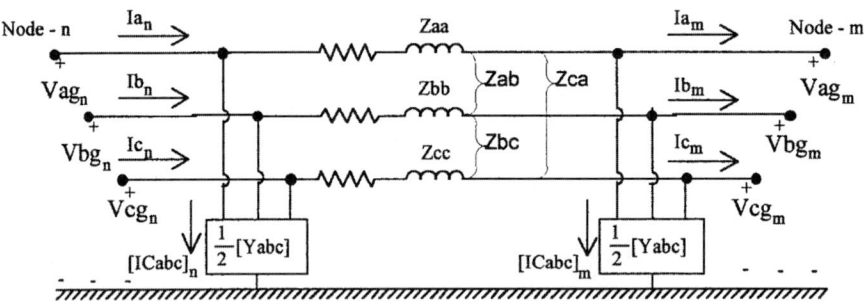

FIGURE 6.13 Three-phase line segment model.

Example 6

Determine the shunt admittance of the single-phase tape shielded cable of Example 3 in the section "Line Impedance".

$$R_b = \frac{d_s}{24} = \frac{1.084}{24} = 0.0452$$

The diameter of the 1/0 AA phase conductor = 0.368 in.

$$R_a = \frac{d_p}{24} = \frac{0.368}{24} = 0.0153$$

Substitute into Eq. (6.68):

$$Y_{ts} = j\frac{77.586}{\ln\left(\frac{R_b}{R_a}\right)} = j\frac{77.586}{\ln\left(\frac{0.0452}{0.0153}\right)} = j71.8169 \; \mu S/\text{mile}$$

Line Segment Models

Exact Line Segment Model

The exact model of a three-phase line segment is shown in Fig. 6.13.

For the line segment in Fig. 6.13, the equations relating the input (Node n) voltages and currents to the output (Node m) voltages and currents are:

$$[VLG_{abc}]_n = [a] \cdot [VLG_{abc}]_m + [b] \cdot [I_{abc}]_m \qquad (6.69)$$

$$[I_{abc}]_n = [c] \cdot [VLG_{abc}]_m + [d] \cdot [I_{abc}]_m \qquad (6.70)$$

where

$$[a] = [U] - \frac{1}{2} \cdot [Z_{abc}] \cdot [Y_{abc}] \qquad (6.71)$$

$$[b] = [Z_{abc}] \qquad (6.72)$$

$$[c] = [Y_{abc}] - \frac{1}{4} \cdot [Z_{abc}] \cdot [Y_{abc}]^2 \qquad (6.73)$$

Distribution Systems

$$[d] = [U] - \frac{1}{2} \cdot [Z_{abc}] \cdot [Y_{abc}] \tag{6.74}$$

In Eqs. (6.71) through (6.74), the impedance matrix $[Z_{abc}]$ and the admittance matrix $[Y_{abc}]$ are defined earlier in this document.

Sometimes it is necessary to determine the voltages at node-m as a function of the voltages at node-n and the output currents at node-m. The necessary equation is:

$$[VLG_{abc}]_m = [A] \cdot [VLG_{abc}]_n - [B] \cdot [I_{abc}]_m \tag{6.75}$$

where

$$[A] = \left([U] + \frac{1}{2} \cdot [Z_{abc}] \cdot [Y_{abc}] \right)^{-1} \tag{6.76}$$

$$[B] = \left([U] + \frac{1}{2} \cdot [Z_{abc}] \cdot [Y_{abc}] \right)^{-1} \cdot [Z_{abc}] \tag{6.77}$$

$$[U] = \begin{bmatrix} 1 & 0 & 0 \\ 0 & 1 & 0 \\ 0 & 0 & 1 \end{bmatrix} \tag{6.78}$$

In most cases the shunt admittance is so small that it can be neglected. When this is the case, the $[a]$, $[b]$, $[c]$, $[d]$, $[A]$, and $[B]$ matrices become:

$$[a] = [U] \tag{6.79}$$

$$[b] = [Z_{abc}] \tag{6.80}$$

$$[c] = [0] \tag{6.81}$$

$$[d] = [U] \tag{6.82}$$

$$[A] = [U] \tag{6.83}$$

$$[B] = [Z_{abc}] \tag{6.84}$$

When the shunt admittance is neglected, Eqs. (6.69), (6.70), and (6.75) become:

$$[VLG_{abc}]_n = [VLG_{abc}]_m + [Z_{abc}] \cdot [I_{abc}]_m \tag{6.85}$$

$$[I_{abc}]_n = [I_{abc}]_m \tag{6.86}$$

$$[VLG_{abc}]_m = [VLG_{abc}]_n - [Z_{abc}] \cdot [I_{abc}]_m \tag{6.87}$$

It is usually safe to neglect the shunt capacitance of overhead line segments for segments less than 25 miles in total length and for voltages less than 25 kV. For underground lines, the shunt admittance should be included.

If an accurate determination of the voltage drops down a line segment is to be made, it is essential that the phase impedance matrix $[Z_{abc}]$ be computed based on the actual configuration and spacings of the overhead or underground lines. No assumptions should be made, such as transposition. The reason for this is best demonstrated by an example.

Example 7

The phase impedance matrix for the line configuration in Example 1 was computed to be:

$$[z_{abc}] = \begin{bmatrix} 0.4576 + j1.0780 & 0.1560 + j0.5017 & 0.1535 + j0.3849 \\ 0.1560 + j0.5017 & 0.466 + j1.0482 & 0.1580 + j0.4236 \\ 0.1535 + j0.3849 & 0.1580 + j0.4236 & 0.4615 + j1.0651 \end{bmatrix} \text{ Ohms/mile}$$

Assume that a 12.47-kV substation serves a load 1.5 miles from the substation. The metered output at the substation is balanced 10,000 kVA at 12.47 kV and 0.9 lagging power factor. Compute the three-phase line-to-ground voltages at the load end of the line and the voltage unbalance at the load.

The line-to-ground voltages and line currents at the substation are:

$$[VLG_{abc}] = \begin{bmatrix} 7200\underline{/0} \\ 7200\underline{/-120} \\ 7200\underline{/120} \end{bmatrix} \quad [I_{abc}]_n = \begin{bmatrix} 463\underline{/-25.84} \\ 463\underline{/-145.84} \\ 463\underline{/94.16} \end{bmatrix}$$

Solve Eq. (6.85) for the load voltages:

$$[VLG_{abc}]_m = [VLG_{abc}]_n - 1.5 \cdot [z_{abc}] \cdot [I_{abc}]_n = \begin{bmatrix} 6761.10\underline{/2.32} \\ 6877.7\underline{/-122.43} \\ 6836.33\underline{/117.21} \end{bmatrix}$$

The voltage unbalance at the load using the NEMA definition is:

$$V_{unbalance} = \frac{\max(V_{deviation})}{V_{avg}} \cdot 100 = 0.937\%$$

The point of Example 1 is to demonstrate that even though the system is perfectly balanced at the substation, the unequal mutual coupling between phases results in a significant voltage unbalance at the load. Significant because NEMA requires that induction motors be derated when the voltage unbalance is 1% or greater.

Approximate Line Segment Model

Many times the only data available for a line segment will be the positive and zero sequence impedances. An approximate three-phase line segment model can be developed by applying the "reverse impedance transformation" from symmetrical component theory.

Using the known positive and zero sequence impedances, the "sequence impedance matrix" is given by:

$$[Z_{seq}] = \begin{bmatrix} Z_0 & 0 & 0 \\ 0 & Z_+ & 0 \\ 0 & 0 & Z_+ \end{bmatrix} \tag{6.88}$$

Distribution Systems

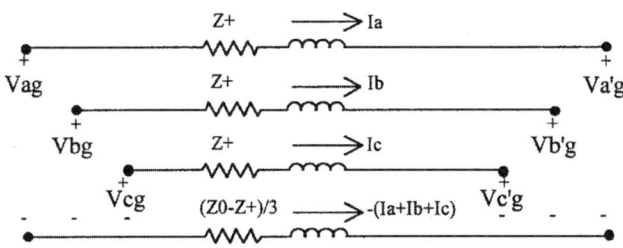

FIGURE 6.14 Approximate line segment model.

The "reverse impedance transformation" results in the following "approximate phase impedance matrix."

$$[Z_{approx}] = [A_s] \cdot [Z_{seq}] \cdot [A_s]^{-1} = \frac{1}{3} \cdot \begin{bmatrix} (2 \cdot Z_+ - Z_0) & (Z_0 - Z_+) & (Z_0 - Z_+) \\ (Z_0 - Z_+) & (2 \cdot Z_+ - Z_0) & (Z_0 - Z_+) \\ (Z_0 - Z_+) & (Z_0 - Z_+) & (2 \cdot Z_+ - Z_0) \end{bmatrix} \quad (6.89)$$

Notice that the approximate phase impedance matrix is characterized by the three diagonal terms being equal and all mutual terms being equal. This is the same result that is achieved if the line is assumed to be transposed. Substituting the approximate phase impedance matrix into Eq. (6.85) results in:

$$\begin{bmatrix} V_{an} \\ V_{bn} \\ V_{cn} \end{bmatrix}_n = \begin{bmatrix} V_{an} \\ V_{bn} \\ V_{cn} \end{bmatrix}_m + \frac{1}{3} \cdot \begin{bmatrix} (2 \cdot Z_+ - Z_0) & (Z_0 - Z_+) & (Z_0 - Z_+) \\ (Z_0 - Z_+) & (2 \cdot Z_+ - Z_0) & (Z_0 - Z_+) \\ (Z_0 - Z_+) & (Z_0 - Z_+) & (2 \cdot Z_+ - Z_0) \end{bmatrix} \cdot \begin{bmatrix} I_a \\ I_b \\ I_c \end{bmatrix}_n \quad (6.90)$$

Eq. (6.90) can be expanded and an equivalent circuit for the approximate line segment model can be developed. This approximate model is shown in Fig. 6.14.

The errors made by using this approximate line segment model are demonstrated in Example 8.

Example 8

For the line of Example 7, the positive and zero sequence impedances were determined to be:

$$Z_+ = 0.3061 + j0.6270 \text{ Ohms/mile}$$

$$Z_0 = 0.7735 + j1.9373 \text{ Ohms/mile}$$

The sequence impedance matrix is:

$$[z_{seq}] = \begin{bmatrix} 0.7735 + j1.9373 & 0 & 0 \\ 0 & 0.3061 + j0.6270 & 0 \\ 0 & 0 & 0.3061 + j0.6270 \end{bmatrix}$$

Performing the reverse impedance transformation results in the approximate phase impedance matrix.

$$[z_{approx}] = [A_s] \cdot [z_{seq}] \cdot [A_s]^{-1} = \begin{bmatrix} 0.4619 + j1.0638 & 0.1558 + j0.4368 & 0.1558 + j0.4368 \\ 0.1558 + j0.4368 & 0.4619 + j1.0638 & 0.1558 + j0.4368 \\ 0.1558 + j0.4368 & 0.1558 + j0.4368 & 0.4619 + j1.0638 \end{bmatrix}$$

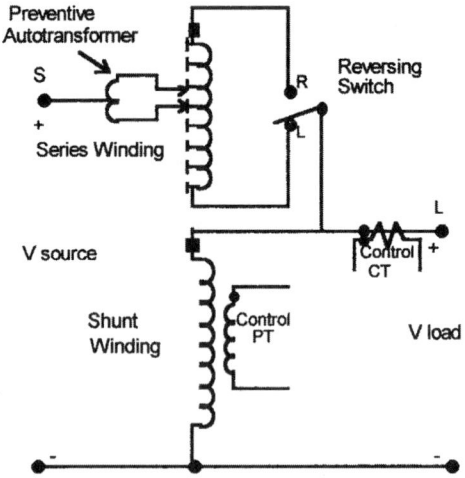

FIGURE 6.15 Step voltage regulator.

Note in the approximate phase impedance matrix that the three diagonal terms are equal and all of the mutual terms are equal.

Use the approximate impedance matrix to compute the load voltage and voltage unbalance as specified in Example 1.

$$\left[VLG_{abc}\right]_m = \left[VLG_{abc}\right]_n - 1.5 \cdot \left[z_{approx}\right] \cdot \left[I_{abc}\right]_n = \begin{bmatrix} 6825.01/-2.51 \\ 6825.01/-122.51 \\ 6825.01/\underline{117.49} \end{bmatrix}$$

Note that the voltages are computed to be balanced. In the previous example it was shown that when the line is modeled accurately, there is a voltage unbalance of almost 1%.

Step-Voltage Regulators

A step voltage regulator consists of an autotransformer and a load tap changing mechanism. The voltage change is obtained by changing the taps of the series winding of the autotransformer. The position of the tap is determined by a control circuit (line drop compensator). Standard step regulators contain a reversing switch enabling a ±10% regulator range, usually in 32 steps. This amounts to a 5/8% change per step or 0.75 volt change per step on a 120-volt base.

A typical step voltage regulator is shown in Fig. 6.15. The tap changing is controlled by a control circuit shown in the block diagram of Fig. 6.16.

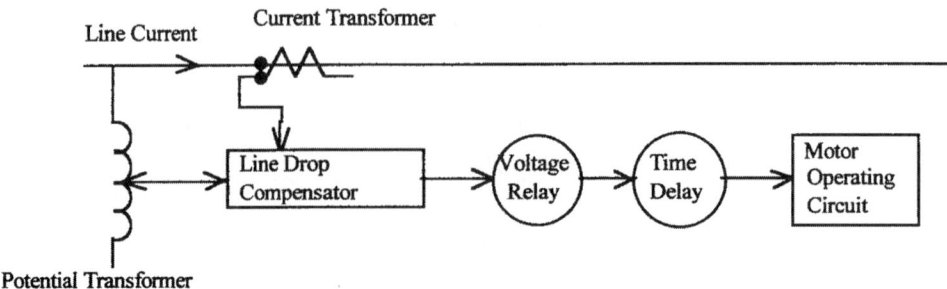

FIGURE 6.16 Regulator control circuit.

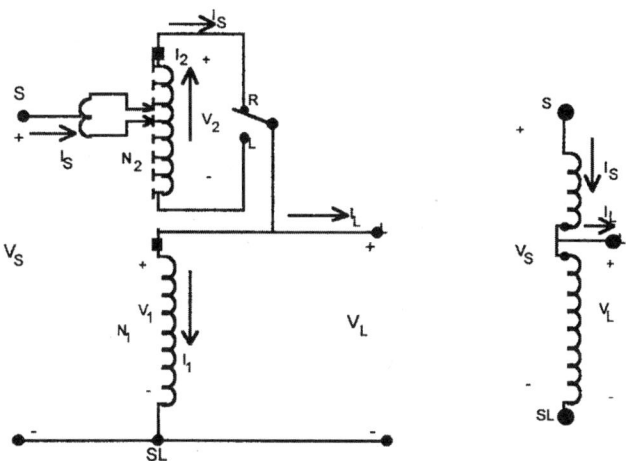

FIGURE 6.17 Voltage regulator in the raise position.

The control circuit requires the following settings:

1. Voltage Level — The desired voltage (on 120-volt base) to be held at the "load center." The load center may be the output terminal of the regulator or a remote node on the feeder.
2. Bandwidth — The allowed variance of the load center voltage from the set voltage level. The voltage held at the load center will be plus or minus one-half the bandwidth. For example, if the voltage level is set to 122 volts and the bandwidth set to 2 volts, the regulator will change taps until the load center voltage lies between 121 volts and 123 volts.
3. Time Delay — Length of time that a raise or lower operation is called for before the actual execution of the command. This prevents taps changing during a transient or short time change in current.
4. Line Drop Compensator — Set to compensate for the voltage drop (line drop) between the regulator and the load center. The settings consist of R and X settings in volts corresponding to the *equivalent* impedance between the regulator and the load center. This setting may be zero if the regulator output terminals are the "load center."

The rating of a regulator is based on the kVA transformed, not the kVA rating of the line. In general this will be 10% of the line rating since rated current flows through the series winding which represents the ±10% voltage change.

Voltage Regulator in the Raise Position

Figure 6.17 shows a detailed and abbreviated drawing of a regulator in the raise position.

The defining voltage and current equations for the regulator in the raise position are as follows:

Voltage Equations *Current Equations*

$$\frac{V_1}{N_1} = \frac{V_2}{N_2} \qquad N_1 \cdot I_1 = N_2 \cdot I_2 \tag{6.91}$$

$$V_S = V_1 - V_2 \qquad I_L = I_S - I_1 \tag{6.92}$$

$$V_L = V_1 \qquad I_2 = I_S \tag{6.93}$$

$$V_2 = \frac{N_2}{N_1} \cdot V_1 = \frac{N_2}{N_1} \cdot V_L \qquad I_1 = \frac{N_2}{N_1} \cdot I_2 = \frac{N_2}{N_1} \cdot I_S \tag{6.94}$$

$$V_S = \left(1 - \frac{N_2}{N_1}\right) \cdot V_1 \qquad I_L = \left(1 - \frac{N_2}{N_1}\right) \cdot I_S \qquad (6.95)$$

$$V_S = a_R \cdot V_L \qquad I_L = a_R \cdot I_S \qquad (6.96)$$

$$a_R = 1 - \frac{N_2}{N_1} \qquad (6.97)$$

Equations (6.96) and (6.97) are the necessary defining equations for modeling a regulator in the raise position.

Voltage Regulator in the Lower Position

Figure 6.18 shows the detailed and abbreviated drawings of a regulator in the lower position. Note in the figure that the only difference between the lower and raise models is that the polarity of the series winding and how it is connected to the shunt winding is reversed.

The defining voltage and current equations for a regulator in the lower position are as follows:

Voltage Equations — *Current Equations*

$$\frac{V_1}{N_1} = \frac{V_2}{N_2} \qquad N_1 \cdot I_1 = N_2 \cdot I_2 \qquad (6.98)$$

$$V_S = V_1 + V_2 \qquad I_L = I_S - I_1 \qquad (6.99)$$

$$V_L = V_1 \qquad I_2 = -I_S \qquad (6.100)$$

$$V_2 = \frac{N_2}{N_1} \cdot V_1 = \frac{N_2}{N_1} \cdot V_L \qquad I_1 = \frac{N_2}{N_1} \cdot I_2 = \frac{N_2}{N_1} \cdot (-I_S) \qquad (6.101)$$

$$V_S = \left(1 + \frac{N_2}{N_1}\right) \cdot V_1 \qquad I_L = \left(1 + \frac{N_2}{N_1}\right) \cdot I_S \qquad (6.102)$$

$$V_S = a_R \cdot V_L \qquad I_L = a_R \cdot I_S \qquad (6.103)$$

$$a_R = 1 + \frac{N_2}{N_1} \qquad (6.104)$$

Equations (6.97) and (6.104) give the value of the effective regulator ratio as a function of the ratio of the number of turns on the series winding (N_2) to the number of turns on the shunt winding (N_1). The actual turns ratio of the windings is not known. However, the particular position will be known. Equations (6.97) and (6.104) can be modified to give the effective regulator ratio as a function of the tap position. Each tap changes the voltage by 5/8% or 0.00625 per unit. Therefore, the effective regulator ratio can be given by:

$$a_R = 1 \mp 0.00625 \cdot Tap \qquad (6.105)$$

In Eq. (6.105), the minus sign applies to the "raise" position and the positive sign for the "lower" position.

Distribution Systems

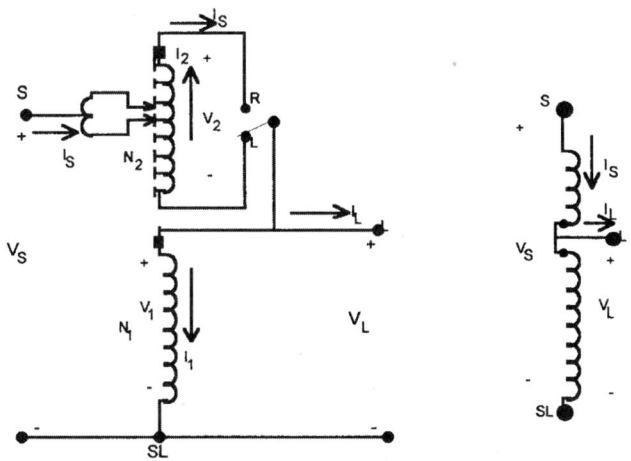

FIGURE 6.18 Regulator in the lower position.

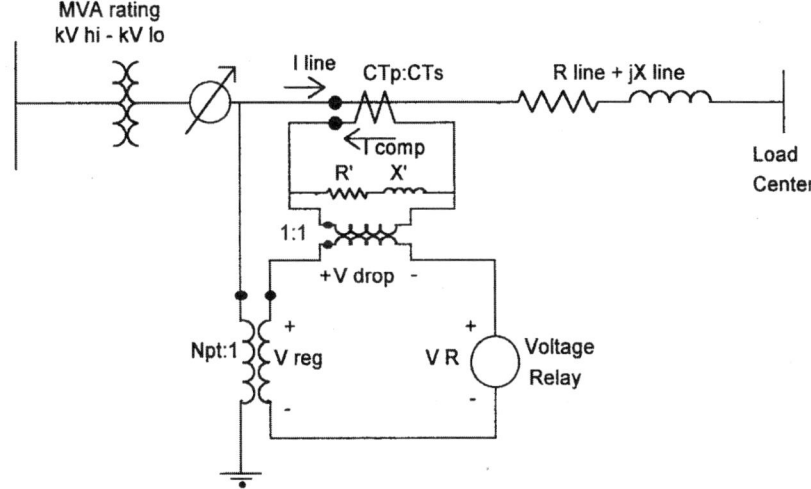

FIGURE 6.19 Line drop compensator circuit.

The Line Drop Compensator

The changing of taps on a regulator is controlled by the "line drop compensator." Figure 6.19 shows a simplified sketch of the compensator circuit and how it is connected to the circuit through a potential transformer and a current transformer.

The purpose of the line drop compensator is to model the voltage drop of the distribution line from the regulator to the "load center." Typically the compensator circuit is modeled on a 120 volt base. This requires the potential transformer to transform rated voltage (line-to-neutral or line-to-line) down to 120 volts. The current transformer turns ratio ($CT_p:CT_s$) where the primary rating (CT_p) will typically be the rated current of the feeder. The setting that is most critical is that of R′ and X′. These values must represent the equivalent impedance from the regulator to the load center. Knowing the equivalent impedance in Ohms from the regulator to the load center (R_{line_ohms} and X_{line_ohms}), the required value for the compensator settings are calibrated in volts and determined by:

$$R'_{volts} + jX'_{volts} = \left(R_{line\_Ohms} + jX_{line\_ohms}\right) \cdot \frac{Ct_p}{N_{pt}} \quad \text{volts} \quad (6.106)$$

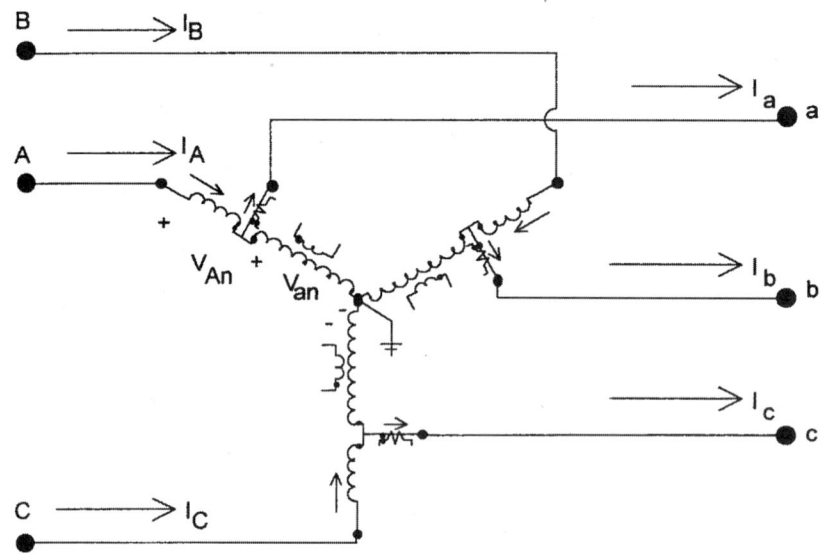

FIGURE 6.20 Wye connected regulators.

The value of the compensator settings in ohms are determined by:

$$R'_{ohms} + jX'_{ohms} = \frac{R'_{volts} + jX'_{volts}}{Ct_S} \quad \text{Ohms} \tag{6.107}$$

It is important to understand that the value of $R_{line\_ohms} + jX_{line\_ohms}$ is not the impedance of the line between the regulator and the load center. Typically the load center is located down the primary main feeder after several laterals have been tapped. As a result, the current measured by the CT of the regulator is not the current that flows all the way from the regulator to the load center. The proper way to determine the line impedance values is to run a power-flow program of the feeder without the regulator operating. From the output of the program, the voltages at the regulator output and the load center are known. Now the "equivalent" line impedance can be computed as:

$$R_{line} + jX_{line} = \frac{V_{regulator\_output} - V_{load\_center}}{I_{line}} \quad \text{ohms} \tag{6.108}$$

In Eq. (6.108), the voltages must be specified in system volts and the current in system amps.

Wye Connected Regulators

Three single-phase regulators connected in wye are shown in Fig. 6.20.

In Fig. 6.20 the polarities of the windings are shown in the "raise" position. When the regulator is in the "lower" position, a reversing switch will have reconnected the series winding so that the polarity on the series winding is now at the output terminal.

Regardless of whether the regulator is raising or lowering the voltage, the following equations apply:

Voltage Equations

$$\begin{bmatrix} V_{An} \\ V_{Bn} \\ V_{Cn} \end{bmatrix} = \begin{bmatrix} a_{R\_a} & 0 & 0 \\ 0 & a_{R\_b} & 0 \\ 0 & 0 & a_{R\_c} \end{bmatrix} \cdot \begin{bmatrix} V_{an} \\ V_{bn} \\ V_{cn} \end{bmatrix} \tag{6.109}$$

Distribution Systems

Equation (6.109) can be written in condensed form as:

$$[VLN_{ABC}] = [aRV_{abc}] \cdot [VLN_{abc}] \tag{6.110}$$

Also:

$$[VLN_{abc}] = [aRV_{ABC}] \cdot [VLN_{ABC}] \tag{6.111}$$

where

$$[aRV_{ABC}] = [aRV_{abc}]^{-1} \tag{6.112}$$

Current Equations

$$\begin{bmatrix} I_A \\ I_B \\ I_C \end{bmatrix} = \begin{bmatrix} \dfrac{1}{a_{R\_a}} & 0 & 0 \\ 0 & \dfrac{1}{a_{R\_b}} & 0 \\ 0 & 0 & \dfrac{1}{a_{R\_c}} \end{bmatrix} \cdot \begin{bmatrix} I_a \\ I_b \\ I_c \end{bmatrix} \tag{6.113}$$

Or:

$$[I_{ABC}] = [aRi_{abc}] \cdot [I_{abc}] \tag{6.114}$$

Also:

$$[I_{abc}] = [aRI_{ABC}] \cdot [I_{ABC}] \tag{6.115}$$

where

$$[aRI_{ABC}] = [aRI_{abc}]^{-1} \tag{6.116}$$

where $0.9 \leq a_{R\_abc} \leq 1.1$ in 32 steps of 0.625%/step (0.75 volts/step on 120 volt base).

Note: The effective turn ratios (a_{R_a}, a_{R_b}, and a_{R_c}) can take on different values when three single-phase regulators are connected in wye. It is also possible to have a three-phase regulator connected in wye where the voltage and current are sampled on only one phase and then all three phases are changed by the same value of a_R (number of taps).

Closed Delta Connected Regulators

Three single-phase regulators can be connected in a closed delta as shown in Fig. 6.21. In the figure, the regulators are shown in the "raise" position.

The closed delta connection is typically used in three-wire delta feeders. Note that the potential transformers for this connection are monitoring the load side line-to-line voltages and the current transformers are monitoring the load side line currents.

Applying the basic voltage and current Eqs. (6.91) through (6.97) of the regulator in the raise position, the following voltage and current relations are derived for the closed delta connection.

$$\begin{bmatrix} V_{AB} \\ V_{BC} \\ V_{CA} \end{bmatrix} = \begin{bmatrix} a_{R\_ab} & 1-a_{R\_bc} & 0 \\ 0 & a_{R\_bc} & 1-a_{R\_ca} \\ 1-a_{R\_ab} & 0 & a_{R\_ca} \end{bmatrix} \cdot \begin{bmatrix} V_{ab} \\ V_{bc} \\ V_{ca} \end{bmatrix} \tag{6.117}$$

Equation (6.115) in abbreviated form can be written as:

$$[VLL_{ABC}] = [aRVD_{abc}] \cdot [VLL_{abc}] \tag{6.118}$$

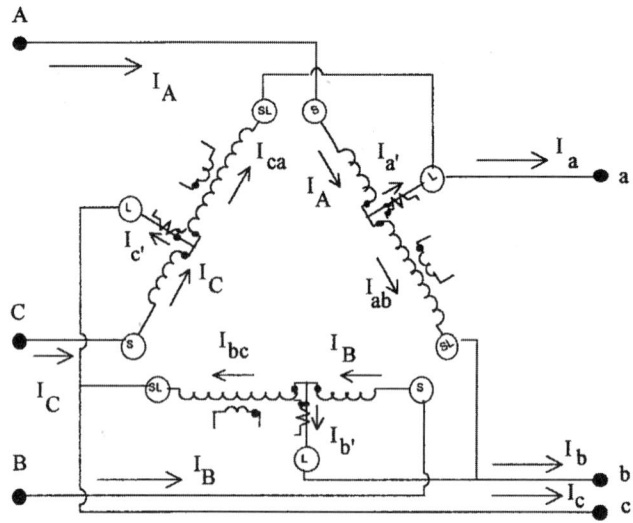

FIGURE 6.21 Delta connected regulators.

When the load side voltages are known, the source side voltages can be determined by:

$$[VLL_{abc}] = [aRVD_{ABC}] \cdot [VLL_{ABC}] \tag{6.119}$$

where

$$[aRVD_{ABC}] = [aRVD_{abc}]^{-1} \tag{6.120}$$

In similar manner the relationships between the load side and source side line currents are given by:

$$\begin{bmatrix} I_a \\ I_b \\ I_c \end{bmatrix} = \begin{bmatrix} a_{R\_ab} & 0 & 1-a_{R\_ca} \\ 1-a_{R\_ab} & a_{R\_bc} & 0 \\ 0 & 1-a_{R\_bc} & a_{R\_ca} \end{bmatrix} \cdot \begin{bmatrix} I_A \\ I_B \\ I_C \end{bmatrix} \tag{6.121}$$

Or:

$$[I_{abc}] = [AID_{ABC}] \cdot [I_{ABC}] \tag{6.122}$$

Also:

$$[I_{ABC}] = [AID_{abc}] \cdot [I_{abc}] \tag{6.123}$$

where

$$[IAD_{abc}] = [IAD_{ABC}]^{-1} \tag{6.124}$$

The closed delta connection can be difficult to apply. Note in both the voltage and current equations that a change of the tap position in one regulator will affect voltages and currents in two phases. As a result, increasing the tap in one regulator will affect the tap position of the second regulator. In most cases the bandwidth setting for the closed delta connection will have to be wider than that for wye connected regulators.

Open Delta Connection

Two single-phase regulators can be connected in the "open" delta connection. Shown in Fig. 6.22 is an open delta connection where two single-phase regulators have been connected between phases AB and CB.

Two other open connections can also be made where the single-phase regulators are connected between phases BC and AC and also between phases CA and BA.

Distribution Systems

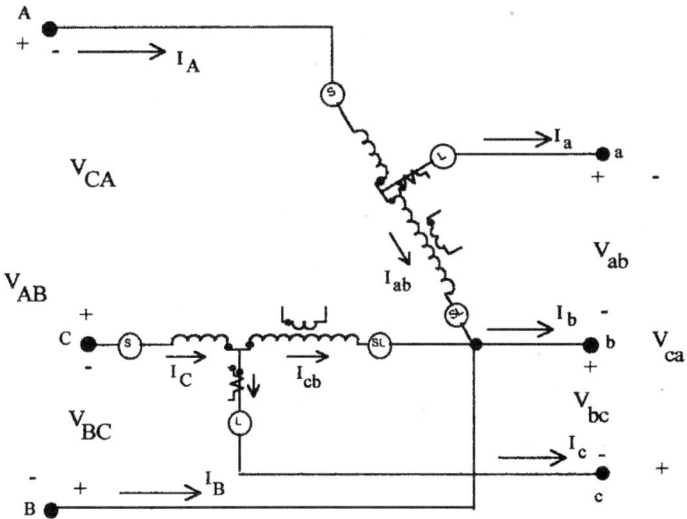

FIGURE 6.22 Open delta connection.

The open delta connection is typically applied to three-wire delta feeders. Note that the potential transformers monitor the line-to-line voltages and the current transformers monitor the line currents. Once again, the basic voltage and current relations of the individual regulators are used to determine the relationships between the source side and load side voltages and currents.

For all three open connections, the following general equations will apply:

$$[VLL_{ABC}] = [aRV_{abc}] \cdot [VLL_{abc}] \tag{6.125}$$

$$[VLL_{abc}] = [aRV_{ABC}] \cdot [VLL_{ABC}] \tag{6.126}$$

$$[I_{ABC}] = [aRI_{abc}] \cdot [I_{abc}] \tag{6.127}$$

$$[I_{abc}] = [aRI_{ABC}] \cdot [I_{ABC}] \tag{6.128}$$

The matrices for the three open connections are defined as follows:

Phases AB & CB

$$[aRV_{abc}] = \begin{bmatrix} a_{R\_A} & 0 & 0 \\ 0 & a_{R\_C} & 0 \\ -a_{R\_A} & -a_{R\_C} & 0 \end{bmatrix} \tag{6.129}$$

$$[aRV_{ABC}] = \begin{bmatrix} \dfrac{1}{a_{R\_A}} & 0 & 0 \\ 0 & \dfrac{1}{a_{R\_C}} & 0 \\ -\dfrac{1}{a_{R\_A}} & -\dfrac{1}{a_{R\_C}} & 0 \end{bmatrix} \tag{6.130}$$

$$[aRI_{abc}] = \begin{bmatrix} \dfrac{1}{a_{R\_A}} & 0 & 0 \\ -\dfrac{1}{a_{R\_A}} & 0 & -\dfrac{1}{a_{R\_C}} \\ 0 & 0 & \dfrac{1}{a_{R\_C}} \end{bmatrix} \tag{6.131}$$

$$[aRI_{ABC}] = \begin{bmatrix} a_{R\_A} & 0 & 0 \\ -a_{R\_A} & 0 & a_{R\_C} \\ 0 & 0 & a_{R\_C} \end{bmatrix} \tag{6.132}$$

Phases BC & AC

$$[aRV_{abc}] = \begin{bmatrix} 0 & -a_{R\_B} & -a_{R\_A} \\ 0 & a_{R\_B} & 0 \\ 0 & 0 & a_{R\_A} \end{bmatrix} \tag{6.133}$$

$$[aRV_{ABC}] = \begin{bmatrix} 0 & -\dfrac{1}{a_{R\_B}} & -\dfrac{1}{a_{R\_A}} \\ 0 & \dfrac{1}{a_{R\_B}} & 0 \\ 0 & 0 & \dfrac{1}{a_{R\_A}} \end{bmatrix} \tag{6.134}$$

$$[aRI_{abc}] = \begin{bmatrix} \dfrac{1}{a_{R\_A}} & 0 & 0 \\ 0 & \dfrac{1}{a_{R\_B}} & 0 \\ -\dfrac{1}{a_{R\_A}} & -\dfrac{1}{a_{R\_B}} & 0 \end{bmatrix} \tag{6.135}$$

$$[aRI_{ABC}] = \begin{bmatrix} a_{R\_A} & 0 & 0 \\ 0 & a_{R\_B} & 0 \\ -a_{R\_A} & -a_{R\_B} & 0 \end{bmatrix} \tag{6.136}$$

Phases CA & BA

$$[aRV_{abc}] = \begin{bmatrix} a_{R\_B} & 0 & 0 \\ -a_{R\_B} & 0 & -a_{R\_C} \\ 0 & 0 & a_{R\_C} \end{bmatrix} \tag{6.137}$$

Distribution Systems

$$[aRV_{ABC}] = \begin{bmatrix} \dfrac{1}{a_{R\_B}} & 0 & 0 \\ -\dfrac{1}{a_{R\_B}} & 0 & -\dfrac{1}{a_{R\_C}} \\ 0 & 0 & \dfrac{1}{a_{R\_C}} \end{bmatrix} \quad (6.138)$$

$$[aRI_{abc}] = \begin{bmatrix} 0 & -\dfrac{1}{a_{R\_B}} & -\dfrac{1}{a_{R\_C}} \\ 0 & \dfrac{1}{a_{R\_B}} & 0 \\ 0 & 0 & \dfrac{1}{a_{R\_C}} \end{bmatrix} \quad (6.139)$$

$$[aRI_{ABC}] = \begin{bmatrix} 0 & -a_{R\_B} & -a_{R\_C} \\ 0 & a_{R\_B} & 0 \\ 0 & 0 & a_{R\_C} \end{bmatrix} \quad (6.140)$$

Generalized Equations

The voltage regulator models used in power-flow studies are generalized for the various connections in a form similar to the ABCD parameters that are used in transmission line analysis. The general form of the power-flow models in matrix form are:

$$[V_{ABC}] = [a] \cdot [V_{abc}] + [b] \cdot [I_{abc}] \quad (6.141)$$

$$[I_{ABC}] = [c] \cdot [V_{abc}] + [d] \cdot [I_{abc}] \quad (6.142)$$

$$[V_{abc}] = [A] \cdot [V_{ABC}] - [B] \cdot [I_{abc}] \quad (6.143)$$

Depending upon the connection, the matrices $[V_{ABC}]$ and $[V_{abc}]$ can be either line-to-line or line-to-ground. The current matrices represent the line currents regardless of the regulator connection. For all voltage regulator connections, the generalized constants are defined as:

$$[a] = [aRV_{abc}] \quad (6.144)$$

$$[b] = [0] \quad (6.145)$$

$$[c] = [0] \quad (6.146)$$

$$[d] = [aRI_{abc}] \quad (6.147)$$

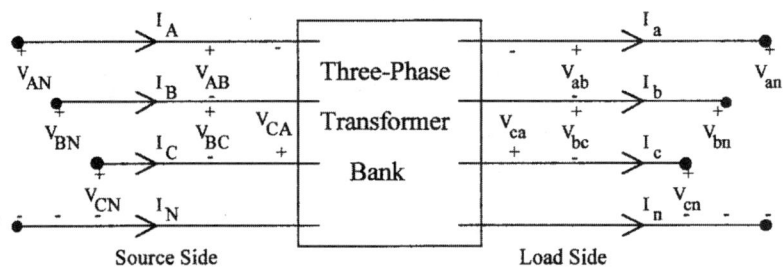

FIGURE 6.23 General transformer bank.

$$[A] = [aRV_{ABC}] \tag{6.148}$$

$$[B] = [0] \tag{6.149}$$

Transformer Bank Connections

Unique models of three-phase transformer banks applicable to radial distribution feeders have been developed (Kersting, 1999). Models for the following three-phase connections are included in this document:

- Delta-Grounded Wye
- Grounded Wye-Delta
- Ungrounded Wye-Delta
- Grounded Wye-Grounded Wye
- Delta-Delta

Figure 6.23 defines the various voltages and currents for the transformer bank models.

The models can represent a step-down (source side to load side) or a step-up (source side to load side) transformer bank. The notation is such that the capital letters **A,B,C,N** will always refer to the **source** side of the bank and the lower case letters **a,b,c,n** will always refer to the **load** side of the bank. It is assumed that all variations of the wye-delta connections are connected in the "American Standard Thirty Degree" connection. The standard is such that:

Step-down connection
 V_{AB} leads V_{ab} by 30°
 I_A leads I_a by 30°

Step-up connection
 V_{ab} leads V_{AB} by 30°
 I_a leads I_A by 30°

Generalized Equations
The models to be used in power-flow studies are generalized for the connections in a form similar to the ABCD parameters that are used in transmission line analysis. The general form of the power-flow models in matrix form are:

$$[V_{ABC}] = [a] \cdot [V_{abc}] + [b] \cdot [I_{abc}] \tag{6.150}$$

$$[I_{ABC}] = [c] \cdot [V_{abc}] + [d] \cdot [I_{abc}] \tag{6.151}$$

$$[V_{abc}] = [A] \cdot [V_{ABC}] - [B] \cdot [I_{abc}] \tag{6.152}$$

Distribution Systems

In Eqs. (6.150), (6.151), and (6.152), the matrices $[V_{ABC}]$ and $[V_{abc}]$ can be either line-to-line voltages (delta connection) or line-to-ground voltages (wye connection). The current matrices represent the line currents regardless of the transformer winding connection.

Common Variable and Matrices

All transformer models will use the following common variable and matrices:

- Transformer turns ratio: $\quad a_T = \dfrac{V_{rated\_source}}{V_{rated\_load}}$ (6.153)

where V_{rated_source} = transformer winding rating on the source side
V_{rated_load} = transformer winding rating on the load side

Note that the transformer "winding" ratings may be either line-to-line or line-to-neutral, depending upon the connection. The winding ratings can be specified in actual volts or per-unit volts using the appropriate base line-to-neutral voltages.

- Source to load matrix voltage relations: $\quad [V_{ABC}] = [AV] \cdot [V_{abc}]$ (6.154)

The voltage matrices may be line-to-line or line-to-neutral voltages depending upon the connection.

- Load to source matrix current relations: $\quad [I_{abc}] = [AI] \cdot [I_{ABC}]$ (6.155)

The current matrices may be line currents or delta currents depending upon the connection.

- Transformer impedance matrix: $\quad [Zt_{abc}] = \begin{bmatrix} Zt_a & 0 & 0 \\ 0 & Zt_b & 0 \\ 0 & 0 & Zt_c \end{bmatrix}$ (6.156)

The impedance elements in the matrix will be the per-unit impedance of the transformer windings on the load side of the transformer whether it is connected in wye or delta.

- Symmetrical component transformation matrix: $\quad [A_S] = \begin{bmatrix} 1 & 1 & 1 \\ 1 & a^2 & a \\ 1 & a & a^2 \end{bmatrix}$ (6.157)

where $a = 1\underline{/120}$

- Phase shift matrix: $\quad [T] = \begin{bmatrix} 1 & 0 & 0 \\ 0 & t & 0 \\ 0 & 0 & t^* \end{bmatrix}$ (6.158)

where $\quad t = \dfrac{1}{\sqrt{3}} \cdot \underline{/-30}$

- Matrix to convert line-to-line voltages to equivalent line-to-neutral voltages:

$$[W] = [A_S] \cdot [T] \cdot [A_S]^{-1} = \dfrac{1}{3} \cdot \begin{bmatrix} 2 & 1 & 0 \\ 0 & 2 & 1 \\ 1 & 0 & 2 \end{bmatrix}$$ (6.159)

Example: $[VLN] = [W] \cdot [VLL]$

- Matrix to convert delta currents into line currents:

$$[DI] = \begin{bmatrix} 1 & 0 & -1 \\ -1 & 1 & 0 \\ 0 & -1 & 1 \end{bmatrix} \qquad (6.160)$$

Example: $[I_{abc}] = [DI] \cdot [ID_{abc}]$

- Matrix to convert line-to-ground or line-to-neutral voltages to line-to-line voltages:

$$[D] = \begin{bmatrix} 1 & -1 & 0 \\ 0 & 1 & -1 \\ -1 & 0 & 1 \end{bmatrix} \qquad (6.161)$$

Example: $[VLL_{abc}] = [D] \cdot [VLN_{abc}]$

The matrices $[a]$, $[b]$, $[c]$, $[d]$, $[A]$, and $[B]$ [see Eqs. (6.150), (6.151), and (6.152)] for each connection are defined at the end of this section.

The Per-Unit System

All transformer models were developed so that they can be applied using either "actual" or "per-unit" values of voltages, currents, and impedances. When the per-unit system is used, all per-unit voltages (line-to-line and line-to-neutral) use the line-to-neutral base as the base voltage. In other words, for a balanced set of three-phase voltages, the per-unit line-to-neutral voltage magnitude will be 1.0 at rated voltage and the per-unit line-to-line voltage magnitude will be the square root of 3 (1.732). In similar fashion, all currents (line currents and delta currents) are based on the base line current. Again, a square root of 3 relationship will exist between the line and delta currents under balanced conditions. The base "line" impedance will be used for all line impedances and for wye and delta connected transformer impedances. There will be different base values on the two sides of the transformer bank.

Base values are computed following the steps listed below:

- Select a base three-phase kVA_{base} and the rated line-to-line voltage, $kVLL_{source}$, on the source side as the base line-to-line voltage.
- Based upon the voltage ratings of the transformer bank, determine the "rated" line-to-line voltage, $kVLL_{load}$, on the load side.
- Determine the "transformer ratio", n_T, as:

$$n_T = \frac{kVLL_{source}}{kVLL_{load}} \qquad (6.162)$$

- The "source" side base values are computed as:

$$kVLN_S = \frac{kVLL_S}{\sqrt{3}} \qquad (6.163)$$

$$I_S = \frac{kVA_{base}}{\sqrt{3} \cdot kVLL_{source}} \qquad (6.164)$$

Distribution Systems

$$Z_S = \frac{kVLL_{source}^2 \cdot 1000}{kVA_B} \tag{6.165}$$

- The load side base values are computed by:

$$kVLN_L = \frac{kVLN_S}{n_T} \tag{6.166}$$

$$I_L = n_T \cdot I_S \tag{6.167}$$

$$Z_L = \frac{Z_S}{n_T^2} \tag{6.168}$$

Thevenin Equivalent Circuit

The study of short circuit studies that occur on the load side of a transformer bank requires the three-phase Thevenin equivalent circuit referenced to the load-side terminals of the transformer. In order to determine this equivalent circuit, the Thevenin equivalent circuit up to the primary terminals of the "feeder" transformer must be determined. A block diagram of the total system is shown in Fig. 6.24.

In Fig. 6.24 the system voltage source will typically be a balanced set of per-unit voltages such that:

$$[Eth_{ABC}] = \begin{bmatrix} E_{AN} \\ E_{BN} \\ E_{CN} \end{bmatrix} = \begin{bmatrix} 1.0\underline{/0} \\ 1.0\underline{/-120} \\ 1.0\underline{/120} \end{bmatrix} \quad \text{per-unit} \tag{6.169}$$

The Thevenin equivalent impedance from the source to the primary terminals of the feeder transformer is given by:

$$[Zth_{ABC}] = [Zsys_{ABC}] + [Zsub_{ABC}] + [ZeqS_{ABC}] \tag{6.170}$$

The values of the source side Thevenin equivalent circuit will be the same regardless of the type of connection of the feeder transformer. The three-phase Thevenin equivalent circuit referenced to the load side of the feeder transformer is shown in Fig. 6.25.

For each three-phase transformer connection, unique values of the matrices $[Eth_{abc}]$ and $[Zth_{abc}]$ are defined as functions of the source side Thevenin equivalent circuit. These definitions are shown for each transformer connection below.

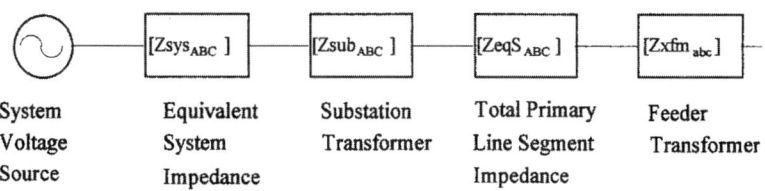

FIGURE 6.24 Total system.

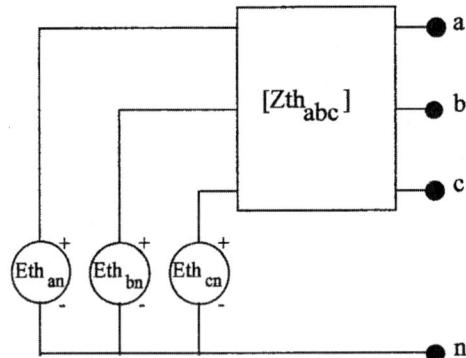

FIGURE 6.25 Three-phase Thevenin equivalent circuit.

Matrix Definitions

Delta-Grounded Wye

Power flow equations:

$$[VLL_{ABC}] = [a] \cdot [VLG_{abc}] + [b] \cdot [I_{abc}]$$

$$[I_{ABC}] = [c] \cdot [VLG_{abc}] + [d] \cdot [I_{abc}]$$

$$[a] = [AV0]$$

$$[b] = [AV0] \cdot [ZNt_{abc}]$$

$$[c] = [0]$$

$$[d] = [DY]$$

$$[VLG_{abc}] = [A] \cdot [VLL_{ABC}] - [B] \cdot [I_{abc}]$$

$$[A] = [AV]^{-1}$$

$$[B] = [ZNt_{abc}]$$

Thevenin equations:

$$[Eth_{abc}] = [AV]^{-1} \cdot [D] \cdot [Eth_{ABC}]$$

$$[Zth_{abc}] = [AV]^{-1} \cdot [D] \cdot [Zth_{ABC}] \cdot [DY] + [Zt_{abc}]$$

The matrices used for the step-down connection are:

$$[AV0] = \frac{a_T}{3} \cdot \begin{bmatrix} 1 & -2 & 1 \\ 1 & 1 & -2 \\ -2 & 1 & 1 \end{bmatrix} \quad [AV]^{-1} = \begin{bmatrix} 0 & 0 & -\frac{1}{a_T} \\ -\frac{1}{a_T} & 0 & 0 \\ 0 & -\frac{1}{a_T} & 0 \end{bmatrix}$$

$$[DY] = \begin{bmatrix} \dfrac{1}{a_T} & -\dfrac{1}{a_T} & 0 \\ 0 & \dfrac{1}{a_T} & -\dfrac{1}{a_T} \\ -\dfrac{1}{a_T} & 0 & \dfrac{1}{a_T} \end{bmatrix} \quad [D] = \begin{bmatrix} 1 & -1 & 0 \\ 0 & 1 & -1 \\ -1 & 0 & 1 \end{bmatrix}$$

$$[ZNt_{abc}] = \begin{bmatrix} Zt_a & 0 & 0 \\ 0 & Zt_b & 0 \\ 0 & 0 & Zt_c \end{bmatrix}$$

The matrices used for the step-up connection are:

$$[AV0] = \dfrac{a_T}{3} \cdot \begin{bmatrix} 2 & -1 & -1 \\ -1 & 2 & -1 \\ -1 & -1 & 2 \end{bmatrix} \quad [AV]^{-1} = \begin{bmatrix} \dfrac{1}{a_T} & 0 & 0 \\ 0 & \dfrac{1}{a_T} & 0 \\ 0 & 0 & \dfrac{1}{a_T} \end{bmatrix}$$

$$[DY] = \begin{bmatrix} \dfrac{1}{a_T} & 0 & -\dfrac{1}{a_T} \\ -\dfrac{1}{a_T} & \dfrac{1}{a_T} & 0 \\ 0 & -\dfrac{1}{a_T} & \dfrac{1}{a_T} \end{bmatrix} \quad [D] = \begin{bmatrix} 1 & -1 & 0 \\ 0 & 1 & -1 \\ -1 & 0 & 1 \end{bmatrix}$$

$$[ZNt_{abc}] = [Zt_{abc}] = \begin{bmatrix} Zt_a & 0 & 0 \\ 0 & Zt_b & 0 \\ 0 & 0 & Zt_c \end{bmatrix}$$

Grounded Wye-Delta

Power flow equations:

$$[VLG_{ABC}] = [a] \cdot [VLL_{abc}] + [b] \cdot [I_{abc}]$$

$$[I_{ABC}] = [c] \cdot [VLL_{abc}] + [d] \cdot [I_{abc}]$$

$$[a] = [EV] \cdot [AV]$$

$$[b] = [EV] \cdot [AV] \cdot [ZNt_{abc}] \cdot [H1]$$

$$[c] = [H2] \cdot [EV] \cdot [AV]$$

$$[d] = [H1] - [H2] \cdot [EV] \cdot [AV] \cdot [ZNt_{abc}] \cdot [H1]$$

$$[VLL_{abc}] = [A] \cdot [VLG_{ABC}] - [B] \cdot [I_{abc}]$$

$$[A] = [AV]^{-1} - [ZNt_{abc}] \cdot [H2]$$

$$[B] = [ZNt_{abc}] \cdot [H1]$$

Thevenin equations:

$$[Eth_{abc}] = [Eth_{ABC}] \underline{/\pm 30}, \quad +30 \text{ for step-up}, \quad -30 \text{ for step-down}$$

$$[Zth_{abc}] = [W] \cdot \left([K1] \cdot [K]^{-1} \cdot [ZeqS_{ABC}] + [ZNt_{abc}]\right) \cdot [H1]$$

Sub-matrices used for the step-down and the step-up connections are:

$$[ZNt_{abc}] = [Zt_{abc}] \cdot [AI]$$

$$[K1] = [AV]^{-1} - [ZNt_{abc}] \cdot [H2]$$

$$[K] = [U] + [Zth_{ABC}] \cdot [H2]$$

$$[EV] = [U] - [AV] \cdot [ZNt_{abc}] \cdot [H2]$$

Matrices used for the step-down connection are:

$$[AV] = \begin{bmatrix} a_T & 0 & 0 \\ 0 & a_T & 0 \\ 0 & 0 & a_T \end{bmatrix} \quad [AI] = \begin{bmatrix} a_T & 0 & 0 \\ 0 & a_T & 0 \\ 0 & 0 & a_T \end{bmatrix}$$

$$[DY] = \begin{bmatrix} a_T & 0 & -a_T \\ -a_T & a_T & 0 \\ 0 & -a_T & a_T \end{bmatrix} \quad [U] = \begin{bmatrix} 1 & 0 & 0 \\ 0 & 1 & 0 \\ 0 & 0 & 1 \end{bmatrix}$$

$$[Zt_{abc}] = \begin{bmatrix} Zt_{ab} & 0 & 0 \\ 0 & Zt_{bc} & 0 \\ 0 & 0 & Zt_{ca} \end{bmatrix}$$

$$[H1] = \frac{1}{a_T \cdot (Zt_{ab} + Zt_{bc} + Zt_{ca})} \cdot \begin{bmatrix} Zt_{ca} & -Zt_{bc} & 0 \\ Zt_{ca} & Zt_{ab} + Zt_{cb} & 0 \\ -Zt_{ab} - Zt_{bc} & -Zt_{bc} & 0 \end{bmatrix}$$

$$[H2] = \frac{1}{a_T \cdot (Zt_{ab} + Zt_{bc} + Zt_{ca})} \cdot \begin{bmatrix} 1 & 1 & 1 \\ 1 & 1 & 1 \\ 1 & 1 & 1 \end{bmatrix}$$

Distribution Systems

The matrices used for the step-up connection are:

$$[AV] = \begin{bmatrix} 0 & 0 & -a_T \\ -a & 0 & 0 \\ 0 & -a & 0 \end{bmatrix} \quad [AI] = \begin{bmatrix} 0 & -a_T & 0 \\ 0 & 0 & -a_T \\ -a_T & 0 & 0 \end{bmatrix}$$

$$[DY] = \begin{bmatrix} a_T & -a_T & 0 \\ 0 & a_T & -a_T \\ -a_T & 0 & a_T \end{bmatrix} \quad [U] = \begin{bmatrix} 1 & 0 & 0 \\ 0 & 1 & 0 \\ 0 & 0 & 1 \end{bmatrix}$$

$$[Zt_{abc}] = \begin{bmatrix} Zt_{ab} & 0 & 0 \\ 0 & Zt_{bc} & 0 \\ 0 & 0 & Zt_{ca} \end{bmatrix}$$

$$[H1] = \frac{1}{a_T \cdot (Zt_{ab} + Zt_{bc} + Zt_{ca})} \cdot \begin{bmatrix} Zt_{ab} + Zt_{bc} & Zt_{bc} & 0 \\ -Zt_{ca} & Zt_{bc} & 0 \\ -Zt_{ca} & -(Zt_{ab} + Zt_{ca}) & 0 \end{bmatrix}$$

$$[H2] = \frac{1}{a_T \cdot (Zt_{ab} + Zt_{bc} + Zt_{ca})} \cdot \begin{bmatrix} -1 & -1 & -1 \\ -1 & -1 & -1 \\ -1 & -1 & -1 \end{bmatrix}$$

Ungrounded Wye-Delta

Power flow equations:

$$[VLN_{ABC}] = [a] \cdot [VLL_{abc}] + [b] \cdot [I_{abc}]$$

$$[I_{ABC}] = [c] \cdot [VLL_{abc}] + [d] \cdot [I_{abc}]$$

$$[a] = [AV]$$

$$[b] = [ZDt_{abc}]$$

$$[c] = [0]$$

$$[d] = [H]$$

$$[VLL_{abc}] = [A] \cdot [VLN_{ABC}] - [B] \cdot [I_{abc}]$$

$$[A] = [AV]^{-1}$$

$$[B] = [ZNt_{abc}] \cdot [H]$$

Thevenin equations:

$$[Eth_{abc}] = [Eth_{ABC}]/\pm 30, \quad +30 \text{ for step-up}, \; -30 \text{ for step-down}$$

$$[Zth_{abc}] = [W] \cdot \left([AV]^{-1} \cdot [W0] \cdot [ZeqS_{ABC}] + [ZNt_{ABC}]\right) \cdot [H]$$

The matrices used used for the step-down connection are:

$$[AV] = \begin{bmatrix} a_T & 0 & 0 \\ 0 & a_T & 0 \\ 0 & 0 & a_T \end{bmatrix} \quad [AI] = \begin{bmatrix} a_T & 0 & 0 \\ 0 & a_T & 0 \\ 0 & 0 & a_T \end{bmatrix}$$

$$[DY] = \begin{bmatrix} a_T & 0 & -a_T \\ -a_T & a_T & 0 \\ 0 & -a_T & a_T \end{bmatrix} \quad [H] = \frac{1}{3 \cdot a_T} \cdot \begin{bmatrix} 1 & -1 & 0 \\ 1 & 2 & 0 \\ -2 & -1 & 0 \end{bmatrix}$$

$$[ZNt_{abc}] = \begin{bmatrix} a_T \cdot Zt_{ab} & 0 & 0 \\ 0 & a_T \cdot Zt_{bc} & 0 \\ 0 & 0 & a_T \cdot Zt_{ca} \end{bmatrix}$$

$$[ZDt_{abc}] = \frac{a_T}{3} \cdot \begin{bmatrix} Zt_{ab} & -Zt_{ab} & 0 \\ Zt_{bc} & 2 \cdot Zt_{bc} & 0 \\ -2 \cdot Zt_{ca} & -Zt_{ca} & 0 \end{bmatrix}$$

The matrices used for the step-up connection are:

$$[AV] = \begin{bmatrix} 0 & 0 & -a_T \\ -a_T & 0 & 0 \\ 0 & -a_T & 0 \end{bmatrix} \quad [AI] = \begin{bmatrix} 0 & -\dfrac{1}{a_T} & 0 \\ 0 & 0 & -\dfrac{1}{a_T} \\ -\dfrac{1}{a_T} & 0 & 0 \end{bmatrix}$$

$$[DY] = \begin{bmatrix} a_T & -a_T & 0 \\ 0 & a_T & -a_T \\ -a_T & 0 & a_T \end{bmatrix} \quad [H] = \frac{1}{3 \cdot a_T} \cdot \begin{bmatrix} 2 & 1 & 0 \\ -1 & 1 & 0 \\ -1 & -2 & 0 \end{bmatrix}$$

$$[ZDt_{abc}] = \frac{a_T}{3} \cdot \begin{bmatrix} 2 \cdot Zt_{ca} & Zt_{ca} & 0 \\ -Zt_{ab} & Zt_{ab} & 0 \\ -Zt_{bc} & -2 \cdot Zt_{bc} & 0 \end{bmatrix}$$

Distribution Systems

$$[ZNt_{abc}] = \begin{bmatrix} 0 & -a_T \cdot Zt_{ab} & 0 \\ 0 & 0 & -a_T \cdot Zt_{bc} \\ a_T \cdot Zt_{ca} & 0 & 0 \end{bmatrix}$$

The Grounded Wye-Grounded Wye Connection

Power flow equations:

$$[VLG_{ABC}] = [a] \cdot [VLG_{abc}] + [b] \cdot [I_{abc}]$$

$$[I_{ABC}] = [c] \cdot [VLG_{abc}] + [d] \cdot [I_{abc}]$$

$$[a] = [AV]$$

$$[b] = [ZNt_{abc}]$$

$$[c] = [0]$$

$$[d] = [AI]^{-1}$$

$$[VLG_{abc}] = [A] \cdot [VLG_{ABC}] - [B] \cdot [I_{ABC}]$$

$$[A] = [AV]^{-1}$$

$$[B] = [AV]^{-1} \cdot [ZNt_{abc}]$$

Thevenin equations

$$[Eth_{abc}] = [AV]^{-1} \cdot [Eth_{ABC}]$$

$$[Zth_{abc}] = [AV]^{-1} \cdot \left([ZeqS_{ABC}] \cdot [AV]^{-1} + [ZNt_{abc}] \right)$$

The matrices used are:

$$[AV] = [AI] = \begin{bmatrix} a_T & 0 & 0 \\ 0 & a_T & 0 \\ 0 & 0 & a_T \end{bmatrix} \quad [AV]^{-1} = [AI]^{-1} \cdot \begin{bmatrix} \dfrac{1}{a_T} & 0 & 0 \\ 0 & \dfrac{1}{a_T} & 0 \\ 0 & 0 & \dfrac{1}{a_T} \end{bmatrix}$$

$$[Zt_{abc}] = \begin{bmatrix} Zt_a & 0 & 0 \\ 0 & Zt_b & 0 \\ 0 & 0 & Zt_c \end{bmatrix} \quad [ZNt_{abc}] = \begin{bmatrix} a_T \cdot Zt_a & 0 & 0 \\ 0 & a_T \cdot Zt_b & 0 \\ 0 & 0 & a_T \cdot Zt_c \end{bmatrix}$$

Delta-Delta

Power flow equations:

$$[VLL_{ABC}] = [a] \cdot [VLL_{abc}] + [b] \cdot [I_{abc}]$$

$$[I_{ABC}] = [c] \cdot [VLL_{abc}] + [d] \cdot [I_{abc}]$$

$$[a] = [AV]$$

$$[b] = [Zt_{abc}] \cdot [G1]$$

$$[c] = [0]$$

$$[d] = [AI]^{-1}$$

$$[VLL_{abc}] = [A] \cdot [VLL_{ABC}] - [B] \cdot [I_{abc}]$$

$$[A] = [AV]^{-1}$$

$$[B] = [Zt_{abc}] \cdot [G1]$$

Thevenin equations:

$$[Eth_{abc}] = [W] \cdot [AV]^{-1} \cdot [D] \cdot [W] \cdot [ELL_S]$$

$$[Zth_{abc}] = [W] \cdot \left([AV]^{-1} \cdot [D] \cdot [ZeqS_{ABC}] \cdot [AV]^{-1} + [Zt_{abc}] \cdot [G1] \right)$$

The matrices used are:

$$[AV] = \begin{bmatrix} a_T & 0 & 0 \\ 0 & a_T & 0 \\ 0 & 0 & a_T \end{bmatrix} \quad [AI] = \begin{bmatrix} a_T & 0 & 0 \\ 0 & a_T & 0 \\ 0 & 0 & a_T \end{bmatrix}$$

$$[D] = \begin{bmatrix} 1 & -1 & 0 \\ 0 & 1 & -1 \\ -1 & 0 & 1 \end{bmatrix} \quad [W] = \frac{1}{3} \cdot \begin{bmatrix} 2 & 1 & 0 \\ 0 & 2 & 1 \\ 1 & 0 & 2 \end{bmatrix}$$

$$[Zt_{abc}] = \begin{bmatrix} Zt_{ab} & 0 & 0 \\ 0 & Zt_{bc} & 0 \\ 0 & 0 & Zt_{ca} \end{bmatrix}$$

$$[G1] = \frac{1}{Zt_{ab} + Zt_{bc} + Zt_{ca}} \cdot \begin{bmatrix} Zt_{ca} & -Zt_{bc} & 0 \\ Zt_{ca} & Zt_{ab} + Zt_{ca} & 0 \\ -Zt_{ab} - Zt_{bc} & -Zt_{bc} & 0 \end{bmatrix}$$

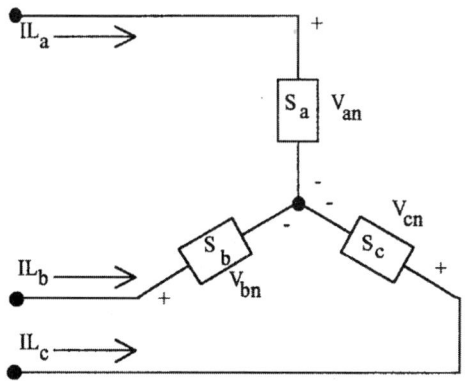

FIGURE 6.26 Wye connected load.

Load Models

Loads can be represented as being connected phase-to-phase or phase-to-neutral in a four-wire wye systems or phase-to-phase in a three-wire delta system. The loads can be three-phase, two-phase, or single-phase with any degree of unbalance and can be modeled as:

- Constant real and reactive power (constant PQ)
- Constant current
- Constant impedance
- Any combination of the above

The load models developed in this document are used in the iterative process of a power-flow program. All models are initially defined by a complex power per phase and either a line-to-neutral (wye load) or a line-to-line voltage (delta load). The units of the complex power can be in volt-amperes and volts or per-unit volt-amperes and per-unit volts.

For both the wye and delta connected loads, the basic requirement is to determine the load component of the line currents coming into the loads. It is assumed that all loads are initially specified by their complex power (S = P + jQ) per phase and a line-to-neutral or line-to-line voltage.

Wye Connected Loads

Figure 6.26 shows the model of a wye connected load.

The notation for the specified complex powers and voltages are as follows:

$$\text{Phase a: } |S_a|\underline{/\theta_a} = P_a + jQ_a \text{ and } |V_{an}|\underline{/\delta_a} \tag{6.171}$$

$$\text{Phase b: } |S_b|\underline{/\theta_b} = P_b + jQ_b \text{ and } |V_{bn}|\underline{/\delta_b} \tag{6.172}$$

$$\text{Phase c: } |S_c|\underline{/\theta_c} = P_c + jQ_c \text{ and } |V_{cn}|\underline{/\delta_c} \tag{6.173}$$

1. Constant Real and Reactive Power Loads

$$IL_a = \left(\frac{S_a}{V_{an}}\right)^* = \frac{|S_a|}{|V_{an}|}\underline{/\delta_a - \theta_a} = |IL_a|\underline{/\alpha_a} \tag{6.174}$$

$$IL_b = \left(\frac{S_b}{V_{bn}}\right)^* = \frac{|S_b|}{|V_{bn}|}/\delta_b - \theta_b = |IL_b|/\alpha_b$$

$$IL_c = \left(\frac{S_c}{V_{cn}}\right)^* = \frac{|S_c|}{|V_{cn}|}/\delta_c - \theta_c = |IL_c|/\alpha_c$$

In this model the line-to-neutral voltages will change during each iteration until convergence is achieved.

2. Constant Impedance Loads

The "constant load impedance" is first determined from the specified complex power and line-to-neutral voltages according to Eq. (6.175).

$$Z_a = \frac{|V_{an}|^2}{S_a^*} = \frac{|V_{an}|^2}{|S_a|}/\theta_a = |Z_a|/\theta_a$$

$$Z_b = \frac{|V_{bn}|^2}{S_b^*} = \frac{|V_{bn}|^2}{|S_b|}/\theta_b = |Z_b|/\theta_b \qquad (6.175)$$

$$Z_c = \frac{|V_{cn}|^2}{S_c^*} = \frac{|V_{cn}|^2}{|S_c|}/\theta_c = |Z_c|/\theta_c$$

The load currents as a function of the "constant load impedances" are given by Eq. (6.176).

$$IL_a = \frac{V_{an}}{Z_a} = \frac{|V_{an}|}{|Z_a|}/\delta_a - \theta_a = |IL_a|/\alpha_a$$

$$IL_b = \frac{V_{bn}}{Z_b} = \frac{|V_{bn}|}{|Z_b|}/\delta_b - \theta_b = |IL_b|/\alpha_b \qquad (6.176)$$

$$IL_c = \frac{V_{cn}}{Z_c} = \frac{|V_{cn}|}{|Z_c|}/\delta_c - \theta_c = |IL_c|/\alpha_c$$

In this model the line-to-neutral voltages in Eq. (6.176) will change during each iteration until convergence is achieved.

3. Constant Current Loads

In this model the magnitudes of the currents are computed according to Eq. (6.174) and then held constant while the angle of the voltage (δ) changes during each iteration. This keeps the power factor of the load constant.

$$IL_a = |IL_a|/\delta_a - \theta_a$$

$$IL_b = |IL_b|/\delta_b - \theta_b \qquad (6.177)$$

$$IL_c = |IL_c|/\delta_c - \theta_c$$

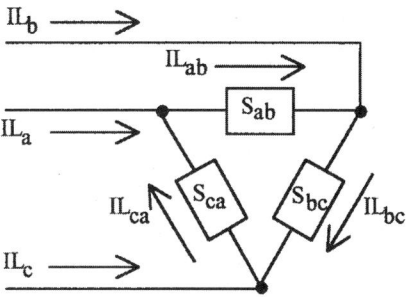

FIGURE 6.27 Delta connected load.

4. Combination Loads
   Combination loads can be modeled by assigning a percentage of the total load to each of the above three load models. The total line current entering the load is the sum of the three components.

Delta Connected Loads

Figure 6.27 shows the model of a delta connected load.

The notation for the specified complex powers and voltages are as follows:

$$\text{Phase ab: } |S_{ab}|\underline{/\theta_{ab}} = P_{ab} + jQ_{ab} \quad \text{and} \quad |V_{ab}|\underline{/\delta_{ab}} \tag{6.178}$$

$$\text{Phase bc: } |S_{bc}|\underline{/\theta_{bc}} = P_{bc} + jQ_{bc} \quad \text{and} \quad |V_{bc}|\underline{/\delta_{bc}} \tag{6.179}$$

$$\text{Phase ca: } |S_{ca}|\underline{/\theta_{ca}} = P_{ca} + jQ_{ca} \quad \text{and} \quad |V_{ca}|\underline{/\delta_{ca}} \tag{6.180}$$

1. Constant Real and Reactive Power Loads

$$\begin{aligned} IL_{ab} &= \left(\frac{S_{ab}}{V_{ab}}\right)^* = \frac{|S_{ab}|}{|V_{ab}|}\underline{/\delta_{ab} - \theta_{ab}} = |IL_{ab}|\underline{/\alpha_{ab}} \\ IL_{bc} &= \left(\frac{S_{bc}}{V_{bc}}\right)^* = \frac{|S_{bc}|}{|V_{bc}|}\underline{/\delta_{bc} - \theta_{bc}} = |IL_{bc}|\underline{/\alpha_{bc}} \\ IL_{ca} &= \left(\frac{S_{ca}}{V_{ca}}\right)^* = \frac{|S_{ca}|}{|V_{ca}|}\underline{/\delta_{ca} - \theta_{ca}} = |IL_{ca}|\underline{/\alpha_{ca}} \end{aligned} \tag{6.181}$$

In this model the line-to-line voltages will change during each iteration until convergence is achieved.

2. Constant Impedance Loads
   The "constant load impedance" is first determined from the specified complex power and line-to-neutral voltages according to Eq. (6.182).

$$Z_{ab} = \frac{|V_{ab}|^2}{S_{ab}^*} = \frac{|V_{ab}|^2}{|S_{ab}|}\underline{/\theta_{ab}} = |Z_{ab}|\underline{/\theta_{ab}} \tag{6.182}$$

$$Z_{bc} = \frac{|V_{bc}|^2}{S_{bc}^*} = \frac{|V_{bc}|^2}{|S_{bc}|}/\theta_{bc} = |Z_{bc}|/\theta_{bc}$$

$$Z_{ca} = \frac{|V_{ca}|^2}{S_{ca}^*} = \frac{|V_{ca}|^2}{|S_{ca}|}/\theta_{ca} = |Z_{ca}|/\theta_{ca}$$

The load currents as a function of the "constant load impedances" are given by Eq. (6.173).

$$IL_{ab} = \frac{V_{ab}}{Z_{ab}} = \frac{|V_{anb}|}{|Z_{ab}|}/\delta_{ab} - \theta_{ab} = |IL_{ab}|/\alpha_{ab}$$

$$IL_{bc} = \frac{V_{bc}}{Z_{bc}} = \frac{|V_{bc}|}{|Z_{bc}|}/\delta_{bc} - \theta_{bc} = |IL_{bc}|/\alpha_{bc} \qquad (6.183)$$

$$IL_{ca} = \frac{V_{ca}}{Z_{ca}} = \frac{|V_{ca}|}{|Z_{ca}|}/\delta_{ca} - \theta_{ca} = |IL_{ca}|/\alpha_{ca}$$

In this model the line-to-neutral voltages in Eq. (6.183) will change during each iteration until convergence is achieved.

3. Constant Current Loads

   In this model the magnitudes of the currents are computed according to Eq. (6.181) and then held constant while the angle of the voltage (δ) changes during each iteration. This keeps the power factor of the load constant.

$$IL_{ab} = |IL_{ab}|/\delta_{ab} - \theta_{ab}$$

$$IL_{bc} = |IL_{bc}|/\delta_{bc} - \theta_{bc} \qquad (6.184)$$

$$IL_{ca} = |IL_{ca}|/\delta_{ca} - \theta_{ca}$$

4. Combination Loads

   Combination loads can be modeled by assigning a percentage of the total load to each of the above three load models. The total delta current for each load is the sum of the three components.

The line currents entering the delta connected load are determined by:

$$\begin{bmatrix} IL_a \\ IL_b \\ IL_c \end{bmatrix} = \begin{bmatrix} 1 & 0 & -1 \\ -1 & 1 & 0 \\ 0 & -1 & 1 \end{bmatrix} \cdot \begin{bmatrix} IL_{ab} \\ IL_{bc} \\ IL_{ca} \end{bmatrix} \qquad (6.185)$$

In both the wye and delta connected loads, single-phase and two-phase loads are modeled by setting the complex powers of the missing phases to zero. In other words, all loads are modeled as three-phase loads and by setting the complex power of the missing phases to zero, the only load currents computed using the above equations will be for the non-zero loads.

Distribution Systems

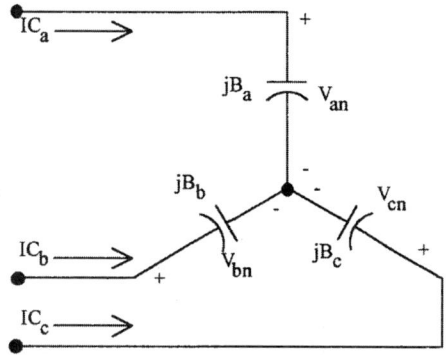

FIGURE 6.28 Wye connected capacitor bank.

Shunt Capacitor Models

Shunt capacitor banks are commonly used in a distribution system to help in voltage regulation and to provide reactive power support. The capacitor banks are modeled as constant susceptances connected in either wye or delta. Similar to the load model, all capacitor banks are modeled as three-phase banks with the kVAr of missing phases set to zero for single-phase and two-phase banks.

Wye Connected Capacitor Bank

A wye connected capacitor bank is shown in Fig. 6.28.

The individual phase capacitor units are specified in kVAr and kV. The constant susceptance for each unit can be computed in either Siemans or per-unit. When per-unit is desired, the specified kVAr of the capacitor must be divided by the base single-phase kVAr and the kV must be divided by the base line-to-neutral kV.

The susceptance of a capacitor unit is computed by:

$$B_{actual} = \frac{kVAr}{kV^2 \cdot 1000} \quad \text{Siemans} \tag{6.186}$$

$$B_{pu} = \frac{kVAr_{pu}}{V_{pu}^2} \quad \text{Per-unit} \tag{6.187}$$

where

$$kVAr_{pu} = \frac{kVAr_{actual}}{kVA_{single\_phase\_base}} \tag{6.188}$$

$$V_{pu} = \frac{kV_{actual}}{kV_{line\_to\_neutral\_base}} \tag{6.189}$$

The per-unit value of the susceptance can also be determined by first computing the actual value [Eq. (6.186)] and then dividing by the base admittance of the system.

With the susceptance computed, the line currents serving the capacitor bank are given by:

$$\begin{aligned} IC_a &= jB_a \cdot V_{an} \\ IC_b &= jB_b \cdot V_{bn} \\ IC_c &= jB_c \cdot V_{cn} \end{aligned} \tag{6.190}$$

Delta Connected Capacitor Bank

A delta connected capacitor bank is shown in Fig. 6.29.

Equations (6.186) through (6.189) can be used to determine the value of the susceptance in actual Siemans and/or per-unit. It should be pointed out that in this case, the kV will be a line-to-line value of the voltage. Also, it should be noted that in Eq. (6.189), the base line-to-neutral voltage is used to compute the per-unit line-to-line voltage. This is a variation from the usual application of the per-unit system where the actual line-to-line voltage would be divided by a base line-to-line voltage in order to get the per-unit line-to-line voltage. That is not done here so that under normal conditions, the per-unit line-to-line voltage will have a magnitude of $\sqrt{3}$ rather than 1.0. This is done so that KCL at each node of the delta connection will apply for either the actual or per-unit delta currents.

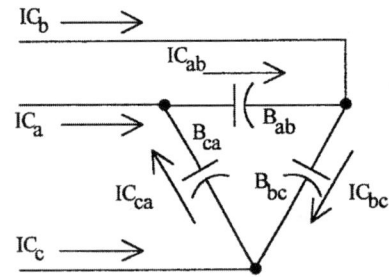

FIGURE 6.29 Delta connected capacitor bank.

The currents flowing in the delta connected capacitors are given by:

$$IC_{ab} = jB_{ab} \cdot V_{ab}$$
$$IC_{bc} = jB_{bc} \cdot V_{bc} \quad (6.191)$$
$$IC_{ca} = jB_{ca} \cdot V_{ca}$$

The line currents feeding the delta connected capacitor bank are given by:

$$\begin{bmatrix} IC_a \\ IC_b \\ IC_c \end{bmatrix} = \begin{bmatrix} 1 & 0 & -1 \\ -1 & 1 & 0 \\ 0 & -1 & 1 \end{bmatrix} \cdot \begin{bmatrix} IC_{ab} \\ IC_{bc} \\ IC_{ca} \end{bmatrix} \quad (6.192)$$

Analysis

Power Flow Analysis

The power-flow analysis of a distribution feeder is similar to that of an interconnected transmission system. Typically what will be known prior to the analysis will be the three-phase voltages at the substation and the complex power of all of the loads and the load model (constant complex power, constant impedance, constant current or a combination). Sometimes, the input complex power supplied to the feeder from the substation is also known.

In the areas of this section entitled "Line Segment Models", "Step-Voltage Regulators", and "Transformer Bank Connections", phase frame models were presented for the series components of a distribution feeder. In the areas entitled "Load Models" and "Shunt Capacitor Models", models were presented for the shunt components (loads and capacitor banks). These models are used in the "power-flow" analysis of a distribution feeder.

A power-flow analysis of a feeder can determine the following by phase and total three-phase:

- Voltage magnitudes and angles at all nodes of the feeder
- Line flow in each line section specified in kW and kVAr, amps and degrees or amps and power factor
- Power loss in each line section
- Total feeder input kW and kVAr
- Total feeder power losses
- Load kW and kVAr based upon the specified model for the load

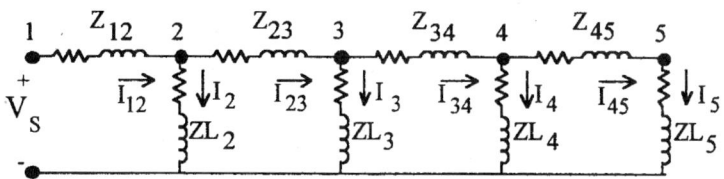

FIGURE 6.30 Linear ladder network.

Because the feeder is radial, iterative techniques commonly used in transmission network power-flow studies are not used because of poor convergence characteristics (Trevino, 1970). Instead, an iterative technique specifically designed for a radial system is used. The "ladder iterative technique" (Kersting and Mendive, 1976) will be presented here.

The Ladder Iterative Technique

Linear Network

A modification of the "ladder" network theory of linear systems provides a robust iterative technique for power-flow analysis. A distribution feeder is non-linear because most loads are assumed to be constant kW and kVAr. However, the approach taken for the linear system can be modified to take into account the non-linear characteristics of the distribution feeder.

For the ladder network in Fig. 6.30 it is assumed that all of the line impedances and load impedances are known along with the voltage at the source (V_s). The solution for this network is to assume a voltage at the most remote load (V_5). The load current I_5 is then determined as:

$$I_5 = \frac{V_5}{ZL_5} \qquad (6.193)$$

For this "end node" case, the line current I_{45} is equal to the load current I_5. The voltage at node 4 (V_4) can be determined using Kirchhoff's Voltage Law:

$$V_4 = V_5 + Z_{45} \cdot I_{45} \qquad (6.194)$$

The load current I_4 can be determined and then Kirchhoff's Current Law applied to determine the line current I_{34}.

$$I_{34} = I_{45} + I_4 \qquad (6.195)$$

Kirchhoff's Voltage Law is applied to determine the node voltage V_3. This procedure is continued until a voltage (V_1) has been computed at the source. The computed voltage V_1 is compared to the specified voltage V_s. There will be a difference between these two voltages. The ratio of the specified voltage to the compute voltage can be determined as:

$$Ratio = \frac{V_S}{V_1} \qquad (6.196)$$

Since the network is linear, all of the line and load currents and node voltages in the network can be multiplied by the Ratio for the final solution to the network.

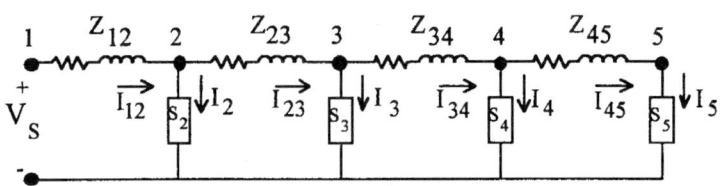

FIGURE 6.31 Non-linear ladder network.

Non-Linear Network

The linear network of Fig. 6.30 is modified to a non-linear network by replacing all of the constant load impedances by constant complex power loads as shown in Fig. 6.31.

The procedure outlined for the linear network is applied initially to the non-linear network. The only difference being that the load current at each node is computed by:

$$I_n = \left(\frac{S_n}{V_n}\right)^* \tag{6.197}$$

The "forward sweep" will determine a computed source voltage V_1. As in the linear case, this first "iteration" will produce a voltage that is not equal to the specified source voltage V_s. Because the network is non-linear, multiplying currents and voltages by the ratio of the specified voltage to the computed voltage will not give the solution. The most direct modification to the ladder network theory is to perform a "backward sweep." The backward sweep commences by using the specified source voltage and the line currents from the "forward sweep." Kirchhoff's Voltage Law is used to compute the voltage at node 2 by:

$$V_2 = V_s - Z_{12} \cdot I_{12} \tag{6.198}$$

This procedure is repeated for each line segment until a "new" voltage is determined at node 5. Using the "new" voltage at node 5, a second "forward sweep" is started that will lead to a "new" computed voltage at the source.

The forward and backward sweep process is continued until the difference between the computed and specified voltage at the source is within a given tolerance.

General Feeder

A typical distribution feeder will consist of the "primary main" with laterals tapped off the primary main, and sublaterals tapped off the laterals, etc. Figure 6.32 shows an example of a typical feeder.

The ladder iterative technique for the feeder of Fig. 6.32 would proceed as follows:

1. Assume voltages (1.0 per-unit) at the "end" nodes (6, 8, 9, 11, and 13).
2. Starting at node 13, compute the node current (load current plus capacitor current if present).
3. With this current, apply Kirchhoff's Voltage Law (KVL) to calculate the node voltages at 12 and 10.
4. Node 10 is referred to as a "junction" node since laterals branch in two directions from the node. This feeder goes to node 11 and computes the node current. Use that current to compute the voltage at node 10. This will be referred to as "the most recent voltage at node 10."
5. Using the most recent value of the voltage at node 10, the node current at node 10 (if any) is computed.
6. Apply Kirchhoff's Current Law (KCL) to determine the current flowing from node 4 towards node 10.
7. Compute the voltage at node 4.
8. Node 4 is a "junction node." An end node downstream from node 4 is selected to start the forward sweep toward node 4.

Distribution Systems

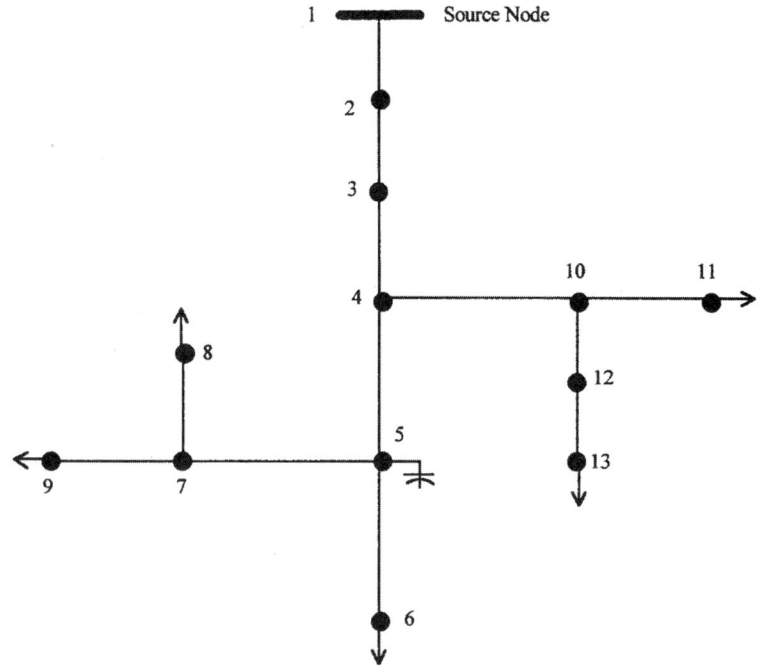

FIGURE 6.32 Typical distribution feeder.

9. Select node 6, compute the node current, and then compute the voltage at "junction node" 5.
10. Go to downstream end node 8. Compute the node current and then the voltage at junction node 7.
11. Go to downstream end node 9. Compute the node current and then the voltage at junction node 7.
12. Compute the node current at node 7 using the most recent value of node 7 voltage.
13. Apply KCL at node 7 to compute the current flowing on the line segment from node 5 to node 7.
14. Compute the voltage at node 5.
15. Compute the node current at node 5.
16. Apply KCL at node 5 to determine the current flowing from node 4 toward node 5.
17. Compute the voltage at node 4.
18. Compute the node current at node 4.
19. Apply KCL at node 4 to compute the current flowing from node 3 to node 4.
20. Calculate the voltage at node 3.
21. Compute the node current at node 3.
22. Apply KCL at node 3 to compute the current flowing from node 2 to node 3.
23. Calculate the voltage at node 2.
24. Compute the node current at node 2.
25. Apply KCL at node 2.
26. Calculate the voltage at node 1.
27. Compare the calculated voltage at node 1 to the specified source voltage.
28. If not within tolerance, use the specified source voltage and the forward sweep current flowing from node 1 to node 2 and compute the new voltage at node 2.
29. The backward sweep continues using the new upstream voltage and line segment current from the forward sweep to compute the new downstream voltage.
30. The backward sweep is completed when new voltages at all end nodes have been completed.
31. This completes the first iteration.
32. Now repeat the forward sweep using the new end voltages rather than the assumed voltages as was done in the first iteration.

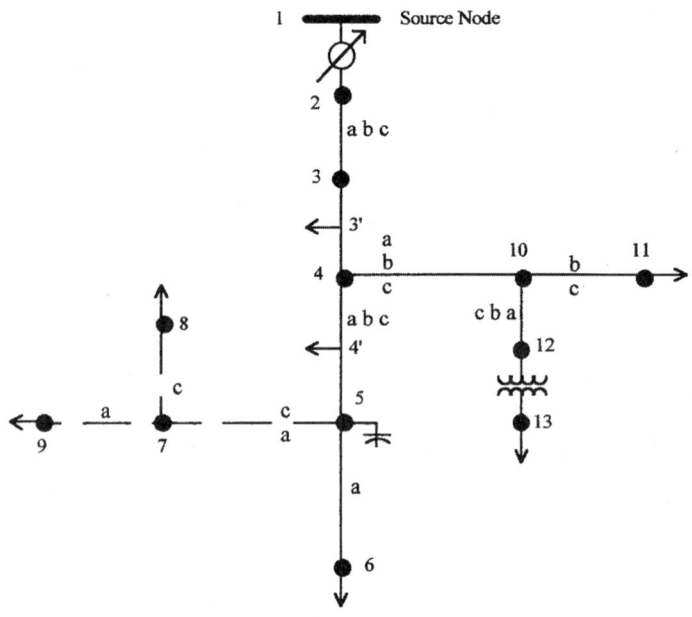

FIGURE 6.33 Unbalanced three-phase distribution feeder.

33. Continue the forward and backward sweeps until the calculated voltage at the source is within a specified tolerance of the source voltage
34. At this point the voltages are known at all nodes and the currents flowing in all line segments are known. An output report can be produced giving all desired results.

The Unbalanced Three-Phase Distribution Feeder

The previous section outlined the general procedure for performing the ladder iterative technique. This section will address how that procedure can be used for an unbalanced three-phase feeder.

Figure 6.33 is the one-line diagram of an unbalanced three-phase feeder.

The topology of the feeder in Fig. 6.33 is the same as the feeder in Fig. 6.32. Figure 6.33 shows more detail of the feeder however. The feeder in Fig. 6.33 can be broken into the "series" components and the "shunt" components.

Series Components

The "series" components of a distribution feeder are:

- Line segments
- Transformers
- Voltage regulators

Models for each of the series components have been developed in prior areas of this section. In all cases, models (three-phase, two-phase, and single-phase) were developed in such a manner that they can be generalized. Figure 6.34 shows the "general model" for each of the series components.

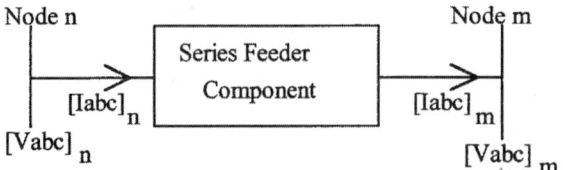

FIGURE 6.34 Series feeder component.

Distribution Systems

The general equations defining the "input" (Node n) and "output" (Node m) voltages and currents are given by:

$$[V_{abc}]_n = [a] \cdot [V_{abc}]_m + [b] \cdot [I_{abc}]_m \tag{6.199}$$

$$[I_{abc}]_n = [c] \cdot [V_{abc}]_m + [d] \cdot [I_{abc}]_m \tag{6.200}$$

The general equation relating the "output" (Node m) and "input" (Node n) voltages is given by:

$$[V_{abc}]_m = [A] \cdot [V_{abc}]_n - [B] \cdot [I_{abc}]_m \tag{6.201}$$

In Eqs. (6.199), (6.200), and (6.201), the voltages are line-to-neutral for a four-wire wye feeder and "equivalent" line-to-neutral for a three-wire delta system. For transformers and voltage regulators, the voltages are line-to-neutral for terminals that are connected to a four-wire wye and line-to-line when connected to a three-wire delta.

Shunt Components
The shunt components of a distribution feeder are:

- Spot loads
- Distributed loads
- Capacitor banks

"Spot" loads are located at a node and can be three-phase, two-phase, or single-phase and connected in either a wye or a delta connection. The loads can be modeled as constant complex power, constant current, constant impedance, or a combination of the three.

"Distributed" loads are located at the mid-section of a line segment. A distributed load is modeled when the loads on a line segment are uniformly distributed along the length of the segment. As in the spot load, the distributed load can be three-phase, two-phase, or single-phase and connected in either a wye or a delta connection. The loads can be modeled as constant complex power, constant current, constant impedance, or a combination of the three. To model the distributed load, a "dummy" node is created in the center of a line segment with the distributed load of the line section modeled at this dummy node.

Capacitor banks are located at a node and can be three-phase, two-phase, or single-phase and can be connected in a wye or delta. Capacitor banks are modeled as constant admittances.

In Fig. 6.33 the solid line segments represent overhead lines while the dashed lines represent underground lines. Note that the phasing is shown for all of the line segments. In the area of this section entitled "Line Impedances", the application of Carson's equations for computing the line impedances for overhead and underground lines was presented. There it was pointed out that two-phase and single-phase lines are represented by a three-by-three matrix with zeros set in the rows and columns of the missing phases.

In the area of this section entitled "Line Admittances", the method for the computation of the shunt capacitive susceptance for overhead and underground lines was presented. Most of the time the shunt capacitance of the line segment can be ignored; however, for long underground segments, the shunt capacitance should be included.

The "node" currents may be three-phase, two-phase, or single-phase and consist of the sum of the load current at the node plus the capacitor current (if any) at the node.

Applying the Ladder Iterative Technique
The previous section outlined the steps required for the application of the ladder iterative technique. For the general feeder of Fig. 6.33 the same outline applies. The only difference is that Eq. (6.199) and (6.200) are used for computing the node voltages on the "forward sweep" and Eq. (6.201) is used for computing

the downstream voltages on the "backward sweep." The $[a]$, $[b]$, $[c]$, $[d]$, $[A]$, and $[B]$ matrices for the various series components are defined in the following areas of this section:

- Line Segments: Line Segment Models
- Voltage Regulators: Step-Voltage Regulators
- Transformer Banks: Transformer Bank Connections

The node currents are defined in the following area:

- Loads: Load Models
- Capacitors: Shunt Capacitor Models

Final Notes

Line Segment Impedances

It is extremely important that the impedances and admittances of the line segments be computed using the exact spacings and phasing. Because of the unbalanced loading and resulting unbalanced line currents, the voltage drops due to the mutual coupling of the lines become very important. It is not unusual to observe a voltage rise on a lightly loaded phase of a line segment that has an extreme current unbalance.

Power Loss

The real power losses of a line segment must be computed as the difference (by phase) of the input power to a line segment minus the output power of the line segment. It is possible to observe a negative power loss on a phase that is lightly loaded compared to the other two phases. Computing power loss as the phase current squared times the phase resistance does not give the actual real power loss in the phases.

Load Allocation

Many times the input complex power (kW and kVAr) to a feeder is known because of the metering at the substation. This information can be either total three-phase or for each individual phase. In some cases the metered data may be the current and power factor in each phase.

It is desirable to have the computed input to the feeder match the metered input. This can be accomplished (following a converged iterative solution) by computing the ratio of the metered input to the computed input. The phase loads can now be modified by multiplying the loads by this ratio. Because the losses of the feeder will change when the loads are changed, it is necessary to go through the ladder iterative process to determine a new computed input to the feeder. This new computed input will be closer to the metered input, but most likely not within a specified tolerance. Again, a ratio can be determined and the loads modified. This process is repeated until the computed input is within a specified tolerance of the metered input.

Short Circuit Analysis

The computation of short-circuit currents for unbalanced faults in a normally balanced three-phase system has traditionally been accomplished by the application of symmetrical components. However, this method is not well-suited to a distribution feeder that is inherently unbalanced. The unequal mutual coupling between phases leads to mutual coupling between sequence networks. When this happens, there is no advantage to using symmetrical components. Another reason for not using symmetrical components is that the phases between which faults occur is limited. For example, using symmetrical components, line-to-ground faults are limited to **phase a to ground.** What happens if a single-phase lateral is connected to phase **b** or **c**? This section will present a method for short circuit analysis of an unbalanced three-phase distribution feeder using the phase frame (Kersting, 1980).

General Theory

Figure 6.35 shows the unbalanced feeder as modeled for short-circuit calculations.

In Fig. 6.35, the voltage sources E_a, E_b, and E_c represent the Thevenin Equivalent line-to-ground voltages at the faulted bus. The matrix [ZTOT] represents the Thevenin Equivalent impedance matrix at the faulted bus. The fault impedance is represented by Z_f in Fig. 6.35.

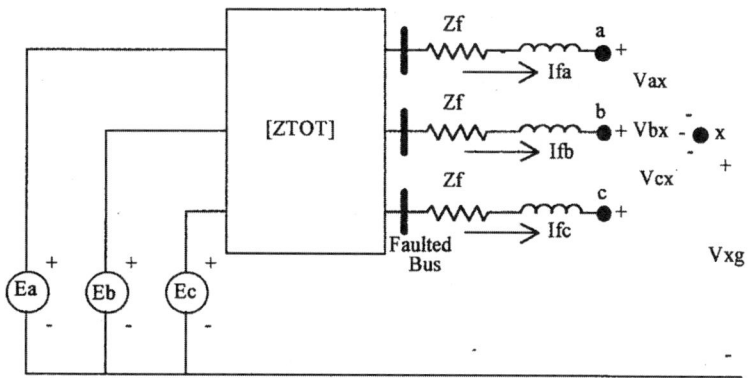

FIGURE 6.35 Unbalanced feeder short-circuit analysis model.

Kirchhoff's Voltage Law in matrix form can be applied to the circuit of Fig. 6.35.

$$\begin{bmatrix} E_a \\ E_b \\ E_c \end{bmatrix} = \begin{bmatrix} Z_{aa} & Z_{ab} & Z_{ac} \\ Z_{ba} & Z_{bb} & Z_{bc} \\ Z_{ca} & Z_{cb} & Z_{cc} \end{bmatrix} \cdot \begin{bmatrix} If_a \\ If_b \\ If_c \end{bmatrix} + \begin{bmatrix} Z_f & 0 & 0 \\ 0 & Z_f & 0 \\ 0 & 0 & Z_f \end{bmatrix} \cdot \begin{bmatrix} If_a \\ If_b \\ If_c \end{bmatrix} + \begin{bmatrix} V_{ax} \\ V_{bx} \\ V_{cx} \end{bmatrix} + \begin{bmatrix} V_{xg} \\ V_{xg} \\ V_{xg} \end{bmatrix} \quad (6.202)$$

Equation (6.202) can be written in compressed form as:

$$[E_{abc}] = [ZTOT] \cdot [If_{abc}] + [ZF] \cdot [If_{abc}] + [V_{abcx}] + [V_{xg}] \quad (6.203)$$

Combine terms in Eq. (6.203).

$$[E_{abc}] = [ZEQ] \cdot [If_{abc}] + [V_{abcx}] + [V_{xg}] \quad (6.204)$$

where:
$$[ZEQ] = [ZTOT] + [ZF] \quad (6.205)$$

Solve Eq. (6.204) for the fault currents:

$$[If_{abc}] = [YEQ] \cdot [E_{abc}] - [YEQ] \cdot [V_{abcx}] - [YEQ] \cdot [V_{xg}] \quad (6.206)$$

where
$$[YEQ] = [ZEQ]^{-1} \quad (6.207)$$

Since the matrices $[YEQ]$ and $[E_{abc}]$ are known, define:

$$[IP_{abc}] = [YEQ] \cdot [E_{abc}] \quad (6.208)$$

Substituting Eq. (6.208) into Eq. (6.206) results in the expanded Eq. (6.209).

$$\begin{bmatrix} If_a \\ If_b \\ If_c \end{bmatrix} = \begin{bmatrix} IP_a \\ IP_b \\ IP_c \end{bmatrix} - \begin{bmatrix} Y_{aa} & Y_{ab} & Y_{ac} \\ Y_{ba} & Y_{bb} & Y_{bc} \\ Y_{ca} & Y_{cb} & Y_{cc} \end{bmatrix} \cdot \begin{bmatrix} V_{ax} \\ V_{bx} \\ V_{cx} \end{bmatrix} - \begin{bmatrix} Y_{aa} & Y_{ab} & Y_{ac} \\ Y_{ba} & Y_{bb} & Y_{bc} \\ Y_{ca} & Y_{cb} & Y_{cc} \end{bmatrix} \cdot \begin{bmatrix} V_{xg} \\ V_{xg} \\ V_{xg} \end{bmatrix} \quad (6.209)$$

Performing the matrix operations in Eq. (6.209):

$$If_a = IP_a - (Y_{aa} \cdot V_{ax} + Y_{ab} \cdot V_{bx} + Y_{ac} \cdot V_{cx}) - Y_a \cdot V_{xg}$$
$$If_b = IP_b - (Y_{ba} \cdot V_{ax} + Y_{bb} \cdot V_{bx} + Y_{bc} \cdot V_{cx}) - Y_b \cdot V_{xg} \quad (6.210)$$
$$If_c = IP_a - (Y_{ca} \cdot V_{ax} + Y_{cb} \cdot V_{bx} + Y_{cc} \cdot V_{cx}) - Y_c \cdot V_{xg}$$

where

$$Y_a = Y_{aa} + Y_{ab} + Y_{ac}$$
$$Y_b = Y_{ba} + Y_{bb} + Y_{bc} \quad (6.211)$$
$$Y_c = Y_{ca} + Y_{cb} + Y_{cc}$$

Equations (6.210) become the general equations that are used to simulate all types of short circuits. Basically there are three equations and seven unknowns (If_a, If_b, If_c, V_{ax}, V_{bx}, V_{cx}, and V_{xg}). The other three variables in the equations (IP_a, IP_b, and IP_c) are functions of the total impedance and the Thevenin voltages and are therefore known. In order to solve Eq. (6.210), it will be necessary to specify four of the seven unknowns. These specifications are functions of the type of fault being simulated. The additional required four knowns for various types of faults are given below:

Three-phase faults

$$V_{ax} = V_{bx} = V_{cx} = 0$$
$$I_a + I_b + I_c = 0 \quad (6.212)$$

Three-phase-to-ground faults

$$V_{ax} = V_{bx} = V_{cx} = V_{xg} = 0 \quad (6.213)$$

Line-to-line faults (assume **i-j** fault with phase **k** unfaulted)

$$V_{ix} = V_{jx} = 0$$
$$If_k = 0 \quad (6.214)$$
$$If_i + If_j = 0$$

Line-to-line-to-ground faults (assume **i-j** to ground fault with **k** unfaulted)

$$V_{ix} = V_{jx} = V_{xg} = 0$$
$$V_{kx} = \frac{IP_k}{Y_{kk}} \quad (6.215)$$

Line-to-ground faults (assume phase **k** fault with phases **i** and **j** unfaulted)

$$V_{kx} = V_{xg} = 0$$
$$If_i = If_j = 0 \quad (6.216)$$

Notice that Eqs. (6.214), (6.215), and (6.216) will allow the simulation of line-to-line, line-to-line-to-ground, and line-to-ground faults for all phases. There is no limitation to **b-c** faults for line-to-line and **a-g** for line-to-ground as is the case when the method of symmetrical components is employed.

References

Carson, J. R., Wave propagation in overhead wires with ground return, *Bell Syst. Tech. J.*, 5, 1926.

Glover, J. D. and Sarma, M., *Power System Analysis and Design*, 2nd ed., PWS Publishing Company, Boston, 1994, chap. 5.

Kersting, W. H., Distribution system short circuit analysis, *25th Intersociety Energy Conversion Conference*, Reno, Nevada, August 1980.

Kersting, W. H., *Milsoft Transformer Models — Theory*, Research Report, Milsoft Integrated Solutions, Inc., Abilene, TX, 1999.

Kersting, W. H. and Mendive, D. L., An application of ladder network theory to the solution of three-phase radial load-flow problems, IEEE Conference Paper presented at the *IEEE Winter Power Meeting*, New York, January 1976.

Kron, G., Tensorial analysis of integrated transmission systems, Part I: The six basic reference frames, *AIEE Trans.*, 71, 1952.

Trevino, C., Cases of difficult convergence in load-flow problems, IEEE Paper n. 71-62-PWR, presented at the IEEE Summer Power Meeting, Los Angeles, CA, 1970.

6.3 Power System Operation and Control

George L. Clark and Simon W. Bowen

Implementation of Distribution Automation

The implementation of "distribution automation" within the continental U.S. is as diverse and numerous as the utilities themselves. Particular strategies of implementation utilized by various utilities have depended heavily on environmental variables such as size of the utility, urbanization, and available communication paths. The current level of interest in distribution automation is the result of:

- The maturation of technologies within the past 10 years in the areas of communication and RTUs/PLCs.
- Increased performance in host servers for the same or lower cost; lower cost of memory.
- The threat of deregulation and competition as a catalyst to automate.
- Strategic benefits to be derived (e.g., potential of reduced labor costs, better planning from better information, optimizing of capital expenditures, reduced outage time, increased customer satisfaction).

While not meant to be all-inclusive, this section on distribution automation attempts to provide some dimension to the various alternatives available to the utility engineer. The focus will be on providing insight on the elements of automation that should be included in an scalable and extensible system. The approach will be to describe the elements of a "typical" distribution automation system in a simple manner, offering practical observations as required.

For the electric utility, justification for automating the distribution system, while being highly desirable, was not readily attainable based on a cost/benefit ratio due to the size of the distribution infrastructure and cost of communication circuits. Still there have been tactical applications deployed on parts of distribution systems that were enough to keep the dream alive. The development of the PC (based on the Intel architecture) and VME systems (based on the Motorola architecture) provided the first low-cost SCADA master systems that were sized appropriately for the small co-ops and municipality utilities. New SCADA vendors then entered the market targeting solutions for small to medium-sized utilities.

Eventually the SCADA vendors who had been providing transmission SCADA took notice of the distribution market. These vendors provided host architectures based on VAX/VMS (and later Alpha/OpenVMS) platforms and on UNIX platforms from IBM and Hewlett-Packard. These systems were required for the large distribution utility (100,000–250,000 point ranges). These systems often resided on company-owned LANs with communication front-end processors and user interface attached either locally on the same LAN or across a WAN.

Distribution SCADA History

Supervisory Control And Data Acquisition (SCADA) is the foundation for the distribution automation system. The ability to remotely monitor and control electric power system facilities found its first application within the power generation and transmission sectors of the electric utility industry. The ability to significantly influence the utility bottom line through the effective dispatch of generation and the marketing of excess generating capacity provided economic incentive. The interconnection of large power grids in the midwestern and the southern U.S. (1962) created the largest synchronized system in the world. The blackout of 1965 prompted the U.S. Federal Power Commission to recommend closer coordination between regional coordination groups (Electric Power Reliability Act of 1967), and gave impetus to the subsequent formation of the National Electric Reliability Council (1970). From that time (1970) forward, the priority of the electric utility has been to engineer and build a highly reliable and secure transmission infrastructure. Transmission SCADA became the base for the large Energy Management Systems that were required to manage the transmission grid. Distribution SCADA languished during this period.

In the mid-1980s, EPRI published definitions for distribution automation and associated elements. The industry generally associates distribution automation with the installation of automated distribution line devices, such as switches, reclosers, sectionalizers, etc. The author's definition of distribution automation encompasses the automation of the distribution substations and the distribution line devices. The automated distribution substations and the automated distribution line devices are then operated as a system to facilitate the operation of the electric distribution system.

SCADA System Elements

At a high level, the elements of a distribution automation system can be divided into three main areas:

- SCADA application and server(s)
- DMS applications and server(s)
- Trouble management applications and server(s)

Distribution SCADA

As was stated in the introduction, the Supervisory Control And Data Acquisition (SCADA) system is the heart of Distribution Management System (DMS) architecture. A SCADA system should have all of the infrastructure elements to support the multifaceted nature of distribution automation and the higher level applications of a DMS. A Distribution SCADA system's primary function is in support of distribution operations telemetry, alarming, event recording, and remote control of field equipment. Historically, SCADA systems have been notorious for their lack of support for the import, and more importantly, the export of power system data values. A modern SCADA system should support the engineering budgeting and planning functions by providing access to power system data without having to have possession of an operational workstation. The main elements of a SCADA system are:

- Host equipment
- Communication infrastructure (network and serial communications)
- Field devices (in sufficient quantity to support operations and telemetry requirements of a DMS platform)

Distribution Systems

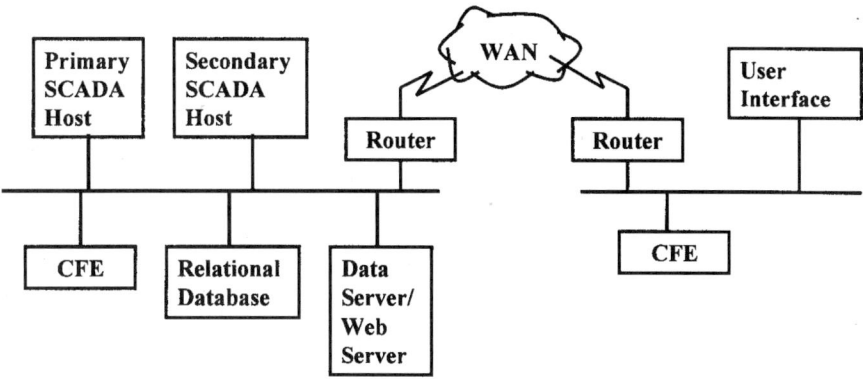

FIGURE 6.36 DA system architecture.

Host Equipment

The authors feel that the essential elements of a distribution SCADA host are:

- Host servers (redundant servers with backup/failover capability).
- Communication front-end nodes (network based).
- Full graphics user interfaces.
- Relational database server (for archival of historical power system values) and data server/Web server (for access to near real time values and events).

The elements and components of the typical distribution automation system are illustrated in Fig. 6.36.

Host Computer System

SCADA Servers

As SCADA has proven its value in operation during inclement weather conditions, service restoration, and daily operations, the dependency on SCADA has created a requirement for highly available and high-performance systems. Redundant server hardware operating in a "live" backup/failover mode is required to meet the high availability criteria. High-performance servers with abundant physical memory, RAID hard disk systems, and interconnected by 10/100 baseT switched Ethernet are typical of today's SCADA servers.

Communication Front-End (CFE) Processors

The current state of host to field device communications still depends heavily on serial communications. This requirement is filled by the CFE. The CFE can come in several forms based on bus architecture (e.g., VME or PCI) and operating system. Location of the CFE in relation to the SCADA server can vary based on requirement. In some configurations the CFE is located on the LAN with the SCADA server. In other cases, existing communications hubs may dictate that the CFE reside at the communication hub. The incorporation of the WAN into the architecture requires a more robust CFE application to compensate for less reliable communications (in comparison to LAN). In general the CFE will include three functional devices: a network/CPU board, serial cards, and possibly a time code receiver. Functionality should include the ability to download configuration and scan tables. The CFE should also support the ability to dead band values (i.e., report only those analog values that have changed by a user-defined amount). CFE, network, and SCADA servers should be capable of supporting worst-case conditions (i.e., all points changing outside of the dead band limits), which typically occur during severe system disturbances.

Full Graphics User Interface

The current trend in the user interface (UI) is toward a full graphics (FG) user interface. While character graphics consoles are still in use by many utilities today, SCADA vendors are aggressively moving their platforms to a full graphics UI. Quite often the SCADA vendors have implemented their new full graphics user interface on low-cost NT workstations using third-party applications to emulate the X11 window system. Full graphic displays provide the ability to display power system data along with the electric distribution facilities in a geographical (or semigeographical) perspective. The advantage of using a full graphics interface becomes evident (particularly for distribution utilities) as SCADA is deployed beyond the substation fence where feeder diagrams become critical to distribution operations.

Relational Databases, Data Servers, and Web Servers

The traditional SCADA systems were poor providers of data to anyone not connected to the SCADA system by an operational console. This occurred due to the proprietary nature of the performance (in memory) database and its design optimization for putting scanned data in and pushing display values out. Power system quantities such as: bank and feeder loading (MW, MWH, MQH, and ampere loading), and bus volts provide valuable information to the distribution planning engineer. The availability of event (log) data is important in postmortem analysis. The use of relational databases, data servers, and Web servers by the corporate and engineering functions provides access to power system information and data while isolating the SCADA server from nonoperations personnel.

Host to Field Communications

Serial communications to field devices can occur over several mediums: copper wire, fiber, radio, and even satellite. Telephone circuits, fiber, and satellites have a relatively high cost. New radio technologies offer good communications value. One such technology is the Multiple Address Radio System (MAS). The MAS operates in the 900 MHz range and is omnidirectional, providing radio coverage in an area with radius up to 20–25 miles depending on terrain. A single MAS master radio can communicate with many remote sites. Protocol and bandwidth limit the number of remote terminal units that can be communicated with by a master radio. The protocol limit is simply the address range supported by the protocol. Bandwidth limitations can be offset by the use of efficient protocols, or slowing down the scan rate to include more remote units. Spread-spectrum and point-to-point radio (in combination with MAS) offers an opportunity to address specific communication problems. At the present time MAS radio is preferred (authors' opinion) to packet radio (another new radio technology); MAS radio communications tend to be more deterministic providing for smaller timeout values on communication no-responses and controls.

Field Devices

Distribution Automation (DA) field devices are multi-featured installations meeting a broad range of control, operations, planning, and system performance issues for the utility personnel. Each device provides specific functionality, supports system operations, includes fault detection, captures planning data and records power quality information. These devices are found in the distribution substation and at selected locations along the distribution line. The multifeatured capability of the DA device increases its ability to be integrated into the electric distribution system. The functionality and operations capabilities complement each other with regard to the control and operation of the electric distribution system. The fault detection feature is the "eyes and ears" for the operating personnel. The fault detection capability becomes increasingly more useful with the penetration of DA devices on the distribution line.

The real-time data collected by the SCADA system is provided to the planning engineers for inclusion in the radial distribution line studies. As the distribution system continues to grow, the utility makes annual investments to improve the electric distribution system to maintain adequate facilities to meet the increasing load requirements. The use of the real-time data permits the planning engineers to optimize the annual capital expenditures required to meet the growing needs of the electric distribution system.

The power quality information includes capturing harmonic content to the 15th harmonic and recording Percent Total Harmonic Distortion (%THD). This information is used to monitor the performance of the distribution electric system.

Modern RTU

Today's modern RTU is modular in construction with advanced capabilities to support functions that heretofore were not included in the RTU design. The modular design supports installation configurations ranging from the small point count required for the distribution line pole-mounted units to the very large point count required for large bulk-power substations and power plant switchyard installations. The modern RTU modules include analog units with 9 points, control units with 4 control pair points, status units with 16 points, and communication units with power supply. The RTU installation requirements are met by accumulating the necessary number of modern RTU modules to support the analog, control, status, and communication requirements for the site to be automated. Packaging of the minimum point count RTUs is available for the distribution line requirement. The substation automation requirement has the option of installing the traditional RTU in one cabinet with connections to the substation devices or distributing the RTU modules at the devices within the substation with fiberoptic communications between the modules. The distributed RTU modules are connected to a data concentrating unit which in turn communicates with the host SCADA computer system.

The modern RTU accepts direct AC inputs from a variety of measurement devices including line-post sensors, current transformers, potential transformers, station service transformers, and transducers. Direct AC inputs with the processing capability in the modern RTU supports fault current detection and harmonic content measurements. The modern RTU has the capability to report the magnitude, direction, and duration of fault current with time tagging of the fault event to 1-millisecond resolution. Monitoring and reporting of harmonic content in the distribution electric circuit are capabilities that are included in the modern RTU. The digital signal processing capability of the modern RTU supports the necessary calculations to report %THD for each voltage and current measurement at the automated distribution line or substation site.

The modern RTU includes logic capability to support the creation of algorithms to meet specific operating needs. Automatic transfer schemes have been built using automated switches and modern RTUs with the logic capability. This capability provides another option to the distribution line engineer when developing the method of service and addressing critical load concerns. The logic capability in the modern RTU has been used to create the algorithm to control distribution line switched capacitors for operation on a per phase basis. The capacitors are switched on at zero voltage crossing and switched off at zero current crossing. The algorithm can be designed to switch the capacitors for various system parameters, such as voltage, reactive load, time, etc. The remote control capability of the modern RTU then allows the system operator to take control of the capacitors to meet system reactive load needs.

The modern RTU has become a dynamic device with increased capabilities. The new logic and input capabilities are being exploited to expand the uses and applications of the modern RTU.

PLCs and IEDs

Programmable Logic Controllers (PLCs) and Intelligent Electronic Devices (IEDs) are components of the distribution automation system, which meet specific operating and data gathering requirements. While there is some overlap in capability with the modern RTU, the authors are familiar with the use of PLCs for automatic isolation of the faulted power transformer in a two-bank substation and automatic transfer of load to the unfaulted power transformer to maintain an increased degree of reliability. The PLC communicates with the modern RTU in the substation to facilitate the remote operation of the substation facility. The typical PLC can support serial communications to a SCADA server. The modern RTU has the capability to communicate via an RS-232 interface with the PLC.

IEDs include electronic meters, electronic relays, and controls on specific substation equipment, such as breakers, regulators, LTC on power transformers, etc. The IEDs also have the capability to support serial communications to a SCADA server. However, the authors' experience indicates that the IEDs are typically reporting to the modern RTU via an RS-232 interface or via status output contact points. As its communicating capability improves and achieves equal status with the functionality capability, the IED has the potential to become an equal player in the automation communication environment.

However, in the opinion of the authors, the limited processing capability for supporting the communication requirement, in addition to its functional requirements (i.e., relays, meters, etc.), hampers the widespread use of the IEDs in the distribution automation system.

Substation

The installation of the SCADA technology in the DA substation provides for the full automation of the distribution substation functions and features. The modular RTU supports the various substation sizes and configuration. The load on the power transformer is monitored and reported on a per-phase basis. The substation low-side bus voltage is monitored on a per phase basis. The distribution feeder breaker is fully automated. Control of all breaker control points is provided including the ability to remotely set up the distribution feeder breaker to support energized distribution line work. The switched capacitor banks and substation regulation are controlled from the typical modular RTU installation. The load on the distribution feeder breaker is monitored and reported on a per-phase basis as well as on a three-phase basis. This capability is used to support the normal operations of the electric distribution system and to respond to system disturbances. The installation of the SCADA technology in the DA substation eliminates the need to dispatch personnel to the substation except for periodic maintenance and equipment failure.

Line

The DA distribution line applications include line monitoring, pole-mounted reclosers, gang-operated switches equipped with motor operators, switched capacitor banks, and pad-mounted automatic transfer switchgear. The modular RTU facilitates the automation of the distribution line applications. The use of the line post sensor facilitates the monitoring capability on a per-phase basis. The direct AC input from the sensors to the RTU supports monitoring of the normal load, voltage and power factor measurements, and also the detection of fault current. The multifeatured distribution line DA device can be used effectively to identify the faulted sections of the distribution circuit during system disturbances, isolate the faulted sections, and restore service to the unfaulted sections of the distribution circuit. The direct AC inputs to the RTU also support the detection and reporting of harmonics to the 15th harmonic and the %THD per phase for voltage and current.

Tactical and Strategic Implementation Issues

As the threat of deregulation and competition emerges, retention of industrial and large commercial customers will become the priority for the electric utility. Every advantage will be sought by the electric utility to differentiate itself from other utilities. Reliable service, customer satisfaction, fast storm restorations, and power quality will be the goals of the utility. Differing strategies will be employed based on the customer in question and the particular mix of goals that the utility perceives will bring customer loyalty.

For large industrial and commercial customers, where the reliability of the electric service is important and outages of more than a few seconds can mean lost production runs or lost revenue, tactical automation solutions may be required. Tactical solutions are typically transfer schemes or switching schemes that can respond independently of operator action, reporting the actions that were initiated in response to loss of preferred service and/or line faults. The requirement to transfer source power, or reconfigure a section of the electric distribution system to isolate and reconnect in a matter of seconds is the primary criteria. Tactical automation based on local processing provides the solution.

In cases where there are particularly sensitive customer requirements, tactical solutions are appropriate. When the same requirements are applied to a large area and/or customer base, a strategic solution based on a distribution management platform is preferred. This solution requires a DMS with a system operational model that reflects the current configuration of the electric distribution system. Automatic fault isolation and restoration applications that can reconfigure the electric distribution system, require a "whole system" model in order to operate correctly and efficiently.

So while tactical automation requirements exist and have significant impact and high profile, goals that target system issues (reduction of system losses, voltage programs, storm restoration) require a strategic solution.

Distribution Systems

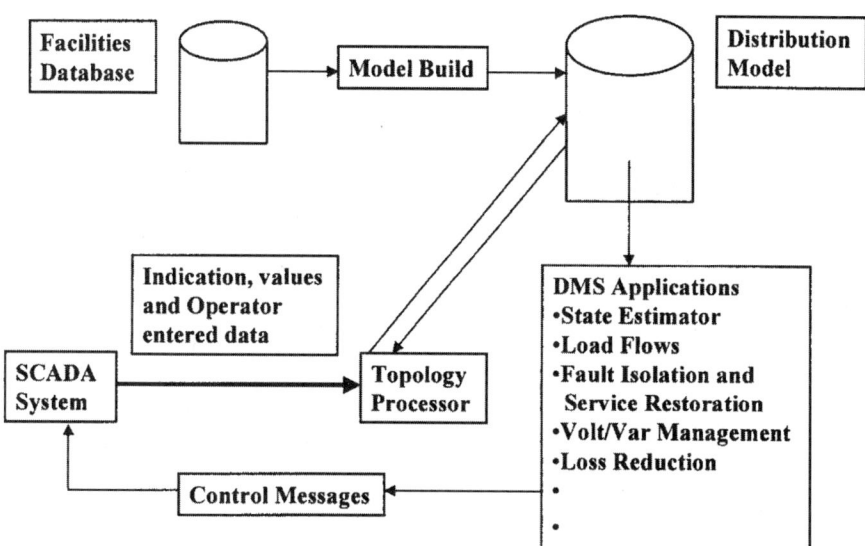

FIGURE 6.37 A DMS platform with SCADA interface.

Distribution Management Platform

Distribution Management System

A functional Distribution Management System (DMS) platform should be fully integrated with the distribution SCADA system. The SCADA-DMS interface should be fully implemented with the capability of passing data [discrete indication (status) and values (analog)] bidirectionally. The SCADA interface should also support device control. Figure 6.37 details the components of a DMS.

Trouble Management Platform

Trouble Management System

In addition to the base SCADA functionality and high-level DMS applications, the complete distribution automation system will include a Trouble Management System (TMS). Trouble Management Systems collect trouble calls received by human operators and Interactive Voice Recorders (IVR). The trouble calls are fed to an analysis/prediction engine that has a model of the distribution system with customer to electrical address relationships. Outage prediction is presented on a full graphics display that overlays the distribution system on CAD base information. A TMS also provides for the dispatch and management of crews, customer callbacks, accounting, and reports. A SCADA interface to a TMS provides the means to provide confirmed (SCADA telemetry) outage information to the prediction engine. Figure 6.38 shows a typical TMS.

Practical Considerations

Choosing the Vendor

Choosing a Platform Vendor

In choosing a platform (SCADA, DMS, TMS) vendor there are several characteristics that should be kept in mind (these should be considered as rule of thumb based on experience of what works and what does not). Choosing the right vendor is as important as choosing the right software package.

Vendor characteristics that the authors consider important are:

- A strong "product" philosophy. Having a strong product philosophy is typically a chicken and egg proposition. Which came first, the product or the philosophy? Having a baseline SCADA application

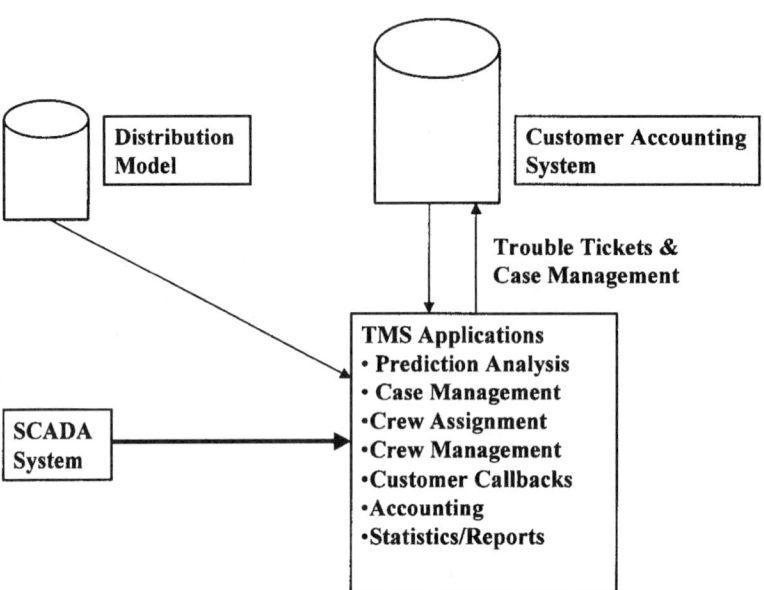

FIGURE 6.38 A TMS platform with SCADA interface.

can be a sign of maturity and stability. Did the platform vendor get there by design or did they back into it? Evidence of a product philosophy include a baseline system that is in production and enhancements that are integrated in a planned manner with thorough testing on the enhancement and regression testing on the product along with complete and comprehensive documentation.
- A documented development and release path projected three to five years into the future.
- By inference from the first two bullets, a vendor who funds planned product enhancements from internal funds.
- A strong and active User Group that is representative of the industry and industry drivers.
- A platform vendor that actively encourages its User Group by incentive (e.g., dedicating part of its enhancement funding to User Group initiatives).
- A vendor that is generally conservative in moving its platform to a new technology; one that does not overextend its own resources.

Other Considerations
- As much as possible, purchase the platform as an off-the-shelf product (i.e., resist the urge to ask for customs that drive your system away from the vendor's baseline).
- If possible, maintain/develop your own support staff.

All "customization" should be built around the inherent capabilities and flexibility of the system (i.e., do not generate excessive amounts of new code). Remember, you will have to reapply any code that you may have developed to every new release; or worse, you will have to pay the vendor to do it for you.

Standards

Internal Standards
The authors highly recommend the use of standards (internal to your organization) as a basis for ensuring a successful distribution automation or SCADA program. Well-documented construction standards that specify installation of RTUs, switches, and line sensors with mechanical and electrical specifications will ensure consistent equipment installations from site to site. Standards that cover nontrivial, but often overlooked issues can often spell the difference between acceptance and rejection by operational users

and provide the additional benefit of having a system that is "maintainable" over the 10–20 year (or more) life of a system. Standards that fall in this category include standards that cover point-naming conventions, symbol standards, display standards, and the all-important Operations Manual.

Industry Standards

In general, standards fall into two categories: standards that are developed by organizations and commissions (e.g., EPRI, IEEE, ANSI, CCITT, ISO, etc.) and *de facto* standards that become standards by virtue of widespread acceptance. As an example of what can occur, the reader is invited to consider what has happened in network protocols over the recent past with regard to the OSI model and TCP/IP.

Past history of SCADA and automation has been dominated by the proprietary nature of the various system vendor offerings. Database schemas and RTU communication protocols are exemplary of proprietary design philosophies utilized by SCADA platform and RTU vendors. Electric utilities that operate as part of the interconnected power grid have been frustrated by the lack of ability to share power system data between dissimilar energy management systems. The same frustration exists at the device level; RTU vendors, PLC vendors, electronic relay vendors, and meter vendors each having their own product protocols have created a "Tower of Babel" problem for utilities. Recently several communications standards organizations and vendor consortiums have proposed standards to address these deficiencies in intersystem data exchange, intrasystem data exchange (corporate data exchange), and device level interconnectivity. Some of the more notable examples of network protocol communication standards are ICCP (Inter Control Center Protocol), UCA (Utility Communication Architecture), CCAPI (Control Center Applications Interface), and UIB (Utility Integration Bus). For database schemas, EPRI's CIM (Common Information Model) is gaining supporters. In RTU, PLC, and IED communications, DNP 3.0 has also received much industry press.

In light of the number of standards that have appeared (and then disappeared) and the number of possibly competing "standards" that are available today, the authors, while acknowledging the value of standards, prefer to take (and recommend) a cautious approach to standards. A wait-and-see posture may be an effective strategy. Standards by definition must have proven themselves over time. Difficulties in immediately embracing new standards are due in part to vendors having been allowed to implement only portions of a standard, thereby nullifying the hopefully "plug-and-play" aspect for adding new devices. Also, the trend in communication protocols has been to add functionality in an attempt to be all-inclusive, which has resulted in an increased requirement on bandwidth. Practically speaking, utilities that have already existing infrastructure may find it economical to resist the deployment of new protocols. In the final analysis, as in any business decision, a "standard" should be accepted only if it adds value and benefit that exceeds the cost of implementation and deployment.

Deployment Considerations

The definition of the automation technology to be deployed should be clearly delineated. This definition includes the specification of the host systems, the communication infrastructure, the automated end-use devices, and the support infrastructure. This effort begins with the development of a detailed installation plan that takes into consideration the available resources. The pilot installation will never be any more than a pilot project until funding and manpower resources are identified and dedicated to the enterprise of implementing the technologies required to automate the electric distribution system. The implementation effort is best managed on an annual basis with stated incremental goals and objectives for the installation of automated devices. With the annual goals and objectives identified, then the budget process begins to ensure that adequate funding is available to support the implementation plan. To ensure adequate time to complete the initial project tasks, the planning should begin 18 to 24 months prior to the budget year. During this period, the identification of specific automation projects is completed. The initial design work is commenced with the specification of field automation equipment (e.g., substation RTU based on specific point count requirements and distribution line RTU). The verification of the communication to the selected automation site is an urgent early consideration in order to minimize the cost of achieving effective remote communications. As the installation year approaches, the associated automation equipment (e.g., switches, motor-operators, sensors, etc.) must be verified to ensure that adequate supplies are stocked to support the implementation plan.

The creation of a SCADA database and display is on the critical path for new automated sites. The database and display are critical to the efficient completion of the installation and checkout tasks. Data must be provided to the database and display team with sufficient lead time to create the database and display for the automated site. The database and display are subsequently used to checkout the completed automated field device. The Point Assignment (PA) sheet is a project activity that merits serious attention. The PA sheet is the basis for the creation of the site-specific database in the SCADA system. The PA sheet should be created in a consistent and standard fashion. The importance of an accurate database and display cannot be overemphasized. The database and display form the basis for the remote operational decisions for the electric distribution system using the SCADA capability. Careful coordination of these project tasks is essential to the successful completion of the annual automation plans.

Training is another important consideration during the deployment of the automation technology. The training topics are as varied as the multidisciplined nature of the distribution automation project. Initial training requirements include the system support personnel in the use and deployment of the automation platform, the end-user (operator) training, and installation teams. Many utilities now install new distribution facilities using energized line construction techniques. The automated field device adds a degree of complexity to the construction techniques to ensure adherence to safe practices and construction standards. These training issues should be addressed at the outset of the planning effort to ensure a successful distribution automation project.

Support Organization

The support organization must be as multidisciplined as the distribution automation system is multi-featured. The support to maintain a deployed distribution automation system should not be underestimated. Functional teams should be formed to address each discipline represented within the distribution automation system. The authors recommend forming a core team that is made up of representation from each area of discipline or area of responsibility within the distribution automation project. These areas of discipline include the following:

- Host SCADA system
- User interface
- Communication infrastructure
- Facilities design personnel for automated distribution substation and distribution line devices
- System software and interface developments
- Installation teams for automated distribution substation and distribution line devices
- End-users (i.e., the operating personnel)

The remaining requirement for the core team is project leadership with responsibility for the project budget, scheduling, management reports, and overall direction of the distribution automation project. The interaction of the various disciplines within the distribution automation team will ensure that all project decisions are supporting the overall project goals. The close coordination of the various project teams through the core team is essential to minimizing decision conflict and maximizing the synergy of project decisions. The involvement of the end-user at the very outset of the distribution automation project planning cannot be overemphasized. The operating personnel are the primary users of the distribution automation technology. The participation of the end-user in all project decisions is essential to ensure that the distribution automation product meets business needs and improves the operating environment in the operating centers. One measure of good project decisions is found in the response of the end-user. When the end-user says, "I like it," then the project decision is clearly targeting the end-user's business requirements. With this goal achieved, then the distribution automation system is in position to begin meeting other corporate business needs for real-time data from the electric distribution system.

7
Electric Power Utilization

Andrew Hanson
ABB Power T&D Company

7.1 Metering of Electric Power and Energy *John V. Grubbs* 7-1

7.2 Basic Electric Power Utilization — Loads, Load Characterization and Load Modeling
    Andrew Hanson 7-12

7.3 Electric Power Utilization: Motors *Charles A. Gross* 7-18

7

Electric Power Utilization

John V. Grubbs
Alabama Power Company

Andrew Hanson
ABB Power T&D Company

Charles A. Gross
Auburn University

7.1 Metering of Electric Power and Energy 7-1
The Electromechanical Meter • Blondel's Theorem • The Electronic Meter • Special Metering • Instrument Transformers • Measuring kVA

7.2 Basic Electric Power Utilization — Loads, Load Characterization and Load Modeling..............................7-12
Basic Load Characterization • Composite Loads and Composite Load Characterization • Composite Load Modeling • Other Load-Related Issues

7.3 Electric Power Utilization: Motors...................................7-18
Some General Perspectives • Operating Modes • Motor, Enclosure, and Controller Types • System Design

7.1 Metering of Electric Power and Energy

John V. Grubbs

Electrical metering deals with two basic quantities: *energy* and *power*. Energy is equivalent to work. Power is the rate of doing work. Power applied (or consumed) for any length of time is energy. In mathematical terms, power integrated over time is energy. The basic electrical unit of energy is the watthour. The basic unit of power is the watt. The watthour meter measures energy (in watthours), while the wattmeter measures the rate of energy, power (in watthours per hour or simply watts). For a constant power level, power multiplied by time is energy. For example, a watthour meter connected for two hours in a circuit using 500 watts (500 watthours per hour) will register 1000 watthours.

The Electromechanical Meter

The electromechanical watthour meter is basically a very specialized electric motor, consisting of

- A *stator* and a *rotor* that together produce torque
- A *brake* that creates a counter torque
- A *register* to count and display the revolutions of the rotor

Single Stator Electromechanical Meter

A two-wire single stator meter is the simplest electromechanical meter. The single stator consists of two electromagnets. One electromagnet is the potential coil connected between the two circuit conductors. The other electromagnet is the current coil connected in series with the load current. Figure 7.1 shows the major components of a single stator meter.

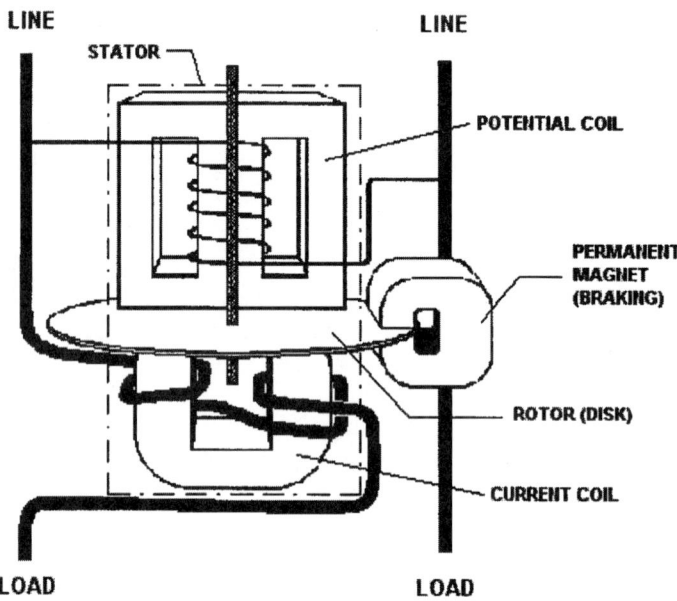

FIGURE 7.1 Main components of electromechanical meter.

The electromagnetic fields of the current coil and the potential coil interact to generate torque on the rotor of the meter. This torque is proportional to the product of the source voltage, the line current, and the cosine of the phase angle between the two. Thus, the torque is also proportional to the power in the metered circuit.

The device described so far is incomplete. In measuring a steady power in a circuit, this meter would generate constant *torque* causing steady acceleration of the rotor. The rotor would spin faster and faster until the torque could no longer overcome friction and other forces acting on the rotor. This ultimate speed would not represent the level of power present in the metered circuit.

To address these problems, designers add a permanent magnet whose magnetic field acts on the rotor. This field interacts with the rotor to cause a *counter torque* proportional to the speed of the rotor. Careful design and adjustment of the magnet strength yields a meter that rotates at a *speed* proportional to power. This speed can be kept relatively slow. The product of the rotor speed and time is revolutions of the rotor. The revolutions are proportional to energy consumed in the metered circuit. One revolution of the rotor represents a fixed number of watthours. The revolutions are easily converted via mechanical gearing or other methods into a display of watthours or, more commonly, kilowatthours.

Blondel's Theorem

Blondel's theorem of polyphase metering describes the measurement of power in a polyphase system made up of an arbitrary number of conductors. The theorem provides the basis for correctly metering power in polyphase circuits. In simple terms, Blondel's theorem states that the total power in a system of (N) conductors can be properly measured by using (N) wattmeters or watt-measuring elements. The elements are placed such that one current coil is in each of the conductors and one potential coil is connected between each of the conductors and some common point. If this common point is chosen to be one of the (N) conductors, there will be zero voltage across one of the measuring element potential coils. This element will register zero power. *Therefore, the total power is correctly measured by the remaining (N – 1) elements.*

In application, this means that to accurately measure the power in a four-wire three-phase circuit (N = 4), the meter must contain (N – 1) or three measuring elements. Likewise, for a three-wire three-phase circuit

(N = 3), the meter must contain two measuring elements. There are meter designs available that, for commercial reasons, employ less than the minimum number of elements (N − 1) for a given circuit configuration. These designs depend on *balanced* phase voltages for proper operation. Their accuracy suffers as voltages become unbalanced.

The Electronic Meter

Since the 1980s, meters available for common use have evolved from (1) electromechanical mechanisms driving mechanical, geared registers to (2) the same electromechanical devices driving electronic registers to (3) totally electronic (or solid state) designs. All three types remain in wide use, but the type that is growing in use is the solid state meter.

The addition of the electronic register to an electromechanical meter provides a digital display of energy and demand. It supports enhanced capabilities and eliminates some of the mechanical complexity inherent in the geared mechanical registers.

Electronic meters contain no moving mechanical parts — rotors, shafts, gears, bearings. They are built instead around large-scale integrated circuits, other solid state components, and digital logic. Such meters are much more closely related to computers than to electromechanical meters.

The operation of an electronic meter is very different than that described in earlier sections for an electromechanical meter. Electronic circuitry samples the voltage and current waveforms during each electrical cycle and converts these samples to digital quantities. Other circuitry then manipulates these values to determine numerous electrical parameters, such as kW, kWh, kvar, kvarh, kQ, kQh, power factor, kVA, rms current, rms voltage.

Various electronic meter designs also offer some or all of the following capabilities:

- **Time of use (TOU)**. The meter keeps up with energy and demand in multiple daily periods. (See section on Time of Use Metering.)
- **Bi-directional**. The meter measures (as separate quantities) energy delivered to and received from a customer. This feature is used for a customer that is capable of generating electricity and feeding it back into the utility system.
- **Loss compensation**. The meter can be programmed to automatically calculate watt and var losses in transformers and electrical conductors based on defined or tested loss characteristics of the transformers and conductors. It can internally add or subtract these calculated values from its measured energy and demand. This feature permits metering to be installed at the most economical location. For instance, we can install metering on the secondary (e.g., 4 kV) side of a customer substation, even when the contractual service point is on the primary (e.g., 110 kV) side. The 4 kV metering installation is much less expensive than a corresponding one at 110 kV. Under this situation, the meter compensates its secondary-side energy and demand readings to simulate primary-side readings.
- **Interval data recording**. The meter contains solid state memory in which it can record up to several months of interval-by-interval data. (See section on Interval Data Metering.)
- **Remote communications**. Built-in communications capabilities permit the meter to be interrogated remotely via telephone, radio, or other communications media.
- **Diagnostics**. The meter checks for the proper voltages, currents, and phase angles on the meter conductors. (See section on Site Diagnostic Meter.)
- **Power quality**. The meter can measure and report on momentary voltage or current variations and on harmonic conditions.

Note that many of these features are available only in the more advanced (and expensive) models of electronic meters.

As an example of the benefits offered by electronic meters, consider the following two methods of metering a large customer who is capable of generating and feeding electricity back to the utility. In this example, the metering package must perform these functions:

Measure kWh delivered to the customer
Measure kWh received from the customer
Measure kvarh delivered
Measure kvarh received
Measure kW delivered
Measure kW received
Compensate received quantities for transformer losses
Record the measured quantities for each demand interval

Method A. (2) kW/kWh electromechanical meters with pulse generators (one for delivered, one for received)
(2) kWh electromechanical meters with pulse generators (to measure kvarh)
(2) Phase shifting transformers (used along with the kWh meters to measure kvarh)
(2) Transformer loss compensators
(1) Pulse data recorder

Method B. (1) Electronic meter

Obviously, the electronic installation is much simpler. In addition, it is less expensive to purchase and install and is easier to maintain.

Benefits common to most solid state designs are high accuracy and stability. Another less obvious advantage is in the area of error detection. When an electromechanical meter develops a serious problem, it may produce readings in error by any arbitrary amount. An error of 10%, 20%, or even 30% can go undetected for years, resulting in very large over- or under-billings. However, when an electronic meter develops a problem, it is more likely to produce an obviously bad reading (e.g., all zeroes; all 9s; a demand 100 times larger than normal; or a blank display). This greatly increases the likelihood that the error will be noticed and reported soon after it occurs. The sooner such a problem is recognized and corrected, the less inconvenience and disruption it causes to the utility and to the customer.

Multifunction Meter

Multifunction or *extended function* refers to a meter that can measure reactive or apparent power (e.g., kvar or kVA) in addition to real power (kW). By virtue of their designs, many electronic meters inherently measure the quantities and relationships that define reactive and apparent power. It is a relatively simple step for designers to add meter intelligence to calculate and display these values.

Voltage Ranging and Multiform Meter

Electronic meter designs have introduced many new features to the watthour metering world. Two features, typically found together, offer additional flexibility, simplified application, and opportunities for reduced meter inventories for utilities.

- *Voltage ranging* – Many electronic meters incorporate circuitry that can sense the voltage level of the meter input signals and adjust automatically to meter correctly over a wide range of voltages. For example, a meter with this capability can be installed on either a 120 volt or 277 volt service.
- *Multiform* – Meter form refers to the specific combination of voltage and current signals, how they are applied to the terminals of the meter, and how the meter uses these signals to measure power and energy. For example, a Form 15 meter would be used for self-contained application on a 120/240 volt 4-wire delta service, while a Form 16 meter would be used on a self-contained 120/208 volt 4-wire wye service. A *multiform* 15/16 meter can work interchangeably on either of these services.

Site Diagnostic Meter

Newer meter designs incorporate the ability to measure, display, and evaluate the voltage and current magnitudes and phase relationships of the circuits to which they are attached. This capability offers important advantages:

- At the time of installation or reinstallation, the meter analyzes the voltage and current signals and determines if they represent a recognizable service type.
- Also at installation or reinstallation, the meter performs an initial check for wiring errors such as crossed connections or reversed polarities. If it finds an error, it displays an error message so that corrections can be made.
- Throughout its life, the meter continuously evaluates voltage and current conditions. It can detect a problem that develops weeks, months, or years after installation, such as tampering or deteriorated CT or VT wiring.
- Field personnel can switch the meter display into diagnostic mode. It will display voltage and current magnitudes and phase angles for each phase. This provides a quick and very accurate way to obtain information on service characteristics.

If a diagnostic meter detects any error that might affect the accuracy of its measurements, it will lock its display in error mode. The meter continues to make energy and demand measurements in the background. However, these readings cannot be retrieved from the meter until the error is cleared. This ensures the error will be reported the next time someone tries to read the meter.

Special Metering

Demand Metering

What is Demand?

Electrical energy is commonly measured in units of kilowatthours. Electrical power is expressed as kilowatthours per hour or, more commonly, kilowatts.

Demand is defined as power averaged over some specified period. Figure 7.2 shows a sample power curve representing instantaneous power. In the time interval shown, the integrated area under the power curve represents the energy consumed during the interval. This energy, divided by the length of the interval (in hours) yields "demand." In other words, the demand for the interval is that value of power that, if held constant over the interval, would result in an energy consumption equal to that energy the customer actually used.

Demand is most frequently expressed in terms of real power (kilowatts). However, demand may also apply to reactive power (kilovars), apparent power (kilovolt-amperes), or other suitable units. Billing for demand is commonly based on a customer's maximum demand reached during the billing period.

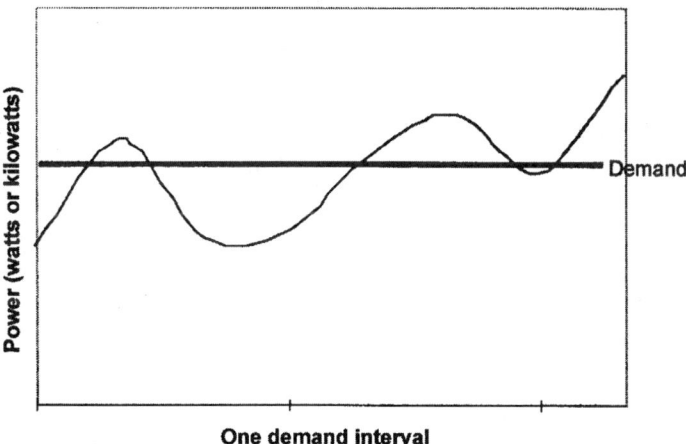

FIGURE 7.2 Instantaneous power vs. demand.

Why is Demand Metered?

Electrical conductors and transformers needed to serve a customer are selected based on the expected maximum demand for the customer. The equipment must be capable of handling the maximum levels of voltages and currents needed by the customer. A customer with a higher maximum demand requires a greater investment by the utility in equipment. Billing based on energy usage alone does not necessarily relate directly to the cost of equipment needed to serve a customer. Thus, energy billing alone may not equitably distribute to each customer an appropriate share of the utility's costs of doing business.

For example, consider two commercial customers with very simple electricity needs. Customer A has a demand of 25 kW and operates at this level 24 hours per day. Customer B has a maximum demand of 100 kW but operates at this level only 4 hours per day. For the remaining 20 hours of the day, "B" operates at a 10 kW power level.

$$\text{"A" uses } 25 \text{ kW} \times 24 \text{ hr} = 600 \text{ kWh per day}$$

$$\text{"B" uses } (100 \text{ kW} \times 4 \text{ hr}) + (10 \text{ kW} \times 20 \text{ hr}) = 600 \text{ kWh per day}$$

Assuming identical billing rates, each customer would incur the same energy costs. However, the utility's equipment investment will be larger for Customer B than for Customer A. By implementing a charge for demand as well as energy, the utility would bill Customer A for a maximum demand of 25 kW and Customer B for 100 kW. "B" would incur a larger total monthly bill, and each customer's bill would more closely represent the utility's cost to serve.

Integrating Demand Meters

By far the most common type of demand meter is the integrating demand meter. It performs two basic functions. First, it measures the *average* power during each *demand interval*. (Common demand interval lengths are 15, 30, or 60 min.) The meter makes these measurements interval-by-interval throughout each day. Second, it retains the maximum of these interval measurements.

The demand calculation function of an electronic meter is very simple. The meter measures the energy consumed during a demand interval, then multiplies by the number of demand intervals per hour. In effect, it calculates the energy that would be used if the rate of usage continued for one hour. The following table illustrates the correspondence between energy and demand for common demand interval lengths.

TABLE 7.1 Energy/Demand Comparisons

Demand Interval	Intervals per Hour	Energy During Demand Interval	Resulting Demand
60 min	1	100 kWh	100 kW
30 min	2	50 kWh	100 kW
15 min	4	25 kWh	100 kW

After each measurement, the meter compares the new demand value to the stored *maximum demand*. If the new value is greater than that stored, the meter replaces the stored value with the new one. Otherwise, it keeps the previously stored value and discards the new value. The meter repeats this process for each interval. At the end of the billing period, the utility records the maximum demand, then resets the stored *maximum demand* to zero. The meter then starts over for the new billing period.

Time of Use Metering

A time of use (TOU) meter measures and stores energy (and perhaps demand) for multiple periods in a day. For example, a service rate might define one price for energy used between the hours of 12 noon and 6 P.M. and another rate for that used outside this period. The TOU meter will identify the hours from 12 noon until 6 P.M. as "Rate 1." All other hours would be "Rate 2." The meter will maintain separate measurements of Rate 1 energy (and demand) and Rate 2 energy (and demand) for the entire billing period. Actual TOU service rates can be much more complex than this example, including features such as

- more than two periods per day,
- different periods for weekends and holidays, and
- different periods for different seasons of the year.

A TOU meter depends on an internal clock/calendar for proper operation. It includes battery backup to maintain its clock time during power outages.

Interval Data Metering

The standard method of gathering billing data from a meter is quite simple. The utility reads the meter at the beginning of the billing period and again at the end of the billing period. From these readings, it determines the energy and maximum demand for that period. This information is adequate to determine the bills for the great majority of customers. However, with the development of more complex service rates and the need to study customer usage patterns, the utility sometimes wants more detail about how a customer uses electricity. One option would be to read the meter daily. That would allow the utility to develop a day-by-day pattern of the customer's usage. However, having someone visit the meter site every day would quickly become very expensive. What if the meter could record usage data every day? The utility would have more detailed usage data, but would only have to visit the meter when it needed the data, for instance, once per month. And if the meter is smart enough to do that, why not have it record data even more often, for instance every hour?

In very simple terms, this is what *interval data metering* does. The interval meter includes sufficient circuitry and intelligence to record usage multiple times per hour. The length of the recording interval is programmable, often over a range from 1 to 60 minutes. The meter includes sufficient solid state memory to accumulate these interval readings for a minimum of 30 days at 15-minute intervals. Obviously, more frequent recording times reduce the days of storage available.

A simple kWh/kW recording meter typically records one set of data representing kWh. This provides the detailed usage patterns that allow the utility to analyze and evaluate customer "load profiles" based on daily, weekly, monthly, or annual bases. An extended function meter is commonly programmed to record two channels of data, e.g., kWh and kvarh. This provides the additional capability of analyzing customers' power factor patterns over the same periods. Though the meter records information in energy units (kWh or kvarh), it is a simple matter to convert this data to equivalent demand (kW or kvar). Since demand represents energy per unit time, simply divide the energy value for one recorder interval by the length of the interval (in hours). If the meter records 16.4 kWh in a 30-minute period, the equivalent demand for that period is 16.4 kWh/(0.5 hours) = 32.8 kW.

A sample 15-minute interval load shape for a 24-hour period is shown in the graph in Fig. 7.3. The minimum demand for that period was 10.5 kW, occurring during the interval ending at 04:30. The maximum demand was 28.7 kW, occurring during the interval ending at 15:15, or 3:15 P.M.

Pulse Metering

Metering pulses are signals generated in a meter for use outside the meter. Each pulse represents a discrete quantity of the metered value, such as kWh, kVAh, or kvarh. The device receiving the pulses determines the energy or demand at the meter by counting the number of pulses occurring in some time interval. A pulse is indicated by the transition (e.g., open to closed) of the circuit at the meter end. Pulses are commonly transmitted on small conductor wire circuits. Common uses of pulses include providing signals to

- customer's demand indicator
- customer's energy management system
- a *totalizer* (see section on Totalized Metering)

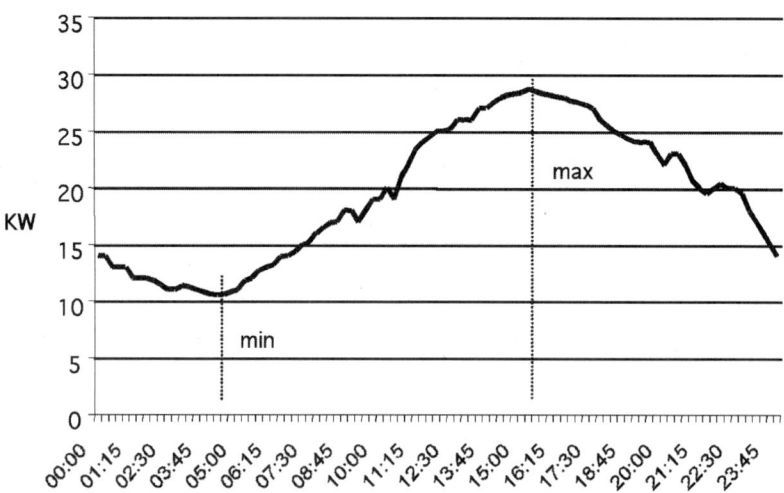

FIGURE 7.3 Graph of interval data.

- a metering data recorder
- a telemetering device that converts the pulses to other signal forms (e.g., telephone line tones or optical signals) for transmission over long distances

Pulse metering is installed when customer service requirements, equipment configurations, or other special requirements exceed the capability of conventional metering. Pulse metering is also used to transmit metered data to a remote location.

Recording Pulses

A meter pulse represents a quantity of energy, not power. For example, a pulse is properly expressed in terms of watthours (or kWh) rather than watts (or kW). A pulse recorder will associate time with pulses as it records them. If set up for a 15-minute recording interval, the recorder counts pulses for 15 min, then records that number of pulses. It then counts pulses for the next 15 min, records that number, and so on, interval after interval, day after day. It is a simple matter to determine the number of pulses recorded in a chosen length of time. Since the number of pulses recorded represents a certain amount of energy, simply divide this energy by the corresponding length of time (in hours) to determine average power for that period.

Example: For a metering installation, we are given that each pulse represents 2400 watthours or 2.4 kWh. In a 15-minute period, we record 210 pulses. What is the corresponding energy (kWh) and demand (kW) during this 15-minute interval?

Total energy in interval = 2.4 kWh per pulse × 210 pulses
= 504 kWh
Demand = Energy/Time = 504 kWh/0.25 hour
= 2016 kW

Often, a customer asks for the demand value of a pulse, rather than the energy value. The demand value is dependent on demand interval length. The demand pulse value is equal to the energy pulse value divided by the interval length in hours.

For the previous example, the kW pulse value would be:

2.4 kWh per pulse/0.25 hours = 9.6 kW per pulse

and the resulting demand calculation is:

Demand = 9.6 kW per pulse × 210 pulses
= 2016 kW

Electric Power Utilization

Remember, however, that a pulse demand value is meaningful only for a specific demand interval. In the example above, counting pulses for any period other than 15 minutes and then applying the kW pulse value will yield incorrect results for demand.

Pulse Circuits

Pulse circuits commonly take two forms (Fig. 7.4):

- *Form A*, a two-wire circuit where a switch toggles between closed and open. Each transition of the circuit (to open or to closed) represents one pulse.
- *Form C*, a three-wire circuit where the switch flip-flops. Each transition (from closed on one side to closed on the other) represents one pulse.

FIGURE 7.4 Pulse circuits.

Use care in interpreting pulse values for these circuits. The value will normally be expressed per *transition*. With Form C circuits, a transition is a change from closed on the first side to closed on the second side. Most receiving equipment interprets this properly. However, with Form A circuits, the transition is defined as a change from open to closed or from closed to open. An initially open Form A circuit that closes, then opens has undergone two (2) transitions. If the receiving equipment counts only circuit closures, it will record only half of the actual transitions. This is not a problem if the applicable pulse value of the Form A circuit is *doubled* from the rated pulse weight per transition. For example, if the value of a Form A meter pulse is 3.2 kWh per transition, the value needed for a piece of equipment that only counted circuit closures would be 3.2 × 2 = 6.4 kWh per pulse.

Totalized Metering

Totalized metering refers to the practice of combining data to make multiple service points look as if they were measured by a single meter. This is done by combining two or more sets of data from separate meters to generate data equivalent to what would be produced by a single "virtual meter" that measured the total load. This combination can be accomplished by:

- Adding recorded interval data from multiple meters, usually on a computer
- Adding (usually on-site) meter pulses from multiple meters by a special piece of metering equipment known as a totalizer
- Paralleling the secondaries of current transformers located in multiple circuits and feeding the combined current into a conventional meter (this works only when the service voltages and ratios of the current transformers are identical)
- Using a multi-circuit meter, which accepts the voltage and current inputs from multiple services

Totalized demand is the sum of the *coincident* demands and is usually less than the sum of the individual peak demands registered by the individual meters. Totalized energy equals the sum of the energies measured by the individual meters.

Table 7.2 illustrates the effects of totalizing a customer served by three delivery (and metering) points. It presents the customer's demands over a period of four demand intervals and illustrates the difference in the maximum totalized demand compared to the sum of the individual meter maximum demands.

TABLE 7.2 Example of Totalized Meter Data

Interval	Meter A	Meter B	Meter C	Totalized (A+B+C)
1	**800**	600	700	2100
2	780	650	**740**	2170
3	750	**700**	500	1950
4	780	680	720	**2180**

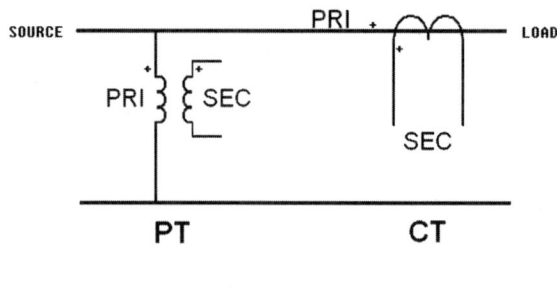

FIGURE 7.5 Instrument transformer symbols.

The peak kW demand for each meter point is shown in bold. The sum of these demands is 2240 kW. However, when summed interval-by-interval, the peak of the sums is 2180 kW. This is the *totalized demand*. The difference in the two demands, 60 kW, represents a cost savings to the customer. It should be clear why many customers with multiple service points desire to have their demands totalized.

Instrument Transformers

Instrument transformers is the general name for members of the family of current transformers (CTs) and voltage transformers (VTs) used in metering. They are high-accuracy transformers that convert load currents or voltages to other (usually smaller) values by some fixed ratio. Voltage transformers are also often called potential transformers (PTs). The terms are used interchangeably in this section. CTs and VTs are most commonly used in services where the current and/or voltage levels are too large to be applied directly to the meter.

A current transformer is rated in terms of its nameplate primary current as a ratio to five amps secondary current (e.g., 400:5). The CT is not necessarily limited to this nameplate current. Its maximum capacity is found by multiplying its nameplate rating by its *rating factor*. This yields the total current the CT can carry while maintaining its rated accuracy and avoiding thermal overload. For example, a 200:5 CT with a rating factor of 3.0 can be used and will maintain its rated accuracy up to 600 amps. Rating factors for most CTs are based on open-air outdoor conditions. When a CT is installed indoors or inside a cabinet, its rating factor is reduced.

A voltage transformer is rated in terms of its nameplate primary voltage as a ratio to either 115 or 120 volts secondary voltage (e.g., 7200:120 or 115000:115). These ratios are sometimes listed as an equivalent ratio to 1 (e.g., 60:1 or 1000:1).

Symbols for a CT and a PT connected in a two-wire circuit are shown in Fig. 7.5.

Measuring kVA

In many cases, a combination watthour demand meter will provide the billing determinants for small- to medium-sized customers served under rates that require only real power (kW) and energy (kWh). Rates for larger customers often require an *extended function* meter to provide the additional reactive or apparent power capability needed to measure or determine kVA demand. There are two common methods for determining kVA demand for billing.

1. **Actual kVA**. This method directly measures actual kVA, a simple matter for electronic meters.
2. **Average Power Factor kVA**. This method approaches the measurement of kVA in a more round-about fashion. It was developed when most metering was done with mechanical meters that could directly measure only real energy and power (kWh and kW). With a little help, they could measure kvarh. Those few meters that could measure actual kVA were very complex and demanded frequent maintenance. The Average Power Factor (APF) method of calculating kVA addressed these limitations. It requires three (3) pieces of meter information:

Electric Power Utilization

- Total real energy(kWh)
- Maximum real demand(kW)
- Total reactive energy(kvarh)

These can be measured with two standard mechanical meters. The first meter measures kWh and kW. With the help of a special transformer to shift the voltage signals 90° in phase, the second mechanical meter can be made to measure kvarh.

APF kVA is determined by calculating the customer's "average power factor" over the billing period using the total kWh and kvarh for the period. This APF is then applied to the maximum kW reading to yield APF kVA. An example of this calculation process follows.

Customer: XYZ Corporation

Billing determinants obtained from the meter:

kWh	981,600
kvarh	528,000
kW	1412

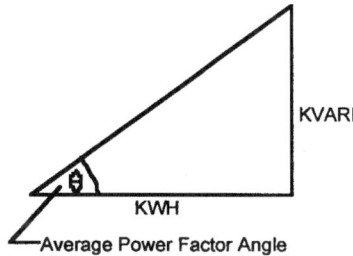

$$\tan(\theta) = \frac{KVARH}{KWH} = \frac{528000}{981600} = 0.5378$$

$$\theta = 28.275°$$

$$APF = \cos(\theta) = 0.881$$

$$KVA\ \text{demand} = \frac{KW\ \text{demand}}{APF} = \frac{1412}{0.881}$$

$$= 1603\ KVA$$

FIGURE 7.6 Calculation of kVA demand using the Average Power Factor method.

Defining Terms

Class: The class designation of a watthour meter represents the maximum current at which the meter can be operated continuously with acceptable accuracy and without excessive temperature rise. Examples of common watthour meter classes are:

Self-contained — Class 200, 320, or 400
Transformer rated — Class 10 or 20

Test amperes (TA): The test amperes rating of a watthour meter is the current that is used as a base for adjusting and determining percent registration (accuracy). Typical test current ratings and their relations to meter class are:

Class 10 and 20 — TA 2.5
Class 200 — TA 30

Self-contained meter: A self-contained meter is one designed and installed so that power flows from the utility system *through* the meter to the customer's load. The meter sees the total load current and full service voltage.

Transformer rated meter: A transformer rated meter is one designed to accept *reduced* levels of current and/or voltage that are directly proportional to the service current and voltage. The primary windings of current transformers and/or voltage transformers are placed in the customer's service and see the total load current and full service voltage. The transformer rated meter connects into the secondary windings of these transformers.

Meter element: A meter element is the basic energy and power measurement circuit for one set of meter input signals. It consists of a current measurement device and a voltage measurement device for one phase of the meter inputs. Usually, a meter will have one less element than the number of wires in the circuit being metered. That is, a 4-wire wye or delta circuit will be metered by a 3-element meter; a 3-wire delta circuit will be metered by a 2-element meter, although there are numerous exceptions.

CT PT ratio: A number or factor obtained by multiplying the current transformer ratio by the potential transformer ratio. Example: If a meter is connected to 7200:120 volt PTs (60:1) and 600:5 CTs (120:1), the CT PT ratio is $60 \times 120 = 7200$. A metering installation may have current transformers but no potential transformer in which case the CT PT ratio is just the CT ratio.

Meter multiplier: Also called the dial constant or kilowatthour constant, this is the multiplier used to convert meter kWh readings to actual kWh. The meter multiplier is the CT PT ratio. For a self-contained meter, this constant is 1.

Further Information

Further information and more detail on many of the topics related to metering can be found in the *Handbook for Electricity Metering*, published by Edison Electric Institute. This authoritative book provides extensive explanations of many aspects of metering, from fundamentals of how meters and instrument transformers operate, to meter testing, wiring, and installation.

7.2 Basic Electric Power Utilization — Loads, Load Characterization and Load Modeling

Andrew Hanson

Utilization is the "end result" of the generation, transmission, and distribution of electric power. The energy carried by the transmission and distribution system is turned into useful work, light, heat, or a combination of these items at the utilization point. Understanding and characterizing the utilization of electric power is critical for proper planning and operation of power systems. Improper characterization of utilization can result of over or under building of power system facilities and stressing of system equipment beyond design capabilities. This section describes some of the basic concepts used to characterize and model loads in electric power systems.

The term *load* refers to a device or collection of devices that draw energy from the power system. Individual loads (devices) range from small light bulbs to large induction motors to arc furnaces. The term *load* is often somewhat arbitrarily applied, at times being used to describe a specific device, and other times referring to an entire facility and even being used to describe the lumped power requirements of power system components and connected utilization devices downstream of a specific point in large-scale system studies.

Basic Load Characterization

A number of terms are used to characterize the magnitude and intensity of loads. Several such terms are defined and uses outlined below.

Energy — Energy use (over a specified period of time) is a key identifying parameter for power system loads. Energy use is often recorded for various portions of the power system (e.g., homes, businesses, feeders, substations, districts). Utilities report aggregate system energy use over a variety of time frames (daily, weekly, monthly, and annually). System energy use is tied directly to sales and thus is often used as a measure of the utility or system performance from one period to another.

Demand — Loads require specific amounts of energy over short periods of time. Demand is a measure of this energy and is expressed in terms of power (kilowatts or Megawatts). Instantaneous demand is the peak instantaneous power use of a device, facility, or system. Demand, as commonly referred to in utility discussions, is an integrated demand value, most often integrated over 10, 15, or 30 min. Integrated demand values are determined by dividing the energy used by the time interval of measurement or the demand interval.

$$\text{Demand} = \frac{\text{Energy Use Over Demand Interval}}{\text{Demand Interval}} \tag{7.1}$$

Integrated demand values can be much lower than peak instantaneous demand values for a load or facility.

Demand Factor — Demand factor is a ratio of the maximum demand to the total connected load of a system or the part of the system under consideration. Demand factor is often used to express the expected diversity of individual loads within a facility prior to construction. Use of demand factors allows facility power system equipment to be sized appropriately for the expected loads.

$$\text{Demand Factor} = \frac{\text{Maximum Demand}}{\text{Total Connected Load}} \tag{7.2}$$

Load Factor — Load factor is similar to demand factor and is calculated from the energy use, the demand, and the period of time associated with the measurement.

$$\text{Load Factor} = \frac{\text{Energy Use}}{\text{Demand} \times \text{Time}} \tag{7.3}$$

A high load factor is typically desirable, indicating that a load or group of loads operates near its peak most of the time, allowing the greatest benefit to be derived from any facilities installed to serve the load.

Composite Loads and Composite Load Characterization

It is impractical to model each individual load connected to a power system to the level of detail at which power is delivered to each individual utilization device. Loads are normally lumped together to represent all of the "downstream" power system components and individual connected loads. This grouping occurs as a result of metering all downstream power use from a certain point in the power system, or as a result of model simplification in which effects of the downstream power system and connected loads are represented by a single load in system analysis.

Coincidence and Diversity

Although individual loads vary unpredictably from hour to hour and minute to minute, an averaging effect occurs as many loads are examined in aggregate. This effect begins at individual facilities (home, commercial establishment, or industrial establishment) where all devices are seldom if ever in operation at the same instant. Progressing from an individual facility to the distribution and transmission systems, the effect is compounded, resulting in somewhat predictable load characteristics.

Diversity is a measure of the dispersion of the individual loads of a system under observation over time. Diversity is generally low in individual commercial and industrial installations. However, at a feeder level, diversity is a significant factor, allowing more economical choices for equipment since the feeder needs to supply power to the aggregate peak load of the connected customers, not the sum of the customer individual (noncoincident) peak loads.

Groups of customers of the same class (i.e., residential, commercial, industrial) tend to have an aggregate peak load per customer that decreases as the number of customers increases. This tendency is termed *coincidence* and has significant impact on the planning and construction of power systems (Willis, 1997). For example, load diversity would allow a feeder or substation to serve a number of customers whose individual (noncoincident) peak demands may exceed the feeder or substation rating by a factor of two or more.

$$\text{Coincidence Factor} = \frac{\text{Aggregate Demand for a Group of Customers}}{\text{Sum of Individual Customer Demands}} \quad (7.4)$$

Note that there is a minor but significant difference between coincidence (and its representation as a coincidence factor) and the demand factor discussed above. The coincidence factor is based on the *observed* peak demand for individuals and groups, whereas the demand factor is based on the *connected* load.

Load Curves and Load Duration

Load curves and load duration curves graphically convey very detailed information about the characteristics of loads over time. Load curves typically display the load of a customer class, feeder, or other portion of a power system over a 24-hour period. Load duration curves display the cumulative amount of time that load levels are experienced over a period of time.

Load curves represent the demand of a load or groups of load over a period of time, typically 24 hours. The curves provide "typical" load levels for a customer class on an hour-by-hour or minute-by-minute basis. The curves themselves represent the demand of a certain class of customers or portion of the system. The area under the curve represents the corresponding energy use over the time period under consideration. Load curves provide easily interpreted information regarding the peak load duration as well as the variation between minimum and maximum load levels. Load curves provide key information for daily load forecasts allowing planners and operators to ensure system capacity is available to meet customer needs. Three sample load curves (for residential, commercial, and industrial customer classes) are shown in Fig. 7.7 through Fig. 7.9.

Load curves can also be developed on a feeder or substation basis, as a composite representation of the load profile of a portion of the system.

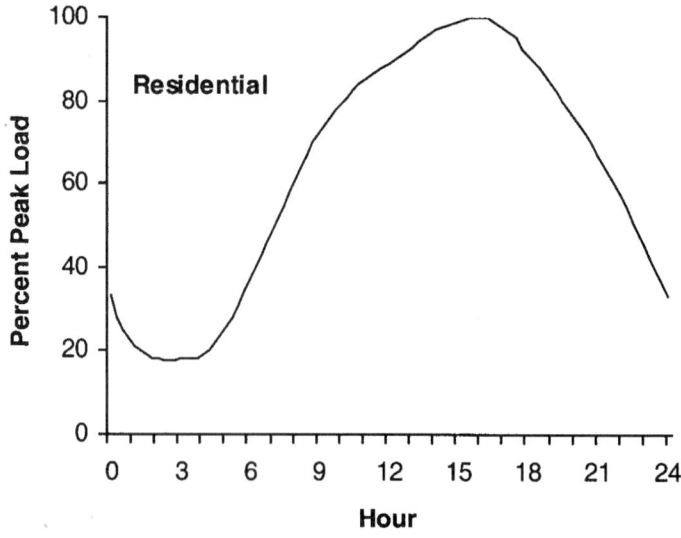

FIGURE 7.7 Residential load curve.

Electric Power Utilization

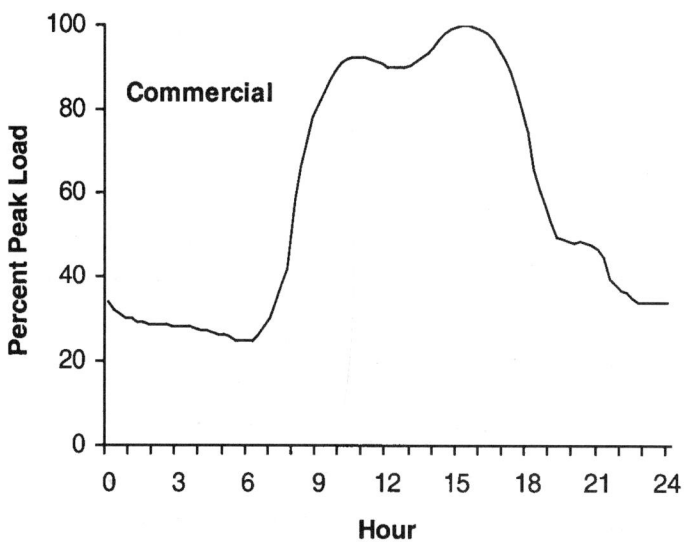

FIGURE 7.8 Commercial load curve.

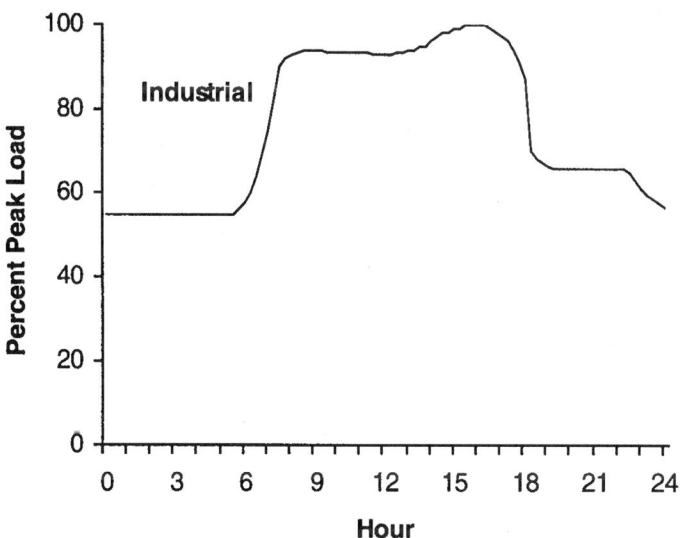

FIGURE 7.9 Industrial load curve.

Load duration curves quickly convey the duration of the peak period for a portion of a power system over a given period of time. Load duration curves plot the cumulative amount of time that load levels are seen over a specified time period. The information conveyed graphically in a load duration curve, although more detailed, is analogous to the information provided by the load factor discussed above. A sample load duration curve is shown in Fig. 7.10.

Load duration curves are often characterized by very sharp ascents to the peak load value. The shape of the remainder of the curves vary based on utilization patterns, size, and content of the system for which the load duration curve is plotted.

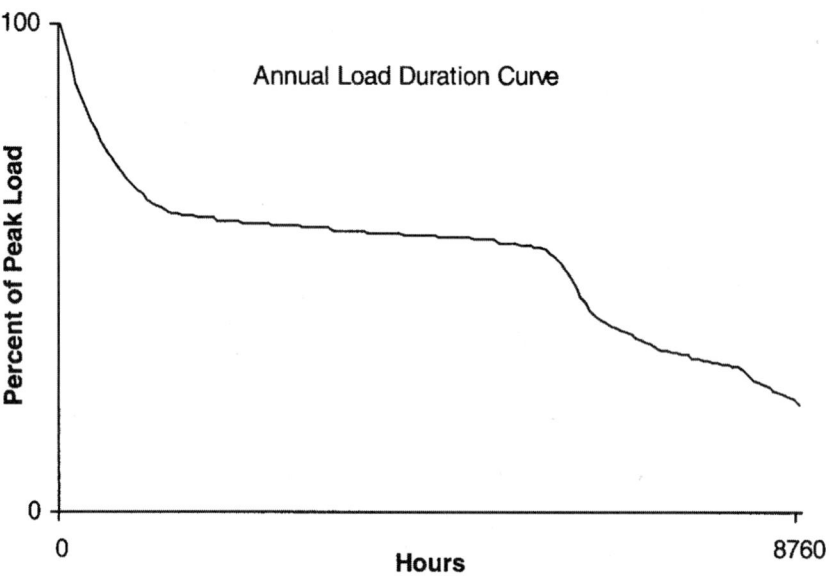

FIGURE 7.10 Annual load duration curve.

Composite Load Modeling

Load models can generally be divided into a variety of categories for modeling purposes. The appropriate load model depends largely on the application. For example, for switching transient analyses, simple load models as combinations of time-invariant circuit elements (resistors, inductors, capacitors) and/or voltage sources are usually sufficient. Power flow analyses are performed for a specific operating point at a specific frequency, allowing loads to be modeled primarily as constant impedance or constant power. However, midterm and extended term transient stability analyses require that load voltage and frequency dependencies be modeled, requiring more complex aggregate load models. Two load models are discussed below.

Composite loads exhibit dependencies on frequency and voltage. Both linear (Elgerd, 1982; Gross, 1986) and exponential models (Arrillaga and Arnold, 1990) are used for addressing these dependencies.

Linear Voltage and Frequency Dependence Model — The linear model provides excellent representation of load variations as frequency and voltages vary by small amounts about a nominal point.

$$P = P_{nominal} + \frac{\partial P}{\partial |\overline{V}|}\Delta|\overline{V}| + \frac{\partial P}{\partial f}\Delta f \tag{7.5}$$

$$Q = Q_{nominal} + \frac{\partial Q}{\partial |\overline{V}|}\Delta|\overline{V}| + \frac{\partial Q}{\partial f}\Delta f \tag{7.6}$$

where $P_{nominal}$, $Q_{nominal}$ are the real and reactive power under nominal conditions,

$\frac{\partial P}{\partial |\overline{V}|}, \frac{\partial P}{\partial f}, \frac{\partial Q}{\partial |\overline{V}|}, \frac{\partial Q}{\partial f}$ are the rates of change of real and reactive power with respect to voltage magnitude and frequency, and

$\Delta|V|, \Delta f$ are the deviations in voltage magnitude and frequency from nominal values.

Electric Power Utilization

The values for the partial derivatives with respect to voltage and frequency can be determined through analysis of metered load data recorded during system disturbances or in the case of very simple loads, through calculations based on the equivalent circuit models of individual components.

Exponential Voltage and Frequency Dependence Model — The exponential model provides load characteristics useful in midterm and extended term stability simulations in which the changes in system frequency and voltage are explicitly modeled in each time step.

$$P = P_{nominal} |\overline{V}|^{pv} f^{pf} \tag{7.7}$$

$$Q = Q_{nominal} |\overline{V}|^{qv} f^{qf} \tag{7.8}$$

where $P_{nominal}$, $Q_{nominal}$ are the real and reactive power of the load under nominal conditions

$|V|$ is the voltage magnitude in per unit

f is the frequency in per unit

pv, pf, qv, and qf are the exponential modeling parameters for the voltage and frequency dependence of the real and reactive power portions of the load, respectively

Other Load-Related Issues

Cold Load Pickup

Following periods of extended service interruption, the advantages provided by load diversity are often lost. The term *cold load pickup* refers to the energization of the loads associated with a circuit or substation following an extended interruption during which much of the diversity normally encountered in power systems is lost.

For example, if a feeder suffers an outage, interrupting all customers on the feeder during a particularly cold day, the homes and businesses will cool to levels below the individual thermostat settings. This situation eliminates the diversity normally experienced, where only a fraction of the heating will be required to operate at any given time. Once power is restored, the heating at all customer locations served by the feeder will attempt to operate to bring the building temperatures back to levels near the thermostat settings. The load experienced by the feeder following reenergization can be far in excess of the design loading due to lack of load diversity.

Cold load pickup can result in a number of adverse power system reactions. Individual service transformers can become overloaded under cold load pickup conditions, resulting in loss of life and possible failure due to overheating. Feeder load levels can exceed protective device ratings/settings, resulting in customer interruptions following initial service restoration. Additionally, the heavily loaded system conditions can result in conductors sagging below their designed minimum clearance levels, creating safety concerns.

Harmonics and Other Nonsinusoidal Loads

Electronic loads that draw current from the power system in a nonsinusoidal manner represent a significant portion of the load connected to modern power systems. These loads cause distortions of the generally sinusoidal characteristics traditionally observed. Harmonic loads include power electronic based devices (rectifiers, motor drives, switched mode power supplies, etc.) and arc furnaces. More details on power electronics and their effects on power system operation can be found in the power electronics section of this handbook.

References

Arrillaga, J. and Arnold, C. P., *Computer Analysis of Power Systems*, John Wiley & Sons, West Sussex, 1990.
Elgerd, O. I., *Electric Energy Systems Theory: An Introduction*, 2nd ed., McGraw Hill Publishing Company, New York, 1982.
Gross, C. A., *Power System Analysis*, 2nd ed., John Wiley & Sons, New York, 1986.
1996 National Electric Code, NFPA 70, Article 100, Batterymarch Park, Quincy, MA.
Willis, H. L., *Power Distribution Planning Reference Book*, Marcel-Dekker, Inc., New York, 1997.

Further Information

The references provide a brief treatment of loads and their characteristics. More detailed load characteristics for specific industries can be found in specific industry trade publications. For example, specific characteristics of loads encountered in the steel industry can be found in Fruehan, R. J., Ed., *The Making, Shaping and Treating of Steel*, 11th ed., AISE Steel Foundation, Pittsburgh, Pennsylvania, 1998.

The quarterly journals *IEEE Transactions on Power Systems* and *IEEE Transactions on Power Delivery* contain numerous papers on load modeling, as well as short and long term load forecasting. Papers in these journals also track recent developments in these areas.

Information on load modeling for long term load forecasting for power system planning can be found the following references respectively:

Willis, H. L., *Spatial Electric Load Forecasting*, Marcel-Dekker, Inc., New York, 1996.
Stoll, H. G., *Least Cost Electric Utility Planning*, John Wiley & Sons, New York, 1989.

7.3 Electric Power Utilization: Motors

Charles A. Gross

A major application of electric energy is in its conversion to mechanical energy. Electromagnetic, or "EM" devices designed for this purpose are commonly called "motors." Actually the machine is the central component of an integrated system consisting of the source, controller, motor, and load. For specialized applications, the system may be, and frequently is, designed as a integrated whole. Many household appliances (e.g., a vacuum cleaner) have in one unit, the controller, the motor, and the load. However, there remain a large number of important stand-alone applications that require the selection of a proper motor and associated control, for a particular load. It is this general issue that is the subject of this section.

The reader is cautioned that there is no "magic bullet" to deal with all motor-load applications. Like many engineering problems, there is an artistic, as well as a scientific dimension to its solution. Likewise, each individual application has its own peculiar characteristics, and requires significant experience to manage. Nevertheless, a systematic formulation of the issues can be useful to a beginner in this area of design, and even for experienced engineers faced with a new or unusual application.

Some General Perspectives

Consider the general situation in Fig. 7.11a. The flow of energy through the system is from left to right, or from electrical source to mechanical load. Also, note the positive definitions of currents, voltages, speed, and torques. These definitions are collectively called the "motor convention," and are logically used when motor applications are under study. Likewise, when generator applications are considered, the sign conventions of Fig. 7.11b (called generation convention) will be adopted. This means that variables will be positive under "normal" conditions (motors operating in the motor mode, generators in the generator mode), and negative under some abnormal conditions (motors running "backwards," for example). Using motor convention:

$$T_{dev} - (T_m + T_{RL}) = T_{dev} - T'_m = J(d\omega_{rm}/dt) \tag{7.9}$$

Electric Power Utilization

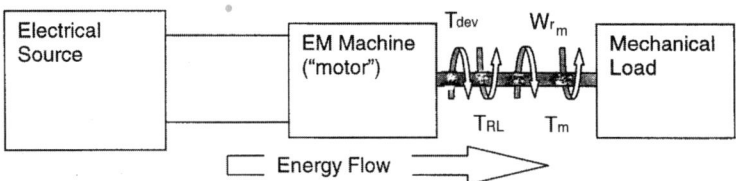

a. the EM rotational machine; motor convention

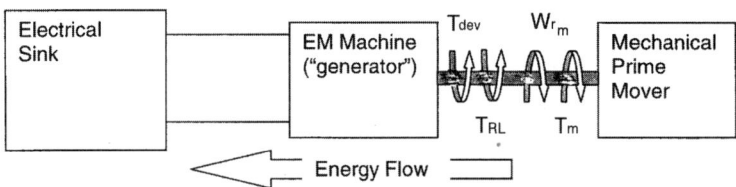

b. the EM rotational machine; generator convention

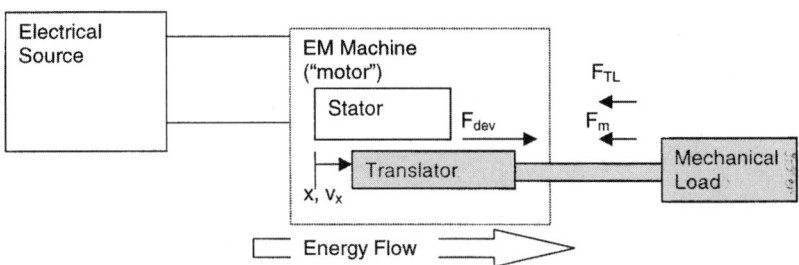

c. the EM translational machine; motor convention

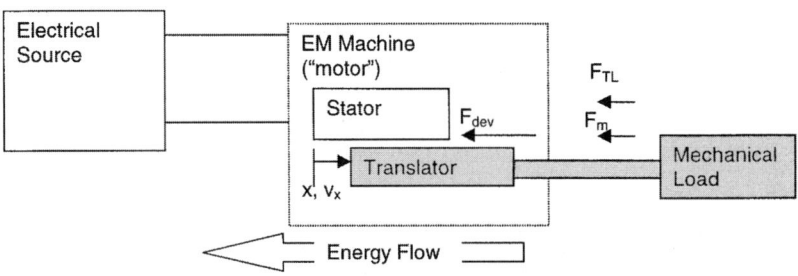

d. the EM translational machine; generator convention

FIGURE 7.11 Motor and generator sign conventions for EM machines.

where T_{dev} = EM torque, produced by the motor, Nm
T_m = torque absorbed by the mechanical load, including the load losses and that used for useful mechanical work, Nm
T_{RL} = rotational loss torque, internal to the motor, Nm
T'_m = $T_m + T_{RL}$ = equivalent load torque, Nm
J = mass polar moment of inertia of all rotating parts, kg-m²
ω_{rm} = angular velocity of rotating parts, rad/s

Observe that whenever $T_{dev} > T'_m$, the system accelerates; if $T_{dev} < T'_m$, the system decelerates. The system will inherently seek out the equilibrium condition of $T_{dev} = T'_m$, which will determine the running speed. In general, the steady state running speed for any motor-load system occurs at the intersection of the motor and load torque-speed characteristics, i.e., where $T_{dev} = T'_m$. If $T_{dev} > T'_m$, the system is accelerating; for $T_{dev} < T'_m$, the system decelerates. Thus, torque-speed characteristics for motors and loads are necessary for the design of a speed (or position) control system.

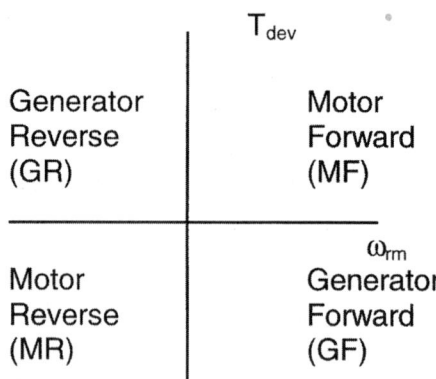

FIGURE 7.12 Operating modes.

The corresponding system powers are:

$P_{dev} = T_{dev}\, \omega_{rm}$ = EM power, converted by the motor into mechanical form, W
$P_m = T_m\, \omega_{rm}$ = power absorbed by the mechanical load, including the load losses and that used for useful mechanical work, W
$P_{RL} = T_{RL}\, \omega_{rm}$ = rotational power loss, internal to the motor, W

Operating Modes

Equation (7.9) implies that torque and speed are positive. Consider positive speed as "forward," meaning rotation in the "normal" direction, which should be obvious in a specific application. "Reverse" is defined to mean rotation in the direction opposite to "forward," and corresponds to $\omega_{rm} < 0$. Positive EM torque is in the positive speed direction. Using motor convention, first quadrant operation means that (1) speed is positive ("forward") and (2) T_{dev} is positive (also forward), and transferring energy from motor to load ("motoring"). There are four possible operating modes specific to the four quadrants of Fig. 7.12. In any application, a primary consideration is to determine which of these operating modes will be required.

Motor, Enclosure, and Controller Types

The general types of enclosures, motors, and controllers are summarized in Tables 7.3, 7.4, and 7.5.

System Design

The design of a proper motor-enclosure-controller system for a particular application is a significant engineering problem requiring engineering expertise and experience. The following issues must be faced and resolved.

Load Requirements

1. The steady-state duty cycle with torque-speed (position) requirements at each load step.
2. What operating modes are required.
3. Dynamic performance requirements, including starting and stopping, and maximum and minimum accelerations.
4. The relevant torque-speed (position) characteristics.
5. All load inertias (J).
6. Coupling options (direct drive, belt-drive, gearing).
7. Reliability of service. How critical is a system failure?
8. Future modifications.

Environmental Requirements

1. Ambient atmospheric conditions (pressure, temperature, humidity, content)
2. Indoor, outdoor application
3. Wet, dry location
4. Ventilation
5. Acceptable acoustical noise levels
6. Electrical/mechanical hazards to personnel
7. Accessibility for inspection and maintenance

Electrical Source Options

1. DC-AC
2. If AC, single- and/or three-phase
3. Voltage level
4. Frequency
5. Capacity (kVA)
6. Protection options
7. Power quality specifications

TABLE 7.3 General Enclosure Types[a]

Types
Open
Drip-proof
Splash-proof
Semi-guarded
Weather protected
Type I
Type II
Totally enclosed
Nonventillated
Fan-cooled
Explosion-proof
Dust-ignition-proof
Water-proof
Pipe-ventilated
Water-cooled
Water-air-cooled
Air-to-air-cooled
Air-over-cooled

[a] See NEMA Standard MG 1.1.25–1.1.27 for definitions.

Preliminary System Design

Based on the information compiled in the steps above, select an appropriate enclosure, motor type, and controller. In general, the enclosure entries, reading from top to bottom in Table 7.3, are from simplest (and cheapest) to most complex (and expensive). Select the simplest enclosure that meets all the environmental constraints. Next, select a motor and controller combination from Tables 7.4 and 7.5. This requires personal experience and/or consulting with engineers with experience relevant to the application.

In general, DC motors are expensive and require more maintenance, but have excellent speed and position control options. Single-phase AC motors are limited to about 5 kW, but may be desirable in locations where three-phase service is not available and control specifications are not critical.

Three-phase AC synchronous motors are not amenable to frequent starting and stopping, but are ideal for medium and high power applications which run at essentially fixed speeds. Three-phase AC cage rotor induction motors are versatile and economical, and will be the preferred choice for most applications, particularly in the medium power range. Three-phase AC wound rotor induction motors are expensive, and only appropriate for some unusual applications.

The controller must be compatible with the motor selected; the best choice is the most economical that meets all load specifications. If the engineer's experience with the application under study is lacking, two or more systems should be selected.

System Ratings

Based on the steps above, select appropriate power, voltage, and frequency ratings. For cyclic loads, the power rating may tentatively be selected based on the "rms horsepower" method (calculating the rms power requirements over the load cycle).

System Data Acquisition

Request data from at least two vendors on all systems selected in the steps above, including:

- circuit diagrams
- performance test data
- equivalent circuit values, including inertia constants
- cost data
- warranties and guarantees

TABLE 7.4 General Motor Types[a]

Type
DC motors (commutator devices)
Permanent magnet field
Wound field
Series
Shunt
Compound
AC motors
Single-phase
Cage rotor
Split phase
Resistance-start
Capacitor start
Single capacitor (start-run)
Capacitor start/capacitor run
Shaded pole
Wound rotor
Repulsion
Repulsion start/induction run
Universal
Synchronous
Hysteresis
Three-phase
Synchronous
Permanent magnet field
Wound field
Induction
Cage rotor
NEMA Design A,B,C,D,F
Wound rotor

[a] See NEMA Standard MG 1.1.1-1.1.21 for definitions.

Engineering Studies

Perform the following studies using data from the system data acquisition step above.

1. Steady state performance. Verify that each candidate system meets all steady state load requirements.
2. Dynamic performance. Verify that each system meets all dynamic load requirements.
3. Load cycle efficiency. Determine the energy efficiency over the load cycle.
4. Provide a cost estimate for each system, including capital investment, maintenance, and annual operating costs.
5. Perform a power quality assessment.

Based on these studies, select a final system design.

Final System Design

Request a competitive bid on the final design from appropriate vendors. Select a vendor based on cost, expectation of continuing technical support, reputation, warranties, and past customer experience.

Field Testing

Whenever practical, customer and vendor engineers should design and perform field tests on the installed system, demonstrating that it meets or exceeds all specifications. If multiple units are involved, one proto-unit should be installed, tested, and commissioned before delivery is made on the balance of the order.

TABLE 7.5 General Motor Controllers

Type
DC motor controllers
Electromechanical
Armature starting resistance; rheostat field control
Power electronic drive
Phase converters: 1, 2, 4 quadrant drives
Chopper control: 1, 2, 4 quadrant drives
AC motor controllers
Single-phase
Electromechanical
Across-the-line: protection only
Step-reduced voltage
Power electronic drive
Armature control: 1, 2, 4 quadrant drives
Three-phase induction
Cage rotor
Electromechanical
Across-the-line: protection only
Step-reduced voltage
Power electronic drive (ASDs)
Variable voltage source inverter
Variable current source inverter
Chopper voltage source inverter
PWM voltage source inverter
Vector control
Wound rotor
Variable rotor resistance
Power electronic rotor power recovery
Three-phase synchronous
Same as cage rotor induction
Brushless DC control

Further Information

The design of a properly engineered motor-controller system for a particular application requires access to several technical resources, including standards, the technical literature, manufacturers' publications, textbooks, and handbooks. The following section provides a list of references and resource material that the author recommends for work in this area. In many cases, more recent versions of publications listed are available and should be used.

Organizations

American National Standards Institute (ANSI), 1430 Broadway, New York, NY 10018.
Institute of Electrical and Electronics Engineers (IEEE), 445 Hoes Lane, Piscataway, NJ 08855.
International Organization for Standardization (ISO) 1, rue de Varembe, 1211 Geneva 20, Switzerland.
American Society for Testing and Materials (ASTM), 1916 Race Street, Philadelphia, PA 19103.
National Electrical Manufacturers Association (NEMA), 2101 L Street, NW, Washington, D.C. 20037.
National Fire Protection Association (NFPA), Batterymarch Park Quincy, MA 02269.
The Rubber Manufacturers Association, Inc., 1400 K Street, NW, Suite 300, Washington, D.C. 20005.
Mechanical Power Transmission Association, 1717 Howard Street, Evanston, IL 60201.

Standards

NEMA MG 1-1987, *Motors and Generators.*
NEMA MG 2-1983, *Safety Standard for Construction and Guide for Selection, Installation and Use of Electric Motors and Generators.*

NEMA MG 3-1984, *Sound Level Prediction for Installed Rotating Electrical Machines.*
NEMA MG 13-1984, *Frame Assignments for Alternating-Current Integral-horsepower Induction Motors.*
ANSI/NFPA 70-1998, *National Electrical Code.*
IEEE Std 1-1969, *General Principles for Temperature Limits in the Rating of Electric Equipment.*
IEEE Std 85-1980, *Test Procedure for Airborne Sound Measurements on Rotating Electric Machinery.*
ANSI/IEEE Std 100-1984, *IEEE Standard Dictionary of Electrical and Electronics Terms.*
IEEE Std 112-1984, *Standard Test Procedure for Potyphase Induction Motors and Generators.*
IEEE Std 113-1985, *Guide on Test Procedures for DC Machines.*
ANSI/IEEE Std 114-1984, *Test Procedure for Single-Phase Induction Motors.*
ANSI/IEEE Std 115-1983, *Test Procedures for Synchronous Machines.*
ANSI/IEEE Std 117-1985, *Standard Test Procedure for Evaluation of Systems of Insulating Materials for Random-Wound AC Electric Machinery.*
ANSI/IEEE Std 304-1982, *Test Procedure for Evaluation and Classification of Insulation Systems for DC Machines.*
ISO R-1000, *SI Units and Recommendations for the Use of their Multiples and of Certain Other Units.*

Books (an abridged sample)

Acarnley, P. P., *Stepping Motors,* 2nd ed., Peter Peregrinus, Ltd., London, 1984.
Anderson, L. R., *Electric Machines and Transformers,* Reston Publishing, Reston, VA, 1981.
Bergseth, F. R. and Venkata, S. S., *Introduction to Electric Energy Devices,* Prentice-Hall, Englewood Cliffs, NJ, 1987.
Brown, D. and Hamilton 111, E. P., *Electromechanical Energy Conversion,* Macmillan, New York, 1984.
Chapman, S. J., *Electric Machinery Fundamentals,* McGraw-Hill, New York, 1985.
DC Motors-Speed Controls-Servo Systems — An Engineering Handbook, 5th ed., Electro-Craft Corporation, Hopkins, MN, 1980.
Del Toro, V, *Electric Machinery and Power Systems,* Prentice-Hall, Englewood Cliffs, NJ, 1986.
Electro-Craft Corporation, *DC Motors, Speed Controls, Servo Systems,* 3rd ed., Pergamon Press, Ltd., Oxford, 1977.
Fitzgerald, A. E., Kingsley, Jr., C., and Umans, S. D., *Electric Machinery,* 5th ed., McGraw-Hill, New York, 1990.
Gonen, T., *Engineering Economy for Engineering Managers,* Wiley, New York, 1990.
Kenjo, T. and S. Nagamori, *Permanent-Magnet and Brush-less DC Motors,* Oxford, Claredon, 1985.
Krause, P. C. and Wasynezk, O., *Electromechanical Machines and Devices,* McGraw-Hill, New York, 1989.
Krein, P., *Elements of Power Electronics;* Oxford Press, 1998.
Moha, N., Undeland, and Robbins, *Power Electronics; Converters, Application, and Design,* 2nd ed., John Wiley & Sons, New York, 1995.
Nasar, S. A. and Boldea, I., *Linear Motion Electric Machines,* John Wiley & Sons, New York, 1976.
Nasar, S. A., Ed., *Handbook of Electric Machines,* McGraw-Hill, New York, 1987.
Patrick, D. R. and Fardo, S. W., *Rotating Electrical Machines and Power Systems,* Prentice-Hall, Englewood Cliffs, NJ, 1985.
Ramshaw, R. and Van Heeswijk, R. G., *Energy Conversion: Electric Motors and Generators,* Saunders College Publishing, Orlando, FL, 1990.
Rashid, M. H., *Power Electronics: Circuits, Devices, and Applications,* 2nd ed., Prentice-Hall, Englewood Cliffs, NJ, 1993.
Sarma, M. S., *Electric Machines: Steady-State Theory and Dynamic Performance,* Brown Publishers, Dubuque, IA, 1985.
Smeatson, R. W., Ed., *Motor Application and Maintenance Handbook,* McGraw-Hill, New York, 1969.
Stein, R., and Hunt, W. T., *Electric Power System Components: Transformers and Rotating Machines,* Van Nostrand, New York, 1979.
Veinott, C. G. and Martin, J. E., *Fractional- and Subfractional-Horsepower Electric Motors,* 4th ed., McGraw-Hill, New York, 1986.
Wenick, E. H., ed., *Electric Motor Handbook,* McGraw-Hill, London, 1978.
Bose, B.K., Power *Electronics and AC Drives,* Prentice-Hall, Englewood Cliffs, NJ, 1985.

8
Power System Analysis and Simulation

L.L. Grigsby
Auburn University

Andrew Hanson
ABB Power T&D Company

    8.1   The Per-Unit System   *Charles A. Gross*..8-1

    8.2   Symmetrical Components for Power System Analysis   *Tim A. Haskew*......................8-14

    8.3   Power Flow Analysis   *L.L. Grigsby and Andrew Hanson*...8-34

    8.4   Fault Analysis in Power Systems   *Charles A. Gross*..8-44

8
Power System Analysis and Simulation

Charles A. Gross
Auburn University

Tim A. Haskew
University of Alabama

L. L. Grigsby
Auburn University

Andrew Hanson
ABB Power T&D Company

8.1 The Per-Unit System 8-1
   Impact on Transformers • Per-Unit Scaling Extended to Three-Phase Systems • Per-Unit Scaling Extended to a General Three-Phase System

8.2 Symmetrical Components for Power System Analysis....8-14
   Fundamental Definitions • Reduction to the Balanced Case • Sequence Network Representation in Per-Unit • Power Transformers

8.3 Power Flow Analysis..8-34
   The Power Flow Problem • Formulation of the Bus Admittance Matrix • Formulation of the Power Flow Equations • Bus Classifications • Generalized Power Flow Development • Solution Methods • Component Power Flows

8.4 Fault Analysis in Power Systems 8-44
   Simplifications in the System Model • The Four Basic Fault Types • An Example Fault Study • Further Considerations • Summary

8.1 The Per-Unit System

Charles A. Gross

In many engineering situations, it is useful to scale or normalize quantities. This is commonly done in power system analysis, and the standard method used is referred to as the per-unit system. Historically, this was done to simplify numerical calculations that were made by hand. Although this advantage has been eliminated by using the computer, other advantages remain:

- Device parameters tend to fall into a relatively narrow range, making erroneous values conspicuous.
- The method is defined in order to eliminate ideal transformers as circuit components.
- The voltage throughout the power system is normally close to unity.

Some disadvantages are that component equivalent circuits are somewhat more abstract. Sometimes phase shifts that are clearly present in the unscaled circuit are eliminated in the per-unit circuit.

It is necessary for power system engineers to become familiar with the system because of its wide industrial acceptance and use and also to take advantage of its analytical simplifications. This discussion is limited to traditional AC analysis, with voltages and currents represented as complex phasor values. Per-unit is sometimes extended to transient analysis and may include quantities other than voltage, power, current, and impedance.

The basic per-unit scaling equation is

$$\text{Per-unit value} = \frac{\text{actual value}}{\text{base value}}. \tag{8.1}$$

The base value always has the same units as the actual value, forcing the per-unit value to be dimensionless. Also, the base value is always a real number, whereas the actual value may be complex. Representing a complex value in polar form, the angle of the per-unit value is the same as that of the actual value.

Consider complex power

$$\mathbf{S} = \mathbf{V}\mathbf{I}^* \tag{8.2}$$

or

$$S\angle\theta = V\angle\alpha \; I\angle-\beta$$

where $\mathbf{V}$ = phasor voltage, in volts; $\mathbf{I}$ = phasor current, in amperes.

Suppose we arbitrarily pick a value S_{base}, a real number with the units of volt-amperes. Dividing through by S_{base},

$$\frac{S\angle\theta}{S_{base}} = \frac{V\angle\alpha \; I\angle-\beta}{S_{base}}.$$

We further define

$$V_{base} \, I_{base} = S_{base}. \tag{8.3}$$

Either V_{base} or I_{base} may be selected arbitrarily, but not both. Substituting Eq. (8.3) into Eq. (8.2), we obtain

$$\frac{S\angle\theta}{S_{base}} = \frac{V\angle\alpha(I\angle-\beta)}{V_{base} \, I_{base}}$$

$$S_{pu}\angle\theta = \left(\frac{V\angle\alpha}{V_{base}}\right)\left(\frac{I\angle-\beta}{I_{base}}\right)$$

$$\mathbf{S}_{pu} = V_{pu}\angle\alpha(I_{pu}\angle-\beta)$$

$$\mathbf{S}_{pu} = \mathbf{V}_{pu} \, \mathbf{I}_{pu}^{*} \tag{8.4}$$

The subscript *pu* indicates per-unit values. Note that the form of Eq. (8.4) is identical to Eq. (8.2). This was not inevitable, but resulted from our decision to relate V_{base}, I_{base}, and S_{base} through Eq. (8.3). If we select Z_{base} by

$$Z_{base} = \frac{V_{base}}{I_{base}} = \frac{V_{base}^2}{S_{base}}. \tag{8.5}$$

Convert Ohm's law:

$$\mathbf{Z} = \frac{\mathbf{V}}{\mathbf{I}} \tag{8.6}$$

Power System Analysis and Simulation

into per-unit by dividing by Z_{base}.

$$\frac{Z}{Z_{base}} = \frac{V/I}{Z_{base}}$$

$$Z_{pu} = \frac{V/V_{base}}{I/I_{base}} = \frac{V_{pu}}{I_{pu}}.$$

Observe that

$$Z_{pu} = \frac{Z}{Z_{base}} = \frac{R+jX}{Z_{base}} = \left(\frac{R}{Z_{base}}\right) + j\left(\frac{X}{Z_{base}}\right)$$

$$Z_{pu} = R_{pu} + jX_{pu} \qquad (8.7)$$

Thus, separate bases for R and X are not necessary:

$$Z_{base} = R_{base} = X_{base}$$

By the same logic,

$$S_{base} = P_{base} = Q_{base}$$

Example 1:

(a) Solve for **Z**, **I**, and **S** at Port ab in Fig. 8.1a.
(b) Repeat (a) in per-unit on bases of $V_{base} = 100$ V and $S_{base} = 1000$ V. Draw the corresponding per-unit circuit.

Solution:

(a) $\mathbf{Z}_{ab} = 8 + j12 - j6 = 8 + j6 = 10 \angle 36.9° \, \Omega$

$\mathbf{I} = \dfrac{\mathbf{V}_{ab}}{\mathbf{Z}_{ab}} = \dfrac{100 \angle 0°}{10 \angle 36.9°} = 10 \angle -36.9°$ amperes

$\mathbf{S} = \mathbf{V}\mathbf{I}^* = (100 \angle 0°)(10 \angle -36.9°)^*$

$\qquad = 1000 \angle 36.9° = 800 + j600$ VA

$P = 800$ W $\qquad Q = 600$ var

(b) On bases V_{base} and $S_{base} = 1000$ VA:

$Z_{base} = \dfrac{V_{base}^2}{S_{base}} = \dfrac{(100)^2}{1000} = 10 \, \Omega$

$I_{base} = \dfrac{S_{base}}{V_{base}} = \dfrac{1000}{100} = 10$ A

$$V_{pu} = \frac{100\angle 0°}{100} = 1\angle 0° \text{ pu}$$

$$Z_{pu} = \frac{8+j12-j6}{10} = 0.8+j0.6 \text{ pu}$$

$$= 1.0\angle 36.9° \text{ pu}$$

$$I_{pu} = \frac{V_{pu}}{Z_{pu}} = \frac{1\angle 0°}{1\angle 36.9°} = 1\angle -36.9° \text{ pu}$$

$$S_{pu} = V_{pu} I_{pu}{}^* = (1\angle 0°)(1\angle -36.9°)^* = 1\angle 36.9° \text{ pu}$$

$$= 0.8+j0.6 \text{ pu}$$

Converting results in (b) to SI units:

$$I = (I_{pu})I_{base} = (1\angle -36.9°)(10) = 10\angle -36.9° \text{ A}$$

$$Z = (Z_{pu})Z_{base} = (0.8+j0.6)(10) = 8+j6 \text{ } \Omega$$

$$S = (S_{pu})S_{base} = (0.8+j0.6)(1000) = 800+j600 \text{ W, var}$$

The results of (a) and (b) are identical.

For power system applications, base values for S_{base} and V_{base} are arbitrarily selected. Actually, in practice, values are selected that force results into certain ranges. Thus, for V_{base}, a value is chosen such that the normal system operating voltage is close to unity. Popular power bases used are 1, 10, 100, and 1000 MVA, depending on system size.

Impact on Transformers

To understand the impact of pu scaling on transformer, consider the three-winding ideal device (see Fig. 8.2).
For sinusoidal steady-state performance:

$$V_1 = \frac{N_1}{N_2} V_2 \tag{8.8a}$$

$$V_2 = \frac{N_2}{N_3} V_3 \tag{8.8b}$$

$$V_3 = \frac{N_3}{N_1} V_1 \tag{8.8c}$$

and

$$N_1 I_1 + N_2 I_2 + N_3 I_3 = 0 \tag{8.9}$$

Power System Analysis and Simulation   8-5

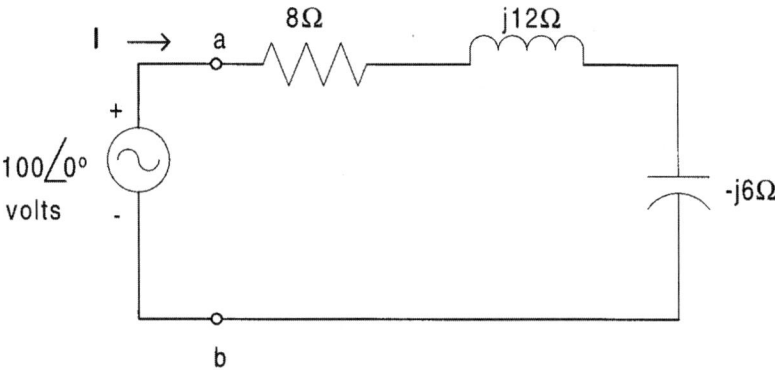

FIGURE 8.1a Circuit with elements in SI units.

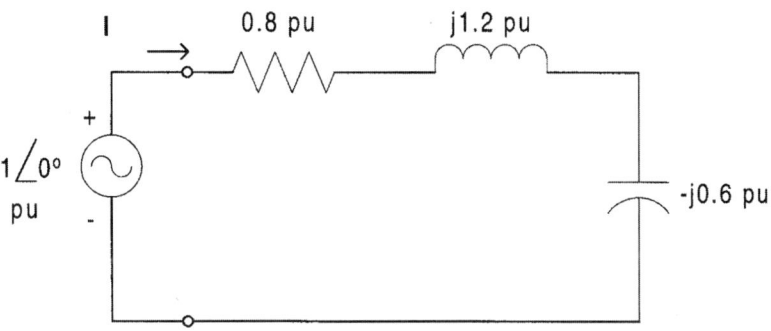

FIGURE 8.1b Circuit with elements in per-unit.

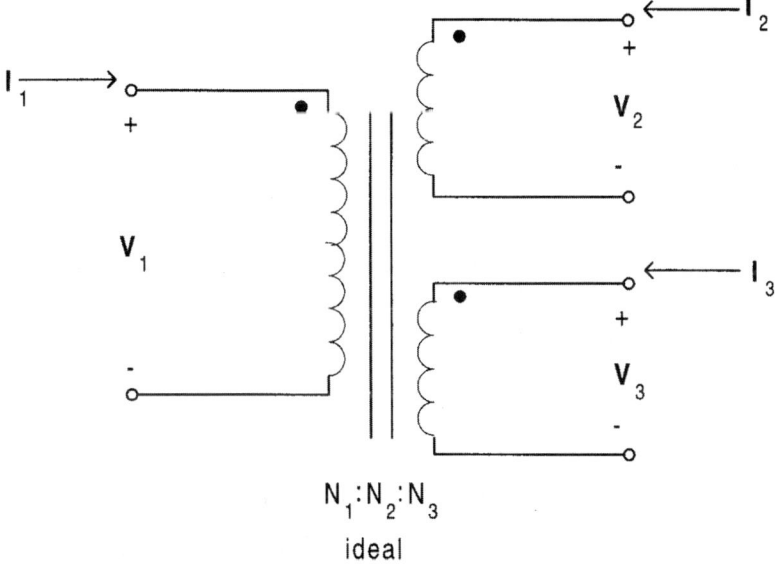

FIGURE 8.2 The three-winding ideal transformer.

Consider the total input complex power **S**.

$$\begin{aligned}
S &= V_1 I_1{}^* + V_2 I_2{}^* + V_3 I_3{}^* \\
&= V_1 I_1{}^* + \frac{N_2}{N_1} V_1 I_2{}^* + \frac{N_3}{N_1} V_1 I_3{}^* \\
&= \frac{V_1}{N_1}\left[N_1 I_1 + N_2 I_2 + N_3 I_3\right]^* \\
&= 0
\end{aligned} \tag{8.10}$$

The interpretation to be made here is that the ideal transformer can neither absorb real nor reactive power. An example should clarify these properties.

Arbitrarily select two base values V_{1base} and S_{1base}. Require base values for windings 2 and 3 to be:

$$V_{2base} = \frac{N_2}{N_1} V_{1base} \tag{8.11a}$$

$$V_{3base} = \frac{N_3}{N_1} V_{1base} \tag{8.11b}$$

and

$$S_{1base} = S_{2base} = S_{3base} = S_{base} \tag{8.12}$$

By definition,

$$I_{1base} = \frac{S_{base}}{V_{1base}} \tag{8.13a}$$

$$I_{2base} = \frac{S_{base}}{V_{2base}} \tag{8.13b}$$

$$I_{3base} = \frac{S_{base}}{V_{3base}} \tag{8.13c}$$

It follows that

$$I_{2base} = \frac{N_1}{N_2} I_{1base} \tag{8.14a}$$

$$I_{3base} = \frac{N_1}{N_3} I_{1base} \tag{8.14b}$$

Recall that a per-unit value is the actual value divided by its appropriate base. Therefore:

$$\frac{V_1}{V_{1base}} = \frac{(N_1/N_2) V_2}{V_{1base}} \tag{8.15a}$$

Power System Analysis and Simulation

and

$$\frac{V_1}{V_{1base}} = \frac{(N_1/N_2)V_2}{(N_1/N_2)V_{2base}} \quad (8.15b)$$

or

$$V_{1pu} = V_{2pu} \quad (8.15c)$$

indicates per-unit values. Similarly,

$$\frac{V_1}{V_{1base}} = \frac{(N_1/N_3)V_3}{(N_1/N_3)V_{3base}} \quad (8.16a)$$

or

$$V_{1pu} = V_{3pu} \quad (8.16b)$$

Summarizing:

$$V_{1pu} = V_{2pu} = V_{3pu} \quad (8.17)$$

Divide Eq. (8.9) by N_1

$$I_1 + \frac{N_2}{N_1}I_2 + \frac{N_3}{N_1}I_3 = 0$$

Now divide through by I_{1base}

$$\frac{I_1}{I_{1base}} + \frac{(N_2/N_1)I_2}{I_{1base}} + \frac{(N_3/N_1)I_3}{I_{1base}} = 0$$

$$\frac{I_1}{I_{1base}} + \frac{(N_2/N_1)I_2}{(N_2/N_1)I_{2base}} + \frac{(N_3/N_1)I_3}{(N_3/N_1)I_{3base}} = 0$$

Simplifying to

$$I_{1pu} + I_{2pu} + I_{3pu} = 0 \quad (8.18)$$

Equations (8.17) and (8.18) suggest the basic scaled equivalent circuit, shown in Fig. 8.3. It is cumbersome to carry the pu in the subscript past this point: no confusion should result, since all quantities will show units, including pu.

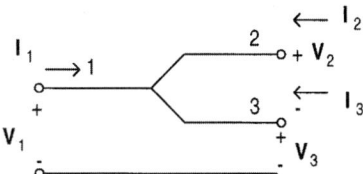

FIGURE 8.3 Single-phase ideal transformer.

Example 2:

The 3-winding single-phase transformer of Fig. 8.1 is rated at 13.8 kV/138kV/4.157 kV and 50 MVA/40 MVA/10 MVA. Terminations are as followings:

    13.8 kV winding: 13.8 kV Source
    138 kV winding: 35 MVA load, pf = 0.866 lagging
    4.157 kV winding: 5 MVA load, pf = 0.866 leading

Using S_{base} = 10 MVA, and voltage ratings as bases,

(a) Draw the pu equivalent circuit.
(b) Solve for the primary current, power, and power, and power factor.

Solution:

(a) See Fig. 8.4.

(b,c) $S_2 = \dfrac{35}{10} = 3.5$ pu $S_2 = 3.5\angle+30°$ pu

$S_3 = \dfrac{5}{10} = 0.5$ pu $S_3 = 0.5\angle-30°$ pu

$V_1 = \dfrac{13.8}{13.8} = 1.0$ pu $V_1 = V_2 = V_3 = 1.0\angle0°$ pu

$I_2 = \left(\dfrac{S_2}{V_2}\right)^* = 3.5\angle-30°$ pu

$I_3 = \left(\dfrac{S_3}{V_3}\right)^* = 0.5\angle+30°$ pu

All values in Per-Unit Equivalent Circuit:

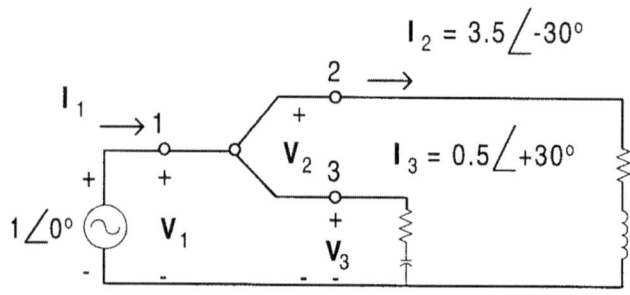

FIGURE 8.4 Per-unit circuit.

$I_1 = I_2 + I_3 = 3.5\angle-30° + 0.5\angle+30° = 3.464 - j1.5 = 3.775\angle-23.4°$ pu

$S_1 = V_1 I_1^* = 3.775\angle+23.4°$ pu

$S_1 = 3.775(10) = 37.75$ MVA; pf = 0.9177 lagging

$I_1 = 3.775\left(\dfrac{10}{0.0138}\right) = 2736$ A

Per-Unit Scaling Extended to Three-Phase Systems

The extension to three-phase systems has been complicated to some extent by the use of traditional terminology and jargon, and a desire to normalize phase-to-phase and phase-to-neutral voltage simultaneously. The problem with this practice is that it renders Kirchhoff's voltage and current laws invalid in some circuits. Consider the general three-phase situation in Fig. 8.5, with all quantities in SI units.

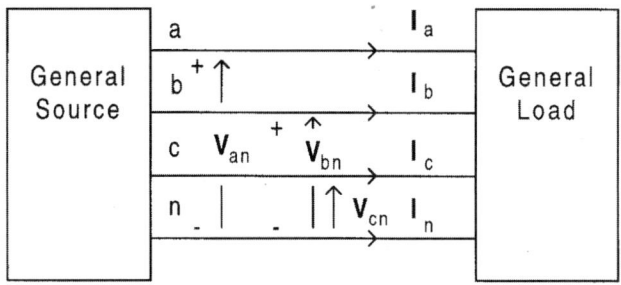

FIGURE 8.5 General three-phase system.

Define the complex operator:

$$\mathbf{a} = 1\angle 120°$$

The system is said to be balanced, with sequence abc, if:

$$\mathbf{V}_{bn} = \mathbf{a}^2 \mathbf{V}_{an}$$

$$\mathbf{V}_{cn} = \mathbf{a} \mathbf{V}_{an}$$

and

$$\mathbf{I}_b = \mathbf{a}^2 \mathbf{I}_a$$

$$\mathbf{I}_c = \mathbf{a} \mathbf{I}_a$$

$$-\mathbf{I}_n = \mathbf{I}_a + \mathbf{I}_b + \mathbf{I}_c = 0$$

Likewise:

$$\mathbf{V}_{ab} = \mathbf{V}_{an} - \mathbf{V}_{bn}$$

$$\mathbf{V}_{bc} = \mathbf{V}_{bn} - \mathbf{V}_{cn} = \mathbf{a}^2 \mathbf{V}_{ab}$$

$$\mathbf{V}_{ca} = \mathbf{V}_{cn} - \mathbf{V}_{an} = \mathbf{a} \mathbf{V}_{ab}$$

If the load consists of wye-connected impedance:

$$Z_y = \frac{\mathbf{V}_{an}}{\mathbf{I}_a} = \frac{\mathbf{V}_{bn}}{\mathbf{I}_b} = \frac{\mathbf{V}_{cn}}{\mathbf{I}_c}$$

The equivalent delta element is:

$$Z_\Delta = 3 Z_Y$$

To convert to per-unit, define the following bases:

$S_{3\phi base}$ = The three-phase apparent base at a specific location in a three-phase system, in VA.
V_{Lbase} = The line (phase-to-phase) rms voltage base at a specific location in a three-phase system, in V.

From the above, define:

$$S_{base} = S_{3\phi base}/3 \tag{8.19}$$

$$V_{base} = V_{Lbase}/\sqrt{3} \tag{8.20}$$

It follows that:

$$I_{base} = S_{base}/V_{base} \tag{8.21}$$

$$Z_{base} = V_{base}/I_{base} \tag{8.22}$$

An example will be useful.

Example 3:

Consider a balanced three-phase 60 MVA 0.8 pf lagging load, sequence abc operating from a 13.8 kV (line voltage) bus. On bases of $S_{3\phi base}$ = 100 MVA and V_{Lbase} = 13.8 kV:

(a) Determine all bases.
(b) Determine all voltages, currents, and impedances, in SI units and per-unit.

Solution:

(a) $S_{base} = \dfrac{S_{3\phi base}}{3} = \dfrac{100}{3} = 33.33$ MVA

$V_{base} = \dfrac{V_{Lbase}}{\sqrt{3}} = \dfrac{13.8}{\sqrt{3}} = 7.967$ kV

$I_{base} = \dfrac{S_{base}}{V_{base}} = 4.184$ kA

$Z_{base} = \dfrac{V_{base}}{I_{base}} = 1.904\ \Omega$

(b) $\mathbf{V}_{an} = 7.967\angle 0°$ kV $\quad (1.000\angle 0°$ pu$)$

$\mathbf{V}_{bn} = 7.967\angle -120°$ kV $\quad (1.000\angle -120°$ pu$)$

$\mathbf{V}_{cn} = 7.967\angle +120°$ kV $\quad (1.000\angle +120°$ pu$)$

$S_a = S_b = S_c = \dfrac{S_{3\phi}}{3} = \dfrac{60}{3} = 20$ MVA $\quad (0.60$ pu$)$

$\mathbf{S}_a = \mathbf{S}_b = \mathbf{S}_c = 16 + j12$ MVA $\quad (0.48 + j0.36$ pu$)$

$\mathbf{I}_a = \left(\dfrac{\mathbf{S}_a}{\mathbf{V}_{an}}\right)^* = 2.510\angle -36.9°$ kA $\quad (0.6000\angle -36.9°$ pu$)$

$\mathbf{I}_b = 2.510\angle-156.9°\,\text{kA}\quad (0.6000\angle-156.9°\,\text{pu})$

$\mathbf{I}_c = 2.510\angle 83.1°\,\text{kA}\quad (0.6000\angle 83.1°\,\text{pu})$

$\mathbf{Z}_Y = \dfrac{\mathbf{V}_{an}}{\mathbf{I}_a} = 3.174\angle+36.9° = 2.539 + j1.904\,\Omega\quad (1.33 + j1.000\,\text{pu})$

$\mathbf{Z}_\Delta = 3\mathbf{Z}_Y = 7.618 + j5.713\,\Omega\quad (4 + j3\,\text{pu})$

$\mathbf{V}_{ab} = \mathbf{V}_{an} - \mathbf{V}_{bn} = 13.8\angle 30°\,\text{kV}\quad (1.732\angle 30°\,\text{pu})$

$\mathbf{V}_{bc} = 13.8\angle -90°\,\text{kV}\quad (1.732\angle -90°\,\text{pu})$

$\mathbf{V}_{ca} = 13.8\angle 150°\,\text{kV}\quad (1.732\angle 150°\,\text{pu})$

Converting voltages and currents to symmetrical components:

$$\begin{bmatrix}\mathbf{V}_0\\ \mathbf{V}_1\\ \mathbf{V}_2\end{bmatrix} = \dfrac{1}{3}\begin{bmatrix}1 & 1 & 1\\ 1 & a & a^2\\ 1 & a^2 & a\end{bmatrix}\begin{bmatrix}\mathbf{V}_{an}\\ \mathbf{V}_{bn}\\ \mathbf{V}_{cn}\end{bmatrix} = \begin{bmatrix}0\,\text{kV}\\ 7.967\angle 0°\,\text{kV}\\ 0\,\text{kV}\end{bmatrix}\quad \begin{pmatrix}0\,\text{pu}\\ 1\angle 0°\,\text{pu}\\ 0\,\text{pu}\end{pmatrix}$$

$\mathbf{I}_0 = 0\,\text{kA}\quad (0\,\text{pu})$

$\mathbf{I}_1 = 2.510\angle -36.9°\,\text{kA}\quad (0.6\angle -36.9°\,\text{pu})$

$\mathbf{I}_2 = 0\,\text{kA}\quad (0\,\text{pu})$

Inclusion of transformers demonstrates the advantages of per-unit scaling.

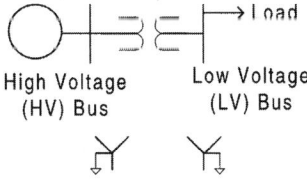

FIGURE 8.6 A three-phase transformer situation.

Example 4:

A 3ϕ 240 kV Y :15 kV Y transformer supplies a 13.8 kV 60 MVA pf = 0.8 lagging load, and is connected to a 230 kV source on the HV side, as shown in Fig. 8.6.

(a) Determine all base values on both sides for $S_{3\phi\text{base}} = 100$ MVA. At the LV bus, $V_{L\text{base}} = 13.8$ kV.
(b) Draw the positive sequence circuit in per-unit, modeling the transformer as ideal.
(c) Determine all currents and voltages in SI and per-unit.

Solution:

(a) Base values on the LV side are the same as in Example 3.
The turns ratio may be derived from the voltage ratings ratios:

$$\frac{N_1}{N_2} = \frac{240/\sqrt{3}}{15/\sqrt{3}} = 16$$

$$\therefore (V_{base})_{HV\ side} = \frac{N_1}{N_2}(V_{base})_{LV\ side} = 16.00(7.967) = 127.5\ \text{kV}$$

$$(I_{base})_{HV\ side} = \frac{S_{base}}{(V_{base})_{HV\ side}} = \frac{33.33}{0.1275} = 261.5\ \text{A}$$

Results are presented in the following chart.

Bus	$S_{3\phi base}$ MVA	$V_{L\ base}$ kV	S_{base} MVA	I_{base} kA	V_{base} kV	Z_{base} ohm
LV	100	13.8	33.33	4.184	7.967	1.904
HV	100	220.8	33.33	0.2615	127.5	487.5

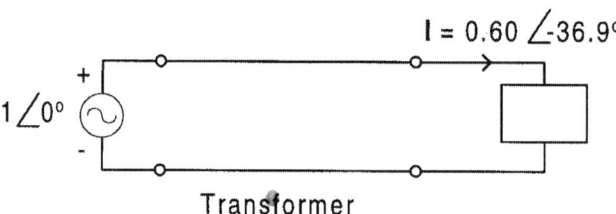

FIGURE 8.7 Positive sequence circuit.

(b) $\mathbf{V}_{LV} = \dfrac{7.967\angle 0°}{7.967} = 1\angle 0°\ \text{pu}$

$S_{1\phi} = \dfrac{60}{3} = 20\ \text{MVA}$

$S_{1\phi} = \dfrac{20}{33.33} = 0.6\ \text{pu}$

(c) All values determined in pu are valid on both sides of the transformer! To determine SI values on the HV side, use HV bases. For example:

$\mathbf{V}_{an} = (1\angle 0°)127.5 = 127.5\angle 0°\ \text{kV}$

$\mathbf{V}_{ab} = (1.732\angle 30°)(127.5) = 220.8\angle 30°\ \text{kV}$

$\mathbf{I}_a = (0.6\angle -36.9°)(261.5) = 156.9\angle -36.9°\ \text{A}$

Example 5:
Repeat the previous example using a 3ϕ 240 kV:15 kV Y↓↑Δ

Solution:
All results are the same as before. The reasoning is as follows.

The voltage ratings are interpreted as line (phase-to-phase) values *independent of connection* (wye or delta). Therefore the turns ratio remains:

$$\frac{N_1}{N_2} = \frac{240\sqrt{3}}{15/\sqrt{3}} = 16$$

As before:

$$(V_{an})_{LV\ side} = 7.967\ kV$$

$$(V_{an})_{HV\ side} = 127.5\ kV$$

However, V_{an} is no longer in phase on both sides. This is a consequence of the transformer model, and not due to the scaling procedure. Whether this is important depends on the details of the analysis.

Per-Unit Scaling Extended to a General Three-Phase System

The ideas presented are extended to a three-phase system using the following procedure.

1. Select a three-phase apparent power base ($S_{3ph\ base}$), which is typically 1, 10, 100, or 1000 MVA. This base is valid at every bus in the system.
2. Select a line voltage base ($V_{L\ base}$), user defined, but usually the nominal rms line-to-line voltage at a user-defined bus (call this the "reference bus").
3. Compute

$$S_{base} = (S_{3ph\ base})/3 \qquad \text{(Valid at every bus)} \qquad (8.23)$$

4. At the reference bus:

$$V_{base} = V_{L\ base}/\sqrt{3} \qquad (8.24)$$

$$I_{base} = S_{base}/V_{base} \qquad (8.25)$$

$$Z_{base} = V_{base}/I_{base} = V_{base}^2/S_{base} \qquad (8.26)$$

5. To determine the bases at the remaining busses in the system, start at the reference bus, which we will call the "from" bus, and execute the following procedure:

   Trace a path to the next nearest bus, called the "to" bus. You reach the "to" bus by either passing over (1) a line, or (2) a transformer.

   (1) **The "line" case:** $V_{L\ base}$ is the same at the "to" bus as it was at the "from" bus. Use Eqs. (8.2), (8.3), and (8.4) to compute the "to" bus bases.

   (2) **The "transformer" case:** Apply $V_{L\ base}$ at the "from" bus, and treat the transformer as ideal. Calculate the line voltage that appears at the "to" bus. This is now the new $V_{L\ base}$ at the "to" bus. Use Eqs. (8.2), (8.3), and (8.4) to compute the "to" bus bases.

Rename the bus at which you are located, the "from" bus. Repeat the above procedure until you have processed every bus in the system.

6. We now have a set of bases for every bus in the system, which are to be used for every element terminated at that corresponding bus. Values are scaled according to:

per-unit value = actual value/base value

where actual value = the actual complex value of S, V, Z, or I, in SI units (VA, V, Ω, A); base value = the (user-defined) base value (real) of S, V, Z, or I, in SI units (VA, V, Ω, A); per-unit value = the per-unit complex value of S, V, Z, or I, in per-unit (dimensionless).

Finally, the reader is advised that there are many scaling systems used in engineering analysis, and, in fact, several variations of per-unit scaling have been used in electric power engineering applications. There is no standard system to which everyone conforms in every detail. The key to successfully using any scaling procedure is to understand how all base values are selected at every location within the power system. If one receives data in per-unit, one must be in a position to convert all quantities to SI units. If this cannot be done, the analyst must return to the data source for clarification on what base values were used.

8.2 Symmetrical Components for Power System Analysis

Tim A. Haskew

Modern power systems are three-phase systems that can be balanced or unbalanced and will have mutual coupling between the phases. In many instances, the analysis of these systems is performed using what is known as "per-phase analysis." In this chapter, we will introduce a more generally applicable approach to system analysis know as "symmetrical components." The concept of symmetrical components was first proposed for power system analysis by C.L. Fortescue in a classic paper devoted to consideration of the general N-phase case (1918). Since that time, various similar modal transformations (Brogan, 1974) have been applied to a variety of power type problems including rotating machinery (Krause, 1986; Kundur, 1994).

The case for per-phase analysis can be made by considering the simple three-phase system illustrated in Fig. 8.8. The steady-state circuit response can be obtained by solution of the three loop equations presented in Eq. (8.27a) through (8.27c). By solving these loop equations for the three line currents, Eq. (8.28a) through (8.28a) are obtained. Now, if we assume completely balanced source operation (the impedances are defined to be balanced), then the line currents will also form a balanced three-phase set. Hence, their sum, and the neutral current, will be zero. As a result, the line current solutions are as presented in Eq. (8.29a) through (8.29c).

$$\bar{V}_a - \bar{I}_a(R_S + jX_S) - \bar{I}_a(R_L + jX_L) - \bar{I}_n(R_n + jX_n) = 0 \quad (8.27a)$$

$$\bar{V}_b - \bar{I}_b(R_S + jX_S) - \bar{I}_b(R_L + jX_L) - \bar{I}_n(R_n + jX_n) = 0 \quad (8.27b)$$

$$\bar{V}_c - \bar{I}_c(R_S + jX_S) - \bar{I}_c(R_L + jX_L) - \bar{I}_n(R_n + jX_n) = 0 \quad (8.27c)$$

$$\bar{I}_a = \frac{\bar{V}_a - \bar{I}_n(R_n + jX_n)}{(R_s + R_n) + j(X_s + X_n)} \quad (8.28a)$$

$$\bar{I}_b = \frac{\bar{V}_b - \bar{I}_n(R_n + jX_n)}{(R_s + R_n) + j(X_s + X_n)} \quad (8.28b)$$

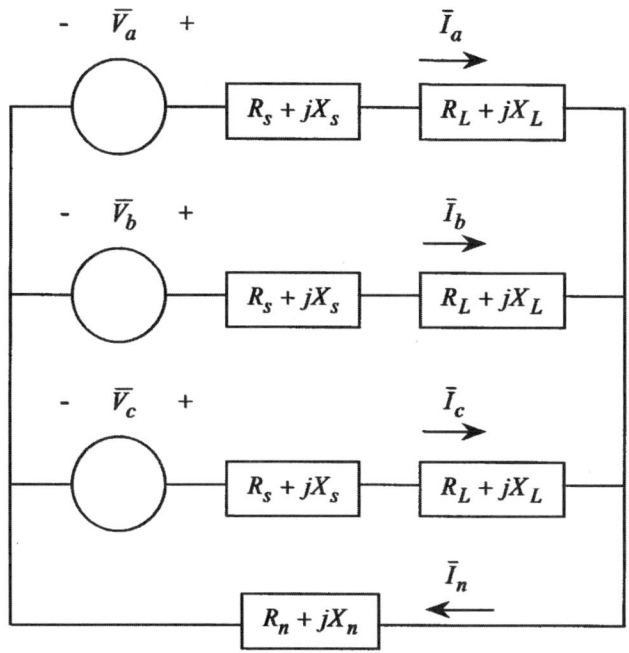

FIGURE 8.8 A simple three-phase system.

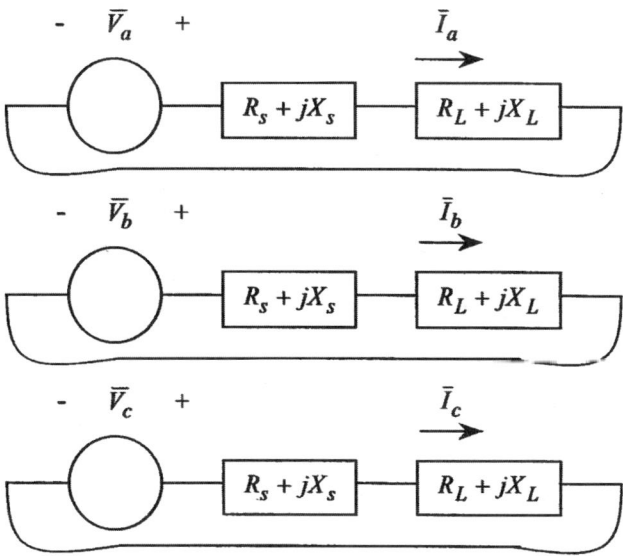

FIGURE 8.9 Decoupled phases of the three-phase system.

$$\bar{I}_c = \frac{\bar{V}_c - \bar{I}_n(R_n + jX_n)}{(R_s + R_n) + j(X_s + X_n)} \quad (8.28c)$$

$$\bar{I}_a = \frac{\bar{V}_a}{(R_s + R_n) + j(X_s + X_n)} \quad (8.29a)$$

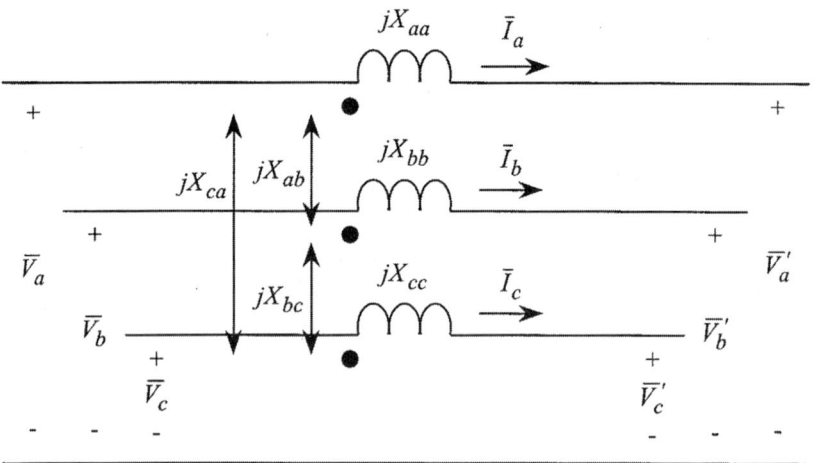

FIGURE 8.10 Mutually coupled series impedances.

$$\bar{I}_b = \frac{\bar{V}_b}{(R_s + R_n) + j(X_s + X_n)} \quad (8.29b)$$

$$\bar{I}_c = \frac{\bar{V}_c}{(R_s + R_n) + j(X_s + X_n)} \quad (8.29c)$$

The circuit synthesis of Eq. (8.29a) through (8.29c) is illustrated in Fig. 8.9. Particular notice should be taken of the fact the response of each phase is independent of the other two phases. Thus, only one phase need be solved, and three-phase symmetry may be applied to determine the solutions for the other phases. This solution technique is the per-phase analysis method.

If one considers the introduction of an unbalanced source or mutual coupling between the phases in Fig. 8.8, then per-phase analysis will not result in three decoupled networks as shown in Fig. 8.9. In fact, in the general sense, no immediate circuit reduction is available without some form of reference frame transformation. The symmetrical component transformation represents such a transformation, which will enable decoupled analysis in the general case and single-phase analysis in the balanced case.

Fundamental Definitions

Voltage and Current Transformation

To develop the symmetrical components, let us first consider an arbitrary (no assumptions on balance) three-phase set of voltages as defined in Eq. (8.30a) through (8.30c). Note that we could just as easily be considering current for the purposes at hand, but voltage was selected arbitrarily. Each voltage is defined by a magnitude and phase angle. Hence, we have six degrees of freedom to fully define this arbitrary voltage set.

$$\bar{V}_a = V_a \angle \theta_a \quad (8.30a)$$

$$\bar{V}_b = V_b \angle \theta_b \quad (8.30b)$$

$$\bar{V}_c = V_c \angle \theta_c \quad (8.30c)$$

We can represent each of the three given voltages as the sum of three components as illustrated in Eq. (8.31a) through (8.31c). For now, we consider these components to be completely arbitrary except for their sum. The 0, 1, and 2 subscripts are used to denote the zero, positive, and negative sequence components of each phase voltage, respectively. Examination of Eq. (8.31a-c) reveals that 6 degrees of freedom exist on the left-hand side of the equations while 18 degrees of freedom exist on the right-hand side. Therefore, for the relationship between the voltages in the abc frame of reference and the voltages in the 012 frame of reference to be unique, we must constrain the right-hand side of Eq. (8.31).

$$\overline{V}_a = \overline{V}_{a_0} + \overline{V}_{a_1} + \overline{V}_{a_2} \tag{8.31a}$$

$$\overline{V}_b = \overline{V}_{b_0} + \overline{V}_{b_1} + \overline{V}_{b_2} \tag{8.31b}$$

$$\overline{V}_c = \overline{V}_{c_0} + \overline{V}_{c_1} + \overline{V}_{c_2} \tag{8.31c}$$

We begin by forcing the a_0, b_0, and c_0 voltages to have equal magnitude and phase. This is defined in Eq. (8.32). The zero sequence components of each phase voltage are all defined by a single magnitude and a single phase angle. Hence, the zero sequence components have been reduced from 6 degrees of freedom to 2.

$$\overline{V}_{a_0} = \overline{V}_{b_0} = \overline{V}_{c_0} \equiv \overline{V}_0 = V_0 \angle \theta_0 \tag{8.32}$$

Second, we force the a_1, b_1, and c_1 voltages to form a balanced three-phase set with positive phase sequence. This is mathematically defined in Eq. (8.33a-c). This action reduces the degrees of freedom provided by the positive sequence components from 6 to 2.

$$\overline{V}_{a_1} = \overline{V}_1 = V_1 \angle \theta_1 \tag{8.33a}$$

$$\overline{V}_{b_1} = V_1 \angle (\theta_1 - 120°) = \overline{V}_1 \bullet 1 \angle{-120°} \tag{8.33b}$$

$$\overline{V}_{c_1} = V_1 \angle (\theta_1 + 120°) = \overline{V}_1 \bullet 1 \angle{+120°} \tag{8.33c}$$

And finally, we force the a_2, b_2, and c_2 voltages to form a balanced three-phase set with negative phase sequence. This is mathematically defined in Eq. (8.34a-c). As in the case of the positive sequence components, the negative sequence components have been reduced from 6 to 2 degrees of freedom.

$$\overline{V}_{a_2} = \overline{V}_2 = V_2 \angle \theta_2 \tag{8.34a}$$

$$\overline{V}_{b_2} = V_2 \angle (\theta_2 + 120°) = \overline{V}_2 \bullet 1 \angle{+120°} \tag{8.34b}$$

$$\overline{V}_{c_2} = V_2 \angle (\theta_2 - 120°) = \overline{V}_2 \bullet 1 \angle{-120°} \tag{8.34c}$$

Now, the right- and left-hand sides of Eq. (8.31a) through (8.31c) each have 6 degrees of freedom. Thus, the relationship between the symmetrical component voltages and the original phase voltages is unique. The final relationship is presented in Eq. (8.35a) through (8.35c). Note that the constant "a" has been defined as indicated in Eq. (8.36).

$$\overline{V}_a = \overline{V}_0 + \overline{V}_1 + \overline{V}_2 \qquad (8.35a)$$

$$\overline{V}_b = \overline{V}_0 + \overline{a}^2 \overline{V}_1 + \overline{a}\,\overline{V}_2 \qquad (8.35b)$$

$$\overline{V}_c = \overline{V}_0 + \overline{a}\,\overline{V}_1 + \overline{a}^2 \overline{V}_2 \qquad (8.35c)$$

$$\overline{a} = 1\angle 120° \qquad (8.36)$$

Equation (8.35) is more easily written in matrix form, as indicated in Eq. (8.37) in both expanded and compact form. In Eq. (8.37), the [T] matrix is constant, and the inverse exists. Thus, the inverse transformation can be defined as indicated in Eq. (8.38). The over tilde (~) indicates a vector of complex numbers.

$$\begin{bmatrix} \overline{V}_a \\ \overline{V}_b \\ \overline{V}_c \end{bmatrix} = \begin{bmatrix} 1 & 1 & 1 \\ 1 & \overline{a}^2 & \overline{a} \\ 1 & \overline{a} & \overline{a}^2 \end{bmatrix} \begin{bmatrix} \overline{V}_0 \\ \overline{V}_1 \\ \overline{V}_2 \end{bmatrix} \qquad (8.37)$$

$$\tilde{V}_{abc} = [\overline{T}] \tilde{V}_{012}$$

$$\begin{bmatrix} \overline{V}_0 \\ \overline{V}_1 \\ \overline{V}_2 \end{bmatrix} = \frac{1}{3} \begin{bmatrix} 1 & 1 & 1 \\ 1 & \overline{a} & \overline{a}^2 \\ 1 & \overline{a}^2 & \overline{a} \end{bmatrix} \begin{bmatrix} \overline{V}_a \\ \overline{V}_b \\ \overline{V}_c \end{bmatrix} \qquad (8.38)$$

$$\tilde{V}_{012} = [\overline{T}]^{-1} \tilde{V}_{abc}$$

Equations (8.39) and (8.40) define an identical transformation and inverse transformation for current.

$$\begin{bmatrix} \overline{I}_a \\ \overline{I}_b \\ \overline{I}_c \end{bmatrix} = \begin{bmatrix} 1 & 1 & 1 \\ 1 & \overline{a}^2 & \overline{a} \\ 1 & \overline{a} & \overline{a}^2 \end{bmatrix} \begin{bmatrix} \overline{I}_0 \\ \overline{I}_1 \\ \overline{I}_2 \end{bmatrix} \qquad (8.39)$$

$$\tilde{I}_{abc} = [\overline{T}] \tilde{I}_{012}$$

$$\begin{bmatrix} \overline{I}_0 \\ \overline{I}_1 \\ \overline{I}_2 \end{bmatrix} = \frac{1}{3} \begin{bmatrix} 1 & 1 & 1 \\ 1 & \overline{a} & \overline{a}^2 \\ 1 & \overline{a}^2 & \overline{a} \end{bmatrix} \begin{bmatrix} \overline{I}_a \\ \overline{I}_b \\ \overline{I}_c \end{bmatrix} \qquad (8.40)$$

$$\tilde{I}_{012} = [\overline{T}]^{-1} \tilde{I}_{abc}$$

Impedance Transformation

In order to assess the impact of the symmetrical component transformation on systems impedances, we turn to Fig. 8.10. Note that the balanced case has been assumed. Kirchhoff's Voltage Law for the circuit dictates equations Eq. (8.41a-c), which are written in matrix form in Eq. (8.42) and even more simply in Eq. (8.43).

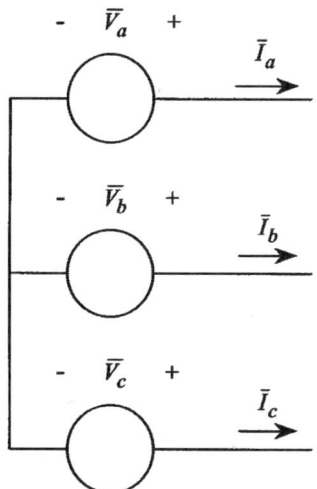

FIGURE 8.11 Three-phase wye-connected source.

$$\bar{V}_a - \bar{V}'_a = jX_{aa}\bar{I}_a + jX_{ab}\bar{I}_b + jX_{ca}\bar{I}_c \tag{8.41a}$$

$$\bar{V}_b - \bar{V}'_b = jX_{ab}\bar{I}_a + jX_{bb}\bar{I}_b + jX_{bc}\bar{I}_c \tag{8.41b}$$

$$\bar{V}_c - \bar{V}'_c = jX_{ca}\bar{I}_a + jX_{bc}\bar{I}_b + jX_{cc}\bar{I}_c \tag{8.41c}$$

$$\begin{bmatrix}\bar{V}_a\\\bar{V}_b\\\bar{V}_c\end{bmatrix} - \begin{bmatrix}\bar{V}'_a\\\bar{V}'_b\\\bar{V}'_c\end{bmatrix} = j\begin{bmatrix}X_{aa} & X_{ab} & X_{ca}\\X_{ab} & X_{bb} & X_{bc}\\X_{ca} & X_{bc} & X_{cc}\end{bmatrix}\begin{bmatrix}\bar{I}_a\\\bar{I}_b\\\bar{I}_c\end{bmatrix} \tag{8.42}$$

$$\tilde{V}_{abc} - \tilde{V}'_{abc} = [\bar{Z}_{abc}]\tilde{I}_{abc} \tag{8.43}$$

Multiplying both sides of Eq. (8.43) by $[\bar{T}]^{-1}$ yields Eq. (8.44). Then, substituting Eq. (8.38) and (8.39) into the result leads to the sequence equation presented in Eq. (8.45). The equation is written strictly in the 012 frame reference in Eq. (8.46) where the sequence impedance matrix is defined in Eq. (8.47).

$$[\bar{T}]^{-1}\tilde{V}_{abc} - [\bar{T}]^{-1}\tilde{V}'_{abc} = [\bar{T}]^{-1}[\bar{Z}_{abc}]\tilde{I}_{abc} \tag{8.44}$$

$$\tilde{V}_{012} - \tilde{V}'_{012} = [\bar{T}]^{-1}[\bar{Z}_{abc}][\bar{T}]\tilde{I}_{012} \tag{8.45}$$

$$\tilde{V}_{012} - \tilde{V}'_{012} = [\bar{Z}_{012}]\tilde{I}_{012} \tag{8.46}$$

$$[\bar{Z}_{012}] = [\bar{T}]^{-1}[\bar{Z}_{abc}][\bar{T}] = \begin{bmatrix}\bar{Z}_{00} & \bar{Z}_{01} & \bar{Z}_{02}\\\bar{Z}_{10} & \bar{Z}_{11} & \bar{Z}_{12}\\\bar{Z}_{20} & \bar{Z}_{21} & \bar{Z}_{22}\end{bmatrix} \tag{8.47}$$

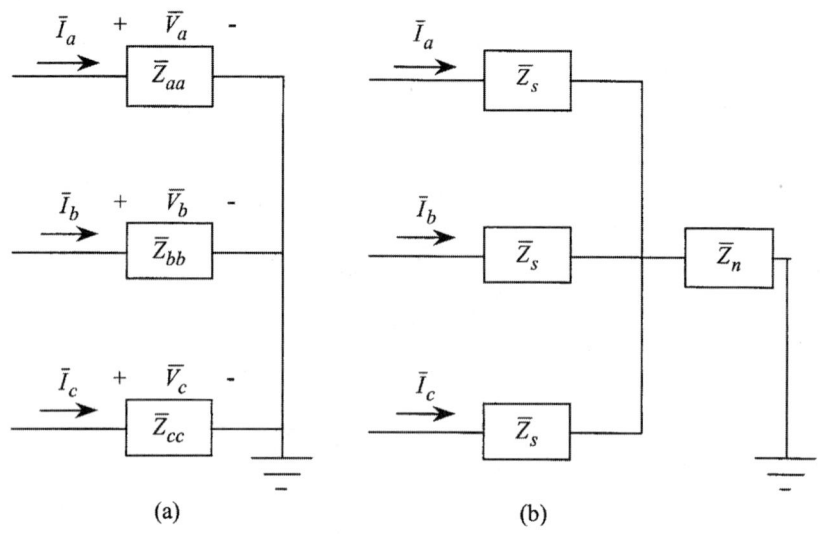

FIGURE 8.12 Three-phase impedance load model.

Power Calculations

The impact of the symmetrical components on the computation of complex power can be easily derived from the basic definition. Consider the source illustrated in Fig. 8.11. The three-phase complex power supplied by the source is defined in Eq. (8.48). The algebraic manipulation to Eq. (8.48) is presented, and the result in the sequence domain is presented in Eq. (8.49) in matrix form and in Eq. (8.50) in scalar form.

$$\bar{S}_{3\phi} = \bar{V}_a \bar{I}_a^* + \bar{V}_b \bar{I}_b^* + \bar{V}_c \bar{I}_c^* = \tilde{V}_{abc}^T \tilde{I}_{abc}^* \tag{8.48}$$

$$\bar{S}_{3\phi} = \tilde{V}_{abc}^T \tilde{I}_{abc}^* = \left\{[\bar{T}]\tilde{V}_{012}\right\}^T \left\{[\bar{T}]\tilde{I}_{012}\right\}^*$$

$$= \tilde{V}_{012}^T [\bar{T}]^T [\bar{T}]^* \tilde{I}_{012}^*$$

$$[\bar{T}]^T [\bar{T}]^* = \begin{bmatrix} 1 & 1 & 1 \\ 1 & \bar{a}^2 & \bar{a} \\ 1 & \bar{a} & \bar{a}^2 \end{bmatrix} \begin{bmatrix} 1 & 1 & 1 \\ 1 & \bar{a} & \bar{a}^2 \\ 1 & \bar{a}^2 & \bar{a} \end{bmatrix}$$

$$= \begin{bmatrix} 3 & 0 & 0 \\ 0 & 3 & 0 \\ 0 & 0 & 3 \end{bmatrix} = 3 \begin{bmatrix} 1 & 0 & 0 \\ 0 & 1 & 0 \\ 0 & 0 & 1 \end{bmatrix}$$

$$\bar{S}_{3\phi} = 3\tilde{V}_{012}^T \tilde{I}_{012}^* \tag{8.49}$$

$$\bar{S}_{3\phi} = 3\left\{\bar{V}_0 \bar{I}_0^* + \bar{V}_1 \bar{I}_1^* + \bar{V}_2 \bar{I}_2^*\right\} \tag{8.50}$$

Note that the nature of the symmetrical component transformation is not one of power invariance, as indicated by the multiplicative factor of 3 in Eq. (8.50). However, this will prove useful in the analysis of

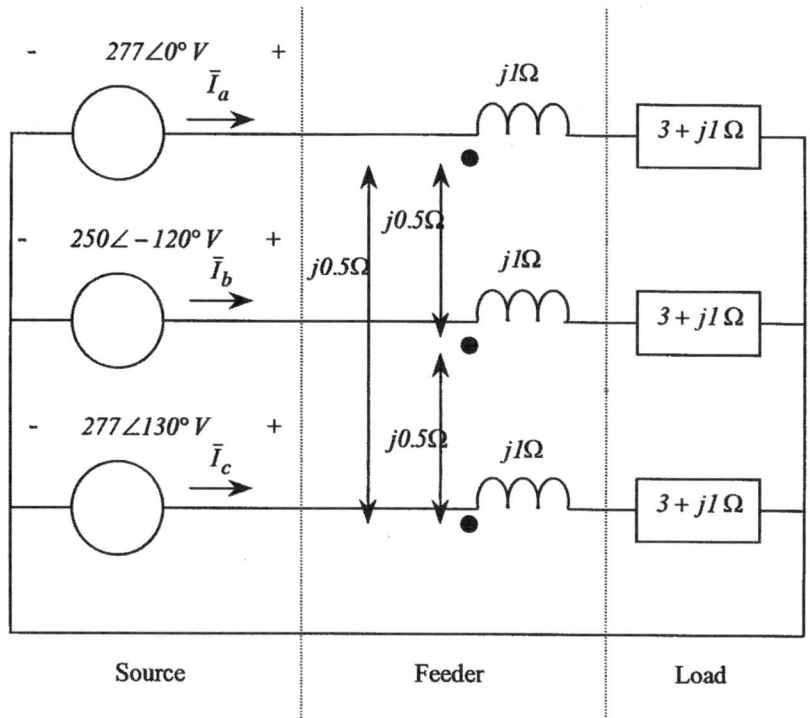

FIGURE 8.13 Power system for Example 1.

balanced systems, which will be seen later. Power invariant transformations do exist as minor variations of the one defined herein. However, they are not typically employed, although the results are just as mathematically sound.

System Load Representation

System loads may be represented in the symmetrical components in a variety of ways, depending on the type of load model that is preferred. Consider first a general impedance type load. Such a load is illustrated in Fig. 8.12a. In this case, Eq. (8.43) applies with $\bar{V}'_{abc} = 0$ due to the solidly grounded Y connection. Therefore, the sequence impedances are still correctly defined by Eq. (8.47). As illustrated in Fig. 8.12a, the load has zero mutual coupling. Hence, the off-diagonal terms will be zero. However, mutual terms may be considered, as Eq. (8.47) is general in nature. This method can be applied for any shunt-connected impedances in the system.

If the load is Δ-connected, then it should be converted to an equivalent Y-connection prior to the transformation (Irwin, 1996; Gross, 1986). In this case, the possibility of unbalanced mutual coupling will be excluded, which is practical in most cases. Then, the off-diagonal terms in Eq. (8.47) will be zero, and the sequence networks for the load will be decoupled. Special care should be taken that the zero sequence impedance will become infinite because the Δ-connection does not allow a path for a neutral current to flow, which is equivalent to not allowing a zero sequence current path as defined by the first row of matrix Eq. (8.40). A similar argument can be made for a Y-connection that is either ungrounded or grounded through an impedance, as indicated in Fig. 8.12b. In this case, the zero sequence impedance will be equal to the sum of the phase impedance and three times the neutral impedance, or, $\bar{Z}_{00} = \bar{Z}_Y + 3\bar{Z}_n$. Notice should be taken that the neutral impedance can vary from zero to infinity.

The representation of complex power load models will be left for the section on the application of balanced circuit reductions to the symmetrical component transformation.

TABLE 8.1 Summary of the Symmetrical Components in the General Case

Quantity	Transformation Equations	
	abc ⇒ 012	012 ⇒ abc
Voltage	$\begin{bmatrix} \bar{V}_0 \\ \bar{V}_1 \\ \bar{V}_2 \end{bmatrix} = \frac{1}{3} \begin{bmatrix} 1 & 1 & 1 \\ 1 & \bar{a} & \bar{a}^2 \\ 1 & \bar{a}^2 & \bar{a} \end{bmatrix} \begin{bmatrix} \bar{V}_a \\ \bar{V}_b \\ \bar{V}_c \end{bmatrix}$ $\tilde{V}_{012} = [\bar{T}]^{-1} \tilde{V}_{abc}$	$\begin{bmatrix} \bar{V}_a \\ \bar{V}_b \\ \bar{V}_c \end{bmatrix} = \begin{bmatrix} 1 & 1 & 1 \\ 1 & \bar{a}^2 & \bar{a} \\ 1 & \bar{a} & \bar{a}^2 \end{bmatrix} \begin{bmatrix} \bar{V}_0 \\ \bar{V}_1 \\ \bar{V}_2 \end{bmatrix}$ $\tilde{V}_{abc} = [\bar{T}] \tilde{V}_{012}$
Current	$\begin{bmatrix} \bar{I}_0 \\ \bar{I}_1 \\ \bar{I}_2 \end{bmatrix} = \frac{1}{3} \begin{bmatrix} 1 & 1 & 1 \\ 1 & \bar{a} & \bar{a}^2 \\ 1 & \bar{a}^2 & \bar{a} \end{bmatrix} \begin{bmatrix} \bar{I}_a \\ \bar{I}_b \\ \bar{I}_c \end{bmatrix}$ $\bar{I}_{012} = [\bar{T}]^{-1} \bar{I}_{abc}$	$\begin{bmatrix} \bar{I}_a \\ \bar{I}_b \\ \bar{I}_c \end{bmatrix} = \begin{bmatrix} 1 & 1 & 1 \\ 1 & \bar{a}^2 & \bar{a} \\ 1 & \bar{a} & \bar{a}^2 \end{bmatrix} \begin{bmatrix} \bar{I}_0 \\ \bar{I}_1 \\ \bar{I}_2 \end{bmatrix}$ $\bar{I}_{abc} = [\bar{T}] \bar{I}_{012}$
Impedance	$[\bar{Z}_{012}] = [\bar{T}]^{-1} [\bar{Z}_{abc}][\bar{T}]$	
Power	$\bar{S}_{3\phi} = \bar{V}_a \bar{I}_a^* + \bar{V}_b \bar{I}_b^* + \bar{V}_c \bar{I}_c^* = \tilde{V}_{abc}^T \tilde{I}_{abc}^*$ $\bar{S}_{3\phi} = 3\{\bar{V}_0 \bar{I}_0^* + \bar{V}_2 \bar{I}_2^* + \bar{V}_3 \bar{I}_3^*\} = 3\tilde{V}_{012}^T \tilde{I}_{012}^*$	

Summary of the Symmetrical Components in the General Three-Phase Case

The general symmetrical component transformation process has been defined in this section. Table 8.1 is a short form reference for the utilization of these procedures in the general case (i.e., no assumption of balanced conditions). Application of these relationships defined in Table 8.1 will enable the power system analyst to draw the zero, positive, and negative sequence networks for the system under study. These networks can then be analyzed in the 012 reference frame, and the results can be easily transformed back into the abc reference frame.

Example 1:

The power system illustrated in Fig. 8.13 is to be analyzed using the sequence networks. Find the following:

(a) three line currents
(b) line-to-neutral voltages at the load
(c) three-phase complex power output of the source

Solution:
The sequence voltages are computed in Eq. (8.51). The sequence impedances for the feeder and the load are computed in Eqs. (8.52) and (8.53), respectively. The sequence networks are drawn in Fig. 8.14.

$$\begin{bmatrix} \bar{V}_0 \\ \bar{V}_1 \\ \bar{V}_2 \end{bmatrix} = \frac{1}{3} \begin{bmatrix} 1 & 1 & 1 \\ 1 & \bar{a} & \bar{a}^2 \\ 1 & \bar{a}^2 & \bar{a} \end{bmatrix} \begin{bmatrix} 255\angle 0° \\ 250\angle -120° \\ 277\angle 130° \end{bmatrix} = \begin{bmatrix} 8.8\angle -171° \\ 267.1\angle 3° \\ 24.0\angle -37° \end{bmatrix} V \qquad (8.51)$$

$$[\bar{Z}_{012}] = [\bar{T}]^{-1} \begin{bmatrix} j1 & j0.5 & j0.5 \\ j0.5 & j1 & j0.5 \\ j0.5 & j0.5 & j1 \end{bmatrix} [\bar{T}] = \begin{bmatrix} j2 & 0 & 0 \\ 0 & j0.5 & 0 \\ 0 & 0 & j0.5 \end{bmatrix} \Omega \qquad (8.52)$$

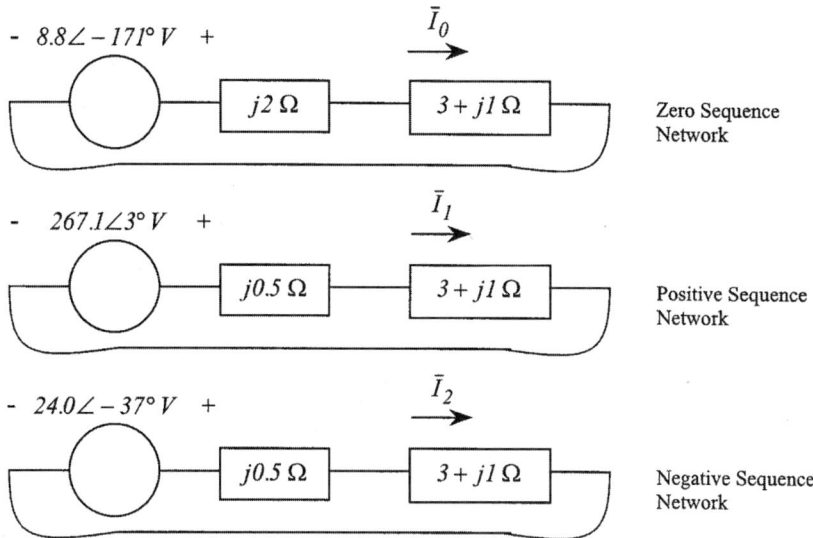

FIGURE 8.14 Sequence networks for Example 1.

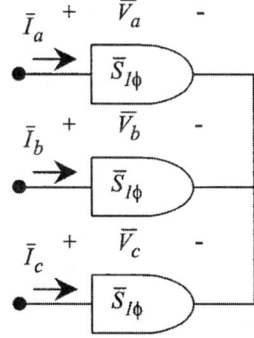

FIGURE 8.15 Balanced complex power load model.

$$[\bar{Z}_{012}] = [T]^{-1} \begin{bmatrix} 3+j1 & 0 & 0 \\ 0 & 3+j1 & 0 \\ 0 & 0 & 3+j1 \end{bmatrix} [T] = \begin{bmatrix} 3+j1 & 0 & 0 \\ 0 & 3+j1 & 0 \\ 0 & 0 & 3+j1 \end{bmatrix} \Omega \quad (8.53)$$

The sequence currents are computed in Eq. (8.54a-c). In Eq. (8.55), the sequence currents and sequence load impedances are used to compute the zero, positive, and negative sequence load voltages.

$$\bar{I}_0 = \frac{8.8\angle -171°}{3+j(1+2)} = 2.1\angle 144° \text{ A} \quad (8.54a)$$

$$\bar{I}_1 = \frac{267.1\angle 3°}{3+j(1+0.5)} = 79.6\angle -24° \text{ A} \quad (8.54b)$$

$$\bar{I}_2 = \frac{24.0\angle -37°}{3+j(1+0.5)} = 7.2\angle -64° \text{ A} \quad (8.54c)$$

$$\begin{bmatrix} \overline{V}_0 \\ \overline{V}_1 \\ \overline{V}_2 \end{bmatrix} = [\overline{Z}_{012}]\tilde{I}_{012} = \begin{bmatrix} 3+j1 & 0 & 0 \\ 0 & 3+j1 & 0 \\ 0 & 0 & 3+j1 \end{bmatrix} \begin{bmatrix} 2.1\angle 144° \\ 79.6\angle -24° \\ 7.2\angle -64° \end{bmatrix}$$
$$= \begin{bmatrix} 6.6\angle 162° \\ 251.7\angle -6° \\ 22.8\angle -46° \end{bmatrix} V \quad (8.55)$$

The three line currents can be computed as illustrated in Eq. (8.56), and the line-to-neutral load voltages are computed in Eq. (8.57). The three-phase complex power output of the source is computed in Eq. (8.58).

$$\begin{bmatrix} \overline{I}_a \\ \overline{I}_b \\ \overline{I}_c \end{bmatrix} = \begin{bmatrix} 1 & 1 & 1 \\ 1 & \overline{a}^2 & \overline{a} \\ 1 & \overline{a} & \overline{a}^2 \end{bmatrix} \begin{bmatrix} 2.1\angle 144° \\ 79.6\angle -24° \\ 7.2\angle -64° \end{bmatrix} = \begin{bmatrix} 83.2\angle -27° \\ 73.6\angle -147° \\ 82.7\angle 102° \end{bmatrix} A \quad (8.56)$$

$$\begin{bmatrix} \overline{V}_a \\ \overline{V}_b \\ \overline{V}_c \end{bmatrix} = \begin{bmatrix} 1 & 1 & 1 \\ 1 & \overline{a}^2 & \overline{a} \\ 1 & \overline{a} & \overline{a}^2 \end{bmatrix} \begin{bmatrix} 6.6\angle 162° \\ 251.7\angle -6° \\ 22.8\angle -46° \end{bmatrix} = \begin{bmatrix} 263.0\angle -9° \\ 232.7\angle -129° \\ 261.5\angle 120° \end{bmatrix} V \quad (8.57)$$

$$\overline{S}_{3\phi} = 3\{\overline{V}_0\overline{I}_0^* + \overline{V}_1\overline{I}_1^* + \overline{V}_2\overline{I}_2^*\} = 57.3 + j29.2 \, kVA \quad (8.58)$$

Reduction to the Balanced Case

When the power system under analysis is operating under balanced conditions, the symmetrical components allow one to perform analysis on a single-phase network in a manner similar to per-phase analysis, even when mutual coupling is present. The details of the method are presented in this section.

Balanced Voltages and Currents

Consider a balanced three-phase source operating with positive phase sequence. The voltages are defined below in Eq. (8.59). Upon computation of Eq. (8.38), one discovers that the sequence voltages that result are those shown in Eq. (8.60).

$$\tilde{V}_{abc} = \begin{bmatrix} V_a \angle \theta_a \\ V_a \angle (\theta_a - 120°) \\ V_a \angle (\theta_a + 120°) \end{bmatrix} \quad (8.59)$$

$$\tilde{V}_{012} = \begin{bmatrix} 0 \\ V_a \angle \theta_a \\ 0 \end{bmatrix} \quad (8.60)$$

In Eq. (8.61), a source is defined with negative phase sequence. The sequence voltages for this case are presented in Eq. (8.62).

$$\tilde{V}_{abc} = \begin{bmatrix} V_a \angle \theta_a \\ V_a \angle (\theta_a + 120°) \\ V_a \angle (\theta_a - 120°) \end{bmatrix} \quad (8.61)$$

$$\tilde{V}_{012} = \begin{bmatrix} 0 \\ 0 \\ V_a \angle \theta_a \end{bmatrix} \tag{8.62}$$

These results are particularly interesting. For a balanced source with positive phase sequence, only the positive sequence voltage is non-zero, and its value is the a-phase line-to-neutral voltage. Similarly, for a balanced source with negative phase sequence, the negative sequence voltage is the only non-zero voltage, and it is also equal to the a-phase line-to-neutral voltage. Identical results can be shown for positive and negative phase sequence currents.

Balanced Impedances

In the balanced case, Eq. (8.42) is valid, but Eq. (8.63a-b) apply. Thus, evaluation of Eq. (8.47) results in the closed form expression of Eq. (8.64a). Equation (8.64b) extends the result of Eq. (8.64a) to impedance rather than just reactance.

$$X_{aa} = X_{bb} = X_{cc} \equiv X_s \tag{8.63a}$$

$$X_{ab} = X_{bc} = X_{ca} \equiv X_m \tag{8.63b}$$

$$[\overline{Z}_{012}] = [\overline{T}]^{-1}[\overline{Z}_{abc}][\overline{T}] = \begin{bmatrix} X_s + 2X_m & 0 & 0 \\ 0 & X_s - X_m & 0 \\ 0 & 0 & X_s - X_m \end{bmatrix}$$

$$= \begin{bmatrix} Z_{00} & 0 & 0 \\ 0 & Z_{11} & 0 \\ 0 & 0 & Z_{22} \end{bmatrix} \tag{8.64a}$$

$$[\overline{Z}_{012}] = \begin{bmatrix} \overline{Z}_s + 2\overline{Z}_m & 0 & 0 \\ 0 & \overline{Z}_s - \overline{Z}_m & 0 \\ 0 & 0 & \overline{Z}_s - \overline{Z}_m \end{bmatrix} = \begin{bmatrix} Z_{00} & 0 & 0 \\ 0 & Z_{11} & 0 \\ 0 & 0 & Z_{22} \end{bmatrix} \tag{8.64b}$$

Balanced Power Calculations

In the balanced case, Eq. (8.58) is still valid. However, in the case of positive phase sequence operation, the zero and negative sequence voltages and currents are zero. Hence, Eq. (8.65) results. In the case of negative phase sequence operation, the zero and positive sequence voltages and currents are zero. This results in Eq. (8.66).

$$\overline{S}_{3\phi} = 3\{\overline{V}_0 \overline{I}_0^* + \overline{V}_1 \overline{I}_1^* + \overline{V}_2 \overline{I}_2^*\}$$
$$= 3\overline{V}_1 \overline{I}_1^* = 3\overline{V}_a \overline{I}_a^* \tag{8.65}$$

$$\overline{S}_{3\phi} = 3\{\overline{V}_0 \overline{I}_0^* + \overline{V}_1 \overline{I}_1^* + \overline{V}_2 \overline{I}_2^*\}$$
$$= 3\overline{V}_2 \overline{I}_2^* = 3\overline{V}_a \overline{I}_a^* \tag{8.66}$$

Examination of Eqs. (8.65) and (8.66) reveals that the nature of complex power calculations in the sequence networks is identical to that performed using per-phase analysis (i.e., the factor of 3 is present). This feature of the symmetrical component transformation defined herein is the primary reason that power invariance is not desired.

Balanced System Loads

When the system loads are balanced, the sequence network representation is rather straightforward. We shall first consider the impedance load model by referring to Fig. 8.12a, imposing balanced impedances, and allowing for consideration of a neutral impedance, as illustrated in Fig. 8.12b. Balanced conditions are enforced by Eq. (8.67a-b). In this case, the reduction is based on Eq. (8.64). The result is presented in Eq. (8.68). Special notice should be taken that the mutual terms may be zero, as indicated on the figure, but have been included for completeness in the mathematical development.

$$\overline{Z}_{aa} = \overline{Z}_{bb} = \overline{Z}_{cc} \equiv \overline{Z}_s \tag{8.67a}$$

$$\overline{Z}_{ab} = \overline{Z}_{bc} = \overline{Z}_{ca} \equiv \overline{Z}_m \tag{8.67b}$$

$$[\overline{Z}_{012}] = \begin{bmatrix} \overline{Z}_s + 2\overline{Z}_m + 3\overline{Z}_n & 0 & 0 \\ 0 & \overline{Z}_s - \overline{Z}_m & 0 \\ 0 & 0 & \overline{Z}_s - \overline{Z}_m \end{bmatrix} = \begin{bmatrix} Z_{00} & 0 & 0 \\ 0 & Z_{11} & 0 \\ 0 & 0 & Z_{22} \end{bmatrix} \tag{8.68}$$

The balanced complex power load model is illustrated in Fig. 8.15. The transformation into the sequence networks is actually defined by the results presented in Eqs. (8.65) and (8.66). In positive phase sequence systems, the zero and negative sequence load representations absorb zero complex power; in negative phase sequence systems, the zero and positive sequence load representations absorb zero complex power. Hence, the zero complex power sequence loads are represented as short-circuits, thus forcing the sequence voltages to zero. The non-zero sequence complex power load turns out to be equal to the single-phase load complex power. This is defined for positive phase sequence systems in Eq. (8.69) and for negative phase sequence systems in Eq. (8.70).

$$\overline{S}_1 = \overline{S}_{1\phi} \tag{8.69}$$

$$\overline{S}_2 = \overline{S}_{1\phi} \tag{8.70}$$

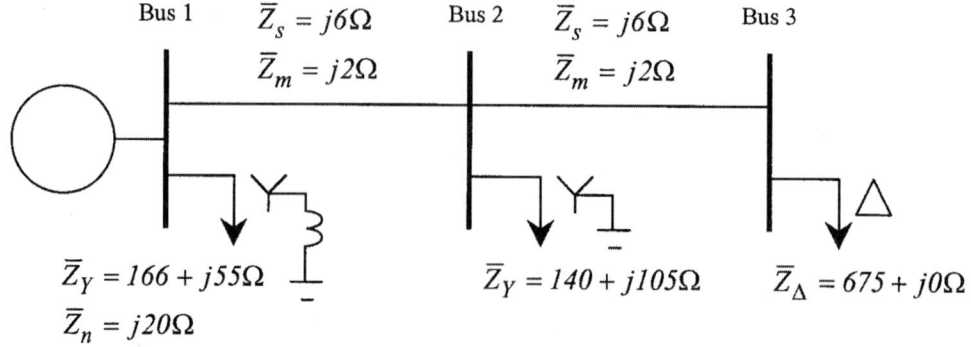

FIGURE 8.16 Balanced power system for Example 2.

Power System Analysis and Simulation

TABLE 8.2 Summary of the Symmetrical Components in the Balanced Case

Quantity	Transformation Equations	
	abc ⇒ 012	012 ⇒ abc
Voltage	Positive Phase Sequence: $\begin{bmatrix}\bar{V}_0\\\bar{V}_1\\\bar{V}_2\end{bmatrix}=[T]^{-1}\begin{bmatrix}\bar{V}_a\\\bar{V}_b\\\bar{V}_c\end{bmatrix}=\begin{bmatrix}0\\\bar{V}_a\\0\end{bmatrix}$	Positive Phase Sequence: $\begin{bmatrix}\bar{V}_a\\\bar{V}_b\\\bar{V}_c\end{bmatrix}=[T]\begin{bmatrix}\bar{V}_0\\\bar{V}_1\\\bar{V}_2\end{bmatrix}=\begin{bmatrix}\bar{V}_1\\a^2\bar{V}_1\\a\bar{V}_1\end{bmatrix}$
	Negative Phase Sequence: $\begin{bmatrix}\bar{V}_0\\\bar{V}_1\\\bar{V}_2\end{bmatrix}=[T]^{-1}\begin{bmatrix}\bar{V}_a\\\bar{V}_b\\\bar{V}_c\end{bmatrix}=\begin{bmatrix}0\\0\\\bar{V}_a\end{bmatrix}$	Negative Phase Sequence: $\begin{bmatrix}\bar{V}_a\\\bar{V}_b\\\bar{V}_c\end{bmatrix}=[T]\begin{bmatrix}\bar{V}_0\\\bar{V}_1\\\bar{V}_2\end{bmatrix}=\begin{bmatrix}\bar{V}_2\\a\bar{V}_2\\a^2\bar{V}_2\end{bmatrix}$
Current	Positive Phase Sequence: $\begin{bmatrix}\bar{I}_0\\\bar{I}_1\\\bar{I}_2\end{bmatrix}=[T]^{-1}\begin{bmatrix}\bar{I}_a\\\bar{I}_b\\\bar{I}_c\end{bmatrix}=\begin{bmatrix}0\\\bar{I}_a\\0\end{bmatrix}$	Positive Phase Sequence: $\begin{bmatrix}\bar{I}_a\\\bar{I}_b\\\bar{I}_c\end{bmatrix}=[T]\begin{bmatrix}\bar{I}_0\\\bar{I}_1\\\bar{I}_2\end{bmatrix}=\begin{bmatrix}\bar{I}_1\\a^2\bar{I}_1\\a\bar{I}_1\end{bmatrix}$
	Negative Phase Sequence: $\begin{bmatrix}\bar{I}_0\\\bar{I}_1\\\bar{I}_2\end{bmatrix}=[T]^{-1}\begin{bmatrix}\bar{I}_a\\\bar{I}_b\\\bar{I}_c\end{bmatrix}=\begin{bmatrix}0\\0\\\bar{I}_a\end{bmatrix}$	Negative Phase Sequence: $\begin{bmatrix}\bar{I}_a\\\bar{I}_b\\\bar{I}_c\end{bmatrix}=[T]\begin{bmatrix}\bar{I}_0\\\bar{I}_1\\\bar{I}_2\end{bmatrix}=\begin{bmatrix}\bar{I}_1\\a\bar{I}_1\\a^2\bar{I}_1\end{bmatrix}$
Impedance	$[\bar{Z}_{012}]=[T]^{-1}[\bar{Z}_{abc}][T]=\begin{bmatrix}\bar{Z}_s+2\bar{Z}_m+3\bar{Z}_n & 0 & 0\\ 0 & \bar{Z}_s-\bar{Z}_m & 0\\ 0 & 0 & \bar{Z}_s-\bar{Z}_m\end{bmatrix}$	
Power	$\bar{S}_{3\phi}=\bar{V}_a\bar{I}_a^*+\bar{V}_b\bar{I}_b^*+\bar{V}_c\bar{I}_c^*=3\bar{V}_a\bar{I}_a^*$ $\bar{S}_{3\phi}=\{\bar{V}_0\bar{I}_0^*+\bar{V}_1\bar{I}_1^*+\bar{V}_2\bar{I}_2^*\}=\begin{cases}3\bar{V}_1\bar{I}_1^* \text{ positive ph. seq.}\\ 3\bar{V}_2\bar{I}_2^* \text{ negative ph. seq.}\end{cases}$	

Summary of Symmetrical Components in the Balanced Case

The general application of symmetrical components to balanced three-phase power systems has been presented in this section. The results are summarized in a quick reference form in Table 8.2. At this point, however, power transformers have been omitted from consideration. This will be rectified in the next few sections.

Example 2:

Consider the balanced system illustrated by the one-line diagram in Fig. 8.16. Determine the line voltage magnitudes at buses 2 and 3 if the line voltage magnitude at bus 1 is 12.47 kV. We will assume positive phase sequence operation of the source. Also, draw the zero sequence network.

Solution:

The two feeders are identical, and the zero and positive sequence impedances are computed in Eqs. (8.71a) and (8.71b), respectively. The zero and positive sequence impedances for the loads at buses 1 and 2 are computed in Eq. (8.72a-b) through (8.73a-b), respectively. The Δ-connected load at bus 3 is converted to an equivalent Y-connection in Eq. (8.74a), and the zero and positive sequence impedances for the load are computed in Eq. (8.74b) and (8.74c), respectively.

$$\bar{Z}_{00_{feeder}} = \bar{Z}_s + 2\bar{Z}_m = j6 + 2(j2) = j10\,\Omega \tag{8.71a}$$

$$\overline{Z}_{11_{feeder}} = \overline{Z}_s - \overline{Z}_m = j6 - j2 = j4\,\Omega \tag{8.71b}$$

$$\begin{aligned}\overline{Z}_{00_{bus1}} &= \overline{Z}_s + 2\overline{Z}_m + 3\overline{Z}_n \\ &= (166 + j55) + 2(0) + 3(j20) = 166 + j115\,\Omega\end{aligned} \tag{8.72a}$$

$$\overline{Z}_{11_{bus1}} = \overline{Z}_s - \overline{Z}_m = (166 + j55) - 0 = 166 + j55\,\Omega \tag{8.72b}$$

$$\begin{aligned}\overline{Z}_{00_{bus2}} &= \overline{Z}_s + 2\overline{Z}_m + 3\overline{Z}_n \\ &= (140 + j105) + 2(0) + 3(0) = 140 + j105\,\Omega\end{aligned} \tag{8.73a}$$

$$\overline{Z}_{11_{bus2}} = \overline{Z}_s - \overline{Z}_m = (140 + j105) - 0 = 140 + j105\,\Omega \tag{8.73b}$$

$$\overline{Z}_{Y_{bus3}} = \frac{\overline{Z}_\Delta}{3} = \frac{675 + j0}{3} = 225 + j0\,\Omega \tag{8.74a}$$

$$\overline{Z}_{00_{bus3}} = \overline{Z}_s + 2\overline{Z}_m + 3\overline{Z}_n = (225 + j0) + 2(0) + 3(\infty) \rightarrow \infty \tag{8.74b}$$

$$\overline{Z}_{11_{bus3}} = \overline{Z}_s - \overline{Z}_m = (225 + j0) - 0 = 225 + j0\,\Omega \tag{8.74c}$$

The zero and positive sequence networks for the system are provided in Figs. 8.17a and b. Note in the zero sequence network, that the voltage at bus 1 has been forced to zero by imposing a short-circuit to reference. For analysis, since the system is balanced, we need only concern ourselves with the positive sequence network. The source voltage at bus 1 is assumed to be the reference with a 0° phase angle. Note that the source voltage magnitude is the line-to-neutral voltage magnitude at bus 1. The positive sequence voltage at bus 2 can be found using the voltage divider, as shown in Eq. (8.75). Note here that the subscript numbers on the voltages denote the bus, not the sequence network. We assume that all voltages are in the positive sequence network. Again using the voltage divider, the positive sequence voltage at bus 3 can be found, as shown in Eq. (8.76). The requested line voltage magnitudes at buses 2 and 3 can be computed from the positive sequence voltages as shown in Eq. (8.77a-b).

$$\overline{V}_2 = 7200\angle 0° \frac{\{(140 + j105)//(225 + j4)\}}{j4 + \{(140 + j105)//(225 + j4)\}} = 7095.9\angle -2°\,V \tag{8.75}$$

$$\overline{V}_3 = 7095.9\angle -2° \frac{225}{225 + j4} = 7094.8\angle -3°\,V \tag{8.76}$$

$$V_{L_2} = \sqrt{3}|\overline{V}_2| = \sqrt{3}(7095.9) = 12{,}290.5\,V \tag{8.77a}$$

$$V_{L_3} = \sqrt{3}|\overline{V}_3| = \sqrt{3}(7094.8) = 12{,}288.6\,V \tag{8.77b}$$

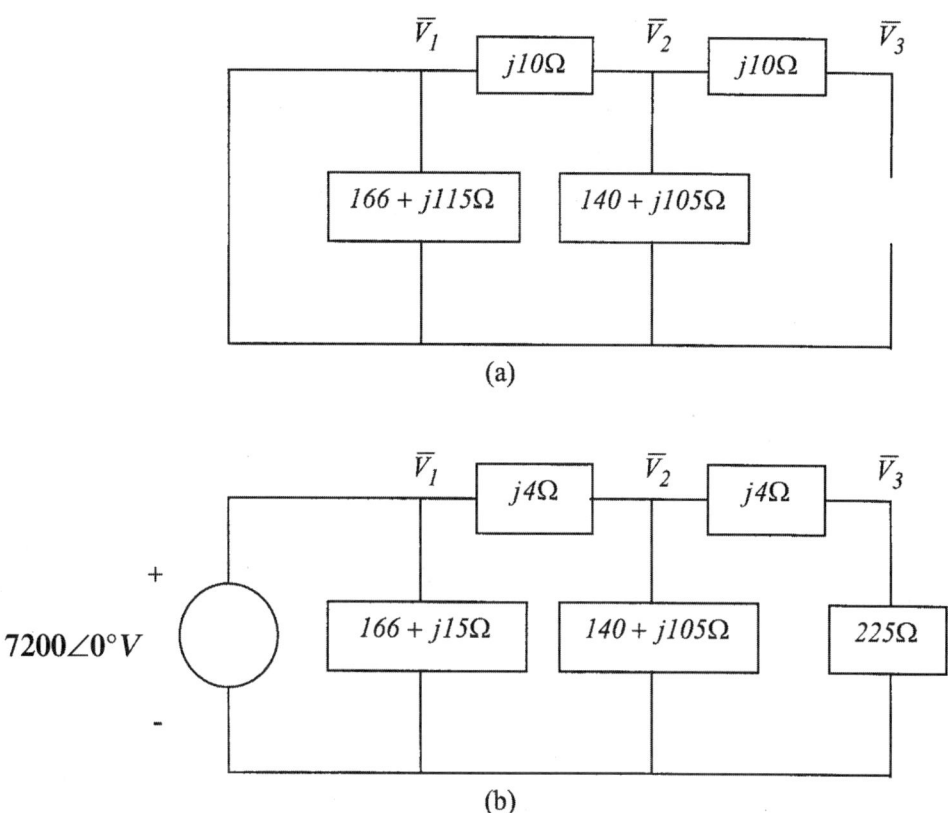

FIGURE 8.17 (a) Zero and (b) positive sequence networks for Example 2.

Sequence Network Representation in Per-Unit

The foregoing development has been based on the inherent assumption that all parameters and variables were expressed in SI units. Quite often, large-scale power system analyses and computations are performed in the per-unit system of measurement (Gross, 1986; Grainger and Stevenson, 1994; Glover and Sarma, 1989). Thus, we must address the impact of per-unit scaling on the sequence networks. Such a conversion is rather straightforward because of the similarity between the positive or negative sequence network and the a-phase network used in per-phase analysis (the reader is cautioned not to confuse the concepts of per-phase analysis and per-unit scaling). The appropriate bases are the same for each sequence network, and they are defined in Table 8.3. Note that the additional subscript "pu" has been added to denote a variable in per-unit; variables in SI units do not carry the additional subscripts.

Power Transformers

For the consideration of transformers and transformer banks, we will limit ourselves to working in the per-unit system. Thus, the ideal transformer in the transformer equivalent circuit can be neglected in the nominal case. The equivalent impedance of a transformer, whether it be single-phase or three-phase, is typically provided on the nameplate in percent, or test data may be available to compute equivalent winding and shunt branch impedances. Developing the sequence networks for these devices is not terribly complicated, but does require attention to detail in the zero sequence case. Of primary importance is the type of connection on each side of the transformer or bank.

The general forms of the per-unit sequence networks for the transformer are shown in Fig. 8.18. Notice should be taken that each transformer winding's impedance and the shunt branch impedance are all

TABLE 8.3 Per-Unit Scaling of Sequence Network Parameters

Quantity	Base Value	Scaling Relationship		
		Zero Sequence	Positive Sequence	Negative Sequence
Voltage	Line-to-Neutral Voltage Base: $V_{LN_{base}} = \dfrac{V_{L_{base}}}{\sqrt{3}}$	$\bar{V}_{0_{pu}} = \dfrac{\bar{V}_0}{V_{LN_{base}}}$	$\bar{V}_{1_{pu}} = \dfrac{\bar{V}_1}{V_{LN_{base}}}$	$\bar{V}_{2_{pu}} = \dfrac{\bar{V}_2}{V_{LN_{base}}}$
Current	Line Current Base: $I_{L_{base}} = \dfrac{S_{3\phi_{base}}}{\sqrt{3}V_{L_{base}}}$	$\bar{I}_{0_{pu}} = \dfrac{\bar{I}_0}{I_{L_{base}}}$	$\bar{I}_{1_{pu}} = \dfrac{\bar{I}_1}{I_{L_{base}}}$	$\bar{I}_{2_{pu}} = \dfrac{\bar{I}_2}{I_{L_{base}}}$
Impedance	Y-Impedance Base: $Z_{Y_{base}} = \dfrac{V_{L_{base}}^2}{S_{3\phi_{base}}}$	$\bar{Z}_{00_{pu}} = \dfrac{\bar{Z}_{00}}{Z_{Y_{base}}}$	$\bar{Z}_{11_{pu}} = \dfrac{\bar{Z}_{11}}{Z_{Y_{base}}}$	$\bar{Z}_{22_{pu}} = \dfrac{\bar{Z}_{22}}{Z_{Y_{base}}}$
Complex Power	Single-Phase Apparent Power Base: $S_{1\phi_{base}} = \dfrac{S_{3\phi_{base}}}{3}$		$\bar{S}_{1\phi_{pu}} = \dfrac{\bar{S}_{1\phi}}{S_{1\phi_{base}}} = \dfrac{\bar{S}_{3\phi}}{S_{3\phi_{base}}}$	

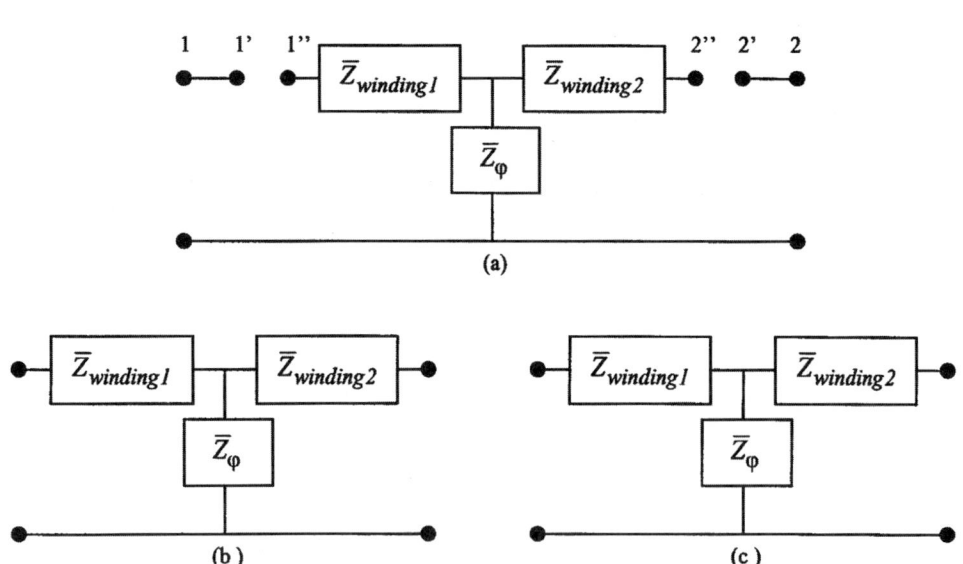

FIGURE 8.18 (a) Zero, (b) positive, and (c) negative sequence transformer networks.

modeled in the circuits. The sequence networks are of the presented form whether a three-phase transformer or a bank of three single-phase transformers is under consideration. Note that the positive and negative sequence networks are identical, and the zero sequence network requires some discussion. The "i^{th} primed" terminals in the zero sequence network are terminated based on the type of connection that is employed for winding i. Details of the termination are presented in Table 8.4.

We must turn our attention to the calculation of the various impedances in the sequence networks as a function of the individual transformer impedances. The zero, positive, and negative sequence impedances are all equal for any transformer winding. Furthermore, the sequence impedances for any transformer winding are equal to the winding impedance expressed in per-unit on the system (not device)

Power System Analysis and Simulation 8-31

TABLE 8.4 Power Transformer Zero Sequence Terminations.

Winding "i" Connection	Connection of Terminals	Schematic Representation
Y	Leave i' and i'' unconnected.	i, i', i'' — $\overline{Z}_e$ — Rest of Network
Y (grounded)	Short i' to i''.	i, i', i'' — $\overline{Z}_e$ — Rest of Network
Y—$\overline{Z}_n$ (grounded)	Connect i' to i'' through $3\overline{Z}_n$.	i, i', $3\overline{Z}_n$, i'' — $\overline{Z}_e$ — Rest of Network
Δ	Short i'' to reference.	i, i', i'' — $\overline{Z}_e$ — Rest of Network

ratings. This is independent of the winding connection (Y or Δ), because of the per-unit scaling. If the sequence networks are to be drawn in SI units, then the sequence impedances for a Δ connection would be 1/3 of the transformer winding impedance. In the case of a three-phase transformer, where the phases may share a common magnetic path, the zero sequence impedance will be different from the positive and negative sequence impedances (Gross, 1986; Blackburn, 1993).

In many cases, a single equivalent impedance is provided on a transformer nameplate. Utilization of this value as a single impedance for the circuit model requires neglecting the shunt branch impedance, which is often justified. If open-circuit test data is not available, or just for the sake of simplicity, the shunt branch of the transformers may be neglected. This leads to the sequence networks illustrated in Fig. 8.19. Here again, care must be taken to place the equivalent transformer impedance in per-unit on the appropriate system bases. Derivation of the equivalent transformer impedance is most appropriately performed in a study focused on power transformers (Gross, 1986; Blackburn, 1993).

Example 3:

Consider the simple power system, operating with positive phase sequence, described by the one-line diagram presented in Fig. 8.20. Compute the line voltage at bus 1, and draw the zero sequence network.

Solution:

We begin by selecting system bases. For simplicity, we choose the system bases to be equal to the transformer ratings. In other words, the system apparent power base is chosen as 750 kVA (three times the single-phase transformer kVA rating), and the line voltage bases at buses 1 and 2 are chosen as 12,470 V (delta side) and 480 V (Y side), respectively. Thus, the transformer impedance provided for the transformer is unaltered when converted to the system bases, as illustrated in Eq. (8.78).

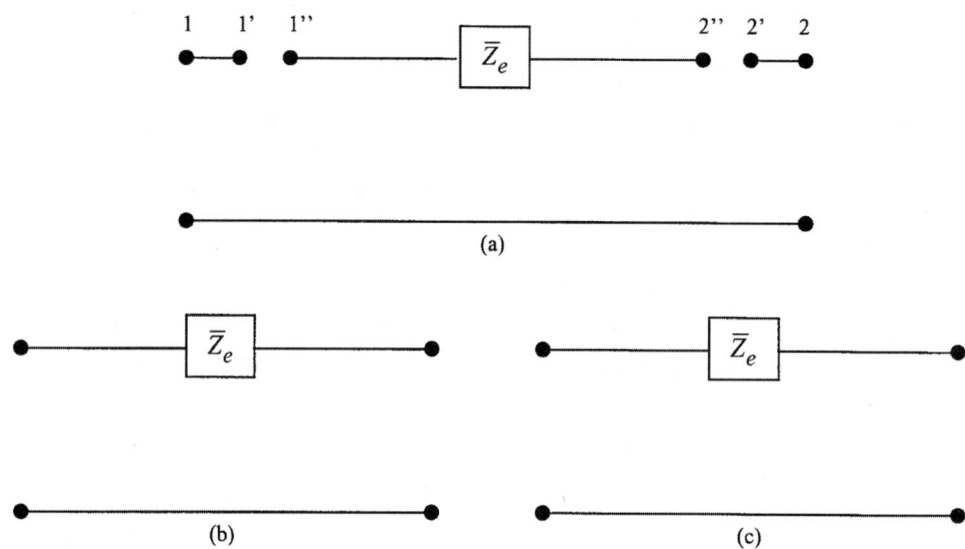

FIGURE 8.19 Reduced (a) zero, (b) positive, and (c) Negative sequence transformer networks.

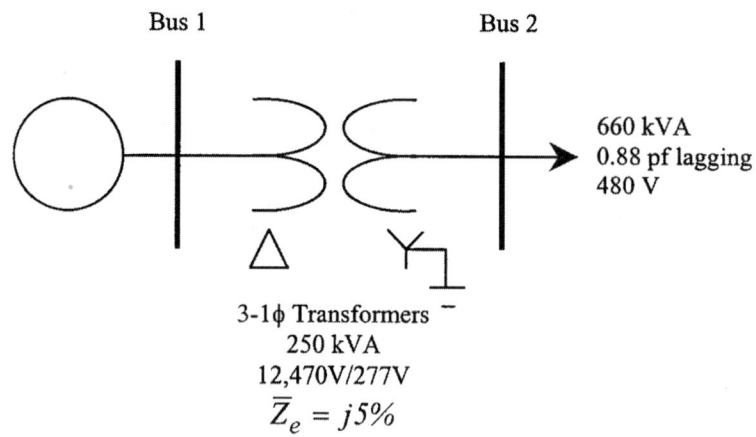

FIGURE 8.20 Power system with a transformer for Example 3.

$$\overline{Z}_e = (j0.05)\frac{Z_{transformer_{base}}}{Z_{Ysystem_{base}}} = (j0.05)\frac{\left(\dfrac{277^2}{250 \times 10^3}\right)}{\left(\dfrac{480^2}{750 \times 10^3}\right)} = j0.05 \quad (8.78)$$

Since balanced conditions are enforced, the load is a non-zero complex power in only the positive sequence network. The positive sequence load value is the single-phase load complex power. In per-unit, the three-phase and single-phase complex powers are equal, as indicated in Eq. (8.79).

$$\overline{S}_{1_{pu}} = \overline{S}_{1\phi_{pu}} = \overline{S}_{3\phi_{pu}} = \frac{\overline{S}_{3\phi}}{S_{3\phi_{base}}} = \frac{666\angle 28.4°}{750} = 0.88\angle 28.4° \quad (8.79)$$

Power System Analysis and Simulation

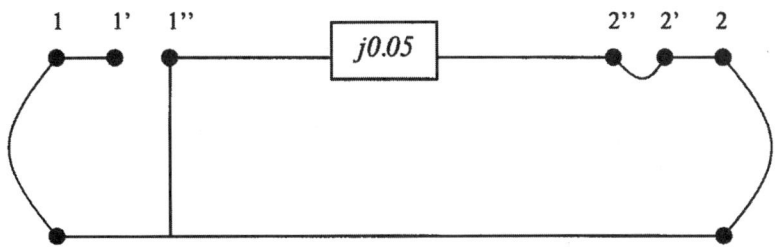

FIGURE 8.21 Zero sequence network for Example 3.

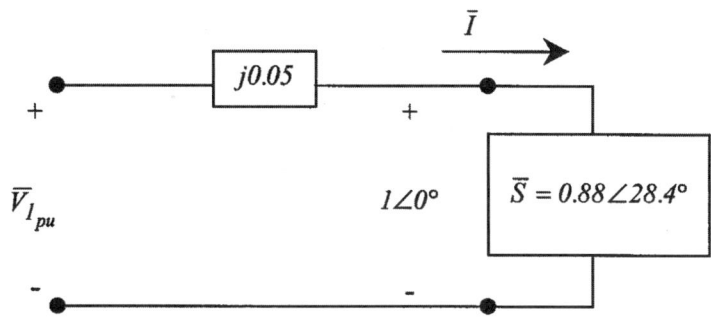

FIGURE 8.22 Positive sequence network for Example 3.

The positive sequence load voltage is the a-phase line-to-neutral voltage at bus 2. If we assume this to be the reference voltage with a zero degree phase angle, then we get 277 ∠ 0°V. In per-unit, this corresponds to unity voltage.

The zero and positive sequence networks are provided in Figs. 8.21 and 8.22, respectively. The line voltage at bus 1 is found by solution of the positive sequence network. The load current is computed from the load voltage and complex power in Eq. (8.80). The positive sequence per-unit voltage at bus 1 is computed in Eq. (8.81). The line voltage at bus 1 is computed from the bus 1 positive sequence voltage in Eq. (8.82). The positive sequence voltage magnitude at bus 1 is the per-unit line-to-neutral voltage magnitude at bus 1. In per-unit, the line and line-to-neutral voltages are equal. Thus, multiplying the per-unit positive sequence voltage magnitude at bus 1 by the line voltage base at bus 1 produces the line voltage at bus 1.

$$\bar{I} = \left(\frac{0.88\angle 28.4°}{1\angle 0°}\right) = 0.88\angle -28.4° \quad (8.80)$$

$$\bar{V}_{1_{pu}} = 1\angle 0° + (0.88\angle -28.4°)(j0.05) = 1.02\angle 2.2° \quad (8.81)$$

$$V_L = \left|\bar{V}_{1_{pu}}\right| V_{L_{base}}(bus1) = 1.02(12,470) = 12,719V \quad (8.82)$$

References

Blackburn, J. L., *Symmetrical Components for Power Systems Engineering*, Marcel Dekker, New York, 1993.
Brogan, W. L., *Modern Control Theory*, Quantum Publishers, Inc., New York, 1974.
Fortescue, C. L., Method of Symmetrical Coordinates Applied to the Solution of Polyphase Networks, *AIEE Transaction*, 37, part 2, 1918.

Glover, J. D. and Sarma, M., *Power System Analysis and Design*, PWS-Kent Publishing Company, Boston, MA, 1989.
Grainger, J. J. and Stevenson, Jr., W. D., *Power System Analysis*, McGraw-Hill, Inc., New York, 1994.
Gross, C. A., *Power System Analysis*, 2nd ed., New York, John Wiley & Sons, New York, 1986.
Irwin, J. D., *Basic Engineering Circuit Analysis*, 5th ed., Prentice-Hall, New Jersey, 1996.
Krause, P. C., *Analysis of Electric Machinery*, McGraw-Hill, New York, 1986.
Kundur, P., *Power System Stability and Control*, McGraw-Hill, Inc., New York, 1994.

8.3 Power Flow Analysis

L. L. Grigsby and Andrew Hanson

The equivalent circuit parameters of many power system components are described in other sections of this handbook. The interconnection of the different elements allows development of an overall power system model. The system model provides the basis for computational simulation of the system performance under a wide variety of projected operating conditions. Additionally, "post mortem" studies, performed after system disturbances or equipment failures, often provide valuable insight into contributing system conditions. This section discusses one such computational simulation, the power flow analysis.

Power systems typically operate under slowly changing conditions which can be analyzed using steady state analysis. Further, transmission systems operate under balanced or near balanced conditions allowing per-phase analysis to be used with a high degree of confidence in the solution. Power flow analysis computationally models these conditions and provides the starting point for most other analyses. For example, the small signal and transient stability effects of a given disturbance are dramatically affected by the "pre-disturbance" operating conditions of the power system. (A disturbance resulting in instability under heavily loaded system conditions may not have any adverse effects under lightly loaded conditions.) Additionally, fault analysis and transient analysis can also be impacted by the "pre-disturbance" operating point of a power system (although, they are usually affected much less than transient stability and small signal stability analysis).

The Power Flow Problem

Power flow analysis is fundamental to the study of power systems forming the basis for other anlayses. Power flow analyses play a key role in the planning of additions or expansions to transmission and generation facilities as well as establishing the starting point for many other types of power system analyses. In addition, power flow analysis and many of its extensions are an essential ingredient of the studies performed in power system operations. In this latter case, it is at the heart of contingency analysis and the implementation of real-time monitoring systems.

The power flow problem (also known as the load flow problem) can be stated as follows:

> For a given power network, with known complex power loads and some set of specifications or restrictions on power generations and voltages, solve for any unknown bus voltages and unspecified generation and finally for the complex power flow in the network components.

Additionally, the losses in individual components and the total network as a whole are usually calculated. Furthermore, the system is often checked for component overloads and voltages outside allowable tolerances.

This section addresses power flow computations for balanced networks (typically applicable to transmission voltage level systems). Positive sequence network components are used for the problem formulation presented here. In the solution of the power flow problem, the network element values are almost always taken to be in per-unit. Likewise, the calculations within the power flow analysis are typically in

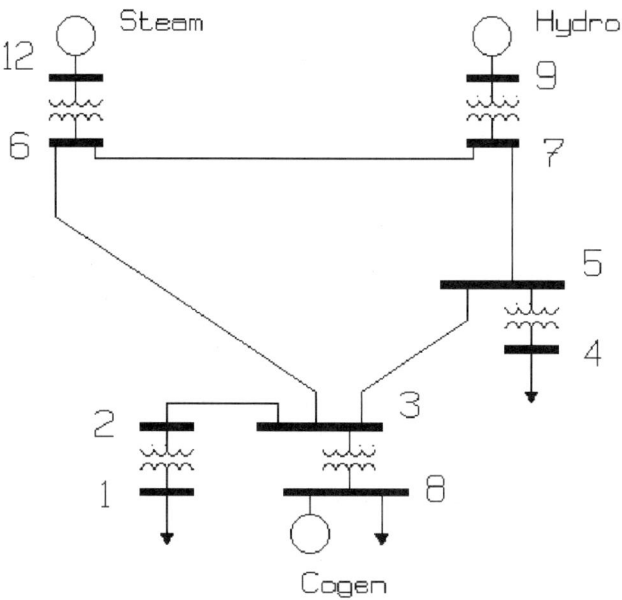

FIGURE 8.23 The one-line diagram of a power system.

per-unit. However, the solution is usually expressed in a mixed format. Solution voltages are usually expressed in per-unit; powers are most often given with kVA or MVA.

The "given network" may be in the form of a system map and accompanying data tables for the network components. More often, however, the network structure is given in the form of a one-line diagram (such as shown in Fig. 8.23).

Regardless of the form of the given network and how the network data is given, the steps to be followed in a power flow study can be summarized as follows:

1. Determine element values for passive network components.
2. Determine locations and values of all complex power loads.
3. Determine generation specifications and constraints.
4. Develop a mathematical model describing power flow in the network.
5. Solve for the voltage profile of the network.
6. Solve for the power flows and losses in the network.
7. Check for constraint violations.

Formulation of the Bus Admittance Matrix

The first step in developing the mathematical model describing the power flow in the network is the formulation of the bus admittance matrix. The bus admittance matrix is an n×n matrix (where n is the number of buses in the system) constructed from the admittances of the equivalent circuit elements of the segments making up the power system. Most system segments are represented by a combination of shunt elements (connected between a bus and the reference node) and series elements (connected between two system buses). Formulation of the bus admittance matrix follows two simple rules:

1. The admittance of elements connected between node k and reference is added to the (k, k) entry of the admittance matrix.
2. The admittance of elements connected between nodes j and k is added to the (j, j) and (k, k) entries of the admittance matrix. The negative of the admittance is added to the (j, k) and (k, j) entries of the admittance matrix.

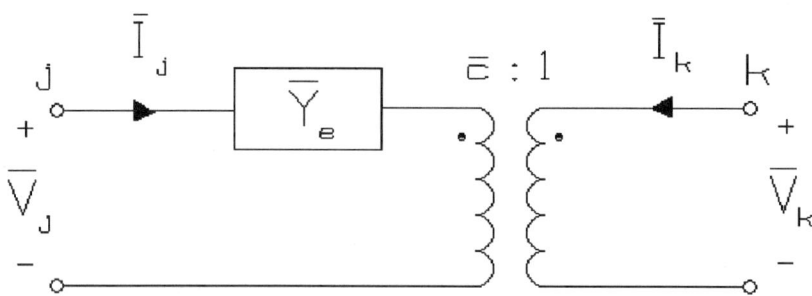

FIGURE 8.24 Off nominal turns ratio transformer.

Off nominal transformers (transformers with transformation ratios different from the system voltage bases at the terminals) present some special difficulties. Figure 8.24 shows a representation of an off nominal turns ratio transformer.

The admittance matrix base mathematical model of an isolated off nominal transformer is:

$$\begin{bmatrix} \bar{I}_j \\ \bar{I}_k \end{bmatrix} = \begin{bmatrix} \bar{Y}_e & -\bar{c}\bar{Y}_e \\ -\bar{c}^*\bar{Y}_e & |\bar{c}|^2\bar{Y}_e \end{bmatrix} \begin{bmatrix} \bar{V}_j \\ \bar{V}_k \end{bmatrix} \qquad (8.83)$$

where $\bar{Y}_e$ is the equivalent series admittance (refered to node j)
$\bar{c}$ is the complex (off nominal) turns ratio
$\bar{I}_j$ is the current injected at node j
$\bar{V}_j$ is the voltage at node j (with respect to reference)

Off nominal transformers are added to the bus admittance matrix by adding the corresponding entry of the isolated off nominal transformer admittance matrix to the system bus admittance matrix.

Formulation of the Power Flow Equations

Considerable insight into the power flow problem and its properties and characteristics can be obtained by consideration of a simple example before proceeding to a general formulation of the problem. This simple case will also serve to establish some notation.

A conceptual representation of a one-line diagram for a four bus power system is shown in Fig. 8.25. For generality, we have shown a generator and a load connected to each bus. The following notation applies:

$$\bar{S}_{G1} = \text{Complex complex power flow into bus 1 from the generator}$$

$$\bar{S}_{D1} = \text{Complex complex power flow into the load from bus 1}$$

Comparable quantities for the complex power generations and loads are obvious for each of the three other buses.

The positive sequence network for the power system represented by the one line diagram of Fig. 8.25 is shown in Fig. 8.26. The boxes symbolize the combination of generation and load. Network texts refer to this network as a five-node network. (The balanced nature of the system allows analysis using only the positive sequence network, reducing each three phase bus to a single node. The reference or ground represents the fifth node.) However, in power systems literature it is usually referred to as a four-bus network or power system.

Power System Analysis and Simulation

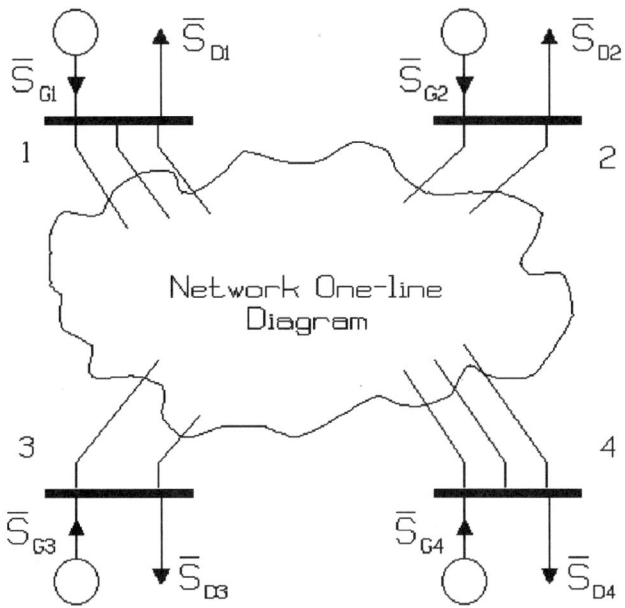

FIGURE 8.25 Conceptual one-line diagram of a four-bus power system.

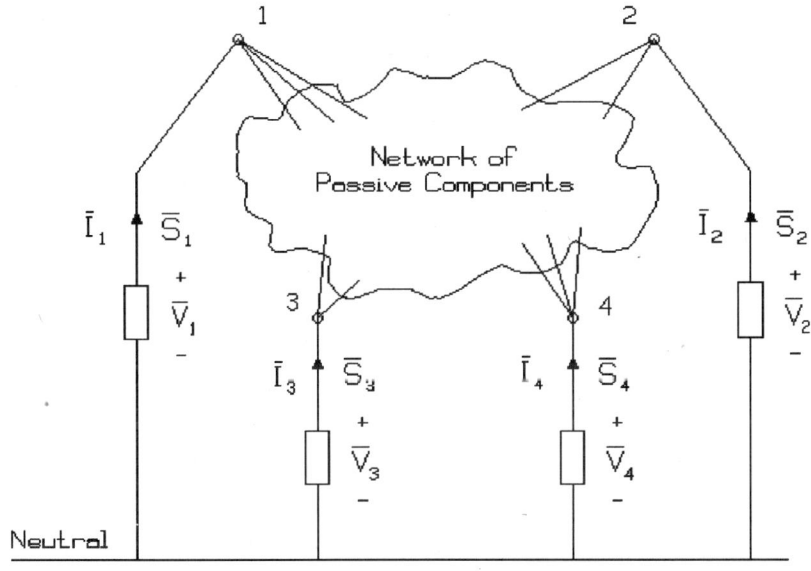

FIGURE 8.26 Positive sequence network for the system of Fig. 8.25.

For the network of Fig. 8.26, we define the following additional notation:

$$\bar{S}_1 = \bar{S}_{G1} - \bar{S}_{D1} = \text{Net complex power injected at bus 1}$$

$$\bar{I}_1 = \text{Net positive sequence phasor current injected at bus 1}$$

$$\bar{V}_1 = \text{Positive sequence phasor voltage at bus 1}$$

The standard node voltage equations for the network can be written in terms of the quantities at bus 1 (defined above) and comparable quantities at the other buses.

$$\bar{I}_1 = \bar{Y}_{11}\bar{V}_1 + \bar{Y}_{12}\bar{V}_2 + \bar{Y}_{13}\bar{V}_3 + \bar{Y}_{14}\bar{V}_4 \qquad (8.84)$$

$$\bar{I}_2 = \bar{Y}_{21}\bar{V}_1 + \bar{Y}_{22}\bar{V}_2 + \bar{Y}_{23}\bar{V}_3 + \bar{Y}_{24}\bar{V}_4 \qquad (8.85)$$

$$\bar{I}_3 = \bar{Y}_{31}\bar{V}_1 + \bar{Y}_{32}\bar{V}_2 + \bar{Y}_{33}\bar{V}_3 + \bar{Y}_{34}\bar{V}_4 \qquad (8.86)$$

$$\bar{I}_4 = \bar{Y}_{41}\bar{V}_1 + \bar{Y}_{42}\bar{V}_2 + \bar{Y}_{43}\bar{V}_3 + \bar{Y}_{44}\bar{V}_4 \qquad (8.87)$$

The admittances in Eqs. (8.84)–(8.87), are the ij^{th} entries of the bus admittance matrix for the power system. The unknown voltages could be found using linear algebra if the four currents $\bar{I}_1 \ldots \bar{I}_4$ were known. However, these currents are not known. Rather, something is known about the complex power and voltage at each bus. The complex power injected into bus k of the power system is defined by the relationship between complex power, voltage, and current given by Eq. (8.88).

$$\bar{S}_k = \bar{V}_k \bar{I}_k^* \qquad (8.88)$$

Therefore,

$$\bar{I}_k = \frac{\bar{S}_k^*}{\bar{V}_k^*} = \frac{\bar{S}_{Gk}^* - \bar{S}_{Dk}^*}{\bar{V}_k^*} \qquad (8.89)$$

By substituting this result into the nodal equations and rearranging, the basic power flow equations for the four-bus system are given as Eqs. (8.90)–(8.93).

$$\bar{S}_{G1}^* - \bar{S}_{D1}^* = \bar{V}_1^*\left[\bar{Y}_{11}\bar{V}_1 + \bar{Y}_{12}\bar{V}_2 + \bar{Y}_{13}\bar{V}_3 + \bar{Y}_{14}\bar{V}_4\right] \qquad (8.90)$$

$$\bar{S}_{G2}^* - \bar{S}_{D2}^* = \bar{V}_2^*\left[\bar{Y}_{21}\bar{V}_1 + \bar{Y}_{22}\bar{V}_2 + \bar{Y}_{23}\bar{V}_3 + \bar{Y}_{24}\bar{V}_4\right] \qquad (8.91)$$

$$\bar{S}_{G3}^* - \bar{S}_{D3}^* = \bar{V}_3^*\left[\bar{Y}_{31}\bar{V}_1 + \bar{Y}_{32}\bar{V}_2 + \bar{Y}_{33}\bar{V}_3 + \bar{Y}_{34}\bar{V}_4\right] \qquad (8.92)$$

$$\bar{S}_{G4}^* - \bar{S}_{D4}^* = \bar{V}_4^*\left[\bar{Y}_{41}\bar{V}_1 + \bar{Y}_{42}\bar{V}_2 + \bar{Y}_{43}\bar{V}_3 + \bar{Y}_{44}\bar{V}_4\right] \qquad (8.93)$$

Examination of Eqs. (8.90)–(8.93) reveals that unless the generation equals the load at every bus, the complex power outputs of the generators cannot be arbitrarily selected. In fact, the complex power output of at least one of the generators must be calculated last since it must take up the unknown "slack" due to the, as yet uncalculated network losses. Further, losses cannot be calculated until the voltages are known. These observations are a result of the principle of conservation of complex power. (i.e., the sum of the injected complex powers at the four system buses is equal to the system complex power losses.)

Further examination of Eqs. (8.90)–(8.93) indicates that it is not possible to solve these equations for the absolute phase angles of the phasor voltages. This simply means that the problem can only be solved to some arbitrary phase angle reference.

In order to alleviate the dilemma outlined above, suppose $\bar{S}_{G4}$ is arbitrarily allowed to float or swing (in order to take up the necessary slack caused by the losses) and that $\bar{S}_{G1}$, $\bar{S}_{G2}$, and $\bar{S}_{G3}$ are specified (other cases will be considered shortly). Now, with the loads known, Eqs. (8.89)–(8.92) are seen as four simultaneous nonlinear equations with complex coefficients in five unknowns $\overline{V}_1$, $\overline{V}_2$, $\overline{V}_3$, $\overline{V}_4$, and $\overline{S}_{G4}$.

The problem of too many unknowns (which would result in an infinite number of solutions) is solved by specifying another variable. Designating bus 4 as the slack bus and specifying the voltage $\overline{V}_4$ reduces the problem to four equations in four unknowns. The slack bus is chosen as the phase reference for all phasor calculations, its magnitude is constrained, and the complex power generation at this bus is free to take up the slack necessary in order to account for the system real and reactive power losses.

The specification of the voltage $\overline{V}_4$, decouples Eq. (8.93) from Eqs. (8.90)–(8.92), allowing calculation of the slack bus complex power after solving the remaining equations. (This property carries over to larger systems with any number of buses.) The example problem is reduced to solving only three equations simultaneously for the unknowns $\overline{V}_1$, $\overline{V}_2$, and $\overline{V}_3$. Similarly, for the case of n buses, it is necessary to solve n – 1 simultaneous, complex coefficient, nonlinear equations.

Systems of nonlinear equations, such as Eqs. (8.90)–(8.92), cannot (except in rare cases) be solved by closed-form techniques. Direct simulation was used extensively for many years; however, essentially all power flow analyses today are performed using iterative techniques on digital computers.

Bus Classifications

There are four quantities of interest associated with each bus:

1. Real Power, P
2. Reactive Power, Q
3. Voltage Magnitude, V
4. Voltage Angle, δ

At every bus of the system, two of these four quantities will be specified and the remaining two will be unknowns. Each of the system buses may be classified in accordance with which of the two quantities are specified. The following classifications are typical:

Slack Bus — The slack bus for the system is a single bus for which the voltage magnitude and angle are specified. The real and reactive power are unknowns. The bus selected as the slack bus must have a source of both real and reactive power, since the injected power at this bus must "swing" to take up the "slack" in the solution. The best choice for the slack bus (since, in most power systems, many buses have real and reactive power sources) requires experience with the particular system under study. The behavior of the solution is often influenced by the bus chosen. (In the earlier discussion, the last bus was selected as the slack bus for convenience.)

Load Bus (P-Q Bus) — A load bus is defined as any bus of the system for which the real and reactive power are specified. Load buses may contain generators with specified real and reactive power outputs; however, it is often convenient to designate any bus with specified injected complex power as a load bus.

Voltage Controlled Bus (P-V Bus) — Any bus for which the voltage magnitude and the injected real power are specified is classified as a voltage controlled (or P-V) bus. The injected reactive power is a variable (with specified upper and lower bounds) in the power flow analysis. (A P-V bus must have a variable source of reactive power such as a generator.)

In all realistic cases, the voltage magnitude is specified at generator buses to take advantage of the generator's reactive power capability. Specifying the voltage magnitude at a generator bus requires a variable specified in the simple analysis discussed earlier to become an unknown (in order to bring the number of unknowns back into correspondence with the number of equations). Normally, the reactive power injected by the generator becomes a variable, leaving the real power and voltage magnitude as the specified quantities at the generator bus.

It was noted earlier that Eq. (8.93) is decoupled, and only Eqs. (8.90)–(8.92) need be solved simultaneously. Although not immediately apparent, specifying the voltage magnitude at a bus and treating the bus reactive power injection as a variable results in retention of, effectively, the same number of complex unknowns. For example, if the voltage magnitude of bus 1 of the earlier four-bus system is specified and the reactive power injection at bus 1 becomes a variable, Eqs. (8.90)–(8.92) again effectively have three complex unknowns. (The phasor voltages $\overline{V}_2$ and $\overline{V}_3$ at buses 2 and 3 are two complex unknowns and the angle δ_1 of the voltage at bus 1 plus the reactive power generation Q_{G1} at bus 1 result in the equivalent of a third complex unknown.)

Bus 1 is called a *voltage controlled bus* since it is apparent that the reactive power generation at bus 1 is being used to control the voltage magnitude. Typically, all generator buses are treated as voltage controlled buses.

Generalized Power Flow Development

The more general (n-bus) case is developed by extending the results of the simple four-bus example. Consider the case of an n-bus system and the corresponding n+1 node positive sequence network. Assume that the buses are numbered such that the slack bus is numbered last. Direct extension of the earlier equations (writing the node voltage equations and making the same substitutions as in the four-bus case) yields the basic power flow equations in the general form.

The Basic Power Flow Equations (PFE)

$$\overline{S}_k^* = P_k - jQ_k = \overline{V}_k^* \sum_{i=1}^{n} \overline{Y}_{ki} \overline{V}_i \tag{8.94}$$

$$\text{for } k = 1, 2, 3, \ldots, n-1$$

and

$$P_n - jQ_n = \overline{V}_n^* \sum_{i=1}^{n} \overline{Y}_{ni} \overline{V}_i \tag{8.95}$$

Equation (8.95) is the equation for the slack bus. Eq. (8.94) represents n-1 simultaneous equations in n-1 complex unknowns if all buses (other than the slack bus) are classified as load buses. Thus, given a set of specified loads, the problem is to solve Eq. (8.94) for the n-1 complex phasor voltages at the remaining buses. Once the bus voltages are known, Eq. (8.95) can be used to calculate the slack bus power.

Bus j is normally treated as a P-V bus if it has a directly connected generator. The unknowns at bus j are then the reactive generation Q_{Gj} and δ_j because the voltage magnitude, V_j, and the real power generation, P_{Gj}, have been specified.

The next step in the analysis is to solve Eq. (8.94) for the bus voltages using some iterative method. Once the bus voltages have been found, the complex power flows and complex power losses in all of the network components are calculated.

Solution Methods

The solution of the simultaneous nonlinear power flow equations requires the use of iterative techniques for even the simplest power systems. Although there are many methods for solving nonlinear equations, only two methods are discussed here.

The Newton-Raphson Method

The Newton-Raphson algorithm has been applied in the solution of nonlinear equations in many fields. The algorithm will be developed using a general set of two equations (for simplicity). The results are easily extended to an arbitrary number of equations.

A set of two nonlinear equations are shown in Eqs. (8.96) and (8.97).

Power System Analysis and Simulation

$$f_1(x_1, x_2) = k_1 \qquad (8.96)$$

$$f_2(x_1, x_2) = k_2 \qquad (8.97)$$

Now, if $x_1^{(0)}$ and $x_2^{(0)}$ are inexact solution estimates and $\Delta x_1^{(0)}$ and $\Delta x_2^{(0)}$ are the corrections to the estimates to achieve an exact solution, Eqs. (8.96) and (8.97) can be rewritten as:

$$f_1\left(x_1^{(0)} + \Delta x_1^{(0)}, x_2^{(0)} + \Delta x_2^{(0)}\right) = k_1 \qquad (8.98)$$

$$f_2\left(x_1^{(0)} + \Delta x_1^{(0)}, x_2^{(0)} + \Delta x_2^{(0)}\right) = k_2 \qquad (8.99)$$

Expanding Eqs. (8.98) and (8.99) in a Taylor series about the estimate yields:

$$f_1\left(x_1^{(0)}, x_2^{(0)}\right) + \left.\frac{\partial f_1}{\partial x_1}\right|^{(0)} \Delta x_1^{(0)} + \left.\frac{\partial f_1}{\partial x_2}\right|^{(0)} \Delta x_2^{(0)} + \text{h.o.t.} = k_1 \qquad (8.100)$$

$$f_2\left(x_1^{(0)}, x_2^{(0)}\right) + \left.\frac{\partial f_2}{\partial x_1}\right|^{(0)} \Delta x_1^{(0)} + \left.\frac{\partial f_2}{\partial x_2}\right|^{(0)} \Delta x_2^{(0)} + \text{h.o.t.} = k_2 \qquad (8.101)$$

where the superscript (0) on the partial derivatives indicates evaluation of the partial derivatives at the initial estimate, and h.o.t. indicates the higher order terms.

Neglecting the higher order terms (an acceptable approximation if $\Delta x_1^{(0)}$ and $\Delta x_2^{(0)}$ are small), Eqs. (8.100) and (8.101) can be rearranged and written in matrix form:

$$\begin{bmatrix} \left.\frac{\partial f_1}{\partial x_1}\right|^{(0)} & \left.\frac{\partial f_1}{\partial x_2}\right|^{(0)} \\ \left.\frac{\partial f_2}{\partial x_1}\right|^{(0)} & \left.\frac{\partial f_2}{\partial x_2}\right|^{(0)} \end{bmatrix} \begin{bmatrix} \Delta x_1^{(0)} \\ \Delta x_2^{(0)} \end{bmatrix} = \begin{bmatrix} k_1 - f_1\left(x_1^{(0)}, x_2^{(0)}\right) \\ k_2 - f_2\left(x_1^{(0)}, x_2^{(0)}\right) \end{bmatrix} \qquad (8.102)$$

The matrix of partial derivatives in Eq. (8.102) is known as the Jacobian matrix and is evaluated at the initial estimate. Multiplying each side of Eq. (8.102) by the inverse of the Jacobian yields an approximation of the required correction to the estimated solution. Since the higher order terms were neglected, addition of the correction terms to the original estimate will not yield an exact solution, but will often provide an improved estimate. The procedure may be repeated, obtaining sucessively better estimates until the estimated solution reaches a desired tolerance. Summarizing, correction terms for the ℓth iterate are given in Eq. (8.103) and the solution estimate is updated according to Eq. (8.104).

$$\begin{bmatrix} \Delta x_1^{(\ell)} \\ \Delta x_2^{(\ell)} \end{bmatrix} = \begin{bmatrix} \left.\frac{\partial f_1}{\partial x_1}\right|^{(\ell)} & \left.\frac{\partial f_1}{\partial x_2}\right|^{(\ell)} \\ \left.\frac{\partial f_2}{\partial x_1}\right|^{(\ell)} & \left.\frac{\partial f_2}{\partial x_2}\right|^{(\ell)} \end{bmatrix}^{-1} \begin{bmatrix} k_1 - f_1\left(x_1^{(\ell)}, x_2^{(\ell)}\right) \\ k_2 - f_2\left(x_1^{(\ell)}, x_2^{(\ell)}\right) \end{bmatrix} \qquad (8.103)$$

$$x^{(\ell+1)} = x^{(\ell)} + \Delta x^{(\ell)} \qquad (8.104)$$

The solution of the original set of nonlinear equations has been converted to a repeated solution of a system of linear equations. This solution requires evaluation of the Jacobian matrix (at the current solution estimate) in each iteration.

The power flow equations can be placed into the Newton-Raphson framework by separating the power flow equations into their real and imaginary parts and taking the voltage magnitudes and phase angles as the unknowns. Writing Eq. (8.103) specifically for the power flow problem:

$$\begin{bmatrix} \Delta \underline{\delta}^{(\ell)} \\ \Delta \underline{V}^{(\ell)} \end{bmatrix} = \begin{bmatrix} \dfrac{\partial \underline{P}}{\partial \underline{\delta}}\bigg|^{(\ell)} & \dfrac{\partial \underline{P}}{\partial \underline{V}}\bigg|^{(\ell)} \\ \dfrac{\partial \underline{Q}}{\partial \underline{\delta}}\bigg|^{(\ell)} & \dfrac{\partial \underline{Q}}{\partial \underline{V}}\bigg|^{(\ell)} \end{bmatrix}^{-1} \begin{bmatrix} \underline{P}(\text{sched}) - \underline{P}^{(\ell)} \\ \underline{Q}(\text{sched}) - \underline{Q}^{(\ell)} \end{bmatrix} \qquad (8.105)$$

The underscored variables in Eq. (8.105) indicate vectors (extending the two-equation Newton-Raphson development to the general power flow case). The (sched) notation indicates the scheduled real and reactive powers injected into the system. $P^{(\ell)}$ and $Q^{(\ell)}$ represent the calculated real and reactive power injections based on the system model and the ℓth voltage phase angle and voltage magnitude estimates. The bus voltage phase angle and bus voltage magnitude estimates are updated, the Jacobian reevaluated, and the mismatch between the scheduled and calculated real and reactive powers evaluated in each iteration of the Newton-Raphson algorithm. Iterations are performed until the estimated solution reaches an acceptable tolerance or a maximum number of allowable iterations is exceeded. Once a solution (within an acceptble tolerance) is reached, P-V bus reactive power injections and the slack bus complex power injection may be evaluated.

Fast Decoupled Power Flow Solution

The fast decoupled power flow algorithm simplifies the procedure presented for the Newton-Raphson algorithm by exploiting the strong coupling between real power and bus voltage phase angles and reactive power and bus voltage magnitudes commonly seen in power systems. The Jacobian matrix is simplified by approximating the partial derivatives of the real power equations with respect to the bus voltage magnitudes as zero. Similarly, the partial derivatives of the reactive power equations with respect to the bus voltage phase angles are approximated as zero. Further, the remaining partial derivatives are often approximated using only the imaginary portion of the bus admittance matrix. These approximations yield the following correction equations:

$$\Delta \underline{\delta}^{(\ell)} = [B']^{-1} \Big[\underline{P}(\text{sched}) - \underline{P}^{(\ell)} \Big] \qquad (8.106)$$

$$\Delta \underline{V}^{(\ell)} = [B'']^{-1} \Big[\underline{Q}(\text{sched}) - \underline{Q}^{(\ell)} \Big] \qquad (8.107)$$

where B′ is an approximation of the matrix of partial derviatives of the real power flow equations with respect to the bus voltage phase angles and B″ is an approximation of the matrix of partial derivatives of the reactive power flow equations with respect to the bus voltage magnitudes. B′ and B″ are typically held constant during the iterative process, eliminating the necessity of updating the Jacobian matrix (required in the Newton-Raphson solution) in each iteration.

The fast decoupled algorithm has good convergence properties despite the many approximations used during its development. The fast decoupled power flow algorithm has found widespread use since it is less computationally intensive (requires fewer computational operations) than the Newton-Raphson method.

Power System Analysis and Simulation

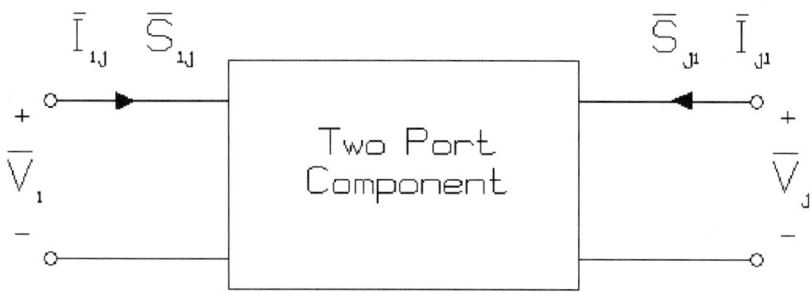

FIGURE 8.27 Typical power system component.

Component Power Flows

The positive sequence network for components of interest (connected between buses i and j) will be of the form shown in Fig. 8.27.

An admittance description is usually available from earlier construction of the nodal admittance matrix. Thus,

$$\begin{bmatrix} \bar{I}_i \\ \bar{I}_j \end{bmatrix} = \begin{bmatrix} \bar{Y}_a & \bar{Y}_b \\ \bar{Y}_c & \bar{Y}_d \end{bmatrix} \begin{bmatrix} \bar{V}_i \\ \bar{V}_j \end{bmatrix} \qquad (8.108)$$

Therefore, the complex power flows and the component loss are:

$$\bar{S}_{ij} = \bar{V}_i \bar{I}_i^* = \bar{V}_i \left[\bar{Y}_a \bar{V}_i + \bar{Y}_b \bar{V}_j \right]^* \qquad (8.109)$$

$$\bar{S}_{ji} = \bar{V}_j \bar{I}_j^* = \bar{V}_j \left[\bar{Y}_c \bar{V}_i + \bar{Y}_d \bar{V}_j \right]^* \qquad (8.110)$$

$$\bar{S}_{loss} = \bar{S}_{ij} + \bar{S}_{ji} \qquad (8.111)$$

The calculated component flows combined with the bus voltage magnitudes and phase angles provide extensive information about the power systems operating point. The pu voltage magnitudes may be checked to ensure operation within a prescribed range. The segment power flows can be examined to ensure no equipment ratings are exceeded. Additionally, the power flow solution may used as the starting point for other analyses.

An elementary discussion of the power flow problem and its solution are presented in this section. The power flow problem can be complicated by the addition of further constraints such as generator real and reactive power limits. However, discussion of such complications is beyond the scope of this section. The references provide detailed development of power flow formulation and solution under additional constraints.

References

Bergen, A. R., *Power Systems Analysis,* Prentice-Hall, Englewood Cliffs, NJ, 1986.
Elgerd, O. I., *Electric Energy Systems Theory — An Introduction,* 2nd ed., McGraw-Hill, New York, 1982.
Glover, J. D. and Sarma, M., *Power System Analysis and Design,* 2nd ed., PWS Publishing, Boston, MA, 1995.
Grainger, J. J. and Stevenson, W. D., *Power System Analysis,* McGraw-Hill, New York, 1994.
Gross, C. A., *Power System Analysis,* 2nd ed., John Wiley & Sons, New York, NY, 1986.

Further Information

The references provide clear introductions to the analysis of power systems. An excellent review of many issues involving the use of computers for power system analysis is provided in July 1974, *Proceedings of the IEEE* (Special Issue on Computers in the Power Industry). The quarterly journal *IEEE Transactions on Power Systems* provides excellent documentation of more recent research in power system analysis.

8.4 Fault Analysis in Power Systems

Charles A. Gross

A **fault** in an electrical power system is the unintentional and undesirable creation of a conducting path (a *short circuit*) or a blockage of current (an *open circuit*). The short-circuit fault is typically the most common and is usually implied when most people use the term *fault*. We restrict our comments to the short-circuit fault.

The causes of faults include lightning, wind damage, trees falling across lines, vehicles colliding with towers or poles, birds shorting out lines, aircraft colliding with lines, vandalism, small animals entering switchgear, and line breaks due to excessive ice loading. Power system faults may be categorized as one of four types: single line-to-ground, line-to-line, double line-to-ground, and balanced three-phase. The first three types constitute severe unbalanced operating conditions.

It is important to determine the values of system voltages and currents during faulted conditions so that protective devices may be set to detect and minimize their harmful effects. The time constants of the associated transients are such that sinusoidal steady-state methods may still be used. The method of symmetrical components is particularly suited to fault analysis.

Our objective is to understand how symmetrical components may be applied specifically to the four general fault types mentioned and how the method can be extended to any unbalanced three-phase system problem.

Note that phase values are indicated by subscripts, a, b, c; sequence (symmetrical component) values are indicated by subscripts 0, 1, 2. The transformation is defined by

$$\begin{bmatrix} \overline{V}_a \\ \overline{V}_b \\ \overline{V}_c \end{bmatrix} = \begin{bmatrix} 1 & 1 & 1 \\ 1 & a^2 & a \\ 1 & a & a^2 \end{bmatrix} \begin{bmatrix} \overline{V}_0 \\ \overline{V}_1 \\ \overline{V}_2 \end{bmatrix} = [T] \begin{bmatrix} \overline{V}_0 \\ \overline{V}_1 \\ \overline{V}_2 \end{bmatrix}$$

Simplifications in the System Model

Certain simplifications are possible and usually employed in fault analysis.

- Transformer magnetizing current and core loss will be neglected.
- Line shunt capacitance is neglected.
- Sinusoidal steady-state circuit analysis techniques are used. The so-called **DC offset** is accounted for by using correction factors.
- Prefault voltage is assumed to be $1\angle 0°$ per-unit. One per-unit voltage is at its nominal value prior to the application of a fault, which is reasonable. The selection of zero phase is arbitrary and convenient. Prefault load current is neglected.

For hand calculations, neglect series resistance is usually neglected (this approximation will not be necessary for a computer solution). Also, the only difference in the positive and negative sequence networks is introduced by the machine impedances. If we select the subtransient reactance X_d'' for the

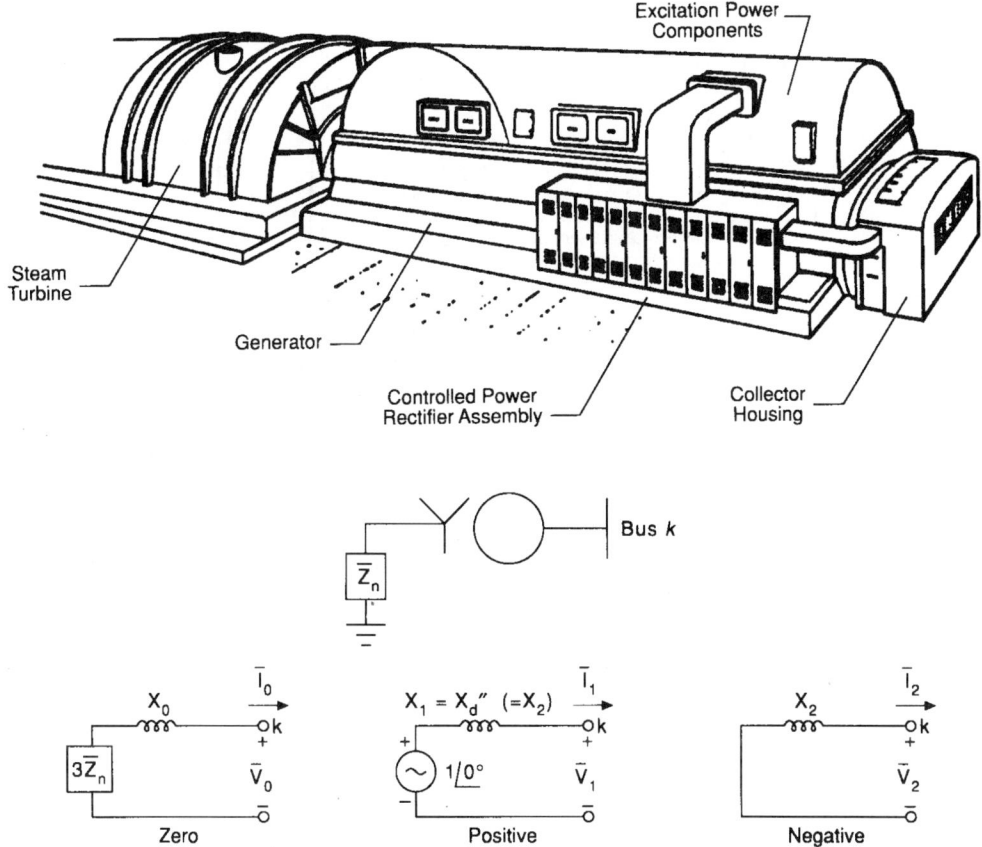

FIGURE 8.28 Generator sequence circuit models.

positive sequence reactance, the difference is slight (in fact, the two are identical for nonsalient machines). The simplification is important, since it reduces computer storage requirements by roughly one-third. Circuit models for generators, lines, and transformers are shown in Figs. 8.28, 8.29, and 8.30, respectively.

Our basic approach to the problem is to consider the general situation suggested in Fig. 8.31(a). The general terminals brought out are for purposes of external connections that will simulate faults. Note carefully the positive assignments of phase quantities. Particularly note that the currents flow *out of* the system. We can construct general *sequence* equivalent circuits for the system, and such circuits are indicated in Fig. 8.31(b). The ports indicated correspond to the general three-phase entry port of Fig. 8.31(a). The positive sense of sequence values is compatible with that used for phase values.

The Four Basic Fault Types

The Balanced Three-Phase Fault

Imagine the general three-phase access port terminated in a fault impedance ($\bar{Z}_f$) as shown in Fig. 8.32(a). The terminal conditions are

$$\begin{bmatrix} \bar{V}_a \\ \bar{V}_b \\ \bar{V}_c \end{bmatrix} = \begin{bmatrix} \bar{Z}_f & 0 & 0 \\ 0 & \bar{Z}_f & 0 \\ 0 & 0 & \bar{Z}_f \end{bmatrix} \begin{bmatrix} \bar{I}_a \\ \bar{I}_b \\ \bar{I}_c \end{bmatrix}$$

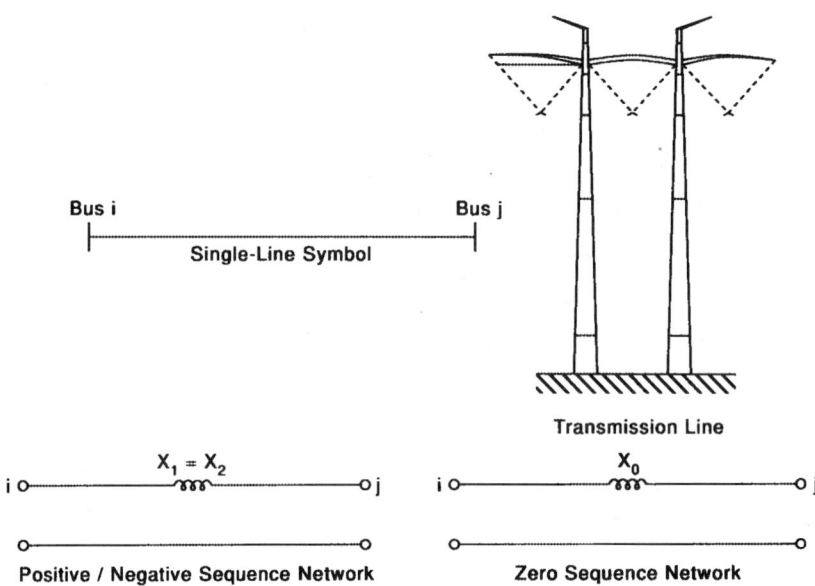

FIGURE 8.29 Line sequence circuit models.

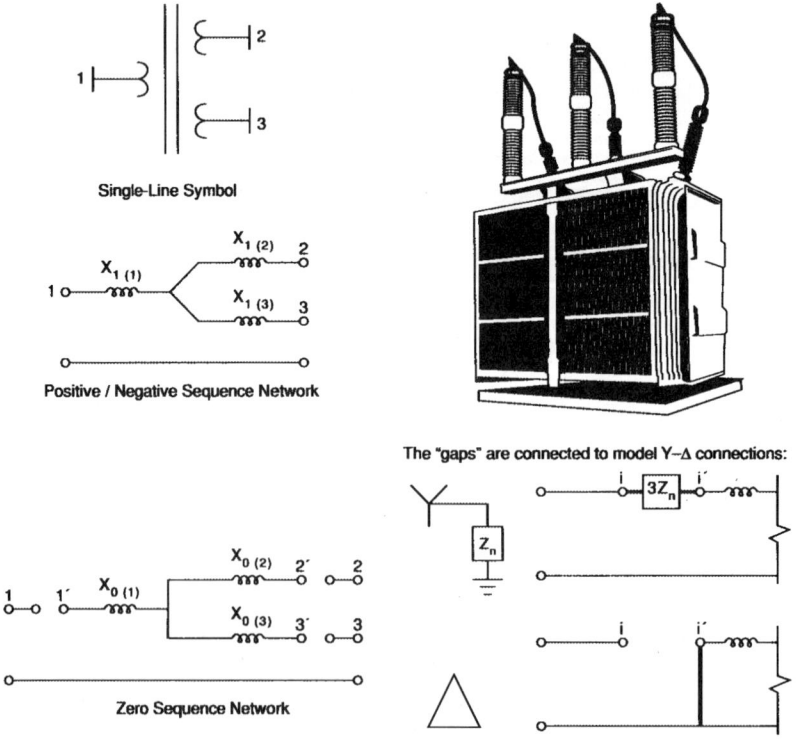

FIGURE 8.30 Transformer sequence circuit models.

Power System Analysis and Simulation

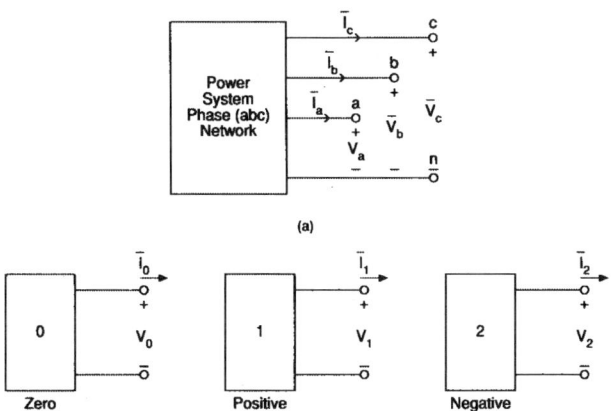

FIGURE 8.31 General fault port in an electric power system. (a) General fault port in phase (*abc*) coordinates; (b) corresponding fault ports in sequence (012) coordinates.

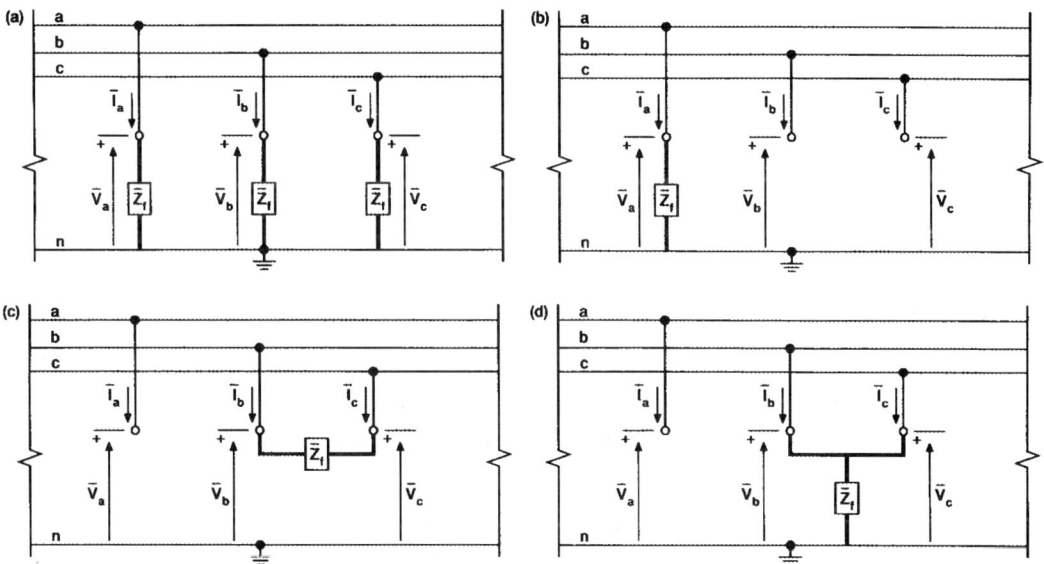

FIGURE 8.32 Fault types. (a) Three-phase fault; (b) single phase-to-ground fault; (c) phase-to-phase fault; (d) double phase-to-ground fault.

Transforming to $[Z_{012}]$,

$$[Z_{012}] = [T]^{-1} \begin{bmatrix} \overline{Z}_f & 0 & 0 \\ 0 & \overline{Z}_f & 0 \\ 0 & 0 & \overline{Z}_f \end{bmatrix} [T] = \begin{bmatrix} \overline{Z}_f & 0 & 0 \\ 0 & \overline{Z}_f & 0 \\ 0 & 0 & \overline{Z}_f \end{bmatrix}$$

The corresponding network connections are given in Fig. 8.33(a). Since the zero and negative sequence networks are passive, only the positive sequence network is nontrivial.

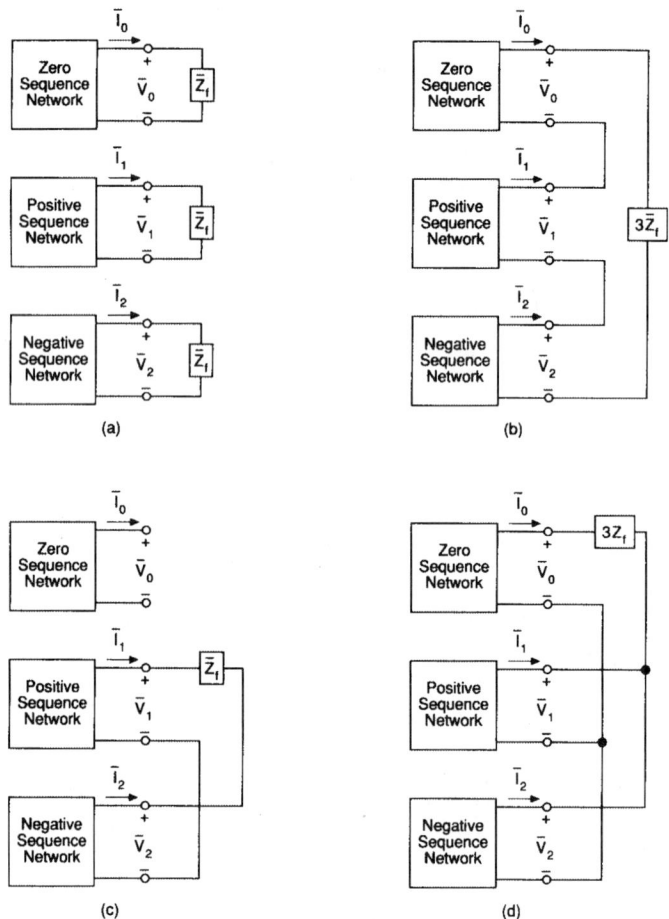

FIGURE 8.33 Sequence network terminations for fault types. (a) Balanced three-phase fault; (b) single phase-to-ground fault; (c) phase-to-phase fault; (d) double phase-to-ground fault.

$$\overline{V}_0 = \overline{V}_2 = 0 \tag{8.112}$$

$$\overline{I}_0 = \overline{I}_2 = 0 \tag{8.113}$$

$$\overline{V}_1 = \overline{Z}_f \overline{I}_1 \tag{8.114}$$

The Single Phase-to-Ground Fault

Imagine the general three-phase access port terminated as shown in Fig. 8.32(b). The terminal conditions are

$$\overline{I}_b = 0 \quad \overline{I}_c = 0 \quad \overline{V}_a = \overline{I}_a \overline{Z}_f$$

Therefore,

$$\overline{I}_0 + a^2 \overline{I}_1 + a \overline{I}_2 = \overline{I}_0 + a \overline{I}_1 + a^2 \overline{I}_2 = 0$$

Power System Analysis and Simulation

or

$$\bar{I}_1 = \bar{I}_2$$

Also,

$$\bar{I}_b = \bar{I}_0 + a^2\bar{I}_1 + a\bar{I}_2 = \bar{I}_0 + (a^2 + a)\bar{I}_1 = 0$$

or

$$\bar{I}_0 = \bar{I}_1 = \bar{I}_2 \tag{8.113}$$

Furthermore, it is required that

$$\begin{aligned}\bar{V}_a &= \bar{Z}_f \bar{I}_a \\ \bar{V}_0 + \bar{V}_1 + \bar{V}_2 &= 3\bar{Z}_f \bar{I}_1\end{aligned} \tag{8.114}$$

In general then, Eqs. (8.113) and (8.114) must be simultaneously satisfied. These conditions can be met by interconnecting the sequence networks as shown in Fig. 8.33(b).

The Phase-to-Phase Fault

Imagine the general three-phase access port terminated as shown in Fig. 8.32(c). The terminal conditions are such that we may write

$$\bar{I}_0 = 0 \qquad \bar{I}_b = -\bar{I}_c \qquad \bar{V}_b = \bar{Z}_f \bar{I}_b + \bar{V}_c$$

It follows that

$$\bar{I}_0 + \bar{I}_1 + \bar{I}_2 = 0 \tag{8.115}$$

$$\bar{I}_0 = 0 \tag{8.116}$$

$$\bar{I}_1 = -\bar{I}_2 \tag{8.117}$$

In general then, Eqs. (8.115), (8.116), and (8.117) must be simultaneously satisfied. The proper interconnection between sequence networks appears in Fig. 8.33(c).

The Double Phase-to-Ground Fault

Consider the general three-phase access port terminated as shown in Fig. 8.32(d). The terminal conditions indicate

$$\bar{I}_a = 0 \qquad \bar{V}_b = \bar{V}_c \qquad \bar{V}_b = (\bar{I}_b + \bar{I}_c)\bar{Z}_f$$

It follows that

$$\bar{I}_0 + \bar{I}_1 + \bar{I}_2 = \bar{0} \tag{8.118}$$

$$\bar{V}_1 = \bar{V}_2 \tag{8.119}$$

and

$$\overline{V}_0 - \overline{V}_1 = 3\overline{Z}_f \overline{I}_0 \qquad (8.120)$$

For the general double phase-to-ground fault, Eqs. (8.118), (8.119), and (8.120) must be simultaneously satisfied. The sequence network interconnections appear in Fig. 8.33(d).

An Example Fault Study

Case: EXAMPLE SYSTEM
Run :
System has data for 2 Line(s); 2 Transformer(s);
4 Bus(es); and 2 Generator(s)

Transmission Line Data

Line	Bus	Bus	Seq	R	X	B	Srat
1	2	3	pos	0.00000	0.16000	0.00000	1.0000
			zero	0.00000	0.50000	0.00000	
2	2	3	pos	0.00000	0.16000	0.00000	1.0000
			zero	0.00000	0.50000	0.00000	

Transformer Data

Trans-former	HV Bus	LV Bus	Seq	R	X	C	Srat
1	2	1	pos	0.00000	0.05000	1.00000	1.0000
	Y	Y	zero	0.00000	0.05000		
2	3	4	pos	0.00000	0.05000	1.00000	1.0000
	Y	D	zero	0.00000	0.05000		

Generator Data

No.	Bus	Srated	Ra	Xd″	Xo	Rn	Xn	Con
1	1	1.0000	0.0000	0.200	0.0500	0.0000	0.0400	Y
2	4	1.0000	0.0000	0.200	0.0500	0.0000	0.0400	Y

Zero Sequence [Z] Matrix

0.0 + j(0.1144)	0.0 + j(0.0981)	0.0 + j(0.0163)	0.0 + j(0.0000)
0.0 + j(0.0981)	0.0 + j(0.1269)	0.0 + j(0.0212)	0.0 + j(0.0000)
0.0 + j(0.0163)	0.0 + j(0.0212)	0.0 + j(0.0452)	0.0 + j(0.0000)
0.0 + j(0.0000)	0.0 + j(0.0000)	0.0 + j(0.0000)	0.0 + j(0.1700)

Positive Sequence [Z] Matrix

0.0 + j(0.1310)	0.0 + j(0.1138)	0.0 + j(0.0862)	0.0 + j(0.0690)
0.0 + j(0.1138)	0.0 + j(0.1422)	0.0 + j(0.1078)	0.0 + j(0.0862)
0.0 + j(0.0862)	0.0 + j(0.1078)	0.0 + j(0.1422)	0.0 + j(0.1138)
0.0 + j(0.0690)	0.0 + j(0.0862)	0.0 + j(0.1138)	0.0 + j(0.1310)

The single-line diagram and sequence networks are presented in Fig. 8.34.

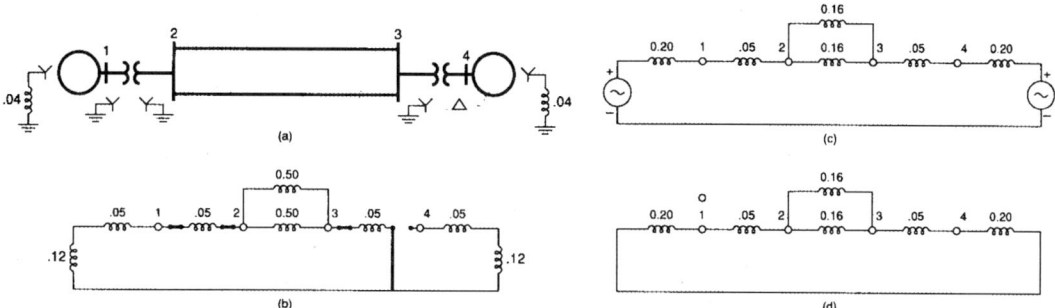

FIGURE 8.34 Example system. (a) Single-line diagram; (b) zero sequence network; (c) positive sequence network; (d) negative sequence network.

Suppose bus 3 in the example system represents the fault location and $\overline{Z}_f = 0$. The positive sequence circuit can be reduced to its Thévenin equivalent at bus 3:

$$E_{T1} = 1.0\underline{/0°} \quad \overline{Z}_{T1} = j0.1422$$

Similarly, the negative and zero sequence Thévenin elements are:

$$\overline{E}_{T2} = 0 \quad \overline{Z}_{T2} = j0.1422$$
$$\overline{E}_{T0} = 0 \quad Z_{T0} = j0.0452$$

The network interconnections for the four fault types are shown in Fig. 8.35. For each of the fault types, compute the currents and voltages at the faulted bus.

Balanced Three-Phase Fault

The sequence networks are shown in Fig. 8.35(a). Obviously,

$$\overline{V}_0 = \overline{I}_0 = \overline{V}_2 = \overline{I}_2 = 0$$

$$\overline{I}_1 = \frac{1\underline{/0°}}{j0.1422} = -j7.032; \quad \text{also } \overline{V}_1 = 0$$

To compute the phase values,

$$\begin{bmatrix} \overline{I}_a \\ \overline{I}_b \\ \overline{I}_c \end{bmatrix} = [T]\begin{bmatrix} \overline{I}_0 \\ \overline{I}_1 \\ \overline{I}_2 \end{bmatrix} = \begin{bmatrix} 1 & 1 & 1 \\ 1 & a^2 & a \\ 1 & a & a^2 \end{bmatrix}\begin{bmatrix} 0 \\ -j7.032 \\ 0 \end{bmatrix} = \begin{bmatrix} 7.032\underline{/-90°} \\ 7.032\underline{/150°} \\ 7.032\underline{/30°} \end{bmatrix}$$

$$\begin{bmatrix} \overline{V}_a \\ \overline{V}_b \\ \overline{V}_c \end{bmatrix} = [T]\begin{bmatrix} 0 \\ 0 \\ 0 \end{bmatrix} = \begin{bmatrix} 0 \\ 0 \\ 0 \end{bmatrix}$$

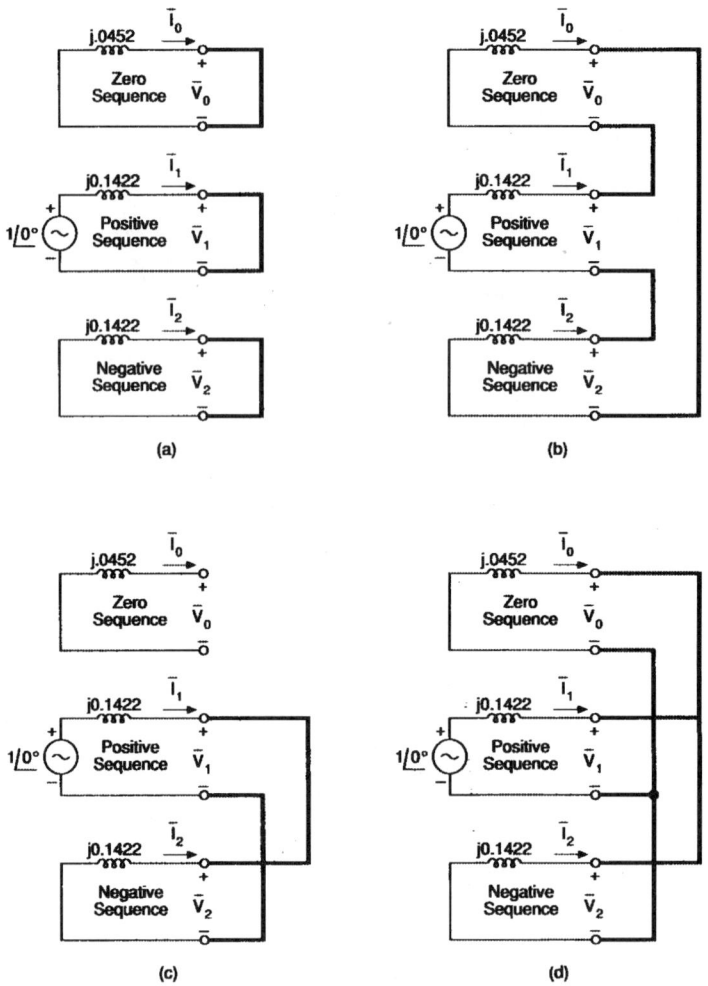

FIGURE 8.35 Example system faults at bus 3. (a) Balanced three-phase; (b) single phase-to-ground; (c) phase-to-phase; (d) double phase-to-ground.

Single Phase-to-Ground Fault

The sequence networks are interconnected as shown in Fig. 8.35(b).

$$\bar{I}_0 = \bar{I}_1 = \bar{I}_2 = \frac{1/0°}{j0.0452 + j0.1422 + j0.1422} = -j3.034$$

$$\begin{bmatrix} \bar{I}_a \\ \bar{I}_b \\ \bar{I}_c \end{bmatrix} = \begin{bmatrix} 1 & 1 & 1 \\ 1 & a^2 & a \\ 1 & a & a^2 \end{bmatrix} \begin{bmatrix} -j3.034 \\ -j3.034 \\ -j3.034 \end{bmatrix} = \begin{bmatrix} -j9.102 \\ 0 \\ 0 \end{bmatrix}$$

The sequence voltages are

$$\overline{V}_0 = -j0.0452(-j3.034) = -1371$$
$$\overline{V}_1 = 1.0 - j0.1422(-j3.034) = 0.5685$$
$$\overline{V}_2 = -j0.1422(-j3.034) = -0.4314$$

The phase voltages are

$$\begin{bmatrix}\overline{V}_a \\ \overline{V}_b \\ \overline{V}_c\end{bmatrix} = \begin{bmatrix}1 & 1 & 1 \\ 1 & a^2 & a \\ 1 & a & a^2\end{bmatrix}\begin{bmatrix}-0.1371 \\ 0.5685 \\ -0.4314\end{bmatrix} = \begin{bmatrix}0 \\ 0.8901\ \underline{/-103.4°} \\ 0.8901\ \underline{/-103.4°}\end{bmatrix}$$

Phase-to-phase and double phase-to-ground fault values are calculated from the appropriate networks [Figs. 8.35(c) and (d)]. Complete results are provided.

Faulted Bus	Phase a	Phase b	Phase c
3	G	G	G

Sequence Voltages

Bus	V0		V1		V2	
1	0.0000/	0.0	0.3939/	0.0	0.0000/	0.0
2	0.0000/	0.0	0.2424/	0.0	0.0000/	0.0
3	0.0000/	0.0	0.0000/	0.0	0.0000/	0.0
4	0.0000/	0.0	0.2000/	−30.0	0.0000/	30.0

Phase Voltages

Bus	Va		Vb		Vc	
1	0.3939/	0.0	0.3939/	−120.0	0.3939/	120.0
2	0.2424/	0.0	0.2424/	−120.0	0.2424/	120.0
3	0.0000/	6.5	0.0000/	−151.2	0.0000/	133.8
4	0.2000/	−30.0	0.2000/	−150.0	0.2000/	90.0

Sequence Currents

Bus to Bus		I0		I1		I2	
1	2	0.0000/	167.8	3.0303/	−90.0	0.0000/	90.0
1	0	0.0000/	−12.2	3.0303/	90.0	0.0000/	−90.0
2	3	0.0000/	167.8	1.5152/	−90.0	0.0000/	90.0
2	3	0.0000/	167.8	1.5152/	−90.0	0.0000/	90.0
2	1	0.0000/	−12.2	3.0303/	90.0	0.0000/	−90.0
3	2	0.0000/	−12.2	1.5152/	90.0	0.0000/	−90.0
3	2	0.0000/	−12.2	1.5152/	90.0	0.0000/	−90.0
3	4	0.0000/	−12.2	4.0000/	90.0	0.0000/	−90.0
4	3	0.0000/	0.0	4.0000/	−120.0	0.0000/	120.0
4	0	0.0000/	0.0	4.0000/	60.0	0.0000/	−60.0

Faulted Bus	Phase a	Phase b	Phase c
3	G	G	G

Phase Currents

Bus to Bus		Ia		Ib		Ic	
1	2	3.0303/	−90.0	3.0303/	150.0	3.0303/	30.0
1	0	3.0303/	90.0	3.0303/	−30.0	3.0303/	−150.0
2	3	1.5151/	−90.0	1.5151/	150.0	1.5151/	30.0
2	3	1.5151/	−90.0	1.5151/	150.0	1.5151/	30.0
2	1	3.0303/	90.0	3.0303/	−30.0	3.0303/	−150.0
3	2	1.5151/	90.0	1.5151/	−30.0	1.5151/	−150.0
3	2	1.5151/	90.0	1.5151/	−30.0	1.5151/	−150.0
3	4	4.0000/	90.0	4.0000/	−30.0	4.0000/	−150.0
4	3	4.0000/	−120.0	4.0000/	120.0	4.0000/	−0.0
4	0	4.0000/	60.0	4.0000/	−60.0	4.0000/	−180.0

Faulted Bus	Phase a	Phase b	Phase c
3	G	0	0

Sequence Voltages

Bus	V0		V1		V2	
1	0.0496/	180.0	0.7385/	0.0	0.2615/	180.0
2	0.0642/	180.0	0.6731/	0.0	0.3269/	180.0
3	0.1371/	180.0	0.5685/	0.0	0.4315/	180.0
4	0.0000/	0.0	0.6548/	−30.0	0.3452/	210.0

Phase Voltages

Bus	Va		Vb		Vc	
1	0.4274/	0.0	0.9127/	−108.4	0.9127/	108.4
2	0.2821/	0.0	0.8979/	−105.3	0.8979/	105.3
3	0.0000/	89.2	0.8901/	−103.4	0.8901/	103.4
4	0.5674/	−61.8	0.5674/	−118.2	1.0000/	90.0

Sequence Currents

Bus to Bus		I0		I1		I2	
1	2	0.2917/	−90.0	1.3075/	−90.0	1.3075/	−90.0
1	0	0.2917/	90.0	1.3075/	90.0	1.3075/	90.0
2	3	0.1458/	−90.0	0.6537/	−90.0	0.6537/	−90.0
2	3	0.1458/	−90.0	0.6537/	−90.0	0.6537/	−90.0
2	1	0.2917/	90.0	1.3075/	90.0	1.3075/	90.0
3	2	0.1458/	90.0	0.6537/	90.0	0.6537/	90.0
3	2	0.1458/	90.0	0.6537/	90.0	0.6537/	90.0
3	4	2.7416/	90.0	1.7258/	90.0	1.7258/	90.0
4	3	0.0000/	0.0	1.7258/	−120.0	1.7258/	−60.0
4	0	0.0000/	90.0	1.7258/	60.0	1.7258/	120.0

Faulted Bus	Phase a	Phase b	Phase c
3	G	0	0

Phase Currents

Bus to Bus		Ia		Ib		Ic	
1	2	2.9066/	−90.0	1.0158/	90.0	1.0158/	90.0
1	0	2.9066/	90.0	1.0158/	−90.0	1.0158/	−90.0
2	3	1.4533/	−90.0	0.5079/	90.0	0.5079/	90.0
2	3	1.4533/	−90.0	0.5079/	90.0	0.5079/	90.0
2	1	2.9066/	90.0	1.0158/	−90.0	1.0158/	−90.0
3	2	1.4533/	90.0	0.5079/	−90.0	0.5079/	−90 0
3	2	1.4533/	90.0	0.5079/	−90.0	0.5079/	−90 0
3	4	6.1933/	90.0	1.0158/	90.0	1.0158/	90.0
4	3	2.9892/	−90.0	2.9892/	90.0	0.0000/	−90.0
4	0	2.9892/	90.0	2.9892/	−90.0	0.0000/	90.0

Faulted Bus	Phase a	Phase b	Phase c
3	0	C	B

Sequence Voltages

Bus	V0		V1		V2	
1	0.0000/	0.0	0.6970/	0.0	0.3030/	0.0
2	0.0000/	0.0	0.6212/	0.0	0.3788/	0.0
3	0.0000/	0.0	0.5000/	0.0	0.5000/	0.0
4	0.0000/	0.0	0.6000/	−30.0	0.4000/	30.0

Phase Voltages

Bus	Va		Vb		Vc	
1	1.0000/	0.0	0.6053/	−145.7	0.6053/	145.7
2	1.0000/	0.0	0.5423/	−157.2	0.5423/	157.2
3	1.0000/	0.0	0.5000/	−180.0	0.5000/	−180.0
4	0.8718/	−6.6	0.8718/	−173.4	0.2000/	90.0

Sequence Currents

Bus to Bus		I0		I1		I2	
1	2	0.0000/	−61.0	1.5152/	−90.0	1.5152/	90.0
1	0	0.0000/	119.0	1.5152/	90.0	1.5152/	−90.0
2	3	0.0000/	−61.0	0.7576/	−90.0	0.7576/	90.0
2	3	0.0000/	−61.0	0.7576/	−90.0	0.7576/	90.0
2	1	0.0000/	119.0	1.5152/	90.0	1.5152/	−90.0
3	2	0.0000/	119.0	0.7576/	90.0	0.7576/	−90.0
3	2	0.0000/	119.0	0.7576/	90.0	0.7576/	−90.0
3	4	0.0000/	119.0	2.0000/	90.0	2.0000/	−90.0
4	3	0.0000/	0.0	2.0000/	−120.0	2.0000/	120.0
4	0	0.0000/	90.0	2.0000/	60.0	2.0000/	−60.0

Faulted Bus	Phase a	Phase b	Phase c
3	0	C	B

Phase Currents

Bus to Bus		Ia		Ib		Ic	
1	2	0.0000/	180.0	2.6243/	180.0	2.6243/	0.0
1	0	0.0000/	180.0	2.6243/	0.0	2.6243/	180.0
2	3	0.0000/	−180.0	1.3122/	180.0	1.3122/	0.0
2	3	0.0000/	−180.0	1.3122/	180.0	1.3122/	0.0
2	1	0.0000/	180.0	2.6243/	0.0	2.6243/	180.0
3	2	0.0000/	−180.0	1.3122/	0.0	1.3122/	180.0
3	2	0.0000/	−180.0	1.3122/	0.0	1.3122/	180.0
3	4	0.0000/	−180.0	3.4641/	0.0	3.4641/	180.0
4	3	2.0000/	−180.0	2.0000/	180.0	4.0000/	0.0
4	0	2.0000/	0.0	2.0000/	0.0	4.0000/	−180.0

Faulted Bus	Phase a	Phase b	Phase c
3	0	G	G

Sequence Voltages

Bus	V0		V1		V2	
1	0.0703/	0.0	0.5117/	0.0	0.1177/	0.0
2	0.0909/	0.0	0.3896/	0.0	0.1472/	0.0
3	0.1943/	−0.0	0.1943/	0.0	0.1943/	0.0
4	0.0000/	0.0	0.3554/	−30.0	0.1554/	30.0

Phase Voltages

Bus	Va		Vb		Vc	
1	0.6997/	0.0	0.4197/	−125.6	0.4197/	125.6
2	0.6277/	0.0	0.2749/	−130.2	0.2749/	130.2
3	0.5828/	0.0	0.0000/	−30.7	0.0000/	−139.6
4	0.4536/	−12.7	0.4536/	−167.3	0.2000/	90.0

Sequence Currents

Bus to Bus		I0		I1		I2	
1	2	0.4133/	90.0	2.4416/	−90.0	0.5887/	90.0
1	0	0.4133/	−90.0	2.4416/	90.0	0.5887/	−90.0
2	3	0.2067/	90.0	1.2208/	−90.0	0.2943/	90.0
2	3	0.2067/	90.0	1.2208/	−90.0	0.2943/	90.0
2	1	0.4133/	−90.0	2.4416/	90.0	0.5887/	−90.0
3	2	0.2067/	−90.0	1.2208/	90.0	0.2943/	−90.0
3	2	0.2067/	−90.0	1.2208/	90.0	0.2943/	−90.0
3	4	3.8854/	−90.0	3.2229/	90.0	0.7771/	−90.0
4	3	0.0000/	0.0	3.2229/	−120.0	0.7771/	120.0
4	0	0.0000/	−90.0	3.2229/	60.0	0.7771/	−60.0

Faulted Bus	Phase a	Phase b	Phase c
3	0	G	G

		Phase Currents					
Bus to Bus		Ia		Ib		Ic	
1	2	1.4396/	−90.0	2.9465/	153.0	2.9465/	27.0
1	0	1.4396/	90.0	2.9465/	−27.0	2.9465/	−153.0
2	3	0.7198/	−90.0	1.4733/	153.0	1.4733/	27.0
2	3	0.7198/	−90.0	1.4733/	153.0	1.4733/	27.0
2	1	1.4396/	90.0	2.9465/	−27.0	2.9465/	−153.0
3	2	0.7198/	90.0	1.4733/	−27.0	1.4733/	−153.0
3	2	0.7198/	90.0	1.4733/	−27.0	1.4733/	−153.0
3	4	1.4396/	−90.0	6.1721/	−55.9	6.1721/	−124.1
4	3	2.9132/	−133.4	2.9132/	133.4	4.0000/	−0.0
4	0	2.9132/	46.6	2.9132/	−46.6	4.0000/	−180.0

Further Considerations

Generators are not the only sources in the system. All rotating machines are capable of contributing to fault current, at least momentarily. Synchronous and induction motors will continue to rotate due to inertia and function as sources of fault current. The impedance used for such machines is usually the transient reactance X'_d or the subtransient X''_d, depending on protective equipment and speed of response. Frequently, motors smaller than 50 hp are neglected. Connecting systems are modeled with their Thévenin equivalents.

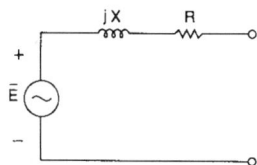

FIGURE 8.36 Positive sequence circuit looking back into faulted bus.

Although we have used AC circuit techniques to calculate faults, the problem is fundamentally transient since it involves sudden switching actions. Consider the so-called DC offset current. We model the system by determining its positive sequence Thévenin equivalent circuit, looking back into the positive sequence network at the fault, as shown in Fig. 8.36. The transient fault current is

$$i(t) = I_{ac}\sqrt{2}\,\cos(\omega t - \beta) + I_{dc}e^{-t/\tau}$$

This is a first-order approximation and strictly applies only to the three-phase or phase-to-phase fault. Ground faults would involve the zero sequence network also.

$$I_{ac} = \frac{E}{\sqrt{R^2 + X^2}} = \text{rms AC current}$$

$$I_{dc}(t) = I_{dc}e^{-t/\tau} = \text{DC offset current}$$

The maximum initial DC offset possible would be

$$\text{Max } I_{DC} = I_{max} = \sqrt{2}I_{AC}$$

The DC offset will exponentially decay with time constant τ, where

$$\tau = \frac{L}{R} = \frac{X}{\omega R}$$

The maximum DC offset current would be $I_{DC}(t)$

$$I_{DC}(t) = I_{DC}e^{-t/\tau} = \sqrt{2}I_{AC}e^{-t/\tau}$$

The *transient rms* current $I(t)$, accounting for both the AC and DC terms, would be

$$I(t) = \sqrt{I_{AC}^2 + I_{DC}^2(t)} = I_{AC}\sqrt{1 + 2e^{-2t/\tau}}$$

Define a multiplying factor k_i such that I_{AC} is to be multiplied by k_i to estimate the interrupting capacity of a breaker which operates in time T_{op}. Therefore,

$$k_i = \frac{I(T_{op})}{I_{AC}} = \sqrt{1 + 2e^{-2T_{op}/\tau}}$$

Observe that the maximum possible value for k_i is $\sqrt{3}$.

Example

In the circuit of Fig. 8.36, $E = 2400$ V, $X = 2\ \Omega$, $R = 0.1\ \Omega$, and $f = 60$ Hz. Compute k_i and determine the interrupting capacity for the circuit breaker if it is designed to operate in two cycles. The fault is applied at $t = 0$.

Solution:

$$I_{ac} \cong \frac{2400}{2} = 1200\ \text{A}$$

$$T_{op} = \frac{2}{60} = 0.0333\ \text{s}$$

$$\tau = \frac{X}{\omega R} = \frac{2}{37.7} = 0.053$$

$$k_i = \sqrt{1 + 2e^{-2T_{op}/\tau}} = \sqrt{1 + 2e^{-0.0067/0.053}} = 1.252$$

Therefore,

$$I = k_i I_{ac} = 1.252(1200) = 1503\ \text{A}$$

The Thévenin equivalent at the fault point is determined by normal sinusoidal steady-state methods, resulting in a first-order circuit as shown in Fig. 8.36. While this provides satisfactory results for the steady-state component I_{AC}, the X/R value so obtained can be in serious error when compared with the rate of decay of $I(t)$ as measured by oscillographs on an actual faulted system. The major reasons for the discrepancy are, first of all, that the system, for transient analysis purposes, is actually high-order, and second, the generators do not hold constant impedance as the transient decays.

Summary

Computation of fault currents in power systems is best done by computer. The major steps are summarized below:

- Collect, read in, and store machine, transformer, and line data in per-unit on common bases.
- Formulate the sequence impedance matrices.
- Define the faulted bus and Z_f. Specify type of fault to be analyzed.
- Compute the sequence voltages.
- Compute the sequence currents.
- Correct for wye-delta connections.
- Transform to phase currents and voltages.

For large systems, computer formulation of the sequence impedance matrices is required. Refer to Further Information for more detail. Zero sequence networks for lines in close proximity to each other (on a common right-of-way) will be mutually coupled. If we are willing to use the same values for positive and negative sequence machine impedances,

$$[Z_1] = [Z_2]$$

Therefore, it is unnecessary to store these values in separate arrays, simplifying the program and reducing the computer storage requirements significantly. The error introduced by this approximation is usually not important. The methods previously discussed neglect the prefault, or load, component of current; that is, the usual assumption is that currents throughout the system were zero prior to the fault. This is almost never strictly true; however, the error produced is small since the fault currents are generally much larger than the load currents. Also, the load currents and fault currents are out of phase with each other, making their sum more nearly equal to the larger components than would have been the case if the currents were in phase. In addition, selection of precise values for prefault currents is somewhat speculative, since there is no way of predicting what the loaded state of the system is when a fault occurs. When it is important to consider load currents, a power flow study is made to calculate currents throughout the system, and these values are superimposed on (added to) results from the fault study.

A term which has wide industrial use and acceptance is the *fault level* or **fault MVA** at a bus. It relates to the amount of current that can be expected to flow out of a bus into a three-phase fault. As such, it is an alternate way of providing positive sequence impedance information. Define

$$\text{Fault level in MVA at bus } i = V_{i_{pu\,nominal}} I_{i_{pu\,fault}} S_{3\phi\,base}$$

$$= (1)\frac{1}{Z_{ii}^1} S_{3\phi\,base} = \frac{S_{3\phi\,base}}{Z_{ii}^1}$$

Fault study results may be further refined by approximating the effect of DC offset.

The basic reason for making fault studies is to provide data that can be used to size and set protective devices. The role of such protective devices is to detect and remove faults to prevent or minimize damage to the power system.

Defining Terms

DC offset: The natural response component of the transient fault current, usually approximated with a first-order exponential expression.

Fault: An unintentional and undesirable conducting path in an electrical power system.

Fault MVA: At a specific location in a system, the initial symmetrical fault current multiplied by the prefault nominal line-to-neutral voltage (×3 for a three-phase system).

Sequence (012) quantities: Symmetrical components computed from phase (*abc*) quantities. Can be voltages, currents, and/or impedances.

References

P. M. Anderson, *Analysis of Faulted Power Systems,* Ames: Iowa State Press, 1973.
M. E. El-Hawary, *Electric Power Systems: Design and Analysis,* Reston, Va.: Reston Publishing, 1983.
M. E. El-Hawary, *Electric Power Systems,* New York: IEEE Press, 1995.
O. I. Elgerd, *Electric Energy Systems Theory: An Introduction,* 2nd ed., New York: McGraw-Hill, 1982.
General Electric, *Short-Circuit Current Calculations for Industrial and Commercial Power Systems,* Publication GET-3550.
C. A. Gross, *Power System Analysis,* 2nd ed., New York: Wiley, 1986.
S. H. Horowitz, *Power System Relaying,* 2nd ed, New York: Wiley, 1995.
I. Lazar, *Electrical Systems Analysis and Design for Industrial Plants,* New York: McGraw-Hill, 1980.
C. R. Mason, *The Art and Science of Protective Relaying,* New York: Wiley, 1956.
J. R. Neuenswander, *Modern Power Systems,* Scranton, Pa.: International Textbook, 1971.
G. Stagg and A. H. El-Abiad, *Computer Methods in Power System Analysis,* New York: McGraw-Hill, 1968.
Westinghouse Electric Corporation, *Applied Protective Relaying,* Relay-Instrument Division, Newark, N.J., 1976.
A. J. Wood, *Power Generation, Operation, and Control,* New York: Wiley, 1996.

Further Information

For a comprehensive coverage of general fault analysis, see Paul M. Anderson, *Analysis of Faulted Power Systems,* New York, IEEE Press, 1995. Also see Chapters 9 and 10 of *Power System Analysis* by C.A. Gross, New York: Wiley, 1986.

9
Power System Protection

Arun Phadke
Virginia Polytechnic Institute

9.1 **Transformer Protection**   *Alex Apostolov, John Appleyard, Ahmed Elneweihi, Robert Haas, and Glenn W. Swift* 9-1

9.2 **The Protection of Synchronous Generators**   *Gabriel Benmouyal* 9-11

9.3 **Transmission Line Protection**   *Stanley H. Horowitz* 9-30

9.4 **System Protection**   *Miroslav Begovic* 9-39

9.5 **Digital Relaying**   *James S. Thorp* 9-44

9.6 **Use of Oscillograph Records to Analyze System Performance**   *John R. Boyle* 9-62

9
Power System Protection

Alex Apostolov
Alstom T&D

John Appleyard
S&C Electric Company

Ahmed Elneweihi
BC Hydro

Robert Haas
Haas Engineering

Glenn W. Swift
APT Power Technologies

Gabriel Benmouyal
Schweitzer Engineering Laboratories, Ltd.

Stanley H. Horowitz
Consultant

Miroslav Begovic
Georgia Institute of Technology

James S. Thorp
Cornell University

John R. Boyle
Power System Analysis

9.1 Transformer Protection...9-1
Types of Transformer Faults • Types of Transformer Protection • Special Considerations • Special Applications • Restoration

9.2 The Protection of Synchronous Generators....................9-11
Review of Functions • Differential Protection for Stator Faults (87G) • Protection Against Stator Winding Ground Fault • Field Ground Protection • Loss-of-Excitation Protection (40) • Current Unbalance (46) • Anti-Motoring Protection (32) • Overexcitation Protection (24) • Overvoltage (59) • Voltage Unbalance Protection (60) • System Backup Protection (51V and 21) • Out-of-Step Protection • Abnormal Frequency Operation of Turbine-Generator • Protection Against Accidental Energization • Generator Breaker Failure • Generator Tripping Principles • Impact of Generator Digital Multifunction Relays

9.3 Transmission Line Protection.......................................9-30
The Nature of Relaying • Current Actuated Relays • Distance Relays • Pilot Protection

9.4 System Protection..9-39
Transient Stability and Out-of-Step Protection • Voltage Stability and Undervoltage Load Shedding • Special Protection Schemes (SPS) • Future Improvements in Control and Protection

9.5 Digital Relaying..9-44
Sampling • Antialiasing Filters • Sigma-Delta A/D Converters • Phasors from Samples • Symmetrical Components • Algorithms

9.6 Use of Oscillograph Records to Analyze System Performance...9-62

9.1 Transformer Protection

Alex Apostolov, John Appleyard, Ahmed Elneweihi, Robert Haas, and Glenn W. Swift

Types of Transformer Faults

Any number of conditions have been the reason for an electrical transformer failure. Statistics show that winding failures most frequently cause transformer faults (ANSI/IEEE, 1985). Insulation deterioration, often the result of moisture, overheating, vibration, voltage surges, and mechanical stress created during transformer through faults, is the major reason for winding failure.

Voltage regulating load tap changers, when supplied, rank as the second most likely cause of a transformer fault. Tap changer failures can be caused by a malfunction of the mechanical switching mechanism, high resistance load contacts, insulation tracking, overheating, or contamination of the insulating oil.

Transformer bushings are the third most likely cause of failure. General aging, contamination, cracking, internal moisture, and loss of oil can all cause a bushing to fail. Two other possible reasons are vandalism and animals that externally flash over the bushing.

Transformer core problems have been attributed to core insulation failure, an open ground strap, or shorted laminations.

Other miscellaneous failures have been caused by current transformers, oil leakage due to inadequate tank welds, oil contamination from metal particles, overloads, and overvoltage.

Types of Transformer Protection

Electrical

Fuse: Power fuses have been used for many years to provide transformer fault protection. Generally it is recommended that transformers sized larger than 10 MVA be protected with more sensitive devices such as the differential relay discussed later in this section. Fuses provide a low maintenance, economical solution for protection. Protection and control devices, circuit breakers, and station batteries are not required.

There are some drawbacks. Fuses provide limited protection for some internal transformer faults. A fuse is also a single phase device. Certain system faults may only operate one fuse. This will result in single phase service to connected three phase customers.

Fuse selection criteria include: adequate interrupting capability, calculating load currents during peak and emergency conditions, performing coordination studies that include source and low side protection equipment, and expected transformer size and winding configuration (ANSI/IEEE, 1985).

Overcurrent Protection: Overcurrent relays generally provide the same level of protection as power fuses. Higher sensitivity and fault clearing times can be achieved in some instances by using an overcurrent relay connected to measure residual current. This application allows pick up settings to be lower than expected maximum load current. It is also possible to apply an instantaneous overcurrent relay set to respond only to faults within the first 75% of the transformer. This solution, for which careful fault current calculations are needed, does not require coordination with low side protective devices.

Overcurrent relays do not have the same maintenance and cost advantages found with power fuses. Protection and control devices, circuit breakers, and station batteries are required. The overcurrent relays are a small part of the total cost and when this alternative is chosen, differential relays are generally added to enhance transformer protection. In this instance, the overcurrent relays will provide backup protection for the differentials.

Differential: The most widely accepted device for transformer protection is called a restrained differential relay. This relay compares current values flowing into and out of the transformer windings. To assure protection under varying conditions, the main protection element has a multislope restrained characteristic. The initial slope ensures sensitivity for internal faults while allowing for up to 15% mismatch when the power transformer is at the limit of its tap range (if supplied with a load tap changer). At currents above rated transformer capacity, extra errors may be gradually introduced as a result of CT saturation.

However, misoperation of the differential element is possible during transformer energization. High inrush currents may occur, depending on the point on wave of switching as well as the magnetic state of the transformer core. Since the inrush current flows only in the energized winding, differential current results. The use of traditional second harmonic restraint to block the relay during inrush conditions may

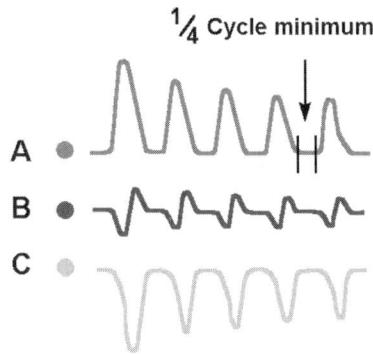

FIGURE 9.1 Transformer inrush current waveforms.

result in a significant slowing of the relay during heavy internal faults due to the possible presence of second harmonics as a result of saturation of the line current transformers. To overcome this, some relays use a waveform recognition technique to detect the inrush condition. The differential current waveform associated with magnetizing inrush is characterized by a period of each cycle where its magnitude is very small, as shown in Fig. 9.1. By measuring the time of this period of low current, an inrush condition can be identified. The detection of inrush current in the differential current is used to inhibit that phase of the low set restrained differential algorithm. Another high-speed method commonly used to detect high-magnitude faults in the unrestrained instantaneous unit is described later in this section.

When a load is suddenly disconnected from a power transformer, the voltage at the input terminals of the transformer may rise by 10–20% of the rated value causing an appreciable increase in transformer steady state excitation current. The resulting excitation current flows in one winding only and hence appears as differential current that may rise to a value high enough to operate the differential protection. A waveform of this type is characterized by the presence of fifth harmonic. A Fourier technique is used to measure the level of fifth harmonic in the differential current. The ratio of fifth harmonic to fundamental is used to detect excitation and inhibits the restrained differential protection function. Detection of overflux conditions in any phase blocks that particular phase of the low set differential function.

Transformer faults of a different nature may result in fault currents within a very wide range of magnitudes. Internal faults with very high fault currents require fast fault clearing to reduce the effect of current transformer saturation and the damage to the protected transformer. An unrestrained instantaneous high set differential element ensures rapid clearance of such faults. Such an element essentially measures the peak value of the input current to ensure fast operation for internal faults with saturated CTs. Restrained units generally calculate an rms current value using more waveform samples. The high set differential function is not blocked under magnetizing inrush or over excitation conditions, hence the setting must be set such that it will not operate for the largest inrush currents expected.

At the other end of the fault spectrum are low current winding faults. Such faults are not cleared by the conventional differential function. Restricted ground fault protection gives greater sensitivity for ground faults and hence protects more of the winding. A separate element based on the high impedance circulating current principle is provided for each winding.

Transformers have many possible winding configurations that may create a voltage and current phase shift between the different windings. To compensate for any phase shift between two windings of a transformer, it is necessary to provide phase correction for the differential relay (see section on Special Considerations).

In addition to compensating for the phase shift of the protected transformer, it is also necessary to consider the distribution of primary zero sequence current in the protection scheme. The necessary filtering of zero sequence current has also been traditionally provided by appropriate connection of auxiliary current transformers or by delta connection of primary CT secondary windings. In microprocessor transformer protection relays, zero sequence current filtering is implemented in software when a delta CT connection would otherwise be required. In situations where a transformer winding can produce zero sequence current caused by an external ground fault, it is essential that some form of zero sequence current filtering is employed. This ensures that ground faults out of the zone of protection will not cause the differential relay to operate in error. As an example, an external ground fault on the wye side of a delta/wye connected power transformer will result in zero sequence current flowing in the current transformers associated with the wye winding but, due to the effect of the delta winding, there will be no corresponding zero sequence current in the current transformers associated with the delta winding, i.e., differential current flow will cause the relay to operate. When the virtual zero sequence current filter is applied within the relay, this undesired trip will not occur.

Some of the most typical substation configurations, especially at the transmission level, are breaker-and-a-half or ring-bus. Not that common, but still used are two-breaker schemes. When a power transformer is connected to a substation using one of these breaker configurations, the transformer protection is connected to three or more sets of current transformers. If it is a three winding transformer or an auto transformer with a tertiary connected to a lower voltage sub transmission system, four or more sets of CTs may be available.

It is highly recommended that separate relay input connections be used for each set used to protect the transformer. Failure to follow this practice may result in incorrect differential relay response. Appropriate testing of a protective relay for such configuration is another challenging task for the relay engineer.

Overexcitation: Overexcitation can also be caused by an increase in system voltage or a reduction in frequency. It follows, therefore, that transformers can withstand an increase in voltage with a corresponding increase in frequency but not an increase in voltage with a decrease in frequency. Operation cannot be sustained when the ratio of voltage to frequency exceeds more than a small amount.

Protection against overflux conditions does not require high-speed tripping. In fact, instantaneous tripping is undesirable, as it would cause tripping for transient system disturbances, which are not damaging to the transformer.

An alarm is triggered at a lower level than the trip setting and is used to initiate corrective action. The alarm has a definite time delay, while the trip characteristic generally has a choice of definite time delay or inverse time characteristic.

Mechanical

There are two generally accepted methods used to detect transformer faults using mechanical methods. These detection methods provide sensitive fault detection and compliment protection provided by differential or overcurrent relays.

Accumulated Gases: The first method accumulates gases created as a by product of insulating oil decomposition created from excessive heating within the transformer. The source of heat comes from either the electrical arcing or a hot area in the core steel. This relay is designed for conservator tank transformers and will capture gas as it rises in the oil. The relay, sometimes referred to as a Buchholz relay, is sensitive enough to detect very small faults.

Pressure Relays: The second method relies on the transformer internal pressure rise that results from a fault. One design is applicable to gas-cushioned transformers and is located in the gas space above the oil. The other design is mounted well below minimum liquid level and responds to changes in oil pressure. Both designs employ an equalizing system that compensates for pressure changes due to temperature (ANSI/IEEE, 1985).

Thermal

Hot Spot-Temperature: In any transformer design, there is a location in the winding that the designer believes to be the *hottest* spot within that transformer (ANSI/IEEE, 1995). The significance of the "hot-spot temperature" measured at this location is an assumed relationship between the temperature level and the rate-of-degradation of the cellulose insulation. An instantaneous alarm or trip setting is often used, set at a judicious level above the full load rated hot-spot temperature (110°C for 65°C rise transformers). [Note that "65°C rise" refers to the full load rated *average* winding temperature rise.] Also, a relay or monitoring system can mathematically integrate the rate-of-degradation, i.e., rate-of-loss-of-life of the insulation for overload assessment purposes.

Heating Due to Overexcitation: Transformer core flux density (B), induced voltage (V), and frequency (f) are related by the following formula.

$$B = k_1 \cdot \frac{V}{f} \qquad (9.1)$$

where K_1 is a constant for a particular transformer design. As B rises above about 110% of normal, that is, when saturation starts, significant heating occurs due to stray flux eddy-currents in the nonlaminated structural metal parts, including the tank. Since it is the voltage/hertz quotient in Eq. (9.1) that defines the level of B, a relay sensing this quotient is sometimes called a "volts-per-hertz" relay. The expressions "overexcitation" and "overfluxing" refer to this same condition. Since temperature rise is proportional to the integral of power with respect to time (neglecting cooling processes) it follows that an inverse-time characteristic is useful, that is, *volts-per-hertz* versus *time*. Another approach is to use definite-time-delayed alarm or trip at specific per unit flux levels.

Heating Due to Current Harmonic Content (ANSI/IEEE, 1993): One effect of nonsinusoidal currents is to cause current rms magnitude (I_{RMS}) to be incorrect if the method of measurement is not "true-rms."

$$I_{RMS}^2 = \sum_{n=1}^{N} I_n^2 \qquad (9.2)$$

where n is the harmonic order, N is the highest harmonic of significant magnitude, and I_n is the harmonic current rms magnitude. If an overload relay determines the I^2R heating effect using the fundamental component of the current only [I_1], then it will underestimate the heating effect. Bear in mind that "true-rms" is only as good as the pass-band of the antialiasing filters and sampling rate, for numerical relays.

A second effect is heating due to high-frequency eddy-current loss in the copper or aluminum of the windings. The winding eddy-current loss due to each harmonic is proportional to the square of the harmonic amplitude and the square of its frequency as well. Mathematically,

$$P_{EC} = P_{EC\ RATED} \cdot \sum_{n=1}^{N} I_n^2 n^2 \qquad (9.3)$$

where P_{EC} is the winding eddy-current loss and $P_{EC-RATED}$ is the rated winding eddy-current loss (pure 60 Hz), and I_n is the nth harmonic current in per-unit based on the fundamental. Notice the fundamental difference between the effect of harmonics in Eq. (9.2) and their effect in Eq. (9.3). In the latter, higher harmonics have a proportionately greater effect because of the n^2 factor. IEEE Standard C57.110-1986 (R1992), *Recommended Practice for Establishing Transformer Capability When Supplying Nonsinusoidal Load Currents* gives two empirically-based methods for calculating the derating factor for a transformer under these conditions.

Heating Due to Solar Induced Currents: Solar magnetic disturbances cause geomagnetically induced currents (GIC) in the earth's surface (EPRI, 1993). These DC currents can be of the order of tens of amperes for tens of minutes, and flow into the neutrals of grounded transformers, biasing the core magnetization. The effect is worst in single-phase units and negligible in three-phase core-type units. The core saturation causes second-harmonic content in the current, resulting in increased *security* in second-harmonic-restrained transformer differential relays, but decreased *sensitivity*. Sudden gas pressure relays could provide the necessary alternative internal fault tripping. Another effect is increased stray heating in the transformer, protection for which can be accomplished using gas accumulation relays for transformers with conservator oil systems. Hot-spot tripping is not sufficient because the commonly used hot-spot simulation model does not account for GIC.

Load Tap-changer Overheating: Damaged current carrying contacts within an underload tap-changer enclosure can create excessive heating. Using this heating symptom, a way of detecting excessive wear is to install magnetically mounted temperature sensors on the tap-changer enclosure and on the main tank. Even though the method does not accurately measure the internal temperature at each location, the *difference* is relatively accurate, since the error is the same for each. Thus, excessive wear is indicated if a relay/monitor detects that the temperature difference has changed significantly over time.

Special Considerations

Current Transformers

Current transformer ratio selection and performance require special attention when applying transformer protection. Unique factors associated with transformers, including its winding ratios, magnetizing inrush current, and the presence of winding taps or load tap changers, are sources of difficulties in engineering a dependable and secure protection scheme for the transformer. Errors resulting from CT saturation and load-tap-changers are particularly critical for differential protection schemes where the currents from more than one set of CTs are compared. To compensate for the saturation/mismatch errors, overcurrent relays must be set to operate above these errors.

CT Current Mismatch: Under normal, non-fault conditions, a transformer differential relay should ideally have identical currents in the secondaries of all current transformers connected to the relay so that no current would flow in its operating coil. It is difficult, however, to match current transformer ratios exactly to the transformer winding ratios. This task becomes impossible with the presence of transformer off-load and on-load taps or load tap changers that change the voltage ratios of the transformer windings depending on system voltage and transformer loading.

The highest secondary current mismatch between all current transformers connected in the differential scheme must be calculated when selecting the relay operating setting. If time delayed overcurrent protection is used, the time delay setting must also be based on the same consideration. The mismatch calculation should be performed for maximum load and through-fault conditions.

CT Saturation: CT saturation could have a negative impact on the ability of the transformer protection to operate for internal faults (dependability) and not to operate for external faults (security).

For internal faults, dependability of the harmonic restraint type relays could be negatively affected if current harmonics generated in the CT secondary circuit due to CT saturation are high enough to restrain the relay. With a saturated CT, 2nd and 3rd harmonics predominate initially, but the even harmonics gradually disappear with the decay of the DC component of the fault current. The relay may then operate eventually when the restraining harmonic component is reduced. These relays usually include an instantaneous overcurrent element that is not restrained by harmonics, but is set very high (typically 20 times transformer rating). This element may operate on severe internal faults.

For external faults, security of the differentially connected transformer protection may be jeopardized if the current transformers' unequal saturation is severe enough to produce error current above the relay setting. Relays equipped with restraint windings in each current transformer circuit would be more secure. The security problem is particularly critical when the current transformers are connected to bus breakers rather than the transformer itself. External faults in this case could be of very high magnitude as they are not limited by the transformer impedance.

Magnetizing Inrush (Initial, Recovery, Sympathetic)

Initial: When a transformer is energized after being de-energized, a transient magnetizing or exciting current that may reach instantaneous peaks of up to 30 times full load current may flow. This can cause operation of overcurrent or differential relays protecting the transformer. The magnetizing current flows in only one winding, thus it will appear to a differentially connected relay as an internal fault.

Techniques used to prevent differential relays from operating on inrush include detection of current harmonics and zero current periods, both being characteristics of the magnetizing inrush current. The former takes advantage of the presence of harmonics, especially the second harmonic, in the magnetizing inrush current to restrain the relay from operation. The latter differentiates between the fault and inrush currents by measuring the zero current periods, which will be much longer for the inrush than for the fault current.

Recovery Inrush: A magnetizing inrush current can also flow if a voltage dip is followed by recovery to normal voltage. Typically, this occurs upon removal of an external fault. The magnetizing inrush is usually less severe in this case than in initial energization as the transformer was not totally de-energized prior to voltage recovery.

Sympathetic Inrush: A magnetizing inrush current can flow in an energized transformer when a nearby transformer is energized. The offset inrush current of the bank being energized will find a parallel path in the energized bank. Again, the magnitude is usually less than the case of initial inrush.

Both the recovery and sympathetic inrush phenomena suggest that restraining the transformer protection on magnetizing inrush current is required at all times, not only when switching the transformer in service after a period of de-energization.

Primary-Secondary Phase-Shift

For transformers with standard delta-wye connections, the currents on the delta and wye sides will have a 30° phase shift relative to each other. Current transformers used for traditional differential relays must be connected in wye-delta (opposite of the transformer winding connections) to compensate for the transformer phase shift.

Phase correction is often internally provided in microprocessor transformer protection relays via software virtual interposing CTs for each transformer winding and, as with the ratio correction, will depend upon the selected configuration for the restrained inputs. This allows the primary current transformers to all be connected in wye.

Turn-to-Turn Faults

Fault currents resulting from a turn-to-turn fault have low magnitudes and are hard to detect. Typically, the fault will have to evolve and affect a good portion of the winding or arc over to other parts of the transformer before being detected by overcurrent or differential protection relays.

For early detection, reliance is usually made on devices that can measure the resulting accumulation of gas or changes in pressure inside the transformer tank.

Through Faults

Through faults could have an impact on both the transformer and its protection scheme. Depending on their severity, frequency, and duration, through fault currents can cause mechanical transformer damage, even though the fault is somewhat limited by the transformer impedance.

For transformer differential protection, current transformer mismatch and saturation could produce operating currents on through faults. This must be taken into consideration when selecting the scheme, current transformer ratio, relay sensitivity, and operating time. Differential protection schemes equipped with restraining windings offer better security for these through faults.

Backup Protection

Backup protection, typically overcurrent or impedance relays applied to one or both sides of the transformer, perform two functions. One function is to backup the primary protection, most likely a differential relay, and operate in event of its failure to trip.

The second function is protection for thermal or mechanical damage to the transformer. Protection that can detect these external faults and operate in time to prevent transformer damage should be considered. The protection must be set to operate before the through-fault withstand capability of the transformer is reached. If, because of its large size or importance, only differential protection is applied to a transformer, clearing of external faults before transformer damage can occur by other protective devices must be ensured.

Special Applications

Shunt Reactors

Shunt reactor protection will vary depending on the type of reactor, size, and system application. Protective relay application will be similar to that used for transformers.

Differential relays are perhaps the most common protection method (Blackburn, 1987). Relays with separate phase inputs will provide protection for three single phase reactors connected together or for a single three phase unit. Current transformers must be available on the phase and neutral end of each winding in the three phase unit.

Phase and ground overcurrent relays can be used to back up the differential relays. In some instances, where the reactor is small and cost is a factor, it may be appropriate to use overcurrent relays as the only protection. The ground overcurrent relay would not be applied on systems where zero sequence current is negligible.

As with transformers, turn-to-turn faults are most difficult to detect since there is little change in current at the reactor terminals. If the reactor is oil filled, a sudden pressure relay will provide good protection. If the reactor is an ungrounded dry type, an overvoltage relay (device 59) applied between the reactor neutral and a set of broken delta connected voltage transformers can be used (ABB, 1994).

Negative sequence and impedance relays have also been used for reactor protection but their application should be carefully researched (ABB, 1994).

Zig-Zag Transformers

The most common protection for zig-zag (or grounding) transformers is three overcurrent relays that are connected to current transformers located on the primary phase bushings. These current transformers must be connected in delta to filter out unwanted zero sequence currents (ANSI/IEEE, 1985).

It is also possible to apply a conventional differential relay for fault protection. Current transformers in the primary phase bushings are paralleled and connected to one input. A neutral CT is used for the other input (Blackburn, 1987).

An overcurrent relay located in the neutral will provide backup ground protection for either of these schemes. It must be coordinated with other ground relays on the system.

Sudden pressure relays provide good protection for turn-to-turn faults.

Phase Angle Regulators and Voltage Regulators

Protection of phase angle and voltage regulators varies with the construction of the unit. Protection should be worked out with the manufacturer at the time of order to insure that current transformers are installed inside the unit in the appropriate locations to support planned protection schemes. Differential, overcurrent, and sudden pressure relays can be used in conjunction to provide adequate protection for faults (Blackburn, 1987; ABB, 1994).

Unit Systems

A unit system consists of a generator and associated step-up transformer. The generator winding is connected in wye with the neutral connected to ground through a high impedance grounding system. The step-up transformer low side winding on the generator side is connected delta to isolate the generator from system contributions to faults involving ground. The transformer high side winding is connected in wye and solidly grounded. Generally there is no breaker installed between the generator and transformer.

It is common practice to protect the transformer and generator with an overall transformer differential that includes both pieces of equipment. It may be appropriate to install an additional differential to protect only the transformer. In this case, the overall differential acts as secondary or backup protection for the transformer differential. There will most likely be another differential relay applied specifically to protect the generator.

A volts-per-hertz relay, whose pickup is a function of the ratio of voltage to frequency, is often recommended for overexcitation protection. The unit transformer may be subjected to overexcitation during generator startup and shutdown when it is operating at reduced frequencies or when there is major loss of load that may cause both overvoltage and overspeed (ANSI/IEEE, 1985).

As with other applications, sudden pressure relays provide sensitive protection for turn-to-turn faults that are typically not initially detected by differential relays.

Backup protection for phase faults can be provided by applying either impedance or voltage controlled overcurrent relays to the generator side of the unit transformer. The impedance relays must be connected to respond to faults located in the transformer (Blackburn, 1987).

Single Phase Transformers

Single phase transformers are sometimes used to make up three phase banks. Standard protection methods described earlier in this section are appropriate for single phase transformer banks as well. If one or both sides of the bank is connected in delta and current transformers located on the transformer bushings are to be used for protection, the standard differential connection cannot be used. To provide proper ground fault protection, current transformers from each of the bushings must be utilized (Blackburn, 1987).

Sustained Voltage Unbalance

During sustained unbalanced voltage conditions, wye-connected core type transformers without a delta-connected tertiary winding may produce damaging heat. In this situation, the transformer case may produce damaging heat from sustained circulating current. It is possible to detect this situation by using either a thermal relay designed to monitor tank temperature or applying an overcurrent relay connected to sense "effective" tertiary current (ANSI/IEEE, 1985).

Restoration

Power transformers have varying degrees of importance to an electrical system depending on their size, cost, and application, which could range from generator step-up to a position in the transmission/distribution system, or perhaps as an auxiliary unit.

When protective relays trip and isolate a transformer from the electric system, there is often an immediate urgency to restore it to service. There should be a procedure in place to gather system data at the time of trip as well as historical information on the individual transformer, so an informed decision can be made concerning the transformer's status. No one should re-energize a transformer when there is evidence of electrical failure.

It is always possible that a transformer could be incorrectly tripped by a defective protective relay or protection scheme, system backup relays, or by an abnormal system condition that had not been considered. Often system operators may try to restore a transformer without gathering sufficient evidence to determine the exact cause of the trip. An operation should always be considered as legitimate until proven otherwise.

The more vital a transformer is to the system, the more sophisticated the protection and monitoring equipment should be. This will facilitate the accumulation of evidence concerning the outage.

History — Daily operation records of individual transformer maintenance, service problems, and relayed outages should be kept to establish a comprehensive history. Information on relayed operations should include information on system conditions prior to the trip out. When no explanation for a trip is found, it is important to note all areas that were investigated. When there is no damage determined, there should still be a conclusion as to whether the operation was correct or incorrect. Periodic gas analysis provides a record of the normal combustible gas value.

Oscillographs, Event Recorder, Gas Monitors — System monitoring equipment that initiates and produces records at the time of the transformer trip usually provide information necessary to determine if there was an electrical short-circuit involving the transformer or if it was a "through-fault" condition.

Date of Manufacture — Transformers manufactured before 1980 were likely not designed or constructed to meet the severe through-fault conditions outlined in ANSI/IEEE C57.109, *IEEE Guide for Transformer Through-Fault Current Duration* (1985). Maximum through-fault values should be calculated and compared to short-circuit values determined for the trip out. Manufacturers should be contacted to obtain documentation for individual transformers in conformance with ANSI/IEEE C57.109.

Magnetizing Inrush — Differential relays with harmonic restraint units are typically used to prevent trip operations upon transformer energizing. However, there are nonharmonic restraint differential relays in service that use time delay and/or percentage restraint to prevent trip on magnetizing inrush. Transformers so protected may have a history of falsely tripping on energizing inrush which may lead system operators to attempt restoration without analysis, inspection, or testing. There is always the possibility that an electrical fault can occur upon energizing which is masked by historical data.

Relay harmonic restraint circuits are either factory set at a threshold percentage of harmonic inrush or the manufacturer provides predetermined settings that should prevent an unwanted operation upon transformer energization. Some transformers have been manufactured in recent years using a grain-oriented steel and a design that results in very low percentages of the restraint harmonics in the inrush current. These values are, in some cases, less than the minimum manufacture recommended threshold settings.

Relay Operations — Transformer protective devices not only trip but prevent reclosing of all sources energizing the transformer. This is generally accomplished using an auxiliary "lockout" relay. The lockout relay requires manual resetting before the transformer can be energized. This circuit encourages manual inspection and testing of the transformer before reenergization decisions are made.

Incorrect trip operations can occur due to relay failure, incorrect settings, or coordination failure. New installations that are in the process of testing and wire-checking are most vulnerable. Backup relays, by design, can cause tripping for upstream or downstream system faults that do not otherwise clear properly.

References

Blackburn, J.L., *Protective Relaying: Principles and Applications,* Marcel Decker, Inc., New York, 1987.
Mason, C.R., *The Art and Science of Protective Relaying,* John Wiley & Sons, New York, 1996.
IEEE Guide for Diagnostic Field Testing of Electric Power Apparatus — Part 1: Oil Filled Power Transformers, Regulators, and Reactors, ANSI/IEEE Std. 62-199S.
Guide for the Interpretation of Gases Generated in oil-Immersed Transformers, ANSI/IEEE C57.104-1991.
IEEE Guide for Loading Mineral Oil-Immersed Transformers, ANSI/IEEE C57.91-1995.
IEEE Guide for Protective Relay Applications to Power Transformers, ANSI/IEEE C37.91-1985.
IEEE Guide for Transformer Through Fault Current Duration, ANSI/IEEE C57.109-1985.
Recommended Practice for Establishing Transformer Capability When Supplying Nonsinusoidal Load Currents, IEEE Std. C57.110-1986(R1992).
IEEE Standard General Requirements for Liquid-Immersed Distribution, Power, and Regulating Transformers, ANSI/IEEE C57.12.00-1993.
Protective Relaying, Theory & Application, ABB, Marcel Dekker, Inc., New York, 1994.
Protective Relays Application Guide, GEC Measurements, Stafford, England, 1975.
Rockefeller, G., et al., Differential relay transient testing using EMTP simulations, paper presented to the 46th annual Protective Relay Conference (Georgia Tech.), April 29–May 1, 1992.
Solar magnetic disturbances/geomagnetically-induced current and protective relaying, *Electric Power Research Institute* Report TR-102621, Project 321-04, August 1993.
Warrington, A.R. van C., *Protective Relays, Their Theory and Practice,* Vol. 1, Wiley, New York, 1963, Vol. 2, Chapman and Hall Ltd., London, 1969.

9.2 The Protection of Synchronous Generators

Gabriel Benmouyal

In an apparatus protection perspective, generators constitute a special class of power network equipment because faults are very rare but can be highly destructive and therefore very costly when they occur. If for most utilities, generation integrity must be preserved by avoiding erroneous tripping, removing a generator in case of a serious fault is also a primary if not an absolute requirement. Furthermore, protection has to be provided for out-of-range operation normally not found in other types of equipment such as overvoltage, overexcitation, limited frequency or speed range, etc.

It should be borne in mind that, similar to all protective schmes, there is to a certain extent a "philosophical approach" to generator protection and all utilities and all protective engineers do not have the same approach. For instance, some functions like overexcitation, backup impedance elements, loss-of-synchronism, and even protection against inadvertant energization may not be applied by some organizations and engineers. It should be said, however, that with the digital multifunction generator protective packages presently available, a complete and extensive range of functions exists within the same "relay": and economic reasons for not installing an additional protective element is a tendancy which must disappear.

The nature of the prime mover will have some definite impact on the protective functions implemented into the system. For instance, little or no concern at all will emerge when dealing with the abnormal frequency operation of hydaulic generators. On the contrary, protection against underfrequency operation of steam turbines is a primary concern.

The sensitivity of the motoring protection (the capacity to measure very low levels of negative real power) becomes an issue when dealing with both hydro and steam turbines. Finally, the nature of the prime mover will have an impact on the generator tripping scheme. When delayed tripping has no detrimental effect on the generator, it is common practice to implement sequential tripping with steam turbines as described later.

The purpose of this article is to provide an overview of the basic principles and schemes involved in generator protection. For further information, the reader is invited to refer to additional resources dealing with generator protection. The ANSI/IEEE guides (ANSI/IEEE, C37.106, C37.102, C37.101) are particularly recommended. The *IEEE Tutorial on the Protection of Synchronous Generators* (IEEE, 1995) is a detailed presentation of North American practices for generator protection. All these references have been a source of inspiration in this writing.

Review of Functions

Table 9.1 provides a list of protective relays and their functions most commonly found in generator protection schemes. These relays are implemented as shown on the single-line diagram of Fig. 9.2.

As shown in the Relay Type column, most protective relays found in generator protection schemes are not specific to this type of equipment but are more generic types.

Differential Protection for Stator Faults (87G)

Protection against stator phase faults are normally covered by a high-speed differential relay covering the three phases separately. All types of phase faults (phase-phase) will be covered normally by this type of protection, but the phase-ground fault in a high-impedance grounded generator will not be covered. In this case, the phase current will be very low and therefore below the relay pickup.

Contrary to transformer differential applications, no inrush exists on stator currents and no provision is implemented to take care of overexcitation. Therefore, stator differential relays do not include harmonic restraint (2nd and 5th harmonic). Current transformer saturation is still an issue, however, particularly in generating stations because of the high X/R ratio found near generators.

TABLE 9.1 Most Commonly Found Relays for Generator Protection

Identification Number	Function Description	Relay Type
87G	Generator phase phase windings protection	Differential protection
87T	Step-up transformer differential protection	Differential protection
87U	Combined differential transformer and generator protection	Differential protection
40	Protection against the loss of field voltage or current supply	Offset mho relay
46	Protection against current imbalance. Measurement of phase negative sequence current	Time-overcurrent relay
32	Anti-motoring protection	Reverse-power relay
24	Overexcitation protection	Volt/Hertz relay
59	Phase overvoltage protection	Overvoltage relay
60	Detection of blown voltage transformer fuses	Voltage balance relay
81	Under- and overfrequency protection	Frequency relays
51V	Backup protection against system faults	Voltage controlled or voltage-restrained time overcurrent relay
21	Backup protection against system faults	Distance relay
78	Protection against loss of synchronization	Combination of offset mho and blinders

The most common type of stator differential is the percentage differential, the main characteristics of which are represented in Fig. 9.3.

For a stator winding, as shown in Fig. 9.4, the restraint quantity will very often be the absolute sum of the two incoming and outgoing currents as in:

$$Irestraint = \frac{|IA\_in| + |IA\_out|}{2}, \quad (9.4)$$

whereas the operate quantity will be the absolute value of the difference:

$$Ioperate = |IA\_in - IA\_out| \quad (9.5)$$

The relay will output a fault condition when the following inequality is verified:

$$Irestraint \geq K \bullet Ioperate \quad (9.6)$$

where K is the differential percentage. The dual and variable slope characteristics will intrinsically allow CT saturation for an external fault without the relay picking up.

An alternative to the percentage differential relay is the high-impedance differential relay, which will also naturally surmount any CT saturation. For an internal fault, both currents will be forced into a high-impedance voltage relay. The differential relay will pickup when the tension across the voltage element gets above a high-set threshold. For an external fault with CT saturation, the saturated CT will constitute a low-impedance path in which the current from the other CT will flow, bypassing the high-impedance voltage element which will not pick up.

Backup protection for the stator windings will be provided most of the time by a transformer differential relay with harmonic restraint, the zone of which (as shown in Fig. 9.2) will cover both the generator and the step-up transformer.

An impedance element partially or totally covering the generator zone will also provide backup protection for the stator differential.

Power System Protection

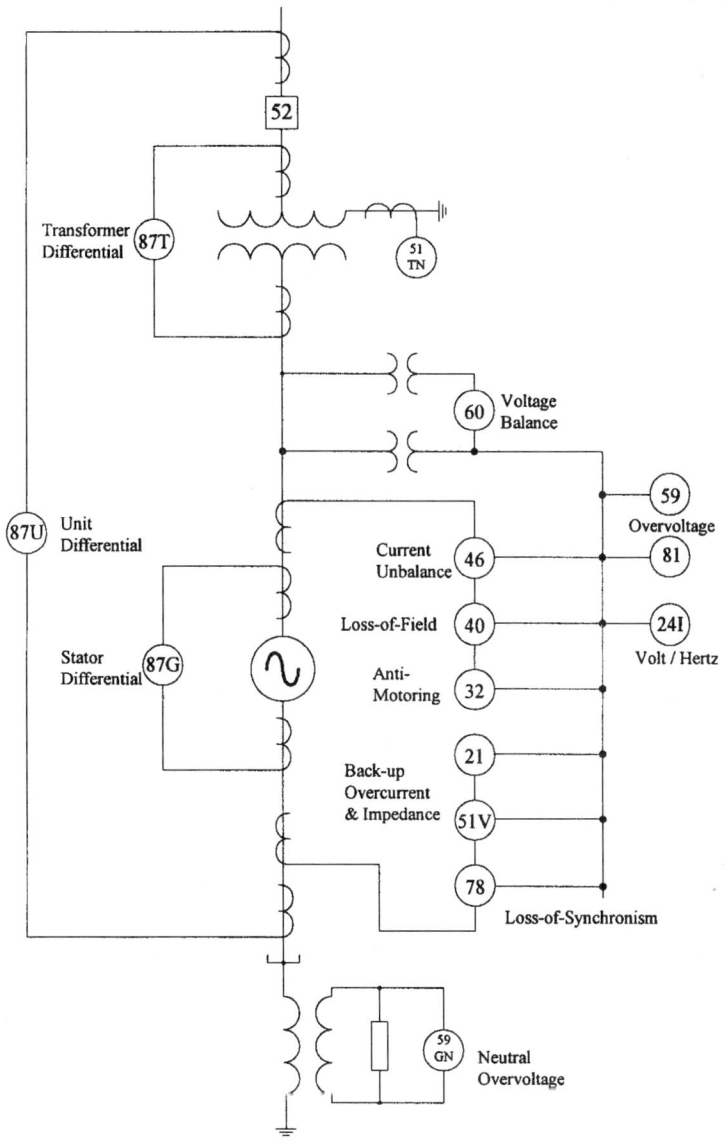

FIGURE 9.2 Typical generator-transformer protection scheme.

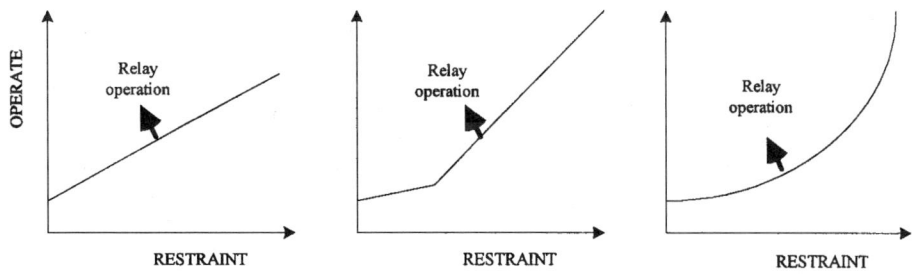

FIGURE 9.3 Single, dual, and variable-slope percentage differential characteristics.

IA_in IA_out

FIGURE 9.4 Stator winding current configuration.

Protection Against Stator Winding Ground Fault

Protection against stator-to-ground fault will depend to a great extent upon the type of generator grounding. Generator grounding is necessary through some impedance in order to reduce the current level of a phase-to-ground fault. With solid generator grounding, this current will reach destructive levels. In order to avoid this, at least low impedance grounding through a resistance or a reactance is required. High-impedance through a distribution transformer with a resistor connected across the secondary winding will limit the current level of a phase-to-ground fault to a few primary amperes.

The most common and minimum protection against a stator-to-ground fault with a high-impedance grounding scheme is an overvoltage element connected across the grounding transformer secondary, as shown in Fig. 9.5.

For faults very close to the generator neutral, the overvoltage element will not pick up because the voltage level will be below the voltage element pick-up level. In order to cover 100% of the stator windings, two techniques are readily available:

1. use of the third harmonic generated at the neutral and generator terminals, and
2. voltage injection technique.

Looking at Fig. 9.6, a small amount of third harmonic voltage will be produced by most generators at their neutral and terminals. The level of these third harmonic voltages depends upon the generator operating point as shown in Fig. 9.6a. Normally they would be higher at full load. If a fault develops near the neutral, the third harmonic neutral voltage will approach zero and the terminal voltage will increase. However, if a fault develops near the terminals, the terminal third harmonic voltage will reach zero and the neutral voltage will increase. Based on this, three possible schemes have been devised. The relays available to cover the three possible choices are:

1. Use of a third harmonic undervoltage at the neutral. It will pick up for a fault at the neutral.
2. Use of a third harmonic overvoltage at the terminals. It will pick up for a fault near the neutral.
3. The most sensitive schemes are based on third harmonic differential relays that monitor the ratio of third harmonic at the neutral and the terminals (Yin et al., 1990).

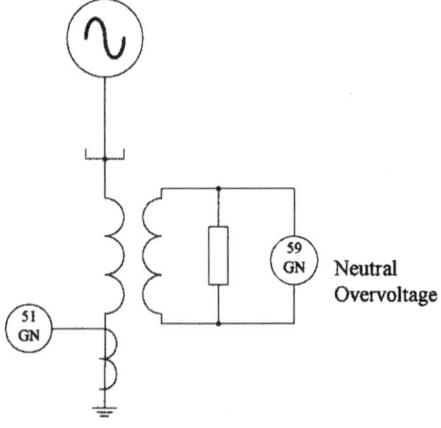

FIGURE 9.5 Stator-to-ground neutral overvoltage scheme.

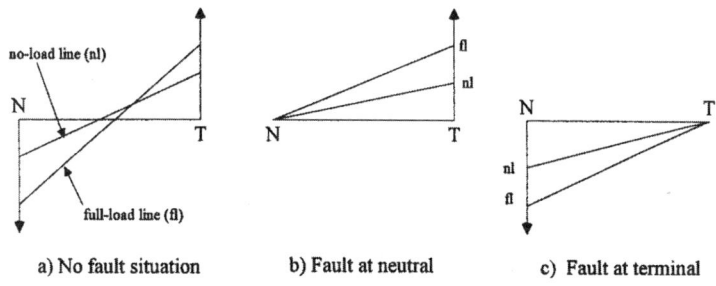

a) No fault situation   b) Fault at neutral   c) Fault at terminal

FIGURE 9.6 Third harmonic on neutral and terminals.

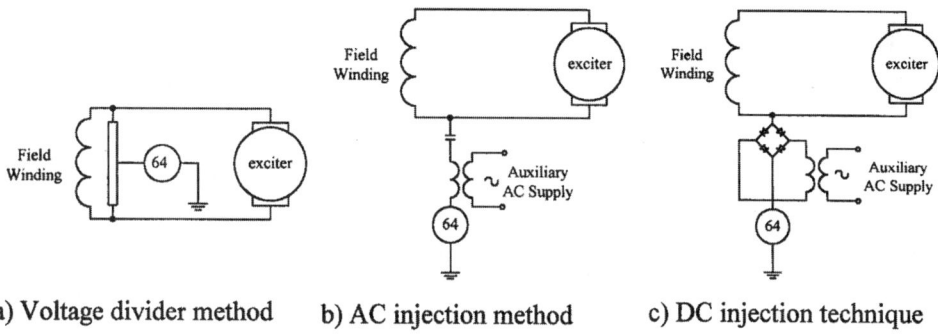

a) Voltage divider method   b) AC injection method   c) DC injection technique

FIGURE 9.7 Various techniques for field-ground protection.

Field Ground Protection

A generator field circuit (field winding, exciter, and field breaker) is a DC circuit that does not need to be grounded. If a first earth fault occurs, no current will flow and the generator operation will not be affected. If a second ground fault at a different location occurs, a current will flow that is high enough to cause damage to the rotor and the exciter. Furthermore, if a large section of the field winding is short-circuited, a strong imbalance due to the abnormal air-gap fluxes could result on the forces acting on the rotor with a possibility of serious mechanical failure. In order to prevent this situation, a number of protecting devices exist. Three principles are depicted in Fig. 9.7.

The first technique (Fig. 9.7a) involves connecting a resistor in parallel with the field winding. The resistor centerpoint is connected the ground through a current sensitive relay. If a field circuit point gets grounded, the relay will pick up by virtue of the current flowing through it. The main shortcoming of this technique is that no fault will be detected if the field winding centerpoint gets grounded.

The second technique (Fig. 9.7b) involves applying an AC voltage across one point of the field winding. If the field winding gets grounded at some location, an AC current will flow into the relay and causes it to pick up.

The third technique (Fig. 9.7c) involves injecting a DC voltage rather than an AC voltage. The consequence remains the same if the field circuit gets grounded at some point.

The best protection against field-ground faults is to move the generator out of service as soon as the first ground fault is detected.

Loss-of-Excitation Protection (40)

A loss-of-excitation on a generator occurs when the field current is no longer supplied. This situation can be triggered by a variety of circumstances and the following situation will then develop:

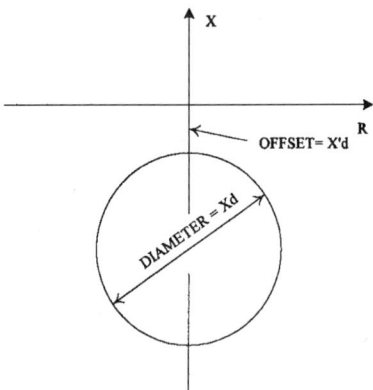

FIGURE 9.8 Loss-of-excitation offset-mho characteristic.

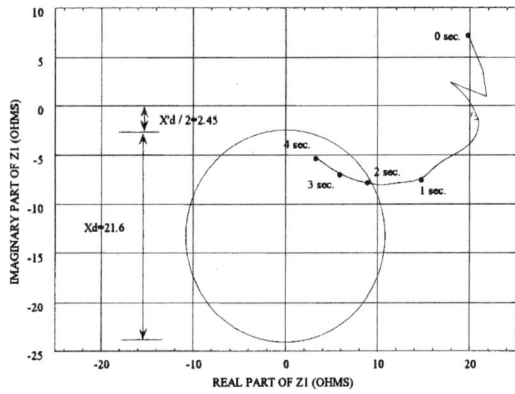

FIGURE 9.9 Loss-of-field positive sequence impedance trajectory.

1. When the field supply is removed, the generator real power will remain almost constant during the next seconds. Because of the drop in the excitation voltage, the generator output voltage drops gradually. To compensate for the drop in voltage, the current increases at about the same rate.
2. The generator then becomes underexcited and it will absorb increasingly negative reactive power.
3. Because the ratio of the generator voltage over the current becomes smaller and smaller with the phase current leading the phase voltage, the generator positive sequence impedance as measured at its terminals will enter the impedance plane in the second quadrant. Experience has shown that the positive sequence impedance will settle to a value between Xd and Xq.

The most popular protection against a loss-of-excitation situation uses an offset-mho relay as shown in Fig. 9.8 (IEEE, 1989). The relay is supplied with generator terminals voltages and currents and is normally associated with a definite time delay. Many modern digital relays will use the positive sequence voltage and current to evaluate the positive sequence impedance as seen at the generator terminal.

Figure 9.9 shows the digitally emulated positive sequence impedance trajectory of a 200 MVA generator connected to an infinite bus through an 8% impedance transformer when the field voltage was removed at 0 second time.

Current Imbalance (46)

Current imbalance in the stator with its subsequent production of negative sequence current will be the cause of double-frequency currents on the surface of the rotor. This, in turn, may cause excessive overheating of the rotor and trigger substantial thermal and mechanical damages (due to temperature effects).

The reasons for temporary or permanent current imbalance are numerous:

- system asymmetries
- unbalanced loads
- unbalanced system faults or open circuits
- single-pole tripping with subsequent reclosing

The energy supplied to the rotor follows a purely thermal law and is proportional to the square of the negative sequence current. Consequently, a thermal limit K is reached when the following integral equation is solved:

$$K = \int_0^t I_2^2 \, dt \qquad (9.7)$$

In this equation, we have:

K = constant depending upon the generator design and size
I_2 = RMS value of negative sequence current
t = time

The integral equation can be expressed as an inverse time-current characteristic where the maximum time is given as the negative sequence current variable:

$$t = \frac{K}{I_2^2} \qquad (9.8)$$

In this expression the negative sequence current magnitude will be entered most of the time as a percentage of the nominal phase current and integration will take place when the measured negative sequence current becomes greater than a percentage threshold.

Thermal capability constant, K, is determined by experiment by the generator manufacturer. Negative sequence currents are supplied to the machine on which strategically located thermocouples have been installed. The temperature rises are recorded and the thermal capability is inferred.

Forty-six (46) relays can be supplied in all three technologies (electromechanical, static, or digital). Ideally the negative sequence current should be measured in rms magnitude. Various measurement principles can be found. Digital relays could measure the fundamental component of the negative sequence current because this could be the basic principle for phasor measurement. Figure 9.10 represents a typical relay characteristic.

Anti-Motoring Protection (32)

A number of situations exist where a generator could be driven as a motor. Anti-motoring protection will more specifically apply in situations where the prime-mover supply is removed for a generator supplying a network at synchronous speed with the field normally excited. The power system will then drive the generator as a motor.

A motoring condition may develop if a generator is connected improperly to the power system. This will happen if the generator circuit breaker is closed inadvertently at some speed less than synchronous speed. Typical situations are when the generator is on turning gear, slowing down to a standstill, or has reached standstill. This motoring condition occurs during what is called "generator inadvertent energization." The protection schemes that respond to this situation are different and will be addressed later in this article.

Motoring will cause adverse effects, particularly in the case of steam turbines. The basic phenomenon is that the rotation of the turbine rotor and the blades in a steam environment will cause windage losses.

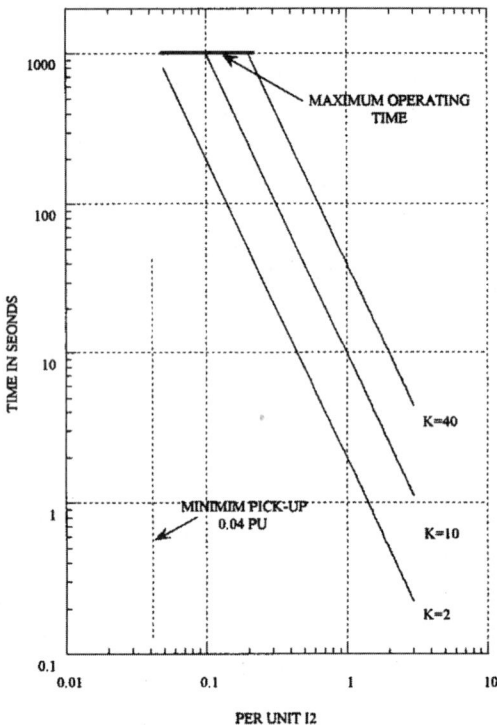

FIGURE 9.10 Typical static or digital time-inverse 46 curve.

Windage losses are a function of rotor diameter, blade length, and are directly proportional to the density of the enclosed steam. Therefore, in any situation where the steam density is high, harmful windage losses could occur. From the preceding discussion, one may conclude that the anti-motoring protection is more of a prime-mover protection than a generator protection.

The most obvious means of detecting motoring is to monitor the flow of real power into the generator. If that flow becomes negative below a preset level, then a motoring condition is detected. Sensitivity and setting of the power relay depends upon the energy drawn by the prime mover considered now as a motor.

With a gas turbine, the large compressor represents a substantial load that could reach as high as 50% of the unit nameplate rating. Sensitivity of the power relay is not an issue and is definitely not critical. With a diesel type engine (with no firing in the cylinders), load could reach as high as 25% of the unit rating and sensitivity, once again, is not critical. With hydroturbines, if the blades are below the tail-race level, the motoring energy is high. If above, the reverse power gets as low as 0.2 to 2% of the rated power and a sensitive reverse power relay is then needed. With steam turbines operating at full vacuum and zero steam input, motoring will draw 0.5 to 3% of unit rating. A sensitive power relay is then required.

Overexcitation Protection (24)

When generator or step-up transformer magnetic core iron becomes saturated beyond rating, stray fluxes will be induced into nonlaminated components. These components are not designed to carry flux and therefore thermal or dielectric damage can occur rapidly.

In dynamic magnetic circuits, voltages are generated by the Lenz Law:

$$V = K \frac{d\phi}{dt} \tag{9.9}$$

Power System Protection

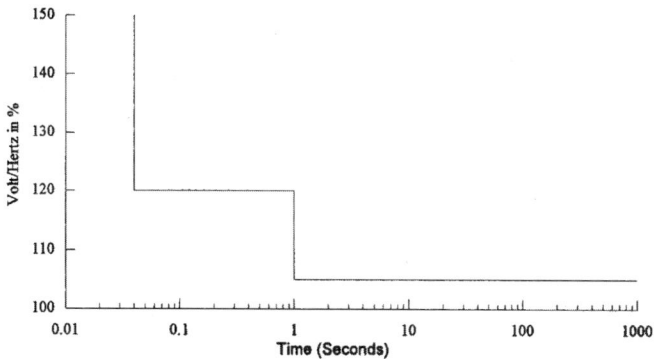

FIGURE 9.11 Dual definite-time characteristic.

Measured voltage can be integrated in order to get an estimate of the flux. Assuming a sinusoidal voltage of magnitude Vp and frequency f, and integrating over a positive or negative half-cycle interval:

$$\phi = \frac{1}{K}\int_0^{T/2} V_p \sin(\omega t + \theta)dt = \frac{V_p}{2\pi f K}\left(-\cos \omega t\right)\Big|_0^{T/2} \quad (9.10)$$

one derives an estimate of the flux that is proportional to the value of peak voltage over the frequency. This type of protection is then called volts per hertz.

$$\phi \approx \frac{V_p}{f} \quad (9.11)$$

The estimated value of the flux can then be compared to a maximum value threshold. With static technology, volts per hertz relays would practically integrate the monitored voltage over a positive or negative (or both) half-cycle period of time and develop a value that would be proportional to the flux. With digital relays, since measurement of the frequency together with the magnitudes of phase voltages are continuously available, a direct ratio computation as shown in Eq. (9.11) would be performed.

ANSI/IEEE standard limits are 1.05 pu for generators and 1.05 for transformers (on transformer secondary base, at rated load, 0.8 power factor or greater; 1.1 pu at no-load). It has been traditional to supply either definite time or inverse-time characteristics as recommended by the ANSI/IEEE guides and standards. Fig. 9.11 represents a typical dual definite-time characteristic whereas Fig. 9.12 represents a combined definite and inverse-time characteristic.

One of the primary requirements of a volt/hertz relay is that it should measure both voltage magnitude and frequency over a broad range of frequency.

Overvoltage (59)

An overvoltage condition could be encountered without exceeding the volt/hertz limits. For that reason, an overvoltage relay is recommended. Particularly for hydro-units, C37-102 recommends both an instantaneous and an inverse element. The instantaneous should be set to 130 to 150% of rated voltage and the inverse element should have a pick-up voltage of 110% of the rated voltage. Coordination with the voltage regulator should be verified.

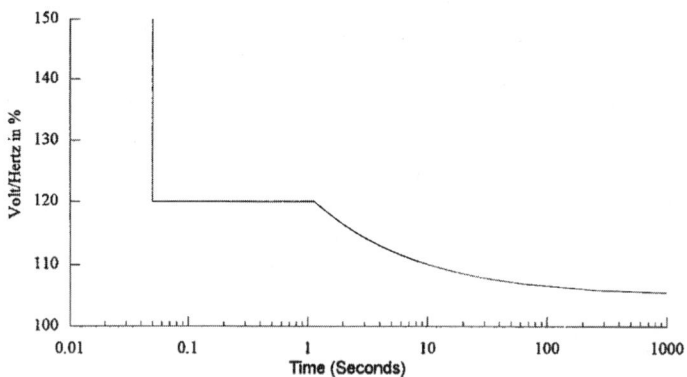

FIGURE 9.12 Combined definite and inverse-time characteristics.

Voltage Imbalance Protection (60)

The loss of a voltage phase signal can be due to a number of causes. The primary cause for this nuisance is a blown-out fuse in the voltage transformer circuit. Other causes can be a wiring error, a voltage transformer failure, a contact opening, a misoperation during maintenance, etc.

Since the purpose of these VTs is to provide voltage signals to the protective relays and the voltage regulator, the immediate effect of a loss of VT signal will be the possible misoperation of some protective relays and the cause for generator overexcitation by the voltage regulator. Among the protective relays to be impacted by the loss of VT signal are:

- Function 21: Distance relay. Backup for system and generator zone phase faults.
- Function 32: Reverse power relay. Anti-motoring function, sequential tripping and inadvertent energization functions.
- Function 40: Loss-of-field protection.
- Function 51V: Voltage-restrained time overcurrent relay.

Normally these functions should be blocked if a condition of fuse failure is detected.

It is common practice for large generators to use two sets of voltage transformers for protection, voltage regulation, and measurement. Therefore, the most common practice for loss of VT signals detection is to use a voltage balance relay as shown in Fig. 9.13 on each pair of secondary phase voltage. When a fuse blows, the voltage relationship becomes imbalanced and the relay operates. Typically, the voltage imbalance will be set at around 15%.

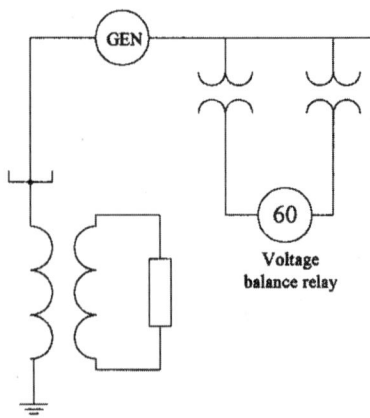

FIGURE 9.13 Example of voltage balance relay.

Power System Protection

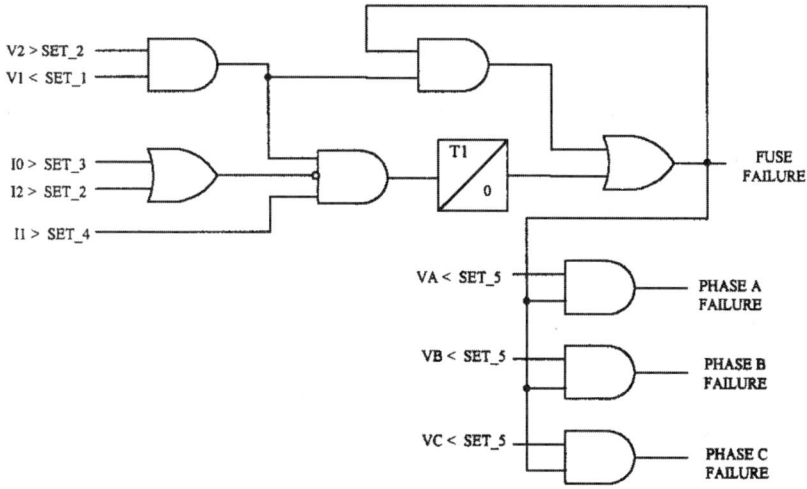

FIGURE 9.14 Symmetrical component implementation of fuse failure detection.

The advent of digital relays has allowed the use of sophisticated algorithms based on symmetrical components to detect the loss of VT signal. When a situation of loss of one or more of the VT signals occurs, the following conditions develop:

- there will be a drop in the positive sequence voltage accompanied by an increase in the negative sequence voltage magnitude. The magnitude of this drop will depend upon the number of phases impacted by a fuse failure.
- in case of a loss of VT signal and contrary to a fault condition, there should not be any change in the current's magnitudes and phases. Therefore, the negative and zero sequence currents should remain below a small tolerance value. A fault condition can be distinguished from a loss of VT signal by monitoring the changes in the positive and negative current levels. In case of a loss of VT signals, these changes should remain below a small tolerance level.

All the above conditions can be incorporated into a complex logic scheme to determine if indeed a there has been a condition of loss of VT signal or a fault. Figure 9.14 represents the logic implementation of a voltage transformer single and double fuse failure based on symmetrical components.

If the following conditions are met in the same time (and condition) during a time delay longer than T1:

- the positive sequence voltage is below a voltage set-value SET_1,
- the negative sequence voltage is above a voltage set-value SET_2,
- there exists a small value of current such that the positive sequence current I1 is above a small set-value SET_4 and the negative and zero sequence currents I2 and I2 do not exceed a small set-value SET_3,

then a fuse failure condition will pick up to one and remain in that state thanks to the latch effect. Fuse failure of a specific phase can be detected by monitoring the level voltage of each phase and comparing it to a set-value SET_5. As soon as the positive sequence voltage returns to a value greater than the set-value SET_1 and the negative sequence voltage disappears, the fuse failure condition returns to a zero state.

System Backup Protection (51V and 21)

Generator backup protection is not applied to generator faults but rather to system faults that have not been cleared in time by the system primary protection, but which require generator removal in order for the fault to be eliminated. By definition, these are time-delayed protective functions that must coordinate with the primary protective system.

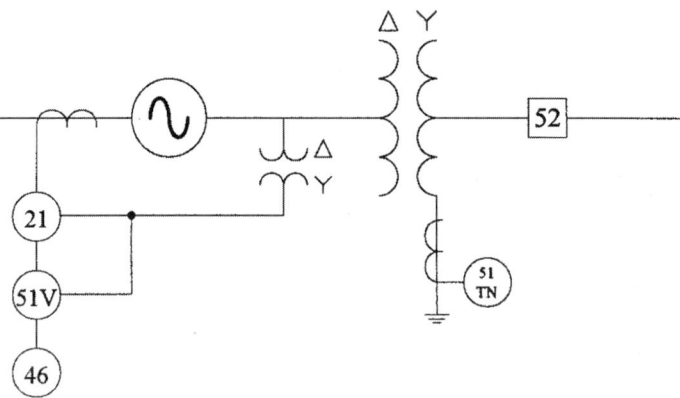

FIGURE 9.15 Backup protection basic scheme.

System backup protection (Fig. 9.15) must provide protection for both phase faults and ground faults. For the purpose of protecting against phase faults, two solutions are most commonly applied: the use of overcurrent relays with either voltage restraint or voltage control, or impedance-type relays.

The basic principle behind the concept of supervising the overcurrent relay by voltage is that a fault external to the generator and on the system will have the effect of reducing the voltage at the generator terminal. This effect is being used in both types of overcurrent applications: the voltage controlled overcurrent relay will block the overcurrent element unless the voltage gets below a pre-set value, and the voltage restraint overcurrent element will have its pick-up current reduced by an amount proportional to the voltage reduction (see Fig. 9.16).

The impedance type backup protection could be applied to the low or high side of the step-up transformer. Normally, three 21 elements will cover all types of phase faults on the system as in a line relay.

As shown in Fig. 9.17, a reverse offset is allowed in the mho element in order for the backup to partially or totally cover the generator windings.

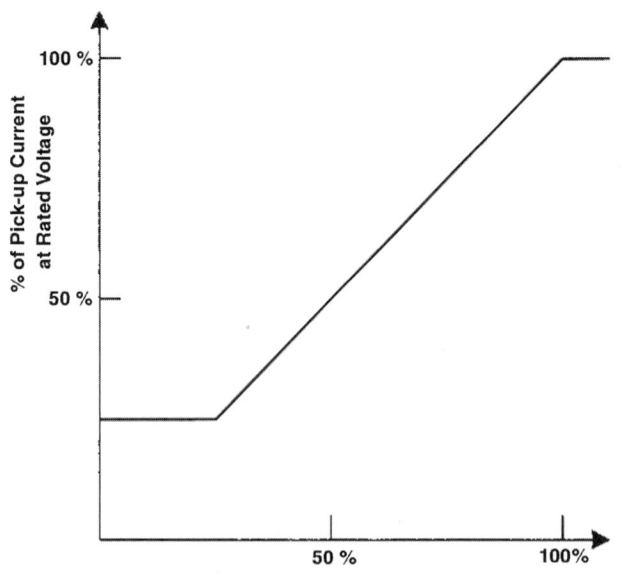

FIGURE 9.16 Voltage restraint overcurrent relay principle.

Power System Protection

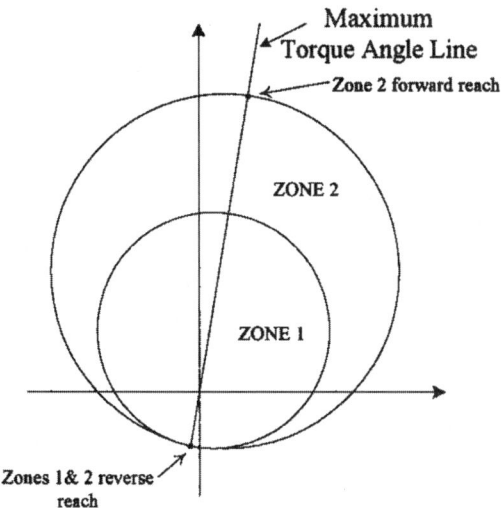

FIGURE 9.17 Typical 21 elements application.

Out-of-Step Protection

When there is an equilibrium between generation and load on an electrical network, the network frequency will be stable and the internal angle of the generators will remain constant with respect to each other. If an imbalance (loss of generation, sudden addition of load, network fault, etc.) occurs, however, the internal angle of a generator will undergo some changes and two situations might develop: a new stable state will be reached after the disturbance has faded away, or the generator internal angle will not stabilize and the generator will run synchronously with respect to the rest of the network (moving internal angle and different frequency). In the latter case, an out-of-step protection is implemented to detect the situation.

That principle can be visualized by considering the two-source network of Fig. 9.18.

If the angle between the two sources is θ and the ratio between the voltage magnitudes is n = E_G/E_S, then the positive sequence impedance seen from location will be:

$$Z_R = \frac{n(Z_G + Z_T + Z_S)(n - \cos\theta - j\sin\theta)}{(n - \cos\theta)^2 + \sin^2\theta} - Z_G. \qquad (9.12)$$

If n is equal to one, Eq. (9.12) simplifies to:

$$Z_R = \frac{n(Z_G + Z_T + Z_S)\left(1 - j\,cotg\dfrac{\theta}{2}\right)}{2} - Z_G \qquad (9.13)$$

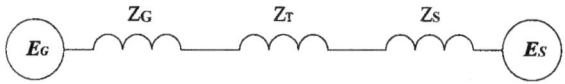

FIGURE 9.18 Elementary two-source network.

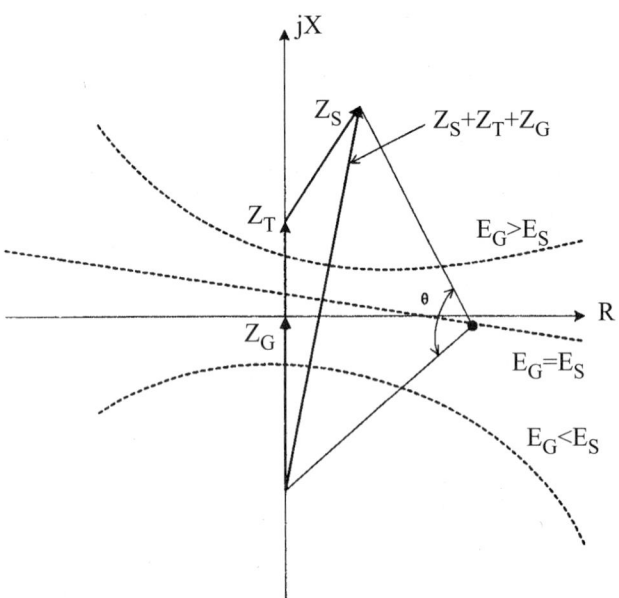

FIGURE 9.19 Impedance locus for different source angles.

The impedance locus represented by this equation is a straight line, perpendicular to and crossing the vector $Z_s + Z_T + Z_G$ at its middle point. If n is different from 1, the loci become circles as shown in Fig. 9.19. The angle θ between the two sources is the angle between the two segments joining Z_R to the base of Z_G and the summit of Z_S. Normally, that angle will take a small value. In an out-of-step condition, it will assume a bigger value and when it reaches 180°, it crosses $Z_s + Z_T + Z_G$ at its middle point.

Normally, because of the machine's inertia, the impedance Z_R moves slowly. The phenomenon can be taken advantage of and an out-of-step condition will very often be detected by the combination a mho relay and two blinders as shown in Fig. 9.20. In this application, an out-of-step condition will be assumed to be detected when the impedance locus enters the mho circle and remains between the two blinders for an interval of time longer than a preset definite time delay. Implicit in this scheme is the fact that the angle between the two sources is assumed to take a large value when Zr crosses the blinders. Implementation of an out-of-step protection will normally require some careful studies and eventually will require some stability simulations in order to determine the nature and the locus of the stable and the unstable swings. One of the paramount requirement of an out-of-step protection is not to trip the generator in case of a stable wing.

Abnormal Frequency Operation of Turbine-Generator

Although it is not a concern for hydraulic generators, the protection against abnormal frequency operation becomes an issue with steam turbine-graters. If the turbine is rotated at a frequency other than synchronous, the blades in the low pressure turbine element could resonate at their natural frequency. Blading mechanical fatigue could result with subsequent damage and failure.

Figure 9.21 (ANSI C37.106) represents a typical steam turbine operating limitation curve. Continuous operation is allowed around 60 Hz. Time-limited zones exist above and below the continuous operation regions. Prohibited operation regions lie beyond.

With the advent of modern generator microprocessor-based relays (IEEE, 1989), there does not seem to be a consensus emerging among the relay and turbine manufacturers, regarding the digital implementation of underfrequency turbine protection. The following points should, however, be taken into account:

Power System Protection

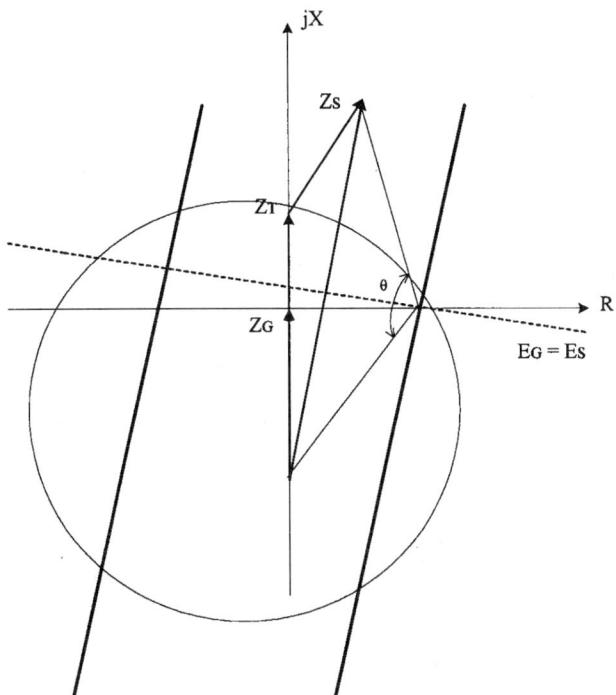

FIGURE 9.20 Out-of-step mho detector with blinders.

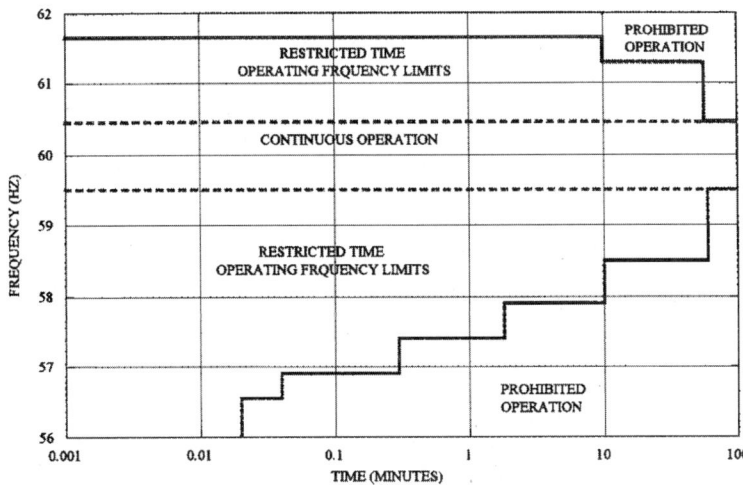

FIGURE 9.21 Typical steam turbine operating characteristic. (Modified from ANSI/IEEE C37.106-1987, Figure 6.)

- Measurement of frequency is normally available on a continuous basis and over a broad frequency range. Precision better than 0.01 Hz in the frequency measurement has been achieved.
- In practically all products, a number of independent over- or under-frequency definite time functions can be combined to form a composite curve.

Therefore, with digital technology, a typical over/underfrequency scheme, as shown in Fig. 9.22, comprising one definite-time over-frequency and two definite-time under-frequency elements is readily implementable.

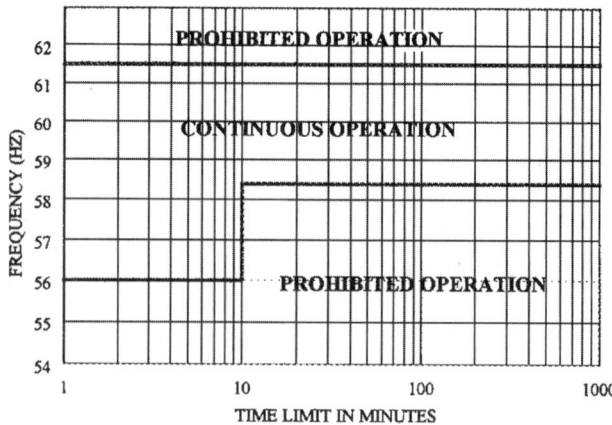

FIGURE 9.22 Typical abnormal frequency protection characteristic.

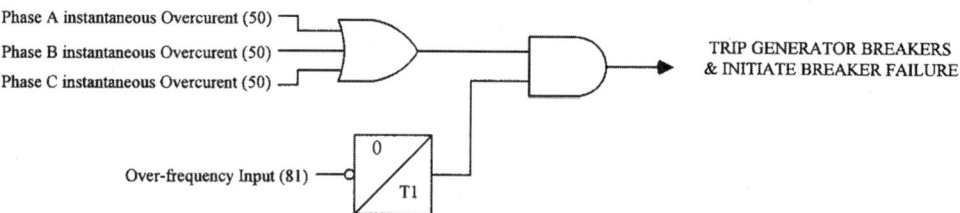

FIGURE 9.23 Frequency supervised overcurrent inadvertent energizing protection.

Protection Against Accidental Energization

A number of catastrophic failures have occurred in the past when synchronous generators have been accidentally energized while at standstill. Among the causes for such incidents were human errors, breaker flashover, or control circuitry malfunction.

A number of protection schemes have been devised to protect the generator against inadvertent energization. The basic principle is to monitor the out-of-service condition and to detect an accidental energizing immediately following that state. As an example, Fig. 9.23 shows an application using an over-frequency relay supervising three single phase instantaneous overcurrent elements. When the generator is put out of service or the over-frequency element drops out, the timer will pick up. If inadvertent energizing occurs, the over-frequency element will pick up, but because of the timer drop-out delay, the instantaneous overcurrent elements will have the time to initiate the generator breakers opening. The supervision could also be implemented using a voltage relay.

Accidental energizing caused by a single or three-phase breaker flashover occurring during the generator synchronizing process will not be detected by the logic of Fig. 9.23. In such an instance, by the time the generator has been closed to the synchronous speed, the overcurrent element outputs would have been blocked.

Generator Breaker Failure

Generator breaker failure follows the general pattern of the same function found in other applications: once a fault has been detected by a protective device, a timer will monitor the removal of the fault. If, after a time delay, the fault is still detected, conclusion is reached that the breaker(s) have not opened and a signal to open the backup breakers will be sent.

Figure 9.24 shows a conventional breaker failure diagram where provision has been added to detect a flashover occurring before the synchronizing of the generator: in addition to the protective relays detecting

Power System Protection

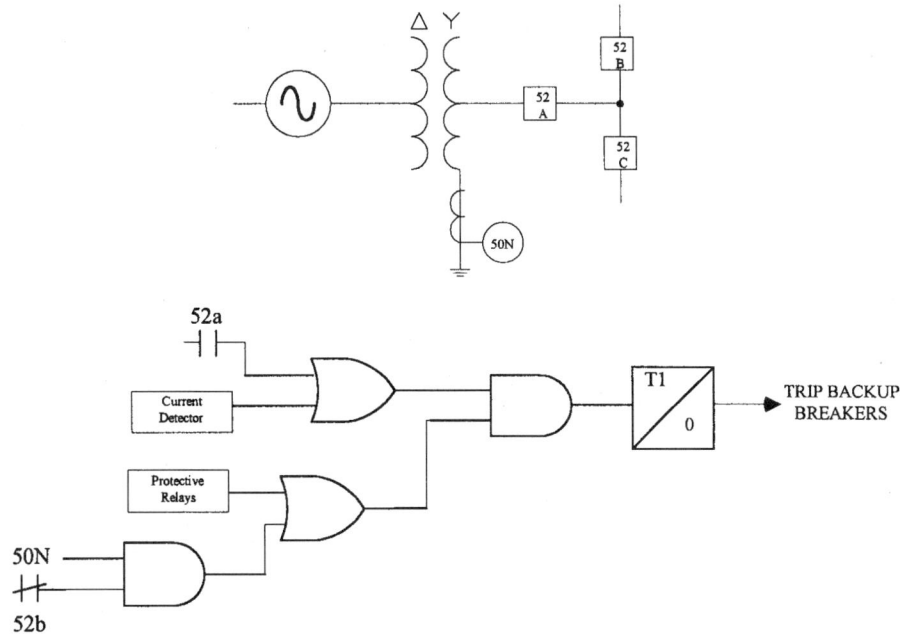

FIGURE 9.24 Breaker failure logic with flashover protection.

a fault, a flashover condition is detected by using an instantaneous overcurrent relay installed on the neutral of the step-up transformer. If this relay picks up and the breaker position contact (52b) is closed (breaker open), then a flashover condition is asserted and breaker failure is initiated.

Generator Tripping Principles

A number of methods for isolating a generator once a fault has been detected are commonly being implemented. They fall into four groups:

- Simultaneous tripping involves simultaneously shutting the prime mover down by closing its valves and opening the field and generator breakers. This technique is highly recommended for severe internal generator faults.
- Generator tripping involves simultaneously opening both the field and generator breakers.
- Unit separation involves opening the generator breaker only.
- Sequential tripping is applicable to steam turbines and involves first tripping the turbine valves in order to prevent any overspeeding of the unit. Then, the field and generator breakers are opened. Figure 9.25 represents a possible logical scheme for the implementation of a sequential tripping function. If the following three conditions are met, (1) the real power is below a negative pre-set threshold SET_1, (2) the steam valve or a differential pressure switch is closed (either condition indicating the removal of the prime-mover), (3) the sequential tripping function is enabled, then a trip signal will be sent to the generator and field breakers.

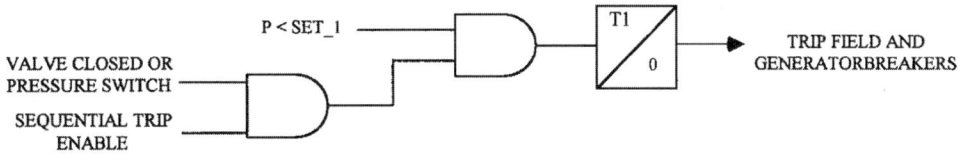

FIGURE 9.25 Implementation of a sequential tripping function.

Impact of Generator Digital Multifunction Relays[1]

The latest technological leap in generator protection has been the release of digital multifunction relays by various manufacturers (Benmouyal, 1988; Yalla, 1992; Benmouyal, 1994; Yip, 1994). With more sophisticated characteristics being available through software algorithms, generator protective function characteristics can be improved. Therefore, multifunction relays have many advantages, most of which stem from the technology on which they are based.

Improvements in Signal Processing

Most multifunction relays use a full-cycle Discrete Fourier Transform (DFT) algorithm for acquisition of the fundamental component of the current and voltage phasors. Consequently, they will benefit from the inherent filtering properties provided by the algorithms, such as:

- immunity from DC component and good suppression of exponentially decaying offset due to the large value of X/R time constants in generators;
- immunity to harmonics;
- nominal response time of one cycle for the protective functions requiring fast response.

Since sequence quantities are computed mathematically from the voltage and current phasors, they will also benefit from the above advantages.

However, it should be kept in mind that fundamental phasors of waveforms are not the only parameters used in digital multifunction relays. Other parameters like peak or rms values of waveforms can be equally acquired through simple algorithms, depending upon the characteristics of a particular algorithm.

A number of techniques have been used to make the measurement of phasor magnitudes independent of frequency, and therefore achieve stable sensitivities over large frequency excursions. One technique is known as frequency tracking and consists of having a number of samples in one cycle that is constant, regardless of the value of the frequency or the generator's speed. A software digital phase-locked loop allows implementation of such a scheme and will inherently provide a direct measurement of the frequency or the speed of the generator (Benmouyal, 1989). A second technique keeps the sampling period fixed, but varies the time length of the data window to follow the period of the generator frequency. This results in a variable number of samples in the cycles (Hart et al., 1997). A third technique consists of measuring the root-mean square value of a current or voltage waveform. The variation of this quantity with frequency is very limited, and therefore, this technique allows measurement of the magnitude of a waveform over a broad frequency range.

A further improvement consists of measuring the generator frequency digitally. Precision, in most cases, will be one hundredth of a hertz or better, and good immunity to harmonics and noise is achievable with modern algorithms.

Improvements in Protective Functions

The following functions will benefit from some inherent advantages of the digital processing capability:

- A number of improvements can be attributed to stator differential protection. The first is the detection of CT saturation in case of external faults that would cause the protection relay to trip. When CT ratios do not match perfectly, the difference can be either automatically or manually introduced into the algorithm in order to suppress the difference.
- It is no longer necessary to provide a Δ-Y conversion for the backup 21 elements in order to cover the phase fault on the high side of the voltage transformer. That conversion can be accomplished mathematically inside the relay.
- In the area of detection of voltage transformer blown fuses, the use of symmetrical components allows identification of the faulted phase. Therefore, complex logic schemes can be implemented where only the protection function impacted by the phase will be blocked. As an example, if a 51V is implemented on all three phases independently, it will be sufficient to block the function only

---

[1]This section was published previously in a modified form in Working Group J-11 of PSRC, Application of multifunction generator protection systems, *IEEE Trans. on PD*, 14(4), Oct. 1999.

on the phase on which a fuse has been detected as blown. Furthermore, contrary to the conventional voltage balance relay scheme, a single VT will suffice when using this modern algorithm.
- Because of the different functions recording their characteristics over a large frequency interval, it is no longer necessary to monitor the frequency in order to implement start-up or shut-down protection.
- The 100% stator-ground protection can be improved by using third-harmonic voltage measurements both at the phase and neutral.
- The characteristic of an offset mho impedance relay in the R-X plane can be made to be independent of frequency by using one of the following two techniques: the frequency-tracking algorithm previously mentioned, or the use of the positive sequence voltage and current because their ratio is frequency-independent.
- Functions which are inherently three-phase phenomena can be implemented by using the positive sequence voltage and current quantities. The loss-of-field or loss-of-synchronism are examples.
- In the reverse power protection, improved accuracy and sensitivity can be obtained with digital technology.
- Digital technology allows the possibility of tailoring inverse volt/hertz curves to the user's needs. Full programmability of these same curves is readily achievable. From that perspective, volt/hertz protection is improved by a closer match between the implemented curve and the generator or step-up transformer damage curve.

Multifunction generator protection packages have other functions that make use of the inherent capabilities of microprocessor devices. These include: oscillography and event recording, time synchronization, multiple settings, metering, communications, self-monitoring, and diagnostics.

References

Benmouyal, G., An adaptive sampling interval generator for digital relaying, *IEEE Trans. on PD*, 4(3), July, 1989.

Benmouyal, G., Design of a universal protection relay for synchronous generators, CIGRE Session, No. 34-09, 1988.

Benmouyal, G., Adamiak, M. G., Das, D. P., and Patel, S. C., Working to develop a new multifunction digital package for generator protection, *Electricity Today*, 6(3), March 1994.

Berdy, J., Loss-of-excitation for synchronous generators, *IEEE Trans. on PAS*, PAS-94(5), Sept./Oct. 1975.

Guide for Abnormal Frequency Protection for Power Generating Plant, ANSI/IEEE C37.106.

Guide for AC Generator Protection, ANSI/IEEE C37.102.

Guide for Generator Ground Protection, ANSI/IEEE C37.101.

Hart, D., Novosel, D., Hu, Y., Smith, R., and Egolf, M., A new tracking and phasor estimation algorithm for generator, *IEEE Trans. on PD*, 12(3), July, 1997.

IEEE Tutorial on the Protection of Synchronous Generators, IEEE Catalog No. 95TP102, 1995.

IEEE Recommended Practice for Protection and Coordination of Industrial and Commercial Power Systems, ANSI/IEEE 242-1986.

Ilar, M. and Wittwer, M., Numerical generator protection offers new benefits of gas turbines, International Gas Turbine and Aeroengine Congress and Exposition, Colone, Germany, June 1992.

Inadvertant energizing protection of synchronous generators, *IEEE Trans. on PD*, 4(2), April 1989.

Wimmer, W., Fromm, W., Muller, P., and Ilar, F., Fundamental Considerations on User-Configurable Multifunctional Numerical Protection, 34-202, CIGRE 1996 Session.

Working Group J-11 of PSRC, Application of multifunction generator protection systems, *IEEE Trans. on PD*, 14(4), Oct. 1999.

Yalla, M. V. V. S., A digital multifunction protection relay, *IEEE Trans. on PD*, 7(1), January 1992.

Yin, X. G., Malik, O. P., Hope, G. S., and Chen, D. S., Adaptive ground fault protection schemes for turbo-generator based on third harmonic voltages, *IEEE Trans. on PD*, 5(2), July, 1990.

Yip, H. T., *An Integrated Approach to Generator Protection*, Canadian Electrical Association, Toronto, March 1994.

9.3 Transmission Line Protection

Stanley H. Horowitz

The study of transmission line protection presents many fundamental relaying considerations that apply, in one degree or another, to the protection of other types of power system protection. Each electrical element, of course, will have problems unique to itself, but the concepts of reliability, selectivity, local and remote backup, zones of protection, coordination and speed which may be present in the protection of one or more other electrical apparatus are all present in the considerations surrounding transmission line protection.

Since transmission lines are also the links to adjacent lines or connected equipment, transmission line protection must be compatible with the protection of all of these other elements. This requires coordination of settings, operating times and characteristics.

The purpose of power system protection is to detect faults or abnormal operating conditions and to initiate corrective action. Relays must be able to evaluate a wide variety of parameters to establish that corrective action is required. Obviously, a relay cannot prevent the fault. Its primary purpose is to detect the fault and take the necessary action to minimize the damage to the equipment or to the system. The most common parameters which reflect the presence of a fault are the voltages and currents at the terminals of the protected apparatus or at the appropriate zone boundaries. The fundamental problem in power system protection is to define the quantities that can differentiate between normal and abnormal conditions. This problem is compounded by the fact that "normal" in the present sense means outside the zone of protection. This aspect, which is of the greatest significance in designing a secure relaying system, dominates the design of all protection systems.

The Nature of Relaying

Reliability

Reliability, in system protection parlance, has special definitions which differ from the usual planning or operating usage. A relay can misoperate in two ways: it can fail to operate when it is required to do so, or it can operate when it is not required or desirable for it to do so. To cover both situations, there are two components in defining reliability:

Dependability — which refers to the certainty that a relay will respond correctly for all faults for which it is designed and applied to operate; and

Security — which is the measure that a relay will not operate incorrectly for any fault.

Most relays and relay schemes are designed to be dependable since the system itself is robust enough to withstand an incorrect tripout (loss of security), whereas a failure to trip (loss of dependability) may be catastrophic in terms of system performance.

Zones of Protection

The property of security is defined in terms of regions of a power system — called zones of protection — for which a given relay or protective system is responsible. The relay will be considered secure if it responds only to faults within its zone of protection. Figure 9.26 shows typical zones of protection with transmission lines, buses, and transformers, each residing in its own zone. Also shown are "closed zones" in which all power apparatus entering the zone is monitored, and "open" zones, the limit of which varies with the fault current. Closed zones are also known as "differential," "unit," or absolutely selective," and open zones are "non-unit," "unrestricted," or "relatively selective."

The zone of protection is bounded by the current transformers (CT) which provide the input to the relays. While a CT provides the ability to detect a fault within its zone, the circuit breaker (CB) provides the ability to isolate the fault by disconnecting all of the power equipment inside its zone. When a CT is part of the CB, it becomes a natural zone boundary. When the CT is not an integral part of the CB,

Power System Protection

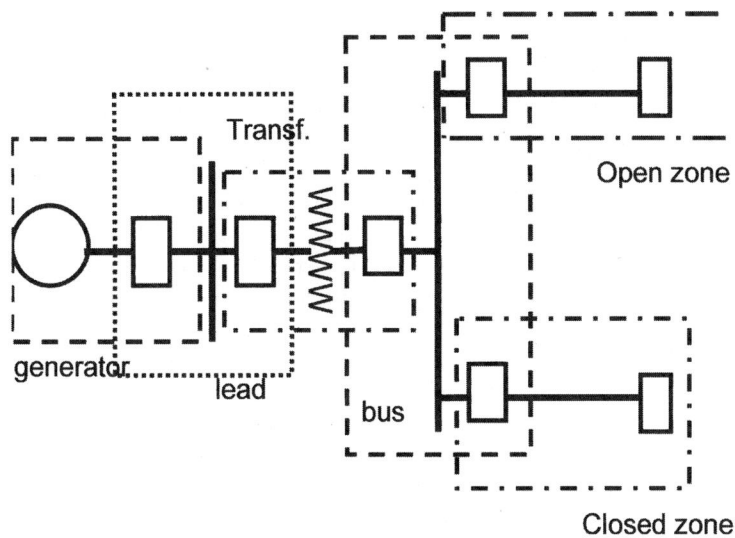

FIGURE 9.26 Closed and open zones of protection. (*Source:* Horowitz, S. H. and Phadke, A. G., *Power System Relaying*, 2nd ed., 1995. Research Studies Press, U.K. With permission.)

special attention must be paid to the fault detection and fault interruption logic. The CTs still define the zone of protection, but a communication channel must be used to implement the tripping function.

Relay Speed

It is, of course, desirable to remove a fault from the power system as quickly as possible. However, the relay must make its decision based upon voltage and current waveforms, which are severely distorted due to transient phenomena that follow the occurrence of a fault. The relay must separate the meaningful and significant information contained in these waveforms upon which a secure relaying decision must be based. These considerations demand that the relay take a certain amount of time to arrive at a decision with the necessary degree of certainty. The relationship between the relay response time and its degree of certainty is an inverse one and is one of the most basic properties of all protection systems.

Although the operating time of relays often varies between wide limits, relays are generally classified by their speed of operation as follows:

1. Instantaneous — These relays operate as soon as a secure decision is made. No intentional time delay is introduced to slow down the relay response.
2. Time-delay — An intentional time delay is inserted between the relay decision time and the initiation of the trip action.
3. High-speed — A relay that operates in less than a specified time. The specified time in present practice is 50 milliseconds (3 cycles on a 60 Hz system).
4. Ultra high-speed — This term is not included in the Relay Standards but is commonly considered to be operation in 4 milliseconds or less.

Primary and Backup Protection

The main protection system for a given zone of protection is called the primary protection system. It operates in the fastest time possible and removes the least amount of equipment from service. On Extra High Voltage (EHV) systems, i.e., 345kV and above, it is common to use duplicate primary protection systems in case a component in one primary protection chain fails to operate. This duplication is therefore intended to cover the failure of the relays themselves. One may use relays from a different manufacturer, or relays based on a different principle of operation to avoid common-mode failures. The operating time and the tripping logic of both the primary and its duplicate system are the same.

It is not always practical to duplicate every element of the protection chain. On High Voltage (HV) and EHV systems, the costs of transducers and circuit breakers are very expensive and the cost of duplicate equipment may not be justified. On lower voltage systems, even the relays themselves may not be duplicated. In such situations, a backup set of relays will be used. Backup relays are slower than the primary relays and may remove more of the system elements than is necessary to clear the fault.

Remote Backup — These relays are located in a separate location and are completely independent of the relays, transducers, batteries, and circuit breakers that they are backing up. There are no common failures that can affect both sets of relays. However, complex system configurations may significantly affect the ability of a remote relay to "see" all faults for which backup is desired. In addition, remote backup may remove more sources of the system than can be allowed.

Local Backup — These relays do not suffer from the same difficulties as remote backup, but they are installed in the same substation and use some of the same elements as the primary protection. They may then fail to operate for the same reasons as the primary protection.

Reclosing

Automatic reclosing infers no manual intervention but probably requires specific interlocking such as a full or check synchronizing, voltage or switching device checks, or other safety or operating constraints. Automatic reclosing can be high speed or delayed. High Speed Reclosing (HSR) allows only enough time for the arc products of a fault to dissipate, generally 15–40 cycles on a 60 Hz base, whereas time delayed reclosings have a specific coordinating time, usually 1 or more seconds. HSR has the possibility of generator shaft torque damage and should be closely examined before applying it.

It is common practice in the U.S. to trip all three phases for all faults and then reclose the three phases simultaneously. In Europe, however, for single line-to-ground faults, it is not uncommon to trip only the faulted phase and then reclose that phase. This practice has some applications in the U.S., but only in rare situations. When one phase of a three-phase system is opened in response to a single phase-to-ground fault, the voltage and current in the two healthy phases tend to maintain the fault arc after the faulted phase is de-energized. Depending on the length of the line, load current, and operating voltage, compensating reactors may be required to extinguish this "secondary arc."

System Configuration

Although the fundamentals of transmission line protection apply in almost all system configurations, there are different applications that are more or less dependent upon specific situations.

Operating Voltages — Transmission lines will be those lines operating at 138 kV and above, subtransmission lines are 34.5 kV to 138 kV, and distribution lines are below 34.5 kV. These are not rigid definitions and are only used to generically identify a transmission system and connote the type of protection usually provided. The higher voltage systems would normally be expected to have more complex, hence more expensive, relay systems. This is so because higher voltages have more expensive equipment associated with them and one would expect that this voltage class is more important to the security of the power system. The higher relay costs, therefore, are more easily justified.

Line Length — The length of a line has a direct effect on the type of protection, the relays applied, and the settings. It is helpful to categorize the line length as "short," "medium," or "long" as this helps establish the general relaying applications although the definition of "short," "medium," and "long" is not precise. A short line is one in which the ratio of the source to the line impedance (SIR) is large (>4 e.g.), the SIR of a long line is 0.5 or less and a medium line's SIR is between 4 and 0.5. It must be noted, however, that the per-unit impedance of a line varies more with the nominal voltage of the line than with its physical length or impedance. So a "short" line at one voltage level may be a "medium" or "long" line at another.

Multiterminal Lines — Occasionally, transmission lines may be tapped to provide intermediate connections to additional sources without the expense of a circuit breaker or other switching device. Such a configuration is known as a multiterminal line and, although it is an inexpensive measure for strengthening the power system, it presents special problems for the protection engineer. The difficulty

Power System Protection 9-33

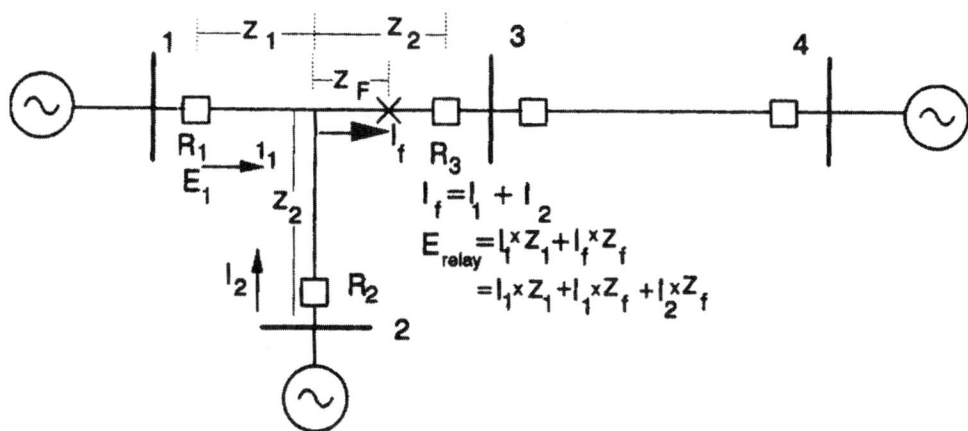

FIGURE 9.27 Effect of infeed on local relays. (*Source:* Horowitz, S. H. and Phadke, A. G., *Power System Relaying*, 2nd ed., 1995. Research Studies Press, U.K. With permission.)

arises from the fact that a relay receives its input from the local transducers, i.e., the current and voltage at the relay location. Referring to Fig. 9.27, the current contribution to a fault from the intermediate source is not monitored. The total fault current is the sum of the local current plus the contribution from the intermediate source, and the voltage at the relay location is the sum of the two voltage drops, one of which is the product of the unmonitored current and the associated line impedance.

Current Actuated Relays

Fuses

The most commonly used protective device in a distribution circuit is the fuse. Fuse characteristics vary considerably from one manufacturer to another and the specifics must be obtained from their appropriate literature. Figure 9.28 shows the time-current characteristics which consist of the minimum melt and total clearing curves.

Minimum melt is the time between initiation of a current large enough to cause the current responsive element to melt and the instant when arcing occurs. Total Clearing Time (TCT) is the total time elapsing from the beginning of an overcurrent to the final circuit interruption; i.e., TCT = minimum melt plus arcing time.

In addition to the different melting curves, fuses have different load-carrying capabilities. Manufacturer's application tables show three load-current values: continuous, hot-load pickup, and cold-load pickup. Continuous load is the maximum current that is expected for three hours or more for which the fuse will not be damaged. Hot-load is the amount that can be carried continuously, interrupted, and immediately reenergized without melting. Cold-load follows a 30-min outage and is the high current that is the result in the loss of diversity when service is restored. Since the fuse will also cool down during this period, the cold-load pickup and the hot-load pickup may approach similar values.

Inverse-Time Delay Overcurrent Relays

The principal application of time-delay overcurrent relays (TDOC) is on a radial system where they provide both phase and ground protection. A basic complement of relays would be two phase and one ground relay. This arrangement will protect the line for all combinations of phase and ground faults using the minimum number of relays. Adding a third phase relay, however, provides complete backup protection, that is two relays for every type of fault, and is the preferred practice. TDOC relays are usually used in industrial systems and on subtransmission lines that cannot justify more expensive protection such as distance or pilot relays.

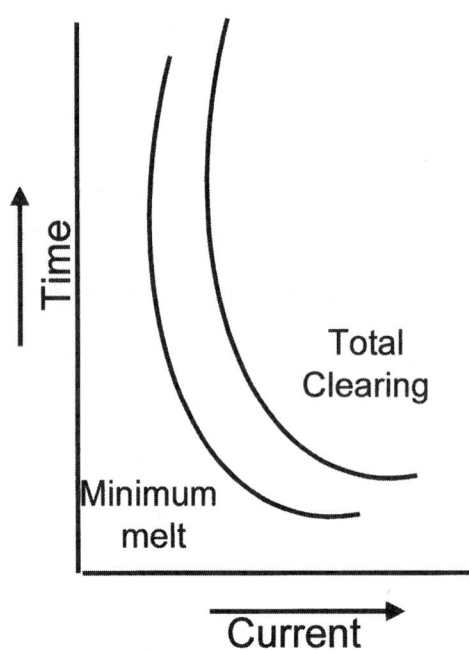

FIGURE 9.28 Fuse time-current characteristic. (*Source:* Horowitz, S. H. and Phadke, A. G., *Power System Relaying*, 2nd ed., 1995. Research Studies Press, U.K. With permission.)

There are two settings that must be applied to all TDOC relays: the pickup and the time delay. The pickup setting is selected so that the relay will operate for all short circuits in the line section for which it is to provide protection. This will require margins above the maximum load current, usually twice the expected value, and below the minimum fault current, usually 1/3 the calculated phase-to-phase or phase-to-ground fault current. If possible, this setting should also provide backup for an adjacent line section or adjoining equipment. The time-delay function is an independent parameter that is obtained in a variety of ways, either the setting of an induction disk lever or an external timer. The purpose of the time-delay is to enable relays to coordinate with each other. Figure 9.29 shows the family of curves of a single TDOC model. The ordinate is time in milliseconds or seconds depending on the relay type; the abscissa is in multiples of pickup to normalize the curve for all fault current values. Figure 9.30 shows how TDOC relays on a radial line coordinate with each other.

Instantaneous Overcurrent Relays

Figure 9.30 also shows why the TDOC relay cannot be used without additional help. The closer the fault is to the source, the greater the fault current magnitude, yet the longer the tripping time. The addition of an instantaneous overcurrent relay makes this system of protection viable. If an instantaneous relay can be set to "see" almost up to, but not including, the next bus, all of the fault clearing times can be lowered as shown in Fig. 9.31. In order to properly apply the instantaneous overcurrent relay, there must be a substantial reduction in short-circuit current as the fault moves from the relay toward the far end of the line. However, there still must be enough of a difference in the fault current between the near and far end faults to allow a setting for the near end faults. This will prevent the relay from operating for faults beyond the end of the line and still provide high-speed protection for an appreciable portion of the line.

Since the instantaneous relay must not see beyond its own line section, the values for which it must be set are very much higher than even emergency loads. It is common to set an instantaneous relay about 125–130% above the maximum value that the relay will see under normal operating situations and about 90% of the minimum value for which the relay should operate.

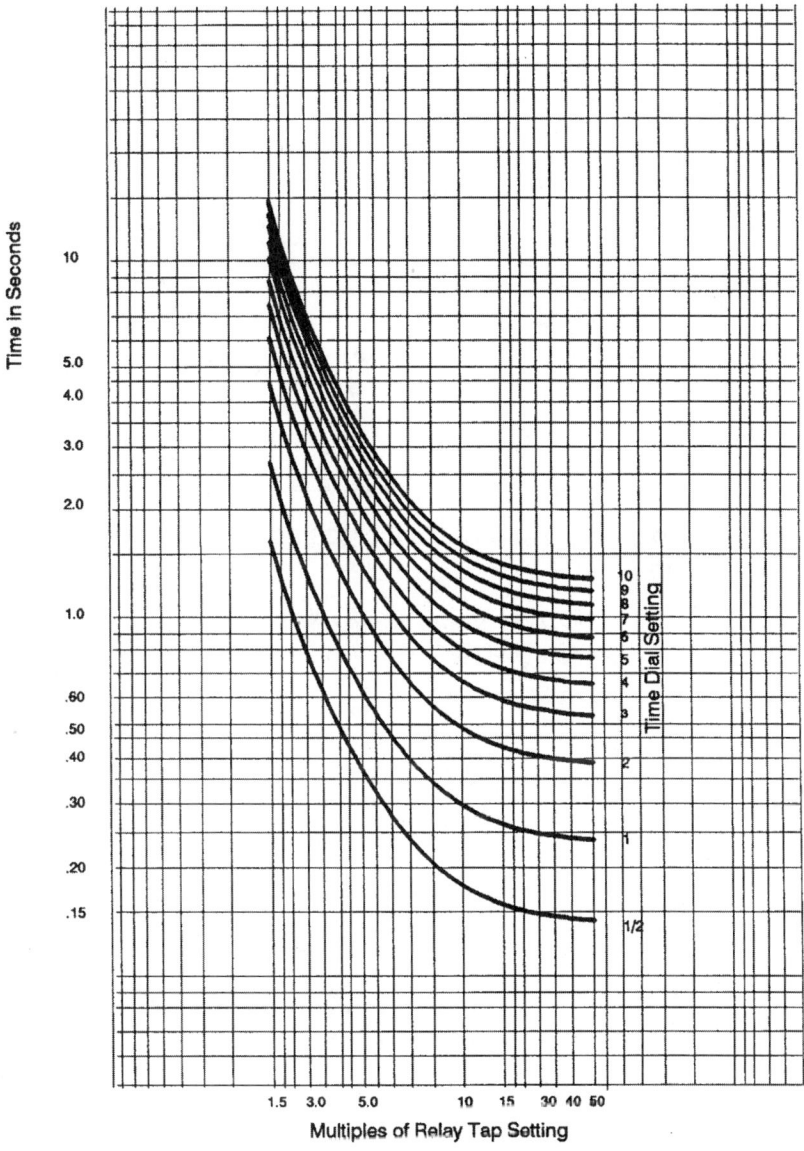

FIGURE 9.29 Family of TDOC time-current characteristics. (*Source:* Horowitz, S. H. and Phadke, A. G., *Power System Relaying*, 2nd ed., 1995. Research Studies Press, U.K. With permission.)

Directional Overcurrent Relays

Directional overcurrent relaying is necessary for multiple source circuits when it is essential to limit tripping for faults in only one direction. If the same magnitude of fault current could flow in either direction at the relay location, coordination cannot be achieved with the relays in front of, and, for the same fault, the relays behind the nondirectional relay, except in very unusual system configurations.

Polarizing Quantities — To achieve directionality, relays require two inputs; the operating current and a reference, or polarizing, quantity that does not change with fault location. For phase relays, the polarizing quantity is almost always the system voltage at the relay location. For ground directional indication, the zero-sequence voltage ($3E_0$) can be used. The magnitude of $3E_0$ varies with the fault

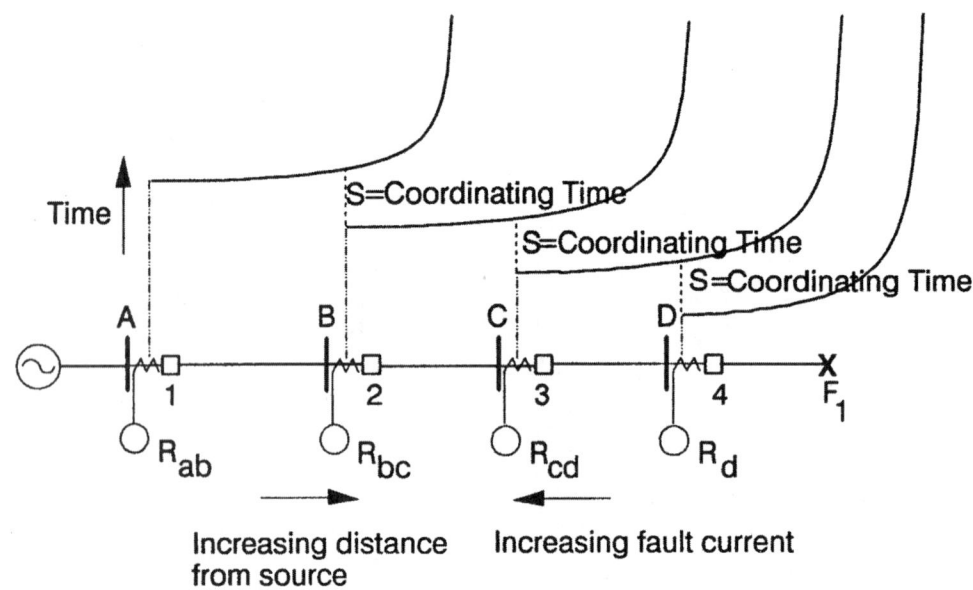

FIGURE 9.30 Coordination of TDOC relays. (*Source:* Horowitz, S. H. and Phadke, A. G., *Power System Relaying*, 2nd ed., 1995. Research Studies Press, U.K. With permission.)

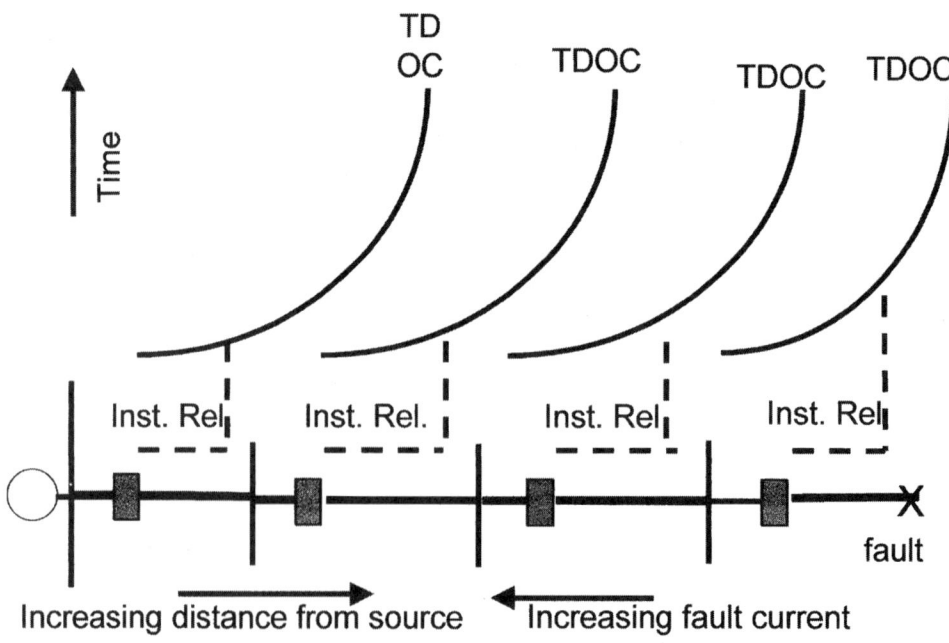

FIGURE 9.31 Effect of instantaneous relays. (*Source:* Horowitz, S. H. and Phadke, A. G., *Power System Relaying*, 2nd ed., 1995. Research Studies Press, U.K. With permission.)

location and may not be adequate in some instances. An alternative and generally preferred method of obtaining a directional reference is to use the current in the neutral of a wye-grounded/delta power transformer.

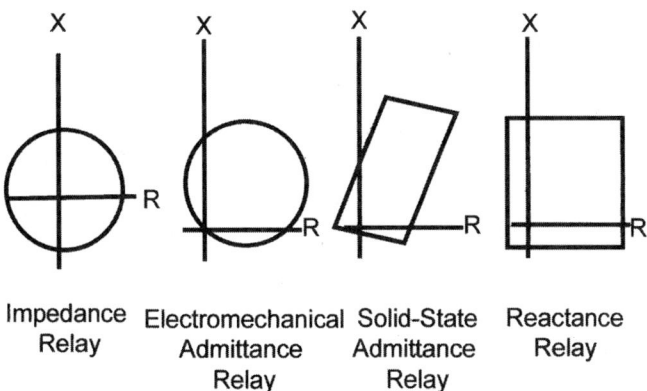

FIGURE 9.32 Distance relay characteristics. (*Source:* Horowitz, S. H. and Phadke, A. G., *Power System Relaying,* 2nd ed., 1995. Research Studies Press, U.K. With permission.)

Distance Relays

Distance relays respond to the voltage and current, i.e., the impedance, at the relay location. The impedance per mile is fairly constant so these relays respond to the distance between the relay location and the fault location. As the power systems become more complex and the fault current varies with changes in generation and system configuration, directional overcurrent relays become difficult to apply and to set for all contingencies, whereas the distance relay setting is constant for a wide variety of changes external to the protected line.

There are three general distance relay types as shown in Fig. 9.32. Each is distinguished by its application and its operating characteristic.

Impedance Relay

The impedance relay has a circular characteristic centered at the origin of the R-X diagram. It is nondirectional and is used primarily as a fault detector.

Admittance Relay

The admittance relay is the most commonly used distance relay. It is the tripping relay in pilot schemes and as the backup relay in step distance schemes. Its characteristic passes through the origin of the R-X diagram and is therefore directional. In the electromechanical design it is circular, and in the solid state design, it can be shaped to correspond to the transmission line impedance.

Reactance Relay

The reactance relay is a straight-line characteristic that responds only to the reactance of the protected line. It is nondirectional and is used to supplement the admittance relay as a tripping relay to make the overall protection independent of resistance. It is particularly useful on short lines where the fault arc resistance is the same order of magnitude as the line length.

Figure 9.33 shows a three-zone step distance relaying scheme that provides instantaneous protection over 80–90% of the protected line section (Zone 1) and time-delayed protection over the remainder of the line (Zone 2) plus backup protection over the adjacent line section. Zone 3 also provides backup protection for adjacent lines sections.

In a three-phase power system, 10 types of faults are possible: three single phase-to-ground, three phase-to-phase, three double phase-to-ground, and one three-phase fault. It is essential that the relays provided have the same setting regardless of the type of fault. This is possible if the relays are connected to respond to delta voltages and currents. The delta quantities are defined as the difference between any two phase quantities, for example, $E_a - E_b$ is the delta quantity between phases a and b. In general, for a multiphase fault between phases x and y,

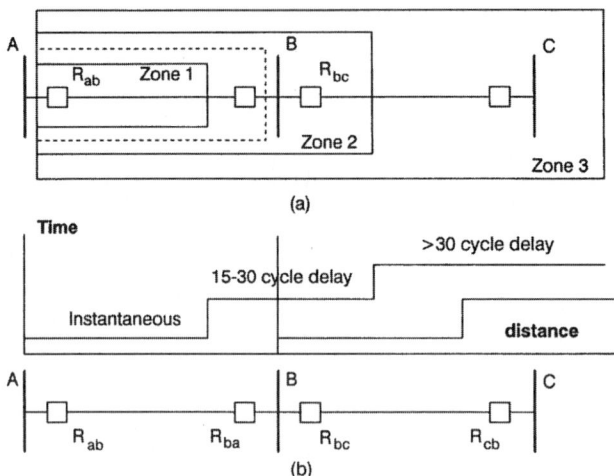

FIGURE 9.33 Three-zone step distance relaying to protect 100% of a line and backup the neighboring line. (*Source:* Horowitz, S. H. and Phadke, A. G., *Power System Relaying*, 2nd ed., 1995. Research Studies Press, U.K. With permission.)

$$\frac{Ex - Ey}{Ix - Iy} = Z1 \tag{9.14}$$

where x and y can be a, b, or c and Z_1 is the positive sequence impedance between the relay location and the fault. For ground distance relays, the faulted phase voltage, and a compensated faulted phase current must be used.

$$\frac{Ex}{Ix + mI_0} = Z_1 \tag{9.15}$$

where m is a constant depending on the line impedances, and I_0 is the zero sequence current in the transmission line. A full complement of relays consists of three phase distance relays and three ground distance relays. This is the preferred protective scheme for high voltage and extra high voltage systems.

Pilot Protection

As can be seen from Fig. 9.33, step distance protection does not offer instantaneous clearing of faults over 100% of the line segment. In most cases this is unacceptable due to system stability considerations. To cover the 10–20% of the line not covered by Zone 1, the information regarding the location of the fault is transmitted from each terminal to the other terminal(s). A communication channel is used for this transmission. These pilot channels can be over power line carrier, microwave, fiberoptic, or wire pilot. Although the underlying principles are the same regardless of the pilot channel, there are specific design details that are imposed by this choice.

Power line carrier uses the protected line itself as the channel, superimposing a high frequency signal on top of the 60 Hz power frequency. Since the line being protected is also the medium used to actuate the protective devices, a blocking signal is used. This means that a trip will occur at both ends of the line unless a signal is received from the remote end.

Microwave or fiberoptic channels are independent of the transmission line being protected so a tripping signal can be used.

Wire pilot channels are limited by the impedance of the copper wire and are used at lower voltages where the distance between the terminals is not great, usually less than 10 miles.

Directional Comparison

The most common pilot relaying scheme in the U.S. is the directional comparison blocking scheme, using power line carrier. The fundamental principle upon which this scheme is based utilizes the fact that, at a given terminal, the direction of a fault either forward or backward is easily determined by a directional relay. By transmitting this information to the remote end, and by applying appropriate logic, both ends can determine whether a fault is within the protected line or external to it. Since the power line itself is used as the communication medium, a blocking signal is used.

Transfer Tripping

If the communication channel is independent of the power line, a tripping scheme is a viable protection scheme. Using the same directional relay logic to determine the location of a fault, a tripping signal is sent to the remote end. To increase security, there are several variations possible. A direct tripping signal can be sent, or additional underreaching or overreaching directional relays can be used to supervise the tripping function and increase security. An underreaching relay sees less than 100% of the protected line, i.e., Zone 1. An overreaching relay sees beyond the protected line such as Zone 2 or 3.

Phase Comparison

Phase comparison is a differential scheme that compares the phase angle between the currents at the ends of the line. If the currents are essentially in phase, there is no fault in the protected section. If these currents are essentially 180° out of phase, there is a fault within the line section. Any communication link can be used.

Pilot Wire

Pilot wire relaying is a form of differential line protection similar to phase comparison, except that the phase currents are compared over a pair of metallic wires. The pilot channel is often a rented circuit from the local telephone company. However, as the telephone companies are replacing their wired facilities with microwave or fiberoptics, this protection must be closely monitored.

9.4 System Protection

Miroslav Begovic

While most of the protective system designs are centered around individual components, system-wide disturbances in power systems are becoming a frequent and challenging problem for electric utilities. The occurrence of major disturbances in power systems requires coordinated protection and control actions to stop the system degradation, restore the normal state, and minimize the impact of the disturbance. Local protection systems are often not capable of protecting the overall system, which may be affected by the disturbance. Among the phenomena that create the power system disturbances are various types of system instability, overloads, and power system cascading (Horowitz and Phadke, 1992; Elmore, 1994; Blackburn, 1987; Phadke and Thorp, 1988; Anderson, 1999).

Power system planning has to account for tight operating margins with less redundancy, because of new constraints placed by restructuring of the entire industry. The advanced measurement and communication technology in wide area monitoring and control are expected to provide new, faster, and better ways to detect and control an emergency (Begovic et al., 1999).

Transient Stability and Out-of-Step Protection

Every time a fault or a topological change affects the power balance in the system, the instantaneous power imbalance creates oscillations between the machines. Stable oscillations lead to transition from one (pre-fault) to another (post-fault) equilibrium point, whereas unstable ones allow machines to oscillate beyond the acceptable range. If the oscillations are large, the stations' auxiliary supplies may undergo severe voltage fluctuations, and eventually trip (Horowitz and Phadke, 1992). Should that

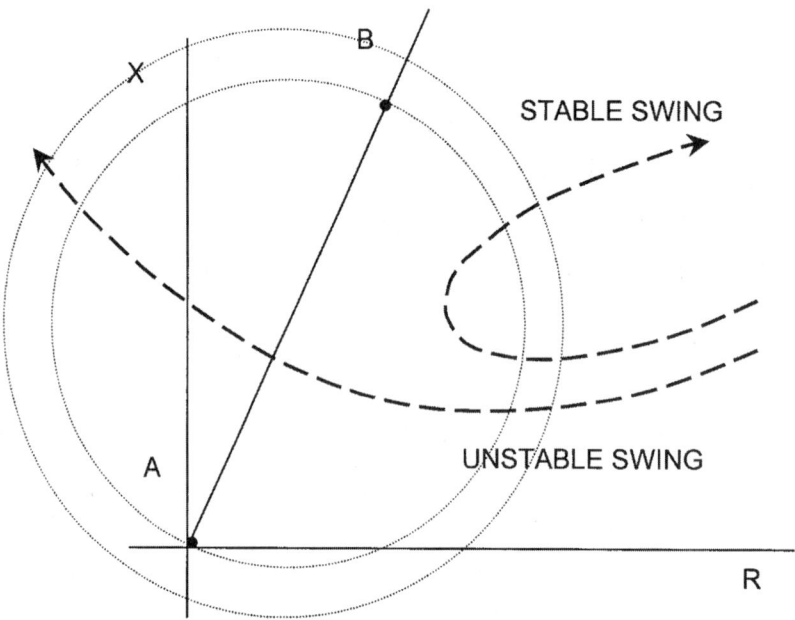

FIGURE 9.34 Trajectories of stable and unstable swings in the impedance plane.

happen, the subsequent resynchronization of the machines might take a long time. It is, therefore, desirable to trip the machine(s) exposed to transient unstable oscillations while the plant auxiliaries remain energized.

The frequency of the transient oscillations is usually between 0.5 and 2 Hz. Since the fault imposes almost instantaneous changes on the system, the slow speed of the transient disturbances can be used to distinguish between the two. For the sake of illustration, let us assume that a power system consists of two machines, A and B, connected by a transmission line. Figure 9.34 represents the trajectories of the stable and unstable swings between the machines, as well as a characteristic of the mho relay covering the line between them, shown in the impedance plane. The stable swing moves from the distant stable operating point towards the trip zone of the relay, and may even encroach on it, then leave again. The unstable trajectory may pass through the entire trip zone of the relay. The relaying tasks are to detect, and then trip (or block) the relay, depending on the situation. Detection is accomplished by out-of-step relays, which have multiple characteristics. When the trajectory of the impedance seen by the relays enters the outer zone (a circle with a larger radius), the timer is activated, and depending on the speed at which the impedance trajectory moves into the inner zone (a circle with a smaller radius), or leaves the outer zone, a tripping (or blocking) decision can be made. The relay characteristic may be chosen to be straight lines, known as "blinders," which prevent the heavy load from being misrepresented as a fault or instability. Another piece of information that can be used in detection of transient swings is that they are symmetrical, and do not create any zero or negative sequence currents.

In the case when power system separation is imminent, out-of-step protection should take place along boundaries that will form islands with matching load and generation. Distance relays are often used to provide an out-of-step protection function, whereby they are called upon to provide blocking or tripping signals upon detecting an out-of-step condition. The most common predictive scheme to combat loss of synchronism is the Equal-Area Criterion and its variations. This method assumes that the power system behaves like a two-machine model where one area oscillates against the rest of the system. Whenever the underlying assumption holds true, the method has potential for fast detection.

Overload and Underfrequency Load Shedding

Outage of one or more power system components due to the overload may result in overload of other elements in the system. If the overload is not alleviated in time, the process of power system cascading may start, leading to power system separation. When a power system separates, islands with an imbalance between generation and load are formed. One consequence of the imbalance is deviation of frequency from the nominal value. If the generators cannot handle the imbalance, load or generation shedding is necessary. A special protection system or out-of-step relaying can also start the separation.

A quick, simple, and reliable way to reestablish active power balance is to shed load by underfrequency relays. The load shedding is often designed as a multistep action, and the frequency settings and blocks of load to be shed are carefully selected to maximize the reliability and dependability of the action. There are a large variety of practices in designing load shedding schemes based on the characteristics of a particular system and the utility practices. While the system frequency is a final result of the power deficiency, the rate of change of frequency is an instantaneous indicator of power deficiency and can enable incipient recognition of the power imbalance. However, change of the machine speed is oscillatory by nature due to the interaction among generators. These oscillations depend on location of the sensors in the island and the response of the generators. The problems regarding the rate-of-change of frequency function are:

- Systems having small inertia may cause larger oscillations. Thus, enough time must be allowed for the relay to calculate the actual rate-of-change of frequency reliably. Measurements at load buses close to the electrical center of the system are less susceptible to oscillations (smaller peak-to-peak values) and can be used in practical applications. Smaller system inertia causes a higher frequency of oscillations, which enables faster calculation of the actual rate-of-change of frequency. However, it causes faster rate-of-change of frequency, and, consequently, a larger frequency drop.
- Even if rate-of-change of frequency relays measure the average value throughout the network, it is difficult to set them properly unless typical system boundaries and imbalance can be predicted. If this is the case (e.g., industrial and urban systems), the rate of change of frequency relays may improve a load shedding scheme (scheme can be more selective and/or faster).

Voltage Stability and Undervoltage Load Shedding

Voltage stability is defined by the System Dynamic Performance Subcommittee of the IEEE Power System Engineering Committee as being the ability of a system to maintain voltage such that when load admittance is increased, load power will increase, and so that both power and voltage are controllable. Also, voltage collapse is defined as being the process by which voltage instability leads to a very low voltage profile in a significant part of the system. It is accepted that this instability is caused by the load characteristics, as opposed to the angular instability that is caused by the rotor dynamics of generators.

The risk of voltage instability increases as the transmission system becomes more heavily loaded. The typical scenario of these instabilities starts with a high system loading, followed by a relay action due to either a fault, a line overload, or hitting an excitation limit.

Voltage instability can be alleviated by a combination of the following remedial measures: adding reactive compensation near load centers, strengthening the transmission lines, varying the operating conditions such as voltage profile and generation dispatch, coordinating relays and controls, and load shedding. Most utilities rely on planning and operation studies to guard against voltage instability. Many utilities utilize localized voltage measurements in order to achieve load shedding as a measure against incipient voltage instability. The efficiency of the load shedding depends on the selected voltage thresholds, locations of pilot points in which the voltages are monitored, locations and sizes of the blocks of load to be shed, as well as the operating conditions that may activate the shedding. The wide variety of conditions that may lead to voltage instability suggests that the most accurate decisions should imply the adaptive relay settings, but such applications are still in the stage of early development.

Special Protection Schemes (SPS)

Increasingly popular over the past several years are the so-called special protection systems, sometimes also referred to as remedial action schemes (Anderson and LeReverend, 1996; McCalley and Fu, 1998). Depending on the power system in question, it is sometimes possible to identify the contingencies or combinations of operating conditions that may lead to transients with extremely disastrous consequences (Tamronglak et al., 1996). Such problems include, but are not limited to, transmission line faults, the outages of lines and possible cascading that such an initial contingency may cause, outages of the generators, rapid changes of the load level, problems with HVDC or FACTS equipment, or any combination of those events.

Among the many varieties of special protection schemes, several names have been used to describe the general category (Elmore, 1994): special stability controls, dynamic security controls, contingency arming schemes, remedial action schemes, adaptive protection schemes, corrective action schemes, security enhancement schemes, etc. In the strict sense of protective relaying, we do not consider any control schemes to be SPS, but only those protective relaying systems that possess the following properties (McCalley and Fu, 1998):

- SPS can be operational ("armed"), or out of service ("disarmed"), in conjunction with the system conditions.
- SPS are responding to very low probability events; hence they are active rarely more than once a year.
- SPS operate on simple, predetermined control laws, often calculated based on extensive offline studies.
- Oftentimes, SPS involve communication of remotely acquired measurement data (SCADA) from more than one location in order to make a decision and invoke a control law.

The SPS design procedure is based on the following (Elmore, 1994):

- **Identification of critical conditions:** On the grounds of extensive offline steady state studies on the system under consideration, a variety of operating conditions and contingencies are identified as potentially dangerous, and those among them that are deemed the most harmful are recognized as the critical conditions. The issue of their continuous monitoring, detection, and mitigation is resolved through offline studies.
- **Recognition triggers:** These are the measurable signals that can be used for detection of critical conditions. Oftentimes, such detection is accomplished through a complicated heuristic logical reasoning, using the logic circuits to accomplish the task: "**If** event A **and** event B occur together, **or** event C occurs, **then**…" Inputs for the decision making logic are called recognition triggers, and can be the status of various relays in the system, sometimes combined with a number of (SCADA) measurements.
- **Operator control:** In spite of extensive simulations and studies done in the process of SPS design, it is often necessary to include human intervention, i.e., to include human interaction in the feedback loop. This is necessary because SPS are not needed all the time, and the decision to arm, or disarm them remains in the hands of an operator.

Among the SPS schemes reported in the literature (Anderson and LeReverend, 1996; McCalley and Fu, 1998), the following are represented:

- Generator Rejection
- Load Rejection
- Underfrequency Load Shedding
- System Separation
- Turbine Valve Control
- Stabilizers
- HVDC Controls

- Out-of-step Relaying
- Dynamic Braking
- Generator Runback
- VAR Compensation
- Combination of schemes

Some of them have already been described in the above text. A general trend continues toward more complex schemes, capable of outperforming the present solutions and taking advantage of the most recent technological developments and advances in systems analysis. Some of the trends are described in the following text (Begovic et al., 1999).

Future Improvements in Control and Protection

Existing protection/control systems may be improved and new protection/control systems may be developed to better adapt to prevailing system conditions during system-wide disturbance. While improvements in the existing systems are mostly achieved through advancement in local measurements and development of better algorithms, improvements in new systems are based on remote communications. However, even if communication links exist, conventional systems that utilize only local information may still need improvement since they are supposed to serve as fallback positions. The increased functions and communication ability in today's SCADA systems provide the opportunity for an intelligent and adaptive control and protection system for system-wide disturbance. This, in turn, can make possible full utilization of the network, which will be less vulnerable to a major disturbance.

Out-of-step relays have to be fast and reliable. The present technology of out-of-step tripping or blocking distance relays is not capable of fully dealing with the control and protection requirements of power systems. Central to the development effort of an out-of-step protection system is the investigation of the multiarea out-of-step situation. The new generation of out-of-step relays has to utilize more measurements, both local and remote, and has to produce more outputs. The structure of the overall relaying system has to be distributed and coordinated through a central control. In order for the relaying system to manage complexity, most of the decisions have to be made locally. Preferably, the relay system is adaptive, in order to cope with system changes. To deal with out-of-step prediction, it is necessary to start with a system-wide approach, find out what sets of information are crucial and how to the process information with acceptable speed and accuracy.

Protection against voltage instability should also be addressed as a part of hierarchical structure. The sound approach for designing the new generation of voltage instability protection is to first design a voltage instability relay with only local signals. The limitations of local signals should be identified in order to be in a position to select appropriate communicated signals. However, a minimum set of communicated signals should always be known in order to design reliable protection, which requires the following: (a) determining the algorithm for gradual reduction of the number of necessary measurement sites with minimum loss of information necessary for voltage stability monitoring, analysis, and control; (b) development of methods (i.e., sensitivity analysis), which should operate *concurrent* with any existing local protection techniques, and possessing superior performance, both in terms of security and dependability.

References

Anderson, P.M., *Power System Protection,* McGraw-Hill and IEEE Press, New York, 1999.

Anderson, P.M., LeReverend, B.K., Industry Experience with Special Protection Schemes, IEEE/CIGRE Committee Report, *IEEE Trans. PWRS* (11), 1166–1179, August 1996.

Begovic, M., Novosel, D., Milisavljevic, M., Trends in power system protection and control, in *Proceedings 1999 HICSS Conference,* Maui, Hawaii, January 4–7, 1999.

Blackburn, L., *Protective Relaying,* Marcel Dekker, New York, 1987.

Elmore, W.A., Editor, *Protective Relaying Theory and Applications,* ABB and Marcel Dekker, New York, 1994.

Horowitz, S.H., Phadke, A.G., *Power System Relaying,* John Wiley & Sons, Inc., New York, 1992.
McCalley, J., Fu. W., Reliability of Special Protection Schemes, IEEE PES paper PE-123-PWRS-0-10-1998.
Phadke, A.G., Thorp, J.S., *Computer Relaying for Power Systems,* John Wiley & Sons, New York, 1988.
Tamronglak, S., Horowitz, S., Phadke, A., Thorp, J., Anatomy of power system blackouts: Preventive relaying strategies, *IEEE Trans. PWRD* (11), 708–715, April 1996.

9.5 Digital Relaying

James S. Thorp

Digital relaying had its origins in the late 1960s and early 1970s with pioneering papers by Rockefeller (1969), Mann and Morrison (1971), and Poncelet (1972) and an early field experiment (Gilcrest et al., 1972; Rockefeller and Udren, 1972). Because of the cost of the computers in those times, a single high-cost minicomputer was proposed by Rockefeller (1969) to perform multiple relaying calculations in the substation. In addition to having high cost and high power requirements, early minicomputer systems were slow in comparison with modern systems and could only perform simple calculations. The well-founded belief that computers would get smaller, faster, and cheaper combined with expectations of benefits of computer relaying kept the field moving. The third IEEE tutorial on microprocessor protection (Sachdev, 1997) lists more then 1100 publications in the area since 1970. Nearly two thirds of the papers are devoted to developing and comparing algorithms. It is not clear this trend should continue. Issues beyond algorithms should receive more attention in the future.

The expected benefits of microprocessor protection have largely been realized. The ability of a digital relay to perform self-monitoring and checking is a clear advantage over the previous technology. Many relays are called upon to function only a few cycles in a year. A large percentage of major disturbances can be traced to "hidden failures" in relays that were undetected until the relay was exposed to certain system conditions (Tamronglak et al., 1996). The ability of a digital relay to detect a failure within itself and remove itself from service before an incorrect operation occurs is one of the most important advantages of digital protection.

The microprocessor revolution has created a situation in which digital relays are the relays of choice because of economic reasons. The cost of conventional (analog) relays has increased while the hardware cost of the most sophisticated digital relays has decreased dramatically. Even including substantial software costs, digital relays are the economic choice and have the additional advantage of having lower wiring costs. Prior to the introduction of microprocessor-based systems, several panels of space and considerable wiring was required to provide all the functions needed for each zone of transmission line protection. For example, an installation requiring phase distance protection for phase-to-phase and three-phase faults, ground distance, ground-overcurrent, a pilot scheme, breaker failure, and reclosing logic demanded redundant wiring, several hundred watts of power, and a lot of panel space. A single microprocessor system is a single box, with a ten-watt power requirement and with only direct wiring, has replaced the old system.

Modern digital relays can provide SCADA, metering, and oscillographic records. Line relays can also provide fault location information. All of this data can be available by modem or on a WAN. A LAN in the substation connecting the protection modules to a local host is also a possibility. Complex multi-function relays can have an almost bewildering number of settings. Techniques for dealing with setting management are being developed. With improved communication technology, the possibility of involving microprocessor protection in wide-area protection and control is being considered.

Sampling

The sampling process is essential for microprocessor protection to produce the numbers required by the processing unit to perform calculations and reach relaying decisions. Both 12 and 16 bit A/D converters

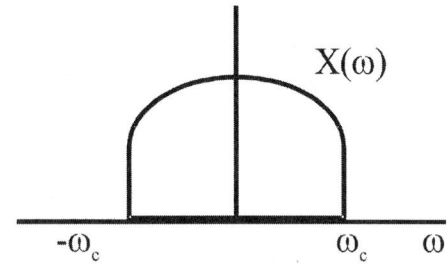

FIGURE 9.35 The Fourier Transform of a band-limited function.

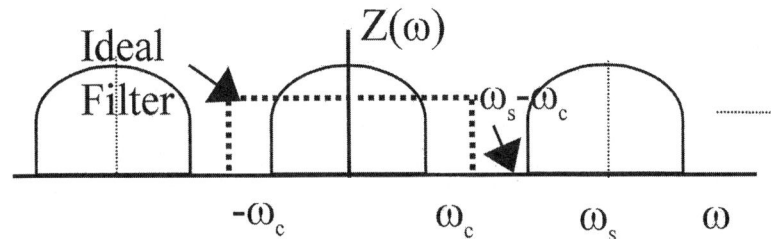

FIGURE 9.36 The Fourier Transform of a sampled version of the signal x(t).

are in use. The large difference between load and fault current is a driving force behind the need for more precision in the A/D conversion. It is difficult to measure load current accurately while not saturating for fault current with only 12 bits. It should be noted that most protection functions do not require such precise load current measurement. Although there are applications, such as hydro generator protection, where the sampling rate is derived from the actual power system frequency, most relay applications involve sampling at a fixed rate that is a multiple of the *nominal* power system frequency.

Antialiasing Filters

ANSI/IEEE Standard C37.90, provides the standard for the Surge Withstand Capability (SWC) to be built into protective relay equipment. The standard consists of both an oscillatory and transient test. Typically the surge filter is followed by an antialiasing filter before the A/D converter. Ideally the signal x(t) presented to the A/D converter x(t) is band-limited to some frequency ω_c, i.e., the Fourier transform of x(t) is confined to a low-pass band less that ω_c such as shown in Fig. 9.35. Sampling the low-pass signal at a frequency of ω_s produces a signal with a transform made up of shifted replicas of the low-pass transform as shown in Fig. 9.36. If $\omega_s - \omega_c > \omega_c$, i.e., $\omega_s > 2\omega_c$ as shown, then an ideal low-pass filter applied to z(t) can recover the original signal x(t). The frequency of twice the highest frequency present in the signal to be sampled is the Nyquist sampling rate. If $\omega_s < 2\omega_c$ the sampled signal is said to be "aliased" and the output of the low-pass filter is not the original signal. In some applications the frequency content of the signal is known and the sampling frequency is chosen to avoid aliasing (music CDs), while in digital relaying applications the sampling frequency is specified and the frequency content of the signal is controlled by filtering the signal before sampling to insure its highest frequency is less than half the sampling frequency. The filter used is referred to as an antialiasing filter.

Aliasing also occurs when discrete sequences are sampled or decimated. For example, if a high sampling rate such as 7200 Hz is used to provide data for oscillography, then taking every tenth sample provides data at 720 Hz to be used for relaying. The process of taking every tenth sample (decimation) will produce aliasing unless a digital antialiasing filter with a cut-off frequency of 360 Hz is provided.

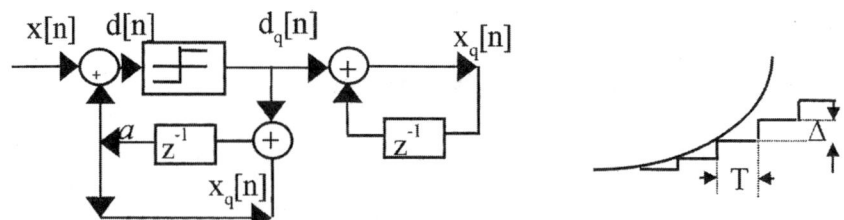

FIGURE 9.37 Delta modulator and error.

Sigma-Delta A/D Converters

There is an advantage in sampling at rates many times the Nyquist rate. It is possible to exchange speed of sampling for bits of resolution. So called Sigma-Delta A/D converters are based on one bit sampling at very high rates. Consider a signal x(t) sampled at a high rate T = 1/fs, i.e., x[n] = x(nT) with the difference between the current sample and α times the last sample given by

$$d[n] = x[n] - \alpha\, x[n-1] \qquad (9.16)$$

If d[n] is quantized through a one-bit quantizer with a step size of Δ, then

$$x_q[n] = \alpha\, x_q[n-1] + d_q[n] \qquad (9.17)$$

The quantization is called delta modulation and is represented in Fig. 9.37. The z^{-1} boxes are unit delays while the one bit quantizer is shown as the box with d[n] as input and $d_q[n]$ as output. The output $x_q[n]$ is a staircase approximation to the signal x(t) with stairs that are spaced at T sec and have height Δ. The delta modulator output has two types of errors: one when the maximum slope Δ/T is too small for rapid changes in the input (shown on Fig. 9.37) and the second, a sort of chattering when the signal x(t) is slowly varying. The feedback loop below the quantizer is a discrete approximation to an integrator with α = 1. Values of α less than one correspond to an imperfect integrator. A continuous form of the delta modulator is also shown in Fig. 9.38. The low pass filter (LPF) is needed because of the high frequency content of the staircase. Shifting the integrator from in front of the LPF to before the delta modulator improves both types of error. In addition, the two integrators can be combined.

The modulator can be thought of as a form of voltage follower circuit. Resolution is increased by oversampling to spread the quantization noise over a large bandwidth. It is possible to shape the quantization noise so it is larger at high frequencies and lower near DC. Combining the shaped noise with a very steep cut-off in the digital low pass filter, it is possible to produce a 16-bit result from the one bit comparator. For example, a 16-bit answer at 20 kHz can be obtained with an original sampling frequency of 400 kHz.

Phasors from Samples

A phasor is a complex number used to represent sinusoidal functions of time such as AC voltages and currents. For convenience in calculating the power in AC circuits from phasors, the phasor magnitude

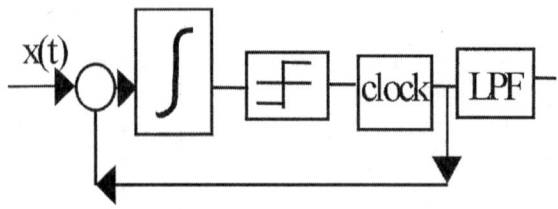

FIGURE 9.38 Signa-Delta modulator.

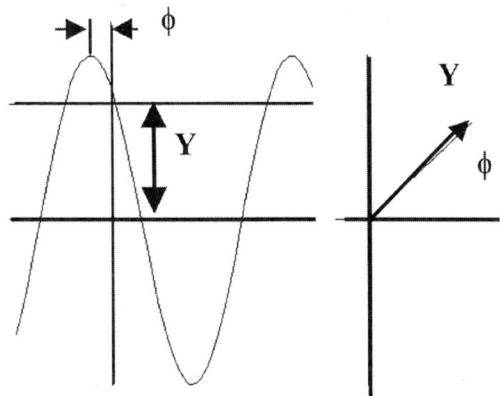

FIGURE 9.39 Phasor representation.

is set equal to the rms value of the sinusoidal waveform. A sinusoidal quantity and its phasor representation are shown in Fig. 9.39, and are defined as follows:

$$\text{Sinusoidal quantity} \qquad \text{Phasor}$$

$$y(t) = Y_m \cos(\omega t + \phi) \qquad Y = \frac{Y_m}{\sqrt{2}} e^{j\phi} \tag{9.18}$$

A phasor represents a single frequency sinusoid and is not directly applicable under transient conditions. However, the idea of a phasor can be used in transient conditions by considering that the phasor represents an estimate of the fundamental frequency component of a waveform observed over a finite window. In case of N samples y_k, obtained from the signal y(t) over a period of the waveform:

$$Y = \frac{1}{\sqrt{2}} \frac{2}{N} \sum_{k=1}^{N} y_k e^{-jk\frac{2\pi}{N}} \tag{9.19}$$

or,

$$Y = \frac{1}{\sqrt{2}} \frac{2}{N} \left\{ \sum_{k=1}^{N} y_k \cos\left(\frac{k2\pi}{N}\right) - j \sum_{k=1}^{N} y_k \sin\left(\frac{k2\pi}{N}\right) \right\} \tag{9.20}$$

Using θ for the sampling angle $2\pi/N$, it follows that

$$Y = \frac{1}{\sqrt{2}} \frac{2}{N} (Y_c - jY_s) \tag{9.21}$$

where

$$Y_c = \sum_{k=1}^{N} y_k \cos(k\theta)$$

$$Y_s = \sum_{k=1}^{N} y_k \sin(k\theta) \tag{9.22}$$

Note that the input signal y(t) must be band-limited to Nω/2 to avoid aliasing errors. In the presence of white noise, the fundamental frequency component of the Discrete Fourier Transform (DFT) given by Eqs. (9.19)–(9.22) can be shown to be a least-squares estimate of the phasor. If the data window is not a multiple of a half cycle, the least-squares estimate is some other combination of Y_c and Y_s, and is no longer given by Eq. (9.21). Short window (less than one period) phasor computations are of interest in some digital relaying applications. For the present, we will concentrate on data windows that are multiples of a half cycle of the nominal power system frequency.

The data window begins at the instant when sample number 1 is obtained as shown in Fig. 9.39. The sample set y_k is given by

$$y_k = Y_m \cos(k\theta + \phi) \tag{9.23}$$

Substituting for y_k from Eq. (9.23) in Eq. (9.19),

$$Y = \frac{1}{\sqrt{2}} \frac{2}{N} \sum_{k=1}^{N} Y_m \cos(k\theta + \phi) e^{-jk\theta} \tag{9.24}$$

or

$$Y = \frac{1}{\sqrt{2}} Y_m e^{j\phi} \tag{9.25}$$

which is the familiar expression Eq. (9.18), for the phasor representation of the sinusoid in Eq. (9.18). The instant at which the first data sample is obtained defines the orientation of the phasor in the complex plane. The reference axis for the phasor, i.e., the horizontal axis in Fig. 9.39b, is specified by the first sample in the data window.

Equations (9.21)–(9.22) define an algorithm for computing a phasor from an input signal. A recursive form of the algorithm is more useful for real-time measurements. Consider the phasors computed from two adjacent sample sets: y_k {k = 1, 2, ···, N} and, y'_k {k = 2, 3, ···, N + 1}, and their corresponding phasors Y^1 and $Y^{2'}$ respectively:

$$Y^1 = \frac{1}{\sqrt{2}} \frac{2}{N} \sum_{k=1}^{N} y_k e^{-jk\theta} \tag{9.26}$$

$$Y^{2'} = \frac{1}{\sqrt{2}} \frac{2}{N} \sum_{k=1}^{N} y_{k+1} e^{-jk\theta} \tag{9.27}$$

We may modify Eq. (9.27) to develop a recursive phasor calculation as follows:

$$Y^2 = Y^{2'} e^{-j\theta} = Y^1 + \frac{1}{\sqrt{2}} \frac{2}{N} (y_{N+1} - y_1) e^{-j\theta} \tag{9.28}$$

Since the angle of the phasor $Y^{2'}$ is greater than the angle of the phasor Y^1 by the sampling angle θ, the phasor Y^2 has the same angle as the phasor Y^1. When the input signal is a constant sinusoid, the phasor calculated from Eq. (9.28) is a constant complex number. In general, the phasor Y, corresponding to the data y_k {k = r, r + 1, r + 2, ···, N + r – 1} is recursively modified into Y^{r+1} according to the formula

$$Y^{r+1} = Y^r e^{-j\theta} = Y^r + \frac{1}{\sqrt{2}} \frac{2}{N} (y_{N+r} - y_r) e^{-j\theta} \qquad (9.29)$$

The recursive phasor calculation as given by Eq. (9.28) is very efficient. It regenerates the new phasor from the old one and utilizes most of the computations performed for the phasor with the old data window.

Symmetrical Components

Symmetrical components are linear transformations on voltages and currents of a three phase network. The symmetrical component transformation matrix S transforms the phase quantities, taken here to be voltages E_ϕ, (although they could equally well be currents), into symmetrical components E_s:

$$E_s = \begin{bmatrix} E_0 \\ E_1 \\ E_2 \end{bmatrix} = SE_\phi = \frac{1}{3} \begin{bmatrix} 1 & 1 & 1 \\ 1 & \alpha & \alpha^2 \\ 1 & \alpha^2 & \alpha \end{bmatrix} \begin{bmatrix} E_a \\ E_b \\ E_c \end{bmatrix} \qquad (9.30)$$

where $(1, \alpha, \alpha^2)$ are the three cube-roots of unity. The symmetrical component transformation matrix S is a similarity transformation on the impedance matrices of balanced three phase circuits, which diagonalizes these matrices. The symmetrical components, designated by the subscripts (0,1,2) are known as the zero, positive, and negative sequence components of the voltages (or currents). The negative and zero sequence components are of importance in analyzing unbalanced three phase networks. For our present discussion, we will concentrate on the positive sequence component E_1 (or I_1) only. This component measures the balanced, or normal voltages and currents that exist in a power system. Dealing with positive sequence components only allows the use of single-phase circuits to model the three-phase network, and provides a very good approximation for the state of a network in quasi-steady state. All power generators generate positive sequence voltages, and all machines work best when energized by positive sequence currents and voltages. The power system is specifically designed to produce and utilize almost pure positive sequence voltages and currents in the absence of faults or other abnormal imbalances. It follows from Eq. (9.30) that the positive sequence component of the phase quantities is given by

$$Y_1 = \frac{1}{3}(Y_a + \alpha Y_b + \alpha^2 Y_c) \qquad (9.31)$$

Or, using the recursive form of the phasors given by Eq. (9.29),

$$Y_1^{r+1} = Y_1^r + \frac{1}{\sqrt{2}} \frac{2}{N} \left[(x_{a,N+r} - x_{a,r}) e^{-jr\theta} + \alpha (x_{b,N+r} - x_{b,r}) e^{-jr\theta} + \alpha^2 (x_{c,N+r} - x_{c,r}) e^{-jr\theta} \right] \qquad (9.32)$$

Recognizing that for a sampling rate of 12 times per cycle, α and α^2 correspond to $\exp(j4\theta)$ and $\exp(j8\theta)$, respectively, it can be seen from Eq. (9.32) that

$$Y_1^{r+1} = Y_1^r + \frac{1}{\sqrt{2}} \frac{2}{N} \left[(x_{a,N+r} - x_{a,r}) e^{-jr\theta} + (x_{b,N+r} - x_{b,r}) e^{j(4-r)\theta} + (x_{c,N+r} - x_{c,r}) e^{j(8-r)\theta} \right] \qquad (9.33)$$

With a carefully chosen sampling rate — such as a multiple of three times the nominal power system frequency — very efficient symmetrical component calculations can be performed in real time. Equations similar to (9.33) hold for negative and zero sequence components also. The sequence quantities can be used to compute a distance to the fault that is independent of fault type. Given the ten possible faults in a three-phase system (three line-ground, three phase-phase, three phase-phase-ground, and three phase), early microprocessor systems were taxed to determine the fault type before computing the distance to the fault. Incorrect fault type identification resulted in a delay in relay operation. The symmetrical component relay solved that problem. With advances in microprocessor speed it is now possible to simultaneously compute the distance to all six phase-ground and phase-phase faults in order to solve the fault classification problem.

The positive sequence calculation is still of interest because of the use of synchronized phasor measurements. Phasors, representing voltages and currents at various buses in a power system, define the state of the power system. If several phasors are to be measured, it is essential that they be measured with a common reference. The reference, as mentioned in the previous section, is determined by the instant at which the samples are taken. In order to achieve a common reference for the phasors, it is essential to achieve synchronization of the sampling pulses. The precision with which the time synchronization must be achieved depends upon the uses one wishes to make of the phasor measurements. For example, one use of the phasor measurements is to estimate, or validate, the state of the power systems so that crucial performance features of the network, such as the power flows in transmission lines could be determined with a degree of confidence. Many other important measures of power system performance, such as contingency evaluation, stability margins, etc., can be expressed in terms of the state of the power system, i.e., the phasors. Accuracy of time synchronization directly translates into the accuracy with which phase angle differences between various phasors can be measured. Phase angles between the ends of transmission lines in a power network may vary between a few degrees, and may approach 180° during particularly violent stability oscillations. Under these circumstances, assuming that one may wish to measure angular differences as little as 1°, one would want the accuracy of measurement to be better than 0.1°. Fortunately, synchronization accuracies of the order of 1 μsec are now achievable from the Global Positioning System (GPS) satellites. One microsecond corresponds to 0.022° for a 60 Hz power system, which more than meets our needs. Real-time phasor measurements have been applied in static state estimation, frequency measurement, and wide area control.

Algorithms

Parameter Estimation

Most relaying algorithms extract information about the waveform from current and voltage waveforms in order to make relaying decisions. Examples include: current and voltage phasors that can be used to compute impedance, the rms value, the current that can be used in an overcurrent relay, and the harmonic content of a current that can be used to form a restraint in transformer protection. An approach that unifies a number of algorithms is that of parameter estimation. The samples are assumed to be of a current or voltage that has a known form with some unknown parameters. The simplest such signal can be written as

$$y(t) = Y_c \cos\omega_0 t + Y_s \sin\omega_0 t + e(t) \tag{9.34}$$

where ω_0 is the nominal power system frequency, Y_c and Y_s are unknown quantities, and e(t) is an error signal (all the things that are not the fundamental frequency signal in this simple model). It should be noted that in this formulation, we assume that the power system frequency is known. If the numbers, Y_c and Y_s were known, we could compute the fundamental frequency phasor. With samples taken at an interval of T seconds,

$$y_n = y(nT) = Y_c \cos n\theta + Y_s \sin n\theta + e(nT) \tag{9.35}$$

Power System Protection

where $\theta = \omega_0 T$ is the sampling angle. If signals other than the fundamental frequency signal were present, it would be useful to include them in a formulation similar to Eq. (9.34) so that they would be included in e(t). If, for example, the second harmonic were included, Eq. (9.34) could be modified to

$$y_n = Y_{1c} \cos n\theta + Y_{1s} \sin n\theta + Y_{2c} \cos 2n\theta + Y_{2s} \sin 2n\theta + e(nT) \tag{9.36}$$

It is clear that more samples are needed to estimate the parameters as more terms are included. Equation (9.36) can be generalized to include any number of harmonics (the number is limited by the sampling rate), the exponential offset in a current, or any known signal that is suspected to be included in the post-fault waveform. No matter how detailed the formulation, e(t) will include unpredictable contributions from:

- The transducers (CTs and PTs)
- Fault arc
- Traveling wave effects
- A/D converters
- The exponential offset in the current
- The transient response of the antialiasing filters
- The power system itself

The current offset is not an error signal for some algorithms and is removed separately for some others. The power system generated signals are transients depending on fault location, the fault incidence angle, and the structure of the power system. The power system transients are low enough in frequency to be present after the antialiasing filter.

We can write a general expression as

$$y_n = \sum_{k=1}^{K} s_k(nT) Y_k + e_n \tag{9.37}$$

If y represents a vector of N samples, and Y a vector of K unknown coefficients, then there are N equations in K unknowns in the form

$$y = SY + e \tag{9.38}$$

The matrix S is made up of samples of the signals s_k.

$$S = \begin{bmatrix} s_1(T) & s_2(T) & \cdots & s_K(T) \\ s_1(2T) & s_2(2T) & \cdots & s_K(2T) \\ \vdots & \vdots & & \vdots \\ s_1(NT) & s_2(NT) & \cdots & s_K(NT) \end{bmatrix} \tag{9.39}$$

The presence of the error e and the fact that the number of equations is larger than the number of unknowns (N > K) makes it necessary to estimate Y.

Least Squares Fitting

One criterion for choosing the estimate $\hat{Y}$ is to minimize the scalar formed as the sum of the squares of the error term in Eq. (9.38), viz.

$$e^T e = (y - SY)^T (y - SY) = \sum_{n=1}^{N} e_n^2 \qquad (9.40)$$

It can be shown that the minimum least squared error [the minimum value of Eq. (9.40)] occurs when

$$\hat{Y} = (S^T S)^{-1} S^T y = By \qquad (9.41)$$

where $B = (S^T S)^{-1} S^T$. The calculations involving the matrix S can be performed off-line to create an "algorithm," i.e., an estimate of each of the K parameters is obtained by multiplying the N samples by a set of stored numbers. The rows of Eq. (9.41) can represent a number of different algorithms depending on the choice of the signals $s_k(nT)$ and the interval over which the samples are taken.

DFT

The simplest form of Eq. (9.41) is when the matrix $S^T S$ is diagonal. Using a signal alphabet of cosines and sines of the first N harmonics of the fundamental frequency over a window of one cycle of the fundamental frequency, the familiar Discrete Fourier Transform (DFT) is produced. With

$$\begin{aligned} s_1(t) &= \cos(\omega_0 t) \\ s_2(t) &= \sin(\omega_0 t) \\ s_3(t) &= \cos(2\omega_0 t) \\ s_4(t) &= \sin(2\omega_0 t) \\ &\vdots \\ s_{N-1}(t) &= \cos(N\omega_0 t/2) \\ s_N(t) &= \sin(N\omega_0 t/2) \end{aligned} \qquad (9.42)$$

The estimates are given by:

$$\hat{Y}_{Cp} = \frac{2}{N} \sum_{n=0}^{N-1} y_n \cos(pn\theta)$$

$$\hat{Y}_{Sp} = \frac{2}{N} \sum_{n=0}^{N-1} y_n \sin(pn\theta) \qquad (9.43)$$

Note that the harmonics are also estimated by Eq. (9.43). Harmonics have little role in line relaying but are important in transformer protection. It can be seen that the fundamental frequency phasor can be obtained as

$$Y = \frac{2}{N\sqrt{2}} (Y_{C1} - jY_{S1}) \qquad (9.44)$$

The normalizing factor in Eq. (9.44) is omitted if the ratio of phasors for voltage and current are used to form impedance.

Power System Protection

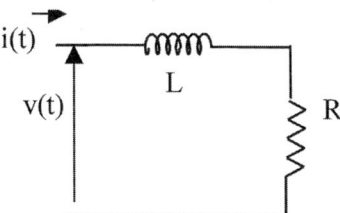

FIGURE 9.40 Model of a faulted line.

Differential Equations

Another kind of algorithm is based on estimating the values of parameters of a physical model of the system. In line protection, the physical model is a series R-L circuit that represents the faulted line. A similar approach in transformer protection uses the magnetic flux circuit with associated inductance and resistance as the model. A differential equation is written for the system in both cases.

Line Protection Algorithms

The series R-L circuit of Fig. 9.40 is the model of a faulted line. The offset in the current is produced by the circuit model and hence will not be an error signal.

$$v(t) = Ri(t) + L\frac{di(t)}{dt} \tag{9.45}$$

Looking at the samples at k, k + 1, k + 2

$$\int_{t_0}^{t_1} v(t)dt = R\int_{t_0}^{t_1} i(t)dt + L(i(t_1) - i(t_0)) \tag{9.46}$$

$$\int_{t_1}^{t_2} v(t)dt = R\int_{t_1}^{t_2} i(t)dt + L(i(t_2) - i(t_1)) \tag{9.47}$$

Using trapezoidal integration to evaluate the integrals (assuming t is small)

$$\int_{t_1}^{t_2} v(t)dt = R\int_{t_1}^{t_2} i(t)dt + L(i(t_2) - i(t_1)) \tag{9.48}$$

$$\int_{t_1}^{t_2} v(t)dt = R\int_{t_1}^{t_2} i(t)dt + L(i(t_2) - i(t_1)) \tag{9.49}$$

R and L are given by

$$R = \left[\frac{(v_{k+1} + v_k)(i_{k+2} - i_{k+1}) - (v_{k+2} + v_{k+1})(i_{k+1} - i_k)}{2(i_k i_{k+2} - i_{k+1}^2)}\right] \tag{9.50}$$

$$L = \frac{T}{2}\left[\frac{(v_{k+2}+v_{k+1})(i_{k+1}+i_k)-(v_{k+1}+v_k)(i_{k+2}+i_{k+1})}{2(i_k i_{k+2}-i_{k+1}^2)}\right] \quad (9.51)$$

It should be noted that the sample values occur in both numerator and denominator of Eqs. (9.50) and (9.51). The denominator is not constant but varies in time with local minima at points where both the current and the derivative of the current are small. For a pure sinusoidal current, the current and its derivative are never both small but when an offset is included there is a possibility of both being small once per period.

Error signals for this algorithm include terms that do not satisfy the differential equation such as the currents in the shunt elements in the line model required by long lines. In intervals where the denominator is small, errors in the numerator of Eqs. (9.50) and (9.51) are amplified. The resulting estimates can be quite poor. It is also difficult to make the window longer than three samples. The complexity of solving such equations for a larger number of samples suggests that the short window results be post processed. Simple averaging of the short-window estimates is inappropriate, however.

A counting scheme was used in which the counter was advanced if the estimated R and L were in the zone and the counter was decreased if the estimates lay outside the zone (Chen and Breingan, 1979). By requiring the counter to reach some threshold before tripping, secure operation can be assured with a cost of some delay. For example, if the threshold were set at six with a sampling rate of 16 times a cycle, the fastest trip decision would take a half cycle. Each "bad" estimate would delay the decision by two additional samples. The actual time for a relaying decision is variable and depends on the exact data.

The use of a median filter is an alternate to the counting scheme (Akke and Thorp, 1997). The median operation ranks the input values according to their amplitude and selects the middle value as the output. Median filters have an odd number of inputs. A length five median filter has an input-output relation between input x[n] and output y[n] given by

$$y[n] = \text{median}\{x[n-2], x[n-1], x[n], x[n+1], x[n+2]\} \quad (9.52)$$

Median filters of length five, seven, and nine have been applied to the output of the short window differential equation algorithm (Akke and Thorp, 1997). The median filter preserves the essential features of the input while removing isolated noise spikes. The filter length rather than the counter scheme, fixes the time required for a relaying decision.

Transformer Protection Algorithms

Virtually all algorithms for the protection of power transformers use the principle of percentage differential protection. The difference between algorithms lies in how the algorithm restrains the differential trip for conditions of overexcitation and inrush. Algorithms based on harmonic restraint, which parallel existing analog protection, compute the second and fifth harmonics using Eq. (9.25) (Thorp and Phadke, 1982). These algorithms use current measurements only and cannot be faster than one cycle because of the need to compute the second harmonic. The harmonic calculation provides for secure operation since the transient event produces harmonic content which delays relay operation for about a cycle.

In an integrated substation with other microprocessor relays, it is possible to consider transformer protection algorithms that use voltage information. Shared voltage samples could be a result of multiple protection modules connected in a LAN in the substation. The magnitude of the voltage itself can be used as a restraint in a digital version of a "tripping suppressor" (Harder and Marter, 1948). A physical model similar to the differential equation model for a faulted line can be constructed using the flux in the transformer. The differential equation describing the terminal voltage, v(t), the winding current, i(t), and the flux linkage $\Lambda(t)$ is:

$$v(t) - L\frac{di(t)}{dt} = \frac{d\Lambda(t)}{dt} \quad (9.53)$$

where L is the leakage inductance of the winding.

Power System Protection

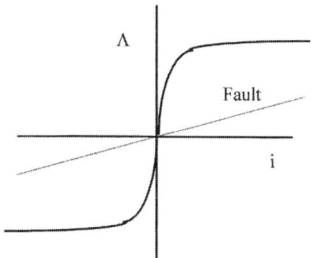

FIGURE 9.41 The flux-current characteristic compared to fault conditions.

Using trapezoidal integration for the integral in Eq. (9.53)

$$\int_{t_1}^{t_2} v(t)\,dt - L\big[i(t_2) - i(t_1)\big] = \Lambda(t_2) - \Lambda(t_1) \qquad (9.54)$$

gives

$$\Lambda(t_2) - \Lambda(t_1) = \frac{T}{2}\big[v(t_2) + v(t_1)\big] - L\big[i(t_2) - i(t_1)\big] \qquad (9.55)$$

or

$$\Lambda_{k+1} = \Lambda_k + \frac{T}{2}\big[v_{k+1} + v_k\big] - L\big[i_{k+1} - i_k\big] \qquad (9.56)$$

Since the initial flux Λ_0 in Eq. (9.56) cannot be known without separate sensing, the slope of the flux current curve is used

$$\left(\frac{d\Lambda}{di}\right)_k = \frac{T}{2}\left[\frac{v_k + v_{k-1}}{i_k - i_{k-1}}\right] - L \qquad (9.57)$$

The slope of the flux current characteristic shown in Fig. 9.41 is different depending on whether there is a fault or not. The algorithm must then be able to differentiate between inrush (the slope alternates between large and small values) and a fault (the slope is always small). The counting scheme used for the differential equation algorithm for line protection can be adapted to this application. The counter increases if the slope is less than a threshold and the differential current indicates trip, and the counter decreases if the slope is greater than the threshold or the differential does not indicate trip.

Kalman Filters

The Kalman filter provides a solution to the estimation problem in the context of an evolution of the parameters to be estimated according to a state equation. It has been used extensively in estimation problems for dynamic systems. Its use in relaying is motivated by the filter's ability to handle measurements that change in time. To model the problem so that a Kalman filter may be used, it is necessary to write a state equation for the parameters to be estimated in the form

$$x_{k+1} = \Phi_k x_k + \Gamma_k w_k \qquad (9.58)$$

$$z_k = H_k x_k + v_k \qquad (9.59)$$

where Eq. (9.58) (the state equation) represents the evolution of the parameters in time and Eq. (9.59) represents the measurements. The terms w_k and v_k are discrete time random processes representing state noise, i.e., random inputs in the evolution of the parameters, and measurement errors, respectively. Typically w_k and v_k are assumed to be independent of each other and uncorrelated from sample to sample. If w_k and v_k have zero means, then it is common to assume that

$$E\{w_k w_j^T\} = Q_k : k = j$$
$$= 0; \ k \neq j \tag{9.60}$$

The matrices Q_k and R_k are the covariance matrices of the random processes and are allowed to change as k changes. The matrix Φ_k in Eq. (9.58) is the state transition matrix. If we imagine sampling a pure sinusoid of the form

$$y(t) = Y_c \cos(\omega t) + Y_s \sin(\omega t) \tag{9.61}$$

at equal intervals corresponding to $\omega \Delta \tau = \Psi$, then the state would be

$$x_k = \begin{bmatrix} Y_C \\ Y_S \end{bmatrix} \tag{9.62}$$

and the state transition matrix

$$\Phi_k = \begin{bmatrix} 1 & 0 \\ 0 & 1 \end{bmatrix} \tag{9.63}$$

In this case, H_k, the measurement matrix, would be

$$H_k = \begin{bmatrix} \cos(k\Psi) & \sin(k\Psi) \end{bmatrix} \tag{9.64}$$

Simulations of a 345 kV line connecting a generator and a load (Gurgis and Brown, 1981) led to the conclusion that the covariance of the noise in the voltage and current decayed in time. If the time constant of the decay is comparable to the decision time of the relay, then the Kalman filter formulation is

$$x = \begin{bmatrix} Y_C \\ Y_S \\ Y_0 \end{bmatrix} \quad \Phi_k = \begin{bmatrix} 1 & 0 & 0 \\ 0 & 1 & 0 \\ 0 & 0 & e^{-\beta t} \end{bmatrix}$$

appropriate for the estimation problem. The voltage was modeled as in Eqs. (9.63) and (9.64). The current was modeled with three states to account for the exponential offset.

and

$$H_k = \begin{bmatrix} \cos(k\Psi) & \sin(k\Psi) & 1 \end{bmatrix} \tag{9.65}$$

The measurement covariance matrix was

$$R_k = K e^{-k\Delta t/T} \tag{9.66}$$

Power System Protection

with T chosen as half the line time constant and different Ks for voltage and current. The Kalman filter estimates phasors for voltage and current as the DFT algorithms. The filter must be started and terminated using some other software. After the calculations begin, the data window continues to grow until the process is halted. This is different from fixed data windows such as a one cycle Fourier calculation. The growing data window has some advantages, but has the limitation that if started incorrectly, it has a hard time recovering if a fault occurs after the calculations have been initiated.

The Kalman filter assumes an initial statistical description of the state x, and recursively updates the estimate of state. The initial assumption about the state is that it is a random vector independent of the processes w_k and v_k and with a known mean and covariance matrix, P_0. The recursive calculation involves computing a gain matrix K_k. The estimate is given by

$$\hat{x}_{k+1} = \Phi_k \hat{x}_k + K_{k+1}\left[z_{k+1} - H_{k+1}\hat{x}_k\right] \tag{9.67}$$

The first term in Eq. (9.67) is an update of the old estimate by the state transition matrix while the second is the gain matrix K_{k+1} multiplying the observation residual. The bracketed term in Eq. (9.67) is the difference between the actual measurement, z_k, and the predicted value of the measurement, i.e., the residual in predicting the measurement. The gain matrix can then be computed recursively. The amount of computation involved depends on the state vector dimension. For the linear problem described here, these calculations can be performed off-line. In the absence of the decaying measurement error, the Kalman filter offers little other than the growing data window. It has been shown that at multiples of a half cycle, the Kalman filter estimate for a constant error covariance is the same as that obtained from the DFT.

Wavelet Transforms

The Wavelet Transform is a signal processing tool that is replacing the Fourier Transform in many applications including data compression, sonar and radar, communications, and biomedical applications. In the signal processing community there is considerable overlap between wavelets and the area of filter banks. In applications in which it is used, the Wavelet Transform is viewed as an improvement over the Fourier Transform because it deals with time-frequency resolution in a different way. The Fourier Transform provides a decomposition of a time function into exponentials, $e^{j\omega t}$, which exist for all time. We should consider the signal that is processed with the DFT calculations in the previous sections as being extended periodically for all time. That is, the data window represents one period of a periodic signal. The sampling rate and the length of the data window determine the frequency resolution of the calculations. While these limitations are well understood and intuitive, they are serious limitations in some applications such as compression. The Wavelet Transform introduces an alternative to these limitations.

The Fourier Transform can be written

$$X(\omega) = \int_{-\infty}^{\infty} x(t) e^{-j\omega t} dt \tag{9.68}$$

The effect of a data window can be captured by imagining that the signal x(t) is windowed before the Fourier Transform is computed. The function h(t) represents the windowing function such as a one-cycle rectangle.

$$X(\omega, t) = \int_{-\infty}^{\infty} x(\tau) h(t-\tau) e^{-j\omega \tau} d\tau \tag{9.69}$$

The Wavelet Transform is written

$$X(s,t) = \int_{-\infty}^{\infty} x(\tau) \left[\frac{1}{\sqrt{s}} h\left(\frac{\tau - t}{s} \right) \right] d\tau \qquad (9.70)$$

where s is a scale parameter and t is a time shift. The scale parameter is an alternative to the frequency parameter of the Fourier Transform. If h(t) has Fourier Transform $H(\omega)$, then h(t/s) has Fourier Transform $H(s\omega)$. Note that for a fixed h(t) that large, s compresses the transform while small s spreads the transform in frequency. There are a few requirements on a signal h(t) to be the "mother wavelet" (essentially that h(t) have finite energy and be a bandpass signal). For example, h(t) could be the output of a bandpass filter. It is also true that it is only necessary to know the Wavelet Transform at discrete values of s and t in order to be able to represent the signal. In particular

$$s = 2^m, \quad t = n2^m \quad m = \ldots, -2, 0, 1, 2, 3, \ldots$$
$$n = \ldots, -2, 0, 1, 2, 3, \ldots$$

where lower values of m correspond to smaller values of s or higher frequencies.

If x(t) is limited to a band B Hz, then it can be represented by samples at $T_S = 1/2B$ sec.

$$x(n) = x(nT_s)$$

Using a mother wavelet corresponding to an ideal bandpass filter illustrates a number of ideas. Figure 9.42 shows the filters corresponding to m = 0,1,2, and 3 and Fig. 9.43 shows the corresponding time functions. Since x(t) has no frequencies above B Hz, only positive values of m are necessary. The structure of the process can be seen in Fig. 9.44. The boxes labeled LPF_R and HPF_R are low and high pass

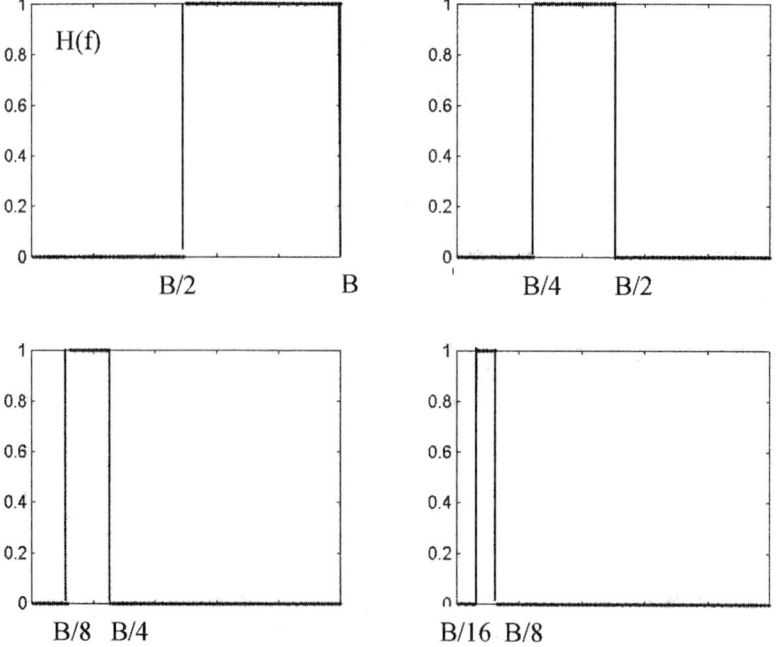

FIGURE 9.42 Ideal bandpass filters corresponding to m = 0, 1, 2, 3.

Power System Protection

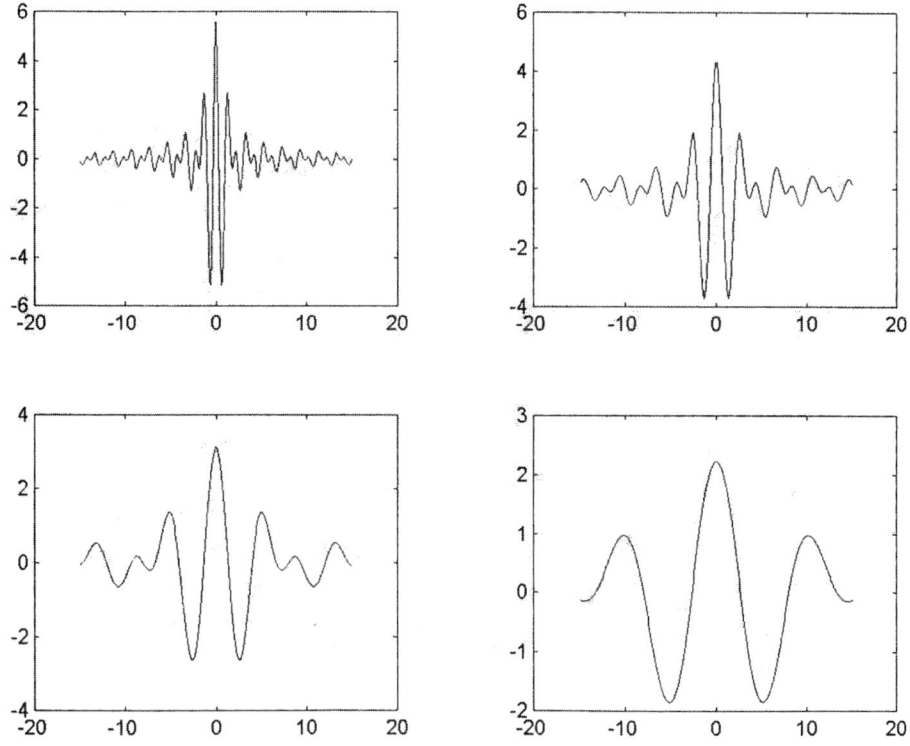

FIGURE 9.43 The impulse responses corresponding to the filters in Fig. 9.42.

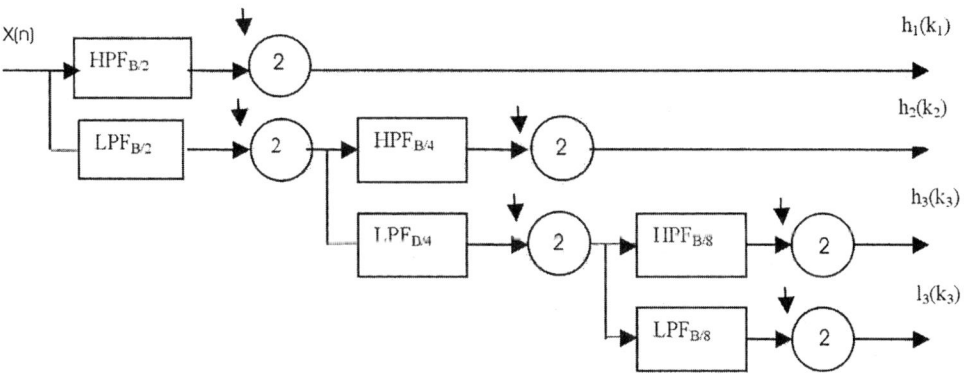

FIGURE 9.44 Cascade filter structure.

filters with cutoff frequencies of R Hz. The circle with the down arrow and a 2 represents the process of taking every other sample. For example, on the first line the output of the bandpass filter only has a bandwidth of B/2 Hz and the samples at T_S sec can be decimated to samples at $2T_S$ sec.

Additional understanding of the compression process is possible if we take a signal made of eight numbers and let the low pass filter be the average of two consecutive samples $(x(n) + x(n + 1))/2$ and the high pass filter to be the difference $(x(n) - x(n + 1))/2$ (Gail and Nielsen, 1999). For example, with

$$x(n) = [-2 \; -28 \; -46 \; -44 \; -20 \; 12 \; 32 \; 30]$$

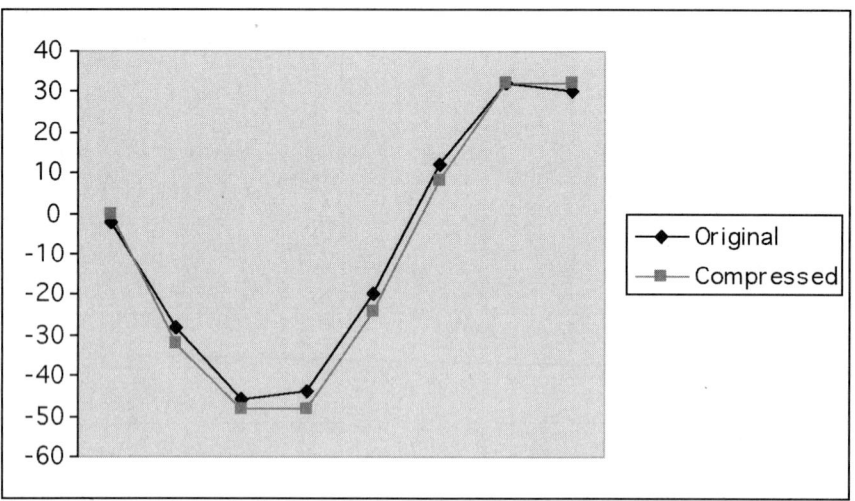

FIGURE 9.45 Original and compressed signals.

we get

$$h_1(k_1) = [13\ -1\ -16\ 1]$$
$$h_2(k_2) = [7\ -8.5]$$
$$h_3(k_3) = [7.75]$$
$$l_3(k_3) = [-0.75]$$

If we truncate to form

$$h_1(k_1) = [16\ 0\ -16\ 0]$$
$$h_2(k_2) = [8\ -8]$$
$$h_3(k_3) = [8]$$
$$l_3(k_3) = [0]$$

and reconstruct the original sequence

$$\tilde{x}(n) = [0\ -32\ -48\ -48\ -24\ 8\ 32\ 32]$$

The original and reconstructed compressed waveform is shown in Fig. 9.45. Wavelets have been applied to relaying for systems grounded through a Peterson coil where the form of the wavelet was chosen to fit unusual waveforms the Peterson coil produces (Chaari et al., 1996).

Neural Networks

Artificial Neural Networks (ANNs) had their beginning in the "perceptron," which was designed to recognize patterns. The number of papers suggesting relay application have soared. The attraction is the use of ANNs as pattern recognition devices that can be trained with data to recognize faults, inrush, or

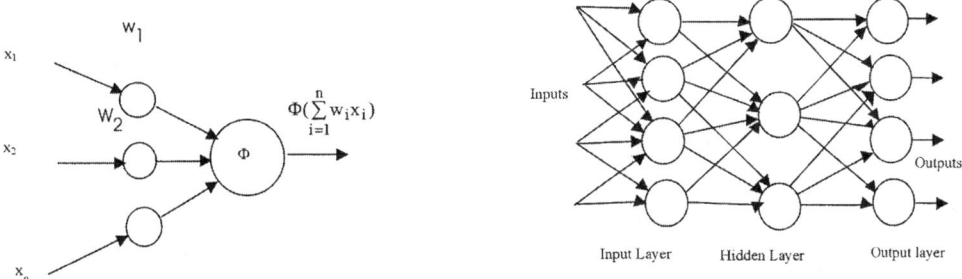

FIGURE 9.46 One neuron and a neural network.

other protection effects. The basic feed forward neural net is composed of layers of neurons as shown in Fig. 9.46.

The function Φ is either a threshold function or a saturating function such as a symmetric sigmoid function. The weights w_i are determined by training the network. The training process is the most difficult part of the ANN process. Typically, simulation data such as that obtained from EMTP is used to train the ANN. A set of cases to be executed must be identified along with a proposed structure for the net. The structure is described in terms of the number of inputs, neuron in layers, various layers, and outputs. An example might be a net with 12 inputs, and a 4, 3, 1 layer structure. There would be 4×12 plus 4×3 plus 3×1 or 63 weights to be determined. Clearly, a lot more than 60 training cases are needed to learn 63 weights. In addition, some cases not used for training are needed for testing. Software exists for the training process but judgment in determining the training sequences is vital. Once the weights are learned, the designer is frequently asked how the ANN will perform when some combination of inputs are presented to it. The ability to answer such questions is very much a function of the breadth of the training sequence.

The protective relaying application of ANNs include high-impedance fault detection (Eborn et al., 1990), transformer protection (Perez et al., 1994), fault classification (Dalstein and Kulicke, 1995), fault direction determination, adaptive reclosing (Aggarwal et al., 1994), and rotating machinery protection (Chow and Yee, 1991).

References

Aggarwal, R. K., Johns, A. T., Song, Y. H., Dunn, R. W., and Fitton, D. S., Neural-network based adaptive single-pole autoreclosure technique for EHV transmission systems, *IEEE Proceedings — C*, 141, 155, 1994.

Akke, M. and Thorp, J. S., Improved estimates from the differential equation algorithm by median post-filtering, *IEEE Sixth Int. Conf. on Development in Power System Protection*, Univ. of Nottingham, UK, March 1997.

Chaari, O, Neunier, M., and Brouaye, F., Wavelets: A new tool for the resonant grounded power distribution system relaying, *IEEE Trans. on Power Delivery*, 11, 1301, July 1996.

Chen, M. M. and Breingan, W. D., Field experience with a digital system with transmission line protection, *IEEE Trans. on Power Appar. and Syst.*, 98, 1796, Sep./Oct. 1979.

Chow, M. and Yee, S. O., Methodology for on-line incipient fault detection in single-phase squirrel-cage induction motors using artificial neural networks, *IEEE Trans. on Energy Conversion*, 6, 536, Sept. 1991.

Dalstein, T. and Kulicke, B., Neural network approach to fault classification for high speed protective relaying, *IEEE Trans. on Power Delivery*, 10, 1002, Apr. 1995.

Eborn, S., Lubkeman, D. L., and White, M., A neural network approach to the detection of incipient faults on power distribution feeders, *IEEE Trans. on Power Delivery*, 5, 905, Apr. 1990.

Gail, A. W. and Nielsen, O. M., Wavelet analysis for power system transients, *IEEE Computer Applications in Power*, 12, 16, January 1999.

Gilcrest, G. B., Rockefeller, G. D., and Udren, E. A., High-speed distance relaying using a digital computer, Part I: System description, *IEEE Trans.*, 91, 1235, May/June 1972.

Girgis, A. A. and Brown, R. G., Application of Kalman filtering in computer relaying, *IEEE Trans. on Power Appar. and Syst.*, 100, 3387, July 1981.

Harder, E. L. and Marter, W. E., Principles and practices of relaying in the United States, *AIEE Transactions*, 67, Part II, 1005, 1948.

Mann, B. J. and Morrison, I. F., Relaying a three-phase transmission line with a digital computer, *IEEE Trans. on Power Appar. and Syst.*, 90, 742, Mar./April 1971.

Perez, L. G., Flechsiz, A. J., Meador, J. L., and Obradovic, A., Training an artificial neural network to discriminate between magnetizing inrush and internal faults, *IEEE Trans. on Power Delivery*, 9, 434, Jan. 1994.

Poncelet, R., The use of digital computers for network protection, *CIGRE*, 32-98, Aug. 1972.

Rockefeller, G. D., Fault protection with a digital computer, *IEEE Trans.*, PAS-88, 438, Apr. 1969.

Rockefeller, G. D. and Udren, E. A., High-speed distance relaying using a digital computer, Part II. Test results, *IEEE Trans.*, 91, 1244, May/June 1972.

Sachdev, M. S. (Coordinator), Advancements in microprocessor based protection and communication, *IEEE Tutorial Course Text Publication*, 97TP120-0, 1997.

Tamronglak, S., Horowitz, S. H., Phadke, A. G., and Thorp, J. S., Anatomy of power system blackouts: Preventive relaying strategies, *IEEE Trans. on Power Delivery*, 11, 708, Apr. 1996.

Thorp, J. S. and Phadke, A. G., A microprocessor based voltage-restraint three-phase transformer differential relay, *Proc. South Eastern Symp. on Systems Theory*, 312, April 1982.

9.6 Use of Oscillograph Records to Analyze System Performance

John R. Boyle

Protection of present-day power systems is accomplished by a complex system of extremely sensitive relays that function only during a fault in the power system. Because relays are extremely fast, automatic oscillographs installed at appropriate locations can be used to determine the performance of protective relays during abnormal system conditions. Information from oscillographs can be used to detect the:

1. Presence of a fault
2. Severity and duration of a fault
3. Nature of a fault (A phase to ground, A – B phases to ground, etc.)
4. Location of line faults
5. Adequacy of relay performance
6. Effective performance of circuit breakers in circuit interruption
7. Occurrence of repetitive faults
8. Persistency of faults
9. Dead time required to dissipate ionized gases
10. Malfunctioning of equipment
11. Cause and possible resolution of a problem

Another important aspect of analyzing oscillograms is that of collecting data for statistical analysis. This would require a review of all oscillograms for every fault. The benefits would be to detect incipient problems and correct them before they become serious problems causing multiple interruptions or equipment damage.

An analysis of an oscillograph record shown in Fig. 9.47 should consider the nature of the fault. Substation **Y** is comprised of two lines and a transformer. The high side winding is connected to ground. Oscillographic information is available from the bus potential transformers, the line currents from breaker **A** on line 1, and the transformer neutral current. An "A" phase-to-ground fault is depicted on line 1. The

Power System Protection

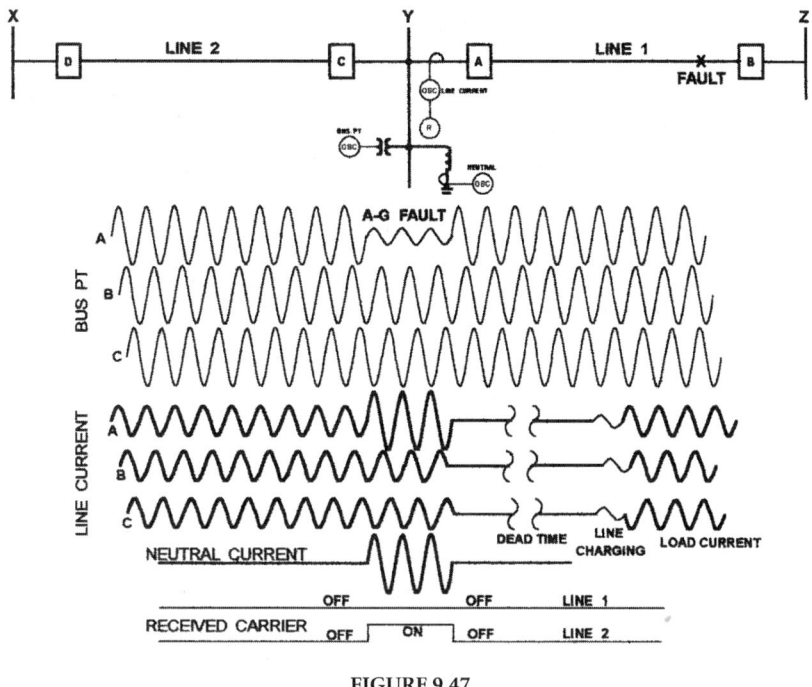

FIGURE 9.47

oscillograph reveals a significant drop in "A" phase voltage accompanied with a rise in "A" phase line 1 current and a similar rise in the transformer neutral current. The "A" phase breaker cleared the fault in 3 cycles (good). The received carrier on line 1 was "off" during the fault (good) permitting high-speed tripping at both terminals (breakers A and B). There is no evidence of AC or DC current transformer (CT) saturation of either the phase CTs or the transformer neutral CT. The received carrier signal on line 2 was "on" all during the fault to block breaker "D" from tripping at terminal "X". This would indicate that the carrier ground relays on the number 2 line performed properly. This type of analysis may not be made because of budget and personnel constraints. Oscillographs are still used extensively to analyze known cases of trouble (breaker failure, transformer damage, etc.), but oscillograph analysis can also be used as a maintenance tool to prevent equipment failure.

The use of oscillograms as a maintenance tool can be visualized by classifying operations as good (A) or questionable (B) as shown in Fig. 9.48. The first fault current waveform (upper left) is classified as A because it is sinusoidal in nature and cleared in 3 cycles. This could be a four or five cycle fault clearing time and still be classified as A depending upon the breaker characteristics (4 or 5 cycle breaker, etc.) The DC offset wave form can also be classified as A because it indicates a four cycle fault clearing time and a sinusoidal waveform with no saturation.

An example of a questionable waveform (B) is shown on the right side of Fig. 9.48. The upper right is one of current magnitude which would have to be determined by use of fault studies. Some breakers have marginal interrupting capabilities and should be inspected whenever close-in faults occur that generate currents that approach or exceed their interrupting capabilities. The waveform in the lower right is an example of a breaker restrike that requires a breaker inspection to prevent a possible breaker failure of subsequent operations.

Carrier performance on critical transmission lines is important because it impacts fast fault clearing, successful high-speed reclosing, high-speed tripping upon reclosure, and delayed breaker failure response for permanent faults upon reclosure, and a "stuck" breaker. In Fig. 9.49 two waveforms are shown that depict adequate carrier response for internal and external faults. The first waveform shows a 3 cycle fault and its corresponding carrier response. A momentary burst of carrier is cut off quickly allowing the breaker to trip in 3 cycles. Upon reclosing, load current is restored. The bottom waveform depicts the

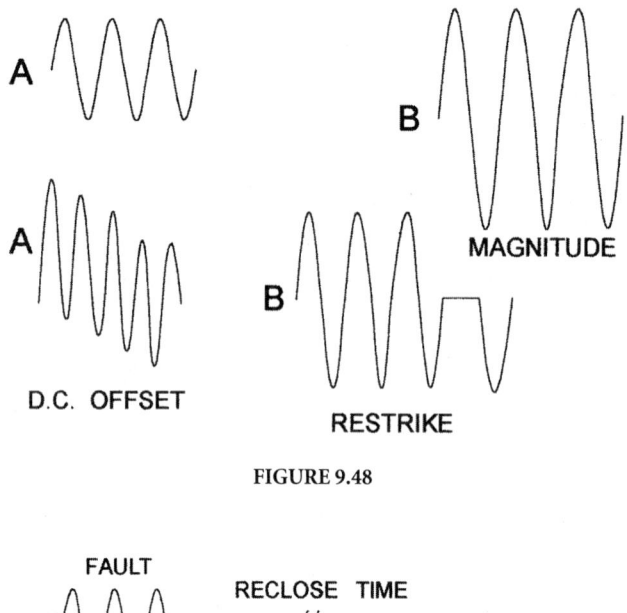

FIGURE 9.48

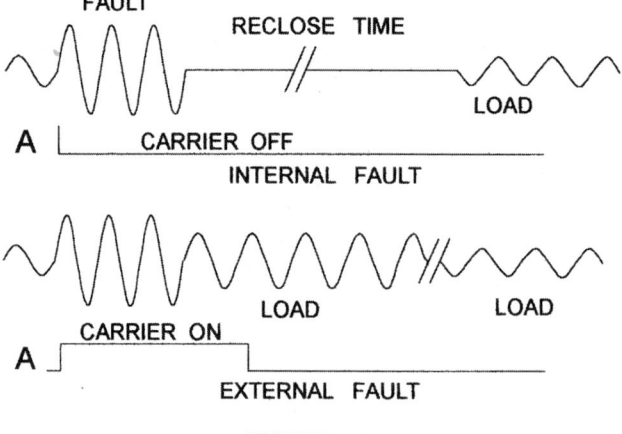

FIGURE 9.49

response of carrier on an adjacent line for the same fault. Note that carrier was "off" initially and cut "on" shortly after fault initiation. It stayed "on" for a few cycles after the fault cleared and stayed "off" all during the reclose "dead" time and after restoration of load current. Both of these waveforms would be classified as "good" and would not need further analysis.

An example of a questionable carrier response for an internal fault is shown in Fig. 9.50. Note that the carrier response was good for the initial 3 cycle fault, but during the reclose dead time, carrier came back "on" and was "on" upon reclosing. This delayed tripping an additional 2 cycles. Of even greater concern is a delay in the response of breaker-failure clearing time for a stuck breaker. Breaker failure initiation is predicated upon relay initiation which, in the case shown, is delayed 2 cycles. This type of "bad" carrier response may go undetected if oscillograms are not reviewed. In a similar manner, a delayed carrier response for an internal fault can result in delayed tripping for the initial fault as shown in Fig. 9.51. However, a delayed carrier response on an adjacent line can be more serious because it will result in two or more line interruptions. This is shown in Fig. 9.52. A fault on line 1 in Fig. 9.47 should be accompanied by acceptable carrier blocking signals on all external lines that receive a strong enough signal to trip if not accompanied by an appropriate carrier blocking signal. Two conditions are shown. A good ("A") block signal and questionable ("B") block signal. The good block signal is shown as one that blocks (comes "on") within a fraction of a cycle after the fault is detected and unblocks (goes "off") a few cycles after the fault is cleared. The questionable block signal shown at the bottom of the waveform in Fig. 9.52

Power System Protection

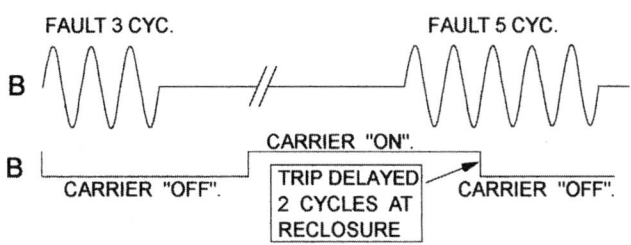

INTERNAL FAULT

FIGURE 9.50

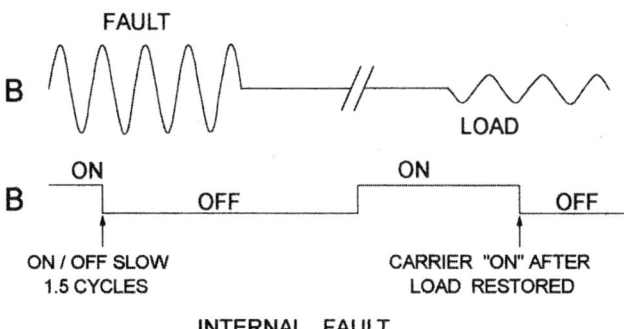

INTERNAL FAULT

FIGURE 9.51

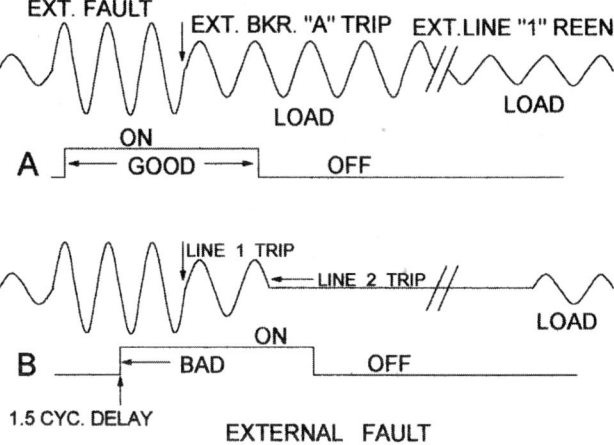

EXTERNAL FAULT

FIGURE 9.52

is late in going from "off" to "on" (1.5 cycles). The race between the trip element and the block element is such that a trip signal was initiated first and breaker "D" tripped 1.5 cycles after the fault was cleared by breaker A in 3 cycles. This would result in a complete station interruption at station "Y."

Impedance relays receive restraint from either bus or line potentials. These two potentials behave differently after a fault has been cleared. This is shown in Fig. 9.53. After breakers "A" and "B" open and the line is deenergized, the bus potential restores to its full value thereby applying full restraint to all

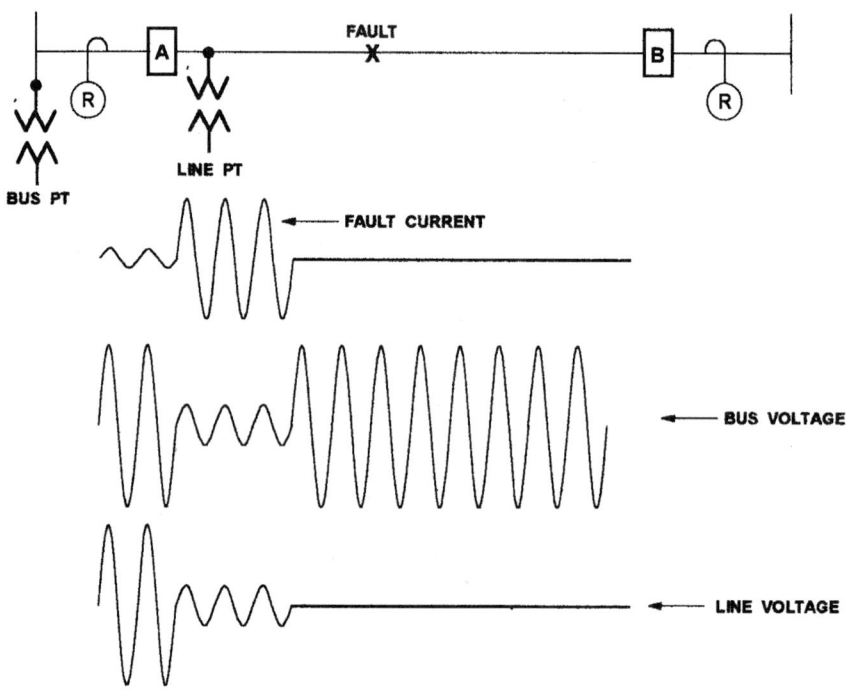

FIGURE 9.53

impedance relays connected to the bus. The line voltage goes to zero after the line is deenergized. Normally this is not a problem because relays are designed to accommodate this condition. However, there are occasions when the line potential restraint voltage can cause a relay to trip when a breaker recloses. This condition usually manifests itself when shunt reactors are connected on the line. Under these conditions an oscillatory voltage will exist on the terminals of the line side potential devices after both breakers "A" and "B" have opened. A waveform example is shown in Fig. 9.54. Note that the voltage is not a 60 Hz wave shape. Normally it is less than 60 Hz depending on the degree of compensation. This oscillatory voltage is more pronounced at high voltages because of the higher capacitance charge on the line. On lines that have flat spacing, the two outside voltages transfer energy between each other that results in oscillations that are mirror images of each other. The voltage on the center phase is usually a constant decaying decrement. These oscillations can last up to 400 cycles or more. This abnormal voltage is applied to the relays at the instant of reclosure and has been known to cause a breaker (for example, "A") to trip because of the lack of coordination between the voltage restraint circuit and the overcurrent monitoring element. Another more prevalent problem is multiple restrikes across an insulator during the oscillatory voltage on the line. These restrikes prevent the ionized gasses from dissipating sufficiently at the time of reclosure. Thus a fault is reestablished when breaker "A" and/or "B" recloses. This phenomena can readily be seen on oscillograms. Action taken might be to look for defective insulators or lengthen the reclose cycle.

The amount of "dead time" is critical to successful reclosures. For example, at 161 kV a study was made to determine the amount of dead time required to dissipate ionized gasses to achieve a 90% reclose success rate. In general, on a good line (clean insulators), at least 13 cycles of dead time are required. Contrast this to 10 cycles dead time where the reclose success rate went down to approximately 50%. Oscillograms can help determine the dead time and the cause of unsuccessful reclosures. Note the dead time is a function of the performance of the breakers at both ends of the line. Figure 9.55 depicts the performance of good breaker operations (top waveform). Here, both breakers trip in 3 cycles and reclose successfully in 13 cycles. The top waveform depicts a slow breaker "A" tripping in 6 cycles. This results in an unsuccessful reclosure because the overall dead time is reduced to 10 cycles. Note, the oscillogram

Power System Protection   9-67

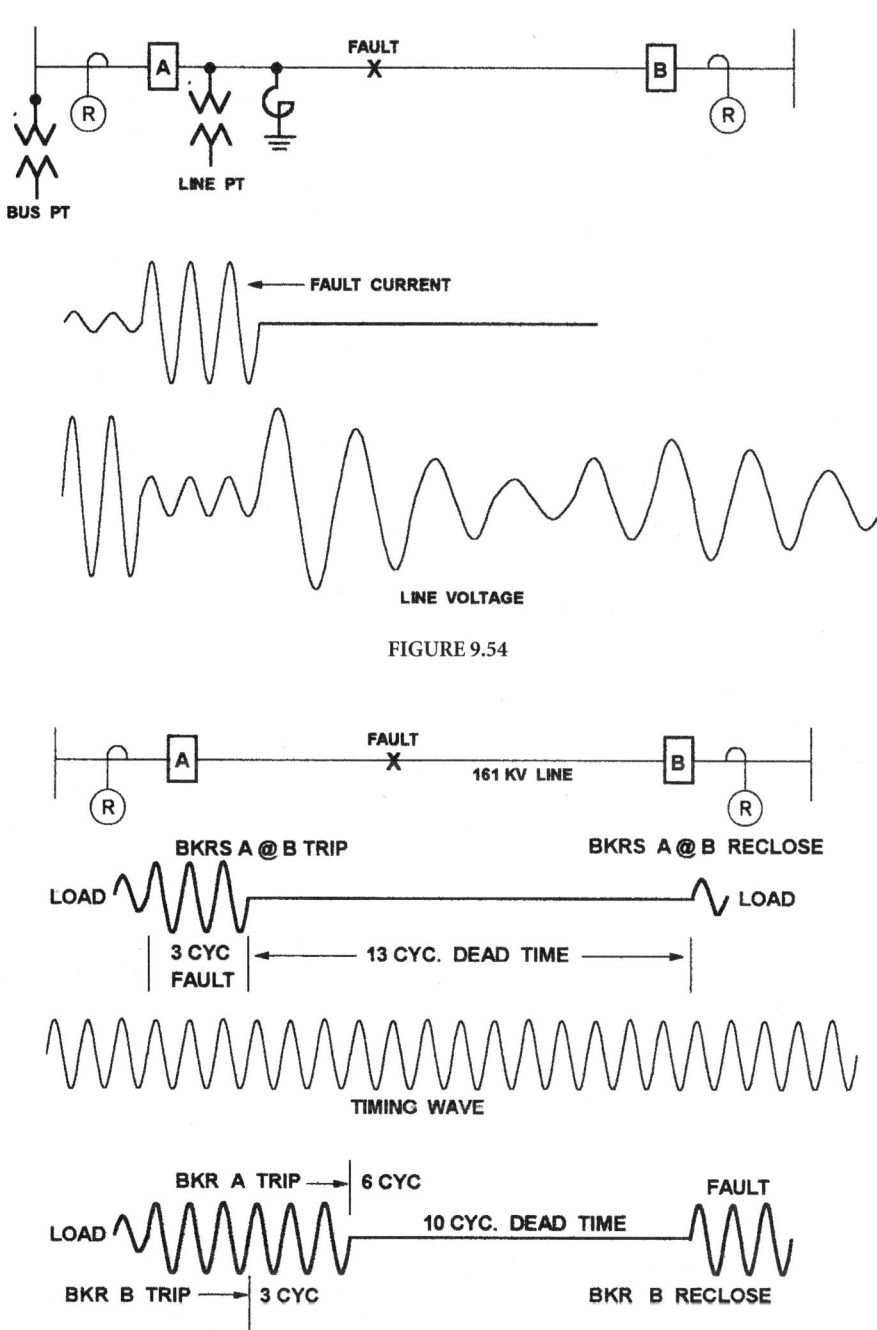

FIGURE 9.54

FIGURE 9.55

readily displays the problem. The analysis would point to possible relay or breaker trouble associated with breaker "A."

Figure 9.56 depicts current transformer (CT) saturation. This phenomenon is prevalent in current circuits and can cause problems in differential and polarizing circuits. The top waveform is an example of a direct current (DC) offset waveform with no evidence of saturation. That is to say that the secondary waveform replicates the primary waveform. Contrast this with a DC offset waveform (lower) that clearly indicates saturation. If two sets of CTs are connected differentially around a transformer and the high

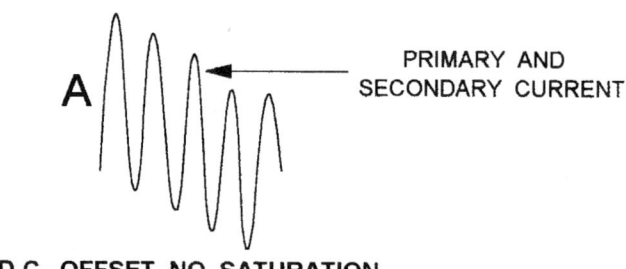
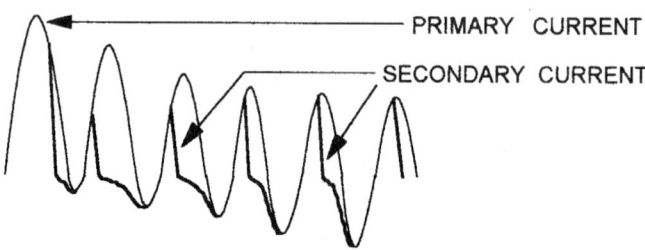

FIGURE 9.56

side CTs do not saturate (upper waveform) and the low side CTs do saturate (lower waveform), the difference current will flow through the operate coil of the relay which may result in deenergizing the transformer when no trouble exists in the transformer. The solution may be the replacement of the offending low side CT with one that has a higher "C" classification, desensitizing the relay or reducing the magnitude of the fault current. Polarizing circuits are also adversely affected by CTs that saturate. This occurs where a residual circuit is compared with a neutral polarizing circuit to obtain directional characteristics and the apparent shift in the polarizing current results in an unwanted trip.

Current reversals can result in an unwanted two-line trip if carrier transmission from one terminal to another does not respond quickly to provide the desired block function of a trip element. This is shown in a step-by-step sequence in Figs. 9.57 through 9.60. Consider a line 1 fault at the terminals of breaker

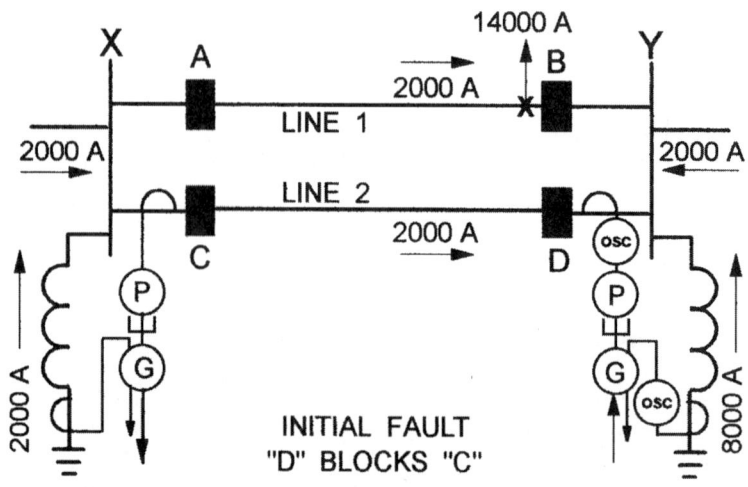

FIGURE 9.57

Power System Protection

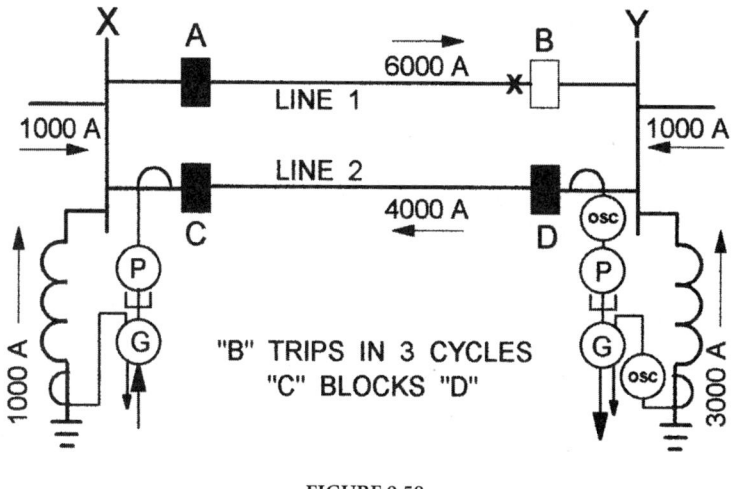

FIGURE 9.58

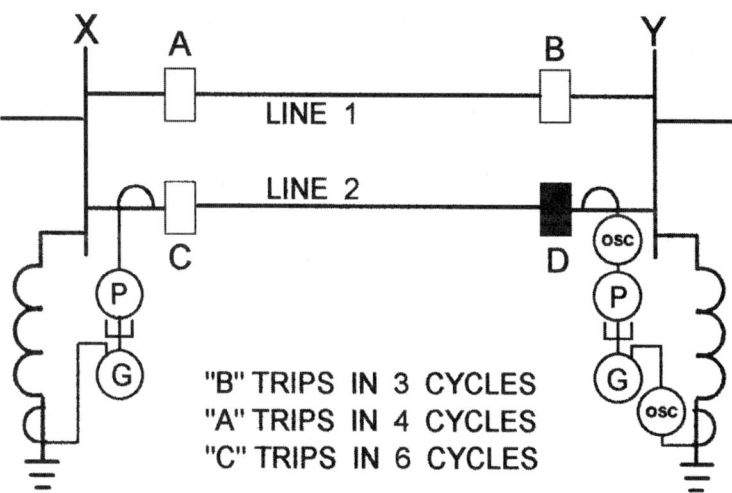

FIGURE 9.59

"B" (Fig. 9.57). For this condition, 2000 amperes of ground fault current is shown to flow on each line from terminal "X" to terminal "Y." Since fault current flow is towards the fault at breakers "A" and "B", **neither** will receive a signal (carrier "off") to initiate tripping. However, it is assumed that both breakers do not open at the same time (breaker "B" opens in 3 cycles and breaker "A" opens in 4 cycles). The response of the relays on line 2 is of prime concern. During the initial fault when breakers "A" and "B" are both closed, a block carrier signal must be sent from breaker "D" to breaker "C" to prevent the tripping of breaker "C." This is shown as a correct "on" carrier signal for 3 cycles in the bottom oscillogram trace in Fig. 9.60. However, when breaker "B" trips in 3 cycles, the fault current in line 2 increases to 4000 amperes and, more importantly, it reverses direction to flow from terminal "Y" to terminal "X." This instantaneous current reversal requires that the directional relays on breaker "C" pickup to initiate a carrier block signal to breaker "D." Failure to accomplish this may result in a trip of breaker "C" if its own carrier signal does not rise rapidly to prevent tripping through its previously made up trip directional elements. This is shown in Fig. 9.59 and oscillogram record Fig. 9.60. An alternate undesirable operation would be the tripping of breaker "D" if its trip directional elements make up before the carrier block signal from breaker "C" is received at breaker "D." The end result is the same (tripping line 2 for a fault on line 1).

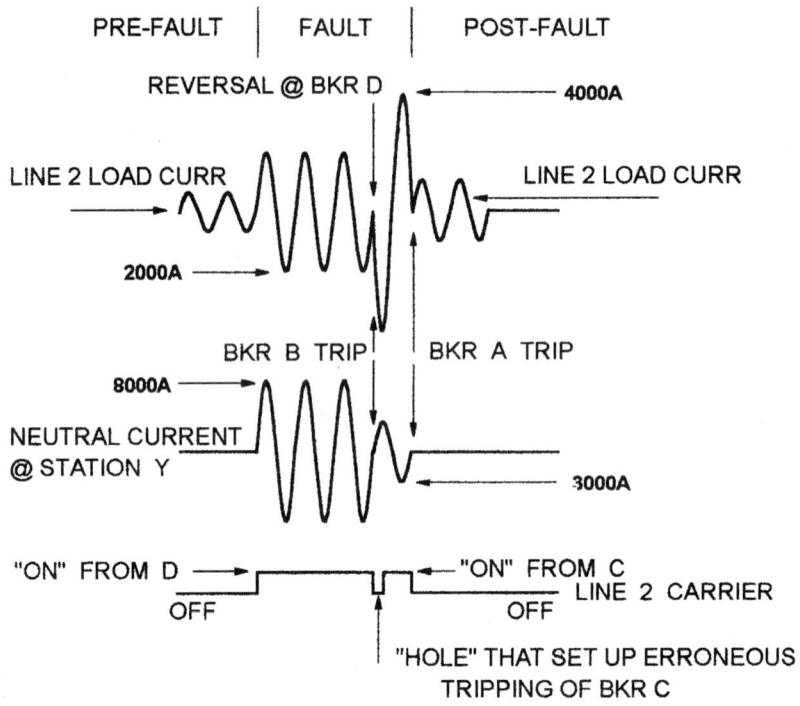

FIGURE 9.60

Restrikes in breakers can result in an explosive failure of the breaker. Oscillogams can be used to prevent breaker failures if the first restrike within the interrupter can be detected before a subsequent restrike around the interrupter results in the destruction of the breaker. This is shown diagrammatically in Fig. 9.61. The upper waveform restrike sequence depicts a ½ cycle restrike that is successfully extinguished within the interrupter. The lower waveform depicts a restrike that goes around the interrupter. This restrike cannot be extinguished and will last until the oil becomes badly carbonized and a subsequent fault occurs between the bus breaker terminal and the breaker tank (ground). In Fig. 9.61 the interrupter bypass fault lasted 18 cycles. Depending upon the rate of carbonization, the arc time could last longer or less before the flashover to the tank. The result would be the same. A bus fault that could have devastating affects. One example resulted in the loss of eight generators, thirteen 161 kV lines, and three 500-kV lines. The reason for the extensive loss was the result of burning oil that drifted up into adjacent busses steel causing multiple bus and line faults that deenergized all connected equipment in the station. The restrike phenomena is a result of a subsequent lightning strikes across the initial fault (insulator). In the example given above, lightning arresters were installed on the line side of each breaker and no additional restrikes or breaker failures occurred after the initial distructive failures.

Oscillography in microprocessor relays can also be used to analyze system problems. The problem in Fig. 9.62 involves a microprocessor differential relay installation that depicts the failure to energize a large motor. The CTs on both sides of the transformer were connected wye-wye but the low side CTs were rolled. The 30° shift was corrected in the relay and was accurately portrayed by oscillography in the microprocessor relay but the rolled CTs produced current in the operate circuit that resulted in an erroneous trip. Note that with the low side CTs rolled, the high and low side currents W1 and W2 are in phase (incorrect). The oscillography output clearly pin-pointed the problem. The corrected connection is shown in Fig. 9.63 together with the correct oscillography (W1 and W2 180° out of phase).

Power System Protection

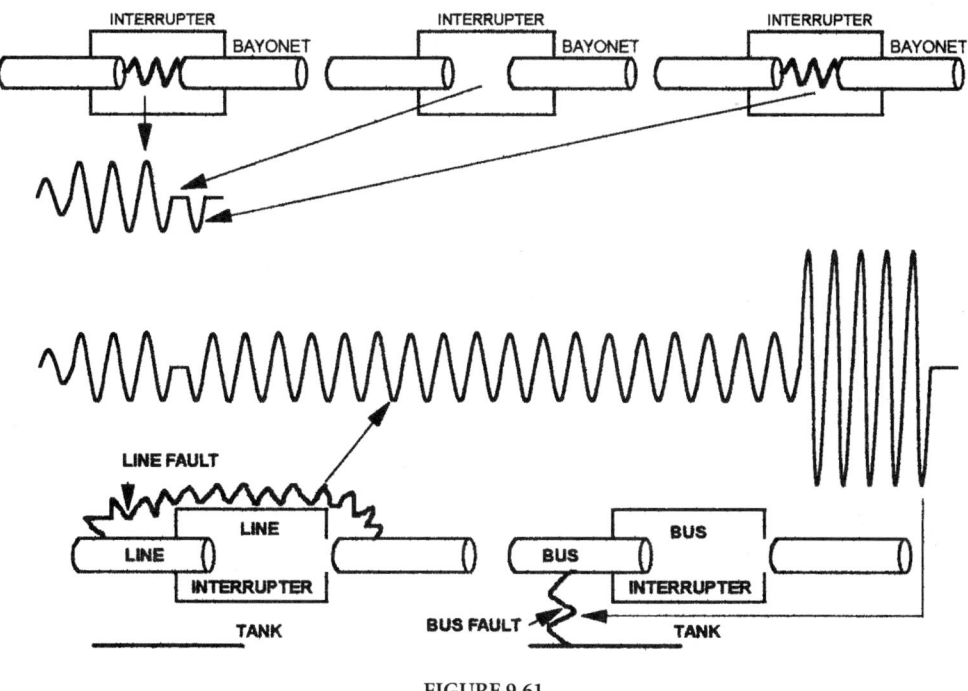

FIGURE 9.61

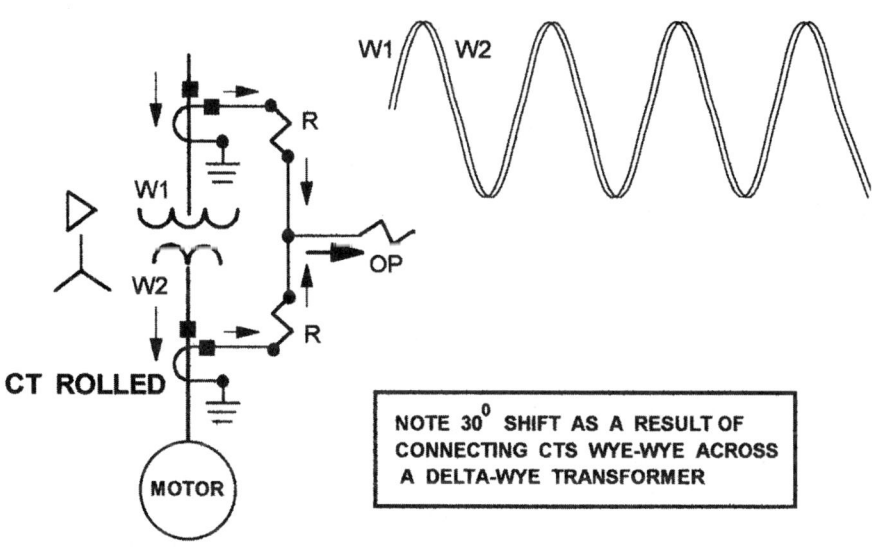

NOTE 30° SHIFT AS A RESULT OF CONNECTING CTS WYE-WYE ACROSS A DELTA-WYE TRANSFORMER

FIGURE 9.62

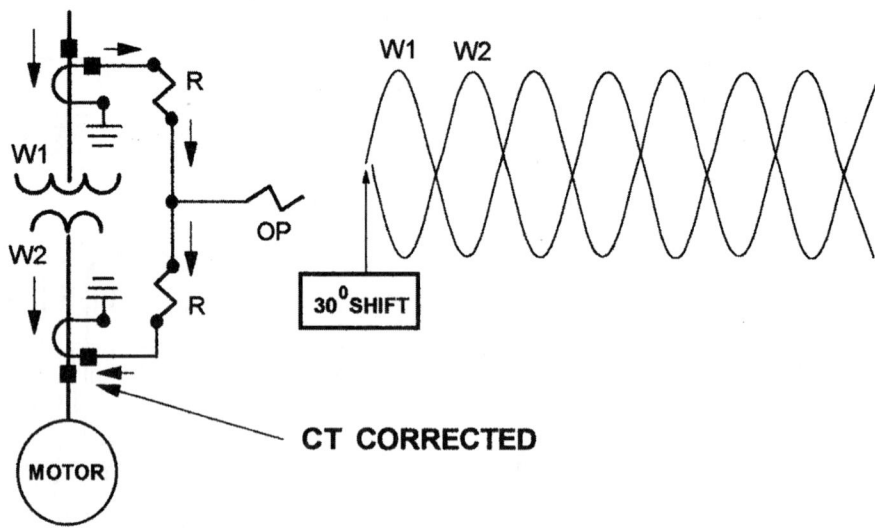

FIGURE 9.63

10
Power System Transients

Pritindra Chowdhuri
Tennessee Technological University

10.1	Characteristics of Lightning Strokes *Francisco de la Rosa*	10-2
10.2	Overvoltages Caused by Direct Lightning Strokes *Pritindra Chowdhuri*	10-8
10.3	Overvoltages Caused by Indirect Lightning Strokes *Pritindra Chowdhuri*	10-21
10.4	Switching Surges *Stephen R. Lambert*	10-36
10.5	Very Fast Transients *Juan A. Martinez-Velasco*	10-41
10.6	Transient Voltage Response of Coils and Windings *Robert C. Degeneff*	10-54
10.7	Transmission System Transients — Grounding *William Chisholm*	10-80
10.8	Insulation Coordination *Stephen R. Lambert*	10-93

10
Power System Transients

10.1	Characteristics of Lightning Strokes 10-2	
	Lightning Generation Mechanism • Parameters of Importance for Electric Power Engineering • Incidence of Lightning to Power Lines • Conclusions	
10.2	Overvoltages Caused by Direct Lightning Strokes 10-8	
	Direct Strokes to Unshielded Lines • Direct Strokes to Shielded Lines • Significant Parameters • Outage Rates by Direct Strokes	
10.3	Overvoltages Caused by Indirect Lightning Strokes 10-21	
	Inducing Voltage • Induced Voltage • Green's Function • Induced Voltage of a Doubly Infinite Single-Conductor Line • Induced Voltages on Multiconductor Lines • Effects of Shield Wires on Induced Voltages • Estimation of Outage Rates Caused by Nearby Lightning Strokes	
10.4	Switching Surges 10-36	
	Transmission Line Switching Operations • Series Capacitor Bank Applications • Shunt Capacitor Bank Applications • Shunt Reactor Applications	
10.5	Very Fast Transients 10-41	
	Origin of VFT in GIS • Propagation of VFT in GIS • Modeling Guidelines and Simulation • Effects of VFT on Equipment	
10.6	Transient Voltage Response of Coils and Windings 10-54	
	Transient Voltage Concerns • Surges in Windings • Determining Transient Response • Resonant Frequency Characteristic • Inductance Model • Capacitance Model • Loss Model • Winding Construction Strategies • Models for System Studies	
10.7	Transmission System Transients — Grounding 10-80	
	General Concepts • Material Properties • Electrode Dimensions • Self-capacitance Electrodes • Initial Transient Response from Capacitance • Ground Electrode Impedance over Perfect Ground • Ground Electrode Impedance over Imperfect Ground • Analytical Treatment of Complex Electrode Shapes • Numerical Treatment of Complex Electrode Shapes • Treatment of Multilayer Soil Effects • Layer of Finite Thickness over Insulator • Treatment of Soil Ionization • Design Recommendations	
10.8	Insulation Coordination 10-93	
	Insulation Characteristics • Probability of Flashover (pfo) • Flashover Characteristics of Air Insulation • Application of Surge Arresters	

Francisco de la Rosa
DLR Electric Power Reliability

Pritindra Chowdhuri
Tennessee Technological University

Stephen R. Lambert
Shawnee Power Consulting, LLC

Juan A. Martinez-Velasco
Universitat Politecnica de Catalunya

Robert C. Degeneff
Rensselaer Polytechnic Institute

William Chisholm
Ontario Hydro Technologies

10.1 Characteristics of Lightning Strokes

Francisco de la Rosa

Lightning, one of Mother Nature's most spectacular events, started to appear significantly demystified after Franklin showed its electric nature with his famous electrical kite experiment in 1752. Although a great deal of research on lightning has been conducted since then, lightning stands nowadays as a topic of considerable interest for investigation (Uman, 1969, 1987). This is particularly true for the improved design of electric power systems, since lightning-caused interruptions and equipment damage during thunderstorms stand as the leading causes of failures in the electric utility industry.

Lightning Generation Mechanism

First Strokes

The wind updrafts and downdrafts that take place in the atmosphere, create a charging mechanism that separates electric charges, leaving negative charge at the bottom and positive charge at the top of the cloud. As charge at the bottom of the cloud keeps growing, the potential difference between cloud and ground, which is positively charged, grows as well. This process will continue until air breakdown occurs. See Fig. 10.1.

The way in which a cloud-to-ground flash develops involves a stepped leader that starts traveling downwards following a preliminary breakdown at the bottom of the cloud. This involves a positive pocket of charge, as illustrated in Fig. 10.1. The stepped leader travels downwards in steps several tens of meters in length and pulse currents of at least 1 kA in amplitude (Uman, 1969). When this leader is near ground, the potential to ground can reach values as large as 100 MV before the attachment process with one of the upward streamers is completed. Figure 10.2 illustrates a case when the downward leader is intercepted by the upward streamer developing from a tree.

It is important to highlight that the terminating point on the ground is not decided until the downward leader is some tens of meters above the ground plane and that it will be attached to one of the growing upward streamers from elevated objects such as trees, chimneys, power lines, and communication facilities. It is actually under this principle that lightning protection rods work, i.e., they have to be strategically located so as to insure the formation of an upward streamer with a high probability of intercepting

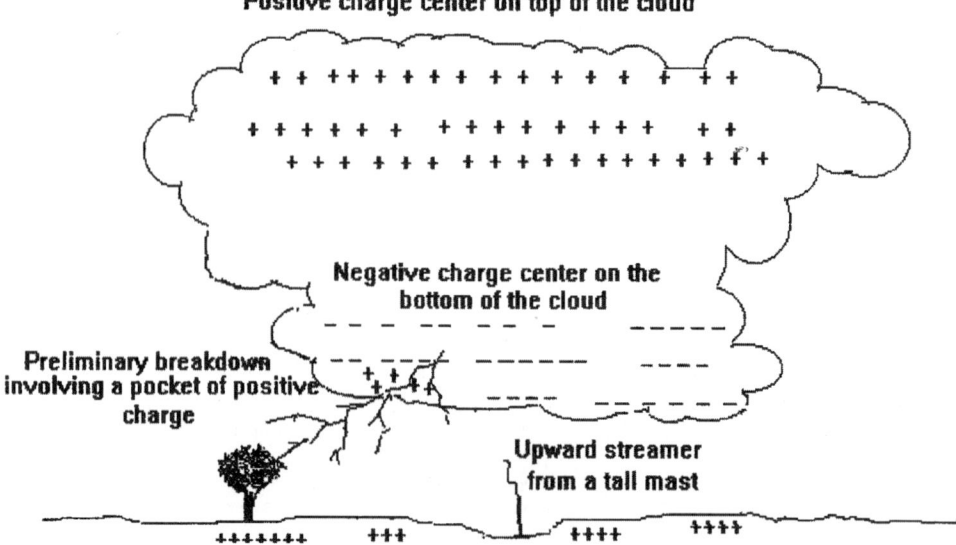

FIGURE 10.1 Separation of electric charge within a thundercloud.

Power System Transients

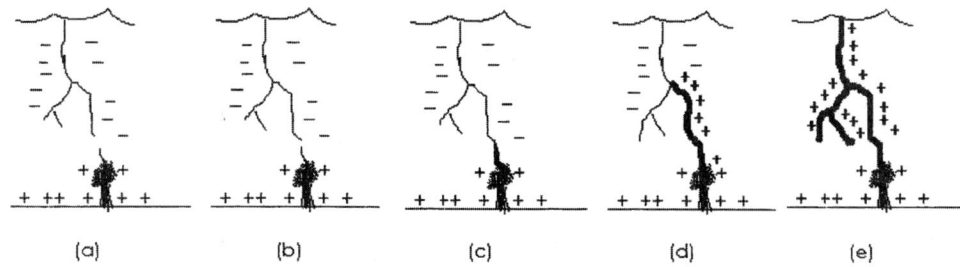

FIGURE 10.2 Attachment between downward and upward leaders in a cloud-to-ground flash.

downward leaders approaching the protected area. For this to happen, upward streamers developing from protected objects within the shielded area have to compete unfavorably with those developing from the tip of the lightning rods.

Just after the attachment process takes place, the charge that is lowered from the cloud base through the leader channel is conducted to ground while a breakdown current pulse, known as the return stroke, travels upward along the channel. The return stroke velocity is around one third the speed of light. The median peak current value associated with the return stroke is reported to be on the order of 30 kA, with rise time and time to half values around 5 and 75 μs, respectively. See Table 10.1 adapted from (Berger et al., 1975).

Associated with this charge transfer mechanism (an estimated 5 C charge is lowered to ground through the stepped leader) are the electric and magnetic field changes that can be registered at close distances

TABLE 10.1 Lightning Current Parameters for Negative Flashes[a]

Parameters	Units	Sample Size	Value Exceeding in 50% of the Cases
Peak current (minimum 2 kA)	kA		
First strokes		101	30
Subsequent strokes		135	12
Charge (total charge)	C		
First strokes		93	5.2
Subsequent strokes		122	1.4
Complete flash		94	7.5
Impulse charge	C		
(excluding continuing current)		90	4.5
First strokes		117	0.95
Subsequent strokes			
Front duration (2 kA to peak)	μs		
First strokes		89	5.5
Subsequent strokes		118	1.1
Maximum di/dt	kA/μs		
First strokes		92	12
Subsequent strokes		122	40
Stroke duration	μs		
(2 kA to half peak value on the tail)		90	75
First strokes		115	32
Subsequent strokes			
Action integral ($\int i^2 dt$)	A²s		
First strokes		91	5.5×10^4
Subsequent strokes		88	6.0×10^3
Time interval between strokes	ms	133	33
Flash duration	ms		
All flashes		94	13
Excluding single-stroke flashes		39	180

[a] Adapted from Berger et al., Parameters of lightning flashes, *Electra* No. 41, 23–37, July 1975.

from the channel and that can last several milliseconds. Sensitive equipment connected to power or telecommunication lines can get damaged when large overvoltages created via electromagnetic field coupling are developed.

Subsequent Strokes

After the negative charge from the cloud base has been transferred to ground, additional charge can be made available on the top of the channel when discharges known as J and K processes take place within the cloud (Uman, 1969). This can lead to some three to five strokes of lightning following the first stroke. A so-called dart leader develops from the top of the channel lowering charges, typically of 1 C, until recently believed to follow the same channel of the first stroke. Studies conducted in the past few years, however, indicate that around half of all lightning discharges to earth, both single- and multiple-stroke flashes, strike ground at more than one point, with the spatial separation between the channel terminations varying from 0.3 to 7.3 km, with a geometric mean of 1.3 km (Thottappillil et al., 1992).

Generally, dart leaders develop no branching and travel downward at velocities of around 3×10^6 m/s. Subsequent return strokes have peak currents usually smaller than first strokes but faster zero-to-peak rise times. The mean inter-stroke interval is about 60 ms, although intervals as large as a few tenths of a second can be involved when a so-called continuing current flows between strokes (this happens in 25–50% of all cloud-to-ground flashes). This current, which is on the order of 100 A, is associated with charges of around 10 C and constitutes a direct transfer of charge from cloud to ground (Uman, 1969).

The percentage of single-stroke flashes presently suggested by CIGRE of 45% (Anderson and Eriksson, 1980), is considerably higher than the following figures recently obtained form experimental results: 17% in Florida (Rakov et al., 1994), 14% in New Mexico (Rakov et al., 1994), 21% in Sri Lanka (Cooray and Jayaratne, 1994) and 18% in Sweden (Cooray and Perez, 1994).

Parameters of Importance for Electric Power Engineering

Ground Flash Density

Ground flash density, frequently referred as GFD or Ng, is defined as the number of lightning flashes striking ground per unit area and per year. Usually it is a long-term average value and ideally it should take into account the yearly variations that take place within a solar cycle — believed to be the period within which all climatic variations that produce different GFD levels occur.

A 10-year average GFD map of the continental U.S. obtained by and reproduced here with permission from Global Atmospherics, Inc. of Tucson, AZ, is presented in Fig. 10.3. Note the considerably large GFD levels affecting the state of Florida, as well as all the southern states along the Gulf of Mexico (Alabama, Mississippi, Louisiana, and Texas). High GFD levels are also observed in the southeastern states of Georgia and South Carolina. To the west, Arizona is the only state with GFD levels as high as 8 flashes/km²/year. The lowest GFD levels (<0.5 flashes/km²/year) are observed in the western states, notably in California, Oregon, and Washington on the Pacific Ocean, in a spot area of Colorado, and in the northeastern state of Maine on the Atlantic Ocean.

It is interesting to mention that a previous (five-year average) version of this map showed levels of around 6 flashes/km²/year also in some areas of Illinois, Iowa, Missouri, and Indiana, not seen in the present version. This is often the result of short-term observations, that do not reflect all climatic variations that take place in a longer time frame.

The low incidence of lightning does not necessarily mean an absence of lightning-related problems. Power lines, for example, are prone to failures even if GFD levels are low when they are installed in terrain with high-resistivity soils, like deserts or when lines span across hills or mountains where ground wire or lightning arrester earthing becomes difficult.

The GFD level is an important parameter to consider for the design of electric power and telecommunication facilities. This is due to the fact that power line performance and damage to power and telecommunication equipment are considerably affected by lightning. Worldwide, lightning accounts for most of the power supply interruptions in distribution lines and it is a leading cause of failures in

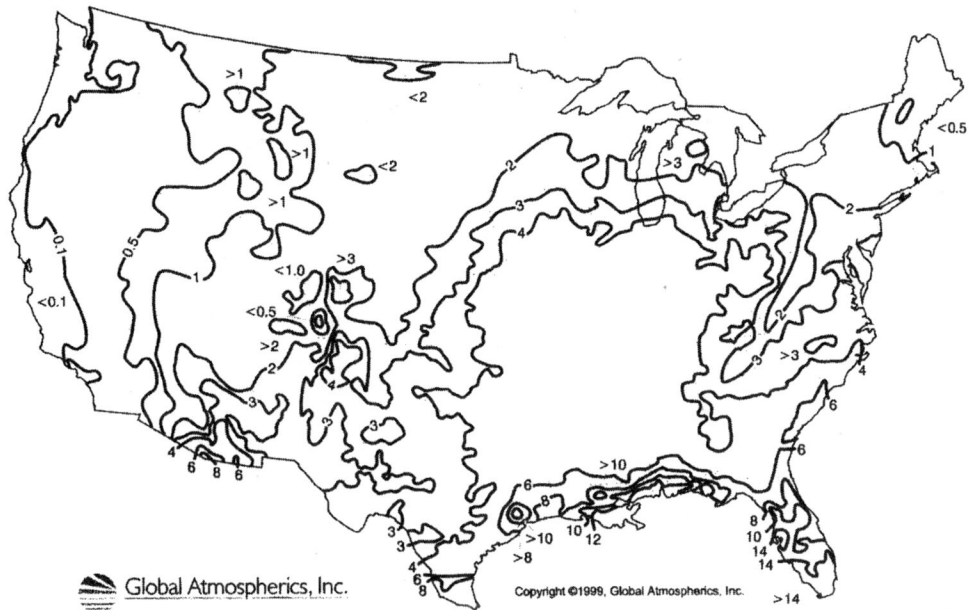

FIGURE 10.3 10-year average GFD map of the U.S. (Reproduced with permission from Global Atmospherics, Inc. of Tucson, AZ.)

transmission systems. In the U.S. alone, an estimated 30% of all power outages are lightning-related on annual average, with total costs approaching one billion dollars (Kithil, 1998).

In De la Rosa et al. (1998), it is discussed how to determine GFD as a function of TD (Thunder Days or Keraunic Level) or TH (Thunder-Hours). This is important where GFD data from lightning location systems are not available. Basically, any of these parameters can be used to get a *rough* approximation of Ground Flash Density. Using the expressions described in Anderson et al. and MacGorman et al. (1984, 1984), respectively:

$$Ng = 0.04 \, TD^{1.25} \; flashes/km^2/year \tag{10.1}$$

$$Ng = 0.054 \, TD^{1.1} \; flashes/km^2/year \tag{10.2}$$

Current Peak Value

Finally, regarding current peak values, first strokes are associated with peak currents around two to three times larger than subsequent strokes. According to De la Rosa et al. (1998), electric field records, however, suggest that subsequent strokes with higher electric field peak values may be present in one out of three cloud-to-ground flashes. These may be associated with current peak values greater than the first stroke peak.

Tables 10.1 and 10.2 are summarized and adapted from (Berger et al., 1975) for negative and positive flashes, respectively. They present statistical data for 127 cloud-to-ground flashes, 26 of them positive, measured in Switzerland. These are the type of lightning flashes known to hit flat terrain and structures of moderate height. This summary, for simplicity, shows only the 50% or statistical value, based on the

TABLE 10.2 Lightning Current Parameters for Positive Flashes[a]

Parameters	Units	Sample Size	Value Exceeding in 50% of the Cases
Peak current (minimum 2 kA)	kA	26	35
Charge (total charge)	C	26	80
Impulse charge (excluding continuing current)	C	25	16
Front duration (2 kA to peak)	µs	19	22
Maximum di/dt	kA/µs	21	2.4
Stroke duration (2 kA to half peak value on the tail)	µs	16	230
Action integral ($\int i^2 dt$)	A²s	26	6.5×10^5
Flash duration	ms	24	85

[a] Adapted from Berger et al., Parameters of lightning flashes, *Electra No. 41*, 23–37, July 1975.

log-normal approximations to the respective statistical distributions. These data are amply used as primary reference in the literature on both lightning protection and lightning research.

The action integral is an interesting concept (i.e., the energy that would be dissipated in a 1-Ω resistor if the lightning current were to flow through it). This is a parameter that can provide some insight on the understanding of forest fires and on damage to power equipment, including surge arresters, in power line installations. All the parameters presented in Tables 10.1 and 10.2 are estimated from current oscillograms with the shortest measurable time being 0.5 µs (Berger and Garbagnati, 1984). It is thought that the distribution of front duration might be biased toward larger values and the distribution of di/dt toward smaller values (De la Rosa et al., 1998).

Incidence of Lightning to Power Lines

One of the most accepted expressions to determine the number of direct strikes to an overhead line in an open ground with no nearby trees or buildings, is that described by Eriksson (1987):

$$N = N_g \left(\frac{28 h^{0.6} + b}{10} \right) \quad (10.3)$$

where

    h is the pole or tower height (m) — negligible for distribution lines
    b is the structure width (m)
    N_g is the Ground Flash Density (flashes/km²/year)
    N is the number of flashes striking the line/100 km/year. For unshielded distribution lines, this is comparable to the fault index due to direct lightning hits. For transmission lines, this is an indicator of the exposure of the line to direct strikes. (The response of the line being a function of overhead ground wire shielding angle on one hand and on conductor-tower surge impedance and footing resistance on the other hand).

Note the dependence of the incidence of strikes to the line with height of the structure. This is important since transmission lines are several times taller than distribution lines, depending on their operating voltage level.

Also important is that in the real world, power lines are to different extents shielded by nearby trees or other objects along their corridors. This will decrease the number of direct strikes estimated by Eq. (10.3) to a degree determined by the distance and height of the objects. In IEEE Std. 1410-1997, a shielding factor is proposed to estimate the shielding effect of nearby objects to the line. An important aspect of this reference work is that objects within 40 m from the line, particularly if equal or higher that 20 m, can attract most of the lightning strikes that would otherwise hit the line. Likewise, the same objects would produce insignificant shielding effects if located beyond 100 m from the line. On the other hand, sectors of lines extending over hills or mountain ridges may increase the number of strikes to the line.

The above-mentioned effects may, in some cases, cancel each other so that the estimation obtained form Eq. (10.3) can still be valid. However, it is recommended that any assessment of the incidence of lightning strikes to a power line be performed by taking into account natural shielding and orographic conditions along the line route. This also applies when identifying troubled sectors of the line for installation of metal oxide surge arresters to improve its lightning performance.

Finally, although meaningful only for distribution lines, the inducing effects of lightning, also described in De la Rosa et al. (1998) and Anderson et al. (1984), have to be considered to properly understand their lightning performance or when dimensioning the outage rate improvement after application of any mitigation action. Under certain conditions, like in circuits without grounded neutral, with low critical flashover voltages, high GFD levels, or located on high resistivity terrain, the number of outages produced by close lightning can considerably surpass those due to direct strikes to the line.

Conclusions

We have tried to present a brief overview of lightning and its effects in electric power lines. It is important to mention that a design and/or assessment of power lines considering the influence of lightning overvoltages has to undergo a more comprehensive manipulation, outside the scope of this limited discussion.

Aspects like the different methods available to calculate shielding failures and backflashovers in transmission lines, or the efficacy of remedial measures are not covered here. Among these, overhead ground wires, metal oxide surge arresters, increased insulation, or use of wood as an arc quenching device, can only be mentioned. The reader is encouraged to look further at the references or to get experienced advice for a more comprehensive understanding of the subject.

References

Anderson, R. B. and Eriksson, A. J., Lightning parameters for engineering applications, *Electra No. 69*, 65–102, March 1980.

Anderson, R. B., Eriksson, A. J., Kroninger, H., Meal, D. V., and Smith, M. A., Lightning and thunderstorm parameters, in *IEE Lightning and Power Systems Conf. Publ. No. 236*, London, 1984.

Berger, K., Anderson, R. B., and Kroninger, H., Parameters of lightning flashes, *Electra No. 41*, 23–37, July 1975.

Berger, K. and Garbagnati, E., Lightning current parameters, results obtained in Switzerland and in Italy, in *Proc. URSI Conf.*, Florence, Italy, 1984.

Cooray, V. and Jayaratne, K. P. S., Characteristics of lightning flashes observed in Sri Lanka in the tropics, *J. Geophys. Res. 99*, 21,051–21,056, 1994.

Cooray, V. and Perez, H., Some features of lightning flashes observed in Sweden, *J. Geophys. Res. 99*, 10,683–10,688, 1994.

Eriksson, A. J., The incidence of lighting strikes to power lines, in *IEEE Trans. on Power Delivery*, PWRD-2(2), 859–870, July 1987.

IEEE Std. 1410-1997, IEEE Guide for Improving the Lightning Performance of Electric Power Distribution Lines, *IEEE PES*, December, 1997, Section 5.

Kithil, R., Lightning protection codes: Confusion and costs in the USA, in *Proc. of the 24th Int'l Lightning Protection Conference*, Birmingham, U.K., Sept 16, 1998.

MacGorman, D. R., Maier, M. W., and Rust, W. D., Lightning strike density for the contiguous United States from thunderstorm duration records, in *NUREG/CR-3759, Office of Nuclear Regulatory Research*, U.S. Nuclear Regulatory Commission, Washington, D.C., 44, 1984.

Rakov, M. A., Uman, M. A., and Thottappillil, R., Review of lightning properties from electric field and TV observations, *J. Geophys. Res. 99*, 10,745–10,750, 1994.

De la Rosa, F., Nucci, C. A., and Rakov, V. A., Lightning and its impact on power systems, in *Proc. Int'l Conf. on Insulation Coordination for Electricity Development in Central European Countries*, Zagreb, Croatia, 1998.

Thottappillil, R., Rakov, V. A., Uman, M. A., Beasley, W. H., Master, M. J., and Shelukhin, D. V., Lightning subsequent stroke electric field peak greater than the first stroke and multiple ground terminations, *J. Geophys. Res.*, 97, 7,503–7,509, 1992.

Uman, M. A. *Lightning*, Dover, New York, 1969, Appendix E.

Uman, M. A., *The Lightning Discharge*, International Geophysics Series, Vol. 39, Academic Press, Orlando, FL, Chapter 1, 1987.

10.2 Overvoltages Caused by Direct Lightning Strokes

Pritindra Chowdhuri

A lightning stroke is defined as a direct stroke if it hits either the tower or the shield wire or the phase conductor. This is illustrated in Fig. 10.4. When the insulator string at a tower flashes over by direct hit either to the tower or to the shield wire along the span, it is called a backflash; if the insulator string flashes over by a strike to the phase conductor, it is called a shielding failure for a line shielded by shield wires. Of course, for an unshielded line, insulator flashover is caused by backflash when the stroke hits the tower or by direct contact with the phase conductor. In the analysis of performance and protection of power systems, the most important parameter which must be known is the insulation strength of the system. It is not a unique number. It varies according to the type of the applied voltage, e.g., DC, AC, lightning, or switching surges. For the purpose of lightning performance, the insulation strength has been defined in two ways: basic impulse insulation level (BIL) and critical flashover voltage (CFO or V_{50}). BIL has been defined in two ways. The statistical BIL is the crest value of a standard (1.2/50-µs) lightning impulse voltage that the insulation will withstand with a probability of 90% under specified conditions. The conventional BIL is the crest value of a standard lightning impulse voltage that the insulation will withstand for a specific number of applications under specified conditions. CFO or V_{50} is the crest value of a standard lightning impulse voltage that the insulation will withstand during 50% of the applications. In this section, the conventional BIL will be used as the insulation strength under lightning impulse voltages. Analysis of direct strokes to overhead lines can be divided into two classes: unshielded lines and shielded lines. The first discussion involves the unshielded lines.

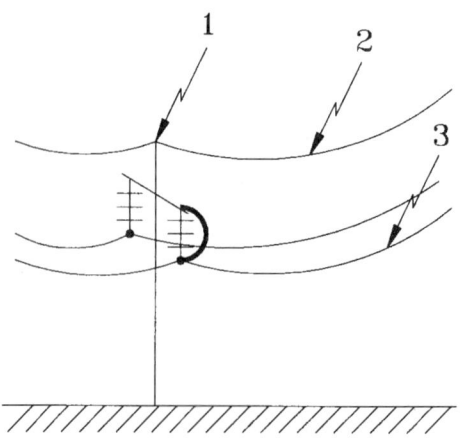

FIGURE 10.4 Illustration of direct lightning strokes to line. (1) backflash caused by direct stroke to tower; (2) backflash caused by direct stroke to shield wire; (3) insulator flashover by direct stroke to phase conductor (shielding failure).

Direct Strokes to Unshielded Lines

If lightning hits one of the phase conductors, the return-stroke current splits into two equal halves, each half traveling in either direction of the line. The traveling current waves produce traveling voltage waves that are given by:

$$V = \frac{Z_o I}{2} \tag{10.4}$$

where I is the return-stroke current and Z_o is the surge impedance of the line, given by $Z_o = (L/C)^{1/2}$, and L and C are the series inductance and capacitance to ground per meter length of the line. These traveling voltage waves stress the insulator strings from which the line is suspended as these voltages arrive at the succeeding towers. The traveling voltages are attenuated as they travel along the line by ground resistance and mostly by the ensuing corona enveloping the struck line. Therefore, the insulators of the towers adjacent to the struck point are most vulnerable. If the peak value of the voltage, given by Eq. (10.4), exceeds the BIL of the insulator, then it might flash over causing an outage. The minimum return-stroke current that causes an insulator flashover is called the critical current, I_c, of the line for the specified BIL. Thus, following Eq. (10.4):

$$I_c = \frac{2 \, BIL}{Z_o} \tag{10.5}$$

Lightning may hit one of the towers. The return-stroke current then flows along the struck tower and over the tower-footing resistance before being dissipated in the earth. The estimation of the insulator voltage in that case is not simple, especially because there has been no concensus about the modeling of the tower in estimating the insulator voltage. In the simplest assumption, the tower is neglected. Then, the tower voltage, including the voltage of the cross arm from which the insulator is suspended, is the voltage drop across the tower-footing resistance, given by $V_{tf} = IR_{tf}$, where R_{tf} is the tower-footing resistance. Neglecting the power-frequency voltage of the phase conductor, this is then the voltage across the insulator. It should be noted that this voltage will be of opposite polarity to that for stroke to the phase conductor for the same polarity of the return-stroke current.

Neglecting the tower may be justified for short towers. The effect of the tower for transmission lines must be included in the estimation of the insulator voltage. For these cases, the tower has also been represented as an inductance. Then the insulator voltage is given by $V_{ins} = V_{tf} + L(dI/dt)$, where L is the inductance of the tower.

However, it is known that voltages and currents do travel along the tower. Therefore, the tower should be modeled as a vertical transmission line with a surge impedance, Z_t, where the voltage and current waves travel with a velocity, v_t. The tower is terminated at the lower end by the tower-footing resistance, R_{tf}, and at the upper end by the lightning channel, which can be assumed to be another transmission line of surge impedance, Z_{ch}. Therefore, the traveling voltage and current waves will be repeatedly reflected at either end of the tower while producing voltage at the cross arm, V_{ca}. The insulator from which the phase conductor is suspended will then be stressed at one end by V_{ca} (to ground) and at the other end by the power-frequency phase-to-ground voltage of the phase conductor. Neglecting the power-frequency voltage, the insulator voltage, V_{ins} will be equal to the cross-arm voltage, V_{ca}. This is schematically shown in Fig. 10.5a. The initial voltage traveling down the tower, V_{to}, is $V_{to}(t) = Z_t I(t)$, where I(t) is the initial tower current which is a function of time, t. The voltage reflection coefficients at the two ends of the tower are given by:

$$a_{r1} = \frac{R_{tf} - Z_t}{R_{tf} + Z_t} \text{ and } a_{r2} = \frac{Z_{ch} - Z_t}{Z_{ch} + Z_t}. \tag{10.6}$$

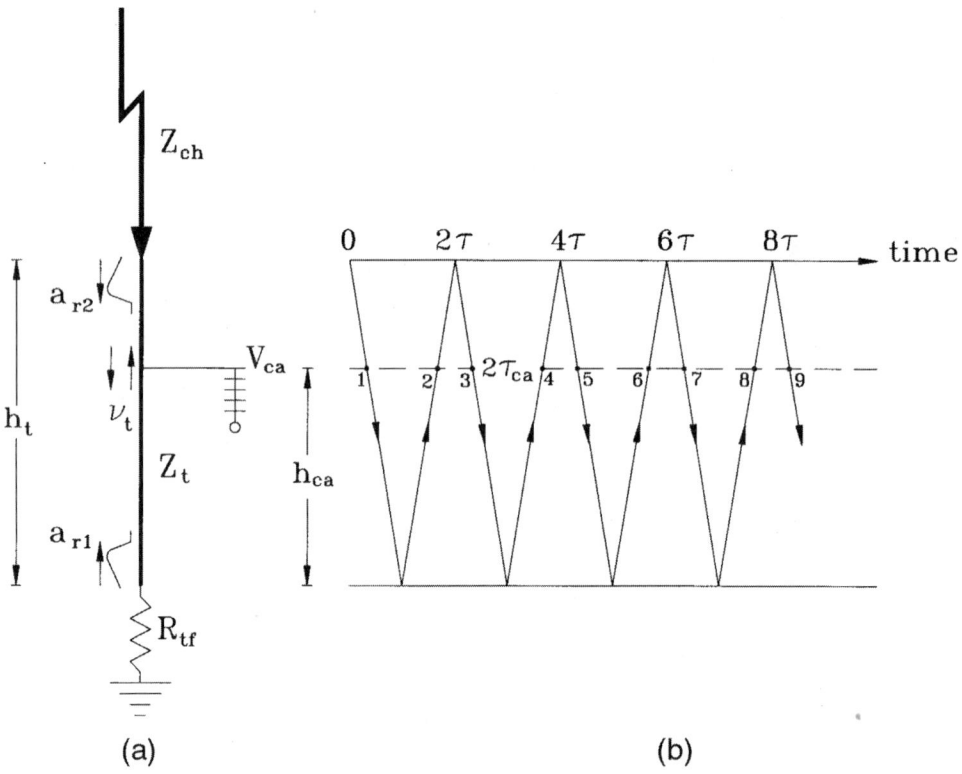

FIGURE 10.5 Lightning channel striking tower top: (a) schematic of struck tower; (b) voltage lattice diagram.

Figure 10.5b shows the lattice diagram of the progress of the multiple reflected voltage waves along the tower. The lattice diagram, first proposed by Bewley (1951), is the space-time diagram that shows the position and direction of motion of every incident, reflected, and transmitted wave on the system at every instant of time. In Fig. 10.5, if the heights of the tower and the cross arm are h_t and h_{ca}, respectively, and the velocity of the traveling wave along the tower is v_t, then the time of travel from the tower top to its foot is $\tau_t = h_t/v_t$, and the time of travel from the cross arm to the tower foot is $\tau_{ca} = h_{ca}/v_t$. In Fig. 10.5b, the two solid horizontal lines represent the positions of the tower top and the tower foot, respectively. The broken horizontal line represents the cross-arm position. It takes $(\tau_t - \tau_{ca})$ seconds for the traveling wave to reach the cross arm after lightning hits the tower top at $t = 0$. This is shown by point 1 on Fig. 10.5b. Similarly, the first reflected wave from the tower foot (point 2 in Fig. 10.5b) reaches the cross arm at $t = (\tau_t + \tau_{ca})$. The first reflected wave from the tower top (point 3 in Fig. 10.5b) reaches the cross arm at $t = (3\tau_t - \tau_{ca})$. The downward-moving voltage waves will reach the cross arm at $t = (2n-1)\tau_t - \tau_{ca}$, and the upward-moving voltage waves will reach the cross arm at $t = (2n-1)\tau_t + \tau_{ca}$, where $n = 1, 2, \cdots, n$. The cross-arm voltage, $V_{ca}(t)$ is then given by:

$$V_{ca}(t) = \sum_{n=1}^{n}(a_{r1}a_{r2})^{n-1}V_{to}\left(t-(2n-1)\tau_t+\tau_{ca}\right)u\left(t-(2n-1)\tau_t+\tau_{ca}\right)+$$

$$a_{r1}\sum_{n=1}^{n}(a_{r1}a_{r2})^{n-1}V_{to}\left(t-(2n-1)\tau_t-\tau_{ca}\right)u\left(t-(2n-1)\tau_t-\tau_{ca}\right) \quad (10.7)$$

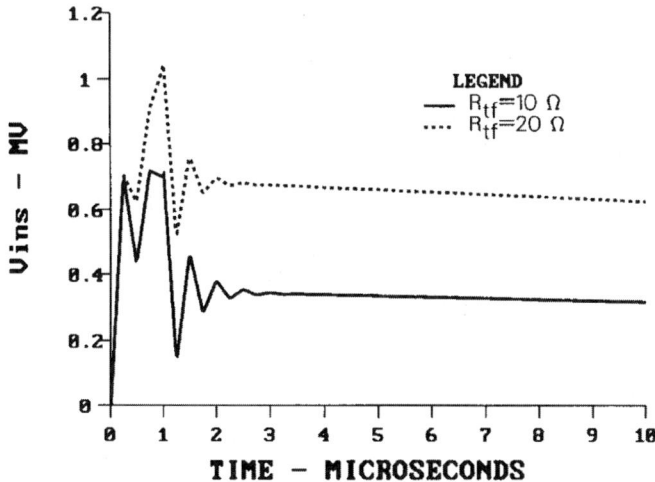

FIGURE 10.6 Profiles of insulator voltage for an unshielded line for a lightning stroke to tower. Tower height = 30 m; cross-arm height = 27.0 m; phase-conductor height = 25.0 m cross-arm width = 2.0 m; return-stroke current = 30 kA @1/50-µs; $Z_t = 100\ \Omega$; $Z_{ch} = 500\ \Omega$.

The voltage profiles of the insulator voltage, $V_{ins}(= V_{ca})$ for two values of tower-footing resistances, R_{tf}, are shown in Fig. 10.6. It should be noticed that the V_{ins} is higher for higher R_{tf} and that it approaches the voltage drop across the tower-footing resistance (IR_{tf}) with time. However, the peak of V_{ins} is significantly higher than the voltage drop across R_{tf}. Higher peak of V_{ins} will occur for (i) taller tower and (ii) shorter front time of the stroke current (Chowdhuri, 1996).

Direct Strokes to Shielded Lines

One or more conductors are strung above and parallel to the phase conductors of single- and double-circuit overhead power lines to shield the phase conductors from direct lightning strikes. These shield wires are generally directly attached to the towers so that the return-stroke currents are safely led to ground through the tower-footing resistances. Sometimes, the shield wires are insulated from the towers by short insulators to prevent power-frequency circulating currents from flowing in the closed-circuit loop formed by the shield wires, towers, and the earth return. When lightning strikes the shield wire, the short insulator flashes over, connecting the shield wire directly to the grounded towers.

For a shielded line, lightning may strike a phase conductor, the shield wire, or the tower. If it strikes a phase conductor but the magnitude of the current is below the critical current level, then no outage occurs. However, if the lightning current is higher than the critical current of the line, then it will precipitate an outage that is called the shielding failure. In fact, sometimes, shielding is so designed that a few outages are allowed, with the objective of reducing the excessive cost of shielding. However, the critical current for a shielded line is higher than that for an unshielded line because the presence of the grounded shield wire reduces the effective surge impedance of the line. The effective surge impedance of a line shielded by one shield wire is given by (Chowdhuri, 1996):

$$Z_{eq} = Z_{11} - \frac{Z_{12}^2}{Z_{22}} \tag{10.8}$$

$$\text{where } Z_{11} = 60\ \ell n \frac{2h_p}{r_p};\ Z_{22} = 60\ \ell n \frac{2h_s}{r_s};\ Z_{12} = 60\ \ell n \frac{d_{p's}}{d_{ps}} \tag{10.9}$$

Here, h_p and r_p are the height and radius of the phase conductor, h_s and r_s are the height and radius of the shield wire, $d_{p's}$ is the distance from the shield wire to the image of the phase conductor in the ground, and d_{ps} is the distance from the shield wire to the phase conductor. Z_{11} is the surge impedance of the phase conductor in the absence of the shield wire, Z_{22} is the surge impedance of the shield wire, and Z_{12} is the mutual surge impedance between the phase conductor and the shield wire.

It can be shown that either for strokes to tower or for strokes to shield wire, the insulator voltage will be the same if the attenuation caused by impulse corona on the shield wire is neglected (Chowdhuri, 1996). For a stroke to tower, the return-stroke current will be divided into three parts: two parts going to the shield wire in either direction from the tower, and the third part to the tower. Thus, lower voltage will be developed along the tower of a shielded line than that for an unshielded line for the same return-stroke current, because lower current will penetrate the tower. This is another advantage of a shield wire. The computation of the cross-arm voltage, V_{ca}, is similar to that for the unshielded line, except for the following modifications in Eqs. (10.6) and (10.7):

1. The initial tower voltage is equal to IZ_{eq}, instead of IZ_t as for the unshielded line, where Z_{eq} is the impedance as seen from the striking point, i.e.,

$$Z_{eq} = \frac{0.5 Z_s Z_t}{0.5 Z_s + Z_t}, \quad (10.10)$$

where $Z_s = 60 \ln(2h_s/r_s)$ is the surge impedance of the shield wire.

2. The traveling voltage wave moving upward along the tower, after being reflected at the tower foot, encounters three parallel branches of impedances, the lightning-channel surge impedance, and the surge impedances of the two halves of the shield wire on either side of the struck tower. Therefore, Z_{ch} in Eq. (10.6) should be replaced by $0.5 Z_s Z_{ch}/(0.5 Z_s + Z_{ch})$.

The insulator voltage, V_{ins}, for a shielded line is not equal to V_{ca}, as for the unshielded line. The shield-wire voltage, which is the same as the tower-top voltage, V_{tt}, induces a voltage on the phase conductor by electromagnetic coupling. The insulator voltage is, then, the difference between V_{ca} and this coupled voltage:

$$V_{ins} = V_{ca} - k_{sp} V_{tt}, \quad (10.11)$$

where $k_{sp} = Z_{12}/Z_{22}$. It can be seen that the electromagnetic coupling with the shield wire reduces the insulator voltage. This is another advantage of the shield wire. To compute V_{tt}, we go back to Fig. 10.5. As the cross arm is moved toward the tower top, τ_{ca} approaches τ_t, and naturally, at tower top $\tau_{ca} = \tau_t$. Then, except the wave 1, the pairs of upward-moving and downward-moving voltages (e.g., 2 and 3, 4 and 5, etc.) arrive at the tower top at the same time. Putting $\tau_{ca} = \tau_t$ in Eq. (10.7), and writing $a_{t2} = 1 + a_{r2}$, we get V_{tt}:

$$V_{tt}(t) = V_{to} u(t) + a_{t2} a_{r1} \sum_{n=1}^{n} (a_{r1} a_{r2})^{n-1} V_{to}(t - 2n\tau_t) u(t - 2n\tau_t). \quad (10.12)$$

From Eq. (10.6), $a_{t2} = 1 + a_{r2} = \dfrac{2 Z_{ch}}{Z_{ch} + Z_t}. \quad (10.13)$

The coefficient, a_{t2}, is called the coefficient of voltage transmission.

When lightning strikes the tower, equal voltages (IZ_{eq}) travel along the tower as well as along the shield wire in both directions. The voltages on the shield wire are reflected at the subsequent towers and arrive

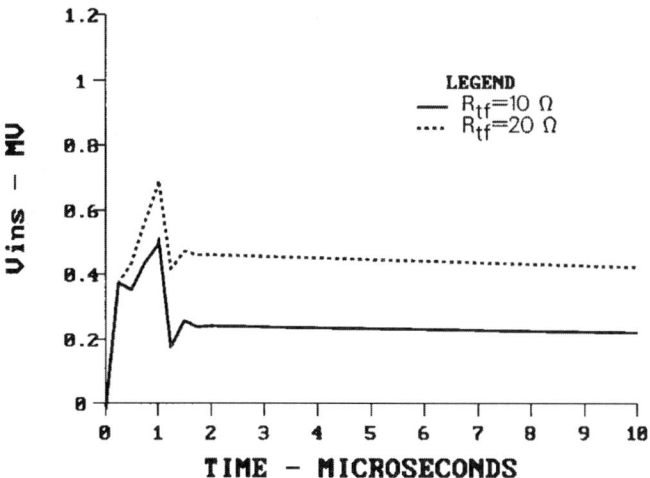

FIGURE 10.7 Profiles of insulator voltage for a shielded line for lightning stroke to tower. Tower height = 30 m; cross-arm height = 27.0 m; phase-conductor height = 25.0 m cross-arm width = 2.0 m; return-stroke current = 30 kA @1/50-μs; Z_t = 100 Ω; Z_{ch} = 500 Ω.

back at the struck tower at different intervals as voltages of opposite polarity (Chowdhuri, 1996). Generally, the reflections from the nearest towers are of any consequence. These reflected voltage waves lower the tower-top voltage. The tower-top voltage remains unaltered until the first reflected waves arrive from the nearest towers. The profiles of the insulator voltage for the same line as in Fig. 10.6 but with a shield wire are shown in Fig. 10.7. Comparing Figs. 10.6 and 10.7, it should be noticed that the insulator voltage is significantly reduced for a shielded line for a stroke to tower. This reduction is possible because (i) a part of the stroke current is diverted to the shield wire, thus reducing the initial tower-top voltage (V_{to} = $I_t Z_t$, I_t < I), and (ii) the electromagnetic coupling between the shield wire and the phase conductor induces a voltage on the phase conductor, thus lowering the voltage difference across the insulator (V_{ins} = $V_{ca} - k_{sp} V_{tt}$).

Shielding Design

Striking distance of the lightning stroke plays a crucial role in the design of shielding. Striking distance is defined as the distance through which a descending stepped leader will strike a grounded object. Whitehead and his associates (1968; 1969) proposed a simple relation between the striking distance, r_s, and the return-stroke current, I, (in kA) of the form:

$$r_s = aI^b \ (m) \tag{10.14}$$

where a and b are constants. The most frequently used value of a is 8 or 10, and that of b is 0.65. Let us suppose that a stepped leader with prospective return-stroke current of I_s, is descending near a horizontal conductor, P, (Fig. 10.8a). Its striking distance, r_s, will be given by Eq. (10.14). It will hit the surface of the earth when it penetrates a plane which is r_s meters above the earth. The horizontal conductor will be struck if the leader touches the surface of an imaginary cylinder of radius, r_s, with its center at the center of the conductor. The attractive width of the horizontal conductor will be ab in Fig. 10.8a. It is given by:

$$ab = 2\omega_p = 2\sqrt{r_s^2 - (r_s - h_p)^2} = 2\sqrt{h_p(2r_s - h_p)} \ \text{for } r_s > h_p \text{ and} \tag{10.15a}$$

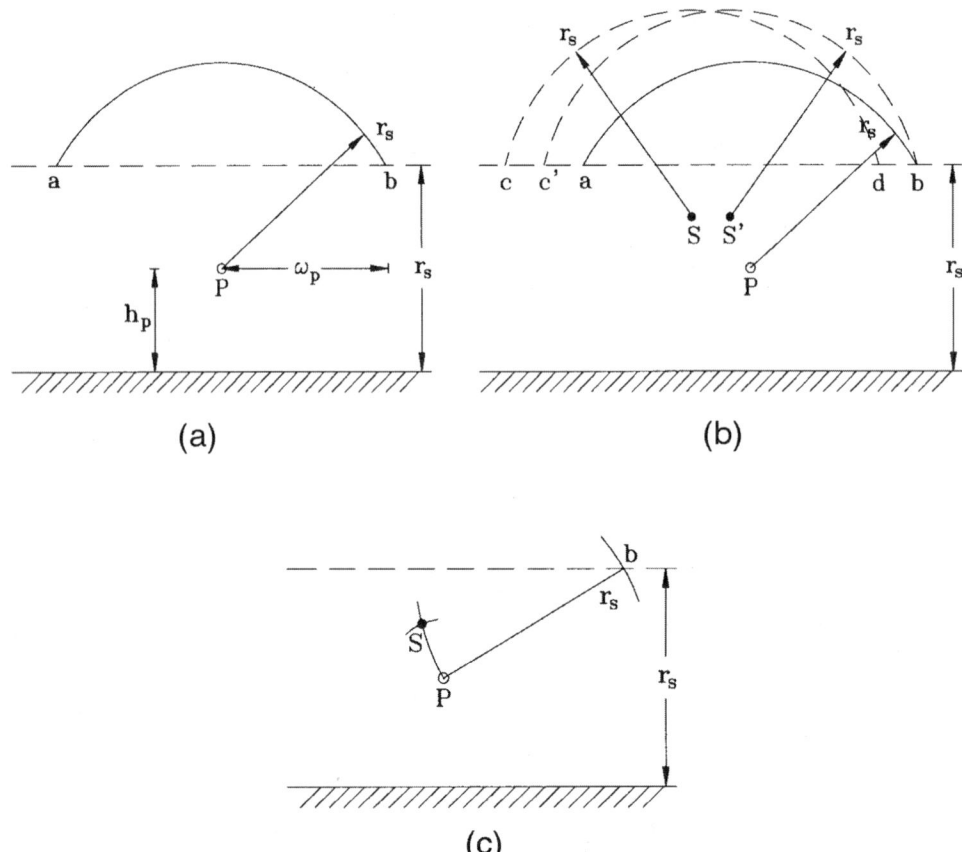

FIGURE 10.8 Principle of shielding: (a) electrogeometric model; (b) shielding principle; (c) placement of shield wire for perfect shielding.

$$ab = 2\omega_p = 2r_s \text{ for } r_s \le h_p, \qquad (10.15b)$$

where h_p is the height of the conductor. For a multiconductor line with a separation distance, d_p, between the outermost conductors, the attractive width will be $2\omega_p + d_p$.

Now, if a second horizontal conductor, S, is placed near P, the attractive width of S will be cd (Fig. 10.8b). If S is intended to completely shield P, then the cylinder around S and the r_s-plane above the earth's surface must completely surround the attractive cylinder around P. However, as Fig. 10.8b shows, an unprotected width, db remains. Stepped leaders falling through db will strike P. If S is repositioned to S' so that the point d coincides with b, then P is completely shielded by S.

The procedure to place the conductor, S, for perfect shielding of P is shown in Fig. 10.8c. Knowing the critical current, I_c, from Eq. (10.5), the corresponding striking distance, r_s, is computed from Eq. (10.14). A horizontal straight line is drawn at a distance r_s above the earth's surface. An arc of radius, r_s, is drawn with P as center, which intersects the r_s-line above earth at b. Then, an arc of radius, r_s, is drawn with b as center. This arc will go through P. Now, with P as radius, another arc is drawn of radius r_{sp}, where r_{sp} is the minimum required distance between the phase conductor and a grounded object. This arc will intersect the first arc at S, which is the position of the shield wire for perfect shielding of P.

Figure 10.9 shows the placement of a single shield wire above a three-phase horizontally configured line for shielding. In Fig. 10.9a, the attractive cylinders of all three phase conductors are contained within the attractive cylinder of the shield wire and the r_s-plane above the earth. However, in Fig. 10.9b where

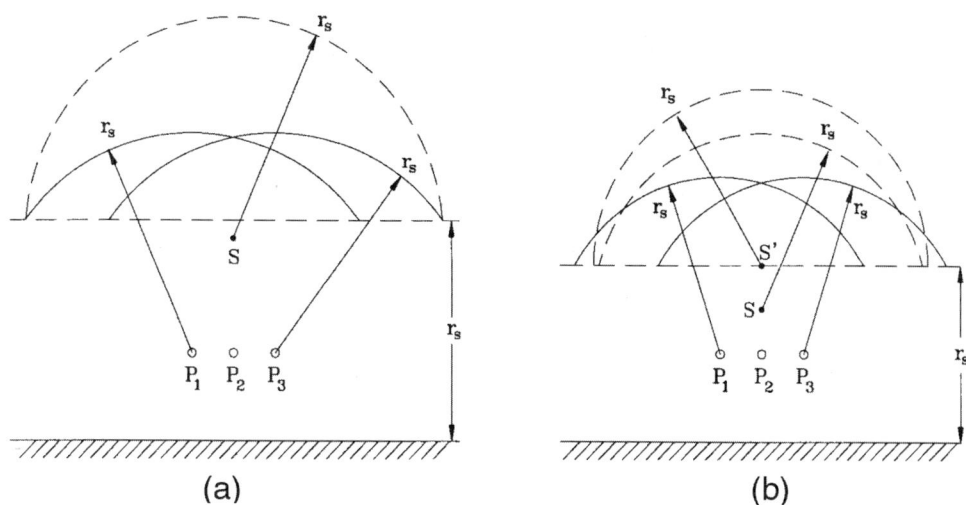

FIGURE 10.9 Shielding of three-phase horizontally configured line by single shield wire: (a) perfect shielding; (b) imperfect shielding.

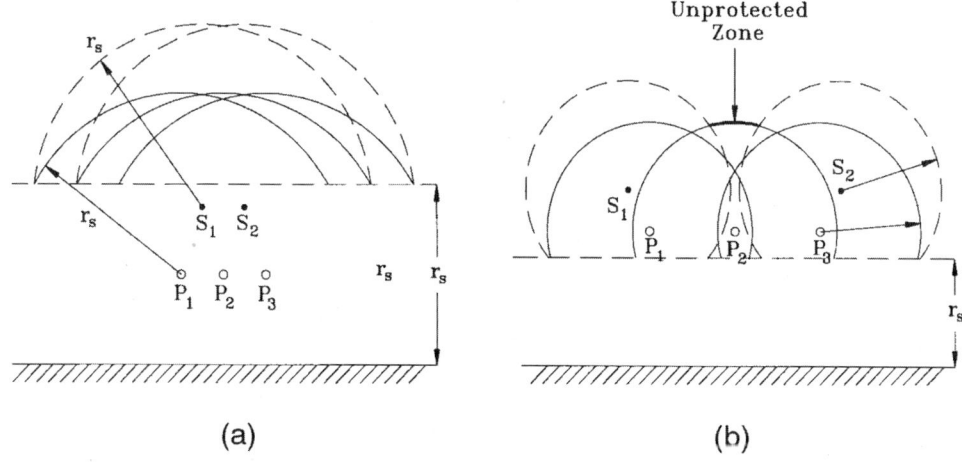

FIGURE 10.10 Shielding of three-phase horizontally configured line by two shield wires: (a) perfect shielding; (b) imperfect shielding.

the critical current is lower, the single shield wire at S cannot perfectly shield the two outer phase conductors. Raising the shield wire helps in reducing the unprotected width, but, in this case, it cannot completely eliminate shielding failure. As the shield wire is raised, its attractive width increases until the shield-wire height reaches the r_s-plane above earth, where the attractive width is the largest, equal to the diameter of the r_s-cylinder of the shield wire. Raising the shield-wire height further will then be actually detrimental. In this case, either the insulation strength of the line should be increased (i.e., the critical current increased) or two shield wires should be used.

Figure 10.10 shows the use of two shield wires. In Fig. 10.10a, all three phase conductors are completely shielded by the two shield wires. However, for smaller I_c (i.e., smaller r_s), part of the attractive cylinder of the middle phase conductor is left exposed (Fig. 10.10b). This shows that the middle phase conductor may experience shielding failure even when the outer phase conductors are perfectly shielded. In that case, either the insulation strength of the line should be increased or the height of the shield wires raised, or both.

Significant Parameters

The most significant parameter in estimating the insulator voltage is the return-stroke current, i.e., its peak value, waveshape, and statistical distributions of the amplitude and waveshape. The waveshape of the return-stroke current is generally assumed to be double exponential where the current rapidly rises to its peak exponentially, and subsequently decays exponentially:

$$I(t) = I_o \left(e^{-a_1 t} - e^{-a_2 t} \right). \tag{10.16}$$

The parameters, I_o, a_1, and a_2 are determined from the given peak, I_p, the front time, t_f, and the time to half value, t_h, during its subsequent decay. However, the return-stroke current can also be simulated as a linearly rising and linearly falling wave:

$$I(t) = \alpha_1 t u(t) - \alpha_2 (t - t_f) u(t - t_f), \tag{10.17}$$

$$\text{where, } \alpha_1 = \frac{I_p}{t_f}, \text{ and } \alpha_2 = \frac{2 t_h - t_f}{2 t_f (t_h - t_f)} I_p. \tag{10.18}$$

I_o, a_1, and a_2 of the double exponential function in Eq. (10.16) are not very easy to evaluate. In contrast, α_1 and α_2 of the linear function in Eq. (10.17) are easy to evaluate as given in Eq. (10.18). The results from the two waveshapes are not significantly different, particularly for lightning currents where t_f is on the order of a few microseconds and t_h is several tens of microseconds. As t_h is very long compared to t_f, the influence of t_h on the insulator voltage is not significant. Therefore, any convenient number can be assumed for t_h (e.g., 50 μs) without loss of accuracy.

The statistical variations of the peak return-stroke curent, I_p, fits the log-normal distribution (Popolansky, 1972). The probability density function, $p(I_p)$, of I_p can then be expressed as:

$$p(I_p) = \frac{1}{\sqrt{2\pi} I_p \sigma_{\ell n I_p}} e^{-0.5 \left(\frac{\ell n I_p - \ell n I_{pm}}{\sigma_{\ell n I_p}} \right)^2}, \tag{10.19}$$

where $\sigma_{\ell n I_p}$ is the standard deviation of $\ell n I_p$, and I_{pm} is the median value of the return-stroke current, I_p. The cumulative probability, P_c, that the peak current in any lightning flash will exceed I_p kA can be determined by integrating Eq. (10.19) as follows:

$$\text{Putting } u = \frac{\ell n I_p - \ell n I_{pm}}{\sqrt{2} \sigma_{\ell n I_p}} \tag{10.20}$$

$$P_c(I_p) = \frac{1}{\sqrt{\pi}} \int_u^\infty e^{-u^2} du = 0.5 \, \text{erfc}(u). \tag{10.21}$$

The probability density function, $p(t_f)$, of the front time, t_f, can be similarly determined by replacing I_{pm} and $\sigma_{\ell n I_p}$ by the corresponding t_{fm} and $\sigma_{\ell n t_f}$ in Eqs. (10.20) and (10.21). Assuming no correlation between I_p and t_f, the joint probability density function of I_p and t_f is $p(I_p, t_f) = p(I_p) p(t_f)$. The equation for $p(I_p, t_f)$ becomes more complex if there is a correlation between I_p and t_f (Chowdhuri, 1996). The statistical parameters (I_{pm}, $\sigma_{\ell nIp}$, t_{fm} and $\sigma_{\ell ntf}$) have been analyzed in (Anderson and Eriksson, 1980; Eriksson, 1986) and are given in (Chowdhuri, 1996):

$$t_{fm} = 3.83 \ \mu s; \ \sigma_{\ell ntf} = 0.553$$

$$\text{For } I_p \leq 20 \text{ kA}: I_{pm} = 61.1 \text{ kA}; \ \sigma_{\ell nIp} = 1.33$$

$$\text{For } I_p > 20 \text{ kA}: I_{pm} = 33.3 \text{ kA}; \ \sigma_{\ell nIp} = 0.605$$

Besides I_p anf t_f, the ground flash density, n_g, is the third significant parameter in estimating the lightning performance of power systems. The ground flash density is defined as the average number of lightning strokes per square kilometer per year in a geographic region. It should be borne in mind that the lightning activity in a particular geographic region varies by a large margin from year to year. Generally, the ground flash density is averaged over ten years. In the past, the index of lightning severity was the keraunic level (i.e., the number of thunder days in a region) because that was the only parameter available. Several empirical equations have been used to relate keraunic level with n_g. However, there has been a concerted effort in many parts of the world to measure n_g directly, and the measurement accuracy has also been improved in recent years.

Outage Rates by Direct Strokes

The outage rate is the ultimate gauge of lightning performance of a transmission line. It is defined as the number of outages caused by lightning per 100 km of line length per year. One needs to know the attractive area of the line in order to estimate the outage rate. The line is assumed to be struck by lightning if the stroke falls within the attractive area. The electrical shadow method has been used to estimate the attractive area. According to the electrical shadow method, a line of height, h_ℓ m, will attract lightning from a distance of $2h_\ell$ m from either side. Therefore, for a 100-km length, the attractive area will be $0.4h_\ell$ km². This area is then a constant for a specific overhead line of given height, and is independent of the severity of the lightning stroke (i.e., I_p). The electrical shadow method has been found to be unsatisfactory in estimating the lightning performance of an overhead power line. Now, the electrogeometric model is used in estimating the attractive area of an overhead line. The attractive area is estimated from the striking distance, which is a function of the return-stroke current, I_p, as given by Eq. (10.14). Although it has been suggested that the striking distance should also be a function of other variables (Chowdhuri and Kotapalli, 1989), the striking distance as given by Eq. (10.14) is being universally used.

The first step in the estimation of outage rate is the determination of the critical current. If the return-stroke current is less than the critical current, then the insulator will not flash over if the line is hit by the stepped leader. If one of the phase conductors is struck, such as for an unshielded line, then the critical current is given by Eq. (10.5). However, for strikes either to the tower or to the shield wire of a shielded line, the critical current is not that simple to compute if the multiple reflections along the tower are considered as in Eqs. (10.7) or (10.12). For these cases, it is best to compute the insulator voltage first by Eqs. (10.7) or by (10.12) for a return-stroke current of 1 kA, then estimate the critical current by taking the ratio between the insulation strength and the insulator voltage caused by 1 kA of return-stroke current of the specified front time, t_f, bearing in mind that the insulator voltage is a function of t_f.

Methods of estimation of the outage rate for unshielded and shielded lines will be somewhat different. Therefore, they are discussed separately.

Unshielded Lines

The vertical towers and the horizontal phase conductors coexist for an overhead power line. In that case, there is a race between the towers and the phase conductors to catch the lightning stroke. Some lightning strokes will hit the towers and some will hit the phase conductors. Figure 10.11 illustrates how to estimate the attractive areas of the towers and the phase conductors.

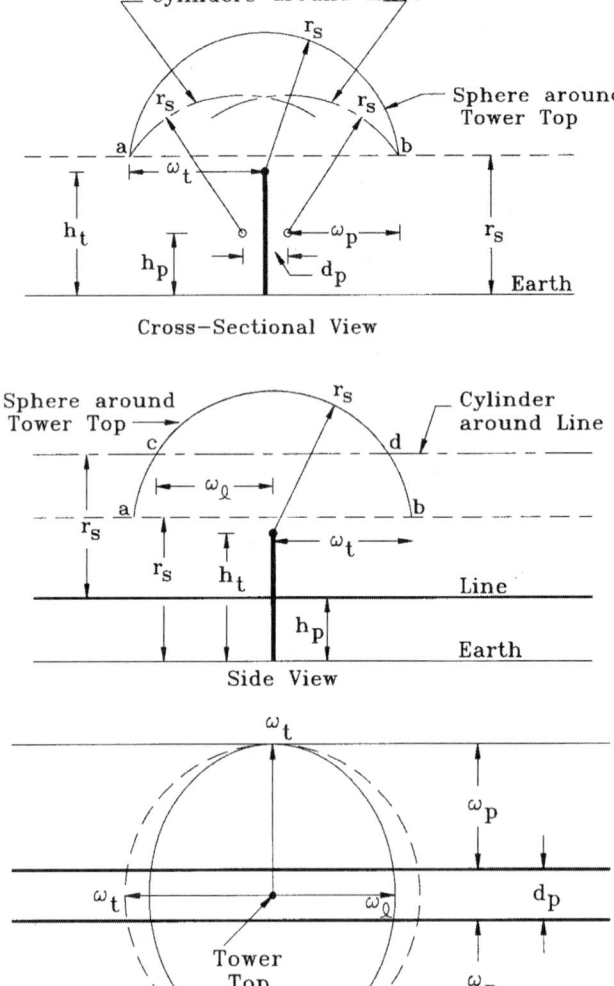

FIGURE 10.11 Attractive areas of tower and horizontal conductors.

The tower and the two outermost phase conductors are shown in Fig. 10.11. In the cross-sectional view, a horizontal line is drawn at a distance r_s from the earth's surface, where r_s is the striking distance corresponding to the return-stroke current, I_s. A circle (cross-sectional view of a sphere) is drawn with radius, r_s, and center at the tip of the tower, cutting the line above the earth at a and b. Two circles (representing cylinders) are drawn with radius, r_s, and centers at the outermost phase conductors, cutting the line above the earth again at a and b. The horizontal distance between the tower tip and either a or b is ω_t. The side view of Fig. 10.11 shows where the sphere around the tower top penetrates both the r_s-plane (a and b) above ground and the cylinders around the outermost phase conductors (c and d). The projection of the sphere around the tower top on the r_s-plane is a circle of radius, ω_t, given by:

$$\omega_t = \sqrt{r_s^2 - (r_s - h_t)^2} = \sqrt{h_t(2r_s - h_t)}. \tag{10.22}$$

The projection of the sphere on the upper surface of the two cylinders around the outer phase conductors will be an ellipse with its minor axis, $2\omega_\ell$, along a line midway between the two outer phase conductors and parallel to their axes; the major axis of the ellipse will be 2_ω, as shown in the plan view of Fig. 10.11. ω_ℓ is given by:

$$\omega_\ell = \sqrt{r_s^2 - \left(r_s - h_t + h_p\right)^2}. \qquad (10.23)$$

If a lightning stroke with return-stroke current I_s or greater, falls within the ellipse, then it will hit the tower. It will hit one of the phase conductors if it falls outside the ellipse but within the width $(2\omega_p + d_p)$; it will hit the ground if it falls outside the width $(2\omega_p + d_p)$. Therefore, for each span length, ℓ_s, the attractive areas for the tower (A_t) and for the phase conductors (A_p) will be:

$$A_t = \pi\omega_t\omega_\ell \text{ and} \qquad (10.24a)$$

$$A_p = \left(2\omega_p + d_p\right)\ell_s - A_t. \qquad (10.24b)$$

The above analysis was performed for the shielding current of the overhead line when the sphere around the tower top and the cylinders around the outer phase conductors intersect the r_s-plane above ground at the same points (points a and b in Fig. 10.11). In this case, $2\omega_t = 2\omega_p + d_p$. The sphere and the cylinders will intersect the r_s-plane at different points for different return-stroke currents; their horizontal segments (widths) can be similarly computed. The equation for ω_t was given above. The equation for ω_p was given in Eq. (10.15). Due to conductor sag, the effective height of a conductor is lower than that at the tower. The effective height is generally assumed as:

$$h_p = h_{pt} - \frac{2}{3}(\text{midspan sag}), \qquad (10.25)$$

where h_{pt} is the height of the conductor at the tower.

The critical current, i_{cp}, for stroke to a phase conductor is computed from Eq. (10.5). It should be noted that i_{cp} is independent of the front time, t_f, of the return-stroke current. The critical current, i_{ct}, for stroke to tower is a function of t_f. Therefore, starting with a short t_f, such as 0.5 μs, the insulator voltage is determined with 1 kA of tower injected current; then, the critical tower current for the selected t_f is determined by the ratio of the insulation strength (e.g., BIL) to the insulator voltage determined with 1 kA of tower injected current. The procedure for estimating the outage rate is started with the lower of the two critical currents (i_{cp} or i_{ct}). If i_{cp} is the lower one, which is usually the case, the attractive areas, A_p and A_t, are computed for that current. If $i_{cp} < i_{ct}$, then this will not cause any flashover if it falls within A_t. In other words, the towers act like partial shields to the phase conductors. However, all strokes with i_{cp} and higher currents falling within A_p will cause flashover. The cumulative probability, $P_c(i_{cp})$, for strokes with currents i_{cp} and higher is given by Eq. (10.21). If there are n_{sp} spans per 100 km of the line, then the number of outages for lightning strokes falling within A_p along the 100-km stretch of the line will be:

$$nfp_o = n_g P_c\left(i_{cp}\right) p\left(t_f\right) \Delta t_f n_{sp} A_p \qquad (10.26)$$

where $p(t_f)$ is the probability density function of t_f, and Δt_f is the front step size. The stroke current is increased by a small step (e.g., 500 A), Δi, $(i = i_{cp} + \Delta i)$, and the enlarged attractive area, A_{p1}, is calculated. All strokes with currents i and higher falling within A_{p1} will cause outages. However, the outage rate for strokes falling within A_p for strokes i_{cp} and greater has already been computed in Eq. (10.26). Therefore, only the additional outage rate, Δnfp, should be added to Eq. (10.26):

$$\Delta \text{nfp} = n_g P_c(i) p(t_f) \Delta t_f n_{sp} \Delta A_p, \qquad (10.27)$$

where $\Delta A_p = A_{p1} - A_p$. The stroke current is increased in steps of Δi and the incremental outages are added until the stroke current is very high (e.g., 200 kA) when the probability of occurrence becomes acceptably low. Then, the front time, t_f is increased by a small step, Δt_f, and the computations are repeated until the probabilty of occurrence of higher t_f is low (e.g., $t_f = 10.5$ μs). In the mean time, if the stroke current becomes equal to i_{ct}, then the outages due to strokes to the tower should be added to the outages caused by strokes to the phase conductors. The total outage rate is then given by:

$$\text{nft} = \text{nfp} + \text{nft} \qquad (10.28a)$$

$$\text{nfp} = \text{nfpo} + n_g n_{sp} \sum P_c(i) p(t_f) \Delta t_f \Delta A_p, \quad \text{and} \qquad (10.28b)$$

$$\text{nft} = \text{nfto} + n_g n_t \sum_{t_f} \sum_i P_c(i) p(t_f) \Delta t_f \Delta A_t. \qquad (10.28c)$$

With digital computers, the total outage rates can be computed within a few seconds.

Shielded Lines

For strokes to the shield wire, the voltage at the adjacent towers will be the same as that for stroke to the tower for the same stroke current. Therefore, there will be only one critical current for strokes to shielded lines, unlike the unshielded lines. The critical current for shielded lines can be computed similar to that for the unshielded lines, except Eq. (10.11) is now used instead of Eq. (10.7).

Otherwise, the computation for shielded lines is similar to that for unshielded lines. The variables h_p and d_p for the phase conductors are replaced by h_s and d_s, which are the shield-wire height and the separation distance between the shield wires, respectively. For a line with a single shield wire, $d_s = 0$. Generally, shield wires are attached to the tower at its top. However, the effective height of the shield wire is lower than that of the tower due to sag. The effective height of the shield wire, h_s, can be computed from Eq. (10.25) by replacing h_{pt} by h_{st}, the shield-wire height at tower.

References

Anderson, R. B. and Eriksson, A. J., Lightning parameters for engineering applications, *Electra*, 69, 65–102, 1980.
Armstrong, H. R. and Whitehead, E. R., Field and analytical studies of transmission line shielding, *IEEE Trans. on Power Appar. and Syst.*, PAS-87, 270-281, 1968.
Bewley, L. V., *Traveling Waves on Transmission Systems*, 2nd ed., John Wiley, New York, 1951.
Brown, G. W. and Whitehead, E. R., Field and analytical studies of transmission line shielding: Part II, *IEEE Trans. on Power Appar. and Syst.*, PAS-88, 617-626, 1969.
Chowdhuri, P., *Electromagnetic Transients in Power Systems*, Research Studies Press, Taunton, U.K. and Taylor and Francis, Philadelphia, PA, 1996.
Chowdhuri, P. and Kotapalli, A. K., Significant parameters in estimating the striking distance of lightning strokes to overhead lines, *IEEE Trans. on Power Delivery* 4, 1970–1981, 1989.
Eriksson, A. J., Notes on lightning parameters, CIGRE Note 33-86 (WG33-01) IWD, 15 July 1986.
Popolansky, F., Frequency distribution of amplitudes of lightning currents, *Electra*, 22, 139–147, 1972.

10.3 Overvoltages Caused by Indirect Lightning Strokes

Pritindra Chowdhuri

A direct stroke is defined as a lightning stroke when it hits either a shield wire, tower, or a phase conductor. An insulator string is stressed by very high voltages caused by a direct stroke. An insulator string can also be stressed by high transient voltages when a lightning stroke hits the nearby ground. An indirect stroke is illustrated in Fig. 10.12.

The voltage induced on a line by an indirect lightning stroke has four components:

1. The charged cloud above the line induces bound charges on the line while the line itself is held electrostatically at ground potential by the neutrals of connected transformers and by leakage over the insulators. When the cloud is partially or fully discharged, these bound charges are released and travel in both directions on the line giving rise to the traveling voltage and current waves.
2. The charges lowered by the stepped leader further induce charges on the line. When the stepped leader is neutralized by the return stroke, the bound charges on the line are released and thus produce traveling waves similar to that caused by the cloud discharge.
3. The residual charges in the return stroke induce an electrostatic field in the vicinity of the line and hence an induced voltage on it.
4. The rate of change of current in the return stroke produces a magnetically induced voltage on the line.

If the lightning has subsequent strokes, then the subsequent components of the induced voltage will be similar to one or the other of the four components discussed above.

The magnitudes of the voltages induced by the release of the charges bound either by the cloud or by the stepped leader are small compared with the voltages induced by the return stroke. Therefore, only the electrostatic and the magnetic components induced by the return stroke are considered in the following analysis. The initial computations are performed with the assumption that the charge distribution along the leader stroke is uniform, and that the return-stroke current is rectangular. However, the result with the rectangular current wave can be transformed to that with currents of any other waveshape by the convolution integral (Duhamel's theorem). It was also assumed that the stroke is vertical and that the overhead line is lossfree and the earth is perfectly conducting. The vertical channel of the return stroke is shown in Fig. 10.13, where the upper part consists of a column of residual charge which is neutralized by the rapid upward movement of the return-stroke current in the lower part of the channel.

Figure 10.14 shows a rectangular system of coordinates where the origin of the system is the point where lightning strikes the surface of the earth. The line conductor is located at a distance y_o meters from the origin, having a mean height of h_p meters above ground and running along the x-direction. The origin of time (t = 0) is assumed to be the instant when the return stroke starts at the earth level.

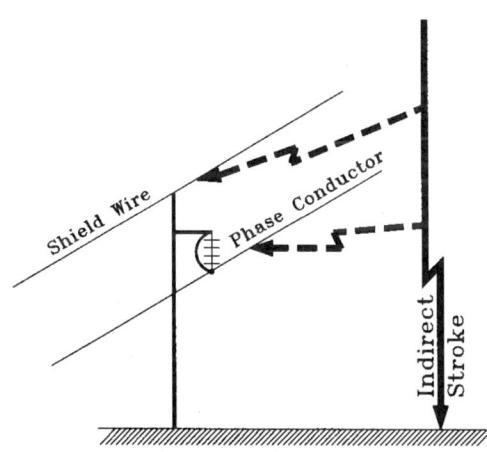

FIGURE 10.12 Illustration of direct and indirect lightning strokes.

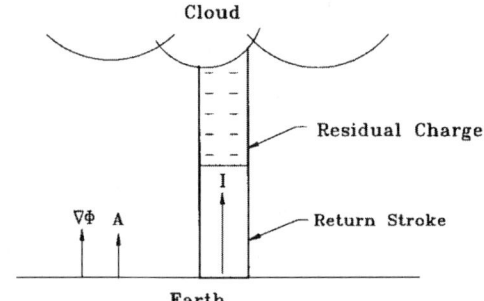

FIGURE 10.13 Return stroke with the residual charge column.

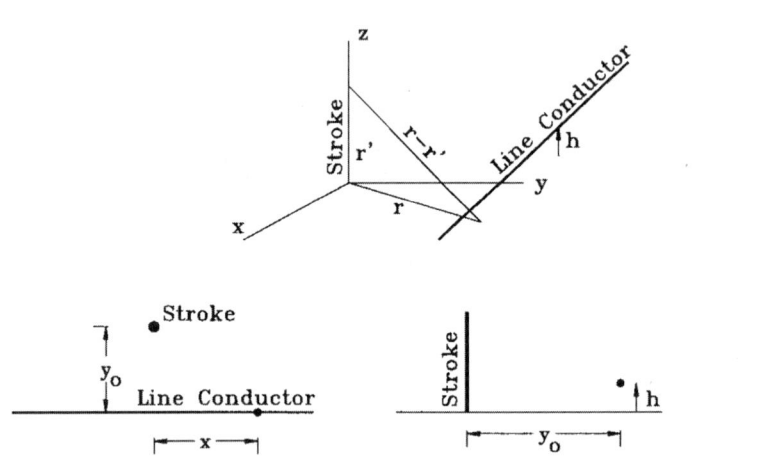

FIGURE 10.14 Coordinate system of line conductor and lightning stroke.

Inducing Voltage

The total electric field created by the charge and the current in the lightning stroke at any point in space is

$$E_i = E_{ei} + E_{mi} = -\nabla\phi - \frac{\partial A}{\partial t}, \qquad (10.29)$$

where ϕ is the *inducing* scalar potential created by the residual charge at the upper part of the return stroke and A is the *inducing* vector potential created by the upward-moving return-stroke current (Fig. 10.13). ϕ and A are called the retarded potentials, because these potentials at a given point in space and time are determined by the charge and current at the source (i.e., the lightning channel) at an earlier time; the difference in time (i.e., the retardation) is the time required to travel the distance between the source and the field point in space with a finite velocity, which in air is $c = 3 \times 10^8$ m/s. These electromagnetic potentials can be deduced from the distribution of the charge and the current in the return-stroke channel. The next step is to find the inducing electric field [Eq. (10.29)]. The *inducing* voltage, V_i, is the line integral of E_i:

$$V_i = -\int_0^{h_p} E_i \cdot dz = -\int_0^{h_p} E_{ei} \cdot dz - \int_0^{h_p} E_{mi} \cdot dz = V_{ei} + V_{mi}. \qquad (10.30)$$

As the height, h_p, of the line conductor is small compared with the length of the lightning channel, the inducing electric field below the line conductor can be assumed to be constant, and equal to that on the ground surface:

$$V_i = \left(\nabla\phi + \frac{\partial A}{\partial t}\right) \cdot h_p. \tag{10.31}$$

The inducing voltage will act on each point along the length of the overhead line. However, because of the retardation effect, the earliest time, t_o, the disturbance from the lightning channel will reach a point on the line conductor would be:

$$t_o = \frac{\sqrt{x^2 + y_o^2}}{c}. \tag{10.32}$$

Therefore, the inducing voltage at a point on the line remains zero until $t = t_o$. Hence,

$$V_i = \psi(x,t)u(t - t_o), \tag{10.33}$$

where $u(t - t_o)$ is the shifted unit step function. The continuous function, $\psi(x,t)$, is the same as Eq. (10.31), and is given, for a negative stroke with uniform charge density along its length, by (Rusck, 1958):

$$\psi(x,t) = -\frac{60 I_o h_p}{\beta}\left[\frac{1-\beta^2}{\sqrt{\beta^2 c^2(t-t_o)^2 + (1-\beta^2)r^2}} - \frac{1}{\sqrt{h_c^2 + r^2}}\right], \tag{10.34}$$

where

I_o = step-function return-stroke current, A
h_p = height of line above ground, m
β = v/c
v = velocity of return stroke
r = distance of point x on line from point of strike, m
h_c = height of cloud charge center above ground, m

The inducing voltage is the voltage at a field point in space with the same coordinates as a corresponding point on the line conductor, but without the presence of the line conductor. The inducing voltage at different points along the length of the line conductor will be different. The overhead line being a good conductor of electricity, these differences will tend to be equalized by the flow of current. Therefore, the actual voltage between a point on the line and the ground below it will be different from the inducing voltage at that point. This voltage, which can actually be measured on the line conductor, is defined as the *induced* voltage. The calculation of the induced voltage is the primary objective.

Induced Voltage

Neglecting losses, an overhead line may be represented as consisting of distributed series inductance L (H/m), and distributed shunt capacitance C (F/m). The effect of the inducing voltage will then be equivalent to connecting a voltage source along each point of the line (Fig. 10.15). The partial differential equation for such a configuration will be:

$$-\frac{\partial V}{\partial x}\Delta x = L\Delta x \frac{\partial I}{\partial t} \quad \text{and} \tag{10.35}$$

$$-\frac{\partial I}{\partial x}\Delta x = C\Delta x \frac{\partial}{\partial t}(V - V_i). \tag{10.36}$$

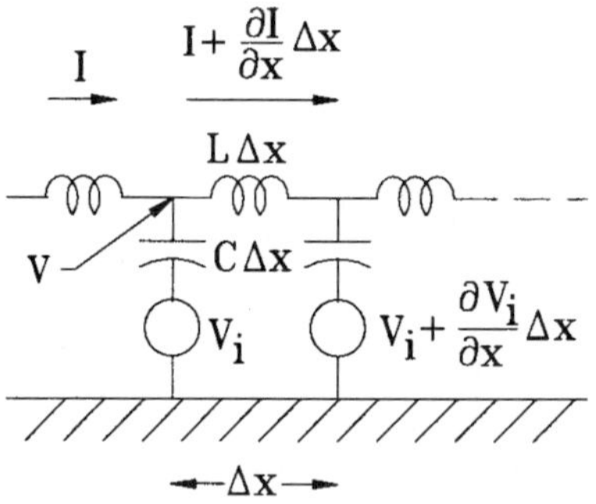

FIGURE 10.15 Equivalent circuit of transmission line with inducing voltage.

Differentiating Eq. (10.35) with respect to x, and eliminating I, the equation for the induced voltage can be written as:

$$\frac{\partial^2 V}{\partial x^2} - \frac{1}{c^2}\frac{\partial^2 V}{\partial t^2} = -\frac{1}{c^2}\frac{\partial^2 V_i}{\partial t^2} = F(x,t), \qquad (10.37)$$

$$\text{where } c = \frac{1}{\sqrt{LC}} = \frac{1}{\sqrt{\epsilon_o \mu_o}} = 3 \times 10^8 \text{ m s}. \qquad (10.38)$$

In Laplace transform, $\dfrac{\partial^2 V(x,s)}{\partial x^2} - \dfrac{s^2}{c^2}V(x,s) = -\dfrac{s^2}{c^2}V_i(x,s) = F(x,s).$ (10.39)

Equation (10.39) is an inhomogeneous wave equation for the induced voltage along the overhead line. It is valid for any charge distribution along the leader channel and any waveshape of the return-stroke current. Its solution can be obtained by assuming F(x,t) to be the superposition of impulses which involves the definition of Green's function (Morse and Feshbach, 1950).

Green's Function

To obtain the voltage caused by a distributed source, F(x), the effects of each elementary portion of the source are calculated and then integrated for the whole source. If G(x;x′) is the voltage at a point x along the line caused by a unit impulse source at a source point x′, the voltage at x caused by a source distribution F(x′) is the integral of G(x;x′)F(x′) over the whole domain (a,b) of x′ occupied by the source, provided that F(x′) is a piecewise continuous function in the domain $a \leq x' \leq b$,

$$V(x) = \int_a^b G(x;x')F(x')dx'. \qquad (10.40)$$

The function G(x;x′), called the Green's function, is, therefore, a solution for a case which is homogeneous everywhere except at one point. Green's function, G(x;x′), has the following properties:

$$\text{(a)} \quad G(x;x'+0) - G(x;x'-0) = 0 \qquad (10.41)$$

$$\text{(b)} \quad \left(\frac{dG}{dx}\right)_{x'+0} - \left(\frac{dG}{dx}\right)_{x'-0} = 1 \qquad (10.42)$$

Power System Transients

(c) G(x;x′) satisfies the homogeneous equation everywhere in the domain, except at the point x = x′, and
(d) G(x;x′) satisfies the prescribed homogeneous boundary conditions.

Green's function can be found by converting Eq. (10.39) to a homogeneous equation and replacing V(x,s) by G(x;x′,s):

$$\frac{\partial^2 G(x;x',s)}{\partial x^2} - \frac{s^2}{c^2} G(x;x',s) = 0. \tag{10.43}$$

The general solution of Eq. (10.43) is given by:

$$G(x;x',s) = A e^{\frac{sx}{c}} + B e^{-\frac{sx}{c}}. \tag{10.44}$$

The constants A and B are found from the boundary conditions and from the properties of Green's function.

Induced Voltage of a Doubly Infinite Single-Conductor Line

The induced voltage at any point, x, on the line can be determined by invoking Eq. (10.40), where $G(x;x') \cdot F(x')$ is the integrand. $F(x')$ is a function of the amplitude and waveshape of the inducing voltage, V_i [Eq. (10.33)], whereas the Green's function, $G(x;x')$ is dependent on the boundary conditions of the line and the properties of Green's function. In other words, it is a function of the line configuration and is independent of the lightning characteristics. Therefore it is appropriate to determine the Green's function first.

Evaluation of Green's Function

As Green's function is finite for $x \to -\infty$ and $x \to +\infty$,

$$G_1 = A e^{\frac{sx}{c}} \text{ for } x < x'; \quad G_2 = B e^{-\frac{sx}{c}} \text{ for } x > x'.$$

From Eq. (10.41): $\quad A e^{\frac{sx'}{c}} = B e^{-\frac{sx'}{c}}$, i.e., $B = A e^{\frac{2sx'}{c}}$.

From Eq. (10.42): $\quad A = -\frac{c}{2s} e^{-\frac{sx}{c}}$; hence, $B = -\frac{c}{2s} e^{\frac{sx}{c}}$.

$$G_1(x;x',s) = -\frac{c}{2s} \exp\left(-\frac{s(x'-x)}{c}\right) \text{ for } x < x' \text{ and} \tag{10.45}$$

$$G_2(x;x',s) = -\frac{c}{2s} \exp\left(-\frac{s(x-x')}{c}\right) \text{ for } x > x'. \tag{10.46}$$

By applying Eq. (10.40):

$$V(x,s) = -\frac{c}{2s}\int_{-\infty}^{x} e^{\frac{s}{c}(x'-x)} F(x',s)dx' - \frac{c}{2s}\int_{x}^{\infty} e^{-\frac{s}{c}(x'-x)} F(x',s)dx' = V_1(x,s) + V_2(x,s). \tag{10.47}$$

Induced Voltage Caused by Return-Stroke Current of Arbitrary Waveshape

The induced voltage caused by return-stroke current, I(t), of arbitrary waveshape can be computed from Eq. (10.39) by several methods. In method I, the inducing voltage, V_i, due to I(t) is found by applying Duhamel's integral (Haldar and Liew, 1988):

$$V_i = \frac{d}{dt}\int_0^t I(t-\tau)V_{istep}(x',\tau)d\tau, \qquad (10.48)$$

where V_{istep} is the inducing voltage caused by a unit step-function current. In other words,

$$V_{istep}(x',\tau) = \psi_o(x',\tau)u(\tau - t_o), \qquad (10.49)$$

where $\psi_o(x',\tau) = \psi(x',\tau)/I_o$, and $\psi(x',\tau)$ is given in Eq. (10.34). Inserting Eq. (10.49) in Eq. (10.48), and taking Laplace transform of V_i in Eq. (10.48):

$$V_i(x',s) = sI(s)\psi_o(x',s)e^{-st_o} \text{ and} \qquad (10.50)$$

$$F(x',s) = -\frac{s^2}{c^2}V_i(x',s) = -\frac{s^3}{c^2}I(s)\psi_o(x',s)e^{-st_o}. \qquad (10.51)$$

Replacing F(x',s) in Eq. (10.47) by Eq. (10.51), the induced voltage, V(x,s) is:

$$V(x,s) = \frac{1}{2c}\left[sI(s)\left\{s\int_{-\infty}^{x}\psi_o(x',s)e^{-s\left(t_o - \frac{x'-x}{c}\right)}dx' + s\int_{x}^{\infty}\psi_o(x',s)e^{-s\left(t_o + \frac{x'-x}{c}\right)}dx'\right\}\right]. \qquad (10.52)$$

Inverting to time domain by convolution integral:

$$V(x,t) = \frac{1}{2c}\int_0^t \frac{d}{dt}I(t-\tau)\left[\frac{d}{d\tau}\int_{-\infty}^{x}\psi_o\left(x',\tau + \frac{x'-x}{c}\right)u\left(\tau - t_o + \frac{x'-x}{c}\right)dx'\right]d\tau +$$

$$\frac{1}{2c}\int_0^t \frac{d}{dt}I(t-\tau)\left[\frac{d}{d\tau}\int_{x}^{\infty}\psi_o\left(x',\tau - \frac{x'-x}{c}\right)u\left(\tau - t_o - \frac{x'-x}{c}\right)dx'\right]d\tau = V_1(x,t) + V_2(x,t) \qquad (10.53)$$

Because of the shifted unit step function in $V_1(x,t)$:

$$\tau \geq t_o - \frac{x'-x}{c}.$$

In the limit, $\tau = t_o - \frac{x_{o1}-x}{c} = \frac{\sqrt{x_{o1}^2 + y_o^2}}{c} - \frac{x_{o1}-x}{c}$

$$\text{or, } x_{o1} = \frac{y_o^2 - (c\tau - x)^2}{2(c\tau - x)}. \qquad (10.54)$$

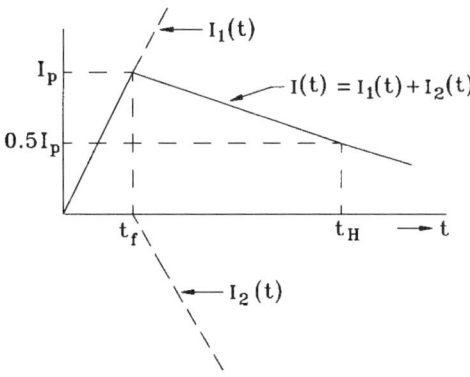

FIGURE 10.16 A linearly rising and falling lightning return-stroke current.

$$\text{Similarly, for } V_2(x,t): \quad x_{o2} = \frac{(c\tau + x)^2 - y_o^2}{2(c\tau + x)}. \tag{10.55}$$

Replacing $-\infty$ by x_{o1} in $V_1(x,t)$, and ∞ by x_{o2} in $V_2(x,t)$ in Eq. (10.53):

$$V_1(x,t) = \frac{1}{2c}\int_0^t \frac{d}{dt}I(t-\tau) \cdot \frac{d}{d\tau}\left\{\int_{x_{o1}}^x \Psi_o\left(x,\tau + \frac{x'-x}{c}\right)dx'\right\}u(\tau - t_o)d\tau \text{ and} \tag{10.56}$$

$$V_2(x,t) = \frac{1}{2c}\int_0^t \frac{d}{dt}I(t-\tau) \cdot \frac{d}{d\tau}\left\{\int_x^{x_{o2}} \Psi_o\left(x,\tau - \frac{x'-x}{c}\right)dx'\right\}u(\tau - t_o)d\tau. \tag{10.57}$$

A lightning return-stroke current can be represented by a linearly rising and linearly falling wave with sufficient accuracy (Fig. 10.16) (Chowdhuri, 1996):

$$I(t) = \alpha_1 t u(t) - \alpha_2 (t - t_f)u(t - t_f) = I_1(t) + I_2(t), \tag{10.58}$$

$$\text{where } \alpha_1 = \frac{I_p}{t_f} \text{ and } \alpha_2 = \frac{2t_H - t_f}{2t_f(t_H - t_f)}I_p. \tag{10.59}$$

It will be evident from Eq. (10.58) that $V_1(x,t)$ in Eq. (10.56) will have two components: one component, $V_{11}(x,t)$, will be a function of $I_1(t)$, and the other component, $V_{21}(x,t)$, will be a function of $I_2(t)$, i.e., $V_1(x,t) = V_{11}(x,t) + V_{21}(x,t)$. Similarly, $V_2(x,t) = V_{12}(x,t) + V_{22}(x,t)$. After integration and simplifying Eq. (10.56), $V_{11}(x,t)$ can be written as:

$$V_{11}(x,t) = -\frac{\alpha_1 h_p}{\beta} \times 10^{-7} u(t - t_o)\left[(1-\beta^2)\ln\frac{f_{11}(\tau=t) \cdot f_{21}(\tau=t_o)}{f_{11}(\tau=t_o) \cdot f_{21}(\tau=t)} + \ln\frac{f_{31}(\tau=t)}{f_{31}(\tau=t_o)}\right] \tag{10.60}$$

where

$$f_{11}(\tau) = m_{11} + \sqrt{m_{11}^2 + a_{11}^2}; \quad f_{21}(\tau) = m_{21} + \sqrt{m_{21}^2 + a_{11}^2}; \quad f_{31}(\tau) = x_{o1} + \sqrt{x_{o1}^2 + y_o^2 + h_c^2};$$

$$m_{11} = x + \beta^2(c\tau - x); \quad m_{21} = x_{o1} + \beta^2(c\tau - x); \quad a_{11}^2 = (1 - \beta^2)\left[y_o^2 + \beta^2(c\tau - x)^2\right].$$

The expression for $V_{21}(x,t)$ is similar to Eq. (10.60), except that α_1 is replaced by $(-\alpha_2)$, and t is replaced by $(t - t_f)$. The computation of $V_2(x,t)$ is similar; namely,

$$V_{12}(x,t) = -\frac{\alpha_1 h_p}{\beta} \times 10^{-7} u(t - t_o)\left[(1-\beta^2)\ell n \frac{f_{12}(\tau=t) \cdot f_{22}(\tau=t_o)}{f_{12}(\tau=t_o) \cdot f_{22}(\tau=t)} - \ell n \frac{f_{32}(\tau=t)}{f_{32}(\tau=t_o)}\right], \quad (10.61)$$

where

$$f_{12}(\tau) = m_{12} + \sqrt{m_{12}^2 + a_{12}^2}; \quad f_{22}(\tau) = m_{22} + \sqrt{m_{22}^2 + a_{12}^2}; \quad f_{32}(\tau) = x_{o2} + \sqrt{x_{o2}^2 + y_o^2 + h_o^2};$$

$$m_{12} = x_{o2} - \beta^2(c\tau + x); \quad m_{22} = x - \beta^2(c\tau + x); \quad a_{12}^2 = (1 - \beta^2)\left[y_o^2 + \beta^2(c\tau + x)^2\right].$$

$V_{22}(x,t)$ can similarly determined by replacing α_1 in Eq. (10.61) by $(-\alpha_2)$, and replacing t by $(t - t_f)$.

The second method of determining the induced voltage, $V(x,t)$, is to solve Eq. (10.47), for a unit step-function return-stroke current, then find the induced voltage for the given return-stroke current waveshape by applying Duhamel's integral (Chowdhuri and Gross, 1967; Chowdhuri, 1989). The solution of Eq. (10.47) for a unit step-function return-stroke current is given by (Chowdhuri, 1989):

$$V_{step}(x,t) = (V_{11} + V_{12} + V_{21} + V_{22})u(t - t_o), \quad (10.62)$$

where

$$V_{11} = \frac{30 h_p (1-\beta^2)}{\beta^2 (ct-x)^2 + y_o^2}\left[\beta(ct-x) + \frac{(ct-x)x - y_o^2}{\sqrt{c^2 t^2 + \frac{1-\beta^2}{\beta^2}(x^2 + y_o^2)}}\right], \quad (10.63)$$

$$V_{12} = \frac{-30 h_p}{\beta}\left[1 - \frac{1}{\sqrt{k_1^2 + 1}} - \beta^2\right]\frac{1}{ct - x}, \quad (10.64)$$

$$V_{21} = \frac{30 h_p (1-\beta^2)}{\beta^2 (ct+x)^2 + y_o^2}\left[\beta(ct+x) - \frac{(ct+x)x + y_o^2}{\sqrt{c^2 t^2 + \frac{1-\beta^2}{\beta^2}(x^2 + y_o^2)}}\right], \quad (10.65)$$

$$V_{22} = \frac{-30 h_p}{\beta}\left[1 - \frac{1}{\sqrt{k_2^2 + 1}} - \beta^2\right]\frac{1}{ct + x}, \quad (10.66)$$

$$k_1 = \frac{2h_c(ct-x)}{y_o^2 + (ct-x)^2}, \tag{10.67}$$

$$k_2 = \frac{2h_c(ct+x)}{y_o^2 + (ct+x)^2}. \tag{10.68}$$

The expressions for the induced voltage, caused by a linearly rising and falling return-stroke current, are given in Appendix I.

The advantage of method II is that once the induced voltage caused by a step-function return-stroke current is computed, then it can be used as the reference in computing the induced voltage caused by currents of any given waveshape by applying Duhamel's integral, thus avoiding the mathematical manipulations for every given waveshape. However, the mathematical procedures are simpler for method I than for method II.

A third method to solve Eq. (10.37) is to apply numerical method which bypasses all mathematical complexities (Agrawal et al., 1980). However, the accuracy of the numerical method strongly depends upon the step size of computation. Therefore, the computation of the induced voltage of long lines, greater than 1 km, becomes impractical.

Induced Voltages on Multiconductor Lines

Overhead power lines are usually three-phase lines. Sometimes several three-phase circuits are strung from the same tower. Shield wires and neutral conductors are part of the multiconductor system. The various conductors in a multiconductor system interact with each other in the induction process for lightning strokes to nearby ground. The equivalent circuit of a two-conductor system is shown in Fig. 10.17. Extending to an n-conductor system, the partial differential equation for the induced voltage, in matrix form, is (Chowdhuri, 1996; Chowdhuri and Gross, 1969; Cinieri and Fumi, 1979; Chowdhuri, 1990):

$$\frac{\partial^2[V]}{\partial x^2} - \frac{1}{c^2}\frac{\partial^2[V]}{\partial t^2} = -[L][C_g]\frac{\partial^2[V_i]}{\partial t^2} = -[M]\frac{\partial^2[V_i]}{\partial t^2}, \tag{10.69}$$

where [L] is an n × n matrix whose elements are:

$$L_{rr} = 2 \times 10^{-7} \ell n \frac{2h_r}{r_r}; \quad L_{rs} = 2 \times 10^{-7} \ell n \frac{d_{r's}}{d_{rs}}.$$

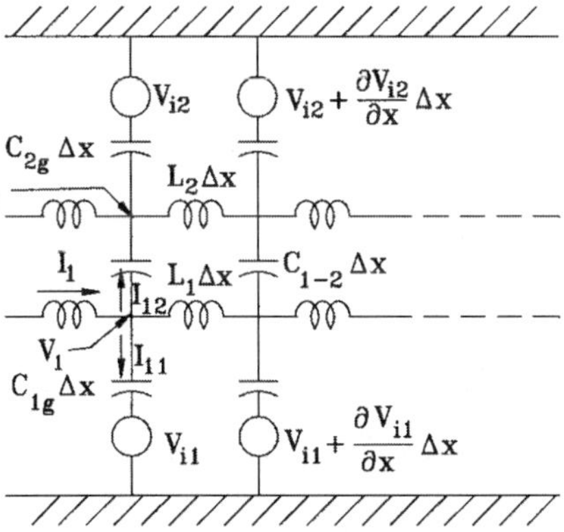

FIGURE 10.17 Equivalent circuit of a two-conductor system.

$[C_g]$ is an n × n diagonal matrix whose elements are, $C_{jg} = C_{j1} + C_{j2} + \cdots + C_{jn}$, where C_{jr} is an element of an n × n matrix, $[C] = [p]^{-1}$ and:

$$P_{rr} = 18 \times 10^9 \ln\frac{2h_r}{r_r}; \quad P_{rs} = 18 \times 10^9 \ln\frac{d_{r's}}{d_{rs}},$$

h_r and r_r are the height above ground and radius of the r-th conductor, $d_{r's}$ is the distance between the image of the r-th conductor below earth and the s-th conductor, d_{rs} is the distance between the r-th and s-th conductors. From Eq. (10.69), for the j-th conductor:

$$\frac{\partial^2 V_j}{\partial x^2} - \frac{1}{c^2}\frac{\partial^2 V_j}{\partial t^2} = -\left(M_{j1}\frac{\partial^2 V_{i1}}{\partial t^2} + \cdots + M_{jj}\frac{\partial^2 V_{ij}}{\partial t^2} + \cdots + M_{jn}\frac{\partial^2 V_{in}}{\partial t^2}\right). \tag{10.70}$$

If the ratio of the inducing voltage of the m-th conductor to that of the j-th conductor is k_{mj} (m = 1,2,…n), then

$$\frac{\partial^2 V_j}{\partial x^2} - \frac{1}{c^2}\frac{\partial^2 V_j}{\partial t^2} = -c^2\left(M_{j1}k_{1j} + \cdots + M_{jj} + \cdots + M_{jn}k_{nj}\right)\frac{1}{c^2}\frac{\partial^2 V_{ij}}{\partial t^2} = \left(M_j c^2\right)F_j(x,t), \tag{10.71}$$

$$\text{where } M_j = M_{j1}k_{1j} + \cdots + M_{jj} + \cdots + M_{jn}k_{nj} \text{ and } F_j(x,t) = \frac{1}{c^2}\frac{\partial^2 V_{ij}}{\partial t^2}. \tag{10.72}$$

If the j-th conductor in its present position existed alone, the partial differential equation of its induced voltage, V_{js}, would have been the same as Eq. (10.37), i.e.,

$$\frac{\partial^2 V_{js}}{\partial x^2} - \frac{1}{c^2}\frac{\partial^2 V_{js}}{\partial t^2} = F_j(x,t). \tag{10.73}$$

Therefore, the ratio of the induced voltage of the j-th conductor in an n-conductor system to that of a single conductor at the same position would be:

$$\frac{V_j}{V_{js}} = M_j c^2. \tag{10.74}$$

The inducing voltage being nearly proportional to the conductor height, and the lateral distance of the stroke point being significantly larger than the separation distance between phase conductors, the presence of other conductors in a horizontally configured line will be minimal. On the other hand, for a vertically configured line, the induced voltage of the highest conductor will be lower than that for the same conductor without any neighboring conductors. Similarly, the lowest conductor voltage will be pulled up by the presence of the neighboring conductors of higher elevation, and the middle conductor will be the least affected by the presence of the other conductors (Chowdhuri, 1996).

Effects of Shield Wires on Induced Voltages

If there are (n + r) conductors, of which r conductors are grounded (r shield wires), then the partial differential equation for the induced voltages of the n number of phase conductors is given by (Chowdhuri, 1996; Chowdhuri and Gross, 1969; Cinieri and Fumi, 1979; Chowdhuri, 1990):

$$\frac{\partial^2[V_n]}{\partial x^2} - \frac{1}{c^2}\frac{\partial^2[V_n]}{\partial t^2} = -[L'][C_{gn}]\frac{\partial^2[V_{in}]}{\partial t^2} = -[M_g]\frac{\partial^2[V_{in}]}{\partial t^2}. \quad (10.75)$$

The matrix [L'] is obtained by partitioning the (n+r)×(n+r) inductance matrix of the (n + r) conductors, and putting $[L'] = [L_{nn}] - [L_{nr}][L_{rr}]^{-1}[L_{rn}]$, where:

$$[L]_{(n+r),(n+r)} = \begin{bmatrix} L_{nn} & L_{nr} \\ L_{rn} & L_{rr} \end{bmatrix}. \quad (10.76)$$

$[C_{gn}]$ is an n × n diagonal matrix each element of which is the sum of the elements of the corresponding row, up to the n-th row, of the original (n + r) × (n + r) capacitance matrix of the (n + r) conductors, $[C] = [p]^{-1}$, where [p] is the matrix of the potential coefficients of the (n + r) conductors. The j-th element of $[C_{gn}]$ is given by:

$$C_{jgn}(j \le n) = \sum_{k=1}^{n+r} C_{jk}. \quad (10.77)$$

From Eq. (10.75), the induced voltage of the j-th conductor is:

$$\frac{\partial^2 V_j}{\partial x^2} - \frac{1}{c^2}\frac{\partial^2 V_j}{\partial t^2} = -c^2\left(M_{gj1}k_{1j} + \cdots + M_{gij} + \cdots + M_{gin}k_{nj}\right)\frac{1}{c^2}\frac{\partial^2 V_{ij}}{\partial t^2} = \left(M_{gj}c^2\right)F_j(x,t). \quad (10.78)$$

Defining the protective ratio as the ratio of the induced voltages on the j-th conductor with and without the shield wires in place:

$$\text{Protective Ratio} = \frac{M_{gj}}{M_j}, \quad (10.79)$$

where M_j is given by Eq. (10.72).

Estimation of Outage Rates Caused by Nearby Lightning Strokes

The knowledge of the following two parameters are essential for estimating the outage rate of an overhead power line: (i) basic insulation level of the line, BIL, and (ii) ground flash density of the region, n_g (number of strokes per km² per year). With this knowledge, the electrogeometric model is constructed to estimate the attractive area (Fig. 10.18). According to the electrogeometric model, the striking distance

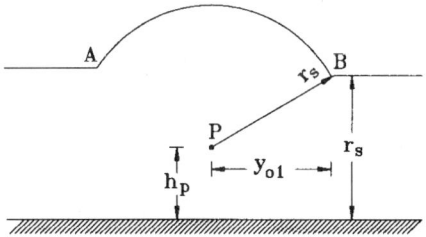

FIGURE 10.18 Electrogeometric model for estimating the least distance of ground strike.

of a lightning stroke is proportional to the return-stroke current. The following relation is used to estimate this striking distance, r_s:

$$r_s = 8 I_p^{0.65} \, (m), \tag{10.80}$$

where I_p is the peak of the return-stroke current. In the cross-sectional view of Fig. 10.18, a horizontal line (representing a plane) is drawn at a distance of r_s meters from the ground plane corresponding to the return-stroke current, I_p. A circular arc is drawn with its center on the conductor, P, and r_s as radius. This represents a cylinder of attraction above the line conductor. The circular arc and the horizontal line intersect at points A and B. The strokes falling between A and B will strike the conductor resulting in direct strokes; those falling outside AB will hit the ground, inducing voltages on the line. The horizontal projection of A or B is y_{o1}, which is given by:

$$y_{o1} = \sqrt{r_s^2 - (r_s - h_p)^2}, \text{ for } r_s > h_p \text{ and} \tag{10.81a}$$

$$y_{o1} = r_s, \text{ for } r_s \leq h_p. \tag{10.81b}$$

y_{o1} is the shortest distance of a lightning stroke of given return-stroke current from the overhead line which will result in a flash to ground.

Analysis of field data shows that the statistical variation of the peak, I_p, and the time to crest, t_f, of the return-stroke current fit lognormal distribution (Anderson and Eriksson, 1980). The probability density function, $p(I_p)$ of I_p then can be expressed as:

$$p(I_p) = \frac{e^{-0.5 f_1}}{I_p \cdot \sigma(\ell n I_p) \cdot \sqrt{2\pi}}, \tag{10.82}$$

$$\text{where } f_1 = \left(\frac{\ell n I_p - \ell n I_{pm}}{\sigma(\ell n I_p)} \right)^2 \text{ and} \tag{10.83}$$

$\sigma(\ell n I_p)$ = standard deviation of $\ell n I_p$, and I_{pm} = median value of I_p. Similarly, the probability density function of t_f can be expressed as:

$$p(t_f) = \frac{e^{-0.5 f_2}}{t_f \cdot \sigma(\ell n t_f) \cdot \sqrt{2\pi}}, \tag{10.84}$$

$$\text{where } f_2 = \left(\frac{\ell n t_f - \ell n t_{fm}}{\sigma(\ell n t_f)} \right)^2. \tag{10.85}$$

The joint probability density function, $p(I_p, t_f)$, is given by:

$$p(I_p, t_f) = \frac{e^{-\frac{0.5}{1-\rho^2} \left(f_1 - 2\rho \sqrt{f_1 \cdot f_2} + f_2 \right)}}{(2\pi)(I_p \cdot t_f)(\sigma(\ell n I_p) \cdot \sigma(\ell n t_f)) \sqrt{1-\rho^2}}, \tag{10.86}$$

where ρ = coefficient of correlation. The statistical parameters of return-stroke current are as follows (Anderson and Eriksson, 1980; Eriksson, 1986):

For $I_p \leq 20$ kA: Median peak current, $I_{pm1} = 61.1$ kA
Log (to base e) of standard deviation, $\sigma(\ell n I_{p1}) = 1.33$
Median time to crest, $t_{fm1} = 3.83$ μs
Log (to base e) of standard deviation, $\sigma(\ell n t_{f1}) = 0.553$

For $I_p > 20$ kA: Median peak current, $I_{pm2} = 33.3$ kA
Log (to base e) of standard deviation, $\sigma(\ell n I_{p2}) = 0.605$
Median time to crest, $t_{fm2} = 3.83$ μs
Log (to base e) of standard deviation, $\sigma(\ell n t_{f2}) = 0.553$
Correlation coefficient, $\rho = 0.47$

To compute the outage rate, the return-stroke current, I_p, is varied from 1 kA to 200 kA in steps of 0.5 kA (Chowdhuri, 1989). The current front time, t_f, is varied from 0.5 μs to 10.5 μs in steps of 0.5 μs. At each current level, the shortest possible distance of the stroke, y_{o1}, is computed from Eq. (10.81). Starting at $t_f = 0.5$ μs, the induced voltage is calculated as a function of time and compared with the given BIL of the line. If the BIL is not exceeded, then the next higher level of current is chosen. If the BIL is exceeded, then the lateral distance of the stroke from the line, y, is increased by Δy (e.g., 1 m), the induced voltage is recalculated and compared with the BIL of the line. The lateral distance, y, is progressively increased until the induced voltage does not exceed BIL. This distance is called y_{o2}. For the selected I_p and t_f, the induced voltage will then exceed the BIL of the line and cause line flashover, if the lightning stroke hit the ground between y_{o1} and y_{o2} along the length of the line. For a 100-km sector of the line, the attractive area, A, will be (Fig. 10.19):

$$A = 0.2(y_{o2} - y_{o1}) \text{ km}^2. \quad (10.87)$$

The joint probability density function, $p(I_p, t_f)$, is then computed from Eq. (10.86) for the selected $I_p - t_f$ combination. If n_g is the ground flash density of the region, the expected number of flashovers per 100 km per year for that particular $I_p - t_f$ combination will be:

$$\text{nfo} = p(I_p, t_f) \cdot \Delta I_p \cdot \Delta t_f \cdot n_g \cdot A, \quad (10.88)$$

where Δi_p = current step, and Δt_f = front time step.

The front time, t_f, is then increased by $t_f - 0.5$ μs to the next step, and nfo for the same current but with the new t_f is computed and added to the previous nfo. Once $t_f = 10.5$ μs is reached, the return-stroke current is increased by $\Delta i_p = 0.5$ kA, and the whole procedure repeated until the limits $I_p = 200$ kA and $t_f = 10.5$ μs are reached. The cumulative nfo will then give the total number of expected line flashovers per 100 km per year for the selected BIL.

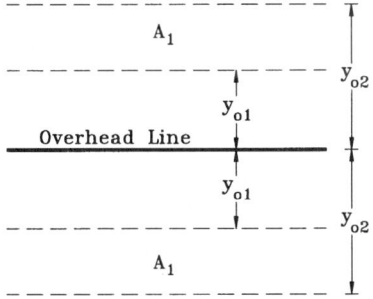

FIGURE 10.19 Attractive area of lightning ground flash to cause line flashover. $A = 2A_1 = 0.2(y_{o2} - y_{o1})$ km².

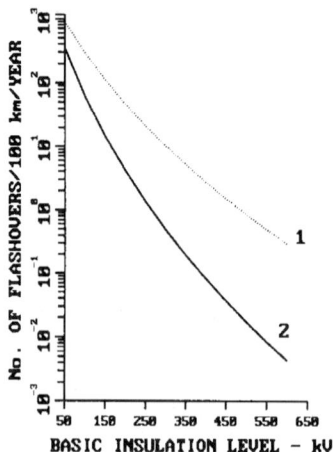

FIGURE 10.20 Flashover rate of overhead line *vs.* BIL. Curve 1: no shield wire: Curve 2: one shield wire. Line height, $h_p = 10$ m; Shield-wire height, $h_{sh} = 11$ m; ground flash density, $n_g = 10/\text{km}^2/\text{year}$.

The lightning-induced outage rates of a 10-m high single conductor are plotted in Fig. 10.20. The effectiveness of the shield wire, as shown in the figure, is optimistic, bearing in mind that the shield wire was assumed to be held at ground potential. The shield wire will not be held at ground potential under transient conditions. Therefore, the effectiveness of the shield wire will be less than the idealized case shown in Fig. 10.20.

References

Agrawal, A. K., Price, H. J., and Gurbaxani, S. H., Transient response of multiconductor transmission lines excited by a nonuniform electromagnetic field, *IEEE Trans. on Electromagnetic Compatibility*, EMC-22, 119, 1980.
Anderson, R. B. and Eriksson, A. J., Lightning parameters for engineering applications, *Electra* 69, 65, 1980.
Chowdhuri, P., Analysis of lightning-induced voltages on overhead lines, *IEEE Trans. on Power Delivery*, 4, 479, 1989.
Chowdhuri, P., *Electromagnetic Transients in Power Systems*, Research Studies Press/John Wiley & Sons, Taunton, U.K./New York, 1996, Chap. 1.
Chowdhuri, P., Estimation of flashover rates of overhead power distribution lines by lightning strokes to nearby ground, *IEEE Trans. on Power Delivery*, 4, 1982–1989.
Chowdhuri, P., Lightning-induced voltages on multiconductor overhead lines, *IEEE Trans. on Power Delivery*, 5, 658, 1990.
Chowdhuri, P. and Gross, E. T. B., Voltage surges induced on overhead lines by lightning strokes, *Proc. IEE* (U.K.), 114, 1899, 1967.
Chowdhuri, P. and Gross, E. T. B., Voltages induced on overhead multiconductor lines by lightning strokes, *Proc. IEE* (U.K.), 116, 561, 1969.
Cinieri, E. and Fumi, A., The effect of the presence of multiconductors and ground wires on the atmospheric high voltages induced on electrical lines, (in Italian), *L'Energia Elettrica*, 56, 595, 1979.
Eriksson, A. J., Notes on lightning parameters for system performance estimation, CIGRE Note 33-86 (WG33- 01) IWD, 15 July 1986.
Haldar, M. K. and Liew, A. C., Alternative solution for the Chowdhuri-Gross model of lightning-induced voltages on power lines, *Proc. IEE* (U.K.), 135, 324, 1988.
Morse, P. M. and Feshbach, H., *Methods of Theoretical Physics*, Vol. 1, McGraw-Hill, New York, 1950, Chap. 7.
Rusck, S., Induced lightning over-voltages on power-transmission lines with special reference to the over-voltage protection of low-voltage networks, *Trans. Royal Inst. of Tech.*, 120, 1, 1958.

Appendix I: Voltage Induced by Linearly Rising and Falling Return-Stroke Current

$$V(x,t) = V_1(x,t)u(t-t_o) + V_2(x,t)u(t-t_{of})$$

where

$$V_1(x,t) = \frac{30\,\alpha_1 h_p}{\beta c}\left[b_o \cdot \ln\frac{f_{12}}{f_{11}} + 0.5\ln(f_{13})\right]; \quad V_2(x,t) = -\frac{30\,\alpha_2 h_p}{\beta c}\left[b_o \cdot \ln\frac{f_{12a}}{f_{11a}} + 0.5\ln(f_{13a})\right]$$

$$b_o = 1-\beta^2; \quad t_{of} = t_o + t_f; \quad t_{tf} = t - t_f$$

$$f_1 = m_1 + (ct-x)^2 - y_o^2; \quad f_2 = m_1 - (ct-x)^2 + y_o^2$$

$$f_3 = m_0 - (ct_o - x)^2 + y_o^2; \quad f_4 = m_o + (ct_o - x)^2 - y_o^2$$

$$f_5 = n_1 + (ct+x)^2 - y_o^2; \quad f_6 = n_1 - (ct+x)^2 + y_o^2$$

$$f_7 = n_o - (ct_o + x)^2 + y_o^2; \quad f_8 = n_o + (ct_o + x)^2 - y_o^2$$

$$f_9 = b_o(\beta^2 x^2 + y_o^2) + \beta^2 c^2 t^2(1+\beta^2); \quad f_{10} = 2\beta^2 ct\sqrt{\beta^2 c^2 t^2 + b_o(x^2 + y_o^2)}$$

$$f_{11} = \frac{c^2 t^2 - x^2}{y_o^2}; \quad f_{12} = \frac{f_9 - f_{10}}{b_o^2 y_o^2}; \quad f_{13} = \frac{f_1 \cdot f_3 \cdot f_5 \cdot f_7}{f_2 \cdot f_4 \cdot f_6 \cdot f_8}$$

$$f_{1a} = m_{1a} + (ct_{tf} - x)^2 - y_o^2; \quad f_{2a} = m_{1a} - (ct_{tf} - x)^2 + y_o^2$$

$$f_{3a} = f_3; \quad f_{4a} = f_4; \quad f_{7a} = f_7; \quad f_{8a} = f_8$$

$$f_{5a} = n_{1a} + (ct_{tf} + x)^2 - y_o^2; \quad f_{6a} = n_{1a} - (ct_{tf} + x)^2 + y_o^2$$

$$f_{9a} = b_o(\beta^2 x^2 + y_o^2) + \beta^2 c^2 t_{tf}^2(1+\beta^2); \quad f_{10a} = 2\beta^2 ct_{tf}\sqrt{\beta^2 c^2 t_{tf}^2 + b_o(x^2 + y_o^2)}$$

$$f_{11a} = \frac{c^2 t_{tf}^2 - x^2}{y_o^2}; \quad f_{12a} = \frac{f_{9a} - f_{10a}}{b_o^2 y_o^2}; \quad f_{13a} = \frac{f_{1a} \cdot f_{3a} \cdot f_{5a} \cdot f_{7a}}{f_{2a} \cdot f_{4a} \cdot f_{6a} \cdot f_{8a}}$$

$$m_o = \sqrt{\left[(ct_o - x)^2 + y_o^2\right]^2 + 4h_c^2(ct_o - x)^2}; \quad m_1 = \sqrt{\left[(ct-x)^2 + y_o^2\right]^2 + 4h_c^2(ct-x)^2}$$

$$n_o = \sqrt{\left[(ct_o + x)^2 + y_o^2\right]^2 + 4h_c^2(ct_o + x)^2} \; ; \; n_1 = \sqrt{\left[(ct + x)^2 + y_o^2\right]^2 + 4h_c^2(ct + x)^2}$$

$$m_{la} = \sqrt{\left[(ct_{tf} - x)^2 + y_o^2\right]^2 + 4h_c^2(ct_{tf} - x)^2} \; ; \; n_{la} = \sqrt{\left[(ct_{tf} + x)^2 + y_o^2\right]^2 + 4h_c^2(ct_{tf} + x)^2}$$

10.4 Switching Surges

Stephen R. Lambert

Switching surges occur on power systems as a result of instantaneous changes in the electrical configuration of the system, and such changes are mainly associated with switching operations and fault events. These overvoltages generally have crest magnitudes which range from about 1 per unit to 3 pu for phase-to-ground surges and from about 2.0 to 4 pu for phase-to-phase surges (in pu on the phase to ground crest voltage base) with higher values sometimes encountered as a result of a system resonant condition. Waveshapes vary considerably with rise times ranging from 50 μs to thousands of μs and times to half-value in the range of hundreds of μs to thousands of μs. For insulation testing purposes, a waveshape having a time to crest of 250 μs with a time to half-value of 2000 μs is often used.

The following addresses the overvoltages associated with switching various power system devices. Possible switching surge magnitudes are indicated, and operations and areas of interest that might warrant investigation when applying such equipment are discussed.

Transmission Line Switching Operations

Surges associated with switching transmission lines (overhead, SF_6, or cable) include those that are generated by line energizing, reclosing (three phase and single phase operations), fault initiation, line dropping (deenergizing), fault clearing, etc. During an energizing operation, for example, closing a circuit breaker at the instant of crest system voltage results in a 1 pu surge traveling down the transmission line and being reflected at the remote, open terminal. The reflection interacts with the incoming wave on the phase under consideration as well as with the traveling waves on adjacent phases. At the same time, the waves are being attenuated and modified by losses. Consequently, it is difficult to accurately predict the resultant waveshapes without employing sophisticated simulation tools such as a transient network analyzer (TNA) or digital programs such as the Electromagnetic Transients Program (EMTP).

Transmission line overvoltages can also be influenced by the presence of other equipment connected to the transmission line — shunt reactors, series or shunt capacitors, static var systems, surge arresters, etc. These devices interact with the traveling waves on the line in ways that can either reduce or increase the severity of the overvoltages being generated.

When considering transmission line switching operations, it can be important to distinguish between "energizing" and "reclosing" operations, and the distinction is made on the basis of whether the line's inherent capacitance retains a trapped charge at the time of line closing (reclosing operation) or whether no trapped charge exists (an energizing operation). The distinction is important as the magnitude of the switching surge overvoltage can be considerably higher when a trapped charge is present; with higher magnitudes, insulation is exposed to increased stress, and devices such as surge arresters will, by necessity, absorb more energy when limiting the higher magnitudes. Two forms of trapped charges can exist — DC and oscillating. A trapped charge on a line with no other equipment attached to the line exists as a DC trapped charge, and the charge can persist for some minutes before dissipating (Beehler, 1964). However, if a transformer (power or wound potential transformer) is connected to the line, the charge will decay rapidly (usually in less than 0.5 sec) by discharging through the saturating branch of the transformer (Marks, 1969). If a shunt reactor is connected to the line, the trapped charge takes on an

Power System Transients

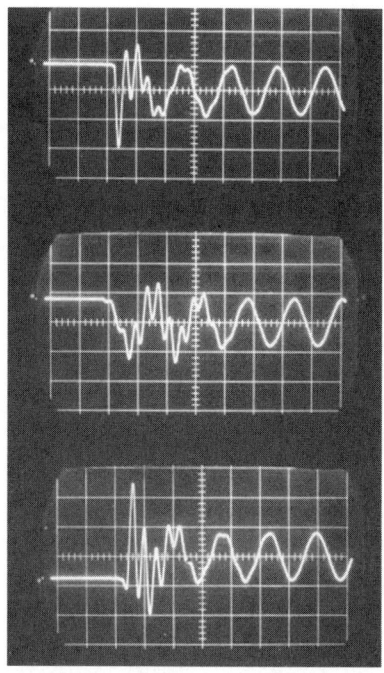

(10 μs/div)

FIGURE 10.21 DC trapped charge.

oscillatory waveshape due to the interaction between the line capacitance and the reactor inductance. This form of trapped charge decays relatively rapidly depending on the Q of the reactor, with the charge being reduced by as much as 50% within 0.5 seconds.

Figures 10.21 and 10.22 show the switching surges associated with reclosing a transmission line. In Fig. 10.21 note the DC trapped charge (approximately 1.0 pu) that exists prior to the reclosing operation (at 20 μs). Figure 10.22 shows the same case with an oscillating trapped charge (a shunt reactor was present on the line) prior to reclosing. Maximum surges were 3.0 for the DC trapped charge case and 2.75 pu for the oscillating trapped charge case (both occurred on phase c).

The power system configuration behind the switch or circuit breaker used to energize or reclose the transmission line also affects the overvoltage characteristics (shape and magnitude) as the traveling wave interactions occurring at the junction of the transmission line and the system (i.e., at the circuit breaker) as well as reflections and interactions with equipment out in the system are important. In general, a stronger system (higher short circuit level) results in somewhat lower surge magnitudes than a weaker system, although there are exceptions. Consequently, when performing simulations to predict overvoltages, it is usually important to examine a variety of system configurations (e.g., a line out of service or contingencies) that might be possible and credible.

Single phase switching as well as three phase switching operations may also need to be considered. On EHV transmission lines, for example, most faults (approximately 90%) are single phase in nature, and opening and reclosing only the faulted phase rather than all three phases, reduces system stresses. Typically, the overvoltages associated with single phase switching have a lower magnitude than those that occur with three phase switching (Koschik et al., 1978).

Switching surge overvoltages produced by line switching are statistical in nature — that is, due to the way that circuit breaker poles randomly close (excluding specially modified switchgear designed to close on or near voltage zero), the instant of electrical closing may occur at the crest of the system voltage, at voltage zero, or somewhere in between. Consequently, the magnitude of the switching surge varies with each switching event. For a given system configuration and switching operation, the surge voltage

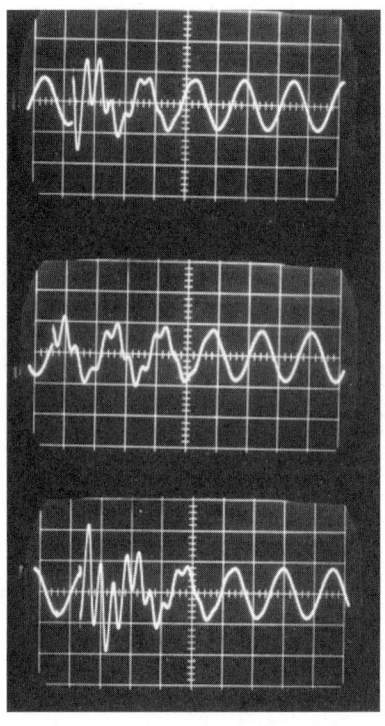

(10 μs/div)

FIGURE 10.22 Oscillating trapped charge.

magnitude at the open end of the transmission line might be 1.2 pu for one closing event and 2.8 pu for the next (Johnson et al., 1964; Hedman et al., 1964), and this statistical variation can have a significantly impact on insulation design (see Section 10.8 on insulation coordination).

Typical switching surge overvoltage statistical distributions (160 km line, 100 random closings) are shown in Figs. 10.23 and 10.24 for phase-to-ground and phase-to-phase voltages (Lambert, 1988), and

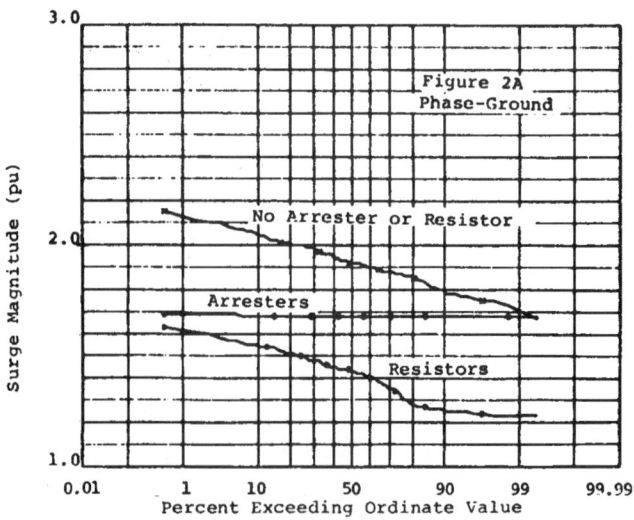

FIGURE 10.23 Phase-to-ground overvoltage distribution.

Power System Transients

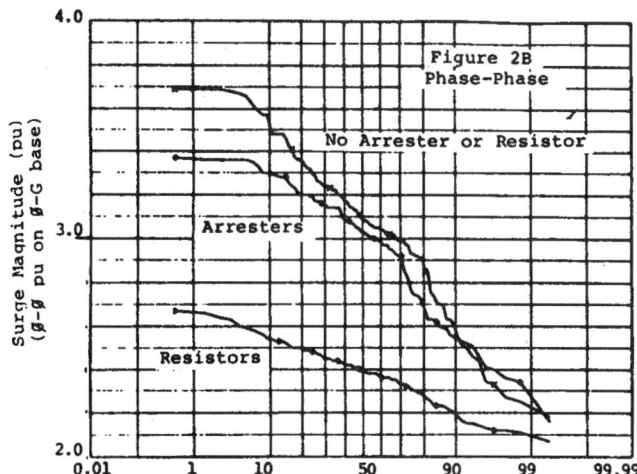

FIGURE 10.24 Phase-to-phase overvoltage distribution.

the surge magnitudes indicated are for the highest that occurred on any phase during each closing. With no surge limiting action (by arresters or circuit breaker preinsertion resistors), phase-to-ground surges varied from 1.7 to 2.15 pu with phase-to-phase surges ranging from 2.2 to 3.7 pu. Phase-to-phase surges can be important to line-connected transformers and reactors as well as to transmission line phase-to-phase conductor separation distances when line uprating or compact line designs are being considered.

Figure 10.23 also demonstrates the effect of the application of surge arresters on phase-to-ground surges, and shows the application of resistors preinserted in the closing sequence of the circuit breaker (400 Ω for 5.56 ms) is even more effective than arresters in reducing surge magnitude. The results shown on Fig. 10.24, however, indicate that while resistors are effective in limiting phase-to-phase surges, arresters applied line to ground are generally not very effective at limiting phase-to-phase overvoltages.

Line dropping (deenergizing) and fault clearing operations also generate surges on the system, although these typically result in phase-to-ground overvoltages having a maximum value of 2 to 2.2 pu. Usually the concern with these operations is not with the phase-to-ground or phase-to-phase system voltages, but rather with the recovery voltage experienced by the switching device. The recovery voltage is the voltage which appears across the interrupting contacts of the switching device (a circuit breaker for example) following current extinction, and if this voltage has too high a magnitude, or in some instances rises to its maximum too quickly, the switching device may not be capable of successfully interrupting.

The occurrence of a fault on a transmission line also can result in switching surge type overvoltages, especially on parallel lines. These voltages usually have magnitudes on the order of 1.8–2.2 pu and are usually not a problem (Kimbark and Legate, 1968; Madzarevic et al., 1977).

Series Capacitor Bank Applications

Installation of a series capacitor bank in a transmission line (standard or thyristor controlled) has the potential for increasing the magnitude of phase-to-ground and phase-to-phase switching surge overvoltages due to the trapped charges that can be present on the bank at the instant of line reclosing. In general, surge arresters limit the phase-to-ground and phase-to-phase overvoltages to acceptable levels; however, one problem that can be serious is the recovery voltage experienced by circuit breakers when clearing faults on a series compensated line. Depending the bank's characteristics and on fault location with respect to the bank's location, a charge can be trapped on the bank, and this trapped charge can add to the surges already being generated during the fault clearing operation (Wilson, 1972). The first circuit breaker to clear is sometimes exposed to excessive recovery voltages under such conditions.

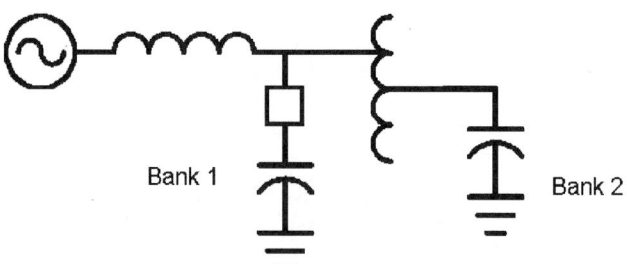

FIGURE 10.25 Voltage magnification circuit.

Shunt Capacitor Bank Applications

Energizing a shunt capacitor bank typically results in maximum overvoltages of about 2 pu or less. However, there are two conditions where significant overvoltages can be generated. One involves a configuration (shown on Fig. 10.25) where two banks are separated by a significant inductance (e.g., a transformer) (Schultz et al., 1959). When one bank is switched, if the system inductance and bank 1 capacitance has the same natural frequency as that of the transformer leakage inductance and the bank 2 capacitance, then a voltage magnification can take place.

Another configuration that can result in damaging overvoltages involves energizing a capacitor bank with a transformer terminated transmission line radially fed from the substation at which the capacitor bank is located (Jones and Fortson, 1985). During bank switching, phase-to-phase surges are imposed on the transformer, and because these are not very well suppressed by the usual phase-to-ground application of surge arresters, transformer failures have been known to result. Various methods to reduce the surge magnitude have included the application of controlled circuit breaker closing techniques (closing near voltage zero), and resistors or reactors preinserted in the closing sequence of the switching devices.

Restriking of the switching device during bank deenergizing can result in severe line-to-ground overvoltages of 3 pu to 5 pu or more (rarely) (Johnson et al., 1955; Greenwood, 1971). Surge arresters are used to limit the voltages to acceptable levels, but at higher system voltages, the energy discharged from the bank into the arrester can exceed the arrester's capability.

Shunt Reactor Applications

Switching of shunt reactors (and other devices characterized as having small inductive currents such as transformer magnetizing currents, motor starting currents, etc.) can generate high phase-to-ground overvoltages as well as severe recovery voltages (Greenwood, 1971), especially on lower voltage equipment such as reactors applied on the tertiary of transformers. Energizing the devices seldom generates high overvoltages, but overvoltages generated during deenergizing, as a result of current chopping by the switching device when interrupting the small inductive currents, can be significant. Neglecting damping, the phase-to-ground overvoltage magnitude can be estimated by:

$$V = i\sqrt{\frac{L}{C}} \qquad (10.89)$$

where i is the magnitude of the chopped current (0 to perhaps as high as 10 A or more), L is the reactor's inductance, and C is the capacitance of the reactor (on the order of a few thousand picofarads). When C is small, especially likely with dry-type reactors often used on transformer tertiaries, the surge impedance term can be large, and hence the overvoltage can be excessive.

To mitigate the overvoltages, surge arresters are sometimes useful, but the application of a capacitor on the terminals of the reactor (or other equipment) have a capacitance on the order of 0.25–0.5 μF is very helpful. In the equation above, note that if C is increased from pF to μF, the surge impedance term is dramatically reduced, and hence the voltage is reduced.

References

Beehler, J. E., Weather, corona, and the decay of trapped energy on transmission lines, *IEEE Trans. on Power Appar. and Syst.* 83, 512, 1964.

Greenwood, A., *Electrical Transients in Power Systems*, John Wiley & Sons, New York, 1971.

Hedman, D. E., Johnson, I. B., Titus, C. H., and Wilson, D. D., Switching of extra-high-voltage circuits, II — surge reduction with circuit breaker resistors, *IEEE Trans. on Power Appar. and Syst.* 83, 1196, 1964.

Johnson, I. B., Phillips, V. E., and Simmons, Jr., H. O., Switching of extra-high-voltage circuits, I — system requirements for circuit breakers, *IEEE Trans. on Power Appar. and Syst.* 83, 1187, 1964.

Johnson, I. B., Schultz, A. J., Schultz, N. R., and Shores, R. R., Some fundamentals on capacitance switching, *AIEE Trans. on Power Appar. and Syst.* PAS-74, 727, 1955.

Jones, R. A. and Fortson, Jr., H. S., Considerations of phase-to-phase surges in the application of capacitor banks, IEEE PES Summer Meeting, 1985, 85 SM 400-7.

Kimbark, E. W. and Legate, A. C., Fault surge versus switching surge: A study of transient overvoltages caused by line-to-ground faults, *IEEE Trans. on Power Appar. and Syst.* PAS-87, 1762, 1968.

Koschik, V., Lambert, S. R., Rocamora, R. G., Wood, C. E., and Worner, G., Long line single-phase switching transients and their effect on station equipment, *IEEE Trans. on Power Appar. and Syst.* PAS-97, 857, 1978.

Lambert, S. R., Effectiveness of zinc oxide surge arresters on substation equipment probabilities of flashover, *IEEE Trans. on Power Delivery* 3(4), 1928, 1988.

Madzarevic, V., Tseng, F. K., Woo, D. H., Niebuhr, W. D., and Rocamora, R. G., Overvoltages on ehv transmission lines due to fault and subsequent bypassing of series capacitors, IEEE PES Winter Meeting, January 1977, F77 237-1.

Marks, L. W., Line discharge by potential transformers, *IEEE Trans. on Power Appar. and Syst.* PAS-88, 293, 1969.

Schultz, A. J., Johnson, J. B., and Schultz, N. R., Magnification of switching surges, *AIEE Trans. on Power Appar. and Syst.* 77, 1418, 1959.

Wilson, D. D., Series compensated lines — voltages across circuit breakers and terminals caused by switching, *IEEE PES Summer Meeting*, 1972, T72 565-0.

10.5 Very Fast Transients

Juan A. Martinez-Velasco

Transient phenomena in power systems are caused by switching operations, faults, and other disturbances, such as lightning strokes. They may appear with a wide range of frequencies that vary from DC to several MHz. A distinction is usually made between slow electromechanical transients and faster electromagnetic transients. The latter type of transients can occur on a time scale that goes from microseconds to several cycles. Due to frequency-dependent behavior of power components and difficulties for developing models accurate enough for a wide frequency range, the frequency ranges are classified into groups, with overlapping between them. An accurate mathematical representation of each power component can generally be developed for a specific frequency range (CIGRE, 1990).

Very Fast Transients (VFT), also known as Very Fast Front Transients, belong to the highest frequency range of transients in power systems. According to the classification proposed by the CIGRE Working Group 33-02, VFT may vary from 100 kHz up to 50 MHz (1990). According to IEC 71-1, the shape of a very fast front overvoltage is "usually unidirectional with time to peak < 0.1 μs, total duration < 3 ms, and with superimposed oscillations at frequency 30 kHz $< f <$ 100 MHz" (1993). In practice, the term VFT is restricted to transients with frequencies above 1 MHz. Several causes can originate these transients in power systems: disconnector operations and faults within gas insulated substations (GIS), switching of motors and transformers with short connections to the switchgear, certain lightning conditions (IEC 71-2, 1996).

This section is exclusively dedicated to explaining the origin, and to analyze the propagation and the effects of VFT in GIS. These transients have a rise time in the range of 4 to 100 ns, and are normally followed by oscillations ranging from 1 to 50 MHz. Their magnitude is in the range of 1.5 to 2 per unit of the line-to-neutral voltage crest, but they can also reach values higher than 2.5 per unit. These values are generally below the BIL of the GIS and connected equipment of lower voltage classes. VFT in GIS are of greater concern at the highest voltages, for which the ratio of the BIL to the system voltage is lower (Yamagata et al., 1996). External VFT can be dangerous for secondary and adjacent equipment.

Origin of VFT in GIS

VFT within a gas-insulated substation (GIS) are usually generated by disconnect switch operations, although other events, such as the closing of a grounding switch or a fault, can also cause VFT.

A large number of pre- or restrikes can occur during a disconnector operation due to the relatively slow speed of the moving contact (Ecklin et al., 1980). Figure 10.26 shows a very simple configuration used to explain the general switching behavior and the pattern of voltages on opening and closing of a disconnector at a capacitive load (Boggs et al., 1982). During an opening operation, sparking occurs as soon as the voltage between the source and the load exceeds the dielectric strength across contacts. After

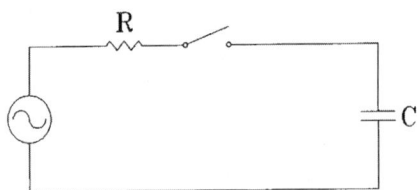

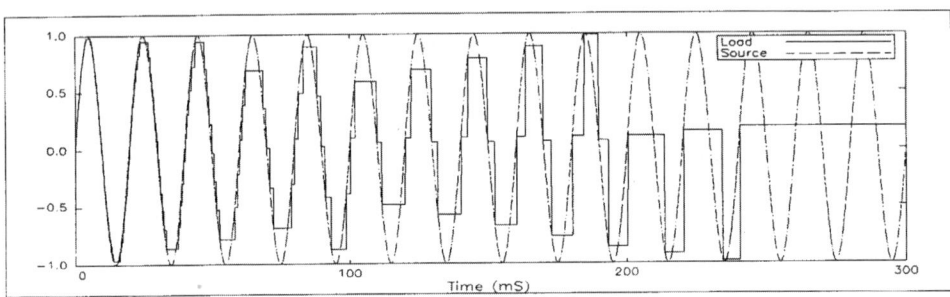

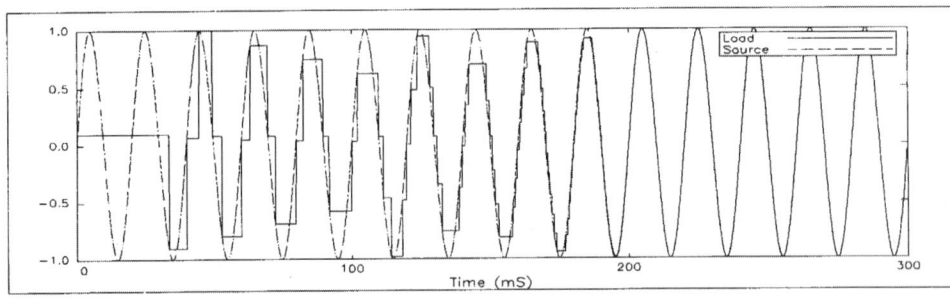

FIGURE 10.26 Variation of load and source side voltages during disconnector switching: (a) scheme of the circuit, (b) opening operation; (c) closing operation.

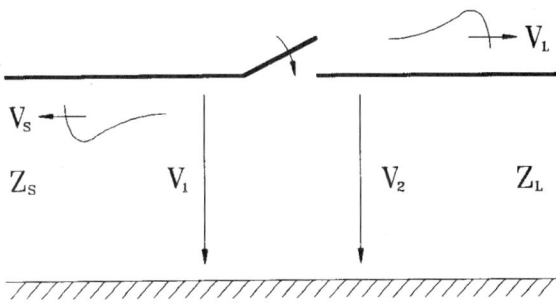

FIGURE 10.27 Generation of very fast transients.

a restrike, a high-frequency current will flow through the spark and equalize the capacitive load voltage to the source voltage. The potential difference across the contacts will fall and the spark will extinguish. The subsequent restrike occurs when the voltage between contacts reaches the new dielectric strength level that is determined by the speed of the moving contact and other disconnector characteristics. The behavior during a closing operation is very similar, and the load side voltage will follow the supply voltage until the contact-make. For a discussion of the physics involved in the restrikes and prestrikes of a disconnect switch operation see (Boggs et al., 1982).

The scheme shown in Fig. 10.27 will be very useful to illustrate the generation of VFT due to a disconnector operation. The breakdown of a disconnector when it is closing originates two surges V_L and V_S which travel outward in the bus duct and back into the source side respectively. The magnitude of both traveling surges is given by

$$V_L = \frac{Z_L}{Z_S + Z_L}(V_1 - V_2) \quad V_S = -V_L \quad (10.90)$$

where Z_S and Z_L are the surge impedances on the source and on the load side respectively. V_1 is the intercontact spark voltage, while V_2 is the trapped charge voltage at the load side.

Steep fronted traveling surges can also be generated in case of a line-to-ground fault, as the voltage collapse at the fault location occurs in a similar way as in the disconnector gap during striking.

Propagation of VFT in GIS

VFT in GIS can be divided into internal and external. Internal transients can produce overvoltages between the inner conductor and the enclosure, while external transients can cause stress on secondary and adjacent equipment. A summary about the propagation and main characteristics of both types of phenomena follows.

Internal Transients

Breakdown phenomena across the contacts of a disconnector during a switch operation or a line-to-ground fault generate very short rise time traveling waves which propagate in either direction from the breakdown location. As a result of the fast rise time of the wave front, the propagation throughout a substation must be analyzed by representing GIS sections as low-loss distributed parameter transmission lines, each section being characterized by a surge impedance and a transit time. Traveling waves are reflected and refracted at every point where they encounter a change in the surge impedance. The generated transients depend on the GIS configuration and on the superposition of the surges reflected and refracted on line discontinuities like breakers, "T" junctions, or bushings. As a consequence of multiple reflections and refractions, traveling voltages can increase above the original values and very high-frequency oscillations occur.

The internal damping of the VFT influencing the highest frequency components is determined by the spark resistance. Skin effects due to the aluminum enclosure can be generally neglected. The main portion of the damping of the VFT occurs by outcoupling at the transition to the overhead line. Due to the traveling wave behavior of the VFT, the overvoltages caused by disconnector switches show a spatial distribution. Normally the highest overvoltage stress is reached at the open end of the load side.

Overvoltages are dependent on the voltage drop at the disconnector just before striking, and on the trapped charge that remains on the load side of the disconnector. For a normal disconnector with a slow speed, the maximum trapped charge reaches 0.5 pu resulting in a most unfavorable voltage collapse of 1.5 pu. For these cases, the resulting overvoltages are in the range of 1.7 pu and reach 2 pu for very specific cases. For a high-speed disconnector, the maximum trapped charge could be 1 pu and the highest overvoltages reach values up to 2.5 pu. Although values larger than 3 pu have been reported, they have been derived by calculation using unrealistic simplified simulation models. The main frequencies depend on the length of the GIS sections affected by the disconnector operation and are in the range of 1 to 50 MHz.

The following examples will be useful to illustrate the influence of some parameters on the frequency and magnitude of VFT in GIS. Figure 10.28 shows two very simple cases, a single bus duct and a "T" junction in which GIS components are modeled as lossless distributed parameter transmission lines. The source side is represented as a step-shaped source in series with a resistance. This is a simplified modeling of an infinite length bus duct. The surge impedance of all bus sections is 50 Ω. For the simplest configuration, the reflections of the traveling waves at both terminals of the duct will produce, when the source resistance is neglected, a pulse-shaped transient of constant magnitude — 2 pu — and constant frequency at the open terminal. The frequency of this pulse can be calculated from the following expression

$$f = \frac{1}{4\tau} \qquad (10.91)$$

where τ is the transit time of the line. If the propagation velocity is close to that of light, the frequency, in MHz, of the voltage generated at the open terminal will be

$$f \approx \frac{75}{d} \qquad (10.92)$$

where d is the duct length, in meters. When a more realistic representation of the source is used, R = 40 Ω, the maximum overvoltage at the open terminal will depend on the voltage at the disconnector just before striking, and on the trapped charge which remains on the load side.

Overvoltages can reach higher values in more complex GIS configurations. The simulations performed for the "T" configuration shown in Fig. 10.28 gave in all cases higher values than in the previous case, where node 4 is the location where the highest overvoltages were originated.

External Transients

Internally generated VFT propagate throughout the GIS and reach bushings where they cause transient enclosure voltages and traveling waves that propagate along the overhead transmission line. An explanation about the generation of external transients and some comments on their main characteristics follow.

Transient Enclosure Voltages

Transient enclosure voltages (TEV), also known as transient ground potential rises (TGPR), are short-duration high-voltage transients that appear on the enclosure of the GIS through the coupling of internal transients to enclosure at enclosure discontinuities. The simplified circuit shown in Fig. 10.29 is used to explain the generation of TEV (Meppelink et al., 1989). At the GIS-air interface, three transmission lines can be distinguished: the coaxial GIS transmission line, the transmission line formed by the bushing conductor and the overhead line, and the GIS enclosure-to-ground transmission line. When an internal wave propagates to the gas-to-air bushing, a portion of the transient is coupled onto the overhead

Power System Transients

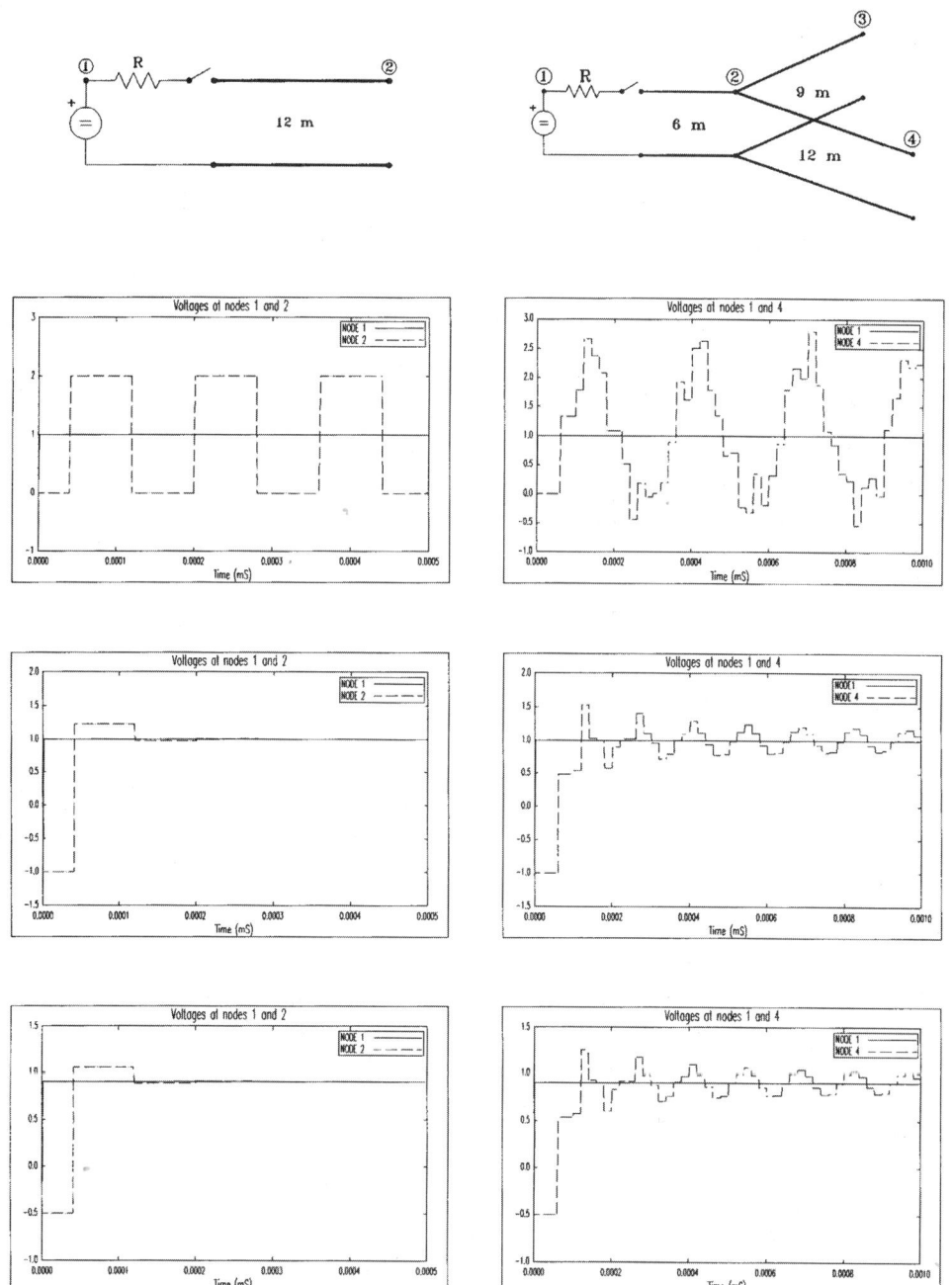

FIGURE 10.28 VFT overvoltages in GIS: (a) scheme of the network, (b) R = 0; V_1 = 1 pu; V_2 = 0, (c) R = 40 Ω; V_1 = 1 pu; V_2 = −1 pu, (d) R = 40 Ω; V_1 = 0.9 pu; V_2 = −0.5 pu.

transmission line, and a portion is coupled onto the GIS enclosure-to-ground transmission line. The wave that propagates along the enclosure-to-ground transmission line is the TEV. The usual location for these voltages is the transition GIS-overhead line at an air bushing, although they can also emerge at visual inspection ports, insulated spacers for CTs, or insulated flanges at GIS/cables interfaces.

TEV waveforms have at least two components; the first one has a short initial rise time and is followed by high-frequency oscillations, in the range of 5 to 10 MHz, determined by the lengths of various sections

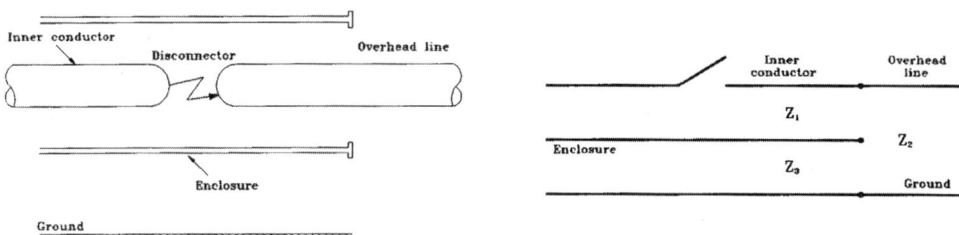

FIGURE 10.29 Generation of TEV: (a) GIS-air transition, (b) single-line diagram.

of the GIS. The second component is of lower frequency, hundreds of kHz, and is often associated with the discharge of capacitive devices with the earthing system. Both components are damped quickly as a result of the lossy nature of the enclosure-to-ground plane transmission mode. TEV generally persists for a few microseconds. The magnitude varies along the enclosure; it can be in the range of 0.1 to 0.3 pu of the system voltage, and reaches the highest magnitude near the GIS-air interface. Mitigation methods include short length leads, low impedance grounding, and the installation of metal-oxide arresters across insulating spacers.

Transients on Overhead Connections
A portion of the VFT traveling wave incident at a gas-air transition is coupled onto the overhead connection and propagates to other components. This propagation is lossy and results in some increase of the waveform rise time. In general, external waveforms have two different characteristics: the overall waveshape that is dictated by lumped circuit parameters, such as the capacitance of voltage transformers or line and earthing inductance, with a rise time in the range of a few hundred nanoseconds; and a fast front portion that is dictated by transmission line effects, with a rise time in the range of 20 ns. A fast rise time of the initial portion is possible as capacitive components, such as bushings, are physically long and distributed and cannot be treated as lumped elements; the magnitude is generally lower than that of internal VFT, it is usually reduced by discontinuities in the transmission path, with a voltage rate-of-rise in the range of 10-30 kV/μs.

Transient Electromagnetic Fields
These fields are radiated from the enclosure and can cause some stress on secondary equipment. Their frequency depends on the GIS arrangement, but are typically in the range of 10 to 20 MHz.

Modeling Guidelines and Simulation

Due to the origin and the traveling nature of VFT, modeling of GIS components makes use of electrical equivalent circuits composed of lumped elements and distributed parameter lines. At very high frequencies, the skin losses can produce an important attenuation; however, these losses are usually neglected, which produces conservative results. Only the dielectric losses in some components (e.g., capacitively graded bushing) need be taken into account. The calculation of internal transients may be performed using distributed parameter models for which only an internal mode (conductor-enclosure) is taken into account, and assuming that the external enclosure is perfectly grounded. If TEV is a concern, then a second mode (enclosure-ground) is to be considered.

The next two sections present modeling guidelines to represent GIS equipment in computation of internal and external transients (Fujimoto et al., 1986; Ogawa et al., 1986; Witzmann, 1987; CIGRE, 1988; Povh, 1996; Fujimoto et al., 1982; Dick et al., 1982). They make use of single-phase models and very simple representations. Depending on the substation layout and the study to be performed, three-phase models for inner conductors (Miri and Binder, 1995) or the outer enclosures (Dick et al., 1982) should be considered. More advanced guidelines have been analyzed and proposed in by Haznadar et al. (1992).

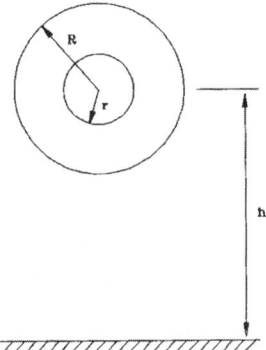

FIGURE 10.30 Coaxial bus duct cross section.

Computation of Internal Transients

A short explanation about the representation of the most important GIS components follows.

Bus Ducts

For frequencies lower than 100 MHz, a bus duct can be represented as a lossless transmission line. The surge impedance and the travel time are calculated from the physical dimensions of the duct. The inductance and the capacitance per unit length of a horizontal single-phase coaxial cylinder configuration, as that shown in Fig. 10.30, are given by the following expressions

$$L'_1 = \frac{\mu_0}{2\pi} \ln \frac{R}{r} \tag{10.93}$$

$$C'_1 = \frac{2\pi\varepsilon}{\ln \frac{R}{r}} \quad (\varepsilon \approx \varepsilon_0) \tag{10.94}$$

from where the following form for the surge impedance is derived

$$Z_1 = \sqrt{\frac{L'_1}{C'_1}} = \frac{\sqrt{\mu_0 \varepsilon}}{2\pi} \ln \frac{R}{r} \approx 60 \ln \frac{R}{r} \tag{10.95}$$

A different approach should be used for vertically oriented bus sections (Miri and Binder, 1995). As for the propagation velocity, empirical corrections are usually needed to adjust its value. Experimental results show that the propagation velocity in GIS ducts is close to 0.95–0.96 of the speed of light (Fujimoto et al., 1986).

Other equipment, such as elbows, can also be modeled as lossless transmission lines.

Surge Arresters

Experimental results have shown that if switching operations in GIS do not produce voltages high enough to cause metal-oxide surge arrester to conduct, then the arrester can be modeled as a capacitance-to-ground. However, when the arrester conducts, the model should take into account the steep front wave effect, since the voltage developed across the arrester for a given discharge current increases as the time to crest of the current increases, but reaches crest prior to the crest of the current. A detailed model must represent each internal shield and block individually, and include the travel times along shield sections, as well as capacitances between these sections, capacitances between blocks and shields, and the blocks themselves.

Circuit Breakers

A closed breaker can be represented as a lossless transmission line whose electrical length is equal to the physical length, with the propagation velocity reduced to 0.95–0.96 of the speed of light. The representation of an open circuit breaker is more complicated due to internal irregularities. In addition, circuit breakers with several chambers contain grading capacitors, which are not arranged symmetrically. The electrical length must be increased above the physical length due to the effect of a longer path through the grading capacitors, while the speed of progression must be decreased due to the effects of the higher dielectric constant of these capacitors.

Gas-to-Air Bushings

A bushing gradually changes the surge impedance from that of the GIS to that of the line. A simplified model may consist of several transmission lines in series with a lumped resistor representing losses; the surge impedance of each line section increases as the location goes up the bushing. If the bushing is distant from the point of interest, the resistor can be neglected and a single line section can be used. A detailed model must consider the coupling between the conductor and shielding electrodes, and include the representation of the grounding system connected to the bushing (Fujimoto and Boggs, 1988; Ardito et al., 1992).

Power Transformers

At very high frequencies, a winding of a transformer behaves like a capacitive network consisting of series capacitances between turns and coils, and shunt capacitances between turns and coils to the grounded core and transformer tank; the saturation of the magnetic core can be neglected, as well as leakage impedances. When voltage transfer has to be calculated, interwinding capacitances and secondary capacitance-to-ground must also be represented. If voltage transfer is not a concern, an accurate representation can be obtained by developing a circuit that matches the frequency response of the transformer at its terminals. Details on the computation of equivalent capacitances have been covered by Chowdhuri (1996).

Spark Dynamics

The behavior of the spark in disconnector operations can be represented by a dynamically variable resistance with a controllable collapse time. In general, this representation does not affect the magnitude of the maximum VFT overvoltages, but it can introduce a significant damping on internal transients (Yanabu et al., 1990).

Computation of TEV

At very high frequencies, currents are constrained to flow along the surface of the conductors and do not penetrate through them. The inside and the outside of a GIS enclosure are distinct, so that transients generated within the substation do not appear on the outside surface of the enclosure until discontinuities in the sheath are encountered. These discontinuities occur at gas-to-air terminations, GIS-cable transitions, or external core current transformers. The modeling of the GIS for computation of TEV must include the effects of the enclosure, the representation of ground straps, and the earthing grid.

Enclosures

A GIS-air termination can be modeled as a junction of three transmission lines, each with its own surge impedance (Fig. 10.29). This equivalent network can be analyzed using lossless transmission line models to determine reflected and transmitted waves. The surge impedance of the enclosure-to-ground transmission line (Fig. 10.29) is derived from the following forms

$$L'_3 \approx \frac{\mu_0}{2\pi} \ln \frac{2h}{R} \qquad (10.96)$$

$$C'_3 \approx \frac{2\pi\varepsilon_0}{\ln \frac{2h}{R}} \qquad (10.97)$$

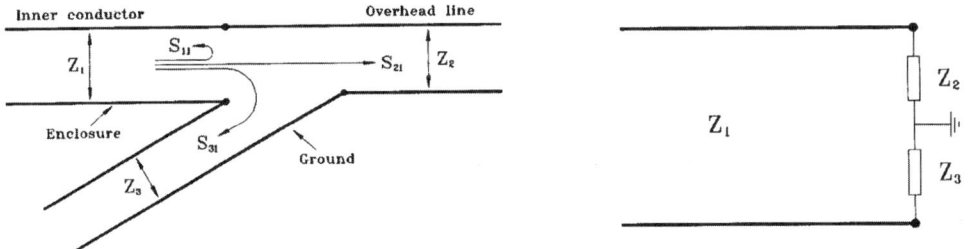

FIGURE 10.31 GIS–air transition. Scattering coefficients: (a) schematic diagram; (b) equivalent circuit.

$$Z_3 = \sqrt{\frac{L'_3}{C'_3}} = \frac{\sqrt{\mu_0 \varepsilon_0}}{2\pi} \ln \frac{2h}{R} \approx 60 \ln \frac{2h}{R} \quad (10.98)$$

The basic mechanism of TEV is defined by the refraction of waves from the internal coaxial bus duct to the enclosure sheath-to-ground system. Figure 10.31 shows the scattering coefficients involved in an air-SF6 transition, and the equivalent circuit to be used for calculating these coefficients. The coefficients S_{ji} represent the refraction of waves from line "i" into line "j".

The coefficient S_{11}, which is also the reflection coefficient at the transition, is given by

$$S_{11} = \frac{(Z_2 + Z_3) - Z_1}{Z_1 + Z_2 + Z_3} \quad (10.99)$$

The refraction coefficient at the transition is then

$$r_t = 1 + S_{11} = \frac{2(Z_2 + Z_3)}{Z_1 + Z_2 + Z_3} \quad (10.100)$$

The magnitude of the transmitted wave onto the outside of the enclosure sheath is given by following scattering coefficient

$$S_{31} = r_t \frac{-Z_3}{Z_2 + Z_3} = -\frac{2Z_3}{Z_1 + Z_2 + Z_3} \quad (10.101)$$

The negative sign means that there is an inversion of the waveform with respect to the internal transient.

Ground Straps

TEV propagates back from the gas-to-air termination into the substation on the transmission line defined by the enclosure and the ground plane. The first discontinuity in the propagation is generally a ground strap. For TEV rise times, most ground straps are too long and too inductive for effective grounding. Ground leads may have a significant effect on the magnitude and waveshape of TEV. This effect can be explained by considering two mechanisms (Fujimoto et al., 1982). First, the ground lead may be seen as a vertical transmission line whose surge impedance varies with height; when the transient reaches the ground strap, a reflected wave is originated that reduces the magnitude of the transmitted wave, with the reduction expressed by the coefficient

$$\frac{2Z_g}{2Z_g + Z_3} \quad (10.102)$$

where Z_g is the surge impedance of the ground strap. As Z_g is usually much larger than Z_3, the attenuation produced by the ground strap will usually be small. Second, when the portion of the wave that propagates down the ground strap meets the low impedance of the ground grid, a reflected wave is produced that propagates back to the enclosure where it will tend to reduce the original wave.

The representation of a ground lead as a constant surge impedance is not strictly correct. It has a continuously varying surge impedance, so that a continuous reflection occurs as a wave propagates down the lead. The ground strap can be divided into sections, each one represented by a surge impedance calculated from the following expression

$$Z_s = 60 \ln \frac{2\sqrt{2}\,h}{r} \qquad (10.103)$$

where r is the lead radius and h is the average height of the section (Fujimoto et al., 1982). However, a constant inductor model may be adequate for straps with travel time less than the surge rise time, while a nonuniform impedance model may be necessary for very long straps.

Earthing Grid

The representation of the earthing grid at TEV frequencies is a very complex task. Furthermore, this grid may not be designed to carry very high frequency currents, as no standards for very high frequency earthing systems are currently available. A simplified modeling may be used by representing the earthing grid as a low value constant resistance.

Advanced models for GIS components in computation of TEV might consider a frequency-dependent impedance for ground straps, a frequency-dependent model for the enclosure-to-ground line (which could take into account earth losses), and the propagation of phase-to-phase modes on the three enclosures (Fujimoto et al., 1982).

Statistical Calculation

The largest VFT stresses under normal operating conditions originate from disconnector operations. The level reached by overvoltages is random by nature. The maximum overvoltage produced by a disconnector breakdown depends on the geometry of the GIS, the measuring point, the voltage prior to the transient at the load side (trapped charge), and the intercontact voltage at the time of the breakdown, as shown in the section on Internal Transients.

Several works have been performed to determine the statistical distribution of VFT overvoltages in a GIS (Boggs et al., 1982; Yanabu et al., 1990; Boggs et al., 1984; Fujimoto et al., 1988). A very simple expression can be used to calculate the transient overvoltages as a function of time t and position s (Boggs et al., 1984; Fujimoto et al., 1988)

$$V(t,s) = V_b * K(t,s) + V_q \qquad (10.104)$$

where $K(t,s)$ is the normalized response of the GIS, V_b is the intercontact spark voltage, and V_q the voltage prior to the transient at the point of interest. As V_b and V_q are random variables, $V(t,s)$ is also random. This equation can be used to estimate worst-case values (Fujimoto et al., 1988).

The performance of a disconnector during an opening operation can be characterized by the pattern of arcing on a capacitive load (Fig. 10.26). A difference in breakdown voltages for the two polarities indicates a dielectric asymmetry. When the asymmetry is large compared to the statistical variance in breakdown voltage, a systematic pattern is originated near the end of the arcing sequence (Boggs et al., 1984). The final trapped charge voltage has a distribution which is very dependent on the asymmetry in the intercontact breakdown voltage.

The dielectric asymmetry of a disconnector is usually a function of contact separation. A disconnector may show a different performance at different operating voltages. A consequence of this performance is that very different stresses will be originated as a result of different operational characteristics.

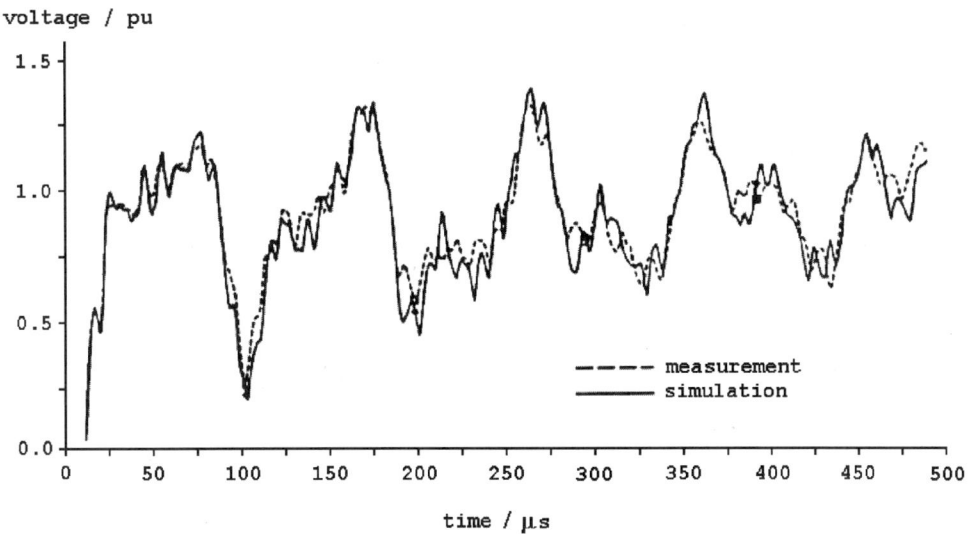

FIGURE 10.32 Comparison of simulation and measurement of disconnect switch induced overvoltages in a 420 kV GIS. (Copyright 1999 IEEE.)

From the results presented in the literature, the following conclusions may be derived (Boggs et al., 1982; Yanabu et al., 1990; Boggs et al., 1984; Fujimoto et al., 1988):

- The value of the trapped charge is mainly dependent on the disconnect switch characteristics: the faster the switch, the greater the mean value that the trapped charge voltage can reach.
- For slow switches, the probability of a re-/prestrike with the greatest breakdown voltage, in the range 1.8–2 pu, is very small; however, due to the great number of re-/prestrikes that are produced with one operation, this probability should not always be neglected.
- The asymmetry of the intercontact breakdown voltage can also affect the trapped charge distribution; in general, both the magnitude and the range of values are reduced if there is a difference in the breakdown voltage of the gap for positive and negative values.

Validation

The results presented in Figs. 10.32 and 10.33 illustrate the accuracy which can be obtained by means of a digital simulation. Figure 10.32 shows a good match between a direct measurement in an actual GIS and a computer result. The simulation was performed including the effects of spacers, flanges, elbows, and other hardware, but neglecting propagation losses (Witzmann, 1987). Figure 10.33 shows that important differences could occur when comparing a measurement and a digital simulation result, although a detailed representation of the GIS was considered. The differences were due to use of low damping equivalent circuits and to limitation of measuring instruments that did not capture very high frequencies (IEEE, 1996).

Effects of VFT on Equipment

The level reached by VFT overvoltages originated by disconnector switching or line-to-ground faults inside a GIS are below the BIL of substation and external equipment (Boersma, 1987). However, aging of the insulation of external equipment due to frequent VFT must be considered. TEV is a low energy phenomenon and it is not considered dangerous to humans; the main concern is in the danger of the surprise-shock effect. External transients can cause interference with or even damage to the substation control, protection, and other secondary equipment (Meppelink et al., 1989; Boersma, 1987). The main effects caused by VFT to equipment and the techniques that can be used to mitigate these effects are summarized below (CIGRE, 1988).

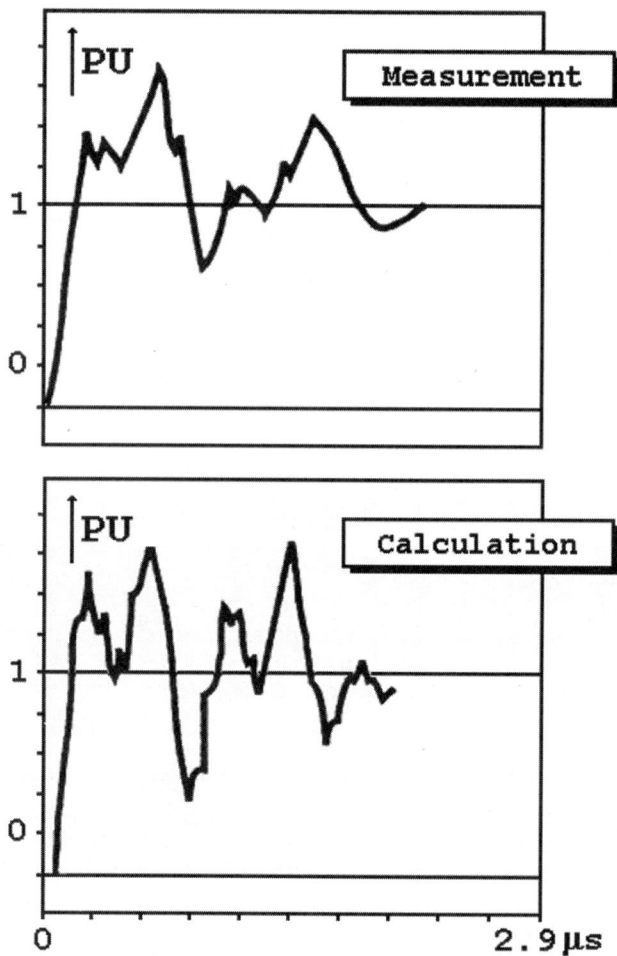

FIGURE 10.33 Measurement and simulation of overvoltages in a 420 kV GIS at closing a switch. (Copyright 1999 IEEE.)

SF6 insulation — Breakdown caused by VFT overvoltages is improbable in a well-designed GIS insulation system during normal operations. The breakdown probability increases with the frequency of the oscillations. In addition, breakdown values can be reduced by insulation irregularities like edges and fissures. However, at ultra high voltage systems, more than 1000 kV, for which the ratio of BIL to the system voltage is lower, breakdown is more likely to be caused. At these levels, VFT overvoltages can be reduced by using resistor-fitted disconnectors (Yamagata et al., 1996).

Transformers — Due to steep fronted wave impulses, direct connected transformers can experience an extremely nonlinear voltage distribution along the high-voltage winding, connected to the oil-SF6 bushings, and high resonance voltages due to transient oscillations generated within the GIS. Transformers can generally withstand these stresses; however, in critical cases, it may be necessary to install varistors to protect tap changers.

Disconnectors and breakers — The insulation system of breakers and switches is not endangered by VFT overvoltages generated in adjacent GIS equipment. Ground faults induced by VFT overvoltages have been observed in disconnectors operations, as residual leader branches can be activated by enhanced field gradient to ground. These faults can be avoided by a proper disconnector design.

Enclosure — TEV can cause sparking across insulated flanges and to insulated busbars of CTs, and can puncture insulation that is intended to limit the spread of circulating currents within the enclosure.

TEV can be minimized with a proper design and arrangement of substation masts, keeping ground leads as short and straight as possible in order to minimize the inductance, increasing the number of connections to ground, introducing shielding to prevent internally generated VFT from reaching the outside of the enclosure, and installing voltage limiting varistors where spacers must be employed.

Bushings — Very few problems have been reported with capacitively graded bushings. High impedances in the connection of the last graded layer to the enclosure should be avoided.

Secondary equipment — TEV may interfere with secondary equipment or damage sensitive circuits by raising the housing potential if they are directly connected, or via cable shields to GIS enclosure by emitting free radiation which may induce currents and voltages in adjacent equipment. Correct cable connection procedures may minimize interference. The coupling of radiated energy may be reduced by mounting control cables closely along the enclosure supports and other grounded structures, grounding cable shields at both ends by leads as short as possible, or using optical coupling services. Voltage limiting devices may have to be installed.

References

Ardito, A., Iorio, R., Santagostino, G., and Porrino, A., Accurate modeling of capacitively graded bushings for calculation of fast transient overvoltages in GIS, *IEEE Trans. on Power Delivery*, 7, 1316, 1992.

Boersma, R., Transient ground potential rises in gas-insulated substations with respect to earthing systems, *Electra*, 110, 47, 1987.

Boggs, S. A., Chu, F. Y., Fujimoto, N., Krenicky, A., Plessl, A., and Schlicht, D., Disconnect switch induced transients and trapped charge in gas-insulated substations, *IEEE Trans. on Power Appar. and Syst.*, 101, 3593, 1982.

Boggs, S.A, Fujimoto, N., Collod, M., and Thuries, E., The modeling of statistical operating parameters and the computation of operation-induced surge waveforms for GIS disconectors, *CIGRE*, Paper No. 13-15, Paris, 1984.

Chowdhuri, P., *Electromagnetic Transients in Power Systems*, RSP-John Wiley, 1996, Chap. 12.

CIGRE Working Group 33/13-09, Very fast transient phenomena associated with gas insulated substations, *CIGRE*, Paper 33-13, Paris, 1988.

CIGRE Joint WG 33/23-12, Insulation co-ordination of GIS: Return of experience, on site tests and diagnostic techniques, *Electra*, 176, 66, 1998.

CIGRE Working Group 33.02, *Guidelines for representation of networks elements when calculating transients*, CIGRE Brochure, 1990.

Dick, E. P., Fujimoto, N., Ford, G. L., and Harvey, S., Transient ground potential rise in gas-insulated substations   problem identification and mitigation, *IEEE Trans. on Power Appar. and Syst.*, 101, 3610, 1982.

Ecklin, A., Schlicht, D., and Plessl, A., Overvoltages in GIS caused by the operation of isolators, in *Surges in High-Voltage Networks*, Ragaller, K., Ed., Plenum Press, New York, 1980, chap. 6.

Fujimoto, N., and Boggs, S. A., Characteristics of GIS disconnector-induced short risetime transients incident on externally connected power system components, *IEEE Trans. on Power Delivery*, 3, 961, 1988.

Fujimoto, N., Chu, F. Y., Harvey, S. M., Ford, G. L., Boggs, S. A., Tahiliani, V. H., and Collod, M., Developments in improved reliability for gas-insulated substations, *CIGRE*, Paper No. 23-11, Paris, 1988.

Fujimoto, N., Dick, E. P., Boggs, S. A., and Ford, G. L., Transient ground potential rise in gas- insulated substations — experimental studies, *IEEE Trans. on Power Appar. and Syst.*, 101, 3603, 1982.

Fujimoto, N., Stuckless, H. A., and Boggs, S. A., Calculation of disconnector induced overvoltages in gas-insulated substations, in *Gaseous Dielectrics IV*, Pergamon Press, 1986, 473.

Haznadar, Z., Carsimamovic, S., and Mahmutcehajic, R., More accurate modeling of gas insulated substation components in digital simulations of very fast electromagnetic transients, *IEEE Trans. on Power Delivery*, 7, 434, 1992.

IEC 71-1, *Insulation Co-ordination — Part 1: Definitions, Principles and Rules*, 1993.

IEC 71-2, *Insulation Co-ordination — Part 2: Application Guide*, 1996.
IEEE TF on Very Fast Transients (Povh, D., Chairman), Modelling and analysis guidelines for very fast transients, *IEEE Trans. on Power Delivery*, 11, 2028, 1996.
Meppelink, J., Diederich, K., Feser, K., and Pfaff, W., Very fast transients in GIS, *IEEE Trans. on Power Delivery*, 4, 223, 1989.
Miri, A. M. and Binder, C., Investigation of transient phenomena in inner- and outer systems of GIS due to disconnector operation, in *Proc. Int. Conf. Power Systems Transients*, Lisbon, 1995, 71.
Ogawa, S., Haginomori, E., Nishiwaki, S., Yoshida, T., and Terasaka, K., Estimation of restriking transient overvoltage on disconnecting switch for GIS, *IEEE Trans. on Power Delivery* 1, 95, 1986.
Witzmann, R., Fast transients in gas insulated substations. Modelling of different GIS components, in *Proc. 5th Int. Symp. HV Engineering*, Braunschweig, 1987.
Yamagata, Y., Tanaka, K., Nishiwaki, S., Takahashi, N., Kokumai, T., Miwa, I., Komukai, T., and Imai, K., Suppression of VFT in 1100 kV GIS by adopting resistor-fitted disconnector, *IEEE Trans. on Power Delivery* 11, 872, 1996.
Yanabu, S., Murase, H., Aoyagi, H., Okubo, H., and Kawaguchi, Y., Estimation of fast transient overvoltage in gas-insulated substation, *IEEE Trans. on Power Delivery*, 5, 1875, 1990.

10.6 Transient Voltage Response of Coils and Windings

Robert C. Degeneff

Transient Voltage Concerns

Normal System Operation

Transformers are normally used in systems to change power from one voltage (or current) to another. This is often driven by a desire to optimize the overall system characteristics, e.g., economics, reliability, and/or performance. To achieve these system goals, a purchaser must specify and a designer must configure the transformer to meet a desired impedance, voltage and power rating, thermal characteristic, short circuit strength, sound level, physical size, and voltage withstand capability. Obviously, many of these goals will produce requirements that are in conflict and prudent compromise will be required. Failure to achieve an acceptable characteristic for any of these makes the overall transformer design unacceptable. Transformer characteristics and the concomitant design process is outlined in Blume et al., (1951); Bean et al., (1959); MIT, (1943); and Franklin, (1983).

Normally a transformer operates under steady voltage excitation. Occasionally a transformer (in fact, all electrical equipment) experiences dynamic and/or transient overvoltages. Often, it is these infrequent transient voltages that establish design constraints for the transformer's insulation system. These constraints often have far reaching effects on the overall equipment design. The transformer must be configured to withstand any abnormal voltages covered in the design specification and realistically expected in service. Often, these constraints have great impact on other design issues and, as such, have significant effect on the overall transformer cost, performance, and configuration. In recent years, engineers have explored the adverse effect of transient voltages on the reliability of transformers (Kogan et al., 1988; Working group 05, 1983; Kogan et al., 1990), and found it to be a major cause of transformer failure.

Sources and Types of Transient Voltage Excitation

The voltages subjected to a transformer's terminals can be broadly classed as steady state and "transient." Normally, the "transient" voltages a transformer experiences are commonly referred to as dynamic, transient, and very fast transients.

The majority of voltages a transformer experiences during its operational life are steady state, e.g., the voltage is within ±10% of nominal and the frequency is within 1% of rated. As power quality issues grow, the effect of harmonic voltages and currents on performance becoming more of an issue. These harmonics are effectively reduced magnitude, steady state voltages and currents at harmonic frequencies (say 2nd to the 50th). These are addressed in great detail in reference IEEE Std. 519 (1999). Strictly speaking, all other voltage excitation are transients, e.g., dynamic, transient, and very fast transient voltages.

Dynamic voltages refer to relatively low frequency (60 to 1500 Hz), damped oscillatory voltage. Magnitudes routinely observed are from 1 to 3 times the systems peak nominal voltage. Transient voltage refers to the class of excitation caused by events like lightning surges, switching events, and line faults causing voltages of the chopped wave form (Degeneff et al., 1982). Normally, these are aperiodic waves. Occasionally, the current chopping of a vacuum breakers will produce transient periodic excitation in the 10 to 200 kHz range (Greenwood, 1994). The term *very fast transient* encompasses voltage excitation with rise times in the range of 50–100 ns and frequencies from 0.5 to 30 MHz. These types of voltages are encountered in gas-insulated stations. The voltages produced within the transformer winding structure by the system it is a part of, must be addressed and understood if a successful insulation design is to be achieved (Narang et al., 1998). Since transient voltages affect system reliability, and that, in turn, system safety and economics, a full understanding of the transient characteristic of a transformer is warranted.

Addressing Transient Voltages Performance

Addressing the issue of transient voltage performance can be divided into three activities: recognition, prediction, and mitigation. By 1950 over 1000 papers had been written to address these issues (Abetti, 1959; 1962; 1964). The first is to appreciate that transient voltage excitation can produce equipment responses different than one would anticipate at first glance. For example, the addition of more insulation around a conductor may in fact make the transient voltage distribution worse and the insulation integrity of the design weaker. Another example would be the internal voltage amplification a transformer experiences when excited near its resonant frequency. The transient voltage distribution is a function of the applied voltage excitation and the shape and material content of the winding being excited. The capability of the winding to withstand the transient voltage is a function of the specific winding shape, the materials voltage versus time characteristic, the past history of the structure, and the statistical nature of the structure's voltage withstand characteristic.

The second activity is to assess or predict the transient voltage within the coil or winding. Today this is generally accomplished using a lumped parameter model of the winding structure and some form of computer solution method that allows the internal transient response of the winding to be computed. Once this voltage distribution is known, its effect on the insulation structure can be computed with a two- or three-dimensional FEM. The resultant voltages, stresses, and creeps are examined in light of the known material and geometrical capability of the system in light of desired performance margins.

The third activity is to establish a transformer structure or configuration that, in light of the anticipated transient voltage excitation and material capability, variability, and statistics nature, will provide acceptable performance margins. Occasionally, nonlinear resistors are used as part of the insulation system to achieve a cost-effective, reliable insulation structure. Additionally, means of limiting the transient excitation include the use of nonlinear resistors, capacitors, and snubbers.

Complex Issue to Predict

The accurate prediction of the transient voltage response of coils and winding has been of interest for almost 100 years. The problem is complex for several reasons. The form of excitation varies greatly. Most large power transformers are unique designs, and as such, each transformer's transient response characteristic is unique. Each has its own impedance versus frequency characteristic. As such, the transient response characteristic of each transformer is different. Generally, the problem is addressed by building a large lumped parameter model of inductances, capacitances, and resistances. Constructing the lumped parameter model is challenging. The resultant mathematical model is ill-conditioned, e.g., the resultant differential equation is difficult to solve. The following sections outline how these challenges are currently addressed.

It should be emphasized that the voltage distribution within the winding is only the first component of the insulation design process. The spatial distribution of the voltages within the winding must be determined and finally the ability of the winding configuration in view of its voltage versus time characteristic must be assessed.

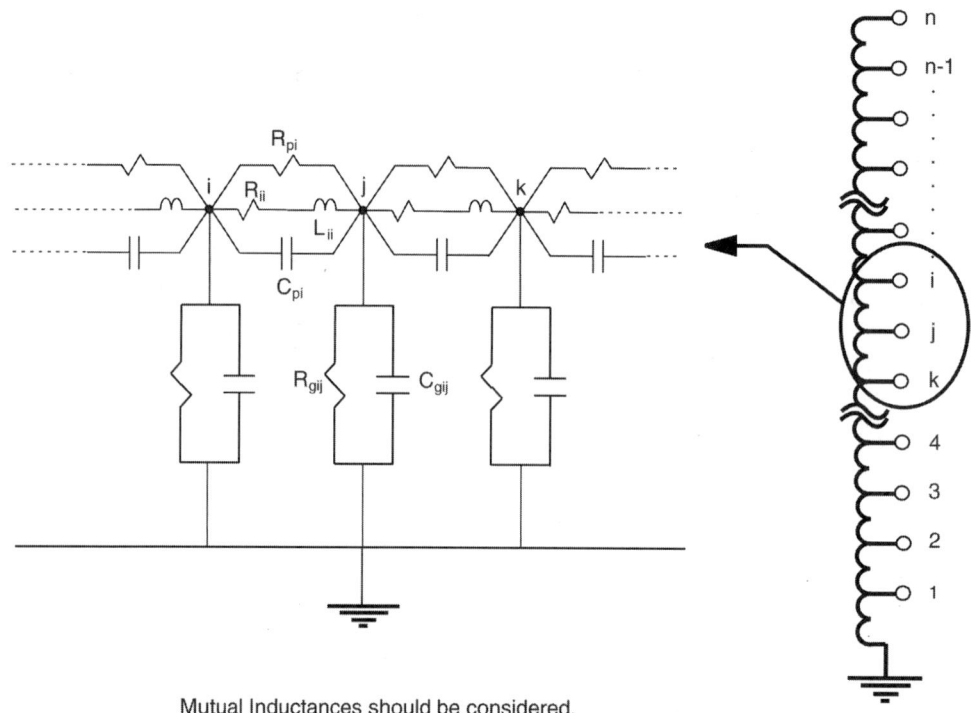

Mutual Inductances should be considered,
i.e. M_{ji} is between segments j and i.

FIGURE 10.34 Sample of section used to model example coil

Surges in Windings

Response of a Simple Coil

Transformer windings are complex structures of wire and insulation. This is the result of many contradictory requirements levied during the design process. In an effort to introduce the basic concepts of transient response, a very simple disk coil was modeled and the internal transient response computed. The coil consisted of 100 identical continuous disk sections of 24 turns each. The inside radius of the coil is 12.55 in., the space between each disk coil is 0.220 in., and the coil was assumed to be in air with no iron core. Each turn was made of copper 0.305 in. in height, 0.1892 in. in the radial direction, with 0.0204 in. of insulation between the turns. For this example, the coil was subjected to a full wave with a 1.0 per-unit voltage. Figure 10.34 provides a sketch of the coil and the node numbers associated with the calculation. For this example, the coil has been subdivided into 50 equal subdivisions with each subdivision a section pair. Figure 10.35 contains the response of the winding as a function of time for the first 200 μsec. It should be clear that the response is complex and a function of both the applied excitation voltage and the characteristics of the coil itself.

Initial Voltage Distribution

If the voltage distribution along the helical shown in Fig. 10.35 is examined at times very close to time zero, it is observed that the voltage distribution is highly nonuniform. For the first few tenths of a microsecond, the distribution is dominated exclusively by the capacitive structure of the coil. This distribution is often referred to as the initial (or short-time) distribution and it is generally highly nonuniform. This initial distribution is shown in Fig. 10.36. For example, examining the voltage gradient over the first 10% of the winding, one sees that the voltage is 82% rather that the anticipated 10% or a rather large enhancement or gradient in some portions of the winding.

Power System Transients

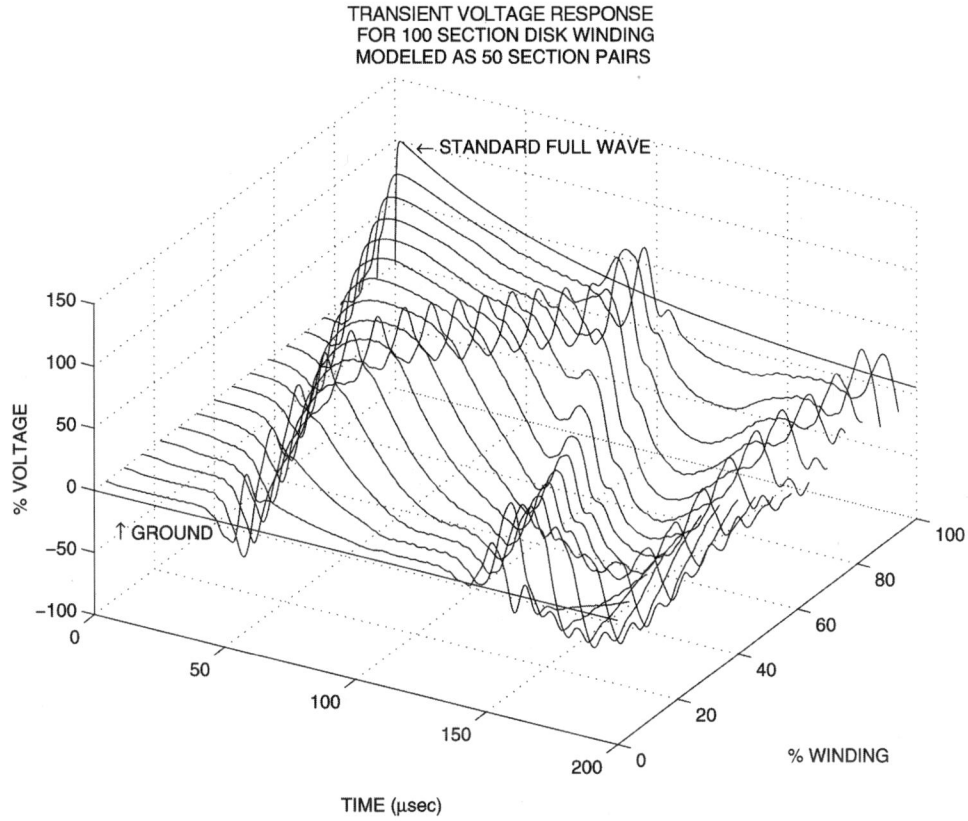

FIGURE 10.35 Voltage versus time for helical winding

The initial distribution shown in Fig. 10.36 is based on the assumption that the coil knows how it is connected, i.e., it requires some current to flow in the winding and this requires some few tens of a ηsecond. The initial distribution can be determined by evaluating the voltage distribution for the windings capacitive network and ignoring both the inductive and resistive components of the transformer. This discussion is applicable for times greater than approximately 0.25 μsec. This is the start of the transient response for the winding.

For times smaller than 0.25 μsec, the distribution is still dictated by capacitance, but the transformers capacitive network is unaware that it is connected. This is addressed in the chapter on very fast transients and in Narang et al. (1998).

Steady-State Voltage Distribution

The steady-state voltage distribution depends primarily on the inductance and losses of the winding's structure. This distribution, referred to by Abetti (1960) as the pseudo-final, is dominated primarily by the self and mutual inductance of the windings and the manner in which the winding is connected. This steady-state voltage distribution is very near (but not identical) to the turns ratio. For the simple winding shown in Fig. 10.34, the distribution is known by inspection and shown in Fig. 10.36, but for more complex windings it can be determined by finding the voltage distribution of the inductive network and ignoring the effect of winding capacitances.

Transient Voltage Distribution

Figure 10.35 shows the transient response for this simple coil. The transient response of the coil is the voltage the coil experiences as the coil transitions between the initial voltage distribution and the steady state distribution. It is the very same idea as pulling a rubber-band away from its stretched (but steady

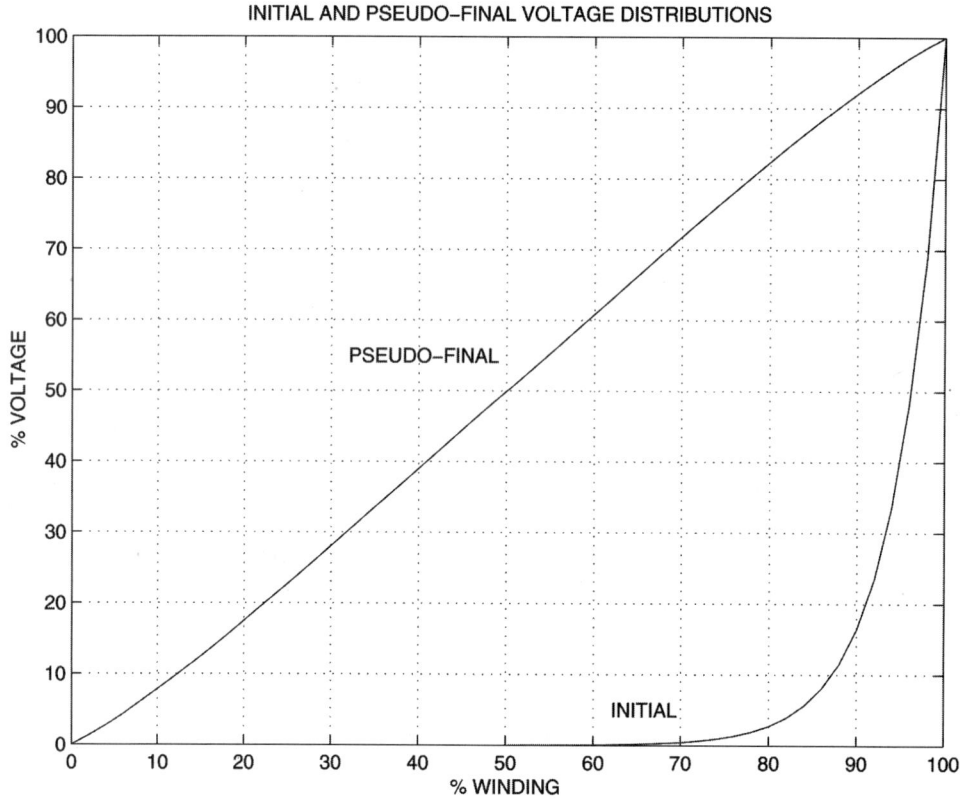

FIGURE 10.36 Initial and pseudo-final voltage distribution

state position) and then letting it go and monitoring its movement in space and time. What is of considerable importance is the magnitude and duration of these transient voltages and the ability of the transformer's insulation structure to consistently survive these voltages.

The coil shown in Fig. 10.34 is a very simple structure. The challenge facing transformer design engineers since the early 1900s has been to determine what this voltage distribution was for a complex winding structure of a commercial transformer design.

Determining Transient Response

History

Considerable research has been devoted to determining the transformer's internal transient voltage distribution. These attempts started at the beginning of the 1900s and have continued at a steady pace for almost 100 years (Abetti, 1959; 1962; 1964). The earliest attempts at using a lumped parameter network to model the transient response was in 1915. Until the early 1960s, these efforts were of limited success due to computational limitations encountered when solving large numbers of coupled stiff differential equations. During this period, the problem was attached in the time domain using either a standing wave approach or the traveling wave method. The problem was also explored in the frequency domain and the resultant individual results combined to form the needed response in the time domain. Abetti introduced the idea of a scale model, or analog for each new design. Then, in the 1960s, with the introduction of the high-speed digital computer, major improvements in computational algorithms, detail, accuracy, and speed were obtained.

In 1956, Rouxel and Waldvogel used an analog computer to calculate the internal voltages in the transformer by solving a system of linear differential equations with constant coefficients resulting from a uniform, lossless, and linear lumped parameter model of the winding where mutual and self inductances were calculated assuming an air core. Later, McWhirter, Fahrnkopf, and Steel (1956) developed a method of determining impulse voltage stresses within the transformer windings through the use of the digital and analog computer applicable to some extent for nonuniform windings. But it was not until 1958 that Dent, Hartill, and Miles recognized the limitations of the analog models and developed a digital computer model in which any degree of nonuniformity in the windings could be introduced and any form of applied input voltage applied. During the mid-1970s, efforts at General Electric (White, 1977; Degeneff et al., 1980; White 1977) focused on building a program to compute transients for core form winding of completely general design. By the end of the 1970s, an adequate linear lossless model of the transformer was available to the industry. However, adequate representation of the effect of the nonlinear core and losses was not available. Additionally, the transformer models used in insulation design studies had little relationship with lumped parameter transformer models used for system studies.

Wilcox improved the transformer model by including core losses in a linear-frequency domain model where self and mutual impedance between winding sections were calculated considering a grain-oriented conductive core with permeability μ_r (1992). In this work, Wilcox modeled the skin effect, the losses associated with the magnetizing impedance, and a loss mechanism associated with the effect of the flux radial component on the transformer core during short circuit conditions. Wilcox applied this modified modal theory to model the internal voltage response on practical transformers. Vakilian and Degeneff (1993) modified White's inductance model (1977) to include the core's saturable characteristics and established a system of O.D.E.s for the linearized lumped parameter LRC network, then used Gear's method to solve for the internal voltage response in time. During the same year, Gutierrez and Degeneff (1993; 1994) presented a transformer model reduction technique as an effort to reduce the computational time required by linear and nonlinear detailed transformer models and to make these models small enough to fit into EMTP. In 1994, de Leon and Semlyen presented a nonlinear three phase transformer model including core and winding losses. This was the first attempt to combine frequency dependency and nonlinearity in a detailed transformer model. The authors used the principle of duality to extend their model to three phase transformers.

Lumped Parameter Model

A device's transient response is a result of the flow of energy between the distributed electrostatic and electromagnetic characteristics of the device. For all practical transformer winding structures, this interaction is quite complex and can only be realistically investigated by constructing a detailed lumped parameter model of the winding structure and then carrying out a numerical solution for the transient voltage response. The most common approach is to subdivide the winding into a number of segments (or groups of turns). The method of subdividing the winding can be complex and, if not addressed carefully, affects the accuracy of the resultant model. The resultant lumped parameter model is composed of inductances, capacitance, and losses. Starting with these inductances, capacitances, and resistive elements, equations reflecting the transformer's transient response can be written in numerous forms. Two of the most common are the basic admittance formulation of the differential equation and the state variable formulation. The admittance formulation is given by Degeneff, 1977:

$$[I(s)] = \left[\frac{1}{s}[\Gamma_n] + [G] + s[C]\right][E(s)] \qquad (10.105)$$

The general state-variable formulation is given by Vakilian (1993) describing the transformer's lumped parameter network at time t:

$$\begin{bmatrix}[L] & 0 & 0 \\ 0 & [C] & 0 \\ 0 & 0 & [U]\end{bmatrix}\begin{bmatrix}[di_e/dt] \\ [de_n/dt] \\ [df_e/dt]\end{bmatrix} = \begin{bmatrix}-[r] & [T]^t \\ -[T] & -[G] \\ -[r] & [T]^t\end{bmatrix}\begin{bmatrix}[i_e] \\ [e_n]\end{bmatrix} - \begin{bmatrix}0 \\ [I_s] \\ 0\end{bmatrix} \qquad (10.106)$$

where the variables in Eqs. (10.105) and (10.106)

$[i_e]$ = Vector of currents in the winding segments
$[e_n]$ = Model's nodal voltages vector
$[f_e]$ = Windings' flux-linkages vector
$[r]$ = Diagonal matrix of windings series resistance
$[T]$ = Windings connection matrix
$[T]^t$ = Transpose of $[T]$
$[C]$ = Nodal capacitance matrix
$[U]$ = Unity matrix
$[I(s)]$ = Laplace transform of current sources
$[E(s)]$ = Laplace transform of nodal voltages
$[\Gamma_n]$ = Inverse Nodal Inductance Matrix = $[T][L]^{-1}[T]^t$
$[L]$ = Matrix of self and mutual inductances
$[G]$ = Conductance matrix, for resistors connected between nodes
$[I_s]$ = Vector of current sources

In a liner representation of an iron core transformer, the permeability of the core is assumed constant regardless of the magnitude of the core flux. This assumption allows the inductance model to remain constant for the entire computation. Equations (10.105) and (10.106) are based on the assumption that the transformer core is linear and the various elements in the model are not frequency dependent. Work in the last decade has addressed both the nonlinear characteristics of the core and the frequency dependent properties of the materials. Much progress has been made, but their inclusion adds considerably to the model's complexity and the computational difficulty. If the core is nonlinear, the permeability changes as a function of the material properties, past history, and instantaneous flux magnitude. Therefore, the associate inductance model is time dependent. The basic strategy for solving the transient response of the nonlinear model in Eq. (10.105) or (10.106) is to linearize the transformer's nonlinear magnetic characteristics at an instant of time based on the flux in the core at that instant.

Two other model formulations should be mentioned. de Leon addressed the transient problem using a model based on the idea of duality (1994). The finite element method has found wide acceptance in solving for electrostatic and electromagnetic field distributions. In some instances it is a very useful in solving for the transient distribution in coils and windings of complex shape.

Frequency Domain Solution

A set of linear differential equations representing the transient response of the transformer can be solved either in the time domain or in the frequency domain. If the model is linear, the resultant solution will be the same for either method. The frequency domain solution requires that the components of the input waveform at each frequency be determined. These individual sinusoidal waves are then applied individually to the transformer and the resultant voltage response throughout the winding is determined. Finally, the total response in the time domain is determined by summing the component responses at each frequency applying superposition. An advantage of this method is that it allows the recognition of frequency-dependent losses to be addressed easily. Disadvantages of this method are that it does not allow the modeling of time-dependent switches, nonlinear resistors like ZnO, or the recognition of nonlinear magnetic core characteristics.

Solution in the Time Domain

The following briefly discusses the solution of Eqs. (10.105) and (10.106). There are numerous methods to solve Eq. (10.105) but it has been found that when solving the stiff differential equation model of a transformer, a generalization of Dommel's method (Dommel, 1969; Degeneff, 1977) works very well. A lossless lumped parameter transformer model containing n nodes has approximately $n(n + 1)/2$ inductors and $3n$ capacitors. Since the total number of inductors far exceeds the number of capacitors in the network, this methodology reduces storage and computational time by representing each capacitor as an inductor in parallel with a current source. The following system of equations results:

$$[\hat{Y}][F(t)] = [I(t)] - [H(t)] \qquad (10.107)$$

where

$[F(t)]$ = nodal integral of the voltage vector
$[I(t)]$ = nodal injected current vector
$[H(t)]$ = Past History current vector

and

$$[\hat{Y}] = \frac{4}{\Delta t^2}[C] + \frac{2}{\Delta t}[G] + [\Gamma_n] \qquad (10.108)$$

The lumped parameter model is composed of capacitances, inductances, and losses computed from the winding geometry, permittivity of the insulation, iron core permeability, and the total number of sections into which the winding is divided. Then the matrix $[\hat{Y}]$ is computed using the integration step size, Δt. At every time step, the above system of equations is solved for the unknowns in the integral of the voltage vector. The unknown nodal voltages, $[E(t)]$, are calculated by taking the derivative of $[F(t)]$. Δt is selected based on the detail of the model and the highest resonant frequency of interest. Normally, Δt is smaller that one-tenth the period of this frequency.

The state variable formulation shown as Eq. (10.106) can be solved using differential equation routines available in IMSL or others based on the work of Gear and Adams (1991). The advantage of these routines is that they are specifically written with the solution of stiff systems of equations in mind. The disadvantage of these routines is that they consume considerable time during the solution.

Accuracy vs. Complexity

Every model of a physical system is an approximation. Even the simplest transformer has a complex winding and core structure and as such possesses an infinite number of resonant frequencies. A lumped parameter model, or for that matter, any model is at best an approximation of the actual device of interest. A lumped parameter model containing a structure of inductances, capacitances, and resistances will produce a resonance frequency characteristic that contains the same number of resonant frequencies as nodes in the model. The transient behavior of a linear circuit (the lumped parameter model) is determined by the location of the poles and zeros of its terminal impedance characteristic. It follows then that a detailed transformer model must possess two independent characteristics to faithfully reproduce the transient behavior of the actual equipment. First, it must include accurate values of R, L, and C, reflecting the transformer geometry. This fact is well appreciated and documented. Second, the transformer must be modeled with sufficient detail to address the bandwidth of the applied waveshape. In a valid model, the highest frequency of interest would have a period at least ten times larger than the travel time of the largest winding segment in the model. If this second characteristic is overlooked, a model can produce results that appear valid, but may have little physical basis. A final issue is the manner in which the transformer structure is subdivided. If care is not taken, the manner in which the model is constructed will itself introduce significant errors and the computation will be mathematically robust but an inaccurate approximation of the physically reality.

Resonant Frequency Characteristic

Definitions

The steady state and transient behavior of any circuit, for any applied voltage, is established by the location of the poles and zeros of the impedance function of the lumped parameter model in the complex plane. The zeros of the terminal impedance function coincide with the natural frequencies of the model, by definition. McNutt (1974) defines terminal resonance as the terminal current maximum and a terminal impedance minimum. In a physical system there are an infinite number of resonances. In a lumped parameter model of a system, there are as many resonances as nodes in the model (or the order of the system). Terminal resonance is also referred to as series resonance (Abetti, 1959; 1954). Terminal antiresonance is defined as a terminal current minimum and a terminal impedance maximum (McNutt et al., 1974). This is also referred to as parallel resonance (Abetti, 1959; 1954). McNutt defines internal resonance as an internal voltage maximum and internal antiresonance as an internal voltage minimum.

Impedance vs. Frequency

The terminal resonances for a system can be determined by taking the square root of the eigenvalues of the system matrix, [A], shown in the state variable representation for the system shown as:

$$[\dot{q}] = [A][q] + [B][u]$$
$$[y] = [C][q] + [D][u]$$
(10.109)

where

[A] = State Matrix
[B] = Input Matrix
[C] = Output Matrix
[D] = Direct Transmission Matrix
[q] = Vector of state variables for system
[$\dot{q}$] = First derivative of [q]
[u] = Vector of input variables
[y] = Vector of output variables

The impedance versus frequency characteristic requires a little more effort. In light of the previous definitions, terminal resonance may be defined as occurring when the reactive component of the terminal impedance is zero. Equivalently, terminal resonance occurs when the imaginary component of the quotient of the terminal voltage divided by the injected terminal current is zero. Recalling that in the Laplace domain, that s is equivalent to $j\omega$ with a system containing n nodes with the excited terminal code j, one can rewrite Eq. (10.105) to obtain:

$$\begin{bmatrix} e_1(s) \\ e_2(s) \\ - \\ e_j(s) \\ - \\ e_n(s) \end{bmatrix} = \begin{bmatrix} Z_{1j}(s) \\ Z_{2j}(s) \\ - \\ Z_{jj}(s) \\ - \\ Z_{nj}(s) \end{bmatrix} [i_j(s)]$$
(10.110)

The voltage at the primary (node j) is in operation form. Rearranging the terminal impedance is given by:

$$Z_t(\omega) = Z_{jj}(j\omega) = Z_{jj}(s) = \frac{e_j(s)}{i_j(s)}$$
(10.111)

In these equations the unknown quantities are the voltage vector and the frequency. It is a simple matter to assume a frequency and solve for the corresponding voltage vector. Solving Eq. (10.111) over a range of frequencies results in the well-known impedance versus frequency plot. Figure 10.37 contains the impedance versus frequency for the example used in Figs. 10.34 and 10.35.

Power System Transients

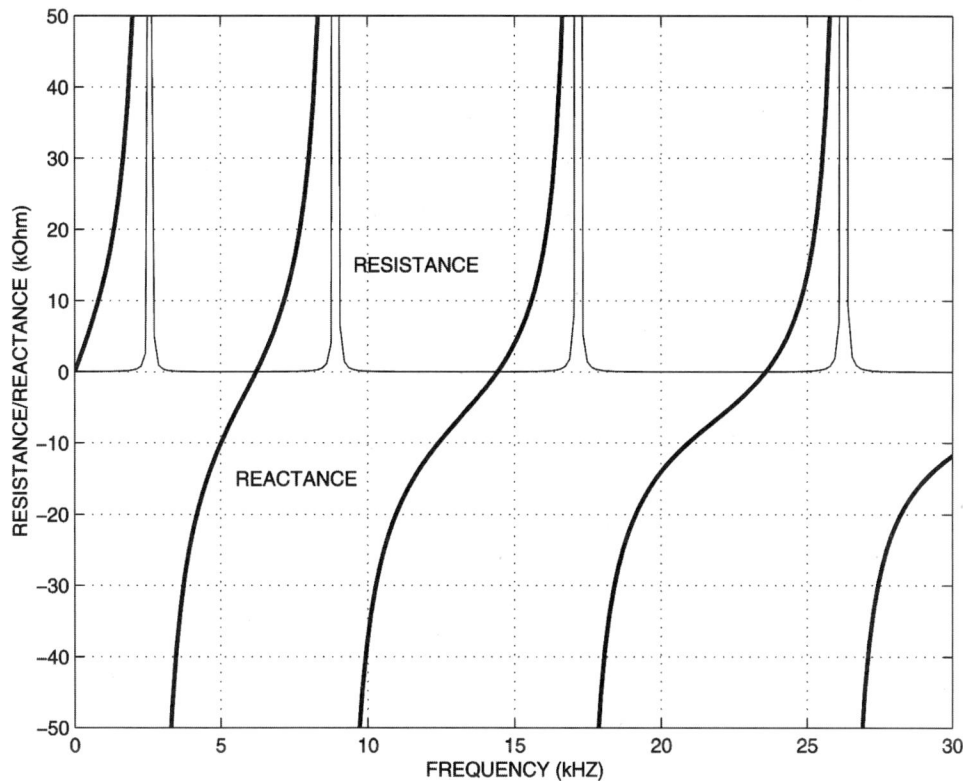

FIGURE 10.37 Terminal impedance for a helical winding

Amplification Factor

The amplification factor or gain function is defined as:

$$[N_{lm,j}] = \frac{\text{Voltage between points } l \text{ and } m \text{ at frequency } \omega}{\text{Voltage applied at input node } j \text{ at frequency } \omega} \quad (10.112)$$

Degeneff (1977) shows that this results in:

$$N_{lm,j} = \frac{Z_{ll}(j\omega) - Z_{mj}(j\omega)}{Z_{jj}(j\omega)} \quad (10.113)$$

It is a simple matter to assume a frequency and solve for the corresponding voltage versus frequency vector. If one is interested in the voltage distribution within a coil at one of the resonant frequencies, this can be found from the eigenvector of the coil at the frequency of interest. If one is interested in the distribution at any other frequency, Eq. (10.113) can be utilized. This is shown in Fig. 10.38.

Inductance Model

Definition of Inductance

Inductance is defined as:

$$L = \frac{d\lambda}{dI} \quad (10.114)$$

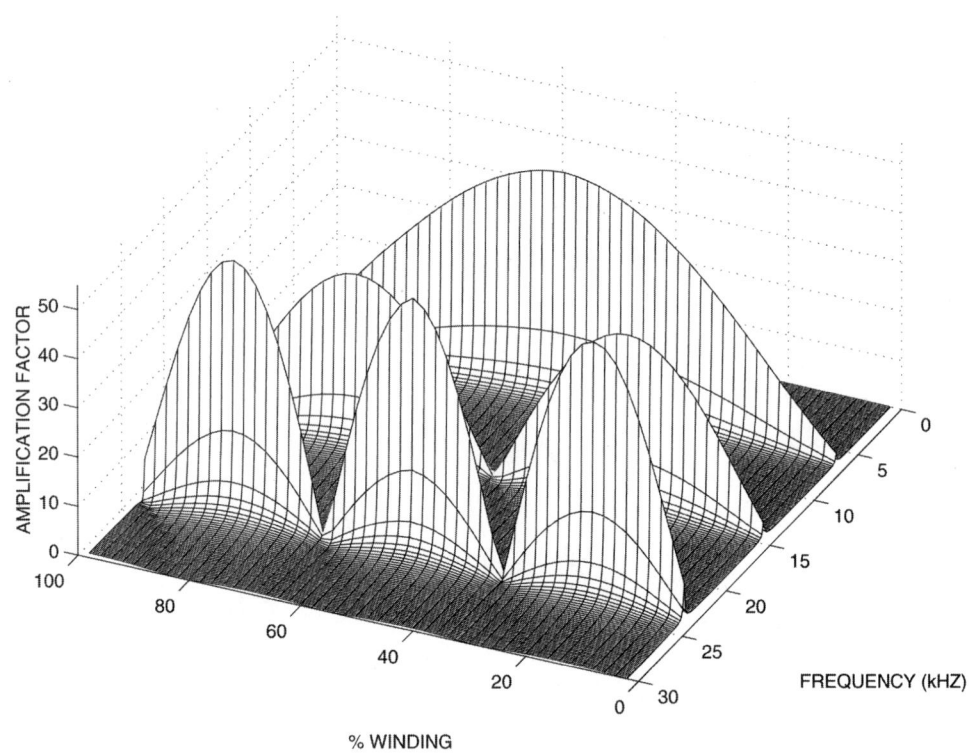

FIGURE 10.38 Amplification factor at 5 from 0 to 30kHz

and if the system is linear:

$$L = \frac{\lambda}{I} \tag{10.115}$$

where

L = inductance, Henrys
λ = flux linkage caused by current I, Weber turns
I = current producing flux linkages, Amperes
$d\lambda$ = first derivative of λ
dI = first derivative of I

L is referred to as self inductance if the current producing the flux is the same current being linked. L is referred to as mutual incidence if the current producing the flux is other than the current being linked (MIT, 1943). Grover (1962) published an extensive work providing expressions to compute the self and mutual inductance in air for a large number of practical conductor and winding shapes.

One of the most difficult phenomena to model is the magnetic flux interaction involving the different winding sections and the iron core. Historically this phenomenon has been modeled by dividing the flux in two components: the common and leakage flux. The common flux dominates when the transformer behavior is studied under open circuit conditions and the leakage flux dominates the transient response when the winding is shorted or loaded heavily. Developing a transformer model capable of representing the magnetic flux behavior for all conditions the transformer will see in factory test and in service requires the accurate calculation of the self and mutual inductances.

Transformer Inductance Model

Until the introduction of the computer there was a lack of practical analytical formulae to compute the self and mutual inductances of coils with an iron core. Rabins (1956) developed expressions to calculate self and mutual inductances for a coil on an iron core based on the assumption of a round core leg and infinite core yokes both of infinite permeability. Fergestad and Heriksen improved Rabins' inductance model in 1974 by assuming an infinite permeable core except for the core leg. In their approach, a set of state variable equations was derived from the classic lumped parameter model of the winding.

White (1977; 1978) derived an expression to calculate the self and mutual inductances in the presence of an iron core with finite permeability under the assumption of an infinitely long iron core. White's inductance model had the advantage that the open circuit inductance matrix could be inverted (White, 1978). White derived an expression for the self and mutual inductance between sections of a transformer winding by solving a two-dimensional problem in cylindrical coordinates for the magnetic vector potential assuming a nonconductive and infinitely-long open core. He assumed that the leakage inductance of an open-core configuration is the same as the closed core.

Starting from the definition of the magnetic vector potential $\vec{B} = \nabla \times \vec{A}$ and $\nabla \cdot \vec{A} = 0$ and using Ampere's law in differential form, $\vec{J}_f = \nabla \times \vec{H}$, White solved the following equation:

$$\nabla^2 \vec{A}(r,z) = -\mu_o \vec{J}(r,z) \tag{10.116}$$

The solution broke into two parts: the air core solution and the change in the solution due to the insertion of the iron core as shown in the following equations:

$$\vec{A}(r,z) = \begin{cases} \vec{A}_o(r,z) + \vec{A}_1(r,z), & 0 \leq r \leq R_c \\ \vec{A}_o(r,z) + \vec{A}_2(r,z), & r > R_c \end{cases}$$

$\vec{A}_o(r, z)$ is the solution when the core is present and $\vec{A}_1(r, z)$ and $\vec{A}_2(r, z)$ are the solutions when the iron core is added. Applying Fourier series to Eq. (10.116), the solution for $\vec{A}_o(r, z)$ was found first and then $\vec{A}_1(r, z)$.

Knowing the magnetic vector potential allows the flux linking a filamentary turn at (r, z) to be determined by recalling $\phi(r, z) = \oint \vec{A}(r, z) \cdot \vec{dl}$. The flux for the filamentary turn is given by:

$$\phi(r,z) = 2\pi r \{\vec{A}_o(r,z) + \vec{A}_2(r,z)\} = \phi_o(r,z) + 2\pi r \vec{A}_2(r,z) \tag{10.117}$$

The flux in air, $\phi_o(r, z)$, can be obtained from known formulae for filaments in air (Grover, 1962); therefore, it is only necessary to obtain the change in the flux linking the filamentary turn due to the iron core. If the mutual inductance, L_{ij} between two coil sections is going to be calculated, then the average flux linking section i needs to be calculated. This average flux is given by

$$\phi_{ave} = \frac{\int_{R_i}^{\bar{R}_i} \int_{Z_i}^{\bar{Z}_i} \phi(r,z) dz dr}{H_i(\bar{R}_i - R_i)} \tag{10.118}$$

Knowing the average flux, the mutual inductance can be calculated using the following expression:

$$L_{ij} = \frac{N_i N_j \phi_{ave}}{I_j} \tag{10.119}$$

White's final expression for the mutual inductance between two coil segments is:

$$L_{ij} = L_{ijo} + 2N_i N_j (1-\nu_r) \mu_o R_c \int_0^\infty \frac{I_0(\omega R_c) I_1(\omega R_c) F(\omega)}{\nu_r + (1-\nu_r)\omega R_c I_1(\omega R_c) K_0(\omega R_c)} d\omega \qquad (10.120)$$

where

$$F(\omega) = \frac{1}{\omega} \left[\frac{1}{\omega(\overline{R}_i - R_i)} \int_{\omega R_i}^{\omega \overline{R}_i} x K_1(x) dx \right] \frac{2}{\omega H_i} \sin\left(\frac{\omega H_i}{2}\right)$$

$$\left[\frac{1}{\omega(\overline{R}_j - R_j)} \int_{\omega R_j}^{\omega \overline{R}_j} x K_1(x) dx \right] \frac{2}{\omega H_j} \sin\left(\frac{\omega H_j}{2}\right) \cos(\omega d_{ij}) \qquad (10.121)$$

L_{ijo} is the air core inductance. $\nu_r = \frac{1}{\mu_r}$ is the relative reluctivity. $I_0(\omega R_c)$, $I_1(\omega R_c)$, $K_0(\omega R_c)$, and $K_1(\omega R_c)$ are modified Bessel functions of first and second kind.

Inductance Model Validity

The ability of the inductance model to accurately represent the magnetic characteristic of the transformer can be assessed by the accuracy with which it reproduces the transformer electrical characteristics, e.g., the short circuit and open circuit inductance, and the pseudo-final (turns ratio) voltage distribution. The short circuit and open circuit inductance of a transformer can be determined by several methods, but the simplest is to obtain the inverse of the sum of all the elements in the inverse nodal inductance matrix, Γ_n. This has been verified (Degeneff, 1978; Degeneff and Kennedy, 1975). The pseudo-final voltage distribution is defined in (Abetti, 1960). It is very nearly the turns ratio distribution and must match what ever voltage distribution the winding arrangement and number of turns dictates. An example of this is contained in (Degeneff and Kennedy, 1975).

Capacitance Model

Definition of Capacitance

Capacitance is defined as:

$$C = \frac{Q}{V} \qquad (10.122)$$

where

C = capacitance between the two plates, farads
Q = charge on one of the capacitor plates, coulombs
L = voltage between the capacitor plates, volts

Snow (1954) published an extensive work on computing the capacitance for unusually shaped conductors. Practically, however, most lumped parameter models of windings are created by subdividing the winding into segments with small radial and axial dimensions and large radiuses where a simple parallel plate formula can be used (1980) to compute both the series and shunt capacitance for a segment. For example:

$$C = \epsilon_o \epsilon_r \frac{RadCir}{nD} \qquad (10.123)$$

Power System Transients

where

ϵ_o = permittivity of freespace, 8.9×10^{-12} coulomb/meters
ϵ_r = relative permittivity between turns
Rad = radial build of the segments turns, meters
Cir = circumference of the mean turn within segment, meters
n = number of turns with the segment
D = separation between turns, meter

In computing these capacitances, the relative permittivity, ϵ_r, of the materials must be recognized. This is a function of the material, moisture content, temperature, and effective age of the material. Clark and Von Hippel (1962; 1964) provide a large database of this type of information. Since most lumped parameter models assume the topology has circular symmetry, if the geometry is unusually complex it may be appropriate to model the system with a three-dimensional FEM. It should be emphasized that all of the above is based on the assumption that the capacitive structure of the transformer is frequency invariant. If the transient model is required to be valid over a very large bandwidth, then the frequency characteristic of dielectric structure must be taken into account.

Series and Shunt Capacitance

In order to construct a lumped parameter model, the transformer is subdivided into segments (or groups of turns). Each of these segments contains a beginning node and an exit node. Between these two nodes there will generally be associated a capacitance, traditionally called the series capacitance. These are the intrasection capacitances. In most cases they are computed using the simple parallel plate capacitance given in Eq. (10.123). An exception to this is the series capacitance of disk winding segments. Expressions for their series capacitance is given in the next section.

Additionally, each segment will have associated with it, capacitances between adjacent sections of turns or to a shield or earth. These are the intersection capacitances. These capacitances are generally referred to as shunt capacitances and are normally divided in half and connected to each end of the appropriate segments. This is an approximation, but if the winding is subdivided into relatively small segments, the approximation is acceptable and the error introduced by the model is small.

Equivalent Capacitance for Disk Windings

This section presents simplified expressions to compute the series capacitance for disk winding section pairs. Since most lumped parameter models are not turn-to-turn models, an electrostatic equivalent of the disk section is used for the series capacitance. It is well known that as the series capacitance of disk winding sections becomes larger with respect to the capacitances to ground, the initial distribution becomes more linear (straight line) and the transient response in general more benign. Therefore, since it is possible to arrange the turns within a disk section in many ways without affecting the section's inductance characteristic or the space or material it requires, the industry has offered many arrangements in an effort to increase this effective series capacitance.

The effective series capacitance of a disk winding is a capacitance, which when connected between the input and output of the disk winding section pair would store the same electrostatic energy the disk section pair would store (between all turns) if the voltage were distributed linearly within the section. This modeling strategy is discussed in detail by Scheich (1965) and Degeneff and Kennedy (1975). Figure 10.39 illustrates the cross-section of three common disk winding configurations. The series capacitance of the continuous disk is given by:

$$C_{continuous} = \frac{2}{3}C_s + \left[\frac{n-2}{n^2}\right]C_t \quad (10.124)$$

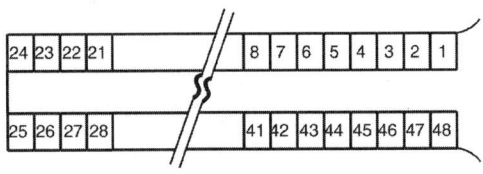

A. Continuous Disk Winding

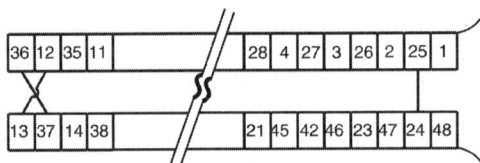

B. Interleaved Disk Winding

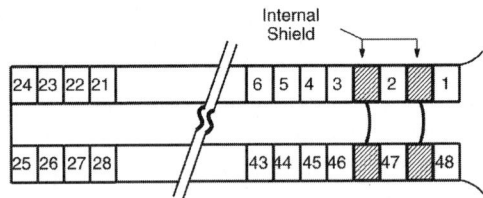

C. Internally Shielded Disk Winding
2 Per Section

FIGURE 10.39 Common disk winding section pairs: A. Continuous; B. Interleaved; and C. Internal Shield.

The series capacitance of the interleaved disk section pair is given by:

$$C_{interleaved} = 1.128 C_s + \left[\frac{n-4}{4}\right] C_t \qquad (10.125)$$

The interleaved disk provides a greater series capacitance than the continuous disk but is more difficult to produce. A winding that has a larger series capacitance than the continuous disk but simpler to manufacture than the interleave is the internally shielded winding. Its series capacitance is given by:

$$C_{internal\ shield} = \frac{2}{3} C_s + \left[\frac{n-2-2n_s}{n^2}\right] C_t + 4 C_{ts} \sum_{i=1}^{n_s} \left[\frac{n_i}{n}\right]^2 \qquad (10.126)$$

where

C_s = Capacitance between sections
C_t = Capacitance between turns
C_{ts} = Capacitance between turn and internal shield
n = Turns in section pair
n_i = Location of shield within section
n_s = Internal shields within section pair

Selecting the disk winding section is often a compromise of electrical performance, economics, and manufacturing preference for a given firm.

Initial Voltage Distribution

The initial voltage distribution can be determined experimentally by applying a voltage wave with a fairly fast rise time (say 0.5 µsec) and measuring the normalized distribution within the winding structure an intermediate time (say 0.3 µsec). The initial distribution can be computed analytically by injecting a current into the excited node and determining the normalized voltage throughout the transformer winding structure. This computational method is outlined in detail by Degeneff and Kennedy (1975). If one is considering a single coil, it is common practice to determine the gradient of the transient voltage near the excited terminal (which is the most severe). This gradient is referred to as α and is found by (Greenwood, 1991):

$$\alpha = \sqrt{\frac{C_g}{C_s}} \tag{10.127}$$

where

α = winding gradient
C_g = capacitance to ground, farads
C_s = effective series capacitance, farads

For the coil shown in Fig. 10.36, the α is on the order of 12.

Loss Model

At steady state, losses are a costly and unwanted characteristic of physical systems. At high frequency, losses produce a beneficial effect in that they reduce the transient voltage response of the transformer by reducing the transient voltage oscillations. In general, the oscillations are underdamped. The effect of damping on the resonant frequency is to reduce the natural frequencies slightly. Losses within the transformer are a result of a number of sources, each source with a different characteristic.

Copper Losses

The losses caused by the current flowing in the winding conductors are referred to as series losses. Series losses are composed of three components: DC losses, skin effect, and proximity effect.

DC Resistance
The conductor's DC resistance is given by:

$$R_{dc} = \rho \frac{l}{A} \tag{10.128}$$

With:

ρ = conductor resistivity, ohm-m
l = length of conductor, m
A = conductor area, m²

ρ is a function of the conductor material and its temperature.

Skin Effect
Lammeraner and Stafl (1966) give an expression for the skin effect in a rectangular conductor. The impedance per unit length of the conductor is given in by:

$$Z = \frac{k}{4h\sigma} \coth kb \quad \Omega/m \tag{10.129}$$

where

$$k = \frac{1+j}{a} \quad (10.130)$$

with

$$a = \sqrt{\frac{2}{\omega\sigma\mu}} \quad (10.131)$$

- h = half the conductor height, m
- b = half the conductor thickness, m
- σ = conductivity of the conductor, S/m
- μ = permeability of the material, H/m
- ω = frequency, rad/s

Defining

$$\xi = b\sqrt{j\omega\sigma\mu} \quad (10.132)$$

and Eq. (10.129) is expressed as

$$Z_{skin} = R_{dc}\xi\coth\xi \quad \Omega/m \quad (10.133)$$

where R_{dc} is the DC resistance per unit length of the conductor. Equation (10.133) is used to calculate the impedance due to the skin effect as a function of frequency.

Proximity Effect

Proximity effect is the increase in losses in one conductor due to currents in other conductors produced by a redistribution of the current in the conductor of interest by the currents in the other conductors. A method of finding the proximity effect losses in the transformer winding consists of finding a mathematical expression for the impedance in terms of the flux cutting the conductors of an open winding section due to an external magnetic field. Since windings in large power transformers are mainly built using rectangular conductors, the problem reduces to the study of eddy current losses in a packet of laminations. Lammeraner and Stafl (1966) provide an expression for the flux as a function of frequency in a packet of laminations. It is given in the following equation:

$$\Phi = \frac{2al\mu}{1+j} H_o \tanh(1+j)\frac{b}{a} \quad (10.134)$$

where l is the conductor length, H_o is the rms value of the magnetic flux intensity, and the remaining variables are the same as defined in Eq. (10.131).

Assuming H_o in Eq. (10.134) represents the average value of the magnetic field intensity inside the conductive region represented by the winding section i and defining L_{ijo} as

$$L_{ijo} = N_i N_j \phi_{ijo} \quad (10.135)$$

where ϕ_{ijo} is the average flux cutting each conductor in section i due to the current I_j, and N is the number of turns in each section, the inductance L_{ij} as a function of frequency is:

Power System Transients

$$L_{ij} = \frac{L_{ijo}}{(1+j)\frac{b}{a}} \tanh\left(1+j\right)\frac{b}{a} \quad \text{H} \tag{10.136}$$

The impedance Z_{proxij} is obtained by multiplying the inductance by the complex variable s. Using the same notation as in Eq. (10.133), the impedance of the conductor due to the proximity effect is given as

$$Z_{prox_{ij}} = s\frac{L_{ij_o}}{\xi}\tanh\xi \quad \Omega \tag{10.137}$$

Core Losses

The effect of eddy currents in the core have been represented in references (de Leon and Semlyen, 1994; Tarasiewicz et al., 1993; Avila-Rosales and Alvarado, 1982) by the well-tested formula:

$$Z = \frac{4N^2 A}{l d^2 \sigma} x \tanh x \tag{10.138}$$

where

$$x = \frac{d\sqrt{j\omega\mu\sigma}}{2} \tag{10.139}$$

and

l = Length of the core limb in meters (axial direction), m
d = Thickness of the lamination, m
μ = Permeability of the material, H/m
N = Number of turns in the coil
A = Total cross-sectional area of all laminations
ω = Frequency, rad/sec

This formula represents the equivalent impedance of a coil wound around a laminated iron core limb. The expression was derived by Avila Rosales and Alvarado (1982) by solving Maxwell's equations assuming the electromagnetic field distribution is identical in all laminations and an axial component of the magnetic flux.

The total hysteresis loss in core volume, V, in which the flux density is uniform everywhere and varying cyclically at a frequency of ω, can be expressed as:

$$P_h = 2\pi\omega\eta V \beta_{max}^n \tag{10.140}$$

with

P_h = Total hysteresis loss in core
η = Constant a function of material
V = Core volume
β = Flux density
n = Exponent dependent upon material, 1.6–2.0
ω = Frequency, rad/sec

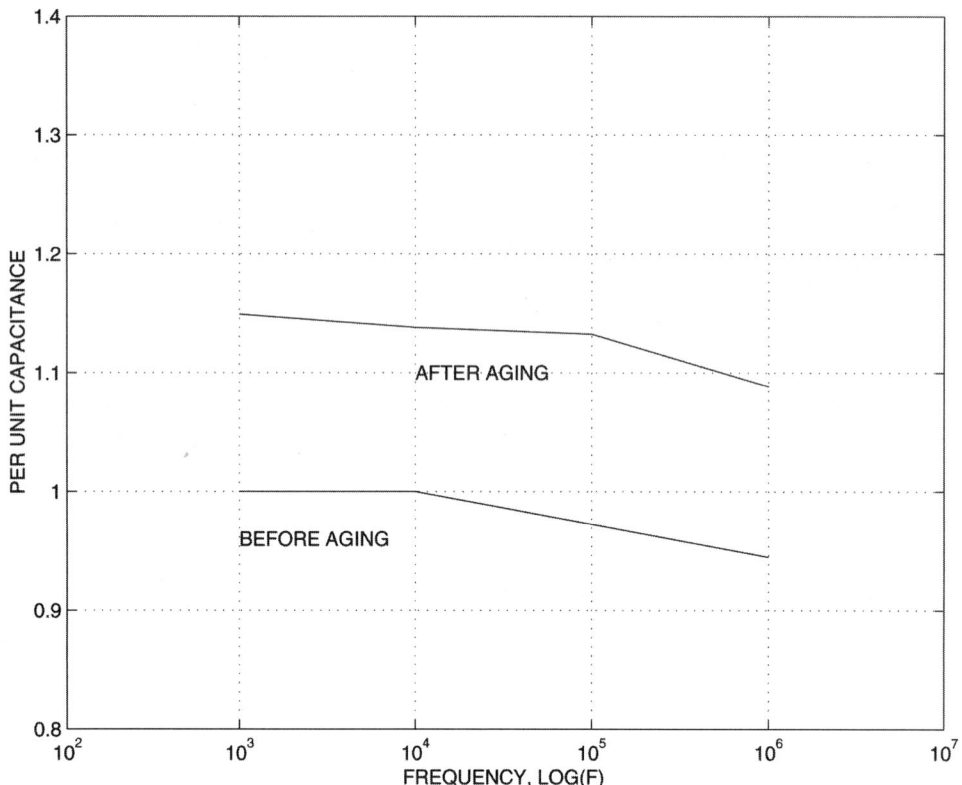

FIGURE 10.40 Oil-soaked paper capacitance and conductance as a function of frequency.

Dielectric Losses

The capacitive structure of a transformer has parallel losses associated with it. At low frequency, the effect of capacitance on the internal voltage distribution can be ignored. As such, the effect of the losses in the dielectric structure can be ignored. However, at higher frequencies the losses in the dielectric system can have a significant effect on the transient response. Batruni et al. (1996) explore the effect of dielectric losses on the impedance versus frequency characteristic of the materials in power transformers.

These losses are frequency dependent and are shown in Fig. 10.40.

Winding Construction Strategies

Design

The successful design of a commercial transformer requires the selection of a simple structure so that the core and coils are easy to manufacture. At the same time, the structure should be as compact as possible to reduce required materials, shipping concerns, and footprint. The form of construction should allow convenient removal of heat, sufficient mechanical strength to withstand forces generated during system faults, acceptable noise characteristics, and an electrical insulation system that meets the system steady state and transient requirements. There are two common transformer structures in use today. When the magnetic circuit is encircled by two or more windings of the primary and secondary, the transformer is referred to as a core-type transformer. When the primary and secondary windings are encircled by the magnetic material, the transformer is referred to as a shell-type transformer.

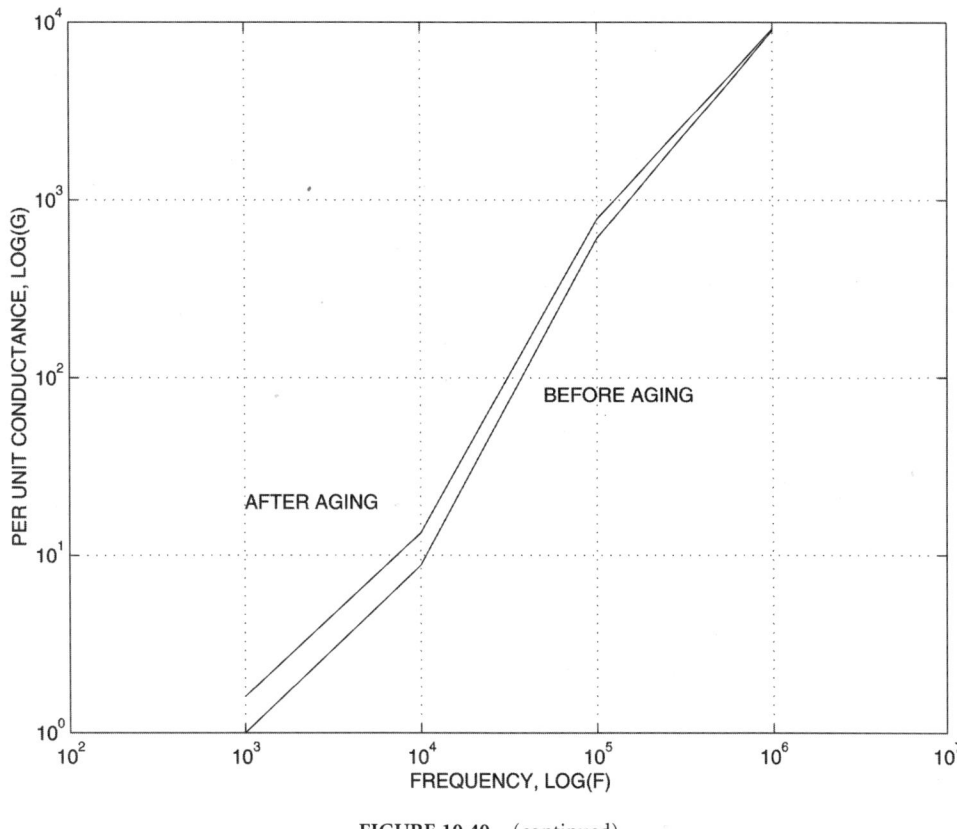

FIGURE 10.40 (continued)

Core-Form

Characteristics of the core-form transformer are a long magnetic path and a shorter mean length of turn. Commonly used core-form magnetic circuits are single phase transformers with a two-legged magnetic path with turns wound around each leg, a three-legged magnetic path with the center leg wound with conductor, and a legged magnetic path with the two interior legs wound with conductors (Bean et al., 1959; MIT, 1943). Three phase core-form designs are generally three legged magnetic cores with all three legs possessing windings and a five-legged core arrangement with the three center legs possessing windings. The simplest winding arrangement has the low voltage winding nearest the core and the high voltage winding wound on top of the low. Normally in the core form construction, the winding system is constructed from helical, layer, or disk type windings. Often the design requirements call for a winding arrangement that is a more complex arrangement, e.g., interleaving high and low voltage windings, interwound taps, windings that are bifurcated have and entry and exit points other than the top or bottom of the coil. All of these variations have, to one degree or another, an effect on the transformer's transient voltage response. To insure an adequate insulation structure, each possible variation must be explored during the design stage to evaluate its effect on the transient overvoltages.

Shell-Form

Shell-form transformer construction features a short magnetic path and a longer mean length of electrical turn. Fink and Beatty (1987) points out that this results in the shell-form transformer having a larger core area and a smaller number of winding turns than the core-form of the same output and performance. Additionally, the shell-form will generally have a larger ratio of steel to copper than an equivalently rated core-form transformer. The most common winding structure for shell-form windings are the primary-secondary-primary (P-S-P) but it is not uncommon to encounter shell form windings of P-S-P-S-P. The winding structure for both the primary and secondary windings are normally of the pancake-type winding structure (Bean et al., 1959).

Proof of Design Concept

The desire of the purchaser is to obtain a transformer at a reasonable price that will achieve the required performance for an extended period of time. The desire of the manufacturer is to construct and sell a product, at a profit, that meets the customer's goals. The specification and purchase contract is the document that combines both the purchaser's requirements and the manufacturer's commitment in a legal format. The specification will typically address the transformer's service condition, rating, general construction, control and protection, design and performance review, testing requirements, and transportation and handling. Since it is impossible to address all issues in a specification, the industry uses standards that are acceptable to purchaser and supplier. In the case of power transformers, the applicable standards would include IEEE C57, IEC 76, and NEMA TR-1.

Standard Winding Tests

ANSI/IEEE C57.12.00 (1993) defines routine and optional tests and testing procedures for power transformers. The following are listed as routine tests for transformers larger than 501 kVA:

1. Winding resistance
2. Winding turns ratio
3. Phase-relationship tests: polarity, angular displacements, phase sequence
4. No-load loss and exciting current
5. Load loss and impedance voltage
6. Low frequency dielectric tests (applied voltage and induced voltage)
7. Leak test on transformer tank

The following are listed as tests to be performed on only one of a number of units of similar design for transformers 501 kVA and larger:

1. Temperature rise tests
2. Lightning-impulse tests (full and chopped wave)
3. Audible sound test
4. Mechanical test from lifting and moving of transformer
5. Pressure tests on tank

Other tests are listed in ANSI/IEEE C57.12.00 (1993) which include, for example, short circuit forces and switching surge impulse tests. Additionally, specific tests may be required by the purchaser based on their applications or field experience.

The variety of transient voltages a transformer may see in its normal useful life are virtually unlimited (Degeneff et al., 1982). It is impractical to proof test each transformer for all conceivable combinations of transient voltage. However, the electrical industry has found that it is possible, in most instances, to assess the integrity of the transformer's insulation systems to withstand transient voltages with the application of a few specific aperiodic voltage waveforms. Figure 10.41 illustrates the full, chopped, and switching surge waveforms. IEEE C57 contains the specific wave characteristics, relationships, and acceptable methods and connections required for these standard tests. Each of these test is designed to test the insulation structure for a different transient condition. The purpose of applying this variety of tests is to substantiate adequate performance of the total insulation system for all the various transient voltages it may see in service.

The insulation integrity for steady state and for dynamic voltages are also assessed by factory tests called out in ANSI/IEEE C57.12.00 (1993). One should not be lulled into thinking that a transformer that has passed all factory voltage tests (both impulse and low frequency) can withstand all transient

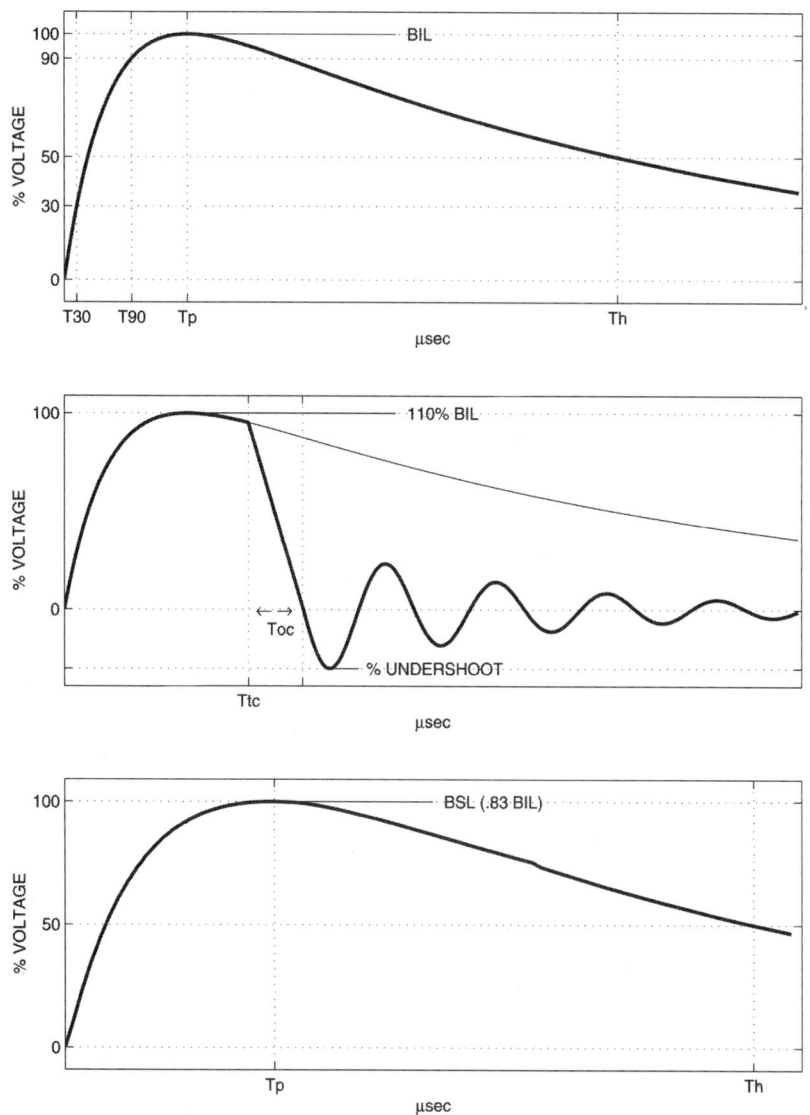

FIGURE 10.41 Standard voltage waveforms for impulse tests: full, chopped, switching surge.

voltages to which the system may subject it. One should always assess the environment the transformer is applied in and determine if there may be unusual transient voltage excitation present in an application that is not covered in the standards (see Degeneff et al., 1982; Greenwood, 1991; McNutt et al., 1974).

Design Margin

The actual level of insulation requested, e.g., BIL/BSL, is determined by recognizing the system within which the transformer will operate, and the arrester protective level. Normally, a minimum protective margin of 15 to 20% between the arrester peak voltage and the transformer capability at three (σ) is established. This is illustrated in Fig. 10.42 for a 230 kV transformer with a 750 kV BIL protected with a 180 kV rated arrester (Balma et al., 1996). The curve designated A in Fig. 10.42 is used to represent the transformer's insulation coordination characteristic (insulation capability) when subjected to aperiodic and oscillatory waveforms. The curve to be used to represent the transformer volt-time insulation

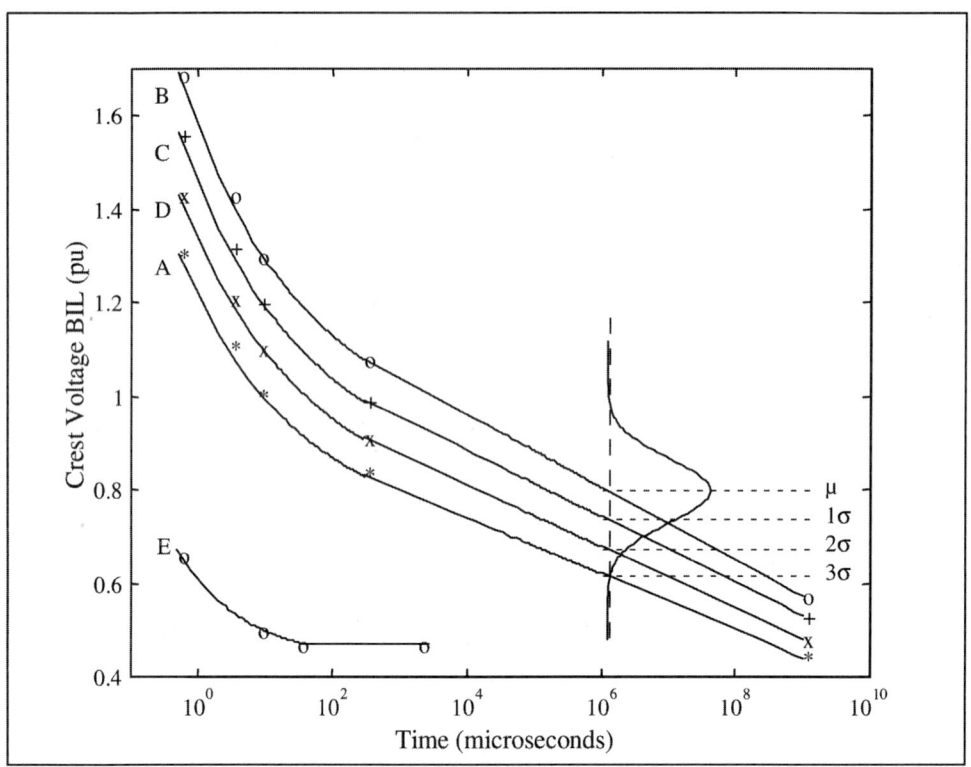

FIGURE 10.42 Voltage-time curve for insulation coordination

coordination characteristic when subjected to aperiodic wave forms with a time to failure between 0.1 and 2000 μsec is to be based on five points (IEEE C62.22-1991). The five points are:

1. Front of Wave Voltage plotted at its time of crest (about 0.5 μsec). If the front of wave voltage is not available, a value of 1.3 times the BIL should be plotted at 0.5 μsec.
2. Chopped Wave Voltage at its time of crest (about 3.0 μsec).
3. Full Wave Voltage (or BIL) plotted about 8 μsec.
4. Switching Surge Voltage (or BSL) plotted about 300 μsec.
5. A point at 2000 μsec where it magnitude is established with the following expression.:

$$logV_{2000} = \frac{logT_{BSL} - logT_{2000}}{m} + logV_{BSL} = \frac{log\frac{300}{2000}}{m} + logV_{BSL}$$

V_{2000} is the voltage at 2000 μsec, T_{2000}, V_{BSL} is the BSL voltage (or 0.83 the BIL), and T_{BSL} equal to 300 μsec. The value of m is established as the inverse of the slope of a straight line drawn on log-log paper from the BSL point to a point established by the peak of the one hour induced test voltage plotted at a time the induced voltage exceeds 90% of its peak value (this would be 28.7% of 3600 sec or 1033.2 sec).

The connection between all points is made with a smooth continuous curve. The first four points in the curve establish an approximate level of insulation voltage capability for which one would anticipate only one insulation failure out of 1000 applications of that voltage level, e.g., at 3 σ the probability of failure is 1.0 − 0.99865 or 0.001. Experience has shown that the standard deviation for transformer insulation structures is on the order of 10 to 15%. Figure 10.42 assumes that σ is 10%. Curve B, or the

50% failure rate curve, is established by increasing the voltage in Curve A by 30%. Therefore, for Curve B, on average the unit would be expected to fail one out of two times if it were subjected to this level of voltage. Curves C and D establish 1 and 2 σ curves, or 16% and 2.3% failure rate curves, respectively. The inserted normal distribution on the right of Fig. 10.42 illustrates this concept. All of this assumes the transformer is new.

Insulation Coordination

In a field installation, an arrester is normally placed directly in front of the transformer to afford it protection from transient voltages produced on the system. Curve E in Fig. 10.42 is a metal oxide arrester protective curve established in a manner similar to that described in IEEE standard C62.2. The curve is specified by three points:

1. The front of wave voltage held by the arrester plotted at 0.5 μs
2. The 8 × 20 μs voltage plotted at 8.0 μs
3. The switching surge voltage plotted in straight line from 30 to 2000 μs

The protective ratio is established by dividing the transformer insulation capability by the arrester protective level for the wave shape of interest. For example, in Fig. 10.42, the protective level for a switching surge is on the order of 177% or $(0.83/0.47) \times 100$.

Additional System Considerations

The standards reflect the growing and learning within the industry and each year they expand in breadth, addressing issues that are of concern to the industry. However, at present the standards are silent with regard to the effects of system voltage on transient response, multi-phase surges, aging or mechanical movements of insulation structures, oscillatory voltage excitation, temperature variations, movement of oil, and loading history. A prudent user will seek similar users and explore their experience base.

Models for System Studies

Model Requirements

The behavior of large power transformers under transient conditions is of interest to both transformer designers and power engineers. The transformer designer employs detailed electrical models to establish a reliable and cost effective transformer insulation structure. The power engineer models not only the transformer, but the system, in order to investigate the effects of power system transients.

Considerable effort has been devoted to computing the transformer's internal transient response. Models of this type may contain several hundred nodes for each phase. This detail is necessary in order to compute the internal response in enough detail to establish an adequate transformer insulation design. The utility engineer usually is not interested in the internal response, but is concerned only with the transformer's terminal response. Even if the transformer's detailed models were available, their use would create system models too large to be effectively used in system studies. Normal practice has been to create a reduced order model of the transformer that represents the terminal response of the transformer. Experience has shown that great care must be taken to obtain a terminal model that provides a reasonable representation of the transformer over the frequency range of interest (Gutierrez et al., 1993; Degeneff, 1977).

The challenge in creating a high fidelity reduced model lies in the fact that as the size of the model is reduced, the number of valid eigenvalues must also decrease. In effect, any static reduction technique will produce a model which is intrinsically less accurate than the more detailed lumped parameters model (Morched et al., 1992; de Leon and Semlyen, 1992).

Reduced Order Model

McNutt (1974) suggested a method of obtaining a reduced order transformer model by starting with the detailed model and appropriately combining series and shunt capacitances. This suggestion was extended by de Leon (1992). This method is limited to linear models and cannot be used to eliminate large proportions of the detailed models without affecting the resulting model's accuracy. Degeneff (1978) proposed a terminal model developed from information from the transformer's nameplate and capacitance measured among the terminals. This model is useful below the first resonant frequency but lacks the necessary accuracy at higher frequencies for system. Dommel (1982) proposed a reduced model for EMTP described by branch impedance or admittance matrix calculated from open and short circuit tests. TRELEG and BCTRAN matrix models for EMTP can be applied only for very low frequency studies. Morched (1992) proposed a terminal transformer model, composed of a synthesized LRC network, where the nodal admittance matrix approximates the nodal admittance matrix of the actual transformer over the frequency range of interest. This method is appropriate only for linear models.

Gutierrez et al. (1993) and Degeneff et al. (1994) present a method for reducing both a detailed linear and nonlinear lumped parameter model to a terminal model with no loss of accuracy. The work starts with Eq. (10.105), then progresses to Eq. (10.108), and then applies Kron reduction to obtain a terminal model of the transformer that retains all the frequency fidelity of the initial transformer lumped parameter model. Gutierrez et al., (1993) and Degeneff et al. (1994) present all the appropriate equations to apply this technique within EMTP.

References

Abetti, P.A., Correlation of forced and free oscillations of coils and windings, *IEEE PAS*, 986–996, December 1959.

Abetti, P.A., Survey and classification of published data on the surge performance of transformers and rotating machines, *AIEE Trans.*, 78, 1403–1414, 1959.

Abetti, P.A., First Supplement to Reference (12), *AIEE Trans.*, 81, 213, 1962.

Abetti, P.A., Second Supplement to Reference (12), *AIEE Trans.*, 83, 855, 1964.

Abetti, P.A., Pseudo-final voltage distribution in impulsed coils and windings, *Trans. AIEE*, 87–91, 1960.

Abetti, P.A., Maginniss, F.J., Fundamental oscillations of coils and windings, *IEEE PAS*, 1–10, February 1954.

Avila-Rosales, J. and Alvarado, L., Nonlinear frequency dependent transformer model for electromagnetic transient studies in power systems, *IEEE Trans. on PAS*, PAS-101, 11, Nov. 1982.

Balma, P.M., Degeneff, R.C., Moore, H.R., and Wagenaar, L.B., The Effects of Long Term Operation and System Conditions on the Dielectric Capability and Insulation Coordination of Large Power Transformers, Paper No. 96 SM 406-9 PWRD, presented at the *Summer Meetings of IEEE/PES*, Denver, CO, 1996.

Batruni, R., Degeneff R., Lebow, M., Determining the Effect of Thermal Loading on the Remaining Useful Life of a Power Transformer from Its Impedance Versus Frequency Characteristics, *IEEE Trans. on Power Delivery*, 11, 3, 1385–1390, July 1996.

Bean, R.L., Crackan, N., Moore, H.R., and Wentz, E., *Transformers for the Electric Power Industry*, Westinghouse Electric Corporation, McGraw-Hill, New York, 1959.

Blume, L.F., Boyajian, A., Camilli, G., Lennox, T.C., Minneci, S., and Montsinger, V.M., *Transformer Engineering*, 2nd ed., John Wiley & Sons, Inc., New York, 416–423, 1951.

Clark, F.M., *Insulating Materials for Design and Engineering Practice*, Wiley, New York, 1962.

Degeneff, R.C., A General Method For Determining Resonances in Transformer Windings, *IEEE Trans. on PAS*, 96, 2, 423–430, March/April 1977.

Degeneff, R.C., A Method for Constructing Terminal Models for Single-Phase n-Winding Transformers, IEEE Paper A78 539-9, Summer Power Meeting, Los Angeles, 1978.

Degeneff, R.C., Reducing Storage and Saving Computational Time with a Generalization of the Dommel (BPA) Solution Method, in *IEEE PICA Conference Proceedings*, Toronto, Canada, May 24–27, 1977, 307–313.

Degeneff, R.C., Simplified Formulas to Calculate Equivalent Series Capacitances for Groups of Disk Winding Sections, General Electric, TIS 75PTD017, August 16, 1976.

Degeneff, R.C., Blalock, T.J., and Weissbrod, C.C., *Transient Voltage Calculations in Transformer Windings*, General Electric Technical Information Series, No. 80PTD006, 1980.

Degeneff, R.C., Gutierrez, M., and Vakilian, M., Nonlinear, Lumped Parameter Transformer Model Reduction Technique, IEEE paper no. 94 SM 409-3 PWRD, 1994.

Degeneff, R.C. and Kennedy, W.N., Calculation of Initial, Pseudo-Final, and Final Voltage Distributions in Coils Using Matrix Techniques, Paper A75-416-8, Summer Power Meeting, San Francisco, CA,

Degeneff, R.C., Neugebaur, W., Panek, J., McCallum, M.E., and Honey, C.C., Transformer Response to System Switching Voltages, *IEEE Trans. on Power Appar. and Syst.*, Pas-101, 6, 1457–1465, June 1982.

Dent, B.M., Hartill, E.R., and Miles, J.G., A method of analysis of transformer impulse voltage distribution using a digital computer, *IEE Proceedings*, 105, Pt. A, 445–459, 1958.

Dommel, H.W., Digital computer solution of electromagnetic transients in single and multiphase networks, *IEEE Trans. on PAS*, 388–399, April 1969.

Dommel, H.W., Dommel, I.I., and Brandwajn, V., Matrix Representation of Three-Phase N-Winding Transformers for Steady State and Transient Studies, *IEEE PAS-1*, 6, June 1982.

Fergestad, P.I. and Henriksen, T., Transient Oscillations in Multiwinding Transformers, *IEEE Trans. on PAS*, 500–507, PAS-93, 1974.

Fink, D.G. and Beaty, H.W., *Standard Handbook for Electrical Engineers*, 12th Edition, McGraw-Hill Book Company, New York, 1987.

FORTRAN Subroutines for Mathematical Applications, Version 2.0, September 1991, MALB-USM-PER-FCT-EN9109-2.0.

Franklin, A.C., *The J & P Transformer Book*, 11th ed., chap. 15, 351–367, 1983.

Greenwood, A., *Electrical Transients in Power Systems*, John Wiley & Sons, Inc., New York, 1991.

Greenwood, A., *Vacuum Switchgear*, Institute of Electrical Engineers, Short Run Press Ltd., Exeter, 1994.

Grover, F.W., *Inductance Calculations — Working Formulas and Tables*, Dover Publications, New York, Inc., 1962.

Gutierrez, M., Degeneff, R.C., McKenny, P.J., and Schneider, J.M., Linear, Lumped Parameter Transformer Model Reduction Technique, IEEE paper no. 93 SM 394-7 PWRD, 1993.

IEEE C62.22-1991, *IEEE Guide for the Application of Metal-Oxide Surge Arrester for Alternating-Current Systems*.

IEEE Std. 519, *IEEE Recommended Practices and Requirements for Harmonic Control in Electrical Power Systems*, IEEE Industry Applications Society and Power Engineering Society, April 12, 1999.

C57, *IEEE Guide and Standards for Distribution, Power, and Regulating Transformers*, 1993.

Kogan, V.I., Fleeman, J.A., Provanzana, J.H., and Shih, C.H., Failure analysis of EHV transformers, *IEEE Trans. in Power Delivery*, 672–683, April 1988.

Kogan, V.I., Fleeman, J.A., Provanzana, J.H., Yanucci, D.A., and Kennedy, W.N., Rationale and implementation of a new 765kV generator step-up transformer specification, CIGRE Paper 12-202, August 1990.

Lammeraner, J. and Stafl, M., *Eddy Currents*, The Chemical Rubber Co. Press, Cleveland, 1966.

de Leon, F., and Semlyen, A., Complete transformer model for electromagnetic transients, *IEEE Transactions on Power Delivery*, 9, 1, 231–239, January 1994.

de Leon, F. and Semlyen, A., Reduced order model for transformer transients, *IEEE Trans. on PWRD*, 7, 1, 361–369, January 1992.

Massachusetts Institute of Technology, Department of Electrical Engineering, *Magnetic Circuits and Transformers*, John Wiley and Sons, 1943.

McNutt, W.J., Blalock, T.J., and Hinton, R.A., Response of Transformer Windings to System Transient Voltages, *IEEE PES Transactions*, 457–467, 1974.

McWhirter, J.H., Fahrnkopf, C.D., and Steele, J.H., Determination of impulse stresses within transformer windings by computers, *AIEE Transactions*, pt. III, 75, 1267–1273, 1956.

Morched, A., Marti, L., and Ottevangers, J., A High Frequency Transformer Model for EMTP, Paper No. 925M 359-0 IEEE 1992 Summer Meeting, Seattle, WA, July 12–16.

Narang, A., Wisenden, D., and Boland, M., *Characteristics of Stress on Transformer Insulation Subjected to Very Fast Transient Voltages*, CEA No. 253 T 784, Canadian Electricity Association, July 1998.

Rabins, Transformer Reactance Calculations with Digital Computers, *AIEE Trans.*, 75, 1, 261–267, July 1956.

Scheich, A., Behavior of Partially Interleave Transformer Windings Subject to Impulse Voltages, *Bulletin Oerlikon*, 389/390, 41–52, 1965.

Snow, C., *Formulas for Computing Capacitance and Inductances,* National Bureau of Standards, Circular 544, Sept. 10, 1954.

Tarasiewicz, E.J., Morched, A.S., Narang, A., and Dick, E.P., Frequency dependent eddy current models for the nonlinear iron cores, *IEEE Trans. on PAS,* 8, 2, 588–597, May 1993.

Vakilian, M., A Nonlinear Lumped Parameter Model for Transient Studies of Single Phase Core Form Transformers, Ph.D. Thesis, Rensselaer Polytechnic Institute, Troy, New York, 1993.

Vakilian, M., Degeneff, R., and Kupferschmid, M., Computing the internal transient voltage response of a transformer with a nonlinear core using Gear's method — Part 1: Theory, *IEEE Trans. on Power Delivery,* 10, 4, 1836–1841, October 1995.

Vakilian M., Degeneff, R., and Kupferschmid, M., Computing the internal transient voltage response of a transformer with a nonlinear core using Gear's method — Part 2: Verification, *IEEE Trans. on Power Delivery,* 10, 2, 702–708, May 1995.

Von Hippel, A., *Dielectric Materials and Applications,* Cambridge, MIT, 1954.

Waldvogel, P. and Rouxel, R., A new method of calculating the electric stresses in a winding subjected to a surge voltage, *Brown Boveri Rev.,* 43, 6, 206–213, June 1956.

White, W.N., *An Examination of Core Steel Eddy Current Reaction Effect on Transformer Transient Oscillatory Phenomena,* General Electric Technical Information Series, No. 77PTD012, April 1977.

White, W.N., *Inductance Models of Power Transformers,* General Electric Technical Information Series, No. 78PTD003, April 1978.

White, W.N., Numerical Transient Voltage Analysis of Transformers and LRC Networks Containing Nonlinear Resistive Elements, *1977 PICA Conference,* 288–294.

Wilcox, D.J., Hurley, W.G., McHale, T.P., and Conton, M., Application of modified modal theory in the modeling of practical transformers, *IEE Proceedings-C,* 139, 6, November 1992.

Wilcox, D.J., Conlon, M., and Hurley, W.G., Calculation of self and mutual impedances for coils on ferromagnetic cores, in *IEE Proceedings,* 135, A, 7, September 1988, pp. 470–476.

An international survey on failures in large power transformers in service, Final Report of Working Group 05 of Study Committee 12 (Transformers), *Electra,* 88, May 1983.

10.7 Transmission System Transients — Grounding

William Chisholm

General Concepts

Electric power systems are often grounded, that is to say "intentionally connected to earth through a ground connection or connections of sufficiently low impedance and having sufficient current-carrying capacity to prevent the buildup of voltages which may result in undue hazard to connected equipment or to persons" (IEEE Std. 100). Grounding affects the dynamic power-frequency voltages of unfaulted phases, and influences the choice of surge protection. Also, the tower-footing impedance is an important specification for estimating the severity of insulator-string transient voltage for a direct lightning strike.

Under AC fault conditions, systems can be grounded by any of three means (IEEE Std. 100):

- **Inductance grounded,** such that the system zero-sequence reactance is much higher than the positive-sequence reactance, and is also greater than the zero-sequence resistance. The ground-fault current then becomes more than 25% of the three-phase fault current.
- **Resistance grounded,** either directly to ground, or indirectly through a transformer winding. The low-resistance-grounded system permits a higher ground-fault current (on the order of 25 A to several kA) for selective relay performance.
- **Resonant grounded,** through a reactance with a value of inductive current that balances the power-frequency capacitive component of the ground-fault current during a single line-to-ground fault. With resonant grounding of a system, the net current is limited so that the fault arc will extinguish itself.

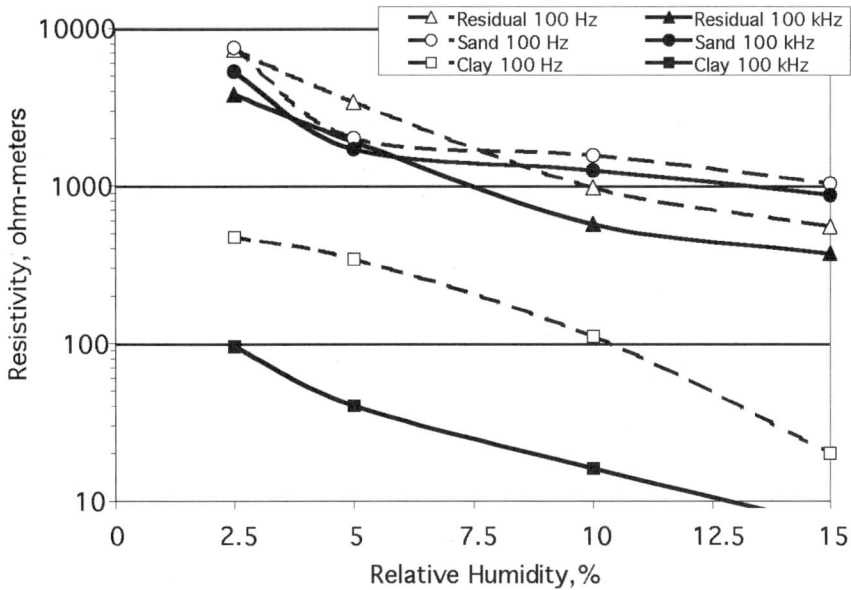

FIGURE 10.43 Change in resistivity with relative humidity and frequency for three typical surface materials. (Filho and Portela, 1987.)

Power system transients have a variety of waveshapes, with spectral energy ranging from the power frequency harmonics up to broadband content in the 300-kHz range, associated with 1-μs rise and fall times of lightning currents and insulator breakdown voltages. With the wide frequency content of transient waveshapes, resonant grounding techniques offer little benefit. Also, resistance grounds that may be effective for power frequency can have an additional inductive voltage rise (L dI/dt) that dominates the transient response. Both resistive and inductive aspects must be considered in the selection of an appropriate ground electrode.

Material Properties

The earth resistivity (ρ, units of Ω-m) dominates the potential rise on ground systems at low frequencies and currents. Near the surface, resistivity changes as a function of moisture, temperature, frequency, and electric field stress. Figure 10.43 shows that this variation can be quite large.

Reconnaissance of earth resistivity, from traditional four-terminal resistance measurements, is a classic tool in geological prospecting (Keller and Frischknecht, 1982). A current I is injected at the outer two locations of an equally spaced line of four probes. A potential difference U then appears between the inner two probes, which are separated by a distance a. The apparent resistivity ρ_a is then defined as:

$$\rho_a = 2\pi a \frac{U}{I}$$

At a given location, several measurements of ρ_a are taken at geometrically-spaced values of a, such as ($a = 1, 2, 3, 5, 10, 20, 30,\ldots 1000$ m). When ρ_a is constant with distance, the assumption of a uniform soil model is justified, and the effective resistivity ρ_e for any electrode size is simply ρ_a. However, in many cases, there are two or more layers of contrasting soil. The most difficult case tends to be a thin conducting top layer (clay, till, sand) over a thick, poorly conducting rock layer ($\rho_1 < \rho_2$). This case will have an increasing value of ρ_a with distance. A conservative interpretation in this case can be simplified: For flat electrodes, the effective resistivity, ρ_e, equals the value of ρ_a observed at a probe spacing of ($a = 2s$), where s is the maximum extent (for example, the radius of a ring electrode).

TABLE 10.3 Values of A, s, and g for Typical Ground Electrodes in Half-Plane

Geometry	A	s	g
Vertical Cylindrical Rod Length l and Radius r	$2\pi r l + \pi r^2 \approx 2\pi r l$	$\sqrt{l^2+r^2} \approx l$	$\sqrt{l^2+2r^2} \approx l$
Solid Conducting Cylinder Length l, Radius r	$2\pi r l + \pi r^2$	$\sqrt{l^2+r^2}$	$\sqrt{l^2+2r^2}$
Buried Circular Disk at Depth h with Radius r	$2\pi r h + \pi r^2$	$\sqrt{h^2+r^2}$	$\sqrt{h^2+2r^2}$
Buried Circular Ring at Depth h with Radius r	$2\pi r h + \pi r^2$	$\sqrt{h^2+r^2}$	$\sqrt{h^2+2r^2}$
Circular Disc on Surface Thickness t, Radius r	$2\pi r t + \pi r^2 \approx \pi r^2$	$\sqrt{t^2+r^2} \approx r$	$\sqrt{t^2+2r^2} \approx \sqrt{2}r$
Oblate (disc-like) Half Spheroid Radius a > thickness b, $\varepsilon = \sqrt{1 - b^2/a^2}$	$\pi a^2 + \pi b^2/(2\varepsilon) \ln((1+\varepsilon)/(1-\varepsilon))$	a	$\sqrt{2a^2+b^2}$
Hemisphere of radius r	$2\pi r^2$	r	$\sqrt{3}*r$
Prolate (tube-like) Half Spheroid Radius a < length b, $\varepsilon = \sqrt{1 - a^2/b^2}$	$\pi a^2 + \pi ab/\varepsilon \arcsin\varepsilon$	b	$\sqrt{2a^2+b^2}$
Rectangular Strip, l long, w wide and t thick	$wt + 2(t+w)l \approx 2lw$	$\sqrt{w^2/4+t^2/4+l^2}$	= s
Two or more Vertical Rods in a Straight Line, total length l, depth h, rod thickness w	$lw + 2(l+w) h \approx 2lh$	$\sqrt{w^2/4+l^2/4+h^2}$	= s
Buried Counterpoise at depth h, thickness w, total length l	$lw + 2(l+w) h \approx 2lh$	$\sqrt{w^2/4+l^2/4+h^2}$	= s
Conducting Box, l long by w wide by h deep	$lw + 2(l+w) h$	$\sqrt{l^2/4+w^2/4+h^2}$	= s
Buried Rectangular Plate, l long by w wide at Depth h	$lw + 2(l+w) h$	$\sqrt{l^2/4+w^2/4+h^2}$	= s
Buried Rectangular Grid l long by w wide at Depth h	$lw + 2(l+w) h$	$\sqrt{l^2/4+w^2/4+h^2}$	= s
Four Vertical Rods of length h on Corners of l by w Rectangle	$lw + 2(l+w) h$	$\sqrt{l^2/4+w^2/4+h^2}$	= s
Four Radial Wires, on Corners l by w at depth h	$lw + 2(l+w) h$	$\sqrt{l^2/4+w^2/4+h^2}$	= s
Surface Plate, l long by w wide by t thick	$lw + 2(l+w) t \approx lw$	$\sqrt{l^2/4+w^2/4+t^2}$	= s

Note: There are really only three equations in Table 10.3; one for a cylinder, one for a spheroid, and one for a box. Different dimensions dominate the s, g, and A terms, depending on the electrode shape.

Electrode Dimensions

Three dimensions are relevant for analysis of electrode response under steady-state and transient conditions. Electrodes with large surface area A in contact with soil will have lower resistance, lower impedance, and less susceptibility to unpredictable effects of soil ionization. For objects with concave features, the area of the smallest convex body that can envelop it is determined. With this model, a tube has the same area as a solid cylinder of the same dimensions. Disk electrodes have two sides. Buried horizontal wires expose area on both sides of the narrow trench. Table 10.3 provides further interpretations.

The second electrode dimension is the 3-dimensional distance from the center of the electrode to its outermost point. For a spheroid in a conducting half-plane, $s=MAX(a,b)$ where a is the maximum cross-section radius and b is the length in the axis of symmetry. Table 10.3 shows that the 3-D extent s of cylinders and prisms are slightly larger than the s for a prolate spheroid of the same depth. The propagation time $\tau = s/c$, calculated from speed-of-light propagation at $c = 3.10^8$ m/s, is used to estimate transient electrode impedance.

The geometric radius of the electrode, $g = \sqrt{R_x^2 + R_y^2 + R_z^2}$ is the third important dimension of the electrode, since it is used to estimate self-capacitance. For long, thin, or rectangular shapes, $g = s$; for a disc, $g = \sqrt{2}\,s$; for a hemisphere, $g = \sqrt{3}\,s$.

Self-capacitance of Electrodes

Electrode capacitance is easily calculated (Chow, 1982; Chow, 1988) and offers an elegant description of grounding response to transients. Also, the self-capacitance C_{self} to infinity of an arbitrary conducting object in full space has a useful dual relation to its steady-state resistance R in a half-space of conducting medium, given by Weber (1950):

$$R = 2\,\varepsilon_o\,\rho/C_{self} \qquad (10.141)$$

where ε_o is the permittivity constant, 8.854×10^{-12} F/m
ρ is the earth resistivity, in Ω-m

The transient impedance of the same arbitrary conducting object can be modeled using the time (τ) it takes to charge up its self-capacitance C_{self}. This time cannot be less than the maximum dimension of the electrode, s, divided by the speed of light. An average surge impedance Z, given by the ratio τ/C, can then be used to relate voltages and currents during any initial surge. The capacitance of an object is approximately (Chow, 1982):

$$C_{self} = \varepsilon_o\, c_f \sqrt{4\pi A} \qquad (10.142)$$

where A is the total surface area of the object, including both sides of disk-like objects
c_f is a correction factor between the capacitance of the object and the capacitance of a sphere with the same surface area. For spheroids, $0.9 < c_f < 1.14$ for $0 < b/a < 8$ (Chow, 1988).

A good estimate for c_f is given by the following expression:

$$c_f = \frac{4\pi g}{\sqrt{4\pi A}\,\ln\frac{4\pi e^{\sqrt{3}} g^2}{3A}} \simeq \frac{3.54\,g}{\sqrt{A}\,\ln\frac{23.7\,g^2}{A}} \qquad (10.143)$$

Equation (10.143) is exact for a sphere, and remains valid for a wide range of electrode shapes, from disc to rod.

Initial Transient Response from Capacitance

Once the capacitance of a conducting electrode has been estimated, its average transient impedance can be computed from the minimum charging time τ. The charging time τ would be the maximum three-dimensional extent s, divided by the speed of light c. The transient impedance of the ground electrode will be seen only during the charging time, and in general it will vary somewhat around the estimated average value.

Figure 10.44 gives the initial transient response of conducting spheroid electrodes.

TABLE 10.4 Transient Impedance of Conducting Electrodes

Shape	Surface Area	3-D Extent s	Capacitance	Travel Time	Transient Impedance
Circular Disc	$2\pi s^2$	s	$0.9\,\pi\varepsilon\,s\,\sqrt{8}$	s/c	47 Ω
Hemisphere	$3\pi s^2$	s	$1.0\,\pi\varepsilon\,s\,\sqrt{12}$	s/c	35 Ω
Long Cylinder	$2\pi r\,s$	s	$3.3\,\pi\varepsilon\,\sqrt{4rs}$	s/c	s/r = 100: 210 Ω
			$7.8\,\pi\varepsilon\,\sqrt{4rs}$		s/r = 1000: 270 Ω
			$20\,\pi\varepsilon\,\sqrt{4rs}$		s/r = 10000: 340 Ω

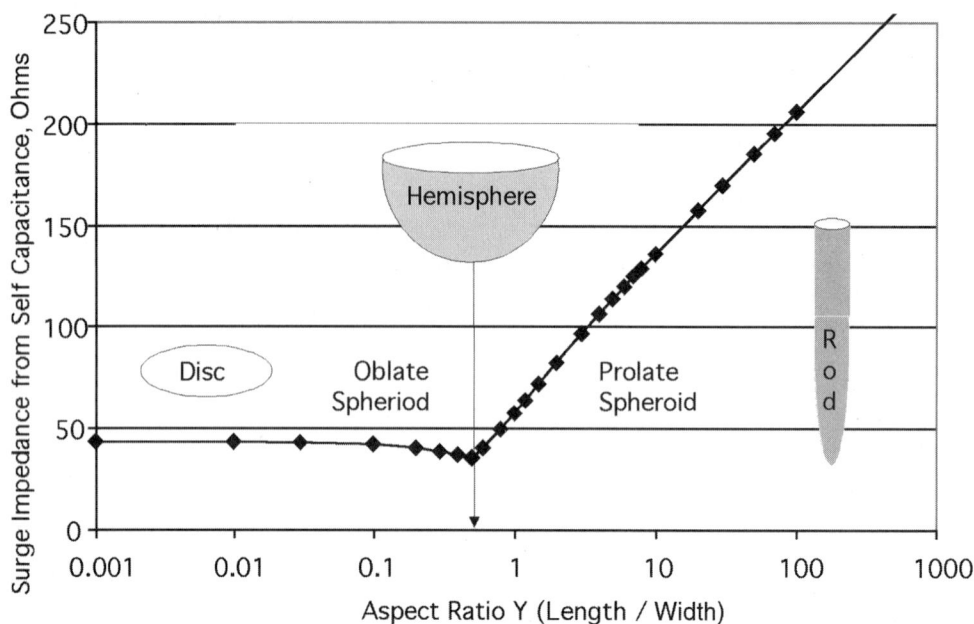

FIGURE 10.44 Relation between transient impedance and aspect ratio: spheroid electrodes in half space.

The main observation from Fig. 10.44 is that wide, flat electrodes will have inherently better transient response than long, thin electrodes. This includes electrodes in both horizontal and vertical planes, so long leads to remote ground electrodes are defective under transient conditions.

For compact electrodes, the response can be lumped into an inductance element ($L_{self} = \tau^2/C_{self}$) by ignoring electromagnetic considerations (lack of return path for inductance integrals). A potential rise on the earth electrode can then be estimated from the simple circuit model:

$$U_{electrode} = RI + \left(\tau^2/C_{self}\right) dI/dt \qquad (10.144)$$

A numerical example for Eq. (10.144) is useful. For ρ = 100 Ω-m, and a disc electrode of 5 m radius, the resistance to remote ground will be 3.2 Ω and τ^2/C_{self} = 0.8 μH. For a typical (median) lightning stroke with I = 30 kA and dI/dt = 24 kA/μs at the peak, the two terms of the peak potential rise will be:

$$(30\text{ kA})(3.2\text{ }\Omega) + (24\text{ kA}/\mu s)(0.8\text{ }\mu H) = 96\text{ kV} + 19\text{ kV}.$$

The inductive term is desirably low in the example, but it can dominate the response of long, thin electrodes in low-resistivity soil. For distributed electrodes, however, the circuit approximation in Eq. (10.144) eventually fails as rate of current rise (dI/dt) increases.

Ground Electrode Impedance over Perfect Ground

The per-unit inductance L of a distributed grounding connection over a conducting plane can be calculated from the surge impedance Z of the wire and its travel time τ:

$$L = Z\tau = \tau^2/C_{self} \qquad (10.145)$$

where

> $Z = 60 \ln(2h/r)$, the surge impedance (Ω) of a wire of radius r over a ground plane at a height h
> $\tau = s/c$, the maximum length of the wire (m) divided by the speed of light (3.0×10^8 m/s).

Ground Electrode Impedance over Imperfect Ground

When the electrode is placed over imperfect ground, the effective return depth of current will increase. Bewley (1963) suggests that the plane for image currents, in his tests of buried horizontal wires, was 61 m (200′) below the earth surface. Some analytical indication of the increase in return depth is given by the normal skin depth, $\delta = 1/\sqrt{\pi f \mu_o / \rho}$, which decreases from 460 m (60 Hz) to 11 m (100 kHz) for 50 Ω-m soil. Deri (1981) proposes a more complete approach, with a frequency-dependent complex depth p, as follows:

$$Z = \frac{j\omega\mu_o}{2\pi} \ln\left(\frac{2(h+p)}{r}\right) \tag{10.146}$$

where

> p is a complex depth, $1/\sqrt{j\omega\mu_o\sigma}$
> ω is 2π times the frequency, Hz
> μ_o is $4\pi \times 10^{-7}$ H/m
> σ is the conductivity, Siemens/m ($1/\rho$)

For good soil with a resistivity of 50 Ω-m, the complex depth is 230 (1 – j) m at 60 Hz and 5.6 (1 – j) m at 100 kHz. The complex depth is related to the normal skin depth by the relation $1/p = (1 + j)\,1/\delta$.

Analytical Treatment of Complex Electrode Shapes

Simple analytical expressions are documented for a variety of regular electrode shapes (see, e.g., Smythe, 1950; Weber, 1950; Keller and Frischknecht, 1982; Sunde, 1949). However, grounding of electrical systems often consists of several interconnected components, making estimation of footing resistance more difficult. The tower foundation can be a single or (more typically) four concrete cylinders, often reinforced with steel. In the preferred case, the steel is bonded electrically, and a grounding connection is brought out of the form before the concrete is poured. In areas with low soil resistivity, four concrete footings can often provide a low tower resistance without supplemental electrodes. In some cases on both transmission and distribution systems, a metal grillage (or pole butt-wrap) is installed at the base after excavation. This deep electrode is more effective than a surface electrode of the same area. Also, grillage and pole-wrap electrodes are protected from vandalism and frost damage.

Supplemental grounding electrodes are often installed during line construction or upgrade. The following approaches are used:

- Horizontal conductors are bonded to the tower, then buried at a practical depth.
- Vertical rods are driven into the soil at some distance from the tower, then bonded to the tower base, again using bare wires, buried at a practical depth.
- Supplemental guy wires are added to the tower (often for higher mechanical rating) and then grounded using rock or soil anchors at some distance away.

Supplemental grounding should be considered to have a finite lifetime of five to twenty years, especially in areas where the soil freezes in winter. Also, auxiliary electrodes such as rock anchors should be designed to carry their share of impulse current, and to withstand the associated traverse forces.

The resistance of an electrode that envelops all contacts can be used to obtain a good estimate of the combined resistance of a complex, interconnected electrode. The resistance of a solid electrode is:

$$R_{geometric} = \frac{\rho}{2\pi s} \ln \frac{2\pi e s^2}{A} \qquad (10.147)$$

where

s is the three-dimensional distance from the center to the furthest point on the electrode
A is the convex surface area that would be exposed if the electrode were excavated
e is the exponential constant, 2.718 (noting $2\pi e \approx 17$)

The resistance can also be based on the geometric radius g rather than the maximum dimension s:

$$R_{geometric} = \frac{\rho}{2\pi g} \ln \frac{11.8 g^2}{A} \qquad (10.48)$$

where

$g = \sqrt{R_x^2 + R_y^2 + R_z^2}$, the geometric radius of the electrode
$R_{x, y, z}$ = maximum x, y, and z dimensions
$11.8 = (2\pi e^{\sqrt{3}})/3$

If the electrode is a wire frame, rather than a solid, then a small correction should be added to the geometric resistance:

$$R_{wire\ frame} = R_{geometric} + \rho_1/l \qquad (10.149)$$

where

l is the total length of the wire frame
ρ_1 is the resistivity of the upper layer of soil (the layer next to the wire)

Numerical Treatment of Complex Electrode Shapes

The effectiveness of supplemental grounding can be conveniently evaluated and visualized using a numerical solution to Laplace's equation in two dimensions. This is readily accomplished using a spreadsheet (such as Excel™) and the following steps:

- Define a rectangular problem space, for example a 50 × 50 box with each cell representing 1 × 1 m.
- Calculate the resistance between nodes, $R_{ab} = \rho/(2\pi\Delta)$ where Δ is the spacing between nodes.
- Highlight the perimeter of the box and set all the values to 0 (the boundary condition).
- Disable iterative calculation (in Excel, at the *Tools…Options…Calculation* menu).
- Fill the interior of the box with the two-dimensional finite-difference operator (Sadiku, 1992). For the Laplace equation, this operator is simply the average of the four adjacent cells, for example cell G5, = 0.25*(G4 + G6 + F5 + H5). Ignore error messages about circular references.
- Calculate the resistances R_f of each individual electrode, using Eq. (10.147) or (10.148). Buried interconnection wire should use the resistance associated with the wire radius and the cell spacing Δ.
- Set the potentials at each electrode location to $(0.25*(G4 + G6 + F5 + H5)/R_{ab} + 1/R_f)/(4/R_{ab} + 1/R_f)$, again using cell G5 as an example.
- Enable iterative calculation (in Excel, at the *Tools…Options…Calculation* menu) and set convergence to a small value.
- Form the sum of all potentials just adjacent to the boundary of the problem space. Each term is proportional to the current density at that point, so the sum (integral), divided by Rab, gives the total current flowing 'out' of the problem space.
- Compute the input current at each electrode, which will be the unit potential minus the node potential, divided by the electrode resistance.

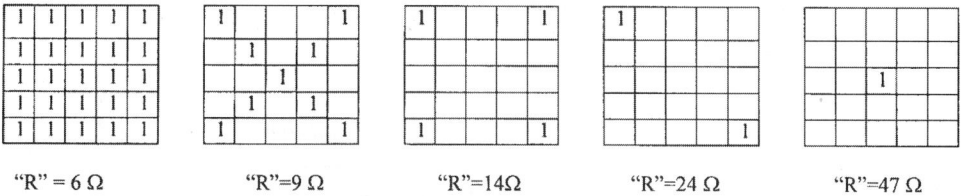

FIGURE 10.45 Numerical solutions of Laplace equation for 3-m rods at 1-m grid spacing.

- Form the sum of all input currents, which should agree with the current at the boundary after iteration.
- Iterate the solution manually (*F9 key*) or automatically until convergence is achieved to a tolerance of about 1 part in 10^5.
- Modify the electrode the shape, size, and configurations as desired and observe changes in resistance.
- Use the three-dimensional plot tool to observe areas of high step and touch potential differences.

There are many approximations and possible errors in the numerical procedure, based on cell size and soil resistivity. However, Fig. 10.45 illustrates the power of the visualization for driven rods:

The relative reduction in resistance is largest when the second electrode is added and additional nearby rods are seen to be less effective at the assumed spacing of $\Delta = 1$m.

Treatment of Multilayer Soil Effects

Generally, the treatment of footing resistance in lightning calculations considers a homogeneous soil with a finite conductivity. This treatment, however, seldom matches field observations, particularly in areas where grounding is difficult. Under these conditions, a thin layer of conducting clay, till, or gravel often rests on top of insulating rock. The distribution of resistivity values for a particular overburden material and condition can be narrow, with standard deviations usually less than 10%. However, the variation of overburden depth with distance can be large. Airborne electromagnetic survey techniques at multiple frequencies in the 10–100 kHz range offer an inexpensive new method of reconnaissance of the overburden parameters of resistivity and depth.

Once a resistivity survey has established an upper-layer resistivity ρ_1, a layer depth d and a lower-layer resistivity ρ_2, the equivalent resistivity ρ_e can be computed. For a disk-like electrode buried just below the surface. ρ_e is given approximately by the following three equations:

$$\rho_e = \rho_1 \frac{1 + C \frac{\rho_2}{\rho_1} \frac{r}{d}}{1 + C \frac{r}{d}}$$

Case 1: Upper Layer (ρ_1) > Lower Layer (ρ_2)

$$C = \frac{1}{1.4 + \left(\frac{\rho_2}{\rho_1}\right)^{0.8}}$$

Case 2: Upper Layer (ρ_1) < Lower Layer (ρ_2)

$$C = \frac{1}{1.4 + \left(\frac{\rho_2}{\rho_1}\right)^{0.8} + \left(\frac{\rho_2}{\rho_1} \frac{r}{d}\right)^{0.5}}$$

These equations provide an empirical fit to the solution of elliptic-integral potentials given by (Zaborsky, 1955).

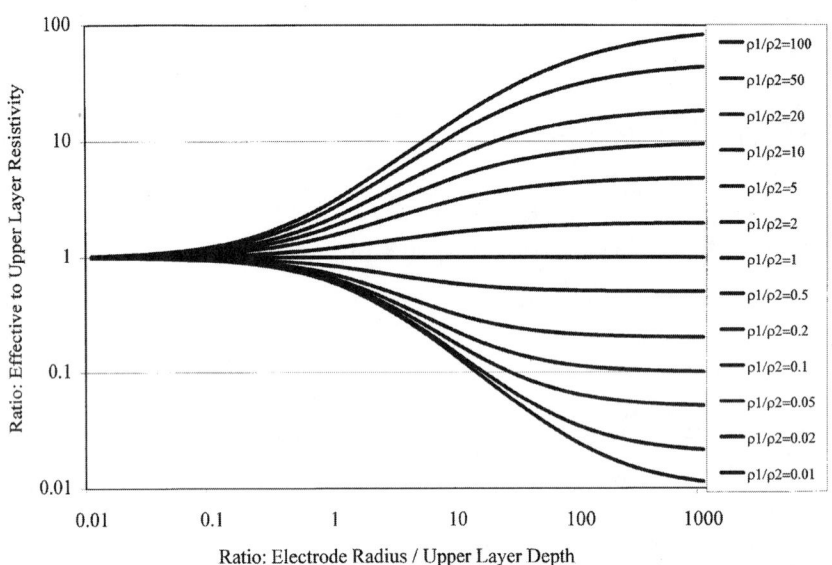

FIGURE 10.46 Relative effective resistivity versus ratio of electrode radius to upper layer depth, with resistivity ratio as a parameter.

Normally, electrode penetration through an upper layer would only be desirable in extreme examples of Case 1 ($\rho_1 \gg \rho_2$). Rather than recomputing the effective resistivity with revised image locations, the effects of the upper layer can be neglected, with the connection through ρ_1 providing only series self-inductance.

Layer of Finite Thickness over Insulator

A simpler two-layer soil treatment is appropriate for Case 2 when $\rho_2 \gg \rho_1$, e.g., (Loyka, 1999). Under these conditions, the following summation ($n = 1,2,3,\ldots\infty$) describes the resistance of a single sphere of radius s in a finitely conducting slab with resistivity ρ_1 and thickness d:

$$R = \rho_1 / \pi \left(1/s + 2\Sigma \left(1 / \sqrt{s^2 + (2nd)^2} \right) \right)$$

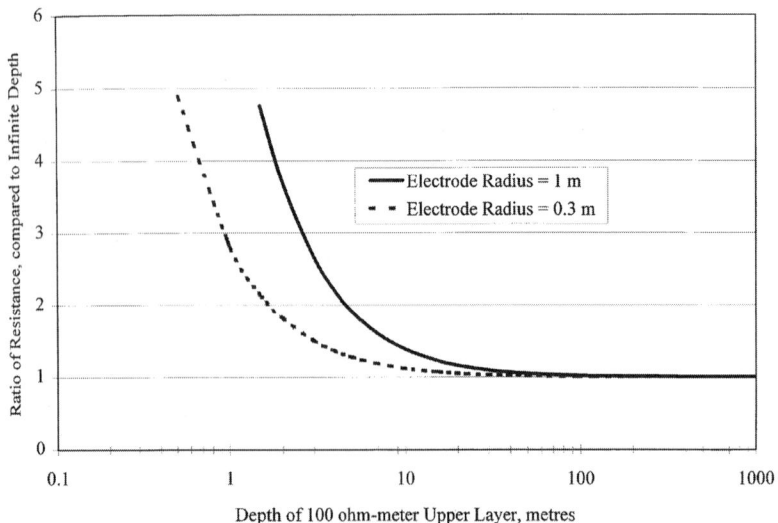

FIGURE 10.47 Asymptotic behavior of spherical electrodes in conducting layer.

Treatment of Soil Ionization

Under high electric fields, the air will ionize and effectively become a conductor. The transient electric fields needed to ionize small volumes of soil, or to flashover across the soil surface are typically between 100 kV/m and 1000 kV/m (Korsuncev, 1958; Oettle, 1988). Considering that the potential rise on a small ground electrode can reach 1 MV, the origins of 10-m furrows around small (inadequate) ground electrodes after lightning strikes becomes clear. Surface arcing activity is unpredictable and may transfer lightning surge currents to unprotected facilities. Thus, power system ground electrodes for lightning protection should have sufficient area and multiplicity to limit ionization.

Korsuncev used similarity analysis to relate dimensionless ratios of s, ρ, resistance R, current I, and critical breakdown gradient Eo as follows:

$$\Pi_1^0 = \frac{1}{2\pi} \ln \frac{2\pi e s^2}{A} \tag{10.150}$$

$$\Pi_2 = \rho I / \left(Eo s^2\right) \tag{10.151}$$

$$\Pi_1 = \mathrm{MIN}\left(\Pi_1^0, 0.26\, \Pi_2^{-0.31}\right) \tag{10.152}$$

$$R = \rho \Pi_1 / s \tag{10.153}$$

where

s is the three-dimensional distance from the center to the furthest point on the electrode, m
I is the electrode current, A
ρ is the resistivity, Ω-m
Eo is the critical breakdown gradient of the soil, usually 300 to 1000 kV/m
R is the resistance of the ground electrode under ionized conditions

The calculation of ionzied electrode resistance proceeds as follows:

- A value of Π_1^0 is calculated from Eq. (10.150). This un-ionized value will range from $\Pi_1^0 = 0.159$ (for a hemisphere) to 1.07 for a 3-m long, 0.01-m radius cylinder.
- A value of Π_2 is calculated from Eq. (10.151). For a 3-m rod at 100 kA in 100 Ω-m soil, with Eo of 300 kV/m, the value of $\Pi_2 - 3.7$ is obtained.
- A value of Π_1 is computed from Π_2 using $0.26\, \Pi_2^{-0.31}$. This value, from Eq. (10.152), represents the fully ionized sphere with the gradient of Eo at the injected current; for the rod example, $\Pi_1 = 0.173$.
- If the ionized value of Π_1 is not greater than Π_1^0, then there is not enough current to ionize the footing, so the un-ionized resistance from Π_1^0 will be seen for the calculation of resistance in Eq. (10.153). For the 3-m rod, ionization reduces the low-current resistance of 36 Ω to 6 Ω at 100 kA.

In two-layer soils with sparse electrodes, ionization effects will tend to reduce the contact resistance [from Eq. (10.149)] without altering the surface area A or characteristic dimension s. This will tend to reduce the influence of ionization.

Design Recommendations

The following advice is especially relevant for transmission towers or other tall structures, where a low-impedance ground is needed to limit lightning-transient overvoltages.

- Choose a wide, flat electrode shape rather than a long, thin shape. Four radial wires of 60 m will be two to eight times more effective than a single radial wire of 240 m under lightning surge conditions.
- Take advantage of natural elements in the structure grounding, such as foundations and guy anchors, by planning for electrical connections and by extending radial wires outwards from these points.
- In rocky areas, use modern airborne techniques to survey resistivity and layer depth using several frequencies up to 100 kHz. Place towers where conductive covering material is deep.
- Provide grounding staff with the tools and techniques to pre-estimate the amount of wire required for target footing impedance values, using simple interpretation of two-layer soil data.
- Near areas where transferred lightning potentials could be dangerous to adjacent objects or systems, use sufficient electrode dimensions to limit ionization, that is, to remain on the Π_1^0 characteristic.

References

Bewley, L.V., *Travelling Waves on Transmission Systems*, Dover, New York, 1963.

Chisholm, W.A. and Janischewskyj, W., Lightning Surge Response of Ground Electrodes, *IEEE Trans. on Power Delivery*, 4(2), April 1989.

Chow, Y.L. and Srivastava, K.D., Non-Uniform Electric Field Induced Voltage Calculations, Final Report for Canadian Electrical Association Contract 117 T 317, February 1988.

Chow, Y.L. and Yovanovich, M.M., The shape factor of the capacitance of a conductor, *J. Appl. Physics*, 52, December 1982.

Deri, A., Tevan, G., Semlyen, A., and Castanheira, A., The Complex Ground Return Plane — A Simplified Model for Homogeneous and Multi-Layer Earth Return, *IEEE Transactions* PAS-100 8, August 1981.

Filho, S.V. and Portela, C.M., Soil Permittivity and Conductivity Behaviour on Frequency Range of Transient Phenomena in Electric Power Systems, Paper 93.06, Proceedings of 5th ISH, August 1987.

IEEE Standard 100, *Dictionary of Electrical and Electronics Terms*, 5th ed., IEEE, New York, 1992.

ANSI/IEEE Standard 80-1986, *IEEE Guide for Safety in AC Substation Grounding*, IEEE/Wiley, New York, 1986.

Keller, G.G. and Frischknecht, F.C., *Electrical Methods in Geophysical Prospecting*, Pergamon, New York, 1982.

Korsuncev, A.V., Application on the theory of similarity to calculation of impulse characteristics of concentrated electrodes, *Elektrichestvo* No. 5, 1958.

Liew, A. and Darveniza, M., Dynamic model of impulse characteristics of concentrated earths, *IEE Proceedings*, 121(2), Feb. 1974.

Loyka, S.L. A simple formula for the ground resistance calculation, *IEEE Transactions on EMC*, 41(2), 152-154, May 1999.

Oettle, E.E., A new general estimation curve for predicting the impulse impedance of concentrated earth electrodes, *IEEE Trans. on Power Delivery*, 3(4), October 1988.

Sadiku, M.N.O., Ed., *Numerical Techniques in Electromagnetics*, CRC Press, Boca Raton, FL, 1992.

Smythe, W.R., *Static and Dynamic Electricity*, McGraw-Hill, New York, 1950.

Sunde, E.D., *Earth Conduction Effects in Transmission Systems*, Van Nostrand, Toronto, 1949.

Tagg, G.F., *Earth Resistances*, George Newnes, London, 1964.

Weber, E., *Electromagnetic Fields Theory and Applications Volume 1 — Mapping of Fields*, Wiley, New York, 1950.

Zaborsky, J., Efficiency of grounding grids with nonuniform soil, *AIEE Transactions*, 1230-1233, December 1955.

Appendix A: Relevant IEEE Grounding Standards

ANSI/IEEE Std 80-1986: *IEEE Guide for Safety in AC Substation Grounding*

Presents essential guidelines for assuring safety through proper grounding at AC substations at all voltage levels. Provides design criteria to establish safe limits for potential differences within a station, under fault conditions, between possible points of contact. Uses a step-by-step format to describe test methods, design and testing of grounding systems. Also provides English translations of three fundamental papers on grounding by Rudenberg, Laurent, and Zeitschrift.

IEEE Std. 81-1983: *IEEE Guide for Measuring Earth Resistivity, Ground Impedance, and Earth Surface Potentials of a Ground System*

The present state of the technique of measuring ground resistance and impedance, earth resistivity, and potential gradients from currents in the earth, and the prediction of the magnitude of ground resistance and potential gradients from scale-model tests are described and discussed. Factors influencing the choice of instruments and the techniques for various types of measurements are covered. These include the purpose of the measurement, the accuracy required, the type of instruments available, possible sources of error, and the nature of the ground or grounding system under test. The intent is to assist the engineer or technician in obtaining and interpreting accurate, reliable data. The test procedures described promote the safety of personnel and property and prevent interference with the operation of neighboring facilities.

IEEE Std. 81.2-1991: *IEEE Guide for Measurement of Impedance and Safety Characteristics of Large, Extended or Interconnected Grounding Systems*

Practical instrumentation methods are presented for measuring the AC characteristics of large, extended, or interconnected grounding systems. Measurements of impedance to remote earth, step and touch potentials, and current distributions are covered for grounding systems ranging in complexity from small grids (less than 900 m^2) with only a few connected overhead or direct-burial bare concentric neutrals, to large grids (greater than 20,000 m^2) with many connected neutrals, overhead ground wires (sky wires), counterpoises, grid tie conductors, cable shields, and metallic pipes. This standard addresses measurement safety; earth-return mutual errors; low-current measurements; power-system staged faults; communication and control cable transfer impedance; current distribution (current splits) in the grounding system; step, touch, mesh, and profile measurements; the foot-equivalent electrode earth resistance; and instrumentation characteristics and limitations.

IEEE Std. 367-1996: *IEEE Recommended Practice for Determining the Electric Power Station Ground Potential Rise and Induced Voltage from a Power Fault*

Information for the determination of the appropriate values of fault-produced power station ground potential rise (GPR) and induction for use in the design of protection systems is provided. Included are the determination of the appropriate value of fault current to be used in the GPR calculation, taking into account the waveform, probability, and duration of the fault current; the determination of inducing currents, the mutual impedance between power and telephone facilities, and shield factors; the vectorial summation of GPR and induction; considerations regarding the power station GPR zone of influence; and communications channel time requirements for noninterruptible services. Guidance for the calculation of power station ground potential rise (GPR) and longitudinal induction (LI) voltages is provided, as well as guidance for their appropriate reduction from worst-case values, for use in metallic telecommunication protection design.

IEEE Std. 524a-1993: *IEEE Guide to Grounding During the Installation of Overhead Transmission Line Conductors—Supplement to IEEE Guide to the Installation of Overhead Transmission Line Conductors*

General recommendations for the selection of methods and equipment found to be effective and practical for grounding during the stringing of overhead transmission line conductors and overhead ground wires are provided. The guide is directed to transmission voltages only. The aim is to present in one document, sufficient details of present day grounding practices and equipment used in effective grounding and to provide electrical theory and considerations necessary to safeguard personnel during the stringing operations of transmission lines.

IEEE Std. 789-1988 (R1994): *IEEE Standard Performance Requirements for Communications and Control Cables for Application in High Voltage Environments*

Requirements are set forth for wires and cables used principally for power system communications and control purposes that are located within electric power stations, installed within the zone of influence of the power station ground potential rise (GPR), or buried adjacent to electric power transmission and distribution lines. The cables can be subjected to high voltages either by conduction or induction coupling, or both. Cable specifications that ensure overall reliability in high-voltage environments are provided. Environmental considerations, operating service conditions, installation practices, and cable design requirements are covered. Design tests, routine production tests, and physical and electrical tests are included.

IEEE Std. 837-1989 (R1996): *IEEE Standard for Qualifying Permanent Connections Used in Substation Grounding*

Directions and methods for qualifying permanent connections used for substation grounding are provided. Particular attention is given to the connectors used within the grid system, connectors used to join ground leads to the grid system, and connectors used to join the ground leads to equipment and structures. The purpose is to give assurance to the user that connectors meeting the requirements of this standard will perform in a satisfactory manner over the lifetime of the installation provided, that the proper connectors are selected for the application, and that they are installed correctly. Parameters for testing grounding connections on aluminum, copper, steel, copper-clad steel, galvanized steel, stainless steel, and stainless-clad steel are addressed. Performance criteria are established, test procedures are provided, and mechanical, current-temperature cycling, freeze-thaw, corrosion, and fault-current tests are specified.

IEEE Std. 1048-1990: *IEEE Guide for Protective Grounding of Power Lines*

Guidelines are provided for safe protective grounding methods for persons engaged in deenergized overhead transmission and distribution line maintenance. They comprise state-of-the-art information on protective grounding as currently practiced by power utilities in North America. The principles of protective grounding are discussed. Grounding practices and equipment, power-line construction, and ground electrodes are covered.

IEEE Std. 1050-1996: *IEEE Guide for Instrumentation and Control Equipment Grounding in Generating Stations*

Information about grounding methods for generating station instrumentation and control (I & C) equipment is provided. The identification of I & C equipment grounding methods to achieve both a suitable level of protection for personnel and equipment is included, as well as suitable noise immunity for signal ground references in generating stations. Both ideal theoretical methods and accepted practices in the electric utility industry are presented.

IEEE Std. 1243-1997: *IEEE Guide for Improving the Lightning Performance of Transmission Lines*

Procedures for evaluating the lightning outage rate of overhead transmission lines at voltage levels of 69 kV or higher are described. Effects of improved insulation, shielding, coupling and grounding on backflashover, and shielding failure rates are then discussed.

IEEE Std. 1313.1-1996: *IEEE Standard for Insulation Coordination—Definitions, Principles, and Rules*

The procedure for selection of the withstand voltages for equipment phase-to-ground and phase-to-phase insulation systems is specified. A list of standard insulation levels, based on the voltage stress to which the equipment is being exposed, is also identified. This standard applies to three-phase AC systems above 1 kV.

IEEE Std. 1410-1997: *IEEE Guide for Improving the Lightning Performance of Distribution Lines*
Procedures for evaluating the lightning outage rate of overhead distribution lines at voltage levels below 69 kV are described. Effects of improved insulation, shielding, coupling, and grounding for direct strokes and induced over-voltage are then discussed.

10.8 Insulation Coordination

Stephen R. Lambert

Insulation coordination is the art of correlating equipment electrical insulation strengths with expected overvoltage stresses so as to result in an acceptable risk of failure while considering economics and operating criteria (McNutt and Lambert, 1992).

Insulation properties can be characterized as self-restoring and non-self-restoring. Self-restoring insulation has the ability to "heal" itself following a flashover, and such insulation media are usually associated with a gas — air, SF_6, etc. Examples include overhead line insulators, station buswork, external bushing surfaces, SF_6 buswork, and even switchgear insulation. With self-restoring insulation, some flashovers are often acceptable while in operation. An EHV transmission line, for example, is allowed to experience occasional line insulator flashovers during switching operations such as energizing or reclosing, or as a result of a lightning flash striking the tower, shield wires, or phase conductors.

Non-self-restoring insulation is assumed to have permanently failed following a flashover, and repairs must be made before the equipment can be put back into service. Insulation such as oil, oil/paper, and solid dielectrics such as pressboard, cross-link polyethylene, butyl rubbers, etc. are included in this insulation class. Any flashover of non-self-restoring insulation, say within a transformer or a cable, is unacceptable as such events usually result in lengthy outages and costly repairs.

The performance level of self-restoring insulation is usually addressed and defined in terms of the probability of a flashover. Thus, for a specific voltage stress, a given piece of insulation has an expected probability of flashover — e.g., a 1 meter conductor-to-conductor gap exposed to a 490 kV switching surge, would be expected to have a 50% chance of flashover; with a 453 kV surge, the gap would be expected to have a 10% chance of flashover, etc. Consequently, when self-restoring insulation is applied, the procedure is to select a gap length that will give the overall desired performance (probability of flashover) as a function of the stress (overvoltages) being applied.

For non-self-restoring insulation, however, any flashover is undesirable and unacceptable, and consequently for application of non-self-restoring insulation, a capability is selected such that the "100%" withstand level (effectively a zero percent chance of flashover) of the insulation exceeds the highest expected stress by a suitable margin.

Insulation Characteristics

Self-restoring (as well as non-self-restoring) insulation has, when exposed to a voltage, a probability of flashover that is dependent on:

- the dielectric material (air, SF_6, oil...)
- the waveshape of the stress (voltage)
- the electrode or gap configuration (rod-rod, conductor to structure...)
- the gap spacing
- atmospheric conditions (for gases)

Probability of Flashover (pfo)

Assuming the flashover characteristics of insulation follow a Gaussian distribution, and this is a good assumption for most insulation media (air, SF_6, oil, oil/paper), the statistical flashover characteristics of insulation can be described by the $V_{50\%}$ or mean value of flashover, and a standard deviation. The $V_{50\%}$ is a function of the rise time of the applied voltage, and when at a minimum, it is usually known as the V_{CFO} or critical flashover voltage.

Consequently, for a given surge level and insulation characteristic, the pfo of a single gap can be described by "p," and can be determined by first calculating the number of standard deviations the stress level is above or below the mean:

$$\#\delta = \frac{V_{stress} - V_{50\%}}{1 \text{ standard deviation}} \quad (10.154)$$

For air insulation, one standard deviation is either 3% of the $V_{50\%}$ for fundamental frequency (50–60 Hz) voltages and for lightning impulses or 6% of the $V_{50\%}$ for switching surge impulses. That the standard deviation is a fixed percentage of the $V_{50\%}$ and is not a function of gap length is very fortuitous and simplifies the calculations. Once the number of standard deviations away from the mean has been found, then by calculation or by entering a table, the probability of occurrence associated with that number of standard deviations is found.

Example:

Assume an insulator has a $V_{50\%}$ of 1100 kV with a standard deviation of 6%, and a switching overvoltage of 980 kV is applied to the insulation. The stress is 1.82 standard deviations below the mean:

$$\#\delta = \frac{980 - 1100}{0.06 \times 1100} \quad (10.155)$$

$$= -1.82 \text{ standard deviations below the mean}$$

By calculation or table, the probability associated with –1.82 standard deviations below the mean (for a normal distribution) is 3.4%. Thus, there is a 3.4% chance of insulation flashover every time the insulation is exposed to a 980 kV surge.

The physics of the flashover mechanism precludes a breakdown or flashover below some stress level, and this is generally assumed to occur at 3.5–4 standard deviations below the mean.

Multiple Gaps per Phase

The probability of flashover, P_n, for n gaps in parallel (assuming the gaps have the same characteristics and are exposed to the same voltage) can be described by the following equation where p is the probability of flashover of one gap. This mathematical expression defines the probability of one or more gaps flashing over, but practically only one gap of the group will flashover as the first gap to flashover reduces the voltage stress on the other gaps.

$$P_n = 1 - (1-p)^n \quad (10.156)$$

Multiple Gaps and Multiple Phases

Analysis of some applications may not only require consideration of multiple gaps in a given phase but also of multiple phases. Consider the pfo analysis of a transmission line; during a switching operation, for example, multiple towers are exposed to surges and at each tower, each of the three phases is stressed (typically by different surge magnitudes). Thus it is important to consider not only the multiple gaps associated with the multiple towers, but also all three phases often need to be considered to determine the overall line probability of flashover. The overall probability of flashover for a given surge, PFO, can be expressed as:

Power System Transients

$$\text{PFO} = 1 - \left(1 - \text{pfo}_{n,a}\right)^g \left(1 - \text{pfo}_{n,b}\right)^g \left(1 - \text{pfo}_{n,c}\right)^g \qquad (10.157)$$

where $\text{pfo}_{n,x}$ is the pfo of the x phase for the given surge, n
g is the number of towers (gaps in parallel)

The simultaneous analysis of all three phases can be important especially when various techniques are used to substantially suppress the surges (Lambert, 1988).

Flashover Characteristics of Air Insulation

Voltage Waveshape

Waveshapes used for testing and for determining the flashover response of insulation have been standardized by various groups and while there is not 100% agreement, the waveshapes used generally conform to the following:

- Fundamental frequency   50 or 60 Hz sine wave (8000 μs rise time)
- Switching impulse   200–250 μs by 2000 μs
- Lightning impulse   1.2 μs by 50 μs

The impulse waveshapes are usually formed by a double exponential having the time to crest indicated by the first number and the time to 50% of the crest on the tail of the wave indicated by the second number. Thus, a lightning impulse would crest at 1.2 μs and following the crest, would fall off to 50% of the crest at 50 μs.

Fundamental frequency characteristics have been published, and typical values are indicated in Fig. 10.48 (Aleksandrov et al., 1962; EPRI, 1982).

Equations have also been published which define the typical responses to positive polarity switching and lightning impulses (Fig. 10.49). Insulation usually has a lower withstand capability when exposed to positive polarity impulses than when exposed to negative impulses; thus, designs are usually based on positive magnitude impulses.

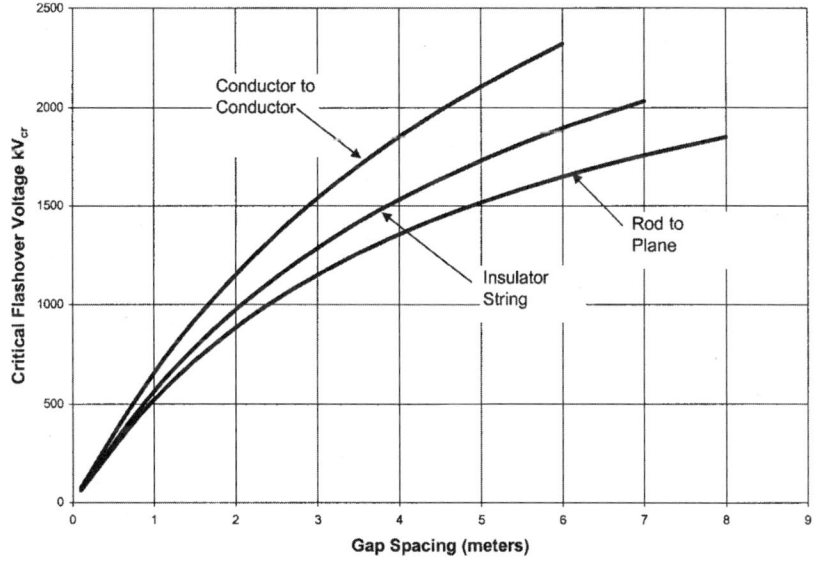

FIGURE 10.48 $V_{50\%}$ for fundamental frequency waveshapes.

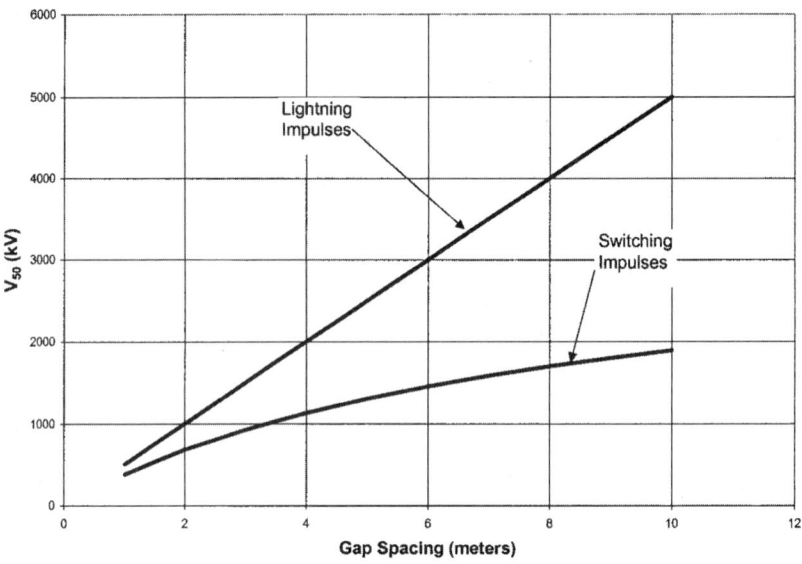

FIGURE 10.49 $V_{50\%}$ for impulses.

For switching surge impulses (Gallet et al., 1976):

$$V_{50\%} = k \frac{3400}{1+\frac{8}{d}} \text{ kV} \qquad (10.158)$$

For lightning impulses:

$$V_{50\%} = k * 500 * d \text{ kV} \qquad (10.159)$$

where k is an electrode factor reflecting the shape of the electrodes (Paris, 1967)
d is the electrode gap spacing in meters

Electrode Configuration

Electrode configuration has a pronounced effect on the $V_{50\%}$ characteristics, and this is reflected as a gap or electrode factor, k (Paris, 1967). Examples of k are:

rod — plane	1.00
conductor — structure	1.30
rod — rod	1.30
conductor — conductor	1.40
conductor — rod	1.65

Effect of Insulator

The presence of an insulator in a gap tends to reduce the gap factor from those given above, mainly due to the terminal electrode configuration (and intermediate flanges for multi-unit column bus support insulators). The reduction increases with increased gap factor and typical correction values may be found in Fig. 10.50. Note that these corrections are subject to variations (Thione, 1984).

Rain has little effect on a gap without an insulator, however, rain does reduce the gap factor when an insulator is present. Reductions as high as 20% have been noted, but in general, a reduction of 4–5% is typical (Thione, 1984).

Power System Transients

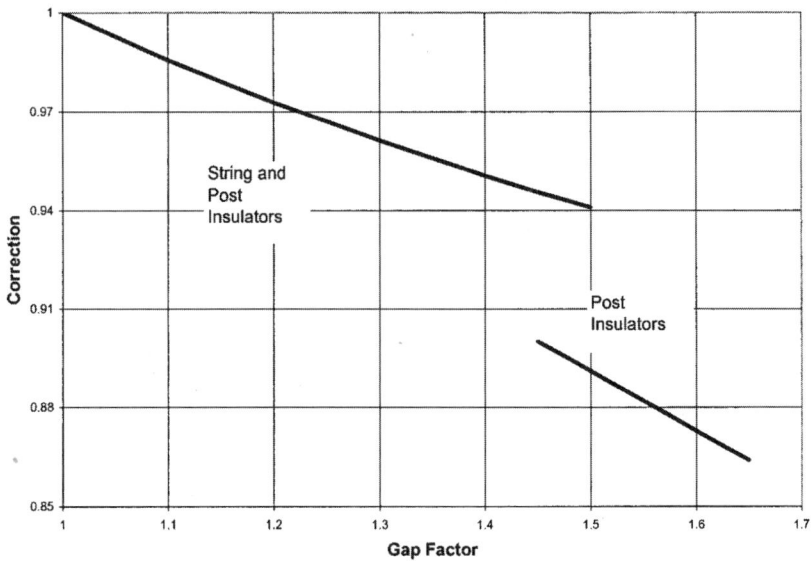

FIGURE 10.50 Gap factor correction for presence of insulator.

Effect of Atmospheric Conditions on Air Insulation

$V_{50\%}$ for gases is affected by temperature, atmospheric pressure, and humidity, and for air the correction can be expressed as:

$$V_{50\%, \text{ambient}} = V_{50\%, \text{STP}} \left(\frac{H_o}{\delta} \right)^n \tag{10.160}$$

where H_o is the humidity correction factor
n is a gap length correction factor
δ is the relative air density correction factor

The correction for temperature and pressure, δ, is known as the RAD (relative air density) correction factor and is expressed by:

$$\delta = \frac{0.386 \, H_{\text{mm of Hg}}}{273.33 + T_{°C}} \tag{10.161}$$

where $H_{\text{mm of Hg}}$ is the atmospheric pressure in mm of Hg
$T_{°C}$ is the temperature in °C

The humidity correction factor, H_o, is given in IEEE 4 (1978) and can be expressed approximately by:

$$H_o \cong 1.1 - 0.00820 * H_{AB} \tag{10.162}$$
$$\cong 1.1 - 0.008071 * VP$$

where H_{AB} is the absolute humidity in g/m³
VP is the vapor pressure in mm of Hg

For switching impulses (and fundamental frequency), the effect of the RAD and humidity on $V_{50\%}$ is, however, a function of the gap length and has less effect on longer gap than on shorter gap lengths. For lengths of 0–1 m, the "n" correction factor is 1.0; from 1–6 m, the correction decreases linearly from

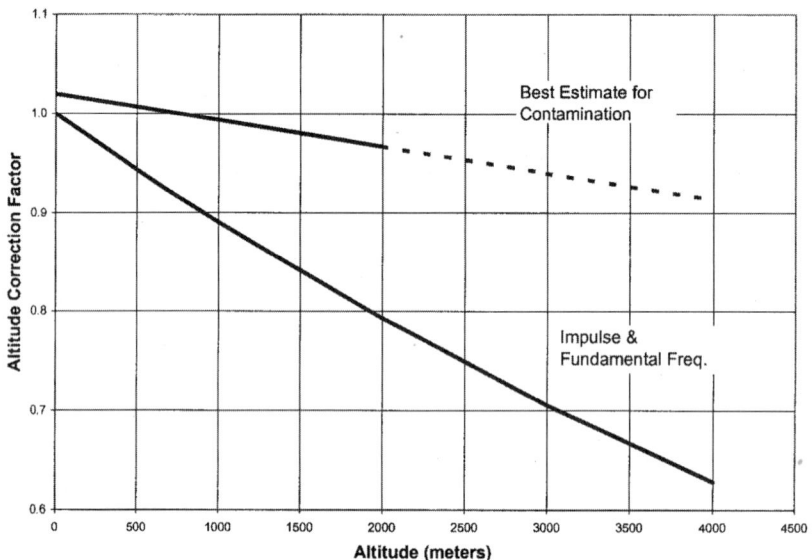

FIGURE 10.51 Altitude correction factors.

1.0 to 0.4; and for lengths greater than 6 m, the factor is 0.4. There is no gap length correction for positive lightning impulses (EPRI, 1982; EPRI, 1975; EEI, 1968). Other approaches for humidity corrections can be found in Menemenlis et al. (1988); Thione (1984); Feser and Pigini (1987).

Altitude

Corrections for altitude are also important as the insulation capability drops off about 10% per 1000 m as shown in Fig. 10.51. There are various equations for the altitude correction factor (ACF), and the following expression is representative of most in use (IEEE 1312, 1999):

$$ACF = e^{-\frac{Ht}{8600}}$$

where Ht is the altitude in meters

Insulator Contamination

Insulator contamination is an important issue for fundamental frequency voltage considerations, and the equivalent salt density, ESDD, approach is extensively used as a design tool. The contamination severity is defined by the ESDD in mg/cm^2, and an insulator creepage distance, in terms of $mm/kV_{rms, phase\ to\ phase}$, can then be selected (IEC 815, 1986). Note that insulator/bushing shed/skirt design has a significant impact on the performance, and some past designs performed poorly due to skirt configuration even though they had large creepage distances. With the ESDD approach, insulators are tested to define their expected performance. Table 10.5 shows the relationship between contamination level, ESDD, and recommended creepage distances.

Altitude also has an effect on the performance of contaminated insulation, and the degradation of capability as a function of altitude may be found in Fig. 10.51 (Mizuno et al., 1997).

Example:

Assume ten identical substation bus support insulators in parallel located in a 500 kV substation located at sea level; this configuration can be described as an air gap, conductor to rod configuration at standard atmospheric conditions. Assume that an overall probability of flashover for the ten gaps of 0.5% is desired when the configuration is exposed to a switching surge of 939 kV (2.3 pu on a 500 kV system). What is the required gap clearance in meters?

TABLE 10.5 Relationship Between Contamination Level, ESDD, and Recommended Creepage Distances

Contamination Level	Example	ESDD (mg/cm²)	Minimum Recommended Creepage Distance (mm/kV$_{rms, phase\ to\ phase}$)
Light	Low industrial activity	0.03–0.06	16
Medium	Industrial activity — some exposure to wind from the sea	0.1–0.2	20
Heavy	Industrial area and areas close to the sea	0.3–0.6	25
Very Heavy	Heavy industrial or sea coast area	>0.6	31

The desired pfo of one gap, p, then should be:

$$\text{pfo}_{10\ \text{gaps}} = 0.005 = 1 - (1-p)^{10}$$

and

$$p = 1 - (1 - 0.005)^{1/10}$$

$$= 0.0005011$$

From tables or calculations for a normal or Gaussian distribution, a probability of 0.0005011 corresponds to 3.29 standard deviations below the mean ($V_{50\%}$). Therefore, the desired $V_{50\%}$ value is:

$$939\ \text{kV} = V_{50\%}(1 - 3.29 * 0.06)$$

and

$$V_{50\%} = 1170\ \text{kV}$$

A standard deviation of 6% is often used for the air gap for switching surge stresses.

With the $V_{50\%}$ of 1170 kV and noting that a conductor to rod gap has a k factor of 1.65 assumed to be reduced to 1.42 due to the presence of the insulator, the desired gap spacing value can be calculated by:

$$1170\ \text{kV} = 1.42 \frac{3400}{1 + \frac{8}{d}}$$

and

$$d = 2.56\ \text{meters}$$

Application of Surge Arresters

Surge arresters are used to limit overvoltages and as a result, allow reductions in the clearances required for self-restoring gaps (e.g., transmission line towers) as well as the capability required for non-self-restoring insulation such as transformer windings. In most applications the proper approach is to determine the minimum arrester rating that can be applied without resulting in damage to the arrester, and then to define the insulation level required so as to result in an acceptable probability of flashover or risk of failure.

For a transmission line application, for example, although the arrester reduces higher magnitude surges to lower levels, the line is still allowed to have a finite, albeit low, probability of flashover for a specific switching operation. Thus, the arrester, by limiting the higher magnitude surges, allows smaller conductor-to-tower clearances.

However, when arresters are used to protect a transformer for example, an insulation level is selected that has a significantly higher capability than the maximum surge allowed by the arrester. This margin between the arrester protective level (lightning or switching surge) is a function of various considerations as well as the conservatism of the person applying the arrester/insulation system.

Today, for new applications, only metal oxide (ZnO) arresters are being applied. Although there are certainly many of the gapped, silicon carbide type arresters still in service and which still perform effectively, in what follows, only metal oxide arresters will be considered to protect insulation. Successful application requires that the arrester survive the electrical environment in which it is placed, and the following arrester capabilities must be carefully considered:

MCOV — maximum fundamental frequency continuous operating voltage applied to the arrester
TOV — temporary fundamental frequency overvoltages to which the arrester may be exposed
Energy — the energy which must be absorbed by the arrester when limiting switching surges

MCOV — The highest system voltage which can be continuously applied to the arrester needs to be determined and the arrester capability, its MCOV rating, should at least be equal to and should usually exceed the highest continuous system voltage by some small margin. For example, if a nominal 345 kV system is never operated above 352 kV, then the maximum continuous voltage that would be expected to be applied to a line to ground arrester would be $352 kV/\sqrt{3} = 203.2$ kV. With today's typical arresters, the next highest available MCOV capability would be 209 kV and is associated with an arrester rated 258 kV.

TOV — On occasion, the fundamental frequency voltage applied to an arrester will exceed the expected MCOV. Examples include fault conditions during which line-to-ground voltages on unfaulted phases can rise significantly (as high as phase-to-phase voltage for ungrounded systems); rise in line voltage when energizing a transmission line (ferranti effect); and voltages that occur during load rejection events — these are usually associated with voltages experienced on a radial transmission line emanating from a generating plant when the load terminal of the line opens unexpectedly.

Energy — When an arrester limits switching surges on a transmission line, it can absorb a significant amount of energy, and it can be important to examine events and determine the energy that could be absorbed. Exceeding the arrester's capability could result in immediate damage to the arrester and failure. It is also important to the arrester's TOV capability as absorbing energy heats the arrester material, and application of a significant temporary overvoltage immediately following absorption of a significant amount of energy could result in thermal runaway and arrester failure.

Following selection of an arrester that would be expected to survive the electrical environment (i.e., the minimum rated arrester), the protective levels of the arrester must be correlated with the insulation capability and acceptable margins between the protective levels and the insulation capability achieved.

The protective level or discharge voltage of an arrester is the voltage magnitude to which the arrester will limit the voltage while discharging a surge, and these levels are a function of the waveshape and rise time of the surge as well as the current magnitude of the discharge. In general, the discharge or protective levels considered for coordination with insulation capability are:

- a 10 kA, 8×20 μs discharge for coordination with the insulation full wave or lightning impulse (BIL) capability, and
- a 0.5–2.0 kA, 36×90 μs discharge for coordination with the switching impulse capability.

There should always be margin between the protective level of the arrester and the insulation capability to allow for uncertainties in arrester protective levels due to surge rise times, discharge currents, and arrester separation distance (faster rise times, higher currents, and longer separation distance or lead lengths generate higher protective levels). Uncertainties in insulation capability include reduced insulation strength due to aging (especially for paper insulation in transformers for example) and limitations of the ability of laboratory dielectric testing to accurately relate to field conditions.

In the author's opinion, a margin of at least 40% is appropriate unless all the uncertainties and the risks are carefully evaluated.

Examples of Surge Arrester Application (Non-Self-restoring Insulation)

34.5 kV System Application

Surge arresters are to be applied line-to-ground at the terminals of a circuit breaker (38 kV rating, 150 kV BIL) used on a solidly grounded 34.5 kV system. The highest expected continuous system voltage is 37 kV, and during fault conditions, the phase-to-ground voltage can rise to 1.4 pu or 27.9 kV. Faults can persist for 20 cycles.

The maximum line-to-ground voltage is $37/\sqrt{3} = 21.4$ kV, and the MCOV of the arrester must meet or exceed this value. An arrester rated 27 kV would be acceptable as it has an MCOV of 22.0 kV. The 1 second TOV capability of the arrester is 31.7 kV, and as this exceeds the 27.9 kV phase-to-ground voltage expected during faults, the 27 kV arrester meets the TOV criteria as well.

A 27 kV arrester has a 10 kA discharge level of 67.7 kV, and thus the margin between the discharge or protective level and the insulation BIL is $(150/67.7 \times 100 - 100)$ or 121%. This margin is obviously more than adequate, and selection of an arrester rated 27 kV would be appropriate.

500 kV System Application

A 500 kV shunt reactor (solidly grounded neutral) is being applied at the end of a 300 km, 500 kV transmission line, and arresters are to be applied line-to-ground on the terminals of the reactor to limit surges to reasonable levels. The reactor is solidly connected to the line and is switched with the line, and the substation at which the reactor resides is at an altitude of 1800 m. The highest expected continuous system voltage is 550 kV. During line switching operations, the circuit breaker at the reactor terminal may not be closed for some period following energizing of the line/reactor from the other terminal, and the phase-to-ground voltage at the reactor can be as high as 1.15 pu for as long as five minutes. Arrester energy requirements were determined (by EMTP or TNA simulations of switching operations) to be well within the capability of an arrester rated 396 kV.

The minimum required MCOV is $550/\sqrt{3} = 317.5$ kV. The minimum required TOV is $1.15 \times 500/\sqrt{3} = 332$ kV for 300 seconds, and for most arresters, such a requirement would correlate with a 1 second TOV rating of 451 kV. An arrester rated 396 kV has a 318 kV MCOV and a 1 second TOV rating of 451 kV; thus, a 396 kV arrester would be the minimum rating that could be used. Of course any arrester rated higher than 396 kV could also be used. The 10 kA lightning (8×20 μs waveform) and switching surge (2 kA, 36×90 μs) discharge levels for a 396 kV and a 420 kV arrester are:

	Discharge Levels	
Rating	10 kA	Switching Surge
396 kV	872 kV	758 kV
420 kV	924 kV	830 kV

BIL values of 1300 kV and 1425 kV for the reactor's *internal insulation* (i.e., insulation not affected by altitude) could be considered as reasonable candidates for a specification. The corresponding switching impulse levels (SIL) would be 1080 kV and 1180 kV, respectively, and the following table indicates the margin between the arrester protective levels and the insulation level.

	1300 kV BIL		1425 kV BIL	
Arrester	396 kV	420 kV	396 kV	420 kV
SIL	42%	30%	56%	42%
BIL	49%	41%	63%	54%

Application of a 420 kV arrester for a 1300 kV BIL insulation level results in margins below 40%, and unless the application is very carefully considered from the point of view of arrester separation distance and lead length, expected maximum discharge current level, wave rise time, etc., a 396 kV arrester would be a better choice. For a 1425 kV BIL, either the 396 kV or the 420 kV arrester would result in sufficient margins.

For *external insulation*, i.e., the reactor bushings, the effect of altitude on the insulation capability needs to be considered. At 1800 m the insulation has only 81% of the withstand capability demonstrated at sea level or 0 m. For example, the SIL of a 1425 kV bushing (1180 kV at sea level) would be reduced to 956 kV at 1800 m (1108 × 0.81 = 956 kV), and application of even a 396 kV arrester would result in a margin of 26% — hardly acceptable.

Assume that a 420 kV arrester was selected to protect the reactor (the arrester itself is rated for application to 3000 m). The switching surge and 10 kA protective levels are 830 kV and 924 kV, respectively. With a desired minimum margin of 40%, and correcting for altitude, the minimum SIL and BIL at sea level (0 m) should be:

$$\text{Minimum SIL} = \frac{830 \times 1.4}{0.81} = 1435 \text{ kV}$$

$$\text{Minimum BIL} = \frac{924 \times 1.4}{0.81} = 1597 \text{ kV}$$

A 1550 kV BIL bushing would have a 1290 kV SIL, and even if one would accept the slightly less than a 36% margin for the BIL, the SIL margin would only be 26%. A 1675 kV BIL bushing would be expected to have a 1390 kV SIL capability, and so the SIL margin would be 36% with a BIL margin of 47%. The next higher rated bushing (1800 kV BIL) would mean applying 800 kV system class bushings, and their increased size and cost would likely not make for a reasonable design. Consequently, specifying a 1675 kV BIL bushing and accepting the slightly reduced SIL margin would be a reasonable compromise.

Effect of Surge Reduction Techniques on Overall PFO

Application of surge arresters to significantly reduce switching surge levels on transmission line and substation insulators can be effective; however, the designer should be aware that the overall pfo of all three phases needs to be considered as it will usually be higher than that found for a single phase by a factor often approaching three. Also for long transmission lines, application of arresters at the line terminals will certainly limit the surges at the terminals but will not limit the surges at other points on the line to the same level. Consequently, the surge distribution along the line may need to be considered (Lambert, 1988; Ribiero et al., 1991).

References

Aleksandrov, G. N., Kizvetter, V. Y., Rudakova, V. M. Tushnov, A. N., The AC flashover voltages of long air gaps and strings of insulators, *Elektrichestvo*, 6, 27, 1962.

EHV Transmission Line Reference Book, Edison Electric Institute, 1968.

Feser, K., Pigini, A., Influence of atmospheric conditions on the dielectric strength of external insulation, *ELECTRA*, No. 112, 83, 1987.

Gallet, G., Bettler, M., Leroy, G., Switching impulse results obtained on the outdoor testing area at Renardieres, *IEEE Trans. on Power Appar. and Syst.*, PAS-95, 2, 580, 1976.

Guide for the Selection of Insulators in Respect of Polluted Conditions, IEC Publication 815, 1986.

IEEE Standard for Insulation Coordination — Part 2, Application Guide, IEEE 1312-1999, Institute of Electrical and Electronic Engineers.

IEEE Standard Techniques for High-Voltage Testing, IEEE 4 — 1978, Institute of Electrical and Electronic Engineers.

Lambert, S. R., Effectiveness of zinc oxide surge arresters on substation equipment probabilities of flashover, *IEEE Transactions on Power Delivery*, 3(4), 1928, 1988.

McNutt, W. J., Lambert, S. R., *Transformer Concepts and Applications Course,* Power Technologies, Inc., Schenectady, NY, 1992.

Menemenlis, C., Carrara, G., Lambeth, P. J., Application of insulators to withstand switching surges in substations, part I: Switching impulse insulation strength, 88 WM 077-0, IEEE/PES Winter Meeting, New York, New York, January 31–February 5, 1988.

Mizuno, Y., Kusada, H., Naito, K., Effect of climatic conditions on contamination flashover voltage of insulators, *IEEE Trans. on Dielectrics and Electrical Insulation*, 4(3), 286, 1997.

Paris, L., Influence of air gap characteristics on line-to-ground switching surge strength, *IEEE Trans. on Power Appar. and Syst.*, PAS-86, 8, 936, 1967.

Ribeiro, J. R., Lambert, S. R., Wilson, D. D., Protection of compact transmission lines with metal oxide arresters, in *CIGRE Leningrad Symposium 1991,* 400–6, S33–91.

Thione, L., Evaluation of the switching impulse strength of external insulation, *Electra*, No. 94, 77, 1984.

Transmission Line Reference Book, 345 kV and Above, 1st ed., Electric Power Research Institute, 1975.

Transmission Line Reference Book, 345 kV and Above/Second Edition, Electric Power Research Institute, 1982.

11
Power System Dynamics and Stability

Richard G. Farmer
Arizona State University

11.1	Power System Stability — Overview *Prabha Kundur*	11-2
11.2	Transient Stability *Kip Morrison*	11-10
11.3	Small Signal Stability and Power System Oscillations *John Paserba, Prabha Kundar, Juan Sanchez-Gasca, and Einar Larsen*	11-20
11.4	Voltage Stability *Yakout Mansour*	11-34
11.5	Direct Stability Methods *Vijay Vittal*	11-42
11.6	Power System Stability Controls *Carson W. Taylor*	11-55
11.7	Power System Dynamic Modeling *William W. Price*	11-72
11.8	Direct Analysis of Wide Area Dynamics *J. F. Hauer, W.A. Mittelstadt, M.K. Donnelly, W.H. Litzenberger, and Rambabu Adapa*	11-82
11.9	Power System Dynamic Security Assessment *Peter W. Sauer*	11-120
11.10	Power System Dynamic Interaction with Turbine-Generators *Richard G. Farmer and Bajarang L. Agrawal*	11-126

11
Power System Dynamics and Stability

Prabha Kundur
Powertech Labs, Inc.

Kip Morrison
Powertech Labs, Inc.

John Paserba
Mitsubishi Electric Power Products, Inc.

Juan Sanchez-Gasca
GE Power Systems

Einar Larsen
GE Power Systems

Yakout Mansour
BC Hydro

Vijay Vittal
Iowa State University

Carson W. Taylor
Carson Taylor Seminars

William W. Price
GE Power Systems

J. F. Hauer
Pacific Northwest National Laboratory

W. A. Mittelstadt
Bonneville Power Administration

M. K. Donnelly
Pacific Northwest National Laboratory

W. H. Litzenberger
Bonneville Power Administration

11.1 Power System Stability — Overview.................................11-2
Basic Concepts • Classification of Power System Stability • Historical Review of Stability Problems • Consideration of Stability in System Design and Operation

11.2 Transient Stability...11-10
Basic Theory of Transient Stability • Methods of Analysis of Transient Stability • Factors Influencing Transient Stability • Transient Stability Considerations in System Design • Transient Stability Considerations in System Operation

11.3 Small Signal Stability and Power System Oscillations..11-20
Nature of Power System Oscillations • Criteria for Damping • Study Procedure • Mitigation of Power System Oscillations • Summary

11.4 Voltage Stability..11-34
Generic Dynamic Load–Voltage Characteristics • Analytical Frameworks • Computational Methods • Mitigation of Voltage Stability Problems

11.5 Direct Stability Methods..11-42
Review of Literature on Direct Methods • The Power System Model • The Transient Energy Function • Transient Stability Assessment • Determination of the Controlling UEP • The BCU (Boundary Controlling UEP) Method • Applications of the TEF Method and Modeling Enhancements

11.6 Power System Stability Controls.............................11-55
Review of Power System Synchronous Stability Basics • Concepts of Power System Stability Controls • Types of Power System Stability Controls and Possibilities for Advanced Control • Dynamic Security Assessment • "Intelligent" Controls • Effect of Industry Restructuring on Stability Controls • Experience from Recent Power Failures • Summary

11.7 Power System Dynamic Modeling..........................11-72
Modeling Requirements • Generator Modeling • Excitation System Modeling • Prime Mover Modeling • Load Modeling • Transmission Device Models • Dynamic Equivalents

Rambabu Adapa
Electric Power Research Institute

Peter W. Sauer
University of Illinois at Urbana

Richard G. Farmer
Arizona State University

Bajarang L. Agrawal
Arizona Public Service Company

11.8 Direct Analysis of Wide Area Dynamics 11-82
  Dynamic Information Needs: The WSCC Breakup of August 10, 1996 • Background • An Overview of WSCC WAMS • Direct Sources of Dynamic Information • Monitor Architectures • Monitor Network Topologies • Networks of Networks • WSCC Experience in Monitor Operations • Database Management in Wide Area Monitoring • Monitor Application Examples • Conclusions

11.9 Power System Dynamic Security Assessment 11-120
  Power System Security Concepts • Dynamic Phenomena • Assessment Methodologies • Summary

11.10 Power System Dynamic Interaction with Turbine-Generators 11-126
  Subsynchronous Resonance • Device Dependent Subsynchronous Oscillations • Supersynchronous Resonance • Device Dependent Supersynchronous Oscillations

11.1 Power System Stability — Overview

Prabha Kundur

This introductory section provides a general description of the power system stability phenomena including fundamental concepts, classification, and definition of associated terms. A historical review of the emergence of different forms of stability problems as power systems evolved and of the developments of methods for their analysis and mitigation is presented. Requirements for consideration of stability in system design and operation are discussed.

Basic Concepts

Power system stability is the ability of the system, for a given initial operating condition, to regain a normal state of equilibrium after being subjected to a disturbance. Stability is a condition of equilibrium between opposing forces; instability results when a disturbance leads to a sustained imbalance between the opposing forces.

The power system is a highly nonlinear system that operates in a constantly changing environment; loads, generator outputs, topology, and key operating parameters change continually. When subjected to a transient disturbance, the stability of the system depends on the nature of the disturbance as well as the initial operating condition. The disturbance may be small or large. Small disturbances in the form of load changes occur continually, and the system adjusts to the changing conditions. The system must be able to operate satisfactorily under these conditions and successfully meet the load demand. It must also be able to survive numerous disturbances of a severe nature, such as a short-circuit on a transmission line or loss of a large generator.

Following a transient disturbance, if the power system is stable, it will reach a new equilibrium state with practically the entire system intact; the actions of automatic controls and possibly human operators will eventually restore the system to normal state. On the other hand, if the system is unstable, it will result in a run-away or run-down situation; for example, a progressive increase in angular separation of generator rotors, or a progressive decrease in bus voltages. An unstable system condition could lead to cascading outages and a shut-down of a major portion of the power system.

The response of the power system to a disturbance may involve much of the equipment. For instance, a fault on a critical element followed by its isolation by protective relays will cause variations in power flows, network bus voltages, and machine rotor speeds; the voltage variations will actuate both generator and transmission network voltage regulators; the generator speed variations will actuate prime mover governors; and the voltage and frequency variations will affect the system loads to varying degrees depending on their individual characteristics. Further, devices used to protect individual equipment may respond to variations in system variables and thereby affect the power system performance. A typical modern power system is thus a very high-order multivariable process whose dynamic performance is influenced by a wide array of devices with different response rates and characteristics. Hence, instability in a power system may occur in many different ways depending on the system topology, operating mode, and the form of the disturbance.

Traditionally, the stability problem has been one of maintaining synchronous operation. Since power systems rely on synchronous machines for generation of electrical power, a necessary condition for satisfactory system operation is that all synchronous machines remain in synchronism or, colloquially, "in step." This aspect of stability is influenced by the dynamics of generator rotor angles and power-angle relationships.

Instability may also be encountered without the loss of synchronism. For example, a system consisting of a generator feeding an induction motor can become unstable due to collapse of load voltage. In this instance, it is the stability and control of voltage that is the issue, rather than the maintenance of synchronism. This type of instability can also occur in the case of loads covering an extensive area in a large system.

In the event of a significant load/generation mismatch, generator and prime mover controls become important, as well as system controls and special protections. If not properly coordinated, it is possible for the system frequency to become unstable, and generating units and/or loads may ultimately be tripped possibly leading to a system blackout. This is another case where units may remain in synchronism (until tripped by such protections as under-frequency), but the system becomes unstable.

Because of the high dimensionality and complexity of stability problems, it is essential to make simplifying assumptions and to analyze specific types of problems using the right degree of detail of system representation. The following subsection describes the classification of power system stability into different categories.

Classification of Power System Stability

Need for Classification

Power system stability is a single problem; however, it is impractical to deal with it as such. Instability of the power system can take different forms and is influenced by a wide range of factors. Analysis of stability problems, including identifying essential factors that contribute to instability and devising methods of improving stable operation is greatly facilitated by classification of stability into appropriate categories. These are based on the following considerations (Kundur, 1994; Kundur and Morrison, 1997):

- The physical nature of the resulting instability related to the main system parameter in which instability can be observed.
- The size of the disturbance considered indicates the most appropriate method of calculation and prediction of stability.
- The devices, processes, and the time span that must be taken into consideration in order to determine stability.

Figure 11.1 shows a possible classification of power system stability into various categories and subcategories. The following are descriptions of the corresponding forms of stability phenomena.

Rotor Angle Stability

Rotor angle stability is concerned with the ability of interconnected synchronous machines of a power system to remain in synchronism under normal operating conditions and after being subjected to a disturbance. It depends on the ability to maintain/restore equilibrium between electromagnetic torque and mechanical torque of each synchronous machine in the system. Instability that may result occurs in the form of increasing angular swings of some generators leading to their loss of synchronism with other generators.

The rotor angle stability problem involves the study of the electromechanical oscillations inherent in power systems. A fundamental factor in this problem is the manner in which the power outputs of

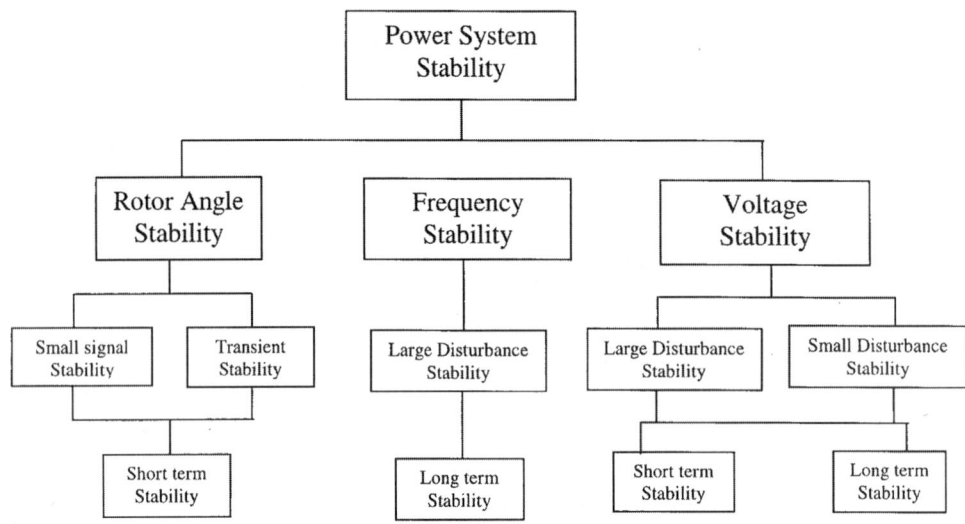

FIGURE 11.1 Classification of power system stability.

synchronous machines vary as their rotor angles change. The mechanism by which interconnected synchronous machines maintain synchronism with one another is through restoring forces, which act whenever there are forces tending to accelerate or decelerate one or more machines with respect to other machines. Under steady-state conditions, there is equilibrium between the input mechanical torque and the output electrical torque of each machine, and the speed remains constant. If the system is perturbed, this equilibrium is upset, resulting in acceleration or deceleration of the rotors of the machines according to the laws of motion of a rotating body. If one generator temporarily runs faster than another, the angular position of its rotor relative to that of the slower machine will advance. The resulting angular difference transfers part of the load from the slow machine to the fast machine, depending on the power-angle relationship. This tends to reduce the speed difference and hence the angular separation. The power-angle relationship, as discussed above, is highly nonlinear. Beyond a certain limit, an increase in angular separation is accompanied by a decrease in power transfer; this increases the angular separation further and leads to instability. For any given situation, the stability of the system depends on whether or not the deviations in angular positions of the rotors result in sufficient restoring torques.

It should be noted that loss of synchronism can occur between one machine and the rest of the system, or between groups of machines, possibly with synchronism maintained within each group after separating from each other.

The change in electrical torque of a synchronous machine following a perturbation can be resolved into two components:

- *Synchronizing torque* component, in phase with a rotor angle perturbation.
- *Damping torque* component, in phase with the speed deviation.

System stability depends on the existence of both components of torque for each of the synchronous machines. Lack of sufficient synchronizing torque results in *aperiodic* or *non-oscillatory instability*, whereas lack of damping torque results in *oscillatory instability.*

For convenience in analysis and for gaining useful insight into the nature of stability problems, it is useful to characterize rotor angle stability in terms of the following two categories:

1. *Small signal* (or *steady state*) *stability* is concerned with the ability of the power system to maintain synchronism under small disturbances. The disturbances are considered to be sufficiently small

that linearization of system equations is permissible for purposes of analysis. Such disturbances are continually encountered in normal system operation, such as small changes in load.

Small signal stability depends on the initial operating state of the system. Instability that may result can be of two forms: (i) increase in rotor angle through a non-oscillatory or aperiodic mode due to lack of synchronizing torque, or (ii) rotor oscillations of increasing amplitude due to lack of sufficient damping torque.

In today's practical power systems, small signal stability is largely a problem of insufficient damping of oscillations. The time frame of interest in small-signal stability studies is on the order of 10 to 20 s following a disturbance.

2. *Large disturbance rotor angle stability or transient stability*, as it is commonly referred to, is concerned with the ability of the power system to maintain synchronism when subjected to a severe transient disturbance. The resulting system response involves large excursions of generator rotor angles and is influenced by the nonlinear power-angle relationship.

Transient stability depends on both the initial operating state of the system and the severity of the disturbance. Usually, the disturbance alters the system such that the post-disturbance steady state operation will be different from that prior to the disturbance. Instability is in the form of aperiodic drift due to insufficient synchronizing torque, and is referred to as *first swing stability*. In large power systems, transient instability may not always occur as first swing instability associated with a single mode; it could be as a result of increased peak deviation caused by superposition of several modes of oscillation causing large excursions of rotor angle beyond the first swing.

The time frame of interest in transient stability studies is usually limited to 3 to 5 sec following the disturbance. It may extend to 10 sec for very large systems with dominant inter-area swings.

Power systems experience a wide variety of disturbances. It is impractical and uneconomical to design the systems to be stable for every possible contingency. The design contingencies are selected on the basis that they have a reasonably high probability of occurrence.

As identified in Fig. 11.1, small signal stability as well as transient stability are categorized as short term phenomena.

Voltage Stability

Voltage stability is concerned with the ability of a power system to maintain steady voltages at all buses in the system under normal operating conditions, and after being subjected to a disturbance. Instability that may result occurs in the form of a progressive fall or rise of voltage of some buses. The possible outcome of voltage instability is loss of load in the area where voltages reach unacceptably low values, or a loss of integrity of the power system.

Progressive drop in bus voltages can also be associated with rotor angles going out of step. For example, the gradual loss of synchronism of machines as rotor angles between two groups of machines approach or exceed 180° would result in very low voltages at intermediate points in the network close to the electrical center (Kundur, 1994). In contrast, the type of sustained fall of voltage that is related to voltage instability occurs where rotor angle stability is not an issue.

The main factor contributing to voltage instability is usually the voltage drop that occurs when active and reactive power flow through inductive reactances associated with the transmission network; this limits the capability of transmission network for power transfer. The power transfer limit is further limited when some of the generators hit their reactive power capability limits. The driving force for voltage instability are the loads; in response to a disturbance, power consumed by the loads tends to be restored by the action of distribution voltage regulators, tap changing transformers, and thermostats. Restored loads increase the stress on the high voltage network causing more voltage reduction. A rundown situation causing voltage instability occurs when load dynamics attempts to restore power consumption beyond the capability of the transmission system and the connected generation (Kundur, 1994; Taylor, 1994; Van Cutsem and Vournas, 1998).

As in the case of rotor angle stability, it is useful to classify voltage stability into the following subcategories:

1. *Large disturbance voltage stability* is concerned with a system's ability to control voltages following large disturbances such as system faults, loss of generation, or circuit contingencies. This ability is determined by the system-load characteristics and the interactions of both continuous and discrete controls and protections. Determination of large disturbance stability requires the examination of the nonlinear dynamic performance of a system over a period of time sufficient to capture the interactions of such devices as under-load transformer tap changers and generator field-current limiters. The study period of interest may extend from a few seconds to tens of minutes. Therefore, long term dynamic simulations are required for analysis (Van Cutsem et al., 1995).
2. *Small disturbance voltage stability* is concerned with a system's ability to control voltages following small perturbations such as incremental changes in system load. This form of stability is determined by the characteristics of loads, continuous controls, and discrete controls at a given instant of time. This concept is useful in determining, at any instant, how the system voltage will respond to small system changes. The basic processes contributing to small disturbance voltage instability are essentially of a steady state nature. Therefore, static analysis can be effectively used to determine stability margins, identify factors influencing stability, and examine a wide range of system conditions and a large number of postcontingency scenarios (Gao et al., 1992). A criterion for small disturbance voltage stability is that, at a given operating condition for every bus in the system, the bus voltage magnitude increases as the reactive power injection at the same bus is increased. A system is voltage unstable if, for at least one bus in the system, the bus voltage magnitude (V) decreases as the reactive power injection (Q) at the same bus is increased. In other words, a system is voltage stable if V-Q sensitivity is positive for every bus and unstable if V-Q sensitivity is negative for at least one bus.

The time frame of interest for voltage stability problems may vary from a few seconds to tens of minutes. Therefore, voltage stability may be either a short-term or a long-term phenomenon.

Voltage instability does not always occur in its pure form. Often, the rotor angle instability and voltage instability go hand in hand. One may lead to the other, and the distinction may not be clear. However, distinguishing between angle stability and voltage stability is important in understanding the underlying causes of the problems in order to develop appropriate design and operating procedures.

Frequency Stability

Frequency stability is concerned with the ability of a power system to maintain steady frequency within a nominal range following a severe system upset resulting in a significant imbalance between generation and load. It depends on the ability to restore balance between system generation and load, with minimum loss of load.

Severe system upsets generally result in large excursions of frequency, power flows, voltage, and other system variables, thereby invoking the actions of processes, controls, and protections that are not modeled in conventional transient stability or voltage stability studies. These processes may be very slow, such as boiler dynamics, or only triggered for extreme system conditions, such as volts/hertz protection tripping generators. In large interconnected power systems, this type of situation is most commonly associated with islanding. Stability in this case is a question of whether or not each island will reach an acceptable state of operating equilibrium with minimal loss of load. It is determined by the overall response of the island as evidenced by its mean frequency, rather than relative motion of machines. Generally, frequency stability problems are associated with inadequacies in equipment responses, poor coordination of control and protection equipment, or insufficient generation reserve. Examples of such problems are reported by Kundur et al. (1985); Chow et al. (1989); and Kundur (1981).

Over the course of a frequency instability, the characteristic times of the processes and devices that are activated by the large shifts in frequency and other system variables will range from a matter of seconds, corresponding to the responses of devices such as generator controls and protections, to several minutes, corresponding to the responses of devices such as prime mover energy supply systems and load voltage regulators.

Although frequency stability is impacted by fast as well as slow dynamics, the overall time frame of interest extends to several minutes. Therefore, it is categorized as a long-term phenomenon in Fig. 11.1.

Comments on Classification

The classification of stability has been based on several considerations so as to make it convenient for identification of the causes of instability, the application of suitable analysis tools, and the development of corrective measures appropriate for a specific stability problem. There clearly is some overlap between the various forms of instability, since as systems fail, more than one form of instability may ultimately emerge. However, a system event should be classified based primarily on the dominant initiating phenomenon, separated into those related primarily with voltage, rotor angle, or frequency.

While classification of power system stability is an effective and convenient means to deal with the complexities of the problem, the overall stability of the system should always be kept in mind. Solutions to stability problems of one category should not be at the expense of another. It is essential to look at all aspects of the stability phenomena, and at each aspect from more than one viewpoint.

Historical Review of Stability Problems

As electric power systems have evolved over the last century, different forms of instability have emerged as being important during different periods. The methods of analysis and resolution of stability problems were influenced by the prevailing developments in computational tools, stability theory, and power system control technology. A review of the history of the subject is useful for a better understanding of the electric power industry's practices with regard to system stability.

Power system stability was first recognized as an important problem in the 1920s (Steinmetz, 1920; Evans and Bergvall, 1924; Wilkins, 1926). The early stability problems were associated with remote power plants feeding load centers over long transmission lines. With slow exciters and noncontinuously acting voltage regulators, power transfer capability was often limited by steady-state as well as transient rotor angle instability due to insufficient synchronizing torque. To analyze system stability, graphical techniques such as the equal area criterion and power circle diagrams were developed. These methods were successfully applied to early systems which could be effectively represented as two machine systems.

As the complexity of power systems increased, and interconnections were found to be economically attractive, the complexity of the stability problems also increased and systems could no longer be treated as two machine systems. This led to the development in the 1930s of the network analyzer, which was capable of power flow analysis of multimachine systems. System dynamics, however, still had to be analyzed by solving the swing equations by hand using step-by-step numerical integration. Generators were represented by the classical "fixed voltage behind transient reactance" model. Loads were represented as constant impedances.

Improvements in system stability came about by way of faster fault clearing and fast acting excitation systems. Steady-state aperiodic instability was virtually eliminated by the implementation of continuously acting voltage regulators. With increased dependence on controls, the emphasis of stability studies moved from transmission network problems to generator problems, and simulations with more detailed representations of synchronous machines and excitation systems were required.

The 1950s saw the development of the analog computer, with which simulations could be carried out to study in detail the dynamic characteristics of a generator and its controls rather than the overall behavior of multimachine systems. Later in the 1950s, the digital computer emerged as the ideal means to study the stability problems associated with large interconnected systems.

In the 1960s, most of the power systems in the U.S. and Canada were part of one of two large interconnected systems, one in the east and the other in the west. In 1967, low capacity HVDC ties were also established between the east and west systems. At present, the power systems in North America form virtually one large system. There were similar trends in growth of interconnections in other countries. While interconnections result in operating economy and increased reliability through mutual assistance, they contribute to increased complexity of stability problems and increased consequences of instability. The Northeast Blackout of November 9, 1965, made this abundantly clear; it focused the attention of the public and of regulatory agencies, as well as of engineers, on the problem of stability and importance of power system reliability.

Until recently, most industry effort and interest has been concentrated on *transient (rotor angle) stability*. Powerful transient stability simulation programs have been developed that are capable of modeling large complex systems using detailed device models. Significant improvements in transient stability performance of power systems have been achieved through use of high-speed fault clearing, high-response exciters, series capacitors, and special stability controls and protection schemes.

The increased use of high response exciters, coupled with decreasing strengths of transmission systems, has led to an increased focus on *small signal (rotor angle) stability*. This type of angle instability is often seen as local plant modes of oscillation, or in the case of groups of machines interconnected by weak links, as interarea modes of oscillation. Small signal stability problems have led to the development of special study techniques, such as modal analysis using eigenvalue techniques (Martins, 1986; Kundur et al., 1990). In addition, supplementary control of generator excitation systems, static Var compensators, and HVDC converters is increasingly being used to solve system oscillation problems. There has also been a general interest in the application of power electronic based controllers referred to as FACTS (Flexible AC Transmission Systems) controllers for damping of power system oscillations (IEEE, 1996).

In the 1970s and 1980s, frequency stability problems experienced following major system upsets led to an investigation of the underlying causes of such problems and to the development of long term dynamic simulation programs to assist in their analysis (Davidson et al., 1975; Converti et al., 1976; Stubbe et al., 1989; Inoue et al., 1995; Ontario Hydro, 1989). The focus of many of these investigations was on the performance of thermal power plants during system upsets (Kundur et al., 1985; Chow et al., 1989; Kundur, 1981; Younkins and Johnson, 1981). Guidelines were developed by an IEEE Working Group for enhancing power plant response during major frequency disturbances (1983). Analysis and modeling needs of power systems during major frequency disturbances was also addressed in a recent CIGRE Task Force report (1999).

Since the late 1970s, voltage instability has been the cause of several power system collapses worldwide (Kundur, 1994; Taylor, 1994; IEEE, 1990). Once associated primarily with weak radial distribution systems, voltage stability problems are now a source of concern in highly developed and mature networks as a result of heavier loadings and power transfers over long distances. Consequently, voltage stability is increasingly being addressed in system planning and operating studies. Powerful analytical tools are available for its analysis (Van Cutsem et al., 1995; Gao et al., 1992; Morison et al., 1993), and well-established criteria and study procedures are evolving (Abed, 1999; Gao et al., 1996).

Clearly, the evolution of power systems has resulted in more complex forms of instability. Present-day power systems are being operated under increasingly stressed conditions due to the prevailing trend to make the most of existing facilities. Increased competition, open transmission access, and construction and environmental constraints are shaping the operation of electric power systems in new ways. Planning and operating such systems require examination of all forms of stability. Significant advances have been made in recent years in providing the study engineers with a number of powerful tools and techniques. A coordinated set of complementary programs, such as the one described by Kundur et al. (1994) makes it convenient to carry out a comprehensive analysis of power system stability.

Consideration of Stability in System Design and Operation

For reliable service, a power system must remain intact and be capable of withstanding a wide variety of disturbances. Owing to economic and technical limitations, no power system can be stable for all possible disturbances or contingencies. In practice, power systems are designed and operated so as to be stable for a selected list of contingencies, normally referred to as "design contingencies" (Kundur, 1994).

Experience dictates their selection. The contingencies are selected on the basis that they have a significant probability of occurrence and a sufficiently high degree of severity, given the large number of elements comprising the power system. The overall goal is to strike a balance between costs and benefits of achieving a selected level of system security.

While security is primarily a function of the physical system and its current attributes, secure operation is facilitated by:

- Proper selection and deployment of preventive and emergency controls.
- Assessing stability limits and operating the power system within these limits.

Security assessment has been historically conducted in an off-line operation planning environment in which stability for the near-term forecasted system conditions is exhaustively determined. The results of stability limits are loaded into look-up tables which are accessed by the operator to assess the security of a prevailing system operating condition.

In the new competitive utility environment, power systems can no longer be operated in a very structured and conservative manner; the possible types and combinations of power transfer transactions may grow enormously. The present trend is, therefore, to use online dynamic security assessment. This is feasible with today's computer hardware and stability analysis software.

Acknowledgment

The classification of power system stability presented in this section is based on the report currently under preparation by a joint CIGRE-IEEE Task Force on Power System Stability Terms, Classification, and Definitions.

References

Abed, A.M., WSCC voltage stability criteria, undervoltage load shedding strategy, and reactive power reserve monitoring methodology, in *Proceedings of the 1999 IEEE PES Summer Meeting*, Edmonton, Alberta, 191, 1999.

Chow, Q.B., Kundur, P., Acchione, P.N., and Lautsch, B., Improving nuclear generating station response for electrical grid islanding, *IEEE Trans.*, EC-4, 3, 406, 1989.

Report of CIGRE Task Force 38.02.14, Analysis and modelling needs of power systems under major frequency disturbances, 1999.

Converti, V., Gelopulos, D.P., Housely, M., and Steinbrenner, G., Long-term stability solution of interconnected power systems, *IEEE Trans.*, PAS 95, 1, 96, 1976.

Davidson, D.R., Ewart, D.N., and Kirchmayer, L.K., Long term dynamic response of power systems — an analysis of major disturbances, *IEEE Trans.*, PAS-94, 819, 1975.

EPRI Report EL-6627, Long-term dynamics simulation: Modeling requirements, Final Report of Project 2473-22, Prepared by Ontario Hydro, 1989.

Evans, R.D. and Bergvall, R.C., Experimental analysis of stability and power limitations, *AIEE Trans.*, 39, 1924.

Gao, B., Morison, G.K., and Kundur, P., Towards the development of a systematic approach for voltage stability assessment of large scale power systems, *IEEE Trans. on Power Systems*, 11, 3, 1314, 1996.

Gao, B., Morison, G.K., and Kundur, P., Voltage stability evaluation using modal analysis, *IEEE Trans.* PWRS-7, 4, 1529, 1992.

IEEE PES Special Publication, FACTS Applications, Catalogue No. 96TP116-0, 1996.

IEEE Special Publication 90TH0358-2-PWR, *Voltage Stability of Power Systems: Concepts, Analytical Tools and Industry Experience*, 1990.

IEEE Working Group, Guidelines for enhancing power plant response to partial load rejections, *IEEE Trans.*, PAS-102, 6, 1501, 1983.

Inoue, T., Ichikawa, T., Kundur, P., and Hirsch, P., Nuclear plant models for medium- to long-term power system stability studies, *IEEE Trans. on Power Systems*, 10, 141, 1995.

Kundur, P., *Power System Stability and Control*, McGraw-Hill, New York, 1994.

Kundur, P., A survey of utility experiences with power plant response during partial load rejections and system disturbances, *IEEE Trans.*, PAS-100, 5, 2471, 1981.

Kundur, P., Morison, G.K., and Balu, N.J., A comprehensive approach to power system analysis, CIGRE Paper 38-106, presented at the 1994 Session, Paris, France.

Kundur, P. and Morison, G.K., A review of definitions and classification of stability problems in today's power systems, Paper presented at the *Panel Session on Stability Terms and Definitions, IEEE PES Winter Meeting*, New York, 1997.

Kundur, P., Rogers, G.J., Wong, D.Y., Wang, L. and Lauby, M.G., A comprehensive computer program package for small signal stability analysis of power systems, *IEEE Trans. on Power Systems*, 5, 1076, 1990.

Kundur, P., Lee, D.C., Bayne, J.P., and Dandeno, P.L., Impact of turbine generator controls on unit performance under system disturbance conditions, *IEEE Trans.* PAS-104, 1262, 1985.

Martins, N., Efficient eigenvalue and frequency response methods applied to power system small-signal stability studies, *IEEE Trans.*, PWRS-1, 217, 1986.

Morison, G.K., Gao, B., and Kundur, P., Voltage stability analysis using static and dynamic approaches, *IEEE Trans. on Power Systems*, 8, 3, 1159, 1993.

Steinmetz, C.P., Power control and stability of electric generating stations, *AIEE Trans.*, XXXIX, 1215, 1920.

Stubbe, M., Bihain, A., Deuse, J., and Baader, J.C., STAG a new unified software program for the study of dynamic behavior of electrical power systems, *IEEE Trans. on Power Systems*, 4, 1, 1989.

Taylor, C.W., *Power System Voltage Stability*, McGraw-Hill, New York, 1994.

Van Cutsem, T., Jacquemart, Y., Marquet, J.N., and Pruvot, P., A comprehensive analysis of mid-term, voltage stability, *IEEE Trans. on Power Systems*, 10, 1173, 1995.

Van Cutsem, T. and Vournas, C., *Voltage Stability of Electric Power Systems*, Kluwer Academic Publishers, Dordrecht, The Netherlands, 1998.

Wilkins, R., Practical aspects of system stability, *AIEE Trans.*, 41, 1926.

Younkins, T.D. and Johnson, L.H., Steam turbine overspeed control and behavior during system disturbances, *IEEE Trans.*, PAS-100, 5, 2504, 1981.

11.2 Transient Stability

Kip Morrison

As discussed in Seciton 11.1, power system stability was recognized as a problem as far back as the 1920s at which time the characteristic structure of systems consisted of remote power plants feeding load centers over long distances. These early stability problems, often a result of insufficient synchronizing torque, were the first emergence of transient instability. As defined in the previous section, *transient stability* is the ability of a power system to remain in synchronism when subjected to large transient disturbances. These disturbances may include faults on transmission elements, loss of load, loss of generation, or loss of system components such as transformers or transmission lines.

Although many different forms of power system stability have emerged and become problematic in recent years, transient stability still remains a basic and important consideration in power system design and operation. While it is true that the operation of many power systems is limited by phenomena such as voltage stability or small-signal stability, most systems are prone to transient instability under certain conditions or contingencies and hence the understanding and analysis of transient stability remain fundamental issues. Also, we shall see later in this section that transient instability can occur in a very short time frame (a few seconds), leaving no time for operator intervention to mitigate problems. It is therefore essential to deal with the problem in the design stage or severe operating restrictions may result.

This section includes a discussion of the basic principles of transient stability, methods of analysis, control and enhancement, and practical aspects of its influence on power system design and operation.

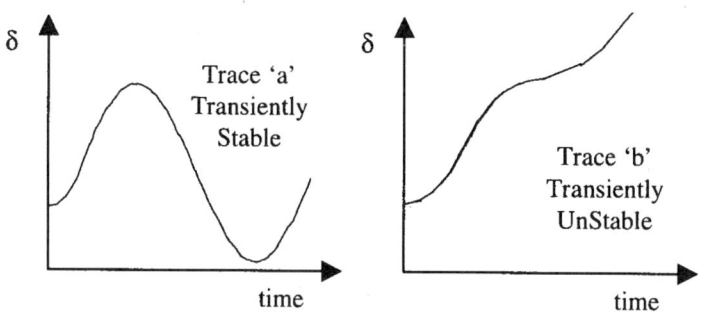

FIGURE 11.2 Typical swing curves.

Basic Theory of Transient Stability

Most power system engineers are familiar with plots of generator rotor angle (δ) versus time as shown in Fig. 11.2. These "swing curves" plotted for a generator subjected to a particular system disturbance show whether a generator rotor angle recovers and oscillates around a new equilibrium point as in trace "a" or whether it increases aperiodically such as in trace "b". The former case is deemed to be transiently stable, and the latter case transiently unstable. What factors determine whether a machine will be stable or unstable? How can the stability of large power systems be analyzed? If a case is unstable, what can be done to enhance stability? These are some of the questions discussed in this section.

Two concepts are essential in understanding transient stability: (i) the swing equation and (ii) the power-angle relationship. These can be used together to describe the equal area criterion, a simple graphical approach to assessing transient stability.

The Swing Equation

In a synchronous machine, the prime mover exerts a mechanical torque T_m on the shaft of the machine and the machine produces an electromagnetic torque T_e. If, as a result of a disturbance, the mechanical torque is greater than the electromagnetic torque, an accelerating torque T_a exists and is given by:

$$T_a = T_m - T_e \tag{11.1}$$

This ignores the other torques caused by friction, core loss, and windage in the machine. T_a has the effect of accelerating the machine which has an inertia J (kg·m²) made up of the inertia of the generator and the prime mover and, therefore,

$$J \frac{d\omega_m}{dt} = T_a = T_m - T_e \tag{11.2}$$

where t is time in seconds and ω_m is the angular velocity of the machine rotor in mechanical rad/s. It is common practice to express this equation in terms of the inertia constant H of the machine. If ω_{0m} is the rated angular velocity in mechanical rad/s, J can be written as:

$$J = \frac{2H}{\omega_{0m}^2} VA_{base} \tag{11.3}$$

Therefore,

$$\frac{2H}{\omega_{0m}^2} VA_{base} \frac{d\omega_m}{dt} = T_m - T_e \tag{11.4}$$

And now, if ω_r denotes the angular velocity of the rotor (rad/s) and ω_0 its rated value, the equation can be written as:

$$2H\frac{d\overline{\omega}_r}{dt} = \overline{T}_m - \overline{T}_e \qquad (11.5)$$

Finally it can be shown that

$$\frac{d\overline{\omega}_r}{dt} = \frac{d^2\delta}{\omega_0 dt^2} \qquad (11.6)$$

where δ is the angular position of the rotor (elec. rad/s) with respect to a synchronously rotating reference frame.

Combining Eqs. (11.5) and (11.6) results in the *swing equation* [Eq. (11.7)], so-called because it describes the swings of the rotor angle δ during disturbances.

$$\frac{2H}{\omega_0}\frac{d^2\delta}{dt^2} = \overline{T}_m - \overline{T}_e \qquad (11.7)$$

An additional term $(-K_D\Delta\overline{\omega}_r)$ may be added to the right side of [Eq. (11.7)] to account for a component of damping torque not included explicitly in T_e.

For a system to be *transiently stable* during a disturbance, it is necessary for the rotor angle (as its behavior is described by the swing equation) to oscillate around an equilibrium point. If the rotor angle increases indefinitely, the machine is said to be *transiently unstable* as the machine continues to accelerate and does not reach a new state of equilibrium. In multimachine systems, such a machine will "pull out of step" and lose synchronism with the rest of the machines.

The Power-Angle Relationship

Consider a simple model of a single generator connected to an infinite bus through a transmission system as shown in Fig. 11.3. The model can be reduced as shown by replacing the generator with a constant voltage behind a transient reactance (classical model). It is well known that there is a maximum power that can be transmitted to the infinite bus in such a network. The relationship between the electrical power of the generator P_e and the rotor angle of the machine δ is given by,

$$P_e = \frac{E'E_B}{X_T}\sin\delta = P_{max}\sin\delta \qquad (11.8)$$

where

$$P_{max} = \frac{E'E_B}{X_T} \qquad (11.9)$$

Equation (11.8) can be shown graphically as Fig. 11.4 from which it can be seen that as the power initially increases, δ increases until reaching 90° when P_e reaches its maximum. Beyond $\delta = 90°$, the power decreases until at $\delta = 180°$, $P_e = 0$. This is the so-called power-angle relationship and describes the transmitted power as a function of rotor angle. It is clear from Eq. (11.9) that the maximum power is a function of the voltages of the generator and infinite bus, and more importantly, a function of the transmission system reactance; the larger the reactance (for example, the longer or weaker the transmission circuits), the lower the maximum power.

Power System Dynamics and Stability

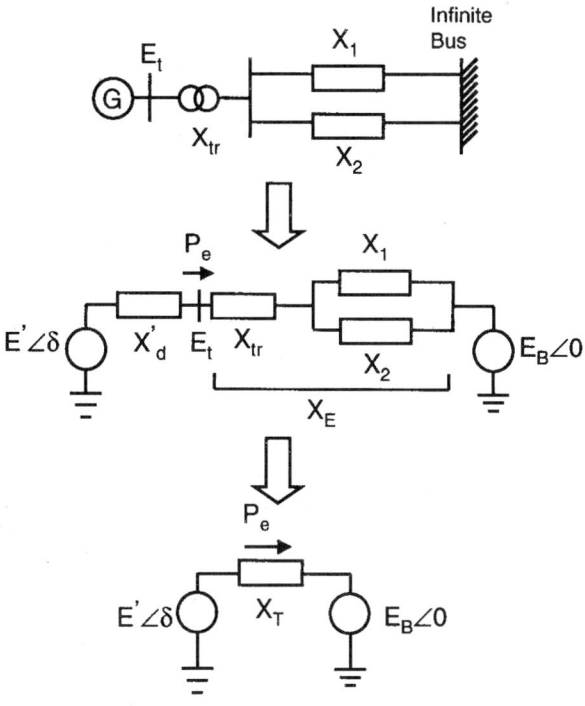

FIGURE 11.3 Single machine system.

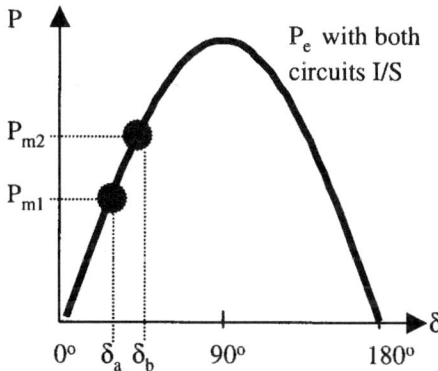

FIGURE 11.4 Power-angle relationship with both circuits in service.

Figure 11.4 shows that for a given input power to the generator P_{m1}, the electrical output power is P_e (equal to P_m) and the corresponding rotor angle is δ_a. As the mechanical power is increased to P_{m2}, the rotor angle advances to δ_b. Figure 11.5 shows the case with one of the transmission lines removed causing an increase in X_T and a reduction P_{max}. It can be seen that for the same mechanical input (P_{m1}), the situation with one line removed causes an increase in rotor angle to δ_c.

The Equal Area Criterion

By combining the dynamic behavior of the generator as defined by the swing equation, with the power-angle relationship, it is possible to illustrate the concept of transient stability using the *equal area criterion*.

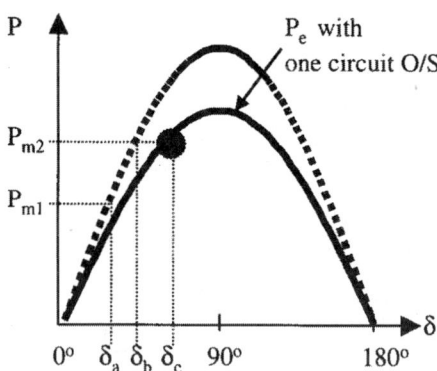

FIGURE 11.5 Power-angle relationship with one circuit out of service.

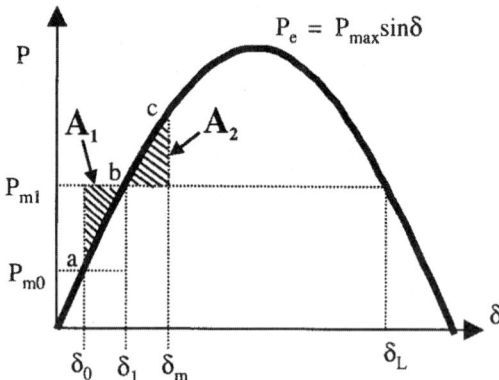

FIGURE 11.6 Equal area criterion for step change in mechanical power.

Consider Fig. 11.6 in which a step change is applied to the mechanical input of the generator. At the initial power P_{m0}, $\delta = \delta_0$ and the system is at operating point "a". As the power is increased in a step to P_{m1} (accelerating power = $P_{m1} = P_e$), the rotor cannot accelerate instantaneously, but traces the curve up to point "b," at which time $P_e = P_{m1}$ and the accelerating power is zero. However, the rotor speed is greater than the synchronous speed and the angle continues to increase. Beyond "b," $P_e > P_m$ and the rotor decelerates until reaching a maximum δ_{max} at which point the rotor angle starts to return towards "b."

As we will see, for a single machine infinite bus system, it is not necessary to plot the swing curve to determine if the rotor angle of the machine increases indefinitely, or if it oscillates around an equilibrium point. The equal area criterion allows stability to be determined using graphical means. While this method is not generally applicable to multi-machine systems, it is a valuable learning aid.

Starting with the swing equation as given by Eq. (11.7) and interchanging per unit power for torque,

$$\frac{d^2\delta}{dt^2} = \frac{\omega_0}{2H}(P_m - P_e) \tag{11.10}$$

Multiplying both sides by $2\delta/dt$ and integrating gives

$$\left[\frac{d\delta}{dt}\right]^2 = \int_{\delta_0}^{\delta} \frac{\omega_0(P_m - P_e)}{H} d\delta \quad \text{or} \quad \frac{d\delta}{dt} = \sqrt{\int_{\delta_0}^{\delta} \frac{\omega_0(P_m - P_e)}{H} d\delta} \tag{11.11}$$

δ_0 represents the rotor angle when the machine is operating synchronously prior to any disturbance. It is clear that for the system to be stable, δ must increase, reach a maximum (δ_{max}), and then change direction as the rotor returns to complete an oscillation. This means that $d\delta/dt$ (which is initially zero) changes during the disturbance, but must, at a time corresponding to δ_{max}, become zero again. Therefore, as a stability criterion,

$$\int_{\delta_0}^{\delta} \frac{\omega_0}{H}\left(P_m - P_e\right)d\delta = 0. \quad (11.12)$$

This implies that the area under the function $P_m - P_e$ plotted against δ must be zero for a stable system, which requires Area 1 to be equal to Area 2. Area 1 represents the energy gained by the rotor during acceleration and Area 2 represents energy lost during deceleration.

Figures 11.7 and 11.8 show the rotor response (defined by the swing equation) superimposed on the power-angle curve for a stable case and an unstable case, respectively. In both cases, a three-phase fault is applied to the system given in Fig. 11.3. The only difference in the two cases is that the fault clearing time has been increased for the unstable case. The arrows show the trace of the path followed by the rotor angle in terms of the swing equation and power-angle relationship. It can be seen that for the stable case, the energy gained during rotor acceleration is equal to the energy dissipated during deceleration ($A_1 = A_2$) and the rotor angle reaches a maximum and recovers. In the unstable case, however, it can be seen that the energy gained during acceleration is greater than that dissipated during deceleration (since the fault is applied for a longer duration), meaning that $A_1 > A_2$ and the rotor continues to advance and does not recover.

Methods of Analysis of Transient Stability
Modeling
The basic concepts of transient stability presented above are based on highly simplified models. Practical power systems consist of large numbers of generators, transmission circuits, and loads.

For stability assessment, the power system is normally represented using a positive sequence model. The network is represented by a traditional positive sequence powerflow model that defines the transmission topology, line reactances, connected loads and generation, and predisturbance voltage profile.

Generators can be represented with various levels of detail, selected based on such factors as length of simulation, severity of disturbance, and accuracy required. The most basic model for synchronous generators consists of a constant internal voltage behind a constant transient reactance, and the rotating inertia constant (H). This is the so-called classical representation that neglects a number of characteristics: the action of voltage regulators, variation of field flux linkage, the impact of the machine physical construction on the transient reactances for the direct and quadrature axis, the details of the prime mover or load, and saturation of the magnetic core iron. Historically, classical modeling was used to reduce computational burden associated with more detailed modeling, which is not generally a concern with today's simulation software and computer hardware. However, it is still often used for machines that are very remote from a disturbance (particularly in very large system models) and where more detailed model data is not available.

In general, synchronous machines are represented using detailed models that capture the effects neglected in the classical model, including the influence of generator construction (damper windings, saturation, etc.), generator controls, (excitation systems including power system stabilizers, etc.), the prime mover dynamics, and the mechanical load. Loads, which are most commonly represented as static voltage and frequency-dependent components, may also be represented in detail by dynamic models that capture their speed torque characteristics and connected loads. There are a myriad of other devices, such as HVDC lines and controls and static Var devices, which may require detailed representation. Finally,

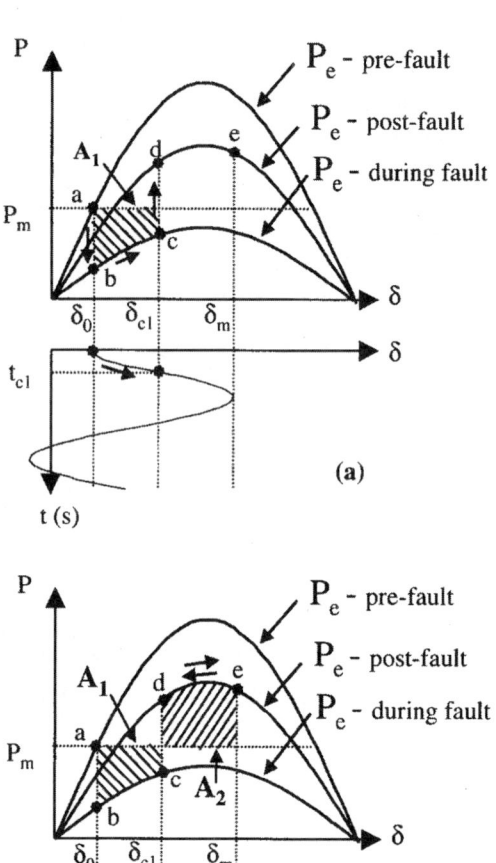

FIGURE 11.7 Equal area criterion for stable case $A_1 = A_2$. (a) Acceleration of rotor. (b) Deceleration of rotor.

system protections are often represented. Models may also be included for line protections (such as mho distance relays), out-of-step protections, loss of excitation protections, or special protection schemes.

Although power system models may be extremely large, representing thousands of generators and other devices producing systems with tens of thousands of system states, efficient numerical methods combined with modern computing power have made time-domain simulation readily available in many commercially available computer programs. It is also important to note that the time frame in which transient instability occurs is usually in the range of 1 to 5 sec, so that simulation times need not be excessively long.

Analytical Methods

To accurately assess the system response following disturbances, detailed models are required for all critical elements. The complete mathematical model for the power system consists of a large number of algebraic and differential equations, including

- Generators stator algebraic equations
- Generator rotor circuit differential equations

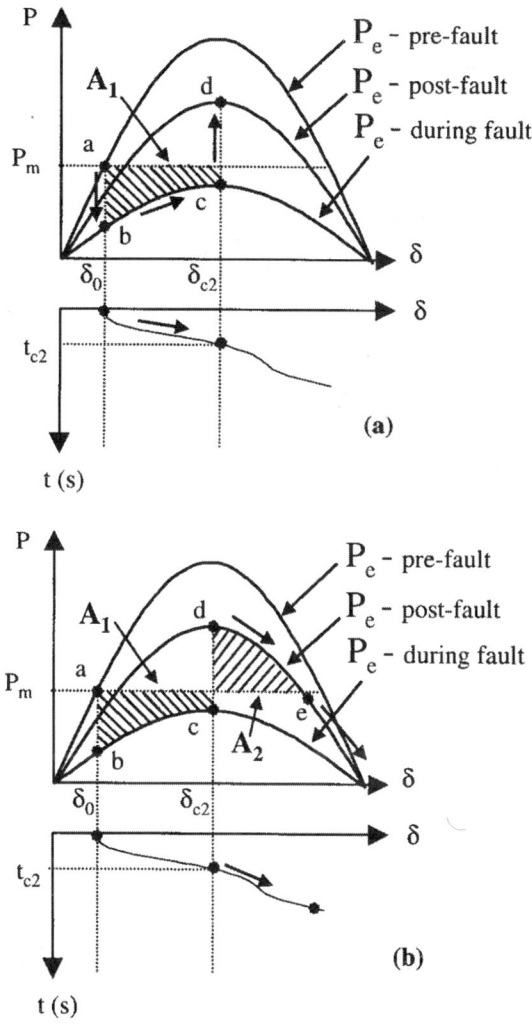

FIGURE 11.8 Equal area criterion for unstable case $A_1 > A_2$. (a) Acceleration of rotor. (b) Deceleration of rotor.

- Swing equations
- Excitation system differential equations
- Prime mover and governing system differential equations
- Transmission network algebraic equations
- Load algebraic and differential equations

While considerable work has been done on *direct methods* of stability analysis in which stability is determined without explicitly solving the system differential equations (see Section 11.5), the most practical and flexible method of transient stability analysis is *time-domain simulation* using step-by-step numerical integration of the nonlinear differential equations. A variety of numerical integration methods are used, including *explicit* methods (such as Euler and Runge-Kutta methods) and *implicit* methods (such as the trapezoidal method). The selection of the method to be used largely depends on the stiffness of the system being analyzed. Implicit methods are generally better suited than explicit methods for systems in which time steps are limited by numerical stability rather than accuracy.

Simulation Studies

Modern simulation tools offer sophisticated modeling capabilities and advanced numerical solution methods. Although simulation tools differ somewhat, the basic requirements and functions are the same.

Input data:
1. Powerflow: Defines system topology and initial operating state.
2. Dynamic data: Includes model types and associated parameters for generators, motors, protections, and other dynamic devices and their controls.
3. Program control data: Specifies such items as the type of numerical integration to use and time-step.
4. Switching data: Includes the details of the disturbance to be applied. This includes the time at which the fault is applied, where the fault is applied, the type of fault and its fault impedance if required, the duration of the fault, the elements lost as a result of the fault, and the total length of the simulation.
5. System monitoring data: This specifies which quantities are to be monitored (output) during the simulation. In general, it is not practical to monitor all quantities because system models are large and recording all voltages, angles, flows, generator outputs, etc., at each integration time step would create an enormous volume. Therefore, it is common practice to define a limited set of parameters to be recorded.

Output data:
1. Simulation log: This contains a listing of the actions that occurred during the simulation. It includes a recording of the actions taken to apply the disturbance and reports on any operation of protections or controls, or any numerical difficulty encountered.
2. Results output: This is an ASCII or binary file that contains the recording of each monitored variable over the duration of the simulation. These results are examined, usually through a graphical plotting, to determine if the system remained stable and to assess the details of the dynamic behavior of the system.

Factors Influencing Transient Stability

Many factors affect the transient stability of a generator in a practical power system. From the small system analyzed above, the following factors can be identified.

- The post-disturbance system reactance as seen from the generator. The weaker the post-disturbance system, the lower P_{max} will be.
- The duration of the fault clearing time. The longer the fault is applied, the longer the rotor will be accelerated and the more kinetic energy will be gained. The more energy that is gained during acceleration, the more difficult it is to dissipate it during deceleration.
- The inertia of the generator. The higher the inertia, the slower the rate of change of angle and the less the kinetic energy gained during the fault.
- The generator internal voltage (determined by excitation system) and infinite bus voltage (system voltage). The lower these voltages, the lower P_{max} will be.
- The generator loading prior to the disturbance. The higher the loading, the closer the unit will be to P_{max}, which means that during acceleration, it is more likely to become unstable.
- The generator internal reactance. The lower the reactance, the higher the peak power and the lower the initial rotor angle.
- The generator output during the fault. This is a function of the fault location and type of fault.

Transient Stability Considerations in System Design

As outlined previously, transient stability is an important consideration that must be dealt with during the design of power systems. In the design process, time-domain simulations are conducted to assess the stability of the system under various conditions and when subjected to various disturbances. Since it is not practical to design a system to be stable under all possible disturbances, design criteria specify the disturbances for which the system must be designed to be stable. The criteria disturbances generally consist of the more statistically probable events which could cause the loss of any system element and typically include three-phase faults cleared in normal time and line-to-ground faults with delayed clearing due to breaker failure. In most cases, stability is assessed for the loss of one element (such as a transformer or transmission circuit) with possibly one element out-of-service predisturbance.

Therefore, in system design, a wide number of disturbances are assessed and if the system is found to be unstable (or marginally stable), a variety of actions can be taken to improve stability. These include the following.

- *Reduction of transmission system reactance*: This can be achieved by adding additional parallel transmission circuits, providing series compensation on existing circuits, and by using transformers with lower leakage reactances.
- *High-speed fault clearing*: In general, two-cycle breakers are used in locations where faults must be removed quickly to maintain stability. As the speed of fault clearing decreases, so does the amount of kinetic energy gained by the generators during the fault.
- *Dynamic braking*: Shunt resistors can be switched in following a fault to provide an artificial electrical load. This increases the electrical output of the machines and reduces the rotor acceleration.
- *Regulate shunt compensation*: By maintaining system voltages around the power system, the flow of synchronizing power between generators is improved.
- *Reactor switching*: The internal voltages of generators, and therefore stability, can be increased by connected shunt reactors.
- *Single pole switching*: Most power system faults are of the single-line-to-ground type. However, in most schemes, this type of fault will trip all three phases. If single pole switching is used, only the faulted phase is removed and power can flow on the remaining two phases, thereby greatly reducing the impact of the disturbance.
- *Steam turbine fast-valving:* Steam valves are rapidly closed and opened to reduce the generator accelerating power in response to a disturbance.
- *Generator tripping*: Perhaps one of the oldest and most common methods of improving transient stability, this approach disconnects selected generators in response to a disturbance. This has the effect of reducing the power that is required to be transferred over critical transmission interfaces.
- *High-speed excitation systems*: As illustrated by the simple examples presented earlier, increasing the internal voltage of a generator has the effect of improving transient stability. This can be achieved by fast-acting excitation systems that can rapidly boost field voltage in response to disturbances.
- *Special excitation system controls:* It is possible to design special excitation systems that can use discontinuous controls to provide special field boosting during the transient period, thereby improving stability.
- *Special control of HVDC links*: The DC power on HVDC links can be rapidly ramped up or down to assist in maintaining generation/load imbalances caused by disturbances. The effect is similar to generation or load tripping.
- *Controlled system separation and load shedding*: Generally considered a last resort, it is often feasible to design system controls that can respond to separate, or island, a power system into areas with balanced generation and load. Some load shedding or generation tripping may also be required in selected islands. In the event of a disturbance, instability can be prevented from propagating and affecting large areas by partitioning the system in this manner. If instability primarily results in generation loss, load shedding alone may be sufficient to control the system.

Transient Stability Considerations in System Operation

While it is true that power systems are designed to be transiently stable, and many of the methods described above may be used to achieve this goal, in actual practice, systems may be prone to instability. This is largely due to uncertainties related to assumptions made during the design process. These uncertainties result from a number of sources, including:

- *Load and generation forecast:* The design process must use forecast information about the amount, distribution, and characteristics of the connected loads, as well as the location and amount of connected generation. These all have a great deal of uncertainty. If the actual system load is higher than planned, the generation output will be higher, the system will be more stressed, and the transient stability limit may be significantly lower.
- *System topology:* Design studies generally assume all elements in service, or perhaps up to two elements out of service. In actual systems, there are usually many elements out of service at any one time due to forced outages (failures) or system maintenance. Clearly, these outages can seriously weaken the system and make it less transiently stable.
- *Dynamic modeling:* All models used for power system simulation, even the most advanced, contain approximations out of practical necessity.
- *Dynamic data:* The results of time-domain simulations depend heavily on the data used to represent the models for generators and the associated controls. In many cases this data is not known (typical data is assumed) or is in error (either because it has not been derived from field measurements or due to changes that have been made in the actual system controls that have not been reflected in the data).
- *Device operation:* In the design process it is assumed that controls and protection will operate as designed. In the actual system, relays, breakers, and other controls may fail or operate improperly.

To deal with these uncertainties in actual system operation, safety margins are used. Operational (short term) time-domain simulations are conducted using a system model that is more accurate (by accounting for elements out on maintenance, improved short-term load forecast, etc.) than the design model. *Transient stability limits* are computed using these models. The limits are generally in terms of maximum flows allowable over critical interfaces, or maximum generation output allowable from critical generating sources. *Safety margins* are then applied to these computed limits. This means that actual system operation is restricted to levels (interface flows or generation) *below* the stability limit by an amount equal to a defined safety margin. In general, the margin is expressed in terms of a percentage of the critical flow or generation output. For example, operation procedure might be to define the *operating limit* as 10% below the stability limit.

References

Elgerd, O. I., *Electric Energy Systems Theory: An Introduction*, McGraw-Hill, New York, 1971.
IEEE Recommended Practice for Industrial and Commercial Power System Analysis, IEEE Std. 399-1997, IEEE 1998.
Kundur, P., *Power System Stability and Control*, McGraw-Hill, New York, 1994.
Stevenson, W. D., *Elements of Power System Analysis*, 3rd ed., McGraw-Hill, New York, 1975.

11.3 Small Signal Stability and Power System Oscillations

John Paserba, Prabha Kundur, Juan Sanchez-Gasca, Einar Larsen

Nature of Power System Oscillations

Historical Perspective

Damping of oscillations has been recognized as important in electric power system operation from the beginning. Indeed before there were any power systems, oscillations in automatic speed controls (governors) initiated an analysis by J.C. Maxwell (speed controls were found necessary for the successful operation of the first steam engines). Aside from the immediate application of Maxwell's analysis, it also had a lasting influence as at least one of the stimulants to the development by E.J. Routh in 1883 of his very useful and widely used method to enable one to determine theoretically the stability of a high-order dynamic system without having to know the roots of its equations (Maxwell analyzed only a second-order system).

Oscillations among generators appeared as soon as AC generators were operated in parallel. These oscillations were not unexpected, and in fact, were predicted from the concept of the power vs. phase-angle curve gradient interacting with the electric generator rotary inertia, forming an equivalent mass-and-spring system. With a continually varying load and some slight differences in the design and loading of the generators, oscillations tended to be continually excited. Particularly in the case of hydro-generators there was very little damping, and so amortisseurs (damper windings) were installed, at first as an option. (There was concern about the increased short-circuit current, and some people had to be persuaded to accept them (Crary and Duncan, 1941).) It is of interest to note that although the only significant source of actual negative damping here was the turbine speed governor (Concordia, 1969), the practical "cure" was found elsewhere. Two points are evident and are still valid. First, automatic control is practically the only source of negative damping, and second, although it is obviously desirable to identify the sources of negative damping, the most effective and economical place to add damping may lie elsewhere.

After these experiences, oscillations seemed to disappear as a major problem. Although there were occasional cases of oscillations and evidently poor damping, the major analytical effort seemed to ignore damping entirely. First using analog, then digital computing aids, analysis of electric power system dynamic performance was extended to very large systems, but still representing the generators (and, for that matter, also the loads) in the simple "classical" way. Most studies covered only a short time period, and as occasion demanded, longer-term simulations were kept in bound by including empirically estimated damping factors. It was, in effect, tacitly assumed that the net damping was positive.

All this changed rather suddenly in the 1960s when the process of interconnection accelerated and more transmission and generation extended over large areas. Perhaps the most important aspect was the wider recognition of the negative damping produced by the use of high-response generator voltage regulators in situations where the generator may be subject to relatively large angular swings, as may occur in extensive networks. (This possibility was already well known in the 1930s and 1940s but had not had much practical application.) With the growth of extensive power systems, and especially with the interconnection of these systems by ties of limited capacity, oscillations reappeared. (Actually, they had never entirely disappeared but instead were simply not "seen".) There are several reasons for this reappearance.

1. For intersystem oscillations, the amortisseur is no longer effective, as the damping produced is reduced in approximately inverse proportion to the square of the effective external-impedance-plus-stator-impedance, and so it practically disappears.
2. The proliferation of automatic controls has increased the probability of adverse interactions among them. (Even without such interactions, the two basic controls, the speed governor and the generator voltage regulator, practically always produce negative damping for frequencies in the power system oscillation range: the governor effect, small, and the AVR effect, large.)
3. Even though automatic controls are practically the only devices that may produce negative damping, the damping of the uncontrolled system is itself very small and could easily allow the continually changing load and generation to result in unsatisfactory tie-line power oscillations.
4. A small oscillation in each generator that may be insignificant may add up to a tie-line oscillation that is very significant relative to its rating.
5. Higher tie-line loading increases both the tendency to oscillate and the importance of the oscillation.

To calculate the effect of damping on the system, the detail of system representation has to be considerably extended. The additional parameters required are usually much less well known than are the generator inertias and network impedances required for the "classical" studies. Further, the total damping of a power system is typically very small and is made up of both positive and negative components.

Thus, if one wishes to get realistic results, one must include all known sources. These sources include: prime movers, speed governors, electrical loads, circuit resistance, generator amortisseurs, generator excitation, and in fact, all controls that may be added for special purposes. In large networks, and particularly as they concern tie-line oscillations, the only two items that can be depended upon to produce positive damping are the electrical loads and (at least for steam-turbine driven generators) the prime mover.

Although it is obvious that net damping must be positive for stable operation, why be concerned about its magnitude? More damping would reduce (but not eliminate) the tendency to oscillate and the magnitude of oscillations. As pointed out above, oscillations can never be eliminated, as even in the best-damped systems, the damping is small, being only a small fraction of the "critical damping." So the common concept of the power system as a system of masses and springs is still valid, and we have to accept some oscillations. The reasons why they are often troublesome are various, depending on the nature of the system and the operating conditions. For example, when at first a few (or more) generators were paralleled in a rather closely connected system, oscillations were damped by the generator amortisseurs. If oscillations did occur, there was little variation in system voltage. In the simplest case of two generators paralleled on the same bus and equally loaded, oscillations between them would produce practically no voltage variation and what was produced would be principally at twice the oscillation frequency. Thus, the generator voltage regulators were not stimulated and did not participate in the activity. Moreover, the close coupling between the generators reduced the effective regulator gain considerably for the oscillation mode. Under these conditions when voltage regulator response was increased (e.g., to improve transient stability), there was little apparent decrease of system damping (in most cases) but appreciable improvement in transient stability. Instability through negative damping produced by increased voltage-regulator gain had already been demonstrated theoretically (Concordia, 1944).

Consider that the system just discussed is then connected to another similar system by a tie-line. This tie-line should be strong enough to survive the loss of any one generator but may be only a rather small fraction of system capacity. Now, the response of the system to tie-line oscillations is quite different from that just described. Because of the high external impedance seen by either system, not only is the positive damping by the generator amortisseurs largely lost, but the generator terminal voltages become responsive to angular swings. This causes the generator voltage regulators to act, producing negative damping as an unwanted side effect. This sensitivity of voltage-to-angle increases as a strong function of initial angle, and thus, tie-line loading. Thus, in the absence of mitigating means, tie-line oscillations are very likely to occur, especially at heavy line loading (and they have on numerous occasions as illustrated in Chapter 3 of CIGRE Technical Brochure No. 111 [1996]). These tie-line oscillations are bothersome, especially as a restriction on the allowable power transfer, as relatively large oscillations are (quite properly) taken as a precursor to instability.

Next, as interconnection proceeds, another system is added. If the two previously discussed systems are designated A and B, and a third system, C, is connected to B, then a chain A-B-C is formed. If power is flowing $A \rightarrow B \rightarrow C$ or $C \rightarrow B \rightarrow A$, the principal (i.e., lowest frequency) oscillation mode is A against C, with B relatively quiescent. However, as already pointed out, the voltages of system B are varying. In effect, B is acting as a large synchronous condenser facilitating the transfer of power from A to C, and suffering voltage fluctuations as a consequence. This situation has occurred several times in the history of interconnected power systems and has been a serious impediment to progress. In this case, note that the problem is mostly in system B, while the solution (or at least mitigation) will be mostly in systems A and C. It would be practically impossible with any presently conceivable controlled voltage support solely in system B to maintain a satisfactory voltage. On the other hand, without system B for the same power transfer, the oscillations would be much more severe. In fact, the same power transfer might not be possible without, for example, a very high amount of series or shunt compensation. If the power transfer is $A \rightarrow B \leftarrow C$ or $A \leftarrow B \rightarrow C$, the likelihood of severe oscillation (and the voltage variations produced by the oscillations) is much less. Further, both the trouble and the cure are shared by all three systems, so effective compensation is more easily achieved. For best results, all combinations of power transfers should be considered.

Aside from this abbreviated account of how oscillations grew in importance as interconnections grew in extent, it may be of interest to mention the specific case that seemed to precipitate the general acceptance of the major importance of improving system damping, as well as the general recognition of the generator voltage regulator as the major culprit in producing negative damping. This was the series of studies of the transient stability of the Pacific Intertie (AC and DC in parallel) on the west coast of the U.S. In these studies, it was noted that for three-phase faults, instability was determined not by severe first swings of the generators but by oscillatory instability of the post fault system, which had one of two parallel AC line sections removed and thus a higher impedance. This showed that damping is important for transient as well as steady-state stability and contributed to a worldwide rush to apply power system stabilizers (PSS) to all generator voltage regulators as a panacea for all oscillatory ills.

But the pressures of the continuing extension of electric networks and of increases in line loading have shown that the PSS alone is often not enough. When we push to the limit, that limit is more often than not determined by lack of adequate damping. When we add voltage support at appropriate points in the network, we not only increase its "strength" (i.e., increased synchronizing power or smaller transfer impedance), but also improve its damping (if the generator voltage regulators have been producing negative damping) by relieving the generators of a good part of the work of voltage regulation and also reducing the regulator gain. This is so whether or not reduced damping was an objective. However, the limit may still be determined by inadequate damping. How can it be improved? There are at least three options:

1. Add a signal (e.g., line current) to the voltage support device control.
2. Increase the output of the PSS (which is possible with the now stiffer system), or do both as found to be appropriate.
3. Add an entirely new device at an entirely new location. Thus the proliferation of controls, which has to be carefully considered.

Oscillations of power system frequency as a whole can still occur in an isolated system, due to governor deadband or interaction with system frequency control, but is not likely to be a major problem in large interconnected systems. These oscillations are most likely to occur on intersystem ties among the constituent systems, especially if the ties are weak or heavily loaded. This is in a relative sense; an "adequate" tie planned for certain usual line loadings is nowadays very likely to be much more severely loaded and, thus, behave dynamically like a weak line as far as oscillations are concerned, quite aside from losing its emergency pick-up capability. There has always been commercial pressure to utilize a line, perhaps originally planned to aid in maintaining reliability, for economical energy transfer simply because it is there. Now, however, there is also "open access" that may force a utility to use nearly every line for power transfer. This will certainly decrease reliability and may decrease damping, depending on the location of added generation.

Power System Oscillations Classified by Interaction Characteristics

Electric power utilities have experienced problems with the following types of subsynchronous frequency oscillations (Kundur, 1994):

- Local plant mode oscillations
- Interarea mode oscillations
- Torsional mode oscillations
- Control mode oscillations

Local plant mode oscillation problems are the most commonly encountered among the above, and are associated with units at a generating station oscillating with respect to the rest of the power system. Such problems are usually caused by the action of the AVRs of generating units operating at high output and feeding into weak transmission networks; the problem is more pronounced with high response excitation systems. The local plant oscillations typically have natural frequencies in the range of 1 to 2 Hz. Their characteristics are well understood and adequate damping can be readily achieved by using supplementary control of excitation systems in the form of power system stabilizers (PSS).

Interarea modes are associated with machines in one part of the system oscillating against machines in other parts of the system. They are caused by two or more groups of closely coupled machines being interconnected by weak ties. The natural frequency of these oscillations is typically in the range of 0.1 to 1 Hz. The characteristics of interarea modes of oscillation are complex and in some respects significantly differ from the characteristics of local plant modes (CIGRE Technical Brochure No. 111, 1996; Kundur, 1994).

Torsional mode oscillations are associated with the turbine-generator rotational (mechanical) components. There have been several instances of torsional mode instability due to interactions with the generating unit excitation and prime mover controls (Kundur, 1994):

- Torsional mode destabilization by excitation control was first observed in 1969 during the application of power system stabilizers on a 555 MVA fossil-fired unit at the Lambton generating station in Ontario. The PSS, which used a stabilizing signal based on speed measured at the generator end of the shaft was found to excite the lowest torsional (16 Hz) mode. The problem was solved by sensing speed between the two LP turbine sections and by using a torsional filter (Kundur et al., 1981; Watson and Coultes, 1973).
- Instability of torsional modes due to interaction with speed governing systems was observed in 1983 during the commissioning of a 635 MVA unit at Pickering "B" nuclear generating station in Ontario. The problem was solved by providing an accurate linearization of steam valve characteristics and by using torsional filters (Lee et al., 1986).
- Control mode oscillations are associated with the controls of generating units and other equipment. Poorly tuned controls of excitation systems, prime movers, static var compensators, and HVDC converters are the usual causes of instability of control modes. Sometimes it is difficult to tune the controls so as to assure adequate damping of all modes. Kundur et al. (1981) describe the difficulty experienced in tuning the power system stabilizers at the Ontario Hydro's Nanticoke generating station in 1979. The stabilizers used shaft speed signals with torsional filters. With the stabilizer gain high enough to stabilize the local plant mode oscillation, a control mode local to the excitation system and the generator field referred to as the "exciter mode" became unstable. The problem was solved by developing an alternative form of stabilizer that did not require a torsional filter (Lee and Kundur, 1986).

Although all of these categories of oscillations are related and can exist simultaneously, the primary focus of this section is on the electromechanical oscillations that affect interarea power flows.

Conceptual Description of Power System Oscillations

As illustrated in the previous subsection, power systems contain many modes of oscillation due to a variety of interactions of its components. Many of the oscillations are due to generator rotor masses swinging relative to one another. A power system having multiple machines will act like a set of masses interconnected by a network of springs and will exhibit multiple modes of oscillation. As illustrated previously in the section "Historical Perspective", in many systems, the damping of these electromechanical swing modes is a critical factor for operating in a secure manner. The power transfer between such machines on the AC transmission system is a direct function of the angular separation between their internal voltage phasors. The torques that influence the machine oscillations can be conceptually split into synchronizing and damping components of torque (de Mello and Concordia, 1969). The synchronizing component "holds" the machines in the power system together and is important for system transient stability following large disturbances. For small disturbances, the synchronizing component of torque determines the frequency of an oscillation. Most stability texts present the synchronizing component in terms of the slope of the power-angle relationship, as illustrated in Fig. 11.9, where K represents the amount of synchronizing torque. The damping component determines the decay of oscillations and is important for system stability following recovery from the initial swing. Damping is influenced by many system parameters, is usually small, and can sometimes become negative in the presence of controls,

Power System Dynamics and Stability

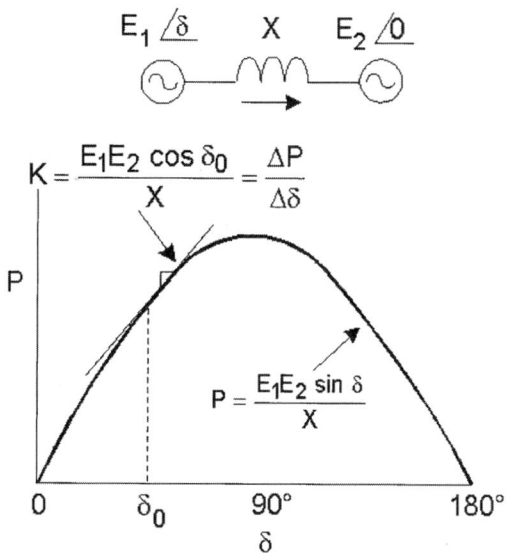

FIGURE 11.9 Simplified power-angle relationship between two AC systems.

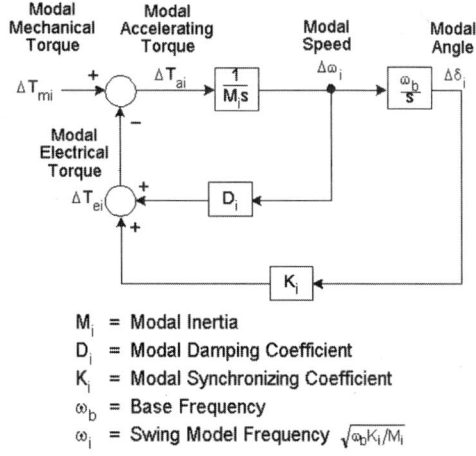

FIGURE 11.10 Conceptual block diagram of a power-swing mode.

(which are practically the only "source" of negative damping). Negative damping can lead to spontaneous growth of oscillations until relays begin to trip system elements or a limit cycle is reached.

Figure 11.10 shows a conceptual block diagram of a power-swing mode, with inertial (M), damping (D), and synchronizing (K) effects identified. For a perturbation about a steady-state operating point, the modal accelerating torque ΔT_{ai} is equal to the modal electrical torque ΔT_{ei} (with the modal mechanical torque ΔT_{mi} considered to be 0). The effective inertia is a function of the total inertia of all machines participating in the swing; the synchronizing and damping terms are frequency dependent and are influenced by generator rotor circuits, excitation controls, and other system controls.

Summary on the Nature of Power System Oscillations

The preceding review leads to a number of important conclusions and observations concerning power system oscillations:

- Oscillations are due to natural modes of the system and therefore cannot be eliminated. However, their damping and frequency can be modified.
- As power systems evolve, the frequency and damping of existing modes change and new ones may emerge.
- The source of "negative" damping is power system controls, primarily excitation system automatic voltage regulators.
- Interarea oscillations are associated with weak transmission links and heavy power transfers.
- Interarea oscillations often involve more than one utility and may require the cooperation of all to arrive at the most effective and economical solution.
- Power system stabilizers are the most commonly used means of enhancing the damping of interarea modes.
- Continual study of the system is necessary to minimize the probability of poorly damped oscillations. Such "beforehand" studies may have avoided many of the problems experienced in power systems (see Chapter 3 of CIGRE Technical Brochure No. 111, 1996).

It must be clear that avoidance of oscillations is only one of many aspects that should be considered in the design of a power system and so must take its place in line along with economy, reliability, operational robustness, environmental effects, public acceptance, voltage and power quality, and certainly a few others that may need to be considered. Fortunately, it appears that many features designed to further some of these other aspects also have a strong mitigating effect in reducing oscillations. However, one overriding constraint is that the power system operating point must be stable with respect to oscillations.

Criteria for Damping

The rate of decay of the amplitude of oscillations is best expressed in terms of the damping ratio ζ. For an oscillatory mode represented by a complex eigenvalue $\sigma \pm j\omega$, the damping ratio is given by:

$$\zeta = \frac{-\sigma}{\sqrt{\sigma^2 + \omega^2}}$$

The damping ratio ζ determines the rate of decay of the amplitude of the oscillation. The time constant of amplitude decay is $1/|\sigma|$. In other words, the amplitude decays to $1/e$ or 37% of the initial amplitude in $1/|\sigma|$ seconds or in $1/(2\pi\zeta)$ cycles of oscillation (Kundur, 1994). As oscillatory modes have a wide range of frequencies, the use of damping ratio rather than the time constant of decay is considered more appropriate for expressing the degree of damping. For example, a 5-s time constant represents amplitude decay to 37% of initial value in 110 cycles of oscillation for a 22 Hz torsional mode, in 5 cycles for a 1 Hz local plant mode, and in one-half cycle for a 0.1 Hz interarea mode of oscillation. On the other hand, a damping ratio of 0.032 represents the same degree of amplitude decay in 5 cycles for all modes.

A power system should be designed and operated so that the following criteria are satisfied for all expected system conditions, including postfault conditions following design contingencies:

1. The damping ratio (ζ) of all system modes oscillation should exceed a specified value. The minimum acceptable damping ratio is system dependent and is based on operating experience and/or sensitivity studies; it is typically in the range 0.03 to 0.05.
2. The small-signal stability margin should exceed a specified value. The stability margin is measured as the difference between the given operating condition and the absolute stability limit ($\zeta = 0$) and should be specified in terms of a physical quantity, such as a power plant output, power transfer through a critical transmission interface, or system load level.

Study Procedure

There is a general need for establishing study procedures and developing widely accepted design and operating criteria with respect to power system oscillations. Tools for the analysis of system oscillations, in addition to determining the existence of problems, should be capable of identifying factors influencing the problem and providing information useful in developing control measures for mitigation.

System oscillation problems are often investigated using nonlinear time-domain simulations as a natural extension to traditional transient stability analysis. However, there are a number of practical problems that limit the effectiveness of using only the time-domain approach:

- The use of time responses exclusively to look at damping of different modes of oscillation could be deceptive. The choice of disturbance and the selection of variables for observing time response are critical. The disturbance may not provide sufficient excitation of the critical modes. The observed response contains many modes, and poorly damped modes may not always be dominant.
- To get a clear indication of growing oscillations, it is necessary to carry the simulations out to 15 s or 20 s or more. This could be time-consuming.
- Direct inspection of time responses does not give sufficient insight into the nature of the oscillatory stability problem; it is difficult to identify the sources of the problem and develop corrective measures.

Spectral estimation (i.e., modal identification) techniques based on Prony analysis may be used to analyze time-domain responses and extract information about the underlying dynamics of the system (Hauer, 1991).

Small-signal analysis (i.e., modal analysis or eigenanalysis) based on linear techniques is ideally suited for investigating problems associated with oscillations. Here, the characteristics of a power system model can be determined for a system model linearized about a specific operating point. The stability of each mode is clearly identified by the system's eigenvalues. Modeshapes and the relationships between different modes and system variables or parameters are identified using eigenvectors (Kundur, 1994). Conventional eigenvalue computation methods are limited to systems up to about 800 states. Such methods are ideally suited for detailed analysis for system oscillation problems confined to a small portion of the power system. This includes problems associated with local plant modes, torsional modes, and control modes. For analysis of interarea oscillations in large interconnected power systems, special techniques have been developed for computing eigenvalues associated with a small subset of modes whose frequencies are within a specified range (Kundur, 1994). Techniques have also been developed for efficiently computing participation factors, residues, transfer function zeros, and frequency responses useful in designing remedial control measures. Powerful computer program packages incorporating the above computational features are now available, thus providing comprehensive capabilities for analyses of power system oscillations (CIGRE Technical Brochure No. 111, 1996; CIGRE Technical Brochure, 2000; Kundur, 1994). For very large interconnected systems, it may be necessary to use dynamic equivalents (Wang et al., 1997). This can only be achieved by developing reduced-order power system models that correctly reflect the significant dynamic characteristics of the interconnected system.

In summary, a complete understanding of power systems oscillations generally requires a combination of analytical tools. Small-signal stability analysis complemented by nonlinear time-domain simulations is the most effective procedure of studying power system oscillations. The following are the recommended steps for a systematic analysis of power system oscillations:

1. Perform an eigenvalue scan using a small-signal stability program. This will indicate the presence of poorly damped modes.
2. Perform a detailed eigenanalysis of the poorly damped modes. This will determine their characteristics and sources of the problem, and assist in developing mitigation measures. This will also identify the quantities to be monitored in time-domain simulations.

3. Perform time-domain simulations of the critical cases identified from the eigenanalysis. This is useful to confirm the results of small-signal analysis. In addition, it shows how system nonlinearities affect the oscillations. Prony analysis of these time-domain simulations may also be insightful (Hauer, 1991).

Mitigation of Power System Oscillations

In many power systems, equipment is installed to enhance various performance issues such as transient, oscillatory, or voltage stability. In many instances, this equipment is power-electronic based, which generally means the device can be rapidly and continuously controlled. Examples of such equipment applied in the transmission system include a static Var compensator (SVC), static compensator (STATCOM), thyristor-controlled series compensation (TCSC) and Unified Power Flow Controller (UPFC). To improve damping in a power system, a supplemental damping controller can be applied to the primary regulator of one of these transmission devices or to generator controls. The supplemental control action should modulate the output of a device in such a way as to affect power transfer such that damping is added to the power system swing modes of concern. This subsection provides an overview on some of the issues that affect the ability of damping controls to improve power system dynamic performance (CIGRE Technical Brochure No. 111, 1996; CIGRE Technical Brochure, 2000; Paserba et al., 1995; Levine, 1995).

Siting

Siting plays an important role in the ability of a device to stabilize a swing mode. Many controllable power system devices are sited based on issues unrelated to stabilizing the network (e.g., HVDC transmission and generators), and the only question is whether they can be utilized effectively as a stability aid. In other situations (e.g., SVC, STATCOM, TCSC, or UPFC), the equipment is installed primarily to help support the transmission system, and siting will be heavily influenced by its stabilizing potential. Device cost represents an important driving force in selecting a location. In general, there will be one location that makes optimum use of the controllability of a device. If the device is located at a different location, a larger-size device may be needed to achieve the desired stabilization objective. In some cases, overall costs may be minimized with nonoptimum locations of individual devices because other considerations must also be taken into account, such as land price and availability, environmental regulations, etc.

The inherent ability of a device to achieve a desired stabilization objective in a robust manner while minimizing the risk of adverse interactions is another consideration that can influence the siting decision. Most often, these other issues can be overcome by appropriate selection of input signals, signal filtering, and control design. This is not always possible, however, so these issues should be included in the decision-making process for choosing a site. For many applications, it will be desirable to apply the devices in a distributed manner. This approach helps maintain a more uniform voltage profile across the network, during both steady-state operation and after transient events. Greater security may also be possible with distributed devices because the overall system is more likely to tolerate the loss of one of the devices.

Control Objectives

Several aspects of control design and operation must be satisfied during both the transient and the steady-state operation of the power system, before and after a major disturbance. These aspects suggest that controls applied to the power system should:

1. Survive the first few swings after a major system disturbance with some degree of safety. The safety factor is usually built into a Reliability Council's criteria (e.g., keeping voltages above some threshold during the swings).
2. Provide some minimum level of damping in the steady-state condition after a major disturbance (post-contingent operation). In addition to providing security for contingencies, some applications will require "ambient" damping to prevent spontaneous growth of oscillations in steady-state operation.
3. Minimize the potential for adverse side effects, which can be classified as follows:

a. Interactions with high-frequency phenomena on the power system, such as turbine-generator torsional vibrations and resonances in the AC transmission network.
   b. Local instabilities within the bandwidth of the desired control action.
4. Be robust so that the control will meet its objectives for a wide range of operating conditions encountered in power system applications. The control should have minimal sensitivity to system operating conditions and component parameters since power systems operate over a wide range of operating conditions and there is often uncertainty in the simulation models used for evaluating performance. Also, the control should have minimum communication requirements.
5. Be highly dependable so that the control has a high probability of operating as expected when needed to help the power system. This suggests that the control should be testable in the field to ascertain that the device will act as expected should a contingency occur. This leads to the desire for the control response to be predictable. The security of system operations depends on knowing, with a reasonable certainty, what the various control elements will do in the event of a contingency.

Closed-Loop Control Design

Closed-loop control is utilized in many power-system components. Voltage regulators, either continuous or discrete, are commonplace on generator excitation systems, capacitor and reactor banks, tap-changing transformers, and SVCs. Modulation controls to enhance power system stability have been applied extensively to generator exciters and to HVDC, SVC, and TCSC systems. A notable advantage of closed-loop control is that stabilization objectives can often be met with less equipment and impact on the steady-state power flows than is generally possible with open-loop controls. While the behavior of the power system and its components is usually predictable by simulation, its nonlinear character and vast size lead to challenging demands on system planners and operating engineers. The experience and intuition of these engineers is generally more important to the overall successful operation of the power system than the many available, elegant control design techniques (Levine, 1995; CIGRE Technical Brochure, 2000).

Typically, a closed-loop controller is always active. One benefit of such a closed-loop control is ease of testing for proper operation on a continuous basis. In addition, once a controller is designed for the worst-case contingency, the chance of a less severe contingency causing a system breakup is lower than if only open-loop controls are applied. Disadvantages of closed-loop control involve primarily the potential for adverse interactions. Another possible drawback is the need for small step sizes, or vernier control in the equipment, which will have some impact on cost. If communication is needed, this could also be a challenge. However, experience suggests that adequate performance should be attainable using only locally measurable signals.

One of the most critical steps in control design is to select an appropriate input signal. The other issues are to determine the input filtering and control algorithm and to assure attainment of the stabilization objectives in a robust manner with minimal risk of adverse side effects. The following subsections discuss design approaches for closed-loop stability controls, so that the potential benefits can be realized on the power system.

Input Signal Selection

The choice of using a local signal as an input to a stabilizing control function is based on several considerations.

1. The input signal must be sensitive to the swings on the machines and lines of interest. In other words, the swing modes of interest must be "observable" in the input signal selected. This is mandatory for the controller to provide a stabilizing influence.
2. The input signal should have as little sensitivity as possible to other swing modes on the power system. For example, for a transmission-line device the control action will benefit only those modes that involve power swings on that particular line. If the input signal was also responsive to local swings within an area at one end of the line, then valuable control range would be wasted in responding to an oscillation that the damping device has little or no ability to control.

3. The input signal should have little or no sensitivity to its own output, in the absence of power swings. Similarly, there should be as little sensitivity to the action of other stabilizing controller outputs as possible. This decoupling minimizes the potential for local instabilities within the controller bandwidth (CIGRE Technical Brochure, 2000).

These considerations have been applied to a number of modulation control designs, which have eventually proven themselves in many actual applications (see Chapter 5 of CIGRE Technical Brochure No. 111 [1996]). For example, the application of PSS controls on generator excitation systems was the first such study that reached the conclusion that speed or power is the best input signal, with frequency of the generator substation voltage being an acceptable choice as well (Larsen and Swann, 1981; Kundur et al., 1989). For SVCs, the conclusion was that the magnitude of line current flowing past the SVC is the best choice (Larsen and Chow, 1987). For torsional damping controllers on HVDC systems, it was found that using the frequency of a synthesized voltage close to the internal voltage of the nearby generator, calculated with locally measured voltages and currents, is best (Piwko and Larsen, 1982). In the case of a series device in a transmission line (such as a TCSC), the considerations listed above lead to the conclusion that using frequency of a synthesized remote voltage to estimate the center-of-inertia of an area involved in a swing mode is a good choice (Levine, 1995). This allows the series device to behave like a damper across the AC line.

Input-Signal Filtering

To prevent interactions with phenomena outside the desired control bandwidth, low-pass and high-pass filtering must be used for the input signal. In certain applications, notch filtering is needed to prevent interactions with certain lightly damped resonances. This has been the case with SVCs interacting with AC network resonances and modulation controls interacting with generator torsional vibrations. On the low-frequency end, the high-pass filter must have enough attenuation to prevent excessive response during slow ramps of power, or during the long-term settling following a loss of generation or load. This filtering must be considered while designing the overall control as it will strongly affect performance and the potential for local instabilities within the control bandwidth. However, finalizing such filtering usually must wait until the design for performance is completed, after which the attenuation needed at specific frequencies can be determined. During the control design work, a reasonable approximation of these filters needs to be included. Experience suggests that a high-pass break near 0.05 Hz (3 s washout time constant), and a double low-pass break near 4 Hz (40 ms time constant) as shown in Fig. 11.11, is suitable for a starting point. A control design that provides adequate stabilization of the power system with these settings for the input filtering has a high probability of being adequate after the input filtering parameters are finalized.

Control Algorithm

Levine (1995) and CIGRE Technical Brochure (2000) present many control design methods that can be utilized to design supplemental controls for power systems. Generally, the control algorithm for damping leads to a transfer function that relates an input signal(s) to a device output. This statement is the starting point for understanding how deviations in the control algorithm affect system performance.

In general, the transfer function of the control (and input-signal filtering) is most readily discussed in terms of its gain and phase relationship versus frequency. A phase shift of 0° in the transfer function

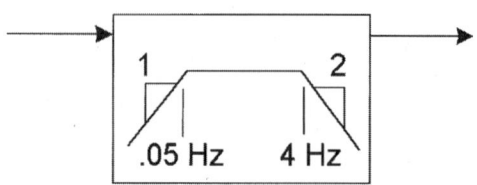

FIGURE 11.11 Initial input signal filtering.

means that the output is proportional to the input, and, for discussion purposes, is assumed to represent a pure damping effect on a lightly damped power swing mode. Phase lag in the transfer function (up to 90°) translates to a positive synchronizing effect, tending to increase the frequency of the swing mode when the control loop is closed. The damping effect will decrease with the sine of the phase lag. Beyond 90°, the damping effect will become negative. Conversely, phase lead is a desynchronizing influence and will decrease the frequency of the swing mode when the control loop is closed. Generally, the desynchronizing effect should be avoided. The preferred transfer function has between 0° and 45° of phase lag in the frequency range of the swing modes that the control is designed to damp.

Gain Selection

After the shape of the transfer function to meet the desired control phase characteristics is designed, the gain of the control is selected to obtain the desired level of damping. To maximize damping, the gain should be high enough to assure full utilization of the controlled device for the critical disturbances, but no higher, so that risks of adverse effects are minimized. Typically, the gain selection is done analytically with root-locus or Nyquist methods. However, the gain must ultimately be verified in the field (see Chapter 8 of CIGRE Technical Brochure No. 111 [1996]).

Control Output Limits

The output of a damping control must be limited to prevent it from saturating the device being modulated. By saturating a controlled device, the purpose of the damping control would be defeated. As a general rule of thumb for damping, when a control is at its limits in the frequency range of interarea oscillations, the output of the controlled device should be just within its limits (Larsen and Swann, 1981).

Performance Evaluation

Good simulation tools are essential to applying damping controls to power transmission equipment for the purpose of system stabilization. The controls must be designed and tested for robustness with such tools. For many system operating conditions, the only feasible means of testing the system is by simulation, so confidence in the power system model is crucial. A typical large-scale power system model may contain up to 15,000 state variables or more. For design purposes, a reduced-order model of the power system is often desirable (Wang et al., 1997). If the size of the study system is excessive, the large number of system variations and parametric studies required becomes tedious and prohibitively expensive for some linear analysis techniques and control design methods in general use today. A good understanding of the system performance can be obtained with a model that contains only the relevant dynamics for the problem under study. The key situations that establish the adequacy of controller performance and robustness can be identified from the reduced-order model, and then tested with the full-scale model. Note that CIGRE Technical Brochure No. 111 (1996) (CIGRE Technical Brochure, 2000) and Kundur (1994) contain information on the application of linear analysis techniques for very large systems.

Field testing is also an essential part of applying supplemental controls to power systems. Testing needs to be performed with the controller open-loop, comparing the measured response at its own input and the inputs of other planned controllers, against the simulation models. Once these comparisons are acceptable, the system can be tested with the control loop closed. Again, the test results should have a reasonable correlation with the simulation program. Methods have been developed for performing such testing of the overall power system to provide benchmarks for validating the full-system model. Such testing can also be done on the simulation program to help arrive at the reduced-order models (Hauer, 1991; Kamwa et al., 1993) needed for the advanced control design methods (Levine, 1995; CIGRE Technical Brochure, 2000). Methods have also been developed to improve the modeling of individual components. These issues are discussed in great detail in Chapters 6 and 8 of CIGRE Technical Brochure No. 111 (1996).

Adverse Side Effects

Historically in the power industry, each major advance in improving system performance has created some adverse side effects. For example, the addition of high-speed excitation systems over 40 years ago caused the destabilization known as the "hunting" mode of the generators. The fix was power system stabilizers, but it took over 10 years to learn how to tune them properly and there were some unpleasant surprises involving interactions with torsional vibrations on the turbine-generator shaft (Larsen and Swann, 1981).

HVDC systems were also found to interact adversely with torsional vibrations [the subsynchronous torsional interaction (SSTI) problem], especially when augmented with supplemental modulation controls to damp power swings. Similar SSTI phenomena exist with SVCs, although to a lesser degree than with HVDC. Detailed study methods have since been established for designing systems with confidence that these effects will not cause trouble for normal operation (Piwko and Larsen, 1982; Bahrman et al., 1980). Another potential adverse side effect is with SVC systems that can interact unfavorably with network resonances. This side effect caused a number of problems in the initial application of SVCs to transmission systems. Design methods now exist to deal with this phenomenon, and protective functions exist within SVC controls to prevent continuing exacerbation of an unstable condition (Larsen and Chow, 1987).

As the available technologies continue to evolve, such as the current industry focus on Flexible AC Transmission Systems (FACTS), new opportunities arise for power system performance improvement. FACTS devices introduce capabilities that may be an order of magnitude greater than existing equipment applied for stability improvement. Therefore, it follows that there may be much more serious consequences if they fail to operate properly. Robust operation and noninteraction of controls for these FACTS devices are critically important for stability of the power system (CIGRE Technical Brochure, 2000; Clark et al., 1995).

Summary

In summary, this section on small signal stability and power system oscillations shows that power systems contain many modes of oscillation due to a variety of interactions among components. Many of the oscillations are due to synchronous generator rotors swinging relative to one another. The electromechanical modes involving these masses usually occur in the frequency range of 0.1 to 2 Hz. Particularly troublesome are the interarea oscillations, which are typically in the frequency range of 0.1 to 1 Hz. The interarea modes are usually associated with groups of machines swinging relative to other groups across a relatively weak transmission path. The higher frequency electromechanical modes (1 to 2 Hz) typically involve one or two generators swinging against the rest of the power system or electrically close machines swinging against each other.

These oscillatory dynamics can be aggravated and stimulated through a number of mechanisms. Heavy power transfers, in particular, can create interarea oscillation problems that constrain system operation. The oscillations themselves may be triggered through some event or disturbance on the power system or by shifting the system operating point across some steady-state stability boundary where growing oscillations may be spontaneously created. Controller proliferation makes such boundaries increasingly difficult to anticipate. Once started, the oscillations often grow in magnitude over the span of many seconds. These oscillations may persist for many minutes and be limited in amplitude only by system nonlinearities. In some cases they cause large generator groups to lose synchronism where part or all of the electrical network is lost. The same effect can be reached through slow cascading outages when the oscillations are strong and persistent enough to cause uncoordinated automatic disconnection of key generators or loads. Sustained oscillations can disrupt the power system in other ways, even when they do not produce network separation or loss of resources. For example, power swings, which are not always troublesome in themselves, may have associated voltage or frequency swings that are unacceptable. Such concerns can limit power transfer even when oscillatory stability is not a direct concern.

Information presented in this section addressing power system oscillations included:

- Nature of oscillations
- Criteria for damping
- Study procedure
- Mitigation of oscillations by control

As to the priority of selecting devices and controls to be applied for the purpose of damping power system oscillations, the following summarizing remarks can be made.

1. Carefully tuned power system stabilizers (PSS) on the major generating units affected by the oscillations should be considered first. This is because of the effectiveness and relatively low cost of PSSs.
2. Supplemental controls added to devices installed for other reasons should be considered second. Examples include HVDC installed for the primary purpose of long-distance transmission or power exchange between asynchronous regions and SVC installed for the primary purpose of dynamic voltage support.
3. Augmentation of fixed or mechanically switched equipment with power-electronics, including damping controls can be considered third. Examples include augmenting existing series capacitors with a thyristor-controlled portion (TCSC).
4. The fourth priority for consideration is the addition of a new device in the power system for the primary purpose of damping.

References

Bahrman, M.P., Larsen, E.V., Piwko, R.J. and Patel, H.S., Experience with HVDC turbine-generator torsional interaction at Square Butte, *IEEE Trans. on PAS*, 99, 966–975, 1980.

CIGRE Task Force 38.01.07 on Power System Oscillations, *Analysis and Control of Power System Oscillations,* CIGRE Technical Brochure No. 111, December 1996.

CIGRE Task Force 38.02.16, *Impact of the Interaction among Power System Controllers,* CIGRE Technical Brochure, 2000.

Clark, K., Fardanesh, B. and Adapa, R., Thyristor controlled series compensation application study — control interaction considerations, *IEEE Trans. on Power Delivery*, 1031–1037, April 1995.

Concordia, C., Steady-state stability of synchronous machines as affected by voltage regulator characteristics, *AIEE Transactions*, 63, 215–220, 1944.

Concordia, C., Effect of prime-mover speed control characteristics on electric power system performance, *IEEE Trans. on PAS*, 88/5, 752–756, 1969.

Crary, S.B. and Duncan, W.E., Amortisseur windings for hydrogenerators, *Electrical World*, 115, 2204–2206, June 28, 1941.

deMello, F.P. and Concordia, C., Concepts of synchronous machine stability as affected by excitation control, *IEEE Trans. on PAS*, 88, 316–329, 1969.

Hauer, J.F., Application of Prony analysis to the determination of model content and equivalent models for measured power systems response, *IEEE Trans. on Power Systems*, 1062–1068, August 1991.

Kamwa, I., Grondin, R., Dickinson, J. and Fortin, S. A minimal realization approach to reduced-order modeling and modal analysis for power system response signals, *IEEE Trans. on Power Systems*, 8, 3, 1020–1029, 1993.

Kundur, P., *Power System Stability and Control*, McGraw-Hill, New York, 1994.

Kundur, P., Klein, M., Rogers, G.J. and Zywno, M.S., Application of power system stabilizers for enhancement of overall system stability, *IEEE Trans. on Power Systems*, 4, 614–626, May 1989.

Kundur, P., Lee, D.C. and Zein El-Din, H.M., Power system stabilizers for thermal units: Analytical techniques and on site validation, *IEEE Trans. on PAS*, 100, 81–85, January 1981.

Larsen E.V. and Chow, J.H., SVC Control Design Concepts for System Dynamic Performance, Application of Static Var Systems for System Dynamic Performance, *IEEE Special Publication No. 87TH1087-5-PWR on Application of Static Var Systems for System Dynamic Performance*, 36–53, 1987.

Larsen, E.V. and Swann, D.A., Applying power system stabilizers, Parts I, II and III, *IEEE Trans. on PAS*, 100, 3017–3046, 1981.

Lee, D.C., Beaulieu, R.E. and Rogers, G.J., Effects of governor characteristics on turbo-generator shaft torsionals, *IEEE Trans. on PAS*, 104, 1255–1261, June 1985.

Lee, D.C. and Kundur, P., Advanced excitation controls for power system stability enhancement, CIGRE Paper 38-01, Paris, 1986.

Levine, W.S., editor, *The Control Handbook*, CRC Press, Boca Raton, FL, 1995.

Paserba, J.J., Larsen, E.V., Grund, C.E. and Murdoch, A., Mitigation of inter-area oscillations by control, *IEEE Special Publication 95-TP-101 on Interarea Oscillations in Power Systems*, 1995.

Piwko, R.J. and Larsen, E.V., HVDC System control for damping subsynchronous oscillations, *IEEE Trans. on PAS*, 101, 7, 2203–2211, 1982.

Wang, L., Klein, M., Yirga, S., and Kundur, P. Dynamic reduction of large power systems for stability studies, *IEEE Trans. on Power Systems*, PWRS-12, 2, 889–895, May 1997.

Watson W. and Coultes, M.E., Static exciter stabilizing signals on large generators — Mechanical problems, *IEEE Trans. on PAS*, 92, 205–212, January/February 1973.

11.4 Voltage Stability

Yakout Mansour

Voltage stability refers to the ability of a power system to maintain its voltage profile under the full spectrum of its operating scenarios so that both voltage and power are controllable at all times.

Voltage instability of radial distribution systems has been well recognized and understood for decades (Venikov, 1970; 1980) and was often referred to as load instability. Large interconnected power networks did not face the phenomenon until late 1970s and early 1980s.

Most of the early developments of the major HV and EHV networks and interties faced the classical machine angle stability problem. Innovations in both analytical techniques and stabilizing measures made it possible to maximize the power transfer capabilities of the transmission systems. The result was increasing transfers of power over long distances of transmission. As the power transfer increased, even when angle stability was not a limiting factor, many utilities have been facing a shortage of voltage support. The result ranged from post contingency operation under reduced voltage profile to total voltage collapse. Major outages attributed to this problem were experienced in the northeastern part of the U.S., France, Sweden, Belgium, Japan, along with other localized cases of voltage collapse (Mansour, 1990). Accordingly, voltage stability imposed itself as a governing factor in both planning and operating criteria of a number of utilities. Consequently, major challenges in establishing sound analytical procedures, quantitative measures of proximity to voltage instability, and margins have been facing the industry for the last two decades.

Voltage instability is associated with relatively slow variations in network and load characteristics. Network response in this case is highly influenced by the slow-acting control devices such as transformer on-load tap changers, automatic generation control, generator field current limiters, generator overload reactive capability, under-voltage load shedding relays, and switchable reactive devices. The characteristics of such devices as to how they influence the network response to voltage variations are generally understood and well covered in the literature. On the other hand, electric load response to voltage variation has only been addressed more recently, even though it is considered the single most important factor in voltage instability.

Generic Dynamic Load-Voltage Characteristics

While it might be possible to identify the voltage response characteristics of a large variety of individual equipment of which a power network load is comprised, it is not practical or realistic to model network load by individual equipment models. Thus, the aggregate load model approach is much more realistic.

Field test results as reported by Hill (1992) and Xu et al. (1996) indicate that typical response of an aggregate load to step-voltage changes is of the form shown in Fig. 11.12. The response is a reflection of the collective effects of all downstream components ranging from OLTCs to individual household loads. The time span for a load to recover to steady-state is normally in the range of several seconds to minutes, depending on the load composition. Responses for real and reactive power are qualitatively similar. It can be seen that a sudden voltage change causes an instantaneous power demand change. This change defines the transient characteristics of the load and was used to derive static load models for angular stability

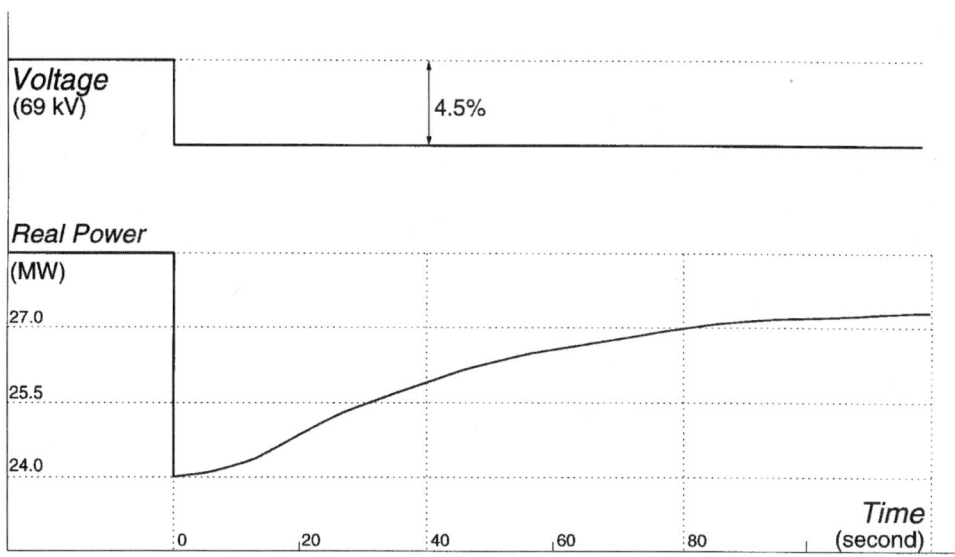

FIGURE 11.12 Aggregate load response to a step voltage change.

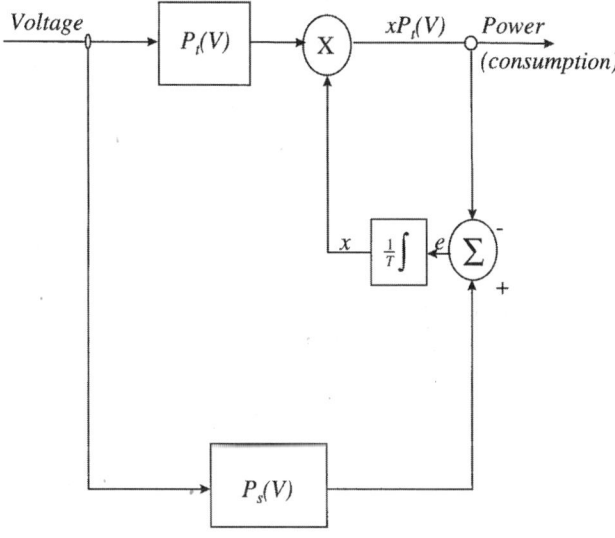

FIGURE 11.13 A generic dynamic load model.

studies. When the load response reaches steady-state, the steady-state power demand is a function of the steady-state voltage. This function defines the steady-state load characteristics known as voltage-dependent load models in load flow studies.

The typical load-voltage response characteristics can be modeled by a generic dynamic load model proposed in Fig. 11.13. In this model (Xu et al., 1993), x is the state variable. $P_t(V)$ and $P_s(V)$ are the transient and steady-state load characteristics, respectively, and can be expressed as:

$$P_t = V^a \quad \text{or} \quad P_t = C_2 V^2 + C_1 V + C_o$$

$$P_s = P_o V^a \quad \text{or} \quad P_s = P_o \left(d_2 V^2 + d_1 V + d_o \right)$$

where V is the per-unit magnitude of the voltage imposed on the load. It can be seen that, at steady-state, state variable x of the model is constant. The input to the integration block, $E = P_s - P$, must be zero and, as a result, the model output is determined by the steady-state characteristics $P = P_s$. For any sudden voltage change, x maintains its predisturbance value initially because the integration block cannot change its output instantaneously. The transient output is then determined by the transient characteristics $P - xP_t$. The mismatch between the model output and the steady-state load demand is the error signal e. This signal is fed back to the integration block that gradually changes the state variable x. This process continues until a new steady-state ($e = 0$) is reached. Analytical expressions of the load model, including real (P) and reactive (Q) power dynamics, are:

$$T_p \frac{dx}{dt} = P_s(V) - P, P = xP_t(V)$$

$$T_q \frac{dy}{dt} = Q_s(V) - Q, Q = yQ_t(V)$$

$$P_t(V) = V^a, P_s(V) = P_o V^a; \quad Q_t(V) = V^\beta, Q_s(V) = Q_o V^\beta$$

Analytical Frameworks

The slow nature of the network and load response associated with the phenomenon made it possible to analyze the problem in two frameworks: (1) long-term dynamic framework in which all slow-acting devices and aggregate bus loads are represented by their dynamic models (the analysis in this case is done through dynamic simulation of the system response to a contingency or load variation), or (2) steady-state framework (e.g., load flow) to determine if the system can reach a stable operating point following a particular contingency. This operating point could be a final state or a midpoint following a step of a discrete control action (e.g., transformer tap change).

The proximity of a given system to voltage instability is typically assessed by indices that measure one or a combination of:

- Sensitivity of load bus voltage to variations in active power of the load.
- Sensitivity of load bus voltage to variations in injected reactive power at the load bus.
- Sensitivity of the receiving end voltage to variations in sending end voltage.
- Sensitivity of the total reactive power generated by generators, synchronous condensers, and SVS to variations in load bus reactive power.

Computational Methods

Load Flow Analysis

Consider a simple two-bus system of a sending end source feeding a $P - Q$ load through a transmission line. The family of curves shown in Fig. 11.14 is produced by maintaining the sending end voltage constant while the load at the receiving end is varied at a constant power factor and the receiving end voltage is calculated. Each curve is calculated at a specific power factor and shows the maximum power that can be transferred at this particular power factor. Note that the limit can be increased by providing more reactive support at the receiving end [limit (2) vs. limit (1)], which is effectively pushing the power factor of the load in the leading direction. It should also be noted that the points on the curves below the limit line Vs characterize unstable behavior of the system where a drop in demand is associated with a drop in the receiving end voltage leading to eventual collapse. Proximity to voltage instability is usually measured by the distance (in PU power) between the operating point on the P–V curve and the limit of the same curve.

Another family of curves similar to that of Fig. 11.15 can be produced by varying the reactive power demand (or injection) at the receiving end while maintaining the real power and the sending end voltage constant. The relation between the receiving end voltage and the reactive power injection at the receiving

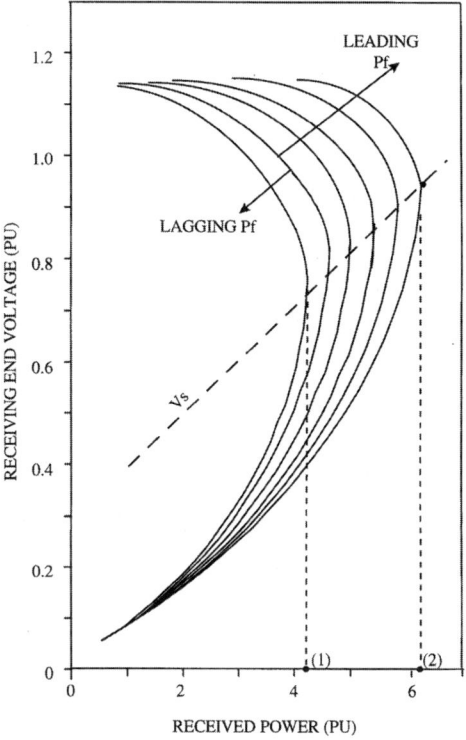

FIGURE 11.14 Pr-Vr characteristics.

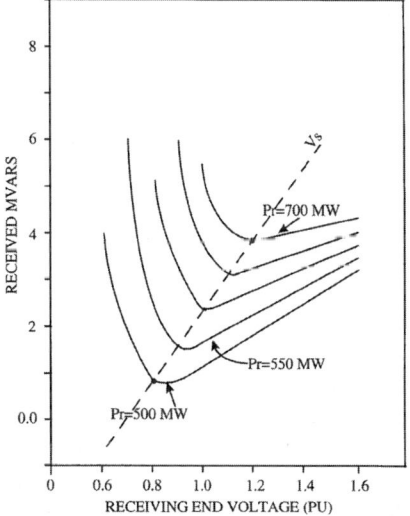

FIGURE 11.15 Q-Vr characteristics.

end is plotted to produce the so called Q–V_r curves of Fig. 11.15. The bottom of any given curve characterizes the voltage stability limit. Note that the behavior of the system on the right side of the limit is such that an increase in reactive power injection at the receiving end results in a receiving end voltage rise while the opposite is true on the left side because of the substantial increase in current at the lower voltage, which, in turn, increases reactive losses in the network substantially. The proximity to voltage instability is measured as the difference between the reactive power injection corresponding to the operating point and the bottom of the curve. As the active power transfer increases (upwards in Fig. 11.15), the reactive power margin decreases as does the receiving end voltage.

The same family of relations in Figs. 11.14 and 11.15 can be and have been used to assess the voltage stability of large power systems. The P–V curves can be calculated using load flow programs. The demand of load center buses are increased in steps at a constant power factor while the generators' terminal voltages are held at their nominal value. The P–V relation can then be plotted by recording the MW demand level against a central load bus voltage at the load center. It should be noted that load flow solution algorithms diverge past the limit and do not produce the unstable portion of the P–V relation. The Q–V relation, however, can be produced in full by assuming a fictitious synchronous condenser at a central load bus in the load center. The Q–V relation is then plotted for this particular bus as a representative of the load center by varying the voltage of the bus (now converted to a voltage control bus by the addition of the synchronous condenser) and recording its value against the reactive power injection of the synchronous condenser. If the limits on the reactive power capability of the synchronous condenser is made very high, the load flow solution algorithm will always converge at either side of the Q–V relation.

Sequential Load Flow Method

The P–V and Q–V relations produced results corresponding to an end state of the system where all tap changers and control actions have taken place in time and the load characteristics were restored to a constant power characteristics. It is always recommended and often common to analyze the system behavior in its transition following a disturbance to the end state. Aside from the full long-term time simulation, the system performance can be analyzed in a quasidynamic manner by breaking the system response down into several time windows, each of which is characterized by the states of the various controllers and the load recovery (Mansour, 1993). Each time window can be analyzed using load flow programs modified to reflect the various controllers' states and load characteristics. Those time windows (Fig. 11.16) are primarily characterized by:

1. Voltage excursion in the first second after a contingency as motors slow, generator voltage regulators respond, etc.
2. The period 1 to 20 sec when the system is quiescent until excitation limiting occurs
3. The period 20 to 60 sec when generator over excitation protection has operated
4. The period 1 to 10 min after the disturbance when LTCs restore customer load and further increase reactive demand on generators
5. The period beyond 10 min when AGC, phase angle regulators, operators, etc. come into play

Voltage Stability as Affected by Load Dynamics

Voltage stability may occur when a power system experiences a large disturbance such as a transmission line outage. It may also occur if there is no major disturbance but the system's operating point shifts slowly towards stability limits. Therefore, the voltage stability problem, as other stability problems, must be investigated from two perspectives, the large-disturbance stability and the small-signal stability.

Large-disturbance voltage stability is event-oriented and addresses problems such as postcontingency margin requirement and response of reactive power support. Small-signal voltage stability investigates the stability of an operating point. It can provide such information as to the areas vulnerable to voltage collapse. In this section, the principle of load dynamics affecting both types of voltage stabilities is analyzed by examining the interaction of a load center with its supply network. Key parameters influencing voltage stability are identified. Since the real power dynamic behavior of an aggregate load is similar to its reactive power counterpart, the analysis is limited to reactive power only.

Large-Disturbance Voltage Stability

To facilitate explanation, assume that the voltage dynamics in the supply network are fast as compared to the aggregate dynamics of the load center. The network can then be modeled by three quasi-steady-state

Power System Dynamics and Stability   11-39

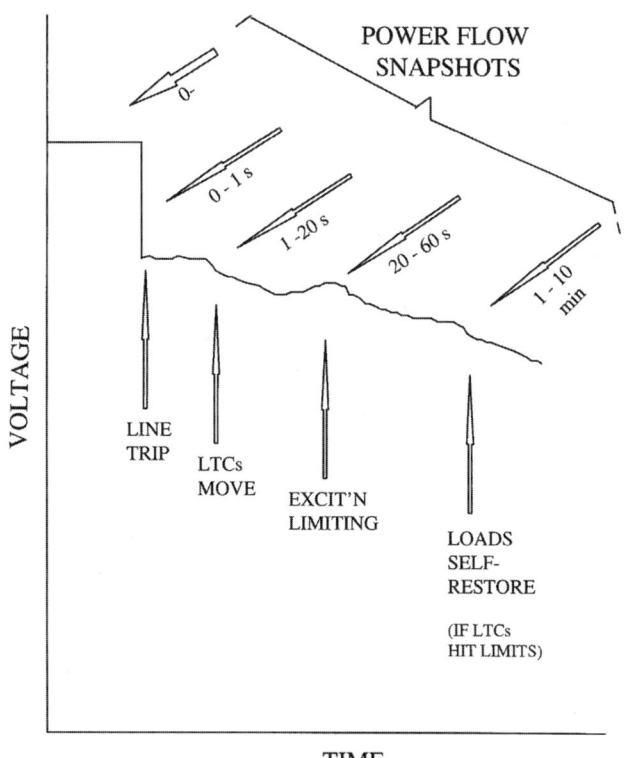

FIGURE 11.16 Breaking the system response down into time periods.

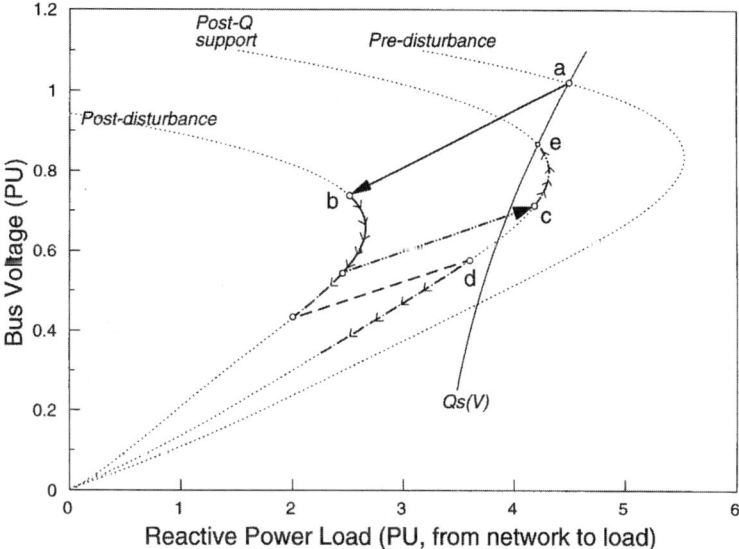

FIGURE 11.17 Voltage dynamics as viewed from V-Q plane.

V–Q characteristics (QV curves), predisturbance, postdisturbance and postdisturbance-with-reactive-support, as shown in Fig. 11.17. The load center is represented by a generic dynamic load. This load-network system initially operates at the intersection of the steady-state load characteristics and the predisturbance network V–Q curve, point a.

The network experiences an outage that reduces its reactive power supply capability to the postdisturbance V–Q curve. The aggregate load responds (see section on Generic Dynamic Load-Voltage Characteristics) instantaneously with its transient characteristics ($\beta = 2$, constant impedance in this example) and the system operating point jumps to point b. Since, at point b, the network reactive power supply is less than load demand for the given voltage:

$$T_q \frac{dy}{dt} = Q_s(V) - Q(V) > 0$$

The load dynamics will try to draw more reactive power by increasing the state variable y. This is equivalent to increasing the load admittance if $\beta = 2$ or the load current if $\beta = 1$. It drives the operating point to a lower voltage. If the load demand and the network supply imbalance persists, the system will continuously operate on the intersection of the postdisturbance V–Q curve and the drifting transient load curve with a monotonically decreasing voltage.

If reactive power support is initiated shortly after the outage, the network is switched to the third V–Q curve. The load responds with its transient chracteristics and a new operating point is formed. Depending on the switch time of reactive power support, the new operating point can be either c, for fas response, or d, for slow response. At point c, power supply is greater than load demand ($Q_s(V) - Q(V) < 0$). The load then draws less power by decreasing its state variable, and as a result, the operating voltage is increased. This dynamic process continues until the power imbalance is reduced to zero, namely a new steady-state operating point is reached (point e). On the other hand, for the case with slow response reactive support, the load demand is always greater than the network supply. A monotonic voltage collapse is the ultimate end. A numerical solution technique can be used to simulate the above process. Equations for the simulation are:

$$T_q \frac{dy}{dt} = Q_s(V) = Q(t); \quad Q(t) = yQ_t(V)$$

$$Q(t) = Network(V_s t)$$

where the function Network $V_s t$ consists of three polynomials each representing one V–Q curve. Figure 11.17 shows the simulation results in V–Q coordinates. The load voltage as a function of time is plotted in Fig. 11.18. The results demonstrate the importance of load dynamics for explaining the voltage stability problem.

Small-Signal Voltage Stability

The voltage characteristics of a power system can be analyzed around an operating point by linearizing the load flow equations around the operating point and analyzing the resulting sensitivity matrices. Recent breakthroughs in the computational algorithms made those techniques efficient and helpful in analyzing large-scale systems, taking into account virtually all the important elements affecting the phenomenon. In particular, singular value decomposition and modal techniques should be of particular interest to the reader and are thoroughly described by Mansour (1993); Lof et al. (IEEE Paper, 1992); Lof et al. (1992); and Gao et al. (1992).

Mitigation of Voltage Stability Problems

The following methods can be used to mitigate voltage stability problems.

Must-Run Generation. Operate uneconomic generators to change power flows or provide voltage support during emergencies or when new lines or transformers are delayed.

Series Capacitors. Use series capacitors to effectively shorten long lines, thus decreasing the net reactive loss. In addition, the line can deliver more reactive power from a strong system at one end to one experiencing a reactive shortage at the other end.

Shunt Capacitors. Though the heavy use of shunt capacitors can be part of the voltage stability problem, sometimes additional capacitors can also solve the problem by freeing "spinning reactive

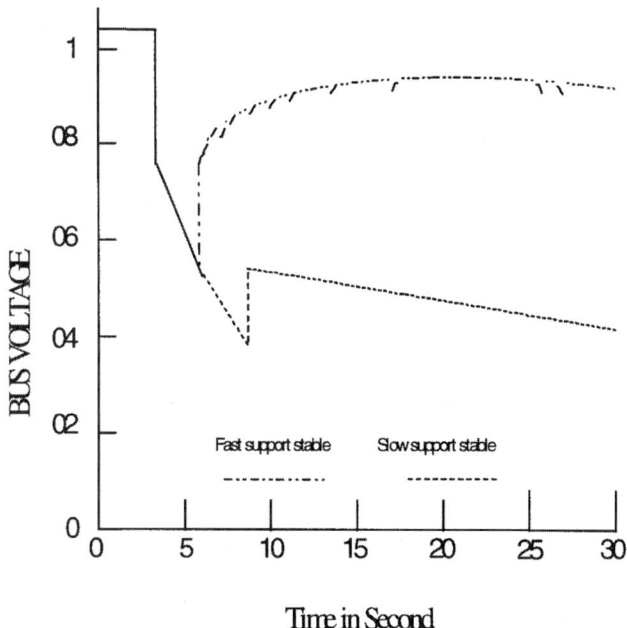

FIGURE 11.18 Simulation of voltage collapse.

reserve" in generators. In general, most of the required reactive power should be supplied locally, with generators supplying primarily active power.

Static Var Compensators (SVC). SVCs, the modern counterpart to the synchronous condenser, are effective in controlling voltage and preventing voltage collapse, but have very definite limitations that must be recognized. Voltage collapse is likely in systems heavily dependent on SVCs when a disturbance exceeding planning criteria takes SVCs to ceiling.

Operate at Higher Voltages. Operating at higher voltage may not increase reactive reserves, but does decrease reactive demand. As such, it can help keep generators away from reactive power limits, and thus help operators maintain control of voltage. The comparison of receiving end Q–V curves for two sending end voltages shows the value of higher voltages.

Undervoltage Load Shedding. A small load reduction, even 5 to 10%, can make the difference between collapse and survival. Manual load shedding is used today for this purpose (some utilities use distribution voltage reduction via SCADA), though it may be too slow to be effective in the case of a severe reactive shortage. Inverse time-undervoltage relays are not widely used, but can be very effective. In a radial load situation, load shedding should be based on primary side voltage. In a steady-state stability problem, the load shed in the receiving system will be most effective even though voltages may be lowest near the electrical center (though shedding load in the vicinity of the lowest voltage may be more easily accomplished, and will be helpful).

Lower Power Factor Generators. Where new generation is close enough to reactive-short areas or areas that may occasionally demand large reactive reserves, a .80 or .85 power factor generator may sometimes be appropriate. However, shunt capacitors with a higher power factor generator having reactive overload capability, may be more flexible and economic.

Use Generator Reactive Overload Capability. Generators should be used as effectively as possible. Overload capability of generators and exciters may be used to delay voltage collapse until operators can change dispatch or curtail load when reactive overloads are modest. To be most useful, reactive overload capability must be defined in advance, operators trained in its use, and protective devices set so as not to prevent its use.

References

Gao, B., Morison, G.K., Kundur, P., Voltage Stability Evaluation Using Modal Analysis, *IEEE Trans. on Power Systems,* 7, 1529–1542, November, 1992.

Hill, D. J., *Nonlinear Dynamic Load Models with Recovery for Voltage Stability Studies,* IEEE Paper #92WM102-4 PWRS.

Lof, P.-A., Andersson, G., Hill, D. J., *Voltage Stability Indices for Stressed Power Systems,* IEEE Paper #92WM101-6 PWRS.

Lof, P.-A., Smed, T., Andersson, G., Hill, D.J., Fast Calculation of a Voltage Stability Index, *IEEE Trans. on Power Systems,* T-PS7, 54–64, February, 1992.

Mansour, Y., Editor, Suggested Techniques for Voltage Stability Analysis, IEEE Special Publication #93TH0620-5-PWR.

Mansour, Y., Editor, *Voltage Stability of Power Systems: Concepts, Analytical Tools, and Industry Experience,* IEEE Special Publication # 90TH0358-2-PWR, 1990.

Venikov, V., *Transient Processes in Electrical Power Systems,* Mir Publishers, Moscow, 1970 and 1980.

Xu, W., Mansour, Y., *Voltage Stability Analysis Using Generic Dynamic Load Models,* IEEE Paper #93WM185-9 PWRS.

Xu, W., Vaahedi, E., Mansour, Y., Tamby, J., *Voltage Stability Load Parameter Determination from Field Tests on B. C. Hydro's System,* IEEE Paper #96SM536-3 PWRS.

11.5 Direct Stability Methods

Vijay Vittal

Direct methods of stability analysis determine the transient stability (as defined in Section 11.1 and described in Section 11.2) of power systems without explicitly obtaining the solutions of the differential equations governing the dynamic behavior of the system. The basis for the method is Lyapunov's second method, also known as Lyapunov's direct method, to determine stability of systems governed by differential equations. The fundamental work of A. M. Lyapunov (1857-1918) on stability of motion was published in Russian in 1893, and was translated into French in 1907 (Lyapunov, 1907). This work received little attention and for a long time was forgotten. In the 1930s, Soviet mathematicians revived these investigations and showed that Lyapunov's method was applicable to several problems in physics and engineering. This revival of the subject matter has spawned several contributions that have led to the further development of the theory and application of the method to physical systems.

The following example motivates the direct methods and also provides a comparison with the conventional technique of simulating the differential equations governing the dynamics of the system. Figure 11.19 shows an illustration of the basic idea behind the use of the direct methods. A vehicle, initially at the bottom of a hill, is given a sudden push up the hill. Depending on the magnitude of the push, the vehicle will either go over the hill and tumble, in which case it is unstable, or the vehicle will climb only part of the way up the hill and return to a rest position (assuming that the vehicle's motion will be damped), i.e., it will be stable. In order to determine the outcome of disturbing the vehicle's equilibrium for a given set of conditions (mass of the vehicle, magnitude of the push, height of the hill, etc.), two different methods can be used:

1. Knowing the initial conditions, obtain a time solution of the equations describing the dynamics of the vehicle and track the position of the vehicle to determine how far up the hill the vehicle will travel. This approach is analogous to the traditional time domain approach of determining stability in dynamic systems.
2. The approach based on Lyapunov's direct method would consist of characterizing the motion of the dynamic system using a suitable Lyapunov function. The Lyapunov function should satisfy certain sign definiteness properties. These properties will be addressed later in this subsection. A natural choice for the Lyapunov function is the system energy. One would then compute the

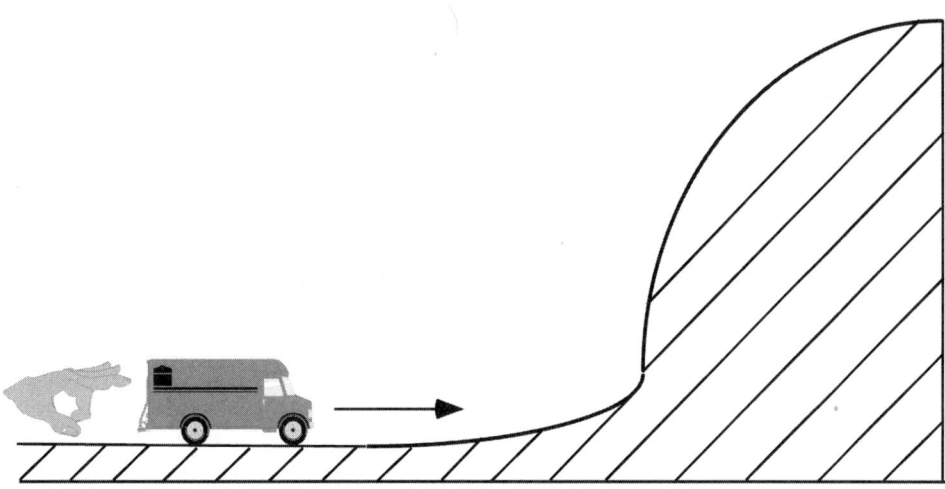

FIGURE 11.19

energy injected into the vehicle as a result of the sudden push, and compare it with the energy needed to climb the hill. In this method, there is no need to track the position of the vehicle as it moves up the hill.

These methods are simple to use if the calculations involve only one vehicle and one hill. The complexity increases if there are several vehicles involved as it becomes necessary to determine (a) which vehicles will be pushed the hardest, (b) how much of the energy is imparted to each vehicle, (c) which direction will they move, and (d) how high a hill must they climb before they will go over the top.

The simple example presented here is analogous to analyzing the stability of a one-machine-infinite-bus power system. The approach presented here is identical to the well-known equal area criterion (Kimbark, 1948; Anderson and Fouad, 1994) which is a direct method for determining transient stability for the one-machine-infinite-bus power system. For a more detailed discussion of the equal area criterion and its relationship to Lyapunov's direct method refer to Pai (1981), chap. 4; Pai (1989), chap. 1; Fouad and Vittal (1992), chap. 3.

Review of Literature on Direct Methods

In the review presented here, we will deal only with work relating to the transient stability analysis of multimachine power systems. In this case the simple example presented above becomes quite complex. Several vehicles which correspond to the synchronous machines are now involved. It also becomes necessary to determine (a) which vehicles will be pushed the hardest, (b) what portion of the disturbance energy is distributed to each vehicle, (c) in which directions the vehicles move, and (d) how high a hill must the vehicles climb before they will go over.

Energy criteria for transient stability analysis were the earliest of all direct methods of multimachine power system transient stability assessment. These techniques were extensions of the equal area criterion to power systems with more than two generators represented by the classical model (Anderson and Fouad, 1994, chap. 2). Researchers from the Soviet Union conducted early work in this area (1930s and 1940s). There were very few results on this topic in Western literature during the same period. In the 1960s the application of Lyapunov's direct method to power systems generated a great deal of activity in the academic community. In most of these investigations, the classical power system model was used. The early work on energy criteria dealt with two main issues: (a) characterization of the system energy, and (b) the critical value of the energy.

Several excellent references that provide a detailed review of the development of the direct methods for transient stability exist. Ribbens-Pavella (1971) and Fouad (1975) are early review papers and provide a comprehensive review of the work done in the period 1960–1975. Detailed reviews of more recent work are conducted in Bose (1984), Ribbens-Pavella and Evans (1985), Fouad and Vittal (1988), and Chiang et al. (1995). The following textbooks provide a comprehensive review and also present detailed descriptions of the various approaches related to direct stability methods: Pai (1981), Pai (1989), Fouad and Vittal (1992), Ribbens-Pavella (1971), and Pavella and Murthy (1994). These references provide a thorough and detailed review of the evolution of the direct methods. In what follows, a brief review of the field and the evolutionary steps in the development of the approaches are presented.

Gorev (1971) first proposed an energy criteria based on the lowest saddle point or unstable equilibrium point (UEP). This work influenced the thinking of power system direct stability researchers for a long time. Magnusson (1947) presented an approach very similar to that of Gorev's and derived a potential energy function with respect to the (posttransient) equilibrium point of the system. Aylett (1958) studied the phase-plane trajectories of multimachine systems using the classical model. An important aspect of this work is the formulation of the system equations based on the intermachine movements. In the period that followed, several important publications dealing with the application of Lyapunov's method to power systems appeared. These works largely dealt with the aspects of obtaining better Lyapunov function, and determining the least conservative estimate of the domain of attraction. Gless (1966) applied Lyapunov's method to the one machine classical model system. El-Abiad and Nagappan (1966) developed a Lyapunov function for multimachine system and demonstrated the approach on a four machine system. The stability results obtained were conservative, and the work that followed this largely dealt with improving the Lyapunov function. A sampling of the work following this line of thought is presented in Willems (1968), Pai et al. (1970), and Ribbens-Pavella (1971). These efforts were followed by the work of Tavora and Smith (1972) dealing with the transient energy of a multimachine system represented by the classical model. They formulated the system equations in the Center of Inertia (COI) reference frame and also in the internode coordinates which is similar to the formulation used by Aylett (1958). Tavora and Smith obtained expressions for the total kinetic energy of the system and the transient kinetic energy, which the authors say determines stability. This was followed by work of Gupta and El-Abiad (1976), which recognized that the UEP of interest is not the one with the lowest energy, but rather the UEP closest to the system trajectory. Uyemura et al. (1996) made an important contribution by developing a technique to approximate the path-dependent terms in the Lyapunov functions by path-independent terms using approximations for the system trajectory.

The work by Athay, Podmore, and colleagues (Athay et al., 1979) is the basis for the transient energy function (TEF) method used today. This work investigated many issues dealing with the application of the TEF method to large power systems. These included:

1. COI formulation and approximation of path-dependent terms.
2. Search for the UEP in the direction of the faulted trajectory.
3. Investigation of the Potential Energy Boundary Surface (PEBS).
4. Application of the technique to power systems of practical sizes.
5. Preliminary investigation of higher-order models for synchronous generators.

This work was followed by the work at Iowa State University by Fouad and colleagues (1981), which dealt with the determination of the correct UEP for stability assessment. This work also identified the appropriate energy for system separation and developed the concept of corrected kinetic energy. Details regarding this work are presented in Fouad and Vittal (1992).

The work that followed largely dealt with developing the TEF method into a more practical tool, and with improving its accuracy, modeling features, and speed. An important development in this area was the work of Bergen and Hill (1981). In this work the network structure was preserved for the classical model. As a result, fast techniques that incorporated network sparsity could be used to solve the problem. A concerted effort was also carried out to extend the applicability of the TEF method to realistic systems. This included improvements in modeling features, algorithms, and computational efficiency. Work related to the large-scale demonstration of the TEF method is found in Carvalho et al. (1986). The work dealing with extending the applicability of the TEF method is presented in Fouad et al. (1986). Significant contributions to this aspect of the TEF method can also be found in Padiyar and Sastry (1987), Padiyar and Ghosh (1989), and Abu-Elnaga et al. (1988).

In Chiang (1985), Chiang et al. (1987), and Chiang et al. (1988), a significant contribution was made to provide an analytical justification for the stability region for multimachine power systems, and a systematic procedure to obtain the controlling UEP was also developed. Zaborsky et al. (1988) also provide a comprehensive analytical foundation for characterizing the region of stability for multimachine power systems.

With the development of a systematic procedure to determine and characterize the region of stability, a significant effort was directed toward the application of direct methods for online transient stability assessment. This work, reported in Waight et al. (1994) and Chadalavada et al. (1997), has resulted in an online tool which has been implemented and used to rank contingencies based on their severity. Another online approach implemented and being used at B.C. Hydro is presented in Mansour et al. (1995). A recent effort with regard to classifying and ranking contingencies quite similar to the one presented in Chadalavada et al. (1997) is described in Chiang et al. (1998).

Some recent efforts (Ni and Fouad, 1987; Hiskens et al., 1992; Jiang et al., 1995) also deal with the inclusion of FACTS devices in the TEF analysis.

The Power System Model

The classical power system model will now be presented. It is the "simplest" power system model used in stability studies and is limited to the analysis of first swing transients. For more details regarding the model, the reader is referred to Anderson and Fouad (1994), Fouad and Vittal (1992), Kundur (1994), and Sauer and Pai (1998). The assumptions commonly made in deriving this model are:

For the synchronous generators

1. Mechanical power input is constant.
2. Damping or asynchronous power is negligible.
3. The generator is represented by a constant EMF behind the direct axis transient (unsaturated) reactance.
4. The mechanical rotor angle of a synchronous generator can be represented by the angle of the voltage behind the transient reactance.

The load is usually represented by passive impedances (or admittances), determined from the predisturbance conditions. These impedances are held constant throughout the stability study. This assumption can be improved using nonlinear models. See Fouad and Vittal (1992), Kundur (1994), and Sauer and Pai (1998) for more details. With the loads represented as constant impedances, all the nodes except the internal generator nodes can be eliminated. The generator reactances and the constant impedance loads are included in the network bus admittance matrix. The generators' equations of motion are then given by

$$M_i \frac{d\omega_i}{dt} = P_i - P_{ei}$$
$$\frac{d\delta_i}{dt} = \omega_i \qquad i = 1, 2, \ldots, n \tag{11.13}$$

where

$$P_{ei} = \sum_{\substack{j=1 \\ j \neq i}}^{n} \left[C_{ij} \sin(\delta_i - \delta_j) + D_{ij} \cos(\delta_i - \delta_j) \right] \tag{11.14}$$

$$P_i = P_{mi} - E_i^2 G_{ii}$$
$$C_{ij} = E_i E_j B_{ij}, \quad D_{ij} = E_i E_j G_{ij}$$

P_{mi} = Mechanical power input
G_{ii} = Driving point conductance
E_i = Constant voltage behind the direct axis transient reactance
ω_i, δ_i = Generator rotor speed and angle deviations, respectively, with respect to a synchronously rotating reference frame
M_i = Inertia constant of generator
$B_{ij}(G_{ij})$ = Transfer susceptance (conductance) in the reduced bus admittance matrix

Equation (11.13) is written with respect to an arbitrary synchronous reference frame. Transformation of this equation to the inertial center coordinates not only offers physical insight into the transient stability problem formulation in general, but also removes the energy associated with the motion of the inertial center which does not contribute to the stability determination. Referring to Eq. (11.13), define

$$M_T = \sum_{i=1}^{n} M_i$$

$$\delta_0 = \frac{1}{M_T} \sum_{i=1}^{n} M_i$$

then,

$$M_T \dot{\omega}_0 = \sum_{i=1}^{n} P_i - P_{ei} = \sum_{i=1}^{n} P_i - 2 \sum_{i=1}^{n-1} \sum_{j=i+1}^{n} D_{ij} \cos \delta_{ij} \tag{11.15}$$

$$\dot{\delta}_0 = \omega_0$$

The generators' angles and speeds with respect to the inertial center are given by

$$\begin{aligned} \theta_i &= \delta_i - \delta_0 \\ \tilde{\omega}_i &= \omega_i - \omega_0 \end{aligned} \quad i = 1, 2, \ldots, n \tag{11.16}$$

and in this coordinate system the equations of motion are given by

$$M_i \dot{\tilde{\omega}}_i = P_i - P_{mi} - \frac{M_i}{M_T} P_{COI}$$
$$\dot{\theta}_i = \tilde{\omega}_i \quad i = 1, 2, \ldots, n \tag{11.17}$$

Review of Stability Theory

A brief review of the stability theory applied to the TEF method will now be presented. This will include a few definitions, some important results, and an analytical outline of the stability assessment formulation.

The definitions and results that are presented are for differential equations of the type shown in Eqs. (11.13) and (11.17). These equations have the general structure given by

$$\dot{x}(t) = f(t, x(t)) \tag{11.18}$$

The system described by Eq. (11.18) is said to be *autonomous* if $f(t, x(t)) \equiv f(x)$, i.e., independent of t and is said to be nonautonomous otherwise.

A point $x_0 \in R^n$ is called an *equilibrium point* for the system [Eq. (11.18)] at time t_0 if $f(t, x_0) \equiv 0$ for all $t \geq t_0$.

An equilibrium point x_e of Eq. (11.18) is said to be an isolated equilibrium point if there exists some neighborhood S of x_e which does not contain any other equilibrium point of Eq. (11.18).

Some precise definitions of stability in the sense of Lyapunov will now be presented. In presenting these definitions, we consider systems of equations described by Eq. (11.18), and also assume that Eq. (11.18) possesses an isolated equilibrium point at the origin. Thus, $f(t, 0) = 0$ for all $t \geq 0$.

The equilibrium $x = 0$ of Eq. (11.18) is said to be *stable* in the sense of Lyapunov, or simply stable if for every real number $\varepsilon > 0$ and initial time $t_0 > 0$ there exists a real number $\delta(\varepsilon, t_0) > 0$ such that for all initial conditions satisfying the inequality $\|x(t_0)\| = \|x_0\| < \delta$, the motion satisfies $\|x(t)\| < \varepsilon$ for all $t \geq t_0$.

The symbol $\|.\|$ stands for a norm. Several norms can be defined on an n-dimensional vector space. Refer to Miller and Michel (1983) and Vidyasagar (1978) for more details. The definition of stability given above is unsatisfactory from an engineering viewpoint, where one is more interested in a stricter requirement of the system trajectory to eventually return to some equilibrium point. Keeping this requirement in mind, the following definition of asymptotic stability is presented.

The equilibrium $x = 0$ of Eq. (11.18) is *asymptotically stable* at time t_0 if

1. $x = 0$ is stable at $t = t_0$

2. For every $t_0 \geq 0$, there exists an $\eta(t_0) > 0$ such that $\underset{t \to \infty}{Lim}\|x(t)\| \to 0$ whenever $\|x(t)\| < \eta$
   (ATTRACTIVITY)

This definition combines the aspect of stability as well as attractivity of the equilibrium. The concept is local, because the region containing all the initial conditions that converge to the equilibrium is some portion of the state space. Having provided the definitions pertaining to stability, the formulation of the stability assessment procedure for power systems is now presented. The system is initially assumed to be at a predisturbance steady-state condition governed by the equations

$$\dot{x}(t) = f^p(x(t)) \quad -\infty < t \leq 0 \tag{11.19}$$

The superscript p indicates predisturbance. The system is at equilibrium, and the initial conditions are obtained from the power flow solution. At $t = 0$, the disturbance or the fault is initiated. This changes the structure of the right-hand sides of the differential equations, and the dynamics of the system are governed by

$$\dot{x}(t) = f^f(x(t)) \quad 0 < t \leq t_{cl} \tag{11.20}$$

where the superscript f indicates faulted conditions. The disturbance or the fault is removed or cleared by the protective equipment at time t_{cl}. As a result, the network undergoes a topology change and the right-hand sides of the differential equations are again altered. The dynamics in the postdisturbance or postfault period are governed by

$$\dot{x}(t) = f(x(t)) \quad t_{cl} < t \leq \infty \tag{11.21}$$

The stability analysis is done for the system in the postdisturbance period. The objective is to ascertain asymptotic stability of the postdisturbance equilibrium point of the system governed by Eq. (11.21). This is done by obtaining the domain of attraction of the postdisturbance equilibrium and determining if the initial conditions of the postdisturbance period lie within this domain of attraction or outside it. The domain of attraction is characterized by the appropriately determined value of the transient energy function. In the literature survey presented previously, several approaches to characterize the domain of attraction were mentioned. In earlier approaches (El-Abiad and Nagappan, 1966; Tavora and Smith, 1972), this was done by obtaining the unstable equilibrium points (UEP) of the postdisturbance system and determining the one with the lowest level of potential energy with respect to the postdisturbance equilibrium. This value of potential energy then characterized the domain of attraction. In the work that followed, it was found that this approach provided very conservative results for power systems. In Gupta and El-Abiad (1976), it was recognized that the appropriate UEP was dependent on the fault location, and the concept of closest UEP was developed. An approach to determine the domain of attraction was also presented by Kakimoto and colleagues (1978; 1981) based on the concept of the potential energy boundary surface (PEBS). For a given disturbance trajectory, the PEBS describes a "local" approximation of the stability boundary. The process of finding this local approximation is associated with the determination of the stability boundary of a lower dimensional system (see Fouad and Vittal [1992], chap. 4 for details). It is formed by joining points of maximum potential energy along any direction originating from the postdisturbance stable equilibrium point. The PEBS constructed in this manner is orthogonal to the equipotential curves. In addition, along the direction orthogonal to the PEBS, the potential energy achieves a local maximum at the PEBS. In Athay et al. (1979), several simulations on realistic systems were conducted. These simulations, together with the synthesis of previous results in the area led to the development of a procedure to determine the correct UEP to characterize the domain of attraction. The results obtained were much improved, but in terms of practical applicability there was room for improvement. The work presented in Fouad et al. (1981) and Carvalho et al. (1986) made several important contributions to determining the correct UEP. The term *controlling UEP* was established, and a systematic procedure to determine the controlling UEP was developed. This will be described later. In Chiang et al. (1985; 1987; 1988), a thorough analytical justification for the concept of the controlling UEP and the characterization of the domain of attraction was developed. This provides the analytical basis for the application of the TEF method to power systems. These analytical results in essence show that the stability boundary of the postdisturbance equilibrium point is made up of the union of the stable manifolds of those unstable equilibrium points contained on the stability boundary. The boundary is then approximated locally using the energy function evaluated at the controlling UEP. The conceptual framework of the TEF approach is illustrated in Fig. 11.20.

The Transient Energy Function

The TEF can be derived from Eq. (11.17) using first principles. Details of the derivation can be found in Pai (1981), Pai (1989), Fouad and Vittal (1992), Athay et al. (1979). For the power system model considered in Eq. (11.17), the TEF is given by

$$V = \frac{1}{2}\sum_{i=1}^{n} M_i \tilde{\omega}_i^2 - \sum_{i=1}^{n} P_i\left(\theta_i - \theta_i^{s2}\right) - \sum_{i=1}^{n-1}\sum_{j=i+1}^{n}\left[C_{ij}\cos\left(\theta_{ij} - \theta_{ij}^{s2}\right) - \int_{\theta_i^{s2}+\theta_j^{s2}}^{\theta_i+\theta_j} D_{ij}\cos\theta_{ij}\,d\left(\theta_i + \theta_j\right)\right] \quad (11.22)$$

where $\theta_{ij} = \theta_i - \theta_j$.

The first term on the right-hand side of Eq. (11.22) is the kinetic energy. The next three terms represent the potential energy. The last term is path dependent. It is usually approximated (Uyemura et al., 1996; Athay et al., 1979) using a straight line approximation for the system trajectory. The integral between two points θ^a and θ^b is then given by

$$I_{ij} = D_{ij}\frac{\theta_i^b - \theta_i^a + \theta_j^b - \theta_j^a}{\theta_{ij}^b - \theta_{ij}^a}\left(\sin\theta_{ij}^b - \sin\theta_{ij}^a\right). \quad (11.23)$$

In Fouad et al. (1981), a detailed analysis of the energy behavior along the time domain trajectory was conducted. It was observed that in all cases where the system was stable following the removal of a

Power System Dynamics and Stability

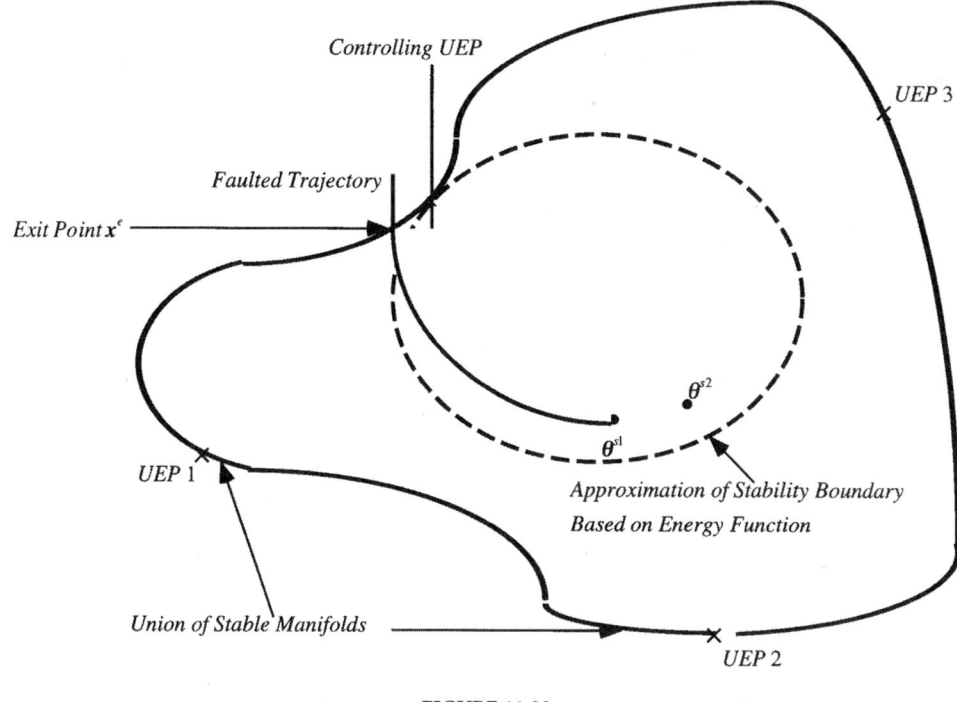

FIGURE 11.20

disturbance, a certain amount of the total kinetic energy in the system was not absorbed. This indicates that not all the kinetic energy created by the disturbance, contributes to the instability of the system. Some of the kinetic energy is responsible for the intermachine motion between the generators and does not contribute to the separation of the severely disturbed generators from the rest of the system. The kinetic energy associated with the gross motion of k machines having angular speeds $\tilde{\omega}_1, \tilde{\omega}_2, \ldots, \tilde{\omega}_k$ is the same as the kinetic energy of their inertial center. The speed of the inertial center of that group and its kinetic energy are given by

$$\tilde{\omega}_{cr} = \sum_{i=1}^{k} M_i \tilde{\omega}_i \bigg/ \sum_{i=1}^{k} M_i \quad (11.24)$$

$$V_{KE_{cr}} = \frac{1}{2}\left[\sum_{i=1}^{k} M_i\right]\left(\tilde{\omega}_{cr}\right)^2 \quad (11.25)$$

The disturbance splits the generators of the system into two groups: the critical machines and the rest of the generators. Their inertial centers have inertia constants and angular speeds $M_{cr}, \tilde{\omega}_{cr}$ and $M_{sys}, \tilde{\omega}_{sys}$, respectively. The kinetic energy causing the separation of the two groups is the same as that of an equivalent one-machine-infinite-bus system having inertia constant M_{eq} and angular speed $\tilde{\omega}_{eq}$ given by

$$M_{eq} = \frac{M_{cr} \times M_{sys}}{M_{eq} + M_{sys}}$$

$$\tilde{\omega}_{eq} = \left(\tilde{\omega}_{cr} - \tilde{\omega}_{sys}\right) \quad (11.26)$$

and the corresponding kinetic energy is given by

$$V_{KE_{corr}} = \frac{1}{2} M_{eq} \left(\tilde{\omega}_{eq} \right)^2 \quad (11.27)$$

The kinetic energy term in Eq. (11.22) is replaced by Eq. (11.27).

Transient Stability Assessment

As described previously, the transient stability assessment using the TEF method is done for the final postdisturbance configuration. The stability assessment is done by comparing two values of the transient energy V. The value of V is computed at the end of the disturbance. If the disturbance is a simple fault, the value of V at fault clearing V_{cl} is evaluated.

The other value of V that largely determines the accuracy of the stability assessment is the critical value of V, V_{cr}, which is the potential energy at the controlling UEP for the particular disturbance being investigated.

If $V_{cl} < V_{cr}$, the system is stable, and if $V_{cl} > V_{cr}$, the system is unstable. The assessment is made by computing the energy margin ΔV given by

$$\Delta V = V_{cr} - V_{cl} \quad (11.28)$$

Substituting for V_{cr} and V_{cl} from Eq. (11.22) and invoking the linear path assumption for the path dependent integral between the conditions at the end of the disturbance and the controlling UEP, we have

$$\Delta V = -\frac{1}{2} M_{eq} \tilde{\omega}_{eq}^{cl\,2} - \sum_{i=1}^{n} P_i \left(\theta_i^u - \theta_i^{cl} \right)$$
$$- \sum_{i=1}^{n-1} \sum_{j=i+1}^{n} \left[C_{ij} \left(\cos\theta_{ij}^u - \cos\theta_{ij}^{cl} \right) \right] - D_{ij} \frac{\theta_i^u - \theta_i^{cl} + \theta_j^u - \theta_j^{cl}}{\left(\theta_{ij}^u - \theta_{ij}^{cl} \right)} \left(\sin\theta_{ij}^u - \sin\theta_{ij}^{cl} \right) \quad (11.29)$$

where $(\theta^{cl}, \tilde{\omega}^{cl})$ are the conditions at the end of the disturbance and $(\theta^u, \mathbf{0})$ represents the controlling UEP. If ΔV is greater than zero the system is stable, and if ΔV is less than zero, the system is unstable. A qualitative measure of the degree of stability (or instability) can be obtained if ΔV is normalized with respect to the corrected kinetic energy at the end of the disturbance (Fouad et al., 1981).

$$\Delta V_n = \Delta V / V_{KE_{corr}} \quad (11.30)$$

For a detailed description of the computational steps involved in the TEF analysis, refer to Fouad and Vittal (1992), chap. 6.

Determination of the Controlling UEP

A detailed description of the rationale in developing the concept of the controlling UEP is provided in Fouad and Vittal (1992), section 5.4. A criterion to determine the controlling UEP based on the normalized energy margin is also presented. The criterion is stated as follows. The postdisturbance trajectory approaches (if the disturbance is large enough) the controlling UEP. This is the UEP with the lowest normalized potential energy margin. The determination of the controlling UEP involves the following key steps:

1. Identifying the correct UEP.
2. Obtaining a starting point for the UEP solution close to the exact UEP.
3. Calculation of the exact UEP.

Identifying the correct UEP involves determining the advanced generators for the controlling UEP. This is referred to as the mode of disturbance (MOD). These generators generally are the most severely disturbed generators due to the disturbance. The generators in the MOD are not necessarily those that lose synchronism. The computational details of the procedure to identify the correct UEP and obtain a starting point for the exact UEP solution are provided in Fouad and Vittal (1992), section 6.6. An outline of the procedure is provided below:

1. Candidate modes to be tested by the MOD test depend on how the disturbance affects the system. The selection of the candidate modes is based on several disturbance severity measures obtained at the end of the disturbance. These severity measures include kinetic energy and acceleration. A ranked list of machines is obtained using the severity measures. From this ranked list, the machines or group of machines at the bottom of the list are included in the group forming the rest of the system and $V_{KE_{corr}}$ is calculated. In a sequential manner, machines are successively added to the group forming the rest of the system and $V_{KE_{corr}}$ is calculated and stored.
2. The list of $V_{KE_{corr}}$ calculated above is sorted in descending order and only those groups within 10% of the maximum $V_{KE_{corr}}$ in the list are retained.
3. Corresponding to the MOD for each of the retained groups of machines in step 2, an approximation to the UEP corresponding to that mode is constructed using the postdisturbance stable equilibrium point. For a given candidate mode, where machines i and j are contained in the critical group, an estimate of the approximation to the UEP for an n-machine system is given by $\left[\theta_{ij}^u\right]^T = \left[\theta_1^{s2}, \theta_2^{s2}, ..., [\pi-\theta_i^{s2}], ..., [\pi-\theta_j^{s2}], ..., \theta_n^{s2}\right]$. This estimate can be further improved by accounting for the motion of the COI, and using the concept of the PEBS to maximize the potential energy along the ray drawn from the estimate and the postdisturbance stable equilibrium point θ^{s2}.
4. The normalized potential energy margin for each of the candidate modes is evaluated at the approximation to the exact UEP, and the mode corresponding to the lowest normalized potential energy margin is then selected as the mode of the controlling UEP.
5. Using the approximation to the controlling UEP as a starting point, the exact UEP is obtained by solving the nonlinear algebraic equation given by

$$f_i = P_i - P_{mi} - \frac{M_i}{M_T} P_{COI} = 0 \quad i = 1, 2, ..., n \tag{11.31}$$

The solution of these equations is a computationally intensive task for realistic power systems. Several investigators have made significant contributions to determining an effective solution. A detailed description of the numerical issues and algorithms to determine the exact UEP solution are beyond the scope of this handbook. Several excellent references that detail these approaches are available. These efforts are described in Fouad and Vittal (1992), section 6.8.

The BCU (Boundary Controlling UEP) Method

The BCU method (Chiang et al., 1985, 1987, 1988) provides a systematic procedure to determine a suitable starting point for the controlling UEP solution. The main steps in the procedure are as follows:

1. Obtain the faulted trajectory by integrating the equations

$$M_i \dot{\tilde{\omega}}_i = P_i^f - P_{ei}^f - \frac{M_i}{M_T} P_{COI}^f$$

$$\dot{\theta}_i = \tilde{\omega}_i, \quad i = 1, 2, ..., n \tag{11.32}$$

Values of θ obtained from Eq. (11.32) are substituted in the postfault mismatch equation given by Eq. (11.31). The exit point x^e is then obtained by satisfying the condition $\sum_{i=1}^{n} -f_i \tilde{\omega}_i = 0$.

2. Using θ^e as the starting point, integrate the associated gradient system equations given by

$$\dot{\theta}_i = P_i - P_{ei} - \frac{M_i}{M_T} P_{COI}, \quad i = 1, 2, \ldots, n-1$$

$$\theta_n = -\sum_{i=1}^{n-1} M_i \theta_i / M_n$$
(11.33)

At each step of the integration, evaluate $\sum_{i=1}^{n} |f_i| = F$ and determine the first minimum of F along the gradient surface. Let θ^* be the vector of rotor angles at this point.
3. Using θ^* as a starting point in Eq. (11.31), obtain the exact solution for the controlling UEP.

Applications of the TEF Method and Modeling Enhancements

The preceding subsections have provided the important steps in the application of the TEF method to analyze the transient stability of multimachine power systems. In this subsection, a brief mention of the applications of the technique and enhancements in terms of modeling detail and application to realistic power systems is provided. Inclusion of detailed generator models and excitation systems in the TEF method are presented in Athay et al. (1979), Fouad et al. (1986), and Waight et al. (1994). The sparse formulation of the system to obtain more efficient solution techniques is developed in Bergen and Hill (1981), Abu-Elnaga et al. (1988), and Waight et al. (1994). The application of the TEF method for a wide range of problems including dynamic security assessment are discussed in Fouad and Vittal (1992), chaps. 9–10; Chadalavada et al. (1997); and Mansour et al. (1995). The availability of a qualitative measure of the degree of stability or instability in terms of the energy margin makes the direct methods an attractive tool for a wide range of problems. The modeling enhancements that have taken place and the continued development in terms of computational efficiency and computer hardware, make direct methods a viable candidate for online transient stability assessment (Waight et al., 1994; Chadalavada et al., 1997; Mansour et al., 1995). This feature is particularly effective in the competitive market environment to calculate operating limits with changing conditions. There are several efforts underway dealing with the development of direct methods and a combination of time simulation techniques for online transient stability assessment. These approaches take advantage of the superior modeling capability available in the time simulation engines, and use the qualitative measure provided by the direct methods to derive preventive and corrective control actions and estimate limits. This line of investigation has great potential and could become a vital component of energy control centers in the near future.

References

Abu-Elnaga, M. M., El-Kady, M. A., and Findlay, R. D., Sparse formulation of the transient energy function method for applications to large-scale power systems, *IEEE Trans. Power Syst.*, PWRS-3, 4, 1648–1654, November 1988.
The All Union Institute of Scientific and Technological Information and the Academy of Sciences of the USSR (in Russia). *Criteria of Stability of Electric Power Systems*. Electric Technology and Electric Power Series, Moscow, 1971 (in Russian).
Anderson, P. M., and Fouad, A. A., *Power System Control and Stability*, IEEE Press, New York, 1994.
Athay, T., Sherkat, V. R., Podmore, R., Virmani, S., and Puech, C., *Transient Energy Stability Analysis, Systems Engineering For Power: Emergency Operation State Control — Section IV*, U.S. Department of Energy Publication No. CONF-790904-PL, 1979.
Aylett, P. D., The energy-integral criterion of transient stability limits of power systems, in *Proceedings of Institution of Electrical Engineers*, 105C, 8, London, September 1958, 527–536.

Bergen, A. R., and Hill, D. J., A structure preserving model for power system stability analysis, *IEEE Trans. Power Appar. Syst.*, PAS-100, 1, 25–35, January 1981.

Bose, A., Chair, IEEE Committee Report, Application of direct methods to transient stability analysis of power systems, *IEEE Trans. Power Appar. Syst.*, PAS-103, 7, 1629–1630, July 1984.

Carvalho, V. F., El-Kady, M. A., Vaahedi, E., Kundur, P., Tang, C. K., Rogers, G., Libaque, J., Wong, D., Fouad, A. A., Vittal, V., Rajagopal, S., Demonstration of large scale direct analysis of power system transient stability, Electric Power Research Institute Report EL-4980, December 1986.

Chadalavada, V., V. Vittal, et al., An on-line contingency filtering scheme for dynamic security assessment, *IEEE Trans. Power Syst.*, 12, 1, 153–161, February 1997.

Chiang, H.-D., A theory-based controlling UEP method for direct analysis of power system transient stability, in *Proceedings of the 1989 International Symposium on Circuits and Systems*, 3, 65–69, 1985.

Chiang, H-D., Chiu, C. C., and Cauley, G., Direct stability analysis of electric power systems using energy functions: Theory, application, and perspective, *IEEE Proceedings*, 83, 11, 1497–1529, November 1995.

Chiang, H. D., Wang, C. S., and Li, H., Development of BCU classifiers for on-line dynamic contingency screening of electric power systems, Paper No. PE-349, IEEE Power Engineering Society Summer Power Meeting, San Diego, CA, July 1998.

Chiang, H.-D., Wu, F. F., and Varaiya, P. P., Foundations of the direct methods for power system transient stability analysis, *IEEE Trans. Circuits and Syst.*, 34, 160–173, February 1987.

Chiang, H.-D., Wu, F. F., and Varaiya, P. P., Foundations of the potential energy boundary surface method for power system transient stability analysis, *IEEE Trans. Circuits and Syst.*, 35, 6, 712–728, June 1988.

El-Abiad, A. H., and Nagappan, K., Transient stability regions of multi-machine power systems, *IEEE Trans. Power Appar. Syst.*, PAS-85, 2, 169–178, February 1966.

Fouad, A. A., Stability theory-criteria for transient stability, in *Proceedings of the Engineering Foundation Conference on System Engineering for Power, Status and Prospects*-NIT Publication No. Conf.-750867, August 1975.

Fouad, A. A., Kruempel, K. C., Mamandur, K. R. C., Stanton, S. E., Pai, M. A., and Vittal, V., Transient stability margin as a tool for dynamic security assessment, EPRI Report EL-1755, March 1981.

Fouad, A. A., Vittal, V., Ni, Y. X., Pota, H. R., Nodehi, K., and Oh, T. K., Extending application of the transient energy function method, Report EL-4980, Palo Alto, CA, EPRI, 1986.

Fouad, A. A., and Vittal, V., *Power System Transient Stability Analysis Using the Transient Energy Function Method*, Prentice-Hall, Inc., New Jersey, 1992.

Fouad, A. A., and Vittal, V., The transient energy function method, *International Journal of Electric Power and Energy Systems*, 10, 4, 233–246, October 1988.

Gless, G. E., Direct method of Liapunov applied to transient power system stability, *IEEE Trans. Power Appar. Syst.*, PAS-85, 2, 159–168, February 1966.

Gupta, C. L., and El-Abiad, A. H., Determination of the closest unstable equilibrium state for Lyapunov's method in transient stability studies, *IEEE Trans. Power Appar. Syst.*, PAS-95, 1699–1712, September/October 1976.

Hiskens, I. A., et al., Incorporation of SVC into energy function method, *IEEE Trans. Power Syst.*, PWRS, 7, 133–140, February 1992.

Jing, C., et al. Incorporation of HVDC and SVC models in the Northern States Power Co. (NSP) network for on-line implementation of direct transient stability assessment, *IEEE Trans. on Power Syst.*, 10, 2, 898–906, May 1995.

Kakimoto, N., and Hayashi, M., Transient stability analysis of multimachine power systems by Lyapunov's direct method, in *Proceedings of 20th Conference on Decision and Control*, San Diego, CA, 1981.

Kakimoto, N., Ohsawa, Y., and Hayashi, M., Transient stability analysis of electric power system via Lure-Type Lyapunov function, Parts I and II, *Transactions IEE of Japan*, 98, 516, 1978.

Kakimoto, N., Ohsawa, Y., and Hayashi, M., Transient stability analysis of large-scale power systems by Lyapunov's direct method, *IEEE Trans. Power App. Syst.*, 103(1), 160–167, January 1978.

Kimbark, E. W., *Power System Stability*, I, John Wiley & Sons, New York, 1948.

Kundur, P., *Power System Stability and Control*, McGraw-Hill, New York, 1994.

Lyapunov, M. A., Problème Général de la Stabilité du Mouvement, *Ann. Fac. Sci. Toulouse*, 9, 203–474, 1907: (French, translation of the original paper published in 1893 in *Comm. Soc. Math. Kharkow*; reprinted as Vol. 17 in *Annals of Mathematical Studies*, Princeton, 1949).

Magnusson, P. C., Transient energy method of calculating stability, *AIEE Trans.*, 66, 747–755, 1947.

Mansour, Y., Vaahedi, E., Chang, A. Y., Corns, B. R., Garrett, B. W., Demaree, K., Athay, T., and Cheung, K., B. C. Hydro's on-line transient stability assessment (TSA): Model development, analysis, and post-processing, *IEEE Trans. Power Syst.*, 10, 1, 241–253, Feb. 1995.

Miller, R. K., and A. N. Michel, *Ordinary Differential Equations*, Academic Press, New York, 1983.

Ni, Y.-X., and A. A. Fouad, A simplified two terminal HVDC model and its use in direct transient stability assessment, *IEEE Trans. Power Syst.*, PWRS-2, 4, 1006–1013, November 1987.

Padiyar, K. R., and Ghosh, K. K., Direct stability evaluation of power systems with detailed generator models using structure preserving energy functions, *Int. J. Electr. Power and Energy Syst.*, 11, 1, 47–56, Jan. 1989.

Padiyar, K. R., and Sastry, H. S. Y., Topological energy function analysis of stability of power systems, *Int. J. of Electr. Power and Energy Syst.*, 9, 1, 9–16, Jan. 1987.

Pai, M. A., *Energy Function Analysis for Power System Stability*, Kluwer Academic Publishers, Boston, 1989.

Pai, M. A., *Power System Stability*, North-Holland Publishing Co., Amsterdam, 1981.

Pai, M. A., Mohan, A., Gopala Rao, J., Power system transient stability regions using Popov's method, *IEEE Trans. Power Appar. Syst.*, PAS-89, 5, 788–794, May/June 1970.

Pavella, M., and Murthy, P. G., *Transient Stability of Power Systems: Theory and Practice*, John Wiley & Sons, Inc., New York, 1994.

Ribbens-Pavella, M., Critical survey of transient stability studies of multi-machine power systems by Lyapunov's direct method, in *Proceedings of 9th Annual Allerton Conference on Circuits and System Theory*, Oct. 1971.

Ribbens-Pavella, M., Transient stability of multi-machine power systems by Lyapunov's direct method, IEEE Winter Power Meeting Conference Paper, 1971.

Ribbens-Pavella, M., and Evans, F. J., Direct methods for studying of the dynamics of large scale electric power systems — a survey, *Automatica*, 21, 1, 1–21, 1985.

Sauer, P. W., and M. A. Pai, *Power System Dynamics and Stability*, Prentice Hall, New York, 1998.

Tavora, Carlos J., and Smith, O. J. M., Characterization of equilibrium and stability in power systems, *IEEE Trans. Power Appar. Syst.*, PAS-72, 1127–1130, May/June.

Tavora, Carlos J., and Smith, O. J. M., Equilibrium analysis of power systems, *IEEE Trans. Power Appar. Syst.*, PAS-72, 1131–1137, May/June.

Tavora, Carlos J., and Smith, O. J. M., Stability analysis of power systems, *IEEE Trans. Power Appar. Syst.*, PAS-72, 1138–1144, May/June.

Uyemura, K., Matsuki, J., Yamada, I., and Tsuji, T., Approximation of an energy function in transient stability analysis of power systems, *Electrical Engineering in Japan*, 92, 6, 96–100, November/December.

Vidyasagar, M., *Nonlinear Systems Analysis*, Prentice-Hall, Englewood Cliffs, NJ, 1978.

Waight, J. G., et al., Analytical methods for contingency selection and ranking for dynamic security analysis, Report TR-104352, Palo Alto, September 1994.

Willems, J. L., Improved Lyapunov function for transient power-system stability, in *Proceedings of the Institution of Electrical Engineers*, 115, 9, London, 1315–1317, September 1968.

Zaborszky, J., Huang, G., Zheng, B., and Leung, T.-C., On the phase-portrait of a class of large nonlinear dynamic systems such as the power system, *IEEE Trans. on Automatic Control*, 32, 4–15, January 1988.

11.6 Power System Stability Controls[1]

Carson W. Taylor

Power system synchronous or angle instability phenomenon limits power transfer, especially where transmission distances are long. This is well recognized and many methods have been developed to improve stability and increase allowable power transfers.

The synchronous stability problem has been fairly well solved by fast fault clearing, thyristor exciters, power system stabilizers, and a variety of other stability controls such as generator tripping. Fault clearing of severe short circuits can be less than three cycles (50 ms for 60 Hz frequency), and the effect of the faulted line outage on generator acceleration and stability may be greater than that of the fault itself.

Nevertheless, requirements for more intensive use of available generation and transmission, more onerous load characteristics, greater variation in power schedules, and industry restructuring pose new concerns. Recent large-scale power failures have heightened the concerns.

In this section we describe the state-of-the art of power system angle stability controls. Controls for voltage stability are described in another section of this chapter and in other literature (CIGRE Brochure No. 101, 1995; CIGRE Brochure No. 128, 1998; IEEE THO 596-7 PWR, 1993; Taylor 1994; Van Cutsem and Vournas, 1998).

We emphasize controls employing relatively new technologies that have actually been implemented by electric power companies, or that are seriously being considered for implementation. The technologies include applied control theory, power electronics, microprocessors, signal processing, transducers, and communications.

Power system stability controls must be effective and robust. Effective in an engineering sense means "cost-effective." Control robustness is the capability to operate appropriately for a wide range of power system operating and disturbance conditions.

Review of Power System Synchronous Stability Basics

Many publications, for example CIGRE Brochure No. 111 (1996), Kundur (1994), and IEEE Discrete Supplementary Controls Task Force (1978), describe the basics, which are briefly reviewed here. Power generation is largely by synchronous generators, which may be interconnected over thousands of kilometers in very large power systems. Thousands of generators must operate in synchronism during normal and disturbance conditions. Loss of synchronism of a generator or group of generators with respect to another group of generators is termed *instability* and could result in expensive widespread power blackouts.

The essence of synchronous stability is balance of individual generator electrical and mechanical torques as described by Newton's second law applied to rotation:

$$J \frac{d\omega}{dt} = T_m - T_e$$

where J is the moment of inertia of the generator and prime mover, ω is speed, T_m is mechanical prime mover torque, and T_e is electrical torque related to generator electric power output. The generator speed determines the generator rotor angle changes relative to other generators. Figure 11.21 shows the basic "swing equation" block diagram relationship for a generator connected to a power system.

The conventional equation form and notation are used. The block diagram is explained as follows:

---

[1]This section is adapted from Chapter 1 of the CIGRE report, *Advanced Angle Stability Controls* (1999), which is based on U.S. government work not covered by copyright.

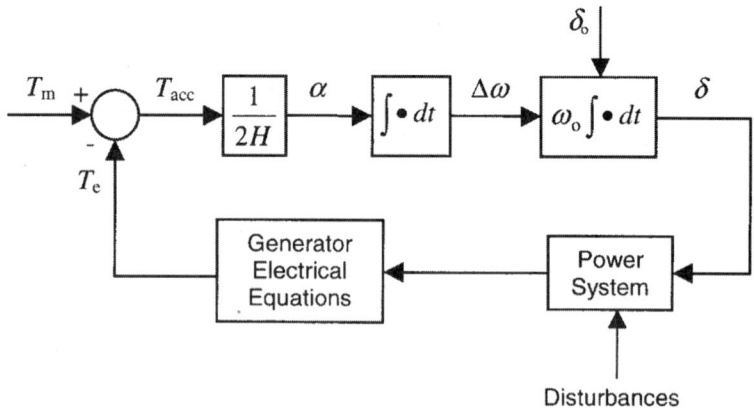

FIGURE 11.21 Block diagram of generator electromechanical dynamics.

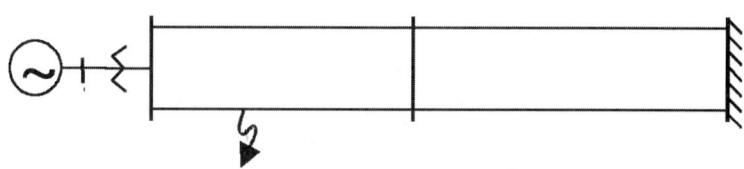

FIGURE 11.22 Remote power plant to large system. Short-circuit location is shown.

- The inertia constant, H, is proportional to the moment of inertia and is the kinetic energy at rated speed divided by the generator MVA rating. Units are MW-seconds/MVA, or seconds.
- T_m is mechanical torque in per unit. As a first approximation it is assumed to be constant. It is, however, influenced by speed controls (governors) and prime mover and energy supply system dynamics.
- ω_o is rated frequency in radians/second.
- δ_o is predisturbance rotor angle in radians relative to a reference generator.
- The power system block comprises the transmission network, loads, power electronic devices, and other generators/prime movers/energy supply systems with their controls. The transmission network is generally represented by algebraic equations. Loads and generators are represented by algebraic and differential equations.
- Disturbances include short circuits, and line and generator outages. A severe disturbance is a three-phase short circuit near the generator. This causes electric power and torque to be zero, with accelerating torque equal to T_m. (Although generator current is very high for the short circuit, the power factor, active current, and active power are close to zero.)

The generator electrical equations block represents the internal generator dynamics. Figure 11.22 shows a simple conceptual model: a remote generator connected to a large power system by two parallel transmission lines with an intermediate switching station. With some approximations adequate for a second or more following a disturbance, the Fig. 11.23 block diagram is realized. The basic relationship between power and torque is $P = T\omega$. Since speed changes are quite small, power is considered equal to torque in per unit. The generator representation is a constant voltage, E', behind a reactance. The transformer and transmission lines are represented by inductive reactances. Using the relation $S = E'I^*$, the generator electrical power is the well-known relation:

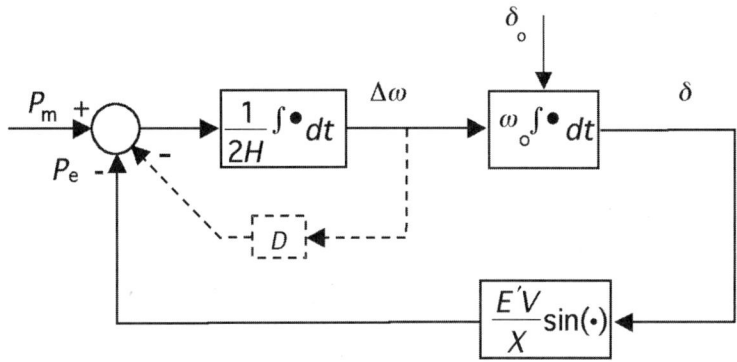

FIGURE 11.23

$$P_e = \frac{E'V}{X}\sin\delta$$

where V is the large system (infinite bus) voltage and X is the total reactance from the generator internal voltage to the large system. The above equation approximates characteristics of a detailed, large-scale model, and illustrates that the power system is fundamentally a highly nonlinear system for large disturbances.

Figure 11.24a shows the relation graphically. The predisturbance operating point is at the intersection of the load or mechanical power characteristic and the electrical power characteristic. Normal stable operation is at δ_o. For example, a small increase in mechanical power input causes an accelerating power that increases δ to increase P_e until accelerating power returns to zero. The opposite is true for the unstable operating point at $\pi - \delta_o$. δ_o is normally less than 45°.

During normal operation, mechanical and electrical torques are equal and a generator runs at close to 50 or 60 Hz rated frequency. If, however, a short circuit occurs (usually with removal of a transmission line), the electric power output will be partially blocked from reaching loads (momentarily) and the generator (or group of generators) will accelerate with increase in generator speed and angle. If the acceleration relative to other generators is too great, synchronism will be lost. Loss of synchronism is an unstable, runaway situation with large variations of voltages and currents that will normally cause protective separation of a generator or a group of generators. Following short-circuit removal, the electrical torque and power developed as angle increases will decelerate the generator. If deceleration reverses angle swing prior to $\pi - \delta'_o$, stability can be maintained at new operating point δ'_o (Fig. 11.24). If the swing is beyond $\pi - \delta'_o$, accelerating power/torque again becomes positive resulting in runaway increase in angle and speed, and instability. Figure 11.24a illustrates the equal area stability criterion for "first swing" stability. If the decelerating area (energy) above the mechanical power load line is greater than the accelerating area below the load line, stability is maintained.

Stability controls increase stability by decreasing the accelerating area or increasing the decelerating area. This may be done by either increasing the electrical power–angle relation, or by decreasing the mechanical power input.

For small disturbances, the block diagram, Fig. 11.23, can be linearized. The block diagram would then be that of a second-order differential equation oscillator. For a remote generator connected to a large system, the oscillation frequency is 0.8 to 1.1 Hz.

Figure 11.23 also shows a damping path (dashed, damping power/torque in phase with speed) that represents mechanical or electrical damping mechanisms in the generator, turbine, loads, and other devices. Mechanical damping arises from the turbine torque-speed characteristic, friction and windage, and components of prime mover control in phase with speed. At an oscillation frequency, the electrical

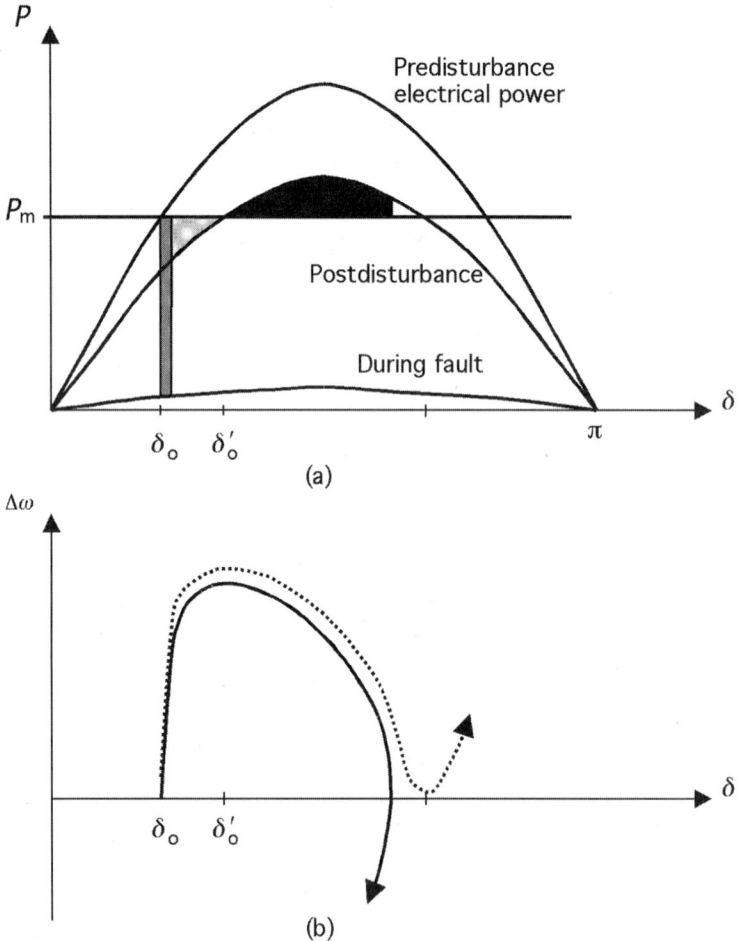

FIGURE 11.24 (a) Power angle curve and equal area criterion. Dark shading for acceleration energy during fault. Light shading for additional acceleration energy because of line outage. Black shading for deceleration energy. (b) Angle-speed phase plane. Dotted line is for unstable case.

power can be resolved into a component in phase with angle (synchronizing power) and a quadrature (90° leading) component in phase with speed (damping power). Controls, notably generator automatic voltage regulators with high gain, can introduce negative damping at some oscillation frequencies. (In any feedback control system, high gain combined with time delays can cause positive feedback and instability.) For stability, the net damping must be positive for both normal conditions and for large disturbances with outages.

Stability controls may also be added to improve damping. In some cases, stability controls are designed to improve both synchronizing and damping torques of generators.

The above analysis can be generalized to large systems. For first swing stability, synchronous stability between two critical groups of generators is of concern. For damping, many oscillation modes are present, all of which require positive damping. The low frequency modes (0.1–0.8 Hz) are most difficult to damp. These modes represent interarea oscillations between large portions of a power system.

Concepts of Power System Stability Controls

Figure 11.25 shows the general structure for analysis of power system stability and for development of power system stability controls.

Power System Dynamics and Stability

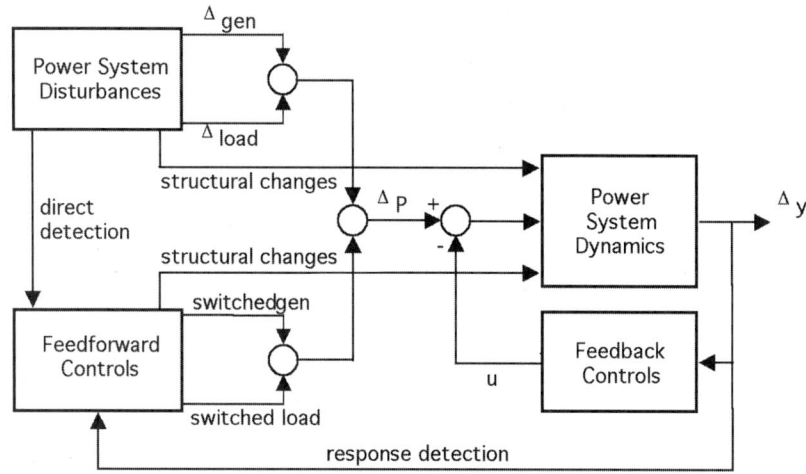

FIGURE 11.25 General power system structure showing stability controls. (From Hauer, J. F., Robustness Issues in Stability Control of Large Electric Power Systems, in *32nd IEEE Conf. Decision and Control*, San Antonio, TX, Dec. 15–17, 1993. With permission.)

Stability problems typically involve disturbances such as short circuits, with subsequent removal of faulted elements. Generation or load may be lost, resulting in generation–load imbalance and frequency excursions. These disturbances stimulate power system electromechanical dynamics. Improperly designed or tuned controls may contribute to stability problems; as mentioned, one example is negative damping torques caused by generator automatic voltage regulators.

Because of power system synchronizing and damping forces (including the feedback controls shown on Fig. 11.25), stability is maintained for most disturbances and operating conditions.

Feedback Controls

The most important feedback (closed-loop) controls are the generator excitation controls (automatic voltage regulator often including power system stabilizer). Other feedback controls include prime mover controls, controls for reactive power compensation such as static var systems, and special controls for HVDC links. These controls are generally linear, continuously active, and based on local measurements.

There are, however, interesting possibilities for very effective discontinuous feedback controls with microprocessors facilitating implementation. Discontinuous controls have certain advantages over continuous controls. Continuous feedback controls are potentially unstable. In complex power systems, continuously controlled equipment may cause adverse modal interactions (Hauer, 1989). Modern digital controls, however, can be discontinuous, and take no action until variables are out-of-range. This is analogous to biological systems (that have evolved over millions of years) that operate on the basis of excitatory stimuli (Studt, 1998).

Bang-bang discontinuous control can operate several times to control large amplitude oscillations, providing time for linear continuous controls to become effective.

If stability is a problem, generator excitation control including power system stabilizers should be high performance.

Feedforward Controls

Also shown on Fig. 11.25 are specialized feedforward (open-loop) controls that are a powerful stabilizing force for severe disturbances and for highly stressed operating conditions. Short circuit or outage events can be directly detected to initiate preplanned actions such as generator or load tripping, or reactive power compensation switching. These controls are rule-based, with rules developed from simulations (i.e., pattern recognition). These "event-based" controls are very effective since rapid control action prevents electromechanical dynamics from becoming stability threatening.

"Response-based" feedforward controls are also possible. These controls initiate stabilizing actions for arbitrary disturbances that cause significant "swing" of measured variables.

Feedforward controls such as generator or load tripping can ensure a postdisturbance equilibrium with sufficient region of attraction. With fast control action the region of attraction can be small compared to requirements with only feedback controls.

Feedforward controls have been termed discrete supplementary controls (IEEE, 1978), special stability controls (IEEE, 1996), special protection systems, remedial action schemes, and emergency control systems (Djakov et al., 1998).

Feedforward controls are very powerful. Although the reliability of special stability controls is often an issue (IEEE/CIGRE, 1996), adequate reliability can be obtained by design. Generally, controls are required to be as reliable as primary protective relaying. Duplicated or multiple sensors, redundant communications, and duplicated or voting logic are common (Dodge et al., 1990). Response-based controls are often less expensive than event-based controls because fewer sensors and communications paths are needed.

Undesired operation by some feedforward controls are relatively benign, and controls can be "trigger happy." For example, infrequent misoperation or unnecessary operation of HVDC fast power change, reactive power compensation switching, and transient excitation boosting may not be very disruptive. Misoperation of generator tripping (especially of steam-turbine generators), fast valving, load tripping, or controlled separation, however, are disruptive and costly.

Synchronizing and Damping Torques

Power system electromechanical stability means that synchronous generators and motors must remain in synchronism following disturbances — with positive damping of rotor angle oscillations ("swings"). For very severe disturbances and operating conditions, loss of synchronism (instability) occurs on the first forward swing within about one second. For less severe disturbances and operating conditions, instability may occur on the second or subsequent swings because of a combination of insufficient synchronizing and damping torques at synchronous machines.

Effectiveness and Robustness

Power systems have many electromechanical oscillation modes, and each mode can potentially become unstable. Lower frequency interarea modes are the most difficult to stabilize. Controls must be designed to be effective for one or more modes, and must not cause adverse interactions for other modes.

There are recent advances in robust control theory, especially for linear systems. For real nonlinear systems, emphasis should be on knowing uncertainty bounds and on sensitivity analysis using detailed nonlinear, large-scale simulation. For example, the sensitivity of controls to different operating conditions and load characteristics must be studied. Online simulation using actual operating conditions reduces uncertainty, and can be used for control adaptation.

Actuators

Actuators may be mechanical or power electronic. There are tradeoffs between cost and performance. Mechanical actuators (circuit breakers) are lower cost, and are usually sufficiently fast for electromechanical stability (e.g., two-cycle opening time, five-cycle closing time). They have restricted operating frequency and are generally used for feedforward controls.

Circuit breaker technology and reliability have improved in recent years (CIGRE Task Force 13.00.1, 1995; Brunke et al., 1994). Bang-bang control (up to perhaps five operations) for interarea oscillations with periods of two seconds or longer is feasible (Furumasu and Hasibar, 1992). Traditional controls for mechanical switching have been simple relays, but advanced controls can approach the sophistication of controls of, for example, thyristor-switched capacitor banks.

Power electronic phase control or switching using thyristors has been widely used in generator exciters, HVDC, and static var compensators. Newer devices, especially gate-turn-off thyristors, now have voltage and current ratings sufficient for high-power transmission applications. Advantages of power electronic actuators are very fast control, unrestricted switching frequency, and minimal transients.

For economy, existing actuators should be used to the extent possible. These include generator excitation and prime mover equipment, HVDC equipment, and circuit breakers. For example, infrequent generator tripping may be cost-effective compared to new power electronic actuated equipment.

Reliability Criteria

Experience shows that instability incidents are usually not caused by three-phase faults near large generating plants that are typically specified in deterministic reliability criteria. Rather they are the result of a combination of unusual failures and circumstances. The three-phase fault reliability criterion is often considered an *umbrella* criterion for less predictable disturbances involving multiple failures such as single-phase short circuits with "sympathetic" tripping of unfaulted lines. Of main concern are multiple *related* failures involving lines on the same right-of-way or with common terminations.

Types of Power System Stability Controls and Possibilities for Advanced Control

Stability controls are of many types, including:

- Generator excitation controls
- Prime mover controls, including fast valving
- Generator tripping
- Fast fault clearing
- High-speed reclosing, and single-pole switching
- Dynamic braking
- Load tripping and modulation
- Reactive power compensation switching or modulation (series and shunt)
- Current injection by voltage source inverter devices (STATCOM, UPFC, SMES, battery storage)
- Fast phase angle control
- HVDC link supplementary controls
- Adjustable-speed (doubly fed) synchronous machines
- Controlled separation and underfrequency load shedding

We will summarize these controls. Chapter 17 of Kundur (1994) provides considerable additional information. Torizuka and Tanaka (1998) describe the use of many of these controls in Japan.

Excitation Control

Generator excitation controls are a basic stability control. Thyristor exciters with high ceiling voltage provide powerful and economical means to ensure stability for large disturbances. Modern automatic voltage regulators and power system stabilizers are digital, facilitating additional capabilities such as adaptive control and special logic (IEEE, 1997; Bollinger et al., 1993; Hajagos and Gerube, 1998; Arcidiancone et al., 1998).

Excitation control is almost always based on local measurements. Therefore, full effectiveness may not be obtained for interarea stability problems where the normal local measurements are not sufficient. Line drop compensation (Rubenstein and Walkley, 1957; Dehdashti et al., 1988) is one method to increase the effectiveness (sensitivity) of excitation control, and improve coordination with static var compensators that normally control transmission voltage with small droop.

Several forms of discontinuous control have been applied to keep field voltage near ceiling levels during the first forward interarea swing (Kundur, 1994; Lee and Kundur, 1986; Taylor et al., 1993). Referring to

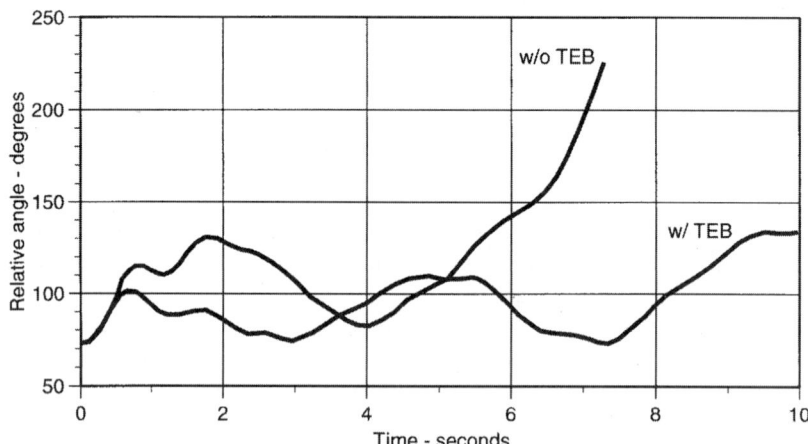

FIGURE 11.26 Rotor angle swing of Grand Coulee Unit 19 in Pacific Northwest relative to the San Onofre nuclear plant in Southern California. The effect of transient excitation boosting (TEB) at the Grand Coulee Third Power Plant following bipolar outage of the Pacific HVDC Intertie (3100 MW) is shown. (From Taylor, C. W., et al., Transient excitation boosting at Grand Coolee Third Power Plant, *IEEE Trans. on Power Syst.*, 8, 1291–1298, August 1993. With permission.)

the above discussion of angle measurement for stability control, the control described in Kundur (1994) and Lee and Kundur (1986) computes change in rotor angle locally from the power system stabilizer (PSS) speed change signal. The control described by Taylor et al. (1993) is a feedforward control that injects a decaying pulse into the voltage regulators at a large power plant following direct detection of a large disturbance. Figure 11.26 shows simulation results using this Transient Excitation Boosting — TEB.

Prime Mover Control Including Fast Valving

Fast power reduction (fast valving) at accelerating sending-end generators is an effective means of stability improvement. Use has been limited, however, because of the coordination required between characteristics of the electrical power system, the prime mover and prime mover controls, and the energy supply system (boiler).

Digital prime mover controls facilitate addition of special features for stability enhancement. Digital boiler controls, often retrofitted on existing equipment, may improve the feasibility of fast valving.

Fast valving is potentially lower cost than tripping of turbo-generators. Kundur (1994) and Bhatt (1996) describe investigations and recent implementations of fast valving. Sustained fast valving may be necessary for a stable postdisturbance equilibrium.

Generator Tripping

Generator tripping is an effective (cost-effective) control especially if hydro units are used. Tripping of fossil units, especially gas- or oil-fired units, may be attractive if tripping to house load is possible and reliable. Gas-turbine and combined-cycle plants constitute a large percentage of new generation. Occasional tripping of these units is feasible and can become an attractive stability control in the future.

Most generator tripping controls are event-based (based on outage of generating plant outgoing lines or outage of tie lines). Several advanced response-based generator tripping controls, however, have been implemented.

The Automatic Trend Relay (ATR) has been implemented at the Colstrip generating plant in eastern Montana (Stigers et al., 1997). The plant consists of two 330 MW units and two 700 MW units. The microprocessor-based controller measures rotor speed and generator power, and computes acceleration and angle. Tripping of 16–100% of plant generation is based on eleven trip algorithms involving acceleration, speed, and angle changes. Because of the long distance to Pacific Northwest load centers, the ATR has operated many times, both desirably and undesirably. There are proposals to use voltage angle

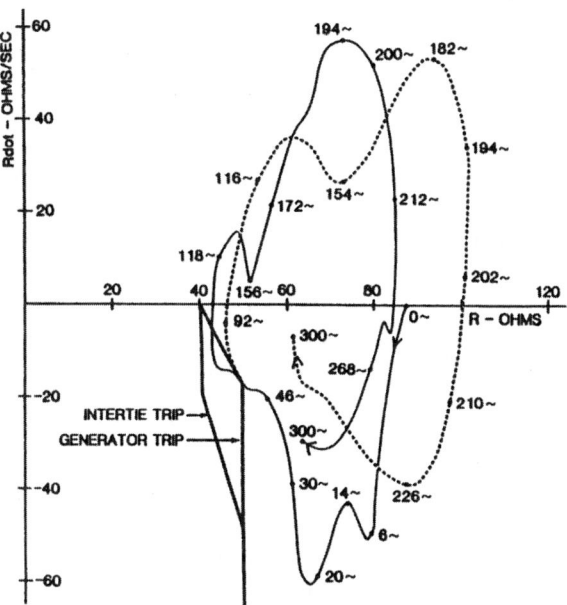

FIGURE 11.27 R–Rdot phase plane for loss of Pacific HVDC Intertie (2000 MW). Solid trajectory is without additional generator tripping. Dashed trajectory is with additional 600 MW of generator tripping initiated by the R–Rdot controller generator trip switching line. (From Haner et al., Experience with the R-Rdot out-of-step relay, *IEEE Trans. on Power Delivery*, PWRD-1, 2, 35–39, April 1986. With permission.)

measurement information (Colstrip 500-kV voltage angle relative to Grand Coulee and other Northwest locations) to adaptively adjust ATR settings, or as additional information for trip algorithms. Another possibility is to provide speed or frequency measurements from Grand Coulee and other locations to base algorithms on speed difference rather than only Colstrip speed (Kosterev et al., 1998).

A Tokyo Electric Power Company stabilizing control predicts generator angle changes and decides the minimum number of generators to trip (Matsuzawa et al., 1995). Local generator electric power, voltage, and current measurements are used to estimate angles. The control has worked correctly for several actual disturbances.

The Tokyo Electric Power Company is also developing an emergency control system that uses a predictive prevention method for step-out of pumped storage generators (Kojima et al., 1997; Imai et al., 1998). In the new method, the generators in TEPCO's network, which swing against their local pumped storage generators after serious faults, are treated as an external power system. The parameters in the external system, such as angle and moment of inertia, are estimated using local online information, and the behavior of local pumped storage generators is predicted based on equations of motion. Control actions (the number of generators to be tripped) are determined based on the prediction.

Haner et al. (1986) describe response-based generator tripping using a phase-plane controller. The controller is based on the apparent resistance — rate-of-change of apparent resistance (R–Rdot) phase plane, which is closely related to an angle difference — speed difference phase plane between two areas. The primary use of the controller is for controlled separation of the Pacific AC Intertie. Figure 11.27 shows simulation results where 600 MW of generator tripping reduces the likelihood of controlled separation.

Fast Fault Clearing, High Speed Reclosing, and Single-Pole Switching

Clearing time of close-in faults can be less than three cycles using conventional protective relays and circuit breakers. Typical EHV circuit breakers have two-cycle opening time. One-cycle breakers have been developed (Berglund et al., 1974), but special breakers are seldom justified. High magnitude short circuits may be detected as fast as one-fourth cycle by nondirectional overcurrent relays. Ultra high-speed traveling wave relays are also available (Esztergalyos et al., 1978). With such short clearing times, and considering that most EHV faults are single-phase, the removed transmission lines or other elements may be the major contributor to generator acceleration. This is especially true if nonfaulted equipment is removed by "sympathetic" relaying.

High speed reclosing is an effective method of improving stability and reliability. Reclosing is before the maximum of the first forward angular swing, but after 30–40 cycle time for arc extinction. During a lightning storm, high-speed reclosing keeps the maximum number of lines in service. High speed reclosing is effective when unfaulted lines trip because of relay misoperations.

Unsuccessful high-speed reclosing into a permanent fault can cause instability and can also compound the torsional duty imposed on turbine-generator shafts. Solutions include reclosing only for single-phase faults, and reclosing from the weak end with hot-line checking prior to reclosing at the generator end. Communication signals from the weak end indicating successful reclosing can also be used to enable reclosing at the generator end (Behrendt, 1996).

Single-pole switching is a practical means to improve stability and reliability in extra high voltage networks where most circuit breakers have independent pole operation (IEEE Committee, 1986; Belotelov et al., 1998). Several methods are used to ensure secondary arc extinction. For short lines, no special methods are needed. For long lines, the four-reactor scheme (Knutsen, 1962; Kimbark, 1964) is most commonly used. High speed grounding switches may be used (Hasibar et al., 1981). A hybrid reclosing method used by Bonneville Power Administration employs single-pole tripping, but with three-pole tripping on the backswing followed by rapid three-pole reclosure; the three-pole tripping ensures secondary arc extinction (IEEE Committee, 1986).

Single-pole switching may necessitate positive sequence filtering in stability control input signals.

For advanced stability control, signal processing and pattern recognition techniques may be developed to detect secondary arc extinction (Finton et al., 1996; Djuric and Terzija, 1995). Reclosing into a fault is avoided and single-pole reclosing success is improved.

High speed reclosing or single-pole switching may not allow increased power transfers because deterministic reliability criteria generally specifies permanent faults. Nevertheless, fast reclosing provides "defense-in-depth" for frequently occurring single-phase temporary faults and false operation of protective relays. The probability of power failures because of multiple line outages is greatly reduced.

Dynamic Braking

Shunt dynamic brakes using mechanical switching have been used infrequently (Kundur, 1994). Normally the insertion time is fixed. One attractive method not requiring switching is neutral-to-ground resistors in generator step-up transformers. Braking automatically results for ground faults, which are most common.

Often generator tripping, which helps ensure a post-disturbance equilibrium, is a better solution.

Thyristor switching of dynamic brakes has been proposed. Thyristor switching or phase control minimizes generator torsional duty (Bayer et al., 1996), and can be a subsynchronous resonance countermeasure (Donnelly et al., 1993).

Load Tripping and Modulation

Load tripping is similar in concept to generator tripping but is at the receiving end to reduce deceleration of receiving-end generation. Interruptible industrial load is commonly used. For example, Taylor et al. (1981) describe tripping of up to 3000 MW of industrial load following outages during power import conditions.

Rather than tripping large blocks of industrial load, it may be possible to trip low priority commercial and residential load such as space and water heaters, or air conditioners. This is less disruptive and the consumer may not even notice brief interruptions. The feasibility of this control depends on implementation of direct load control as part of demand side management, and on the installation of high-speed communication links to consumers with high-speed actuators at load devices. Although unlikely because of economics, appliances such as heaters could be designed to provide frequency sensitivity by local measurements.

Load tripping is also used for voltage stability. Here the communication and actuator speeds are generally not as critical.

It is also possible to modulate loads such as heaters to damp oscillations (Samuelsson and Eliasson, 1997; Kamwa et al., 1998; Dagle, 1997).

Clearly, load tripping or modulation of small loads will depend on the economics, and the development of fast communications and actuators.

Reactive Power Compensation Switching or Modulation

Controlled series or shunt compensation improves stability, with series compensation generally being the most powerful. For switched compensation, either mechanical or power electronic switches may be used. For continuous modulation, thyristor phase control of a reactor (TCR) is used. Mechanical switching has the advantage of lower cost. The operating times of circuit breakers are usually adequate, especially for interarea oscillations. Mechanical switching is generally single insertion of compensation for synchronizing support. In addition to previously mentioned advantages, power electronic control has advantages in subsynchronous resonance performance.

For synchronizing support, high-speed series capacitor switching has been used effectively on the North American Pacific AC intertie for over 25 years (Kimbark, 1966). The main application is for full or partial outages of the parallel Pacific HVDC intertie (event-driven control using transfer trip over microwave radio). Series capacitors are inserted by circuit breaker opening; operators bypass the series capacitors some minutes after the event. Response-based control using an impedance relay was also used for some years, and new response-based controls are being investigated.

Thyristor-based series compensation switching or modulation has been developed, with several installations in service or planned (Cristi et al., 1992; Piwko et al., 1994; Zhou et al., 1988). Thyristor-controlled series compensation (TCSC) allows significant time-current dependent increase in series reactance over nominal reactance. With appropriate controls, this increase in reactance can be a powerful stabilizing force.

Thyristor-controlled series compensation was chosen for the 1020 km, 500-kV intertie between the Brazilian North–Northeast networks and the Brazilian Southeast network (Gama et al., 1998). The TCSCs at each end of the intertie are modulated using line power measurements to damp low-frequency (0.12 Hz) oscillations. Figure 11.28, from commissioning field tests (Gama, 1999), shows the powerful stabilizing benefits of TCSCs.

Zhou et al. (1998) describe a TCSC application in China for integration of a remote power plant using two parallel 500-kV transmission lines (1300 km). Transient stability simulations indicate that 25% thyristor-controlled compensation is more effective than 45% fixed compensation. Several advanced TCSC control techniques are promising. The state-of-the-art is to provide both transient stability and damping control modes.

Zhou and Liang (1999) surveyed TCSC stability controls, providing 85 references.

For synchronizing support, high speed switching of shunt capacitor banks is also effective. Again on the Pacific AC intertie, four 200 MVAr shunt banks are switched for HVDC and 500-kV AC line outages (Furumasu and Hasibar, 1992); new response-based controls are being investigated.

High speed mechanical switching of shunt banks as part of a static var system is common. For example, the Forbes SVS near Duluth, Minnesota, includes two 300 MVAr 500-kV shunt capacitor banks (Sybille et al., 1996). Generally, it is effective to augment power electronic controlled compensation with fixed or mechanically switched compensation.

Static var compensators are applied along interconnections to improve synchronizing and damping support. Voltage support at intermediate points allow operation at angles above 90°. SVCs are modulated to improve oscillation damping. One study (CIGRE, 1996; Larsen and Chow, 1987) showed line current magnitude to be the most effective input signal. Synchronous condensers can provide similar benefits,

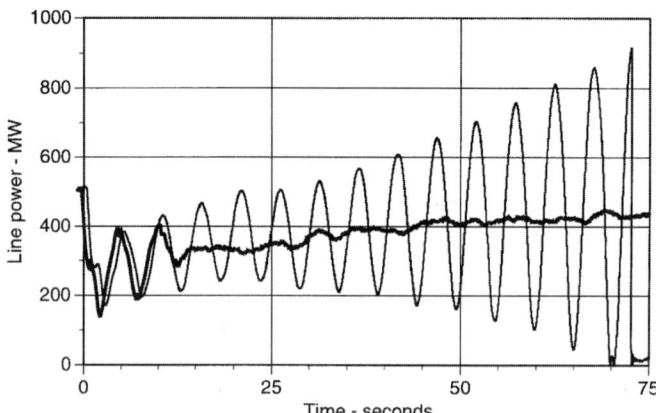

FIGURE 11.28 Effect of TCSCs for trip of a 300 MW generator in the North–Northeast Brazilian network. Results are from commissioning field tests in March 1999. The thin line without TCSC power oscillation damping shows interconnection separation after 70 seconds. The thick line with TCSC power oscillation damping shows rapid oscillation damping.

but are not competitive today with power electronic control. Available SVCs in load areas may be used to indirectly modulate load to provide synchronizing or damping forces.

Digital control allows many new control strategies. Gains in supervision and optimization adaptive control are common. For series or shunt power electronic devices, control mode selection allows bang-bang control, synchronizing versus damping control, and other nonlinear and adaptive strategies.

Current Injection by Voltage Source Inverters

Advanced power electronic controlled equipment employ gate turn-off thyristors. Reactive power injection devices include static compensator (STATCOM), static synchronous series compensator (SSSC), and unified power flow controller (UPFC). CIGRE Task Force 38.01.07 (1996) describes the use of these devices for oscillation damping.

As with conventional thyristor-based equipment, it is often effective for voltage source inverter control to also direct mechanical switching.

Voltage source inverters may also be used for real power series or shunt injection. Superconducting magnetic energy storage (SMES) or battery storage is the most common. For angle stability control, injection of real power is more effective than reactive power.

For transient stability improvement, SMES can be smaller MVA size and lower cost than a STATCOM. SMES is less location-dependent than a STATCOM.

Fast Voltage Phase Angle Control

Voltage phase angles, and thereby rotor angles, can be directly and rapidly controlled by power electronic controlled series compensation (discussed above) or phase shifting transformers. This provides powerful stability control. Although one type of thyristor-controlled phase shifting transformer was developed over fifteen years ago (Stemmler and Güth, 1982), high cost has presumably prevented installations. Fang and MacDonald (1998) describe an application study.

The unified power flow controller incorporates GTO-thyristor phase shifting and series compensation control, and one installation (not a transient stability application) is in service (Rahman et al., 1997).

One concept employs power electronic series or phase shifting equipment to control angles across an interconnection within a small range (Christensen, 1997). On a power–angle curve, this can be visualized as keeping high synchronizing coefficient (slope of power–angle curve) during disturbances.

Bonneville Power Administration developed a novel method for transient stability by high speed 120° phase rotation of transmission lines between networks losing synchronism (Cresap et al., 1981). This

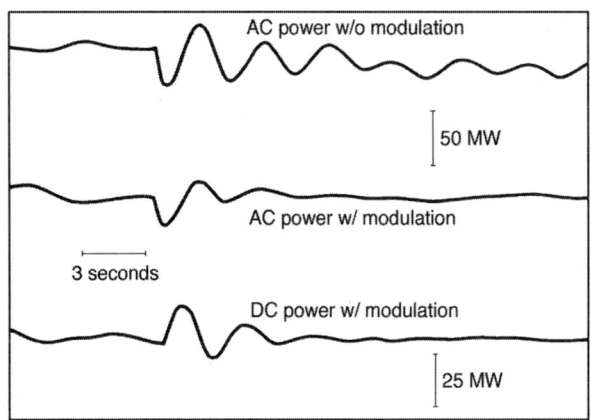

FIGURE 11.29 System response to Pacific AC Intertie series capacitor bypass with and without DC modulation. (From Cresap et al., Operating experience with modulation of the Pacific HVDC intertie, *IEEE Trans. on Power Appar. and Syst.*, PAS-98, 1053–1059, July/August 1978. With permission.)

technique is very powerful (perhaps too powerful!) and raises reliability and robustness issues, especially in the usual case where several lines form the interconnection. It has not been implemented.

HVDC Link Supplementary Controls

HVDC DC links are installed for power transfer reasons. In contrast to the above power electronic devices, the available HVDC converters provide the actuators so that stability control is inexpensive. For long distance HVDC links within a synchronous network, HVDC modulation can provide powerful stabilization, with active and reactive power injections at each converter. Control robustness, however, is a concern.

HVDC link stability controls are described in CIGRE (1996), IEEE (1991), and reports by Cresap et al. (1978). The Pacific HVDC Intertie modulation control implemented in 1976 is unique in that a remote input signal from the parallel Pacific AC Intertie was used (Cresap et al., 1978). Figure 11.29 shows commissioning test results.

Adjustable Speed (Doubly Fed) Synchronous Machines

CIGRE (1996) summarizes stability benefits of adjustable speed synchronous machines that have been developed for pumped storage applications in Japan. Fast digital control of excitation frequency enables direct control of rotor angle.

Controlled Separation and Underfrequency Load Shedding

For very severe disturbances and failures, maintaining synchronism may not be possible or cost effective. Controlled separation based on out-of-step detection or parallel path outages mitigates the effects of instability. Stable islands are formed, but underfrequency load shedding may be required in islands that were importing power.

Advanced controlled separation schemes are described in Ohura et al. (1990), Haner et al. (1986), Taylor et al. (1983), and Centeno et al. (1997). Recent proposals advocate use of voltage phase angle measurements for controlled separation.

Dynamic Security Assessment

Control design and settings, along with transfer limits, are usually based on off-line simulation (time and frequency domain), and on field tests. Controls must then operate appropriately for a variety of operating conditions and disturbances.

Recently, however, online dynamic (or transient) stability/security assessment software has been developed. State estimation and online power flow determine the base operating conditions. Simulation of potential disturbances is then based on actual operating conditions, reducing uncertainty of the control environment. Dynamic security assessment is presently used to determine arming levels for generator tripping controls (Mansour et al., 1995; Ota et al., 1996).

With today's computer capabilities, hundreds or thousands of large-scale simulations may be run each day to provide an organized database of system stability properties. Security assessment is made efficient by techniques such as fast screening and contingency selection, and smart termination of strongly stable or unstable cases. Parallel computation is straightforward using multiple workstations for different simulation cases; common initiation may be used for the different contingencies.

In the future, dynamic security assessment may be used for control adaptation to current operating conditions. Another possibility is stability control based on neural network or decision tree pattern recognition. Dynamic security assessment provides the database for pattern recognition techniques. Pattern recognition may be considered data compression of security assessment results.

Industry restructuring requiring near real-time power transfer capability determination may accelerate the implementation of dynamic security assessment, facilitating advanced stability controls.

"Intelligent" Controls

Mention has already been made of rule-based controls and pattern recognition based controls.

As a possibility, Chiang and Wong (1995) describes a sophisticated self-organizing neural fuzzy controller (SONFC) based on the speed–acceleration phase plane. Compared to the angle–speed phase plane, control tends to be faster and both final states are zero (using angle, the postdisturbance equilibrium angle is not known in advance). The controllers are located at generator plants. Acceleration and speed can be easily measured/computed using, for example, the techniques developed for power system stabilizers.

The SONFC could be expanded to incorporate remote measurements. Dynamic security assessment simulations could be used for updating/retraining of the neural network fuzzy controller. The SONFC is suitable for generator tripping, series or shunt capacitor switching, HVDC control, etc.

Effect of Industry Restructuring on Stability Controls

Industry restructuring will have much impact on power system stability. New, frequently changing power transfer patterns cause new stability problems. Most stability and transfer capability problems must be solved by new controls and new substation equipment, rather than by new transmission lines.

Different ownership of generation, transmission, and distribution makes necessary power *system* engineering more difficult. New power industry standards along with ancillary services mechanisms are being developed. Controls such as generator or load tripping, fast valving, higher than standard exciter ceilings, and power system stabilizers may be ancillary services. In large interconnections, independent grid operators or security coordination centers may facilitate dynamic security assessment and centralized stability controls.

Experience from Recent Power Failures

Recent cascading power outages demonstrated the impact of control and protection failures, the need for "defense-in-depth," and the need for advanced stability controls.

The July 2, 1996, and August 10, 1996, power failures (WSCC, 1996; Taylor and Erickson, 1997; Hauer et al., 1997; Kosterev et al., 1999) in western North America showed need for improvements and innovations in stability control areas such as:

- Fast insertion of reactive power compensation, and fast generator tripping using response-based controls.
- Special HVDC and SVC control.

- Power system stabilizer design and tuning.
- Controlled separation.
- Power system modeling and data validation for control design.
- Adaptation of controls to actual operating conditions.

Summary

Power system angle stability can be improved by a wide variety of controls. Some methods have been used effectively for many years, both at generating plants and in transmission networks. New control techniques and actuating equipment are promising.

We provide a broad survey of available stability control techniques with emphasis on implemented controls, and on new and emerging technology.

References

Arcidiancone, V., Corsi, S., Ottaviani, G., Togno, S., Baroffio, G., Raffaelli, C., and Rosa, E., The ENEL's experience on the evolution of excitation control systems through microprocessor technology, *IEEE Trans. on Energy Conversion*, 13, 3, 292–299, September 1998.

Bayer, W., Habur, K., Povh, D., Jacobson, D. A., Guedes, J. M. G., and Marshall, D. A., Long Distance Transmission with Parallel AC/DC Link From Cahora Bassa (Mozambique) to South Africa and Zimbabwe, CIGRÉ paper 14-306, 1996.

Behrendt, K. C., Relay-to-Relay Digital Logic Communication for Line Protection, Monitoring, and Control, in *Proceedings of the 23rd Annual Western Protective Relay Conference*, Spokane, Washington, October 1996.

Belotelov, A. K., Dyakov, A. F., Fokin, G. G., Ilynichnin, V. V., Leviush, A. I., and Strelkov, V. M., Application of Automatic Reclosing in High Voltage Networks of the UPG of Russia under New Conditions, CIGRÉ paper 34-203, 1998.

Berglund, R. O., Mittelstadt, W. A., Shelton, M. L., Barkan, P., Dewey, C. G., and Skreiner, K. M., One-cycle fault interruption at 500 kV: System benefits and breaker designs, *IEEE Trans. on Power Appar. Syst.*, PAS-93, 1240–1251, September/October 1974.

Bhatt, N. B., Field experience with momentary fast turbine valving and other special stability controls employed at AEP's Rockport Plant, *IEEE Trans. on Power Syst.*, 11, 1, 155–161, February 1996.

Bollinger, K. E., Nettleton, L., Greenwood-Madsen, T., and Salyzyn, M., Experience with digital power system stabilizers at steam and hydro generating plants, *IEEE Trans. on Energy Conversion*, 8, 2, June 1993.

Brunke, J. H., Esztergalyos, J. H., Khan, A. H., and Johnson, D. S., Benefits of Microprocessor-Based Circuit Breaker Control, CIGRÉ paper 23/13-10, 1994.

Centeno, V., Phadke, A. G., Edris, A., Benton, J., Gaudi, M., and Michel, G., An adaptive out of step relay, *IEEE Trans. on Power Delivery*, 12, 1, 61–71, January 1997.

Chang, H.-C., and Wang, M.-H., Neural network-based self-organizing fuzzy controller for transient stability of multimachine power systems, *IEEE Trans. on Energy Conversion*, 10, 2, 339–347, June 1995.

Christensen, J. F., New control strategies for utilizing power system network more effectively, *Electra*, 173, 5–16, August 1997.

Christl, N., Sadek, K., Hedin, R., Lützelberger, P., Krause, P. E., Montoya, A. H., McKenna, S. M., and Torgerson, D., Advanced Series Compensation (ASC) with Thyristor Controlled Impedance, CIGRÉ paper 14/37/38-05, 1992.

CIGRÉ Task Force 13.00.1, Controlled switching — A state-of-the-art survey, *Electra*, 163, 65–97, December 1995.

CIGRÉ TF 38.02.17, *Advanced Angle Stability Controls*, 2000 (in press).

CIGRÉ TF 38.01.06 *Load Flow Control in High Voltage Systems Using FACTS Controllers*, October 1995.

CIGRÉ TF 38.01.07, *Analysis and Control of Power System Oscillations*, Brochure 111, December 1996.
CIGRÉ TF 38.02.12, *Criteria and Countermeasures for Voltage Collapse*, CIGRÉ Brochure 101, October 1995. Summary in *Electra*, October 1995.
CIGRÉ WG 34.08, *Protection against Voltage Collapse*, CIGRÉ Brochure No. 128, 1998. Summary in *Electra*, No. 179, pp. 111–126, August 1998.
Cresap, R. L., Scott, D. N., Mittelstadt, W. A., and Taylor, C. W., Damping of Pacific AC Intertie Oscillations via Modulation of the Parallel Pacific HVDC Intertie, CIGRÉ paper 14-05, 1978.
Cresap, R. L., Scott, D. N., Mittelstadt, W. A., and Taylor, C. W., Operating experience with modulation of the Pacific HVDC Intertie, *IEEE Trans. on Power Appar. Syst.*, PAS-98, 1053–1059, July/August 1978.
Cresap, R. L., Taylor, C. W., and Kreipe, M. J., Transient stability enhancement by 120-degree phase rotation, *IEEE Trans. on Power Appar. Syst.*, PAS-100, 745–753, February 1981.
Dagle, J., Distributed-FACTS: End-Use Load Control for Power System Dynamic Stability Enhancement, in EPRI Conference, *The Future of Power Delivery in the 21st Century*, La Jolla, California, 18–20 November 1997.
Dehdashti, A. S., Luini, J. F., and Peng, Z., Dynamic voltage control by remote voltage regulation for pumped storage plants, *IEEE Trans. on Power Syst.*, 3, 3, 1188–1192, August 1988.
Djakov, A. F., Bondarenko, A., Portnoi, M. G., Semenov, V. A., Gluskin, I. Z., Kovalev, V. D., Berdnikov, V. I., and Stroev, V. A., The Operation of Integrated Power Systems Close to Operating Limits with the Help of Emergency Control Systems, CIGRÉ paper 39-109, 1998.
Djuric, M. B., and Terzija, V. V., A new approach to the arcing faults detection for fast autoreclosure in transmission systems, *IEEE Trans. on Power Delivery*, 10, 4, 1793–1798, October 1995.
Dodge, D., Doel, W., and Smith, S., Power system stability control using fault tolerant technology, *ISA Instrumentation in Power Industry*, paper 90-1323, 33, 33rd Power Instrumentation Symposium, May 21–23, 1990.
Donnelly, M. K., Smith, J. R., Johnson, R. M., Hauer, J. F., Brush, R. W., and Adapa, R., Control of a dynamic brake to reduce turbine-generator shaft transient torques, *IEEE Trans. on Power Syst.*, 8 1, 67–73, February 1993.
Esztergalyos, J. H., Yee, M. T., Chamia, M., and Lieberman, S., The Development and Operation of an Ultra High Speed Relaying System for EHV Transmission Lines, CIGRÉ paper 34-04, 1978.
Fang, Y. J., and Macdonald, D. C., Dynamic quadrature booster as an aid to system stability, *IEE Proc.—Gener. Transm. Distrib.*, 145, 1, 4147, January 1998.
Fitton, D. S., Dunn, R. W., Aggarwal, R. K., Johns, A. T., and Bennett, A., Design and implementation of an adaptive single pole autoreclosure technique for transmission lines using artificial neural networks, *IEEE Trans. on Power Delivery*, 11, 2, 748–756, April 1996.
Furumasu, B. C., and Hasibar, R. M., Design and Installation of 500-kV Back-to-Back Shunt Capacitor Banks, *IEEE Trans. on Power Delivery*, 7, 2, 539–545, April 1992.
Gama, C., Brazilian North–South Interconnection — Control Application and Operating Experience with a TCSC, in *Proceedings of 1999 IEEE/PES Summer Meeting*, July 1999.
Gama, C., Leoni, R. L., Gribel, J., Fraga, R., Eiras, M. J., Ping, W., Ricardo, A., Cavalcanti, J., and Tenório, R., Brazilian North–South Interconnection — Application of thyristor controlled series compensation (TCSC) to damp inter-area oscillation mode, CIGRÉ paper 14-101, 1998.
Hajagos, L. M., and Gerube, G. R., Utility experience with digital excitation systems, *IEEE Trans. on Power Syst.*, 13, 1, 165–170, February 1998.
Haner, J. M., Laughlin, T. D., and Taylor, C. W., Experience with the R-Rdot out-of-step relay, *IEEE Trans. on Power Delivery*, PWRD-1, 2, 35–39, April 1986.
Hasibar, R. M., Legate, A. C., Brunke, J. H., and Peterson, W. G., The application of high-speed grounding switches for single-pole reclosing on 500-kV power systems, *IEEE Trans. on Power Appar. Syst.*, PAS-100, 4, 1512–1515, April 1981.
Hauer, J. F., Robust damping controls for large power systems, *IEEE Control Systems Magazine*, January 1989.
Hauer, J. F., Robustness Issues in Stability Control of Large Electric Power Systems, *32nd IEEE Conference on Decision and Control*, San Antonio, Texas, December 15–17, 1993.

Hauer, J., Trudnowski, D., Rogers, G., Mittelstadt, W., Litzenberger, W., and Johnson, J., Keeping an eye on power system dynamics, *IEEE Computer Applications in Power*, 10, 1, 26–30, January 1997.

IEEE Committee Report, HVDC controls for system dynamic performance, *IEEE Trans. on Power Syst.*, 6, 2, 743–752, May 1991.

IEEE Committee Report, Single-pole switching for stability and reliability, *IEEE Trans. on Power Syst.*, PWRS-1, 25–36, May 1986.

IEEE/CIGRÉ Committee Report (Anderson, P. M. and LeReverend, B. K.), Industry experience with special protection schemes, *IEEE Trans. on Power Syst.*, 11, 3, 1166–1179, August 1996.

IEEE Digital Excitation Applications Task Force, Digital excitation technology — A review of features, functions and benefits, *IEEE Trans. on Energy Conversion*, 12, 3, September 1997.

IEEE Discrete Supplementary Control Task Force, A description of discrete supplementary controls for stability, *IEEE Trans. on Power Appar. Syst.*, PAS-97, 149–165, January/February 1978.

IEEE FACTS Working Group, *FACTS Applications*, IEEE/PES 96TP116-0.

IEEE Power System Relaying Committee WG K12, *System Protection and Voltage Stability*, 93 THO 596-7 PWR, 1993.

IEEE Special Stability Controls Working Group, Annotated bibliography on power system stability controls: 1986–1994, *IEEE Trans. on Power Syst.*, 11, 2, 794–800, August 1996.

Imai, S., Syoji, T., Yanagihashi, K., Kojima, Y., Kowada, Y., Oshida, H., and Goda, T., Development of predictive prevention method for mid-term stability problem using only local information, *Transactions of IEE Japan*, 118-B, 9, 1998.

Kamwa, I., Grondin, R., Asber, D., Gingras, J. P., and Trudel, G., Active power stabilizers for multimachine power systems: Challenges and prospects, *IEEE Trans. on Power Syst.*, 13, 4, 1352–1358, November 1998.

Kimbark, E. W., Improvement of system stability by switched series capacitors, *IEEE Trans. on Power Appar. Syst.*, PAS-85, 2, 180–188, February 1966.

Kimbark, E. W., Suppression of ground-fault arcs on single-pole switched lines by shunt reactors, *IEEE Trans. on Power Appar. Syst.*, PAS-83, 3, 285–290, March 1964.

Knutsen, N., Single-Phase Switching of Transmission lines Using Reactors for Extinction of the Secondary Arc, CIGRÉ paper 310, 1962.

Kojima, Y., Taoka, H., Oshida, H., and Goda, T., On-line modeling for emergency control systems, *IFAC/CIGRE Symposium on Control of Power Systems and Power Plant*, 627–632, 1997.

Kosterev, D. N., Esztergalyos, J., and Stigers, C. A., Feasibility study of using synchronized phasor measurements for generator dropping controls in the Colstrip System, *IEEE Trans. on Power Syst.*, 13, 3, 755–762, August 1998.

Kosterev, D. N., Taylor, C. W., and Mittelstadt, W. A., Model validation for the August 10, 1996 WSCC system outage, *IEEE Trans. on Power Syst.*, 14, 3, 967–979, August 1999.

Kundur, P., *Power System Stability and Control*, McGraw-Hill, New York, 1994.

Larsen, E. V., and Chow, J. H., SVC control design concepts for system dynamic performance, *Application of Static Var Systems for System Dynamic Performance*, IEEE Special Publication 87TH1087-5-PWR, 36–53, 1987.

Lee, D. C., and Kundur, P., Advanced Excitation Controls for Power System Stability Enhancement, CIGRÉ paper 38-01, 1986.

Mansour, Y., Vaahedi, E., Chang, A. Y., Corns, B. R., Garrett, B. W., Demaree, K., Athay, T., and Cheung, K., B. C. Hydro's on-line Transient Stability Assessment (TSA) model development, analysis, and post-processing, *IEEE Trans. on Power Syst.*, 10, 1, 241–253, February 1995.

Matsuzawa, K., Yanagihashi, K., Tsukita, J., Sato, M., Nakamura, T., and Takeuchi, A., Stabilizing control system preventing loss of synchronism from extension and its actual operating experience, *IEEE Trans. on Power Syst.*, 10, 3, 1606–1613, August 1995.

Ohura, Y., Suzuki, M., Yanagihashi, K., Yamaura, M., Omata, K., Nakamura, T., Mitamura, S., and Watanabe, H., A predictive out-of-step protection system based on observation of the phase difference between substations, *IEEE Trans. on Power Delivery*, 5, 4, 1695–1704, November 1990.

Ota, H., Kitayama, Y., Ito, H., Fukushima, N., Omata, K., Morita, K., and Kokai, Y., Development of transient stability control system (TSC System) based on on-line stability calculation, *IEEE Trans. on Power Syst.*, 11, 3, 1463–1472, August 1996.

Piwko, R. J., Wegner, C. A., Furumasu, B. C., Damsky, B. L., and Eden, J. D., The Slatt Thyristor-Controlled Series Capacitor Project — Design, Installation, Commissioning and System Testing, CIGRÉ paper 14-104, 1994.

Rahman, M., Ahmed, M., Gutman, R., O'Keefe, R. J., Nelson, R. J., and Bian, J., UPFC application on the AEP system: Planning considerations, *IEEE Trans. on Power Syst.*, 12, 4, 1695–1701, November 1997.

Rubenstein, A. S., and Walkley, W. W., Control of reactive KVA with modern amplidyne voltage regulators, *AIEE Transactions*, 961–970, December 1957.

Samuelsson, O., and Eliasson, B., Damping of electro-mechanical oscillations in a multimachine system by direct load control, *IEEE Trans. on Power Syst.*, 12, 4, 1604–1609, November 1997.

Stemmler, H., and Güth, G., The thyristor-controlled static phase shifter—A new tool for power flow control in AC transmission systems, *Brown Boveri Review*, 69, 3, 73–78, March 1982.

Stigers, C. A., Woods, C. S., Smith, J. R., and Setterstrom, R. D., The acceleration trend relay for generator stabilization at Colstrip, *IEEE Trans. on Power Delivery*, 12, 3, 1074–1081, July 1997.

Studt, T., Computer Scientists search for ties to biological intelligence, *R&D Magazine*, 77–78, October 1998.

Sybille, G., Giroux, P., Dellwo, S., Mazur, R., and Sweezy, G., Simulator and field testing of forbes SVC, *IEEE Trans. on Power Delivery*, 11, 3, 1507–1514, July 1996.

Taylor, C. W., *Power System Voltage Stability*, McGraw-Hill, New York, 1994.

Taylor, C. W., and Erickson, D. C., Recording and analyzing the July 2 cascading outage, *IEEE Computer Applications in Power*, 10, 1, 26–30, January 1997.

Taylor, C. W., Haner, J. M., Hill, L. A., Mittelstadt, W. A., and Cresap, R. L., A new out-of-step relay with rate of change of apparent resistance augmentation, *IEEE Trans. on Power Appar. Syst.*, PAS-102, 3, 631–639, March 1983.

Taylor, C. W., Mechenbier, J. R., and Matthews, C. E., Transient excitation boosting at Grand Coulee Third Power Plant, *IEEE Trans. on Power Syst.*, 8, 3, 1291–1298, August 1993.

Taylor, C. W., Nassief, F. R., and Cresap, R. L., Northwest power pool transient stability and load shedding controls for generation–load imbalances, *IEEE Trans. on Power Appar. Syst.*, PAS-100, 7, 3486–3495, July 1981.

Torizuka, T., and Tanaka, H., An outline of power system technologies in Japan, *Electric Power Systems Research*, 44, 1–5, 1998.

Van Cutsem, T., and Vournas, C., *Voltage Stability of Electric Power Systems*, Kluwer Academic Publishers, Dordrecht, the Netherlands, 1998.

WSCC reports on July 2, 1996 and August 10, 1996 outages, available at www.wscc.com.

Zhou, X., et al., Analysis and control of Yimin–Fentun 500 kV TCSC system, *Electric Power Systems Research* 46, 157–168, 1998.

Zhou, X., and Liang, J., Overview of control schemes for TCSC to enhance the stability of power systems, *IEE Proc.-Gener. Transm. Distrib.*, 146, 2, 125–130, March 1999.

11.7 Power System Dynamic Modeling

William W. Price

Modeling Requirements

Analysis of power system dynamic performance requires the use of computational models representing the nonlinear differential-algebraic equations of the various system components. While scale models or analog models are sometimes used for this purpose, most power system dynamic analysis is performed

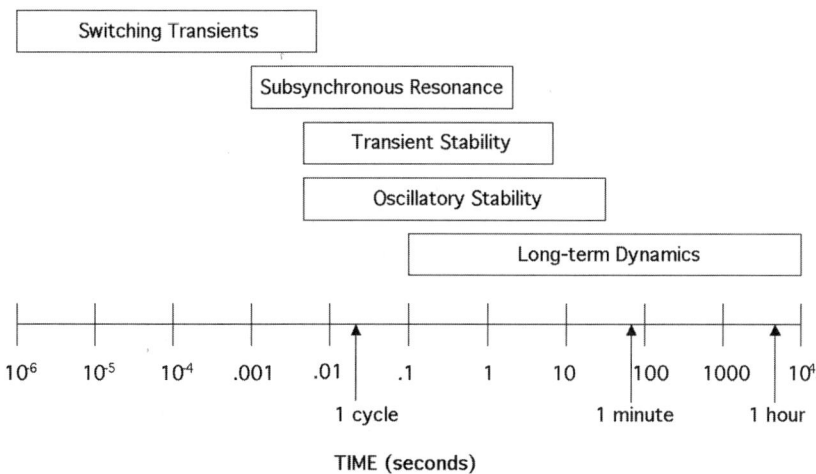

FIGURE 11.30 Time scale of power system dynamic phenomena.

with digital computers using specialized programs. These programs include a variety of models for generators, excitation systems, governor-turbine systems, loads, and other components. The user is therefore concerned with selecting the appropriate models for the problem at hand and determining the data to represent the specific equipment on his or her system. The focus of this article is on these concerns.

The choice of appropriate models depends heavily on the time scale of the problem being analyzed. Figure 11.30 shows the principal power system dynamic performance areas displayed on a logarithmic time scale ranging from microseconds to days. The lower end of the band for a particular item indicates the smallest time constants that need to be included for adequate modeling. The upper end indicates the approximate length of time that must be analyzed. It is possible to build a power system simulation model that includes all dynamic effects from very fast network inductance/capacitance effects to very slow economic dispatch of generation. However, for efficiency and ease of analysis, normal engineering practice dictates that only models incorporating the dynamic effects relevant to the particular performance area of concern be used.

This section focuses on the modeling required for analysis of power system stability, including transient stability, oscillatory stability, voltage stability, and frequency stability. For this purpose, it is normally adequate to represent the electrical network elements (transmission lines and transformers) by algebraic equations. The effect of frequency changes on the inductive and capacitive reactances is sometimes included, but is usually neglected, since for most stability analysis the frequency changes are small. The modeling of the various system components for stability analysis purposes is discussed in the remainder of this section. For greater detail, the reader is referred to (Kundur, 1994) and the other references cited below.

Generator Modeling

The model of a generator consists of two parts: the acceleration equations of the turbine-generator rotor, and the generator electrical flux dynamics.

Rotor Mechanical Model

The acceleration equations are simply Newton's Second Law of Motion applied to the rotating mass of the turbine-generator rotor, as shown in block diagram form in Fig. 11.31. The following points should be noted:

1. The inertia constant (H) represents the stored energy in the rotor in MW-seconds, normalized to the MVA rating of the generator. Typical values are in the range of 3 to 15, depending on the type and size of the turbine-generator. If the inertia (J) of the rotor is given in kg-m/sec, H is computed as follows:

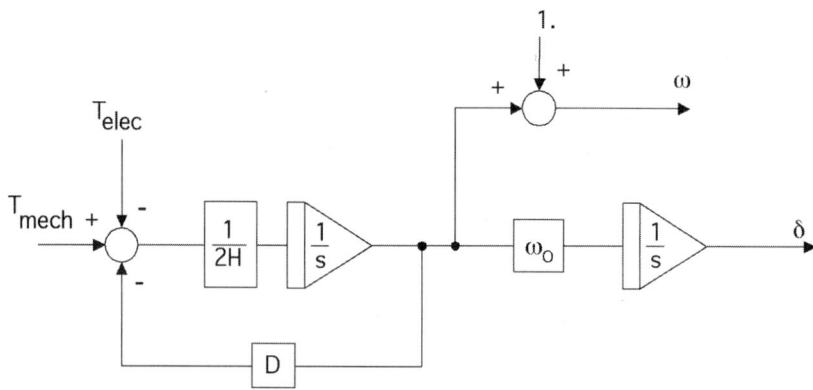

FIGURE 11.31 Generator rotor mechanical model.

$$H = 5.48 \times 10^{-9} \frac{J(RPM)^2}{MVARating} \text{ MW-sec/MVA}$$

2. Sometimes the mechanical power and electrical power are used in this model instead of the corresponding torques. Since power equals torque multiplied by rotor speed, the difference is small for operation close to nominal speed. However, there will be some effect on the damping of oscillations (*IEEE Trans.*, Feb. 1999).
3. Most models include the damping factor (D) shown in Fig. 11.31. It is used to model oscillation damping effects that are not explicitly represented elsewhere in the system model. The selection of a value for this parameter has been the subject of much debate (*IEEE Trans.*, Feb. 1999). Values from 0 to 4 or higher are sometimes used. The recommended practice is to avoid the use of this parameter by including sources of damping in other models, e.g., generator amortisseur and eddy current effects, load frequency sensitivity, etc.

Generator Electrical Model

The equivalent circuit of a three-phase synchronous generator is usually rendered as shown in Fig. 11.32. The three phases are transformed into a two-axis equivalent, with the direct (d) axis in phase with the rotor field winding and the quadrature (q) axis 90 electrical degrees ahead. For a more complete discussion of this transformation and of generator modeling, see IEEE Std. 1110-1991. In this equivalent circuit, r_a and L_ℓ represent the resistance and leakage inductance of the generator stator, L_{ad} and L_{aq} represent the mutual inductance between stator and rotor, and the remaining elements represent rotor windings or equivalent windings. This equivalent circuit assumes that the mutual coupling between the rotor windings and between the rotor and stator windings are the same. Additional elements can be added (IEEE Std. 1110-1991) to account for unequal mutual coupling, but most models do not include this since the data is difficult to obtain and the effect is small.

The rotor circuit elements may represent either physical windings on the rotor or eddy currents flowing in the rotor body. For solid-iron rotor generators such as steam-turbine generators, the field winding to which the DC excitation voltage is applied, is normally the only physical winding. However, additional equivalent windings are required to represent the effects of eddy currents induced in the body of the rotor. Salient-pole generators, typically used for hydro-turbine generators, have laminated rotors with lower eddy currents. However, these rotors often have additional amortisseur (damper) windings embedded in the rotor.

Data for generator modeling are usually supplied as synchronous, transient, and subtransient inductances and open circuit time constants. The relationships between these parameters and the equivalent network elements are shown in Table 11.1. Note that the inductance values are often referred to as

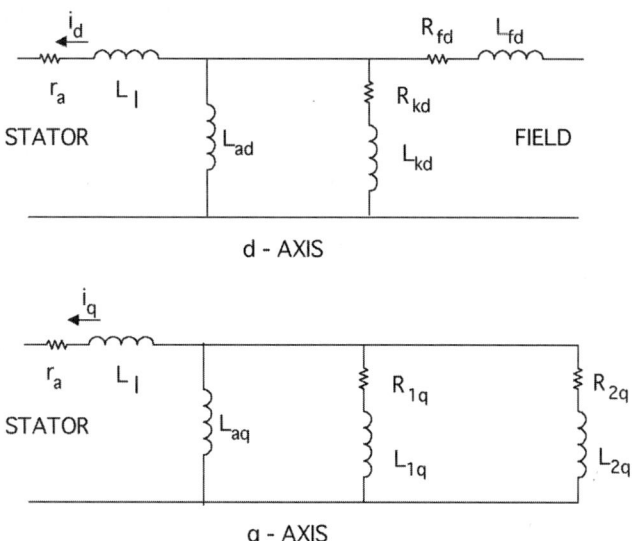

FIGURE 11.32 Generator equivalent circuit.

TABLE 11.1 Generator Parameter Relationships

	d-axis	q-axis
Synchronous inductance	$L_d = L_\ell + L_{ad}$	$L_q = L_\ell + L_{aq}$
Transient inductance	$L'_d = L_\ell + \dfrac{L_{ad} L_{fd}}{L_{ad} + L_{fd}}$	$L'_q = L_\ell + \dfrac{L_{aq} L_{1q}}{L_{aq} + L_{1q}}$
Subtransient inductance	$L''_d = L_\ell + \dfrac{L_{ad} L_{fd} L_{kd}}{L_{ad} L_{fd} + L_{ad} L_{kd} + L_{fd} L_{kd}}$	$L''_q = L_\ell + \dfrac{L_{aq} L_{1q} L_{2q}}{L_{aq} L_{1q} + L_{aq} L_{2q} + L_{1q} L_{2q}}$
Transient open circuit time constant	$T'_{do} = \dfrac{L_{ad} + L_{fd}}{\omega_0 R_{fd}}$	$T'_{qo} = \dfrac{L_{aq} + L_{1q}}{\omega_0 R_{1q}}$
Subtransient open circuit time constant	$T''_{do} = \dfrac{L_{ad} L_{fd} + L_{ad} L_{kd} + L_{fd} L_{kd}}{\omega_0 R_{kd}(L_{ad} + L_{fd})}$	$T''_{qo} = \dfrac{L_{aq} L_{1q} + L_{aq} L_{2q} + L_{1q} L_{2q}}{\omega_0 R_{2q}(L_{aq} + L_{1q})}$

reactances. At nominal frequency, the per unit inductance and reactance values are the same. However, as used in the generator model, they are really inductances, which do not change with changing frequency.

These parameters are normally supplied by the manufacturer. Two values are often given for some of the inductance values, a saturated (rated voltage) and unsaturated (rated current) value. The unsaturated values should be used, since saturation is usually accounted for separately, as discussed below.

For salient-pole generators, one or more of the time constants and inductances may be absent from the data, since fewer equivalent circuits are required. Depending on the program, either separate models are provided for this case or the same model is used with certain parameters set to zero or equal to each other.

Saturation Modeling

Magnetic saturation effects may be incorporated into the generator electrical model in various ways. The data required from the manufacturer is the open-circuit saturation curve, showing generator terminal voltage vs. field current. If the field current is given in amperes, it can be converted to per unit by dividing by the field current at rated terminal voltage on the air-gap (no saturation) line. (This value of field current is sometimes referred to as AFAG.) Often the saturation data for a generator model is input as only two points on the saturation curve, e.g., at rated voltage and 120% of rated voltage. The model then automatically fits a curve to these points.

The open circuit saturation curve characterizes saturation in the d-axis only. Ideally, saturation of the q-axis should also be represented, but the data for this is difficult to determine and is usually not provided. Some models provide an approximate representation of q-axis saturation based on the d-axis saturation data (IEEE Std. 1110-1991).

Excitation System Modeling

The excitation system provides the DC voltage to the field winding of the generator and modulates this voltage for control purposes. There are many different configurations and designs of excitation systems. Stability programs usually include a variety of models capable of representing most systems. These models normally include the IEEE standard excitation system models, described in IEEE Standard 421.5 (1992). Reference should be made to that document for a description of the various models and typical data for commonly used excitation system designs.

The excitation system consists of several subsystems, as shown in Fig. 11.33. The excitation power source provides the DC voltage and current at the levels required by the generator field. The excitation power may be provided by a rotating exciter, either a DC generator or an AC generator (alternator) and rectifier combination, or by rectifiers supplied from the generator terminals (or other AC source). Excitation systems with these power sources are often classified as "DC," "AC," and "static," respectively. The maximum (ceiling) field voltage available from the excitation power source is an important parameter. Depending on the type of system, this ceiling voltage may be affected by the magnitude of the field current or the generator terminal voltage, and this dependency must be modeled since these values may change significantly during a disturbance.

The automatic voltage regulator (AVR) provides for control of the terminal voltage of the generator by changing the generator field voltage. There are a variety of designs for the AVR, including various means of ensuring stable response to transient changes in terminal voltage. The speed with which the field voltage can be changed is an important characteristic of the system. For the "DC" and most of the "AC" excitation systems, the AVR controls the field of the exciter. Therefore, the speed of response is limited by the exciter's time constant. The speed of response of excitation systems is characterized according to IEEE Standard 421.2 (1990).

A power system stabilizer (PSS) is frequently, but not always, included in an excitation system. It is designed to modulate the AVR input in such a manner as to contribute damping to intermachine oscillations. The input to the PSS may be generator rotor speed, electrical power, or other signals. The PSS usually is designed with linear transfer functions whose parameters are tuned to produce positive damping for the range of oscillation frequencies of concern. It is important that reasonably correct values be used for these parameters. The output of the PSS is limited, usually to ±5% of rated generator terminal voltage, and this limit value must be included in the model.

The excitation system includes several other subsystems designed to protect the generator and excitation system from excessive duty under abnormal operating conditions. Normally, these limiters and protective modules do not come into play for analysis of transient and oscillatory stability. However, for longer-term simulations, particularly related to voltage instability, overexcitation limiters (OEL) and under-excitation limiters (UEL) may need to be modeled. While there are many designs for these limiters, typical systems are described in *IEEE Trans.* (Dec. and Sept., 1995).

Prime Mover Modeling

The system that drives the generator rotor is often referred to as the prime mover. The prime mover system includes the turbine (or other engine) driving the shaft, the speed control system, and the energy supply system for the turbine. The following are the most common prime mover systems:

Power System Dynamics and Stability

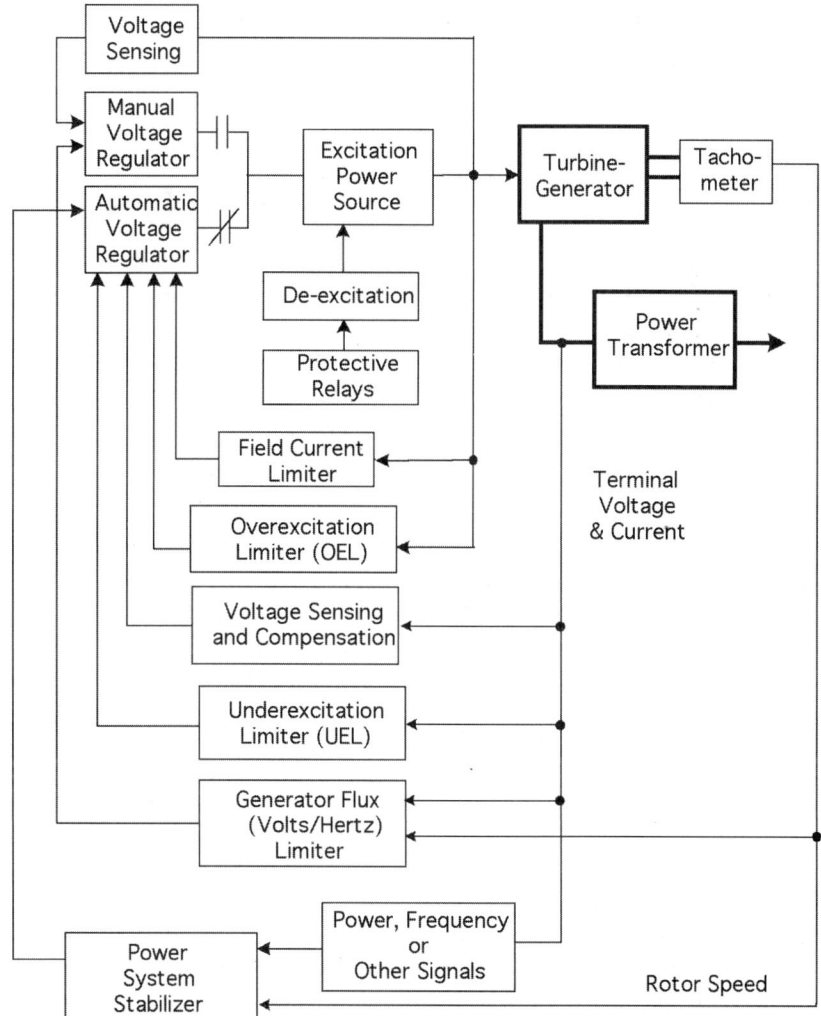

FIGURE 11.33 Excitation system model structure.

- Steam turbine
  - Fossil fuel (coal, gas, or oil) boiler
  - Nuclear reactor
- Hydro turbine
- Combustion turbine (gas turbine)
- Combined cycle (gas turbine and steam turbine)

Other less common, and generally smaller prime movers include wind turbine, geothermal steam turbine, solar thermal steam turbine, and diesel engine.

For analysis of transient and oscillatory stability, greatly simplified models of the prime mover are sufficient since, with some exceptions, the response times of the prime movers to system disturbances are slow compared with the time duration of interest, typically 10 to 20 sec or less. For simple transient stability analysis of only a few seconds duration, the prime mover model may be omitted altogether by assuming that the mechanical power output of the turbine remains constant. An exception is for a steam-turbine system equipped with "fast valving" or "early valve actuation" (EVA). These systems are designed to reduce turbine power output rapidly for nearby faults by quickly closing the intercept valves between the high-pressure and low-pressure turbine sections (Younkins et al., 1987).

For analysis of disturbances involving significant frequency excursions, the turbine and speed control (governor) systems must be modeled. Simplified models for steam and hydro turbine-governor systems are given in *IEEE Trans.* (Dec. 1973; Feb. 1992) and these models are available in most stability programs. Models for gas turbines and combined cycle plants are less standard, but typical models have been described in several references (Rowan, 1983; Hannett and Khan, 1993; *IEEE Trans.*, Aug. 1994).

For long-term simulations involving system islanding and large frequency excursions, more detailed modeling of the energy supply systems may be necessary. There are a great many configurations and designs for these systems. Models for typical systems have been published (*IEEE Trans.*, May 1991). However, detailed modeling is often less important than incorporating key factors that affect the plant response, such as whether the governor is in service and where the output limits are set.

For a fossil fuel steam plant, the coordination between the speed control and steam pressure control systems has an important impact on the speed with which the plant will respond to frequency excursions. If the governor directly controls the turbine valves (boiler-follow mode), the power output of the plant will respond quite rapidly, but may not be sustained due to reduction in steam pressure. If the governor controls fuel input to the boiler (turbine-follow mode), the response will be much slower but can be sustained. Modern coordinated controls will result in an intermediate response to these two extremes. The plant response will also be slowed by the use of "sliding pressure" control, in which valves are kept wide open and power output is adjusted by changing the steam pressure.

Hydro plants can respond quite rapidly to frequency changes if the governors are active. Some reduction in transient governor response is often required to avoid instability due to the "nonminimum phase" response characteristic of hydro turbines, which causes the initial response of power output to be in the opposite of the expected direction. This characteristic can be modeled approximately by the simple transfer function: $(1 - sT_w)/(1 + sT_w/2)$. The parameter T_w is called the water starting time and is a function of the length of the penstock and other physical dimensions. For high-head hydro plants with long penstocks and surge tanks, more detailed models of the hydraulic system may be necessary.

Gas (combustion) turbines can be controlled very rapidly, but are often operated at maximum output (base load), as determined by the exhaust temperature control system, in which case they cannot respond in the upward direction. However, if operated below base load, they may be able to provide output in excess of the base load value for a short period following a disturbance, until the exhaust temperature increases to its limit. Typical models for gas turbines and their controls are found in Rowan (1983) and *IEEE Trans.* (Feb. 1993).

Combined cycle plants come in a great variety of configurations, which makes representation by a typical model difficult (*IEEE Trans.*, 1994). The steam turbine is supplied from a heat recovery steam generator (HRSG). Steam is generated by the exhaust from the gas turbine(s), sometimes with supplementary firing. Often the power output of the steam turbine is not directly controlled by the governor, but simply follows the changes in gas turbine output as the exhaust heat changes.

Load Modeling

For dynamic performance analysis, the transient and steady-state variation of the load P and Q with changes in bus voltage and frequency must be modeled. Accurate load modeling is difficult due to the complex and changing nature of the load and the difficulty in obtaining accurate data on its characteristics. Therefore, sensitivity studies are recommended to determine the impact of the load characteristics on the study results of interest. This will help to guide the selection of a conservative load model or focus attention on where load modeling improvements should be sought.

For most power system analysis purposes, "load" refers to the real and reactive power supplied to lower voltage subtransmission or distribution systems at buses represented in the network model. In addition to the variety of actual load devices connected to the system, the "load" includes the intervening distribution feeders, transformers, shunt capacitors, etc. and may include voltage control devices, including automatic tap-changing transformers, induction voltage regulators, automatically switched capacitors, etc.

For transient and oscillatory stability analysis, several levels of detail can be used, depending on the availability of information and the sensitivity of the results to the load modeling detail. *IEEE Trans.* (May 1993 and Aug. 1995) discuss recommended load modeling procedures. A brief discussion is given below.

1. **Static load model** — The simplest model is to represent the active and reactive load components at each bus by a combination of constant impedance, constant current, and constant power components, with a simple frequency sensitivity factor, as shown in the following formula:

$$P = P_0 \left[P_1 \left(\frac{V}{V_0}\right)^2 + P_2 \left(\frac{V}{V_0}\right) + P_3 \right] \left(1 + L_{DP}\Delta f\right)$$

$$Q = Q_0 \left[Q_1 \left(\frac{V}{V_0}\right)^2 + Q_2 \left(\frac{V}{V_0}\right) + Q_3 \right] \left(1 + L_{DQ}\Delta f\right)$$

If nothing is known about the characteristics of the load, it is recommended that constant current be used for the real power and constant impedance for the reactive power, with frequency factors of 1 and 2, respectively. This is based on the assumption that typical loads are about equally divided between motor loads and resistive (heating) loads.

Most stability programs provide for this type of load model, often called a ZIP model. Sometimes an exponential function of voltage is used instead of the three separate voltage terms. An exponent of 0 corresponds to constant power, 1 to constant current, and 2 to constant impedance. Intermediate values or larger values can be used if available data so indicates. The following, more general model, permitting greater modeling flexibility, is recommended in *IEEE Trans.* (August 1995):

$$P = P_0 \left[K_{PZ}\left(\frac{V}{V_0}\right)^2 + K_{PI}\left(\frac{V}{V_0}\right) + K_{PC} + K_{P1}\left(\frac{V}{V_0}\right)^{n_{PV1}}\left(1 + n_{PF1}\Delta f\right) + K_{P2}\left(\frac{V}{V_0}\right)^{n_{PV2}}\left(1 + n_{PF2}\Delta f\right) \right]$$

$$Q = Q_0 \left[K_{QZ}\left(\frac{V}{V_0}\right)^2 + K_{QI}\left(\frac{V}{V_0}\right) + K_{QC} + K_{Q1}\left(\frac{V}{V_0}\right)^{n_{QV1}}\left(1 + n_{QF1}\Delta f\right) + K_{Q2}\left(\frac{V}{V_0}\right)^{n_{QV2}}\left(1 + n_{QF2}\Delta f\right) \right]$$

2. **Induction motor dynamic model** — For loads subjected to large fluctuations in voltage and/or frequency, the dynamic characteristics of the motor loads become important. Induction motor models are usually available in stability programs. Except in the case of studies of large motors in an industrial plant, individual motors are not represented. But one or two motor models representing the aggregation of all of the motors supplied from a bus can be used to give the approximate effect of the motor dynamics (Nozari et al., 1987). Typical motor data are given in the General Electric Company *Load Modeling Reference Manual* (1987).
3. **Detailed load model** — For particular studies, more accurate modeling of certain loads may be necessary. This may include representation of the approximate average feeder and transformer impedance as a series element between the network bus and the bus where the load models are connected. For long-term analysis, the automatic adjustment of transformer taps may be represented by simplified models. Several load components with different characteristics may be connected to the load bus to represent the composition of the load.

Load modeling data can be acquired in several ways, none of which are entirely satisfactory, but contribute to the knowledge of the load characteristics:

1. **Staged testing of load feeders** — Measurements can be made of changes in real and reactive power on distribution feeders when intentional changes are made in the voltage at the feeder, e.g., by changing transformer taps or switching a shunt capacitor. The latter has the advantage of providing an abrupt change that may provide some information on the dynamic response of the load as well as the steady-state characteristics. This approach has limitations in that only a small range of voltage can be applied, and the results are only valid for the conditions (time of day, season, temperature, etc.) when the tests were conducted. This type of test is most useful to verify a load model determined by other means.
2. **System disturbance monitoring** — Measurements can be made of power, voltage, and frequency at various points in the system during system disturbances, which may produce larger voltage (and possibly frequency) changes than can be achieved during staged testing. This requires installation and maintenance of monitors throughout the system, but this is becoming common practice on many systems for other purposes. Again, the data obtained will only be valid for the conditions at the time of the disturbance, but over time many data points can be collected and correlated.
3. **Composition-based modeling** — Load models can also be developed by obtaining information on the composition of the load in particular areas of the system. Residential, commercial, and various types of industrial loads are composed of various proportions of specific load devices. The characteristics of the specific devices are generally well-known (General Electric Company, 1987). The mix of devices can be determined from load surveys, customer SIC classifications, and from typical compositions of different types of loads (General Electric Company, 1987).

Transmission Device Models

For the most part, the elements of the transmission system, including overhead lines, underground cables, and transformers, can be represented by the same algebraic models used for steady-state (power flow) analysis. Lines and cables are normally represented by a pi-equivalent with lumped values for the series resistance and inductance and the shunt capacitance. Transformers are normally represented by their leakage inductance, resistance, and tap ratio. Transformer magnetizing inductance and eddy current (no-load) losses are sometimes included.

Other transmission devices that require special modeling include high-voltage direct current (HVDC) systems (Kundur, 1994) and power electronic (PE) devices. The latter includes static Var compensators (SVC) (*IEEE Trans.*, Feb. 1994) and a number of newer devices (TCSC, STATCON, UPFC, etc.) under the general heading of flexible AC transmission systems (FACTS) devices. Many of these devices have modulation controls designed to improve the stability performance of the power system. It is therefore important that these devices and their controls be accurately modeled. Due to the developmental nature of many of these technologies and specialized designs that are implemented, the modeling usually must be customized to the particular device.

Dynamic Equivalents

It is often not feasible or necessary to include the entire interconnected power system in the model being used for a dynamic performance study. A certain portion of the system that is the focus of the study, the "study system," is represented in detail. The remainder of the system, the "external system," is represented by a simplified model that is called a dynamic equivalent. The requirements for the equivalent depend on the objective of the study and the characteristics of the system. Several types of equivalents are discussed below.

1. **Infinite bus** — If the external system is very large and stiff, compared with the study system, it may be adequate to represent it by an infinite bus, that is, a generator with very large inertia and

very small impedance. This is often done for studies of industrial plant power systems or distribution systems that are connected to higher voltage transmission systems.
2. **Lumped inertia equivalent** — If the external system is not infinite with respect to the study system but is connected at a single point to the study system, a simple equivalent consisting of a single equivalent generator model may be used. The inertia of the generator is set approximately equal to the total inertia of all of the generators in the external area. The internal impedance of the equivalent generator should be set equal to the short-circuit (driving point) impedance of the external system viewed from the boundary bus.
3. **Coherent machine equivalent** — For more complex systems, especially when interarea oscillations are of interest, some form of coherent machine equivalent should be used. In this case, groups of generators in the external system are combined into single lumped inertia equivalents if these groups oscillate together for interarea modes of oscillation. Determination of such equivalents requires specialized calculations for which software is available (Price et al., 1978; Price et al., 1996).

References

Kundur, P., *Power System Stability and Control*, McGraw-Hill, New York, 1994.
Damping representation for power system stability analysis, *IEEE Trans.*, PWRS-14, 151–157, Feb. 1999.
IEEE Standard 1110-1991, *IEEE Guide for Synchronous Generator Modeling Practices in Stability Analysis*, 1991.
IEEE Standard 421.5-1992, *IEEE Recommended Practice for Excitation System Models for Power System Stability Studies*, 1992.
IEEE Standard 421.2-1990, *IEEE Guide for Identification, Testing, and Evaluation of the Dynamic Performance of Excitation Control Systems*, 1990.
Recommended models for overexcitation limiting devices, *IEEE Trans.*, EC-10, 706–713, Dec. 1995.
Underexcitation limiter models for power system stability studies, *IEEE Trans.*, EC-10, 524–531, Sept. 1995.
Younkins, T.D., Kure-Jensen, J., et al., Fast valving with reheat and straight condensing steam turbines, *IEEE Trans.*, PWRS-2, 397–404, May 1987.
Dynamic models for steam and hydro turbines in power system studies, *IEEE Trans.*, PAS-92, 1904–1915, Dec. 1973.
Hydraulic turbine and turbine control models for system dynamic studies, *IEEE Trans.*, PWRS-7, 167–179, Feb. 1992.
Rowan, W.I., Simplified mathematical representations of heavy-duty gas turbines, *ASME Trans. (Journal of Engineering for Power)*, 105(1), 865–869, Oct. 1983.
Hannett, L.N., Khan, A., Combustion turbine dynamic model validation from tests, *IEEE Trans.*, PWRS-8, 152–158, Feb. 1993
Dynamic models for combined cycle plants in power systems, *IEEE Trans.*, PWRS-9, 1698–1708, August 1994.
Dynamic models for fossil fueled steam units in power system studies, *IEEE Trans.*, PWRS-6, 753–761, May 1991.
Load representation for dynamic performance analysis, *IEEE Trans.*, PWRS-8, 472–482, May 1993.
Standard load models for power flow and dynamic performance simulation, *IEEE Trans.*, PWRS-10, August 1995.
Nozari, F., Kankam, M.D., Price, W.W., Aggregation of induction motors for transient stability load modeling, *IEEE Trans.*, PWRS-2, 1096–1103, Nov. 1987.
General Electric Company, *Load Modeling for Power Flow and Transient Stability Computer Studies-Load Modeling Reference Manual*, EPRI Final Report EL-5003, 2, Jan. 1987.
Kundur, P., *Power System Stability and Control*, Section 10.9 Modelling of HVDC systems, McGraw-Hill, New York, 1994.

Static Var compensator models for power flow and dynamic performance simulation, *IEEE Trans.*, PWRS-9, 229–240, Feb. 1994.

Price, W.W., Hargrave, A.W., Hurysz, B.J., Chow, J.H., Hirsch, P.M., Large-scale system testing of a power system dynamic equivalencing program, *IEEE Trans.*, PWRS-13, 768–774, Aug. 1998.

Price, W.W., Hurysz, B.J., Chow, J.H., Hargrave, A.W., Advances in power system dynamic equivalencing, *Proc. Fifth Symposium of Specialists in Electric Operational and Expansion Planning (V SEPOPE)*, 155–169, Recife, Brazil, May 1996.

11.8 Direct Analysis of Wide Area Dynamics

J. F. Hauer, W. A. Mittelstadt, M. K. Donnelly, W. H. Litzenberger, and Rambabu Adapa

The material to follow deals with the direct analysis of power system dynamic performance. By "direct" we mean that the analysis is performed on the physical system, and that any use of system models is secondary. Many of the tools and procedures are as applicable to simulated response as to measured response, however. Comparison of the results thus obtained is strongly recommended as a means to test model validity.

The resources needed for direct analysis of a large power system represent significant investments in measurement systems, mathematical tools, and staff expertise. New market forces in the electricity industry require that the "value engineering" of such investments be considered very carefully. Many guidelines for this can be found in collective utility experience of the Western Systems Coordinating Council (WSCC), in western North America.

Dynamic Information Needs: The WSCC Breakup of August 10, 1996

Large power systems are very rich in information that can be developed from direct measurements of dynamic behavior. Progressive electrical utilities are developing comprehensive data acquisition facilities. The emerging critical path challenge is to extract essential information from this data, and to distribute the pertinent information where and when it is needed. Otherwise, system control centers will be progressively inundated by potentially valuable data that they are not yet able to fully utilize.

New factors are rapidly compounding this problem. Utility restructuring promises to sharply increase the need for measurement-based information while shrinking the time frame in which it must be produced and distributed. In addition, financial pressures dictate that cost recovery for the requisite technology investments be prompt and low risk.

These issues were brought into sharp and specific focus by the massive breakup experienced by the Western North American Power System on August 10, 1996. The mechanism of failure (though perhaps not the cause) was a transient oscillation under conditions of high power transfer on long paths that had been progressively weakened through a series of seemingly routine transmission line outages. Later analysis of monitor records, as in Fig. 11.34, provides many indications of potential oscillation problems. Verbal accounts also suggest that less direct indications of a weakened system were observed by system operators for some hours, but that there had been no means for interpreting them. It is very likely that, buried within the measurements already at hand, lay the information that system behavior was abnormal and that the system itself was vulnerable.

Utility restructuring, through several mechanisms, is making it impossible to predict system vulnerabilities as accurately or as promptly as the increasingly volatile market demands. It is quite possible that standard planning models could not have predicted the August 10 breakup, even if the conditions leading up to it had been known in full detail. This situation has deep roots and many ramifications (WSCC Work Group, 1990; Hauer and Hunt, 1996; Stahlkopf and Wilhelm, 1997; Kosterev et al., 1999; Hauer and Taylor, 1998; Taylor, 1999). Correcting the situation will require many years of concerted effort on many fronts.

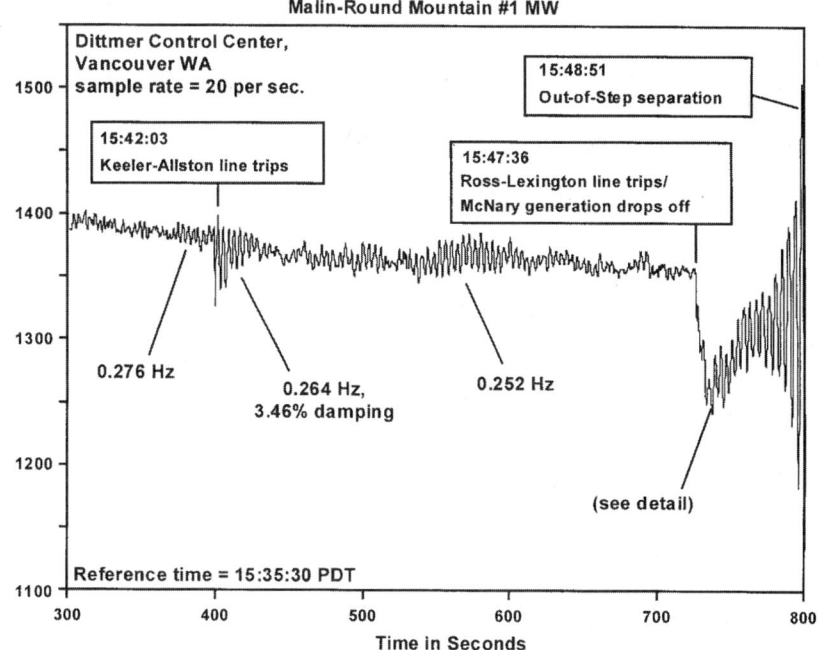

FIGURE 11.34 Oscillation buildup for the WSCC breakup of August 10, 1996.

An interim solution is to reinforce capabilities for predicting system vulnerability with the capability to detect and recognize its symptoms as evidenced in dynamic measurements. This can also be viewed as a form of measurement-based dynamic security assessment (DSA). In the Western Power System, the technology and infrastructure that this requires are being developed as extensions of the DOE/EPRI Wide Area Measurement System (WAMS) Project and related efforts (Hauer et al., 1997; Hauer et al., Jan. 1999; Hauer et al., May 1999).

Background

Comprehensive monitoring of a large power system is a long step beyond the monitoring of local devices or even regional performance. Developing a suitable investment plan calls for close attention to emerging information needs and information technologies. Even more important — perhaps decisively so — are the interutility practices and infrastructure through which dynamic information resources are collectively reinforced and operated.

Timeliness of the information is becoming an increasingly important consideration. The pace of utility restructuring strongly encourages more aggressive use of power delivery assets, which translates into progressive encroachment upon customary operating margins. This can greatly increase the need for direct evidence concerning the proximity and nature of safe operating limits.

The Western North America power system provides a useful example of the agencies at work. Driven by stability considerations, and in an earlier regulatory environment, the member utilities of the Western North America power system have made significant progress in the development of monitor facilities for examining system behavior. This development, coordinated through the WSCC, is a collective response to their shared needs for measurement-based information about system characteristics, model fidelity, and operational performance (WSCC Work Group, 1990; Hauer and Hunt, 1996; Kosterev et al., 1999).

Utility restructuring is now carrying these and all utilities toward the future in large abrupt steps. Infrastructure development for acquiring wide-area dynamic information started in response to technical needs in a cost-based environment. Its value in the new price-based environment is likely to be much higher through direct services such as

- **Real-time determination of transmission capacities** — assuming necessary progress with mathematical tools.
- **Early detection of system problems** — this enables cost reductions through performance-based maintenance, and provides a safety check on network loading.
- **Refinement of planning, operation, and control processes** — essential to best use of transmission assets.

Such an infrastructure can also provide indirect benefits as an enterprise network (*IEEE Spectrum*, July 1991; Grenier and Metes, 1992), or "people net." It is a proven means through which a wide range of technical skills can be accessed and shared among the utilities as a virtual technology staff. This may be very important to the future power system, especially in situations where difficult stability problems constrain the use of transmission assists.

An Overview of WSCC WAMS

Collectively, in response to the needs that utilities share in operating their facilities, these efforts are leading toward an integrated dynamic information network that spans the entire WSCC system. WSCC WAMS is evolving as a hierarchical network of dynamic monitors, plus the information tools and general infrastructure necessary for effective use of the acquired data. Data sources are of many kinds, and they may be located anywhere in the power system. This is also true for those who need the data, or those who need various kinds of information extracted from the data.

It would not be practical or sufficient to just collect all WAMS data at one location and then permit users to retrieve it as needed. Data volume and information demands call for distributed storage and management. Central to this is a WAMS Information Manager having enough "intelligence" to route data and manage archives on the basis of the information contained. This issue is pursued in a later section.

Direct Sources of Dynamic Information

There are a variety of means by which dynamic information can be extracted from a large power system. These include

- Disturbance analysis
- Ambient noise measurements
    - spectral signatures
    - open-loop/closed-loop spectral comparisons
    - correlation analysis
- Direct tests with
    - low level noise inputs
    - mid-level inputs with special waveforms
    - high level pulse inputs
    - network switching

Each has its own merits, disadvantages, and technical implications (Hauer et al., 1999; Hauer, 1995; Phadke, 1993; EPRI, Dec. 1993; Hauer et al., 1995; Hauer, 1996). For comprehensive results, at best cost, a sustained program of direct power system analysis will draw upon all of these in combinations that are tailored to the circumstances at hand.

Some of these operations, such as network switching tests, are outside the usual scope of power system monitoring. All of them represent required functionalities for the WAMS infrastructure, however, and for the monitor facilities within it. *Monitoring is a subset of measurement operations.* Even so, it is the monitor facilities that provide the backbone for the dynamic information infrastructure.

From a functional standpoint, wide-area monitoring for a large power system involves the following general functions:

- **Disturbance monitoring**: characterized by large signals, short event records, moderate bandwidth, and straightforward processing. Highest frequency of interest is usually in the range of 2 Hz to perhaps 5 Hz. Operational priority tends to be very high.
- **Interaction monitoring**: characterized by small signals, long records, higher bandwidth, and fairly complex processing (such as correlation analysis). Highest frequency of interest ranges to 20–25 Hz for rms quantities, but may be substantially higher for direct monitoring of phase voltages and currents. Operational priority is variable with the application and usually less than for disturbance monitoring.
- **System condition monitoring**: characterized by large signals, very long records, very low bandwidth. Usually performed with data from SCADA or other EMS facilities. Highest frequency of interest is usually in the range of 0.1 Hz to perhaps 2 Hz. Core processing functions are simple, but associated functions such as state estimation and dynamic or voltage security analysis can be very complex. Operational priority tends to be very high.

These functions are all quite different in their objectives, priorities, technical requirements, and information consumers. At many utilities they are supported by separate staff structures and by separate data networks.

What is a Monitor?

The power system contains many devices that can serve as monitors for some processes and purposes. This document narrows the field somewhat, through the following definition:

A *monitor* is any device that automatically records power system data, either selectively or continuously, according to some mechanism that permits the data to be retrieved later for analysis and display.

With exceptions as noted, present attention is further restricted to monitors that record one or more of the following:

- Dynamic "swing" interactions among generators and loads through an interconnecting electrical transmission system.
- The performance of specific facilities involved in swing interactions. This includes generators, loads, and control systems.

This focus excludes most digital fault recorders and SCADA systems, since their typical record lengths and data rates are (respectively) not adequate for capturing swing dynamics. Network condition monitors might be included, depending upon the specifics of the device.

Overall, *monitor facilities* often contain a wide range of recording devices that are not swing monitors. The ubiquitous (and underutilized) digital fault recorder, or DFR, is a particulary good example of this. The facilities may also include devices that have little or no direct recording capability, but that do provide essential "intelligence" to overall facility performance. It is becoming established practice to classify all of the aforementioned devices as *intelligent electronic devices*, or IEDs (Smith, 1996; Carolsfeld, 1997). Drawing upon this terminology, just about any modern measurement system reduces to an IED network.

Monitor System Functionalities

The primary rule in power system measurements is to *record good data, and keep them safe!* These are two distinct functions and both call for close attention. If monitor facilities are used to support direct tests, then an even more important rule enters the picture, which is to *do no harm to the power system*. In both cases it is the measurement of low-level interactions that define the most rigorous functionality requirements to the overall measurement system. This is due, in part, to long recording periods plus the need to resolve very small changes in large signals (Hauer, 1995).

There is a more demanding need, however — the evaluation of system dynamic performance in nearly real time (NRT). This is particularly critical during staged tests when a close balance must be maintained

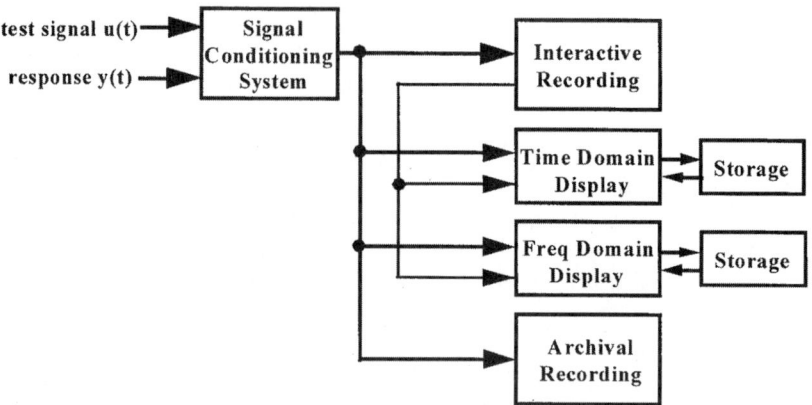

FIGURE 11.35 Information functions in measurement of low level interactions.

between system security and the quality of test results. It is also important during routine monitor operations as a means for identifying important data and for generation of operator alerts.

Figure 11.35, with the test signal deleted for the case of passive monitoring, represents core information functions that are needed in both situations. The functions are:

- **Signal conditioning**, to assure the measurement quality and prompt observability of important signal features.
- **Archival recording**, to assure safekeeping of important results. This should be as comprehensive as possible, and include all phases of testing that involve switching operations or application of test signals.
- **Interactive recording,** to permit prompt examination of data that cannot be fully assessed in real time. This also provides backup recording of high priority signals.
- **Time domain display,** to permit frequent review of signal waveforms for evidence of data quality and emerging trouble on the power system.
- **Frequency domain display**, to permit frequent review of signal spectra for evidence of data quality and possible trouble on the power system.

The NRT displays permit appropriate actions to be taken by the test team or monitor operations staff. In some cases the analysis tools underlying these displays are also used in event detection logic (EDL) to trigger automatic functions such as accessory data capture, information routing, or operator alerts.

Figure 11.35 represents a paradigm that is fundamental to low level measurements of power system dynamics. Not all monitors are interaction monitors, and many interaction monitors lack some of the functions shown. *The required functionality resides in the overall measurement system*. Figure 11.35 will reappear in many expanded forms throughout the sections to follow.

Event Detection Logic

Data recording in an individual monitor is either selective or continuous. In selective recording, data "snapshots" are collected upon command of "trigger logic." This requires that dynamic event signatures be detected in real time — otherwise useful data is lost. Continuous recording requires very similar logic, but uses it to sort records already captured according to their information content.

Triggers for initiating data capture can be classified in several ways. A "manual vs. automatic" classification provides

- manual
  - local
  - remote

- automatic
  - pre-scheduled
  - internal event detection
  - cross-trigger (from another monitor or IED)

Another classification is

- external
  - local manual
  - remote manual
  - prescheduled (clock initiated)
  - cross-trigger (from another monitor or IED)
- internal event detection

This highlights the fact that EDL is fundamentally different from some other triggers. EDL is also a core issue in monitor design and operation, whether recording is selective or continuous.

There are four basic factors involved in detecting the onset of a dynamic event. They are

- magnitude
- persistence
- frequency content
- context

A simple disturbance trigger might examine just magnitude and persistence, in tests of form "Do the latest M samples each exceed threshold $T(M)$?" (Hauer and Vakili, 1990). It is useful to think of the context factor as adjusting such thresholds to system conditions, such as network stress or the operational status of key system resources.

A partial list of signatures through which events can be detected, and perhaps recognized, includes the following:

- Steps or swings in tieline power.
- Large change, or rate of change, in bus voltage or frequency.
- Sustained or poorly damped oscillations, perhaps in conjunction with some other event.
- Large increase in system noise level.
- Increase of system activity in some critical frequency band.
- Unusual correlation or phasing between fluctuations in two given signals.

The tools needed to extract useful signature information from measured data range from straightforward heuristics to very advanced methods of signal analysis. Recognition of the underlying events calls for pattern recognition logic to match extracted signatures against known event templates.

Monitor Architectures

The vast majority of dynamic monitors only capture data for disturbances that are strongly observable in the monitor inputs. The "trigger to archive" or "snapshot" monitor in Fig. 11.36 is typical. A circulating prehistory buffer retains the most recently acquired data, assuring that a certain amount of information will be provided about system conditions before the disturbance is detected. Trigger logic causes the disturbance to be recorded as one or more data snapshots in the motor archives. The device shown also has the following accessory features:

- External trigger inputs for initiating data capture.
- Flags and alarms sent out to indicate the triggering condition.

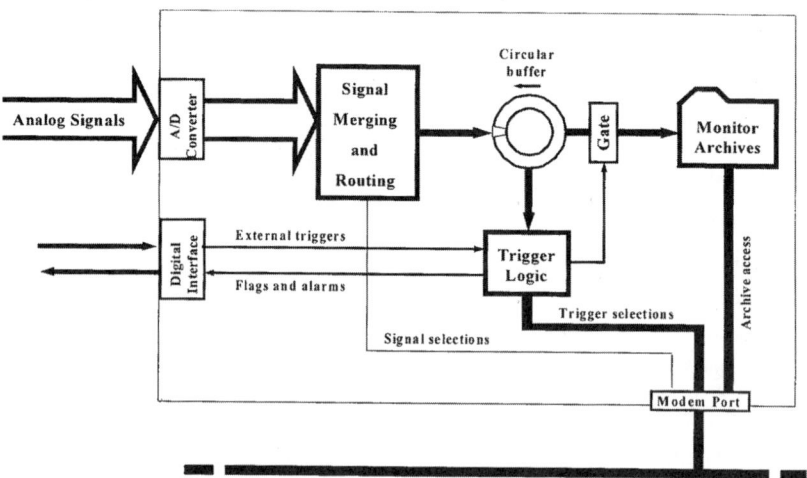

FIGURE 11.36 A simple "snapshot" disturbance monitor.

- External control of monitor settings via a modem port.
- Retrieval of archive data via a modem port.

The triggered monitor uses a rather conservative architecture that is necessary and appropriate when substantial amounts of data must be captured at high rates (as in digital fault recording). The reliance upon triggered recording — together with a general tendency toward short records and even shorter prehistories–are serious handicaps in wide area measurements, however.

WSCC experience suggests that triggered data capture does not provide an adequate basis for wide-area measurements. Even rather large events may not be sensed by trigger logic that is remote from the site of the disturbance. Records for a cascading failure that develops slowly, from some fairly small initiating event, are very unlikely to present a comprehensive view of the mechanism by which the small failure propagated into a very large one.

Figure 11.37 illustrates the point. The record there, collected on BPA's earlier Power System Disturbance Monitor, indicates peak-to-peak 0.7 Hz swings of roughly 900 MW on the Pacific AC Intertie (PACI). However, as is usual with triggered monitors, it failed to capture the all-important interval during which

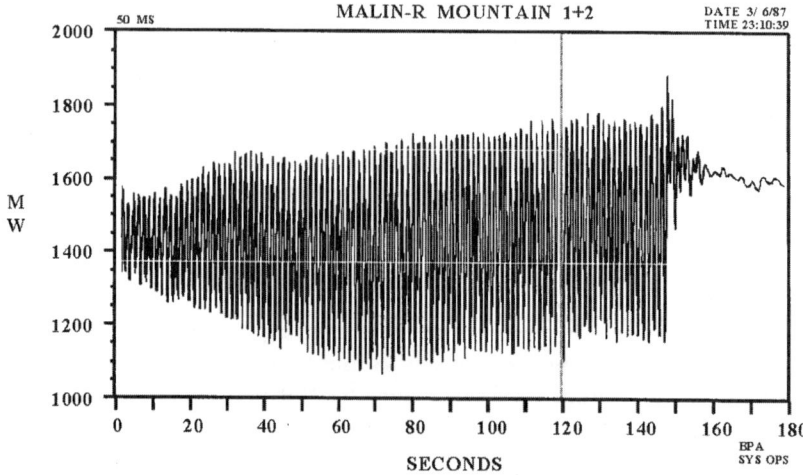

FIGURE 11.37 Western system oscillations of March 6, 1987 (sum for two parallel circuits).

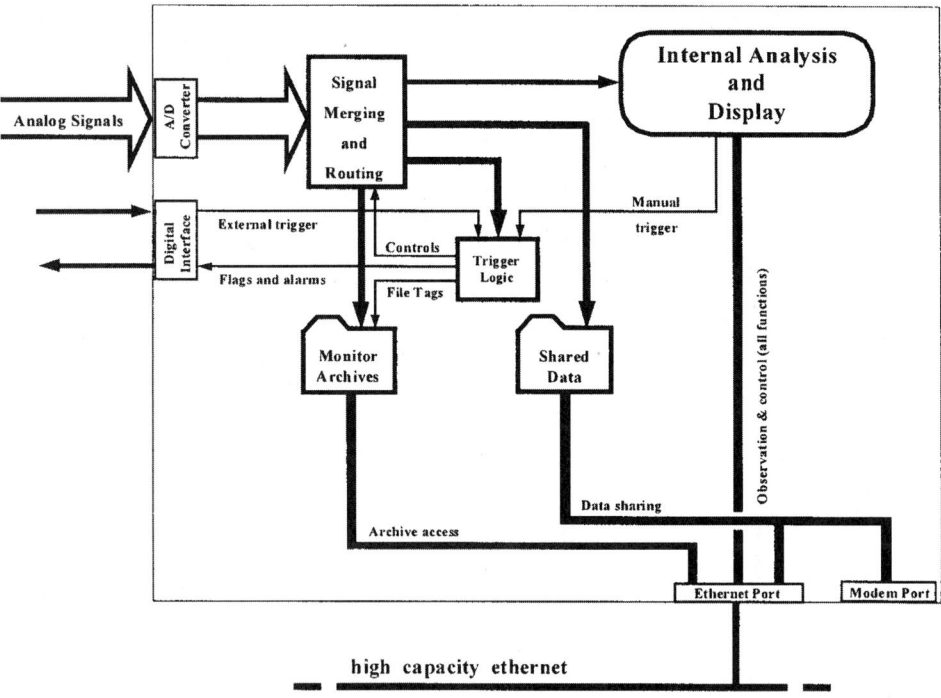

FIGURE 11.38 Basic architecture for a continuously recording monitor.

the oscillations started. Without this, whatever indications there may have been to warn system operators of pending trouble remain unknown.

Some triggered monitors can support interaction measurements under manual control, provided that their storage capabilities and analog to digital (A/D) resolution are sufficient. Their functionality in this mode of operation is similar to that of a high-quality tape recorder. The more "intelligent" functions expected of a monitor are lost, and sorting through the data can become a very laborious manual exercise. Monitors that are explicitly designed to operate in a continuous recording mode are a far better option.

Figure 11.38 shows a basic "stream to archive" monitor that, typically, will maintain a continuous data record for periods ranging to several weeks. While triggers are used, their role is rather different than for a triggered monitor. Data capture in a continuous monitor *does not* depend upon trigger logic. Instead, trigger logic is used to:

- **tag the acquired data** in ways indicating its likely value. Factors in this are:
  - the magnitude, duration, and frequency content of dynamic activity.
  - the context in which the data were acquired — e.g., event specific external triggers and system stress levels (from condition monitoring).
- **generate external flags or alarms** when data values reach critical levels.

The value tags will also determine such matters as the priority, routing, archiving level, compression, and retention time for the associated data. In a fully developed monitor system, the value tags will be amended by an Archive Scanner that automatically reviews and restructures collective monitor archives.

This continuous monitor also has some accessory features beyond those in Fig. 11.36. These are

- Real-time analysis and display, with an option for manual triggering.
- NRT data sharing through the computer network.

The next architecture takes the functionality progression several steps further.

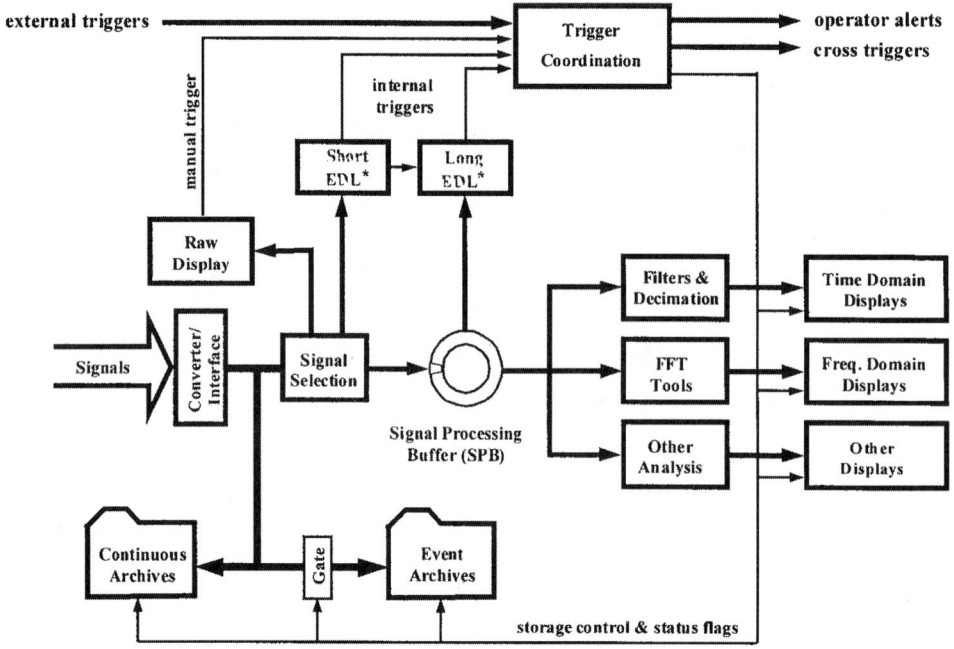

FIGURE 11.39 A continuous monitor combining short and long EDL.

Figure 11.39 shows a configuration that is representative of a mature interactions monitor that is operating in a manned substation or in a system control center. Absence of an indicated perimeter for the monitor reflects the fact that this might well be a locally networked facility, containing several computers or other IEDs plus linkages to the energy management system (EMS).

The indicated triggers are both external and internal, manual and automatic. The internal automatic triggers are classified as short or long (fast or slow), depending upon length of the data segment needed by the associated event detection logic. Short EDL can work with a short block of recent data and is usually sufficient for disturbance monitoring.

A distinguishing feature in this architecture is the signal processing buffer (SPB) used for advanced triggers (in the long EDL) and in special displays. SPB functionality is essential for extracting interaction signatures, and for presenting those signatures to operations staff for their interpretation and review. At hardware level, however, this functionality can be distributed among one or more buffers internal to the monitor itself plus external buffers for shared access to the record stream at file level.

A next step in monitor refinement is to enhance the EDL and trigger coordination functions of Fig. 11.39 through artificial intelligence. Figure 11.40 represents a Diagnostic Event Scanner (DES) suitable for this purpose, and for the Archive Scanner mentioned earlier.

Monitor Network Topologies

With respect to their architecture, monitor facilities consist of

- **Central monitors** that continually scan signals from transducers, and similar instruments, that are communicated from remote sites.
- **Distributed monitor networks,** which are found in a variety of general forms
- **Local monitors,** for which remote access is usually weak and manually initiated.

BPA operates a centralized monitor system, based at the Dittmer Control Center, that is interlaced with an expanding network of remote monitors and secondary recording devices. Many of the remote monitors provide local or regional surveillance over some important parts of the control system. The

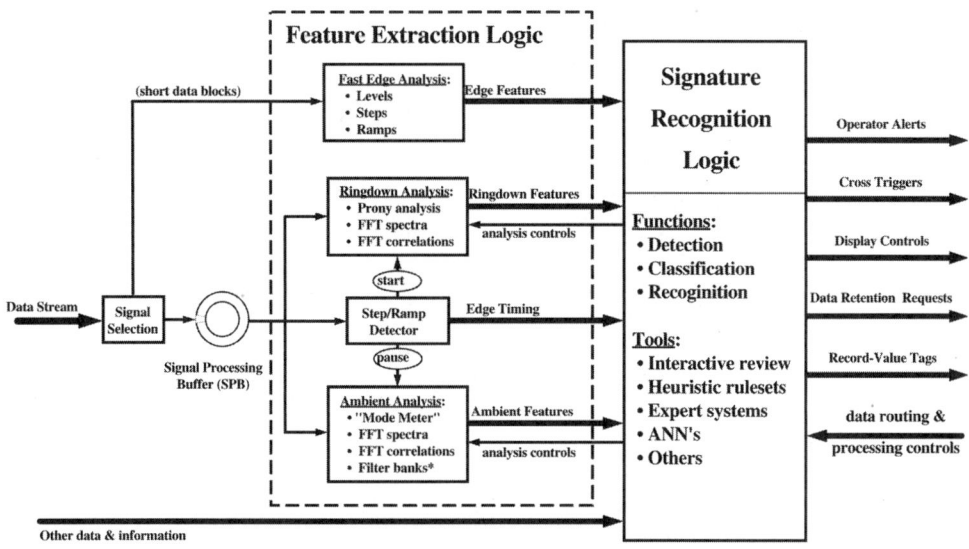

FIGURE 11.40 Diagnostic Event Scanner (DES) within a continuous monitor.

general strategy is to collect essential signals (including alarms) from such sites on the central monitors, at modest data rates, and to perform comprehensive high-speed monitoring locally. In most cases, the high-speed data is immediately available to site operators, but is selectively forwarded to the control center or to system analysts.

BPA has three central monitors at or near the Dittmer Control Center (Fig. 11.41). The newest of these, based upon BPA's rapidly evolving Phasor Data Concentrator (PDC), is discussed a bit later. The other two, the Power System Analysis Monitor (PSAM) and the Dittmer PPSM, are indicated in Fig. 11.42.

The PSAM and PPSM have access, in parallel, to several hundred analog signals. Most of these signals represent power flowing within the BPA service area and contain useful information at frequencies up to perhaps 2 Hz. Signals associated with control projects are an increasingly common exception to this. They ordinarily use 20 Hz channels and the corresponding transducers, if any, tend to be conventional electronic units modified for either a 2 Hz or 20 Hz bandwidth.

In this arrangement, the 2 Hz signals present a comprehensive view of interarea "swing" dynamics visible in BPA interchanges, plus information about voltages, reactive flow, important loads, and automatic generation control (AGC). The 20 Hz signals convey a necessary minimum of essential information about controller behavior. They also facilitate alignment and cross analysis of central monitor records with more comprehensive local recordings. The total communications burden for this mixture of central plus distributed monitoring is much less than for a fully centralized monitor system. Furthermore, much of the data can be moved with adequate speed and reliability using general purpose computer networking technology.

This deployment of remote monitors on a general network represents an evolutionary step in the gradual transition from analog technology to digital. Direct replacement of BPA's present point-to-point analog communications by digital channels of comparable bandwidth and resolution (about 14 bits, after filtering) would be needlessly expensive at this time. Best use of digital technology will call for new architectures, which should readily accommodate the rapid product enhancements so characteristic of the "information age."

The third monitor system at the Dittmer Control Center is entirely digital. The system consists of multiple Phasor Measurement Units (PMUs) that are linked together by one or more PDC units. The PMUs are synchronized digital transducers that stream their data, in real time, to the Dittmer PDC(s). The general functions and topology for this network resemble those for the Dittmer PSAM and PPSM. Data quality for the phasor technology appears to be very high, however, and secondary processing of the acquired phasors can provide a broader range of signal types.

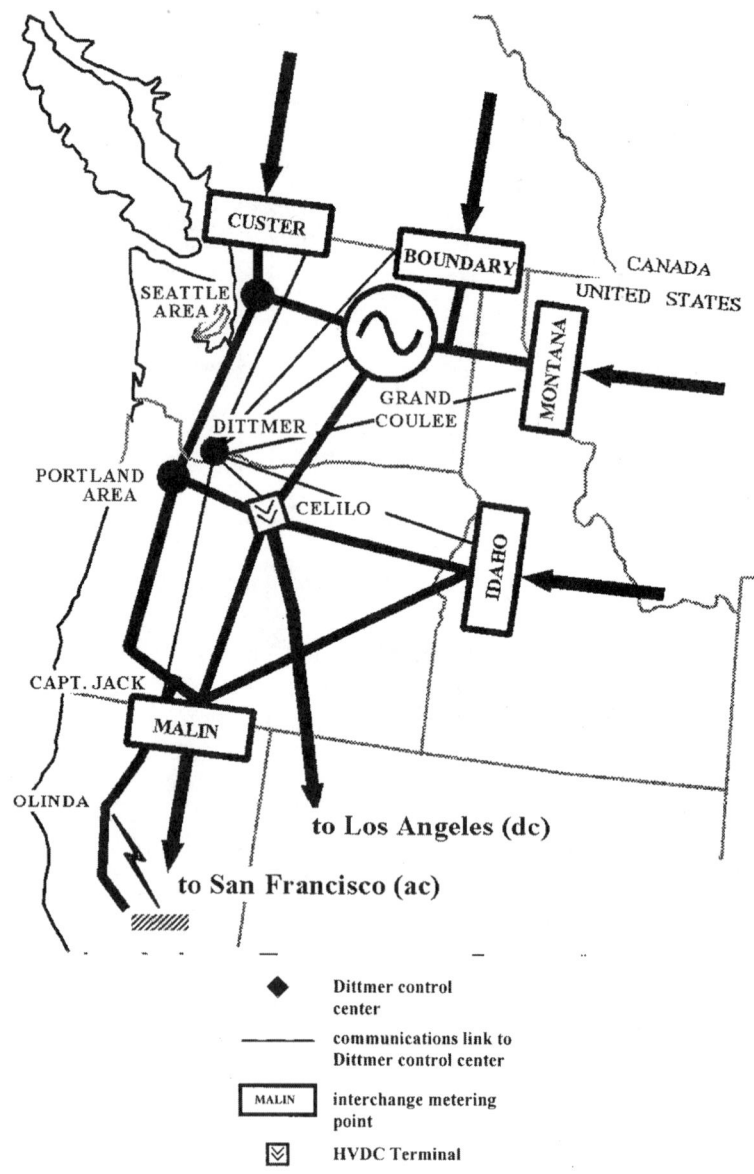

FIGURE 11.41 BPA central monitor coverage.

Phasor networks provide their best value in applications that are mission-critical, and that involve truly wide area measurements. Both factors encourage real-time data links among regional phasor networks. This can be accomplished at both the PMU and the PDC levels. Connecting a PMU to multiple PDC units is straightforward and has already been done. Selective forwarding of PDC signals to other PDC units seems feasible and is under very active development. A copy of the BPA PDC became operational at Southern California Edison (SCE) facilities in mid-1998, and another copy became operational at WAPA in November 1998. The resulting network is evolving toward a topology of the sort indicated in Fig. 11.43.

Central monitor systems, though very effective, are not a complete answer:

- They do not serve information needs at local or regional level.
- The dedicated communications can be very expensive.

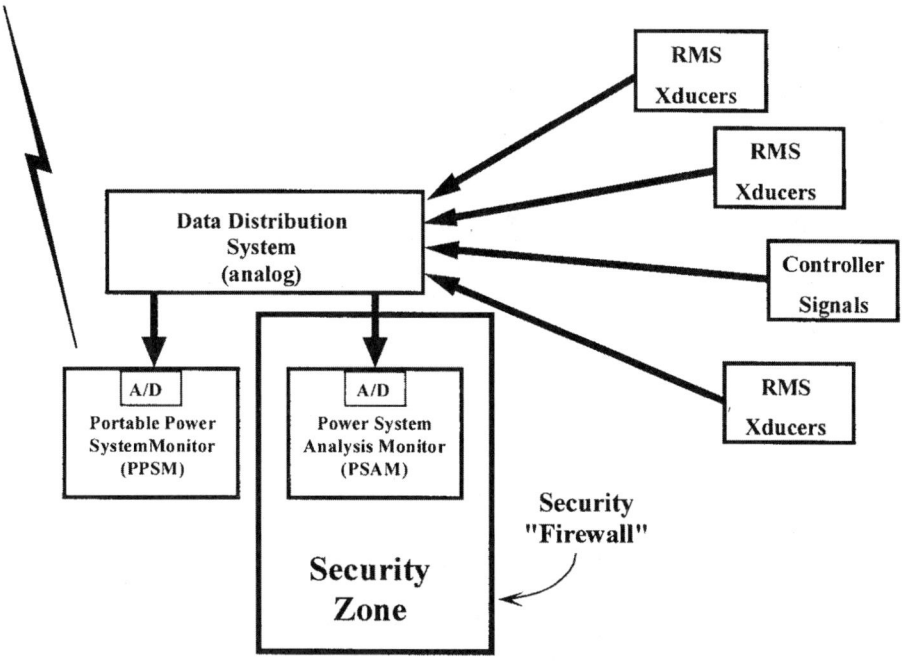

FIGURE 11.42 Topology of BPA central monitor system for analog signals.

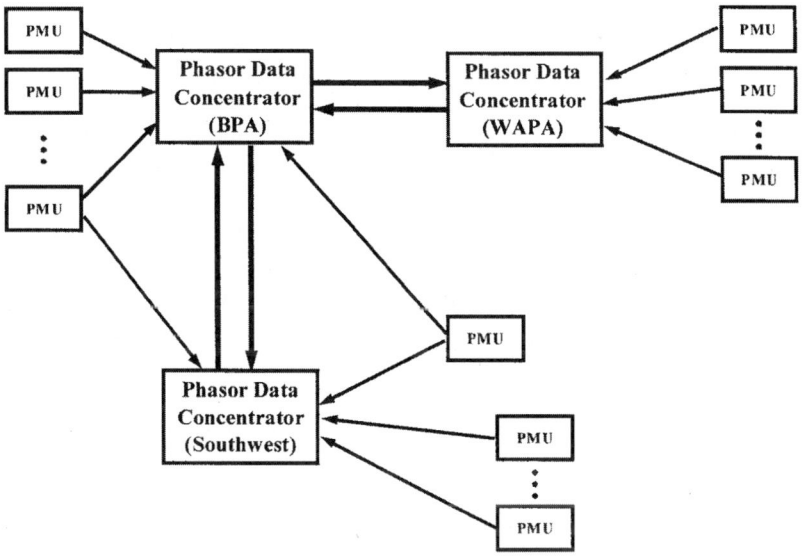

FIGURE 11.43 Partial topology of the emerging WSCC phasor measurements network.

- Communication failures may cause important data to be lost.
- Data rates and data volumes for some dynamic processes are so high that continuous transmission to central facilities is not practical. This is especially likely for high performance control systems.
- Under normal circumstances, a lot of data are too mundane to merit the costs of continuous transmission. Transmission should be selective, and based upon information value.

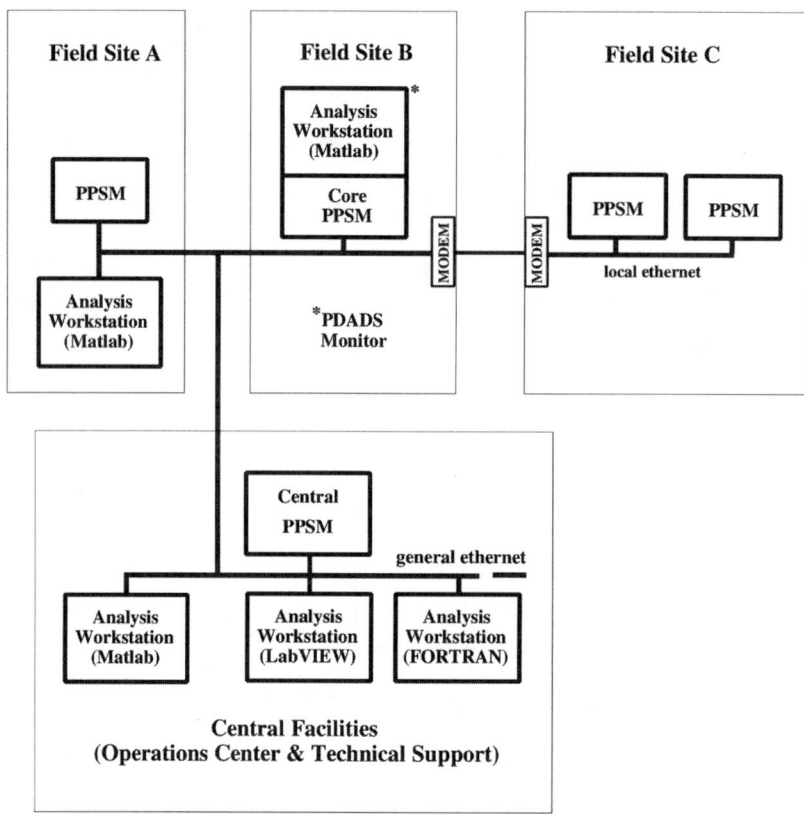

FIGURE 11.44 A distributed network for dynamic information.

In many situations, better performance and better "value engineering" can be obtained through distributed monitor networks, in which local storage and selective messaging minimize reliance upon dedicated communications.

Distributed monitor networks can take many forms, ranging from dial-up access among sites to NRT communication of data and information on full-time computer networks. A general computer network permits a much wider range of functions and technologies than those of the central monitors represented in Figs. 11.42 and 11.43. It is particularly valuable for broadening the staff support base for monitor operations.

Figure 11.44 shows a dynamic information network that represents the topology and functionalities associated with distributed regional monitoring. The nomenclature there is based upon BPA/PNNL elements of the WAMS technology package. A "monitor" in this context generally means a *measurements workstation*, consisting of

- a "core monitor" or *data capture unit* (DCU).
- one or more *analysis toolsets*. Most of this functionality is applied to data already captured but, as indicated in Figs. 11.35 and 11.38, some may be used in near real time.

The analysis toolsets, when operated separately from the DCU, provide an *analysis workstation*. The following comments apply to the network of Fig. 11.44:

- All monitor units and most workstations can communicate with one another.
- PPSM data capture (in the DCU) follows the logic shown in Fig. 11.38. Continuous recording is customary but optional.

- PPSM units normally include LabVIEW® analysis station capabilities suitable for initial analysis of captured data. For in-depth analysis and design, this is augmented or replaced by a more advanced package based upon Matlab® (plus imbedded FORTRAN®).
- The diagram for Field Site B shows a PPSM variant in which the LabVIEW analysis tool has been replaced by a more extensive Matlab toolset for combined analysis and design. The result is a Portable Dynamic Analysis and Design System (PDADS).
- Analysis in the FORTRAN environment is often performed on larger computers that do not communicate readily with the PPSM or other workstations. Then the communications may be limited to data transfers only.

In point of fact, there are two monitor types shown in Fig. 11.44. The standard PPSM is entirely a LabVIEW-based virtual instrument (Santori, 1990; Schoukens, 1993). It finds its best uses in field analysis or in predefined environments where "pushbutton" analysis is appropriate. The other monitor is an extended PPSM that uses the standard LabVIEW DCU, while turning to Matlab to support its more advanced functions (Hauer et al., 1997; Matlab Reference Guide).

This network topology is suitable when security requirements are modest. Dial-in modems can be particularly inviting points for external attack, however, and they may not be suitable for applications that need a high level of data security (Jones and Skelton, 1999).

Networks of Networks

A well-evolved monitor system will necessarily involve a mixture of technologies, data sources, functions, operators, and data consumers. In broad terms,

- **Required functionalities** are determined by who must see what, when, and in what form.
- **System configuration** is strongly influenced by geography, ownership, selected technology, and the technology already in service (legacy systems).
- **Investment value** is strongly enhanced through:
    - selective use of IEDs to supervise, integrate, or replace legacy systems.
    - organization of IEDs (including monitors) into local, regional, and wide area networks appropriate to their functions and technologies.

The choices that a particular utility will make are strongly colored by its operating and business requirements and more generally by the value it places upon information.

Overall, the forces at work strongly favor wide area measurement systems that evolve as "networks of networks." There are a lot of advantages to this. Interleaving networks that have different topologies and different base technologies can make the overall network much more reliable, while broadening the alternatives for value engineering. It also permits networks to be operated on the basis of ownership. The ability of a utility to retain data until it is no longer sensitive (delayed release) will almost certainly prove necessary for information sharing in the new power system.

Figure 11.45 shows interleaved networks of PMU, PPSM, and DSM units (plus analysis workstations). These devices are proprietary to Macrodyne, BPA/PNNL, and Power Technologies Inc. In this case,

- The PMU network has dedicated microwave communications in addition to dial-up links.
- The DSM "network" consists of individual units, accessed from a DSM base unit via dial-up links.
- The PPSM network serves in several roles. In addition to advanced monitor functions, it also provides the PMU and DSM units with:
    - alternate communication paths, via the computer network.
    - local, high volume archiving.
    - analysis and display functions available both on site and through remote teleoperation.

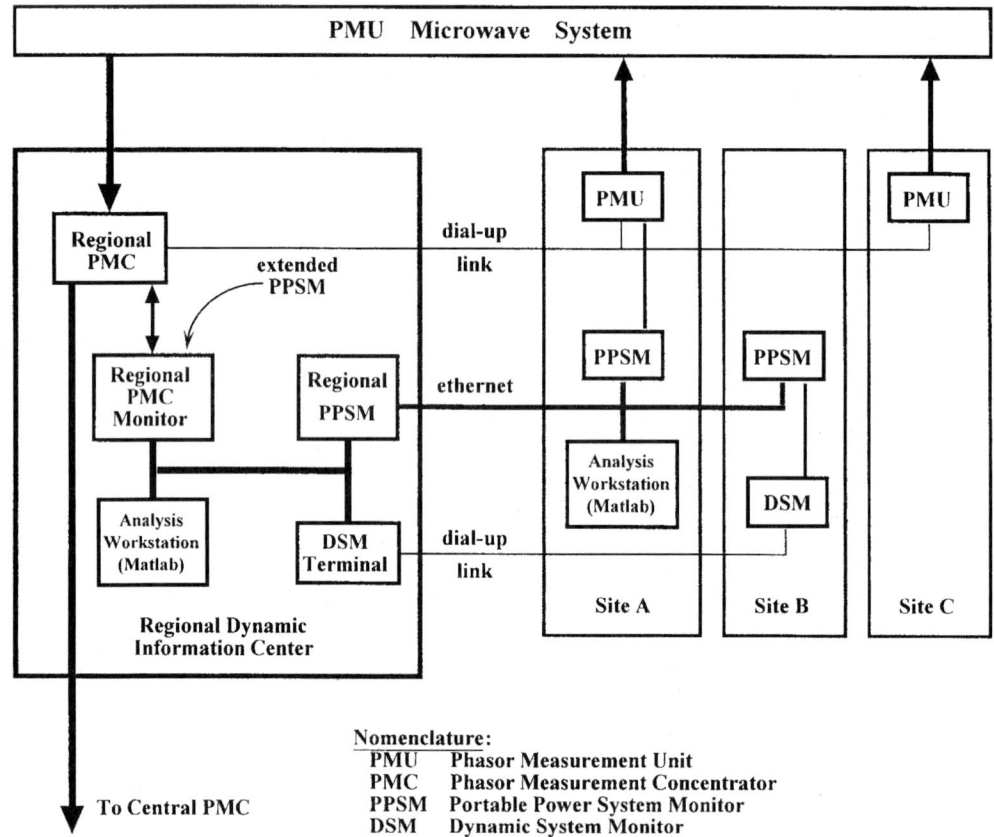

FIGURE 11.45 Interleaved networks of PMU, PPSM, and DSM units (plus analysis workstations).

Typical network details are shown in Figs. 11.46 and 11.47. Note that Fig. 11.46 indicates yet another layer of networking, for a digital transducer network. This could range from a basic PMU/PMT configuration to the highly versatile Integrated Object Network produced by Power Measurement, Ltd. (Carolsfeld, 1997).

Figure 11.47 represents a local measurement network developing for the 500 kV Thyristor Controlled Series Capacitor (TCSC) that was installed at BPA's Slatt substation under an EPRI FACTS project (Piwko et al., 1995; Hauer et al., 1996; Trudnowski et al., 1996). The functionalities there are highly desirable for any large control project. Guidelines for network organization are that:

- Any monitor (DFR, DSM, PPSM), through the local area network (LAN), triggers data collection on the others according to its own rules.
- The PPSM can supervise and control all other monitors or major instruments. All measurements are available to the PPSM for local display and analysis, for archiving, or for routing to other locations on the wide area network (WAN).
- The WAN permits remote observation and control of all devices in the measurement LAN through the PPSM.
- The PPSM can extract real time data from any other PPSM on the WAN and incorporate that data into its own processing.
- Every monitor, and the majority of major instruments can be accessed through a telephone connection as a (lower performance) backup to WAN failure.

Power System Dynamics and Stability

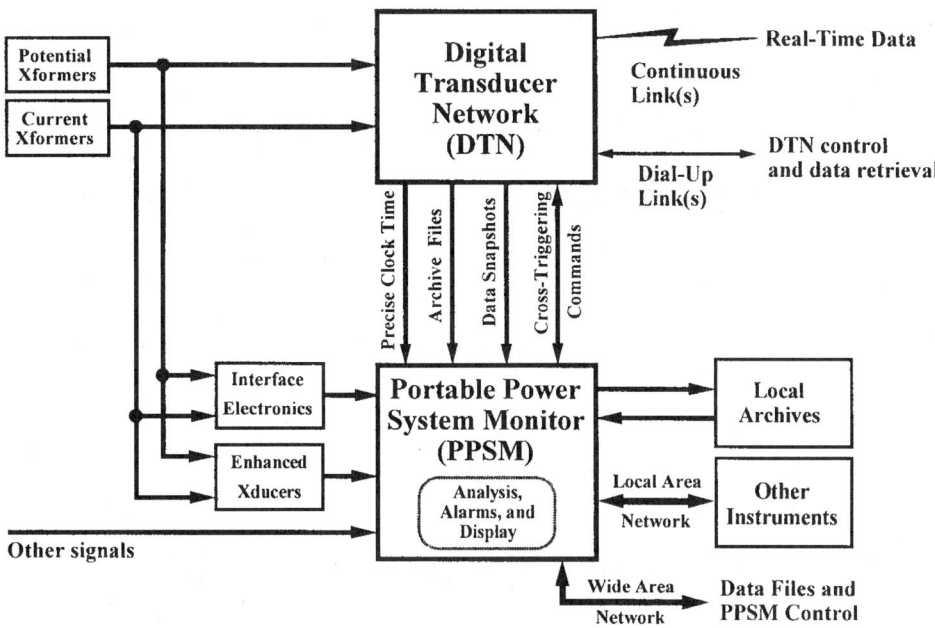

FIGURE 11.46 PPSM interconnections to local transducers.

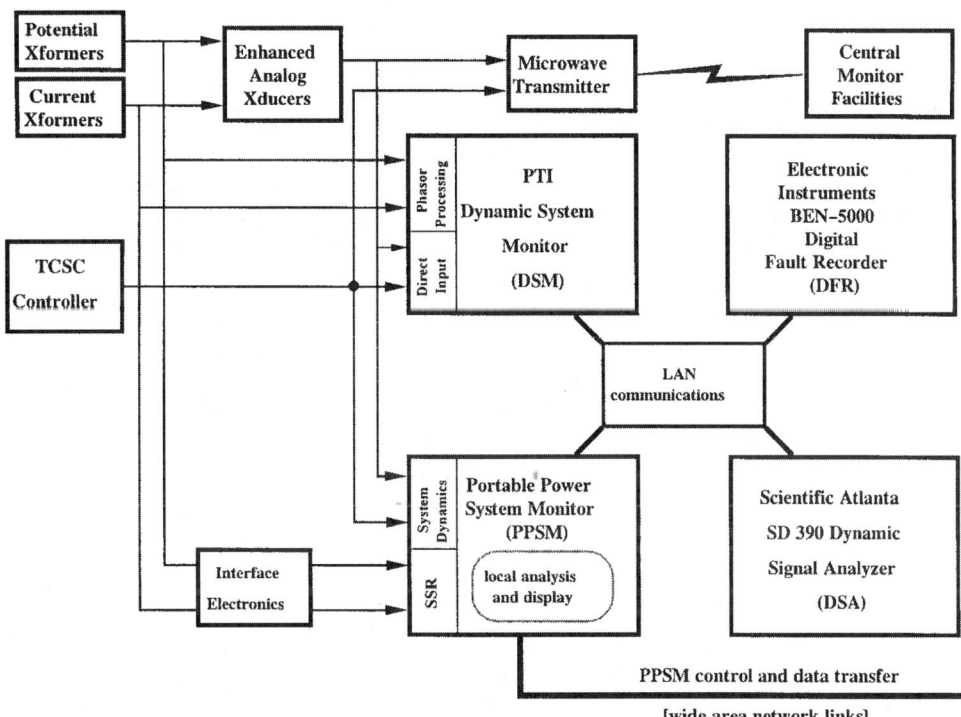

FIGURE 11.47 Monitor organization local to the Slatt TCSC.

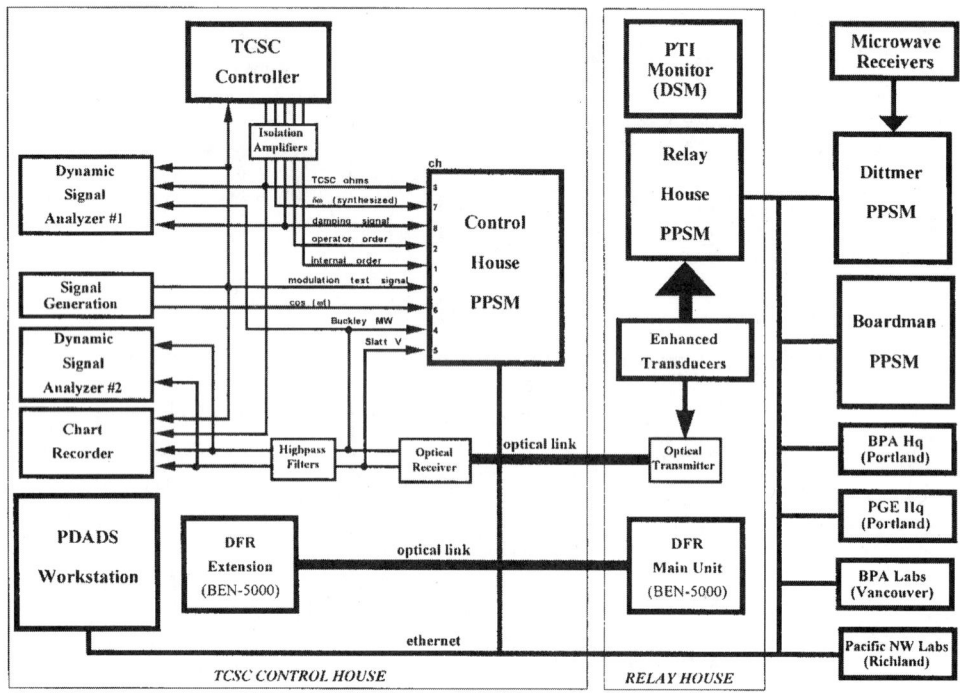

FIGURE 11.48 Slatt measurement network for TCSC modulation tests of June 6–7, 1994.

In addition to its usual functions, the DFR is used for waveform recording in controller performance tests. The DSA, by itself, is not quite a monitor. It does have good recording capabilities and it is readily networked. It is quite clearly a rather advanced IED and a valuable addition to overall monitor facilities.

Figure 11.48 provides an expanded view of the Slatt measurement network and of the regional network containing it. Information facilities of this kind are particularly valuable for certification and operation of major control systems such as the TCSC. For example, control engineers can perform wide area performance tests without the cost and delay of special purpose communication links. Staged tests can be supported remotely, with a minimum of travel, and technical staff can time-share that support across several projects while remaining in their normal work areas. *Workteam functionality* is a major requirement source in the proper design of dynamic information networks.

The final network to consider in this "network of networks," then, is the "people net." As indicated in Fig. 11.48, the TCSC Project draws upon remote technical support from three institutions (BPA, PNNL, and Portland General Electric) at four different locations. Some of this support was provided interactively during commissioning tests (Hauer et al., 1999), and all of it is available as needed.

Developing the proper interfaces between the "people net" and the monitor network(s) is critical to value engineering of the dynamic information system. The information users determine *who must know what, when, and in what form*. The information providers must deal with *who must do what, when, and with what tools*. Various aspects of these questions are treated below. The very important issue of placing a value on the information itself is reserved for separate discussion.

WSCC Experience in Monitor Operations

A competent monitor network is the "backbone" of the dynamic information infrastructure and it is a fundamental requirement for wide area control (CIGRE Task Force 38.01.07, 1996; CIGRE Task Force

38.02.17, 1999; CIGRE Task Force 38.03.17, 1999). We should expect (or hope) that most of measurement functions will be mundane ones, performed unobtrusively under routine operating conditions. However, the network must stand ready to provide mission-critical services with little or no warning. Some examples of high value support are

- Early warnings of trouble arising on the system or in specific equipment.
- Integrated recording and information sharing for major system disturbances.
- Real-time recording and analysis during tests of controller performance, or of wide area system dynamics.
- Recording of anomalous system behavior that is too intermittent for scheduled examination.

The reader is advised that these are performance objectives, and that they might not be fully mature operational realities.

The WSCC utilities, individually and collectively, have been comparatively aggressive in their development of monitor facilities (Hauer et al., 1999; Clark et al., 1992; Taylor et al., 1996; Taylor and Erikson, 1997; Martin, 1992; Agee and Girgis, 1996). Operation of those facilities has revealed a number of general problems:

- *Of the triggered monitors installed on the system,* expect no more than 50% to provide good records for a major disturbance. Leading problems are:
  - failure to detect the event (i.e., to trigger data capture)
  - failure to trigger soon enough, or to "retrigger" often enough during protracted disturbances.
  - inadequate data storage.
  - overwriting of stored data before they have been downloaded to secure archives.
  - monitor out of service.
  - monitor failure from loss of supply power.
- *Of the continuous monitors installed on the system,* expect about 90% to provide good records for a major disturbance. Leading problems are:
  - monitor out of service.
  - monitor failure from loss of supply power.
- *The value of a particular event record* may be higher for some other utility than for its owner. This situation may not be recognized for several days after the event, in which case the record may well have been deleted from the data system.
- *Determination of predisturbance conditions,* though fundamental to subsequent model studies, is seriously hampered by sparse monitor coverage plus failure to retain relevant EMS data.
- *Operations staff are very cautious* about high-level staged tests, and becoming more so (see Hauer and Hunt, 1996 and comments by Scottish Power in CIGRE Task Force 38.01.07, 1996). Better use must be made of chance disturbances, plus low-level ("non-intrusive") tests and measurements.
- *Very few triggered monitors* are designed for low level tests and measurements. These are the defining tasks for a continuous interactions monitor such as BPA's PPSM and PDC.
- *Necessary tools and skills* for conducting staged tests and extracting dynamic information from measured data, are not evenly distributed among the utilities. Some mechanism is needed for sharing these resources (and their costs) to meet shared utility needs.
- *An emergency response plan* is needed to assure safe retention and prompt integration of measured data following major disturbances. It is *not realistic* to task primary operations staff with this function, and it is one that should be automated anyway.

The WAMS effort and the WAMS technologies are directly rooted in this experience.

Database Management in Wide Area Monitoring

Deployment of new monitors, and the proliferation of IEDs in general, are overcoming many of the problems in acquiring raw data. It is clear that this emerging abundance is producing a new generation of challenges in monitor operations. Chief among them are

- Timely extraction and routing of information resident in the data.
- Avoiding premature deletion of valuable data, but without inundating data facilities.

EPRI and BPA have initiated exploratory research into a generic WAMS Information Manager to deal with such matters. A key element in this is a WAMS Database Manager (DBM). Conceptually, its functions include the following:

- Automatic Routing:
  - Operator alerts
  - Cross triggers to local and remote monitors
  - Event-driven control of local displays
  - Data retention requests to local and remote monitors
- Servicing of Staff Requests:
  - Data transfers
  - Special data operations and displays
  - External triggering of local or remote monitors
  - Special log entries
- Background Directory Operations:
  - Exchanges among DBM units
  - Integration, annotation, and indexing
  - Posting on EMS, OASIS, WWW
- Background Data Operations:
  - Launching and supervision of the Archive Scanner
  - Content based compression and archiving
  - Logging of events and summary features
- Utility Functions:
  - File merging and compression
  - Hardcopy generation

The monitors involved in this include SCADA, DFR, and other IED units in addition to the usual swing monitors.

Placing a Value on Information

It is apparent that the utilities face a massive investment in information technology. Planning these investments encounters a very basic question: just how do you place a value on information? A partial answer is that the value of information is precisely that of the decisions derived from it.

Figure 11.49, another paradigm, is useful for expanding upon this statement.

Decision Time Scales

- Automatic Control:
  - Protection: milliseconds
  - Stability: milliseconds to seconds
  - AGC: seconds to minutes

Power System Dynamics and Stability

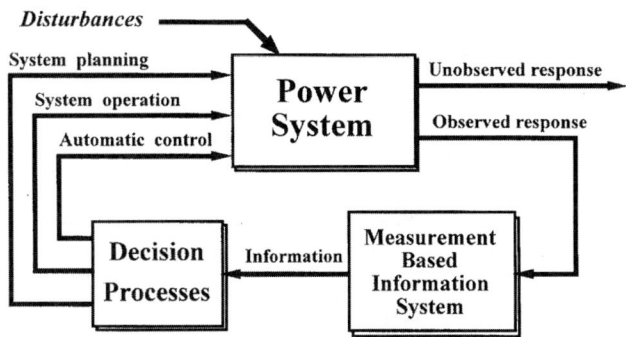

FIGURE 11.49 The cycle of measurement, information, and decisions.

- System Operation:
  - minutes to weeks
- System Planning:
  - Operations: hours to weeks
  - Expansion: weeks to years

Decision processes in a power system range from the very rapid ones preprogrammed into protective control equipment to the very slow ones associated with expansion planning. In all cases, the decisions are derived, with varying degrees of immediacy, from measurement-based information. In some cases, the information is encapsulated in a model, or perhaps in operating policies. In others, the data is processed immediately, e.g., as a controller input or as a signal to system operators.

Accumulated over time, information provides a knowledge base that permeates utility practices and, often enough, those of the industry. Such long-term effects, together with the many paths by which information enters utility decision processes, will defeat any direct attempt to place a value upon it. More constructive results follow from considerations of affordability and risk management.

Some suggestions in the matter are these:

- Consider information as an insurance policy against operational uncertainty:
  - How much insurance is enough?
  - How much risk is too much?
- Distinguish between value, cost, and affordability.
- Consider all cost elements, especially lead time and staff demands.

Another factor, one that may preempt many of these considerations, is regulatory mandates issued at various levels of government. It is very likely that an infrastructure for developing and exchanging dynamic information will be found necessary for assuring power system reliability and, thereby, the public interest.

Monitor Placement

There can be no simple rule for determining what to monitor or how to configure particular monitor systems. WSCC recommendations in (WSCC Work Group, 1990) are still a reasonable guide with respect to functionalities. They state that the overall monitor facilities should support the following applications:

- analysis of system disturbances and oscillation incidents
- performance evaluation for major control systems
- early detection of poor damping or other unusual system behavior
- validation of computer models

These same guidelines also state that the overall facilities should provide a comprehensive report of line flows, bus conditions, and control actions for

- major and/or critical interchanges and load areas
- major control systems, especially those for HVDC line flow
- major and/or critical generation projects

Regarding technical performance, the guidelines emphasize long-term recording capabilities with an overall signal resolution equivalent to 14–16 binary digits and sample rates at or above 20 samples per second. Some general guidelines on technology choices are provided in Hauer et al. (1999) and Hauer et al. (1995) and in Appendix A of this section. The underlying perspective there is that the rationale for power system performance monitors should parallel that for flight recorders on commercial aircraft.

Surveillance of major interactions is fundamental to the objectives of wide area monitoring, and to the broader objectives of a dynamic information infrastructure. Figure 11.50 indicates the more important interaction paths in the WSCC system, plus "index" generators where the associated dynamic modes are particularly observable and/or controllable.

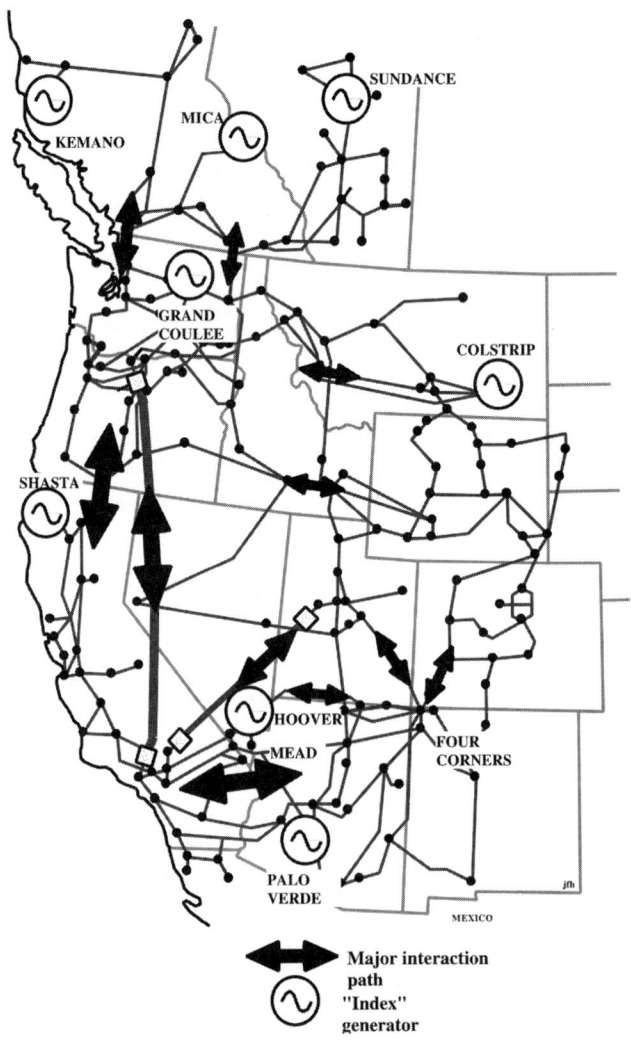

FIGURE 11.50 Major interaction paths for the western North America power system.

Monitoring may well call for different technologies in different regions. The western power system has an underlying loop structure, with important radial extensions into British Columbia, Alberta, and eastern Montana. Remote imports into the Los Angeles area often approach 10% of system capacity. Power flow along the western side of the loop is concentrated on a few major interties. Along the east side of the loop power flow is diffused across a web of more numerous and weaker lines, eventually concentrating along a major transmission corridor that approaches Los Angeles from the east.

Monitor needs on the west side of the system have the following characteristics:

- The dominant problem is interarea "swing" dynamics and related control systems.
- Measurements of dynamic state are the key indicator for system performance.
- Fairly good signals for the needed quantities are available from enhanced conventional transducers for power, frequency, and voltage magnitude.
- Critical information is concentrated at a fairly small number of sites.
- substantial monitor facilities are already in place.

However, on the eastern side,

- The dominant problem is overloading of weak links through power surges or through inadvertent "loop" flows.
- Measurements of static state (powerflow variables) are the key indicators for system performance.
- Competent measurement of complex voltage and current requires precise time synchronization technology that is just now reaching maturity (Phadke, 1993; Hauer, 1996; Hauer et al., 1999).
- Critical information is broadly dispersed across the network, and the appropriate measurement points are not obvious.
- Monitor facilities are in the early stages of development.

The natural "state variables" for the west side are, for the most part, those associated with generator and control system dynamics. The natural state variables on the east side are those associated with powerflow on the network.

Access to Monitor Data

In-depth analysis of monitor data is often done at locations well removed from the original point of data collection, and in some other computing environment. Overall monitor facilities must accommodate many different users for the data collected. Some users may need regular access to records from a number of monitors and monitor types. It is highly desirable that:

- overall data access be provided within a consistent, well integrated framework.
- the users process the data with the same tools that they use in other work of the same general nature.

Figure 11.51 indicates three of the more common analysis environments for monitor data. The LabVIEW environment is typical of normal field operations, the FORTRAN/C environment is typical for system planning, and the Matlab environment is typical for advanced analysis and for control system design. Figure 11.52 shows the various means of access to PDC data at BPA. Special Matlab toolsets indicated there include the BPA/PNNL Power System Identification (PSI) Toolbox, of which the Power System Monitor (PSM) Tools are a subset. Toolsets coded in Visual Basic are also in use, but not indicated in this figure.

Monitor Application Examples

This section presents a cross-section of PPSM data collected on the WSCC system. The examples are organized according to the conditions under which they were obtained, i.e., during ambient conditions, system disturbances, and staged tests.

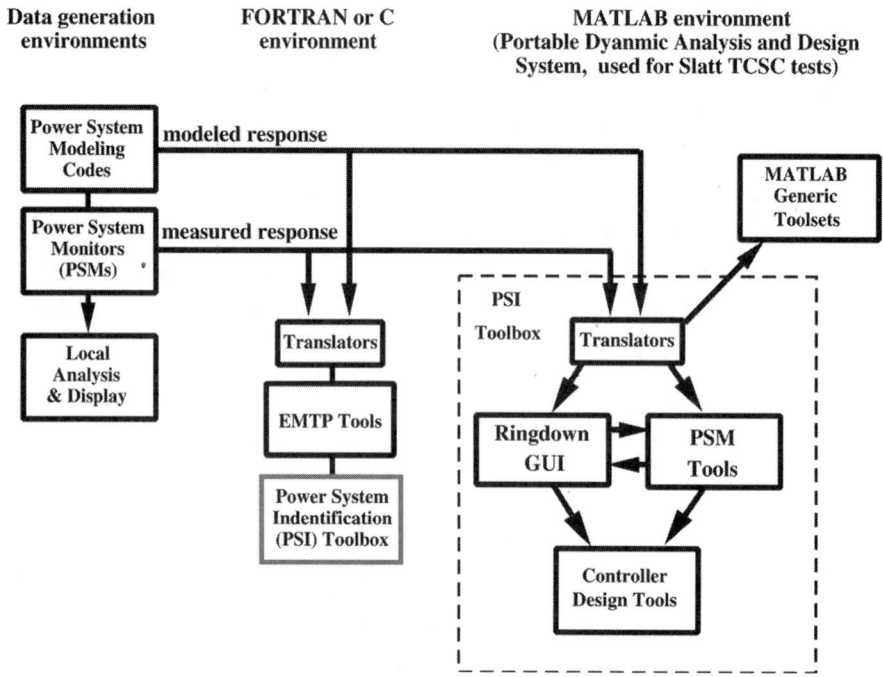

FIGURE 11.51 Typical analysis environments for power system response records.

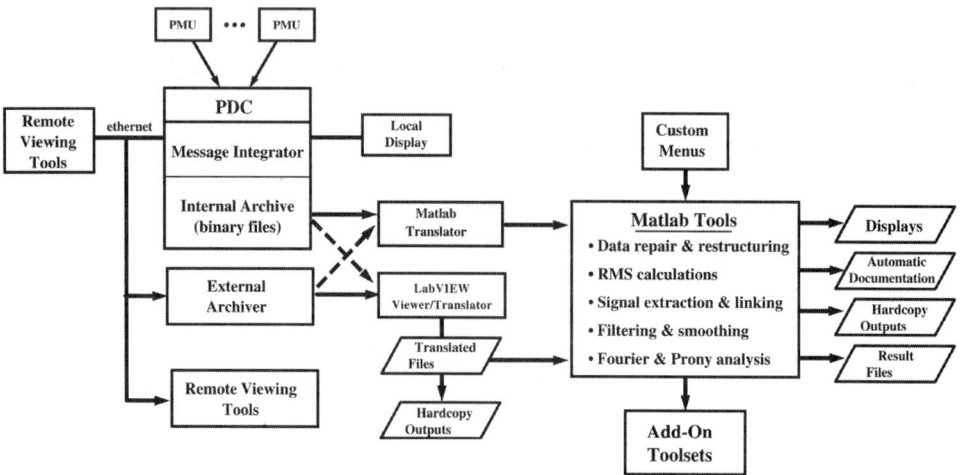

FIGURE 11.52 Surrounding network for the Phasor Data Concentrator at BPA's Dittmer Control Center.

Examples for Ambient Conditions

Major modes for the WSCC system are conspicuous in ambient noise activity under just about all conditions. The activity shown in Fig. 11.53, on BPA interchanges into Canada and California, is typical. Power flows there are represented as north to south; slow trends have been removed by a highpass filter set at 0.1 Hz, and lowpass filtering at 1.0 Hz was used to reduce local-mode activity plus extraneous noise. Figs. 11.54 and 11.55 show that ambient swings are strong, highly coherent, and sharply focused near 0.29 Hz. (See Brigham [1988], Bendat and Piersol [1980], Farmer and Agrawal [1983], and Trudnowski et al. [1999] for analysis principles.) Those on the Canada interchanges (Ingledow and Boundary)

Power System Dynamics and Stability

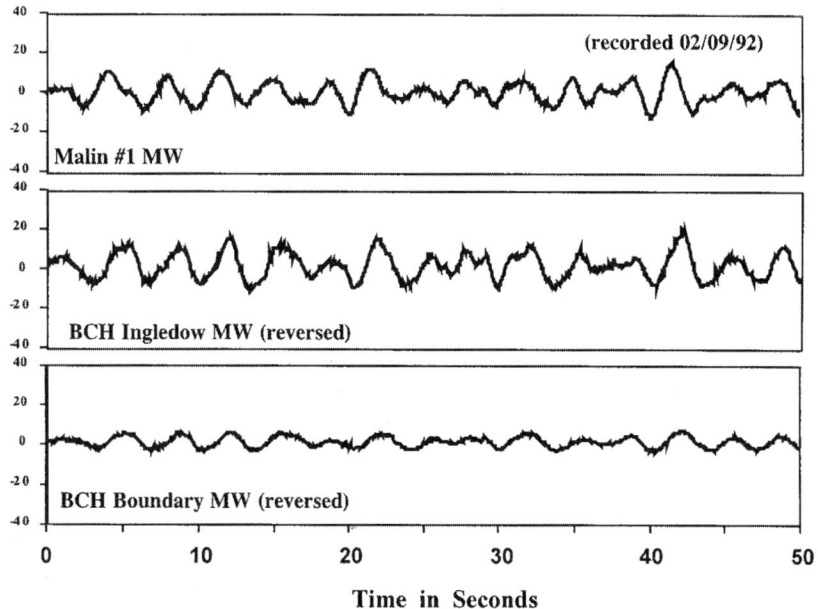

FIGURE 11.53 Signals for ambient noise on major BPA interchanges, 02/09/92.

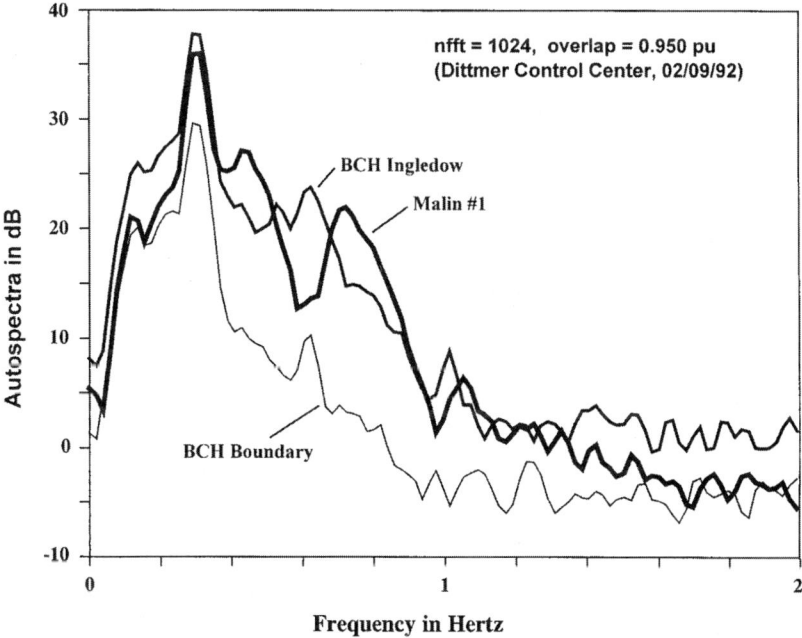

FIGURE 11.54 Spectra for ambient noise on major BPA interchanges, 02/09/92.

are in phase, and slightly offset from those with California (at Malin). Such results are a constant reminder that the "AC Intertie" mode associated with the Pacific AC Intertie (PACI) is very much alive.

A next issue is whether warning signs such as those conspicuous in Fig. 11.34 can be extracted from prior ambient activity. Comparison of Figs. 11.53 and 11.56 indicates that such warnings, if present, are not obvious in the time domain records.

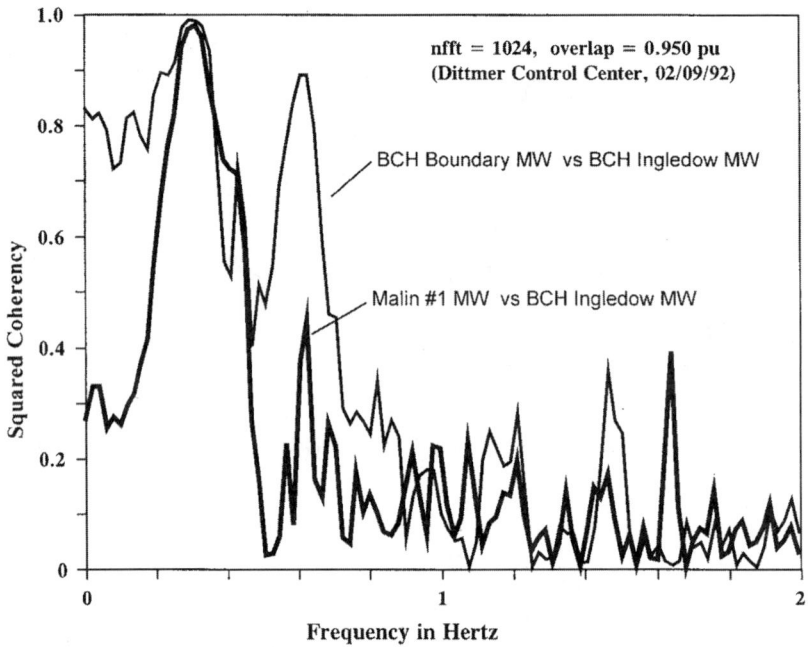

FIGURE 11.55 Coherencies for ambient noise on major BPA interchanges, 02/09/92.

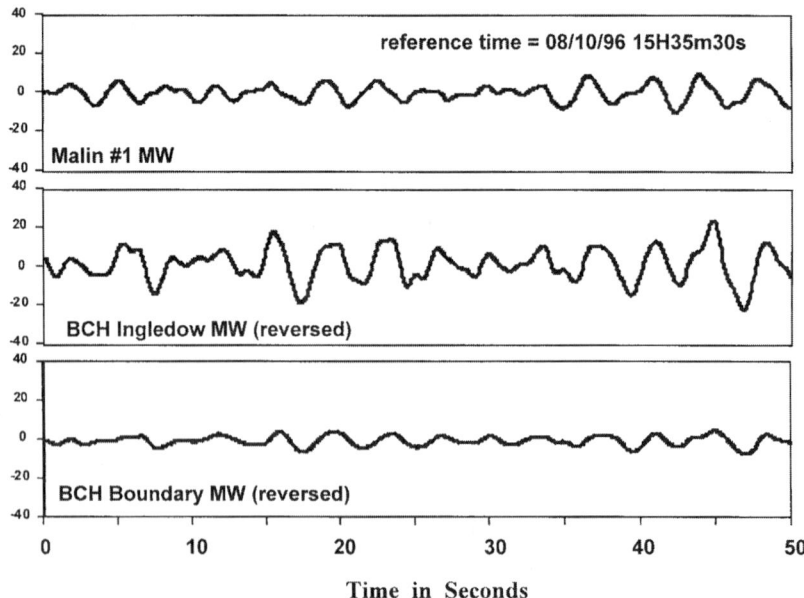

FIGURE 11.56 Sample of ambient noise on Major BPA Interchanges, starting 400 sec before Keeler-Allston line trip on August 10, 1996.

Figures 11.57 and 11.58 show Fourier analysis results for 30 minutes prior to tripping of the Keeler-Allston line on August 10, 1996 (see also Fig. 11.34). At first inspection, the highly structured spectrum in Fig. 11.58 is very similar to what is usually observed. Closer inspection, with more subtle tools, may show otherwise. The abundance of data collected for this very important disturbance may contain a number of hidden clues concerning proximity to the limits of safe operation.

Power System Dynamics and Stability   11-107

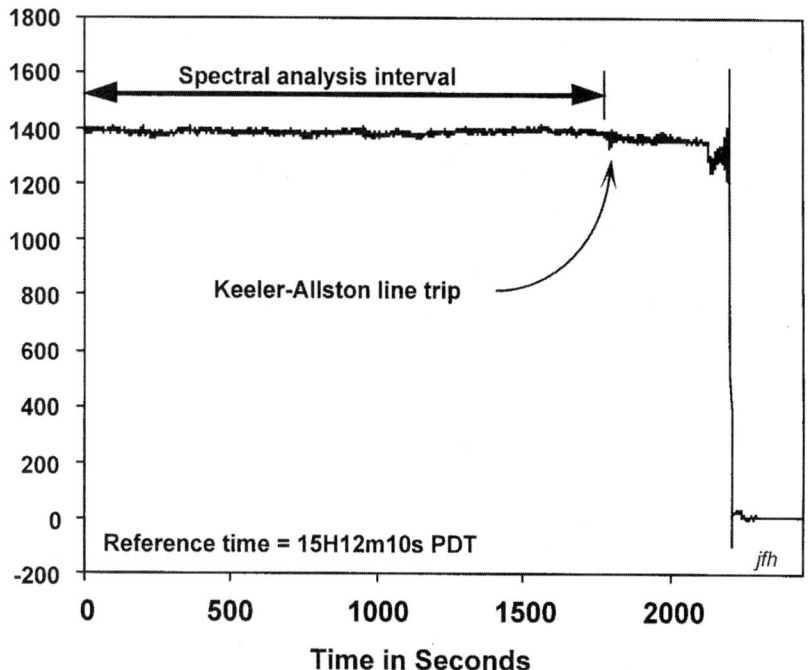

FIGURE 11.57 Record for prehistory analysis of conditions preceding Keeler-Allston line trip.

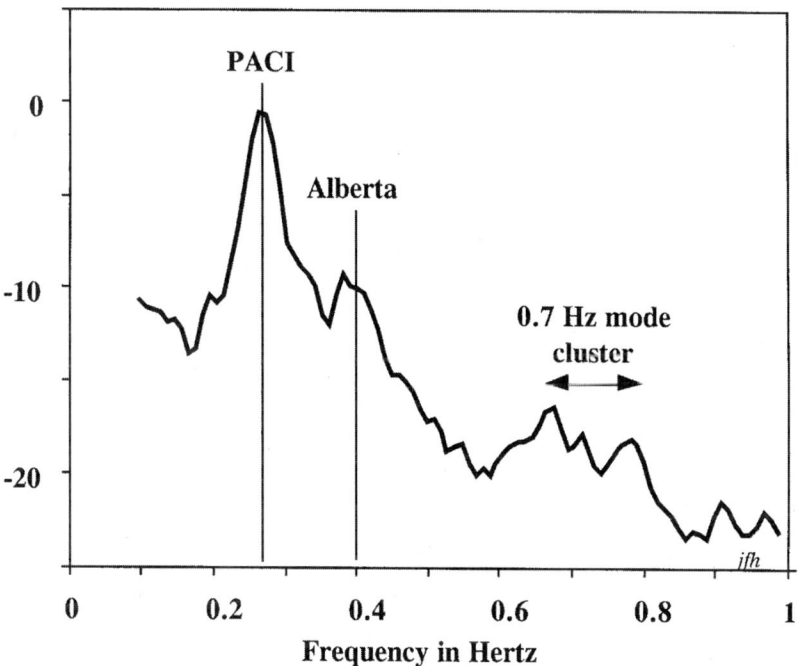

FIGURE 11.58 Prehistory analysis for conditions preceding Keeler-Allston line trip.

Figures 11.59 and 11.60 show an example of wide area correlation involving two widely spaced continuous monitors. The signals represent bus frequency at Kyrene substation in Phoenix, AZ, and bus frequency at Tacoma substation (near Seattle, WA, but recorded on the central PPSM at the Dittmer Control Center). The substations are about 1000 miles apart.

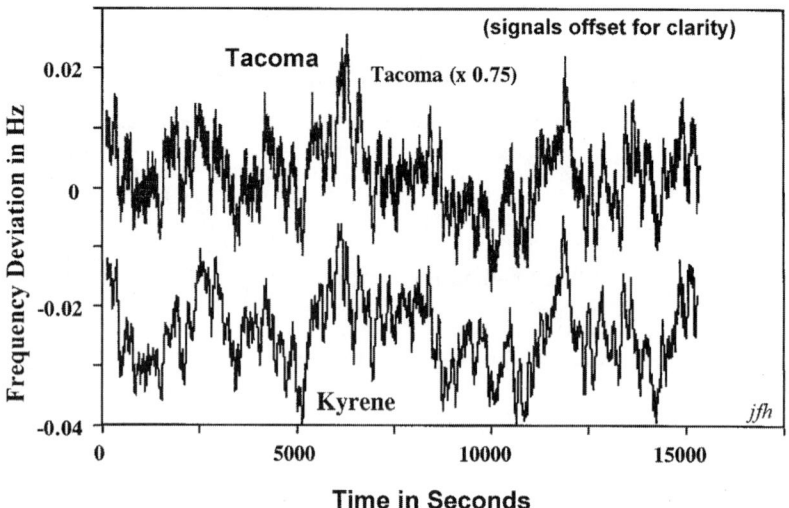

FIGURE 11.59 Ambient fluctuations in bus frequency, Tacoma and Kyrene substations (Tacoma, WA and Phoenix, AZ), April 15, 1996.

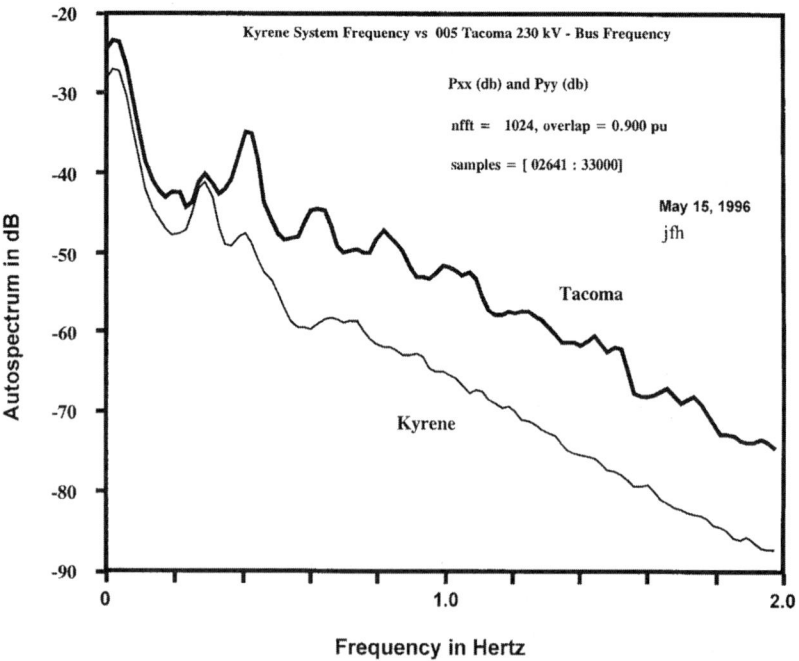

FIGURE 11.60 Autospectra for ambient fluctuations in bus frequency, Tacoma and Kyrene substations (Tacoma, WA and Phoenix, AZ), April 15, 1996.

The signals were obtained under ambient conditions from ordinary frequency transducers and without direct synchronization. Even so, the activity peaks (Fig. 11.60) and the correlations between them (Fig. 11.61) provide dynamic signature information up to roughly 0.8 Hz.

The records used in this analysis are 25 minutes long, at 20 samples per second. In other applications, such as the commissioning of major feedback controllers, the recording periods would span many hours and much of the correlation analysis would be done at the test control site in real time.

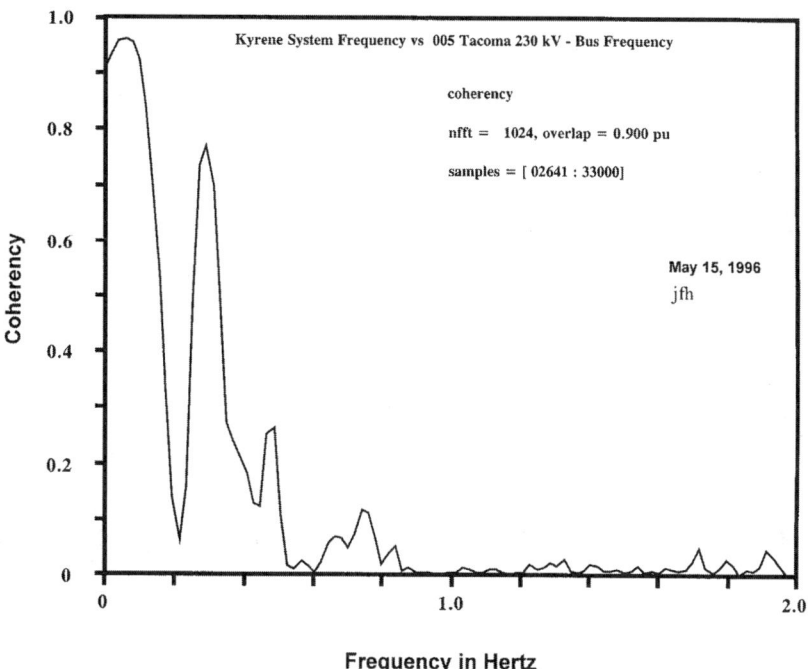

FIGURE 11.61 Coherency of ambient fluctuations in bus frequency, Tacoma and Kyrene substations (Tacoma, WA and Phoenix, AZ), April 15, 1996.

Examples from Major Disturbances

"Ringdown" signals from major disturbances are a good source of information concerning the behavior of oscillatory dynamics and of the control systems that affect them. Generator trips units at the Palo Verde plant (near Phoenix, AZ) are particularly useful for determining current frequency and damping of the PACI mode.

Figure 11.62 shows limiting cases among five trips during the winter of 1992–1993. Implicit in this figure we find the following information:

- Frequency and damping of the PACI mode vary considerably during normal conditions.
- Damping of the PACI mode can reach values low enough to be considered potentially dangerous.
- Frequency content of the ringdown signal is dominated by the PACI mode. This "mode shape" information
  - provides an excellent modeling benchmark for time domain simulations and for eigenanalyis.
  - indicates that Palo Verde is a good location for metering the effectiveness of PACI damping controls.
  - suggests that Palo Verde itself may be a good point for damping the PACI mode.

Figure 11.63 is just one among many examples gathered through collective efforts of the WSCC system oscillation work groups and, more recently, through the WAMS Project.

The events of August 10, 1996, provided ample opportunity (and incentive) to examine the PACI mode under more extreme conditions. Development of these conditions and the subsequent separation are shown in Figs. 11.63 and 11.64. Tripping of McNary generation seems to have been a major factor (Kosterev et al., 1999; Taylor, 1999) in the breakup. Whether this major loss of system support actually caused the oscillations, or merely made bad conditions worse, cannot be determined without very exacting studies based upon fully realistic models. This work is still in progress.

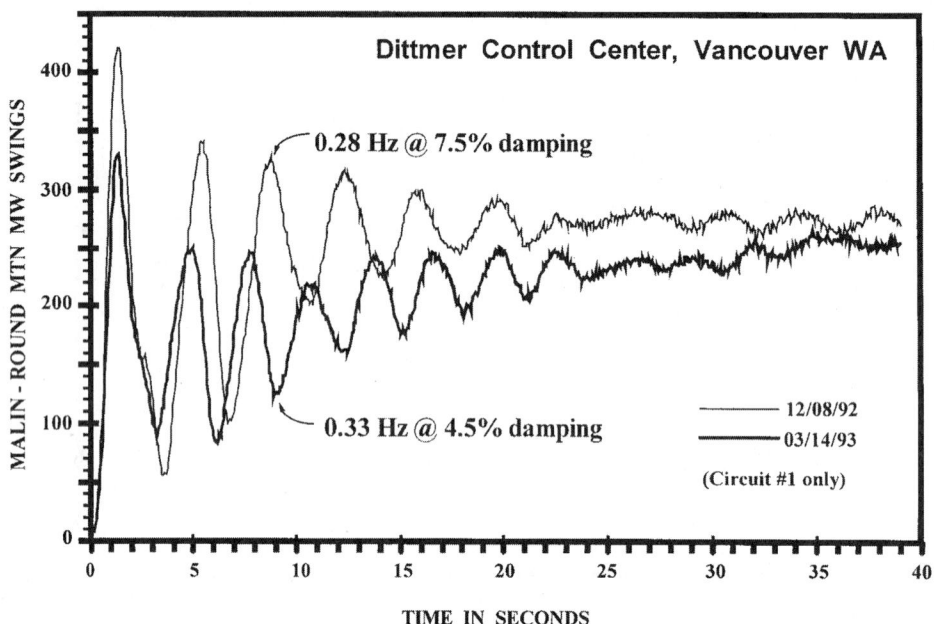

FIGURE 11.62 AC Intertie Response to Palo Verde Unit Trips, 12/08/92 vs. 03/14/93.

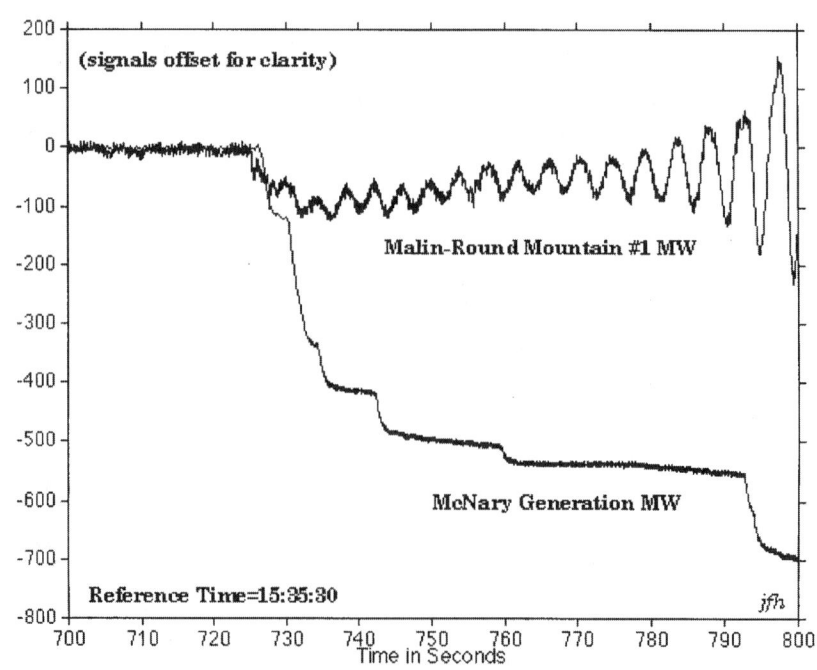

FIGURE 11.63 McNary plant generation during WSCC breakup of August 10, 1996.

On December 8, 1998, a major portion of the electrical services to San Francisco were lost through a chain of unintended breaker actions. All three PDC units collected good data (as did the Dittmer PPSM). Table 11.2 describes the data obtained, and Fig. 11.65 is a sample of the bus angles and frequencies. The primary rms signals indicated there are those conventionally produced in the PSI Toolbox, which was used to align and process the PDC records.

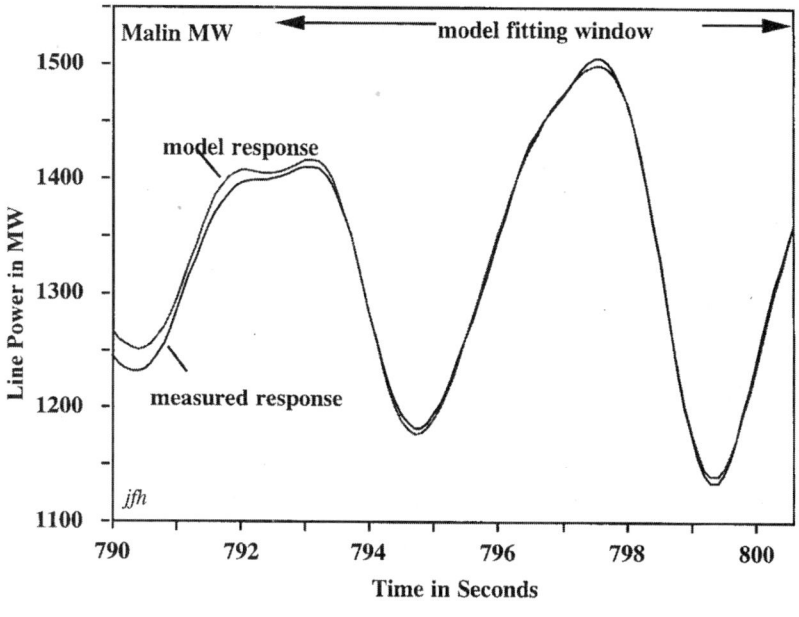

FIGURE 11.64 Oscillation modes just prior to final separation on August 10, 1996.

Major disturbances trigger a number of actions among the operating utilities. A high priority is to determine what happened, why it happened, and how to avoid it in future operations. Key steps in this include the following:

- Integration of operating records
  - initiating and transdisturbance events
  - predisturbance activity and conditions
  - system performance and consequences
- Search for warning signs
  - dynamic signatures in prior swing activity
  - information from model-based stability tools (DSA, voltage stability)
- Support for countermeasures
  - engineering review
  - model development and analysis
  - operator alerts and guidelines

A general process and toolsets for this are indicated in Fig. 11.66 (see appendices of Hauer [1987] plus Kundur et al. [1990], Hauer [1982], Hauer et al. [1990], Hauer [1990], Kamwa et al. [1996], Trudnowski et al. [1999]). Table 11.3, which tracks behavior of the PACI mode before and during the August 10 breakup, provides good eigenvalue benchmarks for development of the necessary models.

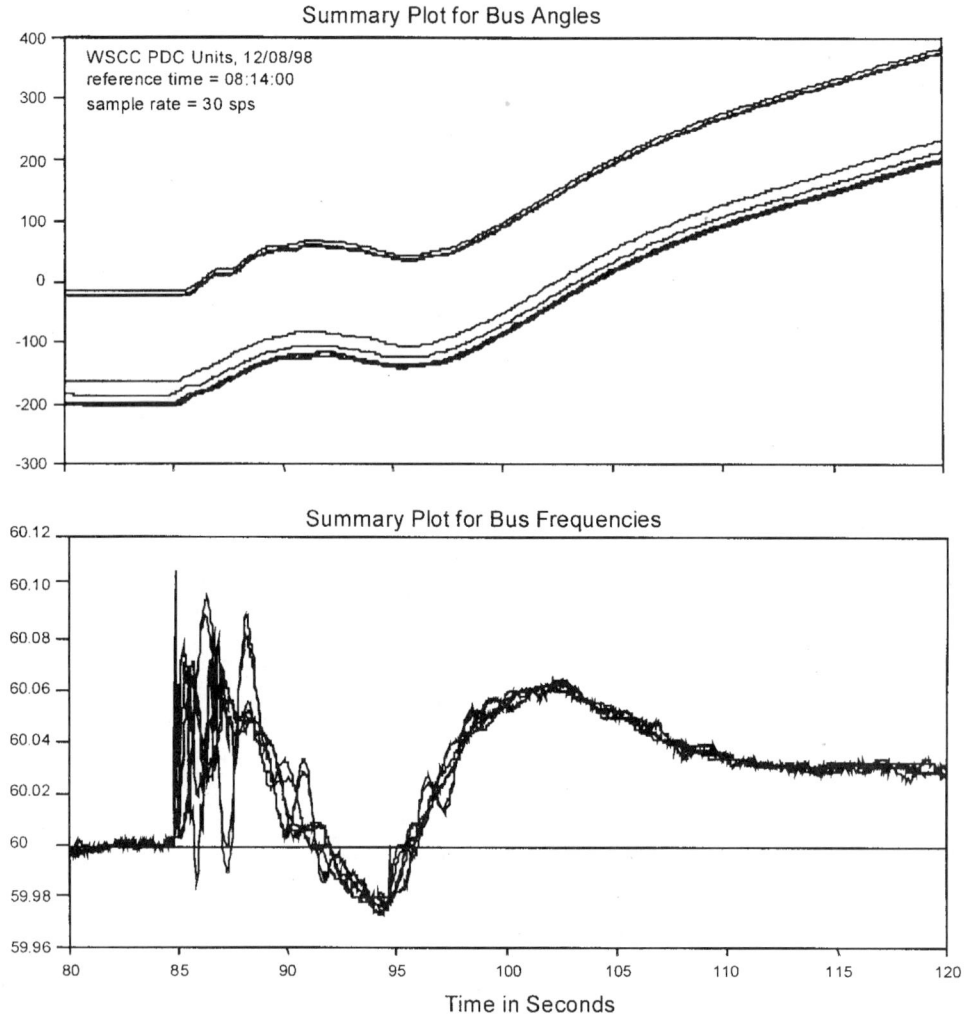

FIGURE 11.65 Bus angle and frequency swings for San Francisco trip on December 8, 1998. (Combined data for WSCC phasor measurement system.)

TABLE 11.2 Data Collected by the WSCC Phasor Measurements Network for the San Francisco Load Trip of December 8, 1998

Organization	PDC Units	PMUs	Phasors	Primary RMS Signals
BPA	1	7	45	153
SCE	1	4	24	86
WAPA	1	3	15	32
WSCC Total	3	14	84	271

Examples From Staged Tests

Figures 11.67 and 11.68 summarize results from 1991 test insertions of BPA's 1400 MW Chief Joseph dynamic brake. They demonstrate that the 500 kV connection to Alberta strongly influences overall system dynamics, and that its operational status may be a factor in tuning of future stability controls such as modulation of HVDC, TCSC, or SVC equipment.

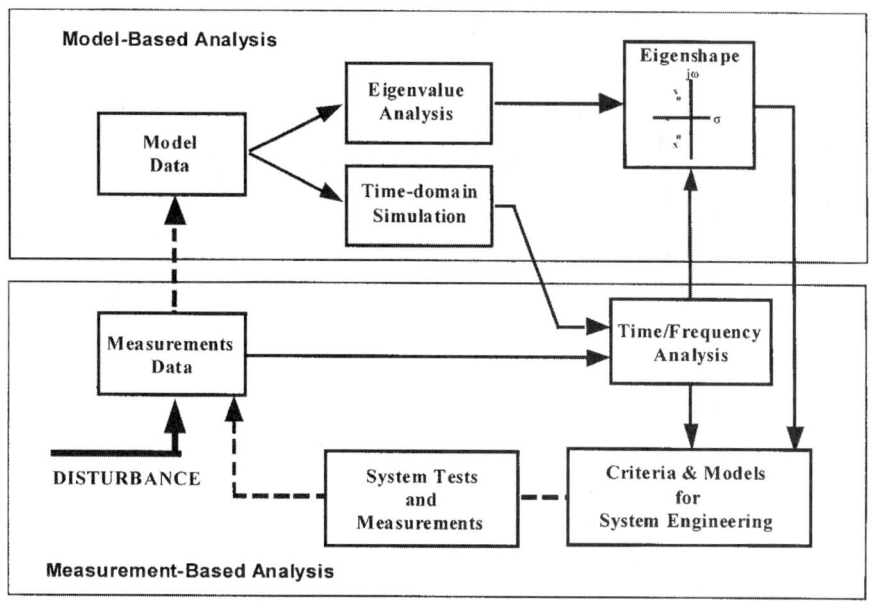

FIGURE 11.66 Integrated use of measurement and modeling tools.

TABLE 11.3 Observed Behavior of the PACI Mode

PACI Mode before August 10, 1996		
Date/Event	Frequency	Damping
12/08/92 (Palo Verde trip)	0.28 Hz	7.5%
03/14/93 (Palo Verde trip)	0.33 Hz	4.5%
07/11/95 (brake insertion)	0.28 Hz	10.6%
07/02/96 (system breakup)	0.22 Hz	1.2%
PACI Mode on August 10, 1996		
Time/Event	Frequency	Damping
10:52:19 (brake insertion)	0.285 Hz	8.4%
14:52:37 (John Day-Marion)	0.264 Hz	3.7%
15:18 (ringing)	0.276 Hz	
15:42:03 (Keeler-Allston)	0.264 Hz	3.5%
15:45 (ringing)	0.252 Hz	
15:47:40 (oscillation start)	0.238 Hz	−3.1%
15:48:50 (oscillation finish)	0.216 Hz	−6.3%

Figure 11.69 was produced during staged tests of the Slatt TCSC. For subsynchronous resonance (SSR) tests, a TCSC modulation signal was produced at the nearby Boardman generator and transmitted to Slatt. PPSM acquisition rates at these sites were set to 400 sps and 900 sps, respectively.

The basic procedure was to sinusoidally modulate the TCSC at a selected shaft torsional frequency, remove the stimulus while switching the TCSC to the desired operating condition, and then record the ringdown in shaft speed. Figure 11.70 is typical. Prony analysis produced values that agree closely with those produced by the General Electric test team with other instruments (Piwko et al., 1995; Hauer et al., 1996; Trudnowski et al., 1996). Figure 11.71 shows composite autospectra for sequential testing of the four shaft modes.

In April 1999, BPA resumed the testing of WSCC dynamics by modulation of the Pacific HVDC Intertie. These had been suspended for more than a decade, due to lack of immediate need and to lack of a suitable

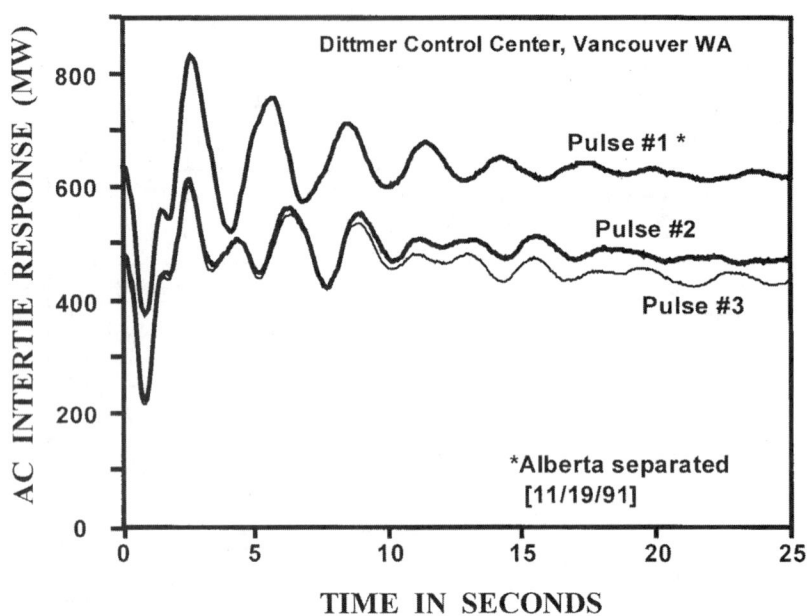

FIGURE 11.67 PACI response to energization of the 1400 MW Chief Joseph dynamic brake.

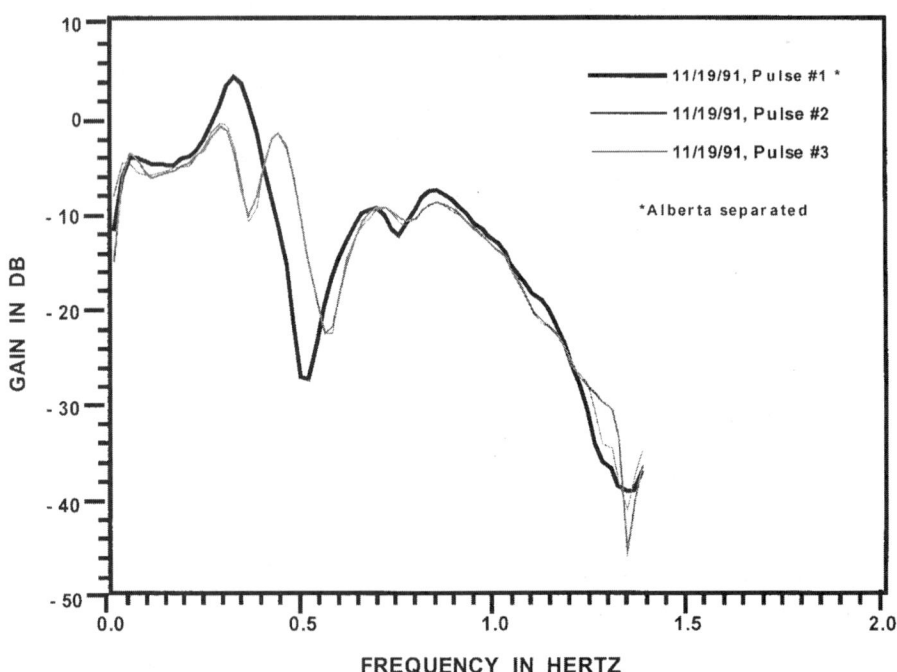

FIGURE 11.68 Effect of Alberta 500 kV connection upon PACI response.

measurement system. The breakup of August 1996 and progressive integration of the WSCC phasor measurement system significantly altered these factors.

For this new test program, the objective is to safely examine the frequency and damping of a single mode with a direct testing procedure that is minimally intrusive and that can be automated. The adopted procedure is similar to the shaft ringdown tests illustrated in Fig. 11.70.

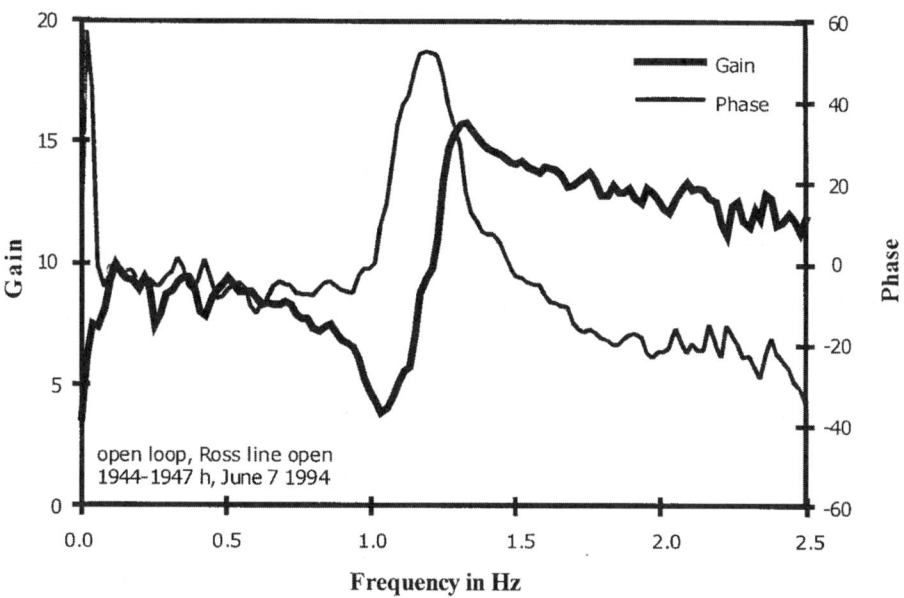

FIGURE 11.69 Frequency response of Slatt-Buckley line MW to TCSC modulation signal.

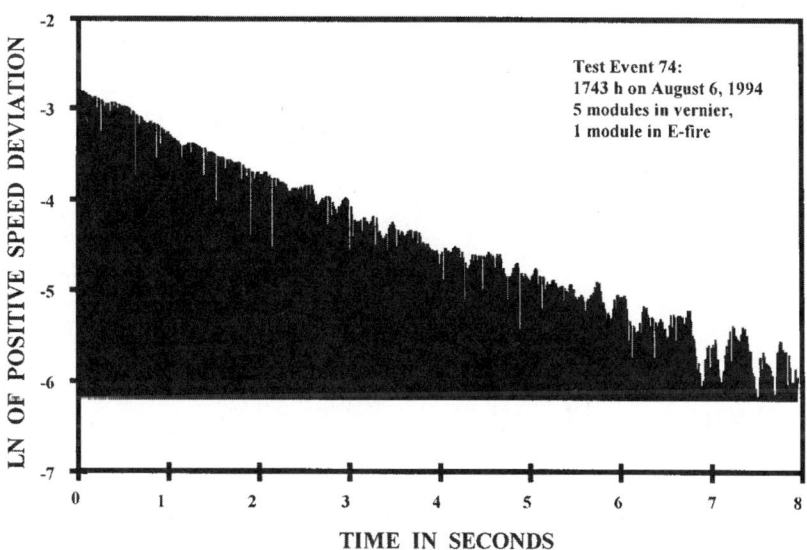

FIGURE 11.70 Boardman generator shaft speed ringdown, TCSC decoupled from shaft.

Figure 11.72 shows typical response for a two different applications of a two-cycle HVDC probing signal, with a base frequency of 0.33 Hz and a level of ±125 MW. If the level of probing were much lower, then some kind of time averaging would be needed during modal analysis. In the present case, spatial averaging across multiple signals should be done when Prony analysis is applied to the ringdown. Another option is to perform "extended" Prony analysis, using input/output tools that extract information from the forced response as well as the ringdown response (Kamwa et al., 1996; Juang, 1994; Van Overshee and De Moor, 1996). These tools require an accurate record of the input signal. Though rarely available in the general case, such records are available for well-instrumented tests.

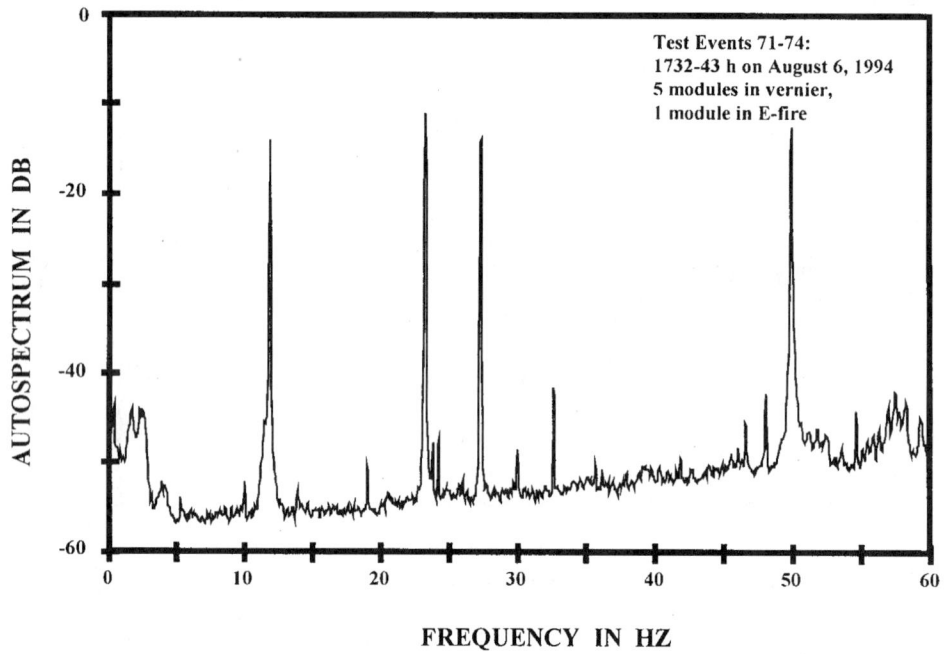

FIGURE 11.71 Composite autospectrum for Boardman generator shaft speed deviations (rear sensor).

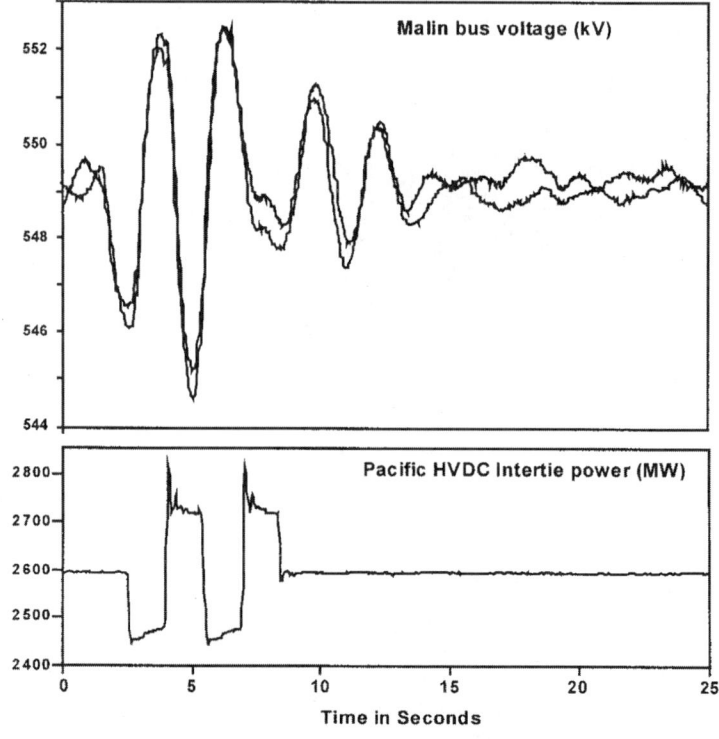

FIGURE 11.72 Power system response to two applications of two cycle probing signal.

Power System Dynamics and Stability

TABLE 11.4 Modal Parameters Identified at Various Stages of the Probing Signal Test on April 27, 1999

	Mode 1			Mode 2		
Test	Frequency Hz	Damping Ratio %	Damping neper/sec	Frequency Hz	Damping Ratio %	Damping neper/sec
B6 — probe	0.303	10.15%	0.193	0.416	9%	0.235
B7 — brake	0.295	9.3%	0.172	0.417	7.9%	0.207
B8 — probe	0.296	10.02%	0.186	0.416	8%	0.209

The results shown in Table 11.4 were produced with a tool of this class, one that is contained in the Matlab System Identification Toolbox®. It was selected on the basis of immediate convenience rather than full comparisons against similar products or algorithms of the same kind, and its use here is exploratory. To this point, the results produced with this tool have been internally consistent, and consistent with those produced with Fourier and Prony methods (see CIGRE Task Force 38.03.17 [1999], chap. 7). Sharper trade-offs among the test procedure, the analysis methods, and the design of the probing signal will evolve with as this BPA project continues.

Conclusions

The material here advances two key points, that

- Timely access to dynamic information is essential to successful grid operation in the emerging power system.
- Lack of such information is a particularly acute threat to system reliability during the transition to a deregulated industry.

The presentation examines these needs and the technology options for meeting them, drawing upon WSCC experience as a guide to the challenges that many other utilities and systems will encounter.

It is reasoned that the dynamic information infrastructure will draw upon a hierarchical monitor network that provides integrated and fairly comprehensive measurements of wide-area dynamics. In a large power system, this is a long step beyond the monitoring of local devices or even regional performance. Reliable data capture, for example, implies subnetworks of continuously recording monitors plus "intelligent" information tools that can cope with the very high data volumes. Even more important — perhaps decisively so — are the interutility practices and infrastructure through which dynamic information resources are collectively reinforced and operated.

Appendix A — A One-Page Perspective on Wide Area Monitoring

1. **Purpose:** the role of monitor facilities is to *avoid* system disturbances, not just to record them. Measured information is a major input to the many decision processes that planning and operation involve.
2. **Value:** information has the same value as the decisions based upon it.
3. **Level of investment:** consider measured information as insurance against operational uncertainty. How much insurance is enough? How much information is too much?
4. **Monitor applications:** in addition to event recording, monitor facilities should support
   a. direct tests and analysis of the power system.
   b. real-time system operations (operator alerts, cross triggers, etc.)
5. **Monitor recording options:**
   a. *evaluate data in real time*, using event detection logic (EDL) to trigger "snapshot" recordings of special activity or conditions.
   b. *record all data continuously*, then apply EDL to archived records.

6. **EDL considerations:**
   a. *EDL tuneup* is ongoing: automatic detection of significant abnormal system behavior requires comprehensive examination and knowledge of normal behavior.
   b. *Real-time EDL* rarely detects the onset of distant events. "Snapshot" monitors often miss the conditions contributing to major disturbances (e.g., cascading outages).
   c. *EDL for scanning continuously recorded archives* can be simplified and/or desensitized for
      - use in real time, as a refinement to the "triggers" now in use.
      - scanning and characterization of snapshot archives.
7. **Value engineering:**
   a. Functionality is the key issue: Who needs to see what, where, when, in what form? Why? What decisions hinge upon the information?
   b. *Staff costs* (engineering, operations, maintenance) are dominant in power system monitoring. Compared to these, hardware is free — configure the hardware to minimize staff costs.
   c. *Continuous recording*, in combination with a competent *archive scanner*, provides the best value engineering for wide area measurements of power system dynamics.
   d. *The archive scanner* required for continuous recording can and should be designed for the overall monitor facility. Better integration and more timely access increases information value while lowering staff costs.
   e. *Ease of use* hastens workflow and reduces direct staff costs, but generally requires that data be translated into multiple work environments.
   f. *User participation* is essential to development and operation of high performance monitor facilities. It is highly desirable that custom data interfaces and EDL be directly accessible for modification by the user, and that this requires a minimum of special skills. Refinement of measurement tools is part of the system engineering process.

References

Agee, J. C., and Girgis, G. K., Integration of Machine Condition and Power System Monitoring at Grand Coulee Powerplant, in *Proceedings of the Sixty-Third Annual International Conference of Doble Clients* — 1996, Doble Engineering Company, Watertown, MA, Section 7–4.

Bendat, J. S., and Piersol, A. G., *Engineering Applications of Correlation and Spectral Analysis*, John Wiley, New York, 1980.

Brigham, E. O., *The Fast Fourier Transform and Its Applications*. Prentice-Hall, Englewood Cliffs, NJ, 1988.

Carolsfeld, R., To measure is to control, *Electrical Business*, 14–15, January 1997.

CIGRE Task Force 38.01.07, *Control of Power System Oscillations*. CIGRE Technical Brochure, December 1996.

CIGRE Task Force 38.02.17, *Advanced Angle Stability Controls*. CIGRE Technical Brochure, to be published in 1999.

CIGRE Task Force 38.03.17, *Impact of the Interaction Among Power System Controls*. CIGRE Technical Brochure, to be published in 1999.

Clark, H. K., Gupta, R. K., Loutan, C., and Sutphin, D. R., Experience with dynamic system monitors to enhance system stability analysis, *IEEE Trans. Power Systems*, PWRS-7, 693–701, May 1992.

Evaluation of Low Frequency System Response: Study Results and Recommendations. Report of the WSCC 0.7 Hz Oscillation Ad Hoc Work Group to the WSCC Technical Studies Subcommittee, September 1990.

Farmer, R. G., and Agrawal, B. L., State-of-the-art techniques for power system stabilizer tuning, *IEEE Trans. PAS*, 699–709, March 1983.

Grenier, R., and Metes, G., *Enterprise Networking: Working Together Apart*, Digital Press, 1992.

Hauer, J. F., BPA Experience in the Measurement of Power System Dynamics, *Inter-Area Oscillations in Power Systems*, IEEE Publication 95 TP 101, 158–163.

Hauer, J. F., Power system identification by fitting structured models to measured frequency response, *IEEE Trans. PAS*, 915–923, April 1982.

Hauer, J. F., Reactive Power Control as a Means for Enhanced Interarea Damping in the Western U. S. Power System—A Frequency-Domain Perspective Considering Robustness Needs, *Application of Static Var Systems for System Dynamic Performance*, IEEE Publication 87TH0187-5-PWR, 79–92.

Hauer, J. F., The Use of Prony Analysis to Determine Modal Content and Equivalent Models for Measured Power System Response, *Eigenanalysis and Frequency Domain Methods for System Dynamic Performance*, IEEE Publication 90TH0292-3-PWR, 105–115.

Hauer, J. F., Validation of phasor calculation in the macrodyne PMU for California-Oregon transmission project tests of March 1993, *IEEE Trans. Power Delivery*, 11, 1224–1231, July 1996.

Hauer, J. F., Demeure, C. J., and Scharf, L. L., Initial results in Prony analysis of power system response signals, *IEEE Trans. Power Systems*, 80–89, February 1990.

Hauer, J. F., Erickson, D. C., Trudnowski, D. J., and Donnelly, M. K., Value Engineering A Dynamic Information Technology Package For Power System Applications, in 1995 *Fault and Disturbance Analysis/Precise Measurements in Power Systems Conference*, Arlington, VA, November 8–10, 1995.

Hauer, J. F., Hughes, F. J., Trudnowski, D. J., Rogers, G. J., Pierre, J. W., Scharf, L. L., and Litzenberger, W. H., A Dynamic Information Manager for Networked Monitoring of Large Power Systems. EPRI Report TR-112031, May 1999.

Hauer, J. F., and Hunt, J. R., in association with the WSCC System Oscillations Work Groups, Extending the Realism of Planning Models for the Western North America Power System, ub *V Symposium of Specialists in Electric Operational and Expansion Planning (SEPOPE)*, Recife (PE) Brazil, May 19-24, 1996.

Hauer, J. F., Mittelstadt, W. A., Litzenberger, W. H., Clemans, C., Hamai, D., and Overholt, P., *Wide Area Measurements for Real-Time Control and Operation of Large Electric Power Systems — Evaluation and Demonstration of Technology for the New Power System*. Prepared for U.S. Department of Energy Under BPA Contracts X5432-1, X9876-2; January 1999. This report and associated attachments are available on compact disk.

Hauer, J. F., Mittelstadt, W. A., Piwko, R. J., Damsky, B. L., and Eden, J. D., Modulation and SSR tests performed on the BPA 500 kV thyristor controlled series capacitor unit at Slatt Substation, *IEEE Trans. Power Systems*, 11, 801–806, May 1996.

Hauer, J. F., and Taylor, C. W., Information, reliability, and control in the new power system, in *Proc. 1998 American Control Conference*, Philadelphia, PA., June 24–26, 1998.

Hauer, J. F., Trudnowski, D. J., Rogers, G. J., Mittlestadt, W. A., Litzenberger, W. H., and Johnson, J. M., Keeping an eye on power system dynamics, *IEEE Computer Applications in Power*, 50–54, October 1997.

Hauer, J. F., and Vakili, F., An oscillation trigger for power system monitoring, *IEEE Trans. Power Systems*, 74–79, February 1990.

Jones, D. A., and Skelton, R. L., The next threat to grid reliability — data security, *IEEE Spectrum*, 36, 6, 46–48, June 1999.

Juang, J. N., *Applied System Identification*, Prentice-Hall, Englewood Cliffs, NJ, 1994.

Kamwa, I., Trudel, G., and Gerin-Lajoie, L., Low-order black-box models for control system design in large power systems, *IEEE Trans. Power Systems*, 11, 1, 303–311, February 1996.

Kosterev, D. N., Taylor, C. W., and Mittelstadt, W. A., Model validation for the August 10, 1996 WSCC System outage, *IEEE Trans. Power Systems*, 14, 3, 967–979, August 1999.

Kundur, P., Rogers, G. J., Wong, D. Y., Wang, L., and Lauby, M. G., A comprehensive computer program package for small signal stability analysis of power systems, *IEEE Trans. Power Systems*, 1076–1083, Nov. 1990.

Martin, K., Phasor Measurement System Test, in *Proc. BPA 1992 Engineering Symposium*, Portland, OR, March 31–April 1 1992, 689–704.

MATLAB — High-Performance Numeric Computation and Visualization Software (Reference Guide). The Math Works, Inc.: Natick, MA.

Phadke, A. G., Synchronized phasor measurements in power systems, *IEEE Computer Applications in Power Systems,* 10–15, April 1993.

Piwko, R. J., Wegner, C. A., Kinney, S. J., and Eden, J. D., Subsynchronous Resonance Performance Tests of the Slatt Thyristor-Controlled Series Capacitor, Paper 95 SM 402-8 PWRD. IEEE/PES 1995 Summer Meeting, Portland, OR.

Real-Time Phasor Measurement for Monitoring and Control. EPRI Report TR-103640S, December 1993.

Santori, M., An instrument that isn't really, *IEEE Spectrum,* 36–39, August 1990.

Schoukens, J., Vanden Bossche, M., and Pintelon, R., Expert opinion: The key to better measurements is software, *IEEE Spectrum,* 55, January 1993.

Smith, H. L., Substation automation problems and possibilities, *IEEE Computer Applications in Power,* 9, 4, 33–36, October 1996.

Special report: Concurrent engineering, *IEEE Spectrum,* 22–37, July 1991.

Stahlkopf, K. E., and Wilhelm, M. R., Tighter controls for busier systems, *IEEE Spectrum,* 34, 4, 48–52, April 1997.

Taylor, C. W., Improving grid behavior, *IEEE Spectrum,* 36, 6, 40–45, June 1999.

Taylor, C. W., and Erickson, D. C., Recording and analyzing the July 2 cascading outage, *IEEE Computer Applications in Power,* 10, 1, 26–30, January 1997.

Taylor, C. W., Mechenbier, J. R., and Burns, J. W., The December 14, 1994 Breakup of the Western North American Power System: Failures, Successes, and Lessons, in *V Symposium of Specialists in Electric Operational and Expansion Planning* (SEPOPE), Recife (PE) Brazil, May 19-24, 1996.

Trudnowski, D. J., Donnelly, M. K., and Hauer, J. F., Estimating damping effectiveness of BPA's thyristor controlled series capacitor by applying time and frequency domain methods to measured response, *IEEE Trans. Power Systems,* 11, 761–766, May 1996.

Trudnowski, D. J., Hauer, J. F., Pierre, J., Litzenberger, W. H., and Maratukulam, D. J., Using the Coherency Function to Detect Large-Scale Dynamic System Modal Observability, in *Proceedings of the* 1999 *American Control Conference,* San Diego, CA, June 1999, 2886–2890.

Trudnowski, D. J., Johnson, J. M., and Hauer, J. F., Making Prony analysis more accurate using multiple signals, *IEEE Trans. Power Systems,* 14, 1, 226–231, February 1999.

Van Overshee, P., and De Moor, B., *Subspace Identification for Linear Systems.* Kluwer Academic Publishers, Dordrecht, The Netherlands, 1996.

11.9 Power System Dynamic Security Assessment

Peter W. Sauer

Power System Security Concepts

Definitions and Historical Perspective

Power system security in the context of this chapter is concerned with the technical performance and quality of service when a disturbance causes a change in system conditions. Strictly speaking, every small change in load is a disturbance that causes a change in system conditions; however, this topic focuses on what could be called "large changes" in system conditions. These changes are referred to as "contingencies." Most commonly, contingencies result in relay operations that are designed to protect the system from faults or abnormal conditions. Typical relay operations result in the loss of a line, transformer, generator, or major load.

When changes occur, the various components of the power system respond and hopefully reach a new equilibrium condition that is acceptable according to some criteria. Mathematical analysis of these responses and the new equilibrium condition is called security analysis. If the analysis evaluates only the expected postdisturbance equilibrium condition (steady-state operating point), this is called Static Security Assessment (SSA). If the analysis evaluates the transient performance of the system as it progresses

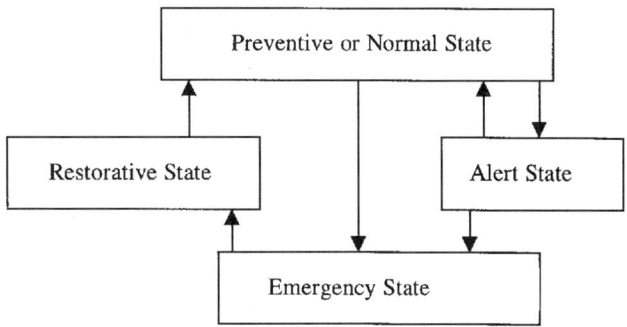

FIGURE 11.73 Operating states.

after the disturbance, this is called Dynamic Security Assessment (DSA). DSA has been formally defined by the Institute of Electrical and Electronics Engineers (IEEE), Power Engineering Society's (PES), Working Group on DSA as follows:

> Dynamic Security Assessment is an evaluation of the ability of a certain power system to withstand a defined set of contingencies and to survive the transition to an acceptable steady-state condition.

Security analysis is a fairly new concept in electric power system analysis. The early power systems were often separate and isolated regions of generators and loads. As systems became larger and more interconnected, the possibility of disturbances propagating long distances increased. The Northeast U.S. blackout of November 1965 started a major emphasis on the reliability and security of electric power systems. The benchmark paper by Tom Dy Liacco (1967) introduced the concept of the Preventive (Normal), Emergency, and Restorative Operating states and their associated controls. The Preventive state is the normal state wherein the system is stable with all components within operating constraints. The Emergency state arises when the system begins to lose stability, or when component operating constraints are violated. The Restorative state is when service to some customers has been lost, usually due to progression through the emergency state and the operation of protective devices. A significant extension added the Alert state between the Preventative (or Normal) state and the Emergency state as shown in Fig. 11.73 (Cihlar et al., 1969). This is a significant extension because it introduced the concept of a "potential emergency." If the occurrence of a likely contingency causes instability or operation with constraint violations, the system is said to be in the Alert state and classified as "insecure."

An extensive report of Dynamic Security Assessment practices in North America summarizes the initial impetus to the activity of the past 10 years (IEEE PES Working Group on DSA, 1988).

Criteria for Security

In terms of operating states, a system is said to be secure if it is in the normal state and will remain in the normal state following all likely contingencies. If a system is in the normal state but will not remain in the normal state following all likely contingencies, then it is reclassified into the Alert state and considered "insecure." The first key criterion here is the concept of "remain in the normal state." SSA can be used to quickly determine if a system is insecure by simply looking at the static outcome of each contingency. However, to be fully secure, DSA must be used to determine if the associated dynamics of each contingency are acceptable. For example, while the voltage levels of the postcontingency system may be normal (as determined by SSA), it is possible that the transient voltage dips during the disturbance may be unacceptable. Furthermore, SSA cannot easily determine if the postcontingency system is stable, or can even be reached due to the transients of the contingency.

The second key criterion is the definition of "likely contingencies." The list of likely contingencies varies from control area to control area depending on operating practice. In most cases, the list consists of single outages such as the loss of a line, transformer, or generator. In other cases, the list may include more complex contingencies that are known to occur with some frequency, and may include a sequence of events such as a fault on for a specified time followed by relay clearing.

Assessment and Control

The adoption of security concepts for electric power systems clearly separates the two functions of assessment and control. Assessment is the analysis necessary to determine the outcome of a "likely" contingency (possibly including all existing automatic controls). Control is the operator intervention or automatic action that might be designed for use to avoid the contingency entirely, or to remedy unacceptable postcontingency conditions. When the controls are implemented, they may then become a part of the assessment analysis through a modification of the contingency description.

Preventative control is the action taken to maneuver the system from the Alert state back to the Normal state. This type of control may be slow, and may be guided by extensive analysis. Emergency control is the action taken when the system has already entered the Emergency state. This type of control must be fast and guided by predefined automatic remedial schemes. Restorative control is the action taken to return the system from the Restorative state to the Normal state. This type of control may be slow, and may be guided by analysis and predefined remedial schemes.

Dynamic Phenomena

Phenomena of Interest

While there are numerous phenomena that are of interest in dynamic analysis, typical DSA programs focus primarily on two phenomena — voltage transients and system stability. The voltage transients are important because they must remain within acceptable limits to avoid further damage or loss of equipment. Normally, the effects of under-overvoltage transients are not included in the large-scale programs that are used for DSA. That is, the automatic tripping and relaying associated with under-overvoltage are not normally modeled as part of the simulation that is being used for DSA. Since these possible actions are not explicitly modeled, the programs simply monitor voltage levels as they progress during a transient.

One of the most basic concepts of system stability is the issue of maintaining synchronous operation of the AC generators. This is usually referred to as "transient stability," and is discussed in Section 11.2 of this chapter. Current DSA programs focus primarily on this type of stability and its associated constraint on operations.

As generator electromechanical dynamics progress during a disturbance, the system voltages and currents can change markedly. These changes can impact voltage-sensitive loads and result in conditions that may be unacceptable even though the generators remain in synchronism. In severe cases, voltage levels can reach points where recovery to nominal levels is impossible. Such voltage collapse conditions normally result in further deterioration of the system and additional relay action or loss of synchronism (Taylor, 1994). The extent to which such phenomena can be detected in DSA programs depends on the level of modeling detail for control systems, relays, and loads.

Time Scales

Power system dynamics include a very wide time-scale classification (Kundur, 1994). These can be loosely described by six categories as shown in Table 11.5.

In order to analyze this wide range of time-scale behavior, considerable care must be given both to efficient modeling and numerical techniques. The majority of current DSA programs consider the dynamics ranging from hundredths of seconds to tens of seconds (the electromechanical dynamics). The challenge of modeling this time range involves properly including the effects of the faster phenomena without explicitly including their fast transients.

Transient Stability

In alternating current (AC) systems, the generators must all operate in synchronism in steady state. When a fault occurs on the system, the electrical power output of some generators (usually those near the fault)

Power System Dynamics and Stability

TABLE 11.5 Power System Time Scales

Category	Time Scale
Lightning propagation	Microseconds to milliseconds
Switching surges	Microseconds to tenths of seconds
Electrical transients	Milliseconds to seconds
Electromechanical	Hundredths of seconds to tens of seconds
Mechanical	Tenths of seconds to hundreds of seconds
Thermal	Seconds to thousands of seconds

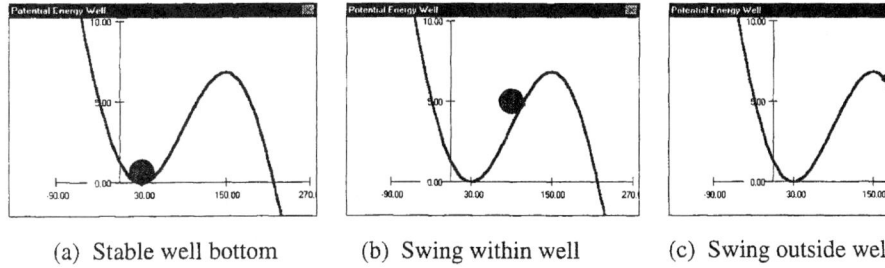

(a) Stable well bottom   (b) Swing within well   (c) Swing outside well

FIGURE 11.74 Transient stability and the energy well.

will tend to decrease. Since the turbine power input does not change instantaneously to match this, these generators will accelerate above the nominal synchronous speed. At the same time, the electrical power output of other generators may increase, resulting in deceleration below the nominal synchronous speed. As a fundamental property of rotating equipment, the generators must all reverse their trends before the energy imbalances become so large that return to synchronous operation is impossible. Transient stability analysis focuses on this phenomenon, which can be visualized through a ball resting in a potential energy well as shown in Fig. 11.74(a). In steady state, the ball is at rest (signifying all generators in synchronism) in the well bottom. Clearly, any small, temporary displacement of the ball will result in a return to the "stable" well bottom. However, if the disturbance is large enough that the ball is pushed over the well boundary, it will not return to the same well bottom. While it may come to rest at some other point, transient stability analysis is concerned with the detection of when the ball will leave the initial well boundary. That is, a fault will cause the ball to move up the side of the well, and must be cleared soon enough in time that the ball never leaves the well [Fig. 11.74(b)]. When the fault remains on the system too long, the ball picks up sufficient kinetic energy to carry it over the well boundary [Fig. 11.74(c)].

When generators accelerate or decelerate with respect to each other, their speed deviations (and corresponding angle deviations) constitute swings. If two or more generators swing apart in speed and then reverse, their return to synchronism could be considered "first-swing stable" if the analysis concludes at the point of return. In many cases, this is sufficient to ensure that the system will remain in synchronism for all time after the return. However, in other cases, the system dynamics may be such that the loss of synchronism does not occur until generators have experienced multiple swings. Deciding when to stop a simulation and declare the result either stable or unstable remains a challenge in DSA analysis.

Modeling

In order to perform computer simulation of the dynamics that may arise during and after a contingency, it is necessary to formulate mathematical equations which capture the fundamental transients. For the phenomena and time scales of interest in current DSA, ordinary differential equations are considered sufficient. Since the primary dynamics of interest are the electromechanical transients (shaft speeds), there will be two differential equations for each generator modeled. In addition, there may be many associated dynamics and controls that influence the electromechanical transients. Finally, there are the quasi-steady-state approximations of the remaining faster and slower dynamics that enter the model as algebraic equations. The resulting mathematical equations have the form given in Eqs. (11.34)–(11.37).

$$\frac{d\delta}{dt} = \omega - \omega_s \quad (11.34)$$

$$\frac{d\omega}{dt} = f(\omega, x, y) \quad (11.35)$$

$$\frac{dx}{dt} = g(\omega, x, y) \quad (11.36)$$

$$0 = h(\delta, x, y) \quad (11.37)$$

In addition, the algebraic equations may need to be modified during simulation to reflect the changes that occur in the network topology as time progresses between a fault application and subsequent clearing. Since it is difficult to guarantee the existence of a "y" solution for the algebraic equations as the dynamic states evolve, this combination of differential and algebraic equations poses theoretical as well as numerical challenges for DSA. Details of the composition of these mathematical models is given in Sauer and Pai (1998), Kundur (1994), and Anderson and Fouad (1993).

Assessment Methodologies

Off-line DSA

In off-line DSA analysis, detailed time-domain stability analysis is performed for all credible contingencies and a variety of operating conditions. In most cases, this off-line analysis is used to determine limits of power transfers across important system interfaces. These limits then are used in an operating environment that is hopefully not significantly different from those conditions considered. Since the analysis is performed off-line, there is not a severe restriction on computation time and therefore detailed analysis can be done for a wide range of conditions and contingencies. These studies include numerical integration of the models discussed above for a certain proposed power transfer condition and for a list of contingencies typically defined by a faulted location and specified fault clearing time (based on known relay settings).

The trajectories of the simulation are analyzed to see if voltage transients are acceptable, and to see if transient stability is maintained for the specified fault clearing time. If the results for one level of power transfer are acceptable for all credible contingencies, the level of proposed power transfer is increased and the analysis is repeated. This process continues until the level of power transfer reaches a point where the system cannot survive all of the credible contingencies. The maximum allowable transfer level is then fixed at the last acceptable level, or reduced by some small amount to provide a margin that would account for changes in conditions when the actual limit is in force.

Online DSA

Online DSA is used to supplement (or update) off-line DSA to consider current operating conditions. A basic online DSA framework includes essentially two steps. The first involves a rapid screening process to limit the number of contingencies that must be evaluated in detail. This rapid screening process might consist of some direct method that avoids long numerical integration times (Pai, 1989; Fouad and Vittal, 1992; Pavella and Murthy, 1993; Chadalavada et al., 1997). In addition to giving fast stability evaluation, these methods inherently include a mechanism for assessing the severity of a contingency. That is, if a system is determined to be stable, the direct methods also provide an indication of "how stable" the system is. This indication usually takes the form of an "energy margin." For example, in reference to the ball motion of Fig. 11.74, the maximum swing of the ball up the side of the energy well could be used to quantify how "close" the ball was to leaving the well.

Most of these methods still require some numerical integration to simulate the impact of a major disturbance and then predict stability or compute a margin to instability. Computation of the margin usually requires the simulation to force the system into an instability, perhaps either by using a sustained fault, or reapplication of the fault after scheduled clearing (Vaahedi et al., 1996).

This first step includes a decision process regarding which contingencies must be studied in greater detail. Those that are judged to be "sufficiently stable" need not be studied further. Those that are considered "marginal" must be studied further. This process includes a ranking strategy that is usually based on the energy margin computed in the direct method. Additional criteria involving artificial intelligence approaches can also be used to aid the decision process (El-Kady et al., 1990).

The second step involves traditional time-domain simulation that includes extensive numerical integration to reveal swing trajectories and voltage variations. This is performed on a small subset of contingencies that were judged to be "marginal" according to the screening process of step one.

In online studies, the time for computation is a severe constraint in addition to the challenge of interpretation and quantification of results. Typical performance goals for online DSA programs are to process 30 contingencies (each having 10 seconds of simulated time) for a 2000 bus, 250 generator system in 10 minutes (Ejebe, 1998).

Summary

Dynamic Security Assessment (DSA) is concerned with the ability of an electric power system to survive a major disturbance. It must assess the quality of the transient behavior as well as stability. DSA is performed both in off-line and online environments and is computationally intensive due to the numerical integration involved in evaluating the transient behavior of the system during major disturbances. Several recent and ongoing projects have addressed the computational issue through screening techniques that provide rapid analysis of stability outcomes and stability margins (Demaree et al., 1994; CIGRE Task Force 38.03.13, 1997). Research in this area is continuing as the need for DSA to evaluate Available Transfer Capability (ATC) becomes stronger in the restructured industry.

References

Anderson, P.M., and Fouad, A.A., *Power System Control and Stability*, Iowa State University Press, Ames, IA, 1977, Reprinted by IEEE Press, 1993.

Chadalavada, V., Vittal, V., Ejebe, G.C., Irisarri, G.D., Tong, J., Pieper, G., and McMullen, M., An On-Line Contingency Filtering Scheme for Dyanmic Security Assessment, *IEEE Trans. on Power Syst.*, 12, 1, 153–159, February, 1997.

Cihlar, T.C., Wear, J.H., Ewart, D.N., and Kirchmayer, L.K., Electric Utility System Security, in *Proceedings of the American Power Conference*, 31, 1969, 891–908.

Demaree, K., Athay, T. Chung, K. Mansour, Y., Vaahedi, E. Chang, A.Y., Corns, B.R., and Garett, B.W., An On-line Dynamic Security Analysis System Implementation, *IEEE Trans. on Power Syst.*, 9, 4, 1716–1722, Nov. 1994.

Ejebe, G.G., On-Line Dynamic Security Assessment, Slides presented to the 1998 IEEE PES Working Group on Dynamic Security Assessment Meeting, February, 1998, Tampa, FL, (Available from the author).

El-Kady, M.A., Fouad, A.A. Liu, C.C., and Venkataraman, S., Use of Expert Systems in Dynamic Security Assessment of Power Systems, in *Proceedings 10th PSCC*, Graz, Austria, 1990.

Fouad, A.A. (Chairman. IEEE PES Working Group on DSA), Dynamic Security Assessment Practices in North America, *IEEE Trans. on Power Syst.*, 3, 3, 1310–1321, August, 1988.

Fouad, A.A., and Vittal, V., *Power System Transient Stability Analysis Using the Transient Energy Function Method*, Prentice-Hall, Englewood Cliffs, NJ, 1992.

Kundur, P., *Power System Stability and Control*, McGraw Hill, New York, 1994.

Dy Liacco, T.E., The Adaptive Reliability Control System, *IEEE Trans. on Power Appar. Syst.*, PAS-86, 5, 517–531, May 1967.

Meyer, B. (Convener) and Nativel, G. (Secretary), New Trends and Requirements for Dynamic Security Assessement, CIGRE Task Force 38.02.13 report, 21 rue d'Artois-R-75008 Paris, France, December 1997.

Pai, M.A., *Energy Function Analysis for Power System Stability*, Kluwer, Boston, MA, 1989.

Pavella, M., and Murthy, P.G., *Transient Stability of Power Systems. Theory and Practice*, John Wiley, New York, 1993.

Sauer, P.W., and Pai, M.A., *Power System Dynamics and Stability*, Prentice-Hall, Upper Saddle River, NJ, 1998.

Taylor, C.W., *Power System Voltage Stability*, McGraw-Hill, New York, 1994.

Vaahedi, E., Mansour, Y., Chang, A.Y., Corns. B.R., and Tse, E.K., Enhanced Second Kick Methods for On-Line Dynamic Security Assessment, *IEEE Trans. on Power Syst.*, 11, 4, 1976–1982, November, 1996.

11.10 Power System Dynamic Interaction with Turbine-Generators

Richard G. Farmer and Bajarang L. Agrawal

Turbine-generators for power production are generally connected to electric power systems that transport the generated power and energy to the user. The power system can range from a single transformer to a complex system. A complex system may contain hundreds of power lines at various voltage levels and hundreds of transformers, turbine-generators, and loads. When the power system and its components are in the normal state, the generators produce sinusoidal voltages at synchronous frequency (60 Hz in the U.S.). The voltages cause currents to flow at synchronous frequency through the power system to the loads. The only current flowing in the generator rotor is the direct current in the generator field. The mechanical torques on the turbine-generator are constant and unidirectional. The system is said to be in synchronism and there is no dynamic interaction between the power system and the turbine-generators.

At other times, the system and its components are disturbed, thereby causing a periodic exchange of energy between the components of the power system. If there is a periodic exchange of energy between a turbine-generator and the power system, we will refer to this energy exchange as *Power System Dynamic Interaction with a Turbine-Generator*. When this occurs, the generator creates oscillating torques on the shafts of the turbine-generator. If the frequency of these torques is equal to, or near one of the natural mechanical frequencies of the turbine-generator or its components, excessive mechanical stress may occur in the turbine-generator. In addition, excessive voltage and current may occur in the generator and power system. Turbine-generator components known to be affected by such interaction are shafts, turbine blades, and generator retaining rings.

There have been several dramatic events resulting from power system dynamic interactions with turbine-generators, including significant turbine-generator damage. Analysis of these events has made the power engineering community aware of the potential for even more extensive turbine-generator damage from power system dynamic interaction. For these reasons, methods have been developed to identify and analyze the potential for power system dynamic interaction and countermeasures have been developed to control such interaction.

This section addresses the types of power system dynamic interaction with turbine-generators that have been identified as potentially hazardous. For each type of interaction, there is a discussion of known events, physical principles, analytic methods, possible countermeasures, and references. The types of interactions to be addressed are:

- Subsynchronous resonance
- Device dependent subsynchronous oscillations
- Supersynchronous resonance
- Device dependent supersynchronous oscillations

Subsynchronous Resonance

Series capacitors have been used extensively since 1950 as a very effective means of increasing the power transfer capability of a power system that has long (150 miles or more) transmission lines. Series capacitors provide a capacitive reactance in series with the inherent inductive reactance of a transmission line, thereby reducing the effective inductive reactance. Series capacitors significantly increase transient and steady state stability limits, in addition to being a near perfect means of var and voltage control. One transmission project, consisting of 1000 miles of 500 kV transmission lines, estimates that the application of series capacitors reduced the project cost by 25%. Until about 1971, it was generally believed that up to 70% series compensation could be used in any transmission line with little or no concern. However, in 1971 it was learned that series capacitors can create an adverse interaction between the series compensated electrical system and the spring-mass mechanical system of the turbine-generators. This effect is called *subsynchronous resonance* (SSR) since it is the result of a resonant condition that has a natural frequency below the fundamental frequency of the power system (Anderson and Farmer, 1996).

Known SSR Events

In 1970, and again in 1971, a 750-MW cross compound Mohave turbine-generator in southern Nevada experienced shaft damage. The damage occurred when the system was switched so that the generator was radial to the Los Angeles area on a 176-mile, series compensated 500 kV transmission line. The shaft damage occurred in the slip ring area of the high-pressure turbine-generator. Metallurgical analysis showed that the shaft had experienced high cycle fatigue leading to plasticity. Fortunately, the plant operators were able to shut the unit down before there was a shaft fracture. In each case, the turbine-generator had to be taken out of service for several months for repairs (Hall and Hodges, 1976). Intensive investigation in the electric power industry led to the conclusion that the Mohave events were caused by a subsynchronous resonance condition referred to as *torsional interaction*. Torsional interaction created sustained torsional oscillations in the second torsional mode, which has a stress concentration point in the slip ring area of the affected turbine-generator. Two other types of SSR have been identified. These are referred to as *induction generator effect* and *torque amplification*. There have been no reported events of induction generator effect or torque amplification.

Subsynchronous Resonance Terms and Definitions

A set of terms and definitions has been developed so engineers can communicate clearly using consistent terminology. Following are definitions for the most commonly used terms. These are consistent with the proposed terms and definitions presented in (*IEEE Trans.*, June 1985).

Subsynchronous: Electrical or mechanical quantities associated with frequencies below the synchronous frequency of a power system.

Supersynchronous: Electrical or mechanical quantities associated with frequencies above the synchronous frequency of a power system.

Subsynchronous Resonance: The resonance between a series capacitor compensated electric system and the mechanical spring-mass system of the turbine-generator at subsynchronous frequencies.

Self-Excitation: The sustainment or growth of response of a dynamic system without externally applied excitation.

Induction Generator Effect: The effect of having subsynchronous positive sequence currents in the armature of a synchronously rotating generator.

Torsional Interaction: Self-excitation of the combined mechanical spring-mass system of a turbine-generator and a series capacitor compensated electric network when the subsynchronous rotor motion developed damping torque is negative and greater in magnitude than the mechanical damping torque of the rotor.

Torque Amplification: The amplification of turbine-generator shaft torque caused by subsynchronous resonance or unfavorable timing of switching events in the electric network.

Subsynchronous Oscillation: The exchange of energy between the electric network and the mechanical spring-mass system of the turbine-generator at subsynchronous frequencies.

Torsional Mode Frequency: A natural frequency of the mechanical spring-mass system of the turbine-generator in torsion.

Torsional Damping: A measure of the decay rate of torsional oscillations.

Modal Model: The mathematical spring-mass representation of the turbine-generator rotor corresponding to one of its mechanical natural torsional frequencies.

Torsional Mode Shape: The relative angular position or velocity at any instant of time of the individual rotor masses of a turbine-generator unit during torsional oscillation at a natural frequency.

Subsynchronous Resonance (SSR) Physical Principles

For this discussion the simplest possible system will be considered with a single turbine-generator connected to a single series compensated transmission lines as shown in Fig. 11.75. The turbine-generator has only two masses connected by a shaft acting as a torsional spring. There are damping elements between the two masses and each mass has a damping element. The electrical system of Fig. 11.75 has a single resonant frequency, f_{er}, and the mechanical spring-mass system has a single natural frequency, f_n. It must be recognized that the electrical system may be a complex grid with many series compensated lines resulting in numerous resonance frequencies f_{er1}, f_{er2}, f_{er3}, etc. Likewise, the turbine-generator may have several masses connected by shafts (springs) resulting in several natural torsional frequencies (torsional modes) f_{n1}, f_{n2}, f_{n3}, etc. Even so, the system of Fig. 11.75 is adequate to present the physical principles of SSR.

SSR is a phenomenon that results in significant energy exchange between the electric system and a turbine-generator at one of the natural frequencies of the turbine-generator below the synchronous frequency, f_o. When the electric system of Fig. 11.75 is series compensated, there will be one subsynchronous natural frequency, f_{er}. For any electric system disturbance, there will be armature current flow in the three phases of the generator at frequency f_{er}. The positive sequence component of these currents will produce a rotating magnetic field at an angular electrical speed of $2\pi f_{er}$. Currents are induced in the rotor winding due to the relative speed of the aforementioned rotating field and the speed of the rotor. The resulting rotor current will have a frequency of $f_r = f_o - f_{er}$. A subsynchronous rotor current creates Induction Generator Effect as will be discussed later in this section. The armature magnetic field, rotating at an angular frequency of f_{er}, interacts with the rotor's DC field, rotating at an angular frequency of f_o, to develop an electromagnetic torque component on the generator rotor at an angular frequency of $f_o - f_{er}$. This torque component contributes to torsional interaction and torque amplification which will be discussed further later in this section.

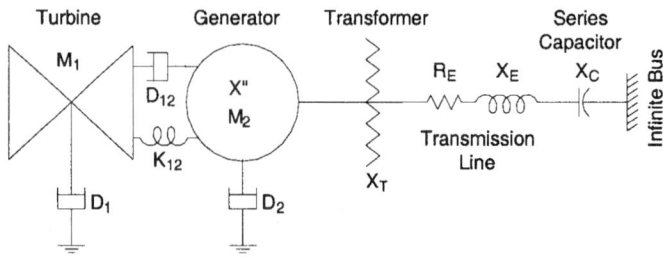

FIGURE 11.75 Turbine-generator with series compensated transmission line. (*Source:* IEEE Committee Report, Terms, definitions and symbols for subsynchronous resonance, *IEEE Transactions*, PAS-104, June 1985. With permission.)

Induction Generator Effect

Induction generator effect involves only the electric system and the generator (does not involve turbines). For an induction machine, the effective rotor resistance as seen from the armature and external power system is given by the following equations:

$$R'_r = \frac{R_r}{s} \quad (11.38)$$

$$s = \frac{f_{er} - f_o}{f_{er}} \quad (11.39)$$

where R'_r = Apparent rotor resistance viewed from the armature
R_r = Rotor resistance
s = Slip
f_{er} = Frequency of the subsynchronous component of current in the armature
f_o = Synchronous frequency

Combining Eqs. (11.38) and (11.39) yields

$$R'_r = \frac{R_r f_{er}}{f_{er} - f_o} \quad (11.40)$$

Since f_{er} is subsynchronous, it will always be less than f_o. Therefore, the effective generator resistance as viewed from the armature circuit will always be negative. If this equivalent resistance exceeds the sum of the positive armature resistance and system resistance at the resonant frequency f_{er}, the armature currents can be sustained or growing. This is known as induction generator effect (Anderson and Farmer, 1996; Kilgore et al., 1977).

Torsional Interaction

Torsional interaction involves both the electrical and the mechanical systems. Both systems have one or more natural frequencies. The electrical system natural frequency is designated f_{er} and the mechanical spring-mass system natural frequency is designated f_n. Generator rotor oscillations at a natural torsional frequency, f_n, induce armature voltage components of subsynchronous frequency, $f_{en}^- = f_o - f_n$, and supersynchronous frequency, $f_{en}^+ = f_o + f_n$. When the frequency of the subsynchronous component of armature voltage, f_{en}^-, is near the electric system natural frequency, f_{er}, the resulting subsynchronous current flowing in the armature is phased to produce a rotor torque that reinforces the initial rotor torque at frequency f_n. If the resultant torque exceeds the inherent damping torque of the turbine-generator for mode n, sustained or growing oscillations can occur. This is known as torsional interaction. For a more detailed mathematical discussion of torsional interaction, see Kilgore et al. (1977) and Anderson et al. (1990).

Torque Amplification

When there is a major disturbance in the electrical system, such as a short circuit, there are relatively large amounts of electrical energy stored in the transmission line inductance and series capacitors. When the disturbance is removed from the system, the stored energy will be released in the form of current flowing at the electrical system resonant frequency, f_{er}. If all, or a portion of the current flows through a generator armature, the generator rotor will experience a subsynchronous torque at a frequency $f_o - f_{er}$. If the frequency of this torque corresponds to one of the torsional modes of the turbine-generator spring-mass system, the spring-mass system will be excited at that natural torsional frequency and cyclic shaft torque can grow to the endurance limit in seconds. This is referred to as torque amplification. For more in-depth treatments of torque amplification, see Joyce et al. (1978) and *IEEE Trans.* (Sept./Oct. 1978).

SSR Control

If series capacitors are to be applied, or seriously considered, it is essential that SSR control be thoroughly investigated. The potential for subsynchronous resonance must be evaluated and the need for countermeasures determined. When a steam-driven turbine-generator is connected directly to a series compensated line, or a grid containing series compensated lines, a potential SSR problem exists. There are three types of series-capacitor applications for which SSR would not be expected. The first occurs when the turbine-generator includes a hydraulic turbine. In this case, the ratio of generator mass to turbine mass is relatively high, resulting in larger modal damping and modal inertia than exists for steam turbine-generators (Anderson et al., 1984). The second type of series capacitor application that is generally free from SSR concerns has turbine-generators connected to an uncompensated transmission system that is overlaid by a series compensated transmission system. The California-Oregon transmission system is of this type with a 500 kV system that has 70% series compensation overlaying an uncompensated 230 kV transmission system. Turbine-generators are connected to the 230 kV system. Extensive study of this system has failed to identify any potential SSR problems. The third type involves series-capacitor-compensation levels below 30%. There have been no potential SSR problems identified for compensation levels below 30%.

For those series capacitor applications that are identified as having potential SSR problems, an SSR countermeasure will be required. Such countermeasures can range from a simple operating procedure to equipment costing millions of dollars. Numerous SSR countermeasures have been proposed and several have been applied (*IEEE Trans.*, Sept/Oct 1980). Fortunately, for every series capacitor installation investigated, an effective SSR countermeasure has been identified.

An orderly approach to planning and providing SSR control has been proposed (Anderson and Farmer, 1996). This includes the five steps presented below.

Screening Studies

Screening studies need to be made to determine the potential SSR problems for every turbine-generator near a series capacitor installation. These studies will probably need to be conducted using estimated data for torsional damping and modal frequencies for the turbine-generator unless the turbine-generator is in place and available for testing. Accurate modal frequencies and damping can only be obtained from tests, although manufacturers will usually provide their best estimate. The most popular analytic tool for screening studies is the Frequency Scanning Technique. This technique can provide an approximate assessment of the potential and severity for the three types of SSR; Induction Generator Effect, Torsional Interaction, and Torque Amplification (Agrawal and Farmer, 1979). To conduct the frequency scan studies, the positive sequence model for the power system is required. Generator impedance as a function of frequency is needed and may be estimated. The best estimates for turbine-generator torsional damping and modal frequencies are required. If the screening study is conducted using estimated data for the turbine-generator, data sensitivity should be examined.

Accurate Studies

If screening studies indicate any potential SSR problem, additional studies are required using the most accurate data as it becomes available from the manufacturer and from tests. The frequency scan program may be adequate for assessment of induction generator effect and torsional interaction, but an eigenvalue study is desirable if large capital expenditures are being considered for self-excitation countermeasures. If the screening studies show any potential for torque amplification, detailed studies must be conducted to calculate the shaft torque levels to be expected and the probability of occurrence. The manufacturer can provide an estimated spring-mass model for the turbine-generator, which can be used for the initial torque amplification studies. The studies can be updated as more accurate data becomes available from tests. The well-known electromagnet transient program (EMTP) is usually used for these studies.

SSR Interim Protection

If series capacitors are to be energized prior to acquiring accurate data from turbine-generator tests, and the above studies indicate a potential SSR problem, interim protection must be provided. Such protection might consist of reduced levels of series compensation, operating procedures to avoid specific levels of series compensation, and/or relays to take the unit off-line in the event an SSR condition is detected.

SSR Tests

Some SSR testing will be required unless the studies discussed above show no potential for the hazards of SSR. The torsional modal frequencies of the spring-mass system can probably be measured through monitoring during normal turbine-generator and system operation. To measure modal damping, it is necessary to operate the turbine-generator at varying load levels while stimulating the spring-mass system. Testing will be discussed in more detail later in this section.

Countermeasure Requirements

When accurate studies have been completed using the turbine-generator model obtained from tests, SSR countermeasures can be selected. The countermeasure selection involves an analysis of the acceptable fatigue life expenditure and turbine-generator tripping. See Computing Fatigue Life Expenditure (later in this section) for a discussion of fatigue life expenditure (FLE). Implementation of the selected countermeasures requires careful coordination. If the countermeasures involve hardware, the effectiveness of the hardware should be determined by testing. Countermeasures will be presented in more detail later in this section.

SSR Analysis

SSR analysis involves the identification of all system and generator operating conditions that result in SSR hazards and the determination of the severity by calculating the negative damping or shaft torque amplification. The primary computer programs used in the industry for SSR analysis are Frequency Scanning, Eigenvalue, and Transient Torque (EMTP). Some program validation has been made in the industry by comparing the results of these analytic methods with test results (IEEE Committee Report, Feb. 1989).

Frequency Scanning

The frequency scanning technique involves the determination of the driving point impedance over the frequency range of interest, as viewed from the neutral of the generator being studied (Agrawal and Farmer, 1979). For frequency scanning, the following modeling is required:

- A positive sequence model of the power system, including series compensation, as viewed from the generator terminals.
- The generator being studied is represented by its induction generator equivalent impedance as a function of slip. This can generally be obtained from the generator manufacturer. If not, an approximation is presented by Kilgore et al. (Nov./Dec. 1977). Other generators in the system are generally modeled by their short-circuit equivalent. Load is generally neglected.

Figure 11.76 is a typical output from a frequency-scanning program. The plots consist of the reactance and resistance as a function of frequency as viewed from the generator neutral. In addition, the 60Hz complements of the modal frequencies have been superimposed and labeled by mode number. The use of frequency scanning to evaluate the three types of SSR will be presented below.

Induction Generator Effect — Frequency scanning is an excellent tool for analysis of induction generator effect. Induction generator effect is indicated when the frequency scan shows that the reactance crosses zero at frequencies corresponding to negative resistance. Such points can be identified by inspection from frequency scan plots.

Torsional Interaction — When a resonant frequency of the electrical system, as viewed from the generator neutral, corresponds to the 60 Hz complement of one of the turbine-generator modal frequencies, negative damping of the turbine-generator exists. If this negative damping exceeds the positive modal

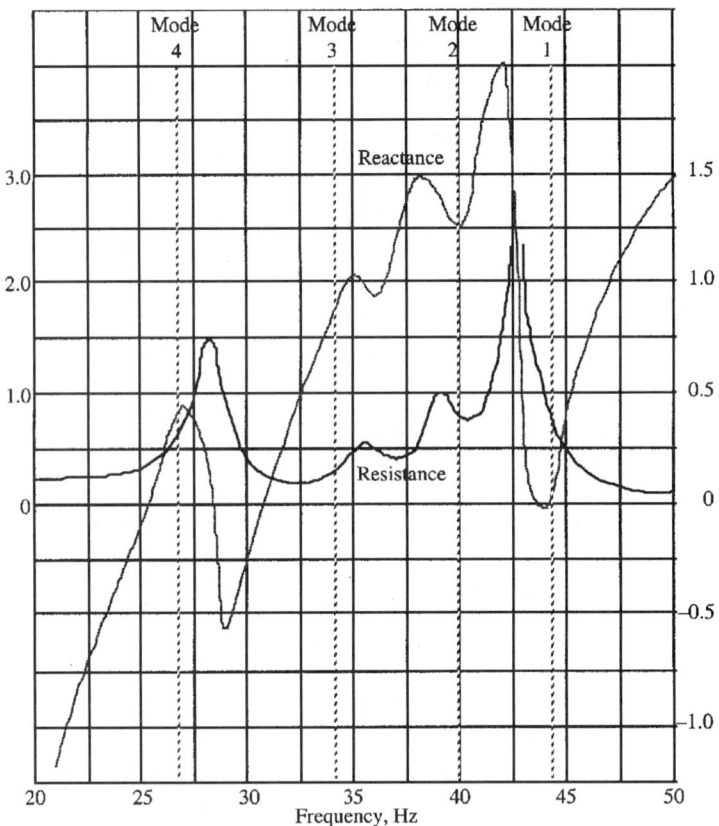

FIGURE 11.76 Frequency scan for the Navajo Project generator connected to the 500 kV system. (*Source:* Anderson, P. M. and Farmer, R. G., Subsynchronous Resonance: *Series Compensation of Power Systems*, PBLSH!, San Diego, 1996. With permission.)

damping of the turbine-generator, sustained or growing shaft torque would be experienced. Such negative damping can be approximated from frequency scanning results according to Kilgore et al. (Nov./Dec. 1977).

Using the method of Kilgore et al. (Nov./Dec. 1977), the amount of negative damping for torsional mode n is directly related to the conductance, G_n, for that mode, and can be calculated by the following approximate formula

$$\Delta\sigma_n = \frac{60 - f_n}{8 f_n H_n} G_n \qquad (11.41)$$

where $\Delta\sigma_n$ = Negative damping for mode n in rad/sec
H_n = Equivalent p.u. stored energy for a pure modal oscillation
See Agrawal and Farmer (1979).
G_n = p.u. conductance of the electrical system including the generator on the generator MVA base at $(60 - f_n)$ Hz.
$G_n = \dfrac{R_n}{R_n^2 + X_n^2}$
R_n = Resistance from frequency scan at $(60 - f_n)$ Hz
X_n = Reactance from frequency scan at $(60 - f_n)$ Hz

Power System Dynamics and Stability

Equation (11.41) neglects the damping due to the supersynchronous components of current. This is generally negligible. Equation 6.4 in Anderson and Farmer (1996) includes the supersynchronous effect. Agrawal and Farmer (1979) include a sample calculation for H_n.

The existence and severity of torsional interaction can now be determined by comparing the negative damping, $\Delta\sigma_n$, determined from frequency scanning for Mode n, with the natural mechanical damping of the turbine-generator for Mode n. In equation form, this is

$$\sigma_{net} = \sigma_n - \Delta\sigma_n \quad (11.42)$$

where σ_{net} = Net torsional damping for Mode n
σ_n = Turbine-generator damping for Mode n
$\Delta\sigma_n$ = Negative damping for Mode n due to torsional interaction

If the net damping, σ_{net}, is negative, torsional interaction instability for Mode n is indicated at the operating condition being studied. From the same frequency scan case, $\Delta\sigma_n$ can be calculated for all other active modes and then compared with the natural damping, σ_n, for the corresponding mode. This provides an indication of the severity of torsional interaction for the operating condition (case) being studied. This process should be repeated for all credible operating conditions that are envisioned.

The natural torsional frequencies and modal damping for the turbine-generator will only be known accurately if the machine has been tested. If estimated data is being used, the possible variations should be accounted for. The simplest way to account for variations in modal frequency is to apply margin. One way is to calculate the maximum conductance for Eq. (11.41) within a frequency range. Agrawal and Farmer (1979) suggests a frequency range of ±1 Hz of the predicted modal frequency. Experience has shown that estimated modal damping can vary from the measured damping by as much as 5 to 1. Therefore, a potential torsional interaction problem should be assumed if a reduction of the estimated natural damping by a factor of 5 results in net negative damping. In such cases, the SSR problem severity can only be finalized when modal damping has been measured.

The frequency scanning technique, as used to calculate negative damping, has been validated through comparison with test results. There has been reasonable correlation, as shown in Agrawal and Farmer (1979) and the IEEE Committee Report (1989), when the turbine-generator model parameters are accurate. Frequency scanning is a cost-effective means to study induction generator effect and torsional interaction. The results must be used with care. If the study results indicate that positive damping exists for all system conditions, but there are large reactance dips (Agrawal and Farmer, 1979), tests should be conducted to validate the study results prior to making a final decision not to implement any countermeasures. Also, if frequency-scanning studies indicate an SSR problem for which countermeasures are required, it is prudent to validate the studies by tests prior to committing to costly countermeasures or series compensation reduction (Anderson and Farmer, 1996).

Torque Amplification — Frequency scanning cannot be used to quantify the torque to be expected for a specific disturbance but it is a very good tool for determining the potential for torque amplification problems and the system configurations that need to be investigated in detail using EMTP. Agrawal and Farmer (1979) suggest that, if a frequency scan case shows a significant reactance dip within ±3 Hz of the 60 Hz complement of a modal frequency of the turbine generator, torque amplification might be expected. This provides an excellent screening tool for developing a list of EMTP cases to be studied. The frequency scan results in Fig. 11.76 suggest potential torque amplification for Mode 1 and Mode 2. The largest reactance dip is near Mode 1, but is slightly detuned. The reactance dip for Mode 2 is smaller but is nearly perfectly tuned. The system configuration represented by Fig. 11.76 was studied using EMTP and found to have serious torque amplification problems. See Farmer et al. (July/Aug. 1977).

Eigenvalue Analysis

Eigenvalue analysis for SSR is straightforward for torsional interaction and induction generator effect since they can be analyzed by linear methods (Anderson and Farmer, 1996). The approach follows:

1. Model the power system by its positive sequence model.
2. Model the generator electrical circuits.
3. Model the turbine-generator spring-mass system with zero damping.
4. Calculate the eigenvalues of the interconnected systems.
5. Analyze the eigenvalues that correspond to the subsynchronous modes of the turbine-generator spring-mass system to evaluate torsional interaction.
6. Analyze the eigenvalues that correspond to the electric system resonant frequency to evaluate induction generator effect.

The eigenvalues to be analyzed for torsional interaction can be identified by comparing the imaginary part of each eigenvalue with the modal frequencies of the spring-mass system. The corresponding real part of the eigenvalue is a quantitative indication of the damping for that mode. If the eigenvalue has a negative real part, positive damping is indicated. If it has a positive real part, negative damping is indicated. The real part of the eigenvalue is a direct measure of the positive or negative damping for each mode. Adding the calculated damping algebraically to the inherent modal damping results in the net modal damping for the system. An unstable eigenvalue, which is only related to the electrical system, may indicate an induction generator effect problem. For a mathematical treatment of modeling for eigenvalue analysis, see Anderson et al. (1990).

Transient Analysis

Transient analysis is required to determine the potential for SSR torque amplification. The well-known EMTP program is very well suited for such an analysis (Gross and Hall, July/Aug. 1978). There are various versions of the program. Bonneville Power Administration (BPA) developed the program and has upgraded it through the years. A version referred to as ATP is in the public domain and is available from BPA. EMTP provides for detailed modeling of those elements required for assessing the severity of SSR torque amplification. This includes the power system, the generators in the system, and the mechanical model of the turbine-generator being studied.

EMTP Power System Model — All elements of the power system are modeled by the three phases, neutral, and ground. The data for the model can generally be provided in the form of phase components or symmetrical components. Special features of series capacitors can be modeled, including capacitor protection by gap flashing or nonlinear resistors. Load is generally neglected.

EMTP Generator Model — The synchronous generator being studied is represented in EMTP by a two-axis Park's equivalent with several rotor circuits on the direct and quadrature axes. The input data can be in the form of either winding data or conventional stability data. The generator data can be obtained from the manufacturer in the form of conventional stability data. All generators in the system, other than the study generator, can generally be represented by a voltage source and impedance without affecting the study accuracy. For a detailed treatment of generator modeling for SSR analysis, refer to (Anderson et al., 1990).

EMTP Turbine-Generator Mechanical Model — The turbine-generator mechanical model in EMTP consists of lumped masses, spring constants, and dampers. The masses and spring constants determine the natural torsional frequencies, which can be verified from tests as soon as the generator goes into service. Experience has shown that the calculated and measured natural torsional frequencies are within ±1 Hz. Damping calculations are not nearly as accurate and variations between measured and calculated modal damping of 5 to 1 have been observed. For EMTP studies, this is further complicated by the fact that damping calculations and damping measurements are in terms of modal damping for each mode, whereas the EMTP mechanical model requires damping values for each mass and between adjacent masses. For torque amplification studies, mechanical damping is not a critical factor. The peak shaft torque would be expected to only vary by about 10% over a range of damping from zero to maximum (Anderson and Farmer, 1996).

Critical Factors for Torque Amplification — The most important use of EMTP for SSR analysis is to find the peak shaft torque that is to be expected when series capacitors are applied. It is necessary to understand that the major torque amplification events due to SSR will occur either during a power system fault or after the clearing of a power system fault. The energy stored in series capacitors during a fault will be discharged as subsynchronous current that can flow in a generator armature, creating amplified subsynchronous torque. The peak shaft torque to be expected depends on many factors. Experience has shown that the dominant factors that should be varied during a torque amplification study are electric system tuning, fault location, fault clearing time, and capacitor control parameters. For a detailed discussion on system tuning and faults, see Anderson and Farmer (1996). For information on capacitor controls, see the IEEE Committee Report (June, 1982).

Computing Fatigue Life Expenditure — When the torque of a turbine-generator shaft exceeds a certain minimum level (Endurance Limit), fatigue life is expended from the shaft during each torsion cycle. The machine manufacturer can generally furnish the fatigue life expenditure (FLE) per cycle corresponding to shaft torque magnitude for each shaft. When plotted, this is referred to as an S-N curve. EMTP can then be used to predict the FLE for a specific system disturbance. One method requires the complete simulation and FLE calculation of an event over approximately 30 seconds, which may be costly if numerous scenarios are to be investigated. An alternate simplified method requires some approximation. For this method, EMTP studies are conducted to find the peak shaft torque that will occur for a given scenario. Since the peak shaft torques generally occur within 0.5 seconds, EMTP simulation and FLE calculation time is minimized. It is assumed that after the shaft torque has peaked, it will decay at a rate corresponding to the mechanical damping of one of the excited modes. The FLE for the simulated event can then be calculated from knowledge of the peak torque, the decay rate, and the S-N curve. This gives conservative estimates of FLE. It is important to recognize that FLE for each incident is cumulative. When the accumulated FLE reaches 100%, the shaft is expected to experience cracks at its surface, but not gross failure. For more detail on computing FLE, see Anderson and Farmer (1996).

Data for SSR Analysis

Data requirements for SSR analysis consist of System Data and Turbine-Generator Data.

System Data — System data for eigenvalue and frequency scanning studies is generally of the same form as the positive sequence data used for power flow, short circuit, and power system stability studies. The data may require refinement to account for the resistance variations with frequency and for system equivalents. The classical short circuit equivalent may not be adequate when the equivalent system includes series capacitors. In such cases, an RLC equivalent might be developed. It should be checked with the frequency-scanning program to determine if the equivalent reasonably approximates the driving point impedance of the system it is to represent over the frequency range of interest (10 to 50 Hz). Large load centers may also require a special equivalent. For one outstanding case where the apparent impedance was actually measured over the frequency range of 15 to 45 Hz, it was found that the Phoenix, Arizona, load must be modeled to provide a good equivalent (Agrawal et al., May 1989). In that case, it was found that the following load model could form an accurate equivalent:

- 60% of the total load consists of induction motor load with x_d'' of 0.135 per unit
- 40% of the load is purely resistive

The validity of such a model for other locations has not been determined.

For torque amplification studies using EMTP, the system data requirements are much more extensive since all three phases and ground are represented. In EMTP the series capacitors can be modeled in detail, including the capacitor protective equipment. For more detail on system data for SSR analysis, see Anderson and Farmer (1996).

Turbine-Generator Data — The IEEE SSR Working Group has developed a set of Recommended SSR Data items that should be furnished by the turbine-generator manufacturer. This is generally the minimum data required for SSR studies. Following is a description of the three types of data:

1. **Generator Electrical Model:** Resistance and reactance as a function of frequency for the generator as viewed from the generator terminals. This should include armature and rotor circuits.
2. **Turbine-Generator Mechanical Model:**
    a. The inertia constant for each turbine element, generator, and exciter.
    b. The spring constants for each shaft connecting turbine elements, generator, and exciter.

c. The natural torsional frequencies and mode shapes as determined for the mechanical model defined by items a and b.
d. The modal damping as a function of load corresponding to the mechanical model defined by items a and b.
3. **Life Expenditure Curves:** For each shaft connecting the turbine elements, generator, and exciter, a plot of the life expended per transient incident as a function of the peak oscillating torque, or an S-N curve showing torque versus number of cycles to crack initiation or crack propagation. The manufacturers should provide all assumptions made in the preparation of these curves.

For more detail on turbine-generator modeling, see Anderson and Farmer (1996) and Anderson et al. (1990).

SSR Countermeasures

If series capacitors are to be used and SSR analysis shows that the hazards of SSR are to be expected, countermeasures must be provided, even if the probability of an SSR event is low. Such countermeasures may not completely eliminate turbine-generator shaft fatigue life expenditure (FLE). Even so, prudent countermeasure selections can probably limit the FLE of any shaft to less than 100% over the expected life of the turbine-generator. A strategy for SSR countermeasure selection should be formulated during the SSR analysis stage so that it can be used as a guide for the studies to be conducted. Farmer and Agrawal (1981) present one utility's guidelines that were developed for countermeasure selection, including the required SSR studies.

Numerous SSR countermeasures have been studied (IEEE Committee Report, Feb. 1992) and twelve, or so, have been applied. Following is a list of the countermeasures known to have been applied with references for each. These are separated into Unit-Tripping and Non-Unit Tripping types.

Unit-Tripping SSR Countermeasures

The following countermeasures will cause the generator to be electrically separated from the power system when a hazardous condition is detected.

- Torsional Motion Relay (Bowler et al., Sept./Oct. 1978; Ahlgren et al., 1982)
- Armature Current Relay (Sun et al., July 1981; Farmer and Agrawal, May 1981)
- Unit-Tripping Logic Schemes (Perez, 1981)

Non-Unit Tripping SSR Countermeasures

The following SSR countermeasures will provide SSR protection without electrically separating the generator from the power system.

- Static Blocking Filter (Farmer et al., July/Aug. 1977; Bowler et al., July/Aug. 1978)
- Dynamic Stabilizer (Ramey et al., Dec. 1981; Ramey et al., 1981; Kimmel et al., Jan. 1984)
- Excitation System Damper (Bowler and Baker, 1981; Bowler and Lawson, 1981)
- Turbine-Generator Modifications (Anderson and Farmer, 1996)
- Pole Face Amortisseur Windings (IEEE Committee Report, Sept./Oct. 1980; Farmer et al., July/Aug. 1977)
- Series Capacitor Bypassing (IEEE Committee Report, June 1982)
- Coordinated Series Capacitor Control With Loading (IEEE Committee Report, Sept./Oct. 1980)
- Operating Procedures (Anderson and Farmer, 1996)
- Thyristor Controlled Series Capacitors (Anderson and Farmer, 1996; IEEE Committee Report, Apr. 1992; Kinney et al., 1993; Hauer et al., 1994)

Fatigue Damage and Monitoring

Fatigue damage of turbine-generator shafts is certainly undesirable, but it may not be practical to completely avoid it. Therefore, it is important to understand the consequences of fatigue damage, and to know how to quantify any fatigue damage experienced so that gross shaft failure is avoided (IEEE Committee Report, June 1985; Ahlgren et al., 1982).

The consequences of high cycle fatigue and low cycle fatigue differ. In the case of high cycle fatigue, where purely elastic deformation occurs, there is no permanent deformation and no irreparable damage. It is said that 100% FLE occurs when cracks are initiated at the stress concentration points on the shaft surfaces. When this point is reached, cracks will be propagated as additional torsional stresses above the endurance limit are experienced. This does not mean that shaft failure will occur when 100% FLE is reached. On the contrary, the ultimate strength of the shaft in torsion is not significantly reduced. It does mean that cracks will be expected to increase in number and size if appropriate action is not taken. Fortunately, the total shaft integrity can be effectively restored by machining the shaft surface to remove the cracks. Cracks can be identified by visual inspection at stress concentration points on the shaft. Even so, it may be very costly to shut a unit down for a visual inspection following an incident suspected to result in significant FLE. For this reason, torsional monitoring techniques have been developed to provide a permanent history of torque experienced by each turbine-generator shaft. The most likely phenomenon leading to high cycle fatigue is torsional interaction.

In the case of low cycle fatigue, for which plastic deformation occurs, the consequences may be quite different from those in high cycle fatigue. When plastic deformation occurs, there is irreversible shaft deformation in torsion (kink). This can result in a bending moment being applied to the shaft each revolution. If the unit continues to operate in this condition, shaft failure in the bending mode may occur. If a monitor detects low cycle fatigue and there is a corresponding increase in lateral vibration, the shaft should be inspected. Whether there should be an immediate unit shutdown for such an incident is subject to judgment. The most likely phenomenon leading to low cycle fatigue is torque amplification.

Shaft torque monitoring techniques have been developed that will provide a permanent history of the approximate torque experienced by each turbine-generator shaft. This information is extremely useful in making a decision following a unit trip by SSR relay action or an unusual event such as out-of-phase synchronization. The options are:

1. Take the unit off-line and inspect the shaft.
2. Inspect the shaft at the next scheduled outage.
3. Synchronize, load the unit, and continue to operate without interruption.

A wrong decision could cause significant shaft damage or an unnecessary unit outage. Several methods have been developed for monitoring shaft torque as reported in the literature (Ahlgren et al., 1982; Walker et al., 1984; Joyce and Lambrecht, 1979; Stein and Fick, Mar./Apr. 1980; Ramey et al., Sept. 1980; Agrawal et al., Aug. 1982).

SSR Testing

The analytic methods and corresponding software for SSR analysis can be perfect but has no value unless the required data for the electric system, generators, and turbine-generator spring-mass-damping system is available and is reasonably accurate. It has been found from tests that the torsional frequencies are usually within one hertz of that predicted by the manufacturer. This implies that the spring-mass model data is reasonably accurate. The turbine-generator manufacturers estimate torsional damping, but testing has shown that damping predictions, when compared with site tests, may have large variations. Therefore, little confidence can be placed in predicted damping and accurate torsional damping values can only be obtained from tests.

SSR tests can vary in their complexity, depending on their purpose, availability of turbine-generator rotor motion monitoring points, type of generator excitation system, power system configuration, and other factors. The minimum and simplest tests are those used to identify the natural torsional frequencies of a turbine-generator. Tests to measure torsional damping are more difficult, particularly at high loading. Various types of tests may be devised to test the effectiveness of countermeasures.

Torsional Mode Frequency Tests
The objective of these tests is to perform a spectrum analysis of rotor motion or shaft strain in torsion at points that respond to all active modes of interest. Rotor motion signals can be obtained by demodulating

the output of a proximity probe mounted adjacent to a toothed wheel on the rotor. Shaft strain is obtainable from strain gauges fixed to the shaft (Walker and Schwalb, 1976). With the use of digital spectrum analyzers, natural torsional frequencies can be measured by merely recording the appropriate signals during normal operation of the turbine-generator unit without any special switching (Anderson and Farmer, 1996).

Modal Damping Tests

The most successful methods for measuring damping is to excite the spring-mass system by some means and then measure the natural decay rate following removal of the stimulus. Two methods have been used to excite the torsional modes. These methods are referred to as the "Impact Method" and the "Steady-State Method."

The Impact Method requires the application of an electrical torque transient to the turbine-generator being tested. The transient must be large enough to allow the decay rate of each modal response to be measured during ring down. Rotor motion is generally the preferred signal, but shaft stress has also been used successfully. Since the transient excites all modes, a series of narrow-band and band-reject filters are applied to the signal to separate the response into the modal components of interest. Switching series capacitor banks or line switching can create the required transient. Synchronizing the generator to the power system may also provide adequate stimulus. Such tests are described in Walker and Schwalb (1976) and Walker et al. (Sept./Dec. 1975).

The Steady-State Method uses a sinusoidal input signal to the voltage regulator of the excitation system, which produces a sinusoidal component of generator field voltage. For some types of excitation systems, this can create a sinusoidal response of the generator rotor. The frequency of the signal is varied to obtain a pure modal response of rotor motion or shaft strain. When a steady-state condition with pure mode stimulus has been obtained, the stimulus is removed and the decaying modal oscillation is recorded and plotted. The decay rate is a measure of the modal damping. This process is repeated for each torsional mode of interest. The Steady-State Method is the preferred method since pure modes can be exited without system switching. This method is only applicable to generators whose excitation system has sufficient gain and speed of response to produce a significant torque from the voltage regulator input signal. Such tests are reported in detail in Walker and Schwalb (1976) and Agrawal and Farmer (Nov. 1988).

The damping measured from either of the two test methods is the net damping of the coupled mechanical and electrical systems. Depending on the system configuration during the damping tests, the measured damping may include positive or negative damping due to interaction of the mechanical and electrical systems. This effect can be calculated from eigenvalue studies or from frequency scanning studies in conjunction with the interaction Eq. (11.41). To obtain the true mechanical damping, the measured damping must be corrected to account for the interaction in accordance with the following equation.

$$\sigma_n = \sigma_{meas} \pm \Delta\sigma_n \qquad (11.43)$$

where σ_n = Mechanical modal damping for Mode n
σ_{meas} = Measured damping from tests
$\Delta\sigma_n$ = Positive or negative damping due to interaction

It is usually important to have measured torsional damping of all active subsynchronous modes as a function of load, ranging from no load to full load. It is often more difficult to obtain full load damping because modal response decreases as damping increases and damping generally increases with load. It may be impossible to obtain adequate torsional excitation at full load. See Fig. 11.77 for results of such tests reported by Walker and Schwalb (1976).

Countermeasure Tests

Testing the effectiveness of any countermeasure to be applied is important, but may not be feasible. For example, if a countermeasure is to limit loss of shaft life for the most severe transient, it is not reasonable

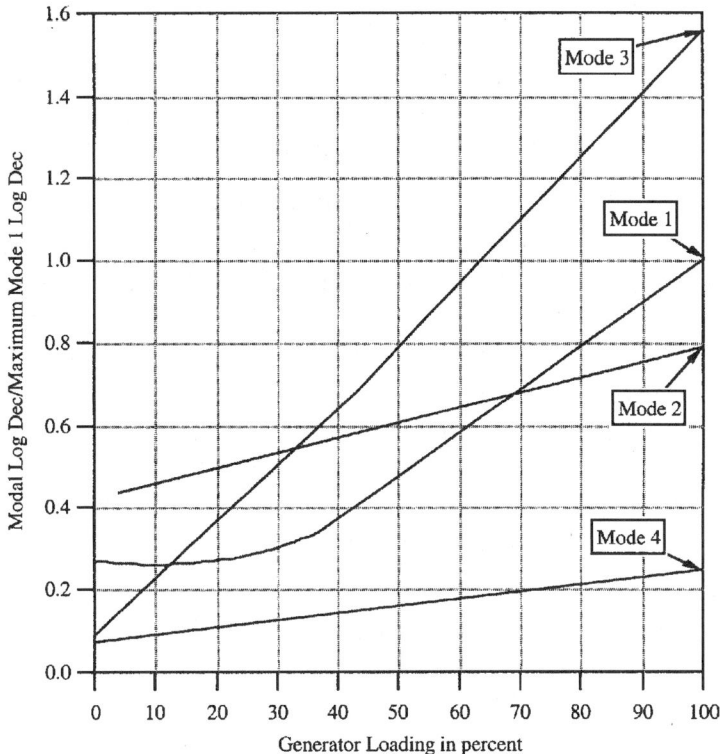

FIGURE 11.77 Variations of modal damping as a function of generator load for Navajo generators. (*Source:* Anderson, P. M. and Farmer, R. G., Subsynchronous Resonance: *Series Compensation of Power Systems*, PBLSH!, San Diego, 1996. With permission.)

to conduct such a test. Tests for effectiveness of torsional interaction countermeasures are practical and should be made whenever possible. One method is to conduct damping tests, as described in the section on Modal Damping Tests, with the countermeasure in service and the system configured to yield significant negative damping due to torsional interaction. Such tests are described by Bowler et al. (July/Aug. 1978) and Tang and Young (1981).

If SSR relays are to be applied, it may be possible to initiate a unit trip by SSR relay action under controlled conditions to verify proper operation. This has been accomplished, at least at one plant, by reducing the relay settings to a very sensitive level, and then causing rotor oscillations by the Steady-State Method described above. The stimulus can be increased to a level of sustained modal oscillations that will cause the relay to pick up. For the reduced setting, the oscillations are kept below the endurance limit. Such a test provides confidence in both the capabilities of the relay to initiate a unit trip and the correct wiring of the circuits from the relay output to the circuit breaker trip coils.

Summary

Consideration must be given to the potential for SSR whenever series capacitors are to be applied. Even so, the ability to analyze and control SSR for the extreme problems encountered has been clearly demonstrated over the last 15 years. Various countermeasures for SSR control have been developed and successfully applied. In many cases, the sole SSR protection can be provided by relays. Monitoring has a place in the SSR field to provide a permanent history of the torques experienced by the shafts and the cumulative shaft life expenditure. Such information can be used to schedule shaft inspection and maintenance, as required, to maintain shaft integrity. Continuous monitoring of SSR countermeasure performance by modern digital equipment can also be cost-effective.

If potential SSR problems are identified when series capacitor applications are considered, there is a clear course established by the utility industry. Analytical methods are available for either cursory or detailed analysis. Countermeasure selection guidelines used by others are available. Testing methods have been developed that vary from simple monitoring to sophisticated signal processing and system switching. SSR can be controlled, thus making it possible to benefit from the distinct advantages of series capacitors.

Device Dependent Subsynchronous Oscillations

Device Dependent Subsynchronous Oscillations is a category of interaction between turbine-generator torsional systems and power system components. Such interaction with turbine-generators has been observed with DC converter controls, variable speed motor controllers, and power system stabilizers. There is potential for such interaction for any wide bandwidth power device located near a turbine-generator.

HVDC Converter Controls

In 1977 tests were conducted to determine the interaction of the Square Butte HVDC converter in North Dakota with the Milton Young #2 turbine-generator. It was found that high-gain power modulation control destabilized the first torsional mode of the turbine-generator at 11.5 Hz. It was also found that the basic HVDC controls created growing torsional oscillations of the turbine-generator in the first torsional mode for a specific system configuration. Investigation of this phenomenon in the power industry shows that any HVDC converter has the potential for creating subsynchronous torsional oscillations in turbine-generators that are connected to the HVDC converter. The potential reduces as the impedance between the two increases. The HVDC system appears as a load to the turbine-generator. The load would be positively damped for crude firing angle control. Successful converter operation requires sophisticated firing angle control. This sophisticated control may make the converter appear as a negatively damped load in the range of 2 to 20 Hz. The potential problem of HVDC converter control interaction with turbine-generators can be investigated by eigenvalue analysis. If negative damping is expected, the problem may be solved by retuning converter controls. Also, a subsynchronous damping controller has been conceptually designed as reported in Bahrman et al. (1980). Mortensen et al. (Jan. 1981) describe a field test and analysis of interaction between a turbine-generator and a HVDC system.

Variable Speed Motor Controllers

In 1979 and 1980, a European fossil fired power plant experienced subsynchronous oscillations of a 775 MW, 3000 RPM turbine-generator. The plant was equipped with variable speed drives for the boiler feedwater pumps. The pump drives are equipped with six pulse subsynchronous converter cascades. For such a converter, the load power to the motors has a component at six times the motor slip frequency. At specific load levels, the feedwater pump speed is such that the load has a component whose frequency corresponds to the 50 Hz complement of one of the natural torsional frequencies of the turbine generator. Under these conditions, the pump load acts as a continuous torsional stimulus of the turbine-generator. Fatigue life expenditure could occur under such conditions, depending on the magnitude of the torsional oscillations. A torsional stress monitor detected the event discussed above. Modeling and analysis in EMTP or similar programs could probably predict such an event, but there is no record of such an analysis. The countermeasure applied to the above problem controls feedwater pump speed to avoid speeds that would excite the natural torsional modes of the turbine-generator (Lambrecht and Kolig, Oct. 1982).

Power System Stabilizers

In 1969, a 500 MW unit was commissioned at the Lambton Generating Station. A power system stabilizer (PSS) was added some time later to provide positive damping for the local mode of about 1.67 Hz. The PSS derived its input signal from rotor motion at a point adjacent to the generator mass. When the PSS was initially tested, sustained 16.0 Hz torsional oscillations of the generator were observed. 16.0 Hz corresponds to the first torsional mode of the turbine-generator mechanical system (Watson and Coultes, Jan./Feb. 1973). From analysis and simulation, it was determined that if the torsional oscillations were allowed to continue, severe turbine-generator shaft damage would occur. It was also learned that any small generator rotor motion at the first torsional mode (16.0 Hz) creates a 16.0 Hz signal input to the PSS. The gain and phase of the PSS and the excitation system creates an oscillating torque on the generator at 16.0 Hz, which reinforces the initiating 16.0 Hz oscillation. This type of problem can be analyzed using either eigenvalue or EMTP-type computer programs, which have provisions for modeling the turbine-generator mechanical system. The essence of the problem can be analyzed manually or using any software that will provide the data for Bode plots.

There are various countermeasures that can be applied to deal with the PSS problem. The countermeasures used at Lambton consisted of moving the rotor motion sensing location to a point of the spring-mass system, which has no torsional motion, or provides positive damping at the active torsional modes. In addition, a 16.35 Hz notch filter was included in the PSS to drastically reduce the gain for the first torsional mode. Others use a high order low pass filter or a wide-band band-reject filter in the PSS loop to insure that torsional oscillations are not generated by the PSS.

Other

In general, any device that controls or responds rapidly to power or speed variations in the subsynchronous frequency range is a potential source for excitation of subsynchronous oscillations. The technical literature includes the effect of governor characteristics on turbine-generator shaft torsionals (Lee et al., June, 1985) and subsynchronous torsional interactions with static var compensators (Piwko et al., Nov. 1990).

Supersynchronous Resonance

The term Supersynchronous Resonance (SPSR) is used here to refer to a torsional resonant condition of a turbine-generator mechanical system at a frequency greater than the frequency corresponding to rated turbine speed and power system rated frequency. Such a resonant condition can be excited from the power system. There have been at least three incidents of turbine blade failure that contributed to the excitation of turbine-generator torsional modes that are very near to 120 Hz.

Known SPSR Events

In 1985, a turbine-generator outside the U.S. experienced the failure of eight blades in the last stage of an 1800-rpm low-pressure turbine with 43-in. last-stage blades. The blades failed at the root attachments to the rotor disk due to high cycle fatigue. A one-year outage was required to repair the unit. In 1993, a turbine-generator in the U.S. experienced the failure of two blades in the next to last row of an 1800-rpm low-pressure turbine with 38-in. last-stage blades. The blades failed at the dovetails on the rotor disk. A 49-day outage was required to repair the unit. The turbine-generator units for both incidents were from the same manufacturer and both have relatively long turbine blades on 1800-rpm low-pressure turbines. A similar event occurred in the 1970s to an 1800-rpm turbine-generator from a different manufacturer (Raczkowski and Kung, 1979).

SPSR Physical Principles

Long turbine blades, such as the 38-in. and 43-in. blades on 1800 rpm low-pressure turbines, often have a natural vibration frequency near 120 Hz when coupled to the rotor disc. A blade-disc with a natural frequency near 120 Hz may be excited by torsional oscillations near 120 Hz (Kung, and LaRosa, 1978). Although turbines are designed to avoid 120 Hz natural torsional frequencies (torsional modes) with at least 0.5 Hz margin, calculations are not exact. The following scenario can contribute to turbine blade failure due to high cycle fatigue: 60 Hz negative sequence current flows in the generator armature due to either unbalanced loads, untransposed lines, or unbalanced faults. The resulting magnetic flux interacting with the field flux results in a 120 Hz electromagnetic torque applied to the turbine-generator mechanical system. This will excite 120 Hz torsional oscillations if there is a 120 Hz torsional mode. 120 Hz torsional oscillations, at points along the shaft where long turbine blades are attached, can excite blade vibration if the blade-disc natural frequency is approximately 120 Hz. Continuous blade vibration, or numerous events, will initiate cracks at the stress concentration points and finally blades will fail.

For the above scenario, there is a torsional mode within 0.5 Hz of 120 Hz. Turbine-generator designers have made efforts to avoid torsional frequencies near 120 Hz but have not always had the technology to accurately calculate the frequencies for the higher torsional modes near 120 Hz. The 1993 blade failure has been attributed to an undetected torsional mode within 0.5 Hz of 120 Hz. For the 1985 blade failure, there were no natural modes within 0.5 Hz of 120 Hz, but the turbine-generator was operating in a relatively small power system whose frequency varied significantly. These frequency variations, in conjunction with negative sequence generator current, excited the torsional modes that were 1 to 2 Hz away from 120 Hz.

Tests and experience have shown that generators experience continuous negative sequence current ranging from 1 to 3%. Of course, much higher negative sequence currents occur during unbalanced fault conditions. Therefore, if the blade-disc natural frequencies are near 120 Hz, it is essential that there are no natural torsional frequencies between 119.5 and 120.5 Hz, under any conditions. The turbine-generator manufacturer calculates the blade-disc natural frequencies and the torsional natural frequencies. Unfortunately, the calculated frequencies may not be sufficiently accurate to determine if blade failure is to be expected. The turbine-generator natural frequencies can be accurately determined from tests. An off-line test has been devised that will accurately show the natural torsional frequencies at no load. This is called a ramp test and consists of monitoring torsional strain at critical points while negative-sequence current flows in the generator armature circuit and the turbine-generator speed is accelerated. The negative sequence current is induced by shorting two generator terminals and controlling field voltage with a separate power supply. The ramp test and other tests are described in Raczkowski and Kung (1978) and Evans et al. (1985). Using accurate test data, an analytic model can be developed by an iterative process. The resulting analytic model can be used to find appropriate countermeasures.

SPSR Countermeasures

The countermeasures that have been applied to avoid turbine blade failure caused by SPSR, involve either moving natural torsional frequencies away from 120 Hz or changing the shapes of modes near 120 Hz, so that they are not excited by electrical torque applied to the generator rotor. This has been successfully accomplished by several methods. One is to braze the tie wires on all last-stage blades. This modification may increase torsional frequencies and it alters the participation of individual blades in the oscillation. A second countermeasure involves adding a mass ring at an appropriate location along the torsional spring-mass system. This modification may reduce the critical torsional frequencies and it will change the mode shapes. Other methods include machining critical sections along the shaft system and changing the generator pole face slotting to move frequencies and modify mode shapes. Tests to determine the natural torsional frequencies following modifications should be made to verify the analytic model (Raczkowski and Kung [1978] and Evans et al. [1985]). A relay has been proposed to alarm or trip the turbine-generator for combinations of negative sequence current and off-nominal frequency operation deemed to be excessive.

Device-Dependent Supersynchronous Oscillations

There has been a series of events that resulted in turbine-generator damage due to supersynchronous resonance stimulated by a power system device. This type of interaction is referred to as Device Dependent Supersynchronous Oscillations (DDSPSO).

Known DDSPSO Events

The Comanche Unit 2 near Pueblo, Colorado went into service in 1975 and during the period of 1987 to 1994, the unit suffered generator damage. In 1987, there was a crack in the generator shaft. In 1993, there were two failures of the rotating exciter. In 1994, there was a retaining ring failure resulting in serious rotor and stator damage (Andorka and Yohn, 1996). All have been attributed to the same phenomena.

DDSPSO Physical Principles

Comanche Unit 2 is about 3 miles from 2 – 60 MVA steel mill arc furnaces. The arc furnaces have a static var compensator (SVC) for flicker control. It has been found that the SVC had a control loop instability that caused negative sequence current to flow in the armature of Comanche 2 at a frequency near 55 Hz. The instability resulted in a 5 Hz amplitude modulation of the 60 Hz SVC current. This modulation created upper and lower sidebands of 55 Hz and 65 Hz in all three phases, but in reverse rotation. The 65 Hz component did not appear outside the SVC delta winding but a 55 Hz negative sequence component flowed in the generator armature. The frequency of this component varied between 54 Hz and 58 Hz, depending on the steel mill operating conditions. This produced a component of electromagnetic torque in the frequency range of 114 to 118 Hz. The natural torsional frequency for Mode 6 of Comanche 2 was about 118 Hz prior to the retaining ring failure. The mode shape for Mode 6 shows large displacement at the two ends of the generator. Therefore, stimulus from the SVC created torsional oscillations that were sufficiently large and sustained to result in high cycle fatigue of the generator shaft, rotating exciter, and retaining ring before the root cause of the problem was found.

DDSPSO Countermeasure

Extensive testing was performed to determine natural modal frequencies for the turbine-generator, the components of armature current, and the arc furnace and SVC stimulus. Once the root cause of the problem was determined, it was a simple matter to retune the control circuit of the SVC (Andorka and Yohn, 1996).

References

Agrawal, B. L., Demcko, J. A., Farmer, R. G., and Selin, D. A., Apparent impedance measuring system (AIMS), in *IEEE Trans.*, PWRS-4, 2, 575–82, May 1989.

Agrawal, B. L., Demcko, J. A., Farmer, R. G., and Selin, D. A., Shaft torque monitoring using conventional digital fault recorders, in *IEEE Trans. on Power Syst.*, 7, 3, 1211–17, Aug. 1992.

Agrawal, B. L. and Farmer, R. G., Effective damping for SSR analysis of parallel turbine generators, in *IEEE Trans.*, PWRS-3, 4, 1441–48, Nov. 1988.

Agrawal, B. L., and Farmer, R. G., Use of frequency scanning technique for subsynchronous resonance analysis, in *IEEE Trans.*, PAS-98, 341–349, Mar./Apr. 1979.

Ahlgren, L., Walve, K., Fahlen, N., and Karlsson, S., Countermeasures against oscillatory torque stresses in large turbogenerators, in CIGRE, Paris, 1982.

Anderson, P. M., Agrawal, B. L., and Van Ness, J. E., *Subsynchronous Resonance in Power Systems*, IEEE Press, New York, 1990.

Anderson, G., Atmuri, R., Rosenqvist, R., and Torseng, S., Influence of hydro units generator-to-turbine ratio on damping of subsynchronous oscillations, in *IEEE Trans.*, PAS-103, 2352–61, 4, Aug. 1984.

Anderson, P. M. and Farmer, R. G., Subsynchronous Resonance, in *Series Compensation of Power Systems*, PBLSH!, San Diego, 1996, chap. 6.

Andorka, M. and Yohn, T., Vibration induced retaining ring failure due to steel mill — power plant electromechanical interaction, in *IEEE 1996 Summer Power Meeting Panel Session on Steel-Making, Inter-Harmonics and Generator Torsional Impacts*, July 1996, Denver, Colorado.

Bahrman, M. P., Larsen E. V., Piwko, R. J. and Patel, H. S., Experience with HVDC — turbine-generator interaction at SquareButte, in *IEEE Trans. on PA&S*, PAS-99, 966–975, 1980.

Bowler, C. E. J., and Baker, D. H., Concepts of Countermeasures for subsynchronous supplementary torsional damping by excitation modulation, in IEEE Special Publication, *Symnposium on Countermeasures for Subsynchronous Resonance*, IEEE Pub. 81TH0086-9-PWR, 1981, 64–69.

Bowler, C. E. J., Baker, D. H., Mincer, N. A. and Vandiveer, P. R., Operation and test of the Navajo SSR protective equipment, in *IEEE Trans.*, PAS-95, 1030–35, July/Aug. 1978.

Bowler, C. E. J., Demcko, J. A., Menkoft, L., Kotheimer, W. C., and Cordray, D., The Navajo SMF type subsynchronous resonace relay, in *IEEE Trans.*, PAS-97, 5, 1489–95, Sep./Oct. 1978.

Bowler, C. E. J. and Lawson, R. A., Operating experience with supplemental excitation damping controls, in ibid., 27–33.

Evans, D. G., Giesecke, H. D., Willman, E. C. and Moffitt, S. P., Resolution of torsional vibration issues for large turbine geneators, in *Proceedings of the American Power Conference*, 57, 1985.

Farmer, R. G. and Agrawal, B. L., Application of subsynchronous oscillation relay — Type SSO, in *IEEE Trans.*, PAS-100, 5, 2442–51, May 1981.

Farmer, R. G. and Agrawal, B. L., Guidelines for the selection of subsynchronous resonance countermeasures, in IEEE Special Publication, *Symposium on Countermeasures for subsynchronous resonance*, IEEE Pub. 81TH0086-9-PWR, 1981, p. 81–85.

Farmer, R. G., Schwalb A. L., and Katz, E., Navajo Project report on subsynchronous resonance analysis and solution, in *IEEE Trans.*, PAS-96, 4, 1226–1232, July/Aug. 1977.

Gross, G., and Hall, M. C., Synchronous machine and torsional dynamics simulation in the computation of electro-magnetic transients, in *IEEE Trans.*, PAS-97, 1074–86, Jul./Aug. 1978.

Hall, M. C. and Hodges D. A., Experience with 500-kV subsynchronous resonance and resulting turbine generator shaft damage at Mohave Generating Station, in IEEE PES Special Publication, *Analysis and Control of Subsynchronous Resonance*, IEEE Publication 76 CH 1066-0-PWR, 1976, 22–29.

Hauer, J. F., Eden, J. D., Donnelly, M. K., Trudnowski, D. J. and Piwko, R. J., Test results and initial operating experience for the BPA 500 kV thyristor controlled series capacitor unit at Slatt Substation, Part II — modulatiion, SSR, and performance monitoring, in BPA Internal Report, 1994.

IEEE Committee Report, Comparison of SSR calculations and test results, in *IEEE Trans.*, PWRS-4, 336–44, 1, Feb. 1989.

IEEE Committee Report, Countermeasures to subsynchronous resonance problems, in *IEEE Trans.*, PAS-99, 1810–18, Sept./Oct. 1980.

IEEE Committee Report, in IEEE Special Publication, *Current Activity in Flexible AC Transmission Systems*, IEEE Publication 92TH0465-5 PWR, April 1992.

IEEE Committee Report, Reader's guide to subsynchronous resonance, in *IEEE Trans.*, PWRS-7, 1, 150–157, Feb. 1992.

IEEE Committee Report, Series capacitor controls and settings as a countermeasure to subsynchronous resonance, in *IEEE Trans.*, PAS-101, 1281–87, June 1982.

IEEE Committee Report, Terms, definitions and symbols for Subsynchronous Resonance, in *IEEE Trans.*, PAS-104, 1326–34, June 1985.

Joyce, J. S., Kulig, T., and Lambrecht, D., Torsional fatigue of turbine-generator shafts caused by different electrical system faults and switching operations, in *IEEE Trans.*, PAS 97, 1965–77, Sept./Oct. 1978.

Joyce, J. S., and Lambrecht, D., Monitoring the fatigue effects of electrical disturbances on steam turbine-generators, in *Proceedings of American Power Conference*, 41, 1979, 1153–62.

Kilgore, L. A., Elliott, L. C. and Taylor, E. T., The prediction and control of self-excited oscillations due to series capacitors in power systems, in *IEEE Trans.*, PAS-96, 1840–46, Nov./Dec. 1977.

Kilgore, L. A., Ramey, D. G., and Hall, M. C., Simplified transmission and generation system analysis procedures for subsynchronous resonance. In *IEEE Trans.*, PAS-96, 1840–1846, Nov./Dec. 1977.

Kimmel, D. S., Carter, M. P., Bednarek, J. N. and Jones, W. H., Dynamic stabilizer on-line experience, in *IEEE Trans.*, PAS-103, 1, 198–212, Jan. 1984.

Kinney, S. J., Mittelstadt W. A. and Suhrbier, R. W., Test results and initial operating experience for the BPA 500 kV thyristor controlled series capacitor unit at Slatt Substation, Part I — design, operation and fault test results, in BPA Internal Report, 1993.

Kung, G. C. and LaRosa, J. A., Response of turbine-generators to electrical disturbances, Presented at the Steam Turbine–Generator Technology Symposium, Charlotte, North Carolina, Oct. 4-5, 1978.

Lambrecht, D. and Kulig, T., Torsional performance of turbine generator shafts especially under resonant excitation, in *IEEE Trans.*, PAS-101, 10, Oct. 1982.

Lee, D. C., Beaulieu, R. E. and Rogers, G. J., Effect of governor characteristics on turbo-generator shaft torsionals, in *IEEE Trans. on PA & S*, PAS-104, 1255–1259, June 1985.

Mortensen, K., Larsen, E. V. and Piwko, R. J., Field test and analysis of torsional interaction between the Coal Creek turbine-generator and the CU HVDC system, in *IEEE Trans.*, PAS-100, 336–344, Jan. 1981.

Perez, A. J., Mohave Project subsynchronous resonance unit tripping scheme, in IEEE Special Publication, *Symposium on Countermeasures for subsynchronous resonance*, IEEE Pub. 81TH0086-9-PWR, 1981, p. 20–22.

Piwko, R. J., Rostamkolai, N., Larsen, E. V., Fisher, D. A., Mobarak, M. A. and Poitras, A. E., Subsynchronous torsional interactions with static var compensators — concepts and practical implications, in *IEEE Trans. on Power Syst.*, 1324–1332, November 1990.

Raczkowski, C. and Kung, G. C., Turbine-generator torsional frequencies — field reliability and testing, in *Proceedings of the American Power Conference*, 1979.

Ramey, D. G., Demcko, J. A., Farmer, R. G., and Agrawal, B. L., Subsynchronous resonance tests and torsional monitoring system verification at the Cholla station, in *IEEE Trans.*, PAS-99, 5, 1900–7, Sep. 1980.

Ramey, D. G., Kimmel, D. S., Dorney, J. W. and Kroening, F. H., Dynamic stabilizer verification tests at the San Juan Station, in *IEEE Trans.*, PAS-100, 5011–19, Dec. 1981.

Ramey, D. G., White, I. A., Dorney, J. H. and Kroening, F. H., Application of dynamic staabilizer to solve an SSR problem, in *Proceedings of American Power Conference*, 43, 1981, 605–9.

Stein, J. and Fick H., The torsional stress analyzer for continuously monitoring turbine generators, in *IEEE Trans.*, PAS-99, 2, 703–10, Mar./Apr. 1980.

Sun, S. C., Salowe, S., Taylor, E. R., and Mummert, C. R., A subsynchronous oscillation relay — Type SSO, in *IEEE Trans.*, PAS-100, 3580–89, July 1981.

Tang, J. F. and Young, J. A., Operating experience of Navajo static blocking filter, in IEEE Special Publication, *Symposium on Countermeasures for Subsynchronous Resonance*, IEEE Pub. 81TH0086-9-PWR, 1981, 23–26.

Walker, D. N., Bowler, C. E. J., Jackson, R. L., and Hodges, D. A., Results of the subsynchronous resonance tests at Mohave, in *IEEE Trans.*, PAS-94, 5, 1878–89, Sep./Dec. 1975.

Watson, W. and Coultes, M. E., Static exciter stabilizing signals on large generators — mechanical problems, in *IEEE Trans. on PA & S*, PAS-92, 204–211, Jan./Feb. 1973.

Walker, D. N., Plaacek, R. J., Bowler, C. E. J., White, J. C., and Edmonds, J. S., Turbine benerator shaft torsional fatigue and monitoring, in *CIGRE*, paper 11-07, 1984.

Walker, D. N. and Schwalb, A. L., Results of subsynchronous resonance test at Navajo, in IEEE PES Special Publication, *Analysis and Control of Subsynchronous Resonance*, IEEE Publication 76 CH 1066-0-PWR, 1976, 37–45.

12
Power System Operation and Control

Bruce F. Wollenberg
University of Minnesota

12.1 Energy Management *K. Neil Stanton, Jay C. Giri, and Anjan Bose* 12-1

12.2 Generation Control: Economic Dispatch and Unit Commitment
Charles W. Richter, Jr. 12-10

12.3 State Estimation *Danny Julian* 12-27

12.4 Optimal Power Flow *M. E. El-Hawary* 12-38

12.5 Security Analysis *Nouredine Hadjsaid* 12-53

12
Power System Operation and Control

K. Neil Stanton
Stanton Associates

Jay C. Giri
ALSTOM ESCA Corporation

Anjan Bose
Washington State University

Charles W. Richter, Jr.
ALSTOM ESCA Corporation

Danny Julian
ABB Power T&D Company

M. E. El-Hawary
Dalhousie University

Nouredine Hadjsaid
Institut National Polytechnique de Grenoble (INPG)

12.1 Energy Management 12-1
Power System Data Acquisition and Control • Automatic Generation Control • Load Management • Energy Management • Security Control • Operator Training Simulator

12.2 Generation Control: Economic Dispatch and Unit Commitment 12-10
Economic Dispatch • The Unit Commitment Problem • Summary of Economical Generation Operation

12.3 State Estimation 12-27
State Estimation Problem • State Estimation Operation • Example State Estimation Problem

12.4 Optimal Power Flow 12-38
Conventional Optimal Economic Scheduling • Conventional OPF Formulation • OPF Incorporating Load Models • SCOPF Including Load Modeling • Operational Requirements for Online Implementation • Conclusions

12.5 Security Analysis...12-53
Definition • Time Frames for Security-Related Decisions • Models • Determinist vs. Probabilistic • Security under Deregulation

12.1 Energy Management

K. Neil Stanton, Jay C. Giri, and Anjan Bose

Energy management is the process of monitoring, coordinating, and controlling the generation, transmission, and distribution of electrical energy. The physical plant to be managed includes generating plants that produce energy fed through transformers to the high-voltage transmission network (grid), interconnecting generating plants, and load centers. Transmission lines terminate at substations that perform switching, voltage transformation, measurement, and control. Substations at load centers transform to subtransmission and distribution levels. These lower-voltage circuits typically operate radially, i.e., no normally closed paths between substations through subtransmission or distribution circuits. (Underground cable networks in large cities are an exception.)

Since transmission systems provide negligible energy storage, supply and demand must be balanced by either generation or load. Production is controlled by turbine governors at generating plants, and automatic generation control is performed by control center computers remote from generating plants. Load management, sometimes called demand-side management, extends remote supervision and control to subtransmission and distribution circuits, including control of residential, commercial, and industrial loads.

FIGURE 12.1 Manitoba Hydro Control Center in Winnipeg, Manitoba, Canada. (Photo used with permission of ALSTOM ESCA Corporation.)

Events such as lightning strikes, short circuits, equipment failure, or accidents may cause a system fault. Protective relays actuate rapid, local control through operation of circuit breakers before operators can respond. The goal is to maximize safety, minimize damage, and continue to supply load with the least inconvenience to customers. Data acquisition provides operators and computer control systems with status and measurement information needed to supervise overall operations. Security control analyzes the consequences of faults to establish operating conditions that are both robust and economical.

Energy management is performed at control centers (see Fig. 12.1), typically called system control centers, by computer systems called *energy management systems* (EMS). Data acquisition and remote control is performed by computer systems called *supervisory control and data acquisition* (SCADA) systems. These latter systems may be installed at a variety of sites including system control centers. An EMS typically includes a SCADA "front-end" through which it communicates with generating plants, substations, and other remote devices.

Figure 12.2 illustrates the applications layer of modern EMS as well as the underlying layers on which it is built: the operating system, a database manager, and a utilities/services layer.

Power System Data Acquisition and Control

A SCADA system consists of a master station that communicates with remote terminal units (RTUs) for the purpose of allowing operators to observe and control physical plants. Generating plants and transmission substations certainly justify RTUs, and their installation is becoming more common in distribution substations as costs decrease. RTUs transmit device status and measurements to, and receive control commands and setpoint data from, the master station. Communication is generally via dedicated circuits operating in the range of 600 to 4800 bits/s with the RTU responding to periodic requests initiated from the master station (polling) every 2 to 10 s, depending on the criticality of the data.

The traditional functions of SCADA systems are summarized:

- Data acquisition: Provides telemetered measurements and status information to operator.
- Supervisory control: Allows operator to remotely control devices, e.g., open and close circuit breakers. A "select before operate" procedure is used for greater safety.
- Tagging: Identifies a device as subject to specific operating restrictions and prevents unauthorized operation.

Power System Operation and Control

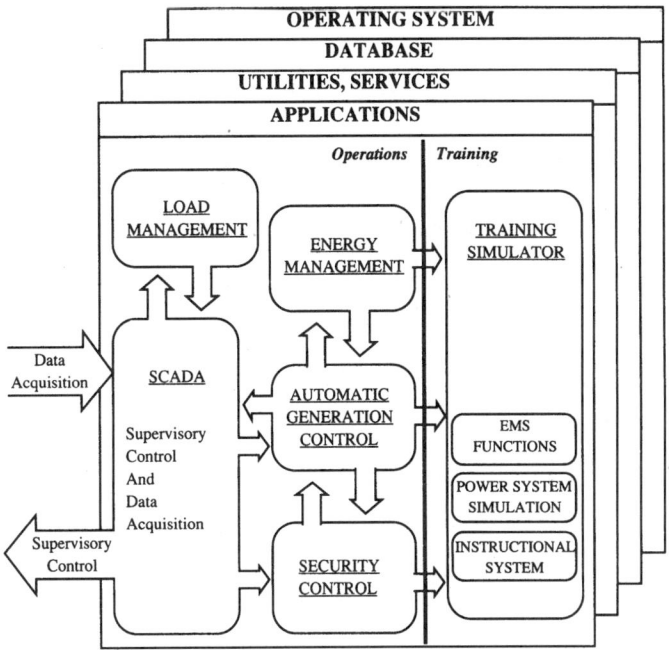

FIGURE 12.2 Layers of a modern EMS.

- Alarms: Inform operator of unplanned events and undesirable operating conditions. Alarms are sorted by criticality, area of responsibility, and chronology. Acknowledgment may be required.
- Logging: Logs all operator entry, all alarms, and selected information.
- Load shed: Provides both automatic and operator-initiated tripping of load in response to system emergencies.
- Trending: Plots measurements on selected time scales.

Since the master station is critical to power system operations, its functions are generally distributed among several computer systems depending on specific design. A dual computer system configured in primary and standby modes is most common. SCADA functions are listed below without stating which computer has specific responsibility.

- Manage communication circuit configuration
- Downline load RTU files
- Maintain scan tables and perform polling
- Check and correct message errors
- Convert to engineering units
- Detect status and measurement changes
- Monitor abnormal and out-of-limit conditions
- Log and time-tag sequence of events
- Detect and annunciate alarms
- Respond to operator requests to:
  - Display information
  - Enter data
  - Execute control action
  - Acknowledge alarms

- Transmit control action to RTUs
- Inhibit unauthorized actions
- Maintain historical files
- Log events and prepare reports
- Perform load shedding

Automatic Generation Control

Automatic generation control (AGC) consists of two major and several minor functions that operate online in realtime to adjust the generation against load at minimum cost. The major functions are load frequency control and economic dispatch, each of which is described below. The minor functions are reserve monitoring, which assures enough reserve on the system; interchange scheduling, which initiates and completes scheduled interchanges; and other similar monitoring and recording functions.

Load Frequency Control

Load frequency control (LFC) has to achieve three primary objectives, which are stated below in priority order:

1. To maintain frequency at the scheduled value
2. To maintain net power interchanges with neighboring control areas at the scheduled values
3. To maintain power allocation among units at economically desired values

The first and second objectives are met by monitoring an error signal, called *area control error* (ACE), which is a combination of net interchange error and frequency error and represents the power imbalance between generation and load at any instant. This ACE must be filtered or smoothed such that excessive and random changes in ACE are not translated into control action. Since these excessive changes are different for different systems, the filter parameters have to be tuned specifically for each control area. The filtered ACE is then used to obtain the proportional plus integral control signal. This control signal is modified by limiters, deadbands, and gain constants that are tuned to the particular system. This control signal is then divided among the generating units under control by using participation factors to obtain *unit control errors* (UCE).

These participation factors may be proportional to the inverse of the second derivative of the cost of unit generation so that the units would be loaded according to their costs, thus meeting the third objective. However, cost may not be the only consideration because the different units may have different response rates and it may be necessary to move the faster generators more to obtain an acceptable response. The UCEs are then sent to the various units under control and the generating units monitored to see that the corrections take place. This control action is repeated every 2 to 6 s.

In spite of the integral control, errors in frequency and net interchange do tend to accumulate over time. These time errors and accumulated interchange errors have to be corrected by adjusting the controller settings according to procedures agreed upon by the whole interconnection. These accumulated errors as well as ACE serve as performance measures for LFC.

The main philosophy in the design of LFC is that each system should follow its own load very closely during normal operation, while during emergencies, each system should contribute according to its relative size in the interconnection without regard to the locality of the emergency. Thus, the most important factor in obtaining good control of a system is its inherent capability of following its own load. This is guaranteed if the system has adequate regulation margin as well as adequate response capability. Systems that have mainly thermal generation often have difficulty in keeping up with the load because of the slow response of the units.

The design of the controller itself is an important factor, and proper tuning of the controller parameters is needed to obtain "good" control without "excessive" movement of units. Tuning is system-specific, and although system simulations are often used as aids, most of the parameter adjustments are made in the field using heuristic procedures.

Economic Dispatch

Since all the generating units that are online have different costs of generation, it is necessary to find the generation levels of each of these units that would meet the load at the minimum cost. This has to take into account the fact that the cost of generation in one generator is not proportional to its generation level but is a nonlinear function of it. In addition, since the system is geographically spread out, the transmission losses are dependent on the generation pattern and must be considered in obtaining the optimum pattern.

Certain other factors have to be considered when obtaining the optimum generation pattern. One is that the generation pattern provide adequate reserve margins. This is often done by constraining the generation level to a lower boundary than the generating capability. A more difficult set of constraints to consider are the transmission limits. Under certain real-time conditions it is possible that the most economic pattern may not be feasible because of unacceptable line flows or voltage conditions. The present-day economic dispatch (ED) algorithm cannot handle these security constraints. However, alternative methods based on optimal power flows have been suggested but have not yet been used for real-time dispatch.

The minimum cost dispatch occurs when the incremental cost of all the generators is equal. The cost functions of the generators are nonlinear and discontinuous. For the equal marginal cost algorithm to work, it is necessary for them to be convex. These incremental cost curves are often represented as monotonically increasing piecewise-linear functions. A binary search for the optimal marginal cost is conducted by summing all the generation at a certain marginal cost and comparing it with the total power demand. If the demand is higher, a higher marginal cost is needed, and vice versa. This algorithm produces the ideal setpoints for all the generators for that particular demand, and this calculation is done every few minutes as the demand changes.

The losses in the power system are a function of the generation pattern, and they are taken into account by multiplying the generator incremental costs by the appropriate penalty factors. The penalty factor for each generator is a reflection of the sensitivity of that generator to system losses, and these sensitivities can be obtained from the transmission loss factors.

This ED algorithm generally applies to only thermal generation units that have cost characteristics of the type discussed here. The hydro units have to be dispatched with different considerations. Although there is no cost for the water, the amount of water available is limited over a period, and the displacement of fossil fuel by this water determines its worth. Thus, if the water usage limitation over a period is known, say from a previously computed hydro optimization, the water worth can be used to dispatch the hydro units.

LFC and the ED functions both operate automatically in realtime but with vastly different time periods. Both adjust generation levels, but LFC does it every few seconds to follow the load variation, while ED does it every few minutes to assure minimal cost. Conflicting control action is avoided by coordinating the control errors. If the unit control errors from LFC and ED are in the same direction, there is no conflict. Otherwise, a logic is set to either follow load (permissive control) or follow economics (mandatory control).

Reserve Monitoring

Maintaining enough reserve capacity is required in case generation is lost. Explicit formulas are followed to determine the spinning (already synchronized) and ready (10 min) reserves required. The availability can be assured by the operator manually, or, as mentioned previously, the ED can also reduce the upper dispatchable limits of the generators to keep such generation available.

Interchange Transaction Scheduling

The contractual exchange of power between utilities has to be taken into account by the LFC and ED functions. This is done by calculating the net interchange (sum of all the buy and sale agreements) and adding this to the generation needed in both the LFC and ED. Since most interchanges begin and end on the hour, the net interchange is ramped from one level to the new over a 10- or 20-min period straddling the hour. The programs achieve this automatically from the list of scheduled transactions.

Load Management

SCADA, with its relatively expensive RTUs installed at distribution substations, can provide status and measurements for distribution feeders at the substation. Distribution automation equipment is now available to measure and control at locations dispersed along distribution circuits. This equipment can monitor sectionalizing devices (switches, interruptors, fuses), operate switches for circuit reconfiguration, control voltage, read customers' meters, implement time-dependent pricing (on-peak, off-peak rates), and switch customer equipment to manage load. This equipment requires significantly increased functionality at distribution control centers.

Distribution control center functionality varies widely from company to company, and the following list is evolving rapidly.

- Data acquisition: Acquires data and gives the operator control over specific devices in the field. Includes data processing, quality checking, and storage.
- Feeder switch control: Provides remote control of feeder switches.
- Tagging and alarms: Provides features similar to SCADA.
- Diagrams and maps: Retrieves and displays distribution maps and drawings. Supports device selection from these displays. Overlays telemetered and operator-entered data on displays.
- Preparation of switching orders: Provides templates and information to facilitate preparation of instructions necessary to disconnect, isolate, reconnect, and reenergize equipment.
- Switching instructions: Guides operator through execution of previously prepared switching orders.
- Trouble analysis: Correlates data sources to assess scope of trouble reports and possible dispatch of work crews.
- Fault location: Analyzes available information to determine scope and location of fault.
- Service restoration: Determines the combination of remote control actions that will maximize restoration of service. Assists operator to dispatch work crews.
- Circuit continuity analysis: Analyzes circuit topology and device status to show electrically connected circuit segments (either energized or deenergized).
- Power factor and voltage control: Combines substation and feeder data with predetermined operating parameters to control distribution circuit power factor and voltage levels.
- Electrical circuit analysis: Performs circuit analysis, single-phase or three-phase, balanced or unbalanced.
- Load management: Controls customer loads directly through appliance switching (e.g., water heaters) and indirectly through voltage control.
- Meter reading: Reads customers' meters for billing, peak demand studies, time of use tariffs. Provides remote connect/disconnect.

Energy Management

Generation control and ED minimize the current cost of energy production and transmission within the range of available controls. Energy management is a supervisory layer responsible for economically scheduling production and transmission on a global basis and over time intervals consistent with cost optimization. For example, water stored in reservoirs of hydro plants is a resource that may be more valuable in the future and should, therefore, not be used now even though the cost of hydro energy is currently lower than thermal generation. The global consideration arises from the ability to buy and sell energy through the interconnected power system; it may be more economical to buy than to produce from plants under direct control. Energy accounting processes transaction information and energy measurements recorded during actual operation as the basis of payment for energy sales and purchases.

Energy management includes the following functions:

- System load forecast: Forecasts system energy demand each hour for a specified forecast period of 1 to 7 days.
- Unit commitment: Determines start-up and shut-down times for most economical operation of thermal generating units for each hour of a specified period of 1 to 7 days.
- Fuel scheduling: Determines the most economical choice of fuel consistent with plant requirements, fuel purchase contracts, and stockpiled fuel.
- Hydro-thermal scheduling: Determines the optimum schedule of thermal and hydro energy production for each hour of a study period up to 7 days while ensuring that hydro and thermal constraints are not violated.
- Transaction evaluation: Determines the optimal incremental and production costs for exchange (purchase and sale) of additional blocks of energy with neighboring companies.
- Transmission loss minimization: Recommends controller actions to be taken in order to minimize overall power system network losses.
- Security constrained dispatch: Determines optimal outputs of generating units to minimize production cost while ensuring that a network security constraint is not violated.
- Production cost calculation: Calculates actual and economical production costs for each generating unit on an hourly basis.

Security Control

Power systems are designed to survive all probable contingencies. A contingency is defined as an event that causes one or more important components such as transmission lines, generators, and transformers to be unexpectedly removed from service. Survival means the system stabilizes and continues to operate at acceptable voltage and frequency levels without loss of load. Operations must deal with a vast number of possible conditions experienced by the system, many of which are not anticipated in planning. Instead of dealing with the impossible task of analyzing all possible system states, security control starts with a specific state: the current state if executing the real-time network sequence; a postulated state if executing a study sequence. Sequence means sequential execution of programs that perform the following steps:

1. Determine the state of the system based on either current or postulated conditions.
2. Process a list of contingencies to determine the consequences of each contingency on the system in its specified state.
3. Determine preventive or corrective action for those contingencies which represent unacceptable risk.

Real-time and study network analysis sequences are diagramed in Fig. 12.3.

Security control requires topological processing to build network models and uses large-scale AC network analysis to determine system conditions. The required applications are grouped as a network subsystem that typically includes the following functions:

- Topology processor: Processes real-time status measurements to determine an electrical connectivity (bus) model of the power system network.
- State estimator: Uses real-time status and analog measurements to determine the "best" estimate of the state of the power system. It uses a redundant set of measurements; calculates voltages, phase angles, and power flows for all components in the system; and reports overload conditions.
- Power flow: Determines the steady-state conditions of the power system network for a specified generation and load pattern. Calculates voltages, phase angles, and flows across the entire system.
- Contingency analysis: Assesses the impact of a set of contingencies on the state of the power system and identifies potentially harmful contingencies that cause operating limit violations.

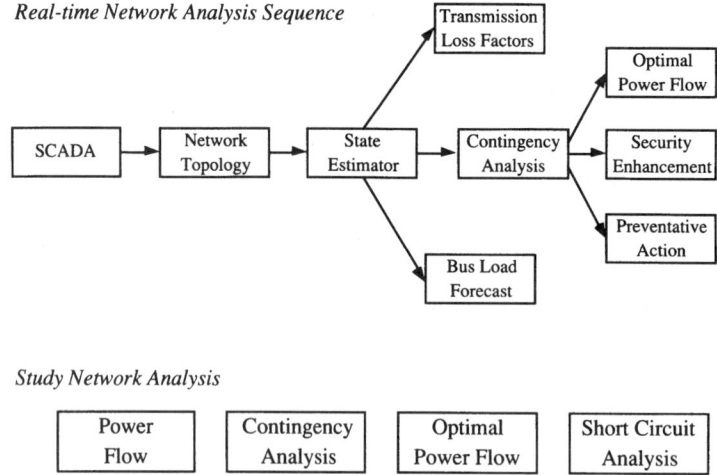

FIGURE 12.3 Real-time and study network analysis sequences.

- Optimal power flow: Recommends controller actions to optimize a specified objective function (such as system operating cost or losses) subject to a set of power system operating constraints.
- Security enhancement: Recommends corrective control actions to be taken to alleviate an existing or potential overload in the system while ensuring minimal operational cost.
- Preventive action: Recommends control actions to be taken in a "preventive" mode before a contingency occurs to preclude an overload situation if the contingency were to occur.
- Bus load forecasting: Uses real-time measurements to adaptively forecast loads for the electrical connectivity (bus) model of the power system network.
- Transmission loss factors: Determines incremental loss sensitivities for generating units; calculates the impact on losses if the output of a unit were to be increased by 1 MW.
- Short-circuit analysis: Determines fault currents for single-phase and three-phase faults for fault locations across the entire power system network.

Operator Training Simulator

Training simulators were originally created as generic systems for introducing operators to the electrical and dynamic behavior of power systems. Today, they model actual power systems with reasonable fidelity and are integrated with EMS to provide a realistic environment for operators and dispatchers to practice normal, every-day operating tasks and procedures as well as experience emergency operating situations. The various training activities can be safely and conveniently practiced with the simulator responding in a manner similar to the actual power system.

An operator training simulator (OTS) can be used in an investigatory manner to recreate past actual operational scenarios and to formulate system restoration procedures. Scenarios can be created, saved, and reused. The OTS can be used to evaluate the functionality and performance of new real-time EMS functions and also for tuning AGC in an off-line, secure environment.

The OTS has three main subsystems (Fig. 12.4).

Energy Control System

The energy control system (ECS) emulates normal EMS functions and is the only part of the OTS with which the trainee interacts. It consists of the supervisory control and data acquisition (SCADA) system, generation control system, and all other EMS functions.

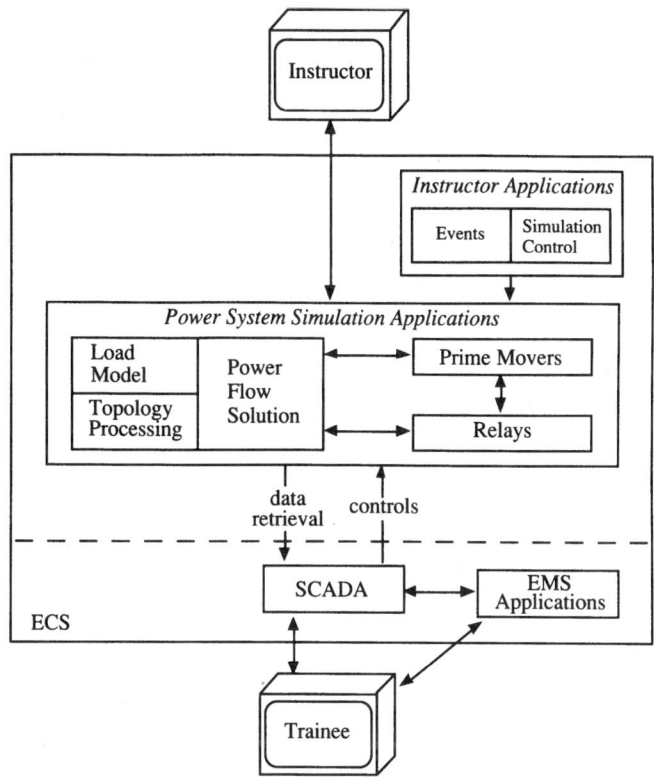

FIGURE 12.4 OTS block diagram.

Power System Dynamic Simulation

This subsystem simulates the dynamic behavior of the power system. System frequency is simulated using the "long-term dynamics" system model, where frequency of all units is assumed to be the same. The prime-mover dynamics are represented by models of the units, turbines, governors, boilers, and boiler auxiliaries. The network flows and states (bus voltages and angles, topology, transformer taps, etc.) are calculated at periodic intervals. Relays are modeled, and they emulate the behavior of the actual devices in the field.

Instructional System

This subsystem includes the capabilities to start, stop, restart, and control the simulation. It also includes making savecases, retrieving savecases, reinitializing to a new time, and initializing to a specific real-time situation.

It is also used to define event schedules. Events are associated with both the power system simulation and the ECS functions. Events may be deterministic (occur at a predefined time), conditional (based on a predefined set of power system conditions being met), or probabilistic (occur at random).

References

Application of Optimization Methods for Economy/Security Functions in Power System Operations, IEEE tutorial course, IEEE Publication 90EH0328-5-PWR, 1990.
Distribution Automation, IEEE Power Engineering Society, IEEE Publication EH0280-8-PBM, 1988.
C. J. Erickson, *Handbook of Electrical Heating,* IEEE Press, 1995.
Energy Control Center Design, IEEE tutorial course, IEEE Publication 77 TU0010-9 PWR, 1977.

Fundamentals of Load Management, IEEE Power Engineering Society, IEEE Publication EH0289-9-PBM, 1988.
Fundamentals of Supervisory Controls, IEEE tutorial course, IEEE Publication 91 EH0337-6 PWR, 1991.
M. Kleinpeter, *Energy Planning and Policy,* New York: Wiley, 1995.
Special issue on computers in power system operations, *Proc. IEEE,* 75, 12, 1987.
W. C. Turner, *Energy Management Handbook,* Fairmont Press, 1997.

Further Information

Current innovations and applications of new technologies and algorithms are presented in the following publications:

- *IEEE Power Engineering Review* (monthly)
- *IEEE Transactions on Power Systems* (bimonthly)
- *Proceedings of the Power Industry Computer Application Conference* (biannual)

12.2 Generation Control: Economic Dispatch and Unit Commitment

Charles W. Richter, Jr.

An area of power system control having a large impact on cost and profit is the optimal scheduling of generating units. A good schedule identifies which units to operate, and the amount to generate at each online unit in order to achieve a set of economic goals. These are the problems commonly referred to as the unit commitment (UC) problem, and the economic dispatch calculation, respectively. The goal is to choose a control strategy that minimizes losses (or maximizes profits), subject to meeting a certain demand and other system constraints. The following sections define EDC, the UC problem, and discuss methods that have been used to solve these problems. Realizing that electric power grids are complex interconnected systems that must be carefully controlled if they are to remain stable and secure, it should be mentioned that the tools described in this chapter are intended for steady-state operation. Short-term (less than a few seconds) changes to the system are handled by dynamic and transient system controls, which maintain secure and stable operation, and are beyond the scope of this discussion.

Economic Dispatch

Economic Dispatch Defined

An *economic dispatch calculation* (EDC) is performed to *dispatch,* or schedule, a set of online generating units to collectively produce electricity at a level that satisfies a specified demand in an economical manner. Each online generating unit may have many characteristics that make it unique, and which must be considered in the calculation. The amount of electricity demanded can vary quickly and the schedule produced by an EDC should leave units able to respond and adapt without major implications to cost or profit. The electric system may have limits (e.g., voltage, transmission, etc.) that impact the EDC and hence should be considered. Generating units may have prohibited generation levels at which resonant frequencies may cause damage or other problems to the system. The impact of transmission losses, congestion, and limits that may inhibit the ability to serve the load in a particular region from a particular generator (e.g., a low-cost generator) should be considered. The market structure within an operating region and its associated regulations must be considered in determining the specified demand, and in determining what constitutes economical operation. An independent system operator (ISO) tasked with maximizing social welfare would likely have a different definition of "economical" than does a generation company (GENCO) wishing to maximize its profit in a competitive environment. The EDC must consider all of these factors and develop a schedule that sets the generation levels in accordance with an economic objective function.

Factors to Consider in the EDC

The Cost of Generation

Cost is one of the primary characteristics of a generating unit that must be considered when dispatching units economically. The EDC is concerned with the short-term operating cost, which is primarily determined by fuel cost and usage. Fuel usage is closely related to generation level. Very often, the relationship between power level and fuel cost is approximated by a quadratic curve: $F = aP^2 + bP + c$. c is a constant term that represents the cost of operating the plant, b is a linear term that varies directly with the level of generation, and a is the term that accounts for efficiency changes over the range of the plant output. A quadratic relationship is often used in the research literature. However, due to varying conditions at certain levels of production (e.g., the opening or closing of large valves may affect the generation cost [Walters and Sheblé, 1992]), the actual relationship between power level and fuel cost may be more complex than a quadratic equation. Many of the long-term generating unit costs (e.g., costs attributed directly to starting and stopping the unit, capital costs associated with financing the construction) can be ignored for the EDC, since the decision to switch on, or *commit*, the units has already been made. Other characteristics of generating units that affect the EDC are the minimum and maximum generation levels at which they may operate. When binding, these constraints will directly impact the EDC schedule.

The Price

The price at which an electric supplier will be compensated is another important factor in determining an optimal economic dispatch. In many areas of the world, electric power systems have been, or still are, treated as a natural monopoly. Regulations allow the utilities to charge rates that guarantee them a nominal profit. In competitive markets, which come in a variety of flavors, price is determined through the forces of supply and demand. Economic theory and common sense tell us that if the total supply is high and the demand is low, the price is likely to be low, and vice versa. If the price is consistently below a GENCO's average total costs, the company may soon be bankrupt.

The Quantity Supplied

The amount of electric energy to be supplied is another fundamental input for the EDC. Regions of the world having regulations that limit competition often require electric utilities to serve all electric demand within a designated service territory. If a consumer switches on a motor, the electric supplier must provide the electric energy needed to operate the motor. In competitive markets, this *obligation to serve* is limited to those with whom the GENCO has a contract. Beyond its contractual obligations, the GENCO may be willing (if the opportunity arises) to supply additional consumer demand. Since the consumers have a choice of electric supplier, a GENCO determining the schedule of its own online generating units may choose to supply all, none, or only a portion of that additional consumer demand. The decision is dependent on the objective of the entity performing the EDC (e.g., profit maximization, improving reliability, etc.).

EDC and System Limitations

A complex network of transmission and distribution lines and equipment are required to move the electric energy from the generating units to the consumer loads. The secure operation of this network depends on bus voltage magnitudes and angles being within certain tolerances. Excessive transmission line loading can also affect the security of the power system network. Since superconductivity is a relatively new field, lossless transmission lines are expensive and are not commonly used. Therefore, some of the energy being transmitted over the system is converted into heat and is consequently lost. The schedule produced by the EDC directly affects losses and security; hence, constraints ensuring proper system operation must be considered when solving the EDC problem.

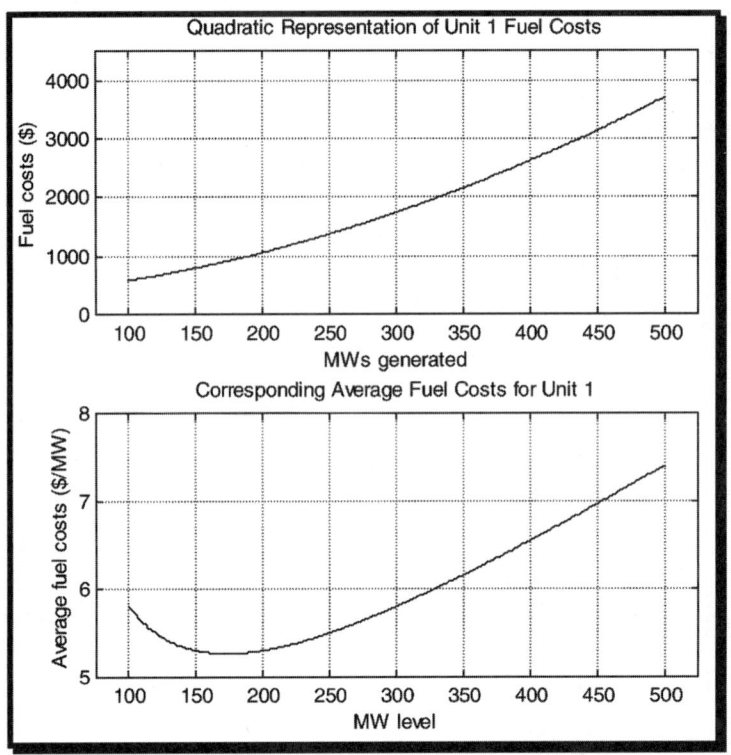

FIGURE 12.5 Relationship between fuel input and power output.

The Objective of EDC

In a regulated, vertically integrated, monopolistic environment, the obligated-to-serve electric utility performs the EDC for the entire service area by itself. In such an environment, providing electricity in an "economical manner" means minimizing the cost of generating electricity, subject to meeting all demand and other system operating constraints. In a competitive environment, the way an EDC is done can vary from one market structure to another. For instance, in a decentralized market, the EDC may be performed by a single GENCO wishing to maximize its expected profit given the prices, demands, costs, and other constraints described above. In a power pool, a central coordinating entity may perform an EDC to centrally dispatch generation for many GENCOs. Depending on the market rules, the generation owners may be able to mask the cost information of their generators. In this case, bids would be submitted for various price levels and used in the EDC.

The Traditional EDC Mathematical Formulation

Assuming operation under a vertically integrated, monopolistic environment, we must meet all demand, D. We must also consider minimum and maximum limits for each generating unit, P_i^{min} and P_i^{max}. We will assume that the fuel costs of the ith operating plant may be modeled by a quadratic equation as shown in Eq. (12.1), and shown graphically in Fig. 12.5. Note that the average fuel costs are also shown in Fig. 12.5.

$$F_i = a_i P_i^2 + b_i P_i + c_i \quad \text{(fuel costs of ith generator)} \tag{12.1}$$

Thus, for N online generating units, we can write a Lagrangian equation, L, which describes the total cost and associated demand constraint, D.

$$L = F_T + \lambda\left(D - \sum_{i=1}^{N} P_i\right) = \sum_{i=1}^{N}\left(a_i P_i^2 + b_i P_i + c_i\right) + \lambda \cdot \left(D - \sum_{i=1}^{N} P_i\right)$$

$$F_T = \sum_{i=1}^{N} F_i \qquad \text{(Total fuel cost is a summation of costs for all online plants)} \qquad (12.2)$$

$$P_i^{\min} \leq P_i \leq P_i^{\max} \quad \text{(Generation must be set between the min and max amounts)}$$

Additionally, note that c_i is a constant term that represents the cost of operating the ith plant, b_i is a linear term that varies directly with the level of generation, P_i, and a_i are terms that account for efficiency changes over the range of the plant output.

In this example, the objective will be to minimize the cost of supplying demand with the generating units that are online. From calculus, a minimum or a maximum can be found by taking the $N + 1$ derivatives of the Lagrangian with respect to its variables, and setting them equal to zero. The shape of the curves is often assumed well behaved — monotonically increasing and convex — so that determining the second derivative is unnecessary.

$$\frac{\partial L}{\partial P_i} = 2a_i P_i + b_i - \lambda = 0 \Rightarrow \lambda = 2a_i P_i + b_i \qquad (12.3)$$

$$\frac{\partial L}{\partial \lambda} = \left(D - \sum_{1}^{N} P_i\right) = 0 \qquad (12.4)$$

λ_i is the commonly used symbol for the "marginal cost" of the i-th unit. At the margin of operation, the marginal cost tells us how many additional dollars the GENCO will have to spend to increase the generation by an additional MW. The marginal cost curve is an positively sloped line if a quadratic equation is being used to represent the fuel curve of the unit. The higher the quantity being produced, the greater the cost of adding an additional unit of the goods being produced. Economic theory says that if a GENCO has a set of plants and it wants to increase production by one unit, it should increase production at the plant that provides the most benefit for the least cost. The GENCO should do this until that plant is no longer providing the greatest benefit for a given cost. At that point it finds the plant now giving the highest benefit-to-cost ratio and increases its production. This is done until all plants are operating at the same marginal cost. When all unconstrained online plants have the same marginal cost, λ (i.e., $\lambda_1 = \lambda_2 = \ldots = \lambda_i = \ldots = \lambda_{\text{SYSTEM}}$), then the cost is at a minimum for that amount of generation. If there were binding constraints, it would prevent the GENCO from achieving that scenario.

If a constraint is binding on a particular unit (e.g., P_i becomes $P_i^{\max}$ when attempting to increase production), the marginal cost of that unit is considered to be infinite. No matter how much money is available to increase plant production by one unit, it cannot do so. (Of course, in the long term, things may be done that can reduce the effect of the constraint, but that is beyond the scope of this discussion.)

EDC Solution Techniques

There are many ways to obtain the optimum power levels that will achieve the objective for the EDC problem being considered. For very simple situations, one may solve the solution directly; but when the number of constraints that introduce nonlinearities to the problem grows, iterative search techniques become necessary. Wood and Wollenberg (1996) describe many such methods of calculating economic dispatch, including the graphical technique, the lambda-iteration method, and the first- and second-order gradient methods. Another method that works well, even when fuel costs are not modeled by a simple quadratic equation, is the genetic algorithm.

TABLE 12.1 Generator Data and Solution for EDC Example

Unit Number	Unit Parameters					Solution		
	P_{min}	P_{max}	A	B	C	P_i (MW)	$/MW ($\lambda_i$)	Cost $/hour
1	100	500	.01	1.8	300	233.2456	6.4649	1263.90
2	50	300	.012	2.24	210	176.0380	6.4649	976.20
3	100	400	.006	2.35	290	342.9094	6.4649	1801.40
4	100	500	.008	2.5	340	247.8070	6.4649	1450.80

In highly competitive scenarios, each inaccuracy in the model can result in losses to the GENCO. A very detailed model might include many nonlinearities, (e.g., valve-point loading, prohibited regions of operation, etc.). Such nonlinearities may mean that it is not possible to calculate a derivative. If the relationship is not well-behaved, there may be no proof that the solution can ever be optimal. With greater detail in the model comes an increase in the amount of time to perform the EDC. Since the EDC is performed quite frequently (on the order of every few minutes), and because it is a real-time calculation, the solution technique should be quick. Since an inaccurate solution may produce a negative impact on the company profits, the solution should also be accurate.

An Example of Cost Minimizing EDC

To illustrate how the EDC is solved via the graphical method, an example is presented here. Assume that a GENCO needs to supply 1000 MW of consumer demand, and that Table 12.1 describes the system on-line units that it is dispatching in a traditional, i.e., vertically integrated, monopolistic environment. Figure 12.6 shows the marginal costs of each of the units over their entire range. It also shows an aggregated marginal cost curve that could be called the system marginal cost curve. This aggregated system curve was created by a horizontal summation of the four individual graphs. Once the system curve is created, one simply finds the desired power level (i.e., 1000 MW) along the x-axis. Follow it up to the curve, and then look to the left. On the y-axis, the system marginal cost can be read. Since no limits were reached, each of the individual λ_is is the same as the system λ. The GENCO can find the λ_i

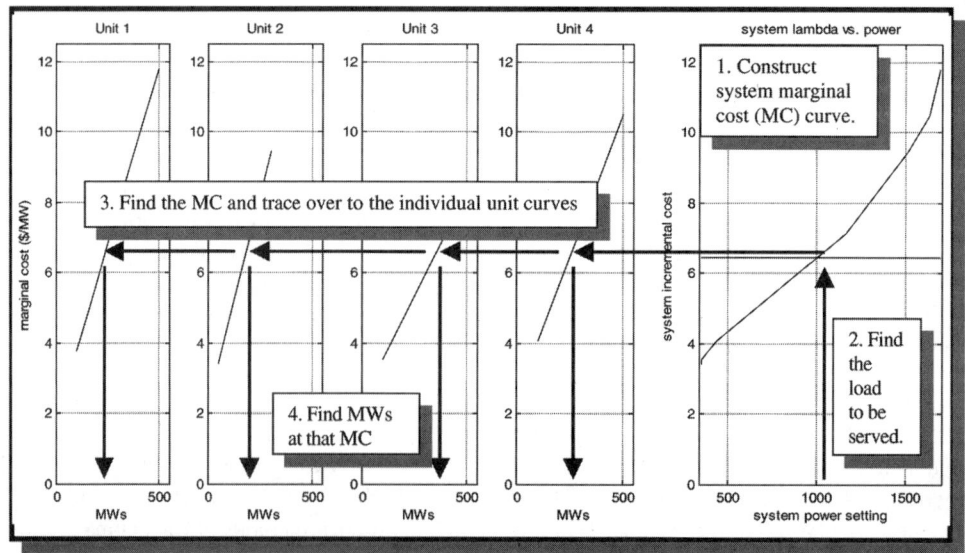

FIGURE 12.6 Unit and aggregated marginal cost curves for solving EDC with the graphical method.

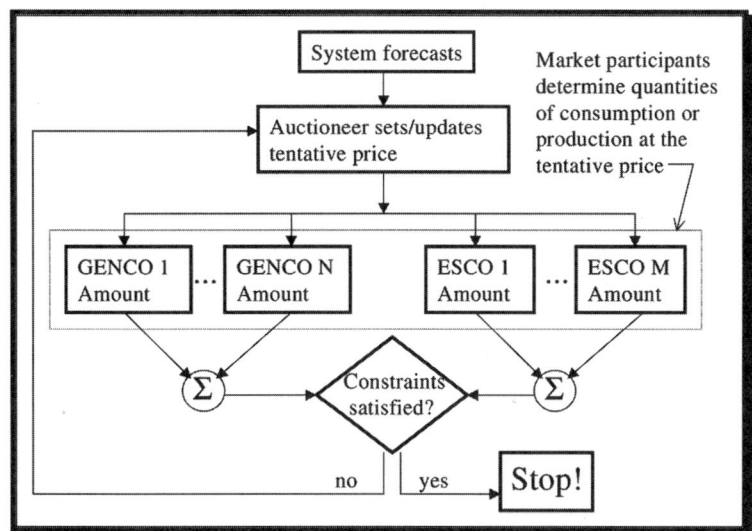

FIGURE 12.7 Economic dispatch and/or unit commitment as an auction.

on each of the unit curves and draw a line straight down from the point where the marginal cost, λ, crosses the curve to find its power level. The generation levels of each online unit are easily found and the solution is shown in the right-hand columns of Table 12.1. The procedure just described is the graphical method of EDC. If the system marginal cost had been above the diagonal portion of an individual unit curve, then we simply set that unit at its P^{max}.

EDC and Auctions

Competitive electricity markets vary in their operating rules, social objectives, and in the mechanism they use to allocate prices and quantities to the participants. Commonly, an auction is used to match buyers with sellers and to achieve a price that is considered fair. Auctions can be sealed bid, open outcry, ascending ask English auctions, descending ask Dutch auctions, etc. Regardless of the solution technique used to find the optimal allocation, the economic dispatch is essentially performing the same allocation that an auction would. Suppose an auctioneer were to call out a price, and ask the participating/online generators how much power they would generate at that level. The reply amounts could be summed to determine the production level at that price. If all of the constraints, including demand, are met, then the most economical dispatch has been achieved. If not, the auctioneer adjusts the price and asks for the amounts at the new price. This procedure is repeated until the constraints are satisfied. Prices may ascend as in the English auction, or they may descend as in the Dutch auction. See Fig. 12.7 for a graphical depiction of this process. For further discussion on this topic, the interested reader is referred to Sheblé (1999).

The Unit Commitment Problem

Unit Commitment Defined

The *unit commitment* (UC) problem is defined as the scheduling of a set of generating units to be on, off, or in stand-by/banking mode for a given period of time to meet a certain objective. For a power system operated by a vertically integrated monopoly, committing units is performed centrally by the utility, and the objective is to minimize costs subject to supplying all demand (and reserve margins). In a competitive environment, each GENCO must decide which units to commit, such that profit is maximized, based on the number of contracted MW; the additional MWhr it forecasts that it can profitably wrest from its competitors in the spot market; and the prices at which it will be compensated.

```
UC Schedule
Hour         1 2 3 4 5 6 ... T
Gen#1:       1 1 1 1 1 1 ... 0
Gen#2:       0 0 0 1 1 1 ... 1
Gen#3:       1 1 1 0 0 0 ... 1
  ....
Gen#N:       1 1 1 1 1 1 ... 0

0=unit off-line    1=unit on-line
```

FIGURE 12.8 A typical unit commitment schedule.

A UC schedule is developed for N units and T periods. A typical UC schedule might look like the one shown in Fig. 12.8. Since uncertainty in the inputs becomes large beyond one week into the future, the UC schedule is typically developed for the following week. It is common to consider schedules that allow unit-status change from hour to hour, so that a weekly schedule is made up of 168 periods. In finding an optimal schedule, one must consider fuel costs, which can vary with time, start-up and shut-down costs, maximum ramp rates, the minimum up-times and minimum down-times, crew constraints, transmission limits, voltage constraints, etc. Because the problem is discrete, the GENCO may have many generating units, a large number of periods may be considered, and because there are many constraints, finding an optimal UC is a complex problem.

Factors to Consider in Solving the UC Problem

The Objective of Unit Commitment

The objective of the unit commitment algorithm is to schedule units in the most economical manner. For the GENCO deciding which units to commit in the competitive environment, economical manner means one that maximizes its profits. For the monopolist operating in a vertically integrated electric system, economical means minimizing the costs.

The Quantity to Supply

In systems with vertically integrated monopolies, it is common for electric utilities to have an obligation to serve all demand within their territory. Forecasters provide power system operators an estimated amount of power demanded. The UC objective is to minimize the total operational costs subject to meeting all of this demand (and other constraints they may be considering).

In competitive electric markets, the GENCO commits units to maximize its profit. It relies on spot and forward bilateral contracts to make part of the total demand known *a priori*. The remaining share of the demand that it may pick up in the spot market must be predicted. This market share may be difficult to predict since it depends on how its price compares to that of other suppliers.

The GENCO may decide to supply less demand than it is physically capable of. In the competitive environment, the obligation to serve is limited to those with whom the GENCO has a contract. The GENCO may consider a schedule that produces less than the forecasted demand. Rather than switching on an additional unit to produce one or two unsatisfied MW, it can allow its competitors to provide that 1 or 2 MW that might have substantially increased its average costs.

Compensating the Electricity Supplier

Maximizing profits in a competitive environment requires that the GENCO know what revenue is being generated by the sale of electricity. While a traditional utility might have been guaranteed a fixed rate of return based on cost, competitive electricity markets have varying pricing schemes that may price electricity at the level of the last accepted bid, the average of the buy, ask, and sell offer, etc. When submitting offers to an auctioneer, the GENCO's offer price should reflect its prediction market share,

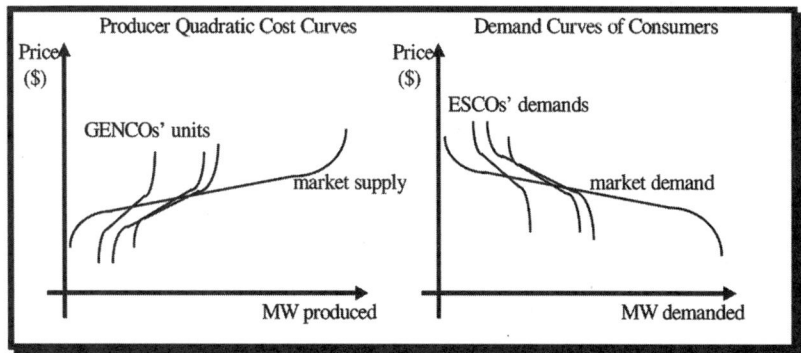

FIGURE 12.9 Treating the market as an additional generator and/or load.

since that determines how many units they have switched on, or in banking mode. GENCOs recovering costs via prices set during the bidding process will note that the UC schedule directly affects the average cost, which indirectly affects the offering price, making it an essential input to any successful bidding strategy.

Demand forecasts and expected market prices are important inputs to the profit-based UC algorithm; they are used to determine the expected revenue, which in turn affects the expected profit. If a GENCO produces two UC schedules each having different expected costs and different expected profits, it should implement the one that provides for the largest profit, which will not necessarily be the one that costs the least. Since prices and demand are so important in determining the optimal UC schedule, price prediction and demand forecasts become crucial. An easy-to-read description of the cost-minimizing UC problem and a stochastic solution that considers spot markets has been presented in Takriti, Krasenbrink, and Wu (1997).

The Source of Electric Energy

A GENCO may be in the business of electricity generation, but it should also consider purchasing electricity from the market, if it is less expensive than its own generating unit(s). The existence of liquid markets gives energy trading companies an additional source from which to supply power that may not be as prevalent in monopolistic systems. See Fig. 12.9. To the GENCO, the market supply curve can be thought of as a pseudo-unit to be dispatched. The supply curve for this pseudo-unit represents an aggregate supply of all of the units participating in the market at the time in question. The price forecast essentially sets the parameters of the unit. This pseudo-unit has no minimum uptime, minimum downtime, or ramp constraints; there are no direct start-up and shutdown costs associated with dispatching the unit.

The liquid markets that allow the GENCO to schedule an additional pseudo unit, also act as a load to be supplied. The total energy supplied should consist of previously arranged bilateral or multilateral contracts arranged through the markets (and their associated reserves and losses). While the GENCO is determining the optimal unit commitment schedule, the energy demanded by the market (i.e., market demand) can be represented as another DISTCO or ESCO buying electricity. Each entity buying electricity should have its own demand curve. The market demand curve should reflect the aggregate of the demand of all the buying agents participating in the market.

Mathematical Formulation for UC

The mathematical formulation for UC depends upon the objective and the constraints that are considered important. Traditionally, the monopolist cost-minimization UC problem has been formulated (Sheblé, 1985):

$$\text{Minimize } F = \sum_{n}^{N} \sum_{t}^{T} \left[\left(C_{nt} + MAINT_{nt} \right) \cdot U_{nt} + SUP_{nt} \cdot U_{nt} \left(1 - U_{nt}\right) + SDOWN_{nt} \cdot \left(1 - U_{nt}\right) \cdot U_{nt-1} \right] \quad (12.5)$$

subject to the following constraints:

$$\sum_{n}^{N}\left(U_{nt}\cdot P_{nt}\right)=D_{t} \qquad \text{(demand constraint)}$$

$$\sum_{n}^{N}\left(U_{nt}\cdot P\max_{n}\right)\geq D_{t}+R_{i} \qquad \text{(capacity constraint)}$$

$$\sum_{n}^{N}\left(U_{nt}\cdot Rs\max_{n}\right)\geq R_{t} \qquad \text{(system reserve constraint)}$$

When formulating the profit-maximizing UC problem for a competitive environment, the obligation-to-serve is gone. The demand constraint changes from an equality to an inequality ($\leq$). In the formulation presented here, we lump the reserves in with the demand. Essentially we are assuming that buyers are required to purchase a certain amount of reserves per contract. In addition to the above changes, formulating the UC problem for the competitive GENCO changes the objective function from cost minimization to profit maximization as shown in Eq. (12.6) below. The UC solution process is shown in block diagram form in Fig. 12.10.

$$\text{Max } \Pi = \sum_{n}^{N}\sum_{t}^{T}\left(P_{nt}\cdot fp_{t}\right)\cdot U_{nt} - F \qquad (12.6)$$

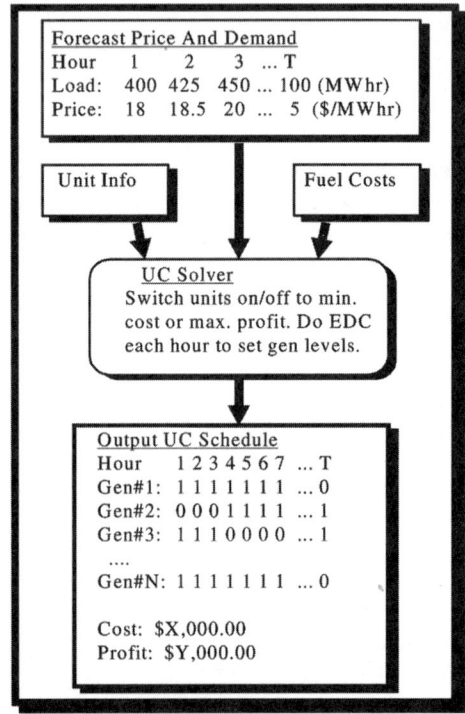

FIGURE 12.10 Block diagram of the UC solution process.

subject to:

$$D_t^{contracted} \leq \sum_{n}^{N} (U_{nt} \cdot P_{nt}) \leq D'_t \quad \text{(demand constraint w/out obligation-to-serve)}$$

$$Pmin_n \leq P_{nt} \leq Pmax_n \quad \text{(capacity limits)}$$

$$|P_{nt} - P_{n,t-1}| \leq Ramp_n \quad \text{(ramp rate limits)}$$

where individual terms are defined as follows:

U_{nt} = up/down time status of unit n at time period t (U_{nt} = 1 unit on, U_{nt} = 0 unit off)
P_{nt} = power generation of unit n during time period t
D_t = load level in time period t
D'_t = forecasted demand at period t (includes reserves)
$D_t^{contract}$ = contracted demand at period t (includes reserves)
fp_t = forecasted price/MWhr for period t
R_t = system reserve requirements in time period t
C_{nt} = production cost of unit n in time period t
SUP_{nt} = start-up cost for unit n, time period t
$SDOWN_{nt}$ = shut-down cost for unit n, time period t
$MAINT_{nt}$ = maintenance cost for unit n, time period t
N = number of units
T = number of time periods
$Pmin_n$ = generation low limit of unit n
$Pmax_n$ = generation high limit of unit n
$Rsmax_n$ = maximum contribution to reserve for unit n

Although it may happen in certain cases, the schedule that minimizes cost is not necessarily the schedule that maximizes profit. Providing further distinction between the cost-minimizing UC for the monopolist and the profit maximizing competitive GENCO is the obligation-to-serve; the competitive GENCO may choose to generate less than the total consumer demand. This allows a little more flexibility in the UC schedules. In addition, our formulation assumes that prices fluctuate according to supply and demand. In cost-minimizing paradigms, it is assumed that leveling the load curve helps to minimize the cost. When maximizing profit, the GENCO may find that under certain conditions, it may profit more under a non-level load curve. The profit depends not only on cost, but also on revenue. If revenue increases more than the cost does, the profit will increase.

The Importance of EDC to the UC Solution

The economic dispatch calculation (EDC) is an important part of UC. It is used to assure that sufficient electricity will be available to meet the objective each hour of the UC schedule. For the monopolist in a vertically integrated environment, EDC will set generation so that costs are minimized subject to meeting the demand. For the price-based UC, the price-based EDC adjusts the power level of each online unit until each has the same incremental cost (i.e., $\lambda_1 = \lambda_2 = \ldots = \lambda_i = \ldots = \lambda_T$). If a GENCO is operating in a competitive framework that requires its bids to cover fixed, start-up, shutdown, and other costs associated with transitioning from one state to another, then the incremental cost used by EDC must embed these costs. We shall refer to this modified marginal cost as a pseudo λ. The competitive generator will generate if the pseudo λ is less than or equal to the competitive price. A simple way to allocate the fixed and transitional costs that result in a $/MWhr figure is shown in Eq. (12.7):

$$\lambda_t = fp_t - \frac{\sum_t \sum_n (\text{transition costs}) + \sum_t \sum_n (\text{fixed costs})}{\sum_t^T \sum_n^N P_{nt}} \quad (12.7)$$

Other allocation schemes that adjust the marginal cost/price according to the time of day or price of power would be just as easy to implement and should be considered in building bidding strategies. Transition costs include start-up, shutdown, and banking costs, and fixed costs (present for each hour that the unit is on), which would be represented by the constant term in the typical quadratic cost curve approximation. For the results presented later in this chapter, we approximate the summation of the power generated by the forecasted demand.

The competitive price is assumed to be equal to the forecasted price. If the GENCO's supply curve is indicative of the system supply curve, then the competitive price will correspond to the point where the demand and supply curves cross. EDC sets the generation level corresponding to the point where the GENCO's supply curve crosses the demand curve, or to the point where the forecasted price is equal to the supply curve, whichever is lower.

Solution Methods

Solving the UC problem to find an optimal solution can be difficult. The problem has a large solution space that is discrete and nonlinear. As mentioned above, solving the UC problem requires that many economic dispatch calculations be performed. One possible way to determine the optimal schedule is to do an exhaustive search. Exhaustively considering all possible ways that units can be switched on or off for a small system can be done, but for a reasonably sized system this would take too long. Solving the UC problem for a realistic system generally involves using methods like Lagrangian relaxation, dynamic programming, genetic algorithms, or other heuristic search techniques. The interested reader may find many useful references regarding cost-minimizing UC for the monopolist in Sheblé and Fahd (1994) and Wood and Wollenberg (1996). Another heuristic technique that has shown much promise and that offers many advantages (e.g., time-to-solution for large systems and ability to simultaneously generate multiple solutions) is the genetic algorithm.

A Genetic-Based UC Algorithm

The Basics of Genetic Algorithms

A genetic algorithm (GA) is a search algorithm often used in nonlinear discrete optimization problems. The development of GAs was inspired by the biological notion of evolution. Initially described by John Holland, they were popularized by David Goldberg who described the basic genetic algorithm very well (Goldberg, 1989). In a GA, data, initialized randomly in a data structure appropriate for the solution to the problem, evolves over time and becomes a suitable answer to the problem. An entire population of candidate solutions (data structures with a form suitable for solving for the problem being studied) is "randomly" initialized and evolves according to GA rules. The data structures often consist of strings of binary numbers that are mapped onto the solution space for evaluation. Each solution (often termed a creature) is assigned a fitness — a heuristic measure of its quality. During the evolutionary process, those creatures having higher fitness are favored in the parent selection process and are allowed to procreate. The parent selection is essentially a random selection with a fitness bias. The type of fitness bias is determined by the parent selection method. Following the parent selection process, the processes of crossover and mutation are utilized and new creatures are developed that ideally explore a different area of the solution space. These new creatures replace less fit creatures from the existing population. Figure 12.11 shows a block diagram of the general GA.

GA for Price-Based UC

The algorithm presented here solves the UC problem for the profit maximizing GENCO operating in the competitive environment (Richter et al., 1999). Research reveals that various GAs have been used by

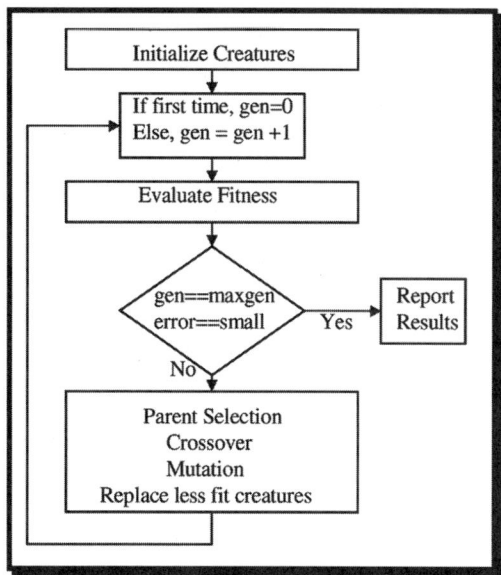

FIGURE 12.11 A simple genetic algorithm.

many researchers in solving the UC problem (Kondragunta, 1997; Kazarlis et al., 1995). However, the algorithm presented here is a modification of a genetic-based UC algorithm for the cost-minimizing monopolist described in Maifeld and Sheblé (1996). Most of the modifications are to the fitness function, which no longer rewards schedules that minimize cost, but rather those that maximize profit. The intelligent mutation operators are preserved in their original form. The schedule format is the same. The algorithm is shown in block diagram format in Fig. 12.12.

The algorithm first reads in the contract demand and prices, the forecast of remaining demand, and forecasted spot prices (which are calculated for each hour by another routine not described here). During the initialization step, a population of UC schedules is randomly initialized. See Fig. 12.13. For each member of the population, EDC is called to set the level of generation of each unit. The cost of each schedule is calculated from the generator and data read in at the beginning of the program. Next, the fitness (i.e., the profit) of each schedule in the population is calculated. "Done?" checks to see whether the algorithm as either cycled through for the maximum number of generations allowed, or whether other stopping criteria have been met. If done, then the results are written to a file; if not done, the algorithm goes to the reproduction process.

During reproduction, new schedules are created. The first step of reproduction is to select parents from the population. After selecting parents, candidate children are created using two-point crossover as shown in Fig. 12.14. Following crossover, standard mutation is applied. Standard mutation involves turning a randomly selected unit on or off within a given schedule.

An important feature of the previously developed UC-GA (Maifeld and Sheblé, 1996) is that it spends as little time as possible doing EDC. After standard mutation, EDC is called to update the profit only for the mutated hour(s). An hourly profit number is maintained and stored during the reproduction process, which dramatically reduces the amount of time required to calculate the profit over what it would be if EDC had to work from scratch at each fitness evaluation. In addition to the standard mutation, the algorithm uses two "intelligent" mutation operators that work by recognizing that, because of transition costs and minimum uptime and downtime constraints, 101 or 010 combinations are undesirable. The first of these operators would purge this undesirable combination by randomly changing 1s to 0s or vice versa. The second of these intelligent mutation operators purges the undesirable combination by changing 1 to 0 or 0 to 1 based on which of these is more helpful to the profit objective.

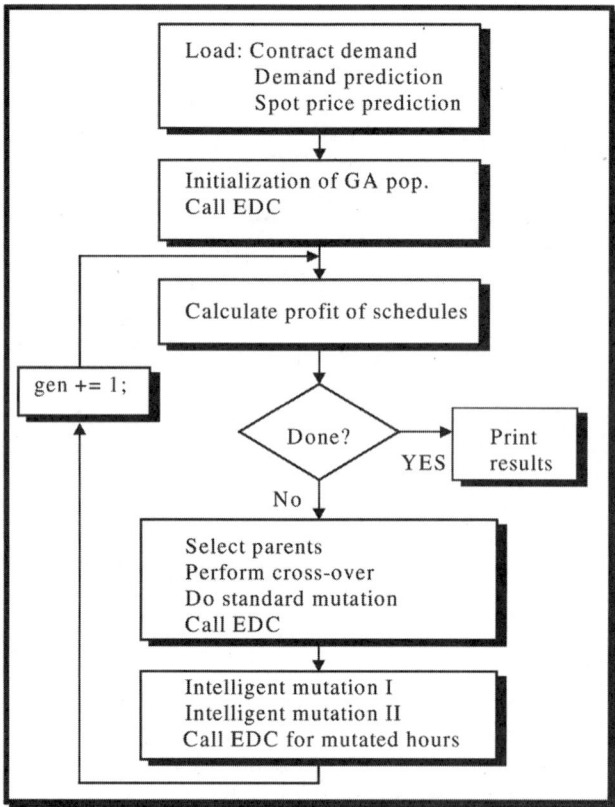

FIGURE 12.12 GA-UC block diagram.

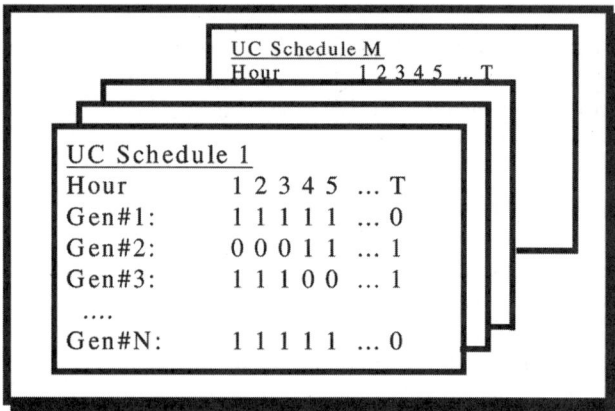

FIGURE 12.13 A population of UC schedules.

Price-Based UC-GA results

The UC-GA is run on a small system so that its solution can be easily compared to a solution by exhaustive search. Before running the UC-GA, the GENCO needs to first get an accurate hourly demand and price forecast for the period in question. Developing the forecasted data is an important topic, but beyond the scope of our analysis. For the results presented in this section, the forecasted load and prices are taken to be those shown in Table 12.2. In addition to loading the forecasted hourly price and demand, the

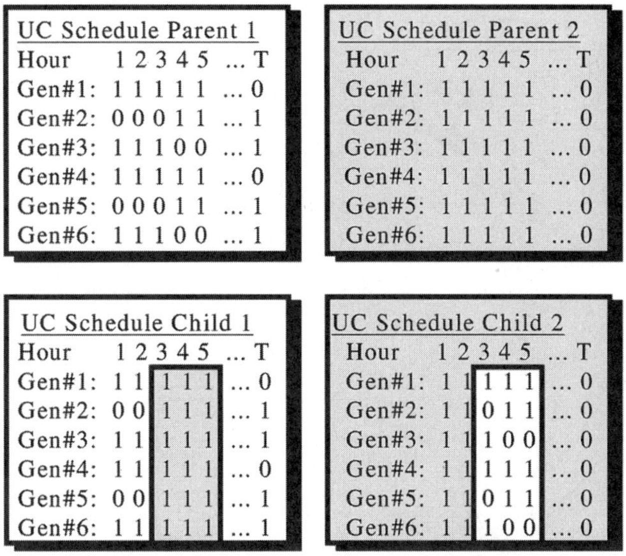

```
┌─────────────────────────────┐  ┌─────────────────────────────┐
│ UC Schedule Parent 1        │  │ UC Schedule Parent 2        │
│ Hour    1 2 3 4 5 ... T     │  │ Hour    1 2 3 4 5 ... T     │
│ Gen#1:  1 1 1 1 1 ... 0     │  │ Gen#1:  1 1 1 1 1 ... 0     │
│ Gen#2:  0 0 0 1 1 ... 1     │  │ Gen#2:  1 1 1 1 1 ... 0     │
│ Gen#3:  1 1 1 0 0 ... 1     │  │ Gen#3:  1 1 1 1 1 ... 0     │
│ Gen#4:  1 1 1 1 1 ... 0     │  │ Gen#4:  1 1 1 1 1 ... 0     │
│ Gen#5:  0 0 0 1 1 ... 1     │  │ Gen#5:  1 1 1 1 1 ... 0     │
│ Gen#6:  1 1 1 0 0 ... 1     │  │ Gen#6:  1 1 1 1 1 ... 0     │
└─────────────────────────────┘  └─────────────────────────────┘

┌─────────────────────────────┐  ┌─────────────────────────────┐
│ UC Schedule Child 1         │  │ UC Schedule Child 2         │
│ Hour    1 2 3 4 5 ... T     │  │ Hour    1 2 3 4 5 ... T     │
│ Gen#1:  1 1 |1 1 1| ... 0   │  │ Gen#1:  1 1 |1 1 1| ... 0   │
│ Gen#2:  0 0 |1 1 1| ... 1   │  │ Gen#2:  1 1 |0 1 1| ... 0   │
│ Gen#3:  1 1 |1 1 1| ... 1   │  │ Gen#3:  1 1 |1 0 0| ... 0   │
│ Gen#4:  1 1 |1 1 1| ... 0   │  │ Gen#4:  1 1 |1 1 1| ... 0   │
│ Gen#5:  0 0 |1 1 1| ... 1   │  │ Gen#5:  1 1 |0 1 1| ... 0   │
│ Gen#6:  1 1 |1 1 1| ... 1   │  │ Gen#6:  1 1 |1 0 0| ... 0   │
└─────────────────────────────┘  └─────────────────────────────┘
```

FIGURE 12.14 Two-point crossover on UC schedules.

TABLE 12.2 Forecasted Demand and Prices for 2-Generator Case

Hour	Load Forecast (MWhr)	Price Forecast ($/MWhr)	Hour	Load Forecast (MWhr)	Price Forecast ($/MWhr)
1	285	25.87	8	328	8.88
2	293	23.06	9	326	9.12
3	267	19.47	10	298	8.88
4	247	18.66	11	267	25.23
5	295	21.38	12	293	26.45
6	292	12.46	13	350	25.00
7	299	9.12	14	350	24.00

TABLE 12.3 Unit Data for 2-Generator Case

	Generator 0	Generator 1
Pmin (MW)	40	40
Pmax (MW)	180	180
A (constant)	58.25	138.51
B (linear)	8.287	7.955
C (quadratic)	7.62e-06	3.05e-05
Bank cost ($)	192	223
Start-up cost ($)	443	441
Shut-down cost ($)	750	750
Min-uptime (hr)	4	4
Min-downtime (hr)	4	4

UC-GA program needs to load the parameters of each generator to be considered. We are modeling the generators with a quadratic cost curve (e.g., $A + B(P) + C(P)^2$), where P is the power level of the unit. The data for the 2-generator case is shown in Table 12.3.

In addition to the 2-unit cases, a 10-unit, 48-hour case is included in this chapter to show that the GA works well on larger problems. While dynamic programming quickly becomes too computationally expensive to solve, the GA scales up linearly with number of hours and units. Figure 12.15 shows the

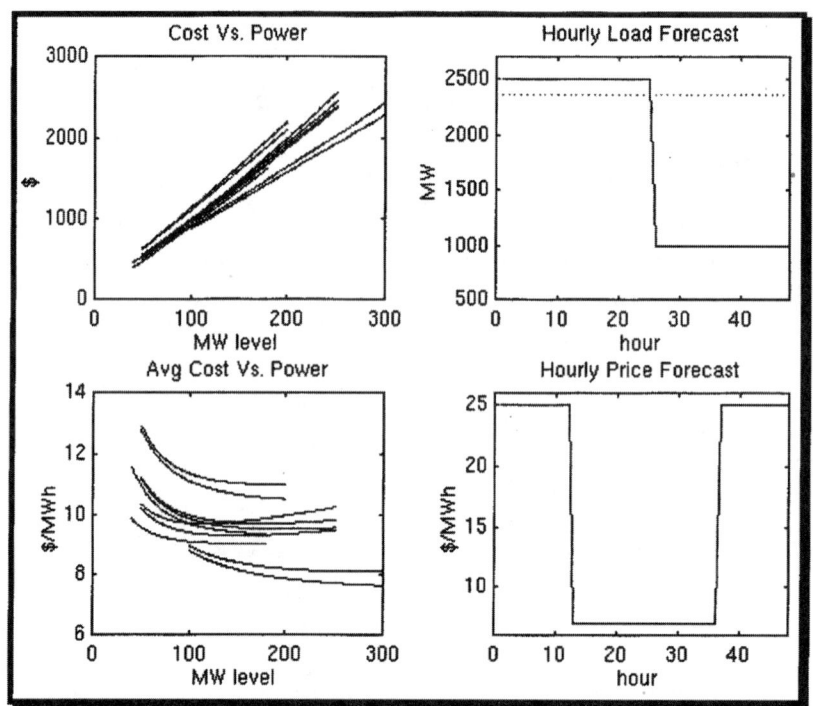

FIGURE 12.15 Data for 10-unit, 48-hour case.

TABLE 12.4 GA Control Parameters

Parameter	Setpoint	Parameter	Setpoint
# of Units	2	System reserve (%)	10
# of Hours	10	Children per generation	10
Popsize	20	UC schedules to keep	1
Generations	50	Random number seed	0.20

costs and average costs (without transition costs) of the 10 generators, as well as the hourly price and load forecasts for the 48 hours. The data was chosen so that the optimal solution was known *a priori*. The dashed line in the load forecast represents the maximum output of the 10 units.

Before running the UC-GA, the user specifies the control parameters shown in Table 12.4, including the number of generating units and number of hours to be considered in the study. The "popsize" is the size of the GA population. The execution time varies approximately linearly with the popsize. The number of generations indicates how many times the GA will go through the reproduction phase. System reserve is the percentage of reserves that the buyer must maintain for each contract. Children per generation tells us how much of the population will be replaced each generation. Changing this can affect the convergence rate. If there are multiple optima, faster convergence can trap the GA in a local suboptimal solution. "UC schedules to keep" indicates the number of schedules to write to file when finished. There is also a random number seed that is set between 0 and 1.

In the 2-generator test cases, the UC-GA was run for the units listed in Table 12.3, and for the forecasted loads and prices listed in Table 12.2. The parameters listed in Table 12.4 were adjusted accordingly. To ensure that the UC-GA is finding optimal solutions, an exhaustive search was performed on some of the smaller cases. Table 12.5 shows the time to solution in seconds for the UC-GA and the exhaustive search methods. For small cases, the exhaustive search was performed and solution time compared to that of

TABLE 12.5 Comparing UC-GA with Exhaustive Search

No. of Generators in Schedule	No. of Hours in Schedule	GA Finds Optimal Solution?	Solution Time for GA (s)	Solution Time Exhaustive Search (s)
2	10	Yes	0.5	674
2	12	Yes	2	6482
2	14	Yes	10	(estimated) 62340
10	48	Yes	730	(estimated) 2E138

TABLE 12.6 The Best UC-GA Schedules of the Population

	Best Schedule for 2-Unit, 10-Hour Case
Unit 1	1111100000
Unit 2	0000000000
Cost	$17,068.20
Profit	$2,451.01

	Best Schedule for 2-Unit, 12-Hour Case
Unit 1	111111000011
Unit 2	000000000000
Cost	$24,408.50
Profit	$4,911.50

	Best Schedule Found by UC-GA for 10-Unit, 48-Hour Case
Unit 1	111111111111000000000000000000000000111111111111
Unit 2	111111111111000000000000000000000000000000000000
Unit 3	111111111111000000000000000000000000000000000000
Unit 4	111111111111000000000000000000000000000000000000
Unit 5	111111111111000000000000000000000000000000000000
Unit 6	111111111111000000000000000000000000000000000000
Unit 7	111111111111000000000000000000000000111111111111
Unit 8	111111111111000000000000000000000000000000000000
Unit 9	111111111111000000000000000000000000111111111111
Unit 10	111111111111000000000000000000000000111111111111
Cost	$325,733.00
Profit	$676,267.00

the UC-GA. Since the exhaustive search solution times were estimated to be prohibitively lengthy, the latter cases were not compared against exhaustive search solutions. Cases with known optimal solutions were used to verify that the UC-GA was, in fact, working for the large cases.

Table 12.6 shows the optimal UC schedules found by the UC-GA for selected cases. Figure 12.16 shows the maximum, minimum and average fitnesses (profit) during each generation of the UC-GA on the 2-generator, 14-hour/period case. The best individual of the population climbs quite rapidly to near the optimal solution. Half of the population is replaced each generation; often the child solutions are poor solutions, hence the minimum fitness tends to remain low over the generations, which is typical for GA optimization.

In the schedules shown in Table 12.6, it may appear as though minimum up- and downtime constraints are being violated. When calculating the cost of such a schedule, the algorithm ensures that the profit is based on a valid schedule by considering a zero surrounded by ones to be a banked unit, and so forth. In addition, note that only the best solution of the population for each of the cases is shown. The existence of additional valid solutions, which may have been only slightly suboptimal in terms of profit, is one of the main advantages of using the GA. It gives the system operator the flexibility to choose the best schedule from a group of schedules to accommodate things like forced maintenance.

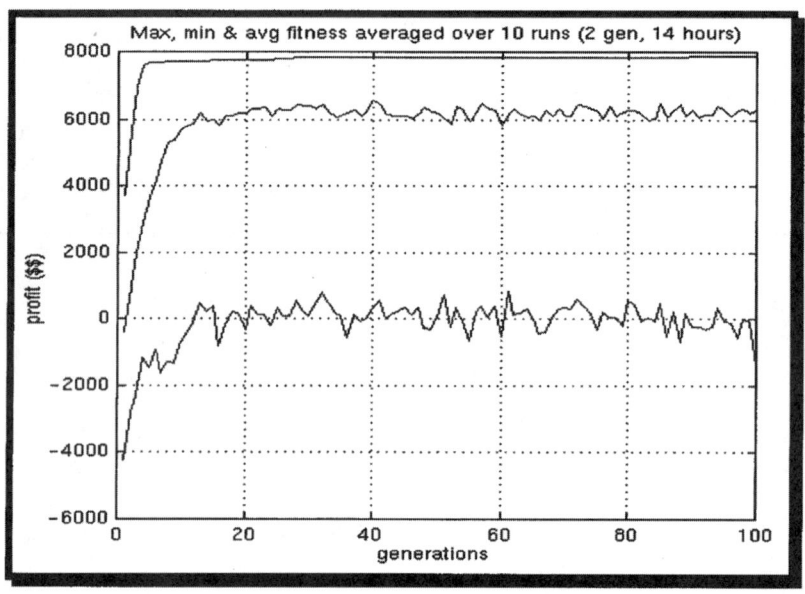

FIGURE 12.16 Max., min., and avg. fitness vs. GA generations for the 2-generator, 14-hour case.

Unit Commitment and Auctions

Regardless of the market framework, the solution method, and who is performing the UC, an auction can model and achieve the optimal solution. As mentioned previously in the section on EDC, auctions (which come in many forms, e.g., Dutch, English, sealed, double-sided, single-sided, etc.) are used to match buyers with sellers and to achieve a price that is considered fair. An auction can be used to find the optimal allocation, and the unit commitment algorithm essentially performs the same allocation that an auction would. Suppose an auctioneer was to call out a price, or a set of prices that is predicted for the schedule period. The auctioneer would then ask all generators how much power they would generate at that level. The generator must consider which units to switch on, and at what level to produce and sell. The reply amounts could be summed to determine the production level at that price. If all of the constraints, including demand, are met, then the most economical combination of units operating at the most economical settings has been found. If not, the auctioneer adjusts the price and asks for the amounts at the new price. This procedure is repeated until the constraints are satisfied. Prices may ascend as in the English auction, or they may descend as in the Dutch auction. See Fig. 12.7 for a graphical depiction of this process. For further discussion on this topic, the interested reader is referred to Sheblé (1999).

Summary of Economical Generation Operation

Since the introduction of electricity supply to the public in the late 1800s, people in many parts of the world have grown to expect an inexpensive reliable source of electricity. Providing that electric energy economically and efficiently requires the generation company to carefully control their generating units, and to consider many factors that may affect the performance, cost, and profitability of their operation. The unit commitment and economic dispatch algorithms play an important part in deciding how to operate the electric generating units around the world. The introduction of competition has changed many of the factors considered in solving these problems. Furthermore, advancements in solution techniques offer a continuum of candidate algorithms, each having its own advantages and disadvantages. Research continues to push these algorithms further. This chapter has provided the reader with an introduction to the problems of determining optimal unit commitment schedules and economic dispatches. It is by no means exhaustive, and the interested reader is strongly encouraged to see the references at the end of the chapter for more details.

References

Goldberg, D., *Genetic Algorithms in Search, Optimization and Machine Learning*. Addison-Wesley Publishing Company, Inc., Reading, MA, 1989.

Kazarlis, S. A., Bakirtzis, A. G., and Petridis, V., A Genetic Algorithm Solution to the Unit Commitment Problem, *1995 IEEE/PES Winter Meeting*, 152-9 PWRS, New York, 1995.

Kondragunta, S., Genetic algorithm unit commitment program, M.S. Thesis, Iowa State University, Ames, IA, 1997.

Maifeld, T., and Sheblé, G., Genetic-Based unit commitment, *IEEE Trans. on Power Syst.*, 11, 1359, August 1996.

Richter, C., and Sheblé, G., A Profit-Based Unit Commitment GA for the Competitive Environment, accepted for *IEEE Trans. on Power Syst.*, publication forthcoming.

Sheblé, G., *Computational Auction Mechanisms for Restructured Power Industry Operation*. Kluwer Academic Publishers, Boston, MA, 1999.

Sheblé, G., Unit Commitment for Operations, Ph.D. Dissertation, Virginia Polytechnic Institute and State University, March, 1985.

Sheblé, G., and Fahd, G., Unit commitment literature synopsis, *IEEE Trans. on Power Syst.*, 9, 128–135, February 1994.

Takriti, S., Krasenbrink, B., and Wu, L. S.-Y., Incorporating Fuel Constraints and Electricity Spot Prices into the Stochastic Unit Commitment Problem, IBM Research Report: RC 21066, Mathematical Sciences Department, T.J. Watson Research Center, Yorktown Heights, New York, December 29, 1997.

Walters, D. C., and Sheblé, G. B., Genetic Algorithm Solution of Economic Dispatch with Valve Point Loading, *1992 IEEE/PES Summer Meeting*, 414-3, New York, 1992.

Wood, A., and Wollenberg, B., *Power Generation, Operation, and Control*. John Wiley & Sons, New York, NY, 1984.

12.3 State Estimation

Danny Julian

An online AC power flow is a valuable application when determining the critical elements affecting power system operation and control such as overloaded lines, credible contingencies, and unsatisfactory voltages. It is the basis for any real-time security assessment and enhancement applications.

AC power flow algorithms calculate real and reactive line flows based on a multitude of inputs with generator bus voltages, real power bus injections, and reactive power bus injections being a partial list. This implies that in order to calculate the line flows using a power flow algorithm, all of the input information (voltages, real power injections, reactive power injections, etc.) must be known *a priori* to the algorithm being executed.

An obvious way to implement an online AC power flow is to telemeter the required input information at every location in the power system. This would require not only a large number of **remote terminal units** (RTUs), but also an extensive communication infrastructure to telemeter the data to the **SCADA** system, both of which are costly. Although the generator bus voltages are usually readily available, the injection data is frequently what is lacking. This is because it is much easier and cheaper to monitor the net injection at a bus than to measure separate injections directly.

Also, this approach presents weaknesses for the online AC power flow that are due to meter accuracy and communication failure. An online power flow relying on a specific set of measurements could become unusable or give erroneous results if any of the predefined measurements became unavailable due to communication failure or due to misoperation of measurement devices. This is not a desirable outcome of an online application designed to alert system operators to unsecure conditions.

Given the above obstacles of utilizing an online AC power flow, work was conducted in the late 1960s and early 1970s (Schweppe and Wildes, Jan. 1970) into developing a process of performing an online power flow using not just the limited data needed for the classical AC power flow algorithm, but using all available measurements. This work led to the **state estimator**, which uses not only the aforementioned voltages but other telemetered measurements such as real and reactive line flows, circuit breaker statuses, and transformer tap settings.

State Estimation Problem

State estimators perform a statistical analysis using a set of m imperfect redundant data telemetered from the power system to determine the state of the system. The state of the system is a function of n **state variables**: bus voltages and relative phase angles, and tap changing transformer positions. Although the state estimation solution is not a "true" representation of the system, it is the "best" possible representation based on the telemetered measurements.

Also, it is necessary to have the number of measurements greater than the number of states ($m \geq n$) to yield a representation of the complete state of the system. This is known as the observability criterion. Typically, m is two to three times the value of n, allowing for a considerable amount of redundancy in the measurement set.

Underlying Assumptions

Telemetered measurements usually are corrupted since they are susceptible to noise. Even when great care is taken to ensure accuracy, unavoidable random noise enters into the measurement process, which distorts the telemetered values.

Fortunately, statistical properties associated with the measurements allow certain assumptions to be made to estimate the true measured value. First, it is assumed the measurement noise has an expected value, or average, of zero. This assumption implies the error in each measurement has equal probability of taking on a positive or negative value. It is also assumed that the expected value for the square of the measurement error is normal and has a standard deviation of σ, and the correlation between measurements is zero (i.e., independent).[1] A variable is said to be normal (or Gaussian) if its probability density function has the form

$$f(v) = \frac{1}{\sigma\sqrt{2\pi}} e^{-\frac{v^2}{2\sigma^2}}. \qquad (12.8)$$

This distribution is also known as the bell curve due to its symmetrical shape resembling a bell as can be seen in Fig. 12.17. The normal distribution is used for the modeling of measurement errors since it is the distribution that results when many factors contribute to the overall error.

Figure 12.17 also illustrates the effect of standard deviation on the normal density function. Standard deviation, σ, is a measure of the spread of the normal distribution about the mean (μ) and gives an indication of how many samples fall within a given interval around the mean. A large standard deviation implies there is a high probability the measurement noise will take on large values. Conversely, a small standard deviation implies there is a high probability the measurement noise will take on small values.

Measurement Representations

Since a measurement is not exact, it can be expressed with an error component of the form

$$z = z_T + v \qquad (12.9)$$

---

[1] In practice, measurements i and j are not necessarily independent since one measurement device may measure more than one value. Therefore, if the measurement device is bad, probably both measurements i and j are bad also.

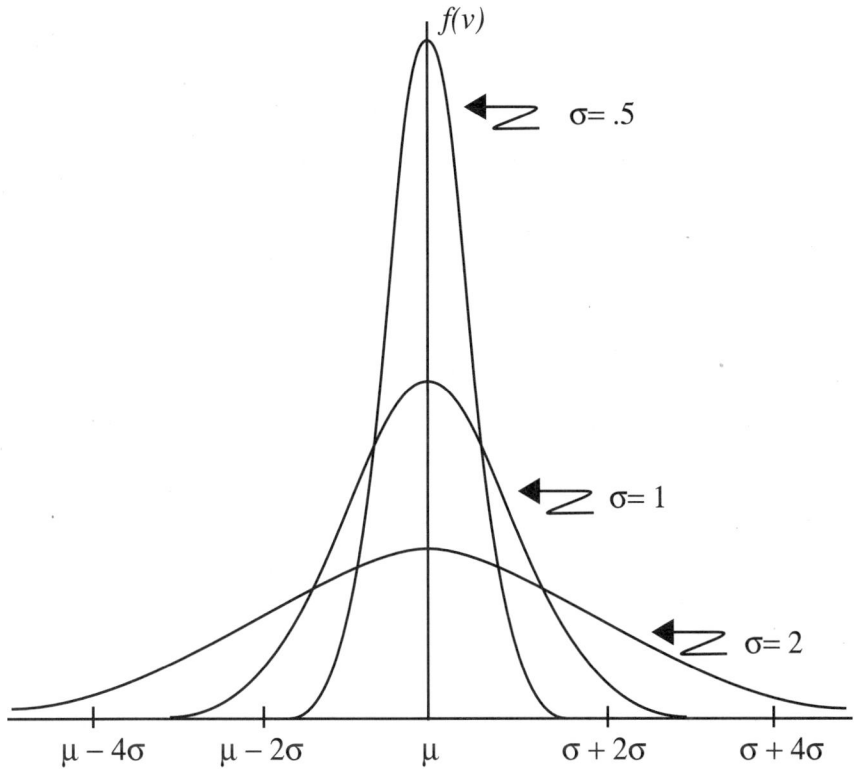

FIGURE 12.17 Normal probability distribution curve with a mean of μ.

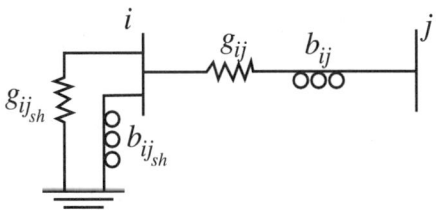

FIGURE 12.18 Transmission line representation.

where z is the measured value, z_T is the true value, and v is the measurement error that represents uncertainty in the measurement. In general, the measured value, as expressed in Eq. (12.9), can be related to the states, x, by

$$z = h(x) + v \qquad (12.10)$$

where $h(x)$ is a vector of nonlinear functions relating the measurements to the state variables. An example of the $h(x)$ vector can be shown using the transmission line in Fig. 12.18.

Assuming real and reactive power measurements are being made at bus i in Fig. 12.18, the equations for line flow from bus i to j need to be determined as

$$P_{ij} = \left|\tilde{V}_i\right|^2 \left(g_{ij} + g_{ij_{sh}}\right) - \left|\tilde{V}_i\right|\left|\tilde{V}_j\right|\left[g_{ij}\cos(\delta_{ij}) + b_{ij}\sin(\delta_{ij})\right] \qquad (12.11)$$

$$Q_{ij} = |\tilde{V}_i|^2 (b_{ij} + b_{i_{sh}}) - |\tilde{V}_i||\tilde{V}_j|[g_{ij}\sin(\delta_{ij}) + b_{ij}\cos(\delta_{ij})] \qquad (12.12)$$

where $|\tilde{V}_i|$ is the magnitude of the voltage at bus i, $|\tilde{V}_j|$ is the magnitude of the voltage at bus j, δ_{ij} is the phase angle difference between bus i and bus j, g_{ij} and b_{ij} are the conductance and susceptance of line i-j, respectively, and $g_{i_{sh}}$ and $b_{i_{sh}}$ are the shunt conductance and susceptance at bus i, respectively.

Using Eqs. (12.11) and (12.12), Eq. (12.10) can now be rewritten as[2]

$$\bar{z} = \bar{h}(x) + \bar{v}$$

$$= \begin{bmatrix} |\tilde{V}_i|^2 (g_{ij} + g_{i_{sh}}) - |\tilde{V}_i||\tilde{V}_j|[g_{ij}\cos(\delta_{ij}) + b_{ij}\sin(\delta_{ij})] \\ |\tilde{V}_i|^2 (b_{ij} + b_{i_{sh}}) - |\tilde{V}_i||\tilde{V}_j|[g_{ij}\sin(\delta_{ij}) + b_{ij}\cos(\delta_{ij})] \end{bmatrix} + \begin{bmatrix} v_{P_{ij}} \\ v_{Q_{ij}} \end{bmatrix} \qquad (12.13)$$

which expresses the measurements entirely in terms of network parameters (which are assumed known) and system states (bus voltage and phase angle).

Solution Methods

The solution to the state estimation problem has been addressed by a broad class of techniques (Filho et al., Aug. 1990) and differs from power flow algorithms in two modes:

1. certain input data are either missing or inexact, and/or
2. the algorithm used for the calculation may entail approximations and approximate methods designed for high speed processing in the online environment.

In this section, two different solution methods to the state estimation problem will be introduced and described.

Weighted Least Squares

The most common approach to solving the state estimation problem is using the method of weighted least squares (WLS). This is accomplished by identifying the values of the state variables that minimize the performance index, J (the weighted sum of square errors):

$$J = \bar{e}^T R^{-1} \bar{e} \qquad (12.14)$$

where the weighting factor, R, is the diagonal covariance matrix of the measurements and is defined as

$$E[\bar{vv}^T] = R = \begin{bmatrix} \sigma_1^2 & 0 & 0 & 0 & 0 \\ 0 & \sigma_2^2 & 0 & 0 & 0 \\ 0 & 0 & \cdots & 0 & 0 \\ 0 & 0 & 0 & \cdots & 0 \\ 0 & 0 & 0 & 0 & \sigma_m^2 \end{bmatrix}. \qquad (12.15)$$

By defining the error, e, in Eq. (12.14) as the difference between the true measured value, z, and the estimated measured value, $\hat{z}$,

$$\bar{e} = \bar{z} - \hat{\bar{z}} \qquad (12.16)$$

---
[2] The superscript ¯ represents a vector.

a new form for the performance index can be written as

$$J = \left(\bar{z} - \bar{h}(x)\right)^T R^{-1} \left(\bar{z} - \bar{h}(x)\right) \tag{12.17}$$

As shown in Eqs. (12.15) and (12.17), the weights are defined by the inverse of the measurements variances. As a result, measurements of a higher quality have smaller variances that correspond to their weights having higher values, while measurements with poor quality have smaller weights due to the correspondingly higher variance values.

In order to minimize the performance index, J, a first-order necessary condition must hold, namely:

$$\left.\frac{\partial J}{\partial \bar{x}}\right|_{x^k} = 0 \tag{12.18}$$

Evaluating Eq. (12.17) at the necessary condition gives the following:

$$H(x^k)^T R^{-1} \left(\bar{z} - \bar{h}(x)\right) = 0 \tag{12.19}$$

where $H(x)$ represents the $m \times n$[3] measurement Jacobian matrix evaluated at iteration k:

$$H(x) = \begin{bmatrix} \frac{\partial h_1}{\partial x_1} & \frac{\partial h_1}{\partial x_2} & \cdots & \frac{\partial h_1}{\partial x_n} \\ \frac{\partial h_2}{\partial x_1} & \frac{\partial h_2}{\partial x_2} & \cdots & \frac{\partial h_2}{\partial x_n} \\ \cdots & \cdots & \cdots & \cdots \\ \cdots & \cdots & \cdots & \cdots \\ \frac{\partial h_m}{\partial x_1} & \frac{\partial h_m}{\partial x_2} & \cdots & \frac{\partial h_m}{\partial x_n} \end{bmatrix}_{x^k} \tag{12.20}$$

A linearized relationship between the measurements and the state variables is then found by expanding the Taylor series expansion of the function $\bar{h}(x)$ around a point x^k:

$$\bar{h}(x^k) = \bar{h}(x^k) + \Delta \bar{x}^k \frac{\partial \bar{h}(x^k)}{\partial \bar{x}} + \text{higher order terms.} \tag{12.21}$$

This set of equations can be solved using an iterative approach such as Newton Raphson's method. At the $(k+1)^{th}$ iteration, the refreshed values of the state variables can be obtained from their values in the previous iteration by:

$$\bar{x}^{k+1} = \bar{x}^k + \left(H(x^k)^T R^{-1} H(x^k) \right)^{-1} H(x^k)^T R^{-1} \left(\bar{z} - \bar{h}(x^k)\right). \tag{12.22}$$

At convergence, the solution $\bar{x}^{k+1}$ corresponds to the weighted least squares estimates of the state variables. Convergence can be determined either by satisfying

---

[3] m represents the number of measurements; n represents the number of states.

$$\max\left(\bar{x}^{k+1} - \bar{x}^k\right) \leq \varepsilon \tag{12.23}$$

or

$$J^{k+1} - J^k \leq \varepsilon \tag{12.24}$$

where ε is some predetermined convergence factor.

Linear Programming

Another solution method that addresses the state estimation problem is linear programming. Linear programming is an optimization technique that serves to minimize a linear objective function subject to a set of constraints:

$$\min\left\{\bar{c}^T \bar{x}\right\}$$
$$\text{s.t. } A\bar{x} = \bar{b} \tag{12.25}$$
$$\bar{x} \geq 0$$

There are many different techniques associated with solving linear programming problems including the simplex and interior point methods.

Since the objective function, as expressed in Eq. (12.17), is quadratic in terms of the unknowns (states), it must be rewritten in a linear form. This is accomplished by first rewriting the measurement error, as expressed in Eq. (12.10), in terms of a positive measurement error, v_p, and a negative measurement error, v_n:

$$\begin{aligned}\bar{z} &= \bar{h}(x) + \bar{v} \\ &= \bar{h}(x) + \bar{v}_p - \bar{v}_n\end{aligned} \tag{12.26}$$

Restricting the positive and negative measurement errors to only nonnegative values insures the problem is bounded. This was not a concern in the weighted least squares approach since a quadratic function is convex and is guaranteed to contain a global minimum.

Using the new definition of a measurement described in Eq. (12.26) and the inverse of the diagonal covariance matrix of the measurements for weights as described in the weighted least squares approach, the objective function can now be written as:

$$J = R^{-1}\left(\bar{v}_p + \bar{v}_n\right) \tag{12.27}$$

The constraints are the equations relating the state vector to the measurements as shown in Eq. (12.26). Once again, since $h(x)$ is nonlinear, it must be linearized around a point x^k by expanding the Taylor series, as was performed previously in the weighted least squares approach. The solution to the state estimation problem can then be determined by solving the following linear program:

$$\min\left\{R^{-1}\left(\bar{v}_p + \bar{v}_n\right)\right\}$$
$$\text{s.t. } \Delta\bar{z}^k - H\left(x^k\right)\Delta\bar{x}^k + \bar{v}_p - \bar{v}_n = 0 \tag{12.28}$$
$$\bar{v}_p \geq 0$$
$$\bar{v}_n \geq 0$$

where $H(x^k)$ represents the $m \times n$ measurement Jacobian matrix evaluated at iteration k as defined in Eq. (12.20).

State Estimation Operation

State estimators are typically executed either periodically (i.e., every 5 min), on demand, or due to a status change such as a breaker operation isolating a line section. To illustrate the relationship of the state estimator with respect to other **EMS** applications, a simple depiction of an EMS is shown below in Fig. 12.19:

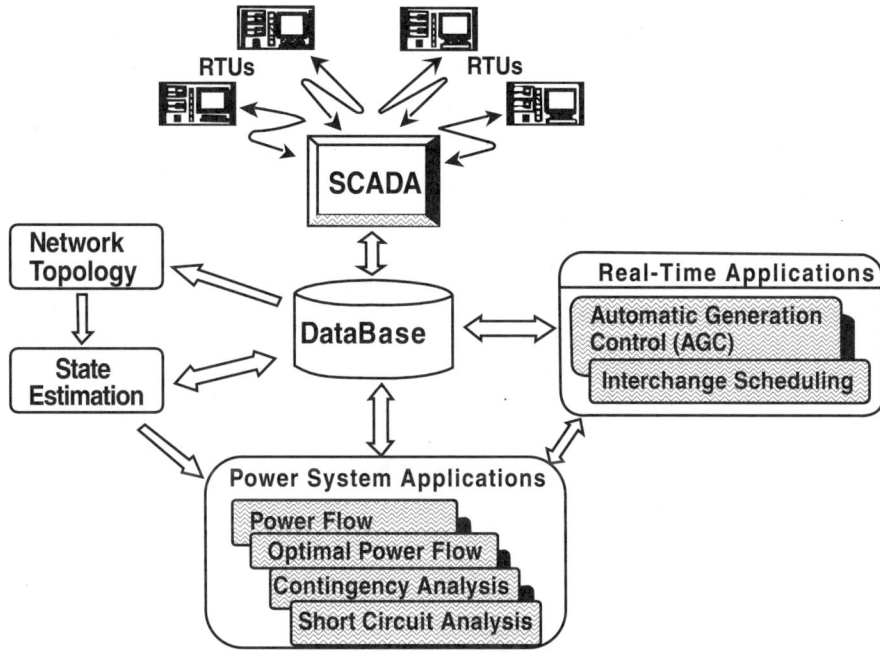

FIGURE 12.19 Simple depiction of an EMS.

As shown, the state estimator receives inputs from the supervisory control and data acquisition (SCADA) system and the network topology assessment applications and stores the state of the system in a central location (i.e., database). Power system applications, such as contingency analysis and optimal power flow, can then be executed based on the state of the system as computed by the state estimator.

Network Topology Assessment

Before the state estimator is executed in realtime, the topology of the network is determined. This is accomplished by a system or **network configurator** that establishes the configuration of the power system network based on telemetered breaker and switch statuses. The network configurator normally addresses questions like:

- Have breaker operations caused individual buses to either be split into two or more isolated buses, or combined into a single bus?
- Have lines been opened or restored to service?

The state estimator then uses the network determined by the network configurator, which consists only of energized (online) lines and devices, as a basis for the calculations to determine the state of the system.

Error Identification

Since state estimators utilize telemetered measurements and network parameters as a foundation for their calculations, the performance of the state estimator depends on the accuracy of the measured data as well as the parameters of the network model. Fortunately, the use of all available measurements introduces a favorable secondary effect caused by the redundancy of information. This redundancy provides the state estimator with more capabilities than just an online AC power flow; it introduces the ability to detect "bad" data. Bad data can come from many sources, such as:

- approximations,
- simplified model assumptions,
- human data handling errors, or
- measurement errors due to faulty devices (e.g., transducers, current transformers).

Telemetered Data

The ability to detect and identify bad measurements is an extremely useful feature of the state estimator. Without the state estimator, obviously wrong telemetered measurements would have little chance of being identified. With the state estimator, operation personnel can have a greater confidence that telemetered data is not grossly in error.

Data is tagged as "bad" when the estimated value is unreasonably different from the measured/telemetered value obtained from the RTU. As a simple example, suppose a bus voltage is measured to be 1.85 pu and is estimated to be 0.95 pu. In this case, the bus voltage measurement could be tagged as bad. Once data is tagged as bad, it should be removed from the measurement set before being utilized by the state estimator.

Most state estimators rely on a combination of preestimation and postestimation schemes for detection and elimination of bad data. Preestimation involves gross bad data detection and consistency tests. Data is identified as bad in preestimation by the detection of gross measurement errors such as zero voltages or line flows that are outside reasonable limits using network topology assessment. Consistency tests classify data either as valid, suspect, or raw for use in postestimation analysis by using statistical properties of related measurements. Measurements are classified as valid if they pass a consistency test that separates measurements into subsets based on a consistency threshold. If the measurement fails the consistency test, it is classified as suspect. Measurements are classified as raw if a consistency test cannot be made and they cannot be grouped into any subset. Raw measurements typically belong to nonredundant portions of the complete measurement set.

Postestimation involves performing a statistical analysis (e.g., hypothesis testing using chi-square tests) on the normalized measurement residuals. A normalized residual is defined as

$$r_i = \frac{z_i - h_i(x)}{\sigma_i} \qquad (12.29)$$

where σ_i is the i-th diagonal term of the covariance matrix, R, as defined in Eq. (12.15). Data is identified as bad in postestimation typically when the normalized residuals of measurements classified as suspect lie outside a predefined confidence interval (i.e., fail the chi-square test).

Parameter Data

In parameter error identification, network parameters (i.e., admittances) that are suspicious are identified and need to be estimated. The use of faulty network parameters can severely impact the quality of state estimation solutions and cause considerable error. A requirement for parameter estimation is that all parameters be identifiable by measurements. This requirement implies the lines under consideration have associated measurements, thereby increasing the size of the measurement set by l, where l is the number of parameters to be estimated. Therefore, if parameter estimation is to be performed, the observability criterion must be augmented to become $m \geq n + l$.

Unobservability

By definition, a state variable is unobservable if it cannot be estimated. Unobservability occurs when the observability criterion is violated ($m < n$) and there are insufficient redundant measurements to determine the state of the system. Mathematically, the matrix $H(x^k)^T R^{-1} H(x^k)$ of Eq. (12.22) becomes singular and cannot be inverted.

The obvious solution to the unobservability problem is to increase the number of measurements. The problem then becomes where and how many measurements need to be added to the measurement set. Adding additional measurements is costly since there are many supplementary factors that must be addressed in addition to the cost of the measuring device such as RTUs, communication infrastructure, and software data processing at the EMS. A number of approaches have been suggested that try to minimize the cost while satisfying the observability criterion (Baran et al., Aug. 1995; Park et al., Aug. 1998).

Another solution to address the problem of unobservability is to augment the measurement set with pseudomeasurements to reach an observability condition for the network. When adding pseudomeasurements to a network, the equation of the pseudomeasured quantity is substituted for actual measurements. In this case, the measurement covariance values in Eq. (12.15) associated with these measurements should have large values that allow the state estimator to treat the pseudomeasurements as if they were measured from a very poor metering device.

Example State Estimation Problem

This section provides a simple example to illustrate how the state estimation process is performed. The WLS method, as previously described, will be applied to a sample system.

System Description

A sample three-bus system is shown below in Fig. 12.20:

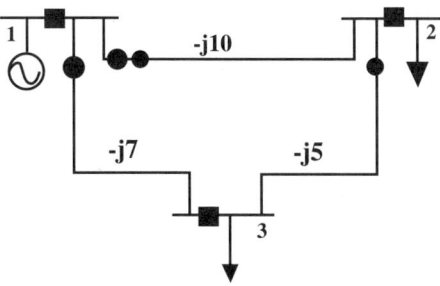

where:

- ■ ⇒ Voltage Measurement (V)
- ● ⇒ Real Power Measurement (MW)
- • ⇒ Reactive Power Measurement (MVAr)

FIGURE 12.20 Sample three-bus power flow system.

Bus 1 is assumed to be the reference bus with a corresponding angle of zero. All other relevant system data is given in Table 12.7.

WLS State Estimation Process

First, the states (x) are defined as the angles at bus 2 and bus 3 and the voltage magnitudes at all buses[4]:

---

[4] The angle at bus one is not chosen as a state since it is designated as the reference bus.

TABLE 12.7 Sample System Data

Measurement Type	Measurement Location	Measurement Value (pu)	Measurement Covariance (σ)
$|\tilde{V}|$	Bus 1	1.02	0.05
$|\tilde{V}|$	Bus 2	1.0	0.05
$|\tilde{V}|$	Bus 3	0.99	0.05
P	Bus 1 – Bus 2	1.5	0.1
Q	Bus 1 – Bus 2	0.2	0.1
P	Bus 1 – Bus 3	1.0	0.1
Q	Bus 2 – Bus 3	0.1	0.1

$$\bar{x} = \begin{bmatrix} \delta_2 \\ \delta_3 \\ |\tilde{V}_1| \\ |\tilde{V}_2| \\ |\tilde{V}_3| \end{bmatrix}$$

This gives a total of seven measurements and five states that satisfy the observability criterion requiring more measurements than states.

Using the previously defined equations for the WLS state estimation procedure, the following can be determined:

$$R = \begin{bmatrix} \sigma_i^2 \end{bmatrix}$$

$$= \begin{bmatrix} (.05)^2 & 0 & 0 & 0 & 0 & 0 & 0 \\ 0 & (.05)^2 & 0 & 0 & 0 & 0 & 0 \\ 0 & 0 & (.05)^2 & 0 & 0 & 0 & 0 \\ 0 & 0 & 0 & (.1)^2 & 0 & 0 & 0 \\ 0 & 0 & 0 & 0 & (.1)^2 & 0 & 0 \\ 0 & 0 & 0 & 0 & 0 & (.1)^2 & 0 \\ 0 & 0 & 0 & 0 & 0 & 0 & (.1)^2 \end{bmatrix}$$

$$\hat{\bar{z}} = h(\bar{x})$$

$$= \begin{bmatrix} x_3 \\ x_4 \\ x_5 \\ 10 x_3 x_4 \sin x_1 \\ 10 x_3^2 - 10 x_3 x_4 \cos x_1 \\ 7 x_3 x_5 \sin x_2 \\ 5 x_4^2 - 5 x_4 x_5 \cos(x_1 - x_2) \end{bmatrix}$$

$$H(x) = \left[\frac{\partial \bar{h}}{\partial x}\right]$$

$$= \begin{bmatrix} 0 & 0 & 1 & 0 & 0 \\ 0 & 0 & 0 & 1 & 0 \\ 0 & 0 & 0 & 0 & 1 \\ 10x_3x_4 \cos x_1 & 0 & 10x_4 \sin x_1 & 10x_3 \sin x_1 & 0 \\ 10x_3x_4 \sin x_1 & 0 & 20x_3 - 10x_4 \cos x_1 & -10x_3 \cos x_1 & 0 \\ 0 & 7x_3x_5 \cos x_2 & 7x_5 \sin x_2 & 0 & 7x_3 \sin x_2 \\ 5x_4x_5 \sin(x_1 - x_2) & -5x_4x_5 \sin(x_1 - x_2) & 0 & 10x_4 - 5x_5 \cos(x_1 - x_2) & -5x_4 \cos(x_1 - x_2) \end{bmatrix}$$

Using zero as an initial guess for the states representing voltage angles (x_1 and x_2) and the measured voltages as given in Table 12.7 for the states representing voltage magnitudes (x_3, x_4, and x_5):

$$\begin{bmatrix} x_1^0 \\ x_2^0 \\ x_3^0 \\ x_4^0 \\ x_5^0 \end{bmatrix} = \begin{bmatrix} 0 \\ 0 \\ 1.02 \\ 1.00 \\ 0.99 \end{bmatrix},$$

the state values at the first iteration are determined by Eq. (12.22) to be

$$\begin{bmatrix} x_1^1 \\ x_2^1 \\ x_3^1 \\ x_4^1 \\ x_5^1 \end{bmatrix} = \begin{bmatrix} 0.147 \\ 0.142 \\ 1.022 \\ 1.003 \\ 0.984 \end{bmatrix}.$$

After four iterations, the state estimation process converges to the final states:

$$\begin{bmatrix} x_1 \\ x_2 \\ x_3 \\ x_4 \\ x_5 \end{bmatrix} = \begin{bmatrix} 0.147 \\ 0.143 \\ 1.016 \\ 1.007 \\ 0.987 \end{bmatrix}.$$

Using the solved voltages and angles from the state estimation process, the line flows and bus injections can now be calculated. With the state of the system now known, other applications such as contingency analysis and optimal power flow may be performed. Notice, the state estimation process results in the state of the system, just as when performing a power flow but without *a priori* knowledge of bus injections.

Defining Terms

Remote Terminal Unit (RTU): Hardware that telemeters systemwide data from various field locations (i.e., substations, generating plants) to a central location.

State estimator: An application that uses a statistical process in order to estimate the state of the system.

State variable: The quantity to be estimated by the state estimator, typically bus voltage and angle.
Network configurator: An application that determines the configuration of the power system based on telemetered breaker and switch statuses.
Supervisory Control and Data Acquisition (SCADA): A computer system that performs data acquisition and remote control of a power system.
Energy Management System (EMS): A computer system that monitors, controls, and optimizes the transmission and generation facilities with advanced applications. A SCADA system is a subset of an EMS.

References

Schweppe, F.C., Wildes, J., Power System Static-State Estimation I,II,III, *IEEE Trans. on Power Appar. Syst.*, 89, 120–135, January 1970.

Filho, M.B.D.C. et al., Bibliography on power system state estimation (1968-1989), *IEEE Trans. on Power Syst.*, 5, 3, 950–961, August 1990.

Baran, M.E. et al., A meter placement method for state estimation, *IEEE Trans. on Power Syst.*, 10, 3, 1704–1710, August 1995.

Park, Y.M et al., Design of reliable measurement system for state estimation, *IEEE Trans. on Power Syst.*, 3, 3, 830–836, August 1998.

12.4 Optimal Power Flow

M. E. El-Hawary

An Optimal Power Flow (OPF) function schedules the power system controls to optimize an objective function while satisfying a set of nonlinear equality and inequality constraints. The equality constraints are the conventional power flow equations; the inequality constraints are the limits on the control and operating variables of the system. Mathematically, the OPF can be formulated as a constrained nonlinear optimization problem. This section reviews features of the problem and some of its variants as well as requirements for online implementation.

Optimal scheduling of the operations of electric power systems is a major activity, which turns out to be a large-scale problem when the constraints of the electric network are taken into account. This document deals with recent developments in the area emphasizing optimal power flow formulation and deals with conventional optimal power flow (OPF), accounting for the dependence of the power demand on voltages in the system, and requirements for online implementation.

The OPF problem was defined in the early 1960s (Burchett et al., Feb. 1982) as an extension of conventional economic dispatch to determine the optimal settings for control variables in a power network respecting various constraints. OPF is a static constrained nonlinear optimization problem, whose development has closely followed advances in numerical optimization techniques and computer technology. It has since been generalized to include many other problems. Optimization of the electric system with losses represented by the power flow equations was introduced in the 1960s (Carpentier, 1962; Dommel and Tinney, Oct. 1968). Since then, significant effort has been spent on achieving faster and robust solution methods that are suited for online implementation, operating practice, and security requirements.

OPF seeks to optimize a certain objective, subject to the network power flow constraints and system and equipment operating limits. Today, any problem that involves the determination of the instantaneous "optimal" steady state of an electric power system is referred to as an Optimal Power Flow problem. The optimal steady state is attained by adjusting the available controls to minimize an objective function subject to specified operating and security requirements. Different classes of OPF problems, designed for special-purpose applications, are created by selecting different functions to be minimized, different sets of controls, and different sets of constraints. All these classes of the OPF problem are subsets of the general problem. Historically, different solution approaches have been developed to solve these different classes of OPF. Commercially available OPF software can solve very large and complex formulations in a relatively short time, but may still be incapable of dealing with online implementation requirements.

There are many possible objectives for an OPF. Some commonly implemented objectives are:

- fuel or active power cost optimization,
- active power loss minimization,
- minimum control-shift,
- minimum voltage deviations from unity, and
- minimum number of controls rescheduled.

In fuel cost minimization, the outputs of all generators, their voltages, LTC transformer taps and LTC phase shifter angles, and switched capacitors and reactors are control variables. The active power losses can be minimized in at least two ways (Happ and Vierath, July, 1986). In both methods, all the above variables are adjusted except for the active power generation. In one method, the active power generation at the swing bus is minimized while keeping all other generation constant at prespecified values. This effectively minimizes the total active power losses. In another method, an actual expression for the losses is minimized, thus allowing the exclusion of lines in areas not optimized.

The behavior of the OPF solutions during contingencies was a major concern, and as a result, security constrained optimal power flow was introduced in the early 1970s. Subsequently, online implementations became a new thrust in order to meet the challenges of new deregulated operating environments.

Conventional Optimal Economic Scheduling

Conventional optimal economic scheduling minimizes the total fuel cost of thermal generation, which may be approximated by a variety of expressions such as linear or quadratic functions of the active power generation of the unit. The total active power generation in the system must equal the load plus the active transmission losses, which can be expressed by the celebrated Kron's loss formula. Reserve constraints may be modeled depending on system requirements. Area and system spinning, supplemental, emergency, or other types of reserve requirements involve functional inequality constraints. The forms of the functions used depend on the type of reserve modeled. A linear form is evidently most attractive from a solution method point of view. However, for thermal units, the spinning reserve model is nonlinear due to the limit on a unit's maximum reserve contribution. Additional constraints may be modeled, such as area interchange constraints used to model network transmission capacity limitations. This is usually represented as a constraint on the net interchange of each area with the rest of the system (i.e., in terms of limits on the difference between area total generation and load).

The objective function is augmented by the constraints using a Lagrange-type multiplier lambda, λ. The optimality conditions are made up of two sets. The first is the problem constraints. The second set is based on variational arguments giving for each thermal unit:

$$\frac{\partial F_i}{\partial P_i} = \lambda \left[1 - \frac{\partial P_L}{\partial P_i} \right] \quad i = 1, \ldots, N \tag{12.30}$$

The optimality conditions along with the physical constraints are a set of nonlinear equations that requires iterative methods to solve. Newton's method has been widely accepted in the power industry as a powerful tool to solve problems such as the load flow and optimal load flow. This is due to its reliable and fast convergence, known to be quadratic.

A solution can usually be obtained within a few iterations, provided that a reasonably good initial estimate of the solution is available. It is therefore appropriate to employ this method to solve the present problem.

Conventional OPF Formulation

The optimal power flow is a constrained optimization problem requiring the minimization of:

$$f = (x,u) \tag{12.31}$$

subject to

$$g(x,u) = 0 \tag{12.32}$$

$$h(x,u) \leq 0 \tag{12.33}$$

$$u^{min} \leq u \leq u^{max} \tag{12.34}$$

$$x^{min} \leq x \leq x^{max} \tag{12.35}$$

Here $f(x,u)$ is the scalar objective function, $g(x,u)$ represents nonlinear equality constraints (power flow equations), and $h(x,u)$ is the nonlinear inequality constraint of vector arguments x and u. The vector x contains dependent variables consisting of bus voltage magnitudes and phase angles, as well as the MVAr output of generators designated for bus voltage control and fixed parameters such as the reference bus angle, noncontrolled generator MW and MVAr outputs, noncontrolled MW and MVAr loads, fixed bus voltages, line parameters, etc. The vector u consists of control variables including:

- real and reactive power generation
- phase-shifter angles
- net interchange
- load MW and MVAr (load shedding)
- DC transmission line flows
- control voltage settings
- LTC transformer tap settings

Examples of equality and inequality constraints are:

- limits on all control variables
- power flow equations
- generation/load balance
- branch flow limits (MW, MVAr, MVA)
- bus voltage limits
- active/reactive reserve limits
- generator MVAr limits
- corridor (transmission interface) limits

The power system consists of a total of N buses, N_G of which are generator buses. M buses are voltage controlled, including both generator buses and buses at which the voltages are to be held constant. The voltages at the remaining (N − M) buses (load buses), must be found.

The network equality constraints are represented by the load flow equations:

$$P_i(V,\delta) - P_{gi} + P_{di} = 0 \tag{12.36}$$

$$Q_i(V,\delta) - Q_{gi} + Q_{di} = 0 \tag{12.37}$$

Two different formulation versions can be considered.

(a) *Polar Form:*

$$P_i(V,\delta) = |V_i| \sum_{1}^{N} |V_j||Y_{ij}| \cos(\delta_i - \delta_j - \phi_{ij}) \quad (12.38)$$

$$Q_i(V,\delta) = |V_i| \sum_{1}^{N} |V_j||Y_{ij}| \sin(\delta_i - \delta_j - \phi_{ij}) \quad (12.39)$$

$$Y_{ij} = |Y_{ij}|\underline{/\phi_{ij}} \quad (12.40)$$

where

- P_i = Active power injection into bus i.
- Q_i = Reactive power injection into bus i.
- $|\tilde{V}_i|$ = Voltage magnitude of bus i.
- δ_i = Angle at bus i.
- $|\tilde{Y}_{ij}|, \phi_{ij}$ = Magnitude and angle of the admittance matrix.
- P_{di}, Q_{di} = Active and reactive load on bus i.

(b) *Rectangular Form:*

$$P_i(e,f) = e_i \left[\sum_{1}^{N} (G_{ij}e_j - B_{ij}f_j) \right] + f_i \left[\sum_{1}^{N} (G_{ij}f_j + B_{ij}e_j) \right] \quad (12.41)$$

$$Q_i(e,f) = f_i \left[\sum_{1}^{N} (G_{ij}e_j - B_{ij}f_j) \right] - e_i \left[\sum_{1}^{N} (G_{ij}f_j + B_{ij}e_j) \right] \quad (12.42)$$

- e_i = Real part of complex voltage at bus i.
- f_i = Imaginary part of the complex voltage at bus i.
- G_{ij} = Real part of the complex admittance matrix.
- B_{ij} – Imaginary part of the complex admittance matrix.

The control variables vary according to the objective being minimized. For fuel cost minimization, they are usually the generator voltage magnitudes, generator active powers, and transformer tap ratios. The dependent variables are the voltage magnitudes at load buses, phase angles, and reactive generations.

Application of Optimization Methods to OPF

Various optimization methods have been proposed to solve the optimal power flow problem, some of which are refinements on earlier methods. These include:

1. Generalized Reduced Gradient (GRG) method.
2. Reduced gradient method.
3. Conjugate gradient methods.
4. Hessian-based method.
5. Newton's method.
6. Linear programming methods.
7. Quadratic programming methods.
8. Interior point methods.

Some of these techniques have spawned production OPF programs that have achieved a fair level of maturity and have overcome some of the earlier limitations in terms of flexibility, reliability, and performance requirements.

Generalized Reduced Gradient Method

The Generalized Reduced Gradient method (GRG), due to Abadie and Carpentier (1969), is an extension of the Wolfe's reduced gradient method (Wolfe, 1967) to the case of nonlinear constraints. Peschon in 1971 and Carpentier in 1973 used this method for OPF. Others have used this method to solve the optimal power flow problem since then (Lindqvist et al., 1984; Yu et al., 1986).

Reduced Gradient Method

A reduced gradient method was used by Dommel and Tinney (1968). An augmented Lagrangian function is formed. The negative of the gradient $\partial L/\partial u$ is the direction of steepest descent. The method of reduced gradient moves along this direction from one feasible point to another with a lower value of f, until the solution does not improve any further. At this point an optimum is found, if the Kuhn-Tucker conditions (1951) are satisfied. Dommel and Tinney used Newton's method to solve the power flow equations.

Conjugate Gradient Method

In 1982, Burchett et al. used a conjugate gradient method, which is an improvement on the reduced gradient method. Instead of using the negative gradient ∇f as the direction of steepest descent, the descent directions at adjacent points are linearly combined in a recursive manner.

$$\Gamma_k = -\nabla f + \beta_k \Gamma_{k-1} \qquad \beta_0 = 0 \qquad (12.43)$$

Here, r_k is the descent direction at iteration "k."

Two popular methods for defining the scalar value β_k are the Fletcher-Reeves method (Carpentier, June 1973) and the Polak-Ribiere method (1969).

Hessian-Based Methods

Sasson (Oct. 1969) discusses methods (Fiacco and McCormick, 1964; Lootsma, 1967; Zangwill, 1967) that transform the constrained optimization problem into a sequence of unconstrained problems. He uses a transformation introduced by Powell and Fletcher (1963). Here, the Hessian matrix is not evaluated directly. Instead, it is built indirectly starting initially with the identity matrix so that at the optimum point it becomes the Hessian itself.

Due to drawbacks of the Fletcher-Powell method, Sasson et al. (1973) developed a Hessian load flow with an extension to OPF. Here, the Hessian is evaluated and solved unlike in the previous method. The objective function is transformed as before to an unconstrained objective. An unconstrained objective is formed. All equality constraints and only the violating inequality constraints are included. The sparse nature of the Hessian is used to reduce storage and computation time.

Newton OPF

Newton OPF has been formulated by Sun et al. (1984), and later by Maria et al. (Aug. 1987). An augmented Lagrangian is first formed. The set of first derivatives of the augmented objective with respect to the control variables gives a set of nonlinear equations as in the Dommel and Tinney method. Unlike in the Dommel and Tinney method where only a part of these are solved by the N-R method, here, all equations are solved simultaneously by the N-R method.

The method itself is quite straightforward. It is the method of identifying binding inequality constraints that challenged most researchers. Sun et al. use a multiply enforced, zig-zagging guarded technique for some of the inequalities, together with penalty factors for some others. Maria et al. used an LP-based technique to identify the binding inequality set. Another approach is to use purely penalty factors. Once the binding inequality set is known, the N-R method converges in a very few iterations.

Linear Programming-Based Methods

LP methods use a linear or piecewise-linear cost function. The dual simplex method is used in some applications (Bentall, 1968; Shen and Laughton, Nov. 1970; Stott and Hobson, Sept./Oct. 1978; Wells, 1968). The network power flow constraints are linearized by neglecting the losses and the reactive powers, to obtain the DC load flow equations. Merlin (1972) uses a successive linearization technique and repeated application of the dual simplex method.

Due to linearization, these methods have a very high speed of solution, and high reliability in the sense that an optimal solution can be obtained for most situations. However, one drawback is the inaccuracies of the linearized problem. Another drawback for loss minimization is that the loss linearization is not accurate.

Quadratic Programming Methods

In these methods, instead of solving the original problem, a sequence of quadratic problems that converge to the optimal solution of the original problem are solved. Burchett et al. use a sparse implementation of this method. The original problem is redefined as simply, to minimize,

$$f(x) \tag{12.44}$$

subject to:

$$g(x) = 0 \tag{12.45}$$

The problem is to minimize

$$g^T p + \frac{1}{2} p^T H p \tag{12.46}$$

subject to:

$$Jp = 0 \tag{12.47}$$

where

$$p = x - x_k \tag{12.48}$$

Here, g is the gradient vector of the original objective function with respect to the set of variables "x." "J" is the Jacobian matrix that contains the derivatives of the original equality constraints with respect to the variables, and "H" is the Hessian containing the second derivatives of the objective function and a linear combination of the constraints with respect to the variables. x_k is the current point of linearization. The method is capable of handling problems with infeasible starting points and can also handle ill-conditioning due to poor R/X ratios. This method was later extended by El-Kady et al. (May 1986) in a study for the Ontario Hydro System for online voltage/var control. A nonsparse implementation of the problem was made by Glavitsch (Dec. 1983) and Contaxis (May, 1986).

Interior Point Methods

The projective scaling algorithm for linear programming proposed by N. Karmarkar is characterized by significant speed advantages for large problems reported to be as much as 50:1 when compared to the simplex method (Karmarkar, 1984). This method has a polynomial bound on worst-case running time that is better than the ellipsoid algorithms. Karmarkar's algorithm is significantly different from Dantzig's simplex method. Karmarkar's interior point rarely visits too many extreme points before an optimal point is found. The IP method stays in the interior of the polytope and tries to position a current solution as the "center of the universe" in finding a better direction for the next move. By properly choosing the step lengths, an optimal solution is achieved after a number of iterations. Although this IP approach requires more computational time in finding a moving direction than the traditional simplex method, better moving direction is achieved resulting in less iterations. Therefore, the IP approach has become a major rival of the simplex method and has attracted attention in the optimization community. Several variants of interior points have been proposed and successfully applied to optimal power flow (Momoh, 1992; Vargas et al., 1993; Yan and Quintana, 1999).

OPF Incorporating Load Models

Load Modeling

The area of power systems load modeling has been well explored in the last two decades of the twentieth century. Most of the work done in this area has dealt with issues in stability of the power system. Load modeling for use in power flow studies has been treated in a few cases (Concordia and Ihara, 1982; IEEE Committee Report, 1973; IEEE Working Group Report, 1996; Iliceto et al., 1972; Vaahedi et al., 1987). In stability studies, frequency and time are variables of interest, unlike in power flow and some OPF studies. Hence, load models for use in stability studies should account for any load variations with frequency and time as well. These types of load models are normally referred to as dynamic load models. In power flow, OPF studies neglecting contingencies, and security-constrained OPF studies using preventive control, time, and frequency, are not considered as variables. Hence, load models for this type of study need not account for time and frequency. These load models are static load models.

In security-constrained OPF studies using corrective control, the time allowed for certain control actions is included in the formulation. However, this time merely establishes the maximum allowable correction, and any dynamic behavior of loads will usually end before any control actions even begin to function. Hence, static load models can be used even in this type of formulation.

Static Load Models

Several forms of static load models have been proposed in the literature, from which the exponential and quadratic models are most commonly used. The exponential form is expressed as:

$$P_m = a_p V^{b_b} \quad (12.49)$$

$$Q_m = a_q V^{bq} \quad (12.50)$$

The values of the coefficients a_p and a_q can be taken as the specified active and reactive powers at that bus, provided the specified power demand values are known to occur at a voltage of 1.0 per unit, measured at the network side of the distribution transformer. A typical measured value of the demand and the network side voltage is sufficient to determine approximately the values of the coefficients, provided the exponents are known. The range of values reported for the exponents vary in the literature, but typical values are 1.5 and 2.0 for b_p and b_q, respectively.

Conventional OPF Studies Including Load Models

Incorporation of load models in OPF studies has been considered in a couple of cases (El-Din et al., 1989; Vaahedi and El-Din, May 1989) for the Ontario Hydro energy management system. In both cases, loss minimization was considered to be the objective. It is concluded by Vaahedi and El-Din (1989) that the modeling of ULTC operation and load characteristics is important in OPF calculations.

The effects of load modeling in OPF studies have been considered for the case where the generator bus voltages are held at prespecified values (Dias and El-Hawary, 1989). Since the swing bus voltage is held fixed at all times (and also the generator bus voltages in the absence of reactive power limit violations), the average system voltage is maintained in most cases. Thus, an increase in fuel cost due to load modeling was noticed for many systems that had a few (or zero) reactive limit violations, and a decrease for those with a noticeable number of reactive limit violations. Holding the generator bus voltages at specified values restricts the available degrees of freedom for OPF and makes the solution less optimal.

Incorporation of load models in OPF studies minimizing fuel cost (with all voltages free to vary within bounds) can give significantly different results when compared with standard OPF results. The reason for this is that the fuel cost can now be reduced by lowering the voltage at the modeled buses along with all other voltages wherever possible. The reduction of the voltages at the modeled buses lowers the power demand of the modeled loads and will thus give the lower fuel cost. When a large number of loads are modeled, the total fuel cost may be lower than the standard OPF. However, a lowering of the fuel cost via a lowering of the power demand may not be desirable under normal circumstances, as this will automatically decrease the total revenue of the operation. This can also give rise to a lower net revenue if the decrease in the total revenue is greater than the decrease in the fuel cost. This is even more undesirable. What is needed is an OPF solution that does not decrease the total power demand in order to achieve a minimum fuel cost. The standard OPF solution satisfies this criterion. However, given a fair number of loads that are fed by fixed tap transformers, the standard OPF solution can be significantly different from the practically observed version of this solution.

Before attempting to find an OPF solution incorporating load models that satisfies the required criterion, we deal with the reason for the problem. In a standard OPF formulation, the total revenue is constant and independent of the solution. Hence, we can define net revenue R_N, which is linearly related to the total fuel cost F_C by the formula:

$$R_N = -F_C + constant \qquad (12.51)$$

The constant term is the total revenue dependent on the total power demand and the unit price of electricity charged to the customers. From this relationship we see that a solution with minimum fuel cost will automatically give maximum net revenue. Now, when load models are incorporated at some buses, the total power demand is not a constant, and hence the total revenue will also not be constant. As a result,

$$R_N = -F_C + R_T \qquad (12.52)$$

where "R_1" is the total demand revenue and is no longer a constant.

If instead of minimizing the fuel cost, we now maximize the net revenue, we will definitely avoid the difficulties encountered earlier. This is equivalent to minimizing the difference between the fuel cost and the total revenue. Hence we see that, in the standard OPF, the required maximum net revenue is implied, and the equivalent minimum fuel cost is the only function that enters the computations.

Security Constrained OPF Including Load Models

A conventional OPF result can have optimal but insecure states during certain contingencies. This can be avoided by using a security constrained OPF. Unlike in the former, for a security constrained OPF, we can incorporate load models in a variety of ways. For example, we can consider the loads as independent of voltage for the intact system, but dependent on the voltage during contingencies. This can be justified by saying that the voltage deviations encountered during a standard OPF and modeled OPF are small compared to those that can be encountered during contingencies. Since the total power demand for the intact system is not changed, fuel cost comparisons between this case and a standard SCOPF seem more reasonable. We can also incorporate load models for the intact system as well as during contingencies, while minimizing the fuel cost. However, we then encounter the problem discussed in the previous section regarding net earnings. Another approach is to incorporate load models for the intact case as well as during contingencies, while minimizing the total fuel cost minus the total revenue.

Inaccuracies of Standard OPF Solutions

It was stated earlier that the standard OPF (or standard security constrained OPF) solution can give results not compatible with practical observations (i.e., using the control variable values from these solutions) when a fair number of loads are fed by fixed tap transformers. The discrepancies between the simulated and observed results will be due to discrepancies between the voltage at a bus feeding a load through a fixed tap transformer, and the voltage at which the specified power demand for that load occurs. The observed results can be simulated approximately by performing a power flow incorporating load models. The effects of load modeling in power flow studies have been treated in a few cases (Dias and El-Hawary, 1990; El-Hawary and Dias, Jan. 1987; El-Hawary and Dias, 1987; El-Hawary and Dias, July 1987). In all these studies, the specified power demand of the modeled loads was assumed to occur at a bus voltage of 1.0 per unit. The simulated modeled power flow solution will be same as the practically observed version only when exact model parameters are utilized.

SCOPF Including Load Modeling

Security constrained optimal power flow (abbreviated SCOPF) takes into account outages of certain transmission lines or equipment (Alsac and Stott, May/June 1974; Schnyder and Glavitsch, 1987). Due to the computational complexity of the problem, more work has been devoted to obtaining faster solutions requiring less storage, and practically no attention has been paid to incorporating load models in the formulations. A SCOPF solution is secure for all credible contingencies or can be made secure by corrective means. In a secure system (level 1), all load is supplied, operating limits are enforced, and no limit violations occur in a contingency. Security level 2 is one where all load is supplied, operating limits are satisfied, and any violations caused by a contingency can be corrected by control action without loss of load. Level 1 security is considered in Dias and El-Hawary (Feb. 1991).

Studies of the effects of load voltage dependence in PF and OPF (Dias and El-Hawary, Sept. 1989) concluded that for PF incorporating load models, the standard solution gives more conservative results with respect to voltages in most cases. However, exceptions have been observed in one test system. Fuel costs much lower than those associated with the standard OPF are obtained by incorporating load models with all voltages free to vary within bounds. This is due to the decrease in the power demand by the reduction of the voltages at buses whose loads are modeled. When quite a few loads are modeled, the minimum fuel costs may be much lower than the corresponding standard OPF fuel cost with a significant decrease in power demand.

A similar effect can be expected when load models are incorporated in security constrained OPF studies. The decrease in the power demand when load models are incorporated in OPF studies may not be desirable under normal operating conditions. This problem can be avoided in a security constrained OPF by incorporating load models during contingencies only. This not only gives results that are more comparable with standard OPF results, but may also give lower fuel costs without lowering the power demand of the intact system. The modeled loads are assumed to be fed by fixed tap transformers and are modeled using an exponential type of load model.

In Dias and El-Harawy (1990), some selected buses were modeled using an exponential type of load model in three cases. In the first, the specified load at modeled buses is obtained with unity voltage. In the second case, the transformer taps have been adjusted to give all industrial-type consumers 1.0 per unit at the low-voltage panel when the high-side voltage corresponds to the standard OPF solution. In the third case, the specified power demand is assumed to take place when the high-side voltages correspond to the intact case of the standard security constrained OPF solution. It is concluded that a decrease in fuel cost can be obtained in some instances when load models are incorporated in security constrained OPF studies during contingencies only. In situations where a decrease in fuel cost is obtained in this manner, the magnitude of decrease depends on the total percentage of load fed by fixed tap transformers and the sensitivity of these loads to modeling. The tap settings of these fixed tap transformers influence the results as well. An increase in fuel cost can also occur in some isolated cases. However, in either case, given accurate load models, optimal power flow solutions that are more accurate than the conventional OPF solutions can be obtained. An alternate approach for normal OPF as well as security constrained OPF is also suggested.

Influence of Fixed Tap Transformer Fed Loads

A standard OPF assumes that all loads are independent of other system variables. This implies that all loads are fed by ULTC transformers that hold the load-side voltage to within a very narrow bandwidth sufficient to justify the assumption of constant loads. However, when some loads are fed by fixed tap transformers, this assumption can result in discrepancies between the standard OPF solution and its observed version. In systems where the average voltage of the system is reasonably above 1.0 per unit (specifically where the loads fed by fixed tap transformers have voltages greater than the voltage at which the specified power demand occurs), the practically observed version of the standard OPF solution will have a higher total power demand, and hence a higher fuel cost, and total revenue, and net revenue. Conversely, where such voltages are lower than the voltage at which the specified power demand occurs, the total power demand, fuel cost, total and net revenues will be lower than expected. For the former case, the system voltages will usually be slightly less than expected, while for the latter case they will usually be slightly higher than expected.

The changes in the power demand at some buses (in the observed version) will alter the power flows on the transmission lines, and this can cause some lines to deliver more power than expected. When this occurs on transmission lines that have power flows near their upper limit, the observed power flows may be above the respective upper limit, causing a security violation. Where the specified power demand occurs at the bus voltages obtained by a standard OPF solution, the observed version of the standard OPF solution will be itself, and there will ideally be no security violations in the observed version.

Most of the above conclusions apply to security constrained OPF as well (Dias and El-Hawary, Nov. 1991). However, since a security constrained OPF solution will in general have higher voltages than its normal counterpart (in order to avoid low voltage limit violations during contingencies), the increase in power demand, and total and net revenues will be more significant while the decrease in the above quantities will be less significant. Also, the security violations due to line flows will now be experienced mainly during contingencies, as most line flows will now usually be below their upper limits for the intact case. For security constrained OPF solutions that incorporate load models only during contingencies, the simulated and observed results will mainly differ in the intact case. Also, with loads modeled during contingencies, the average voltage is lower than for the standard security constrained OPF solution and hence there will be more cases with a decrease in the power demand, fuel cost, and total and net revenues in the observed version of the results than for its standard counterpart.

Operational Requirements for Online Implementation

The most demanding requirements on OPF technology are imposed by online implementation. It was argued that OPF, as expressed in terms of smooth nonlinear programming formulations, produces results that are far too approximate descriptions of real-life conditions to lead to successful online implementations. Many OPF formulations do not have the capability to incorporate all operational considerations into the solutions. Moreover, some operating practices are occasionally incompatible with such OPF formulations. Consequently, many proposed "theoretical optimal solutions" are of little value to the operators who are almost constantly presented with simultaneous events that are outside the scope of OPF definition. These limitations, if properly addressed, do not have to prevent OPF programs from being used in practice, especially when the operational optimal solution may also not be known. Papalexopoulos (1996) offers some of the requirements that need to be met so that OPF applications are useful to, and usable by, the dispatchers in online applications.

Speed Requirements

Fast OPF programs designed for online application are needed because under normal conditions, the state of the power system changes continuously and can change abruptly during emergency conditions. The changes involve the evolution of bus active and reactive power generation and loads with time, control variables moving to and off their limits as time changes, and topology changes due to switching operations and other planned or forced outages. The need for fast OPF solutions is especially true when an excessive amount of calculations due to modeling of contingency constraints or repeated OPF runs is involved.

In general, an online OPF calculation should have been completed before the state of the power system has changed to another state that is appreciably different from the earlier state. Determining the optimal execution frequency to maximize the benefits of the computations depends on the specific situation and is limited by finite computing resources. It may be preferable to develop incrementally correct and flexible algorithms to offer fast and more frequent scheduling. This leads us to conclude that conventional formulations and algorithms characterized with quadratic convergence that give very accurate and "mathematically optimal" solutions, but neglect operational realities are not appropriate for online implementation. Fast and frequent scheduling requires "hot start" OPF capabilities developed to take advantage of the optimal status of previously optimized operating points. The hot start option is significant when the rate of change of system state is small and previously optimized points are still "relevant" to the current operating conditions.

Robustness of OPF Solutions with Respect to Initial Guess Point

An OPF program needs to produce consistent solutions and thus must not be sensitive to the selected initial guess used. In addition, changes in the OPF solutions between operating states need to be consistent with the changes in the power system operating constraints. The OPF solutions will never be exactly the same when starting from different initial guess points because the solution process is iterative. Any differences should be within the tolerances specified by the convergence criterion, and of a magnitude that would be considered insignificant to the operator. First-order OPF solution methods were not well received because noticeably different solutions could be obtained when an OPF algorithm was initialized from different initial guess points, with only one (or even none) of the solutions actually constituting a local optimum. Theoretically, if the objective function and the feasible region can be shown to be convex, then the optimal solution will be unique (Gill et al., 1981). Unfortunately, the complexity of the nonlinear equations and inequality constraints involved in OPF problems make it untenable to rigorously prove convexity. If multiple local minima actually exist, then additional computational or heuristic methods must be used to resolve the issue.

A normally feasible OPF solution space may become nonconvex (thus leading to multiple OPF solutions) due to two considerations. The first is due to use of discontinuous techniques to model specific operating practices and preferences, and the second is due to modeling of local controls. The conventional power flow problem with local control capability, whose implicit objective is feasible with respect to a limited set of inequalities, does not have a unique solution. Nevertheless, solutions of the same problem from different starting conditions usually match quite closely. Occasionally, different initial guess solutions can lead to different solutions. This takes place when two or more feasible voltage levels can satisfy nonlinear loads. OPF applications, however, should be able to overcome this type of ambiguity.

Discrete Modeling

Discrete control is widely used in the electric network. For example, transformers are used for voltage control, shunt capacitors and reactors are switched on or off to correct voltage profiles and to reduce active power transmission losses, and phase shifters are used to regulate the MW flows of transmission lines. An efficient and effective OPF discretization procedure is needed to assist the operators in utilizing discrete controls in a realistic and optimal or near-optimal manner. Discrete elements to be included in the OPF formulation are branch switching; prohibited zones of generator cost curves; and priority sequence levels for unfeasibility handling. OPF algorithms designed for online applications should be able to appropriately handle the discrete aspects of the problem.

Using both discrete and continuous controls converts the OPF into a mixed discrete-continuous optimization problem. A possible accurate solution using a method such as mixed-integer nonlinear programming would be orders of magnitude slower than ordinary nonlinear programming methods (Gill et al., 1981). Linear programming-based OPF algorithms allow substantial recognition of discrete controls by setting the cost curve segment break points at discrete control steps. However, most methods that solve for a nonseparable objective function by nonlinear programming methods do not properly model discrete controls.

Current OPF algorithms treat all controls as continuous variables during the initial solution process. Once the continuous solution is obtained, each discrete variable is moved to the nearest discrete setting. This produces acceptable solutions, assuming that the step sizes for the discrete controls are sufficiently small, which is usually the case for transformer taps and phase shifter angles (Papalexopoulos et al., 1996). Approximate solutions that can produce near-optimal results appear to be a reasonable alternative to rigorous solution methods. One such scheme (Liu et al., 1991) uses penalty functions for discrete controls. The object is to penalize the continuous approximations of discrete control variables for movements away from their discrete steps. This scheme is well suited for Newton-based OPF algorithms. The scheme consists of a set of rules to determine the timing of introduction and criteria of updating the penalties in the optimization process. This heuristic algorithm is of limited scope. Substantially more work is needed to effectively resolve all problems associated with the discrete nature of controls and other discrete elements of the OPF problem.

Detecting and Handling Infeasibility

As the requirements for satisfactory system operation increase, the region of feasible solutions that satisfy all constraints simultaneously may become too small. In this case, there is a need to establish criteria to prioritize the constraints. For OPF applications, this means that when a feasible solution cannot be found, it is still very important for the algorithm to suggest the "best optimal" engineering solution in some sense, even though it is infeasible. This is even more critical for OPF applications that incorporate contingency constraints.

There are several approaches to deal with this problem. In one approach, all power flow equations are satisfied and only the soft constraints that truly cause the bottlenecks are allowed to be violated using a least squares approximation process. An LP approach introduces a weighted slack variable for each binding constraint. If a constraint can be enforced, the slack variable will be reduced to zero and the constraint will be satisfied. The constraints causing infeasibility will have non-zero slack variables whose magnitudes are proportional to the amounts they need to be relaxed to achieve feasibility. Usually, all binding constraints of a particular type are modeled as if they have identical infeasibility characteristics. That is, all slack variables corresponding to these binding constraints share the same cost curve, and their sensitivities are scaled by a weighting factor associated with the type of the corresponding constraint. Using Newton's method, if the OPF does not converge in the first specified set of iterations, the constraint weighting factors, corresponding to the penalty functions associated with the load bus voltage limits and the branch flow limits, will be reduced successively until a solution is reached. This normally results in all constraints being met except for those load bus voltage and branch flow limits that contribute to infeasibility. Special care should be taken in selecting the proper weighting factors to avoid numerical problems and produce acceptable solutions.

Another approach develops hierarchical rules that operate on the controls and constraints of the OPF problem. The rules introduce discontinuous changes in the original OPF formulation. These changes include using a different set of control/constraint limits, expansion of the control set by class or individually, branch switching, load shedding, etc. They are usually implemented in a predefined priority sequence to be consistent with utility practices. The decision as to when to proceed to the next priority level of modifications to achieve feasibility is critical, especially when it involves radial overloads, normally overloaded constraints and constraints known to have "soft" limits. The selection of a final optimal solution among all the others in the set is achieved with the implementation of a "preference index." An application of the preference index approach that minimizes postcontingency line overloads due to generator outages is given in (Yokoyama et al., 1988).

Consistency of OPF Solutions with Other Online Functions

Online OPF is implemented in either study or closed-loop mode. In study mode, the OPF solutions are presented as recommendations to the operator. In closed-loop mode, control actions are implemented in the system via the SCADA system of the EMS (*IEEE Trans.*, June, 1983). In closed-loop mode, OPF is triggered by a number of events, including an operator request, the execution of the real-time sequence and security analysis, structural change, large load change, etc. A major concern for an OPF in closed-loop mode is the design of its interface with the other online functions, which are executed at different frequencies. Some of these functions are unit commitment, economic dispatch (ED), real-time sequence, security analysis, automatic generation control (AGC), etc. To reduce the discrepancy between ideal and realistic OPF solutions, emphasis should be placed on establishing consistency between these functions and static optimal solutions produced by OPF. This requires proper interfacing and integration of OPF with these functions. The integration design should be flexible enough to allow OPF formulation modifications consistent with the ever dynamic and sometimes ill-defined security problem definition.

Ineffective "Optimal" Rescheduling

Production-grade OPF algorithms use all available control actions to obtain an optimal solution, but for many applications it is not practical to execute more than a limited number of control actions. The OPF problem then becomes one of selecting the best set of actions of a limited size out of a much larger set of possible actions. The problem was identified but no concrete remedies were offered. It is not possible to select the best and most effective set of a given size from existing OPF solutions that use all controls to solve each problem. The control actions cannot be ranked and the effectiveness of an action is not related to its magnitude. Each control facility participates in both minimization of the objective function and enforcement of the constraints. Separation of the two effects for evaluation purposes is not feasible. The problem is difficult to define analytically and existing conventional technologies are not adequate. It is important to note that emerging computational intelligence tools such as fuzzy reasoning and neural networks may offer some resolution. The problem of ineffective rescheduling is related to but is not identical to the "minimum number of controls" objective. It is also closely linked to the problem of discrete control variables, since methods that recognize the discrete nature of some control facilities tend to decrease the number of control actions by keeping inefficient discrete controls at their initial settings.

OPF-Based Transmission Service Pricing

OPF programs are capable of computing marginal costs. Information about the optimal states with respect to changes, such as load variations, operating limit changes, or constraint parameter changes, can be used in many practical applications. Specifically, the sensitivities of the production cost of generation with respect to changes in the bus active power injections are called Bus Incremental Costs (BICs). BICs can be used as nodal prices for pricing transmission services, as they reflect the transmission loss and the congestion components for transferring power from one point to another. In a lossless network with no binding constraints, all BICs should be equal. However, when an operating limit is reached, the congestion component takes effect and all BICs in the network can be different. This means that nodal price differences across uncongested lines can be much larger than marginal losses. Extensive experience has shown that it is possible for power to flow from a bus with higher nodal price to a bus with lower nodal price, resulting in negative transmission charges. Failure to properly account for this effect can lead to unacceptable incentives for transmission users. The same applies in the case of transmission reinforcements to mitigate congestion. If as a result of the upgrades, the incremental transmission rights (positive or negative) are not accounted for properly, similar distortions are possible.

Conclusions

A review of recent developments in optimal economic operation of electric power systems with emphasis on the optimal power flow formulation was given. We dealt with conventional formulations of economic dispatch, conventional optimal power flow, and accounting for the dependence of the power demand on voltages in the system. Challenges to OPF formulations and solution methodologies for online application were also outlined.

References

Abadie, J. and Carpentier, J., Generalization of the Wolfe reduced gradient method to the case of nonlinear constraints, *Optimization*, R. Fletcher, Ed., Academic Press, New York, 1969, 37–47.

Alsac, O. and Stott, B., Optimal load flow with steady-state security, *IEEE Trans. Power App. Syst.*, PAS-93, 745–751, May/June 1974.

Fiacco, A.V. and McCormick, G.P., Computational algorithm for the sequential unconstrained minimization technique for nonlinear programming, *Management Sci.*, 10, 601–617, 1964.

Benthall, T.P., Automatic load scheduling in a multiarea power system, *Proc. IEE*, 115, 592–596, April 1968.

Burchett, R.C., Happ, H.H., Vierath, D.R., and Wirgau, K.A., Developments in optimal power flow, *IEEE Trans. Power App. and Syst.*, PAS-101, 406–414, February 1982.

Burchett, R.C., Happ, H.H., and Vierath, D.R., Quadratically convergent optimal power flow, *IEEE Trans. Power App. and Syst.*, PAS-103, 3267–3275, November 1984.

Carpentier, J., Contribution a l'etude du dispatching economique, *Bull. Soc. Francaise Electriciens*, 8, 431–447, 1962.

Carpentier, J., Differential Injections Method: A general method for secure and optimal load flows, in *IEEE PICA Conference Proceedings* Minneapolis, MN, June 1973, 255–262.

Concordia, C. and Ihara, S., Load representation in power system stability studies, *IEEE Trans. Power App. and Syst.*, PAS-101, 1982.

Contaxis, G.C., Delkis, C., and Korres, G., Decoupled optimal power flow using linear or quadratic programming, *IEEE Trans. Power Syst.*, PWRS-1, 1–7, May 1986.

Dias, L.G. and El-Hawary, M.E., A comparison of load models and their effects on the convergence of Newton power flows, *Inter. J. of Electr. Power and Energy Syst.*, 12, 3–8, 1990.

Dias, L.G. and El-Hawary, M.E., Effects of active and reactive modelling in optimal load flow studies, *IEE Proceedings*, 136, Part C, 259–263, September 1989.

Dias, L.G. and El-Hawary, M.E., Effects of load modeling in security constrained OPF studies, *IEEE Trans. on Power Systems*, 6, 1, 87–93, February 1991.

Dias, L.G. and El-Hawary, M.E., Security constrained OPF: influence of fixed tap transformer loads, *IEEE Trans. on Power Systems*, 6, 1366–1372, November 1991.

Dommel, H.W. and Tinney, W.F., Optimal power flow solutions, *IEEE Trans. Power App. and Syst.*, PAS-87, 1866–1876, October 1968.

El-Din, H.M.Z., Burns, S.D., and Graham, C.E., Voltage/var control with limited control actions, in *Proceedings of the Canadian Electrical Association Spring Meeting*, Toronto, Ontario, March 1989.

El-Hawary, M.E., Power system load modeling and incorporation in load flow solutions, in *Proceedings of the Third Large Systems Symposium*, University of Calgary, June 1982.

El-Hawary, M.E. and Dias, L.G., Bus sensitivity to model parameters in load-flow studies, *IEE Proceedings*, 134, Part C, 302–305 July 1987.

El-Hawary, M.E. and Dias, L.G., Incorporation of load models in load flow studies: form of model effects, *IEE Proceedings*, 134, Part C, 27–30 January 1987.

El-Hawary, M.E. and Dias, L.G., Selection of buses for detailed modeling in load flow studies, *Electric Machines and Power Syst.*, 12, 83–92, 1987.

El-Kady, M.A., Bell, B.D., Carvalho, V.F., Burchett, R.C., Happ, H.H., and Vierath, D.R., Assessment of real-time optimal voltage control, *IEEE Trans. Power Systems*, PWRS-1, 98–107, May 1986.

Fletcher, R. and Powell, M.J.D., A rapidly convergent descent method for minimization, *Computer Journal*, 6, 163–168, 1963.

Gill, P.E., Murray, W., and Wright, M.H., *Practical Optimization*, Academic Press, New York, 1981.

Glavitsch, H. and Spoerry, M., Quadratic loss formula for reactive dispatch, *IEEE Trans. Power App. and Syst.*, PAS-102, 3850–3858, December 1983.

Happ, H.H. and Vierath, D.R., The OPF for operations planning and for use on line, in *Proceedings of the Second International Conference on Power Systems Monitoring and Control*, July 1986, 290–295.

IEEE Committee Report, System load dynamic simulation, effects and determination of load constants, *IEEE Trans. Power App. and Syst.*, PAS-92, 600–609, 1973.

IEEE Current Operating Problems Working Group Report, On-line load flows from a system operator's viewpoint, *IEEE Trans. on Power Appar. Syst.*, PWRS-102, 1818–1822, June 1983.

IEEE Working Group Report, The effect of frequency and voltage on power system loads, Paper 31 CP 66-64, IEEE Winter Meeting, New York, 1966.

Iliceto, F., Ceyhan, A., and Ruckstuhl, G., Behavior of loads during voltage dips encountered in stability studies, *IEEE Trans. Power App. and Syst.*, PAS-91, 2470–2479, 1972.

Karmarkar, N., New polynomial-time algorithm for linear programming, *Combinatorica*, 4, 1984, 00.373-395.

Kuhn, H.W. and Tucker, A.W., Nonlinear programming, in *Proc. of Second Berkeley Symposium on Mathematical Statistics and Probability*, University of California, Berkeley, 1951, 481–492.

Lindqvist, A., Bubenko, J.A., and Sjelvgren, D., A generalized reduced gradient methodology for optimal reactive power flows, in *Proceedings of the 8th PSCC*, Helsinki, Finland, 1984.

Liu, E., Papalexopoulos, A.D., and Tinney, W.F., Discrete shunt controls in a Newton optimal power flow, *IEEE Winter Power Meeting 1991*, Paper 91WM 041-4 PWRS.

Lootsma, F.A., Logarithmic programming: a method of solving nonlinear programming problems, *Philips Res. Repts.*, 22, 329–344, 1967.

Maria, G.A. and Findlay, J.A., A Newton optimal power flow program for Ontario hydro EMS, *IEEE Trans. Power Syst.*, PWRS-2, 576–584, Aug. 1987.

Merlin, A., On optimal generation planning in large transmission systems (The Maya Problem), in *Proc. of 4th PSCC*, Grenoble, paper 2.1/6, September 1972.

Momoh, J.A., Application of Quadratic Interior Point Algorithm to Optimal Power Flow, EPRI Final report RP 2473-36 II, March 1992.

Papalexopoulos, A. D., Challenges to an on-line OPF implementation, in *IEEE Tutorial Course Optimal Power Flow: Solution Techniques, Requirements, and Challenges*, Publication 96 TP 111-0, IEEE Power Engineering Society, 1996.

Papalexopoulos, A.D., Imparato, C.F., and Wu, F.F., Large-scale optimal power flow: effects of initialization, decoupling and discretization, *IEEE Trans. on Power Appar. Syst.*, PWRS-4, 748–759, May 1989.

Peschon, J., Bree, D.W., and Hajdu, L.P., Optimal solutions involving system security, in *Proc. of the 7th PICA Conf.*, Boston, MA, 1971, 210–218.

Polak, E. and Ribiere, G., Note sur la convergence de methods de directions conjugees, *Rev. Fr. Inform. Rech. Operation*, 16-R1, 35–43, 1969.

Sasson, A.M., Combined use of the Powell and Fletcher-Powell nonlinear programming methods for optimal load flows, *IEEE Trans. Power App. and Syst.*, PAS-88, 1530–1537, October 1969.

Sasson, A.M., Viloria, F., and Aboytes, F., Optimal load flow solution using the Hessian matrix, *IEEE Trans. Power App. and Syst.*, PAS-92, 31–41, Jan/Feb 1973.

Schnyder, G. and Glavitsch, H., Integrated security control using an optimal power flow and switching concepts, in *IEEE PICA Conf. Proc.*, Montreal, Canada, May 18–22, 1987, 429–436.

Shen, C.M. and Laughton, M.A., Power system load scheduling with security constraints using dual linear programming, *IEE Proceedings*, 117, 2117–2127, Nov. 1970.

Stott, B., Alsac, O., and Monticelli, A. J., Security analysis and optimization, *Proc. IEEE*, 75, 1623–1644, Dec. 1987.

Stott, B. and Hobson, E., Power system security control calculations using linear programming. Parts 1 and 2, *IEEE Trans. Power App. and Syst.*, PAS-97, 1713–1731, Sept./Oct. 1978.

Sun, D.I., Ashley, B., Brewer, B., Hughes, A., and Tinney, W.F., Optimal power flow by Newton approach, *IEEE Trans. Power App. and Syst.*, PAS-103, 2864–2880, Oct. 1984.

Vaahedi, E. and El-Din, H.M.Z., Considerations in applying optimal power flow to power systems operation, *IEEE Trans. Power Syst.*, PWRS-4, 694–703, May 1989.

Vaahedi, E., El-Kady, M., Libaque-Esaine, J.A., and Carvalho, V.F., Load models for large scale stability studies from end user consumption, *IEEE Trans. Power Syst.*, PWRS-2, 864–871, 1987.

Vargas, L.S., Quintana, V.H., and Vannelli, A., A tutorial description of an interior point method and its applications to security-constrained economic dispatch, *IEEE Trans. on Power Systems*, 8, 1315–1323, 1993.

Wells, D.W., Method for economic secure loading of a power system, *IEE Proceedings*, 115, 1190–1194, Aug. 1968.

Wolfe, P., Methods of nonlinear programming, *Nonlinear Programming*, J. Abadie, Ed., North Holland, Amsterdam, 1967, 97–131.

Yan, X. and Quintana, V. H., Improving an interior-point-based OPF by dynamic adjustments of step sizes and tolerances, *IEEE Trans. on Power Systems*, 14, 709–717, 1999.

Yokoyama, R., Bae, S.H., Morita, T., and Sasaki, H., Multiobjective optimal generation dispatch based on. probability security criteria, *IEEE Trans. on Power Appar. Syst.*, PWRS-3, 317–324, 1988.

Yu, D.C., Fagan, J.E., Foote, B., and Aly, A.A., An optimal load flow study by the generalized reduced gradient approach, *Electric Power Systems Research*, 10, 47–53, 1986.

Zangwill, W.I., Non-linear programming via penalty functions, *Management Sci.*, 13, 344–358, 1967.

12.5 Security Analysis

Nouredine Hadjsaid

The power system as a single entity is considered the most complex system ever built. It consists of various equipment with different levels of sophistication, complex and nonlinear loads, various generations with a wide variety of dynamic responses, a large-scale protection system, a wide-area communication network, and numerous control devices and control centers. This equipment is connected with a large network (transformers, transmission lines) where a significant amount of energy transfer often occurs. This system, in addition to the assurance of good operation of its various equipment, is characterized by an important and simple rule: electricity should be delivered to where it is required in due time and with appropriate features such as frequency and voltage quality. Environmental constraints, the high cost of transmission investments and low/long capital recovery, and the willing of utilities to optimize their network for more cost effectiveness makes it very difficult to expand or oversize power systems. These constraints have pushed power systems to be operated close to their technical limits, thus reducing security margins.

On the other hand, power systems are continuously subjected to random and various disturbances that may, under certain circumstances, lead to inappropriate or unacceptable operation and system conditions. These effects may include cascading outages, system separation, widespread outages, violation of emergency limits of line current, bus voltages, system frequency, and loss of synchronism (Debs and Benson, 1975). Furthermore, despite advanced supervisory control and data acquisition systems that help the operator to control system equipment (circuit breakers, on-line tap changers, compensation and control devices, etc.), changes can occur so fast that the operator may not have enough time to ensure system security. Hence, it is important for the operator not only to maintain the state of the system within acceptable and secure operating conditions but also to integrate preventive functions. These functions should allow him enough time to optimize his system (reduction of the probability of occurrence of abnormal or critical situations) and to ensure recovery of a safe and secure situation.

Even though for small-scale systems the operator may eventually, on the basis of his experience, prevent the consequences of most common outages and determine the appropriate means to restore a secure state, this is almost impossible for large systems. It is therefore essential for operators to have at their disposal, efficient tools capable of handling a systematic security analysis. This can be achieved through the diagnosis of all contingencies that may have serious consequences. This is the concern of **security analysis**.

The term contingency is related to the possibility of losing any component of the system, whether it is a transmission line, a transformer, or a generator. Another important event that may be included in this definition concerns busbar faults (bus split). This kind of event is, however, considered rare but with (serious) dangerous consequences. Most power systems are characterized by the well-known $N - 1$ security rules where N is the total number of system components. This rule is the basic requirement for the planning stage where the system should be designed in order to withstand (or to remain in a normal state) any single contingency. Some systems also consider the possibility of $N - 2/k$ (k is the number of contingencies), but mostly for selected and specific cases.

Definition

The term security as defined by NERC (1997) is the ability of the electric systems to withstand sudden disturbance such as electric short-circuits or unanticipated loss of system elements. (See Appendix A).

Security analysis is usually handled for two time frames: static and dynamic. For the static analysis, only a "fixed picture" or a snapshot of the network is considered. The system is supposed to have passed the transient period successfully or be dynamically stable. Therefore, the monitored variables are line flows and bus voltages. Hence, all voltages should be within a predefined secure range, usually around ±5% of nominal voltage (for some systems, such as distribution networks, the range may be wider). In fact, if bus voltages drop below a certain level, there will be a risk of voltage collapse in addition to high losses. On the other hand, if bus voltages are too high compared to nominal values, there will be equipment degradation or damage. Furthermore, overload of transmission lines may be followed by unpredictable line tripping that accelerates the degradation of the voltage profile.

Line flows are related to circuit overload (lines and transformers) and should keep below a maximum limit, usually settled according to line thermal limits. The dynamic security is related to loss of synchronism (transient stability) and oscillatory swings or dynamic instability. In that case the evolution of essential variables are monitored based upon a required time frame (transient period).

Normally, system security is analyzed differently whether it is considered for planning studies or for monitoring and operational purposes. The difference comes from the type of action that should be initiated in case of expected harmful contingencies. However, for both stages, all variables should remain within the bounded domain defining or determining system normal state (Fink, 1978).

Time Frames for Security-Related Decision (McCaulley et al., 1999)

There are generally three different time frames for security-related decisions. In operations, the decision-maker is the operator, who must continuously monitor and operate his system economically in such a way that the normal state is appropriately preserved (maintained). For this purpose, he has specific tools for diagnosing his system and operating rules that allow the required decisions to be made in due time. In operational planning, the operating rules are developed recognizing that the bases for the decision are reliability/security criteria specifying minimum operating requirements, which define acceptable performance for the credible contingencies. In facility planning, the planner must determine the best way to reinforce the transmission system, based on reliability/security criteria for system design, which generally adhere to the same disturbance-performance criteria specified by minimum operating requirements.

One may think that since these systems are designed to operate "normally" or in "a secure state" for a given security rule $(N - k)$, there is nothing to worry about during operations. The problem is that, during the planning stage and for a set of given economical constraints, a number of assumptions are made for operating conditions that concern topology, generation, and consumption. Since there may be several years between the planning stage and the operations, the uncertainties in the system's security may be very significant. Therefore, security analysis is supplemented by operational planning and operations studies.

Power System Operation and Control 12-55

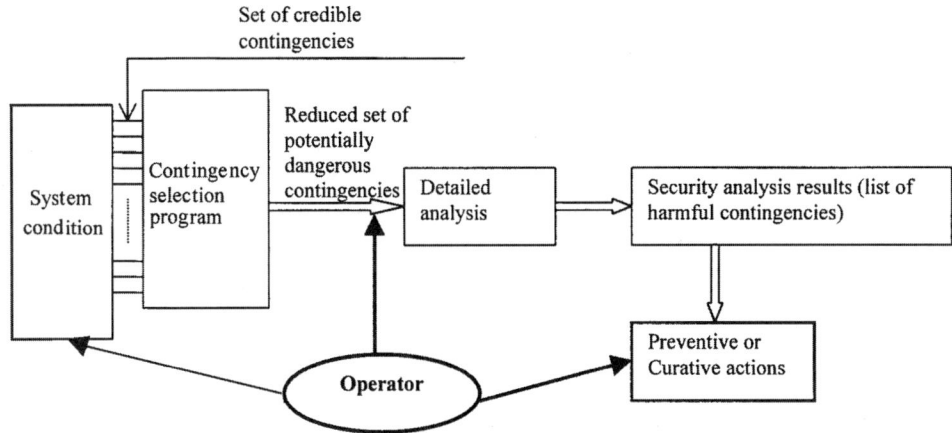

FIGURE 12.21 Contingency analysis procedure.

The decision following any security analysis can be placed in one of two categories: preventive or corrective actions. For corrective actions, once a contingency or an event is determined as potentially dangerous, the operator should be confident that in case of that event, he will be able to correct the system by means of appropriate actions on system conditions (generation, load, topology) in order to keep the system in a normal state and even away from the insecure region. The operator should also prepare a set of preventive actions that may correct the effect of the expected dangerous event.

In operations, the main constraint is the time required for the analysis of the system's state and for the required decision to be made following the security analysis results. The security analysis program should be able to handle all possible contingencies, usually on the N – 1 basis or on specific N – 2. For most utilities, the total time window considered for this task is between 10 min and 30 min. Actually for this time window, the system's state is considered as constant or quasi-constant allowing the analysis to be valid within this time frame. This means that changes in generation or in consumption are considered as negligible.

For large systems, this time frame is too short even with very powerful computers. Since it is known that only a small number of contingencies may really cause system violations, it has been realized that it is not necessary to perform a detailed analysis on all possible contingencies, which may be on the order of thousands. For this purpose, the operator may use his engineering judgment to select those contingencies that are most likely to cause system violation. This procedure has been used (and is still in use) for many years in many control centers around the world. However, as system conditions are characterized by numerous uncertainties, this approach may not be very efficient especially for large systems.

The concept of contingency selection has arisen in order to reduce the list of all possible contingencies to only the potentially harmful. The selection process should be very fast and accurate enough to identify dangerous cases (Hadjsaid, 1992). This process has existed for many years, and still is a major issue in all security studies for operations whether for static or dynamic and transient purposes.

Models

The static security analysis is mainly based on load flow equations. Usually, active/angle and reactive/voltage problems are viewed as decoupled. The active/angle subproblem is expressed as (Stott and Alsac, 1974):

$$\Delta\theta = \left[dP/d\theta\right]^{-1} \Delta P \qquad (12.53)$$

where $\Delta\theta$ is a vector of angular changes with a dimension of Nb – 1 (Nb = number of buses), ΔP a vector of active injection changes (Nb – 1) and $[dP/d\theta]$ is a part of the Jacobian matrix. In the DC approach, this Jacobian is approximated by the B′ (susceptance) matrix representing the imaginary part of the Ybus matrix. This expression is used to calculate the updated angles following a loss of any system component. With appropriate numerical techniques, it is straightforward to update only necessary elements of the equation. Once the angles are calculated, the power flows of all lines can be deducted. Hence, it is possible to check for line limit violation.

Another approach that has been, and still is used in many utilities for assessing the impact of any contingency on line flows is known as shift factors. The principle used recognizes that the outage of any line will result in a redistribution of the power previously flowing through this line on all the remaining lines. This distribution is mainly affected by the topology of the network. Hence, the power flow of any line ij following an outage of line km can be expressed as (Galiana, 1984) (see Appendix B for more details):

$$P_{ij/km} = P_{ij} + \alpha_{ij/km} * P_{km} \qquad (12.54)$$

where $P_{ij/km}$ is the active power flow on line ij after the outage of line km
P_{ij}, P_{km} is the active power previously flowing respectively on line ij and km (before the outage)
$\alpha_{ij/km}$ is the shift factor for line ij following the outage of line km

Equation (12.54) shows that the power flow of line ij ($P_{ij/km}$) when line km is tripped, is determined as the initial power flow on line ij (P_{ij}) before the outage of line km plus a proportion of the power flow previously flowing on line km. This proportion is defined by the terms $\alpha_{ij/km} * P_{km}$.

The shift factors are determined in a matrix form. The important features of these factors are the simplicity of computing and their dependency on network topology. Therefore, if the topology does not change, the factors remain constant for any operating point. The main drawback of these factors is that they are determined on the basis of DC approximation and the shift factor matrix should be updated for any change in the topology. In addition, for some complex disturbances such as bus split, updating these factors becomes a complicated task.

A similar method based on reactive power shift factors has been developed. Interested readers may refer to Ilic-Spong and Phadke (1986) and Taylor and Maahs (1991) for more details.

The reactive/voltage subproblem can be viewed as (Stott and Alsac, 1974):

$$\Delta V = \left[dQ/dV \right]^{-1} \Delta Q \qquad (12.55)$$

where ΔV is the vector of voltages change (Nb − Ng, Ng is the number of generators)
ΔQ is the vector of reactive power injections change (Nb − Ng, Ng is the number of generators)
[dQ/dV] is the Jacobean submatrix

In the well-known FDLF (Fast Decoupled Load Flow) model (Stott and Alsac, 1974), the Jacobian submatrix is replaced by the B″ (susceptance) matrix representing the imaginary part of the Ybus matrix with a dimension of Nb − Ng, where Ng is the number of voltage regulated (generator) buses. In addition, the vector ΔQ is replaced by $\Delta Q/V$.

Once bus voltages are updated to account for the outage, the limit violations are checked and the contingency effects on bus voltages can be assessed.

The most common framework for the contingency analysis is to use approximate models for the selection process, such as the DC model, and use the AC power flow model for the evaluation of the actual impact of the given contingency on line flows and bus voltages.

Concerning the dynamic security analysis, the framework is similar to the one in static analysis in terms of selection and evaluation. The selection process uses simplified models, such as Transient Energy Functions (TEF), and the evaluation one uses detailed assessing tools such as time domain simulations. The fact that the dynamic aspect is more related to transient/dynamic stability technique makes the process much more complicated than for the static problem. In fact, in addition to the number of contingencies to be analyzed, each analysis will require detailed stability calculations with an appropriate network and system component model such as the generator model (park, saturation, etc.), exciter (AVR: Automatic Voltage Regulator; PSS: Power System Stabilizer), governor (nuclear, thermal, hydroelectric, etc.), or loads (non-linear, constant power characteristics, etc.). In addition, integration and numerical solutions are an important aspect for these analyses.

Determinist vs. Probabilistic

The basic requirement for security analysis is to assess the impact of any possible contingency on system performance. For the purpose of setting planning and operating rules that will enable the system to be operated in a secure manner, it is necessary to consider all credible contingencies, different network configurations, and different operating points for given performance criteria. Hence, in the deterministic approach, these assessments may involve a large number of computer simulations even if there is a selection process at each stage of the analysis. The decision in that case is founded on the requirement that each outage event in a specified list, the contingency set, results in system performance that satisfies the criteria of the chosen performance evaluation (Fink and Carlsen, 1978). To handle these assessments for all possible situations by an exhaustive study is generally not reasonable. Since the resulting security rules may lead to the settlement and schedule of investment needs as well as operating rules, it is important to optimize the economical impact of security measures that have to be taken in order to be sure that there is no unnecessary or unjustified investment or operating costs. This has been the case for many years, since the emphasis was on the most severe, credible event leading to overly conservative solutions.

One way to deal with this problem is the concept of the probability of occurrence (contingencies) in the early stage of security analysis. This can be jointly used with a statistical approach (Schlumberger et al., 1999) that allows the generation of appropriate scenarios in order to fit more with the reality of the power system from the technical point of view as well as from the economical point view.

Security under Deregulation

With deregulation, the power industry has pointed out the necessity to optimize the operations of their systems leading to less investment in new facilities and pushing the system to be exploited closer to its limits. Furthermore, the open access has resulted in increased power exchanges over the interconnections. In some utilities, the number of transactions previously processed in one year is now managed in one day. These increased transactions and power exchanges have resulted in increased parallel flows leading to unpredictable loading conditions or voltage problems. A significant number of these transactions are non-firm and volatile. Hence, the security can no longer be handled on a zonal basis but rather on large interconnected systems.

Appendix A

The current NERC basic reliability requirement from NERC Policy 2- transmission (Pope, 1999) is:

Standards

1. Basic reliability requirement regarding single contingencies: All control areas shall operate so that instability, uncontrolled separation, or cascading outages will not occur as a result of the most severe single contingency.

    1.1 Multiple contingencies: Multiple outages of credible nature, as specified by regional policy, shall also be examined and, when practical, the control areas shall operate to protect against instability, uncontrolled separation, or cascading outages resulting from these multiple outages.

    1.2 Operating security limits: Define the acceptable operating boundaries

2. Return from Operating security limit violation: Following a contingency or other event that results in an operating security limit violation, the control area shall return its transmission system to within operating security limits soon as possible, but no longer than 30 minutes.

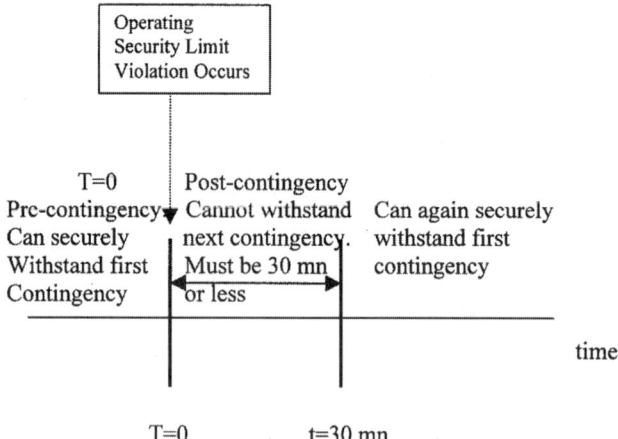

FIGURE 12.22 Current NERC basic reliability requirement. (Pope, J. W., Transmission Reliability under Restructuring, in *Proceedings of IEEE SM 1999*, Edmonton, Alberta, Canada, 162–166, July 18–22, 1999. With permission.)

Appendix B

Shift factor derivation (Galiana, 1984)
Consider a DC load flow for a base case:

$$[B']\underline{\theta} = \underline{P}$$

where θ is the vector of phase angles for the base case
[B′] is the susceptance matrix for the base case
P is the vector of active injections for the base case

Suppose that the admittance of line jk is reduced by ΔY_{jk} and the vector ΔP is unchanged, then:

$$\left[[B'] - \Delta Y_{jk}\, \underline{e}_{jk}\, \underline{e}_{jk}^{\,T}\right]\underline{\theta} = \underline{P}$$

where $\underline{e}_{jk}$ is the vector (Nb − 1) containing 1 in the position j, −1 in the position k and 0 elsewhere
T is the Transpose

Now we can compute the power flow on an arbitrary line lm when line jk is outaged:

$$P_{lm/jk} = Y_{lm}(\theta_l - \theta_m) = Y_{lm}\, \underline{e}_{lm}^{\,T}\, \underline{\theta}$$

$$= Y_{lm}\, \underline{e}_{lm}^{\,T}\left[[B'] - \Delta Y_{jk}\, \underline{e}_{jk}\, \underline{e}_{jk}^{\,T}\right]^{-1} P$$

By using the matrix inversion lemma, we can compute:

$$P_{lm/jk} = Y_{lm}\, \underline{e}_{lm}^{\,T}\left[[B']+\left([B']^{-1}\underline{e}_{jk}\, \underline{e}_{jk}^{\,T}[B']^{-1}\right)\Big/\left((\Delta Y_{jk})-1-\underline{e}_{jk}^{\,T}[B']^{-1}\underline{e}_{jk}\right)\right]P$$

Finally:

$$P_{lm/jk} = P_{lm} + \alpha_{jk/jk} * P_{jk}$$

where

$$\alpha_{jk/jk} = Y_{lm} * \left(\Delta Y_{jk}/Y_{jk}\right) * \left(e_{lm}^T \left[B'\right]^{-1} e_{jk}\right) \Big/ \left(1 - \Delta Y_{jk} e_{jk}^T \left[B'\right]^{-1} e_{jk}\right)$$

References

Debs, A. S. and Benson, A. R., Security Assessment of Power Systems, in *System Engineering for Power: Status and Prospects,* Henniker, New Hampshire, 144-178, Aug. 17–22, 1975.

Fink, L. and Carlsen, K., Operating Under Stress and Strain, *IEEE Spectrum,* 15, 48–53, March, 1978.

Galiana, F. D., Bound estimates of the severity of line outages in power system contingency analysis and ranking, *IEEE Trans. on Power Appar. and Syst.,* PAS-103(9), 2612–2624, September 1984.

Hadjsaid, N., Benahmed, B., Fandino, J., Sabonnadiere, J.-Cl., and Nerin, G., Fast contingency screening for voltage-reactive considerations in security analysis, *IEEE Winter Meeting,* 1992 WM 185-9 PWRS.

Ilic-Spong, M. and Phadke, A., Redistribution of reactive power flow in contingency studies, *IEEE Trans. on Power Syst.,* PWRS-1(3), 266–275, August 1986.

McCaulley, J. D., Vittal, V., and Abi-Samra, N., An overview of risk based security assessment, in *Proceedings of IEEE SM'99,* Edmonton, Alberta, Canada, 173–178, July 18–22, 1999.

The North American Reliability Council, NERC Planning Standards, approved by NERC Board of Trustees, September, 1997.

Pope, J. W., Transmission reliability under restructuring, in *Proceedings of IEEE SM'99,* Edmonton, Alberta, Canada, 162–166, July 18–22, 1999.

Schlumberger, Y., Lebrevelec, C., and De Pasquale, M., Power system security analysis: New approaches used at EDF, in *Proceedings of IEEE SM'99,* Edmonton, Alberta, Canada, 147–151, July 18–22, 1999.

Stott, B. and Alsac, O., Fast decoupled load flow, *IEEE Trans. on Power Appar. and Syst.,* PAS-93, pp. 859-869, May/June 1974.

Taylor, D. G. and Maahs, L. J., A reactive contingency analysis algorithm using MW and MVAR distribution factors, *IEEE Trans. on Power Syst.,* 6, 349–355, February 1991.

13
Power System Planning (Reliability)

Gerald B. Sheblé
Iowa State University

13.1   **Planning**   *Gerald B. Sheblé* 13-1

13.2   **Short-Term Load and Price Forecasting with Artificial Neural Networks**
      Alireza Khotanzad 13-16

13.3   **Transmission Plan Evaluation — Assessment of System Reliability**
      N. Dag Reppen and James W. Feltes 13-26

13.4   **Power System Planning**   *Hyde M. Merrill* 13-40

13.5   **Power System Reliability**   *Richard E. Brown* 13-51

13
Power System Planning (Reliability)

Gerald B. Sheblé
Iowa State University

Alireza Khotanzad
Southern Methodist University

N. Dag Reppen
Niskayuna Power Consultants, LLC

James W. Feltes
Power Technologies

Hyde M. Merrill
Merril Energy, LLC

Richard E. Brown
ABB Power T&D Company

13.1 Planning 13-1
   Defining a Competitive Framework
13.2 Short-Term Load and Price Forecasting with
   Artificial Neural Networks..13-16
   Artificial Neural Networks • Short-Term Load Forecasting •
   Short-Term Price Forecasting
13.3 Transmission Plan Evaluation — Assessment of
   System Reliability 13-26
   Bulk Power System Reliability and Supply Point Reliability •
   Methods for Assessing Supply Point Reliability • Probabilistic
   Reliability Assessment Methods • Application Examples
13.4 Power System Planning..13-40
   Planning Entities • Arenas • The Planning Problem •
   Planning Processes
13.5 Power System Reliability..13-51
   NERC Regions • System Adequacy Assessment • System
   Security Assessment • Probabilistic Security Assessment •
   Distribution System Reliability • Distribution Reliability
   Indices • Storms and Major Events • Component Reliability
   Data • Utility Reliability Problems • Reliability Economics •
   Annual Variations in Reliability

13.1 Planning

Gerald B. Sheblé

Capacity expansion decisions are made daily by government agencies, private corporations, partnerships, and individuals. Most decisions are small relative to the profit and loss sheet of most companies. However, many decisions are sufficiently large to determine the future financial health of the nation, company, partnership, or individual. Capacity expansion of hydroelectric facilities may require the commitment of financial capital exceeding the income of most small countries. Capacity expansion of thermal fossil fuel plants is not as severe, but does require a large number of financial resources including bank loans, bonds for long-term debt, stock issues for more working capital, and even joint-venture agreements with other suppliers or customers to share the cost and the risk of the expansion. This section proposes several mathematical optimization techniques to assist in this planning process. These models and methods are tools for making better decisions based on the uncertainty of future demand, project costs, loan costs, technology change, etc. Although the material presented in this section is only a simple model of the process, it does capture the essence of real capacity expansion problems.

This section relies on a definition of electric power industry restructuring presented in (Sheblé, 1999). The new environment within this work assumes that the vertically integrated utility has been segmented into a horizontally integrated system. Specifically, GENCOs, DISTCOs, and TRANSCOs exist in place of the old. This work does not assume that separate companies have been formed. It is only necessary that comparable services are available for anyone connected to the transmission grid.

As can be concluded, this description of a deregulated marketplace is an amplified version of the commodity market. It needs polishing and expanding. The change in the electric utility business environment is depicted generically below. The functions shown are the emerging paradigm. This work outlines the market organization for this new paradigm.

Attitudes toward restructuring still vary from state to state and from country to country. Many electric utilities in the U.S. have been reluctant to change the status quo. Electric utilities with high rates are very reluctant to restructure since the customer is expected to leave for the lower prices. Electric utility companies in regions with low prices are more receptive to change since they expect to pick up more customers. In 1998, California became the first state in the U.S. to adopt a competitive structure, and other states are observing the outcome. Several states on the eastern coast of the U.S. have also restructured. Some offer customer selection of supplier. Some offer markets similar to those established in the United Kingdom, Norway, and Sweden, but not Spain. Several countries have gone to the extreme competitive position of treating electricity as a commodity as seen in New Zealand and Australia. As these markets continue to evolve, governments in all areas of the world will continue to form opinions on what market, operational, and planning structures will suit them best.

Defining a Competitive Framework

There are many market frameworks that can be used to introduce competition between electric utilities. Almost every country embracing competitive markets for its electric system has done so in a different manner. The methods described here assume an electric marketplace derived from commodities exchanges like the Chicago Mercantile Exchange, Chicago Board of Trade, and New York Mercantile Exchange (NYMEX) where commodities (other than electricity) have been traded for many years. NYMEX added electricity futures to their offerings in 1996, supporting this author's previous predictions (Sheblé, 1991; 1992; 1993; 1994) regarding the framework of the coming competitive environment. The framework proposed has similarities to the Norwegian-Sweden electric systems. The proposed structure is partially implemented in New Zealand, Australia, and Spain. The framework is being adapted since similar structures are already implemented in other industries. Thus, it would be extremely expensive to ignore the treatment of other industries and commodities. The details of this framework and some of its major differences from the emerging power markets/pools are described in Sheblé (1999).

These methods imply that the ultimate competitive electric industry environment is one in which retail consumers have the ability to choose their own electric supplier. Often referred to as retail access, this is quite a contrast to the vertically integrated monopolies of the past. Telemarketers are contacting consumers, asking to speak to the person in charge of making decisions about electric service. Depending on consumer preference and the installed technology, it may be possible to do this on an almost real-time basis as one might use a debit card at the local grocery store or gas station. Real-time pricing, where electricity is priced as it is used, is getting closer to becoming a reality as information technology advances. Presently, however, customers in most regions lack the sophisticated metering equipment necessary to implement retail access at this level.

Charging rates that were deemed fair by the government agency, the average monopolistic electric utility of the old environment met all consumer demand while attempting to minimize their costs. During natural or man-made disasters, neighboring utilities cooperated without competitively charging for their assistance. The costs were always passed on to the rate payers. The electric companies in a country or continent were all members of one big happy family. The new companies of the future competitive environment will also be happy to help out in times of disaster, but each offer of assistance will be priced recognizing that the competitor's loss is gain for everyone else. No longer guaranteed a rate of return, the entities participating in the competitive electric utility industry of tomorrow will be profit driven.

Preparing for Competition

Electric energy prices recently rose to more than $7500/MWh in the Midwest (1998) due to a combination of high demand and the forced outage of several units. Many midwestern electric utilities bought energy at that high price, and then sold it to consumers for the normal rate. Unless these companies thought they were going to be heavily fined, or lose all customers for a very long time, it may have been more fiscally responsible to terminate services.

Under highly competitive scenarios, the successful supplier will recover its incremental costs as well as its fixed costs through the prices it charges. For a short time, producers may sell below their costs, but will need to make up the losses during another time period. Economic theory shows that eventually, under perfect competition, all companies will arrive at a point where their profit is zero. This is the point at which the company can break even, assuming the average cost is greater than the incremental cost. At this ideal point, the best any producer can do in a competitive framework, ignoring fixed costs, is to bid at the incremental cost. Perfect competition is not often found in the real world for many reasons. The prevalent reason is *technology change*. Fortunately, there are things that the competitive producer can do to increase the odds of surviving and remaining profitable.

The operational tools used and decisions made by companies operating in a competitive environment are dependent on the structure and rules of the power system operation. In each of the various market structures, the company goal is to maximize profit. Entities such as commodity exchanges are responsible for ensuring that the industry operates in a secure manner. The rules of operation should be designed by regulators prior to implementation to be complete and "fair." *Fairness* in this work is defined to include noncollusion, open market information, open transmission and distribution access, and proper price signals. It could call for maximization of social welfare (i.e., maximize everyone's happiness) or perhaps maximization of consumer surplus (i.e., make customers happy).

Changing regulations are affecting each company's way of doing business and to remain profitable, new tools are needed to help companies make the transition from the old environment to the competitive world of the future. This work describes and develops methods and tools that are designed for the competitive component of the electric industry. Some of these tools include software to generate bidding strategies, software to incorporate the bidding strategies of other competitors, and updated common tools like economic dispatch and unit commitment to maximize profit.

Present View of Overall Problem

This work is motivated by the recent changes in regulatory policies of interutility power interchange practices. Economists believe that electric pricing must be regulated by free market forces rather than by public utilities commissions. A major focus of the changing policies is "competition" as a replacement for "regulation" to achieve economic efficiency. A number of changes will be needed as competition replaces regulation. The coordination arrangements presently existing among the different players in the electric market would change operational, planning, and organizational behaviors.

Government agencies are entrusted to encourage an open market system to create a competitive environment where generation and supportive services are bought and sold under demand and supply market conditions. The open market system will consist of generation companies (GENCOs), distribution companies (DISTCOs), transmission companies (TRANSCOs), a central coordinator to provide independent system operation (ISO), and brokers to match buyers and sellers (BROCOs). The interconnection between these groups is shown in Fig. 13.1.

The ISO is independent and a dissociated agent for market participants. The roles and responsibilities of the ISO in the new marketplace are yet not clear. This work assumes that the ISO is responsible for coordinating the market players (GENCOs, DISTCOs, and TRANSCOs) to provide a reliable power system functions. Under this assumption, the ISO would require a new class of optimization algorithms to perform price-based operation. Efficient tools are needed to verify that the system remains in operation

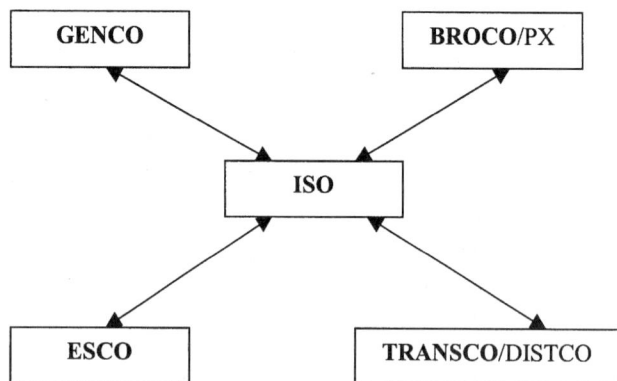

FIGURE 13.1 New organizational structure.

with all contracts in place. This work proposes an energy brokerage model for all services as a novel framework for price-based optimization. The proposed foundation is used to develop analysis and simulation tools to study the implementation aspects of various contracts in a deregulated environment.

Although it is conceptually clean to have separate functions for the GENCOs, DISTCOs, TRANSCOs, and the ISO, the overall mode of real-time operation is still evolving. Presently, two possible versions of market operations are debated in the industry. One version is based on the traditional power pool concept (POOLCO). The other is based on transactions and bilateral transactions as presently handled by commodity exchanges in other industries. Both versions are based on the premise of price-based operation and market-driven demand. This work presents analytical tools to compare the two approaches. Especially with the developed auction market simulator, POOLCO, multilateral, and bilateral agreements can be studied.

Working toward the goal of economic efficiency, one should not forget that the reliability of the electric services is of the utmost importance to the electric utility industry in North America. In the words of the North American Electric Reliability Council (NERC), reliability in a bulk electric system indicates *"the degree to which the performance of the elements of that system results in electricity being delivered to customers within accepted standards and in the amount desired. The degree of reliability may be measured by the frequency, duration, and magnitude of adverse effects on the electric supply."* The council also suggests that reliability can be addressed by considering the two basic and functional aspects of the bulk electric system — adequacy and security. In this work, the discussion is focused on the adequacy aspect of power system reliability, which is defined as the static evaluation of the system's ability to satisfy the system load requirements. In the context of the new business environment, market demand is interpreted as the system load. However, a secure implementation of electric power transactions concerns power system operation and stability issues:

1. *Stability issue*: The electric power system is a nonlinear dynamic system comprised of numerous machines synchronized with each other. Stable operation of these machines following disturbances or major changes in the network often requires limitations on various operating conditions, such as generation levels, load levels, and power transmission changes. Due to various inertial forces, these machines, together with other system components, require extra energy (reserve margins and load following capability) to safely and continuously actuate electric power transfer.
2. *Thermal overload issue*: Electrical network capacity and losses limit electric power transmission. Capacity may include real-time weather conditions as well as congestion management. The impact of transmission losses on market power is yet to be understood.
3. *Operating voltage issues*: Enough reactive power support must accompany the real power transfer to maintain the transfer capacity at the specified levels of open access.

In the new organizational structure, the services used for supporting a reliable delivery of electric energy (e.g., various reserve margins, load following capability, congestion management, transmission losses, reactive power support, etc.) are termed supportive services. These have been called "ancillary services" in the past. In this context, the term "ancillary services" is misleading since the services in question are not ancillary but *closely bundled* with the electric power transfer as described earlier. The open market system should consider all of these supportive services as an integral part of power transaction.

This work proposes that supportive services become a competitive component in the energy market. It is embedded so that no matter what reasonable conditions occur, the (operationally) centralized service will have the obligation and the authority to deliver and keep the system responding according to adopted operating constraints. As such, although competitive, it is burdened by additional goals of ensuring reliability rather than open access only. The proposed pricing framework attempts to become economically efficient by moving from cost-based to price-based operation and introduces a mathematical framework to enable all players to be sufficiently informed in decision-making when serving other competitive energy market players, including customers.

Economic Evolution

Some economists speculate that regional commodity exchanges within the U.S. would be oligopolistic in nature (having a limited numbers of sellers) due to the configuration of the transmission system. Some postulate that the number of sellers will be sufficient to achieve near-perfect competition. Other countries have established exchanges with as few as three players. However, such experiments have reinforced the notion that collusion is all too tempting, and that market power is the key to price determination, as it is in any other market. Regardless of the actual level of competition, companies that wish to survive in the deregulated marketplace must change the way they do business. They will need to develop bidding strategies for trading electricity via an exchange.

Economists have developed theoretical results of how variably competitive markets are supposed to behave under varying numbers of sellers or buyers. The economic results are often valid only when aggregated across an entire industry and frequently require unrealistic assumptions. While considered sound in a macroscopic sense, these results may be less than helpful to a particular company (not fitting the industry profile) that is trying to develop a strategy that will allow it to remain competitive.

Generation companies (GENCOs), energy service companies (ESCOs), and distribution companies (DISTCOs) that participate in an energy commodity exchange must learn to place effective bids in order to win energy contracts. Microeconomic theory states that in the long term, a hypothetical firm selling in a competitive market should price its product at its marginal cost of production. The theory is based on several assumptions (e.g., all market players will behave rationally, all market players have perfect information) that may tend to be true industry-wide, but might not be true for a particular region or a particular firm. As shown in this work, the normal price offerings are based on average prices. Markets are very seldom perfect or in equilibrium.

There is no doubt that deregulation in the power industry will have many far-reaching effects on the strategic planning of firms within the industry. One of the most interesting effects will be the optimal pricing and output strategies generator companies (GENCOs) will employ in order to be competitive while maximizing profits. This case study presents two very basic, yet effective means for a single generator company (GENCO) to determine the optimal output and price of their electrical power output for maximum profits.

The first assumption made is that switching from a government regulated, monopolistic industry to a deregulated competitive industry will result in numerous geographic regions of oligopolies. The market will behave more like an oligopoly than a purely competitive market due to the increasing physical restrictions of transferring power over distances. This makes it practical for only a small number of GENCOs to service a given geographic region.

Market Structure

Although nobody knows the exact structure of the emerging deregulated industry, this research predicts that regional exchanges (i.e., electricity mercantile associations [EMAs]) will play an important role. Electricity trading of the future will be accomplished through bilateral contracts and EMAs where traders

Level					
1	FERC	SPUC	SPUC	…………	SPUC
2			NERC		
3			ICA/ISO/RTO		
4	GENCO	ESCO	TRANSCO	DISTCO	EMA
5			MARKCO		BROCO

FIGURE 13.2 Business environmental model.

bid for contracts via a double auction. The electric marketplace used in this section has been refined and described by various authors. Fahd and Sheblé (1992a) demonstrated an auction mechanism. Sheblé (1994b) described the different types of commodity markets and their operation, outlining how each could be applied in the evolved electric energy marketplace. Sheblé and McCalley (1994e) outlined how spot, forward, future, planning, and swap markets can handle real-time control of the system (e.g., automatic generation control) and risk management. Work by Kumar and Sheblé (1996b) brought the above ideas together and demonstrated a power system auction game designed to be a training tool. That game used the double auction mechanism in combination with classical optimization techniques.

In several references (Kumar, 1996a, 1996b; Sheblé 1996b; Richter 1997a), a framework is described in which electric energy is only sold to distribution companies (DISTCOs), and electricity is generated by generation companies (GENCOs) (see Fig. 13.2). The North American Electric Reliability Council (NERC) sets the reliability standards. Along with DISTCOs and GENCOs, energy services companies (ESCOs), ancillary services companies (ANCILCOs), and transmission companies (TRANSCOs) interact via contracts. The contract prices are determined through a double auction. Buyers and sellers of electricity make bids and offers that are matched subject to approval of the independent contract administrator (ICA), who ensures that the contracts will result in a system operating safely within limits. The ICA submits information to an independent system operator (ISO) for implementation. The ISO is responsible for physically controlling the system to maintain its security and reliability.

Fully Evolved Marketplace

The following sections outline the role of a horizontally integrated industry. Many curious acronyms have described generation companies (IPP, QF, Cogen, etc.), transmission companies (IOUTS, NUTS, etc.), and distribution companies (IOUDC, COOPS, MUNIES, etc.). The acronyms used in this work are described in the following sections.

Horizontally Integrated

The restructuring of the electric power industry is most easily visualized as a horizontally integrated marketplace. This implies that interrelationships exist between generation (GENCO), transmission (TRANSCO), and distribution (DISTCO) companies as separate entities. Note that independent power producers (IPP), qualifying facilities (QF), etc. may be considered as equivalent generation companies. Nonutility transmission systems (NUTS) may be considered as equivalent transmission companies. Cooperatives and municipal utilities may be considered as equivalent distribution companies. All companies are assumed to be coordinated through a regional Transmission Corporation (or regional transmission group).

Federal Energy Regulatory Commission (FERC)

FERC is concerned with the overall operation and planning of the national grid, consistent with the various energy acts and public utility laws passed by Congress. Similar federal commissions exist in other government structures. The goal is to provide a workable business environment while protecting the economy, the customers, and the companies from unfair business practices and from criminal behavior. GENCOs, ESCOs, and TRANSCOs would be under the jurisdiction of FERC for all contracts impacting interstate trade.

State Public Utility Commission (SPUC)
SPUCs protect the individual state economies and customers from unfair business practices and from criminal behavior. It is assumed that most DISTCOs would still be regulated by SPUCs under performance-based regulation and not by FERC. GENCOs, ESCOs, and TRANSCOs would be under the jurisdiction of SPUCs for all contracts impacting intrastate trade.

Generation Company (GENCO)
The goal for a generation company, which has to fill contracts for the cash and futures markets, is to package production at an attractive price and time schedule. One proposed method is similar to the classic decentralization techniques used by a vertically integrated company. The traditional power system approach is to use Dantzig-Wolfe decomposition. Such a proposed method may be compared with traditional operational research methods used by commercial market companies for a "make or buy" decision.

Transmission Company (TRANSCO)
The goal for transmission companies, which have to provide services by contracts, is to package the availability and the cost of the integrated transportation network to facilitate transportation from suppliers (GENCOs) to buyer (ESCOs). One proposed method is similar to oil pipeline networks and energy modeling. Such a proposed method can be compared to traditional network approaches using optimal power flow programs.

Distribution Company (DISTCO)
The goal for distribution companies, which have to provide services by contracts, is to package the availability and the cost of the radial transportation network to facilitate transportation from suppliers (GENCOs) to buyers (ESCOs). One proposed method is similar to distribution outlets. Such proposed methods can be compared to traditional network approaches using optimal power flow programs. The disaggregation of the transmission and the distribution system may not be necessary, as both are expected to be regulated as monopolies at the present time.

Energy Service Company (ESCO)
The goal for energy service companies, which may be large industrial customers or customer pools, is to purchase power at the least cost when needed by consumers. One proposed method is similar to the decision of a retailer to select the brand names for products being offered to the public. Such a proposed method may be compared to other retail outlet shops.

Independent System Operator (ISO)
The primary concern is the management of operations. Real-time control (or nearly real-time) must be completely secure if any amount of scheduling is to be implemented by markets. The present business environment uses a fixed combination of units for a given load level, and then performs extensive analysis of the operation of the system. If markets determine schedules, then the unit schedules may not be fixed sufficiently ahead of realtime for all of the proper analysis to be completed by the ISO.

Regional Transmission Organization (RTO)
The goal for a regional transmission group, which must coordinate all contracts and bids among the three major types of players, is to facilitate transactions while maintaining system planning. One proposed method is based on discrete analysis of a Dutch auction. Other auction mechanisms may be suggested. Such proposed methods are similar to a warehousing decision on how much to inventory for a future period. As shown later in this work, the functions of the RTG and the ISO could be merged. Indeed, this should be the case based on organizational behavior.

Independent Contract Administrator (ICA)
The goal for an Independent Contract Administrator is a combination of the goals for an ISO and an RTG. Northern States Power Company originally proposed this term. This term will be used in place of ISO and RTG in the following to differentiate the combined responsibility from the existing ISO companies.

Time Horizon (Months)						
0..............1	2	....	12	....	18	....	360
Spot Market	Forward Market					
	Swap Market (Market to Market Contracts)					
		Futures Market			Planning Market	

FIGURE 13.3 Interconnection between markets.

Electric Markets

Competition may be enhanced through the various markets: cash, futures, planning, and swap. The cash market facilitates trading in spot and forward contracts. This work assumes that such trading would be on an hourly basis. Functionally, this is equivalent to the interchange brokerage systems implemented in several states. The distinction is that future time period interchange (forward contracts) are also traded.

The futures market facilitates trading of futures and options. These are financially derived contracts used to spread risk. The planning market facilitates trading of contracts for system expansion. Such a market has been proposed by a west coast electric utility. The swap market facilitates trading between all markets when conversion from one type of contract to another is desired. It should be noted that multiple markets are required to enable competition between markets.

The structure of any spot market auction must include the ability to schedule as far into the future as the industrial practice did before deregulation. This would require extending the spot into the future for at least six months, as proposed by this author (Sheblé, 1994). Future month production should be traded for actual delivery in forward markets. Future contracts should be implemented at least 18 months into the future if not 3 years. Planning contracts must be implemented for at least 20 years into the future, as recently offered by TVA, to provide an orderly, predictable expansion of the generation and transmission systems. Only then can timely addition of generation and transmission be assured. Finally, a swap market must be established to enable the transfer of contracts from one period (market) to another.

To minimize risk, the use of option contracts for each market should be implemented. Essentially, all of the players share the risk. This is why all markets should be open to the public for general trading and subject to all rules and regulations of a commodity exchange. Private exchanges, not subject to such regulations, do not encourage competition and open price discovery.

The described framework (Sheblé, 1996b) allows for cash (spot and forward), futures, and planning markets as shown in Fig. 13.3. The *spot market* is most familiar within the electric industry (Schweppe, 1988). A seller and a buyer agree (either bilaterally or through an exchange) upon a price for a certain amount of power (MW) to be delivered sometime in the near future (e.g., 10 MW from 1:00 p.m. to 4:00 p.m. tomorrow). The buyer needs the electricity, and the seller wants to sell. They arrange for the electrons to flow through the electrical transmission system and they are happy. A *forward contract* is a binding agreement in which the seller agrees to deliver an amount of a particular product in a specified quality at a specified time to the buyer. The forward contract is further into the future than is the spot market. In both the forward and spot contracts, the buyer and seller want physical goods (e.g., the electrons). A *futures contract* is primarily a financial instrument that allows traders to lock in a price for a commodity in some future month. This helps traders manage their risk by limiting potential losses or gains. Futures contracts exist for commodities in which there is sufficient interest and in which the goods are generic enough that it is not possible to tell one unit of the good from another (e.g., 1 MW of electricity of a certain quality, voltage level, etc.). A futures *option contract* is a form of insurance that gives the option purchaser the right, but not the obligation, to buy (sell) a futures contract at a given price. For each options contract, there is someone "writing" the contract who, in return for a premium, is obligated to sell (buy) at the strike price (see Fig. 13.3). Both the options and the futures contracts are financial instruments designed to minimize risk. Although provisions for delivery exist, they are not convenient (i.e., the delivery point is not located where you want it to be located). The trader ultimately cancels his position in the futures market, either with a gain or loss. The physicals are then purchased on the spot market to meet demand with the profit or loss having been locked in via the futures contract.

Power System Planning (Reliability)

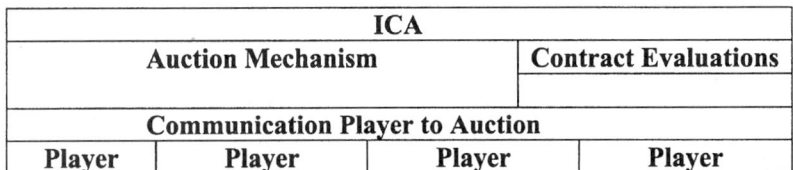

FIGURE 13.4 Computerized markets.

Time Line Into Future (Hours)				Information
0..744				Level
GENCO	ESCO	MARKCO	BROCO	Player
Bids	Bids	....	Bids	Communication
Spot	Forward	Futures	Planning	Markets
Swap				Market
ICA/ISO/RTO				Coordination
GENCO	ESCO	MARKCO	BROCO	Players
...	...	...	...	...

FIGURE 13.5 Electric market.

A *swap* is a customized agreement in which one firm agrees to trade its coupon payment for one held by another firm involved in the swap. Finally, a *planning market* is needed to establish a basis for financing long term projects like transmission lines and power plants (Sheblé, 1993).

Computerized Auction Market Structure

Auction market structure is a computerized market, as shown in Fig. 13.4. Each of the agents has a terminal (PC, workstation, etc.) connected to an auctioneer (auction mechanism) and a contract evaluator. Players generate bids (buy and sell) and submit the quotation to the auctioneer. A bid is a specified amount of electricity at a given price. The auctioneer binds bids (matching buyers and sellers) subject to approval of the contract evaluation. This is equivalent to the pool operating convention used in the vertically integrated business environment.

The contract evaluator verifies that the network can remain in operation with the new bid in place. If the network cannot operate, then the match is denied. The auctioneer processes all bids to determine which matches can be made. However, the primary problem is the complete specification of how the network can operate and how the agents are treated comparably as the network is operated closer to limits. The network model must include all constraints for adequacy and security.

The major trading objectives are hedging, speculation, and arbitrage. Hedging is a defense mechanism against loss and/or supply shortages. Speculation is assuming an investment risk with a chance for profit. Arbitrage is crossing sales (purchases) between markets for riskless profit. This work assumes that there are four markets commonly operated: forward, futures, planning, and swaps (Fig. 13.5).

Forward Market: The forward contracts reflect short term future system conditions. In the forward market, prices are determined at the time of the contract but the transactions occur at some future time. Optimization tools for short term scheduling problems can be enhanced to evaluate trading opportunities in the forward market. For example, short term dispatching algorithms, such as economic unit commitment dispatch, can be used to estimate and earn profit in the forward market.

Futures Market: A futures market creates competition because it unifies diverse and scattered local markets and stabilizes prices. The contracts in the futures market are risky because price movements over time can result in large gains or losses. There is a link between forward markets and futures markets that restricts price volatility. *Options* (options contracts) allow the agent to exercise the right to activate a contract or cancel it. Claims to buy are called "call" options. Claims to sell are called "put" options.

A more detailed discussion of an electric futures contract is discussed in Sheblé (1994b). The components include trading unit, trading hours, trading months, price quotation, minimum price fluctuation, maximum daily price fluctuation, last trading day, exercise of options, option strike prices, delivery, delivery period, alternate delivery procedure, exchange of futures for, or in connection with, physicals, quality specifications, and customer margin requirements.

Swap Market: In the swap market, contract position can be closed with an exchange of physical or financial substitutions. The trader can find another trader who will accept (make) delivery and end the trader's delivery obligation. The acceptor of the obligation is compensated through a price discount or a premium relative to the market rate.

The financial drain inflicted on traders when hedging their operations in the futures market is slightly higher than the one inflicted through direct placement in the forward market. An optimal mix of options, forward commitments, futures contracts, and physical inventories is difficult to assess and depends on hedging, constraints imposed by different contracts, and the cost of different contracts. A clearinghouse such as a swap market handles the exchange of various energy instruments.

Planning Market: The growth of transmission grid requires transmission companies to make contracts based on the expected usage to finance projects. The planning market would underwrite equipment usage subject to the long term commitments to which all companies are bound by the rules of network expansion to maintain a fair marketplace. The network expansion would have to be done to maximize the use of transmission grid for all agents. Collaboration would have to be overseen and prohibited with a sufficiently high financial penalty. The growth of the generation supply similarly requires such markets. However, such a market has been started with the use of franchise rights (options) as established in recent Tennessee Valley Authority connection contracts. This author has published several papers outlining the need for such a market. Such efforts are not documented in this work.

Capacity Expansion Problem Definition

The capacity expansion problem is different for an ESCO, GENCO, TRANSCO, DISTCO, and ANSILCO. This section assumes that the ICA will not own equipment but will only administer the contracts between players. The capacity expansion problem is divided into the following areas: generation expansion, transmission expansion, distribution expansion, and market expansion. ESCOs are concerned with market expansion. GENCOs are concerned with generation expansion. TRANSCOs are concerned with transmission expansion. DISTCOs are concerned with distribution expansion. ANSILCOs are concerned with supportive devices expansion. This author views ancillary services as a misnomer. Such services are necessary supportive services. Thus, the term "supportive" will be used instead of ancillary. Also, since supportive devices are inherently part and parcel of the transmission or distribution system, these devices will be assumed into the TRANSCO and DISTCO functions without loss of generality. Thus, ANSILCOs are not treated separately.

Based on the above idealized view of the marketplace, the following generalizations are made. GENCOs are concerned with the addition of capacity to meet market demands while maximizing profit. Market demands include bilateral contracts with the EMA as well as bilateral contracts with ESCOs or with the ICA. ESCOs are concerned with the addition of capacity of supplying customers with the service desired to maintain market share. ESCOs are thus primarily concerned with the processing of information from marketplace to customer. However, ESCOs are also concerned with additional equipment supplied by DISTCOs or TRANSCOs to provide the level of service required by some customers. ESCOs are thus concerned with all aspects of customer contracts and not just the supply of "electrons."

The ICA is concerned with the operation of the overall system subject to the contracts between the buyers and the sellers and between all players with ICA. The overall goal of the ICA is to enable any customer to trade with any other customer with the quick resolution of contract enforcement available through mercantile associations. The ICA maintains the reliability of the network by resolving the unexpected differences between the contracts, real operation, and unplanned events. The ICA has the authority, through contracts, to buy generation services, supportive services, and/or transmission services, or to curtail contracts if the problems cannot be resolved with such purchases as defined in these contracts.

Thus, the ICA has the authority to connect or disconnect generation and demand to protect the integrity of the system. The ICA has the authority to order new transmission or distribution expansion to maintain the system reliability and economic efficiency of the overall system. The economic efficiency is determined by the price of electricity in the cash markets on a periodic basis. If the prices are approximately the same at all points in the network, then the network is not preventing customers from getting to the suppliers. Similarly, the suppliers can get to the buyers. Since all buyers and suppliers are protected from each other through the default clauses of the mercantile agreement, it does not matter which company deals with other companies as the quick resolution of disputes is guaranteed. This strictness of guarantee is the cornerstone of removing the financial uncertainty at the price of a transaction fee to cover the costs of enforcement.

The goal of each company is different but the tools are the same for each. First, the demand must be predicted for future time periods sufficiently into the future to maintain operation financially and physically. Second, the present worth of the expansion projects has to be estimated. Third, the risks associated with each project and the demand-forecast uncertainty must be estimated. Fourth, the acceptable value at risk acceptable for the company has to be defined. Fifth, the value at risk has to be calculated. Sixth, methods of reducing the value at risk have to be identified and evaluated for benefits. Seventh, the overall portfolio of projects, contracts, strategies, and risk has to be assessed. Only then can management decide to select a project for implementation.

The characteristics of expansion problems include:

1. The cost of equipment or facilities should exhibit economies of scale for the same risk level baring technology changes.
2. Time is a primary factor since equipment has to be in place and ready to serve the needs as they arise. Premature installation results in idle equipment. Delayed installation results in lost market share.
3. The risk associated with the portfolio of projects should decrease as time advances.
4. The portfolio has to be revalued at each point when new information is available that may change the project selection, change the strategy, or change the mix of contracts.

The capital expansion problem is often referred to as the "capital budgeting under uncertainty" problem (Aggarwal, 1993). Thus, capital expansion is an exercise in estimating the present net value of future cash flows and other benefits as compared to the initial investment required for the project given the risk associated with the project(s). The key concept is the uncertainty and thus the risk of all business ventures. Uncertainties may be due to estimation (forecasting) and measurement errors. Such uncertainties can be reduced by the proper application of better tools. Another approach is to investment in information technology to coordinate the dissemination of information. Indeed, information technology is one key to the appropriate application of capital expansion.

Another uncertainty factor is that the net present value depends on market imperfections. Market imperfections are due to competitor reactions to each other's strategies, technology changes, and market rule changes (regulatory changes). The options offered by new investment are very hard to forecast. Also the variances of the options to reduce the risk of projects are critical to proper selection of the right project. Management has to constantly revalue the project, change the project (including termination), integrate new information, or modify the project to include technology changes.

Estimates have often been biased by management pressure to move ahead, to not investigate all risks, or to maintain strategies that are not working as planned. Uncertainties in regulations and taxes are often critical for the decision to continue.

There are three steps to any investment plan: investment alternative identification, assessment, selection and management of the investment as events warrant.

The remaining sections outline the necessity of each step by simple models of the problem to be solved at each step. Since simple problems are given, linear programming solution techniques may be used to solve them. Indeed, the theory of optimization can yield valuable insight as to the importance of further investigations. The inclusion of such models is beyond the scope of this work.

Capacity expansion is one aspect of capital budgeting. Marketing and financial investments are also capital budgeting problems. Often, the capacity expansion has to be evaluated not only on the projects merits, but also the merits of the financing bundled with the project.

Other Sections on Planning

The following sections on planning deal with the overall approach as described by Dr. H. Merrill and include sections on forecasting, power system planning, transmission planning, and system reliability. Forecasting demand is a key issue for any business entity. Forecasting for a competitive industry is more critical than for a regulated industry. Transmission planning is discussed based on probabilistic techniques to evaluate the expected advantages and costs of present and future expansion plans. Reliability of the supply is covered, including transmission reliability. The most interesting aspect of the electric power industry is the massive changes presently occurring. It will be interesting to watch as the industry adapts to regulatory changes and as the various market players find their corporate niche in this new framework.

References

R. Aggarwal, *Capital Budgeting Under Uncertainty*, Prentice-Hall, Englewood Cliffs, NJ, 1993.

M. L. Baughman, J. W. Jones, and A. Jacob, Model for Evaluating the Economics of Cool Storage Systems, *IEEE Trans. Power Syst.*, 8(2), May 1993.

W. G. Bently and J. C. Evelyn. "Customer Thermal Energy Storage: A Marketing Opportunity for Cooling Off Electric Peak Demand," *IEEE Trans. on Power Systems*, 1(4), 973-979, 1987.

R. Billinton and G. Lian, "Monte Carlo Approach to Substation Reliability Evaluation," *IEE Proc.*, 140(2), 147-152, 1991.

R. Billinton and L. Wenyuan, "Hybrid Approach for Reliability Evaluation of Composite Generation and Transmission Systems Using Monte-Carlo Simulation and Enumeration Technique," *IEE Proc.*, 138(3), 233-241, 1991.

R. Billinton and W. Li, "A Monte Carlo Method for Multi-Area Generation System Reliability Assessment," *IEEE Trans. Power Syst.*, 7(4), 1487-1492, 1992.

R. Billinton and L. Gan, "Monte Carlo Simulation Model for Multiarea Generation System Reliability Studies," *IEE Proc.*, 140(6), 532-538, 1993.

B. R. Binger and E. Hoffman, *Microeconomics with Calculus*. Scott, Foresman and Company, Glenview, 1988.

Lynn E. Bussey, The Economic Analysis of Industrial Projects, Prentice-Hall, Englewood Cliffs, NJ, 1981.

H. P. Chao, "Peak Load Pricing and Capacity Planning with Demand and Supply Uncertainty," *Bell J. Econ.*, 14(1), 179-190, 1983.

C. S. Chen and J. N. Sheen, "Cost Benefit Analysis of a Cooling Energy Storage System," *IEEE Trans. Power Syst.*, 8(4), 1993.

W. Chu, B. Chen, and C. Fu, "Scheduling of Direct Load Control to Minimize Load Reduction for a Utility Suffering from Generation Shortage," *IEEE/PES Winter Meeting*, 1993, Columbus, OH.

J. S. Clayton, S. R. Erwin, and C. A. Gibson. "Interchange Costing and Wheeling Loss Evaluation by Means of Incrementals," *IEEE Trans. Power Syst.*, 5(3), 759-765, 1990.

R. E. Clayton and R. Mukerji, "System Planning Tools for the Competitive Market," *IEEE Computer Appl. Power*, 50, 1996.

A. I. Cohen, J. W. Patmore, D. H. Oglevee, R. W. Berman, L. H. Ayers, and J. F. Howard, "An Integrated System for Load Control," *IEEE Trans. Power Syst.*, PWRS-2(3), 1987.

H. G. Daellenbach, *Systems and Decision Making*, John Wiley & Sons, New York, 1994.

B. Daryanian, R. E. Bohn, and R. D. Tabors, "Optimal Demand-side Response to Electricity Spot Prices for Storage-type Customers," *IEEE Trans. Power Syst.*, 4(3), 897-903, 1989.

A. K. David and Y. Z. Li, "Effect of Inter-temporal Factors on the Real-time Pricing of Elasticity," *IEEE Trans. Power Syst.*, 8(1), 1993.

J. T. Day, "Forecasting Minimum Production Costs with Linear Programming," *IEEE Trans. Power Apparatus Syst.*, PAS-90(2), 814-823, 1971.

S. Dekrajangpetch and G. B. Sheblé, "Alternative Implementations of Electric Power Auctions," in *Proc. 60th Am. Power Conf.*, 60-1, 394-398, 1998.

J. K. Delson, X. Feng, and W.C Smith, "A Validation Process for Probabilistic Production Costing Programs," *IEEE Trans. Power Syst.*, 6(3), 1326-1336, 1991.

G. Fahd and G. Sheblé, "Optimal Power Flow of Interchange Brokerage System Using Linear Programming," *IEEE Trans. Power Syst.*, T-PWRS, 7(2), 497-504, 1992.

G. Fahd, Dan Richards, and Gerald B. Sheblé, "The Implementation of an Energy Brokerage System Using Linear Programming," *IEEE Trans. Power Syst.*, T-PWRS, 7(1), 90-96, 1992.

George Fahd, "Optimal Power Flow Emulation of Interchange Brokerage Systems Using Linear Programming," Ph.D. dissertation, Auburn University, 1992.

H. H. Happ, Report on Wheeling Costs, Case 88-E-238, The New York Public Service Commission, Feb. 1990.

B. F. Hobbs and R. E. Schuler, "An Assessment of the Deregulation of Electric Power Generation Using Network Models of Imperfect Spatial Markets," *Papers of the Regional Science Association*, 57, 75-89, 1985.

W. W. Hogan, "A Market Power Model with Strategic Interaction in Electricity Networks," *Energy J.*, 18(4), 107-141, 1997.

S. R. Huang and S. L. Chen, "Evaluation and Improvement of Variance Reduction in Monte-Carlo Production Simulation," *IEEE Trans. Energy Conv.*, 8(4), 610-619, 1993.

M. Ilic, F. Galiana, and L. Fink, *Power Systems Restructuring: Engineering and Economics*, Kluwer Academic Publishers, Norwell, MA, 1998.

K. Kelley, S. Henderson, P. Nagler, and M. Eifert, Some Economic Principles for Pricing Wheeled Power. National Regulatory Research Institute, August 1987.

B. A. Krause and J. McCalley "Bulk Power Transaction Selection in a Competitive Electric Energy System with Provision of Security Incentives," *Proc. 26th Ann. North Am. Power Symp.*, Manhattan, Kansas, September 1994, 126-136.

J. Kumar and G. B. Sheblé, "A Decision Analysis Approach to Transaction Selection Problem in a Competitive Electric Market," *Electric Power Syst. Res. J.*, 1997.

J. Kumar and G. B. Sheblé, "A Framework for Transaction Selection Using Decision Analysis Based upon Risk and Cost of Insurance," *Proc. 29th North Am. Power Symp.*, Kansas State University, KS, 1994, 548-557.

J. Kumar and G. B. Sheblé, "Transaction Selection Using Decision Analysis Based Upon Risk and Cost of Insurance," *IEEE Winter Power Meeting*, 1996.

J. Kumar Electric Power Auction Market Implementation and Simulation, Ph.D. dissertation, Iowa State University, 1996.

C. N. Kurucz, D. Brandt, and S. Sim, "A Linear Programming Model for Reducing System Peak Through Customer Load Control Programs," *IEEE PES Winter Meeting*, 96 WM 239-9 PWRS, Baltimore, MD, 1996.

K. D. Le, R. F. Boyle, M. D. Hunter, and K. D. Jones, "A Procedure for Coordinating Direct-Load-Control Strategies to Minimize System Production Cost," *IEEE Trans. Power App.. Syst.*, PAS-102(6), 1983.

S. H. Lee and C. L. Wilkins, "A Practical Approach to Appliance Load Control Analysis: A Water Heater Case Study," *IEEE Trans. Power App.. Syst.*, 7(4), 1992.

F. N. Lee, "Three-Area Joint Dispatch Production Costing," *IEEE Trans. Power Syst.*, 3(1), 294-300, 1988.

T. Y. Lee and N. Chen, "The Effect of Pumped Storage and Battery Energy Storage Systems on Hydrothermal Generation Coordination," *IEEE Trans. Energy Conv.*, 7(4), 631-637, 1992.

T. Y. Lee and N. Chen, "Optimal Capacity of the Battery Storage System in a Power System," *IEEE Trans. Energy Conv.*, 8(4), 667-673, 1993.

T. Y. Lee and N. Chen, "Effect of Battery Energy Storage System on the Time-of-Use Rates Industrial Customers," *IEE Proc.: Generator Transmission Distribution*, 141(5), 5521-528, 1994.

A. P. Lerner, "Monopoly and the Measurement of Monopoly Power," *Rev. Econ. Stud.*, 1, 157-175, 1934.

M. Lin, A. Breipohl, and F. Lee, "Comparison of Probabilistic Production Cost Simulation Methods" *IEEE Trans. Power Syst.,* 4(4), 1326-1333, 1989.

J. McCalley and G. B. Sheblé, "Competitive Electric Energy Systems: Reliability of Bulk Transmission and Supply," tutorial paper presented at the *Fourth International Conference of Probabilistic Methods Applied to Power Systems,* 1994.

J. McCalley, A. Fouad, V. Vittal, A. Irizarry-Rivera, R. Farmer, and B. Agarwal, "A Probabilistic Problem in Electric Power System Operation: The Economy-Security Tradeoff for Stability-Limited Systems," *Proc. Third Intl. Workshop on Rough Sets and Soft Computing,* November 10-12, 1994, San Jose, CA.

H. M. Merril and A. J. Wood, "Risk and Uncertainty in Power System Planning," *10th Power Syst. Comp. Conf.,* PSCC, Graz, Austria, August 1990.

H. M. Merrill, "Have I Ever Got a Deal for You. Economic Principles in Pricing of Services," IEEE SP 91EH0345-9-PWR, pp. 1-8, 1991.

V. Miranda, "Power System Planning and Fuzzy Sets: Towards a Comprehensive Model Including All Types of Uncertainties," *Proc. PMAPSí94,* Rio de Janeiro, Brazil, September 1994.

V. Miranda and L. M. Proença, "A General Methodology for Distribution Planning Under Uncertainty, Including Genetic Algorithms and Fuzzy Models in a Multi-criteria Environment," *Proc. Stockholm Power Tech, SPT'95,* Stockholm, Sweden, June 18-22, 832-837, 1995.

R. E. Mortensen, and K. P. Haggerty, "Dynamics of Heating and Cooling Loads: Models, Simulation, and Actual Utility Data," *IEEE Trans. Power Syst.,* 5(1), 253-248, 1990.

K.-H. Ng and G. B. Sheblé, "Direct Load Control — A Profit-based Load Management Using Linear Programming," *IEEE Trans. Power Syst.,* 13(2), 1998.

K.-H. Ng Reformulating Load Management Under Deregulation, Master's thesis, Iowa State University, Ames, May 1997.

R. P. O'Neill and C. S. Whitmore, "Network Oligopoly Regulation: An Approach to Electric Federalism," *Electricity and Federalism Symp.,* June 24, 1993 (Revised March 16, 1994).

S. S. Oren, P. Spiller, P. Variya, and F. Wu, "Nodal Prices and Transmission Rights: A Critical Appraisal," University of California at Berkeley Research Report, December 1994.

S. S. Oren, "Economic Inefficiency of Passive Transmission Rights in Congested Electricity Systems with Competitive Generation," *Energy J.,* 18(1), 63-83, 1997.

H. R. Outhred, "Principles of a Market-Based Electricity Industry and Possible Steps Toward Implementation in Australia," *Intl. Conf. Adv. Power Syst. Control, Operation and Management,* Hong Kong, Dec. 7-10, 1993.

B. J. Parker, E. Denzinger, B. Porretta, G. J. Anders, and M. S. Mirsky, "Optimal Economic Power Transfers," *IEEE Trans. Power Syst.,* 4(3), 1167-1175, 1989.

C. Parker and J. Stremel, "A Smart Monte Carlo Procedure for Production Costing and Uncertainty Analysis," *Proc. Am. Power Conf.,* 58(II), 897-900, 1996.

V. Pereira, B. G. Gorenstin, and Morozowski Fo, "Chronological Probabilistic Production Costing and Wheeling Calculations with Transmission Network Modeling," *IEEE Trans. Power Syst.,* 7(2), 885-891, 1992.

D. Post, *Electric Power Interchange Transaction Analysis and Selection,* Master's thesis, Iowa State University, Ames, 1994.

D. Post, S. Coppinger, and G. Sheblé, "Application of Auctions as a Pricing Mechanism for the Interchange of Electric Power," *IEEE Trans. Power Syst.,* 10(3), 1580-1584, 1995.

M. V. Rakic and Z. M. Markovic, "Short Term Operation and Power Exchange Planning of Hydro-thermal Power Systems," *IEEE Trans. Power Syst.,* 9(1), 1994.

N. S. Rau, "Certain Considerations in the Pricing of Transmission Service," *IEEE Trans. Power Syst.,* 4(3), 1133-1139, 1989.

C. Richter and G. Sheblé, "Genetic Algorithm Evolution of Utility Bidding Strategies for the Competitive Marketplace," *1997 IEEE/PES Summer Meeting,* PE-752-PWRS-1-05-1997, New York: IEEE, 1997.

C. Richter and G. Sheblé, "Building Fuzzy Bidding Strategies for the Competitive Generator," *Proc. 1997 North Am. Power Symp.,* 1997.

C. Richter and G. Sheblé, "Bidding Strategies that Minimize Risk with Options and Futures Contracts," in *Proc. 1998 Am. Power Conf., Session 25, Open Access II-Power Marketing, Paper C*, 1998.

S. Roy, "Goal-programming Approach to Optimal Price Determination for Inter-area Energy Trading," *Intl. J. Energy Res.*, 17, 847-862, 1993.

P. Rupanagunta, M. L. Baughman, and J. W. Jones, "Scheduling of Cool Storage Using Non-linear Programming Techniques," *IEEE Trans. Power Syst.*, 10(3), 1995.

T. Russel, "Working with an Independent Grid in the UK — A Generator's View," *Proc. 24th Ann. North Am. Power Symp.*, Manhattan, Kansas, 270-275, September 1992.

F. C. Schweppe, M. C. Caramanis, R. D. Tabors, and R. E. Bohn, *Spot Pricing of Electricity*, Kluwer Academic Publishers, Boston, MA, 1988.

G. B. Sheblé, "Electric Energy in a Fully Evolved Marketplace," *Proc. 26th Ann. North Am. Power Symp.*, Manhattan, Kansas, pp. 81-90, September 1994.

G. B. Sheblé, "Simulation of Discrete Auction Systems for Power System Risk Management," *Proc. 27th Ann. Frontiers of Power Conf.*, Oklahoma State University, Stillwater, OK, pp. I.1-I.9, 1994.

G. Sheblé and G. Fahd, "Unit Commitment Literature Synopsis," *IEEE Trans. Power Syst.*, 9, 128-135, 1994.

G. Sheblé and J. McCalley, "Discrete Auction Systems for Power System Management," presented at the *1994 National Science Foundation Workshop*, Pullman, WA, 1994.

G. Sheblé, "Priced Based Operation in an Auction Market Structure," *IEEE Trans. on Power Systems*, 11(4), 1770-1777, 1996.

Sheblé, G. B., *Computational Auction Methods for Restructured Power System Industry Operation*, Kluwer Academic Press, Boston, MA, 1999.

D. Shirmohammadi, P. R. Gribik, T. K. Law, J. H. Malinowski, and R. E. O'Donnell, "Evaluation of Transmission Network Capacity Use for Wheeling Transactions," *IEEE Trans. Power Syst.*, 4(4), 1405-1413, 1989.

J. Skeer, "Highlights of the International Energy Agency Conference on Advanced Technologies for Electric Demand-side Management," *Proc. Adv. Technol. Electric Demand-Side Management*, International Energy Agency, Sorrento, Italy, 1991.

V. L. Smith, "Electric Power Deregulation: Background and Prospects," *Contemporary Policy Issues*, 6, 14-24, 1988.

S. Smith, "Linear Programming Model for Real-time Pricing of Electric Power Service," *Operations Res.*, 41, 470-483, 1993.

R. L. Sullivan, *Power System Planning*, McGraw-Hill, New York, 1977.

A. Svoboda and S. Oren, "Integrating Price-based Resources in Short-term Scheduling of Electric Power Systems," *IEEE Trans. Energy Conv.*, 9, 760-769, 1994.

R. D. Tabors, "Transmission System Management and Pricing: New Paradigms and International Comparisons," Paper WM110-7 presented at the *IEEE/PES Winter Meeting*, T-PWRS, February 1994.

A. Vojdani, C. Imparto, N. Saini, B. Wollenberg, and H. Happ, "Transmission Access Issues," presented at the *1995 IEEE/PES Winter Meeting*, 95 WM 121-4 PWRS, IEEE, New York, 1994.

L. Wang, "Approximate Confidence Bounds on Monte Carlo Simulation Results for Energy Production," *IEEE Trans. Power Syst.*, 4(1), 69-74, 1989.

C. Wang and J. R. McDonald, *Modern Power System Planning*, McGraw-Hill, New York, 1994.

D. C. Wei and N. Chen, "Air-Conditioner Direct Load Control by Multi-pass Dynamic Programming," *IEEE Trans. Power Syst.*, 10(1), 1995.

H. L. Willis, *Spatial Electric Load Forecasting*, Marcel Dekker, New York, 1996, 14-17.

W. E. Winston and C. A. Gibson, "Geographical Load Shift and its Effect on Interchange Evaluation," *IEEE Trans. Power Syst.*, 3(3), 865-871, 1988.

A. J. Wood and B. F. Wollenberg, *Power Generation, Operation, and Control*, 2nd ed., John Wiley & Sons, New York, 1996.

F. Wu and P. Varaiya, "Coordinated Multi-lateral Trades For Electric Power Networks: Theory and Implementation," University of California at Berkeley Research Report, June 1995.

13.2 Short-Term Load and Price Forecasting with Artificial Neural Networks[1]

Alireza Khotanzad

Artificial Neural Networks

Artificial neural networks (ANN) are systems inspired by research into how the brain works. An ANN consists of a collection of arithmetic computing units (nodes or neurons) connected together in a network of interconnected layers. A typical node of an ANN is shown in Fig. 13.6. At the input side, there are a number of so-called "connections" that have a weight of "W_{ij}" associated with them. The input denoted by X_i gets multiplied by W_{ij} before reaching node j via the respective connection. Inside the neuron, all the individual inputs are first summed up. The summed inputs are passed through a nonlinear single-input, single-output function "S" to produce the output of the neuron. This output in turn is propagated to other neurons via corresponding connections.

While there are a number of different ANN architectures, the most widely used one (especially in practical applications) is the multilayer feed-forward ANN, also known as a multilayer perceptron (MLP), shown in Fig. 13.7. An MLP consists of n input nodes, h so called "hidden layer" nodes (since they are not directly accessible from either input or output side), and m output nodes connected in a feed-forward fashion. The input layer nodes are simple data distributors whereas neurons in the hidden and output layers have an S-shaped nonlinear transfer function known as the "sigmoid activation function," $f(z) = 1/1 + e^{-z}$ where z is the summed inputs.

For hidden layer nodes, the output is:

$$H_j = \frac{1}{1+\exp\left(-\sum_{i=1}^{n} W_{ij} X_i\right)}$$

where H_j is the output of the jth hidden layer node, $j = 1, ..., h$, and X_i represents the ith input connected to this hidden node via W_{ij} with $i = 1, ..., n$.

The output of the kth output node is given by

$$Y_k = \frac{1}{1+\exp\left(-\sum_{j=1}^{h} W_{jk} H_j\right)}$$

where Y_k is the output of the kth output layer node with $k = h + 1, ..., m$, and W_{jk} representing connection weights from hidden to output layer nodes.

One of the main properties of ANNs is the ability to model complex and nonlinear relationships between input and output vectors through a learning process with "examples." During learning, known input-output examples, called the training set, are applied to the ANN. The ANN learns by adjusting or adapting the connection weights through comparing the output of the ANN to the expected output. Once the ANN is trained, the extracted knowledge from the process resides in the resulting connection weights in a distributed manner.

---

[1] This work was supported in part by the Electric Power Research Institute and 1997 Advanced Technology Program of the State of Texas.

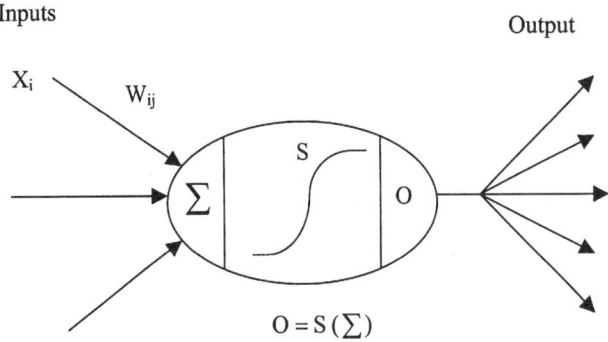

FIGURE 13.6 Model of one node of an ANN.

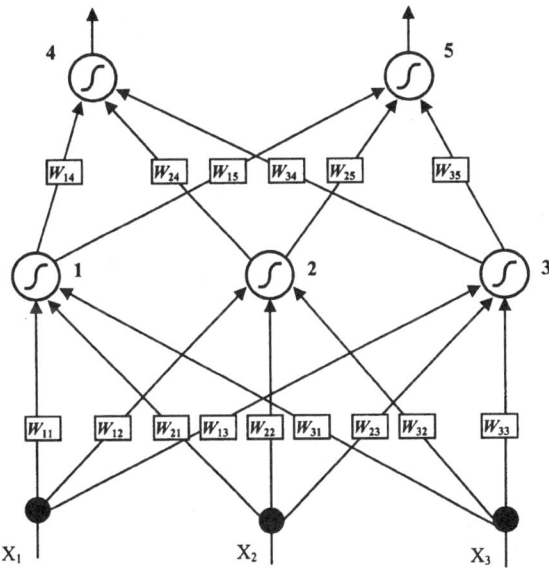

FIGURE 13.7 An example of an MLP with 3 input, 3 hidden, and 2 output nodes.

A trained ANN can generalize (i.e., produce the expected output) if the input is not exactly the same as any of those in the training set. This property is ideal for forecasting applications where some historical data exists but the forecast indicators (inputs) may not match up exactly with those in the history.

Error Back-Propagation Learning Rule

The MLP must be trained with historical data to find the appropriate values for W_{ij} and the number of required neurons in the hidden layer. The learning algorithm employed is the well-known error back-propagation (BP) rule (Rumelhart and McClelland, 1986). In BP, learning takes place by adjusting W_{ij}. The output produced by the ANN in response to inputs is repeatedly compared with the correct answer. Each time, the W_{ij} values are adjusted slightly in the direction of the correct answers by back-propagating the error at the output layer through the ANN according to a gradient descent algorithm.

To avoid overtraining, the cross-validation method is used. The training set is divided into two sets. For instance, if three years of data is available, it is divided into a two-year and a one-year set. The first set is used to train the MLP and the second set is used to test the trained model after every few hundred passes over the training data. The error on the validation set is examined. Typically this error decreases as the number of passes over the training set is increased until the ANN is overtrained, as signified by a rise in this error. Therefore, the training is stopped when the error on the validation set starts to increase.

This procedure yields the appropriate number of epochs over the training set. The entire three years of data is then used to retrain the MLP using this number of epochs.

In a forecasting application, the number of input and output nodes is equal to the number of utilized forecast indicators and the number of desired outputs, respectively. However, there is no theoretical approach to calculate the appropriate number of hidden layer nodes. This number is determined using a similar approach for training epochs. By examining the error over a validation set for a varying number of hidden layer nodes, a number yielding the smallest error is selected.

Adaptive Update of the Weights During Online Forecasting

A unique aspect of the MLPs used in the forecasting systems described in this section is the adaptive update of the weights during online operation. In a typical usage of an MLP, it is trained with the historical data and the weights of the trained MLP are then treated as fixed parameters. This is an acceptable procedure for many applications. However, if the modeled process is a nonstationary one that can go through rapid changes, e.g., variations of electric load due to weather swings or seasonal changes, a tracking mechanism with sensitivity to the recent trends in the data can aid in producing better results.

To address this issue, an adaptive weight adjustment strategy that takes place during online operation is utilized. The MLP is initially trained using the BP algorithm; however, the trained weights are not treated as static parameters. During online operation, these weights are adaptively updated on a sample-by-sample basis. Before forecasting for the next instance, the forecasts of the past few samples are compared to the actual outcome (assuming that actual outcome for previous forecasts have become available) and a small scale error BP operation is performed with this data. This mini-training with the most recent data results in a slight adjustment of the weights and biases them toward the recent trend in data.

Short-Term Load Forecasting

The daily operation and planning activities of an electric utility requires the prediction of the electrical demand of its customers. In general, the required load forecasts can be categorized into short-term, mid-term, and long-term forecasts. The short-term forecasts refer to hourly prediction of the load for a lead time ranging from one hour to several days out. The mid-term forecasts can either be hourly or peak load forecasts for a forecast horizon of one to several months ahead. Finally, the long-term forecasts refer to forecasts made for one to several years in the future.

The quality of short-term hourly load forecasts has a significant impact on the economic operation of the electric utility since many decisions based on these forecasts have significant economic consequences. These decisions include economic scheduling of generating capacity, scheduling of fuel purchases, system security assessment, and planning for energy transactions. The importance of accurate load forecasts will increase in the future because of the dramatic changes occurring in the structure of the utility industry due to deregulation and competition. This environment compels the utilities to operate at the highest possible efficiency, which, as indicated above, requires accurate load forecasts. Moreover, the advent of open access to transmission and distribution systems calls for new actions such as posting the available transmission capacity (ATC), which will depend on the load forecasts.

In the deregulated environment, utilities are not the only entities that need load forecasts. Power marketers, load aggregators, and independent system operators (ISO) will all need to generate load forecasts as an integral part of their operation.

This section describes the third generation of an artificial neural network (ANN) hourly load forecaster known as ANNSTLF (Artificial Neural Network Short-Term Load Forecaster). ANNSTLF, developed by Southern Methodist University and PRT, Inc. under the sponsorship of the Electric Power Research Institute (EPRI), has received wide acceptance by the electric utility industry and is presently being used by over 40 utilities across the U.S. and Canada.

Application of the ANN technology to the load forecasting problem has received much attention in recent years (Bakirtzis et al., 1996; Dillon et al., 1991; Ho et al., 1992; Khotanzad et al., 1998; Khotanzad et al., 1997; Khotanzad et al., 1996; Khotanzad et al., 1995; Lee et al., 1992; Lu et al., 1993; Mohammed et al., 1995; Papalexopoulos et al., 1994; Park et al., 1991; Peng et al., 1993). The function learning property of ANNs enables them to model the correlations between the load and such factors as climatic conditions, past usage pattern, the day of the week, and the time of the day, from historical load and weather data. Among the ANN-based load forecasters discussed in published literature, ANNSTLF is the only one that is implemented at several sites and thoroughly tested under various real-world conditions.

A noteworthy aspect of ANNSTLF is that a single architecture with the same input-output structure is used for modeling hourly loads of various size utilities in different regions of the country. The only customization required is the determination of some parameters of the ANN models. No other aspects of the models need to be altered.

ANNSTLF Architecture

ANNSTLF consists of three modules: two ANN load forecasters and an adaptive combiner (Khotanzad et al., 1998). Both load forecasters receive the same set of inputs and produce a load forecast for the same day, but they utilize different strategies to do so. The function of the combiner module is to mix the two forecasts to generate the final forecast.

Both of the ANN load forecasters have the same topology with the following inputs:

- 24 hourly loads of the previous day
- 24 hourly weather parameters of the previous day (temperatures or effective temperatures, as discussed later)
- 24 hourly weather parameters forecasts for the coming day
- Day type indices

The difference between the two ANNs is in their outputs. The first forecaster is trained to predict the regular (base) load of the next day, i.e., the 24 outputs are the forecasts of the hourly loads of the next day. This ANN will be referred to as the "Regular Load Forecaster (RLF)." On the other hand, the second ANN forecaster predicts the *change* in hourly load from yesterday to today. This forecaster is named the "Delta Load Forecaster (DLF)."

The two ANN forecasters complement each other because the RLF emphasizes regular load patterns whereas the DLF puts stronger emphasis on yesterday's load. Combining these two separate forecasts results in improved accuracy. This is especially true for cases of sudden load change caused by weather fronts. The RLF has a tendency to respond slowly to rapid changes in load. On the other hand, since the DLF takes yesterday's load as the basis and predicts the changes in that load, it has a faster response to a changing situation.

To take advantage of the complimentary performance of the two modules, their forecasts are adaptively combined using the recursive least squares (RLS) algorithm (Proakis et al., 1992). The final forecast for each hour is obtained by a linear combination of the RLF and DLF forecasts as:

$$\hat{L}_{k+1}(i) = \alpha_B(i)\hat{L}_{k+1}^{RLF}(i) + \alpha_C(i)\hat{L}_{k+1}^{DLF}(i), \quad i = 1, \ldots, 24$$

The $\alpha_B(i)$ and $\alpha_C(i)$ coefficients are computed using the RLS algorithm. This algorithm produces coefficients that minimize the weighted sum of squared errors of the past forecasts denoted by J,

$$J = \sum_{k=1}^{N} \beta^{N-k} \left[L_k(i) - \hat{L}_k(i) \right]^2$$

where $L_k(i)$ is the actual load at hour i, N is the number of previous days for which load forecasts have been made, and β is a weighting factor in the range of $0 < \beta \leq 1$ whose effect is to de-emphasize (forget) old data.

The block diagram of the overall system is shown in Fig. 13.8.

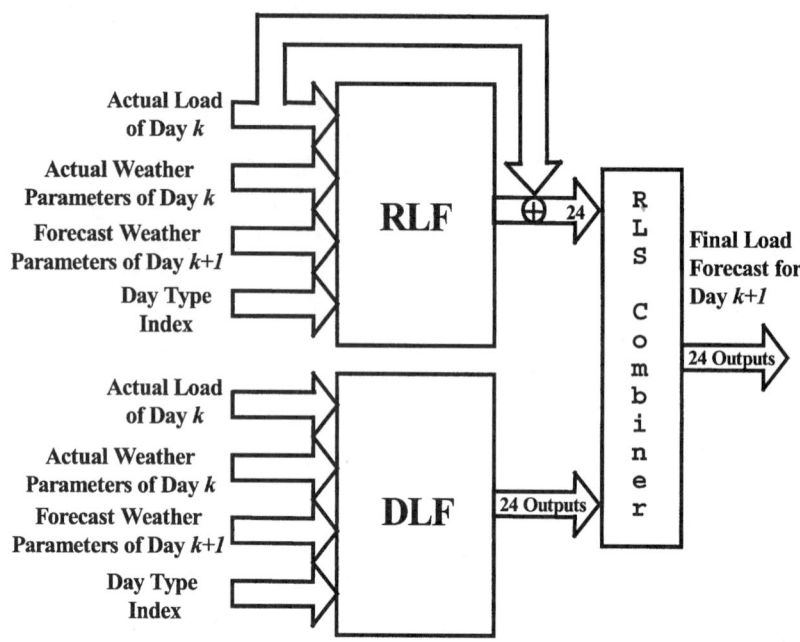

FIGURE 13.8 Block diagram of ANNSTLF.

Humidity and Wind Speed

Although temperature (T) is the primary weather variable affecting the load, other weather parameters, such as relative humidity (H) and wind speed (W), also have a noticeable impact on the load. The effects of these variables are taken into account through transforming the temperature value into an effective temperature, T_eff, using the following transformation:

$$T\_eff = T + \alpha * H$$

$$T\_eff = T - \frac{W * (65° - T)}{100}$$

Holidays and Special Days

Holidays and special days pose a challenge to any load forecasting program since the load of these days can be quite different from a regular workday. The difficulty is the small number of holidays in the historical data compared to the typical days. For instance, there would be three instances of Christmas Day in a training set of three years. The unusual behavior of the load for these days cannot be learned adequately by the ANNs since they are not shown many instances of these days.

It was observed that in most cases, the profile of the load forecast generated by the ANNs using the concept of designating the holiday as a weekend day, does resemble the actual load. However, there usually is a significant error in predicting the peak load of the day. The ANNSTLF package includes a function that enables the user to reshape the forecast of the entire day if the peak load forecast is changed by the user. Thus, the emphasis is placed on producing a better peak load forecast for holidays and reshaping the entire day's forecast based on it.

The holiday peak forecasting algorithm uses a novel weighted interpolation scheme. This algorithm will be referred to as "Reza algorithm" after the author who developed it (Khotanzad et al., 1998). The general idea behind the Reza algorithm is to first find the "close" holidays to the upcoming one in the

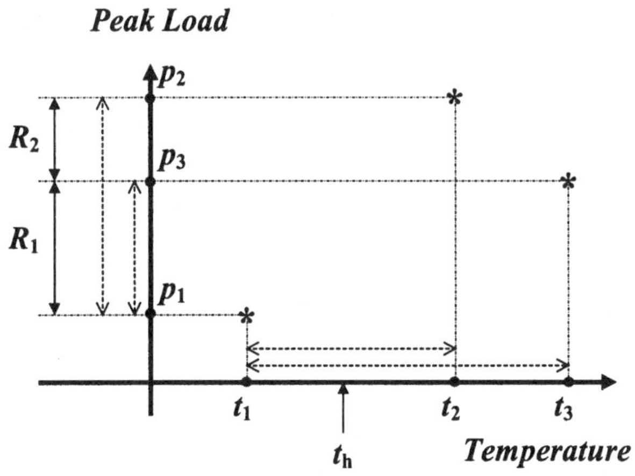

Power System Planning (Reliability)

Peak Load

FIGURE 13.9 Example of peak load vs. temperature at peak load for a three-holiday database.

historical data. The closeness criterion is the temperature at the peak-load hour. Then, the peak load of the upcoming holiday is computed by a novel weighted interpolation function described in the following.

The idea is best illustrated by an example. Let us assume that there are only three holidays in the historical data. The peak loads are first adjusted for any possible load growths. Let (t_i, p_i) designate the i-th peak-load hour temperature and peak load, respectively. Fig. 13.9 shows the plot of p_i vs. t_i for an example case.

Now assume that t_h represents the peak-load hour temperature of the upcoming holiday. t_h falls in between t_1 and t_2 with the implication that the corresponding peak load, p_h, would possibly lie in the range of $[p_1, p_2] = R_1 + R_2$. But, at the same time, t_h is also between t_1 and t_3 implying that p_h would lie in $[p_1, p_3] = R_1$. Based on this logic, p_h can lie in either R_1 or $R_1 + R_2$. However, note that R_1 is common in both ranges. The idea is to give twice as much weight to the R_1 range for estimating p_h since this range appears twice in pair-wise selection of the historical data points.

The next step is to estimate p_h for each nonoverlapping interval, R_1 and R_2, on the y axis, i.e., $[p_1, p_3]$ and $[p_3, p_2]$.

For $R_1 = [p_1, p_3]$ interval:

$$\hat{p}_{h1} = \frac{p_3 - p_1}{t_3 - t_1} * (t_h - t_1) + p_1$$

For $R_2 = [p_3, p_2]$ interval:

$$\hat{p}_{h2} = \frac{p_2 - p_3}{t_2 - t_3} * (t_h - t_3) + p_3$$

If any of the above interpolation results in a value that falls outside the respective range, R_i, the closest p_i, i.e., maximum or minimum of the interval, is used instead.

The final estimate of p_h is a weighted average of $\hat{p}_{h1}$ and $\hat{p}_{h2}$ with the weights decided by the number of overlaps that each pair-wise selection of historical datapoints creates. In this case, since R_1 is visited twice, it receives a weighting of two whereas the interval R_2 only gets a weighting coefficient of one.

$$\hat{p}_h = \frac{w_1 * \hat{p}_{h1} + w_2 * \hat{p}_{h2}}{w_1 + w_2} = \frac{2 * \hat{p}_{h1} + 1 * \hat{p}_{h2}}{2 + 1}$$

TABLE 13.1 Utility Information for Performance Study

Utility	No. Days in Testing Period	Weather Variable	Location
1	141	T	Canada
2	131	T	South
3	365	T,H,W	Northeast
4	365	T	East Coast
5	134	T	Midwest
6	365	T	West Coast
7	365	T,H	Southwest
8	365	T,H	South
9	174	T	North
10	275	T,W	Midwest

TABLE 13.2 Summary of Performance Results in Terms of MAPE

Utility	MAPE OF	Days-Ahead						
		1	2	3	4	5	6	7
1	All hours	1.91	2.29	2.53	2.71	2.87	3.03	3.15
	Peak	1.70	2.11	2.39	2.62	2.73	2.94	3.10
2	All hours	2.72	3.44	3.63	3.77	3.79	3.83	3.80
	Peak	2.64	3.33	3.46	3.37	3.42	3.52	3.40
3	All hours	1.89	2.25	2.38	2.45	2.53	2.58	2.65
	Peak	1.96	2.26	2.41	2.49	2.60	2.69	2.82
4	All hours	2.02	2.37	2.51	2.58	2.61	2.65	2.69
	Peak	2.26	2.59	2.69	2.83	2.85	2.93	2.94
5	All hours	1.97	2.38	2.61	2.66	2.65	2.65	2.74
	Peak	2.03	2.36	2.49	2.37	2.49	2.51	2.55
6	All hours	1.57	1.86	1.99	2.08	2.14	2.17	2.18
	Peak	1.82	2.25	2.38	2.50	2.61	2.62	2.63
7	All hours	2.29	2.79	2.90	3.00	3.05	3.10	3.18
	Peak	2.42	2.78	2.90	2.98	3.07	3.17	3.28
8	All hours	2.22	2.91	3.15	3.28	3.39	3.45	3.50
	Peak	2.38	3.00	3.12	3.29	3.40	3.45	3.52
9	All hours	1.63	2.04	2.20	2.32	2.40	2.41	2.50
	Peak	1.83	2.25	2.36	2.51	2.54	2.64	2.78
10	All hours	2.32	2.97	3.25	3.38	3.44	3.52	3.56
	Peak	2.15	2.75	2.93	3.08	3.16	3.27	3.27
Average	All hours	2.05	2.53	2.72	2.82	2.89	2.94	2.99
	Peak	2.12	2.57	2.71	2.80	2.89	2.97	3.03

Performance

The performance of ANNSTLF is tested on real data from ten different utilities in various geographical regions. Information about the general location of these utilities and the length of the testing period are provided in Table 13.1.

In all cases, three years of historical data is used to train ANNSTLF. Actual weather data is used so that the effect of weather forecast errors do not alter the modeling error. The testing is performed in a blind fashion meaning that the test data is completely independent from the training set and is not shown to the model during its training.

One-to-seven-day-ahead forecasts are generated for each test set. To extend the forecast horizon beyond one day ahead, the forecast load of the previous day is used in place of the actual load to obtain the next day's load forecast.

The forecasting results are presented in Table 13.2 in terms of mean absolute percentage error (MAPE) defined as:

Power System Planning (Reliability)

TABLE 13.3 Training and Test Periods for the Price Forecaster Performance Study

Database	Training Period	Test Period	MAE of Day-Ahead Hourly Price Forecasts ($)
CALPX	Apr 23, 98–Dec 31, 98	Jan 1, 99–Mar 3, 99	1.73
PJM	Apr 2, 97–Dec 31, 97	Jan 2, 98–Mar 31, 98	3.23

$$MAPE = \frac{100}{N} \sum_{i=1}^{N} \frac{|Actual(i) - Forecast(i)|}{Actual(i)}$$

with N being the number of observations. Note that the average MAPEs over ten utilities as reported in the last row of Table 13.3 indicate that the third-generation engine is quite accurate in forecasting both hourly and peak loads. In the case of hourly load, this average remains below 3% for the entire forecast horizon of seven days ahead, and for the peak load it reaches 3% on the seventh day. A pictorial example of one-to-seven-day-ahead load forecasts for utility 2 is shown in Fig. 13.10.

As pointed out earlier, all the weather variables (T or T_eff) used in these studies are the actual data. In online usage of the model, weather forecasts are used. The quality of these weather forecasts vary greatly from one site to another. In our experience, for most cases, the weather forecast errors introduce approximately 1% of additional error for one-to-two-days out load forecasts. The increase in the error for longer range forecasts is more due to less accurate weather forecasts for three or more days out.

Short-Term Price Forecasting

Another forecasting function needed in a deregulated and competitive electricity market is prediction of future electricity prices. Such forecasts are needed by a number of entities such as generation and power system operators, wholesale power traders, retail market and risk managers, etc. Accurate price forecasts enable these entities to refine their market decisions and energy transactions leading to significant economic advantages. Both *long-term* and *short-term* price forecasts are of importance to the industry. The long-term forecasts are used for decisions on transmission augmentation, generation expansion, and distribution planning whereas the short-term forecasts are needed for daily operations and energy trading decisions. In this work, the emphasis will be on short-term hourly price forecasting with a horizon extending up to the next 24 hours.

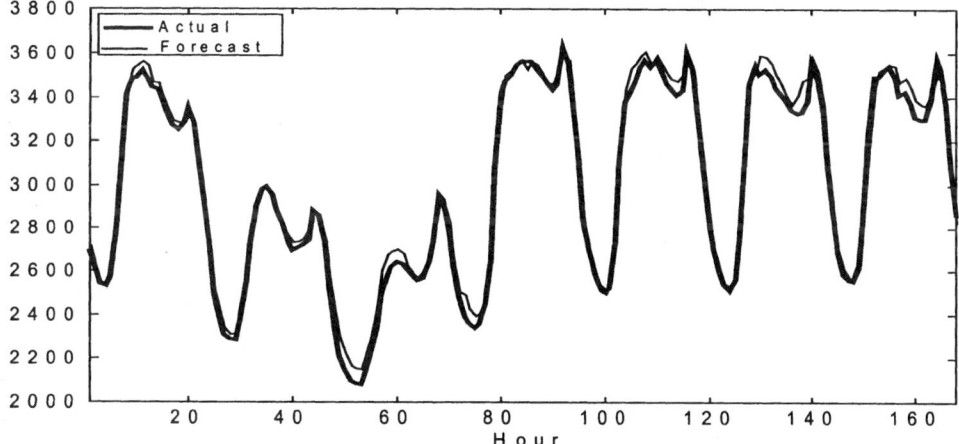

FIGURE 13.10 An example of a one-to-seven-day-ahead load forecast.

In general, energy prices are tied to a number of parameters such as future demand, weather conditions, available generation, planned outages, system reserves, transmission constraints, market perception, etc. These relationships are nonlinear and complex and conventional modeling techniques cannot capture them accurately. In a similar manner to load forecasting, ANNs could be utilized to "learn" the appropriate relationships. Application of ANN technology to electricity price forecasting is relatively new and there are few published studies on this subject (Szkuta et al., 1998).

The adaptive BP MLP forecaster described in the previous section is used here to model the relationship of hourly price to relevant forecast indicators. The system is tested on data from two power pools with good performance.

Architecture of Price Forecaster

The price forecaster consists of a single adaptive BP MLP with the following inputs:

- Previous day's hourly prices
- Previous day's hourly loads
- Next day's hourly load forecasts
- Next day's expected system status for each hour

The expected system status input is an indicator that is used to provide the system with information about unusual operating conditions such as transmission constraints, outages, or other subjective matters. A bi-level indicator is used to represent typical vs. atypical conditions. This input allows the user to account for his intuition about system condition and helps the ANN better interpret sudden jumps in price data that happen due to system constraints.

The outputs of the forecaster are the next day's 24 hourly price forecasts.

Performance

The performance of the hourly price forecaster is tested on data collected from two sources, the California Power Exchange (CALPX) and the Pennsylvania-New Jersey-Maryland Independent System Operator (PJM). The considered price data are the Unconstrained Market Clearing Price (UMCP) for CALPX, and Market Clearing Price (MCP) for PJM. The average of Locational Marginal Prices (LMP) uses a single MCP for PJM. The training and test periods for each database are listed in Table 13.3. Testing is performed in a blind fashion, meaning that the test data is completely independent from the training set and is not shown to the model during its training. Also, actual load data is used in place of load forecast.

The day-ahead forecast results are presented in the first column of Table 13.4 in terms of mean absolute error (*MAE*) expressed in dollars. This measure is defined as:

$$MAE = \frac{100}{N} \sum_{i=1}^{N} \left| Actual\ Price(i) - Forecast\ Price(i) \right|$$

with N being the total number of hours in the test period.

To put these results in perspective, the sample mean and standard deviation of hourly prices in the test period are also listed in Table 13.4. Note the correspondence between MAE and the standard deviation of data, i.e., the smaller standard deviation results in a lower MAE and vice versa.

TABLE 13.4 Results of Performance Study for the Test Period

Database	MAE of Day-Ahead Hourly Price Forecasts ($)	Sample Mean of Actual Hourly Prices ($)	Sample Standard Deviation of Actual Hourly Prices ($)
CALPX	1.73	19.98	5.45
PJM	3.23	17.44	7.67

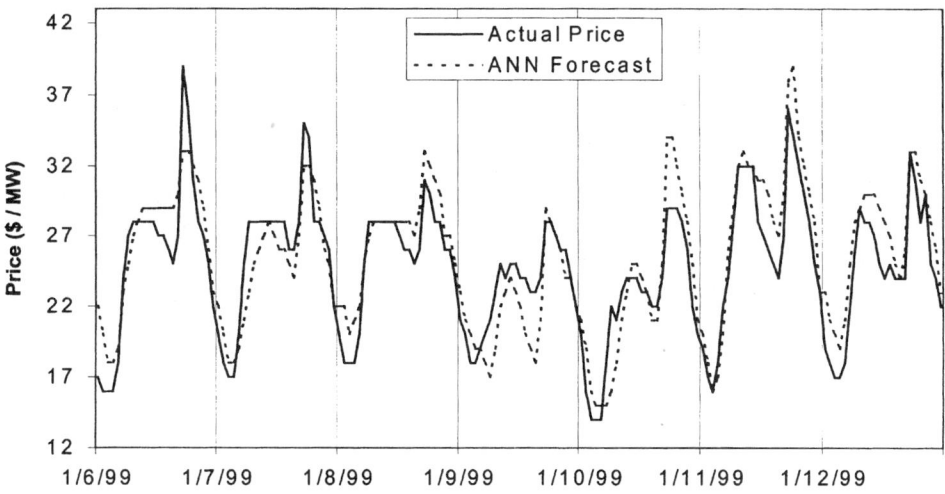

FIGURE 13.11 An example of the ANN price forecaster performance for CALPX price data.

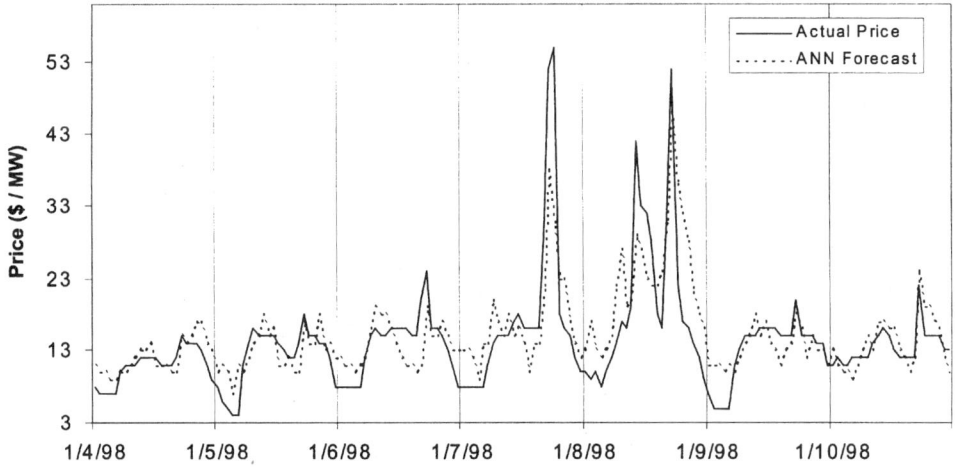

FIGURE 13.12 An example of the price forecaster performance for PJM price data.

Figures 13.11 and 13.12 show a representative example of the performance for each of the databases. It can be seen that the forecasts closely follow the actual data.

References

Bakirtzis, A.G., et. al, A neural network short term load forecasting model for the Greek power system, *IEEE Trans. PWRS*, 11, 2, 858–863, May, 1996

Dillon, T.S., Sestito, S., and Leung, S., Short term load forecasting using an adaptive neural network, *Electrical Power and Energy Systems,* 13, 4, Aug. 1991.

Ho, K., Hsu, Y., and Yang, C., Short term load forecasting using a multi-layer neural network with an adaptive learning algorithm, *IEEE Trans. PWRS*, 7, 1, 141–149, Feb. 1992.

Khotanzad, A., Afkhami-Rohani, R., and Maratukulam, D., ANNSTLF — Artificial neural network short-term load forecaster-generation three, *IEEE Trans. on Power Syst.*, 13, 4, 1413–1422, November, 1998.

Khotanzad, A., Afkhami-Rohani, R., Lu, T.L., Davis, M.H., Abaye, A., and Maratukulam, D.J., ANNSTLF — A neural network-based electric load forecasting system, *IEEE Trans. on Neural Networks*, 8, 4, 835–846, July, 97.

Khotanzad, A., Davis, M.H., Abaye, A., and Martukulam, D.J., An artificial neural network hourly temperature forecaster with applications in load forecasting, *IEEE Trans. PWRS*, 11, 2, 870-876, May 1996.

Khotanzad, A., Hwang, R.C., Abaye, A., and Maratukulam, D., An adaptive modular artificial neural network hourly load forecaster and its implementation at electric utilities, *IEEE Trans. PWRS*, 10, 3, 1716–1722, Aug. 1995.

Lee, K.Y., Cha, Y.T., and Park, J.H., Short-term load forecasting using an artificial neural network, *IEEE Trans. PWRS*, 7, 1, 124-132, Feb. 1992.

Lu, C.N., Wu, N.T., and Vemuri, S., Neural network based short term load forecasting, *IEEE Trans. PWRS*, 8, 1, 336–342, Feb. 1993.

Mohammed, O., Park, D., Merchant, R., et. al, Practical experiences with an adaptive neural network short-term load forecasting system, *IEEE Trans. PWRS*, 10, 1, 254–265, Feb. 1995.

Papalexopolos, A.D., Hao, S., and Peng, T.M., An implementation of a neural network based load forecasting model for the EMS, *IEEE Trans. PWRS*, 9, 4, 1956–1962, Nov. 1994.

Park, D.C., El-Sharkawi, M.A., Marks, R.J., Atlas, L.E., and Damborg, M.J., Electric load forecasting using an artificial neural network, *IEEE Trans. PWRS*, 442–449, May 1991.

Peng, T.M., Hubele, N.F., and Karady, G.G., Advancement in the application of neural networks for short-term load forecasting, *IEEE Trans. PWRS*, 8, 3, 1195–1202, Feb. 1993.

Proakis, J.G., Rader, C.M., Ling, F., and Nikias, C.L., *Advanced Digital Signal Processing*, Macmillan Publishing Company, New York, NY, 1992, 351–358.

Rumelhart, D.E. and McClelland, J.L., *Parallel Distributed Processing*, Vol. 1, MIT Press, Cambridge, 1986.

Szkuta, B.R., Sanabria, L.A., and Dillon, T.S., Electricity price short-term forecasting using artificial neural networks, *IEEE Trans. PWRS*, 14, 3, 851–857, Aug. 1999.

13.3 Transmission Plan Evaluation — Assessment of System Reliability

N. Dag Reppen and James W. Feltes

Bulk Power System Reliability and Supply Point Reliability

Transmission systems must meet performance standards and criteria that ensure an acceptable level of quality of electric service. Service quality means continuity of supply and constancy of voltage waveform and power system frequency. Frequency is typically not an issue in large interconnected systems with adequate generation reserves. Similarly, voltage quality at the consumer connection is typically addressed at the distribution level and not by reinforcing the transmission system. This leaves continuity of power supply as the main criterion for acceptable transmission system performance.

Requirements for continuity of supply are traditionally referred to as power system reliability. Reliability criteria for transmission systems must address both local interruptions of power supply at points in the network as well as widespread interruptions affecting population centers or entire regions. Local and widespread interruptions are typically caused by different types of events and require different evaluation approaches.

Additional transmission facilities will virtually always increase reliability, but this remedy is constrained by the cost of new facilities and environmental impacts of new construction. Reliability objectives, therefore, must be defined explicitly or implicitly in terms of the value of reliable power supply to the consumer and to society at large. Reflecting the different concerns of local interruptions and widespread interruptions, reliability objectives are different for the bulk transmission system than for the local area transmission or subtransmission systems supplying electric power to electric distribution systems. These two aspects of power system reliability will be referred to as bulk power system reliability (Endrenyi et al., 1982, Parts 1 and 2) and supply point reliability.

Bulk Transmission Systems Reliability is Evaluated Using Deterministic Reliability Criteria

A distinguishing characteristic of bulk transmission systems is that severe disturbances arising in them can have widespread impact. Major failures of bulk transmission systems have resulted in interruption of thousands of MW of load and interruption of service to millions of customers. Three important characteristics of reliable bulk transmission system performance are:

1. Low risk of widespread shutdown of the bulk transmission system,
2. Confinement of the extent of bulk transmission system shutdown when it occurs, and
3. Rapid restoration of operation following shutdown of the bulk transmission system.

Most interconnected systems have reliability criteria and design standards that explicitly aim at limiting the risk of widespread shutdowns or blackouts. Such criteria may call for transmission reinforcements or limitations of power transfers across the system. The two other characteristics are addressed by sharpening operating command and control functions and improving control and communication facilities. Therefore, transmission system plans are typically evaluated with respect to reliability criteria that are aimed at limiting the risk of system shutdowns.

The U.S. National Electric Reliability Council (NERC), formed in response to the 1965 Northeast blackout, has developed basic design criteria aimed at reducing the risk of "instability and uncontrolled cascading" that may lead to system blackouts. The various regional reliability councils have interpreted these requirements in various ways and produced additional criteria and guides to address this problem (NERC, 1988). Deterministic criteria for bulk power systems will typically include the following requirements:

1. Test criteria for simulated tests aimed at avoiding overload cascading and instability, including voltage collapse. These test criteria specify in generic form:
   a. the system conditions to be tested: e.g., peak load conditions, lines or generators assumed out on maintenance, transfer levels
   b. the type of failure that initiates a disturbance: e.g., type and location of short circuit
   c. assumptions to be applied regarding the operation of protection systems and other control systems
   d. the allowable limits of system response: line and transformer loading limits, high and low voltage limits, and criteria for stable operation

   The system must be reinforced to meet these criteria.
2. Requirements to test extreme contingencies such as the simultaneous outage of two or more parallel lines or the loss of entire substations. These tests are made to determine and understand the vulnerability of the system to such events. When critical extreme contingencies are identified, steps should be taken to minimize the risk of occurrence of such events.
3. Criteria and guides for protection system design to reduce the risk of critical protection initiated disturbances and for protection misoperation that may aggravate a serious system condition.

Evaluations of the system response to specified severe but rare types of failure events are labeled deterministic. The likelihood of the event specified is not considered, except in a qualitative way when the criteria were created. Since only a small subset of all potentially critical events can be tested, the tests are sometimes referred to as "umbrella" tests. A system that passes these selected tests is believed to have a degree of resiliency that will protect it not only for the specific disturbances simulated, but also for a multitude of other disturbances of similar type and severity.

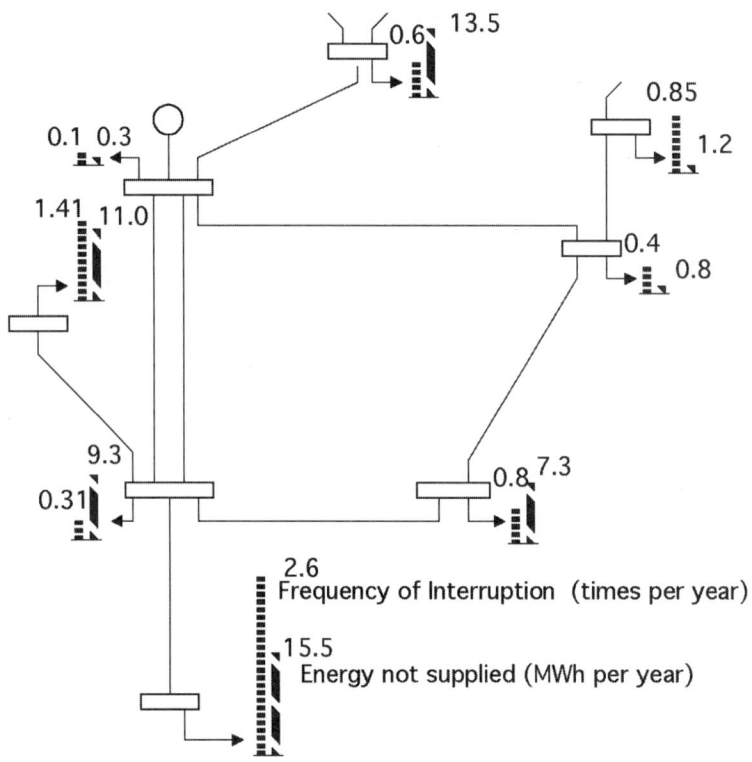

FIGURE 13.13 Prediction of supply point reliability.

Supply Point Reliability is Evaluated Using Either Deterministic or Probabilistic Reliability Criteria

Reliability objectives at the local area transmission or subtransmission level focus on the reliability of supply to specific supply points as shown in Fig. 13.13. Statistically, the reliability of supply may be expressed in terms of the frequency of occurrence of load interruptions, the amount of load interrupted, and the duration of the interruptions. Frequency of interruptions and MWh not served over a period such as a year are commonly used measures for the observed or predicted reliability of power supply to a particular node in a transmission system. Probabilistic reliability methods are required to predict reliability in these terms (Billinton and Allan, 1984; Endrenyi, 1978; Salvaderi et al., 1990). These methods will typically consider more likely events rather than the more extreme and very rare events that can lead to system shutdown. This is justified since system shutdown occurrences are not frequent enough to significantly impact the reliability measures calculated.

While it is practical to perform probabilistic calculations to assess supply point reliability, deterministic simulation tests are also commonly used. As a minimum, deterministic criteria call for load flow testing of all single line and single transformer outages. This is referred to as single contingency testing or N-1 testing. For each of these outages, no line or transformer shall exceed its emergency rating, and no voltage shall violate specified high and low emergency voltage limits. Violation of these criteria calls for system reinforcements. Exceptions are typically made for supply points with low peak demand where it is judged to be too expensive to provide for redundant service. Some utilities use a peak load criterion such as 25 MW, above which redundant transmission connections to a supply point are called for.

Methods for Assessing Supply Point Reliability

Supply point reliability may be assessed in four different ways in order of increasing complexity:

1. **Deterministic:** System alternatives must meet criteria specifying allowable system response to specified contingencies.
2. **Probabilistic — System Trouble:** System alternatives must meet criteria specified in terms of probabilistic reliability indices reflecting risk of unacceptable system response.
3. **Probabilistic — Consumer Impact:** Same as (2), but criteria are specified in terms of consumer impact such as risk of supply interruption or risk of load curtailment.
4. **Cost/Benefit Analysis:** This approach is based on the concept that the proper level of service reliability should be defined by the balance of incremental worth of service reliability improvement and incremental cost of providing that improvement. The approach is also referred to as "effectiveness analysis" or "value based" reliability assessment.

The limitation of the deterministic approach (1) is that it considers only the initial system problems for a few contingencies. These contingencies have typically been selected by committee based on a mixture of judgment, tradition, and experience. If the selected contingencies do not cover all important reliability concerns, the resulting system may be unreliable. If the selected contingencies put undue emphasis on severe but rare events, an unnecessarily expensive system alternative may be selected.

The probabilistic approach (2) aims at eliminating the dependency on judgment in the selection of contingencies by attempting to look at all significant contingencies. In addition, it weighs the importance of the results for each contingency according to the severity of the system problems caused by each contingency and the frequency of occurrence of each contingency.

Approach (3) looks deeper into the problem, in that it is concerned with the impact on the consumer. However, the criteria used to define an acceptable level of reliability are still judgmental. For example, how many interruptions per year would be acceptable or what percentage of total MWh demand is it acceptable to interrupt or curtail? In the cost/benefit approach (4), the criterion for acceptable reliability is implicit in the methodology used.

Reliability Measures — Reliability Indices

Reliability can be measured by the frequency of events having unacceptable impacts on the system or on the consumer, and by the severity and duration of the unacceptable impacts. Thus, there are three fundamental components of reliability measures:

1. Frequency of unacceptable events,
2. Duration of unacceptable events, and
3. Severity of unacceptable events.

From these, other measures, such as probability of unacceptable events, can be derived. An expectation index, such as the loss of load expectation (LOLE) index commonly used to measure the reliability of a generating system is, in its nature, a probability measure. While probability measures have proved useful in generation reliability assessment, they may not be as meaningful in assessing the reliability of a transmission system or a combined generation/transmission system. It is, for example, important to differentiate between 100 events which last 1 sec and 1 event which lasts 100 sec. Since probability measures cannot provide such differentiation, it is often necessary to apply frequency and duration measures when assessing the reliability of transmission systems.

Probabilistic reliability measures or indices can express the reliability improvements of added resources and reinforcements quantitatively. However, several indices are required to capture various reliability aspects. There are two major types of indices: system indices and consumer or load indices (Guertin et al., 1978; Fong et al., 1989). The former concerns itself with system performance and system effects, the latter with the impact on the consumer. The reliability cost measure used in cost/benefit analysis may be classified as a consumer index.

System Indices

Indices suitable for transmission system reliability evaluation may be divided into system problem indices and load curtailment indices.

System problem indices measure frequency, duration, probability, and severity of system problems. Some examples:

- Frequency of circuit overloads (overloads/year),
- Average duration of circuit overloads (hours), and
- Probability of circuit overloads.

Load curtailment indices measure severity in terms of load interrupted or curtailed. The salient characteristic of these indices is that the severity of any event, regardless of the system problems resulting from the event, is expressed in terms of load curtailment. From the three fundamental reliability measures (frequency, duration, and load curtailment), a series of derived reliability indices may be defined as illustrated by the following examples.

Basic Annual Indices

- Frequency of load curtailment $\quad F = \sum_{i} F_i \left(yr^{-1} \right)$

- Hours of load curtailment $\quad D = \sum_{i} F_i D_i \left(h\ yr^{-1} \right)$

- Power curtailed $\quad C = \sum_{i} F_i C_i \left(MW\ yr^{-1} \right)$

- Energy curtailed $\quad E = \sum_{i} F_i D_i C_i \left(MWh\ yr^{-1} \right)$

where

F_i = frequency of event i (yr^{-1})
D_i = duration of event i (h)
C_i = MW load curtailed for event i (MW)
i = all events for which $C_i > 0$

Energy curtailment (E), expressed in MWh not served, is often referred to as *Energy Not Served* (ENS), *Expected Energy Not Served* (EENS), or *Expected Unserved Energy* (EUE).

Load curtailment indices are sometimes normalized to system size. Two commonly used indices are:

- Power interruption index $\quad C_N = C/CMX$ (yr^{-1})
- Energy curtailment index $\quad E_N = E/CMX$ (h yr^{-1})

where CMX = peak load for system, area, or bus.

$E_N \times 60$ is referred to as system minutes, the equivalent number of minutes per year of total system shutdown during peak load conditions.

Cost of Interruptions to Consumers

The fact that a sudden interruption of very short duration can have a significant impact and that an outage of 4 h may have a significantly more severe impact than two outages of 2 h each, illustrates the limitations of simple aggregated reliability measures such as MWh not served. This is an important

Power System Planning (Reliability)

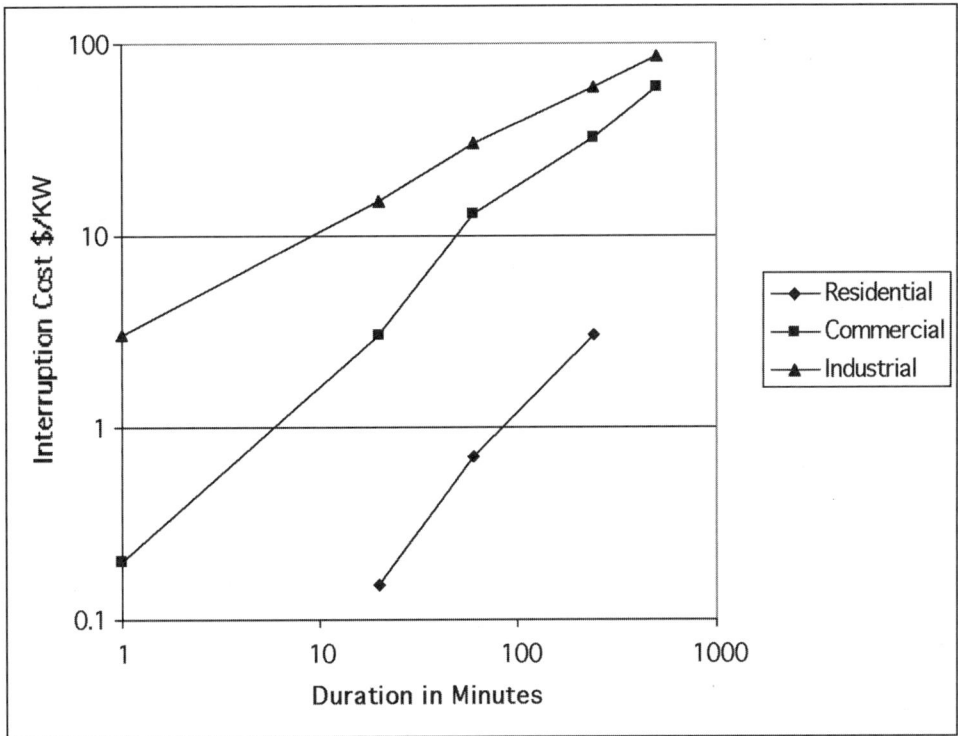

FIGURE 13.14 Illustration of customer damage functions for residential, commercial, and industrial load for process-oriented industrial load. The cost of very short duration outages may be much higher than shown here.

limitation since the various transmission reinforcement options considered may have dramatically different impacts as far as interruption durations are concerned. Since it is difficult to use a multiparameter measure when comparing reinforcement alternatives, a single aggregate measure is much preferred as long as it includes the main reliability factors of concern. The concept of cost to consumers of unreliability expressed in dollars per year has emerged as a practical measure of reliability when comparing transmission reinforcement alternatives. As a measure, reliability cost has the additional important advantage that it can be aggregated with installation cost and operating cost to arrive at a minimum "total cost" design in a cost/benefit analysis.

Conceptually, the annual reliability cost for a group of customers is the aggregated worth the customers put on avoiding load interruptions. In some cases the costs are tangible, allowing reliable dollar cost estimates; in other instances the impacts of interruptions are intangible and subjective, but still real in the eyes of the consumer. Surveys aimed at estimating what consumers would be willing to pay, either in increased rates or for backup service, have been used in the past to estimate the intangible costs of load interruptions. The results of these investigations may be expressed as *Customer Damage Functions*, (CDF), as illustrated in Fig. 13.14 (Mackay and Berk, 1978; Billinton et al., 1983).

Customer damage functions can be used to estimate the dollar cost of any particular load interruption given the amount of load lost and the duration of the interruption. If a customer damage function can be assigned for each supply point, then a cost of interrupted load may be determined.

Outage Models

Generation and transmission outages may be classified in two categories — forced outages and scheduled outages. While forced outages are beyond the control of the system operators, scheduled outages can usually be postponed if necessary to avoid putting the system in a precarious state. These two outage categories must, therefore, be treated separately (Forrest et al., 1985).

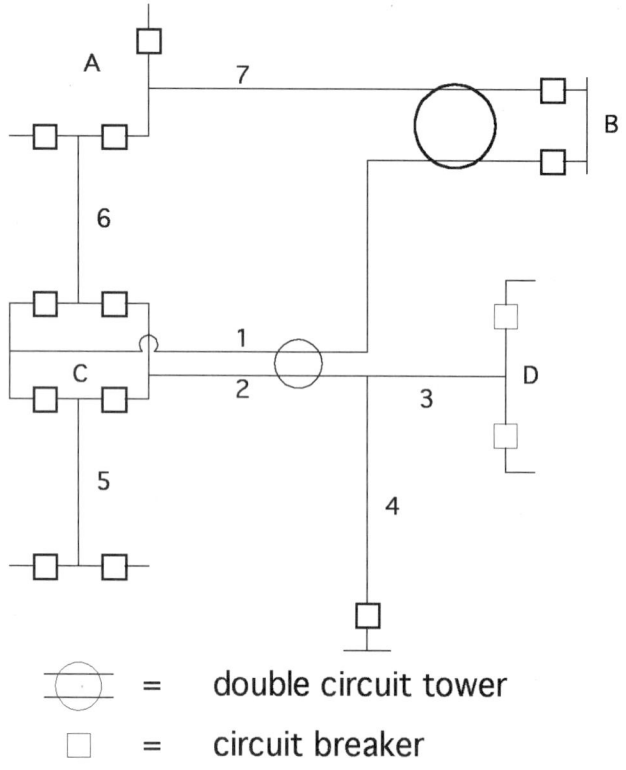

FIGURE 13.15 Sample system illustrating outage types.

Forced Outage Models

The variety and characteristics of forced outage events may be illustrated with reference to Fig. 13.15. Transmission lines and transformers that can be isolated from the system by the opening of circuit breakers are referred to as "elements."

Three categories of forced outage events are recognized:

1. **Single Component Outage Event** — The outage event involves only one element. For example, a fault on circuit 1, cleared by circuit breakers in a normal manner, would only affect circuit 1.
2. **Common Mode Outage Event** — This is a multiple element outage event where a single initiating cause results in multiple element outages where the outages are not consequences of each other. For example, a single lightning stroke taking out both circuits of the double circuit line exiting substation B would be a common mode outage. This event results in the simultaneous outage of circuits 1 and 7.
3. **Substation Related Outage Events** — This is a multiple element outage event that depends on the protection system response to a fault on a component in the substation or on an element connected to the substation. Examples of substation related outage events are:
   a. Stuck breaker — if the breaker common to circuits 1 and 6 is stuck, a fault on either circuits 1 and 6 would result in both circuits out.
   b. Tapped circuits — a fault on circuit 2 would result in circuits 2, 3 and 4 going out together.
   c. Breaker fault – if there is a fault on the breaker common to circuits 1 and 6, both circuits 1 and 6 would be outaged.
   d. Bus section fault — a fault on the bus section in substation B would outage circuits 1 and 7.

A common mode outage event may be combined with substation related outage events. For example, a common mode failure of circuits 1 and 2 would result in an outage event encompassing circuits 1, 2,

Power System Planning (Reliability)

3, and 4. Two or more independent outage events from either of the three outage categories may overlap in time, creating more complex outages. Accurate tools for the prediction of reliability measures include most if not all of these outage types.

Probabilistic Reliability Assessment Methods

Probabilistic reliability assessment tools falls in one of two categories (Endrenyi et al., 1982):

1. The contingency enumeration method
2. The Monte Carlo method

In general, the contingency enumeration method is capable of looking at severe and rare events such as transmission events in great detail, but cannot practically look at many operating conditions. In contrast, the Monte Carlo methods are capable of looking at operating conditions in great detail (Noferi et al., 1975). However, from a computational standpoint, it is not possible to capture with precision, the impact of infrequent but severe transmission contingencies. Thus, the two methods are capturing different aspects of the reliability problem.

Contingency Enumeration Approach

The contingency enumeration approach to reliability analysis includes the systematic selection and evaluation of disturbances, the classification of each disturbance according to failure criteria, and the accumulation of reliability indices. Contingency enumeration techniques are structured so as to minimize the number of disturbances that need to be investigated in detail. This is achieved by testing, to the extent possible, only those disturbances that are sufficiently severe to cause trouble and sufficiently frequent to impact the risk indices to be computed.

The contingency enumeration approach is structured as shown in Fig. 13.16. For a specific predisturbance condition, a contingency is selected and tested to determine whether the contingency causes any immediate system problem such as a circuit overload or a bus voltage out of limits. If it does not, a new contingency is selected and tested.

The occurrence of a system problem may by itself be logged as a failure. However, in many cases, it will be possible to adjust generation or phase shifters to relieve overloads and to adjust generator voltages

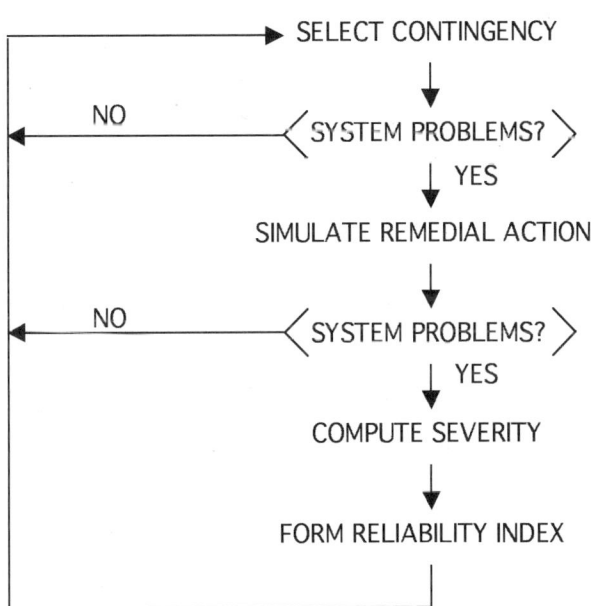

FIGURE 13.16 Contingency enumeration approach.

or transformer taps to bring bus voltages back within range. It is, therefore, of interest to determine whether it is possible to eliminate a system problem by such corrective actions. A failure is logged when corrective actions, short of curtailing consumer loads, are insufficient to eliminate the system problems. The severity of such system problems may be assessed by computing the amount and location of load curtailment necessary to eliminate the problem. In this way, it is possible to compute supply point reliability indices that measure the frequency, duration, and amount of expected load curtailment.

Monte Carlo Approach

Monte Carlo methods (Oliveira et al., 1989) may be sequential or nonsequential. The sequential approach simulates the occurrences of random events through time, recognizing the statistical properties of the various types of events. Typically, the time functions of load and planned generation schedules are established for a period of a year. Starting at the beginning of the year, a sequence of forced shutdown and restoration of transmission and generating equipment is then determined based on random sampling in accordance with the statistical characteristics of the equipment failure processes. The response of the power system during equipment outages is simulated by power flow solutions. Whenever a system condition violating predefined failure criteria is encountered, the occurrence and characteristic of this failure is recorded. At the end of one year of simulation, parameters describing the "observed" reliability of the system can be determined. These parameters may include frequency of equipment overload, frequency of voltage violations, MWh not served, average duration and severity of specified types of failures, etc. This process is illustrated in Fig. 13.17.

One year of simulation constitutes one particular sample scenario governed by the random properties of equipment failure. In order to obtain a measure of the inherent reliability of the system, it is necessary to repeat the simulation over the annual period many times and calculate the reliability measures as the mean of the results obtained over the repeated annual simulations. For reliable systems, several hundred annual simulations may be required to obtain convergence in the reliability measures calculated.

When using the sequential approach, it is possible to model time dependencies between key variables. This allows elaborate modeling of energy limited resources such as hydro plants and pumped hydro. It is also possible to simulate environmental effects such as the occurrence of lightning storms that may impact the failure rate of transmission equipment. A brute-force sequential Monte Carlo simulation would be prohibitively time-consuming when applied to large systems. Practical techniques rely on special sampling techniques and acceptable approximations in power system modeling.

If time dependencies are not essential, the nonsequential approach may be used. In this case, hours of simulation may be selected at random rather than in sequence. For a specific hour, a precontingency state is established including bus loads and matching generation dispatch. When it can be used, the nonsequential approach is typically much faster than the sequential approach.

Comparison of Contingency Enumeration and Monte Carlo Simulation

From the preceding discussion, it is clear that the Monte Carlo method differs from the contingency enumeration method in the way power system states including load, generation dispatch, and component outages are selected. The actual network solution and corrective action models used may be the same or similar for both methods. The major advantage of the Monte Carlo method is the ease with which comprehensive statistics of power system states can be included. This makes the method suitable for computing period reliability indices such as annual indices (CIGRE, 1992).

The Monte Carlo method may not be suitable for estimating the probability or frequency of occurrence of infrequent events. The contingency enumeration method may, therefore, be a more practical approach in system design. In comparing system alternatives to strengthen a local area, the Contingency Enumeration approach will provide consistent and real differences in reliability indices computed for specific situations. Unless the reliability of the alternatives are far apart, it would be very time-consuming and perhaps impractical to obtain acceptable differences in reliability by means of the Monte Carlo method. One way to mitigate this problem is to remove time-consuming calculations from the inner loop of the Monte Carlo calculations. In one approach, which is used to assess the reliability of supply to load centers,

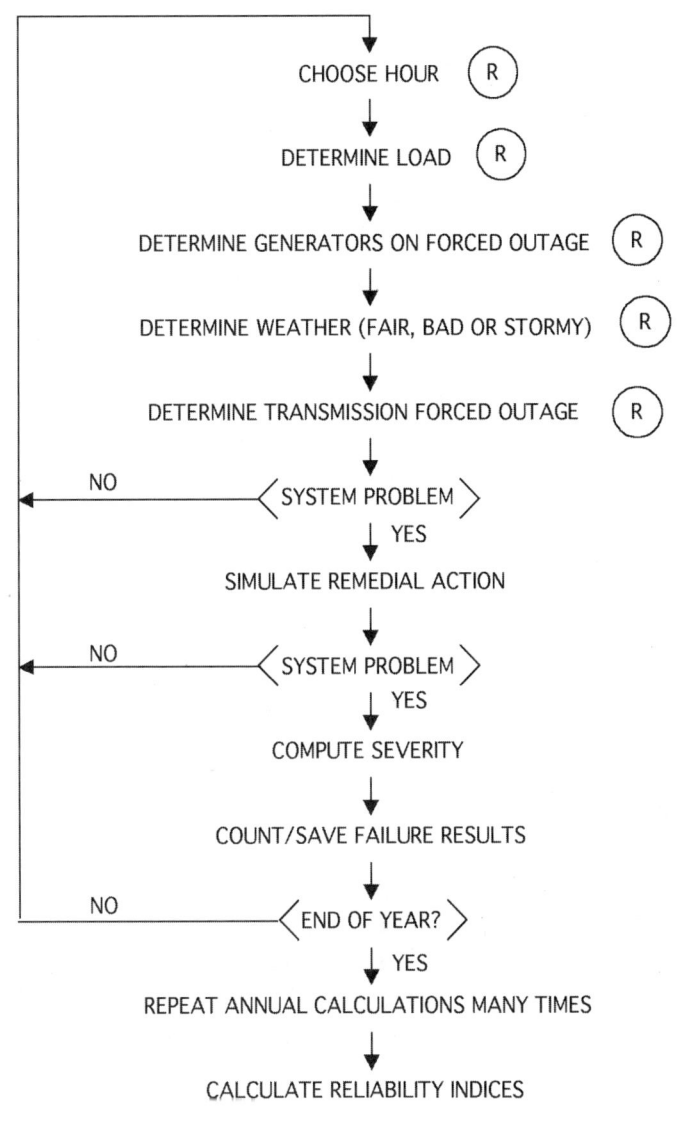

FIGURE 13.17 Possible computational sequence for Monte Carlo method.

the impact of rare transmission failures are obtained from precomputed look-up tables of transmission import limits. Using this approach and various sampling techniques, several thousand years of operation can be simulated in minutes.

Application Examples

The techniques described above are presently used for transmission planning by major utilities. The following examples illustrate some of these methods. The first example uses contingency enumeration while the second example uses Monte Carlo techniques.

Calculation of the Reliability of Electric Power Supply to a Major Industrial Complex

Contingency enumeration techniques were used to assess the reliability of the power supply to a major manufacturing complex (Reppen et al., 1990). In this analysis, the reliability concerns encompassed system events and conditions that are capable of disturbing or shutting down all or portions of the manufacturing processes. The events of concern included initial interruptions, sustained interruptions, overloads, voltage violations, voltage collapse, and overload cascading. The reliability effects of possible system reinforcements in the immediate local power supply area and in the main grid supplying this area were evaluated by a comprehensive probabilistic reliability analysis.

The characteristics of the power system were radial feeds to the plant with provisions for automatic and manual switchover to alternative supply in case of loss of voltage on primary supply feeders, an extensive local 132 kV system, and a regional 300 kV and 420 kV transmission system. Contingency enumeration methods allow detailed modeling of the network, including the modeling of automatic responses of the power system to disturbances such as special relaying schemes for line tripping, generation runback, and load transfer.

Three typical categories of outages — single element outages, independent overlapping outages of multiple elements, and dependent multiple element outages — were considered in the reliability studies. The term "element" encompasses generating units as well as "transmission elements" such as transmission lines, transformers, capacitor banks, and static var devices. The reliability computations included network analysis of outages, classification of failure events according to type and severity, and calculation of reliability indices. Reliability indices representing the predicted frequency of each of the types of failure events were computed as well as load interruption and energy curtailment indices. The indices computed are referred to as annualized indices reflecting the reliability level that would be experienced if the precontingency condition considered should exist for an entire year.

The full analysis included assessment of existing power supply conditions, impact of system reconfiguration on the reliability of supply, reliability effects of system reinforcements, and impact of conditions in the main grid. Here we will concentrate on the reliability effects of system reinforcements. Two reinforcements were analyzed: construction of a new 132 kV line completing a loop at some distance from the plant and construction of a 300 kV ring connecting several of the power supply buses to the plant.

Figure 13.18 presents the results of the investigation using the energy curtailment index defined earlier. The energy curtailment index aggregates the expected loss of energy on an annualized basis. The results indicate that reinforcement A (the remote line) has no significant effect on the reliability of power supply to the plant. Reinforcement B (the 300 kV ring) provides a substantial overall improvement, although there is no significant improvement in the energy curtailment index for the winter case.

While some of the suggested means of improving reliability could have been predicted prior to the analysis, the relative effectiveness of the various actions would not be apparent without a formal reliability analysis. Performing a reliability analysis of this type gives excellent insights into the dominating failure phenomena that govern system performance. The detailed contingency information available promotes understanding of the way systems fail, while the reliability indices computed provide the perspective necessary to make appropriate system design decisions.

Local Area Reliability

The second example considers the task of improving the reliability of electric power to a local area such as a city, major industrial complex, or other load center (Reppen, 1998). In simple terms, the reliability of the supply is a function of the following parameters:

- The load in the area as it fluctuates over time.
- The maximum amount of power that can be imported from the main grid. This import limit varies by maintenance and forced outage of transmission and generating equipment.
- Maximum available local generation at any particular time as it is affected by generation maintenance and forced outages.

Power System Planning (Reliability)

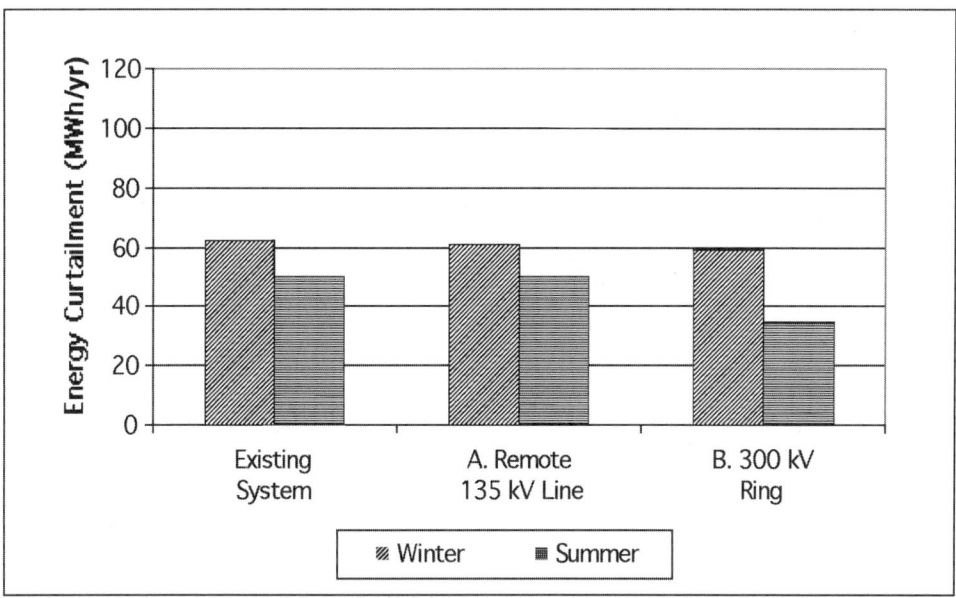

FIGURE 13.18 Benefit of system reinforcements. Annualized energy curtailment (MWh/yr). Sustained interruptions only.

At any particular time, load curtailment will occur if the load exceeds the maximum import capability into the area plus the maximum generation available in the area. Therefore, reliability of supply to consumers in the area can be measured in terms of statistics of load curtailment. Popular load curtailment measures include frequency of interruptions and energy not served (MWh/year). In addition, the expected annual customer interruption cost can be predicted using Monte Carlo techniques that simulate system conditions repeatedly over a time period to develop reliability measures by aggregating and averaging the impacts of individual load curtailment events. This allows the use of interruption cost functions (customer damage function) to estimate the expected annual cost of load interruptions.

Accepting calculated annual interruption costs as a realistic measure of the economic impact on the consumer, one might declare a system reinforcement alternative to be justified from a reliability standpoint if the reduction in interruption cost is greater than the net cost of investment and operation. While such a criterion may not necessarily be appropriate in all cases, it should provide a relevant benchmark in most environments.

Figure 13.19 shows key components of the power supply picture for a small city with a peak load of 210 MW. The city is supplied by two generators totaling 150 MW, and by a 138 kV double circuit transmission line from the main grid. Prime reinforcement options are as follows:

1. Add a new single circuit transmission line as indicated in Fig. 13.19.
2. Add one gas turbine generator of size to be determined.
3. Add two identical gas turbine generators of size to be determined.

Key questions of interest are:

1. Can the line addition be justified on the basis of savings in customer interruption cost?
2. What generator capacities will produce the same reliability improvements as the line addition?
3. What size generators can be justified on the basis of customer interruption?

Figure 13.20 shows results obtained from the Monte Carlo calculations along with the annual fixed charges for investment cost and net annual operating costs. Significant observations that can be made from Fig. 13.20 are:

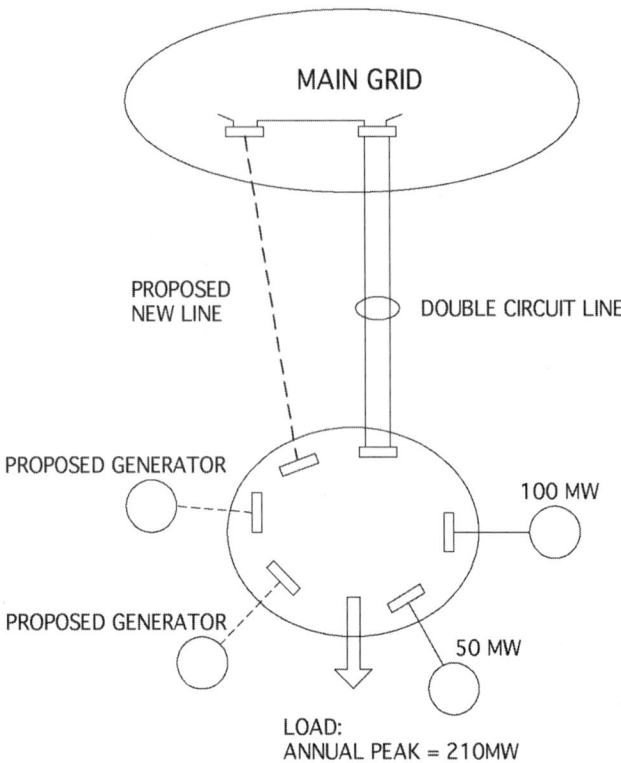

FIGURE 13.19 Power supply configuration for a small city.

- The additions of a new 138 kV line, a 50 MW generator, or two 25 MW generators have approximately the same interruption cost savings. However, the reason for this is that all three alternatives are an effective overkill, reducing the interruption cost to almost 0. If load is anticipated to grow, the line addition will be the better performer at a lower cost.
- One 15 MW to 25 MW generator would give much improved performance and at a cost which can be justified (marginally) based on reliability worth as expressed by the interruption cost curves.
- The line addition is clearly the most cost-effective alternative.

The first set of calculations compared the benefit of reinforcement by transmission or generation. While such additions are typically the most powerful reinforcements from the standpoint of improved reliability, they are typically also the most expensive. In addition, when dealing with small local systems, the natural or cost-effective line and generation additions are often more than what's needed to satisfy reliability needs for the next few years. This is particularly true for low load growth scenarios. Thus, there is a need for less expensive alternatives that typically will have smaller incremental reliability benefits than the addition of transmission lines and generators. The results of four such alternatives are shown in Fig. 13.20:

1. Improve reliability performance of existing generators.
2. Improve grounding of transmission lines.
3. Introduce live line maintenance.
4. Interruptible load contracts.

All of the short term measures except live line maintenance can be justified as the annual interruption cost is greater than the total expense associated with the reliability improvement. Also, live line maintenance and transmission line grounding have too small an impact to be of interest.

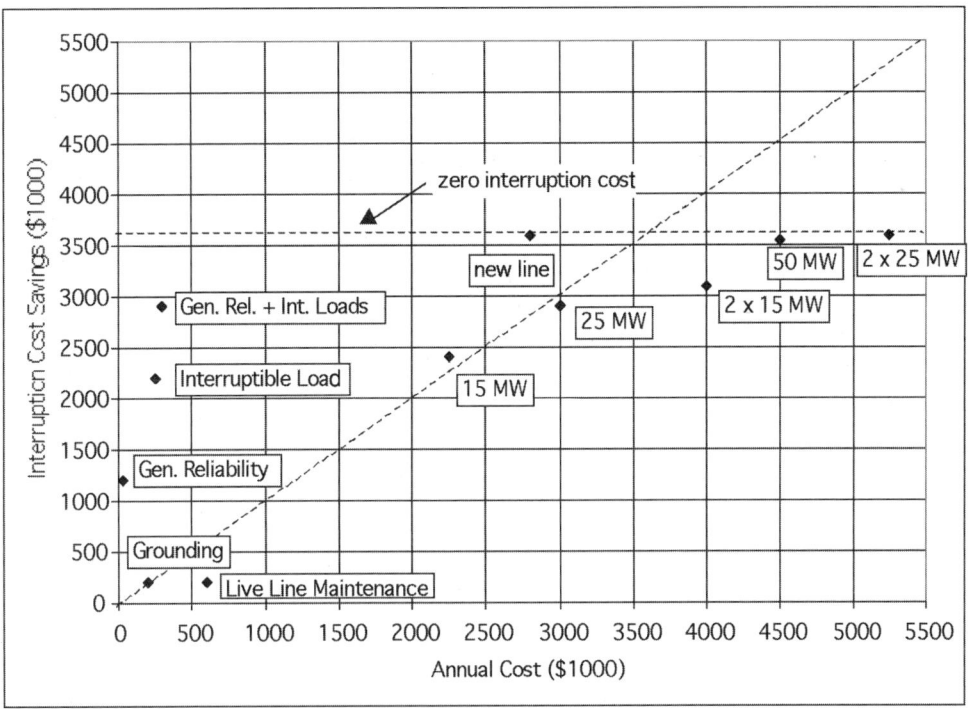

FIGURE 13.20 Annual savings in cost of interruption vs. annual combined investment and operating costs of transmission and generation reinforcements and short term measures. Measures in the upper left triangle can be justified on account of savings in interruption cost.

The most cost effective short term measure comes from improvements in the reliability performance of the 100 MW generator, closely followed by interruptible load contracts. If both of these short term measures are taken, the reliability improvement matches that obtainable from the addition of a 25 MW gas turbine generator and at a much lower cost.

This example illustrates how it is possible to use Monte Carlo reliability calculations to predict and compare the benefit-cost trade-off of transmission and generation reinforcements and various short term measures.

References

Billinton, R., Allan, R. N., Power-system reliability in perspective, *IEE J. Electron. Power*, 30, 231–236, March 1984.

Billinton, R., Wacker, G., Wojczynski, E., Comprehensive bibliography on electrical service interruption costs, *IEEE Trans. Power Appar. Syst.*, PAS-102, 6, 1831–1837, June 1983.

CIGRE Task Force 38.03.10 — 1992, *Power System Reliability Analysis Volume 2 Composite Power System Reliability Evaluation*, 1992.

Discussion of Regional Council Planning Reliability Criteria and Assessment Procedures, A Reference Document by the North American Electric Reliability Council, June 1988.

Endrenyi, J., *Reliability Modeling in Electric Power Systems*, John Wiley & Sons, Ltd., New York, 1978.

Endrenyi, J., Albrecht, P. F., Billinton, R., Marks, G. E., Reppen, N. D., Salvadori, L., Bulk Power System Reliability Assessment — Why and How? Part 1: Why? IEEE Paper 82WM 147-7, presented at the *Winter Power Meeting*, New York, NY, February 1–5, 1982.

Endrenyi, J., Albrecht, P. F., Billinton, R., Marks, G. E., Reppen, N. D., Salvadori, L., Bulk Power System Reliability Assessment — Why and How? Part 2: How? IEEE Paper 82WM 148-5, presented at the *Winter Power Meeting*, New York, NY, February 1–5, 1982.

Fong, C. C., Billinton, R., Gunderson, R. O., O'Neill, P. M., Raksany, J., Schneider, Jr., A. W., Silverstein, B., Bulk system reliability — Measurement and indices, *IEEE Trans. Power Appar. Syst.*, 4, 3, 829–835, Aug. 1989.

Forrest, D. W., Albrecht, P. F., Allan, R. N., Bhavaraju, M. P., Billinton, R., Landgren, G. L., McCory, M. F., Reppen, N. D., Proposed terms for reporting and analyzing outages of electrical transmission and distribution facilities, *IEEE Trans. Power Appar. Syst.*, PAS-104, 2, 337–348, Feb. 1985.

Guertin, M. B., Albrecht, P. F., Bhavaraju, M. P., Billinton, R., Jorgensen, G. E., Karas, A. N., Masters, W. E., Patton, A. D., Reppen, N. D., Spence, R. P., Reliability indices for use in bulk power supply adequacy evaluation, *IEEE Trans. Power Appar. Syst.*, PAS-97, 4, 1097–1103, July/Aug. 1978.

Mackay, E. M., Berk, L.H., Costs of Power Interruptions to Industry Survey Results, CIGRE, Paper 32-07, Aug. 30–Sept. 7, 1978.

Noferi, P. L., Paris, L., Salvaderi, L., Monte Carlo Methods for Power System Reliability Evaluations in Transmission or Generation Planning, in *Proceedings*, 1975 *Reliability and Maintainability Symposium*, Washington, 1975.

Oliveira, G. C., Pereira, M. V. F., Cunha, S. H. F., A Technique for Reducing Computational Effort in Monte-Carlo Based Composite Reliability Evaluation, IEEE 1989 WM 174-4 PWRS.

Overview of Planning Reliability Criteria of the Regional Reliability Councils of NERC, A Reference Document by the North American Electric Reliability Council, 1988.

Reppen, N. D., Balancing Investments and Operating Costs with Customer Interruption Costs to Give Increased Reliability, presented at the IEE Colloquium on Tools and Techniques for Dealing with Uncertainty, London, U.K., January 27, 1998.

Reppen, N. D., Carlsen, T., Glende, I., Bostad, B., Lam, B. P., Calculation of the Reliability of Electric Power Supply to a Major Industrial Complex, presented at the *10th Power Systems Computation Conference (PSCC)*, Graz, Austria, August 19–24, 1990.

Salvaderi, L., Allan, R., Billinton, R., Endrenyi, J., Mc Gillis, D., Lauby, M., Manning, P., Ringlee, R., State of the Art of Composite-System Reliability Evaluation, CICRE Session, paper 38-104, Paris, August 26–September 1, 1990.

13.4 Power System Planning

Hyde M. Merrill

Power system planning is the recurring process of studying and determining what facilities and procedures should be provided to satisfy and promote appropriate future demands for electricity. The electric power system as planned should meet or balance societal goals. These include availability of electricity to all potential users at the lowest possible cost, minimum environmental damage, high levels of safety and reliability, etc. Plans should be technically and financially feasible. Plans also should achieve the objectives of the entity doing the planning, including minimizing risk.

The *electric power system* is a force-at-a-distance energy-conversion system. It consists of three principal elements:

- Current- and voltage-producing, transmitting, and consuming hardware,
- Control and protective devices, and
- Planning, operating, commercial, and regulatory practices and procedures.

These definitions are very different from would have appeared on these pages 25 years ago. They no doubt will seem quaint 25 years hence. At this writing, the electric power industry worldwide is experiencing its most dramatic changes in two generations. These changes affect planning, but this section is intended as a practical exposition, not as a history lesson or a prophesy. We therefore will focus on how planning is or should be done today, avoiding flights of fantasy into the past or future (Sullivan, 1977; Stoll et al., 1989; Kahn, 1988; Ringlee, 1989).

Planning considers:

- Options,
- Uncertainties, and
- Attributes.

Options are the choices available to the planner. Uncertainties are parameters whose values are not known precisely or cannot be forecast without error. Attributes are measures of "goodness." Stakeholder objectives are expressed in terms of attributes. Physical, economic, and institutional realities determine how different options and uncertainties affect the attributes.

The planning problem is to identify and choose among options, in the presence of uncertainties, so as to maximize or minimize (as the case may be) the attributes.

Planning Entities

Planners generally are trained as engineers, economists, civil servants, businessmen, or mathematicians. They do power system planning for the following entities:

- Vertically integrated utilities owning generation, transmission, and distribution systems.
- Transmission companies, independent system operators (ISO), and regional transmission organizations (RTO). Transmission companies own transmission assets; the latter do not, but may have some responsibility for their planning.
- Pools or combinations of vertically integrated utilities.

Other organizations do planning studies and higher-level power sector planning. A step removed from the operation and management of the power system, their interest is in seeing that it meets society's goals:

- Various levels of government.
- International development banks.

Still other organizations do power system planning studies, but without system responsibility. They wish to understand the market for various services and how they might compete in it — its economics and technical requirements for entry.

- Independent power producers (IPP) or nonutility generators (NUG). These include qualifying facilities (QF as defined by the U.S. Public Utilities Regulatory Policy Act of 1978) and exempt wholesale generators (EWG as defined by the U.S. Energy Policy Act of 1992). These are subject to less stringent regulation than are utilities. They neither enjoy monopoly protection nor have an obligation to provide electricity at cost-based tariffs.
- Large industrial users.
- Commercial middlemen who buy and sell electrical energy services.
- Investors.

All of these are supported by independent purveyors of planning information. Consultants with specialized analytic skills also do planning studies.

Arenas

Planning is done in several arenas, distinguished by the planning horizon and by the types of options under consideration. These arenas include:

- **Long-term vs. short-term planning**. Economists distinguish these by whether capital investment options are considered. For engineers, long-term planning has a distant horizon (perhaps 30 years for generation and half of that for transmission). Short-term planning considers about 5 years. Operations planning is for as short as a few hours and is not treated here.

- **Generation vs. transmission vs. least-cost planning.** Generation and transmission planning focus on supply options. Least-cost planning includes demand-side options for limiting or shaping load.
- **Products and services.** Some entities provide power (kW) and energy (kWh). Others plan the transmission system. Others provide for auxiliary services (voltage and power control, electrical reserves, etc.). Still others plan for diversified services like conservation and load management.

Other arenas require engineering and economic skills, but are within the purview of a book on business or policy rather than an engineering handbook.

- **Competitive markets.** Strategic planning is particularly concerned with financial and business plans in competitive markets.
- **Sector evolution.** Defining the form of the future power sector, including the relationships between competitive forces, regulation, and the broadest social objectives, is a particularly vital planning function.

The Planning Problem

Options

Power generation, transformer, transmission system, substation, protection, and operation and control options are discussed in other chapters of this handbook. Other options are discussed below.

Planning and Operating Standards or Criteria

Planning and operating criteria have a dual nature: they are both attributes and options. Here we will emphasize the fact that they are options, subject to change. Though they have no intrinsic value, standards or criteria are important for several reasons. Their consistent application allows independent systems to interconnect electrically in symmetrical relationships that benefit all. Criteria can also eliminate the need for planners to ask constantly, "How much reliability, controllability, etc., do I need to provide?" Criteria include:

- Maximum acceptable loss-of-load probability (LOLP) or expected unserved demand, minimum required reserve margins, and similar generation planning standards,
- What constitutes a single contingency (transmission systems are often designed to withstand "any" single contingency) and whether particular single contingencies are excluded because they are unlikely or expensive to forestall,
- Permissible operating ranges (voltages, power flows, frequency, etc.) in the normal or preventative state, the emergency state, and the restorative state, and
- How criteria are to be measured or applied.

Most power systems in industrialized nations are designed and operated so that:

1. With all elements in service, power flows, voltages, and other parameters are within normal ranges of the equipment,
2. The system remains stable after any single contingency, and
3. Power flows, voltages, and other operating parameters are within emergency ranges following any single contingency.

For financial and economic reasons, developing countries choose weaker criteria.

Demand Management

Demand-side planning often is tied to generation planning because it affects the power and energy that the power plants will need to provide. There is no perfect classification scheme for demand-side options. Some overlapping classifications are:

- Indirect load control vs. direct load control by the bulk system operator,
- Power (kW) or energy (kWh) modification or both, and
- Type of end-use targeted.

TABLE 13.5 Appliances and Sectors under Direct Utility Control, U.S. — 1983

Appliance or Sector	Number Controlled	Percent of Total Controlled
Electric water heaters	648,437	43%
Air conditioners	515,252	34%
Irrigation pumps	14,261	1%
Space heating	50,238	3%
Swimming pool pumps	258,993	17%
Other	13,710	1%
Total	**1,500,891**	**100%**
Residential	1,456,212	97%
Commercial	29,830	2%
Industrial	588	—
Agricultural	14,261	1%

Source: New Electric Power Technologies: Problems and Prospects for the 1990s, Washington, D.C.: U.S. Congress, Office of Technology Assessment, OTA-E-246, July 1985.

Table 13.5 shows the type of load under direct utility control in the U.S. early in the 1980s, when enthusiasm for demand-side options was especially high.

One of the most effective examples of load control was reported by a German utility. Typical off-peak winter demand was less than 70% of the peak for the same day. An indirect program promoted storage space heaters that use electricity at night, when demand is low, to heat ceramic bricks. During the day, air forced among the bricks transfers the heat to the living space. Within five years the program was so popular that direct control was added to avoid creating nighttime peaks. The winter daily load shape became practically flat.

Market and Strategic Options

Market and strategic options are also important. These range from buying a block of power from a neighboring utility to commodity trading in electricity futures to mergers, divestitures, and acquisitions.

Uncertainties

Uncertainty can seldom be eliminated. Planning and forecasting are linked so that even if the forecasts are wrong, the plans are right (Bjorkland, 1987).

Models of Uncertainty (Schweppe, 1973)

Probabilistic models, where different outcomes are associated with different probabilities, are valid if the probability structure is known. The events involved must occur often enough for the law of large numbers to apply, or else the probabilities will have little relationship to the frequencies of the outcomes. Generation planners have excellent probabilistic reliability models. (Generation and transmission reliability evaluation are treated in more detail in a separate section.)

Unknown-but-bounded (set theoretic) models are used when one or both of the conditions above are not met. For instance, transmission planners design to withstand any of a set of single contingencies, usually without measuring them probabilistically.

Demand Growth

Planners forecast the use of energy (MWh) for a period (e.g., a year) first. They divide this by the hours in the period to calculate average demand, and divide again by the projected load factor (average MW demand/peak MW demand) to forecast peak demand. Three techniques are used most often to forecast energy.

Extrapolation — Exponential growth (e.g., 4%/year) appears as a straight line on semi-log paper. Planners plot past loads on semi-log paper and use a straight edge to extrapolate to the future.

Econometric models — Econometric models quantify relationships between such parameters as economic activity and population and use of electricity. The simplest models are linear or log-linear:

$$D_i = f(P, GDP, \text{etc.}) = k_1 D_{i-1} + k_2 P_i + k_3 GDP_i + \ldots \qquad (13.1)$$

D_i = Demand or log(demand) in period i
P_i = Population or log(population) in period i
GDP_i = Gross domestic product or log(gross domestic product) or some measure of local economic activity
k_1, k_2, k_3, etc. are coefficients

Econometric models are *developed* in a trial-and-error process. Variants of Eq. (13.1) are hypothesized and least squares (regression) analysis is used to find values of coefficients that make Eq. (13.1) fit historical data. Econometric models are *used* by first forecasting population, economic activity, etc., and from them calculating future energy demand using Eq. (13.1).

End-use models — First, the number of households is forecast. Then the per-household penetration of various appliances is projected. The average kWh used by each appliance is estimated and is multiplied by the two previous numbers. The results are summed over all types of appliances.

Performance — Extrapolation became suspect after U.S. load forecasts in the 1970s were consistently too high. Econometric modeling is more work but is more satisfying. End-use modeling requires considerable effort but gives the most accurate forecasts of residential load.

Real drivers — One fundamental driver for per-capita load growth is the replacement by electricity of other forms of energy use. The second is the creation of new uses of energy that are uniquely satisfied by electricity.

During the decades when U.S. electric demand grew at over 7% per year, the demand for all forms of energy (of which electricity is a part) grew at about 2% per year. This obviously could not continue: the two cannot cross. The growth of electricity demand began to drop off about 1955, declining noticably in the 1970s and thereafter. The drop in load growth was attributed to the oil crises of the 1970s. Post-1973 conservation played a part, but by then electricity had captured about all the market share it was going to get by replacement and creation of new demands for energy.

In developing countries, both fundamental drivers are limited by the ability of the electric companies to finance the necessary generation and distribution infrastructure, which is very expensive. Demand is also limited by their ability to generate. In industrialized countries, availability of capital and power plant performance are not constraining. In all countries, elasticity reduces demand if electricity is costly. This effect is much stronger in countries with low per-capita income and for energy-intensive industrial load.

Fuel and Water

In the near term, strikes, weather, and natural disasters can interrupt production or delivery. Fuel inventories and the ability to redispatch provide good hedges. In the intermediate term, government action can make fuel available or unavailable. For instance, in 1978, the U.S. Congress forbade burning natural gas by utilities, perceiving that there was a shortage. The shortage became a glut once the U.S. natural gas market was deregulated. In the long term, any single source of fuel is finite and will run out. British coal, which had fueled the industrial revolution, was shut down in the 1990s because it had been worked out.

The more important fuel uncertainties, however, are in price. For instance, Fig. 13.21 shows that the price of crude oil doubled in 1974 and again in 1979. Recognizing the high variability, in 1983 the U.S. Department of Energy forecast a fuzzy band instead of a single trajectory (U.S. Dept. of Energy, 1983). It is interesting that within a year of the publication of this projection, the price of oil had dropped below the low limit of the band, and it has remained there until this writing. Planners must consider extreme possibilities for all uncertainties.

Brazil, Norway, the Pacific Northwest, Quebec, and a number of developing countries are highly dependent on hydropower. Systems usually are planned and operated so that there will not be a shortfall unless one of the worst hydrological years in recorded history recurs.

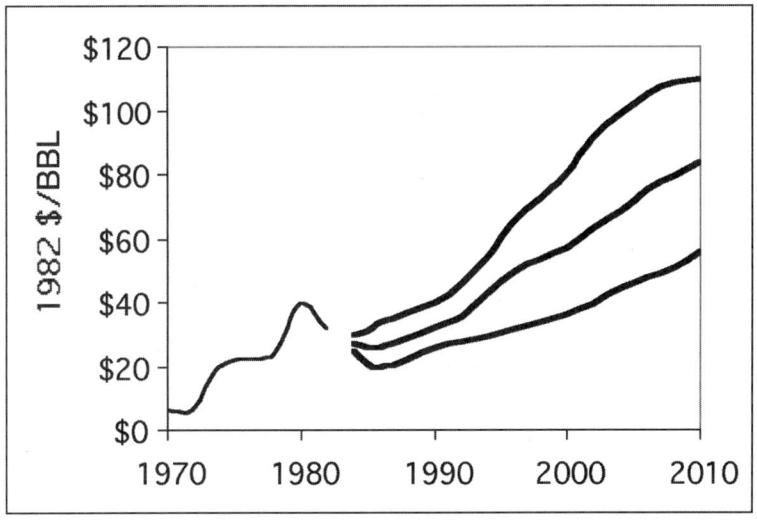

FIGURE 13.21 World oil price projections, 1983.

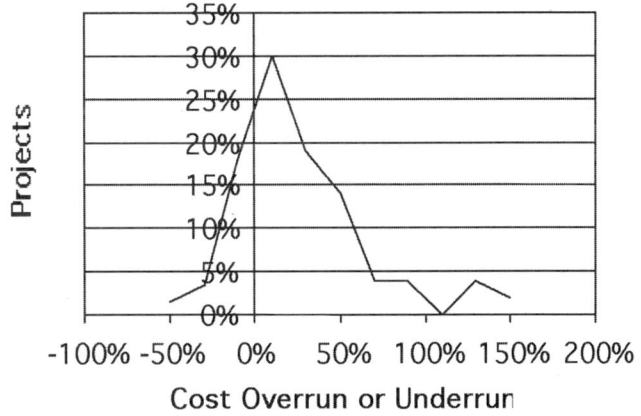

FIGURE 13.22 Budget vs. actual costs, power projects in developing countries.

Construction

Three major construction uncertainties are: How long will it take to build? How much will it cost? Will the project be completed?

A World Bank study of 41 hydro projects revealed that 37% experienced a schedule slip of 30% or more, including 17% with schedule slips of 60%–100% (Crousillat, 1989).

Figure 13.22 shows the range of actual versus budgeted cost for a number of World Bank-financed projects. The distribution is not symmetrical — overruns are much more frequent than under-budget projects. Some of the worst cases in the Fig. 13.22 data occurred during periods of unexpected high inflation (Crousillat, 1989). A 1983 report projected that the cost of some 40 U.S. nuclear plants scheduled for completion by 1990 would be close to normally distributed, with the least expensive costing a bit under $2000/kW and the most expensive three times higher, at $6000/kW (U.S. Dept. of Energy, 1989).

Possibly the most expensive nuclear plant ever built, the Shoreham Plant on Long Island, was completed at a cost of some $16 billion U.S. It was shut down by the state before producing a single kWh of commercial energy.

Technology

New technologies are generally less certain than mature technologies in their cost, construction time, and performance.

Even mature technologies may have important uncertainties. For example, transmission transfer capability is an important measure of transmission system capability. It is usually expressed as a single number, but it is actually a time-varying random variable.

Demand Management

Demand management programs are risky, in part because of uncertainty in the public's response to them. The two major uncertainties are:

- What fraction of eligible customers will respond to a particular program?
- How much will the average customer change his use of electricity?

These uncertainties are affected strongly by the design of the program, the incentives offered, how it is marketed, etc. Carrying out a carefully designed pilot program can reduce the uncertainty. The pilot program should be done in the region of the future commercial program.

Markets and Capital Recovery

For many years, vertically integrated utilities were guaranteed the recovery of all costs, including capital invested, plus a modest but sure profit. The customer paid this and absorbed the market uncertainties. At this writing, the regulated monopoly, cost-recovery market is being replaced in many states and countries by a more competitive market. Some market risks are being transferred from customers to utilities, power marketers, generating companies, speculators, and others.

This creates new uncertainties. For example, in competitive generation markets it is not known which potential generating units will be built. This affects both transmission and generation planning.

Regulation

For the foreseeable future, government will play a key role. The uncertainty in what governments will do propagates into uncertainties in profitability of various players, in market entry, in prices, etc.

For example, in the 1980s, U.S. state and federal governments encouraged utilities to implement demand-side programs. Program costs, and in some cases costs of foregone sales, were recovered through tariffs. The government interest later switched to competitive markets. These markets do not have such a convenient mechanism for encouraging demand-side management. As a result, demand-side programs became less attractive.

Severe Events

High-risk, low-probability events usually are not considered by standard planning practices. For example, transmission planners design so that the system will withstand any single contingency. Planning procedures, criteria, and methods generally ignore several simultaneous or near-simultaneous contingencies. The power system is not designed to withstand them — whether or not it does is happenstance.

In January 1998 an ice storm of unprecedented magnitude struck the northeastern U.S. and Quebec. Ice on transmission lines greatly exceeded design standards. Many towers collapsed. All lines feeding Montreal, and all lines south and east of the city, were on the ground. The government later announced a high-risk, low-probability standard: the system should be designed and operated to prevent loss of more than half of the Montreal load should such an event recur.

Attributes

Attributes measure "goodness" in different ways, from different perspectives. Each stakeholder has objectives; they are expressed in terms of attributes.

Customers of various kinds (residential, commercial, industrial, etc.):

- Cost of electricity
- Other costs absorbed by the customer
- Quality of service (reliability, voltage control, etc.)

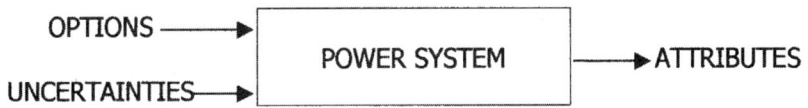

FIGURE 13.23 Options, uncertainties, and attributes.

Investors in various providers of energy and services

- New capital required
- Net income, earnings per share, and other measures of income
- Cash flow, coverage ratios, and other measures of cash use and replacement

Employees

- Security
- Promotion opportunities
- Salaries
- Healthiness and safety of working conditions

Taxpayers

- Tax revenues
- Expenditures from public funds

Neighbors (environmentalists, visitors, local inhabitants, competitors, etc.)

- Emissions or thermal discharges
- Community disruption
- Employment opportunities
- Rights-of-way and other intrusions
- Flooding
- Measures of market power

The list of attributes given above is not complete, and different attributes are important for different studies. Deciding on the planning objectives and the attributes for a given study is an important initial step in power system planning.

Some attributes are measured using complex computer models. For others, approximate or ad hoc models may be adequate or may be the best that is available. The planner calculates how the options and the uncertainties (Fig. 13.23) affect the attributes.

Standards or criteria are surrogates for some attributes that are difficult or impossible to compute.

Planning Processes

Setting Standards or Criteria

Planning objectives often conflict. For example, maximizing reliability and minimizing environmental impacts generally conflict with minimizing costs.

Since all attributes cannot be measured in dollars, achieving the right trade-off can be a difficult socio-technico-economic-institutional problem. Doing this every day would burden the planning process. Having standards avoids having to revisit such judgments continually. For instance, once it is decided that (say) 20% generation reserve provides adequate reliability at an acceptable cost, the planner accepts 20% as a standard and designs to meet it. Testing whether a particular plan meets the reserve criterion is easy; the planner can concentrate on other issues.

Standards should be examined from time to time. If society becomes poorer or richer, its pocketbook may speak for lower or higher standards of service. Changes in technology may justify a change in standards — for instance, development of better scrubbers may make it reasonable to insist on reduced SO_2 or NOx emissions. Increased reliance on electricity may require more reliability: a proposal to shut off the power throughout the U.S. for one minute to salute Edison's death was quashed. Had he died in 1900 instead of 1931, it might have been practical.

Assessment

Forecasts and Projections

Not all uncertainties create risk for every planning study. Those that do for a particular study are identified. Forecasts and projections are developed for these uncertainties.

System State

The state assessment begins with an evaluation of the technical and economic attributes of the present and future power system. Does it and will it satisfy established technical standards? Is it economical? Does it meet other objectives?

Chapter 8 of this handbook, "Power System Analysis and Simulation," describes the analytical tools available to planners. It also describes how these tools are used. The phenomena analyzed are described in Chapters 10 through 12.

Generation Planning

Chapter 2 of this handbook, "Electric Power Generation: Conventional Methods," describes generation planning options and their characteristics. The generation planner does a preliminary selection from among them, recognizing any special features of his planning problem.

Planners measure how the various options would alleviate deficiencies discovered in the assessment step. The effects of various options or combinations of options and the effects of uncertainties on other attributes are also measured.

In particular, planners compute reserve margin or other measures of reliability. They simulate the operation of the system to measure operating cost and to determine if the operation is within acceptable ranges of other parameters.

Transmission Planning

Traditional transmission options and new technologies are described in Chapters 3, 4, 11, 14, and 15 of this handbook. Like the generation planner, the transmission planner makes a preliminary selection based on the needs and development pattern of his system.

For instance, for technical and commercial reasons, a given system will use only a few distinct voltage classes. So a system whose existing transmission consists of 138-kV, 345-kV, and 765-kV equipment will rarely add new circuits at 230 kV or 400 kV, even though these may be popular elsewhere.

Transmission planners then identify a set of specific options and measure how these options in various combinations, along with the important uncertainties, affect the attributes. Load flow, short circuit, and stability analyses are performed to determine if voltages and currents are within acceptable bounds under various system states, and if the system will remain stable for all contingencies. How often and how much the operation of the generation system will be constrained by transmission limitations is an important consideration.

Least-Cost Planning

Least-cost planning is also known as integrated resource planning or integrated demand/supply planning. It considers supply-side options (generally generation options) on a level playing field with demand-side options (generally conservation, indirect load shifting, or direct load control). These options include incentives to encourage utilities and consumers to change energy consumption patterns.

As with generation planning and transmission planning, a preliminary selection weeds out options that are clearly not of interest in a particular area.

Power System Planning (Reliability)

The least-cost planning process includes computing values of key attributes for various options and uncertainties.

Making Choices

A key question in generation, transmission, and least-cost planning is: How is one plan selected over another? A few distinctive approaches will be described.

Minimize Revenue Requirements

The planner selects the best option from the ratepayer's perspective. He selects the plan that will minimize the ratepayer's cost of electricity while satisfying reliability, environmental, and other criteria.

The ratepayer's cost — an attribute — is the revenue that the utility will have to collect to recover all operating and capital costs and to earn a commission-approved return on unrecovered investor capital:

$$RR_i(O,U) = FC_i(O,U) + VC_i(O,U) \quad (13.2)$$

$RR_i(O,U)$ = Revenue requirements in period i
$FC_i(O,U)$ = Fixed costs in period i
$VC_i(O,U)$ = Variable costs in period i
O = A selection of the various options
U = Realizations or values of the various uncertainties

Fixed costs are independent of how much or how little a piece of equipment is used. Depreciation (recovery of investors' capital), interest on debt, and profit (return on unrecovered capital) are typical fixed costs.

Variable costs — fuel cost — for example, are related to how much a piece of equipment is used. These costs include all system costs, not just the cost of the individual option. For instance, old plants may run less when a new plant is built. The variable cost includes fuel cost for all plants.

To apply this traditional method, the planner must know his company's return rate, which is set by the regulator. In a closely related method, the market defines the cost of electricity and Eq. (13.2) is solved for the internal rate of return (IRR). The option selected is the one that maximizes the IRR, the investor's profit.

Cost-Benefit Analysis

If the benefits exceed the costs, a project is worth doing.

Typically, costs are incurred first, and benefits come later. A dollar of benefit later is not worth the same as a dollar of cost today. Present worth analysis is a way to compare dollars at different times. The basic equation is:

$$P = S/(1+i)^n \quad (13.3)$$

In Eq. (13.3), P is the present worth or equivalent value today of an amount S, n years in the future, with i the discount rate or annual cost of capital.

Cost-benefit analysis is also used to rank mutually exclusive projects — the one with the highest benefit/cost ratio wins.

Multi-Objective Decision Analysis

Utility Function Methods

Table 13.6 compares two options for a new power plant in Utah (Keeney et al., 1981). Which choice is better? The attributes are combined in a utility function of the form:

$$U(x) = k_1 \text{Economics} + k_2 \text{Environment} + \ldots + k_7 \text{Feasibility} \quad (13.4)$$

TABLE 13.6   Attributes: Nuclear Plant vs. Coal Plant

	Wellington Coal Plant	Green River Nuclear Plant
Economics ($/MWh)	60.7	47.4
Environment (corridor-miles)	532.6	500.8
Public disbenefits ($ × 000,000)	15.0	22.6
Tax revenues ($ × 000,000/year)	3.5	1.0
Health lost (equivalent years)	446.7	6.3
Public attitudes	0.33	−1.0
Feasibility	60.0	37.0

From Keeney, R. L. et al., *Decision Framework for Technoloy Choice*, Report EA-2153, Electric Power Research Institute, Palo Alto, CA, 1981. With permission.

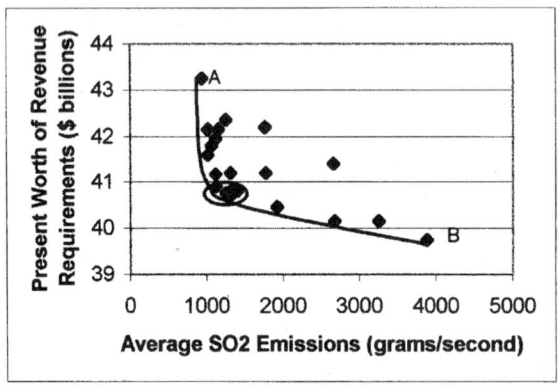

FIGURE 13.24   Trade-off: cost vs. SO_2 emissions.

In Eq. (13.4), x takes on one of two values, "coal" or "nuclear." [The actual functional form for a particular study may be more complicated than Eq. (13.4).] The coefficients k_i reflect the relative importance of each attribute. These coefficients convert the different attributes to a common measure. The choice that minimizes the utility function wins. In this study, for the values of coefficients selected U(coal) was $131.4/MWh; U(nuclear) was $162.9/MWh.

This approach has many variants. Uncertainties can be included, making U(x) a random variable. Work has been done to develop methods for determining the decision-maker's values (the coefficients) and risk tolerance.

Trade-Off Analysis

Trade-off analysis measures each attribute in its natural units, without reducing them all to a common measure, and seeks reasonable compromises at points of saturation or diminishing return. A good compromise will not necessarily optimize any of the attributes, but will come close to optimizing all of them.

For example, Fig. 13.24 shows 22 plans examined in an energy strategy study. The plans in region A minimize SO_2 emissions, but are very costly. The plans in region B are cheap but have high emissions. The plans at the knee of the trade-off curve are at the point of diminishing returns for one attribute against the other. Significant reductions in emissions can be had at little cost by moving from B to the knee. Going beyond the knee toward A will not reduce SO_2 much more but will increase the cost significantly. The plans at the knee come close to minimizing both cost and emissions.

Trade-off analysis can be done graphically for two-attribute problems. More than two attributes cannot be graphed easily but can be analyzed mathematically (Crousillat, 1993).

Risk

Risk is the hazard due to uncertainty. Risk is also associated with decisions. Without uncertainties and alternatives, there is no risk.

System planning, engineering, and operating procedures have evolved to reduce the risks of widespread or local service interruptions. Another section of this handbook describes methods for modeling and enhancing reliability.

Not all risks are included in reliability analysis, however. Much talk about risk is directed to financial risks. Other important risks are not quantified in dollars.

One measure of risk is *robustness*, the likelihood that a particular decision will not be regretted. *Exposure* is the possible loss (in terms of an attribute) under adverse realizations of uncertainties. *Regret* is the difference between the value of an attribute for a particular set of decisions and realizations of uncertainties, and the value of the attribute for optimized decisions with perfect foreknowledge of the uncertainties.

Planners develop *hedges*, options that increase robustness or decrease exposure or regret. Building small generating units instead of large ones is an example; an insurance policy is another (De la Torre et al., 1999).

References

Bjorklund, G. J., Planning for uncertainty at an electric utility, *Public Utilities Fortnightly*, October 15, 1987.
Crousillat, E., *Incorporating Risk and Uncertainty in Power System Planning*, I&ED Energy Series paper # 17, The World Bank, Washington D.C., 1989.
Crousillat, E. O., Dörfner, P., Alvarado, P., Merrill, H. M., Conflicting objectives and risk in power system planning, *IEEE Trans. Power Systems*, 8(3), Aug. 1993.
Energy Projections to the Year 2010, U.S. Dept of Energy, Office of Policy, Planning and Analysis, report DOE/PE-0029/2, October 1983.
The Future of Electric Power in America: Economic Supply for Economic Growth, U.S. Dept of Energy, Office of Policy, Planning and Analysis, report DOE/PE-0045, June 1983.
Kahn, E., *Electric Utility Planning & Regulation*, American Council for an Energy-Efficient Economy, Washington, 1988. This text presents power system planning in a broad social context and is close in spirit to this chapter.
Keeney, R. L., Beley, J. R., Fleischauer, P., Kirkwood, C. W., Sicherman, A., *Decision Framework for Technology Choice*, Report EA-2153, Electric Power Research Institute, Palo Alto, 1981.
Ringlee, R. J., Ed., special section on electric utility systems planning, *Proc. IEEE*, 77(6), June 1989. This special section contains several excellent papers on engineering issues in planning.
Schweppe, F. C., *Uncertain Dynamic Systems*, Prentice-Hall, Englewood Cliffs, NJ, 1973, chap. 3.
Stoll, H. G., Garver, L. J. (sic), Jordan, G. A., Price, W. H., Sigley, R. F., Jr., Szczepanski, R. S., Tice, J. B., *Least-Cost Electric Utility Planning*, John Wiley & Sons, New York, 1989. Its title notwithstanding, this is a general text on planning.
Sullivan, R. L., *Power System Planning*, McGraw-Hill, New York, 1977. This text covers some of the mechanics of planning.
De la Torre, T., Feltes, J. W., Gómez, T., and Merrill, H. M., Deregulation, privatization, and competition: transmission planning under uncertainty, *IEEE Trans. Power Systems*, 14(2), May 1999.

13.5 Power System Reliability

Richard E. Brown

The electric power industry began in the late 1800s as a component of the electric lighting industry. At this time, lighting was the only application for electricity, and homes had other methods of illumination if the electricity supply was interrupted. Electricity was essentially a luxury item and reliability was not an issue.

As electricity became more common, new applications began to appear. Examples include electric motors, electric heating, irons, and phonographs. People began to grow accustomed to these new electric appliances, and their need for reliable electricity increased. This trend culminated with the invention of the radio. No nonelectrical appliance could perform the same function as a radio. If a person wanted to listen to the airwaves, electricity was required. As radio sales exploded in the 1920s, people found that reliable electricity was a necessity. By the late 1930s, electricity was regarded as a basic utility (Philipson and Willis, 1999).

As electric utilities expanded and increased their transmission voltage levels, they found that they could improve reliability by interconnecting their system to neighboring utilities. This allowed connected utilities to "borrow" electricity in case of an emergency. Unfortunately, a problem on one utility's system could now cause problems to other utilities. This fact was made publicly evident on November 9, 1965. On this day, a major blackout left cities in the northeastern U.S. and parts of Ontario without power for several hours. Homes and businesses had become so dependent on electricity that this blackout was crippling. Action was needed to help prevent such occurrences from happening in the future.

NERC Regions

The North American Electric Reliability Council (NERC) was formed in 1968 as a response to the 1965 blackout. By this time, reliability assessment was already a mature field and was being applied to many types of engineered systems (Billinton and Allan, 1988; Ramakumar, 1993). NERC's mission is to promote the reliability of the North America's bulk power system (generation and transmission). It reviews past events; monitors compliance with policies, standards, principles, and guides; and assesses future reliability for various growth and operational scenarios. NERC provides planning recommendations and operating guidelines, but has no formal authority over electric utilities.

Since most of the transmission infrastructure in the U.S. and Canada is interconnected, bulk power reliability must look at systems larger than a single utility. The territory covered by NERC is far too large to study and manage as a whole, and is divided into ten regions. These NERC regions are: East Central Area Reliability Coordination Agreement (ECAR), Electric Reliability Council of Texas (ERCOT), Florida Reliability Coordinating Council (FRCC), Mid-Atlantic Area Council (MAAC), Mid-Atlantic Interconnected Network (MAIN), Mid-Continent Area Power Pool (MAPP), Northeast Power Coordinating Council (NPCC), Southeastern Electric Reliability Council (SERC), Southwest Power Pool (SPP), and the Western Systems Coordinating Council (WSCC). The geographic territories assigned to the ten NERC regions are shown in Fig. 13.25.

Even though there are ten NERC regions, there are only four major transmission grids in the U.S. and Canada: the area associated with the Western Systems Coordinating Council, the area associated with the Electric Reliability Council of Texas, Quebec, and the Eastern United States. These are usually referred to as the Western Interconnection, the ERCOT Interconnection, the Quebec Interconnection, and the Eastern Interconnection. Each of these grids is highly interconnected within their boundaries, but only has weak connections to the other grids. The geographic territories associated with these four interconnections are shown in Fig. 13.25.

NERC looks at two aspects of bulk power system reliability: system adequacy and system security. A system must have enough capacity to supply power to its customers (adequacy), and it must be able to continue supplying power to its customers if some unforeseen event disturbs the system (security). Each of these two aspects of reliability is further discussed below.

System Adequacy Assessment

System adequacy is defined as the ability of a system to supply all of the power demanded by its customers (Billinton and Allan, 1988). Three conditions must be met to ensure system adequacy. First, its available generation capacity must be greater than the demanded load plus system losses. Second, it must be able to transport this power to its customers without overloading any equipment. Third, it must serve its loads within acceptable voltage levels.

System adequacy assessment is probabilistic in nature (Allan et al., 1994; Schilling et al., 1989). Each generator has a probability of being available, P_A, a probability of being available with a reduced capacity,

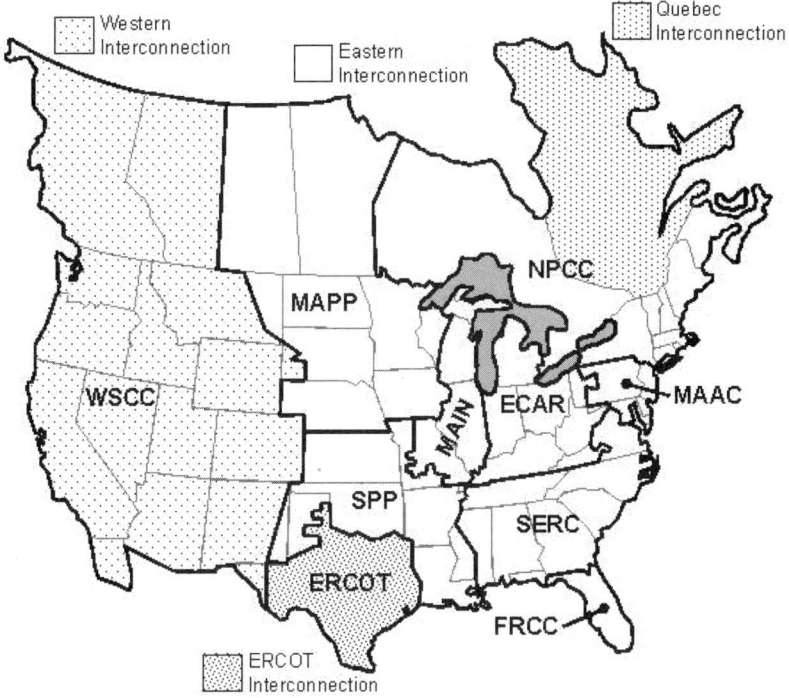

FIGURE 13.25 NERC regions.

TABLE 13.7 Generator State Probabilities

Generator State		
Generator 1	Generator 2	Probability
Available	Available	$P_{A1}\, P_{A2}$
Available	Reduced	$P_{A1}\, P_{R2}$
Available	Unavailable	$P_{A1}\, P_{U2}$
Reduced	Available	$P_{R1}\, P_{A2}$
Reduced	Reduced	$P_{R1}\, P_{R2}$
Reduced	Unavailable	$P_{R1}\, P_{U2}$
Unavailable	Available	$P_{U1}\, P_{A2}$
Unavailable	Reduced	$P_{U1}\, P_{R2}$
Unavailable	Unavailable	$P_{U1}\, P_{U2}$

P_R, and a probability of being unavailable, P_U. This allows the probability of all generator state combinations to be computed. A simple two-generator example is shown in Table 13.7. There are nine possible generator state combinations, and the probability of being in a particular combination is the product of the individual generator state probabilities. In general, if there are n generators and x possible states for each generator, the number of possible generator state combinations is:

$$\text{Generator State Combinations} = x^n \qquad (13.5)$$

In addition to generator state combinations, loading behavior must be known. Information is found by looking at historical load bus demand in recent years. For the best accuracy, 8760 hour peak demand curves are used for each load bus. These correspond to hourly peak loads for a typical year. To reduce

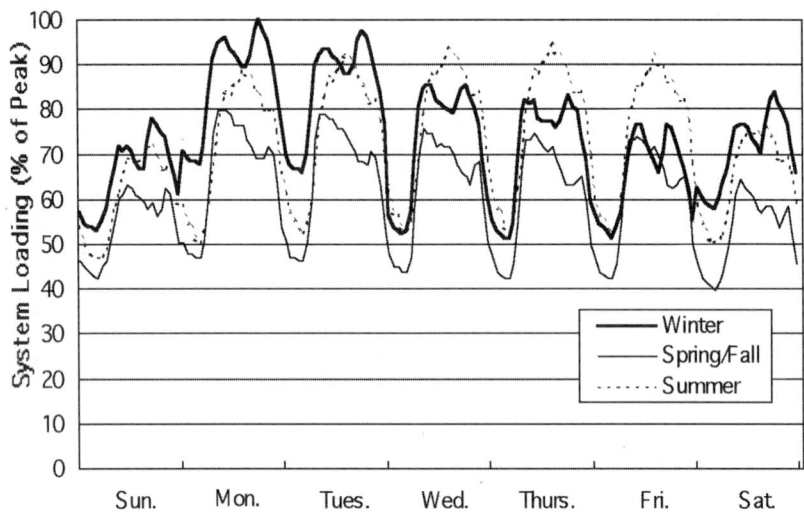

FIGURE 13.26 Weekly load curves by season.

computational and data requirements, it is usually acceptable to reduce each set of 8760-hour load curves to three weekly load curves (168 hours each). These correspond to typical weekly load patterns for winter conditions, spring/autumn conditions, and summer conditions. Weekly load curves can be scaled up or down to represent temperatures that are above or below normal. Sample weekly load curves for a winter peaking load bus are shown in Fig. 13.26.

To perform an adequacy assessment, each generation state combination is compared with all hourly loading conditions. For each combination of generation and loading, a power flow is performed. If the available generation cannot supply the loads or if any constraints are violated, the system is inadequate and certain loads must be shed. After all generation/load combinations are examined, the adequacy assessment is complete.

An adequacy assessment produces the following information for each load bus: (1) the combinations of generation and loading that result in load interruptions, and (2) the probability of being in each of these inadequate state combinations. From this information, it is simple to compute the expected number of interruptions for each load bus, the expected number of interruption minutes for each load bus, and the expected amount of unserved energy for each load bus. These load bus results can then be aggregated to produce the following system indices:

- *LOLE* (Loss of Load Expectation) — the expected number of hours per year that the system will have to shed load.
- *EENS* (Expected Energy Not Served) — the expected number of megawatt hours per year that a system will not be able to supply.

System adequacy assessment assumes that the transmission system is available. This may not always be the case. A classic example is the 1965 blackout, which was initiated by the unexpected loss of a transmission line. To address such events, system security assessment in required.

System Security Assessment

System security is defined as the ability of a power system to supply all of its loads in the event of one or more contingencies (a contingency is an unexpected event such as a system fault or a component outage). This is divided into two separate areas: static security assessment and dynamic security assessment.

Static security assessment determines whether a power system is able to supply peak demand after one or more pieces of equipment (such as a line or a transformer) are disconnected. The system is tested by

Power System Planning (Reliability)

removing a piece (or multiple pieces) of equipment from the normal power flow model, rerunning the power flow, and determining if all bus voltages are acceptable and all pieces of equipment are loaded below emergency ratings. If an unacceptable voltage or overload violation occurs, load must be shed for this condition and the system is *insecure*. If removing any single component will not result in the loss of load, the system is *N-1 Secure*. If removing any X arbitrary components will not result in the loss of load, the system is *N-X Secure*. N refers to the number of components on the system and X refers to the number of components that can be safely removed.

Performing a static security assessment can be computationally intensive. For example, an N-2 assessment on a modest system with 5000 components (1500 buses, 500 transformers, and 3000 lines) will require more than 25 million power flows to be performed. For this reason, contingency ranking methods are often used. These methods rank each contingency based on its likelihood of resulting in load curtailment. Contingencies are examined in order of their contingency ranking, starting with the most severe. If a prespecified number of contingencies are tested and found to be secure, it is assumed that contingencies with less severe rankings are also secure and do not need to be examined.

Static security assessment is based on steady state power flow solutions. For each contingency, it assumes that the system protection has properly operated and the system has reached a steady state. In fact, the power system may not actually reach a steady state after it has been disturbed. Checking whether a system will reach a steady state after a fault occurs is referred to *as dynamic security assessment* (also referred to as *transient security assessment*).

When a fault occurs, the system is less able to transfer power from synchronous generators to synchronous motors. Since the instantaneous power input has not changed, generators will begin to speed up and motors will begin to slow down (analogous to the chain slipping while riding a bicycle). This increases the rotor angle difference between generators and motors. If this rotor angle exceeds a critical value, the system will become unstable and the machines will not be able to regain synchronism. After the protection system clears the fault, the rotor angle difference will still increase since the power transfer limits of the system are still less than the prefault condition. If the fault is cleared quickly enough, this additional increase will not cause the rotor angle difference to exceed the critical angle and the system will return to a synchronous state (ABB Power, 1997).

An example of a transient stability test is shown in Fig. 13.27. This shows the rotor angle difference between a synchronous generator and a synchronous motor during a fault sequence. When the fault occurs, the rotor angle begins to increase. If the fault is not cleared, the rotor angle quickly exceeds the critical angle. If the fault is cleared at 0.3 sec, the rotor angle still increases beyond the critical value. The

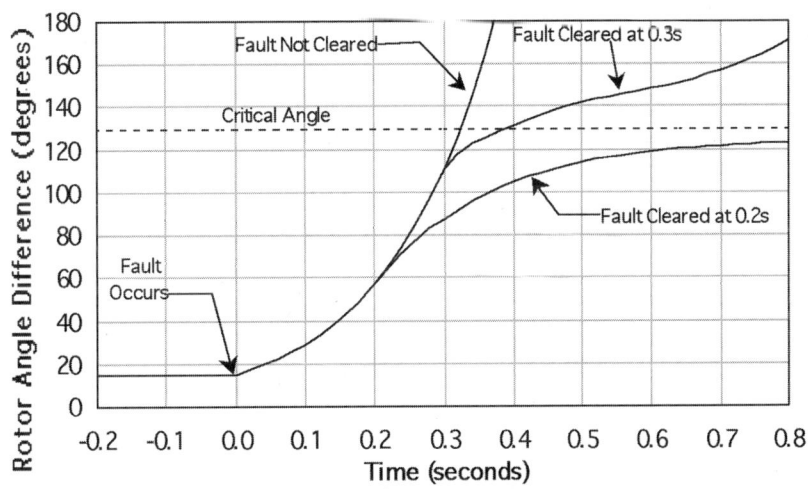

FIGURE 13.27 Dynamic security assessment.

system is dynamically stable for this fault if it is cleared in 0.2 sec. The rotor angle will still increase after the fault occurs, but will stabilize below the critical value.

A dynamic security assessment will consist of many transient stability tests that span a broad range of loading conditions, fault locations, and fault types. To reduce the number of tests required, contingency rankings (similar to static security assessment) can be used.

Probabilistic Security Assessment

Although the "N-1 Criterion" remains popular, it has received much criticism since it treats unlikely events with the same importance as more frequent events. Using the N-1 Criterion, large amounts of money may be spent to reinforce a system against a very rare event. From a reliability perspective, this money *might* be better spent in other areas such as replacing old equipment, decreasing maintenance intervals, adding automation equipment, adding crews, and so on. To make such value judgments, both the impact of each contingency and its probability of occurrence must be considered (Endrenyi, 1978). This is referred to as *probabilistic security assessment*. To do this type of assessment, each piece of equipment needs at least two fundamental pieces of information: the *failure rate* of the equipment (usually denoted λ, in failures per year), and the mean time to repair of the equipment (usually denoted $MTTR$, in hours).

Performing a probabilistic security assessment is similar to a performing a standard static security assessment. First, contingencies are ranked and simulated using a power flow. If a contingency results in the loss of load, information about the number and size of interrupted loads, the frequency of the contingency, and the repair time of the contingency are recorded. This allows quantities such as *EENS* (Expected Energy Not Served) to be easily computed. If contingency i causes kW_i amount of kilowatts to be interrupted, then *EENS* is equal to:

$$EENS = \sum_i kW_i \lambda_i MTTR_i \qquad (13.6)$$

It is important to note that this is the *EENS* due to contingencies, and is separate from the *EENS* due to generation unavailability. It is also important to note that this formula assumes that $\lambda_i MTTR_i$ is small when compared to one year. If this is not the case, a component will experience fewer failures per year than its failure rate and the equation must be adjusted accordingly.

Distribution System Reliability

The majority of customer reliability problems stem from distribution systems. For a typical residential customer with 90 min of interrupted power per year, between 70 and 80 minutes will be attributable to problems occurring on the distribution system that it is connected to (Billinton and Jonnavitihula, 1996). This is largely due to radial nature of most distribution systems, the large number of components involved, the sparsity of protection devices and sectionalizing switches, and the proximity of the distribution system to end-use customers.

Since reliability means different things to different people, it is necessary to address the definition of "distribution system reliability" in more detail. In distribution systems, reliability primarily relates to equipment outages and customer interruptions:

- **Outage** — when a piece of *equipment* is deenergized.
- **Momentary interruption** — when a *customer* is deenergized for less than a few minutes.
- **Sustained interruption** — when a *customer* is deenergized for more than a few minutes.

Customers do not, in the strictest sense, experience power outages. Customers experience power interruptions. If power is restored within a few minutes, it is considered a momentary interruption. If not, it is considered a sustained interruption. The precise meaning of "a few minutes" varies from utility to utility, but is typically between 1 and 5 min. The IEEE defines a momentary interruption based on 5 min. (Note: some references classify interruptions into four categories rather than two. Instantaneous interruptions last a few seconds, momentary interruptions last a few minutes, temporary interruptions last a few hours, and sustained interruptions last many hours.)

On a historical note, momentary interruptions used to be considered a "power quality issue" rather than a "reliability issue." It is now generally agreed that momentary interruptions are an aspect of reliability since (1) momentary interruptions can cause substantial problems to all types of customers, and (2) many trade-offs must be made between momentary interruptions and sustained interruptions during system planning, operation, and control. It can also be observed that customer voltage sags, typically considered a power quality issue, are slowly becoming a reliability issue for similar reasons.

Distribution system reliability is not dependent solely upon component failure characteristics. It is also dependent upon how the system responds to component failures. To understand this, it is necessary to understand the sequence of events that occurs after a distribution system fault.

Typical Sequence of Events after an Overhead Distribution Fault

1. The fault causes high currents to flow from the source to the fault location. These high currents may result in voltage sags for certain customers. These sags can occur on all feeders that have a common coupling at the distribution substation.
2. An instantaneous relay trips open the feeder circuit breaker at the substation. This causes the entire feeder to be deenergized. A pause allows the air around the fault to deionize, and then a reclosing relay will close the circuit breaker. If no fault current is detected, the fault has cleared itself and all customers on the feeder have experienced a momentary interruption.
3. If the fault persists, time overcurrent protection devices are allowed to clear the fault. If the fault is on a fused lateral, the fuse will blow and customers on the lateral will be interrupted. If the feeder breaker trips again, the reclosing relay will repeat the reclosing process a preset number of times before locking out. After the feeder breaker locks out, all customers on the feeder will be interrupted.
4. The electric utility will receive trouble calls from customers with interrupted power. It will dispatch a crew to locate the fault and isolate it by opening up surrounding sectionalizing switches. It may also attempt to reconfigure the distribution system in an attempt to restore power to as many customers as possible while the fault is being repaired. Fault isolation can be very fast if switches are motor operated and remotely controlled, but switching usually takes between 15 and 60 min.
5. The crew repairs the faulted equipment and returns the distribution system to its normal operating state.

As can be seen, *a fault on the distribution system will impact many different customers in many different ways*. In general, the same fault will result in voltage sags for some customers, momentary interruptions for other customers, and varying lengths of sustained interruptions for other customers, depending on how the system is switched and how long the fault takes to repair.

Distribution system reliability assessment methods are able to predict distribution system reliability based on system configuration, system operation, and component reliability data (Brown et al., 1996). This ability is becoming increasingly important as the electric industry becomes more competitive, as regulatory agencies begin to regulate reliability, and as customers begin to demand performance guarantees. The most common reliability assessment methods utilize the following process: (1) they simulate a system's response to a contingency, (2) they compute the reliability impact that this contingency has on each customer, (3) the reliability impact is weighted by the probability of the contingency occurring, and (4) steps 1–3 are repeated for all contingencies. Since this process results in the reliability that each customer can expect, new designs can be compared, existing systems can be analyzed, and reliability improvement options can be explored.

Distribution Reliability Indices

Utilities typically keep track of customer reliability by using reliability indices. These are average customer reliability values for a specific area. This area can be the utility's entire service area, a particular geographic region, a substation service area, a feeder service area, and so on. The most commonly used reliability indices give each customer equal weight. This means that a large industrial customer and a small residential customer will each have an equal impact on computed indices. The most common of these *customer reliability indices* are: System Average Interruption Frequency Index (SAIFI), System Average Interruption Duration Index (SAIDI), Customer Average Interruption Duration Index (CAIDI), and Average System Availability Index (ASAI) (IEEE Working Group, 1998). Notice that these indices are redundant. If SAIFI and SAIDI are known, both CAIDI and ASAI can be calculated. Formulae for these indices are:

$$SAIFI = \frac{Total\ Number\ of\ Customer\ Interruptions}{Total\ Number\ of\ Customers\ Served} \quad \text{per year} \tag{13.7}$$

$$SAIDI = \frac{\sum Customer\ Interruption\ Durations}{Total\ Number\ of\ Customers\ Served} \quad \text{hours per year} \tag{13.8}$$

$$CAIDI = \frac{\sum Customer\ Interruption\ Durations}{Total\ Number\ of\ Customer\ Interruptions} = \frac{SAIDI}{SAIFI} \quad \text{hours per interruption} \tag{13.9}$$

$$ASAI = \frac{Customer\ Hours\ Service\ Availability}{Customer\ Hours\ Service\ Demand} = \frac{8760 - SAIDI}{8760} \quad \text{per unit} \tag{13.10}$$

Some less commonly used reliability indices are not based on the total number of customers served. The Customer Average Interruption Frequency Index (CAIFI) and the Customer Total Average Interruption Duration Index (CTAIDI) are based upon the number of customers that have experienced one or more interruptions in the relevant year. The Average System Interruption Frequency Index (ASIFI) and the Average System Interruption Duration Index (ASIDI) are based upon the connected kVA of customers (these are sometimes referred to as load-based indices). Formulae for these indices are:

$$CAIFI = \frac{Total\ Number\ of\ Customer\ Interruptions}{Customers\ Experiencing\ 1\ or\ more\ Interruptions} \quad \text{per year} \tag{13.11}$$

$$CTAIDI = \frac{\sum Customer\ Interruption\ Durations}{Customers\ Experiencing\ 1\ or\ more\ Interruptions} \quad \text{hours per year} \tag{13.12}$$

$$ASIFI = \frac{Connected\ kVA\ Interrupted}{Total\ Connected\ kVA\ Served} \quad \text{per year} \tag{13.13}$$

$$ASIDI = \frac{Connected\ kVA\ Hours\ Interrupted}{Total\ Connected\ kVA\ Served} \quad \text{hours per year} \tag{13.14}$$

Power System Planning (Reliability)

As momentary interruptions become more important, it becomes necessary to keep track of indices related to momentary interruptions. Since the duration of momentary interruptions is of little consequence, a single frequency related index, the Momentary Average Interruption Frequency Index (MAIFI), is all that is needed. MAIFI, like SAIFI, weights each customer equally (there is currently no load-based index for momentary interruptions). The formula for MAIFI is:

$$MAIFI = \frac{\text{Total Number of Customer Momentary Interruptions}}{\text{Total Number of Customers Served}} \quad \text{per year} \quad (13.15)$$

The precise application of MAIFI varies. This variation is best illustrated by an example. Assume that a customer experiences three recloser operations followed by a recloser lockout, all within a period of one minute. Some utilities would not count this event as a momentary interruption since the customer experiences a sustained interruption. Other utilities would count this event as three momentary interruptions and one sustained interruption. Similarly, if a customer experiences three recloser operations within a period of one minute with power being restored after the last reclosure, some utilities would count the event as three momentary interruptions and other utilities would could the event as a single momentary interruption.

Storms and Major Events

When electric utilities compute reliability indices, they often exclude interruptions caused by "storms" and "major events." The definition of a major event varies from utility to utility, but a typical example is when more than 10% of customers experience an interruption during the event. The event starts when the notification of the first interruption is received and ends when all customers are restored service.

In nonstorm conditions, equipment failures are independent events — the failure of one device is completely independent of another device. In contrast, major events are characterized by common-mode failures. This means that a common cause is responsible for all equipment failures. The result is that many components tend to fail at the same time. This puts a strain on utility resources, which can only handle a certain number of concurrent failures (Brown et al., 1997). The most common causes of major events are wind storms, ice storms, and heat waves.

Wind storms refer to linear winds that blow down trees and utility poles. The severity of wind storms is dependent upon sustained wind speed, gust speed, wind direction, and the length of the storm. Severity is also sensitive to vegetation management and the time elapsed since the last wind storm. Since a wind storm will tend to blow over all of the weak trees, a similar storm occurring a few months later may have little impact. A U.S. map showing wind speeds for the worst expected storm in 50 years is shown in Fig. 13.28.

Ice storms refer to ice buildup on conductors. This has four major effects: (1) it places a heavy physical load on the conductors and support structures, (2) it increases the cross-sectional area that is exposed to the wind, (3) ice can break off and cause a conductor to jump into the phase wires located above it, and (4) galloping. Galloping occurs when ice buildup assumes a teardrop shape and acts as an airfoil. During high winds, this can cause conductors to swing wildly and with great force. Ice can also cause problems by accumulating in trees, causing limbs to break off, and causing entire trunks to fall over into power lines.

Heat waves are extended periods of exceedingly hot weather. This hot weather causes electricity demand to skyrocket due to air-conditioning loads. At the same time, conductors cannot carry as much electricity since they cannot transfer heat as effectively to their surroundings. This combination of heavy loading and conductor de-rating can cause overhead wires to become overloaded and sag to dangerous levels. Overloaded cables will cause insulation to lose life. In a worst-case scenario, the maximum power transfer capabilities of the system can be approached, resulting in a voltage collapse condition. Humidity exacerbates the impact of heat waves since it causes air conditioners to consume more energy.

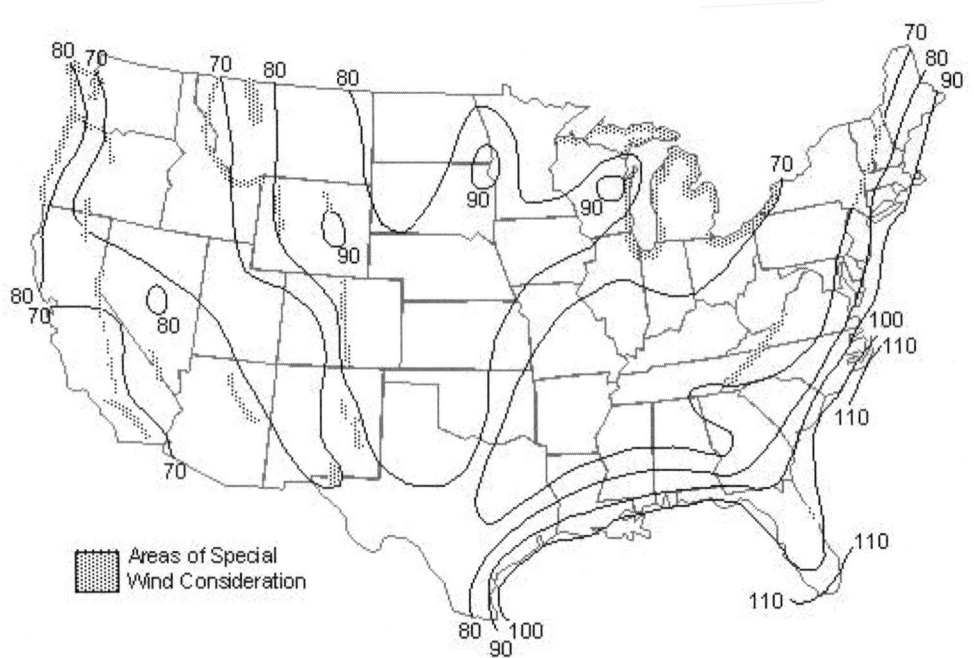

FIGURE 13.28 50-year wind storm (sustained wind speed in mi/hr).

Component Reliability Data

For a reliability model to be accurate, component reliability data must be representative of the system being modeled. Utilities recognize this and are increasing their efforts to keep track of component failure rates, failure modes, repair times, switching times, and other important reliability parameters. Unfortunately, reliability statistics vary widely from utility to utility and from country to country. The range of equipment reliability data that can be found in published literature is shown in Table 13.8.

TABLE 13.8 Equipment Reliability Data

Component	Failure Rate (per year)	MTTR (hours)
Substation Equipment		
Power transformers	0.015–0.07	15–480
Circuit breakers	0.003–0.02	6–80
Disconnect switches	0.004–0.16	1.5–12
Air insulated buswork	0.002–0.04	2–13
Overhead Equipment		
Transmission lines[a]	0.003–0.140	4–280
Distribution lines[a]	0.030–0.180	4–110
Switches/fused cutouts	0.004–0.014	1–4
Pole mounted transformer	0.001–0.004	3–8
Underground Equipment		
Cable[a]	0.005–0.04	3–30
Padmount switches	0.001–0.01	1–5
Padmount transformers	0.002–0.003	2–6
Cable terminations/joints	0.0001–0.002	2–4

[a] Failure rates for lines and cable are per mile.

Because component reliability is very system specific, it is beneficial to calibrate reliability models to historical reliability indices. In this process, component reliability parameters are adjusted until historical reliability indices match computed reliability indices (Brown et al., 1998). The amount that each parameter is adjusted should depend on the confidence of the original value and the sensitivity of the reliability indices to changes in this value. To illustrate, consider an overhead distribution system. A reliability model of this system is created using component reliability data from published literature. Unfortunately, the reliability indices that the model produces do not agree with the historical performance of the system over the past few years. To fix this, the failure rate and repair times of overhead lines (along with other component parameters) can be adjusted until predicted reliability matches historical reliability.

Utility Reliability Problems

To gain a broader understanding of power system reliability, it is necessary to understand the root causes of system faults and system failures. A description of major failure modes is now provided.

Underground Cable

A major reliability concern pertaining to underground cables is electrochemical treeing. Treeing occurs when moisture penetration in the presence of an electric field reduces the dielectric strength of cable insulation. When the dielectric strength is degraded sufficiently, transients caused by lightning or switching can result in dielectric breakdown. Electrochemical treeing usually affects extruded dielectric cable such as cross-linked polyethylene (XLPE) and ethylene-propylene rubber (EPR), and is largely attributed to insulation impurities and bad manufacturing. To reduce failures related to electrochemical treeing, a utility can install surge protection on riser poles (transitions from overhead to underground), can purchase tree-retardant cable, and can test cable reels before accepting them from the manufacturer.

Existing cable can be tested and replaced if problems are found. One way to do this is to apply a DC voltage withstand test (approximately 3 times nominal RMS voltage). Since cables will either pass or not pass this test, information about the state of cable deterioration cannot be determined. Another popular method for cable testing is to inject a small signal into one end and check for reflections that will occur at partial discharge points. Other methods are measuring the power factor over a range of frequencies (dielectric spectroscopy), analyzing physical insulation samples in a lab for polymeric breakdown (degree of polymerization), and using cable indentors to test the hardness of the insulation.

Not all underground cable system failures are due to cable insulation. A substantial percentage occurs at splices, terminations, and joints. Major causes are due to water ingress and poor workmanship. Heat shrink covers can be used to waterproof these junctions and improve reliability.

The last major reliability concern for underground cable is dig-ins. This is when excavation equipment cuts through one or more cables. To prevent dig-ins, utilities should encourage the public to have cable routes identified before initiating site excavation. In extreme cases where high reliability is required, utilities can place cable in concrete-encased duct banks.

Transformer Failures

Transformers are critical links in power systems, and can take a long time to replace if they fail. Through-faults cause extreme physical stress on transformer windings, and are the major cause of transformer failures. Overloads rarely result in transformer failures, but do cause thermal aging of winding insulation.

When a transformer becomes hot, the insulation on the windings slowly breaks down and becomes brittle over time. The rate of thermal breakdown approximately doubles for every 10°C. 10°C is referred to as the "Montsinger Factor" and is a rule of thumb describing the Arrhenius theory of electrolytic dissociation. Because of this exponential relationship, transformer overloads can result in rapid transformer aging. When thermal aging has caused insulation to become sufficiently brittle, the next fault current that passes through the transformer will mechanically shake the windings, a crack will form in the insulation, and an internal transformer fault will result.

Extreme hot-spot temperatures in liquid-filled transformers can also result in failure. This is because the hot spot can cause free bubbles that reduce the dielectric strength of the liquid. Even if free bubbles

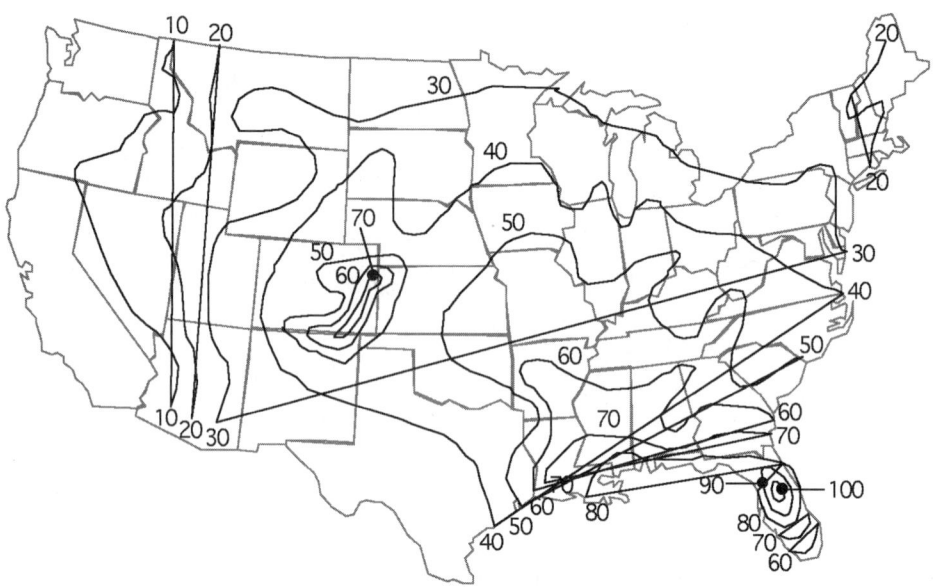

FIGURE 13.29 Number of thunderstorm days per year.

are not formed, high temperatures will increase internal tank pressure and may result in overflow or tank rupture.

Many transformers are fitted with load tap changers (LTCs) for voltage regulation. These mechanically moving devices have historically been prone to failure and can substantially reduce the reliability of a transformer (Willis, 1997). Manufacturers have addressed this problem and new LTC models using vacuum technology have succeeded in reducing failure rates.

Lightning

A lightning strike occurs when the voltage generated between a cloud and the ground exceeds the dielectric strength of the air. This results in a massive current stroke that usually exceeds 30,000 amps. To make matters worse, most strokes consist of multiple discharges within a fraction of a second. Lightning is the major reliability concern for utilities located in high keraunic areas (Burke, 1994). An isokeraunic map for the U.S. is shown in Fig. 13.29.

Lightning can affect power systems through direct strikes (the stroke contacts the power system) or through indirect strikes (the stroke contacts something in close proximity and induces a traveling voltage wave on the power system). Lightning can be protected against by having a high system BIL, by using shield wires, by using surge arrestors to clamp voltages across equipment, and by having a low impedance ground. Direct strikes are virtually impossible to protect against on a distribution system.

Tree Contact

Trees continuously grow, can fall over onto conductors, can drop branches onto conductors, can push conductors together, and can serve as gateway for animals. This is why many utilities spend more on tree trimming than on any other preventative maintenance activity.

When a tree branch bridges two conductors, a fault does not occur immediately. This is because a moist tree branch has a substantial resistance. A small current begins to flow and starts to dry out the wood fibers. After several minutes, the cellulose will carbonize, resistance will be greatly reduced, and a short circuit will occur. Branches brushing against a single phase conductor typically *do not* result in system faults.

Faults due to tree contact can be reduced by using tree wire. This is overhead wire with an insulated jacket similar to cable. Tree wire can be effective, but faults tend to result in conductor burndown since they will not motor (move themselves along the conductor) like faults on bare conductor.

Birds

Birds are the most common cause of animal faults on both transmission systems and air insulated substations. Different types of birds cause different types of problems, but they can generally be classified as nesting birds, roosting birds, raptors, and woodpeckers.

Nesting birds commonly build their homes on transmission towers and in substations. Nesting materials can cause faults, and bird excrement can contaminate insulators. Nesting birds also attract predators such as raccoons, snakes, and cats. These predators can be a worse reliability problem than the birds themselves.

Roosting birds use electrical equipment to rest on or to search for prey. They can be electrocuted by bridging conductors with their wings, and their excrement can contaminate insulators. To prevent birds from roosting, anti-roosting devices can be placed on attractive sites. For locations that cater to thousands of roosting birds, more extreme deterrent methods such as pyrotechnics can be used.

Raptors are birds of prey such as eagles, hawks, ospreys, owls, and vultures. Reliability problems are similar to other roosting and nesting birds, but special consideration may be required since most raptors are protected by the federal government.

Woodpeckers peck holes in wood with their beaks as they search for insects. This does not harm trees (the bark regenerates), but can cause devastating damage to utility poles. This can be prevented by using steel poles, by using repellent, or by tricking a woodpecker into believing that there is already a resident woodpecker (woodpeckers are quite territorial).

Squirrels

Squirrels are a reliability concern for all overhead distribution systems near wooded areas. Squirrels will not typically climb utility poles, but will leap onto them from nearby trees. They cause faults by bridging grounded equipment with phase conductors. Squirrel problems can be mitigated by cutting down nearby access trees or by installing animal guards on insulators.

Snakes

Snakes are major reliability concerns in both substations and underground systems. They can squeeze through very small openings, can climb almost anything, and have the length to easily span phase conductors. Snakes are usually searching for food (birds in substations and mice in underground systems), and removing the food supply can often remove the snake problem. Special "snake fences" are also available.

Insects

It is becoming more common for fire ants to build nests in pad mounted equipment. Their nesting materials can cause short circuits, the ants can eat away at conductor insulation, and they make equipment maintenance a challenge.

Bears, Bison, and Cattle

These large animals do not typically cause short circuits, but degrade the structural integrity of poles by rubbing on guy wires. Bears can also destroy wooden poles by using them as scratching posts, and black bears can climb wooden utility poles. These problems can be addressed by placing fences around poles and guy wire anchors.

Mice, Rats, and Gophers

These rodents cause faults by gnawing through the insulation of underground cable. They are the most common cause of animal-related outages on underground equipment. To make matters worse, they will attract snakes (also a reliability problem). Equipment cabinets should be tightly sealed to prevent these small animals from entering. Ultrasonic devices can also be used to keep rodents away (ultrasonic devices will not keep snakes away).

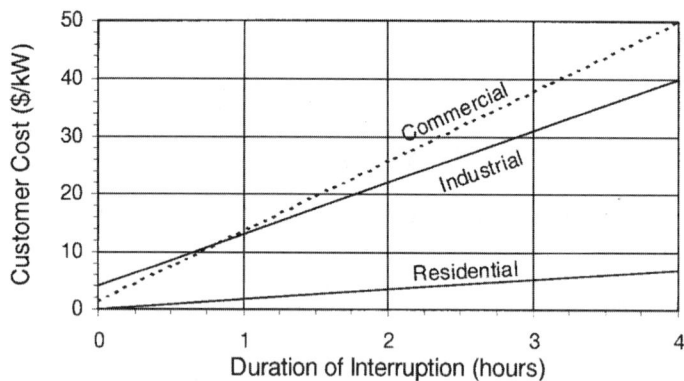

FIGURE 13.30 Typical U.S. customer interruption costs (1999 dollars).

Vandalism

Vandalism can take many different forms, from people shooting insulators with rifles to professional thieves stealing conductor wire for scrap metal. Addressing these reliability problems will vary greatly from situation to situation.

Reliability Economics

When a power interruption occurs, both the utility and the interrupted customers are inconvenienced. The utility must spend money to fix the problem, will lose energy sales during the interruption, and may be sued by disgruntled customers. From the customer perspective, batch processes may be ruined, electronic devices may crash, production may by lost, retail sales may be lost, and inventory (such as refrigerated food) may be ruined.

When a customer experiences an interruption, there is an amount of money that it would be willing to pay to have avoided the interruption. This amount is referred to as the customer's incurred cost of poor reliability, and consists of a base cost plus a time dependent cost. The base cost is the same for all interruptions, relates to electronic equipment shutdown and interrupted processes, and is equivalent to the cost of a momentary interruption. The time dependent cost relates to lost production and extended inconvenience, and reflects that customers would prefer interruptions to be shorter rather than longer.

The customer cost of an interruption varies widely from customer to customer and from country to country. Other important factors include the time of year, the day of the week, the time of day, and whether advanced warning is provided. Specific results are well documented by a host of customer surveys (Billinton et al., 1983; Tollefson et al., 1991; Tollefson et al., 1994; IEEE Std. 493-1990). For planning purposes, it is useful to aggregate these results into a few basic customer classes: commercial, industrial, and residential. Since larger customers will have a higher cost of reliability, results are normalized to the peak kW load of each customer. Reliability cost curves for typical U.S. customers are shown in Fig. 13.30.

Average customer cost curves tend to be linear and can be modeled as an initial cost plus a first order, time dependent cost. Specific customer cost curves may be extremely nonlinear. For example, a meat packing warehouse depending upon refrigeration may be unaffected by interruptions lasting many hours. At a certain point, the meat will begin to spoil and severe economic losses will quickly occur. After the meat spoils, additional interruption time will harm this particular customer much more.

Annual Variations in Reliability

Power system reliability varies from year to year. In a lucky year, a system may have a SAIDI of 30 min. The next year, this exact same system may experience a SAIDI of 8 h. This type of variation is inevitable and must be considered when comparing reliability indices. It is also important to note that the variance of reliability indices will tend to be less for areas serving more customers. Individual customer reliability will tend to be the most volatile, followed by feeder reliability, substation reliability, regional reliability, and so forth.

The importance of annual reliability variance will grow as utilities become subject to performance-based rates and as customer reliability guarantees become more common. These types of contracts expose utilities to new risks that must be understood and managed. Since performance-based contracts penalize and reward utilities based on reliability, annual variations must be understood for fair contracts to be negotiated and managed.

Contractual issues concerning service reliability are becoming important as the electric industry becomes more competitive. Customers can choose between suppliers, and realize that there is a trade-off between reliability and rates. Some customers will demand poor reliability at low rates, and other customers will demand high reliability at premium rates. To address the wide variation in customer needs, utilities can no longer be suppliers of energy alone, but must become suppliers of both energy and reliability. Power system reliability is now a *bona fide* commodity with explicit value for utilities to supply and explicit value that customers demand.

References

Allan, R. N., Billinton, R., Breipohl, A. M., and Grigg, C. H., Bibliography on the application of probability methods in power system reliability evaluation, *IEEE Trans. on Power Syst.*, 9, 1, 41–49, Feb. 1994.

Billinton, R. and Allan, R. N., *Reliability Assessment of Large Electric Power Systems*, Kluwer Academic Publishers, Dordrecht, The Netherlands, 1988.

Billinton, R. and Allan, R., *Reliability Evaluation of Engineering Systems: Concepts and Techniques* (2nd edition) Plenum Press, New York, 1992.

Billinton, R. and Jonnavitihula, S., A Test System for Teaching Overall Power System Reliability Assessment, *IEEE Trans. on Power Syst.*, 11, 4, 1670–1676, Nov. 1996.

Billinton, R., Wacker, G., and Wojczynski, E., Comprehensive bibliography on electrical service interruption costs, *IEEE Trans. on Power Appar. Syst.*, PAS-102, 6, 1831–1837, June 1983.

Brown, R. E., Gupta, S., Christie, R. D., Venkata, S. S., and Fletcher, R. D., Distribution system reliability analysis using hierarchical Markov modeling, *IEEE Trans. on Power Delivery*, 11, 4, 1929–1934, Oct. 1996.

Brown, R. E., Gupta, S., Christie, R. D., Venkata, S. S., and Fletcher, R. D., Distribution system reliability: momentary interruptions and storms, *IEEE Trans. on Power Delivery*, 12, 4, 1569–1575, October 1997.

Brown, R. E. and Ochoa, J. R., Distribution system reliability: default data and model validation, *IEEE Trans. on Power Syst.*, 13, 2, 704–709, May 1998.

Burke, J. J., *Power Distribution Engineering*, Marcel Dekker, Inc., New York, 1994.

Electrical Transmission and Distribution Reference Book, ABB Power T&D Company, Inc., Raleigh, NC, 1997.

Endrenyi, J., *Reliability in Electric Power Systems*, John Wiley & Sons, Ltd., New York, 1978.

IEEE Recommended Practice for the Design of Reliable Industrial and Commercial Power Systems, IEEE Std. 493-1990.

IEEE Working Group on System Design, *Trial Use Guide for Power Distribution Reliability Indices*, Report P1366, IEEE, 1998.

Philipson, L. and Willis, H. L., *Understanding Electric Utilities and De-regulation*, Marcel Dekker, Inc., New York, 1999.

Ramakumar, R., *Engineering Reliability: Fundamentals and Applications*, Prentice-Hall, Inc., Englewood Cliffs, NJ, 1993.

Schilling, M. T., Billinton, R., Leite da Silva, A. M., and El-Kady, M. A., Bibliography on composite system reliability (1964–1988), *IEEE Trans. on Power Syst.*, 4, 3, 1122–1132, Aug. 1989.

Tollefson, G., Billinton, R., Wacker, G., Chan, E., and Aweya, J., A Canadian customer survey to assess power system reliability worth, *IEEE Trans. on Power Syst.*, 9, 1, 443–450, Feb. 1994.

Tollefson, G., Billinton, R., and Wacker, G., Comprehensive bibliography on reliability worth and electric service consumer interruption costs 1980–1990, *IEEE Trans. on Power Syst.*, 6, 4, 1508–1514, Nov. 1991.

Willis, H. L., *Power Distribution Planning Reference Book*, Marcel Dekker, Inc., New York, 1997.

14
Power Electronics

Mark Nelms
Auburn University

14.1 Power Semiconductor Devices *Kaushik Rajashekara* 14-1

14.2 Uncontrolled and Controlled Rectifiers *Mahesh M. Swamy* 14-8

14.3 Inverters *Michael Giesselmann* 14-37

14.4 Active Filters for Power Conditioning *Hirofumi Akagi* 14-44

14
Power Electronics

Kaushik Rajashekara
Delphi Automotive Systems

Mahesh M. Swamy
Yaskawa Electric America

Michael Giesselmann
Texas Tech University

Hirofumi Akagi
Tokyo Institute of Technology

14.1 Power Semiconductor Devices 14-1
Thyristor and Triac • Gate Turn-Off Thyristor (GTO) • Reverse-Conducting Thyristor (RCT) and Asymmetrical Silicon-Controlled Rectifier (ASCR) • Power Transistor • Power MOSFET • Insulated-Gate Bipolar Transistor • MOS-Controlled Thyristor (MCT)

14.2 Uncontrolled and Controlled Rectifiers 14-8
Uncontrolled Rectifiers • Controlled Rectifiers • Conclusion

14.3 Inverters 14-37
Fundamental Issues • Single Phase Inverters • Three Phase Inverters • Multilevel Inverters • Line Commutated Inverters

14.4 Active Filters for Power Conditioning 14-44
Harmonic-Producing Loads • Theoretical Approach to Active Filters for Power Conditioning • Classification of Active Filters • Integrated Series Active Filters • Practical Applications of Active Filters for Power Conditioning

14.1 Power Semiconductor Devices

Kaushik Rajashekara

The modern age of power electronics began with the introduction of thyristors in the late 1950s. Now there are several types of power devices available for high-power and high-frequency applications. The most notable power devices are gate turn-off thyristors, power Darlington transistors, power MOSFETs, and insulated-gate bipolar transistors (IGBTs). Power semiconductor devices are the most important functional elements in all power conversion applications. The power devices are mainly used as switches to convert power from one form to another. They are used in motor control systems, uninterrupted power supplies, high-voltage DC transmission, power supplies, induction heating, and in many other power conversion applications. A review of the basic characteristics of these power devices is presented in this section.

Thyristor and Triac

The thyristor, also called a silicon-controlled rectifier (SCR), is basically a four-layer three-junction *pnpn* device. It has three terminals: anode, cathode, and gate. The device is turned on by applying a short pulse across the gate and cathode. Once the device turns on, the gate loses its control to turn off the device. The turn-off is achieved by applying a reverse voltage across the anode and cathode. The thyristor symbol and its volt-ampere characteristics are shown in Fig. 14.1. There are basically two classifications of thyristors: converter grade and inverter grade. The difference between a converter-grade and an inverter-grade thyristor is the low turn-off time (on the order of a few microseconds) for the latter. The converter-grade thyristors are slow type and are used in natural commutation (or phase-controlled) applications. Inverter-grade thyristors are used in forced commutation applications such as DC-DC choppers and

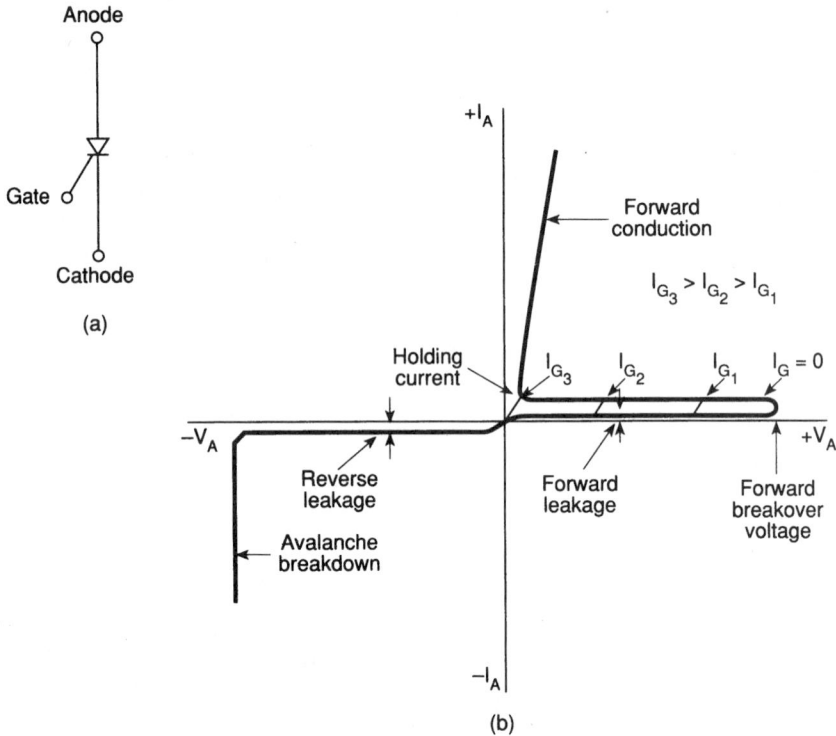

FIGURE 14.1 (a) Thyristor symbol and (b) volt-ampere characteristics. (*Source:* B.K. Bose, *Modern Power Electronics: Evaluation, Technology, and Applications*, p. 5. © 1992 IEEE.)

DC-AC inverters. The inverter-grade thyristors are turned off by forcing the current to zero using an external commutation circuit. This requires additional commutating components, thus resulting in additional losses in the inverter.

Thyristors are highly rugged devices in terms of transient currents, *di/dt*, and *dv/dt* capability. The forward voltage drop in thyristors is about 1.5 to 2 V, and even at higher currents of the order of 1000 A, it seldom exceeds 3 V. While the forward voltage determines the on-state power loss of the device at any given current, the switching power loss becomes a dominating factor affecting the device junction temperature at high operating frequencies. Because of this, the maximum switching frequencies possible using thyristors are limited in comparison with other power devices considered in this section.

Thyristors have I^2t withstand capability and can be protected by fuses. The nonrepetitive surge current capability for thyristors is about 10 times their rated root mean square (rms) current. They must be protected by snubber networks for *dv/dt* and *di/dt* effects. If the specified *dv/dt* is exceeded, thyristors may start conducting without applying a gate pulse. In DC-to-AC conversion applications, it is necessary to use an antiparallel diode of similar rating across each main thyristor. Thyristors are available up to 6000 V, 3500 A.

A triac is functionally a pair of converter-grade thyristors connected in antiparallel. The triac symbol and volt-ampere characteristics are shown in Fig. 14.2. Because of the integration, the triac has poor reapplied *dv/dt*, poor gate current sensitivity at turn-on, and longer turn-off time. Triacs are mainly used in phase control applications such as in AC regulators for lighting and fan control and in solid-state AC relays.

Gate Turn-Off Thyristor (GTO)

The GTO is a power switching device that can be turned on by a short pulse of gate current and turned off by a reverse gate pulse. This reverse gate current amplitude is dependent on the anode current to be turned off. Hence there is no need for an external commutation circuit to turn it off. Because turn-off

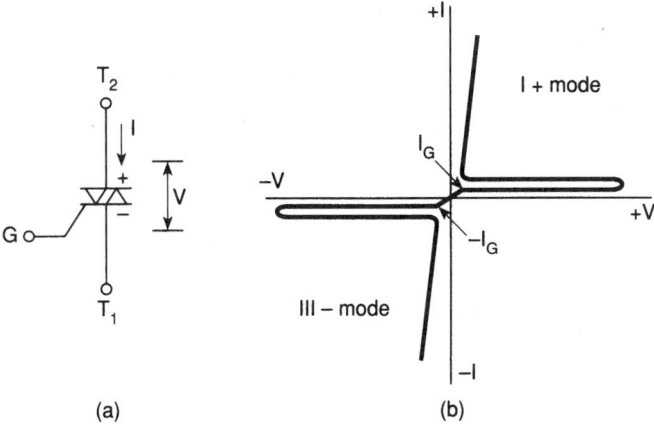

FIGURE 14.2 (a) Triac symbol and (b) volt-ampere characteristics. (*Source:* B.K. Bose, *Modern Power Electronics: Evaluation, Technology, and Applications*, p. 5. © 1992 IEEE.)

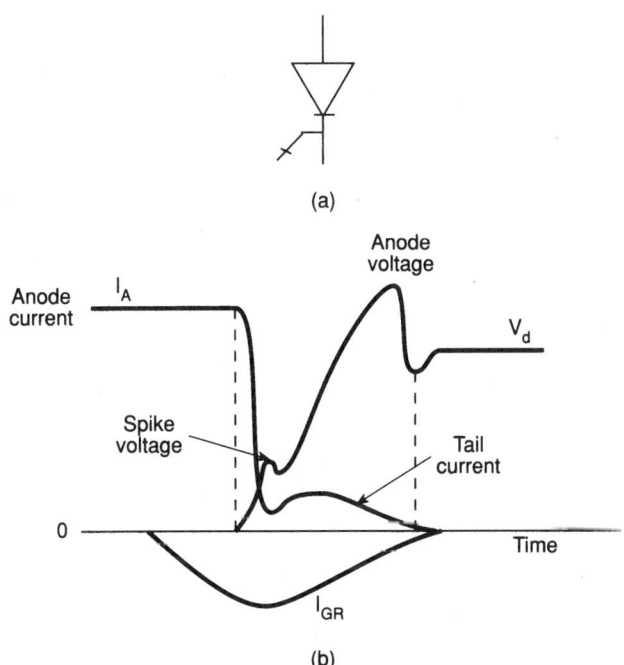

FIGURE 14.3 (a) GTO symbol and (b) turn-off characteristics. (*Source:* B.K. Bose, *Modern Power Electronics: Evaluation, Technology, and Applications*, p. 5. © 1992 IEEE.)

is provided by bypassing carriers directly to the gate circuit, its turn-off time is short, thus giving it more capability for high-frequency operation than thyristors. The GTO symbol and turn-off characteristics are shown in Fig. 14.3.

GTOs have the I^2t withstand capability and hence can be protected by semiconductor fuses. For reliable operation of GTOs, the critical aspects are proper design of the gate turn-off circuit and the snubber circuit. A GTO has a poor turn-off current gain of the order of 4 to 5. For example, a 2000-A peak current GTO may require as high as 500 A of reverse gate current. Also, a GTO has the tendency to latch at temperatures above 125°C. GTOs are available up to about 4500 V, 2500 A.

Reverse-Conducting Thyristor (RCT) and Asymmetrical Silicon-Controlled Rectifier (ASCR)

Normally in inverter applications, a diode in antiparallel is connected to the thyristor for commutation/freewheeling purposes. In RCTs, the diode is integrated with a fast switching thyristor in a single silicon chip. Thus, the number of power devices could be reduced. This integration brings forth a substantial improvement of the static and dynamic characteristics as well as its overall circuit performance.

The RCTs are designed mainly for specific applications such as traction drives. The antiparallel diode limits the reverse voltage across the thyristor to 1 to 2 V. Also, because of the reverse recovery behavior of the diodes, the thyristor may see very high reapplied dv/dt when the diode recovers from its reverse voltage. This necessitates use of large RC snubber networks to suppress voltage transients. As the range of application of thyristors and diodes extends into higher frequencies, their reverse recovery charge becomes increasingly important. High reverse recovery charge results in high power dissipation during switching.

The ASCR has similar forward blocking capability to an inverter-grade thyristor, but it has a limited reverse blocking (about 20–30 V) capability. It has an on-state voltage drop of about 25% less than an inverter-grade thyristor of a similar rating. The ASCR features a fast turn-off time; thus it can work at a higher frequency than an SCR. Since the turn-off time is down by a factor of nearly 2, the size of the commutating components can be halved. Because of this, the switching losses will also be low.

Gate-assisted turn-off techniques are used to even further reduce the turn-off time of an ASCR. The application of a negative voltage to the gate during turn-off helps to evacuate stored charge in the device and aids the recovery mechanisms. This will, in effect, reduce the turn-off time by a factor of up to 2 over the conventional device.

Power Transistor

Power transistors are used in applications ranging from a few to several hundred kilowatts and switching frequencies up to about 10 kHz. Power transistors used in power conversion applications are generally *npn* type. The power transistor is turned on by supplying sufficient base current, and this base drive has to be maintained throughout its conduction period. It is turned off by removing the base drive and making the base voltage slightly negative (within $-V_{BE(max)}$). The saturation voltage of the device is normally 0.5 to 2.5 V and increases as the current increases. Hence, the on-state losses increase more than proportionately with current. The transistor off-state losses are much lower than the on-state losses because the leakage current of the device is of the order of a few milliamperes. Because of relatively larger switching times, the switching loss significantly increases with switching frequency. Power transistors can block only forward voltages. The reverse peak voltage rating of these devices is as low as 5 to 10 V.

Power transistors do not have I^2t withstand capability. In other words, they can absorb only very little energy before breakdown. Therefore, they cannot be protected by semiconductor fuses, and thus an electronic protection method has to be used.

To eliminate high base current requirements, Darlington configurations are commonly used. They are available in monolithic or in isolated packages. The basic Darlington configuration is shown schematically in Fig. 14.4. The Darlington configuration presents a specific advantage in that it can considerably increase the current switched by the transistor for a given base drive. The $V_{CE(sat)}$ for the Darlington is generally more than that of a single transistor of similar rating with corresponding increase in on-state power loss. During switching, the reverse-biased collector junction may show hot-spot breakdown effects that are specified by reverse-bias safe operating area (RBSOA) and

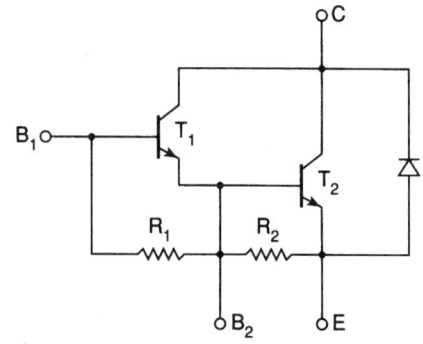

FIGURE 14.4 A two-stage Darlington transistor with bypass diode. (*Source:* B.K. Bose, *Modern Power Electronics: Evaluation, Technology, and Applications*, p. 6. © 1992 IEEE.)

forward-bias safe operating area (FBSOA). Modern devices with highly interdigited emitter base geometry force more uniform current distribution and therefore considerably improve secondary breakdown effects. Normally, a well-designed switching aid network constrains the device operation well within the SOAs.

Power MOSFET

Power MOSFETs are marketed by different manufacturers with differences in internal geometry and with different names such as MegaMOS, HEXFET, SIPMOS, and TMOS. They have unique features that make them potentially attractive for switching applications. They are essentially voltage-driven rather than current-driven devices, unlike bipolar transistors.

The gate of a MOSFET is isolated electrically from the source by a layer of silicon oxide. The gate draws only a minute leakage current on the order of nanoamperes. Hence, the gate drive circuit is simple and power loss in the gate control circuit is practically negligible. Although in steady state the gate draws virtually no current, this is not so under transient conditions. The gate-to-source and gate-to-drain capacitances have to be charged and discharged appropriately to obtain the desired switching speed, and the drive circuit must have a sufficiently low output impedance to supply the required charging and discharging currents. The circuit symbol of a power MOSFET is shown in Fig. 14.5.

Power MOSFETs are majority carrier devices, and there is no minority carrier storage time. Hence, they have exceptionally fast rise and fall times. They are essentially resistive devices when turned on, while bipolar transistors present a more or less constant $V_{CE(sat)}$ over the normal operating range. Power dissipation in MOSFETs is $I_d^2 R_{DS(on)}$, and in bipolars it is $I_C V_{CE(sat)}$. At low currents, therefore, a power MOSFET may have a lower conduction loss than a comparable bipolar device, but at higher currents, the conduction loss will exceed that of bipolars. Also, the $R_{DS(on)}$ increases with temperature.

An important feature of a power MOSFET is the absence of a secondary breakdown effect, which is present in a bipolar transistor, and as a result, it has an extremely rugged switching performance. In MOSFETs, $R_{DS(on)}$ increases with temperature, and thus the current is automatically diverted away from the hot spot. The drain body junction appears as an antiparallel diode between source and drain. Thus, power MOSFETs will not support voltage in the reverse direction. Although this inverse diode is relatively fast, it is slow by comparison with the MOSFET. Recent devices have the diode recovery time as low as 100 ns. Since MOSFETs cannot be protected by fuses, an electronic protection technique has to be used.

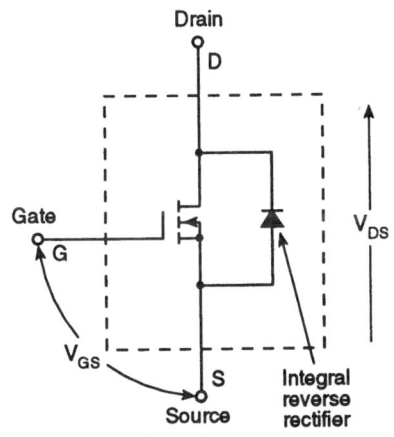

FIGURE 14.5 Power MOSFET circuit symbol. (*Source:* B.K. Bose, *Modern Power Electronics: Evaluation, Technology, and Applications*, p. 7. © 1992 IEEE.)

With the advancement in MOS technology, ruggedized MOSFETs are replacing the conventional MOSFETs. The need to ruggedize power MOSFETs is related to device reliability. If a MOSFET is operating within its specification range at all times, its chances for failing catastrophically are minimal. However, if its absolute maximum rating is exceeded, failure probability increases dramatically. Under actual operating conditions, a MOSFET may be subjected to transients — either externally from the power bus supplying the circuit or from the circuit itself due, for example, to inductive kicks going beyond the absolute maximum ratings. Such conditions are likely in almost every application, and in most cases are beyond a designer's control. Rugged devices are made to be more tolerant for over-voltage transients. Ruggedness is the ability of a MOSFET to operate in an environment of dynamic electrical stresses, without activating any of the parasitic bipolar junction transistors. The rugged device can withstand higher levels of diode recovery dv/dt and static dv/dt.

Insulated-Gate Bipolar Transistor (IGBT)

The IGBT has the high input impedance and high-speed characteristics of a MOSFET with the conductivity characteristic (low saturation voltage) of a bipolar transistor. The IGBT is turned on by applying a positive voltage between the gate and emitter and, as in the MOSFET, it is turned off by making the gate signal zero or slightly negative. The IGBT has a much lower voltage drop than a MOSFET of similar ratings. The structure of an IGBT is more like a thyristor and MOSFET. For a given IGBT, there is a critical value of collector current that will cause a large enough voltage drop to activate the thyristor. Hence, the device manufacturer specifies the peak allowable collector current that can flow without latch-up occurring. There is also a corresponding gate source voltage that permits this current to flow that should not be exceeded.

Like the power MOSFET, the IGBT does not exhibit the secondary breakdown phenomenon common to bipolar transistors. However, care should be taken not to exceed the maximum power dissipation and specified maximum junction temperature of the device under all conditions for guaranteed reliable operation. The on-state voltage of the IGBT is heavily dependent on the gate voltage. To obtain a low on-state voltage, a sufficiently high gate voltage must be applied.

In general, IGBTs can be classified as punch-through (PT) and nonpunch-through (NPT) structures, as shown in Fig. 14.6. In the PT IGBT, an N+ buffer layer is normally introduced between the P+ substrate and the N− epitaxial layer, so that the whole N− drift region is depleted when the device is blocking the off-state voltage, and the electrical field shape inside the N− drift region is close to a rectangular shape. Because a shorter N− region can be used in the punch-through IGBT, a better trade-off between the forward voltage drop and turn-off time can be achieved. PT IGBTs are available up to about 1200 V.

High voltage IGBTs are realized through a nonpunch-through process. The devices are built on an N− wafer substrate which serves as the N− base drift region. Experimental NPT IGBTs of up to about 4 KV have been reported in the literature. NPT IGBTs are more robust than PT IGBTs, particularly under short circuit conditions. But NPT IGBTs have a higher forward voltage drop than the PT IGBTs.

The PT IGBTs cannot be as easily paralleled as MOSFETs. The factors that inhibit current sharing of parallel-connected IGBTs are (1) on-state current unbalance, caused by $V_{CE}(sat)$ distribution and main circuit wiring resistance distribution, and (2) current unbalance at turn-on and turn-off, caused by the switching time difference of the parallel connected devices and circuit wiring inductance distribution. The NPT IGBTs can be paralleled because of their positive temperature coefficient property.

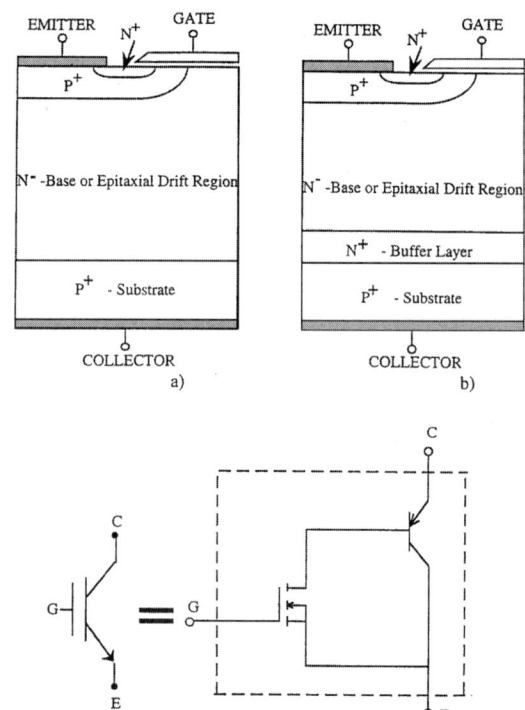

FIGURE 14.6 (a) Nonpunch-through IGBT, (b) punch-through IGBT, (c) IGBT equivalent circuit.

MOS-Controlled Thyristor (MCT)

The MCT is a new type of power semiconductor device that combines the capabilities of thyristor voltage and current with MOS gated turn-on and turn-off. It is a high power, high frequency, low conduction

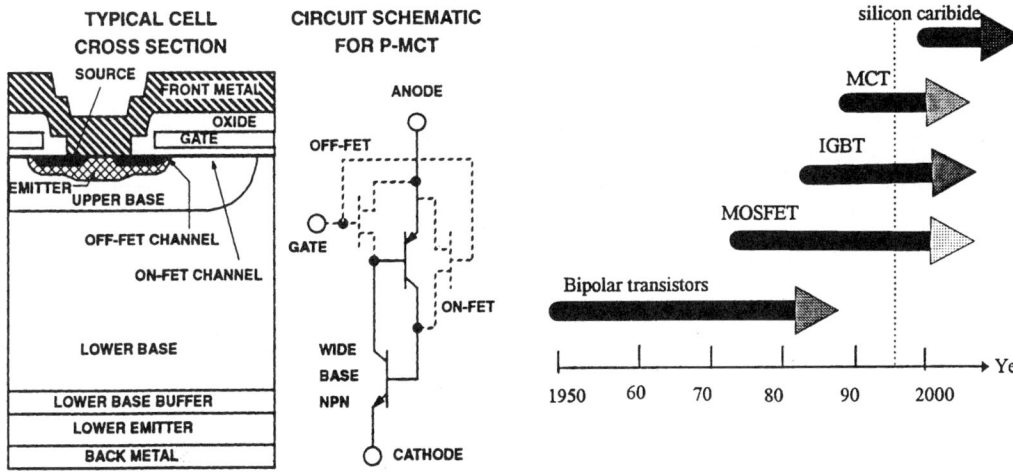

FIGURE 14.7 (*Source:* Harris Semiconductor, *User's Guide of MOS Controlled Thyristor.* With permission.)

FIGURE 14.8 Current and future power semiconductor devices development direction. (*Source:* A.Q. Huang, Recent Developments of Power Semiconductor Devices, VPEC Seminar Proceedings, pp. 1–9. With permission.)

drop and a rugged device, which is more likely to be used in the future for medium and high power applications. A cross-sectional structure of a *p*-type MCT with its circuit schematic is shown in Fig. 14.7. The MCT has a thyristor type structure with three junctions and PNPN layers between the anode and cathode. In a practical MCT, about 100,000 cells similar to the one shown are paralleled to achieve the desired current rating. MCT is turned on by a negative voltage pulse at the gate with respect to the anode, and is turned off by a positive voltage pulse.

The MCT was announced by the General Electric R & D Center on November 30, 1988. Harris Semiconductor Corporation has developed two generations of p-MCTs. Gen-1 p-MCTs are available at 65 A/1000 V and 75A/600 V with peak controllable current of 120 A. Gen-2 p-MCTs are being developed at similar current and voltage ratings, with much improved turn-on capability and switching speed. The reason for developing a p-MCT is the fact that the current density that can be turned off is 2 or 3 times higher than that of an n-MCT; but n-MCTs are the ones needed for many practical applications. Harris Semiconductor Corporation is in the process of developing n-MCTs, which are expected to be commercially available during the next one to two years.

The advantage of an MCT over IGBT is its low forward voltage drop. N-type MCTs will be expected to have a similar forward voltage drop, but with an improved reverse bias safe operating area and switching speed. MCTs have relatively low switching times and storage time. The MCT is capable of high current densities and blocking voltages in both directions. Since the power gain of an MCT is extremely high, it could be driven directly from logic gates. An MCT has high *di/dt* (of the order of 2500 A/μs) and high *dv/dt* (of the order of 20,000 V/μs) capability.

The MCT, because of its superior characteristics, shows a tremendous possibility for applications such as motor drives, uninterrupted power supplies, static VAR compensators, and high power active power line conditioners.

The current and future power semiconductor devices developmental direction is shown in Fig. 14.8. High-temperature operation capability and low forward voltage drop operation can be obtained if silicon is replaced by silicon carbide material for producing power devices. The silicon carbide has a higher band gap than silicon. Hence, higher breakdown voltage devices could be developed. Silicon carbide devices have excellent switching characteristics and stable blocking voltages at higher temperatures. But the silicon carbide devices are still in the very early stages of development.

References

B.K. Bose, *Modern Power Electronics: Evaluation, Technology, and Applications,* New York: IEEE Press, 1992.

Harris Semiconductor, *User's Guide of MOS Controlled Thyristor.*

A.Q. Huang, Recent Developments of Power Semiconductor Devices, in *VPEC Seminar Proceedings,* September 1995, 1–9.

N. Mohan and T. Undeland, *Power Electronics: Converters, Applications, and Design,* John Wiley & Sons, New York, 1995.

J. Wojslawowicz, Ruggedized transistors emerging as power MOSFET standard-bearers, *Power Technics Magazine,* January 1988, 29–32.

Further Information

B.M. Bird and K.G. King, *An Introduction to Power Electronics,* Wiley-Interscience, New York, 1984.

R. Sittig and P. Roggwiller, *Semiconductor Devices for Power Conditioning,* Plenum, New York, 1982.

V.A.K. Temple, Advances in MOS controlled thyristor technology and capability, *Power Conversion,* 544–554, Oct. 1989.

B.W. Williams, *Power Electronics, Devices, Drivers and Applications,* John Wiley, New York, 1987.

14.2 Uncontrolled and Controlled Rectifiers

Mahesh M. Swamy

Rectifiers are electronic circuits that convert bidirectional voltage to unidirectional voltage. This process can be accomplished either by mechanical means like in the case of DC machines employing commutators or by static means employing semiconductor devices. Static rectification is more efficient and reliable compared to rotating commutators. This section covers rectification of electric power for industrial and commercial use. In other words, we will not be discussing small signal rectification that generally involves low power and low voltage signals. Static power rectifiers can be classified into two broad groups. They are (1) uncontrolled rectifiers and (2) controlled rectifiers. Uncontrolled rectifiers make use of power semiconductor diodes while controlled rectifiers make use of thyristors (SCRs), gate turn-off thyristors (GTOs), and MOSFET-controlled thyristors (MCTs).

Rectifiers, in general, are widely used in power electronics to rectify single-phase as well as three-phase voltages. DC power supplies used in computers, consumer electronics, and a host of other applications typically make use of single-phase rectifiers. Industrial applications include, but are not limited to, industrial drives, metal extraction processes, industrial heating, power generation and transmission, etc. Most industrial applications of large power rating typically employ three-phase rectification processes.

Uncontrolled rectifiers in single-phase as well as in three-phase circuits will be discussed, as will controlled rectifiers. Application issues regarding uncontrolled and controlled rectifiers will be briefly discussed within each section.

Uncontrolled Rectifiers

The simplest uncontrolled rectifier use can be found in single-phase circuits. There are two types of uncontrolled rectification. They are (1) half-wave rectification and (2) full-wave rectification. Half-wave and full-wave rectification techniques have been used in single-phase as well as in three-phase circuits. As mentioned earlier, uncontrolled rectifiers make use of diodes. Diodes are two-terminal semiconductor devices that allow flow of current in only one direction. The two terminals of a diode are known as the anode and the cathode.

Mechanics of Diode Conduction

The anode is formed when a pure semiconductor material, typically silicon, is doped with impurities that have fewer valence electrons than silicon. Silicon has an atomic number of 14, which according to Bohr's atomic model means that the K and L shells are completely filled by 10 electrons and the remaining 4 electrons occupy the M shell. The M shell can hold a maximum of 18 electrons. In a silicon crystal, every atom is bound to four other atoms, which are placed at the corners of a regular tetrahedron. The bonding, which involves sharing of a valence electron with a neighboring atom is known as covalent bonding. When a Group 3 element (typically boron, aluminum, gallium, and indium) is doped into the silicon lattice structure, three of the four covalent bonds are made. However, one bonding site is vacant in the silicon lattice structure. This creates vacancies or *holes* in the semiconductor. In the presence of either a thermal field or an electrical field, electrons from a neighboring lattice or from an external agency tend to migrate to fill this vacancy. The vacancy or *hole* can also be said to move toward the approaching electron, thereby creating a mobile hole and hence current flow. Such a semiconductor material is also known as lightly doped semiconductor material or *p type*. Similarly, the cathode is formed when silicon is doped with impurities that have higher valence electrons than silicon. This would mean elements belonging to Group 5. Typical doping impurities of this group are phosphorus, arsenic, and antimony. When a Group 5 element is doped into the silicon lattice structure, it oversatisfies the covalent bonding sites available in the silicon lattice structure, creating excess or loose electrons in the valence shell. In the presence of either a thermal field or an electrical field, these loose electrons easily get detached from the lattice structure and are free to conduct electricity. Such a semiconductor material is also known as heavily doped semiconductor material or *n type*.

The structure of the final doped crystal even after the addition of *acceptor* impurities (Group 3) or *donor* impurities (Group 5), remains electrically neutral. The available electrons balance the net positive charge and there is no charge imbalance.

When a *p*-type material is joined with an *n*-type material, a *p-n* junction is formed. Some loose electrons from the *n*-type material migrate to fill the holes in the *p*-type material and some holes in the *p*-type migrate to meet with the loose electrons in the *n*-type material. Such a movement causes the *p*-type structure to develop a slight negative charge and the *n*-type structure to develop some positive charge. These slight positive and negative charges in the *n*-type and *p*-type areas, respectively, prevent further migration of electrons from *n*-type to *p*-type and holes from *p*-type to *n*-type areas. In other words, an energy barrier is automatically created due to the movement of charges within the crystalline lattice structure. Keep in mind that the combined material is still electrically neutral and no charge imbalance exists.

When a positive potential greater than the barrier potential is applied across the *p-n* junction, then electrons from the *n*-type area migrate to combine with the holes in the *p*-type area, and vice versa. The *p-n* junction is said to be *forward-biased*. Movement of charge particles constitutes current flow. Current is said to flow from the anode to the cathode when the potential at the anode is higher than the potential at the cathode by a minimum threshold voltage also known as the junction barrier voltage. The magnitude of current flow is high when the externally applied positive potential across the *p-n* junction is high.

When the polarity of the applied voltage across the *p-n* junction is reversed compared to the case described above, then the flow of current ceases. The holes in the *p*-type area move away from the *n*-type area and the electrons in the *n*-type area move away from the *p*-type area. The *p-n* junction is said to be *reverse-biased*. In fact, the holes in the *p*-type area get attracted to the negative external potential and similarly the electrons in the *n*-type area get attracted to the positive external potential. This creates a depletion region at the *p-n* junction and there are almost no charge carriers flowing in the depletion region. This phenomenon brings us to the important observation that a *p-n* junction can be utilized to force current to flow only in one direction, depending on the polarity of the applied voltage across it. Such a semiconductor device is known as a *diode*. Electrical circuits employing diodes for the purpose of making the current flow in a unidirectional manner through a load are known as *rectifiers*. The voltage-current characteristic of a typical power semiconductor diode along with its symbol is shown in Fig. 14.9.

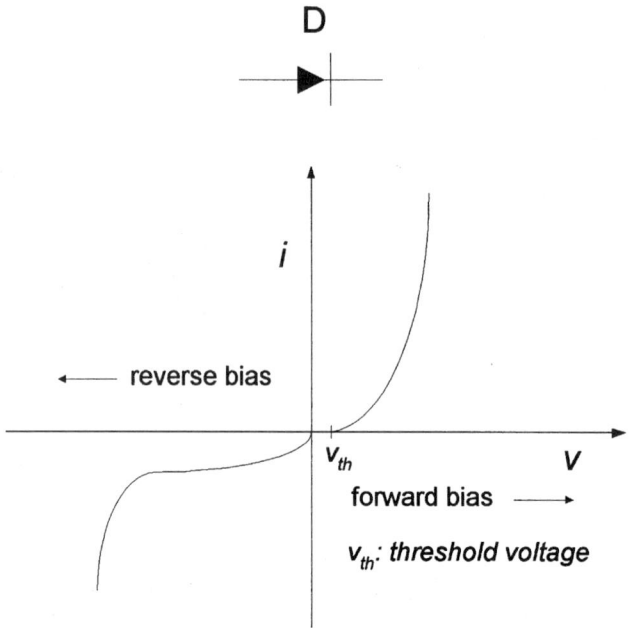

FIGURE 14.9 Typical v-i characteristic of a semiconductor diode and its symbol.

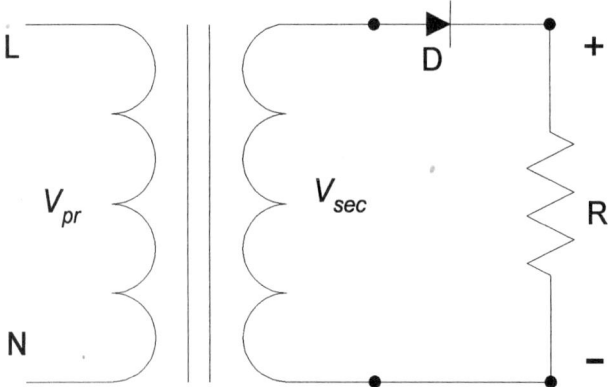

FIGURE 14.10 Electrical schematic of a single-phase half-wave rectifier circuit feeding a resistive load. Average output voltage is V_o.

Single-Phase Half-Wave Rectifier Circuits

A single-phase half-wave rectifier circuit employs one diode. A typical circuit, which makes use of a half-wave rectifier, is shown in Fig. 14.10.

Power Electronics   14-11

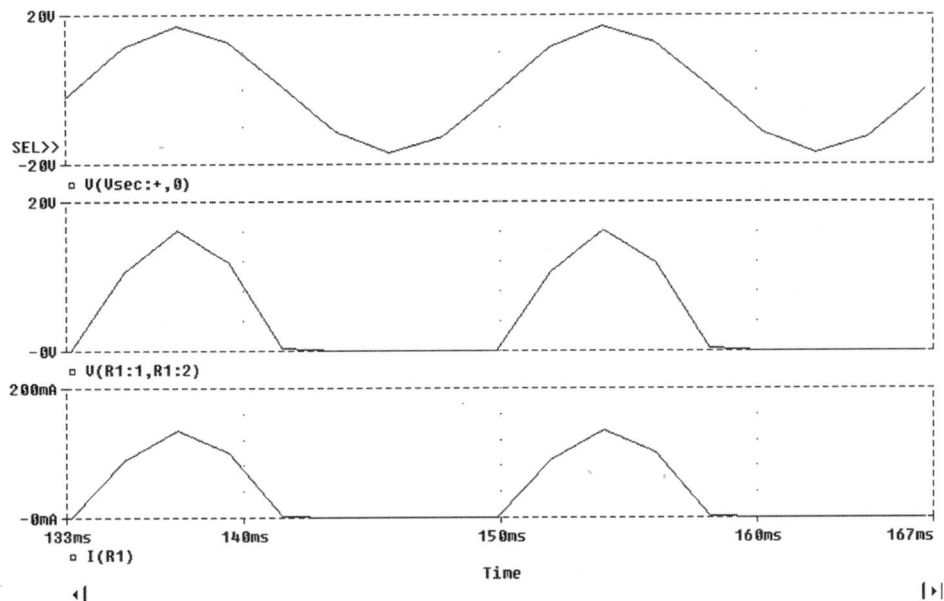

FIGURE 14.11 Typical waveforms at various points in the circuit of Fig. 14.10. For a purely resistive load, $V_o = \sqrt{2} * V_{sec}/\pi$.

A single-phase AC source is applied across the primary windings of a transformer. The secondary of the transformer consists of a diode and a resistive load. This is typical since many consumer electronic items including computers utilize single-phase power.

Typically, the primary side is connected to a single-phase AC source, which could be 120 V, 60 Hz, 100 V, 50 Hz, 220 V, 50 Hz, or any other utility source. The secondary side voltage is generally stepped down and rectified to achieve low DC voltage for consumer applications. The secondary voltage, the voltage across the load resistor, and the current through it is shown in Fig. 14.11.

As one can see, when the voltage across the anode-cathode of diode D1 in Fig. 14.10 goes negative, the diode does not conduct and no voltage appears across the load resistor R. The current through R follows the voltage across it. The value of the secondary voltage is chosen to be 12 VAC and the value of R is chosen to be 120 Ω. Since, only one-half of the input voltage waveform is allowed to pass onto the output, such a rectifier is known as a *half-wave rectifier*. The voltage ripple across the load resistor is rather large and, in typical power supplies, such ripples are unacceptable. The current through the load is discontinuous and the current through the secondary of the transformer is unidirectional. The AC component in the secondary of the transformer is balanced by a corresponding AC component in the primary winding. However, the DC component in the secondary does not induce any voltage on the primary side and hence is not compensated for. This DC current component through the transformer secondary can cause the transformer to saturate and is not advisable for large power applications. In order to smooth the output voltage across the load resistor R and to make the load current continuous, a smoothing filter circuit comprised of either a large DC capacitor or a combination of a series inductor and shunt DC capacitor is employed. Such a circuit is shown in Fig. 14.12.

The resulting waveforms are shown in Fig. 14.13. It is interesting to see that the voltage across the load resistor has very little ripple and the current through it is smooth. However, the value of the filter components employed is large and is generally not economically feasible. For example, in order to get a voltage waveform across the load resistor R, which has less than 6% peak-peak voltage ripple, the value of inductance that had to be used is 100 mH and the value of the capacitor is 1000 μF. In order to improve the performance without adding bulky filter components, it is a good practice to employ full-wave

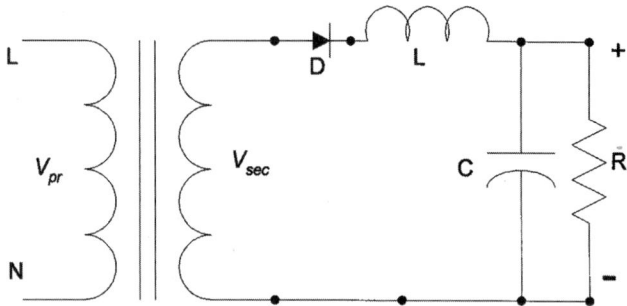

FIGURE 14.12 Modified circuit of Fig. 14.10 employing smoothing filters.

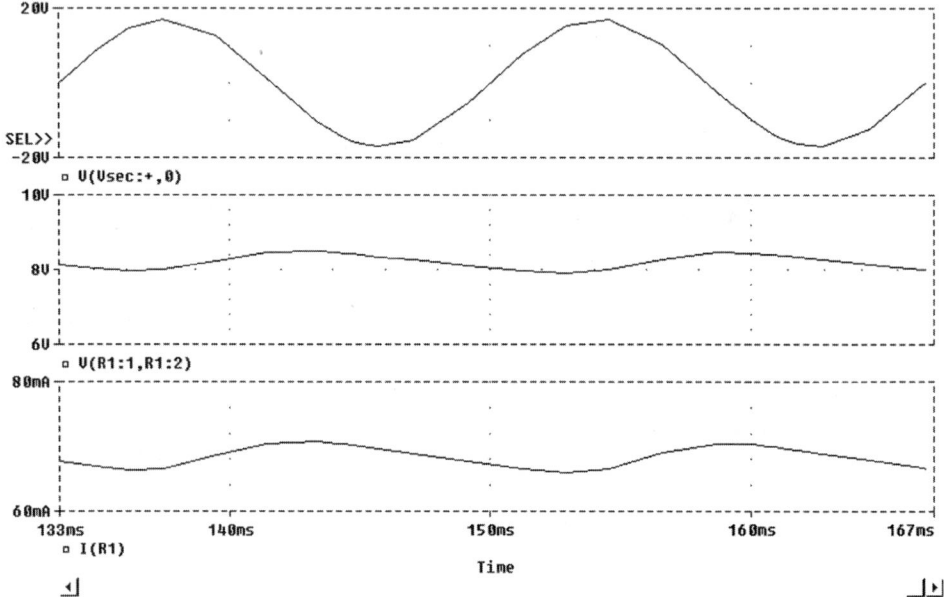

FIGURE 14.13 Voltage across load resistor R and current through it for the circuit in Fig. 14.12.

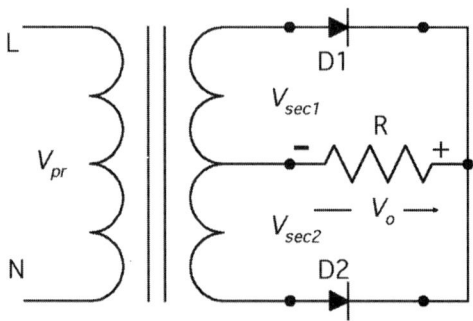

FIGURE 14.14 Electrical schematic of a single-phase full-wave rectifier circuit. Average output voltage is V_o.

rectifiers. The circuit in Fig. 14.10 can be easily modified into a full-wave rectifier. The transformer is changed from a single secondary winding to a center-tapped secondary winding. Two diodes are now employed instead of one. The new circuit is shown in Fig. 14.14.

Power Electronics

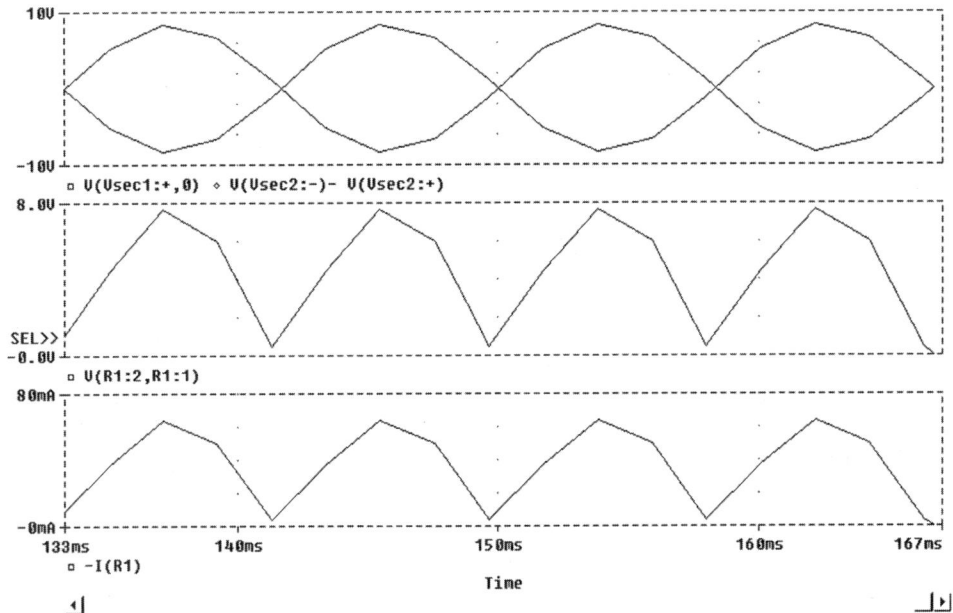

FIGURE 14.15 Typical waveforms at various points in the circuit of Fig. 14.14. For a purely resistive load, $V_o = 2 \ast \sqrt{2} \ast V_{sec}/\pi$.

Full Wave Rectifiers

The waveforms for the circuit of Fig. 14.14 are shown in Fig. 14.15. The voltage across the load resistor is a full-wave rectified voltage. The current has subtle discontinuities but can be improved by employing smaller size filter components. A typical filter for the circuit of Fig. 14.14 may include only a capacitor. The waveforms obtained are shown in Fig. 14.16.

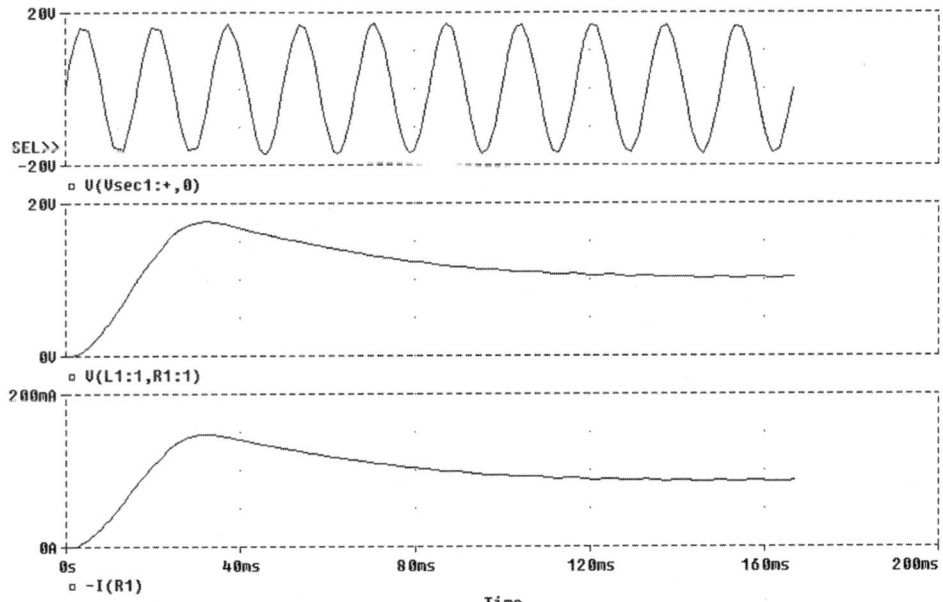

FIGURE 14.16 Voltage across the load resistor and current through it with the same filter components as in Fig. 14.12. Notice the conspicuous reduction in ripple across R.

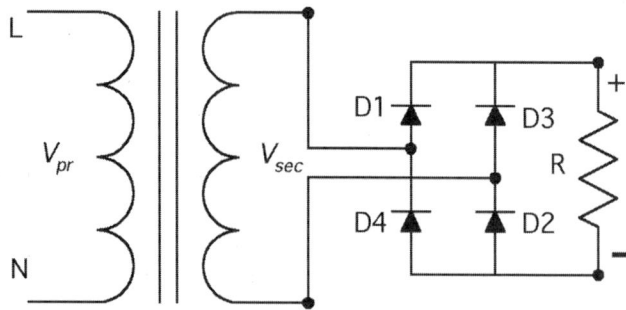

FIGURE 14.17 Schematic representation of a single-phase full-wave H-bridge rectifier.

Yet another way of reducing the size of the filter components is to increase the frequency of the supply. In many power supply applications similar to the one used in computers, a high frequency AC supply is achieved by means of switching. The high frequency AC is then level translated via a ferrite core transformer with multiple secondary windings. The secondary voltages are then rectified employing a simple circuit as shown in Fig. 14.10 or Fig. 14.12 with much smaller filters. The resulting voltage across the load resistor is then maintained to have a peak-peak voltage ripple of less than 1%.

Full-wave rectification can be achieved without the use of center-tap transformers. Such circuits make use of 4 diodes in single-phase circuits and 6 diodes in three-phase circuits. The circuit configuration is typically referred to as the *H-bridge* circuit. A single-phase full-wave H-bridge topology is shown in Fig. 14.17. The main difference between the circuit topology shown in Figs. 14.14 and 14.17 is that the H-bridge circuit employs 4 diodes while the topology of Fig. 14.14 utilizes only two diodes. However, a center-tap transformer of a higher power rating is needed for the circuit of Fig. 14.14. The voltage and current stresses in the diodes in Fig. 14.14 are also greater than that occurring in the diodes of Fig. 14.17.

In order to comprehend the basic difference in the two topologies, it is interesting to compare the component ratings for the same power output. To make the comparison easy, let both topologies employ very large filter inductors such that the current through R is constant and ripple-free. Let this current through R be denoted by I_{dc}. Let the power being supplied to the load be denoted by P_{dc}. The output power and the load current are then related by the following expression:

$$P_{dc} = I_{dc}^2 * R.$$

The rms current flowing through the first secondary winding in the topology in Fig. 14.14 will be $I_{dc}/\sqrt{2}$. This is because the current through a secondary winding flows only when the corresponding diode is forward biased. This means that the current through the secondary winding will flow only for one half cycle. If the voltage at the secondary is assumed to be V, the VA rating of the secondary winding of the transformer in Fig. 14.14 will be given by:

$$VA_1 = V * I_{dc}/\sqrt{2}$$

$$VA_2 = V * I_{dc}/\sqrt{2}$$

$$VA = VA_1 + VA_2 = \sqrt{2} * V * I_{dc}$$

This is the secondary-side VA rating for the transformer shown in Fig. 14.14.

For the isolation transformer shown in Fig. 14.17, let the secondary voltage be V and the load current be of a constant value I_{dc}. Since, in the topology of Fig. 14.17, the secondary winding carries the current I_{dc} when diodes D1 and D2 conduct and as well as when diodes D3 and D4 conduct, the rms value of the secondary winding current is I_{dc}. Hence, the VA rating of the secondary winding of the transformer shown in Fig. 14.17 is V^*I_{dc}, which is less than that needed in the topology of Fig. 14.14. Note that the primary VA rating for both cases remains the same since in both cases the power being transferred from the source to the load remains the same.

When diode D2 in the circuit of Fig. 14.14 conducts, the secondary voltage of the second winding V_{sec2} (=V) appears at the cathode of diode D1. The voltage being blocked by diode D1 can thus reach 2 times the peak secondary voltage (=2^*V_{pk}) (Fig. 14.15). In the topology of Fig. 14.17, when diodes D1 and D2 conduct, the voltage V_{sec} (=V), which is same as V_{sec2} appears across D3 as well as across D4. This means that the diodes have to withstand only one times the peak of the secondary voltage, V_{pk}. The rms value of the current flowing through the diodes in both topologies is the same. Hence, from the diode voltage rating as well as from the secondary VA rating points of view, the topology of Fig. 14.17 is better than that of Fig. 14.14. Further, the topology in Fig. 14.17 can be directly connected to a single-phase AC source and does not need a center-topped transformer. The voltage waveform across the load resistor is similar to that shown in Figs. 14.15 and 14.16.

In many industrial applications, the topology shown in Fig. 14.17 is used along with a DC filter capacitor to smooth the ripples across the load resistor. The load resistor is simply a representative of a load. It could be an inverter system or a high-frequency resonant link. In any case, the diode rectifier-bridge would see a representative load resistor. The DC filter capacitor will be large in size compared to an H-bridge configuration based on three-phase supply system. When the rectified power is large, it is advisable to add a DC link inductor. This can reduce the size of the capacitor to some extent and reduce the current ripple through the load. When the rectifier is turned on initially with the capacitor at zero voltage, a large amplitude of charging current will flow into the filter capacitor through a pair of conducting diodes. The diodes D1~D4 should be rated to handle this large surge current. In order to limit the high inrush current, it is a normal practice to add a charging resistor in series with the filter capacitor. The charging resistor limits the inrush current but creates a significant power loss if it is left in the circuit under normal operation. Typically, a contactor is used to short-circuit the charging resistor after the capacitor is charged to a desired level. The resistor is thus electrically nonfunctional during normal operating conditions. A typical arrangement showing a single-phase full-wave H-bridge rectifier system for an inverter application is shown in Fig. 14.18.

The charging current at time of turn-on is shown in a simulated waveform in Fig. 14.19. Note that the contacts across the soft-charge resistor are closed under normal operation. The contacts across the soft-charge resistor are initiated by various means. The coil for the contacts could be powered from the input AC supply and a timer or it could be powered on by a logic controller that senses the level of voltage across the DC bus capacitor or senses the rate of change in voltage across the DC bus capacitor. A simulated waveform depicting the inrush with and without a soft-charge resistor is shown in Figs. 14.19(a) and (b), respectively.

For larger power applications, typically above 1.5 kW, it is advisable to use a higher power supply. In some applications, two of the three phases of a three-phase power system are used as the source powering

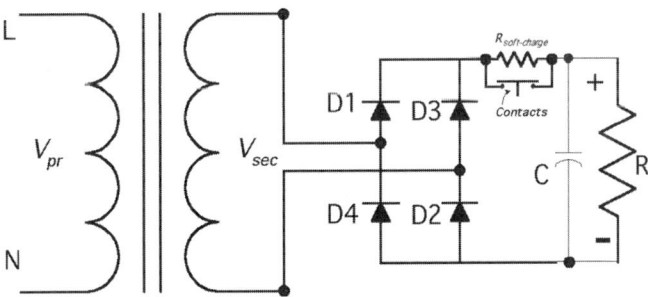

FIGURE 14.18 Single-phase H-bridge circuit for use with power electronic circuits.

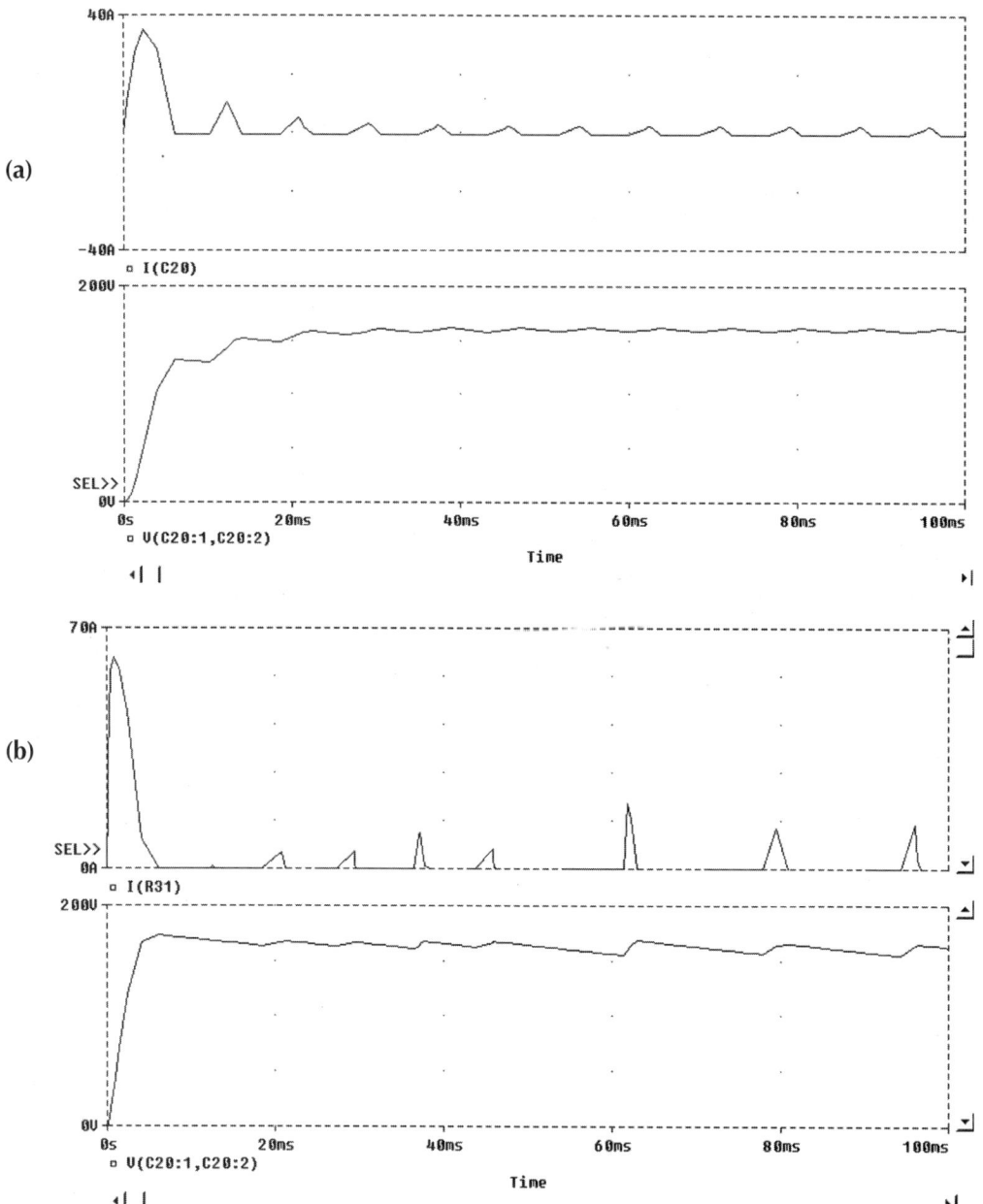

FIGURE 14.19 (a) Charging current and voltage across capacitor for a typical value of soft-charge resistor of 2Ω. The DC bus capacitor is about 1000 µF. The load is approximately 200 Ω. (b) Charging current and voltage across capacitor for no soft charge resistor. The current is limited by the system impedance and by the diode forward resistance. The DC bus capacitor is about 1000 µF. The load is approximately 200 Ω.

the rectifier of Fig. 14.17. The line-line voltage could be either 240 VAC or 480 VAC. Under those circumstances, one may go up to 10 kW of load power before adopting a full three-phase H-bridge configuration. Beyond 10 kW, the size of the capacitor becomes too large to achieve a peak-peak voltage ripple of less than 5%. Hence, it is advisable then to employ three-phase rectifier configurations.

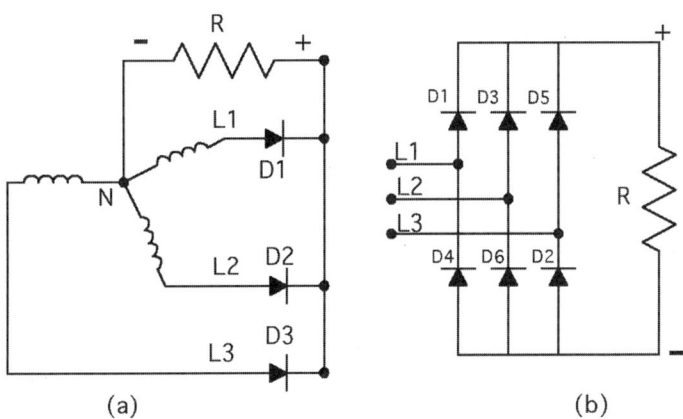

FIGURE 14.20 Schematic representation of three-phase rectifier configurations: (a) half-wave rectifier needing a neutral point, N; and (b) full-wave rectifier.

Three-Phase Rectifiers (Half-Wave and Full-Wave)

Similar to the single-phase case, there exist half-wave and full-wave three-phase rectifier circuits. Again, similar to the single-phase case, the half-wave rectifier in the three-phase case also yields DC components in the source current. The source has to be large enough to handle this. Therefore, it is not advisable to use three-phase half-wave rectifier topology for large power applications. The three-phase half-wave rectifier employs three diodes while the full-wave H-bridge configuration employs six diodes. Typical three-phase half-wave and full-wave topologies are shown in Fig. 14.20.

In the half-wave rectifier shown in Fig. 14.20(a), the shape of the output voltage and current through the resistive load is dictated by the instantaneous value of the source voltages, L1, L2, and L3. These source voltages are phase shifted in time by 120 electrical degrees, which corresponds to approximately 5.55 msec for a 60 Hz system. This means that if one considers the L1 phase to reach its peak value at time t_1, the L2 phase will achieve its peak 120 electrical degrees later (t_1 + 5.55 msec), and L3 will achieve its peak 120 electrical degrees later than L2 (t_1 + 5.55 msec + 5.55 msec). Since all three phases are connected to the same output resistor R, the phase that provides the highest instantaneous voltage is the phase that appears across R. In other words, the phase with the highest instantaneous voltage reverse biases the diodes of the other two phases and prevents them from conducting, which consequently prevents those phase voltages from appearing across R. Since a particular phase is connected to only one diode in Fig. 14.20(a), only three pulses, each of 120° duration, appear across the load resistor, R. Typical output voltage across R for the circuit of Fig. 14.20(a) is shown in Fig. 14.21(a).

A similar explanation can be provided to explain the voltage waveform across a purely resistive load in the case of the three-phase full-wave rectifier shown in Fig. 14.20(b). The output voltage that appears across R is the highest instantaneous line-line voltage and not simply the phase voltage. Since there are six such intervals, each of 60 electrical degrees duration in a given cycle, the output voltage waveform will have six pulses in one cycle [Fig. 14.21(b)]. Since a phase is connected to two diodes (diode pair), each phase conducts current out and into itself, thereby eliminating the DC component in one complete cycle.

The waveform for a three-phase full-wave rectifier with a purely resistive load is shown in Fig. 14.21(b). Note that the number of humps in Fig. 14.21(a) is only three in one AC cycle, while the number of humps in Fig. 14.21(b) is six in one AC cycle.

In both the configurations shown in Fig. 14.20, the load current does not become discontinuous due to three-phase operation. Comparing this to the single-phase half-wave and full-wave rectifier, one can say that the output voltage ripple is much lower in three-phase rectifier systems compared to single-phase rectifier systems. Hence, with the use of moderately sized filters, three-phase full-wave rectifiers

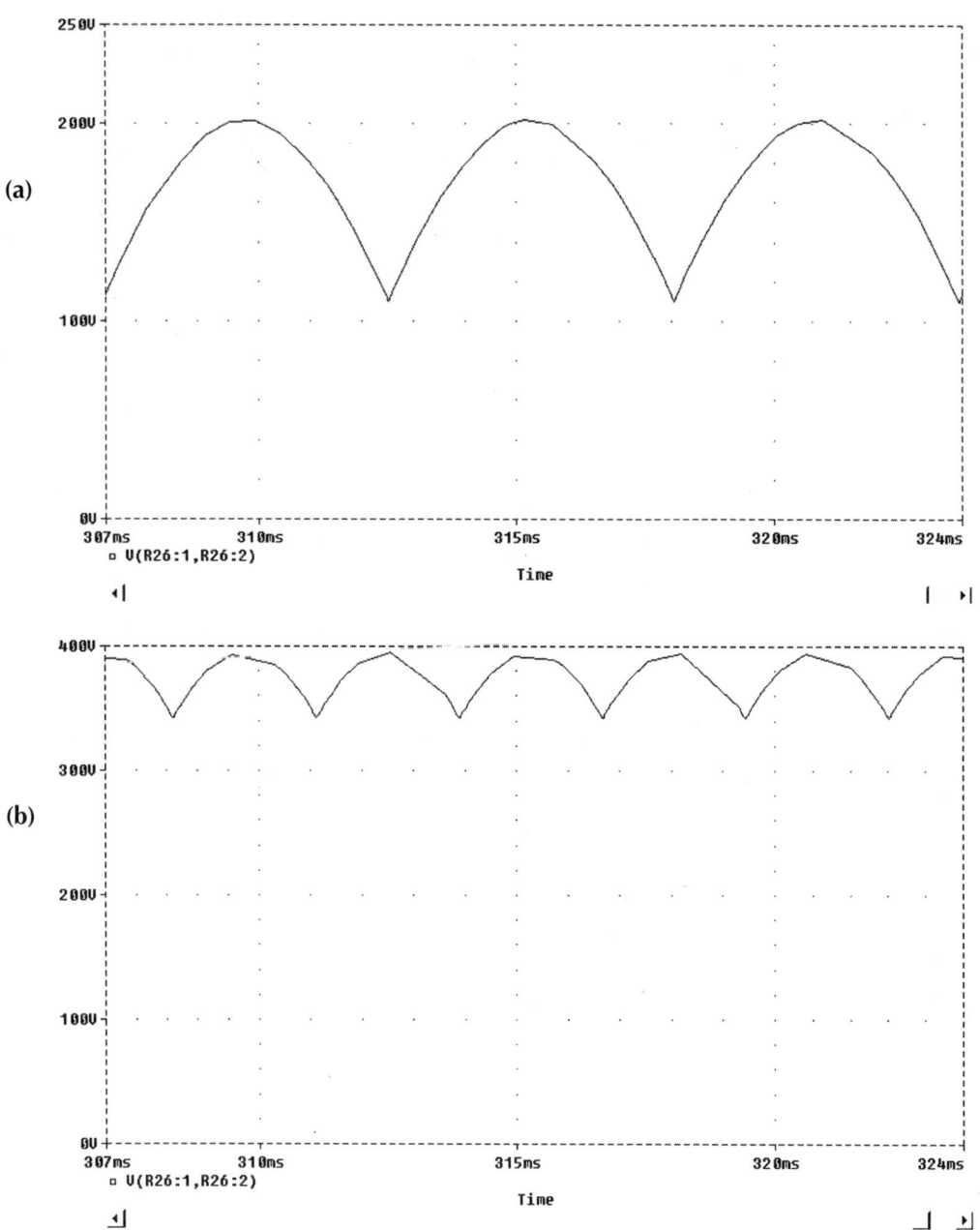

FIGURE 14.21 (a) Typical output voltage across a purely resistive network for the half-wave rectifier shown in Fig. 14.12(a). (b) Typical output voltage across a purely resistive network for the full-wave rectifier shown in Fig. 14.12(b).

can be operated at hundred to thousands of kilowatts. The only limitation would be the size of the diodes used and power system harmonics, which will be discussed next. Since there are six humps in the output voltage waveform per electrical cycle, the three-phase full-wave rectifier shown in Fig. 14.20(b) is also known as a six-pulse rectifier system.

Average Output Voltage

In order to evaluate the average value of the output voltage for the two rectifiers shown in Fig. 14.20, the output voltages in Figs. 14.21(a) and (b) have to be integrated over a cycle. For the circuit shown in Fig. 14.20(a), the integration yields the following:

$$V_o = \frac{3}{2\pi} \int_{\pi/6}^{5\pi/6} \sqrt{2} V_{L-N} \sin(wt) d(wt)$$

$$V_o = \frac{3 * \sqrt{3} * \sqrt{2} * V_{L-N}}{2 * \pi}$$

Similar operations can be performed to obtain the average output voltage for the circuit shown in Fig. 14.20(b). This yields:

$$V_o = \frac{3}{\pi} \int_{\pi/3}^{2\pi/3} \sqrt{2} V_{L-L} \sin(wt) d(wt)$$

$$V_o = \frac{3 * \sqrt{2} * V_{L-L}}{\pi} = \frac{3 * \sqrt{2} * \sqrt{3} * V_{L-N}}{\pi}$$

In other words, the average output voltage for the circuit in Fig. 14.20(b) is twice that for the circuit in Fig. 14.20(a).

Influence of Three-Phase Rectification on the Power System

Events over the last several years have focused attention on certain types of loads on the electrical system that result in power quality problems for the user and utility alike. When the input current into the electrical equipment does not follow the impressed voltage across the equipment, then the equipment is said to have a nonlinear relationship between the input voltage and input current. All equipment that employs some sort of rectification (either 1-ph or 3-ph) are examples of nonlinear loads. Nonlinear loads generate voltage and current harmonics that can have adverse effects on equipment designed for operation as linear loads. Transformers that bring power into an industrial environment are subject to higher heating losses due to harmonic generating sources (nonlinear loads) to which they are connected. Harmonics can have a detrimental effect on emergency generators, telephones, and other electrical equipment. When reactive power compensation (in the form of passive power factor improving capacitors) is used with nonlinear loads, resonance conditions can occur that may result in even higher levels of harmonic voltage and current distortion, thereby causing equipment failure, disruption of power service, and fire hazards in extreme conditions.

The electrical environment has absorbed most of these problems in the past. However, the problem has now reached a magnitude where Europe, the U.S., and other countries have proposed standards to responsibly engineer systems considering the electrical environment. IEEE 519-1992 and IEC 1000 have evolved to become a common requirement cited when specifying equipment on newly engineered projects.

Why Diode Rectifiers Generate Harmonics

The current waveform at the inputs of a three-phase full-wave rectifier is not continuous. It has multiple zero crossings in one electrical cycle. The current harmonics generated by rectifiers having DC bus capacitors are caused by the pulsed current pattern at the input. The DC bus capacitor draws charging current only when it gets discharged due to the load. The charging current flows into the capacitor when the input rectifier is forward biased, which occurs when the instantaneous input voltage is higher than

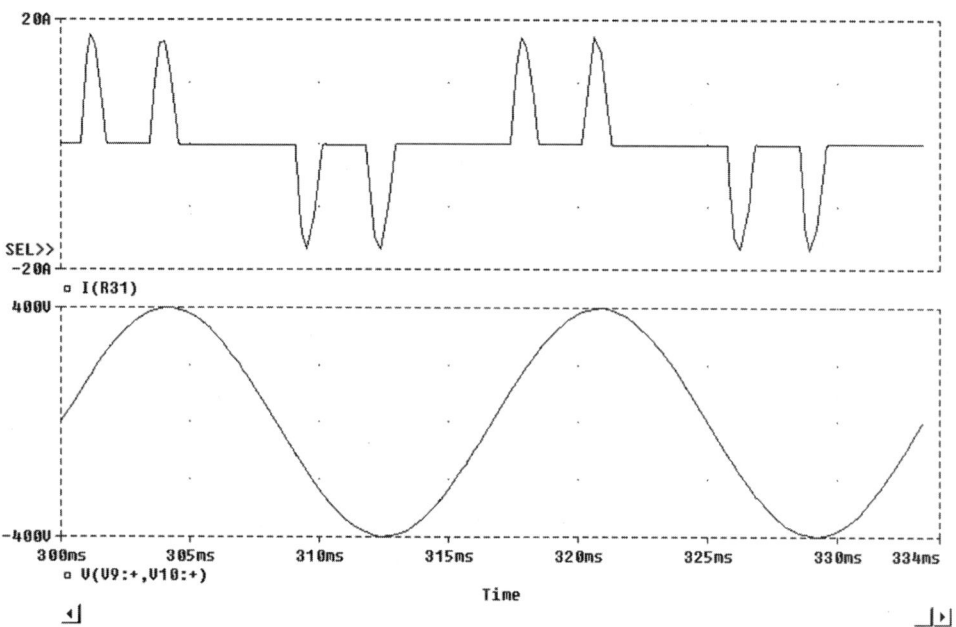

FIGURE 14.22 Typical pulsed current waveform as seen at input of a three-phase diode rectifier with DC capacitor filter. The lower trace is input line-line voltage.

the steady-state DC voltage across the DC bus capacitor. The pulsed current drawn by the DC bus capacitor is rich in harmonics due to the fact that it is discontinuous as shown in Fig. 14.22. Sometimes there are also voltage harmonics that are associated with three-phase rectifier systems. The voltage harmonics generated by three-phase rectifiers are due to the flat-topping effect caused by a weak AC source charging the DC bus capacitor without any intervening impedance. The distorted voltage waveform gives rise to voltage harmonics that could lead to possible network resonance.

The order of current harmonics produced by a semiconductor converter during normal operation is termed characteristic harmonics. In a three-phase, six-pulse rectifier with **no DC bus capacitor**, the characteristic harmonics are nontriplen odd harmonics (e.g., 5th, 7th, 11th, etc.). In general, the characteristic harmonics generated by a semiconductor recitifier are given by:

$$h = kq \pm 1$$

where h is the order of harmonics; k is any integer, and q is the pulse number of the semiconductor rectifier (six for a six-pulse rectifier). When operating a six-pulse rectifier system with a DC bus capacitor (as in voltage source inverters, or VSI), one may start observing harmonics of orders other than those given by the above equation. Such harmonics are called *noncharacteristic* harmonics. Though of lower magnitude, these also contribute to the overall harmonic distortion of the input current. The per-unit value of the characteristic harmonics present in the theoretical current waveform at the input of the semiconductor converter is given by $1/h$, where h is the order of the harmonics. In practice, the observed per-unit value of the harmonics is much greater than $1/h$. This is because the theoretical current waveform is a rectangular pattern made up of equal positive and negative halves, each occupying 120 electrical degrees. The pulsed discontinuous waveform observed commonly at the input of a three-phase full-wave rectifier system depends greatly on the impedance of the power system, the size of the DC bus capacitors, and the level of loading of the DC bus capacitors. Total harmonic current distortion is defined as:

Power Electronics

$$THD_I = \frac{\sqrt{\sum_{n=2}^{n=\infty} I_n^2}}{I_1}$$

where I_1 is the rms value of the fundamental component of current; and I_n is the rms value of the nth harmonic component of current.

Harmonic Limits Based on IEEE Std. 519-1992

The IEEE Std. 519-1992 relies strongly on the definition of the point of common coupling or PCC. The PCC from the utility viewpoint will usually be the point where power comes into the establishment (i.e., point of metering). However, IEEE Std. 519-1992 also suggests that **"within an industrial plant, the point of common coupling (PCC) is the point between the nonlinear load and other loads"** (IEEE Std. 519-1992). This suggestion is crucial since many plant managers and building supervisors feel that it is equally, if not more important to keep the harmonic levels at or below acceptable guidelines within their facility. In view of the many recently reported problems associated with harmonics within industrial plants, it is important to recognize the need for mitigating harmonics at the point where the offending equipment is connected to the power system. This approach would minimize harmonic problems, thereby reducing costly downtime and improving the life of electrical equipment. If one is successful in mitigating individual load current harmonics, then the total harmonics at the point of the utility connection will in most cases meet or exceed the IEEE recommended guidelines. In view of this, it is becoming increasingly common for specifiers to require nonlinear equipment suppliers to adopt the procedure outlined in IEEE Std. 519-1992 to mitigate the harmonics to acceptable levels at the point of the offending equipment. For this to be interpreted equally by different suppliers, the intended PCC must be identified. If the PCC is not defined clearly, many suppliers of offending equipment would likely adopt the PCC at the utility metering point, which would not benefit the plant or the building, but rather the utility.

Having established that it is beneficial to adopt the PCC to be the point where the nonlinear equipment connects to the power system, the next step is to establish the short circuit ratio. Short circuit ratio calculations are key in establishing the allowable current harmonic distortion levels. For calculating the short circuit ratio, one has to determine the available short circuit current at the input terminals of the nonlinear equipment. The short-circuit current available at the input of nonlinear equipment can be calculated by knowing the value of the short-circuit current available at the secondary of the utility transformer supplying power to the establishment (building) and the series impedance in the electrical circuit between the secondary of the transformer and the nonlinear equipment. **In practice, it is common to assume the same short circuit current level as at the secondary of the utility transformer feeding the nonlinear equipment.** The next step is to compute the fundamental value of the rated input current into the nonlinear equipment (three-phase full-wave rectifier in this case). An example is presented here to recap the above procedure. A widely used industrial equipment item that employs a three-phase full-wave rectifier is the voltage source inverter (VSI). These are used for controlling speed and torque of induction motors. Such equipment is also known as an Adjustable Speed Drive (ASD) or Variable Frequency Drive (VFD).

A 100-hp ASD/motor combination connected to a 480-V system being fed from a 1500-kVA, three-phase transformer with impedance of 4% is required to meet IEEE Std. 519-1992 at its input terminals. The rated current of the transformer is 1500*1000/($\sqrt{}$(3)*480), and is calculated to be 1804.2 A. The short circuit current available at the secondary of the transformer is equal to the rated current divided by the per unit impedance of the transformer. This is calculated to be: 45,105.5 A. The short circuit ratio, which is defined as the ratio of the short circuit current at the PCC to the fundamental value of the nonlinear current is computed next. NEC amps for 100-hp, 460-V is 124 A. Assuming that the short circuit current at the ASD input is practically the same as that at the secondary of the utility transformer, the short-circuit

TABLE 14.1 Current Distortion Limits for General Distribution Systems

	(120 V through 69,000 V) Maximum Harmonic Current Distortion in percent of I_L Individual Harmonic Order (Odd Harmonics)[a]					
I_{sc}/I_L	<11	$11 \leq h \leq 17$	$17 \leq h \leq 23$	$23 \leq h \leq 35$	$35 \leq h$	TDD[b]
<20[c]	4.0	2.0	1.5	0.6	0.3	5.0
20 < 50	7.0	3.5	2.5	1.0	0.5	8.0
50 < 100	10.0	4.5	4.0	1.5	0.7	12.0
100 < 1000	12.0	5.5	5.0	2.0	1.0	15.0
>1000	15.0	7.0	6.0	2.5	1.4	20.0

[a] Even harmonics are limited to 25% of the odd harmonic limits above.
[b] TDD is Total Demand Distortion and is defined as the harmonic current distortion in % of maximum demand load current. The maximum demand current could either be a 15-minute or a 30-minute demand interval.
[c] All power generation equipment is limited to these values of current distortion, regardless of actual I_{sc}/I_L; where I_{sc} is the maximum short circuit current at PCC and I_L is the maximum demand load current (fundamental frequency) at PCC.
Source: IEEE Std. 519-1992.

ratio is calculated to be: 45,105.5/124, which equals 363.75. On referring to IEEE Std. 519-1992, Table 10.3 (IEEE Std. 519-1992), the short circuit ratio falls in the 100–1000 category. For this ratio, the total demand distortion (TDD) at the point of ASD connection to the power system network is recommended to be 15% or less. For reference, see Table 14.1.

Harmonic Mitigating Techniques

Various techniques of improving the input current waveform are discussed below. The intent of all techniques is to make the input current more continuous so as to reduce the overall current harmonic distortion. The different techniques can be classified into four broad categories:

1. Introduction of line reactors and/or DC link chokes
2. Passive filters (series, shunt, and low pass broadband filters)
3. Phase multiplication (12-pulse, 18-pulse rectifier systems)
4. Active harmonic compensation

The following paragraphs will briefly discuss the available technologies and their relative advantages and disadvantages. The term three-phase line reactor or just reactor is used in the following paragraphs to denote three-phase line inductors.

Three-Phase Line Reactors

Line reactors offer a significant magnitude of inductance that can alter the way the current is drawn by a nonlinear load such as a rectifier bridge. The reactor makes the current waveform less discontinuous, resulting in lower current harmonics. Since the reactor impedance increases with frequency, it offers larger impedance to the flow of higher order harmonic currents. Therefore, it is instrumental in impeding higher frequency current components while allowing the fundamental frequency component to pass through with relative ease.

On knowing the input reactance value, one can estimate the expected current harmonic distortion. A table illustrating the typically expected input current harmonics for various amounts of input reactance is shown in Table 14.2.

Input reactance is determined by the accumulated impedance of the AC reactor, DC link choke (if used), input transformer, and cable impedance. To maximize the input reactance while minimizing AC voltage drop, one can combine the use of both AC-input reactors and DC link chokes. One can approximate the total effective reactance and view the expected harmonic current distortion from Table 14.2. The effective impedance value in percent is based on the actual loading and is:

TABLE 14.2 Percent Harmonics vs. Total Line Impedance

Harmonic	Total Input Impedance							
	3%	4%	5%	6%	7%	8%	9%	10%
5th	40	34	32	30	28	26	24	23
7th	16	13	12	11	10	9	8.3	7.5
11th	7.3	6.3	5.8	5.2	5	4.3	4.2	4
13th	4.9	4.2	3.9	3.6	3.3	3.15	3	2.8
17th	3	2.4	2.2	2.1	0.9	0.7	0.5	0.4
19th	2.2	2	0.8	0.7	0.4	0.3	0.25	0.2
%THID	44	37	35	33	30	28	26	25
True rms	1.09	1.07	1.06	1.05	1.05	1.04	1.03	1.03

$$Z_{eff} = \frac{\sqrt{3}*2*\pi*f*L*I_{act(fnd.)}}{V_{L-L}}*100$$

where $I_{act(fnd.)}$ is the fundamental value of the actual load current and V_{L-L} is the line-line voltage. The effective impedance of the transformer as seen from the nonlinear load is:

$$Z_{eff,\,x-mer} = \frac{Z_{eff,\,x-mer}*I_{act(fnd.)}}{I_r}$$

where $Z_{eff,x-mer}$ is the effective impedance of the transformer as viewed from the nonlinear load end; Z_{x-mer} is the nameplate impedance of the transformer; and I_r is the nameplate rated current of the transformer.

On observing one conducting period of a diode pair, it is interesting to see that the diodes conduct only when the instantaneous value of the input AC waveform is higher than the DC bus voltage by at least 3 V. Introducing a three-phase AC reactor in between the AC source and the DC bus makes the current waveform less pulsating because the reactor impedes sudden change in current. The reactor also electrically differentiates the DC bus voltage from the AC source so that the AC source is not clamped to the DC bus voltage during diode conduction. This feature practically eliminates flat topping of the AC voltage waveform caused by many ASDs when operated with weak AC systems.

DC Link Choke
Based on the above discussion, it can be noted that any inductor of adequate value placed between the AC source and the DC bus capacitor of the ASD will help in improving the current waveform. These observations lead to the introduction of a DC link choke, which is electrically present after the diode rectifier and before the DC bus capacitor. The DC link choke performs very similar to the three-phase line inductance. The ripple frequency that the DC link choke has to handle is six times the input AC frequency for a six-pulse ASD. However, the magnitude of the ripple current is small. One can show that the effective impedance offered by a DC link choke is approximately half of that offered by a three-phase AC inductor. In other words, a 6% DC link choke is equivalent to a 3% AC inductor from an impedance viewpoint. This can be mathematically derived equating AC side power flow to DC side power flow as follows:

$$P_{ac} = \frac{3*V_{L-N}^2}{R_{ac}}; \quad P_{ac} = P_{dc}$$

V_{L-N} is the line-neutral voltage at the input to the three-phase rectifier.

$$P_{dc} = \frac{V_{dc}^2}{R_{dc}}; \quad V_{dc} = \frac{3*\sqrt{3}*\sqrt{2}*V_{L-N}}{\pi}; \quad \text{Hence, } R_{dc} = 2*\left(\frac{9}{\pi^2}\right)R_{ac}$$

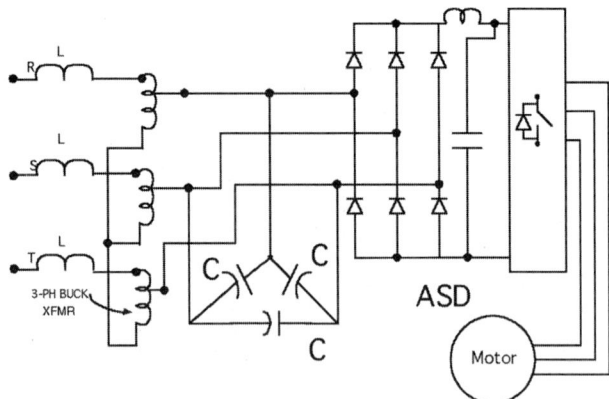

FIGURE 14.23 Schematic representation of a low-pass broadband harmonic filter connected to an ASD with diode rectifier front end. U.S. Patent 5,444,609.

Since $9/\pi^2$ is approximately equal to 1, the ratio of DC impedance to AC impedance can be said to be approximately 1:2. The DC link choke is less expensive and smaller than a three-phase line reactor and is often included inside an ASD. However, as the derivation shows, one has to keep in mind that the effective impedance offered by a DC link choke is only half its numerical impedance value when referred to the AC side. DC link chokes are electrically after the diode bridge and so they do not offer any significant spike or overvoltage surge protection to the diode bridge rectifiers. It is a good engineering practice to incorporate both a DC link choke and a three-phase line reactor in an ASD for better overall performance.

Passive Filters

Passive filters consist of passive components like inductors, capacitors, and resistors arranged in a predetermined fashion either to attenuate the flow of harmonic components through them or to shunt the harmonic component into them. Passive filters can be of many types. Some popular ones are series passive filters, shunt passive filters, and low-pass broadband passive filters. Series and shunt passive filters are effective only in the narrow proximity of the frequency at which they are tuned. Low-pass broadband passive filters have a broader bandwidth and attenuate almost all harmonics above their cutoff frequency. However, applying passive filters requires good knowledge of the power system because passive filter components can interact with existing transformers and power factor correcting capacitors and could create electrical instability by introducing resonance into the system. Some forms of low-pass broadband passive filters do not contribute to resonance but they are bulky, expensive, and occupy space. A typical low-pass broadband filter structure popularly employed by users of ASDs is shown in Fig. 14.23.

Phase Multiplication

As discussed previously, the characteristic harmonics generated by a full-wave rectifier bridge converter is a function of the pulse number for that converter. A 12-pulse converter will have the lowest harmonic order of 11. In other words, the 5th, and the 7th harmonic orders are theoretically nonexistent in a 12-pulse converter. Similarly, an 18-pulse converter will have harmonic spectrum starting from the 17th harmonic and upwards. The lowest harmonic order in a 24-pulse converter will be the 23rd. The size of the passive harmonic filter needed to filter out the harmonics reduces as the order of the lowest harmonic in the current spectrum increases. Hence, the size of the filter needed to filter the harmonics out of a 12-pulse converter is much smaller than that needed to filter out the harmonics of a 6-pulse converter. However, a 12-pulse converter needs two 6-pulse bridges and two sets of 30° phase shifted AC inputs. The phase shift is achieved either by using an isolation transformer with one primary and two phase-shifted secondary windings or by using an autotransformer that provides phase-shifted outputs. Many different autotransformer topologies exist and the choice of a topology over the other involves a compromise between ease of construction, performance, and cost. An 18-pulse converter would need three 6-pulse diode bridges and three sets of 20° phase shifted inputs; similarly, a 24-pulse converter

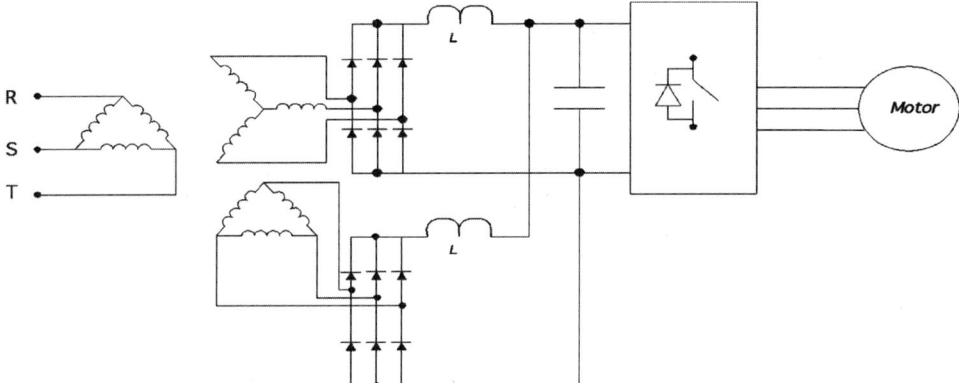

FIGURE 14.24 Schematic of a 12-pulse converter employing a three-winding transformer. Note that the input transformer has to be sized for rated power operation.

would need four 6-pulse diode bridges and four sets of 15° phase-shifted inputs. The transformers providing the phase-shifted outputs for multipulse converters have to be properly designed to handle circulating harmonic flux.

A typical 12-pulse structure is shown in Fig. 14.24. In one electrical cycle, the DC voltage will have 12 humps and hence the name 12-pulse rectifier.

Active Harmonic Compensation

Most passive techniques discussed above aim to cure the harmonic problems once nonlinear loads have created them. However, motor-drive manufacturers are developing rectification techniques that do not generate low-order harmonics. These drives use active front ends. Instead of using diodes as rectifiers, the active front-end ASDs make use of active switches like IGBTs along with parallel diodes. Power flow through a switch becomes bidirectional and can be manipulated to recreate a current waveform that linearly follows the applied voltage waveform.

Apart from the active front ends, there also exist shunt active filters used for actively introducing a current waveform into the AC network, which, when combined with the harmonic current, results in an almost perfect sinusoidal waveform.

One of the most interesting active filter topologies for use in retrofit applications is the combination of a series active filter along with shunt tuned passive filters. This combination is also known as the hybrid structure.

Most active filter topologies are complicated and require active switches and control algorithms that are implemented using digital signal processing (DSP) chips. The active filter topology also needs current and voltage sensors and corresponding analog-to-digital (A/D) converters. This extra hardware increases the cost and component count, reducing the overall reliability and robustness of the design. Manufacturers of smaller power equipment like computer power supplies, lighting ballast, etc. have successfully employed active circuits, employing boost converter topologies.

Controlled Rectifiers

Controlled rectifier circuits make use of devices known as "thyristors." A thyristor is a four-layer (p-n-p-n), three-junction device that conducts current only in one direction similar to a diode. The last (third) junction is utilized as the control junction and consequently the rectification process can be initiated at will provided the device is favorably biased and the load is of favorable magnitude. The operation of a thyristor can be explained by assuming it to be made up of two transistors connected back-to-back as shown in Fig. 14.25.

Let α_1 and α_2 be the ratio of collector to emitter currents of transistors Q1 and Q2, respectively. In other words:

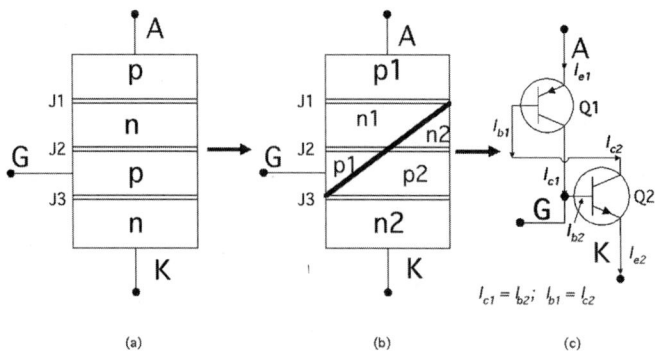

FIGURE 14.25 Virtual representation of a thyristor to explain its operation.

$$\alpha_1 = \frac{I_{c1}}{I_{e1}}; \quad \alpha_2 = \frac{I_{c2}}{I_{e2}};$$

Also, from Fig. 14.25: $I_{e1} = I_{e2} = I_A$ where I_A is the anode current flowing through the thyristor. From transistor theory, the value of I_{e2} is equal to $I_{c2} + I_{b2} + I_{lkg}$; where I_{lkg} is the leakage current crossing the $n1$-$p2$ junction. From Fig. 14.25, $I_{b2} = I_{c1}$. Hence, the anode current can be rewritten as:

$$I_A = I_{c1} + I_{c2} + I_{lkg}$$

Substituting the collector currents by the product of ratio α and emitter current, the anode current becomes:

$$I_A = (\alpha_1 * I_{e1}) + (\alpha_2 * I_{e2}) + I_{lkg}$$

$$I_A = (\alpha_1 + \alpha_2) I_A + I_{lkg}$$

$$I_A = \frac{I_{lkg}}{1 - (\alpha_1 + \alpha_2)}$$

If the ratios of the collector current to base current (gain) of the transistors are assumed to be β_1 and β_2, respectively, then the relationship between to β_1, β_2 and α_1, α_2 can be written as:

$$\alpha_1 = \frac{\beta_1}{1+\beta_1}; \quad \alpha_2 = \frac{\beta_2}{1+\beta_2}$$

Substituting for α_1 and α_2 in the expression for I_A yields the following expression:

$$I_A = \frac{(1+\beta_1)(1+\beta_2) I_{lkg}}{1 - \beta_1 \beta_2}.$$

If the values of α_1 and α_2 are low (low gains), then the anode current is low and comparable to the leakage current. Under this condition, the thyristor is said to be in its OFF state. However, if the effective gain of the transistor is such that the product of the gains are close to 1 (i.e., sum of the ratios of α_1 and

α_2 are close to 1), then there is a large increase in anode current and the thyristor is said to be in conduction. External circuit conditions can be changed to influence the product of the gains ($\beta_1\beta_2$). Some techniques of achieving this are briefly discussed next.

Increasing applied voltage: On applying a voltage across the anode to cathode terminals of the thyristor (anode being more positive than the cathode), one can see that junctions J1 and J3 in Fig. 14.25 are forward biased while junction J2 is reverse biased. The thyristor does not conduct any current and is said to be in a blocking state. On increasing the applied voltage, minority carriers in junction J2 (i.e., holes in $n1$, $n2$ and electrons in $p1$, $p2$) start acquiring more energy and hence start to migrate. In the process, these holes could dislodge more holes. Recombination of the electrons and holes also occur, which creates more motion. If the voltage is increased beyond a particular level, the movement of holes and electrons becomes great and junction J2 ceases to exist. The product of the gains of the two transistors in the two-transistor model is said to achieve values close to unity. This method of forcing current to flow through the thyristor is not recommended since junction J2 gets permanently damaged and the thyristor ceases to block forward voltage. Hence, this method is a destructive method.

High dv/dt: As explained earlier, junction J2 is the forward blocking junction when a forward voltage is applied across anode to cathode of a thyristor. Any p-n junction behaves like a depletion region when it is reverse-biased. Since J2 is reverse-biased, this junction behaves like a depletion region. Another way of looking at a depletion region is that the boundary of the depletion region has abundant holes and electrons while the region itself is depleted of charged carriers. This characteristic is similar to that of a capacitor. If the voltage across the junction (J2) changes very abruptly, then there will be rapid movement of charged carriers through the depleted region. If the rate of change of voltage across this junction (J2) exceeds a predetermined value, then the movement of charged carriers through the depleted region is so high that junction J2 is again annihilated. After this event, the thyristor is said to have lost its capability to block forward voltage and even a small amount of forward voltage will result in significant current flow, limited only by the load impedance. This method is destructive too, and is hence not recommended.

Temperature: Temperature affects the movement of holes and electrons in any semiconductor device. Increasing the temperature of junction J2 will have a very similar effect. More holes and electrons will begin to move, causing more dislodging of electrons and holes from neighboring lattice. If a high temperature is maintained, this could lead to an avalanche breakdown of junction J2 and again render the thyristor useless since it would no longer be able to block forward voltage. Increasing temperature is yet another destructive method of forcing the thyristor to conduct.

Gate current injection: If a positive voltage is applied across the gate to cathode of a thyristor, then one would be forward biasing junction J3. Charged carriers will start moving. The movement of charged carriers in junction J3 will attract electrons from n2 region of the thyristor (Fig. 14.25). Some of these electrons will flow out of the gate terminal but there would be ample of electrons that could start crossing junction J2. Since electrons in p2 region of junction J2 are minority carriers, these can cause rapid recombination and help increase movement of minority carriers in junction J2. By steadily increasing the forward biasing potential of junction J3, one could potentially control the depletion width of junction J2. If a forward biasing voltage is applied across anode to cathode of the thyristor with its gate to cathode favorably biased at the same time, then the thyristor can be made to conduct current. This method achieves conduction by increasing the leakage current in a controlled manner. The gain product in the two-transistor equivalent is made to achieve a value of unity in a controlled manner and the thyristor is said to turn ON. This is the only recommended way of turning ON a thyristor. When the gate-cathode junction is sufficiently forward biased, the current through the thyristor depends on the applied voltage across the anode-cathode and the load impedance. The load impedance and the externally applied anode-cathode voltage should be such that the current through the thyristor is greater than a minimum current known as *latching current*, I_l. Under such a condition, the thyristor is said to have *latched ON*. Once it has latched ON, the thyristor remains ON. In other words, even if the forward biasing voltage across the gate-cathode terminals is removed, the thyristor continues to conduct. Junction J2 does not exist during the ON condition. The thyristor reverts to its blocking state only when the current through it falls below a minimum threshold value known as *holding current*, I_h. Typically, holding current is lower than latching current ($I_h < I_l$). There are two ways of achieving this. They are either (1) increase the load impedance to such a value that the thyristor current falls below I_h or (2) apply reverse-biasing voltage across the anode-cathode of the thyristor.

An approximate v-i characteristic of a typical thyristor and its symbol are shown in Fig. 14.26.

Since the thyristor allows flow of current only in one direction like a diode and the instant at which it is turned ON can be controlled, the device is a key component in building a controlled rectifier unit. One can replace the diode in all the circuits discussed so far with the thyristor. Because of its controllability, the instant at which the thyristor conducts can be delayed to alter the average and rms output voltages. By doing so, one can choose to control the output voltage and power of a rectifier circuit. Hence, rectifiers that employ thyristors are also known as silicon controlled rectifiers or SCR.

A typical single-phase, R-L rectifier circuit with one thyristor as the rectifier is shown in Fig. 14.27. The figure also shows the relevant circuit waveforms. The greatest difference between this circuit and its diode counterpart is also shown for comparison. Both circuits conduct beyond π radians due to the presence of the inductor L since the average voltage across an inductor is zero. If the value of the circuit components and the input supply voltage are the same in both cases, the duration for which the current flows into the output R-L load depends on the values of R and L. In the case of the diode circuit, it does not depend on anything else; while in the case of the thyristor circuit, it also depends on the instant the thyristor is given a gate trigger.

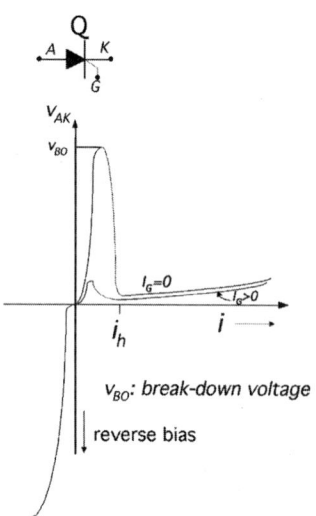

FIGURE 14.26 v-i characteristic of a thyristor along with its symbol.

From Fig. 14.27, it is interesting to note that the energy stored in the inductor during the conduction interval can be controlled in the case of a thyristor is such a manner so as to reduce the conduction interval and thereby alter (reduce) the output power. Both the diode and the thyristor show reverse recovery phenomenon. The thyristor, like the diode, can block reverse voltage applied across it repeatedly, provided the voltage is less than its breakdown voltage.

Gate Circuit Requirements

The trigger signal should have voltage amplitude greater than the minimum gate trigger voltage of the thyristor being turned ON. It should not be greater than the maximum gate trigger voltage, either. The gate current should likewise be in between the minimum and maximum values specified by the thyristor manufacturer. Low gate current driver circuits can fail to turn ON the thyristor. The thyristor is a current controlled switch and so the gate circuit should be able to provide the needed turn ON gate current into the thyristor. Unlike the bipolar transistor, the thyristor is not an amplifier and so the gate current requirement does not absolutely depend on the voltage and current rating of the thyristor. Sufficient gate trigger current will turn ON the thyristor and current will flow from the anode to the cathode provided that the thyristor is favorably biased and the load is such that the current flowing is higher than the latching current of the thyristor. In other words, in single phase AC to DC rectifier circuits, the gate trigger will turn ON the thyristor only if it occurs during the positive part of the AC cycle (Fig. 14.27). Any trigger signal during the negative part of the AC cycle will not turn ON the thyristor and the thyristor will remain in blocking state. Keeping the gate signal ON during the negative part of the AC cycle does not typically damage a thyristor.

Single-Phase H-bridge Rectifier Circuits with Thyristors

Similar to the diode H-bridge rectifier topology, there exist SCR-based rectifier topologies. Because of their unique ability to be controlled, the output voltage and hence the power can be controlled to desired levels. Since the triggering of the thyristor has to be synchronized with the input sinusoidal voltage in

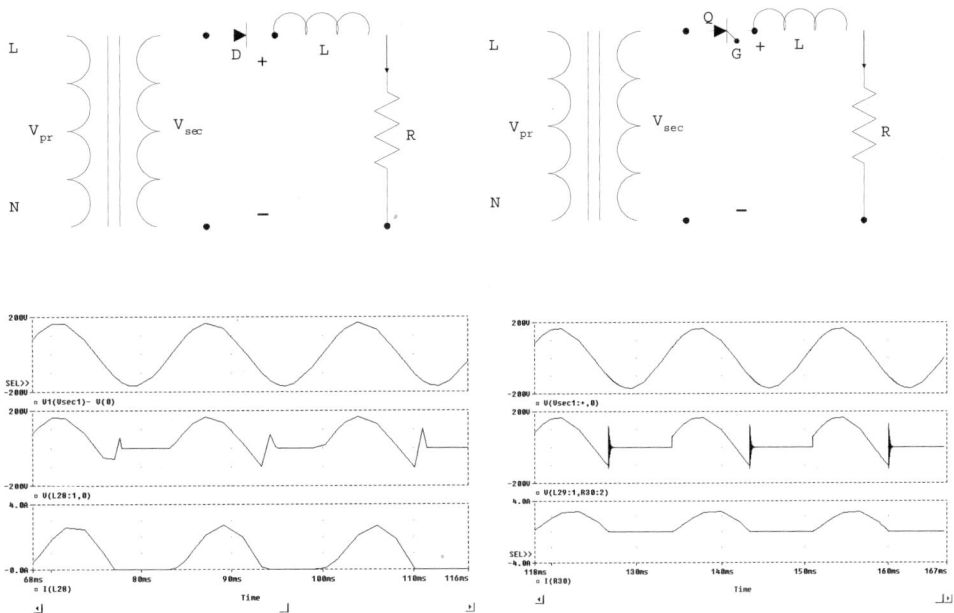

FIGURe 14.27 Comparing a single diode rectifier circuit with a single thyristor rectifier circuit. Note that the thyristor conduction is delayed deliberately to bring out the differences.

an AC to DC rectifier circuit, one can achieve a soft-charge characteristic of the filter capacitor. In other words, there is no need for employing soft-charge resistor and contactor combination as is required in single-phase and three-phase AC to DC rectifier circuits with DC bus capacitors.

In controlled AC to DC rectifier circuits, it is important to discuss control of resistive, inductive, and resistive-inductive load circuits. DC motor control falls into the resistive-inductive load circuit. DC motors are still an important part of the industry. However, the use of DC motors in industrial applications is declining rapidly. Control of DC motors are typically achieved by controlled rectifier circuits employing thyristors. Small motors of less than 3 kW (approximately 5 hp) rating can be controlled by single-phase SCR circuits while larger ratings require three-phase versions. A typical single-phase H-bridge SCR-based circuit for the control of a DC motor is shown in Fig. 14.28. Typical output waveforms are shown in Fig. 14.29. The current in the load side can be assumed continuous due to the large inductance of the armature of the DC motor.

In Fig. 14.28, V_f is the field voltage, which is applied externally and generally is independent of the applied armature voltage. Such a DC motor is known as a separately excited motor. I_a is the armature current while I_f is the field current. The output of the controlled rectifier is applied across the armature. Since the output voltage can be controlled, one can effectively control the armature current. Since the

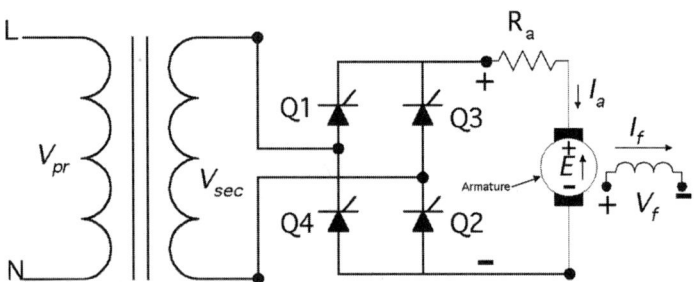

FIGURE 14.28 Single-phase DC motor control circuit for controlling a separately excited DC motor. R_a indicates equivalent armature resistance and E is the back emf.

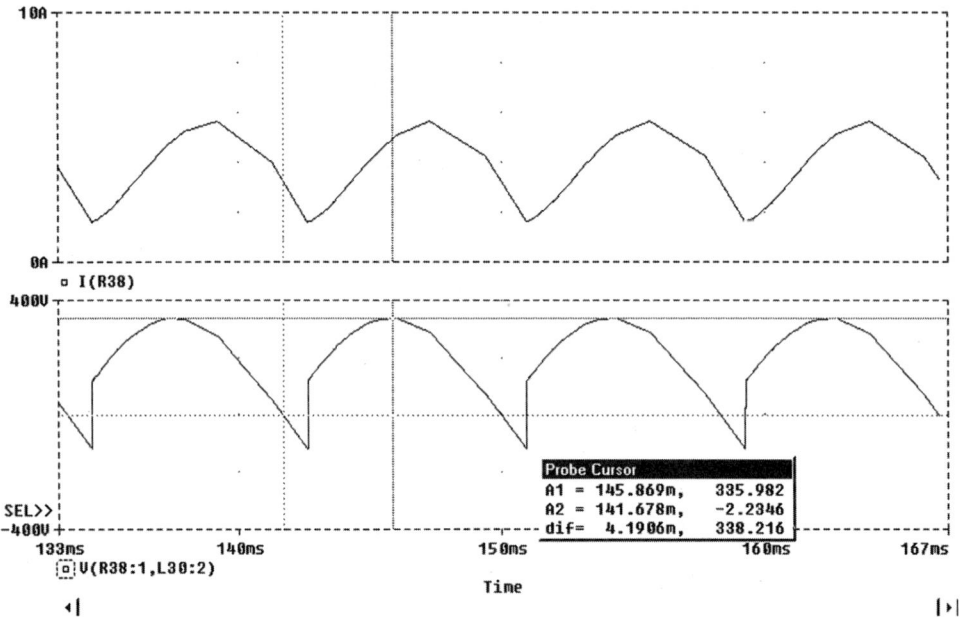

FIGURE 14.29 Armature current and output voltage of AC to DC rectifier employed to control a DC motor.

torque produced by a DC motor is directly proportional to the armature current, the torque developed can thus be controlled.

$$T = K\phi I_a;$$

where K is the motor constant and depends on the number of armature conductors, number of poles, and type of winding employed in the DC machine. ϕ is flux produced by the field and is proportional to the field current, I_f. Hence, the torque produced by a DC machine can be rewritten as $T = K(K_1 I_f)I_a$. By keeping the field current constant, the torque then becomes directly proportional to the armature current, which is controlled by controlling the output voltage of the AC to DC controlled rectifier. In the circuit shown in Fig. 14.28, it is interesting to note that the current I_a, cannot flow in the opposite direction. Hence, the motor cannot generate negative torque. In order to make the motor run in the opposite direction, the direction of the field has to be changed. Speed control within the base speed can also be accomplished by controlling the armature voltage as is shown below.

$$E = K\phi\omega = K(K_1 I_f)\omega;$$

ω is the speed of the armature in radians/sec. The back emf, E, is the difference between the output DC voltage of the AC to DC controlled rectifier and the drop across the equivalent armature resistance. Hence, E can be rewritten as:

$$E = V_a - (I_a R_a); \quad \omega = \frac{V_a - (I_a R_a)}{K K_1 I_f}$$

For control of speed beyond base speed, the field current has to be altered. Hence, it can be shown that controlling the armature current can control the speed and torque produced by a DC machine. Controlling the output DC voltage can control the armature current. Because of the large inductance of the armature circuit, the current through it can be assumed to be continuous for a practical operating region. The average output voltage of a single-phase AC to DC rectifier circuit for continuous current operation is given by (referring to Fig. 14.29):

$$V_o = \frac{1}{\pi} \int_{\alpha}^{\pi+\alpha} \left(\sqrt{2} * V_{rms} \right) d(wt) = \frac{2 * \sqrt{2} * V_{rms} * \cos(\alpha)}{\pi};$$

for continuous current condition. By controlling the triggering angle, α, one can control the average value of the output voltage, V_O. If armature current control is the main objective (to control output torque), then one can configure the controller of Fig. 14.28 with a feedback loop. The measured current can be compared with a set reference and the error can be used to control the triggering angle, α. Since the output voltage and hence the armature current are not directly proportional to α but to $cos(\alpha)$, the above method will yield a nonlinear (co-sinusoidal) relationship between the output voltage and control angle, α. However, one could choose to use the error signal to control $cos(\alpha)$ instead of α. This would then yield a linear relationship between the output voltage and cos of control angle, α.

It is interesting to note from the equation for the output average voltage that the output average voltage can become negative if the triggering angle is greater than 90 electrical degrees. This leads us to the topic of regeneration. AC to DC controlled rectifiers employing thyristors and having large inductance on the DC side can be made to operate in the regeneration mode by simply delaying the trigger angle. This is quite beneficial in overhauling loads like cranes. When the load on a hook of the crane has to be lifted up, electrical energy is supplied to the motor. The voltage across the motor is positive and the current through the armature is positive. Positive torque is generated and hence the load moves up. When the load is to be brought down, the load starts to rotate the motor in the opposite direction due to gravity. The voltage at the terminals becomes negative since speed is negative. The thyristors are gated at an angle greater than 90 electrical degrees to match the generated (negative) voltage of the DC motor. Since current through the thyristors cannot go negative, current is forced to flow into the DC motor in the positive direction. The large inductance of the motor helps to maintain the positive direction of current through the armature. Positive torque is still produced since the direction of current is still positive and the field remains unchanged. In other words, the motor develops positive torque and tries to move the load up against gravity but the gravity is pushing the motor down. The product of current through the motor and the voltage across it is negative, meaning that the motor is not consuming energy, and on the contrary, is producing electrical energy — the kinetic energy due to the motor's downward motion is partly converted to electrical energy by the field and armature. This energy produced by the motor is routed out to the supply via the appropriately gated thyristors. Conversion of kinetic energy to electrical energy acts like a ***dynamic-brake*** and slows the rapid downward descent of the load.

A typical crane is required to operate in all four quadrants (Fig. 14.30). In the first quadrant, the motor develops positive torque and the motor runs in the positive direction, meaning its speed is positive — the product of torque and speed is power, and so positive electric power is supplied to the motor from the AC to DC rectifier. When the crane with a load is racing upward, close to the end of its travel, the

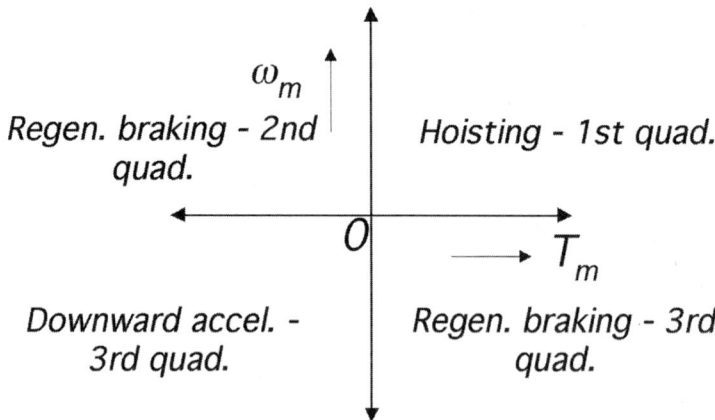

FIGURE 14.30 Four-quadrant operation of a crane or hoist.

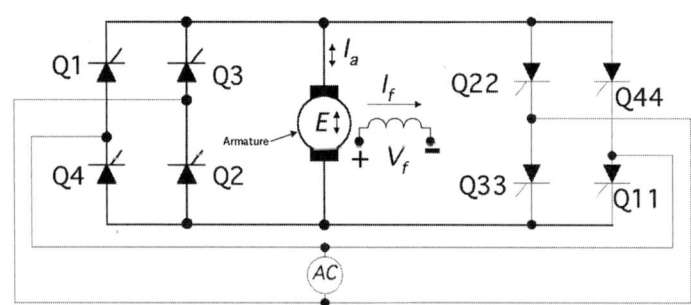

FIGURE 14.31 Two rectifier-bridge arrangements for four-quadrant operation of DC motor.

AC to DC controlled rectifier is made to stop powering the motor. The rectifier generates practically no voltage. The inertia of the load moving upward generates a voltage in the form of a back emf. This voltage is fed into a second rectifier bridge arranged in the opposite direction as shown in Fig. 14.31. The second bridge is turned ON to let the generated voltage across the still upwardly mobile motor flow into the utility, thereby converting the inertial motion to electric power. In the second quadrant, speed remains positive but torque becomes negative, since the current through the motor flows in the opposite direction into the second rectifier bridge arrangement (Fig. 14.31). The product of speed and torque is negative, meaning that the motor behaves like a generator during this part of the travel.

The third quadrant, as explained earlier, occurs at the beginning of the lowering action. Both torque and speed are negative and so the product of torque and speed is positive. Power is applied to the motor to overcome static friction and accelerate the rotating parts of the mechanism to move the load downward. In this case, the direction of armature current through the motor is opposite to that in quadrant 1, and the electrical power needed by the motor is supplied by the second rectifier bridge arrangement (Fig. 14.31).

The mechanical load and motor arrangement goes into the fourth quadrant of operation for the larger part of the downward motion. This is the period during which the motor resists the tendency of the load to accelerate downward by developing positive torque. Since motion is downward, speed is negative and the product of torque and speed is negative. This means the motor behaves like a generator as explained earlier.

Since the thyristors cannot conduct in the opposite direction, a new inverter section had to be provided to enable the four-quadrant operation needed in cranes and hoists. The method by which unidirectional electrical power was routed to the bidirectional AC utility lines is known as inversion (opposite of rectification). Since no external means of switching OFF the thyristors was employed, the process of inversion was achieved by natural commutation. Such an inverter is also known as a line commutated inverter.

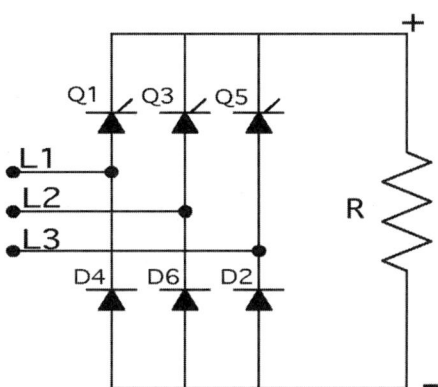

FIGURE 14.32 A typical three-phase semiconverter. Rarely employed in modern industry.

Three-Phase Controlled AC to DC Rectifier Systems

The observations made so far for the single-phase controlled AC to DC rectifiers can be easily extended to three-phase versions. An important controlled rectification scheme that was not mentioned in the single-phase case is the semiconverter circuit. In Fig. 14.28, if the thyristors Q2 and Q4 are replaced by diodes (D2 and D4), then the circuit of Fig. 14.28 is converted into a semiconverter circuit. Such a circuit does not have the potential to provide regeneration capability and hence is of limited use. However, in dual converter applications, especially in three-phase versions, there are a few instances where a semiconverter can be employed to reduce cost. A typical three-phase semiconverter circuit will consist of three thyristors and three diodes arranged in an H-bridge configuration as shown in Fig. 14.32.

Three-phase dual converter schemes similar to the one shown in Fig. 14.31 are still employed to operate large steel mills, hoists, and cranes. However, the advent of vector controlled AC drives has drastically changed the electrical landscape of the modern industry. Most DC motor applications are being rapidly replaced by AC motors with field-oriented control schemes. DC motor application in railway traction has also seen significant reduction due to the less expensive and more robust AC motors.

However, there are still a few important applications where three-phase controlled rectification (inversion) is the most cost-effective solution. One such application is the regenerative converter module that many inverter-drive manufacturers provide as optional equipment to customers with overhauling loads. Under normal circumstances, during the motoring mode of operation of an AC drive, the regenerative unit does not come into the circuit. However, when the DC bus voltage tends to go higher than a predetermined level due to overhauling of the load, the kinetic energy of the load is converted into electrical energy and is fed back into the AC system via a six-pulse thyristor-based inverter-bridge. One such scheme is shown in Fig. 14.33.

Average Output Voltage

In order to evaluate the average value of the output voltage for a three-phase full-bridge converter, the process of integrating the output voltage similar to the one in Fig. 14.21(b) has to be undertaken. For the circuit shown in Fig. 14.20(b), where the diodes are replaced by thyristors, the integration yields the following:

$$V_o = \frac{3}{\pi} \int_{\alpha+(\pi/3)}^{\alpha+(2\pi/3)} \sqrt{2} V_{L-L} \sin(wt) d(wt)$$

$$V_o = \frac{3 * \sqrt{2} * V_{L-L} * \cos(\alpha)}{\pi} = \frac{3 * \sqrt{2} * \sqrt{3} * V_{L-N} * \cos(\alpha)}{\pi}$$

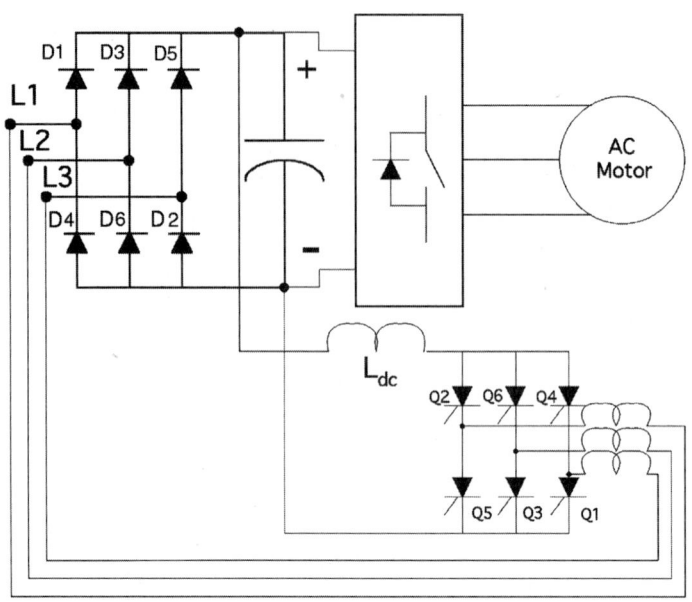

FIGURE 14.33 Use of six-pulse thyristor bridge in the inverter mode to provide regeneration capability to an existing AC drive system.

The average output voltage for the circuit in Fig. 14.20(b) with the diodes being replaced by thyristors is only different in the *cosine* of the triggering angle, α. If the triggering angle is zero, the circuit performs similar to a three-phase diode rectifier and the average output voltages become the same.

HVDC Transmission Systems

One area where it is difficult to replace the use of high voltage, high current carrying thyristors is high voltage DC (HVDC) transmission systems. When a large amount of power is to be transported over long distances, or under water, it has been found that high voltage DC transmission is more economical. HVDC systems are in reality back-to-back rectifier systems. The sending end rectifier system consists typically of 12- or 24-pulse thyristor bridges while the receiving end consists of a similar configuration but in the opposite direction. The receiving end 12- or 24-pulse bridge operates in the inverter mode while the sending end operates in the rectifier mode. 12-pulse configuration is achieved by cascading two 6-pulse bridges in series while 24-pulse configuration needs four 6-pulse bridges cascaded in series. Typical advantages of high voltage DC transmission over high voltage AC transmission is briefly listed below:

1. No stability problems due to transmission line length since no reactive power needs to be transmitted.
2. No limitation of cable lengths for underground cable or submarine cable transmission due to the fact that no charging power compensation need be done.
3. AC power systems can be interconnected employing a DC tie without reference to system frequencies, short circuit power, etc.
4. High-speed control of DC power transmission is possible due to the fact that the control angle, α, has a relatively short time constant.
5. Fault isolation between receiving end and sending end can be dynamically achieved due to fast efficient control of the high voltage DC link.
6. Employing simple control logic can change energy flow direction very fast. This can help in meeting peak demands at either the sending or the receiving station.
7. High reliability of thyristor converter and inverter stations makes this mode of transmission a viable solution for transmission lengths typically over 500 km.
8. The right-of-way needed for high voltage DC transmission is much lower than that of AC transmission of the same power capacity.

Power Electronics

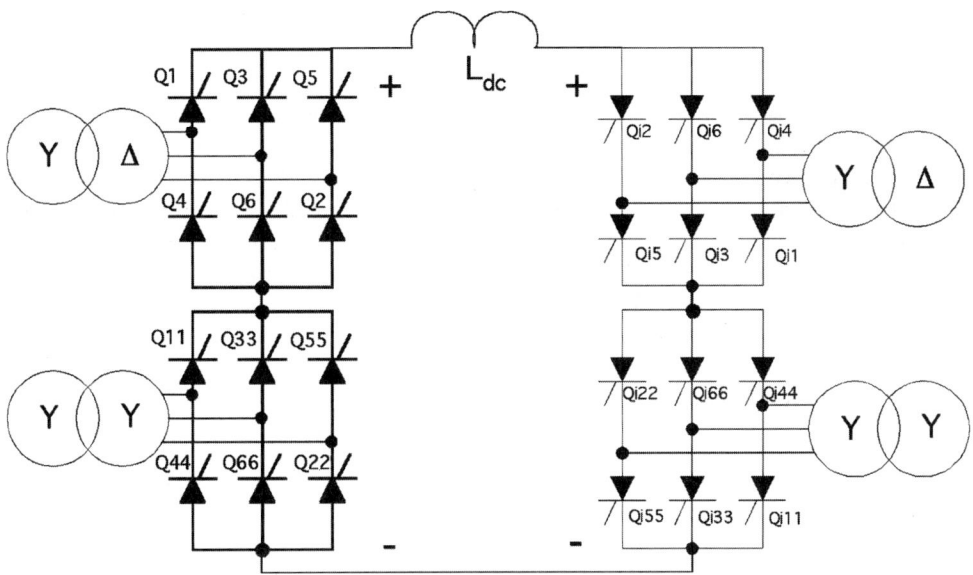

FIGURE 14.34 Schematic representation of a bipolar HVDC system employing 12-pulse rectification/inversion scheme.

The advantages of DC transmission over AC should not be misunderstood and DC should not be considered as a general substitute for AC power transmission. In a power system, it is generally believed that both AC and DC should be considered as complementary to each other, so as to bring about the integration of their salient features to the best advantage in realizing a power network that ensures high quality and reliability of power supply. A typical rectifier-inverter system employing a 12-pulse scheme is shown in Fig. 14.34.

Typical DC link voltage can be as high as 400 kV to 600 kV. Higher voltage systems are also in use. Typical operating power levels are over 1000 MW. There are a few systems transmitting close to 3500 MW of power through two bipolar systems. Most thyristors employed in large HVDC transmission systems are liquid cooled to improve their performance.

Power System Interaction with Three-phase Thyristor AC to DC Rectifier Systems

Similar to the diode rectifiers, the thyristor based AC to DC rectifiers also suffer from low order current harmonics. In addition to current harmonics, there is a voltage notching phenomenon occurring at the input terminals of an AC to DC thyristor based rectifier system. The voltage notching is a very serious problem. Since thyristors are generally slower to turn ON and turn OFF compared to power semiconductor diodes, there are nontrivial durations during which an outgoing thyristor and an incoming thyristor remain in conduction, thereby creating a short-circuit across the power supply phases feeding the corresponding thyristors. Thyristors used in rectifiers are generally known as phase control type thyristors and have typical turn OFF times of 50 to 100 μsec. Thyristors employed in inverter circuits typically are faster and have turn OFF times in the 10 to 50 μsec range.

Notching can create major disturbances in sensitive electronic equipment that rely on the zero-crossing of the voltage for satisfactory operation. Multiple pseudo zero-crossings of the voltage waveform can occur due to the notching effect of thyristor based rectifier systems. Notching can create large magnitudes of currents to flow into power-factor correcting capacitors, thereby potentially causing permanent damage to them. IEEE Std. 519-1992 in the U.S. has strict regulations regarding the depth of the notch as well as the duration of the notch. AC line inductors in series with the supply feeding power to the three-phase bridge help to minimize the notching effect on the power system. The theory behind this phenomenon is discussed next.

When an external inductance is added in front of a three-phase AC to DC rectifier employing thyristors, the duration of commutation increases. In other words, the time period for which the outgoing thyristor remains in conduction along with the incoming thyristor increases. This overlap period causes the average output voltage to reduce because during this period, the output voltage is composed of two shorted phases and a healthy phase. The extent of reduction in the output voltage depends on the duration of overlap in electrical degrees. The duration of overlap in electrical degrees is commonly represented by µ. The overlap duration is directly proportional to the value of the external inductance used. If no external line inductor is used, then this duration will depend on the existing inductance of the system including the wiring inductance. In order to compute the factors influencing the overlap duration, a simple model can be assumed. Assume that the line is comprised of inductance L in each phase. Let the DC load current be I_{dc} and let it be assumed that this current does not change during the overlap interval. The current in the incoming thyristor is zero at start and by the end of the overlap interval, it rises to I_{dc}. Based on this assumption, the relationship between current and voltage can be expressed as:

$$v_{ab} = \sqrt{2} * V_{L-L} * \sin(wt) = 2 * L * (di/dt)$$

$$\sqrt{2} * V_{L-L} * \int_{\alpha+(\pi/3)}^{\alpha+(\pi/3)+\mu} \sin(wt) d(t) = 2 * L * \int_{0}^{I_{dc}} di$$

$$I_{dc} = \frac{\sqrt{2} * V_{L-L} * (\cos(\alpha+\pi/3) - \cos(\alpha+\pi/3+\mu))}{2wL} = \frac{\sqrt{2} * V_{L-L} * \sin(\alpha+\pi \ 3+\mu/2) * \sin(\mu/2)}{wL}$$

For small values of overlap angle µ, $sin(\mu/2) = \mu/2$ and $sin(\alpha + \pi/3 + (\mu/2)) = sin(\alpha + \pi/3)$. Rearranging the above equation yields:

$$\mu = \frac{2wL * I_{dc}}{\sqrt{2} * V_{L-L} * \sin(\alpha+\pi/3)}.$$

From the above expression, it is interesting to note the following:

1. If the inductance L in the form of either external inductance or leakage inductance of transformer or lead length is large, the overlap duration will be large.
2. If the load current, I_{dc} is large, the overlap duration will be large.
3. If the delay angle is small, then the inductance will store more energy and so the duration of overlap will be large. The minimum value of delay angle α is 0° and the maximum value typically is 60°.

The average output voltage will reduce due to the overlap angle as mentioned before. In order to compute the average output voltage with a certain overlap angle, the limits of integration have to be changed. This exercise yields the following:

$$V_o = \frac{3}{\pi} \int_{\alpha+\mu+(\pi \ 3)}^{\alpha+\mu+(2\pi/3)} \sqrt{2} V_{L-L} \sin(wt) d(wt)$$

$$V_o = \frac{3 * \sqrt{2} * V_{L-L} * \cos(\alpha+\mu)}{\pi} = \frac{3 * \sqrt{2} * \sqrt{3} * V_{L-N} * \cos(\alpha+\mu)}{\pi}$$

Thus, it can be seen that the overlap angle has an equivalent effect of advancing the delay angle, thereby reducing the average output voltage. From the discussions in the previous paragraphs on notching, it is interesting to note that adding external inductance increases the duration of the overlap and reduces the average value of the output DC voltage. However, when viewed from the AC source side, the notching effect is conspicuously reduced and in some cases not observable. Since all other electrical equipment in the system will be connected to the line side of the AC inductor (in front of a thyristor based AC to DC rectifier), these equipment will not be affected by the notching phenomenon of thyristors. The external inductance also helps limit the circulating current between the two thyristors during the overlap duration.

Conclusion

Uncontrolled and controlled rectifier circuits have been discussed in this chapter. An introduction to the theory of diode and thyristor conduction has been presented to explain the important operating characteristics of these devices. Rectifier topologies employing both diodes and thyristors and their relative advantages and disadvantages have been discussed. Use of a dual thyristor bridge converter to achieve four-quadrant operation of a DC motor has been discussed. The topic of high-voltage DC (HVDC) transmission has been briefly introduced. Power quality issues relating to diode and thyristor-based rectifier topologies has also been addressed. To probe further into the various topics briefly discussed in this section, the reader is encouraged to refer to the references listed below.

References

Dewan, S. B. and Straughen, A., *Power Semiconductor Circuits,* John Wiley & Sons, New York, 1975.
Hoft, R. G., *Semiconductor Power Electronics,* Van Nostrand Reinhold Company, New York, 1986.
IEEE Recommended Practices and Requirements for Harmonic Control in Electrical Power Systems, IEEE Std. 519-1992.
Laughton M. A. and Say, M. G., Eds., *Electrical Engineer's Reference Book –14th edition,* Butterworths, 1985.
Sen, P. C., *Principles of Electric Machines and Power Electronics,* John Wiley & Sons, New York, 1997.
Passive Harmonic Filter Systems for Variable Frequency Drives, U.S. Patent 5,444,609, 1995.

14.3 Inverters

Michael Giesselmann

Inverters are used to create single or polyphase AC voltages from a DC supply. In the class of polyphase inverters, three-phase inverters are by far the largest group. A very large number of inverters are used for adjustable speed motor drives. The typical inverter for this application is a "hard-switched" voltage source inverter producing Pulse-Width Modulated (PWM) signals with a sinusoidal fundamental (Holtz, 1992). Recently research has shown detrimental effects on the windings and the bearings resulting from unfiltered PWM waveforms and recommend the use of filters (Cash and Habetler, 1998; Von Jouanne et al., 1996). A very common application for single-phase inverters are so-called "Uninterruptable Power Supplies" (UPS) for computers and other critical loads. Here, the output waveforms range from square wave to almost ideal sinusoids. UPS designs are classified as either "off-line" or "on-line". An off-line UPS will connect the load to the utility for most of the time and quickly switch over to the inverter if the utility fails. An online UPS will always feed the load from the inverter and switch the supply of the DC bus instead. Since the DC bus is heavily buffered with capacitors, the load sees virtually no disturbance if the power fails.

In addition to the very common hard-switched inverters, active research is being conducted on soft-switching techniques. Hard-switched inverters use controllable power semiconductors to connect an output terminal to a stable DC-bus. On the other hand, soft switching inverters have an oscillating intermediate circuit and attempt to open and close the power switches under zero-voltage and or zero-current conditions.

A separate class of inverters are the line commutated inverters for multimegawatt power ratings, that use Thyristors (also called Silicon Controlled Rectifiers, SCRs). SCRs can only be turned "on" on command. After being turned on, the current in the device must approach zero in order to turn the device off. All other inverters are self commutated, meaning that the power control devices can be turned on and off. Line commutated inverters need the presence of a stable utility voltage to function. They are used for DC links between utilities, ultra long distance energy transport, and very large motor drives (Ahmed, 1999; Barton, 1994; Mohan et al., 1995; Rashid, 1993; Tarter, 1993). However, the latter application is more and more taken over by modern hard-switched inverters including multilevel inverters (Brumsickle et al., 1998; Tolbert et al., 1999).

Modern inverters use **I**solated **G**ate **B**ipolar **T**ransistors (IGBTs) as the main power control devices (Mohan et al., 1995). Besides IGBTs, power MOSFETs are also used especially for lower voltages, power ratings, and applications that require high efficiency and high switching frequency. In recent years, IGBTs, MOSFETs, and their control and protection circuitry have made remarkable progress. IGBTs are now available with voltage ratings of up to 3300 V and current ratings up to 1200 A. MOSFETs have achieved on-state resistances approaching a few milliohms. In addition to the devices, manufacturers today offer customized control circuitry that provides for electrical isolation, proper operation of the devices under normal operating conditions, and protection from a variety of fault conditions (Mohan et al., 1995). In addition, the industry provides good support for specialized passive devices such as capacitors and mechanical components such as low inductance bus-bar assemblies to facilitate the design of reliable inverters. In addition to the aforementioned inverters, a large number of special topologies are used. A good overview is given by Gottlieb (1984).

Fundamental Issues

Inverters fall in the class of power electronics circuits. The most widely accepted definition of a power electronics circuit is that the circuit is actually processing electric energy rather than information. The actual power level is not very important for the classification of a circuit as a power electronics circuit. One of the most important performance considerations of power electronics circuits, like inverters, is their energy conversion efficiency. The most important reason for demanding high efficiency is the problem of removing large amounts of heat from the power devices. Of course, the judicious use of energy is also paramount, especially if the inverter is fed from batteries such as in electric cars. For these reasons, inverters operate the power devices, which control the flow of energy, as switches. In the ideal case of a switching event, there would be no power loss in the switch since either the current in the switch is zero (switch open) or the voltage across the switch is zero (switch closed) and the power loss is computed as the product of both. In reality, there are two mechanisms that do create some losses, however; these are on-state losses and switching losses (Bird et al., 1993; Kassakian et al., 1991; Mohan et al., 1995; Rashid, 1993). On-state losses are due to the fact that the voltage across the switch in the on state is not zero, but typically in the range of 1–2 V for IGBTs. For power MOSFETs, the on-state voltage is often in the same range, but it can be substantially below 0.5 V due to the fact that these devices have a purely resistive conduction channel and no fixed minimum saturation voltage like bipolar junction devices (IGBTs). The switching losses are the second major loss mechanism and are due to the fact that, during the turn on and turn off transition, current is flowing while voltage is present across the device. In order to minimize the switching losses, the individual transitions have to be rapid (tens to hundreds of nanoseconds) and the maximum switching frequency needs to be carefully considered.

In order to avoid audible noise being radiated from motor windings or transformers, most modern inverters operate at switching frequencies substantially above 10 kHz (Bose, 1992; 1996).

Single Phase Inverters

Figure 14.35 shows the basic topology of a full bridge inverter with single-phase output. This configuration is often called an H-bridge, due to the arrangement of the power switches and the load. The inverter can deliver and accept both real and reactive power. The inverter has two legs, left and right. Each leg

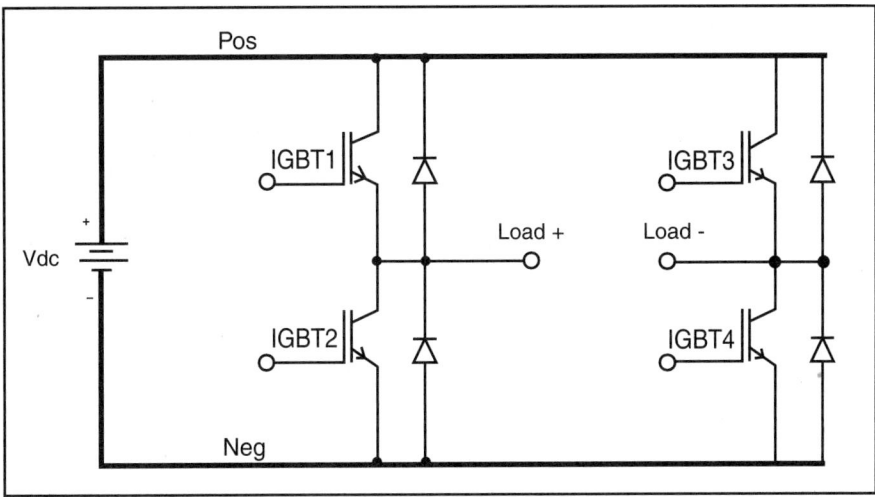

FIGURE 14.35 Topology of a single-phase, full bridge inverter.

consists of two power control devices (here IGBTs) connected in series. The load is connected between the mid-points of the two phase legs. Each power control device has a diode connected in antiparallel to it. The diodes provide an alternate path for the load current if the power switches are turned off. For example, if the lower IGBT in the left leg is conducting and carrying current toward the negative DC bus, this current would "commutate" into the diode across the upper IGBT of the left leg, if the lower IGBT is turned off. Control of the circuit is accomplished by varying the turn on time of the upper and lower IGBT of each inverter leg, with the provision of never turning on both at the same time, to avoid a short circuit of the DC bus. In fact, modern drivers will not allow this to happen, even if the controller would erroneously command both devices to be turned on. The controller will therefore alternate the turn on commands for the upper and lower switch, i.e., turn the upper switch on and the lower switch off, and vice versa. The driver circuit will typically add some additional blanking time (typically 500–1000 ns) during the switch transitions to avoid any overlap in the conduction intervals.

The controller will hereby control the duty cycle of the conduction phase of the switches. The average potential of the center-point of each leg will be given by the DC bus voltage multiplied by the duty cycle of the upper switch, if the negative side of the DC bus is used as a reference. If this duty cycle is modulated with a sinusoidal signal with a frequency that is much smaller than the switching frequency, the short-term average of the center-point potential will follow the modulation signal. "Short-term" in this context means a small fraction of the period of the fundamental output frequency to be produced by the inverter. For the single phase inverter, the modulation of the two legs are inverse of each other such that if the left leg has a large duty cycle for the upper switch, the right leg has a small one, etc. The output voltage is then given by Eq. (14.1) in which m_a is the modulation factor. The boundaries for m_a are for linear modulation. Values greater than 1 cause overmodulation and a noticeable increase in output voltage distortion.

$$V_{ac1}(t) = m_a \cdot V_{dc} \cdot \sin(\omega_1 \cdot t) \qquad 0 \leq m_a \leq 1 \qquad (14.1)$$

This voltage can be filtered using a LC low pass filter. The voltage on the output of the filter will closely resemble the shape and frequency of the modulation signal. This means that the frequency, wave-shape, and amplitude of the inverter output voltage can all be controlled as long as the switching frequency is at least 25–100 times higher than the fundamental output frequency of the inverter (Holtz, 1992). The actual generation of the PWM signals is mostly done using microcontrollers and **D**igital **S**ignal **P**rocessors (DSPs) (Bose, 1987).

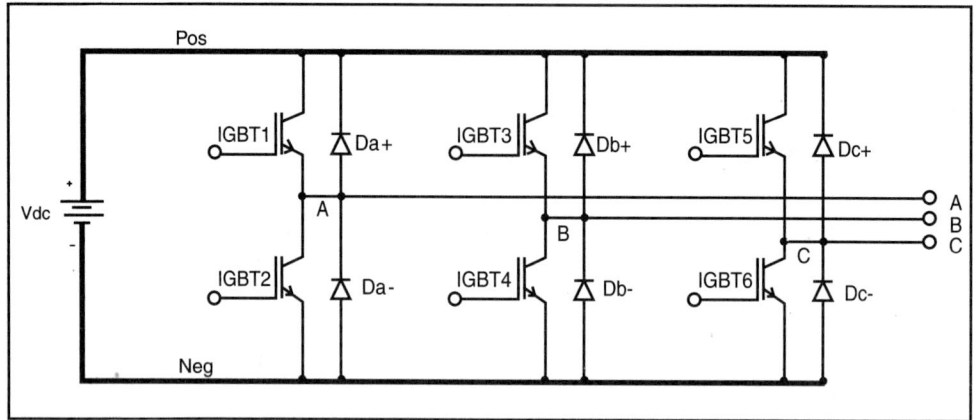

FIGURE 14.36 Topology of a three-phase inverter.

Three Phase Inverters

Figure 14.36 shows a three-phase inverter, which is the most commonly used topology in today's motor drives. The circuit is basically an extension of the H-bridge style single-phase inverter, by an additional leg. The control strategy is similar to the control of the single-phase inverter, except that the reference signals for the different legs have a phase shift of 120° instead of 180° for the single-phase inverter. Due to this phase shift, the odd triplen harmonics (3rd, 9th, 15th, etc.) of the reference waveform for each leg are eliminated from the line-to-line output voltage (Mohan et al., 1995; Novotny and Lipo, 1996; Rashid, 1993; Shepherd and Zand, 1979). The even numbered harmonics are canceled as well if the waveforms are pure AC, which is usually the case. For linear modulation, the amplitude of the output voltage is reduced with respect to the input voltage of a three phase rectifier feeding the DC bus by a factor given by Eq. (14.2).

$$\frac{3}{(2 \cdot \pi)} \cdot \sqrt{3} = 82.7\% \tag{14.2}$$

To compensate for this voltage reduction, the fact of the harmonics cancellation is sometimes used to boost the amplitudes of the output voltages by intentionally injecting a 3rd harmonic component into the reference waveform of each phase leg (Mohan et al., 1995).

Figure 14.37 shows the typical output of a three-phase inverter during a startup transient into a typical motor load. This figure was created using circuit simulation. The upper graph shows the pulse-width modulated waveform between phases A and B, whereas the lower graph shows the currents in all three phases. It is obvious that the motor acts as a low-pass filter for the applied PWM voltage and the current assumes the waveshape of the fundamental modulation signal with very small amounts of switching ripple.

Like the single-phase inverter based on the H-bridge topology, the inverter can deliver and accept both real and reactive power. In many cases, the DC bus is fed by a diode rectifier from the utility, which cannot pass power back to the AC input. The topology of a three-phase rectifier would be the same as shown in Fig. 14.36 with all IGBTs deleted.

A reversal of power flow in an inverter with a rectifier front end would lead to a steady rise of the DC bus voltage beyond permissible levels. If the power flow to the load is only reversing for brief periods of time, such as to brake a motor occasionally, the DC bus voltage could be limited by dissipating the power in a so-called brake resistor. To accommodate a brake resistor, inverter modules with an additional 7th IGBT (called "brake-chopper") are offered. This is shown in Fig. 14.38. For long-term regeneration, the

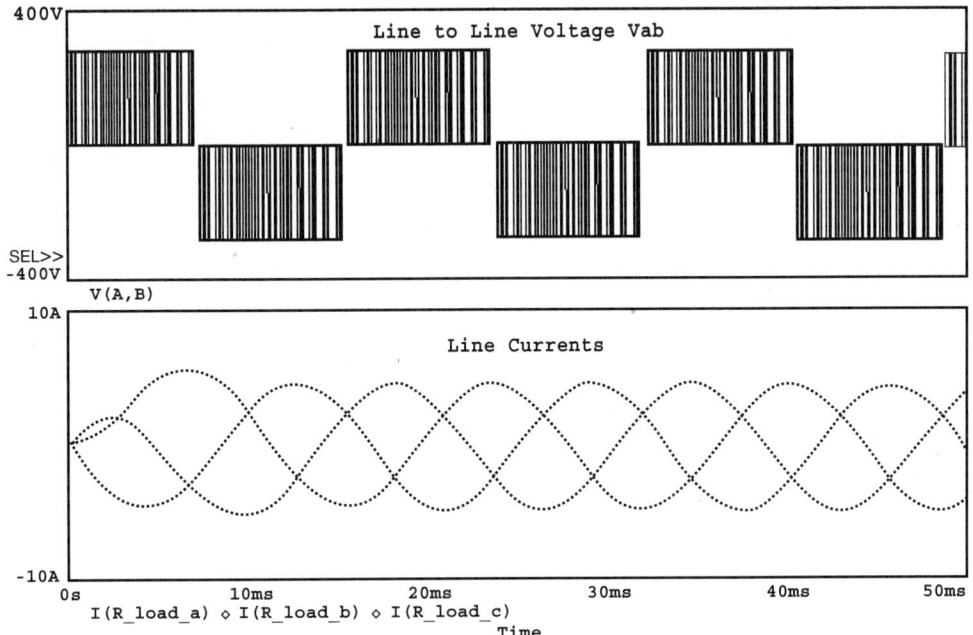

FIGURE 14.37 Typical waveforms of inverter voltages and currents.

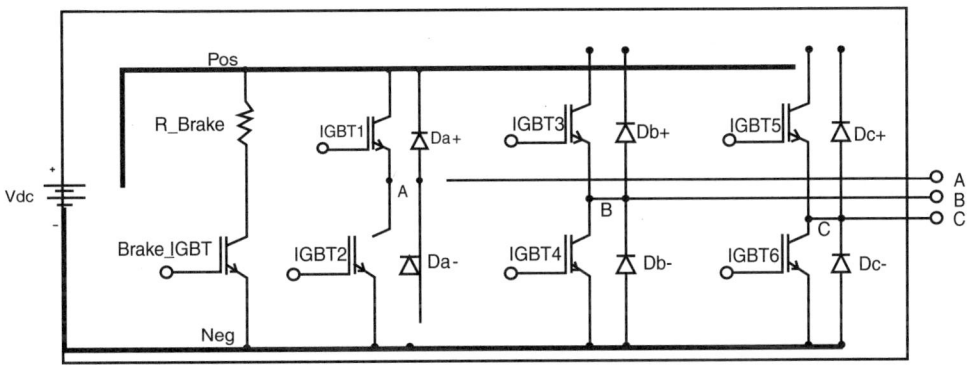

FIGURE 14.38 Topology of a three-phase inverter with brake-chopper IGBT.

rectifier can be replaced by an additional three-phase converter (Mohan et al., 1995). This additional converter is often called a controlled synchronous rectifier. The additional converter including its controller is of course much more expensive than a simple rectifier, but with this arrangement bidirectional power flow can be achieved. In addition, the interface toward the utility system can be managed such that the real and reactive power that is drawn from or delivered to the utility can be independently controlled. Also, the harmonics content of the current in the utility link can be reduced to almost zero. The topology for an arrangement like this is shown in Fig. 14.39.

The inverter shown in Fig. 14.36 provides a three-phase voltage without a neutral point. A fourth leg can be added to provide a four-wire system with a neutral point. Likewise four, five, or n-phase inverters can be realized by simply adding the appropriate number of phase legs.

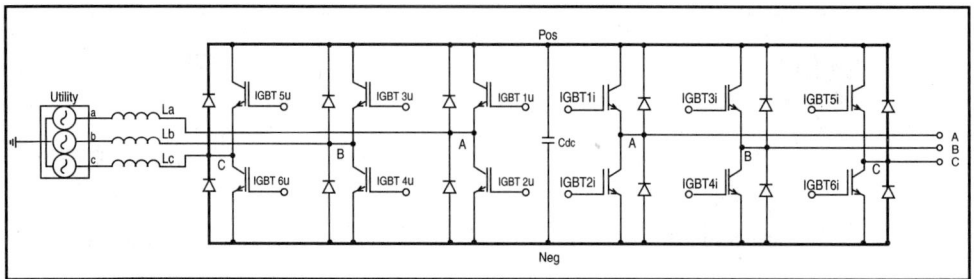

FIGURE 14.39 Topology of a three-phase inverter system for bidirectional power flow.

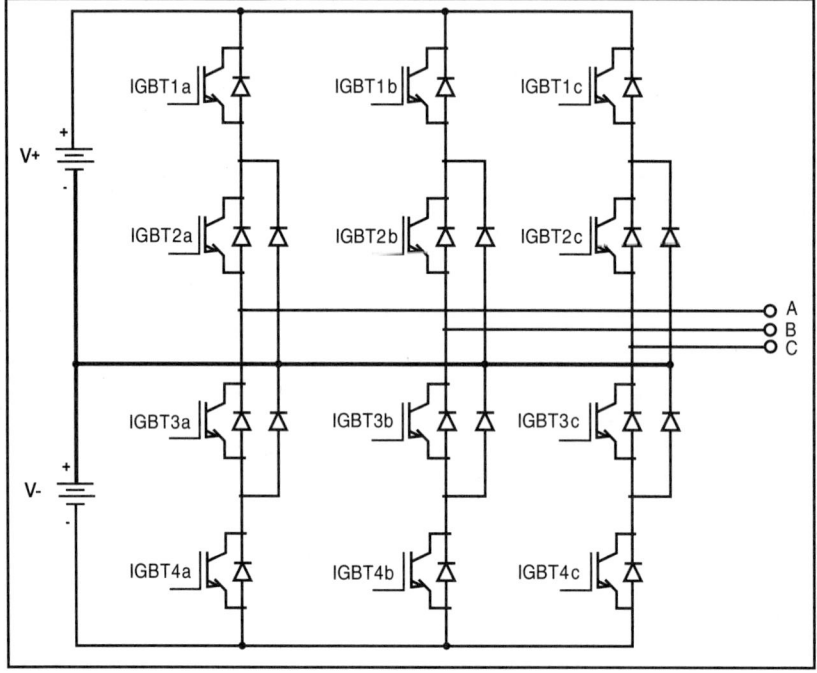

FIGURE 14.40 Topology of a three-level inverter.

As in single-phase inverters, the generation of the PWM control signals is done using modern microcontrollers and DSPs. These digital controllers are typically not only controlling just the inverter, but through the controlled synthesis of the appropriate voltages, motors and attached loads are controlled for high-performance dynamic response. The most commonly used control principle for superior dynamic response is called field-oriented or vector control (Bose, 1987; Bose, 1996; DeDonker and Novotny, 1988; Lorenz and Divan, 1990; Trzynadlowski, 1994).

Multilevel Inverters

Multilevel inverters are a class of inverters where a DC source with several tabs between the positive and negative terminal is present. The two main advantages of multilevel inverters are the higher voltage capability and the reduced harmonics content of the output waveform due to the multiple DC levels. The higher voltage capability is due to the fact that clamping diodes are used to limit the voltage stress

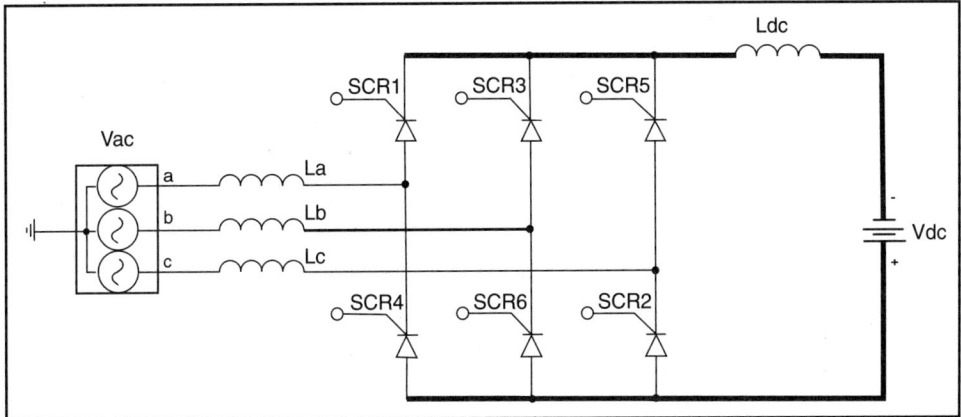

FIGURE 14.41 Line commutated converter in inverter mode.

on the IGBTs to the voltage differential between two tabs on the DC bus. Figure 14.40 shows the topology of a three-level inverter. Here, each phase leg consists of four IGBTs in series with additional antiparallel and clamping diodes. The output is again at the center-point of the phase leg. The output of each phase can be connected to the top DC bus, the center connection of the DC supply, or the negative DC bus. This amounts to three distinct voltage levels for the voltage of each phase, which explains the name of the circuit. It turns out that the resulting line-to-line voltage has five distinct levels in a three-phase inverter.

Line Commutated Inverters

Figure 14.41 shows the topology of a line commutated inverter. In Fig. 14.41 the SCRs are numbered according to their firing sequence. The circuit can operate both as a rectifier and an inverter. The mode of operation is controlled by the firing angle of the SCRs in the circuit (Ahmed, 1999; Barton, 1994; Mohan et al., 1995). The reference value for the firing angle α is the instant when the voltage across each SCR becomes positive, i.e., when an uncontrolled diode would turn on. This time corresponds to 30° past the positive going zero crossing of each phase. By delaying the turn-on angle α more than 90° past this instant, the polarity of the average DC bus voltage reverses and the circuit enters the inverter mode. The DC source in Fig. 14.41 shows the polarity of the DC voltage for inverter operation. The firing delay angle corresponds to the phase of the utility voltage. The maximum delay angle must be limited to less than 180°, to provide enough time for the next SCR in the sequence to acquire the load current. Equation (14.3) gives the value of the DC output voltage of the converter as a function of the delay angle α and the DC current I_{dc}, which is considered constant.

$$V_{dc} = \frac{3}{\pi} \cdot \left(\sqrt{2} \cdot V_{LL} \cdot \cos(\alpha) - \omega \cdot L_S \cdot I_{dc} \right) \qquad (14.3)$$

V_{LL} is the rms value of the AC line-to-line voltage, ω is the radian frequency of the AC voltage, and L_s is the value of the inductors La, Lb, and Lc in Fig. 14.41. Line commutated inverters have a negative impact on the utility voltage and a relatively low total power factor. Equation (14.4) gives an estimate of the total power factor of the circuit shown in Fig. 14.41 for constant DC current and negligible AC line reactors.

$$PF = \frac{3}{\pi} \cdot \cos(\alpha) \qquad (14.4)$$

References

Ahmed A., *Power Electronics for Technology,* Prentice-Hall, Upper Saddle River, NJ, 1999.
Barton, T. H., *Rectifiers, Cycloconverters, and AC Controllers,* Oxford University Press Inc., New York, 1994.
Bird, B. M., King, K.G., Pedder, D. A. G., *An Introduction to Power Electronics,* 2nd Edition, John Wiley & Sons, Inc., New York, 1993.
Bose, B. K., *Modern Power Electronics, Evolution, Technology, and Applications,* IEEE Press, Piscataway, NJ, 1992.
Bose, B. K., *Microcomputer Control of Power Electronics and Drives,* IEEE Press, Piscataway, NJ, 1987.
Bose, B. K., *Power Electronics and Variable Frequency Drives,* IEEE Press, Piscataway, NJ, 1996.
Brumsickle, W.E., D.M. Divan, T.A. Lipo, Reduced switching stress in high-voltage IGBT inverters via a three-level structure, *IEEE-APEC* 2, 544–550, February 1998.
Cash, M.A., and Habetler, T.G., Insulation failure prediction in induction machines using line-neutral voltages, *IEEE Trans. on Ind. Appl.,* 34, 6, 1234–1239, Nov/Dec 1998.
De Donker, R., and Novotny, D. W., The universal field-oriented controller, in *Conf. Rec. IEEE-IAS* 1988, 450–456.
Gottlieb, I. M., *Power Supplies, Switching Regulators, Inverters and Converters,* TAB Books, Inc., Blue Ridge Summit, PA, 1984.
Holtz, J., Pulsewidth modulation — A survey, *IEEE Trans. on Ind. Electr.,* 39, 5, 410–420, 1992.
Kassakian, J. G., Schlecht, M. F., Verghese, G. C., *Principles of Power Electronics,* Addison-Wesley Publishing Company, Reading, MA, 1991.
Lorenz, R.D., and Divan, D.M., Dynamic analysis and experimental evaluation of delta modulators for field oriented induction machines, *IEEE Trans. Ind. Appl.,* 26, 2, 296–301, 1990.
Mohan, N., Undeland, T., Robbins, W., *Power Electronics, Converters, Applications, and Design,* 2nd ed., John Wiley & Sons, New York, 1995.
Novotny, D. W., and Lipo, T. A., *Vector Control and Dynamics of AC Drives,* Oxford Science Publications, New York, 1996.
Rashid, M. H., *Power Electronics, Circuits, Devices, and Applications,* Prentice-Hall, Englewood Cliffs, NJ, 1993.
Shepherd, W., and Zand, P., *Energy Flow and Power Factor in Nonsinusoidal Circuits,* Cambridge University Press, London and New York, 1979.
Tarter, R. E., *Solid State Power Conversion Handbook,* John Wiley & Sons, Inc., New York, 1993.
Tolbert, L.M., Peng, F.Z., and Habetler, T.G., Multilevel converters for large electric drives, *IEEE Trans. Ind. Appl.*. 35, 1, 36–44, Jan/Feb 1999.
Trzynadlowski, A. M., *The Field Orientation Principle in Control of Induction Motors,* Kluwer Academic Publishers, Inc., Dordrecht, The Netherlands, 1994.
Von Jouanne, A., Rendusara, D., Enjeti, P., Gray, W., Filtering techniques to minimize the effect of long motor leads on PWM inverter fed AC motor drive systems, *IEEE Trans. Ind. Appl.,* July/Aug. 1996, 919–926.

14.4 Active Filters for Power Conditioning

Hirofumi Akagi

Much research has been performed on active filters for power conditioning and their practical applications since their basic principles of compensation were proposed around 1970 (Bird et al., 1969; Gyugyi and Strycula, 1976; Kawahira et al., 1983). In particular, recent remarkable progress in the capacity and switching speed of power semiconductor devices such as insulated-gate bipolar transistors (IGBTs) has spurred interest in active filters for power conditioning. In addition, state-of-the-art power electronics technology has enabled active filters to be put into practical use. More than one thousand sets of active filters consisting of voltage-fed pulse-width-modulation (PWM) inverters using IGBTs or gate-turn-off (GTO) thyristors are operating successfully in Japan.

Active filters for power conditioning provide the following functions:

- reactive-power compensation,
- harmonic compensation, harmonic isolation, harmonic damping, and harmonic termination,
- negative-sequence current/voltage compensation,
- voltage regulation.

The term "active filters" is also used in the field of signal processing. In order to distinguish active filters in power processing from active filters in signal processing, the term "active power filters" often appears in many technical papers or literature. However, the author prefers "active filters for power conditioning" to "active power filters," because the term "active power filters" is misleading to either "active filters for power" or "filters for active power." Therefore, this section takes the term "active filters for power conditioning" or simply uses the term "active filters" as long as no confusion occurs.

Harmonic-Producing Loads

Identified Loads and Unidentified Loads

Nonlinear loads drawing nonsinusoidal currents from utilities are classified into identified and unidentified loads. High-power diode/thyristor rectifiers, cycloconverters, and arc furnaces are typically characterized as identified harmonic-producing loads because utilities identify the individual nonlinear loads installed by high-power consumers on power distribution systems in many cases. The utilities determine the point of common coupling with high-power consumers who install their own harmonic-producing loads on power distribution systems, and also can determine the amount of harmonic current injected from an individual consumer.

A "single" low-power diode rectifier produces a negligible amount of harmonic current. However, multiple low-power diode rectifiers can inject a large amount of harmonics into power distribution systems. A low-power diode rectifier used as a utility interface in an electric appliance is typically considered as an unidentified harmonic-producing load. Attention should be paid to unidentified harmonic-producing loads as well as identified harmonic-producing loads.

Harmonic Current Sources and Harmonic Voltage Sources

In many cases, a harmonic-producing load can be represented by either a harmonic current source or a harmonic voltage source from a practical point of view. Figure 14.42(a) shows a three-phase diode rectifier with a DC link inductor L_d. When attention is paid to voltage and current harmonics, the rectifier can be considered as a harmonic current source shown in Fig. 14.42(b). The reason is that the load impedance is much larger than the supply impedance for harmonic frequency ω_h, as follows:

$$\sqrt{R_L^2 + (\omega_h L_d)^2} \gg \omega_h L_S.$$

Here, L_S is the sum of supply inductance existing upstream of the point of common coupling (PCC) and leakage inductance of a rectifier transformer. Note that the rectifier transformer is disregarded from Fig. 14.42(a). Figure 14.42(b) suggests that the supply harmonic current i_{Sh} is independent of L_S.

Figure 14.43(a) shows a three-phase diode rectifier with a DC link capacitor. The rectifier would be characterized as a harmonic voltage source shown in Fig. 14.43(b) if it is seen from its AC terminals. The reason is that the following relation exists:

$$\frac{1}{\omega_h C_d} \ll \omega_h L_S.$$

This implies that i_{Sh} is strongly influenced by the inductance value of L_S.

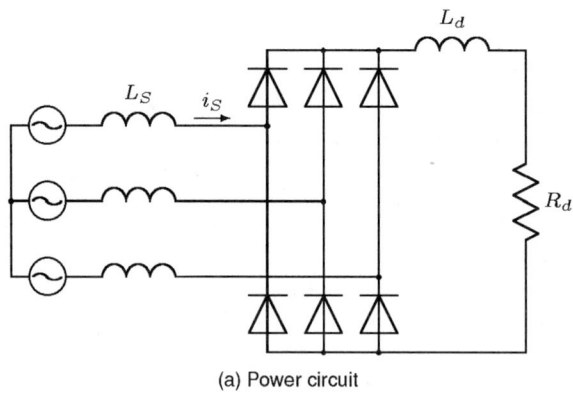

(a) Power circuit

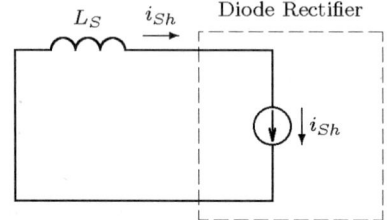

(b) Equivalent circuit for harmonic on a per-phase base

FIGURE 14.42 Diode rectifier with inductive load.

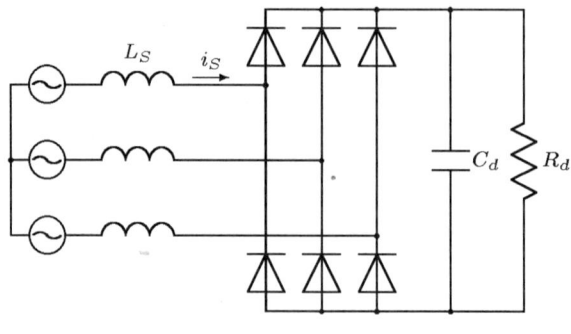

(a) Power circuit

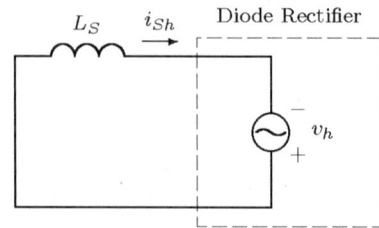

(b) Equivalent circuit for harmonic on a per-phase base

FIGURE 14.43 Diode rectifier with capacitive load.

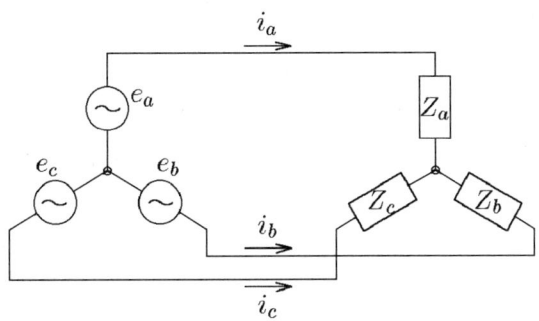

FIGURE 14.44 Three-phase three-wire system.

Theoretical Approach to Active Filters for Power Conditioning
The Akagi-Nabae Theory

The theory of instantaneous power in three-phase circuits is referred to as the "Akagi-Nabae theory" (Akagi et al., 1983; Akagi et al., 1984). Figure 14.44 shows a three-phase three-wire system on the *a-b-c* coordinates, where no zero-sequence voltage is included in the three-phase three-wire system. Applying the theory to Fig. 14.44 can transform the three-phase voltages and currents on the *a-b-c* coordinates into the two-phase voltages and currents on the α-β coordinates, as follows:

$$\begin{bmatrix} e_\alpha \\ e_\beta \end{bmatrix} = \sqrt{\frac{2}{3}} \begin{bmatrix} 1 & -1/2 & -1/2 \\ 0 & \sqrt{3}/2 & -\sqrt{3}/2 \end{bmatrix} \begin{bmatrix} e_a \\ e_b \\ e_c \end{bmatrix} \quad (14.5)$$

$$\begin{bmatrix} i_\alpha \\ i_\beta \end{bmatrix} = \sqrt{\frac{2}{3}} \begin{bmatrix} 1 & -1/2 & -1/2 \\ 0 & \sqrt{3}/2 & -\sqrt{3}/2 \end{bmatrix} \begin{bmatrix} i_a \\ i_b \\ i_c \end{bmatrix}. \quad (14.6)$$

As is well known, the instantaneous real power either on the *a-b-c* coordinates or on the α-β coordinates is defined by

$$p = e_a i_a + e_b i_b + e_c i_c = e_\alpha i_\alpha + e_\beta i_\beta. \quad (14.7)$$

To avoid confusion, *p* is referred to as three-phase instantaneous real power. According to the theory, the three-phase instantaneous imaginary power, *q*, is defined by

$$q = e_\alpha i_\beta - e_\beta i_\alpha. \quad (14.8)$$

The combination of the above two equations bears the following basic formulation:

$$\begin{bmatrix} p \\ q \end{bmatrix} = \begin{bmatrix} e_\alpha & e_\beta \\ -e_\beta & e_\alpha \end{bmatrix} \begin{bmatrix} i_\alpha \\ i_\beta \end{bmatrix}. \quad (14.9)$$

Here, $e_\alpha \cdot i_\alpha$ or $e_\beta \cdot i_\beta$ obviously means instantaneous power in the α-phase or the β-phase because either is defined by the product of the instantaneous voltage in one phase and the instantaneous current in the same phase. Therefore, p has a dimension of [W]. Conversely, neither $e_\alpha \cdot i_\beta$ nor $e_\beta \cdot i_\alpha$ means instantaneous power because either is defined by the product of the instantaneous voltage in one phase and the instantaneous current in the other phase. Accordingly, q is quite different from p in dimension and electric property although q looks similar in formulation to p. A common dimension for q should be introduced from both theoretical and practical points of view. A good candidate is [IW], that is, "imaginary watt."

Equation (14.9) is changed into the following equation:

$$\begin{bmatrix} i_\alpha \\ i_\beta \end{bmatrix} = \begin{bmatrix} e_\alpha & e_\beta \\ -e_\beta & e_\alpha \end{bmatrix} \begin{bmatrix} p \\ q \end{bmatrix} \tag{14.10}$$

Note that the determinant with respect to e_α and e_β in Eq. (14.9) is not zero. The instantaneous currents on the α-β coordinates, i_α and i_β, are divided into two kinds of instantaneous current components, respectively:

$$\begin{bmatrix} i_\alpha \\ i_\beta \end{bmatrix} = \begin{bmatrix} e_\alpha & e_\beta \\ -e_\beta & e_\alpha \end{bmatrix}^{-1} \begin{bmatrix} p \\ 0 \end{bmatrix} + \begin{bmatrix} e_\alpha & e_\beta \\ -e_\beta & e_\alpha \end{bmatrix}^{-1} \begin{bmatrix} 0 \\ q \end{bmatrix}$$

$$\equiv \begin{bmatrix} i_{\alpha p} \\ i_{\beta p} \end{bmatrix} + \begin{bmatrix} i_{\alpha q} \\ i_{\beta q} \end{bmatrix} \tag{14.11}$$

Let the instantaneous powers in the α-phase and the β-phase be p_α and p_β, respectively. They are given by the conventional definition as follows:

$$\begin{bmatrix} p_\alpha \\ p_\beta \end{bmatrix} = \begin{bmatrix} e_\alpha i_\alpha \\ e_\beta i_\beta \end{bmatrix} = \begin{bmatrix} e_\alpha i_{\alpha p} \\ e_\beta i_{\beta p} \end{bmatrix} + \begin{bmatrix} e_\alpha i_{\alpha q} \\ e_\beta i_{\beta q} \end{bmatrix} \tag{14.12}$$

The three-phase instantaneous real power, p, is given as follows, by using Eqs. (14.11) and (14.12):

$$p = p_\alpha + p_\beta = e_\alpha i_{\alpha p} + e_\beta i_{\beta p} + e_\alpha i_{\alpha q} + e_\beta i_{\beta q}$$

$$= \frac{e_\alpha^2}{e_\alpha^2 + e_\beta^2} p + \frac{e_\beta^2}{e_\alpha^2 + e_\beta^2} p + \frac{-e_\alpha e_\beta}{e_\alpha^2 + e_\beta^2} q + \frac{e_\alpha e_\beta}{e_\alpha^2 + e_\beta^2} q \tag{14.13}$$

The sum of the third and fourth terms on the right-hand side in Eq. (14.13) is always zero. From Eqs. (14.12) and (14.13), the following equations are obtained:

$$p = e_\alpha i_{\alpha p} + e_\beta i_{\beta p} \equiv p_{\alpha p} + p_{\beta p} \tag{14.14}$$

$$0 = e_\alpha i_{\alpha q} + e_\beta i_{\beta q} \equiv p_{\alpha q} + p_{\beta q}. \tag{14.15}$$

Inspection of Eqs. (14.14) and (14.15) leads to the following essential conclusions:

- The sum of the power components, $p_{\alpha p}$ and $p_{\beta p}$, coincides with the three-phase instantaneous real power, p, which is given by Eq. (14.7). Therefore, $p_{\alpha p}$ and $p_{\beta p}$ are referred to as the α-phase and β-phase instantaneous active powers.

- The other power components, $p_{\alpha q}$ and $p_{\beta q}$, cancel each other and make no contribution to the instantaneous power flow from the source to the load. Therefore, $p_{\alpha q}$ and $p_{\beta q}$ are referred to as the α-phase and β-phase instantaneous reactive powers.
- Thus, a shunt active filter without energy storage can achieve instantaneous compensation of the current components, $i_{\alpha q}$ and $i_{\beta q}$ or the power components, $p_{\alpha q}$ and $p_{\beta q}$. In other words, the Akagi-Nabae theory based on Eq. (14.9) exactly reveals what components the active filter without energy storage can eliminate from the α-phase and β-phase instantaneous currents, i_α and i_β or the α-phase and β-phase instantaneous real powers, p_α and p_β.

Energy Storage Capacity

Figure 14.45 shows a system configuration of a shunt active filter for harmonic compensation of a diode rectifier, where the main circuit of the active filter consists of a three-phase voltage-fed PWM inverter and a DC capacitor, C_d. The active filter is controlled to draw the compensating current, i_{AF}, from the utility, so that the compensating current cancels the harmonic current flowing on the AC side of the diode rectifier with a DC link inductor.

Referring to Eq. (14.10) yields the α-phase and β-phase compensating currents,

$$\begin{bmatrix} i_{AF\alpha} \\ i_{AF\beta} \end{bmatrix} = \begin{bmatrix} e_\alpha & e_\beta \\ -e_\beta & e_\alpha \end{bmatrix}^{-1} \begin{bmatrix} p_{AF} \\ q_{AF} \end{bmatrix}. \tag{14.16}$$

Here, p_{AF} and q_{AF} are the three-phase instantaneous real and imaginary power on the AC side of the active filter, and they are usually extracted from p_L and q_L. Note that p_L and q_L are the three-phase instantaneous real and imaginary power on the AC side of a harmonic-producing load. For instance, when the active filter compensates for the harmonic current produced by the load, the following relationships exist:

$$p_{AF} = -\tilde{p}_L, \quad q_{AF} = -\tilde{q}_L. \tag{14.17}$$

Here, $\tilde{p}_L$ and $\tilde{q}_L$ are AC components of p_L and q_L, respectively. Note that the DC components of p_L and q_L correspond to the fundamental current present in i_L and the AC components to the harmonic current. In general, two high-pass filters in the control circuit extract $\tilde{p}_L$ from p_L and $\tilde{q}_L$ from q_L.

The active filter draws p_{AF} from the utility, and delivers it to the DC capacitor if no loss is dissipated in the active filter. Thus, p_{AF} induces voltage fluctuation of the DC capacitor. When the amplitude of p_{AF} is assumed to be constant, the lower the frequency of the AC component, the larger the voltage fluctuation (Akagi et al., 1984; Akagi et al., 1986). If the period of the AC component is one hour, the DC capacitor

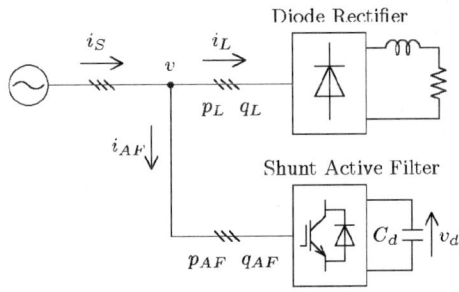

FIGURE 14.45 Shunt active filter.

has to absorb or release electric energy given by integration of p_{AF} with respect to time. Thus, the following relationship exists between the instantaneous voltage across the DC capacitor, v_d and p_{AF}:

$$\frac{1}{2}C_d v_d^2(t) = \frac{1}{2}C_d v_d^2(0) + \int_0^t p_{AF} dt. \quad (14.18)$$

This implies that the active filter needs an extremely large-capacity DC capacitor to suppress the voltage fluctuation coming from achieving "harmonic" compensation of $\tilde{p}_L$. Hence, the active filter is no longer a harmonic compensator, and thereby it should be referred to as a "DC capacitor-based energy storage system," although it is impractical at present. In this case, the main purpose of the voltage-fed PWM inverter is to perform an interface between the utility and the bulky DC capacitor.

The active filter seems to "draw" q_{AF} from the utility, as shown in Fig. 14.45. However, q_{AF} makes no contribution to energy transfer in the three-phase circuit. No energy storage, therefore, is required to the active filter, independent of q_{AF}, whenever $p_{AF} = 0$.

Classification of Active Filters

Various types of active filters have been proposed in technical literature (Moran, 1989; Grady et al., 1990; Akagi, 1994; Akagi and Fujita, 1995; Fujita and Akagi, 1997; Aredes et al., 1998). Classification of active filters is made from different points of view (Akagi, 1996). Active filters are divided into AC and DC filters. Active DC filters have been designed to compensate for current and/or voltage harmonics on the DC side of thyristor converters for high-voltage DC transmission systems (Watanabe, 1990; Zhang et al., 1993) and on the DC link of a PWM rectifier/inverter for traction systems. Emphasis, however, is put on active AC filter in the following because the term "active filters" refers to active AC filters in most cases.

Classification by Objectives: Who is Responsible for Installing Active Filters?

The objective of "who is responsible for installing active filters" classifies them into the following two groups:

- Active filters installed by *individual consumers* on their own premises in the vicinity of one or more identified harmonic-producing loads.
- Active filters being installed by *electric power utilities* in substations and/or on distribution feeders.

Individual consumers should pay attention to current harmonics produced by their own harmonic-producing loads, and thereby the active filters installed by the individual consumers are aimed at compensating for current harmonics.

Utilities should concern themselves with voltage harmonics, and therefore active filters will be installed by utilities in the near future for the purpose of compensating for voltage harmonics and/or of achieving "harmonic damping" throughout power distribution systems or "harmonic termination" of a radial power distribution feeder. The section titled "Practical Applications of Active Filters for Power Conditioning" describes a shunt active filter intended for installation by electric power utilities on the end bus of a power distribution line.

Classification by System Configuration

Shunt Active Filters and Series Active Filters

A standalone shunt active filter shown in Fig. 14.45 is one of the most fundamental system configurations. The active filter is controlled to draw a compensating current, i_{AF} from the utility, so that it cancels current harmonics on the AC side of a general-purpose diode/thyristor rectifier (Akagi et al., 1990; Peng et al., 1990; Bhattacharya et al., 1998) or a PWM rectifier for traction systems (Krah and Holtz, 1994). Generally, the shunt active filter is suitable for harmonic compensation of a current harmonic source such as diode/thyristor rectifier with a DC link inductor. The shunt active filter has the capability of damping harmonic resonance between an existing passive filter and the supply impedance.

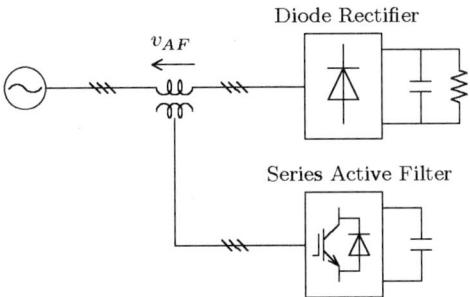

FIGURE 14.46 Series active filter.

TABLE 14.3 Comparison of Shunt Active Filters and Series Active Filters

	Shunt Active Filter	Series Active Filter
System configuration	Figure 14.45	Figure 14.46
Power circuit of active filter	Voltage-fed PWM inverter **with** current minor loop	Voltage-fed PWM inverter **without** current minor loop
Active filter acts as	Current source: i_{AF}	Voltage source: v_{AF}
Harmonic-producing load suitable	Diode/thyristor rectifiers with **inductive** loads, and cycloconverters	Large-capacity diode rectifiers with **capacitive** loads
Additional function	Reactive power compensation	AC voltage regulation
Present situation	Commercial stage	Laboratory stage

Figure 14.46 shows a system configuration of a series active filter used alone. The series active filter is connected in series with the utility through a matching transformer, so that it is suitable for harmonic compensation of a voltage harmonic source such as a large-capacity diode rectifier with a DC link capacitor. The series active filter integrated into a diode rectifier with a DC common capacitor is discussed in section V. Table 14.3 shows comparisons between the shunt and series active filters. This concludes that the series active filter has a "dual" relationship in each item with the shunt active filter (Akagi, 1996; Peng, 1998).

Hybrid Active/Passive Filters

Figures 14.47, 14.48, and 14.49 show three types of hybrid active/passive filters, the main purpose of which is to reduce initial costs and to improve efficiency. The shunt passive filter consists of one or more tuned LC filters and/or a high-pass filter. Table 14.4 shows comparisons among the three hybrid filters in which the active filters are different in function from the passive filters. Note that the hybrid filters are applicable to any current harmonic source, although a harmonic-producing load is represented by a thyristor rectifier with a DC link inductor in Figs. 14.47, 14.48, and 14.49.

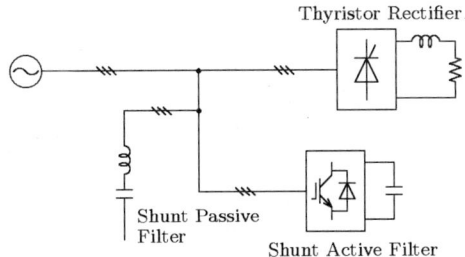

FIGURE 14.47 Combination of shunt active filter and shunt passive filter.

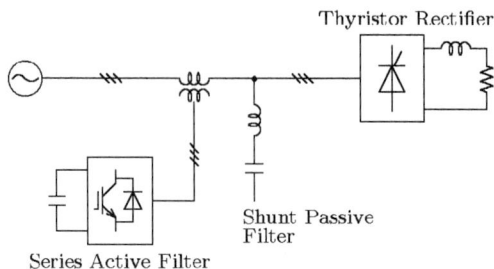

FIGURE 14.48 Combination of series active filter and shunt passive filter.

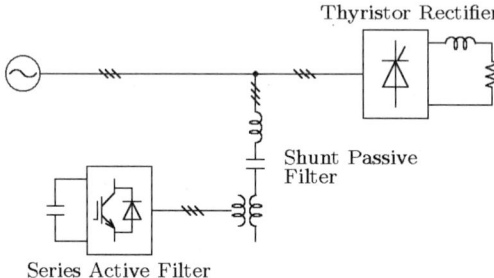

FIGURE 14.49 Series active filter connected in series with shunt passive filter.

TABLE 14.4 Comparison of Hybrid Active/Passive Filters

	Shunt Active Filter Plus Shunt Passive Filter	Series Active Filter Plus Shunt Passive Filter	Series Active Filter Connected in Series with Shunt Passive Filter
System configuration	Figure 14.47	Figure 14.48	Figure 14.49
Power circuit of active filter	• Voltage-fed PWM inverter **with** current minor loop	• Voltage-fed PWM inverter **without** current minor loop	• Voltage-fed PWM inverter **with** or **without** current minor loop
Function of active filter	• Harmonic compensation	• Harmonic isolation	• Harmonic isolation or harmonic compensation
Advantages	• General shunt active filters applicable • Reactive power controllable	• Already existing shunt passive filters applicable • No harmonic current flowing through active filter	• Already existing shunt passive filters applicable • Easy protection of active filter
Problems or issues	• Share compensation in frequency domain between active filter and passive filter	• Difficult to protect active filter against overcurrent • No reactive power control	• No reactive power control
Present situation	• Commercial stage	• A few practical applications	• Commercial stage

Such a combination of a shunt active filter and a shunt passive filter as shown in Fig. 14.47 has already been applied to harmonic compensation of naturally-commutated twelve-pulse cycloconverters for steel mill drives (Takeda et al., 1987). The passive filters absorbs 11th and 13th harmonic currents while the active filter compensates for 5th and 7th harmonic currents and achieves damping of harmonic resonance between the supply and the passive filter. One of the most important considerations in system design is to avoid competition for compensation between the passive filter and the active filter.

The hybrid active filters, shown in Fig. 14.48 (Peng et al., 1990; Peng et al., 1993; Kawaguchi et al., 1997) and in Fig. 14.49 (Fujita and Akagi, 1991; Balbo et al., 1994; van Zyl et al., 1995), are right now on the commercial stage, not only for harmonic compensation but also for harmonic isolation between

Power Electronics

supply and load, and for voltage regulation and imbalance compensation. They are considered prospective alternatives to pure active filters used alone. Other combined systems of active filters and passive filters or LC circuits have been proposed in Bhattacharya et al. (1997).

Classification by Power Circuit

There are two types of power circuits used for active filters; a voltage-fed PWM inverter (Akagi et al., 1986; Takeda et al., 1987) and a current-fed PWM inverter (Kawahira et al., 1983; van Schoor and van Wyk, 1987). These are similar to the power circuits used for AC motor drives. They are, however, different in their behavior because active filters act as nonsinusoidal current or voltage sources. The author prefers the voltage-fed to the current-fed PWM inverter because the voltage-fed PWM inverter is higher in efficiency and lower in initial costs than the current-fed PWM inverter (Akagi, 1994). In fact, almost all active filters that have been put into practical application in Japan have adopted the voltage-fed PWM inverter as the power circuit.

Classification by Control Strategy

The control strategy of active filters has a great impact not only on the compensation objective and required kVA rating of active filters, but also on the filtering characteristics in transient state as well as in steady state (Akagi et al., 1986).

Frequency-Domain and Time-Domain

There are mainly two kinds of control strategies for extracting current harmonics or voltage harmonics from the corresponding distorted current or voltage; one is based on the Fourier analysis in the frequency-domain (Grady et al., 1990), and the other is based on the Akagi-Nabae theory in the time-domain. The concept of the Akagi-Nabae theory in the time-domain has been applied to the control strategy of almost all the active filters installed by individual high-power consumers over the last ten years in Japan.

Harmonic Detection Methods

Three kinds of harmonic detection methods in the time-domain have been proposed for shunt active filters acting as a current source i_{AF}. Taking into account the polarity of the currents i_S, i_L and i_{AF} in Fig. 14.45 gives

$$\text{load-current detection:} \quad i_{AF} = -i_{Lh}$$

$$\text{supply-current detection:} \quad i_{AF} = K_S \cdot i_{Sh}$$

$$\text{voltage detection:} \quad i_{AF} = K_V \cdot v_h.$$

Note that load-current detection is based on feedforward control, while supply-current detection and voltage detection are based on feedback control with gains of K_S and K_V, respectively. Load-current detection and supply-current detection are suitable for shunt active filters installed in the vicinity of one or more harmonic-producing loads by individual consumers. Voltage detection is suitable for shunt active filters that will be dispersed on power distribution systems by utilities, because the shunt active filter based on voltage detection is controlled in such a way to present infinite impedance to the external circuit for the fundamental frequency, and to present a resistor with low resistance of $1/K_V$ [Ω] for harmonic frequencies (Akagi et al., 1999).

Supply-current detection is the most basic harmonic detection method for series active filters acting as a voltage source v_{AF}. Referring to Fig. 14.46 yields

$$\text{supply current detection:} \quad v_{AF} = G \cdot i_{Sh}.$$

The series active filter based on supply-current detection is controlled in such a way to present zero impedance to the external circuit for the fundamental frequency and to present a resistor with high resistance of $G\,[\Omega]$ for the harmonic frequencies. The series active filters shown in Fig. 14.46 (Fujita and Akagi, 1997) and Fig. 14.48 (Peng et al., 1990) are based on supply current detection.

Integrated Series Active Filters

A small-rated series active filter integrated with a large-rated double-series diode rectifier has the following functions (Fujita and Akagi, 1997):

- harmonic compensation of the diode rectifier,
- voltage regulation of the common DC bus,
- damping of harmonic resonance between the communication capacitors connected across individual diodes and the leakage inductors including the AC line inductors,
- reduction of current ripples flowing into the electrolytic capacitor on the common DC bus.

System Configuration

Figure 14.50 shows a harmonic current-free AC/DC power conversion system described below. It consists of a combination of a double-series diode rectifier of 5 kW and a series active filter with a peak voltage and current rating of 0.38 kVA. The AC terminals of a single-phase H-bridge voltage-fed PWM inverter are connected in "series" with a power line through a single-phase matching transformer, so that the combination of the matching transformers and the PWM inverters forms the "series" active filter. For small to medium-power systems, it is economically practical to replace the three single-phase inverters with a single three-phase inverter using six IGBTs. A small-rated high-pass filter for suppression of switching ripples is connected to the AC terminals of each inverter in the experimental system, although it is eliminated from Fig. 14.50 for the sake of simplicity.

The primary windings of the Y-Δ and Δ-Δ connected transformers are connected in "series" with each other, so that the combination of the three-phase transformers and two three-phase diode rectifiers forms the "double-series" diode rectifier, which is characterized as a three-phase twelve-pulse rectifier. The DC terminals of the diode rectifier and the active filter form a common DC bus equipped with an electrolytic capacitor. This results not only in eliminating any electrolytic capacitor from the active filter, but also in reducing current ripples flowing into the electrolytic capacitor across the common DC bus.

Connecting only a commutation capacitor C in parallel with each diode plays an essential role in reducing the required peak voltage rating of the series active filter.

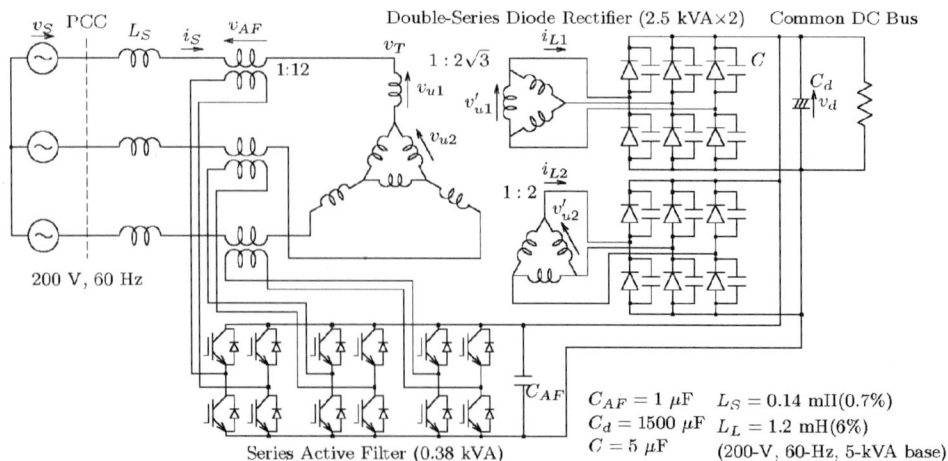

FIGURE 14.50 The harmonic current-free ac/dc power conversion system.

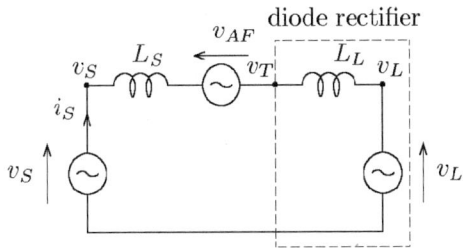

FIGURE 14.51 Single-phase equivalent circuit.

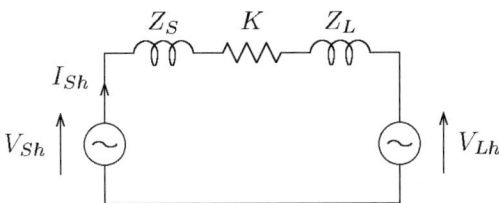

FIGURE 14.52 Single-phase equivalent circuit with respect to harmonics.

Operating Principle

Figure 14.51 shows an equivalent circuit for the power conversion system on a per-phase basis. The series active filter is represented as an AC voltage source v_{AF}, and the double-series diode rectifier as the series connection of a leakage inductor L_L of the transformers with an AC voltage source v_L. The reason for providing the AC voltage source to the equivalent model of the diode rectifier is that the electrolytic capacitor C_d is directly connected to the DC terminal of the diode rectifier, as shown in Fig. 14.50.

The active filter is controlled in such a way as to present zero impedance for the fundamental frequency and to act as a resistor with high resistance of $K\ [\Omega]$ for harmonic frequencies. The AC voltage of the active filter, which is applied to a power line through the matching transformer, is given by

$$v_{AF}^* = K \cdot i_{Sh} \tag{14.19}$$

where i_{Sh} is a supply harmonic current drawn from the utility. Note that v_{AF} and i_{Sh} are instantaneous values. Figure 14.52 shows an equivalent circuit with respect to current and voltage harmonics in Fig. 14.51. Referring to Fig. 14.52 enables derivation of the following basic equations:

$$I_{Sh} = \frac{V_{Sh} - V_{Lh}}{Z_S + Z_L + K} \tag{14.20}$$

$$V_{AF} = \frac{K}{Z_S + Z_L + K}(V_{Sh} - V_{Lh}) \tag{14.21}$$

where V_{AF} is equal to the harmonic voltage appearing across the resistor K in Fig. 14.51.

If $K \gg Z_S + Z_L$, Eqs. (14.20) and (14.21) are changed into the following simple equations.

$$I_{Sh} \approx 0 \tag{14.22}$$

$$V_{AF} \approx V_{Sh} - V_{Lh}. \tag{14.23}$$

Equation (14.22) implies that an almost purely sinusoidal current is drawn from the utility. As a result, each diode in the diode rectifier continues conducting during a half cycle. Equation (14.23) suggests that the harmonic voltage V_{Lh}, which is produced by the diode rectifier, appears at the primary terminals of the transformers in Fig. 14.50, although it does not appear upstream of the active filter or at the utility-consumer point of common coupling (PCC).

Control Circuit

Figure 14.53 shows a block diagram of a control circuit based on hybrid analog/digital hardware. The concept of the Akagi-Nabae theory (Akagi, 1983; Akagi, 1984) is applied to the control circuit implementation. The p-q transformation circuit executes the following calculation to convert the three-phase supply current i_{Su}, i_{Sv}, and i_{Sw} into the instantaneous active current i_p and the instantaneous reactive current i_q.

$$\begin{bmatrix} i_p \\ i_q \end{bmatrix} = \sqrt{\frac{2}{3}} \begin{bmatrix} \cos \omega t & \sin \omega t \\ -\sin \omega t & \cos \omega t \end{bmatrix} \cdot \begin{bmatrix} 1 & -1/2 & -1/2 \\ 0 & \sqrt{3}/2 & -\sqrt{3}/2 \end{bmatrix} \begin{bmatrix} i_{Su} \\ i_{Sv} \\ i_{Sw} \end{bmatrix}. \tag{14.24}$$

The fundamental components in i_{Su}, i_{Sv}, and i_{Sw} correspond to DC components in i_p and i_q, and harmonic components to AC components. Two first-order high-pass-filters (HPFs) with the same cut-off frequency of 10 Hz as each other extract the AC components $\tilde{i}_p$ and $\tilde{i}_q$ from i_p and i_q, respectively. Then, the p-q transformation/inverse transformation of the extracted AC components produces the following supply harmonic currents:

$$\begin{bmatrix} i_{Shu} \\ i_{Shv} \\ i_{Shw} \end{bmatrix} = \sqrt{\frac{2}{3}} \begin{bmatrix} 1 & 0 \\ -1/2 & \sqrt{3}/2 \\ -1/2 & -\sqrt{3}/2 \end{bmatrix} \cdot \begin{bmatrix} \cos \omega t & -\sin \omega t \\ \sin \omega t & \cos \omega t \end{bmatrix} \begin{bmatrix} \tilde{i}_p \\ \tilde{i}_q \end{bmatrix}. \tag{14.25}$$

Each harmonic current is amplified by a gain of K, and then it is applied to the gate control circuit of the active filter as a voltage reference v_{AF}^* in order to regulate the common DC bus voltage, v_{AFf}^* is divided by the gain of K, and then it is added to $\tilde{i}_p$.

The PLL (phase locked loop) circuit produces phase information ωt which is a 12-bit digital signal of 60×2^{12} samples per second. Digital signals, $\sin \omega t$ and $\cos \omega t$, are generated from the phase information, and then they are applied to the p-q (inverse) transformation circuits. Multifunction in the transformation circuits is achieved by means of eight multiplying D/A converters. Each voltage reference, v_{AF}^* is compared with two repetitive triangular waveforms of 10 kHz in order to generate the gate signals for

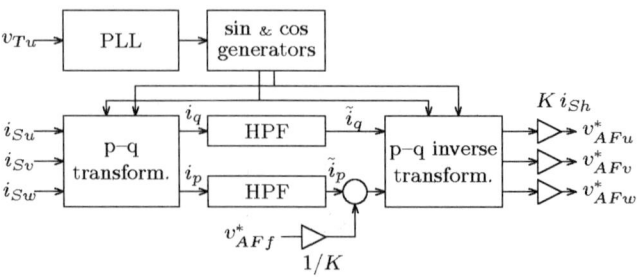

FIGURE 14.53 Control circuit for the series active filter.

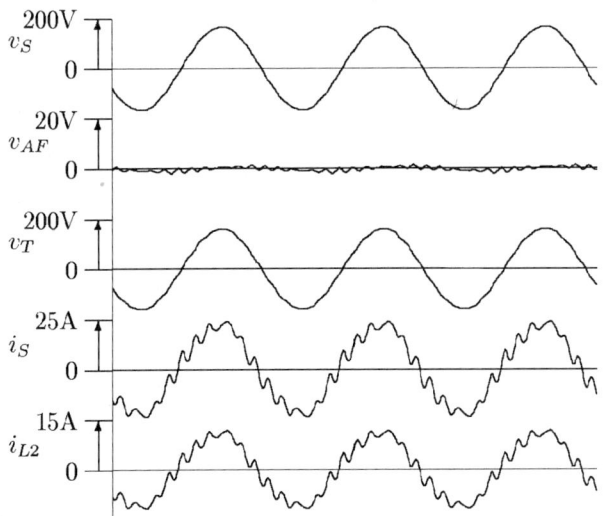

FIGURE 14.54 Experimental waveforms before starting the series active filter.

the IGBTs. The two triangular waveforms have the same frequency, but one has polarity opposite to the other, so that the equivalent switching frequency of each inverter is 20 kHz, which is twice as high as that of the triangular waveforms.

Experimental Results

In the following experiment, the control gain of the active filter, K, is set to 27 Ω, which is equal to 3.3 p.u. on a 3ϕ 200-V, 15-A, 60-Hz basis. Equation (14.20) suggests that the higher the control gain, the better the performance of the active filter. An extremely high gain, however, may make the control system unstable, and thereby a trade-off between performance and stability exists in determining an optimal control gain. A constant load resistor is connected to the common DC bus, as shown in Fig. 14.50.

Figures 14.54 and 14.55 show experimental waveforms, where a 5-μF commutation capacitor is connected in parallel with each diode used for the double-series diode rectifier. Table 14.5 shows the THD of i_S and the ratio of each harmonic current with respect to the fundamental current contained in i_S. Before starting the active filter, the supply 11th and 13th harmonic currents in Fig. 14.54 are slightly magnified due to resonance between the commutation capacitors C and the AC line and leakage inductors, L_S and L_L. Nonnegligible amounts of 3rd, 5th, and 7th harmonic currents, which are so-called "non-characteristic current harmonics" for the three-phase twelve-pulse diode rectifier, are drawn from the utility.

Figure 14.55 shows experimental waveforms where the peak voltage of the series active filter is imposed on a limitation of ±12 V inside the control circuit based on hybrid analog/digital hardware. Note that the limitation of ±12 V to the peak voltage is equivalent to the use of three single-phase matching transformers with turn ratios of 1:20 under the common DC link voltage of 240 V. After starting the active filter, a sinusoidal current with a leading power factor of 0.96 is drawn because the active filter acts as a high resistor of 27 Ω, having the capability of compensating for both voltage harmonics V_{Sh} and V_{Lh}, as well as of damping the resonance. As shown in Fig. 14.55, the waveforms of i_S and v_T are not affected by the voltage limitation, although the peak voltage v_{AF} frequently reaches the saturation or limitation voltage of ±12 V.

The required peak voltage and current rating of the series active filter in Fig. 14.55 is given by

$$3 \times 12^V / \sqrt{2} \times 15^A = 0.38 \text{ kVA}, \qquad (14.26)$$

which is only 7.6% of the kVA-rating of the diode rectifier.

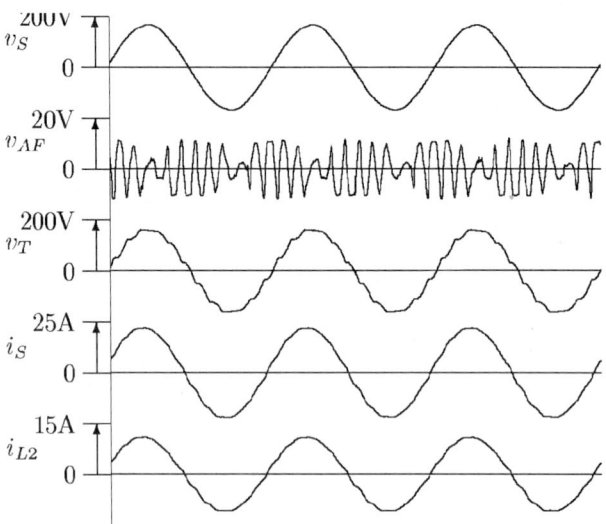

FIGURE 14.55 Experimental waveforms after starting the series active filter.

TABLE 14.5 Supply Current THD and Harmonics Expressed as the Harmonic-to-Fundamental Current Ratio [%], Where Commutation Capacitors of 5 µF are Connected

	THD	3rd	5th	7th	11th	13th
Before (Fig. 14.54)	16.8	5.4	2.5	2.2	12.3	9.5
After (Fig. 14.55)	1.6	0.7	0.2	0.4	0.8	1.0

The harmonic current-free AC-to-DC power conversion system has both practical and economical advantages. Hence, it is expected to be used as a utility interface with large industrial inverter-based loads such as multiple adjustable speed drives and uninterruptible power supplies in the range of 1–10 MW.

Practical Applications of Active Filters for Power Conditioning

Present Status and Future Trends

Shunt active filters have been put into practical applications mainly for harmonic compensation, with or without reactive-power compensation. Table 14.6 shows ratings and application examples of shunt active filters classified by compensation objectives.

Applications of shunt active filters are expanding, not only into industry and electric power utilities but also into office buildings, hospitals, water supply utilities, and rolling stock. At present, voltage-fed

TABLE 14.6 Shunt Active Filters on Commercial Base in Japan

Objective	Rating	Switching Devices	Applications
Harmonic compensation with or without reactive/negative-sequence current compensation	10 kVA ~ 2 MVA	IGBTs	Diode/thyristor rectifiers and cycloconverters for industrial loads
Voltage flicker compensation	5 MVA ~ 50 MVA	GTO thyristors	Arc furnaces
Voltage regulation	40 MVA ~ 60 MVA	GTO thyristors	Shinkansen (Japanese "bullet" trains)

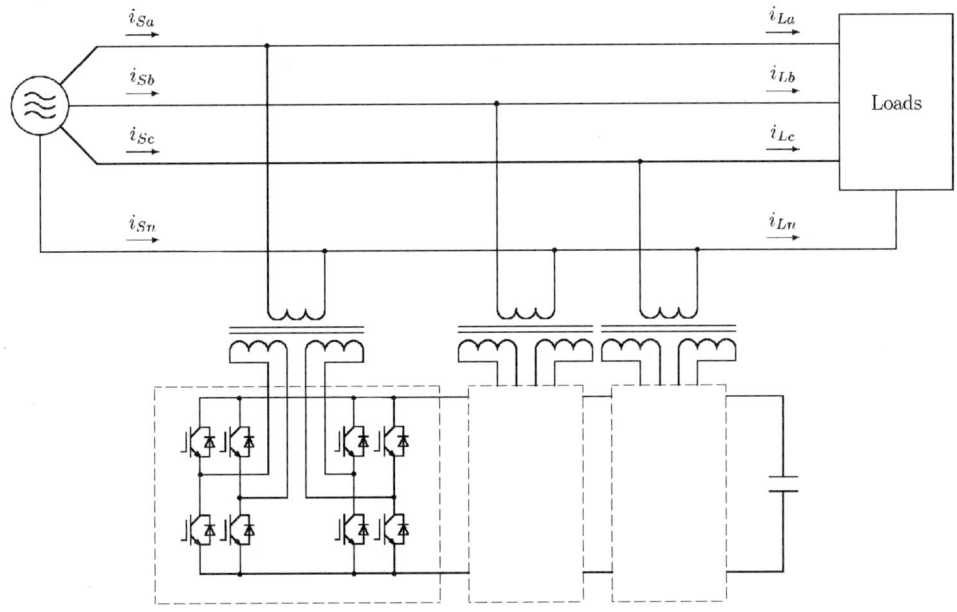

FIGURE 14.56 Shunt active filter for three-phase four-wire system.

PWM inverters using IGBT modules are usually employed as the power circuits of active filters in a range of 10 kVA to 2 MVA, and DC capacitors are used as the energy storage components.

Since a combined system of a series active filter and a shunt passive filter was proposed in 1988 (Peng et al., 1990), much research has been done on hybrid active filters and their practical applications (Bhattacharya et al., 1997; Aredes et al., 1998). The reason is that hybrid active filters are attractive from both practical and economical points of view, in particular, for high-power applications. A hybrid active filter for harmonic damping has been installed at the Yamanashi test line for high-speed magnetically-levitated trains (Kawaguchi et al., 1997). The hybrid filter consists of a combination of a 5-MVA series active filter and a 25-MVA shunt passive filter. The series active filter makes a great contribution to damping of harmonic resonance between the supply inductor and the shunt passive filter.

Shunt Active Filters for Three-Phase Four-Wire Systems

Figure 14.56 depicts the system configuration of a shunt active filter for a three-phase four-wire system. The 300-kVA active filter developed by Meidensha has been installed in a broadcasting station (Yoshida et al., 1998). Electronic equipment for broadcasting requires single-phase 100-V AC power supply in Japan, and therefore the phase-neutral rms voltage is 100 V in Fig. 14.56. A single-phase diode rectifier is used as an AC-to-DC power converter in an electronic device for broadcasting. The single-phase diode rectifier generates an amount of third-harmonic current that flows back to the supply through the neutral line. Unfortunately, the third-harmonic currents injected from all of the diode rectifiers are in phase, thus contributing to a large amount of third-harmonic current flowing in the neutral line. The current harmonics, which mainly contain the 3rd, 5th, and 7th harmonic frequency components, may cause voltage harmonics at the secondary of a distribution transformer. The induced harmonic voltage may produce a serious effect on other harmonic-sensitive devices connected at the secondary of the transformer.

Figure 14.57 shows actually measured current waveform in Fig. 14.56. The load currents, i_{La}, i_{Lb}, and i_{Lc}, and the neutral current flowing on the load side, i_{Ln}, are distorted waveforms including a large amount of harmonic current, while the supply currents, i_{Sa}, i_{Sb}, and i_{Sc}, and the neutral current flowing on the supply side, i_{Sn}, are almost sinusoidal waveforms with the help of the active filter.

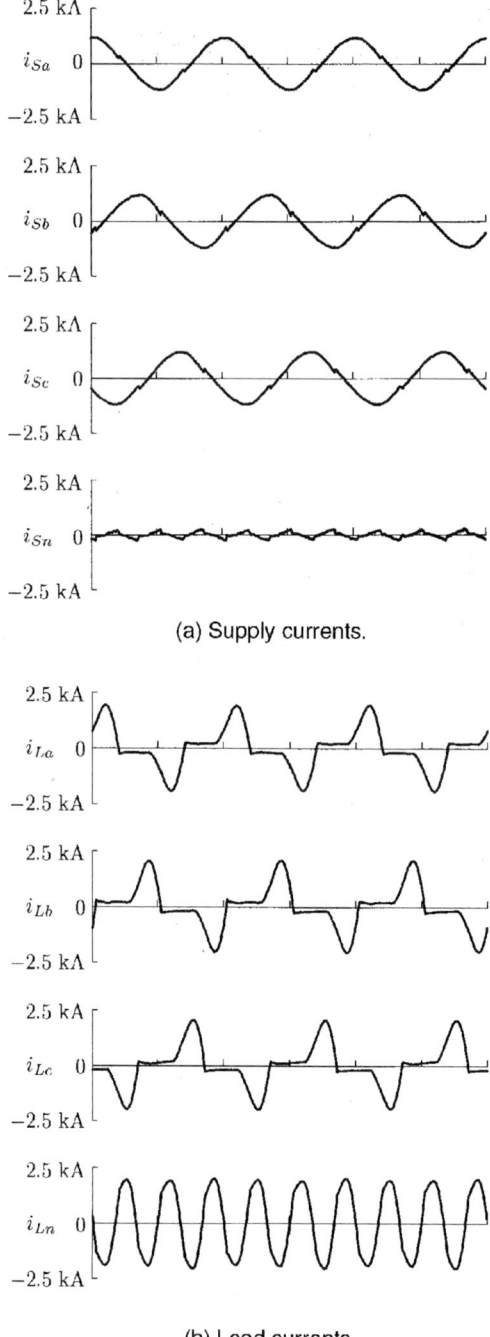

FIGURE 14.57 Actual current waveforms.

The 48-MVA Shunt Active Filter for Compensation of Voltage Impact Drop, Variation, and Imbalance

Figure 14.58 shows a power system delivering electric power to the Japanese "bullet trains" on the Tokaido Shinkansen. Three shunt active filters for compensation of fluctuating reactive current/negative-sequence current have been installed in the Shintakatsuki substation by the Central Japan Railway Company (Iizuka

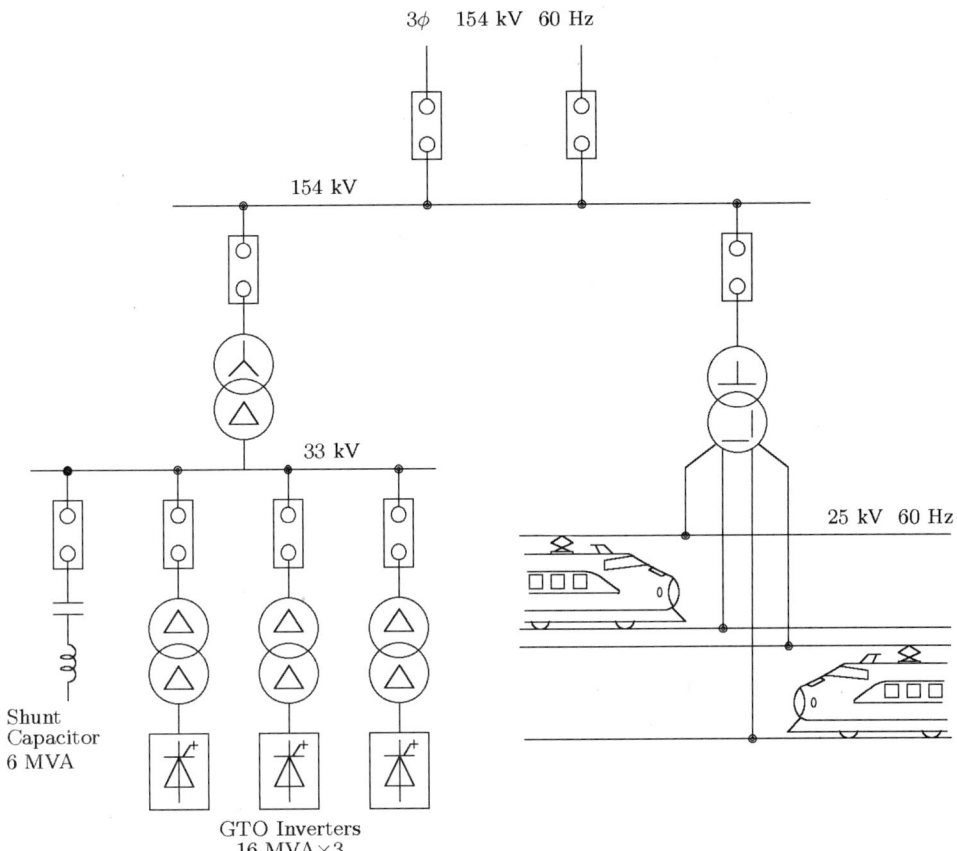

FIGURE 14.58 The 48-MVA shunt active filter installed in the Shintakatsuki substation.

et al., 1995). The shunt active filters, manufactured by Toshiba, consist of voltage-fed PWM inverters using GTO thyristors, each of which is rated at 16 MVA. A high-speed train with maximum output power of 12 MW draws unbalanced varying active and reactive power from the Scott transformer, the primary of which is connected to the 154-kV utility grid. More than twenty high-speed trains pass per hour during the daytime. This causes voltage impact drop, variation, and imbalance at the terminals of the 154-kV utility system, accompanied by a serious deterioration in the power quality of other consumers connected to the same power system. The purpose of the shunt active filters with a total rating of 48 MVA is to compensate for voltage impact drop, voltage variation, and imbalance at the terminals of the 154-kV power system, and to improve the power quality. The concept of the instantaneous power theory in the time-domain has been applied to the control strategy for the shunt active filter.

Figure 14.59 shows voltage waveforms on the 154-kV bus and the voltage imbalance factor before and after compensation, measured at 14:20–14:30 on July 27, 1994. The shunt active filters are effective not only in compensating for the voltage impact drop and variation, but also in reducing the voltage imbalance factor from 3.6 to 1%. Here, the voltage imbalance factor is the ratio of the negative to positive-sequence component in the three-phase voltages on the 154-kV bus. At present, several active filters in a range of 40 MVA to 60 MVA have been installed in substations along the Tokaido Shinkansen (Takeda et al., 1995).

Acknowledgment

The author would like to thank Meidensha Corporation and Toshiba Corporation for providing helpful and valuable information of the 300-kVA active filter and the 48-MVA active filter.

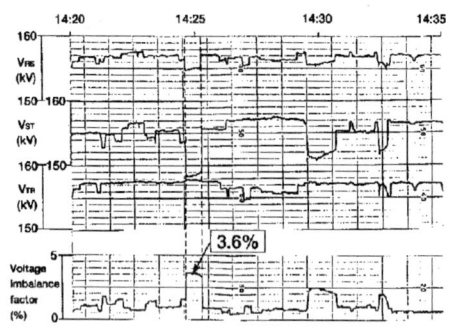
(a) Before compensation.

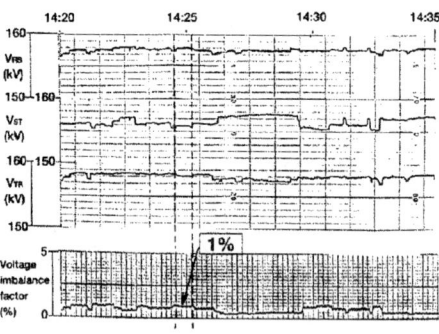
(b) After compensation.

FIGURE 14.59 Installation effect.

References

Akagi, H., New trends in active filters for power conditioning, *IEEE Trans. Ind. Appl.*, 32, 6, 1312–1322, 1996.

Akagi, H., Trends in active power line conditioners, *IEEE Trans. Power Electronics*, 9, 3, 263–268, 1994.

Akagi, H., and Fujita, H., A new power line conditioner for harmonic compensation in power systems, *IEEE Trans. Power Deliv.*, 10, 3, 1570–1575, 1995.

Akagi, H., Fujita, H., and Wada, K., A shunt active filter based on voltage detection for harmonic termination of a radial power distribution line, *IEEE Trans. Ind. Appl.*, 35, 3, 638–645, 1999.

Akagi, H., Kanazawa, Y., and Nabae, A., Generalized theory of the instantaneous reactive power in three-phase circuits, in *Proceedings of the 1983 International Power Electronics Conference*, Tokyo, Japan, 1983, 1375–1386.

Akagi, H., Kanazawa, Y., and Nabae, A., Instantaneous reactive power compensators comprising switching devices without energy storage components, *IEEE Trans. Ind. Appl.*, 20, 3, 625–630, 1984.

Akagi, H., Nabae, A., and Atoh, S., Control strategy of active power filters using multiple voltage-source PWM converters, *IEEE Trans. Ind. Appl.*, 22, 3, 460–465, 1986.

Akagi, H., Tsukamoto, Y., and Nabae, A., Analysis and design of an active power filter using quad-series voltage-source PWM converters, *IEEE Trans Ind. Appl.*, 26, 1, 93–98, 1990.

Aredes, M., Heumann, K., and Watanabe, E. H., A universal active power line conditioner, *IEEE Trans. Power Delivery*, 13, 2, 545–551, 1998.

Balbo, N., Penzo, R., Sella, D., Malesani, L., Mattavelli, P., and Zuccato, A., Simplified hybrid active filters for harmonic compensation in low voltage industrial applications, in *Proceedings of the 1994 IEEE/PES International Conference on Harmonics in Power Systems*, 1994, 263–269.

Bhattacharya, S., Frank, T. M., Divan, D., and Banerjee, B., Active filter system implementation, *IEEE Industry Applications Magazine*, 4, 5, 47–63, 1998.

Bhattacharya, S., Cheng, P., and Divan, D., Hybrid solutions for improving passive filter performance in high power applications, *IEEE Trans. Ind. Appl.*, 33, 3, 732–747, 1997.

Bird, B. M., Marsh, J. F., and McLellan, P. R., Harmonic reduction in multiple converters by triple-frequency current injection, *IEE Proceedings*, 116, 10, 1730–1734, 1969.

Fujita, H., and Akagi, H., An approach to harmonic current-free AC/DC power conversion for large industrial loads: The integration of a series active filter and a double-series diode rectifier, *IEEE Trans. Ind. Appl.*, 33, 5, 1233–1240, 1997.

Fujita, H., and Akagi, H., A practical approach to harmonic compensation in power systems–Series connection of passive and active filters, *IEEE Trans. Ind. Appl.*, 27, 6, 1020–1025, 1991.

Fujita, H., and Akagi, H., The unified power quality conditioner: The integration of series- and shunt-active filters, *IEEE Trans. Power Electronics*, 13, 2, 315–322, 1998.

Grady, W. M., Samotyj, M. J., and Noyola, A. H., Survey of active power line conditioning methodologies, *IEEE Trans. Power Deliv.*, 5, 3, 1536–1542, 1990.

Gyugyi, L., and Strycula, E. C., Active AC power filters, in *Proceedings of the 1976 IEEE/IAS Annual Meeting*, 1976, 529–535.

Iizuka, A., Kishida, M., Mochinaga, Y., Uzuka, T., Hirakawa, K., Aoyama, F., and Masuyama, T., Self-commutated static var generators at Shintakatsuki substation, in *Proceedings of the 1995 International Power Electronics Conference*, Yokohama, Japan, 1995, 609–614.

Kawaguchi, I., Ikeda, H., Ogihara, Y., Syogaki, M., and Morita, H., Novel active filter system composed of inverter bypass circuit for suppression of harmonic resonance at the Yamanashi maglev test line, in *Proceedings of the IEEE-IEEJ/IAS Power Conversion Conference*, 175–180, 1997.

Kawahira, H., Nakamura, T., Nakazawa, S., and Nomura, M., Active power filters, in *Proceedings of the 1983 International Power Electronics Conference*, Tokyo, Japan, 1983, 981–992.

Krah, J. O., and Holtz, J., Total compensation of line-side switching harmonics in converter-fed AC locomotives, in *Proceedings of the 1994 IEEE/IAS Annual Meeting*, 1994, 913–920.

Lêe, T-N., Pereira, M., Renz, K., and Vaupel, G., Active damping of resonances in power systems, *IEEE Trans. Power Deliv.*, 9, 2, 1001–1008, 1994.

Moran, S., A line voltage regulator/conditioner for harmonic-sensitive load isolation, in *Proceedings of the 1989 IEEE/IAS Annual Meeting*, 947–951, 1989.

Peng, F. Z., Application issues of active power filters, *IEEE Industry Application Magazine*, 4, 5, 21–30, 1998.

Peng, F. Z., Akagi, H., and Nabae, A., Compensation characteristics of the combined system of shunt passive and series active filters, *IEEE Trans. Ind. Appl.*, 29, 1, 144–152, 1993.

Peng, F. Z., Akagi, H., and Nabae, A., A new approach to harmonic compensation in power systems—A combined system of shunt passive and series active filters, *IEEE Trans. Ind. Appl.*, 26, 6, 983–990, 1990.

Peng, F. Z., Akagi, H., and Nabae, A., A study of active power filters using quad-series voltage-source PWM converters for harmonic compensation, *IEEE Trans. Power Electronics*, 5, 1, 9–15, 1990.

Van Schoor, G., and van Wyk, J., A study of a system of a current-fed converters as an active three-phase filter, in *Proceedings of the 1987 IEEE/PELS Power Electronics Specialist Conference*, 482–490, 1987.

Takeda, M., Murakami, S., Iizuka, A., Kishida, M., Mochinaga, Y., Hase, S., and Mochinaga, H., Development of an SVG series for voltage control over three-phase unbalance caused by railway load, in *Proceedings of the 1995 International Power Electronics Conference*, Yokohama, Japan, 1995, 603–608.

Takeda, M., Ikeda, K., and Tominaga, Y., Harmonic current compensation with active filter, in *Proceedings of the 1987 IEEE/IAS Annual Meeting*, 808–815, 1987.

Watanabe, E. H., Series active filter for the DC side of HVDC transmission systems, in *Proceedings of the 1990 International Power Electronics Conference*, Tokyo, Japan, 1024–1030, 1990.

Yoshida, T., Nakagawa, G., Kitamura, H., and Iwatani, K., Active filters (multi-functional harmonic suppressors) used to protect the quality of power supply from harmonics and reactive power generated in loads, *Meiden Review*, 262, 5, 13–17, 1998 (in Japanese).

Zhang, W., Asplund, G., Åberg, A., Jonsson, U., and Lööf, O., Active DC filter for HVDC system—A test installation in the Konti-Skan at Lindome converter station, *IEEE Trans. Power Deliv.*, 8, 3, 1599–1605, 1993.

Van Zyl, A., Enslin, J. H. R., and Spée, R., Converter based solution to power quality problems on radial distribution lines, in *Proceedings of the 1995 IEEE/IAS Annual Meeting*, 2573–2580, 1995.

15
Power Quality

S.M. Halpin
Mississippi State University

15.1	Introduction *S.M. Halpin*	15-1
15.2	Wiring and Grounding for Power Quality *Christopher J. Melhorn*	15-2
15.3	Harmonics in Power Systems *S.M. Halpin*	15-16
15.4	Voltage Sags *M.H.J. Bollen*	15-24
15.5	Voltage Fluctuations and Lamp Flicker in Power Systems *S.M. Halpin*	15-42
15.6	Power Quality Monitoring *Patrick Coleman*	15-49

15
Power Quality

S. M. Halpin
Mississippi State University

Christopher J. Melhorn
EPRI PEAC Corporation

M. H. J. Bollen
Chalmers University of Technology

Patrick Coleman
Alabama Power Company

15.1 Introduction 15-1
15.2 Wiring and Grounding for Power Quality....................15-2
 Definitions and Standards • Reasons for Grounding • Typical Wiring and Grounding Problems • Case Study
15.3 Harmonics in Power Systems.......................................15-16
15.4 Voltage Sags 15-24
 Voltage Sag Characteristics • Equipment Voltage Tolerance • Mitigation of Voltage Sags
15.5 Voltage Fluctuations and Lamp Flicker in Power Systems..15-42
15.6 Power Quality Monitoring 15-49
 Selecting a Monitoring Point • What to Monitor • Selecting a Monitor • Summary

15.1 Introduction

S. M. Halpin

Electric power quality has emerged as a major area of electric power engineering. The predominant reason for this emergence is the increase in sensitivity of end-use equipment. This chapter is devoted to various aspects of power quality as it impacts utility companies and their customers and includes material on (1) grounding, (2) voltage sags, (3) harmonics, (4) voltage flicker, and (5) long-term monitoring. While these five topics do not cover all aspects of power quality, they provide the reader with a broad-based overview that should serve to increase overall understanding of problems related to power quality.

Proper grounding of equipment is essential for safe and proper operation of sensitive electronic equipment. In times past, it was thought by some that equipment grounding as specified in the U.S. by the National Electric Code was in contrast with methods needed to insure power quality. Since those early times, significant evidence has emerged to support the position that, in the vast majority of instances, grounding according to the National Electric Code is essential to insure proper and trouble-free equipment operation, and also to insure the safety of associated personnel.

Other than poor grounding practices, voltage sags due primarily to system faults are probably the most significant of all power quality problems. Voltage sags due to short circuits are often seen at distances very remote from the fault point, thereby affecting a potentially large number of utility customers. Coupled with the wide-area impact of a fault event is the fact that there is no effective preventive for all power system faults. End-use equipment will, therefore, be exposed to short periods of reduced voltage which may or may not lead to malfunctions.

Like voltage sags, the concerns associated with flicker are also related to voltage variations. Voltage flicker, however, is tied to the likelihood of a human observer to become annoyed by the variations in the output of a lamp when the supply voltage amplitude is varying. In most cases, voltage flicker considers (at least approximately) periodic voltage fluctuations with frequencies less than about 30–35 Hz that are small in size. Human perception, rather than equipment malfunction, is the relevant factor when considering voltage flicker.

For many periodic waveform (either voltage or current) variations, the power of classical Fourier series theory can be applied. The terms in the Fourier series are called harmonics; relevant harmonic terms may have frequencies above or below the fundamental power system frequency. In most cases, nonfundamental frequency equipment currents produce voltages in the power delivery system at those same frequencies. This voltage distortion is present in the supply to other end-use equipment and can lead to improper operation of the equipment.

Harmonics, like most other power quality problems, require significant amounts of measured data in order for the problem to be diagnosed accurately. Monitoring may be short- or long-term and may be relatively cheap or very costly and often represents the majority of the work required to develop power quality solutions.

In summary, the power quality problems associated with grounding, voltage sags, harmonics, and voltage flicker are those most often encountered in practice. It should be recognized that the voltage and current transients associated with common events like lightning strokes and capacitor switching can also negatively impact end-use equipment. Because transients are covered in a separate chapter of this book, they are not considered further in this chapter.

15.2 Wiring and Grounding for Power Quality

Christopher J. Melhorn

Perhaps one of the most common problems related to power quality is wiring and grounding. It has been reported that approximately 70 to 80% of all power quality related problems can be attributed to faulty connections and/or wiring. This section describes wiring and grounding issues as they relate to power quality. It is not intended to replace or supercede the National Electric Code (NEC) or any local codes concerning grounding.

Definitions and Standards

Defining grounding terminology is outside the scope of this section. There are several publications on the topic of grounding that define grounding terminology in various levels of detail. The reader is referred to these publications for the definitions of grounding terminology.

The following is a list of standards and recommended practice pertaining to wiring and grounding issues. See the section on References for complete information.

National Electric Code Handbook, 1996 edition.
IEEE Std. 1100-1999. *IEEE Recommended Practice for Powering and Grounding Electronic Equipment.*
IEEE Std. 142-1991. *IEEE Recommended Practice for Grounding Industrial and Commercial Power Systems.*
FIPS-94 Publication
Electrical Power Systems Quality

The National Electric Code

NFPAs *National Electrical Code Handbook* pulls together all the extra facts, figures, and explanations readers need to interpret the 1999 NEC. It includes the entire text of the Code, plus expert commentary, real-world examples, diagrams, and illustrations that clarify requirements. Code text appears in blue type and commentary stands out in black. It also includes a user-friendly index that references article numbers to be consistent with the Code.

Several definitions of grounding terms pertinent to discussions in this article have been included for reader convenience. The following definitions were taken from various publications as cited.

From the *IEEE Dictionary* — *Std. 100*

Grounding: A conducting connection, whether intentional or accidental, by which an electric circuit or equipment is connected to the earth, or to some conducting body of relatively large extent that serves in place of the earth. It is used for establishing and maintaining the potential of the earth (or of the conducting body) or approximately that potential, on conductors connected to it; and for conducting ground current to and from the earth (or the conducting body).

Green Book (IEEE Std. 142) Definitions:

Ungrounded System: A system, circuit, or apparatus without an intentional connection to ground, except through potential indicating or measuring devices or other very high impedance devices.

Grounded System: A system of conductors in which at least one conductor or point (usually the middle wire or neutral point of transformer or generator windings) is intentionally grounded, either solidly or through an impedance.

NEC Definitions:

Refer to Figure 15.1.

Bonding Jumper, Main: The connector between the grounded circuit conductor (neutral) and the equipment-grounding conductor at the service entrance.

Conduit/Enclosure Bond: (bonding definition) The permanent joining of metallic parts to form an electrically conductive path which will assure electrical continuity and the capacity to conduct safely any current likely to be imposed.

Grounded: Connected to earth or to some conducting body that serves in place of the earth.

Grounded Conductor: A system or circuit conductor that is intentionally grounded (the grounded conductor is normally referred to as the neutral conductor).

Grounding Conductor: A conductor used to connect equipment or the grounded circuit of a wiring system to a grounding electrode or electrodes.

Grounding Conductor, Equipment: The conductor used to connect the noncurrent-carrying metal parts of equipment, raceways, and other enclosures to the system grounded conductor and/or the grounding electrode conductor at the service equipment or at the source of a separately derived system.

Grounding Electrode Conductor: The conductor used to connect the grounding electrode to the equipment-grounding conductor and/or to the grounded conductor of the circuit at the service equipment or at the source of a separately derived system.

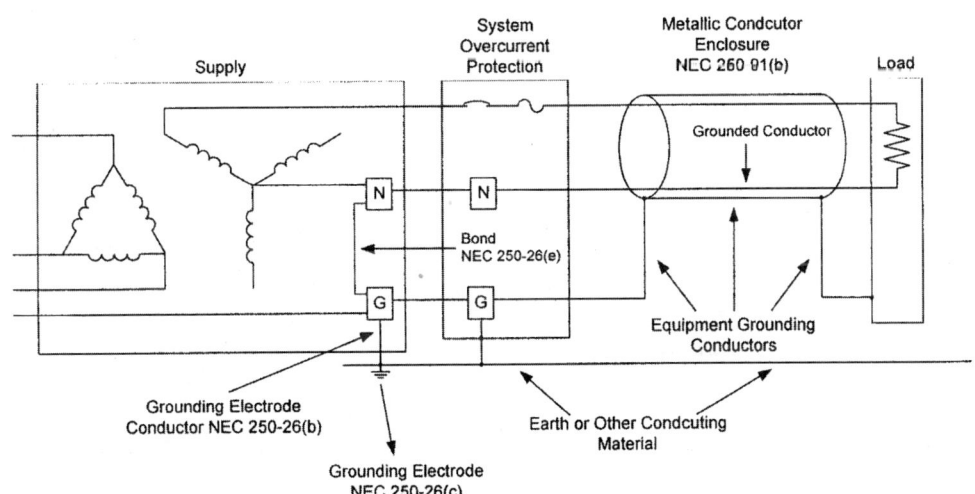

FIGURE 15.1 Terminology used in NEC definitions.

Grounding Electrode: The grounding electrode shall be as near as practicable to and preferably in the same area as the grounding conductor connection to the system. The grounding electrode shall be: (1) the nearest available effectively grounded structural metal member of the structure; or (2) the nearest available effectively grounded metal water pipe; or (3) other electrodes (Section 250-81 & 250-83) where electrodes specified in (1) and (2) are not available.

Grounding Electrode System: Defined in NEC Section 250-81 as including: (a) metal underground water pipe; (b) metal frame of the building; (c) concrete-encased electrode; and (d) ground ring. When these elements are available, they are required to be bonded together to form the grounding electrode system. Where a metal underground water pipe is the only grounding electrode available, it must be supplemented by one of the grounding electrodes specified in Section 250-81 or 250-83.

Separately Derived Systems: A premises wiring system whose power is derived from generator, transformer, or converter windings and has no direct electrical connection, including a solidly connected grounded circuit conductor, to supply conductors originating in another system.

Reasons for Grounding

There are three basic reasons for grounding a power system: personal safety, protective device operation, and noise control. All three of these reasons will be addressed.

Personal Safety

The most important reason for grounding a device on a power system is personal safety. The safety ground, as it is sometimes called, is provided to reduce or eliminate the chance of a high touch potential if a fault occurs in a piece of electrical equipment. Touch potential is defined as the voltage potential between any two conducting materials that can be touched simultaneously by an individual or animal.

Figure 15.2 illustrates a dangerous touch potential situation. The "hot" conductor in the piece of equipment has come in contact with the case of the equipment. Under normal conditions, with the safety ground intact, the protective device would operate when this condition occurred. However, in Fig. 15.2, the safety ground is missing. This allows the case of the equipment to float above ground since the case of the equipment is not grounded through its base. In other words, the voltage potential between the equipment case and ground is the same as the voltage potential between the hot leg and ground. If the operator would come in contact with the case and ground (the floor), serious injury could result.

In recent years, manufacturers of handheld equipment, drills, saws, hair dryers, etc. have developed double insulated equipment. This equipment generally does not have a safety ground. However, there is

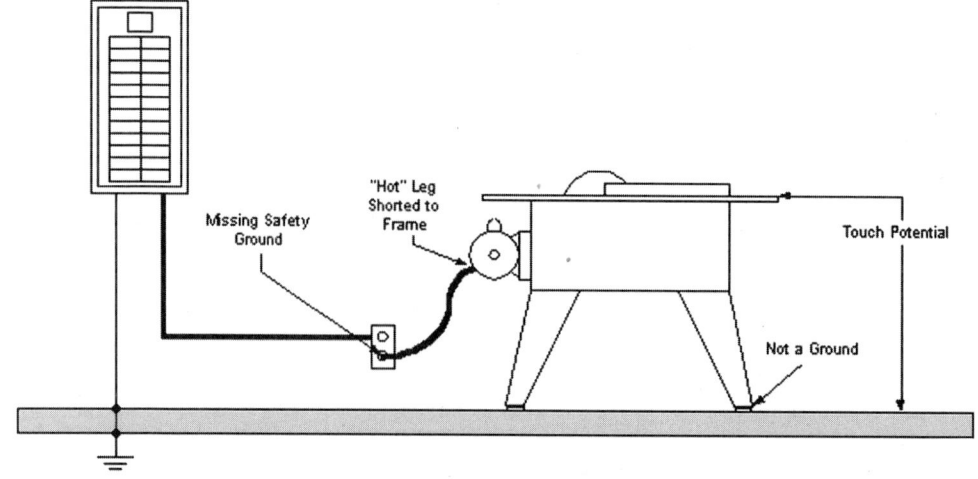

FIGURE 15.2 Illustration of a dangerous touch potential situation.

TABLE 15.1 Example Ground Impedance Values

Protective Device Rating	Voltage to Ground 120 Volts	Voltage to Ground 277 Volts
20 Amps	1.20 Ω	2.77 Ω
40 Amps	0.60 Ω	1.39 Ω
50 Amps	0.48 Ω	1.11 Ω
60 Amps	0.40 Ω	0.92 Ω
100 Amps	0.24 Ω	0.55 Ω

never any conducting material for the operator to contact and therefore there is no touch potential hazard. If the equipment becomes faulted, the case or housing of the equipment is not energized.

Protective Device Operation

As mentioned in the previous section, there must be a path for fault current to return to the source if protective devices are to operate during fault conditions. The National Electric Code (NEC) requires that an effective grounding path must be mechanically and electrically continuous (NEC 250-51), have the capacity to carry any fault currents imposed on it without damage (NEC 250-75). The NEC also states that the ground path must have sufficiently low impedance to limit the voltage and facilitate protective device operation. Finally, the earth cannot serve as the equipment-grounding path (NEC-250-91(c)).

The formula to determine the maximum circuit impedance for the grounding path is:

$$\text{Ground Path Impedance} = \frac{\text{Maximum Voltage to Ground}}{\text{Overcurrent Protection Rating} \times 5}$$

Table 15.1 gives examples of maximum ground path circuit impedances required for proper protective device operation.

Noise Control

Noise control is the third main reason for grounding. Noise is defined as unwanted voltages and currents on a grounding system. This includes signals from all sources whether it is radiated or conducted. As stated, the primary reason for grounding is safety and is regulated by the NEC and local codes. Any changes to the grounding system to improve performance or eliminate noise control must be in addition to the minimum NEC requirements.

When potential differences occur between different grounding systems, insulation can be stressed and circulating currents can be created in low voltage cables (e.g., communications cables). In today's electrical environment, buildings that are separated by large physical distances are typically tied together via a communication circuit. An example of this would be a college campus that may cover several square miles. Each building has its own grounding system. If these grounding systems are not tied together, a potential difference on the grounding circuit for the communication cable can occur. The idea behind grounding for noise control is to create an equipotential grounding system, which in turn limits or even eliminates the potential differences between the grounding systems. If the there is an equipotential grounding system and currents are injected into the ground system, the potential of the whole grounding system will rise and fall and potential differences will not occur.

Supplemental conductors, ground reference grids, and ground plates can all be used to improve the performance of the system as it relates to power quality. Optically isolated communications can also improve the performance of the system. By using the opto-isolators, connecting the communications to different ground planes is avoided. All improvements to the grounding system must be done in addition to the requirements for safety.

Separation of loads is another method used to control noise. Figure 15.3 illustrates this point. Figure 15.3 shows four different connection schemes. Each system from left to right improves noise control.

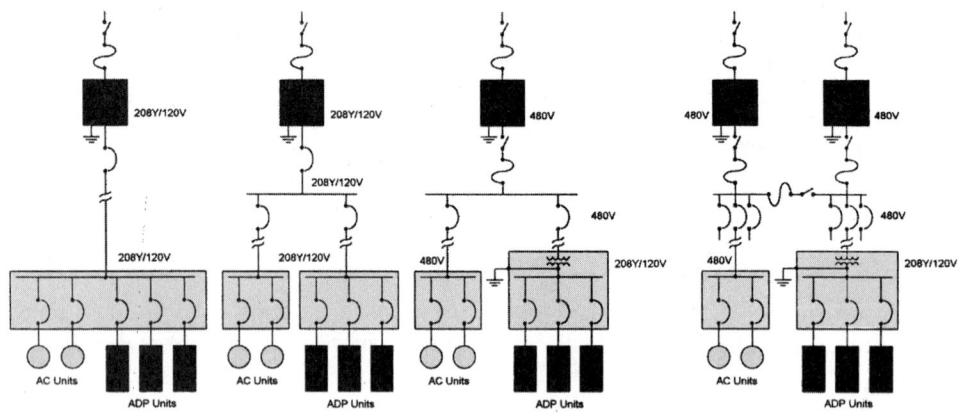

FIGURE 15.3 Separation of loads for noise control.

TABLE 15.2 Typical Wiring and Grounding Problems and Causes

Wiring Condition or Problem Observed	Possible Cause
Impulse, voltage drop out	Loose connections
Impulse, voltage drop out	Faulty breaker
Ground currents	Extra neutral-to-ground bond
Ground currents	Neutral-to-ground reversal
Extreme voltage fluctuations	High impedance in neutral circuit
Voltage fluctuations	High impedance neutral-to-ground bonds
High neutral to ground voltage	High impedance ground
Burnt smell at the panel, junction box, or load	Faulted conductor, bad connection, arcing, or overloaded wiring
Panel or junction box is warm to the touch	Faulty circuit breaker or bad connection
Buzzing sound	Arcing
Scorched insulation	Overloaded wiring, faulted conductor, or bad connection
Scorched panel or junction box	Bad connection, faulted conductor
No voltage at load equipment	Tripped breaker, bad connection, or faulted conductor
Intermittent voltage at the load equipment	Bad connection or arcing

As seen in Figure 15.3, the best case would be the complete separation (system on the far right) of the ADP units from the motor loads and other equipment. Conversely, the worst condition is on the left of Fig. 15.3 where the ADP units are served from the same circuit as the motor loads.

Typical Wiring and Grounding Problems

In this section, typical wiring and grounding problems, as related to power quality, are presented. Possible solutions are given for these problems as well as the possible causes for the problems being observed on the grounding system. (See Table 15.2.)

The following list is just a sample of problems that can occur on the grounding system.

- Isolated grounds
- Ground loops
- Missing safety ground
- Multiple neutral-to-ground bonds
- Additional ground rods
- Insufficient neutral conductors

Power Quality

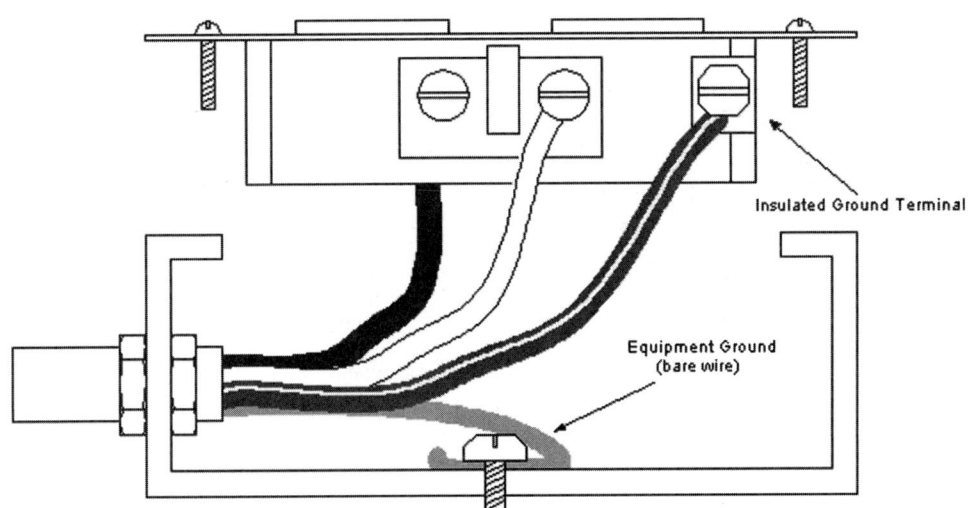

FIGURE 15.4 Properly wired isolated ground circuit.

Insulated Grounds

Insulated grounds in themselves are not a grounding problem. However, improperly used insulated grounds can be a problem. Insulated grounds are used to control noise on the grounding system. This is accomplished by using insulated ground receptacles, which are indicated by a "Δ" on the face of the outlet. Insulated ground receptacles are often orange in color. Figure 15.4 illustrates a properly wired insulated ground circuit.

The 1996 NEC has this to say about insulated grounds.

NEC 250-74. Connecting Receptacle Grounding Terminal to Box. An equipment bonding jumper shall be used to connect the grounding terminal of a grounding-type receptacle to a grounded box.

Exception No. 4. Where required for the reduction of electrical noise (electromagnetic interference) on the grounding circuit, a receptacle in which the grounding terminal is purposely insulated from the receptacle mounting means shall be permitted. The receptacle grounding terminal shall be grounded by an insulated equipment grounding conductor run with the circuit conductors. This grounding conductor shall be permitted to pass through one or more panelboards without connection to the panelboard grounding terminal as permitted in Section 384-20, Exception so as to terminate within the same building or structure directly at an equipment grounding conductor terminal of the applicable derived system or source.

(FPN): Use of an isolated equipment grounding conductor does not relieve the requirement for grounding the raceway system and outlet box.

NEC 517-16. Receptacles with Insulated Grounding Terminals. Receptacles with insulated grounding terminals, as permitted in Section 250-74, Exception No. 4, shall be identified; such identification shall be visible after installation.

(FPN): Caution is important in specifying such a system with receptacles having insulated grounding terminals, since the grounding impedance is controlled only by the grounding conductors and does not benefit functionally from any parallel grounding paths.

The following is a list of pitfalls that should be avoided when installing insulated ground circuits.

- Running an insulated ground circuit to a regular receptacle.
- Sharing the conduit of an insulated ground circuit with another circuit.
- Installing an insulated ground receptacle in a two-gang box with another circuit.

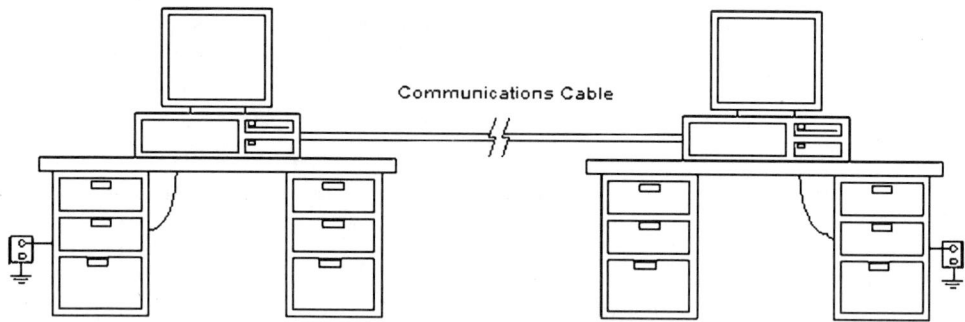

FIGURE 15.5 Circuit with a ground loop.

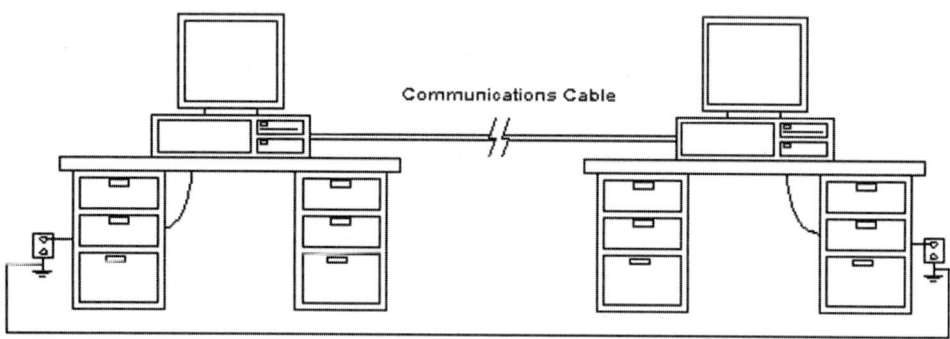

FIGURE 15.6 Grounding electrodes must be bonded together.

- Not running the insulated ground circuit in a metal cable armor or conduit.
- Do not assume that an insulated ground receptacle has a truly insulated ground.

Ground Loops

Ground loops can occur for several reasons. One is when two or more pieces of equipment share a common circuit like a communication circuit, but have separate grounding systems (Fig. 15.5).

To avoid this problem, only one ground should be used for grounding systems in a building. More than one grounding electrode can be used, but they must be tied together (NEC 250-81, 250-83, and 250-84) as illustrated in Fig. 15.6.

Missing Safety Ground

As discussed previously, a missing safety ground poses a serious problem. Missing safety grounds usually occur because the safety ground has been bypassed. This is typical in buildings where the 120-volt outlets only have two conductors. Modern equipment is typically equipped with a plug that has three prongs, one of which is a ground prong. When using this equipment on a two-prong outlet, a grounding plug adapter or "cheater plug" can be employed provided there is an equipment ground present in the outlet box. This device allows the use of a three-prong device in a two-prong outlet. When properly connected, the safety ground remains intact. Figure 15.7 illustrates the proper use of the cheater plug.

If an equipment ground is not present in the outlet box, then the grounding plug adapter should not be used. If the equipment grounding conductor is present, the preferred method for solving the missing safety ground problem is to install a new three-prong outlet in the outlet box. This method insures that the grounding conductor will not be bypassed. The NEC discusses equipment grounding conductors in detail in Section 250 — Grounding.

Power Quality 15-9

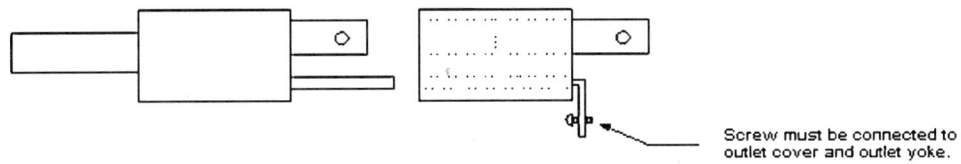

FIGURE 15.7 Proper use of a grounding plug adapter or "cheater plug."

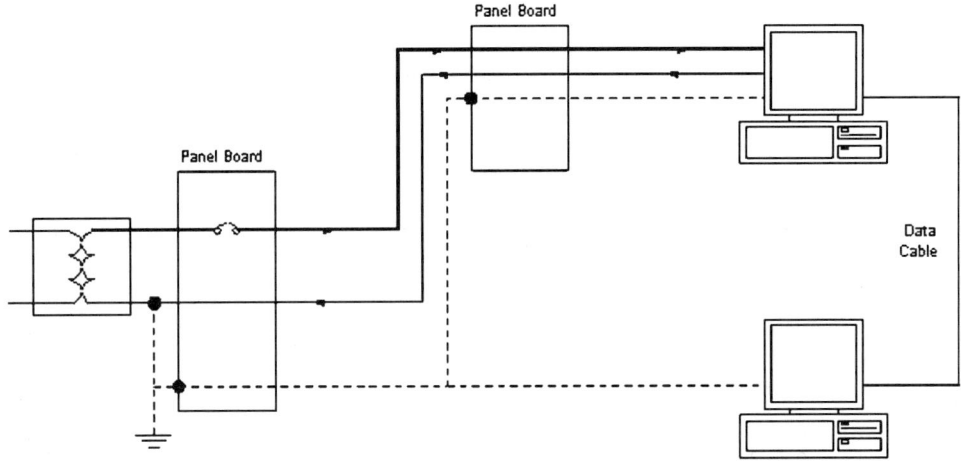

FIGURE 15.8 Neutral current flow with one neutral-to-ground bond.

Multiple Neutral to Ground Bonds

Another misconception when grounding equipment is that the neutral must be tied to the grounding conductor. Only one neutral-to-ground bond is permitted in a system or sub-system. This typically occurs at the service entrance to a facility unless there is a separately derived system. A separately derived system is defined as a system that receives its power from the windings of a transformer, generator, or some type of converter. Separately derived systems must be grounded in accordance with NEC 250-26.

The neutral should be kept separate from the grounding conductor in all panels and junction boxes that are downline from the service entrance. Extra neutral to ground bonds in a power system will cause neutral currents to flow on the ground system. This flow of current on the ground system occurs because of the parallel paths. Figures 15.8 and 15.9 illustrate this effect.

As seen in Fig. 15.9, neutral current can find its way onto the ground system due to the extra neutral-to-ground bond in the secondary panel board. Notice that not only will current flow in the ground wire for the power system, but currents can flow in the shield wire for the communication cable between the two PCs.

If the neutral-to-ground bond needs to be reestablished (high neutral-to-ground voltages), this can be accomplished by creating a separately derived system as defined above. Figure 15.10 illustrates a separately derived system.

Additional Ground Rods

Additional ground rods are another common problem in grounding systems. Ground rods for a facility or building should be part of the grounding system. The ground rods should be connected where all the building grounding electrodes are bonded together. Isolated grounds can be used as described in the

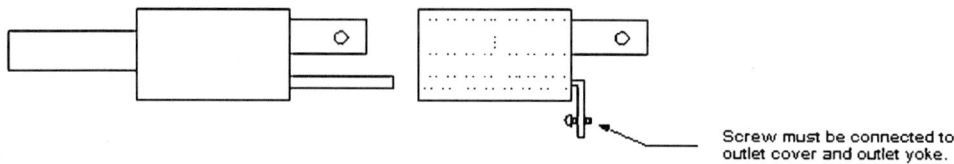

FIGURE 15.7 Proper use of a grounding plug adapter or "cheater plug."

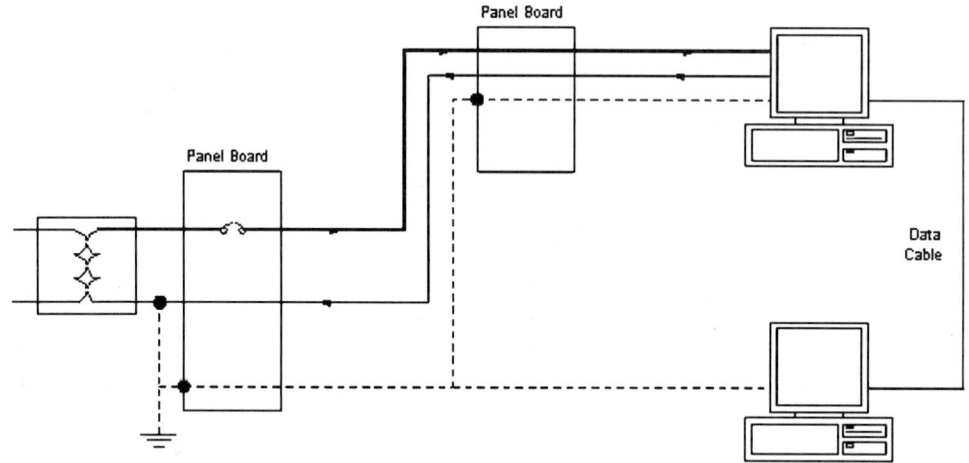

FIGURE 15.8 Neutral current flow with one neutral-to-ground bond.

Multiple Neutral to Ground Bonds

Another misconception when grounding equipment is that the neutral must be tied to the grounding conductor. Only one neutral-to-ground bond is permitted in a system or sub-system. This typically occurs at the service entrance to a facility unless there is a separately derived system. A separately derived system is defined as a system that receives its power from the windings of a transformer, generator, or some type of converter. Separately derived systems must be grounded in accordance with NEC 250-26.

The neutral should be kept separate from the grounding conductor in all panels and junction boxes that are downline from the service entrance. Extra neutral-to-ground bonds in a power system will cause neutral currents to flow on the ground system. This flow of current on the ground system occurs because of the parallel paths. Figures 15.8 and 15.9 illustrate this effect.

As seen in Fig. 15.9, neutral current can find its way onto the ground system due to the extra neutral-to-ground bond in the secondary panel board. Notice that not only will current flow in the ground wire for the power system, but currents can flow in the shield wire for the communication cable between the two PCs.

If the neutral-to-ground bond needs to be reestablished (high neutral-to-ground voltages), this can be accomplished by creating a separately derived system as defined above. Figure 15.10 illustrates a separately derived system.

Additional Ground Rods

Additional ground rods are another common problem in grounding systems. Ground rods for a facility or building should be part of the grounding system. The ground rods should be connected where all the building grounding electrodes are bonded together. Isolated grounds can be used as described in the

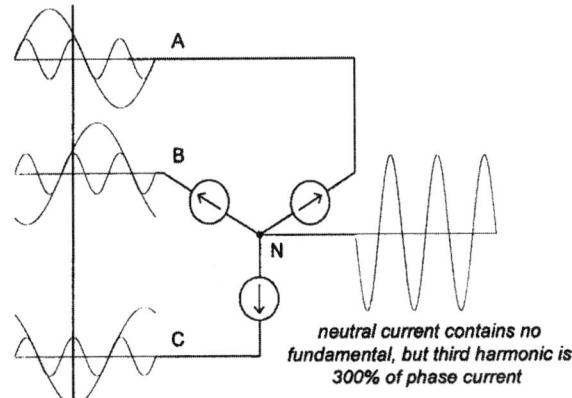

FIGURE 15.11 A balanced three-phase system.

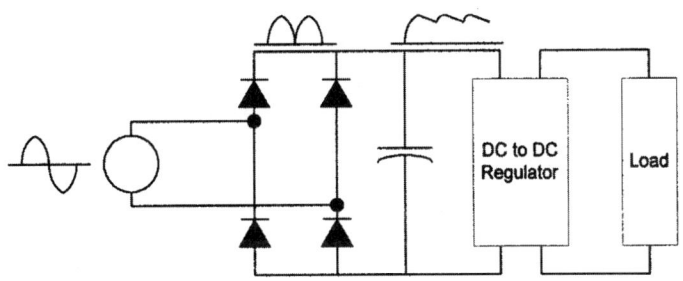

FIGURE 15.12 The basic one-line for a SMPS.

However, PCs, laser printers, and other pieces of electronic office equipment all use the same basic technology for receiving the power that they need to operate. Figure 15.12 illustrates the typical power supply of a PC. The input power is generally 120 volts AC, single phase. The internal electronic parts require various levels of DC voltage (e.g., ±5, 12 volts DC) to operate. This DC voltage is obtained by converting the AC voltage through some type of rectifier circuit as shown. The capacitor is used for filtering and smoothing the rectified AC signal. These types of power supplies are referred to as switch mode power supplies (SMPS).

The concern with devices that incorporate the use of SMPS is that they introduce triplen harmonics into the power system. Triplen harmonics are those that are odd multiples of the fundamental frequency component (h = 3, 9, 15, 21, …). For a system that has balanced single-phase loads as illustrated in Fig. 15.13, fundamental and third harmonic components are present. Applying Kirchoff's current law at node N shows that the fundamental current component in the neutral must be zero. But when loads are balanced, the third harmonic components in each phase coincide. Therefore, the magnitude of third harmonic current in the neutral must be three times the third harmonic phase current.

This becomes a problem in office buildings when multiple single-phase loads are supplied from a three-phase system. Separate neutral wires are run with each circuit, therefore the neutral current will be equivalent to the line current. However, when the multiple neutral currents are returned to the panel or transformer serving the loads, the triplen currents will add in the common neutral for the panel and this can cause over heating and eventually even cause failure of the neutral conductor. If office partitions are used, the same, often undersized neutral conductor is run in the partition with three-phase conductors. Each receptacle is fed from a separate phase in order to balance the load current. However, a single

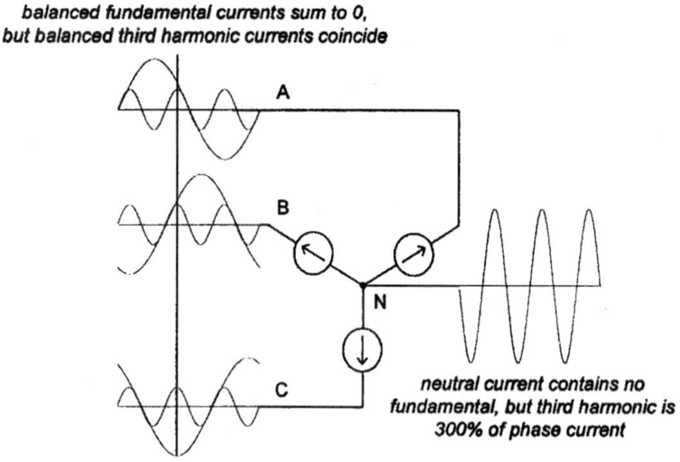

FIGURE 15.13 Balanced single-phase loads.

TABLE 15.3 Summary of Wiring and Grounding Issues

Summary Issues
Good power quality and noise control practices do not conflict with safety requirements.
Wiring and grounding problems cause a majority of equipment interference problems.
Make an effort to put sensitive equipment on dedicated circuits.
The grounded conductor, neutral conductor, should be bonded to the ground at the transformer or main panel, but not at other panel down line except as allowed by separately derived systems.

neutral is usually shared by all three phases. This can lead to disastrous results if the partition electrical receptacles are used to supply nonlinear loads rich in triplen harmonics.

Under the worst conditions, the neutral current will never exceed 173% of the phase current. Figure 15.13 illustrates a case where a three-phase panel is used to serve multiple single-phase SMPS PCs.

Summary

As discussed previously, the three main reasons for grounding in electrical systems are:

1. Personal safety
2. Proper protective device operation
3. Noise control

By following the guidelines found below, the objectives for grounding can be accomplished.

- All equipment should have a safety ground. A safety ground conductor
- Avoid load currents on the grounding system.
- Place all equipment in a system on the same equipotential reference.

Table 15.3 summarizes typical wiring and grounding issues.

Case Study

This section presents a case study involving wiring and grounding issues. The purpose of this case study is to inform the reader on the procedures used to evaluate wiring and grounding problems and present solutions.

Power Quality

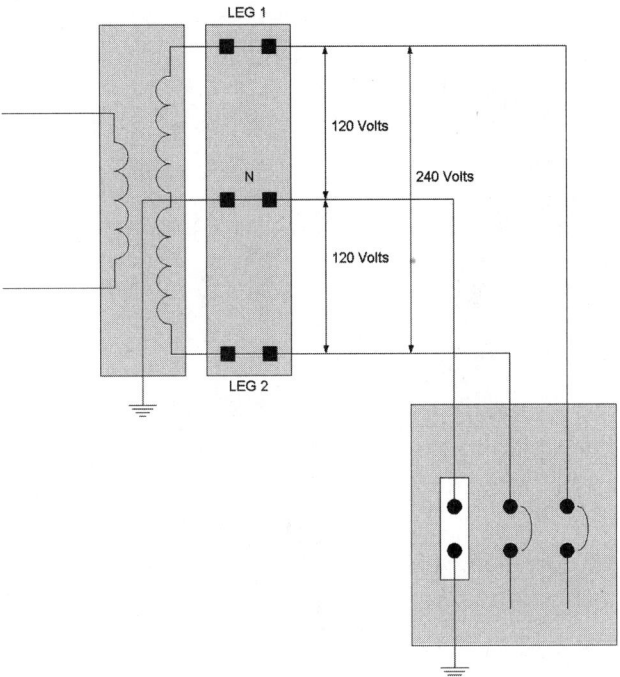

FIGURE 15.14 Split-phase system serving a residential customer.

Case Study — Flickering Lights

This case study concerns a residential electrical system. The homeowners were experiencing light flicker when loads were energized and deenergized in their homes.

Background

Residential systems are served from single-phase transformers employing a spilt secondary winding, often referred to as a single-phase three-wire system. This type of transformer is used to deliver both 120-volt and 240-volt single-phase power to the residential loads. The primary of the transformer is often served from a 12 to 15 kV distribution system by the local utility. Figure 15.14 illustrates the concept of a split-phase system.

When this type of service is operating properly, 120 volts can be measured from either leg to the neutral conductor. Due to the polarity of the secondary windings in the transformer, the polarity of each 120-volt leg is opposite the other, thus allowing a total of 240 volts between the legs as illustrated. The proper operation of this type of system is dependent on the physical connection of the neutral conductor or center tap of the secondary winding. If the neutral connection is removed, 240 volts will remain across the two legs, but the line-to-neutral voltage for either phase can be shifted, causing either a low or high voltage from line to neutral.

Most loads in a residential dwelling, i.e., lighting, televisions, microwaves, home electronics, etc., are operated from 120 volts. However, there are a few major loads that incorporate the use of the 240 volts available. These loads include electric water heaters, electric stoves and ovens, heat pumps, etc.

The Problem

In this case, there were problems in the residence that caused the homeowner to question the integrity of the power system serving his home. On occasion, the lights would flicker erratically when the washing machine and dryer were operating at the same time. When large single-phase loads were operated, low power incandescent light bulb intensity would flicker.

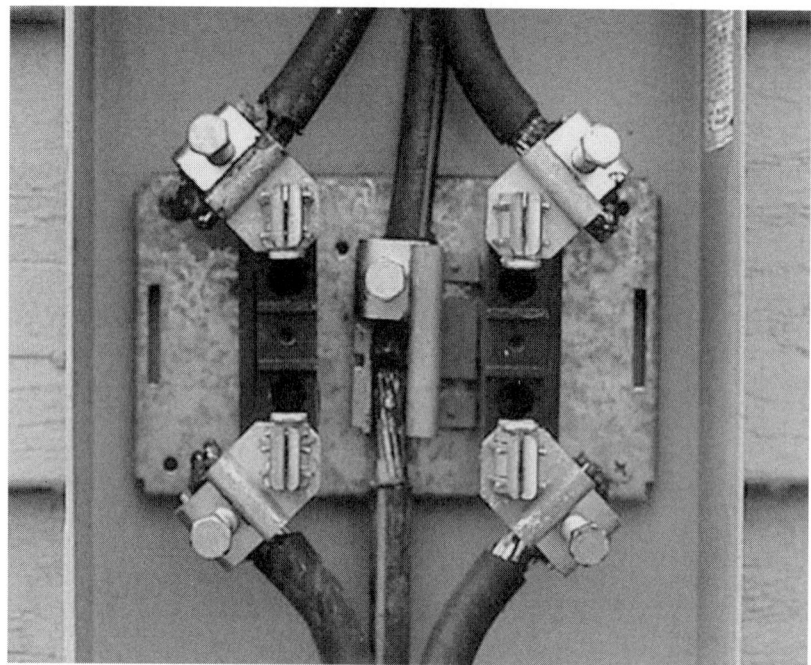

FIGURE 15.15 Actual residential meter base. Notice the missing neutral clamp on load side of meter.

Measurements were performed at several 120-volt outlets throughout the house. When the microwave was operated, the voltage at several of the 120-volt outlets would increase from 120 volts nominal to 128 volts. The voltage would return to normal after the microwave was turned off. The voltage would also increase when a 1500-Watt space heater was operated. It was determined that the voltage would decrease to approximately 112 volts on the leg from which the large load was served. After the measurements confirmed suspicions of high and low voltages during heavy load operation, finding the source of the problem was the next task at hand.

The hunt began at the service entrance to the house. A visual inspection was made of the meter base and socket after the meter was removed by the local utility. It was discovered that one of the neutral connectors was loose. While attempting to tighten this connector, the connector fell off of the meter socket into the bottom of the meter base (see Fig. 15.15). Could this loose connector have been the cause of the flickering voltage? Let's examine the effects of the loose neutral connection.

Figures 15.16 and 15.17 will be referred to several times during this discussion. Under normal conditions with a solid neutral connection (Fig. 15.16), load current flows through each leg and is returned to the source through the neutral conductor. There is very little impedance in either the hot or the neutral conductor; therefore, no appreciable voltage drop exists.

When the neutral is loose or missing, a significant voltage can develop across the neutral connection in the meter base, as illustrated in Fig. 15.17. When a large load is connected across Leg 1 to N and the other leg is lightly loaded (i.e., Leg 1 to N is approximately 10 times the load on Leg 2 to N), the current flowing through the neutral will develop a voltage across the loose connection. This voltage is in phase with the voltage from Leg 1 to N' (see Fig. 15.17) and the total voltage from Leg 1 to N will be 120 volts. However, the voltage supplied to any loads connected from Leg 2 to N' will rise to 128 volts, as illustrated in Fig. 15.17. The total voltage across the Leg 1 and Leg 2 must remain constant at 240 volts. It should be noted that the voltage from Leg 2 to N will be 120 volts since the voltage across the loose connection is 180° out of phase with the Leg 2 to N' voltage.

Therefore, with the missing neutral connection, the voltage from Leg 2 to N' would rise, causing the light flicker. This explains the rise in voltage when a large load was energized on the system.

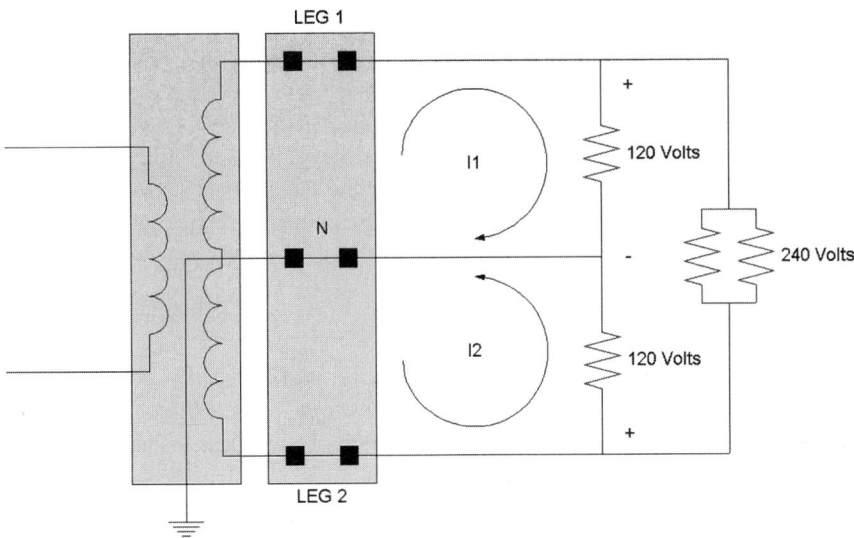

FIGURE 15.16 The effects of a solid neutral connection in the meter base.

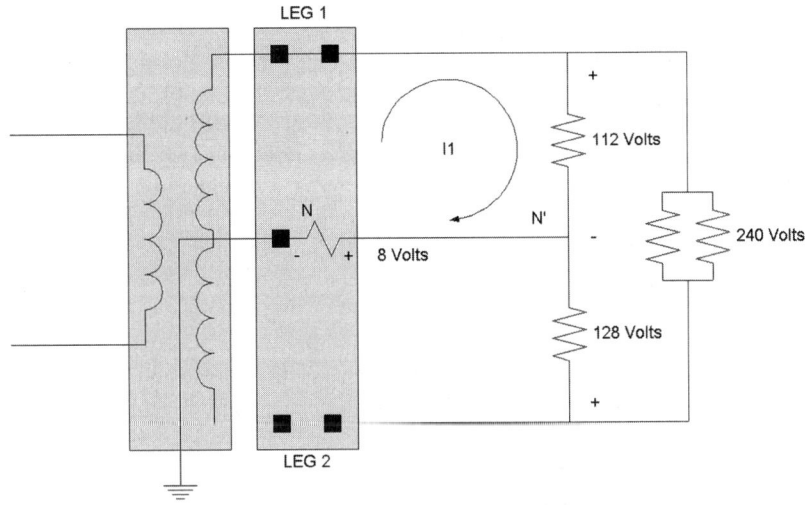

FIGURE 15.17 The effects of a loose neutral connection in the meter base.

The Solution
The solution in this case was simple — replace the failed connector.

Conclusions
Over time, the neutral connector had become loose. This loose connection caused heating, which in turn caused the threads on the connector to become worn, and the connector failed. After replacing the connector in the meter base, the flickering light phenomena disappeared.

On systems of this type, if a voltage rise occurs when loads are energized, it is a good indication that the neutral connection may be loose or missing.

References

Dugan, R. C. et al., *Electrical Power Systems Quality*, McGraw-Hill, New York, 1995.
FIPS-94 Publication.
IEEE Std. 142-1991. *IEEE Recommended Practice for Grounding Industrial and Commercial Power Systems*, The Institute of Electrical and Electronics Engineers, New York, New York, 1991.
IEEE Std. 1100-1999. *IEEE Recommended Practice for Powering and Grounding Electronic Equipment*, The Institute of Electrical and Electronics Engineers, New York, New York, 1999.
Melhorn, Christopher J., Coping with non-linear computer loads in commercial buildings — Part I, *emf-emi control* 2, 5, September/October, 1995.
Melhorn, Christopher J., Coping with non-linear computer loads in commercial building — Part II, *emf-emi control* 2, 6, January/February, 1996.
Melhorn, Chris, Flickering Lights — A Case of Faulty Wiring, *PQToday*, 3, 4, August 1997.
National Electrical Code Handbook, National Fire Protection Agency, Quincy, MA, 1996 edition.
Understanding the National Electric Code, 1993 Edition, Michael Holt, Delmar Publishers, Inc., 1993.

15.3 Harmonics in Power Systems

S. M. Halpin

Power system harmonics are not a new topic, but the proliferation of high-power electronics used in motor drives and power controllers has necessitated increased research and development in many areas relating to harmonics. For many years, high-voltage direct current (HVDC) stations have been a major focus area for the study of power system harmonics due to their rectifier and inverter stations. Roughly two decades ago, electronic devices that could handle several kW up to several MW became commercially viable and reliable products. This technological advance in electronics led to the widespread use of numerous converter topologies, all of which represent nonlinear elements in the power system.

Even though the power semiconductor converter is largely responsible for the large-scale interest in power system harmonics, other types of equipment also present a nonlinear characteristic to the power system. In broad terms, loads that produce harmonics can be grouped into three main categories covering (1) arcing loads, (2) semiconductor converter loads, and (3) loads with magnetic saturation of iron cores. Arcing loads, like electric arc furnaces and florescent lamps, tend to produce harmonics across a wide range of frequencies with a generally decreasing relationship with frequency. Semiconductor loads, such as adjustable-speed motor drives, tend to produce certain harmonic patterns with relatively predictable amplitudes at known harmonics. Saturated magnetic elements, like overexcited transformers, also tend to produce certain "characteristic" harmonics. Like arcing loads, both semiconductor converters and saturated magnetics produce harmonics that generally decrease with frequency.

Regardless of the load category, the same fundamental theory can be used to study power quality problems associated with harmonics. In most cases, any periodic distorted power system waveform (voltage, current, flux, etc.) can be represented as a series consisting of a DC term and an infinite sum of sinusoidal terms as shown in Eq. (15.1) where ω_0 is the fundamental power frequency.

$$f(t) = F_0 + \sum_{i=1}^{\infty} \sqrt{2} F_i \cos(i\omega_0 t + \theta_i) \tag{15.1}$$

A vast amount of theoretical mathematics has been devoted to the evaluation of the terms in the infinite sum in Eq. (15.1), but such rigor is beyond the scope of this section. For the purposes here, it is reasonable to presume that instrumentation is available that will provide both the magnitude F_i and the phase angle θ_i for each term in the series. Taken together, the magnitude and phase of the ith term completely describe the ith harmonic.

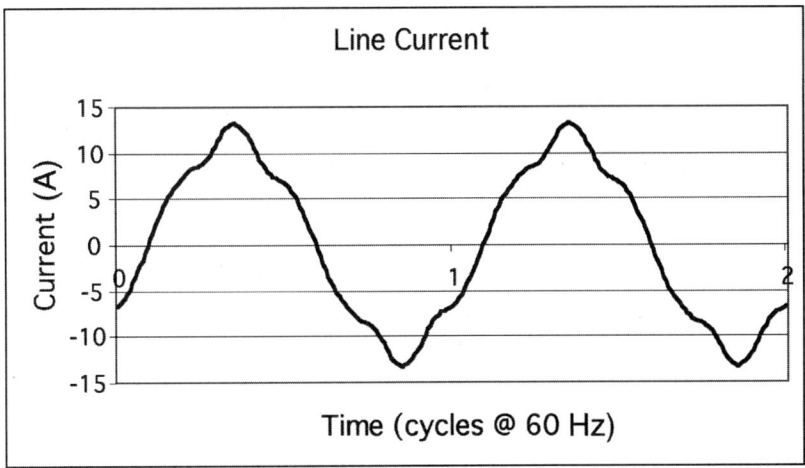

FIGURE 15.18 Current waveform.

It should be noted that not all loads produce harmonics that are integer multiples of the power frequency. These noninteger multiple harmonics are generally referred to as interharmonics and are commonly produced by arcing loads and cycloconverters. All harmonic terms, both integer and noninteger multiples of the power frequency, are analytically treated in the same manner, usually based on the principle of superposition.

In practice, the infinite sum in Eq. (15.1) is reduced to about 50 terms; most measuring instruments do not report harmonics higher than the 50th multiple (2500–3000 Hz for 50–60 Hz systems). The reporting can be in the form of a tabular listing of harmonic magnitudes and angles or in the form of a magnitude and phase spectrum. In each case, the information provided is the same and can be used to reproduce the original waveform by direct substitution into Eq. (15.1) with satisfactory accuracy. As an example, Fig. 15.18 shows the (primary) current waveform drawn by a small industrial plant. Table 15.4 shows a table of the first 31 harmonic magnitudes and angles. Figure 15.19 shows a bar graph magnitude spectrum for this same waveform. These data are widely available from many commercial instruments; the choice of instrument makes little difference in most cases.

TABLE 15.4 Current Harmonic Magnitudes and Phase Angles

Harmonic #	Current (A_{rms})	Phase (deg)	Harmonic #	Current (A_{rms})	Phase (deg)
1	8.36	−65	2	0.01	−167
3	0.13	43	4	0.01	95
5	0.76	102	6	0.01	8
7	0.21	−129	8	0	−148
9	0.02	−94	10	0	78
11	0.08	28	12	0	−89
13	0.04	−172	14	0	126
15	0	159	16	0	45
17	0.02	−18	18	0	−117
19	0.01	153	20	0	22
21	0	119	22	0	26
23	0.01	−76	24	0	143
25	0	0	26	0	150
27	0	74	28	0	143
29	0	50	30	0	−13
31	0	−180			

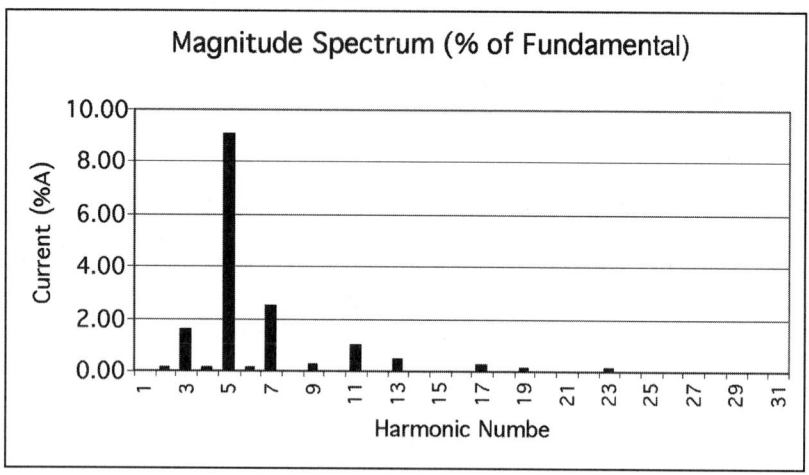

FIGURE 15.19 Harmonic magnitude spectrum.

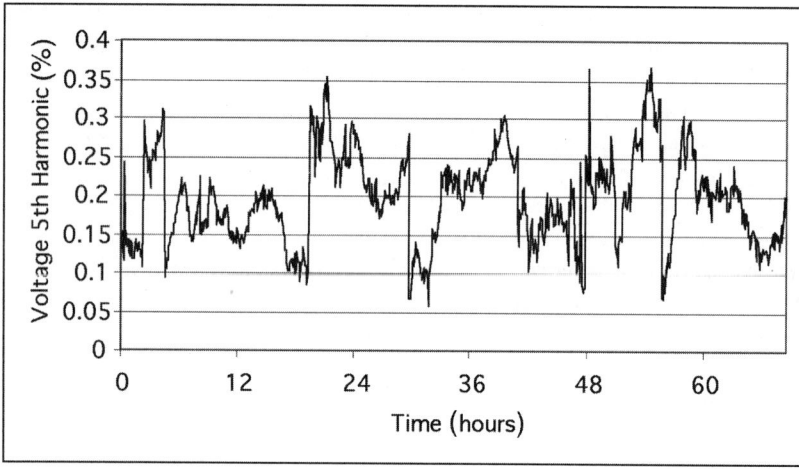

FIGURE 15.20 Example of time-varying nature of harmonics.

A fundamental presumption when analyzing distorted waveforms using Fourier methods is that the waveform is in steady state. In practice, waveform distortion varies widely and is dependent on both load levels and system conditions. It is typical to assume that a steady-state condition exists at the instant at which the measurement is taken, but the next measurement at the next time could be markedly different. As examples, Figs. 15.20 and 15.21 show time plots of 5th harmonic voltage and the total harmonic distortion, respectively, of the same waveform measured on a 115 kV transmission system near a five MW customer. Note that the THD is fundamentally defined in Eq. (15.2), with 50 often used in practice as the upper limit on the infinite summation.

$$\text{THD}(\%) = \frac{\sqrt{\sum_{i=2}^{\infty} F_i^2}}{F_1} *100\% \qquad (15.2)$$

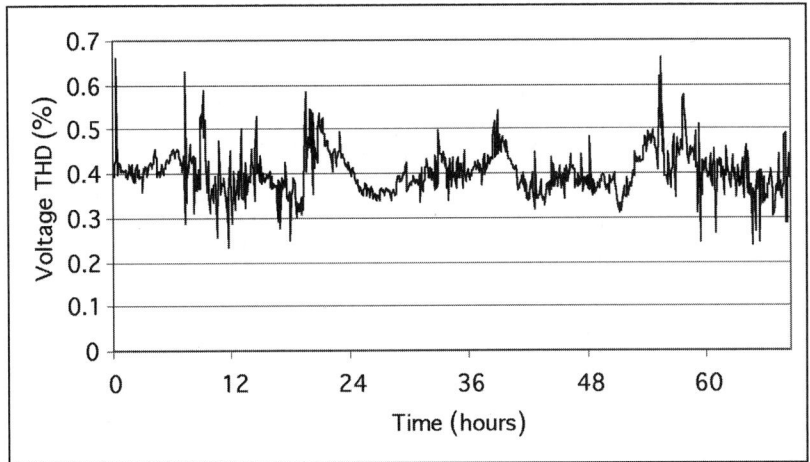

FIGURE 15.21 Example of time-varying nature of voltage THD.

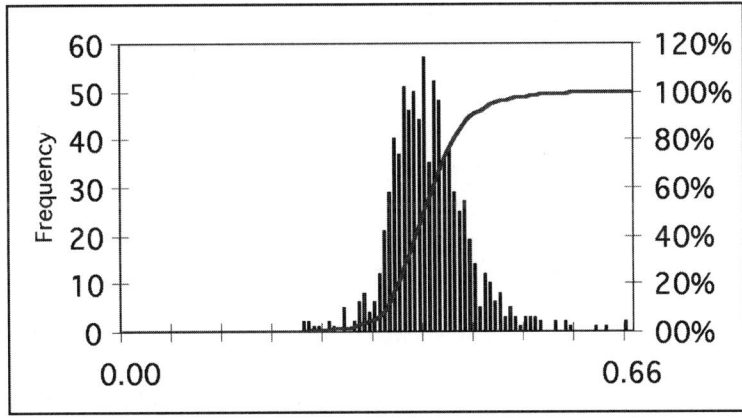

FIGURE 15.22 Probabilistic representation of voltage THD.

Because harmonic levels are never constant, it is difficult to establish utility-side or manufacturing-side limits for these quantities. In general, a probabilistic representation is used to describe harmonic quantities in terms of percentiles. Often, the 95th and 99th percentiles are used for design or operating limits. Figure 15.22 shows a histogram of the voltage THD in Fig. 15.21, and also includes a cumulative probability curve derived from the frequency distribution. Any percentile of interest can be readily calculated from the cumulative probability curve.

Both the Institute of Electrical and Electronics Engineers (IEEE) and the International Electrotechnical Commission (IEC) recognize the need to consider the time-varying nature of harmonics when determining harmonic levels that are permissible. Both organizations publish harmonic limits, but the degree to which the various limits can be applied varies widely. Both IEEE and IEC publish "system-level" harmonic limits that are intended to be applied from the utility point-of-view in order to limit power system harmonics to acceptable levels. The IEC, however, goes further and also publishes harmonic limits for individual pieces of equipment.

The IEEE limits are covered in two documents, IEEE 519-1992 and IEEE 519A (draft). These documents suggest that harmonics in the power system be limited by two different methods. One set of harmonic limits is for the harmonic current that a user can inject into the utility system at the point

TABLE 15.5 IEEE-519 Harmonic Current Limits

I_{SC}/I_L [a]	h < 11	11 ≤ h < 17	17 ≤ h < 23	23 ≤ h < 35	35 ≤ h	TDD
\multicolumn{7}{c}{V_{supply} ≤ 69kV}						
<20[b]	4.0	2.0	1.5	0.6	0.3	5.0
20–50	7.0	3.5	2.5	1.0	0.5	8.0
50–100	10.0	4.5	4.0	1.5	0.7	12.0
100–1000	12.0	5.5	5.0	2.0	1.0	15.0
>1000	15.0	7.0	6.0	2.5	1.4	20.0
\multicolumn{7}{c}{69kV < V_{supply} ≤ 161 kV}						
<20[b]	2.0	1.0	0.75	0.3	0.15	2.5
20–50	3.5	1.75	1.25	0.5	0.25	4.0
50–100	5.0	2.25	2.0	1.25	0.35	6.0
100–1000	6.0	2.75	2.5	1.0	0.5	7.5
>1000	7.5	3.5	3.0	1.25	0.7	10.0
\multicolumn{7}{c}{V_{supply} > 161 kV}						
<50	2.0	1.0	0.75	0.3	0.15	2.5
≥50	3.5	1.75	1.25	0.5	0.25	4.0

Note: Even harmonics are limited to 25% of the odd harmonic limits above. Current distortions that result in a DC offset, e.g., half wave converters, are not allowed.

[a] I_{SC} = maximum short-circuit current at PCC. I_L = maximum demand load current (fundamental frequency component) at PCC.

[b] All power generation equipment is limited to these values of current distortion, regardless of actual I_{SC}/I_L.

where other customers are or could be (in the future) served. (Note that this point in the system is often called the point of common coupling, or PCC.) The other set of harmonic limits is for the harmonic voltage that the utility can supply to any customer at the PCC. With this two-part approach, customers insure that they do not inject an "unreasonable" amount of harmonic current into the system, and the utility insures that any "reasonable" amount of harmonic current injected by any and all customers does not lead to excessive voltage distortion.

Table 15.5 shows the harmonic current limits that are suggested for utility customers. The table is broken into various rows and columns depending on harmonic number, short circuit to load ratio, and voltage level. Note that all quantities are expressed in terms of a percentage of the maximum demand current (I_L in the table). Total demand distortion (TDD) is defined to be the rms value of all harmonics, in amperes, divided by the maximum (12 month) fundamental frequency load current, I_L, with this ratio then multiplied by 100%.

The intent of the harmonic current limits is to permit larger customers, who in concept pay a greater share of the cost of power delivery equipment, to inject a greater portion of the harmonic current (in amperes) that the utility can absorb without producing excessive voltage distortion. Furthermore, customers served at transmission level voltage have more restricted injection limits than do customers served at lower voltage because harmonics in the high voltage network have the potential to adversely impact a greater number of other users through voltage distortion.

Table 15.6 gives the IEEE 519-1992 voltage distortion limits. Similar to the current limits, the permissible distortion is decreased at higher voltage levels in an effort to minimize potential problems for the majority of system users. Note that Tables 15.5 and 15.6 are given here for illustrative purposes only; the reader is strongly advised to consider additional material listed at the end of this section prior to trying to apply the limits.

The IEC formulates similar limit tables with the same intent: limit harmonic current injections so that voltage distortion problems are not created; the utility will correct voltage distortion problems if they exist and if all customers are within the specified harmonic current limits. Because the numbers suggested

TABLE 15.6 IEEE 519-1992 Voltage Harmonic Limits

Bus voltage at PCC (V_{L-L})	Individual Harmonic Voltage Distortion (%)	Total Voltage Distortion — THD_{Vn} (%)
$V_n \leq 69$ kV	3.0	5.0
69 kV $< V_n \leq$ 161 kV	1.5	2.5
$V_n >$ 161 kV	1.0	1.5

Note: High-voltage systems can have up to 2.0% THD where the cause is an HVDC terminal that will attenuate by the time it is tapped for a user.

by the IEC are similar (but not identical) to those given in Tables 15.5 and 15.6, the IEC tables for system-level harmonic limits given in IEC 1000-3-6 are not repeated here.

While the IEEE harmonic limits are designed for application at the three-phase PCC, the IEC goes further and provides limits appropriate for single-phase and three-phase individual equipment types. The most notable feature of these equipment limits is the "mA per W" manner in which they are proposed. For a wide variety of harmonic-producing loads, the steady-state (normal operation) harmonic currents are limited by prescribing a certain harmonic current, in mA, for each watt of power rating. The IEC also provides a specific waveshape for some load types that represents the most distorted current waveform allowed. Equipment covered by such limits include personal computers (power supplies) and single-phase battery charging equipment.

Even though limits exist, problems related to harmonics often arise from single, large "point source" harmonic loads as well as from numerous distributed smaller loads. In these situations, it is necessary to conduct a measurement, modeling, and analysis campaign that is designed to gather data and develop a solution. As previously mentioned, there are many commercially available instruments that can provide harmonic measurement information both at a single "snapshot" in time as well as continuous monitoring over time. How this information is used to develop problem solutions, however, can be a very complex issue.

Computer-assisted harmonic studies generally require significantly more input data than load flow or short circuit studies. Because high frequencies (up to 2–3 kHz) are under consideration, it is important to have mathematically correct equipment models and the data to use in them. Assuming that this data is available, there are a variety of commercially available software tools for actually performing the studies.

Most harmonic studies are performed in the frequency domain using sinusoidal steady-state techniques. (Note that other techniques, including full time-domain simulation, are sometimes used for specific problems.) A power system equivalent circuit is prepared for each frequency to be analyzed (recall that the Fourier series representation of a waveform is based on harmonic terms of known frequencies), and then basic circuit analysis techniques are used to determine voltages and currents of interest at that frequency. Most harmonic producing loads are modeled using a current source at each frequency that the load produces (arc furnaces are sometimes modeled using voltage sources), and network currents and voltages are determined based on these load currents. Recognize that at each frequency, voltage and current solutions are obtained from an equivalent circuit that is valid at that frequency only; the principle of superposition is used to "reconstruct" the Fourier series for any desired quantity in the network from the solutions of multiple equivalent circuits. Depending on the software tool used, the results can be presented in tabular form, spectral form, or as a waveform as shown in Table 15.4 and Figs. 15.18 and 15.19, respectively. An example voltage magnitude spectrum obtained from a harmonic study of a distribution primary circuit is shown in Fig. 15.23.

Regardless of the presentation format of the results, it is possible to use this type of frequency-domain harmonic analysis procedure to predict the impact of harmonic producing loads at any location in any power system. However, it is often impractical to consider a complete model of a large system, especially when unbalanced conditions must be considered. Of particular importance, however, are the locations of capacitor banks.

When electrically in parallel with network inductive reactance, capacitor banks produce a parallel resonance condition that tends to amplify voltage harmonics for a given current harmonic injection.

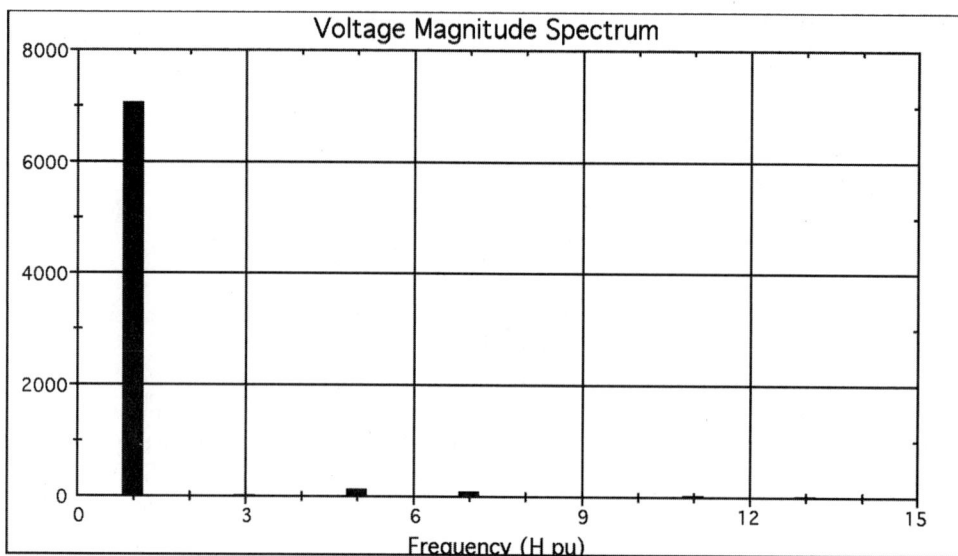

FIGURE 15.23 Sample magnitude spectrum results from a harmonic study.

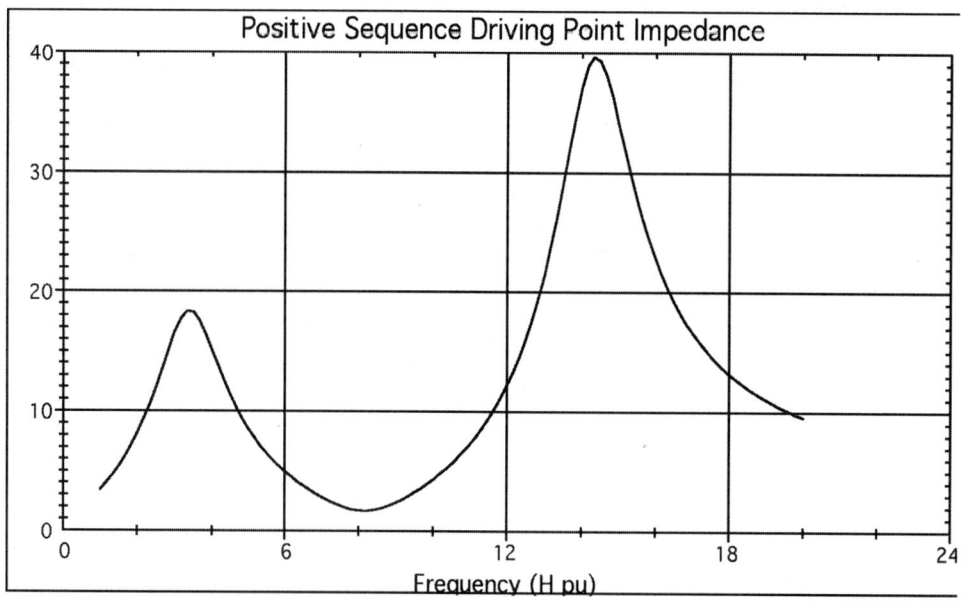

FIGURE 15.24 Sample frequency scan results.

When electrically in series with network inductive reactance, capacitor banks produce a series resonance condition that tends to amplify current harmonics for a given voltage distortion. In either case, harmonic levels far in excess of what are expected can be produced. Fortunately, a relatively simple calculation procedure called a frequency scan, can be used to indicate potential resonance problems. Figure 15.24 shows an example of a frequency scan conducted on the positive sequence network model of a distribution circuit. Note that the distribution primary included the standard feeder optimization capacitors.

A frequency scan result is actually a plot of impedance vs. frequency. Two types of results are available: driving point and transfer impedance scans. The driving point frequency scan shown in Fig. 15.24 indicates how much voltage would be produced at a given bus and frequency for a one-ampere current

Power Quality

injection at that same location and frequency. Where necessary, the principle of linearity can be used to scale the one-ampere injection to the level actually injected by specific equipment. In other words, the driving point impedance predicts how a customer's harmonic producing load could impact the voltage at that load's terminals. Local maximums, or peaks, in the scan plot indicate parallel resonance conditions. Local minimums, or valleys, in the scan plot indicate series resonance.

A transfer impedance scan predicts how a customer's harmonic producing load at one location can impact voltage distortions at other (possibly very remote) locations. In general, to assess the ability of a relatively small current injection to produce a significant voltage distortion (due to resonance) at remote locations (due to transfer impedance) is the primary goal of every harmonic study.

Should a harmonic study indicate a potential problem (violation of limits, for example), two categories of solutions are available: (1) reduce the harmonics at their point of origin (before they enter the system), or (2) apply filtering to reduce undesirable harmonics. Many methods for reducing harmonics at their origin are available; for example, using various transformer connections to cancel certain harmonics has been extremely effective in practice. In most cases, however, reducing or eliminating harmonics at their origin is effective only in the design or expansion stage of a new facility. For existing facilities, harmonic filters often provide the least-cost solution.

Harmonic filters can be subdivided into two types: active and passive. Active filters are only now becoming commercially viable products for high-power applications and operate as follows. For a load that injects certain harmonic currents into the supply system, a DC to AC inverter can be controlled such that the inverter supplies the harmonic current for the load, while allowing the power system to supply the power frequency current for the load. Figure 15.25 shows a diagram of such an active filter application.

For high power applications or for applications where power factor correction capacitors already exist, it is typically more cost effective to use passive filtering. Passive filtering is based on the series resonance principle (recall that a low impedance at a specific frequency is a series-resonant characteristic) and can be easily implemented. Figure 15.26 shows a typical three-phase harmonic filter (many other designs are also used) that is commonly used to filter 5th or 7th harmonics.

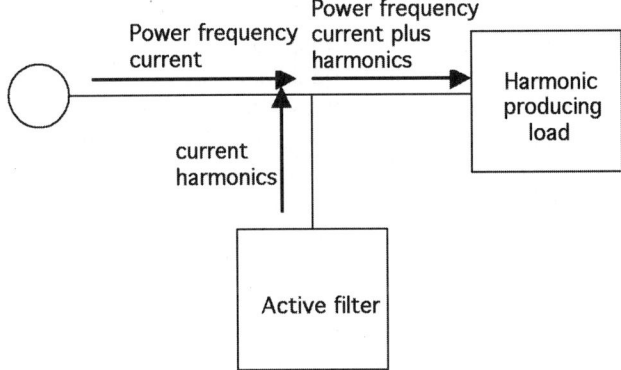

FIGURE 15.25 Active filter concept diagram.

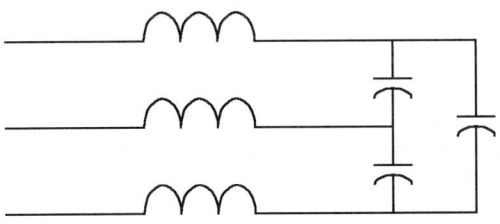

FIGURE 15.26 Typical passive filter design.

It should be noted that passive filtering cannot always make use of existing capacitor banks. In filter applications, the capacitors will typically be exposed continuously to voltages greater than their ratings (which were determined based on their original application). 600 V capacitors, for example, may be required for 480 V filter applications. Even with the potential cost of new capacitors, passive filtering still appears to offer the most cost effective solution to the harmonic problem at this time.

In conclusion, power system harmonics have been carefully considered for many years and have received a significant increase in research and development activity as a direct result of the proliferation of high-power semiconductors. Fortunately, harmonic measurement equipment is readily available, and the underlying theory used to evaluate harmonics analytically (with computer assistance) is well understood. Limits for harmonic voltages and currents have been suggested by multiple standards-making bodies, but care must be used because the suggested limits are not necessarily equivalent.

Regardless of which limit numbers are appropriate for a given application, multiple options are available to help meet the levels required. As with all power quality problems, however, accurate study on the "front end" usually will reveal possible problems in the design stage, and a lower-cost solution can be implemented before problems arise.

The material presented here is not intended to be all-inclusive. The suggested reading provides further documents, including both IEEE and IEC standards, recommended practices, and technical papers and reports that provide the knowledge base required to apply the standards properly.

Further Information

Arrillaga, J., Bradley, D., and Bodger, P., *Power System Harmonics*, John Wiley, New York, 1985.

Mohan, N., Undeland, T. M., and Robbins, W. P., *Power Electronics: Converters, Applications, and Design*, John Wiley, New York, 1989.

Heydt, G. T., *Electric Power Quality*, Stars in a Circle Publications, 1991.

IEEE Standard 519-1992: *Recommended Practices and Requirements for Harmonic Control in Electrical Power Systems*, IEEE Press, April 1993.

Dugan, R. C., McGranaghan, M. F., and Beaty, H. W., *Electrical Power Systems Quality*, McGraw-Hill, New York, 1996.

P519A Task Force of the Harmonics Working Group and SCC20-Power Quality, *Guide for Applying Harmonic Limits on Power Systems* (draft), IEEE, May 1996.

IEC 61000-3-2, *Electromagnetic compatibility (EMC) — Part 3-2: Limits — Limits for harmonic current emissions (equipment input current <= 16 A per phase)*, Ed. 1.2 b:1998.

IEC 61000-3-6 TR3, *Electromagnetic compatibility (EMC) — Part 3: Limits — Section 6: Assessment of emission limits for distorting loads in MV and HV power systems — Basic EMC Publication*, Ed. 1.0 b:1996.

IEC 61000-4-7, *Electromagnetic compatibility (EMC) — Part 4: Testing and measurement techniques-Section 7: General guide on harmonics and interharmonics measurements and instrumentation, for power supply systems and equipment connected thereto*, Ed. 1.0 b:1991.

UIE, *Guide to Quality of Electrical Supply for Industrial Installations, Part 3: Harmonics*, 1998.

IEEE Harmonics Modeling and Simulation Task Force, IEEE Special Publication #98-TP-125-0: *IEEE Tutorial on Harmonics Modeling and Simulation*, IEEE Press, 1998.

15.4 Voltage Sags

M. H. J. Bollen

Voltage sags are short duration reductions in rms voltage, mainly caused by short circuits and starting of large motors. The interest in voltage sags is due to the problems they cause on several types of equipment. Adjustable-speed drives, process-control equipment, and computers are especially notorious for their sensitivity (Conrad et al., 1991; McGranaghan et al., 1993). Some pieces of equipment trip when

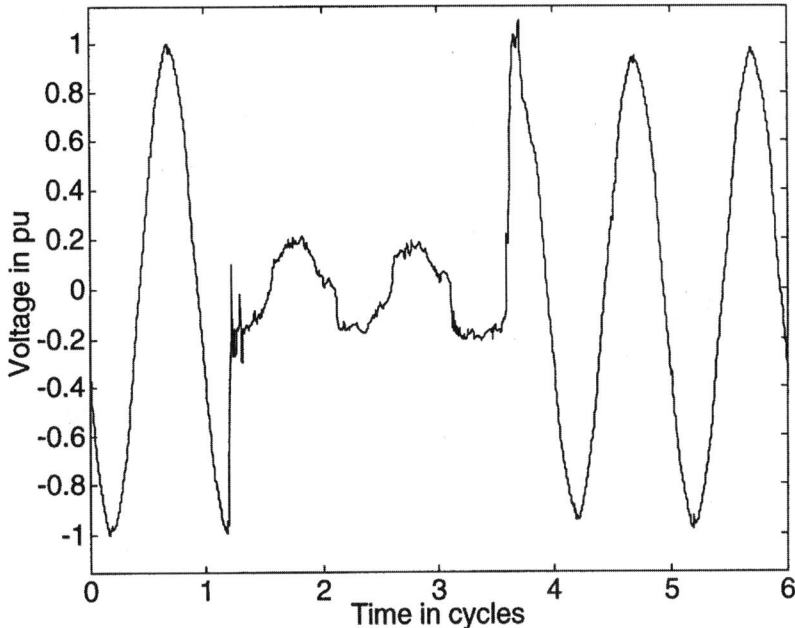

FIGURE 15.27 A voltage sag — voltage in one phase in time domain.

the rms voltage drops below 90% for longer than one or two cycles. Such a piece of equipment will trip tens of times a year. If this is the process-control equipment of a paper mill, one can imagine that the costs due to voltage sags can be enormous. A voltage sag is not as damaging to industry as a (long or short) interruption, but as there are far more voltage sags than interruptions, the total damage due to sags is still larger. Another important aspect of voltage sags is that they are hard to mitigate. Short interruptions and many long interruptions can be prevented via simple, although expensive measures in the local distribution network. Voltage sags at equipment terminals can be due to short-circuit faults hundreds of kilometers away in the transmission system. It will be clear that there is no simple method to prevent them.

Voltage Sag Characteristics

An example of a voltage sag is shown in Fig. 15.27.[1] The voltage amplitude drops to a value of about 20% of its pre-event value for about two and a half cycles, after which the voltage recovers again. The event shown in Fig. 15.27 can be characterized as a voltage sag down to 20% (of the pre-event voltage) for 2.5 cycles (of the fundamental frequency). This event can be characterized as a voltage sag with a magnitude of 20% and a duration of 2.5 cycles.

Voltage Sag Magnitude — Monitoring

The magnitude of a voltage sag is determined from the rms voltage. The rms voltage for the sag in Fig. 15.27 is shown in Fig. 15.28. The rms voltage has been calculated over a one-cycle sliding window:

$$V_{rms}(k) = \sqrt{\frac{1}{N} \sum_{i=k-N+1}^{i=k} v(i)^2} \qquad (15.3)$$

---

[1]The datafile containing these measurements was obtained from a Website with test data set up for IEEE project group P1159.2: http://grouper.ieee.org/groups/1159/2/index.html.

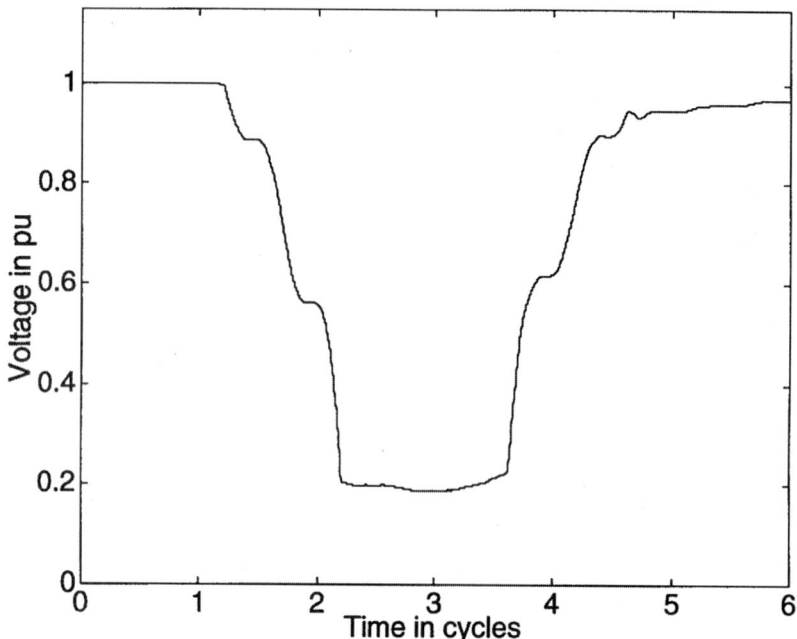

FIGURE 15.28 One-cycle rms voltage for the voltage sag shown in Fig. 15.27.

with N the number of samples per cycle, and $v(i)$ the sampled voltage in time domain. The rms voltage as shown in Fig. 15.28 does not immediately drop to a lower value, but takes one cycle for the transition. This is due to the finite length of the window used to calculate the rms value. We also see that the rms value during the sag is not completely constant and that the voltage does not immediately recover after the fault.

There are various ways of obtaining the sag magnitude from the rms voltages. Most power quality monitors take the lowest value obtained during the event. As sags normally have a constant rms value during the deep part of the sag, using the lowest value is an acceptable approximation.

The sag is characterized through the remaining voltage during the event. This is then given as a percentage of the nominal voltage. Thus, a 70% sag in a 230-V system means that the voltage dropped to 161 V. The confusion with this terminology is clear. One could be tricked into thinking that a 70% sag refers to a drop of 70%, thus a remaining voltage of 30%. The recommendation is therefore to use the phrase "a sag down to 70%." Characterizing the sag through the actual drop in rms voltage can solve this ambiguity, but this will introduce new ambiguities like the choice of the reference voltage.

Origin of Voltage Sags

Consider the distribution network shown in Fig. 15.29, where the numbers (1 through 5) indicate fault positions and the letters (A through D) loads. A fault in the transmission network, fault position 1, will cause a serious sag for both substations bordering the faulted line. This sag is transferred down to all customers fed from these two substations. As there is normally no generation connected at lower voltage levels, there is nothing to keep up the voltage. The result is that all customers (A, B, C, and D) experience a deep sag. The sag experienced by A is likely to be somewhat less deep, as the generators connected to that substation will keep up the voltage. A fault at position 2 will not cause much voltage drop for customer A. The impedance of the transformers between the transmission and the subtransmission system are large enough to considerably limit the voltage drop at high-voltage side of the transformer. The sag experienced by customer A is further mitigated by the generators feeding into its local transmission substation. The fault at position 2 will, however, cause a deep sag at both subtransmission substations and thus for all customers fed from here (B, C, and D). A fault at position 3 will cause a short or long

Power Quality

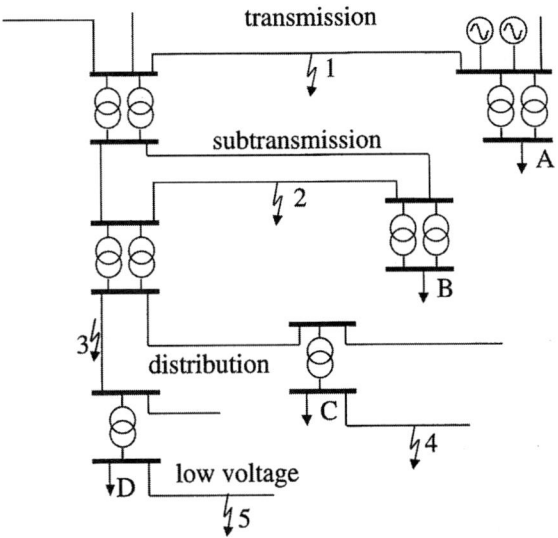

FIGURE 15.29 Distribution network with load positions (A through D) and fault positions (1 through 5).

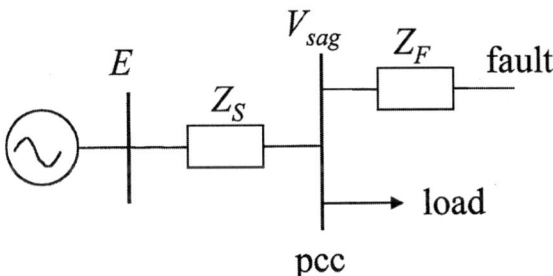

FIGURE 15.30 Voltage divider model for a voltage sag.

interruption for customer D when the protection clears the fault. Customer C will only experience a deep sag. Customer B will experience a shallow sag due to the fault at position 3, again due to the transformer impedance. Customer A will probably not notice anything from this fault. Fault 4 causes a deep sag for customer C and a shallow one for customer D. For fault 5, the result is the other way around: a deep sag for customer D and a shallow one for customer C. Customers A and B will not experience any significant drop in voltage due to faults 4 and 5.

Voltage Sag Magnitude — Calculation

To quantify sag magnitude in radial systems, the voltage divider model, shown in Fig. 15.30, can be used, where Z_S is the source impedance at the point-of-common coupling; and Z_F is the impedance between the point-of-common coupling and the fault. The point-of-common coupling (pcc) is the point from which both the fault and the load are fed. In other words, it is the place where the load current branches off from the fault current. In the voltage divider model, the load current before, as well as during the fault is neglected. The voltage at the pcc is found from:

$$V_{sag} = \frac{Z_F}{Z_S + Z_F} \qquad (15.4)$$

where it is assumed that the pre-event voltage is exactly 1 pu, thus $E = 1$. The same expression can be derived for constant-impedance load, where E is the pre-event voltage at the pcc. We see from Eq. (15.4)

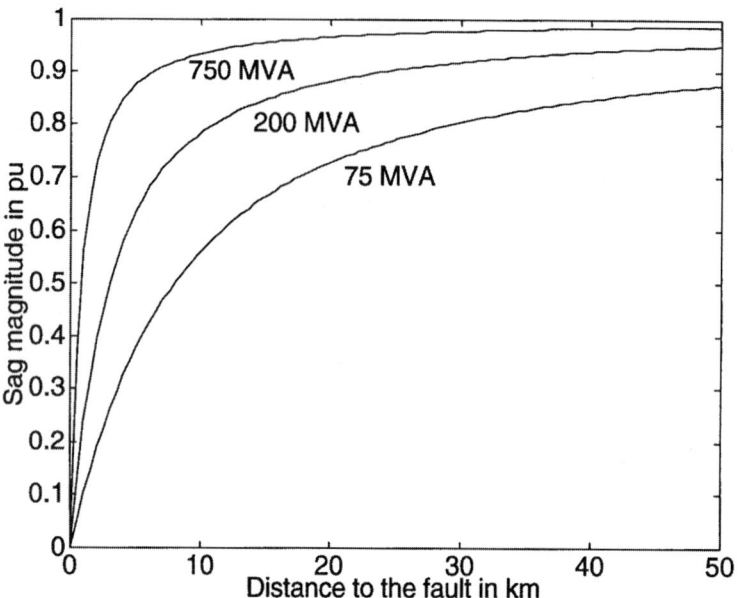

FIGURE 15.31 Sag magnitude as a function of the distance to the fault.

that the sag becomes deeper for faults electrically closer to the customer (when Z_F becomes smaller), and for weaker systems (when Z_S becomes larger).

Equation (15.4) can be used to calculate the sag magnitude as a function of the distance to the fault. Therefore, we write $Z_F = zd$, with z the impedance of the feeder per unit length and d the distance between the fault and the pcc, leading to:

$$V_{sag} = \frac{zd}{Z_S + zd} \qquad (15.5)$$

This expression has been used to calculate the sag magnitude as a function of the distance to the fault for a typical 11 kV overhead line, resulting in Fig. 15.31. For the calculations, a 150-mm² overhead line was used and fault levels of 750 MVA, 200 MVA, and 75 MVA. The fault level is used to calculate the source impedance at the pcc and the feeder impedance is used to calculate the impedance between the pcc and the fault. It is assumed that the source impedance is purely reactive, thus Z_S = j 0.161Ω for the 750 MVA source. The impedance of the 150 mm² overhead line is $z = 0.117 + j\ 0.315$ Ω/km.

Propagation of Voltage Sags

It is also possible to calculate the sag magnitude directly from fault levels at the pcc and at the fault position. Let S_{FLT} be the fault level at the fault position and S_{PCC} at the point-of-common coupling. The voltage at the pcc can be written as:

$$V_{sag} = 1 - \frac{S_{FLT}}{S_{PCC}} \qquad (15.6)$$

This equation can be used to calculate the magnitude of sags due to faults at voltage levels other than the point-of-common coupling. Consider typical fault levels as shown in Table 15.7. This data has been used to obtain Table 15.8, showing the effect of a short circuit fault at a lower voltage level than the pcc. We can see that sags are significantly "damped" when they propagate upwards in the power system. In a sags study, we typically only have to take faults one voltage level down from the pcc into account. And

TABLE 15.7 Typical Fault Levels at Different Voltage Levels

Voltage Level	Fault Level
400 V	20 MVA
11 kV	200 MVA
33 kV	900 MVA
132 kV	3000 MVA
400 kV	17,000 MVA

TABLE 15.8 Propagation of Voltage Sags to Higher Voltage Levels

Fault at:	Point-of-Common Coupling at:				
	400 V	11 kV	33 kV	132 kV	400 kV
400 V	—	90%	98%	99%	100%
11 kV	—	—	78%	93%	99%
33 kV	—	—	—	70%	95%
132 kV	—	—	—	—	82%

even those are seldom of serious concern. Note, however, that faults at a lower voltage level may be associated with a longer fault-clearing time and thus a longer sag duration. This especially holds for faults on distribution feeders, where fault-clearing times in excess of one second are possible.

Critical Distance

Equation (15.5) gives the voltage as a function of distance to the fault. From this equation we can obtain the distance at which a fault will lead to a sag of a certain magnitude V. If we assume equal X/R ratio of source and feeder, we get the following equation:

$$d_{crit} = \frac{Z_S}{z} \times \frac{V}{1-V} \qquad (15.7)$$

We refer to this distance as the critical distance. Suppose that a piece of equipment trips when the voltage drops below a certain level (the critical voltage). The definition of critical distance is such that each fault within the critical distance will cause the equipment to trip. This concept can be used to estimate the expected number of equipment trips due to voltage sags (Bollen, 1998). The critical distance has been calculated for different voltage levels, using typical fault levels and feeder impedances. The data used and the results obtained are summarized in Table 15.9 for the critical voltage of 50%. Note how the critical distance increases for higher voltage levels. A customer will be exposed to much more kilometers

TABLE 15.9 Critical Distance for Faults at Different Voltage Levels

Nominal Voltage	Short-Circuit Level	Feeder Impedance	Critical Distance
400 V	20 MVA	230 mΩ/km	35 m
11 kV	200 MVA	310 mΩ/km	2 km
33 kV	900 MVA	340 mΩ/km	4 km
132 kV	3000 MVA	450 mΩ/km	13 km
400 kV	10000 MVA	290 mΩ/km	55 km

of transmission lines than of distribution feeder. This effect is understood by writing Eq. (15.7) as a function of the short-circuit current I_{flt} at the pcc:

$$d_{crit} = \frac{V_{nom}}{zI_{flt}} \times \frac{V}{1-V} \qquad (15.8)$$

with V_{nom} the nominal voltage. As both z and I_{flt} are of similar magnitude for different voltage levels, one can conclude from Eq. (15.8) that the critical distance increases proportionally with the voltage level.

Voltage Sag Duration

It was shown before, the drop in voltage during a sag is due to a short circuit being present in the system. The moment the short circuit fault is cleared by the protection, the voltage starts to return to its original value. The duration of a sag is thus determined by the fault-clearing time. However, the actual duration of a sag is normally longer than the fault-clearing time.

Measurement of sag duration is less trivial than it might appear. From a recording the sag duration may be obvious, but to come up with an automatic way for a power quality monitor to obtain the sag duration is no longer straightforward. The commonly used definition of sag duration is the number of cycles during which the rms voltage is below a given threshold. This threshold will be somewhat different for each monitor but typical values are around 90% of the nominal voltage. A power quality monitor will typically calculate the rms value once every cycle.

The main problem is that the so-called post-fault sag will affect the sag duration. When the fault is cleared, the voltage does not recover immediately. This is mainly due to the reenergizing and reacceleration of induction motor load (Bollen, 1995). This post-fault sag can last several seconds, much longer than the actual sag. Therefore, the sag duration as defined before, is no longer equal to the fault-clearing time. More seriously, different power quality monitors will give different values for the sag duration. As the rms voltage recovers slowly, a small difference in threshold setting may already lead to a serious difference in recorded sag duration (Bollen, 1999).

Generally speaking, faults in transmission systems are cleared faster than faults in distribution systems. In transmission systems, the critical fault-clearing time is rather small. Thus, fast protection and fast circuit breakers are essential. Also, transmission and subtransmission systems are normally operated as a grid, requiring distance protection or differential protection, both of which allow for fast clearing of the fault. The principal form of protection in distribution systems is overcurrent protection. This requires a certain amount of time-grading, which increases the fault-clearing time. An exception is formed by systems in which current-limiting fuses are used. These have the ability to clear a fault within one half-cycle. In overhead distribution systems, the instantaneous trip of the recloser will lead to a short sag duration, but the clearing of a permanent fault will give a sag of much longer duration.

The so-called magnitude-duration plot is a common tool used to show the quality of supply at a certain location or the average quality of supply of a number of locations. Voltage sags due to faults can be shown in such a plot, as well as sags due to motor starting, and even long and short interruptions. Different underlying causes lead to events in different parts of the magnitude-duration plot, as shown in Fig. 15.32.

Phase-Angle Jumps

A short circuit in a power system not only causes a drop in voltage magnitude, but also a change in the phase angle of the voltage. This sudden change in phase angle is called a "phase-angle jump." The phase-angle jump is visible in a time-domain plot of the sag as a shift in voltage zero-crossing between the pre-event and the during-event voltage. With reference to Fig. 15.30 and Eq. (15.4), the phase-angle jump is the argument of V_{sag}, thus the difference in argument between Z_F and $Z_S + Z_P$. If source and feeder impedance have equal X/R ratio, there will be no phase-angle jump in the voltage at the pcc. This is the case for faults in transmission systems, but normally not for faults in distribution systems. The latter may have phase-angle jumps up to a few tens of degrees (Bollen, 1999; Bollen et al., 1996).

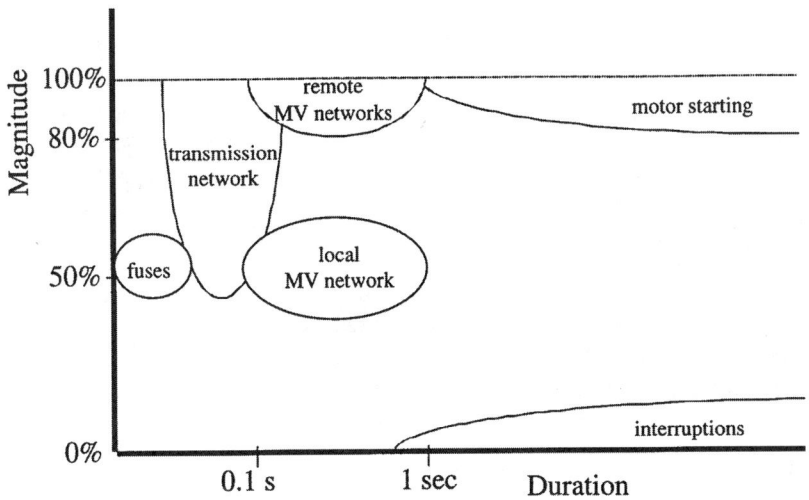

FIGURE 15.32 Sags of different origin in a magnitude-duration plot.

Figure 15.30 shows a single-phase circuit, which is a valid model for three-phase faults in a three-phase system. For nonsymmetrical faults, the analysis becomes much more complicated. A consequence of nonsymmetrical faults (single-phase, phase-to-phase, two-phase-to-ground) is that single-phase load experiences a phase-angle jump even for equal X/R ratio of feeder and source impedance (Bollen, 1999; Bollen, 1997).

To obtain the phase-angle jump from the measured voltage waveshape, the phase angle of the voltage during the event must be compared with the phase angle of the voltage before the event. The phase angle of the voltage can be obtained from the voltage zero-crossings or from the argument of the fundamental component of the voltage. The fundamental component can be obtained by using a discrete Fourier transform algorithm. Let $V_1(t)$ be the fundamental component obtained from a window $(t-T,t)$, with T one cycle of the power frequency, and let $t = 0$ correspond to the moment of sag initiation. In case there is no chance in voltage magnitude or phase angle, the fundamental component as a function of time is found from:

$$V_1(t) = V_1(0)e^{j\omega t} \qquad (15.9)$$

The phase-angle jump, as a function of time, is the difference in phase angle between the actual fundamental component and the "synchronous voltage" according to Eq. (15.9):

$$\phi(t) = \arg\{V_1(t)\} - \arg\{V_1(0)e^{j\omega t}\} = \arg\left\{\frac{V_1(t)}{V_1(0)}e^{-j\omega t}\right\} \qquad (15.10)$$

Note that the argument of the latter expression is always between −180° and +180°.

Three-Phase Unbalance

For three-phase equipment, three voltages need to be considered when analyzing a voltage sag event at the equipment terminals. For this, a characterization of three-phase unbalanced voltage sags is introduced. The basis of this characterization is the theory of symmetrical components. Instead of the three-phase voltages or the three symmetrical components, the following three (complex) values are used to characterize the voltage sag (Bollen and Zhang, 1999; Zhang and Bollen, 1998):

- The "characteristic voltage" is the main characteristic of the event. It indicates the severity of the sag, and can be treated in the same way as the remaining voltage for a sag experienced by a single-phase event.
- The "PN factor" is a correction factor for the effect of the load on the voltages during the event. The PN factor is normally close to unity and can then be neglected. Exceptions are systems with a large amount of dynamic load, and sags due to two-phase-to-ground faults.
- The "zero-sequence voltage," which is normally not transferred to the equipment terminals, rarely affects equipment behavior. The zero-sequence voltage can be neglected in most studies.

Neglecting the zero-sequence voltage, it can be shown that there are two types of three-phase unbalanced sags, denoted as types C and D. Type A is a balanced sag due to a three-phase fault. Type B is the sag due to a single-phase fault, which turns into type D after removal of the zero-sequence voltage. The three complex voltages for a type C sag are written as follows:

$$V_a = F$$
$$V_b = -\frac{1}{2}F - \frac{1}{2}jV\sqrt{3} \qquad (15.11)$$
$$V_c = -\frac{1}{2}F + \frac{1}{2}jV\sqrt{3}$$

where V is the characteristic voltage and F the PN factor. The (characteristic) sag magnitude is defined as the absolute value of the characteristic voltage; the (characteristic) phase-angle jump is the argument of the characteristic voltage. For a sag of type D, the expressions for the three voltage phasors are as follows:

$$V_a = V$$
$$V_b = -\frac{1}{2}V - \frac{1}{2}jF\sqrt{3} \qquad (15.12)$$
$$V_c = -\frac{1}{2}V + \frac{1}{2}jF\sqrt{3}$$

Sag type D is due to a phase-to-phase fault, or due to a single-phase fault behind a Δy-transformer, or a phase-to-phase fault behind two Δy-transformers, etc. Sag type C is due to a single-phase fault, or due to a phase-to-phase fault behind a Δy-transformer, etc. When using characteristic voltage for a three-phase unbalanced sag, the same single-phase scheme as in Fig. 15.30 can be used to study the transfer of voltage sags in the system (Bollen, 1999; Bollen, 1997).

Equipment Voltage Tolerance

Voltage Tolerance Requirement

Generally speaking, electrical equipment prefers a constant rms voltage. That is what the equipment has been designed for and that is where it will operate best. The other extreme is zero voltage for a longer period of time. In that case the equipment will simply stop operating completely. For each piece of equipment there is a maximum interruption duration, after which it will continue to operate correctly. A rather simple test will give this duration. The same test can be done for a voltage of 10% (of nominal), for a voltage of 20%, etc. If the voltage becomes high enough, the equipment will be able to operate on it indefinitely. Connecting the points obtained by performing these tests results in the so-called "voltage-tolerance curve" (Key, 1979). An example of a voltage-tolerance curve is shown in Fig. 15.33: the requirements for IT-equipment as recommended by the Information Technology Industry Council (ITIC, 1999).

Power Quality

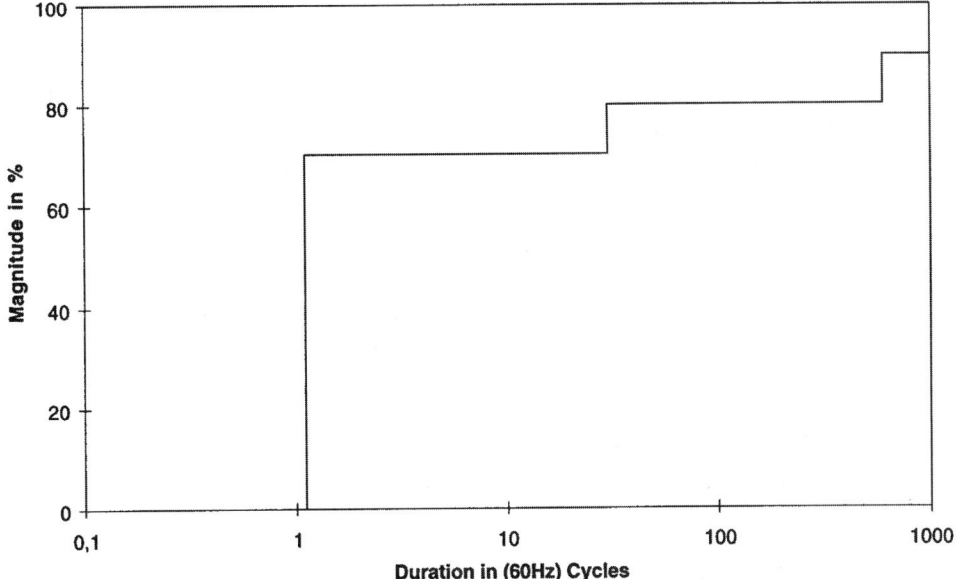

FIGURE 15.33 Voltage-tolerance requirement for IT equipment.

Strictly speaking, one can claim that this is not a voltage-tolerance curve as described above, but a requirement for the voltage tolerance. One could refer to this as a voltage-tolerance requirement and to the result of equipment tests as a voltage-tolerance performance. We see in Fig. 15.33 that IT equipment has to withstand a voltage sag down to zero for 1.1 cycle, down to 70% for 30 cycles, and that the equipment should be able to operate normally for any voltage of 90% or higher.

Voltage Tolerance Performance

Voltage-tolerance (performance) curves for personal computers are shown in Fig. 15.34. The curves are the result of equipment tests performed in the U.S. (EPRI, 1994) and in Japan (Sekine et al., 1992). The shape of all the curves in Fig. 15.34 is close to rectangular. This is typical for many types of equipment, so that the voltage tolerance may be given by only two values, maximum duration and minimum voltage, instead of by a full curve. From the tests summarized in Fig. 15.34 it is found that the voltage tolerance of personal computers varies over a wide range: 30–170 ms, 50–70% being the range containing half of the models. The extreme values found are 8 ms, 88% and 210 ms, 30%.

Voltage-tolerance tests have also been performed on process-control equipment: PLCs, monitoring relays, motor contactors. This equipment is even more sensitive to voltage sags than personal computers. The majority of devices tested tripped between one and three cycles. A small minority was able to tolerate sags up to 15 cycles in duration. The minimum voltage varies over a wider range: from 50% to 80% for most devices, with exceptions of 20% and 30%. Unfortunately, the latter two both tripped in three cycles (Bollen, 1999).

From performance testing of adjustable-speed drives, an "average voltage-tolerance curve" has been obtained. This curve is shown in Fig. 15.35. The sags for which the drive was tested are indicated as circles. It has further been assumed that the drives can operate indefinitely on 85% voltage. Voltage tolerance is defined here as "automatic speed recovery, without reaching zero speed." For sensitive production processes, more strict requirements will hold (Bollen, 1999).

Single-Phase Rectifiers

The sensitivity of most single-phase equipment can be understood from the equivalent scheme in Fig. 15.36. The power supply to a computer, process-control equipment, consumer electronics, etc. consists of a single-phase (four-pulse) rectifier together with a capacitor and a DC/DC converter. During

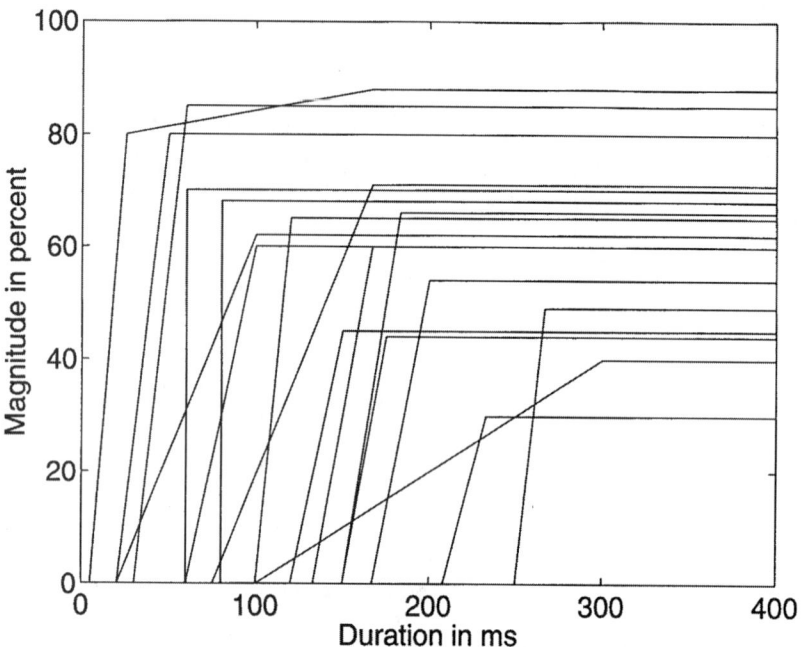

FIGURE 15.34 Voltage-tolerance performance for personal computers.

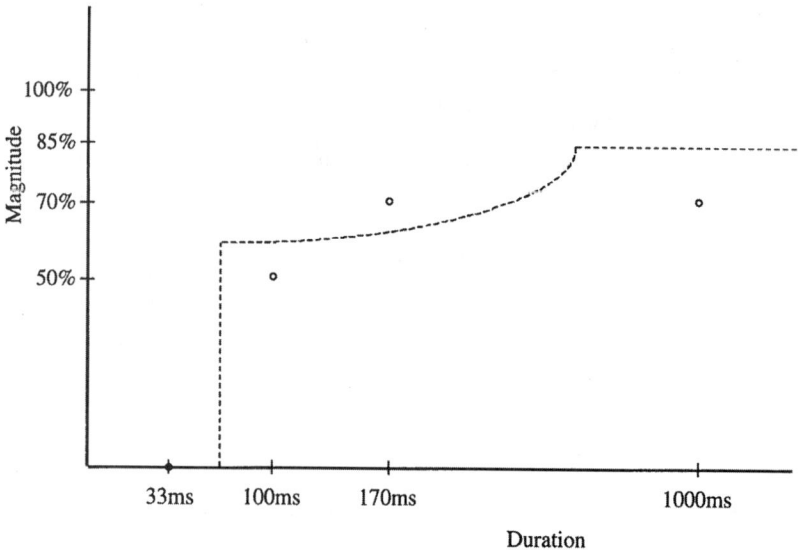

FIGURE 15.35 Average voltage-tolerance curve for adjustable-speed drives.

normal operation the capacitor is charged twice a cycle through the diodes. The result is a DC voltage ripple:

$$\varepsilon = \frac{PT}{2V_0^2 C} \qquad (15.13)$$

with P the DC bus active-power load, T one cycle of the power frequency, V_0 the maximum DC bus voltage, and C the size of the capacitor.

Power Quality

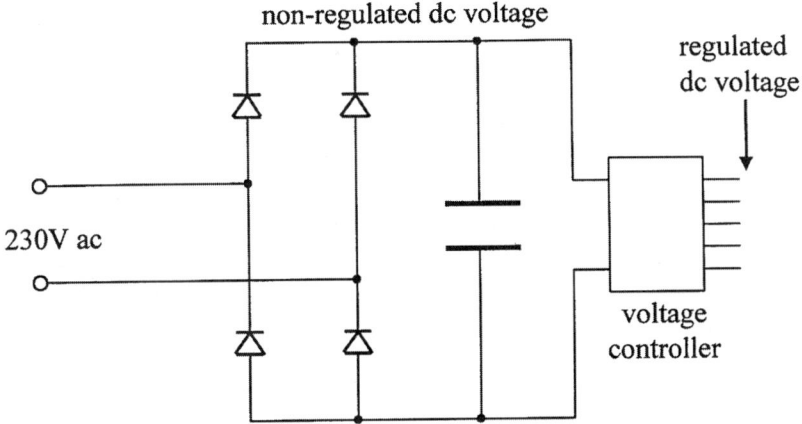

FIGURE 15.36 Typical power supply to sensitive single-phase equipment.

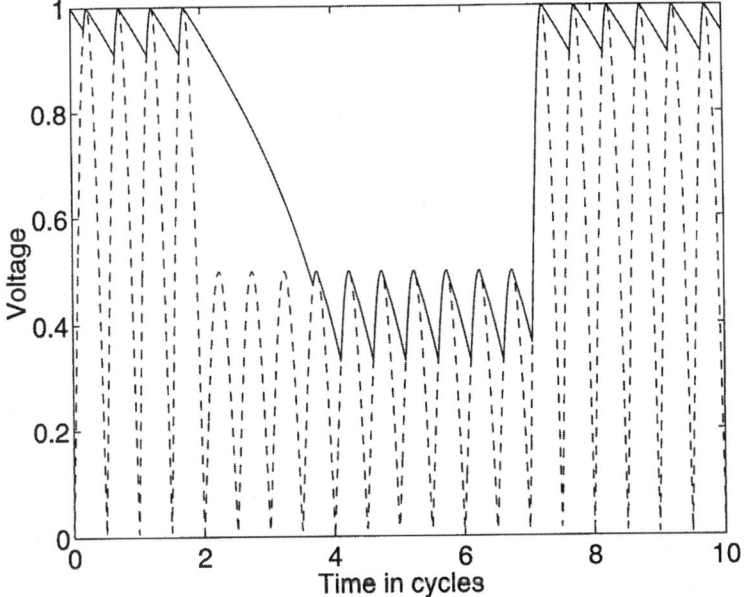

FIGURE 15.37 Absolute value of AC voltage (dashed) and DC bus voltage (solid line) for a sag down to 50%.

During a voltage sag or interruption, the capacitor continues to discharge until the DC bus voltage has dropped below the peak of the supply voltage. A new steady state is reached, but at a lower DC bus voltage and with a larger ripple. The resulting DC bus voltage for a sag down to 50% is shown in Fig. 15.37, together with the absolute value of the supply voltage. If the new steady state is below the minimum operating voltage of the DC/DC converter, or below a certain protection setting, the equipment will trip. During the decaying DC bus voltage, the capacitor voltage $V(t)$ can be obtained from the law of conservation of energy:

$$\frac{1}{2}CV^2 = \frac{1}{2}CV_0^2 - Pt \quad (15.14)$$

where a constant DC bus load P has been assumed. From Eq. (15.14) the voltage as a function of time is obtained:

$$V(t) = \sqrt{V_0^2 - \frac{2P}{C}t} \qquad (15.15)$$

Combining this with Eq. (15.13) gives the following expression:

$$V(t) = V_0 \sqrt{1 - 4\varepsilon \frac{t}{T}} \qquad (15.16)$$

The larger the DC ripple in normal operation, the faster the drop in DC bus voltage during a sag. From Eq. (15.16) the maximum duration of zero voltage t_{max} is calculated for a minimum operating voltage V_{min}, resulting in:

$$t_{max} = \frac{1 - \left(\frac{V_{min}}{V_0}\right)^2}{4\varepsilon} T \qquad (15.17)$$

Three-Phase Rectifiers

The performance of equipment fed through three-phase rectifiers becomes somewhat more complicated. The main equipment belonging to this category is formed by AC and DC adjustable-speed drives. One of the complications is that the operation of the equipment is affected by the three voltages, which are not necessarily the same during the voltage sag. For non-controlled (six pulse) diode rectifiers, a similar model can be used as for single-phase rectifiers. The operation of three-phase controlled rectifiers can become very complicated and application-specific (Bollen, 1996). Therefore, only noncontrolled rectifiers will be discussed here. For voltage sags due to three-phase faults, the DC bus voltage behind the (three-phase) rectifier will decay until a new steady state is reached at a lower voltage level, with a larger ripple. To calculate the DC bus voltage as a function of time, and the time-to-trip, the same equation as for the single-phase rectifier can be used.

For unbalanced voltage sags, a distinction needs to be made between the two types (C and D), as introduced in the section on Three-Phase Unbalance. Figure 15.38 shows AC and DC side voltages for a sag of type C with $V = 0.5$ pu and $F = 1$. For this sag, the voltage drops in two phases where the third phase stays at its presag value. Three capacitor sizes are used (Bollen and Zhang, 1999); a "large" capacitance is defined as a value that leads to an initial decay of the DC voltage equal to 10%, which is 433 µF/kW for a 620 V drive. In the same way, "small" capacitance corresponds to 75% per cycle initial decay, and 57.8 µF/kW for a 620 V drive. It turns out that even for the small capacitance, the DC bus voltage remains above 70%. For the large capacitance value, the DC bus voltage is hardly affected by the voltage sag. It is easy to understand that this is also the case for type C sags with an even lower characteristic magnitude V (Bollen, 1999; Bollen and Zhang, 1999).

Figure 15.39 shows the equivalent results for a sag of type D, again with $V = 0.5$ and $F = 1$. As all three AC voltages show a drop in voltage magnitude, the DC bus voltage will drop even for a large capacitor. But the effect is still much less than for a three-phase (balanced) sag.

The effect of a lower PN factor ($F < 1$) is that even the highest voltage shows a drop for a type C sag, so that the DC bus voltage will always show a small drop. Also for a type D sag, a lower PN factor will lead to an additional drop in DC bus voltage (Bollen and Zhang, 1999).

Mitigation of Voltage Sags

From Fault to Trip

To understand the various ways of mitigation, the mechanism leading to an equipment trip needs to be understood. The equipment trip is what makes the event a problem; if there are no equipment trips,

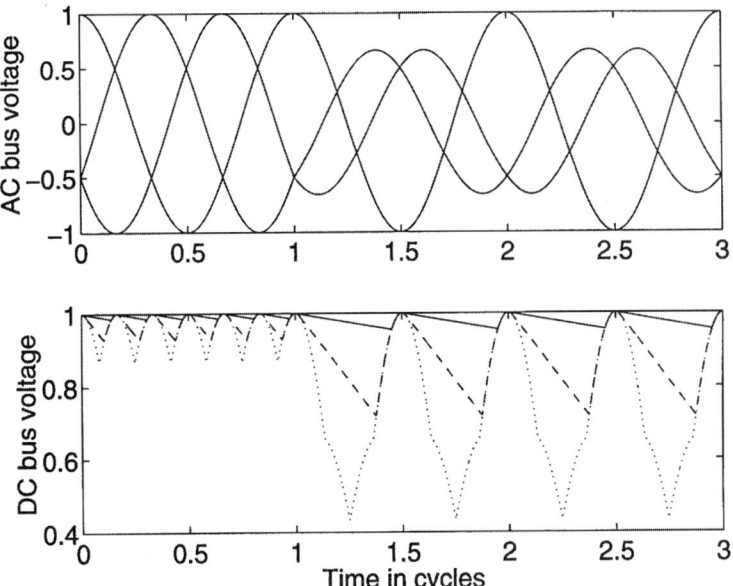

FIGURE 15.38 AC and DC side voltages for a three-phase rectifier during a sag of type C.

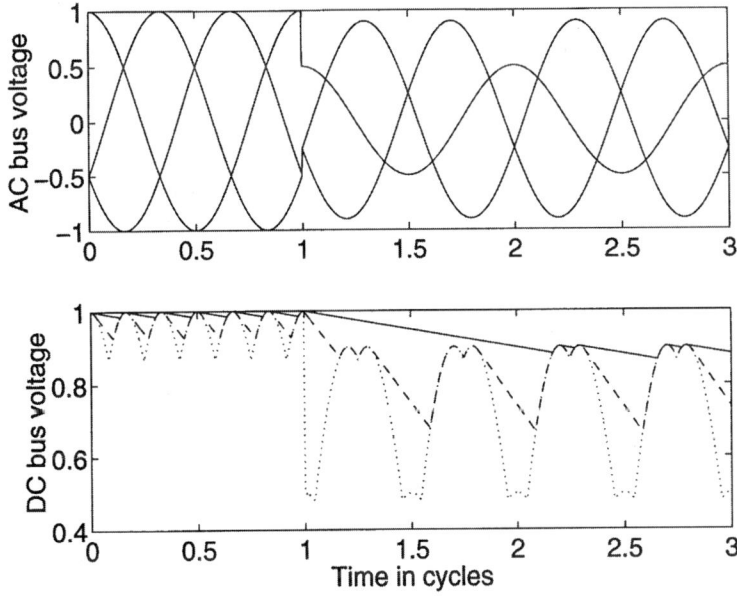

FIGURE 15.39 AC and DC side voltages for a three-phase rectifier during a sag of type D.

there is no voltage sag problem. The underlying event of the equipment trip is a short-circuit fault. At the fault position, the voltage drops to zero, or to a very low value. This zero voltage is changed into an event of a certain magnitude and duration at the interface between the equipment and the power system. The short-circuit fault will always cause a voltage sag for some customers. If the fault takes place in a radial part of the system, the protection intervention clearing the fault will also lead to an interruption. If there is sufficient redundancy present, the short circuit will only lead to a voltage sag. If the resulting event exceeds a certain severity, it will cause an equipment trip.

Based on this reasoning, it is possible to distinguish between the following mitigation methods:

- Reducing the number of short-circuit faults.
- Reducing the fault-clearing time.
- Changing the system such that short-circuit faults result in less severe events at the equipment terminals or at the customer interface.
- Connecting mitigation equipment between the sensitive equipment and the supply.
- Improving the immunity of the equipment.

Reducing the Number of Faults

Reducing the number of short-circuit faults in a system not only reduces the sag frequency, but also the frequency of long interruptions. This is thus a very effective way of improving the quality of supply and many customers suggest this as the obvious solution when a voltage sag or interruption problem occurs. Unfortunately, most of the time the solution is not that obvious. A short circuit not only leads to a voltage sag or interruption at the customer interface, but may also cause damage to utility equipment and plant. Therefore, most utilities will already have reduced the fault frequency as far as economically feasible. In individual cases, there could still be room for improvement, e.g., when the majority of trips are due to faults on one or two distribution lines. Some examples of fault mitigation are:

- Replace overhead lines by underground cables.
- Use special wires for overhead lines.
- Implement a strict policy of tree trimming.
- Install additional shielding wires.
- Increase maintenance and inspection frequencies.

One has to keep in mind, however, that these measures can be very expensive, especially for transmission systems, and that their costs have to be weighted against the consequences of the equipment trips.

Reducing the Fault-Clearing Time

Reducing the fault-clearing time does not reduce the number of events, but only their severity. It does not do anything to reduce to number of interruptions, but can significantly limit the sag duration. The ultimate reduction of fault-clearing time is achieved by using current-limiting fuses, able to clear a fault within one half-cycle. The recently introduced static circuit breaker has the same characteristics: fault-clearing time within one half-cycle. Additionally, several types of fault-current limiters have been proposed that do not actually clear the fault, but significantly reduce the fault current magnitude within one or two cycles. One important restriction of all these devices is that they can only be used for low- and medium-voltage systems. The maximum operating voltage is a few tens of kilovolts.

But the fault-clearing time is not only the time needed to open the breaker, but also the time needed for the protection to make a decision. To achieve a serious reduction in fault-clearing time, it is necessary to reduce any grading margins, thereby possibly allowing for a certain loss of selectivity.

Changing the Power System

By implementing changes in the supply system, the severity of the event can be reduced. Here again, the costs may become very high, especially for transmission and subtransmission voltage levels. In industrial systems, such improvements more often outweigh the costs, especially when already included in the design stage. Some examples of mitigation methods especially directed toward voltage sags are:

- Install a generator near the sensitive load. The generators will keep up the voltage during a remote sag. The reduction in voltage drop is equal to the percentage contribution of the generator station to the fault current. In case a combined-heat-and-power station is planned, it is worth it to consider the position of its electrical connection to the supply.

- Split buses or substations in the supply path to limit the number of feeders in the exposed area.
- Install current-limiting coils at strategic places in the system to increase the "electrical distance" to the fault. The drawback of this method is that this may make the event worse for other customers.
- Feed the bus with the sensitive equipment from two or more substations. A voltage sag in one substation will be mitigated by the infeed from the other substations. The more independent the substations are, the more the mitigation effect. The best mitigation effect is by feeding from two different transmission substations. Introducing the second infeed increases the number of sags, but reduces their severity.

Installing Mitigation Equipment

The most commonly applied method of mitigation is the installation of additional equipment at the system-equipment interface. Also recent developments point toward a continued interest in this way of mitigation. The popularity of mitigation equipment is explained by it being the only place where the customer has control over the situation. Both changes in the supply as well as improvement of the equipment are often completely outside of the control of the end user. Some examples of mitigation equipment are:

- *Uninterruptable power supply (UPS)*. This is the most commonly used device to protect low-power equipment (computers, etc.) against voltage sags and interruptions. During the sag or interruption, the power supply is taken over by an internal battery. The battery can supply the load for, typically, between 15 and 30 minutes.
- *Static transfer switch*. A static transfer switch switches the load from the supply with the sag to another supply within a few milliseconds. This limits the duration of a sag to less than one half-cycle, assuming that a suitable alternate supply is available.
- *Dynamic voltage restorer (DVR)*. This device uses modern power electronic components to insert a series voltage source between the supply and the load. The voltage source compensates for the voltage drop due to the sag. Some devices use internal energy storage to make up for the drop in active power supplied by the system. They can only mitigate sags up to a maximum duration. Other devices take the same amount of active power from the supply by increasing the current. These can only mitigate sags down to a minimum magnitude. The same holds for devices boosting the voltage through a transformer with static tap changer.
- *Motor-generator sets*. Motor-generator sets are the classical solution for sag and interruption mitigation with large equipment. They are obviously not suitable for an office environment but the noise and the maintenance requirements are often no problem in an industrial environment. Some manufacturers combine the motor-generator set with a backup generator; others combine it with power-electronic converters to obtain a longer ride-through time.

Improving Equipment Voltage Tolerance

Improvement of equipment voltage tolerance is probably the most effective solution against equipment trips due to voltage sags. But as a short-time solution, it is often not suitable. In many cases, a customer only finds out about equipment performance after it has been installed. Even most adjustable-speed drives have become off-the-shelf equipment where the customer has no influence on the specifications. Only large industrial equipment is custom-made for a certain application, which enables the incorporation of voltage-tolerance requirements in the specification.

Apart from improving large equipment (drives, process-control computers), a thorough inspection of the immunity of all contactors, relays, sensors, etc. can significantly improve the voltage tolerance of the process.

Different Events and Mitigation Methods

Figure 15.32 showed the magnitude and duration of voltage sags and interruptions resulting from various system events. For different events, different mitigation strategies apply.

Sags due to short-circuit faults in the transmission and subtransmission system are characterized by a short duration, typically up to 100 ms. These sags are very hard to mitigate at the source and improvements in the system are seldom feasible. The only way of mitigating these events is by improvement of the equipment or, where this turns out to be unfeasible, installing mitigation equipment. For low-power equipment, a UPS is a straightforward solution; for high-power equipment and for complete installations, several competing tools are emerging.

The duration of sags due to distribution system faults depends on the type of protection used — ranging from less than a cycle for current-limiting fuses up to several seconds for overcurrent relays in underground or industrial distribution systems. The long sag duration also enables equipment to trip due to faults on distribution feeders fed from other HV/MV substations. For deep long-duration sags, equipment improvement becomes more difficult and system improvement easier. The latter could well become the preferred solution, although a critical assessment of the various options is certainly needed.

Sags due to faults in remote distribution systems and sags due to motor starting should not lead to equipment tripping for sags down to 85%. If there are problems, the equipment needs to be improved. If equipment trips occur for long-duration sags in the 70–80% magnitude range, changes in the system have to be considered as an option.

For interruptions, especially the longer ones, equipment improvement is no longer feasible. System improvements or a UPS in combination with an emergency generator are possible solutions here.

References

Bollen, M.H.J., Characterization of voltage sags experienced by three-phase adjustable-speed drives, *IEEE Trans. on Power Delivery,* 12, 4, 1666–1671, Oct. 1997.

Bollen, M.H.J., The influence of motor reacceleration on voltage sags, *IEEE Trans. on Ind. Appl.,* 31, 4, 667–674, July 1995.

Bollen, M.H.J., Method of critical distances for stochastic assessment of voltage sags, *IEE Proceedings — Generation, Transmission and Distribution,* 145, 1, 70–76, Jan. 1998.

Bollen, M.H.J., *Solving Power Quality Problems, Voltage Sags and Interruptions,* IEEE Press, 1999.

Bollen, M.H.J., Wang, P., Jenkins, N., Analysis and consequences of the phase jump associated with a voltage sag, in *Power Systems Computation Conference,* Dresden, Germany, Aug. 1996.

Bollen, M.H.J., Zhang, L.D., Analysis of voltage tolerance of adjustable-speed drives for three-phase balanced and unbalanced sags, in *IEEE Industrial and Commercial Power Systems Technical Conference,* Sparks, Nevada, May 1999. Scheduled to appear in *IEEE Transactions on Industry Applications,* May/June 2000.

Conrad, L., Little, K., Grigg, C., Predicting and preventing problems associated with remote fault-clearing voltage dips, *IEEE Trans. on Ind. Appl.,* 27, 1, 167–172, Jan. 1991.

Information Technology Industry Council, Interteq, http://www.itic.com (1999).

Key, T.S., Diagnosing power-quality related computer problems, *IEEE Trans. on Ind. Appl.,* 15, 4, 381–393, July 1979.

McGranaghan, M.F., Mueller, D.R., Samotej, M.J., Voltage sags in industrial power systems, *IEEE Trans. on Ind. Appl.,* 29, 2, 397–403, March 1993.

Morgan, L., Power Quality Event Characterization (1159.2), Duke Power, (updated Dec. 29, 1999), http://grouper.ieee.org/groups/1159/21index.html.

PQTN Brief 7: Undervoltage Ride-through Performance of Off-the-Shelf Personal Computers, EPRI Power Electronics Application Centre, Knoxville, TN, 1994.

Sekine, Y., Yamamoto, T., Mori, S., Saito, N., Kurokawa, H., Present state of momentary voltage dip interferences and the countermeasures in Japan, *Int. Conf. on Large Electric Networks (CIGRE),* 34th Session, Paris, France, Sept. 1992.

Zhang, L.D., Bollen, M.H.J., A method for characterizing unbalanced voltage dips (sags) with symmetrical components, *IEEE Power Engineering Review,* 18, 7, 50–52, July 1998.

Further Information

IEEE Recommended Practice for Monitoring Electric Power Quality, IEEE Std. 1159-1995, IEEE, New York, 1995.

European standard EN-50160, *Voltage Characteristics of Electricity Supplied by Public Distribution Systems*, CENELEC, Brussels, Belgium, 1994.

IEEE Recommended Practice for Powering and Grounding Sensitive Electronic Equipment, IEEE Std. 1100-1992.

IEEE Recommended Practice for the Design of Reliable Industrial and Commercial Power Systems (The Gold Book), IEEE Std. 493-1997.

IEEE Recommended Practice for Evaluating Electric Power System Compatibility with Electronic Process Equipment, IEEE Std. 1346-1998.

IEC 61000-4-11, *Electromagnetic Compatibility (EMC) Voltage Dips, Short Interruptions and Voltage Variations Immunity Tests*.

Bollen, M.H.J., Tayjasajant, T., Yalcinkaya, G., Assessment of the number of voltage sags experienced by a large industrial customer, *IEEE Trans. on Ind. Appl.*, 33, 6, 1465–1471, Nov. 1997.

Bollen, M.H.J., Fast assessment methods for voltage sags in distribution systems, *IEEE Trans. on Ind. Appl.*, 31, 6, 1414–1423, Nov. 1996.

Collins, E.R., Morgan, R.L., A three-phase sag generator for testing industrial equipment, *IEEE Trans. on Power Delivery*, 11, 1, 526–532, Jan. 1996.

Conrad, L.E., Bollen, M.H.J., Voltage sag coordination for reliable plant operation, *IEEE Trans. on Ind. Appl.*, 33, 6, 1459–1464, Nov. 1997.

Diliberti, T.W., Wagner, V.E., Staniak, J.P., Sheppard, S.L., Orfloff, T.L., Power quality requirements of a large industrial user: a case study, *IEEE Industrial and Commercial Power Systems Technical Conference*, Detroit, MI, May 1990, 1–4.

Dorr, D.S., Point of utilization power quality study results, *IEEE Trans. on Ind. Appl.*, 31, 4, 658–666, July 1995.

Dorr, D.S., Hughes, M.B., Gruzs, T.M., Jurewicz, R.E., McClaine, J.L., Interpreting recent power quality surveys to define the electrical environment, *IEEE Trans. on Ind. Appl.*, 33, 6, 1480–1487, Nov. 1997.

Dorr, D.S., Mansoor, A., Morinec, A.G., Worley, J.C., Effects of power line voltage variations on different types of 400-W high-pressure sodium ballasts, *IEEE Trans. on Ind. Appl.*, 33, 2, 472–476, March 1997.

Dugan, R.C., McGranaghan, M.F., Beaty, H.W., *Electric Power Systems Quality*, McGraw-Hill, New York, 1996.

Gunther, E.W., Mehta, H., A survey of distribution system power quality — Preliminary results, *IEEE Trans. on Power Delivery*, 10, 1, 322–329, Jan. 1995.

Holtz, J., Lotzhat, W., Stadfeld, S., Controlled AC drives with ride-through capacity at power interruption, *IEEE Trans. on Ind. Appl.*, 30, 5, 1275–1283, Sept. 1994.

Kojovic, L.J., Hassler, S., Application of current limiting fuses in distribution systems for improved power quality and protection, *IEEE Trans. on Power Delivery*, 12, 2, 791–800, April 1997.

Koval, D.O., Leonard, J.J., Rural power profiles, *IEEE Trans. on Ind. Appl.*, 30, 2, 469–75, March–April 1994.

Koval, D.O., R.A. Bocancea, M.B. Hughes, Canadian national power quality survey: Frequency of industrial and commercial voltage sags, *IEEE Trans. on Ind. Appl.*, 35, 5, 904–910, Sept. 1998.

Lamoree, J., Mueller, D., Vinett, P., Jones, W., Samotyj, M., Voltage sag analysis case studies, *IEEE Trans. on Ind. Appl.*, 30, 4, 1083–1089, July 1994.

Mansoor, A., Collins, E.R., Morgan, R.L., Effects of unsymmetrical voltage sags on adjustable-speed drives, *7th IEEE Int. Conf on Harmonics and Quality of Power (ICHPQ)*, Las Vegas, NV, Oct. 1996, 467–472.

Middlekauff, S.W., Collins, E.R., System and customer impact: considerations for series custom power devices, *IEEE Trans. on Power Delivery*, 13, 1, 278–282, Jan. 1998.

Pilay, P., Ed., Motor drive/power systems interactions, *IEEE Industry Applications Society Tutorial Course*, Oct. 1997.

Pumar, C., Amantegui, J., Torrealday, J.R., Ugarte, C., A comparison between DC and AC drives as regards their behavior in the presence of voltage dips: new techniques for reducing the susceptibility of AC drives, *Int. Conf. on Electricity Distribution (CIRED)*, June 2–5 1997, Birmingham, U.K., 9/1-5.

Rioual, P., Pouliquen, H., Louis, J.-P., Regulation of a PWM rectifier in the unbalanced network state using a generalized model, *IEEE Trans. on Power Electronics*, 11, 3, 495–502, May 1996.

Schwartzenberg, J.W., DeDoncker, R.W., 15 kV medium voltage transfer switch, *IEEE Industry Applications Society Annual Meeting*, Oct. 1995, Orlando, FL, 2515–2520.

Smith, R.K., Slade, P.G., Sarkozi, M., Stacey, E.J., Bonk, J.J., Mehta, H., Solid state distribution current limiter and circuit breaker: application requirements and control strategies, *IEEE Trans. on Power Delivery*, 8, 3, 1155–1164, July 1993.

Strangas, E.G., Wagner, V.E., Unruh, T.D., Variable speed drives evaluation test, *IEEE Industry Applications Society Annual Meeting*, Oct. 1996, San Diego, CA, 2239–2243.

Wang, P., Jenkins, N., Bollen, M.H.J., Experimental investigation of voltage sag mitigation by an advanced static VAr compensator, *IEEE Trans. on Power Delivery*, 13, 4, 1461–1467, Oct. 1998.

Woodley, N., Sarkozi, M., Lopez, F., Tahiliani, V., Malkin, P., Solid-state 13-kV distribution class circuit breaker: planning, development and demonstration, *IEE Conf on Trends in Distribution Switchgear*, Nov. 1994, London, U.K., 163–167.

Woodley, N.H., Morgan, L., Sundaram, A., Experience with an inverter-based dynamic voltage restorer, *IEEE Trans. on Power Delivery*, 14, 3, 1181–1186, July 1999.

Yalçinkaya, G., Bollen, M.H.J., Crossley, P.A., Characterisation of voltage sags in industrial distribution systems, *IEEE Trans. on Ind. Appl.*, 34, 4, 682–688, July 1998.

Zhang, L.D., Bollen, M.H.J., A method for characterisation of three-phase unbalanced dips (sags) from recorded voltage waveshapes, *International Telecommunications Energy Conference (INTELEC)*, Copenhagen, Denmark, June 1999.

van Zyl, A., Enslin, J.H.R., Spée, R., Converter-based solution to power quality problems on radial distribution lines, *IEEE Trans. on Ind. Appl.*, 32, 6, 1323–1330, Nov. 1996.

15.5 Voltage Fluctuations and Lamp Flicker in Power Systems

S. M. Halpin

Voltage flicker is a problem that has existed in the power industry for many years. Many types of end-use equipment can create voltage flicker, and many types of solution methods are available. Fortunately, the problem is not overly complex, and it can often be analyzed using fairly simple methods. In many cases, however, solutions can be expensive. Perhaps the most difficult aspect of the voltage flicker problem has been the development of a widely accepted definition of just what "flicker" is and how it can be quantified in terms of measurable quantities.

To electric utility engineers, voltage flicker is considered in terms of magnitude and rate of change of voltage fluctuations. To the utility customer, however, flicker is considered in terms of "my lights are flickering." The necessary presence of a human observer to "see" the change in lamp (intensity) output in response to a change in supply voltage is the most complex factor for which to account. Significant research, dating back to the early 20th century, has been devoted to establishing an accurate correlation between voltage changes and observer perceptions. This correlation is essential so that a readily measurable quantity, supply voltage, can be used to predict a human response.

The early work regarding voltage flicker considered voltage flicker to be a single-frequency modulation of the power frequency voltage. Both sinusoidal and square wave modulations were considered as shown mathematically in Eqs. (15.18) and (15.19), with most work concentrating on square wave modulation.

Power Quality

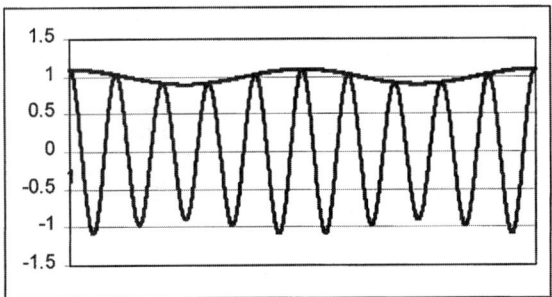

FIGURE 15.40   Sinusoidal voltage flicker.

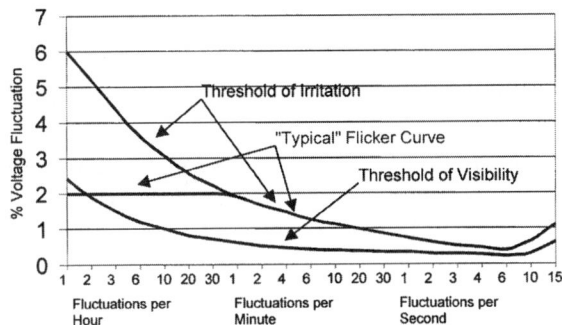

FIGURE 15.41   Typical flicker curves.

$$v(t) = \sqrt{2} V_{rms} \cos(\omega t)\{1.0 + V \cos(\omega_m t)\} \qquad (15.18)$$

$$v(t) = \sqrt{2} V_{rms} \cos(\omega t)\{1.0 + V \text{square}(\omega_m t)\} \qquad (15.19)$$

Based on Eqs. (15.18) and (15.19), the voltage flicker magnitude can be expressed as a percentage of the root-mean-square (rms) voltage, where the term "V" in the two equations represents the percentage. While both the magnitude of the fluctuations ("V") and the "shape" of the modulating waveform are obviously important, the frequency of the modulation is also extremely relevant and is explicitly represented as ω_m. For sinusoidal flicker [given by Eq. (15.18)], the total waveform appears as shown in Fig. 15.40 with the modulating waveform shown explicitly. A similar waveform can be easily created for square-wave modulation.

To correlate the voltage change percentage, V, at a certain frequency, ω_m, with human perceptions, early research led to the widespread use of what is known as a flicker curve to predict possible observer complaints. Flicker curves are still in widespread use, particularly in the U.S. A typical flicker curve is shown in Fig. 15.41 and is based on tests conducted by the General Electric Company. It is important to realize that these curves are developed based on square wave modulation. Voltage changes from one level to another are considered to be "instantaneous" in nature, which may or may not be an accurate representation of actual equipment-produced voltage fluctuations.

The curve of Fig. 15.41 requires some explanation in order to understand its application. The "threshold of visibility" corresponds to certain fluctuation magnitude and frequency pairs that represent the borderline above which an observer can just perceive lamp (intensity) output variations in a 120 V, 60 Hz, 60 W incandescent bulb. The "threshold of irritation" corresponds to certain fluctuation magnitude and frequency pairs that represent the borderline above which the majority of observers would be irritated

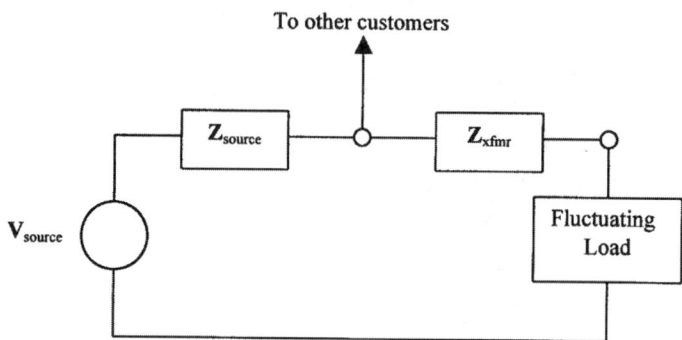

FIGURE 15.42 Example circuit for flicker calculations.

by lamp (intensity) output variations for the same lamp type. Two conclusions are immediately apparent from these two curves: (1) even small percentage changes in supply voltage can be noticed by persons observing lamp output, and (2) the frequency of the voltage fluctuations is an important consideration, with the frequency range from 6–10 Hz being the most sensitive.

Most utility companies do not permit excessive voltage fluctuations on their system, regardless of the frequency. For this reason, a "typical" utility flicker curve will follow either the "threshold of irritation" or the "threshold of visibility" curve as long as the chosen curve lies below some established value (2% in Fig. 15.41). By requiring that voltage fluctuations not exceed the "borderline of visibility" curve, the utility is insuring conservative criteria that should minimize potential problems due to voltage fluctuations.

For many years, the generic flicker curve has served the utility industry well. Fluctuating motor loads like car shredders, wood chippers, and many others can be fairly well characterized in terms of a duty cycle and a maximum torque. From this information, engineers can predict the magnitude and frequency of voltage changes anywhere in the supplying transmission and distribution system. Voltage fluctuations associated with motor starting events are also easily translated into a point (or points) on the flicker curve, and many utilities have based their motor starting criteria on this method for many years. Other loads, most notably arcing loads, cannot be represented as a single flicker magnitude and frequency term. For these types of loads, utility engineers typically presume either worst-case or most-likely variations for analytical evaluations.

Regardless of the type of load, the typical calculation procedure involves either basic load flow or simple voltage division calculations. Figure 15.42 shows an example positive sequence circuit with all data assumed in per-unit on consistent bases.

For fluctuating loads that are best represented by a constant power model (arc furnaces and load torque variations on a running motor), basic load flow techniques can be used to determine the full-load and no-load (or "normal condition") voltages at the "critical" or "point of common coupling" bus where other customers might be served. For fluctuating loads that are best represented by a constant impedance model (motor starting), basic circuit analysis techniques readily provide the full-load and no-load ("normal condition") voltages at the critical bus. Regardless of the modeling and calculation procedures used, equations similar to Eq. (15.20) can be used to determine the percentage voltage change for use in conjunction with a flicker curve. Of course, accurate information regarding the frequency of the assumed fluctuation is absolutely necessary. Note that Eq. (15.20) represents an over-simplification and should therefore not be used in cases where the fluctuations are frequent enough to impact the average rms value (measured over several seconds up to a minute). Other more elaborate formulas are available for these situations.

$$\% \text{ Voltage Change} = \left(1.0 - \frac{V_{\text{full load}}}{V_{\text{normal}}}\right) * 100\% \qquad (15.20)$$

Power Quality

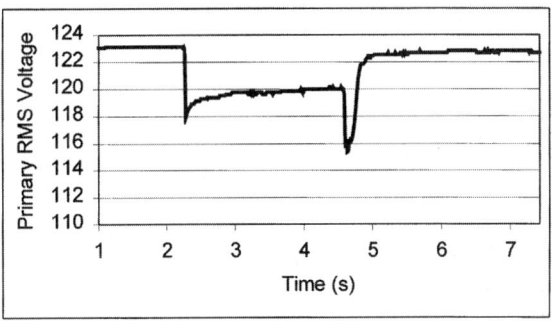

FIGURE 15.43 Poorly timed motor starter voltage fluctuation.

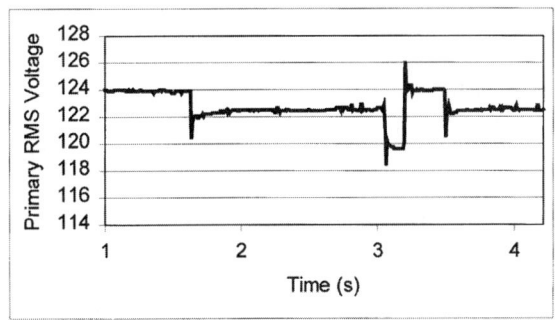

FIGURE 15.44 Adaptive-var compensator effects.

From a utility engineer's viewpoint, the decision to either serve or deny service to a fluctuating load is often based on the result of Eq. (15.20) [or a more complex version of Eq. (15.20)] including information about the frequency at which the calculated change occurs. From this simplified discussion, several questions arise:

1. How are fluctuating loads taken into account when the nature of the fluctuations is not constant in magnitude?
2. How are fluctuating loads taken into account when the nature of the fluctuations is not constant in frequency?
3. How are static compensators and other high response speed mitigation devices included in the calculations?

As examples, consider the rms voltage plots (on 120 V bases) shown in Figs. 15.43 and 15.44. Figure 15.43 shows an rms plot associated with a poorly timed two-step reduced-voltage motor starter. Figure 15.44 shows a motor starting event when the motor is compensated by an adaptive-var compensator. Questions 1–3 are clearly difficult to answer for these plots, so it would be very difficult to apply the basic flicker curve.

In many cases of practical interest, "rules of thumb" are often used to answer approximately these and other related questions so that the simple flicker curve can be used effectively. However, these assumptions and approaches must be conservative in nature and may result in costly equipment modifications prior to connection of certain fluctuating loads. In modern environment, it is imperative that end-users operate at the least total cost. It is equally important that end-use fluctuating loads not create problems for other users. Due to the conservative and approximate nature of the flicker curve methodology, there is often significant room for negotiation, and the matter is often not settled considering only engineering results.

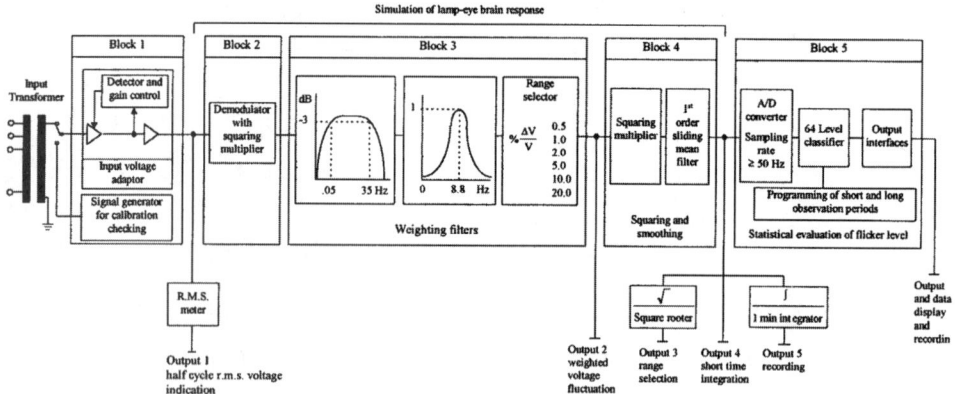

FIGURE 15.45 Flicker meter block diagram.

For roughly three decades, certain engineering groups have recognized the limitations of the flicker curve methods and have developed alternative approaches based on an instrument called a flicker meter. This work, driven strongly in Europe by the International Union for Electroheat (UIE) and the International Electrotechnical Commission (IEC), appears to offer solutions to many of the problems with the flicker curve methodology. Many years of industrial experience have been obtained with the flicker meter approach, and its output has been well-correlated with complaints of utility customers. At this time, the Institute of Electrical and Electronics Engineers (IEEE) is working toward adopting the flicker meter methodology for use in North America.

The flicker meter is a continuous time measuring system that takes voltage as an input and produces three output indices that are related to customer perception. These outputs are: (1) instantaneous flicker sensation, P_{inst}, (2) short-term flicker severity, P_{st}, and (3) long-term flicker severity, P_{lt}. A block diagram of an analog flicker meter is shown in Fig. 15.45.

The flicker meter takes into account both the physical aspects of engineering (how does the lamp [intensity] output vary with voltage?) and the physiological aspects of human observers (how fast can the human eye respond to light changes?). Each of the five basic blocks in Fig. 15.45 contribute to one or both of these aspects. While a detailed discussion of the flicker meter is beyond the scope of this section, the function of the blocks can be summarized as follows.

Blocks 1 and 2 act to process the input voltage signal and to partially isolate only the modulating term in Eqs. (15.18) or (15.19). Block 3 completes the isolation of the modulating signal through complex filtering and applies frequency-sensitive weighting to the "pure" modulating signal. Block 4 models the physiological response of the human observer, specifically the short-term memory tendency of the brain to correlate the voltage modulating signal with a human perception ability. Block 5 performs statistical analysis on the output of Block 4 to capture the cumulative effects of fluctuations over time.

The instantaneous flicker sensation is the output of Block 4. The short- and long-term severity indices are the outputs of Block 5. P_{inst} is available as an output quantity on a continuous basis, and a value of 1.0 corresponds with the threshold of visibility curve in Fig. 15.41. A single P_{st} value is available as an output every ten minutes, and a value of 1.0 corresponds to the threshold of irritation curve in Fig. 15.41. Of course, a comparison can only be made for certain inputs.

For square wave modulation, Fig. 15.46 shows a comparison of the "irritation level" given by IEEE Std. 141 (Red Book) and that level predicted by the flicker meter to be "irritating" (P_{st} = 1.0). For these comparisons, the lamp type used is a 120 V, 60 Hz, 60 W incandescent bulb. Note that the flicker curve taken from IEEE Std. 141 is essentially identical to the "borderline of irritation" curve given in Fig. 15.41.

As Fig. 15.46 clearly demonstrates, the square wave modulation voltage fluctuations that lead to irritation are nearly identical as predicted by either a standard flicker curve or a flicker meter.

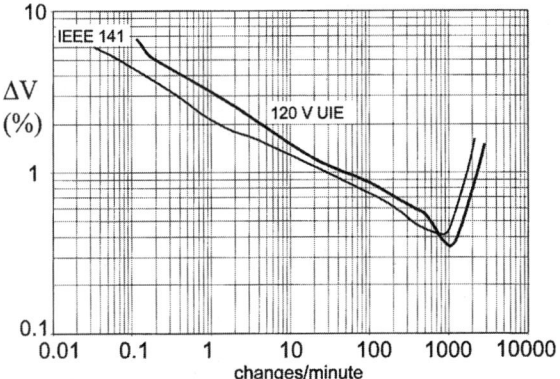

FIGURE 15.46 Threshold of irritation flicker curve and $P_{st} = 1.0$ curve from a flicker meter.

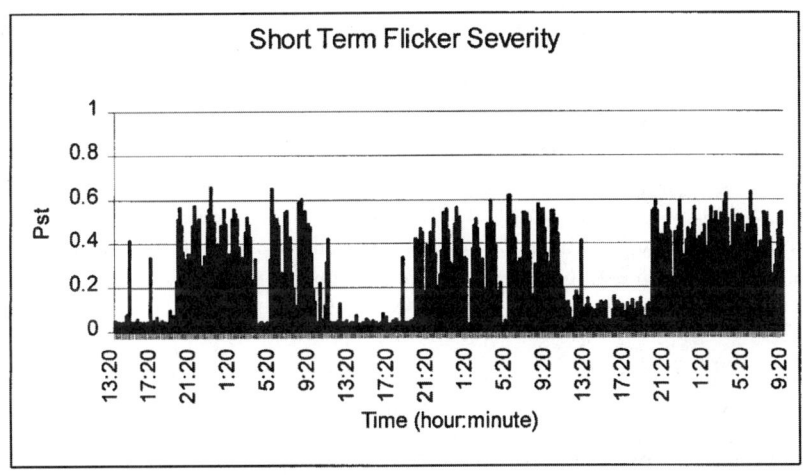

FIGURE 15.47 Short term flicker severity example plot.

The real advantage of the flicker meter methodology lies in that fact that the continuous time measurement system can easily predict possible irritation for arbitrarily complex modulation waveforms. As an example, Fig. 15.47 shows a plot of P_{st} over a three-day period at a location serving a small electric arc furnace. (Note: In this case, there were no reported customer complaints and P_{st} was well below the irritation threshold value of 1.0 during the entire monitoring period.)

Due to the very random nature of the fluctuations associated with an arc furnace, the flicker curve methodology cannot be used directly as an accurate predictor of irritation levels because it is appropriate only for the "sudden" voltage fluctuations associated with square wave modulation. The trade-off required for more accurate flicker prediction, however, is that the inherent simplicity of the basic flicker curve is lost.

For the basic flicker curve, simple calculations based on circuit and equipment models in Fig. 15.42 can be used. Data for these models is readily available, and time-tested assumptions are widely known for cases when exact data are not available. Because the flicker meter is a continuous-time system, continuous-time voltage input data is required for its use. For existing fluctuating loads, it is reasonable to presume that a flicker meter can be connected and used to predict whether or not the fluctuations are irritating. However, it is necessary to be able to predict potential flicker problems prior to the connection of a fluctuating load well before it is possible to measure anything.

There are three possible solutions to the apparent "prediction" dilemma associated with the flicker meter approach. The most basic approach is to locate an existing fluctuating load that is similar to the one under consideration and simply measure the flicker produced by the existing load. Of course, the engineer is responsible for making sure that the existing installation is nearly identical to the one proposed. While the fluctuating load equipment itself might be identical, supply system characteristics will almost never be the same.

Because the short-term flicker severity output of the flicker meter, P_{st}, is linearly dependent on voltage fluctuation magnitude over a wide range, it is possible to linearly scale the P_{st} measurements from one location to predict those at another location where the supply impedance is different. (In most cases, voltage fluctuations are directly related to the supply impedance; a system with 10% higher supply impedance would expect 10% greater voltage fluctuation for the same load change.) In evaluations where it is not possible to measure another existing fluctuating load, other approaches must be used.

If detailed system and load data are known, a time-domain simulation can be used to generate a continuous-time series of voltage data points. These points could then be used as inputs to a simulated flicker meter to predict the short-term flicker severity, P_{st}. This approach, however, is usually too intensive and time-consuming to be appropriate for most applications. For these situations, "shape factors" have been proposed that predict a P_{st} value for various types of fluctuations.

Shape factors are simple curves that can be used to predict, without simulation or measurement, the P_{st} that would be measured if the load were connected. Different curves exist for different "shapes" of voltage variation. Curves exist for simple square and triangular variations, as well as for more complex variations such as motor starting. To use a shape factor, an engineer must have some knowledge of (1) the magnitude of the fluctuation, (2) the shape of the fluctuation, including the time spent at each voltage level if the shape is complex, (3) rise time and fall times between voltage levels, and (4) the rate at which the shape repeats. In some cases, this level of data is not available, and assumptions are often made (on the conservative side). It is interesting to note that the extreme of the conservative choices is a rectangular fluctuation at a known frequency; which is exactly the data required to use the basic flicker curve of Fig. 15.41.

Using either the flicker curve for simple evaluations or the flicker meter methodology for more complex evaluations, it is possible to predict if a given fluctuating load will produce complaints from other customers. In the event that complaints are predicted, modifications must be made prior to granting service. The possible modifications can be made either on the utility side or on the customer (load) side (or both), or some type of compensation equipment can be installed.

In most cases, the most effective, but not least cost, ways to reduce or eliminate flicker complaints are to either (1) reduce the supply system impedance of the whole path from source to fluctuating load, or (2) serve the fluctuating load from a dedicated and electrically remote (from other customers) circuit. In most cases, utility revenue projections for customers with fluctuating loads do not justify such expenses, and the burden of mitigation is shifted to the consumer.

Customers with fluctuating load equipment have two main options regarding voltage flicker mitigation. In some cases, the load can be adjusted to the point that the frequency(ies) of the fluctuations are such that complaints are eliminated (recall the frequency-sensitive nature of the entire flicker problem). In other cases, direct voltage compensation can be achieved through high-speed static compensators. Either thyristor-switched capacitor banks (often called adaptive var compensators or AVCs) or fixed capacitors in parallel with thyristor-switched reactors (often called static var compensators or SVCs) can be used to provide voltage support through reactive compensation in about one cycle. For loads where the main contributor to a large voltage fluctuation is a large reactive power change, reactive compensators can significantly reduce or eliminate the potential for flicker complaints. In cases where voltage fluctuations are due to large real power changes, reactive compensation offers only small improvements and can, in some cases, make the problem worse.

In conclusion, it is almost always necessary to measure/predict flicker levels under a variety of possible conditions, both with and without mitigation equipment and procedures in effect. In very simple cases, a basic flicker curve will provide acceptable results. In more complex cases, however, an intensive measurement, modeling, and simulation effort may be required in order to minimize potential flicker complaints.

While this section has addressed the basic issues associated with voltage flicker complaints, prediction, and measurement, it is not intended to be all-inclusive. A number of relevant publications, papers, reports, and standards are given for further reading, and the reader should certainly consider these documents carefully in addition to what is provided here.

Further Information

Xenis, C.P. and Perine, W., Slide rule yields lamp flicker data, *Electrical World*, October 1937.

Seebald, R.C., Buch, J.F., and Ward, D.J., Flicker Limitations of Electric Utilities, *IEEE Trans. on Power Appar. Syst.*, PAS-104, 9, September 1985.

Sakulin, M. and Key, T.S., UIE/IEC Flicker Standard for Use in North America: Measuring Techniques and Practical Applications, in *Proceedings of PQA'97*, March 1997.

IEEE Standard 141-1993: *Recommended Practice for Power Distribution in Industrial Plants*, IEEE, 1993.

Bergeron, R., Power Quality Measurement Protocol: CEA Guide to Performing Power Quality Surveys, CEA Report 220 D 771, May 1996.

IEC Publication 868, *Flickermeter-Functional and Design Specifications*, 1986.

IEC 61000-4-15, *Flickermeter-Functional and Design Specifications*, 1997-11.

UIE WG on Disturbances, Connection of Fluctuating Loads, 1998.

UIE WG on Disturbances, Flicker Measurement and Evaluations: 2nd Revised Edition, 1992.

UIE WG on Disturbances, Guide to Quality of Electrical Supply for Industrial Installations, Part 5: Flicker and Voltage Fluctuations, 1999.

IEC 1000-3-3, *Electromagnetic Compatibility (EMC) Part 3: Limits — Part 3: Limitation of Voltage Fluctuations and Flicker in Low-Voltage Supply Systems for Equipment with Rated Current ≤ 16 A*, 1994.

IEC 1000-3-5, *Electromagnetic Compatibility (EMC) Part 3: Limits — Part 5: Limitation of Voltage Fluctuations and Flicker in Low-Voltage Supply Systems for Equipment with Rated Current > 16 A*, 1994.

IEC 1000-3-7, *Electromagnetic Compatibility (EMC) Part 3: Limits — Part 7: Assessment of Emission Limits for Fluctuating Loads in MV and HV Power Systems*, 1996.

IEC 1000-3-11, *Electromagnetic Compatibility (EMC) Part 3: Limits — Part 11: Limitation of Voltage Changes, Voltage Fluctuations, and Flicker in Public Low Voltage Supply Systems with Rated Current ≤ 75 A and Subject to Conditional Connection*, 1996.

15.6 Power Quality Monitoring

Patrick Coleman

Many power quality problems are caused by inadequate wiring or improper grounding. These problems can be detected by simple examination of the wiring and grounding systems. Another large population of power quality problems can be solved by spotchecks of voltage, current, or harmonics using hand held meters. Some problems, however, are intermittent and require longer-term monitoring for solution.

Long-term power quality monitoring is largely a problem of data management. If an RMS value of voltage and current is recorded each electrical cycle, for a three-phase system, about 6 gigabytes of data will be produced each day. Some equipment is disrupted by changes in the voltage waveshape that may not affect the rms value of the waveform. Recording the voltage and current waveforms will result in about 132 gigabytes of data per day. While modern data storage technologies may make it feasible to record every electrical cycle, the task of detecting power quality problems within this mass of data is daunting indeed.

Most commercially available power quality monitoring equipment attempts to reduce the recorded data to manageable levels. Each manufacturer has a generally proprietary data reduction algorithm. It is critical that the user understand the algorithm used in order to properly interpret the results.

Selecting a Monitoring Point

Power quality monitoring is usually done to either solve an existing power quality problem, or to determine the electrical environment prior to installing new sensitive equipment. For new equipment, it is easy to argue that the monitoring equipment should be installed at the point nearest the point of connection of the new equipment. For power quality problems affecting existing equipment, there is frequently pressure to determine if the problem is being caused by some external source, i.e., the utility. This leads to the installation of monitoring equipment at the service point to try to detect the source of the problem. This is usually not the optimum location for monitoring equipment. Most studies suggest that 80% of power quality problems originate within the facility. A monitor installed on the equipment being affected will detect problems originating within the facility, as well as problems originating on the utility. Each type of event has distinguishing characteristics to assist the engineer in correctly identifying the source of the disturbance.

What to Monitor

At minimum, the input voltage to the affected equipment should be monitored. If the equipment is single phase, the monitored voltage should include at least the line-to-neutral voltage and the neutral-to-ground voltages. If possible, the line-to-ground voltage should also be monitored. For three-phase equipment, the voltages may either be monitored line to neutral, or line to line. Line-to-neutral voltages are easier to understand, but most three-phase equipment operates on line-to-line voltages. Usually, it is preferable to monitor the voltage line to line for three-phase equipment.

If the monitoring equipment has voltage thresholds which can be adjusted, the thresholds should be set to match the sensitive equipment voltage requirements. If the requirements are not known, a good starting point is usually the nominal equipment voltage plus or minus 10%.

In most sensitive equipment, the connection to the source is a rectifier, and the critical voltages are DC. In some cases, it may be necessary to monitor the critical DC voltages. Some commercial power quality monitors are capable of monitoring AC and DC simultaneously, while others are AC only.

It is frequently useful to monitor current as well as voltage. For example, if the problem is being caused by voltage sags, the reaction of the current during the sag can help determine the source of the sag. If the current doubles when the voltage sags 10%, then the cause of the sag is on the load side of the current monitor point. If the current increases or decreases 10–20% during a 10% voltage sag, then the cause of the sag is on the source side of the current monitoring point.

Sensitive equipment can also be affected by other environmental factors such as temperature, humidity, static, harmonics, magnetic fields, radio frequency interference (RFI), and operator error or sabotage. Some commercial monitors can record some of these factors, but it may be necessary to install more than one monitor to cover every possible source of disturbance.

It can also be useful to record power quantity data while searching for power quality problems. For example, the author found a shortcut to the source of a disturbance affecting a wide area by using the power quantity data. The recordings revealed an increase in demand of 2500 KW immediately after the disturbance. Asking a few questions quickly led to a nearby plant with a 2500 KW switched load that was found to be malfunctioning.

Selecting a Monitor

Commercially available monitors fall into two basic categories: line disturbance analyzers and voltage recorders. The line between the categories is becoming blurred as new models are developed. Voltage recorders are primarily designed to record voltage and current stripchart data, but some models are able to capture waveforms under certain circumstances. Line disturbance analyzers are designed to capture voltage events that may affect sensitive equipment. Generally, line disturbance analyzers are not good voltage recorders, but newer models are better than previous designs at recording voltage stripcharts.

Power Quality

In order to select the best monitor for the job, it is necessary to have an idea of the type of disturbance to be recorded, and an idea of the operating characteristics of the available disturbance analyzers. For example, a common power quality problem is nuisance tripping of variable speed drives. Variable speed drives may trip due to the waveform disturbance created by power factor correction capacitor switching, or due to high or low steady state voltage, or, in some cases, due to excessive voltage imbalance. If the drive trips due to high voltage or waveform disturbances, the drive diagnostics will usually indicate an overvoltage code as the cause of the trip. If the voltage is not balanced, the drive will draw significantly unbalanced currents. The current imbalance may reach a level that causes the drive to trip for input overcurrent. Selecting a monitor for variable speed drive tripping can be a challenge. Most line disturbance analyzers can easily capture the waveshape disturbance of capacitor switching, but they are not good voltage recorders, and may not do a good job of reporting high steady state voltage. Many line disturbance analyzers cannot capture voltage unbalance at all, nor will they respond to current events unless there is a corresponding voltage event. Most voltage and current recorders can easily capture the high steady state voltage that leads to a drive trip, but they may not capture the capacitor switching waveshape disturbance. Many voltage recorders can capture voltage imbalance, current imbalance, and some of them will trigger a capture of voltage and current during a current event, such as the drive tripping off.

To select the best monitor for the job, it is necessary to understand the characteristics of the available monitors. The following sections will discuss the various types of data that may be needed for a power quality investigation, and the characteristics of some commercially available monitors.

Voltage

The most commonly recorded parameter in power quality investigations is the RMS voltage delivered to the equipment. Manufacturers of recording equipment use a variety of techniques to reduce the volume of the data recorded. The most common method of data reduction is to record Min/Max/Average data over some interval. Figure 15.48 shows a strip chart of rms voltages recorded on a cycle-by-cycle basis. Figure 15.49 shows a Min/Max/Average chart for the same time period. A common recording period is 1 week. Typical recorders will use a recording interval of 2–5 minutes. Each recording interval will produce three numbers: the rms voltage of the highest 1 cycle, the lowest 1 cycle, and the average of every cycle during the interval. This is a simple, easily understood recording method, and it is easily implemented

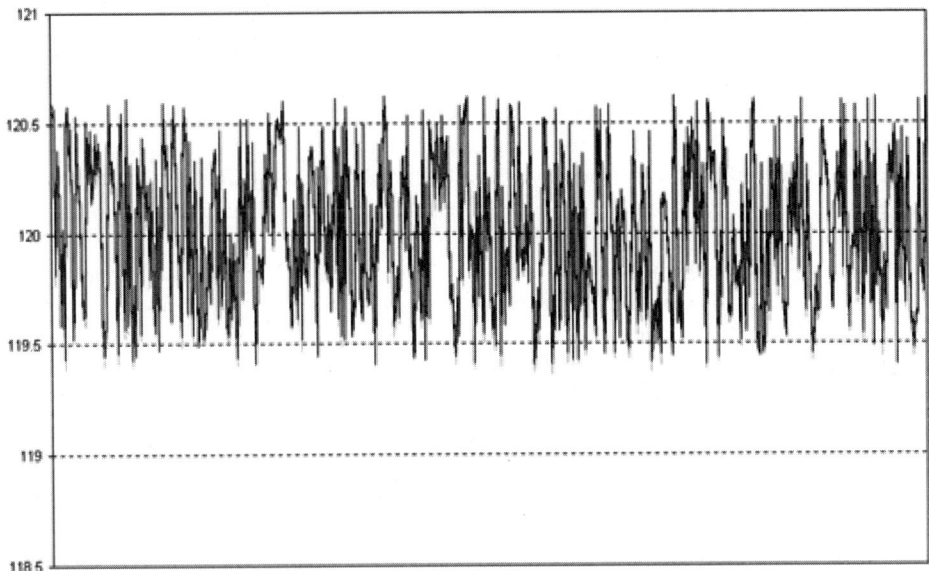

FIGURE 15.48 RMS voltage stripchart, taken cycle by cycle.

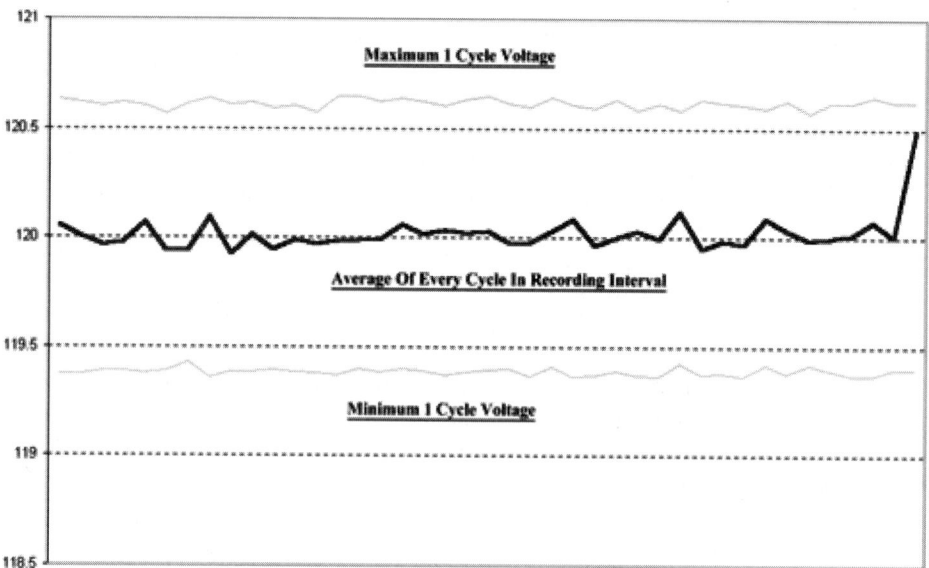

FIGURE 15.49 Min/Max/Average stripchart, showing the minimum single cycle voltage, the maximum single cycle voltage, and the average of every cycle in a recording interval. Compare to the Fig. 15.48 stripchart data.

by the manufacturer. There are several drawbacks to this method. If there are several events during a recording interval, only the event with the largest deviation is recorded. Unless the recorder records the event in some other manner, there is no time-stamp associated with the events, and no duration available. The most critical deficiency is the lack of a voltage profile during the event. The voltage profile provides significant clues to the source of the event. For example, if the event is a voltage sag, the minimum voltage may be the same for an event caused by a distant fault on the utility system, and for a nearby large motor start. For the distant fault, however, the voltage will sag nearly instantaneously, stay at a fairly constant level for 3–10 cycles, and almost instantly recover to full voltage, or possibly a slightly higher voltage if the faulted section of the utility system is separated. For a nearby motor start, the voltage will drop nearly instantaneously, and almost immediately begin a gradual recovery over 30–180 cycles to a voltage somewhat lower than before. Figure 15.50 shows a cycle-by-cycle recording of a simulated adjacent feeder fault, followed by a simulation of a voltage sag caused by a large motor start. Figure 15.51 shows a Min/Max/Average recording of the same two events. The events look quite similar when captured by the Min/Max/Average recorder, while the cycle-by-cycle recorder reveals the difference in the voltage recovery profile.

Some line disturbance analyzers allow the user to set thresholds for voltage events. If the voltage exceeds these thresholds, a short duration stripchart is captured showing the voltage profile during the event. This short duration stripchart is in addition to the long duration recordings, meaning that the engineer must look at several different charts to find the needed information.

Some voltage recorders have user-programmable thresholds, and record deviations at a higher resolution than voltages that fall within the thresholds. These deviations are incorporated into the stripchart, so the user need only open the stripchart to determine, at a glance, if there are any significant events. If there are events to be examined, the engineer can immediately "zoom in" on the portion of the stripchart with the event.

Some voltage recorders do not have user-settable thresholds, but rather choose to capture events based either on fixed default thresholds or on some type of significant change. For some users, fixed thresholds are an advantage, while others are uncomfortable with the lack of control over the meter function. In units with fixed thresholds, if the environment is normally somewhat disturbed, such as on a welder

Power Quality

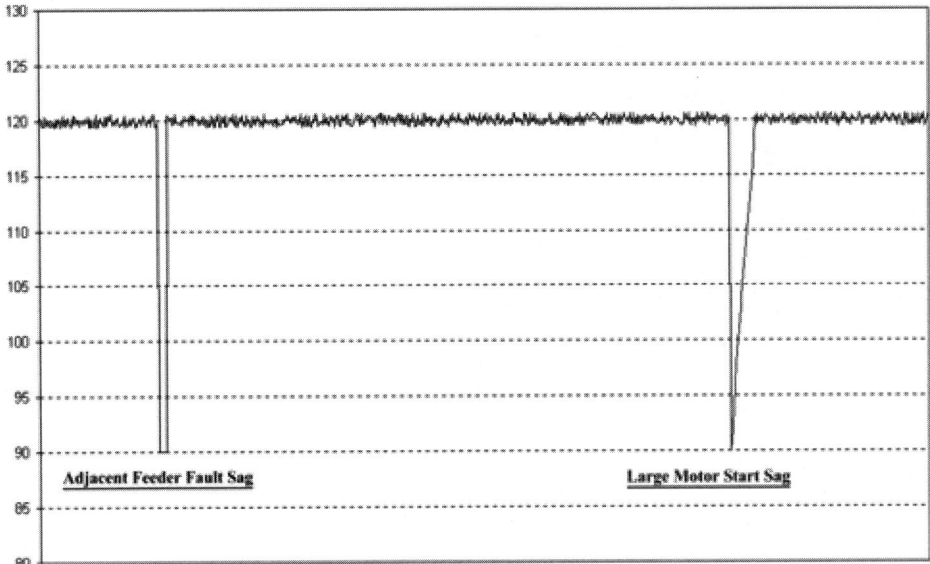

FIGURE 15.50 Cycle-by-cycle rms stripchart showing two voltage sags. The sag on the left is due to an adjacent feeder fault on the supply substation, and the sag on the right is due to a large motor start. Note the difference in the voltage profile during recovery.

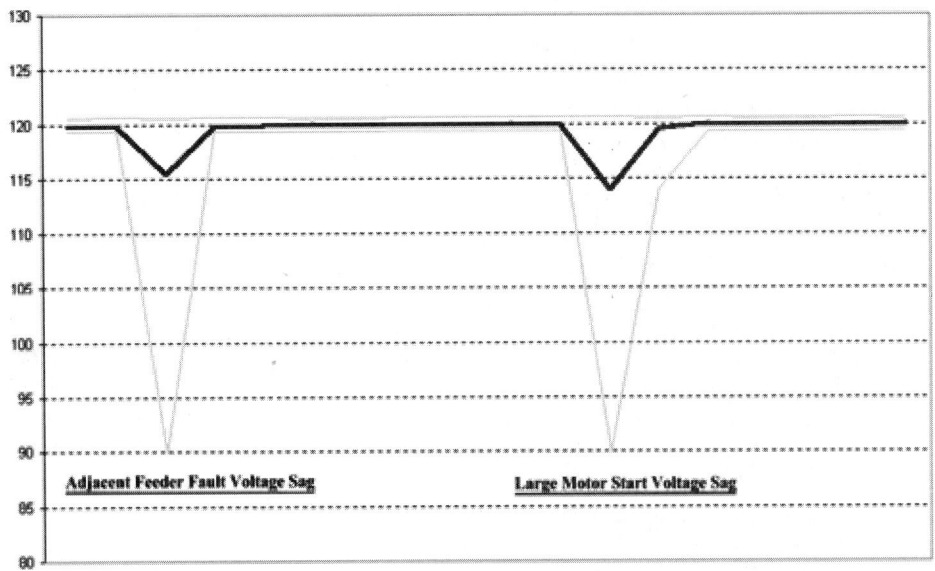

FIGURE 15.51 Min/Max/Average stripchart of the same voltage sags as Fig. 15.50. Note that both sags look almost identical. Without the recovery detail found in Fig. 15.50, it is difficult to determine a cause for the voltage sags.

circuit at a motor control center, the meter memory may fill up with insignificant events and the monitor may not be able to record a significant event when it occurs. For this reason, monitors with fixed thresholds should not be used in electrically noisy environments.

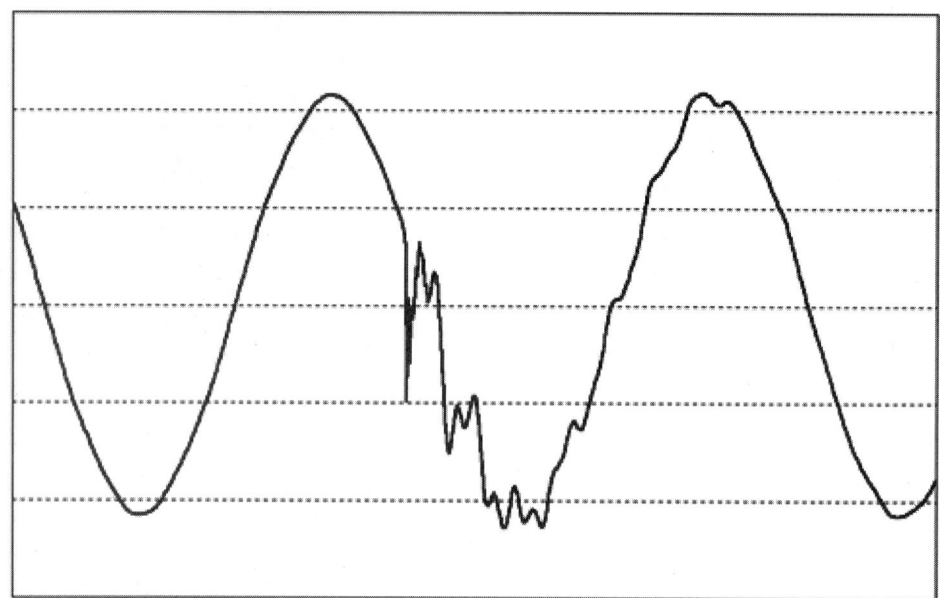

FIGURE 15.52 Typical voltage waveform disturbance caused by power factor correction capacitor energization.

Voltage Waveform Disturbances

Some equipment can be disturbed by changes in the voltage waveform. These waveform changes may not significantly affect the rms voltage, yet may still cause equipment to malfunction. An rms-only recorder may not detect the cause of the malfunction. Most line disturbance analyzers have some mechanism to detect and record changes in voltage waveforms. Some machines compare portions of successive waveforms, and capture the waveform if there is a significant deviation in any portion of the waveform. Others capture waveforms if there is a significant change in the rms value of successive waveforms. Another method is to capture waveforms if there is a significant change in the voltage total harmonic distortion (THD) between successive cycles.

The most common voltage waveform change that may cause equipment malfunction is the disturbance created by power factor correction capacitor switching. When capacitors are energized, a disturbance is created that lasts about 1 cycle, but does not result in a significant change in the rms voltage. Figure 15.52 shows a typical power factor correction capacitor switching event.

Current Recordings

Most modern recorders are capable of simultaneous voltage and current recordings. Current recordings can be useful in identifying the cause of power quality disturbances. For example, if a 20% voltage sag (to 80% of full voltage) is accompanied by a small change in current (plus or minus about 30%), the cause of the voltage sag is usually upstream (toward the utility source) of the monitoring point. If the sag is accompanied by a large increase in current (about 100%), the cause of the sag is downstream (toward the load) of the monitoring point. Figure 15.53 shows the rms voltage and current captured during a motor start downstream of the monitor. Notice the large current increase during starting and the corresponding small decrease in voltage.

Some monitors allow the user to select current thresholds that will cause the monitor to capture both voltage and current when the current exceeds the threshold. This can be useful for detecting over- and under-currents that may not result in a voltage disturbance. For example, if a small, unattended machine is tripping off unexpectedly, it would be useful to have a snapshot of the voltage and current just prior to the trip. A threshold can be set to trigger a snapshot when the current goes to zero. This snapshot can be used to determine if the input voltage or current was the cause of the machine trip.

Power Quality

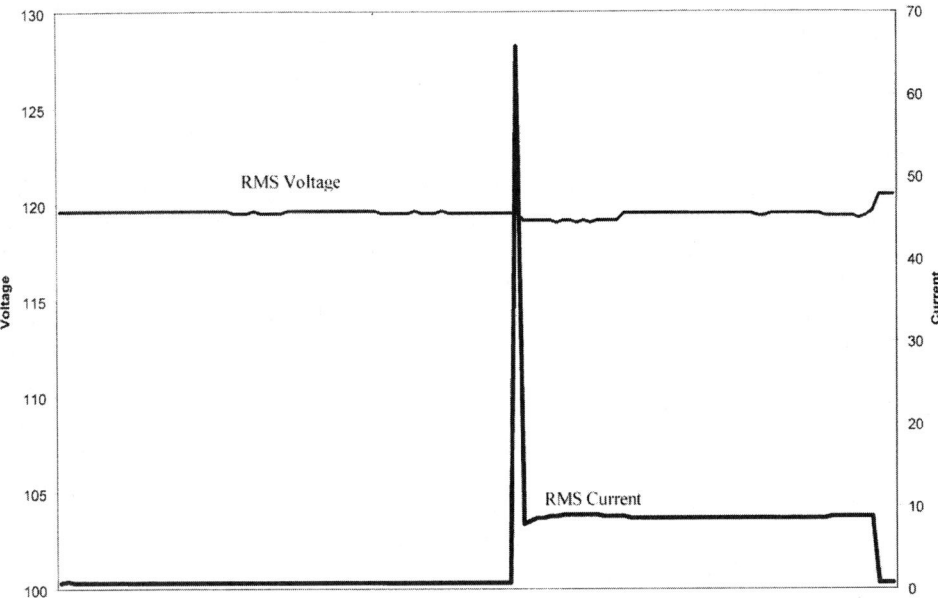

FIGURE 15.53 RMS stripcharts of voltage and current during a large current increase due to a motor start downstream of the monitor point.

Current Waveshape Disturbances

Very few monitors are capable of capturing changes in current waveshape. It is usually not necessary to capture changes in current waveshape, but in some special cases this can be useful data. For example, inrush current waveforms can provide more useful information than inrush current rms data. Figure 15.54 shows a significant change in the current waveform when the current changes from zero to

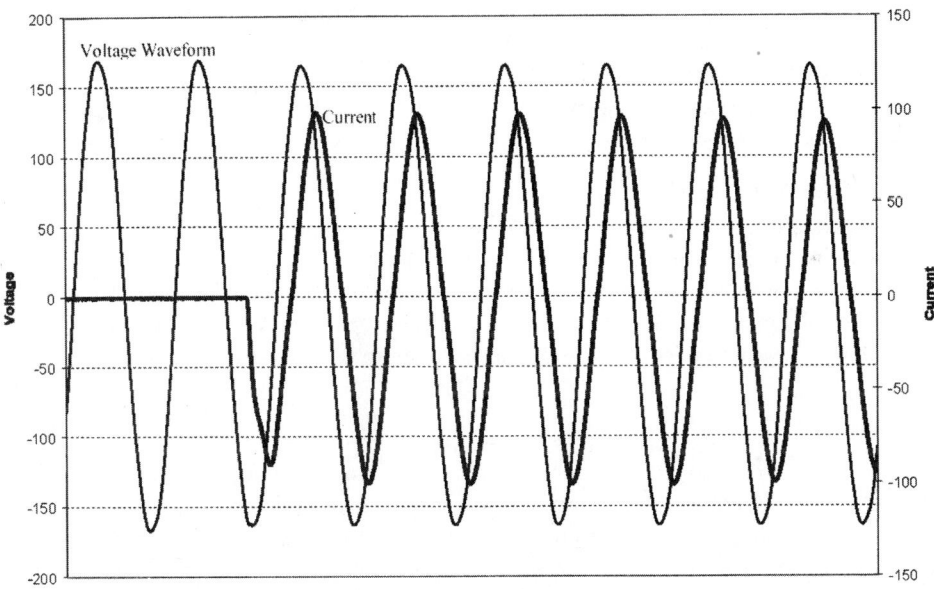

FIGURE 15.54 Voltage and current waveforms for the first few cycles of the current increase illustrated in Fig. 15.53.

nearly 100 amps peak. The shape of the waveform, and the phase shift with respect to the voltage waveform, confirm that this current increase was due to an induction motor start. Figure 15.54 shows the first few cycles of the event shown in Fig. 15.53.

Harmonics

Harmonic distortion is a growing area of concern. Many commercially available monitors are capable of capturing harmonic snapshots. Some monitors have the ability to capture harmonic stripchart data. In this area, it is critical that the monitor produce accurate data. Some commercially available monitors have deficiencies in measuring harmonics. Monitors generally capture a sample of the voltage and current waveforms, and perform a Fast Fourier Transform to produce a harmonic spectrum. According to the Nyquist Sampling Theorem, the input waveform must be sampled at least twice the highest frequency that is present in the waveform. Some manufacturers interpret this to mean the highest frequency of interest, and adjust their sample rates accordingly. If the input signal contains a frequency that is above the maximum frequency that can be correctly sampled, the high frequency signal may be "aliased," that is, it may be incorrectly identified as a lower frequency harmonic. This may lead the engineer to search for a solution to a harmonic problem that does not exist. The aliasing problem can be alleviated by sampling at higher sample rates, and by filtering out frequencies above the highest frequency of interest. The sample rate is usually found in the manufacturer's literature, but the presence of an antialiasing filter is not usually mentioned in the literature.

Flicker

Some users define flicker as the voltage sag that occurs when a large motor starts. Other users regard flicker as the frequent, small changes in voltage that occur due to the operation of arc furnaces, welders, chippers, shredders, and other varying loads. Nearly any monitor is capable of adequately capturing voltage sags due to occasional motor starts. The second definition of flicker is more difficult to monitor. In the absence of standards, several manufacturers have developed proprietary "flicker" meters. In recent years, an effort has been made to standardize the definition of "flicker," and to standardize the performance of flicker meters. At the time of this writing, several monitor manufacturers are attempting to incorporate the standardized flicker function into their existing products.

High Frequency Noise

Sensitive electronic equipment can be susceptible to higher frequency signals imposed on the voltage waveform. These signals may be induced on the conductors by sources such as radio transmitters or arcing devices such as fluorescent lamps, or they may be conductively coupled by sources such as power line carrier energy management systems. A few manufacturers include detection circuitry for high frequency signals imposed on the voltage waveform.

Other Quantities

It may be necessary to find a way to monitor other quantities that may affect sensitive equipment. Examples of other quantities are temperature, humidity, vibration, static electricity, magnetic fields, fluid flow, and air flow. In some cases, it may also become necessary to monitor for vandalism or sabotage. Most power quality monitors cannot record these quantities, but other devices exist that can be used in conjunction with power quality monitors to find a solution to the problem.

Summary

Most power quality problems can be solved with simple hand-tools and attention to detail. Some problems, however, are not so easily identified, and it may be necessary to monitor to correctly identify the problem. Successful monitoring involves several steps. First, determine if it is really necessary to monitor. Second, decide on a location for the monitor. Generally, the monitor should be installed close to the affected equipment. Third, decide what quantities need to be monitored, such as voltage, current, harmonics, and power data. Try to determine the types of events that can disturb the equipment, and select a meter that is capable of detecting those types of events. Fourth, decide on a monitoring period. Usually, a good first choice is at least one business cycle, or at least 1 day, and more commonly, 1 week. It may be necessary to monitor until the problem recurs. Some monitors can record indefinitely by discarding older data to make space for new data. These monitors can be installed and left until the problem recurs. When the problem recurs, the monitoring should be stopped before the event data is discarded.

After the monitoring period ends, the most difficult task begins — interpreting the data. Modern power quality monitors produce reams of data during a disturbance. Data interpretation is largely a matter of experience, and Ohm's law. There are many examples of disturbance data in books such as *The BMI Handbook of Power Signatures, Second Edition*, and the *Dranetz Field Handbook for Power Quality Analysis*.

Index

A

AAA conductors. *See* Conductors
AAC conductors. *See* Conductors
Abadie studies, 12-42
ABB Power, 4-178, 13-55
Abed studies, 11-8
Abetti studies, 10-55, 10-62, 10-66
Above ground cable systems, 4-48 to 62
Absolutely selective zones, 9-30 to 31
Abu-Elnaga studies, 11-44, 11-52
AC. *See* Alternating current (AC)
Acceptor impurities, 14-9
Accessory equipment, power transformers, 3-25 to 29
Access roads, 5-86
Accidental energization, 9-26
Accidental ground circuit, 5-92 to 95
Accredited Standards Committee (ANSI) C57
   American National Standards, 3-250 to 253
   dry-type transformers standards list, 3-262 to 263
   liquid-filled transformers standards list, 3-266 to 269
Accredited Standards Organization (ASO), 3-253
Accuracy
   economic dispatch calculations, 12-14, 12-19 to 20
   genetic algorithm (GA), 12-22
   instrument transformers classes, 3-108 to 110
   optimal power flow (OPF), 12-46
   sagging, 4-119
   subsynchronous resonance studies, 11-130
   synchronization, 9-50

ACE Satellite (Advanced Composition Explorer), 4-161 to 163
ACF. *See* Altitude correction factor (ACF)
Acid, 3-204 to 205, 3-208
AC Intertie. *See* Pacific AC Intertie (PACI)
Acoustics. *See also* Sound levels
   abatement methods, 5-82 to 83
   corona discharges, 4-141 to 145
   geomagnetically induced currents (GIC), 4-153
   load tap changers, 3-287 to 288
   power transformers, 3-277 to 278
   substation automation systems, 5-43 to 44
   substation buildings, 5-38
   substation design, 5-81 to 83
   transformer sound levels, 3-170
ACRC conductors. *See* Aluminum conductors
ACSR conductors. *See* Aluminum conductors
Action integral, lightning, 10-6
Active filters
   applications, 14-58 to 61
   control circuit, 14-56
   energy storage capacity, 14-49 to 50
   general concepts, 14-44 to 61
   harmonic-producing loads, 14-45
   hybrid active/passive filters, 14-50 to 51
   identified loads, 14-45
   installation responsibility, 14-50
   integrated series, 14-54 to 58
   models, 14-47 to 50
   operating principle, 14-55
   shunt and series type, 14-50 to 51

type of, 14-50 to 54
unidentified loads, 14-45
Active harmonic compensation, 14-25
Actual kVA, 7-10 to 11
Actuators, 11-60 to 61
Adams, Gear and, studies, 10-61
Adhesive bonding, 3-35
Adjustable series compensation (ASC), 4-177
Adjustable speed drive (ASD), 14-21
Admittance
   buses matrix, 8-35 to 36
   concentric neutral cables, 6-27
   length, line (cables), 4-30
   models, 6-25 to 28, 10-59
   overhead lines, 6-25 to 26
   relays, 9-37
ADSS. *See* All Dielectric Self-Supporting Cable (ADSS)
Advanced Composition Explorer (ACE Satellite), 4-161 to 163
Advance-retard switch (ARS), 3-188, 3-197
AEP. *See* American Electric Power
Aeroderivative turbines, 2-30 to 31
Aesthetics, proposed substation sites, 5-78 to 81
AFAG (saturation modeling), 11-75 to 76
AFC (alkaline fuel cell), *1-12*
AGC. *See* Automatic generation control (AGC)
Agee and Girgis studies, 11-99
Aggarwal studies, 9-61, 13-11
Agrawal and Farmer studies, 11-130 to 139
Agrawal studies, 10-29
Ahlgren studies, 11-137

I-1

Ahmed studies, 14-38, 14-43
Air-blast breakers, 5-81
Air core reactors, 5-84 to 85
Air cores, 3-4, 3-81
Air flow, monitoring, 15-56
Air insulated substations (AIS)
   bus design, 5-81
   bus/switching configurations, 5-18 to 23
   local control cabinet (LCC), 5-13
   sulfur hexaflouride (SF6), 5-2 to 18
   vs-gas-insulated substations, 5-2
Air insulation, 10-95 to 96
Air magnetic circuit breakers, 5-26
Air type bushings
   air-to-air, 3-171 to 172
   air-to-oil, 3-171
   insulated, 3-176, 3-178 to 179
AISI 400 series stainless steel, 3-34
Akagi and Fujita studies, 14-50 to 54
Akagi-Nabae theory, 14-47 to 49, 14-53, 14-56
Akagi studies, 14-47 to 56
Akke and Thorp studies, 9-54
Alabama, 1-8, 10-4
Alarms, 5-14, 5-75
Albertson and Van Baelen studies, 4-163
Aleksandrov studies, 10-95
Algebraic equations. *See* Algorithms; Models
Algorithms. *See also* Analysis and assessment; Models; specific methods
   control, damping, 11-30 to 31
   Dantzig simplex method, 12-43 to 44
   Dantzig-Wolfe decomposition, 13-7
   de-bouncing, 5-50
   dynamic programming, 12-20
   economic dispatch, 12-5, 12-12-14
   ellipsoid, 12-43
   first and second order gradient methods, 12-13 to 14
   genetic algorithm (GA), 12-14, 12-20 to 26
   graphical technique, 12-13
   holiday peak forecasting, 13-20 to 21
   Karmarkar's, 12-43
   Lagrangian equations, 12-12 to 13, 12-20, 12-39, 12-42
   lamba-iteration method, 12-13
   least squares fitting, 9-51 to 52
   linear programming, 12-32, 12-43, 12-49
   line protection, 9-53
   load flow analysis, 11-36 to 38

Newton-Raphson method, 8-40 to 42, 12-31, 12-39
   parameter estimation, 9-50 to 51
   power flow, 12-30
   power flow analysis, 8-34 to 43
   quadratic programming methods (OPF), 12-43
   recursive least squares (RLS), 13-19
   relays, 9-50 to 61
   Reza, 13-20 to 21
   sequential load flow, 11-38 to 39
   special days peak forecasting, 13-20 to 21
   state estimation, 12-30
   transformer protection, 9-54 to 55
   weighted least squares (WLS) method, 12-30 to 32, 12-35 to 37
Aliasing, 9-45
Alkaline fuel cell (AFC), *1-12*
Allan, Billinton and, studies, 13-28, 13-52
Allan studies, 13-52
All Dielectric Self-Supporting Cable (ADSS), 4-44
Allegheny Power System, 4-154 to 155
Allowable stroke current, 5-114 to 115
Alsac, Stott and, studies, 12-56 to 57
Alsac and Stott studies, 12-46
Alternating current (AC)
   AC to DC rectifiers, 14-33, 14-35 to 37
   calculating faults, 8-44 to 59
   corona, 4-137
   corona acoustics, 4-145
   disadvantages, 4-169
   frequency converters, 5-29 to 30
   inductance grounding, 10-80
   Pacific AC Intertie, 11-63 to 67
   power flow, 12-27
   resistance grounding, 10-80
   resonance grounding, 10-80
   transient stability, 11-122 to 123
Altitude, 4-29
Altitude correction factor (ACF), 10-98
Aluminum, 3-18, 5-3
Aluminum Association, 4-102
Aluminum Company of America, 4-102, 4-104
Aluminum conductors
   all aluminum (AAC), 4-106
   all aluminum alloy (AAAC), 4-106
   aluminum-alloy reinforced (ACAR), 4-106
   steel-reinforced (ACSR), 4-106
Alvarado, Avila-Rosales and, studies, 10-71

American Electric Power (AEP), 5-32, 5-61
American National Standards Institute (ANSI)
   acceptance process, 3-250 to 253
   ANSI/UL 506 "Standard for Safety for Specialty Transformers," 3-256 to 257
   ANSI/UL 1446 "Standard for Safety for Systems of Insulating Materials-General," 3-257, 3-261
   ANSI/UL 1562 "Standard for Safety for Transformers, Distribution, Dry-Type-Over 600 Volts," 3-265
   ANSI/UL 1561 "Standard for Saftey for Dry-Type General Purpose and Power Transformers," 3-260 to 261
American Society of Civil Engineers (ASCE), 5-125
American Society of Mechanical Engineers Hydro Power Technical Committee, 2-2
Amorphous silicon cells, *1-15*
Ampacity, transmission line conductors, 4-66 to 67
Ampere's law, 10-65, 67 to 68
Amplification factor, 10-63
Amplification (SSR), 11-129, 11-133 to 135
Analog data acquisition, 5-44 to 49
Analog models. *See* Models
Analysis and assessment. *See* Algorithms; Models
   cost/benefit analysis, 13-29, 13-49 to 50
   data from monitoring, 15-57
   distorted waveforms, 15-18
   distribution system, 13-57
   dynamic security, 13-55
   Eigenvalue, 11-133 to 134
   electrode response, 10-82 to 83
   forecasts and projections, 13-48
   load flow, 11-36 to 38
   methodologies, 11-124 to 125
   methods, system response assessment, 11-16 to 17
   multi-objective decision, 13-49 to 50
   network topology, 12-33
   off-line dynamic system, 11-124
   on-line dynamic system, 11-124 to 125
   power flow systems, 8-34 to 43
   probabilistic reliability, 13-33 to 39
   probabilistic security, 13-56
   risk, 5-137

Index

security assessment, 11-120 to 125, 12-54 to 58
sequential load flow, 11-38 to 39
short circuits, 6-64 to 67
software for, 11-68
stability, transient, 11-16 to 17, 11-50
state estimation, 12-27 to 37, 13-48
static security, 13-54 to 56
Static Security Assessment (SSA), 11-120 to 121
subsynchronous resonance, 11-131 to 136
symmetrical component transformation, 8-14 to 33
system adequacy, 13-52 to 54
system security, 13-54 to 56
system state, 13-48
trade-off, 13-50
transient conditions, 8-34
transients, 10-82 to 83
transient security, 13-55
transmission plan evaluation, 13-26 to 39
using transient energy function (TEF), 11-50
voltage stability, 11-36
wide area dynamics, 11-82 to 117
ANCILCOs. *See* Ancillary services companies (ANCILCOs)
Ancillary services companies (ANCILCOs), 2-34, 13-6, 13-10
Anderson and Eriksson studies, 4-167, 5-111, 10-4, 10-17, 10-32 to 33
Anderson and Farmer studies, 11-127 to 138
Anderson and Fouad studies, 11-43, 11-45
Anderson and LeReverend studies, 9-42
Anderson studies, 4-165, 5-29, 5-105 to 106, 5-108, 5-118, 9-39, 10-5, 10-7
Andorka and Yohn studies, 11-142, 11-142 to 143
Angular displacement, 3-42, 3-130
Animals, 5-88 to 91, 9-2, 13-63
ANN. *See* Artificial neural networks (ANN)
ANNSTLF. *See* Artificial neural network short-term load forecaster (ANNSTLF)
ANSI. *See* American National Standards Institute (ANSI)
ANSI C57. *See* Accredited Standards Committee (ANSI) C57
Antiresonance, 10-62
Aperiodic instability, 11-4

APF. *See* Average power factor (APF) kVA
Ap index, 4-157 to 159
Applications
  active filters, 14-58 to 61
  commercial, 4-57 to 59, 6-72
  distributed utilities, 2-34 to 35
  distribution transformers, 3-51
  dry type transformers, 3-64
  dynamic load modeling, 6-2 to 3
  flickering lamps, 15-13 to 15
  grounding, 15-12 to 15
  high voltage, 4-48 to 62
  line, distribution automation, 6-72
  load modeling, 6-2 to 10
  load tap changers, 3-188 to 190
  monitoring, 11-103 to 117
  on-line monitoring technologies, 3-272 to 288
  photovoltaic (PV) cells, 1-17 to 19
  probabilistic reliability assessment methods, 13-35 to 39
  PV-powered, 1-18 to 20
  reactors, 3-81 to 103
  residential, 4-57 to 59
  shunt capacitor banks, 4-170 to 174
  snap shot load modeling, 6-2
  static load modeling, 6-2
  step-voltage regulators, 3-69 to 74, 3-80
  substation automation systems, 5-41 to 53
  switching, equipment, 5-24
  voltage sags, 15-26 to 27
  wind power, 1-2 to 4
  wiring, 15-12 to 15
Applied voltage tests, 3-49, 3-222
Approximate line segment models, 6-30 to 32
Arbitrary waveforms, 10-26 to 29
Arbutus conductor, 4-103
Archival recording, 11-86
Arcidiancone studies, 11-61
Arcing, 3-208, 15-16
Arcing switches, 3-185 to 187, 3-196
Arc suppression reactors (Peterson Coils), 3-95 to 97, 9-60
Ardito studies, 10-48
Aredes studies, 14-50, 14-59
Arizona, 4-2 to 7, 4-170, 10-4, 11-107, 11-109
Arnold, Arrillaga and, studies, 7-16
Aromatic oils, 3-206 to 207
Arresters
  distribution transformers, 3-50
  field installations, 10-77
  internal transients, 10-47
  metal oxide, 10-46, 10-77, 10-100
  surge, 5-10, 10-99 to 102

transmission systems, 4-4
Arrhenius reaction, 3-154
Arrillaga and Arnold studies, 7-16
Artificial neural networks (ANN)
  digital relaying, 9-60 to 61
  price forecasting, 13-23 to 25
  short term loads, 13-23
Artificial neural network short-term load forecaster (ANNSTLF), 13-18 to 23
ASAI. *See* Average system availability index (ASAI)
ASC. *See* Adjustable series compensation (ASC)
ASCE. *See* American Society of Civil Engineers (ASCE)
ASCR. *See* Asymmetrical silicon-controlled rectifier (ASCR)
ASD. *See* Adjustable speed drive (ASD)
ASIDI. *See* Average system interruption duration index (ASIDI)
ASIFI. *See* Average system interruption frequency index (ASIFI)
Askarels, 3-33
ASO. *See* Accredited Standards Organization (ASO)
Asplund studies, 5-29, 5-39
Asymmetrical silicon-controlled rectifier (ASCR), 14-4
ATC. *See* Available transfer capability (ATC); Available transmission capacity (ATC)
Athay studies, 11-44, 11-48, 11-52
Atmospheric pressure, 10-95 to 96
ATR. *See* Automatic trend relay (ATR)
Attributes (goodness), 13-46 to 47
Auctions, 12-15
  computerized, 13-9 to 10
  Dutch, 12-26
  English, 12-26
  market simulators, 13-4
  unit commitment, 12-26
Audible noise. *See* Acoustics
Auroral electrojet current, 4-150 to 151, 4-159, 4-162 to 163
Automatic closing, distributed utilities, 2-32 to 33
Automatic generation control (AGC)
  energy management, 12-4 to 5
  monitors, 11-91
  optimal power flow (OPF), 12-50
Automatic trend relay (ATR), 11-62
Automatic voltage regulator (AVR), 11-76, 12-58
Automation technology
  distribution systems, 6-67 to 76

substation systems, 5-41 to 53
Automotive applications. *See* Electric vehicles
Autotransformers
  connections, 3-130, 3-133
  load tap changers, 3-190
  protection, 9-4
  step-voltage regulators, 3-74
Auxiliary losses, 3-158
Auxiliary systems, 2-10, 2-21
Auxiliary transformers, 2-24
Available transfer capability (ATC), 11-125
Available transmission capacity (ATC), 13-18
Average power factor (APF) kVA, 7-10 to 11
Average system availability index (ASAI), 13-58
Average system interruption duration index (ASIDI), 13-58
Average system interruption frequency index (ASIFI), 13-58
Average Voltage Voltmeter Method transformer testing, 3-225
Avila-Rosales and Alvarado studies, 10-71
Axial forces
  power transformers, 3-15, *3-16*
Aylett studies, 11-44

B

Backflashover, 4-168
Background investigation, transformers, 3-243
Back-propagation (BP), 13-17, 13-24
Bahrman studies, 11-140
Bakirtzis studies, 13-18
Balanced case, reduction to, 8-24 to 28
Balanced three-phase faults, 8-45 to 48, 8-51 to 52
Balbo studies, 14-52
Balma studies, 10-75
Bandwidth
  control circuit, 6-33
  load tap changing, 3-137 to 139
  step-voltage regulators, 78 to 79
Bang-bang discontinuous control, 11-59
Barrel windings, 3-21
Barrier insulators, 5-14
Barriers, animal damage, 5-90
Barton studies, 14-43
Bar-type current transformers, 3-122
Basic annual indices, 13-30
Basic Impulse Insulation Level (BIL)
  applied voltage tests, 3-222

design margin, 10-75 to 76
dry type transformers, 3-66
instrument transformers, 3-110 to 111
insulation strength, 3-222, 10-8
lightning, 5-114 to 115, 10-8 to 9, 10-31 to 33
load tap changers, 3-201
power transformers, 3-12
surge arresters application, 10-101 to 102
switching impulse tests, 3-220 to 221
very fast transients, 10-42
Bassetto studies, 3-155
Batteries, 1-9 to 10
Battery chargers, 5-84 to 85
Bayer studies, 11-64
Bayonets, 3-49
BCT (Current Transformer over Bushing), 3-122
BCTRAN matrix models, 10-78
BCU method. *See* Boundary controlling UEP method
Bean studies, 10-54, 10-73 to 74
Bearings and couplings, synchronous machinery, 2-15
Beatty, Fink and, studies, 4-97 to 99, 10-73
Beck, Wagner, McCann studies, 5-109
Beehler studies, 10-36
Behrendt studies, 11-64
Belotelov studies, 11-64
Bengtsson studies, 3-277 to 278, 3-284, 3-287
Benmouyal studies, 9-28
Benson, Debs and, studies, 12-54
Bentall studies, 12-43
Bentonite clay, 5-68
Beranek studies, 5-38
Bergen and Hill studies, 11-44, 11-52
Berger and Floret studies, 5-123
Berger and Garbagnati studies, 10-6
Berger studies, 4-167, 10-3, 10-5
Bergovic studies, 9-39, 9-42
Berk, Mackay and, studies, 13-31
Bervall, Evans and, studies, 11-7
Bewley studies, 10-10
Bhattacharya studies, 14-50, 14-53, 14-59
Bhatt studies, 11-62
BICs. *See* Bus incremental costs (BICs)
Bi-directional meters, 7-3
Biegelmeier and Lee studies, 5-96
BIL. *See* Basic Impulse Insulation Level (BIL)
Billing. *See* Metering
Billinton and Allan studies, 13-28, 13-52

Billinton and Jonnavitihula studies, 13-56
Billinton studies, 13-31, 13-64
Binder, Miri and, studies, 10-46 to 47
Birds, damage from, 5-90, 13-63
Bird studies, 14-38, 14-44
Bjorkland and Johnsson studies, 5-39
Bjorkland studies, 13-43
Blackburn studies, 8-31, 9-39
Blackouts. *See* Outages
Black starts, 2-1, 2-21
Block box (current transformers), 3-117 to 118
Blocking schemes, directional comparison, 9-39
Blocks, stringing, 4-116
Blondel's theorem, 7-2
Blume studies, 10-54
Body current limits, 5-95 to 96
Boersma studies, 10-51
Boggs studies, 10-42 to 43, 10-50 to 51
Bollen, Zhang and, studies, 15-31, 15-36
Bollen studies, 15-29 to 36
Bollinger studies, 11-61
Bolted (grip), 4-116
Boltzmann, Stefen-, studies, 4-67
Bonneville Power Administration (BPA)
  centralized monitor system, 11-90 to 91
  Dittmer Control Center, 11-90 to 91
  M&E insulator ratings, 4-29
  Power System Disturbance Monitor (PSDM), 11-88
  reclosing method, 11-64
  Slatt-Buckley 500-kV line, 4-178 to 179, 11-96, 11-113
  transient stability method (novel), 11-66, 11-134
Bortnik studies, 4-129
Bose studies, 14-38 to 41
Boundary controlling UEP method, 11-51
Bowler studies, 11-139
BP. *See* Back-propagation (BP)
Brakes, dynamic, 11-64, 11-112, 14-31
Branches. *See* Trees
Brayton Cycle systems, 2-31
Brazil, 11-65, 13-44
Breakdown streamers, 4-137
Breaker-and-a-half configuration, 5-22, 9-4
Breakers
  actuators, mechanical, 11-60
  air insulated substations, 5-5
  air magnetic, 5-26

Index

breaker-and-a-half configuration, 5-22, 9-4
bulk oil, 5-26 to 27
bypass, 4-177
classifications, 5-26
dead tank design, 5-26 to 27
distribution transformers, 3-50
electric and magnetic fields (EMF), 5-84 to 85
failures, 9-26, 9-70
gas, 5-27
gas-insulated substations, 5-5, 5-13
hydroelectric generators, 2-7
internal transients, 10-48
live-tank design, 5-27
ratings, 5-27 to 28
reactor applications, 3-99
restrikes, 9-70
seismic events, 5-133
stuck, 9-63
substation noise source, 5-81
sulfur hexaflouride (SF6), 5-27
switching equipment, 5-26, 5-26 to 28
thermal generating plants, 2-23 to 24
transmission systems, 4-4
vacuum, 5-27
very fast transients (VFT), 10-52
Breingan, Chen and, studies, 9-54
Brogan studies, 8-14
Broker companies (BROCOs), 13-3
Brown, Gurgis and, studies, 9-56
Brown studies, 13-57, 13-59
Brumsickle studies, 14-38
Bubble evolution, 3-156
Buildings, 5-37 to 38, 5-80, 5-134 to 135
Bulk oil circuit breakers, 5-26 to 27
Bulk power system reliability, 13-26 to 27
Bullet trains, Japan, 14-60
Bullwheel, 4-116
Bundle conductors
  capacitance, 4-80 to 82
  corona performance, 4-147 to 148
  software for, 4-148
  spiraling effect, 4-64 to 66
Burchett studies, 12-38, 12-42
Burdens, instrument transformers, 3-107
Buried, direct-, installations, 3-43
Buried cables, 4-59 to 60
Buried contact (thin silicon) cells, 1-14
Burning, body current limits, 5-95
Burst coronas, 4-136
Busbar faults, 12-55
Bus ducts, 10-47

Buses
  admittance matrix, 8-35 to 36
  air insulated substations, 5-18 to 23
  busbar, 12-55
  electric and magnetic fields (EMF), 5-84 to 85
  proposed substation sites, 5-80 to 81
  ring, 5-21, 9-4
  seismic events, 5-125
  sulfur hexaflouride (SF6), 5-8
  thermal generating plants, 2-20 to 21, 2-24
  tiebreakers, 5-19 to 21
  tie reactors, 3-86, 3-100 to 101
  voltages, regulating, 3-137 to 139
  voltage stability, 11-5 to 6
Bushings
  animal damage, 5-89
  circuit breakers, 5-27
  Current Transformer over Bushings (BCT), 3-122
  current transformers, 5-45 to 47
  electrical, 3-171 to 184
  gas-to-air, internal transients, 10-48
  on-line maintenance, 3-282 to 284
  porcelain, 4-176
  potential transformers, 5-46 to 47
  power factor (Coso), 3-283 to 284
  SF6-to-air, 5-8 to 10
  SF6-to-oil, 5-9
  transformer failure, 9-2
  transformer installation, 3-237, 3-242
  very fast transients (VFT), 10-52
Bus incremental costs (BICs), 12-50
Bypass breaker, 4-177

C

Cabinet, controls, 3-239
Cables
  Carson's equation, 6-12
  catenary, 4-89 to 97, 4-113 to 114
  concentric neutral, 6-20 to 22, 6-27
  electric and magnetic fields (EMF), 5-84 to 85
  fire hazard, 5-135
  gas-insulated substations, 5-9
  insulated power, 4-48 to 62
  length, 4-30, 4-90 to 91, 9-32 to 33
  modeling, 6-19 to 22
  status points, 5-50
  substation bus design, 5-81
  substation oil source risk, 5-63
  tape shielded, 6-22 to 24, 6-27 to 28
  thermal generating plants, 2-25
  transducers outputs, 5-47

underground, 13-61
Cadmium telluride cells, 1-16
CAES (Compressed air energy storage), 1-8
Cahill studies, 4-119
CAIDI. *See* Customer average interruption duration index (CAIDI)
CAIFI. *See* Customer average interruption frequency index (CAIFI)
Calculations. *See* Algorithms; Models
California, 1-10
California-Oregon transmission system, 11-130
California Power Exchange (CALPX), 13-24
Canada
  Hydro-Quebec, 4-143 to 144, 4-143 to 145, 4-151
  Manitoba, 5-30
  Nelson River, 5-30
  Ontario Hydro, 11-8, 11-24, 12-44
  seismic qualifications methods, 5-129
  transmission systems, 4-7
  wind power installed capacity, 1-2
Canvas list, American National Standards, 3-250
Capacitance
  cables, 4-53, 4-53 to 54
  dissipation factor (Tano), 3-213
  electrode response, 10-83
  graded bushings, 3-173 to 175
  instrument transformers, 3-280 to 282
  models, 10-66 to 72
  on-line, bushing, 3-283 to 284
  series, modeling, 10-67
  shunt, modeling, 6-25 to 28, 10-67
  transmission line conductors, 4-75 to 82
Capacitative coupled voltage transformers (CCVT), 3-116 to 117
Capacitive reactance, transmission line conductors, 4-75 to 82
Capacitor banks
  harmonics, 15-22
  modeling, 6-63
  series, 4-175 to 180
  shunt, 4-170 to 174, 6-57 to 58
  switchable, 6-4
Capacitors
  correction type, 3-141 to 142
  inrush/outrush reactors, 3-88 to 89
  ultra type, 2-31
  voltage transformers, 5-46 to 47
Capacity expansion, 13-1 to 12
Carlsen, Fink and, studies, 12-58

Carlsson studies, 5-39
Carolsfeld studies, 11-85, 11-96
Carpenter studies, 5-123
Carpentier studies, 12-38, 12-42
Cars, electric. See Electric vehicles
Carson's equations, 6-11 to 15, 6-63
Carton and Peyrot studies, 4-17 to 18
Carvalho studies, 11-44, 11-48
Cascading, power system, 9-41
Cash and Habetler studies, 14-37
Cast insulation bushings, 3-176
Catch basins, 5-69 to 72
Catenary cables, 4-89 to 97, 4-113 to 114
CCVT (capacitative coupled voltage transformers), 3-116 to 117
Cellulose insulation, 3-154 to 155, 3-204 to 205
Centeno studies, 11-67
Center of inertia (COI), 11-44
Central Japan Railway Company, 14-60
Ceramic insulators, 4-36 to 41
Ceramic superconductors, 1-8
CFE. See Communication front-end (CFE) processors
CFO. See Critical flashover (CFO)
Chaari studies, 9-60
Chadalavada studies, 11-44, 11-52, 11-124
Change-overs, 3-188, 3-196 to 200
Characteristics
 air insulation, flashover, 10-95 to 99
 cellulose insulation, 3-204 to 205
 frequency resonance, 10-61 to 63
 generic dynamic load-voltage, 11-34 to 36
 insulation, 10-93
 lightning strokes, 10-1 to 7
 loads, 7-12 to 13
 paper insulation, 3-204 to 205
 voltage sag, 15-25 to 32
Charbonneau, Lacroix and, studies, 4-129
Charging mechanism, lightning, 10-2 to 4
Chemical energy, 1-10 to 13
Chemical repellants, animal damage, 5-91
Chen and Breingan studies, 9-54
Chew studies, 3-150 to 153
Chiang and Wong studies, 11-68
Chiang studies, 11-44 to 45, 11-48, 11-51
Chief Joseph substation, 11-112
Children per generation, 12-24
China, 1-2, 11-65
Chow, Larsen and, studies, 11-30, 11-65

Chow and Yee studies, 9-61
Chowdhuri and Gross studies, 10-27 to 30
Chowdhuri and Kotapalli studies, 10-17
Chowdhuri studies, 10-11 to 16, 10-27 to 33, 10-48
Chow studies, 10-83, 11-6
Christofersen studies, 5-38
Chrysler, 1-9
Chu studies, 3-275
CIGS (copper indium (gallium) diselenide), 1-15 to 16
Cihlar studies, 11-121
Cinieri and Fumi studies, 0-29 to 30
Circuit breakers. See Breakers
Circuits. See also Short circuits
 control, 5-52, 6-33
 equivalent, 3-5 to 6, 3-45
 interruption, 2-24
 iron core, 3-5 to 10
 lighting, 2-22
 open, 8-44
 open conditions, 3-119
 per units scaling, 8-1 to 14
 pulse, metering, 7-9
Circuit switches, 5-26
Circular windings, 3-18 to 20
Circulating current paralleling, 3-145 to 147
Claiborne, Oommen and, studies, 3-34
Claiborne studies, 3-33
Clamping systems, 3-180
Clarke, modes of, 4-141
Clark studies, 11-32, 11-99
Classifications
 active filters, 14-50 to 54
 air insulators, 3-178 to 179
 buses, 8-39 to 40
 bushings tests, 3-181 to 184
 circuit breakers, 5-26
 data capture triggers, 11-86 to 7
 distribution transformers, 3-58
 dry type transformers, 3-63 to 64
 electrical bushings, 3-171 to 176
 energy storage technologies, 1-7 to 10
 fuel cells, 1-11 to 13
 generator sizes, 2-32
 hydroelectric plants, 2-2
 loads, 6-1 to 2
 load tap changer ratings, 3-193 to 196
 power system oscillations, 11-23 to 24
 power system stability, 11-3 to 7
 power transformers, 3-12 to 14, 3-13 to 14
 stability controls, 11-61

substation automation architecture, 5-53
transformer tests, 3-210
voltage stability, 11-6
voltage transformers, 3-113 to 114
Class II power transformers, testing, 3-223
Cleanliness
 gas-insulated substations, 5-17
 substation buildings, 5-37 to 38
Climatology, magneto-telluric, 4-157 to 161
Clipping in conductors, 4-45, 4-46
Clipping offsets, 4-119 to 120
Closed-loop controls, 11-29, 11-59
Closed zones, 9-30 to 31
Coarse change-over selector, 3-188
Cochrane studies, 5-38
Cogen generation companies, 13-6
Coherent machine equivalent, 11-81
COI. See Center of inertia (COI)
Coils. See also Windings
 hydroelectric generators, 2-5
 orientation, power transformers, 3-20
 Petersen, 3-95 to 97, 9-60
 as storage medium, 1-8
 surges, 10-56 to 58
 synchronous machinery stator, 2-14
 transient voltage response, 10-54 to 78
Coincidence factor, 7-13 to 14
Cold-loads, 7-17, 9-33
Cold resistance, 3-232 to 233
Collection pits, oil, 5-67
Colorado, 10-4, 11-142 to 143
Colstrip generating plant, 11-62 to 63
Comanche Unit 2, 11-142 to 143
Combination metering units, 3-129
Combustion turbines, 1-8, 2-30 to 31
Commercial applications, 4-57 to 59, 6-72
Common mode outage event, 13-32
Communication, 2-32 to 33, 6-70 to 72
Communication front-end (CFE) processors, 6-69
Communities
 residential distribution layout, 4-57 to 59
 single family, 1-19
 substation considerations, 5-76 to 87
Comparison blocking schemes, directional, 9-39
Compensation, high-voltage windings, 3-114
Compensation to land owners, 1-6 to 7

Index

Complex electrode shapes, 10-86 to 87
Component based load modeling, 6-3
Composite insulators, 4-32 to 35, 4-39 to 40
Composite loads, 6-9 to 10, 7-13 to 17
Composition-based modeling, 11-80
Compressed air energy storage (CAES), 1-8
Compression turbines, 1-8
Computational methods. *See* Algorithms; Models
Computers. *See* Technology
Concentric configurations, cables, 4-49 to 50
Concentric neutral cables, 6-20 to 22, 6-27
Concordia, de Mello and, studies, 11-24
Concordia and Ihara studies, 12-44
Concordia studies, 11-21 to 22
Condensers, synchronous, 2-12 to 13
Conductor blocks, 4-44
Conductors
   all aluminum (AAC), 4-106
   all aluminum alloy (AAAC), 4-106
   aluminum-alloy reinforced (ACAR), 4-106
   Arbutus conductor, 4-103
   cable, 4-49 to 50
   catenary cable, slack/sag, 4-91, 4-113 to 114
   configuration, 4-148
   design parameters, 3-177 to 180
   Drake, 4-90, 4-97 to 102, 4-105 to 113
   Graphic Method for Sag Tension Calculations for ACSR and Other Conductors, 4-102 to 103
   grounding, 10-80 to 90, 15-3
   images, 6-13
   installation, 4-115 to 120
   insulation, distribution transformers, 3-32
   joining, distribution transformers, 3-32 to 33
   line, selection, 4-148
   prestressing, 4-119
   sag/tension, 4-89 to 128
   selection, 4-148
   solid, capacitance, 4-76 to 77
   steel-reinforced (ACSR), 4-106
   stroke current magnitude, 5-106 to 107
Conductors, Guide to the Installation of Overhead Transmission Lines, 4-115
Conduits, cables in, 4-59 to 60
Configurations

breaker-and-a-half, 5-22, 9-4
bus/switching, 5-18 to 23
concentric, cables, 4-49 to 50
conductors, 4-148
double bus, double breaker, 5-19 to 21
double bus, single breaker, 5-21
electrode, voltage characteristics, 10-96
single bus, 5-19 to 21
software for bundled conductors, 4-148
transient voltages, 10-55
transmission line protection, 9-32 to 33
windings, faults, 9-3
Conjugate gradient method (OPF), 12-42
Connections. *See also* Delta connections; Wye connections
   cable, 5-9
   current transformers, 3-121
   direct transformer, 5-9
   distribution transformers, 3-39 to 42
   high-voltage windings, 3-114
   step-voltage regulators, 3-71 to 72, 3-74, 3-80
Conrad studies, 15-24
Constant PO
   delta connections, 6-55
   wye connections, 6-53 to 54
Construction
   current transformers, 3-122 to 125
   distribution transformers, 3-31 to 35, 3-35
   planning uncertainties, 13-45
   power transformers, 3-17 to 25
   substations, 5-3 to 17, 5-86 to 87
   transmission lines, 4-42 to 48
   voltage transformers, 3-115
Contact bouncing, 5-50
Contact life, load tap changers, 3-195
Contacts, monitoring substation automation systems, 5-49 to 50
Contact wear model, 3-286
Containment pits, oil, 5-67
Contamination, insulation media, 3-209
Contaxis studies, 12-43
Contingency enumeration approach, 13-33 to 39
Continuous-load, defined, 9-33
Continuous power frequency voltages, 4-24 to 25
Control building, fire hazard, 5-135
Control hierarchy, hydroelectric plants, 2-9 to 10

Control inputs, 3-136 to 137
Controlled rectifiers, 14-25 to 37
Controlled separation schemes, 11-67
Controllers
   power electronics substations, 5-32 to 36
   substation automation systems, 5-51 to 52
   variable speed motor, 11-140
Controlling UEP, 11-48 to 52
Control mode oscillations, 11-24
Controls
   gas-insulated substations, 5-11 to 13
   stability, power system, 11-55 to 69
   system oscillation mitigation, 11-28 to 29
   testing transformers, 3-214
   transformer installation, 3-239
   transmission systems, 4-4
Converters, 9-46
Converti studies, 11-8
Cookson and Farrish studies, 5-3
Coolants, 3-33 to 34
Coolers, 3-239
Cooling, 5-36
Cooling classes
   dry type transformers, 3-63 to 64
   loading power transformers, 3-149 to 151
   power transformers, 3-13
Cooling fan noise, 3-164
Cooperative utilities (COOPS), 13-6
COOPS. *See* Cooperative utilities (COOPS)
Cooray and Jayaratne studies, 10-4
Cooray and Perez studies, 10-4
Copper
   losses models, 10-69 to 71
   pilot channels, 9-38
   power transformer windings, 3-18
   substations, 5-3, 5-5
Copper indium (gallium) diselenide cells (CIGS), 1-15 to 16
Core-forms
   construction, 3-17 to 18
   distribution transformers, 3-36
   winding strategies, 10-73
Cores. *See also* No-load losses
   air, 3-4
   construction, 3-17
   distribution transformers, 3-32, 3-36
   instrument transformers, 3-104 to 107
   iron/steel, 3-4 to 7
   losses model, 10-71
   meggar readings, 3-235
   nickel-iron, 3-104 to 105
   power transformers, 3-13

reactors, 3-81
transformer sound levels, 3-163 to 164, 3-169
Corona
   discharge modes, 4-129 to 137
   effect on overhead lines, 4-137 to 144
   losses, 4-137 to 138, 4-143 to 145
   rings, suspension (composite) insulators, 4-34
Coronal hole activity, 4-157
Correction capacitors, 3-141 to 142
Coso (insulation power factor), 3-213
Cost and economics. *See also* Economic dispatch (ED)
   actuators, 11-61
   air insulated substations, 5-17 to 18, 5-19 to 23
   autotransformers, 3-133
   batteries, 1-9 to 10
   breaker-and-a-half configurations, 5-22
   cadmiumtelluride cells, 1-16
   capacitor banks, 4-175 to 176
   compressed air energy storage, 1-8
   cost/benefit analysis, 13-29
   current transformers, 3-129
   distribution automation, 6-67
   dry type transformers, 3-67
   energy management, 12-6 to 7
   gallium arsenide cells, 1-15
   gas-insulated substations, 5-17 to 18, 5-28
   geomagnetically induced current, 4-155 to 156
   high speed grounding switches, 5-25
   hydroelectric power, 2-1
   induction generators, 2-4
   interruptions to consumers, 13-30 to 31
   least cost planning, 13-48 to 50
   line conductor selection, 4-148
   load tap changer maintenance, 3-203
   load tap changers, 3-284
   microturbines, 2-30
   non-self restoring insulation, 10-93
   oil containment options, 5-65 to 66
   on-line monitoring system, 3-270
   phosphoric acid fuel cells, 1-12
   power transformers, 3-15 to 17
   price forecasting, 13-16 to 25
   reactive power compensation, 4-169
   reliability, 64
   series capacitors, 11-127
   shielding systems, lightning, 5-119

silicon cells, 1-14
substation automation systems, 5-45
superconducting magnetic energy storage, 1-8
transformer installation, 3-241
transmission line conductors, 4-64 to 65
utility-interactive PV systems, 1-18
various wind turbines, 1-3
wind power plants, 1-2
Coulombs, 10-66
Coultes, Watson and, studies, 11-24, 11-141
Countermeasures, subsynchronous resonance, 11-131, 11-136 to 137, 11-139
Countermeasures, supersynchronous resonance, 11-142
Couplers, linear, 3-125
Crary and Duncan studies, 11-21
Crawler tractor, 4-116
Creep elongation, 4-104, 4-118
Cresap studies, 11-66 to 67
Crisi studies, 11-65
Critical distance, voltage sags, 15-29 to 30
Critical flashover (CFO), 5-114, 10-7, 10-8. *See also* Flashover voltage
Cross-arm voltage (V_{ca}), 10-12
Cross-bonded cables shields, 4-56
Cross-connected current transformers paralleling, 3-145
Crousillat studies, 13-45
Crude oils. *See* Oils
CSP[1] Transformers, 3-50
CT. *See* Current transformers (CT)
CTAIDI. *See* Customer total average interruption duration index (CTAIDI)
CT (current transformers), 3-117 to 118
Culverts, 5-69 to 72
Current
   balanced, 8-24 to 25
   balance paralleling, 3-145 to 147
   carrying capacity, transmission line conductors, 4-66 to 67
   delta connections, 6-56
   grid, 5-102
   harmonic sources, 14-45 to 46
   imbalance, 9-16 to 17
   injection, 11-66
   input, load tap changing, 3-137
   limited fuses, distribution transformers, 3-49, 3-50
   limiting reactors, 3-82 to 88
   mismatch, 9-6

peak value, lightning, 10-5 to 6
recordings, 15-51 to 55
reversals, 9-68 to 69
symmetrical component transformation, 8-16 to 18
wye connections, 6-54
Current transformer over bushings (BCT), 3-122
Current transformers (CT), 3-117 to 118
   current mismatch, 9-6
   line drop compensator modeling, 6-35-36
   load tap changing, 3-137
   metering, 7-10 to 11
   pockets, bushings, 3-180
   protection methods, 9-6
   saturation, 9-6, 9-67 to 72
   seismic events, 5-133
   substation automation systems, 5-45 to 47
   transmission systems, 4-4
   zones of protection, 9-30 to 31
Curves
   flicker, 15-43 to 46
   life expenditure, 11-136
   loads, 7-14 to 16
   melting, 9-33
   swing, 11-11 to 12
   tripping, 5-26
   vee, synchronous machinery, 2-15
Customer average interruption duration index (CAIDI), 13-58
Customer average interruption frequency index (CAIFI), 13-58
Customer total average interruption duration index (CTAIDI), 13-58
Cyclicality, 3-153 to 154
Cyclone fences, animal damage, 5-90

D

Dagle studies, 11-65
Daily operation records, 9-9 to 10
Dakin studies, 3-154
Dalstein and Kulicke studies, 9-61
Dalziel and Lee studies, 5-95 to 96
Damping
   control algorithms, 11-30 to 31
   criteria for, 11-26
   modal tests (SSR), 11-138-139
   oscillations, 11-20 to 33, 11-58
   power system synchronous stability, 11-57 to 58
   reactor, 4-177
   torque, 11-4, 11-60
Dantzig simplex method, 12-43 to 44

Index

Dantzig-Wolfe decomposition, 13-7
Darlington configuration, 14-4
Dart leaders, lightning, 10-4
Data. *See also* Supervisory Control And Data Acquisition
  analog, 5-44 to 49
  capture triggers, 11-86 to 90
  continuous monitoring, 11-89
  Data Acquisition, Processing and Control System Subcommittee (IEEE), 5-60 to 61
  load modeling, 11-80
  monitoring functions, 11-85 to 86
  motor-enclosure-controller systems, 7-21
  parameter, 12-34
  subsynchronous resonance, 11-1369
  telemetered, 12-34
  trigger logic, 11-89
  value tags, 11-89
  wind speed collection sets, 1-5
Data Acquisition Units (DAU), 3-271
Database Manager (DBM), 11-100 to 101
Data capture unit (DCU), 11-94 to 95
Data concentrator, 5-55
Data warehouse, 5-55 to 56, 5-60
Date of manufacture, 9-10
DAU. *See* Data Acquisition Units (DAU)
Davidson studies, 11-8
Dawson studies, 4-134
DBM. *See* Database Manager (DBM)
DBPC (di-tertiary butyl paracresol), 3-207
DBP (di-tertiary butylphenol), 3-207
DC. *See* Direct current (DC)
(DCT) Direct current transformers, 3-126
DCU. *See* Data capture unit (DCU)
DDSPSO. *See* Device Dependent Supersynchronous Oscillations (DDSPSO)
Dead-end spans, 4-107
Dead-front padmounted transformers, 3-59, 3-61 to 62
Dead tank circuit breakers, 5-26 to 27
"Dead time," 9-67
De-bouncing algorithms, 5-50
Debris removal, 5-86
Debs and Benson studies, 12-54
Decision processes, 11-101
De-energized Tap Changer (DETC), 3-23 to 25
Defibrillation, 5-95
Degeneff, Vakilian and, studies, 10-59

Degeneff and Kennedy studies, 10-66, 10-67
Degeneff studies, 10-55, 10-59, 10-60, 10-63, 10-66, 10-74 to 75, 10-77 to 78
Dehdashti studies, 11-61
Dehydrating (desiccant) breathers, 3-26
Delta, sigma-, AD converters, 9-46
Delta connections
  current transformers, 3-121
  distribution transformers, 3-40
  double wound transformers, 3-132 to 133
  load tap changers, 3-189
  models, 6-37 to 57
  step-voltage regulators, 3-71 to 72, 3-74, 3-80
  symmetrical component transformation, 8-21
Delta load forecaster (DLF), 13-18
Delta modulation, 9-46
Delta var paralleling, 3-146
Demand management, 13-42 to 43
Demand metering, 7-5 to 6
Demaree studies, 11-125
Denmark, 1-1 to 2
Dent studies, 10-59
Department of Energy (DOE)
  electric vehicle applications, 1-9
  studies, 13-45
  wind turbines, 1-1 to 2
De-Q'ing, 3 to 89
Deregulation. *See also* Regulation
  price forecasting, 13-23 to 24
  security analysis, 12-58
  tactical automation, 6-72
  utility company competition, 13-2 to 12
Desiccant (dehydrating) breathers, 3-26
Design
  buses, 5-80 to 81
  closed-loop control, 11-29
  commercial distribution cable lines, 4-57 to 59
  control circuits, 5-52
  empirical, 5-108 to 110
  fire protection, 5-137
  gas-insulated substations, 5-15
  grounding system, 5-98 to 104
  grounding tips, 10-89 to 90
  lightning protection, 5-105 to 123
  loading power transformers, 3-149
  meters, 7-1 to 11
  motor-enclosure-controller systems, 7-20 to 23
  power systems, 11-18 to 19
  power system stability, 11-8 to 9

sag-tension parameters, 4-112 to 115
seismic event protection, 5-126
shielding, 10-13 to 16
substations, 5-23
system oscillation mitigation, 11-28 to 29
transient stability, 11-18 to 19
transmission line components, 4-17 to 22
winding construction, 10-72 to 77
Detailed load model, 11-79
DETC (De-energized Tap Changer), 3-23 to 25
Determinist method approach, 12-58, 13-27 to 29
DeTourreil studies, 4-41
Device Dependent Supersynchronous Oscillations (DDSPSO), 11-142 to 143
Dew point test, 3-235
DFT. *See* Fourier series
DGA. *See* Dissolved gas analysis (DGA)
Diagnostic function of oil, 3-208
Diagnostic meters, 7-3 to 5
Dianostic tests, transformers, 3-244 to 245
Dias and El-Hawary studies, 12-44, 12-46
Dick studies, 10-46
Dielectric gas (SF6), 5-2 to 18
Dielectric properties
  bushings tests, 3-182
  disconnector asymmetry, 10-50 to 51
  moisture in oil, 3-156
  requirements, load tap changers, 3-195
  transformer failures, 3-247 to 248
  transients tests, 3-215
  withstand testing, 3-214 to 223
Diesel generators, 1-3, 2-21
Differential schemes
  equations, 9-53 to 55
  phase comparison, 9-39
  pilot wires, 9-39
  restrained relay, 9-2
  stator faults, protection, 9-11 to 14
  zones, 9-30 to 31
Digital multifunction relays, 9-28 to 29
Digital relays, 9-44 to 61
Digital signal processors (DSPs), 14-25, 14-39
Dillon studies, 13-18
Diode conduction mechanics, 14-9
Direct analysis, wide area dynamics, 11-82 to 117
Direct-buried cables, 4-59 to 60

Direct-buried distribution transformers, 3-43
Direct current (DC)
  AC to DC rectifiers, 14-33, 14-35 to 37
  corona acoustics, 4-145
  frequency converters, 5-29 to 30
  generator field circuit protection, 9-14
  geomagnetically induced currents (GIC), 4-150
  interposing relays, 5-51
  link choke, 14-23 to 24
  offset, fault analysis, 8-57 to 58
  offset waveforms, 9-67
  Pacific HVDC Intertie, 4-8, 11-23, 11-113
  resistance model, 10-69
  synchronous machinery, 2-15
  thermal generating plants, 2-23, 2-25
  transformers (DCT), 3-126
  transformer sound levels, 3-170
  transmission system lines, 4-7 to 8
  trapped charges, 10-36 to 39
Directional comparison blocking schemes, 9-39
Directional overcurrent relays, 9-35 to 36
Direct lightning strokes, 10-8 to 20
Direct stability methods, 11-42 to 52
Direct transformer connections, 5-9
Discharge control systems, 5-69 to 72
Discharge current limiting reactors, 3-89
Discharge lamps, 6-9
Disconnectors, 10-52
Disconnect switch operations
  gas-insulated substations, 5-7
  load break, 5-7, 5-24 to 25
  switching equipment, 5-24 to 25
  transmission systems, 4-4
  very fast transients development, 10-42
Disc windings, 3-21 to 23, *3-27 to 28*, 10-67
Dismantling process, 3-245 to 247
Dissolved gas analysis (DGA)
  geomagnetically induced current, 4-155
  load tap changers, 3-288
  power transformers, 3-272 to 275
  transformer installation, 3-241 to 242
Distance relays, 9-37 to 38
DISTCOs. *See* Distribution companies (DISTCOs)
Distributed loads, modeling, 6-63
Distributed monitor networks, 11-90
Distributed utilities (DU), 2-27 to 35

Distribution automation, 6-67 to 76
Distribution companies (DISTCOs), 12-17, 13-3 to 12
Distribution feeders
  analysis techniques, 6-58 to 67
  modeling, 6-1 to 58
  series components, 6-62 to 63
  shunt components, 6-63
Distribution lines, transmission system, 4-2 to 12
Distribution Management System (DMS), 6-68, 6-73 to 76
Distribution SCADA, 6-67 to 76
Distribution systems
  analysis, 6-58 to 67
  modeling, 6-11 to 58
  operation and control, 6-67 to 76
  power system loads, 6-1 to 10
Distribution transformers, 3-30 to 51
Disturbances
  geomagnetic, 4-150 to 163
  monitoring, 11-82 to 117
  rotor angle stability, 11-5
  security assessment, 11-120 to 125
  system monitoring, 11-80
  voltage stability, 11-39 to 40
Ditching, oil containment, 5-67
Di-tertiary butyl paracresol (DBPC), 3-207
Di-tertiary butylphenol (DBP), 3-207
Dittmer Control Center, 11-90 to 91, 11-107
Diurnal cycle, 1-4
Divah, Lorenz and, studies, 14-42
Diversity factor, 7-13 to 14
Djakov studies, 11-60
Djuric and Terzija studies, 11-64
DLF. *See* Delta load forecaster (DLF)
DMS. *See* Distribution Management System (DMS)
Dodge studies, 11-60
DOE. *See* Department of Energy (DOE)
Dommel and Tinney studies, 12-38, 12-42
Dommel studies, 10-60, 10-78
Domun studies, 3-278
De Donker and Novotny studies, 14-42
Donnelly studies, 11-64
Donor impurities, 14-9
Double bus, double breaker configuration, 5-19 to 21
Double bus, single breaker configuration, 5-21
Double phase-to-ground faults, 8-49 to 50
Double reversing change-over selector, 3-188

Double wound transformers, 3-132 to 133
Doubly infinite single-conductor line, 10-25 to 29
Downstream power system, 7-13
Draft tube water depression, 2-11
Drainage systems, 5-69 to 72
Drake conductor, 4-90, 4-97 to 102, 4-105 to 113
Dreyfuss studies, 4-16 to 17
Drum pullers, 4-116
Dry-type reactors, 3-81 to 82
Dry-type transformers
  design, 3-36, 3-63 to 67
  sound levels, 3-170
  standards, 3-255 to 268
DSM. *See* Dynamic System Monitor (DSM)
DSPs. *See* Digital signal processors (DSPs)
DU. *See* Distributed utilities (DU)
Duality, model, 10-60
Dual sensing links, 3-49, 3-50
Ducts, cables in, 4-59 to 60
Duhamel's theorem, 10-21, 10-26
Duncan, Crary and, studies, 11-21
Duplex connections, 3-41
Duplex reactors, 3-87 to 88
Duplex transformers, 3-37
Duration curves, load, 7-14 to 16
Dutch auctions, 12-15
Duval and Lamarre studies, 3-278
DVR. *See* Dynamic voltage restorer (DVR)
Dwellings, single family, 1-19
Dynamic brakes, 11-64, 11-112, 14-31
Dynamic load-voltage, generic, 11-34 to 36
Dynamic modeling, power systems, 11-72 to 81
Dynamic reactive compensation (TCR), 3-92 to 93
Dynamic System Monitor (DSM), 11-95 to 96
Dynamic voltage restorer (DVR), 15-39
Dynamic voltages, 10-55

E

Early value actuation (EVA), 11-77
Earthing grid, 10-50
Earthquakes. *See* Seismic considerations
E assembly, 3-17, *3-19*
Eborn studies, 9-61
Ecklin studies, 10-42
Economic dispatch (ED). *See also* Cost and economics

Index

accuracy of solution, 12-14
algorithms, 12-12 to 14
application, 12-14 to 15
auctions, 12-15
automatic generation control, 12-5
cost/price, 12-11
defined, 12-10
objectives, 12-12
optimal power flow (OPF), 12-50
quantity supplied, 12-11
solution techniques, 12-13
system limitation, 12-11
Economics. *See* Cost and economics
Ecostar vehicle, 1-10
ED. *See* Economic dispatch (ED)
Eddy losses
  cables, 4-55
  distribution transformers, 3-44
  model, 10-71
  thermal response prediction, 3-152
  transformer testing, 3-225, 3-228
Edison, electric shut-down salute to, 13-48
Edison cells, 1-9 to 10
Edison Electric Institute (EEI), 4-16
EENS. *See* Expected energy not served (EENS)
Efficiency. *See* Cost and economics; Performance
Ehrenberg studies, 4-89, 4-107
EHV. *See* Extra-high-voltage (EHV)
Eigenvalue analysis, 11-133 to 134
Ejebe studies, 11-125
E-Kady studies, 12-43
El-Abiad, Gupta and, studies, 11-44, 11-48
El-Abiad and Nagappan studies, 11-44, 11-48
El-Din, Vaahedi and, studies, 12-44
El-Din studies, 12-44
Electrical bushings, 3-171 to 184
Electrical insulation, oil function, 3-207
Electrical interlock paralleling, 3-146
Electrical resistance, cables, 4-52 to 54, 4-62 to 63
Electrical signals, 3-276 to 277
Electrical source options, 7-21
Electrical studies, thermal generating plants, 2-26
Electric art furnace series reactors, 3-95
Electric energy storage, 5-29
Electric fences, substation animal barrier, 5-90
Electric generation
  combustion turbines, 2-30 to 31
  distributed utilities, 2-27 to 35
  fuel cells, 2-28
  hydroelectric, 2-1 to 11
  microturbines, 2-28 to 30
  photovoltaics, 1-14 to 19, 2-31
  storage technologies, 1-7 to 13, 2-31
  synchronous machinery, 2-12 to 20
  thermal generating plants, 2-20 to 26
  thermal generation plants, 2-20 to 26
  wind power, 1-1 to 7
Electricity futures (stock market), 13-2, 13-8 to 10
Electricity mercantile associations (EMAs), 13-5 to 6, 13-10
Electric Power Reliability Act of 1967, 6-68
Electric Power Research Institute (EPRI)
  artificial neural network short-term load forecaster, 13-18
  battery storage development, 1-9
  hydroelectric power plants, 2-1
  insulator failure, 4-39
  Slatt-Buckley 500-kV line, 4-178 to 179, 11-96, 11-113
  Utility Substation Communication Initiative, 5-57 to 58, 5-61
  voltage tolerance performance, 15-33
Electric power utilization
  loads, 7-12 to 17
  metering, 7-1 to 11
  motors, 7-18 to 23
Electric shock, 5-95 to 96
Electric vehicles, 1-7, 1-9 to 11, 2-28
Electrodes
  grounding, 15-3 to 4
  response, 10-82 to 83
  transmission system transients, 10-80 to 90
Electrogeometric model (EGM), 5-110 to 121
Electromagnetic fields (EMF)
  community considerations, 5-84 to 85
  Electromagnetic Transients Program (EMTP), 10-36, 10-78, 11-130, 11-134 to 136
  interference, 4-138 to 141
  transients, 10-46
Electromechanical dynamics. *See* Rotational inertia
Electromechanical meters, 7-1 to 2
Electronic meters, 7-3 to 5
Electronics
  active filters, 14-44 to 61
  active filters, type of, 14-50 to 54
  asymmetrical silicon-controlled rectifiers (ASCRs), 14-4
  controlled rectifiers, 14-25 to 37
  gate turn-off thyristors (GTOs), 14-2 to 3
  harmonic-producing loads, 14-45 to 46
  insulated-gate bipolar transistors (IGBTs), 14-6, 14-38 to 43, 44 to 61
  integrated series active filters, 14-54 to 58
  inverters, 14-37 to 43
  line commutated inverters, 14-43
  metal oxide semiconductor field-effect transistor (MOSFET), 14-5 to 6, 14-38
  models, active filters, 14-47 to 50
  MOS-controlled thyristors (MCT), 14-6 to 7
  multilevel inverters, 14-42 to 43
  reverse-conduction thyristors (RCTs), 14-4
  semiconductor devices, 14-1 to 7
  single phase inverters, 14-38 to 39
  three phase inverters, 14-40 to 42
  thyristors, 14-1 to 4, 14-6 to 7
  transistors, 14-4 to 6
  triac, 14-1 to 2
  uncontrolled rectifiers, 14-8 to 25
Eleftherion studies, 3-278
Elgerd studies, 7-16
El-Hawary, Dias and, studies, 12-44, 12-46
Eliasson, Samuelsson and, studies, 11-65
El-Kady studies, 11-125
Elmore studies, 9-39, 9-42
Elongation, transmission lines, 4-97 to 99, 4-103 to 104, 4-118
EMAs. *See* Electricity mercantile associations (EMAs)
Emergency facilities, 5-63
Emergency services, 2-35
Emergency system state, 11-121 to 122
EMF. *See* Electromagnetic fields (EMF)
Empirical curves lightning design, 5-108 to 110
EMS, Energy management system (EMS)
EMTP. *See* Electromagnetic fields
Enclosures
  dry type transformers, 3-64 to 65
  padmounted transformers, 3-63
  proposed substation sites, 5-80
  vault and subsurface transformers, 3-58
  very fast transients (VFT), 10-52
Endangered animals, damage from, 5-89

End fittings, suspension insulators, 4-33
Endrenyi studies, 13-27 to 28, 13-33, 13-56
End user interface satisfaction, 6-76
Energization, accidental, 9-26
Energy
  active filters, 14-49 to 50
  arrester capabilities, 10-100 to 101
  electric storage, 5-29
  flow direction, 7-18
  management, 12-1 to 9
  sound levels, 3-160 to 170
  storage, types of, 1-8
  stored sources, seismic events, 5-133
  transmission systems, 4-4 to 12
  wind generators, 1-4
Energy, Department of. *See* Department of Energy (DOE)
Energy control system simulation, 12-8
Energy management system (EMS)
  general concepts, 12-1 to 9
  monitor architecture, 11-90
  state estimation, 12-33
Energy service companies (ESCOs)
  functions, 13-5 to 6, 13-10
  role, 13-7
  unit commitment, 12-17
English auctions, 12-15
Environment
  impact on thermal generating plants, 2-20
  insulation, 4-27 to 29
  microturbines, 2-30
  motor-enclosure-controller systems, 7-21
  substation automation systems, 5-43
  substation equipment, 5-28
EPRI. *See* Electric Power Research Institute (EPRI)
Equal area criterion, 11-13 to 15
Equilibrium. *See* Stability
Equipment
  external insulation, 4-23
  lightning impulse tests, 3-217 to 219
  oil containment pits, 5-67
  oil-filled, risk, 5-63 to 64
  substation noise source, 5-81 to 83
  switching, substations, 5-23 to 28
  tension stringing, 4-116 to 117
  transmission line construction, 4-44
  very fast transients (VFT), 10-51 to 53

voltage tolerance, 15-32 to 36, 15-39
Equivalent circuits
  distribution transformers, 3-45
  iron core transformers, 3-5 to 6
Equivalent Salt Deposit Density (ESDD), 3-178 to 179, 4-28, 10-98 to 99
Erection technique (transmission line structure), 4-16
Erikson, Taylor and, studies, 11-99
Eriksson, Anderson and, studies, 4-167, 5-111, 10-4, 10-32 to 33
Eriksson studies, 3-277, 10-6, 10-17, 10-32 to 33
ESCOs. *See* Energy service companies (ESCOs)
ESDD. *See* Equivalent Salt Deposit Density (ESDD)
Essential services, thermal generating plants, 2-22
Esters, distribution transformers, 3-33 to 34
Esztergalyos studies, 11-64
Euler's equation, 3-162, 11-17
Europe, 1-9
EV. *See* Electric vehicles
EVA. *See* Early value actuation (EVA)
Evaluations. *See* Analysis and assessment
Evans, Ribbens-Pavella and, studies, 11-43 to 44
Evans and Bergvall studies, 11-7
Even-ordered harmonics. *See* Harmonics
Events. *See also* specific events
  detection logic, 11-86 to 87
  event-based controls, 11-59
  indices exclusions, 13-59
  planning uncertainties, 13-46
  recorders, transformers, 9-10
Everyday tension, 4-104
EV Plus vehicle, 1-10
EWG. *See* Exempt wholesale generators (EWG)
Exact line segment models, 6-28 to 30
Excitation
  control, 11-61
  function of, 4-139
  hydroelectric plants system, 2-5, 2-8
  loss-of-, protection, 9-15 to 16
  system modeling, 11-76
  thermal generating plants system, 2-25
  transient sources, 10-54 to 55
Exempt wholesale generators (EWG), 13-41
Expansion cap, 3-179 to 180

Expected energy not served (EENS), 13-54, 13-56
Explicit methods, assessment, 11-17
Exponential voltage and frequency dependence model, 7-17
Extended function meters, 7-4
External inductance, 69 to 70
External insulators, 4-23 to 29
External transients, 10-44 to 46. *See also* Transient conditions
Extra-high-voltage (EHV)
  corona discharges, 4-137 to 144
  faults, 10-36
  transmission systems, 4-4 to 8, 4-129

F

FA cooling, 3-149 to 150
FACTS. *See* Flexible AC Transmission System (FACTS)
Fahd, Shelbe and, studies, 12-20
Fahd and Shelbe studies, 13-6
Fahrnkopf studies, 10-59
Failures
  instrument transformers, 3-279
  insulators, 4-35 to 41
  rolling sphere method, 5-121
  transformer investigation, 3-242 to 249
  transformers, 13-61 to 62
  transient suppressors, 5-44
Fang and MacDonald studies, 11-66
Fans, 1-18, 3-164
Farad, 10-66
Faraday's Law, 3-117 to 118
Fardanesh studies, 5-31
Far field locations, sound measurements, 3-169
Farmer, Agrawal and, studies, 11-130 to 139
Farmer, Anderson and, studies, 11-127 to 138
Farrish, Cookson and, studies, 5-3
Fast decoupled power flow method, 8-42
Fast power reduction (fast valving), 11-62
Fast valving (fast power reduction), 11-62, 11-77
Fatigue damage (SSR), 11-137 to 138
Fatigue life expenditure (FLE), 11-135 to 138
Faults
  analysis, 8-34
  busbar, 12-55
  clearing-time, 11-63 to 64
  currents, 2-32 to 33, 3-14
  fast fault clearing, 11-63 to 64
  ground, 9-22

Index

I-13

instrument transformer failure, 3-280
insulation stress, 4-25
locations in insulated cables, 4-60 to 62
low current winding, 9-3
mechanical protection, 9-4
monitoring, 5-32 to 33
overhead distribution, 13-57
phase, 9-22
power systems analysis, 8-44 to 59
reactor applications, 3-99
sensing links, 3-49, 3-50
short circuit analysis, 6-66
stator, differential protection, 9-11 to 14
switching surges, 10-39
system, 9-21 to 22
thermal protection methods, 9-4 to 6
through, 9-7
transformer types, 9-1 to 2
turn-to-turn, 9-7
types, 8-45 to 59
voltage sags, 15-36 to 36
Fault-sensing links, 3-49
FBSOA. *See* Forward-bias safe operating area (FBSOA)
Federal Energy Regulatory Commission (FERC), 13-60
Feedback controls. *See* Closed-loop controls
Feeders, staged testing, 11-80
Feedforward controls. *See* Open-loop controls
Fences, animal damage, 5-90
Fergestad and Herkisen studies, 10-65
Ferranti rise, 3-214 to 215
Ferroresonance
  dielectric withstand testing, 3-215
  distribution transformers, 3-40 to 41
  voltage transformers, 3-114 to 115
Feser and Pigini studies, 10-98
Feser studies, 3-278
Fiacco and McCormick studies, 12-42
Fiber optics, 3-151 to 152, 9-38
Fibrillation, 5-95 to 96
Field circuit protection, 9-14
Field devices
  distribution automation, 6-70 to 72
  sound measurements, 3-169 to 170
  substation automation, 5-55
Field-oriented dynamic response, 14-42
Fifteenth harmonics. *See* Harmonics
Fifth harmonics. *See* Harmonics
Filter reactors, 3-94
Filters
  antialiasing, 9-45
  harmonics, 3-94, 5-29 to 30
  Kalman, 9-55 to 57
  low pass filters (LPF), 9-46
  reactors, 3-94
Final sag/tension, 4-105
Finite difference approximation, 3-163
Finite element method model, 10-60
Fink and Beatty studies, 4-97 to 99, 10-73
Fink and Carlsen studies, 12-58
Fink studies, 12-55
Finton studies, 11-64
Fire protection. *See also* Protection
  hydrogenerators, 2-7
  National Fire Protection Association, 1-17
  oil containment, 5-67 to 69
  substation buildings, 5-38
  substation design, 5-85
  substation operations, 5-87
  substations, 5-134 to 139
  thermal generating plants, 2-22
"First house protection," 142 to 143
First strokes, lightning, 10-2 to 3
Fixed angles lightning design, 5-108
Fixed tap transformers, 12-47
Flanges, bushings, 3-179
Flashover voltage
  air insulation characteristics, 10-95 to 99
  backflashover, 4-168
  critical flashover (CFO), 5-114, 10-7, 10-8
  insulation stress, 4-28
  insulator failure, 4-36 to 41
  probability of (pfo), 10-93 to 95
FLE. *See* Fatigue life expenditure (FLE)
Flecther-Reeves method, 12-42
Fletcher-Powell method, 12-42
Flexible AC Transmission System (FACTS)
  controllers, 5-30 to 32
  high voltage power electronic substations, 5-29
  power system oscillations, 11-8, 11-32
  reactive power compensators, 4-178
  transient energy function (TEF), 11-45
Flickering. *See also* Grounding
  application, 15-13 to 15
  curves, 15-43 to 46
  general concepts, 15-42 to 49
  meter methodology, 15-47 to 49
  monitoring, 15-56

static var compensator operation, 4-173
Floret, Berger and, studies, 5-123
Florida, 10-4
Florida Solar Energy Center, 1-17
Flow blocking systems, 5-73
Flow rate, hydroelectric plants, 2-2 to 3
Fluctuations
  power quality, 15-42 to 49
  protection, 9-2, 9-4
Fluid flow, monitoring, 15-56
Fluorescent lamps, 6-4, 6-9
Flux
  current transformers, 3-120 to 121, 3-125
  geomagnetically induced currents (GIC), 4-154 to 155
  iron/steel core transformers, 3-4 to 7
  modeling, 10-64 to 66
  power transformers, 3-29
  transmission line conductors, 65 to 67
Flywheel energy storage, 1-8, 2-31
FOA cooling, 3-149 to 150
Foata studies, 3-288
Fong studies, 13-29
"Fooler" controls, paralleling, 3-143, 3-145
Forbes SVS, 11-65
Forced circulation, power transformers, 3-13
Forced outage models, 13-32
Ford Motor Company, 1-9, 1-10
Form A pulse circuit, 7-9
Form C pulse circuit, 7-9
Formulations. *See* Algorithms; Models
Forrest studies, 13-31
Fortescue studies, 8-14
Fortson, Jones and, studies, 10-40
Forward-bias safe operating area (FBSOA), 14-5
Fossil fuel plants, 11-78, 11-140
Fouad, Anderson and, studies, 11-43, 11-45
Fouad, Ni and, studies, 11-44
Fouad and Vittal studies, 11-43 to 45, 11-48 to 52, 11-124
Fouad studies, 11-43 to 52
Fourier series. *See also* Algorithms
  active filters, 14-53
  antialiasing filters, 9-45
  digital multifunction relays, 9-28
  Discrete Fourier Transform (DTF), 9-28, 9-48, 9-51 to 52
  fifth harmonic measurement, 9-3
  Fourier Transform vs-Wavelet Transform, 9-57 to 58

inductance models, 10-65
phasors, 9-48
transformer losses, 3-44
Four secondary bushings, 3-40
Fox studies, 3-278
Framing, transmission line construction, 4-45
Francis and Kaplan turbine, 2-8
Franklin studies, 10-54
Frequencies
  converters, 5-29 to 30
  effect, transmission lines, 4-64
  resonance characteristics, 10-61 to 63
  scanning technique (SSR), 11-131
  stability, 11-6 to 7
  turbine-generators, abnormal operation, 9-23 to 24
  variable frequency drive (VFD), 14-21
Frequency-domain application
  active filters, 14-53
  harmonics, 15-21
  monitoring, 11-86
  transient voltage response, 10-60
Frischknecht, Keller and, studies, 10-81
Frost damage, 10-85
Fruth and Fuhr studies, 3-276
FTIR sensing. *See* Infrared spectroscopy (FTIR) sensing
Fu, McCalley and, studies, 9-42
Fuel cells
  distributed utilities, 2-28
  gas sensors, 3-275
  non-conventional power generation, 1-10 to 13
  sensors, 3-282
Fuel cost minimization, 12-39, 12-41, 12-45 to 46
Fuhr, Fruth and, studies, 3-276
Fujimoto studies, 10-46 to 47, 10-49 to 51
Fujita and Akagi studies, 14-50 to 54
Full graphics user interface, 6-69 to 70
Full-tension stringing, 4-43, 4-115 to 116
Fully insulated connections, 3-39
Fumi, Cinieri and, studies, 10-29
Furumasu and Hasibar studies, 11-65
Fuses
  capacitor units, 4-176
  current limited, 3-49, 3-50
  distribution circuit protective device, 9-33
  power, 5-25 to 26
  ratings, 5-25
  transformer fault protection method, 9-2

Futures, electricity (stock market), 13-2, 13-8 to 10
Fuzzy controller, self-organizing neural, (SONFC), 11-68

G

GA. *See* Genetic algorithm
GaAs (gallium arsenide), 1-15
Gail and Nielsen studies, 9-59
Gain function, 10-63
Gain selection, system oscillation mitigation, 11-31
Galiana studies, 12-57
Gallium arsenide cells (GaAs), 1-15
Galloway studies, 3-45
Gama studies, 11-65
Gao studies, 11-6, 11-8, 11-40
Gaps, 10-94 to 95, 10-99
Garbagnati, Berger and, studies, 10-6
Gary studies, 4-139 to 141
Gas
  circuit breakers, 5-27
  handling systems, 5-14
  monitors, transformers, 9-10
  on-line analysis, 3-282
  sensors, development, 3-275
  transformer installation, 3-238
Gas-in-oil analysis. *See* Dissolved gas analysis (DGA)
Gas-insulated bushings, 3-176, 3-179
Gas-insulated substations (GIS)
  alarms, 5-14
  breakers, 5-5, 5-13
  bus design, 5-81
  cables, 5-9
  controls, 5-11 to 13
  design, 5-15
  disconnect switch operations, 5-7
  equipment installation, 5-28
  grounding, 5-15 to 16
  ground switches, 5-7
  installation considerations, 5-17, 5-28
  interlocks, 5-17
  maintenance, 5-17
  modeling, 10-46
  monitoring, 5-14, 5-17
  multipoint grounding, 5-16
  overvoltages, 10-44
  power frequency withstand voltage test, 5-17
  principles, 5-2 to 18
  protection, 5-17
  service life, 5-5
  temperatures, 5-14
  testing, 5-16
  very fast transients, 10-41 to 53
  vs-air insulated substation (AIS), 5-2

Gas-to-air bushings, internal transients, 10-48
Gate circuits requirements, 14-28
Gate current injection, 14-27
Gate turn-off thyristors (GTOs), 11-60, 14-2 to 3
Gaulard and Gibbs, 3-31
Gauss, 5-84
Gaussian distribution, 10-93, 10-99
Gauss's law, 4-76
Gay, Tarrant and, studies, 1-16
Gear and Adams studies, 10-61
GENCOs. *See* Generation companies (GENCOs)
General Electric Company
  flicker curves, 15-43
  load modeling, 11-80
  nameplate ratings, 3-149
  prime mover modeling, 11-78
  thyristor controlled series compensators (TCSC), 4-178
Generalized reduced gradient method (OPF), 12-42
General Motors, 1-9
Generated acoustic power density, 4-141
Generation, children per, 12-24
Generation companies (GENCOs), 12-11 to 22, 13-3 to 12
Generation stations, transmission systems, 4-3
Generators
  backup protection, 9-21 to 22
  classifications, 2-32
  coherent machine equivalent, 11-81
  diesel, 1-3, 2-21
  digital multifunction protection, 9-28 to 29
  dynamic interaction, power system, 11-126 to 143
  equal area criterion, 11-13 to 15
  excitation control, 11-61
  field circuit protection, 9-14
  grounding, 9-14
  hunting mode, 11-31
  hydraulic, 9-11
  hydroelectric, 2-5, 2-7
  induction, 2-4
  infinite bus, 11-80
  lumped inertia equivalent, 11-81
  must-run generation, 11-40
  power angle relationship, 11-12 to 15
  power factor, 11-40
  power system models, 11-73 to 76
  prime mover modeling, 11-76 to 78
  protection for, 9-8
  reactive overload capability, 11-40

Index

swing curves, 11-11 to 12
synchronous, 2-12 to 13, 9-11 to 29
transient energy function (TEF) effect, 11-49
transient stability, 11-11 to 18
tripping, 9-27, 11-60, 11-62 to 63
turbine-generators, 9-23 to 24
wind, 1-1 to 7
Generic dynamic load-voltage, 11-34 to 36
Generic Object Models for Substation and Feeder Equipment (GOMSFE), 5-61
Genetic algorithm (GA)
  accuracy, 12-22
  basic principles, 12-20
  economic dispatch, 12-14
  price-based unit commitment, 12-20 to 25
Geomagnetically induced currents (GIC), 4-150 to 165, 9-5
Geometric Mean Distances (GMD), 6-17, 6-20 to 24
Germany
  compressed air energy storage, 1-8
  load control, 13-43
  wind power installed capacity, *1-2*
  zinc air batteries, 1-10
Gerube, Hajagos and, studies, 11-61
GFD. *See* Ground flash density (GFD or Ng)
GHCC. *See* Global Hydrology and Climate Center (GHCC)
Ghosh, Padiyar and, studies, 11-44
GIC. *See* Geomagnetically induced currents (GIC)
Gill studies, 12-48 to 49
Gilman and Whitehead studies, 5-111, 5-114 to 115
Girgis, Agee and, studies, 11-99
GIS. *See* Gas insulated substations (GIS)
Glass insulators, 4-30 to 32
Glavitsch, Schynder and, studies, 12-46
Glavitsch studies, 12-43
Glaze ice. *See* Icing
Gless studies, 11-44
Global Atmospherics, Inc-, 4-166, 10-4
Global Hydrology and Climate Center (GHCC), 4-166
Glodjo studies, 3-275
Glover and Sarma studies, 6-25, 8-29
Glow (corona modes), 4-133, 4-136 to 137, 4-145
GMD. *See* Geometric Mean Distances (GMD)
Goldberg studies, 12-20
Golde studies, 5-123

GOMSFE. *See* Generic Object Models for Substation and Feeder Equipment (GOMSFE)
Goosen studies, 3-184
Gorev studies, 11-44
Gottlieb studies, 14-38
Governor system, 2-8
GPR. *See* Ground potential rise (GPR)
Grady studies, 14-50, 14-53
Graetz bridge converters, 5-29
Grainger and Stevenson studies, 8-29
Grand Coulee, 11-63
Graphic Method for Sag Tension Calculations for ACSR and Other Conductors, 4-102 to 103
Great Barrington system, 3-31
Great Geomagnetic Storm, March 1989, 4-151
Greenhouse gases, 5-3
Green's function, 10-24 to 26
Greenwood studies, 10-40, 10-55, 10-69, 10-75
Grenier and Metes studies, 11-84
GRG. *See* Generalized reduced gradient method (OPF)
Grid resistance, 5-101 to 102
Griffin studies, 3-272, 3-274
Grips, 4-116
Gross, Chowdhuri and, studies, 10-27 to 30
Gross and Hall studies, 11-134
Gross studies, 7-16, 8-21, 8-29, 8-31
Ground, stator-to, fault protection, 9-14
Grounded systems, thermal generating plants, 2-21 to 22, 2-25
Ground faults, 9-22
Ground flash density (GFD or Ng)
  parameters, 5-108, 10-4 to 5
  protection, 4-165 to 166
Ground grid, electromagnetic fields (EMF), 5-84 to 85
Grounding. *See also* Wiring
  applications, 15-12 to 15
  electrode properties, 10-84 to 85
  gas-insulated substations, 5-15 to 16
  general concepts, 10-80 to 81
  generator, 9-14
  harmonics, 15-11 to 12
  high speed switches, 5-25
  insulated, 15-7 to 8
  leads, 10-49 to 50
  loops, 15-8
  low-resistance, 2-22
  multilayers of soil, 10-87 to 88

multiple neutral to ground bonds, 15-9
neutral conductor, 15-10, 15-10 to 12
office equipment, 15-10 to 11
power quality, 15-2 to 15
problems, 15-6 to 12
reasons for, 15-4 to 6
rods, 15-9 to 10
safety ground, missing, 15-8
standards, 15-2 to 4
substation design, 5-85, 5-92-104
transmission line construction, 4-45
transmission system transients, 10-80 to 90
Ground potential rise (GPR), 5-96, 5-102
Ground straps, 10-49 to 50
Ground switches
  gas-insulated substations, 5-7
Grove, William, 1-10, 2-28
Grover studies, 10-64 to 65
Guertin studies, 13-29
Guha studies, 1-15
Guided liquid flow, 3-20
Guide to the Installation of Overhead Transmission Line Conductors, 4-115
Gulf of Mexico, states along, 10-4
Gupta and El-Abiad studies, 11-44, 11-48
Gurgis and Brown studies, 9-56
Guth, Stemmler and, studies, 11-66
Gutierrez studies, 10-59, 10-77 to 78
Gyugyi studies, 14-44

H

Habetler, Cash and, studies, 14-37
Hadjsaid studies, 12-56
Hajagos and Gerube studies, 11-61
Haldar and Liew studies, 10-26
Half-cycle saturation, 4-152, 4-159
Hall, Gross and, studies, 11-134
Hall and Hodges studies, 11-127
Hall-effect devices, 3-126
Halogenated fluids, 3-33
Haner studies, 11-63, 11-67
Hannett and Khan studies, 11-78
Happ and Vierath studies, 12-39
Harder and Marter studies, 9-54
Harlow studies, 5-61
Harmonic-producing loads, 14-45 to 46
Harmonics
  active, compensation, 14-25
  active filters, 14-44 to 61
  analyzing, 15-18
  arcing loads, 15-16

capacitor banks, 15-22
capturing, 6-70
corona discharges, 4-141
detection method, 14-53 to 54
diode rectifiers, 14-19 to 25
distortion, 14-20, 15-56
distribution transformers, 3-44 to 45
dry type transformers, 3-66 to 67
eleventh (11th), 14-52, *15-17*
even-ordered, 3-44 to 45, 4-152 to 153
fifteenth (15th), 6-70, *15-17*
fifth (5th), 9-3, 14-52, *15-17*
filters, 3-94, 5-29 to 30, 15-23 to 24
geomagnetically induced current, 4-152 to 153
grounding, 15-11 to 12
impedance, 15-22 to 23
limits, 15-19 to 21
loads, 7-17, 14-45 to 46
magnitudes and angles, 15-17
mitigation, 14-22 to 24
noncharacteristic, 14-20
noninteger multiple, 15-17
nontriplen odd, 14-20
odd-ordered, 3-44, 4-152 to 153, 15-11 to 12
parameter estimation algorithms, 9-51
phase multiplication, 14-24 to 25
power quality, 15-16 to 24
saturated magnetic element loads, 15-16
second (2nd), 9-3, *15-17*
semiconductor converter loads, 15-16
seventh (7th), 14-52, *15-17*
standard limits, 14-21
STATCOM, 4-179
stator differential relays, 9-11
steady state conditions, 15-18
third (3rd), 3-132, 4-141, 9-14, *15-17*
thirteenth (13th), 14-52, *15-17*
Total Harmonic Distortion (THD), 6-70, 15-54
transformer protection method, 9-5
transformer sound levels, 3-163 to 164, 3-169 to 170
triplen currents, 15-11 to 12
Harris, Pohlman and, studies, 4-16 to 17
Harris Semiconductor Corporation, 14-7
Hartill studies, 10-59
Hart studies, 9-28
Hasibar, Furumasu and, studies, 11-65

Haslbar studies, 11-64
Hauer and Hunt studies, 11-82 to 83, 11-99
Hauer and Taylor studies, 11-82
Hauer and Vakili studies, 11-87
Hauer studies, 11-27 to 31, 11-59 to 68, 11-84 to 113
Hayman studies, 3-43
Hazards, 5-87, 5-134 to 135
Haznadar studies, 10-46
H-bridge, single-phase, thyristors, 14-28 to 32, 14-38 to 40
Head, hydroelectric plants, 2-2 to 3
Heat. *See* Temperatures
Heat recovery steam generator (HRSG), 11-78
Heat run tests, 3-231 to 233
Heavy frame turbines, 2-30 to 31
Hedman studies, 10-38
Helical windings, 3-21, *3-27*
Helicopters, transmission lines, 4-43
Heriksen, Fergestad and, studies, 10-65
Hessian based methods (OPF), 12-42
Heydt studies, 4-166
Hierarchy control
  hydroelectric plants, 2-9 to 10
High speed grounding switches, 5-25
High speed reclosing (HSR), 9-32, 11-63 to 64
High-temperature hydrocarbons (HTHCs), 3-33
High-temperature superconducting (HTS) transformer, 3-29 to 30
High-temperature superconductors, 1-4, 1-8
High-voltage direct current (HVDC) systems
  actuators, 11-60 to 61
  closed-loop controls, 11-59
  converter controls, 11-140
  link supplementary controls, 11-67
  misoperation, 11-60
  power electronics, 5-29 to 30
  reactors, 3-94 to 95
  system oscillation mitigation, 11-28 to 29, 11-32 to 33
  testing, 11-113
  thyristors, 14-34 to 35
  transmission systems, 4-8, 11-23
High-voltage generator (powerformer), 3-29
High-voltage lines, 4-137 to 144
  corona performance, 4-144 to 149
  impedance, 6-16
  transmission systems, 4-6 to 8, 4-129
High-voltage power electronic substations, 5-29 to 39

High-voltages. *See* Voltages
Highway information signs, PV-powered, 1-19
Hill, Bergen and, studies, 11-44, 11-52
Hiskens studies, 11-44
History
  direct stability methods, 11-42 to 45
  distribution transformers, 3-30 to 31, 3-51
  electric power industry, 13-51 to 52
  gas sensors, 3-30 to 31
  genetic algorithms, 12-20
  high-voltage lines, 4-129
  hydroelectric power, 2-1
  lightning study, 4-165
  power system oscillations, 11-20 to 23
  reactors, 3-81
  security, power systems, 11-120 to 121
  spacer replacement, 4-46 to 47
  stability problems, 11-7 to 8
  substation seismic requirements, 5-124 to 125
  Supervisory Control And Data Acquisition (SCADA), 6-68
  transient response, 10-58 to 59
  wind turbines, 1-1 to 2
Hobson, Stott and, studies, 12-43
Hodges, Hall and, studies, 11-127
Hoffman studies, 1-15
Holiday peak forecasting, 13-20 to 21
Holt studies, 5-30
Holtz, Krah and, studies, 14-50
Holtz studies, 14-37, 14-39
Homes, single family. *See* Communities
Honda, 1-10
Horizontal shaft propeller turbine unit, *2-3, 2-4*
Horowitz and Phadke studies, 9-39
Hospitals, 2-35
Host processors, 5-56 to 57, 6-69 to 70
Ho studies, 13-18
Hot-loads, 9-33. *See also* Loads
Hot resistance, 3-232 to 233
Hot start (OPF), 12-48
Hottest-spot temperature. *See also* Temperatures
  distribution transformers, 3-48
  heat run tests, 3-231
  loading power transformers, 3-151 to 152
  transformer protection method, 9-4 to 5
  vaults and subsurface enclosures, 3-58

Index

HRSG. *See* Heat recovery steam generator (HRSG)
HSR. *See* High speed reclosing (HSR)
HTHCs (high-temperature hydrocarbons), 3-33
HTS transformer, 3-29 to 30
Huang studies, 1-15
Human current limits, 5-95 to 96
Humidity. *See* Moisture
Hunt, Hauer and, studies, 11-82 to 83, 11-99
Hunting mode, generators, 11-31
HVDC. *See* High-voltage direct current (HVDC) systems
HVDC Intertie. *See* Pacific HVDC Intertie
Hybrid active/passive filters, power conditioning, 14-50 to 51
Hydroelectric lakes, 1-4
Hydroelectric power generation, 2-1 to 11
Hydrogen gas, oil degradation, 3-208
Hydrophobicity, 4-40
Hydro plants, 11-78
Hydro-Quebec, 4-143 to 144, 4-143 to 145, 4-151
Hysteresis loss, 3-225, 10-71

I

Ianna studies, 4-141, 4-148
ICA. *See* Independent contract administrator (ICA)
Icing. *See also* Moisture; Seasons
 ice storms, 13-59
 insulation stress, 4-28
 loading on conductor lines, 4-93 to 96, 4-100 to 102
 weight span, 4-114
IEC. *See* International Electrotechnical Commission (IEC)
IED. *See* Intelligent electronic devices (IED)
IEEE. *See* Institute of Electrical and Electronic Engineers (IEEE)
IGBT. *See* Insulated-gate bipolar transistors (IGBTs)
Igneous rock, geomagnetic storms, 4-151
Ihara, Concordia and, studies, 12-44
III-V technology, 1-15
Iizuka studies, 14-60
Iliceto studies, 12-44
Illinois, 10-4
Imai studies, 11-63
Impact method (SSR), 11-138
Impedance
 arcing switch operation, 3-200
 auxiliary transformers, 2-24

balanced, 8-25 to 26
delta connections, 6-55 to 56
distributed utilities, 2-33
distribution transformers, 3-36
duplex reactors, 3-88
ground electrodes, 10-84 to 85
harmonics, 15-22 to 23
line, modeling, 6-11 to 24
line segments, 6-64
load loss measurements, 3-226 to 227
load tap changing transformers, 3-145
mutual, 6-11 to 12, 6-14, 6-17
neutral grounding reactors, 3-87
paralleling, 3-147 to 148
phase, matrix, 6-14 to 32
phase reactors, 3-85
power transformers, 3-14 to 15, 8-29 to 31
primitive, 6-14
relays, 9-7 to 8, 9-37, 9-65 to 66
reverse transformation, 6-30 to 32
self, 6-11 to 12, 6-14, 6-17
sequence, 6-15 to 19
symmetrical component transformation, 8-18 to 19, 8-21
triangle, 3-47
volts, 3-47
wye connections, 6-54
Imperfect ground, electrodes, 10-84 to 85
Implicit methods, assessment, 11-17
Improved Design Criteria of OHTLs Based on Reliability Concepts, 4-21
Impulse tests, 3-49, 4-61
Independent contract administrator (ICA), 13-7, 13-10 to 11
Independent power producers (IPP), 13-6, 13-41
Independent system operators (ISOs)
 economic dispatch, 12-10
 functions, 13-3 to 4
 load forecasting, 13-18
 Pennsylvania-New Jersey-Maryland Independent System Operator (PJM), 13-24
 planning entities, 13-41
 role, 13-7
Indexes. *See* Indices
India, *1-2*
Indiana, 10-4
Indices, 4-162, 13-29 to 30, 13-58 to 59
Indirect lightning strokes, 10-21 to 34
Indoor substations, fire hazard, 5-135
Induced voltage

distribution transformers, 3-49
indirect lightning strokes, 10-23 to 31
tests, 3-222 to 223
Inducing voltages, 10-22 to 24
Inductance
 cables, 4-53, 4-62 to 63
 external, 69 to 70
 grounding, 10-80
 internal, 67 to 69
 mathematical definition, 10-63 to 64
 models, 10-63 to 64
 single-phase currents, 70 to 72
 three-phase currents, 4-72 to 75, 4-75
 transmission line conductors, 67 to 75
Induction generator effect, 11-129, 11-131
Induction generators, 2-4 to 7
Induction motors
 dynamic model, 11-79
 thermal generating plants, 2-24 to 25
Inertia
 hydroelectric plants, 2-7
 microturbines, 2-29
Infinite bus, 11-80
Information Technology Industry Council (ITIC) studies, 15-32
Information value, 11-100 to 101
Infrared spectroscopy (FTIR) sensing, 3-275
Initial inrush, 9-6 to 7
Initial sag/tension, 4-105
Initial voltage distribution, 10-56 to 57, 10-69
Inoue studies, 11-8
Input signals, 11-29 to 30
Inrush current, power transformers, 3 29
Inspection
 helicopters, 4-47 to 48
 hydroelectric plants, 2-11
 receiving transformers, 3-235 to 237
 structural, 4-47
 tower lines, 4-47
 transmission line structure, 4-16
Instability
 angle, 11-55
 non-oscillatory, 11-4
 synchronous, 11-55
 voltages, 9-41, 9-43
Installation considerations. *See also* Location guidelines
 conductors, 4-115 to 120
 current transformers, 3-126 to 128

distribution automation, 6-76 to 77
distribution transformers, 3-42 to 43
gas-insulated substations, 5-17, 5-28
monitor location, 15-50
submersible transformers, 3-55 to 58
surface operable transformers, 3-55 to 58
temporary use of current transformers, 3-122
vaults, 3-42, 3-52 to 55
Installed wind power capacity, 1-2
Instantaneous overcurrent relays, 9-34
Institute of Electrical and Electronic Engineers (IEEE)
  Conductors, Guide to the Installation of Overhead Transmission Lines, 4-115
  Data Acquisition, Processing and Control System Subcommittee, 5-60 to 61
  dry-type transformers standards list, 3-262 to 263
  295 "Electronics Power Transformers," 3-258
  Green Book, 15-3
  liquid-filled transformers standards list, 3-266 to 269
  Power Engineering Society Substations Committee, 5-60 to 61
  Seismic Design for Substations, Recommended Practice for, 5-124 to 125
  System Dynamic Performance Subcommittee, 9-43
Instructional system simulation, 12-9
Instrument transformers
  description, 3-104 to 129
  metering, 7-10 to 11
  paralleling, 3-147
  substation automation systems, 5-45 to 47
Insulated-gate bipolar transistors (IBGTs), 14-6
Insulated-gate bipolar transistors (IGBTs), 14-6, 14-38 to 43, 14-44 to 61
Insulation
  air, 10-95 to 96
  animal outages, 5-91
  condition testing, 3-212 to 214
  coordination, 10-77, 10-93 to 102
  coordination techniques, 10-93 to 102

instrument transformers systems, 3-110 to 111
lightning, 4-168
loading power transformers, 3-154 to 156
power transformer classes, 3-12 to 13
strength, 10-8
transformer failure, 3-247 to 248, 9-1 to 2
types for transformers, 3-204 to 209
Insulation power factor (Coso)
  insulation condition, 3-213
  power transformer failures, 3-278
Insulators
  air classifications, 3-178 to 179
  animal damage, 5-89
  barrier, 5-14
  bushings, 3-178 to 180
  ceramic, 4-36 to 41
  composite, 4-32 to 35, 4-39 to 42
  contamination, 10-98 to 99
  corona rings, 4-34
  effect of, 10-96
  end fittings, 4-33
  external, 4-23 to 29
  failure, 4-35 to 41
  flashover voltage, 4-36 to 41
  glass, 4-30 to 32
  internal, 4-23
  leakage current, 4-37
  long rod, 4-32
  M&E ratings, 4-29
  moisture, effect on, 4-37
  nonceramic, 4-32 to 35, 4-38
  performance, 4-41 to 42
  pollution, 4-36 to 41, 4-38
  porcelain, 4-30 to 32, 4-35 to 36
  post-type, 4-32, 4-35
  rods, fiberglass-reinforced plastic, 4-34 to 35
  salt spray, 4-28, 4-36
  strings, 4-30 to 31, 5-115
  suspension, 4-33, 4-34, 4-107
  transmission system, 4-23 to 42
  voltage, 10-17
  washing, 4-47
Integrated Object Network, 11-96
Integrated series active filters, 14-54 to 58
Integrating demand meters, 7-6
Intelligent electronic devices (IED)
  analog data sources, 5-48
  controlling capability, 5-52
  distribution automation, 6-71 to 72, 6-75
  Trial Use Recommended Practice for Data Communications Between Intelligent Electronic

Devices and Remote Terminal Units in a Substation, 5-61
Interaction monitoring, 11-85
Interactive recording, 11-86
Interactive Voice Recorders (IVR), 6-73
Interarea mode oscillations, 11-24, 11-32, 11-58
Interchange transaction scheduling, 12-5 to 6
Interconnected star connections, 3-132 to 134
Interface considerations
  distributed utilities, 2-32 to 33
  distribution automation, 6-69 to 70
  end-user satisfaction, 6-76
  RS-232, 6-71
  SCADA-DMS, 6-73
  substation automation, 5-58 to 60
Interference, 5-38
Interim protection (SSR), 11-131
Interior installations, distribution transformers, 3-43
Interior point method (OPF), 12-43 to 44
Interlocks
  gas-insulated substations, 5-17
  oil spills, 5-75
Internal inductance, 67 to 69
Internal insulators, 4-23
Internal transients, 10-43 to 44, 10-47 to 48. *See also* Transient conditions
International Electrotechnical Commission (IEC)
  76-1, 3-268
  functions and standards, 3-253 to 255
  harmonics limits, 15-19 to 21
  TC96, 3-257 to 258
  Technical Committee 14 Power Transformers, 3-261 to 62, 3-264
International System of Units (SI), per unit system, 8-14
Interrupting arcs, sulfur hexaflouride (SF6), 5-2
Interrupting times, circuit breakers, 5-27
Interties. *See* Pacific AC Intertie (PACI); Pacific HVDC Intertie
Interval data metering, 7-3
Inverse-time delay overcurrent relays, 9-33 to 34
Inverters
  distributed utilities, 2-34
  general concepts, 14-37 to 43
  issues, 14-38

Index

line commutated, 14-43
multilevel, 14-42 to 43
single-phase current, 14-38 to 39
three-phase current, 14-40 to 42
Ionization, soil, 10-89
Ions, 4-130 to 137
IOUDC distribution companies, 13-6
IOUTS transmission companies, 13-6
Iowa, 10-4
IPP. *See* Independent power producers (IPP); independent power producers (IPP)
Iravani, Zao and, studies, 5-39
Ireland, 1-2
I^2R losses
   ampacity, 4-66
   current-carrying capacity, 4-66
   distribution transformers, 3-44
   power transformers, 3-15
   thermal response prediction, 3-152
   transformer testing, 3-226, 3-228
Iron cores. *See also* No-load losses
   dry type reactors, 3-81
   transformers, 3-4 to 7
Irwin studies, 8-21
Islanding, distributed utilities, 2-32 to 33
ISO. *See* Independent system operators (ISOs)
Isokeraunic maps, 5-107
Isolated phase, 2-25
Isolation devices, substation animal outages, 5-91
Isoparaffin, 3-206 to 207
ITIC. *See* Information Technology Industry Council (ITIC) studies
IVR. *See* Interactive Voice Recorders (IVR)

J

Jacobian matrix, 8-41 to 42, 12-43, 12-56 to 57
Japan
   active filters applications, 14-44 to 61
   bullet trains, 14-60
   Central Japan Railway Company, 14-60
   lead-acid batteries, 1-9
   power system stability controls, 11-61
   Shintakatsuki substation, 14-60
   Tokaido Shinkansen, 14-60
   Tokyo Electric Power Company (TEPCO), 11-63
   Yamanashi test line, 14-59
Jayaratne, Cooray and, studies, 10-4

Jiang studies, 11-44
Johnson, Younkins and, studies, 11-8
Johnson studies, 10-38, 10-40
Johnsson, Bjorkland and, studies, 5-39
Jones and Fortson studies, 10-40
Jones and Skelton studies, 11-95
Jonnavitihula, Billinton and, studies, 13-56
Jordan, Trinh and, studies, 4-133 to 134
Joselyn studies, 4-157
Joyce studies, 11-130
Juang studies, 11-115
Juette studies, 4-139, 4-141

K

Kahn studies, 13-40
Kakimoto studies, 11-48
Kalman filters, 9-55 to 57
Kamwa studies, 11-31, 11-65, 11-115
Kansas, 1-5
Kaplan turbine, 2-8
Kappenman studies, 4-162 to 163
Karmarkar studies, 12-43
Kassakian studies, 14-38
Kawaguchi studies, 14-52, 14-59
Kawahira studies, 14-44, 14-53
Kayenta studies, 5-39
Kazarlis studies, 12-21
Keeler-Allston line, 11-107
Keller and Frischknecht studies, 10-81, 10-85
Kelly studies, 3-154 to 156
Kennedy, Degeneff and, studies, 10-66, 10-67
Kentucky, 5-32
Keraunic levels, 5-107, 10-5, 10-17
Kersting and Mendive studies, 6-59
Key studies, 15-32
Khan, Hannett and, studies, 11-78
Khotanzad studies, 13-18 to 19
Kilgore studies, 11-129 to 133
Kilowatt hour costs
   compressed air energy storage, 1-8
   various wind turbines, 1-3
   wind power plants, 1-2
Kilowatts vs kilowatthours, defined, 7-5
Kimbark and Legate studies, 10-39
Kimbark studies, 11-43, 11-64, 11-65
K index, 4-162
Kirchoff's current law
   ladder iterative technique, 6-60 to 61
   per unit scaling, 8-9
Kirchoff's voltage law
   impedance, 8-18

   ladder iterative technique, 6-59 to 60
   per unit scaling, 8-9
   short circuit analysis, 6-65
Kithil studies, 10-5
Knight studies, 5-31
Knutsen studies, 11-64
Kogan studies, 10-54
Kojima studies, 11-63
Kolig, Lambrecht and, studies, 11-140
Kondragunta studies, 12-21
Korsunecy studies, 10-89
Koschik studies, 10-36
Kosterev studies, 11-63, 11-68, 11-82 to 83, 11-109
Kotopalli, Chowdhuri and, studies, 10-17
Krah and Holtz studies, 14-50
Kramer and Ruff studies, 3-191
Krasenbrink, and Wu, Takriti, studies, 12-17
Krause studies, 8-14
Krider studies, 4-166
   arresters, distribution transformers, 3-50
   bushings, 3-182
   impulse tests, 3-215 to 220
   overvoltages, insulation stress, 4-27
   transmission system protection, 4-165 to 168
Krishnayya studies, 4-129, 5-38
Kron's reduction method
   distribution system modeling, 6-14 to 25
   optimal power flow, 12-39
   transient voltage reduced order model, 10-78
Kuhn-Tucker studies, 12-42
Kulicke, Dalstein and, studies, 9-61
Kumar studies, 13-6
Kundur, Lee and, studies, 11-24, 11-61 to 62
Kundur studies, 2-7 to 8, 8-14
   excitation control, 11-61 to 62
   power system classifications, 11-5 to 6, 11-23 to 31
   power system model, 11-45
   power system stability, 11-3, 11-8
   shunt dynamic braking, 11-64
   transmission devices, 11-80
Kung, Raczkowski and, studies, 11-141 to 142
Kung and de laRosa studies, 11-141
KVA
   measurements, 7-10 to 11
   ratings defined, 3-12
KWh. *See* Kilowatt hour costs
Kyrene substation, 11-107

L

LabVIEW, 11-95, 11-103
Lachman studies, 3-284
Lacroix and Charbonneau studies, 4-129
Ladder iterative technique, 6-59 to 64
Lagrangian equations
  economic dispatch calculations (EDC), 12-12 to 13
  optimal power flow, 12-39, 12-42
  unit commitment, 12-20
Lamarre, Duvall and, studies, 3-278
Lambert, McNutt and, studies, 10-93
Lambert studies, 10-38, 10-95
Lambrecht and Kolig studies, 11-140
Lambton Generating Station, 11-140 to 141
Lammasniemi studies, 1-15
Lammeraner and Stafl studies, 10-69 to 70
Lamp flicker. *See also* Grounding
  application, 15-13 to 15
  flicker curves, 15-43 to 46
  general concepts, 15-42 to 49
  meter methodology, 15-47 to 49
  monitoring, 15-56
  static var compensator operation, 4-173
LAN. *See* Local area networks (LAN)
Land rights, wind turbines, 1-5 to 7
Laplace transform, 10-24 to 26, 10-62, 10-86
Large disturbances. *See* Disturbances
Larsen, Piwko and, studies, 11-30
Larsen and Chow studies, 11-30, 11-65
Larsen and Swann studies, 11-30 to 32
Latching devices, 5-52
Laughton, Shen and, studies, 12-43
Laurentian shield, 4-151
Layer windings, 3-21, *3-26*
Lead-acid batteries, 1-9
Leader, lightning, 4-166 to 167
Leads, ground, 10-49 to 50
Leakage current, insulator failure, 4-37
Leakage resistance, iron core transformers, 3-8 to 9
Least cost planning, 13-48 to 50
Least squares fitting, 9-51 to 2
Lee, Biegelmeier and, studies, 5-96
Lee, Dalziel and, studies, 5-95 to 96
Lee and Kundur studies, 11-24, 11-61 to 62
Lee studies, 5-112, 5-115, 11-24, 11-141, 13-18
Legate, Kimbark and, studies, 10-39
Lemke studies, 3-276
Length, line (cables)
  catenary, 4-90 to 91
  shunt admittance, 4-30
  transmission distances, 11-55
  transmission line configurations, 9-32 to 33
Lenz law, 9-18 to 19
De Leon and Semlyen studies, 10-59 to 60, 10-71
De Leon models, 10-77
LeReverend, Anderson and, studies, 9-42
Level spans, catenary cables, 4-89 to 90
Levine studies, 11-28 to 30
LFC. *See* Load frequency control (LFC)
Dy Liacco studies, 11-121
Liebfried studies, 3-284
Liew, Haldar and, studies, 10-26
Life cycle, batteries, 1-9 to 10
Life expectancy, power transformers, 3-12
Lighting
  circuits, thermal generating plants, 2-22
  proposed substation sites, 5-80
  substation safety/security, 5-85
Lightning
  action integral, 10-6
  charging mechanism, 10-2 to 4
  connections, 5-10
  current peak value, 10-5 to 6
  dart leaders, 10-4
  direct lightning strokes, 10-8 to 20
  first strokes, 10-2 to 3
  grounding, 5-104 to 123
  indirect lightning strokes, 10-21 to 34
  outages, 10-17 to 20
  restrike phenomena, 9-70
  return strokes, 10-3
  shielded lines, 10-9 to 11
  stepped leaders, 10-2, 10-13 to 14, 10-21
  strike incidence, 10-6
  stroke characteristics, 10-1 to 7
  unshielded lines, 10-9 to 11
  utility reliability problems, 13-62
"Likely contingencies," 11-121 to 122
Linck studies, 5-121
Lindqvist studies, 12-42
Linear arrangement, load tap changers, 3-188
Linear couplers, current transformers, 3-125
Linear programming
  optimal power flow, 12-43, 12-49
  state estimation, 12-32
Linear transformations, 9-49 to 50
Linear voltage and frequency dependence model, 7-16
Line-commutated inverters
  distributed utilities, 2-34
  general concepts, 14-43
Line conductors, selection, 4-148
Line construction, 4-9 to 12, 4-42 to 48
Line disturbance analyzers, 15-50 to 54
Line drop
  control circuit, 6-33
  modeling, 6-35-36
  Resistive and Reactive Compensation control settings, 3-79 to 80, 3-139 to 143
  switching surges, 10-39
Line energization, insulation stress, 4-25
Line parameters, transmission system, 4-62 to 82
Line protection algorithms, 9-53
Lines. *See* Cables
Line segments
  impedances, 6-64
  models, 6-28 to 32
Line structures, transmission system, 4-13 to 22
Line traps, electric and magnetic fields (EMF), 5-84 to 85
Link choke, direct current, 14-23 to 24
Linnet studies, 6-14
Lipo, Nowotny and, studies, 14-40
Liquid-filled transformers
  dry-type comparison, 3-36
  loading, 3-48
  standards, 3-256, 3-266-269
Liquid-immersed transformers, 3-269-288. *See also* Submersible transformers
Liquid level indicator, 3-25
Liquid nitrogen, 3-30
Liquid preservation systems, 3-26 to 29
Liquid temperature indicator, 3-26
Lithium batteries, 1-10
Liu studies, 12-49
Live-front padmounted transformers, 3-59, 3-60 to 61
Live-tank circuit breakers, 5-27
Load break disconnect switches, 5-7, 5-24 to 25
Load bus (P-Q bus), 8-39
Load current failures, 3-278
Load curtailment indices, 13-30
Load flow
  analysis, 11-38 to 39
  problem, 8-34 to 43

Index

thermal generating plants, 2-26
Load frequency control (LFC), 12-4 to 5
Loading recommendations
 distribution transformers, 3-48
 power transformers, 3-149 to 159, 3-158 to 159
 tap changers, 3-184 to 204
 transmission lines, 4-17 to
Load losses
 distribution transformers, 3-44, 3-48
 iron core transformers, 3-9
 power transformers, 3-15
 transformer testing, 3-226 to 227
Loads
 arcing, 15-16
 characteristics, 7-12 to 13
 cold, 7-17, 9-33
 composite, 7-13 to 17
 continuous, 9-33
 curves, 7-14 to 16
 delta connections, 6-56
 demand factor, 7-13
 frequency control, 12-4 to 5
 harmonic-producing, 14-45, 14-45 to 46
 hot, 9-33
 load factor, 7-13
 management, 12-6
 models, 6-53 to 56, 11-78 to 90
 modulation, 11-64 to 65
 noise, 3-163 to 164
 optimal power flow models, 12-44 to 46
 regulating voltage, 3-139
 rejection, 4-25
 saturated magnetic elements, 15-16
 semiconductor converter, 15-16
 shedding, 9-41, 11-40, 11-67
 short term, 13-16 to 25
 thermostatically controlled, modeling, 6-9
 tripping, 11-64 to 65
 utilization, 7-12 to 17
 voltage stability, 11-34 to 41
 wye connections, 6-55
Load shedding, 9-41, 11-40, 11-67
Load tap changers (LTCs)
 control circuits, 6-33
 extended control, 3-142-143
 immersed in oil, 3-184 to 204
 on-line maintenance, 3-284 to 288
 paralleling transformers, 3-135 to 159
 refurbishment/replacement, 3-203
 step-voltage regulators, 3-68, 3-74 to 77, 142 to 143
 transformer failure, 9-2

transformer installation, 3-239
transformer protection method, 9-6
turns ratio adjustment, 3-23 to 25
Local area networks (LAN), 5-57 to 58, 6-69
Local backup, EHV systems, 9-32
Local control cabinet (LCC), 5-11 to 13, 5-133
Local plant mode oscillations, 11-23
Location guidelines
 distribution transformers, 3-42 to 43
 hydroelectric plants, 2-2
 power transformers, 3-12
 wind turbines, 1-4 to 5, 1-5 to 6
Lock-in-step paralleling, 3-146
Lockout relays, 9-10
Loeb studies, 4-129
Lof studies, 11-40
Logs, daily
 substation automation, 5-59
 transformers, 9-9 to 10
LOLE. *See* Loss of load expectation (LOLE)
LOLP, loss-of-load probability
Long rod insulators, 4-32
Loop-style padmounted transformers, 3-58
Lootsma studies, 12-42
Lorenz and Divah studies, 14-42
Loss compensation meters, 7-3
Losses
 ampacity, 4-66
 cables, 4-55
 corona, 4-137 to 138, 4-143 to 145
 current-carrying capacity, 4-66
 distribution transformers, 3-44, 3-48
 iron core transformers, 3-9
 power transformers, 3-15
 steady state model, 10-69
 substation converters/controllers, 5-36
 thermal response prediction, 3-152
 total hysteresis, 10-71
 transformer testing, 3-224 to 225, 3-225, 3-226, 3-226 to 227, 3-228
Loss-of-excitation protection, 9-15 to 16
"Loss of life," power transformers, 3-12, 3-15, 3-158
Loss-of-load expectation (LOLE), 13-29, 13-54
Loss-of-load probability, 13-42
Louisiana, 10-4
Low current winding faults, 9-3
Lower end shield bushings, 3-181
Lower position, regulators, 6-34

Low-frequency tests
 bushings, 3-182
 dielectric transformer testing, 3-221 to 223
Low pass filters (LPF), 9-46
Low-resistance grounding, thermal generating plants, 2-22
LPF. *See* Low pass filters (LPF)
Lummis, Pohlman and, studies, 4-16 to 17
Lumped inertia equivalent, 11-81
Lumped parameter model, 10-59 to 61, 10-66 to 67, 10-78
Lu studies, 13-18
Lyapunov studies, 11-42 to 44, 11-47

M

MacGorman studies, 4-165, 10-5
Mackay and Berk studies, 13-31
MacLaurin expansion, 4-89
Madzarevic studies, 10-39
MAE. *See* Mean absolute error (MAE)
Magentic energy storage, wind turbines, 1-4, 1-8
Magnetic Blowout, 5-51
Magnetic circuit heating, failure, 3-248
Magnetic flux
 current transformers, 3-120 to 121, 3-125
 iron/steel core transformers, 3-4 to 7
 modeling, 10-64 to 66
 power transformers, 3-29
 transmission line conductors, 65 to 67
Magnetic storms, 4-150 to 165
Magnetizing inrush, 9-6 to 7, 9-10
Magnetomotiveforce, 3-106
Magneto-telluric climatology, 4-157 to 161
Magnusson studies, 11-44
Maifeld and Shelbe studies, 12-21
MAIFI. *See* Momentary average interruption frequency index (MAIFI)
Maine, 10-4
Maintenance
 air insulated substations, 5-19 to 23
 composite insulators, 4-40 to 42
 gas-insulated substations, 5-17
 immersed transformers, 3-270
 load tap changers, 3-202 to 203
 substations, 5-87
 thermal generating plants, 2-26
 transmission line conductors, 4-64 to 66
 transmission lines, 4-42 to 48

transmission line structure, 4-16
transmission system, 4-42 to 48
vault and subsurface transformers, 3-58
Manholes
    cable installation, 4-60
    flow blocking systems, 5-73
Manitoba Hydro. *See* Canada
Mann and Morrison studies, 9-44
Mansour studies, 11-34, 11-38, 11-40, 11-44, 11-52, 11-68
MAPE. *See* Mean absolute percentage error (MAPE)
Maria studies, 12-42
Market, stock, 13-2, 13-8 to 10, 13-46
Marks studies, 10-36
Marter, Harder and, studies, 9-54
Martin studies, 11-99
MAS. *See* Multiple Address Radio System (MAS)
Master/follower (slave) paralleling, 3-146
Masts
    shielding, 5-109 to 110
    stroke current magnitude, 5-106 to 107
Matlab, 11-95
Matrix definitions, models, 6-46 to 57
Mauthe studies, 5-3
Maxwell's equations, 5-84, 10-71, 11-21
Maynard studies, 4-162
McCalley, Shelbe and, studies, 13-6
McCalley and Fu studies, 9-42
McCann, Beck, Wagner studies, 5-109
McCaulley studies, 12-55 to 56
McClelland, Rumelhart and, studies, 13-17
McCormick, Fiacco and, studies, 12-42
McDonald, Fang and, studies, 11-66
McDonald and Saxton studies, 5-61
MCFC (molten carbonate fuel cell), 1-12, 1-13, 2-28, 2-29
McGranaghan studies, 15-24
McNutt
    system study models, 10-77
McNutt and Lambert studies, 10-93
McNutt studies, 3-155 to 156, 10-61 to 62, 10-75
MCO, arrester capabilities, 10-100
MCT. *See* MOS-controlled thyristors (MCT)
McWhirter studies, 10-59
M&E. *See* Mechanical and electrical (M&E) value
Meadow Brook Substation, 4-154 to 155
Mean absolute error (MAE), 13-24 to 25

Mean absolute percentage error (MAPE), 13-22 to 23
Mean time to repair (MTTR), 13-56
Measurement based load modeling, 6-3
Measurements. *See also* Per unit system
    corona, 4-138
    impedance, load loss, 3-226 to 227
    kVA, 7-10 to 11
    metering, 7-1 to 11
    relative capacitance, 3-280 to 282
    sound levels, 3-160 to 170, 3-233
    telemetered, 12-27, 12-34
    Transformer Loss Measurement Guide, 3-227
    Winding Resistance Measurements, 3-226
Mechanical and electrical (M&E) value, 4-29
Mechanical life, load tap changers, 3-195
Mechanical protection, 9-4
Mechanical stress, insulation, 4-29
Mechanical tests, bushings, 3-183
Medium power transformers, 3-149 to 150
Meggar readings, 3-235, 4-61
Mehraban studies, 5-31
Meidensha Corporation, 14-61
De Mello and Concordia studies, 11-24
Melting curves, 9-33
Mendive, Kersting and, studies, 6-59
Menemenlis studies, 10-98
Meppelink studies, 10-44, 10-51
Mercury
    monitoring substation automation systems, 5-50
    vapor lamps, modeling, 6-9
Merlin studies, 12-43
Merrill studies, 13-12
Mesh voltage, 5-96, 5-98 to 100
Metallic sheath, cables, 4-51
Metallic wires (pilot wires), 9-39
Metal oxide arresters, 10-46, 10-77, 10-100
Metal oxide semiconductor field-effect transistor (MOSFET), 14-5 to 6, 14-38
Metal oxide varistors (MOV)
    series capacitor banks, 4-176 to 177, 4-179
    substation automation systems, 5-43 to 44
Metering
    accuracy and stability, 7-4
    current transformers, 3-129, 5-46
    digital relays, 9-44
    net, 1-3

revenue accuracy, 5-45
types of, 7-1 to 11
Metes, Grenier and, studies, 11-84
Mexico, 5-129
Micrometerology, transmission line system, 4-18, 4-20
Microprocessor type, 9-3 to 4
Microturbines, 2-28 to 30
Microwave channels, 9-38
Miles studies, 10-59
Minnesota, 11-65
Miri and Binder studies, 10-46 to 47
Mississippi, 10-4
Missouri, 10-4
Mitigation
    harmonics, 14-22 to 24
    lightning, 4-168
    power flow control reactor overload, 3-89 to 91
    power system oscillations, 11-28 to 32
    substation animal damage, 5-90
    transient voltages, 10-55
    voltage sags, power quality, 15-36 to 40
    voltage stability problems, 11-40 to 41
Mixed single-and three-phase loads, 3-38
Mizuno studies, 10-98
MLP. *See* Multilayer perceptron (MLP)
Mobile equipment, substation oil source risk, 5-63
MOD. *See* Mode of disturbance (MOD)
Modal damping tests, 11-138 to 139
Models. *See also* Algorithms; Analysis and assessment
    active filters, 14-47 to 50
    admittance formulation, 10-59
    Akagi-Nabae theory, 14-47 to 49, 14-53, 14-56
    approximate line segments, 6-30 to 32
    cables, 6-19 to 22
    capacitance, 10-66 to 72
    capacitor banks models, 6-63
    closed delta connections, 6-37 to 38
    complex electrode shapes, 10-86 to 87
    component loads, 6-3
    composite loads, 6-9 to 10, 7-16 to 17
    contact wear, load tap changers, 3-286
    copper losses, 10-69
    current, 6-36 to 37
    Darlington configuration, 14-4

Index

delta connections, 6-37 to 57
detailed load, 11-79
discharge lamps, 6-9
distributed loads, 6-63
distribution feeders, 6-1 to 58, 6-62 to 63
distribution transformers, 3-45-48
duality, 10-60
dynamic, 6-2 to 3, 11-72 to 81
dynamic system assessment (DSA), 11-123 to 124
econometric, 13-43 to 44
electrogeometric (EGM), 5-110 to 121
Electromagnetic Transients Program (EMTP), 11-135
end use, 13-44
exact line segment, 6-28 to 30
excitation systems, 11-76
exponential, 6-4 to 6, 6-8, 7-17
extrapolation, 13-43 to 44
finite element method, 10-60
fluorescent lamps, 6-9
frequency-domain application, 10-60
gas-insulated substations, 10-46
generators, 11-73 to 76, 11-135 to 136
Generic Object Models for Substation and Feeder Equipment (GOMSFE), 5-61
harmonics distortion, 14-20
inductance, 10-63 to 64
induction motor dynamic, 11-79
linear voltage and frequency dependence model, 7-16
line drop, 6-35-36
line impedance, 6-11 to 24, 6-63
line segment, 6-28 to 32
load flow analysis, 11-36 to 38
loads, 6-2 to 10, 6-53 to 56, 11-78 to 80
load windows, 6-9 to 10
lumped parameter, 10-59 to 61, 10-78
matrix definitions, 6-46 to 57
measurement based, loads, 6-3
mercury vapor lamps, 6-9
open delta connections, 6-38 to 41
optimal power flow, 12-40 to 44, 12-48 to 49
outages, 13-31 to 33
overhead lines, 6-14
per unit system, 6-44 to 45
polynomial, 6-6 to 8
power flow, 6-42, 7-58 to 64, 12-40 to 44
power system loads, 6-1 to 10, 6-2 to 10

Power System Object Model (PSOM), 5-61
power systems, 11-45 to 46, 11-72 to 81, 11-135
prime mover, 11-76 to 78
protective relays, 6-9
reactive power loads, 6-4
real drivers, 13-44
reduced order, 10-77 to 78
regulators, step-voltage, 6-32 to 41
rotor mechanical, 11-73 to 74
sag magnitude, voltage, 15-27 to 28
sag/tension in conductor lines, 4-102 to 107
saturation, 11-75 to 76
sequential load flow, 11-38 to 39
short circuit analysis, 7-64 to 67
shunt admittance, 6-25 to 28
snap shot load, 6-2
sodium vapor lamps, 6-9
solar activity, 4-161 to 163
spot loads, 6-63
Stanley, 3-31
state estimation, 12-27 to 33
state-variable formulation, 10-59
static loads, 6-2, 11-79, 12-44
steady state losses, 10-69
step-voltage regulators, 6-32 to 41
system studies, 10-77
thermostatically controlled loads, 6-9
Thevenin equivalent, 6-45 to 46
time-domain application, 10-60 to 61
transformer bank connections, 6-42 to 52
transformers, 6-42 to 52
transient enclosure voltage (TEV), 10-48 to 50
transient stability, 11-15 to 16
transmission devices, 11-80
underground lines, 6-14, 6-19
very fast transients (VFT), 10-46 to 51
voltage divider, 15-27
voltage regulators, 6-9
voltages, 6-36 to 37
voltage sag magnitude, 15-27 to 28
weighted least squares (WLS) method, 12-30 to 32, 12-35 to 37
wye connections, 6-36 to 57
Mode of disturbance (MOD), 11-51
Modes of Clarke, 4-141
Mohammed studies, 13-18
Mohan studies, 14-38 to 43
Moisture
air insulation, effect on, 10-95 to 96
dew point test, 3-235

effect on insulation, 3-204 to 205
gap effect, 10-96
instrument transformer failure, 3-280
insulation stress, 4-28
insulator failure, 4-37
monitoring, 15-56
in oil, power transformers, 3-275 to 276
substation automation systems, 5-43
substation buildings, 5-37 to 38
substation drainage system, 5-69 to 72
transformer failure, 3-248 to 249, 9-2
Molten carbonate fuel cell (MCFC), *1-12*, 1-13, 2-28, *2-29*
Momentary average interruption frequency index (MAIFI), 13-59
Momoh studies, 12-44
Monitoring
air insulated substations, 5-17
applications, 11-103 to 117
architectures, 11-87 to 90
archival recording, 11-86
central monitors, 11-90
continuous, 11-89
database managers, 11-100 to 101
defined, 11-85
distributed monitor networks, 11-90
disturbance, 11-85
fatigue (SSR), 11-137 to 138
frequency-domain application, 11-86
functions, 11-85 to 86
gas-insulated substations, 5-14, 5-17
interaction, 11-85
interactive recording, 11-86
line disturbance analyzers, 15-50 to 54
local monitors, 11-90
networks, 11-90 to 100
oil spill detection, 5-75
on-line, instrument transformers, 3-279 to 282
on-line, of liquid-immersed transformers, 3-269 to 288
on-line systems, 3-269 to 288
placement location, 11-101 to 103, 15-50
power quality, 15-49 to 57
power transformers accessories, 3-25 to 29
reserve, 12-5
sag magnitude, voltage, 15-25 to 26
signal conditioning, 11-86

split-core type current transformers, 3-122
stream-to-archive, 11-89
substation automation systems, 5-49 to 51
system condition, 11-85
time-domain application, 11-86
voltage recorders, 15-50 to 54
wide area dynamics, 11-84 to 117
Montana, 11-62
Monte Carlo Simulation, 5-111, 13-33 to 39
Montoya studies, 5-31
De Moor, Van Overshee and, studies, 11-115
Moore studies, 3-156
Moran studies, 14-50
Moreau studies, 4-139 to 141
Morison studies, 11-8
Morrison, Mann and, studies, 9-44
Morshuis studies, 3-276
Mortensen studies, 11-140
MOS-controlled thyristors (MCT), 14-6 to 7
MOSFET. *See* Metal oxide semiconductor field-effect transistor (MOSFET)
Moskowitz, Fthenakis and studies, 1-16
Motor current index, 3-284
Motor-enclosure-controller systems, 7-20 to 23
Motoring, 9-17 to 18
Motors
   control centers, 2-24
   electric power utilization, 7-18 to 23
   synchronous, 2-12 to 13
   thermal generating plants, 2-24 to 25
   variable speed controllers, 11-140
Mountainous shielding, 10-7
Mousa and Srivastava studies, 5-112, 5-120 to 121
Mousa studies, 5-106, 5-111 to 113
MOV. *See* Metal oxide varistors (MOV)
Moving contact velocity, 5-7
MTTR. *See* Mean time to repair (MTTR)
Multiconductor lines, induced voltage, 10-29 to 30
Multicrystal silicon cells, 1-14
Multiform meters, 7-4
Multifunction generator protection, 9-28 to 29
Multifunction meters, 7-4
Multigrounded shields, cables, 4-55 to 56

Multilayer perceptron (MLP), 9-60 to 61, 13-16
Multilayers of soil, grounding, 10-87 to 88
Multilevel inverters, 14-42 to 43
Multiple Address Radio System (MAS), 6-70
Multiple coarse change-over selector, 3-189
Multiple gaps and phases, 10-94 to 95
Multiple shielding electrodes, 5-118 to 119
Multipoint grounding, gas-insulated substations, 5-16
Multiterminal transmission lines, 9-32 to 33
Multiwinding transformers, 3-130, 3-133
Municipal utilities (MUNIES), 13-6
MUNIES. *See* Municipal utilities (MUNIES)
Murthy, Pavella and, studies, 11-44, 11-124
Must-run generation, 11-40
Mvar's, 3-91
Myers studies, 3-31, 3-275

N

Naggapan, El-Abiad and, studies, 11-44, 11-48
Nameplate ratings, 3-149 to 150, 3-154, 3-158
Nanticoke generating station, 10-24
Napthenes, 3-206 to 207
Narang studies, 10-55, 10-57
NASA
   fuel cells, 1-10, 2-28
   solar wind monitoring, 4-161 to 163
National Earthquake Hazard Reduction Program (NEHRP), 5-126, 5-129
National Electrical Code (NEC), 1-17 to 18
National Electrical Manufacturing Association (NEMA)
   ST-20 "Dry-Type Transformers for General Applications," 3-258 to 259
   St-1 "Specialty Transformers (Except General Purpose Type)", 3-256
   TP-1 "Guide for Determining Energy Efficiency for Distribution Transformers," 3-259, 3-263 to 264, 3-267 to 268
   TP-2 "Standard Test Method for Measuring the Energy Consumption of Distribution Transformers," 3-260, 3-264 to 265, 3-268 to 269
   TR-1 "1993 (R-1999) Transformers Regulators and Reactors," 3-269
National Electrical Safety Code (NESC)
   conductor loads, 4-93, 4-96
   transmission line structure, 4-13 to 14
National Electric Reliability Council. *See* North American Electric Reliability Council (NERC)
National Fire Protection Association, 1-17. *See also* Fire protection
National Renewable Energy Laboratory, 1-17
Natural shielding, 10-7
Nearly real time (NRT), 11-85 to 86, 11-89
Needle nozzles, 2-4, 2-7
Negative corona modes, 4-130 to 133
Negative pulseless glow, 4-133. *See also* Glow (corona modes)
Negative reactance paralleling, 3-145
Neglect series resistance, 8-44
NEHRP. *See* National Earthquake Hazard Reduction Program (NEHRP)
Nelson River converter station, 5-30
NEMA. *See* National Electrical Manufacturing Association (NEMA)
NERC. *See* North American Electric Reliability Council (NERC)
Nerve damage, body current limits, 5-95
NESC. *See* National Electrical Safety Code (NESC)
Netherlands, *1-2*
Net metering, 1-3
Networks, lightning detection, 5-108
Network transformers
   installation, 3-53
   preferred locations, 3-42
Neural fuzzy controller, self-organzing (SONFC), 11-68
Neural networks, artificial (ANN)
   digital relaying, 9-60 to 61
   price forecasting, 13-23 to 25
   short term loads, 13-23
Neutral grounding reactors, 3-87
Neutrals
   double wound transformers, 3-132
   grounding transformers, 3-133 to 134

Index

load tap changing transformers, 3-145
thermal generating plants, 2-25
New Mexico, lightning, 10-4
Newton OPF, 12-42
Newton-Raphson algorithm
  nonlinear power flow, 8-40 to 42
  optimal power flow, 12-39, 12-42
  state estimation, 12-31
Newton's Second Law
  rotor mechanical model, 11-73
  sound energy measurements, 3-162
  synchronous stability, 11-55
Ng. *See* Ground flash density (GFD or Ng)
Ni and Fouad studies, 11-44
Nicad batteries, 1-9 to 10
Nickel batteries, 1-9 to 10
Nickel-iron core material, 3-104-105
Nielsen, Gail and, studies, 9-59
Nitrogen, liquid, 3-30
NOAA, 4-162
Noferi studies, 13-33
Noise. *See* Sound levels
No-load losses
  distribution transformers, 3-44, 3-46 to 48
  iron core transformers, 3-7 to 8
  power transformers, 3-15
  transformer testing, 3-224 to 225
Nonceramic insulators, 4-32 to 35, 4-38
Non-conventional electric generation
  advanced energy techologies, 1-7 to 13
  photovoltaics, 1-14 to 19, 2-31
  wind power, 1-1 to 7
Non-directed flow cooling, 3-149 to 150
Noninteger multiple harmonics, 15-17
Non-linear resistors, 3-201
Non-oscillatory instability, 11-4
Nonpunch-through IGBTs (NPT IGBTs), 14-6
Non-rechargeable batteries, 1-9
Non-salient poles, synchronous machinery, 2-14-15
Non-self restoring insulation, 10-93 to 102
Nonsinusoidal loads, 7-17
Non-synchronous ties (DC lines), 4-7
Nonuniform field gap, 4-129 to 130
Non-unit zones, 9-30 to 31
Nonutility transmission systems (NUTS), 13-6
Normal system state, 11-121 to 122
North American Electric Reliability Council (NERC)

cascading prevention standards, 13-28
importance of reliability, 13-4
reliability standards, 13-6
security defined, 12-55
Northeast Blackout, November 9, 1965, 11-8, 11-121, 13-27, 13-52
Novotny, De Donker and, studies, 14-42
Nowotny and Lipo studies, 14-40
Nozari studies, 11-79
Nozzle. *See* Needle nozzle
NPT IGBTs. *See* Nonpunch-through IGBTs (NPT IGBTs)
NRT. *See* Nearly real time (NRT)
Numerical treatments. *See* Algorithms; Models
NUTS. *See* Nonutility transmission systems (NUTS)
Nyquist methods, 9-45 to 46, 11-31

O

OA cooling, 3-149 to 150
OCF. *See* Overload Capacity Factor (OCF)
Odd-ordered harmonics. *See* Harmonics
OEL. *See* Overexcitation limiters (OEL)
Oettle studies, 10-89
Office equipment, grounding, 15-10 to 11
Off-line dynamic system assessment, 11-124
Offset-mho relay, 9-16
Ogawa studies, 10-46
OHGWs. *See* Overhead ground wires (OHGWs)
Ohm's Law
  current transformers, 3-118
  instruments transformers, 3-106
  per unit system, 8-2 to 3
  transformer testing, 3-229
OHTL. *See* Overhead transmission lines (OHTL)
Ohura studies, 11-67
Oil conservators, transformer installation, 3-238
Oil-flow relays, 3-200 to 201
Oil rise, heat run tests, 3-231
Oils
  composition, 3-206 to 207
  containment, 5-62 to 75
  degradation, 3-208
  distribution transformers, 3-31, 3-33 to 34
  immersed reactors, 3-81 to 82
  impregnated bushings, 3-176

insulated (filled) bushings, 3-176, 3-179
insulating material, 3-205 to 208
power transformer failures, 3-278
reactors, 3-81
reservoirs, 3-179 to 180
spill prevention, 5-66 to 75
thermal response predictions, 3-152 to 153
transformer installation, 3-240 to 241
Oil stop valve
  flow blocking systems, 5-73
Oil-water separator system, 5-72 to 73
Oliveira studies, 13-34
125-V DC systems, thermal generating plants, 2-23
One-line diagrams, 2-21
On-line gas analysis, instrument transformers, 3-282
On-line monitoring systems, 3-269 to 288
Onset streamers, 4-135 to 136
Ontario Hydro. *See also* Canada
  Nanticoke generating station, 11-24
  optimal power flow studies, 12-44
  studies, 11-8
Oommen and Claiborne studies, 3-34
Oommen studies, 3-156, 3-276
Open circuits conditions, current transformers, 3-119
Open delta connections, modeling, 6-38 to 41
Open-loop controls, 11-59 to 60
Open systems, substation automation, 5-53 to 54
Open zones, 9-30 to 31
Operations
  air insulated substations, 5-17
  distributed utilities, 2-33
  dry type transformers, 3-65
  gas-insulated substations, 5-17
  power systems, 11-8 to 9, 11-20
  step-voltage regulators, 3-74 to 78
  substations, 5-87
  system states, 11-121 to 122
Operator Training Simulator programs, 6-9 to 10
OPF. *See* Optimal Power Flow
OPT-GW. *See* Optical Power Telecommunications-Ground Wire
Optical Power Telecommunications-Ground Wire (OPT-GW), 4-44
Optical Transient Detector (OTD), 4-166
Optimal power flow (OPF)
  closed-loop mode, 12-50

discrete control, 12-48 to 49
economic scheduling, 12-39, 12-50
hot start, 12-48
infeasibility, 12-49
initial guess point, 12-48
load models, 12-44 to 46
online implementation, 12-47 to 51
pricing, 12-50
principles, 12-38 to 39
security, 12-45
soft limits, 12-50
software, 12-39
solution models, 12-40 to 44
speed requirement, 12-48
study mode, 12-50
Supervisory Control And Data Acquisition (SCADA), 12-50
Oregon, 10-4
Oregon, California-, transmission system, 11-130
Orientation of coils, power transformers, 3-20
Orographical shielding, 10-7
Orrell studies, 5-112, 5-115, 5-121
Oscillations
control mode, 11-24
damping, 11-20 to 33, 11-58
description, power system, 11-24 to 25
Device Dependent Supersynchronous Oscillations (DDSPSO), 11-142 to 143
general concepts, 10-25 to 26
interarea mode, 11-24, 11-32, 11-58, 11-60
local plant mode, 11-23
mitigation, 11-28 to 32
power systems, 11-20 to 33, 11-60
prime mover modeling, 11-77
stability, small signal, 11-2 to 9
study procedures, 11-27 to 28
subsynchronous, 11-23 to 24, 11-140 to 141
supersynchronous, 11-142 to 143
torsional mode, 11-24
transient, 9-39 to 41, 10-45 to 46
trapped charges, 10-36 to 39
Oscillographs
aliasing, 9-45
digital relays, 9-44
fault analysis, 8-58
power systems, 9-62 to 72
transformers, 9-10
Osterdorp studies, 4-17, 4-20 to 21
Ota studies, 11-68
Outages
animal-caused, 5-88 to 91
Comanche Unit 2, 11-142 to 143

direct lightning strokes, 10-17 to 20
models, 13-31 to 33
nearby lightning strokes, 10-31 to 34
non-self restoring insulation, 5-88 to 91, 10-93
Northeast Blackout, November 9, 1965, 11-8, 11-121, 13-27, 13-52
planning uncertainties, 13-46
San Francisco, December 8, 1998, 11-110
supersynchronous resonance (SPSR) events, 11-141
Western North American Power System, August 10, 1996, 11-68, 11-82 to 117
Western North America Power System, July 2, 1996, 11-68
Outdoor valves, 5-38
Out-of-step protection, 9-23 to 24, 9-39 to 41
Out-of-step relays, 9-43
Overcurrent blocking device, 3-201
Overcurrents
directional, relays, 9-35 to 36
flucutation protection, 9-2
instantaneous, relays, 9-34
inverse-time delay, relays, 9-33 to 34
relays, 9-2, 9-7 to 8
thyristor, 5-34
transformer fault protection method, 9-2
Overend and Smith studies, 4-119
Overexcitation limiters (OEL), 11-76
Overexcitation protection
synchronous generators, 9-18 to 19
transformers, 9-4 to 5
Overflux, 9-4
Overhead conductors, transmission lines, 4-82
Overhead connections, transients, 10-46
Overhead ground wires (OHGWs)
stroke current magnitude, 5-106 to 107
transmission line structure, 4-13, 4-18, 4-20
Overhead installations, distribution transformers, 3-42
Overhead lines
modeling, 6-14
shunt admittance, 6-25 to 26
Overhead Transmission Line Conductors, Guide to the Installation of, 4-115

Overhead transmission lines (OHTL), 4-13 to 22, 4-64 to 66
Overload Capacity Factor (OCF), 4-13 to 14
Overload mitigation. *See* Mitigation
Overload shedding, 9-43
Overreaching relays, 9-39
Overvoltages. *See also* Surges
direct lightning strokes, 10-8 to 20
gas-insulated substations, 10-44
indirect lightning strokes, 10-21 to 34
instrument transformer failure, 3-280
protection, 4-175
shielded line calculation, 4-167 to 168
synchronous generators, 9-19
Oxidation
inhibitors, 3-207
transformer failure, 3-248
Oxygen
effect on insulation, 3-204 to 205
effect on oils, 3-208
transformer failure, 3-249

P

PA. *See* Pulse accumulators (PA)
PACI. *See* Pacific AC Intertie (PACI)
Pacific AC Intertie (PACI)
mode, 11-109 to 111
monitoring applications, 11-104
power system stability control, 11-63 to 67
Pacific HVDC Intertie
testing, 11-113
transmission systems, 4-8, 11-23
Packet radio, 6-70
Padiyar and Ghosh studies, 11-44
Padiyar and Sastry studies, 11-44
Padmounted transformers, 3-43, 3-58 to 63
PAFC (phosphoric acid fuel cell), 1-12 to 13, 2-28, *2-29*
Pai, Sauer and, studies, 11-45
Pairs studies, 10-95
Pai studies, 11-43, 11-48, 11-124
Palmer studies, 3-32
Palo Verde Nuclear Power Plant, 11-109
Pancake-type winding structure
description, 3-20, *3-24 to 25*
shell-form construction, 3-17 to 18, 10-74
Papalexopoulos studies, 12-47, 12-49, 13-19
Paper insulation
characteristics, 3-204 to 205

Index

loading power transformers, 3-154 to 155
Paraffin, 3-206 to 207
Paralleling transformers
 connections, 3-134 to 135
 load tap changing control, 3-135 to 148
 step-voltage regulators, 3-142 to 147
Parallel resonance, 10-62
Parameter estimation, relaying algorithms, 9-50 to 51
Park studies, 13-19
Partial discharge (PD)
 effect on oils, 3-208
 instrument transformer failure, 3-280
 instrument transformers, 3-282
 power transformers, 3-276 to 278
 testing insulated cables, 4-61
 transformer testing, 3-222 to 223
Paserba studies, 11-28
Passive filters, harmonic mitigation, 14-24
Pavella and Murthy studies, 11-44, 11-124
PCC. *See* Point of common coupling (PCC)
PCM (pulse code modulated), 1-17
PCU (power conditioning unit), 1-17 to 18
PD. *See* Partial discharge (PD)
PDADS. *See* Portable Dynamic Analysis and Design System (PDADS)
PDSM. *See* Power System Disturbance Monitor (PSDM)
Peek's experimental law, 4-146
Pelton turbine, 2-8
PEM. *See* Polymer Electrolyte Membrane Fuel Cell (PEMFC)
Peng studies, 13-19, 14-50 to 54, 14-59
Pennsylvania-New Jersey-Maryland Independent System Operator (PJM), 13-24
Perceptron, multilayer (MLP), 9-60 to 61, 13-16
Pereira studies, 5-29
Perez, Cooray and, studies, 10-4
Perez studies, 9-61
Perfect ground, electrodes, 10-84 to 85
Performance
 artificial neural network short-term load forecasting, 13-22 to 23
 carrier, importance, 9-63

corona, 4-144 to 149
distribution transformers, 3-45 to 48
insulators, 4-41 to 42
monitoring substation automation systems, 5-49 to 50
online dynamic system assessment programs, 11-125
oscillograph analysis, 9-62 to 72
power transformers, 3-15
seismic qualifications methods, 5-128 to 133
site seismic qualifications, 5-126 to 133
substation automation systems, 5-45
synchronous machinery, 2-15 to 20
system oscillation mitigation, 11-31
tolerance, voltage, 15-33
transformer testing, 3-224 to 233
transient voltages, 10-55
voltage transformers, 3-112 to 113
Performance factor (PF), 5-130
Permits, substations, 5-85 to 86
Per phase analsis, 8-14
Per unit system
 modeling, 6-44 to 45
 power systems, 8-1 to 14
 power transformers, 8-29 to 33
 scaling, 8-1 to 4
 sequence network representation, 8-29 to 33
 susceptance, 6-57 to 58
 transients, 8-1
 variations, 8-14
Petersen Coils, 3-95 to 97, 9-60
Peyrot, Carton and, studies, 4-17 to 18
PF. *See* Performance factor (PF)
Phadke, Horowitz and, studies, 9-39
Phadke, Thorp and, studies, 9-54
Phadke and Thorp studies, 9-39
Phadke studies, 11-103
Phase angles
 jumps, 15-30 to 31
 regulators protection for, 9-8
Phase comparison, 9-39
Phase conductors, lightning strikes, 10-9 to 11
Phase faults, 9-22
Phase impedance matrix, 6-14 to 32
Phase multiplication, harmonics, 14-24 to 25
Phase reactors
 current limiting reactors, 3-85 to 86
 vs-bus tie reactors, 3-100 to 101
Phase relation tests, 3-211 to 212

Phase-resolved PD analysis (PRPDA), 3-276
Phase reversing, hydroelectric plants, 2-11
Phases, multiple, 10-94 to 95
Phase shift, primary-secondary, 9-7
Phase-shifting transformers (PST), 3-191 to 192, 3-197
Phase spacing, 5-24
Phase-to-phase faults, 8-49
Phasing diagrams, thermal generating plants, 2-26
Phasing of inputs, 3-137
Phasor Data Concentrator (PDC), 11-91 to 92
Phasor Measurement Units (PMU), 11-91 to 98
Phasors
 power system state, 9-50
 from samples, 9-46 to 50
Philipson and Willis studies, 13-52
Phoenix, Arizona, 4-2 to 7, 11-107, 11-109
Phosphoric acid fuel cell (PAFC), 1-12 to 13, 2-28, *2-29*
Photomultiplier study, 4-136
Photovoltaic (PV) cells, 1-14 to 19, 2-31
Picocolumbs (pC), 3-223
Pierce studies, 3-149 to 153
Pigini, Feser and, studies, 10-98
Pilot channels, 9-38 to 39
Pilot lines, 4-44
Pilot wires (metallic wires), 9-39
Pits, oil containment, 5-67
Piwko and Larsen studies, 11-30
Piwko studies, 5-31, 11-65, 11-96, 11-113, 11-141
PJM. *See* Pennsylvania-New Jersey-Maryland Independent System Operator (PJM)
Planning
 arenas, 13-41 to 42
 entities, 13-41
 generation, 13-48
 least cost, 13-48 to 50
 power system, 13-40 to 51
 problems, 13-42 to 47
 processes, 13-47 to 51
 transmission, 13-48
Plastic, equipment pits, 5-68
Platform selection guidelines, 6-73 to 76
PMU. *See* Phasor Measurement Units (PMU)
Pocketbook (grip), 4-116
Podmore studies, 11-44
Pohlman and Harris studies, 4-16 to 17

Pohlman and Lummis studies, 4-16 to 17
Point discharge phenomenon, 5-123
Point of common coupling (PCC), 14-21, 15-28
Polak-Ribiere method, 12-42
Polarity
 current transformers, 3-121
 distribution transformers, 3-42
 instrument transformers, 3-107
 paralleling, 3-147
 single-phase transformers, 3-130
 tests, transformers, 3-211 to 212
Pollution
 batteries, 1-9 to 10
 fuel cells, 1-13
 insulation stress, 4-28 to 29
 insulator failure, 4-36 to 41, 4-38
Polyaromatic oils, 3-206 to 207
Polychlorinated biphenyls (PCBs), 3-33
Polymer Electrolyte Membrane Fuel Cell (PEMFC), 1-11 to 12, 2-29. See also Proton Exchange Membrane Fuel Cell (PEFC)
Polyphase metering, 7-2
Poncelet studies, 9-44
Ponds, collecting, 5-67
POOLCO. See Power pool concept (POOLCO)
Popolansky studies, 10-9
Popsize (GA population), 12-24
Porcelain
 insulator failure, 4-35 to 36
 insulators, 4-30 to 32
 seismic events, 5-125
Porrino studies, 5-30
Portable Dynamic Analysis and Design System (PDADS), 11-95
Portable Power System Monitor (PPSM), 11-91 to 98
Portland cement, 4-30, 4-35
Portugal, 1-2
Position determination, load tap changers, 3-286
Positive corona modes, 4-133 to 137
Positive pressure system, transformer installation, 3-239
Postal trucks, zinc air batteries, 1-10
Postdisturbance equilibrium condition, 11-120 to 121
Post mortem studies, 8-34
Post-type insulators, 4-32, 4-35
Potentials, retarded, 10-22
Potential transformers (PT), 4-4
Povh studies, 10-46
Power angle relationship, 11-12 to 15

Power cables, transmission system, 4-48 to 62. See also Cables
Power calculations
 balanced, 8-25
 symmetrical component transformation, 8-20
Power conditioning unit (PCU), 1-17 to 18
Power electronics, high voltage substations, 5-29 to 39
Power Engineering Society Substations Committee (IEEE), 5-60 to 61
Power factor generators, voltage stability, 11-40
Power flow
 analysis, 6-58 to 64
 calculated component, 8-43
 control reactors, 3-89 to 91
 economic scheduling, 12-39, 12-50
 equations, 6-46 to 52, 8-36 to 43
 modeling, 6-42
 optimal, 12-38 to 51
 program inadequacy, 6-11
 systems analysis, 8-34 to 43
Powerformer, 3-29
Power frequency withstand voltage test, 5-17
Power fuses, 5-25 to 26
Power line balance, 3-101 to 102
Power loads, reactive models, 6-4
Power plants, 1-2, 1-11
Power pool concept (POOLCO), 13-4
Power quality
 equipment voltage tolerance, 15-32 to 36
 flickering, 4-172, 15-13 to 15, 15-42 to 49, 15-56
 flucutations, 15-42 to 49
 grounding, 15-2 to 15
 harmonics, 15-16 to 24
 meters, 7-3
 mitigation of voltage sags, 15-36 to 40
 monitoring, 15-49 to 57
 voltage sags, 15-24 to 40
 wiring, 15-2 to 15
Power reduction (fast valving), 11-62, 11-77
Power System Analysis Monitor (PSAM), 11-91 to 98
Power System Disturbance Monitor (PSDM), 11-88
Power System Identification (PSI) Toolbox, 11-103
Power System Monitor (PSM) Tools, 11-103
Power System Object Model (PSOM), 5-61
Power systems

artificial neural networks, 13-16 to 25
balanced, 8-26
capacity expansion, 13-1 to 12
coils, transient voltage response, 10-54 to 78
competition, 13-1 to 12
complexity, 12-54
controls, stability, 11-55 to 69
design, 11-8 to 9
digital relaying, 9-44 to 61
direct lightning strokes, 10-8 to 20
dynamic interaction, 11-126 to 143
dynamic modeling, 11-72 to 81
dynamic simulation, 12-9
economic dispatch, 12-10 to 26
energy management, 12-1 to 9
factors affecting stability, 11-18
fault analysis, 8-44 to 59
generation control, 12-10 to 26
generators, 11-126 to 143
grounding, 10-80 to 90
history, 13-51 to 52
indirect lightning strokes, 10-21 to 34
insulation coordination, 10-93 to 102
lightning stroke characteristics, 10-1 to 7
load modeling, 6-1 to 10
modeling, 11-45 to 46, 11-72 to 81
operations, 11-8 to 9, 11-20
oscillations, 11-20 to 33
oscillographs records, 9-62 to 72
per unit system, 8-1 to 14
phasors, 9-50
planning, 13-40 to 51
power flow, optimal, 12-38 to 51
power flow analysis, 8-34 to 43
price forecasting, 13-23 to 25
protection, 9-30 to 39
reliability, 13-26 to 39, 13-51 to 65
security analysis and assessment, 11-120 to 125, 12-54 to 58
short term loads, 13-16 to 25
small signal stability, 11-20 to 33
stability, 11-2 to 9, 11-55 to 69, 13-51 to 65
state estimation, 12-27 to 37
switching surges, 10-36 to 40
symmetrical components, 8-14 to 33
synchronous generators protection, 9-11 to 29
transformer protection, 9-1 to 10
transmission line protection, 9-30 to 39
turbines, 11-126 to 143
unit commitment, 12-10 to 26

Index

very fast transients (VFT), 10-41 to 53
windings, transient voltage response, 10-54 to 78
Power system stabilizers (PSS)
  excitation system, 11-76
  oscillation damping, 11-23, 11-33
  security analysis model, 12-58
Power transformers
  description, 3-11 to 30
  geomagnetically induced currents (GIC), 4-152 to 157
  internal transients, 10-48
  load tap changing, 3-23, 3-135
  on-line monitoring technologies, 3-272
  per unit system, 8-29 to 33
  power fuses, 5-25 to 26
  standards, 3-255, 3-258, 3-260 to 262
  system studies, 10-77
PPSM. *See* Portable Power System Monitor (PPSM)
P-Q bus (load bus), 8-39
Prediction, transient voltages, 10-55
Pre-disturbance operation point, 8-34
Prescribed contours, sound measurements, 3-167
Pressure
  instrument transformers, 3-282
  levels, sound, 3-160 to 170
  relays, 9-4
  relief devices, power transformers, 3-25
  sensing (releasing) relays, 3-201
  system, positive, 3-239
  vessel requirements, 5-3
Prestressing conductor, 4-119
Preventative system state, 11-121 to 122
Price forecasting, 13-23 to 25. *See also* Cost and economics
Price studies, 6-6 to 7, 11-81
Primary batteries, 1-9
Primary-secondary phase shift, 9-7
Primary winding material, instrument transformers, 3-112
Prime mover modeling, 11-76 to 78
Primitive impedance, 6-14
Proakis studies, 13-19
Probabilistic method approach, 12-58, 13-28 to 29, 13-33 to 39
Probability of flashover (pfo), 10-93 to 95
Problems
  grounding, 15-6 to 12
  insulated cables, 4-60 to 62

lightning, utility reliability, 13-62
planning, 13-42 to 47
reliability, 13-61 to 64
transformer failure investigation, 3-242 to 249
transformer investigation, 3-242 to 249
voltage stability, 11-40 to 41
wiring, 15-6 to 12
Production tests. *see* Testing
Programmable logic controllers (PLCs), 6-71, 6-75
Prony analysis, 11-115 to 117
Protected animals, damage from, 5-89
Protection. *See also* Fire protection
  air insulated substations, 5-17, 5-19 to 23
  algorithms for, 9-50 to 61
  from animals, 5-88 to 91, 9-2, 13-63
  anti-motoring, 9-17 to 18
  cables, 4-52
  current transformers, 3-120
  distributed utilities, 2-32 to 33
  distribution transformers, 3-49 to 50
  dry type transformers, 3-66
  electrical types of, 9-2 to 4
  fire, substation buildings, 5-38
  "first house," 142 to 143
  gas-insulated substations, 5-17
  grounding, 15-2 to 15
  hydroelectric plants, 2-10
  lightning, 4-165 to 168
  load tap changer devices, 3-200 to 201
  loss-of-excitation, 9-15 to 16
  mechanical types of, 9-4
  metal oxide varistor system, 4-176 to 177
  microprocessor transformers, 9-3 to 4
  multiple voltages, 5-119
  oil containment, fire, 5-68 to 69
  out-of-step, 9-23 to 24, 9-39 to 41
  overexcitation, 9-18 to 19
  overhead conductors, 4-89
  padmounted transformers, 3-62 to 63
  power transformers, 3-25 to 29, 5-25 to 26
  remedial action schemes, 9-42 to 43
  substation fires, 5-134 to 139
  substation lightning strokes, 5-104 to 123
  substation operations, 5-85, 5-87
  thermal generating plants, 2-22, 2-26
  thermal types of, 9-4

transformer, 9-1 to 10
vault usage, 3-52 to 54
Protective relays, modeling, 6-9
Protocols, local area networks, 5-57
Proton Exchange Membrane Fuel Cell (PEFC), 2-28. *See also* Polymer Electrolyte Membrane Fuel Cell (PEMFC)
Proximity effects
  current transformers, 3-125
  losses model, 10-70 to 71
PRPDA. *See* Phase-resolved PD analysis (PRPDA)
PRT, Inc-, 13-18
PSAM. *See* Power System Analysis Monitor (PSAM)
PSI Toolbox. *See* Power System Identification (PSI) Toolbox
PSM Tools. *See* Power System Monitor (PSM) Tools
PSOM. *See* Power System Object Model (PSOM)
PSS. *See* Power system stabilizers (PSS)
PT. *See* Potential transformers (PT)
PT IGBTs. *See* Punch-through IGBTs (PT IGBTs)
Public concerns, transmission line structure, 4-16. *See also* Communities
Pueblo, Colorado, 11-142 to 143
Pullers, transmission line construction, 4-44
Pulley wheels, 4-116 to 117
Pulse, 7-7 to 9
Pulse accumulators (PA), 5-48 to 49
Pulse code modulated (PCM), 1-17
Pulse metering, 7-7 to 9
Pulse-width-modulated (PWM) signals, 14-37 to 43, 14-44 to 61
Pumped-storage power plants, 2-10 to 11
Pump/fan operation, power transformers, 3-279
Pump motor starting, hydroelectric plants, 2-11
Pump noise, 3-164
Punch-through IGBTs (PT IGBTs), 14-6
Purefoy, Ms-, 3-249
P-V bus (voltage controlled bus), 8-39 to 40
PVC pipe, substation animal barrier, 5-90
PV (photovoltaic) cells, 1-14 to 19, 2-31
PWM. *See* Pulse-width-modulated (PWM) signals

Q

QF. *See* Qualifying facilities
Quadratic programming methods (OPF), 12-43
Qualifying facilities (QF), 13-6, 13-41
Quantification of sound levels, 3-160 to 170, 3-233
Quintana, Yan and, studies, 12-44

R

Rabins studies, 10-65
Racoons, substation damage, 5-90
Raczkowski and Kung studies, 11-141 to 142
Radial forces, power transformers, 3-15, *3-16*
Radial-style padmounted transformers, 3-58
Radiating surfaces, sound measurements, 3-165 to 166
Radiators, transformer installation, 3-238
Radioactivity, lightning rods, 5-123
Radios
 frequency noise source, 5-81
 interference, 4-140, 4-143, 5-38
 Multiple Address Radio System (MAS), 6-70
 packet, 6-70
Raether studies, 4-129
Rahman studies, 11-66
Rain. *See* Moisture
Raised block function, 3-143
Raise position (voltage regulators), 6-33 to 34
Rakov studies, 10-4
Ramakumar studies, 13-52
Rashid studies, 14-38, 14-40
Ratings
 capacitor banks, 4-175 to 176
 circuit breakers, 5-26 to 28
 current transformers, 3-119, 3-120, 7-10
 hydroelectric plants, 2-4
 load tap changers, 3-193 to 196
 motor-enclosure-controller systems, 7-21
 phase reactors, 3 to 85
 power fuses, 5-25
 power transformers, 3-12 to 14
 shunt reactors, 3 to 91
 thyristor controlled reactors, 3 to 93
 voltage transformers, 3-113 to 114, 7-10
Ratio Correction Factor (RCF)
 CT/VT units, 3-129
 instrument transformers, 3-108 to 110
Ratio tests, 3-211 to 212
RBSOA. *See* Reverse-bias safe operating area (RBSOA)
RCF (Ratio Correction Factor)
 CT/VT units, 3-129
 instrument transformers, 3-108 to 110
RCT. *See* Reverse-conduction thyristor (RCTs)
Reactance
 capacitivem, transmission line conductors, 4-75 to 82
 load tap changer type, 3-186 to 187
 relays, 9-37
Reactive capability curves, 2-16 to 17, *2-19*
Reactive currents
 balance paralleling, 3-146
 power balance, 3-102 to 103
 power compensation, 4-169 to 180
Reactive overload capability, 11-40
Reactive power
 compensation switching, 11-65 to 66
 cost and economics, compensation, 4-169
 delta connections, 6-55
 Flexible AC Transmission System (FACTS), 4-178
 load models, 6-4
 modulation, 11-65 to 66
 shunting, 4-169 to 174
 Static Compensators (STATCOM), 4-179 to 180, 11-66
 Static Var Compensators (SVC), 4-172 to 174, 11-65 to 66
 thyristor controlled series compensators (TCSC), 4-178 to 179
 transmission system, 4-169 to 180
 wye connections, 6-53
Reactors
 damping, 4-177
 general concepts, 3-81 to 103
Real-time sequence, optimal power flow (OPF), 12-50
Rechargeable batteries, 1-9 to 10
Reclosures
 automatic, 9-32
 "dead time," 9-67
Recognition, transient voltages, 10-55
Recovery inrush, 9-7
Rectangular windings, 3-18 to 20
Rectifiers
 alternating current to direct current, 14-33, 14-35 to 37
 asymmetrical silicon-controlled (ASCR), 14-4
 average output voltage, 14-19
 controlled, 14-25 to 37
 diode conduction mechanics, 14-9
 full wave, 14-13 to 16
 single-phase currents, 15-33 to 36
 single-phase half wave circuits, 14-10 to 12
 three-phase, 14-17 to 19
 three-phase currents, 15-36
Recuperator, microturbines, 2-29
Recursive least squares (RLS) algorithm, 13-19
Reduced gradient method (OPF), 12-42
Reduced order models, 10-77 to 78
Reduction, fast power (fast valving), 11-62, 11-77
Reduction, voltage, 4-6
Reduction to the balanced case, 8-24 to 28
Reel stands, 4-44
Refurbishment/replacement, load tap changers, 3-203
Regional transmission organizations (RTOs), 13-7, 13-41
Regular load forecaster (RLF), 13-19
Regulation. *See also* Deregulation
 distribution transformers, 3-47
 noise limits by government, 5-82
 planning uncertainties, 13-46
 step-voltage, 3-68 to 80, 3-156 to 157
 utility companies, 13-3
Regulators, step-voltage, modeling, 6-32 to 41
Relational databases, 6-70
Relative capacitance measurements, 3-280 to 282
Relatively selective zones, 9-30 to 31
Relative polarity, instrument transformers, 3-107
Relative tan delta measurements, 3-280 to 282
Relays
 admittance, 9-37
 algorithms, 9-50 to 61
 automatic trend relay (ATR), 11-62
 bus/switching configurations, 5-19 to 23
 capacitor bank protection, 4-177
 coordination studies, 2-26
 current actuated, 9-33 to 36
 digital, 9-44 to 61
 directional overcurrent, 9-35 to 36
 distance, 9-37 to 38
 fundamentals, 9-30 to 33
 generator digital multifunction protection, 9-28 to 29
 impedance, 9-7 to 8, 9-37, 9-65 to 66

Index

instantaneous overcurrent, 9-34
interposing, 5-51
interposing, direct currents, 5-51
inverse-time delay overcurrent, 9-33 to 34
lockout, 9-10
offset-mho, 9-16
oil-flow, 3-200 to 201
operations, restoration, 9-10
out-of-step, 9-43
overcurrent, 9-7 to 8
overcurrent protection method, 9-2
overreaching, 9-39
overvoltage, 9-19
pressure, protection method, 9-4
pressure sensing (releasing), 3-201
protective, modeling, 6-9
reactance, 9-37
restrained differential, 9-2
speed, 9-31
stator differential, 9-11
sudden pressure, 3-26
transmission line protection, 9-30 to 39
volts-per-hertz, protection method, 9-9
zones of protection, 9-30 to 31
Reliability
 air insulated substations, 5-19 to 23
 animal aspects, 5-88 to 91
 bulk power system, 13-26 to 27
 component, data, 13-60 to 61
 determinist approach, 13-27 to 29
 distribution system, 13-56
 economics, 64
 importance of, 13-4
 power systems, 13-26 to 39, 13-51 to 65
 probabilistic method approach, 13-28 to 29, 13-33
 relays, 9-30
 seasonal, 13-64 to 65
 security defined, 12-55
 stability controls, 11-61
 standards, 13-6
 supply point, 13-26, 13-28 to 33
 transmission line system, 4-18 to 20
 transmission plan evaluation, 13-26 to 39
 utility problems, 13-61 to 64
Remanence. *See* Residual magnetism
Remedial action protection schemes, 9-42 to 43
Remote backup, EHV systems, 9-32
Remote communication meters, 7-3
Remote terminal units (RTU)
 energy management, 12-2 to 4

load management, 12-6
state estimation, 12-27
Supervisory Control And Data Acquisition (SCADA), 6-71 to 75, 12-27
Renz studies, 5-31
Reppen studies, 13-36
Required response spectra (RRS), 5-126 to 133
Reserve monitoring, 12-5
Residential distribution layout, 4-57 to 59
Residual magnetism
 current transformers, 3-120 to 121, 3-125
 iron/steel core transformers, 3-4 to 7
 power transformers, 3-29
Resin-bonded paper insulated bushings, 3-176
Resistance
 cables, 4-52 to 54, 4-62 to 63
 grid, 5-101 to 102
 grounding, 10-80
 transmission lines, 4-64 to 66
Resistance-type load tap changers, 3-185 to 186
Resistivity, soil, 10-80 to 90
Resonance
 frequency characteristics, 10-61 to 63
 grounding, 10-80
 insulation stress, 4-25
 parallel, 10-62
 subsynchronous, 4-177, 11-127 to 140
 supersynchronous (SPSR), 11-141 to 143
 terminal, 10-61 to 63
Response-based controls, 11-60
Response spectra, required (RRS), 5-126 to 133
Restorative system state, 11-121 to 122
Restortation
 insulation, 10-93 to 102
 transformers, 9-9 to 10
Restrike phenomena, 9-70
Retarded potentials, 10-22
Retrofit installations, window-type current transformers, 3-126 to 128
Return strokes
 arbitrary waveshapes, 10-26 to 29
 magnitude, 5-105 to 108
 velocity, 10-3
Revenue metering. *See* Metering
Reverse-bias safe operating area (RBSOA), 14-4

Reverse-conduction thyristor (RCTs), 14-4
Reverse impedance transformation, 6-30 to 32
Reverse power flow, 3-143 to 144
Reverse reactance paralleling, 3-145
Reversing change-over selector, 3-188
Ribbens-Pavella and Evans studies, 11-43 to 44
Ribbens-Pavella studies, 11-43 to 44
Richter studies, 13-6
Right-of-way acquisitions, wind turbines, 1-5 to 7
Rime ice. *See* Icing
Ring buses, 5-21, 9-4
Ringdown signals, 11-109
Ringlee studies, 13-40
Risk analysis. *See* Analysis and Assessment
Rizk studies, 4-166
RLF. *See* Regular load forecaster (RLF)
RLS. *See* Recursive least squares (RLS)
Rms horsepower method, 7-21
Robustness, 11-55
Rock, geomagnetic storms, 4-151
Rockefeller and Udren studies, 9-44
Rockefeller studies, 9-44
Rods
 fiberglass-reinforced plastic, 4-34 to 35
 lightning, 5-123
Rolling sphere method, 5-115 to 118, 5-121
Root locus plot, 11-31
De la Rosa, Kung and, studies, 11-141
De la Rosa studies, 10-5, 10-5 to 7
Rotational inertia, 6-3
Rotor angles
 fast voltage control, 11-66
 stability, 11-3 to 5
 swings, 11-60
 vs-time, 11-11
Rotors
 hydroelectric generator, 2-6
 mechanical model, 11-73 to 74
 prime mover modeling, 11-76 to 78
 synchronous machinery, 2-14 to 15
Routh studies, 11-21
Rouxel and Waldvogel studies, 10-58 to 59
Rowan studies, 11-78
Royalty payments. *See* Compensation to land owners
RS-232 interface, 6-71
RTO. *See* Regional transmission organizations (RTOs)
RTU. *See* remote terminal units (RTU)

Rubber, equipment pits, 5-68
Rubenstein and Walkley studies, 11-61
Ruff, Kramer and, studies, 3-191
Ruling span concept, transmission lines, 4-107 to 112
Rumelhart and McClelland studies, 13-17
Runge-Kutta method, 11-17

S

Sabotage. *See* Vandalism
Sachdev studies, 9-44
Safety
  energization, accidental, 9-26
  ground circuit, accidental, 5-92 to 95
  grounding, 15-4
  hazards to humans, 5-136
  substations, 5-85
Sag
  conductor lines, 4-89 to 128
  voltage, 15-24 to 40
SAIDI. *See* System average interruption duration index (SAIDI)
SAIFI. *See* System average interruption frequency index (SAIFI)
Salient poles
  hydroelectric generators, 2-4
  synchronous machinery, 2-14 to 15
Salt River Project (SRP), 4-170
Salt spray. *See also* Equivalent Salt Deposit Density (ESDD)
  insulator failure, 4-36
  insulator stress, 4-28
Salvaderi studies, 13-28
Sampling, digital relaying, 9-44 to 45
Samuelsson and Eliasson studies, 11-65
Sandia National Laboratories, 1-17
San Francisco outage, December 8, 1998, 11-110
Santori studies, 11-95
Sargent studies, 5-111
Sarma, Glover and, studies, 6-25, 8-29
Sasson studies, 12-42
Sastry, Padiyar and, studies, 11-44
Satellite monitoring, solar conditions, 4-161 to 163
Saturation
  current transformers, 3-118 to 119, 9-6
  synchronous machinery, 2-15, 2-16
Saturation modeling, 11-75 to 76
Sauer and Pai studies, 11-45
Saxton, McDonald and, studies, 5-61

SCADA. *See* Supervison Control And Data Acquisition (SCADA)
Scale models. *See* Models
Scaling, analog data acquisition, 5-48
SCE. *See* Southern California Edison (SCE)
Scheduling
  ineffective optimal rescheduling, 12-50
  interchange transaction, 12-5 to 6
  optimal power flow, 12-38 to 51
Scheich studies, 10-67
Schilling studies, 13-52
Schlumberger studies, 12-58
Schneider studies, 5-30
Van Schoor and van Wyk studies, 14-53
Schoukens studies, 11-95
Schultz studies, 10-40
Schwalb, Walker and, studies, 11-138
Schweppe and Wildes studies, 12-27
Schweppe studies, 13-8, 13-43
Schynder and Glavitsch studies, 12-46
SCOPF. *See* Security contrained optimal power flow
Scott transformers, 3-126, 14-61
Screening studies, 11-130
Screw windings, 3-21
SCRs (silicon-controlled rectifiers). *See* Thyristors
Seasons
  corona performance, 4-144 to 146
  electrodes, frost damage, 10-85
  heat waves, 13-59
  ice storms, 13-59
  impact on conductor loads, 4-93 to 96
  impact on reliability, 13-64 to 65
  insulation stress, 4-28
  load cycles, 3-154, 3-158
  major events, 13-59
  short-term load forecasting, 13-20 to 21
  transformer assembly, 3-237
  wind speeds, 1-4
  wind storms, 13-59
Secondary, primary-, phase shift, 9-7
Secondary batteries, 1-9 to 10
Secondary equipment, very fast transients (VFT), 10-52
Second harmonics, 9-3, 9-11
Sectionalized cable shields, 4-56
Security
  animal deterrents, 5-88 to 91
  assessment software, 11-68
  control, power systems, 12-7 to 8
  decision time frame, 12-55 to 56
  deregulation, 12-58
  determinist approach, 12-58
  models, 12-56 to 58

optimal power flow (OPF), 12-45, 12-50
power systems analysis and assessment, 11-120 to 125, 12-54 to 58
probabilistic method approach, 12-58
shift factors method, 12-57
substations, 5-85, 5-88 to 91
transient conditions, 12-55
transmission line system, 4-20
Security contrained optimal power flow (SCOPF), 12-45 to 47
Seismic considerations
  National Earthquake Hazard Reduction Program (NEHRP), 5-126, 5-129
  Seismic Design for Substations, Recommended Practice for, 5-124 to 125
  substation requirements, 5-124 to 134
Sekine studies, 15-33
Self-commuted inverters, distributed utilities, 2-34
Self-organizing neural fuzzy controller (SONFC), 11-68
Self-restoring gaps, 10-99
Self-restoring insulation, 10-93 to 102
Semiconducting shield cables, 4-51
Semiconductor devices, 14-1 to 7
Semi-tension stringing, 4-43, 4-115 to 116
Semlyen, de Leon and, studies, 10-59 to 60, 10-71
Sensors, immersed transformers, 3-271
Sequences
  impedance, 6-15 to 19
  network representation, 8-29 to 36
  transformer tests, 3-210 to 211
Sequential load flow method, 11-38 to 39
Series capacitance, models, 10-67
Series capacitors
  banks, switching surges, 10-39
  voltage stability, 11-40
Series compensation, 4-174 to 177
Series components, distribution feeders, 6-62
Servers, types of, 6-69 to 70
Service life, gas-insulated substations, 5-5
Setting, transmission line construction, 4-46
Set Voltage control settings
  load tap changing, 3-137
  step-voltage regulators, 78 to 79
Sewage disposal facilities, 5-78, 5-86

Index

SF6 insulation, very fast transients (VFT), 10-52
Shaft stop, microturbines, 2-30
Shaped compact single-core cables, 4-49 to 50
Sheath, metallic, 4-51
Sheaves, 4-116
Shedding loads, 9-41, 11-40
Shelbe, Fahd and, studies, 13-6
Shelbe, Maifeld and, studies, 12-21
Shelbe and Fahd studies, 12-20
Shelbe and McCalley studies, 13-6
Shelbe studies, 12-15, 12-17, 12-26, 13-2 to 10
Shell-forms
  construction, 3-17 to 18
  distribution transformers, 3-36
  winding strategies, 10-73 to 74
Shen and Laughton studies, 12-43
Shepherd and Zand studies, 14-40
Shield bonding methods, cables, 4-54 to 57
Shielded lines, lightning, 4-167 to 168
Shielding, substation lightning strokes, 5-104 to 123
Shielding electrodes, multiple, 5-118 to 119
Shielding systems, lightning, 5-118 to 121
Shield wires, induced voltage, 10-30 to 31, 10-34
Shift factors method, 12-57
Shintakatsuki substation, Japan, 14-60
Shock waves, 5-124 to 134
Short circuits
  analysis, 6-64 to 67
  autotransformers, 3-133
  current limiting reactors, 3-82
  current strength, load tap changers, 3-195
  distribution transformers, 3-47
  fault analysis, 8-44 to 59
  high-voltage windings, 3-114
  iron core transformers, 3-9 to 10
  power transformers, 3-14 to 15
  program inadequacy, 6-11
  reactor applications, 3-98 to 103
  studies, thermal generating plants, 2-26
  synchronous machinery, 2-15, *2-19*
  transformer failure, 3-248
Shorted core reading, 3-235
Shunt capacitors, voltage stability, 11-40
Shunting
  admittance, modeling, 6-25 to 28
  branch impedance, 8-29 to 31

capacitance, transmission lines, 4-62 to 63
capacitance models, 10-67
dynamic brakes, 11-64, 11-112
reactive power compensation, 4-169 to 174
windings, step-voltage regulators, 3-74
Shunt reactors
  banks, switching surges, 10-40
  protection for, 9-8
  steady state reactive compensation, 3-91 to 92
SI. *See* International System of Units (SI)
Si cells. *See* Photovoltaic (PV) cells
Side effects, system oscillation mitigation, 11-31
Siemans computation, susceptance, 6-57
Siemens Corporation, 4-178
Sigma-delta A/D converters, 9-46
Signal processing buffer (SPB), 11-90
Signals
  conditioning, 11-86
  pulse-width-modulated, 14-37 to 43
Silicon-carbide (SiC) elements, 3-201
Silicon cells, 1-14 to 15
Silicon-controlled rectifiers (SCRs). *See* Thyristors
Silicones, distribution transformers, 3-33
Silver plate, substations, 5-5
Simulations
  auction market, 13-4
  dynamic system assessment (DSA), 11-123 to 124
  energy control system, 12-8
  instructional system, 12-9
  Monte Carlo, 5-111, 13-33 to 39
  operator training, 12-8 to 9
  power flow analysis, 8-34 to 43
  power system dynamic, 12-9
  proposed substation sites, 5-78 to 80
  time-domain, 11-17
  transient stability, 11-18
  very fast transients (VFT), 10-46 to 51
  visual, proposed substation sites, 5-78 to 80
Single bus configurations, air insulated substations, 5-19 to 21
Single component outage event, 13-32
Single-core cables, 4-54
Single-crystal silicon cells, 1-14
Single-phase currents

capacitance, line, 4-77 to 78
connections, 3-39, 3-39 to 40
core-form construction, 3-17, *3-18*
distribution transformers, 3-36
half-wave rectifier circuits, 14-10 to 12
H-bridge, thyristors, 14-28 to 32, 14-38 to 40
inverters, 14-38 to 39
motors, thermal generating plants, 2-25
non-parallel transformers, 3-135
padmounted transformers, 3-58, 3-59 to 60
per unit system, transformers, 8-4 to 8
polarity, distribution transformers, 3-42
polarity, transformers, 3-130
protection for transformers, 9-9
rectifiers, 15-33 to 36
submersible transformers, 3-55 to 56
substations, 5-3, 5-16
switches, 5-24
two-wire, line inductance, 70 to 72
vault transformers, 3-53
Single phase-to-ground faults, 8-48 to 49, 8-52 to 57
Single-point grounded shields, cables, 4-54 to 55
Single-pole switching, 11-63 to 64
Single stator electromechanical meters, 7-1 to 2
Site diagnostic meters, 7-4 to 5
Siting. *See also* Installation considerations; Location Guidelines
  air insulated substation cost, 5-17 to 18
  gas-insulated substation cost, 5-17 to 18
  proposed substations, 5-80
  stabiliation importance, 11-28
  underground distribution transformers, 3-58
Six-phrase connections, 12-pulse rectifier systems, 3-134
Skelton, Jones and, studies, 11-95
Skin effect, model, 10-69 to 70
Slack bus, 8-39
Slack stringing, 4-43, 4-115 to 116
Slatt-Buckley substation 500-kV line, 4-178 to 179, 11-96, 11-113. *See also* Thyristor controlled series compensators (TCSC)
Small signal stability, 11-4, 11-20 to 33, 11-40
Smede studies, 5-38

SMES (superconducting magnetic energy storage), 1-4, 1-8
Smith, Overend and studies, 4-119
Smith, Tavora and, studies, 11-44, 11-48
Smith-Putnam machine, 1-1
Smith studies, 11-85
Smythe studies, 10-85
Snakes, damage from, 5-90, 13-63
Snow studies, 10-66
Social objectives (GENCOs), 12-10 to 11, 12-15
Sodium sulfur batteries, 1-10
Sodium vapor lamps, modeling, 6-9
SOFC (solid oxide fuel cell), 1-12, 1-13, 2-28, 2-29
Software and languages. *See also* Technology
   bundle conductor configurations, 4-148
   FORTRAN, 11-95, 11-103
   LabVIEW, 11-95, 11-103
   Matlab, 11-95, 11-103
   optimal power flow, 12-39
   Portable Dynamic Analysis and Design System (PDADS), 11-95
   Power System Identification (PSI) Toolbox, 11-103
   Power System Monitor (PSM) Tools, 11-103
   stability analysis, 11-9
   stability/security assessment, 11-68
   substation automation, 5-56 to 57, 5-58
Soil
   grounding properties, 10-80 to 90
   ionization, 10-89
   multilayers, grounding, 10-87 to 88
   oil containment, 5-66
   proposed substation sites, 5-78
   resistivity, 5-101
   temperature, effect on electrodes, 10-85
Solar activity
   heating, wind speed, 1-4
   power system impact, 4-150 to 165
   transformer protection method, 9-5
"Solar Shingle," 1-15
Solar (Sunspot) Cycles
   19, 4-159
   21, 4-159
   22, 4-150-151, 4-157, 4-159
   23, 4-157
Solid bushings, 3-172 to 173
Solid conductor, capacitance, 4-76 to 77

Solid insulation, vault and subsurface enclosures, 3-58
Solid oxide fuel cell (SOFC), 1-12, 1-13, 2-28, 2-29
Solid state meters. *See* Electronic meters
SONFC. *See* Self-organizing neural fuzzy controller (SONFC)
Sound levels. *See also* Acoustics
   abatement methods, 5-82 to 83
   causes and effects, 3-159 to 170
   cooling fan noise, 3-164
   dry-type transformers, 3-170
   energy, 3-160 to 170
   frequency noise source, 5-81, 15-56
   grounding, 15-15 to 6
   harmonics, 3-163 to 164
   measurements, 3-160 to 170, 3-233
   substation equipment, 5-28
   substations, 5-87
   transformer core, 3-163 to 164, 3-169
   transformers, 3-159 to 170, 3-233
   transformer testing, 3-3-233
   windings, 3-163 to 164, 3-169, 3-233
Southern California Edison (SCE), 11-92
Southern Methodist University, 13-18
Southwire's Wire and Cable Technology Group, 4-119
Spacers (spacer dampers), transmission line construction, 4-45 to 47
Spain, 1-2
Spans, catenary cables, 4-89 to 90
Spark dynamics, internal transients, 10-48
Spark gaps, 3-201
SPB. *See* Signal processing buffer (SPB)
SPCC. *See* Spill Prevention Control and Countermeasure (SPCC)
Special days peak forecasting, 13-20 to 21
Special protection schemes (SPS), 9-42 to 43
Spectra, required response (RRS), 5-126 to 133
Spheroid electrodes, 10-83
Spill Prevention Control and Countermeasure (SPCC), 5-62 to 6, 5-62 to 63
Spiraling/bundle conductor effect, 4-64 to 66
Spiral windings, 3-21
Split core-type current transformers, 3-122

Spot loads, modeling, 6-63
Spotting, tower, 4-115
SPS. *See* Special protection schemes (SPS)
SPSR. *See* Supersynchronous resonance (SPSR)
SPUC. *See* State Public Utility Commission (SPUC)
Square Butt HVDC converter, 11-140
Squirrel cage induction motors, 2-24
Squirrels, damage from, 5-89 to 90, 13-63
Sri Lanka, lightning, 10-4
Srivastava, Mousa and, studies, 5-112, 5-120 to 121
SRP. *See* Salt River Project (SRP)
SSA. *See* Static Security Assessment (SSA)
SSR. *See* Subsynchronous resonance (SSR)
SSTI. *See* Subsynchronous torsional interaction (SSTI)
Stability
   controls, power system, 11-58 to 67
   direct methods, 11 to 17, 11-42 to 52
   factors affecting, 11-18
   frequency, 11-6, 11-8
   metering, 7-4
   power systems, 11-2 to 9, 11-122 to 124, 13-51 to 65
   prime mover modeling, 11-77
   problems, 11-7 to 8
   rotor angle, 11-3 to 5
   small signals, 11-4, 11-20 to 33
   synchronous generators, 11-55 to 58
   theory, 11-46 to 48
   transients, 9-39 to 41, 11-10 to 20
   transients assessment, 11-50, 11-122 to 124
   Transient Stability program, 6-6, 6-9
   voltages, 9-41, 9-43, 11-34 to 41
Stacked cores, distribution transformers, 3-36
Stafl, Lammeraner and, studies, 10-69 to 70
Stahlkopf and Wilhelm studies, 11-82
Stand-alone PV systems, 1-18 to 19
Standards. *See also* National Electric Code (NEC); specific organizations
   bushings, 3-176 to 177
   platform selection guidelines, 6-74 to 75
   transformer testing, 3-209 to 210
Stanley model, 3-31
Star connections, 3-132 to 134

Index

Starting duty, synchronous machinery, 2-16 to 20
Start-up plan, thermal generating plants, 2-26
STATCOM. *See* Static Compensators (STATCOM)
State estimation
 analysis with, 12-27 to 37
 application, 12-35 to 37
 energy management systems (EMS), 12-33
 error identification, 12-34
 network topology assessment, 12-33
 operation, 12-33 to 35
 principles, 12-27
 solution, 12-30 to 33
 unobservability, 12-35
State public utility commission (SPUC), 13-7
State-variable formulation, 10-59
Static Compensators (STATCOM)
 future expectations, 5-39
 power electronics substations, 5-29
 reactive power compensation, 4-179 to 180
 reactive power injection device, 11-66
 system oscillations, 11-28
Static electricity, monitoring, 15-56
Static load model, 11-79
Static Security Assessment (SSA), 11-120 to 121
Static Var Compensators (SVC)
 actuators, 11-60 to 61
 closed-loop control, 11-29
 controllers, 5-31, 5-35
 power electronics substations, 5-29
 reactive power compensation, 4-172 to 174, 11-65 to 66
 system oscillations, 11-28, 11-32 to 33
 transmission systems, 4-4
 voltage stability, 11-40
Statistical analysis, oscillographs, 9-62 to 72
Stators
 faults, 9-11 to 14
 hydroelectric plants, 2-5 to 6
 synchronous machines, 2-13 to 14
Steady state conditions
 analyzing waveforms, 15-18
 electrode response, 10-82 to 83
 losses model, 10-69
 operating point, 11-120 to 121
 optimal power flow, 12-38
 reactive compensation (shunt reactors), 3-91 to 92
 rotor angle stability, 11-4
 voltage distribution, 10-56 to 57

Steady state method (SSR), 11-139
Steam turbines
 graters, 9-23 to 24
 prime mover, 9-11
Steel
 distribution transformers, 3-34
 instruments transformers, 3-104 to 107
 pulling lines, transmission line construction, 4-44
 transformers, 3-4 to 7
Steel studies, 10-59
Stefan-Boltzmann studies, 4-67
Steinmetz studies, 11-7
Stemmler and Guth studies, 11-66
Step-down connections
 delta-grounded wye, 6-46
 grounded wye-delta, 6-48
 ungrounded wye-delta, 6-50
Step-down transformers
 underground distribution transformers, 3-52 to 63
Stepped leaders, 10-2, 10-13 to 14, 10-21
Step-up connections
 delta-grounded wye, 6-47
 ungrounded wye-delta, 6-50
Step-up transformers
 hydroelectric generators, 2-7
 protection for, 9-8
 thermal generating plants, 2-25
Step-voltages
 grounding, 5-98 to 100
 human exposures, 5-96
 modeling, 6-32 to 41
 paralleling with transformers, 3-142 to 148
 regulators, general concepts, 3-68 to 80
Stevenson, Grainger and, studies, 8-29
Stigers studies, 11-62
Stock market, 13-2, 13-8 to 10, 13-46
Stoll studies, 13-40
Stopwatch method, sagging, 4-119
Storage systems
 batteries, 1-9 to 10, 2-31
 compressed air energy storage, 1-8
 flywheel, 1-8, 2-31
 fuel cells, 1-10 to 13
 superconducting magnetic energy storage, 1-8
 wind turbines, 1-4
Storage tanks, substation oil source risk, 5-64
Stored energy sources, seismic events, 5-133
Storms. *See also* Seasons
 ice, 13-59
 magnetic, 4-150 to 165

Stott, Alsac and, studies, 12-46
Stott and Alsac studies, 12-56 to 57
Stott and Hobson studies, 12-43
Stranded bundle conductors, capacitance, 4-80 to 82
Straps, ground, 10-49 to 50
Stray losses
 power transformers, 3-15
 transformer testing, 3-226, 3-228
Streamers
 breakdown, 4-137
 electromagnetic interference, 4-138
 negative, 4-133
 onset, 4-135 to 136
 Trichel, 4-131 to 133
Stream-to-archive monitoring, 11-89
Stress-strain curves, sag/tension numerical models, 4-102 to 104
Stringing
 creep, 4-118 to 119
 methods, 4-115 to 116
 sag tables, 4-112
 transmission lines, 4-42 to 43, 4-45
Strings, insulator, 4-30 to 31
Stroke current, 10-11
Strokes. *See* Lightning
Structures
 lightning strike incidence, 10-6
 stroke current magnitude, 5-106 to 107
 transmission lines, 4-13 to 22
Strycula studies, 14-44
Stubb studies, 11-8
Studies, 10-54
Studt studies, 11-59
Submersible transformers
 installation, 3-55 to 58
 preferred locations, 3-42
Substation related outage events, 13-32
Substations. *See also* specifc substation types
 air insulated, 5-18 to 23
 animal deterrents, 5-88 to 91
 automation systems, 5-41 to 53, 5-53 to 61
 community considerations, 5-76 to 87
 distribution automation, 6-72
 external insulation, 4-23 to 24
 fire protection, 5-134 to 139
 gas-insulated, 5-2 to 18, 5-28
 grounding, 5-85, 5-104 to 123
 lightning, 5-104 to 123
 load curves, 7-14
 oil containment, 5-62 to 75
 power electronics, high-voltage, 5-29 to 39

security, 5-88 to 91
seismic events, 5-124 to 134
switching equipment, 5-23 to 28
Subsynchronous frequency oscillations, 11-23 to 24
Subsynchronous resonance (SSR)
   analysis, 11-131 to 136
   control, 11-130 to 131
   countermeasures, 11-131, 11-136 to 137
   fatique damage, 11-137 to 138
   monitoring, 11-137 to 138
   physical properies, 11-128 to 129
   principles of, 4-77
   testing, 11-138 to 139
Subsynchronous torsional interaction (SSTI), 11-32, 11-127 to 128
Subtransmission lines, 4-9
Subway transformers, 3-53
Sudden pressure relay, power transformers, 3-26
Suitcase (grip), 4-116
Sulfur hexaflouride (SF6), 5-8 to 10
   air insulated substations (AIS), 5-2 to 18
   buses, 5-8
   bushings, 5-9, 5-10
   circuit breakers, 5-27
   dielectric gas, 5-2 to 18
   insulation, very fast transients (VFT), 10-52
   interrupting arcs, 5-2
Sullivan studies, 13-40
Sullivan substation, 4-180
Sunde studies, 10-85
Sunspot cycles. See Solar (Sunspot) cycles
Sun studies, 12-42
Superconducting magnetic energy storage (SMES), 1-4, 1-8
Supersynchronous resonance (SPSR), 11-141 to 143
   countermeasures, 11-141
   physical properies, 11-141 to 142
Supervision Control And Data Acquisition (SCADA)
   optimal power flow (OPF), 12-50
   remote terminal units, 12-27
Supervison Control And Data Acquisition (SCADA)
   data warehouse, 5-55 to 56
   digital relays, 9-44
   distribution automation, 6-67 to 76
   intelligent electronic devices (IED), 5-52
   power systems, 12-2 to 4, 12-6
   substation automation systems, 5-41

Supply point reliability, 13-26, 13-28 to 33
Surface operable installations, 3-55 to 58
Surges. See also Overvoltages
   arresters, 10-99 to 102
   direct lightning strokes, 10-8 to 20
   dry type transformers protection, 3-66
   gas-insulated substations, arresters, 5-10
   indirect lightning strokes, 10-21 to 34
   internal transients, 10-47
   Krider studies, arresters, 3-50
   switching, 10-36 to 40
   thermal generating plants protection, 2-26
   transformer failure, 9-2
   transmission systems, arresters, 4-4
Surges withstand capability (SWC), 9-45
Susceptance, 6-57 to 58
Suspended units, 2-7
Suspension (composite) insulators, 4-33, 4-34
Suspension insulators, 4-107
Sustained voltage unbalance, 9-9
SVC. See Static Var Compensators (SVC)
Sverak equation, 5-95
Sverak studies, 5-100 to 102
Swales, 5-69 to 72
Swann, Larsen and, studies, 11-30 to 32
SWC. See Surges withstand capability (SWC)
Sweden, 1-2
Sweden, lightning, 10-4
Swings (rotor angles), 11-11 to 12, 11-60
Switchable capacitor banks, 6-4
Switches
   control, latching devices, 5-52
   gas-insulated substations, 5-7
   ground, gas-insulated substations, 5-7
   high speed grounding, 5-25
   load break disconnect, 5-7, 5-24 to 25
   switching equipment, 5-24 to 25
Switchgears
   bus design, 5-81
   construction, noise abatement methods, 5-83
   electric and magnetic fields (EMF), 5-84 to 85
   hydroelectric generators, 2-7

thermal generating plants, 2-23 to 24
transmission systems, 4-3 to 4
Switching
   configurations, substations, 5-18 to 23
   controls, 5-13
   equipment, 5-23 to 28
   impulse tests, 3-220 to 221
   insulation stress, 4-25
   noise abatement methods, 5-82 to 83
   single-pole, 11-63 to 64
   surges, 5-10, 10-36 to 40
   voltages insulation stress, 4-26
Switching surges
   series capacitor banks, 10-39
   shunt capacitor banks, 10-40
   shunt reactor banks, 10-40
Switchyard facilities. See Substations
Switzerland, 10-5
Sybille studies, 11-65
Symmetrical components, 9-49 to 50
   balanced case, 8-27 to 28
   power systems, 8-14 to 33
Sympathetic inrush, 9-7
Synchronization accuracies, 9-50
Synchronous condensers
   fire hazard, 5-134 to 135
   future expectations, 5-39
Synchronous generators, 2-4 to 7
   accidental energization, 9-26
   protection methods, 9-11 to 29
   stability basics, 11-55 to 58
Synchronous machinery
   adjustable speed (doubly fed), 11-67
   general concepts, 2-12 to 20
   microturbines, 2-30
   torque, 11-4, 11-60
Synchronous motors
   fault analysis, 8-57
   thermal generating plants, 2-25
Synchronous ties (AC lines), 4-7
Synchronous torque, 11-4, 11-60
System architecture, substation automation, 5-54 to 56
System average interruption duration index (SAIDI), 13-58
System average interruption frequency index (SAIFI), 13-58
System condition monitoring, 11-85
System disturbance monitoring, 11-80
System flow, direction, 7-18
System indices, 13-30
System problem indices, 13-30
System responsibilites, substation automation, 5-54

Index

Szkuta studies, 13-24

T

Tacoma substation, 11-107
Tactical automation, 6-72
Takeda studies, 14-52 to 53, 14-61
Takriti, Krasenbrink, and Wu studies, 12-17
Tamronglak studies, 9-42, 9-44
Tanaka studies, 11-61
Tan delta measurements, relative, 3-280 to 282
Tandem cells, 1-15
Tang and Young studies, 11-139
Tanks
   and cabinet materials, distribution transformers, 3-34 to 35
   heating, 3-13, 3-41
   materials, vaults and subsurface enclosures, 3-57 to 58
Tano (capacitance and dissipation factor), 3-213
Tap changers. *See* Load tap changers (LTCs)
Tape shielded cables, 6-22 to 24, 6-27 to 28
Taps. *See also* Load tap changers (LTCs)
   dry type transformers, 3-63
   selectors, load tap changers, 3-185 to 187
   step-voltage regulators, 3-68, 3-74 to 77
   turns ratio adjustment, 3-23 to 25
Tarasiewicz studies, 10-71
Tarrant and Gay studies, 1-16
Tarter studies, 14-38
Tavora and Smith studies, 11-44, 11-48
Taylor and Erikson studies, 11-99
Taylor series, 8-41, 12-31 to 32
Taylor studies, 11-5, 11-8, 11-55, 11-61 to 62, 11-64 to 65, 11-67, 11-82, 11-99, 11-109
TCF (Transformer Correction Factor)
   CT/VT units, 3-129
   instrument transformers, 3-108 to 110
TCR. *See* Thyristor controlled reactors (TCR)
TCSC. *See* Thyristor controlled series compensators (TCSC)
TCT. *See* Total clearing time (TCT)
TDD. *See* Total demand distortion (TDD)
TDOC. *See* Time-delay overcurrent (TDOC) relays
TEB. *See* Transient excitation boosting (TEB)

Technology. *See also* Software and languages
   actuators, 11-60
   comparison of fuel cells, *1-12*
   computational simulation, 8-34 to 43
   database managers, 11-100 to 101
   distribution automation, 6-67 to 76
   electric and magnetic fields (EMF), 5-84 to 85
   energy management system (EMS), 11-90, 12-1 to 9
   energy storage, 1-7 to 13
   fault analysis, 8-59
   III-V, 1-15
   III-V advances, 1-15
   immersed transformers, 3-271
   microprocessor transformers, 9-3 to 4
   office equipment grounding, 15-10 to 11
   optimal power flow, 12-47 to 51
   planning uncertainties, 13-46
   power system modeling, 11-72 to 73
   power system stability, 11-9
   substation automation, 5-53 to 61
   value of information, 11-100 to 101
Telemetry, SCADA, 6-73
Television interference, 4-140 to 141, 5-38
Temperatures. *See also* Hottest-spot temperature
   air insulation, effect on, 10-95 to 96
   dissipation, effect on oil function, 3-207
   distribution transformers, 3-48
   dry type transformers, 3-65
   electrodes, 10-85
   failures, 3-178
   gas-insulated substations, 5-14
   insulation, 3-204 to 205
   insulation stress, 4-27
   iron core transformers
   loading power transformers, 3-149 to 159
   load tap changers, 3-286 to 287
   monitoring, 15-56
   oil function, 3-207
   oils, 3-208
   padmounted transformers, 3-63
   power transformers, 3-13 to 14, 3-278
   sag change, 4-99 to 100
   step-voltage regulators, 3-68
   substation automation systems, 5-43

   substation buildings, 5-37 to 38
   switching equipment, 5-23 to 24
   thyristor, 5-34, 14-3
   thyristors, 14-27
   transformer protection method, 9-4 to 6
   transmission line resistance, 4-64
   vaults and subsurface enclosures, 3-58
   winding resistance, 3-230
Temporary installations, split-core type current transformers, 3-122
Temporary overload, load tap changers, 3-195
Temporary voltages, insulation stress, 4-25 to 26
Tennessee Valley Authority, Sullivan substation, 4-180
Tension, conductor lines, 4-89 to 128
Tensioners, transmission line construction, 4-44
TEPCO. *See* Tokyo Electric Power Company (TEPCO)
Terminal equipment, generators, 2-7
Terminal resonance, 10-61 to 63
Termination kit, cables, 5-9
Terrain, clipping offsets, 4-119 to 120
Terzija, Djuric and, studies, 11-64
Tesla, 5-84
Testing
   bushings, 3-181 to 184
   corona performances, 4-149
   countermeasures (SSR), 11-139
   distribution transformers, 3-48 to 49
   gas-insulated substations, 5-16
   insulated cables, 4-60 to 62
   motor-enclosure-controller systems, 7-21
   subsynchronous resonance, 11-131, 11-138 to 139
   thermal generating plants, 2-26
   unit commitment genetic algorithm, 12-24 to 25
   windings, 10-74 to 75
TEV (Transient enclosure voltages). *See* Transient conditions
Texas, 10-4
TGPR (Transient ground potential rises). *See* Transient conditions
Thaake studies, 11-84
THD. *See* Total Harmonic Distortion (THD)
Thermal diagnostics, load tap changers, 3-286 to 287
Thermal failure, transformers, 3-248
Thermal faulting, effect on oils, 3-208

Thermal generating plants, 2-20 to 26, 2-22
Thermally upgraded paper, distribution transformers, 3-32
Thermal measurements. *See* Temperatures
Thermal ratings, instrument transformers, 3-111 to 112
Thermal tests, bushings, 3-183 to 184
Thermal time constant, distribution transformers, 3-48
Thermosiphon effect, 3-13
Thermostatically controlled loads, modeling, 6-9
Thevenin equivalent, 5-92 to 95
 delta-delta, 6-52
 delta-grounded wye, 6-46
 fault analysis, 8-51, 8-57 to 58
 grounded wye-delta, 6-48
 grounded wye-grounded wye, 6-51
 models, 6-45 to 46
 short circuit analysis, 6-64
 ungrounded wye-delta, 6-50
Thin silicon (buried contact) cells, 1-14
Third harmonics, 3-132, 4-141, 9-14
Thompson, Elihu, 3-31
Thorp, Akke and, studies, 9-54
Thorp, Phadke and, studies, 9-39
Thorp and Phadke studies, 9-54
Thottappillil studies, 10-4
Three-core type cables, 4-54
Three-phase currents
 AC to DC rectifiers, 14-33, 14-35 to 37
 angular displacement, 3-42, 3-130
 capacitance, line, 4-77 to 78
 connections, 3-40 to 41, 3-130
 core-form construction, 3-17
 distribution transformers, 3-37 to 39
 full-bridge converters, 14-33 to 34
 inductance, line, 4-72 to 75
 inverters, 14-40 to 42
 padmounted transformers, 3-58, 3-60 to 62
 per unit system, transformers, 8-9 to 14
 rectifiers, 14-17 to 19, 14-22 to 23, 15-36
 submersible transformers, 3-57
 substations, 5-3, 5-16
 subway transformers, 3-54
 switches, 5-24
 transposed line, inductance, 4-75
 unbalanced distribution feeder, modeling, 6-62 to 63
 vault transformers, 3-54
 voltage sags, 15-31 to 32

Three secondary bushings
 distribution transformers, 3-39 to 40
"Threshold of irritation," 15-43 to 44
"Threshold of visibility," 15-43 to 44
Through-faults
 instrument transformer failure, 3-280
 power system protection, 9-7
Thrust bearing supports, hydrogenerators, 2-7
Thunder days. *See* Keraunic levels
Thunder hours. *See* Keraunic levels
Thunderstorm days. *See* Keraunic levels
Thyristor controlled reactors (TCR), 3-92 to 93, 4-172, 4-178 to 179
Thyristor controlled series compensators (TCSC). *See also* Slatt-Buckley substation 500-kV line
 Brazil, 11-65
 China, 11-65
 closed-loop control, 11-29
 future expectatons, 5-39
 power electronic substations, 5-29
 reactive power compensation, 4-178 to 179
 system oscillations, 11-28, 11-33
Thyristors
 actuators, 11-60 to 61
 dynamic braking, 11-64
 gate circuits requirements, 14-28
 gate current injection, 14-27
 gate turn off (GTOs), 11-60, 14-2 to 3
 high dv/dt, 14-27
 increasing applied voltage, 14-27
 line cummutated inverters, 14-38
 MOS-controlled (MCT), 14-6 to 7
 reverse-conduction (RCT), 14-4
 semiconductor devices, 14-1 to 4
 single-phase *H-bridge*, 14-28 to 32, 14-38 to 40
 temperatures, 5-34, 14-27
Tickler coil, 3-189
Tie-in resistors, 3-198
Tillman studies, 3-150, 3-156 to 157
Time Delay
 control circuit, 6-33
 load tap changing, 3-139
 overcurrent (TDOC) relays, 9-33 to 34
 step-voltage regulators, 3-79
Time-domain applications, 15-21
 harmonics, 14-53
 monitoring, 11-86
 transient stability, 11-17

transient voltage response, 10-60 to 61
Time of use (TOU) metering, 7-3, 7-6 to 7
Time scales, 11-122
Tinney, Dommel and, studies, 12-38, 12-42
TMS. *See* Trouble management system (TMS)
TNA (Transient network analyzer). *See* Transient conditions
Tokaido Shinkansen, Japan, 14-60
Tokyo Electric Power Company (TEPCO), 11-63
Tolbert studies, 14-38
Tollefson studies, 13-64
Torgerson studies, 5-29
Torizuka studies, 11-61
Torque
 amplification (SSR), 11-129, 11-133 to 135
 damping, 11-12, 11-60
 initial peak, load tap changers, 3-284
 load tap changers, 3-284
 synchronous machines, 11-4, 11-60
De la Torre studies, 13-51
Torsional interaction (SSR), 11-129, 11-131 to 133
Torsional mode frequency tests, 11-138
Torsional mode oscillations, 11-24
Toshiba Corporation, 14-61
Total clearing time (TCT), 9-33
Total demand distortion (TDD), 14-22, 15-20
Total Harmonic Distortion (THD), 6-70, 15-54
Total hyteresis loss, 10-71
Totalized metering, 7-9 to 10
Total owning cost. *See* Cost and economics
Touch voltage, 5-92 to 98
TOV, arrester capabilities, 10-100
"Tower of Babel" problem, 6-75
Tower spotting, 4-115
Toxicity, batteries, 1-9 to 10
Traditional utilities, distributed utilities applications, 2-34
Training simulators, 12-8 to 9
Transducers, 5-47 to 48
Transfer bus, air insulated substations, 5-19 to 21
Transferred voltages, 5-98
Transfer tripping scheme, 9-39
Transformer bank connections, modeling, 6-42 to 52
Transformer Correction Factor (TCF) CT/VT units, 3-129

Index

instrument transformers, 3-108 to 110
Transformer failure, 9-2
Transformer Loss Measurement Guide, 3-227
Transformers
  algorithms, protection, 9-54 to 55
  distribution type, 3-30 to 51
  dry type, 3-36, 3-63 to 67
  electrical bushings, 3-171-, 3-171 to 184
  electric and magnetic fields (EMF), 5-84 to 85
  failures, 13-61 to 62
  fault types, 9-1 to 2
  fixed tap, 12-47
  general distribution design, 3-36 to 42
  generators, 2-7
  geomagnetically induced currents (GIC), 4-152 to 157
  installation, 3-234 to 240
  installation and maintenance, 3-234 to 242
  instrument type, 3-104 to 129
  insulating media, 3-204 to 209
  liquid-immersed on-line monitoring, 3-269-288
  loading, 3-149 to 159
  maintenance, 3-241 to 242
  microprocessor type, 9-3 to 4
  modeling, 6-42 to 52
  on-line monitoring, 3-269 to 288
  optimal power flow, fixed tap, 12-47
  per unit scaling, 8-4 to 8
  per unit system, 8-4 to 8
  power type, 3-11 to 30
  practical design, 3-7 to 10
  problems and failure investigation, 3-242 to 249
  protection, 9-1 to 10
  protection algorithms, 9-54 to 55
  reactors, 3-81 to 103
  sound levels, 3-159 to 170, 3-233
  step-voltage regulators, 3-68 to 80
  stress, 4-153
  testing, 3-209 to 234
  theory and principles, 3-3 to 11
  thermal generating plants, 2-20 to 21, 2-24
  transformer connections, 3-130 to 135
  underground distribution type, 3-52 to 63
  U-S-equipment standards, 3-249 to 269
  very fast transients (VFT), 10-52

Transient conditions. *See also* Transient energy function (TEF)
  analysis, 8-34
  assessment, 8-34, 11-67 to 68, 11-122 to 124
  coils, 10-54 to 78
  dielectric tests, 3-215
  electrodes, 10-82 to 84
  electromagnetic fields, 10-46
  excitation sources, 10-54 to 55
  externally generated, 10-44 to 46
  finite element method model, 10-60
  general concepts, 11-10 to 20
  internally generated, 10-43 to 44, 10-47 to 48
  network analyzer (TNA), 10-36
  Optical Transient Detector (OTD), 4-166
  overhead connections, 10-46
  performance, 10-55
  per-unit system, 8-1
  power systems stability, 11-5
  prime mover modeling, 11-77
  security analysis, 12-5
  sources, 10-54 to 55
  stability, 11-122 to 123
  stability models, 11-15 to 16
  system protection, 9-39 to 41
  transient enclosure voltage (TEV), 10-44 to 45, 10-48 to 50
  Transient excitation boosting (TEB), 11-62
  transient ground potential rises (TGPR), 10-44 to 45
  Transient Stability program, 6-6, 6-9
  transmission system, grounding, 10-80 to 90
  types of, 10-54 to 55
  very fast transients (VFT), 10-41 to 53
  voltage concerns, 10-54 to 78
  windings, 10-54 to 78
Transient energy function (TEF). *See also* Transient conditions
  applications, 11-52
  assessement using, 11-50
  direct stability methods, 11-44 to 50
  Flexible AC Transmission System (FACTS), 11-45
  modeling enhancements, 11-52
  principles, 11-48 to 50
  stabilty theory, 11-46 to 48
Transient excitation boosting (TEB), 11-62
Transistors

insulated-gate bipolar (IGBT), 14-6, 14-38 to 43, 14-44 to 61
  power, 14-4 to 6
Transit method, sagging, 4-119
Transit rides, 3-235
Transmission companies (TRANSCOs)
  capacity expansion, 13-10
  functions, 13-3 to 6
  role, 13-7
Transmission Line Conductors, Guide to the Installation of Overhead, 4-115
Transmission lines
  arresters, 10-99
  configurations, 9-32 to 33
  construction, 4-45
  doubly infinite single-conductor, 10-25 to 29
  lightning strikes on shielded, 10-9 to 11
  lightning strikes on unshielded, 10-9 to 11
  protection, 9-30 to 39
  shielded, 10-20
  unshielded, 10-17 to 20
Transmission plan evaluation, 13-26 to 39
Transmission SCADA, 6-68
Transmission system
  accesories, 4-23 to 42
  distribution lines, 4-2 to 12
  geomagnetic disturbances, 4-150 to 163
  grounding transients, 10-80 to 90
  high voltage applications, 4-48 to 62
  insulators, 4-23 to 42
  lightning protection, 4-165 to 168
  line construction, 4-9 to 12, 4-42 to 48
  line parameters, 4-62 to 82
  line structures, 4-13 to 22
  maintenance, 4-42 to 48
  power cables, 4-48 to 62
  reactive power compensation, 4-169 to 180
  studies, 6-11
  subtransmission lines, 4-9
  systems, 4-129 to 149
Transmission towers, lightning, 4-168
Transorbs, 5-43
Trapezoidal method, 11-17
Trapped charges, 10-36 to 39, 10-44
Travelers, 4-116
Trees, 13-62
TRELEG matrix models, 10-78
Trevino studies, 6-59
Triac, 14-1 to 2
Trichel streamers, 4-131 to 133

Triggered air gap, 4-177
Trigger logic, 11-86 to 90
Trinh and Jordan studies, 4-133 to 134
Trinh and Maruvada studies, 4-148
Trinh studies, 4-129 to 144
Triplen currents, 15-11 to 12
Triplex transformers, 3-37
Tripping considerations
    curves, 5-26
    generators, 9-27, 11-60, 11-62 to 63
    loads, 11-64 to 65
    transfer, 9-39
Trouble management system (TMS), 6-73
Troubleshooting. *See also* Problems
    insulated cables, 4-60 to 62
    transformer investigation, 3-242 to 249
Trudnowski studies, 11-96, 11-113
Trzynadlowski studies, 14-42
T-T connections, 3-41
T2 transformer, paralleling, 3-148
Tucker, Kuhn-, studies, 12-42
Turbines
    abnormal frequency operation, 9-23 to 24
    aeroderivative, 2-30 to 31
    combustion, 2-30 to 31, 11-78
    compression vs-combustion, 1-8
    dynamic interaction, power systems, 11-126 to 143
    Francis and Kaplan unit, 2-8
    gas, 1-4, 2-21, 11-78
    generators, 9-23 to 24, 11-126 to 143
    heavy frame, 2-30 to 31
    horizontal shaft propeller unit, 2-4
    hydroelectric plants, 2-4
    largest wind, 1-1
    microturbines, 2-28 to 31
    Pelton, 2-8
    power systems, 11-126 to 143
    steam, 9-11, 9-23 to 24
    thermal generating plants, 2-25
    utility scale, 1-1 to 2
    vertical shaft, Francis unit, 2-4
    wind, 1-1 to 7
Turns ratio. *See also* Load tap changers (LTCs)
    load tap changing, 3-23 to 25
    paralleling, 3-147
Turn-to-turn faults, 9-7
Tuttle studies, 1-16
TV interference, 4-140 to 141, 5-38
Two-breaker schemes, 9-4
250-V DC systems, thermal generating plants, 2-23

Two secondary bushings, distribution transformers, 3-39 to 40
Two-winding transformers
    applied voltage tests, 3-222
    insulation resistance, 3-214
    step-voltage regulators, 3-74 to 77
Tykeson studies, 5-30
Type A step-voltage regulators, 3-77
Type B step-voltage regulators, 3-78

U

UC. *See* Unit commitment (UC)
UCE. *See* Unit control errors (UCE)
Udren, Rockefeller and, studies, 9-44
Udren, Rockefeller and Udren, 9-44
UEL. *See* Underexcitation limiters (OEL)
UEP. *See* Unstable equilibrium points (UEP)
UL. *See* Underwriters Laboratories (UL)
Ullal studies, 1-16
Ultra-capacitors, 2-31
Ultrasonic devices, substation animal barrier, 5-91
Uman studies, 4-166, 10-2 to 4
Umbrella criterion, 11-61, 13-27
Umbrella units, 2-7
Unbalanced voltages, 5-25 to 26
Uncontrolled rectifiers, 14-8 to 25
Underexcitation limiters (OEL), 11-76
Underfrequency load shedding, 9-43, 11-67
Underground cables (lines)
    high-voltage applications, 4-48 to 62
    modeling, 6-14, 6-19
    reliability problems, 13-61
    shunt admittance, 6-26
    transmission lines, 4-64 to 66
Underground distribution transformers, 3-42, 3-52 to 63
Undervoltage load shedding, 9-43, 11-40
Underwriters Laboratories (UL)
    ANSI/UL 506 "Standard for Safety for Specialty Transformers," 3-256 to 257
    ANSI/UL 1446 "Standard for Safety for Systems of Insulating Materials-General," 3-257, 3-261
    ANSI/UL 1562 "Standard for Safety for Transformers, Distribution,

Dry-Type-Over 600 Volts," 3-265
    photovoltaics, 1-17
Ungrounded shields, cables, 4-54 to 55
Ungrounded systems, thermal generating plants, 2-21 to 22
Unified Power Flow Controllers (UPFC)
    future expectations, 5-39
    power electronics substations, 5-29, 5-31 to 32
    reactive power injection device, 11-66
Uninterruptable power supply (UPS), 14-37, 15-39
Unit commitment (UC)
    algorithms, 12-17 to 20
    auctions, 12-26
    defined, 12-15 to 16
    electric energy source, 12-17
    objective, 12-16
    optimal power flow (OPF), 12-50
    quantity supplied, 12-16
    selling electricity, 12-16 to 17
    solutions, 12-19 to 20
Unit control errors (UCE), 12-4
United Kingdom, 1-2
United States. *See also* specific state
    compressed air energy storage, 1-8
    Department of Energy studies, 13-45
    fused capacitor units, 4-176
    lead-acid batteries, 1-9
    transmission systems, 4-7
    utility-scale turbines, 1-1 to 2
    wind power installed capacity, 1-2
United States Advanced Battery Consortium (USABC), 1-9, 1-10
Unit systems, protection for, 9-8
Unit zones, 9-30 to 31
Unobservability, 12-35
Unrestricted zones, 9-30 to 31
Unshielded transmission lines, direct strokes, 10-17 to 20
Unstable equilibrium points (UEP), 11-48 to 52
UPFC. *See* Unified Power Flow Controllers (UPFC)
Upgrading, transmission line structure, 4-16
Uplift, 4-115
UPS. *See* Uninterruptable power supplies (UPS)
Urine (predator), substation animal barrier, 5-91
USABC (United States Advanced Battery Consortium), 1-9, 1-10

Index

User interface trends, 6-69 to 70
Utility distribution system, non-parallel transformers, 3-135
Utility-interactive PV systems, 1-17 to 18
Utility Substation Communication Initiative, 5-57 to 58, 5-61
Utilization, electric power
  loads, 7-12 to 17
  metering, 7-1 to 11
  motors, 7-18 to 23
UV radiation, insulation stress, 4-28
Uyemura studies, 11-44

V

Vaahedi and El-Din studies, 12-44
Vaahedi studies, 11-125, 12-44
Vacuum circuit breakers, 5-27
Vacuum cycle, transformer installation, 3-240
Vacuum filling system, 3-240
Vacuum interrupters, 3-186 to 187, 3-202
Vacuum processing, 3-35
Vakili, Hauer and, studies, 11-87
Vakilian and Degeneff studies, 10-59
Value of information, 11-100 to 101
Value tags, 11-89
Valves, outdoor, 5-38
Valving, fast, 11-62, 11-77
Van Baelen, Albertson and, studies, 4-163
Vancers studies, 5-38
Van Cutsem and Vournas studies, 11-5, 11-55
Van Cutsem studies, 11-6, 11-8
Vandalism, 9-2, 10-85, 13-64, 15-56
Van Overshee and De Moor studies, 11-115
Vargas studies, 12-44
Variable frequency drive (VFD), 14-21
Vault type transformers
  installation, 3-52 to 55, 3-53
  preferred locations, 3-42
Vector control dynamic response, 14-42
Vee curves, synchronous machinery, 2-15, 2-17
Vehicles, electric. *See* Electric vehicles
Vendors, platform, 6-73 to 76
Venikov studies, 11-34
Ventricular fibrillation, 5-95 to 96
Vertical shaft Francis turbine unit, 2-3, 2-4
Very fast front transients. *See* Transient conditions

Very fast transients. *See* Transient conditions
VFD. *See* Variable frequency drive (VFD)
VFT (Very fast transients). *See* Transient conditions
Vibrations
  hydrogenerators, 2-7
  load tap changers, 3-287 to 288
  monitoring, 15-56
  synchronous machines, 2-13
  transformer failure, 9-2
Vierath, Happ and, studies, 12-39
Visual simulation, proposed substation sites, 5-78 to 80
Vithayathil studies, 5-30, 5-39
Vittal, Fouad and, studies, 11-43 to 45, 11-48 to 52, 11-124
Voltage controlled bus (P-V bus), 8-39 to 40
Voltage ranging meters, 7-4
Voltage recorders, 15-50 to 54
Voltage reduction, transmission systems, 4-6
Voltage regulation
  loading power transformers, 3-156 to 157
  load tap changing, 3-137 to 141
  modeling, 6-9
  protection for, 9-8
  transformer failure, 9-2
Voltages
  accidental ground circuit, 5-92 to 95
  analysis, 11-36
  balanced, 8-24 to 25
  body current limits, 5-95 to 98
  corona modes, 4-129 to 137
  distribution, 10-56 to 58
  distribution, initial, 10-69
  distribution transformers, 3-45 to 48
  dynamic, 10-55
  effect on metering, 7-3
  extended control, 142 to 143
  extra-high-voltage, 4-4 to 8, 4-129, 4-137 to 144, 9-31 to 32
  extra-high-voltage (EHV), 10-36
  fluctuations, protection, 9-4
  generic dynamic load characteristics, 11-34 to 36
  harmonic sources, 14-45 to 46
  high-voltage lines, 4-6 to 8, 4-129
  induced, 10-23 to 31
  inducing, 10-22 to 24
  initial distribution, 10-69
  input, load tap changing, 3-136
  instability, 9-41, 9-43
  instrument transformer failure, 3-280

insulation stress, 4-24 to 27
insulator strings, 5-115
iron core transformers, 3-10 to 11
large disturbances, 11-39 to 40
load dynamics, 11-39 to 40
models, 6-36 to 37
multiple, 5-119
network transformers, 3-53
overvoltage relays, 9-19
phase angle control, 11-66
power compensation, 4-169 to 180
power quality, 15-24 to 40
power transformer failures, 3-278
rectifiers, average output, 14-19
reduction control, 3-142 to 143
reverse power flow, 3-143 to 144
sags, 15-24 to 40
single-crystal silicon cells, 1-14
single-phase padmounted transformers, 3-59 to 60
single-phase vault transformers, 3-53 to 54
small signals, 11-40
stability, 9-41, 9-43, 11-5 to 6, 11-34 to 41
step regulators, 3-68 to 80
sustained unbalanced, 9-9
symmetrical component transformation, 8-16 to 18
synchronous machinery, 2-13
thermal generating plants, 2-21, 2-23 to 24
three-phase full-bridge converters, 14-33 to 34
three-phase padmounted transformers, 3-60 to 62
tolerance requirements, 15-32 to 33, 15-39
transformer failure, 9-2
transmission line configurations, 9-32 to 33
unbalanced, 5-25 to 26
Voltage source inverter (VSI), 14-21
Voltage taps, bushings, 3-180
Voltage tests, insulated cables, 4-61
Voltage transformers (VT)
  imbalance protection, 9-20 to 21
  instrument transformers, 3-112 to 117
  load tap changing, 3-136 to 137
  metering, 7-10 to 11
Volt-ampere-watt test, 3-214
Volts, 10-66
Volts-per-hertz
  overvoltages, 9-19
  relays, protection method, 9-9
Von Hippel studies, 10-67
Von Jouanne studies, 14-37
Vournas, Van Cutsem and, studies, 11-5, 11-55

VSI. *See* Voltage source inverter (VSI)
VT. *See* Voltage transformers (VT)

W

Wagner, McCann, Beck studies, 5-109
Wagner studies, 5-109
Waight studies, 11-44, 11-52
Waldvogel, Rouxel and, studies, 10-58 to 59
Walker and Schwalb studies, 11-138
Walkley, Rubenstein and, studies, 11-61
Walls, noise abatement methods, 5-83
WAMS. *See* Wide Area Measurement System (WAMS) project
WAN. *See* Wide area networks (WAN)
Wang studies, 11-27
WAPA. *See* Western Area Power Administration (WAPA)
Washing insulators, 4-47
Washington, 10-4
Waste treatment plants, distributed utilities, 2-35
Watanabe studies, 14-50
Water. *See also* Moisture
  action, loading power transformers, 3-155 to 156
  oil-water, separator system, 5-72 to 73
  pumping, 1-19, 3-58
  substation construction drainage, 5-86
  supply, proposed substation sites, 5-78
Watson and Coultes studies, 11-24, 11-141
Wavelet transforms, 9-57 to 60
Waveshapes. *See* Harmonics
Weather. *See* Seasons
Weather shed, suspension (composite) insulators, 4-35
Webers measure, 5-84
Weber studies, 10-83, 10-85
Web sites, battery techologies, 1-9
Weighted least squares (WLS) method, 12-30 to 32, 12-35 to 37
Weight span, 4-114
Weiss studies, 5-39
Wells studies, 12-43
Wenham, Green and studies, 1-15
Wenzel studies, 3-277
Western Area Power Administration (WAPA), 11-92
Western North American Power System, 11-82 to 117

Western Systems Coordinating Council (WSCC), 11-82 to 117
Westinghouse, George, 3-30 to 31
Wet snow. *See* Icing; Moisture
Wet switching impulse withstand voltage test, 3-182
Wetting sources, contacts, 5-50
Whitehead, Gilman and, studies, 5-111, 5-114 to 115
Whitehead studies, 4-129, 5-111, 10-13
White studies, 10-59, 10-65
Wicket gates, 2-4, 2-7, 2-10 to 11
Wide area dynamics, analysis, 11-82 to 117
Wide Area Measurement System (WAMS) project, 11-82 to 117
Wide area networks (WAN), 5 to 55, 5 to 57, 6 to 69
Wilcox studies, 10-59
Wildes, Schweppe and, studies, 12-27
Wilhelm, Stahlkopf and, studies, 11-82
Wilkins studies, 11-7
Willems studies, 11-44
Willis, Philipson and, studies, 13-52
Willis studies, 7-14
Wilson studies, 10-39
Wind
  conductor lines, 4-93 to 96
  power, 1-1 to 7
  sag/tension numerical models, 4-103 to 104
  short-term load forecasting, 13-20
Winding Resistance Measurements, 3-226, 3-228 to 232
Windings. *See also* Coils
  barrel, 3-21
  circular, 3-18 to 20
  compensation, high-voltage, 3-114
  configurations, faults, 9-3
  connections, high-voltage windings, 3-114
  construction, 10-72 to 77
  copper, 3-18
  disc, 3-21 to 23, *3-27 to 28*
  disk, models, 10-67
  distribution transformers, 3-32, 3-36
  dry type transformers, 3-64
  elliptical-shaped, 3-20
  heat source, 3-13
  helical, 3-21, *3-27*
  instrument transformers, 3-112
  layer, 3-21, *3-26*
  load tap changers, 3-196 to 200
  low current, faults, 9-3
  magnetizing current, 3-5 to 6

  nameplate ratings, 3-150
  pancake-type, 3-17, 3-20, *3-24 to 25*, 10-74
  paralleling, 3-32, 3-147
  power transformers, 3-18 to 23, 8-29 to 31
  rectangular, 3-18 to 20
  screw, 3-21
  short circuits, high-voltage, 3-114
  sound levels, 3-163 to 164, 3-169, 3-233
  spiral, 3-21
  stator, 9-12 to 14
  surges, 10-56 to 58
  temperatures, 3-26, 3-153
  transformer failures, 3-278, 9-1
  transient voltage response, 10-54 to 78
Window modeling, load, 6-9 to 10
Window-type current transformers construction, 3-122
Wind Power Monthly, 1-2
Wind span, 4-114
Winkelman studies, 4-89, 4-91, 4-107
Wiring. *See also* Grounding
  power quality, 15-2 to 15
  problems, 15-6 to 12
  status points, 5-50 to 51
  transducer standards, 5-47
  transformer controls test, 3-214
Withstand voltage, insulator strings, 5-115
Witzmann studies, 10-46, 10-51
Wolfe studies, 12-42
Wollenberg, Wood and, studies, 12-13, 12-20
Wong, Chiang and, studies, 11-68
Wood and Wollenberg studies, 12-13, 12-20
Working Group studies, 2-8
Wound cores, distribution transformers, 3-36
Wound potential transformers, 5-46 to 47
Wound rotor induction motors, 2-25
Wound-type current transformers, 3-122
Wozniak studies, 2-8
WSCC. *See* Western Systems Coordinating Council (WSCC)
Wu, Takriti, Krasenbrink and, studies, 12-17
Wye connections
  current transformers, 3-121
  distribution transformers, 3-39 to 42
  double wound transformers, 3-132 to 133
  load tap changers, 3-189 to 190

Index

models, 6-37 to 57
regulator models, 6-36
step-voltage regulators, 3-70 to 71, 3-74, 3-80
symmetrical component transformation, 8-21
Van Wyk, van Schoor and, studies, 14-53

X

Xu studies, 11-35

Y

Yalla studies, 9-28
Yamanashi test line, Japan, 14-59
Yanabu studies, 10-48, 10-50 to 51
Yan and Quintana studies, 12-44
Yang studies, 1-15
Yard
    noise abatement methods, 5-83
    surfacing, oil containment, 5-66
Ybus matrix, 12-57
Yee, Chow and, studies, 9-61
Yin studies, 9-14
Yip studies, 9-28
Yohn, Andorka and, studies, 11-142 to 143
Yokoyama studies, 12-50
Yoshida studies, 14-59
Young, Tang and, studies, 11-139
Young studies, 5-111
Younkins and Johnson studies, 11-8
Yu studies, 12-42

Z

Zaborsky studies, 11-44
Zand, Shepherd and, studies, 14-40
Zangwill studies, 12-42
Zao and Iravani studies, 5-39
Zhang and Bollen studies, 15-31, 15-36
Zhang studies, 14-50
Zhou and Liang studies, 11-65
Zhou studies, 11-65
Zig-zag connections
    distribution transformers, 3-41
    double wound transformers, 3-132 to 133
    protection for transformers, 9-8
Zinc air batteries, 1-10
Zinc oxide, 5-10
Zinc-oxide (ZnO) varistors, 3-201
Zones of protection, 9-30 to 31
Zweibel studies, 1-15
Van Zyl studies, 14-52